International Symposium on Civil and Environmental Engineering (ISCEE 2016)

MATEC Web of Conferences Volume 103 (2017)

Melaka, Malaysia
5 - 6 December 2016

Part 1 of 2

Editors:

Zainorizuan Mohd Jaini
Lee Yee Yong
Alvin John Lim Meng Siang
Mohamad Hanifi Othman
Siti Nazahiyah Rahmat
Mohd Shalahuddin Adnan

ISBN: 978-1-5108-3893-2

Printed from e-media with permission by:

Curran Associates, Inc.
57 Morehouse Lane
Red Hook, NY 12571

Some format issues inherent in the e-media version may also appear in this print version.

Printed by Curran Associates, Inc. (2017)

For additional information, please contact EDP Sciences – Web of Conferences at the address below.

EDP Sciences – Web of Conferences
17, Avenue du Hoggar
Parc d'Activité de Courtabœuf
BP 112
F-91944 Les Ulis Cedex A
France

Phone: +33 (0) 1 69 18 75 75
Fax: +33 (0) 1 69 28 84 91

contact-edps@webofconferences.org

Additional copies of this publication are available from:

Curran Associates, Inc.
57 Morehouse Lane
Red Hook, NY 12571 USA
Phone: 845-758-0400
Fax: 845-758-2633
Email: curran@proceedings.com
Web: www.proceedings.com

TABLE OF CONTENTS

PART 1

PART 2

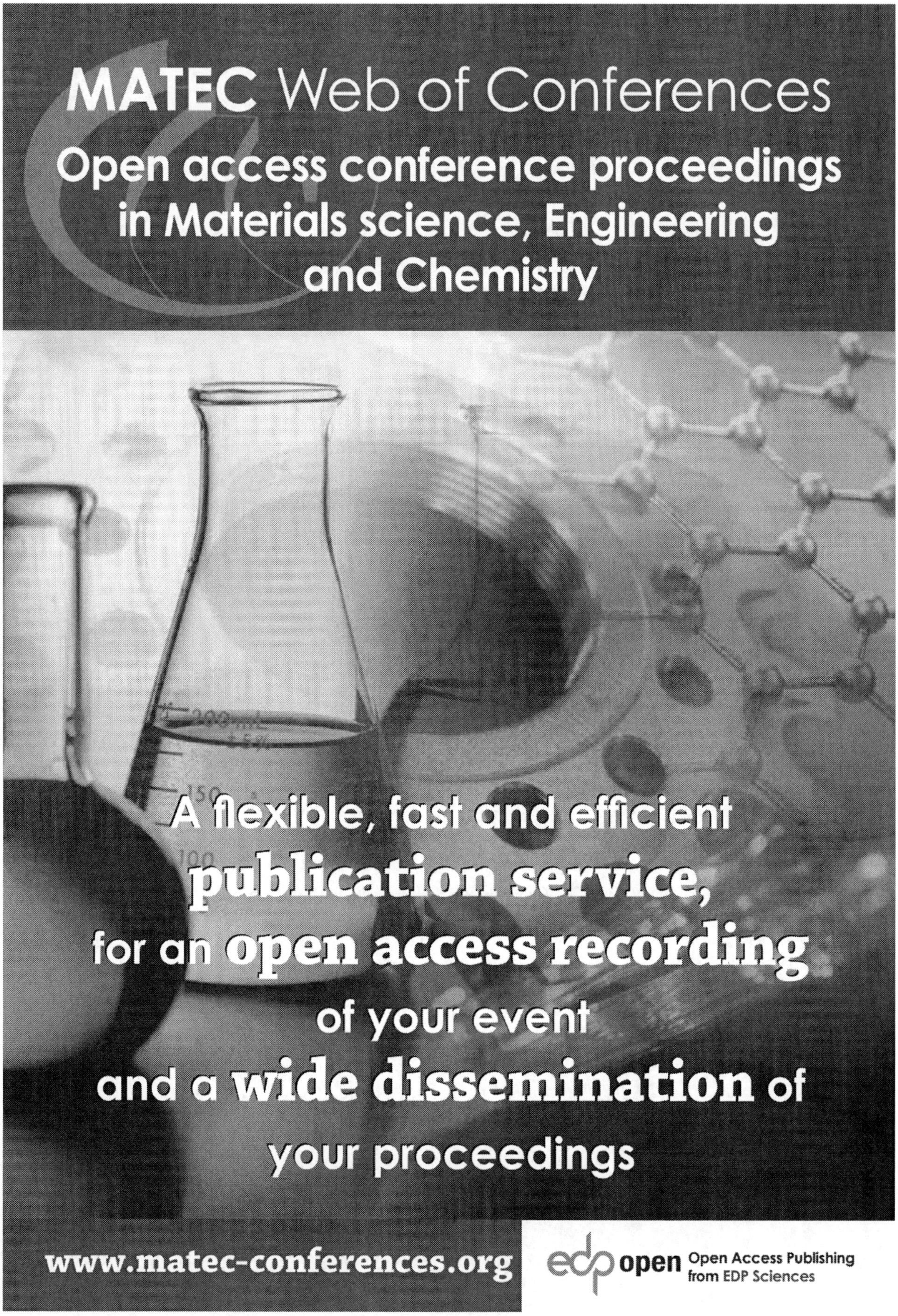
MATEC Web of Conferences
Open access conference proceedings in Materials science, Engineering and Chemistry
A flexible, fast and efficient publication service, for an open access recording of your event and a wide dissemination of your proceedings
www.matec-conferences.org
edp open Open Access Publishing from EDP Sciences

Conference Committees

Organizing Committee

Executive Chairman	:	Assoc. Prof. Dr. Mohd Haziman Wan Ibrahim
Co-Chairman	:	Dr. Kamaruddin Ambak
		Dr. Norwati Jamaluddin
		Assoc. Prof. Dr. Norzila Othman
Deputy Chairman	:	Assoc. Prof. Dr. Aziman Madun
		Dr. Ahmad Zurisman Mohd Ali
		Dr. Mohd Adib Mohammad Razi
Secretary	:	Dr. Shahiron Shahidan
Treasurer	:	Dr. Hazrul Abdul Hamid
Promotions	:	Dr. Basil David Daniel
		Dr. Hartini Kasmin
		Dr. Norhayati Abd Ghafa
		Mrs. Jalilah Md Mohtar
Programme	:	Dr. Nor Amani Filzah Mohd Kamil
		Dr. Nur Hazurina Othman
Technical	:	Sr. Dr. Mustaffa Anjang Ahmad
		Ir. Dr. Raha Abd Rahman
		Mr. Saifullizan Mohd Bukari
		Mrs. Zarina Md Ali

Managing Editor

Dr. Zainorizuan Mohd Jaini
Dr. Norashidah Abd Rahman
Dr. Noorwirdawati Ali
Dr. Faisal Sheikh Khalid
Dr. Goh Wan Inn
Dr. Nor Haslinda Abas
Dr. Lee Yee Yong
Dr. Muhammad Fikri Hasmori
Dr. Siti Hidayah Abu Talib
Dr. Alvin John Lim Meng Siang
Dr. Joewono Prasetijo
Dr. Mohamad Hanifi Othman
Dr. Siti Nazahiyah Rahmat
Dr. Mohd Azlan Yusuf
Dr. Mohd Shalahuddin Adnan
Dr. Mohd Hairul Khamidun

Scientific Committee

Preface

We are very grateful that this international conference is convened today to gather more than one hundred speakers presenting papers pertaining to the field of Civil and Environmental Engineering. ISCEE 2016 Jointly organised by the Centers of Research in the Faculty; Jamilus Research Center for Sustainable Construction (JRC), Micropollutant Research Center (MPRC) and Smart Driving Research Center (SDRC) as a platform for scientists, researchers and engineers to share their knowledge and experiences in creating sustainable and environmental friendly world for the present and the future. We all know that this world is borrowed from the next generation and in other for us to ensure that it is well looked-after, we have to maintain the three important pillars of sustainability i.e. Economics, Environment and Society. These three pillars are the catalyst to ensure that the world is livable, continue to exist and at the same time is developed to enhance the socio-culture of mankind.

The ISCEE 2016 theme, 'Sustainable Infrastructure for Global Awareness and a Better Future' indicates that the needs to prepare the world for future generation to continue the responsibility to make the world a better place to live. It is the desire of all population and generations that mankind will continue to exist and lead the world. It has been portrayed by many in science fiction movies that the apes will rule the world or the world will be destroyed because of our negligence and too engross in the development of artificial intelligence (AI). It is always possible if we do not care to ensure that what exists can be de-existed and what is impossible is always possible. The best example is the telephone evolution. From wired to wireless and from a typical phone to a smart phone. Constructing buildings was considered difficult was limited to certain people. But now, everybody can construct a house with some skills with the application of industrialized building system (IBS). With research works on reducing the usage of natural materials and increasing the use of recycled materials, pollution and negative environmental impacts can be reduced. Thus, it is our duty to ensure that we conduct the necessary steps towards sustainability of the world in whatever discipline and role that we have.

Technically Above the Rest

Sincerely,

Mohd Haziman Wan Ibrahim
Executive Chairman

Kamarudin Ambak
Norwati Jamaluddin
Norzila Othman
Co-Chairman
International Symposium on Civil and Environmental Engineering 2016 (ISCEE 2016)

About ISCEE 2016

The Faculty of Civil and Environmental Engineering, Universiti Tun Hussein Onn Malaysia (UTHM) is proud in organizing the International Symposium on Civil and Environmental Engineering 2016 (ISCEE 2016). Jointly organised by the Centers of Research in the Faculty, i.e. Jamilus Research Center for Sustainable Construction (JRC), Micropollutant Research Center (MPRC) and Smart Driving Research Center (SDRC), ISCEE 2016 aims to bring together experts from the academia and industry to share their research and discuss about current trends, new innovations, challenges and solutions in the field of Civil and Environmental Engineering. ISCEE 2016 is co-organised by Universitas Gadjah Mada, Universitas Muhammadiyah Surakarta, Universitas Brawijaya, the Institute for Infrastructure Engineering and Sustainable Management and Jabatan Kerja Raya.

- Theme -

Sustainable Infrastructure for Global Awareness and a Better Future

- Main Organizer -

UTHM
Universiti Tun Hussein Onn Malaysia

- Co-Organizers -

IIESM
JKR

- Parallel Conference -

International Conference on Sustainable Construction and Structures
(ISuCOS 2016)
International Conference on Sustainable Environment and Water Research
(ICSEWR 2016)
International Conference on Sustainable Infrastructure Engineering
(ICSIE 2016)

Statement of Peer review

In submitting conference proceedings to MATEC Web of Conferences, I certify to the Publisher that I adhere to the **Policy on Publishing Integrity** of the journal in order to safeguard good scientific practice in publishing.

1. All articles have been subjected to peer review administered by the proceedings editors.
2. Reviews have been conducted by expert referees, who have been requested to provide unbiased and constructive comments aimed, whenever possible, at improving the work.
3. Proceedings editors have taken all reasonable steps to ensure the quality of the materials they publish and their decision to accept or reject a paper for publication has been based only on the merits of the work and the relevance to the journal.

Title, date and place of the conference

International Symposium on Civil and Environmental Engineering 2016 (ISCEE 2016)
5-6 December 2016
Novotel Melaka, Melaka, Malaysia

Proceedings editor(s):

Zainorizuan Mohd Jaini
Lee Yee Yong
Alvin John Lim Meng Siang
Mohamad Hanifi Othman
Siti Nazahiyah Rahmat
Mohd Shalahuddin Adnan

Date and editor's signature

13/ 3/ 2017

DOI: 10.1051/matecconf/201710301001

Strength of Binary and Ternary Blended Cement Pastes Containing Palm Oil Fuel Ash and Metakaolin Exposed To Sodium Sulphate

Mohd Hanif Ismail[1,*], *Nor Hazurina* Othman[1] *Ahmad Zurisman* Mohd Ali[1], *Megat Azmi* Megat Johari[2], and *Kamar Shah* Ariffin[3]

[1]Faculty of Civil and Environmental Engineering, Universiti Tun Hussein Onn Malaysia, 86400, Johor, Malaysia
[2]School of Civil Engineering, Universiti Sains Malaysia, 11800, Pulau Pinang, Malaysia
[3]School of Materials and Mineral Resources Engineering, Universiti Sains Malaysia, 11800, Pulau Pinang, Malaysia

Abstract. The compressive strength of binary (BBCP) and ternary blended cement pastes (TBCP) containing Palm Oil Fuel Ash (POFA) and Metakaolin (MK) exposed to 3 % sodium sulphate solution has been studied. The ordinary Portland cement (OPC) was partially replaced with POFA and MK on mass-for-mass basis. All specimens were first cured for 28 days in normal water and subsequently subjected to full immersion in sodium sulphate solution for 150 days for the compressive strength evaluation. The results show that partial replacement of cement with POFA and MK improved the durability of the cement pastes by reducing the damage caused by sulphate attack.

1 Introduction

By using recycled materials in concrete as a filler or additive, it was found that the mechanical properties of the concrete can be improved [1-3]. Furthermore, the use of recycled materials in concrete can reduce environmental pollution and promote recycling in the industry. POFA use as a cement replacement material is an effort to control the pollution of the environment by reducing the use of cement in the construction industry. The use of POFA and MK as pozzolanic material by partially replacing cement has been shown to improve the durability of cement paste, mortar and concrete [4-7]. POFA contains high proportion of silica while MK possesses high content of alumina, hence the utilization of POFA and MK will affect the durability of cement paste, mortar and concrete. This paper will discuss the changes of compressive strength of OPC, BBCP and TBCP containing POFA and MK exposed to 3 % sodium sulphate solution. In addition, the changes in the intensity of calcium hydroxide (CH), calcium silicate hydrate (C-S-H) and ettringite (ETT) will also be discussed. The changes in the intensity of this mineral are

* Corresponding author: mohdhanif@uthm.edu.my

correlated to the changes in the compressive strength of OPC paste as well as the BBCP and TBCP pastes containing POFA and MK.

2 Experiment

2.1 Materials preparation

The ordinary Portland cement was partially replaced by POFA and MK at the rates of 0 %, 20 % and 40 % by mass of total binder. The POFA was treated via heat treatment and ground to a median particle size of approximately 9 µm. While, the Kaolin was turned into Metakaolin also via heat treatment and ground to a median particle size of approximately 8 µm. The water binder (w/b) ratio was kept constant at 0.3 for all mixtures. The proportions of the cement pastes are shown in Table 1. All specimens were removed from their respective moulds 24 hours after casting and cured in water maintained at room temperature for 28 days. Then, all specimens were subjected to full immersion in 3 % sodium sulphate solution up to age of 180 days.

Table 1. Mix proportions of cement paste and pastes containing POFA and MK

Mix	OPC (%)	POFA (%)	MK (%)	w/b
OPC	100	-	-	0.3
M5	95	-	5	0.3
M10	90	-	10	0.3
P20	80	20	-	0.3
P40	60	40	-	0.3
P15M5	80	15	5	0.3
P35M5	60	35	5	0.3
P10M10	80	10	10	0.3
P30M10	60	30	10	0.3

2.2 Compressive strength test

The compressive strength of the hardened cement paste cubes was determined using a 3000 kN automatic compression machine. The test was performed on average of five number of samples of 50 x 50 x 50 mm cement paste cubes at the ages of 60, 90 and 180 days according to ASTM C109.

2.3 X-Ray Diffraction (XRD) analysis

The X-ray scans were performed for 2 theta between 5^0 and 90^0 at 60, 90 and 180 days on crushed pastes sample in powder form using X-ray diffraction (XRD) Advance Bruker DX8 equipment according to BS EN 13925-1:2003. The collected X-ray scans recorded as intensity in unit counts.

Then, the quantitative intensity changes of C-S-H, CH and ETT in the OPC, BBCP and TBCP were observed using XRD analyzer via X'Pert Highscore Plus software via Refinement Method on each testing age. All the XRD data from the X'Pert Highscore Plus software was further analyzed using Microsoft Excel.

3 Results and discussion

3.1 Changes of compressive strength

The compressive strength of cement pastes exposed to normal water and 3 % sodium sulphate solution is given in Table 2. While the changes of compressive strength of cement pastes exposed to 3 % sodium sulphate solution compared to cement paste exposed to normal water is given in Fig. 1. From those Table and Figure, it shows that the compressive strength of OPC, BBCP and TCBP containing POFA and MK increased when exposed to 3 % sodium sulphate solution after 30 (60 days/age) and 60 (90 days/age) days of exposure period but decreased upon a prolonged exposure period of 150 (180 days/age) days.

Table 2. Compressive strength of cement pastes exposed to normal water and 3% sodium sulphate solution

	normal water (MPa)			3% sodium sulphate solution (MPa)		
Mix	60 d	90 d	180 d	60 d	90 d	180 d
OPC	69.5	72.0	75.0	75.0	78.0	71.2
M5	71.0	73.0	76.0	76.0	79.0	72.3
M10	74.0	76.0	78.0	77.0	81.0	73.0
P20	75.0	76.7	78.5	76.8	80.1	73.2
P40	76.0	78.8	81.0	79.0	82.0	79.7
P15M5	73.0	76.0	79.0	78.5	82.0	74.8
P35M5	74.5	77.5	80.0	78.0	81.5	78.0
P10M10	71.5	75.0	79.3	78.3	81.0	74.7
P30M10	72.5	76.0	79.5	80.0	85.0	81.8

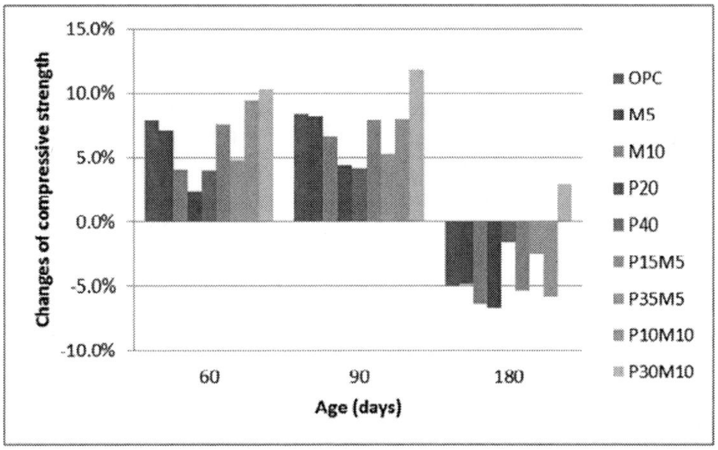

Fig 1. Changes of compressive strength of cement pastes exposed to 3 % sodium sulphate solution compared to cement pastes exposed to normal water.

The increase of strength of OPC cement paste exposed to 3 % Na_2SO_4 solution was influenced by the reaction of sulphate ions with CH to form gypsum [8,9]. Gypsum formation at the beginning lead to denser structure as a result of precipitation of products within voids and micropores [8-10]. The decrease in strength of the OPC cement paste exposed to 3 % Na_2SO_4 solution was affected by the excessive formation of gypsum and ettringite which create internal stress in the cement paste and decrease the C-S-H formation [9].

The increase in strength of the BBCP containing MK was lower than the OPC cement paste but BBCP containing MK has higher durability compared to OPC cement paste for short and long term exposure because the BBCP containing MK has a higher C-S-H content, outcomes of cement hydration and pozzolanic reaction and less formation of gypsum and ettringite than OPC cement paste [11].

It was found that the BBCP containing POFA have higher durability compared to the OPC cement paste and the BBCP containing MK for short and long term exposure. Higher C-S-H contents in the BBCP containing POFA resulting from hydration and pozzolanic reaction and lower formation of gypsum and ettringite than the OPC cement paste and the BBCP containing MK makes the BBCP containing POFA more resistant to sulphate with higher strength value [6].

The increase in strength may be attributed to; 1) continuous hydration of unhydrated cement components and additional reaction of POFA and MK to form more C-S-H leading to an increase in compressive strength 2) The reaction of sulphate ions with hydrated cement component to form gypsum and ettringite. These two reactions led to denser structure as a result of precipitation of products within the voids and micropores in the cement pastes [8, 10]. The compressive strength of cement pastes initially increased, and then began to decrease until the specimens eventually deteriorated after 150 days of exposure. The reaction of sulphate ions with unhydrated cement component to form gypsum and ettringite become more dominant leading to formation of internal stress and microcracks and this decreases the strength [8,10].

3.2 Losses of CH, C-S-H and changes of ettringite intensity

The losses of CH and C-S-H intensity of cement pastes exposed to 3% sodium sulphate soluiton compared to cement pastes exposed to normal water is shown in Fig. 2. Meanwhile Fig. 3 shows the changes of ettringite intensity of cement pastes exposed to 3% sodium sulphate solution compared to cement pastes exposed to normal water.

After being exposed to sodium sulphate solution for 150 days, the P30M10 cement paste shows the highest losses or reduction of CH intensity while, the OPC cement paste shows the lowest reduction of CH intensity after being exposed to sodium sulphate solution for 150 days. The reduction of CH intensity was due to 1) pozzolanic reaction; CH reacts with silica and alumina to form C-S-H 2) the formation of ettringite; CH reacts with sulphate ions to form gypsum then, gypsum reacts with C_3A, CAH and CSAH to form ettringite thus reduces the calcium content in the cement paste. The reduction of CH in the cement paste exposed to sulphate solution weakens the strength of the cement pastes [9].

When compared to the cement pastes exposed to normal water, the reduction in C-S-H intensity of the cement pastes exposed to sodium sulphate solution increased after 30, 60 and 150 days of exposure.

The count of C-S-H for each specimens is declining after both samples were exposed to 3 % sodium sulphate solution. The OPC cement paste shows the highest losses of C-S-H intensity, meanwhile the P30M10 shows the lowest reduction of C-S-H intensity after 150 days of sulphate solution exposure.

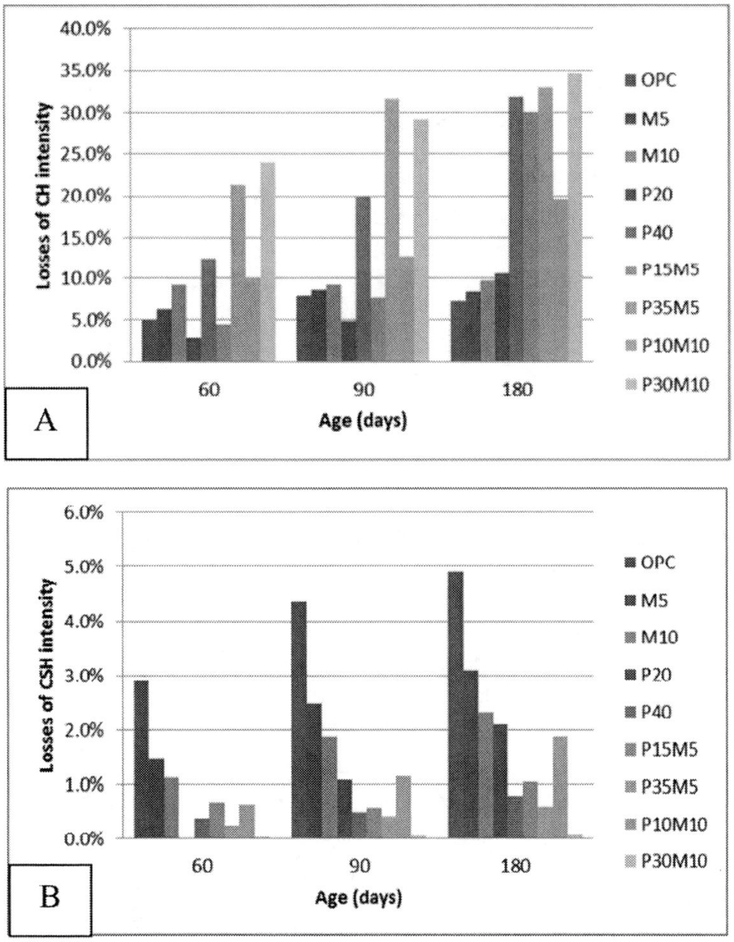

Fig 2. Losses of CH (A) and C-S-H (B) intensity of cement pastes exposed to 3 % sodium sulphate solution compared to cement pastes exposed to normal water.

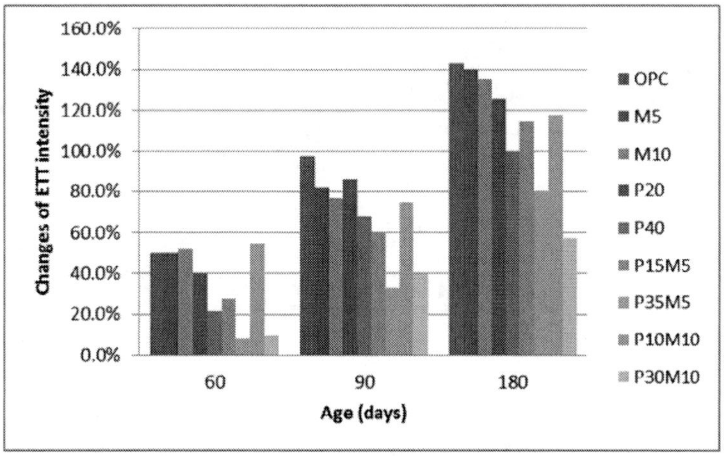

Fig 3. Changes of ettringite intensity of cement pastes exposed to 3 % sodium sulphate solution compared to cement pastes exposed to normal water.

DOI: 10.1051/matecconf/201710301001

The reduction of C-S-H intensity was due to 1) CH reacts with sulphate ions to form gypsum; a reduction of CH contents causes a reduction in the formation of C-S-H 2) formation of ettringite; gypsum reacts with C_3A, CAH and CSAH to form ettringite reduces the calcium contents in cement paste, the reduction of the calcium also leads to reduction of C-S-H contents.

From Fig. 3, the OPC cement paste shows the highest increase in ettringite intensity meanwhile the P30M10 cement paste shows the lowest increase in ettringite intensity after 150 days of exposure. The increased ettringite intensity in cement pastes was due to the presence of sulphate ions in sodium sulphate solution. The calcium hydroxide (CH) reacts with sulphate ion to form gypsum and gypsum reacts with C_3A, CAH and CSAH to form more ettringite. The losses of C-S-H in OPC and P30M10 cement paste can be seen in Fig. 4 and Fig. 5.

The P30M10 cement paste exhibited the highest reduction losses in CH content in sulphate solution, but portrayed a much higher durability compared to the other cement pastes based on the test results in Fig. 1. The P30M10 underwent high formation of C-S-H which contributed to the strength of the P30M10 cement paste. The P30M10 cement paste also recorded the lowest formation of ettringite which also contributed to higher durability performance compared to other cement paste.

The OPC cement paste showed the highest CH content compared to the other cement pastes and suffered a reduction of strength, moreover the OPC cement paste demonstrated lower durability against sulphate attack. High CH contents of the OPC cement paste caused high formation of gypsum and ettringite thus promoting deterioration to the OPC cement paste after 150 days of exposure to sodium sulphate.

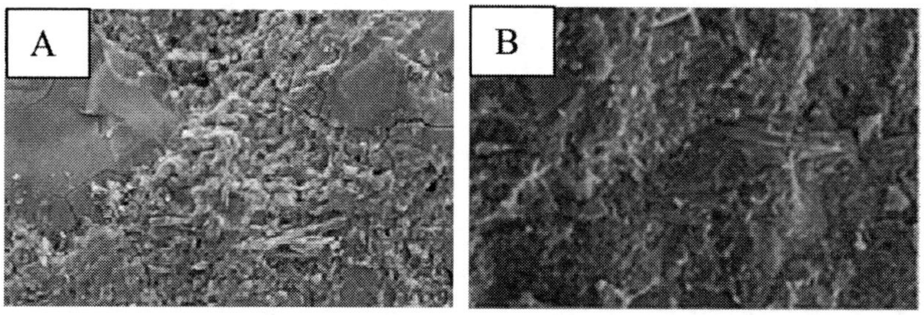

Fig 4. OPC cement paste before (A) and after (B) exposure to 3% sodium sulphate solution.

Fig 5. P30M10 cement paste before (A) and after (B) exposure to 3% sodium sulphate solution.

Overall, high of C-S-H contents contributes to high strength properties and high durability against sulphate attack. The losses of C-S-H contents in the cement paste cause by the formation of gypsum and ettringite which result in high formation of internal stress in the cement paste thus deteriorate the cement paste and reduces the strength of the cement paste. The use of pozzolan lowers the CH content and increases the C-S-H in the cement paste thus reduces the formation of gypsum and ettringite in the cement paste and increases the durability of the cement paste when exposed to sulphate solution.

4 Conclusions

The C-S-H content in the TBCP containing POFA and MK was higher than the OPC cement paste and the BBCP containing POFA and MK resulting from cement hydration and pozzolanic reaction and lower formation of gypsum and ettringite. As a result, the TBCP containing POFA and MK is more durable than the OPC cement paste and the BBCP containing POFA and MK with the higher strength. Furthermore, the P30M10 cement paste is the best and the most suitable cement paste for sulphate environment with the highest resistence against sulphate attack compared to the other tested specimens.

The authors gratefully acknowledge the Universiti Sains Malaysia for providing the needed support including financial support through the Research University (1001/PAWAM/814191) Grant Scheme for undertaking the research work.

References

[1] S.K. Faisal, J.M. Irwan, N. Othman, M.H. Wan Ibrahim, Flexural toughness of ring-shaped waste bottle fibre concrete, MATEC Web of Conference, **47**, 01002, (2016)
[2] S.K. Faisal, J.M. Irwan, N. Othman, W. Ibrahim, R.M Asyraf, M.M.K. Annas, Performance of concrete using light waste PET fibre, Advance Materials Research, **795**, 352-355, (2014)
[3] J.M. Irwan, M.M.K Annas, A.K. Aeslina, R.M. Asyraf, S.K. Faisal, Cracking propagation of reinforced concrete using polyethylene terephtalate (PET) bottles as fine aggregate, Advanced Materials Research, **911**, 474-478, (2014)
[4] J.J. Brooks and M.A. Megat Johari, Effect of metakaolin on creep and shrinkage of concrete, Cement Concrete Composition, **23**, 495-502, (2001)
[5] H.S. Kim, S.H. Lee and H.Y Moon, Strength properties and durability aspects of high strength concrete using Korean metakaolin, Construction and Building Materials, **21**, 1229-1237, (2007)
[6] W. Kroehong, T. Sinsiri, C. Jaturapitakkul, P. Chindaprasirt, Effect of palm oil fuel ash fineness on the microstructure of blended cement paste, Construction and Building Materials, **25**, 4095-4104, (2011)
[7] M.A. Megat Johari,A.M. Zeyad, N. Muhamad Bunnori and K.S. Ariffin, Engineering and transport properties of high-strength green concrete containing high volume of ultrafine palm oil fuel ash, Construction and Building Materials, **30**, 281-288, (2012)
[8] E.E. Hekal, E. Kishar, and H. Mostafa, Magnesium sulphate attack on hardened blended cement pastes under different circumstances, Cement Concrete Research, **32**, 1421-1427, (2002)
[9] M.S. Hasan, *Prediciting life expectancy of concrete septic tanks exposed to sulphate and biogenic sulphuric acid attack*, Master Thesis, RMIT University, Australia, (2009)
[10] P.J. Ramadhansyah, *Properties of concrete containing rice husk ash under aggressive environments subjected to wetting and drying*, PhD Thesis, Universiti Sains Malaysia, Malaysia, (2012)
[11] A.A. Ramezanianpour and H. Bahrami Jovein, Influence of metakaolin as supplementary cementing material on strength and durability of concretes, Construction and Building materials, **30**, 470-479, (2012)
[12] S.M.A. El-Gamal, M.S. Amin and M. Ramadan, *Hydration characteristic and compressive strength of hardened cement paste containing nano-metakaolin*, Housing

and Building National Research Centre, accepted on 4 November 2014 http://ees.elsevier.com/hbrcj
[13] M.W. Hussin and A.S.M.A. Awal, Palm oil fuel ash-a potential pozzolanic material in concrete construction, *Int. Conference on Urban Engineering in Asian Cities in the 21st Century*, Bangkok, Thailand, (1996)
[14] C. Jaturapitakkul, K. Kiattikomol, W. Tangchirapat and T. Saeting, Evaluation of the sulphate resistance of concrete containing palm oil fuel ash, Construction Building Materials, **21**(7), 1399-405, (2007)
[15] J.M. Khatib and S. Wild, Pore size distribution of metakaolin paste, Cement Concrete Research. **26**(10),1545-1553, (1996)
[16] P.K. Mehta, *Concrete: Structure, properties and materials*, Prentice-Hall International New Jersey, USA, (1986)
[17] M.A. Megat Johari, J.J. Brooks, Shahid Kabir and Patrice Rivard, Influence of supplementary cementitious materials on engineering properties of high strength concrete, Construction and Building Materials, **25**, 2639-2648, (2011)

Mechanical Properties of Concrete Using Eggshell Ash and Rice Husk Ash As Partial Replacement Of Cement

Nurul Shahadahtul Afizah Asman[1,*], *Salinah* Dullah[1] , *Janice* Lynn Ayog[1], *Adriana* Amaludin[1], *Hassanel* Amaludin[1], *C. H.* Lim[1] , and *Aslina* Baharum[2]

[1]Faculty of Engineering, Universiti Malaysia Sabah, Jalan UMS, 88400, Kota Kinabalu, Sabah, Malaysia
[2]Faculty of Computer and Informatics, University Malaysia Sabah, Jalan UMS, 88400 Kota Kinabalu, Sabah, Malaysia

Abstract. This research was carried out to determine the optimum percentage of eggshell ash and rice husk ash (RHA) as partial cement replacement. The samples were tested for its mechanical properties by using concrete grade G30 with cube mould (100 mm x 100 mm x 100 mm) and prisms (100 mm x 100 mm x 500 mm). The samples were mixed with eggshell ash and RHA admixture with different proportions (2%:8%, 4%:6%, 6%:4%). Several types of test were conducted towards the samples, which are the slump test, compressive and flexural test. Based on previous researches, the strength of concrete reduced as replaced with eggshells. Most of the researches show the similar trend when partial cement is replaced using eggshell ash. Thus, to increase the strength, an admixture which has pozzolanic reactivity called rice husk ash (RHA) is introduced into mix design which has been proved can help to improve the strength of concrete.

1 Introduction

Concrete is commonly utilized as a material to construct building structures such as beam, column and slab due to its sustainability of carrying large loads. Nevertheless, concrete strength has its limit when it comes to excessive force exerted that may lead to concrete failure. In order to overcome the problem, identification of various researches on discovery of new materials has been made. This research addresses the effects on the strength of concrete by using eggshells as partial cement replacement with proper proportion in concrete mix.

Thereunto, a growing concern worldwide of reducing and reusing waste materials as resources for producing renewable material which have the possibility to be counted on as an alternative or substitute to the existing material has become awareness. This may assist in providing cooperation to any parties involved in reducing the amount of waste as well as to protect the environment. Also, increasing demand for cement in recent days has led to a

* Corresponding author: nurulshahadahtul@gmail.com

boost production of cement in the world. The increase amount was predicted from 2283 million tones in 2005 to about 2836 million tones in the year of 2010 due to emerging economics and developing countries particularly in Asia countries with nearly produced 70% of the world production. The objectives of conducting this research are to determine the optimum percentage of eggshell ash and rice husk ash (RHA) as partial cement replacement. Also, determine the mechanical properties of concrete by comparing the performance of the concrete with the conventional concrete. The scope of this research is limited to partial replacement of cement using recycle wastes such as eggshell and rice husk ash.

2 Effects of eggshell ash and rice husk ash (RHA)

According to Okonkwo et al. [1], eggshell is mainly composed of calcium compounds that is very similar to that of cement and presented as being composed of 93.70% calcium carbonate. Thus, it is an alternative that enables the cement to be replaced with eggshell. It is an advantage where the amount of cement usage can be reduced at the same time saves budget for any project. Calcium oxide is produced when calcium carbonate is incinerated to ash at temperature 500°C and provides as a good accelerator due to extra calcium oxide in it. Calcium oxide also improved the setting time, as minimizing the length of setting time is much desirable in construction industry since it can be interrupted by unpredictable weather especially in the rainy season. Several researches conducted before also discovered the similar trend reduction in strength. For example, cement replacement by using timber waste ash and potential of wood waste ash as an addictive in concrete [2-3]. Assessments made by these researchers, were focused on mechanical strength, durability and shrinkage properties. However, the results indicated that as the number of percentage of cement being replaced, compressive strength of mortar decreased. Not only that, a remarkable decrease of air permeability was observed.

Most of researches state similar reduction in strength of concrete, another material has to be added together into the mix that has give positive impacts to concrete. Jayasankar [4] stated that the combination of fly ash, rice husk ash and eggshell powder was confirmed has equal strength with the conventional concrete. Thus, rich husk ash is chosen as an admixture to be added into the mix at the same time without changing the eggshell ash as partial cement replacement.

Rice Husk Ash (RHA) helps to improve the pozzolanic reactivity of cement. This pozzolanic reactivity refers to the high content of amorphous silica and to a very large surface area governed by the porous structures of the particles. The fineness of the RHA also assists in the reactivity by providing a larger surface area for a better reaction.

Researchers Cheah and Ramli [5] also stated that the improvement in the compressive strength and flexural strength of cement mortar with the use of high calcium wood ash (HCWA) in combination with densified silica fume (DSF) is due to the higher rate of pozzolanic reaction between the amorphous silica mineral of DSF and Portland cement contributed by HCWA. It is similar when eggshell ash, which consists of mainly of calcium to be reacted with RHA that content of pozzolanic amorphous silica.

According to Chao-Lung et al. [6], by adding the ground RHA up to 20%, concrete can be strengthen and enhance the durability properties of concrete. It improved the compressive strength, resistance to electricity and ultrasonic pulse velocity. Furthermore, as the number of curing day increases, higher strength and resistivity are gained.

Meanwhile, according to research done by Sensale [7] stated that by replacing Portland cement by RHA proven to enhance the performance of the concrete. The physical effect is linked to particle size that produces a refinement on the pore structure. But the degree of fineness of RHA has certain level that is allowable to be achieved. During the grinding of

RHA, high degree of fineness should be avoided. This is because the derivation of pozzolanic activity is mainly come from the internal surface area of the particles. It is suggested that at least about 75% of the RHA particles are in a size range at from about 4 μm to about 75 μm [7].

These show that there is a linear relationship as the percentage of RHA added into concrete mix increases, so does the compressive strength compared to conventional concrete. Simultaneously, a durability study of high strength of concrete mixed with part replacement of cement by RHA should be conducted along with its economic aspects.

3 Methodology

Laboratory experiments were conducted by using main ingredients of concrete mix design, cement, fine aggregate, coarse aggregate and water. Meanwhile, eggshell ash and RHA were used as partial substitutes for cement in the mix design. After that, tests were conducted such as slump, compressive strength and flexural strength tests. Then the data for all experiments were recorded for the analysis.

Eggshells were collected from restaurants near the school and incinerated into ash with temperature about 500°C, thoroughly ground and sieved through 75 μm sieve. In the meantime, RHA was obtained from a factory that was already burnt through combustion process.

The mix design is based on grade G30 (1:2:3). The 100 mm x 100 mm x 100 mm cubes and 100 mm x 100 mm x 500 mm prisms were used to observe the strength at 7, 21 and 28 days of curing.

The test specimens were mixed with eggshell ash and RHA admixture added. Partial cement will be replaced using eggshell ash with RHA during mixing the concrete design according to determined proportions. Table 1 shows the proportions of cement that is partially mixed with eggshell ash and RHA for each test.

Laboratory experiments were conducted by using main ingredients of concrete mix design, cement, fine aggregate, coarse aggregate and water. Meanwhile, eggshell ash and RHA were used as partial substitutes for cement in the mix design. After that, tests were conducted such as slump, compressive strength and flexural strength tests. Following that, the data for all experiments were recorded for the analysis.

Eggshells were collected from restaurants near the school and incinerated into ash with temperature about 500°C, thoroughly ground and sieved through 75 μm sieve. In the meantime, RHA was obtained from a factory that was already burnt through combustion process.

The mix design is based on grade G30 (1:2:3). The 100 mm x 100 mm x 100 mm cubes and 100 mm x 100 mm x 500 mm prisms were used to observe the strength at 7, 21 and 28 days of curing.

The test specimens were mixed with eggshell ash and RHA admixture added. Partial cement will be replaced using eggshell ash with RHA during mixing the concrete design according to determined proportions. Table 1 shows the proportions of cement that is partially mixed with eggshell ash and RHA for each test.

Table 1. Proportions of cement, eggshell ash and RHA.

	Cement	Eggshell ash	RHA
Normal concrete G30	100%	-	-
Concrete A	90%	2%	8%
Concrete B	90%	4%	6%
Concrete C	90%	6%	4%

4 Results

4.1 Slump test

Based on ASTM C 143, concretes with replacement eggshell and RHA is more workable compared to normal concrete G30, as shown in Fig. 1. Normal concrete G30 and concrete with 2% eggshell ash and 8% RHA are considered less workable and have high tendency to segregate. Concrete with 4% eggshell height of slump continues to increase as shown by concrete with 6% eggshell ash and 4% RHA replaced with 6% RHA is considered workable in this result.

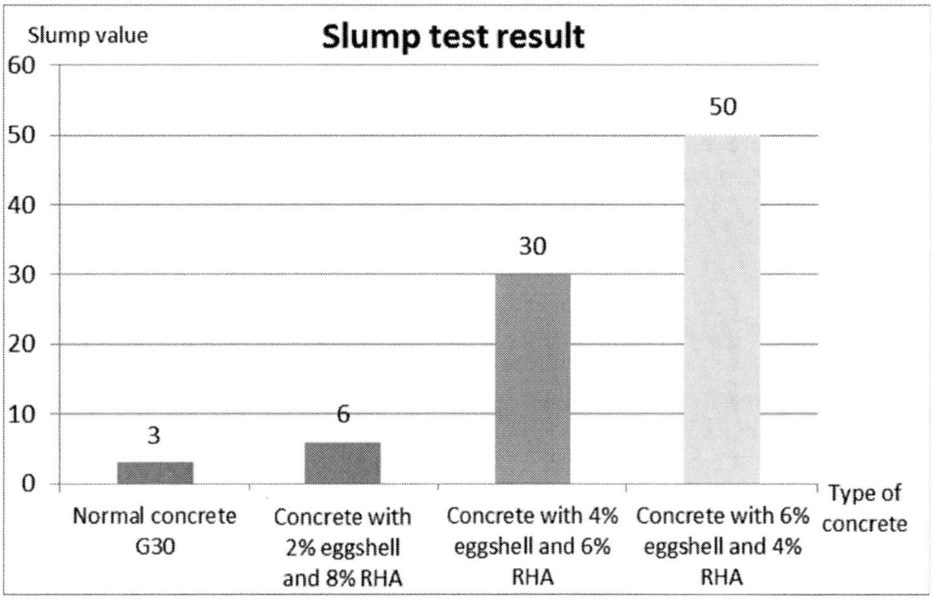

Fig. 1. Workability of Normal Concrete G30, Concrete with Addition of Eggshell and RHA.

4.2 Compressive strength

For compressive strength, replacement of eggshell ash and RHA reduced the strength. Nevertheless, strength of concrete is developed at 28 days of curing. In addition, as the percentage of replacement eggshell ash increased along with reduction in percentage of RHA. The result for flexural strength also almost shows a proportional performance same as compressive strength as shown in Fig. 2. However, for concrete with 6% eggshell ash and 4% RHA the flexural strength reduced.

4.3 Flexural strength

Based on the result it can be considered that concrete with replacement of 4% eggshell ash and 6% RHA is the best optimum usage with a targeted slump between 10mm to 30mm and improved strength development in flexural between 7 to 28 days of curing although its compressive strength is lower that concrete with 6% eggshell ash and 4% RHA (Fig. 3).

DOI: 10.1051/matecconf/201710301002

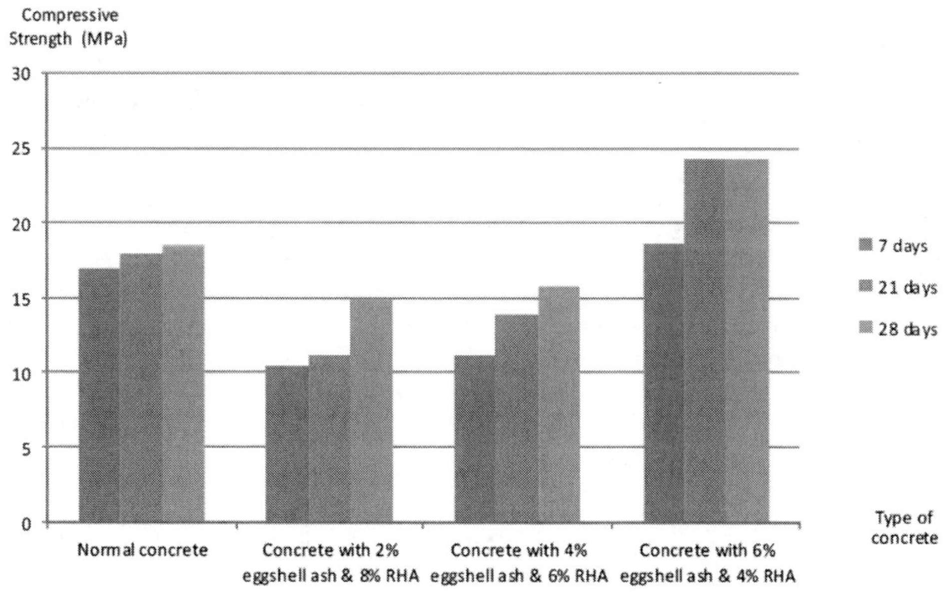

Fig. 2. Compressive strength of normal concrete G30, concrete with addition of eggshell and RHA.

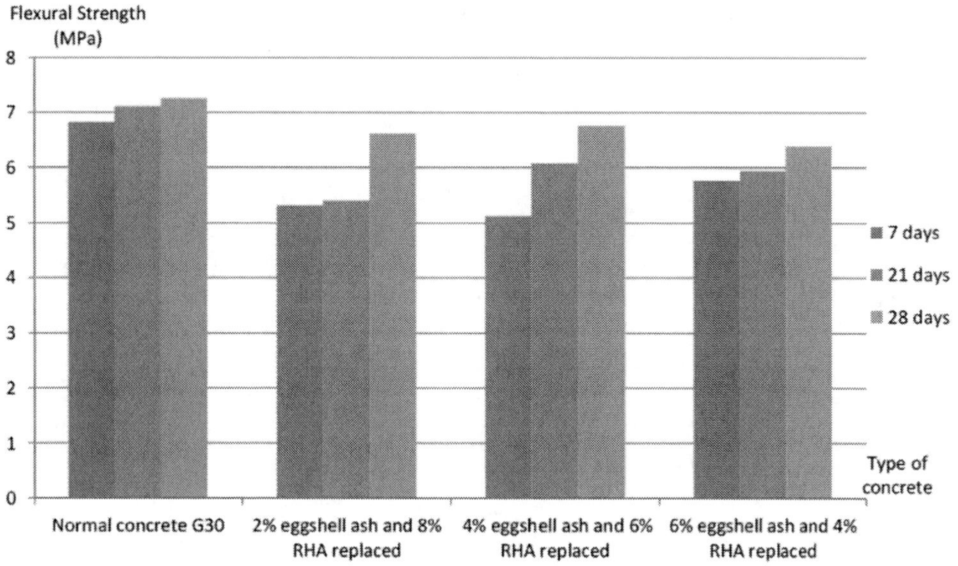

Fig. 3. Flexural strength of normal concrete G30 and concrete with addition of eggshell ash and RHA.

4.4 Comparison with conventional concrete

When it comes to comparing the obtained results between conventional uses based on workability, the concrete with 2% eggshell ash and 8% RHA can be used for road. Meanwhile, concrete with 4% eggshell ash and 6% RHA is suitable for road and light reinforcement. Other than that, 6% eggshell ash and 4% RHA can be applied for flat slab and heavy reinforcement.

5 Conclusion

This research results shows workability in slump test can only be achieved with the mixing of 4% eggshell ash and 6% RHA with is reaching to 30mm slump. Meanwhile, for the normal concrete G30 and concrete with 2% eggshell ash and 8% RHA are less workable with 3mm and 6mm slump respectively. Concrete with 2% eggshell and 8% RHA is classified as very low degree of workability with 6mm slump and can be applied for making road. Meanwhile, Concrete with 4% eggshell ash and 6% RHA is classified as low degree of workability with 30mm slump and can be applied for making road and light reinforcement of concrete. Lastly, concrete with 6% eggshell ash and 4% RHA is classified as medium degree of workability with 50mm slump and can be applied for making road and heavy reinforcement of concrete.

Overall, for compressive and flexural strength of the concrete with eggshell ash and RHA replaced gains lower strength compared to normal concrete using fully OPC. However, the strength developed from day 7 to day 28 due to pozzolanic effect that refers to high content of amorphous silica that gives positive effect to the strength of concrete.

The authors of this paper would like to express their gratitude and sincere appreciation to University Malaysia Sabah for providing the financial support.

References

[1] U.N. Okonkwo, I.C. Odiong, and E.E. Akpabio, The effects of eggshell ash on strength properties of cement-stabilized lateric, Int. J. of Sustainable Construction Engineering and Technology, **13**(1),18-25 (2012)

[2] A.U. Elinwha and Y.A. Mohmood, Ash from timber waste as cement replacement material, **24**(2), 219-222 (2012)

[3] F. Udoeyo, H. Inyang, D. Young, and E. Oparadu, Potential of wood waste ash as an additive in concrete, J. Mater. Civ. Eng., **18**(4), 605–611 (2006)

[4] R. Jayasankar, N. Mahindran, and R. Ilangovan, Studies on concrete using fly ash, rice husk ash and egg shell powder, *Int. J. of Civil and Structural Engineering*, Integrated Publishing Services, 1(3), 362-372, (2010)

[5] C.B. Cheah and M. Ramli., Mechanical strength, durability and drying shrinkage of structural mortar containing HCWA as partial replacement of cement, Construction and Building Materials, **30**, 320-329 (2011)

[6] H.C. Lung, B.L. Tuan, and C.C. Tsun, Effect of rice husk ash on the strength and durability characteristics of concrete, Construction and Building Materials, **25**, 3768-3772 (2011)

[7] G.R. de Sensale, Effect of rice husk ash on durability of cementitious materials, Cement and Concrete Research, **32**,718-725 (2010)

The Effects of Design Strength, Fly Ash Content and Curing Method on Compressive Strength of High Volume Fly Ash Concrete: A Design of Experimental

Mochamad Solikin[1,*], and *Budi* Setiawan[1]

[1]Civil Engineering Department, 57102, Universitas Muhammadiyah Surakarta, Jl. A. Yani Tromol Pos 1 Pabelan Surakarta, Jawa tengah, Indonesia

Abstract. High volume fly ash concrete becomes one of alternatives to produce green concrete as it uses waste material and significantly reduces the utilization of Portland cement in concrete production. Although using less cement, its compressive strength is comparable to ordinary Portland cement (hereafter OPC) and the its durability increases significantly. This paper reports investigation on the effect of design strength, fly ash content and curing method on compressive strength of High Volume Fly Ash Concrete. The experiment and data analysis were prepared using minitab, a statistic software for design of experimental. The specimens were concrete cylinder with diameter of 15 cm and height of 30 cm, tested for its compressive strength at 56 days. The result of the research demonstrates that high volume fly ash concrete can produce comparable compressive strength which meets the strength of OPC design strength especially for high strength concrete. In addition, the best mix proportion to achieve the design strength is the combination of high strength concrete and 50% content of fly ash. Moreover, the use of spraying method for curing method of concrete on site is still recommended as it would not significantly reduce the compressive strength result.

1 Introduction

Concrete, composed of cement, aggregate, water and optional admixture, is the main material for construction and its consumption becomes a second large material of human use after water [1, 2]. Almost all of construction elements need concrete not only for their structural elements such as slab, beam, column and foundation but also for non-structural element such as concrete wall and roof tile. Moreover,

* Corresponding author: msolikin@ums.ac.id

concrete is used for all types of construction such as building, road pavement, and bridge. Therefore, the provision of concrete needs a large amount of material.

The demand of large material to produce concrete raise many ideas of alternative materials as part of replacement of concrete ingredient. This replacement is not only to reduce the cost of production but also to save the environment. The reduction of cost production is due to the fact that the alternative material is usually taken from waste material. Moreover, the use of waste material results in the reduction of energy as there is no need to produce new material, which means that the replacement concrete is environment-friendly material.

One of the most preferable material as cementitious or part of cement replacement is fly ash, a by-product of combustion in coal fire power plant [3, 4]. In 2010, the annual production of coal is 780 million metric tonnes, which further produces 117 million metric tonnes of fly ash or 3.3% of global cement consumption for concrete [5-7] [3-5] . One of ideas to use significant amount of fly ash as cementitious in concrete production is termed as high volume fly ash concrete in which the maximum amount of cement used is 50% as the other binder utilized class C or class F fly ash [8] [6]. The technology has proven to give much advantage to hardened concrete properties and has comparable strength to OPC concrete [9-11].

To get comparable strength to OPC (Ordinary Portland Cement), previous researcher such as Shaikh, Supit and Parker applied high volume fly ash concrete for high design strength concrete in which the total amount of binder is quite higher than its application on normal strength concrete [12-14]. For example: at 56 days testing age, 42 MPa compressive strength of High Volume Fly Ash Concrete is similar to 42 compressive strength of OPC concrete in which the cement content is twice of that of HVFA concrete. In addition, the more fly ash is used in High volume fly ash concrete, the lower the compressive strength becomes [15]. Moreover, the curing regime becomes a factor which gives effect to the compressive strength result of high volume fly ash concrete based on comparison of water immersing and steam curing method [16]. However, daily spraying becomes the most available curing method for high volume fly ash concrete which is cast on site.

Having discussed related literature, it can be noted that the design strength, the amount of cement replacement and the curing method becomes influential factors which can give effects to compressive strength of high volume fly ash concrete. Hence, this paper presents an investigation report of the effect of those factors to the compressive strength of concrete by conducting a series of specimen test. The design of Experimental (DOE) is the method which is used to prepare and analysed the specimens by utilizing Minitab 17, a design of experimental software.

2 Research method

This research employed experimental program to analyse the effects of design strength, fly ash content and curing method on compressive strength of high volume fly ash concrete. The factors to be analysed are shown in Table 1, in which high and low value refer to different methods of application. The table shows that the design experiment for each factor consists of two values, low value and high value. The design strength comprises 25 MPa and 45 MPa design strength; the fly ash content consists of 50% and 70% and the curing method consists of fully immersion and spraying method. Fly ash content of 70% means that the concrete binder is composed of 30% cement and of 70% fly ash. Moreover, in spraying curing method, the concrete specimen was regularly sprayed twice a day for the period of 14 days.

Table 1. Composition of each factor.

No	Factor	Low value	High value
1	Design strength	25 MPa	45 MPa
2	Fly ash content	50%	70%
3	Curing method	Fully immersion	Spraying

The composition of each factor examined were inputted into Minitab 17, a statistical software to conduct design of experimental. The result of mix proportion combination for full factorial design is shown in the following picture:

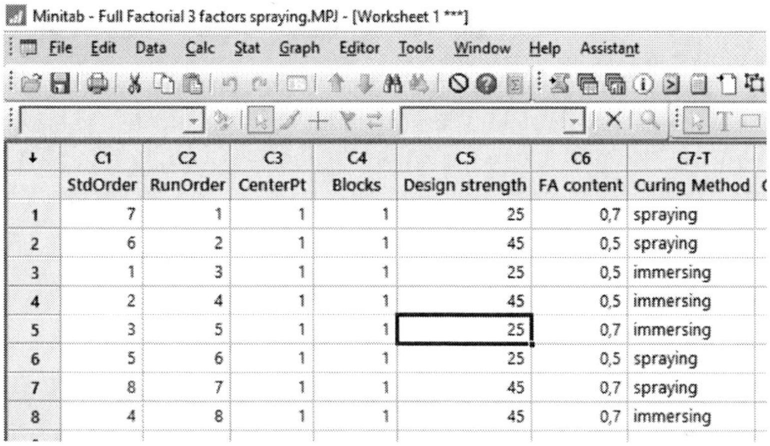

↓	C1 StdOrder	C2 RunOrder	C3 CenterPt	C4 Blocks	C5 Design strength	C6 FA content	C7-T Curing Method
1	7	1	1	1	25	0,7	spraying
2	6	2	1	1	45	0,5	spraying
3	1	3	1	1	25	0,5	immersing
4	2	4	1	1	45	0,5	immersing
5	3	5	1	1	25	0,7	immersing
6	5	6	1	1	25	0,5	spraying
7	8	7	1	1	45	0,7	spraying
8	4	8	1	1	45	0,7	immersing

Fig. 1. Design of experiment combination.

Based on mix proportion combination result from Minitab 17, a series of concrete cylinder specimens were prepared with the diameter of 15 cm and the height of 30 cm. The total number of specimens were 24 concrete cylinders. Besides, the composition of mix proportion for the cylinder was prepared based on ACI mix proportion for normal strength and for high strength. For high volume fly ash concrete, fly ash is utilized as part of cement replacement and the mix proportion is shown in the flowing table.

Deploying ACI design of normal strength concrete of 25 MPa, the mix proportion used 376 kg binder to produce 1 m^3 of concrete. Therefore, HFVA (High volume fly ash) used 188 kg cement and 188 kg fly ash, whereas UHFVA (Ultra High volume fly ash) used 113 kg of cement and 264 kg of fly ash. The mix proportion of fine aggregate and coarse aggregate remained the same for both HFVA and UHFVA. Since High volume fly ash concrete used low water content of 0.35, 0.4% of sika *viscocrete* 1003 to the total binder was used. Similar to the mix proportion of normal design strength, high strength concrete used 537 kg of total binder which was proportionally distributed to Portland cement and fly ash.

Table 2. Mix proportion of concrete.

Mix Design	Cement (kg/m^3)	Fly ash (kg/m^3)	Fine Aggregate (kg/m^3)	Coarse Aggregate (kg/m^3)	Water (kg/m^3)	SP
Normal strength HFA	188.0	188.0	637.0	1,250.0	128	0,4%
Normal strength UHVFA	113.0	264.0	637.0	1,250.0	128	0,4%
High strength HVFA	269.0	269.0	613.0	919.0	188	0,4%
High strength UHVFA	173.0	173.0	613.0	919.0	188	0,4%

Before casting the specimen, the materials for concrete were tested to ensure that all materials satisfied the quality of standard material for concrete. The tests for fine aggregate were clay lumps and water absorption tests. In addition, the coarse aggregate was tested for its abrasion resistance. Visual test for cement and chemical content of fly ash was also conducted.

After all materials were tested, they were mixed in a drum rotating mini mixer to achieve the design slump before it was casted into cylinder mould. On the following day, the specimens were removed from the mould and curing method were applied based on curing method regime. At the age of 56 days, the specimens were tested for its compressive strength.

3 Results and discussion

3.1 Materials tested

The clay lump test and water absorption test for fine aggregate show the result value of 2.86% and 1.63%, both of which satisfy the standard of fine aggregate for concrete, whereas the abrasion test for coarse aggregate results in the value of 21.54%, which is below the maximum degradation of coarse aggregate of 40%.

The fly ash which is used in this experiment was produced by PLTU Tanjung Jati Jepara, Central Java and the chemical content analysis was conducted at Laboratorium Pengujian dan Kalibrasi BARISTAND, Surabaya, East Java. As shown in Table 3, the fly ash used in this research is class F Fly ash as the total amount of $SiO_3 + Al_2O_3 + Fe_2O_3$ is 86.36% [17].

Table 3. Chemical properties of fly ash (mass %).

Chemical analysis	Fly ash PLTU Tanjung jati	ASTM Standard for class F
Silika dioxida, SiO_2	66.50	$SiO_2 + Al_2O_3 + Fe_2O_3$ min. 70%
Aluminium oxida, Al_2O_3	16.20	
Fery oxida, Fe_2O_3	3.66	
TiO_2	0.10	
Magnesium oxida, MgO	21.20	
Ca O	16.40	

3.2 Analysing factor

After testing the compressive strength of all concrete cylinder specimens at their age of 56 days, the data were analysed using MINITAB 17. The analysis results i.e. the main effect factor, the significant factor and the size of effect are shown in the following graphs:

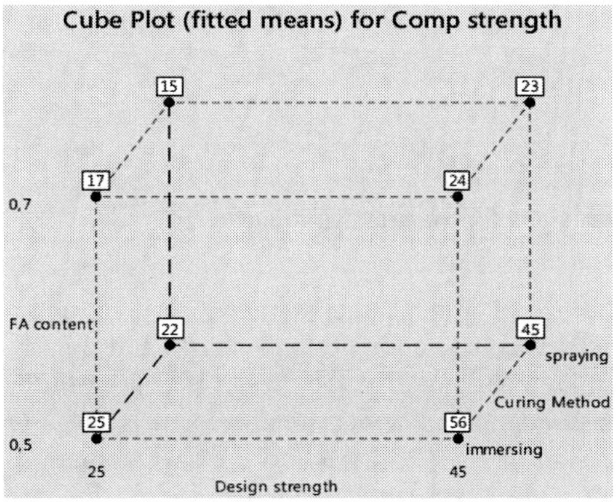

Fig. 2. Cube plot of compressive strength for each mix proportion.

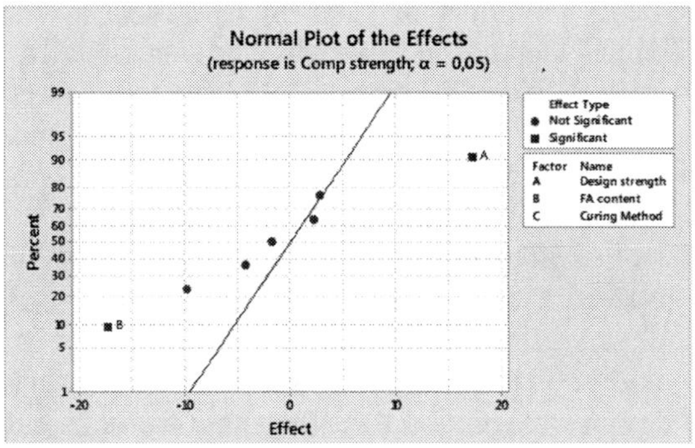

Fig. 3. Significant factor.

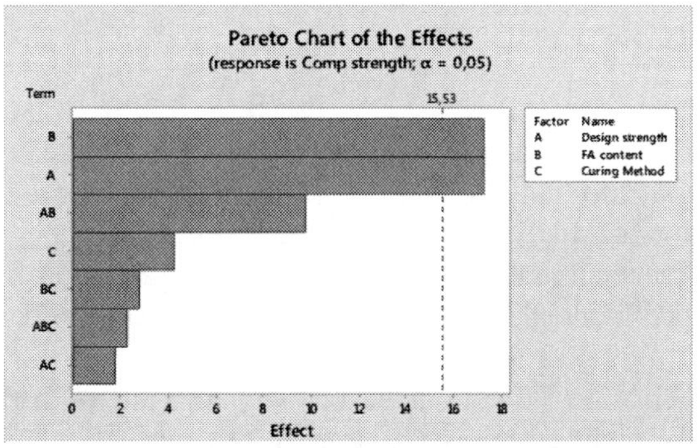

Fig. 4. Size of effect.

 DOI: 10.1051/matecconf/201710301003

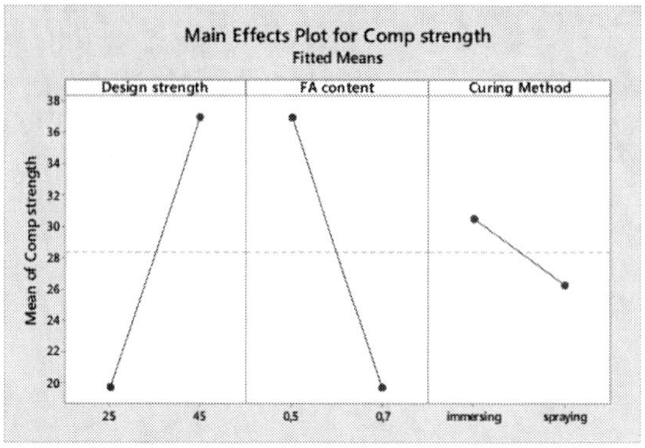

Fig. 5. Average of main plot for compressive strength.

As seen in the above graphs, the factors that give significant effect to the compressive strength of concrete are design strength and fly ash content. The effect can be obtained on the condition that the design strength of the concrete is $f'_c = 45$ MPa and the fly ash content as cement replacement is 50%. On the contrary, the normal strength design of $f'_c = 25$ MPa and the high content of fly ash as cement replacement of 70% are not recommended to be used for the mix proportion, as it is would lead to lower compressive strength.

The higher cement content in high design strength concrete significantly leads to more $Ca(OH)_2$ content, a product of cement hydration, which then reacts with silica, a main chemical content in fly ash, and beneficially results in better hydration process in High Strength concrete. Nevertheless, lower cement content in combination with high level fly ash content in high strength concrete leads to the lower compressive strength of concrete, a condition which confirms the previous research [15].

Those graphs also reveal that curing method of fully immersing and spraying do not significantly give effect on compressive strength result of high volume fly ash concrete. Nevertheless, as the possible curing method on site is spraying water on the concrete, spraying is still adequate as curing method for concrete.

Further, when design strength and fly ash content factors are used together in the mix design of concrete, they would give more significant effect rather than what the curing method role to get higher or lower compressive strength.

4 Conclusion

The result of the research demonstrates that high volume fly ash concrete can produce comparable compressive strength as design strength especially for high strength concrete. In addition, the best mix proportion to achieve the design strength is the combination of high strength concrete and 50% content of fly ash as replacement. Moreover, the use of spraying method for curing method of concrete on site is still recommended, as it would not significantly reduce the compressive strength result.

Further research must be conducted on why high strength design of concrete mix proportion makes the compressive strength of High volume fly ash better rather than normal strength design. Besides, the effect of curing method on the compressive strength of larger dimension concrete needs to be clarified.

This research is made possible because of the *Hibah Bersaing* grant from Directorate Research and Community Service (*Ditlitabmas*) Ministry of Research and Higher Education Indonesia under the contract number of 186.15/A3-III/LPPM/V/2016. The authors wish to acknowledge the support given by technical staff in PT Pioner Beton Surakarta and concrete laboratory at Universitas Muhammadiyah Surakarta as well as the assistance of Qanitah Masykuroh in proofreading.

References

[1] V. Atakan, J. Jain, D. Ravikumar, L. McCandlish and N. DeCristofaro, *Technologies that Reduce Water Use in Cement and Concrete Help Global Industry Address Mounting Concerns of Water Scarcity,* (2016), Retrieved on October 12, 2016 from http://solidiatech.com/wp-content/uploads/2016/08/ Addressing-Water-Scarcity-in-Cement-Concrete-Industries-Solidia Technol ogies-August-2016.pdf

[2] L. J. Murdock, J. D. Dewar and K. M. Brook, *Concrete materials and practice*, Edward Arnold, London, UK, (1991)

[3] S. H. Kosmatka, B. Kerkhoff, and W. C. Panares, *Design and Control of Concrete Mixtures 14th edition*, Portland Cement Association, Illinois, USA, (2003)

[4] E. G. Nawy, *Fundamentals of High Strength High Performance Concrete*, Longman Group Limited, London, UK, (1996)

[5] E. Safitri and Djumari, Kajian Teknis dan Ekonomis Pemanfaatan Limbah Batu Bara (Fly Ash) pada Produksi Paving Block, J. of Media Teknik Sipil, **9**(1), 36-40 (2009)

[6] C. Heidrich, H. J. Feuerborn and A. Weir, Coal Combustion Product: A Global Perspective, *Proc. of 2013 World of Coal Ash (WOCA) Conference*, Lexington, KY, United States, (2013)

[7] Mineral resources program, *Cement*, 2015, Retrieved on August 20, 2016 from https://minerals.usgs.gov/minerals/pubs/commodity/cement/mcs-2015-cemen.pdf

[8] V. M. Malhotra and P. K. Mehta, *High Performance, High-Volume Fly Ash Concrete: materials, mixture proportioning, properties, construction practice, and case histories. 2nd ed.*, Supplementary Cementing Materials for Sustainable Development Inc., Ottawa, Canada, (2005)

[9] A. Bilodeau and V. M. Malhotra, High-Volume Fly Ash System: Concrete Solution for Sustainable Development, ACI Materials J., **97**(1), 41-50 (2000)

[10] M. Solikin, S. Setunge, and I. Patnaikuni, The Influence of Lime Water as Mixing Water on The Compressive Strength Development of High Volume Ultra Fine Fly Ash Mortar, *Proc. of The Sixth International Structural Engineering and Construction Conference (ISEC-6)*, Zurich, Switzerland, (2011)

[11] M. Solikin, S. Setunge, and I. Patnaikuni, Experimental Design Analysis of Ultra Fine Fly Ash, Lime Water, and Basalt Fibre in Mix Proportion of High Volume Fly Ash Concrete, Pertanika J. Science & Technology, **21**(2), 589-600 (2013)

[12] F. U. A. Shaikh, S. W. M. Supit, and P. K. Sarker, A study on the effect of nano silica on compressive strength of high volume fly ash mortars and concretes, J. Materials & Design, **60**(1), 433-442, (2014)

[13] F. U. A. Shaikh and S. W. M. Supit, Compressive strength and durability properties of high volume fly ash (HVFA) concretes containing ultrafine fly ash (UFFA), J. of Construction and Building Materials, **82**(8), 192-205, (2015)

[14] T. F. Prihantoro and M. Solikin, Perkembangan Kuat Tekan Beton Mutu Tinggi dengan Memanfaatkan Teknologi High Volume Fly Ash Concrete, *Proc. of Simposium Nasional Rekayasa Aplikasi Perancangan Industri (RAPI XVI),* Surakarta, Indonesia, (2015)

[15] C.-H. Huang, S.-K. Lin, C.-S. Chang, and H.-J. Chen, Mix proportions and mechanical properties of concrete containing very high-volume of Class F fly ash, J. of Construction and Building Materials, **46**(9), 71-78, (2013)

[16] H. Yazici, S. Aydin, H. Yigiter, and B. Baradan, Effect of steam curing on class C high-volume fly ash concrete mixtures, Cement and Concrete Research, **35**(6), 1122–1127, (2005)

[17] ASTM C 618-03, *Standard specification for fly ash and raw calcined natural pozzolan for use as mineral admixture in portland cement concrete*, Annual Book of ASTM Standard, Philadelphia, United States, (2003)

Reducing Heavy Metal Element from Coal Bottom Ash by Using Citric Acid Leaching Treatment

Ahmad Asyari Yahya [1], *Noorwirdawati* Ali[1,*], *Nur Liyana* Mohd Kamal[2], *Shahiron* Shahidan[1], *Salmia* Beddu[2], *Muhd. Fadhil* Nuruddin[3], and *Nasir* Shafiq[3]

[1]Jamilus Research Center, Faculty of Civil and Environmental Engineering, Universiti Tun Hussein Onn Malaysia, Malaysia
[2]Civil Engineering Department, Universiti Tenaga Nasional, Malaysia
[3]Civil & Environmental Engineering Department, Universiti Teknologi Petronas, Malaysia

Abstract. Coal ash is the residue that is produced during coal combustion for instance fly ash, bottom ash or boiler slag which was primarily produced from the combustion of coal. With growth in coal burning power station, huge amount of coal bottom ash (CBA) considered as hazardous material which are normally disposed in an on-site disposal system without any commercialization purpose. Previous researchers have studied the extraction of silica from agricultural wastes such as palm ash and rice husk ash (RHA) and CBA by using leaching treatment method. In this study, the weaker acid, citric acid solution was used to replace the strong acid in leaching treatment process. Result showed that the heavy metal content such as Copper (Cu), Zinc (Zn) and Lead (Pb) can be decrease. Meanwhile the silica can be extracted up to 44% from coal bottom ash using citric acid leaching treatment under the optimum reaction time of 60 minutes with solution temperature of 60^0C and concentration of citric acid more than 2%.

1 Introduction

Coal bottom ash has a potential to be used in industry. Instead of dumping, the physical and chemical properties of CBA shows that it could be used as a possible applications including as cement replacement, ceramic products and construction materials [1-3]. Previous researchers have proved the existence of silica as a major compound and heavy metal impurities in coal bottom ash [4-5] and therefore, there is a possibility to extract silica from coal bottom ash and at the same time reducing the heavy metal element in coal bottom ash.

Silicon dioxide also known as silica is one of the basic material and the valuable inorganic multipurpose chemical compound. Normally, silica naturally occurred as quartz, sand or flint. It can exist either in gel, crystalline or amorphous forms. It is the most abundant material on the earth's crust. Nowadays, extraction process was used to produce

* Corresponding author: wirda@uthm.edu.my

silica from quartz or sand Sodium silicate, the precursor for silica production is currently manufactured by smelting quartz sand with sodium carbonate at 1300^0C [6].

It is also found that the metallic element such as Copper(Cu), Nickel(Ni), Chromium(Cr), Zinc(Zn) and Lead(Pb) [7, 8] resulting the classification of CBA in Malaysia under the Schedule Waste (SW 104) Environment Quality Act [9]. Even though the utilization of CBA have been practice with positive result, but weathering and erosion over time will cause the heavy metal to leach out. The existence of those elements in the ash could pollute the environment and pose a danger to public health [10] and therefore, an effective treatment need to be done to overcome the problem.

Leaching treatment show a potential route in silica extraction and reducing heavy metal compound. From previous research, strong acid such as sulphuric acid (H_2SO_4), hydrochloric acid (HCL) and nitric acid (HNO_3) solution were used conventionally in leaching treatment in extraction of silica material [7, 11] Other than that, sodium oxide (NaOH) and ammonium citrate solution were also being used in leaching treatment as a silica extraction and to reduce the heavy metal impurities [8, 12]. However, these leaching agents is hazardous to environment and human, the strong acid leaching treatment also had an economic problems since the materials are expensive due to the corrosion resistance to strong acid and special disposal treatment of used strong acid.

The objective of this study was to study the possibility in developing new method in extracting silica compound and reducing heavy metal element by using citric acid leaching method.

2 Materials: Coal bottom ash

Tanjung Bin is one of power plant in Malaysia that used coal to produce electricity. It has been reported that, Tanjung Bin power plant needs about 18 000 ton/ day of coal to generate electricity [13]. Coal bottom ash is one of the example of the coal waste product that are produced by the Tanjung Bin power plant. Due to the large volume of waste material, the large quantity of coal ash will be considerable disposal concern because of the increase requirement for ash storage space. Hence, this will increase the expanses to obtain larger area [13, 14] and it could lead to the environmental problem to the future [7]. Table 1 shows that CBA has about 41% of silicon element and the list heavy metal element that exist in coal bottom ash.

Table 1. Chemical composition of untreated CBA.

Compound	Concentration (%)
SiO_2	41.10
Al_2O_3	14.20
Fe_2O_3	6.88
Pb	0.0031
Cu	0.0103
Ni	0.0015
Zn	0.0267
Cr	0.0034

3 Method and procedures: Leaching process

Generally, leaching refers to the disposal of substance from a solid via a liquid extraction media. The desired component diffuses into the solvent from its natural solid form. In leaching process, there are three important parameters which are temperature, contact time per area and selection of solvent. In order to obtain an optimized solubility and mass transfer, the temperature was adjusted accordingly. There are two categories in leaching which are percolation and solid dispersed. Usually, the solid was crushed into small pieces before being contacted with solvent.

Coal bottom ash that was collected from Tanjung bin power plant was used as raw material to produce silica material and eliminating heavy metal compound. Previous research clarified that the citric acid solution leaching treatment can increase the silica content for palm ash [11] while the heavy metal content can be reduce by extraction and leaching treatment [15].

40g coal bottom ash with average particle size of 75 μm was put into 1000 mL beaker that contained citric acid solution. The beaker was placed on the hot plate magnetic stirrer as shown in Fig. 1. The concentration and temperature of the citric acid solution were selected as the operating parameter. The concentration was controlled from 2% to 5% by changing the mixing ratio between citric acid powder (EMORY, Purity $\geq$ 99.5%) and distilled water.

Fig. 1. Heating and stirring process in CBA leaching treatment.

The solution temperature was changed from 30^{0}C to 70^{0}C with reaction time recorded within 60 minutes. After the acid leaching process, the water rinsing treatment of coal bottom ash was conducted with distilled water at room temperature that functions as a removal of excessive citric acid content in the ash as shown in Fig. 2.

The ash then was filtered by using vacuum filtered (GAST, MODEL:DOA-P504-BN) as shown in Fig. 3 to obtain the retained treated ash. After that, the ash was dried at 60^{0}C for 60 minutes in the oven. It was then combusted at 800^{0}C for 30 minutes in the furnace (Carbolite HTF Furnace, Model 1800). The airflow rate in the combustion was set to 2 ml/s. Then, the treated ash containing metallic compound was tested by using X-ray fluorescence spectroscopy (Model XRF BRUKER S4 PIONEER).

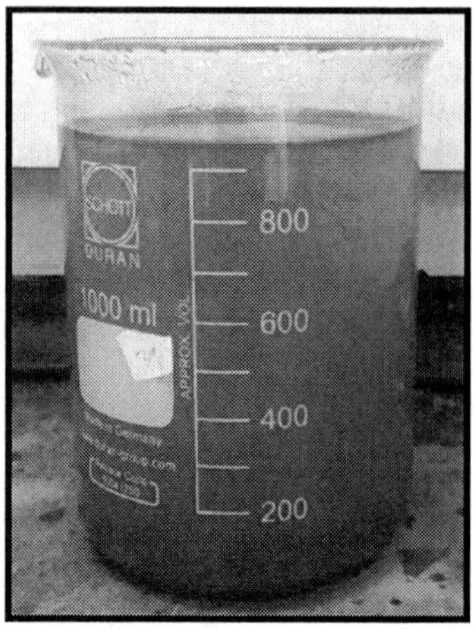

Fig. 2. Water rinsing treatment at room temperature.

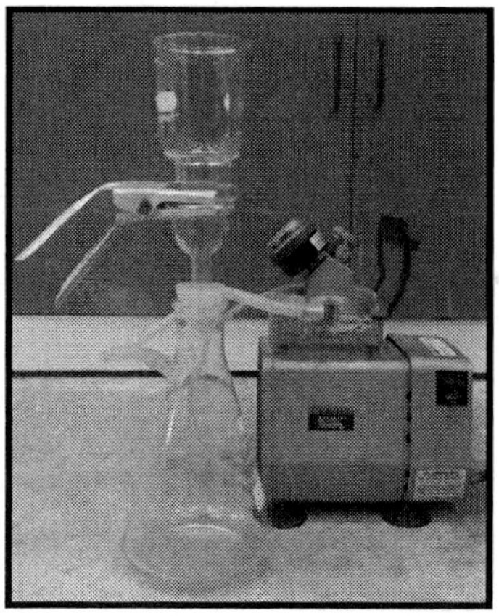

Fig. 3. Image of vacuum filtered (GAST, MODEL: DOA-P504-BN).

4 Results and discussions

In this study, citric acid seems to be as a good alternative to substitute strong acid in leaching process since citric acid reacted as an organic acid which it has less hazardous effect compared to the stronger acid usage. The data was analyses by comparing the result with untreated CBA as shown in table 1. Temperature, time and selection of solvent acted as an important parameter that was considerate in leaching process.

4.1 The effect of solution temperature

Table 2 shows the chemical composition of coal bottom ash after leaching treatment was done under different solution temperature. Some researchers had done leaching experiments at ambient temperature. From the result, the temperature tends to vaporize organic acids and water [16]. The use of high temperature can increase the leaching performance but the use of organic acid with high temperature seen to be limited due to low boiling temperature and decomposition of citric acid. Experimental result from table 2 shows that the metallic content for Pb, Cu, and Zn shows some decrement compared to the untreated CBA specimen. The higher the temperature, the more the percentage of heavy metal was leached out. However, the percentage of Ni and Cr show some increment when higher temperature was used. The silicon oxide (SiO_2) content of treated CBA increase up to 44.5.%. Moreover, the result from table 3 shows the some fluctuation value for the metallic content in treated CBA and the SiO_2 content was increased up to 43% compared to control specimen.

Table 2. The chemical composition of coal bottom ash treated with 2% citric acid by leaching treatment with difference solution temperature.

Solution tempera-ture (^0C)	Sio_2	Al_2O_3	Fe_2O_3	Pb	Cu	Ni	Zn	Cr
30^0C	39.90	15.20	7.16	0.0027	0.0063	0.0077	0.0055	0.0090
40^0C	39.50	14.60	6.88	0.0028	0.0062	0.0131	0.0048	0.0514
50^0C	39.30	14.50	7.77	0.0026	0.0060	0.0126	0.0047	0.0383
60^0C	44.50	15.40	6.10	0.0018	0.0058	0.0073	0.0068	0.0450
Untreated CBA	41.10	14.20	6.88	0.0031	0.0103	0.0015	0.0267	0.0034
	increa-se	increase	mode-rate	decrease	decrease	increase	decrease	incre-ase

Table 3. The chemical composition of coal bottom ash treated with 3% citric acid by leaching treatment with difference solution temperature.

Solution tempera-ture (^0C)	Sio_2	Al_2O_3	Fe_2O_3	Pb	Cu	Ni	Zn	Cr
30^0C	42.60	14.40	6.07	0.0015	0.0056	0.0068	0.0062	0.0096
40^0C	42.20	14.20	6.17	0.0013	0.0054	0.0069	0.0059	0.0171
50^0C	40.80	15.20	7.17	0.0025	0.0059	0.0118	0.0046	0.0357
60^0C	43.40	14.90	5.95	0.0017	0.0055	0.0072	0.0066	0.0113
Untreated CBA	41.10	14.20	6.88	0.0031	0.0103	0.0015	0.0267	0.0034
	incre-ase	moderate	decrease	decrease	decrease	increase	decrease	increase

4.2 The effect of solvent with acid solution

The reaction between solvent and acid solution is an important factor in order to achieve a good result in leaching treatment. Table 4 shows the chemical composition of coal bottom ash after being treated via citric acid solution treatment with different concentration. The temperature of citric acid solution was fixed at 60^0C with reaction time of 60 minutes. From the experimental result at table 6, it shows that the purity of silica (SiO_2) of bottom ash increased about 43 % when citric acid solutions were employed with a concentration of 2% and higher. The heavy metal impurities such as Pb, Cu and Zn show some decrement when the citric acid concentration was increased.

Thus, its shows that citric acid leaching treatment was essential in order to increase the purity of silica (SiO_2) in bottom ash. The percentage of Al_2O_3 compound also seems to be increased compared to untreated CBA sample although the percentage of Al_2O_3 compound decrease with increasing of acid concentration. Apart from that, the leaching treatment help to reduce the heavy metal impurities content compared to untreated CBA except for Ni and

Cr element. The increasing percentage of Cr and Ni may be due to the lack of time reaction during leaching treatment. The color of pulverized coal bottom ash after citric acid leaching treatment was changed from dark grey to brown color as shown in Fig. 3.

Table 4. Chemical composition of coal bottom ash treated with difference concentration of citric acid with temperature of 60^0C.

Acid Concen -tration (%)	**SiO_2**	**Al_2O_3**	**Fe_2O_3**	**Pb**	**Cu**	**Ni**	**Zn**	**Cr**
2	44.50	15.40	6.10	0.0018	0.0058	0.0073	0.0068	0.0450
3	43.40	14.90	5.95	0.0017	0.0055	0.0072	0.0066	0.0113
Untrea-ted CBA	41.10	14.20	6.88	0.0031	0.0103	0.0015	0.0267	0.0034
	increase	increase	decrease	decrease	decrease	increase	decrease	increase

(a) (b)

Fig. 3. Image of coal bottom ash (a) before leaching treatment (b) after leaching treatment.

5 Conclusions

The experimental studies show that there is a possibility to extract silica and at the same time reduce the heavy metal impurities except for Cr and Ni from CBA by using acid leaching method. This could be due to the effect of less time reaction of the treatment. From the result, it shows that the purity of silica was increased up to 43 % with the increasing of acid concentration under extracting temperature of 60^0C and optimum reaction time of 60 minutes. Moreover, although the metallic content for Pb, Cu and Zn are not fully eliminated, but it can be reduce with higher concentration and temperature. But, for the metallic content of Cr and Ni, further studies need to be done to determine the efficient factor of leaching treatment on eliminating the element. Therefore, it can be conclude that by increasing the acid concentration and temperature of the solution may lead to the better result.

This research was supported by Postgraduate Research Grant under Office of Research, Innovation, Commerlization, Consultancy management (ORICC), Universiti Tun Hussein Onn Malaysia (Project Vote No: U444) and Fundamental Research Grant Scheme (FRGS) by Minister of Higher Education (Vote No: FRGS/1/2015/TK06/UNITEN/03/2).

References

[1] S.O. Bada and S. Potgieter-Vermaak, Evaluation and treatment of coal fly ash for adsorption application, Leonardo Electronic J. of Practices and Technologies, 37-48, (2008)

[2] M. Cheriaf, J. C. Rocha, and J. Péra, Pozzolanic properties of pulverized coal combustion bottom ash, Cem. Concr. Res., **2**(9), 1387–1391, (1999)

[3] S. Shahidan, H. B. Koh, A. M. S. Alansi and L. Y. Loon, Strength Development and Water Permeability of Engineered Biomass Aggregate Pervious Concrete, *MATEC Web of Conf.*, 2–7, (2016)

[4] S. Shahidan, I. Isham, and N. Jamaluddin, A Review on Waste Minimization by Adopting in Self Compacting Concrete, *MATEC Web of Conf.*, 1–7, (2016)

[5] M. Tiwari, S. K. Sahu, R. C. Bhangare, P. Y. Ajmal, and G. G. Pandit, Elemental characterization of coal, fly ash, and bottom ash using an energy dispersive X-ray fluorescence technique, Applied Radiation and Isotope, **90**, 53-57, (2014)

[6] U. Kalapathy, A. Proctor and J. Shultz, A simple method for production of pure silica from rice hull ash, Bioresour. Technol., **73**, 257-262, (2000)

[7] C.P. Faizul, C. Abdullah, B. Fazlul, Extraction of Silica from Palm Ash using Organic Acid Leaching Treatment, Key Eng. Mater., **594**, 329–333, (2014)

[8] R. A. Kamarudin, A. S. Matlob, Z. Jubri and Z. Ramli, Extraction of silica and alumina from coal fly ash for the synthesis of zeolites, *Proc. 3rd Int. Conf. on Energy and Environment: Advancement Towards Global Sustainability*, Malacca, Malaysia 456-461(2009)

[9] A. U. Abubakar and K. S. Baharudin, Tanjung Bin Coal Bottom Ash: From Waste to Concrete Material, Adv Mater. Res., **705**, 163-168, (2013)

[10] A. A. Kadir, M. I. H. Hassan, and M. M. A. B. Abdullah, Investigation on Leaching Behaviour of Fly Ash and Bottom Ash Replacement in Self-Compacting Concrete, *IOP Conf. Ser. Mater. Sci. Eng.*, IOP, **133**, 1-12 (2016)

[11] J. Umeda and K. Kondoh, High-purity amorphous silica originated in rice husks via carboxylic acid leaching process, J. Mater. Sci., **43**(22),7084–7090, (2008)

[12] A. Quek, W. Xu, L. Guo, and D. Wu, Heavy Metal Removal from Incineration Bottom Ash through Washing with Rainwater and Seawater, Int. J. Waste Resour., **6**(1) , 2-9 (2016)

[13] A. Muhardi, A. Marto, K. A. Kassim, A. M. Makhtar, L. F. Wei, and Y. S. Lim, Engineering characteristics of Tanjung Bin coal ash, Electron. J. Geotech. Eng., **15**, 1117-1129, (2010)

[14] M. L Jayaranjan, A. P Annachhatre, Precipitation of heavy metals from coal ash leachate using biogenic hydrogen sulfide generated from FGD gypsum., Water Sci. Technol., **2**(67), 311–8, (2013)

[15] N. Ernida, Z. Abidin, M. Haziman, W. Ibrahim, N. Jamaluddin, K. Kamaruddin, A. Farhan Hamzah, The Strength Behavior of Self-Compacting Concrete incorporating Bottom Ash as Partial Replacement to Fine Aggregate, ARPN J. of Eng. & App. Sci., **11**(4), 2570-2575, (2016)

[16] D. B. Sarode, R. N. Jadhav, V. A. Khatik, S. T. Ingle, and S. B. Attarde, Extraction and leaching of heavy metals from thermal power plant fly ash and its admixture, Polish J. Environ. Stud., **19**(6), 1325-1330, (2010)

A Comprehensive Review on the Properties of Coal Bottom Ash in Concrete as Sound Absorption Material

Nurul Izzati Raihan Ramzi Hannan [1], *Shahiron* Shahidan[1,*], *Noorwirdawati* Ali[1], and *Mohamad Zulkhairi* Maarof[2]

[1]Jamilus Research Centre, Faculty of Civil and Environmental Engineering, Universiti Tun Hussein Onn Malaysia, 86400 Parit Raja, Batu Pahat, Johor, Malaysia
[2]Faculty of Engineering Technology, Universiti Tun Hussein Onn Malaysia, 86400 Parit Raja, Batu Pahat, Johor, Malaysia

Abstract. The government is currently implementing policies to increase the usage of coal as fuel for electricity generation. At the same time, the dependency on gas will be reduced. In addition, coal power plants in Malaysia produce large amounts of industrial waste such as bottom ash which is collected in impoundment ponds (ash pond). However, millions of tons of coal ash (bottom ash) waste are collected in ponds near power plant stations. Since bottom ash has been classified as hazardous material that threatens the health and safety of human life, an innovative and sustainable solution has been introduced to reuse or recycle industrial waste such as coal bottom ash in concrete mixtures to create a greener and more sustainable world. Bottom ash has the potential to be used as concrete material to replace fine aggregates, coarse aggregates or both. Hence, this paper provides an overview of previous research which used bottom ash as fine aggregate replacement in conventional concrete. The workability, compressive strength, flexural strength, and sound absorption of bottom ash in concrete are reviewed.

1 Introduction

Major environmental problems around the world are usually due to the disposal of waste materials such as industrial waste, construction waste, household waste and etc. However, industrial waste, also known as production by-products, has been investigated extensively as sustainable alternatives to Portland cement in concrete [1][47]. Recently, waste accumulation has become a major problem to the environment as well as human beings [2]. It is now a global concern to find environmentally friendly solutions for the safe disposal of industrial waste to sustain a cleaner and greener environment.

Coal ash is one of the biggest sources of industrial waste that is produced from power plant stations. Cheriaf et al. [3] stated that 1.2 million metric tons of coal ash has been produced during the combustion of 2.9 million metric tons of coal. Coal is widely used in

* Corresponding author: shahiron@uthm.edu.my

the production of electricity, steel and cement manufacturing and is therefore an important source of energy. In Malaysia, coal has classified as an significant source of fuel [4, 5]. Nowadays, fossil fuels become the foremost basis for energy generation [6]. The demand for fossil fuels has increased due to a high demand for electricity generation, as well as the consumption of fossil fuels by several countries such as USA, EU, India, and China [7-9].

However, based on Bajare et al. [10], several types of coal ash waste have been identified such as bottom ash, fly ash, boiler slag, flue gas, desulfurised material and etc. Fly ash is one of the well-known coal combustion products where it is usually used in concrete mixtures as blended cement, besides fly ash were also known as as supplementary cementitious material for increasing the workability and durability of concrete [10-13][48-49]. On the other hand, other researchers have classified bottom ash as industrial waste that have been produced at the bottom of coal furnace during the coal combustion for electricity generation. It forms from agglomerated ash particles that are glassy and coarse particles that have been produced in a pulverized coal furnace which were fall at the bottom of the furnace due to the properties of bottom ash which is heavier and harder to be carried by flue gas compared to fly ash [10, 14].

Recently, the investigation on reutilizing bottom ash as a construction material is aggressively being conducted among researchers [3, 12, 13, 14]. The usage of bottom ash as construction material has brought many benefits to the industry due to its low cost, low density fill and its suitability to be used as replacement of natural quarried sand or natural coarse aggregates which makes it a more sustainable as construction material. Kalyoncu [15] has conducted a studied and found that the size of bottom ash is similar to coarse and fine-grain, sand-like material which is collected from the bottom of coal-fired boilers. Meanwhile, the physical and chemical properties of bottom ash have the desired properties which make it suitable for use in engineering construction [3, 16, 17, 18]. But, some studies have found that bottom ash contains heavy metal and arsenic. Therefore, the Department of Environment, Malaysia, has classified coal combustion products as hazardous substances because several environmental issues have occurred due to excessive production of waste materials. Massive production of coal ash has negatively affected the environment such as by contaminating underground, adjacent soil and surface water due to limitation of dumping space. Coal ash is still being treated as waste and put in impoundment ponds, silos or landfills.

However, the American Coal Ash Association (ACAA) has lead a survey on the applications of bottom ash and it is found that usually bottom ash can be used for several construction purposes such as transportation, construction, agriculture and so on. It can also be used as soil modification, structural fillers, road base material and etc. [19]. Therefore, this paper aims to review previous research that has been carried out on the effect of bottom ash in concrete as sound absorption material. From the previous research, it had found that the presence of bottom ash in the concrete mixture can be either a partial or total replacement of fine aggregates has effect on the workability, compressive strength, flexural strength and sound absorption in concrete were reviewed to ensure either it is suitable to be replaced with natural aggregate (fine or coarse) in concrete as sound absorption material and also to verify the quality of bottom ash by confirming desired properties of the product.

2 Properties of Coal Bottom Ash

2.1 Chemical properties

A study on the replacement of coal bottom ash as a construction material such as cement or fine aggregate replacement has been carried out from the previous research. But there are some differences that can be seen on the chemical composition of CBA due to the different sources of coal that have been used as shown in Table 1. Based on Table 1, usually bottom ash is classified as a class F as a result of SiO2 + Al2O3 + Fe2O3 that exceeds 70% as described in ASTM C 618. Therefore, due to bottom ash included in class F, it has pozzolanic properties and also has a very low lime (CaO) content which is less than 7%. To produce cementitious mixtures, bottom ash requires a cementing agent such as quicklime or hydrated lime, Portland cement that reacts with water. Meanwhile, the loss on ignition (LOI) of bottom ash is still low which is less than 6% as specified in ASTM C 618. LOI is a method for identifying the mass loss of carbon, sodium and other elements from solid combustion residues by heating. Unburned carbon in coal ash can significantly affect its advantageous applications in concrete mixtures. According to Kurama et al. [20] LOI is defined as the total of unburned carbon in fly ash which is a common approach in cement and concrete applications.

Table 1. Chemical properties of bottom ash according to different locations

Power Plant Station/ Chemical Composition (%)	Spanish power plant [21]	Guru Hargobind Thermal Power Plant Bathinda, India [22]	TNB Electric power Plant Perak, Malaysia [19]	Seocheon coal-fired thermal power plant, South Korea [23]
SiO_2	52.30	56.44	54.80	44.2
Al_2O_3	25.14	29.24	28.50	31.5
Fe_2O_3	9.23	8.44	8.49	8.9
CaO	2.37	0.75	4.20	2.0
MgO	1.84	0.40	0.35	2.6
Na_2O	0.66	0.09	0.08	-
K_2O	3.72	1.29	0.45	-
TiO_2	1.45	3.36	2.71	2.4
P_2O_5	0.25	-	0.28	-
SO_3	0.03	0.24	-	-
LOI	1.07	0.89	2.46	-

2.2 Physical properties

From the laboratory investigations that have been carried out by Singh and Siddique [24], coal bottom ash is found to be a dark grey material by having angular particles shapes. It is also an irregular, porous material which has a rough surface texture. Coal ash resulting from the coal combustion process at 1400-1500°C at the Seocheon thermal power plant station produces bottom ash of different sizes which can be divided into fine and coarse bottom ash [25]. Fig. 1 shows the different types of bottom ash. Abu Bakar et.al [26] found that the size of bottom ash during the sieve analysis test was in the range between gravel and fine sand with a very low percentage of silt-clay sized particles which pass through the 75μm sieve. Therefore, there is a good correlation to show that the size of bottom ash possesses an ideal quality thereby enhancing the quality of concrete especially in terms of strength. Park et al. [27] lead an SEM analysis on the coal bottom ash, and it has been proved that coal bottom ash has a porous surface with many voids. It was also found that the area with voids was not as dense compared to other areas. The small particles could fall out relatively easily, as shown in Fig. 2.

Meanwhile, the percentage of water absorption, specific gravity, and fineness modulus of several types of bottom ash has shown variety results due to the different sources of coal for each thermal power plant station depending on its chemical composition. Table 2 shows the specific gravity, percentage of water absorption and fineness modulus according to different thermal power plant stations in India. From the results, it was found that the specific gravity of bottom ash is between 1.39-2.66 where it differs for each power plant, which is lower than the specific gravity of coarse aggregates (2.51) and fine aggregates (2.62) [28]. According to Muhardi et al. [8], the low specific gravity of bottom ash is due to lower iron oxide content. Bradshaw et al. [29] also specified that higher carbon content in bottom ash results in lower specific gravity. Water absorption of fine aggregate (3.7%) is lower than the percentage of water absorption of bottom ash [30]. Besides that, bottom ash generally has a low specific gravity due to its porous structure and popcorn-like particles which easily degrade under loading or compaction [31].

(a) (b)

Fig. 1. Particle shape of bottom ash (measured in mm); (a) coarse bottom ash and (b) fine bottom ash [26].

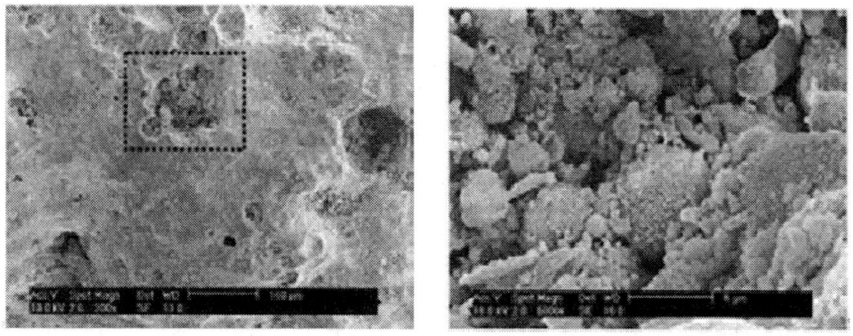

(a) (b)

Fig. 2. SEM analysis of coarse coal bottom ash; (a) 200x and (b) 5000x [27].

Table 2. Physical properties of bottom ash according to different Thermal Power Plants in India

Power plant station/ Physical properties	Guru Hargobind Thermal Power Plant Bathinda India [22]	Panipat Thermal Power Plant Haryana, India [32]	Eklahare Thermal Power Plant Nashik City Maharashtra, India [30]
Specific gravity	1.30	2.66	1.93
Water absorption	31.58	12.00	10.06
Fineness modulus (%)	1.37	3.12	2.7

3 Properties of concrete

3.1 Fresh coal bottom ash concrete

Based on the current environmental issues due to the dumping of industrial waste such as bottom ash, there are various possible solutions that can be done such as reusing industrial waste in the production of concrete by replacing fine aggregates, coarse aggregates or both. Therefore, fresh concrete properties are important for a study and include workability, air content and bulk density. The tests were all conducted according to ASTM.

3.1.1 Workability

The ratio of coarse or fine aggregates in the concrete mixture is one of the factors that will affect the workability of concrete and water demand depending on the properties of the aggregates. The inter particle friction between the aggregates will be decreased due to the physical properties of bottom ash particles which are smaller than natural river sand which is fine and irregularly-shaped.. According to Singh et al. [12], the workability of bottom ash concrete decreases in the concrete mixture in which contains bottom ash for different grades of concrete. The flow characteristic of fresh coal bottom ash concrete is affected due to increased inter particles [24]. However, the workability of concrete can be measured in terms of compaction factor. M. Singh et al. [11] has conducted the test and found that the workability of concrete decreases with the increase in the total amount of the fine aggregates replaced by bottom ash.100% of coal bottom ash as fine aggregate replacement in concrete has resulted in an 86% decrease in terms of slump value. This might be due to the decreased cement content, the use of super plasticizer and higher water cement ratio. Due to the higher workability of bottom ash concrete at a fixed cement content and water cement ratio, there will be a decrease in the compressive strength of concrete [32]. Raju et al. [33] observed that the replacement of bottom ash up to 50% increases water absorption compared to sand particles. Aggregate which are porous will absorb more water to provide workability. Previous research also found that the water absorption of bottom ash affects concrete workability where porous particles of aggregates act as water reservoirs for future hydration of cement [34].

3.2 Hardened concrete

The effects of bottom ash on hardened concrete properties will be extracted from previous research. Hardened properties of bottom ash concrete include compressive strength and flexural strength.

3.2.1Compressive strength

From previous research, several discussions on the compressive strength of bottom ash concrete were found: (i) The compressive strength of the concrete that contains bottom ash as a fine aggregate replacement slightly decreases compared to normal concrete [22][45]. This is confirmed by Kou et al. [35] who found that the compressive strength and drying shrinkage decrease with the increase in fine bottom ash content. The result shows that the strength of concrete decreases when the percentage of bottom ash replacement increases due to the bonding between the aggregates in the mixture. However, according to Kim et at. [25], the compressive strength of fine bottom ash and coarse bottom ash after 7 days and 28 days were around 60 to 70 MPa respectively with various percentages of bottom ash content in concrete. The strength was also affected by the proportion of cement paste which is made using water, binder materials and superplastisizer. Besides that, when using a fixed water cement ratio with different amounts of bottom ash content, the strength of concrete may decrease due to the high initial free water content used in the mixture which reduces bleeding and poorer interfacial bonding between the aggregates and the cement paste. (ii) at the early curing stage, the compressive strength of normal concrete is higher than the mean of compressive strength of bottom ash concrete which is lower [36]. However, a longer curing time will affect the strength of bottom ash concrete which is almost similar to the strength of normal concrete [37]. Due to the replacement of stronger material with delicate material as well as the non-appearance of pozzolanic activity by bottom ash, a decrease in compressive strength of bottom ash concrete may be observed at an early stage. Mathiraja [38] observed that the pozzolanic activity of bottom ash becomes slows down at partial (14 days) curing age and after the period where the bottom ash starts reacting with calcium hydroxide to create C-S-H gel and needles.

Previous research has shown conflicting results regarding the effect of bottom ash concrete on compressive strength. Therefore, further research on bottom ash aggregates is needed to review the effect of aggregate proportion and water absorption that will influence the compressive strength of bottom ash concrete.

3.2.2 Flexural strength

Based on a study by Wongkeo et al. [39], it was found that the flexural strength value for bottom ash lightweight concrete increases when there is a decrease in bottom ash content in Portland cement. The increase of flexural strength is related to the increase of the bulk density of concrete. However, Raju et al. [33] have also conducted a flexural strength test on the bottom ash concrete and found that the flexural strength of concrete decreases when natural fine aggregates are replaced with bottom ash. The flexural strength of bottom ash concrete slightly decreases with the use of fine and coarse bottom ash [26]. Besides that, the difference sizes of bottom ash will affect the flexural strength of bottom ash concrete. From previous research, the modulus of rupture decreases when fine aggregates are totally replaced with fine or coarse bottom ash. Park et al. [27] discovered that the flexural strength of bottom ash concrete tends to reduce all aggregate grading as the bottom ash mixing ratio increases. This means that the fractured section of concrete with bottom ash aggregates shows a flatter surface than normal aggregates [40]. Therefore, it can be concluded that the flexural strength of bottom ash concrete decreases due to the crack propagation in bottom ash concrete which occurs more easily through bottom ash particles compared to normal aggregates which are difficult to penetrate. Consequently, the direction of cracks will affect the normal concrete which made by normal aggregate.

3.3 Acoustic properties

Porous material is usually effective as a sound absorber. Since bottom ash particles are angular, irregular and have a porous structure, they are suitable to be used as sound absorption material.

3.3.1 Sound absorption

Generally, to reduce the noise level, reverberation time, remove echoes as well as to avoid sound from being stuck by concave surface, there are various types of sound absorbing materials that can be used. Coal bottom ash is one of the sound absorptive materials that have the potential to absorb sound that has been produced by transportation or traffic.

Arenas et al. [41] have conducted an investigation on the development of acoustic barriers made of the co-combustion of bottom ash. Meanwhile, 80% of bottom ash used in porous concrete for sound absorbing purposes has shown similar properties as the conventional materials used for sound absorbing applications [21][44][46]. From the research, it was found that the sound absorption coefficient increases when the bottom ash particle size is larger. The grain particle size of bottom ash will influence the acoustic performance and non-acoustic performance (physical and mechanical properties) of the product which contains bottom ash. The larger particle size of bottom ash has resulted in the best sound absorption coefficient due to the higher porosity of particles [21]. Usually, there are several factors that will influence the absorption performance of concrete and previous research has found that acoustic absorption of materials depends on the thickness of the specimen being tested. The acoustic absorption of concrete which has the same composition of bottom ash but varying in thickness will influence the sound absorption coefficient. Thus, the sound absorption coefficient for the thicker specimen slightly decreases the ability of the specimen to absorb sound towards frequency level. The sound absorption coefficient of noise also depends on the materials used. Since bottom ash is a porous material that is used in concrete, the energy of sound will dissipate in the porous structure of bottom ash. In addition, due to the higher porosity of bottom ash, it can potentially cause the air to lost momentum. When a sound wave strikes a material, a portion of the energy is reflected, another portion is absorbed by the material while the rest is transmitted as sound [41, 43]. Changes in direction and viscous drag are also involved in the movement of air particles in the pores. Kucsmarski et al. [42] indicated that less energy is transmitted to the solid structure as more is reflected from the surface when the pore size decreases, this making the material less useful for acoustic absorption.

4 Conclusions

This paper had reviewed previous works that were carried out on the chemical and physical properties, fresh and hardened propertied of bottom ash concrete. The remarks of this literature review can be summarised as follows:

i) From the previous studies conceive that bottom ash has greater potential to be used as one of the construction material, where it has been proved that it is usable as an additive in concrete mixtures to produce durable concrete. Therefore, it has been proved that bottom ash can be used in the concrete mixture either as fine aggregate or coarse aggregate replacement. However, an observation was made where coal bottom ash is suitable to be used as partial or total replacement of fine aggregate (sand) in concrete. A suitable optimum percentage of bottom ash was obtained, where up to 50% of bottom ash without any admixture such as superplasticiser can be used as non-structural concrete such as noise barriers, wall panel and partitions.
ii) The slump test experiment can determine the workability of fresh concrete which refers to true slump values. The workability of concrete has affected due to the presence of bottom ash in the concrete mixture, which it is decreases as the content percentage of bottom ash in concrete mixture increases. Therefore, to obtain the best results in terms of workability, it is important to get the optimum value of mixing.
iii) The addition of bottom ash in the concrete mixture has decreased the mechanical properties of bottom ash concrete strength at early stages. This reduction of strength increases with the increase in bottom ash content. But, after 14 days of curing the strength of bottom ash concrete becomes similar to normal concrete.
iv) Modulus of rupture of concrete decreases when the fine aggregates were replaced with bottom ash. The crack propagation occurs more easily through bottom ash particles.

v) The usage of bottom ash in concrete will also help in protecting the environment especially in terms of noise pollution. The physical properties of bottom ash show that it has a higher porosity and is therefore suitable to be used in concrete as sound absorption material. Nevertheless, from the review, there is very little research focusing on the effect bottom ash with respect to the grain size particle, thickness of specimen, water absorption, air flow resistivity, porosity, density, placement of sound absorptive material, speed of sound and various percentages of bottom ash content in concrete in order to obtain good noise absorption characteristics. Therefore, this warrants further research on this matter.

The authors would like to express their gratitude to Universiti Tun Hussein Onn Malaysia for funding the research under the FRGS Grant Vot 1576.

References

[1] M.C Han, D. Han and J.K. Shin, Use of bottom ash and stone dust to make lightweight aggregate. Construction and Building Materials, **99**, 192–199 (2015)

[2] C. Arenas, C. Leiva, L.F. Vilches and H. Cifuentes, Use of co-combustion bottom ash to design an acoustic absorbing material for highway noise barriers. Waste Management, **33**(11), 2316–2321 (2013)

[3] M. Cheriaf, J.C. Rocha and J. Péra, Pozzolanic properties of pulverized coal combustion bottom ash, Cement and Concrete Reserach, **29**(9),1387– 1391, (1999)

[4] M. Tiwari, S.K. Sahu, R.C. Bhangare, P.Y. Ajmal and G.G. Pandit, Elemental characterization of coal, fly ash, and bottom ash using an energy dispersive X-ray fluorescence technique, Applied Radiation. Isotopes, **90**, 53–7, (2014)

[5] A. Marto, K.A. Kassim, A.M. Makhtar, F.W. Lee and S.L. Yap, Engineering characteristics of Tanjung Bin coal ash, Electronic Journal of Geotechnical Engineering, 1117-1129, (2010)

[6] M.L.D. Jayaranjan, E.D. van Hullebusch and A.P. Annachhatre, Reuse options for coal fired power plant bottom ash and fly ash, Rev. Environ. Sci. Bio/Technology, no. January (2016)

[7] M.H.W. Ibrahim, A.F. Hamzah, N. Jamaluddin, P.J. Ramadhansyah and A.M. Fadzil, Split tensile strength on self-compacting concrete containing coal bottom ash, Procedia - Soc. Behav. Sci., **195**, 2280–2289, (2015)

[8] P. Chindaprasirt, C. Jaturapitakkul, W. Chalee and U. Rattanasak, Comparative study on the characteristics of fly ash and bottom ash geopolymers, Waste Management, **29**(2), 539–543, (2009)

[9] M. Abdul Rahim, N.M. Ibrahim, Z. Idris, Z.M. Ghazaly, S. Shahidan, N.L. Rahim, L.A. Sofri and N.F. Isa, Properties of concrete with different percentange of the rice husk ash (RHA) as partial cement replacement, Material Science Forum, **803**, 288–293, (2014)

[10] D. Bajare, G. Bumanis, and L. Upeniece, Coal combustion bottom ash as microfiller with pozzolanic properties for traditional concrete, Procedia Engineering, **57**, 149–158, (2013)

[11] M.C. Han, D. Han and J.K. Shin, Use of bottom ash and stone dust to make lightweight aggregate, Construction and Building Material, **99**, 192–199, Nov. (2015).

[12] M. Singh and R. Siddique, Effect of coal bottom ash as partial replacement of sand on workability and strength properties of concrete, J. of Cleaner. Productions, **112**, 620–630, (2016)

[13] S. Shahidan, N.M. Bunnori, N. Md Nor and S.R. Basri, Damage severity evaluation on reinforced concrete beam by means of acoustic emission signal and intensity analysis, *IEEE Symposium on Industrial Electronics and Applications*, 337–341, (2011)
[14] E. Menéndez, A.M. Álvaro, M.T. Hernández and J.L. Parra, New methodology for assessing the environmental burden of cement mortars with partial replacement of coal bottom ash and fly ash, J. Environ. Manage, **133**, 275–283, (2014)
[15] J.M. Irwan, S.K.. Faisal, N. Othman, M.H. Wan Ibrahim. R.M. Asyraf and M.M.K. Annas, Performance of concrete using light waste PET fibre, Advanced Materials Research, **795**, 352-355 (2014)
[16] R. Kaminskas, R. Kubiliūtė,. The Effect of coal ash on synthesis and properties of tricalcium silicate, Materials Science (Medžiagotyra), **16**(3), 73-79, (2010)
[17] K.Haldung and K. Mine, Usage of coal combustion bottom ash in concrete mixture, Construction and Building Materials, **22**(9), 1922-1928, (2008)
[18] S. Shahidan, I. Isham and N. Jamaluddin, A review on waste minimization by adopting in self compacting concrete, MATEC Web Conf., **47**, 1–7, (2016)
[19] M. Syahrul, M. Sani, F. Muftah and Z. Muda, The properties of special concrete using washed bottom ash (WBA) as partial sand replacement, Int. J. Sustain. Constr. Eng. Technol., **1**(2), 65–76, (2010)
[20] H. Kurama and M. Kaya, Usage of coal combustion bottom ash in concrete mixture, Construction and Building Material, **22**(9), 1922–1928, (2008)
[21] C. Arenas, C. Leiva, L.F. Vilches and H. Cifuentes, Use of co-combustion bottom ash to design an acoustic absorbing material for highway noise barriers, Waste Management, **33**(11), 2316– 2321, (2013)
[22] M. Singh and R. Siddique, Compressive strength, drying shrinkage and chemical resistance of concrete incorporating coal bottom ash as partial or total replacement of sand, J. Clean. Prod., **112**, 620–630, (2014)
[23] J.G. Jang, H.J. Kim, H.K. Kim and H.K. Lee, Resistance of coal bottom ash mortar against the coupled deterioration of carbonation and chloride penetration, Materials and Design, **93**, 160–167, (2016)
[24] M. Singh and R. Siddique, Effect of coal bottom ash as partial replacement of sand on properties of concrete, J. Clean. Prod., **112**, 620–630, (2013)
[25] H.K. Kim and H.K. Lee, Use of power plant bottom ash as fine and coarse aggregates in high-strength concrete, Constr. Build. Mater, **25**(2), 1115–1122, (2011)
[26] A.U. Abubakar, K.S. Baharudin and I. Technology, Potential use of Malaysian thermal power plants, **3**(2), 25–37, (2012)
[27] S.B. Park, Y.I. Jang, J. Lee and B.J. Lee, An experimental study on the hazard assessment and mechanical properties of porous concrete utilizing coal bottom ash coarse aggregate in Korea, J. Hazard. Mater, **166**(1), 348–355, (2009)
[28] S. Abo-Qudais, Effect of concrete mixing parameters on propagation of ultrasonic waves, Construction and Building Materials, **19**(4), 257–263, (2005)
[29] C.H. Benson and S. Bradshaw, *User guideline for coal bottom ash and boiler slag in green infrastructure construction*, Recycled Materials Resource Center Report, Univerisity of Wisconsin-Madison, USA (2011)
[30] M.P. Kadam and Y.D. Patil, Effect of coal bottom ash as sand replacement on the properties of concrete with different W/C ratio, Int. J. Adv. Technol. Civ. Eng., **2**(1), 45–50 (2013)
[31] C.W Lovell, W.H. Huang and J.E. Lovell, Bottom ash as highway material, *The 70th Annual Meeting of the Transportation Research Board, Washington, D.C*, (2013)
[32] P. Aggarwal, Y. Aggarwal and S.M. Gupta, Effect of bottom ash as replacement of fine, Asian J. Civ. Eng. (Building Housing), **8**(1), 49–62, (2007)

[33] R. Raju, M.M. Paul and K.A. Aboobacker, Strength performance of concrete using bottom ash a fine, Int. J. Res. Eng. Technol., **2**(9), 111–122, (2014)
[34] L.B. Andrade, J.C. Rocha and M. Cheriaf, Evaluation of concrete incorporating bottom ash as a natural aggregates replacement, Waste Manag., **27**(9), 1190–1199, (2007)
[35] S.C. Kou and C.S. Poon, Properties of self-compacting concrete prepared with coarse and fine recycled concrete aggregates, Cem. Concr. Compos, **31**(9), 622–627, (2009).
[36] N. Ghafoori, J. Bucholc, Investigation of lignite based bottom ash for structural concrete, Journal of Materials in Civil Engineering, 128–37, (1996)
[37] M. Singh and R. Siddique, Strength properties and micro-structural properties of concrete containing coal bottom ash as partial replacement of fine aggregate, J. Clean. Prod., **112**, 620–630, (2014)
[38] C. Mathiraja, A study on concrete using bottom ash, manufacturing sand and hybrid steel and coir fibres, IOSR Journal of Mechanical and Civil Engineering, **10**(1), 55-57, (2013)
[39] W. Wongkeo, P. Thongsanitgarn, K. Pimraksa and A. Chaipanich, Compressive strength, flexural strength and thermal conductivity of autoclaved concrete block made using bottom ash as cement replacement materials, **35**, 434–439, (2012).
[40] N. Md Nor, N. Muhamad Bunnori, A. Ibrahim, S. Shahidan and S.N.M. Saliah, An investigation on acoustic wave velocity of reinforced concrete beam in-plane source, *IEEE 7th International Colloquium on Signal Processing and Its Applications,* 19–22 (2011)
[41] C. Arenas, L.F. Vilches, H. Cifuentes, C. Leiva, J. Vale and C. Fernández-pereira, Development of acoustic barriers mainly composed of co-combustion bottom ash, *World Coal Ash Conf.*, (2011)
[42] M.A. Kuczmarski and J.C. Johnston, *Acoustic absorption in porous materials*, NASA Technical Report, Cleveland, Ohio, No. 216995 , 1– 20, (2011)
[43] S.A. Kudus, N. Muhammad Bunnori, S.R. Basri, S. Shahidan, M N.M. Jamil and N. M. Noor, An overview current application of artificial neural network in concrete, Advanced Material Research, **626**, 372–375, (2012)
[44] S. Shahidan, R. Pullin, K.M. Holford, M.B.N and N. Nor, Quantitative evaluation of the relationship between tensile crack and shear movement in concrete beams, Advanced Material Research, **626**, 355–359, (2013)
[45] N.I.R. Ramzi hannan, S. Shahidan, M.Z. Maarof and N. Ali, physical and chemical properties of coal bottom ash (CBA) from Tanjung Bin Power Plant, IOP Conf. Ser. Mater. Sci. Eng., **160**, 12056, (2016)
[46] S. Shahidan, N.I.R.R. Hannan, M.Z. Maarof and A.S. Leman, A comprehensive review on the effectiveness of existing noise barriers commonly used in the railway industry, MATEC Web Conf., **87**(1017), 1–8, (2017)
[47] S. Shahidan, S.S.M. Zuki and N. Jamaluddin, Damage grading system for severity assessment on concrete structure, Case Stud, Constr. Mater., **5**, 79–86, (2016)
[48] S. Shahidan, I. Isham and N. Jamaluddin, A review on waste minimization by adopting in self compacting concrete, MATEC Web Conf., **47**, 1–7, (2016)
[49] S. Shahidan, S.S. Mohd Zuki, M.A. Rahim and N. Ali, amplitude distribution of emission wave for cracking process, MATEC Web Conf., **47**, 2–8, (2016)

Potential Mixture of POFA and SCBA as Cement Replacement in Concrete – A Review

Noorwirdawati Ali[1,*], *Mohd Hazeeq Adhahuddinsa* Mohd Sobri [2], *Josef* Hadipramana[1], *Abdul Aziz* Abdul Samad[1], and *Noridah* Mohamad[1]

[1]Jamilus Research Center, Universiti Tun Hussein Onn Malaysia, 86400 Parit Raja, Johor, Malaysia
[2]Faculty of Civil and Environmental Engineering, Universiti Tun Hussein Onn Malaysia, 86400 Parit Raja, Johor, Malaysia

Abstract. Concrete is an important material used in all kind of building construction and ordinary Portland cement (OPC) is one of an important element in the production of concrete. However, the production of cement causes a problem because of high CO_2 emission to atmosphere. The manufacture of 1 tonnes of cement would produce approximately released 1 tonnes of CO_2. So, the need to search another material that can replace a cement with same properties and environmental friendly are crucial. The suitable material to replace cement has to be a pozzolanic materials. This is because pozzolanic materials has cementitious properties and high silica content. Palm oil fuel ash (POFA) and sugarcane bagasse ash (SCBA) are the material that suitable to replace cement because of high silica content. The use of POFA and SCBA in concrete has been studied by many researcher and it has been proved to improve the mechanical strength of the concrete either in normal concrete, high strength concrete or lightweight concrete. This paper would discuss the overview of the previous study on the cement replacement by POFA and SCBA and the potential of the both materials to be mix together to improve its properties. The chemical element which will be the focus point is SiO_4, MgO, CaO and SO_3, while the physical and mechanical properties such as workability, specific gravity, compressive strength and tensile strength will also be reviewed.

1 Introduction

Nowadays, the population of the world is concern about the cost of building materials that keep increasing from day to day. With the increase in demand, the building materials are not enough for the provision of adequate housing for people in the world and cause the price of the materials keep increasing [1, 2]. Besides that, the use of cement in construction are the most worrying environment concerns because it would release carbon dioxide (CO_2) gas to atmosphere. 5-7% of world CO_2 emissions are caused by this industry and it expected to increase because of an increasing demand of cement [3]. To prevent the problem become worst, many campaign and investigation had been done to search another

* Corresponding author: wirda@uthm.edu.my

material as alternative to reduce the cement usage in construction. Many researchers are interested to use an agricultural and industrial waste as a potential cement materials (CRM) due to amount of waste that keep generated through years [4, 5]. These waste did not produce new CO_2 to atmosphere where CO_2 is only cycled in the atmosphere. As such, it can be stated that the production of energy from biomass is a renewable energy. Today, 9-14% of energy sources of industrialized countries and 20-33% from developing countries are contributed by biomass [6].

Malaysia is popular among the agricultural countries in the world with a major agriculture crops grown such as oil palm, sugarcane, coconut and rubber. From 32.90 million ha of total land of Malaysia, 14.9% of it is applied for agricultural activities while 61% of it are still being a natural forest [7]. As a popular vegetable oil for cooking and food processing, palm oil has produce a huge amount of waste throughout the year. Malaysia is one of the world's largest producer and exporter of palm oil in the world where palm oil plantation covers 4.98 Mha of the land area with about 417 palm oil mills throughout the country. These factors allow it to produce massive amount of palm oil by-product such as empty fruit bunch, oil palm shell and POFA. The amount of these products are increasing every year. It expected in 2020, the number of solid biomass would reach 100 million dry tonnes [7-9]. The other agriculture waste need to be concerned is sugarcane bagasse. Sugarcane bagasse is a fibrous waste that left after extraction and crushing processes [7]. For bagasse ash, it was a by-product that unused in sugar mill industries and can be used to replace a Portland cement due to its pozzolanic properties. Each year, around 410,000 to 620,000 tons of bagasse was produced and the trend are likely to increase every year. It is causes by the fact that the sugarcane is known as one of the agricultural crops and it was an important raw material in the production of ethanol. There are some utilization of bagasse ash such as backfill or fertilizers but the utilization is still minimum and poorly managed. Most of the bagasse ash is dumped in landfills and cause pollution and environmental problem [10].

Supplementary Cementing Materials (SCM) is the farming waste materials that had been utilized and POFA and SCBA are the example of SCM. These materials would achieve a high amount of silica after burning at a high temperature and enables it to show its pozzolanic reactivity. Pozzolanic properties is important for the later age strength development of the concrete [11]. In this study, POFA and SCBA will be mixed to stabilize the chemical content of both materials. Each material has an element that need to added and reduced. The silica content (SiO_4) of POFA and SCBA is quite similar, which is in range of 59.62% to 66.91% for POFA and 50.90% to 65.26% for SCBA, respectively. MgO content in POFA is quite high which is 3.00% to 4.52% and it range quite higher compared to OPC which is just around 1.20% to 4.80%. For POFA, it contained high percentages of sulphur trioxide (SO_3) which is in range of 0.30% to 1.28% compared to 0.04% to 0.21% SO_3 for SCBA [12]. SO_3 is important because it may cause expansion which compensates shrinkage. The mixture with higher SO_3 content would results in lesser shrinkage [13].

2 Cement

2.1 Physical properties

From the finding of previous research, an ordinary Portland cement has a specific surface area of 0.325 m^2/g with a density of 3.11 g/cm^3. The particle size distribution are in the range of 1.2 to 5.0 μm [14]. The specific gravity of the cement mortar with OPC is 3.16 with a fineness and soundness value of 3.10 m^2/kg and 1.61 mm [15]. It was also found that the cement mortars with OPC has a porosity of 12.21% and water absorption of 6.20% [16].

2.2 Mechanical properties

A cement paste has a compressive strength value of 41.5 MPa with a modulus of rupture of 11 MPa. Other than that it has modulus of elasticity of 1270 MPa and toughness value of 340 J/m^2. The value of bulk density of the cement paste is 1.85 g/cm^3 with water absorption and porosity value of 16% and 29% [17]. Results from previous research has recorded that the the compressive and flexural strength of cement is 47 MPa and 9.23 MPa after 28 days while the drying shrinkage of cement at 28 days is 0.076% [18].

2.3 Chemical properties

From the finding of previous research, OPC contained high lime content (CaO) which is around 61.50% to 65.4% followed by silica (SiO_4) in range 20.40% to 21.98%. It is followed by alumina (Al_2O_3) with range 3.68% to 5.27%. Iron (Fe_2O_3) and magnesia (MgO) with range content of 2.27% to 4.19% and 1.20% to 4.80% each [12].

2.4 Cement hydration

Cement hydration is a process when a cement is mixed with water and become a cement paste. It was a complex chemical reaction and these process cannot be reversed. This reaction is influenced by several factors, both internal and external factors. An examples of factors that influenced the hydration process is temperatures and the superplasticizer presence in the mixture. These factor are important because it plays an important role in hydration process and significantly influence the characteristics time. Porosity and pore size of the produced cement based materials also important for hydration process because it influences the final strength and durability of the concrete [19].

There are 5 stages during the hydration process. The stage is namely as initial reaction period, induction period, acceleration period, deceleration period and slow reaction period. The peaks process occur twice which is at initial reaction period and at the end of acceleration period and both of them known as first peak and second peak. The first peak is important for production of calcium sulfoaluminate hydrate (AFt) that formed from ion dissolution between C_3A and calcium sulphate. The second peak is producing C-S-H and calcium hydroxide (CH) from the C_3S hydration that happen during the stage [20].

3 Pozzolanic material

Pozzolan materials is defined by Kanning et al. [21] as a materials that contain high amount of silica and alumina in it chemical structure. These materials may have a little or no portions of cement, so it need to be grinded into a fine size particle because it would form a compound with the cementitious properties when react with calcium hydroxide in a water at ordinary temperatures. In the end, the pozzolanic activity would arises cause by the reaction between amorphous silica and $Ca(OH)_2$ which form a calcium silicate hydrate gel (C-S-H) that important to improve the mechanical properties of concrete.

The first empire used a pozzolanic materials in construction are Roman Empire and it continuously used for ecology and technology purposes. It had been proved that it can produce a strong and durable materials when it mixed with lime and the performance is better compare to Portland cement [22]. In this study, pozzolan is the main element that would be used because POFA and SCBA is well-known as a pozzolanic materials which can be used as cement replacement and has been proved by many researchers. To activate the pozzolan in the concrete mixture, the hydration process of cement is needed to characterize the pozzolanic activity of the paste. Calcium hydroxide was form by the hydration process of almost 28 wt% of ordinary Portland cement. In the end, hydrated calcium silicates are formed from the reaction between calcium hydroxide in water and amorphous SiO_2 in a pozzolanic materials [23].

The pozzolanic activity of the pozzolanic materials are affected by its burning process. The pozzolan is formed from the orthosilicic acid which come from the ground water and later it become an amorphous silica by polymerization in the plant cell. Reactive amorphous silica is formed when the agriculture waste is being burn or used as fuel [24].

4 Properties of concrete with POFA

4.1 Physical properties

POFA has a varying specific gravity, however its value never exceeded 3.0. The specific gravity of POFA would be increase after the grinding process because of a decrease of porosity. An increase of POFA replacement level would decrease the workability of concrete composites [25]. Drying shrinkage of POFA concrete has the same value as OPC concrete when 10% of OPC is replace by POFA. However, the shrinkage increased slightly at 28 days because of the increasing of unground POFA percentage as a cement replacement [25].

However, the water requirement for concrete with POFA is higher compared to normal concrete. So, the production of concrete with POFA will require more water [25]. The workability and density of the concrete also tends to decrease with the increased of POFA replacement [26, 27].

4.2 Mechanical properties

From the finding of Khankhaje et al. [25], the compressive strength of concrete decreased with the increased of percentage replacement of POFA and it depends on the finest of the POFA particles. However, for tensile strength, it shows an improvement where a specimen with 0.4 POFA/cement ratio has a tensile value of approximately 3.52 MPa. This value is 9% higher compared to control mix and 16% higher than a specimen with 1.2 POFA/cement ratio. For modulus of elasticity (MOE), the specimen mixed with POFA either fine or coarse are in range of 25.0 to 28.0 GPa. The value are quite similar to OPC concrete which is 27.5GPa [25].

However, after POFA replaced a cement by 50%, the compressive and tensile strength drop and recorded lower value than normal concrete [25]. Other than that, slow pozzolanic reactivity of POFA would decrease the early strength of concrete due to low C_3S content that only presence in cement [28].

5 Properties of concrete with SCBA

5.1 Physical properties

Previous research has proven that SCBA has a value of specific surface area which is three times higher compare to OPC because SCBA is finer. Therefore, it required more water in concrete production [29]. Since processed SCBA has lower specific gravity compared to OPC, the powder volume needed is higher and it lead to higher water requirement to reach the same workability as normal concrete. The rate of drying shrinkage also increases initially but then it start to slightly decrease [24]. When cement is replaced with SCBA, the volumetric weight of concrete will decreased cause by the low density of the ash [30]. The friction between cement and SCBA particles also increased cause by angularity, irregular shape and high porosity of SCBA [31].

5.2 Mechanical properties

When a 20% of cement is replaced with SCBA in concrete, the compressive strength of concrete shows an increase by 10% compared to normal concrete. However, the strength shows a reduction when the cement is replaced by 15-25% of SCBA [24]. For tensile strength, 20% replacement of bagasse ash shows an increase of tensile strength up to 4.81 MPa. However, at 25% to 30% replacement, the values decrease to 3MPa [12]. Other than that, an addition of sugarcane bagasse fiber are able to increase the modulus of elasticity of the epoxy and it also would improve if the bagasse is mix with glass fibre [32]. Replacement level, composition and fineness of SCBA and cement are the factor that influence the improvement of compressive strength and tensile strength of concrete [29].

5.3 Chemical properties

Similar as POFA as pozzolanic materials, SCBA also has high content of silica (SiO_2) which is in the range 59.90% to 84.1%. Iron (Fe_2O_3) and alumina (Al_2O_3) content in SCBA are between 0.4% to 5.52% and 0.9% to 9.05%, respectively. Other than that, the content of magnesia (MgO) and lime (CaO) are in the range of 1.10% to 2.85% and 2.5% to 10.5% [4, 12].

6 Compatibility of the mixture of POFA and SCBA to replace cement

A materials need to have a compatibility to be mixed to replace a cement either by chemical or mechanical properties. For POFA and SCBA, there are several elements in each material that can be improved. According to Aprianti [12], the silica content (SiO_4) of POFA and SCBA is higher compared to OPC, which in the range of 59.62% to 66.91% for POFA and 50.90% to 65.26% for SCBA. However, the finding of Embong et al. [4] stated that SiO_4 content for SCBA can reach up to 84.1% if it was treated with acid treatment and burnt at suitable temperature. Aprianti [12] also stated that MgO content in POFA was quite high which is around 3.00% to 4.52% and it range higher compared to OPC which is just around 1.20% to 4.80%. According to Aitcin [33], when the MgO content is too high, a detrimental expansive action can occur in hardened concrete. The late hydration of periclase, the crytalline form of MgO in Portland cement clinkers, in the form of brucite $Mg(OH)_2$, results in volumetric expansion that can lead to a severe cracking of concrete. Therefore, SCBA is chosen to be mixed with POFA to stabilize the content of MgO in POFA because MgO content in SCBA is just around 1.10% to 2.85%. The range of lime content (CaO) in SCBA is 4.01% to 10.5% which is higher compared to POFA where it contain CaO in the range of 4.92% to 6.4% [12]. According to Kaewmanee et al. [13], lime is important to early setting and contribute to a higher compressive strength, especially in early ages of concrete. Aprianti [12] found that OPC also contained high CaO content in range of around 61.50% to 65.4%.

For POFA, it contained high percentages of sulphur trioxide (SO_3) which is in the range of 0.30% to 1.28% compared to 0.04% to 0.21% SO_3 for SCBA. OPC itself contained a high value of SO_3 which is in range of 2.19% to 2.95% compared to POFA and SCBA [12]. From the finding of Embong et al. [4], SO_3 content in SCBA can reach 0.50% if it is treated with acid treatment and burnt at suitable temperature. According to Kaewmanee et al. [13], SO_3 is important because it may causes expansion which compensates shrinkage. The mixture with higher SO_3 content would results in less shrinkage. Therefore, POFA can help in increasing the content of SO_3 when mix with SCBA in concrete. Table 1 shows range of chemical content of POFA and SCBA in percentage [4, 12].

Table 1. Range of chemical content of POFA and SCBA in percentage [4, 12]

Chemical Content	POFA	SCBA
Silica / Silicon dioxide (SiO_2)	59.62% - 66.91%	59.90% - 84.1%
Iron / Ferric Oxide (Fe_2O_3)	1.9% - 5.72%	0.4% - 5.52%
Alumina / Aluminium Trioxide (Al_2O_3)	2.5% - 6.44%	0.9% - 9.05%
Lime content / Calcium Oxide (CaO)	4.92% - 6.4%	2.50% - 10.5%%
Sulfuric Anhydride / Sulfur Trioxide (SO_3)	0.30% - 1.28%	0.04% - 0.50%
Magnesia / Magnesium Oxide (MgO)	3.00% - 4.52%	1.10% - 2.85%
Potassium Oxide (K_2O)	4.97% - 7.52%	1.35% - 2.50%
Sodium Oxide (Na_2O)	0.19% - 0.76%	0.30% - 0.92%
LOI	6.57% – 10.02%	0.2% - 19.60%

7 Conclusion

In short, this paper summarized the properties of POFA and SCBA in normal concrete as cement replacement. From the previous research, POFA and SCBA have proven to improve the physical and mechanical properties of concrete. However, each materials has its deficiency when it mixed with concrete in term of their properties especially chemical properties. So, these materials has a potential to be mixed together to get an optimum mixture that would have quite similar or approaching the similar properties of OPC. Consequently, the production of sustainable and green concrete could be realized. From the result, it would lead to waste conservation and at the same time reduce the negative impact on the environment. This would ultimately lead to a more sustainable construction for the construction industry and the same time a cleaner environment for the society to live in.

This research was supported by Short Term Grant, Universiti Tun Hussein Onn Malaysia (Project Vote No: U340).

References

[1] D.A. Adesanya and A.A. Raheem, Development of corn cob ash blended cement, Construction and Building Materials, **23**(1), 347-352 (2009)
[2] S.R. Abdullah, W.R. Wan Zainal Abidin and S. Shahidan, Strength of Concrete Containing Rubber Particle as Partial Cement Replacement, MATEC Web Conf., **47**, 01009 (2016)
[3] K. Hung, U.J. Alengaram, M. Zamin and S. Poh, Feasibility study of high volume slag as cement replacement for sustainable structural lightweight oil palm shell concrete, J. of Cleaner Production, **91**, 297-304 (2015)

[4] R. Embong, N. Shafiq, A. Kusbiantoro and M.F. Nuruddin, Effectiveness of low-concentration acid and solar drying as pre-treatment features for producing pozzolanic sugarcane bagasse ash, J. of Cleaner Production, **112**, 953-962 (2016)
[5] S. Shahidan, I. Isham and N. Jamaluddin, A Review on waste minimization by adopting in self compacting concrete, MATEC Web Conf., **47**, 1-7 (2016)
[6] A. Pereira, J.L. Akasaki, J.L.P. Melges, M.M. Tashima, L. Soriano, M.V. Borrachero, and J. Payá, Mechanical and durability properties of alkali-activated mortar based on sugarcane bagasse ash and blast furnace slag, Ceramics Int., **41**(10), 1-13 (2015)
[7] K.Y. Foo, A vision on the opportunities , policies and coping strategies for the energy security and green energy development in Malaysia. Renewable and Sustainable Energy Reviews, **51**, 1477-1498 (2015)
[8] M.F. Awalludin, O. Sulaiman, R. Hashim, and W.N.A.W. Nadhari, An overview of the oil palm industry in Malaysia and its waste utilization through thermochemical conversion, specifically via liquefaction. Renewable and Sustainable Energy Reviews, **50**, 1469-1484 (2015)
[9] M.E. Rahman, A.L. Boon, A.S. Muntohar, M.N. Hashem Tanim, and V. Pakrashi, Performance of masonry blocks incorporating Palm Oil Fuel Ash. J. of Cleaner Production, **78**, 195-201 (2014)
[10] A. Rerkpiboon, W. Tangchirapat, and C. Jaturapitakkul, Strength, chloride resistance, and expansion of concretes containing ground bagasse ash. Construction and Building Materials, **101**, 983-989 (2015)
[11] K.H. Mo, U. Johnson Alengaram, M.Z. Jumaat, S.P. Yap, and S.C. Lee, Green concrete partially comprised of farming waste residues: A review. J. of Cleaner Production, **117**, 122-138 (2016)
[12] S.E. Aprianti, A Huge Number of Artificial Waste Material can be Supplementary Cementitious Material (SCM) for Concrete Production – A review Part II. J. of Cleaner Production. Elsevier Ltd. **142**, 4178-4194 (2016)
[13] K. Kaewmanee, P. Krammart, T. Sumranwanich, P. Choktaweekarn, and S. Tangtermsirikul, Effect of free lime content on properties of cement-fly ash mixtures. Construction and Building Materials, **38**, 829-836 (2013)
[14] M. S. Kirgiz, Advancements in mechanical and physical properties for marble powder–cement composites strengthened by nanostructured graphite particles. Mechanics of Materials, **92**, 223–234 (2016)
[15] A. Bahurudeen, D. Kanraj, V. Gokul Dev and M. Santhanam, Performance evaluation of sugarcane bagasse ash blended cement in concrete. Cement and Concrete Composites, **59**, 77-88 (2015)
[16] P. Lertwattanaruk and A. Suntijitto, Properties of natural fiber cement materials containing coconut coir and oil palm fibers for residential building applications. Construction and Building Materials, **94**, 664-669 (2015)
[17] G.A.M. Brasileiro, J.A.R. Vieira, and L.S. Barreto, Use of coir pith particles in composites with Portland cement. J. of Environmental Management, **131**, 228-238 (2013)
[18] K. Zheng, J. Zhou and M. Gbozee, Influences of phosphate tailings on hydration and properties of Portland cement. Construction and Building Materials, **98**, 593-601 (2015)
[19] A. Pop and I. Ardelean, Monitoring the size evolution of capillary pores in cement paste during the early hydration via diffusion in internal gradients. Cement and Concrete Research, **77**, 76-81 (2015)
[20] J. Han, K. Wang, J. Shi, and Y. Wang, Mechanism of triethanolamine on Portland cement hydration process and microstructure characteristics. Construction and Building Materials, **93**, 457-462 (2015)

[21] R.C. Kanning, K.F. Portella, M.O.G.P. Bragança, M.M. Bonato, J.C.M. Dos Santos, Banana leaves ashes as pozzolan for concrete and mortar of Portland cement. Construction and Building Materials, **54**, 460-465 (2014)

[22] J.C.B. Moraes, J.L. Akasaki, J.L.P. Melges, J. Monzó, M.V. Borrachero, L. Soriano, M.M. Tashima, Assessment of sugar cane straw ash (SCSA) as pozzolanic material in blended Portland cement: Microstructural characterization of pastes and mechanical strength of mortars. Construction and Building Materials, **94**, 670-677 (2015)

[23] P. Šiler, P. Bayer, T. Sehnal, I. Kolářová, T. Opravil, and F. Šoukal, Effects of high-temperature fly ash and fluidized bed combustion ash on the hydration of Portland cement. Construction and Building Materials, **78**, 181-188 (2015)

[24] A. Bahurudeen and M. Santhanam, Cement & Concrete Composites Influence of different processing methods on the pozzolanic performance of sugarcane bagasse ash. Cement and Concrete Composites, **56**, 32-45 (2015)

[25] E. Khankhaje, M.W. Hussin, J. Mirza, M. Rafieizonooz, M.R. Salim, H.C. Siong and M.N.M. Warid, On blended cement and geopolymer concretes containing palm oil fuel ash. Materials and Design, **89**, 385-398 (2016)

[26] A.S.M.A. Awal, and I.A. Shehu, Evaluation of heat of hydration of concrete containing high volume palm oil fuel ash. Fuel, **105**, 728-731, (2013)

[27] N. Ranjbar, M. Mehrali, U.J. Alengaram, H.S.C. Metselaar and Jumaat, Compressive strength and microstructural analysis of fly ash/palm oil fuel ash based geopolymer mortar under elevated temperatures. Construction and Building Materials, **65**, 114-121 (2014)

[28] M. Momeen Ul Islam, K.H. Mo, U.J. Alengaram and M.Z. Jumaat, Mechanical and fresh properties of sustainable oil palm shell lightweight concrete incorporating palm oil fuel ash, J. of Cleaner Production, **115**, 307-314 (2015)

[29] K. Ganesan, K. Rajagopal, and K. Thangavel, Evaluation of bagasse ash as supplementary cementitious material, Cement and Concrete Composites, **29**(6), 515-524 (2007)

[30] J.C. Arenas-piedrahita, P. Montes-garcía, J.M. Mendoza-rangel and H.Z.L. Calvo, Mechanical and durability properties of mortars prepared with untreated sugarcane bagasse ash and untreated fly ash, Construction and Building Materials, **105**, 69-81, (2016)

[31] G. Sua-iam and N. Makul, Use of increasing amounts of bagasse ash waste to produce self-compacting concrete by adding limestone powder waste, J. of Cleaner Production, **57**, 308-319, (2013)

[32] D. Verma, P.C. Gope, M.K. Maheshwari and R.K. Sharma, Bagasse fiber composites-A review, J. of Materials and Environmental Science, **3**(6), 1079-1092 (2012)

[33] P.C. Aitcin,, *Binders for Durable and Sustainable Concrete,* Taylor & Francis Group, Abingdon, Oxon, United Kingdom, (2008)

A Review: The Effect of Grinded Coal Bottom Ash on Concrete

Nurul Fasihah Basirun[1], *Mohd Haziman* Wan Ibrahim[1,*], *Norwati* Jamaludin[1] and *Ramadhansyah* Putra Jaya[2]

[1]Jamilus Research Centre, Universiti Tun Hussein Onn Malaysia, 86400 Parit Raja, Johor, Malaysia
[2]Faculty of Civil Engineering, Universiti Teknologi Malaysia, 81310 Skudai, Johor, Malaysia

Abstract. This paper offers a review on the use of grinded coal bottom ash (CBA) on the concrete properties as demonstrated by strength test and microstructure test. Amount of CBA from power plant station was disposed in landfill because of the particle shape had a rough particles. By finding an alternative way to gain its surface area by grinding and used as replacement material as cement replacement may give a good side feedback on the strength and morphology of concrete. Most of the prior works studied on the grinded fly ash and grinded rice husk ash. The study on the influence of grinded CBA on the properties of concrete still limited and need more attention Therefore, the review on the effect of grinded CBA on the strength and microstructure of concrete are discussed.

1 Introduction

Concrete is the main choice as building material among all contractors in this world. This is due to concrete is the cheapest than other material and it also can be formed into variety of shapes and sizes. Most of dam, bridge and building all are made by concrete. Burj khalifa is the highest building in this world is one of an example building use concrete as main construction material. It is about 333,000 meters cube concrete is used [1]. As the use of concrete high, the production Portland cements also increase. In 2011, the world production of cement reaches about 3.6 billion tons as the consumption of concrete rises [2]. This is because Portland cement is main ingredients in production concrete and then will mix with sand, water and aggregate [3]. In produce cement powder, the first step are by milling the natural material such as chalk, clay and limestone , then it continues by heating in a kiln. Last part is by uniting it with gypsum [4]. Production for a ton of cement discharged nearly one ton of greenhouse gases, also high emission of CO_2 which making it uneconomical process created and high emission of carbon dioxide (CO_2) and this making it uneconomical [5]. Recently, Malaysia has encouraged the use of "green and recycled by products"[6]. Coal Bottom Ash (CBA) which is the result from the burning of coal is one of those by products. Malaysia has established 4 electrical power stations which composed waste of 80.1 % of fly ash and 19.9% of CBA [7].Thus, by using coal bottom ash (CBA), it suggested that it can reduce the heat in the atmosphere and carbon dioxide (CO_2). Besides,

* Corresponding author: haziman@uthm.edu.my

the uses of recycle product could give an advantage such as in solving environment and economic issues [8]. However, from the literature, it was found that CBA has a large particle size and a porous surface causing in higher water demand and lower compressive strength [9]. Therefore, in order to overcome problem associated with particle shapes of raw CBA, a few researcher made the incorporation of very small size of CBA by increase the surface area [10]. CBA is believed to have good pozzolanic properties if it has high fineness [11]. Previous research reports have investigated on the properties of CBA. However, research studies on the properties of grinded are rarely found. Therefore, the review on the effect of grinded CBA on the strength and microstructure of concrete are discussed.

1.1 Coal bottom ash

Coal is black in color and consists largely of carbon and is made of decomposed plants. According to Pet et al. [12] coal is classified as an organic sedimentary rock. In Malaysia, coal found at 3 geographical provinces in Sarawak, Sabah and Peninsular Malaysia. Coal is transforms electricity by burned in a furnace with a boiler at Power plant. There are four famous electric power plants in peninsular Malaysia that burn the coal namely, Tanjung Bin,Johor (2,100MW), Jimah, Negeri Sembilan (1,400MW), Sultan Salahuddin Abdul Aziz/Kapar, Selangor (2,420 MW) and Sultan Azlan Shah/Mahjung,Perak (2100 MW) power plants[13].

Farhana et al. [13] estimate from 2000 to up to ten years, the use of coal to generate electricity increase by 80%. Fig. 1 shows the system that pictured the process CBA disposed at the bottom [14].According to Abu Bakar et al. [15] in their studied found that the CBA normally found as coarse, porous, light, glassy, granular, and has similar color with cement. Besides, coal bottom ash (CBA) has pozzolanic activity which is attributed to the presence of SiO_2 and Al_2O_3. Its reacts with calcium hydroxide $Ca(OH)_2$ during cement hydration, to form additional Calcium Silicate Hydrate (CSH) and Cacium Aluminate Hydrate (CAH). Therefore with additional Calcium Silicate Hydrate (CSH) and Cacium Aluminate Hydrate (CAH) are essential in forming denser matrix most important to better strength [16].

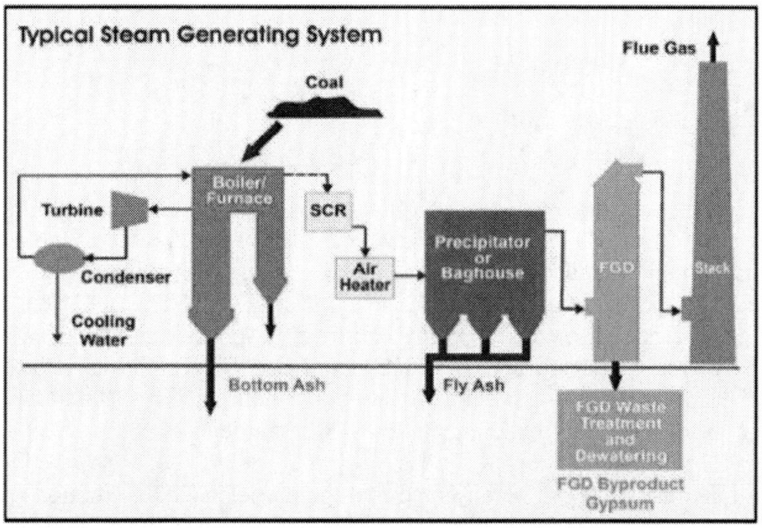

Fig. 1. Power plant that generate CBA (15)

2 Properties of grinded coal bottom ash

2.1 Chemical properties

The chemical analysis of CBA by using X-ray fluorescence (XRF) will reveal that the main chemical compounds include Silicates (SiO_3), Aluminates (Al_2O_3) and Iron oxide (Fe_2O_3) with small amounts of calcium (Ca), magnesium (Mg), and sulfate (SO_3). Table 1 shows the XRF of Ordinary Portland cement, sieve coal bottom ash and grinded CBA. The work of Kim et al. [17] and Jaturapitakkul [18] showed that the sum main chemical that contain in CBA, SiO_2 + Al_2O_3+Fe_2O_3 were exceeds 70% respectively. According to Standard Specification for Coal Fly Ash and Raw or Calcined Natural Pozzolan for Use in Concrete (ASTM C 618) for chemical composition grinded CBAes can be classified as Class F fly ash when the sum of silicates (SiO_3), aluminates (Al_2O_3) and iron oxide (Fe_2O_3) are exceed 70% [12]. The work Kim et al. [19] shows comparison result between sieved CBA and grinded CBA. The tabled shows that carbon content of grinded CBA is higher than sieve CBA. According to Jaturapitakkul [18] have notice an increase in pozzolanic activity with grinding of CBA. In their studied the variation in silica composition as different grinds time which indicates that the silica percentage is increased marginally after grinding [18].

2.2 Particle shape

Fig. 1 shows the SEM photographs of OPC, raw CBA and grinded CBA. The particle shape of Portland cement (Fig. 2(a)) is solid and irregular [21] while according to [18] observation on the raw CBA (Fig. 2(b)) shows that raw CBA is enormous,random shape, and high void. Thus, those materials with are preferred in construct roadway base and as course aggregate replacement under road base course [22]. Bajare et al. [9] states that the problem related to its particles size can be solved when grinding CBA. Besides, in their study also found that pozzolanic activity of CBA increases as the specific surface area increase. In addition, Bajare et al. [9] in their SEM report on 4 minute and 15 minute grinded CBA as in (Fig. 2(c, d)) can be observed that it is texture and shape similarity with the OPC photographs. Cheriaf et al. [23] conclude in their study that the small particles of grinded CBA can decrease the pore. Thus, it can give improvement in the strength of concrete.

2.3 Particle Size Distribution

Reported by Jaturapitakkul et al. [18] is tabulated in Table 3 show that the percentage amount of particles retained on sieve 325(45mm). According to ASTM 618, the allowance for fly ash is less than 34% retained on sieve 325. As shown in table 3, the percentage of grinded CBA and Portland cement type 1 are less than 34%, while raw CBA result in percentage 94.5% which is more than 34% as allowance in ASTM 618. Hence, the major raw CBA particle is course. Based on the fact from [24], CBA greater than 5mm in size suitable for coarse aggregate and less than 5mm can be replace as a fine aggregate. Therefore, the raw CBA unclassified to be as fly ash as in ASTM 618 and the results also suggest that the raw CBA in experimental by [25] not suitable to be used as a pozzolanic material.

From all the results, it can be observed that the particle size distribution are descrease with increase in grinding duration, hence a positive effect was shown in terms of concrete density, as well as improvement in concrete properties.

Table 1. Chemical composition of CBA

Property	Nagaratnamn et al. [20]	Kim et al. [19]	Kim et al. [19]	Jaturapitakkul et al. [18]	Jaturapitakkul et al. [18]
	OPC	**Sieved CBA**	**Grinded CBA**	**Raw CBA**	**Grinded CBA**
SiO_2	20.62	45.74	45.37	46.02	48.12
Al_2O_3	5.22	25.33	25.12	22.31	10.64
Fe_2O_3	3.60	6.86	5.81	10.64	10.55
CaO	64.99	0.99	1.07	11.48	11.65
K_2O	-	3.71	3.87	3.47	3.45
TiO_2	-	3.03	2.84	-	-
MgO	0.07	1.25	1.16	3.47	3.45
Na_2O	0.50	0.70	0.64	0.07	0.07
SO_3	-	-	-	1.52	1.76

Table 2. Physical properties of Portland cement, original CBA and Ground CBA (18)

Sample	Retained on Sieve 325 (%)
Portland cement type 1	4.8
Original CBA	94.5
Grinded bottom ash	2.8

Table 3. Physical proprties of portland cement, original CBA and Ground CBA

Sample	Portland cement type 1	Original CBA	1 hour	2 hour	3 hour	4 hour	6 hour
Specific gravity	3.14	2.00	2.02	2.11	2.14	2.21	2.25

2.4 Specific Gravity

To enhance the performance of CBA Cheriaf et al. [23] in their study conduct ground in a ball mill for 1–6 hour. As shown in Table 4, as grinding time extend to 6 hour, the reading of specific gravity increment from grind at time 1 hour to 6 hour grinding time. The same phenomenon was observed by Payá et al. [26] when grinding fly ash. They concluded that it was due to crushing porous carbon particles. A study by Jaturapitakkul et al. [18] reports that specific gravity of portland cement measured by the pycnometer method was 3.14 while raw CBA was 2.00. Jaturapitakkul et al. [18] concluded in their study that reducing size of bottom ash by different grinding time give better result in microstructure and improve pozzolanic activity.

3 Properties of hardened coal bottom ash concrete

Since the CBA with different grinding time has different properties than normal CBA, it behaves differently in concrete mixes and causes the concrete with replacement CBA to perform different for different grinding time. The properties of CBA on concrete with different grinding time are shown below.

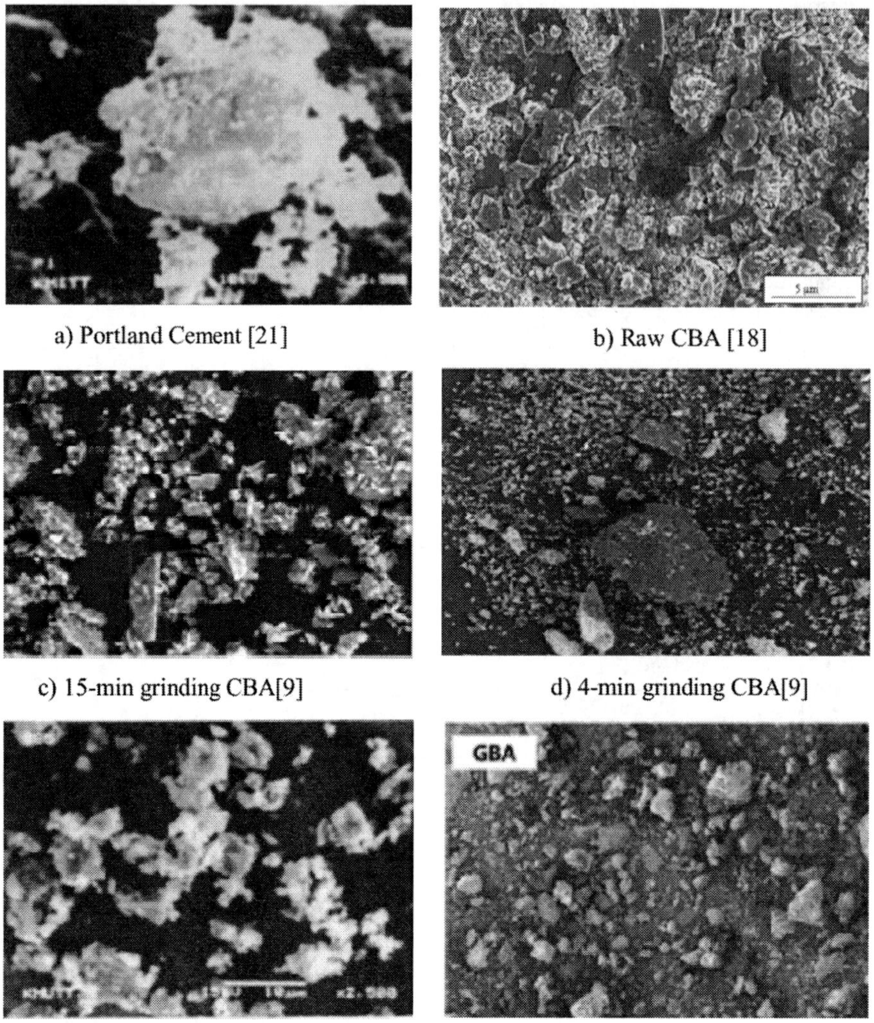

a) Portland Cement [21]

b) Raw CBA [18]

c) 15-min grinding CBA[9]

d) 4-min grinding CBA[9]

e) retained on Sieve 325 is less than 5% by weight [9]

f) 30-min grinding CBA[22]

Fig. 2. Scanning electron micrograph of grinded CBA

3.1 Microstructure

Scanning Electron Microscopy (SEM) was conducted by a few researches to analyze the microstructure. According to Siddique et al. [27] it is well known that, the calcium-silicate-hydrate (CSH) is major present. According to work by Whittaker et al. [28] shown in the SEM images as in Fig. 3(b) their replacement of ordinary portland cement with grinded CBA resulted in an increase in the porosity of the structure. Therefore, in their experimental result conclude that the porosity of the hardened structures may result to a higher water demand. As observe by Singh et al. [29] there were formation of CSH gel in mixture.

As observed by Cheriaf et al. [23] the SEM show a slaglike morphology as shown in Fig. 3. Most of slaglike morphology of grinded CBA unreacted by calcium hydroxide. The grinded CBA was grind in a grinding machine for 6 hour. At up to 28 days, $Ca(OH)_2$ began to be work with the particle in concrete, the reaction lead to the formation of CSH gel and needles as in Fig. 3.

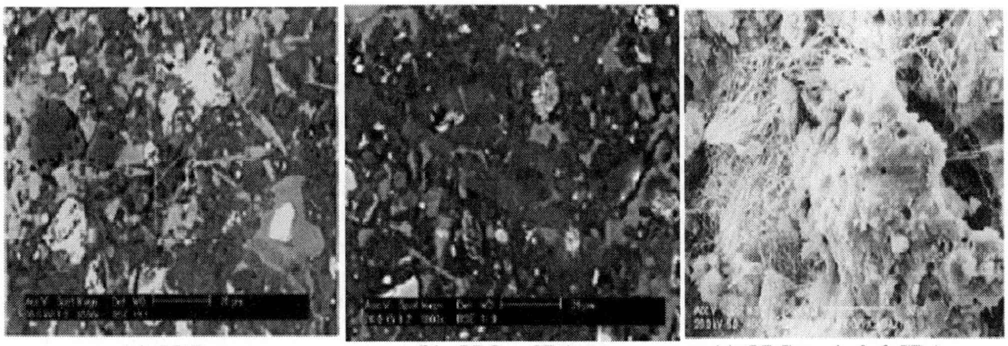

(a) OPC (b) OPC + CBA (c) OPC + grinded CBA

Fig. 3. Image of grinded CBA

3.2 Compressive strength

Compressive strength of CBA is influenced by the grinding time of CBA by high ball mill. Most research recommended that, the grinded CBA has a potential to be a good pozzolanic material by increase in fineness.

According to Kurama et al. [30] the reactive calcium silica in ash will react with lime will form calcium silicate and aluminates hydrates. These hydrates will fill the voids and eliminate lime in concrete. Thus, this will reducing the permeability and enhances the strength of concrete. According to their experimental results, compressive strength increase as amount of grinded CBA replacement up to 10%. At 56 day, about 5% increases in compressive strength compare to the control mixture. Therefore, addition of grinded CBA as cement replacement could gain the requirement properties such pozzolanic, thus improve the strength and durability of concrete.

Differently with the work result by Kim et al. [17], the higher the replacement grinded CBA percentage resulted in lower compressive strength. As shown in table, the result of compressive strength of grinded CBA for 3, 28 and 91 day was replaced with cement in percentage replacement of 15% and 30%. The control cube was obtaining 79 MPa for compressive strength test higher than grinded CBA, 34 and 38 MPa respectively for 15% and 30% replacement levels. The control cube had strength higher than 15 and 30% replacement grinded CBA cube. As observed in Ghafoori et al. [31], due to the angular shape of grinded CBA give an effect on the amount of water and the result of compressive strength at initial curing. At the end 60 days curing, there are 7% increase in strength was recorded and 13.7% increase in strength at 90 day curing.

Bajare et al. [9], CBA with grinding time 15 minutes is 31 MPa higher than 27MPa from CBA grinded at 240 seconds. Jaturapitakkul et al. [18] found similar behavior with tests using 20% replacement of ground bottom ash that performed as well as control. The result of compressive strength at 28 days was 37.1 MPa or 97% of the control concrete. This study use grinded CBA which passin

g on Sieve 325 more than 95%. The compressive strength with 20% replacement CBA increase from 36.6 to 37.1 MPa due to CBA exhibited good pozzolanic after being ground to smaller particle.

Canpolat et al. [32] observed that there poor performance the replacement of cement with bottom ash at initial curing. At the age of 7 days, there was a continuing improvement in the performance of the mixtures. When curing was extended to 28 days, a dramatic increase in the performance of the mixtures was noticed. Most of the compressive strength values of all the cement mixes are higher than control sample.

4 Discussion and conclusion

This paper was prepared by reviewing the different on grinding time of CBA and the effect on concrete in terms of compressive strength. From the review, the properties of concrete containing CBA were improved by decreasing the particle size of CBA. Furthermore, SEM analysis shows that the texture of the cement paste containing CBA with different grinding time was denser and compact compare with raw CBA. Generally, pozzolanic reaction of CBA was directly related to its fineness. Thus, the fineness of CBA is increased with increased in grinding time. From the literature review, by increasing the fineness of CBA in concrete mixture helps to achieve denser and more impermeable cement matrix.

References

[1] S.A. Kallen, *Burj Khalifa, The Tallest Tower in the World*, Norwood House Press, Chicago, (2014)

[2] A.M. Neville and J.J. Brooks, *Concrete Technology*, Pearson Education Limited, Harlow, England, (2003)

[3] P.K. Mehta and P.J.M. Monteiro, *Concrete, Structure, Properties and Materials*, Prentice Hall, New York, (1993)

[4] N. Md Noor, *Physical Performance and Durability Evaluation of Rubberized Concrete, Physical Performance and Durability*, Doctor of Engineering, Kyushu University, Fukuoka, (2014)

[5] S. Adekunle, S.Ahmad, M. Maslehuddin and H.J. Al-Gahtani, Properties of SCC prepared using natural pozzolana and industrial wastes as mineral fillers, Cement and Concrete Composites, **62**, 125–133 (2015)

[6] A.U. Abubakar and K.S. Baharudin, Potential use of Malaysian thermal power plants, **3**(2), 25–37 (2012)

[7] N. Jamaluddin, A.F. Hamzah and M.H.Wan Ibrahim, Fresh properties and flexural strength of SCC CBA, 1–6 (2016)

[8] S. Shahidan, Properties of concrete with difference percentage of the RHA as partial cement replacement, Material Science Forum, **803**, 288-293 (2014)

[9] D. Bajare, G. Bumanis and L. Upeniece, CCBA as microfiller with pozzolanic properties for traditional concrete, Procedia Engineering, **57**, 149-158 (2013)

[10] A.R. Awang, A. Marto and A.M. Makhtar, Advane science influene technology morphological and strength properties of Tanjung Bin coal ash mixtures for applied in geotechnical engineering work, 55–62 , (2012)

[11] N.E. Zainal Abidin, M.H. Wan Ibrahim, N. Jamaluddin, K. Kamaruddin and A.F. Hamzah, The effect of bottom ash on fresh characteristic, compressive strength and water absorption of self-compacting concrete, Applied Mechanics and Materials, **660**, 145–151 (2014)

[12] C.S. Pei, Coal as an energy resource in Malaysia, 399–410 (1993)

[13] A. Farhana and B. Jamaludin, *Energy Mix and Alternatives Energy for Sustainable Development in Malaysia*, Uiversiti Putra Malaysia, 1–9 (2001)

[14] M.S.H. Mohd Sani, F. Muftah and Z. Muda, The prop of special concrete using washed bottom ash (WBA) as partial sand replacement, Int. J. of Sustainable Construction Engineering and Technology, **1**(2), 65-76 (2010)

[15] A.U. Abubakar, K.S. Baharudin and T. Infrastructure, Properties of concrete using Tanjung Bin Power, **3**(2), 56–69 (2012)

[16] P. Filipponi, A. Polettini, R. Pomi and P. Sirini, Physical and mechanical properties of cement-based products containing incineration bottom ash, Waste Management, **23**(2), 145–56 (2003)

[17] H.K. Kim and H.K. Lee, Use of power plant bottom ash as fine and coarse aggregates in high-strength concrete, Construction and Building Materials, **25**(2), 1115–1122 (2011)
[18] C. Jaturapitakkul and R. Cheerarot, Development of bottom ash as pozzolanic material, **0899-1561**, 48–53 (2003)
[19] H.K. Kim, Utilization of sieved and ground coal bottom ash powders as a coarse binder in high-strength mortar to improve workability, Construction and Building Materials, **91**, 57–64 (2015)
[20] B.H. Nagaratnam, M.E. Rahman, A.K. Mirasa, M.A. Mannan and S.O Lame, Workability and heat of hydration of self-compacting concrete incorporating agro-industrial waste, J. of Cleaner Production, **112**, 882–894 (2016)
[21] Z. Zhang, B. Zhang, and P. Yan, Comparative study of effect of raw and densified silica fume in the paste, mortar and concrete, Construction and Building Materials, **105**, 82–93 (2016)
[22] M.H. Wan Ibrahim, A.F. Hamzah, N.Jamaluddin, R. Putrajaya and M.F. Arsha, Split tensile strength on self-compacting concrete containing coal bottom ash, Procedia – Social and Behavioral Sciences, **195**, 2280–2289 (2015)
[23] M. Cheriaf, J. Cavalcante and J. Pera , Pozzolanic properties of pulverized CCBA, Cement and Concrete Research, **29**, 1387–1391 (1999)
[24] R. Siddique, Utilization of coal combustion by-products in sustainable construction material resources, Conservation and Recycling, **54**(12), 1060-1066 (2010)
[25] A. Abdul Kadir and M.I. Hakeem, An overview of fly ash and bottom ash replacement in SCC, *Advanced Materials Engineering and Technology*, Bandung, (2013)
[26] J. Paya, J. Monzo, M.V. Borrachero and E. Peris-Mora, Mechanical treatment of fly ashes, Part I: Physical-chemical characterization of ground fly ashes, Cement and Concrete Research, **25**(7), 1469–1479 (1995)
[27] R. Siddique, P. Aggarwal and Y. Aggarwal, Influence of water/powder ratio on strength properties of self-compacting concrete containing coal fly ash and bottom ash, Construction and Building Materials, **29**, 73–81 (2012)
[28] M. Whittaker, R. Taylor, Q. Li, S. Li, The effect of bottom ash as a partial cement replacement
[29] M. Singh and R. Siddique, Effect of coal bottom ash as partial replacement of sand on workability and strength properties of concrete, J. of Cleaner Production, **112**, 620–630 (2016)
[30] H. Kurama and M. Kaya, Usage of coal combustion bottom ash in concrete mixture, Construction and Building Materials, **22**, 1922-1928 (2008)
[31] N. Ghafoori and J. Bucholc, Investigation of lignite-based bottom ash for structural Concrete, J. Materials in civil Engineering, **8** , 128–137 (1996)
[32] F. Canpolat, K. Y lmaz, M. M. Kose, M. Sumer and M. A. Yurdusev, Use of zeolite, coal bottom ash and fly ash as replacement materials in cement production, Cement and Concrete Research, **34**, 731-735 (2004)

A Preliminary Modified Volumetric Mix Design for Self-Compacting Concrete (SCC) by Utilizing Asphalt Dust Waste (ADW)

Isham Ismail[1], and *Shahiron* Shahidan[1,*]

[1]Jamilus Research Centre, Faculty of Civil and Environmental Engineering, Universiti Tun Hussein Onn Malaysia, 86400 Batu Pahat, Johor, Malaysia

Abstract. The construction of asphalt roads has been commonly used as for infrastructure development since ancient times. The production of asphalt using drum mix plants has created powder waste in the premix batching plant industry. It consists of very fine powders and causesproblems to the environment today. However, powder waste can be successfully and economically utilized in the development of self-compacting concrete (SCC). The aim of this study is to evaluate the fresh properties and the compressive strength of SCC containing asphalt dust waste (ADW). For this purpose, a simple modified volumetric mix design approach based on the Okumura Method was adapted to produce SCC mixes. In order to obtain this SCC, a series of SCC mixes with a specific water/binder ratio (0.3), superplasticizer ratio (2%) as well as different amounts of aggregate, sand and ADW were prepared. Fresh properties of the SCC mix were tested using tests such as slump-flow, J-Ring, L-box and V-funnel which showed satisfactory results for $MD8_{40}$ and $MD8_{50}$. The compressive strength for $MD8_{40}$ and $MD8_{50}$ after 28 days was 40.8 MPa and 34.7 MPa respectively. In conclusion, the simple modified volumetric mix design method by utilizing ADW in SCC can be suggested for the development of SCC.

1 Introduction

Concrete is a common material that has been used for housing, tunnel, foundation, precast elements, high rise buildings and many other types of construction all over the world [1]. In 1986, Self Compacting Concrete (SCC) was introduced by Okamura as a solution for producing durable self-leveling concrete, which can flow and be compacted by its own weight into every corner of the formwork during placing and without requiring the vibrating process for normal vibrated concrete [2].

Through this development, it has become a great opportunity for researchers and the concrete industry to produce more economical and durable SCC by utilizing waste material either for granular or powder replacement in the development of SCC [3-7]. The performance of waste material incorporated in SCC varies and it depends on the physical

* Corresponding author: shahiron@uthm.edu.my

and chemical properties of the materials used [4, 8, 9]. However, there are no specific methods that can be claimed as a standard to produce SCC. It depends on actual situations to obtain high quality SCC with satisfactory properties such as filling ability, passing ability and stability [10-11].

In short, this paper presents a preliminary study based on a modified volumetric mix design method by utilizing ADW to improve the rheological properties in SCC. The results show good performance for fresh state SCC without segregation and bleeding effects. A compressive strength of more than 30 MPa in the hardened state for $MD8_{40}$ and $MD8_{50}$ was also achieved.

2 Modified volumetric mix design for SCC

Generally, there are many methods to produce SCC and it is not restricted to any standard at this moment [12-14]. Researchers from academic institutions, concrete batching plants, precast plants and developers have developed their own mix proportioning methods [13].

In this experimental research, a total of 6 types SCC mixtures were produced. It was a modified method based on recommendations from Okamura's method as well as a EFNARC specifications. This simplified mix design method is based on a volumetric ratio to allocate amounts of cement, aggregate, sand and ADW by percentage from total volume required. Each mix was prepared by fixing 25% of cement content from the total volume, varying amounts of granular (aggregate and sand) and ADW, a water/binder ratio of 0.3 and 2% super plasticiser. The details of the volumetric mix design flow and mix proportion in volumetric ratio are shown in Fig. 1 and Table 1.

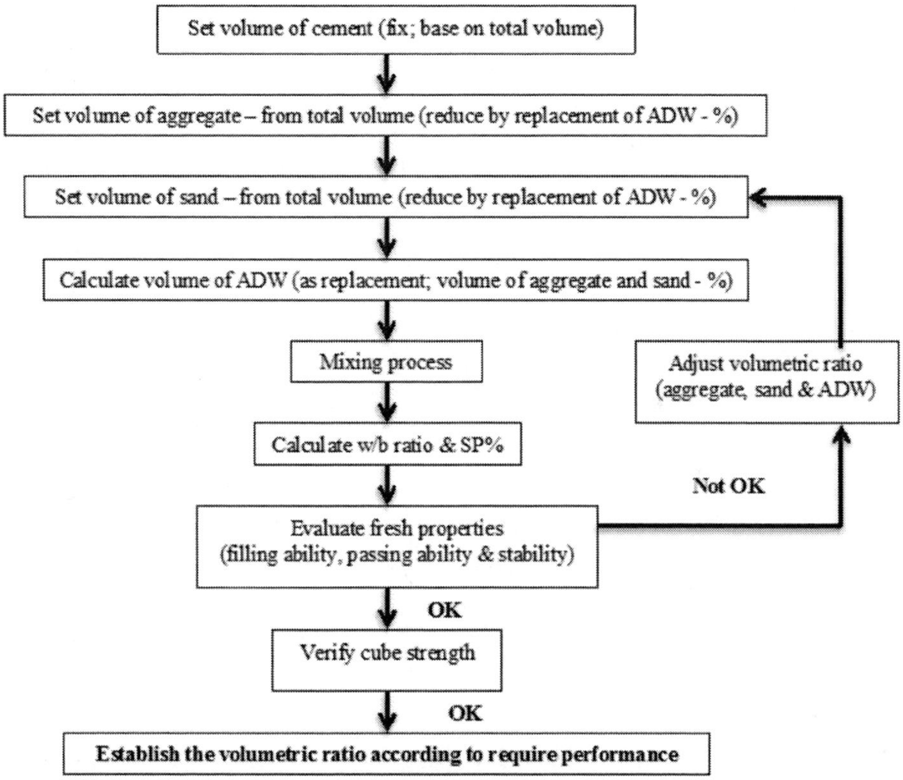

Fig. 1. Modified volumetric mix design procedure.

After several trials, this simplified mix design method obtained satisfactory fresh test criteria. Finally, the detailed actual mix design ratio was calculated as preliminary data for the production of SCC by utilizing ADW as shown in Table 3 and 4.

Table 1. Mix proportion in volumetric ratio.

Mix types	**Main Materials**				**w/b**	**SP**
	Cement	**Aggregate**	**Sand**	**ADW**		
MD8$_0$	25% Total Volume	60% TV	15%TV	0% x (vol. agg. + vol. sand)	0.3	2%
MD8$_{10}$		(60% TV) – (10% x 60%TV)	(15% TV) – (10% x 15%TV)	10% x (vol. agg. + vol. sand)		
MD8$_{20}$		(60% TV) – (20% x 60%TV)	(15% TV) – (20% x 15%TV)	20% x (vol. agg. + vol. sand)		
MD8$_{30}$		(60% TV) – (30% x 60%TV)	(15% TV) – (30% x 15%TV)	30% x (vol. agg. + vol. sand)		
MD8$_{40}$		(60% TV) – (40% x 60%TV)	(15% TV) – (40% x 15%TV)	40% x (vol. agg. + vol. sand)		
MD8$_{50}$		(60% TV) – (50% x 60%TV)	(15% TV) – (50% x 15%TV)	50% x (vol. agg. + vol. sand)		

3 Experimental studies

There are many types of mix design methods and alternative materials for the development of SCC [10-11]. The main principles of developed SCC are to achieve the fresh properties, which are filling ability, passing ability and stability without compromising the strength of the hardened SCC. Therefore, this research mainly focuses on developing a simple volumetric mix design method by utilizing ADW for SCC.

ADW is a powder waste generated from the production of asphalt premix for road construction pavements using drum mix plants. Based on previous studies, it is a potential material to improve the rheological properties of fresh SCC such as Fly Ash, Silica Fume, Granulated Blast Furnace Slag, Ground Glass Filler, Ceramic Powder Waste and Marble Powder Waste [4],[7-8],[15]. SEM photographs for ADW are shown in Fig. 2. ADW has angular shapes with a rough surface texture. Hence, this is an opportunity to use waste material in the development of sustainable concrete products in order to create a better environment for the local community.

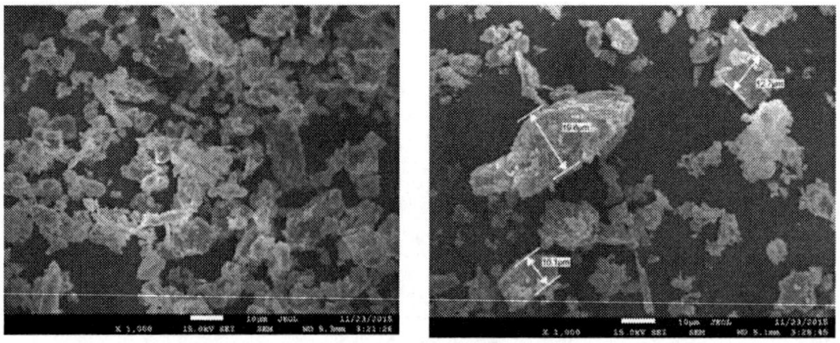

Fig. 2. Angular shape and rough surface texture of ADW from SEM analysis.

However, the performance of ADW in terms of rheological properties is still questionable. Therefore, this preliminary research has used ADW as an alternative powder waste material to produce paste materials and at the same time explore a new approach to develop SCC in a lab by using a modified volumetric mix design ratio. The mixtures were established and evaluated through observation and tests on the fresh properties (slump flow, J Ring, L Box and V Funnel). The performance of hardened concrete strength was measured through a cube test after 7 and 28 days respectively.

3.1 Materials

The production of SCC in this research was done using modified volumetric mix design ratio. The amount of cement, aggregate, sand and ADW will be calculated as mentioned in Table 1 with a suitable water binder ratio (0.3) and superplasticizer content (2%). The chemical properties of cement, sand, aggregate and ADW are shown in Table 2.

Table 2. The chemical properties of cement, sand, aggregate and ADW.

Properties	Cement OPC	Chipping	Sand	ADW
Calcium Oxide, (CaO)	55.60 %	1.85 %	-	2.98 %
Silicon Dioxide, (SiO2)	14.20 %	43.70 %	30.10 %	44.00 %
Aluminum Oxide, (Al2O3)	3.68 %	8.50 %	2.13 %	9.22 %
Iron (III) Oxide, (Fe2O3)	3.49 %	3.38 %	0.30 %	4.24 %
Sulfur Trioxide, (SO3)	3.20 %	-	-	-
Magnesium Oxide, (MgO)	1.48 %	0.49 %		0.62 %
Carbon, (C)	1.00 %	1.00 %	1.00 %	1.00 %
Potassium Oxide, (K2O)	0.48 %		0.19 %	4.69 %
Zirconium Dioxide, (ZrO2)	-	-	0.47 %	-
Titanium Dioxide, (TiO2)	-	0.39 %	0.32 %	0.46 %
Sodium Oxide, (Na2O)	-	-	-	1.00 %
Titanium, (Ti)	0 < LLD	-	-	-
Calcium, (Ca)	-	-	0 < LLD	-
Phosphorus, (P)	-	0 < LLD	-	0 < LLD
(S)	-	-	-	0 < LLD
Zirconium, (Zr)	-	-	-	0 < LLD

Local Ordinary Portland Cement according to MS EN 197-1, CEM I 42.5 N was used in all mixes. The maximum aggregate size and sand grain size was 12 mm and 4 mm respectively. The gradation curves of the granular materials are shown in Fig. 3. The aggregates, sand and ADW were air dried prior to mixing. 2% of superplasticizer Darex Super 20 was used in order to increase the workability of SCC for all mixes. Meanwhile, the water binder ratio used was set to 0.3.

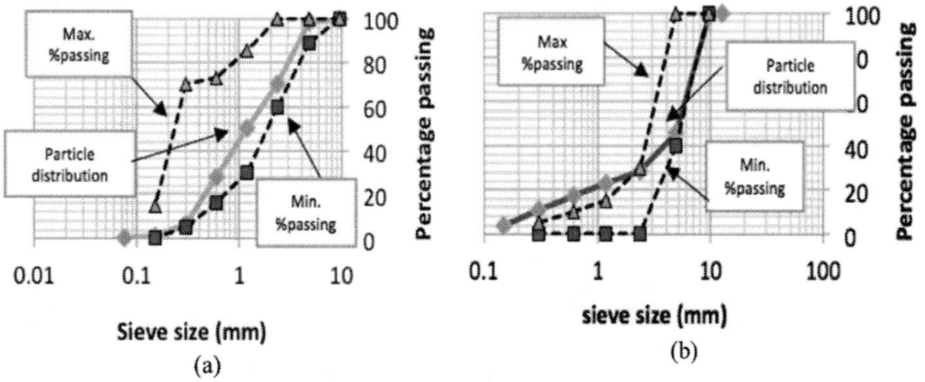

Fig. 3. Particle distribution of the granular materials (a) sand (b) aggregate.

DOI: 10.1051/matecconf/201710301008

3.2 Test equipment

Self-compacting concrete is also known as self-consolidating concrete. It is a highly workable concrete that can flow inside the formwork system and spread through reinforcement under its own weight during the fresh state [2 ,11]. Therefore, these fresh properties should be tested and evaluated for filling ability (V-Funnel test), passing ability (L-Box) and stability (slump flow and J Ring test). Filling ability is the capability of the concrete to flow and fill all spaces or shapes created in the formwork system without using vibration tools. Passing ability is the ability of concrete to flow through and bond to reinforcement bars without any blocking under its own weight. Stability is the performance of internal resistance from the concrete mixture to flow without segregation [16].

3.3 Calculation of modified volumetric mix design method

The aim of developing a procedure using modified volumetric mix design method is to produce a simple understanding on the distribution of cement, aggregate, sand and ADW using locally available materials. This design procedure may require interpolation distribution of materials to achieve target compressive strength after a series of trial mixes [12].

Calculation of Modified Volumetric Ratio (MVR) is based on Volumetric Ratio (VR) and controlled by Weight Ratio (WR) as shown in Table 3 and 4. Conversion of VR to WR is based on the estimation of packing factor in 1 liter of each material. The estimated packing factor for cement is 1.15, 1.33 for aggregates, 1.6 for sand and 1.05 for ADW.

Table 3. Volumetric ratio (VR) for 1000 litre mixture.

1000 L	$MD8_0$		$MD8_{10}$		$MD8_{20}$		$MD8_{30}$		$MD8_{40}$		$MD8_{50}$	
volume	**VR**	**%**	**VR**	**%**	**VR**	**%**	**VR**	**%**	**VR**	**%**	**VR**	**%**
Cement	250	25	250	25	250	25	250	25	250	25	250	25
Agg.	600	60	540	54	480	48	420	42	360	36	300	30
Sand	15	15	135	13.5	120	12	105	10.5	90	9	75	7.5
ADW	0	0	75	7.5	150	15	225	22.5	300	30	375	37.5

Table 4. Weight ratio (WR) for 1000 litre mixture base on volumetric ratio (VR).

1000 L	$MD8_0$		$MD8_{10}$		$MD8_{20}$		$MD8_{30}$		$MD8_{40}$		$MD8_{50}$	
volume	**WR**	**%**	**WR**	**%**	**WR**	**%**	**WR**	**%**	**WR**	**%**	**WR**	**%**
Cement	287.5	21.65	287.5	22.1	287.5	22.5	287.5	23.0	287.5	23.4	287.5	23.9
Agg.	801	60.31	720.9	55.3	640.8	50.1	560.7	44.8	480.6	39.1	400.5	33.3
Sand	239.7	18.05	215.7	16.6	191.7	15.0	167.8	13.4	143.8	11.7	119.8	10.0
ADW	0	0	78.9	6.1	158.0	12.4	236.9	19.0	315.9	25.7	394.9	32.8

4 Experimental results

The fresh properties and compressive strength of developed ADW-SCC were measured as shown in Table 5 and illustrated in Fig. 4 and Fig. 5. Each mixture has a good workability as shown by slump flow values of more than 390 mm. However, only $MD8_{40}$ and $MD8_{50}$ exceeded the minimum slump flow requirement of 500 mm and showed excellent performance during the fresh state in terms of stability, filling ability and passing ability and also compressive strength.

The compressive strength of the developed ADW-SCC against the various amounts of granular materials achieved the minimum target strength (30 MPa) except for $MD8_0$. All of these performances were influenced by a significant amount of paste volume (cement and ADW) present in each mixture as shown in Fig. 6.

Table 5. Fresh properties of ADW-SCC.

	Fresh State				**Compressive strength (MPa)**
	Slump Flow	**J Ring (mm)**	**L box**	**V funnel**	
$MD8_0$	390	-	-	-	25.7
$MD8_{10}$	445	-	-	-	34.7
$MD8_{20}$	405	-	-	-	38.3
$MD8_{30}$	440	-	-	-	46
$MD8_{40}$	812mm, $T_{500} = 2$ s	760mm, $T_{500} = 2$ s, H1 = 15 mm H2 = 9 mm H1/H2 = 1.67	$T_{20} = 1$ s, $T_{40} = 2$ s, H1 = 90 mm H2 = 88 mm H1/H2 = 1.0	$T_{10s} = 12$ s $T_{5m} = 15$ s	40.8
$MD8_{50}$	830mm, $T_{500} = 3$ s	785mm, $T_{500} = 3$ s, H1 = 20 mm H2 = 20 mm, H1/H2 = 1.0	$T_{20} = 2$ s, $T_{40} = 3$ s, H1 = 90 mm H2 = 90 mm H1/H2 = 1.0	$T_{10s} = 11$ s $T_{5m} = 18$ s	34.7

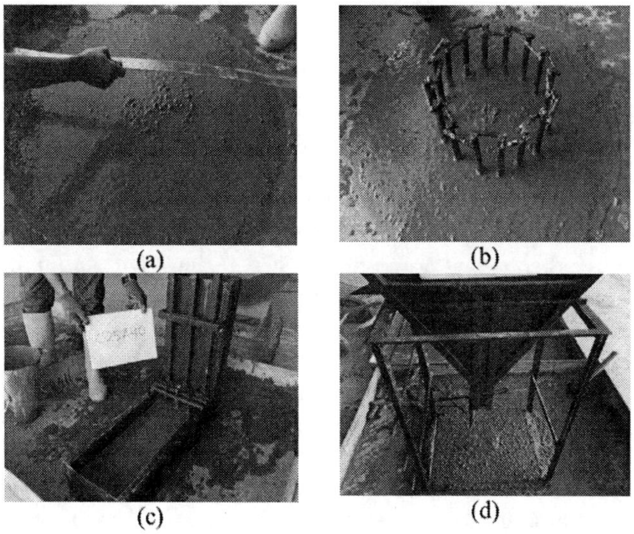

Fig. 4. Fresh state test using (a) slump flow test, (b) J Ring test, (c) L Box test and (d) V Funnel test.

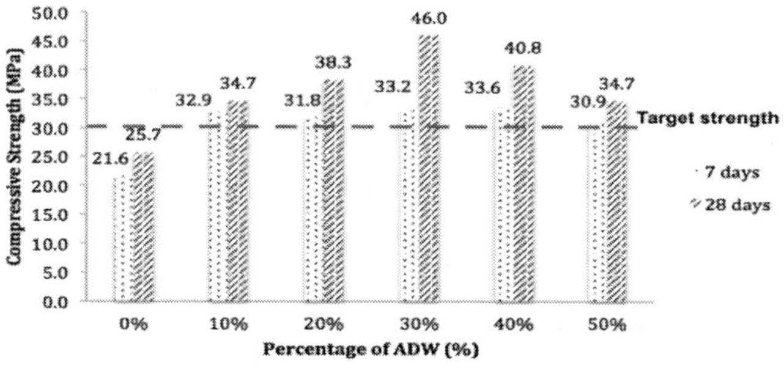

Fig. 5. Compressive strength measured on day 7 and day 28.

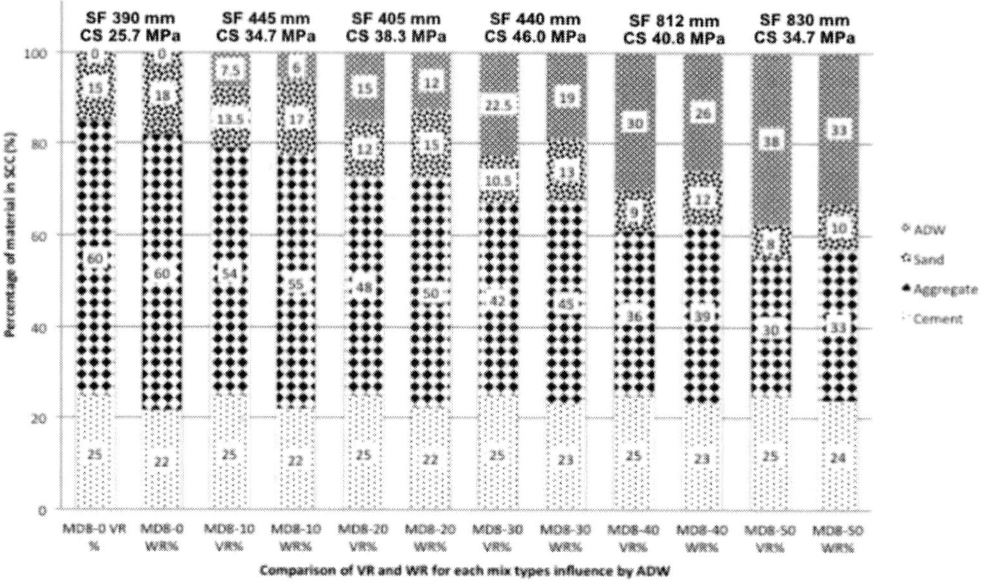

Fig. 6. Distribution of materials in ADW-SCC base on VR (%) and WR (%).

5 Conclusions

From the findings of this experimental modified volumetric mix design for self-compacting concrete by utilizing ADW, the following conclusions can be drawn:

i) The disposal of ADW materials from the premix batching industry is one of the major problems for the environment today. However, this study opens a new field for utilizing ADW in SCC and successfully proven and economically utilized ADW as a paste mixture to improve rheological properties in SCC technology. Furthermore, the physical properties of ADW showed that it is also suitable to be used as additional filler material to improve hardened strength in terms of packing factor.

ii) The distribution of materials is highly useful to produce ADW-SCC with reference to design slump flow (SF), compressive strength (CS) and material distribution. Moreover, w/b ratio and amount of SP can be adjusted to achieve minimum slump flow. Higher w/b ratio and SP (%) will improve the rheological effect. However, lower w/b ratio and SP content is good for fracture toughness.

iii) Further studies on mortar paste volume should be carried out in order to evaluate the hardened performance of ADW-SCC. The amount of paste materials and w/b ratio should be minimized to obtain high performance SCC.

In general, the development of modified volumetric mix design for SCC by utilizing ADW showed the possibility of producing high performance ADW-SCC. It can be concluded that ADW contributes to better stability, filling ability and passing ability of the ADW-SCC. However, further work needs to be done on the effect of ADW in terms of other aspects such as shrinkage, crack pattern, creep and other durability performances.

The authors gratefully acknowledge the Ministry of Higher Education Malaysia (MOHE) for providing financial support for authors and to Universiti Tun Hussein Onn Malaysia (UTHM) for providing the resources needed to complete this paper.

References

[1] P.L. Domone, Self-compacting concrete: An analysis of 11 years of case studies, Cem. Concr. Compos., **28**(2), 197–208, (2006)
[2] H. Okamura and M. Ouchi, Self-compacting concrete, Adv. Concr. Technol., **1**(1), 5–15, (2003)
[3] Y. Sheen, L.-J. Huang, T.H. Sun and D.H. Le, Engineering properties of self-compacting concrete containing stainless steel slags, Procedia Eng., **142**, 79–86, (2016)
[4] M. Benaicha, X. Roguiez, O. Jalbaud, Y. Burtschell and A. H. Alaoui, Influence of silica fume and viscosity modifying agent on the mechanical and rheological behavior of self compacting concrete, Constr. Build. Mater., **84**, 103–110, (2015)
[5] D. Carro López, B. González Fonteboa, J. de Brito, F. Martínez-Abella, I. González-Taboada and P. Silva, Study of the rheology of self-compacting concrete with fine recycled concrete aggregates, Constr. Build. Mater., **96**, 491–501, (2015)
[6] E. Güneyisi, M. Gesoglu, A. Al-Goody and S. İpek, Fresh and rheological behavior of nano-silica and fly ash blended self-compacting concrete, Constr. Build. Mater., **95**, 29–44, (2015)
[7] I. Ismail, N. Jamaluddin and S. Shahidan, A review on performance of waste materials in self compacting concrete (SCC), J. Teknol., **5**, 29–35, (2016)
[8] K. E. Alyamaç and R. Ince, A preliminary concrete mix design for SCC with marble powders, Constr. Build. Mater., **23**(3), 1201–1210, (2009)
[9] H. Y. Aruntas, Workability and mechanical properties of self-compacting concretes containing LLFA , GBFS and MC, **73**, 626–635, (2014)
[10] C. Shi, Z. Wu, K. Lv and L. Wu, A review on mixture design methods for self-compacting concrete, Constr. Build. Mater., **84**, 387–398, (2015)
[11] EFNARC, *Specification and guidelines for self-compacting concrete*, United Kingdom, (2002)
[12] N. Su, K.C. Hsu and H.W. Chai, A simple mix design method for self-compacting concrete, Cem. Concr. Res., **31**(12), 1799–1807, (2001)
[13] C. Selvamony, M. S. Ravikumar, S. U. Kannan and S. B. Gnanappa, Investigations on self-compacted self-curing concrete using limestone powder and clinkers, ARPN J. Eng. Appl. Sci., **5**(3), 1–6, (2010)
[14] M. C. S. Nepomuceno, L. a. Pereira-De-Oliveira, S. M. R. Lopes and R. M. C. Franco, Maximum coarse aggregate's volume fraction in self-compacting concrete for different flow restrictions, Constr. Build. Mater., **113**, 851–856, (2016)
[15] H. Zhao, W. Sun, X. Wu and B. Gao, The properties of the self-compacting concrete with fly ash and ground granulated blast furnace slag mineral admixtures, J. Clean. Prod., **95**, 66–74, (2015)
[16] M.L. Ghambhir, *Concrete technology (theory and practice)*, Fourth Eds, Tata McGrawhill Education Private Limited, (2009)

Estimating Corrosion Initiation Period Due to Chloride Ingress into Reinforced Self-Compacting Concrete Incorporating High Volume Fly Ash

Stefanus Kristiawan[1,*], *Wibowo*[1], *Sunarmasto*[1], *Buntara S* Gan[2] and *Septina* Nurrohmah[1]

[1]SMARTCrete Research Group, Civil Engineering Department, Sebelas Maret University, Jl. Ir. Sutami No. 36A Surakarta 57126, Indonesia
[2]Department of Architecture, College of Engineering, Nihon University, Koriyama, Japan

Abstract. One of the main sources of reinforcement corrosion is the ingress of chloride ion into a concrete structure. The rate of chloride ingress and its consequence to initiate corrosion could determine the service life of the structure. This paper aims to estimate the corrosion initiation period of reinforced self-compacting concrete incorporating high volume fly ash. The governing equation whih characterize the chloride ingress through ionic diffusion is numerically solved using finite difference method. The numerical solutions are used to assess the influence of fly ash content, concrete cover and temperature on the time of corrosion initiation.

1 Introduction

Service life of structure is an important aspect that should be taken into account in the design process. A structure must be designed in such a way to achieve an economical structure at an acceptable reliability throughout its service life in accordance with the needs of the user. The reliability must cover all types of actions that may arise during the expected service life of the structure. These provisions must be met by means of suitable material selection, appropriate design method and detailing, and proper control of design process, production, execution and use [1]. In an attempt to meet the requirements, it is necessary to have a fundamental understanding of the factors that predominantly influence the service life of the structure.

The influencing parameters that determine the service life of structure may be governed by the rate of material degradation. In the case of reinforced concrete structure, the degradation rate could be related to the ingress of aggressive agents from the environment surrounding the concrete structure. Thus, the type of aggressive agents and their concentration are among those of the determinant factors. The degradation rate also depends on the quality of the concrete itself, especially the porosity/permeability of concrete. Taking into account these two factors (environment and concrete quality), many design Codes provide a clause regarding the environmental classification and its

* Corresponding author: s.a.kristiawan@ft.uns.ac.id

corresponding proportion of concrete to satisfy the performance. This clause indicates that the method used to meet the service life requirement is a prescriptive approach [2,3].

The prescriptive approach is developed on the basis of available data of performance of existing structures. This approach has a limitation in its application when new material is introduced since there are no reference data for such material. The limitation encourages designer to move toward the performance-based approach [4-5]. Many researches have been carried out to develop performance-based method for estimating service life of structure by considering particular environmental action, for example chloride ingress into concrete [6-10]. The basic principle of the method is to estimate the service life as the sum of the corrosion initiation and propagation period. The initiation period is the time required for chloride ions to penetrate into the body of concrete and reach a critical concentration at the surface of the embedded reinforcement which could initiate corrosion. The propapation period corresponds to the degradation that follows until the limit state is attained beyond which structure will fail.

The mechanisms by which aggressive ions may penetrate into concrete could be due to one or combination of the following ionic transport mechanisms [10]: pressure-induced water flow, water absorption, water vapour diffusion, wick action, ion diffusion, gas diffusion and pressure-induced gas flow. For chloride ions transport, the predominant mechanism is thought to be the ionic diffusion. Hence, the performance-based methods for estimating corrosion initiation are developed on the basis of calculating chloride ions diffusion causing corrosion initiation. The chloride ions diffusion is governed by the Fick's second law as follows:

$$\frac{\partial C(x,t)}{\partial t} = D\frac{\partial^2 C(x,t)}{\partial x^2} \tag{1}$$

where x, t, $c(x,t)$ *and* D represents depth of chloride penetration, time, concentration of chloride ion at depth x and time t, and chloride ion diffusion coefficient of concrete, respectively. The boundary conditions for solving Equation (1) is:

$$C(x,t)\begin{cases} C_S \text{ if } x=0 \\ Co \text{ if } x\approx\infty \end{cases} \tag{2}$$

where c_O and c_S is an initial concentration of chloride ion in the body of concrete and concentration of chloride ion at the concrete surface, respectively. Assuming homogenous diffusion of concrete property, solution of Equation (1) with boundary condition as in Equation (2) gives:

$$\frac{C_S - C(x,t)}{C_S - C_o} = erf\left(\frac{x}{2\sqrt{Dt}}\right) \tag{3}$$

Equation (3) may be used to estimate the corrosion initiation by substituting x with the depth of embedded reinforcement and $c(x,t)$ with the critical chloride concentration (c_{cr}).

Sun et al. [11] confirmed that it may not appropriate to assume a homogenous characteristic of concrete diffusion along its depth. In fact, many factors affect the ionic transport properties of chloride into concrete i.e. reactivity of chloride ion and its chemical bound with the cement paste constituent, the time dependent property of concrete, the temperature of the environment, etc. Hence, Equation (3) may not correctly predict the corrosion initiation. Adjustment of Equation (1) is, therefore, necessary to accommodate any factors influencing the ionic transport properties of chloride into concrete. Consequently, the exact solution of Equation (1) will not easily be determined. For this reason, numerical solution could be introduced to properly describe the complex ionic transport behaviour. With numerical solution, concrete is divided into several layers along its depth. Initial boundary conditions and any factors affecting the transport properties of the chloride ions could be assigned for each discrete element. An example of such solution is that of Life-365 software, which has been developed for estimating service life of reinforced concrete due to chloride ingress [12]. The software capable to predict the effect of various factors either environmental conditions or concrete ingredients on the service life of reinforced concrete structure. However, the software is not suitable to be used for estimating special type of concrete which is not covered by its database. This paper aims to provide numerical solution for estimating corrosion initiation of special type of concrete i.e. self-compacting concrete incorporating high volume fly ash. The high volume fly ash is defined by the replacement level of cement with fly ash where at least 50% (by weight of total binder) of fly ash is introduced. This special type of concrete is then termed as high volume fly ash-self compacting concrete (HVFA-SCC). The finite difference method is chosen to obtain the numerical solution of Equation (1). The results could be used to estimate the influence fly ash content on the corrosion initiation period of HVFA-SCC. The numerical solution is extended to include the influence of concrete cover and temperature on the corrosion initiation period of the said concrete.

2 Numerical solution with finite difference method

The finite difference formulation of Equation (1) is given as follows:

$$C(t+\Delta t) = C(t) + \left(D\frac{C(x+\Delta x) - 2C(x) + C(x-\Delta x)}{\Delta x^2} \right)\Delta t \tag{4}$$

The influence of the temperature on the coefficient diffusion D is taken into account in the current research by adopting the relation as that of Life-365 model. The influence of fly ash is inserted in the relation as follows:

$$D_{fG(T)} = D_{ref}.FA.\exp\left[\frac{\upsilon}{R} \cdot \left(\frac{1}{\tau_{ref}} - \frac{1}{\tau} \right) \right] \tag{5}$$

where $D_{fa(T)}$ is coefficient diffusion of HVFA-SCC with fly ash content *fa* at absolute temperature T (K). Meanwhile, D_{ref} and T_{ref} is reference coefficient diffusion and reference temperature, respectively. For the current research, the D_{ref} is represented by $D_{50(293)}$ i.e. the coefficient diffusion of HVFA-SCC with 50% fly ash content determined at T_{ref} about 293 K (20°C). U is activation energy of the diffusion process (35000 J/mol) and R is gas constant (8.314472 J/mol.K). FA is ratio of coefficient diffusion of HVFA-SCC at fly ash content higher than 50% to that of D_{ref}. Hence, its value is 1 for HVFA-SCC with fly ash content 50%. For HVFA-SCC with higher fly ash content (more than 50%), Equation (6) may be used to substitute FA of Equation (5).

$$FA = 0.00014 fa^3 - 0.0242 fa^2 + 1.351 fa - 23.8 \tag{6}$$

Equation (6) is derived using data from [13]. The basic mixture of SCC used in this reference was 738 kg of cement, 579 and 703 kg of fine and course aggregate, respectively, 211 kg of water and 7.27 kg of superplasticizer as per m^3. Fly ash contents in the range of 50-70% were introduced to partially substitute cement. The chloride profiles of SCC with various fly ash contents were determined by salt ponding test as specified in AASHTO

T259 where the surface of HVFA-SCCs were exposed to solution of 3% NaCl. This test was carried out under laboratory environment. After 90 days period of exposure, the amount of chloride ingress into HVFA-SCCs were determined along the depth of the penetrations. These chloride profiles were then analyzed to obtain coefficient diffusions and concentrations of chloride ion at the surface of concrete (*Cs*) shown in Table 1 [13]. The relationship of ratio of coefficient diffusion of HVFA-SCC with fly ash content more than 50% to D_{eff} i.e. *FA* versus fly ash content *fa* could be established by regression analysis (Fig. 1) resulting in Equation (6).

Table 1. Coefficient diffusion of HVFA-SCC and *Cs* value at various fly ash contents.

Fly ash content	**50%**	**55%**	**60%**	**65%**	**70%**
Coefficient diffusion (m^2/s)	2.20E-12	1.98E-12	1.90E-12	1.70E-12	2.10E-12
Cs (% by weight of concrete)	0.9963	0.9024	0.8229	0.8064	0.8735

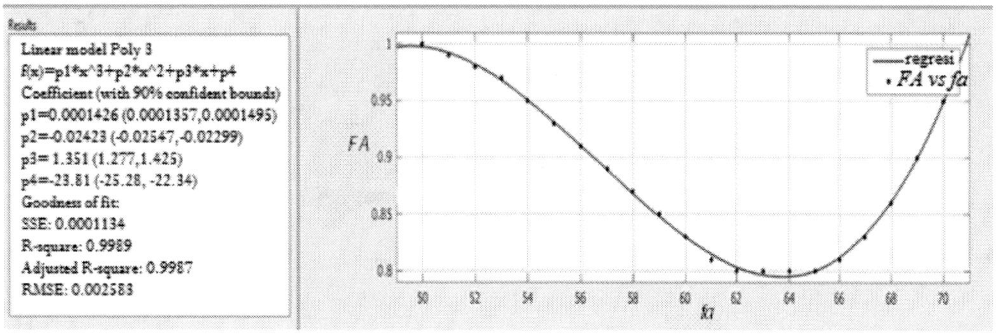

Fig. 1. Relationship of *FA* and *fa*.

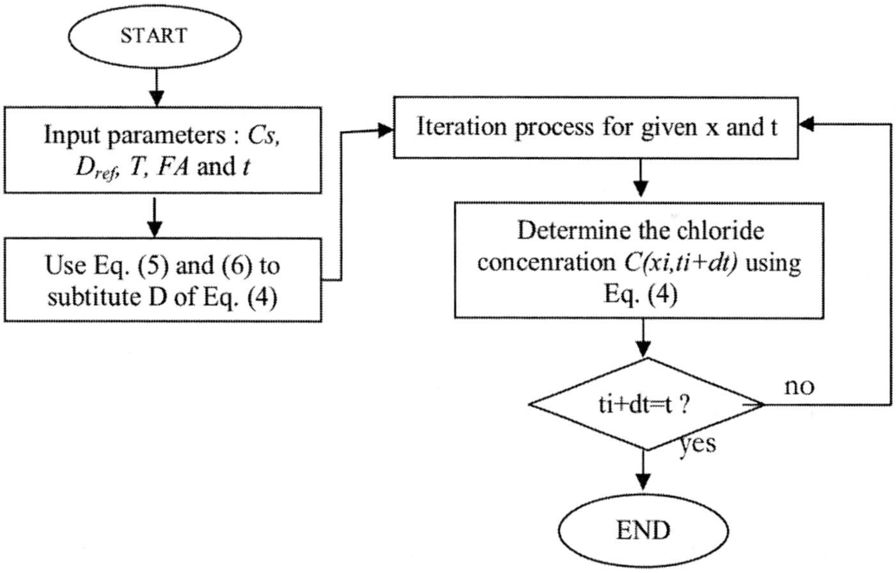

Fig. 2. Flowchart of numerical solution procedure.

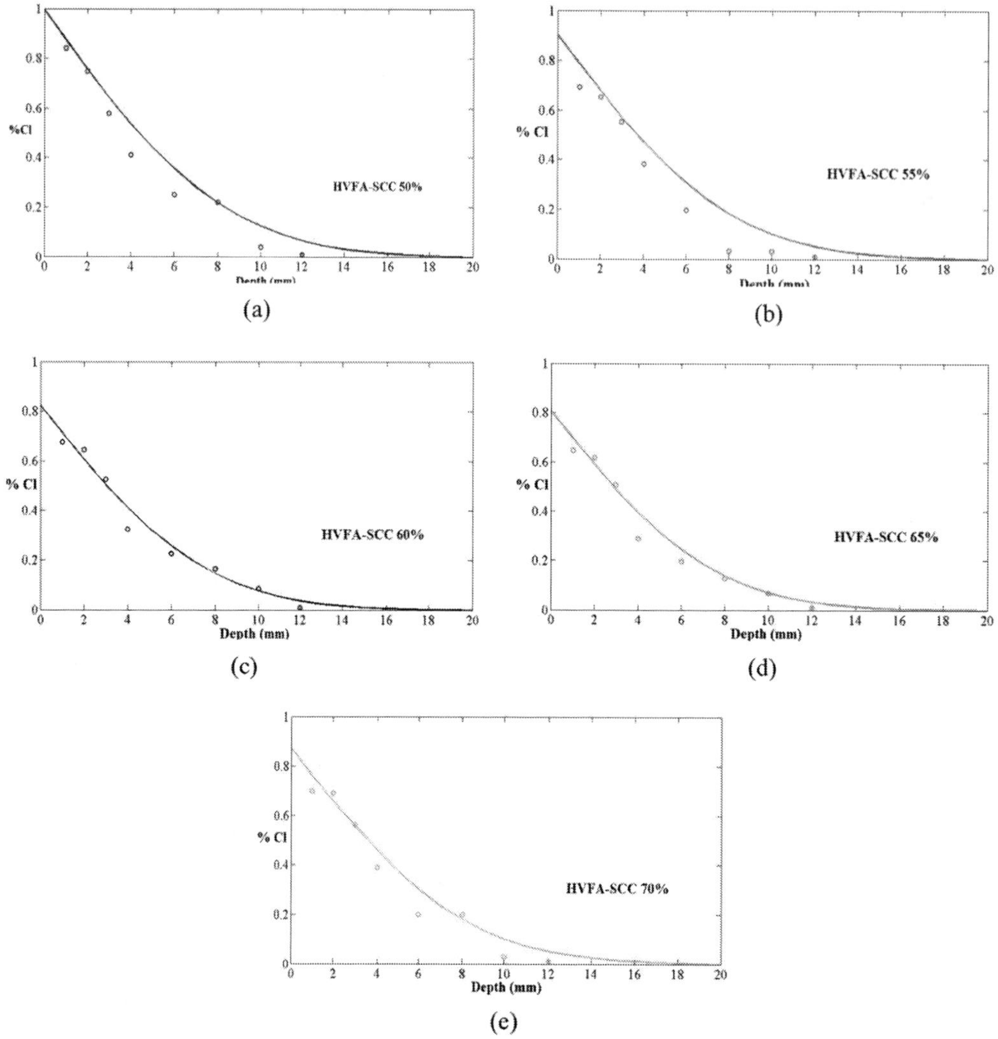

Fig. 3. Comparison of the numerical solution (line) and the experimental data of chloride profiles (dots) for HVFA-SCC with various fly ash contents.

Fig. 2 shows the procedure for obtaining the numerical solution of Equation (4). The required input parameters are D_{ref}, Cs, T, t and fa. The value of Cs refers to Table 1. With the given input parameters, the first step is to substitute FA of Equation (5) with Equation (6) and then uses the $D_{fa(T)}$ of Equation (5) to substitute the coefficient diffusion D of Equation (4). Iteration process is carried out to evaluate Equation (4) for given x and t.

The results of numerical solution are validated with the experimental data of chloride profiles from [13] as shown in Fig. 3. Generally, it can be conveniently concluded that a good agreement exists between the results of numerical solution and the experimental data.

3 Estimating the corrosion initiation of HVFA-SCC

The numerical solution by finite difference method as described in the previous section is now applied to estimate the corrosion initiation period of HVFA-SCC. The critical chloride concentration (c_{cr}) which could possibly initiate the corrosion is in the range of 0.4-1.0 % by weight of cement (or 0.06-0.3% by weight of concrete) [14]. For this study, c_{cr} is assumed to be 0.06% by weight of concrete. The embedded reinforcement is given at a distance of 50 mm from the concrete surface. The environmental condition is represented by the value of chloride concentration at the surface of concrete (*Cs*), where in this case is assumed to be 0.2% by weight of concrete. This value is taken to represent a typical temperate climate marine environment [15]. In reality, there is a period which is required for the chloride concentration to reach its maximum concentration at the concrete surface. The period to build to maximum concentration is omitted in this simulation. Hence, the estimated corrosion initiation of this simulation is based on assumption that *Cs* occurs instantly. Fig. 4. (a) shows the influence of fly ash content on the time of corrosion initiation. Increasing fly ash content from 50% to 60% could prolong the initiation period at about 3 years. The figure also clearly illustrates an optimum fly ash content of 64% that will provide the longest corrosion initiation.

Fig. 4.(b) illustrates the influence of concrete cover on the time of corrosion initiation of HVFA-SCC with fly ash content of 50%. This figure provides a good indication of the important of concrete cover to protect the embedded reinforcement. The effect of increasing concrete cover is to prolong the corrosion initiation period at an accelerating rate. For examples, if concrete cover is inceased by 10 mm from its original value of 40 mm, the corrosion initiation is prolonged by 5 years. Further increase of 10 mm will give an additional 7 years of corrosion initiation period.

The effect of temperature on the corrosion initiation period of HVFA-SCC with fly ash content of 50% is illustrated in Fig. 4. (c). The concrete cover used for this simulation is 40 mm. The figure clearly shows that an increase of environment temperature from 25^{0}C to 35°C will reduce the corrosion initiation from 14 years to 9 years.

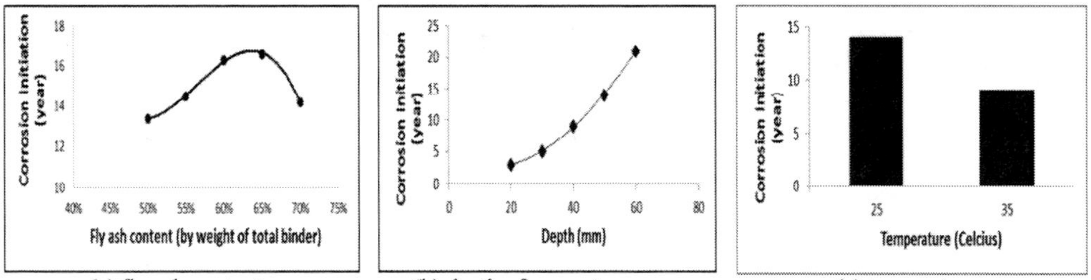

(a) fly ash content (b) depth of concrete cover (c) temperature

Fig. 4. Effect of various parameters on corrosion initiation period of HVFA-SCC.

4 Conclusions

Numerical solution by finite difference method has been adopted to characterize chloride ion diffusion into self-compacting concrete incorporating high volume fly ash. Generally, a good agreement exists between the results of numerical solution and the experimental data. The numerical solutions are then applied to estimate the corrosion initiation period of mentioned concrete and assess the influence of fly ash content, concrete cover and temperature. The optimum fly ash content which yields the longest corrosion initiation period is about 64%. It is also found that concrete cover is essential to protect the embedded reinforcement. An increase of concrete cover will prolong the corrosion initiation period at accelerating rate. Higher temperature will decrease corrosion initiation period.

References

[1] Eurocode 1 General actions-Part 1-1: *Densities, self-weight, imposed loads for buildings,* British Standard, London, United Kingdom, (2002)

[2] Eurocode 2 General actions-Part 1-1: *Design of concrete structures*, British Standard, London, United Kingdom, (2004)
[3] Eurocode 1 General actions-Part 1: *Specification, performance, production and Conformity*, British Standard, London, United Kingdom, (2000)
[4] International Federation for Structural Concrete (*fib*), *Bulletin 34: Model code for service life design,* the International Federation for Structural Concrete (*fib*), Lausanne, Switzerland, (2006)
[5] International Federation for Structural Concrete (*fib*), *Model code for concrete structures 2010*, Wilhelm Ernst & Sohn, Rotherstraße 21, 10245 Berlin, (2013)
[6] I. Khan, R. Francois and Castel, Cem. Conc. Res. **56**, 84 (2014)
[7] U. Angst, A. Ronnquist, B. Elsener, S.K. Larsen and O. Vennesland, Corr. Sci. **53**,177 (2011)
[8] U. Angst, B. Elsener, C.K. Larsen and O. Vennesland, Corr. Sci. **53**, 1451 (2011)
[9] U. Angst, B. Elsener, C.K. Larsen and O. Vennesland, Electrocimica Acta **56**, 5877 (2011)
[10] R.M. Ferrera, *Probability-based durability analysis of concrete structures in marine environment* (PhD Thesis, University of Minho, Portugal, (2004)
[11] Sun G, Zhang Y, Sun W, Liu Z, and Wang C, Constr. Build. Mat. **25 (10)**, (2011)
[12] M.A. Ehlen, *Life-365^{TM} Service Life Prediction Model^{TM} and Computer Program for Predicting the Service Life and Life-Cycle Cost of Reinforced Concrete Exposed to Chlorides Version 2.2*, Life 365TM Consortium III, USA, (2014)
[13] S.A. Kristiawan, S. As'ad, Wibowo, B.S. Gan, D.P. and Sitompul, Effect of high volume fly ash on the chloride penetration resistance of Self-Compacting Concrete (SCC), *Advances in Civil, Architectural, Structural and Constructional Engineering.* CRCPress, 31-34, (2016)
[14] G. Markeset and O. Vennesland, *Critical chloride content in reinforced concrete,* COIN Project Report **6**, SINTEF Building and Infrastructure, Oslo, Norway, (2009)
[15] A. Costa and J. Appleton, Mat. Struc. **32**, 354 (1999)

Rheology of Green Self-Consolidated Concrete (GSCC) with Ternary Blend under Coupled Effects of Temperature and Prolonged Mixing

Samer Al-Martini [1,*], and *Mohammed* Al-Khatib [2]

[1] Department of Civil Engineering, College of Engineering, Abu Dhabi University, Abu Dhabi, UAE
[2] Lab Teaching Assistant of Civil Engineering, College of Engineering, AlHosn University, Abu Dhabi, UAE

Abstract. The fresh quality of a concrete mixture subjected to high temperature and prolonged mixing is adversely affected. These harsh conditions would result in a stiff concrete that is hard to handle and place. This paper investigates the coupled effects of temperature and prolonged mixing on green self-consolidating concrete (GSCC) incorporating a ternary blend of fly ash and silica fume. This study was carried out in the construction material laboratory of Abu Dhabi University under an outdoor environment during the summer of 2014. The SCC mixtures were continuously mixed for 110 minutes at a temperature ranging from 25 to 40 °C to simulate the actual conditions of concrete during transporting to a construction site. The fresh properties were tested using BT2 portable rheometer, slump flow, T_{50} and visual stability index (VSI). The results showed that the ternary blend of silica fume and fly ash enhanced the rheological properties of SCC mixtures under high temperature and prolonged mixing time.

1 Introduction

Concrete mixtures, in United Arab Emirates (UAE) during summer, are usually subjected to both high temperature and prolonged mixing during its transportation to construction sites. These harsh conditions would adversely affect the fresh and hardened properties of concrete. As such, concrete in a transit truck would arrive to the construction site with low quality, and may not meet the job requirements for placement and consolidation. Consequently, such concrete shipment is most likely rejected, resulting in delaying the construction process, wasting materials and affecting the environment.

Several researchers investigated the coupled effects of temperature and prolonged mixing on self-consolidating concrete (SCC). For instance, Al-Martini and Al-Khatib [1], investigated the effect of incorporating fly ash, as partial replacement of cement by mass, on self-consolidating concrete (SCC) under extreme conditions (high temperature and prolonged continuous mixing). They reported that the addition of fly ash alleviated the negative effect of temperature and mixing time on the flowability of SCC mixtures. It was

* Corresponding author: samer.almartini@adu.ac.ae

also documented that SCC mixtures with fly ash had better strength and durability than the reference mixture (i.e., SCC with 100 % cement).

Al-Khatib and Al-Martini [2] studied the effect of incorporating silica fume by mass of cement on SCC mixtures under hot weather and prolonged mixing. The results showed that the slump flow was adversely affected by the temperature and prolonged mixing. Also, the presence of silica fume improved the compressive strength and chloride penetration resistance of SCC mixtures, when compared to the controlled mixtures.

Al-Martini and Nehdi [3] investigated the coupled effects of mixing time and ambient temperature on the slump flow and compressive strength of self-consolidating concrete (SCC). They found that the slump flow decreased with an increased mixing time, while the compressive strength increased with continuous mixing time. The results also revealed that SCC mixtures with slump flow in the range of 400 and 600 mm had superior compressive strength compared with mixtures with slump value lower or higher than that range.

The current research investigates the coupled effects of temperature and prolonged mixing on the rheology of self-consolidating concrete incorporating a ternary blend of supplementary cementitious materials (SCMs), namely fly ash and silica fume, as partial replacement of cement by mass. The mixtures were continuously mixed for 110 minutes under ambient temperature ranging from 25 °C to 40 °C. The main goal of the study is to depict the effect of ternary blend of silica fume and fly ash on SCC mixtures under extreme conditions.

2 Experimental program

Locally available coarse and fine aggregates were used in this study. The coarse aggregates were with 10 mm nominal maximum size. Type I Portland cement, fly ash and silica fume in accordance with ASTM C 150 and ASTM C 618 [4, 5] were used. A second generation polycarboxylate-based high range water reducing admixture (PCH) following the ASTM C 494 type F requirements at a dosage of 1.8% was incorporated in the SCC mixtures [6]. It should be noted that concrete mixtures with no supplementary materials (i.e., 100% cement) were denoted as SCC in the paper, and those incorporating ternary blend were denoted as GSCC.

The concrete was continuously mixed for 110 minutes in a concrete drum mixer. The experiments were conducted during the summer of 2014 in Abu Dhabi. For temperature of 25 °C, concrete was mixed in the construction materials lab with a controlled environment, and for the other tested temperatures (30, 35, and 40 °C) the concrete was mixed out door. The mixing procedure used in this research adopted the one presented by Al-Martini and Nehdi [3]. The coarse and fine aggregates were stored in the same environment of mixing for 24 hours before mixing. The rheology, slump flow, T_{50}, and visual stability index (VSI), were tested at an interval of 30 min with accordance to ASTM C 1611 [7].

3 Rheology test

In this study, BT2 rheometer was used to evaluate the rheological properties of the SCC mixtures investigated. It is a small portable device with an impeller geometry, which can be easily handled and used onsite. Its container can accommodate a concrete sample with a volume up to 20 liters. It needs just one rotational cycle to measure the rheological properties of concrete. The rheometer used in this study measures the torque of concrete at several rotational speeds ranging from 0 to 4 m/s. It should be noted that the BT2 rheometer produces the results in relative values (i.e., relative viscosity (N. s) and relative yield stress (N.m) [8]. The relative yield stress and the relative plastic viscosity are measured using the flow curves produced by the BT2 rheometer. The relative yield stress is the intercept of the fitted data line with the torque axis, and relative viscosity is the slope of that line. The torque of SCC and GSCC mixtures increases linearly with the increase of the flow rate. This relation can be expressed by:

$$T = g + hN \tag{1}$$

where T is the torque (N.m), g is the flow resistance (N.m), h is the viscosity factor (N.s), and N is the rotational speed (m/s).

4 Results and discussion

4.1 Effect of temperature and mixing time on relative viscosity of SCC and GSCC mixtures

Fig. 1 shows the effect of temperature on the relative viscosity of SCC and GSCC mixtures. The ambient temperature ranged from 25 to 45 °C. Figure 1a shows data after short mixing time (20 min). In general, the relative viscosity increased with increasing temperature for all mixtures as shown in Fig. 1 (a). At low and moderate temperatures (25 and 30 °C), GSCC mixtures with ternary blend had higher relative viscosities than that for SCC mixtures with 100% cement, while at high temperatures (35 °C and 40 °C), the GSCC mixtures had lower relative viscosities than that for the SCC mixtures.

The results of the SCC and GSCC mixtures after 110 min of continuous mixing are shown in Fig. 1 (b). The results show that the GSCC mixtures containing ternary blend had lower relative viscosity than that for the SCC mixtures with 100% cement at all temperatures. This indicates that the relative viscosity of the GSCC mixture (with ternary blend) was highly sensitive to the prolonged mixing compared to the SCC mixture (with 100% cement). It can be further observed that the relative viscosity of mixtures after prolonged mixing was higher than that after short mixing for all temperatures.

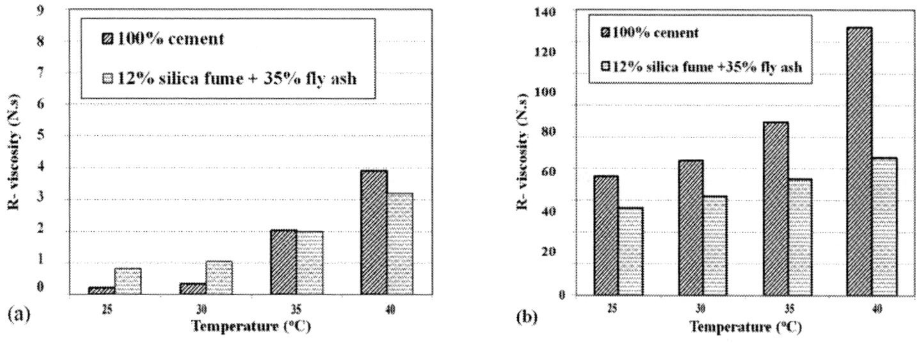

Fig. 1. Relative viscosity of SCC and GSCC mixtures: at (a) 20 min and (b) 110 min

4.2 Effect of temperature and mixing time on relative yield stress of SCC and GSCC mixtures

The relative yield stress values of GSCC mixture incorporating a ternary blend of silica fume and fly ash (35% fly ash and silica fume at 12%) are illustrated in Fig. 2. Data results for control mixtures (SCC with 100% cement) are also presented in the figure. In general, the relative yield stress increased with a higher temperature for short and long continuous mixing. After short mixing time (20 min), the relative yield stress for GSCC mixtures was higher than that of SCC mixture at all temperatures (Fig. 2 (a)). Fig. 2 (b) shows the effect of temperature on the relative yield stress after 110 min of continuous mixing. Similar to the short mixing time, the yield stress values of SCC mixtures were higher than that of GSCC mixture at all mixing temperatures.

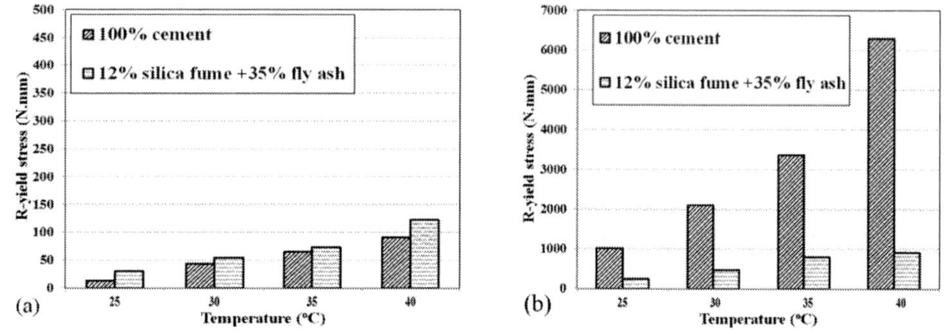

Fig. 2. Relative yield stress of SCC and GSCC mixtures: at (a) 20 min (b) 110 min

4.3 Effect of temperature and mixing time on slump flow, T_{50}, and VSI of SCC and GSCC mixtures

Table 1 shows the results of slump flow, flow rate (T_{50}), and visual stability index (VSI) of SCC and GSCC mixtures at temperature ranging from 25 to 40 °C and prolonged mixing for up to 110 min. The results show that, in general, the slump flow values and flow rate (T_{50}) values decreased with increasing temperature and mixing time, while the stability (VSI) improved.

Table 1. Slump flow, Flow rate (T_{50}), and VSI results of SCC mixtures incorporating ternary blend

Temperature °C		**25**		**30**		**35**		**40**	
Mixtures	Test	20	110	20	110	20	110	20	110
100% CEMENT	Slump flow (mm)	843	295	794	276	691	210	675	343
	T_{50} (s)	1.9	-	4	-	10.9	-	18.1	-
	VSI	(3)	(1)	(3)	(0)	(2)	(0)	(2)	(0)
12%SF+35%FA	Slump flow (mm)	774	550	709	469	676	412	619	396
	T_{50} (s)	9.6	87	14.5	-	18.6	-	29	-
	VSI	(2)	(0)	(1)	(0)	(0)	(0)	(0)	(0)

It can be observed that the presence of the ternary blend of silica fume and fly ash in concrete enhanced the stability of the SCC mixtures, when compared with the 100% cement mixture. After short mixing time (20 min), the slump flow values and the T_{50} values of GSCC mixtures were slightly lower than that of SCC mixtures. However, incorporating the ternary blend of fly ash and silica fumes improved the slump flow and T_{50} after long mixing as evidenced from the higher slump values for the GSCC mixtures after 110 min of continuous mixing. It should be noted that some data results of T_{50} after long mixing time are missing in the table, because those mixtures became stiff, and their slump flow did not reach the 500 mm diameter. Based on these results, it can be argued that ternary blend (30% fly ash and 12% silica fume) is beneficial for SCC mixtures subjected to prolonged mixing time.

5 Conclusion

The coupled effects of temperature and mixing time on the rheological properties of SCC mixtures with 100 % cement and green self-consolidating concrete mixtures (GSCC) incorporating a ternary blend of silica fume and fly ash was investigated. The water to cementitious ratio was kept constant (w/c = 0.36). A constant dosage of polycarboxylate-based high range water reducing admixture was incorporated in all mixtures to maintain adequate initial slump flow. Based on the experimental results, the following can be concluded:

i) The slump flow decreased with increasing the temperature and mixing time.
ii) The combination of fly ash and silica fume improved the stability of SCC mixtures.
iii) The relative viscosity of the SCC mixtures with 100 % cement was more sensitive to the prolonged mixing than the ternary blend mixture (GSCC).
iv) The relative yield stress increased with increasing temperature for all mixing times.
v) It can be argued that ternary blend of fly ash and silica fume is beneficial for SCC mixtures under prolonged mixing time as the slump flow and T_{50} improved.

The authors would like to acknowledge the financial support of Abu Dhabi University (ADU), Grant received in 2015 from the Centre of Excellence on Sustainable Built Environment at ADU (cost center No. 19300078). Thanks are also due to a number of admixtures, additives and aggregates suppliers who contributed materials used in this investigation.

References

[1] S. Al-Martini and M. Al-Khatib, An experimental study on self-consolidating concrete (SCC) under hot weather and hauling time, Key Engineering Materials J., **677**, 3-7 (2016)
[2] M. Al-Khatib and S. Al-Martini, Experimental investigation on the effect of silica fume on green self-consolidating concrete under extreme conditions, *Proc. of Advances and Trends in Engineering Materials and their Applications*, Toronto, (2016)
[3] S. Al-Martini and M. Nehdi, Effects of heat and mixing time on self-compacting Concrete, ICE Construction Materials J., **163,** 175-182 (2010)
[4] ASTM C150 / C150M-16e1, *Standard specification for portland cement,* ASTM International, West Conshohocken, (2016)
[5] ASTM C618-15, *Standard specification for coal fly ash and raw or calcined natural pozzolan for use in concrete*, ASTM International, West Conshohocken, (2015)
[6] ASTM C 494-05a, *Standard specification for chemical admixtures for concrete*, ASTM International, West Conshohocken, (2005)
[7] ASTM C 1611/C 1611M-05, *Standard test method for slump-flow of self consolidating concrete*, ASTM International, West Conshohocken, (2005)
[8] E. Koehler, *Aggregates in Self-Consolidating Concrete*, PhD Thesis, The University of Texas, Austin, (2007)

Mechanical and Fresh State Properties of Medium Strength Self-Compacting Concrete (SCC) Containing Polypropylene Fibres

Anas Alalwany[1], *Noridah* Mohamad[1,*], *A .A. A.* Samad[1], and *W. I.* Goh[1]

[1]Department of Structure and Material Engineering, Faculty of Civil and Environmental Engineering, Universiti Tun Hussein Onn Malaysia, Parit Raja, 86400 Batu Pahat, Johor

Abstract. Due to the rising demand of infrastructure development, concrete has been one of the vital ingredients in this field considering its flexibility to mould any structural shape and design. This paper investigates the effect of adding polypropylene fibres on the mechanical and fresh properties of self-compacting concrete (SCC). The mechanical property tests were conducted on various SCC specimen which include cubes, cylinders and beams to determine its compressive strength, splitting tensile strength and flexural strength of concrete. The workability of self-compacting concrete in terms of viscosity is determined from slump flow test (Diameter and T500), and V-funnel. The results indicated that the addition of polypropylene fibres improved the mechanical properties of SCC at almost every volume fraction considered in this study. It was also observed that the addition of 0.1% fibre significantly enhanced the splitting tensile strength and flexural strength of concrete. Finally, the overall results indicated that introducing fibres to concrete would result in a noticeable decrease in workability and fresh state properties.

1 Introduction

In response to enhance the ordinary concrete, there is a rising need for further development to self-compacting concrete (SCC). In 1988, it was the very first investigation that focused to accomplish concrete structures with high level of durability. Accordingly, a range of researches were conducted, whereas the concrete was used for practical purposes by large construction companies based in Japan [1].

SCC is a non-segregated concrete free of air and spaces that adjusts to any formwork [1]. Without such vibration process, it flows constantly by the means of its weight, thoroughly packing up the formwork and accomplishing full compaction status even with a very packed reinforcement. SCC is utterly one of the most adapted concrete particularly with the use of some chemicals such as super plasticizers and other viscosity modifiers, and minerals i.e. fly ash, slag, silica fume as admixtures in the concrete. Advantages of SCC include faster construction, reduction in manpower, easier placing, uniform and

*Corresponding author: noridah@uthm.edu.my

complete consolidation, better surface finishes, improved durability, increased strength and safer working environment [1].

Polypropylene fibres (PP) were first suggested as an admixture to concrete in 1965, where it was applied in the constructing of anti-blast buildings in the US [2]. The fibre has subsequently been improved further and at present it is used either as short discontinuous fibrillated material for production of fibre reinforced concrete or a continuous mat for production of thin sheet components. It is used in concrete engineering for its enormous advantages such as ductility, calibre, and dispersal so it can pin down any possible cracks [3]. Additionally, it has been proved that adding fibres to concrete enhances formability and bending strength [4].

In this paper, fresh state and mechanical properties of SCC incorporating with PP are thoroughly discussed. The fresh state properties of SCC were determined by using a specific mixture design provided from many trial designs. The mechanical properties of SCC namely; concrete flexural strength, splitting tensile strength and compressive strength were estimated by conducting a compressive strength test on SCC cubes cylinders and beams. All tests were conducted according to the British Standard code [5, 6]. SCC concrete incorporating PP is able to produce concrete with compressive strength higher than conventional concrete.

2 Materials and method

Self-compacting concrete specimens were formed by mixing cement, water, fine aggregate, coarse aggregates, PP and super plasticiser. A Mega Mesh I PP fibre was utilized in this experiment as shown in Fig. 1. Mega Mesh I is used as concrete enhancing material in two mixing designs. PP fibre is considered to offer maximum anchorage that in turn boosts the resistance to pull-out with its three dimensional planes.

Fig. 1. Mega Mesh I.

2.1 Mixture design

A combination of three experiments was tested within the design base on previous work done by Sonebi [7]. 0.2 m^3 of concrete was mixed in each experiment as shown in Table 1

Table 1. Concrete mix design.

PP fibre (kg/m3)	Cement (kg/m3)	FA (kg/m^3)	PP Fiber (kg/m3)	Sand (kg/m3)	Coarse aggregate (kg/ m^3)	SP (kg/ m^3)	Water / (Cement +FA) Ratio
0	270	140	0.00	740	850	2	0.55
0.45	270	140	0.45	740	850	2	0.55
0.9	270	140	0.90	740	850	2	0.55

Note; PP: Polypropylene, FA: Fly ash, SP: Super plasticizers.

2.2 Laboratory tests

The laboratory tests conducted in this study are categorized into two types; namely, fresh state property test and mechanical strength test on SCC cubes and cylindrical specimens containing PP. Fresh concrete testing was selected to investigate the main properties of the fresh mixture in addition to estimate the mixture's passing ability, resistance to segregation and filling ability. Tests were as follows: V funnel, slump flow and slump flow T50. On the other hand, the mechanical strength tests included comprehensive strength tests on the SCC cubes and cylinders to determine the compressive, splitting tensile and flexural strengths for each selected mixture.

3 Results and discussion

3.1 Fresh and physical properties

3.1.1 Slump flow test

From the slump flow test, the fibre distribution was found to be uniform. All mixtures have good fluidity and exhibited self-compacting characteristics. Table 2 demonstrates the data for workability of the fresh SCC as achieved from slump flow tests (Slump flow diameter and T_{500}), T_{5min} for various percentage of PP used.

Table 2. The result of slump flow test and Slump flow T_{500} test.

Sample	Max Diameter,dmax (mm)	Slump Flow Test Perpendicular Diameter,d_{perp}(mm)	Average Diameter, S (mm)	T500 (s)
Mix 1	540	560	550	6
Mix 2	480	420	500	4
Mix 3	360	300	430	2

Results in Table 2 and Fig. 2 show that for the SCC control mixture, the slump flow spread diameter was 550 mm while for T_{500} it achieved 500 mm flow spread. With the presence of 0.9 kg/m^3 PP, the slump flow spread diameter and T_{500} showed decrement up to 430 mm and 2 s, respectively. Mix 2 recorded optimum workability compared to Mix 3. However, the spreading range obtained for Mix 2 was within the acceptable range, which is from 500mm to 700mm [8].

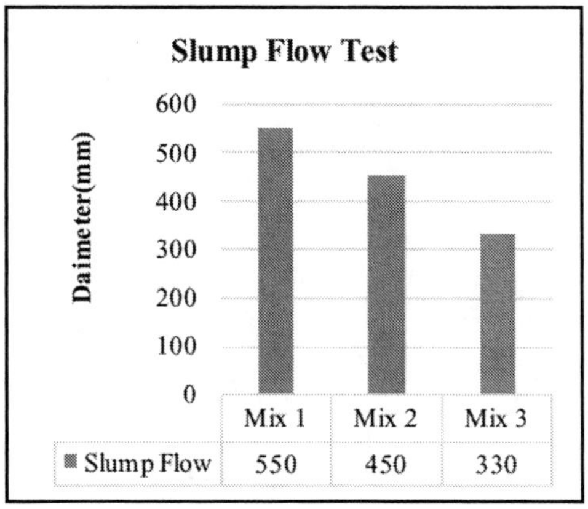

Fig. 2. The bar graph of slump flow result.

3.1.2 V-funnel test

In this experiment, the funnel was filled with the concrete and flow time was accordingly taken for it to flow through the apparatus. The main reason for conducting the V-Funnel test was to assess the stability of SCC mixture [9]. In one study, it was found that a less than 6 second flow time is significantly giving high quality to the SCC [10]. See Table 3 and Fig. 3.

Table 3. The result of V-funnel test.

Experiment	V-Funnel (sec)
Mix 1	5
Mix 2	8
Mix 3	10

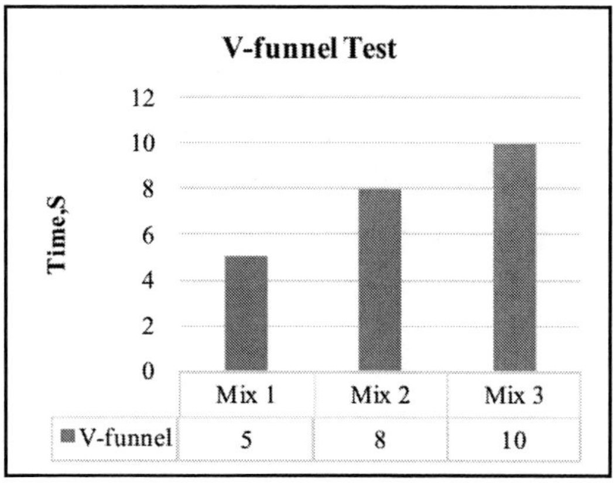

Fig. 3. The bar graph of V-funnel result.

Table 3 shows the result of the V-Funnel test for each selected mixture in this experiment. The measurement of time through V-Funnel test was ranging from 5 to 10 seconds affected by the type of mixture used for each case. A lower flow time of 5 seconds was recorded for the free PPF mixture, while the SCC had highest flow time of 10 seconds. The graphical result is shown in Fig. 3.

DOI: 10.1051/matecconf/201710301011

3.2 Mechanical properties

3.2.1 Compressive strength

Fig. 4 demonstrates the values of compressive strength test for each mixture. Using different PP fibres proportions has indicated vital change in the related results. Increasing the PP fibres in the experiment has shown an enormous increase in the compressive strength values. This change can be justified that the PP fibre is preventing any further extension of cracks as well as reducing the stress extension at the edge of cracks. Another study confirmed the optimistic role of PP fibre in preventing and delaying any extension of cracks [12]. Higher compressive strength attained is due to higher efficiency of PP fibres in bridging the macro-cracks developed in the concrete [13]. The greatest value of compressive strength was attained by the mixture containing 0.9 kg/m^3 PP fibres.

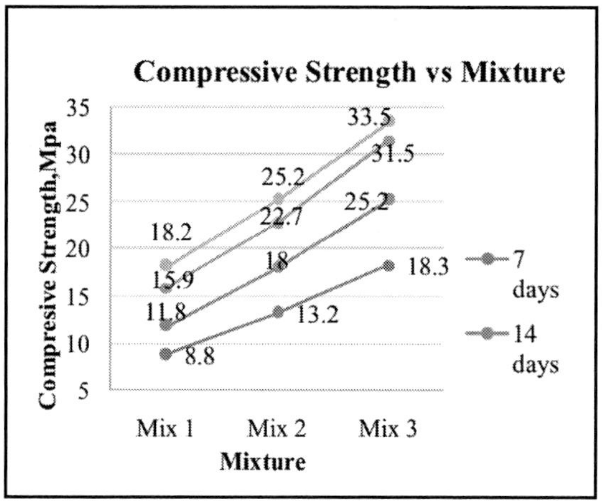

Fig. 4. Compressive strengths values of different PP fibre content.

3.2.2 Splitting tensile and strength flexural strength

The test of splitting tensile was performed as described in the British Standard [14] by using a 150 mm x 300 mm cylinder at the age of 28 days.

Fig. 5. Test of splitting tensile.

Flexural experiment was conducted using 100 mm x 100 mm x 500 mm concrete beams as prescribed in British Standard [15] at the age of 28 days. The flexural strength is calculated by means of constant moment in the central zone of the specimen using two-point loading with a loading rate from the testing machine of 0.04N/mm²/sec.

Fig. 6. Flexural test and specimen.

Three ages were considered for testing in this study, 7, 28 and 56 days. Splitting tensile and flexural strength tests were evaluated in many mixture proportions. Fig 7 and 8 illustrate the values of the selected tests accordingly. It was noticed that increasing the PP fibres proportion has led to an increase in the splitting tensile strength values. This tendency was reported in previous study [12]. It can be noticed that increasing the addition of PP fibres from 0.45 to 0.9 kg/m^3 has affected the splitting tensile strength values from 2.8 to 3.2 Mpa respectively. Results, in general, indicated the improvement of SCC by the adding of PP fibres to the mixture.

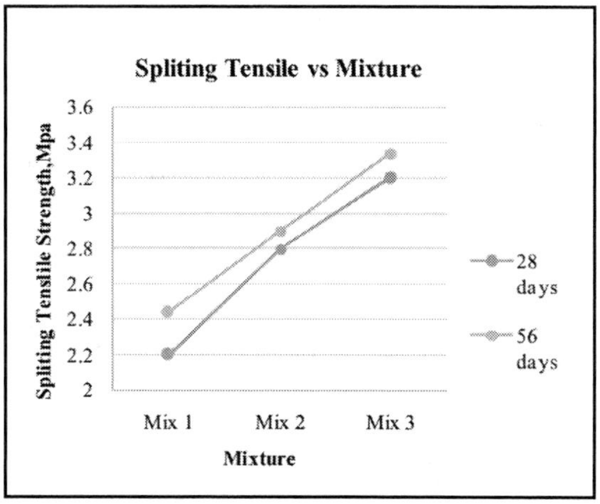

Fig. 7. Tensile strengths of different polypropylene fiber content.

Also, there has been recorded that the addition of PP fibres in the concrete was advantageous and has an enormous role in the flexural strength test. Fig. 8 shows the increased reading of flexural strength test from 2.8 to 3.7 Mpa when adding the PP fibres to the mixture.

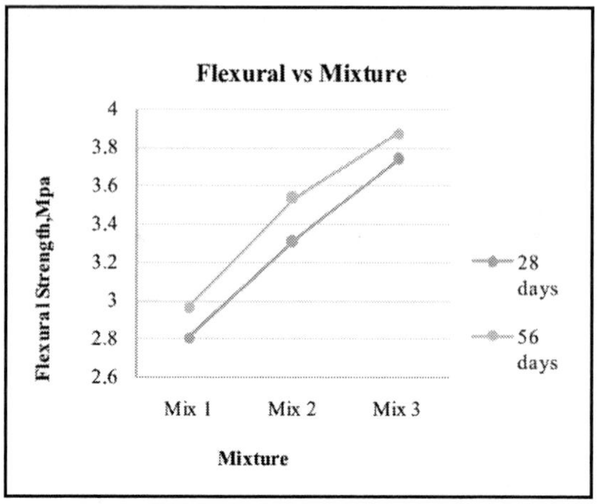

Fig. 8. Flexural strengths of different polypropylene fiber content.

4 Conclusion

The workability of the SCC concrete with PP fibres has shown a considerable decrease with increasing the PP fibres dosage. As tested in the mechanical experiments, adding PP fibres has resulted in much more strong concrete and more crack resistant concrete. The optimum percentage of PP with adequate strength was found to be about 0.9 kg/m^3, as this has been confirmed due to the highest compressive strength value in SCC mixture with 0.9 kg/m^3 of PP fibres which is 31.5Mpa at 28 days. Also, it was noticed that increasing the PP fibre content has led to a noticeable enhancement in mechanical properties of concrete. It is believed that PP fibres has shown its ability at preventing any further extension of cracks as well as reducing the stress extension at the edge of cracks.

The authors gratefully acknowledge the financial support of Higher Ministry of Education and Universiti Tun Hussein Onn Malaysia. (FRGS Vot 1525)

References

[1] H. Okamura, M. Ouchi, Self-compacting concrete. J. of Advance Concrete Technology. Japan Concrete Institute, **1**(5), 5-15, (2003)
[2] T.Ch. Mahdavi, L. Swamy Raju and D. Mathur, Polypropylene fiber reinforced concrete- a review, *Int. Conf. (ACECIM),* 114-119, (2014)
[3] Z. Bayasi and M. McIntyre, Application of fibrillated polypropylene fibers for restraint of plastic shrinkage cracking in silica fume concrete, ACI Materials J., **99**, 337-344 (2002)
[4] S.P. Shah, S.E. Swartz and C. Ouyang, *Fracture mechanics of concrete*, John Wiley & Sons, Inc., New York, USA, (1995)
[5] BS 1881: Part 114, *Determination of density of hardened concrete*, British Standard Association (1983)
[6] BS 1881: Part 116, *Method for determination of compressive strength of concrete cubes*, British Standard Institution (1983)
[7] M. Sonebi, Medium strength self-compacting concrete containing fly ash, Cement and Concrete Research, **34**,1199–1208, (2004)

[8] N. Bouzoubaâ, M. Lachemi, Self-compacting concrete incorporating high volumes of class f fly ash, Cement and Concrete Research, **31**(3), 413-420, (2001)

[9] K. Ozawa, N. Sakata, H. Okamura, Evaluation of self-compact ability of fresh concrete using the funnel test, *Proc. Japan Society of Civil Engineering,* 59–75, (1995)

[10] K. Khayat, A. Ghezal, S. Hadriche, Factorial design model for proportioning self-consolidating concrete, Materials and Structures, **32**(9), 679–686, (1999)

[11] EFNARC-The European Guidelines for Self-Compacting Concrete: Specification, Production, The European Federation of Specialist Construction Chemicals and Concrete Systems, (2005)

[12] V. Afroughsabet and T. Ozbakkaloglu, Mechanical and durability properties of high-strength concrete containing steel and polypropylene fibers, Construction and Building Materials, **38**(3), 389-404, (2015)

[13] M. Hsiea, C. Tua and P. Songb, Mechanical properties of polypropylene hybrid fiber-reinforced concrete, J. Materials Science and Engineering, **27**, 297–306, (2008)

[14] BS 1881: Part 117, British Standard Institution, *Method for determination of compressive strength of concrete cubes*, British Standard Association, (1983)

[15] BS 1881: Part 118, *Method for determination of compressive strength of concrete cube*, British Standard Association (1983)

Strength Properties of Foamed Concrete Containing Crushed Steel Slag as Partial Replacement of Sand with Specific Gradation

Hock Yong Tiong[1*], *Siong Kang* Lim[1], and *Jee Hock* Lim[1]

[1]Lee Kong Chian Faculty of Engineering & Science, Universiti Tunku Abdul Rahman, Bandar Sungai Long, Cheras, 43000, Kajang, Selangor, Malaysia

Abstract. Lightweight construction material, notably foamed concrete, had become more favourable to reduce building weight and cost, accelerate construction process, and ease handling of precast segment. Simultaneously, rapid development had result in price rising of conventional material and environmental issue due to abundant wastes, for instance steel slag. As a consequence, feasibility of steel slag to be incorporated in lightweight foamed concrete for both structural and non-structural purpose is worth to be investigated. This paper is aimed to evaluate the effects of crushed steel slag, as partial replacement of sand with specific gradation, on performance of lightweight foamed concrete (LFC) with density of 1600 kg/m^3 to 1700 kg/m^3 in terms of compressive and tensile strengths. Different steel slag based LFCs were developed by replacing 0, 25, 50, 75 and 100% of steel slag for sand. Different water to cement ratios (w/c) and dosages of super-plasticizer (sp) were adopted to confirm certain workability, strength properties was then studied for ages of 7 and 28 days. The laboratory results showed that lightweight foamed concrete with incorporation of crushed steel slag has decreased strength; however it still achieves structural strength of 17 MPa when replacement level is less than 25% at density of 1600 kg/m^3 to 1700 kg/m^3.

1 Introduction

Lightweight foamed concrete (LFC), as one type of the lightweight concretes (LWC), is a mortar, in which artificial air voids are introduced by suitable foaming agent [1].

LWC possess a density between 300 kg/m^3 and 1850 kg/m^3 based on its type and usage, and can be classified by its purpose to structural lightweight concrete, concrete masonry unit, and insulating concrete [2]. According to ASTM C 330–89 [3], structural lightweight concrete should achieve 28-day cylinder compressive strength (f_{ck}) of 17 MPa, which approximately equivalent to characteristic cube strength (f_{cu}) of 21 MPa.

In the review study by Amran *et* al. [4], foamed concrete has found applied in many area such as structural application, lightweight blocks and pre-cast panel, shock absorbing barriers for traffic, road sub-base, cavity and void filling, repair and rehabilitation, fire,

* Corresponding author: tionghy3@gmail.com

thermal, and acoustic insulation, and many other purposes [1,5,6]. Included in this review, annual market size of foamed concrete is estimated 250,000 – 300,000 m^3 in the UK, 50,000 m^3 in Western Canada, and 250,000 m^3 just in floor heating system in Korea [7-9].

Steel slag is by-product of steel manufacturing. It has been noticed that per year fifty million tons of steel slag is generated from throughout the world [10, 11].

From the research carried out in the Asian country, approximately 10% of steel slag was produced from the production of steel, for instance, in Malaysia about 750 tons of steel were produced daily and approximately 7.5 tons of slag were produced [12].

However, the rate of utilization of steel slag in Malaysia is too low compared to some advanced countries. Albeit, steel slag is still used in some application such as asphalt aggregates for road construction, sub-base, cement stabilization, ground improvement and some other miscellaneous usage [12].

Maslehuddin *et* al. [13] fully replace coarse limestone aggregate (SG=2.51) with coarse steel slag (SG=3.51) in normal concrete and report 5% increment in compressive strength, reduction in absorption and permeability, but reduction in tensile strength.

Kothai and Malathy [14] partially replace fine sand (SG=2.65) with fine steel slag (SG=2.95) in normal concrete and found increment of compressive strength at 5% to 10% when replacement level is between 20% and 40%, but have negligible increment on tensile strength; the gradation show that steel slag used has larger fineness modulus than that of sand.

Falade *et* al. [15] replace sand (SG=2.66) with steel slag (SG=3.47) in foamed concrete and found optimum replacement level of 30% with 20% increment in compressive strength, but the gradation are not specified.

Based on the above literature review, this study is aimed to study the effects of crushed steel slag on strength properties of lightweight foamed concrete (LFC) with density of 1600 kg/m^3 to 1700 kg/m^3 in terms of compressive and tensile strengths at ages of 7 and 28 days, where the gradation of steel slag is same as that of sand, and steel slag is replacing the sand at replacement level of 0%, 25%, 50%, 75% and 100%.

2 Experimental procedures

In this study, strength properties of lightweight foamed concrete with super-plasticizer and crushed steel slag as partial of sand with specific gradation were evaluated.

2.1 Materials

Raw materials used for this study included cement, sand, steel slag, water, foam agent and super-plasticizer (sp).

The ordinary Portland cement complying with Type I Portland Cement in accordance with ASTM C 150-05 [16] was sieved through 300 μm to remove all lumps, subsequently store in moisture-proof container indoor.

The natural river sand was oven-dried and sieved to ensure 100% passing rate for 2.36mm opening. The steel slag, obtained from local hot-roll steel manufacturer, was washed, oven-dried, crushed, and sieved to have equivalent gradation with sand, as shown in Fig. 1. The sand has SG of 2.65 while steel slag has SG of 3.8.

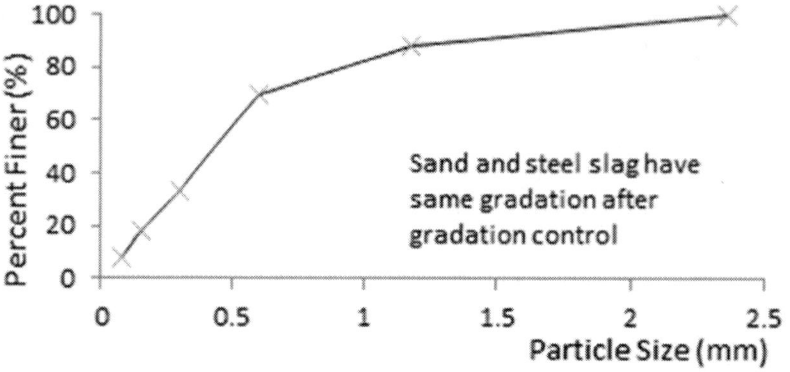

Fig. 1. Gradation of sand and steel slag.

The water used was tap water from the municipal water supply in accordance with ASTM 1602-06 [17]. Petroleum base synthetic detergent was used as foaming agent. Besides, the carboxylic ether type super-plasticizer was used in this study.

2.2 Mix proportions

The foamed concrete in this study was produced by using pre-foaming method, where base mix and pre-formed foam are produced separately and then mix together thoroughly [1,18].

The foam used was dry foam that produced by forcing the foaming agent solution through a series of high density restrictions, and compressed air into mixing chamfer; it is very stable and has size of smaller than 1mm [1,19]. Foam agent and water was mixed in ratio of 1:30 in foam generator before applying compressed air.

The foamed concrete mix proportion and base mix density for different levels of steel slag replacement to sand is presented in Table 1.

Table 1. Mix proportion and details of foamed concrete.

Replacement Level		0%	25%	50%	75%	100%
w/c		0.38	0.373	0.365	0.358	0.35
sp (%)		1	0.875	0.75	0.625	0.5
material per m^3 foamed concrete	cement (kg)	693	695	698	700	702
	sand (kg)	693	522	349	175	0
	steel slag (kg)	0	174	349	525	702
	water (kg)	261	257	253	249	245
	sp (L)	2.63	2.27	1.91	1.56	1.23
base mix density (kg/m^3)		2215	2285	2360	2440	2525
foam required	volume (m^3)	0.255	0.278	0.300	0.323	0.347
	mass (kg)	11.47	12.49	13.52	14.55	15.60

The foam volume required is affected by base mix density and target density. In this study, dosage of super-plasticizer and water to cement ratio were varied to obtain equivalent level of base mix flowability based on observation. Base mix density is mainly affected by replacement level of steel slag due to large difference between specific gravity of steel slag and that of sand. The water to cement ratio has very minor effect on base mix density, for instance, when the water to cement ratio for 0% replacement was change to 0.35, the base mix density, instead of 2215 kg/m³, will be 2250 kg/m³ compare to 2525 kg/m³ at 100% replacement level.

2.3 Testing methods

Testing method used for this study include inverted slump test, compression, split tensile, and flexural tests.

Inverted slump test was performed as per ASTM C 1611-05 [20] to determine the workability of foamed concrete. The slump cone, as per ASTM C 143-08 [21], was placed on a centre of flat base plate with the smaller opening facing down, and filled with foamed concrete. Immediately, the cone was lifted up vertically with a distance of 225 ± 75 mm in 3 ± 1 seconds. The average spread diameter was measured after the concrete stop flowing.

Compressive test was conducted as per BS EN 12390-3 [22]. The concrete cubical specimens with dimension of 100 mm were tested by using a universal compression machine with loading rate of 3 kN/s.

Split tensile test was conducted as per BS EN 12390-6 [23]. The cylindrical specimens with 100 mm diameter and 200 mm length were place horizontally with packing strip and tested by universal compression test machine with loading rate of 1.76 kN/s.

Flexural test was performed using centre-point loading method as per BS EN 12390-5 [24]. The specimens with 40mm x 40mm cross section and total length of 160mm were tested by using testing machine named INSTRON 5582 with effective length of 120mm and loading rate of 0.008 mm/min.

3 Results and discussion

3.1 Workability

Inverted slump test results were shown in Table 2. It was noticed that increment of foam volume has reduced the inverted slump flow of foamed concrete, as shown in Table 2.

Table 2. Workability of foamed concrete.

Replacement Level	0%	25%	50%	75%	100%
Inverted Slump value (mm)	655	680	675	645	555

3.2 Compressive strength

Hardened density, percentage of foam, compressive strength and corresponding compressive strength were tested and evaluated in Table 3.

The relationship between compressive strength and percentage of foam volume, regardless the effect of steel slag, was shown in Fig. 2. Base on the relationship, the corresponding strength at any density between 1600 kg/m³ to 1700 kg/m³ can be calculated. Hence, the corresponding compressive strength at density of 1700 kg/m³ was calculated and shown in Table 3.

The compressive strength, performance index of compressive strength, and corresponding compressive strength were compared, as shown in Fig. 3. Performance index is strength per 1000 kg/m³ density, calculated by strength divided by the density, was used to evaluate the strength performance.

It was noticeable that compressive strength and performance index were influenced by the achieved hardened density, whereas corresponding compressive strength shows an approximate linear transforms, in which, compressive strength was reduced when steel slag replacement level was increased.

The previous researches which indicate increment of strength might have different gradation for sand and steel slag, where steel slag might have larger fineness modulus [14], and have smaller specific gravity [13-15].

No doubt certain amount of fine steel slag is able to improve the strength of concrete [14,15], but it was insufficient to compensate the reduction of strength cause by increment of foam volume due to high specific gravity of steel slag used.

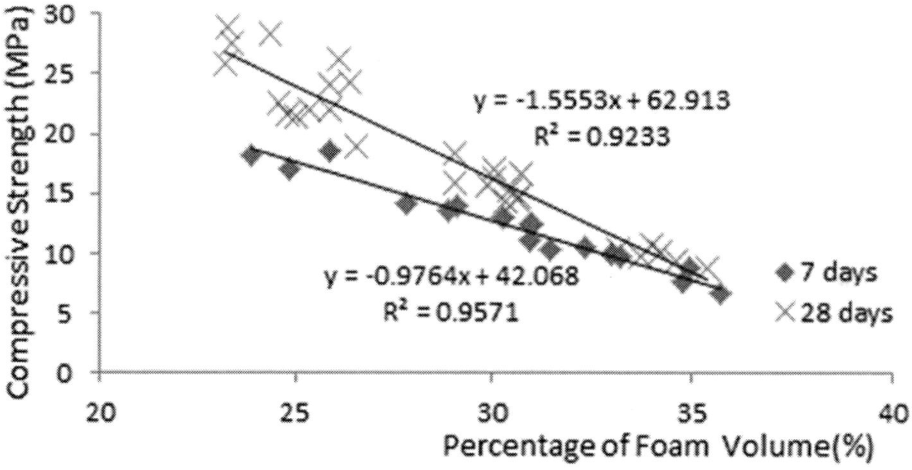

Fig. 2. Relationship between compressive strength and foam volume.

Table 3. Compressive strength, hardened density, and percentage of foam volume

Replacement level		0%	25%	50%	75%	100%
Percentage of Foam in Volume at 1700 kg/m³		23.2	25.6	27.9	30.3	32.7
7 days	Hardened Density (kg/m³)	1647	1622	1634	1654	1637
	Percentage of Foam Volume (%)	25.6	29.0	30.7	32.2	35.2
	Compressive Strength (MPa)	17.0	13.8	12.6	10.4	7.7
	Corresponding Compressive Strength at 1700kg/m³ (MPa)	19.3	17.2	15.4	12.6	9.6
28 days	Hardened Density (kg/m³)	1677	1676	1647	1630	1660
	Percentage of Foam Volume (%)	24.3	26.6	30.2	33.2	34.3
	Compressive Strength (MPa)	24.8	21.9	16.1	10.8	10.1
	Corresponding Compressive Strength at 1700kg/m³ (MPa)	26.4	23.6	19.6	15.3	12.4
7 days / 28 days Corresponding Strength		73%	73%	79%	82%	77%

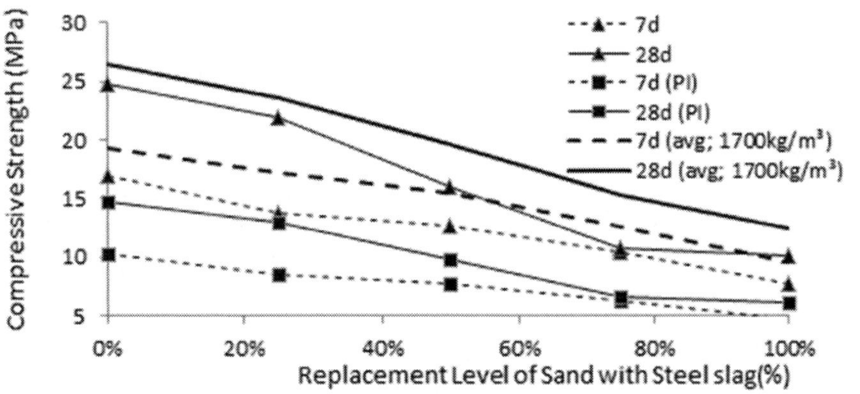

Fig. 3. Comparison between compressive strength, performance index (PI) of compressive strength, corresponding compressive strength and replacement level of steel slag.

3.3 Split tensile strength

Hardened density, percentage of foam, split tensile strength and corresponding split tensile strength were tested and evaluated in Table 4.

The relationship between split tensile strength and percentage of foam volume, was shown in Fig. 4, and the corresponding split tensile strength was calculated based on the relationship and shown in Table 4.

The split tensile strength, performance index of split tensile strength, and corresponding split tensile strength were compared in Fig. 5. It was also noticeable that split tensile strength and performance index were influenced by the achieved hardened density. Therefore, corresponding split tensile strength was estimated for comparison and concluded that, similar to compressive strength, split tensile strength was reduced almost linearly when steel slag replacement level was increased.

Split tensile strength is found at 8% to 10% corresponded to compressive strength in this study. Byun *et* al. had found a similar result [4, 18].

Table 4. Split tensile strength, hardened density, and percentage of foam in volume

Replacement level		0%	25%	50%	75%	100%
Percentage of Foam in Volume at 1700 kg/m^3		23.2	25.6	27.9	30.3	32.7
7 days	Hardened Density (kg/m^3)	1629	1655	1628	1635	1648
	Percentage of Foam Volume (%)	26.4	27.5	31.0	33.0	34.7
	Split Tensile Strength (MPa)	1.6	1.43	1.04	0.91	0.74
	Corresponding Split Tensile Strength at 1700kg/m^3 (MPa)	1.90	1.62	1.35	1.17	0.94
28 days	Hardened Density (kg/m^3)	1659	1637	1644	1647	1631
	Percentage of Foam Volume (%)	25.1	28.3	30.3	32.5	35.4
	Split Tensile Strength (MPa)	2.00	1.61	1.44	1.14	0.79
	Corresponding Split Tensile Strength at 1700kg/m^3 (MPa)	2.21	1.93	1.72	1.39	1.11
7 days / 28 days Corresponding Strength		86%	84%	78%	84%	85%

DOI: 10.1051/matecconf/201710301012

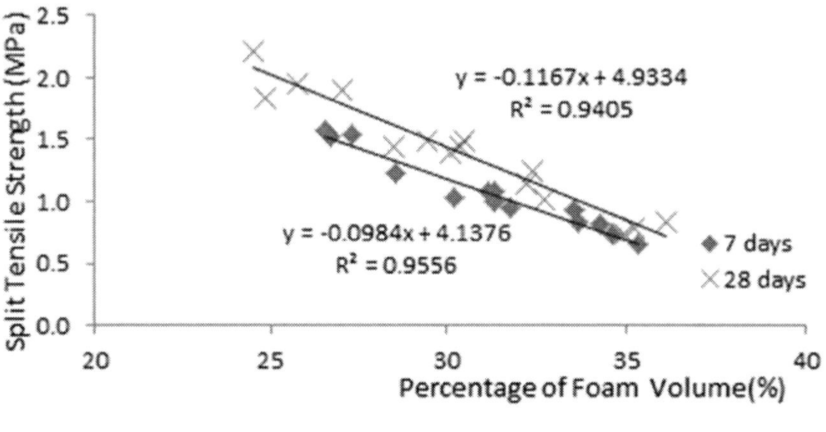

Fig. 4. Relationship between split tensile strength and foam volume.

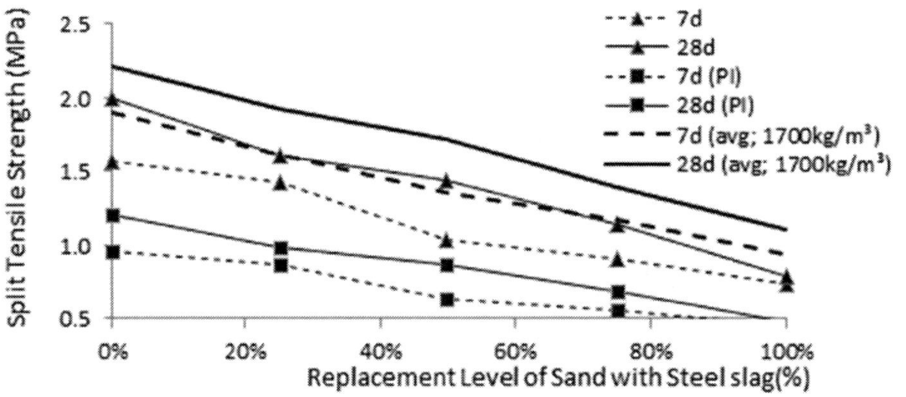

Fig. 5. Comparison between split tensile strength, performance index (PI) of split tensile strength, corresponding split tensile strength and replacement of steel slag

3.4 Flexural strength

Hardened density, flexural strength and performance index of flexural strength were tested and evaluated in Table 5.

The flexural strength and performance index of flexural strength were compared in Fig. 6. Approximate linear reduction of flexural strength performance index by increment of steel slag replacement was observed.

Table 5. Flexural strength, hardened density, and performance index of flexural strength

Replacement level		0%	25%	50%	75%	100%
7 days	Hardened Density (kg/m^3)	1628	1616	1615	1647	1616
	Flexural Strength (MPa)	7.40	6.15	4.44	3.53	3.17
	Performance Index (PI, MPa/1000kg/m^3)	4.51	3.81	2.75	2.14	1.96
28 days	Hardened Density (kg/m^3)	1632	1621	1624	1595	1625
	Flexural Strength (MPa)	7.80	7.23	5.95	5.07	4.04
	Performance Index (PI, MPa/1000kg/m^3)	4.78	4.46	3.66	3.18	2.49
7 days / 28 days Strength		94%	85%	75%	70%	78%

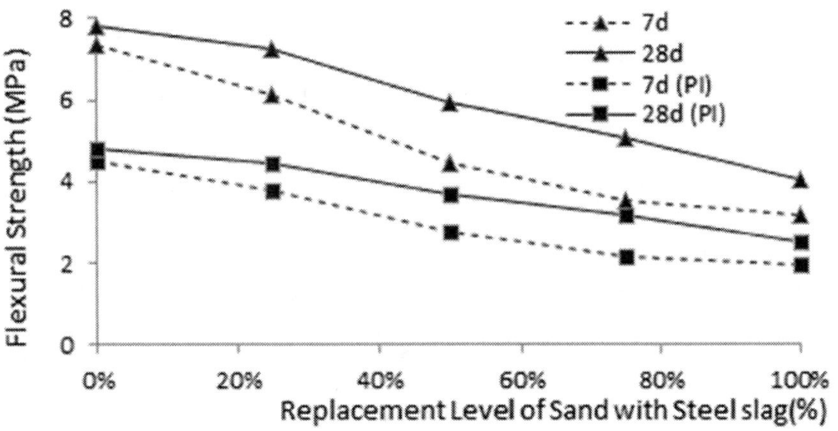

Fig. 6. Comparison between flexural strength, performance index (PI) of flexural strength, and replacement level of steel slag.

Performance index of flexural strength is found at 32% to 48% corresponded to its performance index of compressive strength in this study, which is very different from split tensile strength. Byun *et* al. also found similar result, which is, flexural strength is 20% to 40% of compressive strength, whereas split tensile strength is 8% to 11% of compressive strength [4, 18].

4 Conclusions

Strength properties of foamed concrete with density of 1600 kg/m³ to 1700 kg/m³ was studies, in which sand with specific gravity of 2.65 was partially replaced by steel slag with specific gravity of 3.8 and similar gradation, with different replacement level, namely 0%, 25%, 50%, 75%, and 100%.

The results show that the compressive, split tensile and flexural strengths were reduced almost linearly with increment of replacement level, however, with replacement level of 25% and lesser at density of 1600 kg/m³ to 1700 kg/m³, the achieved 28 days strength fulfil the structural strength of 17 MPa set by ASTM C 330–89 [3].

The split tensile strength is found 8% to 10% of compressive strength, whereas flexural strength is found 32% to 48% of compressive strength.

The support provided by Universiti Tunku Abdul Rahman (UTAR) in the form of a research grant (Vote No. 6200/LG3), and Masteel BhD in the form of raw steel slag for this study is very much appreciated.

References

[1] K. Ramamurthy, E. Kunhanandan Nambiar, and G. Indu Siva Ranjani, A classification of studies on properties of foam concrete, Cement and Concrete Composites, **31** (6), 388-396, (2009)

[2] A. Neville and J. Brooks, *Concrete technology*. Harlow, Essex, UK: Longman Scientific & Technical, 346, 350-351, (1987)

[3] ASTM C 330-89, *Standard Specification for Lightweight Aggregates for Structural Concrete*, ASTM International, Conshohocken, Pennsylvania, United States, (1989)

[4] Y. Amran, N. Farzadnia and A. Abang Ali, Properties and applications of foamed concrete; a review, Construction and Building Materials, **101**, 990-1005, (2015)

[5] N. Uddin, Structural characterization of hybrid fiber reinforced polymer (FRP)-autoclave aerated concrete (AAC) panels, J. of Reinforced Plastics and Composites, **25** (9), 981–999, (2006)

[6] A. Just, and B. Middendorf, Microstructure of high-strength foam concrete, Materials Characterization, **60** (7), 741–748, (2009)

[7] M.R. Jones, and A. McCarthy, *Behaviour and Assessment of Foamed Concrete for Construction Applications*, Thomas Telford, London, UK (2005)

[8] M. Röβler, and I. Odler, Investigations on the relationship between porosity, structure and strength of hydrated Portland cement pastes I. Effect of porosity, Cement and Concrete Research, **15** (2), 320-330 (1985)

[9] K. Yang, K. Lee, J. Song, and M. Gong, Properties and sustainability of alkali-activated slag foamed concrete, J. of Cleaner Production, **68**, 226–233, (2014)

[10] M. Hainin, M.A. Aziz, Z. Ali, R. Putra Jaya, M. El-Sergany, and H. Yaacob, Steel slag as a road construction material, J. Technolology, **73** (4) (2015)

[11] H. Motz, and J. Geiseler, Products of steel slags an opportunity to save natural resources, Waste Management, **21**(3), 285–293, (2001)

[12] E. Oluwasola, M. Hainin and M. A. Aziz, Characteristics and utilization of steel slag in road construction, J. Technology, **70** (7), (2014)

[13] M. Maslehuddin, A. Sharif, M. Shameem, M. Ibrahim, and M. Barry, Comparison of properties of steel slag and crushed limestone aggregate concretes, Construction and Building Materials, **17** (2), 105-112, (2003)

[14] P. Kothai, and R. Malathy, Utilization of steel slag in concrete as a partial replacement material for fine aggregates, IJIRSET Int. J. of Innovative Research In Science, Engineering and Technology, **3** (4), (2014)

[15] F. Falade, E. Ikponmwosa, and B. Ukponu, Structural assessment of foamed concrete containing steel slag as partial replacement of sand, UNILAG J. of Medicine, Science and Technology, **1** (1), 137-145, (2015)

[16] ASTM C 150-05, *Standard Specification for Portland Cement*, ASTM International, Conshohocken, Pennsylvania, United States, (2005)

[17] ASTM C1602-06, *Standard Specification for Mixing Water Used in the Production of Hydraulic Cement Concrete*, ASTM International, Conshohocken, Pennsylvania, United States, (2006)

[18] K. Byun, H. Song, and S. Park, Development of structural lightweight foamed concrete using polymer foam agent, Int. Congress on Polymers in Concrete, **19**, (1998)

[19] D. Aldridge, and T. Ansell, Foamed concrete: production and equipment design, properties, applications and potential. *In: Proc. of one day seminar on foamed concrete: properties, applications and latest technological developments*, Loughborough University, Loughborough, UK, (2001)

[20] ASTM C 1611-05, *Standard Test Method for Slump Flow of Self-Consolidating Concrete*, ASTM International, Conshohocken, Pennsylvania, United States, (2005)

[21] ASTM C 143-08, *Standard Test Method for Slump of Hydraulic-Cement Concrete*, ASTM International, Conshohocken, Pennsylvania, United States, (2008)

[22] BS EN 12390-3, *Testing Hardened Concrete: Compressive Strength of Test Specimens*, British Standard Institution, London, UK, (2002)

[23] BS EN 12390-6, *Testing Hardened Concrete: Tensile Splitting Strength of Test Specimens*, British Standard Institution, London, UK, (2000)

[24] BS EN 12390-5, *Testing Hardened Concrete: Flexural Strength of Test Specimens*, British Standard Institution, London, UK, (2000)

Microstructure and Tensile Strength of Foamed Concrete with Added Polypropylene Fibers

A. R. Hazlin [1], *Afif* Iman[1,*], *N.* Mohamad[1], *W. I.* Goh[1], *L. M.* Sia[1], *A. A. A.* Samad[1], and Noor*wirdawati* Ali[1]

[1]Department of Structure and Material Engineering, Faculty of Civil and Environmental Engineering, Universiti Tun Hussein Onn Malaysia, Parit Raja, 86400 Batu Pahat, Johor, Malaysia

Abstract. An experimental study was conducted to improve the strength of foamed concrete by adding fibres reinforced material, polypropylene (PP) fibres. The aim of the study was to determine the effect of PP fibres on the microstructure and tensile strength of foamed concrete. Foamed concrete with density 1600 kg/m^3 and 1800 kg/m^3 were cast and tested with 0.05%, 0.10% and 0.15% percentage of PP fibres. The findings shows that foamed concrete added by PP fibres gave better performance in tensile strength compared to foamed concrete without PP fibres. With the addition of 0.05% of PP fibres, the tensile strength increased by 35.06% and 40.30% for foamed concrete with densities 1600 kg/m^3 and 1800 kg/m^3, respectively. Through the study of microstructure of foamed concrete, the air voids of foamed concrete with various densities and the interfacial adhesion between concrete and fibres were observed. The foamed concrete added by PP fibres gave better performance in tensile strength compared to foamed concrete without PP fibres.

1 Introduction

Foamed concrete is lightweight and has lower density compared to normal concrete. With its advantage of low density, foamed concrete imposes little stress on the substructure. Among its many advantages is it can be used as a viable solution for reducing load on burdened soil, a particularly important issue in settlement.

However, despite the attractive advantages of foamed concrete, only limited data exists on its performance and guidance on its applications. Foamed concrete has a compressive strength between 1 MPa to 10 MPa [1]. The lack of coarse aggregate in formed concrete makes it prone to any types of shrinkage cracks. Therefore, it has lower strength compare to normal concrete.

Cracks play an important role in concrete as they change concrete structure into permeable element and has a high risk of corrosion. Cracks will reduce the quality of concrete and make the structure out of service [2]. Therefore, it is important to reduce the crack width by adding fibres. PP fibres has been used into admixture of concrete which can

* Corresponding author: noridah@uthm.edu.my

increase the tensile strength and mainly improve the prevention of post cracking that begin at micro cracking The structural of concrete such as shear resistance, bond strength and resistance to cracking depend on the tensile strength; the higher the tensile strength the better the structural properties [3].

In order to minimize the crack and increase the tensile strength of foamed concrete, PP fibres was added as additives. The percentages of PP fibres which was added in this research were 0.05%, 0.10% and 0.15% of the total volume of foamed concrete with densities 1600 kg/m^3 and 1800 kg/m^3.

2 Materials and method

2.1 Materials and mixture composition

The materials used in the foamed concrete admixture are Ordinary Portland cement (OPC), dried sand pass through 3mm, foam and water. The foam agent used to produce foam is SikaAER-50/50. The stable foam was produced with one portion of foam agent and twenty portion of water. In this study, the density of foamed concrete was fixed as 1600kg/m^3 and 1800kg/m^3. The sand/cement and water/cement ratios chosen in this study were 2 and 0.55, respectively.

2.2 Polypropylene fibres properties

Fibrillated PP fibres with 12 mm of length was used in this study. Fibrillated PP consists of interconnected fibres strand that open during mixing [4]. The fibrillated fibres reduce plastic shrinkage and permeability, to increase impact resistance, abrasion resistance fatigue and cohesiveness [5]. Characteristic tensile strength of PP fibres is 300 N/mm^2 with 900 kg/m^3 density [6]. In this study, dosage 0.05%, 0.10% and 0.15% of PP fibres by volume were added into each sample of foamed concrete with density 1600 kg/m^3 and 1800 kg/m^3 respectively.

2.3 Experimental test procedures

OPC, fine sand, water and PP fibres were batched and mixed in a concrete mixer. The foam was added gradually in the mixture according to BS EN 12350-6 [7] until the targeted density was achieved. The mixed foamed concrete was poured into the mould and were cured in room temperature.

Splitting tensile test was performed in accordance with BS-EN-12390-6 [8]. An axial load with a specified rate of loading was applied to cylinder with diameter of 150 mm and height of 300 mm until failure. Mean value obtained from three cylinders were then took as splitting tensile strength for each foamed concrete mixture at age of 28 days.

Microstructural image analysis was carried out by means of Scanning Electron Microscope (SEM). SEM is used for the examination and analysis of microstructural characteristics of solid objects. High resolution can be obtained when a bulk objects are examined [9]. A small piece of crushed foamed concrete from the tested cylinder was used for microstructure analysis.

3 Results and discussion

3.1 Density

It was noticed that most samples had higher density compared to the target density as shown in Table 1 and Table 2. It was difficult to achieve the target density since there were a lot of unpredictable elements during mixing work and environments factors. The production of stable foamed concrete mix is depends on many factors which were, selection of foaming agent, method of foam preparation and addition for uniform air-voids distribution, material section and mixture design strategies, production of foamed concrete, and performance with respect to fresh and hardended state were of great significance [10].

Table 1. Actual density of 1600kg/m^3 foamed concrete samples for cylinder.

Fibres Content	Specimen		Volume (m^3)	Weight (kg)	Density (kg/m^3)	Average (kg/m^3)
0%	AControl	1	0.0053	9.18	1732	1734
		2	0.0053	9.15	1726	
		3	0.0053	9.24	1743	
0.05%	APP01	1	0.0053	9.30	1755	1752
		2	0.0053	9.23	1741	
		3	0.0053	9.33	1760	
0.10%	APP02	1	0.0053	9.25	1745	1753
		2	0.0053	9.28	1751	
		3	0.0053	9.35	1764	
0.15%	APP03	1	0.0053	9.84	1857	1862
		2	0.0053	9.83	1855	
		3	0.0053	9.93	1874	

Table 2. Actual density of 1800kg/m^3 foamed concrete samples for cylinder.

Fibres Content	Specimen		Volume (m^3)	Weight (kg)	Density (kg/m^3)	Average (kg/m^3)
0%	BControl	1	0.0053	10.17	1919	1928
		2	0.0053	10.20	1924	
		3	0.0053	10.29	1941	
0.05%	BPP01	1	0.0053	10.73	2024	2013
		2	0.0053	10.65	2009	
		3	0.0053	10.63	2006	
0.10%	BPP02	1	0.0053	10.48	1977	1970
		2	0.0053	10.45	1972	
		3	0.0053	10.39	1960	
0.15%	BPP03	1	0.0053	10.72	2023	2019
		2	0.0053	10.72	2023	
		3	0.0053	10.66	2011	

3.2 Tensile strength

The results of testing the split tensile strength of foamed concrete with PP fibres addition using cylindrical specimen, diameter 150mm, height 300mm can be seen in Table 3.

Table 3. Result of split tensile strength.

Target Density(kg/m^3)	Average Split Tensile Strength (MPa)			
	Control	PP01	PP02	PP03
1600 (Set A)	0.77	1.04	0.9	1.24
1800 (Set B)	1.34	1.88	1.8	1.7

From Fig. 1, the result obtained for the control specimens shows a huge different in split tensile strength. The result for tensile strength of target density 1600 kg/m^3 foamed concrete is 0.77 MPa, while tensile strength of target density 1800 kg/m^3 foamed concrete is 1.34 MPa. It shows an increment of 0.57 MPa. The increase occurs due to the increasing density and decreasing foam volume. The strength of foamed concrete appeared to increase exponentially with an increase in concrete density or with a decrease in foam volume [11].

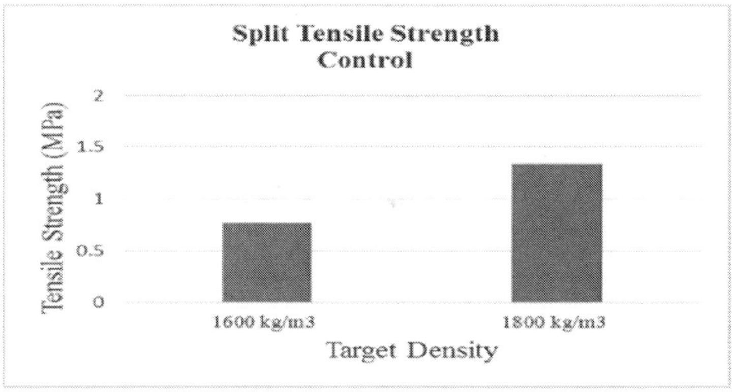

Fig. 1. The tensile strength versus density

From the Fig. 2, we can see that the tensile strength of foamed concrete with target density 1800 kg/m^3 (Set B) is higher compare to foamed concrete with target density 1600 kg/m^3 (Set A). Therefore, the effect of density and foam volume on the strength of foamed concrete can clearly see from this figure. The higher the density of foamed concrete, the higher its strength.

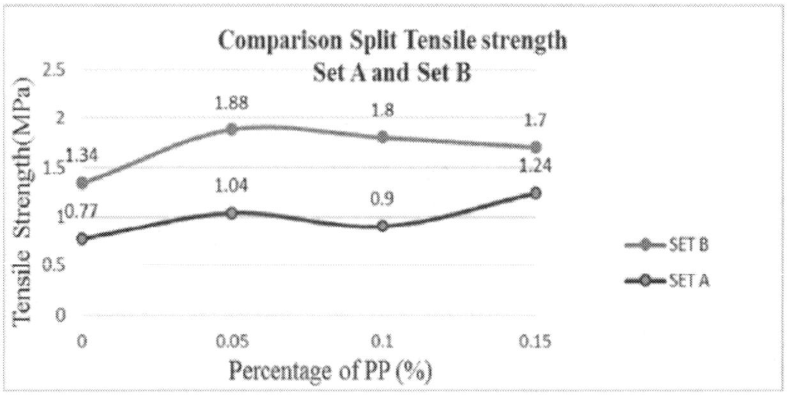

Fig. 2. The tensile strength of foamed concrete with target density 1600 kg/m^3 (set A) and 1800 kg/m^3 (set B) versus percentage of PP fibres.

Both foamed concrete with density 1600 kg/m^3 and 1800 kg/m^3 show the peak reading of tensile strength when 0.05% fibres content was added. However, for foamed concrete with density 1600 kg/m^3, another peak reading of tensile strength was noticed when 0.15% fibres was added. Foamed concrete with density 1600 kg/m^3 and added 0.15% fibres (APP03) was found to have higher strength compared to other specimen in Set A.

From Fig. 2, for foamed concrete with both densities, the PP fibres content for optimum tensile strength is found to be 0.05%. The research carried by Jones & McCarthy (2005) and Chen Bing *et al* (2012) recommended much higher percentage of PP content to be added to achieve the higher strength of foamed concrete which are 0.50% and 0.80%, respectively [11]. However, type of PP fibres used in foamed concrete carried out by previous researcher mostly was monofilament fibres. According to Medina *et al* (2015), the use of 0.07% fibrillated fibres in normal concrete was found to have maximum cracking control. Larger amount of PP fibres did not improve cracking control [12]. Therefore, it was reasonable that the optimum percentage of PP fibres as additive in foamed concrete for both 1600 kg/m^3 and 1800 kg/m^3 density was 0.05%.

3.3 Microstructure

Fig. 3 shows the air voids in foamed concrete without fibres for densities 1600 kg/m^3.and 1800 kg/m^3. Both figure had the same scale which was 35x. The air voids of 1600 kg/m^3 foamed concrete was found to have bigger size compared to 1800 kg/m^3 foamed concrete. The greater the air content the lower the density. This is why the foamed concrete of 1600 kg/m^3 density has the bigger voids size compared to the foamed concrete of 1800 kg/m^3.

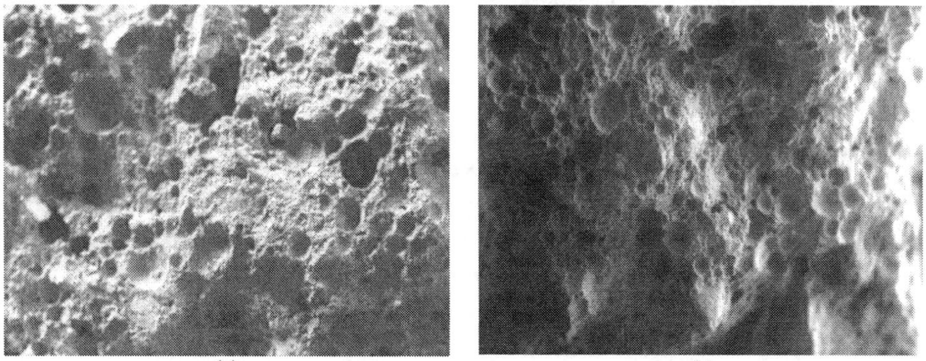

(a) (b)

Fig. 3. Air voids in foamed concrete (35x): (a) 1600 kg/m^3 density - AControl (b) 1800 kg/m^3 – Bcontrol.

Fig. 4 shows the microstructure of foamed concrete with density 1600 kg/m^3 without fibres and with fibres which has scale factor of 300x. From the high resolution image, the microstructure of foamed concrete without fibres was observed to have micro porous and rough surface in the Fig. 4(a). Fig. 4(b) shows the microstructure of foamed concrete with fibres, the interfacial bonding between PP fibres and matrix was observed in the figure within the red circle. The foamed concrete around fibres usually dense due to fibrillated process and interfacial adhesion. The bubbles formed in the matrix of foamed concrete produce micro porous and hence reduce the interfacial bonding in the matrix. Hence, the presence of PP fibres in foamed concrete play role to bridging force crossing crack of matrix and reduce the micro crack [13].

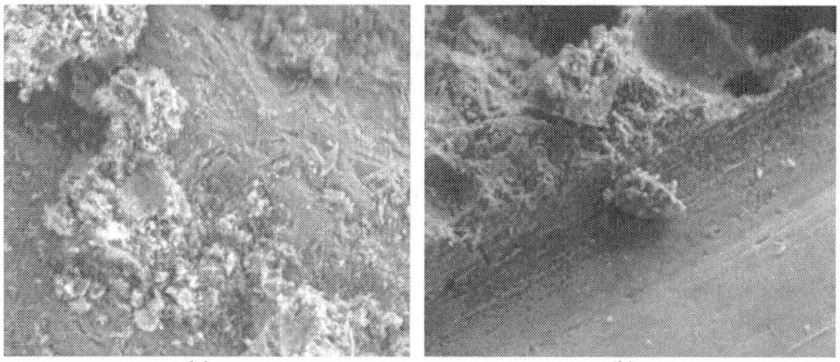

(a) (b)

Fig. 4. Microstructure of foamed concrete (300x) with density 1600 kg/m^3 (a) without fibres – Acontrol (b) with fibres – APP01.

Fig. 5(a) shows the microstructure of the foamed concrete with fibres at the part of failure after the split tensile strength test with the scale of 35x. The PP fibres was being pulled out from foamed concrete and caused the crack in foamed concrete. Fig. 5(b) shows the image of microstructure of the PP fibres after test with the scale of 600x. There are some foamed concrete matrix left on the PP fibres, which can observed in the red circle in Fig. 5(b). It means that there was an interfacial adhesion between the foamed concrete and PP fibres. Through the study of tensile strength of foamed concrete with various percentage of PP fibres, it was found that the 0.05% of PP fibres showed the best interfacial adhesion, which gave the optimum tensile strength.

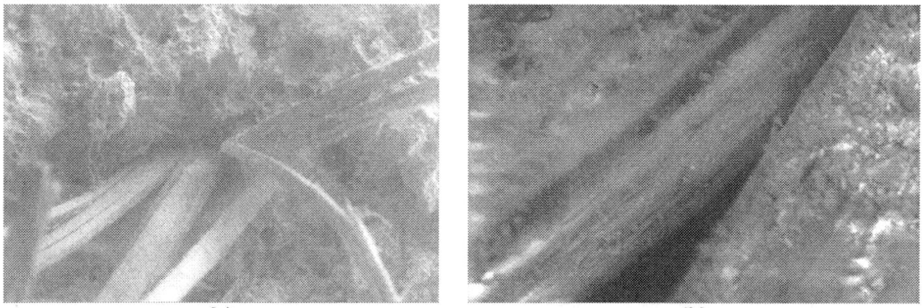

(a) (b)

Fig. 5. (a) Microstructure of the foamed concrete with fibres at the part of failure (35x) (b) Microstructure of the PP fibres after being pull out from foamed concrete (600x).

4 Conclusions

From the results if research, as well as data analysis and discussion, it can be concluded as follows:

i) The tensile strength of target density 1800 kg/m^3 foamed concrete was higher than the target density 1600 kg/m^3 foamed concrete. The higher the density of foamed concrete, the higher its strength.
ii) The tensile strength of foamed concrete with PP fibres was higher than the foamed concrete without PP fibres. The optimum percentage of PP fibres as additive in foamed concrete for both 1600 kg/m^3 and 1800 kg/m^3 density was 0.05%.

iii) Through the study of microstructure of foamed concrete, the air voids of 1600 kg/m^3 foamed concrete was found to have bigger size compared to 1800 kg/m^3 foamed concrete. Besides that, the interfacial adhesion between the foamed concrete and PP fibres was observed.

iv) The actual density of the samples had higher density compared to the target density. It was difficult to achieve the target density since there were a lot of unpredictable elements during mix work and environment factors.

The authors would like to thank the Universiti Tun Hussein Onn Malaysia and Ministry of Higher Education Malaysia (FRGS-VOT 1525) for their financial support for this project.

References

[1] M. Jones and A. McCarthy, Preliminary views on the potential of foamed concrete as a structural material. Magazine of Concrete Research, **57**(1), 21-31, (2005)

[2] A. Neville and J. Brooks, *Concrete technology*. England: Prentice Hall, England, (2001)

[3] A.H. Ameer, H.T. Nicholas and R.D. Andrew, The use of additives to enhance properties of pre-formed foamed concrete, IACSIT International J. of Engineering and Technology, **7**(4), 286-293 (2015)

[4] A. Laning, Primarily used to reduce shrinkage cracking, polypropylene, nylon, and polyester fibers offer other benefits as well, Article on Concrete Construction, The Aberdeen Group, (1992)

[5] J.E. Shoenberger and J. G. Tom, *Polypropylene fibres in portland cement concrete pavements*, Washington, (1992)

[6] Z. Deng and J. Li, tension and impact behaviours of new type fibre reinforced concrete, Computers and Concrete, 4(1), 19-32, (2007)

[7] BS EN 12350-6, *Testing fresh concrete density*, British Standard Institution, London, (2000)

[8] BS EN 12390-6, *Testing hardened concrete part 6: Tensile splitting strength of test specimens*, British Standard Institution, London, (2009)

[9] K.M. Nemati, Fracture analysis of concrete using scanning electron microscopy, Scanning, **19**, 426-43, (1997)

[10] K. Ramamurthy, E. Kunhanandan Nambiar and G. Indu Siva Ranjani, A classification of studies on properties of foam concrete, Cement & Concrete Composites, **31**, 388-396, (2009)

[11] Chen Bing, Wu Zhen and Liu Ning, Experimental research on properties of high-strength foamed concrete, J. Mateias in Civil Engineering, **24**, 113-118, (2012)

[12] N.F. Medina, G. Barluenga and F. Hernandez-Olivares, Combined effect of polypropylene fibres and silica fume to improve the durability of concrete with natural pozzolans blended cement, Construction and Building Materials, **96**, 556-566, (2015)

[13] J. Hadipramana, A. Abdul Samad, A. Ahmad Zaidi, Noridah Mohammad and N. Ali, Contribution of polypropylene fibre in improving strength of foamed concrete, Advanced Materials Research, **626**, 762-768, (2013)

Potential of Hollow Glass Microsphere as Cement Replacement for Lightweight Foam Concrete on Thermal Insulation Performance

Shahiron Shahidan[1,*] , *Eeydzah* Aminuddin [2], *Khairiyah* Mohd Noor[1], *Nurul Izzati Raihan* Ramzi Hannan [1], and *Nur Amira* Saiful Bahari[1]

[1] Jamilus Research Centre, Faculty of Civil and Environmental Engineering, Universiti Tun Hussein Onn Malaysia
[2]Faculty of Civil Engineering, Universiti Teknologi Malaysia, Skudai, Johor, Malaysia

Abstract. Global warming can be defined as a gradual increase in the overall temperature of the earth's atmosphere. A lot of research work has been carried out to reduce that heat inside the residence such as the used of low density products which can reduce the self-weight, foundation size and construction costs. Foamed concrete it possesses high flow ability, low self-weight, minimal consumption of aggregate, controlled low strength and excellent thermal insulation properties. This study investigate the characteristics of lightweight foamed concrete where Portland cement (OPC) was replaced by hollow glass microsphere (HGMs) at 0%, 3%, 6%, 9% by weight. The density of wet concrete is 1000 kg/m3 were tested with a ratio of 0.55 for all water binder mixture. Lightweight foamed concrete hollow glass microsphere (HGMs) produced were cured by air curing and water curing in tank for 7, 14 and 28 days. A total of 52 concrete cubes of size 100mm x 100mm x 100mm and 215mm x102.5mm x 65mm were produced. Furthermore, Scanning Electron Microscope (SEM) and X-ray fluorescence (XRF) were carried out to study the chemical composition and physical properties of crystalline materials in hollow glass microspheres. The experiments involved in this study are compression strength, water absorption test, density and thermal insulation test. The results show that the compressive strength of foamed concrete has reached the highest in 3% of hollow glass microsphere with less water absorption and less of thermal insulation. As a conclusion, the quantity of hollow glass microsphere plays an important role in determining the strength and water absorption and also thermal insulation in foamed concrete and 3% hollow glass microspheres as a replacement for Portland cement (OPC) showed an optimum value in this study as it presents a significant effect than other percentage.

* Corresponding author: shahiron@uthm.edu.my

1 Introduction

In a concrete structure, self-weight represents a very large proportion of the total load on the structure, and clearly there are advantages in reducing the density of concrete [1, 2]. Foamed concrete was widely used and has been used as an insulation material since 1923, where the first comprehensive review on foamed concrete was presented by [23] and a detailed treatment by [24], summarizing the its composition and properties, significant improvements over the past 20 years in production equipment and better quality surfactants (foaming agents) have enabled the use of foamed concrete on a larger scale [3, 4].

Foamed concrete is used for a variety of applications, ranging from thermal insulation and fire protection to void-filling and building elements with successively increasing density and strength requirements, such as, an insulating fill in fire walls or other precast elements, a replacement for soils and backfills, and the construction of cast-in-place piles. Lightweight concrete application in construction reduces building costs, eases construction and has the advantage of being a relatively 'green' building material [5]. Since it can be classified as an ideal material for residential construction which can prevent the heat level for buildings from increasing due to global warming, thermal insulation properties offer the advantage of functionality and being cost effective for house building projects in particular [6]. In addition, a product of low density reduces its own weight, the size of the foundation, and construction costs [7].

According to [9], Hollow Glass Microspheres (HGMs) seem to be a new and innovative product low in density, which can survive extremely high compressive forces, providing compounders and processors with new application opportunities [8-11]. HGMs are being developed in various sizes which possess light weight and high strength under hydrostatic pressure HGMs consist of an outer stiff glass and an inner inert gas, which results in some unique properties, such as being light in weight and having low thermal conductivity. These spheres might be expected to form a good inclusion system for the purpose of increasing the elastic stiffness of a soft light-weight matrix [10, 12, 17]. HGMs modified to replace aggregates as an additive will have a lower density, improved compressive strength and lower thermal insulation. Therefore, it is believed that the compressive strength of the lightweight concrete can be improved by modifying the HGMs into lightweight concrete, where HGMs can be used as cement replacement due to chemical properties. It also have a potential in thermal insulation and the effect can be determined by using difference percentages of HGMs as cement replacement in lightweight foamed concrete. Therefore this research will be mainly emphasized on lightweight foamed concrete using HGMs material for insulation. The objective of this paper is to determine the engineering properties of the lightweight foam concrete contain HGMs material as a cement replacement.

2 Hollow glass microspheres (HGMs)

Low densities of Hollow Glass Microspheres (3MTM Glass Bubble) are often referred to as shells, microballons, shirasuballons, vesicles and microcapsules. These hollow glass microspheres use as a density reducing agent for oil and gas drilling fluids and drilling cements and also can be used as a low-density filler material that is ideal for plastic and rubber parts created from compression moulding, extrusion processes and or other vigorous processing equipment [7]. Hollow spheres are used as lightweight filler in composite materials such as syntactic foam and lightweight concrete. With good thermal insulation properties and resistance to the external impact and help to reduce weight, reduce noise, vibration and harmonics, reduce thermal expansion [9]. Microspheres are usually white or grey in colour, and several manufacturers offer various inorganic hollow microspheres. The wall thicknesses of the particles are generally 10% of their diameter [10, 15, 16]. There is a characteristic of Hollow Glass Microspheres showed in Table 1.

Table 1. Typical Physical Properties of HGMs (source :3M[TM])

Properties	**3M[TM] Glass Bubbles iM30K**
Chemical resistance	chemical properties of 3M glass bubbles resemble those of a soda-lime borosilicate glass
Isostatic crush strength (psi)	28,000
True density (g/cc)	0.60
Packing factor (bulk density to true particle density)	63%
Oil absorption (g oil/100 cc of polymer additive)	33.5
Softening point (°C)	600
Floatation (density)	90% (in volume)
Volatile content (by weight)	0.5% (max)
Alkalinity (miliequaivalent/gram max)	0.5

3 Materials and methods

The material required in this study is foam, cement, fine aggregate and water and Hollow Glass Microsphere (HGMs). The HGMs was tested with SEM and XRF to identify the surface and the chemical composition respectively. The concrete mixes are determined using ratio 2:1 which represent cement to sand and water cement ratio at 0.55% for cement slurry foam production. The cement slurry foamed was poured in cubic moulds which was size 100mm x 100mm x 100mm and curing for 7, 14 and 28 days. The total 53 cubes specimens were tested for compressive strength, water absorption and thermal conductivity at 215mm x 102.5mm x 65mm in volume. The replacement of cement with HGMs is varied from M1:0%, M2:3%, M3:6% and M4:9%. These were done to determine the proportion that would give the most favourable result. The 0% served as control for other sample. In this study, the tests have been carried out to ensure the acceptance of HGMs as replacement cement for binder in lightweight foamed concrete mixture.

4 Results and discussion

4.1 Physical and chemical properties

Fig. 1 shows the SEM image of Hollow Glass Microspheres. The results suggest a smooth clean surface, which would not promote interfacial adhesion between the filler material and polymer matrix. A perfectly spherical shape with an aspect ratio of one makes glass bubbles efficient volume fillers. While Table 2 and Table 3 shows the chemical composition of HGMs and Portland cement.

4.2 Density

Fig. 2 represent the dry densities of foamed concrete fluctuate when the HGMs are added into the foam concrete. It shows that the plastic densities of concrete mortar before being added to the foam concrete were very high. The mixture's density may be slightly lower than normal concrete density due to the HGMs added into the mortar mixtures.

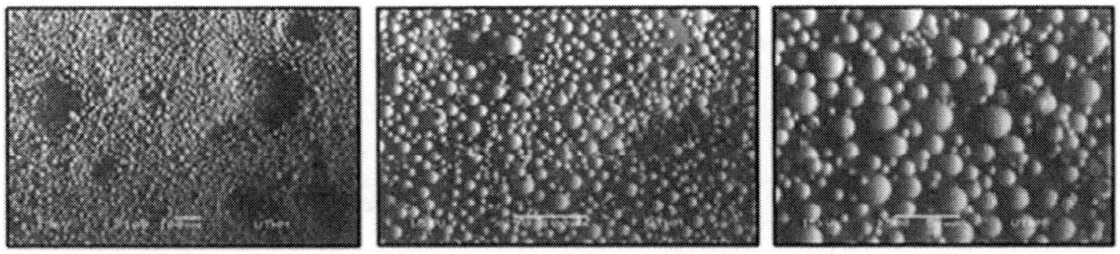

(a) (b) (c)

Fig. 1. SEM of HGMs: (a) 100μm 100x Mag; (b) 100μm 250 x Mag; (c) 50μm 500x Mag.

Table 2. Chemical composition of HGMs

Chemical Composition	Concentration (%)
SiO_2	65.30
CaO	13.40
Na_2O	5.92
C	0.10
P_2O_5	0.59
SO_3	0.27
Al_2O_3	0.26

Table 3. Chemical composition of Portland Cement

Chemical Composition	Concentration (%)
SiO_2	21.06
CaO	57.98
Al_2O_3	6.10
Fe_2O_3	3.08
MgO	2.74
SO_3	2.40
Loss	4.07

After the addition of foam into mortar mixtures, the densities of the mixture decreased significantly. It is because the foam agent occupies the space in the mortar mixtures known as voids. The consistency of the mix to which the foam is added is important, as this consistency affects the stability of the mix. Consistency is significantly reduced when foam is added. [13, 16].

According to Hamidah et al [18], a lower density foamed concrete can achieve a strength equal to that of a higher density foamed concrete by increasing the cement content. The control sample decreased from 1000 kg/m^3 to 943 kg/m^3. This may be due to the drying air content in concrete [19]. The results show that the more HGMs are added to the sample, the more concrete density decreases. This may be because cement and HGMs possess different weight.

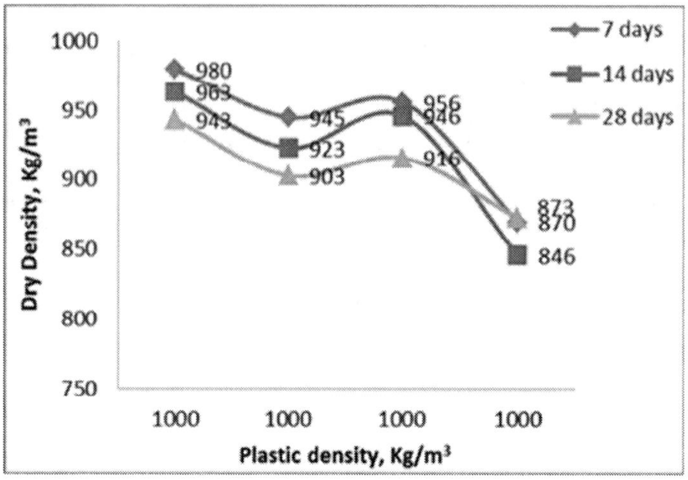

Fig. 2. Comparison between plastic density and dry density of foamed concrete at 7, 14 and 28 days

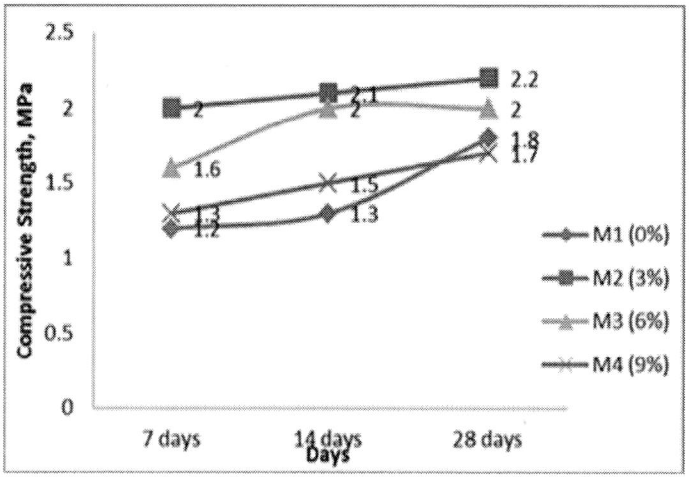

Fig. 3. The compressive strength against the days of curing.

4.3 Compressive strength and water absorption

The comparison of compressive strength of foamed concrete at 7, 14 and 28 days is shown in Fig. 3. The different percentages of HGMs have affected the compressive strength of the concrete. The result show the compressive strength of foamed concrete is fluctuated which is the maximum strength of concrete is M2 where its compressive strength recorded is 2.2 MPa with 3% of HGMs at density 998 kg/m^3, M3 2.1 MPa with 6% at density 916 kg/m^3 and the minimum strength is M4 which is compressive strength 1.7 MPa with 9% of HGMs. These values are quite good compared with compressive strength reported by [16] for foamed concrete (2 to–18MPa at density of 1000 to 1500 kg/m^3) at where the sample were replacement of cement (up to 75% by weight) with both classified and unclassified fly ash.

The compressive strength of foamed concrete in 28 days has increase significantly compare to the compressive strength at 7 and 14 days. It may be due to the different percentage of HGMs used in the foamed concrete and also influenced by the physical properties of HGMs; this is because the strength development with exposure to natural weathering was crucially dependent on the humidity and temperature changes [14, 15, 20]. While, table 4 shows the percentage of water absorbed by HGMs foamed concrete with different curing days. The percentage of water absorbed in the foam concrete was influenced by the different percentage of HGMs used which is M2 where 3% of HGMs used was absorbed less water than other mixture. Meanwhile, M4 with 9% of HGMs achieved the highest absorption. This may be due to the higher percentage of air void and effect of HGMs foamed concrete. Therefore, when the higher percentage of air void occurred, the more the water immersed into the foamed concrete. It can be seen that, water absorption increased when the percentage of HGMs is increased. Fig. 4 show the comparison of percentage of water absorption of foamed concrete at 28 days.

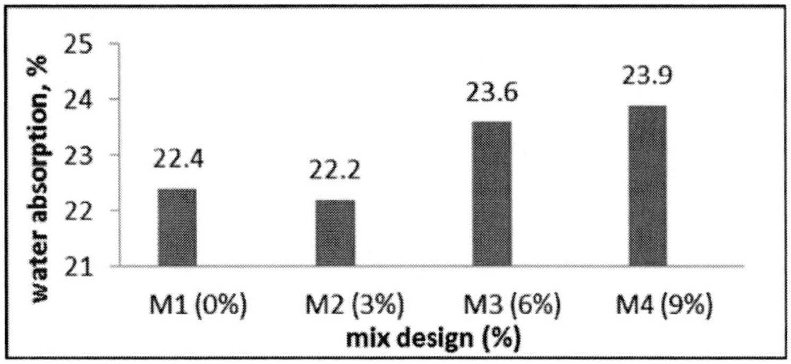

Fig. 4. Percentage of water absorption of foamed concrete at 28 days.

Table 4. water absorption of foamed concrete at 28 days.

Mix Design	Water absorption of foamed concrete for 28 days		
	Oven Dry Density (Kg/m^3)	**Immersed Density Kg/m^3**	**Percentage Absorption (%)**
M1 (0% HGM)	983	1203	22.4
M2 (3% HGM)	930	1136	22.2
M3 (6% HGM)	906	1120	23.6
M4 (9% HGM)	836	1036	23.9

4.4 Thermal conductivity

Table 5 shows the result of the thermal conductivity testing of three different percentages of Hollow Glass Microsphere (HGMs) and the dry density of concrete. After adding The HGMs the result shows that the k-value was decrease. The result of foam concrete that was added by 3% of the HGMs was almost same as the result with the control sample which is 0.25 w/mk. but when the HGMs were added at 6% the result decrease to 0.20 w/mk. from this result, from Figure. 5 (a) it can be see the HGMs in foamed concrete have a potential in thermal insulation for concrete construction. From the previous research, Saygılı and

Baykal [21] reported that the decrease in the thermal conductivity is due to the increase of void ratio that decreased the unit weight of concrete. Since air is the poorest conductor compared to the solid and liquid due to its molecular structure [22], it contributes to the lower thermal conductivity in porous concrete. Also, Ramamurthy et al. [16] reported that foamed concrete has excellent thermal insulating properties due to its cellular microstructure. From the result, it was found that the optimum HGMs content is 3% with the average range of thermal conductivity of 0.2596 W/mK with highest compressive strength value of 2.2 MPa. The thermal conductivity of M2 almost similar as M1(control sample) but for the compressive strength, the M1 and M2 show the large margin which was 2.0 to 2.2 MPa with thermal conductivity below than 0.3 w/mk its quite good result for used HGMs in foamed concrete. Fig. 5 (b) shows the graft the effect of HGMs on mechanical properties and thermal conductivity.

Table 5. Thermal conductivity test result

Mix Design	**Thermal conductivity of different HGMs Percentage for 28 days**		
	M1 (0%)	M2 (3%)	M3 (6%)
Density (Kg/m^3)	970	998	984
Thermal (w/mk)	0.2507	0.2596	0.2029

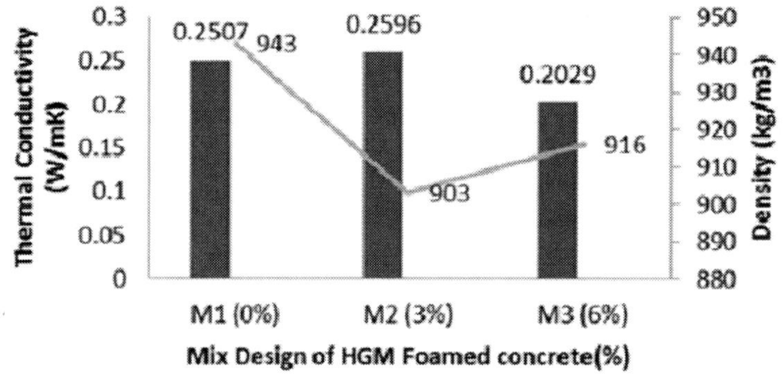

Fig. 5 (a). Thermal conductivity against density of HGMs foamed concrete.

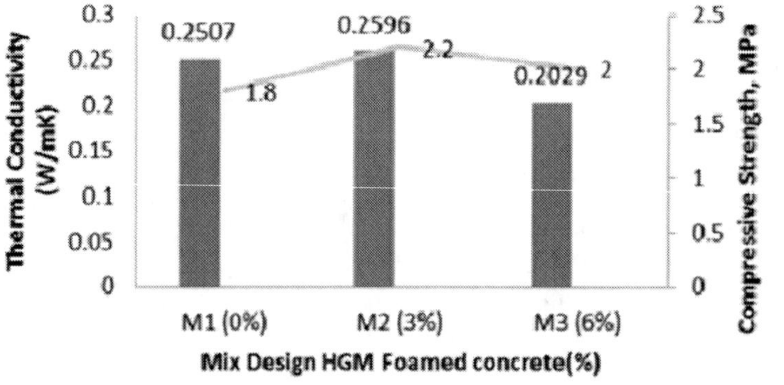

Fig. 5 (b). Effect of HGMs on mechanical properties and thermal conductivity.

5 Conclusions

The conclusion as follows:

- The result gained from this study shows that the Hollow Glass Microsphere foamed concrete has a range density of between 900 kg/m3 and 1000kg/m3.
- The optimum percentage of the usage of HGMs as partial replacement of cement in foamed concrete is 3% due its compressive strength is higher than other percentage of HGMs.
- The percentage of water absorption of Hollow Glass Microsphere foamed concrete on 28 days were between 22.2% and 23.9%.
- The thermal conductivity of HGMs foamed concrete decrease after added the HGMs by percentage 3%, 6% and 9%.

A further research of the properties and the potential of the HGMs should be carried out since the HGMs may contain the chemical material which will infiltrate the water body and hence affecting the ground and environment.

The authors would like to express their deepest appreciation to the Ministry of Higher Education Malaysia, Universiti Tun Hussein Onn Malaysia for funding the research under grant U523.

References

[1] E.K Nambiar, K. Ramamurthy, Influence of filler type on the properties of foam concrete, Cement and Concrete Composites, **28**(5), 475-480, (2006)

[2] S. Shahidan, H.B. Koh, A.M.S. Alansi, L.Y. Loon, Strength development and water permeability of engineered biomass aggregate pervious concrete, MATEC Web Conf., **47**, 2–7, (2016)

[3] S. Shahidan, S.S.M. Zuki, N. Jamaluddin, Damage grading system for severity assessment on concrete structure, Case Stud. Constr. Mater., **5**, 79–86, (2016)

[4] M.A. Rahim, Z. Ghazaly, R.Nurazira, R. Mamat, M.A. Azizan, N.F. Isa, S. Shahidan, Experimental study of slurry infiltrated fiber reinforced concrete, Material Science Forum, **857**, 363–366, (2016)

[5] T.Y. Lo, W.C. Tang, H.Z. Cui, The effects of aggregate properties on lightweight concrete, Building and Environment, **42**, 3025-3029, (2007)

[6] N. Ali, K.Y. Yaacob, S. Shahidan, S.R. Abdullah, Investigation of compressed earth brick containing ceramic waste, ARPN J. Eng. Appl. Sci., **11**(8), 5459–5462, (2016)

[7] J. Alduaij, K. Alshaleh, M.N. Haque, K. Ellaithy, Lightweight concrete in hot coastal area, Cement and Concrete Composites, **21**(5), 453-458, (1999)

[8] S.A. Kudus, N.M. Bunnori, S.R. Basri, S. Shahidan, M.N.M. Jamil, N.M. Noor, An overview current application of artificial neural network in concrete, Advanced Material Research, **626**, 372–375, (2012)

[9] A.S. D'Souza, K. Hendrikson, Innovative high strength glass microspheres for extruded and injection molded plastics, *The 3rd Int. Conf. High Perform. Fillers,* (2007)

[10] K.J. Lee, R.A. Westmann, *Elastic properties of hollow-sphere-reinforced composite*, University of California, Los Angeles, (1969).

[11] St. Paul, Energy and Advanced Materials Division 3M Center, Building 223-6S-04, Retrieve on October 2016 from http:// www.3m.com.

[12] N. Fredrick, P. Lin, J.K. Koech, Y. Wang, Surface modification of hollow glass microspheres, Materials Science and Application, 856-860, (2012)

[13] N. Md Nor, N. Muhamad Bunnori, A. Ibrahim, S. Shahidan, S.N.M. Saliah, An investigation on acoustic wave velocity of reinforced concrete beam in-plane source,

Proc. 2011 IEEE 7th Int. Colloquium on Signal Processing and Its Applications, CSPA, 19–22, (2011).

[14] B.W. Sands, W.C. Howes, Hollow glass microspheres - A cost-effective lightweight filler, Plastics Engineering, **38**(4), 31-33, (1982)

[15] K. Ramamurthy, E.KK. Nambiar, G. Indu Siva Ranjani, A classification of studies on properties of foam concrete, Cement and Concrete Composites, **31**(6), 388-396, (2009)

[16] M.S. Hamidah, I. Azmi, M.R.A. Ruslan, K. Kartini, N.M. Fadhil, Optimisation of foamed concrete mix of different sand-cement ratio and curing conditions, *Proc. of Int. Conf. on the use of foamed concrete in construction*, 37-44, (2005)

[17] S. Shahidan, R. Pullin, K.M. Holford, M.B.N, N. Nor, Quantitative evaluation of the relationship between tensile crack and shear movement in concrete beams, Advanced Material Research, **626**, 355–359, (2013)

[18] E.P. Kearsley, P.J. Wainwright, The effect of high fly ash content on the compressive strength of foamed concrete, Cement and Concrete Research, **31** (1), 105–112, (2001)

[19] S.K. Lim, T.C. Ling, M.W. Hussin, Ground-granulated blast-furnace slag as potential filler in polyester grout: Compressive strength development, ACI Material J., **108**, 120–127, (2011)

[20] M. Abdul Rahim, N.M. Ibrahim, Z. Idris, Z.M. Ghazaly, S. Shahidan, N L. Rahim, L. A. Sofri, N. F. Isa, Properties of concrete with different percentange of the rice husk ash (RHA) as partial cement replacement, Material Science Forum, **803**, 288–293, (2014)

[21] A. Saygılı, G. Baykal, A new method for improving the thermal insulation properties of fly ash, Energy and Buildings, **43**(11), 3236–3242, (2011)

[22] S.C. Ng, K.S. Low, Thermal conductivity of newspaper sandwiched aerated lightweight concrete panel, Energy and Buildings, **42**(12), 2452–2456, (2010)

[23] R.C. Valore, Cellular concrete part 1: Composisiton and method of production, ACI Journal, **50**, 773-796, (1954)

[24] G. Rudnai, *Lightweight concretes*, Budapest, Akademi Kiado, (1963)

Properties of Pervious Concrete Containing Scrap Tyre Tubes

Koh Heng Boon[1,*], *David* Yeoh[1], and *Norwati* Jamaluddin[1]

[1]Department of Structural and Material Engineering, Faculty of Civil and Environmental Engineering, Universiti Tun Hussein Onn Malaysia, 86400, Batu Pahat, Johor, Malaysia

Abstract. There is a huge quantity of waste tyre tubes generated every year due to the increasing of motorcycle user. Therefore, recycling of the waste tyre tubes is become mandatory. The aim of this research was to study the properties of pervious concrete containing scrap tyre tube (STT) rubber particles with percentages of 3%, 5% and 7% of the cement content. The properties studied are void content, compressive strength measured at 7, 14 and 28 days, flexural strength and flow rate which were determined at 28 day. The experimental results showed that, there were increased in void content and flow rate of pervious concrete containing STT. Both compressive strength and flexural strength of pervious concrete containing STT showed a lower value compared to the control mix without STT. The reductions of the mechanical strengths are likely due to the increase of void content. Overall, pervious concrete which contains 7% STT has shown an increment of mechanical strengths and flow rate compared to other STT pervious concrete. Nonetheless, the results indicate that there are potentials for use of STT in pervious concrete, especially for use in pervious concrete applications such as pavements, driveways and parking lots.

1 Introduction

With the ever increasing vehicles there is a growing need for recycling the waste tyre. The waste tyre generated annually in Malaysia was estimated approximately 57,391 tonnes and about 60% of the waste tyres are disposed via unknown routes [1]. Tyre landfilling is responsible for a serious ecological threat. Mainly waste tyres disposal areas contribute to the reduction of biodiversity also the tyres hold toxic and soluble components [2]. Waste tyre management and disposal is a major environmental concern in many countries. Waste tyre management is increasingly becoming a significant environmental, health, and aesthetic problem that is not easily solved. The use of waste tyres as a concrete additive is a possible disposal solution [3].

Pervious concrete can be other name as porous concrete or no fine concrete. It is different from conventional concrete in that it contains no fine aggregate. Combination the uniform coarse aggregate gradations and low water content such as water to cement ratios ranging from 0.25 to 0.35 make for a good concrete mixture with suitable void contents ranging from 11% to 35% [4, 5]. Pervious concrete characteristics differ from conventional

* Corresponding author: koh@uthm.edu.my

concrete in several other ways. Compared to conventional concrete, pervious concrete has a lower compressive strength, higher permeability, and a lower unit weight, approximately 70% of conventional concrete [6].

Based on report published by American Concrete Institute [7], the use of pervious concrete for sustainable construction activities continues to rise due to its several environmental benefits. Most important among these benefits is the reduction of storm water runoff and recharging of groundwater when used as a paving material for parking lots, low volume pavements, and sidewalks. Pervious concretes have been found to be very effective in recharging groundwater and conserving water resources, making it an effective storm water management tool when used as a surface course in large exposed areas such as parking lots. Recently there has been considerable interest in using pervious concrete to mitigate the tyre-pavement interaction noise [8] in concrete pavements.

The use of waste tyres in normal weight concrete has been investigated since many years ago, but to date, research on the use of scrap tyre tubes (STT) in pervious concrete has not been found in Malaysia. Generally, concrete containing waste tyres gained lower compressive strength, but rubberized concrete has some desirable characteristics such as lower density, higher impact and toughness resistance, enhanced ductility and better sound insulation [9]. The aim of this research is to determine the effect of STT use in pervious concrete with regards to the void content, flow rate and mechanical properties. The mechanical properties that have been investigated are compressive strength and flexural strength.

2 Experimental details

2.1 Materials

Constituent materials for pervious concrete mixes included Portland composite cement meeting MS EN 197-1:2007 [10] requirements, crushed granite aggregate with a nominal size of 10 mm, tyre tube-rubber particles processed from scrap tyre tubes (STT) with a nominal size of 6 mm and superplasticizer. STT were collected from the motorcycle workshops in Parit Raja. STT were washed to remove the dust and grease. STT were cut into small irregular shape particles by using plastic granulator machine. The STT and tyre tube-rubber particles are shown in Fig. 1 and Fig. 2, respectively. Tyre tube particles were not pretreated before incorporation into the pervious concrete mixture.

Fig. 1. Scrap tyre tube.

Fig. 2. Processed tyre tube-rubber particles.

2.2 Mix design

Four pervious concrete mixes, as shown in Table 1 were prepared. Since, there is no existing standard method of designing concrete mixes incorporating STT, the method of mix design proposed by Bhutta et al. [11] was employed to design a pervious concrete mix without STT. The mix was designed to have a 28-day target compressive strength of 10 MPa. STT particles were used as the additive with percentages by weight of 3%, 5% and 7% of the cement content. Superplasticiser of 1 litre/100 kg cement was used to increase the workability of the low water cement ratio pervious concrete mixtures.

Table 1. Pervious concrete mixture proportions

Mix ID	Cement (kg/m^3)	Coarse Aggregates (kg/m^3)	Tyre Tube-Rubber Particles (kg/m^3)	Water (litre)	Superplasticiser (litre)	Remark
M0	350	1430	0	105	3.5	Control mix – 0% STT
M1	350	1430	10.5	105	3.5	3% STT
M2	350	1430	17.5	105	3.5	5% STT
M3	350	1430	24.5	105	3.5	7% STT

2.3 Specimens and tests

Concrete cubes of 100 x 100 x 100 mm were used to evaluate the compressive strength and void content. Rectangular prisms size 100 x 100 x 500 mm length was used for evaluating flexural strength. Meanwhile, two types of panel size 300 x 300 x 25 mm thick and 300 x 300 x 25 mm thick were used for the flow test. All specimens were cured under the water until the testing age. The number of specimens used in the tests and the distribution is tabulated in Table 2.

Table 2. Number of specimens

Specimen	Test	Number of specimens		
		7-day	14-day	28-day
100 x 100 x 100 mm cube	Compressive Strength	3	3	3
100 x 100 x 100 mm cube	Void Content	-	-	3
100 x 100 x 500 mm prism	Flexural Strength	-	-	3
300 x 300 x 25 mm thick panel	Flow Rate	-	-	2
300 x 300 x 50 mm thick panel	Flow Rate	-	-	2

Compressive strength test was conducted based on BS EN 12390-3: 2009 [12]. Three cubes were tested for each age. Concrete cubes used for void content are the same cubes used for the 28 day compressive strength test. Void content test (Fig. 3) was carried out at 28 day accordance to BS EN 12390-7: 2009 [13]. Void content of pervious concrete is defined as the percentage of the total volume of void to the total volume of specimen. It was calculated based on the following equation:

$$Void \quad content\ (\%) = 1 - \left[\frac{(W2 - W1)}{vp} \right] x100 \tag{1}$$

where;

W1 is the weight of specimen submerged under water (kg)
W2 is the weight of specimen at a saturated surface dry condition (kg)
v is the volume of specimen (m^3)
ρ is the density of water in (kg/m^3)

Fig. 3. Void content test.

Flexural strength test was carried out based on BS EN 12390-5: 2009 [14]. Three prisms were tested for the flexural strength at 28 days. Four-point flexural strength test is shown in Fig. 4. 300 mm square pervious concrete panels were tested for the vertical and horizontal flow rate at 28 days. Since, there is no existing standard method of testing flow rate for pervious concrete, two apparatus were designed for the test. Fig. 5 and Fig 6. show the schematic diagram for the vertical flow test and apparatus for the horizontal flow test, respectively.

Fig. 4. Four-point flexural strength test.

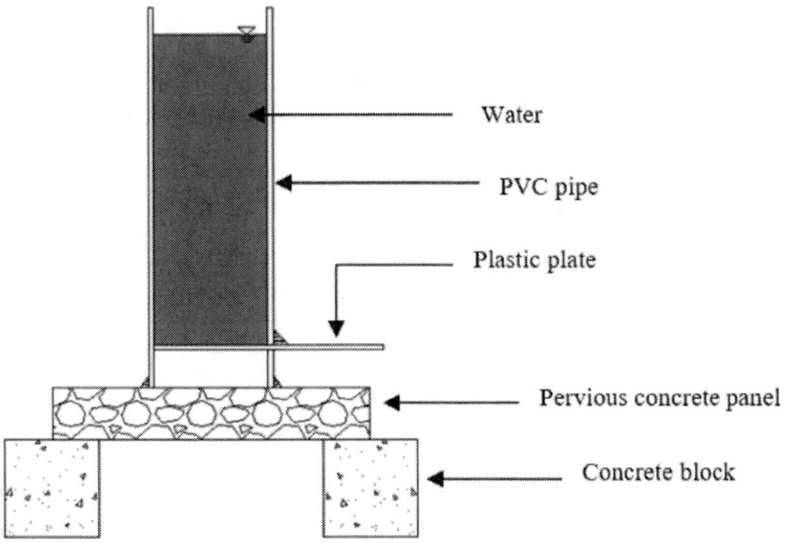

Fig. 5. Schematic diagram of horizontal flow rate test.

Fig. 6. Apparatus for the horizontal flow rate test.

3 Results and discussions

3.1 Void content

The results of void content are shown in Fig. 7. Results showed that pervious concrete containing STT rubber particles achieved higher void content compared to the control specimen. The control specimen (M0) without STT achieved the lowest void content of 28%. The results obtained in this study are similar to the findings of Yukari [15] that void content is slightly higher for self-compacting no fines concrete containing small particles additives. However, the effect of increasing STT particles as shown in the figure, which indicates that there was a reduction of void content. This can be seen from the void content of M1 specimen (3% STT), which recorded the highest value of 31% has reduced gradually to 30% and 29% for M2 (5% STT) and M3 (7% STT) specimens, respectively.

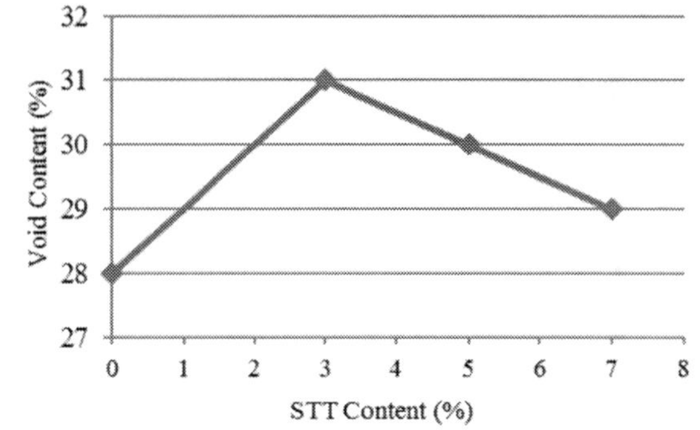

Fig. 7. Void content of pervious concrete.

3.2 Compressive strength and flexural strength

The compressive strength development of 100 mm cubes of M0-M3 mixes are shown in Fig. 8. In order to compare the relative gain of strength at different ages, the compressive strength of the mixes is also presented in Fig. 9. Results showed that all STT mixes gained lower compressive strength for all ages. Among all the pervious concrete containing STT, the compressive strength for M3 is slightly higher than M1 and M2. At 28 day, control specimens (M0) achieved the highest compressive strength of 11.3 MPa followed by M3, which gained 7.6 MPa, M2 achieved 6.35 MPa and M1 achieved the lowest compressive strength with 5.8 MPa. The graphical presentation of the 28 days compressive strength is given in Fig. 10. The reduction of compressive strength for pervious concrete containing STT may be due to the increment of void content and the STT particles are not distributed uniformly in the concrete.

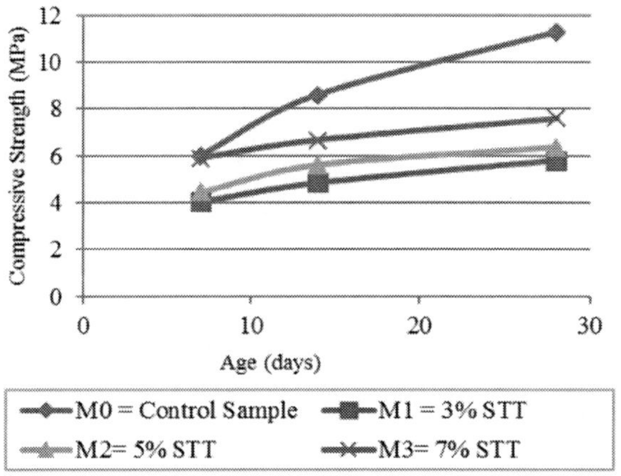

Fig. 8. Compressive strength development of pervious concrete.

Fig. 11 shows the plot of 28 day compressive strength of pervious concrete against void content of this study together with the findings by Tennis et al. [4], Bhutta et al. [11], Schaefer et al. [5] and Ireen Azfar, [16]. Based on the data shown in Fig. 11, the compressive strength of pervious concrete is inversely proportional to the void content.

Results of the 28 day flexural strength are shown in Fig. 12. It can be seen that, flexural strength for all pervious concrete containing STT particles is lower compared to the control specimens. Based on the plot, there is no specific trend in the relationship between the proportion of STT added in pervious concrete and flexural strength. The ratio between compressive strength and flexural strength is presented in Table 3. M1 and M3 specimens have shown a higher value of the ratio between the flexural strength and compressive strength. Pervious concrete containing 7% STT particles gained 3.9% higher than the control specimen (M0). The results obtained here show that, the use of STT more than 7% has the tendency to improve flexural strength of pervious concrete.

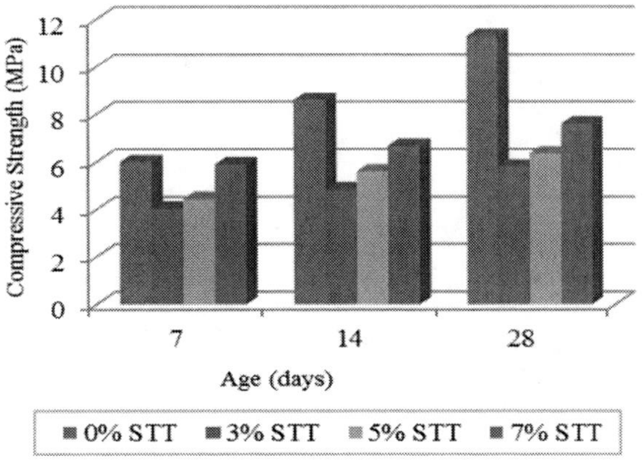

Fig. 9. Compressive strength of pervious concrete.

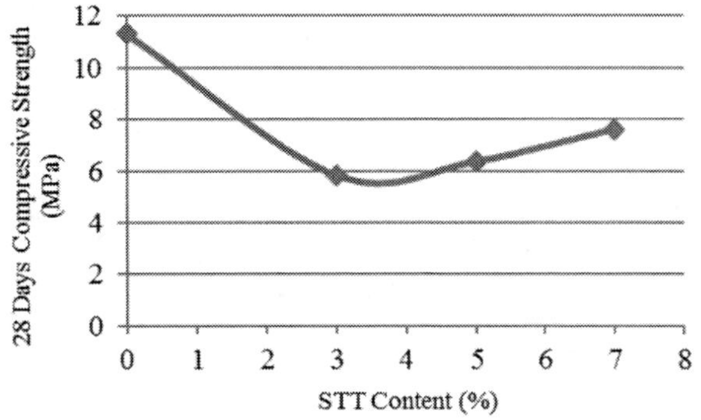

Fig. 10. 28 days compressive strength of pervious concrete.

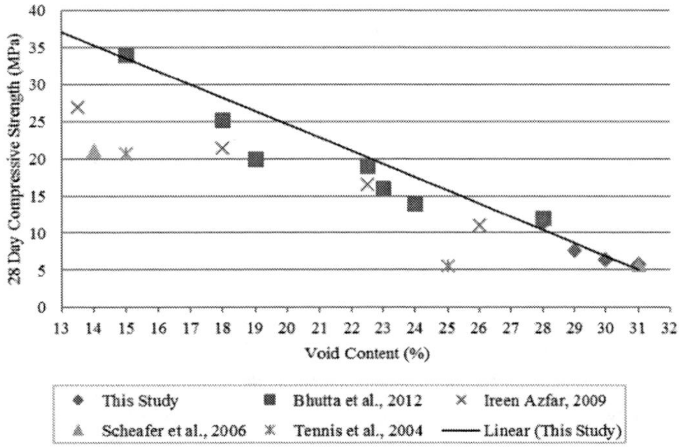

Fig. 11. 28 day compressive strength against void content.

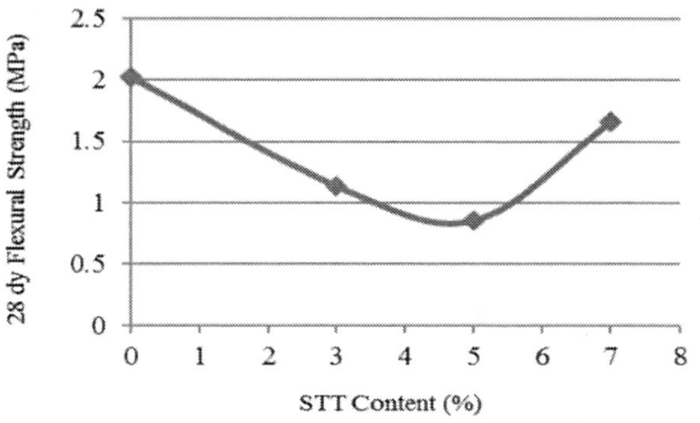

Fig. 12. Flexural strength of pervious concrete.

Table 3. Ratio of compressive strength / flexural strength

Mix ID	28 day compressive strength (MPa)	28 day flexural strength (MPa)	Percentage of flexural strength/ compressive strength (%)
M0	11.3	2.02	17.9
M1	5.8	1.13	19.5
M2	6.35	0.85	13.4
M3	7.6	1.66	21.8

3.3 Flow rate

Fig. 13 shows the vertical flow rate of the pervious concrete. Time to achieve the same water level in horizontal flow test is shown in Fig. 14. As expected, pervious concrete with higher void content achieved higher flow rate. The thicker panel achieved a lower vertical flow rate and horizontal flow. Vertical flow rate of 97.6 to 139.8 L/m^2/s was achieved by the thin panels. Meanwhile, the thicker panels achieved a vertical flow rate of 92.4 to 126.2 L/m^2/s. For the horizontal flow, 22 litre water was used in the test. The time ranging from 3.0 to 5.9 seconds was recorded for achieving the same water level.

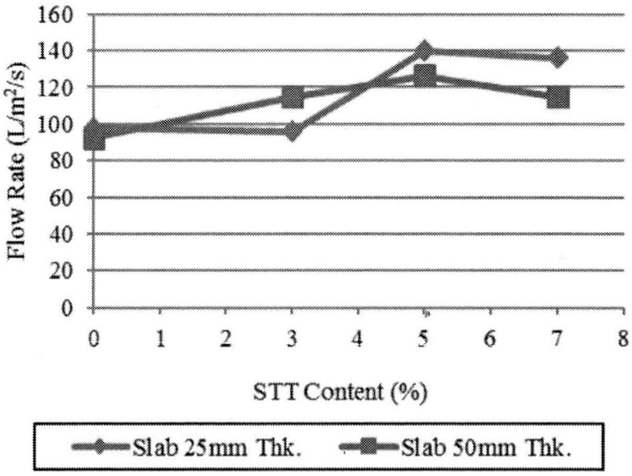

Fig. 13. Vertical flow rate.

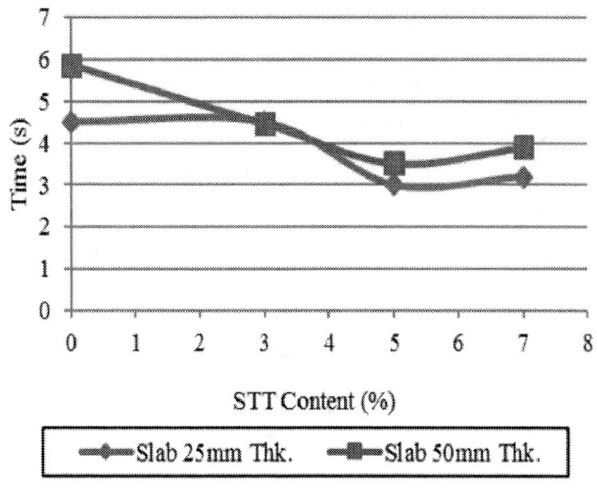

Fig. 14. Time achieved for horizontal flow.

4 Conclusion

Inclusion of scrap tyre tube rubber particles led to increment in void content and reductions in strengths of pervious concrete. However, pervious concrete with coarse rubber particles exhibits an acceptable compressive strength of more than 5 MPa. There is no distinct relationship could be established between the properties (flexural strength and flow rate) and proportion of rubber particles in pervious concrete mixes.

This study has focused on the void content, mechanical strengths and flow rate of pervious concrete containing small quantity of scrap tyres tube rubber particles. The results obtained in this study showing that there are potentials for use of scrap tyre tube rubber particles in pervious concrete. Pervious concrete which contains 7% STT has achieved higher mechanical strengths with better flow rate compared to other STT pervious concrete. However, there is a need for future studies to investigate the properties of pervious concrete with high volume scrap tyres tube rubber particles, and the durability of pervious concrete.

The experiments were performed in the Laboratory of Materials Engineering at Universiti Tun Hussein Onn Malaysia. The pervious concrete samples were prepared by Mr. Ch'ng Han Leong with the help of laboratory technicians. The authors wish to express the acknowledgements to their help.

References

[1] S.K. Thiruvangodan, *Waste tyre management in Malaysia*, PhD Thesis, Universiti Putra Malaysia, Malaysia, (2006)
[2] K. Day, K. Holtze, J. Metcalfe, C. Bishop, B. Dutka, Toxicity of leachate from automobile tyres to aquatic biota, Chemosphere, **27**, 665–75, (1993)
[3] A.R. Khaloo, M. Dehestania, P. Rahmatabadi, Mechanical properties of concrete containing a high volume of tire-rubber particles, Waste Management, **28**, 2472-2482, (2008)
[4] P.D. Tennis, M.L. Leming, D.J. Akers, *Pervious concrete pavements,* EB302.02, Portland Cement Association, Skokie, Illinois, and National Ready Mixed Concrete Association: Silver Spring, Maryland, (2004)

[5] V.R. Schaefer, K. Wang, M.T. Suleimam, J. Kevern, *Mix design development for pervious concrete in cold weather climates*, Center for Transportation Research and Education, Iowa State University, (2006)
[6] A.M. Mulligan, *Attainable compressive strength of pervious concrete paving systems*, MSc Thesis, University of Central Florida Orlando, (2005)
[7] ACI 522R-10, *Pervious concrete*, American Concrete Institute Committee, (2010)
[8] O. Deo, N. Neithalath, Compressive behavior of pervious concretes and a quantification of the influence of random pore structure features, Materials Science and Engineering A. **528**, 402-212, (2010)
[9] R. Siddique, T.R. Naik, Properties of concrete containing scrap-tire rubber – An overview, Waste Management, **24**, 563-569, (2004)
[10] MS EN 197-1, *Cement - Composition, specifications and conformity criteria for common cements*, Department of Standards Malaysia, Malaysia, (2007)
[11] M.A.R. Bhutta, K. Tsuruta, J. Mirza, Evaluation of high-performance porous concrete properties, Construction and Building Materials, 31, pp. 67-73, (2012)
[12] BS EN 12390-3, *Testing hardened concrete. Compressive strength of test specimens,* British Standards Institution, London, (2009)
[13] BS EN 12390-7, *Testing hardened concrete. Density of hardened concrete,* British Standards Institution, London, (2009)
[14] BS EN 12390-5 , *Testing hardened concrete. Flexural strength test specimens,* British Standards Institution, London, (2009)
[15] B.E.M.E. Yukari Aoki, *Development of pervious concrete*, Master Thesis, University of Technology, Sydney, (2009)
[16] B.S. Ireen Azfar, *Evaluation of pervious concrete as a best management practice for stormwater in Edwards Aquifer Region*, Master Thesis, University of Texas, San Antonio, United States, (2009)

Sand Cement Brick Containing Recycled Concrete Aggregate as Fine-Aggregate Replacement

Faisal Sheikh Khalid[1,*], *Herman* Shah Herman[1] , *Nurul Bazilah* Azmi[1], and *Mohd Irwan* Juki[1]

[1]Jamilus Research Centre, Fakulti Kejuruteraan Awam dan Alam Sekitar, Universiti Tun Hussein Onn Malaysia, 86400, Parit Raja, Batu Pahat, Johor, Malaysia

Abstract. Nowadays, the usage amount of the concrete is increasing drastically. The construction industry is a huge consumer of natural consumer. It is also producing the huge wastage products. The usage of concrete has been charged to be not environmentally friendly due to depletion of reserve natural resources, high energy consumption and disposal issues. The conservation of natural resources and reduction of disposal site by reuse and recycling waste material was interest possibilites. The aim of this study is to determine the physical and mechanical properties of sand cement brick containing recycled concrete aggregate and to determine the optimum mix ratio containing recycled concrete aggregate. An experiment done by comparing the result of control specimen using 100% natural sand with recycled concrete aggregate replacement specimen by weight for 55%, 65%, and 75%. The sample was tested under density, compressive strength, flexural strength and water absorption to study the effect of using recycled concrete aggregate on the physical and mechanical properties of bricks. The result shows that the replacement of natural sand by recycled concrete aggregate at the level of 55% provide the highest compressive and flexural strength compared to other percentage and control specimen. However, if the replacement higher than 55%, the strength of brick was decreased for compressive and flexural strength, respectively. The relationship of compressive-flexural strength is determined from statistical analysis and the predicted result can be obtained by using equation $f_{f,RCA} = 0.5375\ (f_c)^{0.3272}$.

1 Introduction

Sand cement brick is a kind commonly used in low- and medium-cost housing development and other commercial constructions in Malaysia. Cement and sand brick is easy to make and inexpensive to produce [1]. However, an issue was production cost of sand cement brick using natural sources, especially in developing areas where manufacturers find it difficult to locate adequate sources of natural aggregate supply. Therefore, the price of natural source such sand has increased, which in turn affected pricing of sand cement bricks

* Corresponding author: faisalsh@uthm.edu.my

[1-2]. Nowadays, many studies focus on the use of waste material in the production of brick concrete.

Therefore, application of recycled materials in the building industry is essential for permanently sustainable development of each county. One possibility is to recycle and reuse waste material concrete. Therefore, recycled concrete aggregate (RCA) as fine aggregate was selected as waste replacement material of sand-aggregate of sand cement brick compared to conventional sand cement brick. Many researchers exhibited that compressive strength of sand cement brick influenced by the physical and mechanical properties of fine RCA. It has been observed from a number of experimental investigations that the compressive strength of concrete using RCA is greatly influenced by the increment in the replacement percentage less than 50% of RCA using the same water-cement ratio compared to concrete using natural aggregates [3-5]. Other researchers show that recycled fine aggregate presented similar [5] or better performance [5-8] with respect to tensile strength. Concrete with RCA at 25% to 50% replacements of sand and aggregates presented sufficient tensile performance compared to normal strength concrete [9-10]. Wojciech et al. [1] claimed that the higher tensile strength of mortar containing fine aggregate is due to the water absorption capacity which is develops a siginificant interfacial bonding between aggregate and matrix. Matias et al. [10] mentioned that the increase in hardened strength is due to the rough surface of recycled fine aggregate which provides better adherence to concrete matrix. Soutsos et al. [11] studied the sand cement brick from waste concrete and bricks to replace coarse and fine aggregates. Southos et al [11] claimed that the brick conatining 60% replacement of coarse aggregate shows presented significant implications.

Ismail and Yaccob [2] claims that destiny of bricks containing 50 % of recycled fine aggregate content presents slightly increase by 1.7 %. The replacement of 50% and 75% containing fine aggregates shows significant effects on the compressive strength of the bricks. However, the flexural strength of the brick decreased when the recycled fine aggregate volume exceeds 50%. Poon et al. [12] declared that the strength of the specimen increased as the percentage of the replacement increased. Not only that, replacement of coarse and fine natural aggregate by recycled aggregates at the level of 25% and 50% only had a little effect compressive strength of the bricks and block. Debieb and Kenai [13] claimed the use of crushed brick as coarse and fine aggregates for mortar concrete. The result shows that the mortar concrete shares similar characteristics with natural-aggregate concrete prepare with recycled aggregate. Their finding found that percentage of 25% and 50% for coarse and fine aggregates, respectively are percentages presents optimum replacement percentages.

As a conclusion, the potential to manufacture brick containing recycled fine aggregate as replacement sand-aggregate material. This study aim to determine the combination of recycled fine aggregate as replacement material to enhance the strength and durability waste brick compared to conventional sand cement brick. Therefore, the use of recycled fine aggregate material in waste brick production will be promoted a sustainable friendly-waste materials and green building.

2 Methodology

The determination of the testing method for optimum mix ratio design of waste brick performance and suitable tools for testing are important to obtain data for analysis purposes. This waste brick was moulded with a mould size 213 mm in length, 101 mm in width, and 63 mm in depth. This study investigated in terms of hardened waste brick to answer the optimum mix ratio of waste RCA. The RCA used in this study were concrete casted in laboratory with strength of 25 MPa. These concrete underwent a further process of mechanised crushing and seiving to produce recycled fine aggregate as follow British

Stadards Instituion [14]. The grain size distributionb of natural sand and recycled fine aggregates as shown in Table 1. Table 2 shows the properties of natural sand and recyled fine aggregate.

Table 1. Sieve analysis of natural and recycled fine aggregate

BS test sieve	Percentage passing by weight	
	Natural sand	Recycled fine aggregate
2.36 mm	93.89	79.41
1.18 mm	77.21	48.56
600	50.2	28.10
425	36.12	19.1
300	22.98	10.5
212	12.91	7.2
150	7.7	6.3

Table 2. Properties of natural and recycled fine aggregate

	Natural sand	Recycled fine aggregate
Dry surface density (gm^3)	2.45	2.55
Water absorption (%)	11.3	4.6

This experiment used recycled fine aggregate as replacement were 55%, 65% and 75% by weight of natural sand [11]. One bacth of bricks had no recycled fine aggregate content as guide to compare the performance of sand cemnet brick containing recycled fine aggregate. The recycled fine aggregates consist of old cement, mortar and natural stone. Design method of the brick design used by BS 6073 [15]. The sand and cement brick design to exceed 7 N/mm^2 The mix ratio design was 1:3. The compressive strength of a material determines its load carrying capacity before failure. British Standards Institution states that the compressive strength of bricks should not be less than 7 N/mm^2. The test was carried out for the sample periods of 7 and 28 days. The sample is placed on compression testing machine. Next, the specimen also tested under flexural strength test. Flexural strength of brick is measure its ability of resists of bending. The sample periods of the specimen that being tested is 28 days.

The water absorption test was conducted to determine the percentage of water absorbed according to the BS 3921:1985, [16] which specify for the testing of bricks. The test also determine the saturation coefficient which is defined as the percentage of pore volume filled in 24 hour of soaking. The test are conducted for 28th days.

3 Result and discussion

The experiment was conducted in Heavy Structures Laboratory Universiti Tun Hussein Onn Malaysia. The results and analysis are divided in two parts, namely physical and mechanical properties. All collected data presented in table and graph to assure for a better and clear understanding of the results.

3.1 Density test

The batch of specimens were divided into 4 parts which contains 0%, 55%, 65% and 75% replacement of the recycled fine aggregate. Each part has 6 sample of the specimen for every percentages of 7 days and 28 days. The density of the specimens are calculated by dividing the weight with the volume of the specimens. Fig. 1 shows the average density of sand cement brick with recycled concrete aggregate. From the result, the average of the density for sand brick with recycled fine aggregate are lower compare to control brick. From the result, the most lower can be seen in bricks with 75% of recycled fine aggregate which reduce the density 4.9% for 7 days sample and 5.4% for 28 days sample compared to control bricks.

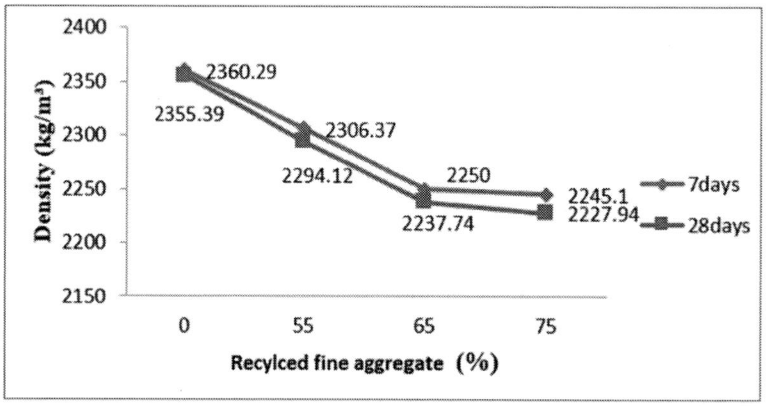

Fig. 1. The average density of sand cement brick with recycled fine aggregate.

3.2 Compressive strength

The compressive strength of the specimens determines its load bearing capacity before failure. Fig. 2 shows the average compressive strength of sand cement brick with fine recycled concrete aggregate.

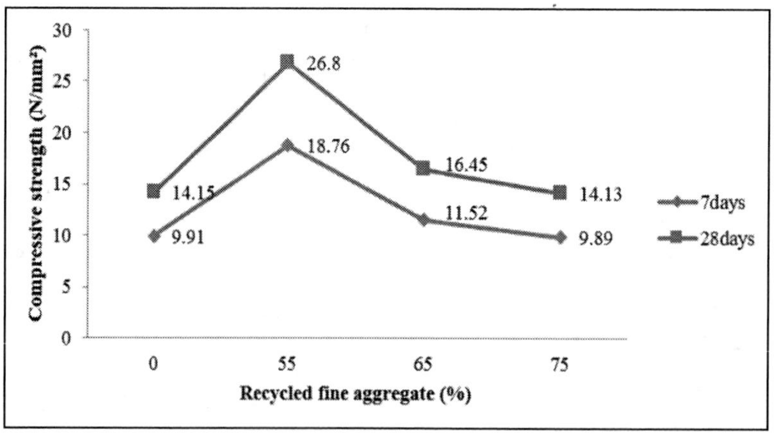

Fig. 2. The average compressive strength of sand cement brick with recycled concrete aggregate.

The specimens period of the sample being tested is 7 days and 28 days with different percentages replacement of recycled fine aggregate. The result shows if the percentage of recycled fine aggregate is added over 55% it will decrease the compressive strength of the brick. From the data collected, it is confirmed that brick with 55% replacement of recycled fine aggregate has good interlocking well graduation between coarse and finer particles of recycled fine aggregate and more homogeneous[2]. The more angular shape roughfer surface texture of recycled fine aggregate are contribute to a higher compressive. Recycled fine aggregate of 55% replacement is within a reasonable range to incorporate into sand cement brick. Therefore, 55% recycled fine aggregate content was selected as the optimum value for natural sand replacment regarding from compressive strength test.

3.3 Flexural strength

The specimen period of the sample is 28 days. The flexural strength of the control brick is 1.34 N/mm². From the result, the flexural strength increases occur when volume dosage rate recycled fine aggregate increases at 55% and achieved strength of 1.58 N/mm^2. But it is gradually decreased when it has more than 55% of volume dosage rate recycled fine aggregate because brick is too porous. It is lead to aggregates are bound not tightly, creating porous and reducing the strength of the brick. Fig. 3 shows the average flexural strength of sand cement brick with recycled fine aggregate. This study presented that the flexural strength of sand cement brick conatining recycled fine aggregate exhibited an increase of 17.9% (contains 55 % of recycled fine aggregate), respectively compared to control sand cement brick. Meanwhile, sand cement brick containing 65% and 75% of recycled fine aggregate presented decreases of 8.2% compared to control specimen.

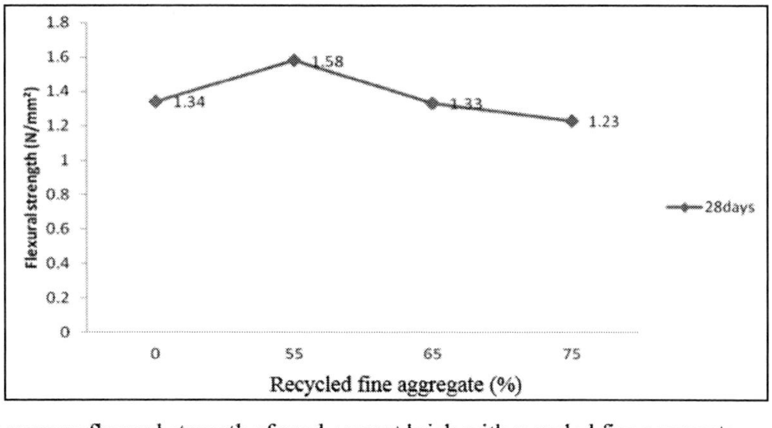

Fig. 3. The average flexural strength of sand cement brick with recycled fine aggregate.

3.4 Water absorption

Fig. 4 shows the results on water absorption with time. Water absorption characteristic of sand cement bricks containing recycled fine aggregate increased the water absorption characteristic of brick. Bricks contentaining 55% of recycled fine aggregate shows similar to the control speciemen. It can be concluede that the density to decrease and the water absorption to increase as the recycled fine aggregate content is increased. The sand cement brick content with 55% recycled fine aggregate show less permeable compared to control speciemen. The sand cement brick containing with high recycled fine aggregate percentage was more porously than the natural sand. It was due the pore is normally discontinuous in a matrix and being completely enveloped by cement paste in brick. The physical size, shape and random distribution of recycled fine aggregates are more coarser than natural sand which is make the packing of mortar matrix becomes less efficient as the recycled aggregate content is increased [2-3].

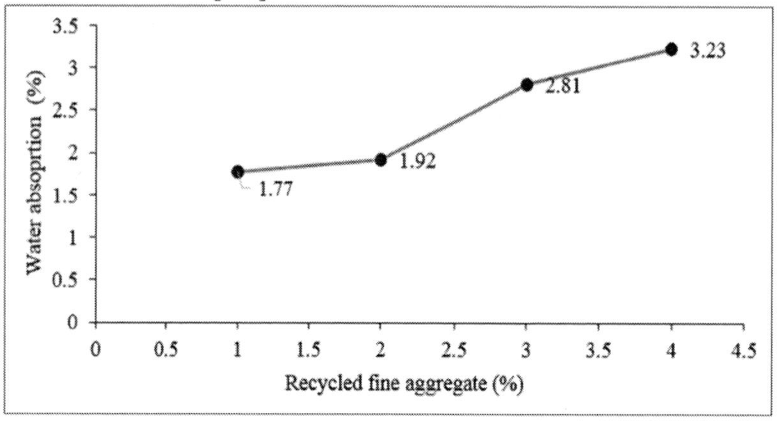

Fig. 4. Water absorption of sand cement brick for 28 days.

3.5 Compressive-flexural strength relationship

The compressive-flexural strength relationship are analysed and presented in the table and graph. The comparison of compressive-flexural relationship is determined by using the equation $f_{f,RCA,} = 0.5375\ (f_c)^{0.3272}$. From the calculation, the relationships agreed with experimental result by 1.00. Table 3 shows Compressive-flexural relationship between experiments and predicted using the equation $f_{f,RCA,} = 0.5375\ (f_c)^{0.3272}$.

Table 3. Compressive-flexural relationship between experiments and predicted using the equation $f_{f,RCA,} = 0.5375\ (f_c)^{0.3272}$.

% RCA	f_c (N/mm²)-comp	f_f (N/mm²)-flexural	Predicted f_f (N/mm²)	Ratio strength of predicted with experimental
0	14.15	1.34	1.28	0.96
55	26.8	1.58	1.58	1.00
65	16.45	1.33	1.34	1.01
75	14.13	1.23	1.28	1.04
Average				1.00

4 Conclusion

As the conclusion, the density of sand bricks containing recycled fine aggregate decreases gradually when the percentages of recycled fine aggregate increases. However, the replacement of recycled fine concrete aggregate at the 55% percentage present significant effects on the compressive strength of the bricks. When the replacements of recycled fine aggregate exceed 55% it will reduce the compressive strength and flexural strength.The replacment of recycled fine aggregate was increased the water absorption compared to control. However, 55 % of replacment presents recycled fine aggregate due to its compressive strength and flexural strength.

This study has been granted by Universiti Tun Hussein Onn Malaysia (UTHM), Malaysia under RAGS Vot. R058.

References

[1] W. Kubissaa, R. Jaskulskia, A. Kopera, and J. Szpetulski, Properties of concretes with natural aggregate improved by RCA addition, Procedia Engineering, **108**, 30-38 (2015)
[2] Ismail and Z. Yaacob, Properties of bricks produced with recycled fine aggregate, Int. J. of Civil, Environmental, Structural, Construction and Architectural Engineering, **4**(7), 670-671, (2010)
[3] L.W. Xiao, Y. Fan, and X. Huang, An overview of study on recycled aggregate concrete in China, Construction and Building Materials, **31,** 364–83, (2012)
[4] McNeil and T. Kang, Recycled concrete aggregates, Int. J. Concrete Structure Materials, **7**(1), 61–9, (2013)
[5] M. Chakradhara Rao, S.K. Bhattacharyya, and S.V. Barai, Influence of field recycled coarse aggregate on properties of concrete, Materials Structure, **44**(1), 205–220, (2011)
[6] P. Saravana Kumar and G. Dhinakaran, Effect of admixed recycled aggregate concrete on properties of fresh and hardened concrete, J. Material Civil Engineering, **24**(4), 494–498, (2012)
[7] S.K. Faisal, J.M. Irwan, N. Othman, and M.H. Wan Ibrahim, Flexural toughness of ring-shaped waste bottle fiber concrete, *The 3rd International Conference on Civil and Environmental Engineering for Sustainability*, MATEC Web of Conf., **47**(1), 1-6, (2016)
[8] J.M. Irwan, S.K. Faisal, N. Othman, and M.H. Wan Ibrahim, Performance of concrete using light waste PET fibre, Advanced Materials Research, **795**, 352-335, (2013)
[9] L. Butler, J.S. West and S.L. Tighe, Effect of recycled concrete coarse aggregate from multiple sources on the hardened properties of concrete with equivalent compressive strength, Construction and Building Materials, **47**, 1292–1301, (2013)
[10] D. Matias, J. de Brito, A. Rosa, and D. Pedro, Mechanical properties of concrete produced with recycled coarse aggregates and influence of the use of superplasticizers, Construction and Building Materilas, **44**,101–109, (2013)
[11] M.N. Soutsos, S.G. Millard, J.H. Bungey, N. Jones, R.G. Tickell, and J. Gradwell. Using recycled demolition waste in concrete building blocks. *Proc. of the Institute of Civil Engineers: Engineering Sustainability*, **157**(3), 139-148, (2004)
[12] C.S. Poon, S.C. Kou and L. Lam, Use of recycled aggregate in molded concrete bricks and blocks, Construction and Building Materals, 16(5), 281-289, (2002)
[13] F. Debieb, and S. Kenai. The use of coarse and fine crushed bricks as aggregate in concrete, Construction and Building Materials, **22**(5), 886–893, (2008)
[14] BS 812, *Methods for determination of particle size distribution, Part 103.1*, British Standards Institution, London, UK, (1985)
[15] BS 6073, *Precast concrete masonry unit, part 1: specification for precast concrete masonry units, part 1*, British Standards Institution, London, UK, (1981)
[16] BS 392, *specification for clay bricks,* Clay Product Standards Committee, British Standard Institution, London, UK, (1985)

Incorporation of Palm Kernel Shell into Fired Clay Brick

Aeslina Abdul Kadir[1,*], *Noor Amira* Sarani [2], and *Syafiqah 'Amirah* Abd Kadir [3]

[1,2,3]Department of Water and Environmental Engineering, Faculty of Civil and Environmental Engineering, Universiti Tun Hussein Onn Malaysia, 86400, Parit Raja, Batu Pahat, Johor

Abstract. Palm oil processing has produce agricultural waste and one of the wastes is palm kernel shell (PKS) which is abundance and creating disposal problems in landfill. Recently, many researchers investigate the incorporation of palm oil waste such as palm fiber, palm oil fuel ash and empty fruit bunch which could have potential as a second raw material in brick manufacture. Therefore, this study is to investigate the physical and mechanical properties of PKS as clay replacement in fired clay brick. Different percentages of PKS (0%, 1%, 5% and 10%) were incorporated into fired clay brick. Manufactured bricks were fired at 1050°C with heating rate of 1°C/min and were tested with dry density, shrinkage, initial rate of suction, water absorption, porosity and compressive strength. XRF result shows that both clay soil and PKS have a highest percentage of Silicon Dioxide (SiO_2) and Iron Oxide (Fe_2O_3). Throughout this study, PKSB1% shows the best result compared to others brick with several improvements on physical and mechanical properties. As the conclusion, PKS could be a potential waste to be utilized as it produces adequate brick with better strength and complied with the requirement.

1 Introduction

Production of palm oil assists Malaysia in terms of economical and also agricultural industry. However, the increased in palm cultivation has led to thousands of tonnes of palm oil waste in Malaysia. Milling process of fresh fruit bunch has produced abundant of palm oil waste such as palm kernel shell (PKS). By sending those agricultural wastes to landfills and disposed through incineration method is not the practical solution because those wastes will release high methane gas during incineration process that caused several environmental problems.

Numerous of researchers had studied incorporation of different types of waste into fired clay brick as an alternative method on maximize the utilization waste product in construction material [1–3]. Furthermore, recycle and reuse of waste product is a worthwhile activity with many beneficial effects to environment and construction field. Based on previous studies, incorporation of waste products into fired clay brick is generously showing the good agreement with others. Therefore, the aim of this study is to

* Corresponding author: aeslina@uthm.edu.my

investigate the possible use of PKS into fired clay brick and its effect on the physical and mechanical effect as it gives positive impact to the environment.

2 Materials and methods

In order to measure the properties of manufactured brick, the materials and methods were discussed in this section.

2.1 Materials

The materials used in this study are PKS and raw clay soil. Upon delivery, raw clay soil and PKS was oven dried for 24 hours at 105°C and then was stored in closed containers. This process is needed in order to ensure that the samples are in dry condition before manufacturing process. The PKS collected from local palm oil mill was ground and sieved passed through 4.75mm sieve to obtain homogenous size.

2.2 Methods

2.2.1 Testing methods for raw clay soils and palm kernel shell

Characterization of chemical composition was tested by using X-ray Fluorescence (XRF) method.

2.2.2 Manufacturing process of control brick and palm kernel shell brick

Once the raw materials have been characterized, raw clay soil was mixed with different percentage of PKS (1%, 5%, and 10%). Necessary amount of water was added to achieve desired consistency of the samples. The mixture then was pressed into mould of 215mm x 115mm x 65mm with pressure of 3000psi. Manufactured samples were oven dried at 105°C for 24 hours and then continued with firing process in furnace at 1050°C using heating rate of 1°C/min. Control brick also was manufactured for control purposed. The manufactured bricks were designed as Control Brick (CB) for the brick without PKS, and PKSB1%, PKSB5% and PKSB10% for the brick with 1%, 5% and 10% of PKS, respectively.

2.2.3 Physical and mechanical properties of manufactured brick

Physical and mechanical properties of fired clay brick were tested such as dry density, shrinkage, initial rate of suction (IRS), water absorption, porosity and compressive strength that has been conducted at laboratory. All the test were carried out according to British Standards and the results are reported in three mean values [4].

3 Results and discussion

3.1 Characterization of manufactured brick

Chemical analysis of raw clay soil and palm kernel shell was carried out to determine the main component of the materials used in this study. The main oxides present in raw clay soil are SiO_2 (55.7%), Al_2O_3 (24.4%) and Fe_2O_3 (4.46%). Meanwhile, other major components observed in palm kernel shell are SiO_2 (22.58%) and Fe_2O_3 (34.37%).

3.2 Physical and mechanical properties

This study is focusing on the different percentages of palm fibre waste to be incorporated into fired clay brick in order to achieve desired performance on physical and mechanical properties.

3.2.1 Dry density

Dry density of the manufactured bricks is presented in Fig. 1 where the values were varied between 1646.93kg/m^3 to 1915.91kg/m^3 depending on the percentage of waste content. Based on the Fig. 1, CB gives the highest number of dry density which was 1915.9kg/m^3 and followed by PKSB1% and PKSB5% with dry density 1844.6kg/m^3 and 1761.5kg/m^3. The lowest density of brick was obtained with the incorporation with 10% of PKS (1646.9kg/m^3). From the results, all the manufactured brick are classified as high gross densities brick as their densities are over than 1000kg/m^3 [5]. Lower density of brick is highly recommended since it could reduce load during construction work besides reduce logistic costs [6–8].

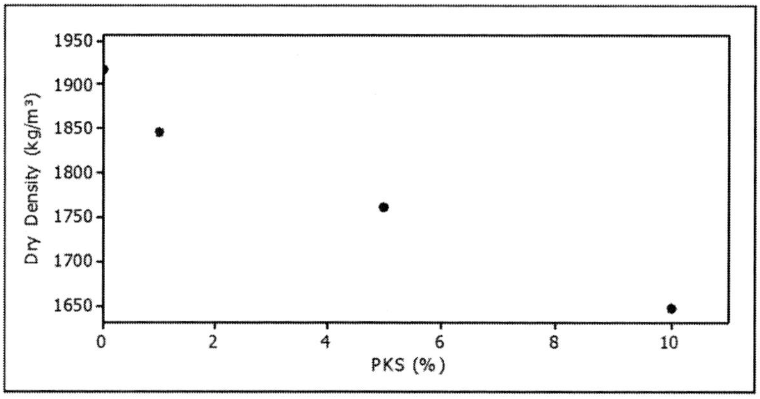

Fig. 1. Dry density of brick.

3.2.2 Shrinkage

Total shrinkage of the manufactured bricks is presented in Fig. 2 where total shrinkage values of the samples were varied within 0.75% to 1.75% depending on the percentage of waste content. As the PKS increased, the value of shrinkage also increased. The lowest was obtained by CB with 0.75%. The result also found that total shrinkage increased with 0.93% (PKSB1%), 1.43% (PKSB5%) and 1.75% (PKSB10%). The requirement has recommended that shrinkage falls within 2.5% to 10% [5]. Unexpectedly the results for all manufactured brick are in good agreement with the requirement.

3.2.3 Initial rate of suction

Initial rate of suction of the manufactured bricks is presented in Fig. 3 where initial rates of suction values of the samples were varied between 2.48kg/m^2.min to 5.61kg/m^2.min depending on the percentage of waste content. As the PKS content increased, the value of initial rate of suction also increased. Based on the result, PKSB10% achieved maximum initial rate of suction which was 5.61kg/m^2.min followed by PKSB5% and PKSB1% which was 3.50kg/m^2.min and 3.29kg/m^2.min respectively. Meanwhile, initial rate of suction for CB was 2.48kg/m^2.min that makes CB obtained the lowest rate among the rest of the brick. However, all manufactured brick shows high initial rate of suction which is over than 2kg/m^2.min. Therefore, brick should be wetted prior 3 hours to 24 hours before laid as suggested by [9].

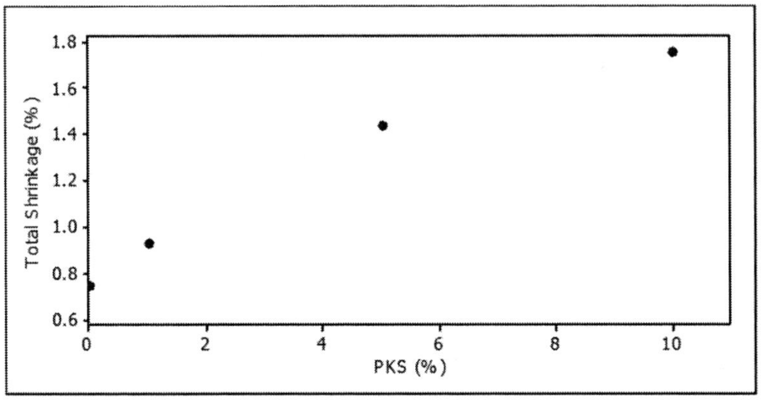

Fig. 2. Total shrinkage of brick.

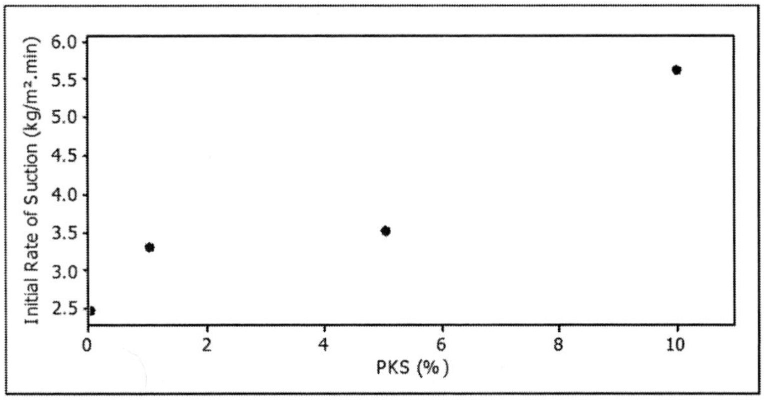

Fig. 3. Initial rate of suction of brick.

3.2.4 Water absorption

Water absorption of the manufactured brick is presented in Fig. 4 where water absorption values of the samples were varied within 12.84% to 17.75% depending on the percentage of waste content. As the PKS content increased, the value of water absorption also increased. Based on the result, CB tends to absorb lowest water absorption with 12.84% compared to others PKSBs. The results also found that increasing 1% to 10% of PKS had increased water absorption of brick with 14.62%, 17.09% and 17.75% respectively. The justification of this behaviour could be related to the organic matter in PKS which eliminate during sintering process thus created numerous pores. According to [4], water absorption of manufactured brick can be classified as Engineering Brick B and brick for Damp Proof Course 2 which only applicable for non-loading application.

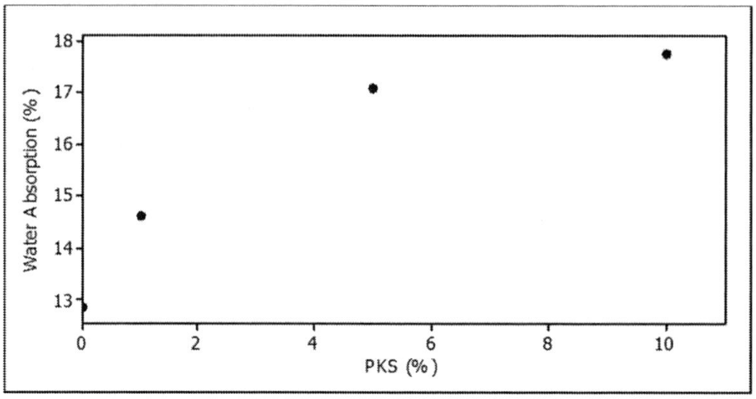

Fig. 4. Water absorption of brick.

3.2.5 Porosity

Porosity of the manufactured bricks is presented in Fig. 5 where porosity values of the samples were varied between 14.65% to 21.35%, depending on the percentage of waste content. Based on the figure, PKSB10% gives the highest porosity (21.34%) followed by PKSB5% and PKSB1% with 17.72% and 16.32% respectively. Meanwhile, CB was obtained lowest porosity with 14.65%. It was acclaimed that high porosity could reduce thermal conductivity of brick as well as beneficial for higher energy saving in building application [10-11].

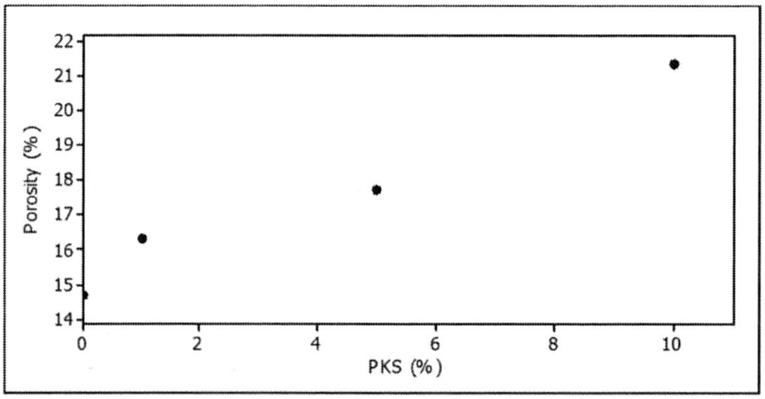

Fig. 5. Porosity of brick.

3.2.6 Compressive strength

Compressive strength of the manufactured bricks is presented in Fig. 6 where compressive strength values of the samples were varied between 6.75MPa to 19.52MPa depending on the percentage of waste content. As the PKS content increased, the value of compressive strength is decreased. From the figure, PKSB1% obtained the highest compressive strength with 24.61% followed by others PKSB with 12.50MPa (PKSB5%) and 6.75MPa (PKSB10%). Meanwhile, compressive strength for CB was recorded as 19.52MPa. It is expected that PKSB10% contained high organic matter thus leaves numerous pores during sintering process which is lead to reduce the strength of brick. According to [4], all the brick were above the minimum requirement and be applied for non loading construction application.

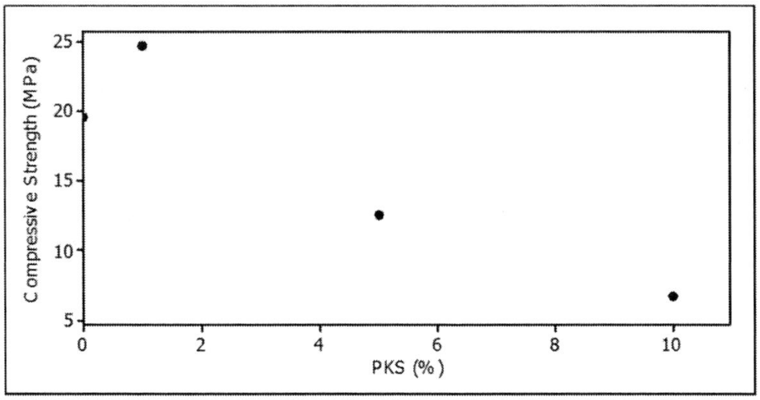

Fig. 6. Compressive strength of brick.

4 Conclusions

PKS can be regarded as a potential addition to raw materials used in the manufacturing of fired clay brick. Incorporation of clayey soil and PKS can be used as building material with acceptable physical and mechanical properties match with the requirement of non-loading application. It was suggested that incorporation of 1% of PKS into fired clay brick could be an alternative method to reduce abundant PKS waste in landfill. Mixing 1% of PKS is recommended as it produced highest compressive strength and lower in density, which are desirable in producing green building product.

The results presented in this paper are part of an ongoing postgraduate research. The authors would like to thank the Faculty of Civil Engineering, UTHM for this study.

References

[1] S.M.S. Kazmi, S. Abbas, M.J. Munir, and A. Khitab, Exploratory study on the effect of waste rice husk and sugarcane bagasse ashes in burnt clay bricks, J. of Building Engineering, **7**, 372–378, (2016)

[2] A.A. Kadir and A. Mohajerani, Effect of heating rate on gas emissions and properties of fired clay bricks and fired clay bricks incorporated with cigarette butts, Applied Clay Science, **104**, 269–276, (2015)

[3] A.A. Kadir, N.A. Sarani, and A.M. Leman, Testing on building material using waste material in fired clay brick, Materials Science Forum, **803**, 330–336, (2014)

[4] BS 3921, *Specification for clay bricks*, British Standard (BS), London, United Kingdom, (1985)

[5] The Brick Industry Association BIA, T*echnical Notes 3A - Brick Masonry Material Properties*, Technical Notes on Brick Construction, The Brick Industry Association, Reston, Virginia, USA, (1992)

[6] A.G. Celik, T. Depci, and A.M. Kılıc, New lightweight colemanite-added perlite brick and comparison of its physicomechanical properties with other commercial lightweight materials, Construction and Building Material, **62**, 59–66, (2014)

[7] P. Panedpojaman and V. Sae-Long, Use of palm oil fuel ash in lightweight brick industry, Academic J. of Science, **2**(2), 163–170, (2013)

[8] A.A. Kadir, A. Mohajerani, F. Roddick and J. Buckeridge, Density, strength, thermal conductivity and leachate characteristics of light-weight fired clay bricks incorporating cigarette butts, *Proc. of the World Academy of Science, Engineering and Technology*, **53**, 1035–1040, (2009)
[9] BS EN 771-1, *Specification for masonry units part 1 : Clay Masonry units*, British Standard, BSI Standards Publication, London, United Kingdom, (2011)
[10] M. Sutcu, Influence of expanded vermiculite on physical properties and thermal conductivity of clay bricks, Ceramics Int., **41**(2), 2819–2827, (2015)
[11] D. Eliche-Quesada, S. Martínez-Martínez, L. Pérez-Villarejo, F.J. Iglesias-Godino, C. Martínez-García, and F.A. Corpas-Iglesias, Valorization of biodiesel production residues in making porous clay brick, Fuel Processing Technology, **103**, 166–173, (2012)

Compressive Strength of Construction Materials Containing Agricultural Crop Wastes: A Review

Nik Nadia Amira Nik Pa[1], *Mohd Haziman* Wan Ibrahim[1,*], *Abdul Halim* Abdul Ghani[1], *Ahmad Zurisman* Mohd Ali[1], and *Mohd Fadzil* Arshad[2]

[1]Jamilus Research Centre , Faculty of Civil and Environmental Engineering, Universiti Tun Hussein Onn Malaysia, 86400, Parit Raja, Batu Pahat, Johor, Malaysia
[2]Institute For Infrustructure Engineering and Sustainable Management, Universiti Teknologi MARA, 40450, Shah Alam, Selangor, Malaysia

Abstract. The growth of industrialization and development of urban life made an increase in demand for cement, concrete and bricks. Exploitation on the non-renewable natural resources for raw materials will keep increased in order to meet the demands for construction materials. At the same time, the problems regarding the agricultural crop wastes such as rice husk, sugarcane bagasse, palm oil fuel waste and elephant grass has become an important issue nowadays. Consequently, there are many researchers who have been studying the viability of using these agricultural wastes as construction materials to meet the industry demands in order to decrease the current use of non-renewable natural resources. This paper reviewing on how agricultural waste could be utilized as replacement materials for construction activities from various researchers. The idea of using agricultural crop wastes was promoted by studying upon their engineering properties. This paper focusing on the compressive strength of the construction materials containing agricultural crop wastes, which was the common parameter considered by most researcher as required by various standards.

1 Introduction

The world's demand for cement until the first half of the 21st century is expected to increase by about 2.5–5.8% every year [1]. Therefore, the exploitation on raw materials from non renewable natural resources were required in large quantities to meet the worldwide demand. At the same time, the cement and concrete productions has been an environmental issue since this industry contributed about 7% of carbon dioxide emission worldwide along with great consumption of energy and aggregate amount [2]. The usage of waste materials for variety of applications in construction industry kept increase due to the growing awareness of environmental issues and resource scarcity. Many industries take a step to find new materials from renewable resources to replace the current used materials derived from non-renewable resources due to strict legislative policies enforcement and in the same

* Corresponding author: haziman@uthm.edu.my

time will protect the environment [3]. The productions and usage of agricultural wastes such as rice husk, sugarcane bagasse and straw, palm oil fuel shell, wheat straw, and elephant grass (*Pennisetum Purpureum*) have been successfully recycled in many products and being a great interest in many publications. The alternatives regarding utilization of agricultural waste to supply the demand for concrete and other construction materials has promote a breakthrough to make the construction industry more environmentally friendly and sustainable [2]. The wide availability of agricultural wastes makes them a suitable and dependable alternative for construction materials such as pozzolans, aggregate, and fibres [4]. Recycling the agricultural wastes will decrease the exploitation on natural resources, contributes to energy saving, and a reduction in the cost of construction materials but still can meet the current demands. It also solves the disposal problem of the wastes which saves landfill space, and hence, helps environmental protection [5, 6].

2 Review of previous research on utilization of agricultural waste materials

The agricultural crop waste materials discussed in this paper were rice husk, elephant grass (*Pennisetum Purpureum*), sugarcane bagasse, and palm oil fuel wastes, as detailed below. Compressive strength of products containing these waste materials was the parameter highlighted in this paper.

2.1 Rice husk

Rice husk is the byproducts which produced after milling process of paddy [7]. Rice husk being disposed and burned in open spaces due to zero commercial value, hence inviting serious environmental problems [8]. In certain country, rice husk was recycled as boiler fuel to generate electricity and steam but contribute about 20% of ash after incineration process [9]. With the proper preparations, rice husks can be processed to be ash and being used as pozzolans and filler in many construction application [10]. High content of amorphous silica (SiO_2) and large surface area contibuted to the high pozzolanic properties and good reactivity of the rice husk ash. In addition, the reactivity of rice husk ash can be improved by increasing its fineness. Recently, many researchers have studied the utilization of rice husk ash as pozzolans and filler in construction material [7,11-21].

Bie et al. [11] conducted a research regarding the mechanical behavior of cement mortar containing RHA burned at different burning conditions in muffle furnace. Cement mortar samples sized of 40 × 40 ×160 mm were prepared by replaced 0%, 5%, 10% and 20% of cement with RHA to obtain their compressive and flexural strength. It was proved that 10% of replacement ratio (by weight) of RHA has the best enhancement effect in compressive and flexural strength of the cement mortar. The result obtained shows that burning conditions can affect the specific surface area and properties of RHA in which 600 °C was the appropriate temperature to produce RHA with large specific surface area due to existence of amorphous silica and also decrease in potassium content. Bakar et al. [7] investigated the effects of RHA fineness on the chemical and physical properties of concrete using eight different fineness grades of RHA. The compressive strength and strength activity index of concrete sample blended with 15% by weight of RHA ground for 90 minutes (median particle size of 9.52μm) has significantly increased. Similarly, Antiohos et al. [12] studied about the effectiveness of rice husk ash as reactive silica and its fineness in mortar and concrete. The results showed that the reactivity of RHA in the mortar and concrete increased when the fineness increased. Hwang et al. [13] studied the strength and durability characteristics of concrete incorporating 0%, 10%, 20% and 30%

replacement ratio of cement by 1 hour ground RHA. The results indicated 20% ground RHA added to concrete has an equivalent values of compressive strength to control samples after 28 days and increased after 91 days of curing period. Eventhough it was possible to produce RHA blended concrete that has similar or better properties than ordinary concrete (without RHA), at least the study agreed that it can reduce the carbon dioxide emission when the level of cement consumptions is decreased. The same pattern were agreed by Shahidan et al [14] in which RHA replacement can increased the compressive strength of concrete after 28 days of curing period due to pozzolanic reactions.

A research conducted by Vishwakarma et al. [15] was a usage of RHA in nano size on the mechanical and microstructural properties of concrete. The RHA was prepared by burned the rice husk at temperature of 600°C to 800°C before being sieved and ground in a ball mill for an hour to get desired nanoparticle size (50 to 70 nanometer). It was found that RHA concrete have higher compressive, tensile and flexural strength than fly ash concrete due to the presence of nanosilica as nanosilica from RHA increased the polymerized layer structure and rough surface texture which increased the bonding between aggregates and cement paste. Meanwhile, a study regarding the usage of 15% of RHA both in natural conditions (without prior treatment or grinding) and ground RHA on the concrete properties was conducted by Zerbino at al. [16]. It was found that natural RHA can reduced the amount of cement content but required an optimization process to obtain acceptable particle size since the raw material contained a residue. Ground RHA provided an excellent characteristics in the fresh state and increased the mechanical properties of concrete even when blended with 25% of RHA. The concrete containing 15 % of natural RHA showed similar mechanical and durability properties with that control concrete however the concrete with ground rice husk ash gave better results. The study proved that the usage of natural RHA in concrete was a good option to dispose this residue even without the prior treatment and grinding process.

Other than that, Kishore et al. [17] studied the strength characteristics of high strength concrete containing RHA and found that the replacement of cement with RHA decreased the compressive strength but achieved the strength target at 10% replacement for both of two different concrete grades (M40 and M50). Meanwhile, Tuan et al. [18] investigated the potentials of the usage of RHA to produce ultra-high performance concrete. It was found that RHA addition which less than 30% does not significantly decreased the compressive strength of ultra-high performance concrete compared to sample containing silica fume. This results was due to the fineness of RHA that has a positive effect on the compressive strength of ultra-high performance concrete and the optimum mean size for RHA was 5.6μm. This study suggested that the cement comsumption can be replaced up to 40% with combination of silica fume and RHA in order to produce ultra high performance concrete.

Chopra et al. [19] investigated the effects of cement replacement with RHA as supplementary cementitious materials in self-compacting concrete (SCC) by replacing cement content with RHA by 0%, 10%, 15% and 20%. The mechanical and durability properties of casted samples being tested after 7, 28 and 56 days of curing period. The results indicated that the compressive and tensile strength increased with the increasing of RHA content. The increasing of compressive strength of concrete was developed from the increasing formation of calcium silicate hydrate (C-S-H) gel which makes its structure dense. The optimum content of RHA replacement was found to be 15%. Hesami et al. [20] studied the usage of RHA as cement replacement material in pervious concrete pavement in order to strengthen pozzolanic cement paste. A concrete mixture with 0%, 2%, 4%, 6%, 8%, 10% and 12% weight percentages of RHA ash were prepared for physical and mechanical properties evaluation. The RHA used in the testing was burned in a furnace for 2 hours which produced black rice husk ash and then it was burned again in a special furnace to decarbonize the carbon content. This process increase the specific surface area and pozzolanic properties of rice husk ash. The results obtained shows that compressive, tensile and flexural of pervious concrete samples mixed with RHA were positively increased and found that 12% of RHA was the optimum content to for concrete pavements. The strength of concrete decreased after the optimum amount of RHA reached. In another study conducted by Jaya et al. [21] found that 40% of RHA as cement replacement in concrete improved resistance to seawater attack and effectively decreased ettringite and gypsum formations.

The possibilities of using rice husk and RHA in production of bricks as alternatives in recycling the waste materials was studied by Sutas et al. [22]. The bricks were produced by adding between rice husk and RHA which varied from 0% to 10% by weight into the clay mixtures. The results indicated that the main component of RHA was made up of silicon oxide in which contributed to the brick's strength while the aluminium and iron oxide changed the brick's color into red after sintering. It was found that the compressive strength and density of the brick can be increased with 2% of RHA addition.

2.2 Pennisetum Purpureum

Pennisetum Purpureum or known as elephant grass or Napier grass is a quick growing plant and has a production of about 40 ton/ha/year [23,24]. The *Pennisetum Purpureum* grass have reasonable energy content and higher calorific value fuels also categorized as biomass materials [25]. At the present, Brazil has been used 100% of *Pennisetum Purpureum* grass as a renewable energy source in several thermoelectric power plant in their country which deposited a large quantities of ashes during direct burning process resulting in environmental problems [23,25]. The extraction of ortho-silicic acid from ground water by the *Pennisetum Purpureum* grass has made the plant contained considerable content of amorphous silica, which made it has pozzolanic properties similar to rice husk and sugarcane bagasse [23,25]. Researchers also conducted the possibility of fiber productions from *Pennisetum Purpureum* grass [26-28].

A laboratory scale study conducted by Nakanishi et al. [29] in which regarding the usage of *Pennisetum Purpureum* ash to manufacture eco-friendly binary cements. *Pennisetum Purpureum* ash were prepared through same process conducted by Nakanishi et al. [25]. The physical and mechanical of blended cement paste with 20% replacement of cement with *Pennisetum Purpureum* ash were evaluated after 28, 60 and 90 days of curing period. The study found that the cement paste containing 20% of *Pennisetum Purpureum* ash have similar strength with the control paste and cement paste containing 20% of silica fume after 90 days of curing period. Similarly, the same facts agreed by Nakanishi et al. [25] in which the main hydrated phases during the pozzolanic reaction in *Pennisetum Purpureum* paste consisted of C-S-H gels similar to silica fume paste. The study agreed that *Pennisetum Purpureum* ash can be cementitious materials as an alternative to traditional pozzolans. Meanwhile, Cordeiro and Sales [23] studied the pozzolanic properties of *Pennisetum Purpureum* ash and its effects on the mechanical properties of concrete. *Pennisetum Purpureum* ash was produced by three different method and laboratory scale tests were conducted to evaluate the physical and chemical properties of the *Pennisetum Purpureum* ash including the pozzolanic and reactivity test on the *Pennisetum Purpureum* ash paste. For the mechanical properties testing, a concrete sample with 20% replacement of cement with *Pennisetum Purpureum* ash were prepared. The results showed no changes in compressive strength, Young's modulus and water absorption rate of concrete containing 20% of *Pennisetum Purpureum* ash when compared to control sample and proved that the pozzolanic activity of *Pennisetum Purpureum* ash maintained the mechanical properties of concrete.

Merta and Tschegg [30] studied the potentials usage of outer core of the *Pennisetum Purpureum* stem as fibres reinforcement for concrete. The study shows that *Pennisetum Purpureum* fibres had tensile strength of about 40 to 60 MPa, which has similar range to wheat straw fibres strength. On the other hand, Haameem et al. [25] investigated the mechanical properties of *Pennisetum Purpureum* (Napier) used as fibre or polyester composites. The extraction of *Pennisetum Purpureum* fibres were done through water retting process and then the fibres were alkali treated using sodium hydroxide (NaOH) solutions. The study found that alkali treated fibres has higher strength than untreated fibres and 10% alkali treated fibres showed highest tensile strength. The tensile and flexural strength of *Pennisetum Purpureum* fibres composites increased with the increasing of fibres volume fractions. It was found that 25% of volume fraction of *Pennisetum Purpureum* fibres showed optimum tensile and flexural strength. Futhermore, the research conducted by Ramaniah et al. [27] and Ridzuan et al. [28], both agreed that *Pennisetum Purpureum* fiber were suitable to be utilized as composites material in many applications due to its properties which were light in weight, have economic value and good thermal insulating properties.

2.3 Sugarcane bagasse

Sugarcane bagasse is a waste produced from extraction of sugarcane for sugar and alcohol production industry [31]. Approximately, 1 kg of sugarcane produced 25% of bagasse and 0.6% of bagasse ash [32]. After combustion process, the sugarcane residue dominated by silicon dioxide (SiO_2) in chemical composition [33]. Disposal problems arises due the high volume of the sugarcane residue, in which turning this ash as a pozzolanic material in the construction sectors would be a suitable disposal way [31].

Tashima et al. [34] studied on the utilization of sugar cane straw ash in blended cement as pozzolans. The microstructural characteristics and mechanical strength were evaluated onto cement paste and cement mortar containing sugarcane straw ash. The results obtained shows a positive development on compressive strength of mortar containing 15% to 30% of sugarcane straw ash. After 28 and 90 days, mortar samples containing 25% and 30% of sugarcane straw ash given same results with the control sample in which, indicated that the sugarcane straw ash has a potential to be used as pozzolanic admixture in blended mortars with high replacement levels. Rukzon and Chindaprasirt [35] carried out the investigation on utilization of finely ground bagasse ash as a pozzolanic material in high strength concrete. The concrete samples were prepared by replacing the cement content with 10%, 20% and 30% of bagasse ash. The concrete samples showed good strength and low porosity when 10% of bagasse ash was used. From the study, it was concluded that the high strength concrete can be produced by replacing the cement with 30% of bagasse ash. On the other hand, an investigation on the strength and durability properties of concrete containing 10%, 20%, 30%, 40% and up to 50% of ground bagasse ash by weight of binder were conducted by Rerkpiboon et al. [36]. The study found that the highest compressive strength of concrete was achieved at 20% of replacement cement by bagasse ash at 90 days age.

Modani and Vyawahare [37] conducted a research on the usage of sugarcane bagasse ash as a partial replacement of fine aggregate in concrete. The concrete samples were casted using untreated sugarcane bagasse ash with replacement ratio of 0%, 10%, 20%, 30% and 40% by volume of fine aggregate. The results indicated that the usage of untreated sugarcane bagasse ash as fine aggregate does not give any decrease in workability and strength of concrete. The compressive strength of concrete containing 10% and 20% of sugarcane bagasse ash increased due to pozzolanic reaction after 28 days of curing period. Lima et al. [38] studied the potentials of compressed earth brick block masonry (CEB) containing sugarcane bagasse through their mechanical properties. Two sets of blocks with

6% and 12% cement addition were prepared while the sugarcane bagasse ratio added to the blocks were 0%, 2%, 4% and 8% for each mixture. The results found that satisfactory compressive strength were achieved by the block containing 12% of cement while the addition of sugarcane bagasse ash did not affect the mechanical strength and water absorption of the CEB. However, the block prepared with 12% of cement and 8% of sugarcane bagasse ash have potentials to be used as non-structure masonry component. Montes-Garcia et al. [39] analyzed the use of lime and sugar cane bagasse ash (SCBA) as chemical stabilizers in compacted soil blocks. The results indicate that blocks manufactured with 10% of lime in combination with 10% of SCBA showed better performance than those containing only lime.

2.4 Palm oil fuel waste

Malaysia, Indonesia, Thailand and Nigeria has become most important agro-industries for oil palm products with Malaysia as the world's largest producer and exporter of palm oil [40]. One of the problems in the palm fruit processing plants is that the process produces large quantities of solid waste resulting in pollution problems [41]. The production of solid waste from the palm oil industry in Malaysia, Indonesia and Thailand was 47, 40 and 3.5 million tons per year, respectively. The solid wastes include oil palm bunches, palm fibres and fruit shells.

Alsubari et al. [42] studied the usage of treated palm oil fuel as supplementary cementitious material (SCM) to produce sustainable self-compacting concrete (SCC). A laboratory scale study were conducted by preparing the concrete mixtures with 0%, 50%, 60% and 70% replacement of cement with treated palm oil fuel ash. The treated palm oil fuel ash SCC has lower compressive strength at early age however, still can be classified as high strength class at 1 day age. SCC blended with 50% to 70% of treated palm oil fuel ash showed similar compressive strength to control SCC and kept increased after 28 days. Tangchirapat et al. [43] investigated the effects of ground palm oil fuel ash utilization as pozzolanic material in high strength concrete by replacing the cement content at 10%, 20% and 30%. The results indicated that ground palm oil fuel ash with high fineness suitable to be used as pozzolans to produce high strength concrete with 20% of replacement. Overall, 20% replacement of ground palm oil fuel ash in high strength concrete mixture has smaller loss in compressive strength compared to high strength concrete made from other type of Portland cement. Muthusamy et al. [44] carried out a study on the production of oil palm shell lightweight aggregate concrete containing palm oil fuel ash. The results indicated that desired compressive strength of oil palm shell lightweight aggregate concrete can be achieved through 20% replacement of palm oil fuel ash. The amount of solid waste disposed by oil palm mill industry to the environment can be reduced by using up this waste as replacement materials in producing oil palm shell lightweight aggregate concrete with the combination of palm oil fuel ash and 100% of oil palm shell. A systematic study on the utilization of ultrafine palm oil fuel ash as a pozzolanic admixtures in green ultra-high performance fiber reinforced concrete was conducted by Aldahdooh et al. [45]. The study indicated that the optimum replacement of Ordinary Portland cement with ultrafine palm oil fuel ash was less than 25% in order to achieve desired compressive strength and agreed that the ultrafine palm oil fuel ash were able to be a good pozzolans in green ultra-high performance fiber reinforced concrete with good engineering properties which can save the cost and contribute to sustainability in concrete industry. As well as the study of lightweight foam concrete incorporating palm oil fuel ash as filler investigated by Lim et al. [46] found the increment in compressive, tensile and flexural strength when 10% to 20% of palm oil fuel ash were replaced in the lightweight foamed concrete compared to sample with 100% of sand filler.

Meanwhile, Muntohar and Rahman [47] made lightweight masonry block from palm oil kernel shell. The masonry block was made from a mixture of Portland cement, standard sand, and palm oil kernel shell and molded into 220 mm × 110 mm × 80 mm specimen size. It was found that the mixtures with same ratio of material produced highest compressive strength however, the compressive strength decreased when the proportion of palm oil kernel shell added to the mixtures increased. The masonry block produced with palm oil kernel shell met the requirement of class B2 and A2 for masonry as followed the building standards.

3 Discussions

In this paper, the discussion centered on usage of agricultural crop wastes in construction materials either as pozzolanic material, filler and fibres composites in different proportions and methods. Different tests were conducted on the product containing agricultural crop wastes by following the various standards to evaluate their properties, but the common performance measure evaluated by the researcher was the compressive strength. The review from different research has agreed that all agricultural wastes has good performance as construction materials whether increased or maintained the compressive strength of the final products. Compressive strength tests will provides the data on safety of the agricultural crop wastes usage in construction materials. In addition, the compressive strength tests will guarantee the quality of the construction materials containing agricultural crop wastes in order to meet specified requirement.

Nevertheless, there some standards and precautions should be followed during the processing of the agricultural crop wastes, for example, these wastes must undergoes various treatment process in which will affecting their effectiveness and properties as pozzolanic ash, fillers or fibre. Among examples of treatments that have been used were chemical treatment using acid solution (hydrocholric acid (HCL)) in which to reduce pottasium content in pozzolans [23, 25, 29], as well as using alkali solution (sodium hydroxide (NaOH)) to treat fibres [26]. The pozzolanic activity of the treated materials has been increased, which increased the compressive strength of construction materials.

The effectiveness of agricultural waste materials also can be preserved through their heating temperature and durations of burning during calcination process and their particle size distribution. In this case, the correct burning procedure was important to produce pozzolanic ash containing amorphous silica with low carbon content [23]. In the context of pozzolanic ash, it was agreed that the finer particles were better in which it will increase the rate of reaction and gain in strength in final products. The finer particles also affected the water cement ratio, workability, creep and shrinkage properties of construction materials. The fineness of materials were depends on their grinding times, which also agreed either can decreased or increased the engineering properties of concrete. It was proved that the strength of concrete increased as the grinding time of RHA increased but then declined after reached the optimum value, as studied by Jaya et al. [48]. As all things considered, the compressive strength of the construction materials were depends on the effectiveness of the agricultural crop wastes used. Equally important, the original agricultural waste materials must be ensured to be free from any contaminants before using it in any product [49].

4 Conclusions

Overall, more research and efforts needs to be done on the environmental, economic, and technical aspects. The government policy related to waste recycling and sustainable development should be strictly implemented in order to promote application of agricultural crop waste materials in construction industries, which can brings this industries to a whole new level.

This research and paper was supported by Postgraduate Research Grantt (GPPS) of Office for Research, Innovation, Commercialization, Consultancy Management (ORICC) Universiti Tun Hussein Onn Malaysia (Project Vot Number U453).

References

[1] K.J. Mun, W.K. Hyoung, C.W. Lee, S.Y. So, and Y.S. Soh, Basic properties of non-sintering cement using phosphor gypsum and waste lime as activator, J. of Constr. Build. Mater., **21**(6), 1342–1350 (2007)

[2] E. Aprianti, P. Shafigh, S. Bahri, and J. Nodeh, Supplementary cementitious materials origin from agricultural wastes – A review, Constr. Build. Mater., **74**, 176-187 (2015)

[3] T. Vaisanen, A. Haapala, R. Lappalainen and L. Tomppo, Utilization of agricultural and forest industry waste and residues in natural fiber-polymer composites: A review, Waste Management, **54**, 62–73 (2016)

[4] P. Shafigh, H. Mahmud, M.Z. Jumaat and M. Zargar, Agricultural wastes as aggregate in concrete mixtures – A review, Constr. Build. Mater., **53**, 110–117 (2014)

[5] S. Raut, R. Ralegaonkar, and S. Mandavgane, Development of sustainable construction material using industrial and agricultural solid waste: A review of waste-create bricks, Constr. Build. Mater., **25**(10), 4037-4042 (2011)

[6] B. Wang, F. Dong, M. Chen, J. Zhu, J. Tan, X. Fu, Y. Wang and S. Chen, Advances in recycling and utilization of agricultural wastes in China: Based on environmental risk, crucial pathways, influencing factors, policy mechanism, *Procedia Environmental Sciences*, **31**, 12-17 (2016)

[7] B.H.A. Bakar, P.J. Ramadhansyah and M.J.M. Azmi, Effects of rice husk ash fineness on the chemical and physical properties of concrete, Magazine of Concrete Research, **63**(5), 313-320 (2011)

[8] E. Rafiee, S. Shahebrahimi, M. Feyzi and M. Shaterzadeh, Optimization of synthesis and characterization of nanosilica produced from rice husk (a common waste material), International Nano Letters; Springer Open J., **2**(29) (2012)

[9] R.S. Bie, X.F Song, Q.Q. Liu, X.Y. Ji and P. Chen, Studies on effect of burning conditions and rice husk ash (RHA) blending amount on the mechanical behavior of cement, J. of Cem. and Concr. Compos., **55**, 162-168 (2014)

[10] R. Jauberthie, F. Rendell, S. Tamba and I.Cisse, Origin of the pozzolanic effect of rice husks, Constr. Build. Mater., **14**, 419-423 (2000)

[11] R.S. Bie, X.F Song, Q.Q. Liu, X.Y. Ji and P. Chen, Studies on effect of burning conditions and rice husk ash (RHA) blending amount on the mechanical behavior of cement, J. of Cem. and Concr. Compos., **55**, 162-168 (2014)

[12] S.K. Antiohos, V.G. Papadakis, and S. Tsimas, Rice husk ash (RHA) effectiveness in cement and concrete as a function of reactive silica and fineness, Cem. and Concr. Research, 61-62, 20-27 (2014)

[13] C. Hwang, L.A. Bui and C. Chen, Effect of rice husk ash on the strength and durability characteristics of concrete, Constr. Build. Mater., **25**(9), 3768-3772 (2011)

[14] S. Shahidan, I. Ismail, and N. Jamaluddin, A Review on Performance of Waste Materials in Self Compacting Concrete (SCC), J. of Technol. (Sciences & Engineering), 78(4), 29–35 (2016)

[15] V. Vishwakarma, D. Ramachandran, N. Anbarasan, and A.M. Rabel, Studies of rice husk ash nanoparticles on the mechanical and microstructural properties of the concrete, *Materials Today: Proceedings*, **3**(6), 1999-2007 (2016)

[16] R. Zerbino, G. Giaccio, and G. Isaia, Concrete incorporating rice-husk ash without processing, Constr. Build. Mater., **25**(1), 371-378 (2011)
[17] R. Kishore, V. Bhikshma and P. Jeevana Prakash, Study on strength characteristics of high strength rice husk ash concrete, Proc. Engineering, **14**, 2666–2672 (2011)
[18] N.V. Tuan, G. Ye, K.V. Breugel, A.L.A. Fraaij, and B.D. Dai, The study of using rice husk ash to produce ultra-high performance concrete, Constr. Build. Mater., **25**(4), 2030-2035 (2011)
[19] D. Chopra, R. Siddique and Kunal, Strength, permeability and microstructure of self-compacting concrete containing rice husk, Biosystems Eng. J., **130**, 72-80 (2015)
[20] S. Hesami, S. Ahmadi, and M. Nematzadeh, Effects of rice husk ash and fiber on mechanical properties of pervious concrete pavement, Constr. Build. Mater., **53**, 680-691 (2014)
[21] R.P. Jaya, M.R. Hainin, B.H. Abu Bakar, M.A. Megat Johari, M.H. Wan Ibrahim, and D.S. Jayanti, Strength and microstructure analysis of concrete containing rice husk ash under seawater attack by wetting and drying cycles, Advances in Cement Research, **26**(3), 145-154 (2014)
[22] J. Sutas, A. Mana and L. Pitak, Effect of rice husk and rice husk ash to properties of bricks, *Proc. Engineering,* **32**, 1061-1067 (2012)
[23] G. Cordeiro and C. Sales, Pozzolanic activity of elephant grass ash and its influence on the mechanical properties of concrete, Cem. and Concr. Compos., **55**, 331-336 (2015)
[24] V. Strezov, T.J. Evans and C. Hayman, Thermal conversion of elephant grass (*Pennisetum Purpureum Schum*) to bio-gas, bio-oil and charcoal, Bioresour. Technol., **99**, 8394-8399 (2008)
[25] E.Y. Nakanishi, M. Frias, S.M. Ramirez, S.F. Santos, M.S. Rodrigues, O. Rodriguez, and H. Savastano Jr, Characterization and properties of elephant grass ashes as supplementary cementing material in pozzolan/$Ca(OH)_2$ pastes, Constr. Build. Mater., **73**, 391-398 (2014)
[26] M. Haameem J.A, M.S. Abdul Majid, M. Afendi, H.F.A. Marzuki, I. Fahmi, and A.G. Gibson, Mechanical properties of Napier grass fibre/polyester composites, J. of Composites Structures, **136**, 1-10 (2016)
[27] K. Ramanaiah, A.V. Ratna Prasad and K. Hema Chandra Reddy, Thermo physical properties of elephant grass fiber-reinforced polyester composites, Materials Letters, **89**, 156–158 (2012)
[28] M.J.M. Ridzuan, M.S. Abdul Majid, M. Afendi, S.N. Aqmariah Kanafiah, J.M. Zahri and A.G. Gibson, Characterisation of natural cellulosic fibre from *Pennisetum purpureum* stem as potential reinforcement of polymer composites, Mater. and Design, **89**, 839–847 (2016)
[29] E.Y. Nakanishi, M. Frias, S.F. Santos, M.S. Rodrigues, R.V. Villa, O. Rodriguez and H.S. Junior, Investigating the possible usage of elephant grass ash to manufacture the eco-friendly binary cements, J. of Cleaner Prod., **116**, 236-243 (2016)
[30] I. Merta and E.K. Tschegg, Fracture energy of natural fibre reinforced concrete, Constr. Build. Mater., **40**, 991-997 (2013)
[31] J.C.B. Moraes, J.L. Akasaki, J.L.P. Melges, J. Monzó, M.V. Borrachero, L. Soriano, J. Payá and M.M. Tashima, Assessment of sugar cane straw ash (SCSA) as pozzolanic material in blended Portland cement: Microstructural characterization of pastes and mechanical strength of mortars, Constr. Build. Mater., **94**, 670–677 (2015)
[32] S.M.S. Kazmi, S. Abbas, M.A. Saleem, M.J. Munir and A. Khitab, Manufacturing of sustainable clay bricks: Utilization of waste sugarcane bagasse and rice husk ashes, Constr. Build. Mater., 120, 29–41 (2016)

[33] N. Ali, N.A. Zainal, M.K. Burhanudin, A.A.A. Samad, N. Mohamad, S. Shahidan and S.R. Abdullah, Physical and Mechanical Properties of Compressed Earth Brick (CEB) Containing Sugarcane Bagasse Ash, *EDP Sciences,* 47 (2016)
[34] M.M. Tashima, J.C.B. Moraes, J.L.P. Melges, J. Monzo, M.V. Borrachero, L. Soriano and J. Paya, Assessment of sugar cane straw ash (SCSA) as pozzolanic material in blended Portland cement: Microstructural characterization of pastes and mechanical strength of mortars, Constr. Build. Mater., **94**, 670-677 (2015)
[35] S. Rukzon and P. Chindaprasirt, Utilization of bagasse ash in high-strength concrete, J. of Mater. and Des., **34**, 45-50 (2012)
[36] A. Rerkpiboon, W. Tangchirapat and C. Jaturapitakkul, Strength, chloride resistance, and expansion of concretes containing ground bagasse ash, Constr. Build. Mater., **101**, 983–989 (2015)
[37] P.O. Modani and M.R. Vyawahare, Utilization of Bagasse Ash as a Partial Replacement of Fine Aggregate in Concrete, Procedia Engineering, **51**(NUiCONE 2012), 25-29 (2013)
[38] S. A. Lima, H. Varum, A. Sales, and V. F. Neto, Analysis of the mechanical properties of compressed earth block masonry using the sugarcane bagasse ash, Constr. Build. Mater., **35**, 829-837 (2012)
[39] R. Alavez-Ramirez, P. Montes-Garcia, J. Martinez-Reyes, D.C. Altamirano-Juarez, and Y. Gochi-Ponce, The use of sugarcane bagasse ash and lime to improve the durability and mechanical properties of compacted soil blocks, Constr. Build. Mater., **34**, 296–305 (2012)
[40] D. Mujah, Compressive strength and chloride resistance of grout containing ground palm oil fuel ash, J. of Cleaner Prod., **112**, 712-722 (2016)
[41] M.R. Hamidian, P. Shafigh, M.Z. Jumaat, U.J. Alengaram and N.H.R. Sulong, A new sustainable composite column using an agricultural solid waste as aggregate, J. of Cleaner Prod., **129**, 282-291 (2016)
[42] B. Alsubari, P. Shafigh and M. Z. Jumaat, Utilization of high-volume treated palm oil fuel ash to produce sustainable self-compacting concrete, J. of Cleaner Prod., **137**, 982-996 (2016)
[43] W. Tangchirapat, C. Jaturapitakkul and P. Chindaprasirt, Use of palm oil fuel ash as a supplementary cementitious material for producing high-strength concrete, Constr. Build. Mater., **23**, 2641–2646 (2009)
[44] K. Muthusamy, N. Zamri, M.A. Zubir, A. Kusbiantoro and S.W. Ahmad, Effect of mixing ingredient on compressive strength of oil palm shell lightweight aggregate concrete containing palm oil fuel ash, Procedia Engineering, **125**, 804-810 (2015)
[45] M.A.A. Aldahdooh, N.M. Bunnori and M.A.M. Johari, Development of green ultra-high performance fiber reinforced concrete containing ultrafine palm oil fuel ash, Constr. Build. Mater., **48**, 379–389 (2013)
[46] S.K. Lim, C.S. Tan, O.Y. Lim and Y.L. Lee, Fresh and hardened properties of lightweight foamed concrete with palm oil fuel ash as filler, Constr. Build. Mater., **46**, 39–47 (2013)
[47] A.S. Muntohar and M.E. Rahman, Lightweight masonry block from oil palm kernel shell, Constr. Build. Mater., **54**, 477–484 (2014)
[48] R.P. Jaya, B.H. Abu Bakar, M.A. Megat Johari, M.H. Wan Ibrahim, Engineering properties of normal concrete Grade 40 containing rice husk ash at different grinding times, Int. J. of Technology, **1**, 10-19 (2011)
[49] L. Zhang, Production of bricks from waste materials – A review, Constr. Build. Mater., **47**, 643–655 (2013)

Utilization of Baggase Waste Based Materials as Improvement for Thermal Insulation of Cement Brick

Eeydzah Aminudin[1,*], *Nur Hafizah* A. Khalid [1], *Nor Ain* Azman[2], *Khairuniza* Bakri[2], *Mohd Fadhil* Md Din[3], *Rozana* Zakaria[1], and *Nur Azmira* Zainuddin[3]

[1]Department of Structures and Materials, Universiti Teknologi Malaysia, 81310, Johor
[2]Department of Building and Construction, Universiti Tun Hussein Onn Malaysia, Batu Pahat
[3]Department of Environmental Engineering, Universiti Teknologi Malaysia, 81310, Johor

Abstract. Building materials having low thermal load and low thermal conductivity will provide thermal comforts to the occupants in building. In an effort to reduce the use of high energy and waste products from the agricultural industry, sugarcane bagasse and banana bagasse has been utilize as an additive in the manufacture of cement brick. The aim of this study is to investigate the insulation and mechanical properties of brick that has been mixed with bagasse and its effectiveness as thermal insulation using heat flow meter. Waste bagasse is being treated using sodium hydroxide (NaOH) and is characterized using SEM and XRF. The samples produced with two different dimensions of 50 mm x 50 mm x 50 mm and 215mm x 102.5mm x 65mm for thermal conductivity test. Next, the sample varies from 0% (control sample), 2%, 4%, 6%, 8% and 10% in order to determine the best mix proportion. The compressive strength is being tested for 7, 14 and 28 days of water curing. Results showed that banana bagasse has lower thermal conductivity compared to sugarcane bagasse used, with compressive strength of 15.6MPa with thermal conductivity 0.6W/m.K.

1 Introduction

In recent years, various studies have been conducted to produce the bricks which having a comparative strength and durable when applying to agricultural waste. In Malaysia, agricultural by-product produces in large quantities especially from oil palm, banana, pineapple, paddy and sugarcane.

Sugarcane bagasse for instance, produces 1 tonnes of sugarcane which generally generate around 280 kg of bagasse for one cycle [1]. This fibrous is remaining by-product after extracted from sugarcane. It is about 54 million dry tonnes of bags are produced annually through the world [1]. Sugarcane bagasse is potentially strong in displacing fossil fuels and can be extensively used in boilers, turbines and furnaces for power generation [2].

* Corresponding author: eeydzah@utm.my

Other applications of sugarcane bagasse are as sources of animal feed, energy, pulp, paper and boards [3].

Another type of by product from agriculture is bananas. Bananas are found abundance in tropical areas particularly. Normally restricted to tropical country, since the banana needs a climate mean temperature about 26.67°C and mean rainfall of 4 in. [6]. The total planted area of banana in Malaysia reached 33,704.2 ha in 2001 [7]. There are 25–80 species of banana in tropical areas, depending on the taxonomist according to [6]. Besides that, this by product fiber bagasse materials is practical to be used since it is renewable, low in cost, density, as well as satisfactory mechanical properties make them attractive and more environmentally friendly to be used as by product not only in the field of construction (partition boards, ceiling panelling), packaging, consumer products but also in transportation (automobiles, railway coaches). Reuse of this wastes as sustainable construction material appears to be viable solution not only to pollution problem but also to the problem of the land-filling and high cost of building materials [8]. Increase of population growth and urbanization, has raise standards of living due to technological innovations which contributed to an increase in the quantity and variety of solid wastes generated by industrial, mining, domestic and agricultural activities [9]. According to [4], improper treatment of sugarcane bagasse causes similar problems. The excess bagasse accumulation presents a waste problem for the sugar industry [2]. Therefore, this study investigated the application of modified sugarcane bagasse with the comparison of banana bagasse as thermal improvement for building brick. The physical and chemical properties of brick containing the sugarcane and banana bagasse are determined and discussed.

2 Literature review

2.1 Development of construction material from agro-waste

Agriculture waste has been used widely especially in construction materials as additive materials in concrete. Marwan [17] produces cement bricks that contain banana fiber as intermixture. Building using the natural fiber brick has become more popular since it is low in cost, readily available raw materials, produces faster and easier construction and economically friendly. Natural fiber bricks offer a competitive alternative to conventional building materials because they utilize local resources that can be both cost effective and energy efficient, and closely follow existing masonry construction practices. It shows that these traditional binders can be replaced by environmentally friendly and sustainable alternatives from unutilized waste [10].

Typically, cement bricks are formed using a mixture (by weight) of angular sand aggregate (40%-70%), clayey soil (30%-60%), cement (4%-10%) and water (8%-12%) [11]. Having a compressive strength in between (4-5 MPa) it has become a basic and universally accepted unit of measurement to specify the quality of the masonry units which restricts their applications to only one-story buildings.

Over 10 million hectares of banana plantation, of an average of 1500 plants per hectare, exist in more than 160 countries globally. It has cause to left over decompose besides creating tons of banana waste, emitting a huge amount of methane gas and carbon dioxide. These emissions have a negative impact on the environment, which increases global warming every year. Every ton of banana waste emits, on average, a half-ton of carbon dioxide per year. This study advances the utility of the cement bricks by adding banana fibers, thereby creating an innovative Green Cement Brick that further improves the mechanical properties of cement brick. Sugarcane bagasse is a fibrous residue produced in large quantity by the sugar industrials. Generally, sugarcane bagasse residue is produced about 32% of every tons of sugarcane after juice extraction [5]. Sugarcane bagasse is a fibre material which is easy to absorb and release water. Sugarcane fibre can absorb water by 78.5%. When the fibre is dry, the water will come out through vaporization process without damaging the cell structure.

The use of fresh untreated fiber in cement brick commonly results in very slow setting and abnormally low strength because of interference with normal setting and hardening processes by carbohydrates, tannins and possibly other substances in the fibers [8]. Pre-treatment is required to disrupt the structure of lingo cellulosic materials during cellulosic ethanol production, because the extensive interactions among cellulose, hemicelluloses and lignin, and the barrier nature of lignin minimize enzyme access to the carbohydrates and result in poor yields of fermentable sugars [9]. Fig. 1, vegetal fibre structure is formed by a central channel called as lumen, responsible for water and nutrients transportation, and by the cell wall. The cell wall of each fibre is composed by several layers such as middle lamella, the thin primary wall, and the secondary wall with is subdivided into external secondary wall (S1), middle secondary wall (S2) and internal secondary wall (S3). These layer are composed of microfibrils oriented into space in defined (angles) form, according to the each layer [13].

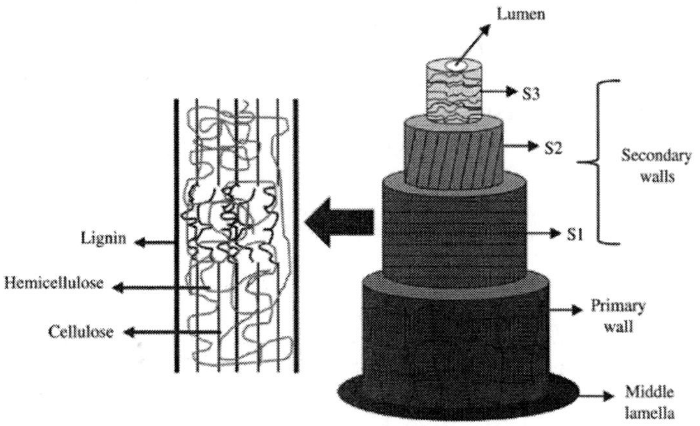

Fig. 1. Schematic representation of plant fiber structure [13].

The primary wall, initially deposited during the growth of cells, consists in a disordered arrangement of cellulose fibrils placed in a matrix of pectin, hemicellulose, lignin and protein [14]. Secondary walls consist of crystalline cellulose microfibrils organized in a spiral arrangement, where the middle layer (s2) determines the fibre mechanical properties. This is consisted by an amount of microfibrils, in a helical conformation of long chains of cellulose. The middle lamella, that is outer layer of cell, is composed predominantly by pectin that acts as cement between fibres [14].

A low thermal conductivity and a low specific heat are required to have a good thermal insulation of buildings. Many studies have shown that, regarding the mechanical behaviour, vegetable fibres are an interesting alternative to asbestos in fibre reinforced cement products manufactured by the Hatcheck process [15,16]. Besides that, the fiber has essentially due to its capability of enhancing the load carrying capacity and ductility of the concrete structures [18]. Study done by [17] shows that percentage of POFA which an agro waste based can lower the thermal conductivity value. Many studies had done which demonstrated the low thermal conductivity of such vegetable/agriculture fibre based composite cement materials but only few resulting from effect of the mixture fibre/matrix.

3 Materials and methods

The materials used in this study are cement, sand, banana trunk fiber, sugarcane bagasse and water. Both fibers are treated using water boiling containing sodium hydroxide (NaOH). Then the fibers are tested with SEM and XRF in order to identify the surface and the chemical composition respectively. For the sample preparation, the mix ratio used was 1:3 which represent cement to sand and water cement ratio of 0.45% for cement brick production. However, both fibers introduced into mortar mix are varied from 0%, 2%, 4%, 6%, 8% and 10%. This is to ensure the optimum results based on the most favourable result. The 0% served as control for other sample. Samples with dimension of 50mm × 50mm × 50mm were produced for compressive strength test. While a dimension of 102.5mm × 215mm × 65mm were used to test the thermal conductivity of samples.

4 Results and Discussions

4.1 Microscopic Analysis (SEM) with the comparison of untreated and treated fiber

Table 1 shows the SEM image of untreated banana trunk fiber and sugarcane fiber with the comparison of treated using NaOH. Comparison between both figures, it shows that surface of treated banana trunk fiber more rough compared to untreated banana trunk fiber. Hence, the bonding between treated fiber and cement would come out with better result in strengthening the brick. On top of that, the presence of pores at the surface of treated banana trunk fiber would result in reducing the thermal conductivity [10].

Table 1. Comparison of microscopic analysis of banana and bagasse fiber before and after treatment.

	Before treatment	After Treatment with NaOH
Microscopic analysis (SEM) of banana trunk fiber	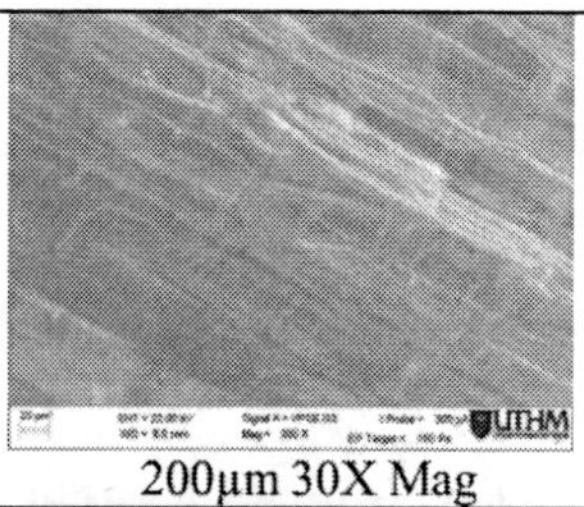 200μm 30X Mag	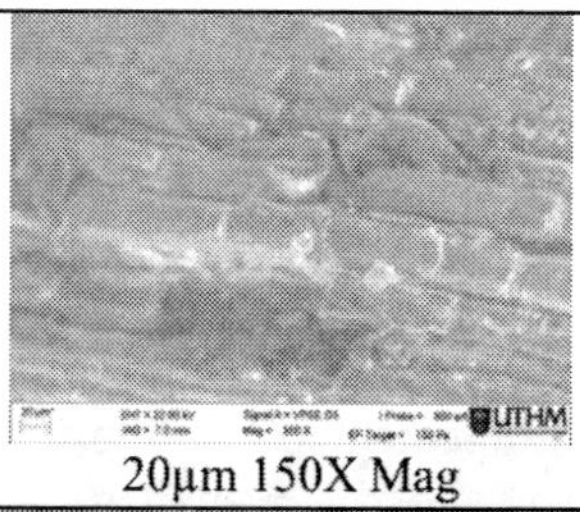 20μm 150X Mag
Microscopic analysis (SEM) of sugarcane bagasse	 200μm 30X Mag	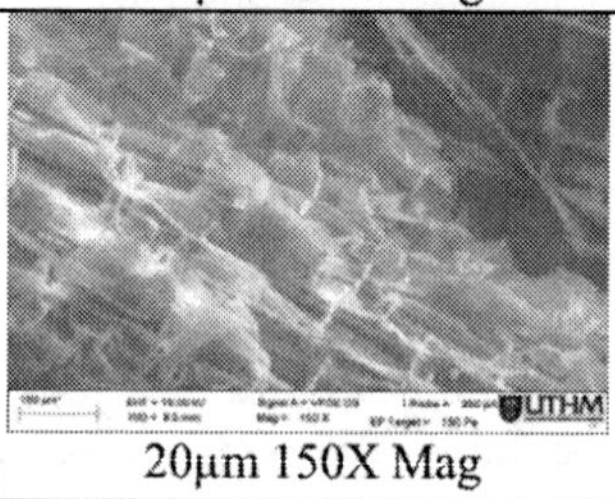 20μm 150X Mag

4.2 Chemical compound of ordinary portland cement, treated banana trunk fiber and sugarcane bagasse

Table 2 shows the XRF result of Ordinary Portland cement, banana trunk fiber and sugarcane bagasse sample respectively. These show the various chemical compounds present in both samples. From Table 2, the top three highest concentration of chemical composition found in Ordinary Portland cement are calcium oxide, CaO with 58.9%, silicon dioxide, $SiO_2$with 16% and aluminium oxide, Al_2O_3 with 4.32%. The other chemical compound found in Ordinary Portland cement such as carbon dioxide, potassium oxide, magnesium oxide, iron oxide and sulphur trioxide ranging from 0.10 to 3.61% of concentration. The percentage of concentration of all chemical compound found in the treated banana trunk fiber sample is quite similar with Ordinary Portland cement. The result shows that the highest percentage of chemical content in banana trunk fiber is potassium oxide with 10.7%. The concentration of the other chemical was from range 0.1% to 3.9%., especially in silicon dioxide, calcium oxide, and ferric oxide. Hence, the overall chemical compounds make the banana trunk fiber and sugarcane bagasse potential to be used as additional fiber in cement brick.

Table 2. Chemical compound of ordinary portland cement, treated banana trunk fiber and sugarcane bagasse.

Chemical compound	Formula	Portland Cement Concentration (%)	Treated Banana Trunk Fiber Concentration (%)	Treated Sugarcane Bagasse Concentration (%)
Carbon dioxide	CO_2	0.10	0.1	0.1
Calcium oxide	CaO	58.90	3.9	2.55
Silicon dioxide	SiO_2	16.00	2.3	0.87
Aluminium oxide	Al_2O_3	4.32	0	0
Sulfur trioxide	SO_3	3.61	0.3	0.51
Iron oxide	Fe_2O_3	3.47	0.7	0.48
Magnesium oxide	MgO	1.81	0	0
Potassium oxide	K_20	0.52	10.7	0.76

4.3 Compressive strength

From the results of sugarcane bagasse fiber as addition in mortar concrete, the results of compressive strength for 28 days of curing are 13.29 MPa for 0%, 15.13 MPa for 2% and 17.23 for 4%. However, the strength started to decrease by 14.42 MPa for 6% of sugarcane bagasse and continuously drop from 8% and 10% which is 13.80 MPa, and 12.01 MPa respectively.This concludes that the additional percentage of more than 4% of sugarcane bagasse will result in decrease of the strength of the sample. It was due to poor workability and less bonding between sugarcane bagasse and mortar. This finding was supported by previous study [12]. The author reported that, even compressive strength is decreased by the incorporation of the waste, but it is producing adequate brick which is comply with the standard for non-load bearing purpose. According to Oyekan, mortar contains about 1% of air voids. Air-entrainment is known to improve the properties mortar in both the fresh and hardened states. In fresh mortar, the minute air bubbles act as small ball-bearings and greatly improve the workability of the mortar. But the more workable mortar loses about 20% of its strength because every 1% of air entrapped produces about 4% loss in strength.

As for banana fiber, the result of compression strength for mortars with 28 days of curing shows that 4% of banana trunk fibers having a higher compressive strength with 15.6 N/mm². It was increased from 14.12 N/mm² and 14.78 N/mm² which were the control sample and 2% of banana trunk fibers. The compression strength was slightly decrease as the content of banana trunk fibers as intermixture in cement brick was increased which were 15.584 N/mm² for 6%, 15.58 N/mm² for 8% and 13.05 N/mm² for 10%. According to Uddin [11], the brick with highest percentage of banana trunk fiber recorded the highest compressive strength. As a conclusion, the workability between fibers and concrete were good as the percentage of fibers increase up to 2% and 4% and has the highest compression strength. Unfortunately, the strength continued to reduce for cement bricks with percentage of 6%, 8% and 10% of banana trunk fibers as intermixture because of the poor bonding between fibers and concrete. Fig. 2 illustrates the compressive strength against percentage with comparison to banana trunk fiber and sugarcane bagasse.

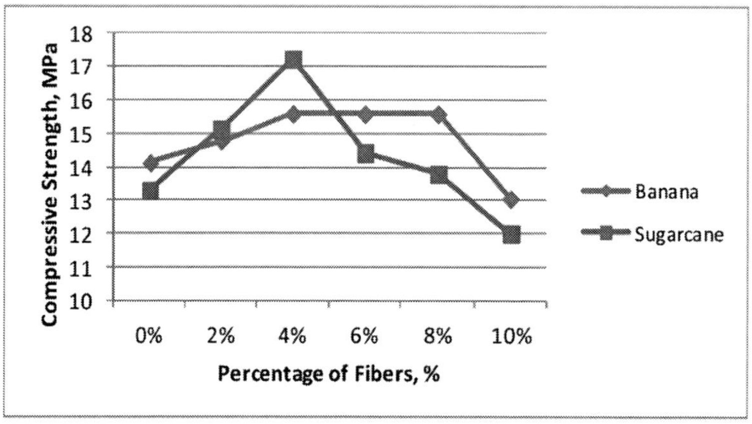

Fig. 2. Compressive strength against percentage with comparison to banana trunk fiber and sugarcane bagasse.

4.4 Correlation between compressive strength and thermal conductivity

The result of thermal conductivity test for the samples of 28th day with different sugarcane bagasse content were shown in Table 3 and Fig. 3 below.

Table 3. Data of thermal conductivity test

Type of fiber during 28days curing age	Thermal conductivity of different sawdust and sugarcane bagasse in different percentage (W/m.K)			
	0%	4%	8%	10%
Sugarcane baggase	0.69	0.68	0.65	0.62
Banana trunk fiber	0.69	0.60	0.63	0.58

Table 3 show the result for thermal conductivity test of all samples. Percentage of sugarcane bagasse found that the graph line for thermal conductivity was lower than the graph line for the compressive strength except for 0% of sugarcane bagasse content. It was obviously seen that the thermal conductivity for 0% of sugarcane bagasse content was higher while, the compressive strength was lower compared to 4% and 8% of sugarcane bagasse content. Even though, the 10% of sugarcane bagasse content gives the best result in reducing thermal conductivity, the strength is less than the required strength so it cannot be used for permanent constructions. Compared with banana fiber thermal conductivity test for cement brick with 10% additive show a good result (0.58 W/m-K) where it is lower compared to the control sample (0.59 W/m-K). Meanwhile, the cement brick with percentage 8% of banana trunk fiber as intermixture recorded the highest thermal conductivity which is 0.63 W/mK and followed by 0.60 W/mK for percentage of 4%.

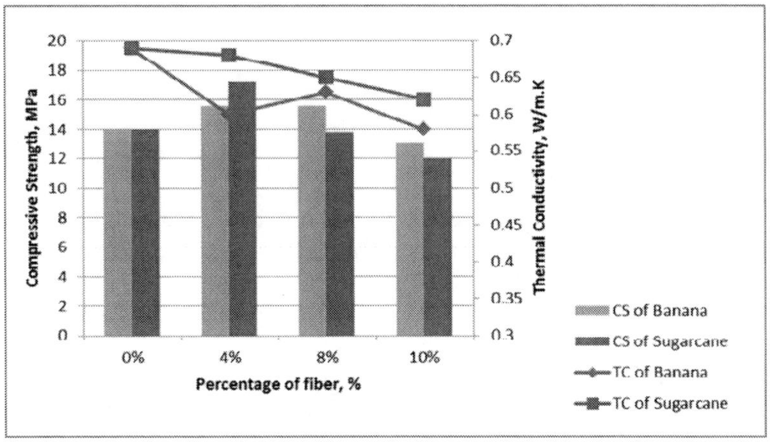

Fig. 3. Relationship between compressive strength (CS) and thermal conductivity (TC).

It can be concluded that the characteristics of banana trunk fiber has a potential to be as thermal improvement in cement brick. In other words, it is suitable to use as intermixture in cement brick and applicable in the construction industry [11]. However, based on both fiber, it can conclude that banana bagasse fiber have a good strength of 15.6MPa with low thermal conductivity 0.6W/m.K comparatively with other percentage and fibers.

5 Conclusion

The chemical composition of sugarcane baggase and banana trunk fiber is suitable and has the potential as an addition in producing cement brick. The bonding between treated both fibers gave better result in strengthening the brick as the treated fibers has a rough surface. Overall, the cement brick with percentage 4% of banana trunk fiber gives the highest value of compressive strength which is 15.59 N/mm² at the age of 28 days of curing. Although the highest value of compressive strength for 28 days of curing is 4%, the optimum mix design is 8% because there is an intersection between compressive strength and thermal conductivity lines. The value at the intersection is compatible with ASTM C207 Type N masonry cement. However, based on both fiber, it can conclude that banana bagasse fiber have a good strength of 15.6MPa with low thermal conductivity 0.6W/m.K comparatively with other percentage and fibers.

The present work benefited from financial support funded by Universiti Teknologi Malaysia (UTM). This project was placed under Vot. number of Q.J130000.2622.12J53 with research title of "Cool Concrete as thermal repellent for mitigating heat island effect" and Vot. number of R.J130000.2722.02K54 with research title of "Reactive Silica Concrete containing eggshell as hybrid concrete supplementary".

References

[1] J.X, Sun, Fractional extraction and structural characterization of sugarcane bagasse hemicelluloses, Carbohydrate Polymers, **56**, 195-204, (2004)

[2] B.C. Suddell and W.J. Evans, The increasing use and application of natural fibre composite materials within the automotive industry, *Proc. of the Seventh Int. Conf. on Woodfiber-Plastic Composites,Forest Products Socety,* Madison, WI, USA, (2003)
[3] K.C.P. Faria, R.F. Gurgel and J.N.F. Holanda, Recycling of sugarcane bagasse ash waste in the production of clay bricks, J. of Environmental Management, **101**, 7-12, (2003)
[4] J. B. Khedari, New lightweight construction materials with low thermal conductivity. Cement and Concrete Composites, **23**, 65-70 (2002).
[5] P.A. Turgut and H.M. Algin, Limestone dust and wood sawdust as brick material, Building and Environment, **42**(9), 3399-3403, (2007)
[6] J.Y. Tock, C.L. Lai, K.T. Lee, K.T, Tan and S. Bhatia, Banana biomass as potential renewable energy resource: A Malaysian case study, Renewable and sustainable energy reviews, **14**(2), 798-805, (2010)
[7] H.P.S. Abdul Khalil, M. Siti Alwani and A.K. Mohd Omar, Chemical composition, anatomy, lignin distribution and cell wall structure of Malaysian plant waste fibers, BioResources, **17**, 220–232, (2006)
[8] A. Pappu, M. Saxena, S. R. Asolekar. Solid wastes generation in India and their recycling potential in building materials, Build Environ, **42**, 2311–2320 (2007)
[9] M.V. Madurwar, R.V. Ralegaonkar and S.A. Mandavgane, Application of agro-waste for sustainable construction materials: A review, Construction and Building Materials, **38**, 872-878, (2013)
[10] T. Rauber, *Natural fibre specialist*, (2007), Retrieved on April 20, 2016 from Bcomp: http://www.bcomp.ch/10-0-natural-fibres.html
[11] M. M. Uddin, Effect of banana fibers on the compressive and flexural, Buildings, **5**, 282-296, (2015)
[12] ASTM Designation C109/C109M-07, Standard Test Method for Compressive Strength of Hydraulic Cement Mortars (Using 2-in. or [50-mm] Cube Specimens), ASTM International, (2008)
[13] W.Y. Li, A.X. Jin, C.F. Liu, R.C. Sun, A.P. Zhang and J.F. Kennedy, Homogeneous modification of cellulose with succine anhydride in ionic liquid using 4-dimethylaminopyridine as a catalyst, Carbohydrate Polymers, **78**(3), 389-395, (2009)
[14] A.K. Pietak, Atomic force microscopy characterization of the surface wettability of natural fibers, Applied surface sciences, **253(7)**, 3627-3635, (2007)
[15] H.J. Savastano, P.G. Warden and R.S.P. Coutts, Performance of low-cost vegetable fibrecement composites under weathering, *Proceedings of CIIB World building congress*, Wellington, New-Zealand, (2001)
[16] H.J. Savastano, P.G. Warden and R.S.P Coutts, Mechanically pulp sisal as reinforcement in cementitious matrices, Cem Concr Compos, **25**, 585–92, (2003)
[17] E. Aminudin, M.F.M. Din, M.W. Hussin, A.H. Abdullah, K. Iwao and Y. Ichikawa, Properties of agro-Industrial aerated concrete as potential thermal insulation for building, *MATEC Web of Conf.*, 1–9, (2016)
[18] S.M.S. Mohsin, M.F. Manaf, N.N. Sarbini and K. Muthusamy, Behaviour of reinforced concrete beams with kenaf and steel hybrid fibre, ARPN J. of Engineering and Applied Sciences, **11**, 5385-5390 (2016)

Utilizing Slurry and Carwash Wastewater as Fresh Water Replacement in Concrete Properties

Shahiron Shahidan[1,*], *Aeslina* Abdul Kadir[2], *Lau* Hai Yee[3], *Nurul Izzati Raihan* Ramzi[4], and *Faisal* Sheikh[5]

[1,3,4,5]Jamilus Research Centre, Faculty of Civil and Environmental Engineering, Universiti Tun Hussein Onn Malaysia
[2]Department of Water and Environmental Engineering, Faculty of Civil and Environmental Engineering, Universiti Tun Hussein Onn Malaysia

Abstract. The high demand for concrete production generates wastewater which causes environmental problems. However, if wastewater is able to be recycled as part of engineering construction materials, many benefits can be reaped. Unfortunately, the use of wastewater in manufacturing concrete is not common. Therefore, this research aims to identify the influence of using slurry water and car wash wastewater on concrete properties, focusing particularly on its mechanical properties. The basic characteristics of wastewater were studied according to USEPA method while the properties of concrete with wastewater were compared according to ASTM C1602 and BS EN 1008 standards. In this paper, the compressive strength, modulus of elasticity and tensile strength were examined in order to determine the mechanical properties of concrete. The wastewater was replaced in the concrete mix from 0% up to 40%. The results indicated that the characteristics of wastewater complied with the BS and ASTM standards. In addition, the results also recommended that the concrete mixture with 20% of wastewater has given the highest compressive strength and modulus of elasticity.

1 Introduction

Recently, construction industry development board (CIDB) of Malaysia has indicated that the number of projects for building and civil engineering work has been recorded the highest numbers compared to electrical and mechanical sectors [1, 2][37]. This growth is due to the greatest investment in construction development projects under the 10th Malaysia Plan from 2011 to 2015 [3]. In line with that, the water demand in construction industries increased dramatically especially Ready Mixed Concrete because it is one of the main suppliers in housing areas [4, 5][36]. Subsequently, previous researchers estimated that approximately 500 to 1500 litres of tap water is required to clean each concrete mixer truck at the end of each day [6-9]. At the same time, slurry water might be produced after

* Corresponding author: shahiron@uthm.edu.my

washing the mixer truck which might lead to environmental pollution [10][35]. Previous studies also mentioned that normally the pH value of slurry water from pH 11 to 12.5 [6-8]. In addition, slurry water is also classified as dangerous wastewater as it can cause surface bodies of water due to alkaline characteristics [4]. Apart from that, typical slurry water that been produced from the ready mixed concrete plants contains dissolved solids such as calcium chloride and sulphate from cement as an admixture [5, 6]. Besides, the investigation on the concrete manufacturing using slurry water from RMC plants also has been conducted [7]. From the studied, it found that the slurry water has a highest content of alkaline and total solids which is exceeded the limit as required in ASTM C94, which may cause the concrete to be more porous and contains a weaker matrix [8, 9].

Reusing alternative car wash wastewater in concrete mixes aims to create a sustainable solution in this industry towards green technology in construction materials. Some researched has used car wash wastewater to manufacture concrete which is the wastewater characteristics include pH value, chloride, alkalinity and sulphate content in wastewater was also determined in this study [11]. The results indicated that the wastewater characteristics were within the ASTM standard limits. The mechanical tests including compressive strength, tensile strength and flexural test were done on day 28. In addition, the percentages of car wash wastewater in the concrete mix ranged from 25% to 100% [11].

Last but not least, there is a lack in literature on the utilization of wastewater in concrete mixing. This research focused mainly on the characteristics of slurry water and car wash wastewater and the effect of the wastewater on concrete properties. Meanwhile, the mechanical tests for concrete were conducted according to ASTM standards as a guidelines, which is also included compressive strength, tensile strength and modulus of elasticity.

2 Materials and method

In this research, the percentages of slurry water used in concrete mixes ranged from 10% to 50% and 10% to 40% for car wash wastewater. The water cement ratio w/c is 0.45 and all the concrete specimens were designed to have a target mean strength of 35 MPa with a slump value between 30mm to 60mm. The concrete mix was designed according to the method set by the Department of Environment (DOE). In addition, the properties of fresh concrete such as unit weight, slump test and the mechanical tests including compressive strength, tensile strength and modulus of elasticity (MOE) were carried out in this study. The concrete specimens tested for compressive strength were cast in moulds with dimensions of 150 x 150 X 150 mm while those tested for tensile strength and MOE were cast in 300 mm x 150mm diameter moulds. Triplicate concrete samples were made for each concrete mechanical testing.

2.1 Materials

2.1.1 Wastewater (slurry and carwash)

Slurry water was collected from three different RMC plants in Batu Pahat, Johor, Malaysia which are Hanson RMC plant, Chin Keng RMC plant and Renggam RMC plant. Before the slurry water was collected, the wash water from concrete mixer truck was discharged into a pit at the RMC plant. On the other hand, car wash wastewater was collected from two different car wash stations in Batu Pahat, Johor, Malaysia which are MJ car wash and Bandar U car wash. Before the wastewater was collected, the washout water from the vehicle was discharged into a pond through a drainage system. The chemical properties of wastewater were analyzed according to the USEPA standard method and compared with the standard specifications [11].

2.1.2 Cement

A local Ordinary Portland cement type CEM 1 42.5N was used in this study. The cement complied with the requirements of MS EN 197-1.

2.1.3 Aggregates

A fine aggregate which is natural sand met the ASTM C33 requirements [12, 13]. The coarse aggregate consisted of crushed rocks and also complied the ASTM C33 specifications. The fine aggregate passed through a 5mm sieve (No 4) and were retained at 75µm (no 200). The coarse aggregate passed through a 20mm sieve and retained at 10mm. All the aggregates were air dried at 20+5^{0}C to obtain saturated surface dry condition.

2.2 Mix proportion of concrete

In this research, all concrete specimens were designed to have a target mean strength of 35MPa with the curing process of 28 days according to the Department of Environment, DOE method.

2.3 Experimental work

2.3.1 Wastewater analysis

The pH value of wastewater was measured by using Hanna HI 8424 digital pH meter. According to the United States Environmental Protection Agency (USEPA) method 300.0, the content of chloride and sulphate in car wash wastewater were analysed. In addition, all the results of wastewater characteristics were compared with ASTM and BS standards.

2.3.2 Concrete testing

The manufacturing and curing process of concrete were according to ASTM standards. However, the concrete slump tests were performed according to ASTM C 143[14].Besides, the tests for unit weight of concrete were conducted according to ASTM C138 [15]. On the other hand, the compressive strength and tensile strength of concrete on day 7 and day 28 were carried out in accordance with ASTM C39 and ASTM C496 respectively [16,17,32]. Meanwhile, the modulus of elasticity of concrete on day 28 was tested according to ASTM C469 [18].

3 Results and discussion

3.1 Design requirement

Table 1 and Table 2 show the results of the characteristics of wastewater and tolerance limit of water mixed in concrete respectively. Based on Table 1 and Table 2, the slurry water and car wash wastewater characteristics completely complied with the standards as well as the typical values suggested by previous researchers except for the pH value for car wash wastewater. Even though this study obtained higher pH values compared to the previous studies, the range of pH values is still in the range of typical car wash wastewater which varies from 6.0 to 9.0.

Table 1. Characteristic of wastewater.

Sources/ Parameters	Parameters		
	pH	Chloride Cl^-,ppm	Sulphate SO_4^{2-},ppm
Slurry water			
Hanson	10.0-11.3	7.5-46.9	108.6-454.9
Chin Keng	11.0-11.9	6.7-28.7	426.0-289.5
Renggam	11.6-12.1	19.9-33.7	69.4-475.8
Carwash Wastewater			
MAHAJU (MJ) Car wash	8-8-9.5	19.4-33.0	12.6-115.5
Bandar U Carwash	9.5-10.6	18.2-25.4	16.1-100.4
Tap water			
Tap water	7.6-8.1	19.4-19.7	13.3-13.4

Table 2. Tolerance limit of water in mixed concrete.

Standards	Tolerance limit, ppm		
	pH	Chloride Cl^-,ppm	Sulphate SO_4^{2-},ppm
ASTM C1602	-	<500	<3000
ASTM C685	-	<500	<3000
BS EN 1008	>4	<500	<2000
BS 3148	-	<5007	<1000
Previous Researchers (slurry water)			
[5-9]	8.1-13.5	20.8-355.0	12.7-1152
Previous Researchers (car wash wastewater)			
[13-14][19][31]	6.5-8.75	9-74.2	21.3-188.0

In addition, the higher pH value in car wash wastewater may be due to chemicals used in pre-soak stages during the car wash process as supported by previous studies by Zaneti et al., [15]. Furthermore, according to BS EN 1008 mentioned that the typical pH value of water for mixing concrete was at least more than four (>4). Based on that, wastewater has a higher potential to be used as replacement in concrete mixes due to its reasonable characteristics.

3.2 Properties of fresh concrete

3.2.1 Slump test

In this study, the concrete slump test was analysed according to ASTM C143 standards and the results were listed in Table 3. From Table 3, the slump values for both types of wastewater were comparable to the control mix. As mentioned in sub topic 2, the target mean slump value varies from 30 mm to 60 mm. These findings definitely proved that the slump value of concrete containing wastewater were in the designated range. In addition, the results show that a higher replacement percentage of wastewater in concrete mix results in greater slump value.

Table 3. Slump Test.

Waste-water Type	CM	SW10 %	SW20 %	SW30 %	SW40 %	SW50 %	CW10 %	CW20 %	CW30 %	CW40 %
Slump value (mm)	32	43	39	40	42	48	42	38	43	45

3.3 Mechanical properties of concrete

3.3.1 Compressive strength

Fig. 1 showed the comparison of compressive strength between slurry water and car wash wastewater mix against the control mix.

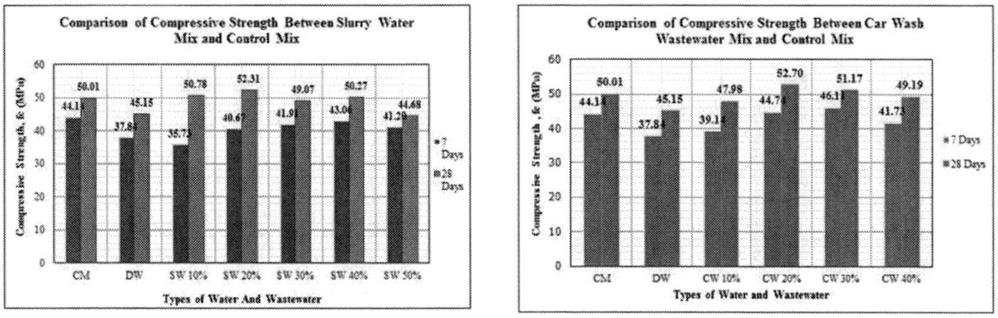

(a) (b)

Fig. 1. Comparison of compressive strength between; (a) slurry water and (b) Car Wash.

According to Fig. 1, all types of slurry water shows the higher strength than the minimum limit of 90% specified in the standards except for concrete specimen with contain 10% of slurry water (SW10%) at only 80.95%. However, on day 28, the concrete specimen that contains 10% to 40% of slurry water has exceeded the limit of the compressive strength, meanwhile by 50% replacement of slurry water in concrete mix was below the minimum limit. This is because the proportion of water for concrete (SW50%) consisted of 50% of slurry water and 50% of tap water, which would cause the pH value to become higher than others. This can be seen from Table 1, where the pH value for slurry water is between 10 to 12.1 compared to tap water which is 7.6 to 8.1. According to the previous studied, the researchers mentioned that the typical range of pH value for water used in concrete manufacturing is between 6.0 to 9.0 [20, 21, 22, 30].

On the other hand, the compressive strength of car wash wastewater concrete was in the range of 88% to 104% of normal concrete. This range was beyond the minimum limit of 90% that was specified in ASTM C1602 except for the 10% car wash wastewater replacement. However, on day 28 all the specimens of concrete mixed with car wash wastewater not only went above the limits of the standard, but also indicated a slight improvement especially for the concrete with 20% and 30% car wash wastewater replacement. This finding may be due to the fine solid particles present in the car wash wastewater could fill the pores or voids between fine aggregates and coarse aggregates so that the strength of concrete could be improved but that would depend on suitable replacement percentages of car wash wastewater in concrete mixes as mentioned and supported by these researchers [23, 24, 29].

In addition, the highest compressive strength for concrete with car wash wastewater was approximately 52.7 MPa compared to concrete with slurry water which is 52.3 MPa. This may be due to the lower sulphate content in car wash wastewater which is 12.6 ppm to 115.5 ppm compared to slurry water with a range between 42.6 ppm to 475.8 ppm. This is because the presence of sulphate content in wastewater may be generated from ash or residue. Thus, the higher sulphate content in wastewater may result in a lot of waste such as ash and residues which may decrease concrete permeability and reduce the strength of the concrete [19].

3.3.2 Modulus of elasticity (MOE)

Consequently, Fig. 2 presented the comparison of modulus of elasticity between slurry water and car wash wastewater mix against control mix.

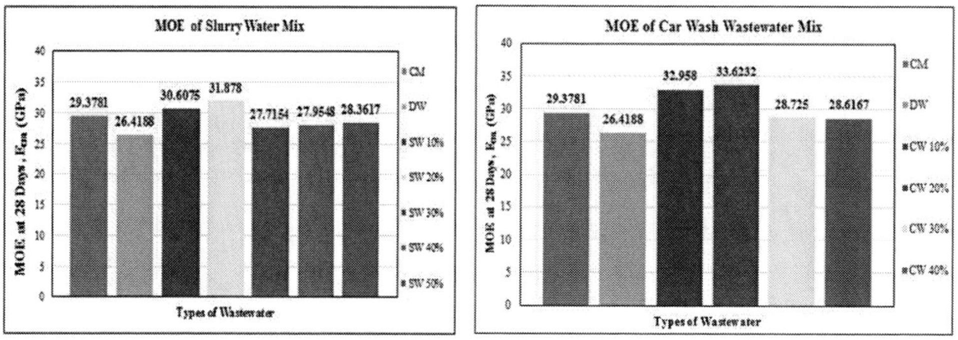

(a) (b)

Fig. 2. MOE of ; (a) Slurry water and (b) Car Wash.

As stated in BS EN 1992, usually the range of modulus of elasticity of concrete (Ecm) for grade 35 is 32 + 6 GPa [25]. Based on Table 7, the results completely achieved the required range of MOE for concrete with a mean strength of 35MPa. In addition, ACI 318-02 and Neville, (1995) also mentioned that the typical MOE value for concrete with a compressive strength of 35 was 30 GPa [26][34]. Therefore, by comparing the data in Table 7 with previous studies, it is clear that the MOE value of slurry water and car wash wastewater complied with the suggested ranges.

On the other hand, the concrete with 20% car wash wastewater replacement demonstrated the highest MOE value at 33.6 GPa compared to slurry water as well as control mix with only 31.9 GPa and 29.3781 GPa respectively. These findings may correlate to the existence of fine solid particles in wastewater which could fill the porosity and voids in between the fine and coarse aggregates in concrete structures. According to previous researchers, the factors affecting the MOE were concrete strength and concrete density [27, 28]. The concrete strength was related to the concrete compressive strength while the concrete density was related to the porosity of concrete. Based on the results in Table 1, the sulphate content in car wash wastewater was between 12.6 ppm to 115.5 ppm compared to tap water which was between 13.3 ppm to 13.4 ppm. Previous researchers revealed that higher sulphate concentration may result in lower concrete impermeability. Thus, it can be summarized that the lower content of sulphate in wastewater may lead to higher MOE of concrete.

3.3.3 Relationship between compressive strength and MOE

According to Fig. 1 and Fig. 2, the modulus of elasticity increased when the concrete compressive strength was increased. In addition, the relationship that can be seen has shown that the concrete mixed with 20% wastewater has achieved the highest compressive strength and modulus of elasticity. Therefore, it can be concluded that the optimum percentage of slurry water and car wash wastewater as a fresh water replacement in concrete mixes is 20%.

4 Conclusion

Based on the results from the experimental work, the following conclusions for the use of car wash wastewater in concrete mixes can be drawn: (a) The pH value for slurry water from ready mixed concrete plants was in the range of 10 to 12 while the pH value for car wash wastewater was approximately 8.8 to 10.6. Both types of wastewater had pH values slightly higher than tap water. However, the content of chloride and sulphate in wastewater were within the ASTM and BS standards requirement; (b) With an increase in the percentage replacement of wastewater in concrete mix has lead to the decrease of compressive strength, and modulus of elasticity of concrete. Generally, the compressive strength of concrete mixed with slurry water is in the range of 81% to 98% for the control mix and 88% to 104% for concrete mixed with car wash wastewater; (c) According to the results for compressive strength and modulus of elasticity, the optimum percentage of car wash wastewater as a fresh water replacement in concrete mixes is 20%.

The authors would like to express their deepest appreciation to the Ministry of Higher Education Malaysia, Universiti Tun Hussein Onn Malaysia for funding the research under grant U523.

References

[1] Z.B. Zaharaton, *Water resources management in Malaysia - the way forward*, Asian Water, Kuala Lumpur, Malaysia, (2004)
[2] S. Shahidan, R. Pullin, K.M. Holford, M.B.N, and N. Nor, Quantitative evaluation of the relationship between tensile crack and shear movement in concrete beams, Advanced Material Research, **626**, 355–359, (2013)
[3] CIDB, *Construction Statistics Quarterly Bulletin*, Construction Industry Development Board Malaysia (CIDB), (Jun 2015)
[4] S. Tsimas and M. Zervaki, Reuse of wastewater from ready-mixed concrete plants. Management of Environmental Quality: An Int. J., **22**(1), 7-17, (2011)
[5] A.M. Wahid and H. Tanaka, Potential implementation of wastewater reclamation and reuse in Malaysian urban area, Int. Sustainability and Civil Engineering J., **1**(1), 75-83, (2012)
[6] G. Asadollahfardi, M. Delnavaz, V. Rashnoiee, and N. Ghonabadi, Use of treated domestic wastewater before chlorination to produce and cure concrete, Construction Build. Materials, **105**, 253–261, (2016)
[7] B. Chatveera, P. Lertwattanaruk and N. Makul, Effect of sludge water from ready-mixed concrete plant on properties and durability of concrete, Cement and Concrete Composites, **28**(5), 441-450, (2006)

[8] S.A. Kudus, N.M. Bunnori, S.R. Basri, S. Shahidan, M.N.M. Jamil and N.M. Noor, An overview current application of artificial neural network in concrete, Advanced Material Research, **626**, 372–375, (2012)
[9] S.A. Chini and W.J. Mbwambo, environmentally friendly solution for the disposal of concrete wash water from ready mixed concrete operations, *in Proc. of CIB W89 Beijing Int. Conf.*, (1996)
[10] R.A. Khan, M.S. Liew and Z. Ghazali, Malaysian construction sector and Malaysia Vision 2020: Developed nation status, Social and Behavioral Sciences, **109**, 507-513, (2014)
[11] ASTM C33, *Specification for concrete aggregates*, Pennsylvania, United States: American Society for Testing and Materials, (2003)
[12] BS 882, *Specification for aggregates from natural sources for concrete*, London: British Standards Institution, (1992).
[13] BS EN 1008, *Mixing water for concrete-specification for sampling, testing and assessing the suitability of water, including water recovered from processes in the concrete industry, as mixing water in concrete*, London: British Standards Institution (2002)
[14] ASTM C143, *Test method for slump of hydraulic cement concrete*, Pennsylvania, United States: American Society for Testing and Materials, (2005)
[15] ASTM C138, *Test method for density unit weight, yield and air content of concrete*, Pennsylvania, United States: American Society for Testing and Materials, (2005)
[16] ASTM C469, *Test method for static modulus of elasticity and poisson ratio of concrete in compression*, Pennsylvania, United States: American Society for Testing and Materials, (2002)
[17] ASTM C39, *Test method for compressive strength of cylindrical concrete specimens*, Pennsylvania, United States: American Society for Testing and Materials, (2005)
[18] ASTM C496, *Test method for splitting tensile strength of cylindrical concrete specimens*, Pennsylvania, United States: American Society for Testing and Materials, (2004)
[19] C. Brown, *Water conservation in the professional car wash industry*, International Car Wash Association, Washington, (2000).
[20] C. Brown, *Water use in the professional car wash industry*, International Car Wash Association, Washington, (2002)
[21] K.S. Al-Jabri, A.H. Al-Saidy, R. Taha and A.J. Al-Kemyani, Effect of using wastewater on the properties of high strength concrete, Procedia Engineering, **14**, 370-376, (2011)
[22] K.J. Kucche, quality of water for making concrete: A review of literature, J. of Scientific and Research Publications, **5**(1), 201-205, (2015)
[23] K.S. Al-Jabri, A.H. Al-Saidy, R. Taha and A.J. Al-Kemyani, Effect of using wastewater on the properties of high strength concrete, Procedia Engineering, **14**, 370-376,(2011)
[24] R. Zaneti, R. Etchepare and J. Rubio, Car wash wastewater reclamation. Full- scale application and upcoming features, Resources, Conservation and Recycling, **55**(11), 953-959, (2011)
[25] K.P. Mehta and P.J. Monteiro, *Concrete: Microstructure, Properties, and Materials*, 4th ed., New York: Mc Graw Hill, (2014)
[26] BS EN 1992, *Design of concrete structure*, London: British Standards Institution, (2004)
[27] A.M. Neville, *Properties of concrete*, Concrete Technology, 4th ed., London: Pearson, (1995)

[28] M.A. Rahim, Z. Ghazaly, R. Nurazira, R. Mamat, M.A. Azizan, N.F. Isa, and S. Shahidan, Experimental study of slurry infiltrated fiber reinforced concrete, Material Science Forum, **857**, 363–366, (2016)

[29] M. Abdul Rahim, N.M. Ibrahim, Z. Idris, Z.M. Ghazaly, S. Shahidan, N.L. Rahim, L. A. Sofri, and N. F. Isa, Properties of concrete with different percentange of the rice husk ash (RHA) as partial cement replacement, Material Science Forum, **803**, 288–293, (2014)

[30] S. Shahidan, N.M. Bunnori, N. Md Nor and S.R. Basri, Damage severity evaluation on reinforced concrete beam by means of acoustic emission signal and intensity analysis, *IEEE Symposium on Industrial Electronics and Applications*, 337–341, (2011)

[31] S. Shahidan, I. Isham and N. Jamaluddin, A Review on waste minimization by adopting in self compacting concrete, MATEC Web Conf., **47**, 1–7, (2016)

[32] J.M. Irwan, S.K. Faisal, N. Othman, M.H. Wan Ibrahim, R.M. Asyraf, M.M.K. Annas, Performance of concrete using light waste PET fibre, Advanced Materials Research, **795**, 352-355, (2014)

[33] R.P. Jaya, B.H. Abu Bakar, M.A.M. Johari, M.H.W. Ibrahim, M.R. Hainin and D.S. Jayanti, Strength and microstructure analysis of concrete containing rice husk ash under seawater attack by wetting and drying cycles, Advances in Cement Research., **26**(3), 145–154, (2014)

[34] S. Shahidan, M.S. Senin, A.A. Kadir, L.H. Yee and N. Ali, properties of concrete mixes with carwash wastewater, MATEC Web Conf., **87**(1018), 1–6, (2017)

[35] S. Shahidan, S.S. M. Zuki and N. Jamaluddin, Damage grading system for severity assessment on concrete structure, Case Stud. Constr. Mater., **5**,79–86, (2016)

[36] S. Shahidan, I. Isham and N. Jamaluddin, A review on waste minimization by adopting in self compacting concrete, MATEC Web Conf., **47**, 1–7, (2016)

[37] S. Shahidan, S.S. Mohd Zuki, M.A. Rahim and N. Ali, Amplitude distribution of emission wave for cracking process, MATEC Web Conf., **47**, 2–8, (2016)

Properties of Sugarcane Fiber on the Strength of the Normal and Lightweight Concrete

Faisal Sheikh Khalid[1,*], *Herman Shah* Herman[1], and *Nurul Bazilah* Azmi[1]

[1]Faculty of Civil and Environment Engineering, Universiti Tun Hussein Onn Malaysia, 86400 Parit Raja, Johor, Malaysia

Abstract. The usage of natural fiber in construction are widely used in building materials engineering. However, using sugarcane fiber waste material as a natural in construction is very precious, because it can increase crack control and ductility, brittle concrete. Furthermore, the usage of sugarcane in construction can reduce of environmental pollution.In this study, a mixture of sugarcane fiber to be used in normal grade concrete and lightweight concrete to determine whether there is an increase in the compressive and tensile strength of the concrete. The objective of this study was to determine the compressive and tensile strength between control concrete and concrete mix with sugarcane fiber. In addition, the optimal volume of sugarcane fiber in the concrete mixture where the percentage of sugarcane fiber used was 0.5%, 1.0% and 1.5%. Compessive strength was tested on days 7 and 28 after curing test is carried out. Meanwhile, the tensile test, has been carried out to measure the tensile strength of sugarcane fiber relations in concrete mixes only at 28 day curing. Result of the testing showed that the optimum value containing admixtures of sugarcane is 0.5%. This percentage get the value of compressive strength is nearest with concrete control and the value of tensile strength is higher than concrete control and also the timing of concrete to cracked getting slower. Therefore, the use of sugarcane fiber suitable for addition that do not exceed 0.5% of the concrete mixture.

1 Introduction

Concrete is made up of three main ingredients which is cement, aggregates and water. However, there were also other materials put into the concrete mix which is known as addictive to increase the strength of the concrete. Concrete has high compressive strength, corrosion and weathering effect are minimal. However, concrete has low tensile strength and easily can be crack due its characteristics which is brittle [1-3].

Nowadays, an increasing demand of concrete in the industries, researchers and scientist are developing alternate binders that are eco friendly and contribute towards waste management especially for agricultural waste such as sugarcane [2].There have many previous experimental have been conducted by using different volume of sugarcane bagasse. Baggase fiber is a by-product of sugarcane industry whose role is sugar, rum or

* Corresponding author: faisalsh@uthm.edu.my

biofuel production. Sugarcane can help to improve the deficient tensile strength of concrete due to its limited tensile strength [2].

Sugarcane baggase have a little impact on the compressive strength of concrete [4, 5]. The increasing of volume of concrete was reduced the compressive strength of the concrete. Olutoge [4] was claimed that baggase ash can increase the strength of concrete when used up to 10% cement and 30% fine aggregates replacement level . However, the flexural strength of concrete is decrease after certain percentage replacement of sugarcane baggase ash. The sugarcane baggase ash concrete had higher compressive strength compare to the concrete without sugarcane baggase ash [5]. The optimal level of sugarcane baggase ash content with 15% replacement. Many researchers show that sugarcane in high performance concrete as a partial replacement of Portland cement improved the rheological properties [6-9]. The method was implemented by grinding treatment that the ash received before its incorporation into the concrete mixtures. Cordeiro et al. [10] found an influence of sugarcane ash on the properties of concretes. Concrete containing sugarcane ash was produced using a vibratory grinding machine in a dry open circuit. Conventional and high-performance concretes were produced using 0% to 20% of these ashes as a cement replacement [11]. It can be concluded that the partial replacement of Portland cement sugarcane ash caused a slight increase in yield stress and a small reduction in plastic viscosity in both the conventional and high-strength concretes. The rheological evaluation of these causes is hardly to determine due to different particle sizes of the ingredients that form the mixture [2, 11].

This paper presents the results of research aimed to investigate the performance of sugaracane baggase as an addition material in normal and lightweight concrete. The utilization of sugarcane bagasse in concrete mix produced desired compressive strength, tensile strength and optimum sugarcane fiber in concrete.

2 Methodology

To achieve the objectives of this study, there are many approaches and methods that should be taken. Methods of work such as literature review, methods of mixing and testing in the laboratory is important to get a more accurate analysis results. The volume of sugarcane are 0.5%, 1.0% and 1.5% and use volume method for the design mix concrete for normal concrete that shown in Table 1 and lightweight concrete that shown in Table 2. The sugarcane fiber was used as shown in Fig. 1.

Fig. 1. Sugarcane fiber.

The sugarcane bagasse will dry under the sun until completely dry for 7 days. Then the bagasse will cut into small relatively uniform strips with estimation 5cm until 10cm. Fig. 2 shows the sugarcane bagasse dry under sun.

Fig. 2. Sugarcane was dried.

Furthermore, bagasse need to be treated first with Sodium Hydroxide NaOH Solution50% dilute for 3 days as in Fig. 3. Purpose of the treatment is to remove the impurities and ensure the sugarcane withstand longer to use in concrete. Then, the treated sugarcane bagasse need to be dried under the sun to ensure it completely dried before adding into concrete mixture. The volume of sugarcane are 0.5%, 1.0% and 1.5% and use volume method for the design mix concrete for normal concrete that shown in Table 1 and lightweight concrete that shown in Table 2.

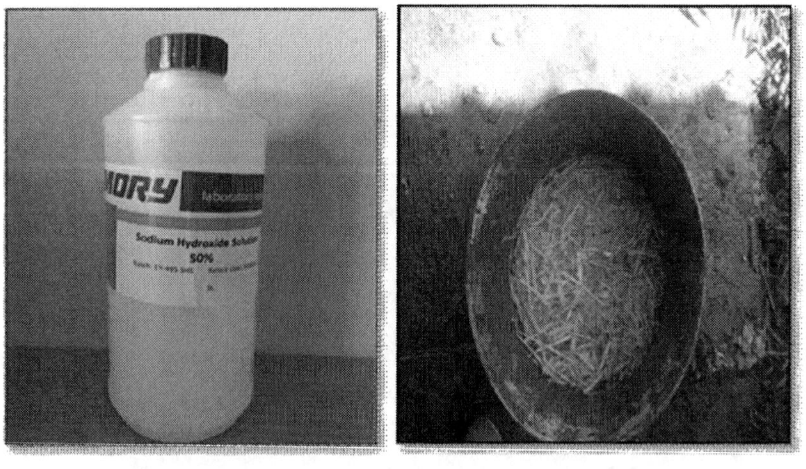

Fig. 3. Sugarcane that have treat in Sodium Hydroxide Solution.

The size of the cube used are 100 mm x 100 mm x 100 mm for the compression test for the tensile test size of cylinder used are 100 mm x 200 mm. There are several tests conducted to achieve the objectives, which is compression and tensile tests of concrete that was done according to standard test [12, 13]. The curing process was carried out for 7 and

28 days in water at room temperature to let the concrete in the moist state as long as possible so that the hardening process happens in moderation in order to achieve maximum strength.

Table 1. Mix design of concrete for 1 m^3.

Water (kg/m^3)	Cement (kg/m^3)	Sand (kg/m^3)	Coarse aggregate (kg/m^3)
209	300	980	805

Table 2. Mix design of lightweight concrete for 1 m^3.

Water (kg/m^3)	Cement (kg/m^3)	Sand (kg/m^3)	Foam (kg/m^3)
175	390	800	235

3 Results and discussion

In this study, there are two types of tests carried out on 72 concrete samples which 32 samples are prepared for normal concrete and another 32 samples are prepared for lightweight concrete. Both of the sample was conducted compressive and tensile tests.

3.1 Compressive strength of the concrete

Fig. 4 shows that the control sample containing 0% sugarcane recorded a reading of 18.00 MPa compressive strength. Samples containing 0.5% additive showed higher readings than the other percentage that is 14.8 MPa for the curing 7 days. The sample containing 1.0% sugarcane representing 11.4 MPa. Sample containing 1.5% sugarcane representing has lowest compressive strength for 7 days curing that is 9.6 MPa.

Meanwhile, compressive strength of concrete for curing period of 28 days after the cube was produced. Based on the figure, the control sample containing 0% sugarcane fiber recorded a reading of compressive strength is 23.80 MPa. For samples containing 0.5% additive showed compressive strength 21.30 MPa meanwhile sample containing 1.0% recorded compressive strength 18.60 MPa. For sample 1.5% additive recorded the lowest compressive strength of concrete for samples 28 days that is 15.30 MPa. The lower compressive strength of fibrous mixtures might be attributed to the presence of fiber bundle (high percentage of fiber) during mixing [14, 15]. These might attributed the voids due tothe addition of high percentage of fiber and the existence of a weak interfacial bonds among the fiber and concrete matrix [9, 10].

The lightweight concrete also was tested with compressive strength and result as shown in Fig. 4. The control sample that containing 0% of sugarcane bagasse was 3.74MPa. Samples containing 0.5% bagasse obtained the highest value of compressive strength which was 6.50MPa. Meanwhile, the value of compressive strength for samples containing 1.0% and 1.5% recorded slightly decline value compared to sample control and 0.5% which were 4.31MPa and 1.54MPa after curing 7 days.

Meanwhile, the result compressive strength of the concrete for 28 days, the control sample obtained value 7.10MPa. The sample containing 0.5% sugarcane bagasse recorded the highest value which is 10.30MPa. However, the samples containing 1.5% sugarcane bagasse showed the lowest value which is 3.80MPa. The compressive strength of the concrete samples for 28 day curing period showed increased high strength and also indicates that the sample has reached maturity in providing more strength.

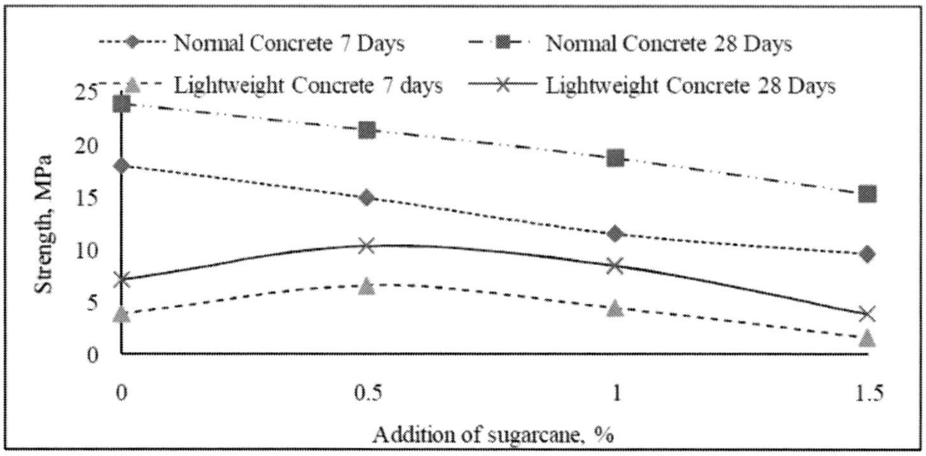

Fig. 4. Compressive Strength of Concrete with Sugarcane Fibre.

3.2 Tensile strength of concrete

The control sample of normal concrete containing 0% sugarcane recorded a reading of 2.10 MPa tensile strength as shown in Fig. 5. For samples containing 0.5% additive showed higher readings than the control sample that is 2.16 MPa for the curing 28 days. The sample containing 1.0% sugarcane representing 2.22 MPa is higher than 0.5%. Sample containing 1.5% sugarcane representing has highest tensile strength for 28 days curing that is 2.29 MPa. For the 28-day tensile the strength, the sample containing 1.5% sugarcane fiber showed a high increase in tensile strength. This indicates that the sample 1.5% has reached a stage of maturity in giving more strength and it also possible to increase other properties of concrete like durability and cracking resistance towards concrete.

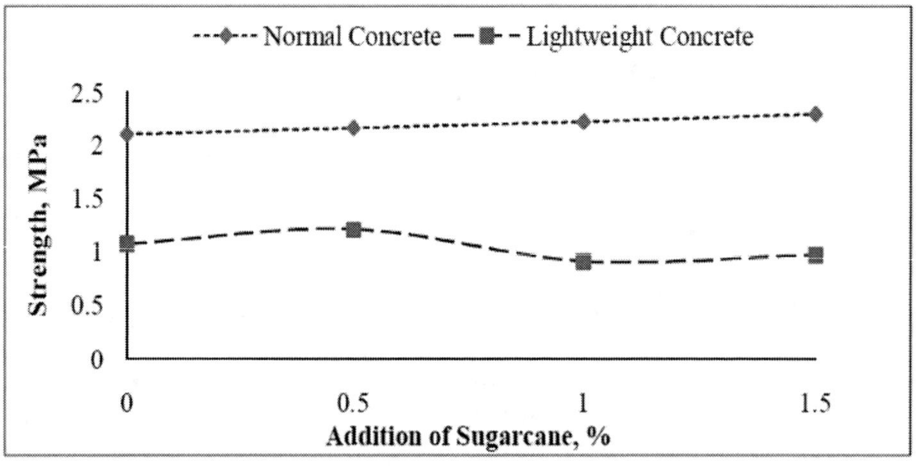

Fig. 5. Tensile strength of the concrete for 28 days.

Meanwhile, the lightweight concrete for control specimen is 1.07MPa. However, the 0.5% of addition of sugarcane bagasse gives maximum tensile strength at 28 days as compared to 1.0% and 1.5% addition of fiber sugarcane bagasse which was 1.21MPa. Concrete with 1.5% sugarcane bagasse slightly decreased by the value of 0.97MPa. The lowest value of tensile strength of this experiment was 1.0% sugarcane bagasse with 0.91MPa. Incorporating of sugarcane bagasse tends to reduce the tensile strength. A decreased of tensile is observed with the increase of fiber proportion. Fig. 5 shows the tensile strength of the concrete for 28 days.

The fiber properties control the strength behavior of concrete that contains fibers. Fibers with higher tensile strength transfer more tensile stress from a cracked zone to the fibers. The tensile strength of fiber used in this research work is significant to create bridge stress between fiber and concrete matrix during tensile test.

4 Conclusions

Based on the analysis of data and discussion that has been carried out, it was shown that concrete containing sugarcane fibers in consistent with the properties of ordinary concrete. Compressive strength of the concrete were decreased when the amount of sugarcane fiber increased. The optimum percentage that give the higher compressive strength is 0.5% either for normal concrete or lightweight concrete and this is not far from the strength of concrete control.Tensile strength of the concrete were increased when the amount of sugarcane fiber increase especially for normal concrete since 1.5% volume of sugarcane is the highest compared with the tensile strength of concrete control. However, optimum tensile strength of the lightweight with sugarcane fibre is 0.5 % . When the sugarcane fibre keep increasing, the tensile strength of the lightweight concrete was reduced.

Physical characteristics of the concrete compressive strength and tensile strength of concrete shows that sugarcane fiber content affects the physical properties of ordinary concrete. The optimum value of sugarcane is 0.5% percent although they did not affect the compressive strength of concrete but in terms of tensile strength percent sugarcane fiber is still beyond the control of the tensile strength of normal concrete. However, the increment of sugarcane fibre in the lightweight concrete reduce the tensile strength of the lightweight concrete. Overall, volume fiber content of 0.5% of sugarcane bagasse was the significantly for mix design for normal concrete and lightweight concrete containing sugarcane.

References

[1] M.S. Frías, E. Villar-Cocina and E. Valencia-Morales, Characterisation of sugar cane straw waste as pozzolanic material for construction: Calcining temperature and kinetic parameters, Waste Management, **27**(53), 3–8, (2007)

[2] D. Govindarajan and G. Jayalakshmi, XRD, FTIR and microstructure studies of calcined sugarcane bagasse ash, Advances in Applied Science Research, **2**, 544–9 (2011a)

[3] J.M. Irwan, S.K. Faisal, N. Othman and M.H. Wan Ibrahim, Performance of concrete using light waste PET fibre, Advanced Materials Research, **795**, 352-335, (2013)

[4] F. Olutoge, O. Ofuyatan, A. Rachael and E. Opaluwa, Strength characteristics of concrete reinforced with sugarcane bagasse fiber, IOSR Journal of Mechanical and Civil Engineering, **12**(3), 68-71, (2015)

[5] M.K. Yashwant and BG. Nareshkumar, An experimental study on bagasse ash as replacement for cement in ligthweight concrete, Int. J. of Latest Trends in Engineering and Technology, **3**(3), 253-260, (2013)

[6] V. Srivastava, P. K. Mehta and S. Nath, Natural fiber in cement and concrete matrices. J. of Environment in Nanotechnology, **2**(3), 63 – 66, (2013)
[7] G.C. Cordeiro, R.D. Toledo-Filho, L.M. Tavares and E.M.R. Fairbairn, Ultrafine grinding of sugar cane bagasse ash for application as pozzolanic admixture in concrete, Cement Concrete Research. **39**,110–5, (2009)
[8] A. Bahurudeen, A.V. Marckson, A. Kishore and M. Santhanam, Development of sugarcane bagasse ash based Portland pozzolana cement and evaluation of compatibility with superplasticizers, Construction Building Materials, **68**, 465– 475, (2014)
[9] M. Frías, E. Villar and H. Savastano, Brazilian sugar cane bagasse ashes from the cogeneration industry as active pozzolans for cement manufacture, Cement Concrete Composite, **33**, 490–496, (2011)
[10] M. Gesoglu, E. Güneyisi, A.H. Nahhab and H. Yazıcı, The effect of aggregates with high gypsum content on the performance of ultra-high strength concretes and Portland cement mortars, Construction Building Materials, **110**, 346–354, (2016)
[11] A.M. Rashad, Metakaolin as cementitious material: history, scours, production and composition – a comprehensive overview, Construction Building Materials, **41**(3), 3 – 18, (2013)
[12] EN 12390-3:2000, *Design of concrete structures: Testing hardened concrete: Compressive strength for test specimens*, British Standard Institution, London, (2000)
[13] EN 12390-6:2000, *Design of concrete structures: Testing of hardened concrete: tensile strength test*, British standard. British Standard Institution, London, (2000)
[14] S. Shahidan, H.B. Koh, A.M.S. Alansi and L.Y. Loon, Strength development and water permeability of engineered biomass aggregate pervious concrete, MATEC Web Conf., **47**, 2–7 (2016)
[15] M. Abdul Rahim, N.M. Ibrahim, Z. Idris, Z.M. Ghazaly, S. Shahidan, N.L. Rahim, L. A. Sofri and N.F. Isa, properties of concrete with different percentage of the rice husk ash (RHA) as partial cement replacement, Material Science Forum, **803**, 288 –293 (2014)

The Effect of Effective Microorganism (EM) Inclusion to the Setting Time of Microbed Cement Paste

Noorli Ismail[1,*], *Hamidah* Mohd Saman[2], *Mohd Faizal* Mohd Jaafar[3], and *Siti Radziah* Abdullah[1]

[1]Faculty of Civil and Environmental Engineering, Universiti Tun Hussein Onn Malaysia, 86400 Parit Raja, Batu Pahat, Johor, Malaysia
[2]Faculty of Civil Engineering, Universiti Teknologi MARA, 40450 Shah Alam, Selangor, Malaysia
[3]Faculty of Civil Engineering and Earth Resources, Universiti Malaysia Pahang, 26600 Pahang, Malaysia

Abstract. The inclusion of Effective Microorganism (EM) lengthen the setting time of microbed cement paste with respect to the control cement paste. However, the investigation on the influence of EM ingredients to the setting time was not performed yet in the earlier study. Therefore, the test of X-Ray Diffraction (XRD) was performed on the EM product used in order to determine their ingredients. Further examination was carried out to analyse the pore characteristics in examining the effect of the pore size distribution to the setting time of microbed cement paste. Potassium Chloride (KCl) was detected in EM liquid using XRD test. It can be concluded, the ingredient of EM which is consist of the molasses influenced the setting time of microbed cement paste. The hardening rate of microbed cement paste was effected by the presence of KCl. The pore size of microbed cement paste recorded the lower pore volume (mL/g) with respect to the control cement paste. The larger intrusion of mercury was recorded at smaller pore diameter in microbed cement paste while control cement paste occupied at the bigger pore diameter. The microbed cement paste (16.5%) recorded the smaller porosity than cement paste without EM (20.3%). The smaller cement pore effects the delayed setting time of microbed cement.

1 Introduction

Increasing of concrete consumption introduces the processed and unprocessed product materials or waste materials in the concrete mix to improve the concrete workability, durability and mechanical properties. The significant outcome by adding these materials are well reported [1-15]. Beginning of the 21th century, other new potential materials have been initiated to enhance the cement based properties is Effective Microorganism (EM) [16-24]. The incorporation of EM in cement paste prolonged the initial and final of setting time. Due to the extended setting time, its effect the process of hydration of microbed

* Corresponding author: noorli@uthm.edu.my

cement paste [21]. The decrement pore size of cement paste particles also lengthen the setting time [25].

However, the study was not published yet in examining the effect of EM ingredients to the delayed setting time of microbed cement paste. Furthermore, the outstanding result performed in previous studies [16-24] initiated the curiosity to carry out further experiments to study the effectiveness of EM in analysing the pore size of microbed cement paste. Therefore, this present study was carried out to investigate the influence of EM inclusion to the setting time of cement paste. Vicat test was performed on the cement paste containing with and without EM. The X-Ray Diffraction (XRD) was carried out in investigating the ingredients of EM liquid while total porosity, total pore area and diameter, their distribution and average were determined to relate the pore characteristics to the setting time of cement paste by performing the Mercury Intrusion Porosimetry (MIP) test.

2 Experimental

The Effective Microorganism (EM) was obtained from Johor Farmers' Organisation, Malaysia. One (1) litre of mixtures consist of distilled water (90%) and EM (10%) was named as Effective Microorganism Solution (EMS). Seven (7) days of fermentation was carried out in plastic bottle to activate the microorganism growth in mixtures. The pH of the mixtures was less than four (4) before being added into the cement paste. The test of Vicat and Mercury Intrusion Porosimetry (MIP) were performed on the microbed and control cement paste while the liquid of EM used was tested using X-Ray Diffraction (XRD). The following subsection describes the preparation of cement paste and testing required.

2.1 Preparation of cement paste

The quantity of water required was calculated from the consistency test which was 28% and 30% recorded in control and microbed cement paste, respectively. The cement paste of 500g was blended with 140ml of distilled water and 150ml of EMS. The size of mould prepared was 50mm and the specimens leave in the room temperature before tested by MIP.

2.2 Consistency and setting time

The cement paste of 500g mixed with 100ml of distilled water and EMS. The plunger of Vicat released at 4min ± 10s after zero time. The penetration was recorded with different water amounts until the distance of plunger and base-plate reached at 6 ± 2 mm. After achieved the consistency of paste, the plunger was replaced by the needle. The needle of Vicat adjusted and reached the surface of the paste. The initial setting time was measured at zero time when the needle released to the time at which the needle penetrates vertically into the paste when the distance reached 6 ± 3 mm between the needle and the base plate. The final setting time determined when the time elapsed was recorded from zero to the time at needle finally penetrates only 0.5 mm into the paste [26].

2.3 X-Ray diffraction (XRD)

XRD test was carried out at Faculty of Pharmacy, Universiti Teknologi MARA (UiTM), Puncak Alam, Selangor with scanning from 2° to 90°. The liquid of EM was put on the specimen's holder before placed at tubular aerosol suspension chamber.

2.4 Mercury intrusion porosimetry (MIP)

The equipment of Micromeritics AutoPore IV was used to obtain data of porosity, pore characteristics and their distribution. The test was performed at the Multipurpose Laboratory, Faculty of Chemical Engineering, Universiti Teknologi MARA (UiTM), Selangor. The crushed of cement paste specimens was tested at the age of 3, 14 and 28 days of curing period. The test was carried out by applying a pressure between 0.1 MPa and 400 MPa.

3 Results and discussion

The results of cement consistency and setting time, XRD and MIP of cement paste with and without Effective Microorganism (EM) was described in the following sections.

3.1 Effective microorganism (EM) inclusion influence the cement consistency and setting time

The inclusion of EM in the cement paste increased the cement consistency and setting time. The cement consistency recorded was 28% and 30% in cement paste without EM and with EM, respectively. In control cement paste, the initial setting time recorded at 45 minutes while final time set at 145 minutes. Initial setting time of microbed cement paste indicated at 55 minutes while final in 170 minutes.

3.2 Effective microorganism (EM) ingredients influence the delayed setting time of microbed cement paste

The analysis of XRD in Fig. 1 shows the liquid of Effective Microorganism (EM) used in this present study contains high potassium chloride (KCl). Generally, KCl is a mineral used as fertilizers by farmers in plantation. This mineral is widely used as main ingredients in sugar which was commonly found in sugarcane [27]. The contents of KCl in EM used was confirmed by the addition of molasses in this present study. The molasses used in this study was produced from the sugarcane. The presence of KCl in EM influence the hardening rate of microbed cement paste which was lengthen the setting time of microbed cement paste [20]. The similar pattern found in the previous study [27-29] also concluded that molasses contents influenced the setting time of cement which was acted as retarder in the cement paste.

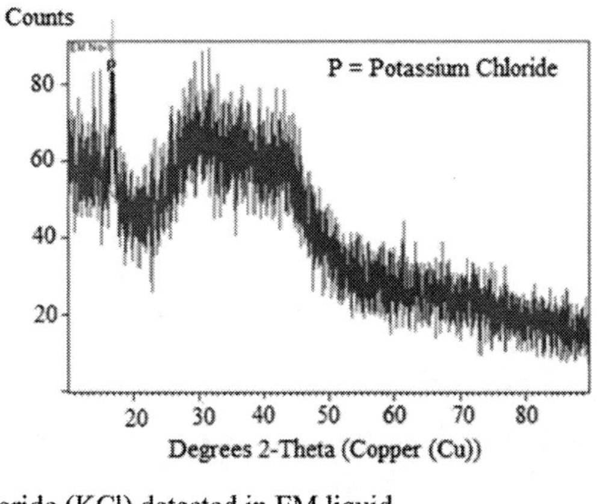

Fig. 1. Potassium chloride (KCl) detected in EM liquid

3.3 Effective microorganism (EM) inclusion influence the pore size and porosity of cement paste

The pore volume (mL/g) against pore size of control and microbed cement paste obtained from Mercury Intrusion Porosimetry (MIP) tests can be referred to Fig. 2 and Fig. 3. The test was performed on the curing day of 28. Generally, the pore diameters of all specimen displays a size more than 1μm. The curves is located at the larger pores region. The pore volume of microbed cement paste slightly lower compared to the cement paste without EM as shown in Fig. 2. In microbed and control cement paste, the larger mercury intrusion presented at pore diameter started from 20μm to 40μm and between 40μm and 50μm, respectively. The smaller of pore size and lower pore volume delayed the setting time of microbed cement. So it is consistent with larger mercury intrusion at range between 40μm and 50μm.

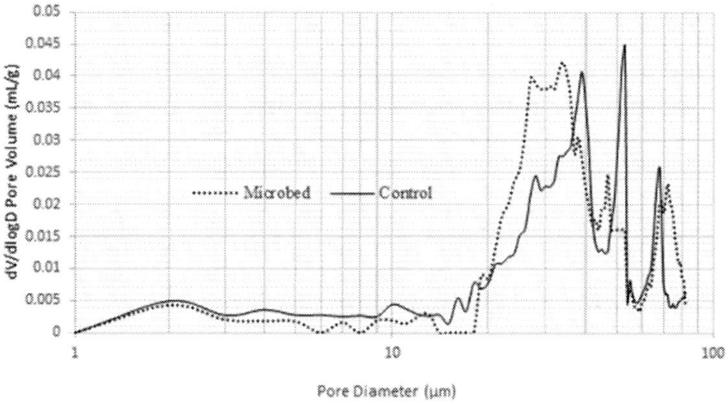

Fig. 2. Pore size distribution of control and microbed cement paste

The pore volume and diameter of microbed cement paste in Fig. 3 shows the highest pore volume found in the specimens of three (3) days with more than 0.12 mL/g. This can be due to the delayed setting time which reduced the hardening rate of the microbed cement paste. The pore diameter decreases with respect to the increment of curing days. In average, the larger mercury intrusion occurred at pore diameter between 20μm to 40μm in all specimens of microbed cement paste.

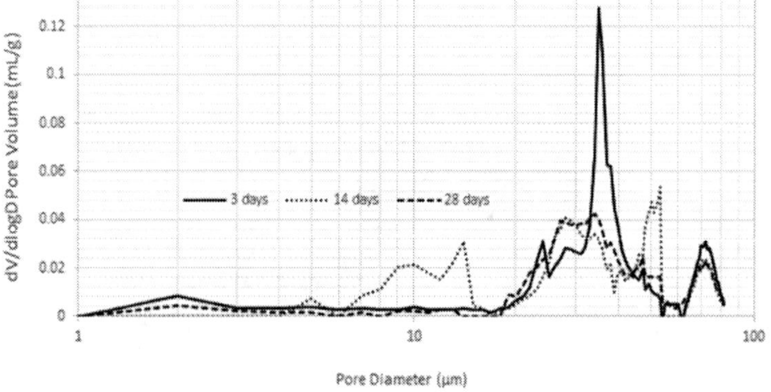

Fig. 3. Pore volume with respect to pore diameter of microbed cement paste

Table 1 presents the microbed cement paste having the reduction of total porosity (16.5%) than the control cement paste (20.3%) while median pore diameter represents the point that located 50% of pore size distribution curve. In microstructure examination found the densification matrix of microbed cement paste was denser and less voids with respect to the control paste without EM [20, 22]. The micrograph image captured was consistent with the result of porosity and median pore diameter. The microbed cement paste presents smaller value of median pore diameter corresponding to control cement paste. The maximum and minimum pore diameter do not show significance effect to the setting time of cement. Rationally, these two (2) data do not correlate to the pore volume of cement paste unless the graph of logarithm was constructed.

Table 1. The pore characteristics of control and microbed cement paste at the age of 28 days of curing

	Control	**Microbed**
Total porosity (%)	20.325	16.492
Average pore diameter (μm)	0.0539	0.0206
Median pore diameter (μm)	0.2424	0.0154
Maximum pore diameter (μm)	319.057	319.163
Minimum pore diameter (μm)	0.00301702	0.003017

4 Conclusion

The ingredient of Effective Microorganism (EM) used as the addition in the cement paste showed the presence of Potassium Chloride (KCl) which was detected using X-Ray Diffraction (XRD) test. Consequently, the setting time and hardening rate of microbed cement paste was effected by the presence of KCl due to the existence of molasses in EM liquid.

The addition of Effective Microorganism (EM) into the cement paste decreased the porosity with 19% corresponding to the control cement paste at the age of 28 days curing. The larger mercury intrusion occurred in microbed cement paste at pore diameter between 20μm and 40μm at the age of 28 days curing. The decrement of pore size delayed the setting time of cement. It can be referred to the result of lower pore volume and smaller pore diameter indicated in microbed of cement paste. The larger pore size recorded when the larger mercury intrusion at range between 40μm and 50μm in cement paste without EM.

References

[1] M. Vitruvius, *The ten books in architecture*, Translator Dover Publication, New York, USA, (1960)
[2] J.P. Adam, *Roman building: Materials and techniques*, Routledge Publication, 11 New Fetter Lane, London, (1999)
[3] N. Bouzoubaa, M.H. Zhang and V.M. Malhotra, Mechanical properties and durability of concrete made with high-volume fly ash blended cements using a coarse fly ash, J. Cement and Concrete Research, **31**(10), 1393–1402, (2001)
[4] D.A. Cengiz, Strength properties of high-volume fly ash roller compacted and workable concrete, and influence of curing condition, J. Cement and Concrete Research, **35**(6), 1112– 1121, (2005)

[5] P. Chindaprasirt, S. Homwuttiwonga and C. Jaturapitakkul, Strength and water permeability of concrete containing palm oil fuel ash and rice husk–bark ash, Construction and Building Materials, **21**(7), 1492-1499, (2007)
[6] K.Y. Foo and B.H. Hameed, Value-added utilization of oil palm ash: A superior recycling of the industrial agricultural waste, J. of Hazardous Materials, **172**(2–3), 523-531, (2009)
[7] S. Rafat, Utilization of silica fume in concrete: Review of hardened properties, J. Conservation and Recycling, **55**(11), 923-932, (2011)
[8] S.H. Adnan, I.A. Rahman, H.M. Saman and F.V. Riza, Pozzolanic properties of micronized biomass silica in enhancing compressive strength and water permeability of concrete, Modern Applied Science, **6**(11), 1-8, (2012)
[9] Z. Jamellodin, M.S. Hamidah, S.H. Adnan, N.S. Mohammad, W.Y. Wan Yusof, Compressive and flexural strength of fine grained mortar containing rice husk ash: A review, Advanced Materials Research, **1051**, 757-762, (2014)
[10] N.E. Zainal Abidin, M.H. Wan Ibrahim, N. Jamaluddin, K. Kamaruddin, A.F. Hamzah, The effect of bottom ash on fresh characteristic, compressive strength and water absorption of self-compacting concrete, Applied Mechanics and Materials, **660**, 145-151, (2014)
[11] M.A. Majid, M.A. Mohd Hishammudin, N.A.A. Hamid, Z. Jamellodin and N. Salleh, Tensile properties of treated and untreated paddy straw fiber using sodium hydroxide strengthened with polypropylene resin, MATEC Web of Conferences, **47**, 01019 (2016)
[12] N. Jamaluddin, A.F. Hamzah, M.H. Wan Ibrahim, R.P. Jaya, M.F. Arshad, N.E.Z. Abidin and N. H. Dahalan, Fresh properties and flexural strength of self-compacting concrete integrating coal bottom ash, MATEC Web of Conferences, **47**, 01010 (2016)
[13] A.F. Hamzah, M.H.W. Ibrahim, N. Jamaluddin, R.P. Jaya, M.F. Arshad and N.E.Z. Abidin, Fresh characteristic and mechanical compressive strength development of self-compacting concrete integrating coal bottom ash as partial fine aggregates replacement, Int. J. of Mechanical and Mechatronics Engineering, **15**(4), 61-67, (2015)
[14] S.N. Mokhatar, J. Hadipramana, S.N. Mohd Isa and M.M. Mustafa, The potential of artificial polyethylene coarse aggregate (APECA) on compressive strength of concrete after exposed by temperatures, MATEC Web of Conferences, 47, 01005 (2016)
[15] S. Shahidan, H.B. Koh, A.M.S. Alansi and L.Y. Loon, Strength development and water permeability of engineered biomass aggregate pervious concrete, MATEC Web of Conferences, **47**, 01007 (2016)
[16] Sato, Nobuyuki, Teruo Higa, S. Sugita and M. Shuya, Some properties of concrete mixed with effective microorganisms and the on-site investigation of the completed structures, In *Proceedings of the 28th International conference, Our world in concrete and structures*, Singapore, (2003)
[17] S. Nobuyuki, S. Masami and S. Syuichi, Investigation on concrete rust preventive agents using microorganisms (mechanisms), In *29th Conference on Our World in Concrete & Structures, Singapore*, 25-26, (2004)
[18] J. M. Yatim, M. Ismail, W.A.W.A. Rahman and V. Y. Chiew, Performance of concrete containing effective microorganisms (EM) under various environments, In *Joint Conference of 7th Asia Pacific Structural Engineering and Construction Conference,* (2009)
[19] Andrew, Tan Cheng Siong, Ibrahim Izni Syahrizal, and Mohd Yatim Jamaluddin, Effective microorganisms for concrete (EMC) admixture–its effects to the mechanical properties of concrete, In *Awam International Conference on Civil Engineering (AICCE'12) Geohazard Information Zonation (GIZ'12),* Penang, (2012)

[20] N. Ismail and M.S. Hamidah, Microstructure examination and strength characteristics of effective microbed cement, *International Conference on Biological, Civil and Environmental Engineering*, Dubai, UAE, (2014)
[21] N. Ismail, M.S. Hamidah, H. Jelani, H. Mansor and M.J. Mohd Faizal, The studies on the effect of molasses in effective microbed cement paste, *International Conference on Agriculture, Environment and Biological Science*, Antalya, Turkey, (2014)
[22] N. Ismail, M.S. Hamidah, H. Yusof and N. A. Adnan, X-ray diffraction (XRD) and thermogravimetric analysis (TGA) on microbed cement paste, *Structure, Materials and Construction Engineering Conference*, Istanbul, Turkey, (2014)
[23] M.N. Isa, M.M. Garba, and A.L. Kawu, Influence of locally made effective microorganisms on the compressive strength of concrete, J. of Multidisciplinary Engineering Science and Technology, **3**(4), 4425-4429, (2016)
[24] M.N. Isa, M.M. Garba and N. Usman, Durability of harden concrete containing locally made effective microorganisms under different environmental conditions, J. of Engineering and Technology, **2**, 1-9, (2015)
[25] P.A. Webb and C. Orr, *Analytical methods in fine particle technology*, Norcross, GA, USA, Micromeritics Instrument Corp., (1997)
[26] BS EN 196-3:2016, *Methods of testing cement. Determination of setting times and soundness*, BSI, (2016)
[27] K. Watanabe, Y. Fukuzawa and S. Kawasaki, Effects of potassium chloride and potassium sulfate on sucrose concentration in sugarcane juice under pot conditions, Sugar Tech, **18**(3), 258–265, (2016)
[28] M.C.G. Juenger and H.M. Jennings, New insights into the effects of sugar on the hydration and microstructure of cement pastes, Cement and Concrete Research, **32**(3), 393–399, (2002)
[29] H. Yildirim and B. Altun, Usage of molasses in concrete as water reducing and retarding admixtures, Indian J. of Engineering and Materials Science, **19**, 421-426, (2012)

Chloride Resistance Behavior on Nano-Metaclayed Ultra-High Performance Concrete

Mohd Faizal Md Jaafar[1,*], *Hamidah* Mohd Saman[2], *Muhd Norhasri* Muhd Sidek[2], *Noorli* Ismail[3], and *Nur Farhayu* Ariffin[1]

[1]Faculty of Civil Engineering and Earth Resources, Universiti Malaysia Pahang, 26600 Gambang, Pahang, Malaysia
[2]Faculty of Civil Engineering, Universiti Teknologi MARA, 40450 Shah Alam, Selangor, Malaysia
[3]Faculty of Civil and Environmental Engineering, Universiti Tun Hussein Onn Malaysia, 86400 Parit Raja, Johor, Malaysia

Abstract. Nowadays, the application of nano materials is getting attention to enhance the conventional concrete properties. The ultrafine particles of nano material also will help reducing the formation of micro pores by acting as a filler agent, produce a very dense concrete and will reduce the growth of micro pores in the UHPC structures. The introduction of nano materials in concrete is to increase the strength and durability of concrete. One of the most important factor affecting concrete durability is chloride penetrability. The purpose of this study was to evaluate the chloride permeability and chloride penetration depth by using rapid chloride permeability test (RCPT) and colorimetric method, respectively. The study was conducted for normal performance concrete (NPC), high performance concrete (HPC), ultra-high performance concrete (UHPC) and a series of UHPC (nano metaclayed-UHPC) incorporating different replacement levels of nano metaclay. All the tests were performed at ages of concretes from 3 up to 365 days. The results showed that the presence of silica fume in HPC and nano metaclay in UHPC reducing the Coulombs passed on the 56-days up to 365-days. The experimental results also revealed the depth of chloride ion penetration for nano metaclayed-UHPC concretes are much lesser than those concretes. As regards to the results, nano metaclay led to noticeable benefit in term of chloride resistance.

1 Introduction

The most important characteristic of concrete is believed to be the one that affecting its durability of concrete. Nowadays, concrete with excellent performance in strength has become one of the most important construction materials. Conversely, it can be misleading to spontaneously assume that such concrete would also possess high performance qualities in marine environment [1]. A lot of deterioration is reported for marine construction since marine environments are very aggressive as well as sea water consists mainly of chlorides and sulphates [2]. As well-known, the major cause for degradation of reinforced concrete

* Corresponding author: faizaljaafar@ump.edu.my

(RC) due to chloride penetration has been the great research effort until today. RC is usually durable, which has resulted in its widespread use for the construction of structures. However, it has become increasingly apparent that attack by aggressive agents such as chloride penetration can significantly affect a structure to deteriorate. Chloride ions can penetrate through the concrete pore network, potentially leading to the corrosion of steel reinforcement in the initiation stage [3]. The corrosion of reinforcing steel in concrete due to chloride penetration in RC structures has received increasing attention in recent years due to its widespread occurrence and the high cost of repair [4, 5].

It is important to note that the main feature of producing superior quality concrete is to minimize the volume of void in the concrete mass to have denser, stronger and more impermeable concrete. The interest in dealing with nano-cement and understanding of the hydration of cement particle to produce high strength concrete will be promised opportunity. Regards to this, the ultra-high performance concrete (UHPC) has been introduced as a new class and one of the most advanced concrete which it has better characteristics in term of strength and durability [6-8]. The use of industrial or agricultural by-product substitutions for cement has greatly contributed to boost up the performance characteristics of the concrete [9-11]. Therefore, the UHPC incorporating material as supplementary cementitious materials (CSM) and nano material has been developed. Previous literatures regard to the improved performance of UHPC containing pozzolanic cement enhanced the strength and increase the durability of concrete are well-documented in Schmidt et al. [12].

Numerous studies outlined that the use of nano-silica in UHPC can promote the hydration of cement to generate more C–S–H gel thus, leads to reduction of capillary pores in concrete [13-16]. Another researcher also found that UHPC contains ultra-fine POFA exhibited the greatest improvement in permeability properties such as porosity, rapid chloride permeability, water permeability and gas permeability [17]. In the same way, the chloride penetration in nanoclayed-UHPC were significantly improved its chloride resistance [18-20]. But, there are still few studies attempting to investigate chloride penetration in UHPC incorporating nano clay. The interest in using nano metaclay in UHPC production has been projected in this present research.

Regards to this, the ability of UHPC to resist chloride ion penetration is a key element for durability of concrete in salt-exposed environment was looked into. The rapid chloride permeability test (RCPT) results have been used as an indicator for concrete durability evaluation. However, several researchers have argued on this test when concrete produce with supplementary cementitious materials. For example, the used of silica fume in concrete mixture can lead to an order of magnitude reduction in Na+, K+, Ca+ and OH- ion concentration in pore solution [21-23]. Therefore, it is not appropriate to use RCPT charge passed alone to evaluate chloride ions penetrability in concrete incorporating SCM [24].

However, Nokken and Hooton [25] argued with this outcome and they found that the effect of using SCM in concrete is small. Therefore, additional method which is less dependent on the electrical conductance and pore solution is required to be validated. As a comparison with previous researches [18, 19], this study aimed to evaluated the chloride penetration profiles in various concrete types which are normal performance concrete (NPC), high performance concrete (HPC), ultra-high performance concrete (UHPC) and a series of UHPC integrating nano metaclay as cement replacement. For that reason, colorimetric method by using silver nitrate ($AgNO_3$) solution was performed other than RCPT.

2 Experimental program

2.1 Materials

In this research, six (6) series of mixtures consists of normal performance concrete (NPC), high performance concrete (HPC), ultra-high performance concrete (UHPC) and UHPC with different percentage of nano metaclay (nano metaclayed-UHPC) were prepared. An ordinary Portland cement (OPC) Type I provided by local supplier was used as a binder for all mixes. The properties of OPC are equivalent to BS EN 197-1:2000 specification. The HPC mixture was prepared by replacing 10% silica fume (SF) from the total weight of OPC. Conversely, the nano metaclayed-UHPC mixture were formulated by replacing the OPC partly with different levels of nano metaclay at 1, 3 and 5% from the total weight of OPC used. The mixtures for nano metaclayed-UHPC were labelled as UHPC-1, UHPC-3 and UHPC-5 respectively. The preparation on calcination process; chemical composition and morphology examination on nano metaclay have been examined in previous studies [19]. The crushed gravel was used as the coarse aggregate with a nominal size of 10 mm. Meanwhile, fine aggregate passing 5 mm was used as sand. In addition, the hyper-superplasticizier (SP) namely Glenium ACE389 SURTEC was added in NPC, HPC, UHPC and nano metaclayed-UHPC mixes to obtain a desired workability [20]. Table 1 displays the mix proportion of concrete for all mixes.

Table 1. Mix proportion for NPC, HPC, UHPC and nano metaclayed UHPC concretes

Concrete Designation	Raw Materials (kg/m^3)							
	OPC	**SF**	**Metaclay**	**Aggregate**	**Sand**	**Water**	**SP**	**w/c**
NPC	380	-	-	995	815	190	2.85	0.5
HPC	416	46	-	1149	619	152	4.20	0.4
UHPC	800	-	0	433	800	160	11.52	0.2
UHPC-1	797	-	8	433	800	160	6.70	0.2
UHPC-3	776	-	24	433	800	160	6.98	0.2
UHPC-5	760	-	40	433	80	160	7.60	0.2

2.2 Preparation of specimens

The concrete cylinder specimens with 100 mm diameter and 200 mm height were prepared for determination chloride penetration and rapid chloride permeability test (RCPT). In the first part, the concrete specimens were cured in water up to 365 days before subjected to RCPT test. The procedures to prepare the sample was called as 2-day preparation. First, the specimens were removed two (2) days earlier from water curing before RCPT to be conducted. Then, the cylindrical specimens were cut into slice samples approximately 50 ±1 mm in thickness using water-cooled diamond saw. After cutting, the samples were left in room temperature before coated with waterproofing coating at the side of samples. The second day of preparation began with the desiccation process to water-saturate the specimens using desiccator chamber for 24 hours. The RCPT was performed after the samples cured for 3, 7, 28, 56, 92, 182 and 365 days.

The second part, the specimens for chloride penetration evaluation were taken out from water curing after cured for 7 days. Before exposed to 3% sodium chloride (NaCl) solution, the specimens were sealed at the top and bottom parts. In this research, waterproofing membrane type T-100 water-based PU bituminous was used to prevent the ingress of chloride ions at the both parts. After the coating membrane dried, the specimens instantly immersed in 3%NaCl solution for chloride exposure. All the specimens were immersed for

3, 7, 28, 56, 92, 182 and 365 days. At each interval of immersion, the concrete specimens were taken out and tested its chloride penetration.

2.3 Testing procedures

In this study, the testing method to determine the chloride penetration can be divided into two (2) comprises of rapid chloride permeability test (RCPT) and colorimetric method ($AgNO_3$ sprayed method). Under chloride permeability of NPC, HPC, UHPC and a series of nano metaclayed-UHPC specimens, the RCPT was adopted. It is widely accepted that the ability of concrete to resist ingress of chloride ions can be resulted by performance the RCPT [7]. The RCPT was conducted for determining the resistance of concrete to chloride ions penetration. The chloride penetration was accelerated into the concrete specimens by applying the direct current of 60 voltage for 6 hours. Then, the concrete permeability levels were classified based on charge passed as specified in standard as tabulated in Table 2. The RCPT procedures were confirming to ASTM C1202-05 specifications [26]. This test was measuring the amount of electrical current that passes through a 50 ±1 mm thick slice of concrete samples. The chloride penetration was accelerated into the samples by applying the direct current of 60- voltage for 6 hours. A sodium chloride solution (NaCl) is applied to one side of the slice and a sodium hydroxide (HCl) solution is applied to the other side. At the end of 6 hours, the samples were removed from the cells and the amounts of coulombs passed through the samples were recorded.

Table 2. Chloride permeability classes based on charge passed [26]

Charge Passed, Coulombs	Permeability Classes
> 4000	High
2000 – 4000	Moderate
1000 – 2000	Low
100 – 1000	Very Low
< 100	Negligible

On the other hand, the chloride penetration depth profile of NPC, HPC, UHPC and a series of nano metaclayed-UHPC samples were determined by chemical method namely silver nitrate colorimetric method. Before the test was carried out, the concrete samples were taken out from 3%NaCl solution after immersed for 3, 7, 28, 56, 92, 182 and 365 days. The exposed samples were left in room temperature to allow it dried for a while. Afterwards, the samples were prepared by splitting the samples axially into two (2) sections using compression machine. Then, the freshly fractured surfaces split (cross-section) of samples was sprayed with 0.1M silver nitrate ($AgNO_3$) solution and left in temperature room until sprayed solution was dried. Subsequently after the sprayed surface was completely dried, white silver chloride precipitation on the surface was visible. While, the area free of chloride ion remains natural or brown in colour. In order to measure the profile depth of the chloride precipitation, the line along the surface that contain chloride was drawn. Then, the penetration depth of the chloride penetrability was measured using digital caliper.

3 Results and discussions

3.1 Rapid chloride permeability test (RCPT)

The results of RCPT is presented in Fig. 1. The figure demonstrates the charge passed through the concrete decreases as the age prolonged for all concretes. The chloride permeability indication can be classified based on the charge passed into concrete as in accordance with ASTM C1202 specifications. At early age, it was revealed that the chloride permeability for NPC concrete can be identified as high level of permeability (> 40000 Coulombs) as compare to those concrete mixes. After matured at 28-days, the chloride penetration in NPC concrete dropped up to 365-days. The Coulombs charge of the NPC concrete is reduced to 2676 Coulombs, indicating moderate chloride penetration resistance. The results also clear that the Coulombs charge through NPC is higher than HPC, UHPC and nano metaclayed-UHPC.

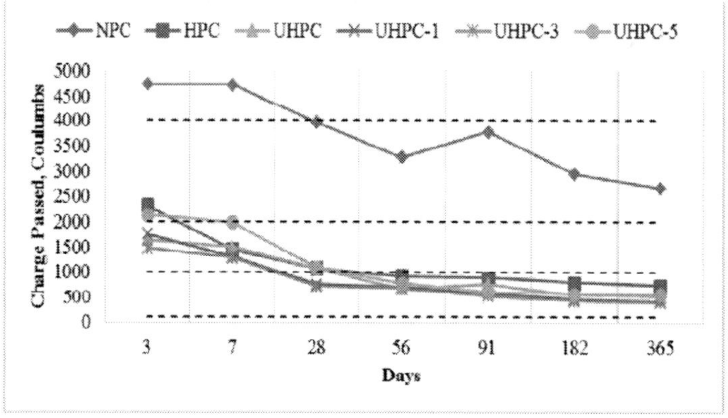

Fig. 1. Charge passed measured through NPC, HPC, plain UHPC and nano metaclayed-UHPC concretes

It was observed that, the presence of silica fume in HPC and nano metaclay in UHPC concretes contribute a benefit in terms of resisting chloride penetration. The incorporation of silica fume and nano metaclay in HPC and UHPC respectively, results in drastic reduction in the coulombs charge. The charge passed for HPC, UHPC and nano metaclay in UHPC concretes are reduced from low penetration to very low penetration of chloride from day-3 to day-56 respectively. In general, it is also revealed that small charge changes in Coulombs passed were observed for HPC, UHPC and a series of nano metaclayed-UHPC concretes after 28-days. This contrasting behavior occurs because of pozzolan-containing concrete increase more rapidly with increase in curing time than do those for the NPC concrete. On the other hand, it is also noted that higher w/c ratio results in higher porosity, so that the total charge passed for NPC concrete is more connected than the pore structure in low w/c ratio concrete.

On the other hand, incorporation of nano metaclay in UHPC concretes had superior chloride resistance other than HPC concrete. The Coulombs passed of the UHPC and a series of nano metaclayed-UHPC slightly decreased after day-56. Among the UHPC concrete, the different is insignificant but the trend is still the same. It was expressed that incorporating 1, 3 and 5% nano metaclay in UHPC can decrease RCPT results up to 76.11%, 72.67% and 73.59% respectively. Previous research also found that the volume of chloride that penetrated into the UHPC is extremely low [7]. These results are consistent with finding from other researchers, where they reported the UHPC have a discontinuously capillary porosity and thus exhibit significantly decreased the chloride permeability [27, 28]. In addition, among the series of nano metaclayed UHPC mixes, UHPC incorporating 1% nano metaclay (UHPC-1) has been proved the lowest chloride penetration compared with other nano metaclayed-UHPC mixes. Incorporating different percentage of nano metaclay in UHPC did not improve greatly the chloride resistance for all the nano metaclayed-UHPC from day-56 up until day-365. However, the reaction of 1% nano metaclay has the anticipated effect on the penetration characteristics of nano metaclayed-

UHPC. It is believed that, the good chloride resistance of nano metaclay blended concretes is due to the refinement of the pore structure. Previous research also stated that nanoclay in UHPC is able to acts as an ultra-filler material and yields a significant reduction in the total charge passed [19].

3.2 Chloride penetration depth profile

The results highlighted the influence of silica fume in HPC and nano metaclay in producing UHPC concretes to chloride penetrability using colorimetric method. The chloride penetration depth profiles with respect to expose days in 3% NaCl solution are graphically presented in Fig. 2 for all concrete mixes. Overall, it was found that the depth of chloride ion penetration in the concrete increased gradually with the time of exposure. Later, the chloride penetration depth decreased after expose for 365-days. As expected, the amount of chloride can be penetrated through NPC concrete recorded highest penetration depth compare to those other concrete series. It was noted that, the depth of the chloride penetration in NPC is significant because of the w/c ratio was high thus, results in higher pore size distribution. It is also clear that w/c ratio is the governing factor for chloride ion penetration and is mostly dependent on concrete pore structure. Fig. 2 also shows the depth of chloride ion penetration was high for NPC and HPC from day-3 to day-91. Later, chloride depth in HPC was higher as compare to plain UHPC and UHPC with 1% and 3% of nano metaclay. In this stage, it is noticed that silica fume had a significant effect on reducing the chloride penetration due to the pozzolanic and micro-filler effect. This finding was also agreed by [29], who found that depth of chloride penetration in concrete made with silica fume was became zero at 91 days.

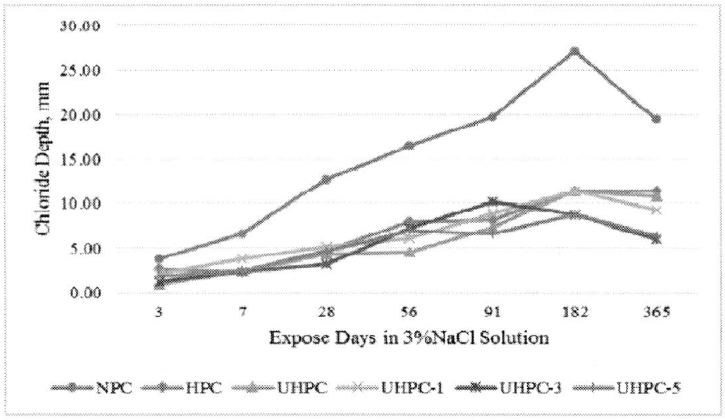

Fig. 2. Chloride penetration depth in NPC, HPC, UHPC and nano metaclayed-UHPC concretes

For UHPC concretes, it is revealed that as the nano metaclay level increases, there is a very marked reduction in the chloride resistance as compare to UHPC alone after 91-days. Due to its ultrafine size, nano metaclay particles fill the voids and reduce the capillary pores. At this stage, the hydration of UHPC concrete with nano metaclay have been also more advanced as a result of the additional hydration and pozzolanic reaction. The denser nano metaclayed UHPC would be less vulnerable to the ingress of the chloride in comparison to those at the age of 56-days. However, it has been observed that UHPC made with 3% nano metaclay improved significantly the chloride resistance at 365-days of exposure. From the present results, it is clearly publicized that nano metaclay in UHPC produces finer pore structure due to formation of the pozzolanic reaction within the capillary pore space. The pore system will also become finer and more segmented with increase in exposure time due to continuing pozzolanic reaction. As well documented in previous studies, the used of nano-silica in concrete will delay and reduce the chloride resistance [13-16].

3.3 Relationship on coulombs passed and chloride penetration depth

In order to investigate the role of nano metaclay in enhancing the chloride resistance of the UHPC concrete, the relationship between Coulombs passed and chloride depth was performed. Fig. 3 displays the relationship between Coulombs passed and chloride penetration depth for series of nano metaclayed-UHPC concretes.

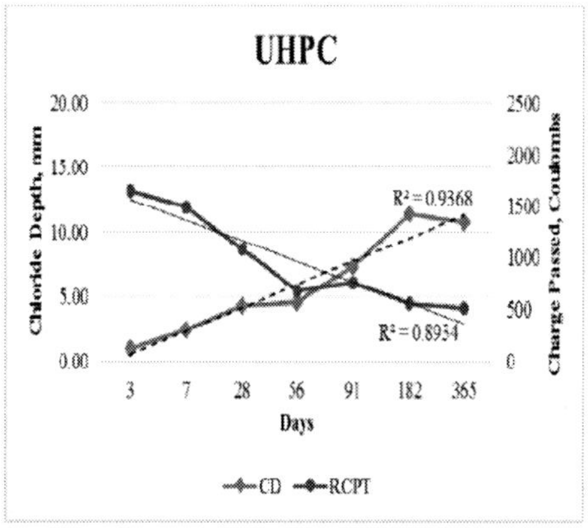

(a)

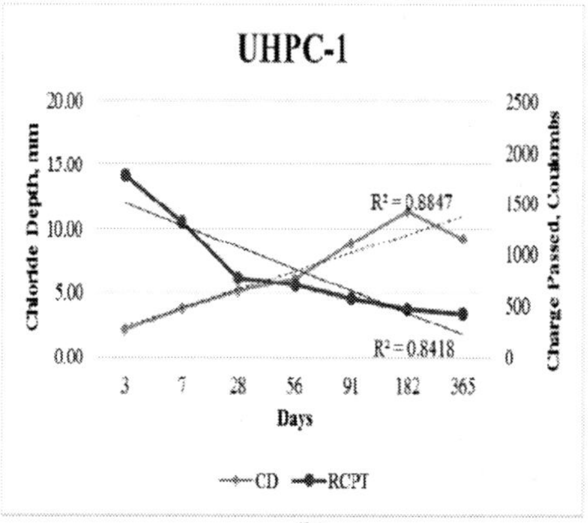

(b)

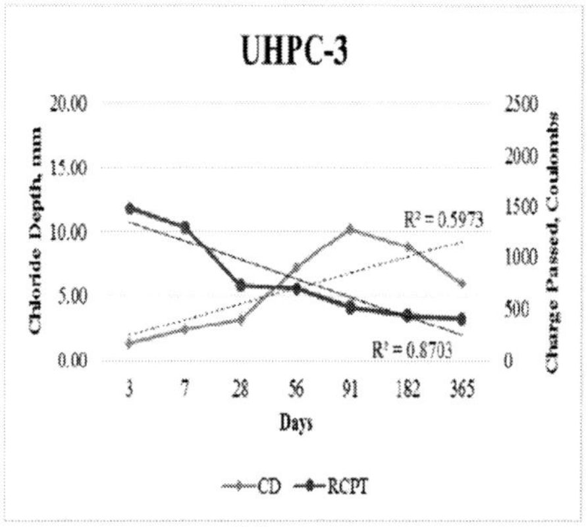

(c)

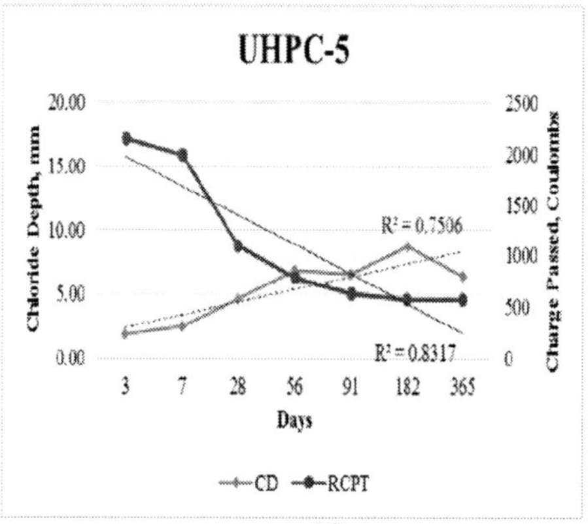

(d)

Fig. 3. Relationship between chloride depth and charge (Coulombs) passed for UHPC and nano metaclayed-UHPC

In general, the performance of chloride resistance of concrete can be improved by obtaining a discontinuous capillary pore structure caused by lowering w/c ratio or the presence of pozzolanic materials. In this case, the w/c ratio used for UHPC and nano metaclayed-UHPC concretes were constant at 0.2. From the graph, it was revealed that the optimum performance of nano metaclay to contribute in the pozzolanic reaction was recorded at day-56 for UHPC-1 and UHPC-5. However, the crossing line for UHPC was plotted just after 56-days and UHPC-3 was obtained nearly 56-days. This can give visual evidence that the degree of hydration in UHPC incorporating nano metaclay appeared considerably at day-56. Therefore, the filler effect and the pore volume refinement occurred at 56-days of age.

In order to verify the real influence of nano metaclay at 56-days, the relationship graph between chloride penetration and charge passed was performed as shown in Fig. 4. It is found that the different percentage of nano metaclay in UHPC has a strong linear with the chloride depth penetration. This mean 77.07% variation in chloride penetration depth profile has been explained by variation in the different levels of nano metaclay replacement and another 22.93% variation can be explained by other factors such as w/c ratio or age of curing. Meanwhile, the relationship between RCPT charge passed and various replacement levels of nano metaclay was not very strong [R-square = 0.6559]. This means that only 65.59% variation in charge passed has been explained by variation in the percentage of nano metaclay and the other of 34.41%can be explained by other factors. The graph indicates the optimum performance of chloride resistance for UHPC incorporating 1% nano metaclay was achieved on the day-56. Previous finding also found that the addition of metakaolin to concrete has a significant effect on to Coulombs passed and chloride penetration depth at early age [30]. Another researcher was also discovered that the use of nano metakaolin in concrete has reduced the capillary in concrete at 28 days of hydration [31].

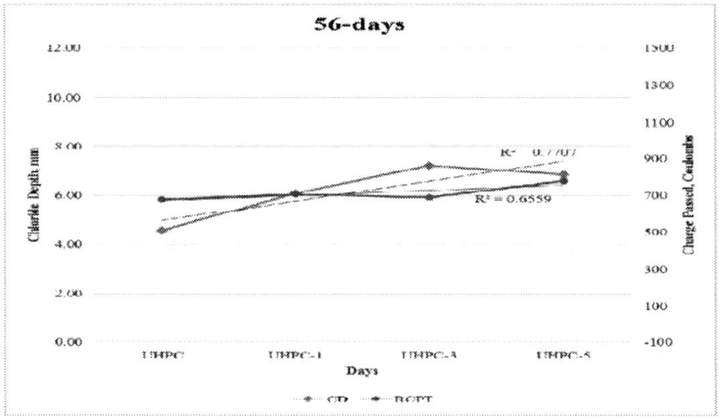

Fig. 4. Relationship on chloride depth and charge passed for UHPC and nano metaclayed-UHPC on the day-56

4 Conclusions

Based on the experimental results of this research, the following conclusions can be drawn:

i) It is deduced that the presence of silica fume and nano metaclay in concrete is effective in reducing Coulombs passed at 56-days up to 365-days. Incorporating nano metaclay in UHPC has a significant effect on the chloride permeability. From the results, 1% nano metaclay is found to be the optimum amount to be used in UHPC production. At this replacement level, the Coulombs passed value significantly reduced up to 76.11%.
ii) The depth of chloride ion penetration for nano metaclayed-UHPC concretes are much lesser than NPC, HPC and UHPC concretes. The capillary chloride penetration gradually decreased towards to 182-days. Contradicted to the results obtained from RCPT, it was found that inclusion of 3% nano metaclay significanty reduced the chloride penetration depth.
iii) The optimum chloride resistance behavior of nano metaclay in UHPC was achieved on the day-56. At this stage, it is noted that the high degree of the hydration process of nano metaclay was developed. It was concluded that the UHPC developed incorporating nano metaclay has significant potential on chloride resistance.

References

[1] O. Kayali and B. Zhu, Corrosion performance of medium-strength and silica fume high-strength reinforced concrete in a chloride solution, Cement and Concrete Composition, **27**, 117-124 (2005)

[2] M. Maes and N. De Belie, Resistance of concrete and mortar against combined attack of chloride and sodium sulphate, Cement and Concrete Composition, **53**, 59-72 (2014)

[3] C. Andrade, R. d'Andrea and N. Rebolledo, Chloride ion penetration in concrete: The reaction factor in the electrical resistivity model, Cement and Concrete Composition, **47**, 41-46 (2013)

[4] Federal Highway Administration, *Building more durable bridges*, FOCUS Federal Highway Administration, Washington DC, United States, (2001)

[5] BRE Centre for Concrete Construction, *Guide to the maintenance, repair and monitoring of reinforced concrete structures*, DME Report 4, Watford, United Kingdom, (2001)

[6] B.E. Graybeal and J.L. Hartmann, Strength and durability of ultra-high performance concrete, *3rd International Symposium on High Performance Concrete*, PCI, Orlando, (2003)
Florida.

[7] B.E. Graybeal and J. Tanesi, Durability of an ultrahigh-performance concrete, J. of Materials in Civil Engineering, **19**(10), 848-854 (2007)

[8] P.R. Prem, B.H. Bharatkumar and N.R. Iyer, Mechanical properties of ultra high performance concrete, Int. Scholarly and Scientific Research and Innovation, **6**(8), 676-685 (2012)

[9] R. Yu, P. Spiesz and H.J.H. Brouwers, Mix design and properties assessment of ultra-high performance fibre reinforced concrete (UHPFRC), Cement and Concrete Research, **56,** 29-39 (2014)

[10] A.S. El-Dieb, Mechanical, durability and microstructural characteristics of ultrahigh-strength self-compacting concrete incorporating steel fibres, Material and Design, **30**, 4286–4292 (2009)

[11] B.A. Tayeh, B.H. Abu Bakar, M.A. Megat Johari and Y.L. Voo, Mechanical and permeability properties of the interface between normal concrete substrate and ultra-high performance fibre concrete overlay, Construction and Building Materials, **36**, 538-548 (2012)

[12] M. Schmidt, E. Fehling, C. Geisenhanslüke, Ultra high performance concrete (UHPC), structural material and engineering series, *Proc. Symposium on Ultra High Performance Concrete*, **3**, Kassel, Germany, (2004)

[13] E. Ghafari, H. Costa, E. Júlio, A. Portugal and L. Durães, The effect of nanosilica addition on flowability, strength and transport properties of ultra-high performance concrete, Materials and Design, **59**, 1-9 (2014)

[14] N. Randl, T. Steiner, S. Ofner, E. Baumgartner and T. Mészöly, Development of UHPC mixtures from an ecological point view, Construction and Building Materials, **67**, 373-378 (2014)

[15] R. Yu, P. Spiesz and H.J.H. Brouwers, Effect of nano-silica on the hydration and microstructure development of ultra-high performance concrete (UHPC) with a low binder amount, Construction and Building Materials, **65**, 140-150 (2014)

[16] M. Alkaysi, S. El-Tawil, Z. Liu and W. Hansen, Effects of silica powder and cement type on durability of ultra high performance concrete (UHPC), Cement and Concrete Composites,**66**(10),47-56 (2016)

[17] A.N. Mohammed, M.A.M. Johari, A.M. Zeyad, B.A. Tayeh and M.O. Yusuf, Improving the engineering and fluid transport properties of ultra-high strength concrete

utilizing ultrafine palm oil fuel ash, J. of Advanced Concrete Technology, **12**, 127-137 (2014)
[18] M.J.M. Faizal, M.S. Hamidah and M.S. Muhd Norhasri, *Strength and chloride content of nanoclayed ultra-high performance concrete*, Structures, Materials and Construction Engineering, DAKAM Publishing (2014)
[19] M.J. Mohd Faizal, M.S. Hamidah, M.S. Muhd Norhasri, I. Noorli and M.P. Mohamad Ezad Hafez, Chloride permeability of nanoclayed ultra-high performance concrete, *Proc. of Int. Civil and Infrastructure Engineering Conf.*, Springer Singapore, 613-625 (2015)
[20] M.J. Mohd Faizal, M.S. Hamidah, M.S. Muhd Norhasri and I. Noorli, Effect of clay as a nanomaterial on corrosion potential of steel reinforcement embedded in ultra-high performance concrete, *Proc. of Int. Civil and Infrastructure Engineering Conf.*, Springer Singapore, 679-689 (2016).
[21] C. Shi, Effect of mixing proportion of concrete on its electrical conductivity and the rapid chloride permeability test (ASTM C1202 or ASSHTO T227) results, Cement and Concrete Research, **34**(3), 537-545 (2004)
[22] M.S. Ahmed, O. Kayali and W. Anderson, Evaluation of binary and ternary blends of pozzolanic materials using the rapid chloride permeability test, J. of Materials in Civil Engineering, **21**(9), 446-453 (2009)
[23] K. Krishnakumar, S. Bhaskar and K. Parthlbah, Evaluation of chloride in OPC concrete by silver nitrate solution spray method, Int. J. of Chemical Technology Research, **6**(5), 2676-2882 (2004)
[24] M.T. Bassuoni, M.L. Nehdi, and T.R. Greenough, Enhancing the reliability of evaluating chloride ingress in concrete using ASTM C1202 rapid chloride penetrability test, J. of ASTM International, **3**(3), 1-13 (2006)
[25] M.R. Nooken and R.D. Hooton, Dependence of rate of absorption on degree of saturation of concrete, Journal of Cememt, Concrete and Aggregates (CCA), **24**(1), 20-24 (2002)
[26] ASTM C1202-07, *Standard test method for the electrical indication of concrete's ability to resist chloride ion penetration*, American Society of Testing and Materials, Pennsylvania, (2007)
[27] C. Vernet, UHPC microstructure and related durability performance laboratory assessment and field experience examples, *Proc. of the Int. Symposium on High Performance Concrete*, Orlando, (2003)
[28] O. Bonneau, C. Vernet, M. Moranville and P.C. Aïtchin, Characterization of the granular packing and percolation threshold of reactive powder concrete, Cement and Concrete Research, **26**(11), 1639-1648 (2000)
[29] S. Ahmad, *Development on test methods for assessment of concrete durability for use in performance-based specification*, Master Thesis, University of Toronto, Canada, (2010)
[30] G. Dhinakaran, S. Thilgavathi and J. Venkataramana, Compressive strength and chloride resistance of metakaolin concrete, KSCE J. of Civil Engineering, **16**(7), 1209-1217 (2012)
[31] H. Shouky, M.F. Kokkata, S.A. Abo-El-Enein, M.S. Morsy, S.S. Shebl, Enhanced physical, mechanical and microstructural properties of lightweight vermiculite cement composites modified with nano metakaolin, Construction and Building Materials, **112**, 276-283 (2016)

Optimum Mix for Pervious Geopolymer Concrete (GEOCRETE) Based on Water Permeability and Compressive Strength

Salaheddin Abdulsalam Arafa[1], *Ahmad Zurisman* Mohd Ali[1,*], *Siti Nazahiyah* Rahmat[2], and *Yee Loon* Lee [3]

[1] Jamilus Research Centre, Universiti Tun Hussein Onn Malaysia, 86400, Batu Pahat, Johor, Malaysia
[2] Faculty of Civil and Environmental Engineering, Universiti Tun Hussein Onn Malaysia, 86400, Batu Pahat, Johor, Malaysia
[3] Faculty of Civil and Environmental Engineering, International Institute of Cambodia, Sangkat Chak Angre Krom, Khan Mean Chey, Phnom Penh, Cambodia

Abstract. The production of ordinary Portland cement (OPC) consumes considerable natural resources and energy, and it also affects the emission of a significant quantity of CO_2 in the atmosphere. This pervious geopolymer concrete study aims to explore an alternative binder without OPC. Pervious geopolymer concretes were prepared from fly ash (FA), sodium silicate ($NaSiO_3$), sodium hydroxide (NaOH) solution, and coarse aggregate (CA). The effects of pervious geopolymer concrete parameters that affect water permeability and compressive strength are evaluated. The FA to CA ratios of 1:6, 1:7,1:8, and 1:9 by weight, CA sizes of 5–10, 10–14, and 14–20 mm, constant $NaSiO_3$/NaOH ratio of 2.5, alkaline liquid to fly ash (AL/FA) ratios of 0.4, 0.5, and 0.6, and NaOH concentrations of 8, 10, and 12 M were the pervious geopolymer concrete mix proportions. The curing temperature of 80 °C for 24 *h* was used. The results showed that a pervious geopolymer concrete with CA of 10 mm achieved water permeability of 2.3 cm/s and compressive strength of 20 MPa with AL/FA ratio of 0.5, NaOH concentration of 10 M, and FA:CA of 1:7. GEOCRETE is indicated to have better engineering properties than does pervious concrete that is made of ordinary Portland cement.

1 Introduction

Concrete is the most used construction material worldwide, and ordinary Portland cement (OPC) is one of its most essential components. At present, Portland cement is an important material for manufacturing concrete. Concrete production fulfills the global demand of over 1.5 billion tons annually. The production of Portland cement releases a significant volume of carbon dioxide (CO_2) to the atmosphere, and it is also energy intensive [1, 2]. For each ton of synthetic Portland cement, one ton of CO_2 is estimated to be emitted into the environment. The process embraces high temperatures (1400–1500 °C) and quarry

* Corresponding author: zurisman@uthm.edu.my

demolition to extract raw materials, and the process causes the release of greenhouse gases such as CO_2 [3]. The costs related with these energy requirements are significant. Thus, further investigation is needed.

In addition to the necessity of large land area for the fly ash disposal, the use of OPC as the primary binder in every building construction is also a well-known environmental issue. Given the continuous growth of construction development, the use of cement materials in construction works is also increasing directly. This condition will increase the demand of OPC production throughout the world. This rising demand contributes roughly 7% of the total greenhouse gas emissions to the atmosphere [2]. Indirectly, this situation causes global warming that negatively affects the planet. Therefore, alternatives must be established to reduce OPC production.

Readily available commercial by-products, such as blast furnace slag and fly ash, have been adopted to meet these demands. By the 2010, roughly 780 billion tons of fly ash were produced [4]. Recycling will improve disposal and concrete manufacturing costs. A considerable electrical power supply is needed for various uses because of modern population growth and rapid development. One of the major sources of power generation is from thermal power plant (TPP). The by-products from these power plants typically contain a fine material type called fly ash and a coarse material type called bottom ash. Disposal of fly ash is a primary problem and requires considerable land area. As the cost and land requirement are increasing, these by-products should be reused for another purpose.

Davidovits [5] proposed that aluminum (Al) and silicon (Si) react with alkaline liquid from a geological source or by-product materials, such as rice husk ash and fly ash, to produce paste. Polymerization process is the chemical reaction referred to in this study, and the term geopolymer is used to describe the paste production. The production of geopolymer concrete does not use Portland cement.

Geopolymer is a type of inorganic polymer that is produced by the reaction of aluminosilicate materials with alkaline liquid [6]. Geopolymer has shown many excellent properties, such as good resistance against acid and sulfate attacks, high early strength, and good performance in high temperature [7-9]. Geopolymer does not use cement, and it reduces the emission of greenhouse gas by 44–64%. Furthermore, certain wastes and by-products, such as blast furnace slag and fly ash, are appropriate sources of aluminosilicate, which are used to produce geopolymer. Given that geopolymers are made from by-product materials, they have lower environmental impact in comparison to other types of concrete binders [10, 11].

The rapid growth of development has caused the imbalance of the storm water management in Malaysia. As natural areas are build up over urbanization storm, water runoff flow and volumes are significantly increased. These conditions are caused by the increase of impermeable land cover and decreased frequency of flooding that occurs downstream of the urban drainage system as well as the increased potential for stream bank erosion because of high flow velocities. Thus, the water quality is degraded in general and contributes to drastic changes in streambeds. The large percentage of impervious surface is noted to contribute to another form of pollution [12].

In this study, the main purpose is to design a concrete that allows water to pass through it (pervious concrete). Besides that, pervious concrete can be used as temporary storage for raining water and pavement. This type of pavement can reduce the skid resistance and provides safety to the users.

Compared to normal concrete, the pervious concrete is a special concrete that contains voids and possesses high water permeability. It has been developed as an environment friendly material that is used in permeable pavement, water purification, thermal insulation, acoustic absorption, and other applications in civil engineering and architecture [13, 14]. Generally, the void content is between 15% and 35% with sizes of connected pore in pervious concrete that ranges from 2 mm to 8 mm in diameter, and the compressive strength is between 2.8 MPa and 28.0 MPa. The high porosity is attained by highly interconnected void content. High void content of pervious concrete allow water to drain naturally in the range of L/min/m^2 81 to 730 L/min/m^2, and this can remove the normal surface-water drainage infrastructure, reduce the runoff from a site, and allow ground water recharge when conventional concrete does not [15]. It can be used in the pedestrian walkways, park areas, tennis courts, greenhouses, areas with light traffic, and other civil engineering and architectural works [15, 16]. Typically, pervious concrete consists of CA, binder, and admixture. Generally, the binding material that is used in pervious concrete is Portland cement, and further investigation on the use of other binding materials, such as geopolymer binders, in pervious concrete is still needed.

Therefore, this study focuses on the use of a fly ash as a source material, sodium silicate and sodium hydroxide solution as an alkaline solution, and CA for making pervious geopolymer concrete. The water permeability and compressive strength of the pervious geopolymer concrete were tested. The obtained data will undoubtedly be beneficial for the future use of fly ash geopolymer in the construction of pervious concrete. Thus, this will lead to the reduction of cement consumption and environmental problems.

2 Properties of materials used

The raw materials for pervious geopolymer concrete mixes are sodium hydroxide, sodium silicates (as alkaline solution), CA, and FA. Fly ash was collected from the Sejingkat power plant in Kuching Sarawak. The chemical composition of fly ash was determined with the use of X-ray (XRD) (SiO_2= 55.7 %, AL_2O_3= 24.22 %, Fe_2O_3= 9.1 %, K_2O =4.42 %, CaO= 2.45%, MgO=1.36%, TiO_2=1.33%, and less than 1% for P_2O_5, BaO, Na_2O, and MnO). Sodium silicate and sodium hydroxide were used as combination of alkaline solution. Sodium hydroxide in pelts forms with 98% purity. This agent was used and purchased from the local chemical supplier, and sodium silicate solution (Na_2O = 16.84%, SiO_2 = 35.01%, and water = 46.37% by mass) was used as alkaline liquid. Dissolving the pelts in water is prepared with sodium hydroxide solution. Tap water that is available in the laboratory was used to prepare the sodium hydroxide solution. The alkaline solution was prepared at least one day before use. Three sizes of CA ranging from either 5 mm to 10 mm, 10 mm to 14 mm, or 14 mm to 20 mm were used in geopolymer pervious concrete mixes. The specific gravity of the natural aggregate was 2.63. The water absorption of the natural aggregate was 2.76% in saturated surface, and a dry condition was used for making pervious geopolymer concrete.

3 Experimental program

The prepare geopolymer concrete mixtures were grouped into four, namely, M1, M2, M3, and M4. M1 group of mixtures was prepared to study the influence of aggregate content on pervious geopolymer concrete properties. M2 group of mixtures was prepared to study the influence of aggregate size on pervious geopolymer concrete properties. M3 group of mixtures was prepared to study the influence of the ratio of alkaline liquid to fly ash (AL/FA) on pervious geopolymer concrete properties. M4 group of mixtures was prepared to study the influence of sodium hydroxide solution concentration on pervious geopolymer concrete properties. An optimum proportion of aggregate content was obtained. Thus, the influences of other parameters were studied by making M2, M3, and M4 groups of mixtures.

4 Mix proportions, mixing, and casting

The optimum pervious geopolymer concrete mix was determined by casting pervious geopolymer concrete cubes in order to obtain the optimum ratios of FA/CA, size of CA, ratio of AL/FA, sodium hydroxide concentration (Molarity) based on compressive strength, and water permeability. Pervious geopolymer concrete was made with FA/CA ratio (1:6, 1:7, 1:7.5, 1:8, and 1:9), the sizes of CA (5–10, 10–14, and 14–20 mm. AL/FA ratio (0.4, 0.5, and 0.6), and sodium hydroxide concentration (8, 10, and 12 M). The $NaSIO_3/NaOH$ ratio was kept constant at 2.5 and cured for 24 h at 80 °C in an oven. The pervious geopolymer concrete mixing was done in a 25 °C controlled room. FA was mixed with alkaline liquid for 5 min in a pan-type mixer. CA was then incorporated and mixed for another minute. After being mixed, the pervious geopolymer concrete was molded in 100x100x100 mm cubes. The specimens were wrapped with a thin plastic sheet to minimize moisture loss before being cured for 24 h at 80 °C in an oven.

5 Testing detail

5.1 Compressive strength

All specimens casted in 100x100x100 mm cubes were molded and tested after 24 h by using a 250 KN capacity loading machine in displacement control at a rate of 1 mm/min until failure [17]. The mean value obtained from the 3 cubes was taken as the cube compressive strength for each pervious geopolymer concrete mix. The compressive strength of pervious geopolymer concrete is calculated by dividing the maximum load attained from the test by the cross-sectional area of the cube.

5.2 Water permeability

Permeability is an important parameter of pervious concrete because the material is designed to be used as drainage layer in pavement structures. The water permeability coefficient of pervious geopolymer concrete was tested by using the constant head method and carried out when a steady state flow was reached [18, 19]. The coefficients of water permeability (k) were the three specimen averages, and they were calculated by following Darcy's law as shown in the following Equation (1).

$$K = \frac{Q}{H} \quad (1)$$

where k is the coefficient of water permeability (cm/s), Q is the quantity of water collected (cm^3) over time t (s), L is the thickness of specimen (cm), H is the water head (h_2–h_1) (cm), and A is the cross-sectional area of the specimen (cm^2).

6 Results and discussions

6.1 Effect of fly ash to aggregate ratio on the permeability and strength of pervious geopolymer concrete

The influence of different FA/CA ratio on the compressive strength 24 h after the temperature curing period and the water permeability of pervious geopolymer concrete are presented in Figure 1. The compressive strength of pervious geopolymer concrete increases with a decrease in aggregate content. For example, the 24 h compressive strength with FA/CA ratios of 1:6, 1:7, 1:8, and 1:9 were 21.6, 19.8, 10.2, and 8.7 MPa, respectively. In addition, the significant increase in water permeability with increasing aggregate content could be observed in Fig. 1.

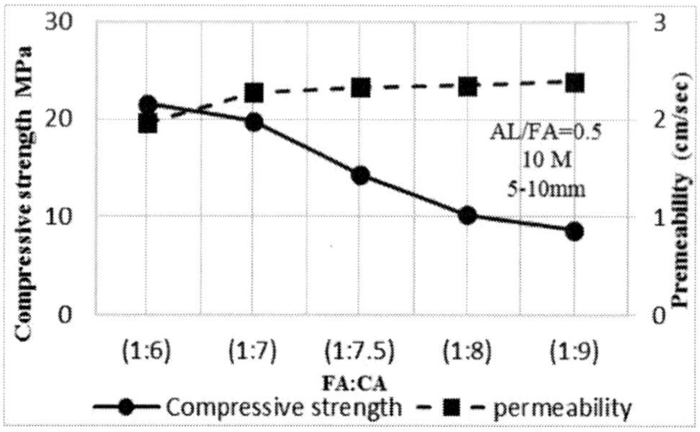

Fig. 1. The relationship between water permeability and compressive strength with different FA/CA percentage.

The optimum aggregate to fly ash ratio is shown in Fig. 1. Water permeability of 2.27 cm/s and compressive strength of 19.8 MPa were obtained, and the optimum with 1:7 aggregate to fly ash ratio was also achieved.

6.2 Effect of aggregate size on strength and permeability of pervious geopolymer concrete

Fig. 2 shows the influence of different aggregate size on the water permeability and compressive strength of pervious geopolymer concrete. The bulk density of aggregate is decreased with the increase in the aggregate size, and this in turn increased the porosity of the concrete. Mahboub et al. (2009) [20] reported that the porosity of pervious concrete can be estimated with the knowledge of the bulk density of the aggregate via the existing linear relationship.

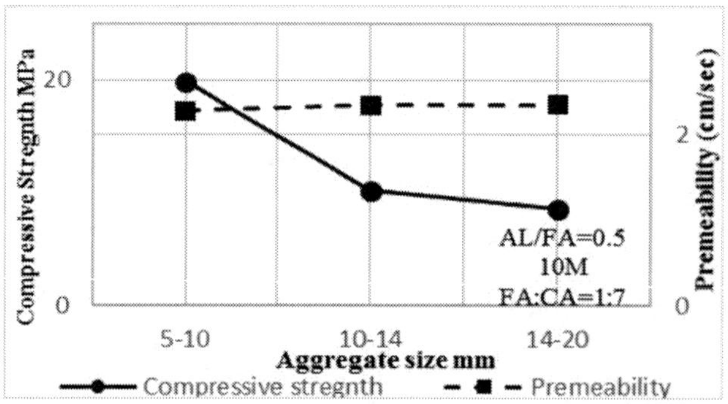

Fig. 2. The relationship between water permeability and compressive strength with different Aggregate sizes

The strength of pervious concrete decreases with the increase in the porosity. The variation in the permeability with water head is observed for pervious concrete with different aggregate sizes. The highest permeability of pervious geopolymer concrete with 14–20 mm aggregate size was 2.35 cm/s. The highest permeability had the highest porosity. The lowest permeability of 2.26 cm/s with the pervious geopolymer concrete with 5–10 mm aggregate size was recorded. The lowest permeability had the lowest porosity.

The relationship between compressive strength and permeability for pervious geopolymer concrete as a function of the water head is shown. As expected, the highest water permeability of pervious geopolymer concrete is when the compressive strength is the lowest. The variation in the permeability with water head is observed for pervious concrete with different aggregate sizes.

6.3 Effect of fly ash to alkaline liquid ratio on strength and permeability of pervious geopolymer concrete

As shown in Fig. 3, the compressive strength of pervious geopolymer concrete increased slightly with an increase in the AL/FA ratio. At AL/FA ratios of 0.40, the pervious geopolymer concrete contained low liquid and was difficult to compact. The pervious geopolymer concrete with high void contents and low compressive strengths was obtained. The compressive strengths of 16.4, 19.8, and 18.7 MPa were obtained for pervious geopolymer concrete with AL/FA ratios of 0.4, 0.5, and 0.6, respectively. With an AL/FA ratio of up to 0.5, the compressive strength increases, and beyond that, it decreases. This was consistent with the study conducted by Sathonsaowaphak et al. (2009) [21]. An increase in the compressive strength of the lignite bottom ash geopolymer mortar with the increase in AL/FA ratios from 0.325 to 0.429 was noted.

The water permeability coefficients of pervious geopolymer concrete were between 2.86 and 2.10 cm/s because of the different void contents. Generally, the water permeability coefficients in this study slightly decreased with increasing AL/FA. For instance, the water permeability coefficients for AL/FA ratios of 0.4, 0.5, and 0.6 were 2.86, 2.32, and 2.10 cm/s, respectively.

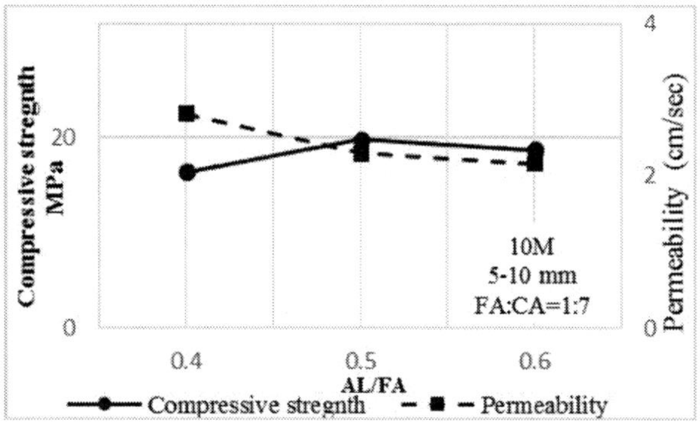

Fig. 3. The relationship between water permeability and compressive strength with different AL/FA ratios

In addition, the significant increase in water permeability with a decrease in the compressive strength is observed as shown in Figure 3. The highest water permeability was 2.86 cm/s with the lowest strength of 16.4, and the highest compressive strength was 19.8 MPa with the lowest water permeability of 2.10 cm/s. These values were obtained from AL/FA ratios of 0.4, 0, and 0.5, respectively. The water permeability increased with a decrease in the compressive strength. However, the good relationship between the compressive strength and water permeability was not because of the narrow void content. The addition of alkali liquid in the mixture increased the paste content, and the excess paste filled the voids. Thus, a relatively dense pervious concrete with low void content, low water permeability, and high strength is observed [22].

6.4 Effect of concentration of sodium hydroxide solution on strength and permeability of pervious geopolymer concrete

The compressive strength and water permeability of pervious geopolymer concrete with different NaOH concentrations but the same paste content are shown in Figure 4. The compressive strength of pervious geopolymer concrete that increased with the increase of NaOH molarity up to a value of 10 is shown in Figure 4, and when NaOH molarity is further increased, the compressive strength decreases. The optimum NaOH concentration in producing a good strength of pervious geopolymer concrete of 19.8 MPa was 10 M.

The water permeability of the pervious geopolymer concrete with different NaOH concentrations and the same paste content did not vary. The water permeability of pervious geopolymer concrete that increased with the increase of NaOH molarity is shown in Figure 4. For instance, the water permeability coefficients for NaOH concentrations of 8 M, 10 M, and 12 M were 2.21 cm/s, 2.28 cm/s, and 2.36 cm/s, respectively. A decreased NaOH concentration in the mixture increased the paste content, and the excess paste filled the voids. Therefore, a relatively dense pervious concrete with low void content and low permeability is observed.

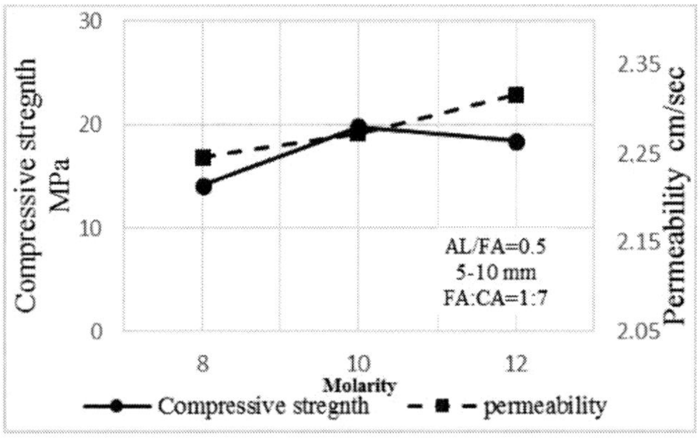

Fig. 4. The relationship between water permeability and compressive strength with different sodium hydroxide concentrations

The increase in water permeability with a decrease in the compressive strength is shown in Figure 4. Water permeability of 2.28 cm/s and compressive strength of 19.8 MPa were obtained with an optimum of 10 M.

7 Conclusion

Based on the compressive strength and water permeability measurement of pervious geopolymer concrete made with class F fly ash, activate alkali, and CA (with different sizes that range from 5 mm to 20 mm), the following conclusions were made:

i) Compressive strength and water permeability of pervious geopolymer concrete depend primarily on the porosity of the concrete; age, binder material type, FA/CA ratio, and CA size showed marginal influence on the strength and water permeability of pervious geopolymer concrete.
ii) This study shows the possibility of obtaining the optimum of compressive strength and water permeability from different FA/CA ratios, different sizes of CA, different AL/FA activator ratios, and different percentage of sodium hydroxide concentrations for pervious geopolymer concrete mixture.
iii) Reduction in the aggregate size decreased the porosity and water permeability of pervious geopolymer concrete, and thus, the strength of pervious concrete is increased.
iv) The FA/CA ratios had a significant effect on pervious geopolymer concrete properties. The increase in the aggregate portion resulted in the increase of pervious geopolymer concrete's porosity and water permeability. The increase was observed with a decrease in the compressive strength and density of the pervious geopolymer concrete.
v) The compressive strength increased with an increase in sodium hydroxide concentration up to a value of 10 M, and when NaOH is further increased, the compressive strength decreases, and it decreases with an increasing alkaline activator/fly ash ratio. In addition, NS/NaOH ratio, AL/FA ratio, and NaOH concentration did not have a significant effect on the water permeability of pervious geopolymer concrete.
vi) Based on the present study, a pervious geopolymer concrete with FA/CA ratio of 1:7, CA of 5–10 mm, NaOH molarity of 10, $Na2SiO_3$/NaOH ratio of 2.5, and AL/FL ratio of 0.5 when cured for 24 h at 80 °C gave optimum ratios for compressive strength of 19.8 MPa after temperature curing for 24 h and water permeability of 2.29 cm/s.
vii) The workability of the pervious geopolymer concrete in fresh state increases with the increase of AL/FL ratio, but it decreases with the increasing sodium hydroxide concentration.

This study was supported by Fundamental Research Grant Scheme (FRGS) (Vote 1523).

References

[1] D. Sumajouw, D. Hardjito, S. Wallah, B. Rangan, Fly ash-based geopolymer concrete: Study of slender reinforced columns, J. of Materials Science, **42**(9), 3124-3130 (2007)
[2] J. Davidovits, Geopolymer, green chemistry and sustainable development solutions, *Proc. of the World Congress Geopolymer*, Geopolymer Institute (2005)
[3] J. Fernandez, M. Ana, A. Palomo, C. Lopez-Hombrados, Engineering properties of alkali-activated fly ash concrete, ACI Materials J., **103**(2),106-112 (2006)
[4] D. Khale, R. Chaudhary, Mechanism of geopolymerization and factors influencing its development, J. of Materials Science **42**(3), 729-746 (2007)
[5] J. Davidovits, Geopolymer chemistry and properties, Geopolymer, **88**, 25-48 (1988)
[6] D. Kong, L. Sanjayan, G. Jay, K. Sagoe-Crentsil, Comparative performance of geopolymers made with metakaolin and fly ash after exposure to elevated temperatures, Cement and Concrete Research, **37**(12),1583-1589 (2007)
[7] D. Kong, L. Sanjayan, Damage behavior of geopolymer composites exposed to elevated temperatures, Cement and Concrete Composites, **30**(10), 986-991 (2008)
[8] A.Z. Mohd Ali, J. Sanjayan, The spalling of geopolymer high strength concrete wall panels and cylinders under hydrocarbon fire, MATEC Web of Conf, **47**, 1-5 (2016)

[9] A. Awal, M.H. Wan Ibrahim, A.Z. Mohd Ali, M. Zakaria, Mechanical properties and thermal behaviour of two-stage concrete containing palm oil fuel ash, International Journal of Geomate, (2017, in press)

[10] S.N. Rahmat, A.Z. Mohd Ali, M.H. Wan Ibrahim, N.A. Alias, Oil and grease (O&G) removal from commercial kitchen waste water using carbonised grass as a key media, *The 9th International Unimas Stem Engineering Conference,* **87**, 01010 (2017)

[11] B. Rangan, B. Vijaya, *Concrete construction engineering handbook*, 2nd eds., CRC Press, New York, United State (2008)

[12] A. Zakaria, N. Azazi, A. Ghani, R. Abdullah, L. Sidek, H. Kassim, A. Ainan, MSMA– A new urban stormwater management manual for Malaysia, *The 6th International Conference on Hydro-Science and Engineering* (2004)

[13] M. Kevern, T. John, V. Schaefer, K. Wang, Evaluation of pervious concrete workability using gyratory compaction, J. of Materials in Civil Engineering, **21**(12), 764-770 (2009)

[14] H. Kim, H. Lee, Acoustic absorption modeling of porous concrete considering the gradation and shape of aggregates and void ratio, J. of Sound and Vibration, **329** (7) 866-879 (2010)

[15] ACI committee 522, *Pervious concrete*, Report No. 522R-10, American Concrete Institute, Detroit, USA, **38** (2010)

[16] J. Tyner, S. Wright, P. Dobbs, Increasing exfiltration from pervious concrete and temperature monitoring, J. of Environmental Management, **90**(8), 2636-2641 (2009)

[17] C 39/C 39M-01, *Standard test method for compressive strength of cylindrical concrete specimens,* American Society for Testing and Materials, USA (2003)

[18] Japan Concrete Institute, Construction and recent applications of porous concrete, *Proc. of the JCI symposium on design*, 1–10 (2004)

[19] S. Park, B. Lee, Y. Jang, A study on the seawater purification characteristics of water-permeable concrete using recycled aggregate, Resour. Conserv. Recy., **(54)**, 658–65 (2010)

[20] K. Mahboub, C. Jonathan, R. Rathbone, T. Robl, B. Davis, Pervious concrete: Compaction and aggregate gradation, ACI Materials J., **106**, (6), 523-528 (2009)

[21] A. Sathonsaowaphak, P. Chindaprasirt, K. Pimraksa, Workability and strength of lignite bottom ash geopolymer mortar, J. of Hazardous Materials, **168**(1), 44-50 (2009)

[22] P. Chindaprasirt, S. Matakana, T. Chareerat, M. Mishima, Y. Yuasa, Cement paste characteristics and porous concrete properties, Constr. Build. Mater., **22,** 894–901 (2008)

Fresh and Hardened State of Polymer Modified Concrete and Mortars – A Review

Nurul Nadrah Aqilah Tukimat[1], *Noor Nabilah* Sarbini[2,*], *Izni Syahrizal* Ibrahim[2], *Chau Khun* Ma[2], and *Khairunisa* Mutusamy[1]

[1]Faculty of Civil Engineering and Earth Resources, Universiti Malaysia Pahang, 26300 Gambang, Pahang, Malaysia
[2]Department of Structures and Materials, Faculty of Civil Engineering, Universiti Teknologi Malaysia, 81310 Johor Bahru, Johor, Malaysia

Abstract. Polymer modified concrete or mortar is an alternative to the advancement of long serving civil engineering material - mortar and concrete. The excellence and promising benefits of modified composites have led to numerous progressive studies of its application. This paper presented a critical review from previous research on the polymer modified concrete and mortar. Both fresh and hardened state behaviours were reviewed as they are important for the development of excellent engineering material. Most of the applications of polymer modified concrete and mortar can be seen in diverse types of polymer such as latex, epoxy and emulsion. The utilization of each type of polymers resulted in different characteristics of composite concrete or mortar. Such applications have contributed to the improvement in terms of workability and mechanical strength, especially at higher grade of composite strength of concrete material.

1 Introduction

The application of polymer modified concrete has been started since a few centuries back. There are three types of polymer concrete, which are polymer concrete, polymer modified concrete and polymer impregnated concrete [1-4]. Polymer concrete can be defined as a composite material with its binder made of synthetic organic polymer, while the polymer-modified concrete consists of the combination of cement and polymer in certain proportion as binder [1]. The preparation for polymer type concrete, especially for polymer concrete and polymer modified concrete, is by mixing the polymer or monomer (in the form of powder or liquid) with aggregates for polymer concrete, and with the addition of cement for polymer modified concrete [2]. On the other hand, polymer impregnated concrete is a special concrete with its preparation needing the injection of low-density monomer into hydrated Portland cement, followed by the thermal catalytic polymerization process [5]. The polymerization process will normally occur during the mixing of concrete.

The use of polymer modified concrete and mortar as construction material might drastically increase the cost of a project; therefore it should be used for special conditions

* Corresponding author: noornabilah@utm.my

or occasions. The benefits of polymer modified concrete and mortar include superior properties, low labour cost and low energy requirements [1]. In details, the advantages of using polymer modified concrete and mortar have disguised its higher cost by providing a construction project with fast curing, excellent strength, durability, excellent damping properties and excellent elastic moduli [1]. These advantages have improved the reliance on conventional construction material, which has disadvantages such as delayed hardening, low tensile strength, large drying shrinkage and low chemical resistance [2].

Recently, the application of polymer in civil engineering fields is expanding. In all of its applications reported on its benefits of repair system for deteriorated reinforced concrete structures, strengthening or retrofitting of existing reinforced concrete structures, adhesive or bonding agents for delamination method, liquid-applied membrane waterproofing system and high grade redispersible polymer powders [6]. This has promoted a very broad alternative in repairing works in a row with other methods of strapping and conventional grouting [7-8]. Besides, it can also be applied as structural members such as flooring and roofing systems due to the development of the polymer itself, which can sustain stresses for heavier role purposes. On the other hand, epoxy latex inside a polymer concrete promises high adhesion, low setting time, post-setting shrinkage, creep resistance, higher acid and alkali resistance [1]. Epoxy is the most expensive polymer material, thus, it is normally used for special applications, such as those needing excellent bonding, or specific to repair works.

This paper reviews the behaviour of polymer modified concrete and mortars during the fresh and hardened state conditions. As in the plain concrete behaviour, the fresh state condition is focussing in the consistency and workability of the composite. Meanwhile, the hardened state condition focused on the mechanical strength of such composite.

2 Fresh state of polymer modified cement

Aggarwal [9] studied the effect of epoxy and acrylic emulsion to water to cement ratio of polymer modified mortar. The epoxy was prepared from bisphenol-A and amino-amide. The selected polymer to cement ratio varied from 10%, 20% and 30%, for both polymers. Through their study, it was found that acrylic decreased water requirement compared to epoxy, and reduced water to cement ratio with further amount of polymer to cement ratio. This is mainly due to the characteristics of polymer, in this case acrylic, which produces higher boundary surfactants due to low surface tension in a composite. This result in higher workability or better flow condition for acrylic compared to epoxy and normal mortar (at the same water content).

The flowability characteristics of polymer type mortar were also studied by Subham and Sing [10]. Two types of polymer were used in their work; styrene-butadine-rubber latex (SBR) and vinyl acetate/ethylene (VAE), which were added to mortar at 5%, 10%, 15% and 20%. A comparison between the two polymers showed that SBR exhibited higher workability compared to VAE. Additionally, the comparison between both polymers with conventional mortar still showed that the polymers had improved the workability. The higher percentage replacement (20%) SBR had improved the workability of polymer modified mortar by 100% from the flow value. This is due to the ability of SBR to improve the flowability of composite through excellent ball bearing action.

El-Dakroury [11] studied the inclusion of polyvinyl alcohol (PVA) with the addition of silica fume to the polymer modified mortar. The silica fume percentage was fixed at 10% cement replacement, while the polymer content of PVA varied from 3%, 5%, 7% and 10% by the weight of cement. The fresh condition of the polymer modified mortar showed that higher addition of PVA retarded the setting time of the polymer modified mortar. This is mainly due to the inclusion of PVA causing limited dissolved Ca^{2+} ion to produce $Ca(OH)_2$

into the solution. Besides, the inclusion of PVA decreased the water absorption of the polymer modified mortar at higher percentage of replacement. This observation was also previously discovered by Barluenga [12], whereby SBR inclusion retarded the initial and final time setting of the composite at higher addition of cement replacement.

The observations are aligned to the one reported by Zivica [13] and Ohama [2]. They stated that the use of polymer modified cement in concrete or mortar causes delayed setting time compared to the conventional one, due to increased polymer to cement ratio. However, this delay does not cause any difficulty or inconvenience to its application [2]. The main reason for the delay is inhibition of cement hydration due to the adsorption of surfactants. Although the flowability of the polymer modified cement is higher compared to the conventional cement, the resistance of the composite to bleeding and segregation is excellent. This benefit is due to the properties of most of the polymer, which are having air-entraining and water-reducing capabilities [2].

Water-soluble polymers help in improving the dry-out phenomena of polymer concrete. The dry-out phenomena from water-soluble polymer improved the viscosity of the water phase in polymer modified cement [2]. The viscosity improves by the existence of a very thin and water-impervious film in PMC. Therefore, water-soluble polymer is normally used in polymer mortar, especially as a binder material. Other previous works studied the effect of cellulose ether (CE) and polyvinyl alcohol (PVA) addition to mortar. Three major observations were made based on; (i) the mobility of pore water and polymers, (ii) volumetric changes, and (iii) reinitiated hydration of the cement. Through their research, three mechanisms that affect the composite are cement hydration, mortar volume change and reversible swelling of the latex films [14-15].

3 Hardened state of polymer modified cement: Mechanical strength and durability

The hardening process for polymer concrete involves the manipulation of polymerization process of the polymers or monomers inside the composite. However, polymer modified concrete has a very complicated relationship as both hydration and polymerization processes will occur inside the composite for the hardening process. Wagner [16] discovered that the cement hydration process precedes the polymerization process for polymer modified concrete, and thus, it should always be ensured that both processes proceed well to form an excellent monolithic composite.

Ohama [2] also reported that the inclusion of polymer latex to modified cement will result in less or no improvement in compressive strength compared to that of the conventional composite. However, he also reported that significant improvement due to such addition can be seen in terms of the flexural or tensile strength of the new composite. In general, there are various factors that can contribute to the strength of the composite. The main factors include the characteristics of polymer and other compositions within the new composite, the mixing proportion of polymer to cement ratio, water-cement ratio, binder-void ratio and the curing method. Table 1 summarizes the findings of previous research on the compressive strength of polymer modified cement. In general, the polymer content that have been used by researchers varied, but limited to less than 25% of cement replacement. The research which reported no improvement on compressive strength are those of Xin [17], Parghi and Shahria Alam [18], and Subham and Sing [10].

To that matter, the research works from Xin [17] and Subham and Sing [10] showed that both normal and high strength concretes gave similar observation due to the addition of polymer. Both used styrene-butadine rubber latex at 5, 10, 15 and 20% polymer content, and showed no improvement in the compressive strength of the polymer modified mortar. The non-improvement can be seen when, the compressive strength have dropped by a maximum of 67% for 20% polymer content in the study by Xin [17]; while for Subham and Singh [10], a maximum of 65% drop was also observed for 20% polymer content. Even though the compressive strength dropped, the workability improved due to higher addition of polymer content. This was also discovered by Xin [17] although they included 0.1% polyester fibre into the modified mortar.

On the other hand, Bhikshma [19] studied styrene-butadine rubber latex polymer concrete. A remarkable improvement was seen for normal and high strength polymer modified concrete. They reported that, at 10% polymer content, the improvement was at a maximum of 50% for normal strength concrete when compared to that of conventional, and 21% for high strength concrete. This shows that the improvement was significant when polymer was used to modify cement inside concrete. Other work from Oussama [20] reported an improvement of up to 207% for 16% polymer content in a modified concrete. The type of polymer that they used was bishenol-A epoxy resin. The significant improvement in compressive strength was also previously reported by Golestaneh [21]. They reported that the improvement was up to 18% for 15% polymer content using bishenol-A epoxy resin in polymer modified concrete with the addition of silica fume as filler.

The flexural strength of polymer modified cement was studied by many researchers as shown in Table 2. Some reported no improvement [17], while others reported marginal improvement [18-19, 21-22], compared to conventional cement. Basically, there are various factors influencing the strength of polymer modified cement. One of the factors is the type and amount of surfactants and antifoamers, and monomer ratio in a copolymer. Wang [22] discovered that the flexural strength of polymer modified mortar (styrene-butadine rubber latex) did not improve for more than 10% polymer content. More than this, polymer content is believed to have thickened the composite while reduced the workability and dropped the strength. Besides, Wang [22] also found that different curing methods severely affected the strength properties. From their work, it was found that the best type of curing method was the mixed method (wet and dry air curing), compared to wet curing alone [22].

In addition, concrete strength also affects the strength of the new polymer modified cement. This was discovered by Bhikshma [19], which found significant improvement on flexural strength for normal grade concrete, compared to high strength concrete. This scenario is closely related to the condition of the concrete. High strength concrete has the characteristic of porosity, compared to normal strength concrete. The porosity caused the addition of polymer does not result to any significant improvement in strength. However, this situation can be improved by using microfiller. This has been addressed previously [23-25]. Microfiller is a very fine powder, which can fill the gaps between aggregates and polymer; such as fly ash and silica fume [4]. When it is added to mortar or concrete, it will fill the gaps between other components and helps to produce denser material.

Despite having uncertainty on the mechanical strengths of polymer modified cement, studies on long term performance by several researchers showed remarkable improvement. Barluenga [12] reported compressive strength improvement of up to 56 days aged. While, optimum polymer content at 10% have showed remarkable improvement of up to 20%. Later, Xin [17] also studied the effect of polymer content on compressive strength for polymer modified mortar up to 90 days. The comparison showed strength improvement between 28 days and 90 days, which was more than 50% for maximum polymer content of 20%.

Table 1. Compressive strength of polymer modified concrete and mortars

Ref	Types of polymer	Compressive strength at 28 days			
		Control strength (N/mm^2)	Polymer content (%)	Compressive strength (N/mm^2)	Difference to control (± %)
[10]	Styrene-butadine rubber latex	23	5	20	-13
			10	14	-39
			15	11	-52
			20	8	-65
[11]	Polyvinyl alcohol	70	3	72	3
			5	90	29
			7	90	29
			10	100	43
[17]	Styrene-butadine rubber latex	55	5	27	-51
			10	28	-49
			15	31	-44
			20	18	-67
[18]	Polyacrylic esther	30	5	27.8	-12
			10	26.5	-17
			15	25	-24
			20	22.7	-33
[19]	Styrene-butadine rubber latex	30	5	35	17
			7.5	40	33
			10	45	50
		72	5	75	7
			7.5	80	14
			10	85	21
[20]	Bisphenol-A as epoxy	30	6	25	-17
			9	36	20
			13	85	183
			16	92	207
[21]	Bisphenol-A as epoxy	73	10	73	0
			15	107	18
			20	91	0
[22]	Styrene-acrylic polymer	28	5	34	21
			10	37	32
			15	40	43
			20	29	4
[26]	Multipurpose bonding agent for concrete	36	9	36	0
			12	38	6
			15	22	-42
	Multipurpose bonding agent for grout	63	9	63	0
			12	72	14
			15	64	-11
	Polyesther resin	55	9	55	0
			12	72	31
			15	63	-13

Table 2. Flexural strength of polymer modified concrete and mortars

Ref	Types of polymer	Polymer content (%)	Flexural strength	
			at 28 days (N/mm^2)	Difference to control (± %)
[12]	Styrene-butadine rubber latex	5	5.3	6
		10	5.35	7
		15	4.9	-2
		20	4.25	-15
		25	5.3	6
[17]	Styrene-butadine rubber latex	5	7	-7
		10	7.5	-9
		15	7.5	-9
		20	7.55	-8
[18]	Polyacrylic esther	5	5.32	33
		10	5.47	37
		15	5.59	40
		20	4.71	18
[19]	Styrene-butadine rubber latex	5	5.5	10
		7.5	6	20
		10	6.3	26
		5	8	14
		7.5	8.2	17
		10	8.2	17
[21]	Bisphenol-A as epoxy	10	14.3	0
		15	22.5	57
		20	10.13	-29
[22]	Styrene-butadine rubber latex	5	8.2	26
		10	8.4	29
		15	8.4	29
		20	8.6	32
[27]	Styrene-acrylic polymer	5	4.8	50
		10	5.4	69
		15	5.6	75
		20	4.2	31

The durability aspect of polymer modified mortar beam was studied by Bureau [28]. An aqueous emulsion of styrene-butadine polymer at 5, 7.5, 10, 12.5, 15 and 20% polymer contents were studied for their durability as a polymer modified mortar. Two main observations made were; less than 10% polymer content influenced the micro-cracks initiated in a composite, while more than 12.5% polymer content consolidated the composite and caused better viscous dissipation of stress redistribution when micro-cracks occur. Later, Pascal [29] also reported on the durability aspect of polymer modified mortar beam. They concluded that the main reason for increased bending capacity in a beam with higher polymer content is due to the continuous polymer network formation [29]. In that condition, the cement hydration process or compressive strength of the mortar had less influence on the bending capacity.

The styrene-butadine polymer network formation was studied by Yang [30]. Through the electro-migration test on a specimen, it was found that the polymer improved the composite against chloride penetration and its permeability. Besides, its network formation produced stiffer composite which led to decreased electric capacitance through the specimen. In general, the addition of polymer altered the internal microstructure and improved through the existence of stronger binding network between aggregate and polymer [30]. Next, this also contributed to reduce the portlandite content while mitigated the carbonation of such composite, and facilitated the chloride binding by the formation of calcium aluminate trisulfate hydrate [30].

The durability of styrene-butadine latex modified concrete at 15% polymer content was studied previously by Shaker [31]. A series of tests including water penetration, sorptivity, abrasion resistance and corrosion resistance were conducted as early at 28 days and up to a long term period of 12 months. Amongst their findings was that polymer modified concrete has the characteristic of denser microstructure due to smaller pores and excellent bonding between inside elements. This characteristic benefits the durability of the composite and gives better water tightness and higher resistance to abrasion and corrosion. This observation was applicable for both short term and long term periods for the specimens.

Redispersible polymer powders (RPW) is added during a dry mixed with other cement and aggregates components, followed by the addition of water [2]. The emulsified RPW will form during wet mixing (with during this time RPW will act like polymer latex). The advantages of using RPW are seen in improving the tensile strength of hardened mortar, and providing flexibility to the inside components [14]. Different types of polymer will give different characteristics. However, the weakness of adding RPW is obvious in terms of the efflorescence of the polymer concrete. Based on a previous study by Silva, the efflorescence process can occur when the acetate polymer undergoes alkaline hydrolysis due to the addition of water [32]. The product of alkaline reacts with Ca^{2+} ion from the cement to form calcium acetate, which is an organic salt causing efflorescence.

On the other hand, Zhu [33] studied the effect of adding vinyl acetate copolymer (VAC) and vinyl chloride terpolymer (VCT) to polymer modified mortar. The addition of 10% of VAC and VCT into polymer modified mortar prolonged the efflorescence time on the surface of the mortar specimen. Extremely, the use of VAC exacerbated the efflorescence of hardened mortar, while VCT exacerbated the efflorescence of fresh mortar. The observation concluded from the research was that VAC increases the soluble salts in hardened mortar and creates larger pores on the surface of the specimen. This condition has promoted the migration of inner salts to outer surface, and then caused the efflorescence. Meanwhile, VCT is a polymer which easily dissolves in the surface water and causes the efflorescence process to initiate during the fresh mortar state. This causes less Ca^{2+} ion being produced at later age and helps to prohibit the migration of K^+ and Na^+ ions, unlike VAC [33].

4 Recent innovation of polymer modified cement

The utilization of polymer waste micronized poly ethylene vinyl acetate (EVA) in mortar was studied by Brancher [34]. Though the compressive strength decreased due to the application, EVA waste contributed to producing more lightweight floor. This lightweight characteristic is due to the rises in void index through the use of EVA waste. Besides, this also contributed to noise attenuation transmitted through the floors and offer better acoustic comfort for residents if used inside a building.

Barbutta [35] reported different types of wastes that have been added as aggregates in polymer modified concrete. A certain proportion of argillaceous, calcareous, marble and fly ashes waste aggregates were added to 12.4% polymer content of epoxy resin with hardener. Scanning electron microscopy analysis showed that homogeneous composite was obtained for polymer modified concrete with calcareous and fly ash, however, non-homogenous composite was obtained from composite with argillaceous and marble wastes. The homogeneous polymer modified concrete with added calcareous and fly ash resulted in increased mechanical strength for the composite. Besides, through the observation, when using the wastes, the reduction in resin usage in a composite is possible without sacrificing its mechanical strength performance [35]. Other than that, properties of certain waste materials such as good thermal insulation as reported by Aminudin [36] may lead to other attractive research in mixing the waste into the polymer concrete or mortar.

The inclusion of discrete fibre reinforcement inside polymer cement is not new and has been addressed before [2]. However, the research and innovation focus in this work have just expanded recently. Kumar and Bari [37] reported that the used of basalt fibre to reinforce polymer modified concrete is possible and could produce excellent composite. The benefits and advantages of basalt fibre such as high tensile strength, non-corrosive, resistance to abrasion, light weight and higher chemical resistance can improve the efficiency of polymer modified concrete [37]. As previously mentioned, the application of wastes in polymer modified cement, basalt fibre shares the same mission in saving the world. The used of environmentally friendly materials and recycled materials could benefit mankind and optimize the material itself. Other recent work had also reported on the use of natural polymer modified cement [38-41]. The efforts of using any environmentally friendly material will benefit the world by sustaining civilization without sacrificing natural sources.

5 Conclusion

In conclusion, polymer modified concrete and mortars with different types of polymer will possess different characteristics during their fresh and hardened states. Most of the works reported on the enhancement of workability during the fresh state and promising strength at higher strength grade. The characteristics of polymer modified concrete and mortars such as high performance, effective adhesive, multi-functional and widely market available cause its usage to be more favourable. The current application with the incorporation of environmental friendly and natural sources as part of the composition shows a good sign of polymer innovation.

The authors would like to acknowledge Universiti Malaysia Pahang and Universiti Teknologi Malaysia and thankful to all personnel that contribute to this paper.

References

[1] O. Figovsky and D. Beilin, *Advanced polymer concretes and compounds*, CRC Press, Taylor & Francis Group, Boca Raton, (2014)
[2] Y. Ohama, *Handbook of polymer-modified concrete and mortars*, Noyes Publications, Park Ridge, New Jersey, USA, (1995)
[3] S.R. Bothra and Y. M. Ghugal, Polymer-Modified concrete: Review, Int. J. of Research in Engineering and Technology, **4**(4), 845-848, (2015)
[4] K.T. Varughese and B.K. Chaturvedi, Fly Ash as fine aggregate in polyester based polymer concrete, Cement and Concrete Composites, **18**(2), 105-108, (1996)
[5] A.S. Momtazi, R.K. Khoshbijari and S.S. Mogharab, Polymers in concrete: Applications and specifications, European Online J. of Natural and Social Sciences, Special Issue on New Trends in Architecture, Civil Engineering and Urban Studies, **3**(3), 62-72, (2015)
[6] M.A. R.Bhutta and Y.Ohama, Recent status of research and development of concrete-polymer composites in Japan, Concrete Research Letters, **1**(4), 125-130, (2010)
[7] C.K. Ma, A.Z. Awang, W. Omar, M. Liang, J. Siow-Wei and M. Azimi, Flexural capacity enhancement of rectangular high-strength concrete columns confined with

post-tensioned steel straps: Experimental investigation and analytical modeling, Structural concrete, **17**(4), 668-676, (2016)

[8] C.K. Ma, A.Z. Awang, R. Garcia, W. Omar and K. Pilakoutas, Behaviour of over-reinforced high-strength concrete beams confined with post-tensioned steel straps–an experimental investigation, Structural Concrete, **17**(5), 768-777, (2016)

[9] L.K. Aggarwal, P.C. Thapliyal and S.R. Karade, Properties of polymer-modified mortars using epoxy and acrylic emulsions, Construction and Building Materials, **21**(2), 379-383, (2007)

[10] P. Subham and Singh, Enhancement of proeperties polymer modified ferrocement, Int. J. of Recent Development in Engineering and Technology, **4**(8), 19-23, (2015)

[11] A. M. El-Dakroury, Mechanical and chemical properties of polyvinyl alcohol modified cement mortar with silica fume used as matrix including radioactive waste, Arab J. of Nuclear Science and Applications, **47**(2), 44-53, (2013)

[12] G. Barluenga and F. Hernandez-Olivares, SBR latex modified mortar rheology and mechanical behaviour, Cement and Concrete Research, **34**(3), 527-535, (2004)

[13] V. Zivica, The properties of cement paste with admixture of polyvinyl acetate emulsion, Bulletin RILEM, **28**, 121-128, (1965)

[14] A. Jenni, L. Holzer, R. Zurbriggen and M. Herwegh, Influence of polymers on microstructure and adhesive strength of cementitious tile adhesive mortars, Cement and Concrete Research, **35**(1), 35-50, (2005)

[15] A. Jenni, R. Zurbriggen, L. Holzer, and M. Herwegh, Changes in microstructures and physical properties of polymer-modified mortars during wet storage, Cement and Concrete Research, **36**(1), 79-90, (2006)

[16] H.B. Wagner, Polymer-Modified hydraulic cements, industrial and engineering chemstry, Product Research and Development, **4**(3), 191-196, (1965)

[17] D. Xin, Z. Xinping, C. Hanlin, W. Dingpeng, and Y. Wei, Mechanical properties of polyester fiber and sbr latex compound-modified mortar for tunnel grouting, Electronic J. of Geotechnical Engineering, **21**(4), 1365-1374, (2016)

[18] A. Parghi and M. Shahria Alam, Effects of curing regimes on the mechanical properties and durability of polymer-modified mortars- an experimental investigation, J. of Sustainable Cement-Based Materials, **5**(5), 324 - 347, (2016)

[19] V. Bhikshma, K. Jagannadha Rao and B. Balaji, An experimental study on behavior of polymer cement concrete, Asian J. of Civil Engineering (Building and Housing), **11**(5), 563-573, (2010)

[20] E. Oussama, G. Elhem, M. Valerie and B.O. Mongi, Mechanical and physical properties of epoxy polymer concrete after exposure to temperatures up to 250°C, Construction and Building Materials, **27**(1), 415-424, (2012)

[21] M. Golestaneh, G. Amini, G.D. Najafpour and M.A. Beygi, Evaluation of mechanical strength of epoxy polymer concrete with silica powder as filler, World Applied Sciences J., **9**(2), 216-220, (2010)

[22] R. Wang, P. Wang and X. Li, Physical and mechanical properties of styrene-butadine rubber emulsion modified cement mortars, Cement and Concrete Research, **35**(5), 900-906, (2005)

[23] R. Bedi, R. Chandra and S. P. Singh, Mechanical properties of polymer concrete, J. of Composites, **2013**, 1-12. (2013)

[24] K.T. Varughese and B.K. Chaturvedi, Fly Ash as fine aggregate in polyester based polymer concrete, Cement and Concrete Composites, **18**(2), 105-108, (1996)

[25] J.P. Gorninski, D.C. Dal Molin and C.S. Kazmierczak, Study on the modulus of elasticity of polymer concrete compounds and comparative assessment of polymer concrete and portland cement concrete, Cement and Concrete Research, **34**(11), 2091-2095, (2004)

[26] H. Abdel-Fattah and M.M. El-Hawary, Flexural behavior of polymer concrete, Construction and Building Material, **13**(5), 253-262, (1999)
[27] M.V.N. Sivakumar, Effect of polymer modification on mechanical and structural properties of concrete – an experimental investigation, Int. J. of Civil and Structural Engineering, **1**(4), 732-740, (2010)
[28] L. Bureau, A. Alliche, Ph. Pilvin and S. Pascal, Mechanical characterization of a styrene-butadine modified mortar, Materials Science and Engineering: A, **308**(1-2), 233-240, (2001)
[29] S. Pascal, A. Alliche and Ph. Pilvin, Mechanical behaviour of polymer modified mortars, Materials Science and Engineering: A, **380**(1-2), 1-8, (2004)
[30] Z. Yang, X. Shi, A. T. Creighton and M. M. Peterson, Effect of styrene-butadine rubber latex on the chloride permeability and microstructure of portland cement mortar, Construction and Building Materials, **23**, 2283-2290, (2009)
[31] F.A. Shaker, A.S. El-Dieb and M.M. Reda, Durability of styrene-butadine latex modified concrete, Cement and Concrete Research, **27**(5), 711-720, (1997)
[32] D.A. Silva, H.R. Roman and P.J.P. Gleize, Evidences of chemical interaction between eva and hydrating portland cement, Cement and Concrete Research, **32**, 1383-1390, (2002)
[33] H. Zhu, P. Wang, R. Wang and G.Zhang, Effects of two redispersible polymer powders on efflorescence of portland cement-based decorative mortar, Materials Science (Medziagotyra), **20**(3), 345-350, (2014)
[34] L.R. Brancher, M.F. de Onunes, A.M.C. Grisa, D.T. Pagnussat and M. Zeni, Acoustic behavior of subfloor lightweight mortars containing micronized poly (Ethylene Vinyl Acetate) (EVA), Materials, **9**(51), 1-9, (2016)
[35] M. Barbuta, M. Rujanu and A.Nicuta, Characterization of polymer concrete with different wastes additions, Procedia Technology, **22**, 407-412, (2016)
[36] E. Aminudin, M.F. Md Din, M.W. Hussin, A H. Abdullah, K. Iwao and Y. Ichikawa, Properties of agro-Industrial aerated concrete as potential thermal insulation for building, MATEC Web of Conferences, **47**, 1–9, (2016)
[37] M. A. Kumar and J. A. Bari, Flexural behaviour of polymer modified basalt fiber reinforced concrete - A review, Int. J. on Applications in Civil and Environmental Engineering, **1**(3), 1-5, (2015)
[38] A.S. Momtazi, R.K. Khoshkbijari and S.S. Mogharab, Polymers in concrete: applications and specifications, Special Issue on New Trends in Architecture, Civil Engineering, and Urban Studies, European Online J. of Natural and Social Sciences, **3**(3), 62-72, (2015)
[39] R.M.I. Retno Susilorini, H. Hardjasaputra, S. Tudjono, G. Hapsari, S,R. Wahyu, G. Hadikusumo and J. Sucipto, The advantage of natural polymer modified mortar with seaweed: green construction material innovation for sustainable concrete, Procedia Engineering, **95**, 419-425, (2014)
[40] C. Niveditha, K. Sarayu, A. Ramachandra Murthy, V. Ramesh Kumar and R.I. Nagesh, Marine algae for cement mortar strengthening, J. of Civil Engineering Research, **4**(2A), 23-25, (2014)
[41] Y.H. Jang, S.O. Han, I.N. Sim and H. Kim, Pretreatment effects of seaweed on the thermal and mechanical properties of seaweed/polypropylene biocomposites, Composites Part A: Applied Science and Manufacturing, **47**, 83-90, (2013)

Contribution of Fineness Level of Fly Ash to the Compressive Strength of Geopolymer Mortar

Firdaus[1,*], *Ishak* Yunus[1], and *Rosidawani*[2]

[1]Civil Engineering Department, Faculty of Engineering, Bina Darma University, Palembang 30264, Indonesia
[2]Civil Engineering Department, Faculty of Engineering, Universitas Sriwijaya, Palembang,30136, Indonesia

Abstract. The development of geopolymers has allowed the flash as the substitution of cement in the application of concrete. Therefore, this will be very useful considering the quite abundant by-product materials from power plants burning coal in South Sumatera. However, the untreated fly ash from the source caused its fineness level unpredictable, whereas the fineness of binder in cementitious material significantly affects the mechanical properties of the harden. Therefore, this study aims to determine the contribution of the fineness level of fly ash to the compressive strength of geopolymer mortar, as well as its excellent composition. Type F fly ash from Tanjung Enim Power Plant was treated by filtering to obtain different fineness levels based on the fall zones of the ash. Activators used in geopolymer mixing were sodium hydroxide (NaOH) and sodium silicate (Na_2SiO_3) with three activator/fly ash ratios which was 0.25, 0.35 and 0.45. The results showed that the fineness level based on fall zone as well as the activator to fly ash ratio significantly influenced the compressive strength of the geopolymer mortar. The compressive strength of the F4-P4 specimen of geopolymer mortar with zone-4 fly ash and an activator ratio of 0.45 achieved 28.2 MPa at 28 days.

1 Introduction

Concrete, as the most favorite material, increases in its usage through the years and requires the availability of seemingly unlimited amounts of each composition material. Consequently, the increasing concrete usage results in high demand for cement. As known, the cement industry is one of the main contributors to CO_2 emissions in the air. One ton of cement produced causes one ton of CO_2 emitted into the air. Carbon dioxide emissions are causing global warming, which causes climate change. For this reason, the research for partial and full cement substitution in concrete has gained a lot of attention in the past decades.

Cementitious materials such fly ash, silica fume, slag, rice husk ash, metakaolin and others are some alternatives that have been proven to be viable as a partial substitution for cement. Fly ash as a waste material of coal combustion is the most available compared to

* Corresponding author: firdaus.dr@gmail.com

other materials. Commonly, fly ash addition to concrete is about 10-15%. However, the use of more than 50% fly ash in concrete has been developed and demonstrated good results in previous research [1, 2]. Although the quality of the concrete did not increase significantly, it still showed excellent mechanical properties.

Utilization of 100% fly ash in substituting cement as a binder has potential in concrete production [3-5]. Fly ash, which combined with a type of activator that contains silica is engineered as an adhesive material through a chemical reaction and produces a new kind of material called a geopolymer. Davidovits [6, 7] pioneered this technique and has presented promising results of geopolymer technology as an alternative for an adhesive cement matrix. Geopolymer binder is derived from natural ingredients and undergoes a polymerization reaction in the hardening process. Natural materials are preferred as a replacement for cement and should contain silica and alumina oxide [7]. The need for a high content of silica and alumina oxide are the key ingredients that undergo polymerization since it will produce a binder in the geopolymer concrete.

Furthermore, the development of geopolymer as an adhesive which can replace cement matrix is a wide open research area. Therefore, optimization of the characteristics of fly ash from different sources in its use as a natural geopolymer ingredient still requires attention. The physical and chemical properties of the final fly ash product affect the geopolymer matrix. Moreover, the type and dosage of activators used can produce a geopolymer matrix has the possibility to optimize this kind of excavated material product. So far, research results have shown that geopolymer concrete has very remarkable properties in strength and performance [3,5,8].

The exploration of geopolymer concrete with a variety of features for application in building construction still has a long way to go. Characteristics of the manufacturing process, particularly the technical properties of strength and durability in aggressive environments, as well as the behavior of structural elements using geopolymer concrete are still being developed. One of the manufacturing processes is a treatment of the fly ash as the primary material of geopolymer. This research results how the treatment of the fly ash in fineness level influences the mechanical properties of the geopolymer mortar.

2 Experimental program

2.1 Filtering of fly ash

The fly ash was treated based on the fall zone parameter, which indicates the fineness of the fly ash. The fall zone parameter was modified with fly ash filtering tool, which is shown in Fig. 1.

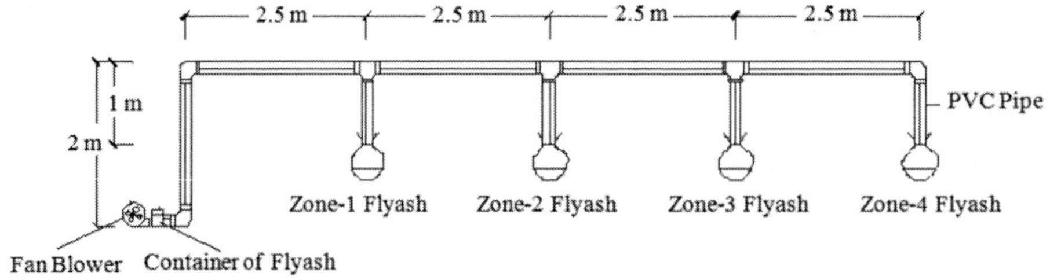

Fig. 1. Fly ash filtering based on fall zone.

The fineness of the fly ash samples was determined by four equally divided fall zones based on distance. The fell fly ash in each zone was used as an ingredient in the mixing of different geopolymer mortar specimens. Fly Ash employed in this work is procured from Tanjung Enim Power Plant in South Sumatera Indonesia and classified as the type F.

2.2 Properties of fly ash

Fly ash inspection included the examination of the result of chemical tests with the X- ray fluorescence (XRF) method to find out the chemical and mineral structures of the fly ash. Moreover, the particle size analysis (PSA) was used to determine the grain size of the fly ash, and the X-ray diffraction (XRD) testing to determine the mineral elements and relative composition of the fly ash. Table 1 gives the XRF analysis results.

Table 1. XRF analysis data of fly ash composition.

Oxides	SiO_2	Fe_2O_3	CaO	Al_2O_3	MgO	SO_3	Na_2O	K_2O	TiO_2	P_2O_5
Zone 0 (%)	43.61	4.59	3.63	20.75	1.00	0.96	2.23	0.51	0.68	0.20
Zone 4 (%)	55.40	5.24	3.71	27.88	1.24	0.35	2.26	0.56	0.79	0.22

Table 1 shows that the fineness level significantly influences the mineral composition of the fly ash. The finest filtered fly ash (zone-4), contains the higher silica and alumina. These materials are indispensable in geopolymer concrete since they are synthetic inorganic compound aluminosilicates [7].

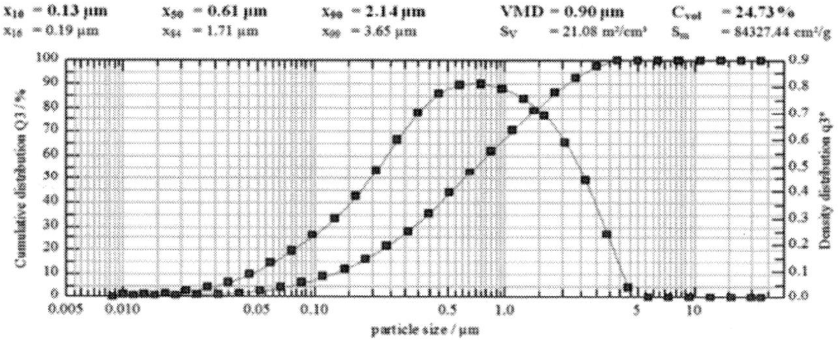

(a) Zone-0 fly ash

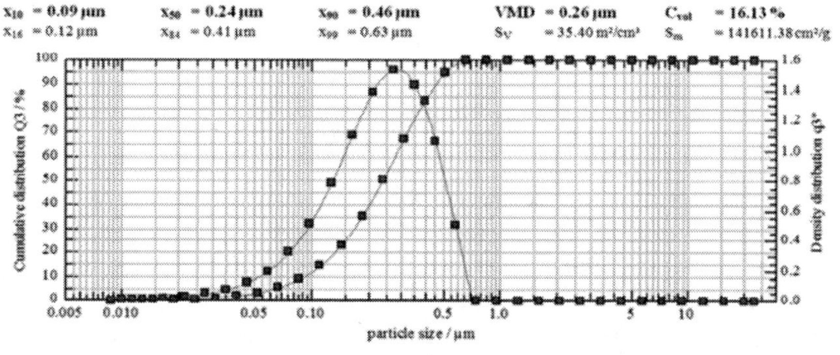

(b) Zone-1 fly ash

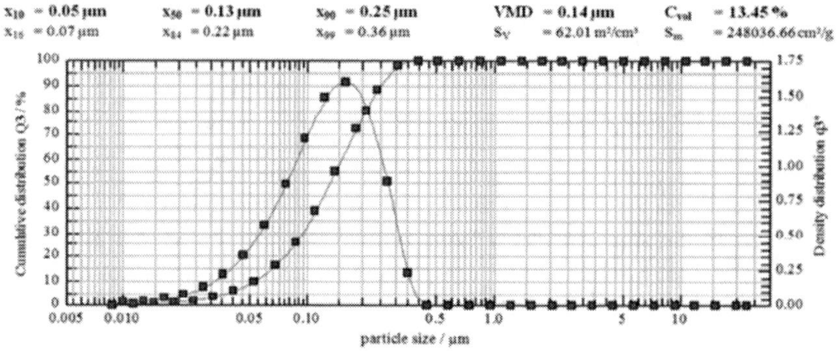

(c) Zone-2 fly ash

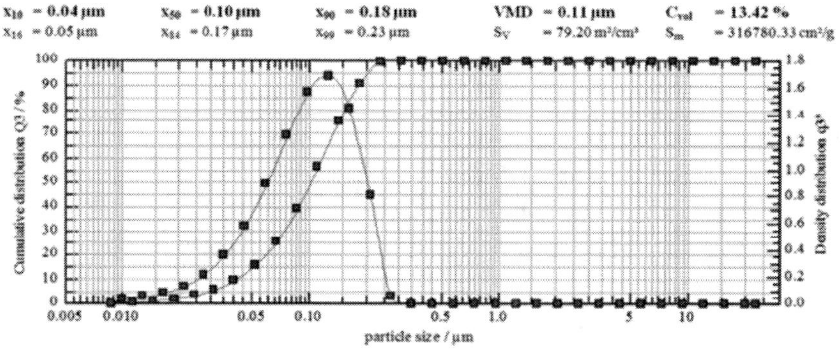

(d) Zone-3 fly ash

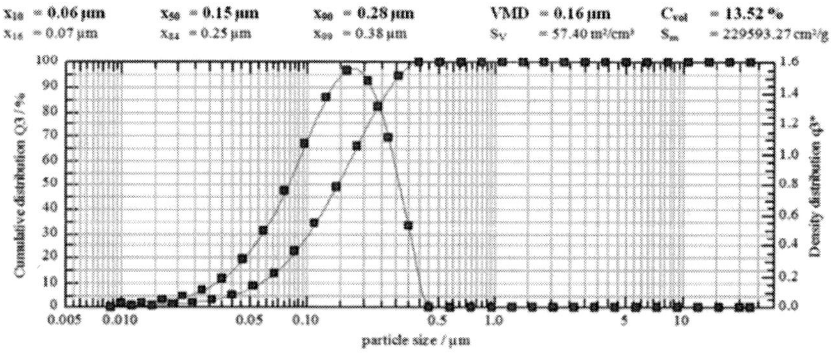

(e) Zone-4 fly ash

Fig. 2. Particle size analysis data of fly ash based on fall zones.

The results of the particle size analysis (PSA) of the fly ash specimens is drawn in Fig. 2. The grain size of unfiltered (original) zone-0 fly ash was 0.13 µm for X_{10} and 3.65 µm for X_{99}. Meanwhile, the grain size was 0.09 µm for X_{10} and 0.63 µm for X_{99}; 0.05 µm for X_{10} and 0.36 µm for X_{99}; 0.04 µm for X_{10} and 0.23 µm for X_{99}; 0.06 µm for X_{10}, and 0.38 µm for zone-1, -2, -3 and -4 fly ash respectively. The X_{10} and X_{99} values of the fly ash from each zone showed that the grain size was fine in the more distant zones. Therefore, this test

 DOI: 10.1051/matecconf/201710301026

show there is a significantly different fineness between untreated fly ash (Zone 0) and the most remote zone (Zone 4).

Fig. 3 (a-e) show the diffraction patterns from the XRD diffraction test of the fly ash. The higher mineral quartz and the finer fly ash produce the higher peak of diffraction. Besides, there was also quartz mineral hematite (Fe_2O_3), $Si0_2$, and $Al_6Si_2O_3$.

Those figures show the diffraction patterns of the fly ash specimens for each zone (0-4). The data clearly demonstrate the difference in composition of the fly ash for each level of fineness. Although all types of the fly ash had an amorphous semi-crystalline form, they differed among other in their degree of fineness.

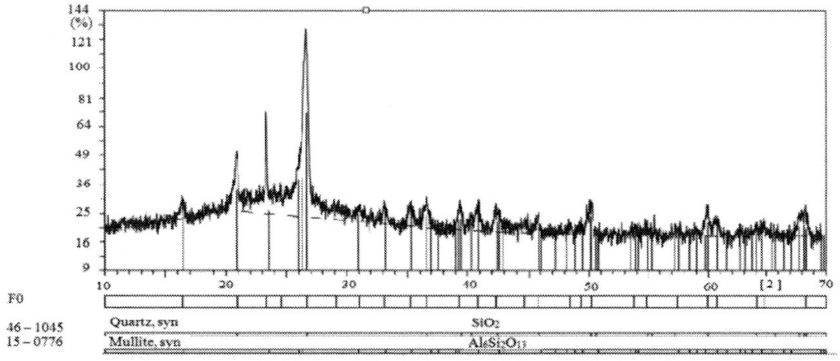

(a) Zone-0 fly ash

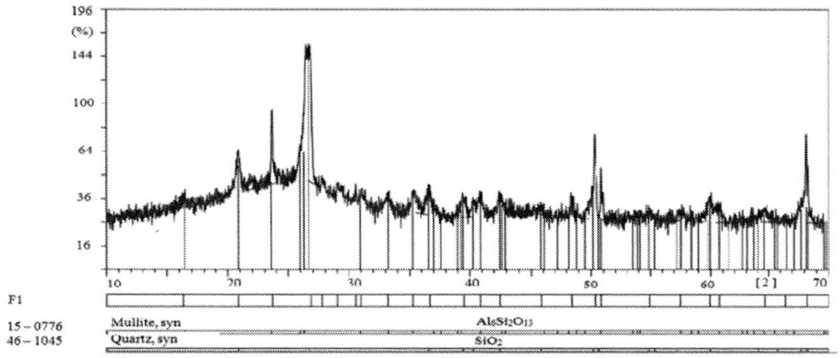

(b) Zone-1 fly ash

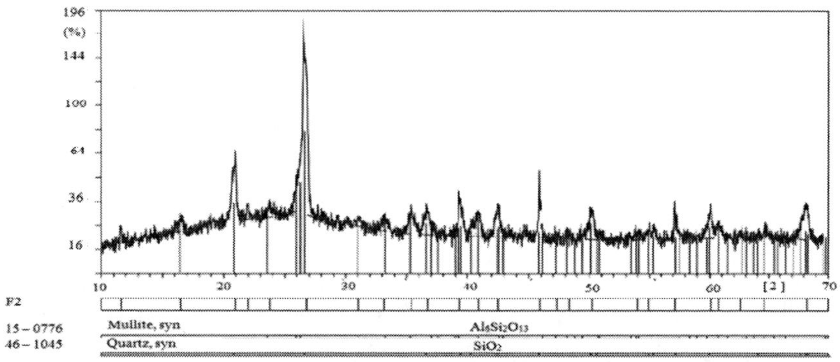

(c) Zone-2 fly ash

DOI: 10.1051/matecconf/201710301026

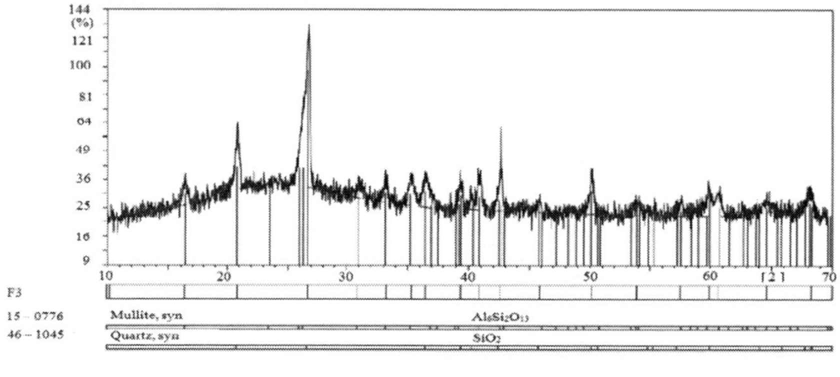

(d) Zone-3 fly ash

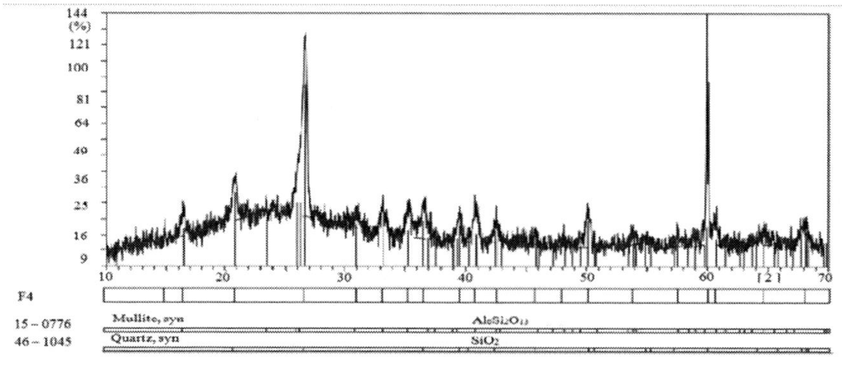

(e) Zone-4 fly ash

Fig. 3. XRD diffraction test of fly ash results based on fall zones.

2.3 Specimens preparation

The variables in this study consisted of five fly ash fineness levels (F0-F4), three activator/fly ash ratios (P2, P3, and P4) and three test ages (7, 21, and 28 days). All of the variables were compared according to the parameter of compressive strength.

Three mixtures of geopolymer mortar were prepared based on the optimum of trial mix design with the compositions which were differentiated by the amount of activator. Sodium hydroxide and sodium silicate were used as an alkaline activator to synthesize the geopolymer with different activator/fly ash ratios which were 0.25, 0.35 and 0.45. Table 2 shows the mix proportion for 1 m^3 mortar using water/fly ash ratio of 0.4.The high range water reducer (HRWR) with HRWR/fly ash ratio of 0.7% was used to improve the workability.

Table 2. Composition of geopolymer mortar.

No.	Material (kg)	Activator ratio (%)		
		0.25	0.35	0.45
1	Fly ash	408	408	408
2	Sand	652	652	652
3	Sodium silicate (water glass)	73	103	131.3
4	Sodium hydroxide (potassium)	28.5	41	52.1
5	Water	163	163	163
6	Superplastizer	2.8	2.8	2.8

3 Results and discussions

Compressive strength tests were conducted on the geopolymer mortar with sodium silicate and sodium hydroxide as an activator at varying ratios to fly ash of 0.25%, 0.35% and 0.45% (P2, P3, and P4) and varying fineness of fly ash (F0, F1, F2, F3, and F4). The compressive test was executed on day 7, 21, and 28, after curing is applied from one day after casting until one day before testing. Table 3 shows the test results of the average compressive strength of the geopolymer mortar specimens.

Table 3. Compressive strength of geopolymer mortar.

Age of Test (day)	Compressive Strength (MPa)														
	F0			F1			F2			F3			F4		
	P2	P3	P4	P2	P3	P4	P2	P3	P4	P2	P3	P4	P2	P3	P4
7	1.5	3.7	4.3	2.7	4.1	6.2	3.9	5.0	8.4	4.8	5.0	10.7	9.7	7.8	14.6
21	2.0	5.2	9.0	3.2	5.3	12.2	3.3	6.3	12.0	4.6	5.9	14.9	10.6	11.8	26.3
28	2.6	5.6	9.5	3.7	6.2	15.5	3.6	6.8	19.2	4.8	6.9	17.5	10.8	12.5	28.2

The curve of specimen F-P2 and F-P3 in Fig. 4 (a) and (b) indicates that there was a slight increase in compressive strength due to the test age. An insignificant increase of all the zone variables for each specimen was shown from the 7th to the 28th day. In contrary, the curve of F-P4 in Fig. 4(c) shows that the compressive strength of each specimen had increased significantly with the testing age. Each figure shows that the compressive strength of the geopolymer mortar from the most distant zone (F4) had the highest compressive strength. There was a significant increase in the compressive strength of the F4 zone compared to the original fly ash (zone 0). Fig. 5 clearly shows the increasing compressive strength on the 28th day of each percentage of activators.

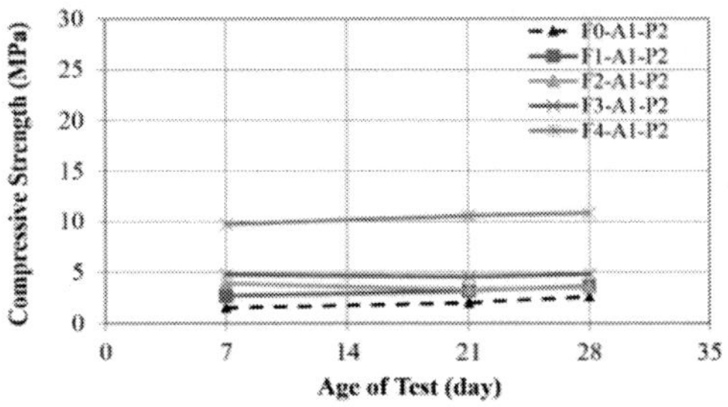

(a) Specimens F-P2

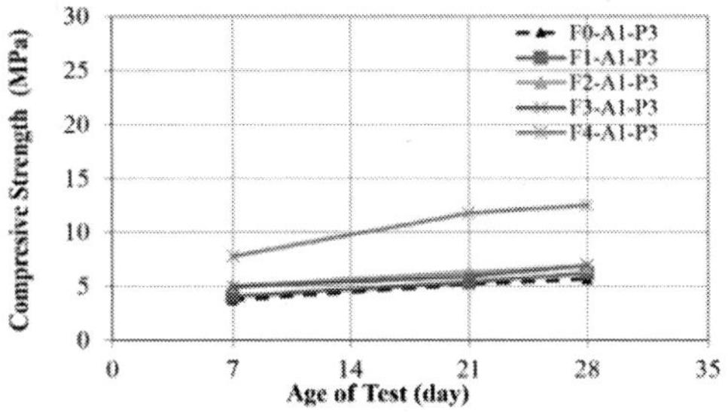

(b) Specimen F-P3

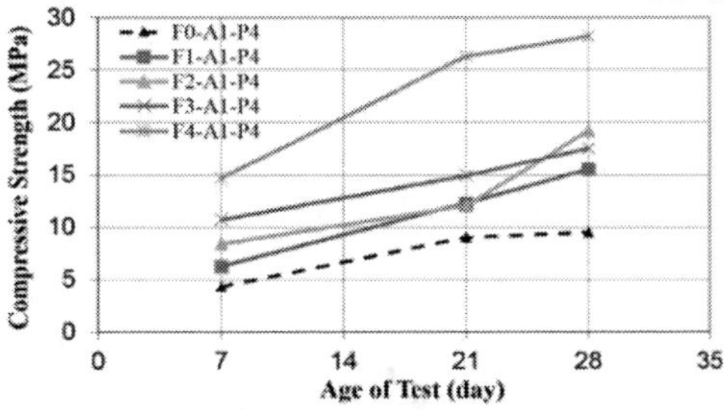

(c) Specimen F-P4

Fig. 4. Compressive strength of specimen F-P2, F-P3 and F-P4.

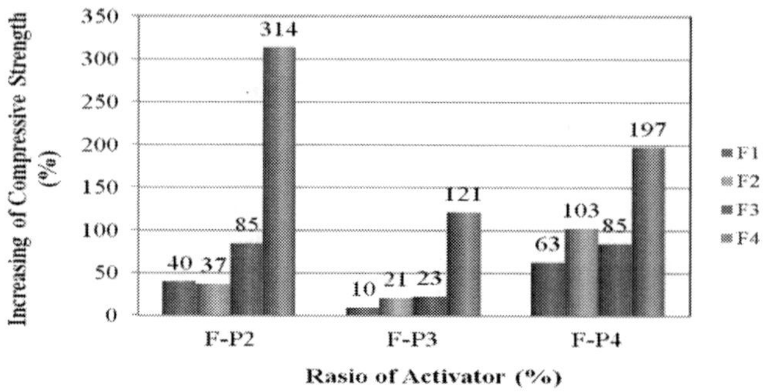

Fig. 5. The Increasing of compressive strength on the 28th day.

From all the compressive strength results perform that the specimens with 0.45% activator (F-P4) showed the highest compressive strength. The specimens with fly ash from each fall zone with 0.45% activator (F-P4) show the significant increasing of compressive strength compared to zone-0. Fly ash fineness based on fall zones contributed significantly to the compressive strength of the geopolymer.

4 Conclusions

i) Fly ash fineness based on fall zones contributed significantly to the compressive strength of the geopolymer mortar.
ii) Geopolymer mortars using fly ash with fineness based on fall zones and mixed with water glass and potassium activators showed a significant increase in compressive strength. The increase in compressive strength of each percentage of activator compared to zone-0 was 197.47%, 121.28% and 313.74% for the for the F4-P4 mixture of the F-P4 specimens, for the F4-P3 mixture of the F-P3 specimens, and an F4-P2 mixture of the F-P2 specimens respectively.
iii) The maximum compressive strengths of the geopolymer mortars based on the fineness level and the composition of the mixture at 28 days were: 28.20 MPa, 12.48 MPa, and 10.84 MPa for F4-P4, F4-P3 and F4-P2 respectively.
iv) The increase in compressive strength compared to fall zone-0 was 313.74% of F4-P2, while the maximum compressive strength was 28.2 MPa for F4-P4.

References

[1] V.M. Malhotra, Introduction: Sustainable development and concrete technology, ACI Concrete Int., **24** (7), 30-34 (2002)
[2] P.K. Mehta, High-performance, high-volume fly ash concrete for sustainable development, *Proc.of the International Workshop on Sustainable Development and Concrete Technology*, Beijing, China, 3-14, (2004)
[3] M.I.A. Aleem and P.D. Arumairaj, Optimum mix for geopolymer concrete, Indian J. of Science and Technology, **5** (3), 2299-2301 (2012)
[4] A.I. Laskar and R. Bhattacharjee, Rheology of flyash based geopolymer concrete, ACI Materials J., **108** (57), 535-543 (2011)
[5] B.S.K. Reddy, J. Varaprasad and K.N.K. Reddy, Strength and workability of low lime fly-ash based geopolymer concrete, Indian J. of Science and Technology, **3**(12), 1188-1189 (2010)

[6] J. Davidovits, Properties of geopolymer cements, *Proc. of the First International Conference on Alkaline Cements and Concretes*, Kiev State Technical University, Kiev, Ukraine, 131-149, (1994)
[7] J. Davidovits. *Geopolymer chemistry and application*, Geopolymer Institute, Saint Quentin, France (2008)
[8] D. Hardjito, B.V. Rangan, Development of flyash-based geopolymer concrete: progress and research needs, *Proc.of the 2nd Asian Concrete Federation Conference*, Bali, Indonesia, 1-5, (2006)

Influence of Ureolytic Bacteria Toward Interlocking Compressed Earth Blocks (ICEB) in Improving Durability of ICEB

M. M. Zamer[1], *J. M.* Irwan [1,*], *Norzila* Othman[2], *S. K.* Faisal[1], *L. H.* Anneza[1], *T.* Teddy[1], and *Abdullah Faisal* Alshalif[1]

[1]Jamilus Research Centre for Sustainable Construction (JRC), Faculty of Civil and Environmental Engineering, Universiti Tun Hussein Onn Malaysia, 86400, Parit Raja, Johor, Malaysia
[2]Micro-Pollutant Research Centre (MPRC), Faculty of Civil and Environmental Engineering, University Tun Hussein Onn Malaysia, 86400, Parit Raja, Johor, Malaysia

Abstract. Interlocking compressed earth blocks (ICEB) are soil stabilized based blocks that allows for mortarless construction. Various studies have been conducted to improve the durability of bricks by using environmental friendly solution. This is because common method used by the construction industries generally involving the use of chemical based substances which will promote pollution to the surrounding. This paper provide the results of Ureolytic Bacteria (UB) in improving the compressive strength and water absorption properties with the percentage of 1%, 3% and 5% UB for 7th ,14th and 28th days of testing. The bacteria were added as partial replacement of limestone water in ICEB. The results of compressive strength and water absorption show that the increment of 15.25% strength and reduction of 15.66% water absorption with 5% of UB on the 28th days of testing compared to the control specimen. Therefore it is hoped that the positive results on using bacteria will continue to improve the durability of the ICEB as one of the environmental friendly solution in order to achieve sustainable construction.

1 Introduction

Interlocking compressed earth block (ICEB) is a cost effective, dry stacked (mortarless) and sustainable construction material which has potential to bring durable and affordable homes to developing countries around the world [1]. Generally ICEB will give illustration of a block that is made up of earth-based material that has been going through a compression process [2]. ICEBs are often chosen because the blocks does not required to be burnt, thus reduce the emission of carbon dioxide. On the other hand, the production from conventional building construction materials pollutes the air, water and land [3]. The production of ICEBs also faster, easier and effective as compared to conventional blocks [4]. These advantages make ICEB practical and preferred construction form.

* Corresponding author: irwan@uthm.edu.my

Previous studies had taken different approach to improve the properties of construction material by introducing the used of bacteria. Researcher such as Willem et al [5] used *Bacillus sphaericus*, Navdeep et al [6] used *Bacillus megaterium*, Abhjit et al [7] used *Bacillus megaterium* and Bernardi et al [8] used *Sporosarcina paseurii*. All bacteria used by previous studies resulted in increasing on compressive strength and reduction on water absorption by comparing control sample and treated sample with bacteria. Positive results from previous studies indicated that the successfulness of using bacteria as an environmental friendly solution in improving the durability of construction material [9]. According to Siddique [10], bacteria are able to promote the precipitation of calcium carbonate ($CaCo_3$) in the form of calcite. The calcite will fill all pores which prevent water ingression. Deposition of calcium carbonate layer on the surface area resulted to decrease water absorption and porosity [10]. Hence it will improve the durability of the material properties.

The understanding on fundamental precipitation calcium carbonate ($CaCo_3$) had been applied in this research with the use of ureolyric bactreria (UB) to improve the durability of ICEB.

2 Methodology

2.1 Enrichment of bacteria

The purpose of enrichment process was to obtain growth curve for the ureolytic bacteria to ensure the survivability of UB in ICEB environment. Two set of enrichment were prepared namely control and treatment specimen. The composition of the treatment enrichment consist of 300ml nutrient broth added with 120ml of 40% urea added also with 1mg of soil substances and lastly 1 cyrogenic bead of ureolytic bacteria. The differences between the compositions of control were the enrichment was done without the soil sample. The method follows the previous study, Irwan [11] by adjusting the pH in alkaline and anaerobic condition were attempt for ensuring the survival of ureolytic bacteria.

2.2 Production of ICEB

The production process of ICEB is a critical element and should be handled very well in order to maintain the quality of the specimen. The materials used to produce the ICEB specimen are soil, sand, cement and limestone water. The addition of bacteria acts as partial replacement of limestone water. Fig. 1 shows the material that has been weight and ready to be mixed. Then the material must first be mixed thoroughly in ICEB mixer. The limestone water and ureolytic bacteria (UB) were added after the mixture are well mix and goes straight to the ICEB mold as shown in Fig. 2. Hydraulic machine was used to compress the mold to form the ICEB specimen. Fig. 3 shows the process of compressing specimen to produce interlocking compressed earth block (ICEB).

2.3 Compressive strength

The compressive strength (*fc*) was determined by using ICEB block (100 x 125 x 250 mm) specimen and the strength are measured for 7th, 14th and 28th days. There are four set batches of ICEB represented by control, 1%, 3% and 5% of UB which was partially replaced of limestone water for ICEB production. All batches were tested in triplicates and the average results of these triplicates are presented and discussed. The test was carried out according to the BS 3921:1985, [12] which specify for the testing of bricks.

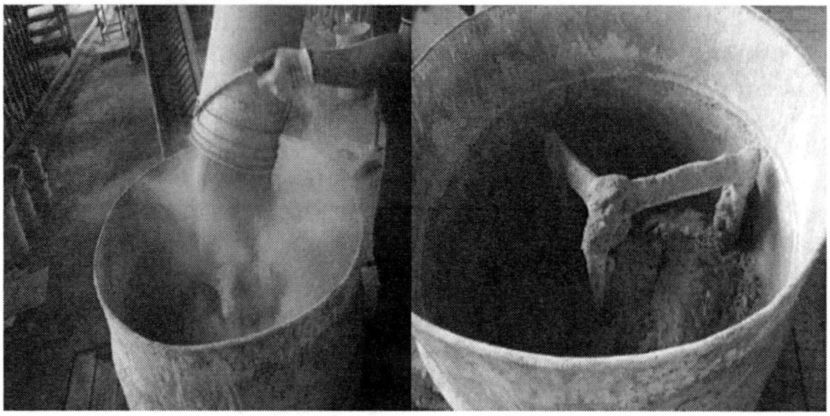

Fig. 1. Mixing of laterite soil, sand and cement for production of ICEB.

Fig. 2. Addition of limestone water and ureolytic bacteria (UB).

Fig. 3. Production of ICEB specimen.

Fig. 4. Compressive strength test of ICEB.

2.4 Water absorption

The water absorption test was conducted to determine the percentage of water absorbed by the bacterial ICEB by monitoring changes in weight among the ICEB with and without bacteria. The test also determined the saturation coefficient which is defined as the percentage of pore volume filled in 24-hour of soaking. The tests are conducted for 7th, 14th and 28th days. The test was carried out also according to the BS 3921-1985, [12] which specify for standard specification for clay masonry units.

Fig. 5. Water absorption test of ICEB at 24-hour submersion.

3 Results and discussion

3.1 Compressive strength

Table 1 and Fig. 4 show the results on compressive strength with time. The compressive strength of ICEB specimen with bacteria increased gradually for 7th, 14th and 28th days of testing compared to the control specimen. The addition of UB with liquid culture of 1%, 3% and 5% in ICEB increased the compressive strength within time compared to control specimen. The patterns of compressive strength increment are the same for all UB addition. The highest compressive strength recorded was 6.35 N/mm^2 at the 28th days of testing for 5% addition of UB.

The positive result indicates the successfulness of UB in deposition of calcium carbonate, $CaCO_3$ for enhancement the strength of ICEB. The increment of compressive strength with added bacteria agreed with Navdeep [6] studies which state that compressive strength can be significantly increased by application of bacteria calcite.

Table 1. Results on compressive strength.

Specimen with UB	Compressive Strength (N/mm^2)		
	7th day	14th day	28th day
0% (Control)	2.24	4.50	5.51
1%	3.05	4.79	6.10
3%	2.79	4.68	6.30
5%	2.76	4.75	6.35

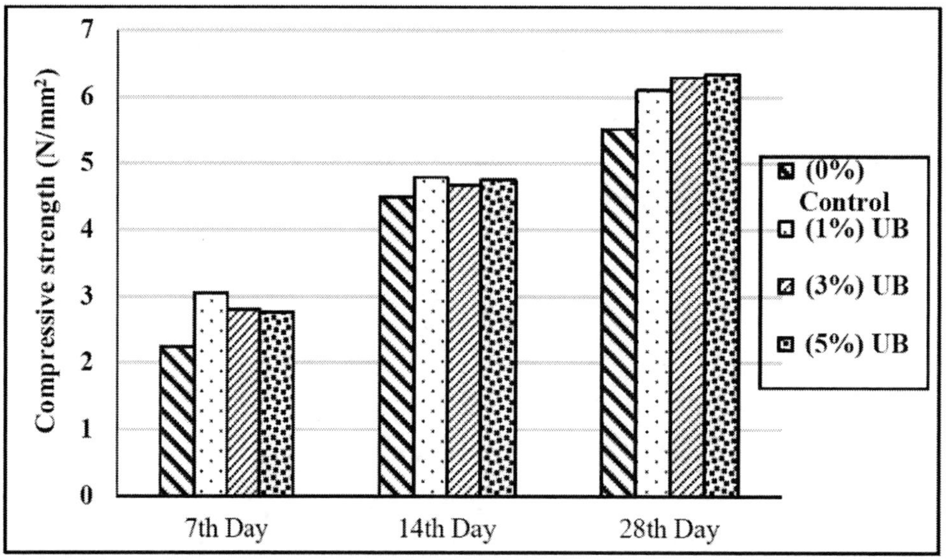

Fig. 4. Relationship between compressive strength and time.

3.2 Water absorption

Table 2 and Fig. 5 show the results on water absorption with time. The percentage of water absorption for 5% addition UB on the 28th days of testing achieved the lowest percentage compared to other ICEB specimen which is 7.70%. The reduction patterns of water absorption percentage are the same for all UB addition (1%, 3% and 5%).

The positive results also indicated the successfulness of UB in depositing calcium carbonate, $CaCo_3$ as calcite in filling the pores to prevent ingress of water absorption. According to Abhjit [7], the calcite crystal act as biosealant by filling the pores which leads to reduction in water absorption, porosity, permeability and enhance the strength of the bricks. Therefore the results on reduction of water absorption collated with studies that was done by Abhjit [7].

DOI: 10.1051/matecconf/201710301027

Table 2. Results on water absorption.

Specimen with UB	Water Absorption (%)		
	7th day	14th day	28th day
(0%) Control	9.81	9.27	9.13
(1%)	9.71	8.77	8.10
(3%)	9.23	8.69	8.05
(5%)	8.73	8.13	7.70

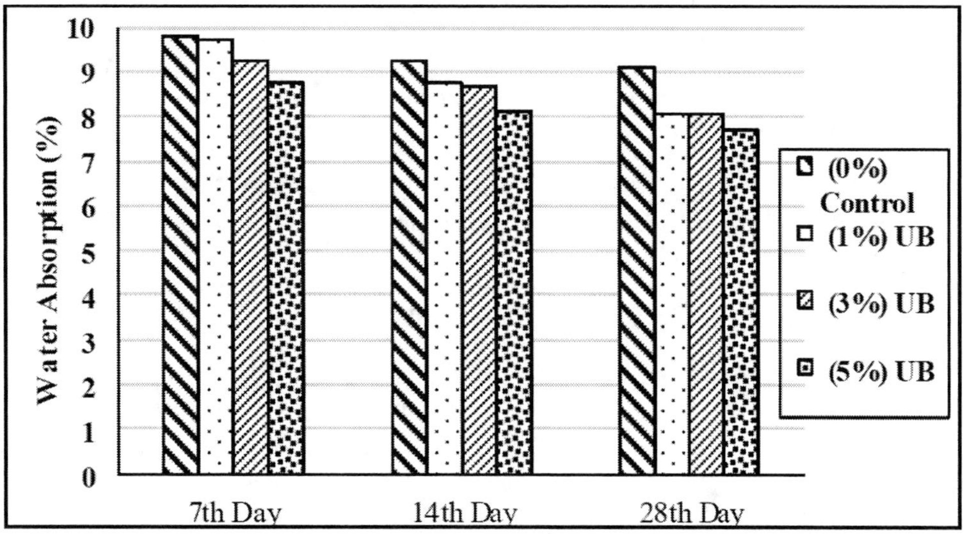

Fig. 5. Relationship between water absorption and time.

4 Conclusion

The results from both compressive strength and water absorption tests show the successfulness of UB in depositing calcium carbonate, $CaCo_3$. Addition of UB in ICEB presents the increment in strength and reduction on absorption of water. The 5% addition of UB in ICEB indicates the best result which achieved highest increment in strength with 15.25% and highest reduction of water absorption with 15.66% at the 28th days of testing. Further study should be conducted through various engineering properties and micro-structure testing in order to determine the influence and morphology of UB in ICEB.

This research was supported by Universiti Tun Hussein Onn Malaysia (UTHM), Malaysia, The Ministry of Higher Education Malaysia through Fundamental Research Grant Scheme (FRGS), Vot. 1577 and collaboration with ICEB industry IRYAS INC. (M) Sdn Bhd.

References

[1] P.T. Laursen, N.A. Herskedal, D.C. Jansen and B. Qu, Interlocking Compressed Earth Blocks Walls: Out-of-Plane Structural Response, *15th World Conference On Earthquake Engineering*, Lisbon, Portugal, 71-81, (2012)

[2] F.V. Riza and I.A. Rahman, The Properties of Compressed Earth-Blocks (CEB) Masonry Blocks, J. Eco-efficient Masonry Bricks and Blocks: Design, Properties and Durability, **17,** 379- 392, (2014)
[3] N. Ali, N.A. Zainal, M.K. Burhanudin, A.A.A. Samad, N. Mohamad, S. Shahidan and S.R. Abdullah, Physical and Mechanical Properties of Compressed Earth Brick (CEB) Containing Sugarcane Bagasse Ash, *MATEC Web of Conferences*, **47,** 1-7, (2016)
[4] F.V. Riza, I.A. Rahman and A.M.A. Zaidi, A brief review of compressed stabilized earth brick (CSEB), *Int. Conference on Science and Social Research*, Kuala Lumpur, Malaysia, 999-1004, (2010)
[5] W. De Muynck, D. Debrouwer, N. De Belie and W. Verstraete, Bacterial carbonate precipitation improves the durability of cementitious materials, J. of Cement and Concrete Research, **38**(7), 1005-1014, (2008)
[6] N.K. Dhami, M.S. Reddy and A. Mukherjee, Improvement in strength properties of ash bricks by bacterial calcite, J. of Ecological Engineering, **39**, 31-35, (2012)
[7] A. Mukherjee, N.K. Dhami, B.V.V Reddy and M.S. Reddey, Bacterial Calcification for Enhancing Performance of Low Embodied Energy Soil-Cement Bricks, *Third Int. Conference on Sustainable Construction Materials and Technology*, Kyoto, Japan, (2013)
[8] D. Bernadi, J.T. Dejong, B.M. Montoya and B.C. Mattinez, Bio-bricks: Biologically cemented sandstone bricks, J. of Construction and Building Material, **55**, 462-469, (2014)
[9] J.M. Irwan, M.M Zamer and N. Othman, A Review on Interlocking Compressed Earth Blocks (ICEB) with Addition of Bacteria, *MATEC Web of Conferences*, **47**, 1-5, (2016)
[10] R. Siddique and N.K. Chahal, Effect of ureolytic bacteria on concrete properties, J. of Construction and Building Materials, **25**(10), 3791-3801, (2011)
[11] J.M. Irwan, L.H. Anneza, N. Othman, T. Husnul and A.F.Alshalif, Isolation and Identification of Concrete Environment bacteria, *IOP Conference Series: Materials Science and Engineering,* **136**(1), 1-9, (2016)
[12] British Standard 3291, *Specification for Clay Bricks*, British Standard Institution, London, United Kingdom, (1985)

Strength and Absorption Rate of Compressed Stabilized Earth Bricks (CSEBs) Due to Different Mixture Ratios and Degree of Compaction

Abd Halid Abdullah[1,*], *Sasitharan* Nagapan[1], *Anna* Antonyova[2], *Kannan* Rasiah[3], *Riduan*Yunus[1], and *Samiullah* Sohu[1]

[1]Faculty of Civil and Environmental Engineering, UniversitiTun Hussein Onn Malaysia, 86400 ParitRaja, BatuPahat, Johor, Malaysia
[2]Faculty of Management, University of Prešov in Prešov, Prešov 08001, Slovak Republic
[3]Department of Mechanical Engineering, Politeknik Port Dickson, Negeri Sembilan, Malaysia

Abstract. Compressed Stabilized Earth Brick (CSEB) is produced by compressing a mixture of water with three main materials such as Ordinary Portland Cement (OPC), soil, and sand. It becomes popularfor its good strength, better insulation properties, and a sustainable product due to its easy production with low carbon emission and less skilled labour required. Different types of local soils usedwill produce CSEB of different physical properties in terms of its strength, durability, and water absorption rate. This study focuses on laterite soil taken from the surrounding local area in Parit Raja, Johor, and CSEB samples are produced based on prototype brick size 100×50×30 mm. The investigations are based on four different degree of compactions (i.e. 1500, 2000, 2500, and 3000 Psi) and three different mix proportion ratios of cement:sand:laterite soil (i.e. 1:1:9, 1:2:8, 1:3:7). A total of 144 CSEB samples have been tested at 7 and 28 days curing periods to determine the compressive strength (BS 3921:1985) and water absorption rate (MS 76:1972). It was found that maximum compressive strength of CSEB was 14.68 N/mm^2 for mixture ratio of 1:3:7 at 2500 Psi compaction. Whereas, the minimum strengthis 6.87 N/mm^2 for 1:1:9mixture ratio at 1500 Psi. Meanwhile, the lowest water absorption was 12.35% for mixture ratio of 1:2:8 at 3000 Psi; while the 1:1:9 mixture ratio at 1500 Psi gave the highest rate of 16.81%. This study affirms that the sand content in the mixture and the degree of compaction would affect the value of compressive strength and water absorption of CSEB.

1 Introduction

Brick is commonly used as building material for wall due to its ability to support live and dead loads, and also serve as heat insulation. It can be categorized as clay bricks, mortar brick, fired or unfired brick, etc. [1]. The prime advantage of utilizing Compressed Stabilized Earth Brick (CSEB) is that it utilizes local materials and reduces transportation costs since the production is in-situ. Other advantages include faster production and easier

* Corresponding author: abdhalid@uthm.edu.my

construction method that generally requires less skilled labour. According to T. Morton [1], a striking contrast between CSEB and conventional brick is lower energy consumed during the production process, and as such less carbon emission. CSEB produces 22 kg CO^2/tonne as compared to that of concrete blocks (143 kg CO^2/tonne), fired clay bricks (200 kg CO^2/tonne) and aerated concrete blocks (280-375 kg CO^2/tonne) respectively [1- 3]. CSEB product is also said to be environmentally friendly and sustainable as it utilizes almost unlimited resource in its natural state that involve no pollution and negligible energy consumption, thus further benefiting the environment by saving biomass fuel [4 - 6].

The main process of producing CSEB involves soil preparation, compaction of mixture, and curing [7]. However, the detail production process involves eight stages including sieving the soil, measuring, mixing, checking the moisture content, moulding, quality control, humid curing and stacking, and curing and stacking [8]. CSEB requires careful and correct selection of the soil to get the best result. After pouring the mixture into the mould, a proper compaction pressure should then be applied. Curing of CSEB usually takes advantage from natural humid where bricks could be stacked immediately after compaction but the strength is gained over time. It is important to prevent rapid drying out; hence, the brick is moist cured under polythene sheet (or wet gunny) in the open air (relative humidity more than 70% is the best condition in ensuring maximum hydration of the used stabilizer) for about 28 days if cement is used as a stabilizer [9, 10].

CSEB made from different types of local soils will give different physical and thermal properties [11, 12]. In this study, CSEB is made from local laterite soil mixed with Ordinary Portland Cement (OPC) as a stabilizer, sand, and water. The primary objective of the study is to determine the optimum compressive strength and water absorption rate of CSEB due to different mixture ratios (i.e. cement:laterite:sand) and degree of compactions.

2 Soil preparation for CSEB

Initially, soil physical testing should be performed to ensure the suitability of laterite soil for CSEB production. The two types of test commonly carried out are Grain Size Analysis Test and Atterberg Limits Test.Grain Size Analysis Test refers to discerning the percentages of particles (by dry mass) within a specified particle size range across all the sizes represented for the sample. The distribution of particle sizes is used to distinguish the maximum particle size and the major portion of the particle sizes, as well as to characterize the soil, such as by the Unified Soil Classification System (USCS) [13]. This test uses a mechanical shaker equipped with various sieve sizes of 5 mm, 2 mm, 1.18 mm, 0.6 mm, 0.425 mm, 0.3 mm, 0.212 mm, 0.15 mm, 0.063 mm, and a pan at the bottom.

Atterberg Limits is the soil classification test to determine the soil's Liquid Limit (LL) and Plastic Limit (PL). The LL and PL are water contents at which the mechanical properties of soil changes. They are applicable to fine-grained soils, and are performed on soils fractions that pass the 0.425 mm (No. 40) sieve. The difference between PL and LL is defined as the Plasticity Index (PI). The LL defines water content at the boundary between liquid and plastic states. It is expressed as a percentage and determining the LL by the multipoint method using the Cone Penetration apparatus. The PL marks the boundary between semi-solid and plastic mechanical behaviour. However, in reality the material slowly transitions between the two. The PL is defined as the water content at which a 3 mm diameter rod of soil begins to crumble. The PI is the difference in water content between LL and PL.

3 CSEB production and sample preparations

A total of 140 CSEB samples size 100×50×30 mm (as shown in Fig. 1) were produced with different mixture ratios and compaction pressures. The process started with preparing all the raw materials such as laterite soil, sand, OPC and water. The particle sizes must be 2 mm for sand and less than 5 mm for laterite soil in order to ensure the binding between all materials when mixed together. The three different mix proportion ratios of cement:sand:laterite soil employed in this study were 1:1:9, 1:2:8, 1:3:7. The added water should not be more than 15% by the ratio. The CSEB mixture was poured into a mould and then compacted using a hydraulic compaction machine.To maintain or avoid the water content in the CSEB samples from draining quickly during curing process, they were placed under covered area or protected from direct sunlight and rain, for two different time periods of 7 days and 28 days.

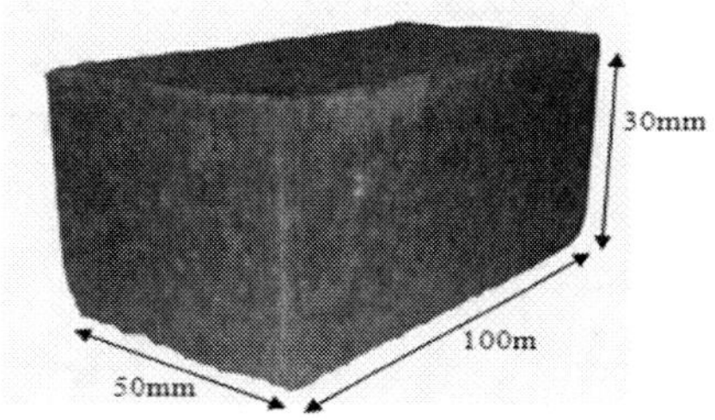

Fig. 1. The finished CSEB.

3.1 Testing compressive strength of CSEB

A total of 3 different samples of the mixture proportions and compaction pressures were taken into account. As a reference, the compressive strength value is referred to in BS 3921:1985 [13] which states that the minimum compressive strength of clay brick is 5 N/mm^2. To obtain accurate values of compressive strength of the surface layer of CSEB, metal plates lining were placed on both its top and bottom sides so that force from the compressive strength Universal Test Machine could be uniformly applied. The values of compressive strength were directly derived from the computer system which presented them in the form of graphs. Then, data were stored and later brought back for analysis.

3.2 Water absorption rate

The samples were tested at 7 and 28 days for the variable mentioned earlier and water absorption rate were taken after 3 days curing period. Based on MS 76:1972 [15], water absorption rate testwhich involved the cooling method where each sample was weighed and dried in an oven for 48 hours at a temperature of 105 °C. It was then soaked in water for 24 hours at room temperature. In order to get the percentage rate of water absorption, the weight of the sample after immersion was divided by the weight of dry sample and then multiplied by 100%.

3.3 Laboratory test results

The results describing the characteristics of the laterite soil due to several laboratory tests are summarized and detailed in Table 1.

Table 1. Identification and Characteristics of the Soil Used

Test Types	Property	
Atterberg Limits	Liquid Limit (LL)	42.5
	Plastic Limit (PL)	27.36
	Plasticity Index (PI)	15.14
Chemical Characteristics	Carbon Dioxide, CO_2 (%)	0.10
	Ferum Trioxide, Fe_2O_3 (%)	36.90
	Silicon Dioxide, SiO_2 (%)	34.20
	Aluminum Oxide, Al_2O_3 (%)	25.40
	Titanium Dioxide, TiO_2 (%)	0.64
	Potassium Oxide, K_2O (%)	1.85
	Sulfur Trioxide, SO_3 (%)	0.27
	Sodium Oxide, Na_2O (%)	0.17
	Magnesium Oxide, MgO (%)	0.21
Normalized Proctor Test	Optimum water content (%)	15
	Optimum cement content (%)	9

3.3.1 Compressive strength

Based on the three mixture ratios, compressive strength tests were performed based on BS 3921:1985[14] to determine the maximum value of CSEBs. The tests were carried out using 3 samples, and the average value of the samples would be considered accordingly as the final value for compressive strength.It can be seen from Fig. 2 that the compressive strength of CSEB samples became higher as their age increased from 7 to 28 days.

3.3.2 Correlation between compressive strength and sand content

Fig. 3 generally shows that as the age of CSEB samples increased from 7 to 28 days regardless of the degree of compactions, the compressive strength would rise as the sand content was added more while reducing the laterite soil content.

For mixture ratio of 1:3:7, the laterite soil content was reduced around 9% and the sand content was added more with the same amount. Correlation between the different percentages of sand content and compaction pressures on the strength can be seen from Figures 3(a) and 3(b) for samples' age of 7 and 28 days. The maximum strength of 14.68% N/mm^2 was obtained for samples with mixture ratio of 1:3:7 subjected to 2500 Psi compaction, while the minimum strength value of 6.87 N/mm^2 for mixture ratio of 1:1:9 with compaction of 1500 Psi. Sand and gravel are also suitable to be used as additive in the soil as they will increase the quality of sand and also improve the particle size distribution, and therefore could affect the value of compressive strength of CSEB [16].

3.3.3 Correlation between compressive strength and degree of compaction

Fig. 4 shows that for all mixture ratios the increased in the degree of compaction pressures in producing CSEB samples would generally lead to the rise in compressive strength despite their ages. Furthermore, the optimum mixture ratio was found to be 1:3:7 as it gave the greatest strength regardless the age of samples. With that mixture ratio, the highest strength values were 12.88 N/mm^2 for 7 days old samples subjected to compaction of 3000 Psi, and 14.68 N/mm^2 for 28 days old samples subjected to compaction of 2500 Psi. In contrast, CSEB samples with mixture ratio of 1:1:9 subjected to 1500 Psi compaction gave the lowest strength values of 7.75 N/mm^2 and 6.87 N/mm^2 at the age of 7 and 28 days, respectively. It can be seen also from Figures 4(a) and 4(b) that the different sand content in the mixture ratios as well as degree of compaction applied could affect the compressive strength of CSEB samples. In general, by increasing the compactions from 5 MPa to 20 MPa, the strength can reach up to 70% [9].Since currently there is no specific standard on the required strength of CSEB, the results of compressive strength values found from this study were compared and referred to the BS 3921:1985 [14]. Minimum strength specified in this standard for clay brick Damp-proof Course 1 is 5 N/mm^2. Thus, all values obtained are beyond the allowable strength specified in the standard.

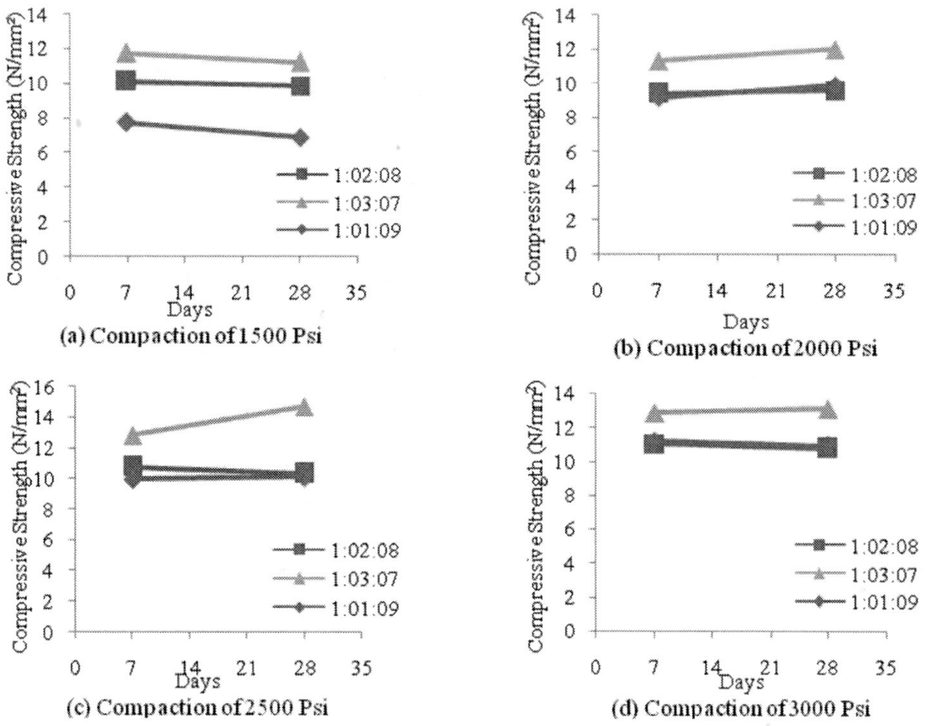

Fig. 2. CSEB strength due to mixture ratios & Compaction vs. Samples' Age (7 & 28 days).

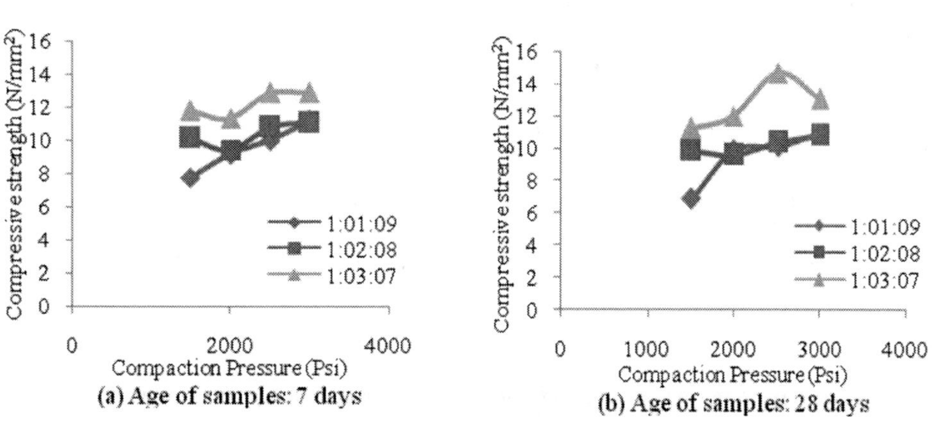

Fig. 3. Effect of sand content on the strength due to age of samples (7 & 28 days).

DOI: 10.1051/matecconf/201710301028

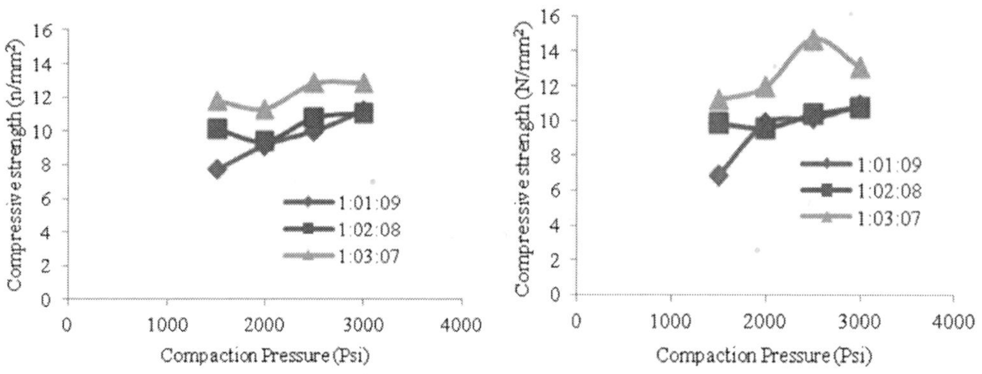

(a) Age of samples: 7 days (b) Age of samples: 28 days

Fig. 4. Strength vs. compaction due to different age of samples (7 & 28 days).

3.3.4 Water Absorption

Water absorption rate for brick is of great concern as it will determine the material's workability due to the effect of weather. The results of rate of water absorption for CSEB samples in this study were compared with that of Load Bearing Bricks Class (1) as specified in the MS 76:1972 [15]. Although currently there is no specific standard of water absorption rate for CSEB, the actual practice in Malaysian construction industry has generally considered the rate of water absorption for bricks should be less than 15%. Lower rate of water absorption indicates that the bricks are better in terms of its quality and resilience to weather effects, particularly when exposed to rain and sunlight.

Fig. 5(a) to Fig. 5(d) show results for water absorption of CSEB samples produced with different mixture ratios as well as subjected to different compaction pressures at the age of 7 and 28 days, respectively. It can be seen that the rate of water absorption rates for CSEB samples at 28 days old were lower as compared to that of 7 days old. This indicates that as the age of CSEB is increasing from 7 to 28 days, the hydration process of cement is getting more complete due to the curing process. Completion of the hydration process would result in the reaction between Calcium Oxide (CaO) with water to produce Calcium Hydroxide $Ca(OH)^2$ which will cover the empty spaces between particles in bricks and making CSEB impervious in nature. This condition could prevent the water absorption into the bricks. The lowest rate of water absorption was found to be 12.35 % for 28 days old CSEB sample with mixture ratio of 1:3:7 subjected to 3000 Psi compaction. Whereas, the highest rate of water absorption was 16.28 % for 28 days CSEB sample with mixture ratio of 1:1:9 subjected to compaction of 1500 Psi. Figure 5 also shows that the rates of water absorption were generally lower than 15% for CSEB samples produced with mixture ratio of 1:3:7 subjected to any degree of compactions (i.e. 1500, 2000, 2500, and 3000 Psi) regardless the samples' ages. Higher rate of water absorption of CSEB may cause swelling of stabilized clay fraction and resulting in losing strength with time. Water absorption and porosity of CSEB would increase with higher clay content, as well as decrease with cement content [2].

3.3.5 Correlation between water absorption and compaction pressure

The different degree of compactions applied in the production of CSEB the optimum water content in the stabilized mixtures. Fig. 6(a) and Fig. 6(b) exhibit that the rates of absorption generally decrease with the increase in compaction pressures. The optimum mixture ratio was found to be 1:3:7 as it generally gave the lower rate of absorption regardless the age of

CSEB samples. Samples with that mixture ratio and compacted by 3000 Psi pressure, the lowest absorption rates were 14.05% and 12.97% for 7 and 28 days, respectively. In contrast, CSEB samples with mixture ratio of 1:1:9 subjected to compaction of 1500 Psi gave the highest absorption rates of 16.81% and 16.03% at 7 and 28 days, respectively.

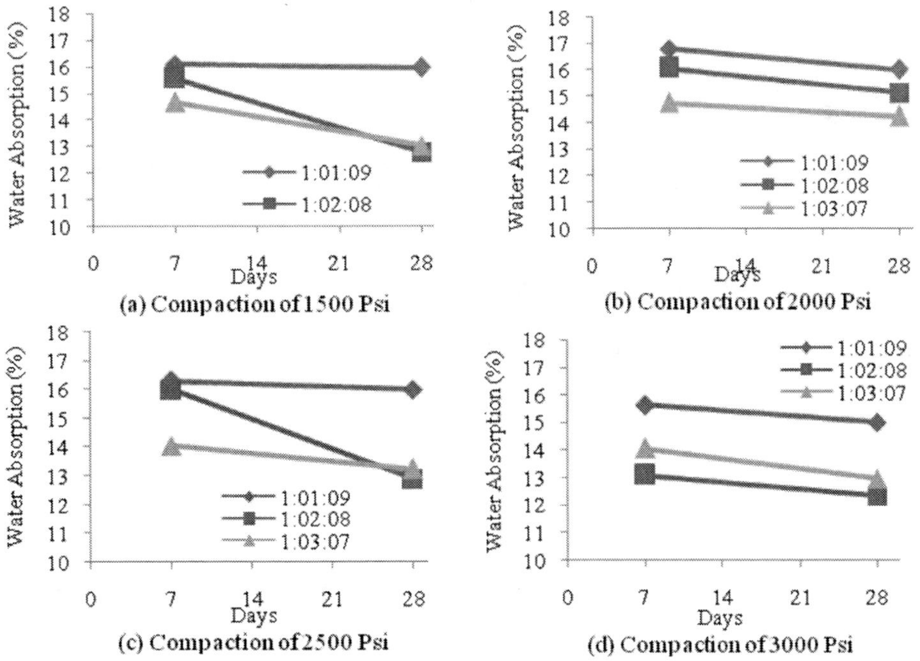

Fig. 5. Absorption due to mixture ratios / Compaction vs. Samples age (7 & 28 days).

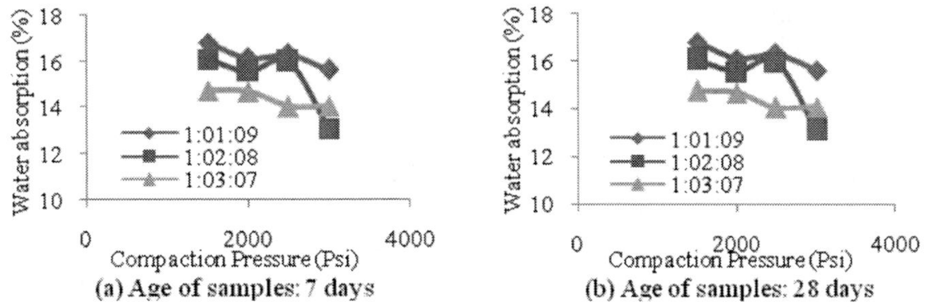

Fig. 6. Absorption vs. Compaction for different age of samples (7 & 28 days).

4 Conclusions

It can be firmly presumed that parameters such as particle size and moisture content of the soil can influence the strength of the CSEB. From the particle size of the soil, we can identify the composition of the soil. The composition of the soil can be adjusted until it falls within the ideal curve limit in particle size distribution to produce high strength CSEB. The mixture ratio plays a vital role in determining the soil-sand-cement strength. Properties of the soil can be the factor of producing the higher strength CSEBs. The quantity of gravel, sand, clay and silt of the soil must be suitable with the ratio of the mixture. Without perfect mixture ratio, the desired strength will not be achieved. Moreover, it can be affirmed that degree of compaction pressures applied in the production of CSEBs greatly affect the compressive strength and rate of water absorption.

We would like to express our appreciation to the staff at the Materials Engineering Laboratory, Universiti Tun Hussein Onn Malaysia (UTHM), for their assistance during the preparation and testing of samples. The authors are alsograteful to the UTHM for giving the financial support under the Short Term Grant (STG) No. U636.

References

[1] T. Morton, *Earth masonry design and construction guidelines*, Construction research communications limited, Berkshire, (2008)

[2] N. Ali, N.A. Zainal, M.K. Burhanudin, A.A.A. Samad, N. Mohamad, S. Shahidan and S.R. Abdullah, Physical and mechanical properties of compressed earth brick (CEB) containing sugarcane bagasse ash, *MATEC Web Conf.*, **47**, 1-7, (2016)

[3] J.C. Morel, A. Mesbah, M. Oggero and P. Walker, Building houses with local material: means to drastically reduce the environmental impact of construction, Building and Environment, **36**, 1119-1126, (2001)

[4] S. Shahidan, H.B. Koh, A.S. Alansi and L.Y. Loon, Strength development and water permeability of engineered biomass aggregate pervious concrete, *MATEC Web Conf.*, **47**, 2–7, (2016)

[5] E.A. Adam and A.R.A. Agib, *Compressed stabilized earth block manufacture in Sudan*, UNESCO, France, (2001)

[6] E. Aminudin, M.F.M. Din, M.W. Hussin, A.H. Abdullah, K. Iwao and Y. Ichikawa, Properties of agro-industrial aerated concrete as potential thermal insulation for building, *MATEC Web Conf.*, **47**, 1-9, (2016)

[7] Gernot Minke, *Building with earth: Design and technology of a sustainable architecture*, Birkhauser Publishers, Switzerland, (2006).

[8] Auroville Earth Institute, *UNESCO Chair Earthen Architecture*, Retrieved on November 15, 2016 from http://www.earth-auroville.com/compressed_stabilised_earth_block_en.php

[9] A. Guettalaet, Durability of lime stabilized earth blocks. Courrier du Savoir, **02**, 61-66, (2002)

[10] P.J. Walker, Strength and erosion characteristics of earth blocks and earth block masonry, Journal of Materials in Civil Engineering, **16**(5), 497-506, (2004).

[11] S.K. Abu Bakar and A.H. Abdullah, Simulation of thermal performance in an office, *Business Engineering & Industrial Applications Colloquium*, IEEE, 318-323, (2012)

[12] A.H. Abdullah, S.K.A. Bakar and I.A. Rahman, Indoor thermal performance of an office building using conventional brick versus interlocking compressed earth brick (ICEB) wall, International Journal of Construction Technology and Management, **1**(1), 22-27, (2013).

[13] John T. Germaine and Amy V. Germaine, *Geotechnical Laboratory Measurements for Engineers*, John Wiley and Sons Inc., United States of America, (2009)

[14] BS 3921:1985. *Specifications for clay brick*, British Standards Institution, London, United Kingdom, (1985)

[15] MS 76:1972. *Specification for bricks and blocks of fired brick earth, clay or shale, Part 2: metric units*, Malaysia Standard, 22,(1972)

[16] German Appropriate Tech. Exchange and CRAterre, *The basics of compressed earth block*, Eschborn: Deutsches Zentrum fürEntwicklungstechnologien-GATE, (1991)

Comparison of Strength Between Laterite Soil and Clay Compressed Stabilized Earth Bricks (CSEBs)

Abd Halid Abdullah[1,*], *Sasitharan* Nagapan[1], *Anna* Antonyova[2], *Kannan* Rasiah[3], *Riduan*Yunus[1], and *Samiullah* Sohu[1]

[1]Faculty of Civil and Environmental Engineering, UniversitiTun Hussein Onn Malaysia, 86400 ParitRaja, BatuPahat, Johor, Malaysia
[2]Faculty of Management, University of Prešov in Prešov, Prešov 08001, Slovak Republic
[3]Department of Mechanical Engineering, Politeknik Port Dickson, Negeri Sembilan, Malaysia

Abstract. Compressed Stabilized Earth Brick (CSEB) is a sustainable material utilizing locally available soils mixed with stabilizers to increase its strength. Two main factors affecting the strength of prototype are its compressive strength and water absorption rate. In this study, the mix proportion ratio for producing CSEB prototypes from laterite soil and clay types was 1:10 (cement:soil). Three different compactions (i.e. 2000, 3000, and 4000 Psi) were applied to produce prototypes with dimension of 100×50×40 mm. The samples were tested at the ages of 7 and 28 days. For laterite soil CSEB, the optimum strength was achieved by the sample subjected to 2000 Psi compaction,withits compressive strength of 9.0 N/mm^2 and water absorption rate of 14.9%. For clay CSEBthe optimum strength observed was the sample subjected to 4000 Psicompaction, with the compressive strength of 5.8 N/mm^2 and water absorption rate of 17%. However, the compressive strength of clay CSEB is still below the level specified in the MS 76:1972, which is over 7 N/mm^2 for the load-bearing fired brick Class (1). This study also revealed that the increased in compression pressure would result in different strength performances for both CSEBs made from laterite soil and clay.

1 Introduction

Until recently, more than half of the world's population have chosen to build their homes of this soil-based structures due to its high durability, low cost, good resistant to fire and weather factors, and aesthetically attractive [1- 5].

Compressed Stabilized Earth Brick (CSEB) is agreen technology in building construction, which has been explored using locally available materials. Soil-based home construction has begun since 1950s in Bogota, Colombia, initially introduced by Raul Ramirez [6]. According to the UNESCO Chair Earthen Architecture [7], clay and laterite soil are suitable to be used in the production of CSEBs, and they are readily available in

* Corresponding author: abdhalid@uthm.edu.my

Malaysia [8]. From using manually controlled pressure in a steel mould in early 1950s, the process of producing CSEBs has undergone many changes. In recent years, hydraulic compression machine has been widely used to improve the quality and also to optimize the cost, effort and time in producing CSEBs [9].

Currently in Malaysia, there is no standard in producing CSEB made from laterite soil and clay. Therefore, this study aims at quantifying the compressive strength and water absorption rate of CSEBs made from laterite soil and clay. The values obtained can be used as references and standards in CSEB production in Malaysia.

2 Overview of CSEB prototype

Fig. 1 shows a CSEB prototype produced with dimensions of 100×50×40 mm. It resembles an Interlocking Compressed Stabilized Earth Brick (ICSEB) sized 100×125×250 mm in terms of strength, mixing ratio, and other attributes. The main purpose of producing this prototype brick is to test its compressive strength and water absorption rate.

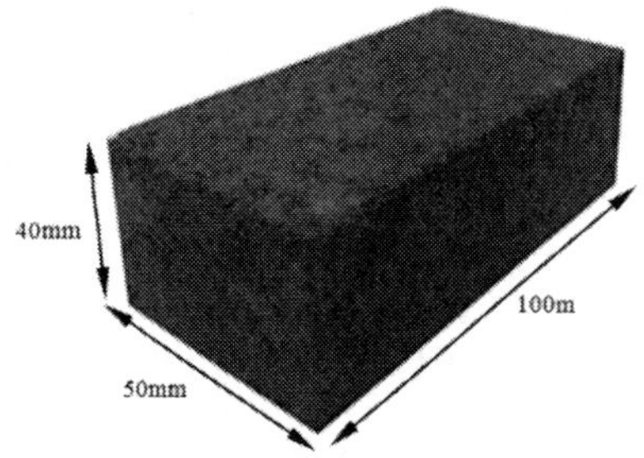

Fig. 1. CSEB prototype dimension.

Soil is the main component for the production of CSEBs [10]. The other ingredients include Ordinary Portland Cement (OPC), sand and water. The properties of the soil and local climatic conditions in a given area should be determined prior to producing CSEB. For instance, dry climatic regions of African continent require different soil than that of tropical climates. Thus, not all types of soil can be used in producing CSEB. Generally soil containing some clay and mud is suitable as it facilitates the adhesion process. The best soil composition contains 15% gravel, 50% sand, 15% silt and 20% clay [10].

Laterite soil consists of clay and iron particles, and characterized by its hardening properties, chemical content and structural evolution. It is porous and soft at high humidity, where exposing it continuously to high ambient temperatures would cause it to harden [11]. In contrast, clay comprises small pores of air and solid particles with no void spaces. It is also soft at high humidity and will quickly harden if continuously exposed to high ambient temperatures. However, clay has a low shear strength and high compressibility level [11].

Sand is also known as the fine aggregates. For CSEB production, the sand should be reasonably fine (i.e. passing 2 mm sieve size) as well as clean and free from any organic material as required by the standard. Additional sand will only be required if the percentage of existing sand content is less than one-third of the overall soil quantity [12].

Cement is a good stabilizer in the production of CSEB since it has adhesive properties that bondminerals' particlestogether. In this study, OPC was chosen due to its suitabilityas admixture in the production of soil-based products. Moreover, this type of cement is also highly suitable for use in a variety of climatic conditions [13].

Water must be clean and free from impurities such as organic and sulphate. The presence of impurities will affect the hydration process particularly during the hardening stage of the bricks. Whereas, during the curing stage, water is needed for a perfect hydration of cement. The amount of water needed in the mixture is between 10 - 20% [11].

3 Preparation of soil for CSEB

Laterite soil and clay were selected to be the main ingredients in the production of CSEB prototypes. Theywere taken at a depth of 1.5 m below ground surface, whereby laterite soil was obtained from a place in Ayer Hitam and clay was acquired from a place in Parit Raja.

The soil samples were analysed and compared with the results found by previous researchers. In particular, analysis results of laterite soil and clay samples were compared with those of Wahab [14] and C.M. Chan [15], respectively. Several tests, which include Particle Distribution Test, Atterberg Limits Test, Chemical Composition Test, and Jar Test, were carried out in accordance to BS 1377:Part 2:1990 [16].

3.1 Particle distribution test (sieve analysis)

Most of the mixing ratiosin producing CSEB prototypes consist of aggregated materials, namely soil and sand, which will particularly affect their durability and performance. Sieve test is crucial for obtaining the rate of aggregate mixture of various sizes.

Based on the AURAM Press 3000 [10], soil containing 15% gravel, 50% sand, 15% silt and 20% clay is considered as having the best soil composition for producing CSEB. Particle Distribution Test on laterite soil found that the percentage of aggregate mixtures of the soil composition were almost similar to those performed by Wahab [14], as outlined in Table 1. These percentages of the composition indicate the suitability of the soil to be used in producing CSEB. For clay, the test results showed that percentage of aggregate mixtures were almost similar to those obtained by C.M. Chan [15]. Apparently, clay composition did not match with that of the best soil composition required for producing CSEB.

Table 1. Soil particle size distribution.

Soil Type	Place taken from (Researcher)	Clay (%)	Sand (%)	Gravel (%)	Mud (%)
Laterite Soil	Ayer Hitam	4	86	10	-
	(Wahab,[14])	7.4	71.5	21.1	-
Clay	Parit Raja	25.8	0.9	-	73.3
	(C.M.Chan, [15])	10.23	0.57	-	89.2

3.2 Atterberg limits test

This experiment is carried out to identify and classify soils samples that have been dried, particularlyto determine the value of liquid limit (LL), plastic limit (PL) and the plasticity index (PI) of the soil samples. The determination of these limits is also used to predict the shear strength and soil sedimentation. Both laterite soil and clay samples had passed through 0.425 mm sieve.

The results of Atterberg limits test as outlined in Table 2 has proven that the laterite soil of Ayer Hitam and clay from Parit Raja to be used in the production of CSEB prototypes in this study are classified as laterite soil and clay. There are obvious similarities in terms of the values for liquid limit (LL), plastic limit (PL) and the plasticity index (PI) of the laterite soil from Ayer Hitam as investigated by Wahab [14], as well as clay from Parit Raja as examined by C.M. Chan [15].

Table 2. Atterberg limits.

Soil Type	Place taken from (Researcher)	Liquid Limit (LL) (%)	Plastic Limit (PL) (%)	Plasticity Index (PI) (%)
Laterite Soil	Ayer Hitam (Wahab, 2008)	43.8	22.5	21.2
		45.0	23.1	21.9
Clay	Parit Raja (C.M.Chan, 2008)	67.5	32.04	35.5
		68	32	35

3.3 Chemical composition test

In order to run this test, samples of clean soil sized 63 µm in the form of dust / dry powder should be prepared. The results of chemical composition analysis, as summarized in Table 3, found that the content of Silicon Dioxide (SiO2) is the highest within the two types of soil samples, which is 42% for laterite soil and 60.6% for clay, respectively. This silica content can help in strengthening the compressive strength of the CSEB prototypes when reacting with cement due topozzolanic process. The high content of lime or Calcium Oxide (CaO) of about 53.6% in cement, would enhance its performance as binding agent between both soil and sand particles.

Table 3. Material chemical composition (weight percentage).

Types of Chemical	Laterite Soil (%)	Clay (%)	OPC (%)
Carbon dioxide, CO_2	0.10	0.10	0.10
Iron (iii) oxide, Fe_2O_3	19.00	6.39	5.10
Silicon dioxide, SiO_2	42.00	60.60	23.40
Aluminium oxide, Al_2O_3	35.00	22.40	10.60
Calcium oxide, CaO	0.29	0.00	53.60
Titanium dioxide, TiO_2	1.01	0.92	0.97
Potassium oxide, K_2O	1.80	2.43	0.71
Sulphur trioxide, SO_3	-	5.63	3.76
Sodium oxide, Na_2O	0.27	0.32	0.27
Magnesium oxide, MgO	0.21	0.88	1.15

3.4 Jar test (sedimentation test)

Soil is a complex building construction material where each soil sample is different from one to another. The Jar test was carried outto determine the composition and the ratio of soil particles. It was observed that the laterite soil samples obtained from Ayer Hitam contains 75% sand and 25% clay. In order to get a good distribution of the particles, the mixture ratio of CSEB prototype should be designed with 1:8:2 (cement:soil:sand). Hence, the existing soil should be added with additional sand as much as 25% of the amount used. Whereas, the clay samples taken from Parit Raja had approximately 15% sand and 85% mud and clay. Since the content of mud and clay is very high, the soil should be modified with mixture ratio of 1:5:5 (cement:soil:sand). This modification is done by adding large quantities of sand to form CSEB prototypes that can withstand compressive stress.

3.5 Raw materials density

The density of the raw materials used in the production of CSEB prototypes is determined in accordance to BS 1881:Part 114:1983 [17], and the results are summarized in Table 4.

Table 4. Density of raw materials.

Material	Density (kg/m^3)
Water	1000
Cement	1095
Laterite Soil	1286
Clay	1095
Sand	1524

3.6 Sampling and testing of CSEB prototype

Three different compactions (i.e. 2000, 3000 and 4000 Psi)were considered in the production of CSEB prototypes. As mentioned earlier, the mixture ratios of cement:soil:sand for both types of soil were 1:8:2 for laterite soil, and 1:5:5 for clay, respectively. The raw materials were mixed together manually based on the required mixture ratiosin producing CSEB prototypes with dimension of 100×50×40 mm. The percentage of water required to produce a good mixture is between 10 - 20% of the weight of the soils. While mixing, a small portion of this mixture was hand gripped to form a soil ball and then dropped to the ground at a height of one metre. The shape and condition of the fallen soil ball would determine the optimum water content needed. The optimum rate of water content is achieved when the fallen soil ball broke into 4 or 5 fractions on the ground. As a result, the optimum percentage of water required for the mixture was found to be 15% for laterite soil and 20% for clay, respectively.

The well-mixed mixture was poured into the mould, and then compacted using a hydraulic compaction machine at three different pressure values of 2000 Psi, 3000 Psi and 4000 Psi. Each prototype sample should be carefully taken out from the mould not to damage its shape at the early stage of production. Each sample should be labelled accordingly to the compaction applied, and subsequently placed in an open area. The curing process was done by spraying each of the prototype samples with water manually every day up to either 7 or 28 days.

CSEB prototype samples were tested for compressive strength and water absorption rate at the age of 7 and 28 days. Universal Testing Machine (UTM) was used for testing compressive strength. For water absorption test, which was based on BS EN 771:Part 1:2003 [18], the cooling method was employed where each prototype sample was weighed and dried in an oven for 24 hours at a temperature of 105 °C. The sample was then soaked in water for 24 hours at room temperature. The percentage rate of water absorption was determined by dividing the weight of the sample after immersion with the weight of dry sample, and then multiplied by 100%. Each physical testing of CSEB prototypes was carried out using 3 samples, and the average value of the samples would be considered accordingly as the final value for both compressive strength and water absorption rate.

4 Results and discussion

4.1 Compressive strength

The CSEB prototype samples produced with different compactions were physically tested through compressive stress test at the age of 7 and 28 days. It was observed that the quantity of laterite soil to be poured into the mould would increase accordingly to the compaction pressures applied. For 2000 Psi compaction, the amount of laterite soil required is 400g. Whereas, 427g and 450g of laterite soil were required for compactions of 3000 Psi and 4000 Psi, respectively.

From Fig. 2(a), it can be seen clearly that the compressive strength of laterite soil CSEB prototype is at its optimum when subjected to 2000 Psi compaction with the strength values increase from 9.0 N/mm^2 for sample age of 7 days to 14.8 N/mm^2 for 28 days old sample. However, as the prototype samples were subjected to compactions of 3000 Psi and 4000 Psi, their compressive strength at 28 days still achieved the predetermined level of strength as stated in MS 76:1972, load-bearing fired brick Class (1), which is at least 7 N/mm^2 [19]. In general, it can be inferred that for all values of compaction pressure applied, the compressive strength of laterite soil CSEB prototypes would be greatly influenced by the age of samples. The increasing age of the samples from 7 days to 28 days would also result in the increase of its compressive strength.

To produce clay CSEB samples, the quantity of clay required to be poured into the mould would increase accordingly to the compaction pressures applied. For 2000 Psi compaction, the amount of clay needed was 410g. Whereas, 437g and 462g of clay were required for compactions of 3000 Psi and 4000 Psi, respectively. Since clay has very high percentage of mud content as confirmed by the particle distribution test, it has to be modified by adding sand to the mixture to form a ratio of 1:5:5. It was also observed that laterite soil content is heavier than clay.

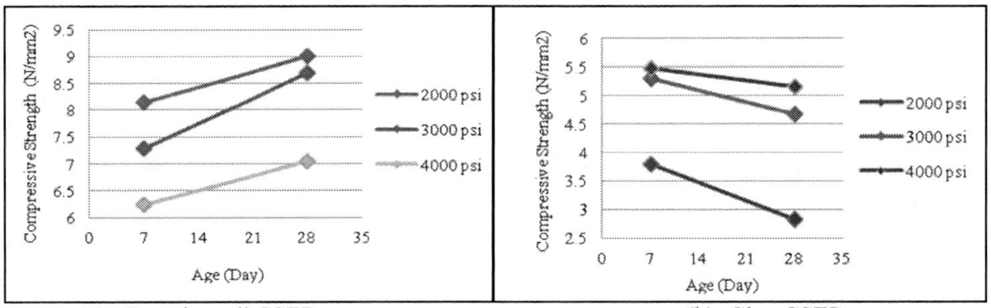

(a) Laterite soil CSEB (b) Clay CSEB

Fig. 2. Compressive strength (N/mm^2) vs.Age of CSEB prototypes (7 and 28 days).

Fig. 2(b) shows the results of the compressive strength of clay CSEB prototype samples. It can be seen clearly that the increased in compaction pressure applied would result in higher compressive strength. The clay CSEB prototype subjected to compaction of 4000 Psi gave the optimum compressive strength value for 7 days sample. Surprisingly, the value of compressive strength lessened to 5.16 N/mm^2 as the age of the sample increased to 28 days. Generally, it can be presumed that for all values of compaction applied, the compressive strength would reduce with the increasing age of the prototype samples from 7 days to 28 days. Additionally, the compressive strength of clay CSEB prototype is generally below the predetermined level of strength as stated in MS 76:1972 for load-bearing fired brick Class (1), which is at least 7 N/mm^2 [19]. This is due to the fact that clay has a very high mud content; and therefore, could not bear the compressive stress imposed.

4.2 Water absorption rate

The results of water absorption rate test on CSEB prototype samples were compared with those of the load bearing fired brick Class (1) based on MS 76:1972 [19]. Although it is not specified in the standard with respect to water absorption rate, the Malaysian construction industry has commonly adopted the absorption rate of less than 15% for this type of brick. Lower absorption rate indicates better quality bricks and higher resistant to the adverse effects of weather.

Fig. 3(a) shows the results of the water absorption rate of laterite soil CSEB prototypes. It can be seen clearly from the graph that the increased in compaction would generally result in lower rates of water absorption. Furthermore, for all values of compaction applied, the quality of laterite soil CSEBs improves as their ages increased from 7 days to 28 days as evidenced by lower percentage rate of water absorption. The rate of water absorption is at its optimum when subjected to 4000 Psi compaction with percentage rates dropped down from 15.4% for sample age of 7 days to 14.5% for 28 days, respectively. This condition occurs due to the fact that higher compaction applied would make the samples become denser, which in turn preventing the absorption of water.

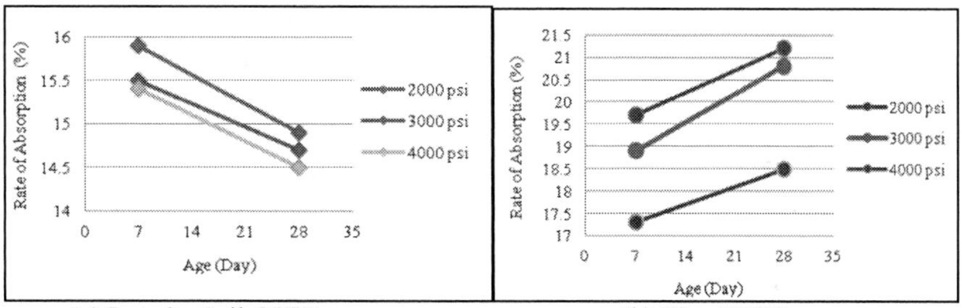

(a) Laterite Soil CSEB (b) Clay CSEB

Fig. 3. Water absorption rate (%) vs.Age of CSEB prototypes (7 and 28 days).

Similarly for clay CSEB prototypes, increasing compaction pressures had resulted in lower water absorption rate as shown in Figure 3(b). Apparently, greater compaction applied means heavier compaction of mixture in the mould resulting in a denser sample that prevents the absorption of water. Thus, increasing the curing age of clay CSEBs would result in higher absorption rate. The optimum absorption rate for clay CSEBs was achieved by the sample subjected to 4000 Psi compaction with the values of 17.3% at the age of 7 days and 18.5% at 28 days, respectively. However, both values of absorption rate still exceed the allowable limit of 15%. The observation on physical properties found that the prototype will decompose at every verticeswhen touched by hand during the transportation process. It can be inferred that the pozzolanic reaction between aggregates within the content of this prototype is very weak.

5 Conclusion

It can be firmly concluded from experimental results that laterite soil is more suitable than clay for CSEB production. It is essential to ensure good distribution of particles during the mixing and compacting processes of producing CSEB. Well-mixed mixture of small and large aggregates can be achieved when the smaller ones filled up the empty spaces in between the larger aggregates. Therefore, mixture ratio of 1:8:2 (cement:soil:sand) was deemed suitable to produce laterite soilCSEBswith 2000 Psi compaction. At this optimum compaction, the aggregate mix remained in its original condition without changing its form. Moreover, all the results of the compressive strength (N/mm^2) and water absorption rate (%) are found to be in accordance with the standards as specified in MS 76:1972 [19].

Our sincere gratitude goes to the staff at Materials Engineering Laboratory, Faculty of Civil and Environmental Engineering, for their tireless assistance during the preparation and testing of samples. I would also like to thank all individuals who have directly or indirectly supported us throughout the course of this study.

References

[1] M.R. Jones and A. McCarthy, Preliminary views on the potential of foamed concrete as a structural materials, Magazine of Concrete Research, **7**(1), 21-31,(2005)

[2] C. Jayasinghe and N. Kamaladasa. *Compressive strength characteristics of cement stabilized rammed earth walls*, Univ. of Moratuwa, Sri Lanka, 1978-1976, (2006)

[3] A. H. Abdullah, S.K.A. Bakar and I.A. Rahman, Indoor thermal performance of an office building using conventional brick versus interlocking compressed earth brick (ICEB) wall, Int. J. of Construction Technology and Management, **1**(1), 22-27, (2013)

[4] S.K. Abu Bakar and A.H. Abdullah, Simulation of thermal performance in an office, *Business Engineering and Industrial Applications Colloquium*, IEEE, 318-323, (2012)

[5] N. Ali, N.A. Zainal, M.K. Burhanudin, A.A.A. Samad, N. Mohamad, S. Shahidan and S.R. Abdullah, Physical and mechanical properties of compressed earth brick (CEB) containing sugarcane bagasse ash, *MATEC Web Conf.*, **47**, 1-7. (2016)

[6] S. Shahidan, H.B. Koh, A.S. Alansi and L.Y. Loon, Strength development and water permeability of engineered biomass aggregate pervious concrete, *MATEC Web Conf.*, **47**, 2-7, (2016)

[7] V. Rigassi and P. OdulCRATerre-EAG, *Compressed earth blocks: Manual of production, Vol. I: Manual of production*, A Publication of The Deutsches Zentrum fur Entwicklungs technologien-GATE, 6, (1985)

[8] UNESCO Chair Earthen Architecture, *Compressed stabilized earth block*, Auroville Earth Institute, (2010).

[9] M.A. Yahaya, *Mengkaji penggunaan tiang kapur terhadap enapan tanah bukan berstruktur kejuruteraan*, Final Year Project Thesis, Universiti Teknologi Malaysia, (2008)

[10] H. Guillaud, T. Joffroy OdulCRATerre-EAG, *Compressed Earth Blocks: Manual of design and construction, Vol. II: Manual of design and construction*, A publication of The Deutsches Zentrum fur Entwicklungs technologien-GATE, 8, (1985)

[11] AURAM Press 3000 (2010), Assessed on 18th October 2016, http://www.aureka.com/earth-construction-equipment/auram-press-3000

[12] M. Shariman, *Pembaikan tanah liat marin dengan pendekatan saliran pugak*, Final Year Project Thesis, UniversitiTeknologi Malaysia, (2008)

[13] A. Bush, *Understanding stabilized earth construction*, Wilson Boulevard, Suite 500 Arlington, Virginia, USA, (1984)

[14] M. Tarmizi, *Pengaruh pasir, kandungan simen dan kaedah pengawetan terhadap konkrit ringan berudara*, Final Year Project Thesis, UniversitiTeknologi Malaysia, (2006)

[15] G.O. Adeyemi and K.A. Wahab, Variability in the geotechnical properties of a lateritic soil from south western Nigeria, Bulletin of Engineering Geology and the Environment,**67**(4), 579-584, (2008)

[16] C.M. Chan and R. Robani, Alternative aggregates from Clay-POC: An exploratory study, UniversitiTun Hussein Onn Malaysia, ICCBT 2008–A–(38), 423-432, (2008).

[17] BS 1377:Part 2:1990, *Methods of test for soils for civil engineering purposes*, British Standard Institute, London, United Kingdom, (1990)

[18] BS 1881: Part 114:1983, *Methods for determination of density of hardened concrete*, British Standard Institute, London, United Kingdom, (1983)

[19] BS EN 771:Part1:2003, *Specification for masonry units. Clay masonry units*, British Standard Institute, London, United Kingdom, (2003)

[20] MS 76:1972, *Specification for bricks and blocks of fired Brick earth, clay or shale, Part 2: Metric units*, Malaysia Standard, 22, Malaysia,(1972)

The Recent Research on Bamboo Reinforced Concrete

Sri Murni Dewi[1,*], and *Devi* Nuralinah[1]

[1]Civil Engineering Department, Brawijaya University, 65145 Malang, Indonesia

Abstract. The paper presents the last research on bamboo reinforced concrete in Brawijaya University Indonesia. Three kinds of structures studied in recent year, the mounting of pegs on reinforcement, the use of lightweight brick to reduce the weight of the beams, and the use the light weight aggregate for bamboo concrete composite frame. All that experiments overcome some problems exist in using bamboo as environmental acceptance structures.

1 Introduction

Bamboo is available everywhere around the world; some regions in the world continue to use bamboo structures to this day. The energy necessary to produce 1 m3 per unit of stress projected in practice for materials commonly used in civil construction has been compared with that of bamboo. It was found that for steel it is necessary to spend 50 times more energy than for bamboo. In the production of one tone steel two tons of CO_2 is produced. In contrast bamboo plant absorbs CO_2 besides producing oxygen. The tensile strength of bamboo is relatively high and can reach 200-300 MPa; this turns the use of bamboo attractive as substitute of steel, especially when considering the relation between tensile resistance and specific weight of bamboo which is six times greater than for the steel.

The potential of bamboo as a substitute for steel reinforcement in reinforced concrete beam has been studied by many researchers (Ghavani [1], Kharee [2], Terai [3], and Dewi [4]). Tensile strength of bamboo reinforcement is almost the same with steel reinforcement, but bamboo material is not as ductile as steel material. All those test results showed that the capacity of flexural beam only 56% of the capacity, if the bamboo reinforcement in beams reached its maximum strength. This low capacity occurs because the bamboo and concrete was not well attached. The recent research head to increase the performance of bamboo reinforced concrete beams by mounted some pegs in the reinforcement. This action aims to strengthen the bond of bamboo and concrete.

As a lightweight material, bamboo is benefit to use as lightweight precast structures. The lightweight structure use for precast beam, to reduce the handling loads. Substitute the steel reinforcement with bamboo is benefit to reduce the weight. For reinforced concrete beams, the concrete in tension area does not require high strength. The concrete can replace

* Corresponding author: srimurnid@ub.ac.id

with light weight brick. The current research related to the replacement the tension area of the beam with lightweight precast brick.

Bamboo reinforced concrete has also been used for roof frame structure or bridge girder. Some study has been done (Dewi [5]), the recent research head to use the light weight aggregate for concrete to reduce the weight of precast frame, and use some pegs into the joints of frame.

2 Bamboo reinforced concrete beam research

Several series of tests on simply supported bamboo reinforced concrete beams have been studied using normal and lightweight aggregates. Ghavani [1] tests 12 cm x 30 cm x300 cm beams with three type reinforcement, steel with reinforcing ratio 0.78%, Bamboo with 3.33% reinforcing ratio, and bamboo with 5% reinforcing ratio. The capacity of steel reinforcement beams were 60 kN. The capacities of bamboo concrete beam with 3.3% reinforcement were 20 kN. The capacities of bamboo concrete beam with 5% reinforcement were 30 kN. Those result equivalent with bamboo reinforcement strength only 35 MPa. It was seen that bamboo concrete beams only has one third capacity.

Karee [2] use 20 cm x 50 cm x 240 cm beams with varying a/d ratio 1,5 – 2, and bamboo reinforcement ratio 1% - 4%.Based on failure load of testing conducted, the equivalent bamboo reinforcement strength varying between 30 MPa-45MPa. The failure loads varied with the compression strength of the concrete, providing a lower failure load for lower compression strengths. Some beams failed in bond-flexure mode, and others failed in shear mode. Those results indicate the requirement of stirrups in the beams.

Terai [3] use 12cm x 25 cm x150 cm beams with varying shear span and shear reinforcement. They use steel stirrups and PP band with varying distance. The maximum equivalent tensile stress reinforcement was 60 – 70 MPa.

From the entire test above, it is found that the potential strength of bamboo not reached yet. To increase the performance of bamboo reinforced concrete beam some treating needs to increase the attached between bamboo and concrete.

As a lightweight reinforcement, bamboo benefit to combine with others lightweight material especially for precast structures. To eluding reduction of strength, the less strength material put in the unsignificant position like tensin area of flexural beam.

Dewi [5] has use bamboo concrete composite for roof frame structure. The loading test of the frame shows that the failure occurs in the joints of the truss and cracks of concrete in a tension bars. The use of pegs in the tension bar at joints area should increase the capacity of the frame.

The other problems of bamboo concrete composite frame is the weight of frame has obstacle for handling. The weight of frame provided from concrete weight despite that the prior strength of the frame comes from bamboo. The solution of those problem is use the lightweight agregate for the concrete parts.

3 Increasing bamboo reinforced concrete beam performance by use pegs along reinforcement

Some research held in Brawijaya University Indonesia, for bamboo reinforced concrete beams. The research related to the effect of adding pegs along the bamboo reinforcement in order to increase the carrying capacity of bamboo reinforced concrete beams. The pegs in the form of bamboo or wood pieces that attached into the reinforcement, is aimed to add the transferred force from the bamboo into the concrete. The addition of pegs along bamboo reinforcement aimed at gaining the transfer of force from the concrete to the bamboo reinforcement, thus the capacity of the beam can be increased.

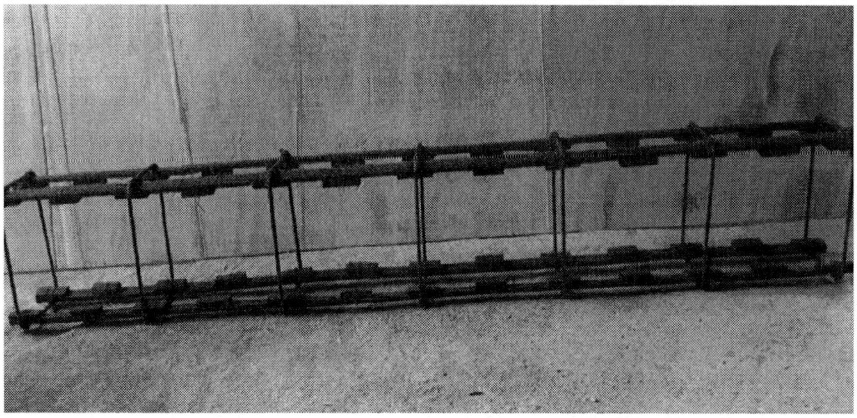

Fig. 1. The pegs along reinforcement.

The research specimen used 18 cm x 28 cm x 160 cm beams. The reinforcement ratio was 0.8% and 1.6 %, and the pegs distance was 6 cm and 12 cm. Two control beams arranged without pegs for reinforcement ratio 0.8% and 1.6 %. The beam loaded in four point load use a spreader beam as shown in Fig. 2.

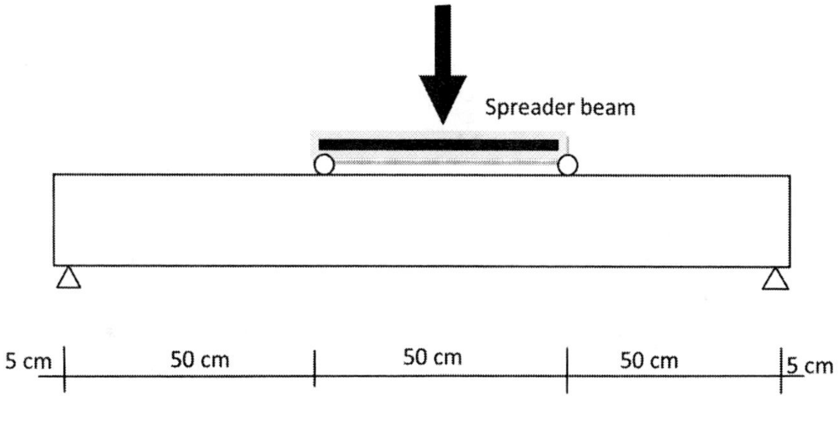

Fig. 2. The loading position.

The load deflection sample for reinforcement ratio 0.8% presented with pegs distance variation in Fig. 3.

The load deflection curve appearance how the inserting pegs increase the capacity and strain energy of the beams. The average tensile stress of the reinforcement also increased from 45 MPa to 90 MPa reached the strength of the bamboo strength. The collapse started from cracked on beam tensile area and the number of cracks is increased, until finally cracked continues at compression area. For all the 24 test objects do not drop the beam reinforcement, but there are some pegs loose and torn.

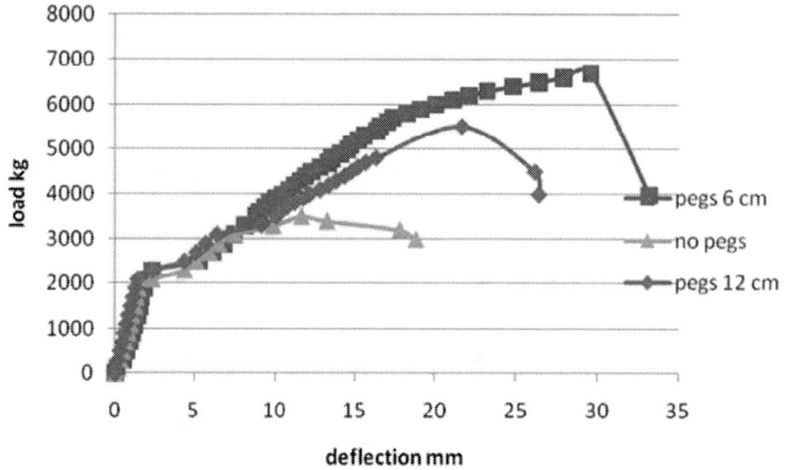

Fig. 3. Load deflection curve with pegs distance variation.

4 Ultilize bamboo and lightweight brick to reduce the weight of precast concrete beam

The lightweight structure needed for precast beam to reduce the handling loads. Replacing steel reinforcement with bamboo is benefit to reduce the weight of buildings. In the case of reinforced concrete beams, area of concrete in tension side is unnecessary . This part can be replaced with a lighter material.. Therefore the concrete at drop area replaced with lightweight brick.

This research was conducted through experiment performed on the beam size 16 cm x 20 cm x 200 cm of bamboo reinforced concrete and light brick. The beam test specimen is shown in Fig. 4.

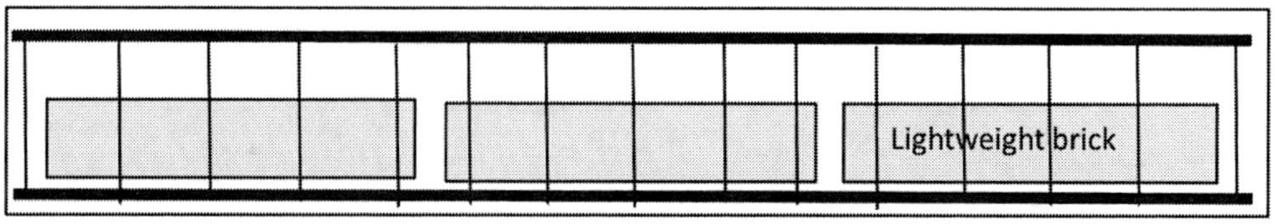

Fig. 4. Bamboo reinforced concrete with lightweight bricks.

The variations of lightweight bricks high were 8.5 cm and 6.5 cm. The beam compared with the ordinary reinforced concrete beam. The use of lightweight brick could reduce 18 % by weight of the beam. The experiments found that the volume weight of the beam varies between 1900 kg /m3 and 2000 kg/m3 and there were no significant decreased of the beam strength. For observations crack patterns, it appeared the relationship between the pattern of cracks with the beam strength, the beam with high strength tend to have more number of cracks.

5 Roof frame from bamboo concrete composite

Bamboo concrete composite has an axial strength compression and tension very high and very appropriate for structure that dominated with axial force likes frame. This composite bamboo concrete frame is useful replace of wood roof frame. The frame should make either in-situ cast concrete or precast concrete.

Fig. 5. Roof frame from bamboo concrete composite.

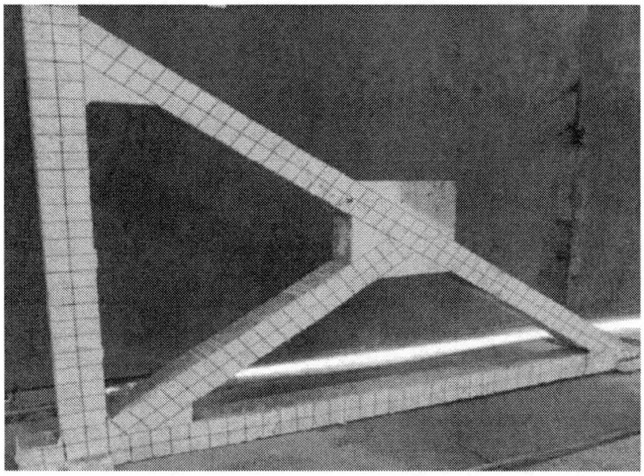

Fig. 6. Precast roof frame from waste brick wall and bamboo reinforcement.

In precast concrete problems, the main obstacle comes from the heavy concrete. The current research is a subtitute the ordinary concrete agregate with lightweight aggregate. For coarse aggregate it were use a waste building brick wall, and for fine aggregate it were substitute a plastic kernel from industrial waste. The composition of those aggregate should reduce the weight by 20%, and only 3% decrease of strength.

6 Conclusion

The use of bamboo for environmentally friendly construction materials, more quickly implemented and added value in terms of cost and environmental sustainability is very interesting to further studied.

The use of pegs along the beam is able to add tensile load bearing capacity reinforcement. Extra pegs reinforcement adds increase effective stress of 45 MPa to 90 MPa.

The use of pegs on reinforcement like hooks on the steel reinforcement will be an opportunity to create a rigid beam-column connection. Another type of pegs was interesting for the next research.

The use of lightweight materials is very advantageous for precast structures and earthquake-resistant structures. The use of construction and industrial waste is benefit in view of environmentally friendly.

The research was held and supported by Ministry of Research Technology and Higher Education Republic of Indonesia.

References

[1] K. Ghavani, Bamboo as reinforcement in structural concrete elements, Cement and Concrete Composite, **27**, 637-649 (2005)

[2] L. Kharee, *Performance evaluation of bamboo reinforced concrete beams*, Master Thesis, University of Texas, Austin, USA, (2005)

[3] M. Terai and K. Minami, Fracture behavior and mechanical properties of bamboo reinforced concrete members, Japan Procedia Engineering,**10**, 2967-2972 (2011)

[4] S.M. Dewi, Bamboo use for earthquake resistance housing, *Int. Conf. on Earthquake and Disaster Mitigation,* Sepuluh November Institute of Technology, Surabaya, Indonesia, (2011)

[5] S.M. Dewi, Roof frame from bamboo concrete composite, J. of Materials Science and Engineering, **B-1**(1), 113-116 (2011)

Experimental Study of Beam-Column Joints Reinforced Concrete with Fiber Concrete and Fly-Ash

Edy Purwanto[1,*], *Bambang* Santosa[1], and *Setiono*[1]

[1]Civil Engineering Department, Faculty of Engineering, Universitas Sebelas Maret, Jl. Ir. Sutami 36A, Surakarta 57126, Indonesia

Abstract. Reinforced concrete elements in the beam - column joint (HBK) plays a very important role for maintaining the structure when subjected to the lateral load, despite the presence of many bars in the area often results in imperfections implementation. The use of fly-ash and fiber in the area of HBK can be one of available solutions; both can increase the strength of concrete and reduce the reinforcement. This study discusses the usage of fly-ash in concrete with proportion of 25% by weight of cement and dramix steel fibers and 10 kg for one m^3 of concrete. This study uses the specimens of HBK with a variation of the concrete at age of 28 and 90 days. The analysis of 28 days HBK results: specimen with normal and with fly ash have average load of 10.53 kN and 6.97 kN where specimen with Fly-ash + additive has 12.65 kN (20.13% higher than normal specimen). At the age of 90 days, HBK with normal concrete, fly ash and fiber concrete can withstand load at 12.73 kN, 10.87 kN, and 13.15 kN respectively, where HBK with fiber + fly-ash reaches at 13.38 kN or has 5.00% more than HBK with normal concrete.

1 Introduction

Elements of the beam column relationship at a reinforced concrete portal play a very important role in accepting the burden of the structure, especially the lateral loads, such as earthquakes. The mechanism requires that the collapse of the structure failure should occur in the area of the beam, or there exists a ductile flexural failure pattern. But in fact, at the time of the earthquake, many parts of the column element, even the beam column relationships, that have failed, resulting a brittle collapse. The weak point in the power beam column relationship can be caused partly by the improper reinforcement detailing and porous concrete casting due to the difficulty of concrete to enter into the area blocked by a number of bars in the area.

To overcome the weakness in the beam column relationship, one of which could be done by increasing the strength of the concrete. It can be done with the addition of steel fibers [1] and the use of fly-ash concrete, where the right use of steel fibers and fly-ash is able to increase the compressive strength and tensile strength of concrete. The use of steel

* Corresponding author: edypurwanto68@staff.uns.ac.id

fibers and fly-ash concrete will also reduce the risk of porous concrete in the area of the beam-column relationship, because the use of stainless steel fibers as micro steel will reduce the amount of reinforcement in the beam-column relationship. Improving the ability of beam-column element of the relationship will enhance the capabilities and behavior of the structure, so it will reduce the risk of destruction of a building and prevents loss of material and human lives [2].

Besides increasing the strength of concrete, the use of fly-ash, which in this case is a waste of coal, will also be very helpful in overcoming the problems of environmental pollution. The presence of waste coal fly-ash at the current amount is very large; it is caused by the development of industries that use energy sources of coal. The problem of global warming has yet to be overcome as well, so that the use of fly-ash in the construction industry will be one of the solutions that can be relied upon. Replacement of Portland cement with fly-ash content of 25% by weight of cement, also greatly reduces the use of cement significantly, especially for the production of 1 ton of Portland cement will also produce 1 ton of CO_2. CO_2 is one of the sources of global warming. Thus reducing the amount of consumption of cement in the construction industry will also reduce the amount of CO_2 in the air.

2 Theoretical

Regional relations beam column is area critical in maintaining the strength and stability of the structure when subjected to the lateral load. As a result of lateral forces acting on the structure, the end of the bending moment on beams at the same joint will rotate in the same direction. This poses a major shear forces on the relationship of column beam. During this time, the area was accounted for strong for their restraint system by reinforcement stirrup. In practice there will be a number of connections of beams and columns as well as the stirrup itself that cause difficulty in implementation. Lack of stirrup bars on the area of main reinforcement that are not unconstrained and forced out due to the high pressure of the concrete core and an error in the installation of reinforcement detailing and implementation of the joint can cause the collapse of a building.

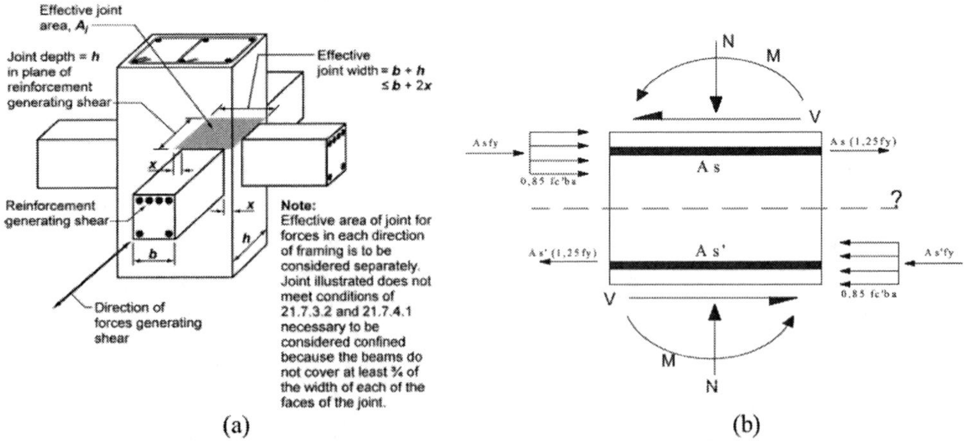

Fig. 1. Effective joint area (a), Calculation of lateral shear force on beam-column joint (b) [3].

From Fig. 1 (b) above, horizontal shear force in the area of relationships beam column (HBK) can be calculated by assuming that the bending elements framed on HBK has reached the capacity, by setting the beam flexural tensile strength of As (1,25fy). Lateral shear force on beam-column joint can be calculated as follows [3][4]:

$$V_u = T_{b1} + T_{s1} + T_{s2} + C_{b2} - V_{col\,1} \quad (1)$$

$$T_{b1} + T_{s1} + T_{s2} = \alpha fy\,(A_{s1} + A_{s\text{-}s1} + A_{s\text{-}s2}) \quad (2)$$

$$C_{b2} = T_{b2} = A_{s2}\alpha fy \quad (3)$$

where ;
Vu : shear force that occurs in joints (kN)
α : 1.25
fy : beam longitudinal reinforcement yield strength (MPa)
Shear strength can be provided by HBK depending on the conditions of confinement that work on HBK. HBK shear strength equation can be calculated as follows:

$$V_{jn} = {}^{c}\sqrt{f'_c}A_j \quad (4)$$

where ;
V_{jn} : nominal joint shear force (kN)
C : 1.7 (HBK bridled at four sides)
: 1.25 (HBK confined on three sides or two opposite sides)
: 1 (for all other relationships)
A_j : effective joint area (mm^2)

Research by the use of fly-ash and steel fiber dramix is expected to improve aspects of strength (f'c) and aspects of ductility.

3 Method

3.1 Material

Additional components in the concrete material are as follows:

i) Fly-ash originating from ready mix concrete plant in the town of Sukoharjo. See Fig. 2 (a).
ii) Steel fibers Dramix RC-80/60 – BN with a length of 60 mm; diameter 0.75 mm; tensile strength of 1225 N / mm^2 and an aspect ratio of length to diameter of 80 . See Fig. 2 (b).
iii) Additive from SikaCim®Concrete, added on fresh concrete at a dosage of 250 mlper sack of cement (50 kg).

(a) Steel fiber Dramix (b) Fly-ash

Fig. 2. Additional materials research.

3.2 Test specimen

There are two types of test specimen, namely :

i) The cylinder speciment, diameter of 150 mm and a height of 300 mm.
ii) Elements of beam-column joint (HBK) with column dimensions of 150 mm x 150 mm height 2000 mm and the beams with dimensions of 150 mm x 200 mm with a length of 1500 mm with a shape as shown in Fig.3. Concrete Repair held only in the meeting of the beam and fields, which are all 500 mm in the central part of the column and 150 mm on the beam in front of the column. The specifications of the test specimen HBK is presented in Table 1 below.

Table 1. Specifications of the test specimens HBK

Sampel	A day of test (day)	Amount	Information
HBK N	28	3	Normal
HBK F	28	3	Fly-ash
HBK AD	28	3	Fly-ash+ Additive
HBK N	90	3	Normal
HBK F	90	3	Fly-ash
HBK S	90	3	Fiber
HBK SF	90	3	Fiber+Fly Ash

3.3 Setting test HBK

Details of the test specimen HBK, setting tools and testing methods of HBK specimen can be seen in the Fig. 3.

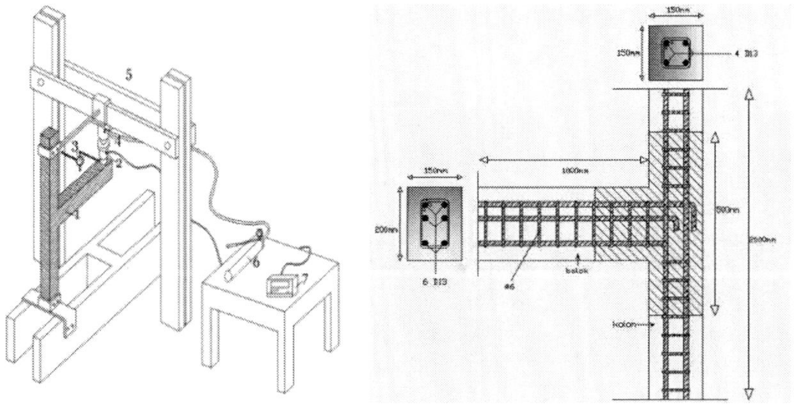

Fig. 3. Setting test and specimens.

4 Result and discussion

4.1 Cylinder

Cylindrical specimen test results on 28th day and 90th day can be seen in the Table 2 and Table 3. From Table 2, it is known that the use of fly-ash concrete with additive at the age of 28 days, can increase the strength of concrete from 18.40 MPa to 23.16 MPa, an increase of 25.86%. Being on a 90th day test from Table 3, the use of fly-ash concrete and fiber will increase from 25.70 MPa to 29.91 MPa, an increase of 16.38%. From of the Table 2 and

Table 3 can also be seen that the use of fly ash on 28th day and 90th day has not provided maximum strength because it is still in the process of reaction.

Table 2. Compressive strength of cylinder on 28th day.

No.	Specimens	Pmax (kN)	f'c (MPa)	f'cr (MPa)	Comparisons (%)
1	S-N1 $_{28}$	340	19,24	18,40	0,00
2	S-N2 $_{28}$	340	18,99		
3	S-N3 $_{28}$	300	16,98		
4	S-F1 $_{28}$	160	8,93	9,56	-48,04
5	S-F2 $_{28}$	150	8,56		
6	S-F3 $_{28}$	200	11,17		
7	S-AD$_{1\text{-}28}$	350	19,59	23,16	25,86
8	S-AD$_{3\text{-}28}$	410	22,31		
9	S-AD$_{6\text{-}28}$	500	27,59		

Table 3. Compressive strength of cylinder on 90th day.

No.	Specimens	Pmax (kN)	f'c (MPa)	f'cr (MPa)	Comparisons (%)
1	S-N1$_{90}$	460	25,46	25,70	0,00
2	S-N2$_{90}$	450	24,91		
3	S-N3$_{90}$	470	26,72		
4	S-F1$_{90}$	480	26,80	25,65	-0,19
5	S-F2$_{90}$	460	25,57		
6	S-F3$_{90}$	450	24,58		
7	S-S1$_{90}$	480	27,53	27,15	4,11
8	S-S2$_{90}$	540	30,83		
9	S-S3$_{90}$	410	23,10		
10	S-SF1$_{90}$	580	31.82	29.91	16.38
11	S-SF2$_{90}$	510	28.86		
12	S-SF3$_{90}$	510	29.05		

4.2 Test of specimens HBK – at 28 days

HBK testing at 28 days presented in the table of load and deflection relationship as shown incFrom Fig. 4 and the obtained value of the maximum load HBK on several conditions as follows:

The average maximum load on HBK-Normal concrete is

$$P_{max\ rt\text{-}rt} = \frac{10.70 + 10.80 + 10.10}{3} = 10.53 \text{ kN} \tag{5}$$

The Average maximum load on HBK-Fly ash is

$$P_{max\ rt\text{-}rt} = \frac{6.90 + 7.00 + 7.01}{3} = 6.97 \text{ kN} \tag{6}$$

The average maximum load HBK-AD (fly-ash + additive) is

$$P_{max\ rt\text{-}rt} = \frac{12.86 + 12.20 + 12.89}{3} = 12.65 \text{ kN} \tag{7}$$

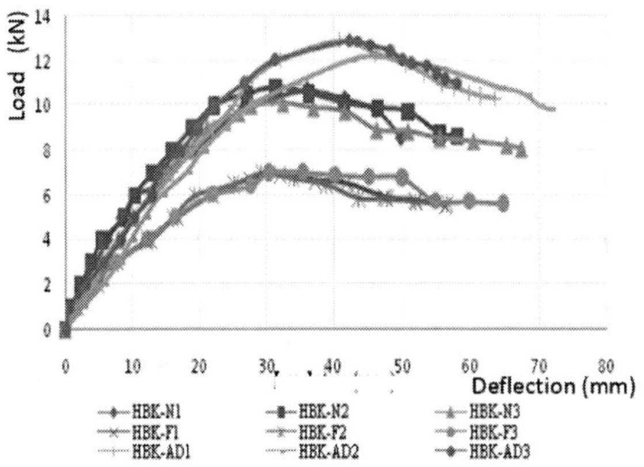

Fig. 4. Relationships of load – deflection of HBK specimens at 28 day.

Then the value of the average load that can be accepted by normal concrete HBK is 10.53 kN and a maximum load average HBK fly ash concrete is 6.97 kN or decreased by 33.81%, while the HBK-Fly ash + additive capable of withstanding the load amounted to 12.65 kN or an increase of 20.13% when compared to normal concrete HBK, HBK while if compared with fly ash concrete increased by 81.49%. Increased ability to withstand the burden shows that the addition of Concrete Additive SikaCim® is able to make fly ash concrete at 28 with more dense so that stronger in the weight-bearing.

4.3 Test of specimens HBK – at 90 days

HBK testing at 90 days presented in the table of load and deflection relationship as shown in Fig. 5, and the calculated mean of the maximum load as follows:

i) The average maximum load on HBK-Normal is 12.73 kN
ii) The average maximum on HBK fly ash concrete is 10. 87 kN

Based on the above calculation, it shows that the capacity load of HBK fly ash concrete is still below normal concrete. The average maximum load - average which can be held by HBK fly ash concrete is 10.873 kN, which is 14.59% below normal HBK concrete that can withstand at 12.730 kN. Increased load obtained HBK fly ash concrete is not maximized due to the slow hardening concrete.

From Fig. 6, it can be seen that the calculated mean of the maximum load average on HBK- concrete fiberis 13.15 kN. So that the fiber-reinforced concrete HBK maximum load increased by 3.27%. The increase in the maximum load of the HBK is a result of the increase of the quality of concrete. The rise of all concrete deflection is caused by the change in style that resulted a shake on the side of the column.

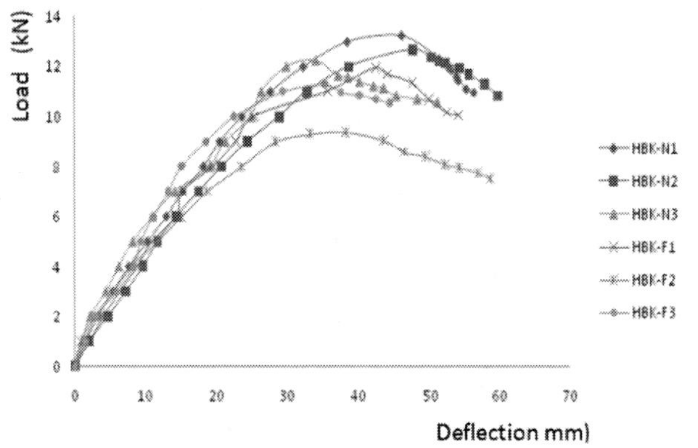

Fig. 5. Relationships of load – deflection of HBKN – HBKF specimens at 90 days.

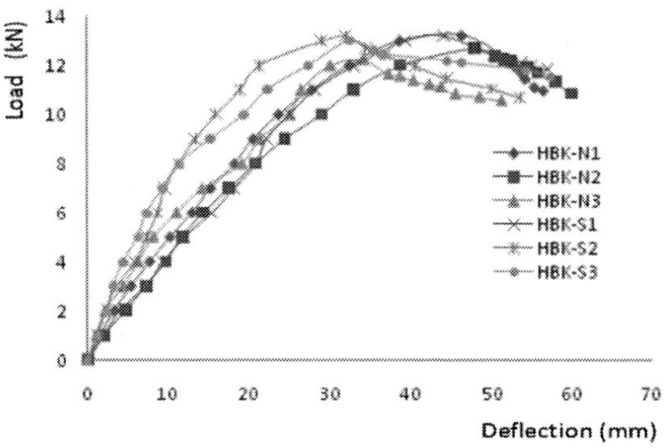

Fig. 6. Relationships of load – displacement of HBKN – HBKS specimens at 90 days.

From Fig. 7, then the calculated mean of the maximum load average on HBK- concrete fiber + fly-ashis 13.38 kN. Based on the above analysis it is known that the average maximum load HBK fiber + fly ash concrete is higher than the normal concrete and HBK test specimens fiber + fly ash concrete experience longer deflection of the test specimen HBK normal concrete. The addition of fly ash + fiber provides maximum expenses increased by 5.08% of normal concrete.

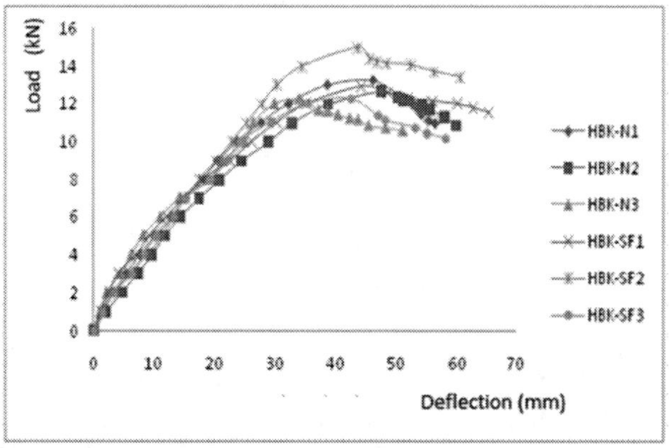

Fig. 7. Relationships of load – displacement of HBKN – HBKSF specimens at 90 days.

4.4 Summary of results

From Table 4, it can be seen that, the use of fly-ash and additives in concrete at 28 days will be able to increase the capacity of the HBK of 20.13% against the normal concrete, while the use of fly-ash and fiber in the concrete life of 90 days will be able to increase capacity by 5.11% compared to normal HBK and increased by 27.07% compared to HBK concrete at 28 days. The use of fly ash at 28 days and 90 days has not been an effective outcome for fly ash is still the process of reaction and should be used additive substances.

Table 4. Summary of testing specimens

Specimen	Load (kN)	Comparisons (%)
Results of the test at 28 day		
HBK N – 28	10,53	0,00
HBK F – 28	6,97	-33,81
HBK AD - 28	12,65	20,13
Results of the test at 90 day		
HBK N – 90	12,73	0,00
HBK F – 90	10,87	-14,61
HBK S – 90	13,15	3,30
HBK SF - 90	13,38	5,11

5 Conclusion

From the testing and analysis conducted drawn conclusions as follows:

i) The use of fly-ash 25% by weight of cement and additive in concrete cylinder at 28 days can increase the strength of concrete from 18.40 MPa to 23.16 MPa, an increase of 23.16%. where at age of 90, the use of fly-ash concrete and fiber will increase from 25.70 MPa to 29.91 MPa, an increase of 16.38%.

ii) The use of steel fibers Dramix and fly-ash in concrete HBK 28 days will improve the ability of concrete 10.53 kN to 12.65 kN, with an increase of 20.13%.

iii) The use of steel fibers Dramix and fly-ash in concrete HBK at age of 90 days will improve the ability of concrete from 12.73 kN to 13.38 kN amounted to an increase of 5.11%. HBK compares to 28 days increased by 27.07%

iv) The use of fly-ash with additive materials and steel fibers effective for increasing load capacity at a time can be a solution to overcome the coal waste in the form of fly-ash.

References

[1] N. Ganesan, P.V. Indira, and R. Abraham, *Steel fibre reinforced high performance concrete beam-column joints subjected to cyclic loading*, ISET J. of Earthquake Technology, **44** (3-4), 445 – 456, (2007)

[2] B. Suhendro, *Concrete fiber local, application concepts and issues*, National Seminar on Construction Materials Engineering, Yogyakarta, Indonesia, (1991)

[3] ACI 318M - 11, *Building code requirements for structural concrete (ACI 318M – 11)*, ACI Standard and Commentary, USA, (2011)

[4] J.C. McCormac, *Reinforced concrete design*, 5th Ed., Vol. II, Erlangga Publisher, Jakarta, Indonesia, (2001)

Performance of Reinforced Concrete Beam with Differently Positioned Replacement Zones of Block Infill under Low Impact Loads

Shahrul Niza Mokhatar[1,*], *Mohamed Muftah* Mustafa[2], *Salih* Saed Rouwab[2], and *Josef* Hadipramana[1]

[1]Jamilus Research Centre, Universiti Tun Hussein Onn Malaysia, Parit Raja-Batu Pahat, 86400, Malaysia
[2]Faculty Civil and Environmental Engineering, Universiti Tun Hussein Onn Malaysia, Parit Raja, Batu Pahat 86400, Johor, Malaysia

Abstract. This paper reveals a study performed on reinforced concrete with artificial aggregate concrete block infill composite beams to innovate a lightweight reinforced concrete utilizing polyethylene (PE) waste materials, such as waste plastic bags. Six beam specimens of normal reinforced concrete (NRC) and different block infill replacement zone positions RCAI (RZ_1) beams containing 100% MAPEA with 50, 95, and 1,000 mm width, height, and length, respectively, were provided for the block infill, whereas RCAI (RZ_2) with different block infill positions containing a 100% MAPEA with 50, 115, and 1000 mm width, height, and length were provided and tested under low impact load. The steel impactor with blunt nose dropped at 0.6 m height which equivalent to 3.5 m/s. The behaviors of the beams were studied relative to the impact force-time and displacement-time histories, the flexural/ bending cracks, and the impact failure. Results show that the overall failure modes of all the beam specimens were successfully recorded. In addition, the residual displacements of the RZ_2 was almost same than those of the RZ_1 and the significantly lower than those of the NRC. In the reinforced concrete beams, less stressed concrete near the neutral axis can be replaced by certain light weight material like waste plastic bags as modified artificial polyethylene aggregates to serve as an artificial aggregate.

1 Introduction

Currently, the reduction of concrete heavy weight has been a challenge to engineers. The weight reduction is necessary because of the increasing demand for large constructions, as well as for economic and aesthetic purposes [1]. Moreover, practitioners have been challenged in producing cost-effective solutions to satisfy this demand. Furthermore, sustainability is another essential concern in the construction industry. Sustainability means

* Corresponding author: shahruln@uthm.edu.my

satisfying the demands of the present generation without compromising the ability of the future generations to meet their needs.

This can be achieved by reducing, reusing, and recycling waste materials; it also means shifting toward green technologies and reducing Carbon dioxide CO2 emissions. Sustainability can also be characterized as the minimized use of resources. As a result, the development of lightweight concrete as an alternative to normal weight concrete has attracted immense interest [2]. At present, lightweight concrete has become most popular among researchers because of its special properties. Furthermore, lightweight concrete can address the weight problem in contrast to normal concrete with its approximate density at about 2,400 kg/m3 [3]. The demand for lightweight concrete in various modern construction applications has been increasing because of its advantages, including its low density that significantly replaced less stressed or unutilized concrete structures by low-cost, sustainable, and environment-friendly alternatives. This can be accomplished by replacing the aggregate in concrete with increased amounts of supplementary cementing materials, or by replacing less stressed or unutilized concrete with other low-cost and environment-friendly materials.

At the same time, Malaysian citizens generate nearly 0.8 kg of garbage daily, which mostly contains polyethylene (PE) materials. Urban areas are estimated to produce 1.5 kg garbage every day. In fact, waste plastic bags are among the major contributors to the increasing number of solid waste production in this country, in which only 1% to 3% of the solid waste materials are recycled. Plastic bags can cause certain inconveniences to society. The aggregate production of plastic waste can produce lightweight concrete characteristics by replacing normal plastic aggregates with lightweight ones. One of the solutions to this solid waste management problem is the conversion of waste plastic bags to modified artificial polyethylene aggregates (MAPEA), which can be utilized on concrete. With its high solidity, good abrasion behavior, and weak thermal conductivity, the use of MAPEA can significantly improve several properties of concrete. In addition, this aggregate has zero impact and crushing value [4]. However, Mokhatar [5] found that polyethylene can be used as an aggregate replacement for reinforced concrete beam (RCB) with artificial aggregate concrete block infill (RCAI). Similarly, the combination of the economy of reinforced block infill beams and the strength of RC beams is attempted in the block infill beams. The bond between the concrete layers and the block infill at its concrete interface are inadequate to ensure that sliding between the two layers is prevented.

In reinforced concrete beams (RCB), the strength of the concrete near the neutral axis is not completely used. The concrete above the neutral axis is less stressed, whereas the concrete below serves as a shear transmitting medium. The partially used RCB concrete can be replaced by blocks, such that, the combination of the economy of the RCB and the strength of the RC beams is attempted in the block-filled beams. The bond between the block and the concrete layers at the block concrete interface should be adequate to prevent slips between the two layers. An experiment was performed on the block-filled RCB based on the maximized stresses at the top and bottom of the beams, as well as the zero stress in the neutral axis. Thus, cheap and light materials can be used near the neutral zone for light weight and economy [6]. The behavior of block-infilled RCB under cyclic loading was investigated through experimentation, and the results were compared with the conventional RCB values. The load carrying capacity of an infilled beam was then observed to be approximately 80% of the conventional RCB [7]. The ratio between the depth of the inserted AAC blocks to the overall depth of the solid section can be a factor contributing to this result [8]. In fact, the experiment on the block-infilled RCB revealed that nearly 30% of concrete was conserved [9]. The analytical and experimental investigations on infilled frames under static and lateral loads suggested that the load-carrying capacity and stiffness of the infilled frame is greater than those of the plain frame [10]. The experiment on the beams with light weight brick core with Ferro cement form observed that the ultimate load of the beams increased when welded wire mesh was used in the fiber cement forms [11].

Impact loading is among the significant factors contributing to the effective use of structural materials. This is predominantly among the factors that focused on the structural elements of both steel and concrete. The impact properties of concrete have attracted interest because of the possibility of impact loading in the predicted service life of a concrete structure. Various situations have been encountered wherein structures are subjected to certain types of impact loading. These situations include various forms of missile impact, gas explosions, construction accidents, vehicle impacts, and pile driving. The impact from an object colliding with a concrete structure can be classified into hard and soft impacts, depending on the relative characteristics of the projectile (or striking object) and the target (or structure). Different forms of damage can occur during a hard impact with concrete structures. The contact zone is subjected to intense dynamic stresses, which can produce crushing, shear failure, and tensile fracturing. These stresses can result in spilling and crater formation, penetration, back-face scabbing, perforation or shear failure, and flexural failure of the target [12].

In summary, the statically loaded RC structure displays dissimilar behavior relative to impact loading structures. Thus, impact loading is among the dynamic factors that should be considered in the analysis and design of structures. The prediction on how high- and low-velocities impact load levels can affect normal RC elements was experimentally tested. The test considered the dynamic (impact) behavior of RC element as studied by Chen et al. [13], Fiquet et al. [14], Kishi et al.[15], and Giraud et al. [16], Fujikake et al. [17], and Saatci et al. [18]. However, these studies were not to fully elucidated the dynamic behavior of RCB with block-infilled. Thus, this paper aims to present the results of the impact response of reinforcement concrete beam RCB with RCAI with differently positioned replacement zones of block infill and to propose a novel lightweight reinforced concrete (RC). The block infill can be placed within of the beam cross - section near the calculated depth of the neutral axis.

2 RCB with RCAI with different block replacement zones (BRZ)

The replacement zone in the RC beam, where concrete near neutral axis may be replaced by block infill was obtained with the use of stress block for the limited state design of the RCB. This block infill replacement zone is shown in Fig. 1. Block replacement zone, RZ_1 can be written as:

$$RZ_1 = d_{max} - d_{min1} \tag{1}$$

$$d_{min1} = \frac{3}{7}X_u + X_{b1} \tag{2}$$

$$X_{b1} = \frac{4}{7}X_u\sqrt{1-\frac{f_{bk}}{f_{ck}}} \tag{3}$$

$$d_{max} = D - d' \text{ where, } D \text{ is a total depth} \tag{4}$$

$$d' = (2 \times \text{cover} + \text{Ø bar}) \tag{5}$$

where d' is the concrete layer thickness necessary to maintain the bond between concrete and steel. Each side of the tension reinforcement is assumed to provide a concrete thickness of not less than the cover provided to satisfy the bond requirement [19]. Block replacement zone, RZ_2 can be expressed by

$$RZ_2 = d_{max} - d_{min2} \tag{6}$$

$$d_{min2} = \frac{3}{7}X_u + X_{b2} \tag{7}$$

$$X_{b2} = \frac{3}{8}(\frac{4}{7}X_u) \tag{8}$$

Where, X_{b1} and X_{b2} are distance between surface block replacement zone to compression force. While, d_{min1} and d_{min2} are distance between surface block replacement zone to top beams. X_u is depth of the neutral axis

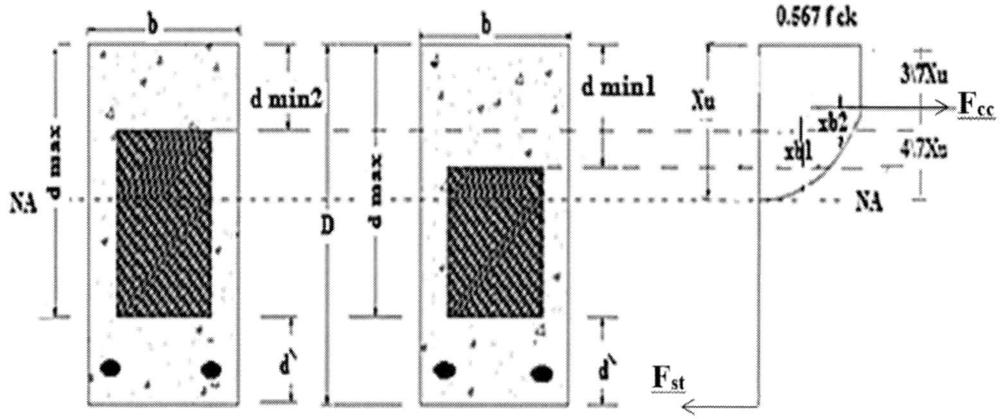

Fig. 1. Different positions of the block infill replacement zone for reinforced concrete infill beam

3 Specimen preparation

The, NRC, RCAI (RZ_1) and RCAI (RZ_2) designs were based on Eurocode 2. All beams possessed a rectangular cross section, 170, 250, and 1,000 mm in width, depth, and length, respectively. Each beam was provided with a bottom longitudinal bar. Two 12 mm bars were used as the bottom steel in each beam. The tensile strength obtained for each bar was 519 MPa. A 6 mm hot-rolled plain round rebar was used for shear reinforcement. The stirrups were vertically rectangular and open at the top. The details of the reinforcement arrangement are shown in Fig. 2. All beams were tested under low impact. The impact test was performed to compare the residual displacements and the loads between the NRC and the RZ_1 and RZ_2 beams. Six beams were prepared, two were used as control samples (NRC) and four were the experimental samples RZ_1 and RCAI RZ_2. RZ_1 beams containing 100% MAPEA with 50, 95, and 1,000 mm width, height, and length, respectively, were provided for the block infill, whereas RZ_2 with different block infill positions containing a 100% MAPEA with 50, 115, and 1000 mm width, height, and length were also provided, as shown in Table 1. The position of the block infill can be placed within the compression region of the beam cross-section when subjected to impact near the calculated depth of the neutral axis.

Table 1. Sample size and volume of the beams

Sample	Size	Volume
Normal Reinforced Concrete, NRC	0.25 m × 0.17 m × 1 m	0.0425 m^3
Beam with RZ_1	0.05 m × 0.095 m × 1 m	0.00475 m^3
Beam with RZ_2	0.05 m × 0.115 m × 1 m	0.00575 m^3

*RZ (1 & 2) = Different positions of the replacement zone of the block

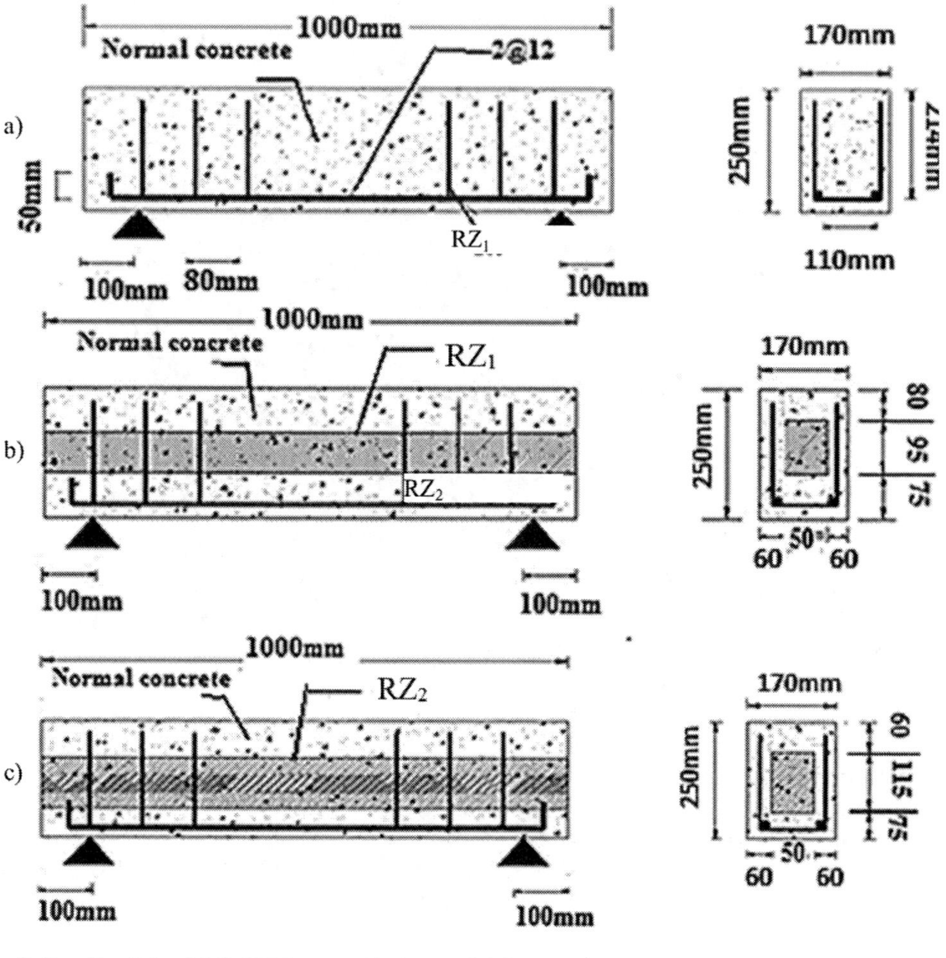

Fig. 2. Details of the a) NRC b) beam with RZ_1 and c) beam with RZ_2

The beam length was 1,000 mm, with a clear span of 800 mm during the preparation for testing. Tests for the NRC, RZ_1, and RZ_2 were conducted to investigate the low impact behavior and failure mechanism of the beams. The test set-up was individually designed to test the beam specimens under impact loading. In this study, approximately 100 kg of drop-weight is acted vertically from a height of 0.6 m, and velocity of the load to be imposed on the specimen was approximately 3.5 m/s. During the impact tests, the steel frame structures and the support conditions should be sufficiently stiff to support the load without any significant deformation. The steel frame, a main structure for the falling weight impact tests, load cell and data logger are shown in Fig. 3.

DOI: 10.1051/matecconf/201710302003

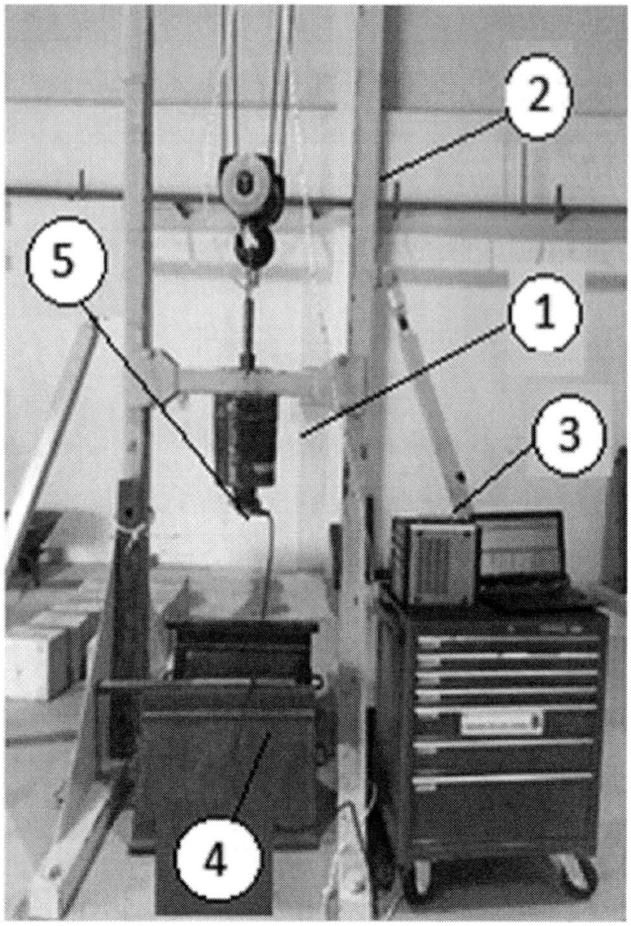

Fig. 3. Main structure of the falling weight system: 1) drop weight, 2) steel frame, 3) data logger, 4) support, and 5) load cell

4 Experimental results and discussions

The main purpose of the experiment is to obtain experimental results to validate the NRC investigation and to classify the ability and accuracy of the proposed RCN, RZ_1, and RZ_2 beams by comparing them to the experimental results. In addition, this study can enrich the understanding on how reinforced concrete structures behave under various impact loads. In this study, the 100 kg steel weight attached to the load cell was acted vertically from a height of 0.6 m. In the low impact loads, the overall failure modes of the specimens were successfully recorded. The typical failure modes of a specimen NRC, RZ_1, and RZ_2 obtained from the impact loading test is shown in Fig. 4. The cracks started to vertically form at the bottom or near the impact point at this drop height. It shows that all beams failed in a bending-failure mode. However, only fine diagonal cracks which started form at the bottom propagating at an angle of approximately 45 degrees upward obtained from specimen NRC. The point at the bottom region of NRC was accomplished from the development of crack pattern and damage. This result suggested that in reinforced concrete beams, the concrete near the neutral axis can be replaced by some light weight concrete beam with RCAI to prevent the formation of a crack.

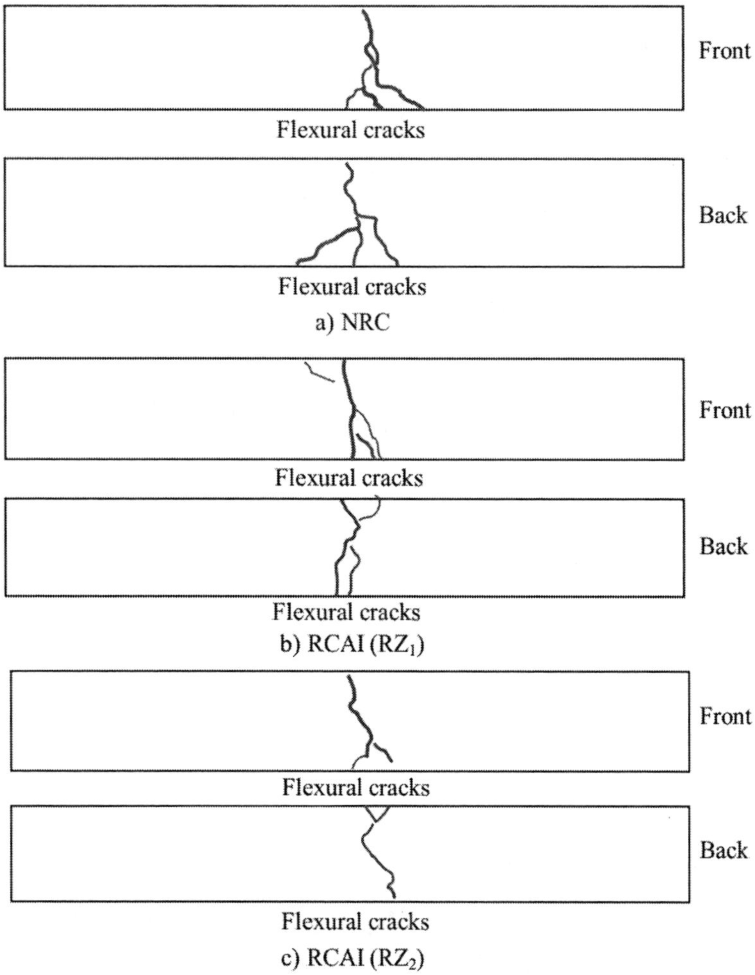

Fig. 4. Crack pattern and failure of the tested beams

The average of the impact force-time and the displacement-time histories of the NRC, RZ_1, and RZ_2 beams are shown in Fig.5. The displacement-time histories during the 100 ms from the beginning of impact are also shown. Fig. 5 indicates that the minimal mid-span displacement of the RZ_2 is 2.54 mm at 10 ms, whereas RZ_1 reported a minimal displacement of 3.10 mm as well. The NRC showed a considerably higher value of displacement value at 4.39 mm relative to the RZ_1 and RZ-2. The displacement basically depended on the stiffness of the beam. Thus, the stiffness of the RZ_1 and the RZ_2 are higher than that of the NRC. From the impact force-time histories shown in Fig. 6, the maximum impact force for the RZ_1 and RZ_2 specimens were approximately 370 and 360 N, respectively, at 10 ms. The NRC has a maximum impact force at 275 N, which is lesser than the above specimens. However, two NRC and the RCAI (RZ_1, RZ_2) specimens behaved differently with regard to the fracture energy or energy absorption. While, the residual energy absorption of the RZ_2 was almost same than those of the RZ_1 and the significantly lower than those of the NRC. The impact loads were nearly the same but the shape of the displacement curve and the periods of the vibrations varied to a certain extent. The difference may be accounted for by the decreased energy loss and absorption of the NRC, which caused by the shorter impact event and larger deflection at failure than those of the RCAI (RZ_1, RZ_2) specimens. Thus, increasing the number of blocks infill placed

DOI: 10.1051/matecconf/201710302003

within the compression region of the beam cross-section when subjected to the impact load, and near the calculated depth of the neutral axis, as presented in Fig. 1, enhances the ability of the RC beam to carry loads at large deflections, thus the energy absorption of the NRC was increased.

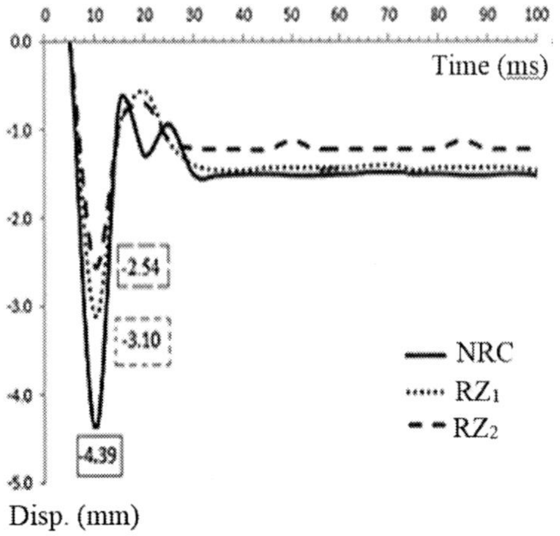

Fig. 5. Mid-span displacement-time histories

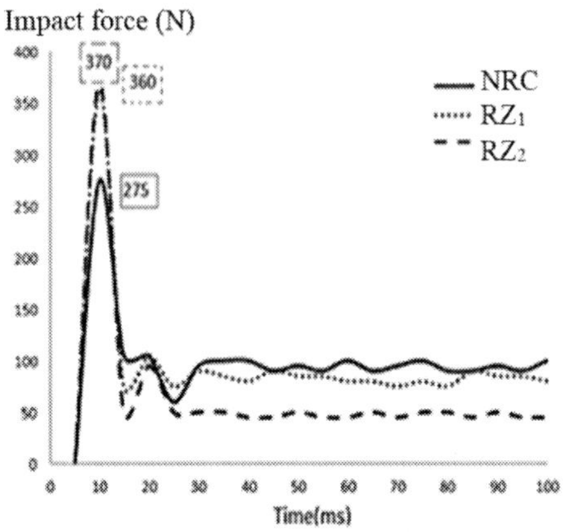

Fig. 6. Impact force-time histories

5 Conclusion

Experimental specimens of the low impact load tests of the NRC beams, and the development of the infill term beams RCAI (RZ_1 and RZ_2) with modified artificial polyethylene aggregates as 100% replacement under a drop height of 0.6 m with impact weight was described. The position of the block replacement zone initiated in the neutral axis, and nearly half of the concrete was replaced by the block infill. The following conclusions are drawn based on the test results:

i) Under low impact loads, insignificant difference in displacement and impact force between the NRC and the RZ_1 and RZ_2 block infill-filled beams were observed. Observed that the average displacement of the NRC a considerably higher value of displacement value relative to the RZ_1 and RZ_2. These results showed that the proposed use of blocks infill produced desirable results under impact loads.
ii) The reduction and economy of light weight beams depended on the percentage of the modified artificial polyethylene aggregates as natural aggregates that replaced the concrete.
iii) Lightweight RC using waste materials as MAPEA was proposed; this necessitate further study.

The authors would like to acknowledge financial and facilities support provided by University Tun Hussein Onn Malaysia under Research Supporting Grant Scheme RSGS Code U100.

References

[1] P. W. Matthew, and D. F. H. Bennett, *Economic long span concrete floors*, British Cement Association, Camberley, Surry, UK, (1990)

[2] K. Ramamurthy, E.K. Kunhanandan Nambiar, and G. Indu Siva Ranjani, Classification of studies on properties of foam concrete, Cement & Concrete Composites, **31**(6), 388–396, (2009)
[3] A.P. Adewuyi, and T. Adegoke, Exploratory study of periwinkle shells as coarse aggregates in concrete works, J. of Engineering and Applied Sciences, **3**(6), 1-5, (2008)
[4] M.M. Mustafa, S.N Mokhatar, and Z.M. Jaini, Properties of different artificial lightweight aggregates and their effect on concrete strength, J. of Engineering and Applied Sciences, **11**(6), 3726- 3730, (2016)
[5] S.N. Mokhatar, Z. M. Jaini, M. K. Burhanudin,M. L. A .Jeni, and M. N. N .Ismail, An experimental study of reinforced concrete beams with artificial aggregate concrete infill under impact loads, *Proc. of the Int. Civil and Infrastructure Engineering Conference,* Springer Singapore, 691-701, (2015)
[6] P. Govindan, and A.R. Santhakumar, Composite action of reinforced cement concrete beams with plain masonry infills, *Proc. of Int. Symposium on Reinforced and Pre-stressed Masonry*, Edinburgh, Scotland, (1984)
[7] P. Govindan, M. Lakshmipathy, and A.R. Santhakumar, Behaviour of infilled beams under cyclic loading, Indian Concrete J., **61** (5), 124-128, (1987)
[8] V.Vimonsatit, A.S. Wahyuni, and H. Nikraz, Reinforced concrete beams with lightweight concrete infill, J. of Scientific Research and Essay, **7**(27), 2370-2379, (2012)
[9] U.B. Choubey, and U. Gupta. Structural response of brick-filled reinforced concrete beams, *Proc. of 4th Int. Seminar on Structural Masonry for Developing Countries,* Madras, India, 226-233, (1992)
[10] S. Arulselvan, K. Subramanian, P.E.B. Pillai, and A.R. Santhakumar, RC infilled frame-RC plane frame interactions for seismic resistance, J. of Applied sciences,**7**(7), 942-959, (2007)
[11] H .F. Ezzat, N.A. Mohamed, B.S. Yousry, and A.A. Ahmed, Permanent ferrocement forms: A viable alternative for construction of concrete beams, *The 30th Conf. on Our World in Concrete & Structures,* Singapore, (2005)
[12] R.S. Ravindrarajah, and M. C. Lyte, Energy absorbing concrete for impact loading, *Proc. of the Int. Conf. on Advances in Concrete Construction,* Hyderabad, India, (2008)
[13] Y.Chen, and I.M .May, Reinforced concrete members under drop-weight impacts, *Proc. of the Institution of Civil Engineers-Structures and Buildings*, **162**(1), 45-56, (2009)
[14] G. Fiquet, and J. Dacquet, Study of the perforation of reinforced concrète slabs by rigid missiles – Experimental study Part II, Nuclear Engineering and Design, **41**(1), 103-120, (1977)
[15] N. Kishi, T. Ohno, H. Mikami, and T .Ando, Effects of boundaries conditions on impact behaviors of reinforced concrete beams subjected to falling-weight impact loads, J. of Structural Engineering, **731**(I-63), 299-316, (2002)
[16] R. Gueraud, A. Sokolovsky, M. Kavyrchine, and M. Astruc, Study of the perforation of reinforced concrete slabs by rigid missiles – General introduction and Experimental Study, Part I, Nuclear Engineering and Design, **41**(1), 91-102, (1977)
[17] K. Fujikake, B. Li, and S. Soeun. Impact response of reinforced concrete beam and its analytical evaluation, J. of Structural Engineering, **135**(8), 938 - 950, (2009)
[18] S. Saatci, and F J. Vecchio. Nonlinear finite element modeling of reinforced concrete structures under impact loads, ACI Structural J., **106**(5), 717-725, (2009)
[19] R .Patel, S.K. Dubey, and K. K. Pathak, Analysis of RC brick filled composite beams using MIF, Procedia Engineering, **51**, 30-34, (2013)

Stress-Strain Relationship of Synthetic Fiber Reinforced Concrete Columns

Rosidawani[1,2,*], *Iswandi* Imran[1], *Ivindra* Pane[1], and *Saptahari* Sugiri[1]

[1]Civil Engineering Department, Faculty of Civil and Environmental Engineering, Institut Teknologi Bandung, 40116 Bandung, Indonesia
[2]Civil Engineering Department, Faculty of Engineering, Universitas Sriwijaya, 30662 Palembang, Indonesia

Abstract. Many empirical confinement models for normal and high strength concrete have been developed. Nevertheless, reported studies in the term of confinement of fiber reinforced concrete are limited. Whereas, the use of fiber reinforced concrete in structural elements has become the subject of the research and has indicated positive experiences. Since the stress-strain relationship of concrete in compression is required for analysis of structural members, the study of the stress-strain relationship for synthetic fiber reinforced concrete is substantial. The aim of the study is to examine the capabilities of the various models available in the literature to predict the actual experimental behavior of synthetic fiber reinforced high-strength concrete columns. The experimental data used are the results of the circular column specimens with the spiral spacing and the volume fraction of synthetic fiber as the test variables. The axial stress-strain curves from the tests are then compared with the various models of confinement from the literature. The performance index of each model is measured by using the coefficient of variation (COV) concept of stress and strain behavior parameter. Among the confinement models, Cusson model shows the closest valid value of the coefficient of variation.

1 Introduction

Many empirical models of compression stress-strain relationship of normal and high strength concrete have been generated by many researchers [1-5]. However, those of models have limitation to fiber reinforced concrete, especially for synthetic fiber reinforced concrete (SNFRC). While, the development of research on synthetic fiber reinforced concrete has been expanded in structural applications. Moreover, this trend indicates positive experiences.

Experimentally compression stress-strain relationship of fiber reinforced concrete gave the significantly different result of the confinement effect from that of no fiber concrete. Consequently, it showed different trends of material behavior. The short and discrete fibers in the concrete mixture modify its typical brittle response to a significant increase in ductility about the no fiber concrete. Furthermore, a previous study has shown that SNFRC

* Corresponding author: rosidawani@gmail.com

could improve the post-peak behavior, ductility, and energy dissipation ability of concrete elements under axial load [6]. It has proven that the combination of synthetic fibers and transverse steel confinement could reduce the amount of confining reinforcement and still behaved in ductile response [6].

The effect of fiber content investigation on the post-peak softening branch of the compressive stress-strain curve of steel fiber reinforced concrete (SFRC) found that an increase in the volumetric ratio of steel fibers leads to a relatively flatter post-peak softening branch of the curve [7]. Accordingly, the same trend was also found on SNFRC [6]. Also, this flatter post-peak softening branch behaved on the relatively same pattern due to the combination of discrete steel or synthetic fibers and reduced transverse steel confinement. Although few published research [8 - 10] have indicated positive field experiences with SNFRC, the ductile behavior of the fiber reinforced concrete which was being concerned with the research interest in the seismic performance of structural elements become an important issue.

Since the stress-strain relationship of concrete in compression is required for analysis of structural members, the study of stress-strain relationship for synthetic fiber reinforced concrete is substantial. Therefore, this paper presents the study, which examines the capabilities of the various models available in the literature to predict the actual experimental behavior of synthetic fiber reinforced high-strength concrete columns.

2 Experimental and analytical program

2.1 Specimens

Investigating the ability of the various models available in the literature to the relative performance of the number of column specimens based on the experimental study [6]. The test specimens used in the circular column specimens with the diameter of 135 mm and the cover of 10 mm contained six longitudinal bars of 10-mm diameter. Those specimens used spiral arrangement for the confinement, which consisted of 6-mm diameter ties spaced at either 30, 45 and 60 mm. The specimen details and instrumentation are shown clearly in Fig. 1.

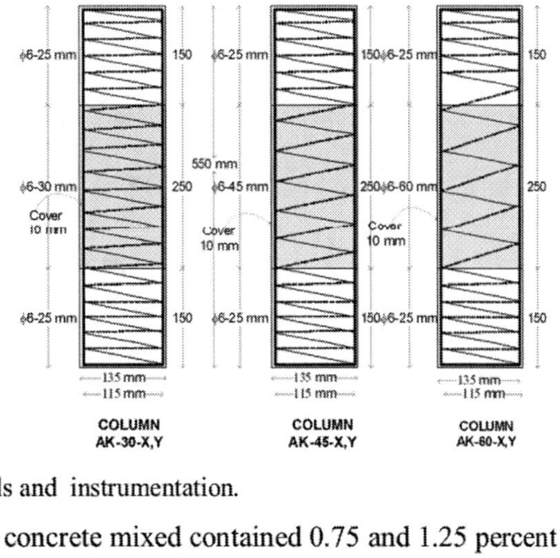

Fig. 1. Specimen details and instrumentation.

Furthermore, the concrete mixed contained 0.75 and 1.25 percent by weight of synthetic fiber and was cast with nominally 70 MPa concrete. The strength of concrete for each series is respectively 64 MPa, 65 MPa and 63 MPa for the 1.25%, 0.75%, 0% volume fraction of fiber inclusion in the concrete mixtures.

2.2 Confinement models

Many researches proposed confinement model for high strength concrete columns, while a few for fiber reinforced concrete. Each model had various limitations. Since limited models available for fiber reinforced concrete, therefore, this investigation used both of fiber reinforced concrete [11,12] and no fiber concrete confinement [1-5] model. The aim is about examining the capabilities of the various models from the literature to predict the actual experimental behavior of synthetic fiber reinforced high-strength concrete columns. Table 1 illustrates those models.

Table 1. Confinement models

Model	Ascending zone	Descending Zone	Peak Stress
	$0 < \varepsilon_c < \varepsilon_{cc}$	$\varepsilon_c > \varepsilon_{cc}$	
Cusson [1]	$f = f'_{cc}\left[\frac{k\frac{\varepsilon_c}{\varepsilon_{cc}}}{k-1+\left(\frac{\varepsilon_c}{\varepsilon_{cc}}\right)^k}\right]$	$f = f'_{cc}exp$ $[k_1(\varepsilon_c - \varepsilon_{cc})^{k_2}]$	$\frac{f'_{cc}}{f'_c} = 1 + 2.1\left(\frac{f_{le}}{f'_c}\right)^{0.7}$
Razvi [2]	$f = f'_{cc}\left[\frac{k\frac{\varepsilon_c}{\varepsilon_{cc}}}{k-1+\left(\frac{\varepsilon_c}{\varepsilon_{cc}}\right)^k}\right]$	$f = f'_{cc} - (\varepsilon_c - \varepsilon_{cc})\cdot$ $\frac{0.15f'_{cc}}{(\varepsilon_{85cc} - \varepsilon_{cc})}$	$f'_{cc} = f'_c + k_1 f_{le}$
Silva [3]	$f = f'_{cc}\left[\frac{k\frac{\varepsilon_c}{\varepsilon_{cc}}}{k-1+\left(\frac{\varepsilon_c}{\varepsilon_{cc}}\right)^k}\right]$	$f = f'_{cc}exp$ $[k_1(\varepsilon_c - \varepsilon_{cc})^{k_2}]$	$f'_{cc} = f_{co} + 4.1f_l$
Assa [4]	$f = f'_{cc}\frac{[\alpha x + (\beta-1)x^2]}{[1+(\alpha-2)x+\beta x^2]}$		$\frac{f'_{cc}}{f'_c} = 1 + 3.36\frac{f_l}{f'_c}$
Li [5]	$0 < \varepsilon_c < \varepsilon_{co}$ $f = E_c\varepsilon_c +$ $\cdot\frac{(f'_c - E_c\varepsilon_{cu})}{\varepsilon_{co}^2}\varepsilon_c^2$ $\varepsilon_{co} \le \varepsilon_c \le \varepsilon_{cc}$ $f = f'_{cc} - \frac{(f'_{cc} - f'_c)}{(\varepsilon_{cc} - \varepsilon_{co})^2}$ $x(\varepsilon_c - \varepsilon_{cc})^2$	$f = f'_{cc} - \beta\frac{f'_{cc}}{\varepsilon_{cc}}$ $x(\varepsilon_c - \varepsilon_{cc})$	$f'_{cc} = f'_c\left[-1.254 + 2.254\right.$ $\left.\sqrt{1 + 7.94\alpha_s\frac{f_l}{f'_c}} - 2\alpha_s\frac{f_l}{f'_c}\right]$ $f'_c \le 52\ MPa$ $\alpha_s = (21.2 - 0.35f'_c)\frac{f_l}{f'_c}$ $f'_c > 52\ MPa$ $\alpha_s = 3.1\frac{f_l}{f'_c}$
Mansur [11]	$f_c = f_{cc}$ $\left[\frac{\beta\left(\frac{\varepsilon_c}{\varepsilon_{cc}}\right)}{\beta-1+\left(\frac{\varepsilon_c}{\varepsilon_{cc}}\right)^\beta}\right]$	$f_c = f_{cc}$ $\left[\frac{k_1\beta\left(\frac{\varepsilon_c}{\varepsilon_{cc}}\right)}{k_1\beta-1+\left(\frac{\varepsilon_c}{\varepsilon_{cc}}\right)^{k_2\beta}}\right]$	$\frac{f_{cc}}{f_o} = 1 + 0.6\left[\rho_s\frac{f_y}{f_o}\right]^{1.23}$
Campione [12]	$\frac{\sigma}{f'_c} = \frac{\beta(\varepsilon/\varepsilon_o)}{\beta-1+(\varepsilon/\varepsilon_o)^\beta}$	$\frac{\sigma}{f'_c} = \eta_d exp$ $\left[-k_d\left(\frac{\varepsilon}{\varepsilon_o} - x_d\right)^2\right]$	$\frac{f'_{cc}}{f'_c} = 1 + 2.1\left[k_e\frac{f_l}{f'_c}\right]^{0.7}$

Cusson [1] developed a confinement model which consists of two-part of stress-strain relationship for ascending and descending formula. This model was formulated based on the experimental study on 50 large-scale, under concentric loading of the concrete compressive strength of the specimens ranged from 60 to 120 MPa, while the ties used yield strength from 400 to 800 MPa.

Razvi [2] proposed the model of the ascending parabolic branch and descending branch up to 20% of the peak stress. The model was determined based on the nearly full-size column specimens of different shapes, sizes, reinforcement configuration, concrete strength, range 414 to 828 MPa and tie yield strength range 400 to 1387 MPa.

Silva [3] developed two formulas for the confinement model which were based on Popovic's model (1973) for ascending model and Fafitis and Shah's model (1985) for descending branch. This model was predicted from 15 circular spirally confined (with yield strength 440 to 560 MPa) normal and high strength concrete columns (35.5 to 125.4 MPa).

Assa [4] proposed confinement model based on 22 150 x 300 mm concrete cylinders which were tested under monotonic concentric compression with 20 to 90 MPa of the strength of concrete. The stress-strain behavior of concrete confined determined by the several types of reinforcement configurations.

Li et al. [5] developed the confinement model from the experimental results based on Murugama's model (1993). The experimental covered forty short column on cross-sectional shapes of cylindrical (240 mm) and square (240 x 240 mm). The parameter of the tests consists of the strength of concrete (35.2 to 82.5 MPa), tie reinforcement yield strength (445 and 1318 MPa), spacing and volumetric ratio.

Mansur [11] proposed the confinement model from an experimental study for high strength plain and fiber concretes confined by transverse ties. This proposed model has reported well agreed with the stress-strain curves generated experimentally. It explained the confinement might also be provided by using a combination of transverse reinforcement and steel fibers. However, it was established the idea in term of incorporating the improvement in ductility due to the confinement in structural design. Furthermore, the experimental covered 180 specimens on cross-sectional shapes of prismatic consist of 75x75x150 mm, 100x100x200 mm, 125 x 125 x 300 mm dimensions with the strength range of 60-120 MPa and transverse ties yield strength of 493 MPa.

Campione [12] developed the model to express the stress-strain relationship in compression of fiber reinforced concrete for both normal and high strength concrete, with and without conventional reinforcement. This model allows one to determine the maximum strength and strain capacity by determining the effective concrete core of the confining devices at rupture.

3 Results and discussions

This paper shows the relative performance of analytical models of stress-strain profile to the experimental results of the confined concrete column specimens. The combination of spirally confinement and volume fraction of synthetic fiber on twenty short column specimens observed experimentally [6]. Therefore, those parameters became the primary interest in predicting the performance of analytical confinement models to the experimental results.

The experimental results show that the ascending branches of the measured stress-strain relationships in all columns tests are almost linear, while the descending branches are flatter than those of no fiber concrete columns. Also, the descending branches of the columns with a closer spiral spacing and higher volume fraction of synthetic fibers are flatter than those with wider spacing and lower volume fraction of synthetic fiber. Seven stress-strain analytical models have been investigated their relative performance to the experimental stress-strain relationships of the specimens to propose the stress-strain models of this behavior. Still, these models apply to high strength concrete on circular sections, and three of models are generated from the fiber reinforced concrete.

The results show that the descending part of the curve of the synthetic fiber reinforced concrete columns are flatter and relatively maintain the stress less than 50% of the peak stress down. These behaviors are significantly different to no fiber concrete specimens. Therefore, the strain associated with the stress of 50% of the peak stress can be considered as an additional curve characterization of the descending portion of the stress-strain models. Consequently, the trend of the descending branch of the analytical models adopts the definition of the corresponding strain of 85% of peak stress and strain of 50% of peak stress to characterize the post-peak region. Fig. 2-Fig. 4 describe those analytical models.

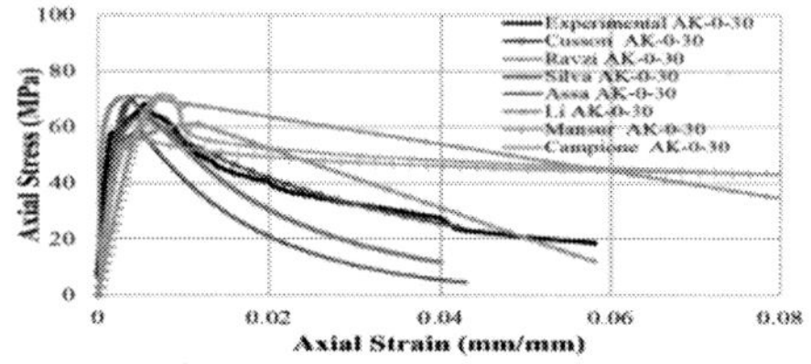

(a) Specimens AK-0-30

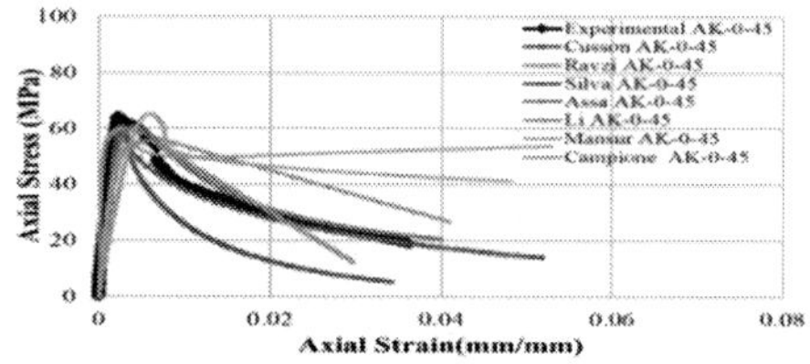

(b) Specimens AK-0-45

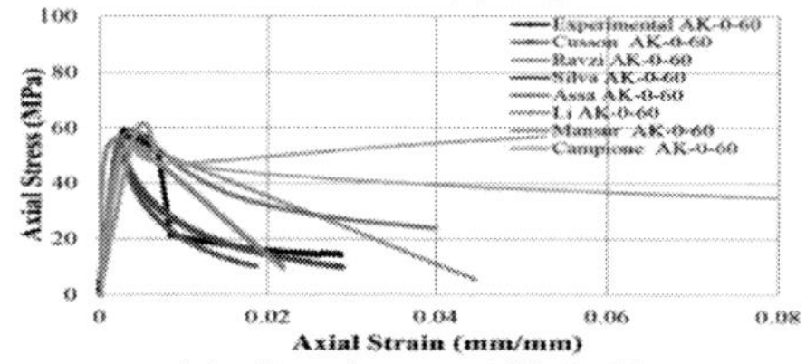

(c) Specimens AK-0-60

Fig. 2. Stress strain curve for specimens AK-0.

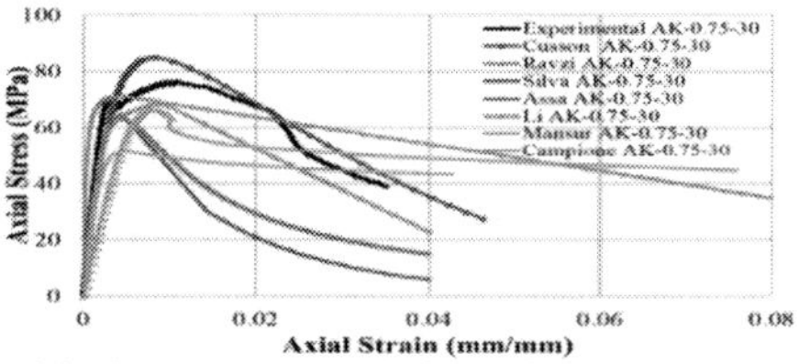

(a) Specimens AK-0.75-30

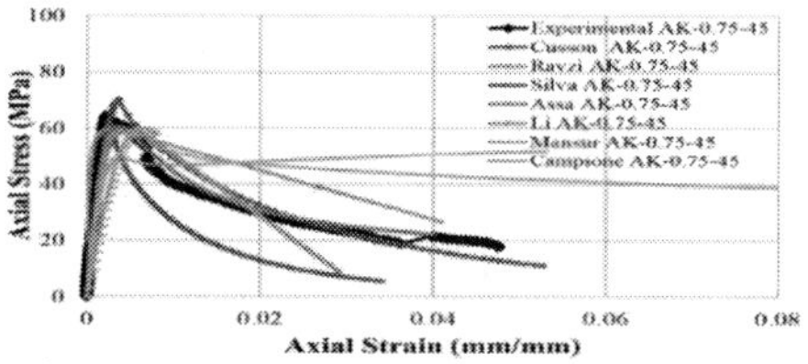

(b) Specimens AK-0.75-45

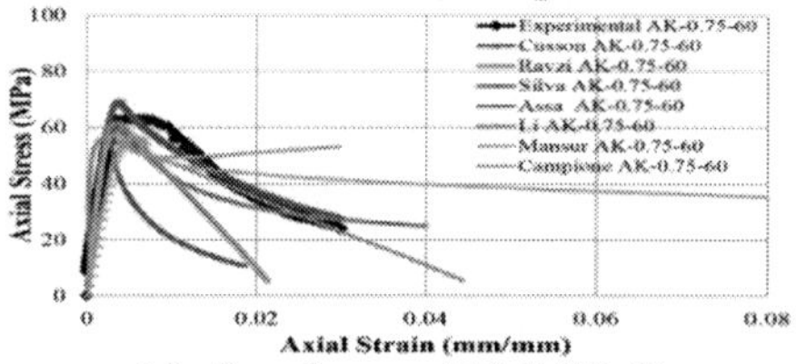

(c) Specimens AK-0.75-60

Fig. 3. Stress strain curve for specimens AK-0.75.

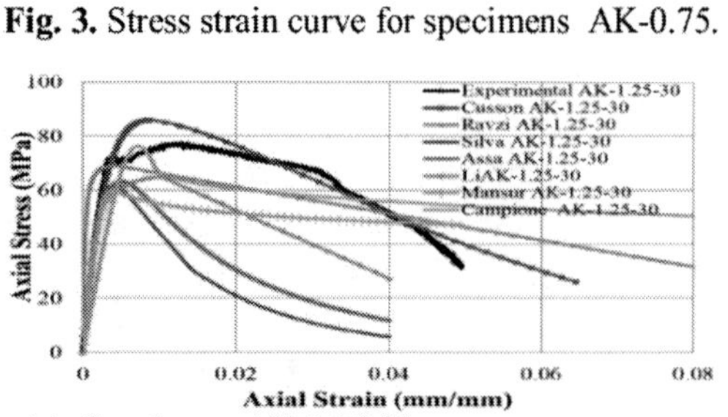

(a) Specimens AK-1.25-30

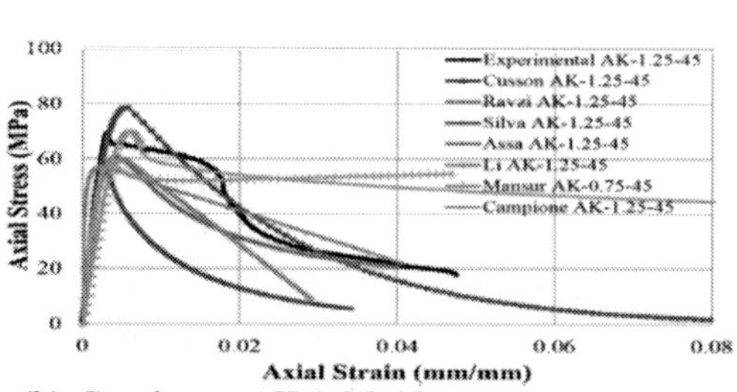

(b) Specimens AK-1.25-45

DOI: 10.1051/matecconf/201710302004

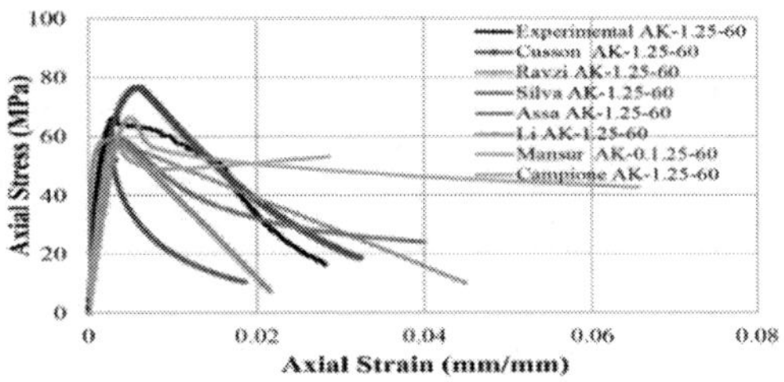

(c) Specimens AK-1.25-60

Fig. 4. Stress-strain curve for specimens AK-1.25.

The graphics in Fig. 2 to Fig. 4 describe the trends of the experimental results. Each picture shows the type of specimens with all confinement models applied to the experimental result. Cusson model shows the best-fitted graphic to the experimental result among others. The graphic trend for each of specimens tends to fit the experimental results. The descending curve retains the trend graph with the experimental data.

Ravzi and Li model give an overview chart opposites between the SNFRC and no fiber concrete specimens. This model describes the slightly underestimate trend chart on SNFRC. Conversely, no fiber concrete specimens illustrated an overestimate trend.

While Silva model provides the same underestimate trend for all the specimens, Assa model gives an illustration that varies in both types of material. Those models show that some are close enough to the experimental results while others remain to underestimate.

Mansur model, which was proposed for the steel fiber reinforced concrete column, shows an opposite trend in the chart between the SNFRC and no fiber concrete specimens. The model which was applied to the no fiber concrete provides an overestimate trend chart, nevertheless, an underestimate trend for the SNFRC.

Like Mansur, Campione model gives the same trend for an overview chart opposite between the SNFRC and no fiber concrete specimens. The overestimate trend charts are provided to the no fiber concrete, while the trend of SNFRC tends to approach with a slight underestimate value and different trend graphs.

Most of the models which are applied to either fiber reinforced concrete or no fiber concrete can define the peak stress (f_{cc}) well. Meanwhile, the peak strain (ε_{cc}) value of all the models on all the specimens is varied. The significant differences are shown in the post-peak response of each model.

The capabilities of the stress-strain models of confined concrete columns to predict the actual behavior are determined by the parameters considered as the peak confined stress, f_{cc}, and the corresponded peak strain, ε_{cc}. Therefore, these parameters are used to show the relative performance of the various models using the ratios of the predicted parameters to the respective experimental parameters for all the specimens. The ratio of peak stress and the corresponded peak strain are shown in Table 2 - Table 5 respectively.

Table 2. Comparison of peak confined stress f_{cc} of no fiber concrete

Specimen	$f_{cc\ exp}$	$f_{cc\ model/exp}$						
		Cusson	Ravzi	Silva	Assa	Li	Mansur	Campione
AK-0-30	68.088	1.023	1.043	0.867	0.928	0.900	0.827	1.049
AK-0-45	64.547	0.935	0.928	0.816	0.937	0.861	0.844	1.011
AK-0-60	59.518	0.958	0.950	0.832	0.994	0.901	0.902	1.038
Average		0.972	0.974	0.838	0.953	0.887	0.858	1.032
COV		5%	6%	3%	4%	3%	5%	2%

Table 3. Comparison of peak confined stress f_{cc} of SNFRC

Specimen	$f_{cc\ exp}$	$f_{cc\ model/exp}$						
		Cusson	Ravzi	Silva	Assa	Li	Mansur	Campione
AK-0.75-30	76.47	1.113	0.932	0.846	0.827	0.894	0.677	0.866
AK-0.75-45	64.55	1.090	0.928	0.863	0.937	0.925	0.774	0.934
AK-0.75-60	63.31	1.097	0.893	0.871	0.934	0.974	0.916	0.898
AK-1.25-30	77.36	1.111	0.894	0.802	0.817	0.840	0.771	0.986
AK-1.25-45	69.49	1.132	0.822	0.802	0.871	0.859	0.847	1.007
AK-1.25-60	66.08	1.160	0.895	0.795	0.895	0.876	0.863	1.005
Average		1.117	0.894	0.830	0.880	0.895	0.808	0.949
COV		2%	4%	4%	6%	5%	10%	6%

Table 4. Comparison of peak confined strain ε_{cc} of no fiber concrete

Specimen	$\varepsilon_{cc\ exp}$	$\varepsilon_{cc\ model/exp}$						
		Cusson	Ravzi	Silva	Assa	Li	Mansur	Campione
AK-0-30	0.0054	0.722	0.841	0.910	0.924	2.143	0.824	1.367
AK-0-45	0.0022	1.219	1.337	1.586	1.923	2.788	1.799	2.695
AK-0-60	0.0028	0.908	0.853	1.052	1.351	1.521	1.296	1.786
Average		0.949	1.011	1.183	1.400	2.151	1.306	1.949
COV		26%	28%	30%	36%	29%	37%	35%

Table 5. Comparison of peak confined strain ε_{cc} of SNFRC

Specimen	$\varepsilon_{cc\ exp}$	$\varepsilon_{cc\ model/exp}$						
		Cusson	Ravzi	Silva	Assa	Li	Mansur	Campione
AK-0.75-30	0.011	0.764	0.423	0.405	0.406	0.855	0.425	0.713
AK-0.75-45	0.002	1.672	1.337	1.531	1.648	2.544	1.848	2.798
AK-0.75-60	0.005	0.831	0.516	0.617	0.731	0.824	0.769	1.246
AK-1.25-30	0.012	0.712	0.351	0.379	0.411	0.840	0.358	0.585
AK-1.25-45	0.003	1.908	0.832	1.113	1.398	1.849	1.279	1.897
AK-1.25-60	0.003	2.035	0.938	0.969	1.269	1.341	1.199	1.631
Average		1.320	0.733	0.836	0.977	1.376	0.980	1.479
COV		47%	51%	54%	55%	51%	58%	56%

Table 6. Comparison of strain at 85% of peak stress (ε_{85cc}) of no fiber concrete

Specimen	$\varepsilon_{85cc\ exp}$	$\varepsilon_{85cc\ model/exp}$						
		Cusson	Ravzi	Silva	Assa	Li	Mansur	Campione
AK-0-30	0.009	0.864	2.879	0.783	1.031	2.200	3.086	1.105
AK-0-45	0.006	0.667	2.163	0.724	1.350	1.742	1.786	1.326
AK-0-60	0.007	0.443	1.388	0.514	1.125	1.101	1.273	0.946
Average		0.658	2.144	0.674	1.169	1.681	2.049	1.126
COV		32%	35%	21%	14%	33%	46%	17%

Table 7. Comparison of strain at 85% of peak stress (ε_{85cc}) of SNFRC

Specimen	$\varepsilon_{85cc\ exp}$	$\varepsilon_{85cc\ model/exp}$						
		Cusson	Ravzi	Silva	Assa	Li	Mansur	Campione
AK-0.75-30	0.022	0.805	1.212	0.324	0.381	0.732	1.292	0.692
AK-0.75-45	0.006	0.975	2.164	0.689	1.198	1.589	0.000	2.151
AK-0.75-60	0.011	0.617	0.824	0.291	0.592	0.586	0.872	1.043
AK-1.25-30	0.032	0.719	0.826	0.234	0.252	0.566	0.766	0.362
AK-1.25-45	0.015	0.658	0.786	0.282	0.544	0.645	0.919	0.572
AK-1.25-60	0.012	0.851	0.887	0.290	0.647	0.597	0.816	0.640
Average		0.771	1.116	0.352	0.602	0.786	0.778	0.910
COV		17%	48%	48%	54%	51%	55%	71%

Table 8. Comparison of strain at 50% of peak stress (ε_{50cc}) of no fiber concrete

Specimen	$\varepsilon_{50cc\ exp}$	$\varepsilon_{50cc\ model/exp}$						
		Cusson	Ravzi	Silva	Assa	Li	Mansur	Campione
AK-0-30	0.027	1	2.882	0.552	0.710	1.495	0	7.081
AK-0-45	0.017	1	2.155	0.576	1.220	1.231	0	8.845
AK-0-60	0.008	1.006	3.261	0.956	2.986	1.895	0	16.563
Average		1.002	2.766	0.695	1.639	1.540	0	10.830
COV		0.3%	20%	33%	73%	22%	N/A	47%

Table 9. Comparison of strain at 50% of peak stress (ε_{50cc}) of SNFRC

Specimen	$\varepsilon_{50cc\ exp}$	$\varepsilon_{50cc\ model/exp}$						
		Cusson	Ravzi	Silva	Assa	Li	Mansur	Campione
AK-0.75-30	0.035	1.000	2.244	0.399	0.522	0.924	0.000	8.704
AK-0.75-45	0.017	1.038	2.155	0.547	1.190	1.123	0.000	13.758
AK-0.75-60	0.023	0.986	1.127	0.308	1.035	0.586	0.000	9.046
AK-1.25-30	0.047	0.998	1.566	0.296	0.406	0.761	0.000	5.951
AK-1.25-45	0.022	1.000	0.443	0.443	0.983	0.907	0.000	10.414
AK-1.25-60	0.020	1.002	1.412	0.364	1.182	0.702	0.000	9.746
Average		1.004	1.491	0.393	0.886	0.834	0.000	9.603
COV		2%	45%	24%	38%	23%	N/A	27%

Since the post-peak behavior of synthetic fiber reinforced concrete columns shows the significant difference compare to those of no fiber one, the parameters of the corresponding strain of the 85% (ε_{cc85}) and 50% (ε_{cc50}) of the peak stress f_{cc} at the descending branch of the curve are also determined. Moreover, these parts become important to convince the aspect of ductility, and the post-peak failure of synthetic fiber reinforced concrete columns. The ratio of those two parameters is illustrated in Table 6 - Table 9.

Definition of the post-peak strain behavior corresponded to the 85% and 50% of the peak stress produces a significant difference among Cusson and other models. Moreover, Cusson model well defines the value of the post-peak strain with small variations on SNFRC and slightly underestimate to the no fiber concrete. Cusson model remarkably good provides the value of the post-peak strain on most of the SNFRC specimens with the COV value below 2% and 20% respectively for ε_{50cc} and ε_{85cc}. Meanwhile, it produces a slight underestimate on the COV value for the no fiber specimens. Cusson model was derived from the steel fiber reinforced concrete specimen column. It shows the remarkably good value on both peak stress (f_{cc}) of the SNFRC and no fiber concrete specimens with 2% and 4.7% COV respectively for SNFRC and no fiber concrete. Peak strain values (ε_{cc}) are also well defined either on SNFRC or no fiber concrete specimens. Although the coefficient of variation produces quite large on both, Cusson model demonstrates the best value compared to the others. Cusson model also shows the best value in the post-peak conditions of the corresponding strain of the 85% and 50% peak stress (ε_{85cc} and ε_{50cc}). Still, the COV value of SNFRC specimens indicates better than that of no fiber concrete specimens.

Ravzi model produces the peak stress, which is relatively close to the experimental result on either SNFRC or no fiber concrete with a good variety. While at the peak and the post-peak strain, Ravzi gives different assessments on both types of specimens. The ε_{cc}, ε_{85cc}, and ε_{50cc} of the no fiber concrete specimens describe much better than those of SNFRC specimen. Overall, all variations in the models to the experimental results tend to produce the overestimate results.

Silva model well generates the peak stress on either SNFRC or no fiber concrete with the great variation (COV). Meanwhile, peak strain on both models represents the value of the test results with considerable variation. This model is not able to describe the post-peak strain value with the consistent pattern for the SNFRC and no fiber concrete as well.

Assa is a model which produces relatively good peak stress on either SNFRC or no fiber concrete with a good variation (COV). Conversely, the peak and the post-peak strain were not able to identify the experimental well, especially for SNFRC specimens.

Li model also produces close enough peak stress with a consistent COV on SNFRC and no fiber concrete specimens. However, peak and post-peak strain are described by the overestimate and underestimate with bad and inconsistent variation.

Mansur and Campione model well characterize the peak stress on the SNFRC and no fiber concrete specimen, while overestimate prediction belongs to the peak and post peak strain. Unfortunately, these two models failed in describing the post strain of ε_{50cc}.

From the seven confinement models which are used in this study, four models of those are equal to the comparison study made by [13], i.e. Ravzi, Silva, Assa and Li models. The same conclusions are obtained that almost all the models can correctly estimate the ascending part of the experimental stress-strain curve. However, the prediction of the descending part shows a wide variation. Also, no models which can accurately predict the complete behavior of high-strength concrete columns. However, those four models obtain better results since the coefficient of variation (COV) acquired is much smaller, especially in the term of strain. This study results, 14%-73% for COV of peak and post-peak strains ratio in predicting experimental values for all column specimens, while 51.31%-93.19% for those of Djauhari et. al.'s study [13].

4 Conclusions

i) The capabilities in predicting the actual compressive test behavior of the various confinement models to the relative performance of the spirally confined no fiber and the synthetic fiber reinforced high-strength concrete columns are defined well for the peak stress, otherwise for the peak and post peak strain.

ii) Most of the models of SNFRC and no fiber concrete correctly estimate the ascending part, while underestimate and overestimate the descending part of the experimental stress-strain curve.

iii) The performance index of each model which was measured by using the coefficient of variation (COV) concept on some behavior parameters such as experimental peak confined strengths f_{cc}, corresponding peak strain ε_{cc}, post-peak strains ε_{85cc}, and ε_{50cc}, from the experimental stress-strain curve show relatively wide variation.

(iv) The coefficient of variation (COV) of each behavior parameter in predicting experimental values for all specimens ranges from 0.3% to 73%. The larger values of the COV show that the models are considered less accurate in predicting the value of the parameters. Among the confinement models, Cusson model shows the closest valid value of the coefficient of variation.

(v) Cussson model well predicted the descending part of the stress-strain curve of the fiber reinforced concrete, which is significantly different from the no fiber concrete.

References

[1] D. Cusson and P. Paultre, Stress-strain model for confined high-strength concrete, J. of Structural Engineering, **121**(3), 468-477, (1995)

[2] S. Razvi and M.Saatcioglu, Confinement model for high-strength concrete, J. of Structural Engineering, **125**(3), 281-289, (1999)

[3] P. Silva, *Effect of concrete strength on axial load response of circular columns*, Master Thesis, McGill University, Montreal, Canada, (2000)

[4] B. Assa, M. Nishiyama and F. Watanabe, New approach for modeling confined concrete I : circular columns, J. of Structural Engineering, **127** (7), 743-750, (2001)

[5] L. Bing, R. Park and H. Tanaka, Stress-strain behavior of high-strength concrete confined by ultra-high and normal-strength transverse reinforcement, ACI Structural J., **98**(3), 395-406, (2001)

[6] Rosidawani, I. Imran, S. Sugiri and I. Pane, Behaviour of macro synthetic fiber reinforced concrete columns under concentric axial compression, Procedia Engineering, **125**, 987-994, (2015)

[7] D.A. Fanella and A.E. Naaman, Stress-strain properties of fiber-reinforced mortar in compression, J. of The American Concrete Institute, **82**(4), 475–483, (1985)

[8] A. Djumbong, D. Cusson and P. Paultre, Experimental study of influence of synthetic fibers on behavior of cylindrical high-strength concrete Columns, *Proc.of the 37th Annual Conf. of the Canadian Society of Civil Engineering,* QC, Canada, 10, (2008)

[9] H. Nguyen. *Shear behavior of polypropylene fiber reinforced concrete beams*, Master Thesis, University of Sherbrooke, Sherbrooke, QC, Canada, (2006)

[10] V. Chernov, H. Zlotnikov and M. Shandalov, Structural synthetic fiber-reinforced concrete-experience with marine applications, Concrete Int., **28**(8), 56-61, (2006)

[11] M.A. Mansur, M.S. Chin, and T.H.Wee, Stress strain relationship of confined high-strength plain and fiber concrete, J. Of Materials in Civil Engineering, **9**(4), 171 -179 (1997)

[12] G. Campione, The effects of fibers on the confinement models for concrete columns Canadian J. of Civil Engineering, **29**(5), 742-750, (2002)

[13] Z. Djauhari, and I. Imran, Study of confinement models for high-strength concrete columns confined by high-strength steel, *Proc.of the Int. Conf. on Earthquake Engineering and Disaster Mitigation*, Jakarta, Indonesia, 518-527, (2008)

 DOI: 10.1051/matecconf/201710302005

Flexural Behaviour of Precast Aerated Concrete Panel (PACP) with Added Fibrous Material: An Overview

Noor Hazlin Abdul Rahim[1], *Noridah* Mohamad[1,*], *Abdul Aziz* Abdul Samad[1], *W. I.* Goh[1], and *Norwati* Jamaluddin[1]

[1]Jamilus Research Center, Faculty of Civil and Environmental Engineering, Universiti Tun Hussein Onn Malaysia, 86400 Parit Raja, Johor, Malaysia

Abstract. The usage of precast aerated concrete panel as an IBS system has become the main alternative to conventional construction system. The usage of this panel system contributes to a sustainable and environmental friendly construction. This paper presents an overview of the precast aerated concrete panel with added fibrous material (PACP). PACP is fabricated from aerated foamed concrete with added Polypropylene fibers (PP). The influence of PP on the mechanical properties of PACP are studied and reviewed from previous research. The structural behaviour of precast concrete panel subjected to flexure load is also reviewed. It is found that PP has significant affects on the concrete mixture's compressive stregth, tensile strength and flexural strength. It is also found that PP manage to control the crack propagation in the concrete panel.

1 Introduction

An innovative new Industrialised Building System (IBS) technology is actively being studied and developed for residential, industrial and commercial building applications. As a developing country, the gap between demand and supply of adequate housing is continuously increasing [1]. IBS is generally preferred due to its prefabrication or precast element which aim to improve the overall construction performance in terms of quality, cost effectiveness, safety and health, waste reduction, efficiency and productivity [2]. Structural components of IBS are manufactured in a factory, on or off site, transported and assembled into a structure with minimal site works [3]. Applications of IBS system in construction has proven to be able to fasten the speed of construction without sacrificng the quality. The IBS system presented in this paper, PACP, is made of precast lightweight aerated concrete with density approximately 80% of the density of normal concrete, in the range of 1440 kg/m^3 to 1840 kg/m^3 [4].

Aerated concrete is a type of lightweight concrete which does not contain coarse aggregates [5]. Generally, aerated concrete is a mixture of cement, sand, water and pre-formed foam. The pre-formed foam is a mixture of foaming agent which either synthetic or

* Corresponding author: noridah@uthm.edu.my

protein based, water and air. It is also known as foamed concrete which have an air content of more than 25% [6]. Aerated lightweight concrete is defined as a type of concrete which consists of expanding agent [7]. It will increase the volume of mixture and give additional qualities such as self-compatibility and lighter weight. Meanwhile, lightweight concrete is a concrete that has low density compared to the normal concrete [8].

The density of aerated concrete can be achieved by controlling a dosage of foam [9]. It is reported that there is a relationship of compressive strength of aerated concrete with its density, water/cement ratio and cement content [10]. Density of aerated concrete has a significant influence on the ultimate strength of concrete mixture. Higher ultimate strength is obtained in appearance of uniform bubbles. Typical mixture details of aerated concrete and properties of aerated concrete are summarized in Table 1 and Table 2.

Table 1. Typical mix details of aerated concrete [10].

Wet density (kg/m^3)	500	900	1300	1700
Dry density (kg/m^3)	360	760	1180	1550
Cement (kg/m^3)	300	320	360	400
Sand (kg/m^3)	-	420	780	1130
Base Mix W/C Ratio	Between 0.5 and 0.6			
Air content (%)	78	62	45	28

Table 2. Properties of aerated concrete [10].

Dry Density (kg/m^3)	Compressive Strength (MPa)	Thermal Conductivity (W/mk)	Modulus of Elasticity (GPa)	Drying Shrinkage (%)
400	0.5-1.0	0.1	0.8-1.0	0.30-0.35
600	1.0-1.5	0.11	1.0-1.5	0.22-0.25
800	1.5-2.0	0.17-0.23	2.0-2.5	0.20-0.22
1000	2.5-3.0	0.23-0.30	2.5-3.0	0.18-0.15
1200	4.5-5.5	0.38-0.40	3.5-4.0	0.11-0.09
1400	6.0-8.0	0.50-0.55	5.0-6.0	0.09-0.07
1600	7.5-10.0	0.62-0.66	10.0-12.0	0.07-0.06

Aghaee & Yazdi [11] investigated the mechanical properties of structural lightweight concrete with added waste steel wires. Four types of mixtures with different percentage of fibers (0%, 0.25%, 0.5%, 0.75%) were tested under compressive, tensile, flexural and impact. The compressive strength increased in mixtures with up to 0.5% of waste steel wires while decreased in mixtures with more than 0.5% added waste. It was found that the specimens were prevented from brittle failure even with the lower dosage of fibers. The higher percentage of waste steel wires showed the maximum flexural strength and maximum energy absorption. Meanwhile, the ultimate impact resistance by higher fibers content increased about thirteen times higher compared to plain lightweight concrete. Thus, the waste steel wires expected to be as reinforcement in structural lightweight concrete due to its strength and mechanical properties.

The investigation on flexural behaviour and ductility of reinforced lightweight concrete beam with different concrete strength and addition of polypropylene fiber were carried out by Nahhas [12]. According to the report, two group of specimens were investigated; namely prismatic concrete beam with different cement dosages (350, 400 and 450kg/m^3) and lightweight reinforced concrete beam with different proportions of fiber (0, 30 and 60kg/m^3). The addition of polypropylene fiber increased the ductility and toughness capacity of prismatic concrete beam. It also increased the bearing strength and ductility found in lightweight reinforced concrete beam. The higher content of cement with higher proportions of polypropylene fiber increased the ductility about 98%. The results indicated that polypropylene fiber tend to control the formation and growth of cracks and finally increased the performance of reinforced concrete beam.

2 Affect of Polypropylene (PP) on mechanical properties of concrete

Fibers are generally used in concrete to control cracking due to plastic and drying shrinkage. Several previous studies conducted on concrete mixture with added PP fibers have indicated that it has some significant effects on the concrete properties. An experimental study was conducted to explore PP fibers effects on compressive, tensile, flexural, shear strength and plastic shrinkage. Compressive strength increased at the age of 28 days with addition of PP fibers at 0.1% to 0.4% but decreased from original with amount of PP fibers more than 0.4% [13]. Higher dosage of PP fibers decreased the strength of concrete matrix due to higher volumes of fibers interfering with the cohesiveness of the concrete matrix. Tensile strength of concrete is only about 10% of its compressive strength. Thus the addition of fibers to a concrete mixture is beneficial to tensile properties of concrete. It reduces propagation of cracks in plastic and hardened state. The tensile strength of concrete increases linearly about 65% to 70% with addition of fibers up to 0.4% after which it decrease. The large amount of fibers reduced the bond strength between concrete ingredients as compared to less proportion of fibers. The flexural strength increased about 80% by adding 0.2% of fibers while shear capacity of concrete have a good agreement with added fibers. The last results on shear cracking is reduced by 83% to 85% by addition of fibers in the range of 0.35% to 0.5%. Overall result showed a notable increase in flexural, tensile and shear strength [13].

It has also been found that utilizing PP fibers in cement matrix has caused a slight enhancement in compressive and flexural strength. However, mechanical strength was decreased by further increasing of fiber content that beyond 0.3%. It is because of poor dispersion of PP fibers in mortar that increase pore volume and creates more micro defects in cement matrix [14].

In a report by Awang et al. [15], a density of 1000 kg/m^3 foamed concrete were used by adding 2 types of fibers namely polypropylene and kenaf. Based on the report, it is shown that addition of fibers had contributed to the peak flexural strength achieved by the specimens before failure. Meanwhile, specimens without fibers experienced immediate failure and broken into 2 parts after the peak flexural strength was reached. This statement is in agreement with the work done by Elsaid et al. where it was stated that when concrete cracks, the fibers prevents the micro-cracks from spreading and breaking into two parts [16].

Ahmad et al. [17] conducted a research on added PP fibers in a concrete mix observed that fibers created bridging across a concrete matrix which help to control shrinkage cracking during plastic stage and matrix micro cracks that occur as the concrete is loaded as illustrated in Fig. 1. It is also found that fibers can prevent the occurrence of large cracks widths. Added PP fibers can control the crack and shrinkage by reducing damages occurred due to shrinkage cracking as shown in Fig. 2.

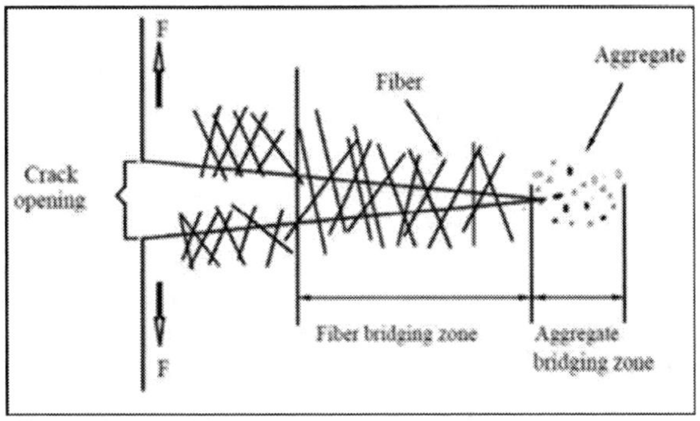

Fig. 1. Fiber bridging across a concrete crack [17].

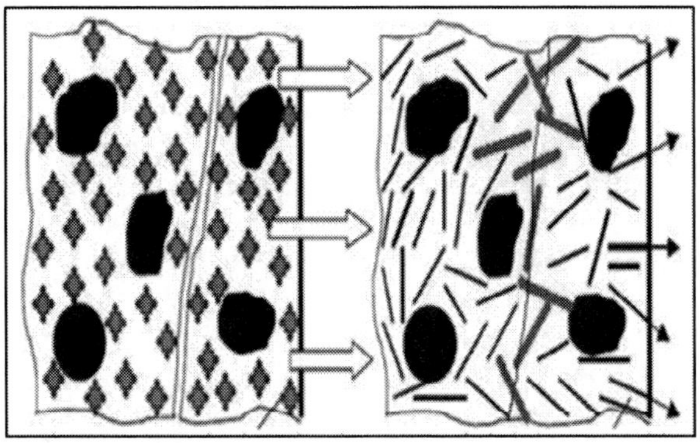

Fig. 2. Reducing the damages of shrinkage cracking [17].

A research was conducted to study the effect of variation of PP fibers ranging from 0.1% to 0.4% together with 0.8% steel fibers on the behaviour of fibrous concrete. It was found that the addition of PP fibers has a slight effect on compressive strength. However, it has a significant affect on the tensile strength [18]. Ramujee [19] studied on compressive strength and splitting tensile strength with different amount of PP fibers varied from 0%, 0.5%, 1%, 1.5% and 2%. The sample with added PP fibers of 1.5% showed better results in comparison with others and reveals that fibers used in concrete were enhanced both compressive and split tensile strength of concrete. It was concluded that the increase of formability and bending strength are the extra advantage of adding the fibers to concrete.

Results from the work by Parveen et al. [18] show that the flexural strength increases with increase in PP fibers due to the additional load taken by fibers that were present in concrete matrix with the optimum value of 0.3%. It is caused by improper mixing of fibers with the matrix takes place due to balling effect of fiber which increase amount of vibration to remove air voids from the mix that causes bleeding problem. The fibrous concrete are more ductile compared to plain concrete. This is caused by the load which was transferred from composites to the fibers at crack surface when the matrix cracked, which prevents the brittle failure of the composites. The 0% and 0.1% of PP fibers behaves like a brittle material which is in elastic energy whereas it become non-linear behaviour with PP fibers more than 0.1%. From this study, it was found that concrete with higher percentage of PP fibers possess higher toughness in plastic energy.

According to Han et al. [20], the evaporation of concrete surface of water is a factor in creating the concrete paste fracture in concrete which to the formation of tension stress since the concrete start to strengthen. Developing such stress during concrete strengthening can lead for crack where the stress is more than concrete strength. The cracks caused by the paste contracting in the concrete are formed in the first hours after pouring the concrete in the formworks and before the concrete reaches its initial strength [21]. It is stated that the best and acceptable method for preventing the cracks formation caused by paste contract is by using fibers [22, 23].

From the previous studies above, it can be concluded that PP has a great potential to be used in concrete mixture as construction material. It was proven that PP fibers added to concrete mixture resulted with slightly increase in compressive strength and significant increase in tensile and flexural strength. It was also proven that added PP fibers have managed to reduce the crack propagation.

3 Behaviour of precast panel under flexure

This paper focuses on the overview of flexure behavior of the precast aerated concrete panel subjected to flexure load. Since it has been proven that PP has great influence in increasing the flexural strength and controlling the crack propagation occurred in concrete panel, the aerated concrete panel studied by the authors is added with PP fibers.

Several previous studies on flexure behavior of precast concrete panel have been studied and investigated by many researchers. Mohamad et al. [24] tested four full scale precast lightweight foam concrete sandwich panel with similar width, height and various thicknesses. The specimens with a double shear steel connector of 6 mm diameter and steel reinforcement of 9 mm diameter were tested under flexure load. It was found that the ultimate flexure load of panel was influenced by compressive strength and thicknesses. The crack pattern showed the crack started at mid span which is in between point of loading and later propagated toward the right and left zones of slab as illustrated in Fig. 3.

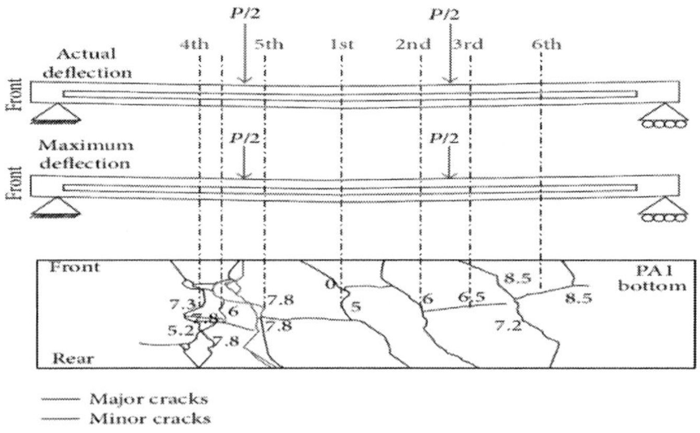

Fig. 3. Crack pattern of Precast Lightweight foamed concrete sandwich panel (PFLP)[24].

Mustaffa et al. [25] investigated structural behaviour of twelve reinforced foamed concrete slab with welded wire mesh under bending. All specimens were tested under four point bending test with varying densities and types of reinforcing steel wire mesh. The higher density of foamed concrete influences the attained ultimate load of slab under test. Mode of failure of all slab panels was found to be similar. The results obtained is found to be reliable to predict ultimate load and structural behaviour of reinforced foamed concrete.

Mugahed et al. [26] conducted experimental and analytical programme of six precast foamed concrete sandwich panels as a one-way working slabs tested under lateral loads. Two dimensional FEA model were carried out by using LUSAS software. Results showed that the increase of aspect ratio has caused decrease on ultimate flexural strength. The crack patterns appeared in one-direction which similar to reports on solid slabs specifically when both concrete wythes act in composite manner. The results obtained from experimental and analytical studies gives a reasonable degree of accuracy.

Benayoune et al. [27] studied FEA of composite precast concrete sandwich panel under flexure by using LUSAS software. Parametric study were conducted with different aspect ratios of slabs. A two dimensional nonlinear model that acting as one-way panel and three dimensional nonlinear model acting as two-way panel with having shear connectors spanning in both directions were simulated. It was found that the placement of shear connectors in both directions gives better load distribution. The ultimate load increased with the increased of shear connectors' number.

Joshani et al. [28] studied the damage model of fracture process of steel concrete composite slabs by using FEA of ABAQUS software. Three dimensional nonlinear model of composite slabs were analysed subjected to four point bending test. Fig. 4 shows damage status in concrete including crack initiation point and propagation path. The results of numerical analysis match the experimental results accurately.

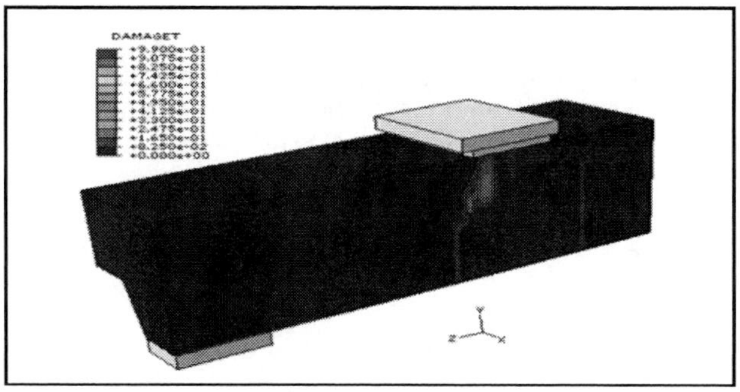

Fig. 4. Damage status at concrete when mid span deflection reached 2.3mm [28].

Mokhatar et al. [29] studied the behaviour of reinforced concrete slabs subjected to impact loading by using ABAQUS software. The response such as time-impact force graph, damage wave propagation, effectiveness of mesh density, effect of projectile size and final crack pattern were verified against existing experimental results. It is shown that the FEA is able to simulate and predict the impact behaviour of structural systems satisfactorily.

Newberry et al. [30] studied pre-stressed concrete sandwich panels subjected to blast loads by using ABAQUS software. FEA used to analysed static and dynamic behavior of foam insulated concrete sandwich wall panels through ultimate capacity. The experimental program used for model development and validation involved both static and dynamic testing of full scale wall panels.

Soronakom et al. [31] studied numerical simulation of full scale elevated slab subjected to a point load at mid span of the central slab. The analysis used concrete damage plasticity model in finite element software ABAQUS with fiber content of 80kg/m^3 and 100kg/m^3. FEA were carried out to calibrate the predicted load deflection response with the averaged experimental response in order to obtain material parameters such as Young's modulus, Poisson ratio and tensile crack relationship. The simulation reasonably agreed with the experimental program.

Abdullah et al. [32] conducted three-dimensional finite element model of steel-concrete composite slab under bending by using ABAQUS software. The quasi static analysis using ABAQUS/Explicit was chosen due to its precise solution where it is dominated by brittle cracking behaviour. A static load is time independent. A dynamic load is time dependent and for which inertial effects cannot be ignored. A quasi-static load is time dependent but is "slow" enough such that inertial effects can be ignored. In ABAQUS, explicit dynamic is used for quasi static analysis. In this analysis, the prescribed displacement was increased slowly enough to eliminate any significant inertia effect. The kinetic energy was chosen as an indicator for the quasi-static response based on the conservation of energy principle. The principle states that the time rate of change of kinetic energy and internal energy for a fixed body of material is equal to the sum of the rate of work done by the external forces. The quasi static response was ensured by keeping the kinetic energy level due to the movement of the model to below 5% of the internal energy at any instance during the time step period. In this way, the external work done by the load is balanced mostly by the internal energy of the whole structure, as is actually happens in static analysis. Limiting the kinetic energy level was attained by adjusting the combination of time step period and the total displacement at the loading nodes accordingly, and monitoring the plots of the kinetic and internal energy with respect to time step period. Longer loading periods can reduce dynamic effects but may increase analysis time significantly. An optimum analysis period was obtained after several trials.

It has conclusively been shown from the review above that FEM has been used for modeling and analysis of structural study due to several limitations such as cost, time and energy required in conducting full-scaled tests. FEA has been proven to be able to predict the ultimate load and structural behaviour of model. The results from previous researches showed a good correlation with the experimental programme.

4 Precast aerated concrete panel with added polypropyle fibers (PACP)

Under flexure load, concrete panel was found to experience crack which started at the mid span and propagated towards its right and left zones. This crack will increase gradually until panel attain its ultimate load and fail. As such, it is critical to find alternative means to control crack propagation in the panel under flexure to avoid abrupt failure.

Therefore, a study on flexural behaviour of PACP tested under four point load as illustrated in Fig. 5, is proposed. The flexural behaviour studied is in the context of its ultimate load, flexural stress, flexural strain, crack patterns, load-deflection profile and the strain distribution on the concrete surface.

PACP is fabricated from cement, sand, preformed foam, water and added with polypropylene fibers, PP. The contribution of PP in the concrete mixture is to strengthen the concrete and to control crack propagation in the concrete against the applied load. It is further strengthened with steel bar reinforcement and stirrups as shown in Fig. 6.

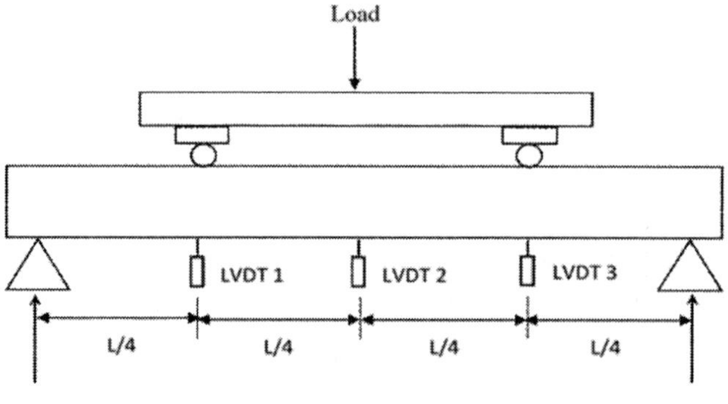

Fig. 5. Test setup specimen with location of strain gauges and LVDT.

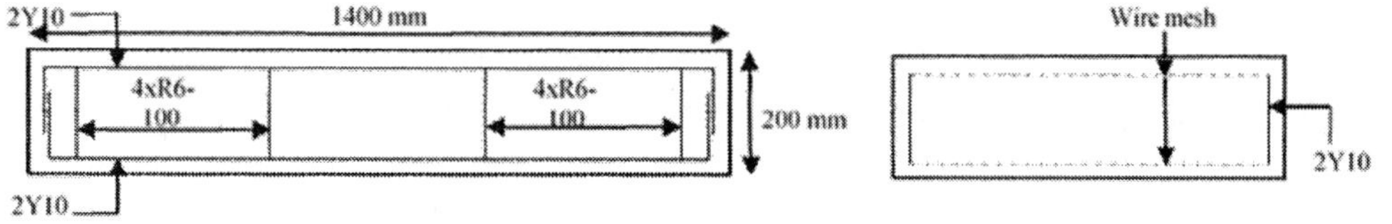

Fig. 6. Steel reinforcement and stirrups in PACP panel.

A three dimensional nonlinear finite element model will be developed by using ABAQUS software to study the PACP's behaviour under flexure loading. The data obtained from experimental results and theoretical calculation will be used as input data in FEM. The results will be analysed for its ultimate flexure load, crack pattern, load-deflection profiles, flexural stress and strain distribution. Parametric study will be conducted by performing series of simulations on both solid and hollow PACP with various heights, overall thicknesses, slenderness ratios and diameters of reinforcement in order to investigate the behaviour of panel is influenced by the size difference. The flexural stress achieved from FEM will be analysed and compared with the values from classical formulae and previous researchers.

5 Conclusion

Precast panel system has been the attractive alternative in the construction industry due to its advantages. Its applications as structural element has led to research on its behaviour under various types of load. PACP discussed in this paper is a precast system which is cast and fabricated from aerated foam concrete with added polypropylene fibres. Its main advantage is greater strength to weight ratio due to the lightweight aerated foam concrete used. Flexural behaviour of PACP subjected to four point bending test will be studied. Under flexure, the cracks initiated and developed in the panel will influence its failure mode. PP fibres are expected to increase PACP's structural integrity by controlling the crack propagation occurred in the slab panel.

The author would like to thank Universiti Tun Hussein Onn Malaysia (UTHM), Batu Pahat, Johor and Higher Ministry Education of Malaysia (FRGS Vot 1525) for its continuous financial support for this research

References

[1] A.T. Waleed, M.S. Jaafar, M. Razali, A. Kadir, A.A.A. Abang, D.N. Trikha and M.S. Najm, Development of an innovative interlocking load bearing hollow block system in Malaysia, Construction and Building Materials, **18**(6), 445-454 (2004)

[2] K.A.M Kamar, M. Alshawi and Z. Hamid, Barriers to industrial building system (IBS): The case of Malaysia, *Proc. of the 9th Int. Postgraduate Research Conferemce*, Salford, United Kingdom, (2009)

[3] Construction Industry Development Board (CIDB), *Construction industry master plan Malaysia 2006-2015*, Malaysia Construction Industry Development Board, Salford, United Kingdom, (2007)

[4] ACI 318-89, *Building code requirements for reinforced concrete* (ACI 318-89) (Revised 1992) and Commentary-ACI 318-89 (Revised 1992), Michigan, American Concrete Institute, (1992)

[5] N. Narayanan and K. Ramamurthy, Structure and properties of aerated concrete: A review, Cement and Concrete Composites, **22**, 321-329 (2000)

[6] D. Aldrigde, *Introduction to foamed concrete: What, why, how? Used of foamed concrete in construction*, Thomas Telford, Scotland, United Kingdom, (2005)
[7] M.L. Zakaria, *Bahan dan binaan*, Dewan Bahasa dan Pustaka, Kuala Lumpur, Malaysia, (1978)
[8] J.M. Scanlon, *Lightweight concrete, Concrete construction handbook*, 4th edition, McGraw-Hill Inc, New York, United States, (1998)
[9] K. Ramamurthy, E.K. Kunhanandan Nambiar and G.A. Indu Siva Ranjani, Classification of studies on properties of foamed concrete. Cement and Concrete Composites, **31**, 388-396 (2006)
[10] British Cement Association, *Foamed concrete composition and properties*, British Cement Association, Camberley, United Kingdom, (1994)
[11] K. Aghaee and M.A. Yazdi, Waste steel wires modified structural lightweight concrete, Materials Research, **17**(4), 958-966 (2014)
[12] M.T. Nahhas, Flexural behavior and ductility of reinforced lightweight concrete beams with polypropylene fiber, J. of Construction Engineering and Management, **1**(1), 4-10 (2013)
[13] S. Ahmed, A.I. Bukhari, I.J. Siddiqui and S.I. Qureshi, A study on properties of polypropylene fiber reinforced concrete, *31st Our World in Concrete and Strcuture*, Singapore, (2006)
[14] A. Sadrmomtazi and A. Fasihi, Influence of polypropylene fibres on the performance of nano-sio2-incorporated mortar, Iranian J. of Science and Technology, Transaction Engineering, **34**, 385-395 (2010)
[15] H. Awang, M.H. Ahmad and M.Z. Al-Mulali, Influence of kenaf and polypropylene fibres on mechanical and durability properties of fibre reinforced lightweight foamed concrete, Engineering Science and Technology, **10**(4), 496-508 (2015)
[16] A. Elsaid, M. Dawood, R. Seracino and C. Bobko, Mechanical properties of kenaf fibre reinforced concrete, Construction and Building Materials, **25**(4), 1991-2001 (2011)
[17] O.A Ahmad and M. Awwad, The effects of polypropylene fibers additions on compressive and tensile strengths of concrete, Int. J. of Civil and Environmental Engineering, **37**(1), 1365-1372 (2015)
[18] Parveen and A. Sharma, Structural behaviour of fibrous concrete using polypropylene fibres, Int. J. of Modern Engineering Research, **3**(3), 1279-1282 (2013)
[19] K. Ramujee, Strength properties of polypropylene fiber reinforced concrete, Int. J. of Inovative Research in Science, Engineering and Technology, **2**(8), 3409-3413 (2013)
[20] C. Han, Y. Hwang, S. Ynag and N. Gowripalan. Performance of spalling resistance of high performance concrete with polypropylene fiber contents and lateral confinement, Cement Concrete Research, **35**, 1747-1753 (2005)
[21] M. Zeiml, D. Leithner, R. Lackner and A.H. Mang, How do polypropylene fibers improve the spalling behavior of in-situ concrete?, Cement and Concrete Research, **36**, 929-942 (2006)
[22] N. Banthia and R. Gupta, Influence of polypropylene fiber geometry on plastic shrinkage cracking in concrete, Cement Concrete Research, **36**, 1263-1267 (2006)
[23] P. Nagarkar, S. Tambe and D.G. Pazare, Study of fibre reinforced concrete, *Proc. of Int. Symposium of fibre reinforced concrete*, Madras, India, (1987)
[24] N. Mohamad, A.I. Khalil, A.A.A. Samad and W.I. Goh, Structural behaviour of precast lightweight foam concrete sandwich panel with double shear truss connectors under flexural load, ISRN Civil Engineering, **2014**, 1- 7 (2014)
[25] N.K. Mustaffa, M.S. Hamidah and H.G. Kwak, Finite element analysis on prediction structural behaviour of reinforced foamed concrete slab, *Proc. of the 10th Int. Conf. on Concrete Engineering and Technology*, Shah Alam, Malaysia, (2009)

[26] Y.H. Mugahed Amran, S.M.R. Rashid, F. Hejazi, N.A. Safiee and A.A. Abang Ali, Structural behavior of laterally loaded precast foamed concrete sandwich panel, Int. J. of Civil, Environmental, Structural Construction and Architectural Engineering, **10**(3), 265-272 (2016)
[27] A. Benayoune, A.A. Abang Ali, A.A. Abdul Samad and D.N. Trikha, Flexural analysis of composite one- and two-way sandwich slabs with truss-shaped connectors, J. The Institution of Engineers Malaysia, **68**(1), 53-60, (2007)
[28] M. Joshani, S.S.R. Koloor and R. Abdullah, Damage mechanic model for fracture process of steel concrete composite slabs, Applied Mechanics and Materials, **165**, 339-345 (2012)
[29] S.N. Mokhatar and R. Abdullah, Computational analysis of reinforced concrete slabs subjected to impact loads, *Int*. J. of Integrated Engineering, **4**(2), 70-76 (2012)
[30] C.M. Newberry, J.M. Hoemann, B.T. Bewick and J.S. Davidson, Simulation of prestressed concrete sandwich panels subjected to blast loads, *2010 Structure Congress*, Orlamdo, Florida, (2010)
[31] C. Soranakom, B. Mobasher and X. Destrée, Numerical simulation of FRC round panel tests and full-scale elevated slabs, American Concrete Institute, **248**, 31-40 (2007)
[32] R. Abdullah, V.P. Paton-Cole, W. Samuel Easterling, Quasi-static analysis of composite slab, Malaysian J. of Civil Engineering, **19**(2), 1-13 (2007)
[33] ASTM Standard E72-80, *Standard test methods of conducting strength tests of panels for building construction,* ASTM International, West Conshohocken, Pennsylvania, (2005)

DOI: 10.1051/matecconf/201710302006

Experimental Investigation of Concrete Filled PVC Tube Columns Confined By Plain PVC Socket

N. Jamaluddin[1], *A. A.* Azeez[1,*], *N.* Abd Rahman[1], *A. N.* Attiyah[2], *M. H.* Wan Ibrahim[1], *N.* Mohamad[1], and *S. H.* Adnan[1]

[1]Jamilus Research Center, Universiti Tun Hussein Onn Malaysia, 86400 Batu Pahat, Johor, Malaysia
[2]Faculty of Civil Engineering, University of Kufa, P.O. Box (21), Najaf Governorate, Iraq

Abstract. This paper presents results of an experimental study for concrete column filled poly vinyl chloride (PVC) tubes confined by plain socket with 5.8 & 6.8 mm thicknesses, 102 mm diameter and 100 mm depth. The total of five concrete filled columns using PVC tubes (CFT PVC) was tested to investigate the columns' behaviour. The column is 700 mm height, 100 mm external diameter and 3.5 mm tube thickness with different thickness of plain socket. The results presented include maximum axial load, plain socket confinement effect, the mode of failure, and lateral PVC strain. The axial load enhancement of PVC-concrete columns confined using plain socket shows an increment of 21.3% up to 55.2% and axial strain from 21% to 40% compared with displacement for control composite columns at 192 kN ultimate load.

1 Introduction

In building construction, the merits of a structure are based on factors such as availability, structural strength, durability, and workability. The properties of the structure materials may differ from each other and there is no single material that can fulfill all structural requirements which resulted in the application of composite structures. Composite columns, particularly composite concrete filled steel tube (CFST) columns, are increasingly used for high-rise building structures, owing to the advantage of combined characteristics of the steel and concrete materials.

PVC has advantages such as low cost, lightweight and is easy to handle and install. It is not affected by corrosion or other forms of degradation; therefore, it is used as an alternative to the metal in many applications where corrosion can compromise functionality and increase maintenance cost. However, the study of concrete-filled PVC tube (CF-PVCT) composite columns are limited even though its advantages are many. The PVC tube does not only protect the core concrete from the corrosion of the atrocious environment, but it is the cheapest material and locally available in abundance. The development of the CFT column using this material could be an achievement for the local construction industries.

* Corresponding author: alyaaabdulrazzaq@yahoo.com

Steel and concrete are the common materials used in the composite CFT columns. The CFT columns are known due to their superior of strength with the ability to withstand the load imposed during the event of Kobe earthquake when their ability accidently encountered. The CFT columns have many constructional advantages, such as high energy absorption, economical formwork and high ductility. These advantages are contributed by the effectively confined concrete core; thus, the realistic prediction of the capacity of these columns must consider the confinement effect provided by the steel tube.

An attempts to use composite columns with plastic tubes can be traced back from the work done by Kurt [1] and recently this column type has gained attention from several researchers [2, 3]. One of the advantages of PVC tube is, being a non-conductor of electricity and immune to the electrochemical based corrosion process. However, not only fewer studies were found on the CFT columns using PVC tubes, its successful design and application requires adequate recognition of its more complex stress/strain and stress/fracture behavior.

The tubes located at the circumferential contribute to strength enhancement of the CFT column. As for concrete filled steel tube columns, CFST, the steel tube not only provide permanent formwork for the concrete core, but its superior strength was contributed from the confinement effect of the steel tube itself that may be expected due to its material properties. Other than that study by Marzouck and Sennah [4] the slenderness ratio also contributes to the columns' strength. Gupta and Verma [5] carried out a study on the length to diameter ratio (D/t) columns effect on such columns. As for PVC, it does not exhibit the simple proportionality between stress and strain that is characteristic of steel material, the PVC capacity to resist fracture is reduced as duration of loading is increased. In addition the PVC tube having a non-stick inner surface resulted in lower friction factors and expected to affect the lateral confining pressure on the concrete core.

In this study, plain sockets were attached to PVC CFT columns with the attention to strengthen such columns. The contribution of the plain socket in confinement effect, load carrying capacity, ductility and durability of the column were investigated.

2 Experimental Details

2.1 Generals

Five specimens of the PVC CFT columns were tested to failure. There were a control specimen, C1 and 100 mm height of plain sockets with different thickness of 5.8 mm and 6.8 mm (Fig. 1) were attached to the rest of the columns (CS1, CS2, CS3 and CS4). Table 1 tabulated the details of the column specimens.

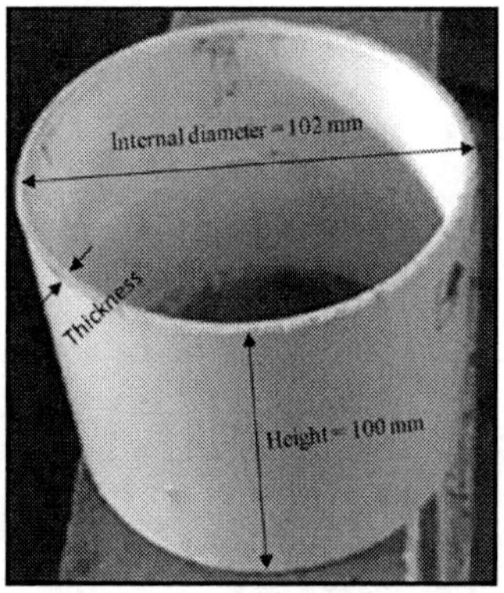

Fig. 1. Details of plain socket

Table 1. Specimens details

Mix ID	Tube Length (m)	Diameter (mm)	Thickness (mm)	Plain socket thickness (mm)
C1				-
CS1				5.8
CS2	700	100	3.5	5.8
CS3				6.8
CS4				6.8

2.2 Materials

2.2.1 Poly Vinyl Chloride (PVC) Tube

The typical physical properties of PVC tubes are shown in Table 2. The PVC tubes were cut into several pieces for the preparation of tensile coupon specimens. The material properties of the PVC element were accordingly determined from mechanical testing on PVC coupons shape as in Fig. 2. The behaviour of PVC described through the stress-strain relationship from the test and the stress-strain graphs are plotted in Fig. 3 with the average of ultimate tensile stress for these coupons is 49.4 MPa.

Table 2. Physical properties of PVC pipes

Parameter	Value
Relative Density	1.42 – 1.48
Elastic tensile modulus	3.0 – 3.3 GPa
Poisson ratio	0.4
Ultimate tensile strength	52 MPa
Elongation at break	50 – 80%
Elastic flexural modulus	2.7 – 3.0 GPa

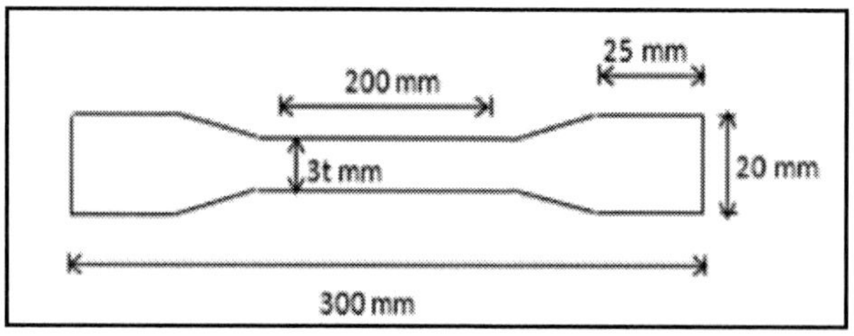

Fig. 2. Coupon scheme

Table 3. Coupon mechanical properties

Property	Value
Average Modulus of elasticity (MPa)	2770
Average ultimate tensile stress (MPa)	49.4

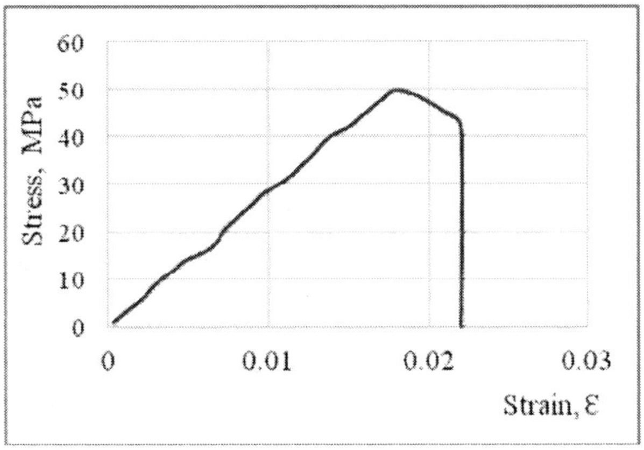

Fig. 3. Stress- strain curve

2.2.2 Properties of concrete

Ready mix concrete with targeted compressive strength of 20 MPa was considered. The average compressive strength results obtained from 150 mm cubes and 100×200 mm cylinders at 28-days of curing are 26.3 MPa and 21.6 MPa, respectively.

2.3 Preparation of the column specimens

The PVC tubes with thickness of 3.5 mm and a diameter of 100 mm were cut to the length of 700 mm as shown in Fig. 4. The tubes were positioned vertically, fixed on the laboratory floor and filled with concrete in approximately 10 cm layers, and each layer was compacted by a steel rod. To prevent the leakage of cement paste from the bottom of the PVC tube, it was covered with nylon. After the tube was filled, the top surface was flush in order to ensure the loading was evenly distributed across the cross section (Fig. 5). The columns were then covered with Hessian and plastic cover for curing. The specimens were moistened with water every day after 24 hours after casting.

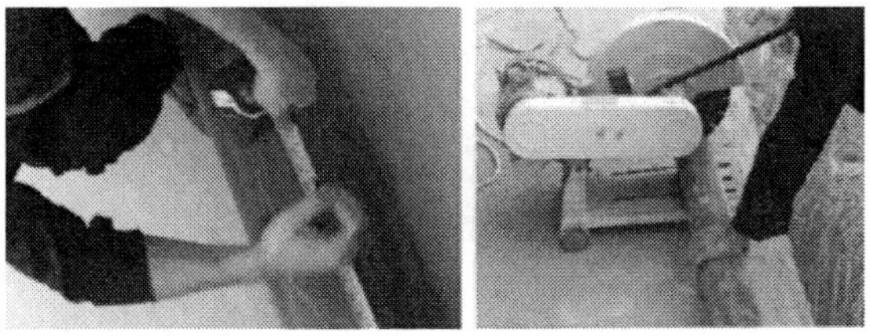

Fig. 4. Preparation of specimens

Fig. 5. Casting the composite columns

2.4 Test setup and procedure

All specimens were tested under axial compression using a universal testing machine with a capacity of 1000 kN. Surface grinding were performed at the top and bottom specimen to remove surface imperfections and maintain uniformity of loading on the surface (Fig. 6). The column specimens were then positioned at the centre of the testing machine to avoid any eccentricity during the testing. The vertical displacement of the lower movable head of the testing machine was measured in relation to the upper head of the testing machine by a dial gauge with magnetic base. Two strain gauges were located at mid height and two-third from the column supports (Fig. 7).The data of loads and displacements were recorded at regular intervals during the tests. The application of the load was continued until the columns failed.

Fig. 6. Preparation of specimens to test

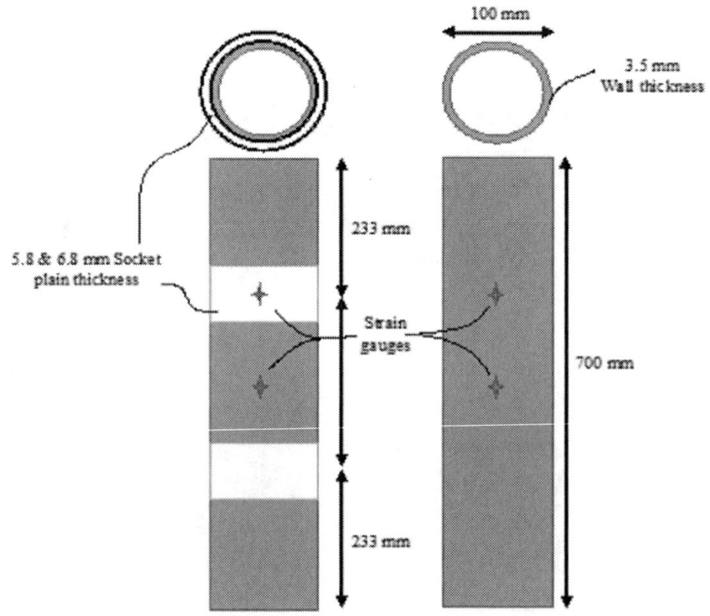

Fig. 7. Schematic diagram of PVC CFT Column

3 Results

3.1 Column strengths

Comparison of the ultimate load made between the control specimen of C1 and PVC CFT columns attached to the plain socket shows the increment range is in between 21% to 55.2% (Table 4). This is expected due to the contribution of the plain socket in confining the concrete core. The PVC tube itself causes the development of a triaxial stress field within the confined concrete, constraining it during dilation and thereby increasing the load carrying capacity. Further increment can be seen for columns with the thicker plain socket.

Table 1. Experimental results

Specimens	Plain socket thickness (mm)	Ultimate load	Enhancement (%)	Maximum displacement (mm)
C1	-	192	-	8.5
CS1	5.8	233	21.3	9.4
CS2	5.8	246	28.1	9.7
CS3	6.8	291	51.5	10.5
CS4	6.8	298	55.2	10.9

3.2 Failure modes

Two types of failure modes were observed in this study; shear failure and outward buckling. For control specimen, C1 the principal failure was due to the shear failure of the concrete core (Fig. 8). This could be seen from the shape of outward buckling of the PVC tube and the failure in the concrete core after the PVC tube was cut out. At the end of the loading stage of the specimen, the C1 column was ruptured due to the concrete core crush and the sudden exploded could be heard from the test without warning. On the other hand, cracks were observed to occur in the plain socket for the rest of the columns (CS1, CS2, CS3 and CS4) which can be considered as the warning before the columns' failure. For these columns, the outward buckling was observed to occur near to the top, bottom, and in between the plain socket (Fig. 9).

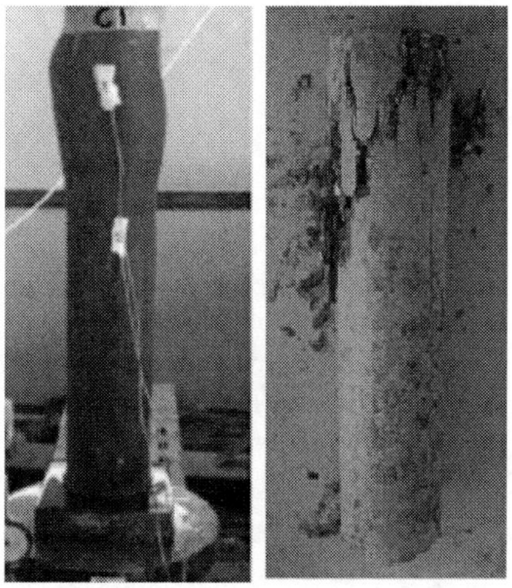

Fig. 8. Specimen C1

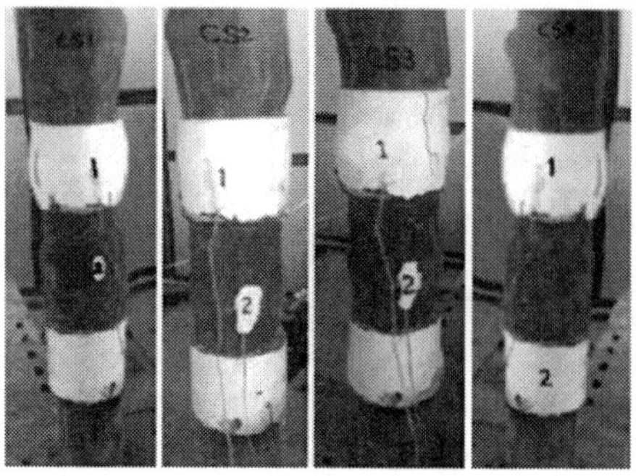

Fig. 9. Failure modes for CFT PVC columns with plain socket

3.3 Strain behavior of confined PVC-concrete composite columns using plain socket

The load-lateral strain graphs of the columns is shown in Fig. 10. From all column specimens, the strain ratio of $\varepsilon_c/\varepsilon_{cs}$ which measured at a two-third column's length is in between 0.77%- 0.93%, whereas at the columns' mid-height, the strain ratio of $\varepsilon_c/\varepsilon_{cs}$ is between 0.75%- 0.93% where; the ε_c is a strain on control specimen, C1 and ε_{cs} is the strain for columns with plain socket. From the figure, it can be observed that the columns with plain socket reached higher load at the similar strain of the control specimen which indicate that the plain socket effectively confining the concrete core at the specific location.

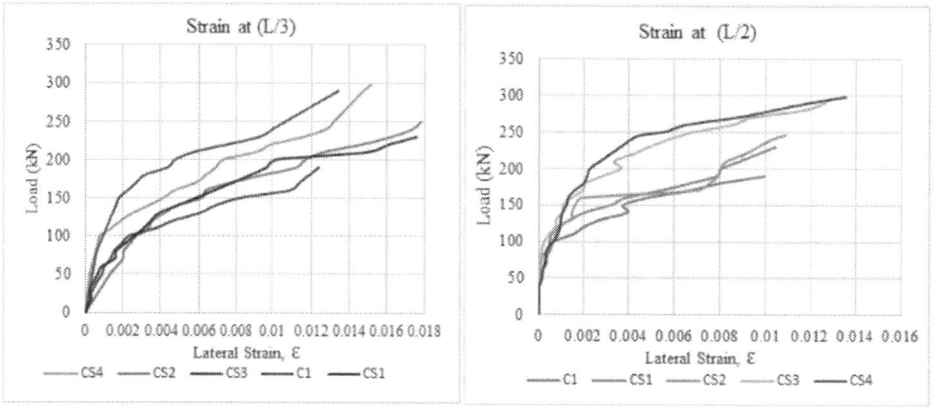

Fig. 10. Lateral strain profile

4. Conclusions

This paper reports the results of an experimental program of CFT columns using plasticized polyvinyl (PVC) tubes. From the experimental results, the plastic pipes were found to be effective in confining the concrete core as evidenced by the increased of ultimate loads. The enhancement in axial load due to confinement by using plain sockets of composite columns is substantial and depends on the level of confinement, in which the ultimate load is increased from 21.3% to 55.2% of the column without the plain socket. It is also evidence that the columns with plain socket reached higher loads compared to the control specimen at similar strain.

The research reported in this paper is a collaboration between the University Tun Hussein Onn Malaysia and University of Kufa, Iraq who are gratefully acknowledged.

References

[1] C.E. Kurt, Concrete filled structural plastic columns, J. of the Structural Division, **104**(ASCE 13478 Proc.), 55-63 (1978)
[2] K.G. Naftary, O.O. Walter and N.M. Geoffrey, Compressive strength characteristics of concrete filled plastic tubes short columns, Int. J. of Science and Research, **3**(9), 2168-74 (2014)
[3] C.M. Usha and H. Eramma, Experimental study on concrete filled, unplasticised poly vinyl chloride (UPVC) tubes, Int. J. of Advanced Technology in Engineering and Science, **2**(7), 2348 – 7550 (2014)
[4] M. Marzouck and K. Sennah, Concrete-filled PVC tubes as compression members: Composite materials in concrete construction, *Proc. of the Int. Congress Challenges of Concrete Construction*, 31-38, Scotland, UK, (2002)
[5] P.K Gupta and V.K. Verma, Study of concrete-filled unplasticized poly-vinyl chloride tubes in marine environment, Int. J. of International Journal of Advanced Structural Engineering, **5**(1), 1-8 (2013)

Effects of Predamaged Level on Confined HSC Columns

Chau-Khun Ma[1,*], *Sofrie Siew-Yung* Chin[1], *Nazirah Mohammad* Apandi[1], *Abdullah Zawawi* Awang[1], and *Wahid* Omar[2]

[1] Department of Structure and Materials, Faculty of Civil Engineering, Universiti Teknologi Malaysia, 81310, Johor Bahru, Malaysia
[2] Office of Chancellery, Universiti Teknologi Malaysia, 81310, Johor Bahru, Malaysia

Abstract. In the design of repair works for damaged concrete, an accurate and representative stress-strain model is of important. The stress-strain model for damaged high strength concrete (HSC) repaired with post-tensioning steel straps confinement yet available, although the confining method has been proven to be effective in improving the performance of non-damaged HSC. A series of experimental test was carried out to investigate the stress-strain relationships of such concrete. A total of 24 HSC cylinders were compressed until certain damaged levels, then repaired by using steel straps. Two important parameters have been identified to have significant effects on the stress-strain relationship of such repaired concrete, namely the confining volumetric ratio and damaged levels. These parameters were incorporated into the development of stress-strain model, which later was shown to correlate well with the experimental results. This paper also has evidenced that existing stress-strain models of damaged concrete are not suitable to be directly applied to the design of repair works using post-tensioning steel straps confinement that produce external lateral stress on damaged columns before subsequent loading applied.

1 Introduction

When a concrete structure was damaged due to major events such as earthquakes, blast or impact, its integrity has been compromised and it can collapse anytime if no immediate actions were taken to restore its capacity [1-6]. An engineer can choose to either replace the affected structure members with new members, demolish the structure for rebuild or emergency repair. Due to the economic consideration, mostly the latter will be chosen as it requires less time, cheaper, easier and most importantly, it is highly effective. Past research works have proven that rapid repairing using confinement is effective in restoring the original strength of the concrete columns [7-10]. Besides, it was also reported that the confinement restored the ductility of damaged concrete as well [10-16]. Despite the effectiveness, the utilisation of such repairing method in the construction industry is particularly limited.

[*] Corresponding author: machaukhun@gmail.com

2 Load-induced damaged specimens

A total of 24 HSC cylinders with diameter of 100 mm and height of 200 mm were prepared in this study. The targeted concrete compressive strength is 50 MPa with mix proportions as shown in Table 1. Another 9 corresponding concrete cubes were prepared alongside with the specimens for the determination of concrete compressive strength. Based on the analysis of the cube tests, the average concrete compressive strength is 50.27 MPa with standard deviation of 4.6 MPa at the age of 28 days after curing. The specimens were then pre-loaded to damage the cylinders to simulate the situations where concrete were damaged at 0%, 50%, 100% and –50% (load reduction of 50% at post-peak region) of axial load capacity. Pre-damage level was determined to justify the behaviour of columns under different confinement configuration. Selected damage level would give better understanding on the columns response in term of strength and ultimate strain. However, the -50% of pre-damaged load was unsuccessful to be achieved as most of the specimens failed immediately after the ultimate load has been reached. Hence, the -50% of pre-damaged levels here also represent any failure load achieved after the ultimate load even it was less than 50%. The details of the specimens are tabulated in Table 2.

Table 1. Mix portions of 1m^3 concrete

Material	Type	Quantity
OPC (kg)	Portland	409
Sand (kg)	River sand	806
Aggregate (kg)	10 mm	1025
Superplasticizer (*l*)	Glenium C380	1.6
Water (kg)	Tap water	162

Table 2a. Details of tested specimens

Specimen ID	Steel Straps layer	Spacing (mm)	Pre-damaged level (%)	Volumetric Ratio (p_v)	Ultimate load (kN)	Ultimate displacement (mm)
C50-0-0 (0)	-	-	-	0	438.3	0.40
C50-30-1 (0)	1	30	-	0.1	394.3	0.60
C50-10-1 (0)	1	10	-	0.15	420.2	0.63
C50-30-2 (0)	2	30	-	0.2	540.4	0.88
C50-30-3 (0)	3	30	-	0.3	533.3	1.40
C50-10-2 (0)	2	10	-	0.3	582.8	2.38
C50-0-0 (+50)	-	-	+50	0	162.6	0.46
C50-30-1 (+50)	1	30	+50	0.1	421.1	0.69
C50-10-1 (+50)	1	10	+50	0.15	420.6	0.85
C50-30-2 (+50)	2	30	+50	0.2	458.7	1.50
C50-30-3 (+50)	3	30	+50	0.3	521.5	1.63
C50-10-2 (+50)	2	10	+50	0.3	563.9	2.64
C50-0-0 (+100)	-	-	+100	0	162.6	0.52
C50-30-1 (+100)	1	30	+100	0.1	286.7	0.76
C50-10-1 (+100)	1	10	+100	0.15	415.5	0.90
C50-30-2 (+100)	2	30	+100	0.2	423.4	1.24
C50-30-3 (+100)	3	30	+100	0.3	512.9	1.55
C50-10-2 (+100)	2	10	+100	0.3	553.8	2.71
C50-0-0 (-50)	-	-	-50	0	149.2	0.61
C50-30-1 (-50)	1	30	-50	0.1	222.3	0.75
C50-10-1 (-50)	1	10	-50	0.15	366.0	1.61

Table 2b. Details of tested specimens

Specimen ID	Steel Straps layer	Spacing (mm)	Pre-damaged level (%)	Volumetric Ratio (p_v)	Ultimate load (kN)	Ultimate displacement (mm)
C50-30-2 (-50)	2	30	-50	0.2	328.3	2.42
C50-30-3 (-50)	3	30	-50	0.3	323.6	2.12
C50-10-2 (-50)	2	10	-50	0.3	404.5	2.95

3 Steel strapping works

In this study, the spacing between the steel straps were fixed into 10 mm (closest spacing) and 30 mm. The spacing were designed not to more than 60mm (center to center). The specimens were confined with 1, 2 and 3 layers steel straps. Confined specimens are shown in Fig. 1. All confined specimens were confined from end to end to prevent concrete crushing near the load platens. Steel strap were install after pre-damaged load to investigate the effectiveness of confinement in repairing works.

Fig. 1. Steel strapping confinement

4 Loading procedure

The compressive strength of the concrete was determined by uniaxial compression test using 'Tinius Olsen' compression machine that has a maximum load capacity of 2,500 kN. All tests were conducted in the Laboratory of Structure and Materials in Universiti Teknologi Malaysia (UTM). The loading is displacement controlled with constant rate of 0.01 mm/s for all specimens (Fig. 2).

During the test, the axial load and vertical strains were recorded by an automatic data acquisition through data logger to monitor the stress-strain behaviour of the specimens. The variations of axial strains of the specimen were measured by two 50 mm omega strain gauges attached at both sides of the specimen. The variations in readings between the two omega strain gauges were monitored closely and the testing was stopped immediately in case the variations exceeding 10%, indicating the specimen was not aligned in the center of the platen. The specimen was then re-positioned and centered again until the variation is less than 10%. This initial testing is very important to avoid any unalign setup which may produce unreliable data. The axial load selected in the pre-testing is choosen to not affect the intergritiy of the specimen. The specimen was monotonically loaded until the specimen failed.

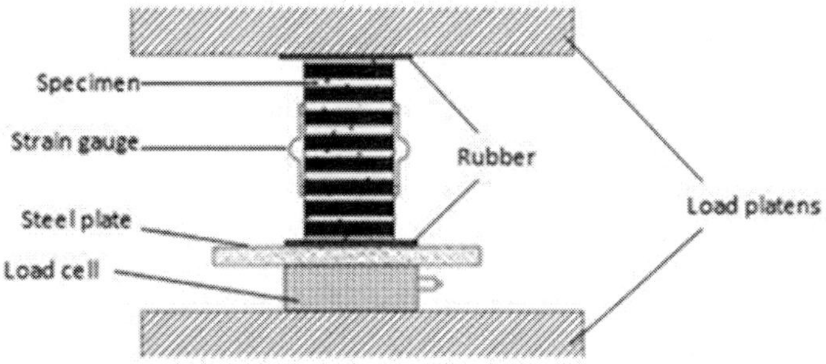

Fig. 2. Compression test setup

5 Infliction of damage

The uniaxial compression testing of concrete cylinders was inflicted by pre-compression of the cylinders to a certain degree. The pre-damaged to the concrete cylinders to a stress that equals a certain percentage of the peak strength 0, +50, 100, -50%, termed as damage level, as shown in Fig. 3. The positive sign indicates a pre-compression load on the ascending branch (before achieving peak strength), whereas the negative sign denotes a load on the descending branch.

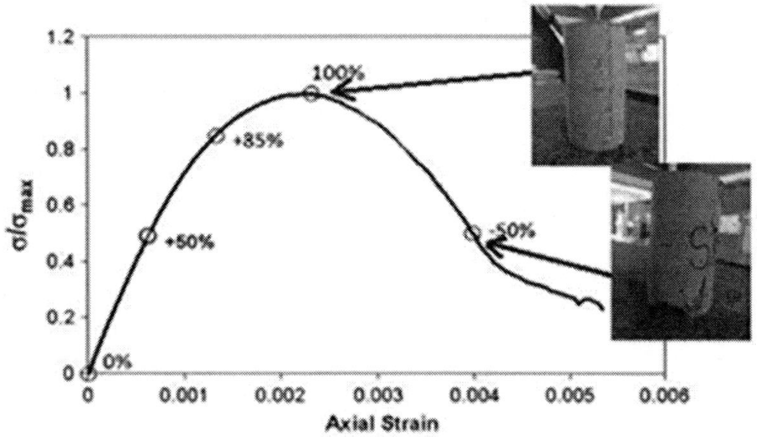

Fig. 3. Definition of damage level

For negative damage levels, the final applied load was calculated and monitored during a test as the actual peak load (observed when the applied load just falls into the descending branch) multiplied by the prescribed percentage. For positive damage levels, the peak load was estimated from the mean peak loads of the specimens in the group that had negative damage levels that had been tested first.

Because of the fast drop in load at the descending branch, the loading rates were carefully examined and selected by testing so that the applied load could be stopped at prescribed loads with reasonable accuracy. For this study, the loading rate will be fixed at 0.01 mm/s. When the pre-damaged load reached a prescribed value, the cylinder was unloaded immediately.

6 Pre-damaged effects on confined HSC columns

The effects of damaged degrees to strength and displacement are as shown in Fig. 4 and Fig. 5 Apparently, a descending relationship exists between ultimate strength and damaged degree. Drastic drop was observed for group $\rho_v = 0$ (unconfined columns). The slope of the trendlines become gentler with the increase of ρ_v. Unexpected large scatter was observed in the relationships of strain versus damaged degree indicating the effect of damaged degree in strain is comparatively negligible. This finding was similar with those reported by Wu et al. in the investigation of repaired concrete using CFRP jacketing [2]. However, no detail explanation was given in their report. Force – displacement relationship of specimen shown in Fig. 5 which represent the effects of various damage level on columns performance after confinement. From the results, damage level at 50% and 100% almost response similary but higher damage level gives lower strength recovery. Volumetric ratio demand of damage columns represented in Fig. 6 that shown higher volumetric ratio is required for undamaged column compared to damaged column which proved that steel strap confinement technique is effective in reducing dilation of concrete column under loading. At damage level of 50% and 100%, equal demand for volumetric ratio is observed.

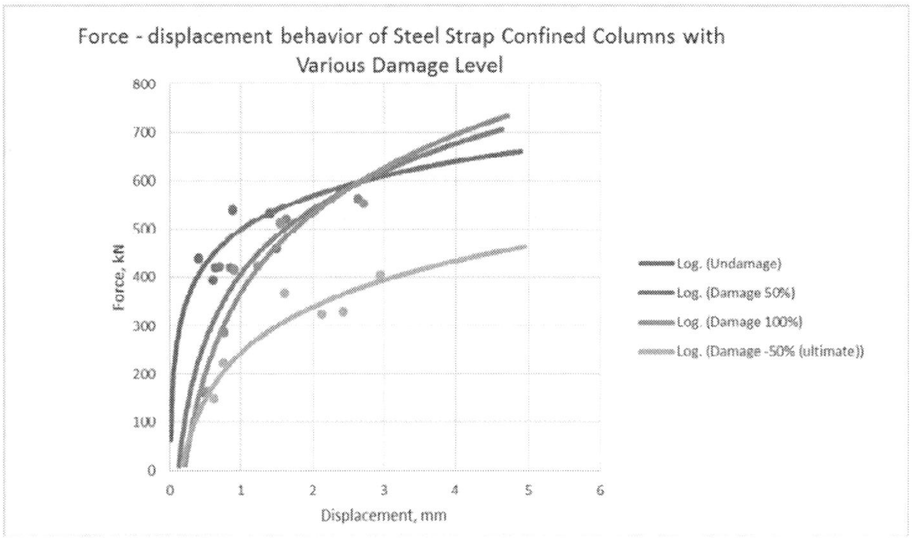

Fig. 4. Force – displacement relationship of repaired columns with various damage levels

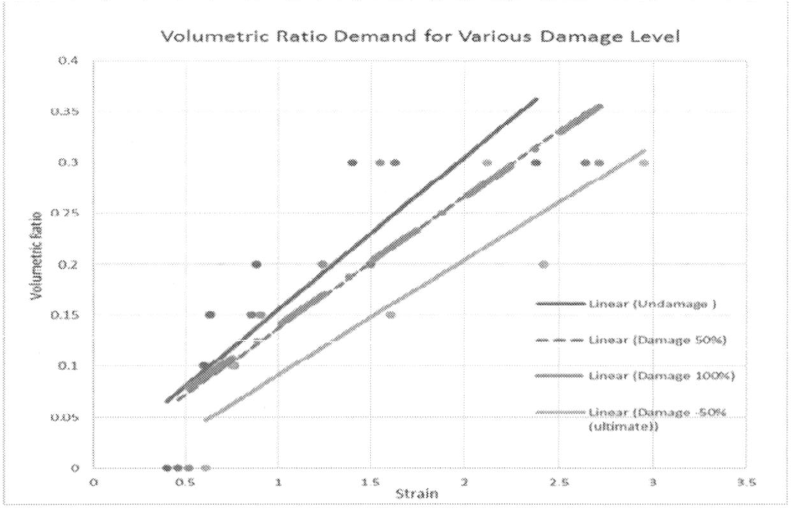

Fig. 5. Volumetric ratio demand for various damage level

DOI: 10.1051/matecconf/201710302007

7 Effects of pre-damaged level of repaired high strength concrete column on deformability

From Fig. 6, with the aid of confinement from steel strap, the deformability of repaired HSC improve with optimum spacing and confinement layer. The optimum spacing for higher deformability of repaired HSC column are 10 mm and 2 layer of steel strap. The deformability can be reflected from stress - strain curve of repaired HSC specimens. For 1 and 2 layer of steel strap confinement, the available spacing are 10 mm and 30 mm respectively whereas for 3 layer of steel strap confinement only available for 30 mm of spacing. The different result from each category are due to effects of pre-damaged level. It is worth noting that under same volumetric ratio, closer spacing (10mm) produce higher ultimate strain of 70%, 61%, 75% and 39% for damage level at 0%, 50%, 100% and -50% respectively.

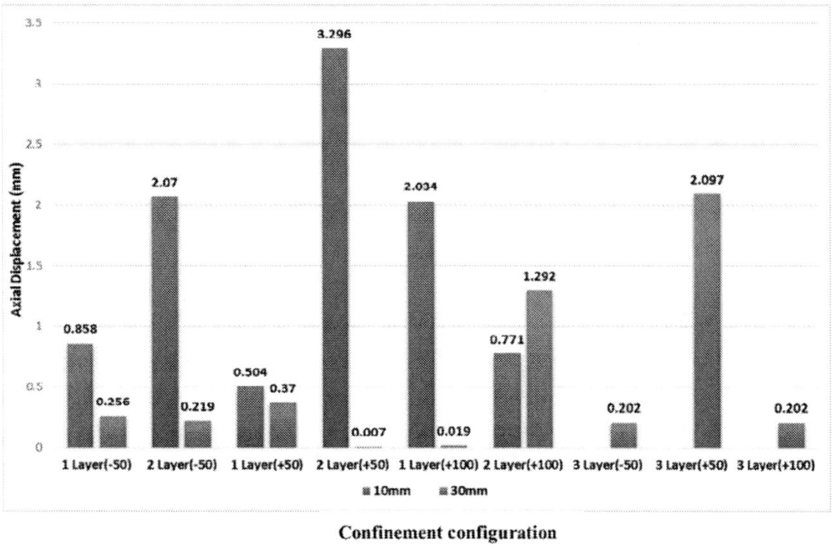

Fig. 6. Deformability of Repaired HSC Columns

8 Summary and conclusion

A series of experimental and analytical studies were performed to investigate the stress-strain of repaired HSC using post-tensioning steel straps confinement. It was evidenced that the pre-damaged degree is an important factor and needed to be considered in estimating the ultimate strength and strain of repaired concrete. In general, it was found that the ultimate strength reduced with the increase of concrete damaged degree. Whilst the ultimate strain increased with damaged degree. Under same volumetric ratio, closer spacing provide better ultimate strain for repaired concrete columns shows the desired for confinement of damage column. Equal volumetric ratio found for damage level at 50% and 100% strength. Force – displacement relationship at each damage level are different but close behaviour observed at 50% and 100% damage level.

This work was funded by Fundamental Research Grant Scheme (FRGS) from Ministry of Higher Education, Malaysia [R.J130000.7822.4F826]. The first author would like to acknowledge that the undergraduate students Amirul Radzi, Melvay Allens and Muhammad Fadhli undertook the tests as their final year projects supervised by the first author.

References

[1] A.G. Tsonos, Effectiveness of CFRP-jackets and RC-jackets in post-earthquake and pre-earthquake retrofitting of beam–column subassemblages, Engineering Structures, **30(**3), 777-793, (2008)

[2] Y.F. Wu, Y.Yun, Y. Wei, and Y. Zhou, Effect of predamage on the stress-strain relationship of confined concrete under monotonic loading, J. of Structural Engineering, **140**(12), 04014093, (2014)

[3] Y.W. Zhou, X.M. Liu, L.L. Sui, F. Xing, and H.J. Zhou, Stress–strain model for fibre reinforced polymer confined load-induced damaged concrete, Materials Research Innovations, **19**(sup6), S6-125, (2015)

[4] M. Panjehpour, N. Farzadnia, R. Demirboga, and Ali, A.A.A. Behavior of high-strength concrete cylinders repaired with CFRP sheets, J. of Civil Engineering and Management, **22**(1), 56-64, (2016)

[5] A. Peled, Confinement of damaged and nondamaged structural concrete with FRP and TRC sleeves, J. of Composites for Construction, **11**(5), 514-522, (2007)

[6] A. Ilki, O. Peker, E. Karamuk, C. Demir, and N. Kumbasar, FRP retrofit of low and medium strength circular and rectangular reinforced concrete columns, J. of Materials in Civil Engineering, **20**(2), 169-188, (2008)

[7] D.S. Gu, G. Wu, Z.S. Wu, and Y.F. Wu, Confinement effectiveness of FRP in retrofitting circular concrete columns under simulated seismic load, J. of Composites for Construction, **14**(5), 531-540, (2010)

[8] R.D. Iacobucci, S.A. Sheikh, and O. Bayrak, Retrofit of square concrete columns with carbon fiber-reinforced polymer for seismic resistance, ACI Structural J., **100**(6), 785-794,(2003)

[9] C.K. Ma, A.Z. Awang, and W. Omar, New theoretical model for SSTT- confined HSC columns, Magazine of Concrete Research, **66**(13), 674-684, (2014)

[10] C.K. Ma, A.Z. Awang, R. Garcia, W. Omar, K. Pilakoutas, and M. Azimi, Nominal curvature design of circular HSC columns confined with post-tensioned steel straps, Structures, **7**, 25-32,(2016)

[11] C.K. Ma, A.Z. Awang, W. Omar, M. Liang, S-W. Jaw, and M. Azimi, Flexural capacity enhancement of rectangular high-strength concrete columns confined with post-tensioned steel straps: experimental investigation and analytical modelling, Structural Concrete, **17**(4), 668-676, (2016).

[12] C.K. Ma, A.Z. Awang, R. Garcia, W. Omar, and K. Pilakoutas, Behaviour of over-reinforced high-strength concrete beams confined with post-tensioned steel straps–an experimental investigation, Structural Concrete, **17**(5), 768-777, (2016)

[13] C.K. Ma, A.Z. Awang and W. Omar, Slenderness effect and upper-bound slenderness limit of SSTT-confined HSC column, Int. J. of Structural Engineering, **5**(3), 279-286, (2014)

[14] C.K. Ma, A.Z. Awang, and W. Omar, Flexural ductility design of confined high-strength concrete columns: Theoretical modelling, Measurement, **78**, 42-48, (2016)

[15] D.E. Lehman, S.E. Gookin, A.M. Nacamuli and J.P. Moehle, Repair of earthquake-damaged bridge columns, ACI Structural J., **98**(3), 233-238, (2001)

[16] D.R. Stoppenhagen, J.O. Jirsa, and L.A. Wyllie, Seismic repair and strengthening of a severely damaged concrete frame, ACI Structural J., **92**(2), 177-187.9.(1995)

Bearing Stress at Failure of Double-Lap Hybrid Joints in Woven Fabric Kenaf Fiber Composite Plates under Quasi-static Loading

Sim Yee Lee[1,*], and *Hilton* Ahmad[1]

[1]Faculty of Civil and Environmental Engineering, Universiti Tun Hussein Onn Malaysia, 86400 Parit Raja, Batu Pahat, Johor, Malaysia

Abstract. The present paper is focused on the bearing stress at failure of double-lap woven fabric kenaf fiber reinforced polymer (KFRP) hybrid bonded-bolted joints in experimental frameworks. The effects of different normalized plate width (plate width/hole diameter, *W/d*), lay-up types and bolt loads were incorporated in current study as specified in testing series. Generally, hybrid joint coupons separated within adhesive layer prior to net-tension failure or bearing/net-tension failure. The bearing stress at failure increased as *W/d* ratio increment, critical *W/d* is given as four and three in clamped and finger tight condition respectively. Lay-up types present insignificant effect to bearing stress at failure due to low volume fiber fraction in kenaf fiber composites. Combination of thicker and clamped conditions plate demonstrated greater bearing stress than equivalent finger-tight (FT) conditions due to higher load transferred from friction, as expected.

1 Introduction

The awareness of renewable plant for massive productions is progressively developed in recent years due to evolution in biotechnology development. Renewable natural fibers able to replace commercially synthetic fiber in production of natural composite materials with medium and low bearing load applications [1]. Generally, natural fiber has greater properties compared to synthetic fiber i.e., higher specific strength [2], low carbon dioxide (CO_2) emissions, biodegradable, good thermal and acoustic insulation. Recent studies investigates the suitability of biodegradable composite materials using natural fibers such as flax, bamboo, pineapple, silk, sisal, jute, kenaf, and ramie for manufacturing sectors. Kenaf fibers is one of the important source of natural fiber extracted from bast and core fibers , previous study reported that the tensile and flexural strength of kenaf reinforced polylactic acid (PLA) composites were approximately 223 MPa and 254 MPa respectively with 70% fiber fraction. It had proved that kenaf fiber showed higher strength compared to other natural fibers in reinforcing PLA [3].

* Corresponding author: simsim1019@hotmail.com

Structures components require joining techniques to assemble different parts and connection is the weakest point when subjected to applied load. Hybrid adhesively bonded-bolted joint (hybrid joint) method is a lap joint technique combines advantages in conventional bolted joints and adhesively bonded joints. Hybrid joints able to decrease high peel stress near plate edges and subsequently increased static strength [4]. The presence of fasteners in hybrid joints assists the continuous operation of the joining structure despite the occurrence of adhesive failure. Observations by Bois *et al.* [5] found that three types of prominent failure modes in hybrid joints, i.e. net-tension failure, adhesive joint failure and bearing failure. However, currently there is no standard for testing method/set-up available as reference for hybrid joints. More recent studies have proved that hybrid joints demonstrated greater static strength than bolted joints or adhesively bonded joints alone. Franco *et al.* [6] had proved that under pre-tension bolt decreased maximum shear and peel stress and therefore increased the static strength of hybrid double-lap with joining GFRP-aluminium plates, approximately similar to the total tensile strength in respective bolted joints and bonded joints. Nabil *et al.* [7] had investigated the strength on bolted joint, adhesively bonded joint and hybrid joint of HexPly M18/1/G939 carbon fiber/epoxy laminates and found that the bolts in hybrid joint relieved exhibited peels stress to delay crack formation within adhesive layer.

However, the research works on kenaf fiber reinforced polymer is relatively limited. It has excellent engineering behaviour such as impact and fatigue resistance due to ability of the crimp to absorb energy as crack developed. Further investigation on these composite types need to be examined for structures applicability, especially in joining structures. Lee *et al.* [8] showed that failure load increased substantially with *W/d* in single-lap hybrid joints on woven fabric kenaf fiber composite plates. Current study broadens the study on bearing stress at failure of hybrid joint double-lap woven fabric KFRP with various parameters. The testing series of current work was designed and mechanical machines were carried out as elaborated in next section.

2 Experimental methodologies

2.1 Testing coupons preparation

The KFRP composite panels were prepared by stacking the plain weave kenaf fiber layer and combined with resin system using hand lay-up fabrication technique under compression. The kenaf fiber yarns have 276 tex (weight in grams of 1000 m of the tow), weaved using rigid plain handloom with tow density of 4 *end per cm* in warp direction and 4 *pick per cm* in weft direction. The term ends and picks describe the number of warp tows and weft tows, respectively, per unit length. Matrix binder system was prepared by mixing thermoset epoxy resin (SP84) and hardener (SP76) with ratio 2:1. Plain weave kenaf fiber layer was placed on aluminium mould and smeared the binder system uniformly with extra care to avoid entrapped voids and bubble formation. These steps were repeated accordingly to the desired stacking sequence and fiber orientations, at least twenty-four hours curing period were allowed to set under the room temperature prior to demolding. Table 1 showed the testing variables tested (lay-up types, end-distance (*e*) to hole diameter ratio, normalized *W/d* and bolt loads) with detailed diagram for one lay-up is given in Fig. 1.

Table 1. Double-lap hybrid joint testing coupon series.

Laminate	Lay-up Orientation	*e/d*	*W/d*	Clamp-up torque
PX2	$(0/90)_S$	6	2, 3, 4, 5	FT, 5Nm
PX4	$(0/90)_{2S}$	6	2, 3, 4, 5	FT, 5Nm
PS2	$(\pm45)_S$	4	2, 3, 4, 5	FT, 5Nm
PQ4	$(0/90/\pm45)_S$	4	2, 3, 4, 5	FT, 5Nm

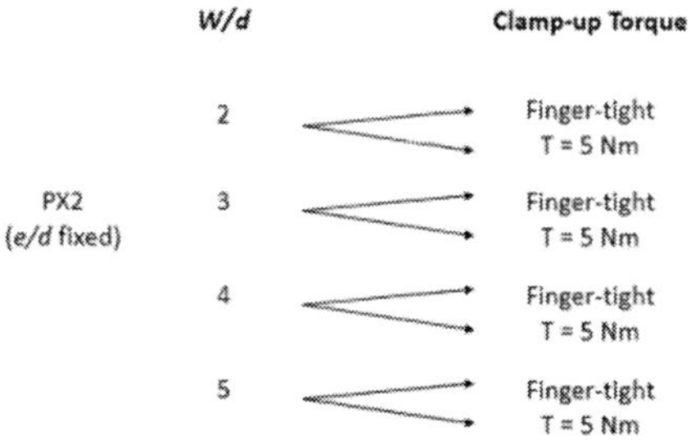

Fig. 1. Testing series carried out on one of the laminate type, PX2

Karakuzu *et al.* [9] concluded that *e/d* ratio was fixed to 6 for cross-ply lay-up and 4 for quasi-isotropic lay-up to eliminate shear-out failure. The hole diameter, *d* used in current work was fixed to 5 mm and drilled by using high-speed drilling bits to reduce hole edge defects. Normalised plate width, *W/d* was ranging between 2 to 5. The fabricated composite panels were sectioned according to the desired testing coupon sizes by using a cooling diamond saw. An adhesive layer (~0.1 mm thick) was applied within the plate overlap region and a M5 size fastening system was then installed under finger-tight/clamped condition by using KTC Digital Torque Wrench GEK 030-C3A immediately prior to mechanical testing to eliminate the bolt relaxation effect. A double-lap hybrid joint configuration ready for mechanical testing is shown in Fig. 2. Finger-tight (FT) condition (~0.5 Nm) was used as the worst case scenario and clamp-up torque with 5 Nm was frequently used in composite joint design [10].

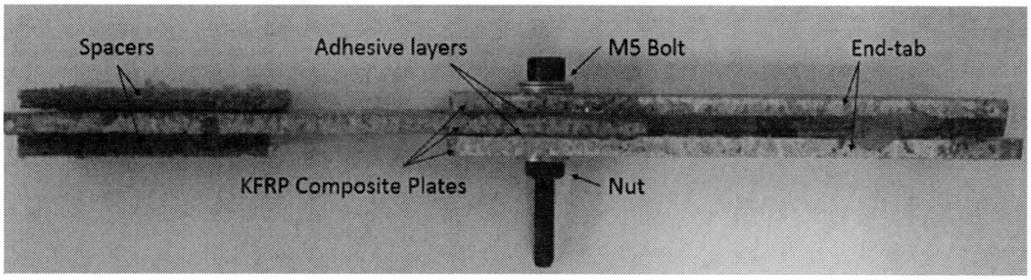

Fig. 2. Photograph of PX4 hybrid double-lap joint.

2.2 Mechanical testing

At least three set of testing joints in each joint designation series were tested and bearing stress at failure values were averaged. ASTM Standard D3039B [11] was referred for the mechanical testing set-up. The quasi-static tests were conducted by using universal testing machine SHIMAZU AG-I with a constant cross-head speed of 0.5 mm/min and 10kN of load cell capacity. The end-tabs were prepared at the coupon edge adhered with aluminium strip prior to testing to avoid slippage during mechanical testing. Spacers with same adjacent plate thickness were prepared to avoid the primary bending as shown in Fig. 2. Bearing stress at failure, σ_b was obtained by using following expression, where P_{max} is the maximum fracture load at failure, *d* is diameter of hole and *t* is plate thickness.

$$\sigma_b = \frac{P_{max}}{dt} \tag{1}$$

3 Results and discussion

3.1 Load-displacement profile

The load-displacement curves were consistent in hybrid joints tested. A Representative load-displacement graph is shown in Fig. 3. From the load-displacement graph shown in Fig. 3, it is clearly presented that there are few stages prior to the ultimate load., as obtained from data-logger, i.e., adhesive failure, bearing contact and plate failure. Plate failure showed net-tension failure in smaller *W/d* coupons and bearing/net-tension failure occurred in larger *W/d* coupons as shown in Fig. 4. At early stage, it showed a linear elastic behaviour before the adhesive failure. As applied load increased, the peel stress near the edges of shear-lap was increased and cohesive cracks initiated and propagated which is also known as cohesive failure. Similar findings by Lee *et al.* [12] where they observed adhesive layer was failed initially, followed by mechanical fastened failure in hybrid joint. Next, the curve showed a constant line where friction load transferred from bolt to joining plates. As the applied load exceeded the bolt load applied ($P > P_{bolt}$), the composite plate were slipped and the bearing contact between bore hole and bolt shaft took place within bearing area. Subsequently, the applied load is borne by individual joining plates until final failure.

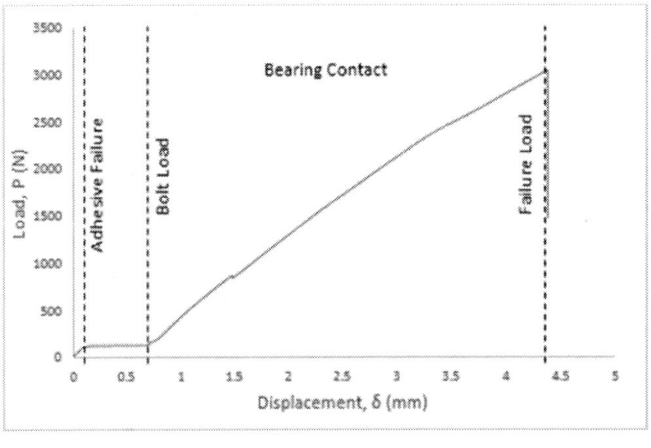

Fig. 3. Load-displacement profile for PQ4 lay-up with finger-tight condition (*W/d* = 3).

3.2 Bearing stress at failure with variation of normalized *W/d*, lay-up and bolt load

Photograph of PQ4 coupons under finger-tight conditions with normalized *W/d* is shown in Fig. 4. It is clearly observed that all testing coupons were failed in net-tension mode. Consistent trend with single-lap non-crimp CFRP hybrid joint was reported by Kelly *et al.* [13], critical *W/d* (transition from net-tension to bearing failure) is given as three.

Compared to catastrophic nature in net-tension failure, bearing failure demonstrated more progressive failure type. Although bearing failure is preferable, but net-tension mode is unavoidable due to geometry constraint and bolt load applied. Current work suggests the critical *W/d* in woven fabric KFRP is larger than five. As expected, bearing stress at failure increased with *W/d* ratio due to larger unloaded region along the net-tension plane.

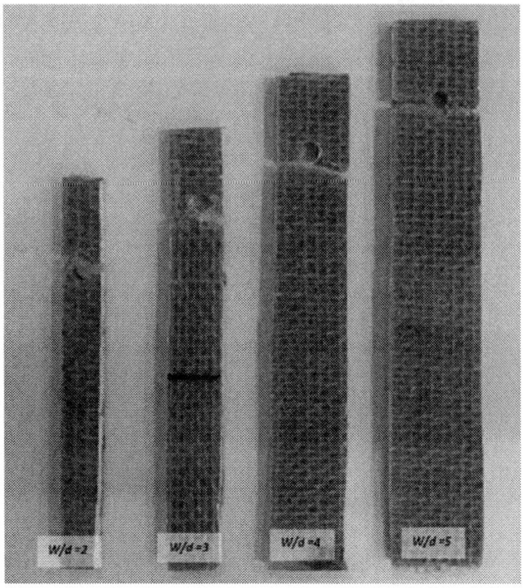

Fig. 4. Photograph of tested coupons for PQ4 lay-up with different *W/d*.

As shown in Fig. 5, both lay-up types showed insignificant effect in bearing stress at failure due to similar volume fiber fraction, V_f of woven fabric kenaf fiber, which is approximately 17%. Although more volume of 0° fiber orientation in cross-ply lay-up, but due to low volume fracture fraction demonstrated insignificant effect to ultimate bearing stress. Therefore, close values of bearing stress at failures with respective to normalized *W/d* in both lay-up types is perhaps not surprising. On the other hands, thicker plate (PX4 and PQ4) showed higher bearing stress at failure than equivalent thinner plates (PX2 and PS2). Generally, thicker coupon is efficient to transfer more load through plate thickness to give higher peak load. Bearing stress increased with plate thickness, consistent with the findings by Gordon Kelly *et al.* [13].

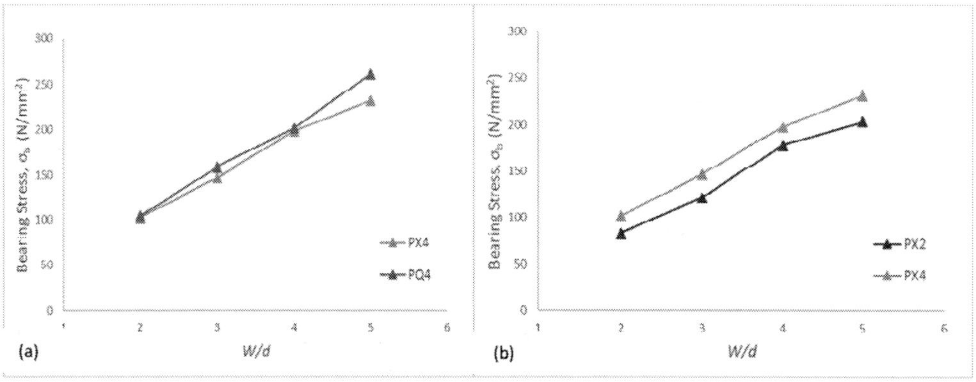

Fig. 5. Effect of (a) lay-up stacking (b) plate thickness on bearing stress at failure with clamped condition.

In application of bolt load, introduction of pre-tension load in bolt able to produce compression in out-of-plane between joining plates. Clamped condition showed more uniform stress distribution along the overlap region and helps to alleviate the peel stress as a result of stress concentration as reported by Franco *et al.* [6]. When applied load is less than bolt load ($P< P_{bolt}$), the load transfer was concentrated within contact region near the bolt head and nut. The amount of applied load, P to surpass the P_{bolt} is smaller under finger-tight condition than under clamped condition, as expected. Thereafter, slippage occurred at the shear plane when $P>P_{bolt}$, both plates were subsequently separated as a result of cohesive cracks and plate splitting along the adhesive layer parallel to applied load. Finally, applied load was borne by the individual plate. Similar findings were reported by Kim, J.-S. *et al.* [14]. Fig. 6 showed the increased bearing stress at failure as a function of bolt load as the frictional load transfer by clamped bolt increased.

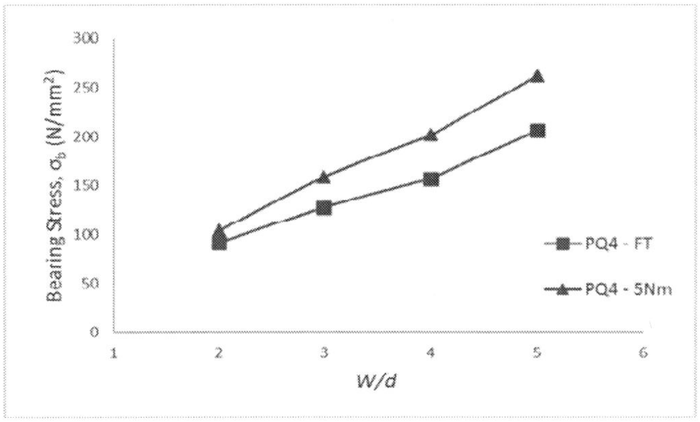

Fig. 6. Effect of bolt load on bearing stress at failure for PS2 and PQ4.

4 Conclusions

All testing coupons exhibited final failure modes in net-tension suggesting critical *W/d* is more than five. Larger normalized plate width, *W/d* showed higher bearing stress at failure due to larger unloaded region along the net-tension plane with larger plate width. Thicker plates and clamped condition able to transfer more load through the plate thickness that gives higher bearing stress at failure. Double-lap joint does not exhibited secondary bending, therefore experimental framework on single-lap joint need to be carried out to broaden the scope of current work.

The authors gratefully acknowledge Universiti Tun Hussein Onn Malaysia (UTHM) and Research Grant Scheme under IGSP U245. This paper was partly sponsored by the Centre for Graduate Studies UTHM.

References

[1] O. Faruk, A. K. Bledzki, H.-P. Fink, and M. Sain, Biocomposites reinforced with natural fibers: 2000 – 2010, Progress in Polymer Science, **37**(11), 1552-1596, (2012)
[2] H. M. Akil, M. F. Omar, A. a. M. Mazuki, S. Safiee, Z. a. M. Ishak and A. Abu Bakar, Kenaf fiber reinforced composites: A review, Material and Design, **32**(8), 4107-4121, (2011)
[3] S. Ochi, Mechanical properties of kenaf fibers and kenaf/PLA composites, Mechanics of Materials, **40**(4-5), 446-452, (2008)

[4] G. Marannano and B. Zuccarello, Numerical experimental analysis of hybrid double lap aluminum-CFRP joints, Composites. Part B Engineering, **71**, 28-39, (2015)
[5] C. Bois, H. Wargnier, J.-C. Wahl and E. L. Goff, An analytical model for the strength prediction of hybrid (bolted/bonded) composite joints, Composite Structures, **97**, 252-260, (2013)
[6] G.D. Franco and B. Zuccarello. Analysis and optimization of hybrid double lap aluminum-GFRP joints, Composite Structures, **116**, 682-693, (2014)
[7] M.C. Nabil, J. Wang, K.C. Wing and P. Chang, Static and fatigue testing bolted, bonded and hybrid step lap joints of thick carbon fibre/epoxy laminates used on aircraft structures, Composite Structures, **142**, 96-106, (2016)
[8] S.Y. Lee and H. Ahmad, Experimental strength of single-lap hybrid joints on woven fabric kenaf fiber composites under quasi static condition, *The 3rd International Conference on Civil and Environmental Engineering for Sustainability*, MATEC Web of Conferences, **47**, Article No. 02003, (2016)
[9] R. Karakuzu, T. Gulem and B. M. Icten, Failure analysis of woven laminated glass-vinylester composites with pin-loaded hole, Composite Structures, **72**, 27-32, (2006)
[10] H. Ahmad, A.D. Crocombe and P.A. Smith, Strength prediction in CFRP woven laminate bolted double-lap joints under quasi-static loading using XFEM, Composites: Part A Applied Science and Manufacturing, **56**, 192-202, (2014)
[11] ASTM D3039/D3039M, *Standard test method for tensile properties of polymer matrix composite materials*, ASTM International, West Conshohocken, United States, (2008)
[12] Y.H. Lee, D.W. Lim, J.H. Choi and M.K. Yoon, Failure load evaluation and prediction of hybrid composite double lap joints, Composite Structures., **92**, 2916-2926, (2010)
[13] G. Kelly and S. Hallstrom, Bearing strength of carbon fibre/epoxy laminates: effects of bolt-hole clearance, Composites: Part B, **35**, 331-343, (2004)
[14] J.S. Kim, J.Y. Lim and W.G. Lee, Joining performance evaluation of different types of gep224 glass/epoxy-to-az31b magnesium alloy single lap joints, Int. J. Precision Engineering and Manufacturing, **16**(6), 1135-1140, (2015)

Concrete-Filled Double Skin Steel Tubular Columns Exposed to ASTM E-119 Fire Curve for 60 and 90 Minutes of Fire

Sharifah Salwa Mohd Zuki[1,2], *Shahiron* Shahidan[2,*], *Choong Kok Keong*[3], *J.* Jayaprakash[3], and *Noorwirdawati* Ali[2]

[1]*School Civil Engineering, Universiti Sains Malaysia, Nibung Tebal 14300 Pulau Pinang, Malaysia*
[2] *Jamilus Research Center, Faculty of Civil and Environmental Engineering, Universiti Tun Hussein Onn Malaysia, 86400 Batu Pahat, Johor, Malaysia*
[3]*Department of Civil Engineering, The University of Notthingham, Malaysia Campus, Semenyih, Selangor, Malaysia*

Abstract: Concrete-filled double skin steel tubular (CFDST) columns with outer steel tubes varying in diameter and thickness were exposed to ASTM E-119 fire curve for 60 and 90 minutes. All specimens (control and heated) failed by outward local buckling of outer steel tube, crushing of concrete and outward and inward buckling of inner steel tube. Residual Strength Index (RSI), secant stiffness and Ductility Index (DI) were calculated in order to compare the performance of control and fire-damaged CFDST columns. It was found that specimens with 90 minutes of fire exposure time possessed highest RSI. Meanwhile, secant stiffness seems to decrease with increased exposure time. And finally, CFDST columns became more ductile after being heated.

1 Introduction

In the event of fire, fire safety is first addressed by sprinklers or smoke detectors which are part of the active fire protection system. However, if the fire becomes larger and this so-called active fire protection system failed to suppress the fire, the passive fire protection system will then take over. The passive fire protection system exists in order to ensure the integrity of the structures for a period of time so that the occupant can safely evacuate the building or until the arrival of fire fighters. The integrity of the structures can be achieved by providing adequate cover to

* Corresponding author: shahiron@uthm.edu.my

reinforcement, adequate fire protection layers, compartmentalization and etc [1-4]. In addition, the structures also need to prevent the fire from spreading and collapsing during fire [2].

There are many known advantages of concrete-filled double skin steel tubular (CFDST) columns over traditional reinforced concrete (RC) columns and concrete-filled steel tubular (CFST) columns in literature. Among them are; 1) higher fire resistance to fire even without fire protection layer on outer steel tube, 2) higher load bearing capacity, 3) prevent the spalling of concrete, 4) eliminate the need for formwork, 5) increase the strength, ductility and energy absorption and 6) reduced column footing [3-9]. However, the usage of CFDST columns is limited to outdoor construction such as transmission towers [7], and bridge piers [8-10] even though the performance of CFDST column is superior to RC and CFST columns. To date, research studies on CFDST columns exposed to fire are limited and can be found in previous research [9-16]. As for the residual strength of CFDST columns after fire, none can be found in the literature. Residual strength of fire-damaged columns is important so that it can be used to assess the damage done by fire. Therefore, engineers can calculate the amount of repair that needs to be done[16-19].

In this study, CFDST columns are subjected to ASTM E-119 fire curve for two different exposure periods, i.e., 60 and 90 minutes. In addition, the diameter and thickness of the outer steel tubes of CFDST columns are varied. The Residual Strength Index, Secant Stiffness and Ductility Index of fire-damaged CFDST columns are calculated and compared with unheated CFDST columns.

2 Experimental program

2.1 Preparation of specimens and material properties

A total of 54 specimens were casted in an upright position as shown in Fig. 1(a). Details of the specimens together with the tensile stress of steel and compressive strength of concrete are presented in Table 1. The overall length of each specimen is 600 mm and for each S1 (represent Series 1 which indicates the outer steel tube diameter of the specimen) there are two different outer steel tube thicknesses, which are 3 mm (represent by C3) and 4 mm (C4) respectively. In addition, the notations 'control', '60' and '90' at the end of the specimen naming convention indicate the un-heated specimen and the specimen heated to 60 and 90 minutes of exposure time, respectively.

The targeted strength of concrete is 30 N/mm^2. For every batch of specimens, six 100×100×100 mm cubes were casted and cured in a concrete laboratory (air-cured). Specimens and cubes were cured in similar conditions until the day of the test for specimens and after 7 and 28 days for the cubes. Three steel coupons were cut from the outer and inner steel tube and tested in order to determine its properties. Both concrete strength after 28 days and tensile stress of outer and inner steel tubes are listed in Table 1. Concrete cubes and steel coupons were tested in accordance with BS EN 12390-3 [20] and ASTM E8/E8M-11 [21] respectively.

2.2 Heating regime

The specimens were heated in a gas-fired fire furnace at the concrete laboratory in Universiti Sains Malaysia (USM). The specimens were heated one at a time as shown in Fig.1(b).The specimens were heated in accordance with ASTM E-119 [22] fire curve until the temperature reached 600°C. The temperature is kept constant for 60 and 90 minutes. After it has reached the targeted exposure time, the furnace will automatically turn off and the lid of the furnace is lifted. The specimens were left to cool down inside the furnace before being taken out of the furnace to completely cool down to room temperature as shown in Fig. 1(c).

The temperature of the furnace for the entire exposure time is measured by four ceramic thermocouples attached to the furnace wall. The temperature of the concrete and inner steel tubes during the heating process is measured by thermocouples. The thermocouple for concrete is inserted during the casting process. Meanwhile, the thermocouple for the inner steel tube is attached to the wall of the tube before the heating process commences. The thermocouple used is thermocouple type K with a length of 1000 mm and a diameter of 3 mm. At the end of the thermocouple, there is a 1000 mm cable that can be attached to a data logger in order to record temperature during the heating process.

2.3 Axial compression test

All specimens were subjected to monotonic axial compression load using a Universal Testing Machine (UTM) with a maximum load capacity of 2000 kN until failure at the Construction Research Institute of Malaysia (CREAM), Kuala Lumpur. Each specimen was equipped with four 10 mm electrical strain gauges located at the mid-height of the specimens. Two strain gauges were placed in horizontal direction, whereas another two were placed in hoop direction. In addition, four Linear Variable Displacement Transducers (LVDTs) were placed at the mid-height of the specimens. During testing, axial load and axial displacement were recorded using UTM. The test arrangement is shown in Fig. 2.

Table 1. Details of tested specimens.

Specimens	Dia. of outer steel tube [mm]	Dia. of inner steel tube [mm]	Compressive strength of concrete after 28 days [N/mm^2]	Tensile stress of outer steel tube [N/mm^2]	Tensile stress of inner steel tube [N/mm^2]	Exposure time [min]
S1-C3-Control	101.6	50.8	38	597	762	-
S1-C3-60	101.6	50.8	38	597	762	60
S1-C3-90	101.6	50.8	38	597	762	90
S1-C4-Control	101.6	50.8	39	561	762	-
S1-C4-60	101.6	50.8	39	561	762	60
S1-C4-90	101.6	50.8	39	561	762	90
S2-C3-Control	127.0	76.2	41	436	449	-
S2-C3-60	127.0	76.2	40	436	449	60

DOI: 10.1051/matecconf/201710302009

S2-C3-90	127.0	76.2	40	436	449	90
S2-C4-Control	127.0	76.2	43	449	449	-
S2-C4-60	127.0	76.2	42	449	449	60
S2-C4-90	127.0	76.2	38	449	449	90
S3-C3-Control	152.4	101.6	40	409	566	-
S3-C3-60	152.4	101.6	40	409	566	60
S3-C3-90	152.4	101.6	41	409	566	90
S3-C4-Control	152.4	101.6	40	430	566	-
S3-C4-60	152.4	101.6	42	430	566	60
S3-C4-90	152.4	101.6	38	430	566	90

(a) (b) (c)

Fig. 1. (a) CFDST columns after casting process; (b) CFDST columns during heating process; (c) CFDST columns after cooling process.

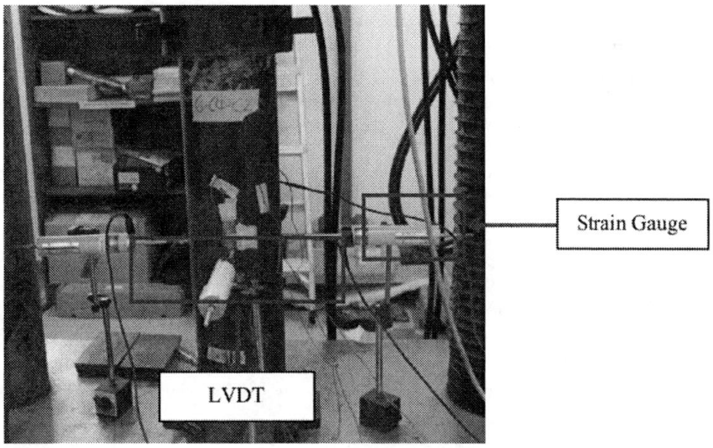

Fig. 2. Test arrangement.

During the curing process, a small amount of shrinkage occurred at the top of the specimens. The gap was filled with high strength epoxy. The epoxy was left overnight to harden. Once hardened, the surface was ground smooth with a grinding wheel for even distribution of stress during the testing process.

3 Result and discussion

3.1 Overall behavior

As seen in Fig. 1(a) and Fig. 1(c), there are no changes in physical appearance of the specimens before and after being exposed to fire. However, the concrete inside CFDST columns changed significantly as shown in Fig. 3. The concrete changed from its normal color to whitish grey. This indicates that the temperature within the concrete has already exceeded 300°C. Researchers often associate a temperature of 300°C as an indicator where concrete starts to lose its strength [23]. Nevertheless, the concrete was still intact due to the outer and inner steel tube, preventing the occurrence of spalling in concrete. However, for specimens exposed to 90 minutes of fire, the resulting concrete was brittle and crumbled under very small pressure as shown in Fig. 3(c). A detailed discussion can be found in [24-26].

All specimens failed by outward local buckling of the outer steel tube, crushing of concrete and outward and inward local buckling of inner steel tube. Outward local buckling of outer steel tube, crushing of concrete and outward and inward local buckling of inner steel tube can be seen in Fig. 3, Fig. 4 and Fig. 5 respectively.

The occurrence of outward local buckling in outer steel tube was due to the presence of concrete that prevented the steel from buckling in an inward direction. As for the inner steel tube, the failure pattern resembles the failure of hollow steel tubes. In addition, the position of local buckling for both outer and inner steel tube corresponded to the position where the crushing of concrete occurs. It was also observed that a longer exposure time leads to more severe buckling and crushing of steel and concrete. Therefore, it can be concluded that a longer exposure time significantly affects the strength of CFDST columns.

(a) (b) (c)

Fig. 3. Condition of concrete (a) control specimen; (b) 60 minutes of fire exposure; (c) 90 minutes of fire exposure.

(a) (b) (c)

Fig. 4. CFDST columns after axial compression test (a) control; (b) 60 minutes of fire exposure; (c) 90 minutes of fire exposure.

(a) (b)

Fig. 5. Failure pattern of inner steel tube (a) top view; (b) side view.

3.2 Residual Strength Index, Secant Stiffness and Ductility Index

Residual Strength Index, RSI is calculated using an equation suggested by Han &Huo [27]. A higher value of RSI means that the damage done by fire is more severe compared to a low value of RSI. Therefore, zero in control specimens means that there is no damage observed among the specimens. The average values of the three specimens in terms of RSI, secant stiffness and Ductility Index, DI are presented in Table 2.

RSI for 60 and 90 minutes of exposure time is shown in Fig. 6. The RSI for 90 minutes of exposure time is higher than 60 minutes exposure time for both outer tube thicknesses, t_0, i.e., 3mm and 4 mm. Higher RSI can be attributed to a longer exposure time. Specimens with a longer exposure time absorb more heat thus increasing the maximum temperature attained by the outer steel tube, concrete and inner steel tube. If the inner steel tube is well protected, the maximum temperature attained will be lower thus it is able to retain its load carrying capacity. As mentioned earlier, the temperature distribution of CFDST columns is discussed in [24-26].

In addition, the specimen with $t_0 = 3$mm is found to have a higher RSI than the specimen with $t_0 = 4$mm . This applies for both 60 and 90 minutes of exposure time. This is because a thicker t_0 slows down the process of heat transfer from steel to concrete. This lowered the maximum temperature attained by concrete and the inner steel tube thereby lowering the RSI value.

From Fig. 6(b), it is interesting to note that the RSI value for S1-C4-60 is negative. The negative value means that the fire-damaged CFDST column in these series gained strength instead of losing its strength after fire exposure. This is the result of a thicker t_0 when compared to S1-C3-60 (RSI = 13%). According to Gunalan & Mahendran (2014)[28, low grade steel loses its strength at a slower rate compared to high grade steel. In addition, yield strength of the steel remains unaffected until it reaches a temperature of around 500 to 600°C[29,30]. With other parameters (exposure time, length, diameter and steel grade) being constant, it can therefore be concluded that specimens with t_0 = 4mm retain more strength than its counterparts thus obtaining a negative value of RSI.

Secant stiffness is calculated by dividing ultimate compressive load with displacement at an ultimate compressive load. Secant stiffness decreases with increased exposure time regardless of t_0. The specimen with 60 minutes of exposure time loses secant stiffness ranging from 11% (S3-C4-60) to 51% (S1-C4-60 and S2-C4-60) from original secant stiffness (Fig. 7). As for the specimen exposed to 90 minutes of exposure time, secant stiffness as high as 64% (S1-C4-90) to as low as 36% (S3-C4-90) is lost. Similar to RSI, a longer exposure time leads to more heat absorption thus resulting in a high reduction in terms of secant stiffness.

The Ductility Index, DI was calculated using an equation suggested by [31]. DI for Series 1 decreases with increased exposure time regardless of t_0. However, for Series 2 and Series 3, DI increases as exposure time increases (Fig. **8**). Interestingly, DI for Series 2 and Series 3 for specimens with t_0 = 4mm exceeded the original DI. This means that the specimens became more ductile after fire exposure. Other researchers like Tao et al. (2008) [32]and Tao & Han (2007) [33]also found out that the ductility of CFST columns after fire exposure increases. This can be attributed to changes in the properties of steel and concrete after being exposed to high temperature.

Table 2. Test results of tested specimens.

Specimens	Residual Strength Index [%]	Secant Stiffness [kN/mm]	Ductility Index, DI
S1-C3-Control	0	32	1.08
S1-C3-60	13	49	1.47
S1-C3-90	16	38	1.23
S2-C3-Control	0	142	1.86
S2-C3-60	4	101	1.35
S2-C3-90	10	84	1.41
S3-C3-Control	0	251	1.40
S3-C3-60	15	131	1.31
S3-C3-90	22	102	1.26
S1-C4-Control	0	108	2.02
S1-C4-60	-2	53	1.63
S1-C4-90	5	39	1.40
S2-C4-Control	0	228	2.04
S2-C4-60	16	111	1.70
S2-C4-90	16	105	2.37
S3-C4-Control	0	201	1.49
S3-C4-60	8	179	1.56
S3-C4-90	15	129	1.80

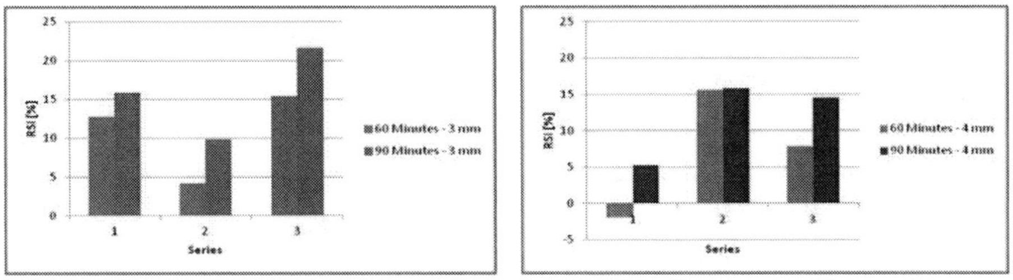

(a) (b)

Fig. 6. RSI of CFDST columns (a) $t_0 = 3mm$; (b) $t_0 = 4mm$.

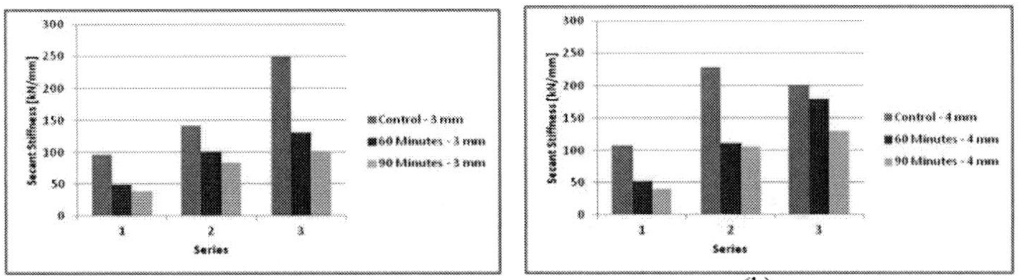

(a) (b)

Fig. 7. Secant stiffness of CFDST columns (a) $t_0 = 3mm$; (b) $t_0 = 4mm$.

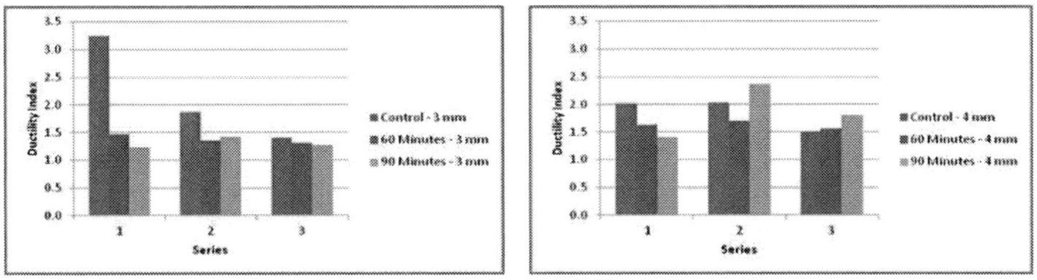

(a) (b)

Fig. 8. DI of CFDST columns (a) $t_0 = 3mm$; (b) $t_0 = 4mm$.

4 Conclusion

From the above discussion, the following conclusions can be drawn:

1) All specimens failed by outward local buckling of outer steel tube, crushing of concrete and outward and inward local buckling of inner steel tube.
2) Local buckling and crushing of concrete is more severe for specimens with 90 minutes of exposure time compared to specimens with 60 minutes of exposure time.
3) Residual Strength Index, RSI for specimens with 90 minutes of exposure time and $t_0 = 3mm$ possessed the highest RSI.
4) Secant stiffness of CFDST columns decreased with increased exposure time regardless of t_0.
5) Ductility Index, DI for Series 2 and Series 3 for specimens with $t_0 = 4mm$ exceeds the original DI.

Authors are thankful for the financial and technical support provided by Universiti Sains Malaysia (USM) and Universiti Tun Hussein Onn Malaysia (UTHM) Vot.

References

[1] J. Purkiss, *Fire Safety Engineering Design of Structures*, Butterworth-Heinemann, Jordan Hill, Oxford (1996)
[2] E.U. Chowdhury, *Behaviour of Fibre Reinforced Polymer Confined*, PhD Thesis, Queen's University, Ontario, Canada, (2009)
[3] J.M. Irwan, S.K. Faisal, N. Othman, M.H. Wan Ibrahim, R.M. Asyraf and M.M.K. Annas, Performance of concrete using light waste PET fibre, Adv. Materials Research, **795**, 352-355, (2014)
[4] S. Shahidan, S.S.M. Zuki and N. Jamaluddin, Damage grading system for severity assessment on concrete structure, Case Studies in Constr. Materials, **5**, 79–86, (2016)
[5] V.K.R. Kodur, T.T. Lie, Experimental Studies on the Fire Resistance of Circular Hollow Steel Columns Filled with Steel-Fibre-Reinforced Concrete, J. of Fire Protection Engineering, **7**, 89-97, (1995)
[6] X. Zhao, R. Grzebieta and M. Elchalakani, Tests of concrete-filled double skin CHS composite stub columns, Steel Comp Struct: Int. J., **2**, 129-146, (2002)
[7] L.H. Han, Y. F. Yang and L. Xu, An experimental study and calculation on the fire resistance of concrete-filled SHS and RHS columns, J. Constr. Steel Res., **59**, 427-452, (2003)
[8] S. Shahidan, I. Isham, and N. Jamaluddin, A Review on Waste Minimization by Adopting in Self Compacting Concrete, MATEC Web of Conf., **47**, 1–7, (2016)
[9] J. Ding, Y.C. Wang, Realistic modelling of thermal and structural behaviour of unprotected concrete filled tubular columns in fire, J. Constr. Steel Res., **64**, 1086-1102, (2008)
[10] L.H. Han, W. Li and R. Bjorhovde, Developments and advanced applications of concrete-filled steel tubular (CFST) structures: Members, J. Constr. Steel Res., **100**, 211-228, (2014)
[11] S. Shahidan, H.B. Koh, A.M.S. Alansi, and L. Y. Loon, Strength Development and Water Permeability of Engineered Biomass Aggregate Pervious Concrete, MATEC Web of Conf., **47**, 1-6, (2016)
[12] X.L. Zhao and R. Grzebieta, Strength and ductility of concrete filled double skin (SHS inner and SHS outer) tubes, Thin-Walled Struct., **40**, 199-213, (2002)
[13] R. Imani, M. Bruneau and G. Mosqueda, Simplified analytical solution for axial load capacity of concrete-filled double-skin tube (CFDST) columns subjected to fire, Engineering Structure, **102**, 156-175, (2015)
[14] H. Lu, X.L. Zhao and L.H. Han, FE modelling and fire resistance design of concrete filled double skin tubular columns, J. Constr. Steel Res., **67**, 1733-1748, (2011)
[15] H. Lu, L.H. Han and X.L. Zhao, Fire performance of self-consolidating concrete filled double skin steel tubular columns: Experiments, Fire Safety J., **45**, 106-115, (2010)

[16] H. Lu, X.L. Zhao and L.H. Han, Testing of self-consolidating concrete-filled double skin tubular stub columns exposed to fire, J. Constr. Steel Res., **66**, 1069-1080, (2010)
[17] H. Yang, L.H. Han, Y.C. Wang, Effects of heating and loading histories on post-fire cooling behaviour of concrete-filled steel tubular columns, J. Constr. Steel Res., **64**, 556-570, (2008)
[18] Y.Yang and L.Han, Fire Resistance of Concrete-Filled Double Skin Steel Tubular Columns, *Proc. of the 4th Int.l Conf. on Adv. in Steel Struc.* Shanghai, China, 1047-1052, (2005)
[19] L. Han, Y. Yang, H.Yang and J. Huo, Residual strength of concrete-filled RHS columns after exposure to the ISO-834 standard fire, Thin-Walled Struct., **40**, 991-1012, (2002)
[20] BSEN 12390-3, *Testing hardened concrete-Part 3: Compressive strength of test specimens*, British Standards, London, United Kingdom, (2009)
[21] ASTM E8/E8M-11, *Standard Test Methods for Tension Testing of Metallic Materials*, West Conshohocken, PA, (2011)
[22] ASTM E119-11, *Standard Test Methods for Fire Tests of Building Construction and Materials*, West Conshohocken, PA, (2010)
[23] N. Short, J. Purkiss, S. Guise, Assessment of fire damaged concrete using colour image analysis, Constr. Build. Mater., **15**, 9-15, (2001)
[24] S.S.M. Zuki, K.K. Choong, J. Jayaprakash and S. Shahidan, Behavior of Fire Exposed Concrete-Filled Double Skin Steel Tubular (CFDST) Columns under Concentric Axial Loads, Appl. Mech. Mater, **773-774,** 938–42, (2015)
[25] S.S.M. Zuki, K.K. Choong, J. Jayaprakash, S. Shahidan, Effect of Diameter on Fire Exposed Concrete-Filled Double Skin Steel Tubular (CFDST) Columns under Concentric Axial Loads, Appl. Mech. Mater., **802**, 130-135, (2015)
[26] S. Shahidan, S.S.M. Zuki, K.K. Choong, J. Jayaprakash, Repaired of Fire-Damaged Concrete-Filled Double Skin Steel Tubular (CFDST) Columns With Fiber Reinforced Polymer (FRP), ARPN J. Eng. Appl. Sci., 11, 3718–372, (2016)
[27] L. Han and J. Huo, Concrete-Filled Hollow Structural Steel Columns after Exposure to ISO-834 Fire Standard, J. Struct. Eng., **129**, 68-78, (2003)
[28] S.Gunalan, M. Mahendran, Structures Experimental investigation of post-fire mechanical properties of cold-formed steels, Thin Walled Struct.,**84**, 241-254, (2014)
[29] M.B. Dwaikat, *Flexural Response Of Reinforced Concrete Beams Exposed To Fire*, PhD Thesis, Michigan State University, (2009)
[30] X. Qiang, F.S.K. Bijlaard and H. Kolstein, Deterioration of mechanical properties of high strength structural steel S460N under steady state fire condition, Mater. Des., **36**, 438-442, (2012)
[31] F. Liu, L. Gardner, H. Yang, Post-fire behaviour of reinforced concrete stub columns confined by circular steel tubes, J. Constr. Steel Res, **102**, 82-103, (2014)
[32] Z. Tao, L.H. Han, J.P. Zhuang, Cyclic performance of fire-damaged concrete-filled steel tubular beam–columns repaired with CFRP wraps, J. Constr. Steel Res., **64**, 37-50, (2008)
[33] Z. Tao, L.H. Han, Behaviour of fire-exposed concrete-filled steel tubular beam columns repaired with CFRP wraps, Thin-Walled Struct.,**45**, 63-76, (2007)

Shear Rehabilitation of RC Deep Beams using NSM CFRP Anchor Bars

Abdul Aziz Abdul Samad[1], *Douread R.* Hassen[1,*], *Noridah* Mohamad[1], *Ali Naji* Attiyah[2], *J.* Jayaprakash[3], and *Priyan* Mendis[4]

[1]Faculty of Civil and Environmental Engineering, Universiti Tun Hussein Onn Malaysia, 86400 Parit Raja, Batu Pahat, Johor, Malaysia
[2]Faculty of Civil Engineering, University of Kufa, Kufa, P.O. Box (21), Najaf Governorate, Iraq
[3]Department of Civil Engineering, University of Nottingham Malaysia Campus, Jalan Broga, 43500 Semenyih, Selangor, Malaysia
[4]Department of Infrastructure Engineering, University of Melbourne, Parkville, 3010 Victoria, Australia

Abstract. This paper presents results of an experimental study for the shear strengthening of reinforced concrete (RC) deep beam using near surface mounted (NSM) carbon fiber reinforced polymer (CFRP) anchor bars. To ensure shear failure, five RC deep beam specimens were cast with nominal shear reinforcement. Shear strengthening of the RC deep beams was conducted by inserting 5 mm diameter by 450 mm length of CFRP bars as anchors with spacing of 100 mm and 150 mm centre to centre from the support of the beam. All RC beams were simply supported and subjected to a four-point bending test with shear span to effective depth ratio of 0.864. The results presented include the ultimate load, CFRP contribution to shear, modes of failure and the load-deflection profile. The shear enhancement of the RC deep beams shows an increment of 17.3% up to 25.5% with decreasing mid-span deflection from 6.4% to 15.1%. In addition, using this technique also increases its flexural beam resistance under the same conditions.

1 Introduction

Near surface mounted (NSM) is one of the most favorable and up-to-date technique for strengthening and repairing damage reinforced concrete structures. Normally NSM technique involves cutting groove along the concrete surface or concrete cover of the structural member and followed by embedding carbon fiber reinforced polymer (CFRP) bars into these grooves. These CFRP bars are than bonded to the concrete grooves by using epoxy resin.

Literature has shown that the NSM technique has been introduced in Europe since the 1950's. In fact, since the last decade, numerous research on NSM has been conducted by various researchers focusing on strengthening of damage structures in flexure using either CFRP laminates, strips or rods [1- 6].

* Corresponding author: dou_444@yahoo.com

However, the authors observed that research on using NSM for shear strengthening for reinforced concrete (RC) deep beams have not been thoroughly investigated [7]. Hence, a new approach to using the NSM CFRP techniques for the strengthening of RC deep beams in shear is presented.

Two NSM techniques were applied concurrently for the strengthening of the RC deep beams. Firstly, within the shear span of the beam, CFRP bars of 5 mm diameter by 450 mm length were inserted vertically (from the bottom upwards) inside the RC deep beam as anchors. The CFRP anchor bars are located adjacent and on each side of the existing steel reinforcement of the RC deep beams. Secondly, additional CFRP longitudinal bars were located in grooves along the soffit of the beam. Five RC deep beams (one control beam and four beams strengthened by the NSM technique) were cast and tested until failure. In this research, the test variables were the spacing of the CFRP anchor bars from the support of the beam. The ultimate load, CFRP contribution to shear, modes of failure and load-deflection profile of the NSM strengthened RC deep beams are presented and discussed in this paper. Subsequently, the experimental results from the NSM strengthened RC deep beams are compared with the control beam.

2 Experimental program

This paper studies the structural performance of applying near surface mounted (NSM) CFRP anchor bars in shear and additional CFRP longitudinal bars in flexure to strengthened RC deep beams with shear discrepancies. Control deep beam (C1) was reinforced in flexure with three 16 mm diameter steel bars at the bottom, two 12 mm diameter steel bars at the top. To ensure that the RC deep beams failed in shear, nominal stirrups of 12 mm diameter at 200 mm centre to centre (c/c) was placed from the support of the beam, as specified by [8]. R1, R2, R3 and R4 are RC deep beam test specimens requiring strengthening using CFRP anchor and longitudinal bars. All four beams were similarly reinforced in shear and flexure as control beam C1. However, for beam specimens R1 and R2, shear strengthening was done by inserting 6mm diameter CFRP anchor bars with spacing of 100mm c/c from the support. Two numbers of 6mm diameter NSM CFRP longitudinal bars was placed along a groove located at the soffit of the beam. Similar CFRP bar arrangements were applied to beam specimens R3 and R4 but with CFRP anchor bar spacing at 150mm c/c. Bonding between CFRP anchor bars and longitudinal bars to the concrete deep beams was achieved by using epoxy resin. Fig. 1 shows detailed description of control beam C1, beam specimens R1 and R2 with CFRP anchor bars at 100mm c/c and beam specimens R3 and R4 with CFRP anchor bars at 150mm c/c.

2.1 Material properties

A ready-mix concrete with target compressive strength of 30 MPa was used. The design slump for the ready-mix concrete was targeted at 75 mm and the maximum aggregate size was selected at 19 mm. Concrete compressive strength was measured from testing six numbers of 150 mm cubes and six numbers of 100 mm by 200 mm cylinders, as specified by [9]. Table 1 shows the average concrete compressive strength which shows good accuracy when average cube strength fcu measures at 30.5 MPa whilst the cylindrical strength achieved 84% of its cube strength at 25.8 MPa.

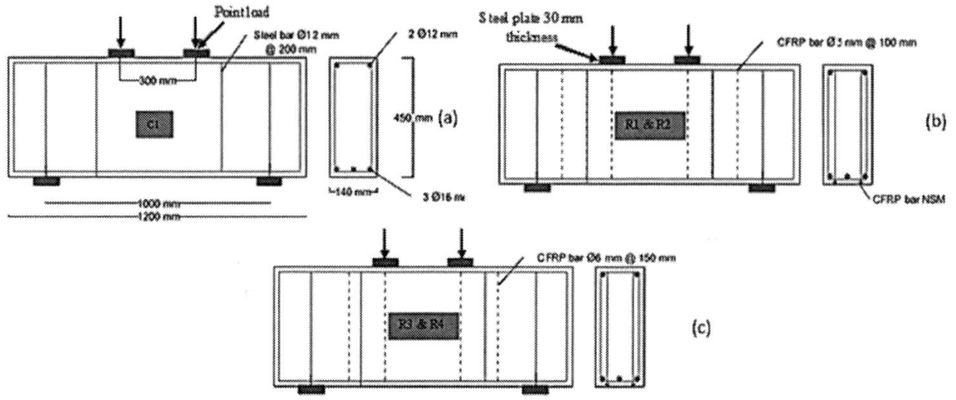

Fig. 1. (a) Control beam C1 (b) Beam specimens R1 and R2 with CFRP anchor bars at 100mm c/c (c) Beam specimens R3 and R4 with CFRP anchor bars at 150 mm c/c.

Table 1. Concrete compression strength at 28 days.

Cubes and Cylinders	Cylinder	Cube
Average Compressive Strength (MPa)	25.8	30.5

The strengthening NSM technique uses 5 mm diameter deformed CFRP bars and the CFRP bar properties are shown in Table 2. The stress–strain relationship for the CFRP bars was obtained by testing the CFRP bars under tension which shows a linear behaviour until a sudden rupture occurred at failure. From the stress-strain curve, a modulus of elasticity at 115 GPa and an ultimate strength of 2300 MPa was recorded. The CFRP strain at rupture was measured at 0.02%. A typical stress-strain curve of the CFRP bars is shown in Fig. 2.

Table 2. CFRP bar properties.

Bar diameter (mm)	Cross sectional area (mm^2)	Tensile Strength (MPa)	Tensile Modulus (GPa)	Ultimate Strain (%)
5	19.63	2300	115	0.02

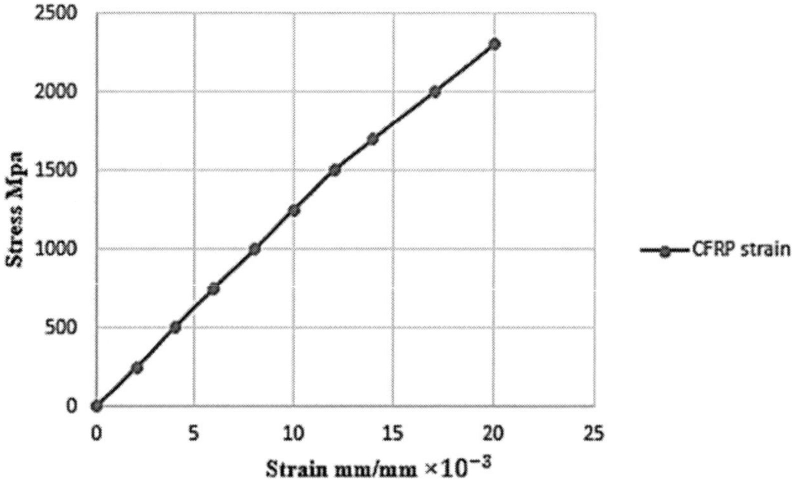

Fig. 2. Stress-strain diagram for CFRP bar.

2.2 Specimen preparation for NSM CFRP anchor bars and longitudinal bars

Preparation of the four strengthened RC deep beams involved: cutting of grooves at approximately 20 mm square cross-section on the flexural zone (soffit) of the RC deep beams, see Fig. 3(a). This was then followed by drilling circular section along the depth of the deep beam (15 mm diameter and up to 450 mm in depth), see Fig. 3(b). The grooves and drilling were cleaned with water jet and air blasted to remove the powdered concrete produced by the cutting process and all the possible loose material. The epoxy paste was prepared by mixing the resin and hardener at 3:1 proportion by volume with a power mixer. The drilled area were then filled with the epoxy paste followed by the insertion of the CFRP bars through the hole. Similar to the procedure used by [2], the epoxy resin was placed to half depth of the grooves. The CFRP bars were then placed in the groove and was lightly pressed. Upon pressing, this forces the paste to flow around the bar and fill completely between the bar and the sides of the groove. The groove was filled with more paste and the surface was leveled. The specimens remained in the laboratory environment for 28 days before testing commences.

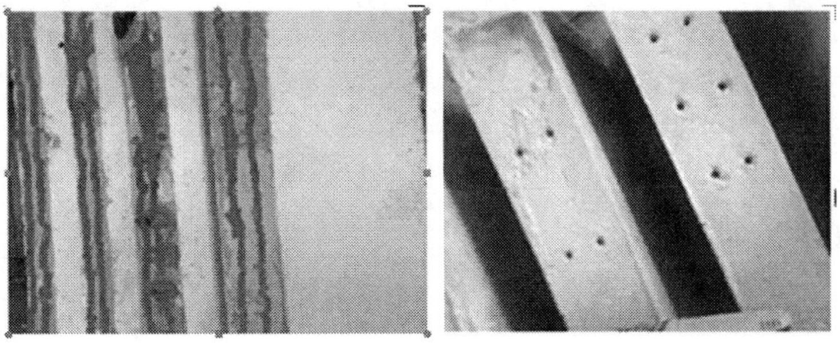

(a) (b)

Fig. 3. (a) Installation of CFRP bars at soffit of beam with epoxy resin (b) Procedure to instill CFRP anchor bars.

2.3 Test setup and test procedure

Beam specimens C1, R1, R2, R3 and R4 were initially tested at 7-days on a four-point bending test as shown in Fig. 1. Load was applied incrementally until first shear crack appeared. At this point, the beams were unloaded and the shear strengthening using NSM CFRP anchor bars within the shear span was conducted. Two longitudinal CFRP bars in flexure was also added to beams R1, R2, R3 and R4. The strengthened RC deep beams were then cured and re-loaded after 28 days until failure. While specimens were reloaded, the vertical deflections at mid-span of the strengthened RC deep beams were measured using digital dial gauges, crack patterns and mode of failure were also recorded.

3 Ultimate and mode of failure

The shear span to effective depth ratio for all RC deep beams was measured at 0.864. Hence, all beams failed consistently under the shear-compression mode of failure as shown in Fig. 4. Table 3 shows the experimental results for all beam specimens at failure. Control beam (C1) failed at an ultimate load of 510 kN and shear failure at 255 kN. For all strengthened RC deep beams, R1 achieved the highest ultimate load and shear load of 641 kN and 320 kN. Beam R2 recorded similar ultimate and shear load at 636 kN and 318 kN respectively. However, when the spacing of CFRP anchor bars increases from 100mm c/c to 150mm c/c, beam R3 and R4 shows lower ultimate and shear load. This shows that the spacing of the anchor bars has significant influence towards the capacity of the strengthened beams. The presence of the CFRP anchor bars has also contributed to the shear strength of the beams. This is clearly seen when the shear enhancement of 17.3% to 25.5% was recorded for all strengthened RC deep beams.

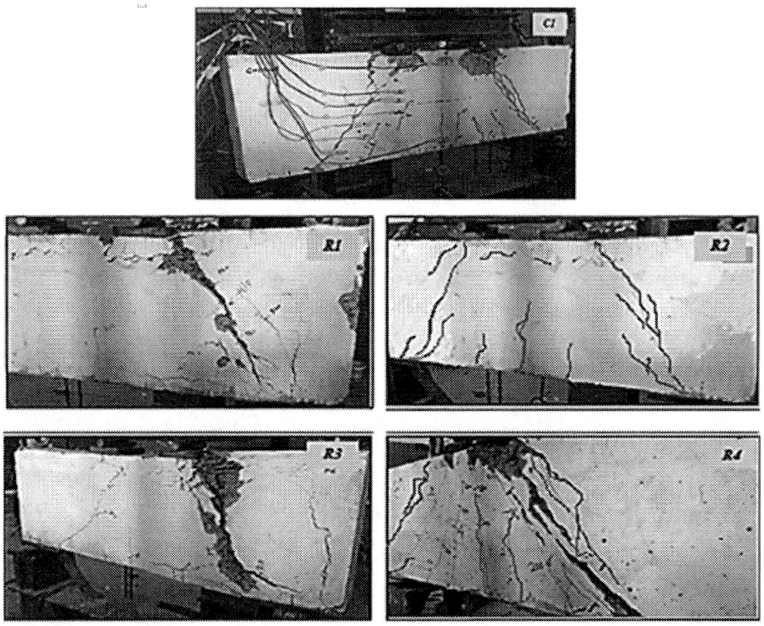

Fig. 4. Failure mode for Beam Specimens C1, R1, R2, R3 and R4.

Table 3. Experimental Results.

Beam	**CFRP Condition from Support**	**Experimental Results**				
		Ultimate Load (kN)	**Shear Force (kN)**	**CFRP Contribu-tion to Shear (kN)**	**Shear Enhancement (%)**	**Modes of Failure**
C1	-	510	255	-	-	Shear - compression
R1	100 mm	641	320	65	25.5%	Shear - compression
R2	100 mm	636	318	63	24.7%	Shear-compression
R3	150 mm	604	307	52	20.4%	Shear-compression
R4	150 mm	598	299	44	17.3%	Shear - compression

4 Mid-span deflection for beam specimens

Table 4 presents the mid-span deflection and its percentage differences at failure. The mid-span deflection for control beam C1 was recorded at 9.8 mm. However, for the strengthened RC deep beams a reduction in deflection was observed at 8.2 mm for beam

R1 and R2, and 9.19 mm and 8.93 mm for beams R3 and R4 respectively. This is a reduction at approximately 6.1% up to 16.3%. The reduction in deflection for the strengthened RC deep beams by NSM CFRP anchor bars and longitudinal bars clearly shows an increment in its inertia value causing the RC deep beams to increase its stiffness behaviour.

Table 4. Mid-span deflection and percentage differences for beam specimens.

Specimens	Shear Force (kN)	Maximum deflection (mm)	Difference in Deflection (mm)	Difference in Deflection (%)
C1	255	9.8	-	-
R1	320	8.2	1.6	16.3
R2	318	8.2	1.6	16.3
R3	307	9.2	0.6	6.1
R4	299	8.9	0.9	9.1

5 Conclusion

This paper presents the research findings of near surface mounted (NSM) CFRP anchor bars in shear and additional CFRP longitudinal bars in flexure for the strengthening of RC deep beams with shear discrepancies. Five RC deep beams (one control beam and four strengthened beams) were tested under a four-point bending test until failure. From the results obtained, the following observation were concluded:

i) All strengthened RC deep beams shows higher ultimate load capacity ranging from 598 kN to 641 kN compared to control beam at 551 kN.
ii) All strengthened RC deep beams shows an enhancement in its shear capacity with an increment of 17.3% to 25.5%.
iii) A lower mid-span deflection measurements were recorded at failure indicating a stiffer response of the strengthened RC deep beams. A deflection reduction by up to 6.1% to 16.3% were achieved compared to the control beam.
iv) The strengthened RC deep beams with NSM CFRP anchor bars with additional longitudinal bars has shown good structural performance. Thus, the NSM technique has good potential for the rehabilitation work for damage RC structures in shear.

The authors would like to acknowledge the support received from Jamilus Research Centre, Faculty of Civil and Environmental Engineering and Centre for Graduate Studies of Universiti Tun Hussein Onn Malaysia.

References

[1] L.D. Lorenzis, A. Nanni and A.L. Tegola, Strengthening of reinforced concrete structures with near surface mounted FRP rods, *Proc. of the Int. Advancing with Composites*, 9-11, (2000)

[2] L.D. Lorenzis and A. Nanni, Shear strengthening of reinforced concrete beams with nea-surface mounted fiber-reinforced polymer rods, Int. J. ACI Structural Journal, **98**(1), 60 – 68 (2001)

[3] R. Parretti and A. Nanni, Strengthening of RC members using near-surface mounted FRP composites, Int. J. Advances in Structural Engineering, **7**(6), 469-483 (2004)

[4] J.A. Barros and S.J. Dias, Near surface mounted CFRP laminates for shear strengthening of concrete beams, Int. J. Cement and Concrete Composites, **28**(3), 276-292, (2006)
[5] J.G. Teng, L.D. Lorenzis, B. Wang, R. Li, T. N. Wong and L. Lam, Debonding failures of RC beams strengthened with near surface mounted CFRP strips, Int. J. of Composites for Construction, **10**(2), 92-105, (2006)
[6] L.D. Lorenzis and J.G. Teng, Near-surface mounted FRP reinforcement: An emerging technique for strengthening structures, Int. J. of Composites Part B: Engineering, **38**(2), 119-143, (2007)
[7] A.K.M.A. Islam, Contribution of NSM CFRP bars in shear strengthening of concrete members, Engineering Structures, **31**(3), 709-714, (2009)
[8] ACI Committee 318-05, *Building Code Requirements for Structural Concrete and Commentary*, American Concrete Institute, Michigan, USA, (2005)
[9] ASTM C39-01, *Standard Test Method for Compressive Strength of Cylindrical Concrete Specimens*, American Society for Testing and Materials, Pennsylvania, USA, (2001)

Innovative Concrete Repairing Technique Using Post Tensioning Steel Straps

Chau-Khun Ma[1,*], *Sofrie* Chin Siew Yung[1], *Nazirah* Apandi[1], *Abdullah* Zawawi Awang[1], and *Wahid* Omar[2]

[1] Department of Structure and Materials, Faculty of Civil Engineering, Universiti Teknologi Malaysia, 81310, Johor Bahru, Malaysia
[2] Office of Chancellery, Universiti Teknologi Malaysia, 81310, Johor Bahru, Malaysia

Abstract. In this paper, innovative technique using low-cost recycled steel straps confinement to repair load-induced damaged high-strength concrete (HSC) columns were studied. This paper explains the effects of repairing technique using post tensioning steel straps. A series of experimental test was carried out to investigate the stress-strain relationships of such concrete. A total of 6 HSC columns were compressed 50% of their ultimate strength, then repaired by using steel straps. The proposed repairing technique significantly improved the performance of damaged concrete columns, in both strength and ductility. It was evidenced from this study that the steel strapping confining technique is effective in repairing of damaged HSC columns but ensured reasonable operating costs.

1 Introduction

Recently, the suitability of using steel straps as confining material has been experimentally investigated [1-5]. The steel straps used in these studies are recycled steel straps which are commonly used as packing strands for exporting good crates. The benefit of using steel straps as confining material, aside from its cost effectiveness, is it allows post-tensioning confining pressure around concrete which ensure fully effective use of the confining material [6-8]. As being reported previously, the passive type of confining materials which are activated by the dilation of the concrete, not reaching their ultimate strain at the ultimate failure. This problem is more severe for high strength concrete (HSC) which generally possess lower Poisson's ratio [9-12]. During the confining works, the steel straps were pulled using tensioner around the concrete up to certain level of strain to ensure even very small dilation of concrete can initiate the confining effects. The confining method using post-tensioned steel straps were evidenced to be effective in increasing the ductility and strength of HSC which is brittle and non-ductile in nature [13].

The current study extends the state-of-art of confining works using steel straps to repair and rehabilitate damaged HSC. It is believed that the proposed model can provide a robust platform for the engineers, academicians, consultants and building pathologists in the

* Corresponding author: machaukhun@gmail.com

DOI: 10.1051/matecconf/201710302011

design of such repair works. The use of steel straps is still limited in the industry although the use of steel straps is effective in repairing structure.

2 Load induced specimen

A total of 6 HSC cylinders with diameter of 100 mm and height of 200 mm were prepared in this study. The targeted concrete compressive strength is 50 MPa with mix proportions as shown in Table 1. Corresponding concrete cubes were prepared alongside with the specimens for the determination of concrete compressive strength. Based on the analysis of the cube tests, the average concrete compressive strength is 50.27 MPa with standard deviation of 4.6 MPa at the age of 28 days after curing. The specimens were then pre-loaded to damage the cylinders to simulate the situations where concrete were damaged at 50% of axial load capacity. The damaged specimens were repaired using two different repairing schemes to investigate the differences in their effectiveness. The first scheme is repair using surface grouting only with mortar that has prescribed mix of 1:1:0.5 (cement: sand: water). No attempts to fill the internal crackings to avoid changes in the specimen properties. Second scheme is repair using surface grouting and then confine using post-tensioned steel straps. The specimens were confined with different spacing and confinement layers of steel straps to achieve different confining volumetric ratio, ρ_v. The different in spacing and layers of confining steel straps will be addressed in confining volumetric ratio hereafter for simplicity in the design which is more favourable in practical. The details of the specimens are tabulated in Table 2.

Table 1. Mix portions of 1m^3 concrete

Material	Type	Quantity
OPC (kg)	Portland	409
Sand (kg)	River sand	806
Aggregate (kg)	10 mm	1025
Superplasticizer (*l*)	Glenium C380	16
Water (kg)	Tap water	162

3 Steel Wrapping Works

In this study, the spacing between the steel straps were fixed into 10 mm and 30 mm. The specimens were confined with 1, 2 and 3 layers steel straps. Steel straps, tensioner used and confined specimens are shown in Fig. 1. All confined specimens were confined from end to end to prevent concrete crushing near the load platens.

4 Load procedure

The compressive strength of the concrete was determined by uniaxial compression test using 'Tinius Olsen' compression machine that has a maximum load capacity of 2,500 kN. All tests were conducted in the Laboratory of structure and materials in Universiti Teknologi Malaysia (UTM). The loading is displacement controlled with constant rate of 0.01 mm/s for all specimens.During the test, the axial load and vertical strains were recorded by an automatic data acquisition through data logger to monitor the stress-strain behaviour of the specimens. The variations of axial strains of the specimen were measured by two 50 mm omega strain gauges attached at both sides of the specimen. The variations in readings between the two omega strain gauges were monitored closely and the testing was stopped immediately in case the variations exceeding 10%, indicating the specimen was not aligned in the center of the platen. The specimen was then re-positioned and centered again until the variation is less than 10%. The specimen was monotonically loaded until the specimen failed. Fig. 2 shows the schematic test setup and instrumentations.

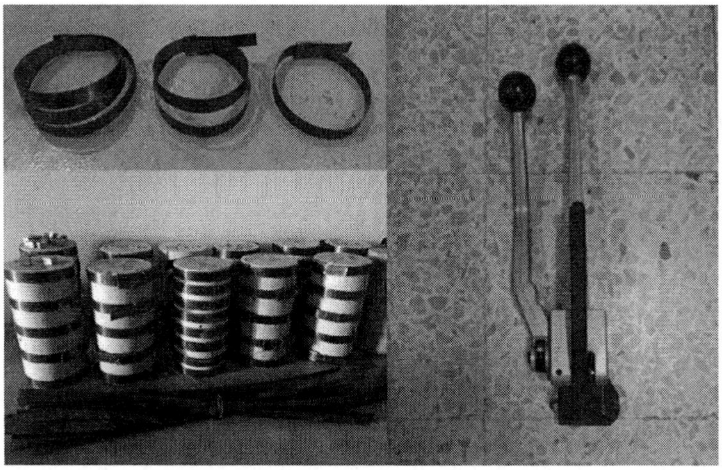

Fig. 1. Steel strapping confinement (a) Steel straps; (b) Confined specimens; (c) Tensioner

Table 2. Details of tested specimens

Specimen ID	Steel Straps layer	Spacing (mm)	Pre-damaged level (%)	Ultimate load (kN)	Ultimate displacement (mm)	Ultimate stress, f_{cu} (N/mm^2)	Degree of damage, δ (%)	Confining volumetric ratio, ρ_v	Confinement pressure, f_l	Repairing/ Strengthening Scheme
C50-0-0 (+50)	-	-	+50	162.6	0.46	50.2	0.05^a	0	0^b	Surface grouting
C50-30-1 (+50)	1	30	+50	421.1	0.69	53.6	0.05^a	0.27	0.05^b	Surface grouting + steel strapping
C50-10-1 (+50)	1	10	+50	420.6	0.85	53.5	0.05^a	0.48	0.09^b	Surface grouting + steel strapping
C50-30-2 (+50)	2	30	+50	458.7	1.50	63.9	0.05^a	0.53	0.10^b	Surface grouting + steel strapping
C50-30-3 (+50)	3	30	+50	521.5	1.63	66.4	0.05^a	0.80	0.15^b	Surface grouting + steel strapping
C50-10-2 (+50)	2	10	+50	563.9	2.64	91.8	0.05^a	0.96	0.18^b	Surface grouting + steel strapping

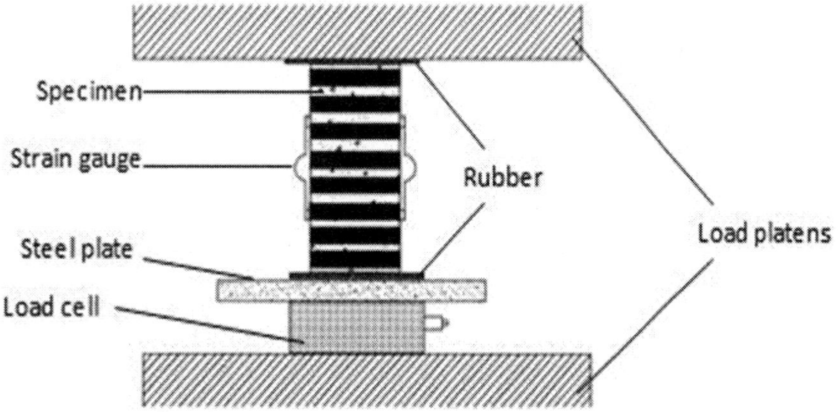

Fig. 2. Compression test setup

5 Test results

5.1 Failure modes

It was observed that all of the specimens failed after the steel straps at approximately mid-height of the specimen snapped. All specimens failed in the same manner with disregard of pre-damaged levels. No apparent surface cracks were observed during the initial stage of testing (< 25% of ultimate load). Minor surface cracks were started to propagate at the mid-span of the columns and continued to implode in both upper and lower directions when the load increased to the second stage (25% - 50% of ultimate load). Thudding sound can be heard at this stage.

The steel straps were observed to prevent the generation of cracks during the third stage (50% -75% of ultimate load). A significant thud sound can be heard when ultimate load is approaching but the columns did not fail due to the confining pressure exerted by the steel straps. At ultimate load, minor concrete crushing and disintegration can be observed at approximately mid-span the repaired columns. The columns still remained intact after the failure. Steel strapping has prevented the columns from crushing and collapsing.

Due to the confinement, the columns did not present abrupt failure as control specimen. Obviously, the specimen confined with higher confining volumetric ratio shows remarkably increase in the ductility. It can also be observed that the concrete cores of specimens with higher pre-damaged level are less intact after failure. This can be due to the pre-damaged has caused existing crack and part of the concrete had already partially disintegrated during the pre-damaging process. Fig.3 shows the failure modes of repaired columns.

Fig. 3. Failure modes: (a) C50-30-1(+50); (b) C50-10-1(+50); (c) C50-30-2(+50)

5.2 Stress-strain relationships

The stress-strain relationships of repairing concrete is shown in Fig. 4. In this figure, the axial stress versus axial strain obtained from the average of the reading from two pi-gauges attached at both sides of the specimens. The figure clearly shows the effectiveness of steel strapping in increasing the performance of damaged concrete. However, it can also be concluded that the existing models which did not considered the pre-damaged levels of concrete are not suitable to be used to predict the stress-strain behaviour of repaired concrete. The accuracy of the existing models in predicting the stress-strain relationships of repaired concrete is further discussed in the following section. The nomenclature of the graph is, C50-30-1(+50) means C50- two 50 mm omega strain gauges attached at both sides of the specimen, 30 mm spacing, 1 layer of steel straps, 50% of pre-damaged level.

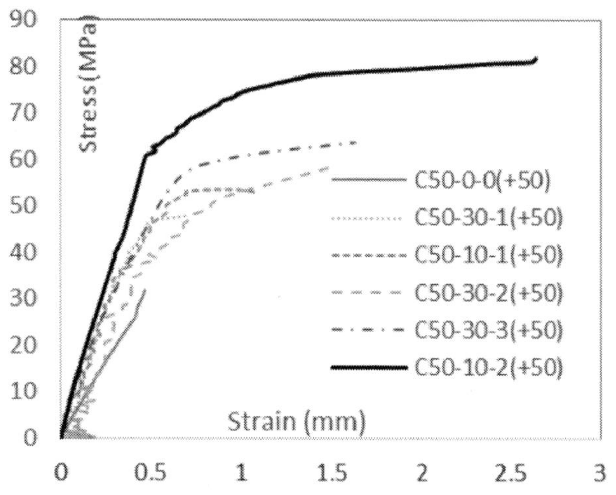

Fig. 4. Load-displacement curves of tested specimens

6 Summary and conclusion

A series of experimental studies were performed to investigate the repaired HSC using post-tensioning steel straps confinement. Along the investigation of the behaviour of repaired HSC, several important conclusions which can be drawn are listed follows:

i) The typical failure mode of repaired HSC indicates that the confined HSC failed at the rupture of steel straps resulting immediate loss of confining pressure. Such failure mode is not associated with the variation of confining volumetric ratio or pre-damaged level.
ii) Damaged degree is an important factor and needed to be considered in estimating the ultimate strength and strain of repaired concrete. In general, it was found that the ultimate strength reduced with the increase of concrete damaged degree. Whilst the ultimate strain increased with damaged degree.
iii) Post-tensioning steel straps confinement increase the rigidity concrete damaged at pre-damaged levels 50%.

This work was funded by Fundamental Research Grant Scheme (FRGS) from Ministry of Higher Education, Malaysia [R.J130000.7822.4F826]. The first author would like to acknowledge that the undergraduate students Amirul Radzi, Melvay Allens and Muhammad Fadhli undertook the tests as their final year projects supervised by the first author.

References

[1] C.K. Ma, A.Z. Awang, and W. Omar, New theoretical model for SSTT- confined HSC columns, Magazine of Concrete Research, **66**(13), 674-684, (2014)
[2] C.K. Ma, A.Z. Awang, R. Garcia, W. Omar, K. Pilakoutas, and M. Azimi, Nominal curvature design of circular HSC columns confined with post-tensioned steel straps, Structures, **7**, 25-32,(2016)
[3] C.K. Ma, A.Z. Awang, W. Omar, M. Liang, S-W. Jaw, and M. Azimi, Flexural capacity enhancement of rectangular high-strength concrete columns confined with post-tensioned steel straps: experimental investigation and analytical modelling, Structural concrete, **17**(4), 668-676, (2016).

[4] C.K. Ma, A.Z. Awang, R. Garcia, W. Omar, and K. Pilakoutas, Behaviour of over-reinforced high-strength concrete beams confined with post-tensioned steel straps–an experimental investigation, Structural Concrete, **17**(5), 768-777, (2016)

[5] Y. Helal, R. Garcia, K. Pilakoutas, M. Guadagnini, and I. Hajirasouliha, Strengthening of short splices in RC beams using post-tensioned metal straps, Materials and Structures, **49**(1-2), 133-147, (2016)

[6] R. Garcia, I. Hajirasouliha, M. Guadagnini, Y. Helal, Y. Jemaa, K. Pilakoutas, P. Mongabure, C. Chrysostomou, N. Kyriakides, A. Iki, and M. Budescu, Full-scale shaking table tests on a substandard RC building repaired and strengthened with post-tensioned metal straps, J. of Earthquake Engineering, **18**(2), 187-213, (2014)

[7] H.P. Lee, A.Z. Awang, and W. Omar, Steel strap confined high strength concrete under uniaxial cyclic compression, Construction and Building Materials, **72**, 48-55, (2014)

[8] C.K. Ma, A.Z. Awang and W. Omar, Slenderness effect and upper-bound slenderness limit of SSTT-confined HSC column, Int. J. of Structural Engineering, **5**(3), 279-286, (2014)

[9] C.K. Ma, A.Z. Awang, and W. Omar, Flexural ductility design of confined high-strength concrete columns: Theoretical modelling, Measurement, **78**, 42-48, (2016)

[10] S.M.S. Mohsin, M.F. Manaf, N.N. Sarbini, and K. Muthusamy, Behaviour of reinforced concrete beams with kenaf and steel hybrid fibre, ARPN J. of Engineering and Applied Sciences, **11**(8), 5385-5390, (2016)

[11] N.N. Sarbini, I.S. Ibrahim, A.A. Saim, M.A.A. Kadir, and S.S.M. Mohsin, Development of shear stress equation contributed by steel fibre in reinforced concrete, ARPN J. of Engineering and Applied Sciences, **11**(8), 5445-5451, (2016)

[12] I.S. Ibrahim, K.S. Elliott, R. Abdullah, A.B.H. Kueh, and N.N. Sarbini, Experimental study on the shear behaviour of precast concrete hollow core slabs with concrete topping, Engineering Structures, **125**, 80-90, 2016.

[13] N.N. Sarbini, I.S. Ibrahim, A.A. Saim, A.B. Abdul-Rahman, N.F. Harun, and N.N. Hasbullah, Shear capacity of composite slab reinforced with steel fibre to that of fabric reinforcement in concrete topping, Materials Research Innovations, **18**(6), 236-240, (2014)

Analytical Prediction on Flexural Response of RC Beams Strengthened with Steel Plates

Rendy Thamrin[1,*]

[1]Civil Engineering Department, Engineering Faculty, Andalas University, Indonesia

Abstract. An analytical prediction on flexural response of reinforced concrete (RC) beams strengthened with steel plates is presented in this paper. The method used in this study is based on theoretical moment-curvature determination which able to determine the complete response of reinforced concrete cross section subjected to monotonic flexural load. A computer program named Reinforced Concrete Cross Section Analysis (RCCSA) was developed and utilized to facilitate the computation process. Two types of flexural strengthening (tension-face and web bonded steel plates) were analyzed using the software and then compared with the experimental result. Result validation using existing available software was also carried out using Respone-2000 software. The comparison shows that the analytical model can be used to predict the flexural response of strengthened reinforced concrete cross section with a good accuracy.

1 Introduction

Structural damage often results after earthquakes occur. When heavy damage is sustained, the building must be demolished. However in many cases repair and strengthening can be a safe and more economical option if carried out appropriately. As regulations and insurance requirements become stricter, there is also a need to strengthen existing structures considered vulnerable to be to future earthquake damage to established standards. To this end a method to adequately calculate the flexural response of the intended strengthening reinforced concrete member is essential.

The development of strengthening methods for reinforced concrete (RC) structures is evolving rapidly as new construction materials are developed. The most recent strengthening innovation in civil construction is the use of fiber reinforced polymer (FRP)[1]. Many types of FRP are now available for strengthening structures including FRP plates and sheets. However, FRP is still expensive compared to traditional steel plates. Steel plates epoxy bonded to the surface of reinforced concrete members have been used for three decades to strengthen and repair reinforced concrete structures [2, 3]. This method appears to be an adequate and economical method for increasing the flexural and shear capacity of existing reinforced concrete members and is still in common use.

Software for structural analysis has also developed rapidly along with improvements in computer hardware. Until now, the finite element method has been the best-known method

* Corresponding author: rendythamrin@gmail.com

available for structural analysis. However, modelling with finite element method is complex and cost consuming for a simple problem such as cross section analysis of reinforced concrete members.

The aim of this study is to develop software that can predict the full flexural response of strengthened reinforced concrete cross section rapidly and accurately. To develop the software, a method based on theoretical moment-curvature determination [4] was applied. Experimental study was also carried out and nine simple supported reinforced concrete beams consisted of three control specimens and six beams with flexural strengthening (bottom and web bonded plated) were tested. The test results are then used to validate the predictions from the software. In addition, existing available software named Respone-2000 (RESP.-2K) [5] was used to verify the predictions from the developed software.

2 Flexural analysis of cross section

Theoretically, cross section analysis of reinforced concrete members is performed based on the relevant material constitutive equations and the assumptions of strain compatibility and equilibrium condition. It is also assumed that the tensile forces are carried by the tensile reinforcement while the compressive forces are carried by the concrete and the compression reinforcement. Two design methods are available in designing reinforced concrete i.e. elastic design and ultimate-strength design. However, because the reinforced concrete structures fail beyond the elastic limit, the ultimate-strength design method is more appropriate to use in this study. Fig. 1 illustrates the ultimate-strength design method using a rectangular reinforced concrete cross section with strain distribution, rectangular stress block, and resultant of internal forces. From this figure, the nominal moment capacity of the cross section is equal to:

$$M_n = Ts.z = Cc.z \tag{1}$$

In this study an analytical computation procedure based on theoretical moment-curvature determination adopted from literature [4] is used. Firstly, the reinforced concrete cross section is divided into a finite number of concrete and reinforcement layers as shown in Fig. 2.

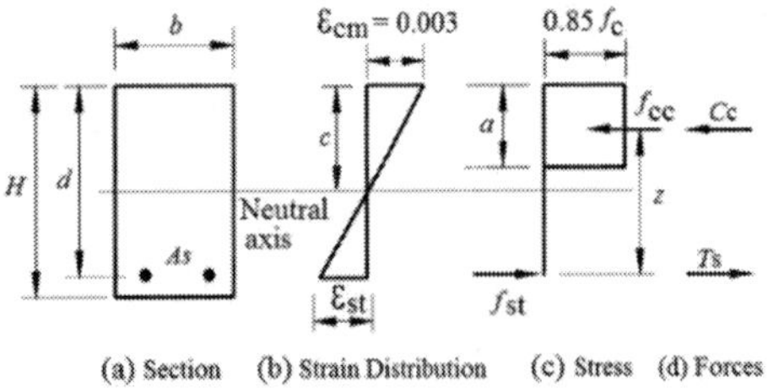

Fig. 1. Stresses and internal forces in a reinforced concrete cross section.

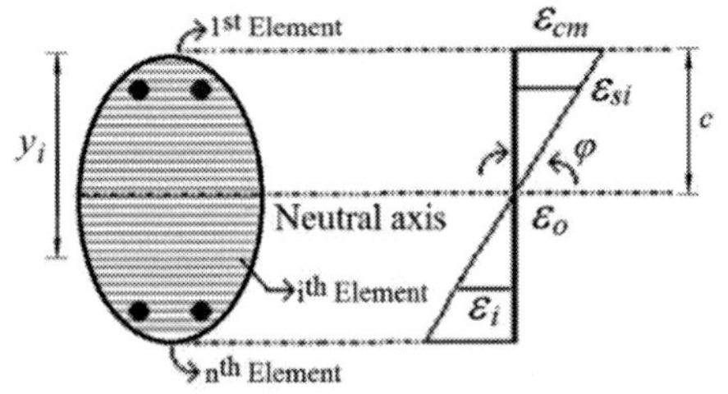

Fig. 2. An arbitrary cross section with assumed strain distribution.

The strain in element i^{th} for each incremental step can be calculated using a given value of curvature, φ, as:

$$\varepsilon_i = \varepsilon_o - (\varphi\, y_i) \tag{2}$$

The corresponding stresses, σ_i, and internal forces, F_i, acting in each steel and concrete element with an area, A_i, are then calculated using the assumed stress-strain relationship of materials and the following equations:

$$\sigma_i = f(\varepsilon_i) \tag{3}$$

$$F_i = A_i\,\sigma_i \tag{4}$$

The equilibrium of internal forces in the cross section is then obtained through an iterative procedure by estimating the axial strain, ε_o, for a given curvature which satisfies the equilibrium conditions written in Equation (5) for each increment of curvature.

$$\sum F_i \approx 0 \tag{5}$$

where F_i is the value of the force of each corresponding element. Finally, internal moment in the cross section can be calculated as:

$$M = \sum F_i y_i \tag{6}$$

Fig. 3 shows the detail of calculation process for a given value of curvature. The input data consists of cross section dimension, reinforcement data, material properties, and given curvature value. By using a given value of curvature and the stress-strain relationship (of concrete, steel reinforcement, and steel plate), the stresses and internal forces acting on each concrete element and reinforcement layer, and the steel plate can be calculated using Equation (3) and Equation (4). The iteration procedure is needed to obtain the neutral axis position which satisfied the equilibrium of the internal forces. This iteration step is started after the calculation of the internal forces for each concrete element and reinforcement layer.

The output for each incremental step is the moment versus curvature for the given value of curvature. The calculation process is stopped when the concrete at the top of compression zone crushed or passed the given limit value of compression strain. In this study, the value of maximum compression strain was assumed to be 0.005. This method was implemented in RCCSA program (FORTRAN language) and a user-friendly interface (VB.Net language) was also developed to facilitate the input data and to visualize the results [6]. Fig. 4 shows the data input page of RCCSA software.

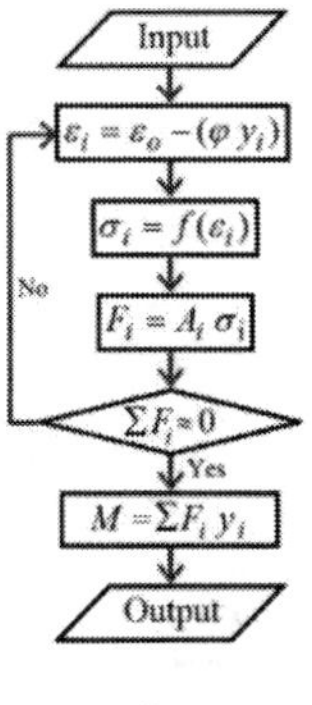

Fig. 3. Algorithm for theoretical moment-curvature determination

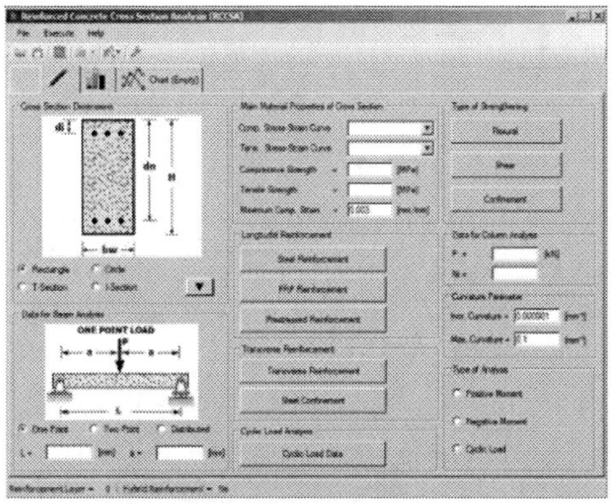

Fig. 4. The interface of RCCSA V4.3.

An analytical load-deflection curve is obtained from moment-curvature distribution corresponding to each incremental step along the beam length and calculated using moment-area theorem. The deflection equation is:

$$\delta = \int_0^{L/2} x\varphi dx \tag{7}$$

where: δ = deflection corresponding to the first moment of the area of the moment-curvature diagram for a half span of a simply supported beam of span L, evaluated with respect to beam support, and φ = the curvature corresponding to each incremental step and obtained from moment-curvature relation calculated as explained previously.

Fig. 5 illustrates the assumed analytical model of strengthened reinforced concrete cross section (bottom and web bonded) with strain distribution, stress distribution and component internal forces. In this assumption, the steel plate acting as reinforcement and connection between the steel plate and the concrete are perfectly bonded. Hence, the strain distribution can be assumed to be linear along the height of the cross section.

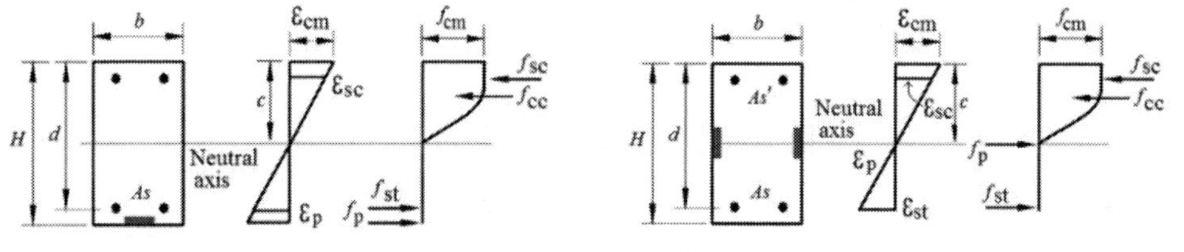

(a) Strain distribution and internal forces in cross section with bottom plate.

(b) Strain distribution and internal forces in cross section with web bonded plate.

Fig. 5. Assumed analytical model of strengthened reinforced concrete cross section.

The stress-strain relationship of concrete in compression used in this study is adopted from literature [7] and shown in Fig. 6(a). This model is a simple empirical equation and leads to efficient programing since only one parameter is needed to obtain the complete curve. The stress-strain relationship of concrete in tension used is linear model up to the tensile strength without a tension stiffening effect as shown in Fig. 6(a). The stress-strain relationship for steel bars employed in this study is adopted from the model proposed by Esmaeily and Xiao [8] and shown in Fig. 6(b). The stress-strain relationship for steel plates used in this study is similar to that used for steel bars as shown in Fig. 6(b).

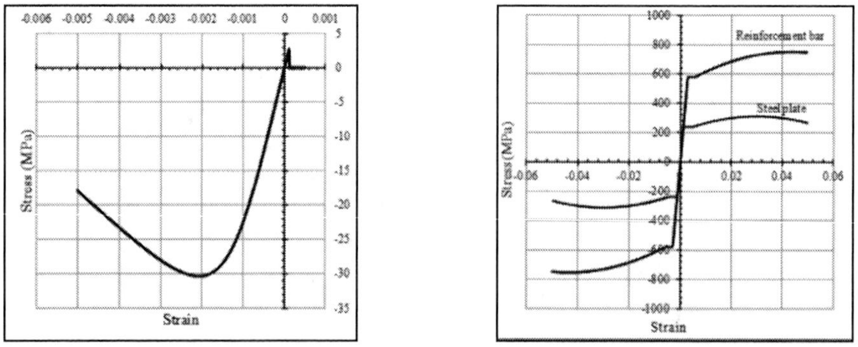

(a) Stress-strain relationship for concrete in compression and tension.

(b) Stress-strain relationship for steel reinforcement and steel plate.

Fig. 6. Stress-strain relationship used in analysis.

3 Experimental study

Nine simply supported reinforced concrete beams with stirrups consisting of three control beams, three beams with a bottom steel plate, and three beams with web bonded steel plates were tested. Fig. 7 shows the configuration of the reinforced concrete beams strengthened using (a) bottom and (b) web bonded steel plates used in this experimental study. Loading position and dimension of the beam are shown schematically in Fig. 7. The total length of all beams was 2300 mm; the clear span length was 2000 mm, the shear span length (*Ls*) was 800 mm and the end anchorage length beyond the support (*La*) was 150 mm. The cross section had dimensions of 125 mm width and 250 mm height.

Fresh concrete was ordered from the ready mix concrete company with the maximum aggregate size of 10 mm and target compressive strength of 30 MPa. Ten concrete cylinders with 150 mm diameter and 300 mm height were casted and tested to obtain the concrete compressive strength. The average concrete cylinder strength obtained from compression tests was 23 MPa at age 28 days.

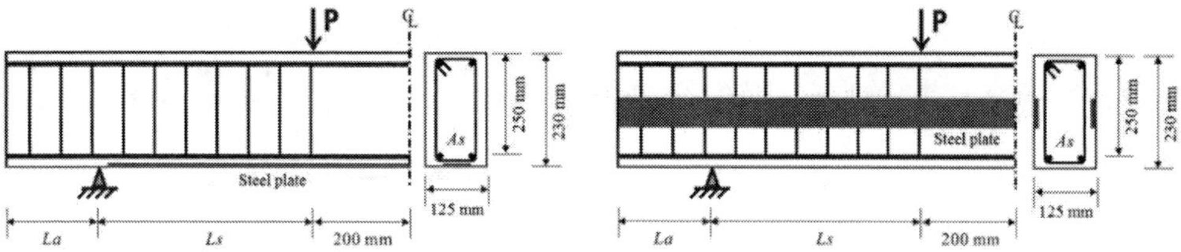

(a) Bottom plate. (b) Web bonded plated.

Fig. 7. Beam dimensions and position of steel plate.

Three ratios of longitudinal reinforcement (0.9%, 1.4% and 2.5%) were used for control specimens and both types of strengthening. Deformed steel bars with 13 mm diameter and 580 MPa yield strength were used as longitudinal reinforcement. The steel stirrups used were closed type with 10 mm diameter and 100 mm spacing with yield strength 391 MPa. Steel plates with 3 mm thickness and 250 MPa yield strength were bonded to the surface of concrete beams. Each of these steel components had a modulus of elasticity of about 200 GPa. The procedure used in applying the steel plates was to first make a groove on the beam surface. This method is identical to the near surface mounted strengthening technique. The steel plates were directly placed inside the grooves and slightly pressed along the whole length of the plate onto the adhesive. In this study, the low price commercially available epoxy adhesive was used to bond the steel plate to the concrete surface. Table 1 lists the details of steel reinforcement and plate used for strengthening.

Table 1. Beam reinforcement and strengthening data.

Specimen	Tensile Reinforcement				Compression Reinforcement				Plate Strengthening			
	dia. (mm)	*N*	*ρ (%)*	*fy* (MPa)	dia. (mm)	*N*	*ρ' (%)*	*fy* (MPa)	Position	Thickness (mm)	Width (mm)	*fy* (MPa)
BK-01		2	0.99						Control specimens without strengthening			
BK-02		3	1.5									
BK-03		5	2.5									
BP-01		2	0.99						Bottom			
BP-02	13	3	1.5	580	13	2	0.99	580	Bottom			
BP-03		5	2.5						Bottom	3	50	240
BT-01		2	0.99						Web			
BT-02		3	1.5						Web			
BT-03		5	2.5						Web			

During the test, the beams were monotonically loaded until failure with two point loads using a 500 kN capacity hydraulic jack. The load was applied by pumping the jack handle with a nearly constant and slow rate. The load was measured using a load cell positioned bellow the hydraulic cylinder and above the spreader beam as shown in Fig. 8(a). The deflections at midspan and at both of the loading points were measured using three linear variable displacement transducers (LVDTs) as shown in Fig. 8(a). The load cell and LVDTs were connected to a data logger to record the data. The cracks occurring during the test were examined and the values of crack loads were marked beside the crack line as shown in Fig. 8(b).

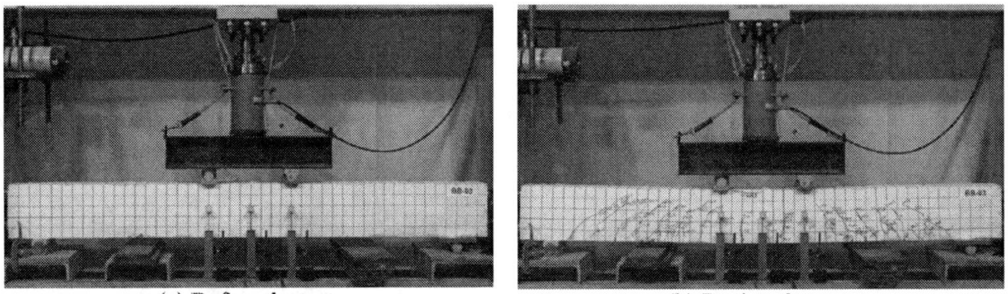

(a) Before the test.　　(b) During the test.

Fig. 8. Experimental set-up showing position of the equipment.

4 Results and discussion

The first crack for all the beams occurred in the midspan zone at an average load level of 8.5 kN. As the load increased the flexural crack appear and developed in the shear span zone. The flexural cracks located in the shear span zone then followed by the growth of diagonal shear cracks. The beams exhibited yielding of longitudinal steel as indicated by the flat part of load-deflection curves shown in Fig. 9 to Fig. 11. The yielding loads of longitudinal reinforcement occurred at different load levels depending on the reinforcement ratio. All the beams failed in flexural failure modes as indicated by crushing of concrete on the surface of the maximum compression zone.

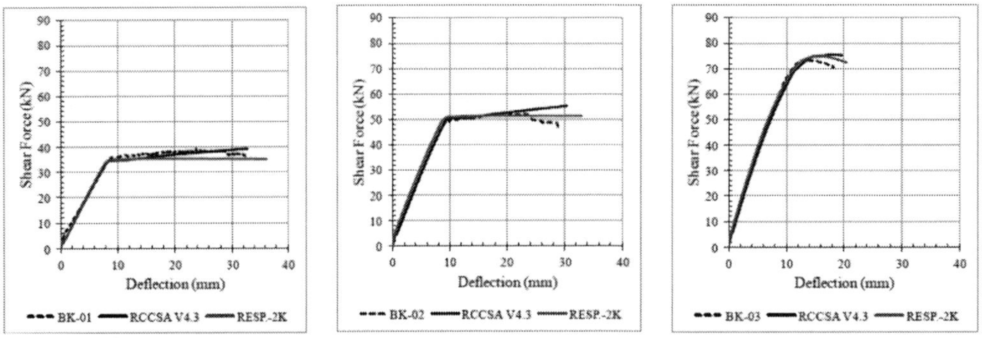

(a)　　(b)　　(c)

Fig. 9. Comparison between analytical and experimental results of beams without strengthening.

Experimental and analytical values of the first crack, yield, and failure loads are tabulated in Table 2. The ratios between experimental and analytical results were calculated to evaluate the accuracy of analytical model (Table 2). The comparison of the analytical (RCCSA V4.3 and RESP.-2K) and experimental midspan load-deflection curves are shown in Fig. 9 to Fig. 11. Fig. 9 shows that analytical calculation (RCCSA V4.3 and RESP.-2K) can accurately predict the general flexural behavior of the beams. However, in the case of the beam with the highest reinforcement ratio (2.5%) and with web bonded plates, the analytical value is slightly smaller than the experimental value as shown in Fig. 11(c).

The deformation behavior of the reinforced concrete beam strengthened with steel plates depends on the bond performance between plates and the concrete surface both for bottom and web bonded members [3]. In three members, labeled BP-01, BP-03 and BT-02 debonding failure occurred before the beam reach its ultimate capacity as shown by the sharp drop in the load-deflection curve in Fig. 10(a), Fig. 10(c), and Fig. 11(b). After debonding occurred the load-deflection behavior of the beam was equivalent to that of the control beams without strengthening.

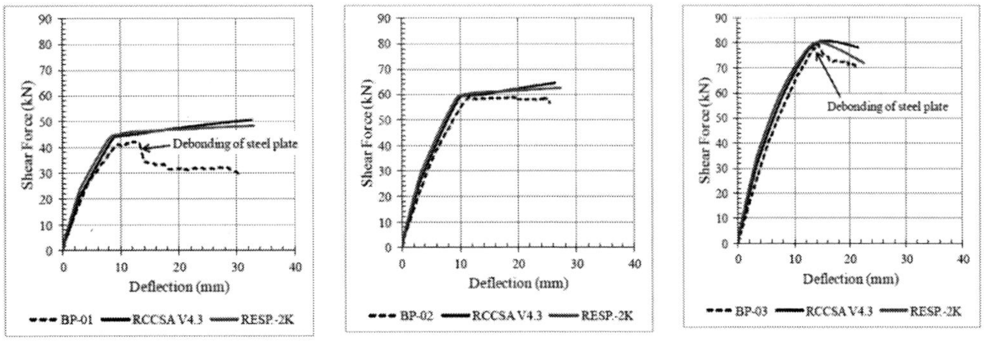

(a) (b) (c)

Fig. 10. Comparison between analytical and experimental results of beams strengthened with bottom plates.

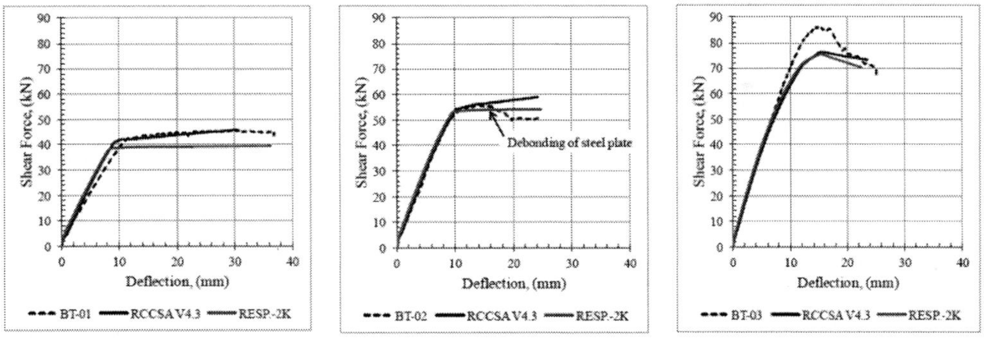

(a) (b) (c)

Fig. 11. Comparison between analytical and experimental results of beams strengthened with web bonded plates.

Both of analytical calculation can predict the general flexural behavior of the beams strengthened with side plated accurately as shown in Fig. 11 apart from one small discrepancy. At the peak load, the analytical result was approximately 10% smaller than the experimental value in the beam with the highest reinforcement ratio (2.5%). The tensile reinforcement of all the beams reached the yield stress of 580 MPa confirming the analytical predictions as shown in figures. The analytical model can accurately demonstrate the general behavior of concrete beams strengthened with steel plates.

As Fig. 9 to Fig. 11 confirm, the analytical model using theoretical moment-curvature determination predicts the flexural response of the beams through all ranges of elastic (before first crack and yielding of tensile reinforcement) and post-peak load. This model is faster to compute than conventional methods and is relatively simple to program.

Table 2. Experimental and analytical beam capacity values with comparison ratio.

Specimen	Experimental (1) First crack load (kN)	Experimental (2) Yield load (kN)	Experimental (3) Failure load (kN)	Analytical (4) First crack load (kN)	Analytical (5) Yield load (kN)	Analytical (6) Failure load (kN)	Ratio (1)/(4)	Ratio (2)/(5)	Ratio (3)/(6)
BK-01	8.9	34.6	38.6	3.9	34.2	39	2.3	1.0	1.0
BK-02	8.5	49.2	51.7	4.1	50	55	2.1	1.0	0.9
BK-03	6	69.8	70.5	4.4	62	75.02	1.4	1.1	0.9
BP-01	8.8	41.1	41.8	6.1	45	52	1.4	0.9	0.8
BP-02	9.3	58.6	58.6	6.3	61	66.9	1.5	1.0	0.9
BP-03	12.4	79.0	79.4	6.4	73	80	1.9	1.1	1.0
BT-01	6.9	42.3	45	5.2	43	47.7	1.3	1.0	0.9
BT-02	7.5	53	55.5	5.8	56	62.0	1.3	0.9	0.9
BT-03	6.5	82	85.9	6.5	68	75	1.0	1.2	1.1

5 Conclusions

The analytical procedure for flexural analysis of strengthened reinforced concrete cross section beams is presented. The procedure adopted is based on the theoretical moment-curvature determination and was implemented into a computer program named RCCSA. Analytical predictions were compared to results of tests on nine reinforced concrete beams. The following conclusions can be noted from this study:

1. The analytical model can be used to predict the flexural behavior of the strengthened beams through all ranges of elastic and post-peak load, is fast to compute and relatively simple to program.
2. All the beams failed in flexural failure modes indicated by crushing of concrete on the surface of maximum compression zone between the two point loads in agreement with the analytical model.
3. The analytical model cannot predict the debonding failure as the parameters regarding the debonding failure load is not yet included in the model.

References

[1] H. Saadatmanesh, and A.M. Malek, Design guidelines for flexural strengthening of RC beams with FRP plates, J. of Composites for Construction, ASCE, **2**(4), 158-164, (1998)

[2] R.N. Swamy, R. Jones, and J.W. Bloxham, Structural behaviour of reinforced concrete beams strengthened by epoxy-bonded steel plates, The Structural Engineer, **65A**(2),59-68, (1987)

[3] D.J. Oehlers, M.S. Mohamed Ali, and W. Luo, Upgrading continuous reinforced concrete beams by gluing steel plates to their tension faces, J. of Structural Engineering, ASCE, **124**(3), 224–32, (1998)

[4] R. Park, and T. Paulay, *Reinforced Concrete Structures*, John Wiley, New York, (1975)

[5] E.C. Bentz, *Response 2000*, retrieved on August 09, 2000 from http://www.ecf.utoronto.ca/~bentz/r2k.htm, (2000)

[6] R. Thamrin, *Reinforced concrete cross section analysis (rccsa) v4.3*, Jurusan Teknik Sipil, Fakultas Teknik, Universitas Andalas, (2014)

[7] T.H. Almusallam, and S.H. Alsayed, Stress-strain relationship of normal, high strength and light weight concrete. Magazine of Concrete Research, **47**(170), 39-44 (1995)

[8] A. Esmaeily, and Y. Xiao, Behavior of Reinforced concrete columns under variable axial loads: Analysis. ACI Structural J., **102**(5), 736-744, (2005)

Shear Strength Prediction for Concrete Beams Reinforced with GFRP Bars

Noor Azlina A. Hamid[1,*], *Rendy* Thamrin[2], *Azmi* Ibrahim[3], *Hanizah* Abdul Hamid[3], *Norhafizah* Salleh[1], *Zalipah* Jamellodin[1], *Masni* A. Majid[1], and *Nur Hafizah* A. Khalid[4]

[1]Faculty of Civil and Environmental Engineering, Universiti Tun Hussein Onn Malaysia, 86400 Parit Raja, Johor, Malaysia
[2]Faculty of Engineering, Universiti Andalas, Padang, Sumatera Barat 25163, Indonesia
[3]Faculty of Civil Engineering, Universiti Teknologi MARA, 40450 Shah Alam, Selangor, Malaysia
[4]Faculty of Civil Engineering, Universiti Teknologi Malaysia, 81310 Skudai, Johor, Malaysia

Abstract. This study presents a shear strength prediction model for concrete beams reinforced with GFRP bars. An empirical equation is developed using multiple regression analysis from the experimental results of 16 RC beams with GFRP bars. The proposed equation involved the parameters that affected the shear strength of beams such as compressive concrete strength, shear span ratio, longitudinal reinforcement ratio and modulus elasticity of the reinforcement. The accuracy of the proposed equation was verified by predicting the available experimental data from the literature. Furthermore, the predictions of shear capacities were compared with the current shear design code of ACI 440.1R-06. As a result, the ACI 440 provides very conservative prediction, while a better prediction is obtained from the shear strength prediction model in the present study.

1 Introduction

Glass fiber-reinforced polymer (GFRP) bars have emerged as an alternative reinforcement material to steel reinforcement due to certain advantages especially in terms of overcoming corrosion and repairing works [1]. In addition, GFRP bars possess higher tensile strength and are lighter than steel bars. However, the brittle behaviour and low modulus of elasticity of GFRP bars remain a matter of concern for concrete structure design.

Recently, the American Concrete Institute [2] has published design guidelines relevant to the use of GFRP bars as the main reinforcement in concrete structures. Accordingly, their design provisions for flexural and shear design followed the conventional reinforced concrete design guidelines. The shear design of reinforced concrete beam considered the contributions of concrete shear strength and shear resistance of stirrups, V_c+V_s, as the total shear resistance, V. However, with respect to different mechanical properties of GFRP, some modifications were made in the calculation of concrete shear strength, V_c of beams reinforced with GFRP bars. The concrete shear strength equation, V_c recommended by the ACI 440 design method [2] consisted of variables such as modulus elasticity of longitudinal

* Corresponding author: azlinah@uthm.edu.my

reinforcement (E_f), longitudinal reinforcement ratio (p_f), concrete compressive strength (f_{cu}), beam width (b_w) and neutral axis depth (c). It was indicated that no shear span to depth ratio (a/d) was included in the equation. Previous studies found that the ACI 440 design method was very conservative and more research is needed with respect to beams with a shear span ratio less than 2.5 [2-4]. Moreover, recent studies [5, 6] have also found that the concrete shear strength estimated by ACI 440 provided the most conservative predictions compared to other existing shear strength model. Consequently, a high amount of reinforcements will be required thereby increasing the cost of building structures.

In this paper, a new concrete shear strength prediction model for concrete beams reinforced with GFRP bars is developed based on the experimental results of this study. Furthermore, the accuracy of the proposed concrete shear strength equations was verified with the test results obtained from the literature. The performance of the proposed equations was also compared with current shear design guidelines of ACI 440 [2].

2 Test Program

2.1 Material properties

As shown in Fig. 1, the sand-coated GFRP bars with a diameter of 16 mm were used as longitudinal reinforcement bars in concrete beams. From the tensile tests on three GFRP samples, the mechanical properties of GFRP bars were obtained. Accordingly, the tensile strength and modulus elasticity of GFRP bars were 766 MPa and 56.6 GPa, respectively. All the tested beams were provided with steel stirrups with a diameter of 8 mm within the shear span. The yield strength and modulus elasticity of stirrups were 454 MPa and 162 MPa respectively. The beams were constructed using two grades of concrete with compressive strengths measuring 24 MPa and 36 MPa respectively.

Fig. 1. Glass fiber reinforced polymer (GFRP) bars.

2.2 Test specimens and instrumentation

This investigation evaluates the contribution of concrete to the shear resistance of concrete beams reinforced with GFRP bars. The test variables consisted of longitudinal reinforcement ratio, shear span ratio and stirrups spacing. As shown in Table 1, a total of sixteen concrete beams were constructed and tested up to failure. The tested beams were divided into two groups: beams with reinforcement ratios of 0.6% and 0.8% (Group 1) and

DOI: 10.1051/matecconf/201710302013

beams with reinforcement ratios of 1.2% and 1.5% (Group 2). All beams measured 200 mm in width with different beam depths of 362 mm (Group 1) and 343 mm (Group 2) as illustrated in Fig. 2. Each group included beams with shear span ratios of 1.5 and 3.0 with steel stirrups spacing of 50 mm and 150 mm. The beams were labelled systematically and can be explained as follows:

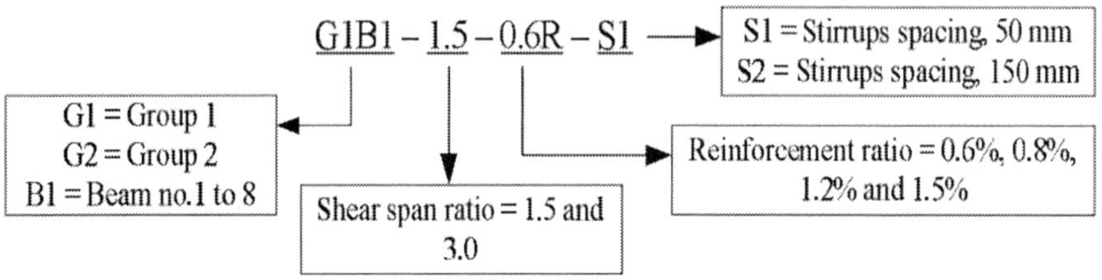

As shown in Fig. 2, all beams were tested over a simply supported with constant clear span of 400 mm. The beams were tested under four point loading. To investigate the stress-strain relationships, two electrical-resistance strain gauges were bonded on the top concrete surface at the middle of beam span. The strain gauges were also bonded on the longitudinal reinforcement bars and stirrups. While, the displacements were measured by using three linear variable displacement transducers (LVDTs) with a 50 mm stroke, that were placed at the mid-span and under the load positions. During testing, all the crack formations on each side of the beam surface were marked.

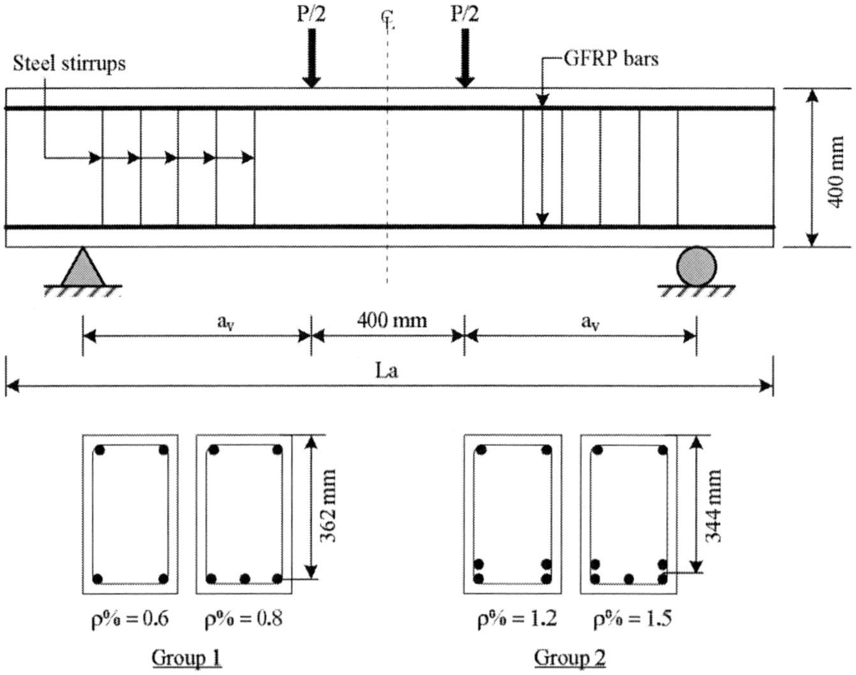

Fig. 2. Details of test specimens and test setup

Table 1(a). Beam details and test results.

Group	Beam	a_v (mm)	L (mm)	f_{cu} (N/mm^2)	Longitudinal reinforcement ratio (%)	Stirrups spacing (mm)
1	G1B1-1.5-0.6R-S1	550	2000	24	0.6	50
	G1B2-1.5-0.8R-S1	550	2000	24	0.8	50
	G1B3-1.5-0.6R-S2	550	2000	24	0.6	150
	G1B4-1.5-0.8R-S2	550	2000	24	0.8	150
	G1B5-3.0-0.6R-S1	1100	3000	24	0.6	50
	G1B6-3.0-0.8R-S1	1100	3000	24	0.8	50
	G1B7-3.0-0.6R-S2	1100	3000	24	0.6	150
	G1B8-3.0-0.8R-S2	1100	3000	24	0.8	150
2	G2B1-1.5-1.2R-S1	550	2000	36	1.2	50
	G2B2-1.5-1.5R-S1	550	2000	36	1.5	50
	G2B3-1.5-1.2R-S2	550	2000	36	1.2	150
	G2B4-1.5-1.5R-S2	550	2000	36	1.5	150
	G2B5-3.0-1.2R-S1	1100	3000	36	1.2	50
	G2B6-3.0-1.5R-S1	1100	3000	36	1.5	50
	G2B7-3.0-1.2R-S2	1100	3000	36	1.2	150
	G2B8-3.0-1.5R-S2	1100	3000	36	1.5	150

Table 1(b). Beam details and test results.

Group	Beam	Test results		
		First diagonal crack load, $V_{c.exp}$ (kN)	Shear load, P/2 (kN)	Failure mode
1	G1B1-1.5-0.6R-S1	41.3	233.2	Flexure
	G1B2-1.5-0.8R-S1	42.7	281.6	Flexure
	G1B3-1.5-0.6R-S2	37.1	139.0	Shear
	G1B4-1.5-0.8R-S2	55.8	181.3	Shear
	G1B5-3.0-0.6R-S1	27.4	99.0	Flexure
	G1B6-3.0-0.8R-S1	24.7	132.1	Flexure
	G1B7-3.0-0.6R-S2	14.4	92.8	Flexure
	G1B8-3.0-0.8R-S2	21.3	125.6	Flexure
2	G2B1-1.5-1.2R-S1	39.0	255.0	Shear
	G2B2-1.5-1.5R-S1	54.0	280.2	Flexure
	G2B3-1.5-1.2R-S2	43.5	179.3	Shear
	G2B4-1.5-1.5R-S2	64.5	219.3	Shear
	G2B5-3.0-1.2R-S1	32.5	151.3	Flexure
	G2B6-3.0-1.5R-S1	30.5	164.4	Flexure
	G2B7-3.0-1.2R-S2	36.0	138.5	Flexure
	G2B8-3.0-1.5R-S2	31.5	140.1	Shear

3 Results and discussion

The test results on first diagonal crack load, shear load and failure mode for each tested beam in this study are listed in Table 1. Typical load-deflection curves between beams with the same shear span ratio and stirrups spacing, but different longitudinal reinforcement ratios of GFRP bars are shown in Fig. 3. It is shown that the shear capacity increases as the longitudinal reinforcement ratios increases. It also revealed that an increase in reinforcement ratio influences the stiffness of the beams. In addition, all beams experienced bilinear load-deflection behaviour after the formation of the first flexural crack within moment region.

Two types of failure were observed namely shear failure and flexure failure. Flexure failure was indicated by excessive concrete crushing at the top concrete sections of the middle beam span. Flexure failure was sudden especially in beams that experienced bar rupture, such in beams G1B1-1.5-0.6R-S1, G1B5-3.0-0.6R-S1, G1B6-3.0-0.8R-S1 and G1B7-3.0-0.6R-S2. The shear failure was exhibited by severe diagonal cracks within the shear span. The failure was also sudden and generally occurred in beams that had stirrups spacing of 150 mm within the shear span ratio of 1.5 such in beams G1B3-1.5-0.6R-S2, G1B4-1.5-0.8R-S2, G2B3-1.5-1.2R-S2 and G2B4-1.5-1.5R-S2.

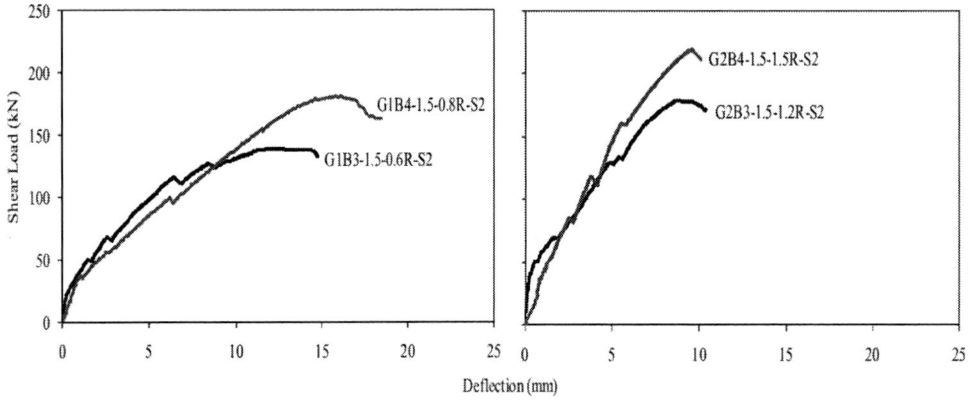

Fig. 3. Typical load-deflection curves of beams with the same shear span ratio of 1.5, stirrups spacing of 150 mm but different reinforcement ratios within 0.6% to 1.5%.

4 Proposed shear design equation

The derivation of the shear design equation, V_c was obtained from the test results of this study. This proposed equation includes the effects of test variables of shear span ratio, reinforcement ratio, modulus of elasticity and concrete compressive strength. The relationship between the test results and variables is plotted in Fig. 4(a) and (b). In the figures, the test results of shear capacity, $V_{c.exp}$ was normalized by the square root of concrete strength, $\sqrt{f_{cu}}$ and the cross section area of the beam, $b_w d$ [7-9]. It was discovered that the normalized shear strength, $V_{c.norm}$ increases as the shear span ratio of (d/a) increases as shown in Fig. 4(a). A linear relationship was also indicated in Fig. 4(b) between normalized shear strength with combination variables of $(p_f E_f f_{cu})^{1/3}$. A similar relationship was reported in the literature for beams reinforced with longitudinal GFRP and CFRP bars. However, the effect of concrete compressive strength was not considered.

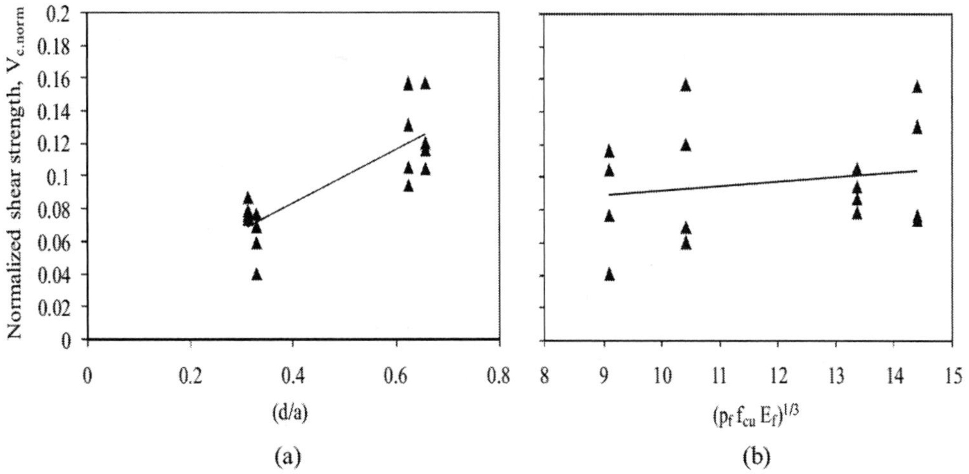

Fig. 4. Relationship between Normalised Shear Strength and Test Variables

Hence, the basic expression for deriving the concrete shear capacity of beams can be written as:

$$V_c = \beta \sqrt{f_{cu}}\, b_w\, d \quad (1)$$

In the above equation, β is the correction factor which incorporated the test variables of (d/a) and $(p_f E_f f_{cu})^{1/3}$ which influenced the shear strength of beams reinforced with GFRP bars. Based on the relationship, the following expression is proposed:

$$\beta = k_1 (p_f E_f f_{cu})^{\frac{1}{3}} + k_2 \left(\frac{d}{a}\right) \quad (2)$$

Using a regression analysis, the constant values of k_1 and k_2 are obtained. By substituting Equation (2) into (1), the proposed shear strength equation can be written as follow:

$$V_{c.pred} = \left[0.00203\, (p_f E_f f_{cu})^{\frac{1}{3}} + 0.153 \left(\frac{d}{a}\right)\right] \sqrt{f_{cu}}\, b_w\, d \quad (3)$$

5 Evaluation of the proposed shear design equation

In order to verify the proposed equation, Equation (3) was compared with the test results of this study and available test results in the literature [3, 6–11] as listed in Table 2. All 42 beams were reinforced with different reinforcement ratios of GFRP and CFRP bars with shear span ratios within 2.3 to 5.0.

In Table 2, the comparison of shear strength predictions according to the proposed equation and current design codes of ACI 440 [2] were compared. The mean value of average (V_{exp}/V_{norm}) for the proposed Equation (3) is 1.12 with a standard deviation of 0.21 and a coefficient of variation of 19%. These low values show that the predictions are close to the real values which provide the most accurate prediction. On the other hand, the predicted shear capacities corresponding to ACI 440 provide the most conservative results compared to the prediction results by Equation (3) with a mean value of 1.57.

Table 2. Verification of proposed shear strength equation.

Sources	Specimens	Types of bar	a/d	b	d	f_{cu}	Longitudinal Reinforcement		$V_{c.exp}$ (kN)	Normalised, $V_{c.norm}$	ACI 440.1R (2006)		Proposed Equation (3)	
							ρ_f (%)	E_f (Mpa)			$V_{c.pred}$ (kN)	$V_{c.norm}/V_{c.pred}$	$V_{c.pred}$ (kN)	$V_{c.norm}/V_{c.pred}$
Present Study	G1B1-1.5-0.6R-S1	GFRP	1.5	200	362	24	0.6	56522	41.3	0.12	0.06	1.92	0.12	0.98
	G1B2-1.5-0.8R-S1	GFRP	1.5	200	362	24	0.8	56522	42.7	0.12	0.07	1.65	0.12	0.99
	G1B3-1.5-0.6R-S2	GFRP	1.5	200	362	24	0.6	56522	37.1	0.10	0.06	1.73	0.12	0.88
	G1B4-1.5-0.8R-S2	GFRP	1.5	200	362	24	0.8	56522	55.8	0.16	0.07	2.16	0.12	1.29
	G1B5-3.0-0.6R-S1	GFRP	3.0	200	362	24	0.6	56522	27.4	0.08	0.06	1.28	0.07	1.12
	G1B6-3.0-0.8R-S1	GFRP	3.0	200	362	24	0.8	56522	24.7	0.07	0.07	0.95	0.07	0.97
	G1B7-3.0-0.6R-S2	GFRP	3.0	200	362	24	0.6	56522	14.4	0.04	0.06	0.67	0.07	0.59
	G1B8-3.0-0.8R-S2	GFRP	3.0	200	362	24	0.8	56522	21.3	0.06	0.07	0.82	0.07	0.84
	G2B1-1.6-1.2R-S1	GFRP	1.6	200	343	36	1.2	56522	39.0	0.09	0.08	1.22	0.12	0.77
	G2B2-1.6-1.5R-S1	GFRP	1.6	200	343	36	1.5	56522	54.0	0.13	0.09	1.53	0.12	1.05
	G2B3-1.6-1.2R-S2	GFRP	1.6	200	343	36	1.2	56522	43.5	0.11	0.08	1.36	0.12	0.86
	G2B4-1.6-1.5R-S2	GFRP	1.6	200	343	36	1.5	56522	64.5	0.16	0.09	1.83	0.12	1.26
	G2B5-3.2-1.2R-S1	GFRP	3.2	200	343	36	1.2	56522	32.5	0.08	0.08	1.02	0.07	1.05
	G2B6-3.2-1.5R-S1	GFRP	3.2	200	343	36	1.5	56522	30.5	0.07	0.09	0.87	0.08	0.96
	G2B7-3.2-1.2R-S2	GFRP	3.2	200	343	36	1.2	56522	36.0	0.09	0.08	1.13	0.07	1.17
	G2B8-3.2-1.5R-S2	GFRP	3.2	200	343	36	1.5	56522	31.5	0.08	0.09	0.89	0.08	0.99
Alam and Hussein (2012)	G-0.5-500	GFRP	2.5	250	455	37.4	0.4	46300	68.0	0.10	0.04	2.44	0.08	1.25
	G-500	GFRP	2.5	250	440	44.7	0.9	46300	77.2	0.10	0.06	1.76	0.09	1.22
	C-0.5-350	CFRP	2.5	250	310	42.4	0.2	144000	58.7	0.12	0.05	2.40	0.08	1.42
	C-0.5-500	CFRP	2.5	250	460	42.4	0.2	144000	70.3	0.09	0.05	1.76	0.08	1.12
	C-500	CFRP	2.5	250	460	42.4	0.5	144000	74.1	0.10	0.07	1.34	0.09	1.10
	C-2.5-500	CFRP	2.5	250	439	42.4	0.7	144000	82.5	0.12	0.09	1.33	0.09	1.24
Ashour and Kara (2014)	B-400-2	CFRP	2.7	200	370	27	0.1	141400	32.9	0.09	0.04	1.93	0.07	1.18
	B-400-4	CFRP	2.7	200	370	27	0.2	141400	36.1	0.09	0.06	1.54	0.08	1.23
Zeidan et al. (2011)	A-I-2.5	CFRP	2.5	150	280	24	0.1	148000	22.5	0.11	0.04	2.51	0.08	1.44
	A-II-2.5	CFRP	2.5	150	280	45	0.1	148000	22.5	0.08	0.04	2.12	0.08	1.01
	A-II-5.0	CFRP	5.0	150	280	49	0.1	148000	12.5	0.04	0.04	1.15	0.05	0.87
	B-II-2.5	CFRP	2.5	150	280	46	0.2	148000	27.5	0.10	0.05	1.86	0.08	1.15
Ashour (2006)	Beam 5	GFRP	2.5	150	269	34	0.9	32000	25.0	0.11	0.05	2.02	0.08	1.31
	Beam 11	GFRP	2.5	150	269	59	1.2	32000	30.0	0.10	0.05	1.83	0.09	1.11
El-Sayed et al. (2006)	GN-2	GFRP	3.1	250	326	44.6	1.2	42000	60.0	0.11	0.07	1.68	0.08	1.44
Melo and Rayol (2002)	Beam 1	GFRP	2.5	150	270	50	0.9	48000	35.0	0.12	0.06	2.05	0.09	1.40
	Beam 2	GFRP	2.5	150	270	50	0.9	48000	35.0	0.12	0.06	2.05	0.09	1.40
	Beam 3	GFRP	2.5	150	270	50	1.6	48000	20.0	0.07	0.08	0.91	0.09	0.75
	Beam 4	GFRP	2.5	150	270	50	2.1	48000	30.0	0.10	0.09	1.22	0.10	1.09
	Beam 5	GFRP	2.5	150	270	50	2.7	48000	30.0	0.10	0.10	1.09	0.10	1.06
	Beam 6	GFRP	2.5	150	270	50	3.3	48000	35.0	0.12	0.10	1.17	0.10	1.20
	Beam 7	GFRP	2.5	150	270	50	3.3	48000	30.0	0.10	0.10	1.00	0.10	1.03
Rendy et al. (2011)	BGN-05	GFRP	2.3	130	210	34	0.6	51500	19.6	0.12	0.06	2.23	0.09	1.42
	BGN-11	GFRP	3.0	130	210	34	0.6	51500	13.6	0.09	0.06	1.55	0.07	1.20
Kim and Jang (2013)	C-2.5-R2-2	CFRP	2.5	150	216	30	0.4	146200	21.6	0.12	0.08	1.53	0.09	1.41
	G-2.5-R1-1	GFRP	2.5	200	216	30	0.3	48200	24.7	0.10	0.04	2.50	0.08	1.36
Total Specimens	42 beams	GFRP and CFRP	2.3 - 5.0	150 - 250	210 - 460	20 - 59	0.11 - 3.26	32000 - 148000			Mean	1.57		1.12
											STDEV	0.51		0.21
											CoV (%)	32		19

6 Conclusions

The following conclusions can be drawn from this study:

(i) Shear strength of the tested beams is proportional to the longitudinal reinforcement ratio and shear span ratio. The stirrups spacing also influence the shear strength of the beams.

(ii) The stiffness of tested beams slightly increases with increasing longitudinal reinforcement ratio after the occurrence of the first flexure crack.

(iii) Proposed Eq. (3) provides a reliable and accurate estimation of the shear strength of beams reinforced with GFRP bars with a mean value of 1.12. The equation accounts for the effect of the longitudinal reinforcement ratio, shear span ratio, modulus elasticity of bars and concrete compressive strength.

(iv) The ACI 440.1R-06 provided very conservative prediction results for all beams reinforced with GFRP bars with a mean value of 1.57.

References

[1] N. Salleh, A. Rahman, M. Sam and J.M. Yatim, Flexural behavior of GFRP RC beam strengthened with carbon fiber reinforced polymer (CFRP) plate: Cracking behavior, Appl. Mech. Mater., **752-753**, 610–616 (2015)

[2] American Concrete Institut (ACI), *Guide for the design and construction of structural concrete reinforced with FRP bars (440-1R-06)*, Farmington Hills, MI (2006)

[3] A.G. Razaqpur, B.O. Isgor, A.M. Asce, S. Greenaway and A. Selley, Concrete contribution to the shear resistance of fiber reinforced polymer reinforced concrete members, J. Compos. Constr., **8**, 452–460 (2004)

[4] N.A.A. Hamid, R. Thamrin, A. Ibrahim and H.A. Hamid, Strain distribution on reinforcement of concrete beams reinforced with glass fiber reinforced polymer (GFRP) bars, Key Eng. Mater., **595**, 812–817 (2014)

[5] A.F. Ashour and I.F. Kara, Size effect on shear strength of FRP reinforced concrete beams, Compos. Part B Eng., **60**, 612–620 (2014)

[6] N.A.A. Hamid, A. Ibrahim, T. Rendy and H.A. Hamid, Effect of longitudinal reinforcement ratio on shear capacity of concrete beams with GFRP bars, *Proc. of the Inter. Civil and Infrastructure Eng.Conference*, Springer, 587–599 (2015)

[7] C.H. Kim and H.S. Jang, Concrete shear strength of normal and lightweight concrete beams reinforced with FRP Bars, J. Compos. Constr., **18**(2), 2–10 (2014)

[8] M.S. Alam and A. Hussein, Unified shear design equation for FRP reinforced concrete members without stirrups, J. Compos. Constr., **17**(5), 575-583 (2013)

[9] A.K. El-sayed, E.F. El-salakawy and B. Benmokrane, Shear strength of FRP-reinforced concrete beams without transverse reinforcement, ACI Struct., **103**(2), 235–243 (2006)

[10] M. Zeidan, M.A. Barakat, Z. Mahmoud and A. Khalifa, Evaluation of concrete shear strength for FRP reinforced beams, *Proc. of the Struct. Congr.*, ASCE, 1816–1826 (2011)

[11] A.F. Ashour, Flexural and shear capacities of concrete beams reinforced with GFRP bars, Constr. Build. Mater., **20**, 1005–1015 (2006)

[12] R. Thamrin, A.A.A. Samad, A. David, Y.E.C. David, N.A.A. Hamid and I. Ali, Experimental study on diagonal shear cracks of concrete beams without stirrups, *Proc. of the fib Symposium PRAGUE*, Czech Concrete Society, 1–8 (2011)

[13] G.S. Melo and J.A. Rayol, Shear resistance of GFRP reinforced concrete beams, *Proc. of the Rehabiliting and Repairing Buildings and Bridges*, ASCE, 47–64 (2002)

Sustainable Shear Behaviour of 2-Span Continuous Reinforced Concrete T-Beams with CFRP Strips

Abdul Aziz Abdul Samad[1,*], *Marwan* B.S. Alferjani[2], *Noorwirdawati* Ali[1], *Noridah* Mohamad[2], *Mohd Hilton* Ahmad[1], *Kong Fah* Tee[3], and *Priyan* Mendis[4]

[1] Jamilus Research Centre, Universiti Tun Hussein Onn Malaysia, 86400 Parit Raja, Johor, Malaysia
[2] Faculty of Civil and Environmental Engineering, Universiti Tun Hussein Onn Malaysia, 86400 Parit Raja, Johor, Malaysia
[3] Department of Engineering Science, University of Greenwich, Kent ME4 4TB, United Kingdom
[4] Department of Infrastructure Engineering, University of Melbourne, Parkville, 3010 Victoria, Australia

Abstract. Reinforced concrete (RC) structures are generally subjected to various load conditions and are susceptible to environmental impact. With time, most of these structures will be subjected to change in their usage. Unexpected increase in load changes may cause severe damage to the structural components, and eventual leading to various types of structural failure. It is noted that shear failure in nature is a sudden failure and should be avoided at all times. To increase its service life due to shear, structures can be strengthened or repaired using CFRP strips wrapped within the shear span (a_v) of the beam. Hence, this paper presents an investigation on the sustainable behaviour of two span continuous RC T-beams strengthened and repaired by wrapping Carbon Fibre Reinforced Polymer (CFRP) strips at intervals along the shear span (a_v) of the beam. The CFRP strips will be orientated in two directions either at 0°/90° or 45°/35°. Five beams, one control beam, two initially strengthened beams and two precracked and repaired beams were cast and cured in room temperature before testing. All T-beams were designed to fail in shear and are subjected to four-point bending test until failure. Upon completion of the test, all beam specimens shows good sustainable behaviour as the ultimate load and shear load shows an increase in capacity by up to 23% with improve ductile behaviour.

1 Introduction

Shear failure of RC beams is mainly caused by the formation of diagonal tension cracks within the web of the beam. This causes the beam to be unstable and consequently causes the structure to fail. In order to resist the shear stresses produced by the applied loads, the shear transfer mechanisms [1] are developed with the beam web as shown in Fig. 1. The

* Corresponding author: azizs@uthm.edu.my

figure clearly show the shear force V of RC beam resisted by the combined action of various forces. Hence, the shear force V can be represented by Equation (1) shown below:

$$V = V_{cc} + V_d + V_{ca} \quad (1)$$

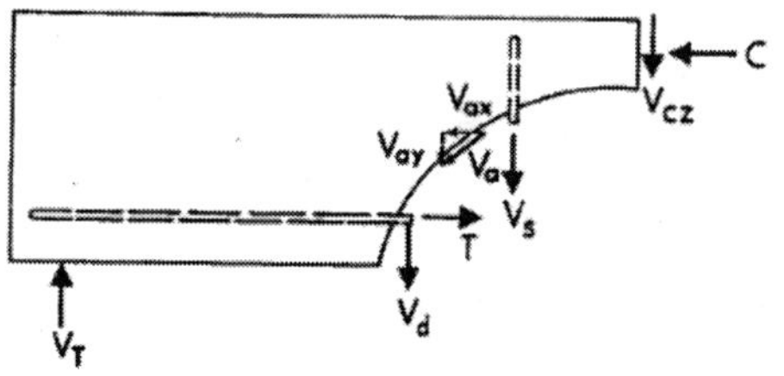

Fig. 1. Shear force transfer mechanism [1]

From Fig. 1, V_{cc} is the un-cracked concrete compression zone, V_d is the dowel action of longitudinal reinforcement and V_{ca} is the vertical component of the force due to aggregate interlock or interface shear transfer. Effectively, the shear failure mechanism of RC beams are governed by a number of factors such as the compressive strength of concrete (f_{cu}), effective depth of the beam (d), maximum aggregate size (d_a), and shear span to depth ratio (a_v/d). It has been reported by Sinha [2] that the behaviour of beams failing in shear can be studied with respect to the different shear span to depth ratio (a_v/d). Three common cases of shear failure (Case I, Case II and Case III) with different a_v/d ratio were outlined by Sinha [2] and are as shown in Fig. 2, Fig. 3 and Fig. 4 respectively.

Case I: Shear compression failure, a_v/d = 1 to 2.5

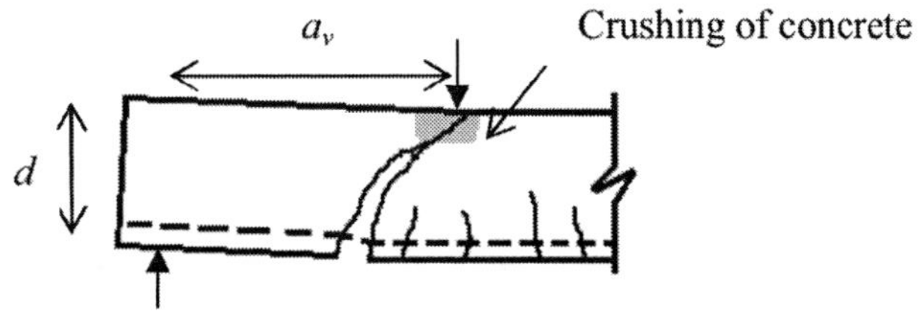

Fig. 2. Shear compression failure [2]

Case II: Shear tension failure, a_v/d = 1 to 2.5

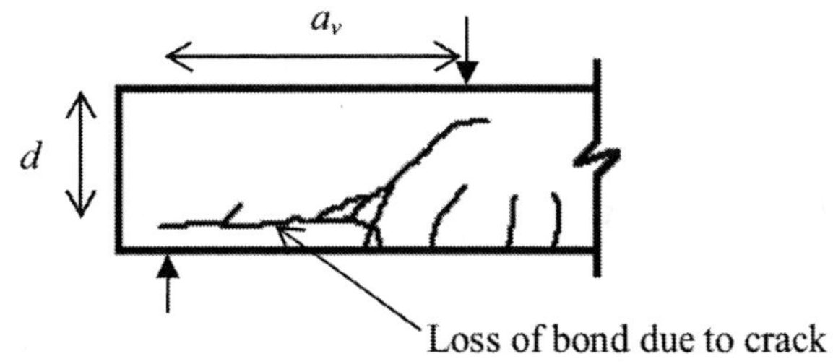

Fig. 3. Shear tension failure [2]

Case III: Diagonal tension failure, $a_v/d \geq 2.5$ to 6

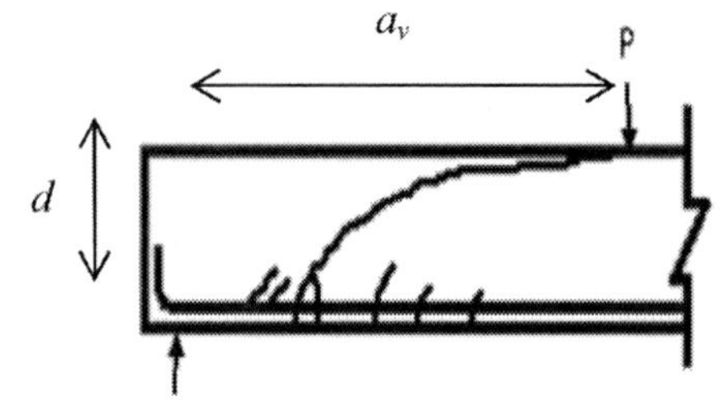

Fig. 4. Diagonal tension failure [2]

To ensure the sustainability and longevity of civil infrastructure, the use of Fibre Reinforced Polymer (FRP) materials for the repair and strengthening of reinforced concrete structures and for other types of construction has become common practice. FRP is a composite material made from strong fibres embedded in a resin matrix, possessing high resistance to corrosion, high strength properties and light weight.

As a material, FRP offers outstanding combination of physical and mechanical properties such as its high tensile strength, it is light weight, it has high stiffness properties, high fatigue strength and has excellent durable properties. Other major characteristics of FRP include its high strength to weight ratio, it has non-conductive, non-corrosive and non-magnetic properties and is resistant to chemical attack [3, 4]. It has also been reported by Alferjani et al. [5], Ali et al. [6] and Shahidan et al. [7] that the application of the FRP composites has been proven to increase the service life contributing to the structures longevity. Another unique property of FRP is that it allows the material to be molded into various shapes and forms, facilitating its flexible use in construction. This permits FRP to be applicable in areas with limited access where traditional techniques would be impractical [8]. However, application of FRP requires the use of additional bonding materials and adhesives, such as epoxy and resins [9], which are relatively expensive compared to traditional materials.

2 Experimental Work

The experimental work consists of testing five 2-span continuous RC T-beams with identical size of 150 mm x 320 mm x 3650 mm, flange width of 400 mm and flange thickness of 120 mm. All T-beams had identical reinforcement detailing, including nominal stirrups and longitudinal reinforcement. Fig. 5 shows the beam size and reinforcement details. All beams were designed to fail in shear and were strengthened and repaired using Carbon Fibre Reinforced Polymer (CFRP) strips from SikaWrap-160 BI C/15 [10], see Fig. 6. The CFRP strips were orientated at, either, 0°/90° (Fig. 7(a)) and 45°/135° (Fig. 7(b)). Due to the limitations of the test frame, the specimens were subjected to a shear span to effective depth ratio (a_v/d) of 1.85. The five beam specimens includes one control beam and four reinforced concrete T-beam. All the specimens are labelled as follows: Control Beam (T1.85-C), Beam 1 (T1.85-U-V1), Beam 2 (T1.85-U-V2), Beam 4 (T1.85-U-S) and Beam 4 (T1.85-L-S). Beam 1 and Beam 2 are initially strengthened beams whilst Beam 3 and Beam 4 are precracked and repaired beams. Observed that Beam 1, Beam 3 and Beam 4 were wrapped with a single layer of CFRP strip whilst Beam 2 with two layers. The investigation comprises of its ultimate load and shear capacity, load-deflection profile, its modes of failure and crack pattern of the specimens. All specimens had similar targeted concrete compressive strength of 30 N/mm^2, main reinforcement tensile strength of 460

N/mm^2 and stirrups tensile strength of 250 N/mm^2. The selected fiber reinforced polymer was a bi-directional CFRP sheet cut at strips of 80mm in width and were spaced at 150mm centre to centre (see Fig. 7(a) and 7(b)). Table 1 shows the material properties and its strength for concrete, steel and CFRP. The application of the CFRP strips was conducted by placing a coat of Sikadur-330 epoxy [9] to the concrete surface, see Fig. 8. The CFRP strips were then placed accordingly within the shear span (a_v) and allowed to cure for one week before being placed in the testing frame. Note that Beam 3 and Beam 4 (as precracked and repaired beams) were preloaded up to 70% of the ultimate load to create the initial shear cracks of the beams. Upon reaching the 70% load, the beams were unloaded followed by the application of the CFRP strips along the initial shear cracks.

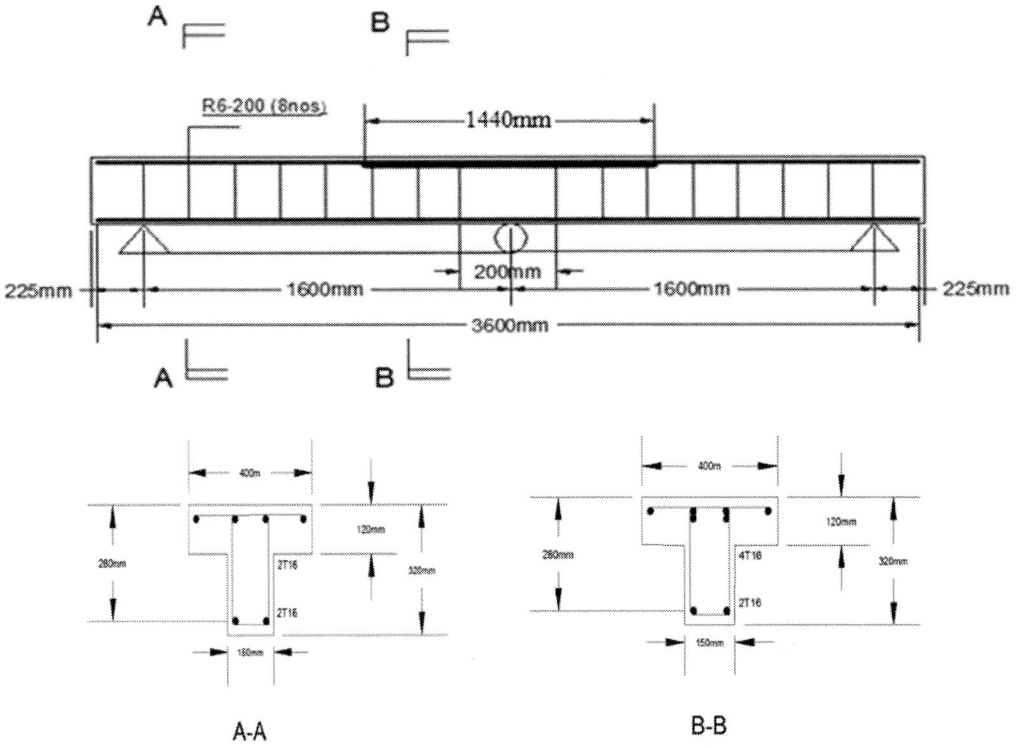

Fig. 5. Reinforcement and cross section details

Fig. 6. SikaWrap-160 BI C/15

Table 1. Material and strength properties for concrete, steel and CFRP

Concrete	7 days		28 days	
Compression Strength, f_{cu} (*MPa*)	26		30	
Steel	16 *mm* diameter		6 *mm* diameter	
Yield Strength, fy (*MPa*)	530		272	
CFRP	Fibre Orientation	Fabric Thickness (*mm*)	Tensile Strength (*MPa*)	Tensile Modulus (*MPa*)
	0°/90°	0.0046/direction	3,800	230,000

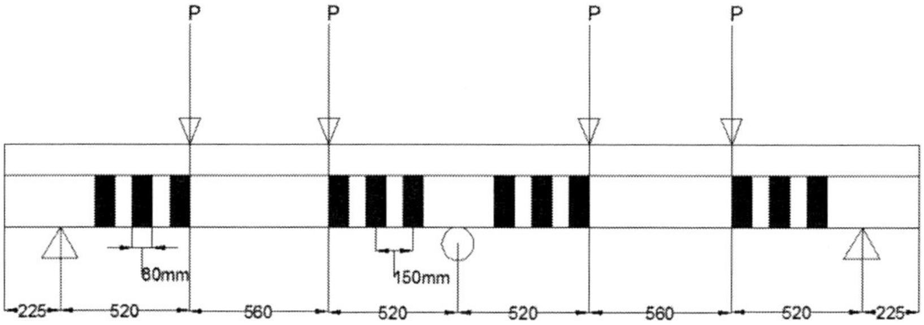

(a) CFRP orientation at 0°/90°

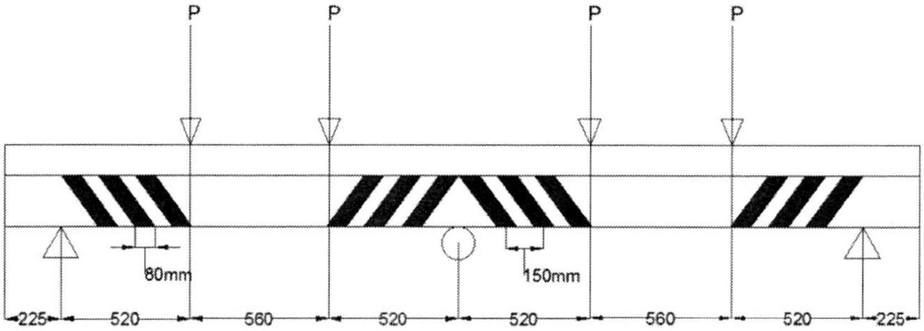

(b) CFRP orientation at 45°/135°

Fig. 7. CFRP orientation of 2 span continuous RC T-beams

Fig. 8. Application of CFRP at shear span (a_v)

3 Results and discussions

3.1 Ultimate load and shear load

Generally at failure, all T-beams either initially strengthened (Beam 1 and Beam 2) or precracked and repaired (Beam 3 and Beam 4) showed higher ultimate load values compared to control beam (T1.85-C). This is clearly indicated in Table 2 where Beam 1, Beam 2, Beam 3 and Beam 4 achieved an ultimate load of 241.2 *kN* up to 247.7 *kN* which compared to 220.4 *kN* for the control beam. This is an increase of approximately 10% to 12%. Likewise, the shear capacity of the beam also showed good enhancement values. By comparison with control beam T1.85-C, Beam 1 and Beam 2 showed shear enhancement of 9.1% and 17.0% respectively whilst Beam 3 and Beam 4 showed enhancement of 6.9% and 13.2%. The higher ultimate load and shear capacity clearly indicates that by externally bonding CFRP strips within the shear span of the T-beams (either initially strengthened or precracked and repaired) ensures its compositeness and its ability to carry more loads.

Table 2. Experimental results (a_v/d=1.85)

Beam Specimens	CFRP Orientation	Ultimate Load (*kN*)	Shear Force (*kN*)	Shear Enhancement (*kN*)	% Shear Enhancement
Control T1.85-C	0°/90°	220.4	73.3	0	0
Beam 1 T1.85-U-V1	0°/90°	247.7	82.4	9.1	12.4
Beam 2 T1.85-U-V2	0°/90°	271.6	90.3	17.0	23.3
Beam 3 T1.85-U-S	0°/90°	241.2	80.2	6.9	9.5
Beam 4 T1.85-L-S	45°/135°	260.1	86.5	13.2	18.0

Note: T = T-beam; 1.85 = shear span to depth ratio; U = CFRP strips at 0°/90° orientation; L = CFRP strips at 45°/135° orientation; V1= initially strengthened, single layer; V2 = initially strengthened, two layers; S = precracked and repaired; C = control beam

3.2 Load-deflection profile

The load-deflection profiles are shown in Fig. 9(a) for Beam 1, Beam 2 (initially strengthened) and Fig. 9(b) for Beam 3, Beam 4 (precracked and repaired beams). Comparison with deflection profile of the control beam (T1.85-C) and findings from Jayaprakash et al. [11, 12] was made for further clarity. By observation, the deflection profile for initially strengthened beams (Beam 1 and Beam 2) showed stiffer responses compared to Jayapakash et al. [11, 12]. However, less displacement values were recorded at failure but still shows an improved ductility behavior compared to control beam. Results show that displacement values for Beam 1 and Beam 2 was recorded at 12 mm and 14 mm respectively.

. For Beam 3 and Beam 4 (precracked and repaired), similar trends and close proximity of deflection profile was observed with Jayaprakash et al. [11, 12]. By comparison with the control beam, the deflection at failure was slightly higher at 11 mm and 13 mm for Beam 3 and Beam 4 respectively indicating an improve ductile behaviour.

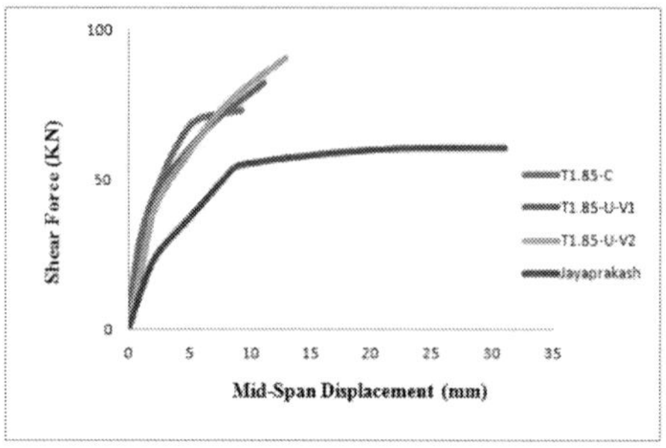

(a) Initially strengthened

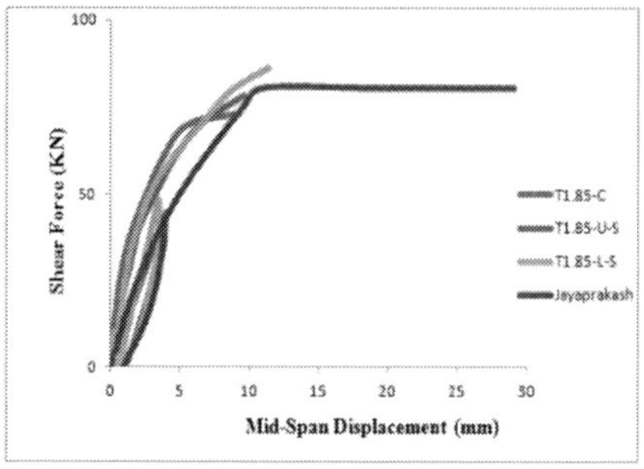

(b) Precracked and repaired

Fig. 9. Load-deflection profile

3.3 Mode of failure and crack pattern

The mode of failure and crack patterns for Control Beam, Beam 1 to Beam 4 is shown in Fig. 10(a) to (e). For control beam (T1.85-C) diagonal shear cracks developed as load was increased until total failure in shear occurs. Similar crck patterns were also observed For Beam 1 to Beam 4. However, upon failure, all strengthened or repaired beams failed by simultaneously rupturing the CFRP strips along its major diagonal shear crack of the beam.

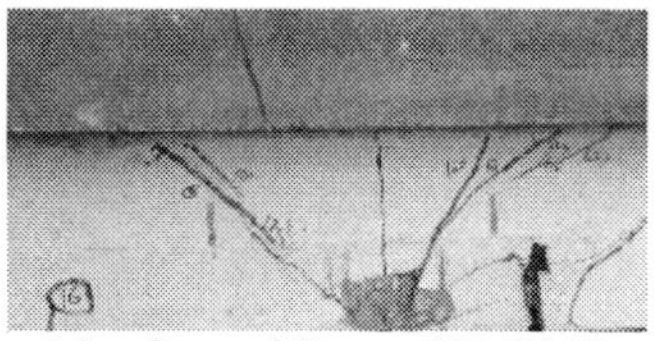

(a) Control Beam (T1.85-C)

(a) Beam 1 (T1.85-U-V1)

(c) Beam 2 (T1.85-U-V2)

(d) Beam 3 (T1.85-U-S)

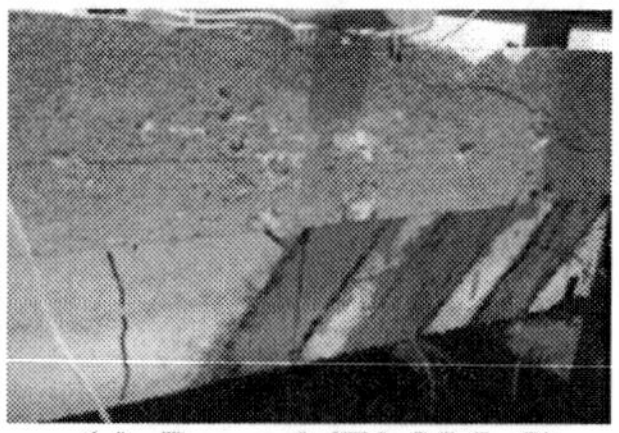

(e) Beam 4 (T1.85-L-S)

Fig. 10. Crack patterns and modes of failure

4 Conclusions

The sustainable behaviour of five full scale 2-span continuous RC T-beams with shear deficiencies were investigated. All beams were tested under a four-point bending test with shear span to effective depth ratio (a_v/d) of 1.85. The T-beams were either initially strengthened or precracked and repaired by wrapping CFRP strips at intervals along the shear span (a_v). Results indicated that the strengthened and repaired T-beams shows good composite responses with increase in ultimate load and shear load of up to 23%. The strengthened and repaired T-beams also shows improve ductility behavior with higher displacement values then the control beam.

The authors acknowledge the funding received from the Fundamental Research Grant Scheme (FRGS) Vot. No. 1573 from the Ministry of Higher Education of Malaysia and the support received from Jamilus Research Centre, Universiti Tun Hussein Malaysia.

References

[1] J.K. Wight and J.G. MacGregor, *Reinforced concrete: Mechanics and design*, 6th ed. Pearson, Upper Saddle River, New Jersey (2012)
[2] S.N. Sinha, *Reinforced concrete design*, 2nd ed. Tata McGraw-Hill (2002)
[3] A. Khalifa and A. Nanni, Improving shear capacity of existing RC T-section beams using CFRP composites, Cement and Concrete Composites, **22**(2), 165-174 (2000)
[4] A. Khalifa and A. Nanni, Rehabilitation of rectangular simply supported RC beams with shear deficiencies using CFRP composites, Construction and Building Materials, **16**, 135-146 (2002)
[5] M.B.S. Alferjani, A.A. Abdul Samad, B.S. Elrawaff and N. Mohamad, Experimental and theoretical investigation shear strengthening of RC precraced continuous T-beams using CFRP strips, Int. J. of Engineering, Transactions B: Applications, **28**(5), 671-676 (2015)
[6] N. Ali, A.A. Abdul Samad, N. Mohamad and J. Jayaprakash, Shear behaviour of precracked continuous beams repaired using externally bonded CFRP strips, Procedia Eng., **53**, 129-144 (2013)
[7] S. Shahidan, S. Salwa, M. Zuki, C. K. Keong and J. Jayaprakash, Repaired of fire-damaged concrete-filled double skin steel tubular (CFDST) columns with fiber reinforced polymer (FRP), ARPN J. Engineering of Applied. Sciences, **11**(6), 3718-3725 (2016)
[8] M.B.S. Alferjani, A.A. Abdul Samad, B.S. Elrawaff, O. Elzaroug and N. Mohamad, Study on shear strengthening of RC continuous T-beams using different layers of CFRP strips, AIP Conference Proceedings, **1660**, 070107 (2015)
[9] Sika Manufacturer's Product Data Sheet, Switzerland, (Supplier: Sika Kimia Sdn. Bhd.), Sika Wrap®-330. 2-party epoxy impregnation resin. Edition 0209/1
[10] Sika Manufacturer's Product Data Sheet, Switzerland, (Supplier: Sika Kimia Sdn. Bhd), Sika Wrap®-160 BI-C/15. Woven carbon fiber fabric for structural strengthening. Edition 11/09/2007
[11] J. Jayaprakash, A.A. Abdul Samad, A. Anvar Abbasovich and A.A. Abang Ali, Shear capacity of precracked and non-precracked reinforced concrete shear beams with externally bonded bi-directional CFRP strips, Construction and Building Materials, **22(6)**, 1148-1165 (2008)
[12] J. Jayaprakash, A.A. Abdul Samad, A.A. Ashrabov and K.K. Choong, Experimental investigation on shear resistance behaviour of RC precracked and non-precracked T-beams using discrete CFRP strips, Int. J. of Integrated Engineering, **1**(2), 1-15 (2009)

Failure Behaviour of Concrete Prisms Strengthened by Various Bond Widths of Carbon Fibre Reinforced Polymer (CFRP)

Syed Mohamad Syahir Syed Mazlan[1,*], *Siti Radziah* Abdullah[1], *Shahiron* Shahidan[1], and *Siti Rahimah* Mohd Noor[1]

[1]Jamilus Research Center, Universiti Tun Hussein Onn Malaysia, 86400 Parit Raja, Johor, Malaysia

Abstract. The use of fibre reinforced polymers (FRP) for external strengthening of concrete structures in various forms such as laminates and sheets has become moderately common. FRP is more effective as it provides a better solution in terms of properties and application. The effectiveness of the FRP strengthening system lies in the bond between concrete and FRP which is affected by various factors such as bond length of FRP, bond width, resin hardening and surface preparation. In certain cases, bond width is a major factor which ensures that the structure can be strengthened properly. This paper focuses mainly on the failure behaviour of concrete prisms strengthened by various bond width ratios of carbon Fibre Reinforced Polymer (CFRP) which are 20mm (0.27), 40mm (0.53), 60mm (0.8) and 75mm (1) respectively. A total of fifteen strengthened prisms measuring 75 mm x 75 mm x 350 mm were cast for a three-point load test. Twelve strengthened prism specimens were externally strengthened with CFRP and three specimens were designated as control. The findings show that a greater width ratio of CFRP will strengthen concrete prisms greatly and will result in different types of failure modes.

1 Introduction

Nowadays, the application of Fiber Reinforced Polymer (FRP) material is increasing in the construction sector especially in strengthening deficient concrete structures [1]. These materials are lightweight, non-corrosive, and exhibit higher tensile strength [2, 3]. FRP can be used to rehabilitate or restore the strength of deteriorated structural members, retrofit or strengthen sound structural members to resist increased loads due to change in use of the structure, or address design or construction errors on a structure [4-6].

Certain advantages of FRP composites including high tensile strength, light weight and ease of application have led engineers all over the world to utilize FRP as an easy-to-use alternative compared to conventional materials for strengthening reinforced concrete structures [7]. Almost all flexural or shear strengthening techniques with FRP rely on the bonding of the composite onto the concrete substrate [8]. The effectiveness of the FRP strengthening system mainly depends on the effective length due to the bond of FRP on the

[*] Corresponding author: syedmsyahir@gmail.com

effective zone [9]. However, temperature sensitivity, concrete quality, surface preparation, width of FRP and bond length are some of the factors that affect the ultimate strength of the specimen between concrete and FRP [10]. In previous studies, it was highlighted that the width ratio of CFRP plays an important role in influencing the debonding failure of CFRP [12-15] The greater width ratio of an FRP plate will develop higher interfacial bond strength of FRP to concrete and a smaller global slip of FRP to concrete [15]. Thus, a greater width ratio of FRP plate will give the greater slope of elastic deformation of the global slip but a lower ductility of FRP to concrete. When the width ratio value is close to 1, it will give the best FRP performance. Other than that, every width ratio will form different types of failure modes as shown in Fig. 1 [14]. Therefore, from previous studies, the result of concrete strengthened with FRP will produce different results based on the types of preparation such as different types of FRP, plies of FRP, bond length of FRP and width ratio of FRP. This study focuses on the effect of different width ratios of CFRP to enhance concrete strength and types of failure modes produced.

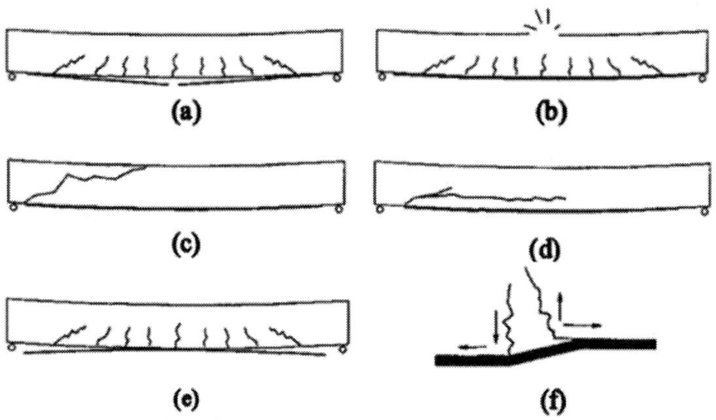

Fig. 1. Failure mode in FRP as a strengthened prism: (a) steel yield and FRP rupture, (b) concrete compression failure, (c) shear failure, (f) debond of layer along rebar, (g) delamination of FRP plate, (h) peeling due to shear cracks.

2 Experimental programs

A total of fifteen (15) strengthened prisms measuring 75 mm x 75 mm x 350 mm were cast for a three-point load test. Twelve (12) strengthened prisms specimens were externally strengthened with CFRP and three (3) specimens were designated as control or non-strengthened prisms.

2.1 CFRP strengthening

Concrete prisms of grade 30 were cured for 28 days. In this study, the type of FRP used was uniaxial CFRP or SIKA WRAP -231C. The mechanical properties of CFRP for tensile strength, modulus of elasticity and tensile strain are 4900 MPa, 230 GPa and 1.7% respectively. The CFRP was attached to the concrete prism as illustrated in Fig. 2 and the prism surface was grinded to obtain a rough surface to ensure perfect bonding and a clean condition [16]. Then the CFRP was attached to the concrete prisms and cured for 7 days at room temperature.

2.2 Specimen characteristics and strengthening conformation

The size of strengthened prisms was 350mm long, 75 mm wide and 75 mm high. The width ratios of the CFRP attached was 20 mm (0.27), 40mm (0.53), 60 mm (0.8) and 75mm (1) respectively. The details on the dimensions and CFRP attachments are illustrated in Fig. 2.

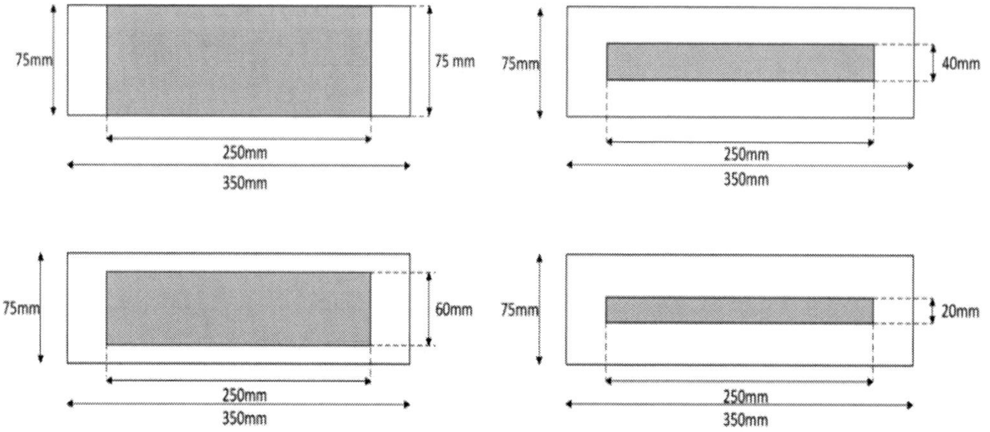

Fig. 2. Bottom view details for the dimensions of CFRP attachment to the strengthened prisms.

2.3 Test Setup

A three-point load test was conducted on strengthened prism specimens to determine the maximum force applied, deflection, and failure behaviour of strengthened prisms after the specimens failed. A test instrumentation setup is presented in Fig. 3. The strengthened prisms specimens were placed onto two-pin supports located at both ends of the specimen. A linear variable displacement transducer (LVDT) was connected to a computer in order to determine the deflection of the strengthened prisms.

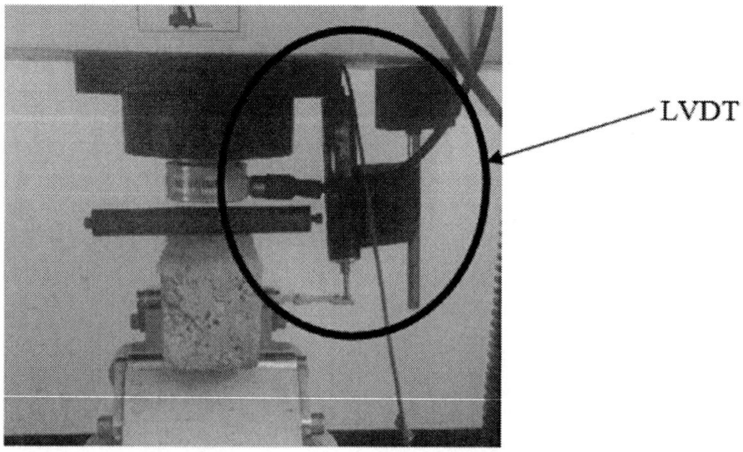

Fig. 3. Strengthened prisms were placed on two-pin supports with a linear variable displacement transducer (LVDT) installed at the centre of the strengthened prisms.

3 Results and analysis

3.1 Effect of width ratio on concrete

The results from the three point load test comprising of ultimate load, average ultimate load and percentage of increase for all tested prisms are presented in Table 1. For control specimens without CFRP, the average ultimate load is 13.06kN. When the CFRP was attached to the concrete prism, it was found that the ultimate load of the concrete increased as the CFRP width ratio increased. For CFRP with a width ratio of 0.27, the percentage of the ultimate load increased by 47.39% compared to the specimen without CFRP. For the specimen attached to CFRP with a width ratio of 0.53, the average ultimate load is 24.6kN which is 89.05% higher than the specimen without CFRP. The trend of ultimate load increases when the width ratio increases and this is graphically presented in Fig. 4. The specimen attached to CFRP with a width ratio of 1 attained the highest average ultimate load of 37.5kN with an increased percentage of 187.23% compared to the specimen without CFRP.

Table 1. Tabulated data of average three point load test

Width Ratio W_p/W_c	Specimen	Three point load test		
		Ultimate load (kN)	Average ultimate load (kN)	Percentage of increase (%)
0	W0	13.64	13.06	0
		11.00		
		14.52		
0.27	W20	18.33	19.25	47.39
		20.61		
0.53	W40	24.81	24.6	89.05
		24.57		
0.80	W60	34.89	33.38	155.67
		32.37		
		32.89		
1.00	W75	36.89	37.50	187.23
		37.73		
		37.89		

From the failure load graph, it can be observed that when the width of ratio increased, the load also increased as shown in figure 4. The failure load for each width ratio resulted in different modes of failure as discussed in [16]. One control prism and four CFRP strengthened concrete prisms experienced different noticeable failure loads respectively. Every prism shows an increase in load between 4kN to 8kN for every width ratio. As the load gradually increased, the prism underwent minor FRP rupture and shear failure respectively [17].

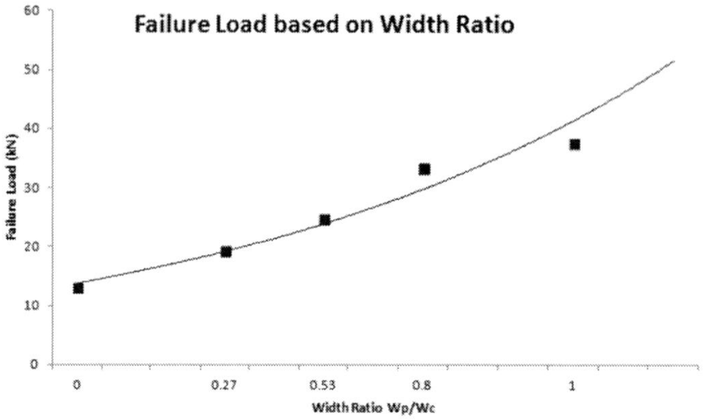

Fig. 4. Failure loads based on the different width ratios.

3.2 Load versus deflection of concrete with various width ratios

The graphs in Fig. 5 show load versus deflection for all the specimens with and without CFRP. It is clear from the figure that the CFRP does not only provide load enhancement but also improve the ductility of the specimen compared to the control specimens. In general, from the load versus deflection graph, it can be observed that an increasing CFRP width ratio leads to an improvement in the ultimate load as well as the stiffness of the strengthened CFRP specimens. This behaviour is noticeable at W75 (width ratio 1) compared to other width ratios and is consistent with the findings by [10].

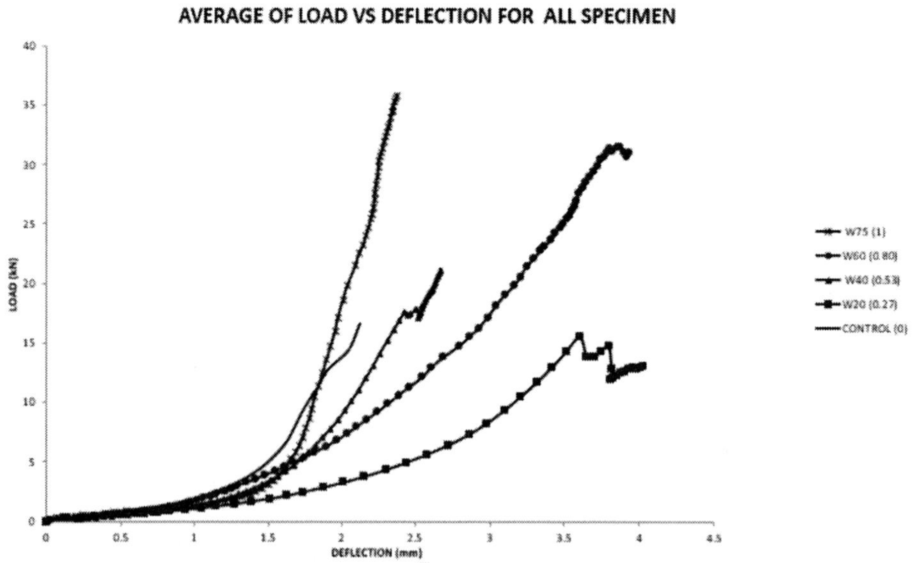

Fig. 5. Load versus deflection for all specimens.

3.3 Failure mode

Table 2 shows the types of failure modes for each specimen with various width ratios. For a control specimen without CFRP, the failure mode is a typical flexural crack. For CFRP width ratios of 0.27 and 0.53, both share a similar failure mode which is CFRP rupture.

Table 2. Types of failure mode for each specimen.

Width Ratio W_p/W_c	Physical Observation	Failure Mode
0	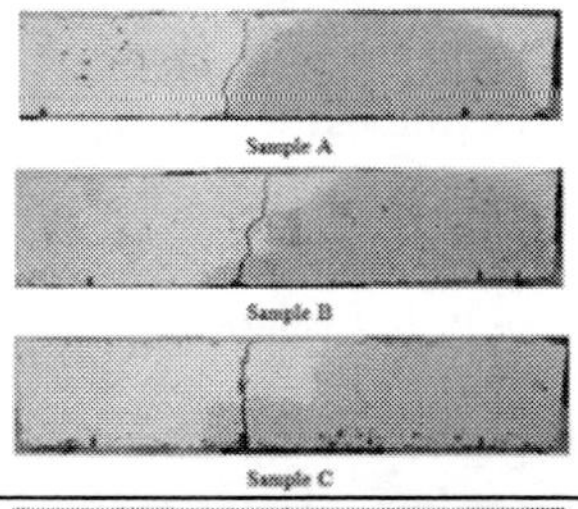	Flexural crack
0.27	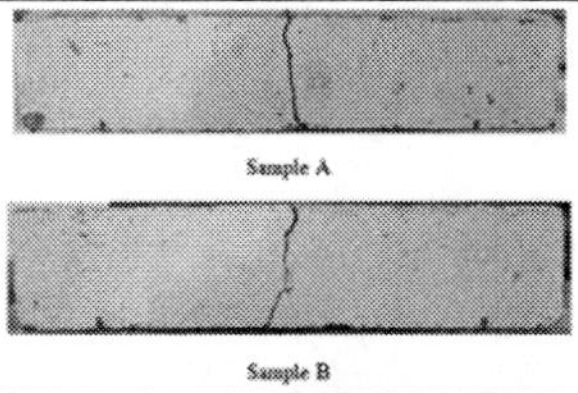	FRP rupture
0.53	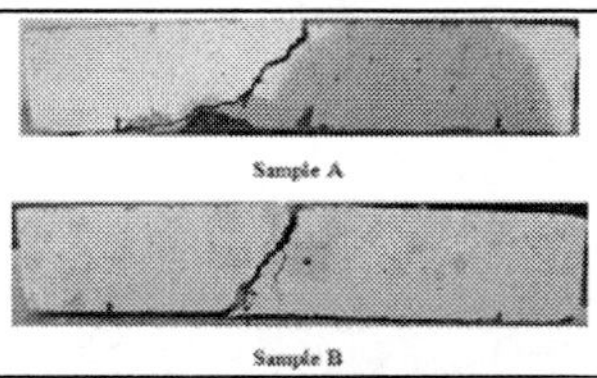	FRP rupture
0.8	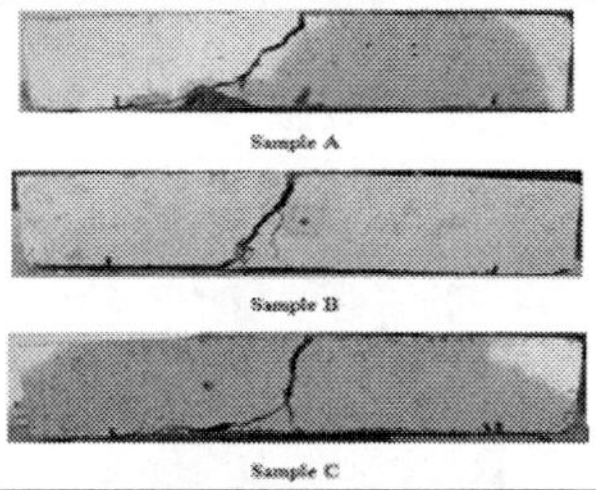	Plate end interfacial debonding (The crack starts to develop at the surface of the concrete prism until it caused the CFRP and concrete to debond)
1	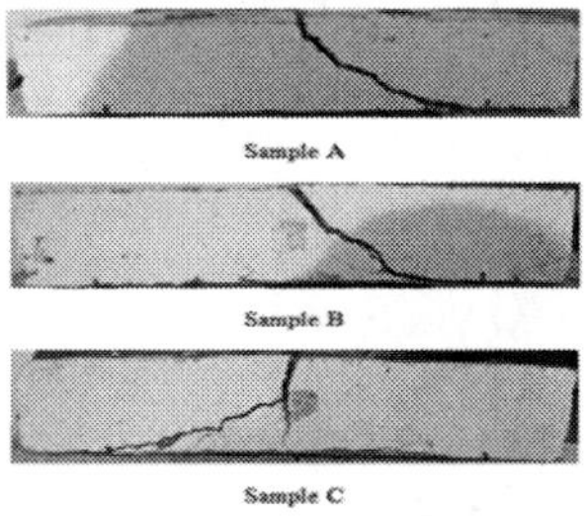	Plate end interfacial debonding (The crack starts to develop at the surface of the concrete prism until it caused the CFRP and concrete to debond)

However, for the crack development of CFRP with a width ratio of 0.53, the crack typically starts from the top and forms a shear crack before CFRP rupture. As the width ratio increases, the failure mode changes from CFRP rupture to plate end interfacial debonding with a thin concrete layer attached to the debonded CFRP sheet [11]. Although the failure mode of specimens attached to CFRP with width ratios of 0.8 and 1 was plate end interfacial debonding, it is proven that the ultimate load is higher as stated in this study (refer Table 1). Similar findings were also reported by Subramaniam et al. [13] indicating that the greater the CFRP width ratio, the higher the debonding strength of CFRP.

4 Conclusion

This is an experimental study on the effect of various CFRP bond widths on concrete prisms. All of the specimens were subjected to a three point load test to determine the ultimate load, deflection and failure mode of the CFRP strengthened concrete prism. It can be concluded that:

i) The ultimate load for CFRP strengthened concrete prisms increased as the CFRP width ratio increases.
ii) Significant load enhancement of more than 80% is achieved when the concrete is strengthened with a CFRP width ratio of 0.5 and above
iii) The increase in CFRP width ratio leads to an increase in the stiffness of the specimen hence resulting in a higher ultimate load.
iv) To avoid plate end debonding failure, it is recommended for future studies to provide an end anchor for the CFRP so that higher ultimate load and ductility can be achieved for the strengthened prisms.

This work was supported by the Jamilus Research Center, Faculty of Civil and Environmental Engineering, University Tun Hussein Onn Malaysia, Center for Graduate Studies, UTHM and the Research Acculturation Grant Scheme (RAGS) Vot. 053.

References

[1] C. Mikami, H.C. Wu and A. Elarbi, Effect of hot temperature on pull-off strength of FRP bonded concrete, Constr. Build. Mater., **91**, 180–186 (2015)

[2] S.R. Abdullah, Fibre reinforced polymer rehabilitation of alkali aggregate reaction-damaged concrete structures, PhD Thesis, Monash University, Australia (2013)

[3] N. Ali, A.A.A. Samad, N. Mohamad and J. Jayaprakash, Shear behaviour of pre-cracked continuous beam repaired using externally bonded CFRP strips, Procedia Eng., **53**, 129–144 (2013)

[4] D. De Domenico, P. Fuschi, S. Pardo and A.A. Pisano, Strengthening of steel-reinforced concrete structural elements by externally bonded FRP sheets and evaluation of their load carrying capacity, Compos. Struct., **118**, 377–384 (2014)

[5] S. Shahidan, S. Salwa, M. Zuki, C.K. Keong and J. Jayaprakash, Repaired of fire-damaged concrete-filled double skin steel tubular (CFDST) columns with fiber reinforced polymer (FRP), ARPN J. Eng. Appl. Sci., **11**(6), 3718–3725, (2016)

[6] S. Shahidan, S.S.M. Zuki and N. Jamaluddin, Damage grading system for severity assessment on concrete structure, Case Stud. Constr. Mater., **5**, 79–86 (2016)

[7] S. Ma, N. M. Bunnori and K.K. Choong, Behavior of reinforced concrete box beam strengthened with CFRP U-wrap strips under torsion, MATEC Web of Conferences, **47**, 02002 (2016)

[8] E. Choi, B.S. Cho, J.S. Jeon and S.J. Yoon, Bond behavior of steel deformed bars embedded in concrete confined by FRP wire jackets, Constr. Build. Mater., **68**, 716–725 (2014)

[9] M. Batikha and F. Alkam, The effect of mechanical properties of masonry on the behavior of FRP-strengthened masonry-infilled RC frame under cyclic load, Compos. Struct., **134**, 513–522 (2015)

[10] H. Mazaheripour and J.A.O. Barros, Tension-stiffening model for FRC reinforced by hybrid FRP and steel bars, **88**, 162–181 (2016)

[11] U.S. Camli and B. Binici, Strength of carbon fiber reinforced polymers bonded to concrete and masonry, Constr. Build. Mater., **21**(7), 1431–46 (2007)

[12] L. De Lorenzis, B. Miller and A. Nanni, Bond of fiber-reinforced polymer laminates to concrete, ACI Mater. J., **98**(3), 256–264 (2001)
[13] K.V. Subramaniam, C. Carloni and L. Nobile, Width effect in the interface fracture during shear debonding of FRP sheets from concrete, Eng. Fract. Mech., **74**(4), 578–594 (2007)
[14] S.K. Woo and Y. Lee, Experimental study on interfacial behavior of CFRP bonded concrete, KSCE J. Civ. Eng., **14**(3), 385–93 (2010)
[15] T. Xu, Z.J. He, C.A. Tang, W.C. Zhu and P.G. Ranjith, Finite element analysis of width effect in interface debonding of FRP plate bonded to concrete, Finite Elem. Anal. Des., **93**(C), 30–41 (2015)
[16] N. Moshiri, A. Hosseini and D. Mostofinejad, Strengthening of RC columns by longitudinal technique, Constr. Build. Mater., **79**, 318-325 (2015)
[17] A. Napoli, G. de Felice, S. De Santis and R. Realfonzo, Bond behaviour of steel reinforced polymer strengthening systems, Compos. Struct., **152**, 499–515 (2016)

Forensic Building: Deterioration and Defect in Concrete Structures

N. Jamaluddin[1], *S. S* Ayop[1,*], *M. H.* Wan Ibrahim[1], *K. H.* Boon[1], *D.* Yeoh[1], *S.* Shahidan[1], *N.* Mohamad[1], *T. N.* Tuan Chik[1], *N. H.* Abd. Ghafar[1], *A. H.* Abdul Ghani[1], and *S.* Shamrul-Mar[1]

[1]Jamilus Research Center, Faculty of Civil and Environmental Engineering, Universiti Tun Hussein Onn Malaysia, 86400 Parit Raja, Johor, Malaysia

Abstract. Forensic building is needed to examine the affected building structure components to assess the structural integrity. This paper highlights some of the studies involved on affected concrete structures in various building types where the non-destructive test (NDT) and destructive tests. The structural integrity can be evaluated based on the extent of deterioration from the experimental results for instance the cause of the corrode reinforcements was due to inadequate concrete cover, whereas the failure of the concrete wall was due to structures' under design which leads to the collapse of the wall. The condition of the floor slab that experience vibration was assessed from the Laser Doppler Vibrator test (LDV). Based on the test results the peak acceleration of the particular floor slab is higher.

1 Introduction

A building is failing when the building loses its ability to perform its function. Building failure can be categorized as physical failure. Examples of physical failures are strength and performance failures due to the inability of the building to function as expected by the established acceptable limit suggested by standard codes. Building a structure carries with it an element of risk; thus, Building Codes establish a standard that structures must meet.

Structural failures can be associated with the choice of materials and shoddy workmanship. This paper highlights some of the assessments made on defect concrete structures and the process involved. The paper is not intended to discredit any parties, but to emphasis on the importance of engineering fundamental and judgement to avoid disastrous event.

Structures will fail when something goes wrong with the design that will affect its strength and stability. The understanding of material properties and behavior are crucial in the design process. Structural failure occurs due to the forces acting on it, either static or dynamic forces. The structures should not be loaded in a way that will cause a strength and system failure.

* Corresponding author: sallehs@uthm.edu.my

 DOI: 10.1051/matecconf/201710302016

Several tests have been conducted on several buildings. The tests were cover-meter test, concrete permeability test, carbonation test, chloride test and half cell potential test for building structure with exposed reinforcements. As for collapse concrete wall the tests involved were coring test and cover-meter test. In the case of vibrated concrete floor, the Laser Doppler Vibrometer test was performed.

2 Structure failures and test results

2.1 Structure with exposed reinforcement

Exposed reinforcements are amongst the severe conditions that could be observed from structures such as at columns and the underside of slab (Fig. 1). The spalling of the concrete parts and rusted steel reinforcement are clearly seen in these structures. These are the examples of distress failures in concrete structures which are caused by corrosion of the steel reinforcement.

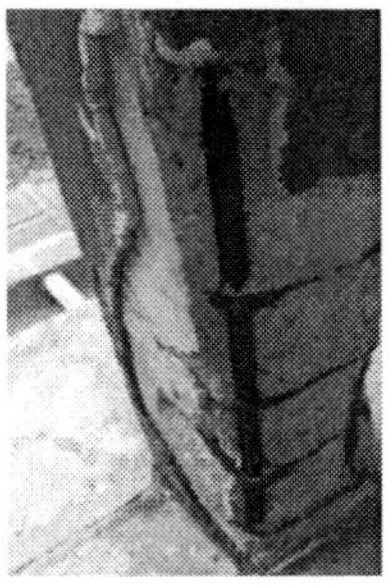
a) Column

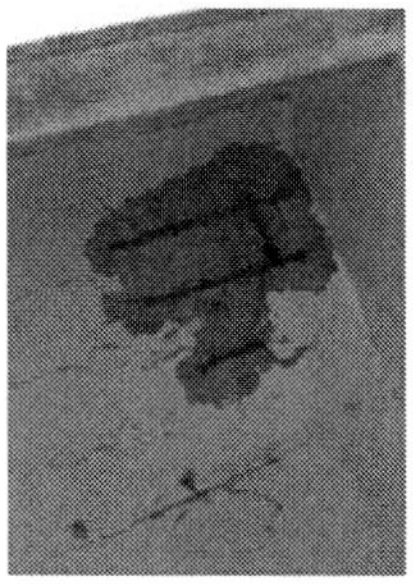
b) Slab

Fig. 1. Structures with exposed reinforcement

An alkaline environment of concrete provides steel reinforcement with corrosion protection [1]. However, the corrosion may take place when the alkalinity of the concrete is reduced or when the chloride concentration in concrete is increased to a certain level [2]. Corrosion of steel reinforcement may cause the concrete cracks and eventually concrete spalling. Rust occupies a greater volume than the steel in concrete, which creates stress in the concrete and eventually cause cracking and spalling of the concrete. Details of the corrosion mechanism have been discussed previously in [3]. In order to investigate the reason of the occurrence of such failure on the particular structures, few tests were conducted as follows;

2.1.1 Cover-meter test

The steel reinforcement can corrode in moist atmospheres, thus the low permeability concrete and adequate concrete cover will help to protect the embedded metals from corrosive materials. A cover-meter test was conducted on the affected structure in order determine the precise concrete cover depth and to determine the exact location of the rebars in concrete. Minimum concrete cover as stated in BS 8110-1-1997 [4], is between 20-50mm. Provision of correct cover depth as per the code recommendations will ensure minimum corrosion of the reinforcement bars. From the test results, the concrete cover was found be varies with the lowest cover is 8mm which may indicate due shoddy construction.

2.1.2 Low permeability concrete

Low permeability can also reduce the corrosion rate. The low permeability concrete could be attained by reducing the ratio of water to cementations materials. Permeability of concrete can be minimized by adopting low water-content ratio, ensuring proper compaction and curing of concrete. However the permeability concrete evaluation, rapid chloride permeability test was not conducted in this case.

2.1.3 Carbonation test

Carbonation is generally a slow process. The amount of carbonation is significantly increased in concrete with a high water-to-cement ratio, low cement content, short curing period, low strength, and highly permeable or porous paste. The carbonation test is carried out to determine the depth of concrete affected due to the combined attack of atmospheric carbon dioxide and moisture and causing a reduction in the level of alkalinity of concrete. Carbonation of concrete also lowers the amount of chloride ions needed to promote corrosion. This test was conducted in accordance with BS EN 14630:2006 [5] and the on-site carbonation tests were conducted in order to obtain real time results.

The test was carried out on the affected structure by spraying the indicator to pre-drilled holes at specific pre-determined locations. The pre-drilled holes using an 18mm diameter drill bit with the total depth of 75 mm as requested by client including the concrete cover were performed. The affected depth is assessed using a solution of phenolphthalein indicator. Fig. 2 shows the carbonation test result. Carbonation test for the pre-drilled holes shows reddish-purple, which indicates that the pH is above 8.6 for un-carbonated condition of such structure. For the phenolphthalein solution which remains colorless, the pH of the concrete should be less than 8.6, suggesting carbonation or fully-carbonated.

Fig. 2. Carbonation test for the pre-drilled holes

2.1.4 Chloride test

Chloride test was carried out in accordance with BS EN 14629:2007 [6]. The test was performed on the dusting powder obtained by drilling from the concrete structure from the specified point which was decided by the client. The concentration of chloride ions was determined from the powder collected at various depths in order to determine the likelihood of corrosion in the reinforcing steel. The concrete was drilled to a depth of 75mm with the dusting powder collected after the first 25mm depth as the first 25mm is considered non-representative. From the chloride test, the chloride contents were determined and compared with limiting values specified for the concrete to assess the risk of corrosion in concrete structure. From the chloride test results (Fig. 3), the chloride content was found to be less than 0.02%, which indicates that the chloride content in current condition is accepted. It should be noted that the results represent the condition of the structure having an adequate thickness cover.

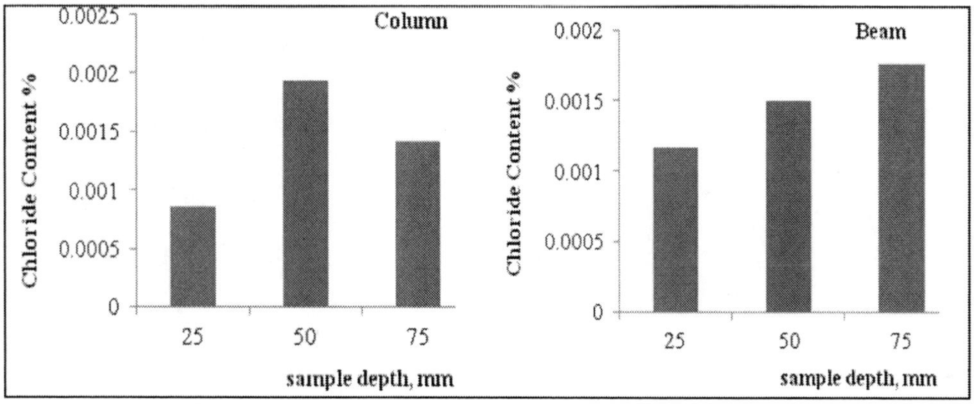

Fig. 3. Results of chloride test

2.1.5 Half cell potential test

Half cell corrosion detection technique was conducted to assess the steel condition in the concrete structure. This technique evaluates the risk of corrosion activity of steel in concrete. The test was conducted by referring to ASTM C876-15 [7] using copper sulphate electrode connected to a voltmeter. Measurements are taken from the pre-drawn grid at 200mm distance. The possible potential electrode values were then compared with the criterion listed in Table 1.

Table 1. Criteria of corrosion according to ASTM C876-91

Open circuit potential (OCP) Values (mV vs. CSE)	Corrosion Condition
< -500	Severe corrosion
< -350	High (<90% of risk corrosion)
-350 to -200	Intermediate corrosion risk
> -200	Low (10% of risk corrosion)

The results of the half cell potential test as shown in Table 2 indicate that the current condition of the reinforcements appeared to be losing its properties.

Table 2. Half cell corrosion results

Location	Indicator
Slab Soffit	Intermediate corrosion risk with certain point has high corrosion (90% risk)
Beam	Low risk corrosion with certain point has intermediate risk.

2.2 Collapse of concrete wall

Another case that involves the structural failure is as shown in Fig.4; the collapsed concrete wall of an effluent tank. Visual inspection during the first visit observed the occurrence of honeycomb along the failure line of the wall concrete.

DOI: 10.1051/matecconf/201710302016

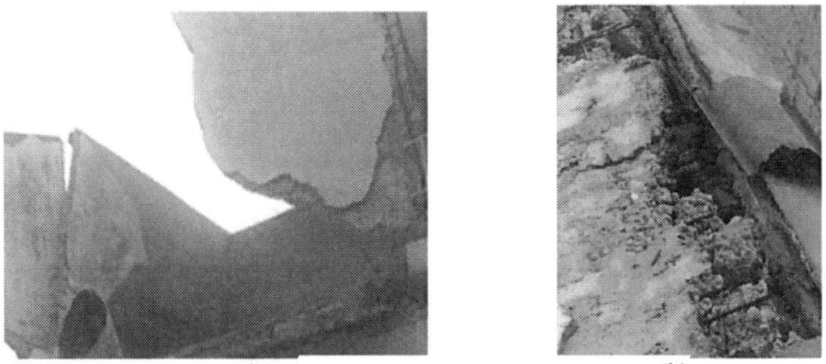

(a) (b)

Fig. 4. Failure of effluent tank concrete wall

It is found that the main reasons for the occurrence of honeycomb are an improper consolidation of the concrete or from using stiff concrete. Junctions in the structure layout are typical spots where honey combs can be observed, which is due to jumbling of reinforcement of between the structures at one place in which special attention is required during concreting and vibrating. The high percentage of large aggregate size may prevent concrete to fill narrow spaces between the reinforcement rods. The presences of honeycomb in the concrete structure not only affect the load bearing capacity but also provide an easy access to water to finds reinforcement rods and start rusting and corrosion. Grout leakage along the structure perimeter.

2.2.1 Coring test

Coring test was conducted on the wall structure of the concrete tank. From the test, the strength of concrete is 27.43 N/mm^2. Considering the condition at the location, for safety reasons the coring sample was taken at the zone where honeycomb was not detected. Thus, due to the presence of a defect in the other zone, one can predict that the concrete strength will further reduce. From the finding, comparison was made in accordance to BS8007:1987 [8] and BS 8110-1-1997 [4] where it is stated that the minimum strength of 35 N/mm^2 is required for liquid retaining structure.

2.2.2 Cover-meter test

The cover-meter test was also carried out on the effluent tank to determine the concrete cover. Findings reveal the minimum concrete cover of 6mm was found on the structure wall. For structure subjected to severe exposure condition, for concrete cover thickness it has different stipulation and protective measures. Relevant measures in terms of work environment and structure type should be taken into consideration. For the concrete wall tank, the minimum concrete cover of 40 mm should be provided.

2.3 Vibrated floor slab

Generally, vibration is not an issue for reinforced concrete floor systems due to inherent mass and stiffness of such systems. However, due to some reasons, the effects of vibration could be the main design issues that need to be addressed. Floor structures are designed for ultimate limit states and serviceability limit state criteria. Floor vibration generally makes people uneasy and creates fear of structural collapse and eventually affects people's sense of well being and ability to carry out tasks. Depending on a number of factors, the vibration can be annoying or worse for the people that are occupying the affected area. Floor vibration can be a greater problem, particularly for long-span floor structure. The main factor of the vibration problems is resonance, which occurs when a load is exerted on a floor at a particular frequency.

A study on a concrete floor slab was carried out on a structure building that has reported to have vibration problems. In this building case, the affected area is 6.5m×10m can be considered as the length of floor span.A preliminary inspection was conducted in the affected area and the followings have been observed:

i) Sense of discomfort for occupants- evident from occupant complaints.
ii) The out-of level condition of the floor- sagging floor
iii) Cracks in the floor slab (Fig. 5) and along the slab boundary

The first action taken in this situation is monitoring the crack development over a period of one week. All existing equipment on the floor space was not removed, but the area was closed for any activities. During the period, the cracks were observed to propagate from 0 to 1.5 inches. It stopped to propagate which indicate that the cracks have reached its equilibrium state.

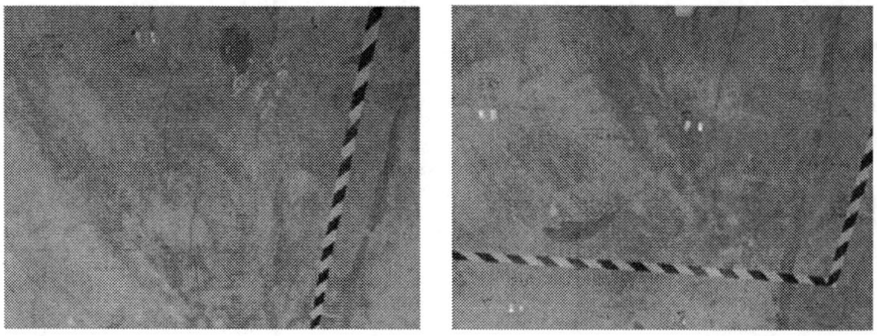

Fig. 5. Cracks on the affected floor slab

The test due to human activity (in this case; walking) was conducted to evaluate the vibration response. For the accepted floor slab response, the peak acceleration of the floor system should be less than or at least equal to the recommended acceleration for a particular occupation.

Peak acceleration is related to the natural frequency (*fn*), the effective weight, and the inherent damping of the floor system. The natural frequency is a measurement of how the floor system will respond to the sources that can cause vibration, and is related to how the occupants will perceive such vibrations.

2.3.1 Laser Doppler Vibrometer test

Vibration waves were captured with Laser Doppler Vibrometer, LDV (Fig. 6) at a designated location on the affected floor area. From the test results, the peak vibration acceleration at the Location of L1 and L6 (Fig. 7) were among the higher as compared to other location with 0.0290 m/s^2 and 0.0219 m/s^2 respectively. The peak vibration acceleration for normal reinforced concrete floors as in previous studies is 0.007 m/s^2 whereas for concrete masonry load bearing system floor, the peak vibration acceleration is 0.00042 m/s^2 [9]. The results indicated that the peak acceleration for the floor is higher as results of people activity in this case walking. Therefore, it has been suggested that further investigation and analyses of natural frequencies, mass, damping for the floor need to be carried out in order to remedy the excessive vibration. Actions are necessary to improve the slab stiffness.

DOI: 10.1051/matecconf/201710302016

Fig. 6. The measurement was recorded based on human's footsteps

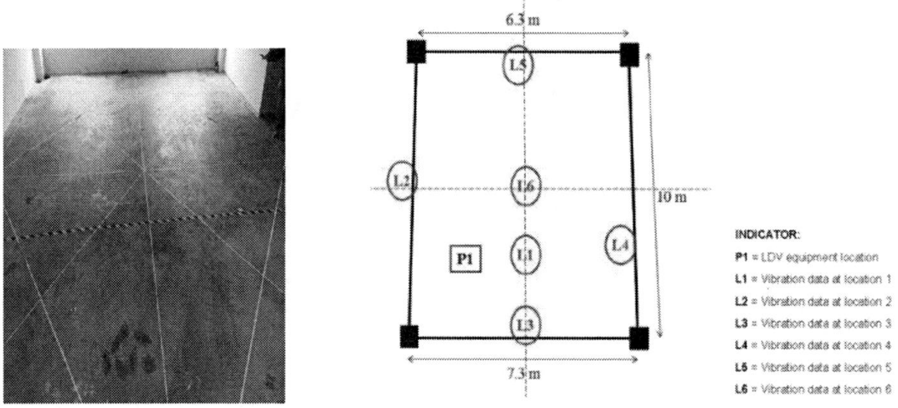

Fig. 7. Mapping for test locations

3 Conclusions

It is impossible to control the causes of possible failures in the structure. However, minimizing these causes will definitely reduce the amount and need for maintenance. With the case of a structure with exposed reinforcement, the principal reason leading to the current failures and defects in such building is potentially associated with deficiencies in the use of materials and construction quality. The corrosion of reinforcing steel bars is a critical concern which reduces significantly the life and durability of concrete structures. The corrosion of the reinforcement was expected due to shoddy construction as the concrete covers were found to vary with the lowest is 8mm.

As for the collapsing concrete wall, from the specific tests which requested by the client, the test results reveal that the structure was under design with the concrete cover and concrete strength are not adequate for the purpose of effluent tank.

In the case of vibrated floor slab, the mass and stiffness of the concrete structure need to be increased as an improvement on the vibration performance of the floors.

References

[1] Portland Cements Association, *Types and causes of concrete deterioration*, (2002), Retrieved on October 10, 2016 from http://www.cement.org/docs/default-

source/fc_concrete_technology/durability/is536-types-and-causes-of-concrete-deterioration.pdf?sfvrsn=4

[2] Portland Cements Association, *Corrosion of embedded metals*, Retrieved on October 10, 2016 from http://www.cement.org/for-concrete-books-learning/concrete-technology/durability/corrosion-of-embedded-materials

[3] S.S. Ayop and J.J. Cairns, Critical study of corrosion damaged concrete structures, Int. J. of Integrated Engineering, **5**(2), 43-50 (2013)

[4] BS 8110-1-1997, *Structural use of concrete. Code of practice for design and construction*, British Standard Institution, London (1997)

[5] BS EN 14630:2006, *Products and systems for the protection and repair of concrete structures*, Test Methods, Determination of Carbonation Depth in Hardened Concrete by the Phenolphthalein Method, British Standards Institute, London, (2006)

[6] BS EN 14629:2007, *Products and systems for the protection and repair of concrete structures*, Test Methods, Determination of Chloride Content in Hardened Concrete, British Standards Institute, London, (2007)

[7] ASTM C876-91, *Standard test method for corrosion potentials of uncoated reinforcing steel in concrete*, ASTM International, West Conshohocken, (1991)

[8] BS 8007:1987, *Code of practice for design of concrete structures for retaining aqueous liquids*, British Standards Institute, London, (1987)

[9] T.N. Tuan Chik, S. Yabi, N.A. Yusoff and M.I. Ghazali, Analysis of the effect of vibration from footfalls on office building, *Proc. of the Int. Civil and Infrastructure Engineering Conf.*, Springer, 3-15 (2014)

Seismic Fragility Curves of Industrial Buildings by Using Nonlinear Analysis

Fadzli Mohamed Nazri[1,*], *Syamimi* Tahar[1], *Siti Nur Aqilah* Saruddin[1], and *Shahiron* Shahidan[2]

[1]School of Civil Engineering, Engineering Campus, Universiti Sains Malaysia, 14300 Nibong Tebal, Penang, Malaysia
[2]Jamilus Research Center-Sustainable Construction, Faculty of Civil and Environmental Engineering, Universiti Tun Hussein Onn Malaysia, 86400, Batu Pahat, Johor, Malaysia

Abstract. This study presents the steel fragility curves and performance curves of industrial buildings of different geometries. The fragility curves were obtained for different building geometries, and the performance curves were developed based on lateral load, which is affected by the geometry of the building. Three records of far-field ground motion were used for incremental dynamic analysis (IDA), and the design lateral loads for pushover analysis (POA). All designs were based on British Standard (BS 5950); however, Eurocode 8 was preferred for seismic consideration in the analysis because BS 5950 does not specify any seismic provision. The five levels of performance stated by FEMA-273, namely, operational phase, immediate occupancy, damage control, life safety, and collapse prevention (CP) were used as main guidelines for evaluating structural performance. For POA, Model 2 had highest base shear, followed by Model 1 and Model 3, even though Model 2 has a smaller structure compared with Model 3. Meanwhile, the fragility curves showed that the probability of reaching or exceeding the CP level of Model 2 is the highest, followed by that of Models 1 and 3.

1 Introduction

Structures need to be well designed and constructed to reduce damage. Industrial structures are commonly composed of prefabricated structures or steel because of their lightweight features and the maximum span between columns must not exceed 10 m. Damage on industrialized structures is commonly caused by brace buckling [1], which can be assessed based on the performance level of drift. Drift is commonly used as a damage parameter to predict the performance of industrialized buildings.

Drift has primary effects caused by the movement of structural elements, non-structural elements, and adjacent structures [2]. Drift can be described by considering the maximum allowable damage state to identify the performance level [3]. The performance level is classified according to FEMA 273 [4] and FEMA-356 [5] guidelines. These guidelines describe four types of performance levels, namely, fully operational (damage is

* Corresponding author: cefmn@usm.my

negligible and facilities continue their operations), operational (minor damage and minor disruption, facilities continue their operations), life safe (damage is moderate to extensive, lives are slightly at risk), and near collapse (structural collapse is impeded, lives are at risk).

Fragility can be used as a tool for predicting structural and non-structural damage [6]. The fragility curve expresses the probability of a structure exceeding or reaching its damage state [7-13], and allows for estimating the structural damage level for a known ground motion index. The current study has two main objectives: (i) to examine the drift of a structure based on different industrial building designs and (ii) to develop the seismic fragility curve for industrial structures based on different geometries.

2 Structural geometry

Three different types of geometry were applied in this study. Geometry is based on the typical types of industrial buildings in Malaysia, as shown in Fig. 1. All models are 6 m in height with different numbers of bays. Table 1 presents the summary of the industrial building models in this study.

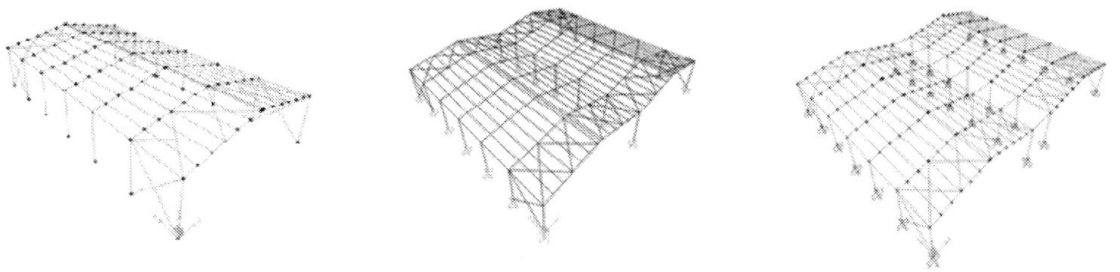

(a) Model 1 (b) Model 2 (c) Model 3

Fig. 1. 3D model with different geometry.

Table 1. Summary of industrial building models.

Frame	nb	Lb [m]	Lc [m]	Beam Section	Column Section
1	1	6	6	UB 533x210x82	UC 356x168x129
2	2	6	6	UB 406x140x46	UC 305x305x198
3	3	6	6	UB 356x171x151	UB 457x152x60

The live loads are 1 kN/m and dead loads are 2 kN/m. The models are assumed suited at soil type A, which is considered stable rock, such as geological formations with least 5 m of weaker material at the surface. Soil type A was used in this study to avoid soil–structure interaction in the analysis. SAP 2000 was used as the main tool to analyze the two types of practice-oriented methods, namely, pushover analysis (POA) and incremental dynamic analysis (IDA).

Several aspects need to be defined to perform POA, including material information, frame dimension, load patterns, load cases, frame hinge properties, and joint constraint [14]. These aspects must be assigned on the model of the existing industrial building for the analysis. All connections must be assigned correctly to avoid error. In this study, models were introduced in uniform distribution.

The selection of ground motion records is important for IDA. An appropriate number of recordings is a significant concern [15-16]. However, most codes [5, 18-19] recommend a minimum of three or seven sets of ground motion records. For this study, three sets of ground motions were used and selected from the PEER NGA website (Table 2).

Table 2. Selected ground motions records

Name Record	Earthquake Location	Year	Magnitude
Chichi	Chichi, Taiwan	1975	7.8
Kobe	Kobe, Japan	1975	6.8
Tehri	Tehri, India	1979	6.8

Far-field (FF) ground motions were chosen based on the following criteria: (i) Joyner–Boore distance is more than 20 km and (ii) magnitude ranges from 7 to 8. These ground motions were then scaled to elastic response spectra. The elastic response spectra were developed from 0 g to 6 g (an increment of 2 g).

To develop a fragility curve, the IDA curve with the displacement or drift of a structure must be obtained. The fragility curve can be plotted using the suggested limit condition in PGA–drift format form [20]. Equation (1), as suggested by Ibrahim and El-Shami [6] was used in this study to develop the fragility curve.

$$P[\mathrm{D/PGA}]=\Phi((\ln(\mathrm{PGA})-\mu)/\sigma) \qquad (1)$$

where ;

D = damage
PGA = Peak Ground Acceleration
Φ = standard normal cumulative distribution
μ = mean
σ = standard deviation of the natural logarithm of PGA

3 Results and discussion

The results obtained from POA and IDA was evaluated by comparing them with the limit state based on PBSD. Five limit states were used, namely, operational phase (OP) with 0.5% drift, immediate occupancy (IO) with 1.0% drift, damage control (DC) with 1.5% drift, life safety (LS) with 2.0% drift, and collapse prevention (CP) with more than 2.5% drift.

3.1 Seismic performance

A base shear versus drift graph can be plotted from the POA results. This performance curve can describe the maximum allowable damage condition for a known seismic hazard. Fig. 2 shows the performance curves for the three models.

Based on Fig. 3, at 0.5% drift, the base shears of Model 2 and 3 were similar, but a small difference was found between the base shears of Model 1 and Model 2 (i.e., 14%). At 1.1% drift, the performance curves of Model 2 and 3 start to diverge. At the CP level, a 16% difference between Model 1 and Model 2 was observed at 2.5% drift. Meanwhile, the difference between Model 2 and Model 3 at 2.5% drift was approximately 21%. Although Model 3 is larger, Model 2 reaches a base shear that is greater than that of Model 3. This difference is probably because the structure of Model 3 is composed of smaller members. The column on Model 3 cannot resist the lateral load exerted on the structure, resulting in a low ability to resist the lateral load applied.

DOI: 10.1051/matecconf/201710302017

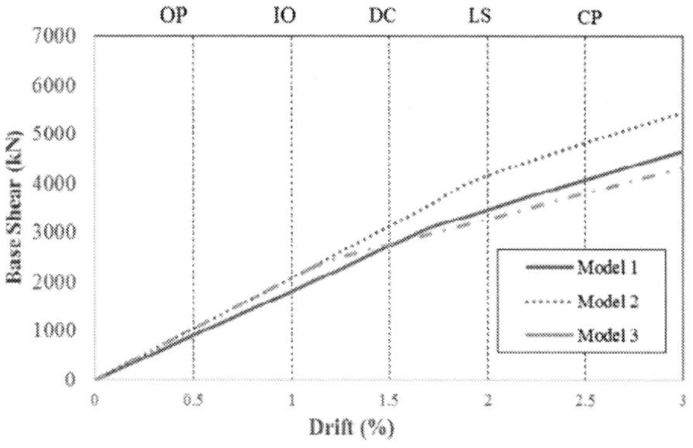

Fig. 2. Seismic performance.

3.2 Hinge deformation

Figure 3 shows the hinge deformation for all models that based on the IDA. The most common failure that occurs in Model 1 and Model 2 is at the column member. For Model 3, failure occurs at the truss and column section. From the hinge deformation result, Model 3 has the worst damage because Model 3 reaches Level 7. The most severe damage occurs at the column, which may be because the structure uses small sections for columns (i.e., UB547×152×60) compared with other models using UC section. Model 2 has a greater damage level than Model 1 because it reached level 5, which is the collapse level. Collapse may occur because of the greater size of the structure of Model 2 than Model 1. In addition, the damage may be due to connection failures, given that Model 2 has no detailed design.

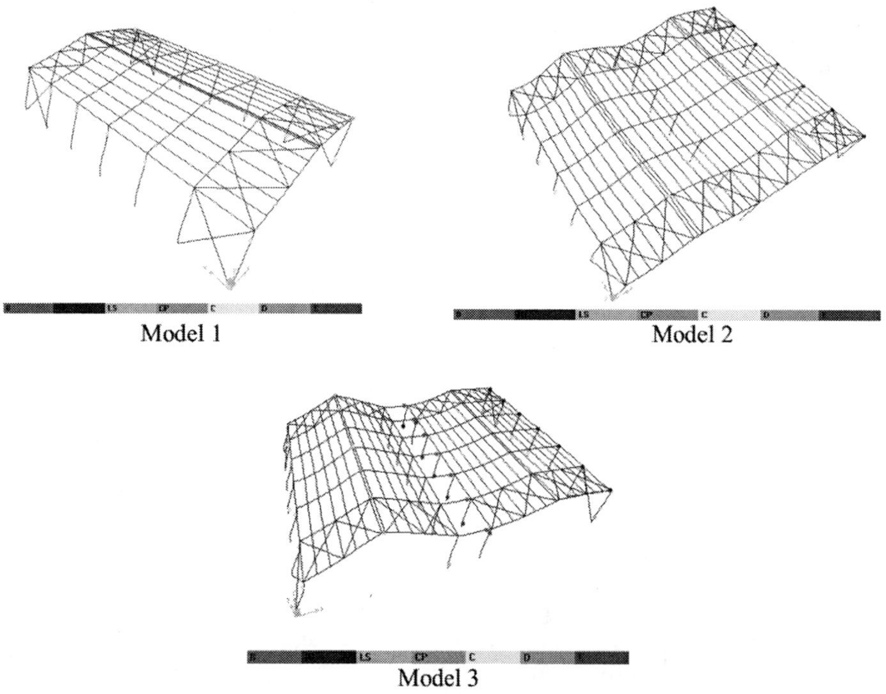

Fig. 3. Hinge deformation

3.3 IDA

IDA was conducted under each ground motion and the scaling PGA increased every 2 g until reaching 6 g. Normally, the analysis would be stopped when PGA reaches 6 g because structures are commonly unstable at 6 g. Every structure has different behavior toward ground motion, and thus the ground motion records have different patterns of IDA curves. An IDA curve shows the relationship between drift and PGA. In this study, drift was only emphasized until 3% to develop the seismic fragility curve; 3% is the maximum drift limit according to FEMA-273. The mean drift was calculated for every PGA to present the average of the IDA curves and was compared with the limit state, as shown in Fig. 4.

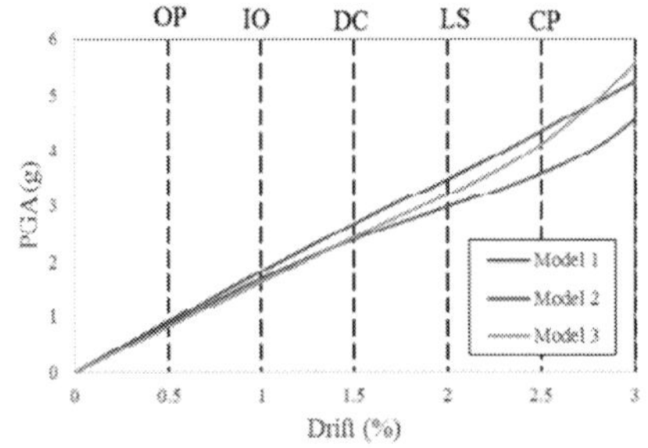

Fig. 4. Mean IDA.

Based on the observation at the CP levels, Model 2 has a better performance than Model 1 and 3, because the members of Model 2 are strong enough to for the ground motion. The larger structural members of Model 2 endow stability when ground motion occurs. The mean difference between Model 2 and 3 is only about 5%. Although Model 3 is larger than Model 2, Model 3 cannot resist the load applied on the structure because the compression members of Model 3 are smaller (UB section). If Model 3 were designed with a UC section at the compression member, then Model 3 would have inferior performance to Model 2. In terms of structural size (whole), Model 3 is the largest, followed by Model 2 and Model 1. Among the models, Model 1 has the lowest performance. The mean difference between Model 1 and Model 2 is about 17%., which might be due to the design and size of the structure. Thus, Model 1 will collapse before Model 2 and Model 3.

3.4 Fragility curve

Two parameters, namely, mean and standard deviation, are needed to develop a fragility curve. The mean and standard deviation of PGA were calculated for every point in the IDA curve, which were across the vertical gridlines of limit state at drifts. The calculated parameters are provided in Table 3. The fragility curves were developed using (1). All seismic fragilities were represented in the damage probability curves, as shown in Fig. 5.

Table 3. Parameters of log-normaldistribution of the models.

Model	**OP**		**IO**		**DC**		**LS**		**CP**	
	μ	σ	μ	σ	μ	σ	μ	σ	μ	σ
1	-0.009	0.700	0.677	0.688	0.689	0.201	0.938	0.253	1.136	0.334
2	-1.332	1.372	0.673	0.871	0.6109	0.0745	0.870	0.0304	1.201	0.131
3	-0.041	0.102	0.607	0.91	0.837	0.361	1.256	0.110	1.512	0.121

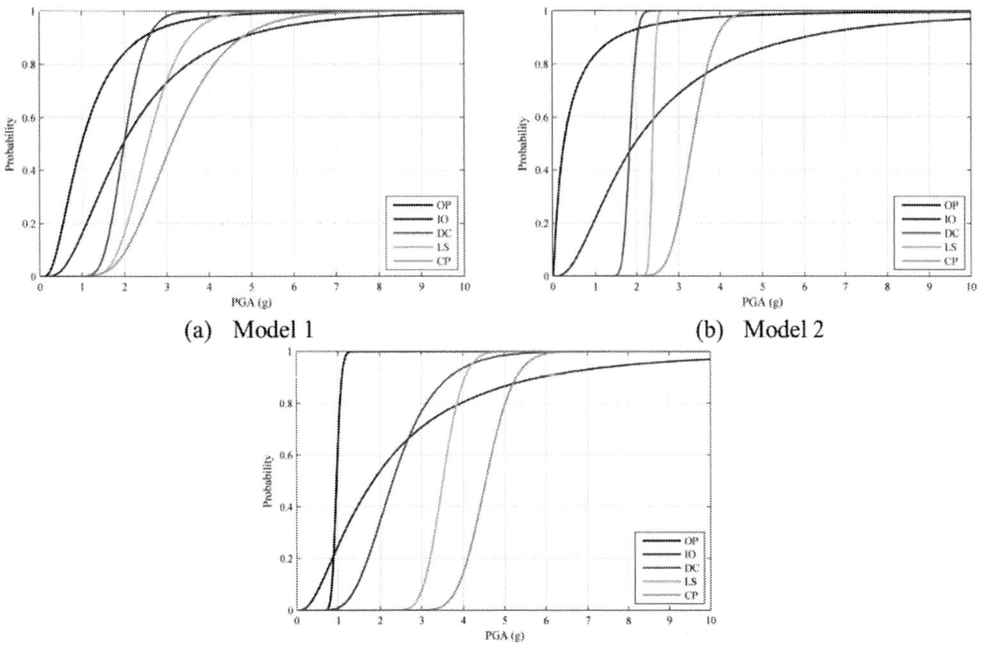

(a) Model 1 (b) Model 2

(c) Model 3

Fig. 5. Seismic fragility curves

As shown in Fig. 4(a), at FF ground motion of 1g, the probabilities of reaching or exceeding the OP and IO levels are 79% and 26%, respectively. At a ground motion of 5 g, the probabilities of reaching or exceeding the CP and IO levels are approximately 92% and 90%, respectively, and the probabilities of exceeding or reaching the OP, LS, and DC levels are almost 100%. The following observations were noticed for Model 2 (Fig. 4(b)): when a PGA of 1 g was exerted, the probabilities of reaching OP and IO levels were 84% and 22%, respectively. When PGA was 5 g, the probability of reaching the DC, LS, and CP levels were 100%. The probabilities of exceeding or reaching the OP and IO levels were 98% and 90%, respectively. Fig 4(c) shows the fragility curve of Model 3. At a PGA of 1 g, the probabilities of exceeding or reaching the OP and IO levels are 60% and 25%, respectively. At a strong ground motion of 5 g, the probabilities of exceeding the OP, IO, DC, LS, and CP levels are 100%, 86%, 98%, 100%, and 78%, respectively.

4 Conclusions

This study develops the fragility curve of industrial buildings based on the practice-oriented methods, namely, POA and IDA. The models were designed based on BS 5950; however, they were analyzed according to Eurocode 8 because BS 5950 does not specify any seismic provision. For the analysis, SAP 2000 software was used as the main tool for analyzing the structure under static load (lateral load) and dynamic load (ground motion). Three sets of ground motion records around Asia were chosen, which were classified as FF. The drift of the structure from POA and IDA were investigated. The following conclusions can be drawn from this study.

i) The performance curves were developed using SAP 2000 software. Lateral load exerted on the structure varies depending on the structure. The structure with a greater mass can resist higher base forces. However, the drift of the structure also depends on the design of the industrial building. For instance, although Model 3 has a larger surface area than Model 2, Model 3 collapses 21% faster than Model 2. Although Model 3 is larger,

Model 2 reaches a base shear that is greater than Model 3. This result is most probably because Model 3 consists of smaller structural members, and the strength of Model cannot resist the load applied to the structure. The different analyses provided different drifts, even for the same model.

ii) The fragility curves were developed for different types of geometry. The fragility curves were analyzed based on five limit states, namely, OP, IO, DC, LS, and CP. By considering the limit states, designers can predict the damage level and possible location of hinge deformation. Comparison of the fragility curves of the three models show that Model 2 has the highest probability of reaching OP and CP levels.

This research was supported by Ministry of Higher Education under the Fundamental Research Grant Scheme (6071321).

References

[1] K. Ozakgul, O. Caglayan and O. Tezer, Investigation of buckled brace system of an existing industrial building, Engineering Failure Analysis, **18**, 455-463, (2011)

[2] G.R. Searer, and S. A. Freeman, Design drift requirement for long period structures, *Proc. of the 13th World Conf. on Earthquake Engineering*, **3292**, (2004)

[3] A. Hokmabadi, B. Fatahi, and B. Samali, Recording inter-storey drifts of structures in time-history approach for seismic design of building frames, Australian J. of Structural Engineering, **13**, 175, (2012)

[4] FEMA-273, *273: NEHRP Guidelines for the seismic rehabilitation of buildings*, Federal Emergency Management Agency, Washington, D.C., United States, (1997)

[5] FEMA-356, *Prestandard and commentary for the seismic rehabilitation of buildings: FEMA-356*, Federal Emergency Management Agency, Washington, D.C., United States, (2000)

[6] Y.E. Ibrahim, and M. M. El-Shami, Seismic fragility curve for mid-rise reinforced concrete frames in kingdom of Saudi Arabia, The IES Journal Part A: Civil and Structural Engineering, **4**(4), 213-223, (2011)

[7] A. Réveillère, P. Gehl, D. Seyedi, and H. Modaressi, Development of seismic fragility curves for mainshock-damaged reinforced-concrete structures, in *Proc. of the 15th World Conf. on Earthquake Engineering*, Lisboa, Portugal, 999, (2012)

[8] E.N. Farsangi, F.H. Rezvani, M. Talebi, and S. A. Hashemi, Seismic risk analysis of steel-mrfs by means of fragility curves in high seismic zones, Advances in Structural Engineering, **17**, 1227-1240, (2014)

[9] C.R. Kumar, K.B. Narayan, and D.V. Reddy, Probabilstic seismic risk evaluation of rc buildings, Int. J. of Research in Engineering and Technology, **3**, 484-495, (2014)

[10] J. Lee, S. Han, and J. Kim, Seismic performance evaluation of apartment buildings with central core, Int. J. of High-Rise Buildings, **3**(1), 9-19, (2014)

[11] V. Silva, H. Crowley, H. Varum, R. Pinho, and R. Sousa, Evaluation of analytical methodologies used to derive vulnerability functions, Earthquake Engineering & Structural Dynamics, **43**, 181-204, (2014)

[12] R. Ahmadi, R. Mulyani, F.M. Nazri, K. Pilakoutas and I. Hajirasouliha, Seismic vulnerability assessment of an industrial building in Peninsular Malaysia, *5th Brunei Int. Conf. on Engineering and Technology (BICET 2014)*, Bandar Seri Begawan, 1-6, (2014)

[13] C. Casotto, V. Silva, H. Crowley, R. Nascimbene, and R. Pinho, Seismic fragility of Italian RC precast industrial structures. Engineering Structures, **94**, 122-136, (2015)

[14] F.M. Nazri, and Y.K. Pang, Seismic performance of moment resisting steel frame subjected to earthquake excitations, Frontiers of Structural and Civil Engineering, **8**, 19-25, (2014)
[15] B.Ö. Ay, and S. Akkar, A procedure on ground motion selection and scaling for nonlinear response of simple structural systems, Earthquake Engineering and Structural Dynamics, **41**, 1693-1707, (2012)
[16] L.H. Najafi, and M. Tehranizadeh, Ground motion selection and scaling in practice. Periodica Polytechnica Civil Engineering, **59**(2), 233 – 248, (2015)
[17] IBC, International Code Council Inc., International Building Code, Falls Church, VA, USA (2000)
[18] Eurocode 8, *Design provisions for earthquake resistance of structures: Part 1-1, General rules - seismic actions and general requirements for structures, BS EN: 1998-1-1: 2004*, British Standards Institution, London, United Kingdom, (2014)
[19] Q. Xue, C.W. Wu, C.C. Chen, and K.C. Chen, The draft code for performance-based seismic design of buildings in Taiwan, Engineering Structures, **30**, 1535-1547, (2008)

Vortex-Induced Vibration Tests of a Marine Growth Wrapped Cylinder at Subcritical Reynolds Number

V. J. Kurian[1,*], *A. M.* Al-Yacouby[1], and *A. R. M.* Khir[1]

[1] Department of Civil and Environmental Engineering, Universiti Teknologi PETRONAS, 32610 Bandar Seri Iskandar, Perak, Malaysia

Abstract. Vortex Induced Vibrations (VIV) may cause great damage to deep water risers. Estimation of accurate hydrodynamic coefficients and response amplitudes for fouled tubular cylinders subjected to VIVs is a complex task. This paper presents the results of an extensive experimental investigation on in-line and cross-flow forces acting on cylinders wrapped with marine growth, subjected to current at Subcritical Reynolds Number. The drag and lift force coefficients have been determined through the use of the Fast Fourier Analysis methods. The different tests were conducted in the offshore engineering laboratory at Universiti Teknologi PETRONAS (UTP), Malaysia. In this study, a cylinder with outer diameter Do = 27 mm, fixed at top as cantilever beam was used. The in-line and cross-flow forces were measured using VIV Force Totaller (VIVFT). VIVFT is a two degree of freedom (2DOF) forces sensor developed by UTP to measure the VIV forces. The tests were conducted for current velocity varied between 0.118 to 0.59 m/s. The test results suggest that the cylinder wrapped with marine growth has shown an overall increase in drag and inertia coefficients as well as on response amplitudes.

1 Introduction

Accurate estimation of VIVs of rough circular cylinder is one of the most difficult task in offshore engineering, due to the complex and non-linear interaction of rough cylinder with the surrounding fluid. VIV of slender cylinder is of practical importance to many fields of engineering. The forces acting on the cylinders experience periodic change due to the vortex shedding. Vibrations resulting from the vortex shedding in the wake of a cylinder exposed to flow can create fatigue damage in production risers used by the offshore oil and gas industry. VIV phenomenon is a function of various parameters, the two main parameters are the Reynolds number and the Strouhal Number, *St.* The Reynolds number is defined as $Re = UD/v$ where $v = 10^{-6}$ m^2/s is the kinematic viscosity of water at room

* Corresponding author: kurian_john@petronas.com.my

temperature and the Strouhal number is defined as $St=f_s D/U$, where U is the fluid velocity, D is the diameter of the cylinder and f_s is the shedding frequency. The fluid-structure coupling is characterized by the reduced velocity $U_r=U/ f_sD$. For instance Feng [1-6] investigated the VIV of circular cylinders. The pioneering experimental investigation conducted to assess the effects of surface roughness on the pressure distribution around a bluf body was conducted by Fage and Warsap [7].

Further, Hover *et al.* [8] conducted experimental study to investigate the effects of thin wires attached on the outer surface and parallel to the axis of a smooth circular cylinder in a steady flow. The reader can also refer to [9-10] for further study on VIVs. However, the effects of marine growth and surface roughness on VIV considering the Malaysian water is not existing. Therefore, the objective of this paper is to assess the effect of marine growth on VIV forces, response amplitudes and the most important flow and structural characteristics governing VIVs, taking into account the marine growth characteristics for the Malaysia offshore locations. According to PTS [11] the marine growth layer thickness to be considered varies from 25 to 127 mm, while the corresponding surface roughness varies from 6.25 to 64 mm, depending on the geographical location and the water depth. Dry unit weight of marine growth recommended for the Malaysian water is 10 kN/m^3 [11].

2 Research methodology

2.1 Details of the wave tank

The VIV model test was conducted in the wave basin of Universiti Teknologi PETRONAS. The wave basin where the tests were carried out is 20 m length × 10 m wide × 1m water depth (d). The wave tank is equipped with 16 wave paddles capable of generating regular, random and multidirectional waves. Besides, the wave tank is equipped with pumps capable of generating currents. However, as the maximum velocity of current in the wave tank was limited to 0.12 m/s at a maximum water depth of 1m, a test section was constructed inside the wave tank to increase the current velocity to meet the subcritical flow regimes requirements. Fig.1a shows a close view of the test section, while Fig. 1b depicts a sectional view of the test section. The response amplitudes of the cylinder due to VIV were measured using OptiTrack motion sensing system, while the current velocity was measured using VECTRINO velocity meter. Due to the limitations of testing facility, a scale factor of 1:55 was adopted and the test results are presented in terms of prototype for engineering applications.

The variation of marine growth thickness with water depth was varied as 3:2:1. The profile of marine growth adopted in this study represents the general pattern of marine growth accumulation on circular cylinders adopted for Malaysian water applications. The surface roughness of marine growth was achieved by wrapping the wet length of the cylinder by sand paper with surface roughness, k = 1mm. The model is connected to a two degree of freedom (2DOF) VIV force sensor specially designed and developed to capture the total VIV forces. The recorded forces have been analysed and the corresponding drag and lift coefficients have been determined using the well-known Morison equation components as described in Section 3.1.

Details of model set up in the wave tank and the VIV force sensor are presented in Fig 2. The VIV experiments were carried out in a test section of 3 m long × 1.5 m. Totally 270 different tests were conducted in the wave tanks. The environmental conditions generated in the wave tank corresponding to the deferent test conditions denoted as T1 to T5 are presented in Table 1.

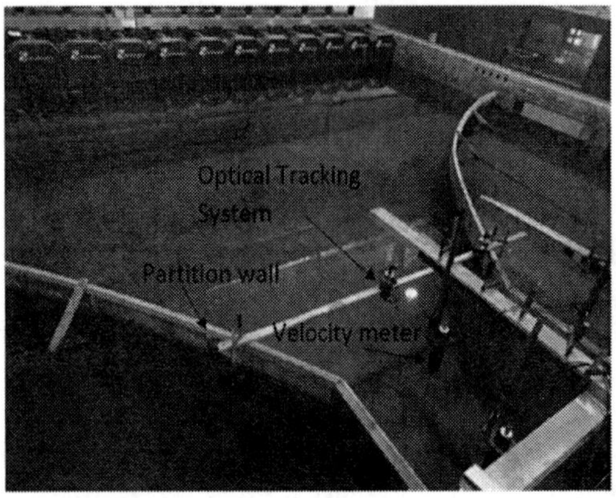

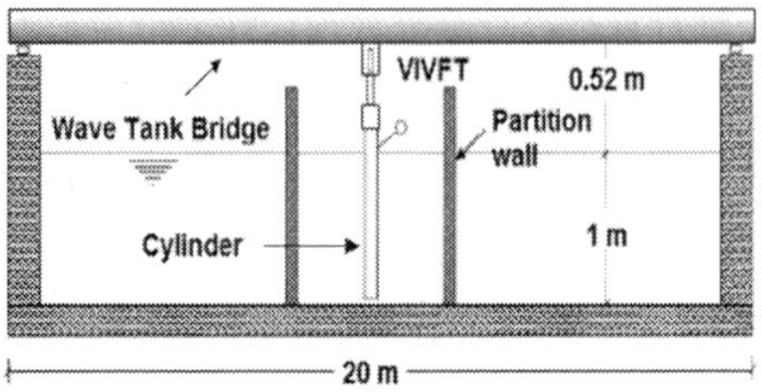

Fig. 1. a) A close view of the wave basin and b) a sectional view of the test section

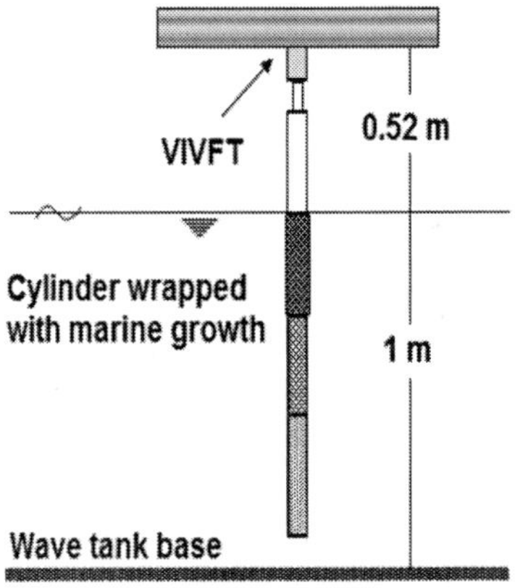

a) Model set up in the wave tank

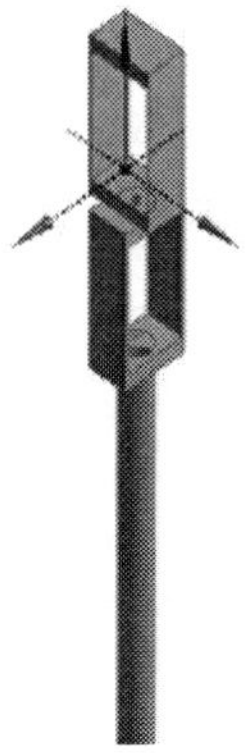

b) Perspective view of a cylinder fitted with VIV force sensor

Fig. 2. Details of model set up and VIV force sensor

Table 1. Variation of current velocity generated in the wave tank

Test Condition	Current Velocity (m/s)
T1	0.12
T2	0.22
T3	0.32
T3	0.41
T4	0.49
T5	0.59

3 Results and discussions

3.1 Drag and lift coefficients

Drag coefficient is a non-dimensional form of the drag force which can be determined as $C_D=F_D/\rho U|U|$ where F_D is the drag force, ρ is the fluid density and U is the flow velocity. Fig. 3a shows the variation of drag coefficient (C_D) with *Re* for the smooth as well as for the marine growth wrapped cylinders. The values of C_D depicted in Fig. 3a shows a slight rise on the drag coefficients as the Re Number increases. The C_D coefficient for the smooth cylinder varied between 0.073 to 0.111, while the C_D values for the rough cylinder varied between 0.370- 0.390. This shows that marine growth increased the surface roughness, the total outer diameter and the added mass, which consequently resulted in comparatively higher drag coefficients. Further, the lift coefficient is an important parameter in VIV of circular cylinders and can be estimated as $C_L=F_L/\rho U|U|$,where F_L is the transverse force. Fig. 3b shows the variation of C_L with Re. The range of C_L for the smooth cylinder varied between 0.073-0.110 while the fouled cylinder has recorded lift coefficients ranging between 0.366-0.385, which indicates that the surface roughness has significantly influenced the values of lift coefficients. The frequency of the lift force on a circular cylinder is consistent with the vortex shedding frequency, and the frequency of the drag force is double that of vortex shedding [9]. The range of drag and lift coefficients are in good agreement with the findings published earlier by [12-13].

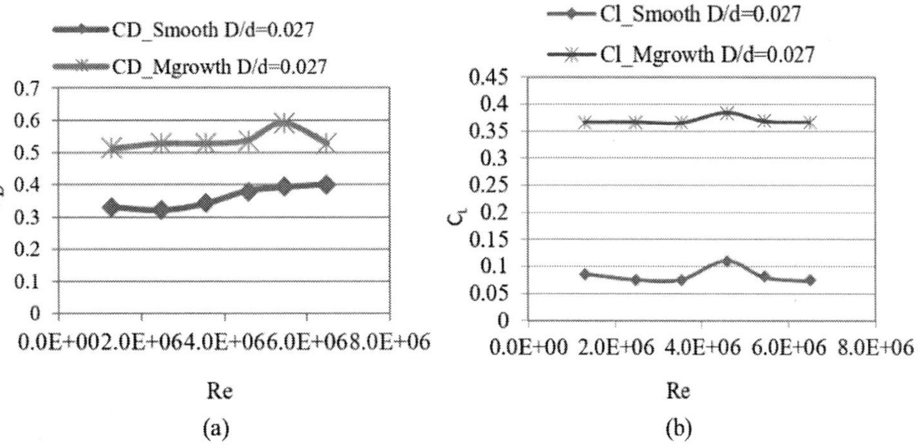

Fig. 3. Variation of drag (a) and lift coefficients with *Re* Number for smooth and rough cylinders with $D/d = 0.027$

3.2 Response Amplitudes Ax/D and Ay/D

Accurate estimation of the fatigue life of production riser experiencing vortex induced vibrations depends upon accurate estimation of the response amplitude and frequencies. However, accurate estimation of the response amplitudes and frequencies are in turn dependent upon several parameters such as surface roughness, hydrodynamic damping etc. Thus, in this section, the effect of marine growths on VIV response amplitude is discussed. The non-dimensional displacement amplitude of the cylinders along the flow directions (*Ax/D*) are depicted in Fig. 4a. The maximum response amplitude for the smooth and the marine growth wrapped cylinder have experienced the maximum response amplitude in the reduced velocity range between 4-7. The plot shows that the rough cylinder has experienced higher displacement although the ultimate response amplitude is in the range of 0.26D. Similarly, Fig.4b depicts the variation of cross-flow displacements of the cylinders (*Ay/D*). From the graph, one can observe that the response amplitude curve showed the initial, intermediate and upper branches which are typically observed for low mass-damping systems. Generally, the fouled cylinder has comparatively experienced higher responses amplitude.

3.3 Response Amplitudes Ax/D and Ay/D

Accurate estimation of the fatigue life of production riser experiencing vortex induced vibrations depends upon accurate estimation of the response amplitude and frequencies. However, accurate estimation of the response amplitudes and frequencies are in turn dependent upon several parameters such as surface roughness, hydrodynamic damping etc. Thus, in this section, the effect of marine growths on VIV response amplitude is discussed. The non-dimensional displacement amplitude of the cylinders along the flow directions (*Ax/D*) are depicted in Fig. 4a. The maximum response amplitude for the smooth and the marine growth wrapped cylinder have experienced the maximum response amplitude in the reduced velocity range between 4-7. The plot shows that the rough cylinder has experienced higher displacement although the ultimate response amplitude is in the range of 0.26D. Similarly, Fig.4b depicts the variation of cross-flow displacements of the cylinders (*Ay/D*). From the graph, one can observe that the response amplitude curve showed the initial, intermediate and upper branches which are typically observed for low mass-damping systems. Generally, the fouled cylinder has comparatively experienced higher responses amplitude.

The maximum response in the cross-flow direction was recorded as *Ay* =1.6D, observed in the reduced velocity interval ranging between 4 -7. The findings of this experimental study can be used to develop empirical models for prediction of drag and lift coefficients and responses for rough cylinders subjected to VIVs.

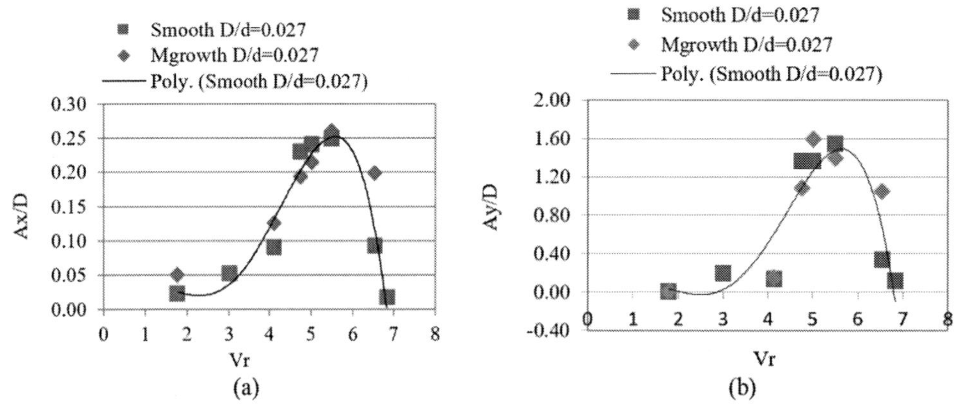

(a) (b)

Fig. 4. Variation of drag (a) and lift coefficients with *Re* for smooth and rough cylinders with *D/d* = 0.027

4 Conclusions

The effect of marine growth on VIV forces and the corresponding response amplitudes have been determined and investigated experimentally in the wave tank using a circular cylinder with relative outer diameter *D/d* = 0.027 mm. From the model tests, the following findings can be concluded:

i) The C_D coefficient for the smooth cylinder varied between 0.073-0.111, while the drag coefficients for the rough cylinder varied between 0.370-0.390.
ii) The range of C_L for the smooth cylinder varied between 0.073-0.110 while the fouled cylinder has recorded lift coefficients ranging between 0.366-0.385, which indicates that the surface roughness has significantly influenced the values of lift coefficients.
iii) Generally, the fouled cylinder has comparatively experienced higher responses amplitude.

The authors would like to gratefully acknowledge their gratitude to Universiti Teknologi PETRONAS for support and encouragement.

References

[1] C. Feng, *The measurement of vortex induced effects in flow past stationary and oscillating circular and D-section cylinders,* Master Thesis, University of British Columbia, Vancouver, Canada (1968)
[2] P.W. Bearman, Vortex shedding from oscillating bluff bodies, Annual Review of Fluid Mechanics, **16**(1),195-222, (1984)
[3] A. Khalak and C. Williamson, Motions, forces and mode transitions in vortex-induced vibrations at low mass-damping, J. of Fluids and Structures, **13**(7), 813-851, (1999)
[4] A. Khalak and C. Williamson, Dynamics of a hydroelastic cylinder with very low mass and damping, J. of Fluids and Structures, **10**(5), 455-472,(1996)
[5] C. Williamson and R. Govardhan, Vortex-induced vibrations, Annu. Rev. Fluid Mech., **36**, 413-455, (2004)
[6] R.D Blevins and C.S. Coughran, Experimental investigation of vortex-induced vibration in one and two dimensions with variable mass, damping, and Reynolds number, J. of Fluids Engineering, **131**(10), 1-7, (2009)
[7] A. Fage and J. Warsap, *The effects of turbulence and surface roughness on the drag of a circular cylinder,* Technical Report R&M 1283, Aeronautical Research Council, London, United Kingdom (1930)
[8] F. Hover, H. Tvedt and M. Triantafyllou, Vortex-induced vibrations of a cylinder with tripping wires, J. of Fluid Mechanics, **448**, 175-195, (2001)
[9] M. Zhao, K. Kaja, Y. Xiang and G. Yan, Vortex-induced vibration (VIV) of a circular cylinder in combined steady and oscillatory flow, Ocean Engineering, **73**, 83-95, (2013)
[10] Y. Gao, D.S. Tan, B. Zhang and S.K. Tan, Experimental study on orbital response and flow behavior behind a freely suspended cylinder, Ocean Engineering, **108**, 439-448. (2015)
[11] PETRONAS Technical Standards (PTS), *Design of fixed offshore structures*, PETRONAS, Kuala Lumpur, Malaysia (2014)
[12] B.M. Sumer and J. Fredsøe, *Hydrodynamics around cylindrical structures*, Advanced Series on Ocean Engineering, World Scientific, (2010)
[13] M. Zdravkovich, Flow around circular cylinders; Vol. 1. Fundamentals, J. of Fluid Mechanics, **350,** 377-378, (2003)

MATEC Web of Conferences **103**, 02019 (2017) DOI: 10.1051/matecconf/201710302019
ISCEE 2016

Vulnerability Assessment of Building Frames Subjected to Progressive Collapse Caused by Earthquake

Fadzli Mohamed Nazri[1,*], *Mohamad Khairi* Adha Zolkifle[1], *Hazrina* Mansor[2], and *Shahiron* Shahidan[3]

[1]School of Civil Engineering, Engineering Campus, Universiti Sains Malaysia, 14300 Nibong Tebal, Penang, Malaysia
[2]Faculty of Civil Engineering, UiTM Shah Alam, 40450 Selangor, Malaysia
[3]Jamilus Research Center, Faculty of Civil and Environmental Engineering, Universiti Tun Hussein Onn Malaysia, 86400 Parit Raja, Johor, Malaysia

Abstract. Progressive collapse is an initial local failure of the structural component and leading to the additional collapse of the building frames. This study investigated the vulnerability of four- and six-storey moment resisting concrete frame (MRCF) buildings subjected to progressive collapse. The four- and six-storey MRCF buildings were designed based on British Standard (BS) and Eurocode (EC). The differences between these two codes were investigated. Nonlinear static analysis, which is also known as pushover analysis (POA), and nonlinear dynamic analysis or incremental dynamic analysis (IDA), were performed for each model to obtain capacity curve and explore vulnerability measures. IDA was conducted using a sample of ground motion from an earthquake that occurred in Ranau, Sabah in 2015. The four-storey building was more vulnerable than the six-storey building.

1 Introduction

Progressive collapse is described as building collapse caused by the loss or failure of a structural load-bearing member because of load hazards. Localized failure facilitates load redistribution to the adjacent member, which then initiates partial or total progressive collapse of a building. Mohamed [1] summarized several events that initiate progressive collapse, namely, gas explosion, blast, foundation failure, accidental impact, fire, and seismic force. According to Tavakoli [2], man-made or natural hazards can induce progressive collapse. Earthquakes may also cause progressive collapse caused by partial or complete failure of critical elements.

According to Byfield [3], the UK Building Regulations [4–5] was established after the incident in Ronan Point apartment in 1968. These regulations ensure a minimum level of structural integrity and modify the American and Canadian codes of practice [6–7]. The

* Corresponding author: cefmn@usm.my

progressive collapse guidelines of the US General Administration [8] and US Department of Defense [9] were introduced after the Oklahoma City bombing and Twin Tower attack in the year of 1995 and 2001. Sasani [10] used SAP2000 to evaluate the three-dimensional model with Bernoulli beam elements; plastic hinges were localized, which are used to model material nonlinearity in beams, columns, and slabs. The approach employs a moment plastic hinge that only accounts for flexural plastic deformation; the plastic hinge is used to model localized plastic hinges. The hinge does not affect the response and structure of the element before yielding.

The progressive collapse of building structures is a complicated mechanical behavior of structural systems with large deformation. However, the lack of experimental technique and numerical simulation for structural systems limits research on this area. This study focuses on parameters that potentially affect progressive collapse, namely, plastic hinges and drifts. Plastic hinges are formed through bending and shear cracking. Building drift is caused by ground motion, such as earthquake. Such a motion could also lead to the deformation and failure of structural member. This study mainly aims (i) to identify the damage measures of building frames that affect the potential of progressive collapse and (ii) to compare the vulnerability of frame designs of buildings based on the British Standard (BS) and Eurocode (EC).

2 Methodology

Pushover analysis (POA) and incremental dynamic analysis (IDA) were conducted to identify the vulnerability of buildings. SAP 2000 software was used as a main tool in order to perform both static an dynamic analyses. The ground motion records from the Ranau earthquake was selected as the main data for the dynamic analysis. Two 3D building frame models of four- and six-storey buildings were analyzed. Two locations of column removal were adopted based on the guideline of the US General Service Administration (GSA) [8]. This analysis involved the following steps. One of the columns at the specific location (Fig. 1) were removed after the structure was designed in a 3D model. Cases 1 and 2 were then subjected to nonlinear analysis. Two cases of column removal were considered in accordance with Section 3.2.9 in the GSA guidelines [8]. Case 1 involves the removal of ground floor internal column at the middle of the building. Case 2 corresponds to the locations of the column removed from the ground floor of the internal middle and external middle sites as shown below.

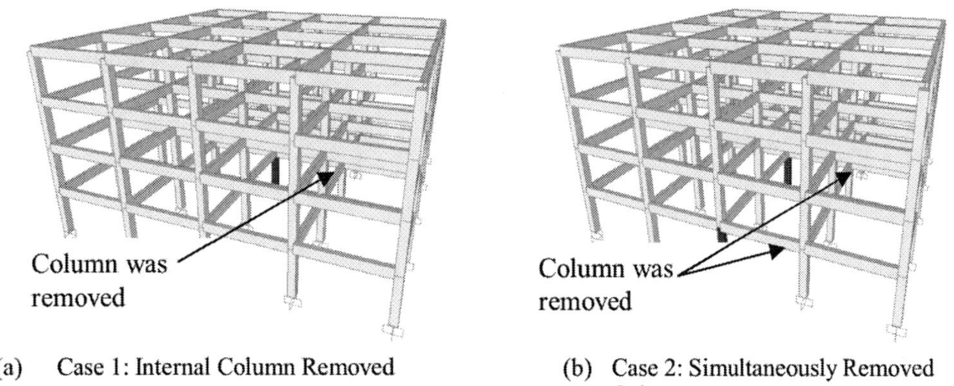

(a) Case 1: Internal Column Removed (b) Case 2: Simultaneously Removed Column

Fig. 1. Column Removal

3 Result and discussions

POA was conducted by applying two load cases to the structure, namely, gravity loads and lateral loads. Gravity load included dead load (DL) and live load (LL), which were applied on the beams. Lateral loads were applied to the height of the building at each joint in a triangular distribution pattern. Four moment-resisting concrete frame (MRCF) models composed of two different storey heights were analyzed. The differences between the two design codes were compared based on force demand.

Fig. 2 (a) shows the maximum base shear of the four-storey BS-based design at approximately 4600 kN and 1.5% drift. The maximum base shear of the six-storey BS was approximately 4100 kN at 1.2% drift. The four-storey MRCF exhibited a higher base force demand than that of the six-storey MRCF. The maximum base force of the four-storey building in the EC-based design was 4300 kN at 1.5% drift, which was higher than the maximum base force of 3900 kN of the six-storey building at 1.2% drift.

BS-based design structure showed a higher base force demand than EC-based design. A four-storey model displayed the strong potential for progressive collapse of both design codes because the base force of the four-storey building is higher than that of the six-storey building. The columns in Case 2 [Fig. 2 (b)] were simultaneously removed. Case 2 shows that the four-storey building with BS-based design exerted a maximum base force of approximately 4200 kN at 1.5% drift. The maximum base force of the six-storey BS was about 3700 kN at 1% drift. By contrast, a slight decrease in base force was observed in the four-storey building designed using EC similar to that in Case 1. The maximum base force of the four-storey building designed using EC was about 3900 kN at 1.5% drift. The maximum base force of the six-storey building designed using EC was 3500 kN at 1% drift.

The comparison of column removal showed that Case 2, which involved simultaneous removal of columns, had a maximum base force lower than that of Case 1, wherein only one internal column was removed. Given the presence of column failure, the load supported by the failed column was transferred to the adjacent member that induced progressive collapse. The role of the structural member of the four-storey building in absorbing load redistribution was lower than that of the six-storey building. Thus, increase in base force rendered the low-rise building susceptible to progressive collapse. According to Tavakoli and Alashti [2], the role of additional elements in resisting progressive collapse increased as the number of storeys increased. This observation was explained by the fact that the weight of structural elements derived from column removal was transferred to other elements. Hoseini [11] also suggested that weight transfer increases pressure to the structure.

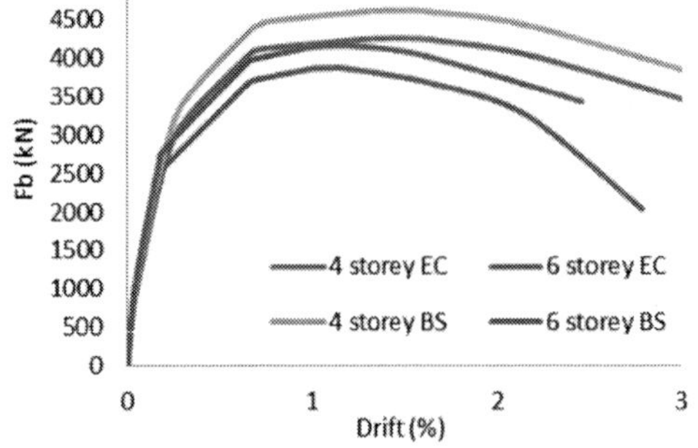

(a) Case 1: Internal column removal

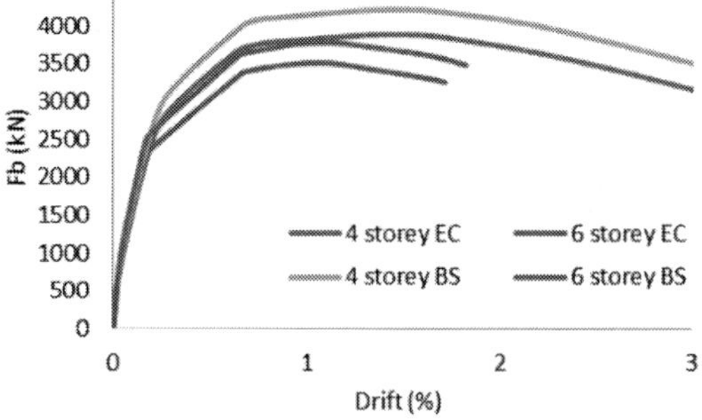

(b) Case 2: Simultaneous column removal

Fig. 2. POA result

IDA was conducted for the four-storey model with Case 2 column removal, which was the worst case according to the POA. Fig. 3 shows the result of IDA analysis of the four-storey building frame subjected to seismic loading. The sample was derived from the ground motion recorded in Kota Kinabalu from the earthquake in Ranau, Sabah. The graph shows that EC-based design recorded a PGA of 7.0 m/s^2 at 3% drift. Compared with the PGA of 6.0 m/s^2 under the BS-based design, only 1.0 m/s^2 was achieved under the EC-based design at 3% drift. The plot for the EC-based design demonstrated a straight upward trend that reached to 10 m/s^2 PGA at about 4.2% drift. PGA for the BS-based design was 7.0 m/s^2 at 4% drift. PGA higher than that of the BS-based design must be attained to displace the building constructed under EC-based design. This requirement may be attributed to the notion that the design procedure considers seismic loading that acts on the building designed based on EC. The difference between the two design codes was about 1.0 m/s^2 PGA. BS-based design showed low performance under seismic loading because of lack of recommended guidelines for seismic design. The stiffness of the structural member was low because of the large displacement under low PGA.

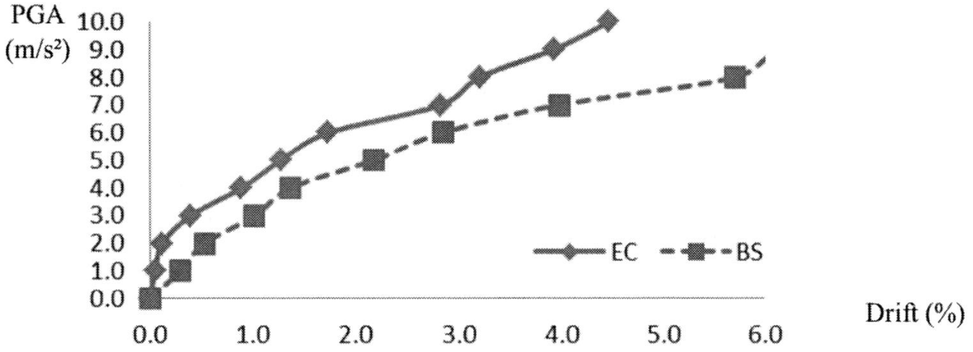

Fig. 3. IDA of the four-storey MRCF for the case of simultaneous column removal

4 Conclusions

This study investigated the behavior of MCRFs and compared the performances of buildings subjected to progressive collapse constructed based on BS and EC designs. Column removal simulated progressive collapse as recommended by GSA guidelines (2013). POA and IDA were performed using SAP2000. The following conclusions were drawn:

i) POA showed that the four-storey building was more vulnerable than the six-storey building. The four-storey building under the EC-based design exhibited the highest peak base force. This result was consistent with Bredean et al. [12], which indicated that low-rise reinforced-concrete buildings were more vulnerable than mid-rise buildings because of the high DCR value of the former.

ii) IDA revealed that the BS-based buildings performed better than EC-based buildings under seismic action. The four-storey MRCF was subjected to seismic loading. Two columns were removed simultaneously. The building designed based on BS recorded a PGA of 6.0 m/s^2 at 3% drift, whereas the building designed by the EC attained a PGA of 7.0 m/s^2.

This research was supported by Universiti Sains Malaysia under Short Term Grant (60313036) and Grant Contract U523.

References

[1] O.A. Mohamed, Assessment of progressive collapse potential in corner floor panels of reinforced concrete buildings, Engineering Structure, **31**, 749–757 (2009)
[2] H.R. Tavakoli and A.R. Alashti, Evaluation of progressive collapse potential of multi-story moment resisting steel frame buildings under lateral loading, Scienctia Iranica, **20**, 1, 77–86 (2013)
[3] M. Byfield, W. Mudalige, C. Morison and E. Stoddart, A review of progressive collapse research and regulations, Proceeding of Institute Civil Engineering, **167**, 8, 447–456 (2014)
[4] British Standards Institute, *Loading for buildings; Part 1: Code of practice for dead and imposed loads*, British Standards Institute, London, UK, (1996)
[5] EN 1990:Eurocode, *Basis of structural design*, British Standards Institution, London, (2000)
[6] Strucutral Engineering Institute. *Minimum design loads for buildings and other structures*, Reston, Va: American Society of Civil Engineers/ Structural Engineering Institute, (2002)
[7] National Building Code of Canada, Part 4 and Commentary C, 1995 edn. National Research Council of Canada, Ottawa, Ontario (1995)
[8] U.S. General Services Administration (GSA), *Alternate path analysis & design guidelines for progressive collapse resistance*, U.S. General Service Administration, (2013)
[9] Department of Defense (DOD), *Interim antiterrorism/force protection construction standards- guidance on structural requirements*,US Department of Defense Document, Washington, DC (2001)
[10] M. Sasani, A. Kazemi, S. Sagiroglu and S. Forest, Progressive collapse resistance of an actual 11-story structure subjected to severe initial damage, J. of Structural Engineering, **137**, 9, 893–902 (2011)
[11] M. Hosseini, N. Fanaie and A.M. Yousefi, Studying the vulnerability of steel moment resistant frames subjected to progressive collapse, Indian J. of Science and Technology, **7**, 3, 335–342 (2014)
[12] L. Bredean, M. Botez and A. Ioani, Progressive collapse risk and robustness of low-rise reinforced concrete buildings, *Proc. of Eleventh International Conference Computer Structural Technology*, Stirlingshire, 1–20 (2012)

Seismic Performance of Precast Polystyrene RC Walls

Ari Wibowo[1,*]

[1]Department of Civil Engineering, Faculty of Engineering, Brawijaya University, Malang, 65149, Indonesia

Abstract. Precast concrete structure such as precast wall is a concept that is growing rapidly these days. However, the earthquake resistance is believed to be one of its drawbacks. Additionally, the large weight of solid elements also increase the building weight significantly which consequently increase the earthquake base shear force as well. Therefore, investigation on the seismic performance of precast concrete wall has been carried out. Three RC wall specimens using wire mesh reinforcement and EPS (Extended Polystyrene System) panel have been tested. This wall was designed as a structural wall that was capable in sustaining lateral loads (in-plane) yet were lightweight to reduce the total weight of the building. Parameter observed was the ratio of height to width (aspect ratio) of wall of 1.0, 1.5 and 2.0 respectively with the aim to study the behaviour of brittle to ductile transition of the wall. Incremental static load tests were conducted until reaching peak load and then followed by displacement control until failure. Several data were measured at every stage of loading comprising lateral load-displacement behaviour, ultimate strength and collapse mechanism. The outcomes showed that precast concrete walls with a steel wire and EPS panel filler provided considerably good resistance against lateral load.

1 Introduction

Shear wall is one of the structural elements of the building functioning in restraining lateral loads such as earthquake excitation and severe wind forces by providing in-plane rigid stiffness of the walls. However, the solid and large weight of walls increase the total weight of building which in turn increase the total earthquake shear force as well, and hence reduce the effectiveness of wall at some extent. Therefore, reducing the self-weight of wall becomes quite important provided that the wall lateral strength capacity does not decrease too much.

One of the idea on minimizing self-weight of the wall is by using Extended Polystyrene (EPS, known as Styrofoam) as panel sandwiched between two steel wires. EPS is regarded as environmentally friendly material since it is fully recyclable and reusable made of plastic. EPS is also fire-proof and commonly used as insulated material and hence increases the fire resistant of the building. In terms of construction method, EPS is better used in

* Corresponding author: ariwibowo@ub.ac.id

precast system rather that cast in-situ due to the simplicity in the manufacturing and on-site construction processes. Investigations on existing precast building showed that such system were interestingly able to provide large lateral drift capacity [1], despite that precast systems are commonly believed to have low resistance capacity against lateral loads such as earthquake excitation. Therefore, the experimental test has been conducted to investigate the overall seismic performance of such walls.

RC walls can be categorised into three groups. RC walls with aspect ratios less or equal than 1.0 are commonly called as short walls which generally develop in-plane strut-and-tie mechanism to resist lateral forces. For aspect ratios of RC walls higher than 2.0, the RC walls will possess low Flexural-to-Shear Strength Ratio (FSSR=M/LV) which tend to develop dominant flexural action and hence are called slender walls. The displacement capacities of high-rise shear walls tend to be higher than short walls with the flexural failures occur at the wall base; whilst on contrary, the failure mode of short walls is mainly dominated by shear mechanism with the crack patterns display shear failure only. For mid-rise RC walls, the seismic behaviour is a combination of both flexure and shear.

For squat RC wall ($H_w/L_w \leq 1.0$), the interaction between shear and flexural capacity is more dependant to each other compared to that of tall RC walls. The behaviour of lateral load-deformation of squat RC walls resembles with that of deep beam, where on contrary, tall RC walls behave in flexural-dominant manner which are quite similar to columns. Another difference between squat and tall RC walls is in the context of gravity load. Low-rise RC walls commonly sustain only small gravity loads which in turn provides less effect on shear strength capacity. It is safe to ignore the axial load in shear strength capacity calculation for squat RC walls. But this case might not be applied for tall RC walls that carry much larger gravity load. For squat RC walls, the requirement of vertical reinforcement may not be necessarily large since squat RC walls possess longer lever arm with small gravity and lateral forces. And thus vertical reinforcement can be distributed uniformly over the full length of the wall, permitting only marginal increment at the vertical edges without losing too much ductility [2]

The purpose of this study was to determine the maximum lateral strength capacity and the lateral load-drift behaviour of EPS RC walls with various aspect ratios when subjected to lateral load. Moreover, the collapse mechanism in the context of the flexural-to-brittle behaviour of EPS walls was of particular interest. Therefore, the specimens were designed by varying ratio of height to width (H_w/L_w) of the wall to investigate the effect of aspect ratio to the collapse mechanism; whilst, other parameters such as vertical and horizontal rebars were set in constant.

2 Experimental test program

Parameters affecting the lateral load-drift behaviour of beam-column elements such as RC columns/walls are axial load ratio, aspect ratio, longitudinal reinforcement ratio and lateral reinforcement ratio. It was observed that for slender RC walls, the ultimate drift capacity increases by decreasing axial load ratio and by increasing transverse steel ratio and aspect ratio. However, the overall trends indicate that some parameters have some interdependence – particularly on the aspect ratio and the longitudinal steel ratio [3]. In this study, the aspect ratio of RC wall has been further investigated.

The experimental tests on three RC wall specimens S1-S3 have been conducted with wall heights of 60 cm, 90 cm and 120 cm, which corresponds to aspect ratio of 1.0, 1.5 and 2.0 respectively. The RC wall specimens were detailed with steel wire (wire mesh) placed on each side of the 4cm thick EPS panel in the core of the concrete wall and connected to each other by ϕ3.0mm wire connectors. The main reinforcement were ϕ2.5 mm galvanized steel wires at 75 mm spacing vertically and horizontally with yield and ultimate tensile strengths of 600 MPa and 680 MPa respectively. The RC wall specimens used two layers of concrete plaster on outer side of walls with thickness of 70 mm (about 1.4 inch) and concrete strength of 15 MPa. The typical RC wall specimens were shown in Fig 1.

Table 1. Basic Property of Column Specimens

Specs.	Width (cm)	Height (cm)	Aspect ratio (L/D)	Reinforcement	
				Vertical	Horisontal
S1	60	60	1.0	ϕ2.5-75	ϕ2.5-75
S2	60	90	1.5	ϕ2.5-75	ϕ2.5-75
S3	60	120	2.0	ϕ2.5-75	ϕ2.5-75

The RC wall specimens were tested using a static load testing procedure to obtain representative lateral load versus displacement curves. The load controlled loading sequence consisted of load increments of 1kN until reaching peak load, and then followed by drift increment of 0.5% until failure.

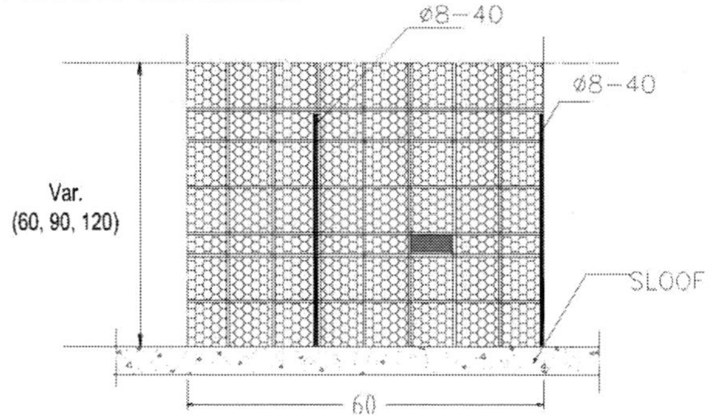

Fig. 1. Geometry and reinforcement details of RC wall specimens

3 Lateral load-displacement behaviour

As shown in Fig. 2a and Table 2, specimen S1 (a=1.0) provided highest peak load but showing quite brittle behaviour with displacement ductility ($\mu=\Delta_u/\Delta_y$) of about 1.7. On the other hand, specimen S3 (a=2.0) could sustain lateral load at the largest ductility of 5.7 but had the least peak load capacity compared to those of other specimens with smaller aspect ratios. Further, specimen S2 (a=1.5) had both peak load and ductility ratio in between of those of specimens S1 and S3. Furthermore, the related drift ($\delta=\Delta/H$) behaviour for all specimens is shown in Fig. 2b.

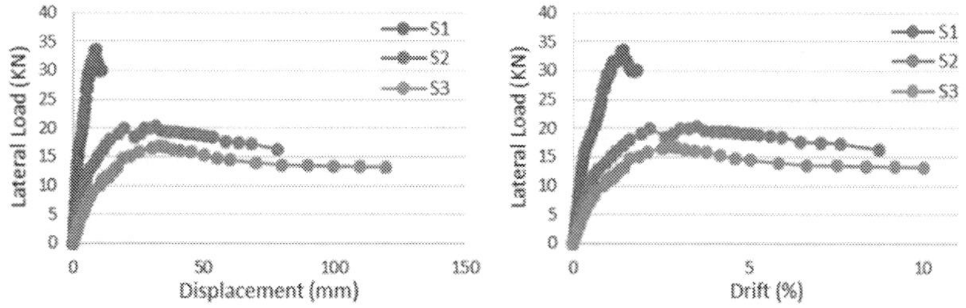

a. Lateral load-displacement behaviour. b. Lateral load–drift behaviour.

Fig. 2. Relationship between lateral load and displacement/drift for all specimens.

 DOI: 10.1051/matecconf/201710302020

Table 2. Peak Load and displacement behaviour of specimens S1-S3.

Specimen	Peak Load, P_u (kN)	Yield Displacement, Δ_y (mm)	Displacement at lateral load failure, Δ_u (mm)	Displacement ductility $\mu=\Delta_u/\Delta_y$
S1	33.6	6.4	10.7	1.7
S2	20.4	15	77	5.1
S3	16.8	21	120	5.7

4 Collapse mechanism

There are three categories of failure commonly defined for RC walls under lateral load, i.e., flexure-critical, shear-critical, and flexure-shear-critical as shown in Fig. 3. A flexure-critical RC wall is one that has sufficient shear strength to develop flexural yielding, whilst a shear-critical wall is one that fails in shear, with the corresponding moment demand significantly less than the moment capacity. A flexure-shear-critical RC wall is one that yields in flexure prior to failing in shear at higher ductility levels.

The failure mechanism can be predicted by calculating the flexural-to-shear strength ratio (*FSSR*) as stated in Equation (1) and using Table 3. It should be noted that the classification system was developed by the ASCE for columns/walls detailed for ductility and hence may not be directly applicable to limited ductile RC walls.

$$FSSR = \frac{M_n}{LV_n} \tag{1}$$

where;

M_n = the nominal moment capacity of a column
V_n = the nominal shear strength of a column
L = wall length

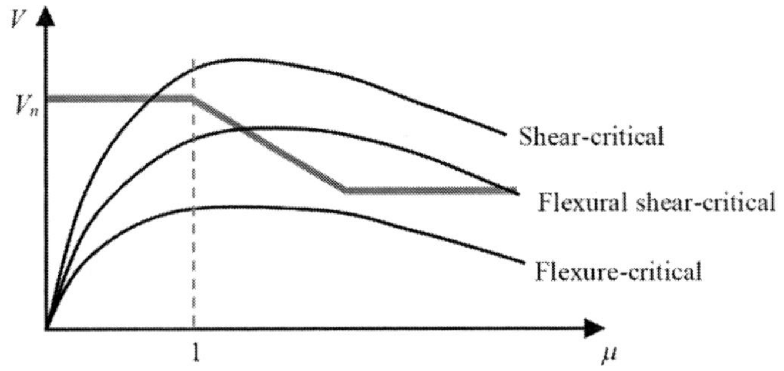

Fig. 3. Shear strength degradation and displacement ductility.

Table 3. Classification of Failure Mechanism (Refer [4], SECTION 6.4.2.2.2).

	Transverse Reinforcement Details		
	ACI conforming details with 135° hooks	**Closed hoops with 90° hooks**	**Other (including lap spliced transverse reinforcement)**
$FSSR \le 0.6$	Flexure	Flexure-shear	Flexure-shear
$0.6 < FSSR \le 1.0$	Flexure-shear	Flexure-shear	Shear
$FSSR > 1.0$	Shear	Shear	Shear

DOI: 10.1051/matecconf/201710302020

All RC wall specimens developed first flexural crack at wall-foundation interface producing gap opening between wall and foundation, which indicated that yield penetration mechanism occurred. Additionally, as the number of load cycle increased towards peak lateral strength, several vertical cracks at bar locations also penetrated deeper into the sloof foundation showing the development of bond slip failure between steel bars and the surrounding concrete. Specimen S1 (a = 1.0) showed shear-dominant cracks with main diagonal shear crack extending from lateral load point at the tip of the wall to the corner of compression zone of concrete at the bottom of the wall specimen. On contrary, the propagation of flexural-shear cracks were localised at the plastic hinge area of specimen S3 (a = 2.0), with no cracks found at the elastic area outside the plastic zone. Whereas, for specimen S2 (a = 1.5) those flexural-shear cracks were observed at a larger extent up to the top of wall specimen. The crack propagation patterns for all specimens are shown in Figure 4.

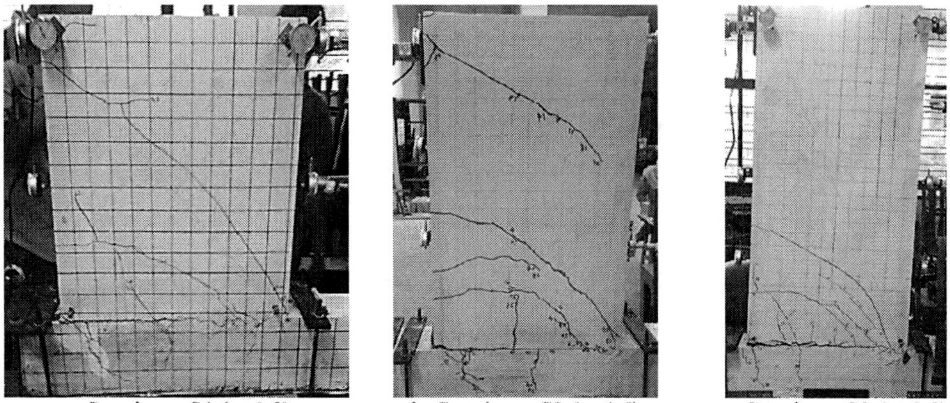

a. Specimen S1 (a=1.0) b. Specimen S2 (a=1.5) c. Specimen S3 (a=2.0)

Fig. 4. Crack propagation over the height of specimens.

The two main shear failure mechanisms of a reinforced concrete elements are shear-compression failure and diagonal tension failure. Shear-compression failure occurs due to arch action causing crushing of concrete along the diagonal strut which usually takes place for cases of RC walls with a very high axial load (i.e. above the balance point) or relatively low shear span ratio (i.e. $a < 2$). For diagonal tension failure, tensile stresses in the concrete governs the failure mechanism by causing the inclined cracks to become unstable and expand through the compression area; where it tends to occur when the axial load is below the balance point and the aspect ratio of the column is larger than 2 ($a > 2$). The tested specimens S1-S3 (aspect ratios $a < 2$ with no axial load applied) were categorized in-between those two conditions, but however, the test results showed that all specimens seems to be dominated by diagonal tension failure. It indicated that axial load ratio provides greater effect than the aspect ratio on shear failure mechanism of short RC walls ($a < 2$).

With the calculated nominal moment and shear capacity of the RC wall of 26.5 kNm and 53.4 kN respectively, the predicted failure mechanisms of RC wall are shear dominated for specimen S1 (a=1.0) and flexure-shear dominated for specimens S2 and S3 which are in good agreement with the test results as shown in Table 4.

The lateral displacement of walls (Δ_{tot}) consists of three components, flexural (Δ_{fl}), yield penetration (Δ_{yp}) and shear deformation (Δ_{sh}) as shown in Fig. 5 (refer [5]). As shown in Fig. 6, specimen S1 with the least aspect ratio of a=1.0 had the largest proportion of shear drift component of about 90% of the total lateral drift, and hence showed the dominant shear behaviour of wall as shown in Fig. 6a. In contrast, the flexural components were the most dominant behaviour for specimens S2 and S3 at around 75%-85% of the total drift due to the large aspect ratio of $a > 1.5$ as shown in Fig. 6b.

$$\Delta_{tot} = \Delta_{fl} + \Delta_{yp} + \Delta_{sh} \tag{2}$$

Table 4. Predicted and Actual Failure Mechanism for all specimens.

Specimen	Aspect ratio	Wall height (m)	FSSR	Failure Mechanism	
				Predicted	Actual
S1	1.0	0.6	0.83	Shear	Shear
S2	1.5	0.9	0.55	Flexure-shear	Flexure shear over wall height
S3	2.0	1.2	0.41	Flexure-shear	Flexure shear concentrated at plastic hinge area

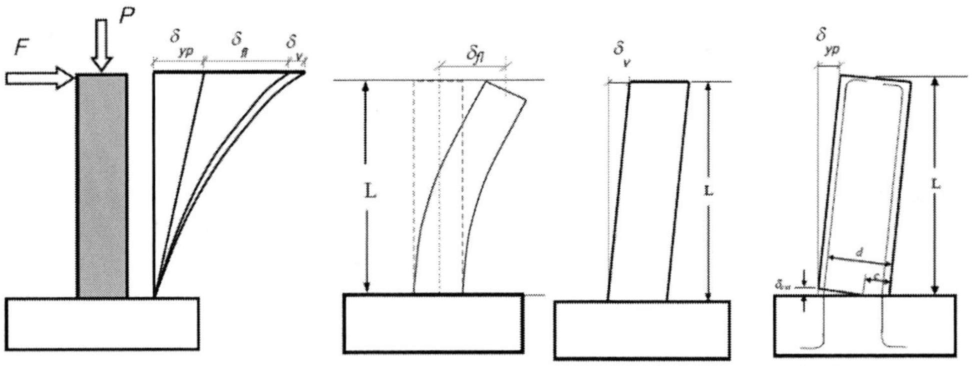

a. Reinforced concrete wall b. Flexural displacement c. Shear displacement d. Yield penetration displacement

Fig. 5. Lateral displacement components of RC wall comprising flexural, shear and yield penetration displacement.

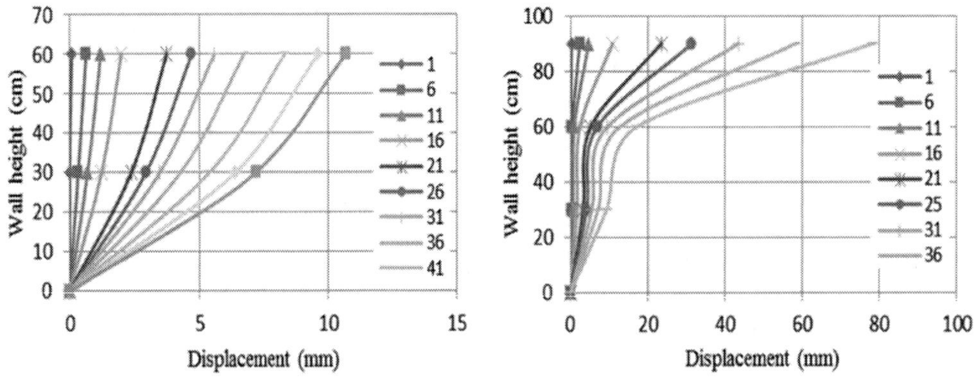

a. Specimen S1 b. Specimen S2

Fig. 6. Total lateral deformation along the RC wall height for each incremental load (note: legends of curves indicate the experimental loading step)

5 Theoretical analysis

The theoretical prediction analysis has been conducted using two models developed by author in a previous journal [5], i.e., detailed and simplified models as shown in Fig. 7 and Table 5. The detailed model predicts the lateral load-drift behaviour of reinforced concrete walls consisting of four stages: cracking, yield, peak, and lateral load failure which were developed using displacement based design methodology. Whereas, the simplified model comprises three stages: cracking, yield and ultimate that were developed based on classic moment-curvature approach. These models are aimed to provide simple procedures for checking lateral load-drift behaviour of non-ductile concrete walls.

The theoretical prediction of lateral load-drift behavior for all specimens S1-S3 are in good agreement with the test results as shown in Fig. 8. Generally, the theoretical models give stiffer results compared with the experimental data which is considered reasonable since the theoretical models are based only on flexural deformation analysis and does not include other deformation components. The comprehensive drift analysis are beyond the scope of this paper.

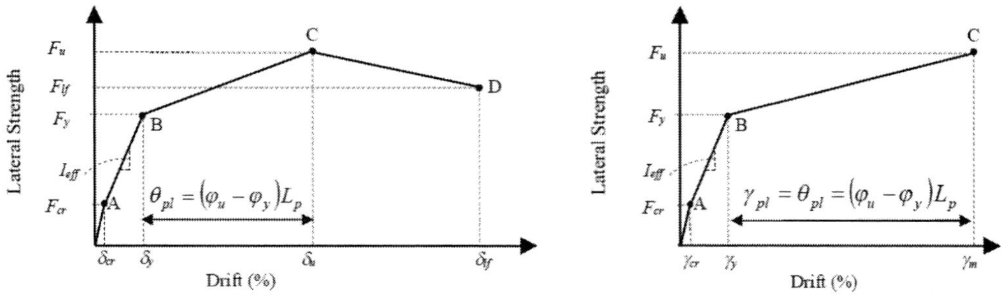

(a). Detailed model (b) Simplified Model

Fig. 7. The detailed and simplified design models of lateral load-drift capacity (refer [5])

Table 5. Analytical model procedures.

Detailed Model	Simplified Model
a. Point A (Cracking) $F_{cr} = \dfrac{M_{cr}}{H_w}$; $\gamma_{cr} = \dfrac{M_{cr}H_w}{3E_cI_g}$ (3) b. Point B (Yield) $F_y = \dfrac{M_y}{H_W}$; $\gamma_y = \dfrac{M_yH_w}{3E_cI_e}$ (4) c. Point C (Peak Strength) $F_u = \dfrac{M_u}{H_w}$; $\gamma_{peak} = \gamma_y + \gamma_{pl.p}$ (5) d. Point D (Ultimate Displacement) $\gamma_u = \dfrac{\gamma_y}{k}\left[(1+k\alpha) - 0.8\dfrac{F_u}{V_u}\right]$ (6)	a. Point A (Cracking) The cracked lateral strength and drift are calculated by assuming cracking drift γ_{cr} = 0.1% or calculated directly using the concrete modulus and cracking strength. b. Point B (Yield) $F_y = \phi F_u$ (7) whilst the yield drift (γ_y) is estimated using $I_{eff} = 0.5I_g$ (8) c. Point C (Ultimate) $\gamma_m = \gamma_y + \gamma_{pl}$ (9) $\gamma_{pl} = \theta_{pl} = \dfrac{W_{cr}}{L_w} = 0.75\dfrac{d_b}{L_w}$ (10)

where ; F_{cr} = cracking load, M_{cr} = cracking moment, H_w = wall height, γ_{cr} = cracking drift, F_y = yield load, M_y = yield moment, γ_y = yield drift, F_u = peak load, M_u = peak moment, γ_{peak} = drift at peak load, γ_u = ultimate drift, γ_{pl} = inelastic drift, α = the drift ductility when the shear strength commences to decline.

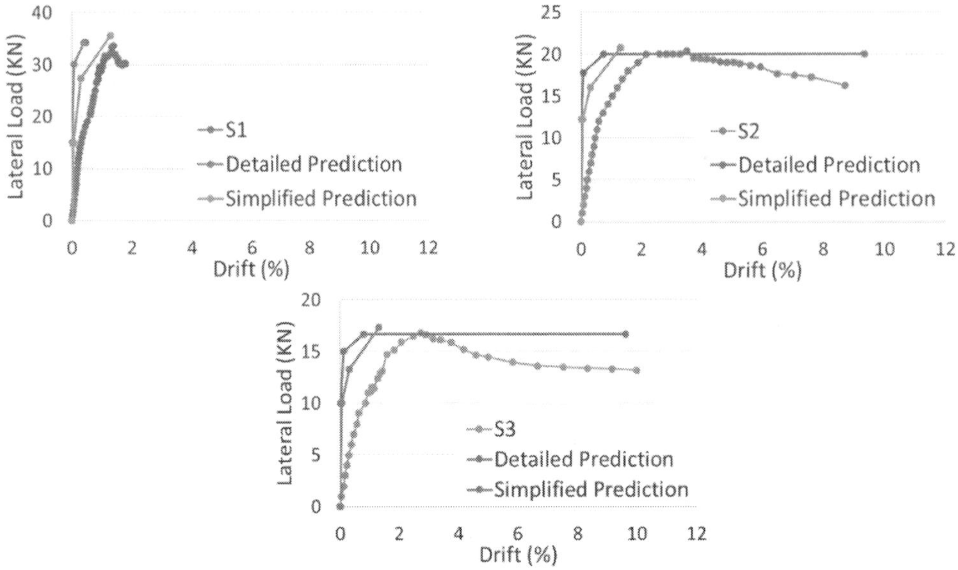

Fig. 8. Comparison of load-drift behaviour curves between experimental test and theoretical prediction results.

6 Summary

Three RC wall specimens have been tested to investigate seismic performance of EPS panel RC wall to investigate brittle-to-failure transition of wall with the outcomes as follows:

i) Squat walls (specimen S1) provided highest peak lateral load but fail in brittle manner. On the other hand, specimens S2 and S3 displayed more ductile behaviour but with smaller peak lateral loads.
ii) Shear-dominated cracks were observed on specimen S1 (a = 1.0) with main diagonal shear crack propagated from tip of wall to the corner of concrete compression zone at the bottom of the wall specimen. Whereas, on contrary, flexural-shear cracks occurred at both specimens S2 and S3. Interestingly, the crack zone of specimen S3 was located at the plastic hinge area of wall; whilst, for specimen S2 (a = 1.5) those cracks were observed at a larger extent up to the top of wall specimen. It indicates that for slender walls, the larger aspect ratio the more localised flexural cracks at plastic hinge area.

References

[1] A. Wibowo, J.L. Wilson, N.T.K. Lam and E.F. Gad, Drift capacity of precast soft storey building in Melbourne, Aust. J. Struct. Eng., **11**, 3, 177-193 (2010)
[2] R. Park and T. Paulay, *Reinforced concrete structures*, John Wiley & Sons, New York, USA, (1975)
[3] A. Wibowo, *Seismic performance of insitu and precast soft storey buildings*, PhD Thesis, Swinburne University of Technology, Melbourne, Australia, (2012)
[4] ASCE, *Seismic rehabilitation of existing buildings,* ASCE/SEI 41-06, American Society of Civil Engineers, Virginia, USA, (2007)
[5] A. Wibowo, J.L. Wilson, N.T.K. Lam and E.F. Gad, Seismic performance of lightly reinforced structural walls for design purposes, Mag. Concrete Res., **65**, 13, 809 –828 (2013)

Diffusion Tests for Chloride in Concrete Using Wind Tunnel Test

Devi Nuralinah[1,*], and *Takumi* Shimomura[2]

[1]Civil Engineering Department, Brawijaya University, 65145 Jl. MT. Haryono 167 Malang, Indonesia
[2]Environmental and Civil Engineering Department, Nagaoka University of Technology, 940-2188 Kamitomioka Machi 1603-1, Nagaoka-Niigata, Japan

Abstract. Diffusion coefficient of chloride in concrete is a predominant factor in chloride penetration into concrete. Two types of diffusion coefficient models are examined here: mean diffusion coefficient *D*, time-dependent diffusion coefficient *D*(*t*). These diffusion coefficients are experimentally derived from the chloride profiles in concrete obtained by the wind tunnel test. It is clearly shown that diffusion coefficients are dependent of water cement ratio but independent of intensity of airborne salt. Time-dependent diffusion coefficients derived based on constant and time-dependent surface chloride are not different. Diffusion coefficient also decreases with increasing of time as explained by other researcher.

1 Introduction

Chloride ingress in concrete is usually expressed by Fick's diffusion equation to practically estimate service-life of concrete structures. In the Standard Specification for Concrete Structures by Japan Society of Civil Engineers, the state in which chloride content at the reinforcing bar portion in concrete reaches the designated threshold value is defined as one of the limit states of structural durability [1]. In the JSCE method, surface chloride content is used as boundary condition for diffusion of chloride in concrete. It is empirically determined based on distance from the shore line. Diffusion coefficient is estimated from concrete properties by means of concrete water-cement ratio and type of cement.

Surface chloride and diffusion coefficient have been investigated in many previous studies. The resistance of concrete to chloride penetration was investigated by the rapid chloride permeability test [2]. The chloride penetration in unsaturated concrete skin was different from that in concrete bulk [3]. Chloride ion diffusion coefficient decreases in long-term chloride exposure [4].

Diffusion coefficient of chloride in concrete is a predominant factor in chloride penetration into concrete. Two types of diffusion coefficient models experimentally derived from the chloride profiles in concrete obtained by the wind tunnel test were examined here: mean diffusion coefficient *D*, and time-dependent diffusion coefficient *D*(*t*) [5,6].

In this paper, relationship between intensity of airborne salt and surface chloride content was investigated based on the test results obtained by the developed testing equipment. On

* Corresponding author: devi@ub.ac.id

the other hand, diffusion coefficients derived from the obtained chloride profile in all specimens are also discussed.

It is well-known that experimentally obtained diffusion coefficient of chlorides in concrete decreases with increasing of exposure time. One of the practical ways to express this tendency in computational model is to regard diffusion coefficient as a function of time, as adopted in many previous studies. If the reason of decreasing of diffusion coefficient is change of pore structure of concrete, time-dependent diffusion coefficient is reasonable.

However, if the true reason for this is that transport of chloride in concrete is not simple diffusion mechanism, then time-dependent diffusion coefficient is not universal way to express the experimental results.

2 Research method

In previous paper, chloride ingress into concrete subjected to airborne salt was developed by wind tunnel test [5-7]. The inside views of that apparatus shown in Fig. 1 has size 1mx1m. The wind path has a length of 12 m in one round. Propeller that generates air flow with diameter of 90 cm was set up in the first floor as shown in Fig. 1(a), as well as driven by an electric motor installed outside of the device. Particles of salt water were produced by putting fine air bubbles into 3% salt water in a bath (Fig. 1(b)). Six concrete specimens whose size is 100 x 100 x 150 m, were prepared with water-cement ratio of 40% and 60%, and cured in water for 28-days. After curing, five surfaces of each specimen were coated with tar epoxy except the exposed surface to ensure one-dimensional chloride ingress into concrete from the exposed surface. Both concrete specimens and gauze specimens were set up and exposed to the airborne salt in the first and the second floor in the wind tunnel as shown respectively in Fig. 1(c) and 1(d). The wind velocity in the tunnel was 1.5 m/s. The measured airborne salt per unit area per unit time, which was expressed in terms of amount of Natrium Chloride was 5 to 68.2 mdd (mg/dm^2/day) depending on the testing position in the wind tunnel.

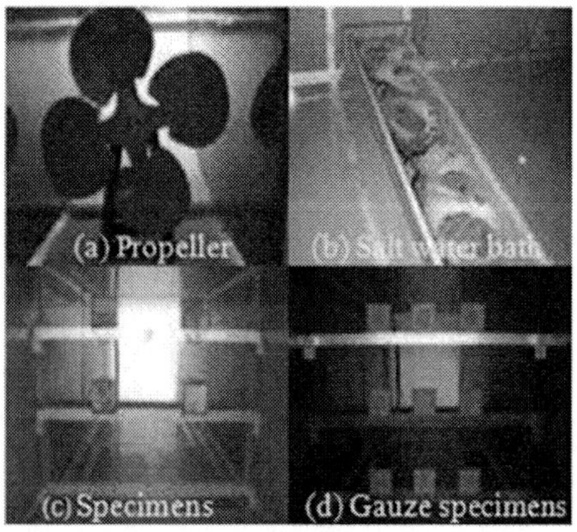

Fig. 1. The inside views of wind tunnel.

Specimens were taken out from the wind tunnel occasionally and chloride content in concrete was measured. From the beginning up to 96 days, samples of concrete powder were taken at the positions of 10 mm, 20 mm, and 40 mm from the exposed surface by drilling from both side surface to the maximum depth of drilled hole. After 96 days, samples were also obtained at the point of 60 and 80 mm from the exposed surface to measure chloride content in deeper portion. Value of chloride content in concrete is expressed in terms of amount of chloride ion per unit volume. After chloride content measurement, specimens were exposed again in the tunnel.

3 Determination of experimental diffusion coefficient

The result of experimental time-dependent chloride profiles in specimen H-40 obtained by wind tunnel test during 240 days is shown as example in Fig. 2.

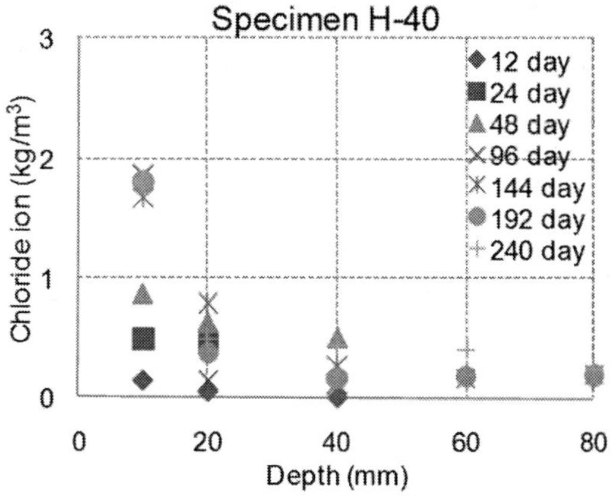

Fig. 2. Time-dependent chloride profiles in concrete specimen H-40.

Transport process of chloride in concrete is evaluated by Fick's diffusion equation coupled-with fixed boundary condition as a standard and most popular method [8-10].

$$\frac{\partial C}{\partial t} = D\frac{\partial^2 C}{\partial x^2} \tag{1}$$

$$C(0,t) = C_0 \tag{2}$$

Integral solution of the governing differential equation of Equation. (1) under the boundary condition Equation. (2) in semi-infinity body is expressed as:

$$C(x,t) = C_0\left\{1 - erf\left(\frac{x}{2\sqrt{Dt}}\right)\right\} \tag{3}$$

where, *erf*(s) is an error function defined as:

$$erf(s) = \frac{2}{\sqrt{\pi}}\int_0^s e^{-\eta^2} d\eta \tag{4}$$

Equation. (3) is often used as practical design equation to estimate service-life of structures since it is convenient to calculate chloride content at the reinforcing bar position in concrete as a function of service time.

Time-dependent and constant diffusion coefficients are derived from the wind tunnel test as examined by other research's group according to following procedure.

i) Transforming Equation. (3), mean diffusion coefficient at the position x_i and at the time t_j is expressed as:

$$D_{i,j}=\frac{1}{t_j\left\{\frac{2}{x_i}erf^{-1}\left(1-\frac{C(x_i,t_j)}{C_0}\right)\right\}^2} \quad (5)$$

where,
$C(x_i,t_j)$: experimentally obtained chloride content in concrete at the position x_i and at the time t_j;
C_0 : surface chloride content. Either mean or time-dependent surface chloride content is used in Equation. (5). Surface chloride content at time t is estimated by extrapolation method

ii) The obtained surface chloride content $C_0(t_j)$ at the sampling time t_j is plotted as a function of time. Then, time-dependent surface chloride content $C_0(t)$ is simulated by a regression curve expressed by Equation. (6).

$$C_0(t)=a(1-e^{-bt}) \quad (6)$$

where;
a, b : parameters for time-dependent surface chloride content determined by least square method.

iii) Then, mean surface chloride during the exposure period is mathematically calculated by averaging $C_0(t_j)$ with respect to time by Equation. (7).

$$C_0=\frac{1}{T}\sum_0^T C_0(t_j)\Delta t_j \quad (7)$$

where,
T : length of time of exposure (day)
$C_0(t_j)$: experimental time-dependent surface chloride content

iv) Series of diffusion coefficient $D_{i,j}$ obtained from the experimental chloride content at each position at the exposure time t_j is averaged with respect to position.

$$D_j=\frac{1}{X}\sum_i D_{i,j}\Delta x_i \quad (8)$$

where,
X : depth of specimen (mm)
Δx_i : distance between measurement positions (mm)

v) Thereafter, regression curve for time-dependent diffusion coefficient $D(t)$ expressed by Equation. (9) is approved to approximate D_j.

$$D(t)=ct^{-d} \quad (9)$$

where;
c and d : parameters for time-dependent diffusion coefficient, which are determined by least square method.

It should be noted that diffusion coefficient obtained by the above procedure is not a theoretical time-dependent diffusion coefficient, because the integral solution Equation. (3)

is based on constant diffusion coefficient. Therefore, diffusion coefficient evaluated by Equation. (5) is considered only as an approximated value of time-dependent diffusion coefficient.

vi) Mean diffusion coefficient in Equation. (10) is mathematically calculated by averaging Equation. (8) with respect to time.

$$D = \frac{1}{T}\sum_{0}^{T} D_j \Delta t_j \tag{10}$$

where;
T : exposure time of specimen
D_j : experimental time-dependent diffusion coefficient as expressed by Eq. (8).

4 Result and discussion

Time-dependent diffusion coefficients of specimens H-40 as example derived from data of 0-240 day under mean and time-dependent surface chloride is shown in Fig. 3 and summarized in Table 1. Two diffusion coefficients both decrease with increasing of time and approach to a same value in long time. Diffusion coefficient based on mean surface chloride $D_1(t)$ and time-dependent surface chloride $D_2(t)$ are close to each other except in early stage of exposure.

Diffusion coefficients based on mean and time-dependent surface chloride in the series of 60% of W/C are relatively greater than those in the series of 40% of W/C. It seems that diffusion coefficients are independent from intensity of airborne salt. Considering these facts, diffusion coefficient can be regarded as material properties independent of environmental condition.

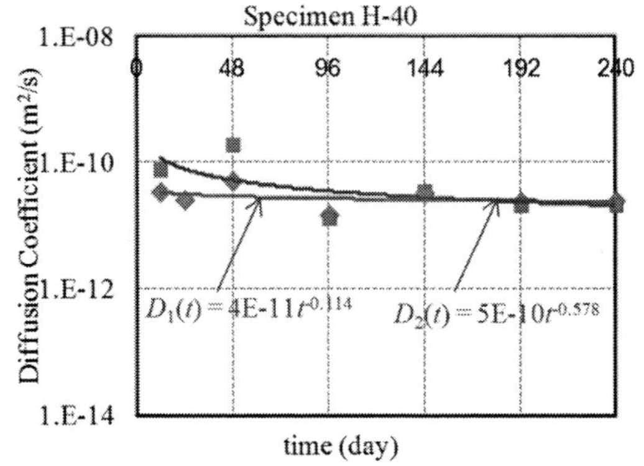

Fig. 3. Time-dependent diffusion coefficient of all specimens derived from 0-240 day data.

Same as many previous studies, experimentally obtained apparent diffusion coefficient of chlorides in concrete in the present study decreases with increasing of exposure time. There are two possible reasons for this. The first is the change of pore structure of concrete during exposure due to cement hydration. The other is that transport phenomena of chloride in concrete cannot be expressed by ideal diffusion equation in reality. If the first reason is dominant, then using time-dependent diffusion coefficient is reasonable. On the other hand, if the second reason is true, using time-dependent diffusion coefficient is not universal way to express the phenomenon because pore structure of concrete does not change. Thus, the relationship between diffusion coefficient and time is affected by environmental condition.

In Fig. 4, time-dependent diffusion coefficients of specimens with high water-cement ratio are relatively higher than those with low water-cement ratio. Diffusion coefficients derived under time-dependent surface chloride are greater than those derived under constant surface chloride in early stage of exposure. However, they almost become same value in long time of exposure.

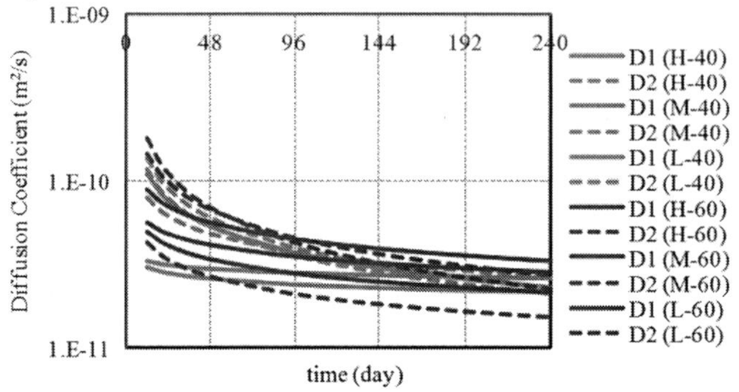

Fig. 4. Time-dependent diffusion coefficient of all specimens derived from 0-240 day data.

Furthermore, mean diffusion coefficient is derived from data of 0-240 day under either mean or time-dependent surface chloride. The results of mean diffusion coefficient are also summarized in Table 1. Relationship between *W/C* and diffusion coefficient D_1, which is averaged $D_1(t)$, is shown in Fig. 5. The influence of intensity of airborne salt on diffusion coefficient is not apparent.

Table 1. Time-dependent and mean diffusion coefficient derived from 0-240 day data.

Specimen	Time-dependent diffusion coefficient (m^2/s)		Mean diffusion coefficient (m^2/s)	
	$D_1(t)$ based on C_0	$D_2(t)$ based on $C_0(t)$	Mean D_1 based on C_0	Mean D_2 based on $C_0(t)$
H-40	$4E\text{-}11t^{-0.114}$	$5E\text{-}10t^{-0.578}$	2.67E-11	4.00E-11
M-40	$4E\text{-}11t^{-0.079}$	$2E\text{-}10t^{-0.368}$	2.87E-11	3.78E-11
L-40	$4E\text{-}10t^{-0.521}$	$6E\text{-}10t^{-0.597}$	3.02E-11	8.45E-11
H-60	$1E\text{-}10t^{-0.231}$	$6E\text{-}10t^{-0.565}$	3.54E-11	6.30E-11
M-60	$1E\text{-}10t^{-0.281}$	$1E\text{-}10t^{-0.343}$	3.11E-11	3.16E-11
L-60	$2E\text{-}10t^{-0.328}$	$1E\text{-}09t^{-0.691}$	4.29E-11	7.52E-11

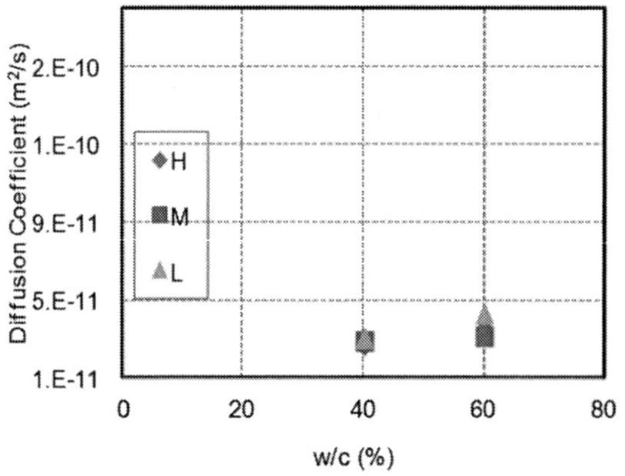

Fig. 5. Relationship between W/C and mean diffusion coefficient D_1 derived from 0-240 day data.

5 Conclusion

Two types of diffusion coefficient, which are mean diffusion coefficient D, time-dependent diffusion coefficient $D(t)$, were derived from the exposure test result by the wind tunnel. The results are concluded as following.

i) It was confirmed that time-dependent diffusion coefficient obtained by the wind tunnel test decreases with increasing of time and approaches to a certain value in long time.

ii) There was no significant difference between time-dependent diffusion coefficient derived based on constant surface chloride and that based on time-dependent surface chloride.

iii) It was found that the derived two types of diffusion coefficient are all dependent of water-cement ratio of concrete but not affected by the intensity of airborne salt that the concrete is exposed.

The authors wish to express their gratitude to Brawijaya University for International Conference funding. Especially thanks to former graduate students of Nagaoka University of Technology for good working in experimental work.

References

[1] JSCE, *Standard specification for concrete structures: Materials and Construction*, Japan Society of Civil Engineers, Tokyo, (2002)
[2] C. Andrade and M.A. Sanjuan, Experimental procedure for the calculation of chloride diffusion coefficients in concrete from migration tests, Int. J. of Advances in Cement Research, **6**(23), 127-134 (1994)
[3] C. Andrade, J.M. Diez and C. Alonso, Mathematical modeling of a concrete surface "skin effect" on diffusion in chloride contaminated media, Int. J. of Advanced Cement Based Materials, **6**(2), 39-44 (1997)
[4] K.M.A. Hossain, Correlations between porosity, chloride diffusivity and electrical resistivity in volcanic pumice-based blended cement pastes, Int. J. of Advances in Cement Research, **17**(1), 29-37 (2005)
[5] D. Nuralinah, K. Kamiura and T. Shimomura, A new laboratory test method for chloride ingress into concrete subjected to airborne salt, *Proc. of the Twelfth Int. Summer Symposium*, Chiba, Japan, 291-294 (2010)
[6] D. Nuralinah, K. Kamiura and T. Shimomura, Analysis of chloride ingress into concrete subjected to airborne salt measured by wind tunnel test, *Proc. of the Japan Concrete Institute (CD-ROM)*, **33**(1), 869-874 (2011)
[7] K. Kamiura, *Development of wind tunnel for simulation of airborne salt*, Master Thesis, Nagaoka University of Technology, Japan, (2009)
[8] G.S. Paul, *Diffusion in solids*, McGraw-Hill, New York (1963)
[9] P. Erwin and M. Leif, *Diffusion of chloride in concrete*, Taylor & Francis Inc., New York (2006)
[10] R.B. Bird, E.S. Warren and N.L. Edwin, *Transport phenomena*, 2nd ed, John Wiley & Sons, Inc., New York (2002)

Thermal Loads Effect on Response of One-Story Reinforced Concrete Frame Buildings in UAE

Sabouni Reem[1,*], and *Sydnaoui* Ikhlass[2]

[1]Civil Engineering Department Chair, Civil Engineering Department. ALHOSN University, Abu Dhabi, UAE
[2]Structural Engineer, Musanada, Abu Dhabi General Services Company, Abu Dhabi, UAE

Abstract. In this paper the effect of temperature change (thermal load) on one-story reinforced concrete frame buildings allocated in the UAE is studied. Sixty eight finite element models with different story heights and slab lengths were developed in ETABS to carry out this study. The effect of temperature change with the variation of story height, slab length and support condition on the lateral displacement and horizontal reactions was studied in this paper. The thermal load results for the fixed support models showed that for different buildings with the same column heights, the external columns' displacements and reactions increased considerably with increasing the slab length. Comparing the horizontal reaction results –due to thermal loads- of the two modeled building heights (3 m and 6 m) showed that the 6 m models have lower reactions at external columns but larger top displacements. The comparison between the lateral load results for fixed and hinged support conditions showed that models with fixed support conditions had lower lateral displacement but higher horizontal reactions due to thermal loads. The effect of the support condition on the lateral displacement was larger for longer slab lengths whereas its effect on the horizontal reaction was larger for smaller slab lengths. The support conditions showed to have larger effect on the shorter column models.

1 Introduction

The UAE is known for its high temperatures especially during the summer period. This high temperature along with the frequent fluctuation in the daily and seasonal temperatures imposes extra loads on structures. The Abu Dhabi International Building Code (2013) (ADIBC) considers temperature loads as one of the main important loads to be verified and analyzed in reinforced concrete structures in the UAE [1]. Seasonal changes in temperature causes overall structural deformation, displacements, in addition to differential stresses in concrete elements. It has to be considered for building serviceability conditions and to eliminate the additional thermal cracks [2]. When studying the effect of temperature change

* Corresponding author: r.arsabouni@alhosnu.ae

on reinforced concrete buildings two important factors should be considered: the nature of the temperature change and the fundamental materials properties of concrete.

In service, thermal effects are related to the difference between the highest temperature during summer and the lowest temperature during winter [2]. The highest temperature registered in the UAE was found to be 50°C (in the western region); whereas the lowest registered temperature was found to be 10°C. The magnitude of the forces developed and the amount of displacements caused by temperature changes are directly related to building length [2]. Contraction and expansion joints are means used to limit the magnitude of forces, displacements and cracking induced by temperature [3]. A common practice in the UAE is that expansion joints spacing shouldn't be more than 120ft (36.6m) considering recognized committees recommendations [1,2,4], when the spacing exceeds 450ft (46m), then a special design including a thermal study has to be performed [2,5].

2 ETABS models input data and conditions

2.1 Concrete properties

The thermal coefficient of concrete is usually affected by many factors such as the type of aggregate, moisture content, mix properties, cement type and age of concrete. For ordinary thermal stresses calculations a thermal coefficient value of 10x10-6\°C can be used when the type of aggregate and concrete degree of saturation are unknown [2, 6]. Accordingly this value was implemented in the used ETABS models (Fig. 1). The rest of the concrete properties are presented in Table 1.

Table 1. Concrete properties used in the model.

Concrete Property	Value
Concrete compressive strength (*fc'*)	40 N/mm^2
Modulus of elasticity (*E*)	30000 MPa
Poisson ratio	0.2
Mass per unit volume	2400 kg/m^3

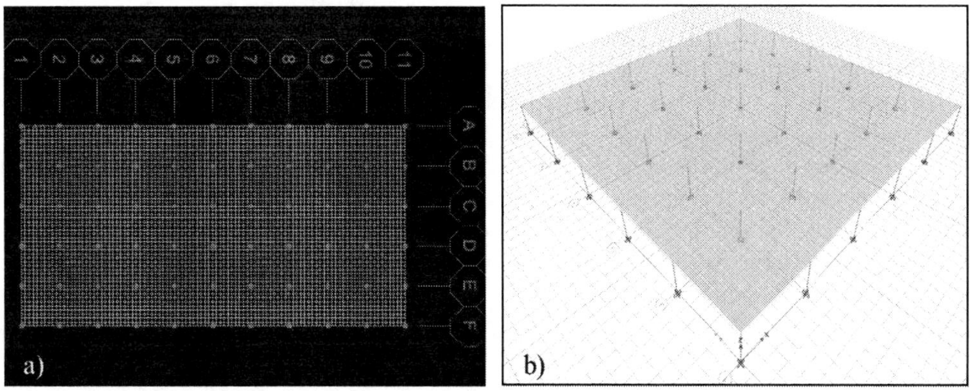

Fig. 1. Typical model: a) Top view slab and b) 3D view.

2.2 ETABS model geometry

Sixty eight ETABS one-story reinforce concrete frame building models were generated in this study (Fig. 1). These models were divided into two sets: fixed support reaction and hinged support reaction and each set consists of two model groups: the first group with column heights of 3.00 m and the second group with column heights of 6 m. Each group consists of 17 different ETABS models. The difference between these models is the concrete slab length which was increased from 50 m to 400 m by increments of 10 m. These models were generated to verify the effect of temperature change with variation in column height, slab length and support conditions on the lateral displacement and horizontal reactions in one-story reinforced concrete frame buildings in UAE. The rest of the model geometry was taken as follows:

i) Slab thickness = 300mm
ii) Columns size = 800 x 800 mm^2
iii) Thermal load = 40 °C
iv) Slab width = 50 m.

Thermal loads (40 C°) were assigned as uniform temperature change at all slab shells elements at upper slab level.

3 Results and analysis

In this section the results of the thermal study conducted on the sixty eight reinforced concrete frame structure models are presented. The results from the first set of models with fixed support conditions were used to study the effect of temperature change with variation in column height and slab length on the lateral displacement and horizontal reactions in one-story reinforced concrete frame buildings in UAE. The results from the second set of models with hinged support conditions were compared to the results of the first set of models to study the influence of the support conditions on the temperature change results.

3.1 Effect of thermal load

For the first set of models (with fixed support) the change in the horizontal reactions and the top deflection at the external columns A1, B1, C1, D1, E1 and F1 with different slab lengths under thermal load are shown in Fig. 2 and Fig. 3 for the 3 m and 6 m high structures, respectively. Due to symmetry in the analyzed models these columns will have the same results as those for columns A11, B11, C11, D11, E11 and F11, respectively. All models were found to be safe under dead and live loads.

These thermal load results show that for the same building the displacements at external columns have different but close values. The results also show that for different buildings with the same column heights, the external columns' displacements increases considerably with increasing the slab length. The external column displacements for 400 m slab was about three times that for the 50 m slab length for the 3 m high column models and about five times for that in the 6 m high column modes. The maximum allowable lateral deflection in columns is (height/180) to avoid damage to external walls [3]. This limit is 16.7 mm and 33.3 mm for the 3 m and 6 m high columns, respectively. The model results showed that this limit was exceeded for slabs with length more than 100 m for the 3 m high column models and for slabs with length more than 200 m for those of the 6 m high column models.

For the same building (in the first model set) the thermal load results showed that the horizontal reactions (parallel to the slab length) at external columns' supports have different but close values. For different buildings with the same column height, the external columns reactions at supports increased considerably with the increase of the slab length. This has a big impact on these columns and its related footing design. The study results showed that the reactions for the 400m long slab was about three times that for 50 m long slab for the 3 m high column models and about nine times for the 6 m high column models. Comparing the results of the two modeled building heights (3 m and 6 m) show that the 6 m models have lower reactions at external columns but larger top displacements due to the thermal loading.

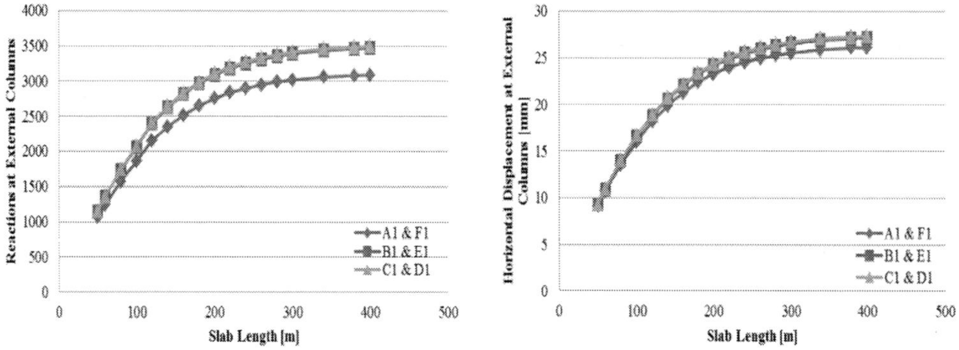

(a) (b)

Fig. 2. Results of thermal study at external columns for 3 m height models and different slab length: a) Reaction and b) Horizontal displacement.

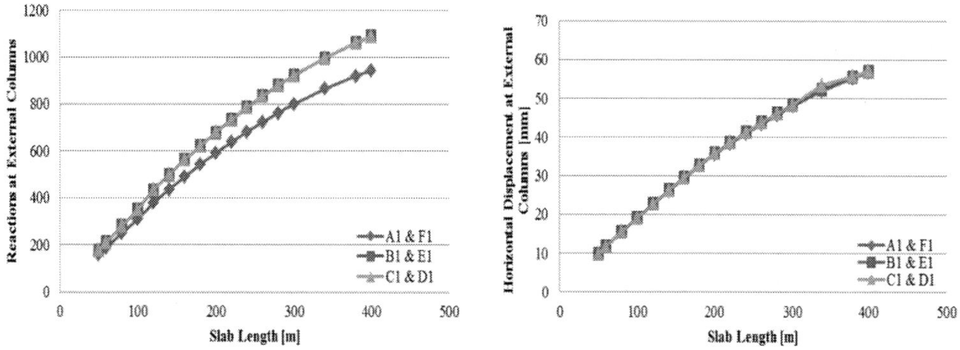

(a) (b)

Fig. 3. Results of thermal study at external columns for 6 m height models and different slab length: a) Reaction and b) Lateral displacement.

3.2 Effect of support conditions on the thermal response

To study the effect of support type on the displacement and reactions of external columns under thermal load a second set of thirty four new models were created. This set of models had the same properties as the first set of models except that the second set had hinged support condition instead of fixed support condition. The displacements under thermal load results for both sets of models are shown in Table 2 and Table 3 for models with column heights (3 m) and (6 m), respectively. The comparison between the results of both sets of models showed that the displacement for the models with hinged support had higher values than that for the fixed support. This increase in the value of displacement increased with the increase of the slab length for each column height. When comparing the results for the 3m high columns models and the 6 m high column models it was found that the support condition had a larger effect on the shorter column length. The ratio of the lateral displacement for the fixed support condition to the hinged support condition for the 6 m column height is larger than 75% for all slab lengths. For the 3 m high column models the ratio of the lateral displacement for the fixed support conditions to the hinged support condition for slab lengths lesser than 160 m were more than 75%. This ratio decreases with the increase of the slab length with a reduction reaching 45% for slab length of 400 m.

Table 2. Lateral deformation for column height (3 m) with fixed and hinged supports.

Slab Length (m)	Fixed columns conditions			Hinged columns conditions		
	A1 & F1 [mm]	B1 & E1 [mm]	C1 & D1 [mm]	A1 & F1 [mm]	B1 & E1 [mm]	C1 & D1 [mm]
50	9.1397	9.258	9.2946	9.8593	9.8544	9.8559
60	10.6902	10.9077	10.9556	11.8096	11.8055	11.8065
80	13.5291	13.9285	13.998	15.6769	15.674	15.6736
100	15.9955	16.5529	16.6606	19.4882	19.4863	19.4845
120	18.1666	18.7634	18.8693	23.1548	23.1542	23.1516
140	19.8894	20.5792	20.6982	26.7737	26.7737	26.7699
160	21.2884	22.0537	22.1834	30.2894	30.286	30.2849
180	22.4085	23.2343	23.3726	33.6918	33.693	33.6867
200	23.2953	24.169	24.314	36.9725	36.9742	36.9668
220	23.9912	24.9024	25.0528	40.1246	40.127	40.1184
240	24.5334	25.4739	25.6284	43.143	43.1459	43.1363
260	24.9536	25.9167	26.0744	46.0239	46.0273	46.0166
280	25.2778	26.2585	26.4186	48.765	48.7688	48.7572
300	25.5271	26.5213	26.6833	51.3654	51.3696	51.3571
340	25.8649	26.8773	27.0419	56.1462	56.1513	56.1371
380	26.0624	27.0855	27.2516	60.3813	60.3872	60.3715
400	26.1277	27.1543	27.3209	62.3026	62.3088	62.2924

Table 3. Lateral deformation for column height (6 m) with fixed and hinged supports.

Slab Length (m)	Fixed columns conditions			Hinged columns conditions		
	A1 & F1 [mm]	B1 & E1 [mm]	C1 & D1 [mm]	A1 & F1 [mm]	B1 & E1 [mm]	C1 & D1 [mm]
50	9.7842	9.8001	9.8057	9.9	9.9	9.9
60	11.691	11.7247	11.7324	11.9	11.9	11.9
80	15.4494	15.5183	15.5299	15.8	15.8	15.8
100	19.1189	19.2221	19.2375	19.7	19.7	19.7
120	22.6826	22.7991	22.8186	23.6	23.6	23.6
140	26.1135	26.2569	26.2801	27.5	27.5	27.5
160	29.4063	29.5755	29.6022	31.4	31.4	31.4
180	32.5508	32.7447	32.7747	35.2	35.2	35.2
200	35.5395	35.7569	35.79	39	39	39
220	38.3674	38.607	38.6431	42.7	42.7	42.7
240	41.0318	41.2923	41.3312	46.5	46.5	46.5
260	43.532	43.8121	43.8538	50.1	50.2	50.1
280	45.8694	46.1679	46.212	53.8	53.8	53.8
300	48.047	48.3625	48.4089	57.4	57.4	57.4
340	51.9406	52.2867	52.3373	64.4	64.4	64.4
380	55.2592	55.6313	55.6854	71.2	71.3	71.2
400	56.7203	57.1039	57.1595	74.6	74.6	74.6

The horizontal reaction under thermal load results for both sets of models are shown in Table 4 and Table 5 for models with column heights (3 m) and (6 m), respectively. The comparison between the horizontal reaction results of both sets of models showed that the reaction for the models with hinged support had lower values than that for the fixed support. The effect of support condition was larger on models with smaller slab lengths for both column heights. The column height was found to be inversely proportional to horizontal reaction values. It is clear that models with 3 m story height and fixed columns supports have the most critical reaction values. Horizontal reaction related to column height 3 m is three to five times bigger than reactions related to columns with height 6 m. Fixed columns reactions for models with 3 m column height are around 15 times the hinged column reaction for model with 3 m column height and slab length 50 m. This ratio decreased for models with longer slabs where, this ratio is around 6 for models with slab length of 400 m.

Table 4. Horizontal reactions for column height (3m) with fixed and hinged supports.

Slab Length (m)	Fixed columns conditions			Hinged columns conditions		
	A1 & F1 [kN]	B1 & E1 [kN]	C1 & D1 [kN]	A1 & F1 [kN]	B1 & E1 [kN]	C1 & D1 [kN]
50	1060.24	1146.31	1151.28	37.38	74.31	74.56
60	1240.91	1352.62	1358.96	45.19	90.16	90.36
80	1571.72	1730.42	1739.33	60.68	121.58	121.68
100	1859.13	2058.64	2069.73	75.95	152.55	152.55
120	2147.02	2393.91	2408.15	111.88	217.19	217.41
140	2351.02	2627.11	2642.98	129.61	252.36	252.49
160	2516.67	2816.47	2833.66	146.83	286.52	286.56
180	2649.3	2968.09	2986.34	163.5	319.58	319.54
200	2754.31	3088.13	3107.21	179.57	351.46	351.33
220	2836.7	3182.32	3202.06	195.01	382.09	381.88
240	2900.9	3255.71	3275.96	209.8	411.42	411.14
260	2950.65	3312.58	3333.23	223.91	439.41	439.06
280	2989.04	3356.46	3377.42	237.34	466.05	465.63
300	3018.56	3390.22	3411.41	250.08	491.31	490.83
340	3058.56	3435.94	3457.45	273.5	537.77	537.17
380	3081.95	3462.67	3484.38	294.24	578.92	578.21
400	3089.68	3471.51	3493.27	303.66	597.59	596.83

Table 5. Horizontal reactions for column height (6m) with fixed and hinged supports.

Slab Length (m)	Fixed columns conditions			Hinged columns conditions		
	A1 & F1 [kN]	B1 & E1 [kN]	C1 & D1 [kN]	A1 & F1 [kN]	B1 & E1 [kN]	C1 & D1 [kN]
50	157.36	176.65	176.92	9.05	17.38	17.47
60	188.25	211.92	212.2	10.95	21.1	21.17
80	249.12	281.43	281.73	14.75	28.52	28.57
100	308.56	349.29	349.61	18.53	35.91	35.94
120	377.2	432.08	432.63	27.25	50.82	50.91
140	434.38	498.2	498.78	31.78	59.44	59.51
160	489.25	561.66	562.26	36.28	68.01	68.05
180	541.65	622.26	622.89	40.75	76.51	76.53
200	591.46	679.86	680.52	45.18	84.94	84.94
220	638.58	734.36	735.04	49.57	93.3	93.27
240	682.98	785.71	786.41	53.91	101.57	101.52
260	724.65	833.9	834.61	58.21	109.75	109.68
280	763.6	878.94	879.68	62.46	117.83	117.74
300	799.89	920.91	921.66	66.65	125.81	125.7
340	864.78	995.95	996.74	74.87	141.45	141.29
380	920.08	1059.9	1060.72	82.84	156.62	156.42
400	944.43	1088.06	1088.89	86.73	164.01	163.8

4 Conclusions

Throughout this research, sixty eight ETABS models of one-story reinforced concrete frame buildings were studied under thermal load from UAE registered values. These models were divided into two sets: fixed support reaction and hinged support reaction and each set consists of two model groups: the first group with column heights of 3 m and the second group with column heights of 6 m. Each group consists of 17 different ETABS models. The difference between these models is the concrete slab length which was increased from 50 m to 400 m by increments of 10 m. The findings of this research confirmed the considerable additional lateral deformations at slab level and horizontal forces at supports due to temperature loads. These deformations and reactions increases considerably with increasing the slab length. For a fixed support condition the horizontal displacements for 400 m slab was about three times that for the 50 m slab length for the 3 m high column models and about five times for that in the 6 m high column modes. Whereas, the reactions for the 400 m long slab was about three times that for 50 m long slab for the 3 m high column models and about nine times for the 6 m high column models.

When comparing the thermal loads results of both fixed and hinged support conditions, the displacement for the models with hinged support condition had higher values than that for the fixed support condition. The ratios of the lateral displacement under thermal loads for the fixed support condition to the hinged support condition for all the 6 m column height models and 3 m column height models with slab lengths lesser than 160 m were larger than 75%. This ratio decreases with the increase of the slab length for the 3 m column height models with a reduction reaching 45% for slab length of 400 m. The comparison between the horizontal reaction results of both sets of models showed that the reaction for the models with hinged support had lower values than that for the fixed support. The effect of support condition was larger on models with smaller slab lengths and shorter column heights. Horizontal reaction related to column height 3m is three to five times bigger than reactions related to columns with height 6m. Fixed columns reactions for models with 3m height are around15 times the hinged column reaction for model with 3m column height and slab length 50m. This ratio will be decreased for models with longer slabs where, this ratio will be around 6 for models with slab length of 400m.

References

[1] ADIBC 2013, *Abu Dhabi international building code*, 1, The Higher Building Code Committee Abu Dhabi, United Arab Emirates, (2013)
[2] ACI Committee 224, *Joints in concrete construction*, 3, ACI, US, (2001)
[3] ACI Committee 209, *Prediction of creep, shrinkage, and temperature effects in concrete structures*, 1, ACI, US, (1997)
[4] SCSE Committee, *Expansion joints in buildings*, 65, National Academy of Sciences, Washington, D.C., US, (2000)
[5] ACI Committee 330, *Guide for design and construction of concrete parking lots*, 1, ACI, US, (1997)
[6] The International Federation for Structural Concrete (Federation international du beton, fib), *fib Model Code for Concrete Structures 2010*, 1, Wiley, Germany, (2010)

Vulnerability Study of Public Buildings Subjected to Earthquake Event

Ismail Rozaina[1,2,*], *Ibrahim* Azmi[2], and *Razali* Norhazlila[2]

[1]Institute for Infrastructure Engineering and Sustainable Management (IIESM), Universiti Teknologi MARA, 40450 Shah Alam, Selangor, Malaysia
[2]Faculty of Civil Engineering, Universiti Teknologi MARA, 40450 Shah Alam, Selangor, Malaysia

Abstract. The present article addresses on the earthquake study due to performance of critical frame reinforced concrete building. The buildings are analysed using Finite Element Modeling (FEM) under different types of analyses including Free Vibration Analysis (FVA), and Time History Analysis (THA) with varies earthquake intensities. The performances of the structure are shown by the yield point at beam-column connections where the internal forces at beam elements exceed the design capacity of the beams. The performance of critical frame reinforced concrete building when subjected to earthquake motion (Near source: Acheh) was addressed. The level of the damage state (Minor, Moderate, Major, Collapse level) has been defined as well. As a result, a greater damage index means that the members yield earlier and the plastic rotation is larger and vice versa. In general, there are no significant damage occurred to the structure. However, some non-structural elements of the building are expected to experience minor damages. These two buildings has no structural response to the earthquake because both Damage Index are less than 1.000. The study indicates that more than 50% of the buildings produced dynamic amplification factors of slightly more than one indicating not much of a dynamic response to the buildings.

1 Introduction

The reinforced concrete frame is easy to get affected due to this earthquake impact. Most of structural damages are results from inability of structural to perform surrounding soil and surface ruptures. It is essential to have a proper design detailing and selection of materials so that the reinforced concrete structure can withstand under severe combination of loadings such as dead load, imposed load, and earthquake loading. Malaysia is located at the low intensity of earthquake region. There are major tectonic plates around Malaysia. Engineers should concern and consider the earthquake impacts of buildings in Malaysian's building design procedure. According to Li [1], the reinforced concrete frame has to meet the requirements of performance-based earthquake engineering, since the influence of damage should be considered when conducting a structural performance assessment.

*Corresponding author: rozaina_fka_uitm@yahoo.com

Moreno [2] present the performance of the building in low to moderate earthquake region. The results show that the buildings expected to have damages due to its poor expected performance and the seismic vulnerability study is very important to predict the damage of the buildings due to earthquake event. R, Ditommaso [3] estimated the fundamental period of reinforced concrete buildings and damage of structural and non-structural element.

IDARC2D performs the nonlinear analysis. Barron [4] found that the flexible model results more displacement and interstorey drift than the rigid frame by get used of the IDARC 2-Dimensional software for reinforced concrete building [5]. The performance normally involves a qualitative process of measurement. Due to the reflection and expectation of the damage state, the values and parameters should be take place in order to measure quantitatively due to the performance level. Table 1 shows the ATC-13 damage level [6] in [7] to define the damage state level which adopted in this study. Table 2 shows the list of buildings analysed in this study.

Table 1. The Structural Engineers Association of California (SEAOC) damage levels [6].

SEAOC Earthquake Level	SEAOC Damage	ATC-13 Damage Factors (State)
Minor	Without any damage	D.F.* = 0 (None) D.F. < 0.01 (Slight)
Moderate	No structural damage, some non-structural damage	0.01 < D.F. < 0.10 (Light) 0.01 < D.F. <= 0.30 (Moderate)
Major	No collapse, some structural damage, non-structural damage considerable	0.30 < D.F. <= 0.60 (Heavy) 0.60 < D.F. < 1.0 (Major)
Collapse	Collapse	D.F. = 1.0 (Destroyed)

Table 2. List of buildings analysed.

No.	Building Name	Story	Height (m)
1	Pejabat Daerah Kuala Terengganu (PDKT)	5	19.20
2	Yayasan Negeri Sembilan (YNS)	13	48.58

2 Free vibration analysis

Free vibration analysis is needed to understand the character of the structures for dynamic impact. According to Brebbia and Nardini [7], free vibration analysis, however, consists of finding only the natural frequencies and modes of the system. The natural periods and mode shapes are the most important factors to determine the dynamic characteristic. By identifying those parameters as well as the periods of the earthquakes, the dynamic characteristic can be specified. The factors are calculated by finding the ratio between the periods of earthquakes and the natural period of the structures. The formula for damping ratio ξ, frequency ratio β and dynamic amplification factors D [8] are shown as below.

$$D = \frac{1}{\sqrt{1-\beta^2+(2\xi\beta)^2}} \tag{1}$$

where;
ξ = C/Cr, (C is structure damping and Cr is critical damping.)
$\beta = \varpi / \omega$ (where ϖ is the frequency of earthquake excitation and ω is the angular frequency of structure.

It shows that the resonance effect at beta = 1. If D is equal to 1 the response of structure in dynamic is equivalent to the static response. However if D is less than 1, there is no structural response to the earthquake load.

Table 3 shows the dynamic characteristic parameter for the public buildings in Malaysia. For dynamic characteristic parameter we need to know the stories of the building, time period T which in second, frequency f in Hz, and angular frequency ω in radian. All of these parameter we can get it from the SAP2000 software. Meanwhile, for dynamic amplification we can calculate the ϖ, β and D from all the value that we gained from the software. Pejabat Daerah Kuala Terengganu consists of 5 stories and the time period of the building is 12.2533 second, the frequency is 0.08161 Hz and the angular frequency is 0.5128 radian. Meanwhile, for Yayasan Negeri Sembilan consists of 13 stories and the time period of the building is 16.0942 second, the frequency is 0.06213 Hz and the angular frequency is 0.3903 radian. All of this value is gained from the SAP2000 software. After that, these values are going to be used to calculate the dynamic amplification.

Table 3. Dynamic characteristic parameter for the public buildings in Malaysia.

Building Name	Story	Time period, T (sec)	Frequency, , f	Angular Frequency, ω (rad)
Pejabat Daerah Kuala Terengganu (PDKT)	5	12.2533	0.08161	0.5128
Yayasan Negeri Sembilan (YNS)	13	16.0942	0.06213	0.3903

Table 4. Dynamic amplification for buildings in Penisular Malaysia.

Site Buildings To Epicenter Acheh	T	$\varpi = \frac{2\pi}{T}$	ω	$\beta = \frac{\varpi}{\omega}$	D = $\frac{1}{\sqrt{1-\beta^2} + (2\xi\beta)^2}$
Pejabat Daerah Kuala Terengganu (PDKT)	3.02	2.08	0.51	4.08	0.064
Yayasan Negeri Sembilan (YNS)	2.36	2.66	0.39	6.82	0.022

Table 4 shows the dynamic amplification for buildings in Peninsular Malaysia. For dynamic amplification parameter we need to know the epicenter of the earthquake, period T which in second, ϖ, ω, β and lastly the dynamic amplification, D. For Pejabat Daerah Kuala Terengganu the time taken, T is 3.02 second, ϖ is 2.08, angular frequency ω is 0.51, β is 4.08 and the dynamic amplification is 0.064. Meanwhile, for Yayasan Negeri Sembilan the time taken, T is 2.36 second, ϖ is 2.66, angular frequency ω is 0.39, β is 6.82 and the dynamic amplification is 0.022. This dynamic amplification is calculated based on the formula that stated in the table below. The results show that the Pejabat Daerah Kuala Terengganu is having a factor of 0.064 due to Acheh earthquake. Whereas the Yayasan Negeri Sembilan is having a factor of 0.022.

3 Dynamic non-linear analysis

3.1 Time history ground motion

The present study the intensities as reported by MMD [8]. The Sumatra–Andaman (Acheh Earthquake) time history record is shown in Fig. 1 with 5000 number of outputs within 0.02 time steps with the maximum acceleration of 0.012g. The Acheh earthquake occurred in December 26, at Indian Ocean near to Acheh with magnitude 9.3 on the Richter scale. The acceleration is simulated to four variation of earthquake intensities; 0.05g, 0.10g, 0.15g, and 0.2g. This study used 5% damped spectral acceleration as adopted in [9-13].

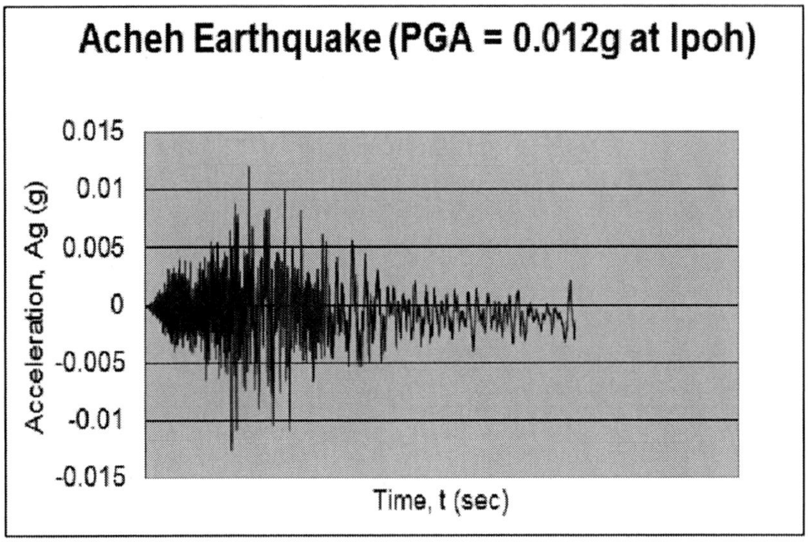

Fig. 1. Time history record of Acheh earthquake for peak ground acceleration 0.012g (December 26, 2004) [10-12].

3.2 Modeling Concepts

IDARC-2D was used to analyze non-linear dynamic to evaluate the vulnerability of the structures based on the damage indices. Fig. 2 outlines the modeling procedures. The modeling concepts applied are regular medium-rise structure and reinforced concrete with moment resisting frame. The dynamic non-linear analysis was performed using computer software IDARC 2D (Damage Analysis). British Standard [14] is referred to determine the loadings applied to the structure in the analysis steps to be applied as nodal loading in the modeling steps as shown in Fig. 4. FORTRAN based program developed by the University of Buffalo called IDARC2D was then used to carry out two dimensional analyses on selected frames from each building. This program is able to compute what is known as "structural damage index" which can be defined as a way of quantifying numerically the seismic damage suffered by buildings. Details of structural members such sizes of columns and beams, their steel reinforcements and the cumulative column axial loads are taken from the results that calculate by manual calculation and entered as input data in IDARC2D. Nodal weights were calculated and were based on tributary areas to the node and frame in question. The models were analyzed using the modified design response spectra. The computed damage indices were tabulated for easy reference and comparison [14-16].

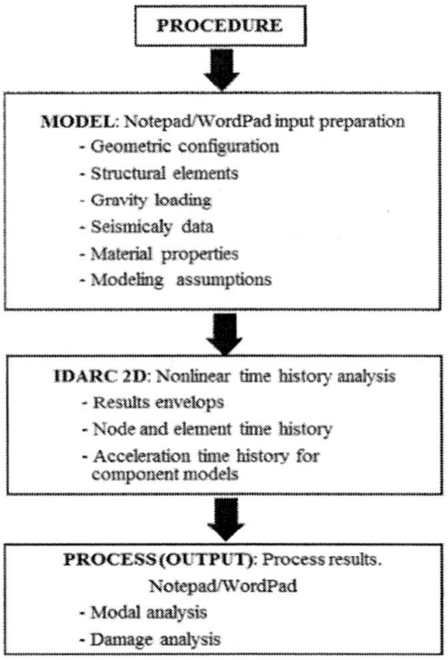

Fig. 2. Flowchart of modeling procedure.

3.3 Plastic Hinge

The term plastic hinge is used to describe the deformation of a section of a frame which consist of beams and columns where the plastic bending occurs. Normally, the plastic hinges for moment resisting frame type of buildings due to structural local failures occur either at beam or column connection. Fig. 3 shows the formation of the damage state location and plastic hinge of the moment resistance frame with four different intensities of ground acceleration. These represent the sequences of damage state of frames under earthquake loading. The study focus on damage level of the building and to know when the building starts yielding. Based on the result, for intensity 0.05g, the damage index is equal to zero and based on Damage Levels as stipulated in ATC-13 [6] the damage level at minor means free from damage. Table 5 and Table 6 show in detail the resultant of dynamic nonlinear analysis under various intensity; 0.05g, 0.10g, 0.15g, and 0.2g. These tables also show the location and time for each yielding point. All buildings has light damage level when the intensity 0.05 g applied to the model.

Table 5. Summarization of first yielding point and damage level for Pejabat Daerah Kuala Terengganu.

Intensity	Floor Level	Location	Time (sec)	Damage Index (Overall)	Damage Level
0.012	-	-	-	-	None
0.05	1	Column	2.29	0.044	Minor slight
0.10	1	Column	2.18	1.000	Collapse
0.15	1	Column	0.14	1.000	Collapse
0.20	1	Column	0.11	1.000	Collapse

Table 6. Summarization of first yielding point and damage level for Yayasan Negeri Sembilan.

Intensity	Floor Level	Location	Time (sec)	Damage Index (Overall)	Damage Level
0.012	-	-	-	-	None
0.05	1	Beam	2.60	0.067	Minor slight
0.10	1	Beam	2.58	0.142	Moderate slight
0.15	1	Beam	0.69	0.168	Moderate slight
0.20	1	Beam	0.52	1.000	Moderate slight

In terms of damage pattern, the frame shows the locations of element structural where it started to yield, crack or collapse.

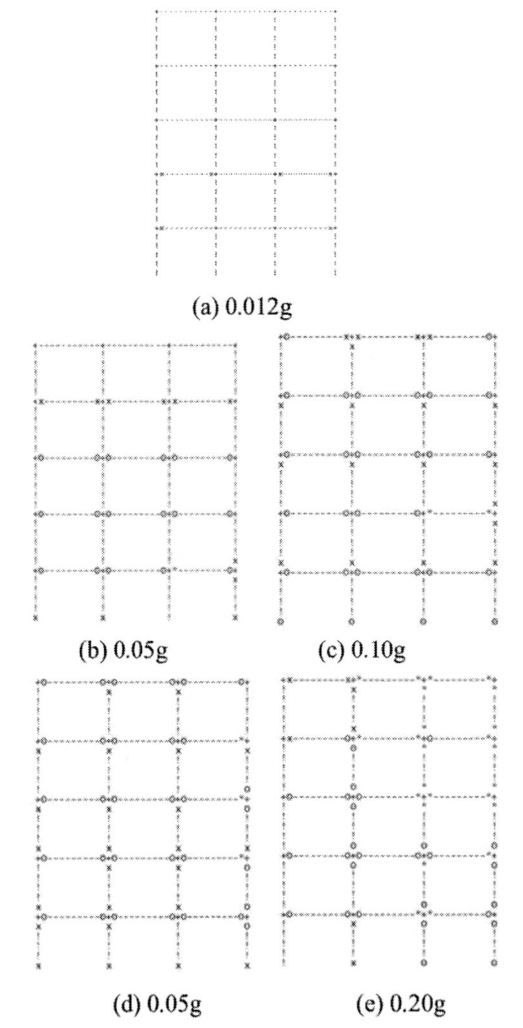

Notation: i = beam, ! = column, x = cracking (for concrete initial yield (for steel), o = plastic hinge developed

Fig. 3. Damage state frame under 0.012g, 0.05g, 0.10g, 0.15g and 0.20g earthquake intensity for Pejabat Daerah Kuala Terengganu.

From the results, the modal frame experienced cracking and yielding at the beam elements at the early stage. In Fig. 3, the 'x' symbol shows the occurrence of crack for concrete frame, the 'o' symbol shows the yield of plastic hinge has developed and the '*' symbol shows the local failure which exceed criteria. Based on Fig. 3, Pejabat Daerah Terenganu has experienced no damage at the early stage which 0.012g intensity applied. For intensity 0.05 the damage pattern show the frame has started experienced yielding and cracking. For example the initial yield of the plastic hinge starts to develop at the beam in the first floor. Apart from the top floor, all beam ends act as plastic hinges, while only the columns show a few hinge behaviour. The beams act as hinges earlier than columns, nearly forming an overall symmetrical beam hinge system. The structure performance measured by the quantitative damage index is more reliable than displacement-based design.

3.4 Performance level of the structure

The program developed by Park [5] is used in this study to provide a measure of the accumulated damage index sustained by the components of the building included the maximum to ultimate deformations ratio. Table 7 shows in detail the summary of overall damage index of the buildings. The overall damage index for the building which can be referred to the SEAOC Damage Level in Table 1 and Table 8. The overall damage index for intensity 0.05g categorized at minor level which no damage occurred for Block K and Police Quarters. Both building experienced no damage during 0.012g intensity and experienced minor (light) level that means no structural damage, some non-structural damage. The reinforced concrete frames was completely collapse for Pejabat Daerah Kuala Terengganu for 0.20g intensity and Yayasan Negeri Sembilan for 0.10g intensity applied.

Table 7. Summarization of damage level.

No.	Building name	Type of building	Overall Structural Damage Index (g)				
			0.012	*0.05*	*0.10*	*0.15*	*0.20*
2.	Pejabat Daerah Kuala Terengganu	Medium-Rise	0	0.067	0.142	0.168	1.000
3.	Yayasan Negeri Sembilan	High-Rise	0	0.044	1.000	1.000	1.000

Table 8. The contour damage level [6].

SEAOC EQ Level	SEAOC Damage	ATC-13			
		None		=	0
Minor	Without any Damage	Slight		<	0.01
	No Structural damage, some non-structural	Light		≤	0.1
Moderate	damage	Moderate		≤	0.3
	No collapse, some structural damage, non-structural damage	Heavy		≤	0.6
Major	considerable	Major		<	1
Collapse	Collapse	Collapse		=	1

4 Conclusions

Most buildings that have been analysed subjected to the Acheh earthquake ground motion, have a damage indexes in the range of 0.0 to 1.000. This low index that the buildings have only light or minor damage level. Local failures are mostly developed at beam connections followed by column connections. In general, there are no significant damage occurred to the structure. However some non-structural elements of the building are expected to experience minor damages. These two buildings has no structural response to the earthquake because both damage index are less than 1.000. From the overall analyses, it can be concluded that the high rise buildings in this study are affected by earthquake load more than the medium rise buildings.

The authors would like to acknowledge that this research has been carried out as part of a project FRGS with File No. 100-RMI/SF 16/6/2 (3/2015) funded by Ministry of Science, Technology and Innovation (MOSTI) and support from Universiti Teknologi MARA (UiTM).

References

[1] A. Adnan, T. L. Y. Patrick, R. Ismail, S.M. Shamsuddin, Artificial neural network application for predicting seismic damage index of buildings in Malaysia, Electronic J. of Structural Engineering, **12**(1), 1-9, (2012)
[2] M. Mardiyono, R. Suryanita, A. Adnan, Intelligent monitoring system on prediction of building damage index using neural network, TELKOMNIKA Indonesian J. of Electrical Engineering, 10(1), 155-164, (2012)
[3] A. Adnan, H. Hendriyawan and M. Irsyam, The effect of the Latest Sumatra earthquake to Malaysian peninsular, J. Kejuruteraan Awam, **14**(2), 12-25, (2002)
[4] T. Balendra and Z. Li, Seismic hazard in Singapore and Malaysia, Seismic hazard of Singapore and Malaysia, Electronic J. Structural Engineering, Special Issue (Earthquake engineering in the low and moderate seismic regions of Southeast Asia and Australia), **8**, 57–63, (2008)
[5] M. F. Jamaludin, *Performance of low rise building due to earthquake*, Bachelor Degree Thesis, Universiti Teknologi MARA, Shah Alam, Malysia, (2011)
[6] M. Hassim, *Analysis of medium rise building due to earthquake*, Bachelor Degree Thesis, Universiti Teknologi MARA, Shah Alam, Malysia, (2011)
[7] C.A. Brebbia and D. Nardini, Dynamic analysis in solid mechanics by an alternative boundary element procedure, Int. J. of Soil Dynamics and Earthquake Engineering **2**(4), 228–33 (1983)
[8] A.K. Chopra, and R.K. Goel, *Capacity-demand-diagram methods for estimating seismic deformation of inelastic structures: SDF systems*, Rep. No. PEER-1999/02, Pacific Earthquake (1999)
[9] M.S.H. Shamsudin, *Performance of low rise concrete building under earthquake loadings*, Bachelor Degree Thesis, Universiti Teknologi MARA, Shah Alam, Malysia (2010)
[10] R. Ismail, I. Azmi and A. Adnan, Application of artificial neural network in damage predeiction of buildings based on seismic zonation in Malaysia, *Proc. of the 27th Conference of Asean Federation of Engineering Organisations (Cafeo 27),* Suntec City Convention Centre, Singapore, (2011).
[11] R. Ismail, I. Azmi and A. Adnan, Seismic damage analysis of reinforced concrete frame of public buildings in Ipoh subjected to Acheh Earthquake event, *Proc. of The Int. Civil and Infrastructure Engineering Conf.*, Springer Singapore , 149 -157 (2015)

[12] R. Ismail, *The application of ANN in seismic evaluation of buildings*, Master Thesis, Universiti Teknologi Malaysia, Shah Alam, Malysia, (2008)
[13] R. Ismail, I. Azmi and A. Adnan, Vulnerability of high-rise buildings in Kuala Lumpur subjected to Acheh earthquake event, *Proc. of the Int. Civil and Infrastructure Engineering Conference*, Springer Singapore, 139 -148 (2015)
[14] Y.J. Park, and A.M. Reinhorn, and S.K. Kunnath, , *IDARC: Inelastic damage analysis of reinforced concrete frame shear wall structures*, Technical Report: National Center for Earthquake Engineering Research, (1990)
[15] Anaxagoras Elenas, Intensity parameters as damage potential descriptors of earthquakes, *III ECCOMAS Thematic Conf. on Computational Methods in Structural Dynamics and Earthquake Engineering*, Corfu, Greece, (2011)
[16] R. Rajabi, M. Barghi, and R. Rajabi, Investigation of Park-Ang damage index model for flexural behavior of reinforced concrete columns, Structural Design of Tall & Special Buildings, **22**(17), pp. 1350–1358 (2013)

Seismic Analysis of Concrete Dam by Using Finite Element Method

Ismail Rozaina[1,2*], *Kamsani* Muhamad Hilfi[2], and *Mohd Nadzri* Nur Insyirah[2]

[1]Institute for Infrastructure Engineering and Sustainable Management (IIESM), Universiti Teknologi MARA, 40450 Shah Alam, Selangor, Malaysia
[2]Faculty of Civil Engineering, Universiti Teknologi MARA, 40450 Shah Alam, Selangor, Malaysia

Abstract. This paper reports a brief study on linear seismic analysis of Sg. Kinta Concrete Dam. The analysis was conducted in order to determine the performance and behaviour of the dam under seismic excitation. The dam was modelled as two-dimensional and developed based on the design drawing that is obtained from Angkasa Consulting Services Sdn. Bhd. The seismic analysis of the dam is conducted using finite element analysis software package LUSAS 14.3 and the dam has been analyse as a plain stress problem with a linear consideration. A set of historic data, with El Centro earthquake acceleration of about 0.50g is used as an earthquake excitation. The natural frequency and mode shape up to fifth mode of the dam has been obtained from the analysis to show the differences of the stress and deformation between each mode. The maximum horizontal and vertical stress of Sg. Kinta dam was found and the distribution of them was discussed in form of contours. The deformation of the dam were also been discussed by comparing the maximum displacement for each mode shaped.

1 Introduction

Dam is a structure that acts as a barrier to hold water or retain the water. Generally, dam has been built since many years ago and there are more than 45000 dams around the world. One of the countries that have lots of dam structure is China with over 22000 of large dams. Most of the dams were built for the irrigation purposes and almost all the large dams were used to produce electricity. There are many dams purpose other than produce electricity such as flood control, supply water to the cities, and river navigation. There are also dams that are built to serve multipurpose providing two or more benefits at once. The Sg. Kinta Dam was operated in August 2007 for the purpose of water supply and it was the first dam that used Roller Compacted Concrete (RCC) Technology in Malaysia. The dam reservoir has 29.9 million m^3storage to supply 227 million litre of water per day (mld) to a new treatment plant and 85 mld to the existing Ulu Kinta treatment plant. The dam is constructed with a height of 85m, 760m long and 70m wide. Sg. Kinta Dam able to provide 639 million litres of water per day to Kinta Valley area.Though, Perak Water Board (LAP) is planning to use Sg. Kinta Dam as electricity generator with the agreement of Tenaga Nasional Berhad (TNB) in their future plan. The finite element method is an approximate

*Corresponding author: rozaina_fka_uitm@yahoo.com

analysis technique which is basically consists of dividing a structure into a number of parts called elements [1]. Lee et al. [2] and Taghizadeh et al. [3] also adopted finite element method to study the performance of their structure as well. This method consists of assuming the continuous function for the solution and obtaining the parameters of the functions in a manner that reduces the error in the solution [4].Norshariza [1] stated that the finite element method is an approximately analysis technique which basically consists of dividing a structure into a number of parts called elements. Each node may have one or more degree of freedom which is defined as independent displacements used to describe the movement of each node. The number and type of elements should be selected so that the deformed shaped of the structure can be effectively represented. Deformations within the elements are assumed to follow predefined functions known as shape functions. The finite element method is the best method used in analysing the internal forces of elements by solving differential equation obtained through their discretization in space dimensions [5].

The most important step in finite element analysis of structure is creating mathematical modelling composed of separated elements which stimulate the real continuous structure. The purpose of this modelling is having a system with limited degree of freedom instead of system infinite degree of freedom [6]. In dam engineering, the finite element analysis is often used to model the physical behaviour of an existing concrete dam. When analysis is completed and results are obtained, large stress tensor fields must be manipulated to recuperate essential engineering quantities: the force and moment resultants acting on various user-defined surfaces [7]. Bathe [8] claimed that finite element method is a numerical method that can be used to solve different kinds of engineering problem in the stable, transient, liner or non-linear cases. De Salvo and Swanson [9] stated that by using the programs with interactive graphical facilities, it is possible to generate finite element models of complex structures with considerable ease and to obtain the results in a convenient, readily assimilated form.

Earthquake is a movement of earth crust which formed along faults and ruptures.Earthquake can be recorded using an instrument called as seismometer. Richter scale is used to represent magnitude of earthquakes. The effects of earthquake can be very severe and can cause major property damage [10]. Braile [11] states that there are four types of seismic wave namely Primary waves (P-waves), Secondary waves (S-waves), Love waves and Rayleigh waves. Seismic design and evaluation is most often based on liner-elastic response-spectrum or time-history analysis procedures, although nonlinear analysis procedures can be used for evaluation of certain nonlinear mechanisms [12]. The equation of displacement of dynamic analysis of dam with n degree of freedom are as follow:

$$M\ddot{U} + C\ddot{U} + KU = F(t) \tag{1}$$

If the displacement is occurred with empty reservoir, the equation of dynamic displacement of dam can be written as:

$$M\ddot{U} + C\dot{r} + Kr = F(t) \tag{2}$$

The absolute and relative displacement of dam nods equation can be shown as:

$$U = r + j.U_g(t) \tag{3}$$

The result of substitution of equation (2) and (3) is:

DOI: 10.1051/matecconf/201710302024

$$M\ddot{r} + C\dot{r} + Kr = - \text{M}.j.\ddot{U}_g(t) \quad (4)$$

The *M, C* and *K* are Mass, damping and stiffness matrix respectively and $\dot{r}$ is applied earthquake acceleration and *r* is the relative displacement of the nodes [13].

Time history is a set of graphical data that shows the intensity of the earthquake and contains acceleration, g of the earthquake event [14]. Seismic analysis of the dam or structure required the time history data. There are several methods exist to input seismic excitation when making the seismic time-history analysis for design: 1) to input the displacement time-history at the base, which is called displacement method; 2) to input the inertia loading calculated from the time-history of support motion acceleration, which is called acceleration method; 3) large mass method [15]. For use in linear dynamic analysis, at least three time-histories (for each component of motion) should be used for each design earthquake. The two basic methods that are commonly used for the seismic dynamic analysis are time-history and response spectrum methods. Naibin J. et al. [15] states that the time-history method is relatively consumed more time, costly and lengthy. On the other hand, the response spectrum method is relatively more rapid, concise, and economical. However, time-history method must be employed when geometrical and/or material non-linearities are present in the structural systems. Nowadays, it is more convenient for using time-history method than before for advancing of computer's hardware and software.

Many researchers Major [16], Schoeber [17], Polyakov [18] and Wyatt [19] stated that earthquake or seismic loads are the major dynamic loads being considered in the analysis and design of dams especially in earthquake prone areas. They are of kinematic origin and owe their existence to vibration caused in the structure by the movement of the earth's surface during an earthquake. They have random characteristics and are regarded as deterministic in practical calculations to simplify the design model [16-20]. Parish et al. [19] shows the deformation of the dam by applying dynamic loading at the base of the foundation layer as a velocity excitation.

2 Research background

This study involved the modelling of Sg. Kinta Concrete Dam located at 12km north east of Ipoh which part of the Greater Ipoh Water Supply Scheme Phase 2. The study comprised of several tasks in finite element modelling and involves a modelling technique for dynamic analysis using a computer program. This research starts with obtaining and understanding the literature review concerned with the related research topic followed by collection of the dam data, understanding the finite element using LUSAS 14.3, building and analyse a model by developing a two dimensions finite element model with the LUSAS 14.3 and finally obtains the result and conclusion based on the analysis.

Sg. Kinta Dam was operated in August 2007 for the purpose of water supply and it was the first dam that used Roller Compacted Concrete (RCC) Technology in Malaysia. The dam is straight axis gravity structure with vertical upstream face and a uniform downstream face at 0.75 in horizontal and 1.0 in vertical from the crest. Figure 1 shows the typical cross section of Sg. Kinta Dam. The material used for RCC is fly ash as replacement of cement to reduce cost, heat of hydration, to increase workability and have long term strength development.

The dam reservoir has 29.9 million m^3 storage to supply 227 million litre of water per day (mld) to a new treatment plant and 85 mld to existing Ulu Kinta treatment plant. The dam is constructed with a height of 85m, 760m long and 70m wide. Sg. Kinta Dam able to provide 639 million litres of water per day to Kinta Valley area. However, Perak Water

Board (LAP) is planning to use Sg. Kinta Dam as electricity generator with the agreement of Tenaga Nasional Berhad (TNB) in their future plan.

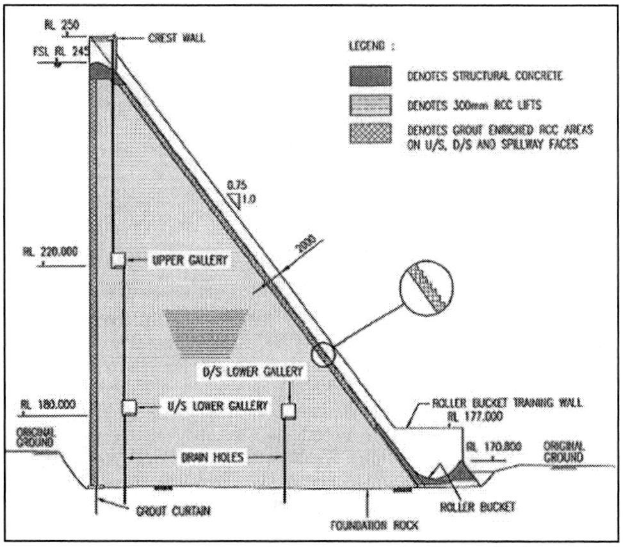

Fig. 1. Typical cross section of Sg. Kinta Dam [21].

3 Modal analysis

The main dynamic analysis is performed by using a modal approach or analysis due to the weight of the dam and horizontal component of the ground motion. The seismic excitation that has been used in the analysis is obtained from the E1 Centro time history data and presented in Fig 2.

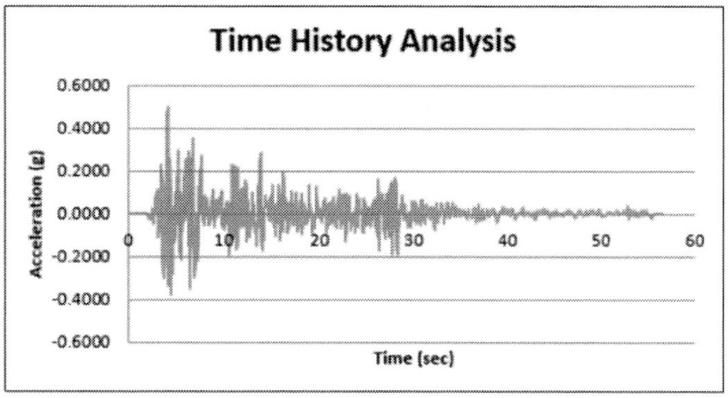

Fig. 2. Time history analysis of E1 centro earthquake [21].

The result obtained from Modal Analysis is the natural frequency and the mode shape of Sg. Kinta Dam under certain seismic excitation or loading. There is no limitation to the mode shape of the dam, but in this study, there are only five different mode shapes of the dam is presented with different natural frequency each. The purpose is just to show the dam deformation under the seismic excitation. Mode shape shows the specific pattern of vibration at a specific frequency at specific period of time. Table 1 shows different mode shape with different period of time.

Table 1. No. of mode shape associate with period.

No. of Mode Shape	Period (sec.)
1	0.32
2	0.24
3	0.17
4	0.16
5	0.13

It is noticed that different mode shape will be associated different frequency. It shows that the period of time is decreased from 0.32 to 0.13 from the mode shape 1 to mode shape 5. There is only a slightly difference in the natural frequency for each mode. This will makes the mode shape or the deformation of the dam change too little for each frequency and will be difficult to see the comparison between each shape.

4 Deformation analysis

The deformation level of each dam occurs due to the assignation of the dynamic loading during the analysis. It can be seen that the deformation of the dam are significantly to the horizontal direction with different displacement each. Fig. 3 shows mode shape 5 with 3.48mm of maximum displacement due to seismic excitation.

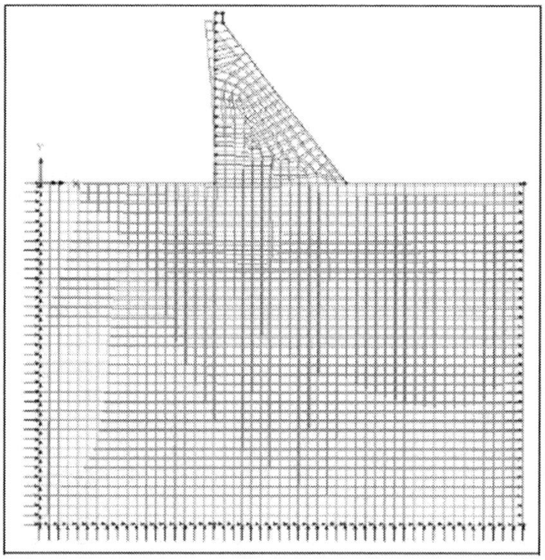

Fig. 3. Mode Shape 5 (Max. displacement = 3.48mm).

5 Stress analysis

The dynamic analysis shows the stress in the dam structures when subjected to the dynamic loading. The purpose of this stress analysis is to determine whether the dam can safely withstand the specified forces. Table 2 shows the normal stress and shear stress of the dam due to the seismic excitation.

Table 2. Maximum normal stress and shear stress.

No. of Mode Shape	Normal Stress (kN/m^2)	Shear StresskN/m^2)
1	194.120	109.008
2	341.909	221.248
3	199.523	97.123
4	2.29304	3.66383
5	436.699	87.1719

The stress of the dam is shown in contours in order to visualize the distribution of the stress. Fig. 4 to Fig. 7 shows the normal and shear stress contours for mode shape 1 and 5 respectively. Fig. 4 shows the maximum stress about 194.12 kN/m^2 at node 1222 and -108.8666 kN/m^2 of minimum stress at node 2 under dynamic loading. 436.699 kN/m^2 of maximum stress at node 1 and -294.538 kN/m^2 at node 2 was shown in Fig. 5.

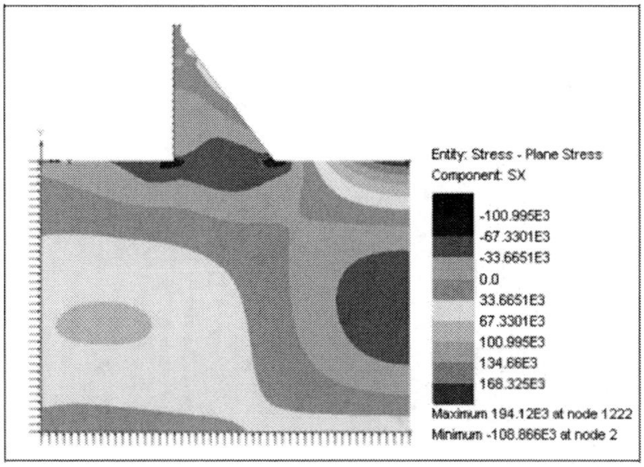

Fig. 4. Mode shape 1 (Normal Stress = 194.12 kN/m^2).

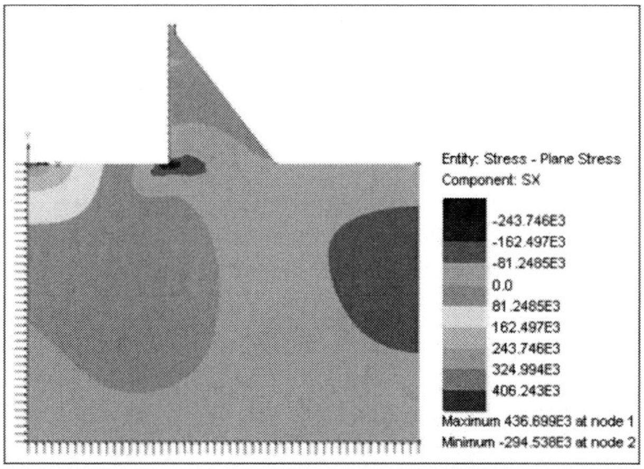

Fig. 5. Mode shape 5 (Normal Stress = 436.699 kN/m^2).

Fig. 6 shows the maximum stress about 109.008 kN/m^2 at node 1272 and -121.955 kN/m^2 of minimum stress at node 1919 under dynamic loading. 87.1719 kN/m^2 of maximum stress at node 1896 and -535.707 kN/m^2 at node 1927 was shown in Fig. 7.

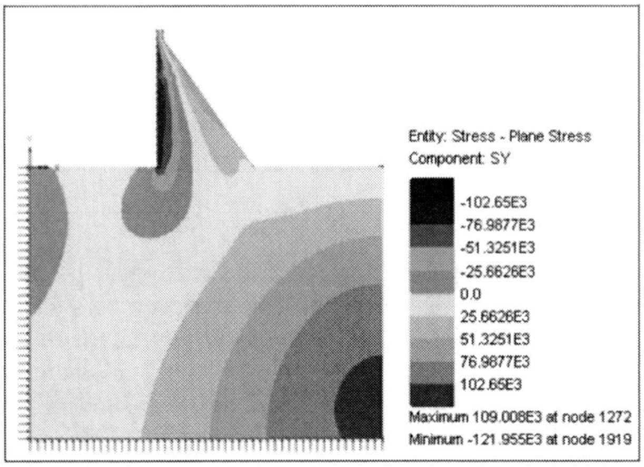

Fig. 6. Mode shape 1 (Shear Stress = 109.008 kN/m^2).

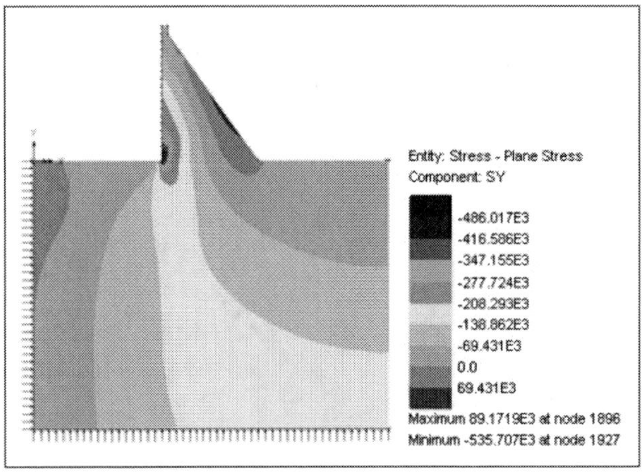

Fig. 7. Mode shape 5 (Shear Stress = 87.1719 kN/m^2).

It is noticed that for each mode shape there are different in the maximum and minimum stress. It is due to the variable of the time history data which is the acceleration of the earthquake. The maximum stresses obtained from the analysis need to be compared with the material allowable capacity in order to determine whether the stress exerted on the dam can with stand the seismic loading. The stress allowable capacity of concrete was around 800 kN/m^2 as shown in Table 3.

Table 3. Allowable capacity.

Normal Stress (Max.)	Shear Stress (Max.)	Allowable Capacity
221.248 kN/m^2	436.699 kN/m^2	800 kN/m^2

It can be seen that the normal stress and shear stress exerted by the dam does not exceeded the allowable capacity of the material.

6 Conclusion

An earthquake analysis of concrete dam, which is referred to as modal analysis and stress analysis has been used in order to find the performance and the behaviour of the dam. The method has been described previously, and the procedure is implemented in a computer program LUSAS 14.3. The analysis of the concrete dam which is Sg. Kinta Dam is considered as linear dynamic analysis and is carried out by utilizing 5 modes of shape for the dam. Based on the dynamic analysis, the performance of the dam can be considered as a satisfactory result due to the acceptable maximum displacement value which is 3.48mm in the mode shape 5. The stress behaviour of the dam was also satisfactorily acceptable because the dam normal and shear stress which is 221.248 kN/m^2 and 436.499 kN/m^2 respectively does not exceed the allowable stress capacity which is 800 kN/m^2. The result obtained from the linear dynamic analysis conducted in this project was also shown that LUSAS 14.3 software was capable in analysing the seismic response.

The authors would like to acknowledge that this research has been carried out as part of a project Science Fund with File No: 100-RMI/SF 16/6/2 (3/2015) funded by Ministry of Science, Technology and Innovation (MOSTI) and support from Faculty of Civil Engineering, Universiti Teknologi MARA (UiTM).

References

[1] M.B. Norshariza, *Modal analysis of concrete bridge decks subjected to free vibration*, Master of Engineering Thesis, Faculty of Civil Engineering, Universiti Teknologi Malaysia, (2005)
[2] J.E. Lee, J.B. Kim, K. Park, Finite element analysis for improvement of folding defects in the forging process of subminiature screws, J. of the Korean Society for Precision Engineering, **32**(6), 509-515 (2015)
[3] M. Taghizadeh, H.R. Ovesy and S.A.M. Ghannadpour, Nonlocal integral elasticity analysis of beam bending by using finite element method, Structural Engineering and Mechanics, **54**(4), 755-769 (2015)
[4] U.S. Dixit, Finite element method: *An introduction, department of mechanical engineering*, Indian Institute of Technology Guwahati-781, **39** (2008)
[5] A.A. Adedeji, Seismic analysis of earth wall gravity dam using decoupled modal approach, Int. Egyptian e-Journal of Engineering and Mathematics: Theory and Application, **5**, 19-34 (2008)
[6] M.A.L. Yanghin and M.A. Hesari, Dynamic analysis of the arch concrete dam under earthquake force with ABAQUS, J. of Applied Sciences, **8**(15), 2648-2658 (2008)
[7] P. McLean, P. Léger and R. Tinawi, Post-processing of finite element stress fields using dual kriging based method for structural analysis of concrete dams, Finite Elements in Analysis and Design, **42**(6), 532-546 (2006)
[8] K.J. Bathe, *Finite Element Procedures*, 1st ed., Prentice Hall Inc., Englewood Cliffs, New Jersey, 735, (1996)
[9] G.J. De Salvo and J.A. Swanson, *ANSYS: Engineering analysis system: User manual*, Swanson Analysis System, (1986)
[10] D. Hyndman, *Natural Hazards and Disasters*, 2nd ed., Brooks.Cole, USA, (2008)
[11] L. Braile, Seismic waves demonstration and animations, Purdue University, 1-15, (2010)

[12] Y. Ghanaat and A.K. Chudgar, Seismic design and evaluation of concrete dam - An Engineering Manual, (2007)
[13] O.C. Zienkiewicz and R.L. Taylor, *The finite element method*, 5th ed., Butterworth-Heinenmann, Oxford, UK, (2000)
[14] K.S. K.Sulaiman, Seismic vulnerability of smart tunnel, Universiti Teknologi Malaysia, (2008)
[15] J. Naibin, M. Qing and Z. Yixiong, *Research on time-history input methodology of seismic analysis*, (2007)
[16] A. Major, *Dynamics in civil engineering*, 2nd ed., Collet's Holding Ltd, London, (1980)
[17] W. Schoeber, Regarding the load bearing behaviour of large dams, Die Wasserwirstchaft, **71**(81/1287), (1981)
[18] S.V. Polyakov, *Design of earthquake resistant structures basic theory of seismic stability*, 2nd ed., Mir Publishers (1985)
[19] Y. Parish and F. Najaei Abadi, Dynamic behaviour of earth dam for variation of earth material stifness, Int. J. of Applied Science, Engineering Technology, **50**, 606-611, (2009)
[20] Y. Ghanaat and B.J. Rasmussen, *Evaluation of seismic stability of claytor dam using linear and nonlinear time history analysis*, (2005)
[21] K. Muhamad Hilfi, Seismic analysis of a concrete dam by using finite element method: A case study of Sungai Kinta dam, Bachelor of Engineering Thesis, Universiti Teknologi MARA, (2010)

An investigation of Crater Diameter on Plain Slab Foamed Concrete Rice Husk Ash (FCRHA) Exposed to Low Impact Loading

Josef Hadipramana[1,*], *Abdul Aziz* Abdul Samad[1], *Shahrul Niza* Mokhatar[1], *Fetra Venny* Riza[1], *Noridah* Mohamad[1] and *Mohd Yusuf* Mohd Wahab[1]

[1] Jamilus Research Centre, Faculty Civil and Environmental Engineering Universiti Tun Hussein Onn Malaysia, 86400 Parit Raja, Johor, Malaysia

Abstract. As sustainable material building and construction, the foamed concrete (FC) in this investigation was modified by adding the Rice Husk Ash (RHA) as sand replacement to increase its strength. Furthermore, this modification material (is called FCRHA) treated on impact loading. This investigation was motivated when the plain slab of FCRHA subjected to small impactor, then the nose impactor over all would penetrate into slab target due to porosity of FCRHA. The experimental produced plain slabs FCRHA and FC (as a control) with 1400 kg/m^3 and 1600 Kg/m^3 of densities. In impact test all plain slabs exposed by 40 mm steel blunt nose impactor with various impact velocities. The result showed the crater which produced by impact loading was not found spalling, scabbing, radial crack and widely cratering. This local damage occurred when porosity of FCRHA took over the impact loading. The nose impactor over all considered have been successful penetrated into slab of FCRHA and FC. Therefore, the diameter of crater equals to diameter of impactor. With this certainty, the prediction penetration depth on plain slab FCRHA (also FC) can be determined in future investigation. In addition, the penetration of impactor on FCRHA with low impact velocity give the same impression on penetration impactor with high impact velocity on FC.

1 Introduction

The application of sustainable material is not sufficient by only using lightweight aerated Foamed Concrete (FC) as construction and building material, yet the used of waste resources becoming popular ideas to reach the expected materials. Since the used of waste materials is cost effective, especially in construction industry. At the same time, the investigation on using of waste Rice Husk Ash (RHA) as construction materials is not something new [1-3] RHA in FC mixes can enhance the strength of FC [4, 5]. RHA increased the stress wall porosity in FC, and along with the increased stress, so as strain in wall porosity, which resulted in FC with RHA (FCRHA) increasing strength [6, 7]. Other than that, the FCRHA gives better impact energy absorption [6, 8]. The RHA act as

* Corresponding author: josef@uthm.edu.my

strengthtener in FC and increase the elasticity of FCRHA, resulting in decreased falling fragment and spalling [9]. When FCRHA exposed to impact loading then, FCRHA is able to spread the impact energy to all slab body and the porosity in FCRHA take over those loading, wherein the impact energy wave changed to tensile or compressive energy (depend on the constraint placed on a slab) taken over by porosity [10].

The basic idea of this investigation presents, that slab of FC or FCRHA decreased the local damage (crater) as wide as touching area of impactor on slab surface, when the slab subjected to impact loading. This local damage occurs, when the porosity of slab takes over the impact energy and prevent widely damage such as radial crack, wide crater, spalling or scabbing [8,9]. Meanwhile, the scabbing, spalling, widely cratering, or radial crack would be created on the rigid concrete, when it is subjected to impact loading [11].

However, previous research [8, 12] observed the crater as local damage on FC and FCRHA by using impactor with large diameter, and the result showed the blunt nose shape not fully penetrated to FC and FCRHA. The penetration depth of FCRHA has more shallow crater than FC due to different strength. This indicates that the local damage and the penetration on FC and FCRHA influenced by the role of impact velocity, target strength, nose shape and size of impactor diameter [13]. Therefore, the local damage in those case could not set at broad of nose shape impactor.

Hence, this investigation offered tracing of local damage in FCRHA should be tested by smaller impactor to prove that local damage trace on FCRHA appeared only as wide as impactor diameter even though exposed by low impact velocity as in high impact velocity on FC [9]. Thereby, the relation between nose shape impactor diameters with impact diameter can be determined.

This investigation provided the plain slabs FC (as a control) and FCRHA with densities 1400 kg/m^2 and 1600 kg/m^2. Those slabs exposed to blunt nose impactor with various low impact velocities. The most important thing in this investigation is create the relation between penetrations of impactor in FCRHA with the area of penetration of impactor on slabs, besides the local damage behaviour observation.

2 Material and experiment

2.1 Material and mix proportion

The RHA obtained from rice manufacture at Muar-Johor Malaysia which un-controlled burning under 700°C during ± 6 hours. The RHA inserted into a mixture concrete to substitute the partly of sand. Where 25% volume of sand replaced by RHA on cement-sand ratios 1:4 by weight. Basically, FCRHA admixture was produced same way as FC, which this experimental, the pre-foaming method was applied. The OPC, sand, water, and RHA mixed evenly as base mixture, further added by stable preformed aqueous foam which made separately. The stable foam was created by 1:20 of foam agent-water ratio.

The base mix of FC and FCRHA produced by 3:5 of water-cement ratio. While, to avoid the absorption of water by RHA in FCRHA, then the water added into a mixture about 1.25 of RHA/water ratio. Finally, the slab target were produced by size of 300mm x 300mm with 200 mm of thickness, where the slabs target have densities about 1400 kg/m^3 and 1600 kg/m^3 that were cured for 28 days.

2.2 Experiment

The primary objective of this experiment was to observe the local damage behavior especially relation between the diameter of impactor and the crater on plain slab surface of

FCRHA that created through low impact velocity. The steel blunt nose impactor (see Figure 1c) with 17 kg of weight and 40mm diameter (d) was used through drop weight impact method [14-16]. This method was applied where, the impactor dropped at various altitudes of 1.2 m, 1.4 m and 1.6 m which equal to 4.8 m/s, 5.2 m/s and 5.6 m/s respectively. A pipe used to guide the fall of impactor (Fig. 1a) that falling perpendicularly on a target. During impact test, the slab target constrained on all sides including up and bottom sides (Fig. 1b) to create the magnitude compressive stress from reflected energy wave [10]. So, the plastic deformation would be expected occurs at constraint area firstly and local damage would be increased. Visually, local damage on surface plain slab was observed for diameter and penetration of impactor.

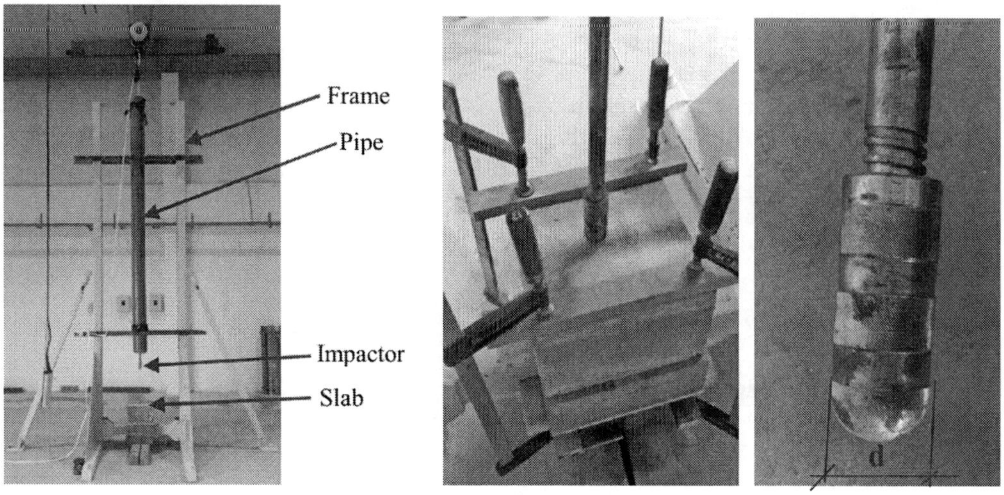

Fig. 1. (a) Drop weight impact frame (b) Slab constrained during test (c) Blunt nose shape of steel impactor

3 Result and discussion

3.1 Local damage behaviour

Local impact occurred in this investigation was crater without radial crack, spalling, scabbing nor wide cratering [11, 17, 18] on FC and FCRHA slab target (see Fig. 2). Since the porosity of FC and FCRHA carried out the impact loading [8]. This impact loading had been transformed to be energy wave, when the impactor initial touched on surface slab target and spread in slab target transversely and longitudinally [10]. In this investigation, all sides of slab target were constrained during impact test. The energy impact that propagate in slab longitudinally is compressive stress wave and it would be enhanced to be double due to no displacement is allowed at side edge (constrained). Furthermore, the increasing stress compressive wave create a plastic compressive wave. Thus, the energy dissipation of plastic compressive wave and plastic deformation should be occurred in constrained area [10]. So that why, in this case usually the constrained area of slab could be failed firstly due to high compression energy wave that generated from impact loading. However, in this experiment the failure of constrained area was not occur to slab target FC and FCRHA (see Fig. 2), since porosity in FC and FCRHA hinder well the compression energy wave [8, 19].

 DOI: 10.1051/matecconf/201710302025

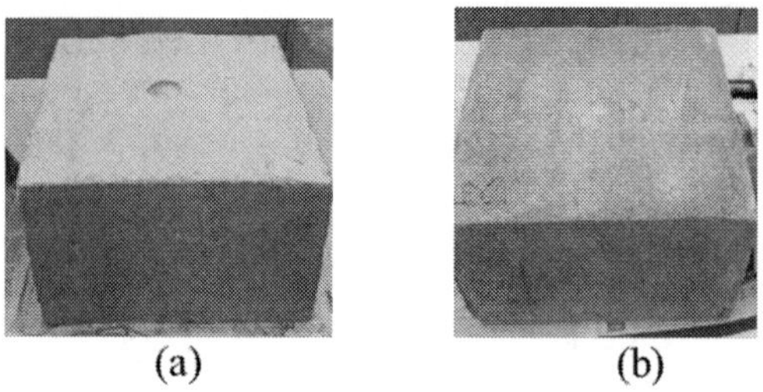

Fig. 2. The local damage as crater on slab surface of a) FC similar with b) FCRHA

Physically, there is difference between FC and FCRHA, FC is seen more porous, while the FCRHA visibly is denser (see Fig. 3). Fig. 3b shows the RHA filled up the porous of FCRHA furthermore, this is caused the FCRHA more strength than FC (see Table 1) [7]. Besides the RHA improved elasticity of porosity in FCRHA [8]. However, this physical condition does not make FCRHA different from RHA when it subjected to impact loading. The crater which created on surface of FCRHA was similar with FC. This indicates that FCRHA was good absorb energy and presence the RHA was not reduce ability absorb energy of FCRHA.

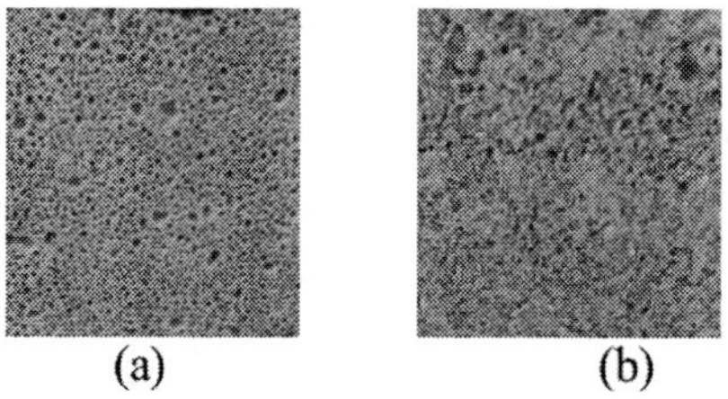

Fig. 3. Appearance of porosity on a) FC and b) FCRHA.

Fig. 4 shows the craters of slab surface of FC (Fig. 4a, 4b and 4c) and FCRHA (Fig. 4d, 4e and 4f) with target density 1400 Kg/m^3. Those slab target generated by impactor with different impact velocities about 4.8 m/s (Fig. 4a and 4d), 5.2 m/s (Fig. 4b and 4e) and 5.6 m/s (Fig. 4c and 4f). All craters clearly show that no radial crack, spalling, or widely cratering created. Nonetheless, the crater of FC for all impact velocities formed the crack on crater beyond, meanwhile the beyond crack on FCRHA visibly vague. This indicates that the elasticity of porosity of FCRHA increased [8, 19].

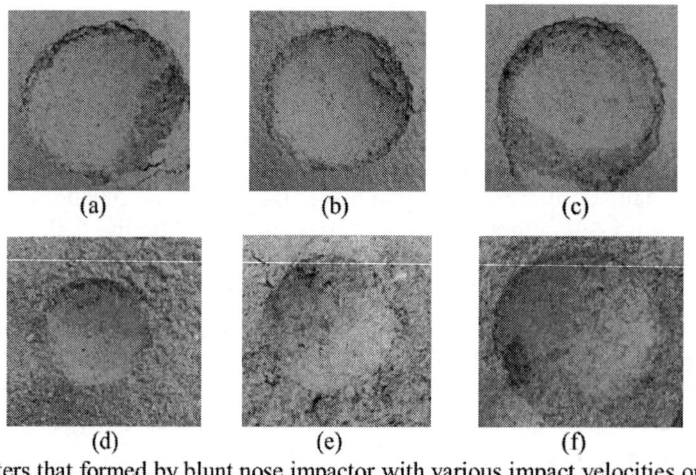

Fig. 4. The craters that formed by blunt nose impactor with various impact velocities on slab target of FC (a, b and c) and FCRHA (d, e and f). All slab target with 1400 kg/m^3 of density

The comparison between slab target of FC and RHA with 1600 kg/m^3 of density was presented in Fig. 5. Fig. 5a, 5b, and 5c show the crater of FC surfaces slab which created by 4.8 m/s, 5.2 m/s, and 5.6 m/s impact velocities, respectively. So that with Fig. 5d, 5e and 5f for FCRHA which were exposed by the same impact velocities as FC, respectively. The result shows no different with 1400 kg/m^3 density of slabs target.

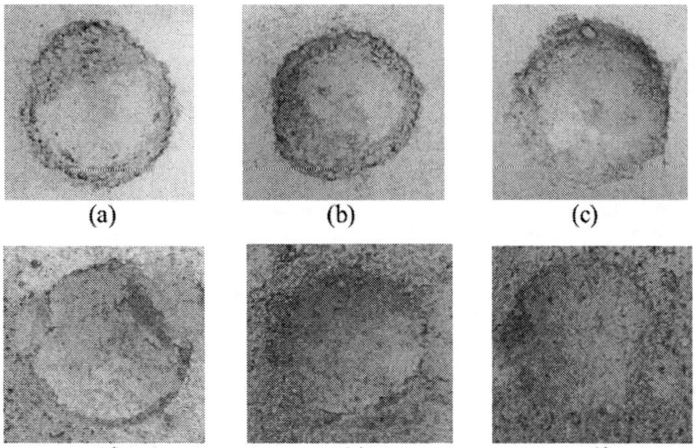

(a) (b) (c)

(d) (e) (f)

Fig. 5. Comparison of local damage on FC (a, b and c) and FCRHA (d, e and f) slab target with 1600 kg/m^3 of density. The local damage created the crater by blunt nose impactor with various velocities

3.2 Penetration values

Basically, penetration depth is essential variable for measuring the localized damage on slab concrete due to impact loading. By the penetration depth the local damage on slab could be determined, even for perforation prediction [20]. Table 1 presents the penetration depth (*Xe*) and diameter of crater (D_C) of each material due to various velocities of impact loading.

The result shows that penetration of FC with 1400 kg/m^3 density was deepest than others. Meanwhile, the diameter of crater which created by impactor shows the impact velocity of 5.6 m/s hit the FC 1400 created the largest crater than others, i.e. 40.7 mm. Actually, the diameter of impactor was 40 mm, it means the nose impactor wholly was success penetrate in slab concrete without cracking, spalling and widely cratering.

Table. 1. Impact test result on FC and FCRHA

Material (density) (kg/m^3)	Compressive strength (f_C) (MPa)	Impact velocity (v) (m/s)	Penetration depth (Xe) (mm)	Diameter of crater (Dc) (mm)
FC (1400)	3.65	4.8	11.2	36.7
		5.2	12.2	38.7
		5.6	14.1	40.7
FC (1600)	4.5	4.8	7.4	30.8
		5.2	8.1	31.7
		5.6	9.3	34.1
FCRHA (1400)	3.57	4.8	7.8	36.3
		5.2	8.1	32.1
		5.6	8.2	32.7
FCRHA (1600)	6.62	4.8	6.1	28.8
		5.2	6.1	29.0
		5.6	6.4	29.0

Relation between penetration depth and impact velocity presents in Fig. 6 and 7. The penetration was not significant for both material FC1400 and FCRHA1400 (see Fig. 6). In the other hand, both materials with 1600 kg/m^3 density, resulted the different penetration depth (see Fig. 7), where both of materials have different compressive strength (see Table 1). From the result indicates that strength of concrete play an important role for penetration depth of concrete beside impact velocity, where the result agrees with previous researchers [13, 17, 20-22].

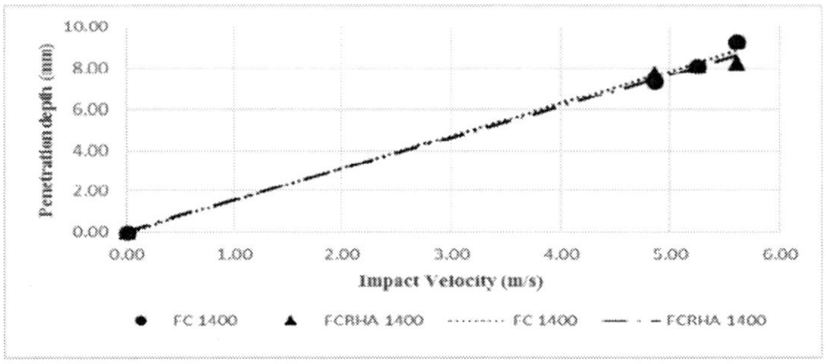

Fig. 6. Penetration depth of impactor on slab of FC and FCRHA with density 1400 kg/m^3

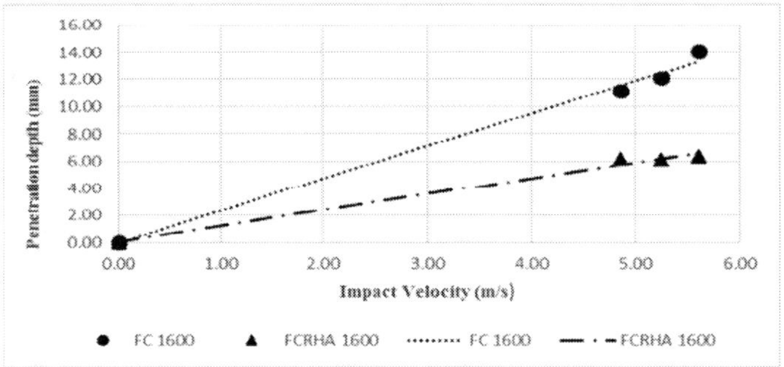

Fig. 7. Penetration depth of impactor on slab of FC and FCRHA with density 1600 kg/m^3

3.3 Diameter of penetration

In the case where the slab concrete could absorbed energy impact without radial crack, spalling, scabbing, nor widely cratering, the diameter crater is important to measure the local damage on slab target. By using different impact velocities on each material target, the result obtained relation between the diameters of the crater on a slab with penetration impactor as whichever seen in Fig. 8. The data trend line created by diameter against penetration is polynomial that will eventually constant at a certain point i.e. the diameter of impactor as seen in curve of FC1400. Meanwhile, other data inclined to follow FC1400, but not reach the value of the impactor diameter. Perhaps it caused by a deficiency of impact energy or the concrete strength was greater than FC1400, which hindering the penetration of impactor [13, 18, 21]. Although, the values of diameter crater each data are different, all penetration of whole nose impactor considered have been successful to penetrate slab of FC and FCRHA. This diameter result gives a contribution for penetration depth prediction of FCRHA and FC by the reason that the crater diameter of FC and FCRHA slab equals to a nose diameter of impactor. Therefore, the result provides the conclusion that penetration of impactor on FCRHA with low impact velocity give the same impression on penetration impactor with high impact velocity on FC [9].

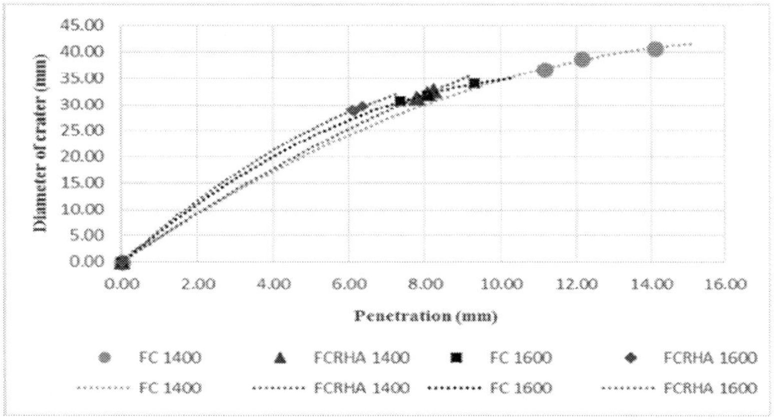

Fig. 8. Diamter of crater for FC and FCRHA with 1400 kg/m^3 and 1600 kg/m^3 due to various impact loading. The diameter crater tending toward a point the diameter of impactor

4 Conclusion

The investigation of diameter crater as local damage on plain slab FC and FCRHA with 1400 kg/m^3 and 1600 kg/m^3 subjected to 4.8 m/s, 5.2 m/s and 5.6 m/s impact velocities has been conducted. The local damage produced crater with no spalling, scabbing radial crack nor widely cratering due to the porosity took over the impact load. The presence of RHA in FCRHA enhanced the wall elasticity of porous, causes the strength of FCRHA increased. The strength of concrete is play important role on penetration beside the impact velocity. So that, penetration of impactor on FCRHA more shallow than FC and cracking in crater beyond seen vague. The objectives of this investigation has been achieved that nose impactor wholly penetrated in plain slab of FCRHA, which its diameter of crater will be used as an orientation on penetration depth prediction and dimension of impact calculation in future investigation.

The authors would like to acknowledge financial and facilities support provided by University Tun Hussein Onn Malaysia under Research Supporting Grant Scheme RSGS Code U100.

References

[1] F.V. Riza, Application of RHA's pozzolanic properties in the making of CEB. Int. J. of Sustainable Construction Engineering and Technology, **2**(2), 32-36 (2011)
[2] S. Singh and D. Kumar, Alternate and low cost construction material: Rice husk ash (RHA). Int. J. of Innovative Research in Advanced Engineering, **1**(6), 214-217 (2014)
[3] M.S. Bhusari, Industrial application of rice husk ash (RHA) as a insulating material: A solution for reduction in land pollution, Int J Adv Engg Tech, **VII**(I), 552-553 (2016)
[4] J. Hadipramana, A.A. . Samad, A.M.A, Zaidi, N. Mohammad and F.V. Riza, Effect of uncontrolled burning rice husk ash in foamed concrete, Advanced Materials Research, **626**, 769-775 (2013)
[5] M.H. Zhang, R. Lastra, V.M. Malhotra, Rice-husk ash paste and concrete: Some aspects of hydration and the microstructure of the interfacial zone between the aggregate and paste, Cement and Concrete Research, **26**(6), 963-977 (1996)
[6] A.M.A. Zaidi, J. Hadipramana, A.A.A. Samad, N. Mohamad and F. V. Riza, Potential of RHA in foamed concrete subjected to dynamic impact loading, Key Engineering Materials, **594-595**, 395-400 (2013)

[7] J. Hadipramana, A.A.A. Samad, A.M.A. Zaidi, N. Mohamad and F.V. Riza, Contribution of RHA granules as filler to improve the impact resistance of foamed concrete. Key Engineering Materials, **594-595**, 93-97 (2013)
[8] J. Hadipramana, A.A.A. Samad, R. Ibrahim, N. Mohamad, and F. Venny Riza, The energy absorption of modified foamed concrete with rice husk ash subjected to impact loading, ARPN Journal of Engineering and Applied Sciences, **11**(12), 7437-7442 (2015)
[9] A.Z.A. Mujahid and Q. M. Li, investigation on penetrating resistance of foamed concrete, Structure and Building, **162**, 77-85 (2009)
[10] G. Lu and T. Yu, *Energy absorption of structures and materials*, Abington Cambridge, England: Woodhead Publishing Limited, 50-51 (2003)
[11] G. Hughes, Hard missile impact on reinforced concrete, Nuclear Engineering and Design, **77**(**1**), 23-35 (1984)
[12] J. Hadipramana, A.A.A. Samad, N. Mohamad, S.N. Mokhatar and F.V. Riza, Penetration depth study from previous empirical formula of modified foamed concrete slab underlow impact load from a non-deformable impactor, Jurnal Teknologi, **78**(5), 147-151 (2016)
[13] Q.M. Li, S.R. Reid, H.M. Wen and A. R. Telford, Local impact effects of hard missiles on concrete targets, Int. J. of Impact Engineering, **32**(1-4), 224-284 (2005)
[14] T. Ando, N. Kishi, H. Mikami and K.G. Matsuoka, Weight falling impact tests on shear-failure type RC beams without stirrups, Structures Under Shock and Impact, **VI**, 579-587 (2000)
[15] K. Toyota, K. Okubo, T. Fujii, T. Oguri and T. Uenoya, Mechanical properties of plain-woven CFRP reinforced by spread fiber tow during and after drop-weight impact, Structures Under Shock and Impact, **IX**, 455-463 (2006)
[16] M.R. Jones and L. Zheng, Energy absorption of foamed concrete from low-velocity impact, ICE Publishing, **65**(4), 209-219 (2012)
[17] M.H. Zhang, V.P. W. Shim, G. Lu and C.W. Chew, Resistance of high-strength concrete to projectile impact, Int. J. of Impact Engineering, **31**(7), 825-841 (2005)
[18] M. Beppu, K. Miwa, M. Itoh, M. Katayama and T. Ohno, Damage evaluation of concrete plates by high-velocity impact, Int. J. of Impact Engineering, **35**(12), 1419-1426 (2008)
[19] L.J. Gibson and M.F. Ashby, *Cellular Solids structure and properties* (second ed.): Cambridge University Press, 175-187 (1997)
[20] Q.M. Li, S.R. Reid and A. M. Ahmad Zaidi, Critical impact energies for scabbing and perforation of concrete target, Nuclear Engineering and Design, **236**(11), 1140-1148 (2006)
[21] D.Z. Yankelevsky, Local response of concrete slabs to low velocity missile impact, Int. J. of Impact Engineering, **19**(4), 331-343 (1997)
[22] Q.M. Li and X.W. Chen, Dimensionless formulae for penetration depth of concrete target impacted by a non-deformable projectile, Int. J. of Impact Engineering, **28**(1), 93-116 (2003)

Estimation of the Scale of Fluctuation for Spatial Variables of RC Structures

S. Hilyati[1], *Z. M.* Nizam[2,3,*], *M. A. A.* Zurisman[3], *Z. A. M.* Hazreek[2], and *A. A. T.* Saiful[2]

[1]Faculty of Civil and Environmental Engineering, Universiti Tun Hussein Onn Malaysia, 86400 Batu Pahat, Johor, Malaysia
[2]Research Centre for Soft Soil, Universiti Tun Hussein Onn Malaysia, 86400 Batu Pahat, Johor, Malaysia
[3]Jamilus Research Centre, Universiti Tun Hussein Onn Malaysia, 86400 Batu Pahat, Johor, Malaysia

Abstract. Dimensional and structural properties of RC structures are non-homogenous due to the quality of workmanship, environmental and material variability. One of the required statistical information for spatial variability analysis of RC structures includes the scale of fluctuation, θ. This paper discusses the estimation of θ for two spatial variables; concrete compressive strength and concrete cover. Methods used to estimate the θ are the Curve fitting method and the Kriging Method. Kriging is an optimal interpolation method which uses the concept of randomness that allows the uncertainty of the predicted values to be calculated. Data measurements for concrete compressive strength and concrete cover were obtained from Peterson (1964) and Public Work Department of Malaysia respectively. The most reliable value for θ of f_{cu} was determined and the value obtained for θ of c was found unreliable due to the insufficient of data points from the available data.

1 Introduction

In the field of the structural reliability of reinforced concrete (RC) structures particularly located in marine environments (eg., for service life predictions), the issue of neglecting the spatial variability of the deterioration parameters across the structure has been raised. It is becoming increasingly difficult to ignore the spatial variability, as most modern structures possess a high degree of structural complexity. Hence, the materials and properties of RC structures are not homogeneous due to the variability of the environment and workmanship [1, 2]. Most of the main parameters in the chloride-induced cracking models are subject to uncertainty. The assumption that materials and dimensional properties of RC structures are homogeneous can lead to non-conservative predictions of failure for RC structures in corrosive environments [3, 4]. Neglecting such sources of uncertainty has a significant impact on the safety and assumed whole life durability performance of the structures [1].

The required statistical information for analysis involving spatial variability includes the mean value (μ), standard deviation (σ) and scale of fluctuation (θ). Data for the first two statistical parameters, μ and σ are available in the literature, however, data for θ is very

* Corresponding author: mnizam@uthm.edu.my

scarce and the insufficient data for θ has been reported by other researchers [1, 2, 5]. The main concern in the spatial variability of the reliability analysis is the lack of data for θ although some studies have included spatial variability for durability aspects such as Mullard and Stewart [2], Vu and Stewart [6], Karimi and Ramachandran [7], and Engelund and Sorensen [8]. However, these studies have been unable to provide accurate mathematical estimation for θ. For example, θ used by Mullard and Stewart [2] was based on engineering judgment which was governed by the correlation function on a particular random field.

There are other important spatial variables that need to be included in the spatial variability of the reliability analysis, including concrete compressive strength, f_{cu} and concrete cover, c. However, data for θ of f_{cu} and c are difficult to find elsewhere. Clearly, due to the limited presence of θ in the literature, θ needs to be further investigated in order to provide more accurate and reliable results for the spatial variability of the reliability analysis in the chloride-induced corrosion problem. This research takes this opportunity to carry out some analysis based on the available data in order to suggest new values of θ for f_{cu} and c.

2 Definition and the development of the scale of fluctuation, θ

The scale of fluctuation, θ is defined as the approximate length over which strong correlation persists in the random field [9-11]. The scale of fluctuation is a parameter used for finding realistically accurate information about the variance of local averages of random fields. The scale of fluctuation of a spatial variable can be estimated from data measurements. For example, a data measurement of concrete compressive strength on a 3 m height of the concrete column. The important information required for calculating θ includes the value of concrete compressive strength along the height of the column and the distances between each measurement. This section summarises the description on the development of θ, including the contribution and recommended values for θ used by previous researchers.

The development for θ was initially supported by Vanmarcke [9] which has proposed various methods for calculating θ for stationary random fields. Another study that contributed to the development of θ by Karimi [12]. This study used historical data from various engineering disciplines to apply to the chloride-induced RC corrosion problem. However, it was found that the probability of the onset of corrosion was underestimated. This is because the spatial variables, C_s and D_{app} were treated as random variables only. The study by Ramachandran et al. [13] has suggested the Kriging method to predict missing data for estimating θ. A recent study done by Connor and Kenshel [1] provides detailed descriptions on the estimation for of θ for C_s and D_{app} from the analysis of experimental data recorded on a bridge in South East Ireland. Two methods were adopted in the study to estimate θ of both spatial variables. The first method is the Curve fitting method used to determine the values for θ through an associated model parameter, known as correlation length, d. The second method is Kriging method to determine the missing data in the specific unmeasured location from the recorded data.

This paper used data from the study by Peterson [14] to determine θ for concrete compressive strength. This research also has an opportunity to access data from the Public work department (PWD) of Malaysia to calculate θ for concrete cover. The purpose of using the concrete cover data from PWD of Malaysia is to study the degree of physical fluctuation of concrete cover on real structures. The new finding for θ were discussed here will serve as a base for further works and possibly be beneficial for providing discussion and information (e.g., limitation and procedures) of θ that is useful for future research.

2.1 The Kriging method

Kriging is a statistical interpolation method which applies a geostatistics method. Geostatistics is defined as the study of phenomena that vary in space and/or time. The Kriging is well known in the fields of mining engineering, geology, hydrology and social science. This method is very appropriate for finding the missing data as Kriging accepts irregular spaced data, calculates an error of estimates which can give an actual measure of the reliability of estimated points and uses the autocorrelation between known data values for estimation of unmeasured values. Kriging itself is an optimal interpolation based on regression against observed *z*-values of surrounding data points, weighted according to spatial values [15].

This method uses the concept of randomness which allows the uncertainty of the predicted values to be calculated. Kriging generates the best linear unbiased estimate at a specific location by employing a semivariogram model. As the estimation of θ uses a curve fitting method, therefore the most suitable interpolation method must be used to estimate the missing data.

Some advantages of Kriging as stated by Geoff Bohling [16], are that 1) it helps to compensate for the effects of data clustering, 2) it can assign individual points within a cluster less weight than isolated data points, 3) it gives an estimate of error (kriging variance), along with estimates of the variable, *Z*, itself (although the error map is basically a scaled version of a map of distance to nearest data point, and therefore not that unique), and availability of estimation error provides basis for stochastic simulation of possible realizations of *Z(u)*.

The basic linear regression estimator *Z*(u)* is

$$Z^*(u)-m(u)=\sum_{\alpha=1}^{n(u)}\lambda_\alpha(u)\left[z(u_\alpha)-m(u_\alpha)\right] \tag{1}$$

In the Equation (1), *Z(u)* is treated as a random field and the aim of this equation is to calculate Kriging weights, λ_α. Kriging calculates a residual at u as the weighted sum of residuals at neighbouring data points. Kriging weights, λ_α, are derived from the covariance function or semivariogram which should characterize the residual component. Kriging weights, λ_α, minimize the variance of the estimator σ^2_E in Equation (2) under the unbiasedness constraint E {Z*(u)-Z(u)}=0.

$$\sigma^2_E(u)=Var\{Z^*(u)-Z(u)\} \tag{2}$$

2.2 Modeling the semivariogram

To model the semivariogram, the empirical semivariogram needs to be replaced with an acceptable semivariogram model. This is due to the kriging algorithm needing access to semivariogram values for lag distances other than those used in the empirical semivariogram. The five most frequently used models are given in Table 1.

The semivariance function $\gamma_{Exp(\tau)}$ used in the equation can be obtained empirically from a set of data points as follows:

$$\lambda_{Exp(\tau)}=\frac{1}{2N(\tau)}\sum_{\alpha=1}^{N(\tau)}\left[z(u_\alpha+\tau)-z(u_\alpha)\right]^2 \tag{3}$$

where $\gamma_{Exp(\tau)}$ denotes the empirical semivariance function and $N(\tau)$ denotes the number of pairs of data that are separated by a lag distance τ.

Table 1. Analytical semivariogram models.

No	Semivariogram name	Semivariogram model
1	Nugget	$g(h)=\begin{cases}0 & if\, h=0\\ c & otherwise\end{cases}$
2	Spherical	$g(h)=\begin{cases}c\cdot\left(1.5\left(\frac{h}{a}\right)-0.5\left(\frac{h}{a}\right)^3\right) & if\, h\leq a\\ c & otherwise\end{cases}$
3	Exponential	$g(h)=c\cdot\left(1-\exp\left(\frac{-3h}{a}\right)\right)$
4	Gaussian	$g(h)=c\cdot\left(1-\exp\left(\frac{-3h^2}{a^2}\right)\right)$
5	Power	$g(h)=c\cdot h^{\omega}\;\; with\, 0<\omega<2$

When the semivariance function is plotted against the corresponding lags, the plot is referred to as the semivariogram. The semivariogram finds the set of weight for estimating the variable value at the location u from values at a set of neighbouring data points. The weight on each data point generally decreases with increasing distance to that point, in accordance with the decreasing data-to-estimation covariances specified in the right-hand vector, k. However, the set of weights is also designed to account for redundancy among the data points, represented in the data point-to-data point covariances in the matrix K. Multiplying k by k-1 (on the left) will downweight points falling in clusters relative to isolated points as the same distance. Stated by Vanmarcke [9], there are two important elements required in order to estimate θ through associated model parameter, *d*. The important elements are 1) sufficient data measurement and 2) an analytical model for $\rho(\tau)$. Once these elements have established, nonlinear regression methods can be used to estimate the parameters of the chosen model (in this case, known as correlation length, *d*). On the other hand, as stated in the study by Connor and Kenshel [1], by curve fitting the analytical model for $\rho(\tau)$ to the correlation coefficients determined from the experimental data $\rho_{Exp(\tau)}$, the value for model parameter *d* that produces the best fit can be obtained, and θ can then be determined based upon its relationship to model parameter *d*. Six types of analytical model (Gaussian ACF) for $\rho(\tau)$ are presented in Table 2.

3 Estimating θ for concrete compressive strength

Data set used to measure θ for concrete compressive strength is obtained from the study done by Peterson [14]. The data from this study found to be significant for estimating θ for concrete compressive strength. An experimental work carried out by Peterson [14] investigates the variation in strength in the longitudinal direction of the column. Data obtained from the study by Peterson [14] shows the variation of f_{cu} from the top of the column to the foot of the column, (f_{cu} in a column varies systematically over its height) and provides the measuring distance and the distance between measured points which are sufficient to produce a reliable value of θ for f_{cu}. The specimens, columns with a size of 0.305 m^2 × 3.05 m tall (12 in.^{2} × 10-ft tall), were removed from the forms 24 hours after casting then totally immersed in a shallow water bath for 25 days. The columns were cut into 0.305 m (12 in.) segments and the ends of the segments were ground smooth. Cores with a diameter of 0.152 m (6 in.) were drilled in a direction parallel to the original column axis and tested when the concrete was 28 days old [17].

Table 2. Analytical semivariogram models.

Type of ACF	ACF Model	θ	Reference
Square exponential (Gaussian)	$\rho(\tau)=\exp\left[-\left(\lvert\tau\rvert/d\right)^2\right]$	$\sqrt{\pi}$ d	Vanmarcke (1983)
Exponential	$\rho(\tau)=\exp\left[-\left(\tau/b\right)\right]$	$2b$	Vanmarcke (1983)
Cosine exponential	$\rho(\tau)=\exp\left(-[\tau]/e\right)\cdot\cos\left([\tau]/e\right)$	e	Kim (2005)
Triangular	$\rho(\tau)=\begin{cases}1-\lvert\tau\rvert/a, & for\ \lvert\tau\rvert\le a\\ 0 & for\ \lvert\tau\rvert>a\end{cases}$	a	Vanmarcke (1983)
Sinusoidal	$\rho(\tau)=\dfrac{\sin\left[-2.2\left(\lvert\tau\rvert/f\right)\right]}{-2.2\left(\lvert\tau\rvert/f\right)}$	f	Gomes and Awruch (2002)
Second-order autoregressive	$\rho(\tau)=\left[1+\left(\lvert\tau\rvert/c\right)\right]\cdot\exp\left(-\lvert\tau\rvert/c\right)$	$4c$	Vanmarcke (1983)

The concrete mixes were named A, B, C, D, E and F. However, only data from concrete mixes A and C were used in this study. As reported by Peterson [14], both concrete mixes (A and C) exhibited adequate cohesion, did not segregate and the concrete was poured in the vertical direction. Each column was divided into nine parts and each part was drilled to investigate the strength of the core cylinders. The variation of f_{cu} from the top of the column to the foot of the column and f_{cu} in a column often varies systematically over its height. The study found that the strength at the bottom is often greater than the strength at the top. The important information required for the study is the concrete compressive strength of the core cylinders with 0.667 m intervals drilled from column A and C as can be seen in Table 3.

3.1 Scale of fluctuation for concrete compressive strength

All the experimental correlation coefficients from column A1-A4 and column C1-C4 are combined in one plot to produce a better fit. The square exponential Gaussian was considered the best fit with experimental correlation coefficients for two columns (Column A and C). Fig. 1 (a) shows a complete set of data after Kriging and Fig. 1 (b) shows

DOI: 10.1051/matecconf/201710302026

experimental correlation coefficient for concrete compressive strength fitted with Gaussian ACF for Column A. Two values of d were chosen and the plot representing each d was superimposed as shown in Fig. 1 (b). The square exponential (Gaussian) ACF with $d = 0.25$ m has a better fit than the square exponential (Gaussian) ACF with $d = 1.13$ m. Therefore, the square exponential (Gaussian) ACF model with $d = 0.25$ m and $\theta = 0.5$ m was chosen to be the best fit with the experimental correlation coefficient for concrete compressive strength for column A.

Table 3. The compressive strength of core cylinders drilled from columns [14].

Column No.	Compressive strength of core cylinders (MPa)								
	1	2	3	4	5	6	7	8	9
A1	14.810	15.000	15.590	14.420	12.940	15.790	14.910	15.590	14.610
A2	21.670	23.630	21.180	23.000	23.830	23.140	23.240	23.340	24.620
A3	17.160	20.200	19.910	18.730	19.120	19.610	19.520	19.220	19.900
A4	19.120	21.380	19.520	20.500	20.300	18.140	21.280	20.890	20.990
C1	28.240	30.890	32.260	28.540	30.700	28.640	32.070	31.190	30.890
C2	30.499	34.324	36.874	34.912	33.147	33.245	36.579	35.206	34.226
C3	28.146	30.891	37.070	37.560	36.187	33.735	37.462	36.972	36.972
C4	31.382	37.070	39.914	40.894	38.639	40.306	40.502	40.796	41.385

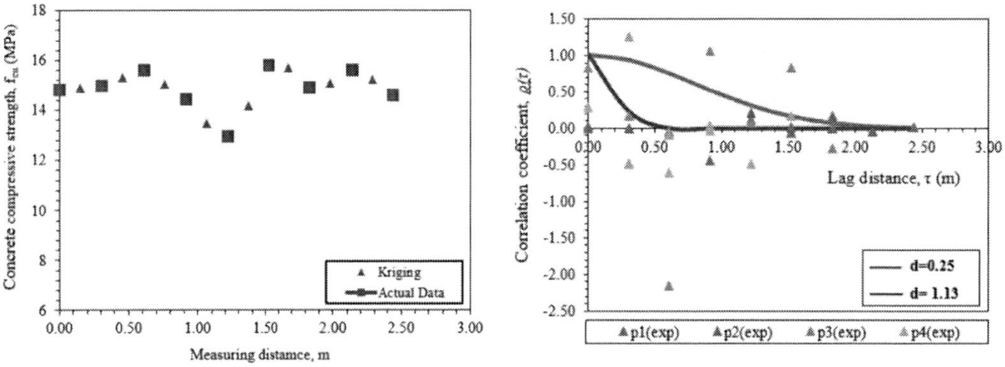

(a) (b)

Fig. 1. (a) A complete set of data after Kriging and (b) Experimental correlation coefficient for concrete compressive strength fitted with Gaussian ACF for Column A.

4 Estimating θ for concrete cover

Data used for estimating θ for concrete cover, c, are obtained from the Public Work Department (PWD) of Malaysia. The data was recorded observations (data measurements) of c on seven piers of a bridge of River Aur, North port of Klang, Selangor, Malaysia. The process of determining the reliable value of θ for concrete cover is the same as the process of determining θ for concrete compressive strength.

4.1 Scale of fluctuation for concrete cover

The experimental correlation coefficient, $\rho_{Exp(\tau)}$ was found not perfectly fit with the chosen ACF model (the square exponential ACF) as can be seen in Fig. 2 (b) and a complete set of data after Kriging shown in Fig. 2 (a). The model nearly fits with the Gaussian ACF model when $d = 0.1$ m. However, with $d = 0.1$ m and $\theta = 0.177$ m is considered very small and this will produce high correlations between elements and concerns in the decomposition of the correlation coefficient matrix. The unreliable values of θ for c is due to the measurements of c were made in less than 1 m. The measuring distance should be at least 3 m in order to have ample information about the degree of fluctuation of the property.

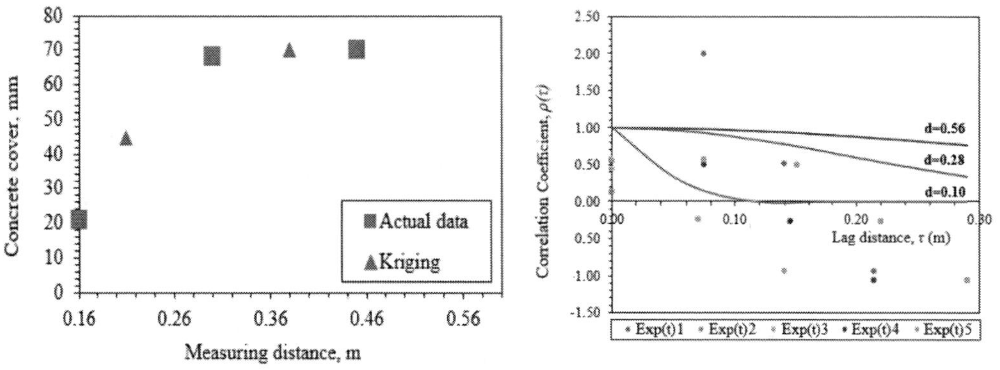

(a) (b)

Fig. 2. (a) A complete set of data after Kriging and **(b)** Experimental correlation coefficient for concrete cover fitted with Gaussian ACF.

5 Conclusion

The purpose of this paper is to determine reliable values for the scale of fluctuation, θ of two important spatial variables: concrete compressive strength, f_{cu} and concrete cover, c. Data from Peterson [14] provides the measuring distance and the distance between measured points which are sufficient to produce a reliable value of θ for f_{cu}. The results of this study support the idea that the increase in the number of observation will increase the accuracy of the value of estimated scale of fluctuation. For estimating missing data, a simple calculation method such as Kriging can be considered as a good starting point when dealing with a continuous record of data, however, the method would not be useful when only a small set of data points is available.

The authors acknowledged the assistances from Prof Mark G. Stewart from the University of Newcastle, Australia, Dr. Lim Char Ching, from the Forensic Unit PWD of Malaysia and Puan Noor Azliliyana bin Kamarohim from IKRAM Infrastructure Asset Management Sdn. Bhd.

References

[1] A.J. O'Connor and O. Kenshel, Experimental evaluation of the scale of fluctuation for spatial variability modeling of chloride-induced reinforced concrete corrosion, J. Bridge Eng., **18**, 3–14, (2013)

[2] M.G. Stewart and J.A. Mullard, Spatial time-dependent reliability analysis of corrosion damage and the timing of first repair for RC structures, Eng. Struct., **29**(7), 1457–1464, (2007)

[3] J.A. Mullard and M.G. Stewart, *Corrosion-induced cover cracking of RC structures: new experimental data and predictive models*, Research Report No. 275.05.09, Centre for Infrastructure Performance and Reliability, the University of Newcastle, New South Wales, Australia, **275**, 1–32, (2009)

[4] M.G. Stewart, Spatial variability of pitting corrosion and its influence on structural fragility and reliability of RC beams in flexure, Struct. Saf., **26**(4), 453–470, (2005)

[5] K.A. Vu and M. G. Stewart, Predicting the likelihood and extent of reinforced concrete corrosion-induced cracking, J. Struct. Eng., **131**(11), 1681–1689, (2005)

[6] K.A. Vu and M.G. Stewart, Structural reliability of concrete bridges including improved chloride-induced corrosion models, Struct. Saf., **22**(4), 313–333, (2000)

[7] A.R. Karimi and K. Ramachandran, *Probalistic estimation of corrosion in bridges due to chlorination*, Application of Statistics and Probability, (2000)

[8] S. Englund and J.D. Sorensen, A probalistic model for chloride ingress and initiation of corrosion in reinforced structures, Structural Safety, **20**, 69-89, (1998)

[9] E.Vanmarcke, *Random fields: Analysis and synthesis*, *Cambridge, Massachusetts,* MIT Press, (1983)

[10] A. Der Kiureghian and J. Ke, The stochastic finite element method in structural reliability, Probabilistic Engineering Mechanics, **3**(2), 83–91, (1988)

[11] A. Haldar and S. Mahadevan, *Reliability assessment using stochastic finite element analysis,* John Wiley & Sons, (2000)

[12] A. R. Karimi, *Probalistic assessment of deterioration and strength of concrete bridge beams and slabs*, Department of Civil Engineering, Imperial College of Science, Technology and Medicine, (2001)

[13] K. Ramachandran, A.R. Karimi and A. Zingoni, Estimation of corrosion initiation time with observed data, *Structural engineering, mechanics and computation,* Elsevier Science, Oxford U.K., 1431-1437, (2001)

[14] N. Peterson, *Strength of concrete in finished structures*, Goteborg: landers boktr, (1964)

[15] P. Goovaerts, *Geostatistics for natural resources evaluation*, Oxford University Press, (1997)

[16] G. Bohling, *Kriging*, C&PE 940, Kansas Geological Survey, Wichita, Kansas, USA, (2005)

[17] F.M. Bartlett and J.G. Macgregor, Variation of in-place concrete strength in structures analysis, ACI Material Journal, **96**, 261–270, (2000)

MATEC Web of Conferences **103**, 02027 (2017) DOI: 10.1051/matecconf/201710302027
ISCEE 2016

A Preliminary Study Application Clustering System in Acoustic Emission Monitoring

Nur Amira Afiza Saiful Bahari[1], *Shahiron* Shahidan[1,*], *Siti Radziah* Abdullah[1] and *Noorwirdawati* Ali[1]

[1]Jamilus Research Center, Faculty of Civil and Environmental Engineering, Universiti Tun Hussein Onn Malaysia, 86400 Batu Pahat, Johor

Abstract. Acoustic Emission (AE) is a non-destructive testing known as assessment on damage detection in structural engineering. It also can be used to discriminate the different types of damage occurring in a composite materials. The main problem associated with the data analysis is the discrimination between the different AE sources and analysis of the AE signal in order to identify the most critical damage mechanism. Clustering analysis is a technique in which the set of object are assigned to a group called cluster. The objective of the cluster analysis is to separate a set of data into several classes that reflect the internal structure of data. In this paper was used *k*-means algorithm for partitioned clustering method, numerous effort have been made to improve the performance of application *k*-means clustering algorithm. This paper presents a current review on application clustering system in Acoustic Emission.

1 Introduction

The Acoustic Emission technique has been widely used in the field civil engineering for structural health monitoring (SHM) [1-6]. Structural health monitoring (SHM) refers to the process of implementation on early damage detection of incipient defects. An acoustic emission (AE) signal is an ultrasonic wave emitted from the sudden release of the strain energy when damage happens, for instance interface debonding, fibre breakage and matrix cracking in composite materials [7-10]. AE signals contain useful information regarding on the damage mechanisms. A major challenge of AE analysis is how to discriminate the AE signals due to the different damage mechanisms [10-13]

Generally accepted ways of discriminating AE signals is cluster analysis. Cluster analysis can be define as generic name for variety of mathematical method, classification numbering in hundreds and that can be used to find out the similarity[14]. The three important things in cluster analysis which are, the selection of AE parameters will be used, choosing the clustering algorithm and validation of the define cluster [15-17].

Many researchers have developed techniques to find out the damage classification by means of AE signal parameters [18-21]. The AE parameters are using to distinguish the type and stage of failure and the integrity of the materials [22-24]. Therefore based on the

* Corresponding author: shahironshahidan@gmail.com

literature, detail definition about clustering method is partition data object such as patterns, entities and unit in some category and it also the analysis of some object into cluster based on the similarity characterization as shown in Fig. 1[9].

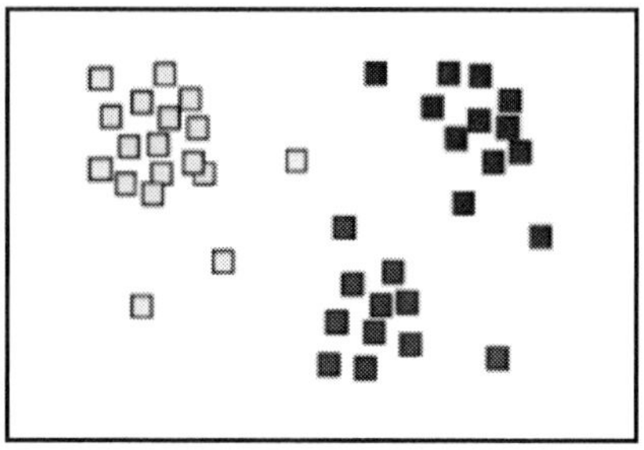

Fig. 1. Cluster based on the similarity.

2 Previous research on AE clustering system

This review paper based on the combining information from available literature journal paper since 2010 till 2016. Table 1a, b and c was shown the previous research focusing on the application of clustering system in Acoustic Emission (AE) and review on AE. The Applications on AE clustering systems are currently well known in term of damage classification. The methods are Neural Network, Artifical Neural Network (ANN), Supervised Clustering and Unsupervised clustering [17, 18, 24]

Table 1. Summary of previous research based on application of clustering in AE.

Year	Ref	Country	Title	Findings
2010	Calabrese et al. [13]	Italy	Use of cluster analysis of acoustic emission signals in evaluating damage severity in concrete structures	Two kinds of unsupervised clustering method was used such as principal component analysis (PCA) and self-organized map (Kahonen map). Combining both method it has been possible to quantify the damage severity and to identify the development of damage itself.
2011	Ono [1]	Los Angeles	Acoustic Emission in material research	Review the progress in method of signal analysis used in Acoustic Emission (AE) as applied to materilas research. The achievment and inadequacy regarding AE from materials during deformation, fracture and other processes are examined systemathically.

Year	Ref	Country	Title	Findings
2013	Pomponi and Vinogradov [14]	Russia	A real-time approach to acoustic emission clustering	New evolutionary clustering algorithm which is adaptive sequential k-means (ASK) has been proposed and prove effective to clasify the AE signal with different emitting sources.
	Yadav and Sharma [15]	India	A review of K-means Algorithm	K-Means firstly proposed by Macqueen in 1967. Three dissimilar modified k-means algorithm are discuss. 1st algorithm remove the limitation of specifying the value of k in advance, 2nd algorithm reduce the computational complexity and also remove dead unit problem. In 3rd algorithm use simple data structure that can be used to store information in each iteration.
	S. Shahidan, et al. [9]	Malaysia	Damage classification in reinforced concrete beam by acoustic emission signal analysis.	The acoustic emission analysis was successfully used to determine the mode of crack and classify the damage level accordance the AE parameters
2014	Li et al. [5]	China	Cluster analysis of acoustic emission signal for 2D and 3D woven glass/epoxy composites.	Peak amplitude and peak frequency were selected as the best cluster definition features from AE parameters by k-means algorithm. This features represent adequately and reproducible way the AE event clustering.
2015	Yang et al. [15]	China	Frequency is the key parameter in discriminating the failure types of thermal barrier coatings: Cluster analysis of acoustic emission signals.	Using the k-means cluster analysis of acoustic emission (AE) signal is the key parameter in discriminate the failure. The findings indicate that the frequency can be applied to identify the AE sources mechanisms.

Year	Ref	Country	Title	Findings
2016	Fallahi et al. [16]	Italy	Supervised and non-supervised AE data classification of nanomodified CFRP during DCB tests.	Detect the damage mechanisms using supervised and non-supervised clustering method. Non-supervised such as k-means and supervised using Neural Network (NN). Result showed, that cluster AE signal are reliable tool to detect damage mechanisms, neural network show 99% of accuracy.
	Farbaniec, Couque, and Dirras [17]	France	Probabilistic improvement of crack propagation monitoring by using acousting emission.	K-means++ algorithm, was used fro clustering with two AE parameters such as peak frequency and number of counts. The AE result indicates is corrected by repetitive tests are useful for monitoring crack ppropagation.
	Assarar et al. [6]	France	Acoustic emission characterization of damage in short Hemp-fiber Reinforced Poklymer (HFRP) composites.	The AE technique combined with scanning electron microscopy was used to identify the microstructurall damage event in overall failure in HFRP composites. The identififcation was made with unsupervised method, k-means. Based on the AE signal induced by interface failure and amplitude range, shown the fibre-matrix interface has a major damage process of HFRP composites.

3 Conclusion

In this review paper, has been made a literature on application clustering in Acoustic emission. AE events can be discriminated in four cluster based on peak amplitude, peak frequency, frequency of centroid and RA value. However, this review have revealed that Peak amplitude (PA) and peak frequency (PF) are the most important parameters in this discrimination. Other than that, this review also found the most favorable to classify the damage mechanisms is used the clustering approach by k-means clustering. The k-means application showed the good results for the clustering of AE signal. Using the k-means cluster analysis of acoustic emission (AE) signal is the key parameter in discriminate the failure. The findings indicate that the frequency can be applied to identify the AE sources mechanisms. The analysis technique proved effective and can be used for the classification of AE data from future tests. It is observed that a lot improvement has been made to the working of k-means algorithm in the past year.

This work was supported by the Faculty of Civil and Environmental Engineering, University Tun Hussein Onn Malaysia, Center for Graduate Studies, UTHM and the Research Grant Contarct Vot U523.

References

[1] K. Ono, Acoustic emission in materials research – A review, J. Acoustic Emission, **29**, 284-308, (2011)
[2] S. Shahidan, S.S. Mohd Zuki, M.A. Rahim, and N. Ali, Amplitude distribution of emission wave for cracking process, MATEC Web of Conf., 2–8, (2016)
[3] T.H. Loutas and V. Kostopoulos, Health monitoring of carbon / carbon, woven reinforced composites. Damage assessment by using advanced signal processing techniques, Part I : Acoustic emission monitoring and damage mechanisms evolution, **69**(2), 265-272, (2009)
[4] N. Godin, S. Huguet, R. Gaertner and L. Salmon, Clustering of acoustic emission signals collected during tensile tests on unidirectional glass/polyester composite using supervised and unsupervised classifiers, NDT E Int., **37**, 253-264, (2004)
[5] L. Li, S.V. Lomov, X. Yan, and V. Carvelli, Cluster analysis of acoustic emission signals for 2D and 3D woven glass / epoxy composites, Composite Structures, **116**, 286-299, (2014)
[6] A.V. Patil, B. Kumar and R.B. Barjibhe, An extensive review on the use of acoustic emission technique for continuous monitoring, Int. Res. J. Eng. and Technology, **3**, 2395-56, (2016)
[7] M. Assarar, D. Scida, W. Zouari, E.H. Saidane, and R. Ayad, Acoustic emission characterization of damage in short hemp-fiber reinforced polypropylene composites, Polymer Composites, **37**(4), 1101-1112, (2014)
[8] M.A.A. Aldahdooh, N.M. Bunnori and M.A. Megat Johari, Damage evaluation of reinforced concrete beams with varying thickness using the acoustic emission technique, Construction and Building Materials, **44**, 812-21, (2012)
[9] S. Shahidan, S.S.M. Zuki, and N. Jamaluddin, Damage grading system for severity assessment on concrete structure, Case Study Construction Materials, **5**, 79–86, (2016)
[10] N. Md Nor, N.M. Bunnori, A. Ibrahim, S. Shahidan, and S.N.M. Saliah, An investigation on acoustic wave velocity of reinforced concrete beam in-plane source, *Proc. IEEE 7th International Colloquium on Signal Processing and Its Applications,* 19-22, (2011)
[11] S. Shahidan, N. M. Bunnori, N. Md Nor, and S.R. Basri, Damage severity evaluation on reinforced concrete beam by means of acoustic emission signal and intensity analysis, *2011 IEEE Symposium on Industrial Electronics and Applications*, 337–341, (2011)
[12] D.D. Doan, E. Ramasso, V. Placet, S. Zhang, L. Boubakar, and N. Zerhouni, An unsupervised pattern recognition approach for AE data originating from fatigue tests on polymer-composite materials, Mechanical System and Signal Processing, **64–65**, 465-478, (2015).

[13] L. Calabrese, G. Campanella and E. Proverbio, Use of cluster analysis of acoustic emission signals in evaluating damage severity in concrete structures, J. Acoustic Emission, **28**, 129-141, (2010)
[14] E. Pomponi and A. Vinogradov, A real-time approach to acoustic emission clustering, Mechanical System and Signal Processing, **40**(2), 791-804, (2013)
[15] J. Yadav and M. Sharma, A Review of K-mean Algorithm, Int. J. of Engineering Trends and Technology, **4**, 2972-2976, (2013)
[16] S. Shahidan, R. Pullin, K.M. Holford, and N. Nor, Quantitative Evaluation of the Relationship between Tensile Crack and Shear Movement in Concrete beams, Adv. Mater. Res., **626**, 355–359, (2013)
[17] S.A. Kudus, N.M. Bunnori, S.R. Basri, S. Shahidan, M.N.M. Jamil, and N.M. Noor, An overview current application of artificial neural network in concrete, Adv. Mater. Res., **626**, 372–375, (2012)
[18] S. Shahidan, S.S. Mohd Zuki, M.A. Rahim, and N. Ali, amplitude distribution of emission wave for cracking process, MATEC Web of Conf., 2–8, (2016)
[19] L. Yang, H.S. Kang, Y.C. Zhou, W. Zhu, C.Y. Cai, and C. Lu, Surface and coatings technology frequency as a key parameter in discriminating the failure types of thermal barrier coatings cluster analysis of acoustic emission signals, Surface and Coating Technology, **264**, 97-104, (2015)
[20] N. Fallahi, G. Nardoni, H. Heidary, R. Palazzetti, X.T. Yan, and A. Zucchelli, Supervised and non-supervised AE data classification of nanomodified CFRP during DCB tests, FME Transaction., **44**(4), 415-21, (2016)
[21] N. Md Nor, N.M. Bunnori, A. Ibrahim, S. Shahidan, and S.N. Mat Saliah, Relationship between acoustic emission signal strength and damage evaluation of reinforced concrete structure: Case studies, *IEEE Symposium on Industrial Electronics and Applications*, 308–313, (2011)
[22] L. Farbaniec, H. Couque, and G. Dirras, Prediction of ductile fracture through small-size notched tensile specimens, *Proc. of the 2015 Annual Conference on Experimental and Applied Mechanics*, 67-72, (2016)
[23] N. Md Nor, N.M. Bunnori, A. Ibrahim, S. Shahidan, and S.N.M. Saliah, An observation of noise intervention into acoustic emission signal on concrete structure, *Proc. IEEE 7th Int. Colloq. Signal Process. Its Appl. CSPA*, 7–10, (2011)
[24] S. Shahidan, N.M.Bunnori, N. Nor, M.A. Rahim, R. Abdullah, and S.S.M. Zuki, Analysis improvements of acoustic emission data in evaluation of concrete beam, ARPN J. Eng. Appl. Sci., **11**(20), 11804–11807, (2016)

Assessment the Strength of Column and Beam at Religious Building Structure by Using Non-Destructive Test

Nor Azizah Adnan[1,*]

[1]Universiti Teknologi Malaysia, Johor Bahru, 81310 Johor, Malaysia

Abstract. The mosque is a sacred place for worship for Muslims. To this date, the traditional mosque needs to be conserved and preserved wood for aesthetic value. Nowadays, most of the mosque materials have a combination of other materials such as concrete, wood and glass. In order to maintain its fit-for-purpose status, non-destructive tests (NDTs) are recommended to be done periodically to identify hidden deteriorations for planning maintenance and repair works without damaging the structure itself. This paper presents two NDTs - Ultrasonic Pulse Velocity tool (PUNDIT) and Schmidt rebound hammer test – done on 38 concrete columns and 29 ground beams of Masjid Kampung Ulu Semerah, Batu Pahat. Recorded readings have shown that most beams and columns had moderate or weak strength. As such, remedial action should be taken to uphold the mosque's structural integrity.

1 Introduction

Referring to Jabatan Agama Islam Johor (JAIJ) and Jabatan Kemajuan Islam Malaysia (JAKIM), Johor has more than eight hundred mosques registered as center perform Friday prayers. Number of one hundred and eighty eight mosques registered as a mosque Friday in Batu Pahat and listed as the highest area in Malaysia peninsular. The geographical this area is located at the west and known for peat soil that is classified into a problematic with high compressibility and low strength. The problem that's need to be faced is about the peat soil sediment. Ideally, construction on peat soil should be avoided, if not, assessments ought to be done for the purpose aforementioned. Assessment of this study is to identify the strength level of column structure based on the results of non-destructive testing (NDT) as preliminary cracks practise. Research on the concrete structure on a mosque is too limited and less attention given. This study took the opportunity to test the ability of concrete to the place of worship that is visited by many people in a certain period. The structural strength of the building provides protection, safety and comfort to its occupants. The stability and strength of a structure depends on the condition of the foundation and underlying soil layers. To ensure that the structure is fit for purpose, tests have to be conducted periodically to safeguard its structural integrity. This is particularly true for structures built on peat soil, which is an organic soil that has high compressibility and low strength.

* Corresponding author: che.nor8253@gmail.com

2 Ultrasonic pulse velocity and rebound hammer

Two types of tests can be done to investigate its structural integrity status, namely Destructive Testing (DT) and Non-Destructive Testing (NDT). J. Helal [1] describe that the tests category in DNT are: (1) Test Hammers Reflection (2) Test UPV (3) The method of penetration resistance (4) Drop resistance method (5) The method of resistance out (6) Methods of test maturity (7) Method absorption test and (8) The method of reinforcement corrosion. All kinds of methods have been based on suitability and testing requirements. The choice of method should be done in order to get the required data reading. A case to note is a series of Table 1 shows findings of previous studies that using NDTs.

Table 1. NDT studies on concrete

Name, Year	Scope Studies	Tools
Masenwat [2] NDT (Non Destructive Testing) in Civil Engineering: Structural Integrity of Concrete Inspection Project	To identify the performance of a concrete structure aged 20 years old included visual inspection, concrete cover meter, half-cell potential, reflection hammer, and UPV. Also carried out the concrete coring, testing carbonation and chloride test. These tests were conducted to get the current reinforcement corrosion rate, the compressive strength of concrete, and the reinforcement position.	1. Ultrasonic Pulse Velocity (UPV/PUNDIT) 2. Half Cell Potential 3. Rebound Hammer Test
I. Lawson [3] Non-Destructive Evaluation of Concrete using Ultrasonic Pulse Velocity	To investigate the relationship between UPV and the compressive strength of concrete test conducted at the age of 2, 7, 15 and 28 days. The UPV and the compressive strength of concrete increases with growth rates at early ages. The UPV results can be determined the uniformity of concrete, detect cracking and voids in concrete, control the quality of concrete and concrete, detect the condition and deterioration of concrete, detect the depth of a surface crack also, determine the strength.	1. PUNDIT
Zia [4] Application Of Non Destructive Test For Structure Evaluation	Visual inspection had found that the effects of fragmentation and significant corrosion on the surface of the concrete. Decision modeling found that the load applied to the structure is heavier than designed. This causes excessive load-bearing structures. In addition, it was found that the steel reinforcement in the concrete having corrosion due to ingress of water into the concrete. The results obtained to determine the appropriate maintenance measures against the structure.	1. PUNDIT 2. Rebound Hammer Test
Pauzi [5] Non destructive Testing (NDT) of concrete containing hematite	Non-destructive ultrasonic and rebound hammer measurements on concrete containing hematite. Rebound Hammer Test show high strength and high density concrete. Pulse velocity but also increase the strength increases when the power is increased again to a constant velocity and purpose to measure the elastic properties of the material and have an affinity with the durability of the material. Recovery pulse velocity should be increased in order to produce good quality rebound.	1. PUNDIT 2. Rebound Hammer Test

This study utilized the NDT method by using Portable Ultrasonic Non-Destructive Digital Indicating Tester (PUNDIT) and Schmidt Rebound Hammer, which is widely used for building conservation works because it can easily detect hidden defects, does not damage important historical structures and the cost effective. Non-destructive testing is a popular method often used in older buildings or historic buildings to determine the strength of the building structure [6]. The published literature has an extensive library of NDT studies done on historical buildings. NDT purpose is to measure the compressive strength of concrete without damaging the original structure surface and value/reading meter referred on the graph to get an estimate compressive strength. The advantage of this test does not cause any damage to the valuable building while giving due engineering data on the current performance or structural integrity status of the building [7].

Fig. 1. Through transmission (direct)

Fig. 2. Diffuse/Scattering or Surface Transmission (indirect)

Table 2. Classification quality of concrete based on the velocity PUNDIT test

Velocity x 10^3 m/s	Quality of Concrete
>4.6	Excellent
3.6 – 4.6	Good
3.0 – 3.6	Impugn
2.1 – 3.0	Poor
<2.1	Very poor

The PUNDIT method uses electrical waves to create a pulse wave that passes through structures in various velocities depending on the concrete strength and size. The transmission transducer has three types of transmission - direct, semi-direct and indirect transmission (Fig. 1 & 2). Direct transmission is the most recommended since it gives the most accurate readings. As for the rebound hammer, it measures the concrete quality through the resulting reflection from the surface of the concrete. The energy absorbed by the hammer recovery allows multiple reflections, and this can be used as a marker to determine the structural strength of concrete. With the reliable readings given by both PUNDIT and Schmidt rebound hammer, appropriate maintenance and repair works can be scheduled and executed to safeguard the structural integrity of the mosque.

The Schmidt Rebound Hammer, should be 12 times [8] the beat (Fig. 3) and hummer head pressed on the concrete surface until the mass spring mass so that the plunger is released and leaves the effect of pressure on the concrete surface (Fig. 4 & 5) and readings shows on slide indicator (Fig. 6) and get the average before rebounding numbers refer to the reference graph (Table 1).

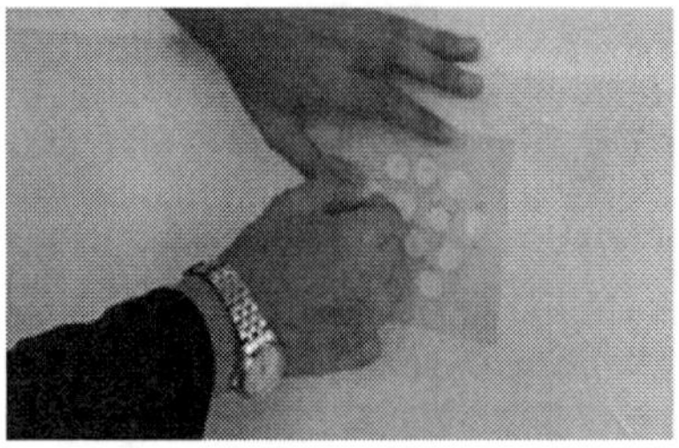

Fig. 3. Points area for rebound beat

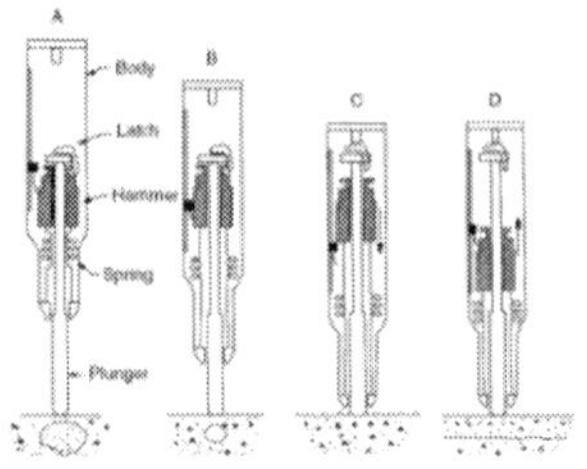

Fig. 4. Standard procedure and guideline

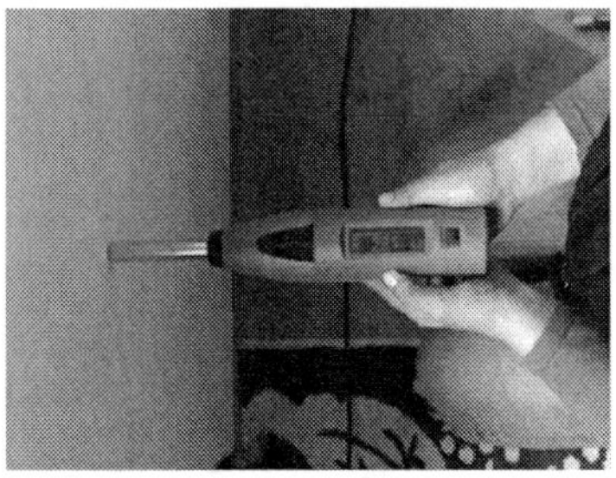

Fig. 5. Hummer head pressed on the wall or concrete surface

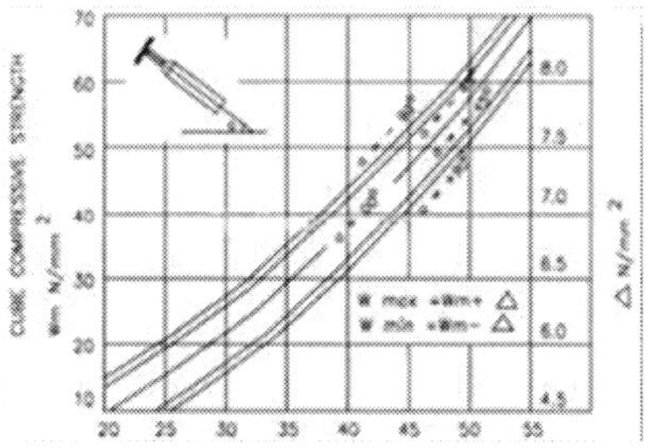

Fig. 6. Graph Strength of Concrete Vs Rebound Number (Yusof, 1995)

Table 3. Concrete quality of Schmidt Rebound Hammer (Hakim, 2002)

Rebound Number Average	Quality of Concrete
>40	Good
30 - 40	Moderate
20 - 30	Poor
<20	Very Poor

3 Methodology

Case study on religious building (mosque) was conduct at Kampung Ulu Semerah, Mukim Air Hitam, Muar, Johore are through site visits, observations, measurement, interviews local committee and perform DNT on structural columns and ground beams of the mosque structure. The purpose of site visit and observation is to know and obtain information relating to the assessment that related to the environment and the strength of the mosque.

Fig. 7. Mosque architectural design

Fig. 8. Interior of the prayer hall

Formally name as Haji Mohammed Mosque was inaugurated in 1999 (Fig. 7 and 8) and geographically stand on soft/peat soil status and the construction costs funded by individuals as donations, but after 5 years of use have shown estranged on joint between pillar-wall and wall-floor than wood strip tenuous effect (Fig. 8). Initial investigations into the mosque found the strength to be confronted the deterioration of columns and beams may affect by soil structure. Observation found that there are several 'tenuous monitor' mounted as regular monitoring for connections is made by other parties. The tenuous between connection timber walls and concrete columns are clearly visible (Fig. 9 and 10). Although the mosque was built in a decade, until now the villagers still use for prayers and perform specific events. However, after a few years there are problems in terms of stability and durability of concrete structures. Raises concerns on safety and security to residents according stability of the mosque.

Fig. 9. Timber floor condition

Fig. 10. Gap between wall and concrete column

Mosque architectural where combined with various materials such timber, concrete, class and metal. Internal space arrangement include 18m x 18m of main prayer hall, 15m x 15m include veranda, teaching and meeting area, 3m x 2.5m of mihrab and 5m x 2.5m of 2 rooms. This mosque was built with four main columns with **10m** heights (0.5m x 0.5cm) from ground level at centre of prayer hall as *suku guru* to support octagonal concrete roof beam of the mosque roof. The suspended floor had assembled with timber stripe on timber floor beam as usually as Malay house concept while that beam is supported by rows of concrete columns with 2m height raised from ground without concrete transverse beams ties. In addition, the brick wall with 3m height stand and tied to 0.58m x 0.38m x 6m height columns are located around the prayer hall, also tied to 1m height timber panel and 1.1m louvered glass windows. The distance between columns to columns is 3m that forming the square of internal space. On top of louvered glass windows have 8cm height timber lintel that tied to glass panel with timber frame and fixed to timber lattice panels or ventilation panel (*sisip angin*) purpose for obtain natural lighting and air ventilation. Although the mosque full designed with glass windows and timber panels but at high-level up to 6m there is no threshold or lintel crossing the walls and more load along the walls then ties tenuous from the columns.

Mosque was measured by regular measuring tape then drawn by AutoCAD to form the layout floor plan. Table 2 shows the layout plan and points position of ground beams; B1-B29 and columns; C1-C38 (Fig. 11). This NDT took about four hours to complete with three readings taken at each location were performed on 29 beams and 38 concrete columns of the mosque. There are four (4) different size for concrete columns which is 0.50m X 0.5m, 0.58m X 0.38m, 0.40m X 0.40m, 0.51m X 0.51m and 0.35m X 0.58m shown in Table 4.

Table 4. Type of columns

Size	Code
0.50m X 0.5m	C1, C2, C3, C4
0.51m X 0.51m	C5, C6, C15, C16, C17, C18, C19, C20
0.58m X 0.38m	C7, C8, C9, C10, C21, C22, C23, C24, C25
0.40m X 0.40m	C11, C12, C13, C14
0.35m X 0.58m	C26, C27, C28, C29,C30, C31, C32
North columns 0.26m	C33, C34, C35, C36, C37, C38

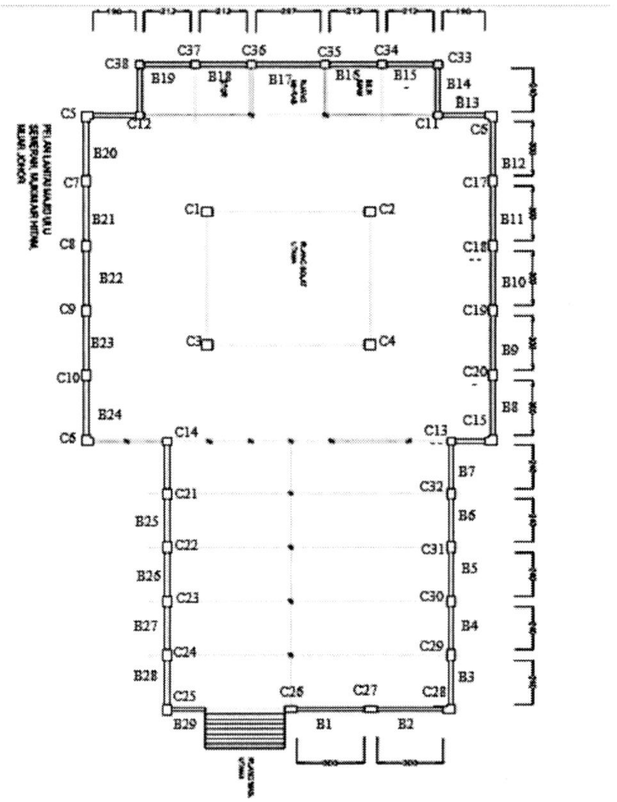

Fig. 11. Location of DNT points at mosque layout plan.

4 Results and discussions

Fig. 12 shows the reading of the results of tests PUNDIT to 38 columns appearing in Masjid Haji Muhamad. The columns position was tested using the method of direct delivery. Columns that have the highest reading speed is column C35 which is 4067m / s, followed by the second highest column is C20, and C38 is the third highest with velocity 3840m / s and 3836m / s respectively. As for the velocity in the range of 3800m / s to 3000m / s is the column C36, C37, C8, C34, C30, C9 and C12. Followed with a velocity in the range of between 2900m / s to 2500m / s is the column C32, C7, C15, C10, C29, C16, C33, C14 and C13. Next to the velocity with values between 2499m / s to 2000 m / s is the column C27, C23, C18, C31, C24, C3 and C21. For columns that have the velocity of 1990m / s to 1000m / s is the column C17, C6, C1, C19, C25, C4, C5, C26, C28, C2 and C22. The very low velocity which is less than 1000 m / s is column C11 by 692 m / s.

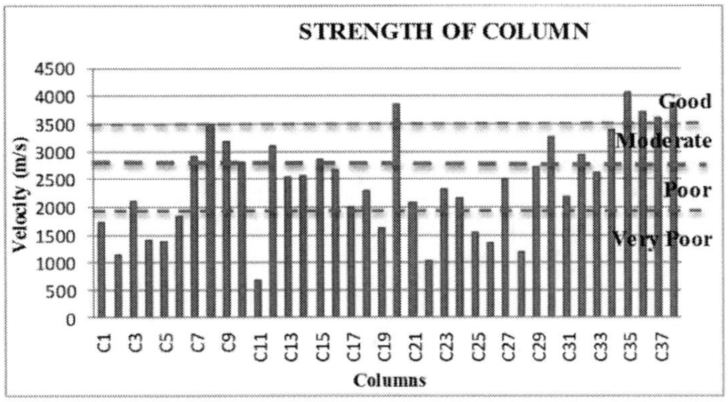

Fig. 12. Test Results PUNDIT for Columns

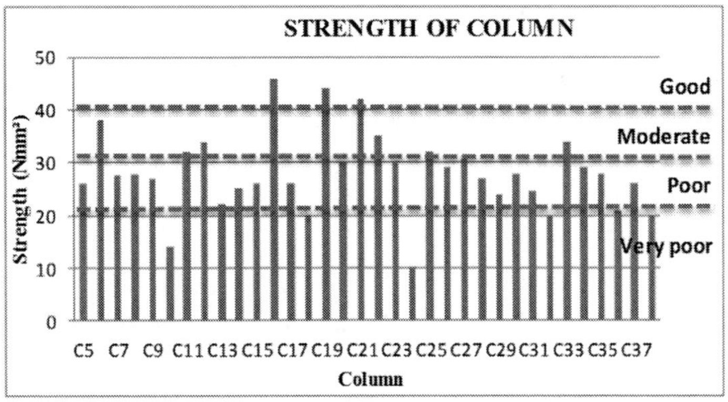

Fig. 13. Schmidt Rebound Hammer test results for columns

Fig. 13 shows the reading of the results of the test hammer Reflection Digischmidt for columns Masjid Haji Muhamad. Based on the graph above, columns C16 had the highest reading of concrete strength of 46N / mm2, followed by columns C19 and C21 with a reading of 44N / mm2 and 42N / mm2. As for the strength of concrete is in the range of 40N / mm2 to 30N / mm2, which is at a moderate level is a columns C6, C22, C12, C33, C11, C25, C27, C20 and C23. Next to the strength of concrete in between the 30N / mm2 to 20N / mm2 is columns C26, C34, C35, C30, C8, C7, C9, C28, C15, C17, C37, C14, C31, C29, C13, C36, C32, C38 and C18. Columns have the lowest concrete strength of less than 20N / mm2 is C10 and C24.

Fig. 14 shows the results of testing for structural ground beams using PUNDIT testing at Masjid Haji Muhammad. Transmission method was been used in this test is indirect method. The ground beam that having the highest value is B23 which velocity of 2320 m / s. While the second highest value is B6 velocity of 2123 m / s. It followed where the velocity readings are within the range of 2000 m / s to 1600 m / s, namely B3, B4, B5, B7, B9, B10, B15, B22, B24, B26 and B29. Next, the readings that obtained between 1500 m / s to 1200m / s is B8, B13, B21, and B27. Ground beam which obtained the lowest reading is B12 which is 968 m / s and followed by B1 of 913 m / s and B17 is 840 m / s. Finally the lowest reading for low velocity is B16 which is 780 m / s.

Fig. 15 shows the readings obtained from Schmidt rebound hammer testing which was conducted on the ground beam at Masjid Haji Muhamad. Based on the results in the graph above, it shows the reading of ground beam for B1 which is the highest concrete strength of 26 N / mm^2 compared with other parts of the ground beam. Ground beam B12 showed the second highest concrete strength of 24 N / mm, followed by B4 and B18. While B3, B5 and B27 have the same strength of 18 N / mm^2. Similarly of B6, B11 and B21 respectively obtained the concrete strength of 16 N / mm^2. While the ground beam of B2 obtaining concrete strength of 14 N / mm^2. Ground beam that have the lowest strength is B7, B8, B9, B10, B13, B16, B26 and B28 in which obtained the same strength of 4 N / mm^2. For the ground beam for B17, B22, B19 and B23 each have the strength of 5 N / mm2 and 9 N / mm^2. From the graph, the strength of concrete beams for B14, B15, B20, B24 and B25 did not give any concrete strength values.

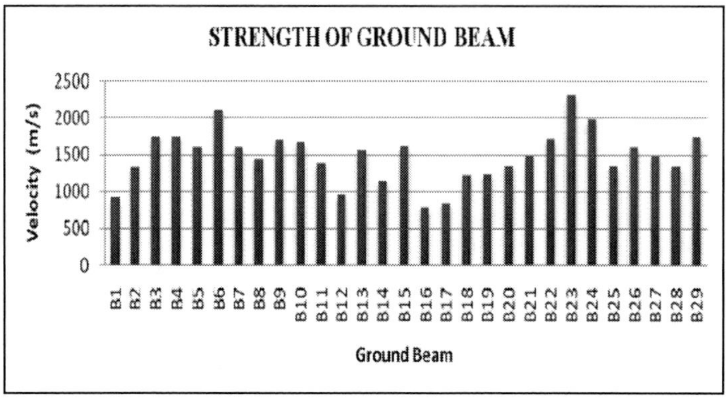

Fig. 14. Test results PUNDIT for ground beams

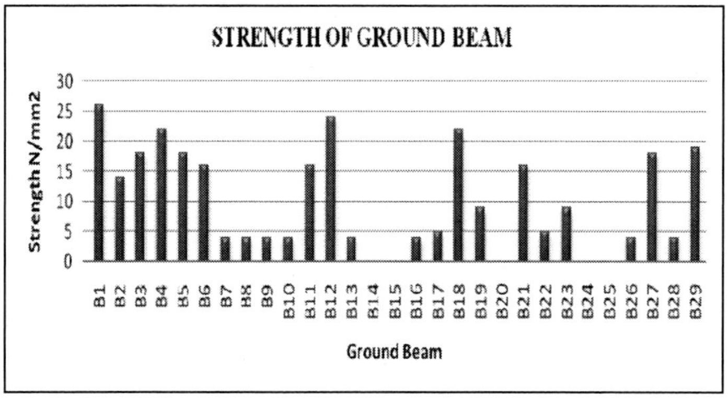

Fig. 15. Schmidt Rebound Hammer Test Results for Ground Beams

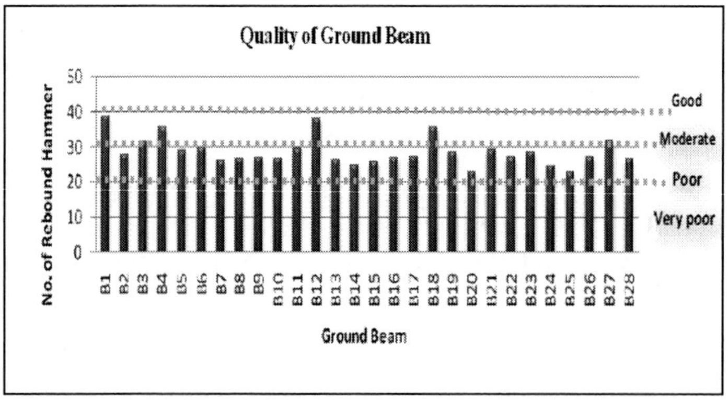

Fig. 16. Schmidt Rebound Hammer test results for quality of ground beams

The Fig. 16 shows the quality of concrete by Schmidt rebound hammer testing. From the graph, it show the highest number of reflection that is B1, where the quality of the concrete is moderate. While the second highest is the moderate quality of the concrete is B18. Under another beam that has a moderate concrete quality is B3. B4, B6, B12, B21, B27 and B29. For the other beams have a poor quality of concrete.

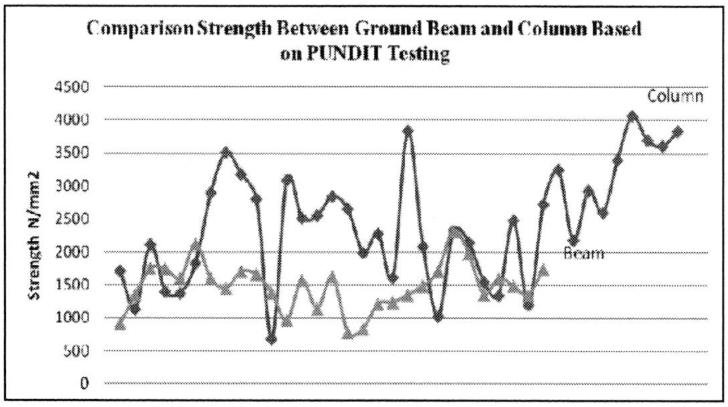

Fig. 17. Comparison strength using PUNDIT testing

According to Fig. 17, it shows the difference of strength for the ground beam and column structures based on PUNDIT testing. From the graph above shows that the structural columns at Masjid Haji Muhammad obtained velocity readings higher than the reading speed of ground beam. Column obtain the highest reading speed which is C20 3840 m / s and it provides a good quality concrete, while the lowest recorded reading speed is column C22 of 1030 m / s which the quality of the concrete is very weak. For the ground beam, the highest reading obtained velocity is 2320 m / s. While for the ground beams of B23 and B16 value is the lowest velocity which is 780 m / s.

Fig. 18 shows the difference in the strength of the structure of columns and ground beam at Masjid Haji Muhamad by using Schmidt rebound hammer testing. Based on the graph (Fig. 18) it to show that the structure of the reading column obtain the high strength compared to strength readings for ground beam structure. The highest reading for the column strength is 46 N / mm^2 for the column C12. While the lowest strength of the column C20 is 10 N / mm^2. Compared with the ground beam structure, the highest reading for the strength of concrete is 34 N / mm^2 for B1 and the lowest value of the concrete strength is 4 N / mm^2.

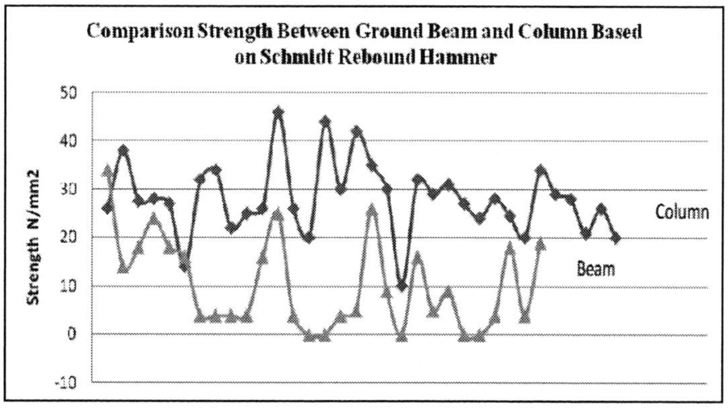

Fig. 18. Comparison strength using Schmidt Rebound Hammer Testing

5 Conclusions

Through testing and data analysis that has been carried out, the conclusion was made based on the test results obtained. This study was conducted to determine the structural strength of wood-based mosque after years of service. Strength level of structure can be determined by using destructive test and non-destructive testing. But for these studies, the selected test is non-destructive testing. This is because these tests can be carried out without disrupting and changing the existing structure. Two methods have been carried out, namely through PUNDIT Test and Schmidt Rebound Hammer. Through both of these tests, data was obtained to determine the structural strength of the Masjid Haji Muhamad. Based on these data, the strength of the structure which are columns and beam are in moderate percentage and weak percentage. Therefore, ongoing review of the appropriate measures for the maintenance and repair needs to be done to ensure the safety of users, so that the mosque will be more permanent.

References

[1] J. Helal, M sofi, P. Mendis, Non-destructive testing of concrete: A Review of method, Electronic Journal of Structure Engineering, **14**(1), 97-105, (2015)

[2] Noor Azreen Masenwat, Kamal Shah Shamsuddin, Mohamad Pauzi Ismail, *NDT (Non Destructive Testing) in Civil Engineering: Structural integrity of concrete inspection project*, NTC 2011: Nuclear Technical Convention, Malaysia, (2011)

[3] I. Lawson, K.A. Danso, H.C. Odoi, C.A. Adjei, F.K. Quashie, I.I. Mumuni, I.S. Ibrahim, Non-destructive evaluation of concrete using ultrasonic pulse velocity, Research Journal of Applied Sciences, Engineering and Technology **3**(6), 499-504, (2011)

[4] M.Z. Rosli, application of non destructive test for structure evaluation, Diss., Universiti Malaysia Pahang, (2010)

[5] N.A.B.M. Pauzi, S. Sani, N. Isa, M.H. Mahmud, *Ujian tidak merosakkan (UTMr) terhadap konkrit mengandungi bijih besi*, Non destructive Testing (NDT) of concrete Containing Hematite, (2007)

[6] B.R. Jayakumar, T. Thavasimuthu, *Practical non-destructive testing*, 2nd Eds., Woodhead Publishing Limited, England, (2002)

[7] A.H. Mohammed, W.M. Wan Mat, *Teknologi penyenggaraan bangunan*, Dewan Bahasa dan Pustaka, Kuala Lumpur, (2002)

[8] K. M. Yusof, *Pengenalan kekuatan dan ketahanlasakan konkrit*, Dewan Bahasa dan Pustaka, Kuala Lumpur, (1995)

Finite Element Simulation of GFRP Reinforced Concrete Beam Externally Strengthened With CFRP Plates

Norhafizah Salleh[1,2,*], *Noor Azlina* Abdul Hamid[1], *Abdul Rahman* Mohd Sam[2], *Jamalludin* Mohd Yatim[2], *Rendy* Thamrin[3], *Nur Hafizah* A. Khalid[2], *Masni* A. Majid[1], and *Zalipah* Jamellodin[1]

[1]Faculty of Civil and Environmental Engineering, Universiti Tun Hussein Onn Malaysia, 86400 Parit Raja, Johor, Malaysia
[2]Faculty of Civil Engineering, Universiti Teknologi Malaysia, 81310 Skudai, Johor, Malaysia
[3]Department of Civil Engineering, Faculty of Engineering, Andalas University, 25163 Sumatera Barat, Indonesia

Abstract. The construction technology now has become more and more advanced allowing the development of new technologies or material to replace the previous one and also solved some of the troubles confronted by construction experts. The Glass Fibre Reinforced Polymer (GFRP) composite is an alternative to replace the current usage of steel as it is rust proof and stronger in terms of stiffness compared to steel. Furthermore, GFRP bars have a high strength-to-weight ratio, making them attractive as reinforcement for concrete structures. However, the tensile behavior of GFRP bars is characterized by a linear elastic stress–strain relationship up to failure and, therefore, concrete elements reinforced with GFRP reinforcement exhibit brittle failure without warning. Design codes encourage over-reinforced GFRP design since it is more progressive and leads to a less catastrophic failure with a higher degree of deformability. Moreover, because of GFRP low modulus of elasticity, GFRP reinforced concrete members exhibit larger deflections and wider cracks width than steel reinforced concrete. This aims of this paper is to developed 2D Finite Element (FE) models that can accurately simulate the respond on an improvement in the deflection of GFRP reinforced concrete beam externally strengthened with CFRP plates on the tension part of beam. The prediction of flexural response according to RCCSA software was also discussed. It was observed that the predicted FE results are given similar result with the experimental measured test data. Base on this good agreement, a parametric study was the performed using the validation FE model to investigate the effect of flexural reinforcement ratio and arrangement of the beams strengthened with different regions of CFRP plates.

* Corresponding author: nhafizah@uthm.edu.my

 DOI: 10.1051/matecconf/201710302029

1 Introduction

Construction industries nowadays face the problem such as durability of the concrete structure due to corrosion. The problem of reinforced steel bar corrosion is one of the most important factors limiting the service life of the structure. Furthermore, repair and retrofit of existing structures are expensive. One of the solutions for this problem is by replacing steel as a reinforcement bar with other material. The suitable alternative material can be used to replace the steel bar is Fiber Reinforced Polymer (FRP). FRP bar can be found with a various type of fiber which is glass, carbon or aramid and they are connected with epoxy resin, polyester or vinyl ester, and therefore be distinguished as GFRP, AFRP, and CFRP products [1]. GFRP is suitable FRP be used to replace the steel due of GFRP give more benefits such as high tensile strength to weight ratio, corrosion free, lightweight, non-magnetic, and non-conductive than the steel bar [2]. GFRP bars also possess mechanical properties different from steel bars, including high tensile strength combined with low elastic modulus and elastic brittle stress–strain relationship [3]. In comparison with steel, GFRP have higher resistance to corrosion, higher tensile capacity and lower weight.

However, concrete members reinforced with GFRP bars exhibit larger deflection and crack width compare with these reinforced with steel due to low modulus of elasticity. Hence, the structure built with the GFRP bar need to have great support to cover the low of modulus of elasticity. To reduce the deflection and crack using GFRP, the beam is being strengthened with Carbon Fiber Reinforcement Polymer (CFRP) as the external reinforcement. CFRP is another type of high-performance fiber available in civil engineering application. CFRP have more potential service life than aramid and glass fibers. Carbon fiber are manufactured by controlled pyrolysis and crystallization of organic precursors at temperatures above 2000°c [4]. This contributes the improvement in reducing the deflection and crack width.

2 Experimental study

This study involved a fabrication and testing of 4 reinforced concrete beam with a size, 200 mm width, 250 mm depth and 2300mm effective length of span and with concrete strength 61.8 MPa. The size beam was selected and design based on the under reinforced (UR) and over reinforced (OR) design and the limitation of testing frame available at the laboratory. The design detail and arrangement of this specimen were shown in Table 1 and Fig. 1. All beams were designed to fail in flexural failure and were test under static load until the beam failed under two point load test.

Table 1. Details of specimens

Group	Specimen	av/d	No of Tensile Reinforce GFRP	Tensile Bar Size (mm)	CFRP Plate Length (mm)	note
UR	CG2	4.85	2	13	-	Un-Strengthenend
	CG2-22				2200	Strengthened
	CG3	4.85	3	13	-	Un-Strengthenend
OR	CG3-22				2200	Strengthened

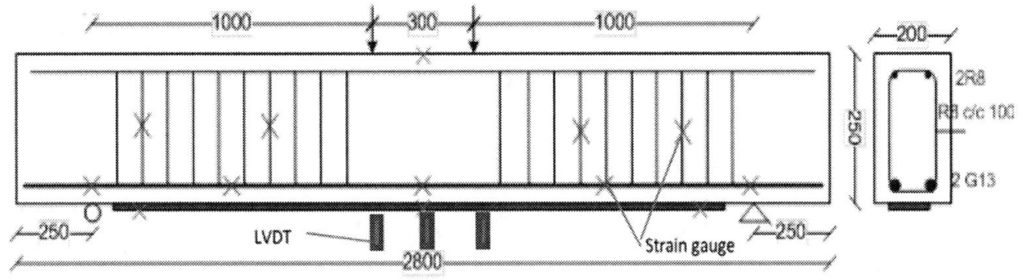

Fig. 1. Arrangement of beam specimen

3 Finite element (FE) analysis- ATENA

In this study, a two-dimensional nonlinear finite element analysis was conducted using the finite element package, ATENA. ATENA is a software for analysis of concrete and reinforced concrete structures. ATENA is a fully windows program with integrated pre-post-processing and finite element solution. The post-processing and pre-processing are improved significantly. In addition, it contains many new materials, solution methods and other capabilities as for instance options for 2D, plane strain, axisymmetric analysis or shell elements [5]. The numerical predictions are compared to the test results.

3.1 FE model

FEA modelled based on measurements made as an experiment before. Fig.2 shows the parameters for the beam has been in and using the FEA. Note that in this figure only half of the beams' length is represented; in fact, using symmetry, and in order to reduce the computational costs, only half of the beams length was modelled [6].

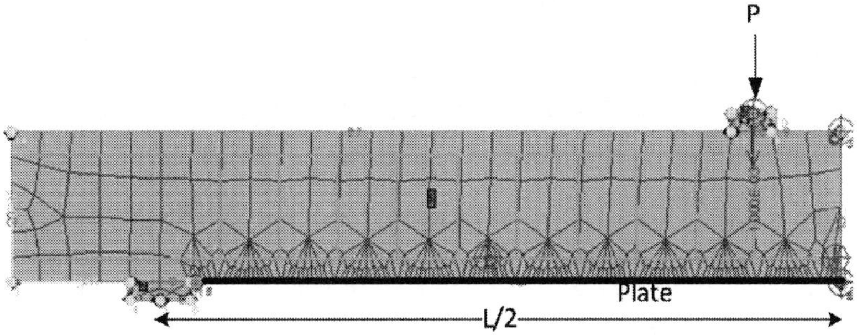

Fig. 2. Geometry of the FE mesh of the beams strengthened with CFRP plate

For descripting the concrete's non-linear behavior, ATENA 2D uses SBETA model, which implements formulations for behaviors as: tension and compression softening, concrete's in tension fracture based on non-linear fracture mechanics, biaxial strength failure criterion, compressive strength reduction after cracking and tension stiffening [5]. SBETA is made by 20 parameters, which can be automatically defined by informing concrete's cubic strength (f_{cu}). For GFRP's behavior, it's possible to consider as linear modelling. The constitutive models used by ATENA 2D are shown in Fig. 3 and Fig. 4. All properties for element are shown in Table 2.

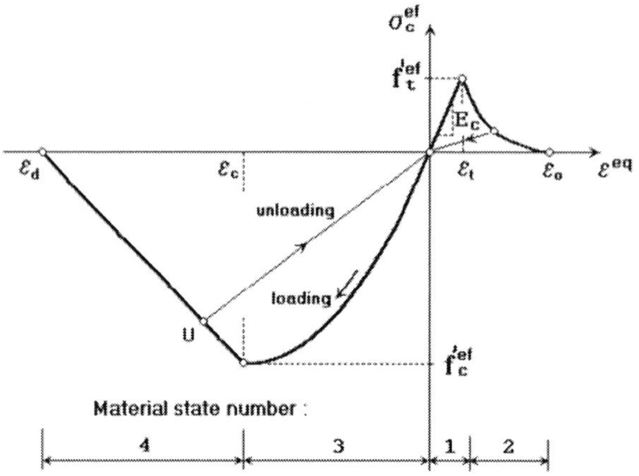

Fig. 3. Uniaxial stress-strain law for concrete [5]

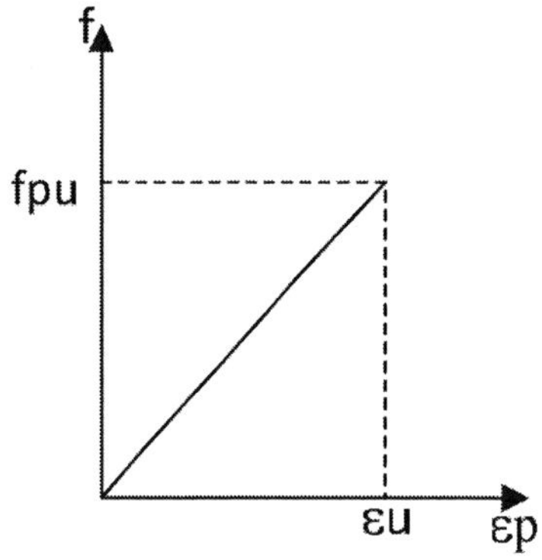

Fig. 4. The linear stress-strain law for GFRP reinforcement

Table 2. Properties of GFRP bar and CFRP plate

Type of Materials	GFRP	CFRP
Nominal size (mm)	13	100x1.2
Cross section area (mm^2)	201	120
Tensile strength (MPa)	981	3100
Elastic modulus (GPa)	43.6	165
Ultimate Elongation (%)	2.29	1.70

4 Numerical model using cross sectional analysis (RCCSA Software)

Numerical model using cross sectional analysis was also carried out. The layered element procedure was used to obtain full flexural response of reinforced concrete cross section. This procedure was implemented into a computer program called Reinforced Concrete Cross Section Analysis (RCCSA) developed by Thamrin [7]. This software also has an ability to predict flexural response of strengthened reinforced concrete cross section.

5 Results and discussion

The load-deflection relationship of experimental, un-strengthened and strengthened beams obtained from the finite element analysis are validated against the experimental results and illustrated in Fig. 5. Comparison of load-deflection curve from the ATENA, RCCSA and experimental shows the same trend of load deflection curves. RCCSA analysis predicts the beam to be stiffer and stronger than ATENA prediction. This is because of the RCCSA using a cross-sectional analytical solution proposed only two kinds of failure, crushing concrete or FRP rupture, can be predicted. Forwarded analysis calculations until crushing concrete or FRP rupture failure modes; thus it appears that the analysis of the cross section (RCCSA) of the beam deflection provides much greater than in experiment and ATENA.

Graph in Fig. 5 shows the increase in load and reduction in deflection occurs for UR beam and beam OR that has been strengthened with CFRP plate. UR beam (CG2-22) load increase of approximately 50%, with a reduction rate of deflection by 40% which makes the stronger beam. Similarly, look at the beam OR where an increase of load rate of 63%, and 33.3% reduction for deflection of the beam (CG3-22). This is shown with the addition of CFRP plate at the tension face of beam can increase the strength and stiffness of GFRP RC beam as in [8] previous research.

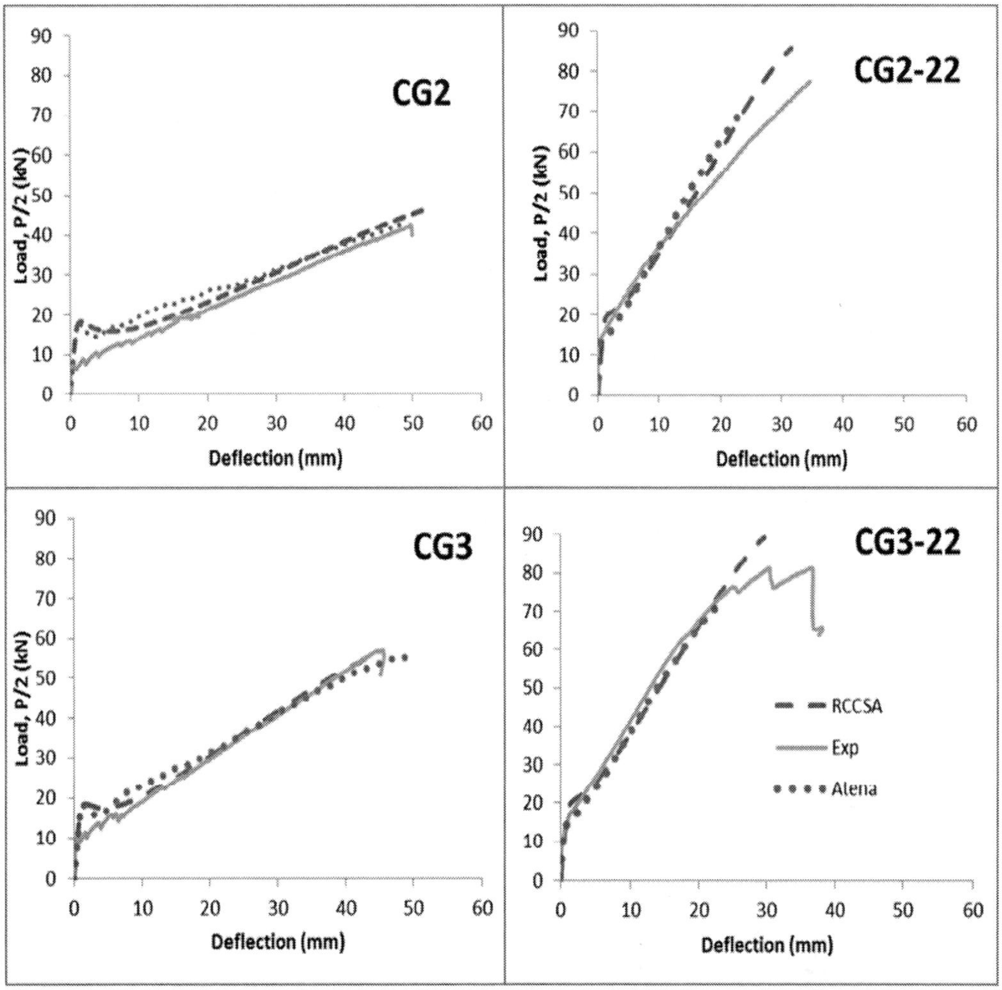

Fig. 5. Load-deflection of beam specimens

6 Conclusion

The general behaviour for FEA capital and RCCSA analysis be shown good agreement with experimental data comparison. The following are the findings that can be produced from studies done:

i) The addition of CFRP on the tension of the beam could fix the maximum value load and maximum deflection that occurs on the beam.
ii) FEA using ATENA can make a simulation for a beam in strengthening with CFRP plate. Expectations for the maximum load of the beam can be obtained correctly.
iii) RCCSA can also produce similar findings with the experimental. Making it easier for the expected results of the experiment will be carried out.
iv) Parametric studies can be made using RCCSA and ATENA, to provide the expected initial findings which will be carried out.

References

[1] I.F. Kara, A.F. Ashour and C. Dundar, Deflection of concrete structures reinforced with FRP bars, Compos. Part B Eng., **44**(1), 375–384 (2013)
[2] A. Rahman, M. Sam and R.N. Swamy, Flexural behaviour of concrete beams reinforced with glass fibre reinforced polymer bars, J. Kejuruteraan Awam, **17**(1), 49–57 (2005)
[3] Ţ. Nicolae, O. Gabriel, I. Dorina and E. Ioana, Fibre reinforced polymer composites as internal and external reinforcements for building elements, Bul. Institutului Politeh. din lasi. Sect. Constr. Arhit., **54**(1), 7 (2008)
[4] L. Van Den Einde, L. Zhao and F. Seible, Use of FRP composites in civil structural applications, Constr. Build. Mater., **17**(6), 389–403, (2003)
[5] V. Červenka, L. Jendele and J. Červenka, *ATENA program documentation part 1 theory*, Červenka Consulting, 1–282 (2012)
[6] J.P. Firmo, M.R.T. Arruda, J.R. Correia and C. Tiago, Flexural behaviour of partially bonded carbon fibre reinforced polymers strengthened concrete beams: Application to fire protection systems design, Mater. Des., **65**, 1064–1074 (2015)
[7] R. Thamrin and T. Kaku, Development length evaluation of reinforced concrete beam with CFRP bars, *Proc. of the Int. Sysmposium on Bond behaviour of FRP in Structure*, 385–392 (2005)
[8] N. Salleh, A.R. Mohd Sam, J. Mohd Yatim and M.F. Osman, Flexural behaviour of reinforced concrete beam with glass fiber reinforced polymer (GFRP) bar strengthened with carbon fiber reinforced polymer (CFRP) plate, Adv. Mater. Res., **1051**, 748–751 (2014)

Fracture Energy of Foamed Concrete: Numerical Modelling Using the Combined Finite-Discrete Element Method

Z. M. Jaini[1,*], *N.* Abd Rahman[1], *R. H. M.* Rum[2], and *M. M.* Haurula[2]

[1]Jamilus Research Center, Universiti Tun Hussein Onn Malaysia, 86400 Parit Raja, Johor, Malaysia
[2]Faculty of Civil and Environmental Engineering, Universiti Tun Hussein Onn Malaysia, 86400 Parit Raja, Johor, Malaysia

Abstract. Foamed concrete is known as lightweight concrete that proven in the low range of densities with good strength. Numerous studies were conducted to investigate its performance in term of mechanical properties. However, these investigations are still not adequate in the requirement of failure and damage mechanisms. This is due to the lack of knowledge on fracture energy that governs the crack resistance of foamed concrete. Therefore, this paper presents the numerical prediction of fracture energy of foamed concrete using the combined finite-discrete element method. The notched beam was modelled in the two-dimensional with unstructured mesh elements and a simple material model of Rotating Crack. The results of numerical prediction signify that various notch depths have an apparent impact toward fracture energy of foamed concrete.

1 Introduction

Fracture energy refers to the most important material property that governs the cracking and failure behaviour, especially for brittle materials. In general, the initiation, growth and propagation of crack are highly associated with fracture energy. Therefore, fracture energy reflects the toughness and crack resistance of material to sustain particular actions. Fracture energy is highly influenced by size effects, mostly dependent on the dimension, geometry and shape of structure, loading condition as well as the present of reinforcements. However, other factors such as strength, particle distribution and mix proportion of materials have also play a major role.

Normal concrete has fracture energy of approximately 75N/m to 100N/m and increases as high as 200N/m for high-performance concrete. Meanwhile, the fracture energy of foamed concrete, including cellular concrete and aerated concrete, was found to be 30% to 50% lower than normal concrete [1]. Void and non-presence of coarse aggregate typically affected the strength and consequently reduced the fracture energy of foamed concrete. However, the strength of foamed concrete also highly depends on its density, which varies from 400kg/m^3 to 1800kg/m^3. Abd Rahman et al. [2], Kozlowski et al. [3] and Jaini et al. [4] suggested that the fracture energy of foamed concrete is within 1.5N/m to 35N/m.

* Corresponding author: rizuan@uthm.edu.my

Current practice to determine fracture energy is through experimental tests such as three-point bending, wedge splitting and disk-shaped compacted tension that require notched specimens. Moreover, due to the advancement in the computational engineering, fracture energy can also be predicted conveniently by using numerical analysis such as finite element method, discrete element method and particle-based method. However, it needs the sophisticated model of constitutive law, representation of softening and crack failure. Therefore, this study aims to investigate the fracture energy of foamed concrete using numerical modelling of the combined finite-discrete element method.

2 Determination of fracture energy

Most of experimental investigations of fracture energy follow the recommendation of Hillerborg model [5] by using the three-point bending test. Based on this model, fracture energy can be determined using the following equation:

$$G_f = \frac{U_0 + m_g \Delta_d}{B(W - a_0)} \tag{1}$$

where U_0 is the area under softening phase of load-displacement profile, m_g is the mass of beam, Δ_d is the displacement of fracture (displacement at the ultimate strength), B is specimen width, W is specimen depth and a_0 is notch depth. However, other models such as introduced by Bazant [6] and Oh et al. [7] can also be adopted to determine fracture energy, where the formula based on the strength and physical properties of concrete.

Numerical modelling is another method to determine fracture energy. This method is very appropriate to predict fracture energy accurately, faster and less tedious. In previous work, Shi and Suzuki [8] conducted numerical modelling of fracture energy using extended fictitious crack model. The model is in two-dimensional based on four-point bending test. Meanwhile, Hanson and Ingraffea [9] also adopted the fictitious model with two-parameter where the non-linear fracture mechanics is taken into account. However, the capability of the numerical modelling is limited to the certain extent of crack opening.

Until recently, numerical modelling of fracture energy of foamed concrete is limited and not well establish. The only available reference is Kozlowski et al. [10] that numerically investigate the fracture energy of foamed concrete using the extended finite element method (XFEM) in Abaqus. In the numerical modelling, the notched beam was simulated under three-point bending test to obtain the fracture processes. It was found that XFEM has successfully modelled the fracture behaviour and provided the decent approximation of load-bearing capacity. This could be a benchmark in the numerical modelling of fracture energy of foamed concrete.

3 Methodology

The numerical modelling is based on the notched beam subjected to the three-point bending test. A series of experimental study, as can be seen in Fig. 1, were conducted to provide reliable information, comparison and appraisal for the numerical modelling.

3.1 Specification of notched beam

The notched beam was modelled with size of 100mm depth (W), 100mm width (B) and 500mm length (L). The schematic design of notched beam is shown in Fig. 2. The notched beam was produced to fulfill the specific density of 1600kg/m^3 with targeted compressive strength of 10MPa to 15MPa. In order to investigate the size effects on fracture energy of

DOI: 10.1051/matecconf/201710302030

foamed concrete, the notch depth was increased based on the depth ratio of $\alpha = a_0/W$ from 0.15 to 0.75 by an increment of 0.15. The specification of notch depth can be referred in Table 1.

Fig. 1. Experimental study of three-point bending test.

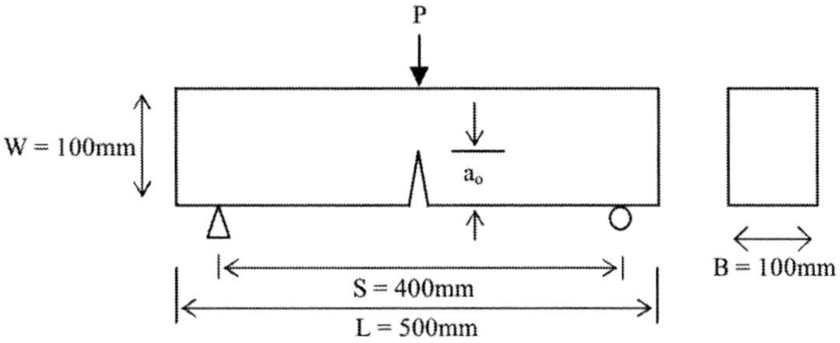

Fig. 2. Dimension of notched beam.

Table 1. Specification of notch depth.

No	Depth Ratio, α	Notch Depth a_o (mm)	Remark
1	0.15	15	
2	0.30	30	Under neutral axis
3	0.45	45	
4	0.60	60	Above neutral axis
5	0.75	75	

3.2 FDEM modelling

The combined finite-discrete element method uses a similar algorithm as in the explicit dynamic finite element method with additional parameters for the discrete element insertion [11]. From the basic equation that governs the response of system is given by:

$$M\ddot{u}(t_n) + C\dot{u}(t_n) + Ku(t_n) = f^{ext}(t_n) \quad (2)$$

Later, the spatially discretized form of the equilibrium equations at time t_n can be rewritten as:

$$M\ddot{u}(t_n) + C\dot{u}(t_n) = f^{ext}(t_n) - f^{\text{int}}(u(t_n)) \quad (3)$$

where M represents the global mass, C is damping and K is stiffness matrices. The f^{int} is the vector of the internal resisting forces, and f^{ext} is the velocity applied forces. The standard notations $\ddot{u}$, $\dot{u}$ and u signify the acceleration, velocity and displacement vectors in the global domain. The subscript n implies a quantity at time t_n. In the context of time integration for nonlinear problems, the governing equation generalises the stiffness dependent elastic forces to include both geometric and constitutive nonlinearities to provide internal forces with the structure [12].

Therefore, the internal force vector has to be updated at each time step as well as at each interaction step during the time integration of the equations of motion. Working with system vectors instead of system matrices, which may be added up by the finite element contributions, for the computation of the state variables u and $\dot{u}$, it is possible to increase the number of degrees of freedom and thus large engineering problem can be treated. The diagonal mass matrix, M is defined by:

$$M = \underset{e=1}{\overset{nelem}{V}} A \sum_{i=1}^{nnode} \int {}^{e}\rho N_i^T [N_1, N_2, ..., N_{nnode}] dV \tag{4}$$

and the diagonal damping matrix, C defining the viscous damping term with a constant factor, α as:

$$C = \alpha M \tag{5}$$

The damping in Equation (5) will damp out rigid body motion and can thus introduce some undesirable viscous system response.

The common case of a varying time step for an explicit nonlinear dynamic analysis is considered, where the integration time steps length, Δt_{n+1} can be identified between the two discrete points in time of t_n and t_{n+1}. A combination of previous and current mid-point velocities form the second order central difference equation that defines the acceleration state of a known point at a known discrete time t_n as:

$$\ddot{u}(t_n) = \frac{\dot{u}_{n+\frac{1}{2}} - \dot{u}_{n-\frac{1}{2}}}{\Delta t_{n+\frac{1}{2}}} \tag{6}$$

Through the direct integration of the acceleration term with respect to the time increment measure, the velocity measure associated with a point, at discrete time t_n can be approximated as:

$$\dot{u}(t_n) = \frac{u_{n+\frac{1}{2}} - u_{n-\frac{1}{2}}}{\Delta t_{n+\frac{1}{2}}} = \frac{u_{n+1} - u_{n-1}}{2\Delta t_{n+\frac{1}{2}}} \tag{7}$$

The mid-point velocity of the previous and current times can also be further determined as a direction combination of the motion terms at time t_{n-1}, t_n and t_{n+1}. The mid-point velocity terms are then defined as:

$$\dot{u}\left(t_{n-\frac{1}{2}}\right) = \frac{u_n - u_{n-1}}{\Delta t_n} \quad ; \quad \dot{u}\left(t_{n+\frac{1}{2}}\right) = \frac{u_{n+1} - u_n}{\Delta t_{n+1}} \tag{8}$$

Therefore, Equation (7) can be written as follows:

$$\dot{u}(t_n) = \frac{\Delta t_{n+1}\dot{u}_{n+\frac{1}{2}} - \Delta t_n \dot{u}_{n-\frac{1}{2}}}{2\Delta t_{n+\frac{1}{2}}} \tag{9}$$

Upon of substitution of Equation (6) and Equation (9) into Equation (3), the velocity at the current time step is determined in term of the previous time step of velocity, displacement and force. The solution to determine this velocity is given by:

$$\dot{u}\left(\Delta t_{n+\frac{1}{2}}\right) = \left[2M + C\Delta t_{n+1}\right]^{-1}\left[D\right] \tag{10}$$

where

$$\left[D\right] = \left(2M - C\Delta t_n\right)\dot{u}_{n-\frac{1}{2}} + 2\Delta t_{n+\frac{1}{2}}\left(f_{n+1}^{\text{int}} - f_{n+1}^{ext}\right) \tag{10}$$

or in global velocity at time t_{n+1} by substitute Equation (3) into Equation (9), yields the following expression:

$$\dot{u}(t_{n+1}) = \frac{2M - C\Delta t_n}{2M - C\Delta t_{n+1}}\dot{u}_{n-\frac{1}{2}} + 2\Delta t_{n-\frac{1}{2}}\frac{f_n^{ext} - f_n^{\text{int}}}{2M + C\Delta t_{n+1}} \tag{11}$$

3.3 Model of notched beam

Using the combined finite-discrete element method, the notched beam was modelled as unstructured three-noded element of the surface domains (refer Fig. 3). The advantage of three-noded element is that the crack can be propagated in irregular directions, either through the boundary element (inter-element) or along failure plane (intra-element). The mesh size was initially determined based on the convergence and critical time step. The surrounding area of notch was discretized with size of 3mm that has similar size as sand, while the rest area adopting size of 5mm to 10mm.

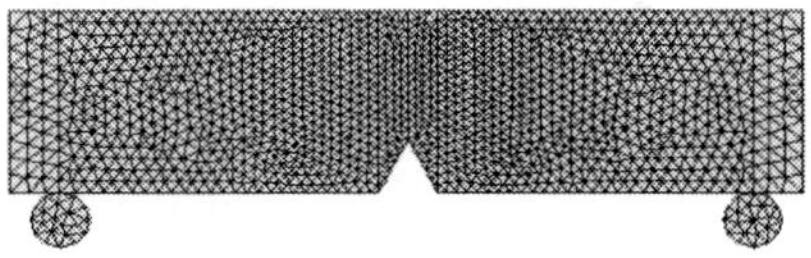

Fig. 3. Unstructured three-noded element of notched beam.

Since the model involves both notched beam and steel support, hence only steel support at left and right sides were constrained. The contact of steel support with notched beam was considered as partially smooth surface. An applied ramp loading was imposed at the mid-span of notched beam with speed of 0.5mm/min. This speed was maintained until notched beam experiences crack formation that leads to totally collapse. The solution algorithm was based on an explicit central difference solution with the time step of 0.2msec, maximum time step of 1s and termination at 200000 steps.

On the other hand, the material properties of foamed concrete and steel support can be referred in Table 2. Foamed concrete was defined as Rotating Crack that only requires elastic properties, tensile strength and strain rate. Meanwhile, the steel support was assigned as fully elastic. Foamed concrete as a quasi-brittle material is represented as continuum from which cracking virtually occurs during the deformation processes. The crack then propagates to all directions when the strength of foamed concrete is degraded. Before experiencing any failure, foamed concrete remains in homogeneous elastic state. After the initial yield, the Rotating Crack represents the anisotropic damage evolution by degrading the elastic modulus in the direction of the major principal stress invariant.

Table 2. Material properties of foamed concrete and steel support

No.	Properties	Foamed Concrete	Steel Support
1	Young's modulus, *E* (GPa)	12	210
2	Poisson's ratio, υ	0.20	0.30
3	Density, ρ (kg/m^3)	1600	7800
4	Tensile strength, σ_t (MPa)	0.56	-
5	Strain rate	0.01	-

4 Results and discussions

The combined finite-discrete element method is based on continuum and fracture failures. Therefore, the effective stress and crack propagation that occur on the notched beam can be simultaneously observed. Effective stress was observed at the initial time and pre-cracking (refer Fig. 4) to understand the deformation behaviour that happen in the different stages.

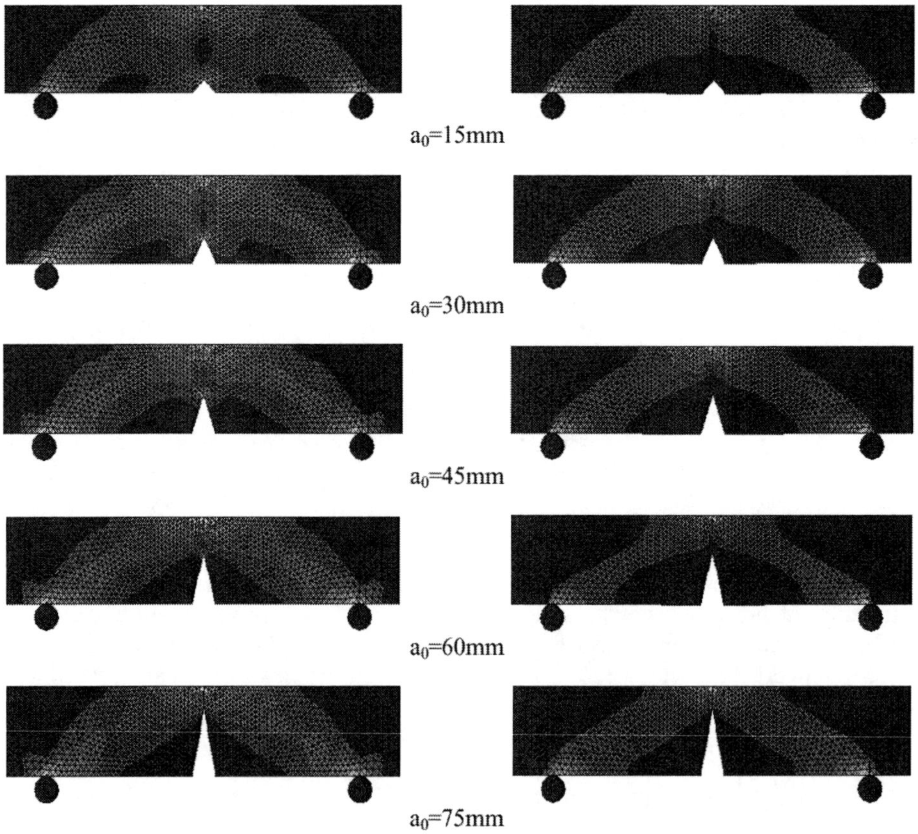

a) initial time b) pre-cracking

Fig. 4. Stress distribution during loaded stage.

At the initial time, no cracking was observed and thus notched beam still in the intact condition. Meanwhile, during which the effective stress exceeded the ultimate strength of foamed concrete, cracks start to initiate and propagate from the notch tip. Under the applied loading, the crack was found propagate very slowly.

Since fracture energy is determined using Hillerborg model, the combined finite-discrete element method was setup to give the load-displacement profile as can be seen in Fig. 5. It was observed from the load-displacement profile that the ultimate load decreases when the notch depth becomes deeper. This subsequently affected the area of softening. Any increment on the notch depth corresponds directly on the reduction of the softening area and hence decrease fracture energy. Based on calculation, fracture energy of foamed concrete is 28.30N/m, 21.28N/m, 10.45N/m, 3.59N/m and 1.37N/m for notch depth of 15mm, 30mm, 45mm, 60mm and 75mm respectively.

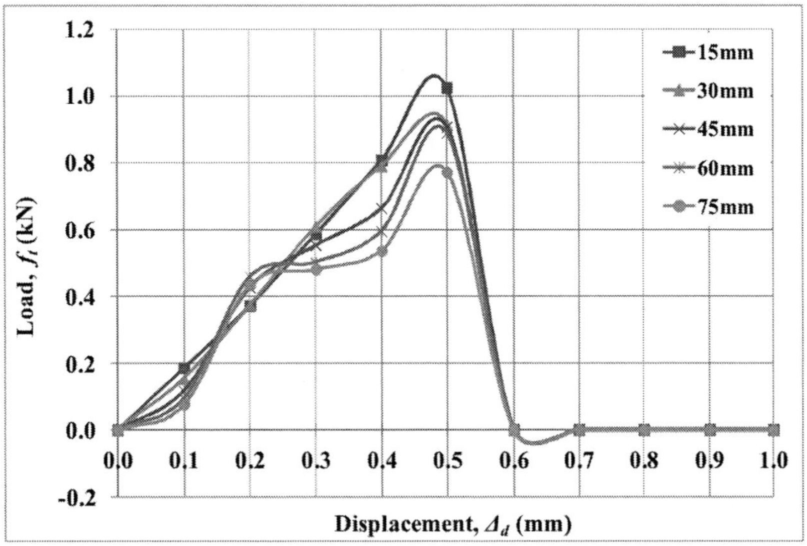

Fig. 5. Load-deflection profile for notched beam subjected three-point bending.

According to Abd Rahman et al. [2], the fracture energy of foamed concrete is about 18N/m to 25N/m for V-notch with the depth of 30mm. It should be noted here that this fracture energy is for notched beam with size 150mm depth (W), 150mm width (B) and 700mm length (L). In the numerical modelling, fracture energy for the similar notch is 21.28N/m. Therefore, the fracture energy that obtained from numerical modelling using similar notch is convincing and favourable. Meanwhile, a comparison of fracture energy from experimental study and numerical modelling is depicted in Fig. 6. Both results were also fitted to produce simple equations based on the notch depth.

Based on the comparison, the fracture energy from numerical modelling is significantly different with experimental study. For notch depth below than 22mm, numerical modelling produces upper bound results. However, numerical modelling underestimates the fracture energy when notch depth increased. The error, although still acceptable, might be caused by the boundary condition between notched beam and steel support which relatively softer than actual. The material properties used in this numerical model are not based on real tests but rather derived from reasonable assumptions. Other than that, it was found that instability of elements occurs during the energy dissipation that disturbs the crack propagation. Table 3 shows the errors of numerical modelling, where notch depth located below than neutral axis has favourable agreement.

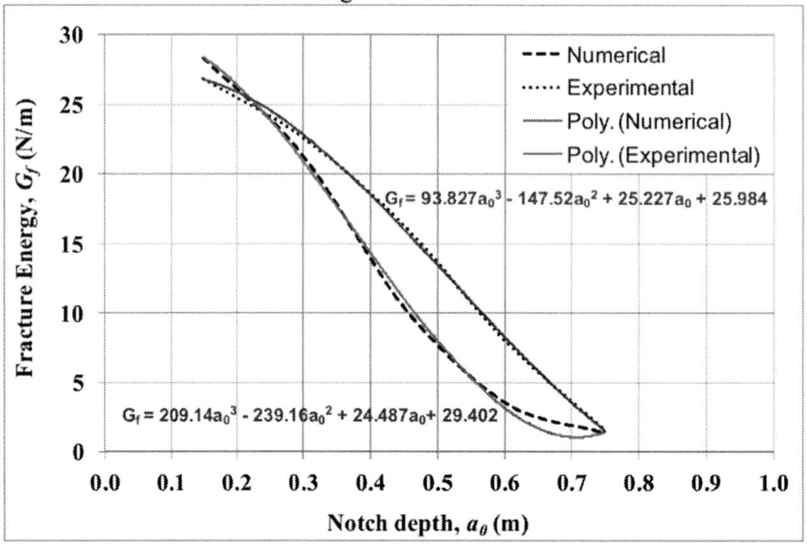

Fig. 6. Comparison of fracture energy between experimental and numerical study.

Table 3. Error in fracture energy of foamed concrete

No.	Notch Depth a_o (mm)	Fracture Energy Experimental $G_{f,exp}$ (N/m)	Fracture Energy Numerical $G_{f,num}$ (N/m)	Error (%)
1	15	26.82	28.30	5.51
2	30	22.59	21.28	5.80
3	45	16.34	10.45	36.05
4	60	8.06	3.59	55.46
5	75	1.56	1.37	12.18

5 Conclusions

Numerical modelling of fracture energy of foamed concrete was performed using the combined finite-discrete element method. The fracture energy was successfully achieved by employing designated material criteria of Rotating-Crack for foamed concrete and elastic model for steel support. The notched beam was modelled using unstructured three-noded mesh element and imposed by loading as similar as three-point bending test. The numerical modelling measures the fracture energy based on Hillerborg model, where the area of load-displacement profile, maximum deflection, mass and dimension of notched beam become dominant parameters.

It was found that the fracture energy of foamed concrete significantly decreased by the notch depth. Although the numerical modelling produces a reliable prediction of fracture energy, the results considerably lower than experimental study. This happen due to identified factors such as the contact interaction between foamed concrete and steel support, material properties and instability on the energy dissipation. However, this approach on numerical modelling of fracture energy of foamed concrete is extremely valuable and the accuracy of results can be further enhanced.

Authors would like to thanks Ministry Education of Malaysia and Universiti Tun Hussein Onn Malaysia for the financial and facilities supports. This study was conducted under Fundamental Research Grant Scheme, FRGS Code 1219.

References

[1] N. Abd Rahman, Z.M. Jaini, N.A. Abd Rahim and S.A. Abd Razak, An experimental study on the fracture energy of foamed concrete using v-notched beams, *Proc. of the Int. Civil and Infrastructure Engineering Conference*, Springer, Singapore, 2015, 97-108
[2] N. Abd Rahman, Z.M. Jaini and N.N.M. Zahir, Fracture energy of foamed concrete by means of the three-point bending test on notched beam specimens, ARPN Journal of Engineering and Applied Sciences,**10**(15), 6562-6570 (2015)
[3] M. Kozlowski, M. Kadela and A. Kukielka, Fracture energy of foamed concrete based on three-point bending test on notched beams, Procedia Engineering, **108**, 349-354 (2015)
[4] Z.M. Jaini, S.N. Mokhatar, A.S.M. Yusof, S. Zulkiply and M.H. Abd Rahman, Effect of pelletized coconut fibre on the compressive strength of foamed concrete, MATEC Web of Conference, **47**, 01013 (2016)
[5] A. Hillerborg, The theoretical basis of a method to determine the fracture energy of concrete, Materials and Structure, **18**(4), 291-296 (1985)
[6] Z.P. Bazant, Concrete fracture models testing and practice, Engineering Fracture Mechanics, **69**, 165-205 (2002)
[7] B.H. Oh, S.Y. Jang and H.K. Byun, Prediction of fracture energy of concrete, KCI Concrete Journal, **11**(3), 211-221 (1999)
[8] Z. Shi and M. Suzuki, Numerical studies on load-carrying capacities of notched concrete beams subjected to various concentrated loads, Construction and Building Materials, **18**(3), 173-180 (2004)
[9] J.H. Hanson and A.R. Ingraffea, Using numerical simulations to compare the fracture toughness values for concrete from the size-effect, two-parameter and fictitious crack model, Engineering Fracture Mechanics, **70**, 1015-1027 (2003)
[10] M. Kozlowski, M. Kadela and M. Gwozdz-Lason, Numerical fracture analysis of foamed concrete beam using XFEM method, Applied Mechanics and Materials, **837**, 183-185 (2016)
[11] Z.M. Jaini, Y.T. Feng, D.R.J. Owen and S.N. Mokhatar, Fracture failure of reinforced concrete slabs subjected to blast loading using the combined finite-discrete element method, Latin American Journal of Solids and Structures, **13**(6), 1086-1106 (2016)
[12] M.G. Cottrell, *The Development of Rational Computational Strategies for the Numerical Modelling of High Velocity Impact*, PhD Thesis, Swansea University, United Kingdom, (2002)

Ranking of Delay Factors for Makkah's Construction Industry

Nashwan Al-Emad[1,*], *Ismail* Abdul Rahman [2], *Sasitharan* Nagapan[1], and *Yaser* Gamil[1]

[1]Faculty of Civil and Environmental Engineering, Universiti Tun Hussein Onn Malaysia, 86400 Parit Raja, Johor, Malaysia
[2]Sariya Contracting Company Ltd, Novotel Business Park, Tower 2, 3rd Floor, Dammam, Khobar Highway, Post Box: 4486, Al Khobar 31952, Kingdom of Saudi

Abstract. This paper presents identification of significant delay factors encountered by Makkah's construction industry using quantitative approach. A structured questionnaire developed based on literature review was verified through pilot study involved selected construction experts. Questionnaire survey was conducted amongst Makkah construction practitioners include contractors, consultants and project management consultancy. The survey managed to collect 100 valid responses which were used to rank the factors using average index approach. Results of the analysis for 10 most significant factors causing construction delay in Makkah construction industry are Difficulties in financing project by contractor, Poor coordination between parties, Shortage of manpower, Delays in producing design documents, Improper planning and scheduling of the project, Delay in progress payments, Low productivity level of labour, Poor communication between parties, Unqualified workforce and Poor contract management. This finding is helpful to Makkah construction's community particularly projects' stakeholders in avoiding potential delay for their future projects.

1 Introduction

Saudi Arabia experiences rapid construction growth in the past decade both in urban and rural areas [1]. There are many challenges faced by the Saudi Arabia construction industry and one of them is project delay [2]. A study conducted by [3] found that around 70% of all public sector construction projects were delayed due to several factors. Hence, this study focused on construction delay factors encountered in Makkah city development.

2 Literature review

Literature review has shown that there are several studies on identifying factors causing delays in construction projects in many parts of the world which also includes Saudi Arabia. Several articles have identified delay factors in construction projects in numerous manners. Based on these articles, 81 common delay factors to construction projects were

* Corresponding author: df080116@gmail.com

extracted for the questionnaire formulation. These factors were verified in pilot study to determine its relevancy to Makkah construction. Outcome of the pilot study was only 37 delay factors are considered relevant to Makkah construction environment.

3 Methodology

Data was collected using questionnaire survey from targeted respondents of the construction community in Makkah city. The questionnaire was designed based on the literature review and validated through the pilot study. Respondents were requested to indicate the level of significance on the identified 37 delay factors using a 5-point likert's significance scale. The measurement scale for this research was adopted from [4] i.e.1 = not significant, 2 = slightly significant, 3 = moderately significant, 4 = significant and 5 = very significant. Questionnaire was distributed to the targeted respondents who are contractors, consultants and project management consultancy (PMC) through site visit and electronic communication. This study managed to secure 100 valid responses from the survey were received back within a stipulated period of one month of data collection.

3.1 Sampling size

Sample size of this study was determined using the following formula [5].

$$SS = \frac{Z^2 \times P\ (1-P)}{C^2} \tag{1}$$

where,

SS = Sample Size
Z = Z value (1.96 for 95% confidence level)
P = percentage picking a choice, expressed as a decimal (0.5 used for sample size needed)
C = margin of error (9 %)

$$SS = \frac{1.96^2 \times 0.5\ (1-0.5)}{0.09^2} = 96.04 \cong \approx 96 \tag{2}$$

Therefore, 100 valid responses used in this study are greater than 96 as minimum samples.

4 Data analysis

The questionnaire form comprises of two parts where the first part is respondents' demographic and the second part is list of delay factors which had been validated by the experts. The demography of the respondents is presented in Table 1.

Table 1 indicates the respondents' demography on their eligibility in giving their expert opinions regarding the delay issues experienced in the construction projects in the city of Makkah. Majority of the respondents are having bachelor degree, working experiences for more than 15 years, holding high position in their company and working in large construction projects.

Table 1. Respondents' demography.

Category	Items	Numbers of respondents	Percentage of respondents
Qualifications	Master	16	16%
	Bachelor	78	78%
	Diploma	6	6%
Working Experience (years)	0-5	16	16%
	06-Oct	32	32%
	Nov-15	33	33%
	16-20	7	7%
	21-25	6	6%
	More than 25	6	6%
Position	Manager	29	29%
	Engineer	60	60%
	Technical Staff	11	11%
Project Contract Amount (Saudi Riyal)	Below SR 10M	3	3%
	SR 10-100M	15	15%
	SR 100M-1B	16	16%
	Above SR 1B	66	66%

4.1 Ranking of delay factors

Before the collected data can be used for further analysis, its stability and consistency were analysed using Cronbach's alpha value. According to [6], if Cronbach alpha value is less than 0.3, the reliability is considered low and data cannot be accepted and if the Cronbach alpha value is more than 0.7 the reliability is considered high. For this research the overall alpha value on the 37 delay factors is 0.942 which consider accepted and reliable. Ranking of the delay factors is based on the average index score of each factor [7]. However, there are possibilities that the factors having same average index score and to resolve this is by introducing standard deviation attained by each factor. Standard deviation (SD) is the amount of variation or dispersion of a set of data values and when SD value is close to 0 it indicates that the data points tend to be very close to the mean (also called the expected value) of the set. Hence this study used two parameters that are average index score and SD to decide the rank of the delay factors. If the factor's average index score is at tie with other factor, then factor with smaller standard deviation should be adjusted for higher ranking. In this case, average index and standard deviation of 37 delay factors was calculated where the average index (AI) values are in the range of 3.98 to 2.79. It was found that 5 sets of factors are at tie ranks (i.e. at numbers 13, 17, 22, 26 and 29) and was resolved by using standard deviation where the ranks of the factors are adjusted accordingly. Results of ranking analysis show 10 most significant factors causing construction delay based on 100 valid responses from experts of Makkah construction industry are presented in Table 2.

The significant delay factors in Table 2 are elaborated in more details in accordance to previous researchers who had carried out similar research in different parts of Saudi Arabia construction industry and also Nashwan Al-Emad's experiences who is working in Makkah city as civil engineer.

Table 2. 10 Most significant factors causing construction delay

Code	Construction Delay Factors	AI	SD	Ranking
F26	Difficulties in financing project by contractor	3.98	1.223	1
F17	Poor coordination between parties	3.83	1.025	2
F33	Shortage of manpower	3.80	1.110	3
F6	Delays in producing design documents	3.75	0.989	4
F3	Improper planning and scheduling of the project	3.72	1.215	5
F27	Delay in progress payments	3.70	1.202	6
F35	Low productivity level of labour	3.67	1.045	7
F18	Poor communication between parties	3.66	0.997	8
F34	Unqualified workforce	3.59	1.147	9
F2	Poor contract management	3.56	1.157	10

4.1.1 Difficulties in financing project by contractor

Difficulty in financing project by contractor is considered as the most significant contributor to construction delay in Makkah city as ranked by all parties with highest average index score of 3.98. Nevertheless, this factor is certainly true for contractors where they are required to start the work once been awarded and requires initial adequate finance for that matter before the payment from the clients can be disbursed. Availability of cash flow is very important for a contractor to carry out the construction activities. Also, most of the main contractors outsource their works to sub-contractors, the failure for contractor to pay to sub-contractors on time resulting in slowing down the progress of the work due to reduce number of workers hired by sub-contractors and un-availability of construction materials. Hence, adequate cash flow and financial stability of contractors is very important in keeping construction progress as planned. This finding concurrently matches with the findings carried out in different parts of Saudi Arabia construction industry by [3, 7-12].

4.1.2 Poor coordination between parties

Poor coordination between parties was ranked as the 2nd significant factor causing construction delay with average index value 3.83 and standard deviation value is 1.025. Coordination is important within multi-participant environment as in most construction projects implementing in Makkah city. First attribute under this factor includes ineffective coordination among construction parties (Client, Consultant, Contractor and stakeholders), lack of coordination between design team and main contractor, lack of coordination between general contractor and subcontractors and also, infective coordination among the contractors' departments. Second attribute related to poor coordination is at design phase between client, stakeholders etc. can lead to serious project delay and is subjected to variation claims by the Contractor, also, non -availability of drawings on time is generally due to lack of coordination between construction site and design office. To prevent this problem, a proper coordination system between project parties should be established to increase the project performance. This result is in line with the results found in Saudi Arabia construction industry done by [13].

4.1.3 Shortage of manpower

Shortage of manpower factor is ranked as the 3rd most significant contributor to construction delay with average index score of 3.80 and standard deviation value is 1.110. Other research works related to Saudi Arabia construction industry [3, 7, 10, 12] also found that shortage of manpower as an important causes of delay in construction. This is certainly true for Saudi Arabia where the locals are not willing to work on construction site. This has resulted to contractors importing labour and therefore reducing their ability to judge the level of skill of their employees. Thus, the quality of labour available is generally quite poor, which leads to low productivity and poor quality of work consequently time delay will be experienced which affects the overall project schedule.

4.1.4 Delays in producing design documents

Delay in producing design documents is the 4th major factor contributing to construction delay in Makkah construction projects with average index score of 3.75 and standard deviation value is 0.989. Poor design and delays in design documents are a common problem that occurs in construction projects. Knowledge and well experience of the designer to all materials, equipment and project specifications is an important attribute for developing a comprehensive design documents system. Besides that, time limitation given to designers during design phase occasionally force the designer to wrap up the necessary design works at a lower quality. If inadequate time is given, the design cannot be developed in a proper manner. This may eventually cause misunderstandings between construction players working on the project. This will automatically give negative effect to projects schedule.

4.1.5 Improper planning and scheduling of the project

In this study improper planning and scheduling of the project was indicated as one of the significant factor causing construction delay in Makkah city. It has an average index value 3.72 and standard deviation value is 0.1.215. Effective planning and scheduling plays a very important role in success of any project. Results of this survey ranked ineffective planning and scheduling of the project as the 5th significant cause of construction delay. Previous researcher e.g., [3, 8, 9, 12, 14] also conducted a research works in different parts of Saudi Arabia and has determined inadequate planning and scheduling of projects as one of the most important causes of project delay. This problem can be overcome by implementing techniques for managing projects with planning and scheduling software packages, as well as contractors should establish a dedicated team for planning, follow-up the progress of the work in daily basis and pending issues.

4.1.6 Delay in progress payments

This the 6th significant factor found by this research with average index and standard deviation values of 3.70 and 1.202 respectively. Progress payment delay is probably the most common source of construction disputes among parties involved. Failure to provide steady monthly progress payment by owner to main contractor will cause work progress delay because there is inadequate cash flow to support construction expenses especially for those contractors who are not financially sound. To reduce this dispute cause, owners should pay progress payment on time and contractors should manage their financial resources and plan cash flow by utilizing progress payment efficiently and effectively .This result in line with the results found by [3, 8, 9, 10, 13, 14] in several parts of Saudi Arabia.

4.1.7 Low productivity level of labour

Low productivity of labour is certainly affecting the activities duration and consequently the total project duration. This factor ranked as the 7th significant factor causing construction delay with average index score 3.670 and standard deviation value 1.045. Low productivity of the labours could be due to several factors including non-availability of materials, tools and equipment, unskilled manpower hired and also labours are not receiving their salaries on time. This has an adverse impact on the project progress. To overcome this problem, it should hire skilled manpower and provide sufficient construction tools and equipment in order to improve and achieve productivity level. This result is in line with the results carried out in Saudi Arabia construction industry by [3, 7, 13].

4.1.8 Poor communication between parties

Poor communication between parties is ranked as the 8th most significant factor contributing to construction delay with average index value 3.66 and standard deviation value is 0. 997. Since there are many parties involved in the project (client, consultant, and contractor) then communication between these parties is very important for project success. This problem invariably leads to rework, and then project will experience delay. Hence, in order to reduce the risk of project delay, it is imperative to establish proper formal communication channels among all project parties during all project phases. This result supported by [8] in his study conducted in the northern part of Saudi Arabia.

4.1.9 Unqualified workforce

In this study, unqualified workforce is indicated as one of the significant contributors to the construction delay in Makkah city with average index score 3.59 and standard deviation value 1.147. Normally, project could experience less progress due to lack of adequate numbers of skilled labour. It will encounter improper execution or rework and this will leads to delay of construction activities and consequently time delay will be experienced. This findings match with the result found by [7, 8, 10] in Saudi Arabia construction industry.

4.1.10 Poor contract management

Poor contract management was ranked by all respondents as the 10th factor causing construction delay with average index and standard deviation values 3.56 and 1.157 respectively. Competence and professionalism of project team plays a key role to a successful project and to prevent the project from facing construction delay. Poor contract management is usually caused by lack of management skills and shortage of professional experts among the project practitioners. As a result of poor contractual skills, the project could experience negative effects to duration of project .This finding in line with the finding found by [7, 9].

5 Conclusions

This paper has presented a descriptive analysis to determine the ranking of delay factors faced by Makkah's construction industry. A total of 37 delay factors were analysed using average index approach and found that 10 most significant factors causing construction delay in Makkah construction industry are Difficulties in financing project by contractor,

Poor coordination between parties, Shortage of manpower, Delays in producing design documents, Improper planning and scheduling of the project, Delay in progress payments, Low productivity level of labour, Poor communication between parties, Unqualified workforce and Poor contract management. It hoped that, these findings will help the authority, researchers and construction players (contractor, consultant and client) to minimize any potential delays occurred in construction industry.

6 Recommendations

Based on the outcome of this study, all Makkah construction practitioners should consider the following points in order to minimize and control construction delays:

i) Contractor should manage his financial resources and plan cash flow by utilizing progress payment effectively and efficiently to avoid any disruption of the project's progress financially.
ii) Client should pay progress payment to the contractor on time to enable the contractor to finance the project.
iii) Proper coordination system between project parties should be established to increase the project performance.
iv) Contractor should establish a dedicated team for planning, follow-up the progress of the work in daily basis and pending issues.
v) It is imperative to establish proper formal communication channels among all project parties throughout project lifecycle.
vi) Contractor should hire competences and professionalism to his organization in order to prevent the project from facing construction delay.
vii) Contractor should hire enough number of skilled labours and motivate them to improve productivity of construction activities.
viii) Producing design documents should be on time, otherwise delay in producing design document will affect slowness of site executions.

Authors acknowledge Universiti Tun Hussein Onn Malaysia for the financial support.

References

[1] N. Samargandi, J. Fidrmuc, and S. Ghosh, *Financial development and economic growth in an oil-rich economy: The case of Saudi Arabia*, Economic Modelling No. 43, Brunel University, London, United Kingdom, (2013)
[2] I.A. Rahman, N. Al-Emad and S. Nagapan, Categorization of Saudi Arabia's construction delay factors using factor analysis technique, *Proc. of the 3rd International Conference on Civil, Offshore and Environmental Engineering*, Kuala Lumpur, Malaysia, (2016)
[3] S.A. Assaf and S. Al-Hejji, Causes of delay in large construction projects, Int. J. of Project Management, **24**(4), 349–357, (2006)
[4] A.H. Memon and I.A. Rahman, SEM-PLS analysis of inhibiting factors of cost performance for large construction projects in Malaysia: perspective of clients and consultants, The Scientific World Journal, (2014)
[5] W.C. Guenther, Sample size formulas for some binomial type problems, Technometrics, **16**(3), 465, (1974)
[6] X. Li and R. Wang, Survey research on relationship among service failure: service recovery and customer satisfaction, *Proc. of the Int. Conf. on Management Science and Engineering*, IEEE, 1121-1123, (2007)

[7] A.H. Memon, I.A. Rahman, M. Akram and N.M. Ali, Significant factors causing time overrun in construction projects of Peninsular Malaysia, Modern Applied Science, 8(4), 16-28, (2014)
[8] I. Mahamid, Micro and macro level of dispute causes in residential building projects: Studies of Saudi Arabia, J. of King Saud University - Engineering Sciences, **1**(28), 12-20, (2014)
[9] A. Albogamy, D. Scott and N. Dawood, *Addressing construction delays in the Kingdom of Saudi Arabia*, PhD Thesis, Teesside University, Middlesbrough, United Kingdom, (2012)
[10] S.A. Assaf, M. Al-Khalil and M. Al-Hazmi, Causes of delay in large building construction projects, J. of Management in Engineering, **11**(2), 45–50, (1995)
[11] M. Al-Khalil and M.A. Al-Ghafly, Delay in public utility projects in Saudi Arabia, Int. J. of Project Management, **17**(2), 101-106, (1999)
[12] A. Al-Kharashi and M. Skitmore, Causes of delays in Saudi Arabian public sector construction projects, Construction Management and Economics, **27**(1), 3-23, (2009)
[13] I. Mahamid, Contributors to schedule delays in public construction projects in Saudi Arabia: Owners' perspective, J. of Construction Project Management and Innovation, **3**(2), 608-619, (2013)
[14] N. Alotaibi, M. Sutrisna and H. Chong, *Managing critical factors causing delays in public construction projects in Kingdom Of Saudi Arabia*, PhD Thesis, Curtin University, Perth, Australia, (2014)

Qualitative Approach on Investigating Failure Factors of Yemeni Mega Construction Projects

Yaser Gamil[1,*], *Ismail* Abdul Rahman[1], *Sasitharan* Nagapan[1], and *Nashwan* Alemad[1]

[1] Faculty of Civil and Environmental Engineering, Universiti Tun Hussein Onn Malaysia, 86400, Parit Raja, Johor, Malaysia

Abstract. Large construction projects in Yemen face many impediments and challenges to be completed successfully. This article presents an investigative cases study of three mega construction projects which are considered failed to achieve their assigned plans and objectives. An exploratory interview was implied in this study via telephone to three construction experts who involved directly in these mega projects at the managerial level. Semi-structured questions based on literature were designed as guide for the interview where the scope was mainly on causes of the mega projects failure. Outcomes from this study found that Yemen construction industry lacks of proper accrediting system to accredit construction stakeholders (contactors, consultant); improper planning for cost and time, lack of effective communication and coordination platforms and lack of advanced technology which are considered amongst the crucial challenges faced by industry. Hence, Yemeni construction industry is in dire needs of an effective platforms and system to administer project communication, information and coordination. Findings of this study can be useful for Yemeni construction community to understand causes of the failure and suggestions provided by experts are also very helpful in creating better quality works.

1 Introduction

Construction industry is complex, fragmented, diversified, schedule driven and involves many stakeholders, commonly faces chronic threats and challenges such as cost and time overruns, poor quality, waste generation etc [1-8]. This is no exception to Yemeni construction industry facing immeasurable challenges and problems where in many cases the projects failed completely to attain its designed plans and goals [9,10]. Among these challenges are poor management, poor planning, poor quality, poor monitoring construction materials cost fluctuation, cost and time overruns and failure of projects [11, 12]. In early 1962, Yemeni construction industry was governed by two bodies named municipal & rural affairs authority and ministry of public works. Then in early 1976, the municipal and rural affairs were placed under the authority of ministry of public work to perform and undertake all the task under this body. However, in 1979 a new ministry of municipal affairs and housing was formed to differ from ministry of public work and forming two distinct entities

* Corresponding author: yaseruthm@yahoo.com

where each of them has a set of functions and responsibilities. Unification of Yemen in 1990 has resulted to the formation of two separate ministries to govern construction industry which; the ministry of construction and development and the ministry of housing and urban planning. Again in 2003, there was a restructuring process of Yemen government which resulted to merge all the construction industry bodies under one discrete body named as ministry of public works and highways to undertake and administer all the construction industries projects [9]. Through the restructuring process of governmental ministries and agencies, the construction industry in Yemen has faced tremendous challenges for 26 years from the reunification of the country in 1990 until 2016 where the industry has defected and faced many development restraints and impediments [10, 11].

In Yemen, an accrediting body which allows construction stakeholders (contactors, consultant) to engage in construction work is the ministry of public works and highways. Unfortunately, this body still lack of proper sets of terms and also enforcements to the stakeholders. This caused to unqualified stakeholders participate in construction work which then results to negative sequences such as poor quality or even to project failure [11,12]. According to Sultan [11], there are 5 major barriers causing deterioration of Yemen construction industry which includes institutional and administrative weakness, red tape, corruption, followed by the lack of infrastructure and also the ineffective law [10]. Sultan [11] added that important factor causes regressive development and deficiencies of Yemen industry is lack of any approved national system of code and standards or specifications where constructions are often over-designed or inappropriate to local needs and priorities [13]. Many cases reported of demolition of structure due to erroneous design, questionable construction practices, or inappropriate material selection. Beside that many of the projects reported failed to achieve time and cost scheduled [11]. For that reason, Yemen government should focus on long-term strategies, plans and policies to an industry which is considered of the main contributor to the national economy to establish new approaches for changing the current state of Yemeni construction industry [13]. Therefore, this study explored qualitatively on the current challenges of Yemen construction projects and alongside on the failure causes of mega construction project. It is to acquire views of current construction industry in particulars mega construction projects which have failed to achieve their plans as a learning scenarios.

2 Methodology of the study

This qualitative study adopted an exploratory interview with construction experts who involved directly in the construction of mega project and have experience for more than 15 years in construction industry. The definition of expert in this study was adopted from a research findings by Fei & Khan [5] which has specified the attributes and tangible criteria of an experts to be as follows (1) age should be older than 30 years (2) years of experience and involvement in construction industry should be more than 10 years (3) the expert should continuously work in a particular company for more than 10 years consequently Investigations on the failure causes of mega construction projects were conducted over telephone, Skype interviews and written answers due to difficulties to access Yemen under current situation. Each expert was given ample time to share their opinions based on his experiences in construction industry. In this exploratory interview, three construction experts were involved where they were selected based on their direct involvement in the mega projects. Therefore, they can be considered as representative of each individual mega project. Questions were designed with reference to previous literatures and findings. The main scope of questions is mainly on the general context of the projects and the causative failures of the projects further, experts were asked to highlight their opinions for preventive measures and future recommendation. Interviews transcriptions are produced in descriptive approach then filtered for relevancy to be in line with the contextual scope where each significant finding is presented in tabulation form and finally recommendations from experts were discussed for future improvements.

3 Interview transcriptions

This investigative study uncovered three mega projects in Yemen which are parliament complex, Sana'a international airport and Sana'a Sa'dah highway where the nominated projects were finally considered as failed. Exploration on these projects were conducted through interview sessions with three experts who are involved in the respective projects via telephone where questions were asked in a broad manner based on their knowledge and experiences working with these projects.

3.1 Sana'a International Airport

Construction of Sana'a international airport was financed by Arab Fund, while the construction was carried out by Chinese Company as the main contractor and an Arabic company as a consultant. A Dutch company was assigned to supervise special work of asphalt strengthening for the runways and supervising the work in the terminal building. Total cost for the project is 500 million USD while cost for consultant was $4 million. For Asphalt work of runways and supervising the work in the terminal building, it cost of $ (2,269,724) dollars.

The expert who gave this feedback has 20 years of working experience in construction industry and holds a bachelor degree in civil engineering. In this project, he was appointed as project coordinator from the inception stage. The project started in 2006 and supposed to end by 2009 but is considered failed where only 40% of the total work project was accomplished during termination of contract. The main causes of project failure are poor planning, poor cost estimation, language difficulty and cultural diversity between construction practitioners, interference of government agencies to interrupt the work and lack of coordination among stakeholders considered very imperative to cause project failure. The expert gave additional factors lead to project failure which are lack of transparency between stakeholders and improper communication that led to the termination of contract. To him, lack of transparency leads to inaccessibility of project information by all stakeholders' hence causing misunderstanding and mistrust among project stakeholders. Also, improper communication caused by ineffective means and lack of technology resulted to ineffective flow of project communication during construction. Therefore, those combinations of factors finally lead to the suspension of contract with main contractors and resulted in the failure of the project.

3.2 Parliament complex

Respondent to provide information for parliament complex project is a former manager of this project. Has being working in Yemeni construction industry for more than 30 years and holds a bachelor degree in civil engineering. This project started early in 2006 and considered failed because only 50% of the work has been accomplished until 2016. According to him, the contract amount for this project was paid in phases not in the form of lump sum. To him, it causes inaccurate planning and execution. Costing of the complex was based on four phases and the expert only managed to provide the cost allocated for phase one which was 24 million USD. According to the expert the project failed due to several reasons which include poor cost planning, bureaucracy, lack of transparency which later causes mistrust among construction parties, improper communication and coordination. Improper communication leads to unnecessary extra works likes redesign, rework and accident on site which also contribute to the failure of the project. According to him that lack of skill labors caused poor quality of work quality and incurred extra cost for more than $20,000 to repair all the defective works. He pointed out that the failure of this project wasn't caused by a sole factor but through combinations of many factors integrated together and subsequently resulted in the suspension of the project. When asked share his opinions regarding the mitigation measures, he implied on providing training for the construction labors regarding the importance of safety measures before the commencement of the project and use of advanced technology to enhance the process of project communication and coordination.

3.3 Sana'a –Sa'dah highway project

Expert who shared information on this failed project is a former site engineer assigned to monitor and coordinate Sana'a Sa'dah highway project. He has more than 15 years of experience in construction and holds bachelor degree in civil engineering. This project was financed through the government's road maintenance fund and the main contractor was the ministry of public work while the subcontractors are local companies. It started in 2000 however only 50% of the work has been accomplished during the termination. Major cause of the project failure is due to employing non-qualified main contactor to take part of this project that led to unsuccessful completion. The main contractor then assigned different sub contactors for each individual road segments of the highway and causing too many contractors for the project. This is due to no standard guidelines and reference/terms to select and validate the legibility of contactor where it was done based on the knowledge of main contractor.

According to him, the progress of work process was regularly interrupted and suspended due to cash flow issue at the same time the subcontractors are not eligible to maintain the work during the interruption of cash flow therefore, since it is a road construction it is susceptible to environmental uncontrolled factors such as raining during season hence once rain happened the embanked road segments need to be re-embanked again due to the erosion caused by rain and that causes cost burden to the subcontractors and finally caused failure of the project.

4 Results and discussion

Reference to the interview transcriptions above, several pertinent points can be extracted and these points are as in Table 1. Table 1 shows summarization of significant results from these cases studies. Sixteen failure factors were extracted for the three mega projects. From these factors, it seems that few of the factors are appeared to be similar for the three projects which are poor planning, lack of accreditation of stakeholders, ineffective government bureaucracy and lack of advance technology to monitor work process. Besides highlighting failures of three mega projects, experts had suggested certain preventive measures and recommendations to enhance the process of construction in mega projects which would be norms and guidelines for future projects. It can be generalized that majority of the preventative measures should be adopted by government agencies. They are related to establishing accrediting and enforcement body to recognize construction stakeholders and also implementation usage of advanced technology in all construction industry facets.

Table 1. Significant information of the three case studies

Items		Projects		
		Parliament complex	**Sana'a international airport**	**Sana'a-Sa'dah highway**
Project components	Project duration	A year for each phase (3 phases)	3 years	2 years
	Total Cost	For first phase of project was 24 million USD	500million USD	16 million USD
	Project stakeholders	• General military division was main contactor • Client was the government	• Main contactor was Chinese company • Financing body was Arab fund • Consultant Yemeni and Dutch (for asphalt and runway construction)	• Main contractor was ministry of public work • Financing body was road maintenance fund
	% of completion	• About 50% of the project work was completed	• 40% of the work project was accomplished	• About 50% completed
	Estimated loss	• About 50 million Riyal	• Approximately more than 20% of the total cost	• About 50% of the total cost of the project
Major causes of projects failure		1. No external consultant to monitor the work 2. Red tape (long time to process orders) 3. Poor financial planning 4. contract conditions were imprecise and unsteady 5. lack of technology	1. Poor planning 2. Poor Financial estimation 3. Poor communication among construction teams 4. Interference of government agencies to interrupt the work 5. Lack of coordination from stakeholders 6. Lack of transparency among construction stakeholders	1. Hiring unqualified subcontractors 2. Excessive use of construction materials due to the lack of unified construction standards and methods. 3. Poor planning and feasibility study 4. Lack of effective management skills 5. Incompetent and inexperienced management of project
Preventive measure		1. Consultant should have been selected from nongovernment body 2. Implication of advanced planning and scheduling methods 3. Associating and implementing advanced technology	1. Improve transparency among construction stakeholders 2. Proper planning and estimation 3. Developing guidelines to administer project information and communication 4. Provide training for labors before the commencement of project	1. Proper selection of subcontractors 2. Proper planning could have prevented the failure 3. Provide skilled managers to train subcontractors in management context 4. Developing guideline to constrain the regulation of accrediting contactors

5 Conclusion

It can be concluded that, Yemen construction industry faces many challenges and problems which requires more attention from the government to improve its policies and strategies. This cases studies managed to uncover most of failure causative factors are relatively due to the governmental policies and lack of accreditation systems of stakeholders. In Sana'a, international airport project, the dominant factor is due to the diversity of stakeholders of the project that caused lack of communication and understanding amongst the construction practitioners. For Parliament complex most of the problems is due to government bureaucracy while in Sana'a-Sa'dah highway, the main issue was the selection of unqualified subcontractors to take part in the project that causes poor quality control and finally resulted in the failure of the project. Hence, Yemen is in dire need of national system of codes, standards and contractual guidelines. Thus, more studies should to be carried out to explore opportunities for improving Yemen construction industry.

Authors acknowledge the financial support provided by Universiti Tun Hussein Onn Malaysia.

References

[1] M.E. Abd El-Razek, H.A. Bassioni and A.M. Mobarak, Causes of delay in building construction projects in Egypt, J. Constr. Eng. and Manage., **134**(11), 831-841, (2008)
[2] W.E. Alaghbari, R.A. Kadir, M.A. Salim and Ernawati, The significant factors causing delay of building construction projects in Malaysia, Eng., Constr. Architect. Manage., **14**(2), 192-206, (2007).
[3] A. Dubois and L. Gadde, The construction industry as a loosely couple system: Implications for productivity and innovation, Constr. Manage. and Econ., **20**(7), 621-631, (2002)
[4] D.W. Chan and M.M. Kumaraswamy, A comparative study of causes of time overruns in Hong Kong construction projects, Int. J. Project Manage., **15**(1), 55-63, (1997).
[5] X. Fei, and T.H. Khan, Identifying Attributes for Expert Construction Project Managers in the Context of China, Int. J. Asian Social Sci., **5**(7), 407-418, (2015).
[6] J.J. Chen, The characteristics and current status of China's construction industry, Constr. Manage. Econ., **16**(6), 711-719, (1998)
[7] P.F. Kaming, P.O. Olomolaiye, G.D. Holt and F.C. Harris, Factors influencing construction time and cost overruns on high-rise projects in Indonesia, Constr. Manage. Econ., **15**(1), 83-94. (1997).
[8] A.H. Memon and I. Abdul Rahman, SEM-PLS Analysis of inhibiting factors of cost performance for large construction projects in Malaysia: Perspective of clients and consultants, The Scientific World J., 165158, (2014)
[9] Ministry of Public works and Highways, Synopsis, Retrieved from: http://www.mpwh-ye.net/en/website/ministry/detail.php?ID=1454&lang=en (2016).
[10] B. Sultan, impediments to the construction industry development of Yemen, J. of Sci. Tech., **18**, (2013).
[11] B.M. Sultan, *The construction industry in Yemen: Towards economic sustainability*, PhD Thesis, Queensland University of Technology, Australia, (2005)
[12] B.M. Sultan and S.L. Kajewski, *The Yemen construction industry: Readying the industry for the successful implementation of sustainability*, Queensland University of Technology, Australia, (2003)
[13] B. Sultan and W.E. Alaghbari, Incompetent construction technologies and resources in the construction industry of Yemen, KICEM J. Constr. Eng. Project Manage., **19**(27), 2233-9582, (2014)

Factor Affecting of BIM Technique in the Construction Firms in Iraq

Hussein M. Hamada[1,*], *Ahmad* Haron[1], *Zahrizan* Zakiria[1], and *Ali M.* Humada[2]

[1]Faculty of Civil Engineering & Earth Resources, University Malaysia Pahang, 26300 Pekan, Pahang, Malaysia
[2]Electricity Production Directorate of Salahaldeen, Ministry of Electricity, 34007, Baiji, Iraq

Abstract. The construction industry in Iraq became much more difficult than before because of complexities in the diagrams and documentations in the construction projects. Building information modeling (BIM) started to appear with entry the foreign companies to build the large projects in Iraq. It is a bode well technique for the construction industry, it has many benefits in all project phases from planning and scheduling to operation and demolition phase. This paper concentrated on the implementation of BIM technology and identify the benefits desired and challenges that reduce adoption this technique in the Iraqi construction industry. A qualitative approach has been used to collect data from two experts in the construction field in Iraq to show the factors affected on using BIM in the Iraqi construction firms. The results indicate to a rare implementation of BIM technology among construction professionals in Iraq. The study recommended encouraging the investment in the Iraqi construction markets to share the knowledge and experiences related to BIM technology between the construction professionals.

1 Introduction

The construction industry in Iraq is still suffering from a lack of experts and competent professionals in the use of modern technology in the construction projects. Therefore, adversely effect on the progress of work in terms of delays in the completion of construction and increased construction costs. Most of the projects in Iraq are still based on the traditional methods in design and planning of projects such as using of 2D graphics and adoption of the critical path method (CPM) to calculate the time needed to complete the projects. Although these methods have been used for many years in the construction projects in the previous years, but it cannot be used in the large and complex projects. Therefore, should adoption of modern techniques, such as BIM technology for being less prone to error in the design and implementation by engineers, and have many benefits that no other. For example, when conduct modifying the properties of any element in the

* Corresponding author: enghu76@gmail.com

construction such as windows, doors or walls, the amendment will appear automatically on all levels in addition to the adjustment in the bill of quantities. The National Building Information Modeling Standards (NBIMS) committee in the USA defines the BIM technology as, a digital representation of physical and functional characteristics of a facility [1-3]. BIM technique is a resource for information required by stakeholders in the construction. projects in order to simplify and facilitate sharing information during the life cycle of the project from planning to demolition phase [3].

2 Problem statement

The complexities and difficulties in interpretation of diagrams and schedules created by the traditional methods such as 2D, and critical path method. These methods lead to errors and conflicts in the implementation of project tasks. Thus, using the traditional methods will lead to increase time and cost of projects, in addition to the effect on the quality of final product. Adoption of new technology in the construction companies has required the knowledge about benefits desired from this technology. Therefore, the aim of this paper is to find out the challenges and benefits of implementation the BIM technique within construction firms in Iraq.

3 Literature review

3.1 Benefits of BIM technology

There are many benefits result from implementation of BIM technology in the construction projects. The most important advantages of BIM technology identified by previous studies such as [4] conducted a survey about the benefits and barriers of BIM technology to the construction companies, more than 400 people participated in this questionnaire survey ,including the contractor, sub-contractor, and project managers. The result of the study proved that most of applications BIM users were to visualization, clash detection, and building design as shown in Fig. 1. Overall, BIM technology assists the project managers and planners to control the cost and time of construction project, in addition, to improving the quality of project model in early phases of project lifetime [5, 6]. Thus, all these advantages will contribute to raising the level of construction productivity [5, 7]. However, many case studies conducted in various projects proved the advantages and benefits of BIM, for instance, [8] implement a case study and said that can be get saving money and time of project if used the BIM technology in the process through minimizing of rework the design documents. While, [9] said that advantages of BIM application by appearing the benefits resulting from the return on investment (ROI) when implementation of BIM in the construction projects.

3.2 Challenges of 4D/BIM technology

There are many challenges and barriers to adoption of BIM technology in the construction firms. The most common challenges to adopting BIM technology within construction firms identified by some researchers such as [4, 10-21]. One of the biggest problems that counter of construction firms, is the difficulty in correct visualize to the construction planning in workspaces through creating complex schedules and difficult in the interpretation [15]. The absence of agreed visualization standards which represent the building components and elements, activities, and resources. Information visualization to 4D data representation as one of the obstacles to implementation the BIM technique in the construction projects [22].

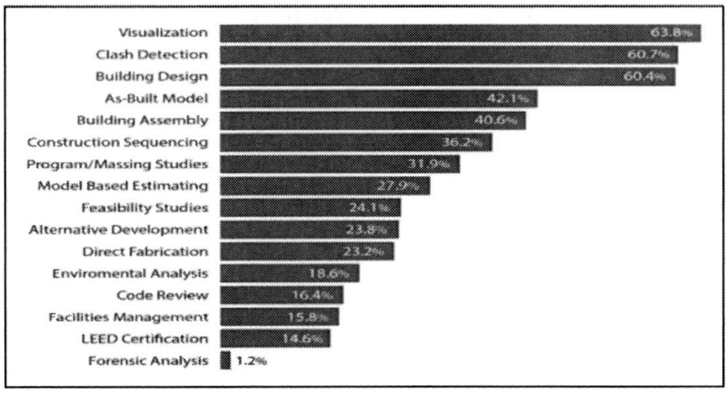

Fig. 1. Benefits of BIM in ranking [4]

3.3 BIM and Iraqi construction industry

Construction sector in Iraq is an important sector by contributing in the providing the houses, hotels, airports, railways and highways in Iraq. Therefore, the construction industry would affect the economy directly. In the last years, the Iraqi construction markets witnessed entry many of construction companies in order to build many projects inside of Iraq. The foreign companies with local companies started to implement the projects by using the BIM technique in their projects. Iraqi construction industry does not have more knowledge and skills to adopt BIM technology in short time for many causes, such as lack of experts in the BIM technology, lack of skills and knowledge for stakeholders involved in the projects, in addition to the cost of software and hardware, and training on it. In a recent survey by [23] proved that 92% of site engineers in the Iraqi companies do not have a professional certificates means that, do not have adequate knowledge and experience in the management of construction projects.

4 Methodology

The method of data collection for this study adopted a qualitative approach, which requires understanding deeply of a problem [24]. The qualitative approach also assist the researcher in finding out thinking of the respondents and what are their answers that effect on the results of research? The respondents in the research have been selected based on the experience in the construction industry and applying the BIM technique in the construction projects as shown in Table 1. All the interview sessions recorded and transcribed after taking approval from participants in the survey. The interview questions open ended were included the objectives following: 1) to find out the current status of BIM within Iraqi construction industry, 2) to collect the projects successful by using BIM technology, 3) to identify the benefits and challenges which lead to the successful projects. The content analysis method has been adopted to find out and validate the results obtained from the interview sessions. After that, the results has been discussed by comparing with the previous studies to find out the similarities among them as in Fig. 2.

Each the respondents R1 and R2, have an experience in the BIM and 4D modeling. In addition, they have high positions in their companies as can be seen in Table 1.

Table 1. Profile of respondents and their experience.

Respondent name	Designation	Experience in construction	Experience in BIM
A	Senior management	More than 20 years	More than 6 years
B	Project manager	More than 20 years	More than 8 years

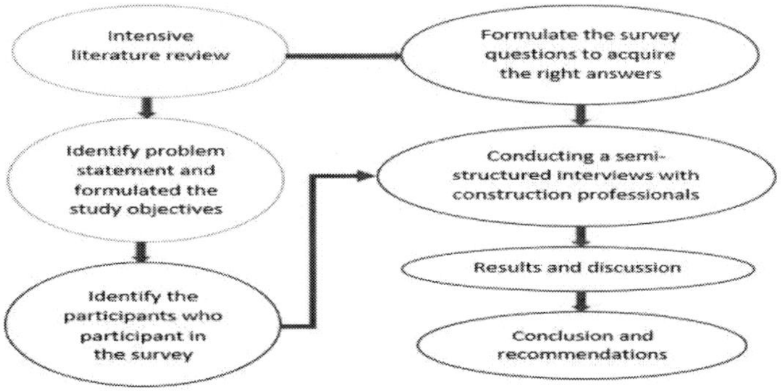

Fig. 2. Research method to data collection.

5 Results and discussion

This chapter will discuss the results obtained from the respondents in the interview sessions, and contains four parts.

5.1 Basra main stadium (by 360 architecture)

There are a number of construction projects implemented by using BIM technique in Iraq, which displays successful implementation and utilization of integrated drawings with schedules in a multidisciplinary environment, such as Basra Main Stadium 65,000 seats and Al Minaa Stadium 30,000 seats [25] located in Basra city in the southern of Iraq. The fabrication process designed by 360 Architecture, Kansas City. The fabrication process consists of 100-foot long glass fiber reinforced plastic (GFRP) as can be seen in Fig. 3 a.

a. Basra main stadium

b. Al Minaa soccer stadium

Fig. 3. (a and b): Basra stadiums by (by 360 architecture)

The construction of Basra sports city was started on 15 July 2009 and completed on 12 October 2013. The city is being funded by the government of Iraq with a budget spending of $550 million, contributed to contain the main stadium with the capacity of 65,000 people.

Table 2. Details of sport city in Basra

No	Items	Details
1	Owner	Ministry of Youth and Sport in Iraq
2	Contractor	Abdullah Al-Jiburi Contracting Company
3	Consultant	360 Architecture, Kansas City
4	Site area	2.5 million-square-foot
5	Contract Period	32 months
6	Start of construction	July 2009
7	Completed of structure	October 2013

5.2 Al Minaa Stadium project

The construction of Al Minaa stadium as shown in Fig. 3b, was started on 2011 and completed on 2013. The stadium also funded by the Iraqi government with a budget $ 86 million, to build the stadium with capacity 30000 people. The contractor was Anwar Soura General Contracting Company, while the consultant and the design team was 360 Architecture in American. The details of Al Minaa stadium as in Table 3 below.

Table 3. Details of Al Minaa stadium in Basra city

No	Items	Details
1	Owner	Ministry of Youth and Sport in Iraq
2	Contractor	Anwar Soura General Contracting Company
3	Consultant	American 360 Architecture
4	Contract Period	24 months
6	Start of construction	2011
7	Completed of structure	2013

The owner of the project is the Ministry of Youth and Sport in Iraq, while the main contractor was Anwar Soura General Contracting Company. R1 represents one of the management of the project, R2 represent one of the contractor staff in the project.

5.3 Benefits of 4D/BIM in the projects

Regarding the benefits of BIM to the project, R1 and R2 said that: Implementation BIM in the construction of sport city in Basra assist the design team on the following points:

i) To share information in the BIM model among the architects and engineers throughout the design phase, in order to meet the design requirements.
ii) Checking logistical and economic constraints a daily in addition to finding out the constructability issues and clashes.
iii) To minimize the time required to fabrication the panels needed to the project by 18 months, therefore adoption of BIM applications to take the design from five models differently.
iv) To achievement the construction and fabrication process of the stadium's skin components at the same time planned.

v) To enable and simplify the smooth interaction among the fabricator and design team, whereas, the project team chose application a digital information in the project [25].
vi) Using BIM software would allow the architects to control the large and complex surface geometry of the project panels.
vii) To do a platform as open communication among all the specialists in the project with the central model.
viii) BIM has the ability to see every detail on a construction project including the wires, the pipes, the columns, beams and collect they in the 3D model before actually starting the construction.

5.4 Challenges of 4D/BIM in the projects

As mentioned by the R1 and R2 in the interview sessions, implementation of BIM in the construction projects faced many challenges, first of all:
i) The high cost to purchase the sophisticated software and hardware related BIM technology such as Tekla and Revit.
ii) Training on the BIM equipment and provide the cost and time needed to acquire the necessary skill and knowledge in this regard. Therefore, most of the BIM companies in Iraq is an international company, thus, they have the ability to implement BIM in their projects.
iii) Compatibility problems between the BIM software, particularly with regard to the structural safety of the construction because of adoption more than an approach in the design the various facilities of the project.

6 Conclusion and recommendations

BIM technology has many benefits, on the other hand, has many challenges as shown in the implementation of sports city in Basra and Al Minaa stadium. All these projects funded by the Iraqi government, and came to enhance the quality of the project by using BIM modeling in the construction industry. In Basra city has been implemented two projects by using BIM technology. All of them designed by the American 360 Architecture. The projects illustrate the benefits and challenges due to use the BIM. some of the benefits identified through the interview sessions with the respondents, such as minimize the time and cost required to the project and getting on the better model of the project, besides clash detection between the project component in an early age of project life. All these factors led to limit the rework for the project tasks. Thus cancel additional time and cost of the project. Many projects have been implemented by application of BIM technology such as an educational al-Najaf hospital in the al-Najaf city, also funded by the Iraqi government. Therefore, the study recommends that the Iraqi government should seek to engage the foreign companies with Iraqi companies in implementing more projects BIM within Iraqi construction industry in order to share the knowledge and experience among construction professionals.

References

[1] H. Penttilä, Describing the changes in architectural information technology to understand design complexity and free-form architectural expression, ITcon, **11** (Special Issue), 395-408 (2006)

[2] B. Ernstrom, D. Hanson, D. Hill, J.J. Clark, M.K. Holder, D.N. Turner and T.W. Barton, *The contractors' guide to BIM,* Associated General Contractors of America, Arlington, VA (2006)
[3] C. Eastman, C.M. Eastman, P. Teicholz and R. Sacks, *BIM handbook: A guide to building information modeling for owners, managers, designers, engineers and contractors*, John Wiley & Sons, New Jersey, Canada (2008)
[4] B. Becerik-Gerber and S. Rice, The perceived value of building information modeling in the US building industry, J. of Information Technology in Construction, **15**, 185-201, (2010)
[5] S. Azhar, A. Nadeem, J. Mok and B. Leung, Building information modeling (BIM): A new paradigm for visual interactive modeling and simulation for construction projects, *Proc. First International Conf. on Construction in Developing Countries***,** Karachi, Pakistan (2008)
[6] M. Fischer and J. Kunz, The scope and role of information technology in construction, *Proceedings-Japan Society of Civil Engineers*, Centre for Integrated Facility Engineering, 1-32 (2004)
[7] P.E. Love, D.J. Edwards, S. Han and Y.M. Goh, Design error reduction: toward the effective utilization of building information modeling, Research in Engineering Design, **22**, 173-187 (2011)
[8] G. Aranda-Mena, J. Crawford, A. Chevez and T. Froese, Building information modelling demystified: Does it make business sense to adopt BIM?, Int. J. of Managing Projects in Business, **2**, 419-434 (2009)
[9] S. Jones, *SmartMarket report, building information modeling transforming design and construction to achieve greater industry productivity*, McGraw-Hill Professional, Bedford, UK (2009)
[10] S.M. Ahmed, H.H. Emam and P. Farrell, Barriers to BIM/4D implementation in Qatar, *The 1st International Conf. of CIB Middle East & North Africa Conf.*, Abu Dhabi, UAE, 533-547 (2014)
[11] Y.H. Ahn, Y.H. Kwak and S.J. Suk, Contractors' transformation strategies for adopting building information modelling, Journal of Management in Engineering, **32**(1), 05015005, (2015).
[12] M. Ali and J. Mohd Noor, Acceptability of Malaysian construction practitioners in adopting 4D, *First International Conf. on Islamic Built Environment*, Bandung, Malaysia (2012)
[13] S. Azhar, M. Khalfan and T. Maqsood, Building information modelling (BIM): Now and beyond, Construction Economics and Building, **12**, 15-28 (2015)
[14] D.M. Brito and E.A. Ferreira, Strategies for representation and analyses of 4D modeling applied to construction project management, Procedia Economics and Finance, **21**, 374-382 (2015)
[15] F. Castronovo, S. Lee, D. Nikolic and J.I. Messner, Visualization in 4D construction management software: A review of standards and guidelines, *Int. Conf. on Computing in Civil and Building Engineering, Computing in Civil and Building Engineering*, 315-322 (2014)
[16] N.N. Dawood and S. Sikka, Measuring the effectiveness of 4D planning as a valuable communication tool', in Messner, International Journal of Information Technology in Construction, **13**, 620-636 (2008)
[17] I. Kaner, R. Sacks, W. Kassian and T. Quitt, Case studies of BIM adoption for precast concrete design by mid-sized structural engineering firms, ITcon, **13** (Special Issue), 303-323 (2008)

[18] M. Kassem, T. Brogden and N. Dawood, BIM and 4D planning: A holistic study of the barriers and drivers to widespread adoption, J. of Construction Engineering and Project Management, **2**, 1-10 (2012)
[19] I. Kiani, A.N. Sadeghifam, S.K. Ghomi and A.K.B. Marsono, Barriers to implementation of Building Information Modeling in scheduling and planning phase in Iran, Australian Journal of Basic and Applied Sciences, **9**, 91-97 (2015)
[20] M. Mohamed Kassem and T. Brogden, BIM and 4D planning: An empirical study of the industry uptake and issues affecting their widespread use, *Proceedings of the CIB W78 2012: 29th International Conf.*, Beirut, Lebanon (2012)
[21] M.H. Tsai, S.C. Kang, and S.H. Hsieh, A three-stage framework for introducing a 4D tool in large consulting firms, Advanced Engineering Informatics, **24**, 476-489 (2010)
[22] R. Mazza, *Introduction to information visualization*, Springer Science & Business Media, London, United Kingdom (2009)
[23] F.M.S. Al-Zwainy, I.A. Mohammed and S.H. Raheem, Investigation and assessment of the project management methodology in Iraqi construction sector, Int. J. of Applied Engineering Research, **11**, 2494-2507 (2016)
[24] S. B. Merriam, *Qualitative research and case study applications in education. Revised and expanded from,* Jossey-Bass Publishers, San Francisco, USA (1998)
[25] J. Schumacher and R. K. Otani, Advanced computational modeling in multidisciplinary design, *Proc., Structures Congress 2012: 20th Analysis and Computation Specialty Conf.*, ASCE, New York, 231-244 (2012)

The Significance of Coordination for Industrialised Building System (IBS) Precast Concrete in Construction Industry

Mohd Khairul Fitri Othman[1,*], *Wan Mohd Nurdden* Wan Muhammad[1], *Nurulhudaya* Abd Hadi[1], and *Mohd Azrai* Azman[1]

[1]Faculty of Architecture, Planning and Surveying, Universiti Teknologi MARA, 94300 Kota Samarahan, Sarawak

Abstract. IBS precast concrete is construction system which is meant to improve the conventional construction process. However IBS precast concrete projects are suffering from serious problems such as cost overrun, delays and less quality of the end product. The absence of coordination is perceived as the reason for this issue. The purpose of this paper is to review the significance of coordination for IBS precast concrete in the construction industry. It if found that the fragmentation which occurs in the construction industry requires continuity of coordination due to the construction activities are intertwined in nature. Coordination is designated to assist stakeholders in completing and complementing each other with the paramount focus of achieving the objective. Proper coordination is required in delivering the desired construction product at the ideal time, cost and quality. As for the findings, the significance of coordination for IBS precast concrete can be seen through the precast concrete construction phases which consist of planning; design; manufacturing; transportation and installation/construction. These phases are meant to complement construction process with the purpose to reduce issues of fragmentation and enhance IBS precast concrete project delivery.

1 Introduction

The construction industry is mostly characterized as a labour intensive industry, which comprises of several wet trade activities and hard labour conditions. Therefore, it is always facing with problems such as delay, cost overruns, low quality, poor safety records as well as environmental impacts [1]. Industrialisation has been viewed as one of a good method for providing solutions for problems mentioned above in construction [2]. It refers to the rationalisation of the construction process by promoting off-site manufacturing [3]. Industrialisation process comprises of investment of equipment, facilities and technology with the objective of maximising production output, minimising labour resource, and improving quality [4].

* Corresponding author: mohd_khairul135@sarawak.uitm.edu.my

One of the principles of IBS in construction is precast concrete. Precast concrete is a construction system in which the concrete will be casted in the mold and then cured in a controlled environment, transported and assembled to produce the structure [5]. The components of precast concrete are produced under laboratory condition using high technology of machinery and computer aided system to meet high quality and durability of the product [6]. By these processes, the quality of precast concrete is known to be more reliable and accurate compared to in-situ concrete construction.

Precast concrete system offers several benefits such as reduce waste; encourage recycling construction waste; reduce construction site cost; improve site safety; enhancing quality under factory production; process standardization; shorten lead time; leading to environmental protection and sustainability; as well as improves construction performance [7]. In addition, precast concrete also contributes to sustainable practices by incorporating integrated design, using materials efficiently, reducing construction waste, site disturbance, and noise [3]. These arguments have clearly recognised IBS precast system as the way forward to improve the conventional concrete system in Malaysia.

The performance of the precast concrete system nevertheless is much dependent on comprehensive coordination between various relevant parties such as clients, architects, engineers, main contractors, subcontractors, suppliers and consultants, as well as the physical flow of materials, services and products. Any delay in delivering precast concrete components to a construction site will delay the construction processes thus increase the costs [8]. Coordination among project participants has been recognized as an important ingredient for the success of many projects [9]. Coordination is almost equivalent in meaning to 'control,' 'planning' or 'management,' but is more descriptive of the relating together of separate activities and their concerted direction towards a common purpose. It is supported by an argument that coordination as one of the principal functions of management, together with planning, organizing, directing and controlling. Moreover, it is also observed that active coordination could minimise, predict and remedy problems caused by design construct lead time, materials availability, manpower, and equipment availability.

2 Coordination in construction industry

The construction industry is critically being suffered from fragmentation issues. Clients, contractors, engineers, quantity surveyors are among the construction parties which directly contact with the construction activities and procedures. Nevertheless, each of the parties is not motivated to collaborate harmonically by self-voluntary [10]. The fragmentation which occurs in the construction industry requires continuity of coordination due to the construction activities are intertwined in nature. Otherwise, it will lead to poor performance of overall construction [11].

Coordination among stakeholders is claimed as a crucial element of the success of construction projects [9]. It is also one of the seven identified management processes besides forecasting, planning, organizing, motivating, controlling and communicating [9]. Coordination is also being defined as closely identical with the definition of control, planning or management. However, coordination is interpreted as more explanatory in collaborating different activities and their intensive focus in an achieving a common purpose [12]. Besides, coordination is also one of the three important functions in the building process together with design and construction [12]. These arguments on defining the coordination have deduced that coordination represents as an indispensable element in managing construction project holistically.

Coordination should be done continuously and it is required at all management levels. It shows that the management responsibility is essential in applying and following up the coordination activities. If any organisation in construction industry fails to coordinate its activities well continuously, the organisation might end up being unsuccessful for a long period [13]. This depiction shows that coordination is the quintessence of management and management functions cannot be achieved by the absence of proper coordination.

The importance of coordination in building construction is highly being recognised by construction players as a result of previous researcher's outcome. Coordination is designated to assist stakeholders in completing and complementing each other with the paramount focus of achieving the objective. The proper coordination activities as tools in managing coordination are required in supporting the desired meeting construction product and performance at the ideal time, cost and quality.

3 Barriers for coordination

Collaborating personnel from a different background is always deemed to contribute to fragmentation issues. The fragmented environment of the construction projects is one of the reasons of the poor performance. This is because most of the construction activities are interdependent and as a remedial action, coordination should take place in a continuous manner [14]. In describing the actual event, relationship based on prejudice between construction stakeholders is one of the industry predicament that influences the coordination process negatively [11]. It can be seen by many activities are performed in the project where if any of the parties come out with a different idea, the project stability is disturbed. The cooperation among the parties is expected to be insufficient and coordination should come into scene to mitigate the problem [10].

The source of the difficulty of construction projects management derives from the interdependent relationship between the parties. It is depicted with the lack of compromise between organisation and technical interdependence with the one who manage the activities [12]. In addition, lack of knowledge is also being argued as the obstacles of coordination. The ability of the parties will be in question while they are refuse to take part in the process. This circumstance is worsened by the communication issues which lead to the breakdown of information flow and failure of coordination process in project management [10].

The main impediment of the coordination application in the construction industry is mainly originated by the nature of being too dependent among project stakeholders and also among the process. Different 'language' and negative perception among the parties have led to communication problem which contributes to failure in achieving project objectives.

4 Coordination phases in precast concrete construction

In implementing a precast concrete system for construction, it is necessary to understand the coordination phases required during off-site and on-site operation to ensure successful implementation. In identifying the precast concrete coordination phases, the principles of the supply chain for precast concrete as being highlighted by Abedi et al. [5] are being utilised. This is due to the precast supply chain phases are intertwined with the coordination activities which being categorised as planning; design; manufacturing/production; transportation; and the installation/assembly and construction.

4.1 Planning

According to Kamar et al., [15] IBS project needs good planning in the aspect of manufacturing, transportation and erection on site. The decision of what component of precast concrete to be produced, what facilities needed to deliver the component and when the installation process will be carried out on the site should be taken into consideration. However, this planning decision should be performed in such a manner that the cost incurred is minimum. Another argument also supports that planning process involves obtaining and providing all documents and reports related to supply chain phase of precast construction [16], which including; design; manufacturing; cost analysis; transportation; installation and construction [17]. This is further supported by Yusof et al., [18] which mentioned that extensive planning and scheduling of activities in advance is crucial in achieving better project delivery and coordination of IBS project besides being recognised as one of the key success factors for IBS project which also apply to IBS precast concrete construction.

4.2 Design

The design of precast concrete component should be conducted thoroughly and carefully to avoid the design changes during construction. The design process should take into account the project's nature, regarding all its characteristics, surrounding environment, and ground conditions rather than only concerned with client need [19]. The most important process in design includes outline technical specification; design sketch; initial cost plan; provisional list of drawings required; and provisional list of annotations for drawings [20]. Kaner et al. [21] stated that all regulations, materials, drawing and appropriate building codes will be established. Furthermore, the quantities, cost estimations, time-scheduling will be prepared by designers and consultants [22]. Besides, the use of computer software such as BIM (Building Information Modelling) will help to enhance design analysis and components production. In the context of IBS precast concrete, standardisation towards process and design is an emphasis in utilising the concept of repetition [18]. These process and measures are important to be well-coordinated in order to convey accurate concept for the manufacturing process.

4.3 Manufacturing

Precast concrete components manufacturing process consist of several sections includes of raw materials selections, [23, 24], batching plant, casting and moulding zone, remoulding and finishing zone, packaging, transportation and the storage yard [25]. These processes manage to act as remedial action to diminish any non-value adding activities such as wastage of raw material; improper handling; long waiting time due to insufficient labour and raw material; double-handling or delivery due to unsatisfied quality or specifications should be avoided [23]. Peng [23], mentioned that manufacturing activities in precast concrete factories could be identified into four phases, includes of site layout management, delivery management of raw materials, production management, and stock management.

4.4 Transportation

Transportation issues are one of the challenges in managing IBS construction. The issues normally associated with size and weight limitations, route restrictions, permitting and the availability of lifting equipment [26]. Transportation management should concentrate on the construction schedule, site layout, cost of plant and the design of plan [27]. Transportation consideration will influence the design of precast components such as size and weight. Based on the study made by Warszawski [28], the length of a volumetric unit should not exceed 12 meter which is the normal vehicle length. The precast components' height, when loaded on the trailer, should not exceed a height limitation of highway ranging from 4.8 to 5.1 meter. The weight of precast concrete component should not exceed 7 tonnes to ensure that a crane can easily erect and transport the material on site [29]. Furthermore, the suitable distance of fabrication plant from development area should be from 50 to 100km, based on the study made by Azman et al., [29]. These considerations are established with the purpose to promote efficient and effective transportation with proper coordination in transferring the components from fabrication plant to site.

4.5 Installation and construction

The reliability of precast concrete construction also depends on the expertise level of a labourer in carrying out the installation task. A thorough supervision, especially in jointing, should be executed to prevent a problem such as water seepage through precast concrete connection [30]. The installation drawings, time-scheduling and instructions which has been produced during a planning phase should be followed accordingly [31]. Furthermore, the delivery process of the precast concrete component to the site should have a continuity and just in time to avoid delay in the installation process and buffer stocks kept on site. As for handling purposes, a tower crane and mobile crane will be used for unloading and placing the component on site as well as for installation [30]. The loading capacity of hoists and cranes for installation of precast concrete components need to be considered based on their weight and size [32, 30]. High accuracy and efficiency when handling IBS components are essential by properly coordinate them during installation and storage [33] with the purpose to avoid a serious problems such as delay and wastage.

5 Discussion

Coordination plays an important function having a considerable effect on the outcome of a construction project. This paper highlighted five phases of coordination based on the critical review of previous researchers regarding construction coordination and precast concrete which includes of planning, design, manufacturing, transportation and installation/ construction.

It is argued that planning phase is the important process where all the documents for all coordination phases would be provided. A variety of reports such as design component, production resources, cost analysis component, installation schedules, bill of quantities, shop drawings and construction drawings are among other reports that will be provided in planning phase. All of these reports are needed to be well-coordinated in order to avoid fragmentation issues in the production of precast concrete.

Meanwhile, design phase also plays a vital process throughout the precast concrete production. In the design process, the designers together with manufacturer and contractor should be involved to avoid constructability issues in future such as delay, late of supply, etc. The technical specification and regulations, types of materials, drawings and the appropriate building codes will be developed. All the drawing produced such as M&E, C&S and architectural drawing should be aligned to avoid disparity. The detailing of structural including joints and connections should also be put into consideration in the design of precast concrete components. All of these processes are required to reduce the errors and discrepancy throughout the precast concrete production, thereby mitigating the delay and incurred cost of rework.

As for manufacturing phase, the precast concrete components are carried out in a factory and systematically done using machine, formwork and other forms of mechanical equipment. The manufacturing process will be categorized into several significant sections consisting of the selection of raw materials, batching plant, moulding zone, finishing zone, packaging, and the storage yard. Besides, crew allocation in every section should be optimised in order to minimise idle times caused by sharing operatives. All of these activities which involved material and human resource should be coordinated wisely to minimise resource wastage and reduce defects and errors in precast concrete production.

The transportation and delivery of precast concrete component should be planned according to the appropriate delivery schedule, quantities, arrival orders, unloading sequences and erection sequence to minimise unnecessary site storage and handling. The transportation of large panels or components is also subject to the restriction of traffic congestion and regulation of the road and transportation authority. The limitations must be taken into consideration when designing and manufacturing a precast concrete system. Where possible, the precast concrete components should be arranged and transported in a manner which they can be easy lifted directly from the lorry for the erection or storage. All these considerations are needed to ensure the precast concrete component can be transported and delivered in efficiently and effectively.

In the installation/construction phase, all the precast components will be installed/assembled based on the appropriate documents which are produced in the planning phase. The installation process should be conducted by the skilled labourers who have expertise in the precast concrete components to avoid poor performance on the structures such as crack, moisture penetration, etc. It is also argued that good integration between all construction members is essential to the successful precast concrete construction. A suitable plant with appropriate loading capacity should be available to use for the site storage and installation of precast components. All these considerations should be well-coordinated to avoid improper assembly of the components thus reduce the defects of the precast concrete construction.

By scrutinising these coordination phases for precast concrete system, it highlights on the apparent gap between the precast concrete system with the significance of associating coordination alongside with the system activities. The coordination activities play important roles in addressing the shortcoming of precast concrete system operation. They also raise up the importance to understand the characteristics of precast concrete system which normally have a relation to the impediment of successful precast concrete system operation. Hence, it is argued that coordination is capable of mediating the deficiencies of precast concrete system.

6 Conclusion

IBS is widely recognised by construction players in Malaysia since its first implementation in the 1960s. Nevertheless, most of the construction project is still carried out in conventional ways due to the absence of coordination in the overall construction process. Coordination is perceived as a paramount element in precast concrete construction. Lack of coordination between design and construction will give adverse effects on construction such as delay and cost overrun. The significance of coordination for IBS precast concrete can be seen through the precast concrete construction phases consist of planning; design; manufacturing; transportation; and installation/construction. These phases are meant to harmonise construction process with the purpose to reduce issues of fragmentation and enhance IBS precast concrete project delivery. Further research is required to identify and verify important coordination activities through precast concrete construction phases and their relationship with precast concrete characteristics.

These researchers would like to express gratitude for the financial support from Kementerian Pendidikan Malaysia (KPM) under the purview of Research Acculturation Grant Scheme (RAGS) 600-RMI/RAGS 5/3 (237/2014) and Research Management Institute (RMI) Universiti Teknologi MARA (UiTM).

References

[1] A.P.C. Chan and T.Y.F. Ma, Materials wastage on commercial projects – contractor's view, *Proc. of the Sixth East Asia-Pacific Conf. on Structure Engineering and Construction*, Taipei, Taiwan, (1998)

[2] L. Koskela, *Application of new production theory in construction*, CIFE Technical Report No. 72, Center for Integrated Facility Engineering, Stanford University, Finland, (1992)

[3] G. Polat, Precast concrete systems in developing vs industrialized countries, J. Civ. Eng. Manag., **16**(1), 85–94, (2010)

[4] W.A. Thanoon, L.W. Peng, M.R. . Kadir, M.S. Jaafar and M.S. Salit, The essential characteristics of industrialised building system, *Int. Conf. Ind. Build. Syst*, Kuala Lumpur, Malaysia, **1999**, 10–11, (2003)

[5] M. Abedi, N.M. Rawai, M.S. Fathi and A.K. Mirasa, Cloud computing as a construction collaboration tool for precast supply chain management, J. Teknol., **70**(7), 1–7, (2014)

[6] A. Neville, Some aspects of sustainability, PCI J., **50**(5), 2–5, (2005)

[7] Y. Chen, G.E. Okudan and D.R. Riley, Sustainable performance criteria for construction method selection in concrete buildings, Autom. Constr., **19**(2), 235–244, (2010)

[8] A.F. Al-Bazi, N. Dawood and J.T. Dean, Improving performance and the reliability of off-site pre-cast concrete production operations using simulation optimisation, J. Inf. Technol. Constr., **15**, 335–356, (2010)

[9] K.N. Jha and K.C. Iyer, Critical determinants of project coordination, Int. J. Proj. Manag., **24**, (4), 314–322, (2006)

[10] W.S. Alaloul, M.S. Liew and N.A.W.A. Zawawi, The characteristics of coordination process in construction projects, *Proc. of the Int. Symposium on Technology Management and Emerging Technologies*, IEEE, 159–164, (2015)

[11] K.H. Tey, M.Y. Aminah, I. Syuhaida and F. W. Lee, A conceptual study of key barriers in construction project coordination, J. Organ. Manag. Stud., **2012**, 333–346, (2012)

[12] G. Higgin and N. Jessop, *Communications in the building industry: The report of a pilot study*, Tavistock Publications, London, United Kingdom, (1965)

[13] M.J. Meixell, M. Nunez and A. Talalayevsky, Activity structures in a project-based environment: A coordination theory perspective, Eng. Manag. IEEE Trans., **53**, 285–296, (2006)

[14] Ö. Wikforss and A. Löfgren, Rethinking communication in construction, *Proc. of the 4th Nordic Conf. on Construction Economics and Organisation*, 337–344, (2007)

[15] K.A.M. Kamar, Z.A. Hamid, M. Zura, M. Zain, A. Hazim, A. Rahim, M.K. Ghani, M. Nor, A. Azman, T.A. Majid, M. Sanusi, S. Ahamad and M.N. Azman, Drivers and barriers of industrialised building system (IBS) roadmaps in Malaysia, Malaysian Constr. Res. J., **9**(1), 1985–3807, (2012)

[16] W.T. Chan and H. Hu, Constraint programming approach to precast production scheduling, J. Constr. Eng. Manag., **128(6)**, 513–521, (2002)

[17] E. Dassori and M. Frasani, Support system for project management and production of concrete precast elements, *Int. Conf. on Advances in Building Technology*, 1689–1696, (2002)
[18] M.R. Yusof, A.H. Nawawi, M.F. Mohammad and M.F. Musa, Key success factors in IBS project management, *Colloquium on Humanities, Science and Engineering*, IEEE, 210–215, (2015)
[19] N.B. Kiong and Z.A. Akasah, Maintenance factor for precast concrete in IBS: A review, *2011 Natl. Postgrad. Conf. - Energy Sustain. Explor. Innov. Minds, NPC 2011*, Kuala Lumpur, Malaysia, (2011)
[20] S.V.N. Biscaya, *Coordination and management of information for construction design projects: A framework for Portugal*, PhD Thesis, University of Salford, United Kingdom, (2012)
[21] I. Kaner, R. Sacks, W. Kassian and T. Quitt, Case studies of BIM adoption for precast concrete design by mid-sized structural engineering firms, Electron. J. Inf. Technol. Constr., **13**(2007), 303–323, (2008)
[22] V. Mlinarić and Z. Sigmund, Problems in large scale precast construction projects, *CIB Jt. Int. Symp. 2009 Constr. Facing Worldw. Challenges*, Dubrovnik, Croatia, 182–188, (2009)
[23] W. Peng and L.S. Pheng, Lean production, value chain and sustainability in precast concrete factory – a case study in Singapore, Lean Constr. J., **2011**, 19–36, (2010)
[24] P.W. and S.P. Low, Lean management and low carbon emissions in precast concrete Factories in Singapore, J. Archit. Eng., **18**, 176–186, (2011)
[25] S. Senaratne and S. Ekanayake, Evaluation of application of lean principles to precast concrete bridge beam production process, J. Archit. Eng., **18**(2), 94–106, (2012)
[26] C.T. Haas and W.R. Fagerlund, *Preliminary research on prefabrication, pre-assembly, modularization and off-site fabrication in construction*, PPMOF Technical Reprot No. 171, The Construction Industry Institute, Austin, Texas, (2002)
[27] I.L. Jabar, F. Ismail and A.A. Mustafa, Issues in managing construction phase of IBS Projects, *Procedia - Soc. Behav. Sci.*, **101**, 81–89, (2013)
[28] A. Warszawski, *Industrialised and automated building systems*, E & FN Spon, London, United Kingdom, (1999)
[29] M.N.A. Azman, M.S.S. Ahamad, T.A. Majid and M.N.S.A. Shah, A study of precast concrete in Malaysia, Concrete, **46**(10), 50–52, (2012)
[30] Y.W. Cheong, H.P. Kwan and A. D. Hariyanto, Quality control in precast production: a case study on tunnel segment manufacture, Dimensi Teknik Arsitektur, **33**(1),153-164, (2005)
[31] A. Rönneblad and T. Olofsson, Application of IFC in design and production of precast concrete constructions, J. Inf. Technol. Constr. (ITcon), **8**, 167–179, (2003)
[32] L.S. Pheng and C.J. Chuan, Just-in-time management in precast concrete construction: a survey of the readiness of main contractors in Singapore, Integr. Manuf. Syst., **12**, 416–429, (2001)
[33] W.M.N.W. Muhammad, M.A. Azman, M.K.F. Othman, N.A. Hadi, S.N.S. Sahimi, and M.F. Mohammad, The insight of industrialised building system (IBS) by Bumiputera construction players, *The 3rd Int. Conf. on Civil and Environmental Engineering for Sustainability* Malacca, Malaysia, (2016)

Exploratory Study on Malaysia Construction Leadership

Nur Ain Ngah Nasaruddin[1,*], and *Ismail Abdul* Rahman[1]

[1]Department of Building and Construction Engineering, Faculty of Civil and Environmental Engineering, Universiti Tun Hussein Onn Malaysia, 86400 Parit Raja, Johor, Malaysia

Abstract. This paper presents an exploratory study on insight issues of leadership quality in Malaysia construction industry. It was conducted using an open interview session which focussed on three main elements which are challenges faced by construction leaders, leader's role to sustain organization's competitiveness and qualities to be an outstanding construction leader. Even though only four senior construction leaders were selected for this qualitative study, but the selection was properly scrutinised based on their expertise as construction leader. During the interview session, leaders were given ample time to express their opinions on the elements with minimum interference. Information revealed was recorded, analysed and synthesized to deduce findings of the study. It was found that significant challenges faced by construction leaders are ensuring project runs smoothly, managing cash flow and ability to manage large number of worker. For ensuring their organization remains competitive, leaders must be creative and persistence in handling construction project and not merely completing the task. To be outstanding, leaders should possess good leadership attributes as problem solver, communicator and motivator in their organizations. This study forms a basis for further exploration on leadership quality for construction industry.

1 Introduction

Rapid growth of construction industry in Malaysia leads to the need for an efficient construction project management teams, especially in terms of leadership. Managing construction project with less spirit of leadership can weaken the organization in its ability to deal challenges in construction industry [1]. According to [2], leadership requires influence to complete missions and to give directions for more well-organized and practical organization. Previous researchers had outlined several issues of leadership in the construction industry such as poor communication, lack of experience, lack of planning, wrongful use of power, lack of adaptability with complex circumstance, lack of resources, lack of strategy management, and lack of superior support [3]. A study found that positive leadership contributed almost 76% to the success of project while negative or poor leadership has contributed to 67% of projects failure [4]. Thus, the need of quality leadership is undeniable to steer construction project success. With appropriate leadership

* Corresponding author: nurain_ngah@yahoo.com

among the construction parties will enable to navigate the construction process in a sustainable manner [5]. Quality leaders able to lead project team in an efficient and effective manner to accomplish the organization and project goals. Hence this exploratory study was intended to gain better understanding of issues on leadership quality in Malaysian construction industry.

2 Methodology

This exploratory study adopted qualitative approach in gathering information from respondents through open interview sessions where the respondents were minimally guided to focus on the identified issues. Guiding questions were prepared based on the literature which comprises of three main elements which are challenges faced by construction leaders, leaders' role to sustain organization's competitiveness and qualities to be outstanding construction leaders. Selection of respondents is crucial in qualitative approach where they should have sound knowledge and experiences on the subject matter to ensure the reliability and validity of their opinions. This study has identified four construction leaders selected from highest G7 grade of CIBD database with working experience for more than 15 years and holding top managerial position. Construction leaders are considered experts of Malaysia construction industry that can contribute reliable information [6].

Experts were contacted to make appointments and to inform the intentions of the interview session for their preparation. Interview sessions were conducted through new technique known as OREO which means Opinion, Reason, Evidence, And Opinion's suggestion [7]. In this technique, experts were required to give opinions on the issues based on their knowledge and experience. Besides giving opinions, they are required to provide reasons/justifications for their opinion and if possible with evidences. Finally, they are required to give suggestions on their opinions. The interview session was conducted in face to face mode where the given information was recorded manually and also captured electronically. Experts were given ample time to voice their opinions on the given topic and gently guided to avoid deviation from the interview's objectives. Information gathered from the interview was analysed and synthesized to produce quality interview transcriptions which elaborate pertinent points in systematic and logical arrangement. These transcriptions were summarized in deductive manner to establish findings.

3 Interviews outcomes

Demography of four experts who participated in this interview are as follows: the first expert has being in the industry for more than 36 years and handled several mega construction projects, the second expert has being in the industry for more than 33 years and also owner of organization, while the third expert having 15 years of working with government mega projects and holding top management position and the final expert having 15 years of working with private mega projects and holding top management position.

3.1 Challenges faced by construction leaders

In the first part of the interview session, experts were requested to voice their opinions regarding challenges faced by construction leaders under the current scenario. The first expert responded by stating three primary challenges facing construction leaders today which are to achieve completion of the targeted construction successfully, ensuring all team

DOI: 10.1051/matecconf/201710303005

members review and respect the project schedule tightly, and managing team member attitude & work ethics correctly. To the second expert, the challenges are to ensure construction project runs smoothly without interference from employee problem, salary disputes, damaged machinery & the uncertainty of building materials prices, and managing cash flow. According to this expert, Malaysia construction industry is facing the shortages of skilled workers and the available foreign workers do not have permit and valid green card to work in Malaysia. Hence, leaders have to address this challenge before entering their workers to work at the construction site. Leaders need to compete for getting workers quickly by giving higher salary to prevent salary disputes among workers. Other important issues are the increases of construction materials prices after project awarded and late payments from the client affecting their cash flow.

According to the third expert, the challenges are to handle the large numbers of worker at one time on the site, handling construction projects in the hectic work area with unstoppable traffic at entry and exit route in the daytime, managing project cash flow with late payment from the client, and ensuring no construction drawing error such as the not accurate of measurement & design errors that will lead to rework. For the fourth expert, construction leaders are facing challenges to ensure the project runs smoothly with the problems of increased of construction material prices, lack of skilled worker on site, employing foreign workers with higher rate of workers levy fees and others. The outcomes for the challenges faced by construction leaders can be summarised as in Table 1.

Table 1. Summary of challenges faced by construction leaders.

First Expert	Second Expert	Third Expert	Forth Expert
Completion of project on time	Ensure project runs smoothly with many problems	Handle large numbers of workers	Ensure project runs smoothly with many problems
Ensure all team members abide with project schedule	Manage project cash flow	Handle project in hectic work area	
Manage team members attitude and work ethics		Manage project cash flow	
		Ensure no construction drawing error	

From Table 1, it can be interpreted that the common challenges according the experts in the order of similarity are ensuring the project runs smoothly within the targeted time, managing cash flow and ability to manage large number of worker. To ensure that the project runs smoothly, several factors are highlighted such as ensuring team member abide the project schedule, work with right attitude, minimise error and ability to work under hectic environment. To avoid delay, time is a crucial factor for construction operation as it will interrupt the sequencing of the construction process. Hence, any challenges must recognize and dealt early by construction leaders who play a vital role in project teams.

3.2 Leaders' role to sustain organization's competitiveness

Second element of the interview is to extract opinions from the experts regarding on how to ensure construction organization stays competitive which means successfully completed the project and also getting new project [8]. The first expert shared experiences in sustaining the organization's competitiveness by effectively utilizing resources to complete the given tasks with high productivity outputs considering cost benefit, workers, machines, and systems. Leader should adopt 'plan the work and work the plan' approach through recognizing, monitoring, and taking immediate actions and this effective approach should be applied as guide for tracking resource, schedule and time.

The second expert uses three approaches comprises of trusting subordinate's credibility in conducting work, evaluating subordinates' performances on construction progress, and aware of subordinates' responsibility to enable works be carried out as scheduled in sustaining the organization. While the third expert stressed that the importance of controlling and monitoring financial status of the project started from the early stage of project implementation. Leader should proactively follows up the work undertaken at the construction site by providing an alternative action/ plan, workaround to meet changing needs and requirements. Leader believes competitive organization must have a 'check and balance' approach among project teams to prevent one person having more power to control over the decision. The final expert sustaining the organization by ensuring accurate estimation on costing and follows the Quality Assessment System for Building Construction Work (Construction Industry Standard; CIS 7: 2006) provided by CIDB. These interview outcomes for role of construction leaders to sustain organization's competitiveness can be summarised as in Table 2.

Table 2. Summary of roles to sustain organization's competitiveness.

First Expert	Second Expert	Third Expert	Forth Expert
Effective use of resources	Trust subordinates credibility	Control and monitor project financial status	Accurate estimation of construction costing
Work according to plan	Evaluate subordinates performances	Proactively follows up work undertaken	Abide Quality Assessment System for Building Construction Work (CIS 7: 2006)
	Awareness of subordinates responsibility	Check and balance among project teams	

From Table 2 shows experts' own approaches to sustain organization's competitiveness. These variety methods are useful to be shared amongst the construction parties especially the contractors to sustain their company competitiveness.

3.3 Qualities to be an outstanding construction leader

Regarding qualities for an outstanding construction leader, the first expert viewed that successful leaders should be innovative to break the project work into daily activities which can speed up productivity. Leader should able to foster good relationship and communication with subordinates by interacting, caring and imparting information effectively to increase cooperation and also improve project completion on-time. Likewise, leader is considered problem solver in tackling issue arises and also making a quick & right decision using resource available for smooth running of construction project. Good leader should be able to delegate work to appropriate worker based on several criteria such as qualification, experience, attitude, and competencies. Leader must act as counsellor in handling disputes at construction site and also motivator to their worker, staff and sub-contractor to work properly for significant impact on productivity. Construction leader needs to work with enthusiasm to make changes for productivity and placing priority to subordinates. Leader should set good examples and inspire their subordinates to emulate like them to work as a team. Leader must show great interest to work, sincere with workers, knowledgeable and hardworking. Strong leader should admit and correct the problem as a symbol of his commitment to the project teams. Outstanding leader should able to evaluate productivity of his previous works for improvement of future project. Finally, he concluded that Malaysia construction industry is vibrant which requires more leaders in the industry.

According to second expert, role of outstanding construction leader is to develop trustworthy among project teams and related parties. Strong leader committed to responsibility and monitor results continuously by evaluating performances of subordinates to achieve targeted progress. Quality leader should able to tolerance with subordinates to enhance teamwork and foster good relationship with subordinates by treating their subordinates fairly, so that, any employee mistakes able to be dealt professionally and solved immediately. Furthermore, leader should be a good problem solver and motivates subordinates to maintain their loyalty to the company. According to this expert, outstanding leaders should have good personal like honest to work and having a good time management to avoid delay.

While the third expert, views good leaders should be proactive in following up work progress and exercising empowerment in organization. Leaders need to have 'check and balance' approach among project teams to prevent one person having more power to control over the decision. Since construction organization in hierarchical structure, quality leader should able to communicate effectively, concerned, understand and listen on workers' welfare to maintaining respect from subordinates. To stay competence, leader should have do self-improvement by gaining knowledge from senior construction leaders, through reading, and actively involved in a construction project. Good leader must be a problem solver using options and resources available in addressing challenges in construction industry. Leader motivates subordinates by contingent reward to show recognition for completing the task early.

The final expert emphasized that good leader should have good communication skill by listening to subordinates to acquire updated information on construction activities. Strong quality leader should able to follow regulations, hardworking, and positive thinking. Leader should also be good problem solver in tackling any issue arises. To this expert, leaders must continually do self-improvement through reading, and actively involved in a construction project. The roles as an outstanding construction leader from all the experts' views can be summarised as in Table 3.

From Table 3, all of the experts agreed that an outstanding construction leader should possess leadership quality as a good problem solver. Besides that, three out of four experts agreed that leader should have good communication skill and also a good motivator in carrying out the construction work.

4 Conclusion

This study has provided valuable insight issues of leadership quality in Malaysia construction industry. It was found ten challenges and roles of construction leaders which are required to sustain organization's competitiveness. To be successful construction leaders, experts gave three most important qualities which are to act as a problem solver, able to communicate well and as a motivator to their team members. This exploratory study's outcome will be a basis for further investigation on leadership characteristics which are required by construction leaders to steer project success.

Table 3. Summary of qualities to be an outstanding construction leader.

First Expert	Second Expert	Third Expert	Forth Expert
Innovative	Trustworthy	Proactive	Good Communication
Good Relationship	Committed	Empowerment	Good Personal
Good Communication	Monitor Result	Good Communication	Problem Solver
Problem Solver	Tolerance	Competence	Committed To Work
Decision Maker	Good Relationship	Problem Solver	
Delegate	Problem Solver	Good Motivator	
Good Counsellor	Good Motivator		
Good Motivator	Good Personal		
Enthusiasm	Good Time Management		
Inspire			
Good Personal			
Competence			
Vision			

This paper was partly sponsored by MyBrain15, Ministry of Education Malaysia and the Centre for Graduate Studies, UTHM. Special thanks to all experts in construction industries for contributing and giving helpful input that made this study possible.

References

[1] C.M.M. Isa, H.M. Saman and C.N. Preece, Determining significant factors influencing Malaysian construction business performance in international markets, J. of Cons. in Devlpg Countries, **20**(2), 1-23 (2015)
[2] K. Gharehbaghi and K. McManus, The construction manager as a leader, Leaders and Mgmt. in Eng., **3**(1), 56-58 (2003)
[3] S.U.R. Toor and S. Ogunlana, Ineffective leadership: Investigating the negative attributes of leaders and organizational neutralizers, Eng., Cons. and Ahc. Mgmt., **16**(3), 254-272 (2009)
[4] S. Nauman and A.M. Khan, Patterns of leadership for effective project management, J. Qual. and Tech. Mgmt., **20**, 1-14 (2008)
[5] N.A.N. Nasaruddin and I.A. Rahman, Leadership quality for Malaysia Construction leader to steer a success construction project, MATEC Web of Conferences, EDP Sciences, **47**, 1-5 (2016)
[6] X. Fei and T. H. Khan, Identifying attributes for expert construction project managers in the context of China, Intl. J. of Asian Soc. Sci., **5**(7), 407-418, (2015)
[7] G. Connell, *Graphic organizers for opinion writing*, (2015), Retrieved on Jun 6, 2016 from http://www.scholastic.com/teachers/top-teaching/2015/03/graphic-organizers-opinion-writing
[8] W. Pasmore, *Developing a leadership strategy: A critical ingredient for organizational success*, Greensboro, North Carolina (2014)

Study on Waste Composition at Taman Pura Kencana, Batu Pahat

Aeslina Abdul Kadir[1,*], *Noor Amira* Sarani [1], *Siti Fatimah* Onthong[1], and *Mohd Ikhmal* Haqeem Hassan[1]

[1]Department of Water and Environmental Engineering, Faculty of Civil and Environmental Engineering, Universiti Tun Hussein Onn Malaysia, 86400, Parit Raja, Batu Pahat, Johor, Malaysia

Abstract. Municipal solid waste management is a major challenge due to the increase in population and the development of a country. The problems also arise when the lifespan of the landfills available are shorten than estimated. The aim of this study is to determine waste composition at Taman Pura Kencana. The waste collection was carried out for 50 houses on a daily basis. The collection and sorting out method was conducted based on Malaysian Standard MS 2505:2012 and the data collected is recorded. The result showed the moisture content was approximately ranging between 25% - 30%. The density for each waste has been calculated and the result was recorded accordingly. The highest density is metal followed by organic waste, glass, napkins, household hazardous waste, textiles, paper, plastic rigid, plastic film, rubber and tetrapek. The waste collected also were categorized and consisted of food waste/organic (43.75%), paper (17.97%), plastics rigid (13.58%), plastic film (10.62%), napkins (4.43%), glass (3.10%), household hazardous waste (1.68%), metal (1.67%), tetrapek (1.34%) , rubber (0.93%), and textiles (0.92%). Results from the analysis illustrated that waste generation in Taman Pura Kencana may influenced by sudden changes in lifestyles, incomes, household size and also increase of population.

1 Introduction

Municipal solid waste management (MSW) becomes a great challenge in development plans throughout the world, especially in rapidly growing cities. One of the most successful countries in transition is Malaysia. Steady economic growth and low unemployment rates driven by stable political conditions and plenty of resources that making Malaysia at same level as developed country [1].

Malaysia is experiencing rapid industrialization and urbanization giving the adverse effects on the environment from the increasing of waste generated. Similar to many other countries, rapid urbanization and industrialization also have changed the characteristics of Malaysian solid waste generated [2]. Furthermore, waste generation rates also increase due to the demand of Malaysian for better quality of life. Nevertheless, waste management is to reduce the amount of waste being produced, thus reducing the disposal costs, the impact on

* Corresponding author: aeslina@uthm.edu.my

the environment, and the impact on human health [3]. The typical solid waste management system practiced in developing country brings many problems such as low collection coverage and irregular collection services, crude open dumping and burning without air and water pollution control and the breeding of vermin and flies. It was reported in 2003, the average amount of municipal solid waste generation per capita per day ranged is between 0.5 to 0.8kg but recently the range has increased 0.5 to 2.5kg, especially in the major cities such as Kuala Lumpur and Petaling Jaya [4]. The national strategic plan on solid waste management estimated that the waste generation rate will increases by 3.59% per year based on population growth projections for the period of 2002 to 2020. Based on that projection, the total waste management in peninsular Malaysia was 23000 tons/day in 2010 and 25000 tons/day in 2012. In 2020, the amount of waste to be generated has been estimated to reach 30000 tons/day [4].

Recently, one of the major concerns is municipal solid waste in residential area. Residential area is potentially exposed to pollution and threat with solid waste disposal. Each house at residential area has the ability to generate solid waste. This situation occurred due to lack of awareness about cleanliness, recycling and effect to the environment, community and existing landfill. Moreover, lacks of solid waste management at residential area also contribute to increase solid waste disposal. Different houses produce different compositions of solid waste based on occupants of house. It is important to identify and evaluate the composition of solid waste at the residential area to determine the most items disposed. Therefore, in this study, the composition of solid waste at Taman Pura Kencana was conducted to determine the most waste disposed from this residential area.

2 Materials and methods

In order to measure the composition of waste produced, the materials and methods were discussed in this section.

2.1 Sampling area

The study was conducted in private properties that covering housing area at Taman Pura Kencana which is limited to 50 units of single-storey houses for a month. However, the sampling size of the collected household reached minimum 200kg of waste weight; therefore it is not necessary to fulfill the requirement for collecting household waste at those 50 units of houses.

2.2 Questionnaire

The survey was conducted to obtain data from each household. The survey is to identify basic information from household and also to get their permission to participate in the study. All the households were briefed regarding the procedure on the waste collection from their house.

2.3 Sampling process

This study is focusing on waste retained and waste discarded which include the waste kept at home for recycling or reuses purpose, as well as the unwanted waste placed at the collection point. Therefore, waste retained and waste discarded generally becomes the main contributor to the total amount of waste generation from a household.

This study also focusing on the sampling of waste as-generated at household, the waste was collected by the student themselves. At the fieldwork, samples were collected by randomly choosing houses in a row within the area of study and these houses were taken for 'as-generated' samples. Then, the residents were provided with two plastics bags labeled organic and non-organic on daily basis to put all their waste into these bags accordingly. Before the collection, the residents were briefed on the procedure and were informed on not to throw any of organic or inorganic recyclable waste during the study out of the house. Time had been arranged between student and household to collect the waste on a daily basis. After the collection process, the organic, non-recyclable and recyclable items were weighed separately and recorded accordingly. Then, the samples were sent for sorting and the procedure was repeated until the required numbers of houses were complete.

2.4 Waste classification and weight of waste composition

The sample waste was sorted and classified into waste composition components as listed. Wide and sufficient area with the flat platform was important for sorting process. Then the wastes were divided accordingly to its composition based on specific description for each category of waste. The description is essential to help recognize the constituents of each waste category.

The total weight of waste composition was determined by weighing all the categories of household waste after the waste was spread on a clean plastic bags and segregation was done by hand. The weight of waste composition then was recorded on the worksheet and illustrated in term of percentage.

2.5 Moisture content

The moisture content of waste composition was determined. Normally, moisture content of solid wastes expressed in percentage or weight. In the wet weight method, the moisture in a sample is expressed as percentage of the wet weight materials meanwhile in dry weight method, it is expressed as percentage of the dry weight material.

Moisture content is the quantity of water contained in a material such as textiles, foods, and wood. Water content is applied in an extensive range of scientific and technical areas and is expressed as a ratio, which can range from zero to the value of the materials porosity at saturation. However, the waste sample in this study was dried in oven at 105°C in order to gain the value for dry weight.

3 Results and discussion

3.1 Survey

3.1.1 Estimation of recyclable waste by weight

Result from the survey in Fig. 1 shown that organic waste dominates the amount of wastes produced with 80kg followed by recyclable wastes which are papers, magazine and plastic rigid with 11.5kg, 11.0kg and 9.95kg accordingly. This trend might be due to daily cooking activities by the occupants. The amount for other wastes such as newspaper (9.7kg), plastic (8.1kg), others (7.0kg), aluminum can (6.7kg), glass bottle (5.9kg) and plastic bag (3.5kg) respectively. Nevertheless, the data might be not too accurate as some of respondents did not answer the questionnaire thoroughly for the first week.

3.1.2 Estimation of recyclable waste by quantity

During the survey process, data analysis in Fig. 2 shows food waste had the highest with 210 followed by paper (197), plastic rigid (175), plastic drinking (150) and napkins (90). The highest number of food waste produced may contribute by residents who are not working and stay at home as housewives. Lower number of waste produced was contribute by aluminium can (79), newspaper (65), and glass bottle (52) produced by residents at Taman Pura Kencana.

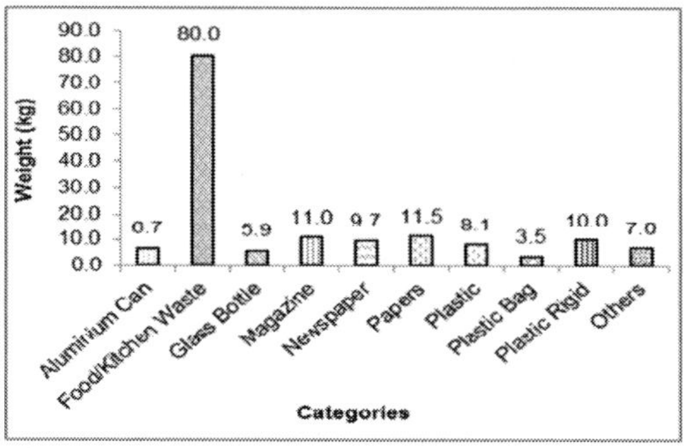

Fig. 1. Estimation of recyclable waste by weight from survey

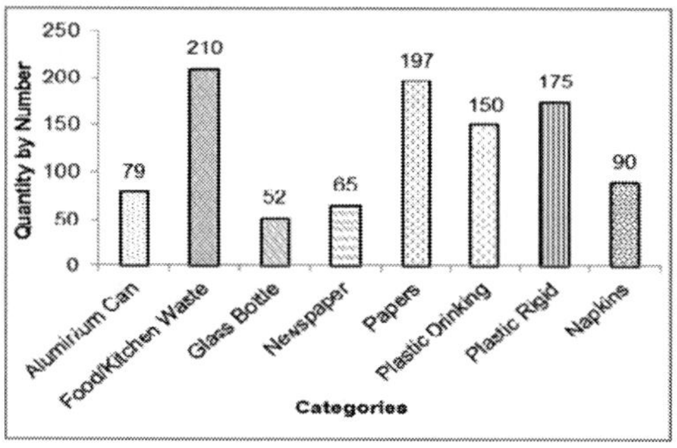

Fig. 2. Estimation of recyclable waste by quantity from survey

3.2 Waste generation and composition

3.2.1 Estimation of organic and inorganic waste generation

Solid waste generation rates offer an estimation of the amount of waste produced by residence over a certain period of time. Refer to Fig. 3 and Fig. 4, comparison between organic and inorganic waste generation for each day is shown. From day 1 to day 7, organic waste was produced almost the same amount of waste and do not have too much differences between each day. The same trend also occurred to inorganic waste where similar data was collected which is 20.26kg, 19.43kg, 20.48kg, 14.6kg, 16.66kg, 19.49kg and 15.04kg respectively.

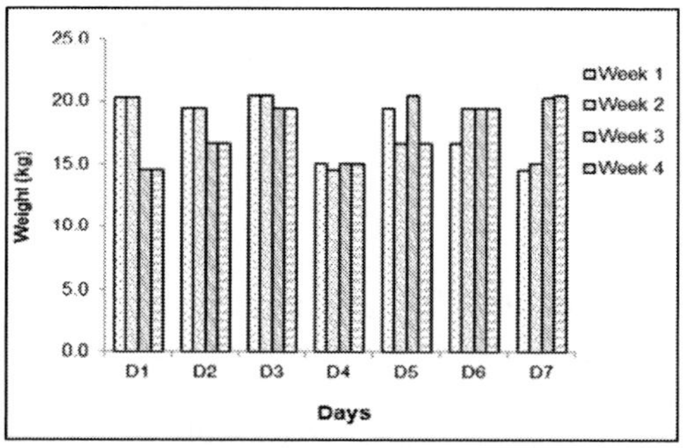

Fig. 3. Waste generation for organic waste

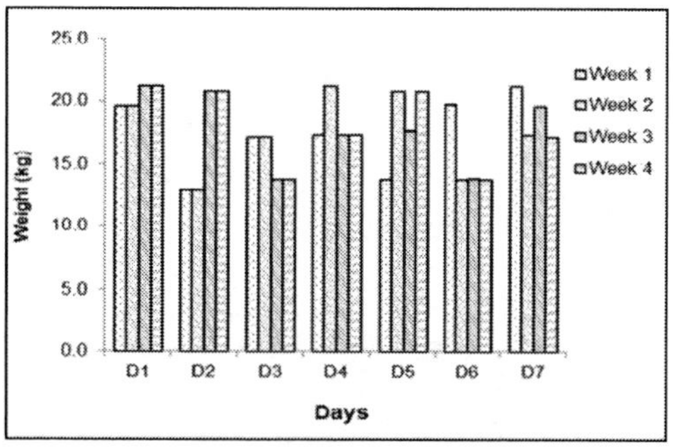

Fig. 4. Waste generation for inorganic waste

3.2.2 Composition of waste disposal for each week

Fig. 5 illustrates the percentage of waste by its category. Based on Fig. 5, it can be concluded that the highest percentage of waste produced at Taman Pura Kencana was food waste which is 47.87%. The second highest waste being produced was paper with 17.41%. Next, plastic rigid, plastic film, metal, napkins and glass which is 11.12%, 9.19%, 5.58%, 3.68% and 2.41% respectively. Tetrapek, textiles and rubber waste are all considered as lowest waste production and each contribute not more than 2%.

Fig. 6 indicates the classification of waste at Taman Pura Kencana according to category by percentage. The pattern of waste production is still the same compared with the sample at first week. Similar to solid waste collected at week 1, food waste is still the highest waste being produced in Taman Pura Kencana which is 46.96% in total for 7 days, followed by paper (18.34%), plastic rigid (14.85%) and plastic film (9.46%). The other categories hold the lowest waste productions which are napkins by 3.92% and glass by 2.95% while other categories obtained less than 2% each.

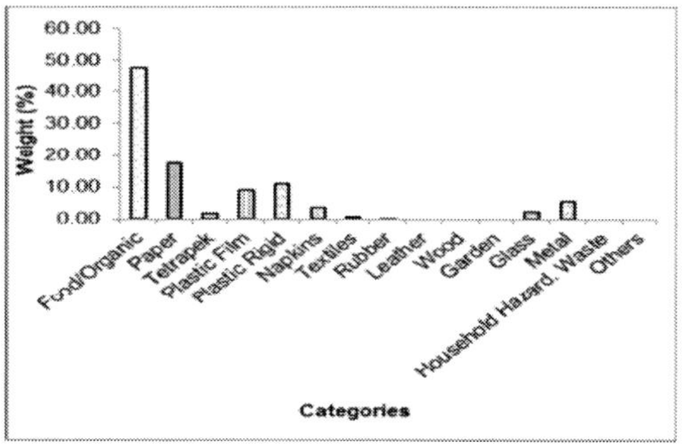

Fig. 5. Composition of waste disposal (Week 1)

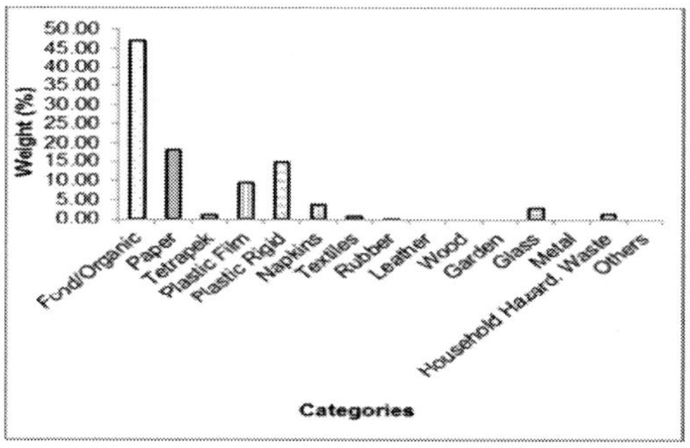

Fig. 6. Composition of waste disposal (Week 2)

Fig. 7 illustrates the percentage of waste by its category. It can be concluded that the highest of waste produce in Taman Pura Kencana was food waste which is 39.52%. the second highest waste being produced was paper (17.91%). Next, plastic rigid and plastic film which is 13.92% and 12.39% respectively. Rubber, tetrapek and textiles waste are all considered as lowest waste generation. Rubber and tetrapek each contribute to 2.02% and 1.49% respectively and textiles contribute less than 1% of the total waste produce in Taman Pura Kencana.

Fig. 8 illustrates the percentage of waste by its category. Based on Fig. 8, it can be concluded that the highest percentage of waste produced at Taman Pura Kencana was food waste which is 40.62%. The second highest waste being produced was paper with 18.22%. Next, plastic rigid, plastic film, napkins and glass which is 14.42%, 11.44%, 4.94%, and 3.88% respectively. Tetrapek, textiles, metal, household hazardous waste, and rubber waste are all considered as lowest waste production and each contribute not more than 3%.

3.3 Moisture content

The results show that values of the moisture content for waste collection in first week are 28.85%, second week 27.40%, third week 26.5% and 26.51% for the fourth week. Usually, the determination of moisture content of wastes may assist in obtaining water content of wastes and thus scientifically quantity the leachate generated in landfill. In addition, moisture contents also the important element during the treatment process of waste incineration and composition.

DOI: 10.1051/matecconf/201710303006

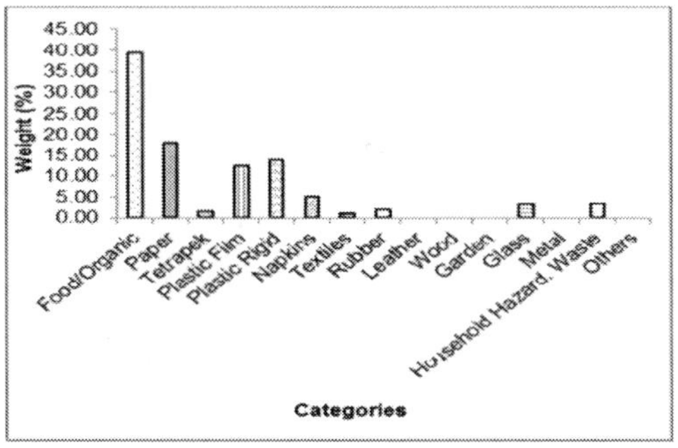

Fig. 7. Composition of waste disposal (Week 3)

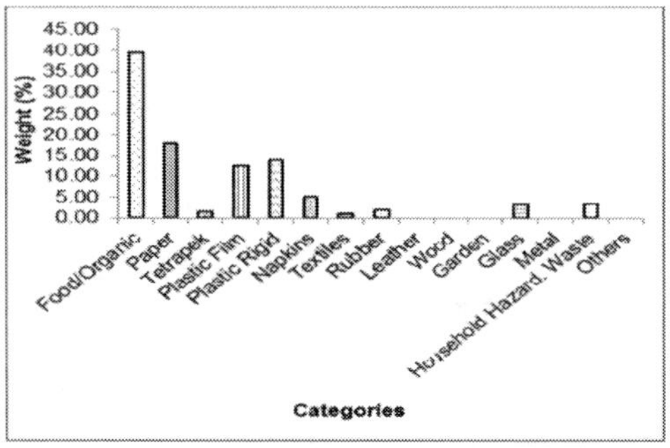

Fig. 8. Composition of waste disposal (Week 4)

4 Conclusions

The waste generation in Taman Pura Kencana may influenced by sudden changes in lifestyles, incomes, household size and also increase of population. These have been explained on the variability between the data of the waste composition. As in Taman Pura Kencana mainly consists of housewives and working person, so their buying power is high to live. In the meantime, information on types of waste is important for separation process, collection and recycling. It would help to understand the influence of socio-economic status and lifestyle on solid waste generation. The results conclude that amount of organic waste is still the highest followed by plastics and paper. The eating habits and family size of the residents of the study area, on the other hand, can influence the household weight significantly. It is suggested to improve proper disposing facilities for recyclable items at each houses, therefore the waste generated is fully under the responsibility of each household's resident. The future research is highly needed in order to gain more reliable information regarding household waste composition and generation rate.

The results presented in this paper are part of an ongoing postgraduate research. The authors would like to thank the Faculty of Civil and Environmental Engineering, UTHM for this study.

References

[1] M.N. Hassan, T.L. Chong, M.M. Rahman, M.N. Salleh, Z. Zakariah, M. Awang and M.N. Yunus, Solid waste management – What's the Malaysian position, *Seminar on Waste to Energy*, Universiti Putra Malaysia, (2000)

[2] G. Tchobanoglous, H. Theisen and S. Vigil, *Integrated solid waste management engineering principles and management issues*, McGraw-Hill, (1993)

[3] A.M.A. Latifah, M.A.A. Samah and M.Z.N. Iyana, Municipal solid waste in Malaysia: Practices and challenge, J. Waste Management, **29**, 2902-2906, (2009)

[4] U.N. Ngoc and H. Schnitzer, Sustainable solutions for solid waste management in Southeast Asian countries, Waste Management, **29**, 1982–1995, (2009)

Overcoming ICT Barriers in IBS Management Process in Malaysia Construction Industry

Peniel Ang Soon Ern[1,*], *Narimah* Kasim[2], *Md Asrul* Nasid Masrom[2], and *Goh* Kai Chen[2]

[1]Department of Civil Engineering Technology, Faculty of Engineering Technology, Universiti Tun Hussein Onn Malaysia, 86400 Parit Raja, Johor, Malaysia
[2]Department of Construction Management, Faculty of Technology Management and Business, Universiti Tun Hussein Onn Malaysia, 86400 Parit Raja, Johor, Malaysia

Abstract. As the result of government's initiative and future outlook in uprising the awareness towards the implementation of sustainable elements in construction industries, the Industrialized Building System (IBS) is one of the approaches that had been introduced as an alternative to conventional building method and produce the new strategy of reducing the waste of materials in construction. The IBS approach is actively promoted through several strategies and incentives as an alternative to conventional building methods. Extensive uptakes of modern Information Communication Technology (ICT) applications are able to support the different IBS processes for effective production. However, it is argued that ICT uptake at the organizational level needs to be regularly improved. Thus, this paper fulfills the need to identify the critical barriers to ICT employment and highlighting the ICT drivers in enhancing the IBS management process. Critical barriers and drivers to ICT uptake were identified through questionnaire survey with industry stakeholders and were analyzed using mean index and critical t-value with the use of SPSS software. The critical drivers are top management support and commitment and improving the availability of the technologies. The critical drivers which were identified are significant in ensuring the enhancement of ICT implementation. Hence, any system in place should avoid the highest ranked barriers or seek to alleviate their impact by focusing on the identified drivers in order to enhance implementation of ICT in IBS process management.

1 Introduction

Information and Communication Technology (ICT) was referred to the combination of manufacturing and services industries that capture, transmit and display data and information electronically [1]. Technological developments have been progressing fast in other industries, but construction for a long period has lagged behind other industries in terms of adoption of new technologies [2]. It is because of many studies having focused on

* Corresponding author: peniel@uthm.edu.my

the potential for harnessing the new ICT technologies for the good of the construction industry all over the world, due to the industry's awareness of the need [3-7].

In this regard, various studies have gone as far as to regard the gap between the construction industry and ICTs as a major problem affecting its whole performance and tried to diagnose the roots of the difficulties in this regard. Thus, a construction technique in which components are manufactured in a controlled environment (on or off site), transported, positioned and assembled into a structure with minimal additional site works, also known as Industrialized Building System (IBS) [8], are not excluded from the enhancement.

In point of fact, extensive use of modern IT tools is able to support the different IBS processes by enabling more accurate documents and hence good conditions for an effective production where errors are discovered early and problems in the manufacturing and assembly phases can be avoided [9]. Significantly, IBS plays an important role in advancing the sustainability of the constructions especially in reducing the rate for materials to be wasted and maintaining the desirable eco-friendly type of materials. With the effectiveness of ICT implementation, IBS uptake will also increase. However, the current practice of ICT implementation in the tracking of IBS components in Malaysia is rather rare. It appears that most innovative system and components are using innovative materials which are based on imported technologies which are obviously more expensive and difficult to purchase by local contractor [10]. Hence, this paper aims at identifying the critical barriers to ICT employment and highlighting the ICT drivers in enhancing the IBS management process especially in Malaysia construction industry.

2 Implementation of ICT in IBS

More sophisticated solutions of ICT based technologies in IBS processes are emerging. Examples include wireless communication, bar-coding and Radio Frequency Identification (RFID) for tagging technologies. A report by Construction Research Institute of Malaysia [11] pointed technology as a main support to impact IBS in the future. 'Building Information Modeling (BIM)' and 'new technology for IBS invented' were classified as the highest ranking scenario of positive impacts on high-high analysis (high impact-high possibilities). In addition, it was stated that further improvement of manufacturing technology is a strong solution to tackle the weakness and threats of IBS adoption.

However, the current practice of ICT uptake in the tracking of IBS components in Malaysia is rather uncommon. There is a lack of effective logistics management in the Malaysian construction industry, where technology intervention plays a very critical role [12]. IBS production process needs a sustained improvement and management in order to ensure proper components are developed and handed over at the right time, in the proper order and without damages.

Among the common ICT applications adopted in the IBS components management process, especially from the management level as indicated by the respondents (R) are the AutoCAD and Microsoft Project software (refer Table 1) [13]. As for the applications employed in the IBS manufacturing yard, there were higher technologies machineries such as the computer controlled batching plants, technology operated moulds, hollow core elematic machine and semi-automated hollow core extruder. Most of the said applications were predominantly used in the design, planning and production stage. However, the usage of the in-house software, Precast Internal System (PIS) was utilised throughout the whole IBS components process management.

The shortcomings of ICT uptake in the construction industry bring forth the need to review possible driver elements for improving ICT uptake in the industry. Several research publications [14-18] indicate that people, process and technology are the three key elements that need to be considered for successful uptake of technologies. Together, these three elements create business value [18].

Table 1. Implementation of existing ICT applications [13].

Ibs process	Existing ict applications	R 1	R 2	R 3	R 4	R 5	R 6
Project management/ coordination	Precast Internal System (PIS)					/	
	MS Excel	/	/	/	/	/	/
	MS Project	/	/	/	/	/	/
	Primavera				/		
Procurement	ACCPAC					/	
Design	AutoCAD	/	/	/	/	/	/
	REVIT						/
	TEKLA				/	/	/
	STAAD. PRO					/	/
	RYAN						/
Computer controlled manufacturing plants	Computer controlled batching plant	/		/			/
	Hollow core plants			/			
	Stencilling manufacturing			/			
	Technology operated mould				/	/	/
	Elematic machine					/	

However, the study further reiterated that "the people, processes, and technology need a leader", just as "an orchestra needs a conductor". The same analogy can be applied to the adoption and uptake of new and innovative technologies within construction companies. The "conductor" in this case is the policy maker. To successfully implement and use any new technology, it requires policy makers to plan and drive policies and strategies. Policies are "written principles or rules to guide decision-making" [19]. All the four elements are considered important for an organisation to effectively employ ICT. A company cannot employ ICT if it satisfies the requirements of just one element and not the others. Hence, all four elements - policy, process, people and technology - need to work hand-in-hand and symbiotically.

Taking this into account, the main elements for ICT uptake for an organisation to employ ICT should comprise of: (a) Policy makers that believes in the technology and takes strategic measures to drive its adoption, uptake and usage in order to mobilise the industry to implementand benefit from it, (b) Processes that enable and support the successful adoption of the technology, (c) People who have adequate skills, understanding of, and belief in, the technology; and finally (d) Technology applications and infrastructures which are adequate. The four elements to ICT uptake will form the theoretical basis to identify critical barriers and critical drivers to the enhancement of ICT uptake in IBS management process.

3 Methodology

This research applied the method of purposive sampling, where the samples are selected based on the knowledge of a population and the purpose of the study [20]. This is a form of non-probability sampling in which it includes specialist knowledge of the research issue.

Apart from the government IBS policy makers, IBS precast manufacturers were the selected respondents which formed the largest population among the other four IBS systems (e.g. IBS formwork, IBS steel systems) in Malaysia (refer Table 2).

Table 2. IBS Manufacturers in Malaysia [21].

IBS type	Total manufacturer
Pre-cast concrete framing, panel and box system	30
Steel framework system	24
Steel framing system	25
Prefabricated timber framing system	4
Block work system	25
Others (on-site)	10
Grand total	118

A total of 41 questionnaires were administered to the entire population of the precast concrete manufacturers and government policy makers, out of which 31 were returned, representing 76% of the entire questionnaires administered. This is however adjudged high response rate from the construction industry, which in general stands at only 25% to 35% [22]. The questionnaire was sent to 33 organizations from the listed companies, of which the majorities are manufacturers under the Registered IBS System Provider (RISP) list in Malaysia. In addition, 8 questionnaires were administered on government policy makers who were direct decision makers in issues pertaining to IBS. The data collection focused on the IBS precast manufacturing process, where ICT applications are more significantly used.

Referring to Fig. 1, the IBS manufacturers and government policy makers were asked to rate the barriers and drivers to ICT uptake in IBS process management from Insignificant to Very Significant range.

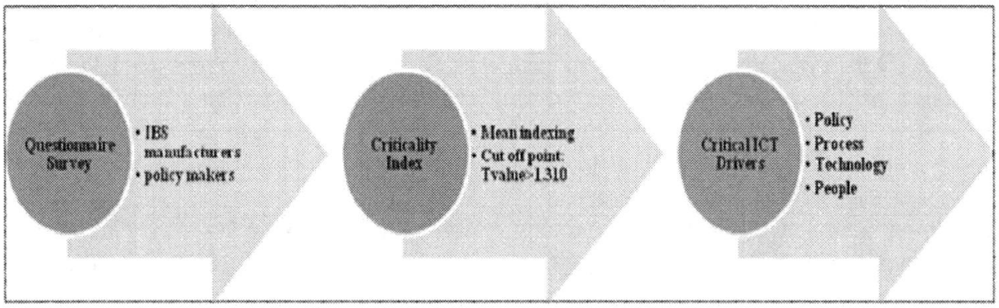

Fig. 1. Flowchart of research methodology.

In order to identify the critical rating from all the sub-components of the ICT barriers-drivers, the mean is fixed at scale '3.5' since ratings above '3' represent 'moderate significant', '4' represent 'significant' and '5' represent 'very significant' [23]. The rankings for the ICT uptake barriers and drivers can be obtained by calculating the critical value of the variables. The mean scores ratings of all proposed ICT uptake barriers and drivers were calculated using Equation (1).

$$\bar{x} = \frac{1(n1) + 2(n2) + 3(n3) + 4(n4) + 5(n5)}{(n1 + n2 + n3 + n4 + n5)} \qquad (1)$$

The null hypothesis (H0: μ1<μ0) against the alternative hypothesis (H1: μ1>μ0) were tested, where μ1 represents the mean of the survey sample population, and μ0 represents the critical rating above 3.5. The decision rule was to reject H0 when the result of the observed t-values (to) Equation (2) was larger than the critical t-value (tc) Equation (3) as shown in Equation (4). The criticality of ICT barriers and drivers in this study was examined using Equation (3) and (4).

$$to = \frac{\bar{x} - \mu 0}{SD / \sqrt{n}} \tag{2}$$

$$tc = t(n - 1, \alpha) \tag{3}$$

$$to > tc \tag{4}$$

The critical t-value, t (30, 0.10) = 1.310 with confidence interval 90% (α= 0.10) was set in this study. T-test analysis was employed to rule out the ICT barrier and ICT driver components which carry the mean value lesser than 3.5. ICT uptake barriers and drivers with t-values that are greater than the critical t-value of 1.310 were shortlisted as the critical barrier-driver to ICT uptake. There are similar studies conducted which applied mean indexing and t-tests. Studies have been done by rating the criticality index for the importance of the sustainability-related cost components [24] and using mean indexing to discuss the importance of evaluation elements for customer satisfaction in project management [25]. Apart from that, t-test analysis has been used in past research in identifying the relative important indicators [26, 27]. The critical barriers and drivers to ICT uptake are identified from the four driver elements.

4 Critical barriers to implement ICT

From the barrier and driver variables identified from the questionnaire survey, further analyses were conducted to shortlist critical ICT barriers and drivers to IBS process management. This was done with the mean indexing, where those with critical t-value was shortlisted. The complete set of rankings will provide an insight in to the effect of different drivers and barriers to construction ICT while identifying their relative importance in ICT implementation decision making.

The summary of the ICT implementation barriers are shown in Table 3. Out of the 34 ICT implementation barriers, there are only six barriers having t-values above the cut-off point of 1.310. The critical barriers, according to the ascending ranking order are high costs or substantial financial commitment in acquiring the technologies; resistance to change; high cost of specialist software; conventional process; conservative attitude of professionals and high cost for training staff. Six of these critical barriers are under the elements of Cost, People and Process. Technology and Policy are not seen as critical barriers to ICT implementation.

People barrier, being one of the top hindrances to ICT implementation has shown the resistance to change and conservative attitude of the professionals. The construction industry is naturally more cautious towards change and slow to adopt new technology. It is suggested that people tend to resist change due to their habits acquired over time, which is a primary constraint to ICT diffusion [28].

The number one barrier to the implementation of ICT was ICT investment restrictions due to budget constraints [29]. Their research finding is in line with the industry IBS stakeholders' opinion that high cost in acquiring technologies is the main barrier for ICT implementation. Another critical barrier in this research is cost of specialist software too high. [30] supported that IT systems are too costly that the organisations are unable to affort IT. High cost investment has been one of the most common barrier in ICT adoption [31-35]. Consequently, many construction case studies found that lack of enough training is one of the main barriers to adopting and using IT/ICT applications [33-35]. This research finding highlights the reason behind the lack of enough training. Number six critical ICT implementation barrier is high cost for training staff. Hence, cost is one element that constraints further development of an organisation to greater ICT uptake. However, [36] stated that cost is not a significant determinant of ICT adoption since most SMEs are aware of the government financial support and incentives. Further, [37] too supported this view for the fact that training cost is not an issue since government agencies have offered and provided a number of necessary training programmes. Despite the studies done, the result from this study shows that cost is a critical barrier to ICT implementation in the IBS process management. The underlying rationale could be caused by the fragmented nature in the construction industry and government not meeting the specific needs in the upgrading of ICT knowledge in the industry. Hence, Cost is one very critical barrier that should be seriously looked into as three of all six critical barriers are under the Cost category.

Table 3. Critical ICT Implementation Barriers for All Industry Stakeholders.

Rank	Barriers	Barrier Elements	T-value (cut-off point>1.310)
1	High costs or substantial financial commitment in acquiring the technologies	Costs	2.855
2	Resistance to change	People	2.299
3	Cost of specialist software too high	Costs	2.219
4	Conventional process	Process	1.58
5	Conservative attitude of professionals	People	1.508
6	High cost for training staff	Costs	1.408

Besides that, People barrier, being one of the top hindrances to ICT implementation has shown the resistance to change and conservative attitude of the professionals. The construction industry is naturally more cautious towards change and slow to adopt new technology. [28] suggested that people tend to resist change due to their habits acquired over time, which is a primary constraint to ICT diffusion. \

ICT is hindered from advancing in the IBS process management due to the conventional process that many IBS players still hold on to. ICT will find difficulty to penetrate into processes that are conventional. Hence, it is imperative that the barriers to ICT implementation be overcome with the critical ICT drivers.

5 Overcoming ICT barriers

Identifying critical drivers to ICT employment will be a strategy to overcome its barriers. Table 4 stated the top ten critical ICT drivers are shortlisted and summarised according to the ascending ranking order. Ten of these critical ICT implementation drivers are under the elements of People, Technology and Policy, where Process is not seen as a driver to ICT implementation. There are two critical drivers under the People element, namely top management support and commitment and devise better training procedures. [38] refers People to adequate skills, understanding of and belief in the technology. Thus, top down commitment is crucial in every organization, so to ensure all decision making are able to be passed down. Strong commitment from top management, including the senior leadership of an organization, is critical to ensuring technological acceptance by creating a supportive context for IT within the organization [39].

Table 4. Top 10 Critical ICT Implementation Drivers for All Industry Stakeholders

Rank	Driver Elements	Drivers	T-value (cut off point>1.310)
1	People	Top management support and commitment	8.469
2	Technology	Improve availability of the technologies	8.088
3	Policy	More cooperation between government and IBS players to promote and implement ICT	7.478
4	People	Devise better training procedures	7.045
5	Technology	Provide adequate ICT support (in-house or external)	6.225
6	Policy	Refine IBS Roadmap to encourage greater ICT uptake	5.903
7	Technology	Allocate adequate resources to implement and upgrade ICT tools	5.491
8	Technology	Consistent commitment and support to use BIM as a primary process and technology	5.479
9	Policy	More ICT course syllabus to start from tertiary level	4.507
10	Policy	Revision of reward systems	4.452

Ten of these critical ICT implementation drivers are under the elements of People, Technology and Policy, where Process is not seen as a driver to ICT implementation. There are two critical drivers under the People element, namely top management support and commitment and devise better training procedures. [38] refers People to adequate skills, understanding of and belief in the technology. Thus, top down commitment is crucial in every organization, so to ensure all decision making are able to be passed down. Strong commitment from top management, including the senior leadership of an organization, is critical to ensuring technological acceptance by creating a supportive context for IT within the organization [39].

The second most critical ICT implementation driver element is Technology. In the latest study of [40] technology plays a moderating role on the relationship between government and a construction organisation readiness level in implementing e-procurement. In order to increase the adoption of ICT in IBS process management, there is a need to first improve the availability of technologies. Studies showed that appropriate training increases the likelihood that IT will be accepted, since training tends to bring about better understanding and more favorable attitudes toward the technology and to result in more frequent use of it [41, 42]. Adequate resources are to be rightfully allocated to implement and upgrade ICT tools whilst consistent commitment to support the use of BIM in the IBS processes. Commitment focuses on end-users who devote themselves to using ICT whereas organisational commitment focuses on top managers who support end-users to use IT/ICT [28]. An organisation's top management also needs to be committed to support and allocate adequate resources for technology investment such as ICT. Commitment is essential to avoid IT project failure [43, 44].

Lastly, Policy is also found to be a critical driver element to greater ICT implementation improvement. [40] suggested that the role of government has a stronger influence on a construction organisation readiness level in implementing ICT. IBS Manufacturers and the Government policy makers agreed that more cooperation is needed between the government and IBS players in order to promote and implement ICT. Other than that, the IBS Roadmap has to be refined and more ICT course syllabus to touch the grass root education. One recurring ICT driver mentioned throughout this study is to revise the reward system [45].

 DOI: 10.1051/matecconf/201710303007

6 Conclusion

The critical drivers which were identified are significant in ensuring the enhancement of ICT implementation. Ten of these critical ICT implementation drivers are under the categories of People, Technology and Policy, where Process is not seen as a driver category to ICT implementation. The critical drivers, according to the ascending ranking order are top management support and commitment; improve availability of the technologies; more cooperation between government and IBS players to promote and implement ICT; devise better training procedures; provide adequate ICT support (in-house or external); refine IBS Roadmap to encourage greater ICT uptake; allocate adequate resources to implement and upgrade ICT tools; consistent commitment and support to use BIM as a primary process and technology; more ICT course syllabus to start from tertiary level and revision of reward systems. Results of the advancement of IBS in the industries could assist in the provision and improvement of higher sustainability to the construction development. These rankings are important to ensure that the industry tackles the nip of the problem instead of focusing on the less critical barrier issues.

This work was supported by RAGS (R059) Universiti Tun Hussein Onn Malaysia (UTHM).

References

[1] Organisation for Economic Co-operation and Development, Measuring the Information Economy Report, Infoeconomy, (2002), Retrieved on May 30, 2003 from http://www.oecd/org/sti/measuring

[2] P.S Brandon, T. Kocaturk, and R. Foundation, Virtual Futures for design, construction and procurement, Blackwell Pub, Malden, MA, (2008)

[3] M. Hannus, M. Blasco, M. Bourdeau, M, Böhms, G. Cooper, F. Garas, T. Hassan, A. Kazi, J. Leinonen, and Y. Rezgui, *Construction ICT roadmap*, Roadcon Project Deliverable Report D, 52, (2003)

[4] A. Adriaanse, H. Voordijk, and G. Dewulf, The use of interorganisational ICT in United States construction projects, Automation in Construction, **19,** 73-83, (2010)

[5] S. Kivrak, G. Arslan, and O. Cagatay, Information technology usage impacts on construction projects'success, (2010)

[6] Y. Rezgui, and A. Zarli, The way to the vision of digital construction: A strategic roadmap, Journal of Construction Engineering and Management, **132**, 767, (2006)

[7] T. Gajendran, and G. Brewer, Cultural consciousness and the effective implementation of information and communication technology, Construction Innovation: Information, Process, Management, **12,** 179-197, (2012)

[8] IBS Roadmap (2003-2010), Construction Industry Development Board (CIDB), Kuala Lumpur, (2003)

[9] J. Lessing, A. Ekholm, and L Stehn, Industrialized Housing - Definition and Loughborough: European Construction Institute, (2005)

[10] *IBS Roadmap Review (Final Report) IBS Centre*, Construction Industry Development Board, Malaysia, Kuala Lumpur, (2007)

[11] CREAM, *The 7th Industrialised Building System (IBS) Roundtable Workshop. Technology Foresight Report*, Construction Research Institute of Malaysia, 1-27, (2011)

[12] K.A.M. Kamar, and Z.I. Hamid, Modernising the Malaysian construction industry through the adoption of industrialised building system (IBS), *The Sixth Int. Conf. on Multi- National Joint Ventures for Construction Works*, Kyoto, Japan, 1-14, (2010)

[13] P.S.E. Ang, and N. Kasim, ICT-readiness in Industrialised Building System (IBS) management processes: Case studies, *1st FPTP Postgraduate Seminar 2013*, Universiti Tun Hussein Onn Malaysia, (2013)
[14] S. Basu, and P. Deshpande, *Wipro's People Processes: A Framework for Excellence*, White paper. (2004) Retrieved on April 23 2011 from http://www.wipro.com/insights/wipropeopleprocesses.htm
[15] C. Goolsby, *Integrated People + Processes + Tools = Best-of- Breed Service Delivery*, Getronics White Paper, (2001) Retrieved on 23 April 2004 http://itpapers.news.com
[16] Fuji Xerox, *Aligning People Processes and Technology*, (2003) Retrieved on April 23 2004 http://www.fujixerox.com.au
[17] B. Larkin, *Aligning People Process and Technology*, (2003) Retrieved on April 23 2004 http://www.paperlesspay.org/articles/Technology.pdf
[18] B. Emmett, IT service management: People process technology, business value, The IT J., Third Quarter, (2002)
[19] Clemson, Definition of Policy, Clemson University - Office of Research Compliance, Definitions of Research Compliance Terms, 2007, Last accessed 14 April 2007, from http: //www.clemson.edu/research/orcSite/orcIRB_DefsP.htm.
[20] A.R. Md Nor, *Statistical methods in reserch*, Pearson, Malaysia, (2009)
[21] *List of Malaysian IBS manufacturers and Registered IBS System Providers (RISP) 2012-2013*, Construction Industry Development Board Malaysia (CIDB), Kuala Lumpur, (2013)
[22] R. Fellows, and A. Liu, *Research methods for construction*, Wiley-Blackwell, West Sussex, (2008)
[23] K.C. Goh, *Developing financial decision support for highway infrastructure sustainability*, PhD Thesis, Queensland University of Technology, (2011)
[24] K.C. Goh, and J. Yang, Importance of sustainability-related cost components in highway infrastructure: perspective of stakeholders in Australia, J. of Infrastructure Systems, **20**(1), (2013)
[25] J.B. Yang, and S.C. Peng, Development of a customer satisfaction evaluation model for construction project management, Building and Environment, **43**(4), 458-468, (2008)
[26] J. Wong, and H. Li, Development of a conceptual model for the selection of intelligent building systems, Building and Environment, **41**(8), 1106-1123, (2006)
[27] Z. Shehu, and A. Akintoye, Major challenges to the successful implementation and practice of programme management in the construction environment: A critical analysis, Int. J. of Project Management, **28**(1), 26-39, (2010)
[28] V. Peansupap, *An exploratory approach to the diffusion of ICT in a project environment*, PhD Thesis, RMIT University, Melbourne, Australia, (2004)
[29] S.L. Kajewski, J.R. Crawford, A., Weippert, P.A. Tilley, S.L. McFallan, T.R. Remmers, and M.A. Haug, *National Perspective on the Status of ICT in the Australian Construction Industry*, (2001)
[30] P. Hawking, A. Stein, D. Wyld, and S. Forster, E-procurement: Is the ugly duckling actually a swan down under?, Asia Pacific J. of Marketing and Logistics, **16**(1), 1-26, (2004)
[31] P.E. Love, Z. Irani, H. Li, E.W. Cheng, and Y.C. Raymond, An empirical analysis of the barriers to implementing e-commerce in small-medium sized construction contractors in the state of Victoria, Australia, Construction Innovation: Information, Process, Management, **1**(1), 31-41, (2001)
[32] L. Marsh, and R. Flanagan, Measuring the costs and benefits of information technology in construction. Engineering, Construction and Architectural Management, **7**(4), 423-35, (2000)

[33] A. D.Songer, R. Young, and K. Davis, Social architecture for sustainable IT implementation in AEC/EPC, *Proc. of IT in Construction in Africa 2001*, Mpumalunga, South Africa, (2001)

[34] P. Stephenson, and S. Blaza, Implementing technological change in construction organisations, *Proc. of the IT in Construction in Africa, Mpumalunga*, South Africa, (2001)

[35] A. Weippert, S.L Kajewski, and P.A. Tilley, Internet-based information and communication systems on remote construction projects: A case study analysis. Construction Innovation, **2**(2), 103-16, (2002)

[36] K.S. Tan, S.C Chong, B. Lin, and U.C Eze, Internet-based ICT adoption: Evidence from Malaysian SMEs, Industrial Management and Data Systems, **109**(2), 224-244, (2009)

[37] S.S. Alama, and M.K.M. Noor, ICT adoption in small and medium enterprises: An empirical evidence of service sectors in Malaysia, Int. J. Bus. Manag., **4**, 112-125, (2009)

[38] K. Ruikar, C.J. Anumba, and P.M. Carrillo, An E-readiness assessment application for construction companies, Automation in Construction, **15**(1), 98-110, (2006)

[39] P. Weill, The relationship between investment in information technology and firm performance: A study of the valve manufacturing sector, Information Systems Research, **3**(4), 307-333, (1992)

[40] Q. Tran, Huang, D. Liu, and B.H.M. Ekram, A construction enterprise's readiness level in implementing eprocurement: A system engineering assessment model, Systems Engineering Procedia, 131-141, (2011)

[41] L. Raymond, *The impact of computer training on the attitudes and usage behavior of small business managers*. Laboratoire en économie et gestion des systèmes de petites dimensions, Dép. d'administration et d'économique, Université du Québec à Trois-Rivières, (1988)

[42] V. Venkatesh, Determinants of perceived ease of use: Integrating control, intrinsic motivation, and emotion into the technology acceptance model, Information Systems Research, **11**(4), 342-365, (2000)

[43] D. Leonard-Barton, and J. J Sviokla, Putting expert systems to work, Harvard Business Review, 91-8, (1988)

[44] R.L. Thompson, C.A. Higgins, and J.M. Howell, Personal computing: Toward a conceptual model of utilization, MIS Quarterly, **15**(1), 125-43, (1991)

[45] R. Mahbub, *An investigation into the barriers to the implementation of automation and robotics technologies in the construction industry*, PhD Thesis, Queensland University of Technology, Australia, (2008)

A Preliminary Review on Economies of Scale (EOS) Towards Industrialized Building System (IBS) Manufacturer

Syazwana Tajul Ariffin[1], *Riduan* Yunus[1,*], *Hairuddin* Mohammad[2], and *Siti Khalijah* Yaman[1]

[1]Jamilus Research Centre, Faculty of Civil and Environmental Engineering, Universiti Tun Hussein Onn Malaysia, 86400 Batu Pahat, Johor, Malaysia
[2]Faculty of Architecture, Planning, and Survey, Universiti Teknologi MARA, Shah Alam, Selangor, Malaysia

Abstract. Industrialized Building System (IBS) is a potential technology to improve productivity of construction industry. Controlled production and minimum generation of construction waste are some of the benefits that can be achieved by replacing conventional construction with IBS. In business, IBS is giving a huge opportunity for manufacturer and supplier to expand their business while contributing to construction development. However, bad strategies will put the company in high risk due to higher initial capital for machines and equipment. Therefore, strategic planning for company's growth, profit maximization, and enhancement of productivity is undeniable to ensure the success of business in construction industry. This preliminary paper is exploring associated factors that affect Economy of Scale (EOS) and their relationships in catalyzing the IBS manufacturer especially precast concrete as the scope of study to continue their business in the construction industry. Thus, a framework of EOS is proposed to assist IBS manufacturers to ensure their company's growth and stability, competitiveness in term of monopoly or an oligopoly, increasing productivity, leading constant returns to scale, and finally increasing the firm's efficiency. The refined EOS's conceptual framework is an important turning point to support the development of decision making tools for IBS manufacturer towards their stability and survival in this highly competitive industry.

1 Introduction

The Malaysia government is greatly promoting the implementation of Industrialized Building System (IBS) to enhance productivity in construction industry through the cooperation of Construction Industry Development Board (CIDB), IBS Centre, and Malaysian Investment Development Authority (MIDA). According to Azman [1], IBS has a potential to improve the quality of constructed projects and eliminate problems which normally faced in conventional method such constructability and higher construction cost.

*Corresponding author: riduan@uthm.edu.my

The success of IBS implementation depends on the supply chain management. Each phase in IBS implementation, starting from design to maintenance and demolition, needs to be examined carefully to improve the deliveries of IBS projects. This study is focusing on manufacturing process which investigates the EOS of a company in sustaining their business in the construction industry. The numbers of IBS manufacturers are increasing over the years. In 2007, the total registered IBS manufacturer is 128 companies and it is increasing to 201 in 2015 [2]. Within the same vein, those increments are translated to competitive advantage in manufacturing which is essential for the industry to maintain uptrend performances due to attainment of economic sustainability, improved equipment efficiency, and reducing production disturbance, all for amplified productivity [3]. Based on a report published by the Economic Planning Unit of Prime Minister's Department [4], during the Eleventh Malaysia Plan (2016-2020), productivity is focused as renewed effort towards national, industry, and enterprise economy.

On the other hand, there are various factors that contributed to the challenges of Malaysia's construction industry especially in the enhancement of productivity. Prior to that, each of construction projects is unique, thus, there are uncertainties which lead to variation in both the prognosis and the actual economic result [5]. However, following a report from Malaysian Investment Development Authority (MIDA) [2], the adoption of IBS in the construction sector is still low and represents only about 15 to 20 per cent of the overall projects in Malaysia despite being in the industry for around 50 years. This percentage is lower than expected notwithstanding huge publicity campaign from the government. The fragmented and disconnected construction supply chains are major causes of the limited uptake as there are only 205 manufacturers that mostly located in the city areas [6].

Subsequently, the manufacturer is struggling to sustain their business due to non-standardization and shortage of demand of IBS components [7]. In most cases, builders have different requests in term of components' specifications, which further require the manufacturers to invest for new molds [7]. This scenario is deemed not sustainable as the manufacturing companies usually have their own economic strategic planning. It is because of the trend of mass production which requires the customization of products and services that eventually leading to cost reduction [8]. Therefore, it is posited that holistic manufacturing strategies must be developed in order to ensure that future productions are able to meet this new demand. Additionally, along the way, a firm needs to organize its operational processes in terms of both technical and economic efficiency to survive in the long run [9]. Thus, scale economies are seen as significant savior which are able to lead for growth towards businesses by reducing average costs and simultaneously increasing levels of productivity appropriately [10]. Here, a positive prospect from the understanding of the concept of EOS might benefit the manufacturing organizations despite of their sizes. Therefore, identifying related factors regarding EOS is deemed paramount and will be a useful reference for the IBS manufacturers' practices. Hence, this preliminary study was carried out by extensive review of established literatures including the authors preceding research [11] that discussed on generic EOS for construction contractor. The qualitative undertaking was helped by a series of thematic process in order to assist the authors on appropriate direction for concluding a reliable and valid finding [12, 13]. Throughout the process, EOS is categorized into internal and external factors, and influenced by apposite supporting factors.

2 Implementation of IBS in Malaysia

The implementation of IBS in Malaysia is still low. Most of industry players have a negative perception and incapable to implement it due to advanced technology which is required in its application [14]. In addition, an emphasis towards contractors' satisfaction level in implementing the IBS is also limited to evaluate subjective performance measurement. Based on comparison of construction costs for the precast wall panels, it was 40% more expensive compared to the cost of a conventional wall, but it promotes less cost for the site labour from RM59 (conventional) to RM23 (precast) per square meter [15].

Previously, Construction Industry Development Board Malaysia (CIDB) [16], highlighted 13 key issues in adopting IBS in term of consultant, contractor, and manufacturer. Examples of problems faced by IBS manufactures which stated in that article are low demand for IBS components, high cost of import duty for IBS manufacturing equipment, expensive installation machinery, sparse location of IBS manufacturer, and costly transportation for components. In addition, Mao et al. [17] identified 24 risk factors in implementing IBS, namely; high cost pressure without economics scale effect, poor cooperation between multi-interface, inappropriate design codes and standards for industrialized building, lack of management practices and experiences, and lack of quality monitoring mechanism in the process of production are some of the high risk factors existed in the process of implementing industrialized building. Therefore, these issues need to be scrutinised in a holistic way as the effective development of firm organization have to ensure the effective economic decisions that may help in minimizing investment, operating, and maintenance expenditures, besides increasing the revenue, savings, and such other positive gains of the organization [18].

3 Economies of scale

Economic of scale exist in several activities in industry. According to Alcorta [19], manufacturing industry was implementing the economies of scale according to product, plant, and firm's dimension. New technologies and de-scaling might give benefit in the level of 'optimal' scales. Scale is a size of output. Economies of scale came with theoretical sense of total cost rises less proportionately than output, and optimal scale occurs at the point where any increase in output no longer reduces but raises unit costs [19]. Economies of scale can be classified into two main types which are internal (within the company) and external, which are relied on the advantages of the other supporting industry as a whole [20, 21, 22].

Pyo and Yang [23] stated that, if the input prices decline as a firm increases its output scale, an economies of scale effect can occur even when there is decreasing or constant returns to scale. Therefore, economies of scale are the relationship of scale of the combination productive factors and quantity of the firm output [24]. A simple economic model is able to develop according to Chamberlinian formulation and modified Dixit Stiglitz to assist the intra and inter industry trade market [25]. According to Pyo & Yang [23], a firm's cost function (C) can be written as in equation (1). A firm's total costs in year (t), where F_t is a function of input prices, Q_t as outputs, and T_t for technology change. Therefore, a firm may achieve positive scale economies even under the condition of decreasing returns to scale when there is a significant decline of factor prices. Meanwhile, the effect of scale economies which is the change of cost with respect to the change of output, can be expressed as in equation (2). The positive EOS ($+EOS_t$) will give economies of scale result where negative EOS ($-EOS_t$) show diseconomies of scale of the operative firm.

$$C_t = C(Q_t F_t T_t) \tag{1}$$

$$\text{EOS}_t = \left(\frac{1}{d \ln C_t / d \ln Q_t} \right) \qquad (2)$$

Apart, by referring to Fig. 1, the short run average cost curve (SRAC) is based on a particular set of fixed costs. As the firm grows, its fixed cost change and the curves that are redrawn are further to the right on the graph. From these, a long run average cost (LRAC) curve can be constructed as it has a same characteristic U-shape as the SRAC [26]. As firms increase quantity output, average costs fall until reaching the optimal marginal cost. Since economies of scale is cost advantages obtained due to size, cost per unit of output are generally decreasing with increasing scale as fixed cost are spread out to more units of output [27].

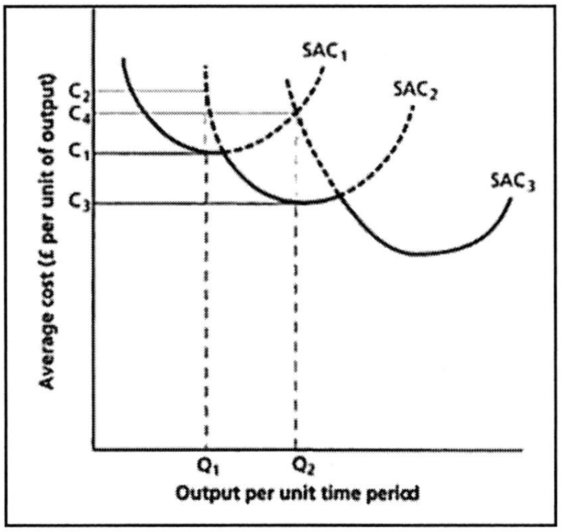

Fig. 1. Economics of scale in short and long run average cost [22].

According to Celli [10][2] that investigates the Small and Medium-Sized (SME) entities familiar with EOS, positive responds were gathered from 59 companies that declared gaining economies of cost by implementing several economies of scale's factors in their productive process. Within organisation, Adam Smith have discussed EOS through division of labour approach, meanwhile Robinson's on economies of massed reserves, management, and large machines [28, 29]. Additionally, internal factors also cover marketing and commercialisation approaches of firm, their workers' specialisation, technical capabilities, and financial portfolios [29, 30]. Meanwhile, from the perspective of external factors, ability of firm to bear business risks, supports to the industry, and supporting infrastructures were some of important determinants to achieve the optimum EOS [22]. Therefore, in order to shed light on each EOS factors, discussions on its impacts towards organisation/firm are paramount, where fair perspectives anchored by appropriate literatures are presented.

There are various impacts of existing EOS in the firm's market. In reality, monopolistic competition is occurred within EOS which in a way faulting fair competitions [31, 32]. These competitions contribute in increment of market size. Monopoly's firm that gain profits will eventually attract competitors that lead to oligopoly and further monopolistic competitive [33]. Melvin [34], stated that main features of monopolistic competition in a trade consist of similar goods but differ marketing strategy from the other firm, where subsequently the firm can raise price and gain benefit from marketing power. The increasing return to scale may also contributed by the advancement of technology (i.e. through internal factor), however, unfortunately, a firm's entry into trades that may in the long-run contributes to zero profit. This possibly occurs if the industry is gaining profit, it will eventually attract new firms which in the end subsidises lower profits. Economies of scale's impacts can also be observed through the changes of growth in size. However the size of firm is not necessary to obtain economies of cost [9]. In another way, productivity will rise as the size of firm are growing, which contributes by internal and external factors such as having specialized workers and higher technology of machineries [35].

Accordingly, Zhi et al. [36] claimed that economies of scale are able to increase firm's productivity given its potential to influence total factor of productivity (TFP). Therefore, long-term economic growth and comprehensive industry level of productivity measure can be determined by referring to an indicator of economies of scale, which is employed construction output per capita. Since firm's productivity is increased, it may lead to their stability in the long run. Besides, constant returns to scale are achieved when the firm experiencing economies of scale where along the way it will grow further with the increase of profitability [37]. This allows costs to be reduced and contributes rise of production as the returns are constant. On the other hand, exploitation in bargaining power and learning on opportunities to reduce costs are able to gain more businesses [34]. Thus, the impacts towards technical and organizational efficiency are very much related to economic sense which implies profit maximizing behaviour of the firm. Higher efficiency and levels of productivity that showed by larger firm is attainable by higher work intensity and the capital size of firm [37]. Therefore, following the previous discussions [11, 12, 13] based on available literatures, a refined framework in visualising EOS are presented as can be perused in Fig. 2.

4 Economies of scale and industrialized building system

IBS approach is not only based on technology solutions, but a holistic process to be implemented [38]. According to Ahmad et al. [39], IBS industry is justified economically when the existing mass production was effectively distributing fixed cost over production of product. In order to achieve economic of scale for manufacturers and buyers, the streamline and standardised IBS component are been highlighted by the CIDB in their IBS Prospectus [40]. In brief, production categories can be divided into craft, mass, and lean production, which came with different operations and outcome as in Table 1. The craft production is known as producing goods with minimal equipment or by hand. Thereafter, mass production is focused on standardised product that is produced in a large scale. Whereby, lean production is a resource utilization towards value and efficiencies of the customer [41,42]. Through preliminary observation, IBS manufacturers in Malaysia are fond towards mass production manufacturing system. These manufacturing systems are able to give advantages in shorter time compared to craft production, where cost reduction and efficiency increment can be enhanced further (i.e. by using economies of scale). Since IBS manufacturing can be improved by economies of scale and automation as it is mass-based production, organisation/firm may draw strategic planning of cost minimisation by utilising all resources for producing goods in the market rather than within a firm, which through the end may lead to change of total cost of production [23]. Likewise, by revisiting the issues in Malaysia's IBS manufacturing (see Chapter 2), strengthening the application of EOS towards IBS manufacturing is seen as inevitable in order to achieve construction industry's vision in the recent Construction Industry Transformation Program (CITP) 2016-2020 (CIDB) [43].

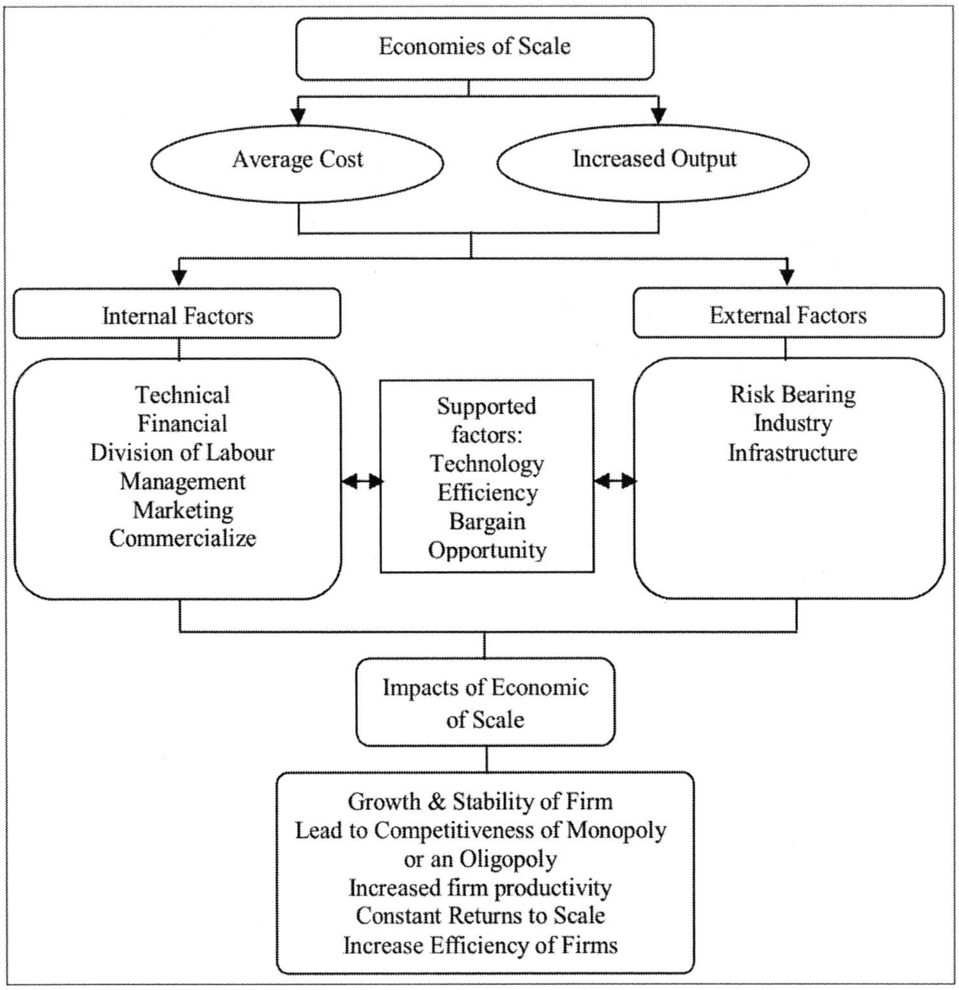

Fig. 2. Refined framework of economies of scale.

Table 1. Categories of production system [42].

Functions	**Craft Production**	**Mass Production**	**Lean Production**
Focus	Task	Product	Customer
Operations	Single Item	Batch and Queue	Synchronized flow and pull
Overall Aim	Mastery of Craft	Reduce cost and increase efficiency	Eliminate waste and add value
Quality	Integration (part of the craft)	Inspection (a second stage after production)	Prevention (built in by design and method)
Business Improvement	Customization	Economies of scale and automation	Flexibility and adaptability
Improvement	Master-Driven (continued improvement)	Expert driven (periodic improvement)	Workforce driven (continued improvement)

5 Conclusion and discussion

In IBS manufacturing, effective strategies will make companies more competitive in the market place, thus creating a lot of spill over effects such as providing employment in particular sector. Unfortunately, the industry was facing a number of shortcomings such as low demand, costly machineries and transportation, location-wise problem, etc. Thus, optimistically, the inclusion of holistic EOS factors (i.e. internal, external, and supports) are expected to counter those deficiencies since given the fact that IBS manufacturing approach is generally adopting mass-based production system, which improvement can be made through EOS. For instances, internal factors of marketing, commercialisation, and financial are very much related to demand and machineries. Whilst, issues on transportation and location might be able to be solved through critical examination of external factors, namely support from surrounding industry and infrastructures.

The impacts from holistic implementation of EOS towards IBS manufacturing firms are very costly to be side-lined. Stability of firm, increasing firm's productivity, and maximising profitability are just some of positive impacts that certainly became objectives of firm's existence. Moreover, following all positive effects will bring (see Figure 2), indirectly, government efforts in nurturing larger uses of IBS components that might be further materialised. Therefore, the authors posit that subsequent research explorations are worth to be taken, where qualitative and quantitative approaches are concerned.

The authors acknowledge and appreciate the financial support provided by the Universiti Tun Hussein Onn Malaysia (UTHM) in Johor, Malaysia through the research university grant.

References

[1] M.N.A. Azman, M.S.S. Ahamad, and N.D. Hilmi, The perspective view of Malaysian industrialized building system (IBS) under IBS precast manufacturing, *4th Int. Engineering Conf. - Toward Engineering 21st Century*, Islamic University of Gaza, 1–13 (2012)

[2] Malaysian Investment Development Authority (MIDA), *Industrialized Building System (IBS) in Malaysia: An Update*, MIDA Newsletter Publication for Global Investors, Malaysia, 3, (December, 2015)

[3] A. Ingemansson and G. S. Bolmsjö, Improved efficiency with production disturbance reduction in manufacturing systems based on discrete-event simulation, J. Manufacturing Technology Management, **15**,(3), 267–279, (2004)

[4] Economic Planning Unit Prime Minister's Department, *Unlocking the potential of productivity*, Srategy Paper 1 No. 9431, Economic Planning Unit, Putrajaya, (2009)

[5] G.M. Winch, Towards a theory of construction as production by projects, Build. Res. Inf., **34**(2), 154–163, (2006)

[6] Construction Industry Development Board (CIDB), IBS Digest, Industrialised Building System (IBS), Kuala Lumpur, Malaysia, (2016)

[7] Construction Industry Development Board (CIDB), Industrialised Building System (IBS), *IBS-Roadmap 2011-2015*, Kuala Lumpur Malaysia, (2010)

[8] D.T. Pham, P.T.N. Pham, and A. Thomas, Integrated production machines and systems – beyond lean manufacturing, J. Manufacturing Technology Management, **19**(6), 695–711, (2008)

[9] M. Celli, Determinants of economies of scale in large businesses, Am. J. Ind. Bus. Management, **3**, 255–261, (2013)

[10] M. Celli, Are EU's Small and medium-sized entities familiar with economies of scale?: Response from the field, World J. Soc. Sci., **5**(3), 101–109, (2015)

[11] S.T. Ariffin, S. Sulaiman, H. Mohammad, S.K. Yaman, and R. Yunus, Factors of Economies of Scale for Construction Contractors, *Int. Congress on Technology, Engineering, and Science*, Kuala Lumpur, Malaysia, (2016)
[12] S.K. Yaman, A.H. Abdullah, H. Mohammad, and F. Hassan, Technical competency of construction manager in Malaysian construction industry, Applied Mechanics and Materials, **773–774**, 1053–1059, (2015)
[13] N. Kariya, Z. Yaakob, M.M. Sairi, H. Mohammad, S. K. Yaman, and N.H. Abas, Investigation of generic house components and their practical ways to be assessed by house buyers during defect liability period in Malaysia, Int. J. Engineering, **29**(10), (2016)
[14] R. Yunus, A.N. Masrom, A.H. Abdullah, and F. Mustakim, conceptual model of contractor satisfaction in the industrialized building system (IBS) implementation, Applied Mechanics and Materials, **773–774**, 828–833, (2015)
[15] T.K. Chan, Mechanisation in construction - A Malaysian perspective, *Proc. of the 19th CIB World Building Congress*, Brisbane, (2013)
[16] Construction Industry Development Board (CIDB), IBS Digest, Industrialised Building System (IBS), Kuala Lumpur, Malaysia, (2015)
[17] C. Mao, L. Shen, L. Luo, and Z. Li, Identification of Risk Factors Influencing the Implementation of Industrialized Building System in China, *Proceedings of the 19th International Symposium on Advancement of Construction Management and Real Estate*, 219–230, (2015)
[18] F. Rangelova, *Fundamentals of economics in sustainable fundamentals of economics in sustainable construction,* Bultest Standard Ltd, Bulgaria, Europe, (2015)
[19] L. Alcorta, *The impact of new technologies on scale in manufacturing industries: Issues and evidence*, Institute for New Technologies, Netherlands, (1994)
[20] A. Wibowo, The contribution of the construction industry to the economy of Indonesia: A systemic approach, *International Seminar CIB 107*, UTM, Malaysia, (2009)
[21] E. Henriksen, K.H.M. Knarvik, and F. S. Steen, *European manufacturing*, London, UK, (2001)
[22] D. Myers, *Construction economics: A new approach*, First Eds., Spon Press, New York, USA, (2004)
[23] M. Pyo and J. Yang, *Optimum allocation between FDI and exporting in the presence of transaction costs and economies of scale: Modeling and examination of MNEs in the automobile industry*, PhD Thesis, The George Washington University, United States, (2008)
[24] G.J. Stigler, The economies of scale, J. Law Economics, **1**(1), 54–71, (1958)
[25] P.R. Krugman, Increasing returns, monopolistic competition, and international trade, J. Int. Economics, **9**(4), 469–479, (1979)
[26] J. Manser, *Economics: A foundation course for the built environment*, E&FN Spoon, London, UK, (1994)
[27] L. Ruddock, *Economics for the modern built environment*, Taylor and Francis, Oxford, (2009)
[28] C.M. Lindsay and M.T. Maloney, Do rate and volume matter?, Transaction Cost Limits to Economies of Scale, Studies in Economics and Finance, **17**(1), 1–14, (1996)
[29] V. Pukelienė and I. Maksvytienė, Economy scale impact on the enterprise competitive advantages, Ekon. mastas ir jo poveikis įmonės konkurenciniams pranašumams., **57**(2), 49–54, (2008)
[30] E. Henriksen, K.H.M. Knarvik, and F. Steen, *Economies of scale in European manufacturing revisited*, Discussion Paper Series No. 2896, Centre for Economic Policy Research, London, United Kingdom, (2001)

[31] A.D. Chandler, *Scale and scope: The dynamics of industrial capitalism*, Harvard University Press, Cambridge, USA, (1990)
[32] M.E. Porter, *The competitive advantage of nations*, Billing and Sons Ltd. Macmillan, Worchester/ London, UK, (1990)
[33] P.R. Krugman and M. Obstfeld, *International economics: Theory and policy*, Sixth eds., World Student Series, United States of America, (2003)
[34] J. R. Melvin, Increasing returns to scale as a determinant of trade, The Canadian J. Economics, **2**(3), 389–402, (1969)
[35] G. A. Carlino, Productivity in Cities: Does City Size Matter?, Business Review, Federal Reserve Bank of Philadelphia, (1987)
[36] M. Zhi, G.B. Hua, S.Q. Wang, and G. Ofori, Total factor productivity growth accounting in the construction industry of Singapore, Construction Management and Economics, **21**(7), 707–718, (2003)
[37] F.C. Chia, *Economic fluctuations and productivity in the Malaysian construction sector*, PhD Thesis, Queensland University of Technology, Queensland, Australia, (2011)
[38] K.A.M. Kamar, Z.A. Hamid, M. Zura, M. Zain, A. Hazim, A. Rahim, M.K. Ghani, M. Nor, A. Azman, T.A. Majid, M. Sanusi, S. Ahamad, and M.N. Azman, Drivers and barriers of industrialised building system (ibs) roadmaps in Malaysia, Malaysian Construction Research J., **9**(1), 1985–3807, (2012)
[39] R. Ahmad, M.F. Mohammad, M.F. Musa, and M.R. Yusof, *The way forward for industrialised building system IBS*, IBS Digest, 23–29, (2015)
[40] Construction Industry Development Board (CIDB), Industrialised Building System (IBS) Roadmap 2011-2015, Kuala Lumpur, Malaysia, (2010)
[41] A. Mossman, Why isn't the UK construction industry going lean with gusto?, Lean Construction J., 24- 36, (2009)
[42] G.F. Jacobs, *Review of lean construction conference proceedings and relationship to the Toyota production system framework*, PhD Thesis, Colorado State University, Colorado, United State, (2010)
[43] Construction Industry Development Board (CIDB), Construction Industry Transformation Programme (CITP) 2016-2020, Kuala Lumpur, Malaysia, (2015)

Stakeholder's Perspective on Risks Allocation in Design and Build Projects in Malaysia

Sabihah Saaidin[1,*], *Intan Rohani* Endut[1], *Siti Akmar* Abu Samah[2], and *Ahmad Ruslan* Mohd Ridzuan[1]

[1]Faculty of Civil Engineering, Universiti Teknologi MARA, 40450 Shah Alam, Selangor, Malaysia
[2]Academy of Language Studies, Universiti Teknologi MARA, 40450 Shah Alam, Selangor, Malaysia

Abstract. Design and Build (DB) projects are prone with high risks. The appropriate risk allocation among contracting parties in Design and Build (DB) projects is an important and essential decision leading to the project performance. The establishment of risk allocation at early stage of the projects will contribute great impact to the project objective. It is required judgement and experiential knowledge of the construction experts to determine the appropriateness of risk allocation to the parties involved in DB projects. Therefore, this study has been carried out to investigate and determine the most responsible parties that should handle the risk in DB projects. A total of 170 responses from stakeholder's (Client, Consultant and Contractor) were received and analysed by using descriptive statistics and Kruskal-Wallis test. Sixty-four risk allocation criteria were used and the results of these analyses shows that 41 risks allocated to the contractor, 9 risks allocated to the client, 6 risks allocated to the consultant, and 8 risks are undecided. There were 17 risks allocations were found with significant differences in between stakeholders on their perceptions. This paper is to help stakeholders define clearly their responsibilities on risks in DB projects and make a better decision in handling it.

1 Introduction

Risks are inherent to all construction projects regardless of the size. Among a variety of risks, a contractor usually takes on many speculative risks especially in design and build (DB) projects. The Design and Build (DB) contract is an innovative procurement method whereby a single contractor is responsible for the design and construction to deliver a construction project for the client. Risk in DB project can vary an incidence between the parties as they wish. Speculative risks can be within or out of the control of a client and contractor. The suitability of design and build project approach must be carefully undertaken by ensuring the contractor is able to manage the risks. They have to be willing and have relevant experiences to control the risk with success; otherwise the contractor will pass these back to the client. To ensure all parties involved with benefit, risks must be identified, assessed and allocated so that responsibility to bear the risk is most appropriate or effective.

[*] Corresponding author: sabihah@ymail.com

The client is now attracted to DB procurement because of several reasons, mainly speed of project time completion, budget cost reduction, simplified contracting and creation of single point responsibility. At the same time, using this kind of contract the contractor is attracted to the high risk. Marques and Berg [1] mentioned that the crucial part of risk management is to identify and allocate risk during planning and construction stage. It is because the contractors in DB project still do not have enough experience and knowledge to identify, allocate and manage risk and effect to the project objective. Therefore, this paper has been carried out to investigate and determine the most responsible parties that should handle the risk in DB project.

2 Risk allocation in design and build projects

Risk allocation is an important issue in DB projects. Proper risk allocation to the contracting parties, mainly the contractor, consultant and client will respond to the best result of the project performance. Usually due to lack of information on a risk allocation system, the clients basically transfer the risk more to the contractor. Formerly, contractors respond these risks by increasing their margin of contingencies sum which the contract price would eventually increase. Generally, risk in DB project cannot be handled by one party; it should be two or share parties to that risk.

Banik and Hannan [2] stated that the allocation of risk comes from the owner and contractor or shared by both parties. A total of eight DB construction project risks was allocated to the owner of which site access/right of way, differing site condition, delayed payment on contract, unidentified utilities, permits and ordinances, change in work, Government Acts and Regulation and tax rate change and Acts of God. Furthermore, nine risks were identified among the twenty-six risks that were allocated to the DB contractors. These were namely defective materials, quality of work, safety and accidents, contractor competence, defective design, labour disputes, establishment of a project cost, actual quantities of work and defensive engineering. Meanwhile, the risk factors shared by both the owner and contractor is contract delay resolution, indemnification and hold harmless, financial failure of any party, change order negotiations and third party delay. Further research by Kartam and Kartam [3] supported the results apportionment of construction risk that also comes from owner and contractor or shared by both parties.

El-Seyegh [4] showed in previous studies that fifty-five percent of the respondents indicated that the risks shared between the client and contractor while thirty-eight percent indicated that the risk allocated to the contractor. Meanwhile, the client allocated the risk only seven percent on their responsibility. It is due to the clients in the UAE that are reluctant to share the risk with contractors. Hence, the results of this situation allow the contractor to bid with higher price in the contract that eventually increases the cost of the project.

3 Research methodology

This study captured the opinions of Malaysian construction industry which is experienced and knowledgeable on the issues of allocation of risk in DB projects. The knowledgeable can defined as a person having or showing ability an understanding and skill gained through experience by handling a complexity of the projects to solve their problems. A research questionnaire was designed to elicit information from respondents namely client, consultant and contractor. The data was received by mailing and personnel meeting. The respondents were done from Project Director, Project Manager, Contract Manager, Design Manager, Construction Manager, Project Engineer, Quantity Surveyor and Site Supervisor.

A comprehensive list of 64 risk factors was identified from the literature review and used in the questionnaire survey. Then, a questionnaire survey was conducted to investigate and determine the perceptions of construction experts on risk allocation with different parties involved in DB projects. The questionnaire was divided into two sections. The first section of the questionnaire solicited demographic information about the respondents. The second section consists of three parties (client, consultant and contractor) and who are most responsible and suitable parties that should handle the risk in DB projects. A total of one hundred seventy (170) useable questionnaires were received from the construction experts in DB projects. The results were analysed by using descriptive statistics (frequency analysis) and Kruskal-Wallis test. Kruskal-Wallis H test is a nonparametric test that can be used to determine the statistically differences between two or more groups of an independent variable.

4 Data analysis and discussion

4.1 Demographic of the respondents

One hundred seventy (170) of the structured questionnaires survey was received from the group of (client, consultant and contractor) involved directly with DB projects. To ensure the validity of the study, the information of respondents was gather from the respondents' working position and years of experience. The results show that the Project Director representing 18% of the respondents, followed by Project Manager at 27%, Construction Manager at 13%, Design Manager at 8%, Contract Manager was represented at 15%, Project Engineer at 12%, Quantity Surveyor at 4% and Site Supervisor was represented at 3% respectively.

Five categories related to the respondents' years of experience were used in this study; < 5 years experiences representing at 9% of the respondents, 6-10 years representing at 15%, 11-15 years representing at 11%, 16-20 years representing at 16% and > 20 years representing at 49% which implies that they are sufficiently experiences and knowledgeable in construction industry to answer the questions.

4.2 Allocation of risk in design and build project

Based on the survey of overall results, the risk allocations in DB project were shown in Table 1, all 64 numbers of risk allocation perceived by respondents were analysed accordingly to fulfil the objective of this study. The criterion for a risk allocation to be appropriated to a particular category (client, contractor and consultant), was that it should get at least a 50% response rate. Those that failed to get such response rate in favour of any category were listed as undecided.

According to the survey, Table 1 shows that a total of 41 risk variables had 64% out of 100% response rate allocated to the contractor. It appears that the respondents agreed that the contractor should be responsible over these risks. It is not surprisingly to note that the contractor is more responsible to the risk that allocated more than 50% of the votes, this is probably due to design and build project delivery system. The client awards with a single entity of contract that is responsible for both design and construction of works. Moreover, the DB concept of combining the designer and contractor into a single entity allows the client to shift the risks mostly to the design-builder. Banik and Hannan [2] supported in their studies that the results apportionment of risk allocated more on contractors.

The risks, lack in availability of equipment, late deliveries of materials, lack in availability of skilled labour, difficulties in availability of materials, supplies of defective materials, incompetent sub-contractors received (97.1%, 96.5%, 96.5%, 93.5%, 92.4%, 95.3%), have the highest response rate in favour of the contractor. These risks should best be controlled by the contractor and it is correct that they be allocated [7]. Kartam and Kartam [3] also agreed in their study the availability of labour, material and equipment is the highest response rate. Incompetent sub-contractor is considered with the responsibility or borne by the contractor. From the hierarchy of organization the sub-contractor is appointed by the contractor. So, the contractor is fully responsible on the performance of sub-contractor. Sub-contractor with poor performance refers to the risk of low quality, delayed completion and disputes with general contractor.

The risks, lack of information on safety at site, lack of training on safety at site, lack of coordination with sub-contractor, lack in preparing method statement or shop drawings by contractor, mistake during construction, lack on supervision of labour and works, contractors lack of staff knowledge/experience, lack of payment (delayed progress payment by contractor to the sub-contractor), and high current workload to contractor received (89.4%, 86.5%, 88.8%, 87.6%, 84.1%, 84.1%, 83.5% 82.9% 80.6%), second highest response rate in favour of the contractor. These risks are mostly related to lack in communication among construction parties. Professional team member experience and competency are considered as a key factor in project success that will also be affecting the cost planning and time management. On the other hand, the most effective way is coordinate closely with supplier and sub-contractor. Adafin [5] suggested that if the project team members were well experienced in the construction of the proposed type of project, this could reflect to the project success. It appears that the contractor now realized that there are the best persons to be responsible to handle and control these risks. The information and training should be provided by the contractor and everybody is responsible when entering the construction site. Although the contractor and client are in the best position to control and enforce compliance with safety regulations at site, safety management requires a top down approach [8].

The risks, lack of teamwork, contractor's detailed design does not meet owner's expectation, insufficient time during request for proposal to sub-contractor, lack of suitable organizational structure, lack of information in drafting request for proposal to sub-contractor and inadequate cash flow by contractor, received (77.6%, 76.5%, 75.9%, 74.7%, 72.9%, 72.9%) third highest response rate in favour of the contractor. The contractor has agreed that these risks come from the management. The contractor considered these risks are allocated to them. Thompson [6] concluded that risk in management is most valuable being implementing at an early stage of the project, which is during design and planning stage. It will be that a major risk might be minimized and avoided. Financial failure must be considered to be the most significant risk category to the contractor or client for the project's objective. Financial risk should consider be borne or responsible by both parties, namely client and contractor.

The risks, insufficient time in preparing tender document to sub-contractor, lack of design-build knowledge/experience/competencies, insufficient time for completion date, lack if quality control and assurance, lack of project manager competency and authority, unforeseen site condition, team members are not participating in technical discussion with owners, insufficient information of site access/right of way and insufficient time to evaluate tender from sub-contractors, received (69.4%, 68.2%, 67.1%, 67.6%, 65.3%, 65.3% 61.2%, 62.4% 60.6%) fourth highest response rate in favour of the contractor. Time management is an important tool for project success. Lack of time management could lead to the project failure time, cost and quality. Hence, the competency, experience and knowledge of project manager drive to project success. The owner who is capable of mobilizing more resources including finances towards site investigations and appointing competent parties to conduct the same should be responsible for the risk of unforeseen site conditions.

The risks, unidentified utilities, delay in commencing work because under-estimated time needed to obtain statutory, complexity of the project, insufficient communication between team member to meet project success, lack of standardised systems during tender evaluation, legal disputes during the construction phase among the parties of the contract, lack of communication of design-builder with end users to meet their requirements, Inflation, tax rate exchange, exchange rate fluctuation/devaluation and catastrophes, received (59.4%, 58.8%, 58.2%, 54.1%, 53.5%, 53.5%, 54.1%, 58.2%, 57.6%, 56.5%, 53.5%), the fifth highest response rate in favour of the contractor. Risks of legal disputes during the construction phase among the parties of the contract could originate due to a mistake or misunderstanding by either party. Hence, this risk should really share the risk. Both parties could hedge against them by including clauses in the contract that enable mediation and arbitration. These methods for resolving disputes are becoming popular in Malaysia as opposed to litigation. Inflation, tax rate exchange, rate fluctuation/devaluation and catastrophes should best be shared instead of being the responsibility of the contractor itself [7]. The contract clauses define the requirement and condition for sharing. These are risks where each party may be able to manage it better under different circumstances and could be specified in the contract.

Meanwhile, a total of nine risk variables had 14% out of 100% response rate should allocated to the client of which client financial capability, lack of payment (delayed progress payment by owner to the contractor), delay in design approval from client, change of original design from client, owner lack of knowledge and experience, poor supervision by the client, change in government policy, change in government regulation and law and political uncertainty. It appears that the clients have realized the futility of shifting the onus of responsibility to the contractors and are ready to take a more proactive role in the management of such risk factors. Poor management may cause risks pertaining to negotiation. The client is in a better position to manage it by specifying clear and fair contract terms and clear workable procedures.

A total of six risk variables had 9% out of 100% response rate should be allocated to the consultant that lack in effectiveness of communication in design, errors of original design, insufficient of original design, insufficient information in contract document, inadequate specification in contract document. The respondents had agreed that the consultant could best manage design, specifications and contract document. Designers may rush to complete the design phase with limited time to the client to start construction early may include a lot of mistakes and defective design occurs. This risk is also related to the quality of design and specification produced by the design professional. This issue might continue during construction phase. The professional designers should improve their design responsibility to fix any deficiency at early planning stage

Lastly, a total of 8 risk variables had 13% out of 100% response rate should be allocated to undecided. The respondents who respond to the survey were undecided on the allocation of the following risks; scope of work is uncertainty, actual quantities differ from the contract quantities, redesign because of over budgeted, bureaucracy in government agencies, slow approval permit by BOMBA department, slow approval permit by local authorities, slow approval permit by town planning department and slow approval permit by public work department. Risks pertaining to permits approval should be shared, with client and contractor [7]. Risk of delay in obtaining permit and approval from the government agencies is categorised as unexpected and avoidable. The good relationship between government and construction parties will run project smoothly. These risks should manage and avoid at early stage of the project by early planning on schedule.

Table 1. Risk allocation in design and build projects.

Risk Description	Risk Allocation (%)			Kruskal – Wallis Test	
	Client	Consultant	Contractor	Chi-square	p-value
Inadequate cash flow by contractor	24.1	2.9	72.9	5.993	0.050
Client financial capability	73.5	1.8	24.7	5.364	0.068
Delay in commencing work because under-estimated time needed to obtain statutory	14.1	27.1	58.8	5.802	0.055
Mistake during construction	2.9	12.9	84.1	1.462	0.481
Lack of payment (delayed progress payment by owner to the contractor)	57.6	2.9	39.4	5.692	0.058
Lack of design/build knowledge / experience / competency	15.9	15.9	68.2	1.939	0.379
Insufficient time for completion date	21.2	11.8	67.1	4.587	0.101
Lack in quality control and assurance	4.1	28.2	67.6	3.193	0.203
Lack of project manager competency and authority	23.5	11.2	65.3	10.786	0.005*
Contractors lack of staffs knowledge / experience	15.3	1.2	83.5	3.246	0.197
Lack of payment (delayed progress payment by contractor to the sub-contractor)	14.7	2.4	82.9	4.056	0.132
Incompetent sub-contractors	4.1	0.6	95.3	4.586	0.101
Lack of teamwork	13.5	8.8	77.6	3.679	0.159
Contractor's detailed design does not meet owner's expectation	8.8	14.7	76.5	0.337	0.845
Lack of coordination with sub-contractor	2.9	8.2	88.8	13.500	0.001*
Complexity of the project	20.6	21.2	58.2	1.820	0.403
Lack of communication of design/builder with end users to meet their requirements	9.4	38.8	51.8	7.253	0.027*
Lack in effectiveness of communication in design	5.3	51.8	42.9	5.223	0.073
Supplies of defective materials	5.9	1.8	92.4	0.579	0.749
Lack in availability of skilled labour	1.8	1.8	96.5	2.118	0.347
Errors of original design	22.4	62.4	15.3	0.931	0.628
Unforeseen site condition	15.3	19.4	65.3	2.485	0.289
Lack of training on safety at site	5.9	7.6	86.5	15.273	0.000*
Lack on supervision of labour and works	4.7	11.2	84.1	10.262	0.006*
Insufficient communication between team member to meet project success	19.4	26.5	54.1	7.443	0.024*
Late deliveries of materials	2.4	1.2	96.5	2.440	0.295
Lack of information on safety at site	2.9	7.6	89.4	26.323	0.000*
Scope of work is uncertainty	27.6	25.9	46.5	1.197	0.550
Unidentified utilities	16.5	24.1	59.4	1.596	0.450
Insufficient of original design	25.9	58.8	15.3	2.486	0.288
Team members are not participating in technical discussions with owners	12.9	25.9	61.2	11.725	0.003*
Lack in availability of equipment	1.8	1.2	97.1	0.802	0.670
Difficulties in availability of materials	3.5	2.9	93.5	0.778	0.678
Delay in design approval from client	52.9	15.3	31.8	4.907	0.086
High current workload to contractor	11.8	7.6	80.6	3.369	0.186
Insufficient information in contract document	24.1	60.6	15.3	3.109	0.211
Inadequate specification in contract	22.4	60.6	17.1	0.218	0.897

document					
Insufficient information of site access/right of way	21.2	16.5	62.4	1.411	0.494
Lack of suitable organizational structure	17.6	7.6	74.7	1.414	0.493
Insufficient time to evaluate tenders from sub-contractor	17.6	21.8	60.6	9.218	0.010*
Change of original design from client	52.4	35.9	11.8	4.297	0.117
Legal disputes during the construction phase among the parties of the contract	27.6	18.8	53.5	2.456	0.293
Lack in preparing method statement or shop drawings by contractor	2.4	10	87.6	0.619	0.734
Exchange rate fluctuation/devaluation	34.1	9.4	56.5	6.340	0.042*
Slow approval permit by BOMBA department	21.8	47.6	30.6	4.555	0.103
Lack of information in drafting request for proposals to sub-contractor	5.9	21.2	72.9	4.393	.111
Insufficient time in preparing tender documents to sub-contractor	7.6	22.9	69.4	9.235	0.010*
Actual quantities differ from the contract quantities	14.1	40	45.9	1.622	0.444
Slow approval permit by local authorities	32.4	38.8	28.8	2.376	0.305
Slow approval permit by town planning department	22.9	45.3	31.8	4.932	0.085
Redesign because of over budgeted	47.1	34.7	18.2	0.109	0.947
Owner lack of knowledge and experience	62.4	4.1	33.5	9.334	0.009*
Insufficient time during request for proposal to sub-contractor	9.4	14.7	75.9	8.934	0.011*
Bureaucracy in government agencies	40	20.6	39.4	4.026	0.134
Slow approval permit by public work department	25.3	43.5	31.2	3.515	0.172
Catastrophes (Act of God)	38.2	8.2	53.5	7.832	0.020*
Poor supervision by the client	59.4	14.1	26.5	9.794	0.007*
Inflation	37.6	4.1	58.2	5.370	0.068
Lack of standardised systems during tender evaluation	17.1	29.4	53.5	22.178	0.000*
Tax rate exchange	35.3	7.1	57.6	5.574	0.062
Change in government policy	55.9	20	24.1	6.361	0.042*
Change in government regulations and law	53.5	19.4	27.1	4.984	0.083
Rigid specifications in contract document	27.6	57.1	15.3	4.028	0.133
Political uncertainty	64.1	8.2	27.6	3.930	0.140

*Indicates that the p-value is less than 0.05

A further analysis was conducted to determine the significance in the differences in the perception of all three parties on risk allocation. Hence, Kruskal-Wallis H test was conducted to determine whether or several independent items come from the same population. The results of the Kruskal-Wallis test for a k-independent sample shows that the respondent groups had different perceptions on 17 out of 64 variables at the 5 per cent ($P<0.05$) significance level, as shown in Table 1; Namely lack of Project Manager Competency and authority, lack of coordination with sub-contractor, lack on supervision of labour and work, insufficient time to evaluate tenders from sub-contractor, insufficient time in preparing tender document to sub-contractor, insufficient time during request for proposal to sub-contractor, lack of standardised systems during tender evaluation, Exchange rate fluctuation/devaluation, catastrophic (Act of God), change in government policy, lack of communication of design-builder with end user to meet their requirements, insufficient communication between team members to meet project success, team members are not participating in technical discussions with owner, Change in government policy, lack of information on safety at site and lack of training on safety at site. With these findings the stakeholders play an important role to improve the ability to identify, allocate and manage risk at early stage of the project design and planning. Kartam and Kartam [3] indicated that the method of transfer and share risks is considered to be ineffective for preventing risks. As a matter of fact, the client or general contractors could exercise to practice of shifting the risks to lower hierarchy of organization.

5 Conclusions

Risk allocations in DB projects are presented. The results have shown that the contractor is the most responsible parties that should handle the risks inherent in DB projects. Namely, the respondents have allocated 41 risks to the contractor, 9 risks to the client, 6 risks to the consultant and 8 risks undecided. While, the results on Kruskal-wallis test shows 17 risks considered as shared risk on respondents perceptions. The clients have realized the futility of shifting the onus of responsibility to the contractors and are ready to take a more proactive role in the management of such risk factors. Risk in design and build projects cannot be entirely eliminated, but it can be better managed by being appropriately allocated. Literally, the contracting parties must be collaborating to reduce risk by providing as much as possible information to the member of the DB at the beginning of the project design and planning. This can be accomplished by providing detailed information on project stakeholders, objective, characteristics, scope and context. Proper risk allocation will help stakeholders to minimize the occurrence of risk and dispute among contracting parties in DB projects.

References

[1] R. Marques and S. Berg, Risks, contracts and private-sector participation in infrastructure. J. Constr. Eng. Manage., 925-932, (2011)

[2] G.C. Banik and F. Hannan. Specialty contractors' perspectives on risk importance and allocation of design-build contract, J. of Construction, 1-11, (2003)

[3] N.A. Kartam and S.A. Kartam, Risk and its management in the Kuwaiti construction industry : A contractors' perspective, Int. J. of Project Management, **19**(6), 325-335, (2001)

[4] S.M. El-Sayegh and M.H. Mansour, Risk assessment and allocation in highway construction project in the UAE, J. Management Eng., 04015004/1-11, (2015)

[5] J. Adafin, J.O.B. Rotimi and S. Wilkinson, Risk impact assessments in project budget development: Architects' perspectives, Architectural Engineering and Design Management, 1-16, (2016)

[6] P.A. Thompson and J.G. Perry, *Engineering construction risks: A Guide to project risk analysis and assessment implications for project client and project managers*, Thomas Telford, London, (1992)

[7] R.J. Smith and W. Gavin, *Risk identification and allocation saving money by improving contracts and contracting practices*, The ASCE Hong Kong International Group and Chartered Institute of Arbitrators, Hong Kong, (1998)

[8] D. Zhao, A.P. McCoy, B.M. Kleiner, TH. Mills and H. Lingard, Stakeholder perceptions of risk in construction, J. of Safety Science, **82,** 111-119, (2016)

A Qualitative Comparison on Guidelines for Construction Workers Accommodation and Facility

Norasyikin Khamis[1,*], *Azeanita* Suratkon[1], *Hairuddin* Mohammad[2], and *Siti Khalijah* Yaman[1]

[1]Faculty of Civil and Environmental Engineering, Universiti Tun Hussein Onn Malaysia, Johor, Malaysia
[2]Faculty of Architecture, Planning and Survey, University of MARA Technology, Shah Alam, Malaysia

Abstract. Construction industry in Malaysia covers almost 1.3 million workers. Thus, realizing the importance of human-centred trait, a long-term plan that is led by the Construction Industry Development Board Malaysia (CIDB) has been designed to improve accommodation and facilities of workers at construction site. In conjunction to that, the Institute for Industrial Research and Standards Malaysia (SIRIM) has developed Malaysia Standard (MS 2593:2015) as the guideline for improvement of accommodation and facilities for workers which is able to contribute to the achievement of the target in the Construction Industry Transformation Programme (CITP). However, a phenomenon that is centred on the level of itscorroboration compared with other similar guidelines has been voiced by some practitioners of construction, especially when the guideline are likely to be the basis of a new related act. Therefore, based on MS 2593:2015, this introductory research is conducted to qualitatively analyse the corroborative level of the guideline compared to its counterpart guidelines. By using a comprehensive document analysis method which anchored to frequency approach and assisted by NVivo v.8 software, the report found out that the Malaysia-based guideline is on par to the restwith some introduction of new variables (especially towards leisure and social aspects). Thus, further research on gauging industry's willingness and existing compliance rates is deemed appropriate in order to inculcate positive impacts towards the objectives of MS 2593:2015, and subsequently contributes to the formation of the forthcoming corresponding act.

1 Introduction

According to Dato' Sri Haji Fadillah Bin Haji Yusof, Minister of Work, the industry has contributed to double-digit growth and maintaining its momentum until recently as new projects and interests were steadily in place [1]. Apart from becoming an important catalyst

*Corresponding author: azeanita@uthm.edu.my

for the nation, its positive spill over effects have nurtured an influx of foreign worker forces in supporting the needs of construction industry [2]. Up to this moment, there are almost 2.1 million workers (minus management staff) pinning on the industry, which large portion of them come from foreign countries [3]. Following the circumstance, providing adequate accommodations and facilities for the workers is paramount, especially towards the responsible parties[4].

In Workers Act 446, those who are responsible for the employees must provide comfortable and adequate accommodations and facilities even if they are foreign workers [5]. Additionally, another similar tone was observed from the Board of Inquiry Act 1990, where responsible parties must ensure that the accommodation of workers has to meet the minimum requirements in accordance with set of standards that is currently available [5]. Lending support to the previous acts, according to a statement made by the Director of Strategic Communications and Public Relations of MRT Corp; Mahmood Abdul Razak, providing systematic and comfortable accommodations and facilities of foreign workers is able to overcome the social problems that currently exist in the local community, which particularly related to problem with drugs within foreign construction workers [6]. Furthermore, issue of accommodation and facilities of construction workers had become a highlight in the recent Construction Industry Transformation Programme (CITP), where improvisation under the purview of CIDB is repeatedly mentioned in order to meet demands of future economy [1].

Thus, noting on the importance of having proper guideline, an effort by the Department of Public Works (JKR) are visualised through their Standard Specifications for Building Works. The exact document describes briefly the need for workers' necessities which can be found in Section A (Preliminary and General Requirements), within a specific Article 28 (Contractor's Temporary Accommodation and Facilities for Workmen Living on Site [7]. Inside, it lists thirteen needs that basically determine the type of material used for workers' accommodation, size of the building and space allocation, amenities and basic facilities to be provided, and person's responsible for keeping and maintaining a list of employees and others shared accommodation sites [7]. Parallel to JKR's guideline, recent publication by the Institute for Industrial Research and Standards Malaysia (SIRIM) on Malaysia Standard of Temporary Construction Site Workers' Amenities and Accommodation (MS 2593:2015) has further strengthened existing measures on national scale. However, since this is a newly published standard, scarcity of in-depth post-study pertaining the interoperability of the standard is very much profound.

These issues do not only lie on the comprehensiveness of the standard per se, but how the standard will behave towards its actual implementation in the construction industry, especially by the imposing organisation (i.e. contractors). Along with future moves by the authorities that intended to gazette an act out of the MS 2593:2015 [4], collective findings are indeed needed to encapsulate their corroborative level, preparedness level, and current tangible milestones. Significant gaps might emanate across the findings which subsequently lend support for any betterments, together with profound constructive spill over effects towards holistic strategies to implement the particular standard. Therefore, in this precise paper, the authors were intrigued to answer the question of 'what?' regarding its corroborative level, which generally attached to the particular standard's fared among its counterparts.

2 Construction site

According to Business & Group (2012), construction sites can be classified asa project's place where there are many dangerous moving machineries and equipment, and contain construction works such as excavation, piling, etc. In other words, it is a place where physical construction works are commenced and utilisation of 4M's (i.e. manpower, machines, materials, and money) is abundance, along with properly placement of hoarding to differentiate its boundary. Therefore, apart from being a workplace for incoming generation [8], compliances with safety requirements are essential in construction site, including the application of safety helmet, safety shoes, and others, where priority is given to safety and health of workers[9].

There are generally three (3) types of work involved in a building concept which are minor construction, major construction, and off-site construction [10]. Minor construction is normally at fixed locations having less than ten employees at a time, with construction period of less than two years. Typically referred as small scale construction, these clusters include newly build building or renovation exercise. On the other hand, major construction is classified under fixed location where there are more than 10 employees working for two consecutive years. Some of the examples are an engineering construction of dam and construction of new housing scheme. Meanwhile, off-site construction is normally categorised as automation-led construction majorly done in large factories, where often associated with prefabricated building components.

3 Construction worker

By and large, construction worker is defined as fully utilisable human resources [11]. According to the Employment Act 1955, Section 2 (1), domestic worker is a worker who is the citizen of this country, and employed with salary as a reward for their contract of service [12]. Indirectly, those definitions are no different to majority of foreign construction workers (e.g. Indonesians, Bangladeshis, Burmese, Vietnamese, etc.) except for their citizenship which in a way differentiate them to indigenous construction workers. Their presence in this country is particularly due to the need of industry in coping with rising number of projects, and to a certain extent, plugging serious lacking of Malaysians workers faced by the construction industry [13]. Although prioritisation and recruitment efforts were given to local workers before any initiative of procuring foreign workers were made[14], issue on shortages of worker is still persisted, hence influxes of them were inevitable [15].Apart from that, workers in the Malaysia's construction industry can be categorised into three (3) broad types, namely; skilled worker, semi-skilled worker, and unskilled worker [14]. However, only semi-skilled and unskilled worker are classed as foreign workers [16].

Skilled workers are defined to be efficient, with exceptional skills, and able to use newer equipment to carry out entrusted works [17]. Basically, five (5) to ten (10) years is required for a person to achieve the status. Meanwhile, semi-skilled workers are defined as workers who have skills in one or more branches of construction trades, but yet to reach similar skills 'level that of skilled worker[17]. Typically, they become assistants to skilled worker, nonetheless there are times when they can produce quality workmanship on par with skilled worker especially when their accumulated years of service are higher. Working based on past experiences and coupled with average skills, semi-skilled foreign workersare commonly associated with the one constantly within the industry and have been residing in Malaysia for minimal period. Last but not least, workers without any skill in any branch of construction trades are called unskilled workers. They usually work within construction site for basic works such as transporting materials, housekeeping, and other general works [17]. Considered as newcomers, unskilled foreign workers are made up of foreigners who came to Malaysia without any basic learning and systematic skills in construction [16]. Normally, they learned by examples of everyday chores, given that proper continuous advanced training is scarce.

4 Worker's accommodation and facility in construction site

Temporary accommodation provides fully equipped shelter to employees who are away from their homes, in a form of comfortable room, dining room, common living room, etc. Meanwhile facilities include communication facilities, recreational supports, medical service, efficient security, etc.[4]. Temporary accommodation and facility should be operated by facility contractor specialist who provides and maintains its standard of care and responsibility through appropriate facility manager. Every effort needs to be executed to reduce the side effects of temporary accommodation and facilities of the area and the surrounding people[4]. Hence, based on typical considerations for initial commencement of such respective move, there are several facets of characteristics need to be pondered, namely; type of workplace, work-related location, type of work to be carried out, number of employees in the workplace, party in charge of work and accommodation, working period(i.e. standby, shifts, etc.), and availability of disable/unfortunate worker [8].

5 Malaysian standards of temporary construction site workers' amenities and accommodation - code of practice (MS 2593:2015)

MS 2593: 2015 was spearheaded by SIRIM since 2014 and was published in December 2015 [4]. The objective is to provide official explanations of the needs for temporary facilities, which include construction workers' welfare, and safety and health of the employees which are deemed appropriate and organised. Furthermore, it also provides reference and guidance towards the provisions in providing reasonable temporary conveniences that are able to be implemented. In MS 2593: 2015, Malaysia is targeting for better accommodation and facilities to the workers at construction site due to various benefits, which through the end will pave direction for respective gazetted act. Within this document, thorough guidelines were presented, where it was divided by ten (10) sections with corresponding sub-sections. For instances; room (with arrangement of beds, wardrobes, lighting, and natural ventilation), sanitary facilities (with toilets, bathrooms, etc.), kitchen and dining place, water supply, toilet (including proper discharged of used water, sewage, and solid particles), drainage, electrical generation, medical facilities and first aid, social and recreational facilities, and signboards availability. Additionally, the guideline is also described in detail on the substantive needs that optimally should be provided in each convenience given.

6 Standards/Guidelines from other countries

There are several related standards from other countries namely Australia, Britain, and the United States of America (USA). Thus, this topic is presented to briefly explain about the particular standards. Standard practice code for Australian sets the practical guidelines in order to achieve the required level of safety at the workplace. Under their safety act (Safety Act, 2008), specific reference to minimum standards of facilities for people involved in construction work is provided [8]. The approved code is designed to be used along with the Work Safety Act and Regulations, but does not have similar legal force where a person or company cannot be prosecuted if they fail to abide to this practice code. The code should be followed unless there are alternative courses of action which able to achieve either similar or better implementation-wise of safety in the workplace. Apart, the code covers guidelines in providing facilities which are based on approved budget which then narrowed down into general provisions, exchanged rooms, dining places, toilet and sanitary, washing place, water supply, and personal closet. Additionally, according to the Australia's Office of Regulatory Services (ORS), relevant inspector can approve additional codes that are relevant to any improvement, or may release Prohibition Notice (PN) to indicate appropriate steps that should be taken to rectify the violation or non-compliance with current acts or the rules of safety [10].

On the other hand, a Guidance Note by the British Councils aimed at providing practical guidance to International Finance Corporation (IFC) (a member of the World Bank Group) and the European Bank for Reconstruction and Development (EBRD) specialists, consultants, and clients on the processes and standards that should be applied to the provision of workers' accommodation in relation to projects funded by them [18]. Running through the particular guideline, applying appropriate standards to the construction and operation of worker housing falls under the performance requirements on worker and also expected working conditions. The Guidance Note also provides examples of good practice approaches that businesses have successfully applied in their past operations. Additionally, a range of different types of workers' accommodation that may be required by various projects and at different stages within projects was documented, these include temporary exploration camps, construction camps, and permanent dormitories. Likewise, the guideline took an initiative to review various international, national, private sector and public sector standards, and other generally applicable guidance. Besides, the guideline promotes implementation and compliances with at least expected minimum benchmark [18].

Separately, a standard provided to Occupational Safety and Health Administration (OSHA) of United States of America aimed to assure safe and healthful working conditions for workers [33]. Apart from delivering guidance focused on workers' accommodation and welfare in construction setting, the said standard (1910.142) became a critical basis for setting up and enforcement of multiple betterment approaches such as training, outreach, education, and any related assistances. Published on 1988 by the United States Departments of Workers, the standard outlaid guidance on implementation of appropriate facilities or conditions which mostly fond towards safety and health of employees, regardless of occupying period in the forms of proper site drainage, shelter, water supply, toilet, laundry facility, handwashing and bathing conveniences, first aid amenities, refuse disposal, etc.

7 Qualitative comparisons on several standards based on Malaysia's MS 2593:2015

Comparison between different documents in term of their latent and manifest values often requires qualitative undertakings. Thus, due to answer prior research question, the authors attempted to impart both values; latent through variables' mapping processes of similar meanings among documents (regardless of sentence's or description's length), and manifest by variables' existence and frequency summary. Given the fact that manual qualitative comparison among the MS 2593:2015 and other standards or guidelines are prone to inconsistency, usage of available software (i.e. NVivo Ver.8) is considered useful. Steered by the works of several previous researchers, namely; Gibbs [19], Richards [20], Grbich [21],Wiltshier [22], Kariya et al. [23], and Ariffin et al. [24], the authors embarked on qualitative analysis based on the following diagram (Fig. 1) which is further described accordingly.

There are 17 variables in MS2593: 2015 which cover the entirety of construction workers' accommodation and facility's segments. These variables were transferred as nodes (using short abbreviations) in Nvivo as shown in Fig. 2. Next, coding exercises of essential words/texts from the MS2593: 2015 were conducted, originally imported into NVivo in MS Word's format, and coded under their respective nodes. Therefore, an appropriate amount of specific text(s) that has been coded to a specific node will reflect the final percentage of the node.

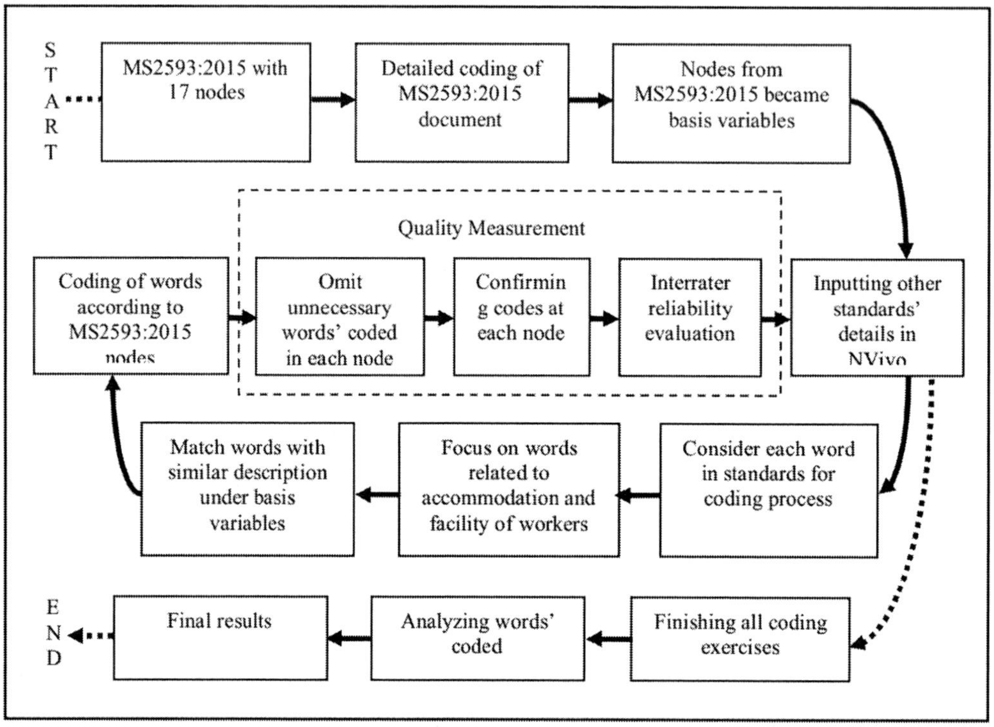

Fig. 1. Overall looping processes through NVivo

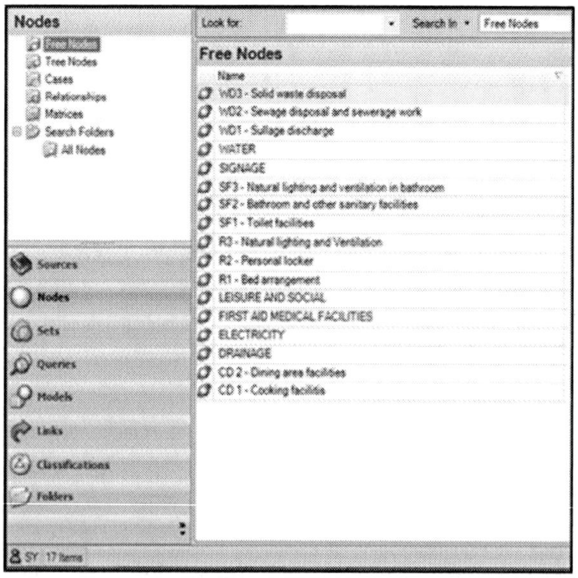

Fig. 2. Screenshot of nodes created in Nvivo based on MS2593: 2015.

Finally, nodes will be the basis of reference for consecutive coding exercises. Besides, guideline documents' obtainment generally relied on online sources, where completeness is of major consideration apart from their suitability. In this preliminary study, the authors were successful in gathering applicable guidelines issued by other countries, namely; the United States of America (1910.142), Britain (Guidance Note), and Australia(Safety Act, 2008). The following figure (Fig. 3) shows the inputted guidelines' folders in NVivo, along with their corresponding documents. Beforehand, importing other documents can be done either through MS Word's or PDF's format. As an internal source in NVivo (Fig. 3), commencement of documents' coding exercises of words is based on predetermined nodes. The process takes all relevant words/texts and was entered into their respective nodes without any editing work.

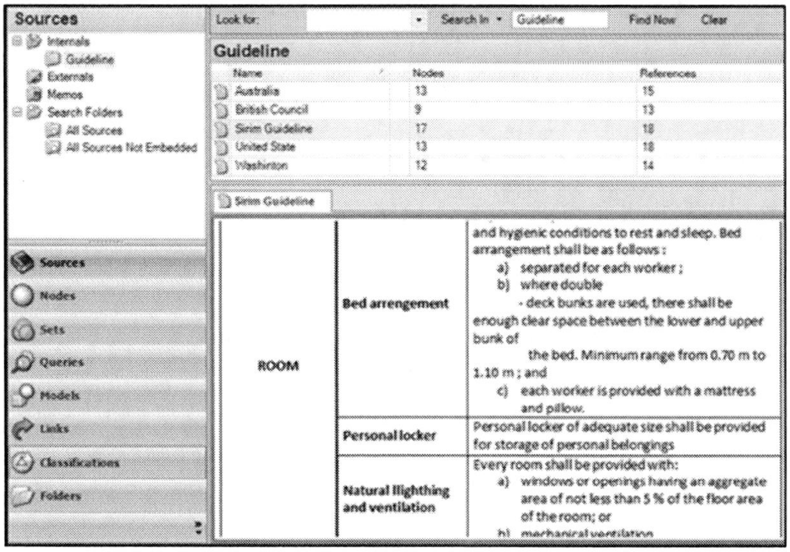

Fig. 3. Guidelines' folder and an example of document's format in NVivo.

7.1 Embedded quality measure

In order to maintain concise and reliable coding exercises, all codes in every single node were scrutinised once more where unnecessary codes were removed (i.e. codes that do not contribute to overall meaning of node). Then, interrater reliability assessment is conducted, which is able to be seen as an indicator for trustworthiness of the coding processes and might as well contributes to its validity [25]. Additionally, the assessment of coding consistency towards either two or more coders is aimed at identifying the reliability for inner coding processes [26 - 28]. It is a crucial evaluation in order to verify reproducibility of coding through previously identified coding steps (Fig. 1). Therefore, driven by the works of Lacobucci [29], Gwet [26], Graham et al. [27], Freelon [30], Hallgren [28], Roberts [31], the research embarked on the particular assessment by selecting a second coder (i.e. senior construction's lecturer). Initially, the judge's coder was briefed and closely recruited for the process towards one document, namely; the US standard (1910.142). The results (Table 1) were in a form of dichotomously coding presence between the two coders within all nodes (i.e. 1 for Yes and 0 for No).

From the analysed data, percentage agreement and Cohen's Kappa are selected as appropriate analysis for coding presence among two coders. As consensus estimates towards nominal data, percentage agreement between two coders is able to evaluate agreement of construct's interpretation [32,37,30]. On the other hand, the application of mathematical improvements for chance agreement between coders through Cohen's Kappa is considered more computationally robust [30, 32, 29, 26, 27]. The analysis was helped by a web-based software created by Freelon [30], and was compared to Graham's rule of thumb (Table 2)[27]. Finally, the results as in Table 3 show that percentage agreement of 88.2% is nearing high benchmark, while Cohen's Kappa of 0.679 is just slightly above the suggested minimum level. Although both results outlaid acceptable outcomes, lower Kappa's however might indicate certain shortcomings during recruitment processes towards the judge's coder. Nevertheless, it is deemed that coding exercises through NVivo are sufficiently reproducible without having to assess all documents [25, 32].

Table 1. Result for coding presence of two coders towards similar document of Washington standard.

Basis Variables	Judge's Coder	The Authors
Cooking Area	Yes	Yes
Dining Area	Yes	Yes
Drainage	No	No
Electricity	Yes	Yes
First Aid Medical Facilities	No	Yes
Leisure And Social	Yes	No
Bed Arrangement	Yes	Yes
Personal Locker	Yes	No
Natural And Ventilation	Yes	Yes
Toilet Facilities	Yes	Yes
Bathroom /Sanitary Facilities	Yes	Yes
Lighting And Ventilation	No	Yes
Signage	Yes	No
Water	Yes	Yes
Sullage Discharge	Yes	No
Sewage Disposal	Yes	Yes
Solid Waste	Yes	Yes

Table 2. Rule of thumb for interpreting interrater result [27]

Interrater reliability	High	Minimum
Percentage agreement	90%	75%
Cohen's Kappa	.81	.61

Table 3. Interrater reliability assessments' results

Interrater reliability	Agreement (%)	Cohen's Kappa	N Agreement	N Disagreement	N Cases	N Decisions
Between coders (i.e. Judge & Authors)	88.2%	0.679	15	2	17	34

7.2 Analysis and extractions

While the latent part was already being partly embedded in the previous section, the final step is the process of analysis and extraction to achieve study's objective. Analysis was divided into several parts including the frequency and percentage of words coded, and their intensities. Those parts are considered important due to the fact that since the integral meaning of numerous codes were dedicated to its particular nodes, subjective issue on similarities of words'/texts' usage is not a concern in this preliminary study, but their comprehensiveness in visualising each node through writings is of major objective. Coupled with detail comparisons between currently applied international standards, therefore, the authors were positive towards exploring the applicability of MS2593: 2015.

To begin with, existence of codes signifies certain consideration that had been given for a particular node, while frequency dictates their comprehensiveness among other documents. As shown in Table 4, it is no surprise that MS2593: 2015 dominate the dichotomous column of codes' existence since they become basis reference of nodes. Following closely are the Australian 2003 and the United States 1910.142 where both have 13 positive nodes. The least can be observed from the British Council 2009 with only 9 nodes. On the other hand, MS2593: 2015 topped the list of frequency of words coded for a single document, with only a difference of ten (10) codes separate Malaysia standard (1043 codes) with the US's (1034 codes). Finally, quite significant gaps persist for the rest of standards in term of their codes' frequency, namely; the Australian 2003 (663 codes) and the British Council 2009 (404 codes). Furthermore, Table 4 also highlighted (i.e. in deeper colour) some variables which has recorded higher frequency of codes for certain variables in few documents. Indirectly it is the evidence that supports particular documents' concentration on workers accommodation and facility's variable. Apart, lighter colour was applied for less concentrated variables.

Table 4. Existence of code(s) and frequency of words coded in every nodes

Basis Variables	**Australia, 2003 [8]**		**British Council, 2009 [18]**		**Malaysia MS2593: 2015 [4]**		**United States, 1910.142 [33]**	
Cooking Area	Yes	32	No	0	Yes	60	Yes	99
Dining Area	Yes	12	No	0	Yes	92	Yes	209
Drainage	Yes	42	No	0	Yes	42	Yes	42
Electricity	Yes	90	No	0	Yes	29	Yes	90
First Aid Medical Facilities	Yes	50	No	0	Yes	17	Yes	50
Leisure and Social	No	0	No	0	Yes	80	No	0
Bed Arrangement	Yes	63	Yes	68	Yes	72	Yes	128
Personal Locker	No	0	Yes	22	Yes	13	No	0
Natural and Ventilation	Yes	69	Yes	35	Yes	28	Yes	64
Toilet Facilities	Yes	38	Yes	68	Yes	262	Yes	31
Bathroom /Sanitary Facilities	Yes	169	Yes	22	Yes	79	Yes	14
Lighting and Ventilation	No	0	Yes	20	Yes	49	Yes	78
Signage	No	0	No	0	Yes	77	No	0
Water	Yes	35	Yes	90	Yes	62	Yes	137
Sullage Discharge	Yes	21	No	0	Yes	25	No	0
Sewage Disposal	Yes	21	Yes	16	Yes	28	Yes	19
Solid Waste	Yes	21	Yes	63	Yes	28	Yes	73
Frequencyof 'Yes" from 17 nodes	13		9		17		13	
Frequency of words coded	663		404		1043		1034	

Next, by variables, intensities of codes were tabulated according to its percentages in horizontal view. The move was intentionally done in order for the authors to further identify emphasis of each standard on observed variables. Based on Table 5, Malaysia's MS2593: 2015, and the US1910.142 was recorded similar achievement for highest percentage of concerning nodes but with dissimilar themes. Specifically; MS2593: 2015: drainage, leisure and social, toilet facility, signage, sullage discharge, and sewage disposal; and the US 1910.142: dining area, drainage, bed arrangement, lighting and ventilation, water, and solid waste facility. Here, all standards showed a number of analogous percentages, precisely for the provision of drainage at workers accommodation, except for the British Council 2009. Nevertheless, for MS2593: 2015, interestingly, there are two (2) variables that achieved full percentage since they are not currently being offered by other standards, namely; leisure and social, and signage. Apart, the remaining standards (i.e. the

British Council 2009) posted lower achievement for highest percentage accurately for the provision of personal locker.

Table 5. Percentage of codes in every variable (i.e. nodes)

Basis Variables	**Australia, 2003 [8]**	**British Council, 2009 [18]**	**Malaysia MS2593: 2015 [4]**	**United States, 1910.142 [33]**	**Total horizontal percentage**
Cooking Area	9.5%	-	17.8%	29.38%	100%
Dining Area	2.61%	-	20.04%	45.53%	
Drainage	33.33%	-	33.33%	33.33%	
Electricity	29.9%	-	9.63%	29.9%	
First Aid Medical Facilities	29.76%	-	10.12%	29.76%	
Leisure and Social	-	-	100%	-	
Bed Arrangement	13.73%	14.81%	15.69%	27.89%	
Personal Locker	-	62.86%	37.14%	-	
Natural and Ventilation	26.54%	13.46%	10.77%	24.62%	
Toilet Facilities	8.19%	14.66%	56.47%	6.68%	
Bathroom /Sanitary Facilities	42.36%	5.51%	19.8%	3.51%	
Lighting and Ventilation	-	9.3%	22.79%	36.28%	
Signage	-	-	100%	-	
Water	7.59%	19.52%	13.45%	29.72%	
Sullage Discharge	45.65%	-	54.35%	-	
Sewage Disposal	20.19%	15.38%	26.92%	18.27%	
Solid Waste	8.14%	24.42%	10.85%	28.29%	
Achievement for highest percentage (darker boxes)	3	1	6	6	

8 Conclusion and way forward

The Malaysia standards (MS2593: 2015) which have 17 noticeable variables on workers accommodation and facility were tested and corroborated towards several other counterpart countries' standards. In seeking its corroborative levelwithout undergoing real life testing, qualitative approach through document analysis was conducted. A thorough steps as in Fig. 1 was implemented to maintain reliable outcomes, backed by positive results of interrater reliability assessments (i.e. percentage agreement and Cohen's Kappa). In a nutshell (i.e. based on Table 4 and Table 5), the authors posited that the MS2593:2015 is on par with other distinguished standards especially from the US 1910.142. Although the MS2593:2015 is seen as focusing on certain variables (i.e. drainage, toilet facility, sullage discharge, and sewage disposal), an introduction of new variables of leisure and social, and signage is applauded to give better values towards construction workers. Indirectly, those initiatives might support government efforts in recruiting more local talents to be involved in the particular industry. Nevertheless, given the fact that MS2593:2015 is comparatively a new guideline, implementation-wise needs to consider collective actions especially towards the enforcement (e.g. the policymaker) and application body (e.g. the contractor). Thus, the authors concluded, findings from the paper has attested the corroborative level of Malaysia's MS2593:2015 which deemed beneficial for the author's subsequent undertakings, and to a certain extent providing an insight towards related stakeholders. However, further research on feedbacks from industries stakeholders, and ground observation of current practices and offerings in term of workers accommodation and facility are inevitable before any subsequent initiatives are introduced (e.g. gazetting act).

The authors would like to thank the Universiti Tun Hussein Onn Malaysia for their assistances throughout the introductory phase of research and sponsoring the inaugural paper.

References

[1] Construction Industry Development Board (CIDB) Malaysia, *Construction Industry Transformation Programme (CITP) 2016-2020*, No. 1. Malaysia: Construction IndustryDevelopment Board (CIDB) Malaysia, (2015)
[2] G. Ofori, Nature of the construction industry , its needs and its development, A Review of Four Decades of Research, **20**(2), 115–135, (2015)
[3] Pook Ah Lek, *The dilemma of having foreign workers in Malaysia*, Sin Chew Daily, Malaysia, (Sept. 2016)
[4] Department of Malaysia Standard, *Temporary construction site workers' amenities and accommodation - Code of practice MS 2593:2015*. Malaysia: Department of Standards Malaysia, (2015)
[5] Lembaga Penyelidikan Undang-undang, *Akta Standard-Standard Minimum Perumahan dan Kemudahan Pekerja,1990 (AKTA 446),* Malaysia, (1990)
[6] I. Rozlin and M.H Ismail, *Kuarters tangani isu pekerja asing*, Sinar Online, (July 15 2015)
[7] Jabatan Kerja Raya Malaysia, *Standard Specifications for Building Works 2014*. Malaysia: Jabatan Kerja Raya Malaysia, (2014)
[8] Government of South Australia, *Guidelines for Workplace Amenities and Accommodation*. Australia: WorkCover Corporation, (2003)
[9] B.Y. Jeong, S. Lee and J. D. Lee, *Workplace Accidents and Work-related Illnesses of Household Waste Collectors,* Saf. Health Work, (2015)
[10] Department of Justice and Community Safety, *Australian Code Of Practice Rail Safety Investigation*, Australia: Department of Justice & Community Safety, **2**, (2010)
[11] Y. Md Junus, *Kajian hak pekerja di tapak bina mengikut peruntukan undang-undang*, Master Thesis , Universiti Teknologi Malaysia, (2002)
[12] M. Government, *Employment Act 1955,* April, **1–120**, (2012)
[13] A. Hamid and B. Singh, Problems faced by contractors in managing foreign workers on construction sites, *2nd Int. Conference on Construction and Project Management*, **15**,131–135, (2011)
[14] Jabatan Imigresen Malaysia, *Pekerja Asing Separa Mahir dan Kurang Mahir,* Jabatan Imigresen News, (2004)
[15] N. Jaffar, A.H. Abdul Tharim and M. N. Shuib, Factors of conflict in construction industry: A literature review, Procedia Engineering, **20**, 193–202, (2011)
[16] O. S. Wei, *Kesan pengurangan buruh asing dalam industri pembinaan. Kajian kes di Johor Bahru,* Master Thesis , Universiti Teknologi Malaysia, Skudai, Malaysia, (2002)
[17] A.M. Ismail, *Penggunaan buruh asing dalam industri pembinaan di Johor,* (2003)
[18] International Finance Corporation and European Bank for Reconstruction and European Bank for Reconstruction and Development, *Workers' accommodation : processes and standards A guidance note by IFC and the EBRD*, European Bank for Reconstruction and Development and International Finance Corporation, London, (2009)
[19] G. Gibbs, *Qualitative data analysis: Explorations with NVivo*, Philadelphia, USA: Open University Press, (2002)

[20] L. Richards, *Handling qualitative data: A practical guide,* London: Sage Publications, London, (2005)
[21] C. Grbich, *Qualitative data analysis: An introduction*, London: SAGE Publication Ltd, London, (2007)
[22] F. Wiltshier, *Researching with NVivo 8*, Forum: Qualitative Social Research, **12**(1), 1-12., (2011)
[23] N. Kariyaa, Z. Yaakobb, M.N.M Sairia, H. Mohammadc, S.K Yaman and R. Yunus, Investigation of generic house components and their practical assessment during defect liability period (DLP) in Malaysia, Int. J. of Engineering, **29**(10), (2016)
[24] S.Ariffin, Z. Yaakobb, S. Sulaiman, H. Mohammad, S.K Yaman, Factors of economies of scale for abstract, I*nt. Congress on Technology, Engineering and Science Conf.*, Kuala Lumpur, Malaysia, (2016)
[25] P.B. Jackson, *Qualitative data analysis with Nvivo*, Second Edi. London: British Library, (2013)
[26] K.L. Gwet, Computing inter-rater reliability and its variance, British J. of Mathematical and Statistical Psychology,**61**,29-48, (2008)
[27] M. Graham, A. Milanowski and J. Miller, *Measuring and Promoting Inter-Rater Agreement of Teacher and Principal Performance Ratings*, (February,2012)
[28] K. Hallgren, Computing inter-rater reliability for observational Data: An overview and tutorial, Tutorials in Quantitative Methods for Psychology, **8(1)**, 23-34, (2012)
[29] D. Iacobucci, Factor analysis, J. Consum. Psychol. Spec. Issue Methodol. Stat. Concerns Exerimental. Behaviour Researcher., **10**(1&2), 75–82, (2001)
[30] D.G. Freelon, ReCal: Intercoder reliability calculation as a web service, Int. J. Internet Sci., **5**(1), 20–33, (2010)
[31] R.E. Stake, *Qualitative research: studying how things work*, New York: The Guilford Press, London, (2010)
[32] S.E. Stemler, A comparison of consensus, consistency, and measurement approaches to estimating interrater reliability, Pract. Assess. Res. Eval., **9**(4), 1–19, (2004)
[33] United States Department of Labor, *US OSHA (2016), General environmental control: temporary labor camps*, Retrieved on 2016 https://www.osha.gov/pls/oshaweb/owadisp.show_document?p_table=STANDARDS &p_id=9791

Development of Occupational Safety and Health Requirement Management System (OSHREMS) Software Using Adobe Dreamweaver CS5 for Building Construction Project

Nor Haslinda Abas[1,*], *Nurjeha* Adman[1], and *Rafikullah* Deraman[1]

[1]Jamilus Research Centre, Universiti Tun Hussein Onn Malaysia, 86400 Parit Raja, Johor, Malaysia

Abstract. The construction industry sector is considered as being risky with frequent and high accident rate. According to Social Security Organization (SOCSO), the construction accidents has arisen from time to time. Construction Industry Development Board (CIDB) has developed the Safety and Health Assessment System in Construction (SHASSIC) for evaluating the performance of a contractor in construction project by setting out the safety and health management and practices, however the requirement checklist provided is not comprehensive. Therefore, this study aims to develop a software system for facilitating OSH in building construction project, namely OSH requirements management system (OSHREMS), using Adobe Dreamweaver CS5 and Sublime Text as PHP editor. The results from a preliminary study which was conducted through interviews showed that, the respondents were only implementing the basic requirements that comply with legislations, with the absence of appropriate and specific guideline in ensuring occupational safety and health (OSH) at the workplace. The tool will be benefits for contractors and other parties to effectively manage the OSH requirements for their projects based on project details.

1 Introduction

Malaysian construction industry is one of the largest industries, which has been instrumental in generating wealth in the development of infrastructure and building with social and economic structures through continued growth. However, it is categorized as dangerous working environment due to high accident and fatality rates, as reported in the Social Security Organization (SOCSO) Annual Report [1-3]. The data collected by the Department of Occupational Safety and Health (DOSH) (as shown in Table 1) further indicates that the construction industry had recorded the highest number of fatalities every year. The construction industry contributes more than one-third of fatalities out of all industries and further proves that safety performance in the construction industry lags behind most other industries [4]. These reports provide clear evidence that the industry is

* Corresponding author: nhaslin@uthm.edu.my

one of the critical sectors in need of a significant and rapid overhaul to its current site safety practices [5].

Table 1. Percentage ratio of number of fatalities in construction industry to the total fatalities reported to DOSH for the years 2007 until 2016 [4, 6]

Year	Number of fatalities in the construction industry	Total number of fatalities reported to DOSH for the respective year	Percentage, %
2007	95	219	43.4
2008	72	230	31.3
2009	62	185	33.5
2010	63	175	36
2011	51	176	29
2012	67	191	35.1
2013	69	185	37.3
2014 (until Nov)	70	184	38
2015	88	214	41.1
2016 (until Aug)	55	135	40.7

It is believed that one of the causes of the increased number of accidents on construction sites is due to the lack of awareness of many parties, particularly employers in providing provisions to ensure safety, health and welfare of the workers. Therefore, it is a responsibility of all parties involved in the construction industry to take preventive action to reduce accidents on the construction site by providing the necessary safety and health of workers at construction site. This study suggests that, one of the ways to avoid accident at construction site is through effective occupational safety and health (OSH) management planning tool. Effective safety can only be achieved when there is a proper management of the interaction between technological systems and people [7]. Even though there are several construction companies has prioritized OSH requirements for their construction projects [8], it is believed that the awareness of contractors (being the management of the project) on OSH management is still lacking [9]. Moreover, it is also believed that many construction companies are not aware of their responsibilities to comply with OSH legislations for effectively managing their construction project.

Realizing the importance of protecting workers from occupational accident and injury, the Government of Malaysia has introduced the necessary legislations related to the industry, in which to be complied by the parties involved in the workplace (including employers and employees). In Malaysia, the main legislations covering health and safety in the workplace are the Occupational Safety and Health Act (OSHA) 1994 [10] and Factory and Machinery Act (FMA) 1967 [11]. The introduction of the comprehensive legislations was in response to the need to cover a more diverse employee base and newer hazards introduced in the workplace [4]. Occupational Safety and Health Act 1194 (Act 514) is the legislations enforced by the Department of Occupational Safety and Health (DOSH) which is a legislative framework in Malaysia construction industry to ensure safety, health and welfare of persons who are at work and to protect other person against safety risk or health in connection with the activities of persons at work. Factory and Machinery Act 1967 (Act 139) is to provide for the control of factories with respect to matters relating to safety, health and welfare of persons therein, the registration and inspection of machinery and for matters connected therewith.

At construction site, the contractor is the most responsible party in ensuring safety and health of the workers. One of the ways the contractor can do to address this is through effective occupational safety and health (OSH) planning tool on a construction site and necessary administrative arrangements before the commencement of construction activity. It is essential for contractors to comply with safety and health requirements stipulated in OSHA 1994 and FMA 1967 (including Building Operations and Works of Engineering Construction (BOWEC) 1986 Regulation) in order to minimize the accident happen on construction worksite [11]. Besides that, even though there exists a tool named Safety and Health Assessment System in Construction (SHASSIC) developed by Construction Industry Development Board [12], this tool is only intended for evaluating the performance of a contractor in construction project, by setting out the safety and health management and practices of contractor for various aspects of the construction work activities. SHASSIC provides a valuable checklist for safety and health compliance to the contractors, but the checklist is not comprehensive which covers three (3) aspects only, i.e. document check, site/workplace inspection and employees interview [4]. Moreover, it is also believed that many construction companies are not aware of their responsibilities to comply with OSH legislations for effectively managing their construction.

Therefore, this study seeks to overcome this problem by developing a system tool that can be used by contractors or other parties to effectively manage or plan the OSH requirements for their project, which comply to related OSH legislation.

2 Development process of the system

The development process consists of several stages, namely i) preliminary survey; ii) identification and validation of OSH requirements based on OSHA 1994, FMA 1967 and BOWEC 1986for building construction project; and iii) preliminary development of decision support system using Adobe Dreamwaver CS5 software.

2.1 Preliminary survey

A preliminary survey was conducted to investigate the issues related to occupational safety and health (OSH) requirements faced by contractors. This task was executed through interviews with five experienced contractors to obtain information about their understandings and implementation of OSH requirements at the workplace. The respondents were those who are the owners of the companies and also who hold the managerial level of the project.

From the interviews, it is found that all contractors did not have a specific guideline or checklist in implementing the safety and health for effective OSH management and planning tool. Most of them implemented general safety precautions as mentioned in the tender document. Other than that, majority of the repsondents are lack of knowledge regarding the requirements of OSH and FMA in their OSH implementation. This showed that there is a strong need to develop a checklist-based-system tool for OSH requirements that comply with OSH legislations based on project details that could help contractors to facilitate OSH requirements for their project effectively, and further improve safety and health performance at the construction site.

2.2 Identification and validation of Occupational Safety and Health (OSH) requirements based on OSHA 1994 and FMA 1967

Next, in order to develop a system of OSH requirements tool for construction projects, in depth study and analysis of the OSH requirements as stipulated in the legislations were executed. The legislations being studied include OSHA 1994, FMA 1967 and BOWEC 1986. The requirements were divided into several categories, which are site management (registration); appointment of safety personnel; safety and health committee; documentation; site planning; concrete work; structural steel; catch platforms; chutes, safety belts and net; ladder and step-ladder; scaffolds; excavation work; piling; and machinery. These requirements serve as the basis for information that underpin the development of the OSHREMS.

In order to ensure the validity of the secondary data for developing the system, the OSH requirements identified need to be validated by an expert panel. Validation process was done through a series of interviews with Johor state DOSH officers whom currently in-charge in the construction safety department. One of the officers' main responsibilities is conducting site inspection and other various OSH compliance for construction sector. From the validation process, there were several requirements identified that need to be improved in terms of naming or addressing the requirements. Whereas, there were also some requirements found irrelevant and need to be excluded, for example, the requirement for timber scaffold. According to one of the panels, the used of timber scaffold is prohibited at construction site.

2.3 Preliminary development of OSHREMS using Adobe Dreamweaver CS5

After obtaining the database of OSH requirements for building construction project, these information were then transferred into a system, using Adobe Dreamweaver CS5 software. Before commencing the development of the system tool, the flow process of the system need to be recognized (as shown in Fig. 1) to facilitate the development process. The OSHREMS commences with the user identification to sign up or sign in to the system, followed by him/her to fill in the project information. Next, the OSHREMS will display the OSH requirements for the specified project, and the user can opt for printing out the OSH requirements listed.

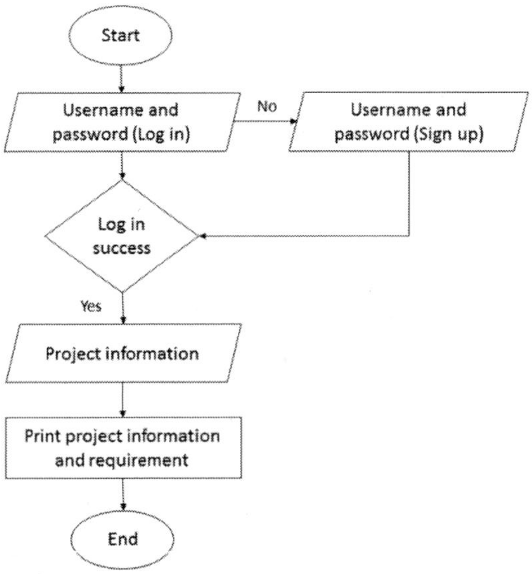

Fig. 1. Flow chart of the system

In developing the OSHREMS, Adobe Dreamweaver was used to create the graphic user interface and to write the codes of the project. The system was operated using MYSQL Server that act as a database, and the system interface was designed using Adobe Dream Weaver CS5 software. This system required some codings to develop the OSH requirement. To develop the interface, some codes for each requirement and interface codes were needed to designing the background and produce the project information interface.

Fig. 2 shows the main menu interface of the OSHREMS. In this interface, the user needs to enter their username and password before click Login. The Reset button is used to clear the username and password at once.

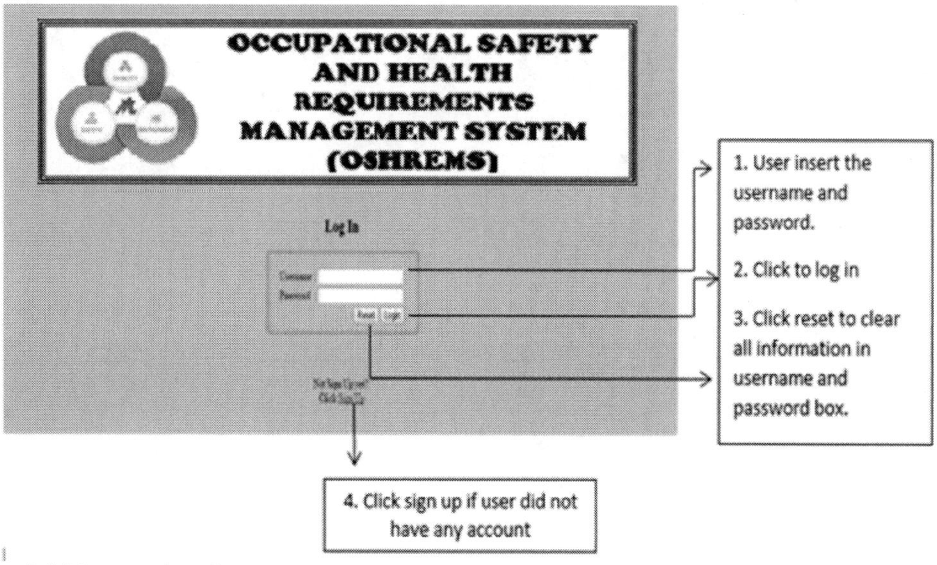

Fig. 2. Main menu interface

Once the user has succesfully logged in to the system, the 'Home interface' will be displayed (Fig. 3). 'Home interface' shows the project information details that the user need to fill. Besides general project information, the user also needs to fill in the information about the plant and scaffolds use (if any) at the construction site. After obtaining the database of OSH requirements for building construction project, based on the information input these information will then be transferred into a system.

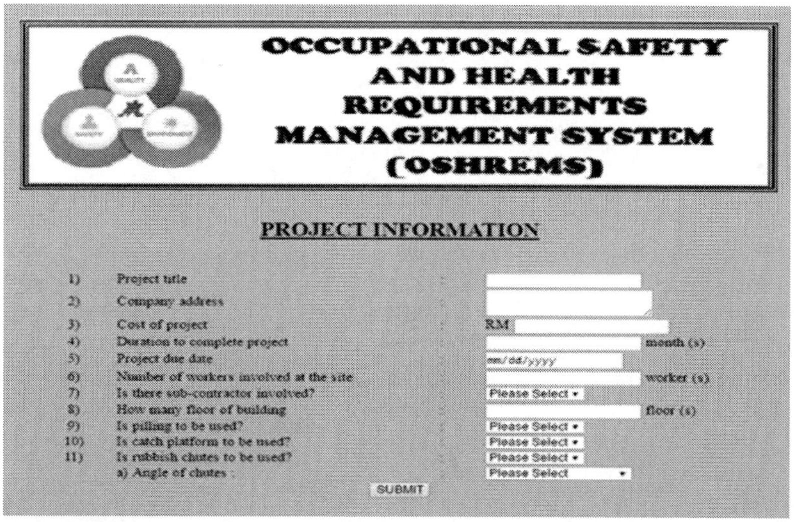

Fig. 3. OSHREMS Home Interface

Once the information of project are determined, the system will identify the requirements of OSHREMS of the project based on the data input and give the output as shown Fig. 4. The user can opt for print the OSH requirement checklist (based on project details) for their project.

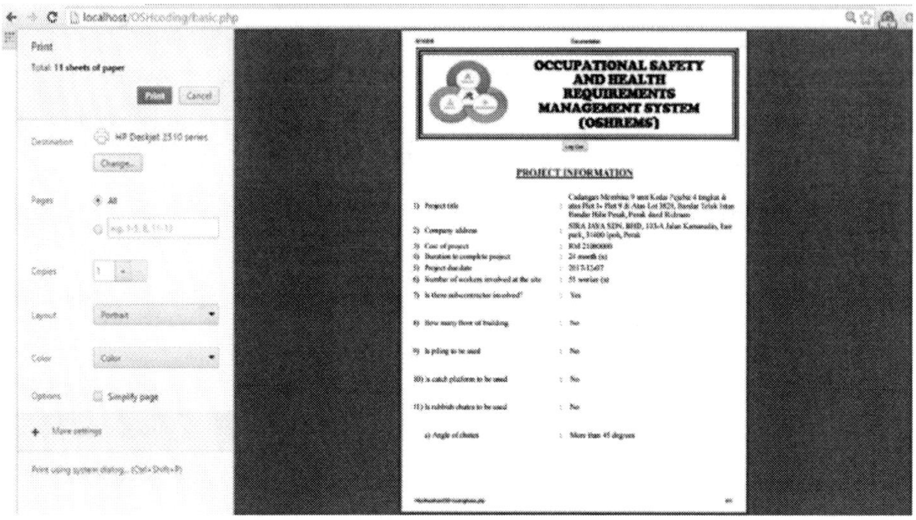

Fig. 4. Example of OSHREMS generated output

2.4 Evaluation of OSHREMS using existing case study

Software validation or, more generally, verification and validation (V&V) is intended to show that a system both conforms to its specification and that it meets the real needs of the user of the application [13]. Therefore, in order to ensure the functionality of the developed system, it was tested using a real case study based on the information of one particular building construction project. The project of "Cadangan Membina 9 unit Kedai Pejabat 4 tingkat di atas Plot1 – Plot 9 di Atas Lot 3829, Bandar Teluk Intan Bandar Hilir Perak, Darul Ridzuan" was selected as a case study. Fig. 5 shows the result from the OSHREMS for the 'Project Information' display as keyed in by the user.

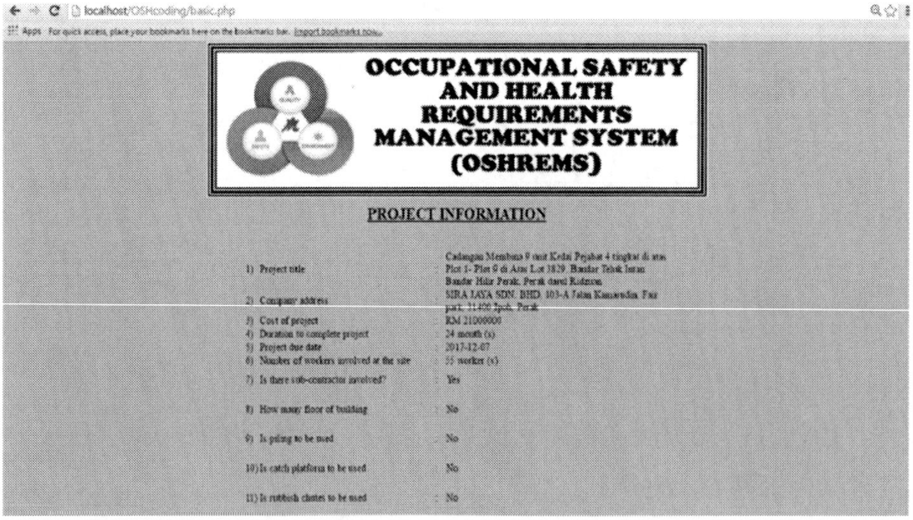

Fig. 5. Example of 'Project Information' display in OSHREMS

With regards to OSH requirements, Fig. 6 shows the OSHREMS output for 'Appointment of Safety Perdonnel' requirements. The OSH requirements shown are based on the information inserted by the user in the project information interface. For example, since the project is more than RM20 million, the appointment of Safety and Health Officer (SHO) is mandatory as stipulated in OSHA 1994 and Occupational Safety and Health (Safety and Health Officer) Order 1997. Safety and health committee is also need to be established because the project involved more than 40 workers (as stipulated under Section 30(1) of OSHA 1994. It is to be noted that this paper only provides an example of part of the OSH requirements needed for a particular building construction project.

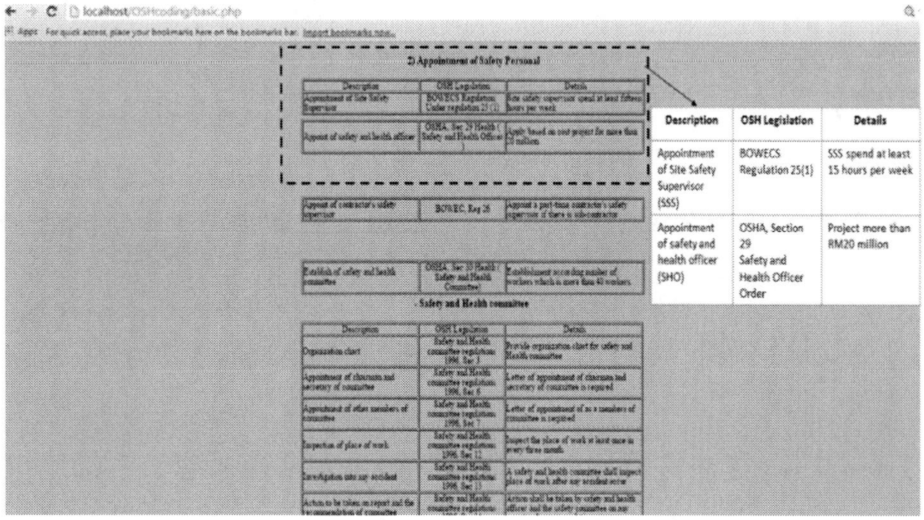

Fig. 6. Example of OSHREMS output for 'Appointment of Safety Personnel' requirement

3. Conclusion

The integration of safety knowledge management and information technology can offer an advanced means for improving safety in the Malaysia construction industry. The purpose of this work is to develop the OSHREMS that can be utilized by contractors and other relevant parties (e.g. client, consultant, local authority, etc.) in the building construction project. The proposed system would pave the way to an innovative safety management approach in construction projects, departing from the traditional mode in Malaysian construction industry. This approach would also add value to the current knowledge-based of construction safety management. The authors believe that OSHREMS tool would be able to provide a guideline that can aid the contractor (and/or other relevant parties) of the building construction project about the legislation-based OSH requirements that need to be fulfilled based on their project information and further enhance safety performance by reducing the number of accidents on sites. The OSHREMS tool is also capable to facilitate OSH management planning that benefits all parties, especially during the planning stage, despites encouraging them to fully implement the OSH requirements that comply to legislations and not only implementing minimum safety requirements. Further research is underway to improve the contents available and test the system in large scale case study.

The results presented in this paper are parts of an ongoing research project. This research was supported by Fundamental Research Grant Scheme (FRGS) (Vot 1579).

References

[1] C. L. Foo, Budgeting For Occupational Safety And Health Management And Its Implementation, *MBAM Annual Safety Conference,* (2005)
[2] SOCSO, *Annual report for 2000*, Social Security Organisation (SOCSO), Kuala Lumpur, Malaysia, (2000)
[3] SOCSO, *Annual report for 2000*, Social Security Organisation (SOCSO), Kuala Lumpur, Malaysia, (2009)
[4] N.H. Abas, *Development of a Knowledge-Based Energy Damage Model for Evaluating Industrialized Building System (IBS) Occupational Health and Safety (OSH) Risk,* PhD Thesis, RMIT University, Melbourne, Australia, (2015)
[5] A.R.A. Hamid, M.Z.A. Majid and B. Singh, Causes of Accidents at Construction Sites, Malaysian Journal of Civil Engineering, **20**(2), 242-259, (2008)
[6] DOSH, *Accident Statistics*, Department of Occupational Safety and Health, (2016), Retrieved on October16, 2016 from http://www.dosh.gov.my
[7] I. F. Mohd Kamar, N.S. Lop, N.M. Salleh, S. Mamter, H.A. Suhaimi, Contractor's Awareness on Occupational Safety and Health (OSH) Management Systems in Construction Industry, *Emerging Technology for Sustainable Development Congress (ETSDC 2014),* EDP Sciences, 1-6, (2014)
[8] B. Fernandez-Muniz, J.M. Montes-Peon, and C. J. Vasquez-Ordas, Safety management system: Development and validation of a multidimensional scale, Journal of Loss Prevention in the Process Industries, **20**(1), 52-68 (2007)
[9] K. Dorjil, B. H. W. Hadikusumo, Safety Management Practices in the Bhutanese Construction Industr, Journal of Construction in Developing Countries, **11**(2), 53-75, (2006)
[10] The Commissioner of Law Revision, *Occupational Safety and Health Act 1994,* The Commissioner of Law Revision, Malaysia, (2006)
[11] The Commissioner of Law Revision, *Factories and Machinery Act 1967,* The Commissioner of Law Revision, Malaysia, (2006)
[12] CIDB, *Construction Industry Standard (CIS) 10:2008 Safety and Health Assessment System in Construction,* Construction Industry Develoment Board (CIDB), Malaysia, (2006)
[13] I. Sommerville, *Software engineering* 9th *Edition*, Pearson Education, Massachusetts, USA, (2011)

Importance Level of Construction Managers' Technical Competency Across Different Contractors' Demographics in Malaysia

Hairuddin Mohammad[1,*], *Padzil Fadzil* Hassan[1], and *Siti Khalijah* Yaman[2]

[1] Faculty of Architecture, Planning, and Survey, University Teknologi MARA, 40450 Shah Alam, Selangor, Malaysia
[2] Faculty of Civil and Environmental Engineering, University of Tun Hussein Onn Malaysia, 86400 Parit Raja, Johor, Malaysia

Abstract. It is undeniable that construction managers play a significant role within construction phase, especially when they are representing construction contractor, one of the key stakeholders within construction project's realm. Reliance towards their technical competency is very much needed for a successfulness of a construction project. Unfortunately, a lot of recurring issues pertaining to their technical incompetency are shrouding the industry. Non-existence of term of reference for construction manager's technical competency profile is dictated as a major setback. Thus, a dedicated and comprehensive research exploration is conducted in order to bridge the gap through establishment of particular competency. Hence, in this instalment, the authors postulated a question of 'how?' which pillars on items' importance level and correlation among different contractors' demographics. Initially, data screening processes over outliers test, reliability test, and normality test were conducted. After that, the importance of technical competency for construction manager was analysed based on their central tendency (i.e. means). Finally, Spearman correlation analysis is directed towards observed ranks from diverse contractors' demographics. In a nutshell, data was free from any extreme outlier and reported high reliability value. However, data was measured as non-normal in nature. Meanwhile, through means, majority of macro items (16 items) were in a range of important (4.00) to very important (5.00). Towards the end, findings for Spearman correlation analysis were mostly leaning towards moderate and strong association's strength across different clusters of diverse contractors' demographics. Nevertheless, there were evidences that major variability of findings for both tests is exhibited through contractors' grades. Therefore, succeeding in-depth quantitative analysis is paramount in order to pin-point significant values onto each individual item.

[*] Corresponding author: hairuddin81@gmail.com

1 Introduction

The Construction Industry Transformation Programme (CITP) which kicked in from 2016 to 2020 was a continuous effort steered by the Construction Industry Development Board (CIDB) Malaysia. Within those documents, one of its major thrust is on alleviating human capital's competency within the industry in order to enhance industry's productivity [1]. Here, issue on incompetency of construction personnel was deemed critical, which unfortunately spread throughout different level of managerial and support's clusters. In line with the conundrum, recurring problems in construction project (i.e. delays, cost overruns, subpar quality, lack of management, etc.) can be observed spawning endlessly in the background, where the contractors are usually to be blamed for such instances [2-5] (to name a few). Knowing the severe impact that will be brought into the industry, change is imminent for indigenous contractors. They must equip themselves with sufficient competency, especially in technical aspects (comprises of knowledge and skills) which become one of the key to construction successfulness, and further differentiate them with other construction practitioners [6,7].

Additionally, the pertaining issue needs not to be seen as isolated as contractor's personnel hierarchy is concerned. Given that organisation's change will incorporate full management's involvement, approach from top-to-down is paramount. Hence, the authors appreciate the fact that construction managers are the key personnel representing contractor's firm that oversee the entire construction processes. Unfortunately, there is a knowledge gap in term of established term of reference for the convenience of education, training, continuous professional development (CPD), and others, which supposed to generically spell out construction manager's technical competency. Furthermore, genuine comments from the construction managers' themselves partly became the driving force for these undertaking.

Therefore, following the previous several presented papers as in Yaman et al. [7], Mohammad et al. [6], and another one on dimensionality of items [18], this instalment proceeds with subsequent analysis on quantitative perspective as these question arose; how important are the generic technical competency of construction manager towards grade, respondents' positions, and years of experience on Malaysia's contractors (minus contractor's category, since they can have more than one category)? The need to preliminarily validate by analysing the importance of the technical competency and summing up generalised perceptions throughout entire country is paramount before subsequent detail analysis is conducted (i.e. to seek items' significant values). Therefore, the succeeding topics were brought forward which encompasses screening processes of collected data, a series of central tendencies' analysis, and simple correlation analysis, which precisely presented in a step-by-step undertakings (n.b. discussions for overall quantitative paradigm can be found in [18]).

2 Data screening processes

First and foremost, after data were collected and inputted into SPSS v.22, summary of respondents' demographics was tabulated as in Table 1. Furthermore, since data fits the cross-sectional survey and majority of questions employed by the Likert-Scale method, careful considerations have been made to preserve their integrity and originality. Therefore, data screening processes has been done through outliers test, reliability test, and normality test, accordingly. It was then followed by common testing for central tendency and correlation as according to classical test theory (CTT). Additionally, to assist the analysis in term of nomenclature, full abbreviations as in Table 2 will be used throughout, whereby screening processes will be considered up to Meso level of items.

Table 1. Summary of collected data

Details	Test/Data		Results
Actual survey	Reliability (α)		0.969
	100% accepted responses (minus incomplete, blank, and unsuitable responses)		373 no.
	Percentage of response (self-administered)		47.35%
	Percentage of response (online)		3.4%
	Category of contractors	Civil engineering	178 no.
		Building construction	219 no.
		Mechanical and electrical	85 no.
	Grade of contractors (the highest number signifies large size contractor)	G1	47 no.
		G2	25 no.
		G3	61 no.
		G4	37 no.
		G5	52 no.
		G6	28 no.
		G7	123 no.
	Position of respondents	Director	131 no.
		Project Manager	68 no.
		Contract Manager	21 no.
		Construction Manager	30 no.
		Site Manager	14 no.
		Site Supervisor	56 no.
		Architect / Engineer / Quantity Surveyor	53 no.
	Years of experience	4 years and below	73 no.
		5 years – 9 years	101 no.
		10 years and above	199 no.

To begin with, outliers are extreme values compared to the rest of data. It is able to show if there is any single item that has extreme recorded data which might disturb the rest of analysis. Outliers can be pre-analysed by looking at mean, 5% trimmed mean (5% TM, by removing 5% of extreme data), and Inter-quartile Range (IQR – above than 3 of IQR is considered Extreme Values) [8, 9]. Therefore, by looking at the following table (Table 3), generally all items (variables) do not have any extreme value. Both values (i.e. mean and 5% trimmed mean) were closer to each other, and consequently attested by lower value of IQR. Hence, subsequent analysis will consider all items without any removal exercise. Apart, through reliability analysis (i.e. Cronbach Alpha, α) where preferred measure of internal consistency is more than 0.7 [10, 11], a value of 0.969 is observed (see Table 1). It shows that those constructs were able to measure their intended purpose, and preliminary evidence that items were closely related as a group and supports high reliability.

Table 2. Full abbreviations used throughout the analysis

Items	Required Technical Competency Detail		Abbreviation		
			Meso Items	Macro Items	Major Items
1	STAFF	Knowledge and skills of staff management (including their development, welfare, laws, etc.).	ST	ST	RESMED
2	MATERIALS 1	Knowledge of construction materials (concrete, steel, wood, soil, etc.).	MT1	MTMED	
3	MATERIALS 2	Knowledge and skills of material management (including their procurement, logistics, supplier, etc.).	MT2		
4	LABOUR	Knowledge and skills of labour management (including their productivity, welfare, laws, etc.).	LB	LB	
5	PLANT 1	Knowledge of construction plant/equipment and their utilization.	PL1	PLMED	
6	PLANT 2	Knowledge and skills of plant management (including their requisition, maintenance, supplier, etc.).	PL2		

7	SUBCONTRACTOR 1	Knowledge and skills of sub-contractor management (including their claims and payments, variations, insurances, etc.).	SU1	SUMED	RESMED
8	SUBCONTRACTOR 2	Knowledge and skills of sub-contractor tendering and bidding (procedures, pretender, bid analysis, quantity take-off, etc.).	SU2		
9	HEALTH & SAFETY 1	Knowledge of health and safety equipment and manual.	HE1	HEMED	OBJMED
10	HEALTH & SAFETY 2	Knowledge and skills of health and safety management (including their practices, compliance, regulation, training, etc.).	HE2		
11	HEALTH & SAFETY 3	Knowledge and skills of risk management (including their assessment and analysis, etc.).	HE3		
12	MONEY	Knowledge and skills of financial and cost management (including their claims and payments, pricing and purchasing practice, etc.).	MO	MO	
13	QUALITY 1	Knowledge of construction specifications, TQM, building codes and standards, etc.	QU1	QUMED	
14	QUALITY 2	Knowledge and skills of quality management (including their administration, assurance, quality control, etc.).	QU2		
15	TIME	Knowledge and skills of time management (including understanding of project scheduling, updating, etc.).	TI	TI	
16	ENVIRONMENT 1	Knowledge and skills of environment management (including their assessment, Environment Management System (EMS), etc.).	EN1	ENMED	
17	ENVIRONMENT 2	Knowledge of green and sustainable construction (including Green Building Index (GBI), green construction techniques, etc.).	EN2		
18	ADMINISTRATION	Knowledge and skills of construction administration (including documentation and record, submissions, plans/drawings, meeting, etc.).	AD	AD	GENMED
19	PRECONSTRUCTION 1	Knowledge and skills of construction site surveying, site layout, temporary structures/work, etc..	PR1	PRMED	
20	PRECONSTRUCTION 2	Knowledge and skills of construction site management (including overall resources management, master programme, construction sequences, Work Breakdown Structure, project start, meetings, etc.).	PR2		
21	PRECONSTRUCTION 3	Knowledge of civil/structural design.	PR3		
22	PRECONSTRUCTION 4	Knowledge of construction systems (including Industrialized Building System (IBS), etc.).	PR4		
23	PRECONSTRUCTION 5	Knowledge of mechanical and electrical systems.	PR5		
24	PRECONSTRUCTION 6	Knowledge of quantity surveying.	PR6		
25	PRECONSTRUCTION 7	Knowledge of Value Engineering (eliminating unnecessary cost which does not contribute to the value of construction).	PR7		
26	PRECONSTRUCTION 8	Knowledge of Constructability (analysis of construction coordination issues associated with various trades).	PR8		
27	PRECONSTRUCTION 9	Knowledge of Lean Construction (application of production management to construction).	PR9		
28	CLOSEOUT & HANDOVER	Knowledge and skills of construction closeout and handover procedures (including their management, commissioning, acceptance, transfer, etc.).	CL	CL	
29	RESPONSIBILITIES 1	Knowledge and skills of construction law and legislation (including authorities processing, etc.).	RS1	RSMED	
30	RESPONSIBILITIES 2	Knowledge and skills of construction businesses (sales, commercial, trades, economic analysis, etc.).	RS2		
31	RESPONSIBILITIES 3	Knowledge and skills of administering and assisting/facilitating client, senior manager and junior manager (including their coordination, liaison of works, orders, etc.).	RS3		
32	COMPUTER & I.T. 1	Knowledge and skills of general computer application (MS Excel, MS Office, Adobe, internet based, etc.).	CO1	COMED	OVEMED
33	COMPUTER & I.T. 2	Knowledge and skills of construction information technology / software application (MS Project, Primavera, CAD, scheduling, estimating, accounting, etc.).	CO2		
34	CONTRACT ADMINISTRATION	Knowledge and skills of construction contracts (including their administration, document, variations, Extension of Time, claims and payments, subcontracts, conflict, etc.).	CT	CT	

Table 3. Outliers result

Items	Mean	5% TM	(IQR)
1	4.4906	4.5481	1
2	4.4933	4.557	1
3	4.445	4.5004	1
4	4.4504	4.5004	1
5	4.2359	4.286	1
6	4.1394	4.1847	1
7	4.4692	4.5243	1
8	4.4772	4.5392	1
9	4.4745	4.5362	1
10	4.4343	4.4915	1
11	4.3834	4.4349	1
12	4.622	4.6971	1
13	4.4692	4.5302	1
14	4.437	4.4915	1
15	4.5523	4.6226	1
16	4.1957	4.2353	1
17	4.1528	4.1936	1
18	4.3861	4.4409	1
19	4.4397	4.5034	1
20	4.4424	4.5034	1
21	4.2279	4.283	1
22	4.1394	4.1906	1
23	4.1153	4.1519	1
24	4.2601	4.3009	1
25	4.1421	4.1728	1
26	4.0831	4.1192	1
27	4.008	4.0536	2
28	4.3432	4.3902	1
29	4.252	4.2949	1
30	4.1287	4.1758	1
31	4.2198	4.2592	1
32	4.1796	4.2204	1
33	4.1072	4.1489	1
34	4.3968	4.4498	1

On the other hand, a normal distribution of data is a symmetric bell-shaped curve defined by two things; the mean (for average) and variance (for variability). In statistical inference, there is a central idea on sample size; when sample size increases, the distribution of data will be approximately normal. Though, in order to reduce any misleading assumption on the distribution of data, the research embarked on checking the normality of data. Additionally, since normality will determine appropriate inference's tests to be conducted afterward (either parametric test or non-parametric test), the research chose histogram analysis, skewness and kurtosis analysis, and Kolmogorov-Smirnov (K-S) and Shapiro-Wilk (S-W) test as the basis for determining normality [12]. In histogram analysis, graphical representation by using histogram that is being superimposed by the theoretical normal distribution curve provides useful information on normality of the data. Thus, each item was subjected to histogram analysis and manual observations were imposed onwards (see Fig. 1 for example). Collectively, all items were deviant to several degrees from normal and outlaid negative skewness which their tails were extending to the left side of the distributions. Therefore, in order to measure their precise deviations, subsequent analysis was conducted (i.e. skewness and kurtosis test).

In skewness analysis, the symmetry of the distribution is related to the skewness [14, 15]. Normal or perfect skewness represents a perfectly symmetric distribution. Scores that clustered to the left with their tail extending to the right are positive skewness. Scores that clustered to the right with their tail extending to the left are negative skewness. Meanwhile, the peak of the distribution is related to the kurtosis [15]. Normal or perfect kurtosis involves a distribution that is in perfect bell-shaped which is not too pointy or flat. Positive kurtosis is indicated by a peak and negative kurtosis is indicated by a flat distribution. For both analysis (i.e. skewness and kurtosis), value of "0" is related to perfectly normal distribution. Thus, farther away from "0" indicates the non-normal distribution, Mutually, as far as the particular analysis is concerned, skewness values ranges from -0.450 to -1.593, and kurtosis values ranges from -.039 to 1.759 (with maximum standard error of 0.252 for both), when there is no observed solid "0" value for both aspects (see Table 4). Unfortunately, since skewness and kurtosis are being analysed separately and rely on arbitrary determination, further test by using Kolmogorov-Smirnov (K-S) and Shapiro-Wilk (S-W) was indeed needed.

Both tests (i.e. K-S and S-W) are established tests for normality that take into account for both Skewness and Kurtosis simultaneously [12]. It was designed to test normality by comparing original data to a normal distribution with the same mean and standard deviation of observed sample. K-S is preferred for large pool of respondents (>2000), while the other way round for S-W (<2000). Generally, hypothesis is made towards the results (by referring to significant values) as follows, H_0 = there is no significant of normality, H_1 = there is significant of normality, and reject H_1 if p-value < 0.05. Therefore, based on Table 5, both tests indicate that all data are non-normal data (i.e. rejecting H_1). For that reason, subsequent inference analysis will rely on non-parametric type of analysis.

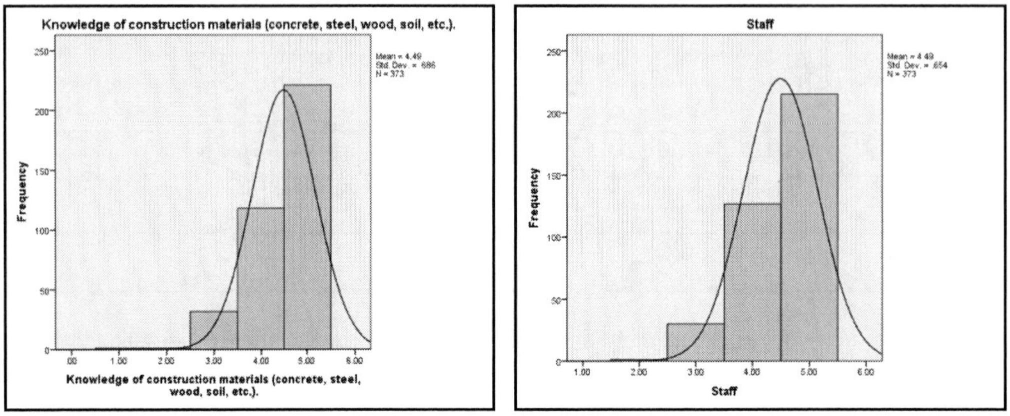

Fig. 1. Examples of histogram analysis

Table 4. Result of skewness and kurtosis analysis

Items	Skewness	Kurtosis
1	-0.979	0.084
2	-1.255	1.509
3	-0.933	0.093
4	-0.767	-0.462
5	-0.768	0.054
6	-0.612	-0.249
7	-0.921	-0.044
8	-1.112	0.459
9	-1.196	1.3
10	-1.074	0.97
11	-0.948	0.399
12	-1.593	1.759
13	-1.053	0.549
14	-0.907	0.059
15	-1.335	1.3
16	-0.622	-0.275
17	-0.591	-0.277
18	-0.903	0.117
19	-1.096	0.66
20	-1.059	0.431
21	-0.753	-0.106
22	-0.69	0.11
23	-0.527	-0.262
24	-0.622	-0.495
25	-0.49	-0.389
26	-0.45	-0.444
27	-0.552	0.048
28	-0.755	-0.28
29	-0.64	-0.254
30	-0.582	-0.32
31	-0.574	-0.192
32	-0.632	-0.039
33	-0.538	-0.425
34	-0.846	0.162

Table 5. Result of Kolmogorov-Smirnov (K-S) and Shapiro-Wilk (S-W) analysis

Items	K-S		S-W	
	Sig.	Statistic	Sig.	Statistic
1	0	0.358	0	0.716
2	0	0.362	0	0.706
3	0	0.34	0	0.736
4	0	0.338	0	0.732
5	0	0.27	0	0.8
6	0	0.241	0	0.822
7	0	0.349	0	0.726
8	0	0.36	0	0.716
9	0	0.355	0	0.716
10	0	0.337	0	0.735
11	0	0.324	0	0.751
12	0	0.427	0	0.621
13	0	0.348	0	0.724
14	0	0.335	0	0.739
15	0	0.387	0	0.679
16	0	0.259	0	0.808
17	0	0.244	0	0.819
18	0	0.318	0	0.756
19	0	0.34	0	0.733
20	0	0.344	0	0.732
21	0	0.263	0	0.802
22	0	0.233	0	0.823
23	0	0.225	0	0.826
24	0	0.274	0	0.795
25	0	0.235	0	0.817
26	0	0.238	0	0.831
27	0	0.241	0	0.844
28	0	0.302	0	0.772
29	0	0.262	0	0.798
30	0	0.23	0	0.824
31	0	0.242	0	0.803
32	0	0.242	0	0.814
33	0	0.235	0	0.827
34	0	0.313	0	0.755

3 Central tendency analysis

In this section, there are two (2) minor inquiries, namely; analysing the importance level of the technical competency and correlational analysis based on contractor's demographics. Most of the works were guided by Creswell [13], and several other literatures that will be specifically cited throughout the writing. Due to the fact that contractors' demographics performed as controlled variables, summary of analysis is needed in order to discover their range of perception level. Therefore, the research only reports on the value of means, standard deviations, and further identifies their ranks according to each demographic cluster. Additionally, in order to give meaningful interpretation, variables were reduced to Macro level with 16 variables by using a function of 'compute variable - median' through an option of Numeric Expression in SPSS. Meanwhile, ranks were identified manually according to their respective mean value and standard deviation (SD).

From Table 6, it is worth noting that most of the contractors postulated the identified technical competency for construction manager is within the range of important (4.00) to very important (5.00). However, there were slight discrepancies on behalf of Grade 5 and Grade 6. For Grade 5 contractors, they positioned the technical competency which related to Plant and Equipment, and Computer and I.T. within the range of slightly important (3.00) to important (4.00). The same goes for Grade 6 contractors with technical competency of Computer and I.T. (see the darken boxes). Meanwhile, following Table 7 shows a summary of means based on respondent's years of experience. It can be observed that from junior to senior construction practitioners, the importance level is ranging from important (4.00) to very important (5.00). The case is quite similar for different respondents' position (see Figure 8), where anomaly is just for technical competency of Computer and I.T. on behalf of Construction Directors (with mean falls between slightly important and important).

As analysed, for 16 items across multiple control variables, there are differences of means which ranges from 3.00 (Likert scale of slightly important) to 4.00 (Likert scale of important), and from 4.00 to 5.00 (Likert scale of very important). Although the former only visualised on several instances, a focus towards contractors' grades is paramount given that most instances occurred within these control variable. However, significant trend is swiftly recognised; especially on the first cluster of technical competency for construction manager (i.e. responsibility – RESMED; such as time and money), where noticeably the respondents (across different control variables) gave extra emphasise since the recorded means of the items were significantly higher and scored greater ranks.

Table 6. Summary of importance's means and ranks of technical competency for construction manager based on contractors' grades

Grade registration under CIDB		ST	LB	MO	TI	AD	CL	CT	MTMED	PLMED	SUMED	HEMED	QUMED	ENMED	PRMED	RSMED	COMED
G1	Mean	4.4894	4.5957	4.6809	4.5319	4.4255	4.4043	4.4043	4.4681	4.2340	4.5745	4.4965	4.4787	4.2979	4.1891	4.2482	4.1064
	SD	.68754	.57708	.59368	.71782	.77304	.68078	.71200	.71021	.91397	.61661	.63654	.73690	.77080	.64868	.70357	.68315
	Rank	**6**	**2**	**1**	**4**	**9**	**10**	**11**	**8**	**14**	**3**	**5**	**7**	**12**	**15**	**13**	**16**
G2	Mean	4.4400	4.6400	4.5200	4.4400	4.2800	4.1200	4.2400	4.5600	4.4400	4.5000	4.4667	4.5200	4.3400	4.2800	4.2267	4.0600
	SD	.65064	.48990	.82260	.76811	.93630	.83267	.59722	.52678	.60069	.55902	.50000	.52994	.64096	.57746	.68530	.61779
	Rank	**8**	**1**	**4**	**9**	**12**	**15**	**13**	**2**	**7**	**5**	**6**	**3**	**10**	**11**	**14**	**16**
G3	Mean	4.5738	4.5082	4.6885	4.6066	4.3607	4.3443	4.4754	4.5656	4.2951	4.5820	4.6011	4.3525	4.2623	4.2313	4.2240	4.2295
	SD	.64444	.59506	.53357	.58534	.63332	.70440	.74401	.59495	.78728	.59986	.51215	.67296	.72805	.62733	.71679	.88776
	Rank	**5**	**7**	**1**	**2**	**9**	**11**	**8**	**6**	**12**	**4**	**3**	**10**	**13**	**14**	**16**	**15**

G4	Mean	4.4054	4.4324	4.7027	4.5676	4.4324	4.2432	4.5676	4.4865	4.2432	4.5000	4.4144	4.5135	4.2162	4.1832	4.2973	4.3649
	SD	.64375	.55480	.57081	.60280	.60280	.72286	.64724	.58317	.69344	.60093	.69124	.55885	.76866	.64215	.65161	.71345
	Rank	**10**	**7**	**1**	**2**	**8**	**14**	**3**	**6**	**13**	**5**	**9**	**4**	**15**	**16**	**12**	**11**
G5	Mean	4.3654	4.2115	4.5577	4.4231	4.4615	4.3269	4.2115	4.2308	**3.8846**	4.3077	4.3269	4.4231	4.0769	4.1923	4.1474	**3.9615**
	SD	.74172	.80041	.66902	.69582	.69906	.70631	.80041	.75057	.86668	.70817	.76836	.67435	.72345	.60902	.71221	.74654
	Rank	**5**	**10**	**1**	**4**	**2**	**6**	**11**	**9**	**16**	**8**	**7**	**3**	**14**	**12**	**13**	**15**
G6	Mean	4.5357	4.5000	4.6786	4.6071	4.2143	4.3571	4.3571	4.4821	4.1786	4.5000	4.4167	4.4286	4.1964	4.1944	4.0595	**3.9643**
	SD	.69293	.74536	.66964	.62889	.83254	.67847	.62148	.68694	.58078	.69389	.68268	.66269	.82034	.62899	.67922	.75680
	Rank	**3**	**5**	**1**	**2**	**11**	**10**	**9**	**6**	**14**	**4**	**8**	**7**	**12**	**13**	**15**	**16**
G7	Mean	4.5285	4.4228	4.5772	4.5935	4.3984	4.3984	4.4228	4.4959	4.1789	4.4309	4.3659	4.4797	4.0732	4.2014	4.1897	4.1829
	SD	.60497	.64029	.67761	.63810	.70988	.75466	.61415	.59882	.74143	.62175	.67672	.60867	.81904	.57636	.65715	.73904
	Rank	**3**	**8**	**2**	**1**	**9**	**10**	**7**	**4**	**15**	**6**	**11**	**5**	**16**	**12**	**13**	**14**

Table 7. Summary of importance's means and ranks of technical competency for construction manager based on respondents' years of experience

Years of construction experience		ST	LB	MO	TI	AD	CL	CT	MTMED	PLMED	SUMED	HEMED	QUMED	ENMED	PRMED	RSMED	COMED
4 years and below	Mean	4.4932	4.5068	4.5616	4.5068	4.4247	4.3425	4.3425	4.3973	4.1644	4.4110	4.4521	4.4247	4.2123	4.1887	4.2100	4.1164
	SD	.70952	.68967	.72622	.74765	.76227	.74943	.67122	.71666	.84185	.76984	.78836	.68044	.82039	.64965	.74847	.77960
	Rank	**4**	**2**	**1**	**3**	**7**	**12**	**11**	**10**	**8**	**9**	**5**	**6**	**13**	**15**	**14**	**16**
5 years - 9 years	Mean	4.4059	4.3069	4.5347	4.4455	4.3465	4.1584	4.3663	4.3564	4.0644	4.3960	4.3696	4.4307	4.1238	4.0968	4.1221	4.1188
	SD	.69539	.67443	.64132	.67045	.75413	.74475	.67413	.65703	.77027	.61366	.65473	.65586	.80127	.65427	.70981	.72852
	Rank	**4**	**10**	**1**	**2**	**9**	**11**	**7**	**8**	**16**	**5**	**6**	**3**	**12**	**15**	**13**	**14**
10 years and above	Mean	4.5327	4.5025	4.6884	4.6231	4.3920	4.4372	4.4322	4.5528	4.2588	4.5352	4.4539	4.4749	4.1859	4.2686	4.2362	4.1658
	SD	.60943	.61030	.60588	.59770	.68692	.69255	.69207	.59116	.74678	.57846	.59927	.61494	.73725	.55490	.63877	.76220
	Rank	**5**	**6**	**1**	**2**	**11**	**9**	**10**	**3**	**13**	**4**	**8**	**7**	**15**	**12**	**14**	**16**

Table 8. Summary of importance's means and ranks of technical competency for construction manager based on respondents' positions

Positions		ST	LB	MO	TI	AD	CL	CT	MTMED	PLMED	SUMED	HEMED	QUMED	ENMED	PRMED	RSMED	COMED
Director	Mean	4.442	4.389	4.702	4.542	4.374	4.328	4.374	4.431	4.072	4.473	4.341	4.416	4.156	4.127	4.157	**3.988**
	Std. Deviation	.70340	.70790	.61637	.65936	.68316	.68436	.69433	.70918	.83002	.66307	.71068	.68262	.77232	.63905	.71531	.75820
	Rank	**4**	**7**	**1**	**2**	**8**	**11**	**9**	**5**	**15**	**3**	**10**	**6**	**13**	**14**	**12**	**16**
Project Manager	Mean	4.529	4.573	4.632	4.661	4.485	4.514	4.588	4.580	4.338	4.573	4.519	4.580	4.279	4.321	4.372	4.279
	Std. Deviation	.63412	.58120	.59612	.53561	.72261	.63464	.57912	.51551	.73526	.52022	.59121	.50822	.75004	.56855	.57743	.67128
	Rank	**8**	**7**	**2**	**1**	**11**	**10**	**3**	**4**	**13**	**6**	**9**	**5**	**16**	**14**	**12**	**15**
Contract Manager	Mean	4.476	4.333	4.761	4.619	4.571	4.571	4.428	4.523	4.238	4.523	4.428	4.571	4.071	4.359	4.238	4.047
	Std. Deviation	.60159	.65828	.43644	.58959	.74642	.74642	.81064	.51177	.80030	.60159	.65949	.74642	.67612	.54081	.63371	.80475
	Rank	**6**	**10**	**1**	**2**	**3**	**3**	**8**	**4**	**12**	**5**	**7**	**3**	**13**	**9**	**11**	**14**
Construction Manager	Mean	4.433	4.566	4.566	4.466	4.233	4.100	4.300	4.450	4.266	4.316	4.300	4.266	4.033	4.003	4.066	4.150
	Std. Deviation	.62606	.50401	.62606	.68145	.72793	.71197	.59596	.67403	.79582	.63631	.76988	.61214	.96431	.61725	.72397	.60387
	Rank	**5**	**1**	**2**	**3**	**11**	**13**	**7**	**4**	**10**	**6**	**8**	**9**	**15**	**16**	**14**	**12**
Site Manager	Mean	4.500	4.571	4.428	4.357	4.285	4.214	4.000	4.642	3.892	4.392	4.357	4.428	4.107	4.134	4.047	4.07
	Std. Deviation	.65044	.51355	.93761	.74495	.61125	.80178	.67937	.60219	.65570	.83617	.82134	.51355	.92359	.50105	.78291	.82874
	Rank	**3**	**2**	**5**	**7**	**9**	**10**	**15**	**1**	**16**	**6**	**8**	**4**	**12**	**11**	**14**	**13**
Site Supervisor	Mean	4.410	4.375	4.357	4.464	4.285	4.232	4.303	4.258	4.142	4.375	4.488	4.401	4.223	4.248	4.142	4.205
	Std. Deviation	.70780	.67588	.79609	.78542	.82494	.87368	.78438	.69407	.78459	.68258	.55764	.69032	.70013	.61299	.73266	.80214
	Rank	**3**	**5**	**7**	**2**	**9**	**12**	**8**	**10**	**16**	**6**	**1**	**4**	**13**	**11**	**15**	**14**
Arch. / Eng. / QS	Mean	4.679	4.471	4.717	4.603	4.434	4.358	4.452	4.584	4.339	4.537	4.572	4.500	4.169	4.283	4.245	4.339
	Std. Deviation	.51041	.63862	.49526	.59935	.69364	.70967	.60657	.53472	.63377	.57050	.53660	.60447	.75284	.56596	.61028	.78919
	Rank	**2**	**8**	**1**	**3**	**10**	**11**	**9**	**4**	**12**	**6**	**5**	**7**	**16**	**14**	**15**	**13**

4 Nonparametric correlation analysis

Due to the fact that all data are considered non-normal, thus, correlation analysis will be conducted through Spearman's Rho analysis. Prior assumption of single-dimensionality of items (i.e. all items contribute to construction manager's technical competency) facilitated the analysis, where their strength of associations was put to test amid diverse clusters in each respective control variables. In line, given that Spearman's Rho mutually relies on ranks of data, findings from the previous analysis were beneficial. Aside from determining the direction of correlations (i.e. positive or negative) among items across different control variables, their observable strength is very much significant in order to describe their degree of associations and further shed some light on subsequent focus towards appropriate control variables. Therefore, based on [16, 17], correlational strength index is defined as weak (at $r \leq 0.5$), moderate (at $0.5 < r < 0.8$), and strong (at $r \geq 0.8$).

Initially, the following table (see Table 9) presented a finding from correlation analysis for diverse contractors' grades. In summary, moderate to strong correlational strength is profound. Nevertheless, there is significantly weak correlation strength that can be detected especially towards G2 contractors towards G4, G5, and G7. On the other hand, as respondents' diverse experience (see Table 10) and positions (see Table 11) is concerned, correlational strength index of moderate to strong dominates the tabulations, except for one; between contract managers and construction managers. All correlations denote positive values which eventually manifested positive linear correlations, where increment values for both cluster of control variables were judged to be perpendicular to each other.

Table 9. Summary of correlation analysis across different contractors' grades

			G1	G2	G3	G4	G5	G6	G7
Spearman's rho	G1	Correlation Coefficient	1.000	.700**	.891**	.712**	.715**	.924**	.756**
		Sig. (2-tailed)	.	.003	.000	.002	.002	.000	.001
		N	16	16	16	16	16	16	16
	G2	Correlation Coefficient	.700**	1.000	.594*	**.476**	**.312**	**.653****	**.482**
		Sig. (2-tailed)	.003	.	.015	.062	.240	.006	.058
		N	16	16	16	16	16	16	16
	G3	Correlation Coefficient	.891**	.594*	1.000	.744**	.682**	.918**	.797**
		Sig. (2-tailed)	.000	.015	.	.001	.004	.000	.000
		N	16	16	16	16	16	16	16
	G4	Correlation Coefficient	.712**	.476	.744**	1.000	.606*	.738**	.815**
		Sig. (2-tailed)	.002	.062	.001	.	.013	.001	.000
		N	16	16	16	16	16	16	16
	G5	Correlation Coefficient	.715**	.312	.682**	.606*	1.000	.721**	.774**
		Sig. (2-tailed)	.002	.240	.004	.013	.	.002	.000
		N	16	16	16	16	16	16	16
	G6	Correlation Coefficient	.924**	.653**	.918**	.738**	.721**	1.000	.903**
		Sig. (2-tailed)	.000	.006	.000	.001	.002	.	.000
		N	16	16	16	16	16	16	16
	G7	Correlation Coefficient	.756**	.482	.797**	.815**	.774**	.903**	1.000
		Sig. (2-tailed)	.001	.058	.000	.000	.000	.000	.
		N	16	16	16	16	16	16	16

Table 10. Summary of correlation analysis across different respondents' years of experience

			4 years and below	5 years - 9 years	10 years and above
Spearman's rho	4 years and below	Correlation Coefficient	1.000	.726**	.756**
		Sig. (2-tailed)	.	.001	.001
		N	16	16	16
	5 years - 9 years	Correlation Coefficient	.726**	1.000	.835**
		Sig. (2-tailed)	.001	.	.000
		N	16	16	16
	10 years and above	Correlation Coefficient	.756**	.835**	1.000
		Sig. (2-tailed)	.001	.000	.
		N	16	16	16

Table 11. Summary of correlation analysis across different respondents' positions

			Director	Project Manager	Contract Manager	Construction Manager	Site Manager	Site Supervisor	Architect and etc.
Spearman's rho	Director	Correlation Coefficient	1.000	.868**	.788**	.806**	.762**	.726**	.879**
		Sig. (2-tailed)	.	.000	.000	.000	.001	.001	.000
		N	16	16	16	16	16	16	16
	Project Manager	Correlation Coefficient	.868**	1.000	.735**	.826**	.565*	.668**	.832**
		Sig. (2-tailed)	.000	.	.001	.000	.023	.005	.000
		N	16	16	16	16	16	16	16
	Contract Manager	Correlation Coefficient	.788**	.735**	1.000	**.469**	.555*	.558*	.711**
		Sig. (2-tailed)	.000	.001	.	.067	.026	.025	.002
		N	16	16	16	16	16	16	16
	Construction Manager	Correlation Coefficient	.806**	.826**	.469	1.000	.688**	.659**	.865**
		Sig. (2-tailed)	.000	.000	.067	.	.003	.006	.000
		N	16	16	16	16	16	16	16
	Site Manager	Correlation Coefficient	.762**	.565*	.555*	.688**	1.000	.671**	.741**
		Sig. (2-tailed)	.001	.023	.026	.003	.	.004	.001
		N	16	16	16	16	16	16	16
	Site Supervisor	Correlation Coefficient	.726**	.668**	.558*	.659**	.671**	1.000	.785**
		Sig. (2-tailed)	.001	.005	.025	.006	.004	.	.000
		N	16	16	16	16	16	16	16
	Architect and etc.	Correlation Coefficient	.879**	.832**	.711**	.865**	.741**	.785**	1.000
		Sig. (2-tailed)	.000	.000	.002	.000	.001	.000	.
		N	16	16	16	16	16	16	16

5 Conclusion and way forward

The authors tried to disseminate critical information through in-depth analysis towards collected data pertaining the importance of technical competency of construction manager.

In this instalment, a focus has been made towards screening processes of quantitative data from questionnaire surveys conducted nationwide. Subsequently, central tendency analysis based on mean towards various control variables was subjected to data, before correlation analysis finalised these preliminary procedures. The main purpose of the analysis is not just to reduce prior assumption regarding the nature of data, but to shed light on substantive control variable to be focused on successive stages.

During screening processes of data, aside from exploring data dimensionality [18], the authors stressed on outliers test, reliability test, and normality test as basis for preliminary analysis. Prior to that, collected responses were sieved thoroughly, where incomplete, blank, and unsuitable responses (such as giving improper responses' signs) were rejected in order to maintain clean and full-bodied data. Following the results of screening undertakings, data were free from outlier and showed high reliability. Apart, normality test was confirmed with non-normal nature of data which typically associated with Likert scale, where data are considered ordinal. As far as the research is concerned, outlier's data will influence subsequent analysis, where apparent variability of findings might appear in the latter stages. More often than not, outlier's data will be treated in the first place by researcher before proceeding towards main analysis. Besides, high reliability value (α) in the paper (see Table 1) is very close to the authors' prior findings [18] through Rasch reliability of item and person. Therefore, the authors were optimist that the internal consistency of data is acceptable. In addition, since data are observed to be ordinal, careful selection of subsequent tests is paramount in order to maintain relevant findings, where reliance towards non-parametric type of analysis is inevitable.

Analysis of central tendency (i.e. mean) outlays comparable findings, where recorded means of items' important level are in the range of important (4.00) to very important (5.00). Only handful cases of lower than important (4.00) can be found especially towards different clusters of contractors' grades. By referring to Table 6, contractors of grade 5 and 6 recorded sub 4.00 for PLMED and COMED, which turned the items into among the lowest ranks. On the other hand, largely, correlation strength of items (which reduced to macro level of 16 items) was within the acceptable range of moderate to strong, despite diverse clusters across multiple control variables. Nevertheless, similarly with findings from previous analysis, weak correlations were mainly identified within several contractors' grades especially towards grade 2 contractors (see Table 9). Although positive correlations were recorded, lower correlation index (at $r \leq 0.5$) signifies questionable degree of association among responses which eventually affect their relationship.

In a nutshell, the findings in this paper echoed previous authors' exploration on central phenomenon in hand as can be perused in Yaman et al. and Mohammad et al. [7, 6]. Due to rigorous processes of initial qualitative endeavours, variability of results throughout these quantitative undertakings was kept at minimum. Nonetheless, further analysis is indeed needed in order to seek significance of items (e.g. their critical *p*-values) especially towards different contractors' grades since the existence of discrepancies eventually posted several latent inquiry, which usually relates to 'why' the circumstances persisted. Moreover, generalisation of findings (e.g. through Kruskall-Wallis H test) across various contractors' grades is also paramount before establishment of technical competency of construction manager can be made.

References

[1] Construction Industry Development Board, *Construction industry transformation programme (CITP) 2016-2020: Driving construction excellence together*, Kuala Lumpur, (2015)

[2] H. Abdul-Rahman, M.A. Berawi, A.R. Berawi, O. Mohamed, M. Othman and I.A. Yahya, Delay mitigation in the Malaysian construction industry, J. of Construction Engineering and Management, **132**(2), 125-133 (2006)
[3] A.R. Ibrahim, M.H. Roy, Z. Ahmed and G. Imtiaz, An investigation of the status of the Malaysian construction industry, Benchmarking, 294-308 (2010)
[4] National Audit Department Malaysia, *Auditor general's report: Government's financial statement, financial management for the year 2013 and activities of the federal ministries/departments and management of the government companies (series three)*, Kuala Lumpur, (2013)
[5] Unit Penyelarasan Pelaksanaan, Jabatan Perdana Menteri, *Pengurusan projek sektor awam. bengkel penggubalan modul dan kaedah pelaksanaan kursus pensijilan pengurusan projek sektor awam*, Kuala Lumpur, (2013)
[6] H. Mohammad, S. Yaman, F. Hassan and Z. Ismail. determining the technical competencies of construction managers in the Malaysia's construction industry, MATEC Web of Conferences, **47**, 04021 (2016)
[7] S. Yaman, A. Abdullah, H. Mohammad and F. Hassan, technical competency of construction manager in Malaysian construction industry, Applied Mechanics and Materials, **773-774**, 1053-1059 (2015)
[8] S. Walfish, A review of statistical outlier methods, Pharmaceutical Technology, **30**(11), 1-5 (2006)
[9] R. Dawson, How significant is a boxplot outlier?, J. of Statistics Education, **19**(2), 1-13 (2011)
[10] J. A. Gliem and R.R. Gliem, Calculating, interpreting, and reporting Cronbach's alpha reliability coefficient for likert-type scales, *Midwest Research to Practice Conf. in Adult, Continuing, and Community Education*, Ohio, (2003)
[11] M. Tavakol and R. Dennick, Making sense of Cronbach's alpha, Int. J. of Medical Education, **2**, 53-55 (2011)
[12] A. Ghasemi and S. Zahediasl, Normality tests for statistical analysis: A guide for non-statisticians, Int. J. of Endocrinology and Metabolism, **10**(2), 486-489 (2012)
[13] J.W. Creswell, *Educational research: Planning, conducting, and evaluating quantitative and qualitative research* (3rd ed.), Pearson, Ohio, (2008)
[14] D.P. Doane and L.E. Seward, Measuring skewness: A forgotten statistic?, J. of Statistics Education, 19(2), 1-18 (2011)
[15] H.Y. Kim, Statistical notes for clinical researchers: Assessing normal distribution (2) using skewness and kurtosis, Restorative Dentistry and Endodontics, 38(1), 52-54 (2013)
[16] P. Singh, P.A. Ghani and T.S Hoon, *Quantitative data analysis for novice researchers*, Primera Publishing, Kuala Lumpur, (2009)
[17] N. Ghazali, S. Yaman and H. Mohammad, Contractors' compliance on occupational safety and health (OSH) policies in Malaysia's construction industry, *Proc. 8th MUCET*, Melaka, (2014)
[18] H. Mohammad, F. Hassan, R. Abd Rashid and S.K. Yaman, Dimensionality analysis of technical competency for Malaysian construction managers, *Int. UNIMAS STEM Engineering Conf.*, Kuching, (2016)

An Investigation Into Design Criteria for Affordable Housing Supply

Abdul Lateef Olanrewaju[1,*], *Lim Tat* Lee[1], *Yeow Tan* Seong[1], *Naoto* Mine[1], and *Chai Teoh* Teng[1]

[1]Faculty of Engineering and Green Technology, Universiti Tunku Abdul Rahman, 31900 Kampar, Perak, Malaysia

Abstract. Affordable housing provision constitutes a very large scheme in any countries due to income distribution and national development. The supply of housing depends on many activities and processes. The purpose of the houses is to meet the requirement of the householders therefore design criteria must address the users' requirements. The establishment of the design criteria constitutes an important activity at the initial phase of the housing development. The design criteria are prepared by the design team in collaboration with many stakeholders especially the householders. Housinsg design criteria are the requirements that must be considered prior to construction of the housing. Lack of adequate information on the design criteria would lead to poor householders' satisfactions, increase in maintenance cost, abandonments and completed but not occupy housing. In Malaysia, many of these consequences are prevalent. However, while information on the house owners' requirements is inconclusive, this current research set out to investigate the design criteria of affordable housing. The increase in the affordable housing gap in Malaysia can be reduced if the designers have a comprehensive understanding of users' requirements and the design criteria. Through a cross sectional survey questionnaire, comprising 25 criteria, 7 criteria were found to be extremely critical. The Kaiser-Meyer-Olkin measure of sampling adequacy indicated that the strength of the relationships among variables was strong (KMO =0.716). Bartlett's test of sphericity, which tests the overall significance of all the correlations within the correlation matrix, was significant χ^2 (325) = 1825.075, $p<0.001$), indicating the data were drawn from the same population and that the criteria were related.Sustainability considerations are now being considered by the providers of affordable housing. Deductively, the results lead to the conclusion that a major factor responsible for the poor homebuyers / homeo ccupiers dissatisfactions is the lack of considerations of the criteria that impact significantly on the buyers/occupiers choice. The findings are significant for developers, urban planners, banks, policy makers and the academia.

* Corresponding author: abdullateef.olanrewaju@ymail.com

1 Introduction

The aim of the study that this paper reports part of is to analysis the supply and demand sides of affordable housing in Malaysia. This current study addresses itself to the design criteria in affordable housing. Olanrewaju, et al. (2016) noted the importance of examining the design criteria and constraints in affordable housing delivery, especially in the contexts of meeting the sustainability requirements [1]. However, the major difference between design criteria and design constraint is that while design criteria are requirements that must be met or take into account during the concept stage of the housing development by the developers/design team, the design constraints are restrictions that keep the design from being able to be the best design. It is noted that while there is information on housing determinants from the householders' perspectives, same cannot be claimed for the supply side. There is a dearth of information on the criteria on which designers based their decisions on housing development in general and to affordable housing in particular. It is likely that a major problem responsible for the imbalance in the supply and demand for affordable housing is the disparities between the design criteria and householders determinants in the affordable housing. If the design criteria and the homebuyers' requirements are not aligned, there are high tendencies that the increase in the housing abandonment and housing project abandonment will continue on abated. There will also be increased in maintenance and refurbishment costs. There will also be increased in home occupiers' dissatisfactions. Basic constraints in affordable housing include the price cap, locations, and quality specifications. For instance, affordable is generally seen to be cheap at least compare to standard houses. This is the problem that the developers have in affordable housing delivery. Hence, there is a constraint on how much the developer will charge for the affordable housing. But since the developers perceive it such, unfortunate though how this may be, there is limitation on the selection of material and components to use, and the quality of labour to employ. This will also affect the site selection. This explains why most affordable housing schemes are located away from the cities and with less social amenities and recreation facilities. This as well explains the poor performances of the affordable housing market. The option for affordable housing is always complicated by the design economic factors. Previously only size and quality standard is considered to influence the cost and the price of building [2]. However, it is now obvious that many different but interrelated criteria do not only influence the housing price, but determine the home buyers' buying/renting decision making. Ultimately, the various criteria have economic implications on home ownerships' or occupier-ships', long term plan. In Malaysia, many housing developments have failed to perform their design functions and too expensive to own and operate on account of their inabilities to consider the various Home buyers' and home users' requirements. Some investigations into the design management are therefore requiring critical investigations. The design criteria is a fusion of home owners requirements and design constraints, hence if the criteria can be identified and ranked and given the respective significance in the solving design problems it should be possible to provide value for money invested in accordance with the basic requirements. Such design will be the best design for the Home buyers' -"designed for home buyers"-. Designing for the buyers is the only feasible solution to ensure Home buyers' satisfactions. Home buyers' satisfactions can be measured in different ways. But today, Home buyers' are concerned about many factors apart from the price of the unit/homes only. For instance, they are concerned about how much they spend on transportation, how far is their home to the workplaces, hospitals, markets, pay for utilities, access to recreations, maintenance and refurbishment costs, future modifications. It is imperative that all stakeholders and especially Home buyers' experiences are well considered in the development of the design criteria. However, it is important to realise that design management is not a pure science rather a social science, implying that knowledge of economics, sociology, psychology, laws, culture and politics will contribute significantly to the development of the design criteria. The purpose of this study was to obtain indications of the design issues obstructing the supply of affordable housing. For this reason, there is a need to be more careful over priorities, so to enhance the delivery of affordable housing supply. Therefore,this study investigates the critical design criteria in order to provide some suggestions and comments of the developers, policy maker and other relevant stakeholders in the delivery of affordable housing supply.

2 Affordable housing and design management

2.1 Affordable housing supply and challenges

To clarify, 'affordability' is an elastic word. At one end, a product is very affordable yet at the other end, it is [totally] unaffordable to the same group of buyers. This makes the issue of affordable housing somewhat complex to understand and explain. Thus 'affordable housing' is defined differently. Often, the term is used quite loosely to represent all types of housing developed by governments, its agencies or in partnership with the developers or social landlords. It will seem that the meaning of affordable housing is more easily defined than understood. The question then arises; at what point is the house that is not affordable to the same buyer was affordable earlier? Effectively, it means household income is a critical issue. Yet, obviously, affordable house carries different meaning to different people. However, to provide context for the definitions and interpretations, housing is 'affordable' if it rental cost, or mortgage repayment cost does not exceed 30 per cent of household income for households in the lowest 40 percent of the income distribution range [3]. For rental purposes, the 30% would include for utility bills, including for electricity, water, gas, sewage and garbage collection. In case of the mortgage, the amount includes the actual consideration, tax, insurance, utilities and maintenance. For instance, cumulatively, monthly expenditure for the rental is 26.7% of the household income [4].However, for the low and middle income earners is 35% [5]. In fact, 2014 inflation in the housing, water, electricity, gas and other fuels category increased to 3.4% (2013: 1.7%), reflecting the upward adjustment in electricity tariffs and a broad-based increase in rents across different types of residential property [6]. Affordable housing is also defined in term of 'median multiplier' [6]. The Demographic compares the median house prices to the median household income to measure affordable housing based on annual international Housing Affordability Survey. The median income in Malaysia, 2016 was estimated at RM5720. As such contextualising the model into the Malaysia scenario, over three years will imply RM205, 000. As such the initial question that possible comes to mind, is what type of house will this amount be able to procure in the cities? The initial answer is none. This is amount is inadequate to buy a house in any of the major cities. This is because less than 50% the housing units built in the year 2015, had a price tag lower than RM 250,000 [7].

"One of Malaysia's longstanding development objectives is the provision of affordable housing for Malaysians in both rural and urban areas, with a focus on lower-income groups" [8]. Therefore, addressing the housing need for the low and middle income earners is very important as they form the basis of the national human capital. The provision of affordable and liveable social housing is very critical, particularly as the government is committed to encouraging greater home ownership among the 'bottom 40% households'. The intense pressure on the provision and maintenance of affordable housing are extreme. By 2020, over 70% of Malaysians are expected to reside in the urban areas. Housing demand and requirements are expected to increase remarkably due to the rapidly growing in population, increase in foreigners (expatriate, students and tourists), migration, changing economic status of the citizens, changes in tastes, and dilapidation of the existing stock. At 2020, the population is estimated to reach 32.4 million. However, there is a high level of insufficiency in home supply. In fact, lack of affordable house is now a source of concern among Malaysians. To illustrate, available statistics show that Malaysia currently has 4,928,883 representing an increase of 1.99% compared to 2014 (i.e. 4,831,791 actually) residences and of the total properties transacted 65% is residential and represents 50.4% in terms of value [9]. That, is home production is approximately 97,092 per annum. However home production has decrease compared to 2014 to 2013, where about 107,000 homes were provided. The plausible implication of these statistics is that there are about 6 persons per home. However, this is inadequate to achieve the above set objectives, countries like Australia and UK have on the average of 2.5 persons per home. Malaysia's growing population is increasing at a rate of about 1.7% a year, that is, in a year approximately 500,000 persons are added. To interpret, in one year, about 400, 000 persons will not have homes to live. This will lead to increase in the prices of home, or many will have to pay unaffordable prices to own or rent a home. Secondly, there will be an increase in maintenance demand. Because, a major factor that determine maintenance demand is the size of the households [10].Lack of affordable housing would lead to other social problems including traffic congestions, air pollution, accidents and urban sprawl. Hence, the existing housing stock would not be part of national gross fixed capital formation that is required in a country to stimulate and supports national development. Therefore policy makers need to take the supply-side imbalance into consideration to provide accommodations that are affordable in all respects. While incomes have increased by 101% since last ten, housing price has increased by corresponding 213% within the same periods. Table 1 contains the median income, since 1997.

Table 1.Household median income distributions (RM), 2006 – 2016 [11]

Period / Area	1997	1999	2002	2004	2007	2009	2012	2014	2016
Malaysia	1,724	1,704	2,049	2,211	2,552	2,841	3,626	4,585	5,720

The government has introduced various schemes to increase home-ownership among Malaysians (e.g. Skim Rumah Pertamaku, Rumah Mesra Rakyat, and Perumahan Rakyat 1 Malaysia). The recently introduce overnight policy rate (OPR) introduced by the Bank Negara is also expected to have some positive impact on home ownerships since payment on the mortgage will be lower. The OPR will see reduction interests on the saving account and deposits are also expected to spur investors to consider investing their savings and deposits in the housing sector. However, with the recent global meltdown cause mainly the housing sector would mean investor will be very careful. Home-ownership was 63% in 2000; this dropped to 55% in 2010 [11]. To interpret, there is an annual decline of about 1% in home-ownership. Home-ownership will thus be less than 50% by 2020 if the housing supply in not revised. While affordable housing is widening, issues pertaining to accessibility to workplaces, sizes, maintenance cost, incessant refurbishments and alterations, modifications, operations, remain major problems that require urgent attention. Many of the problems are related to the design of the housing. Quality degradation was highly due to no fixed design criteria has been set to housing development [12].

2.2 Design management of affordable housing

Housing delivery involves various sequential and sometimes overlapping processes and phases depending on the stakeholders and procurement strategy adopted. The RIBA 2013 Plan of Work identified 8 phases of the building delivery processes comprising of: strategic definition, preparation and brief, concept design, developed design, technical design, construction, handover & close out and in use. The strategic definition, preparation and brief, concept designs are the most critical as they involve the project definition and design phases. Many professional parties involve in the design phases of housing delivery, notably architects, construction managers, quantity surveyors, services engineers and structural engineers. Often, based on the traditional procurement strategy,the architects provide the prime consultancy duties and roles. Irrespective of the procurement strategy and clients' characteristics, the importance of the strategic definition, briefing and design are strategic. It is in these phases that the housing needs, homebuyers' requirements, criteria and constraints of the housing scheme is defined, refined, and adopted. Once the phases are completed, the complete design is produced, and once the complete design is produced, about 70% of the construction cost has been determined. Unfortunately, except for individual private funded dwellings, the requirements of the home buyers are not conclusive at this stage. The future owners are not known. The requirements are only based on the design team's experiences. This often complicates the design management for public housing and housing provided by the developers. However, inadequacies analysis of the home buyers' requirements, and the housing constraints would lead to poor design criteria developments and this in turn leads to variations, cost overrun, delays, poor quality and other problems [17]. Ultimately, this will manifest in homeowners' dissatisfaction during the use phase of the housing. The dissatisfactions can come in terms of high maintenance cost, high utility bills, modification and refurbishment and location problems. The location problems may be due to lack of access to workplaces, children'd schools, recreation and amenities. Some homebuyers are dissatisfied if the house failed to promote social integration. The increase in awareness of pollution, waste, degradation and climate changes has also brought achieving sustainability requirements into the main focused, with many homeoweners, now questioning the sustainability credential of developers and contractors [1].

Design quality is a combination of functionality, building quality and impact of buildings. In most any project type, design quality can be uniquely defined for each facility. Yet, there are some general considerations that apply to most projects which are the efficient use of space, positive environmental impact, secure and safe facilities, energy efficient buildings, accessibility [14]. A good design involves balancing of community values, individual needs, aesthetic judgements, technical requirements. High design quality buildings ensure high productivity of the occupiers. Good design will provide value for money in terms of total cost and cost-in-use [15]. Aside from the great performance that quality design brings, better design housing would sell for higher values and at faster rates than others. Decisions made at the outset about size, form, materials, layout and aesthetics affect all the generations who reside in that house. The design criteria are the requirements that the design team uses as a benchmark to finalise the design. The design criteria are the product of design constraints, homebuyers' requirements, building codes and standards. The determinations of the design criteria affect developers' revenues and profits. The homebuyers' housing requirements on the other hand, are the criteria upon which buyers based their decisions on when completing their buying decisions. The criteria are extremely critical to them because they influence their satisfactions, rents, mortgage payment, maintenance costs, and transportation costs to workplaces, hospitals, market, schools, and other cost operations. It is very critical that both the homebuyers, housing requirements and design are congruently aligned.Therefore, decisions made during before and during the design phase are crucial to homeowners' satisfactions. Therefore, the design team and developer need to evaluate the design criteria for housing development to maximise their profit and increase homebuyers' satisfactions. The design criteria are strategic and should be finalised prior to the construction phase.

3 Outline of resigning method

In general, two types of questions exist in research design, namely the descriptive question and explanatory question. But this is sometime referred to as descriptive research and explanatory research respectively. The distinction between the two is pertinent here for a number of reasons. While the descriptive question is concerned with what is currently going into a 'system' the explanatory question addresses itself to giving insight into why that is going into a system. But, since the essence of this research is to incite the why question, this research is descriptive in nature. In other words, it aims to describe the design criteria that those concerns with affordable housing delivery based their decisions on. There is need to examine these criteria, there seems to be mismatch between users' requirements the affordable housing. This study consists of a literature review and survey. The questionnaires were administered to the respondents attending the ARCHIDEX (International Architecture, Interior Design & Building Exhibition 2016) in the Kuala Lumpur Convention Centre. ARCHIDEX is help yearly and attending by architects and other members of the building design teams (i.e. engineers, quantity surveyors) in Malaysia and around the South East Asia countries. The ARCHIDEX 2016 was held on 20 July, 2016 to 23 July, 2016 and attended by more than 3000 delegates and exhibitors. This survey was conducted 23 (Saturday) July 2016. Altogether, 25 design criteria that are used in various housings across the world were addressed to respondents. The criteria were identified based on a literature review [i.e. 16-20] and the authors' experiences. Respondents were asked based on their current experience, to tick the criticality to which the criteria are considered in the design of affordable housing on a four continuum scale; where 1 denotes extremely critical, and 4 denotes not critical. Two and 3 are located in between. The criticality of the criteria is determined by the mean score. For interpretation, mean score of 1.00- 1.75 denotes extremely critical; 1.76-2.50 denotes very critical; 2.51-3.25 denotes critical and 3.26 - 4.00 denotes not critical. The mode technique was used to analyse the demography of the respondents. Mode was also used to determine the distribution of respondents with respect to the scales.

4 Analysing the results of the survey

4.1 Analysis of respondents' profile

All data gathered adopted Statistical Package for Science Social (SPSS) version 20 for analysis. Altogether 111 completed survey forms were returned and the results of which are contained tables 2, 3 and 4 and Fig. 1. More than 70% of the respondents have obtained university degrees with more than 20% obtained either MSc. or Ph.D. (Table 2). Most of the respondents have either architectural or engineering background. Half of the respondents are either from building developer organisations or private architectural firms (Table3). Many also from the government sector as well as private quantity surveying firms and contractors organisations. Most help strategic positions in their organisations. About 75% have been working for more than five years and more than half have more than 10 years working experience (Fig. 1). More than 90% belong to various notable organisations in the Malaysian construction and housing industries (Table4). 41% of respondents were the members of the Board of architects Malaysia while 28% were members of the Board of Engineers Malaysia. 16% and 7% were members of the Board of Surveyors Malaysia and members of REHDA. On this basis, it is inferred that the respondents have adequate knowledge and skill have the capabilities and abilities to provide unbiased and valid information.

Table 2. Distribution of respondent's qualification

Qualification	Diploma	BSc.	MSc/MBA	Ph.D.	Others
Frequency (%)	24.3	51.4	19.8	1.8	2.7

Table 3. Distribution of respondent's professional background

Professional background	Quantity surveying	Engineering	Architecture	Estate management	Town planning	Others
Frequency (%)	18.9	27.0	40.5	2.7	0.9	9.9

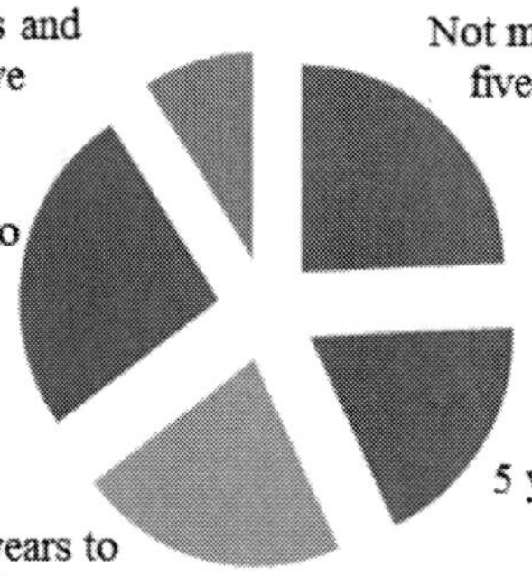

Fig. 1. Distribution of respondent's working experience

Table 4. Distribution of respondent's position

Position	Managing director	Manager	Project manager	Principal partners	Quantity surveyor	Architect	Engineer	Others
Frequency (%)	4.5	9.9	14.4	1.8	15.3	27.9	17.1	9.1

4.2 Analysis design criteria

In order to test the strength of the criteria, both Crunch alpha reliability and Extraction communities were performed. From the results contain in Table 5 all the criteria were very reliable and highly valid except for maintenance cost that obtained a validity score slightly below the recommended score of 0.5. The KMO also produced 0.716 signifying the lack of multicollinarity problems among the variable and that the respondents were drawn from those with similar experience. Since the Bartlett's Test of Sphericity is significant is confirmation that all the criteria are related. A one way t-test was conducted, the results of which were not reported here due to space constraint shows that all the criteria were significant ($p=0.001$), indicating that the population will probably measure all the critical criteria. Therefore, all the criteria are good indicators and appropriate for the aim of the study. Moving forwards, A K-S Test was conducted to tests the null hypothesis that the criteria are normally distributed. We find that the K-S Tests for most of the variable are not significant thus for most of the variables. In other words, the H_0 is accepted. Therefore, the sample is a representative of the population. To determine the degree of criticality of each of the criteria, the mean score for each was computed and contain in Table 5. The mean scores range from 1.57 through 2.98. While the safety of occupiers and security of property was the highest criterion and co-locating of outdoor facilities was the lowest criterion. The standard deviations also indicate that the disparities of measurement were suitable. Using the developed index scale in section 3 above, 7 criteria were extremely critical, 14 criteria were very critical and 8 criteria were measured critical. As may be seen none of the criteria obtained score up to 3.26 signifying that none of the 25 criteria was not critically considered in the design of affordable housing.In general, the results were consistent with our undisclosed hypothesis. It is not surprising that the respondent considered the safety of the building occupiers and security of their properties as major criteria in the affordable housing delivery. The increase in maintenance costs and future medications were also ranked as very critical. This is not unexpected. Researches have revealed that the operational costs of housing far outweigh the capital costs [21-22]. The quality of the environment also emerged as critical considerations. There are many cases of police reports on the behaviours of some owners of adjacent buildings. This is also related to location factor. For instance, if affordable housing is poorly located close to industrial areas, it will always affect the owners of buildings in the surrounding areas. However, the characters of the surrounding building turned out to be a very critical consideration by the design team/ developers in affordable housing supply. But issue relating to the location and surrounding areas of the affordable housing is quite complex. It may be that affordable housing is by policy not sited close to high end price for the negative perception that affordable housing has a negative implication on the property price. But while there are continuing oppositions to locating affordable housing in the cities, based on the review of seventeen studies that have attempted to measure the effect of affordable housing on property values it was found that the extent to which property values are lowered depends many factors [23]. The research indicates that the concentration of affordable housing is only a factor among many that affect property value.

Table 5. Distribution of descriptive statistics of the design criteria

Criteria	Validity (communalities)	Standard deviation	Mean	Ranking
Safety & security of occupier & property	0.725	0.756	1.57	1
Maintenance costs	0.498	0.824	1.58	2
Minimisation of noise & nuisance	0.836	0.776	1.60	3
Extent of required future modification	0.726	0.849	1.69	4
Waste disposal	0.717	0.783	1.69	5
Character of the surrounding buildings	0.720	0.783	1.71	6
Element/component cost	0.729	0.780	1.74	7
Use products & processes adapt to climate change	0.765	0.932	1.85	8
Water bills	0.909	0.793	1.86	9
Energy bills	0.773	0.800	1.87	10
Telephone bills	0.904	0.850	1.88	11
Use of local labour	0.784	1.058	1.92	12
Use of local materials	0.796	1.041	1.92	13

Use products & processes that reduce environmental impacts	0.706	0.873	1.94	14
Variety in housing types	0.643	0.991	2.19	15
A sense of place	0.707	1.026	2.21	16
Co-locating indoor activity areas in the building	0.711	0.937	2.43	17
Cater for senior citizens	0.803	0.965	2.43	18
Connections to local transit service	0.775	0.995	2.49	19
Provide views to the outdoors from indoor rooms	0.882	1.048	2.49	20
Access to recreation facilities by the households	0.870	1.013	2.52	21
Cater for disables	0.850	0.939	2.54	22
Locate stairs near the building main entrance	0.853	0.905	2.60	23
Access to employment opportunities of the household	0.731	1.043	2.71	24
Co-locate outdoor elements	0.531	0.949	2.98	25

Kaiser-Meyer-Olkin Measure of Sampling Adequacy=716; Bartlett's Test of Sphericity= (325)1825.075, Sig. = 001.

This current research founds that issue of waste disposal and cost of the building element was also very critical consideration by the design team. In other words, the design team considers waste management in the delivery of affordable housing. The design team also takes critical analysis of costs of elements in affordable delivery. In other words, the developers are probably very careful not to specify expensive materials and components for affordable housing delivery. This is very interesting because affordable housing is generally defined to mean cheap or low quality housing. Therefore, this finding is not surprising. In general, while some of the lower ranked criteria expected some were not expected. For instance, it is not surprising to find that accesses to employment opportunities of the household are and include rooftop gardens not critical, it is surprising to find that creation of quality neighbourhoods and co-locate outdoor elements were rated low. These findings are rather complicated and difficult to interpret. However, it is crucial that the quality of neighbourhood be considered in locating of affordable housing. This is to ensure social integration and community well-being. The qualities of the neighbourhood also include the availability of good health care facility, schools, and good roads, public parks, playing grounds, market and recreational amenities. But since the qualities of the neighbourhoods are not important, it will not be important to co-locate outdoor elements. It is only when the neighbourhoods are vibrant that sharing and using basic outdoor facilities like sport complex will be useful.

5 Conclusion and recommendations to the design teams and developers

This study has investigated the design criteria for affordable housing. Many criteria actually influence the design teams' decisions. Sustainability issues were revealed to be major determinants. However, ranking the criteria for housing development is a daunting task since development is distinct. Yet this information will guide the stakeholders, especially the design team about which design criteria are uniquely necessary for a development. The theoretical purpose is to ensure that homebuyers are satisfied with the homes and that the developers would be able to make 'sufficient' profit margin. In order to ensure homebuyers'satisfactions are maximised, it is very necessary that the house is designed for that purpose. The concept of 'design for homeowners' should be considered a practical possibility. This will ensure the completed homes meet the requirements of the buyers. To be strictly designed for the homeowners, housing ensures that little amount is spent on energy and waters used and little waste is generated/disposed. The house should be designed with a purpose or promoting occupants and community well-being and social integration. The house should be designed to be constructed and operated with reuse and recycle materials and components. Be readily maintained, refurbished and modified when required. Should use local materials and labour its construction and operation. The salvaged materials on demolition should be recycled and reuse and to be nontoxic. The occupiers will not pay more than required for transportation to workplaces, hospital, parks, and markets on account of them living in the house (location!). The house will not contribute to crime and vandalism rates in the neighbourhoods. And it should not destroy biodiversity. While, this study has been able to provide some preliminary information on the design issues of affordable housing in Malaysia, the readers should be aware that the design management, especially for affordable housing is very complex and sophisticated because various stakeholders, including contractors, urban planners, government, developers, housing associations, banks, service providers (for water, electricity, waste disposal) and insurances are involved. Often due to poor policy formulation and implementations and separation of roles it is difficult to identify the roles of each of the stakeholders. However, while this research has also been able to identify and prioritise the design criteria, there is the need to categorise these criteria for meaningful interpretation. Without categorising the criteria, decision making based ranking alone would not be rigorous. The combine impact of certain criteria will not be obvious and thus lead to poor performance. Therefore, a follow up article will fill this gap, using the factor analysis technique. Furthermore, further researches are useful to improve this research by 1) ways of increasing the sample size and 2) to validity the findings. There is no similar previous study to provide some reference for comparisons. It is equally important to investigate the requirements of the affordable households with respect to what specifically they expected from their houses. Our next research also seeks to develop a Critical Success Factor (CFS) for affordable housing policy makers, urban planners, housing developers, banks and relevant parties.

The material reported in this paper is fully supported by FRGS Grant under No. FRGS/1/2015/TK06/UTAR/02/2. The financial contribution is highly appreciated.

References

[1] A. Olanrewaju, S.T. Tan, L.T. Lee, M. F. Ayob, S.Ang, Investigating the compatibility of affordable housing with sustainability criteria: A conceptual framework, *Proc. of Putrajaya International Built Environment, Technology and Engineering Conf.*, 228-240, Bangi, Malaysia, (2016)
[2] A. Ashworth, Cost studies of buildings. 4th ed., Pearson Education Limited, Harlow, England, (2004)
[3] J. D, Dacquisto, R.D, Rodda, *Housing impact analysis*, U.S. Department of Housing and Urban Development, Government Printing Office, Washington, USA, (2006)
[4] Department of Statistics, *The Malaysian economy in brief*, Malaysia, (2014)
[5] Department of Statistics, *Malaysia report on household expenditure survey*, (2011), Retrieved on March 14, 2014 from http://www.statistics.gov.my/portal

[6] *Demographia 11th annual international housing affordability survey 2015 ratings for metropolitan market*, (2015), Retrieved on March 5, 2015 from http://www.demographia.com/dhi.pdf
[7] *National property information centre change value of transactions by price range for the prinsipal property sub-sectors*, (2015), Retrieved on August 29, 2016 from http://napic.jpph.gov.my/portal
[8] Economic Planning Unit Prime Minister's Department, *Tenth Malaysian Plan 20011-2015*, , Putrajaya, Malaysia, (2010)
[9] National Property Information Centre, *Summary of existing stock for residential, commercial, leisure and industrial property sub-sector as at Q1-Q4 2015* (2015). Retrieved on June 06, 2016 from http://napic.jpph.gov.my/portal
[10]A.A. Olanrewaju, Green maintenance management initiatives for university buildings, Built Environment Journal, **8**(1), 17-24, (2011)
[11]Department of Statistics, *Report on characteristics of household 2010: Population and housing census of Malaysia,* Malaysia, (2014)
[12]E.M. Kamal, H. Hassan, A. Osmadi, Factors Influencing the housing price: Developers' perspective, Int. J. of Social, Behavioral, Educational, Economic, Business and Industrial Engineering, **10**(5), 1603–1609, (2016)
[13]A. Musa, N. Mohd Tawil, A. Che Ani, B. Hassan, An analysis of the selection criteria in purchasing a house in Klang Valley towards the quality affordable housing, Life Science Journal, **12**(3), 51–57, (2015)
[14]J. Neale, *Improving housing quality - unlocking the market,* Royal Institute of British Architects, London, (2009)
[15]A.M. Abdul Rahman, N.M.S. Al-Obaidi, K. Ismail, M.L. YokeMui, Rethinking the Malaysian affordable housing design typology in view of global warning considerations, J. of Sustainable Development, **6**(7), 134–146, (2013)
[16]W.B. Olesen, The philosophy behind EN15251: Indoor environmental criteria for design and calculation of energy performance of buildings, Energy and Buildings, **7**, 740–749, (2007)
[17]I. Sartori, A.GHestnes, Energy use in the life cycle of conventional and low-energy buildings: A review article, Energy and Buildings, **39**(3), 249–257, (2007)
[18]S. Lang, Progress in energy-efficiency standards for residential buildings in China, Energy and Buildings, **36**(12), 1191–1196, (2004)
[19]A. Kaklauskas, K.E. Zavadskas, K.E.S. Raslanas, Multivariant design and multiple criteria analysis of building refurbishments, Energy and Buildings, **37**(4), 361–372, (2005)
[20]Housing Corporation, *Design and quality standards: Strong communities*, (2007), Retrieved on June 25, 2016 from https://www.gov.uk/government/publications/design-and-quality-standards
[21]H.N. Ahuja, M.A. Walsh, *Construction management and engineering: Successful methods in cost engineering*, John Wiley, Canada, (1983)
[22]R. Flanagana, C. Jewell, *Whole life appraisal for construction*, Blackwell Publishing, Oxford, UK, (2005)
[23]T.M. Nguyen, Does affordable housing detrimentally affect property values. A review of the literature, J. of Planning Literature, **20**(1), 15-26, (2005)

MATEC Web of Conferences **103**, 03014 (2017) DOI: 10.1051/matecconf/201710303014
ISCEE 2016

Awareness on Sustainable Affordable Housing Among Homebuyers in Malaysia

Salem Ang[1,*], *AbdulLateef* Olanrewaju[2], *Fah Choy* Chia[1], and *Yeow Seong* Tan[2]

[1]Univerisiti Tunku Abdul Rahman UTAR, LKC FES, Sungai Long Campus, 43000, Kajang, Selangor, Malaysia
[2]Construction Management Department, Univerisiti Tunku Abdul Rahman UTAR, FEGT, Kampar Campus, 31900, Kampar, Perak, Malaysia

Abstract. This paper concerns about the delivery of Sustainable Affordable Housing (SAH) in Malaysia. The built environment, operation and demolishment of housing have the profound impact on our nature environment. SAH can be a solution toward sustainability development because it considered from an environmental friendly, socially enhancing, and economically benefits perfectives towards housing development. This paper aims to explore the awareness of homebuyers in Malaysia towards sustainable affordable housing. This research used the method of unstructured interview among homebuyers toward SAH in Malaysia. Based on the finding of the study, the awareness of homebuyers in SAH are relatively inadequate. Respondents are required the brief explanation on concept and elements of SAH for continue the interview. Based on Porter Five Forces Model, homebuyers' requirements hold the bargaining power of customers to determine the features of housing provided by property provider. The knowledge of homebuyers on SAH is crucial for the sustainable development, because homebuyers are one of the key drivers for implementing the sustainable requirements in housing development.

1 Introduction

Conventional house is generally known to fulfill the basic needs of shelter for human being, whereas modern housing should integrate sustainable criteria into housing [1]. Sustainable housing concept can be rooted to sustainable development, which is defined as the "development that meets the needs of the present without compromising the ability of future generations to meet their needs in the future" [2]. It explained that the resources for our future generation depend on the current resource utilization. The amalgamation between affordable house and sustainability is vital for short term and long term development of a country. It could solve this current social issue, by saving the housing operation cost, buying cost as well as to preserve the environment resources for our future generation. For that reason, this paper concern about the delivery of SAH in Malaysia. The aim of this study is to explore the awareness among homebuyers in Malaysia. Homebuyers are crucial stakeholder in developing the sustainable requirements in housing.

[*] Corresponding author: salemang.my@gmail.com

Literally, homebuyers are the end user of supply chain in housing sector. According to Porter's Five Forces [3], the bargaining power of buyer is able to influence the output of the supplier. Homebuyers' determination on buying a sustainable house influences the decision of developers in providing the types of house.

2 Literature review

According to Department of Statistic Malaysia [4], the median income of Malaysian households are RM 4,585 in the year 2014 with 11.7% compounded annual growth rate (CAGR) compare to RM 3,626 in the year 2012. Assuming we carry the same CAGR percentage of 11.7% to the stretch of 2014 to 2016, this will result in the median income of Malaysian households being RM 5,804 monthly. According to the 'Median Multiple' methodology developed by Demographic International and recommended by the World Bank and the United Nations to evaluate urban housing markets, a house is considered affordable if a household can finance it with less than three times its annual household income (house price-to-income ratio of 3.0 and below) [5]. This suggests that houses priced up to RM 209,000 are considered affordable to a median Malaysian household. However, the developments of affordable house in the regular context focus on social and economic aspects and had neglected the factor of environmental. Sustainable affordable housing, which focuses on social and economic aspect, requires attention on environmental factors as well. This is because housing generally has a profound influence on the social-economic wellbeing of the human society and sustainability of the environmental during its production and consumption [6].

The Malaysia federal and state government has placed efforts in developing housing for low and middle income. The government has introduce housing scheme such as Perumahan Rakyat 1Malaysia (PR1MA), MyHome, Skim Rumah Pertamaku (SRP) and Program Perumahan Rakyat (PPR) and also financial aids such as Mydeposit, 20% stamp duty deduction for first-time house buyers. Despite efforts from the government, the number of affordable houses is severely insufficient. One of the reasons is due to the Malaysia's plans for affordable housing not receiving the required attention and it is not developing well. The plan of building total units of affordable house is 550,700 units between years 1986-2005; however, the completion rate is only 57%. This disappointing achievement is due to a complicated and ambiguous relationship between federal, state and local levels [7].

While there is insufficient affordable housing supply, there are needs to provide sustainable affordable housing because housing has been identified as one of the major resource users in energy and water consumption, raw material consumption and utilization of land [8]. In Malaysia, 90 percent of building is housing [9]. Buildings are at the centre of our economic and social lives, providing shelter, work places and spaces for commerce and leisure. However, according to UNEP [10] construction, operation and demolishment of buildings put a tremendous strain on our environment, buildings being responsible for a significant share of approximately 40% of global energy usage, resource consumption of more than 30% of materials used and 20% of water use and waste generation with 30% of solid waste, and 20% of wastewater. The building sector has the highest potential for sustainability improvements with large-scale reductions in carbon emissions, as well as the efficient use of resources can be achieved at low or no cost.

Despite current efforts from the Malaysian government to curb carbon emissions, Malaysia is currently rank as the 23th out of 216 countries in the world that have the largest amount of carbon emission in the year 2015 [11]. In terms of sectorial percentage, 24% of total carbon dioxide comes from the construction sector in the country [12]. In Malaysia, residential building in construction sector consists of 29.9% of total value of construction work done in the first quarter of 2015 [13]. For that reason, residential housing is one of the factors that contribute to the emission of carbon dioxide in the environment. As displayed in Fig.1, carbon emission per capital in Malaysia increasing from year 1970 to 2011, which resulted from 1.34 to 7.9 metric ton per capital, whereas Singapore's decreasing from 8.78 to 4.32 metric ton per capital during the same period. As shown in Fig.1, Singapore has been reducing their carbon emission since 2006. It is most probably due to the implementation of Green Mark by Building and Construction Authority (BCA). This can be one of the recent indicator or reference for our building construction. The method of Singapore in achieving the carbon reduction should be introduced and compared with the Malaysian GBI in the literature where it could be one of the awareness and implementation for carbon reduction.

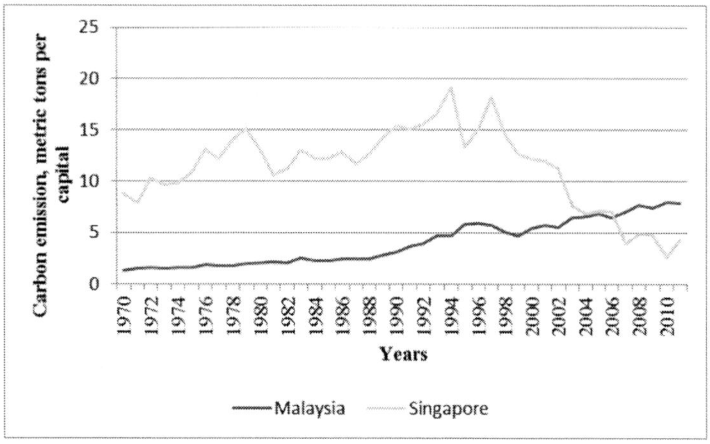

Fig. 1. Carbon emission metric tons per capital between Malaysia and Singapore [14]

Housing has the profound influence not limited to the environment only, but also in impact on the social and economic matter as well. The shortage of houses in Malaysia is decreasing the living standard, especially to the lower income family. According to National Property Information Centre (NAPIC) [15], up to the 4th quarter of 2015, Malaysia has a total of 4.93 million houses. As NAPIC does not track rural houses, assume that only urbanites were taken into account in the survey. This accounts for 70% of 30.8 million populations in Malaysia [16] which equals to 21 million people. Therefore, on average, there are 4.4 persons per household in our country which is the same ratio Australia had back in 1927. Malaysia needs to build 4 million to 7.8 million more houses to match the same ratio as the UK or Australia today [17].

In Malaysia, the green building assessment tool is the Green Building Index (GBI). The GBI rating system are integrated into 6 components which are Energy Efficiency (EE), Indoor Environment Quality (EQ), Material and Resources (MR), Sustainable Site Planning and Management (SM), Water efficiency (WE) and Innovation (IN), and each component carries different points. The allocation of assessment points is categorized by each components based on residential and non-residential groups. GBI rating classification consists of Platinum, follow by Gold, Silver and Certified [18]. Setting an example for the nation, Perdana Putra achieved a Platinum GBI rating; it is one of the iconic buildings of the offices of Malaysia's Prime Minister. In 2010, Perdana Putra was refurbished with efficient technology that reduced the building's energy intensity by 38 percent, its water usage by 40 percent and its CO2 emissions by 30 percent [19]. The same initiative of environmental friendly features can be implemented in affordable housing.

SAH is a new concept to Malaysians its development certainly not common in Malaysia. Most of the stakeholders in the housing sector are not aware of this concept. In addition, there is also inadequate understanding about the idea and benefits of SAH among homebuyers [20]. Hence, SAH plays a crucial role in integrated environmental, social and economic sustainability. SAH is not only about cheap and decent houses. There must be consideration for a boarder range of factors such as access to employment opportunities, energy efficient, interest rates and mortgage availability and so on [21]. The awareness of homebuyers is crucial for exertion in developing SAH by stipulating demands to various stakeholders and partnering with developers, homebuilders, architects, agencies, and non-profits to design, build and deliver the SAH.

In its 2007 report, the Intergovernmental Panel on Climate Change (IPCC) [22] highlighted the building sector as having the greatest potential to reduce GHG emissions, at the lowest cost. With proven and commercially-available technologies, the energy consumption in both new and existing buildings, as well as the related GHG emissions, can be reduced by 30-50% without significantly increasing the investment costs of new construction or renovation projects.

3 Research methodology

The objective of the paper is to identify the awareness of homebuyers in sustainable affordable housing in Malaysia. The qualitative data was collected through a face-to-face unstructured interview, unstructured interview allow respondents to provide their opinion in almost complete freedom [23]. During the MAPEX exhibition, many homebuyers are new to the concept of this study, therefore by face to face interview allows them to ask and express their view freely.

The interview was conducted in Malaysia Property Exhibition (MAPEX) in Ipoh, Perak, Malaysia on the 31st of July 2016. This exhibition was organised by Real Estate and Housing Developers Association (REHDA) and it consisted of 51 booths and 35 developers, several bank and property agencies. This exhibition also consisted of 20,000 visitors. The total properties in this exhibition were worth approximately RM 5 billion [24]. The questions were asked to respondents was not in the order as listed below:

a) Do you know what is sustainable development and sustainable affordable house?
b) What are your concerns in purchasing a house?
c) Do you know any affordable house scheme or financial aids for low and middle income in Malaysia?
d) Feasibility of factoring environmental, social and economic features into housing by introducing sustainable affordable house.
e) Feasibility of purchasing a house with RM 209,000 in Malaysia.
f) Would it affect developers and the government's decision in providing SAH for homebuyers if the demand increasing?

4 Results and discussion

The respondents were homebuyers in the property fair. The participants were asked on their understanding and opinions regarding sustainable affordable housing. During the interview, the respondents were asked within the scope covering sustainable development, affordable housing, sustainable affordable house, environmental, social, economic aspect of housing, affordability issue. Twenty respondents participated in the interview for this research.

The 20 respondent's profiles are displayed in Table 1. The respondents consist of 9 females and 11 males, the varying from 23 years old to 50 years old. The family size varies from 1-5 and 15 of them are looking for their first house to purchase.

Table 1. Respondent's profile

No.	Gentle	Age group	Family size	Purchase for first house
1.	Male	28	3	Yes
2.	Female	30-35	2	Yes
3.	Male	24	2	Yes
4.	Male	34	4	No
5.	Female	23	2	Yes
6.	Male	30	5	Yes
7.	Female	35-40	5	No
8.	Male	50	5	No
9.	Male	24	2	Yes
10.	Male	34	4	Yes
11.	Female	23	5	Yes
12.	Female	32	3	Yes
13.	Male	33	2	No
14.	Female	23	2	Yes
15.	Female	39	5	No
16.	Male	30-35	5	Yes
17.	Male	34	4	No
18.	Male	24	3	Yes
19.	Female	34	3	Yes
20.	Female	23	1	Yes

The first question of the interview for homebuyers was to determine their understanding of the term "sustainable development" and "sustainable affordable house". The outcome of the question shows that all 20 respondents have never come across in regards to sustainable development and sustainable affordable house. In order to continue the survey more specifically, the general concept of incorporating environment, social and economy with house was being elaborated to the respondents. Furthermore, the respondents were being inquired on the integration of SAH features such as strategic location with affordable price, as well as saving utility bills in a long run by harvesting rain water and energy saving technology. All the respondents replied that these housing features are not accessible in Malaysia. From the collected responds of homebuyers, it shows that the understanding of SAH in Malaysia is relatively inadequate. This phenomenon resulted in a weak awareness SAH among homebuyers.

The second question concerned with the homebuyers purchasing a house. As displayed in Table 2, 15 respondents answered that price of the property is their priority concern in purchasing a house, followed by location, safety features, transportation and recreation facilities. Another 5 respondents answered location, follow by price, safety features, recreation facilities, and eligibility for mortgage.

The third question was regarding the understanding of affordable house scheme in Malaysia such as PR1MA, MyHome, Programme Rumah Mesra Rakyat and Peaple's Housing Program (PPR) and also financial aids such as Mydeposit, 20% stamp duty deduction for first time house buyer. 10 respondents implied that they are not aware on the details of the affordable housing scheme, even though these schemes are available near them. Another 10 respondents has no awareness of the schemes. This shows that the publicity of affordable house is not proficient and information is not delivered to those in needs of shelter. Although affordable houses are initiated by the federal and state government, which contains social and economic components, the aspect of environment is often not been put into attention. The fourth question was the availability of purchasing a house with RM 209,000 in Malaysia for personal residence purposes. 16 of the respondents answered it is possible. However, the house will be a house that is aged 20 years and above, and major repair work will be required, which will involve major additional cost. Furthermore, the location of the house would be in an outskirt area, which is approximately 60-80 km from the city.

Table 2. Homebuyer's determinant in buying a house

	First	Second	Third	Fourth	Fifth
15 respondents	Price	Location	Safety	Transportation	Recreation Facilities
5 respondents	Location	Price	Safety	Recreation Facilities	Eligibility for Mortgage

The fifth question is to understand whether developers and government would provide SAH in view of consumer's increase demand request on SAH. All of the respondents mentioned that they have not requested this feature from any parties, and perhaps it would be feasible if more homebuyers suggested and submitted the requests to the developers or government.

The sixth question was regarding the feasibility to factoring environment, social and economic aspects into housing by introducing SAH in Malaysia. All of the respondents answered the concept is not feasible in Malaysia as they perceived that the price of the house would be higher, as higher cost would be required to factor in those aspects into housing.

The awareness and understanding of homebuyers in regards to SAH was relatively poor as shown in the interviewers' responds. According to Porter's Five Forces, the bargaining power of buyer is able to influence the output of supplier; the stipulation of demand is able to dominate the supply requirement. The same principle could be applied to sustainable affordable housing as well. Homebuyer's determination on buying a sustainable house influences the decisions of developers in providing the type of houses. If the awareness and understanding of sustainable affordable housing for homebuyers is developed, it could trigger the initiative of developer to fulfil the customer's needs. The growth in consumers' demand is likely to encourage the voluntarily integrated sustainable features into future developments [25]. Homebuyers do not feel the necessity to incorporate any sustainable concept in their houses, because Malaysia is a country that is sanctified with abundance of natural resources such as water, oil and energy. Thus, there is the lack of exigency of this issue.

5 Conclusion

The aim of this study is to explore the understanding and awareness of homebuyers in Malaysia towards SAH. The study has found out that the awareness in SAH among homebuyers is relatively poor. However, if the awareness and understanding of sustainable affordable housing for homebuyers is developed and the request is computed to the developers, it could trigger developers to fulfil the customers' needs. The growth in consumers' demand is likely to encourage developers to integrate sustainable features into affordable house. According to Olanrewaju and et al [26,27], the Royal Institute of the British Architects (RIBA) Plan of Work 2013 suggests to use in delivery of SAH by various stakeholders. For instance, Malaysia government, architect, developer, engineering designer, manufacturer, buyer, supplier and so on. In this research, we have only focused on the awareness of homebuyers. The formulation of policies and stakeholders engagement required incorporation and feedback. Future research may aim to study the awareness, understanding, commitment, analysis of obstacles and roles of duties of each stakeholder in the delivery of SAH. The SAH would greatly achieve the sustainable development and not only fulfill the needs of current generation, but also preserve the resources for future generation to fulfill their needs as well.

This material is based on research supported by UTARRF under Grant No. IPSR/RMC/UTARRF/2013/O01.

References

[1] P. Suryandari, Transformation concept of sustainable Architecture for middle income house in Indonesia since 1980 till 2012 , Global Journal of Scientific Researches, **1**(2), 52-59, (2013)
[2] World Commission on Environment and Development, *Our common future* , Oxford University Press, Oxford, (1987)
[3] M.E. Porter, How competitive forces shape strategy, Harvard Business Review, March–April, 137–56, (1979)
[4] Department of Statistic Malaysia, *Report of household income and basic amenities survey 2014*, Retrieved on August 20, 2016 from https://www.statistics.gov.my/index.php?r=column/pdfPrev&id=aHhtTHVWNVYzTFBua2dSUlBRL1Rjdz09 , (2015)
[5] Bank Negara Malaysia, *Economic developments in 2015*, Retrieved on August 20, 2016 from http://www.bnm.gov.my/files/publication/ar/en/2015/cp01.pdf, (2015)
[6] E.O. Ibem, D.E. Azuh, Framework for evaluating the sustainability of public housing programmes in developing countries, J. of Sustainable Development and Environmental Protection **1**(3), 24-39, (2011)
[7] T.H. Tan, Sustainability and housing provision in Malaysia, J. of Strategic Innovation and Sustainability, **7**(1), 62-71, (2011)
[8] S.M. Sood, K.H. Chua, Y.P. Leong, Sustainable development in the building sector: A green building framework in Malaysia, Best Practices and Sustainable Development in Construction Paper, WASET, 01-08, University Tenaga Nasional, Malaysia, (2011)
[9] A. Olanrewaju. L.T. Lee, S.Y. Tan, Analysis of sustainable maintenance behaviours in housing operation, Implementing Innovative Ideas in Structural Engineering and Project Management, 895 -900, (2015)
[10] UNEP, Sustainable solutions for social housing, Retrieved on August 27, 2016 from http://www.unep.org/sustainablesocialhousing/pdfs/SUSHI_Guidelines_2013.pdf (2013)
[11] Global Carbon Atlas, CO2 emission, Retrieved on August 23, 2016 from http://www.globalcarbonatlas.org/?q=en/emissions (2015)
[12] M.M. Klufallah, M.F. Nuruddin, M.F. Khamidi, N. Jamaludin, Assessment of carbon emission reduction for buildings projects in Malaysia - A comparative analysis, E3S Web of Conferences, **3**, 01016, (2014)

[13] Department of Statistic Malaysia, *Quarterly construction stat*istic, (2015), Retrieved on August 23, 2016 https://www.statistics.gov.my/index.php?r=column/pdf
[14] Worldbank, *CO2 emissions (metric tons per capita)*, (2015), Retrieved on August 21, 2016 http://data.worldbank.org/indicator/EN.ATM.CO2E.PC?year_high_desc=false
[15] National Property Information Centre, Properties stock as in Quarter 2015, (2016), Retrived on Aug 10, 2016 from http://napic.jpph.gov.my/portal/key-statistics
[16] Worldometer, *Malaysia population*, (2016), Retrieved on Aug 27, 2016 from http://www.worldometers.info/world-population/malaysia-population/
[17] T. Alan, *8 million more houses needed in Malaysia*, The Star, (November 11, 2014)
[18] Green Building Index, *The GBI rating system*, (2013), Retrieved on August 28, 2016 from http://www.greenbuildingindex.org/how-GBI-works2.html
[19] Clean Malaysia, *The rise of green building in Malaysia*, (2015), Retrieved on August 28, 2016 from http://cleanmalaysia.com/2015/08/28/the-rise-of-green-building-in-malaysia/
[20] K.C. Goh, T. Seow, H.H. Goh, *Challenges of implementing sustainability in Malaysian housing industry*, (2013), Retrieved on August 28, 2016 from http://eprints.uthm.edu.my/3964/
[21] M. Emma, M. Vida, Criteria for sustainable housing affordability, Journal of Environmental Engineering, **3**(966–973), (2011)
[22] R.K. Pachauri, A. Reisinger, IPCC fourth assessment report, IPCC, Geneva, (2007)
[23] R. Kumar, *Research methodology: A step-by-step guide for beginners*, SAGE, Los Angeles, United State, (2011)
[24] L. Ivan, *Property expo is back with booths*, The Star, Retrieved on August 15, 2016 from http://www.thestar.com.my/metro/community/2016/07/27/property-expo-is-back-this-years-mapex-to-feature-51-booths/ (July 27, 2016)
[25] M. Osamani, A. O'reilly, Feasibility of zero carbon homes in England by 2016: A house builder's perspective, Building and Environment, **44**(9),1917-1924, (2009)
[26] A. Olanrewaju, Y.S. Tan, T.L. Lee, F. Ayo, S. Ang, Investigate the compatibility of affordable housing with sustainability criteria: a conceptual frame, *Putrajaya International Built Environment Technology and Engineering Conference 2016*, Putrajaya, Malaysia, 228-240, (2016)
[27] Royal Institute of British Architects, *The RIBA Plan of Work 2013*, (2013), Retrieved on August 11, 2016 from https://www.ribaplanofwork.com/

MATEC Web of Conferences **103**, 03015 (2017) DOI: 10.1051/matecconf/201710303015
ISCEE 2016

Head of Households in Terengganu

Izzah Sofea Ibrahim[1,*], *Noor Yasmin* Zainun[1], and *Nurfilzah* Mohd Rawan[1]

[1]Faculty of Civil and Environment Engineering, University Tun Hussein Onn Malaysia, 86400 Parit Raja, Batu Pahat, Johor, Malaysia

Abstract. In the history of construction development, headship rate has been thought of as a key factor in projecting households. By implementing the household formation, the housing demand can be predicted in line with the objectives of the 10th Malaysia Plan (RMK-10) to match the demand and supply of housing in Malaysia. Case study for this paper was in Terengganu, Malaysia. The objective of this study is to identify the headship rate for the year 1980, 2000 and 2010. Data obtained from the Census of Malaysia was analyzed according to their headship rate for the male and female. Then, the graph was plotted according to gender and age group. Results show that at the early group, the headship rate was increased for male while for female headship rate was increased at elderly age group. Therefore, in conclusion, male in Terengganu more likely to own a home at a young age and vice versa for female.

1 Introduction

According to Population and Housing Census of Malaysia 2010, average annual population growth rate from the year 2000 until 2010 increase by 2.17% [1]. The increasing of this population will increased the housing demand as well as in Terengganu district which is the study area. In 2014, the total population at Terengganu was 1.14 million persons, increased to 1.16 million persons in year 2015 and estimated population will increased to 1.18 million persons in the year 2016 [2]. Since the Census was conducted once for 10 years, the headship can classified as stable headship and can be used for 10 years [3]. The previous censuses were conducted in 1970, 1980, 1991, 2000 and 2010. The aim of this paper is to identify the headship rate for the year 1980, 2000 and 2010 in Terengganu. The data later, will used to forecast the household formation in Terengganu but will not discussed in this paper.

The main challenge faced by the government of Terengganu was the issues of property and homes that are affected by supply and demand [4]. If the supply is high but demand is low, the prices will fall. Moreover, National Housing Department reported that 688 units of house was unsold compared 2,236 units that have been built [5]. This imperfection of housing market distorted by external influences such as unethical speculation by housing developers has contribute to the market failure [6].

* Corresponding author: nryasmin@uthm.edu.my

DOI: 10.1051/matecconf/201710303015

2 Headship rate

The headship rate system has now been generalized to produce a system whereby numbers of households are calculated in terms of number of head of households which are identified by age and gender [7]. Thus far, previous studies suggested that headship rate are then the proportion of person in any age, sex and marital status group who head a household, and number of households can simply be produced for a new projected population by multiplying these rates by the numbers in each population group [8]. [9] Reported that, average household size is the inverse of the average household headship rate, which is defined as the number of persons in the population who are heads of households divided by the population, expressed as a rate as defined in Equation (1),

$$\text{Headship rate} = \frac{\text{head}}{\text{population}} \times 100 \qquad (1)$$

Since this definition controls for variations in population size, variations in the age specific headship rates reflect changes in the propensity of the population within each category to form households [10]. The growth in households is a manifestation of individuals young adult preferring to form their own household based on their needs [11].

The term housing need indicator refers to the lack of the number of households that do not have accommodation while the demand for housing is market-driven concept associated with the type and number of the house where users choose a house based on the preferences and ability to pay [12]. Moreover, housing demand is also defined as the number of homes needed to be one residence per household [13]. Considering all of these evidence, it seems that housing demand was referring to the housing market with the combination of two concepts which is needs and demands.

Zainun et al. [14] found the method to forecasting the housing demand to fulfil the needs of citizens and to avoid a mismatch of demand and needs. In Terengganu, 583 units of abandoned or sick projects have been built and by 532 units have been sold. According [15], the problem of land, supply, cost and others are among the factors contributing to the problem of abandoned projects.

3 Scope and limitation

This study was conducted primarily for the purpose of identifying the headship rate on the previous year. 2% Census data for the year 1980, 2000 and 2010 was synthesizing, analyzing and interpreting in the table and graph. The age group was limit within the age of 15-19, 20-24, 25-29, 30-34, 35-39, 40-44, 45-49, 50-54, 60-64, 65-69, 70-74, 75-79 and 80+.

4 Methodology

Headship rate was analysed from the year 1980, 2000 and 2010, the data was obtain from Census of Malaysia [16-18]. Head, population and gender was the most important indicator to be considered in this study [19]. Fig. 1 illustrates the flow of this study. The Census data obtained from the Department of Statistics Malaysia was synthesis by age group. Early group within the age of 15-19 until 80 above was considered based on the marital status [20]. Two group of gender which is male and female were categorised, then the data will be synthesis by head and population within the age group. By using formula that has mention earlier, the headship rate can be identified. Headship rate for male and female in the year 1980, 2000 and 2010 was interpreted in the Table 1 and Table 2.

Example calculation of the headship rate for the male in the year 1980 which is age group 15-19 was shown below:

$$\text{Headship rate} = \frac{20(\text{ male head })}{444(\text{ male population })} \text{ x } 100 = 4.5\%$$

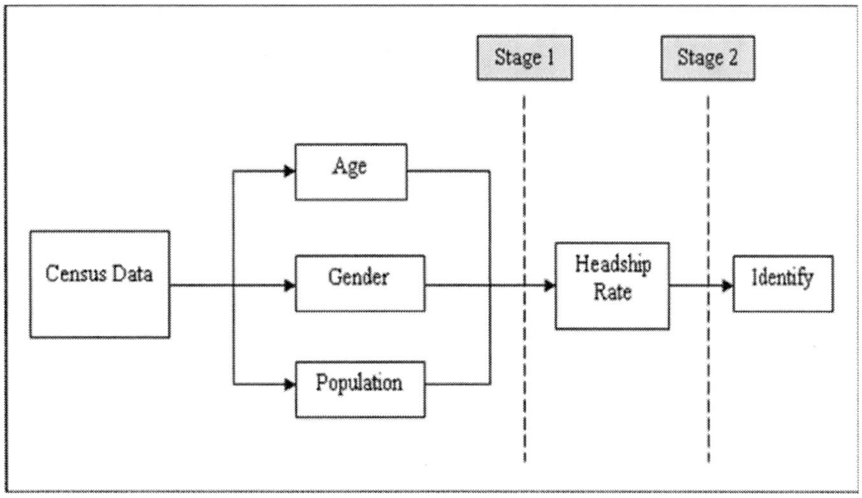

Fig. 1. The flow of this study.

Table 1. Headship rate for male year 1980, 2000 and 2010

Age Group	YEAR								
	1980			2000			2010		
	Male Heads	Male Population	Headship Rate	Male Heads	Male Population	Headship Rate	Male Heads	Male Population	Headship Rate
15-19	20	444	4.50	15	929	1.61	1	1119	0.09
20-24	83	329	25.23	57	582	9.79	69	965	7.15
25-29	154	276	55.80	231	565	40.88	233	814	28.62
30-34	211	263	80.23	377	536	70.34	332	588	56.46
35-39	196	222	88.29	461	550	83.82	398	557	71.45
40-44	193	213	90.61	455	500	91.00	459	549	83.61
45-49	122	129	94.57	396	421	94.06	499	535	93.27
50-54	113	120	94.17	337	351	96.01	449	486	92.39
55-59	104	115	90.43	210	220	95.45	358	375	95.47
60-64	92	98	93.88	173	178	97.19	237	252	94.05
65-69	61	76	80.26	82	91	90.11	182	191	95.29
70-74	44	50	88.00	81	89	91.01	138	158	87.34
75-79	25	32	78.13	57	62	91.94	55	65	84.62
80+	10	16	62.50	43	54	79.63	68	99	68.69

DOI: 10.1051/matecconf/201710303015

Table 2. Headship rate for female year 1980, 2000 and 2010

Age Group	YEAR								
	1980			2000			2010		
	Female Heads	Female Population	Headship Rate	Female Heads	Female Population	Headship Rate	Female Heads	Female Population	Headship Rate
15-19	10	450	2.22	1	936	0.11	0	990	0.00
20-24	23	395	5.82	22	538	4.09	37	902	4.10
25-29	33	344	9.59	21	548	3.83	36	654	5.50
30-34	21	224	9.38	21	571	3.68	27	572	4.72
35-39	33	235	14.04	26	540	4.81	42	572	7.34
40-44	37	210	17.62	32	489	6.54	43	578	7.44
45-49	36	147	24.49	45	395	11.39	62	568	10.92
50-54	25	116	21.55	58	273	21.25	91	461	19.74
55-59	41	128	32.03	54	212	25.47	87	354	24.58
60-64	29	101	28.71	63	188	33.51	92	277	33.21
65-69	24	58	41.38	31	104	29.81	74	202	36.63
70-74	18	55	32.73	41	99	41.41	85	166	51.20
75-79	8	29	27.59	31	55	56.36	28	61	45.90
80+	13	26	50.00	35	85	41.18	50	113	44.25

5 Results and discussion

The trend of the graph on Fig. 2 shows that the headship rate increasing at the early age group and remain constant at the middle of age group then decrease when come to older age group. The higher of headship rate at early group because of the improved material situation of and propensity to leave the parental home of young people [21]. Moreover, this result will impacted to projecting the household formation in Terengganu. Higher increasing of headship rate from the age group of 25-29 to 30-34 was increased by 27.8% for the year 1980 while the lower decreasing headship rate from 75-79 to 80 and above was decreased by 15.9%. Moreover, headship rate for the year 2000, 31.1% was increased from the age group of 20-24 to 25-29 while 30.6% for the year 2010 from the same age group. The lower decreasing headship rate for the year 2000 and 2010 from the same age group which is 75-79 to 80 and above with the value 12.3% and 15.6% respectively.

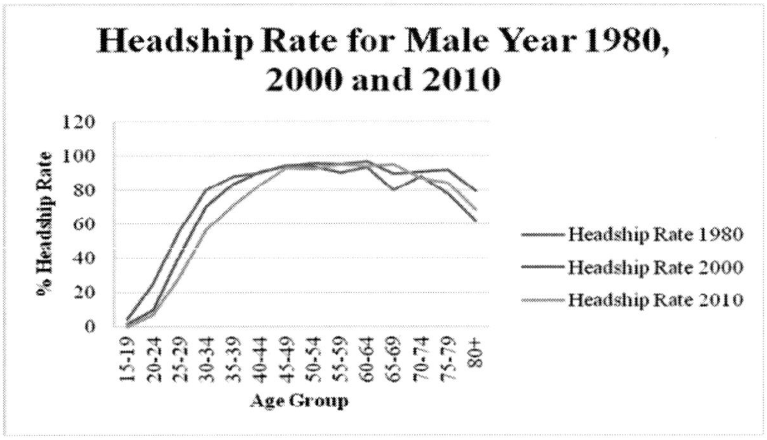

Fig. 2. Headship rate for male year 1980, 2000 and 2010.

Fig. 3 shows the headship rate for female for year 1980, 2000 and 2010 graph was fluctuating. [22] reported that the factor of this fluctuating was income, marriage and education. The higher increasing of headship rate mostly comes from the elderly age group. This results corresponding on the previous research by Moonie (2013) reported that the relatively high headship rate among elderly may be linked in demographic change by transition from high fertility and mortality rate thus possibly a higher number of older person living alone [23]. For the year 1980, age group of 65-69 to 70-74 showed that the highest increasing of headship rate by 14.6% while 5.3% the lowest decreasing from the age group 70-74 to 75-79. Besides that, 14.9% from the age group 70-74 to 75-79 was the highest increasing in headship rate and 15.2% from the age group 75-79 to 80 and above was the lowest decreasing. Last but not least, the higher increasing of headship rate for the year 2010 from the age group of 75-79 to 80 and above was increased by 22.4% while the lower decreasing was 8.6% decrease from the age group of 65-69 to 70-74.

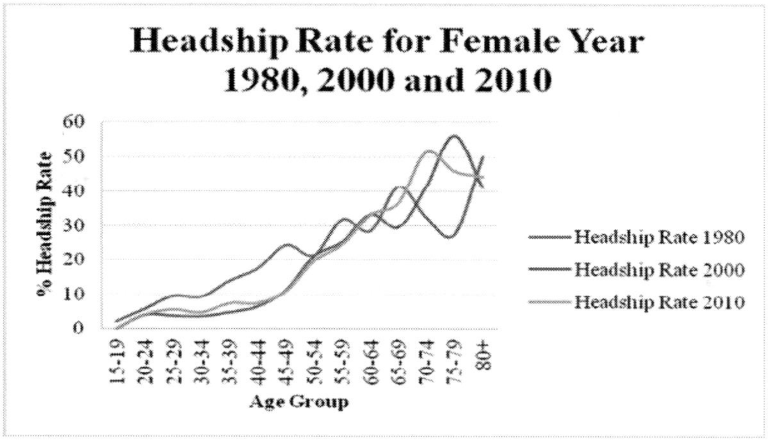

Fig. 3. Headship rate for female year 1980, 2000 and 2010.

6 Conclusion

In a nutshell, headship rate was increased at the early age group for the male while at the elderly age group for the female. This increased because of the factor of marital status and fatality. By considering the headship rate, the household formation can be forecasting accurately based on their demand. This study hoped can be helpful to the government sector and private agencies to construct housing.

This research was collaboration between UTHM and Sirim Berhad under special grant funded by MOSTI. The authors whould like to thank University Tun Hussein Onn Malaysia, Sirim Berhad and MOSTI for financial support and priority necessary infrastructure to carry out research work.

References

[1] Department of Statistics Malaysia, *Population and housing census of Malaysia 2010*, Department of Statistics Malaysia, 77, (2011)
[2] Department of Statistics, *Population quick info*, (2016), Retrieved on July 9, 2016 from http://pqi.stats.gov.my
[3] L. Goodman, R. Pendall, and J. Zhu, *Headship and homeownership: What does the future hold?*, Urban Institute, Washington, (2015)
[4] N. Yatim, *Hartanah di Kuala Terengganu Mahal*, Sinar Harian, 35, (October 8, 2014).

[5] National Housing Department, *Laporan Perangkaan Suku Tahun Pertama 2016*, 1–21 (2016)
[6] A. Rameli, D.F. Johar, and A.P.D.H. Siong, The management of housing supply in malaysia : incorporating market mechanisms in housing planning process, *Int. Conference on Construction Industry 2006*, 1–17, (2006)
[7] United Nations, *Headship rate method*, (1957), Retrieved on May 1, 2016 from http://www.un.org/esa/population/techcoop/SocInd/manual7/chapter4.pdf
[8] J.F. Ermisch, and E. Overton, Minimal household units: A new approach to the analysis of household formation, Population Studies, **39**(1), 33–54, (1985)
[9] F.L. Mackellar, W. Lutz, C. Prinz, and A. Goujon, Population, households, and CO2 emissions, Population and Development Review, **21**(4), 849–865, (1995)
[10] L.B. Smith, Household headship rate, Household Formation and Housing Demand in Canada, **60**(2), 180–188, (1984)
[11] M. Carliner, Headship rates and housing demand, Housing Economics, 8–12, (2013)
[12] S. Heath, Housing demand and need, Social Policy Section, 1–21, (2014)
[13] J.Z. Hansen, P. Stephensen, and J.B. Kristensen, Household formation and housing demand forecasts, (2013), Retrieved on May 5, 2016 from www.dreammodel.dk/SMILE
[14] N.Y. Zainun, I.A. Rahman, and M. Eftekhari, Forecasting low-cost housing demand in pahang, malaysia using artificial neural networks, J. of Mathematic Research, **2**(1), 14–19, (2011)
[15] M.F. Hashim, *Kesan punca masalah projek sakit*, Utusan Malaysia, Kuala Lumpur, (February 16, 2010).
[16] Department of Statistics Malaysia, *2% population and housing census of Malaysia 1980*, Putrajaya, Malaysia, (1980)
[17] Department of Statictics Malaysia, *2% population and housing census of Malaysia 2000*, Putrajaya, Malaysia, (2000)
[18] Department of Statistics Malaysia, *2% population and housing census of Malaysia 2010*, Putrajaya, Malaysia (2010)
[19] S. De Vos, and K. Holden, Measures comparing living arrangements of the elderly: Assessment, Population and Development Review, **14**(4), 688–704, (2007)
[20] F.T.S. Court, Acts Islamic Family Law (Federal Territories) 1984, (1993)
[21] N. Keilman, Recent trends in family and household composition in Europe, Euopean. J. of Population, **3**(3-4), 297–325, (1988)
[22] R.A. Moffitt, Welfare benefits and female headship in US time series, The American Economic Review, **90**(2), 373-377, (2000)
[23] S.C. Moonie, *The projected impact of demographic changes on household size and housing stock in Saint Lucia: 2015-2030*, Master Thesis, The University of the West Indies, Mona, Kingston, Jamaica, (2013)

Post Windstorm Evaluation of Critical Aspects Causing Damage to Rural Houses in the Northern Region of Peninsular Malaysia

Shaharudin Shah Zaini[1,*], *Taksiah* A. Majid[1], *Siti Noratikah* Che Deraman[1], *Farah Alwani* Wan Chik[1], and *Mohd Khairul* Azuan Muhammad[1]

[1]Disaster Research Nexus, School of Civil Engineering, Universiti Sains Malaysia, Engineering Campus, Penang, Malaysia

Abstract. Rural houses are susceptible to roof blown off and severe damage during a windstorm event due to the lack of engineering considerations. The aim of this paper is to conduct a post windstorm evaluation on the damaged rural houses located in the northern region of Peninsula Malaysia. Several activities were involved during the post windstorm survey including site visualization, site measurement and interview. Critical aspects including types of damages, types of houses, gap height, overhang length, roof geometry, roof pitch, roof cladding and terrain category were analysed using a simple bar chart. It is anticipated that the presence of kitchen house influences the overall stability of the rural houses due to the formation of gap height.

1 Introduction

The non-engineered low-rise buildings are defined in various building categories, including residential, institutional, and commercial structures. These types of buildings are most vulnerable to be damaged during windstorm [1, 2, 3]. Typically, they were built with a minimum or no structural engineering design compared to other buildings that use codes and standards [4]. In Malaysia, rural houses are considered as non-engineered building, typically built according to the location and culture. One of the features of the rural houses in Malaysia is the presence of the core house (rumah ibu) and the extension house or kitchen house (rumah dapur) as reported by [5]. Fig. 1(a) shows an example of a rural house comprising a core and kitchen house.

Fig. 1(b) shows a rural house that suffers from roof blown-off and severe damage to the truss system during a windstorm event. Such damages are normally associated to the high uplift pressure generated at the roof of the house. [6-7] concluded that roof is the most damaging and vulnerable part of the house during windstorm as it is always subjected to the largest wind force. Failure of the roofing system is primarily from the weakness of the attachment and roofing material as stated by [8].

* Corresponding author: ceshaharudin@usm.my

The purpose of this paper is to conduct a post windstorm evaluation on the damaged rural houses located in the northern region of Peninsula Malaysia. The aim of this study is to identify the factors that can contribute to the damage of the houses. All the information gathered from this survey can be used for future detail and in-depth analysis on the wind pressure distribution surrounding a rural house using wind tunnel test or computational fluid dynamics.

Fig. 1. (a) Core and kitchen house (b) Severe damage to the roofing system of a rural house.

2 Post windstorm damage survey

This study adopted the site survey method. The post windstorm survey area covered several states in the north bound of Peninsula Malaysia namely Perak, Penang and Kedah. The selected location of the damaged houses was identified via the local news. The data were collected from August 2014 until April 2016. A total of 35 houses were involved in this survey. Only those houses that suffered significant damage to the roof claddings, trusses and walls were considered.

Various activities were involved including site visualization, site measurement and interview. Site visualization involved capturing photos of the damaged houses, their surroundings and the damaged structural elements. On the other hand, the site measurement was performed in order to obtain relevant information such as the dimension of the house, roof pitch, overhang length and gap height. In addition, a sketch for each house was produced to show the overall geometry of the house and the roof. However, the findings will be discussed in other publication. The data from the site survey was presented in the form of vertical bar graph. This type of presentation is more compatible and easier to comprehend as it summarizes the number of the damage occurrences within a particular factor.

3 Results and discussion

3.1 Types of damages

Wind can cause different types of damage to the rural house such as roof blown-off, wall damage and total collapse. The survey data in Fig. 2 shows that 21 houses (60%) experience damage at the roof of the core house only and this number is triple the houses experiencing damage the roof of the kitchen house only. In addition, 3 houses suffer damage at both roofs while another 3 houses suffer damage at the roof and wall of the core house. The survey also recorded 1 house that was totally collapsed due to the windstorm. The results matched the statement provided by [6, 7] where the roof system was found to be the weakest component for a low rise building. It is interesting to note that more number of roof damage was recorded for the core house compared to the kitchen house suggesting that presence of the gap between the two roofs and the orientation of the kitchen house may have influence the wind pressure within the vicinity of the roof area. Fig. 3 shows some examples of the damaged rural houses.

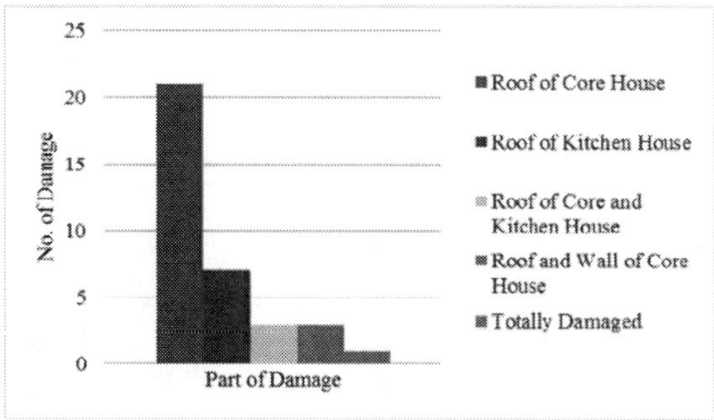

Fig. 2. Types of damages.

(a) (b) (c)

Fig. 3. Example of types of damage (a) roof of core house (b) roof of kitchen house (c) roof and wall of core house.

3.2 Types of houses

The main frame of rural houses was built using two different types of construction materials namely concrete and timber. Fig. 4 shows that 17 single-storey (concrete), 14 double-storey houses (concrete and timber) and 4 elevated and landed (concrete and timber) houses suffer one form of damage as presented earlier in Fig. 3.

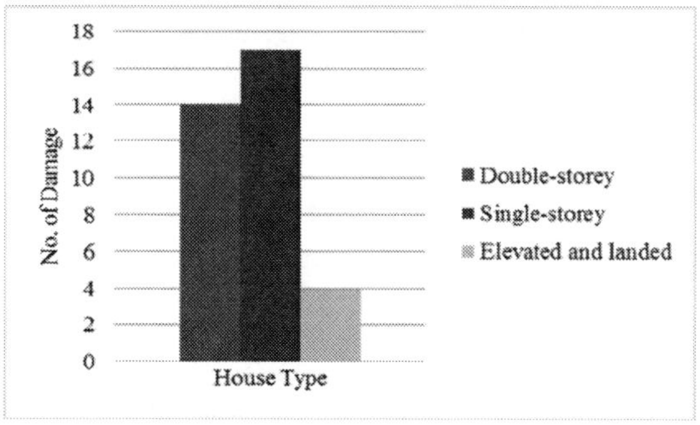

Fig. 4. Number of damage according to the types of houses.

It was thought that double-storey and elevated houses would show more number of damages due to the fact that higher structure generally will be exposed to higher wind speed and pressure. However, this contradicting result can be related to the difference in the gap height of between the core and kitchen house. Some typical examples of the houses are shown in Fig. 5. Obviously, the gap height for the single-story houses were smaller compared to other types of houses in this survey. The construction materials for the main frame were found to be insignificantly influencing the occurrence of damage due to fact that most all damages occurred at roof compared to other parts of the house. However, in general, timber frame is considered to be more susceptible to damage compared to concrete.

(a) (b) (c)

Fig. 5. Severe damage to the roof of (a) single-storey (b) double-storey (c) landed and elevated rural houses.

The survey also revealed that there were no standard dimensions to reflect a rural house. Based on the interview conducted, the size of the house was highly dependent on the available budget and current cost of construction materials. However, using a simple normalisation approach, the best ratio was found to be 3:2:1 representing the length, width and height of the house, respectively. This ratio can be used as a reference for modelling purposes.

3.3 Gap height

Fig. 6. shows the location of the gap between the roof of the core and kitchen house. Currently, there is very limited information in the open literature explaining the influence of the gap on the distribution of pressure surrounding the roof area.

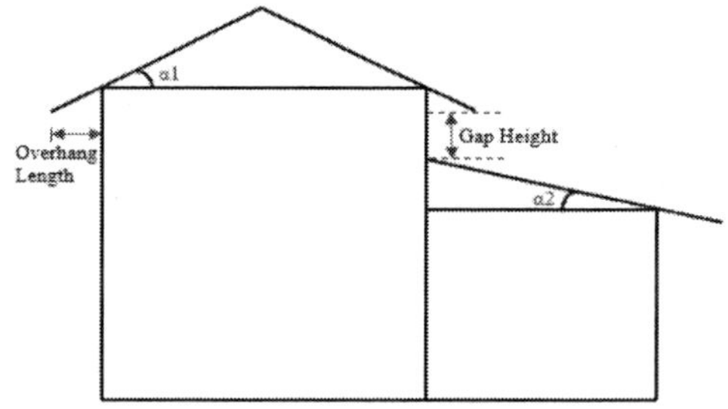

Fig. 6. Schematic diagram showing the location of the gap, the measurement of the gap height between the roof of the core and kitchen house as well as the overhang length.

However, the results in Fig. 7 show a significant trend where rural houses with small gap height are prone to damage (mostly to the roof). This statement is particularly true where 12 houses with gap height less than 0.5 m suffered severe damage compared to the larger ones. However, the actual number of these houses with gap height in this category may be affected because no data was made available for another 15 houses due to unforeseen matters. As the gap height increases, the number of damaged houses reduces significantly. Theoretically, the maximum gap height is achieved when there is no kitchen house being attached to the core house. In this case, the wind force may have directly impinged on the wall of the core house without developing any recirculation zone in the windward direction as shown in the work of [9]. As such, this finding gives an early indication on the sensitivity of the gap height and kitchen house towards the vulnerability of rural houses against windstorm.

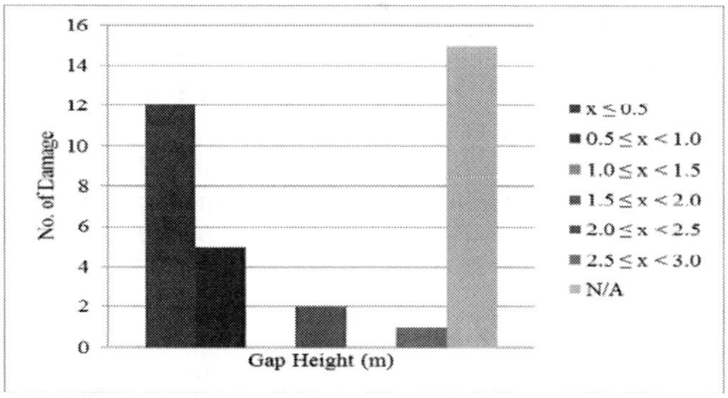

Fig. 7. Number of damages affected by the gap height.

3.4 Overhang length

Roof is built with overhang to protect the wall façade from being moist [10]. In Malaysia, the overhang is made been by extending the roof surface. In this case, the roofing system (cladding and rafter) at the overhang area acts as a cantilever structure. The presence of overhang may initiate damage to the roof especially when there is no adequate support to hold the roofing system from uplift. [11] reported that pressure underneath the overhang was able to account for 50% of the total wind acting on the overhang of a house.

This survey splits the number of damaged houses based on the overhang length of the core and kitchen house as shown in Fig. 8(a) and (b), respectively. To simplify the data, the overhang length is grouped according to the length within a stipulated range. It can be seen that most of the damaged rural houses are built with overhang length in the range of 0.6 to 0.9 m where 15 and 11 houses are affected when considering the overhang length for the core and kitchen house, respectively. The presence of overhang length may create high turbulence and uplift underneath the overhang area. The deflected flow from the windward wall increased the pressure on the lower eave surface while high suction was developed on the upper eave surface immediately after separation had occurred as reported by [10].

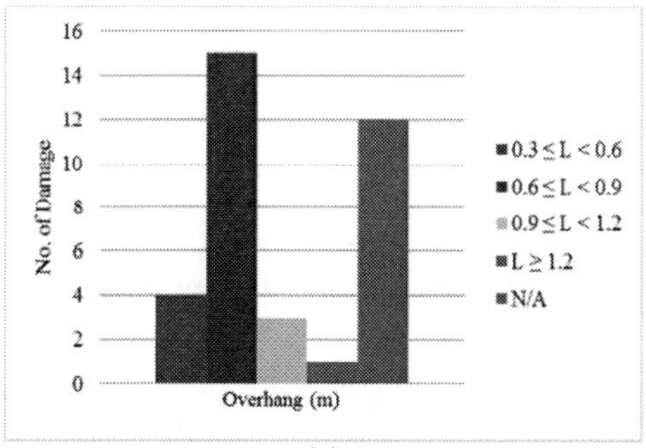

(a)

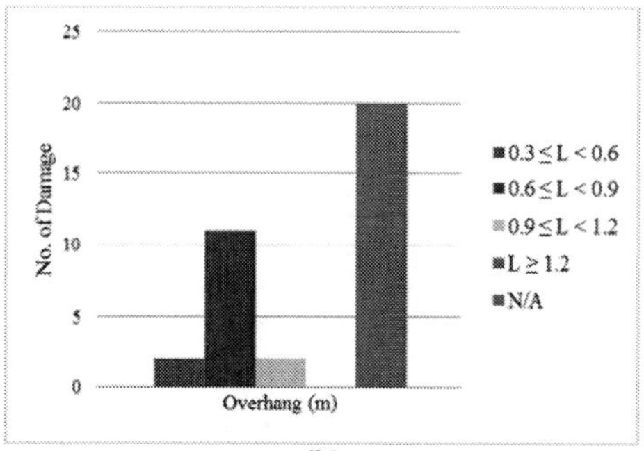

(b)

Fig. 8. Number of damages associated to the overhang length of (a) core house (b) kitchen house.

3.5 Roof geometry

The geometry of the roof may have significant influence on robustness of the roofing system. It is well recognized that roof geometry can change the flow patterns around the house and building [12]. There are four different types of roof geometry that can be commonly found in the rural area namely the gable, dutch-gable, shed and lean-to (skillion) roof as shown in Fig. 9. Based on post windstorm damage survey, 29 core houses were constructed with gable roof geometry. The rest of the roof geometries showed insignificant influence on the damage and less than 3 houses were recorded for each case as shown in Fig. 9(a). [12, 13] proved that gable roof always recorded a high suction on the roof while [14] found that the most critical suctions values were observed near the gable end. On the contrary, 27 kitchen houses were constructed with lean-to roof geometry as shown in Fig. 9(b). It is thought that presence of wall at the far end of the roof may developed turbulence and induced uplift underneath the overhang of the roof for the core house. Fig. 10 shows some examples of the roof geometry.

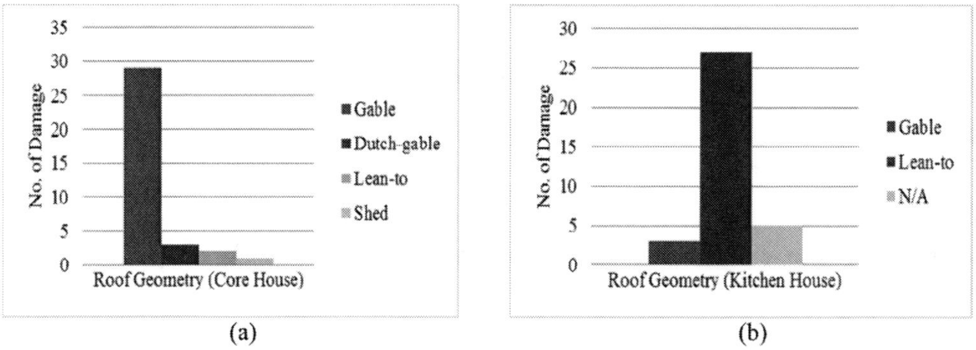

Fig. 9. Number of damages caused by roof geometry for (a) core house (b) kitchen house.

Fig. 10. Example of roof geometry (a) gable; (b) dutch-gable.

3.6 Roof pitch

Fig. 11(a) shows that the roof pitch of the core house for 25 damaged houses is located in the intermediate range between $7° \leq x \leq 27°$. Similarly, the same range of roof pitch for the kitchen house also exhibits the highest number of damaged as shown in Fig. 11(b). Different roof pitches produced different flow patterns on the roof. For wind flow perpendicular to the ridge for intermediate roof, the flow separated at the leading edge, reattached on the upwind slope of roof, then separated again at the ridge. As stated by [15], this range of roof pitch was able to produce worst results and particularly true for gable roof.

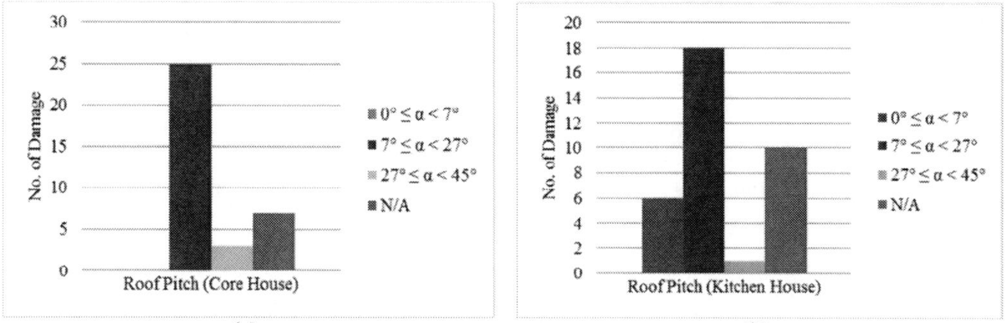

(a) (b)

Fig. 11. Number of damages with respect to the range of the roof pitch for (a) core house and (b) kitchen house.

3.7 Terrain category

Fig. 12 shows that 20 of the damaged houses are located in an open terrain with no or few obstructions, commonly known as Terrain Category 1. These areas normally cover paddy fields and water surfaces. Such areas can largely be found in the part of the mainland of Penang, Kedah and Perak. On the other hand, the other 15 damaged houses are located in an open area with a few scattered obstructions having height generally from 1.5 m to 10.0 m, commonly known as Terrain Category 2. [16] reported that with increase in the surrounding obstructions, the mean wind pressure acting on the building decreased while unsteady pressure increased. Open terrain can cause a direct attack where the wind will directly impinge on the structures. Topographic features such as hills can also speed-up or slow-down the wind flow [7].

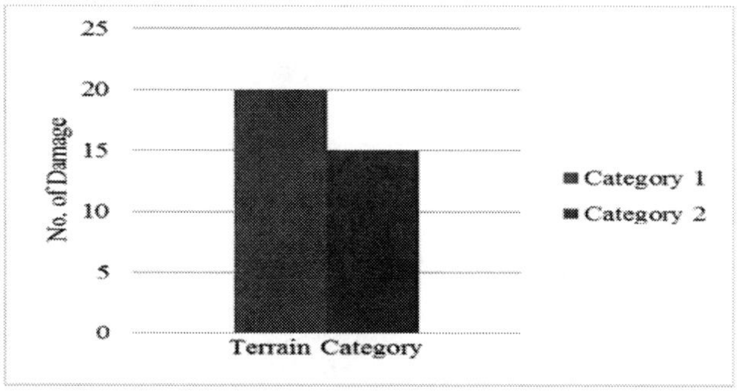

Fig. 12. Number of damages based on the surrounding conditions.

4 Conclusions

Based on the survey, it can be concluded that there are several critical aspects that are related to the damage to of rural houses in the northern part of Peninsula Malaysia. The critical aspects include types of damages, types of houses, gap height, overhang length, roof geometry, roof pitch, roof cladding and terrain categories. However, the degree of influence of each critical aspect or the combination of the critical aspects towards the damage requires further analysis. The presence of kitchen house also appeared to have some form of influences on the overall stability of the rural houses especially by creating the gap height between. As the construction of kitchen house is common for rural houses in Malaysia, this component should be included in the model for future analysis using experimental or numerical approach.

The authors express their gratitude for the financial support of the Fundamental Research Grant Scheme (FRGS) from the Ministry of Higher Education Grant (203.PAWAM.6071317), MyBrain15 Scholarship from the Ministry of Education Malaysia.

References

[1] Y. Uematsu, and N. Isyumov, Wind pressures acting on low-rise buildings, J. Wind Eng & Ind. Aerodyn, 82, 1-25 (1999)
[2] J.D. Holmes, *Wind loading of structures*, Spon Press, London, UK, (2001)
[3] H.W. Tieleman, J. Wind Eng & Ind. Aerodyn., **91**(12-15), 1627-1649 (2003)
[4] M.K.A. Muhammad, T.A. Majid, N.I. Ramli, F.A. Wan Chik, and S.N.C. Deraman, An overview of Non-engineered buildings roofing system failure under wind loads *Proc. of Awam Int. Conf. of Civil Engineering*, Kuala Lumpur, Malaysia, (2015)
[5] S. Said, The grammar of traditional Malay long-roof houses, *Proc. Post-Graduate Seminar,* Universiti Teknologi Malaysia, Skudai, Malaysia, (2007)
[6] Y. Tamura, H. Kikuchi, and K. Hibi, Extreme wind pressure distributions on low-rise building models, J. Wind Eng. Ind. Aerodyn., 89, 1635-1646 (2001)
[7] J. Ginger, D. Henderson, M. Edwards, and J. Holmes, Wind-related disaster risk reduction (WRDRR) activities in asia-pacific region and cooperative actions, *Int. Association for Wind Engineering,* Kanagawa, Japan, (2010)
[8] F.A. Wan Chik, N.I. Ramli, M.K.A. Muhammad, T.A. Majid, and A. Hafiz, The effect of wind load on rural roofing system, *Proc. of the Eight Asia-Pacific Conf. on Wind Engineering*, Research Publishing, Singapore, (2013)
[9] Y. Tominaga, SI. Akabayashi, T. Kitahara, and Y. Arinami, *Build. Environ*, **84**, 204-213 (2015)
[10] K. Narayan, and A. Gairola, Wind pressures on gable roof building with overhang, J. Wind Eng & Ind. Aerodyn., **7**(1), 16-29 (2010)
[11] K.J. Dreher and J.E. Cermak, *Wind loads on a house roof*, Report CER 72-73, KJD-JEC22, Colorado State University, Colorado, (1973)
[12] S. Ahmad and K. Kumar, Effect of geometry on wind pressures on low-rise hip roof buildings, J. Wind Eng & Ind. Aerodyn, **90**(7), 755-779 (2002a)
[13] D. Prasad, T. Uliate, and M. R. Ahmed, Wind loads on low-rise building models with different roof configurations, Int. J. of Fluid Mechanics Research, **36**, 231-243 (2009)
[14] T. Stathopoulos, K. Wang, and H. Wu, Wind pressure provisions for gable roofs of intermediate roof slope, Wind Struct., **4**(2), 119-130, (2001)
[15] S. Ahmad, and K. Kumar, Wind pressures on low-rise hip roof buildings, Wind Struct., **5**(6), 493-514 (2002b)
[16] T.C.E. Ho, D. Surry, and A.G. Davenport, Variability of low building wind loads due to surroundings, J. Wind Eng & Ind. Aerodyn, **38**(2-3), 297-310 (1991)

MATEC Web of Conferences 103, 03017 (2017) DOI: 10.1051/matecconf/201710303017
ISCEE 2016

Conceptualising the Life Expectancy of House Components in Malaysia

Norfarahayu Kariya[1], *Nor Haslinda* Abas[1,*], *Hairuddin* Mohammad[2], and *Siti Khalijah* Yaman[1]

[1]Department of Building and Construction Engineering, Faculty of Civil and Environmental Engineering, Universiti Tun Hussein Onn Malaysia, 86400 Parit Raja, Batu Pahat, Johor, Malaysia
[2]Faculty of Architecture, Planning, & Survey, Universiti Teknologi Mara, 40450 Shah Alam, Selangor, Malaysia

Abstract. House maintenance is crucial to ensure the longevity of house components, in the same time retaining their functionality. Unfortunately, house owners are usually caught off-guarded, especially when house components were already deteriorated without warning. Adding salt to the wound, major house maintenance and rectification generally occur after defect liability period in which the costs incurred are not covered by the developer. The pertaining issue of lack prevention measures towards house components at certain point of time is prevalent. Since there is no established pre-rectification guideline that is able to assist house owners to proactively self-evaluate their house components from time to time, the said problem stays firm. Here, as far as maintenance's guideline is concerned, knowledge gap in terms of life expectancy (LE) of generic house components is well observed. Therefore, the authors embarked on research exploration through literatures on identifying the factors leading to LE for house components and further suggested average LE for generic house components. However, the authors limit the paper within the initial process of research exploration to understand the central phenomenon of this study, which only governed the process of reviewing literatures. In a nutshell, the authors posit that LE was worth explored by embedding a multi-layered thematic analysis which comprised of several interconnected themes between house components and life expectancy of house component. Finally, the authors are optimist that the initial outcome from the literature reviews is paramount for subsequent phases, for instance validation and data gathering phases.

1 Introduction

The average house price in Malaysia have continued to rise from RM138,712 in 2000 to RM317,768 in 2015 [1]. Despite the increment of house price, many house buyers are still looking for new house, where it has become the biggest investment for them. There are several common considerations that will influence the potential house buyers to finalize their purchase decisions like house price, location of the house, the developer's track record,

*Corresponding author: nhaslin@uthm.edu.my

future capital gain, and etc. However, those considerations frequently shadowed the important latent factor, which is building maintenance. Building (particularly house) maintenance is crucial when the house owners need to extend the functionality and longevity of the house [2]. Additionally, due to the augmentation of house and material price every year [3], periodic and pre-maintenance actions by owners are seen to be extremely important. Unfortunately, most house owners are caught off guarded towards the need of maintenance until in the very later stage when deteriorated house components were evident. Since reparation in those stage is generally incurring much larger cost [2,4], the particular situation is able to prolong from bad to worst especially for the ones with limited funds. Adding salt to the wound, major house maintenance and rectification often occurred after defect liability period, where it is the responsibility of the owners to inspect and rectify the defects. Failure to rectify at the specified time may cause injury to the owners and damage to the property [5].

Seeing the above concerning issue, the authors embarked on conceptualising the overall research problem and found out that there is no proper pre-rectification guideline or standard that can assist house owners to self-evaluate their house components from time to time. Thus, preliminary preparations are needed to preserve the components from further and critical damages, and to avoid infectious damages towards other part of components [6]. As house maintenance is important, having initial information for house owners regarding the life expectancy of each house components is paramount. Apart from providing knowledge per se, the move could provide sufficient buffer period for house owner towards future maintenance, and lest paving for establishment of proper guideline regarding the matter. In order to give a fair example from other developed nation, in the UK, there are established guidelines for house owners regarding to maintenance from the pre-stages of construction until occupancy period [7,8]. Features in the guideline include the life expectancy of house components, where a proper study towards it has been conducted on a larger scale [9]. However, these guidelines cannot be directly applied in Malaysia due to several differences including regions, weather and environment; in which the differences could possibly affect the life expectancy of each house components. Realizing the importance of having a preliminary knowledge of house components' life expectancy, the authors therefore intend to determine an appropriate approach of conceptualising the factors which possibly affect the life expectancy of house components and their associated average of life expectancy based on the Malaysia's environment before further establishment are conducted.

2 Malaysia's housing environment

Malaysia is made up of thirteen states and three federal territories across the South China Sea namely; Peninsular Malaysia and East Malaysia with a total landmass of 328,543 square kilometres [10]. Peninsular Malaysia consists of eleven states and two federal territories which is categorised by region namely; Northern region (i.e Perlis, Kedah, Penang, Perak), East Coast region (i.e. Kelantan, Terengganu, Pahang), Central region (i.e. Selangor, federal territories of Kuala Lumpur and Putrajaya) and Southern region (i.e. Negeri Sembilan, Melaka, Johor). While East Malaysia consists of the Malaysian state of Sabah, Sarawak, and the Federal Territory of Labuan. In 2016, total population of Malaysia is estimated at 31.7 million of persons, comprises of 16.4 million of males and 15.3 million of females. Selangor recorded the highest percentage of population in 2016 followed by Sabah and Johor. While Wilayah Persekutuan Labuan and Wilayah Persekutuan Putrajaya are the smallest percentage of population [11]. Located near the equator, Malaysia's climate is categorised as equatorial, being hot and humid throughout the year [12].

The growth of Malaysian construction industry has been increased from year to year and at the same time has contributed to the strengthening of economic development of the country. Generally contributing a moderate growth until 2008, fortunately the contribution of the construction industry is on the rise from 2009 until recently [13]. This phenomenon has enrolled Malaysia in line with other developed countries such as United Kingdom, United States of America, Australia and many more [14]. Apart, recorded value of total construction works done in 2015 is over RM30 million, comprises of residential buildings, non-residential buildings, civil engineering and special trade sectors [13]. Among these sector, civil engineering sector is accountable as a strong contender by 33.9%, followed by non-residential buildings (32.3%), residential buildings (29.1%) and special trade (4.8%) [13].

From micro perspective, through the statistics released by the Ministry of Housing and Local Government (KPKT) and the Department of Valuation and Property Services (JPPH), it was reported that the residential sector continued to lead the property market, accounting for 246,225 (2013) and 247,251 (2014) of units, and Selangor, Kuala Lumpur, and Johor are the highest states with new launches of housing developments in the particular periods. However, few months after the implementation of Goods and Services Tax (GST) in early 2015, this sector showed moderate growth due to various reactions amongst construction players including manufacturers and suppliers, and house buyers [15]. Furthermore, general property prices have increased every year since 2000 (RM138,712) until third quarter of 2015 (RM317,768) [13]. However, the average house price in Malaysia has been slightly reduced to RM315, 287 at the end of fourth quarter 2015. Accordingly, reported house price changes for terraced house is 5.4% at RM278,826, high-rise residential property is 6% at RM296,826, detached house is 4.1% at RM524,260, and semi-detached house is 6% at RM469,823 [13]. Since housing sector becomes one of the main focus of the national agenda (i.e. providing affordable houses for the *Rakyat),* the Government has created various housing programs, schemes, and incentives to help people in various categories of household income, to be able to own a house.

Thus, for the final leg of becoming an advanced country by 2020, Malaysia has adopted the concept of National Blue Ocean Strategy (NBOS) in their five-year strategies of The Eleventh Malaysia Plan, 2016–2020. The Eleventh Malaysia Plan was embedding the Blue Ocean Strategy's tools and frameworks to formulate high-impact and low-cost national strategies that were able to be rapidly executed [16]. Throughout the plan, a theme of "anchoring growth on people" which has six strategic thrusts with six game changers was capsulated. One of the thrust is on improving the wellbeing for all Malaysian, focusing on one of the aspiration's focus area (Focus Area-B), which is providing adequate, quality, and affordable housing to poor, low, and middle income households. Additionally, this aspiration was also considering management and regular standard maintenance of public housing (in Strategy B3) [17].

In Malaysia, basically, most of house accommodations are divided into various categories which are terrace or link house, detached house or bungalows, semi-detached house, long house, flats and apartments, condominiums and shop house [17,18]. In the development of house, there are several parties involved during the construction and after the completion of the house (either from the public sector or the private sector). The key stakeholders are the Ministry of Housing and Local Government (KPKT) with several local authorities, project manager, consultant, contractors, manufacturers and suppliers, and house buyers [20]. Between these parties, there are business agreements in the form of contracts to complete the work in the project, such as; design, engineering, construction, management, and maintenance [20,21]. Meanwhile, private association representing all private housing developers across the Peninsular Malaysia known as Real Estate and Housing Developers' Association Malaysia (REHDA) is recognized as the leading representative body for private property developers, which being involved primarily in advocacy and governance [23].

3 House components

Structure is a complex system which usually made up from a number of frames or members which bears the dead load and live load. Whilst, a frame or members may consist of several components, and might also need a number of sub-components [7,23,24]. The same can be viewed towards a house, which generally needs a collection of components and sub-components. Primarily, structure of house needs several members, which are foundation, column, beam, slab, and truss [26]. Then, to encapsulate all members in becoming envelope-type of structure, floor, wall and roof are constructed and jointed together [26]. Basically, general building works are divided into four different components: structural, architectural, mechanical and electrical (M & E), and external works [27].

Furthermore, in order to view house components in a more comprehensible manner, in-depth literature reviews were conducted to identify house components based on written manuscripts, published sources from all over the world, and Malaysian house components catalogue. Along the way, the research employed a mapping system to outline the major house components which can be assessed using sense perceptions (i.e. vision, hear, touch, smell) since house owner became the concern throughout the research. Lending a support from the works of Kariya et al. [26], the move involved multi-layered thematic analysis as a document analysis tool to collect, filter, and analysed the information based on the research needs [26]. Therefore, Table 1 shows the list of major house components that are available to house owner which are related to Malaysia's environment, and their associated indigenous manufacturers and suppliers of each component, where later on will serve as part of purposive respondents in subsequent stages of research exploration. As can be seen, the main house components consist of twelve (12) main components, namely; staircase, door, window, floor, wall, ceiling, roof and awning, house fittings and fixtures, mechanical and electrical fittings, main gate and fencing, window and door grill, and apron and perimeter drain (citations for each component are shown in Table 1). However, generally they have their own sub-components which at the end will contribute in forming the main components itself. These house components later on become a basis to identify the factors that affect life expectancy of the components and further suggest average life expectancy for each generic house components.

4 Life expectancy (LE)

There are a handful number of sources and guidelines on the life expectancy of different house components around the globe. These include the published guideline in 2006 from National Association of Home Builders (NAHB), along with Bank of America titled "Study of Life Expectancy of Home Components" based on manufacturers, trade associations, and product researchers [9]. According to the definition from Ngo et al. in their article on Establishment of Standards in Property Maintenance - A Pilot Study in Kuala Lumpur, the life expectancy of house component can be defined as the years of life remaining of the components before they break, damage, need further maintenance or need replacement [40]. Their expectancy of life is statistically in the range of number of year's remains of the components starting from their first installation or first built of the components until their last functional usability. The ages between 10 - 12 years appears to be the point of the building's life when wear and tear of the building fabric is increasingly visible, where major repairs and refurbishment become necessary to keep the building economically competitive [40]. In considering the life expectancy of building, three maintenance strategies are taken which are development, implementation, and management [41].

Table 1. House components (minus sub-components) according to sense perception assessment which related to Malaysia's environment.

Component	Sense perception	References
Staircase	Vision	*Authors*: [9], [28], [29], *Manufacturers/suppliers*: Reliance Home Sdn. Bhd., Wira Sky Steel Sdn. Bhd., Supercool Creative Culture Sdn. Bhd., COD Marketing Sdn. Bhd., Weng Hing Stainless Steel Sdn. Bhd., Wira, Wrought Iron & Stainless Steel Sdn. Bhd., Solid Glass & Metal Industries Sdn. Bhd.
Door	Vision Touch Hear	*Authors*: [7,9,27–30], *Manufacturers/suppliers*: Amax Lifestyle Product Sdn. Bhd., Golden Best, Reliance Home Sdn. Bhd., Space Kete Sdn. Bhd., THC Metal Engineering Sdn. Bhd., Tropicale Window & Door Sdn. Bhd., Delux Stainless Steel Sdn. Bhd., Wira Wrought Iron & Stainless Steel Sdn. Bhd., Ga Hing Trading Sdn. Bhd., Honglee Group (M) Sdn. Bhd., Solid Glass & Metal Industries Sdn. Bhd., Vin Art Building Materials Sdn. Bhd.
Window	Vision Touch Hear	*Authors*: [25,27–29,31–33], *Manufacturers/suppliers*: Amax Lifestyle Product Sdn. Bhd., Breezway Window & Door Marketing Sdn. Bhd., Reliance Home Sdn. Bhd., Space Kete Sdn. Bhd., Wira Wrought Iron & Stainless Steel Sdn. Bhd., Honglee Group (M) Sdn. Bhd., Solid Glass & Metal Industries Sdn. Bhd.
Floor	Vision Touch Hear	*Authors*: [7,9,25,28–36], *Manufacturers/suppliers*: Orion Contract Sdn. Bhd., Trocellen S.E.A Sdn. Bhd., Bellezza Ceramiche Sdn Bhd., Decasa Marbles Sdn. Bhd., Gold Rich Stoneworks Sdn. Bhd., Basis Supreme Sdn. Bhd., Coco Vinyl Flooring, Continental Aces Sdn. Bhd., Excel Floor Sdn. Bhd., Floor Depot, Hoss Marketing Sdn. Bhd., I Vanquish Marketing Sdn Bhd, Inova Resources Sdn. Bhd., Lumber Mart Sdn. Bhd., Netwood, Sugawa Lonwood Sdn. Bhd., Swoods Sdn. Bhd.
Wall	Vision Touch Hear Smell	*Authors*: [7,8,25,27–31,34], *Manufacturers/suppliers*: Elegant Haus Wall Coverings, Lamex Décor (M) Sdn Bhd., Metro Furnishing Sdn Bhd., Nippon Paint, Orion Contract Sdn Bhd, Thinkerbell Design Solution Sdn Bhd, Bellezza Ceramiche Sdn Bhd, Decasa Marbles Sdn Bhd, Gold Rich Stoneworks Sdn Bhd.
Ceiling	Vision	*Authors*: [25,29,34,36] *Manufacturers/suppliers*: Teng Plaster Ceiling & Renovation, Ting Fung Plasterceil Sdn Bhd.
Roof	Vision	*Authors*: [8,9,25,28,30–33,35–37], *Manufacturers/suppliers*: The Cascade Team, Wira Sky Steel Sdn. Bhd., De Skylight Roofing Sdn Bhd, Delux Structural Works sdn Bhd, Monier Malaysia Sdn Bhd, Supercool Creative Culture Sdn Bhd, COD Marketing Sdn Bhd, Delux Stainless steel Sdn Bhd, Solid Glass & Metal Industries Sdn Bhd, Vin Art Building Materials Sdn Bhd.
House fittings & fixtures	Vision Touch	*Authors*: [9,25,28,29], *Manufacturers/suppliers*: Reliance Home Sdn. Bhd., Decasa Marbles Sdn Bhd, Gold Rich Stoneworks Sdn Bhd, Bathcare Marketing Sdn Bhd, Duravit & Hansgrohe, Ga Hing Trading Sdn Bhd, Kosa Malaysia Sdn Bhd, Lg Global (VTO Bath), Molecq Sdn Bhd, Solid Glass & Metal Industries Sdn Bhd, The Ultimate Bath Gallery.
M&E fittings	Vision Touch	*Authors*: [9,25,28,29,38]
Main gate & fencing	Vision Touch	*Authors*: [28,29,38], *Manufacturers/suppliers*: Wira Sky Steel Sdn. Bhd., Supercool Creative Culture Sdn Bhd, B.T. Engineering Sdn Bhd, COD Marketing Sdn Bhd, Delux Stainless steel Sdn Bhd, Weng Hing Stainless Steel Sdn Bhd, Wira Wrought Iron & Stainless Steel Sdn. Bhd., Honglee Group (M) Sdn Bhd.
Door & window grill	Vision Touch	*Authors*: [29], *Manufacturers/suppliers*: Wira Sky Steel Sdn. Bhd., B.T. Engineering Sdn Bhd, C.S. Alloy Design (M) Sdn Bhd, Weng Hing Stainless Steel Sdn Bhd (2016), Honglee Group (M) Sdn Bhd.
Drain & perimeter drain	Vision	*Author*: [29], *Manufacturers/suppliers*: Hume Concrete Sdn. Bhd., E-Rete (Malaya) Sdn. Bhd., Oka Corporation Bhd., Lau Tat Sdn. Bhd., Tekun Concrete Manufacturer,

A maintenance strategy is being developed for each building that takes into consideration for its priority, life expectancy, condition, and the surrounding environment. Maintenance and improvement strategies are related to three broad bands of life expectancy which are short life (0-5 years), medium life (5-15 years) and long life (15 years and above)[40].

A study from National Association of Home Builders (NAHB) has concluded that, there are several factors that could affect the life expectancy of house components. These are qualities of products and materials, the quality of their installation (including workmanship and tools), their level of maintenance and replacement, exposure to weather and climatic conditions, and intensity of use. Besides, there are some components to remain functional but become outdated because of changing of styles and technology, or because of product improvements [9]. Therefore, it is understandable that life expectancy of house components may vary in every country depending on the previous affecting factors of their life span. Thus, for that reason, the identification of life expectancy for house components is indeed needed to encapsulate those factors in order to envisage LE holistically.

5 Identifying process of house components' life expectancy

In conducting a research, basically, there are eight (8) types of research design that will be selected by researchers based on their research purpose; which are experimental and quasi-experimental designs, correlational designs, survey designs, grounded theory designs, ethnographic designs, narrative research designs, mixed method design, and action research designs [42]. This study is much likely embarked on pragmatic moves, where mixed method design consisting of qualitative and quantitative inquiry is considered.

Learning from past studies conducted by Kariya et al. [26], Mohammad et al. [43], and several other researchers from developed countries (e.g. UK and US), the first phase of the research (i.e. identifying house components) were adopting qualitative approach through appropriate document analysis (i.e. literature reviews) in order to answer the "what?" question before moving on to the next phase of research. Literatures are derived mostly from books, journals, and legitimate documents which provide established lists of house components, life expectancy's factors, and other valuable information such as suggested average life expectancy. A multi-layered thematic process is embedded in order to capture associated research's key points whilst encapsulate prior identified research's scopes. These several layers of themes were purposely produced in order to collect, filter, and analyse related information [43], particularly on sense perceptions and local housing environment. Then, collected data will be directed to successive phase of research exploration.

Subsequently, due to the nature of research topic that requires further insight on LE's phenomenon and at the same time necessitates large pool of data, the second phase (i.e. life expectancy identification) are embedding both qualitative and quantitative approaches respectively. Qualitative method through purposive semi-structured interviews towards relevant construction practitioners were deemed paramount in order to shed light on certain variables' clusters (i.e. house components, life expectancy's factor, range of life expectancy) and doubled as validation purpose. Meanwhile, the succeeding quantitative method through questionnaire surveys were dedicated to capture purposive respondents' perceptions (i.e. significant factor of life expectancy and life expectancy) and to generalise appropriate outcomes. In order to identify life expectancy's factors, house components were divided into several generic types before differentiation of materials and finishing follows. According to preliminary observations, types of components dictate different life expectancies, and the circumstance are further varied consistent with material selection and finishing applied. On the other hand, life expectancy itself covers an average life span of components and their average periodic maintenance, respectively. After that, quality measure will be imposed at the end of the research before finalising any findings. The following figure (see Fig. 1) shows the thematic processes in a glance.

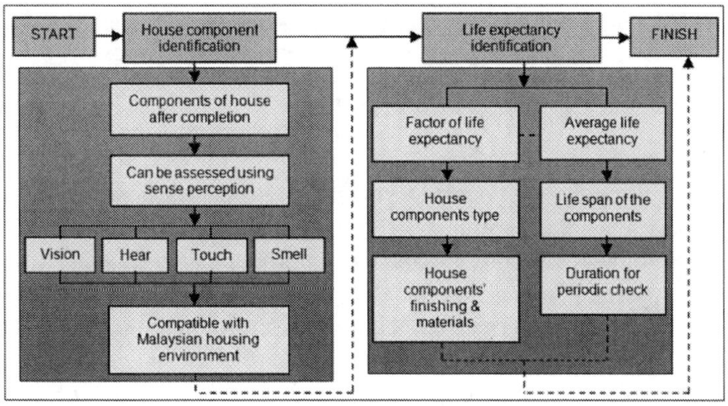

Fig. 1. Multi-layered thematic process.

6 Conclusion and way forward

This paper serves as the basis for preliminary explorations aiming at identifying literatures and references that will benefit the authors in conducting subsequent moves. It is understood that house maintenance is an important facet of management for house owners. Late awareness on house components' defects can be dangerous to the occupants of the house itself. Generally, the ostensive damages on the house components usually occur after the Defect Liability Period (DLP) in which any cost of repairing the damages is borne by the house buyers themselves. Due to that, a study on life expectancy of house components is deemed as needed, in which the findings will become a term of reference for house buyers. Aside from assisting for periodic maintenance in preserving house components from further critical damages on behalf of house owners, the positive outcome may spill over towards other services entity such as the insurance's sector. The research is adopting a mixed-methodology, where the excerpt of research phases can be perused in Figure 1.

In a nutshell, in order to identify house component's life expectancy, several clusters of variable need to be concerned together such as house components (including their materials and finishing), factors that affect the life expectancy, and the average life expectancy itself (including their lifespan and periodic maintenance period). Besides, past studies conducted should be referred to in order to grasp the pragmatic moves taken by them. For instance, in developed country (e.g. UK, US), studies on life expectancy was already conducted, however their findings may differ due to localised housing environment such as climate condition and types of house components. Thus, the authors' mooted house components as listed on Table 1 are able to be assessed using sense perception (i.e. vision, touch, hear, smell), and its compatibility with Malaysia's environment has been identified. Protected and closely covered components such as house foundation and beams, and climate incompatibility such as roof chimney and fireplace were excluded in the list. Therefore, preliminary finding shows that there are twelve (12) main components of house including staircase, door, window, floor, wall, ceiling, roof, house fittings and fixtures, mechanical and electrical fittings, main gate and fencing, door and window grill, and exterior house drain. Along with their appropriate sub-components, the research is optimistically ready for successive explorations. To sum up, it can be concluded that successive tools (i.e. multi-layered thematic process) to analyse related literatures are required in order to avoid any deviation from the research aim and scope. This preliminary action aptly benefited the exploration of central phenomenon in hand, and at the same time capturing apposite pragmatic moves taken by past researchers for subsequent betterment in this research.

The authors would like to convey sincere gratitude to Universiti Tun Hussein Onn Malaysia (UTHM) for sponsoring this paper and research through their in-house research grant.

References

[1] Valuation and Property Services Department, *Property market report*, Ministry of Finance Malaysia, Kuala Lumpur, (2015)
[2] A. Suffian, Some common maintenance problems and building defects: Our experiences, Procedia Enginering, **54**, 101–108, (2013)
[3] C. Liew, N.A. Haron, Factors influencing the rise of house price in Klang Valley, Int. J. of Research in Engineering and Technology, **2**(10), 261–272, (2013)
[4] A.S. Ali, S.N. Kamaruzzaman, R. Sulaiman, Y.C. Peng, Factors affecting housing maintenance cost in Malaysia, J. Facility Management, **8**(4), 285–298, (2010)
[5] Buildings Department Hong Kong, *Building maintenance guidebook*, Hong Kong, (2002)
[6] B. Kayan, Maintaining old building during post gazzeted period - case study: Kuala Lumpur, the Malaysian surveyor, J. Inst. Surv., **40.1**, 15–22, (2006)
[7] R. Coleman, *Residential construction for new home covered by home warranty insurance in British Columbia*, Government of British Columbia, Canada, (2014)
[8] Department of Communities and Local Government, *A decent home: Definition and guidance for implementation*, London, (2006)
[9] Economics Group of NAHB, *Study of life expectancy of home components*, US Department of Housing and Urban Development, USA, (2007)
[10] Worldometers, *Malaysia Population, Dadax*, (2016), Retrieved on October, 2016 from http://www.worldometers.info/world-population/malaysia-population
[11] Department of Statistics Malaysia, *Department of Statistics Malaysia press release*, Kuala Lumpur, (2016)
[12] S.S. Hock, *The population of Malaysia*, Second edi., Institute of Southeast Asian Studies, Singapore, (2007)
[13] Department of Statistics Malaysia, *Department of Statistics Malaysia press release quarterly construction statistics, Fourth Quarter 2015*, Malaysia, (2016)
[14] Investopia, *Top 25 developed and developing countries*, (2016), Retrieved on November 22, 2016 from http://www.investopedia.com/updates/top-developing-countries
[15] Kementerian Kesejahteraan Bandar Perumahan dan Kerajaan Tempatan, *KPKT statistics 2014*, Malaysia, (2014)
[16] Blue Ocean Strategy, *Blue ocean strategy for national developm*ent, (2016), Retrieved on October 11, 2016 from https://www.blueoceanstrategy.com/blue-ocean-strategy-for-national-development
[17] Economic Planning Unit, *Rancangan Malaysia Kesebelas (Eleventh Malaysia Plan): 2016-2020*, Percetakan Nasional Malaysia Berhad, Malaysia, (2015)
[18] Y.H. Lee, *Homebuyers' requirements for housing development within districts in Selangor*, Master Thesis, Universiti Teknologi Malaysia, Malaysia, (2006)
[19] S.A.Q.S.A. Karim, *The significance of owner participation for a successful detached housing development in Malaysia*, PhD Thesis, Queensland Univesity of Technology, Queensland, Australia, (2013)

[20] D.J. Tan, F.E.M. Ghazali, Critical success factors for Malaysian contractors in international construction projects, *Int. Conf. on Engineering, Project and Production Management,* Singapore, 127-138, (2011)
[21] U.K. Zolkafli, N. Zakaria, H. Salleh, The comparative study of international standard form of building contract (FIDIC conditions of contracts for construction - The new red book 1999) and Malaysian's Standard Form of Building Cotract (PWD 203A), *Int. Engineering Education Conf.*, (2011)
[22] H. Singh, *Construction contracts: An overview*, The Ingenieur, 6–20, (2005)
[23] Real Estate and Housing Developers' Association Malaysia, *Official Website of REHDA: Real Estate and Housing Developers' Association Malaysia (REHDA)*, (2000), Retrieved on March 21, 2016 from http://rehda.com
[24] Canadian Association of Home and Property Inspectors, *National Standards of Practice*, Canada, (2012)
[25] Carson Dunlop and Associates Ltd, *Inspection a house: A Guide for buyers, owners and renovators*, Third eds., Dearborn Trade Publishing, USA, (2004)
[26] N. Kariya, Z. Yaakob, M.N.M. Sairi, H. Mohammad, S.K. Yaman, N.H. Abas, Investigation of generic house components and their practical ways to be assessed by house buyers during defect liability period (DLP) in Malaysia, *Int. Congress on Technology, Engineering and Science*, Kuala Lumpur, (2016)
[27] Construction Industry Development Board Malaysia, *Construction industry standard CIS 7:2014 quality assessment system for building construction works*, Kuala Lumpur, (2014)
[28] H. John, C. Geoff, *The technology of building defects*, Taylor and Francis, (1997)
[29] International Association of Certified Home Inspectors, *InterNACHI'S estimated life expectancy chart for Florida homes*, (2006), Retrieved on December 4, 2015 from http://www.nachi.org/florida-life-expectancy.htm
[30] Construction Industry Development Boar, *CIS 7:2006 Quality assessment system for building construction work*, Malaysia, (2006)
[31] M.S. Richard, H. Dave, *Home buyer's checklist,* McGraw, New York, (1993)
[32] B.A. Richardson, *Defects and deterioration in buildings*, 2nd eds. Spon Press, New York, (2001)
[33] Home Builders Association of Northwest Indiana, *How long will it last? The life expectancy of your home's components*, Northwest Indiana, (2015)
[34] G. Peter, *Building surveys*, Butterworth-Heinemann, Italy, (2006)
[35] B. Jay, *Building construction inspection: A guide for architects*, Wiley, USA, (1991)
[36] D. James, A.N. Edward, *Building surveys and reports*, Wiley, New York, (2011)
[37] M.D. Spada, *The home inspection book: A guide for professionals,* South-Western Educational Publishing, Ohio, (2002)
[38] Jabatan kerja Raya Malaysia, *JKR standard specification*, Malaysia, (2014)
[39] American Society of Home Inspectors, *The standard of practice for home inspection and the code of ethics for the home inspection profession*, USA, (2014)
[40] M. Ngo, Establishment of standards in property maintenance - A pilot study in Kuala Lumpur, *Building Miantenance and Modernisation Worlwide,* Singapore, (1990)
[41] R.A. Bates, Maintenance strategy - A Hong Kong prospective, *Proc. of the Int. Symposium on Property Maintenance Management and Modernisation*, Logman Singapore, 28–38, (1990)
[42] J.W. Creswell, *Educational research: Planning, conducting, and evaluating quantitative and qualitative research*, Pearson Education, USA, (2012)
[43] H. Mohammad, S.K. Yaman, F. Hassan, Z. Ismail, Z. Determining the technical competencies of construction managers in the Malaysia's construction industry, MATEC Web of Conf., **47**, 04021, (2016)

Abu Dhabi's New Urban Islands and Shorefront Development

Mohamed El Amrousi[1,*], and *Mohamed* Elhakeem[2]

[1]Department of Architecture, Abu Dhabi University, P.O. Box 59911, UAE
[2]Department of Civil Engineering, Abu Dhabi University, P.O. Box 59911, UAE

Abstract. Abu Dhabi is in the process of urbanizing a group of Islands that surround its northern and eastern coastlines. Al-Lulu, Al-Saadiyat, Al-Maryah, Al-Reem and Yas Islands are all new urban enclaves that were desert islands and marshlands yet, have been developed over the past decade to urban islands that include epic and entertainment centres such as the Abu Dhabi Louvre, Guggenheim Museum, Ferrari World, NYU Abu Dhabi, the Paris Sorbonne Abu Dhabi in addition to iconic and exclusive waterfront residential units. These new islands re-brand the image of the main archipelago of Abu Dhabi that for decades retained a grid street pattern and pragmatic concrete blocks created in the late 1970s. The new urban islands transform Abu Dhabi's image into a multinational modern Arab city seeking to become part of the global city network. Abu Dhabi's new urban islands also act as breakwaters that protect the main archipelago's coastline from erosion resulting from tidal change, because they are designed to include concrete and stone breakwater barriers. This paper represents a cross-disciplinary research between Civil Engineering and Architecture Departments in an attempt to explore the emerging infrastructure and urban expansion of Abu Dhabi from a multi-disciplinary perspective. We also highlight through simulating the effect of breakwaters on wave heights two scenarios for Al-Lulu Island the importance of these new barrier Islands on the urban expansion of Abu Dhabi.

1 Introduction

Abu Dhabi is reshaping much of its coastline through new urban developments and hence changing its image from a city known for its conservative local community to one that is more attractive to expatriate residents. This is achieved through the creation of a new group of urban islands that reshape desert islands and existing marshlands connecting them to the mainland through a series of bridges and supporting them with the necessary infrastructure. These islands also act as protection for the coastline of the main archipelago because they entail the infrastructure through breakwaters and marinas to create barrier islands. These new urban islands are designed to exhibit the richness of Abu Dhabi as the Capital of an oil rich State, this is manifested through the construction of mega museums, iconic buildings and clustered epic centers to create new public spaces. The transition of

* Corresponding author: Mohamed.amrousi@adu.ac.ae

Abu Dhabi from a transit city, which till the 1970's exhibited characteristics of tribal conservatism and remoteness from modernity represents a new phase in the city's urban expansion, infrastructure and socio-cultural establishment (Fig. 1). Abu Dhabi is an archipelago connected to the mainland via two bridges, therefore many of its conservative geographies can be understood if associated with traditional Island communities. This was especially visible in its architecture, urban fabric shaped along modernist design ideologies, traditional community, practices and regulations. In the last decade Abu Dhabi has announced is 2030 Vision that advocates plans to become more open internationally through iconic mega projects and an infrastructure and new spaces of entertainment. Abu Dhabi's urban expansion plan includes a gentrification of the existing building stock in the living city mainly the Down Town area and its Corniche. As its main waterfront promenade the Corniche of Abu Dhabi is undergoing major reconstruction projects that aim to improve its public spaces and communal spaces of gathering. This is exhibited through the creation of public beaches connected to the Downtown via tunnels, the municipality is also creating new entertainment hubs along the Corniche that include cafes, spaces of gathering, centers for sports activities, special lanes for jogging and cycling and shaded seating (Fig. 2). These are all initiatives that are essential in order to maintain the competitiveness of the old downtown especially when juxtaposed with the new upscale facilities emerging on its new urban islands. All major investments are supported by Abu Dhabi's royal family and aim to support international initiatives to improve the quality of life in the city and to transform the city from a transit one to one of long term residency and multinational investment. Part of the internationalization of Abu Dhabi includes Sassaki's plan for the gentrification of the old Port-Mina Zayed that has commenced to transform the old warehouses and shipyards into contemporary museums, and cultural facilities. Sassaki's design incorporates the new Abu Dhabi Media Company headquarters which will be sustained by hotels, retail, and high rise residential units overviewing the Corniche. These visible changes in Abu Dhabi that aim for communal wellness and its transformation to a healthy city are witnessed by its residents especially the new buildings emerging along its coastline overlooking its newly created estuaries that separate between the mainland and its urban Islands. Abu Dhabi's ambitious program of shorefront urban development's aim to serve a larger economic agenda by increasing demand for construction sector services and supplementary investments related to tourism, entertainment and business entrepreneurship. This paper is divided into two sections that complement each other and highlight the cross-disciplinary venues of research possible between architecture and Civil Engineering. The first part addresses Abu Dhabi's shorefront expansion in the form of urban islands that also act as breakwater barrier to protect the mainland/archipelago. The second studies via simulation the shorefront protection system that accompanies the creation of these new urban islands using Lulu Island as a case study.

Fig. 1. Abu Dhabi 1948-1950.

Fig. 2. Abu Dhabi; Marina, the Corniche and Old Port-Al-Mina.

2 Abu Dhabi's urban islands

Al-Lulu, Al-Saadiyat, Al-Maryah, Al-Reem and Yas Islands represent the new urban expansion to the main archipelago of Abu Dhabi, in their planning which is part of the Abu Dhabi 2030 Vision, reducing the effect of marginalization that occurs to old urban centers once new ones appear has been taken into consideration. This is achieved through the visual and physical connectivity infrastructure represented by successive bridges and tunnels, and waterfront promenades that do not obscure the visibility between new and existing. This link is further cemented by signs stating the names of these islands, thereby reducing the effect of marginalization of the old created by the vividness of the new monumental and upscale mega projects separated by water canals. Al-Saadiyat, Al-Maryah, Al-Reem and Yas Islands have evolved with a cultural and environmental perception in mind in that they offer new venues for local and expatriate residents of the city to enjoy international museums, high-end hotels and residences without leaving their homeland [1]. This development is also paralleled with an aspect of environmental sustainability through the plantation of mangrove trees in marsh lands and lagoons created in between these islands. Although the phenomena of creating artificial islands is one that is spreading though the Gulf States, there are different planning ideologies even within the UAE, for example Dubai's artificial Islands differ from those of Abu Dhabi. In terms of construction the Palm Islands in Dubai's reflect on its urban agenda to retain its status as a Star city in the Gulf region. Visual connectivity is reduced to provide privacy its residents. In terms of construction they are fully artificial, created by dredging and backfilling to shape a Palm like waterfront real-estate development that is only visible from top-down aerial views. The Palm (Jumeirah) is protected by a concrete barrier that is in the form of a crescent creating finger islands known as the 'Palm fronds' were canal-estate development offers its residents high-end villas and private beaches, while public or semi-private beaches are scarce on Palm Jumeirah—the most complete of the Palm islands trilogy. Dubai's 'World Islands' further support its policy for dredging and backfilling parts of its waterfronts that are even less connected to its mainland creating a core/periphery relationship were the periphery emphasizes the concept of 'Fantasy Islands'. In Abu Dhabi the new urban islands are very real and accessible to all even though they do offer a degree of exclusivity manifested through their building typologies. The shapes of Abu Dhabi's islands with their smooth coastlines and narrow estuaries created between them highlights the duality of their role as barrier and urban islands that support the image of Abu Dhabi as an emerging modern city with cultural hubs and clustered epic and entertainment centers open to all its residents [2].

Abu Dhabi's new urban islands manifested the need for Abu Dhabi to reshape its image through new forms of real-estate development warranted by an increase in the diversity of socio-cultural communities seeking to reside and invest in Abu Dhabi. Saadiyat has a visible Cultural clustering with projects that have acclaimed international visibility like the Desert Louvre by Jean Nouvel, the Abu Dhabi Guggenheim by Frank Gehry, and the forthcoming Zayed National Museum are all mega projects indicative of the fact that pubic construction projects, are likely to remain a key driver of the construction sector and the economy as a whole [3]. Al-Maryah and Al-Reem Islands were planned to represent the corporate image of the Abu Dhabi through the creation of a new Central Business District, this is especially evident in the multi-scaled and inter-connected group of buildings on Al-Reem Island known as the Gate Towers that is developed by Sorouh. The Abu Dhabi Gate Towers, with their 3533 residential units in four towers as well as retail and leisure facilities that pay tribute to Singapore's Marina Bay Sands development designed by Moshe Safdie, and represent an example of branding of the spaces created by globalization. Accompanying this phenomena there is also a sense of sameness or creation of architectural facsimiles that accompany the globalization of space. The creation of clustered/exclusive residences along newly formulated shorelines, and creating cities that share spaces of sameness is a phenomenon that is being witnessed in the network of global cities such as Dubai, Doha, Singapore and Hong Kong etc., and can be understood as an expansion inscribed by the emergence of new social groups and expatriate communities. Abu Dhabi's interest to seek a share of the capital investment worldwide inscribed on its planners acceptance of changing its initial low profile fabric to one that is branded through monumentality and mega projects [4]. Al-Dar and Sorouh have been merged to create a mega construction company that is able to become a major player as master developer behind Shams Abu Dhabi, a project that aims to create a mixture of commercial, retail and residential space, including the 74-storey Sky Tower and 65- story Sun Tower. Other mega projects are represented by Yas Mall that is currently the largest mall in Abu Dhabi with its 235,000 sq. meters of shopping space and 370 stores 60 of which are international restaurants (a major change for a city that has retained for decades a conservative tribal culture), Yas Marina adds new multi-national public space to Abu Dhabi in terms of waterfront entertainment projects that focus on public spaces of gathering along the Marina shorefront with pedestrian paths that are specifically designed for public uses like a jogging, walking and cycling etc. Here there is no simple demarcation between public and private waterfront landscapes [5]. Yas Island highlights the emergence of high-end clusters of epic and entertainment Centers on a mega scale such as the Ferrari World, Viceroy Hotel and Yas Mall that could have created a juxtaposition of old and new is manifested by the modernist 1970's design of buildings on the mainland and its new offshore developments. Abu Dhabi new expansion via urban islands, changes in the Downtown and Mina areas represent a challenge for urban planners and designers as they attempt to balance between retaining cultural values and the conservative community of the oil rich State and the socio-cultural changes inevitable through the emergence of spaces of entertainment and epic centers in order to avoid the pitfall of loss of identity and marginalization of some communities within the existing/living city. In the Abu Dhabi 2030 Vision retention of Emirati identity and cultural values is highlighted, as special districts for nationals are planned in parallel to the emerging iconic architectures in Abu Dhabi's new shorefront urban expansion (Fig. 3).

Fig. 3. Al-Saadiyat-Abu Dhabi Louvre-Al-Reem Islands-Yas Island Marina.

3 Globalization and urban-island development

The common perception of Islands are usually associated with an image of spaces that escape modernity via isolation spaces that offer its visitors a utopian experience that is embedded in the traditional attire of their communities. However, Al-Saadiyat, Al-Reem Al-Maryah and Yas Islands reflect an alternative perception of space, one that is shaped by new building morphologies that create spaces of emerging epic centers of culture and leisure, this heavy investment in the construction industry manifests Abu Dhabi's interest to emerge on the map of global cities through State sponsored mega-projects. However, Abu Dhabi's investments are not dominated by an attire of exclusivity, they also highlight the importance of culture as main theme, exemplified by the Construction of the Abu Dhabi Louvre, the Abu Dhabi Guggenheim, and the new Sheikh Zayed National Museum that aim to brand Abu Dhabi as a city that values culture and heritage. While there is a clear paradigm shift in Abu Dhabi's urban strategies, which for decades aimed to preserve a tribal-social structure through the exclusion of entertainment and leisure centers, many of the pitfalls of Dubai's rapid urbanization have been avoided especially urban fragmentation. Much has been done to avoid social segregation and to connect the mainland through several bridges and tunnels to the newly developed urban islands of Abu Dhabi. There still is a clear manifestation of global agendas that decontextualize regional-design criteria on both environmental and cultural levels, yet many of the cultural projects aim to create spaces and places of familiarity and interest for the local community and international investors. As outlined by Anna Gospodini, culture and leisure waterfront epic-centers created in Abu Dhabi correspond to manifest a process of urban renewal and generate new building typologies in the post-modern city. These are especially manifested through the adoption of brand names of in institutions of Culture such as the Guggenheim and Louvre, in education such as NYU and the Paris-Sorbonne and healthcare such as the Cleveland Clinic [6]. The global extent of artificial residential waterways created in the last decade highlights the degree of investment in infrastructure by the State and confirms a new interpretation of the modern city as a hybrid group of spaces that includes spaces of the living city and a new social structure created on the spaces of Abu Dhabi's urban which are somehow 'more real' as extensions of city spaces even though they are visually distinct through clear references to international counter structures. Nevertheless they cannot be considered as Peripheral to a Core rather a more upscale and exclusive version of the mainland represent constant evolution of the city [7].

Al-Saadiyat, Al-Maryah, Al-Reem and Yas Islands also manifest a transitional shift in perception of communities living in gulf/desert regions that may have been classified as tribal or conservative ones that adopt an attire and behaviors of new forms of universalism, created by spaces and a new genealogy of buildings. These communal paradigm shift is associated with the city image-branding created by the multiplication of massive structures built with corporate imagery, especially on Al-Reem and Al-Maryah Islands. All of Abu Dhabi's new islands reflect design criteria that caters to spaces mediation between the local residents and the expatriate community (with the exception of Al-Samaliyah Island that was created as an urban museum of local heritage). Despite growing competition for the niche in cultural and entertainment facilities Abu Dhabi's morphological development of Al-Reem Island embraces these economically high-risk ventures. As King outlines globalization, and shared interdependency have become a fact of life through which a global imaginary is becoming an environment and an identity [8]. This may offer an explanation for the simulacra of some of the international iconic buildings such as the Marina Bay Sands Towers and the Gate Towers of Abu Dhabi which both reflect the extensive use of glass facades in hot humid climates in addition to colossal scale and similarity of form. This combination of the spatial reproduction and re-enactment of narratives seeks to create a new image of becoming a global city through building typologies that serve to repackage the image of Abu Dhabi from a conventional Arab city to one that offers the potential of an International lifestyle.

4 The shoreline protection system of Abu Dhabi's islands

In this section of the study we evaluate the design of the current shore protection system of Abu Dhabi City Main Island. Two scenarios were simulated: scenario 1) Prior to 1970, where neither the breakwaters nor the Lulu Island was constructed (Fig. 4A); scenario 2) the shoreline condition after 2003, where Abu Dhabi City shoreline was reshaped and protected by the onshore and offshore breakwaters, and the Lulu Island (Fig. 4B). The Lulu Island assumed to provide additional protection to Abu Dhabi City shoreline by diffracting the wave energy. Another function of the Island was to hide the offshore breakwaters and to be used as a recreation area providing an aesthetic appeal from the shoreline. The Main Island is facing the Arabian Gulf from the north-west side and is exposed to continuous current wave action. Prior to 1970, where neither the breakwaters nor the Lulu Island was constructed (Fig. 4A), the waves were approaching Abu Dhabi City shoreline without any diffraction causing washout of the sand and hence beach erosion.

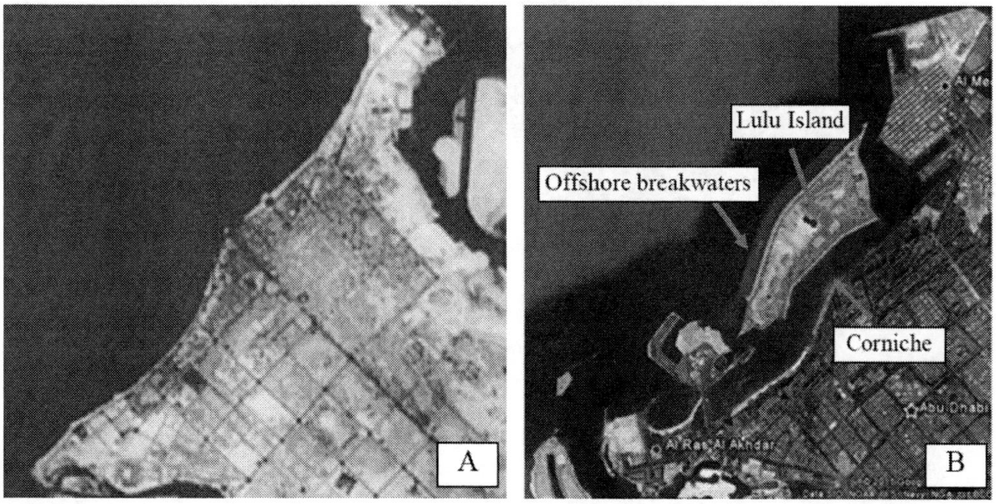

Fig. 4. Abu Dhabi City shoreline: A) prior to 1970; B) after 2003.

In 2003, a shore protection system was developed on this side from the Main Island of Abu Dhabi City (Fig. 4B). The developing of the shore protection system of Abu Dhabi City Main Island went through three stages: 1) reshaping of the shoreline by degrading and damping huge amounts of natural sand; 2) constructing of onshore and offshore breakwaters; and 3) developing of the Lulu Island. Due to the complex geometry of Abu Dhabi City shoreline protection system, a 2D-hydrodynamic model was used in the evaluation of the design. The Costal Graphical Wave (CGWAVE) developed by the US Army Corps of Engineers [9-11] was used to simulate the wave characteristics when they approach the breakwaters and shoreline. The model was calibrated by matching the model wave outputs at the Um Al Dalkh Wave Station, located in the south-west of the model domain, with the measured values of the station. This was done iteratively by changing the wave height at the boundary until the simulated wave height at the station matches the measured value.

Two scenarios were simulated using CGWAVE: scenario 1) Abu Dhabi City shoreline prior to 1970, where neither the breakwaters nor the Lulu Island was constructed. Under this condition, the waves were approaching the shoreline without any diffraction causing washout of the sand and hence beach erosion. In this scenario, we have examined the tidal action in terms of wave height and water surface elevation; and scenario 2) Abu Dhabi City shoreline after 2003 representing the current shoreline condition, where the shoreline was reshaped and protected by onshore and offshore breakwaters, and the Lulu Island. The Lulu Island assumed to provide additional protection to Abu Dhabi City shoreline by diffracting the wave energy. In this scenario, we wanted to examine effectiveness of current shore protection system in providing sufficient protection to the Abu Dhabi City shoreline by damping the wave energy.

In the simulations, we have examined the tidal action in terms of wave height due to wind blowing from the west direction towards Abu Dhabi City shoreline. Fig. 5A shows CGWAVE simulations of scenario 1 for the wave height along Abu Dhabi City shoreline prior to 1970, where neither the breakwaters nor the Lulu Island was constructed.

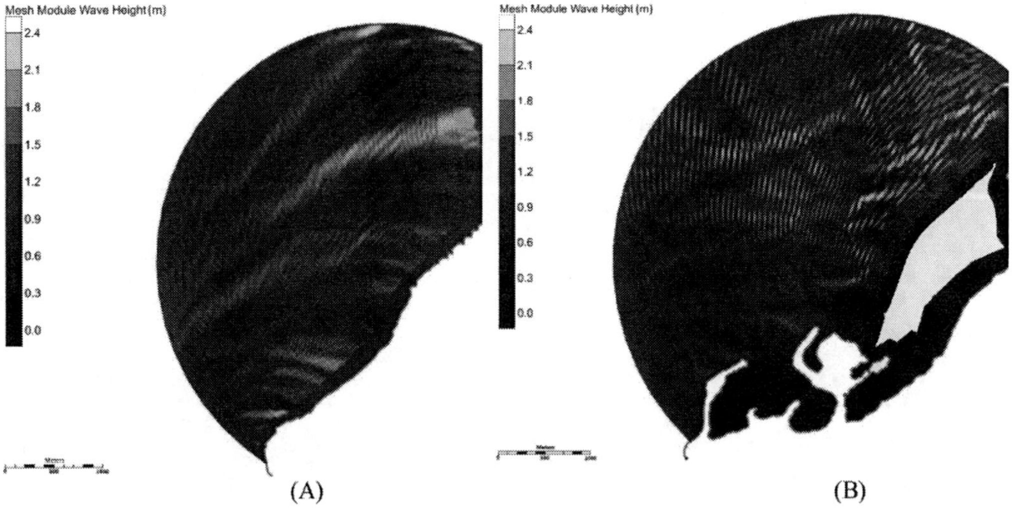

Fig. 5. Predicted wave height at Abu Dhabi City shoreline: (A) scenario 1; (B) scenario 2.

The figure show the simulation for it can be seen that the wave heights vary from 0.1 to 3.0 m. Ocean waves lose part of their energy as they approach the shoreline, but the energy is still great enough for the wave to reach the shoreline at heights that allow them to cause beach erosion. Fig. 5A shows scenario 2 after 2003 representing the current shoreline condition, where the shoreline was reshaped and protected by onshore and offshore breakwaters, and the Lulu Island. It can be seen that the offshore breakwaters dissipate most of the ocean wave energy before they reach the Lulu Island and then the Island completely hinders the waves from reaching to the Abu Dhabi City shoreline. It is also evident that some areas behind the island are completely dead from any wave action due to the offshore breakwaters; the water is almost completely still. The wave heights along the shore and around the Lulu Island are almost zero as can be seen in Fig. 5B. Thus, the proposed design of the shore protection system was successful in controlling beach erosion and flooding by damping the wave energy.

5 Conclusion

Throughout this study we highlight the possibilities of cross-disciplinary research between architecture and Civil Engineering applied to the new urban expansions of Abu Dhabi and its urban sprawl. The study highlights that the creation of new urban Islands have multiple functions such as breakwaters that allow for the creation of new public beaches and spaces of gathering in the living city/archipelago, and act as an integral part of architectural branding, by changing the image of the city from a transit city to one that aims to become part of the global city network similar to Singapore, Hong Kong and Kuala Lumpur etc. Abu Dhabi's urban islands such as Al-Reem, Al-Maryah, Al-Saadiyat and Yas Island represent major investment opportunities in the construction and real estate sector. The State sponsored infrastructure and emerging mega museums and cultural hubs transformed Islands that were primarily desert and marshlands to urban enclaves. Such investments had to be protected from coastal erosion created by waves and tidal changes through the construction of water barriers and breakwater protection systems. Abu Dhabi's 2030 Vision is beginning to materialize to the community as new urban morphologies and iconic buildings rebrand the image of Abu Dhabi. The urban island development represents an alternative solution to that of Dubai's Palm Islands that can only be appreciated from above/aerial views and were constructed artificially by backfilling parts of Dubai's waterfront. Abu Dhabi's new Islands were reshaped to conceal the breakwaters that protect the main archipelago and create new epic-centers and pedestrian promenades along the series of water canals and estuaries. Abu Dhabi's urban expansion that is part of the 2030 Vision manifests and clear sense of comprehensive planning that includes investment in infrastructure in order to support the new spaces of public gathering, culture and leisure.

References

[1] P. Gupta, Futures, fakes and discourses of the gigantic and miniature in 'The World' islands, Dubai, Island Studies J., **10**(2), 181-196 (2015)
[2] K. Zhang and S. Leatherman, Barrier Island population along the U.S. Atlantic and Gulf Coasts, J. of Coastal Research, **27**(2), 356–363 (2011)
[3] D. Ponzini, Large scale development projects and star architecture in the absence of democratic politics: The case of Abu Dhabi, UAE, Cities, **28**, 251–259 (2011)
[4] G. Evans, Hard-branding the cultural city from Prado to Prada, Int. J. of Urban and Regional Research, **27**(2), 417- 440 (2003)
[5] Oxford Business Group, *The Report: UAE: Abu Dhabi 2015*, Oxford Business Group London, (2015)
[6] A. Gospodini, Portraying, classifying and understanding the emerging landscapes in the post-industrial city, Cities, **23**(5), 311–330 (2006)

[7] A. Grydehøj, Constructing a center on the periphery: urbanization and urban design in the island city of Nuuk, Greenland, Island Studies J., **9**(2), 205-222 (2014)
[8] A.D. King, *Spaces of global cultures: Architecture, urbanism, identity*, New York, Routledge, (2004)
[9] Z. Demirbilek, V.G. Panchang, *CGWAVE: A coastal surface-water wave model of the mild-slope equation*, Technical Report CHL-98-26 (US Army Engineer Waterways Experiment Station, Vicksburg, Mississippi, (1998)
[10] US Army Corps of Engineers, *Coastal engineering manual,* Report No. EM 1110-2-1100, US Army Engineers Waterways Experiment Station, Vicksburg, Mississippi, (1995)
[11] B. Xu, V.G. Panchang, Z. Demirbilek, Exterior reflection in elliptic harbor wave models, J. of Waterway, Port, Coastal, an Ocean Engineeing, **122,** 118-126 (1996)

MATEC Web of Conferences **103**, 03019 (2017) DOI: 10.1051/matecconf/201710303019
ISCEE 2016

Building Design Variables Usage as a Tool of Value Engineering During Designing

Nur Sahid [1,*], *Ainomugisha* Safiki[2], and *Mochamad* Solikin[1]

[1]Civil Engineering Department, 57102 UM Surakarta, Indonesia
[2]Civil Engineering Graduate Studies, 57102 UM Surakarta, Indonesia

Abstract. The deployment of construction economics has become more desirable today, mainly due to need to establish how building costs are spent, and be able to come up with the most optimum alternatives. This research article therefore, explored 1) the various costs inputs called design variables used by design consultants, plus their underlying factors, 2) how the practice of value engineering (VE) impacts on projects in Indonesia. A qualitative methodology, was used inform of a questionnaire, designed based on a 5 pointer liker scale approach, and distributed among 30 respondents consisting of consultants and clients in areas of Surakarta. The collected data was processed using statistical method of relative importance index, followed by descriptive analysis inform of bar and pie charts. The results obtained were that building plane shape (index 83.2), was mostly used, which itself depended on external features of building membrane and shape of building site (80.0 each), then other variables were building complexity (82.1); and building façade (77.9), meanwhile the least used was sharing walls (index 62.1), Lastly, VE was found to benefit the industry by producing designs which meet time, cost and quality targets, on the other hand material wastages and loss of confidence was reported once VE was neglected.

1 Introduction

In more recent studies under design economics, it can be seen that there is an attempt to represent as closely as possible how costs are actually spent [1]. In line with that, it was established by [2] and [3] that the level of influence on these building costs goes on declining as project develops through the various stages of its life cycle. These economic studies have led to new phenomena such as building design variables, which have generally been viewed as an attempt to have a representation or form of geometry of a building, project how this form of geometry impacts on costs, as early as possible i.e. before most of the decisions are spent. They have been generally organized in form of a list of alternative design proposals, and quality specifications. Hence in one way or the other helping the building client by making him aware of the likely financial commitments before extensive work is undertaken.

[*]Corresponding author: nur.Sahid@ums.co.id

This whole process or movement towards construction economics has bread into two concepts but which are merely similar i.e. Value management (VM), being about getting the right project, and Value Engineering (VE) as what is done to get the project right.

According to the research that was made in United States of America (USA) by [4], it was indicated that with this application of VE on construction project, a great deal of savings amounting up to between 34-36 % of total costs of a project could be realized. Furthermore, this concept has been greatly viewed to be in line with the famous quote of John Ruskin's (1819–1900), who said, 'It is not the cheaper things in life that we want to possess, but the expensive things that cost less' [5].

In Indonesian perspective development of this VE begun to be realized when *Departemen Pekerjaan Umum (DPU)* released National Standard competence of works *(Standar Kompetensi Kerja Nasional Indonesia) (SKKNI)* to the value engineering professionals. However, this was followed with establishment that this VE concept is not fully realized, that's why from their report in 2008 it led to issuance of bye law *06/PRT/M/2008* on 27 June 2008. The law decreed that in case of discovery of existence of inefficiencies or wastages resulting from unnecessary construction costs, poor type of form of construction, poor cost estimates, and even the method of construction itself, then it becomes incumbent on owners and service providers to do VE. Therefore, it's upon this background that the author sought to undertake study survey so as to achieve the following objectives: 1) firstly to establish the various design variables used by architect consultants in Indonesia, plus underlying factors for their selection, then 2) seeks to find out how the practice of VE impacts of project in Indonesia.

2 Related literature study

2.1 Design variables theory

In trying to meet client requirements and external constraints brought about by matters like statutory requirements, environmental factors and construction process among others, the architects have started to model their designs using parameters called design variables. The building design variables have been suggested as part of the solution by helping in the field of economics especially during modelling on the constraints of cost analysis and forecasts of the project under the concept of providing value for money products or VE. Therefore, from these design variables, design decisions are normally established, as they give solutions of challenges to do with form, time and economy for buildings [6]. According to [7], also it's these variables which form designers' forecasts, because they give the information for forecasting and determining whether value can be achieved at an acceptable cost [8]. The practice is so important in that, clients are able to get reliable cost advice to enable them assess and choose viability of a project when it is still early [9], by indicating how they impact on costs hence enable clients whether to incur that cost or not. The following have been established in previous studies:

2.1.1 Building plan shape

This is generally defined to stand for the spatial feature which defines the outline of the building. It impacts on the areas and sizes of the vertical members like walls their accompanying finishes, windows, partitions and the finishes used, etc, plus the perimeter details which include ground beams, fascias, and the eaves of roofs.

Over the years, studies focusing on the problems of plan shapes and construction cost have been on the increase, resulting in large number of publications. They have generally established that the cost of building construction increased due to increases in external walls, ceilings, floors or the roof [10-14]. Therefore, the various previous researchers all have concluded that, perimeter-to-floor ratio unit construction cost and overall project costs vary with plan shape complexity or irregularity. This being attributed to the fact that a particular shape of the building affects significantly costs of a great number of building elements like foundations, walls, building structure frame, finishes and decorations, roofing, electrical and mechanical services, which later alone also impact on costs of operating and maintaining the building hence overall life cycle costs affected.

2.1.2 Average storey height

This is defined as height from finished floor to next finished floor or from finished floor to ceiling or head room height (bungalow. Generally the following items may experience increases as a result of changes i.e. decrease or an increase in storey height: 1). The amount of heat energy may increase due to volume of building increasing plus the length which supply cables have to run it may also be increased. 2) Longer service and waste pipes may be required. 3) Chances of having expensive roof costs due to cost of hoisting are high. 4) The costs for circulation elements like staircases and lifts also increase as they are vertical elements whose quantity depend on headroom height. 5) Chances are high of cost increases in applying finishes and decorations due to very high heights of ceilings, sometimes calling for additional scaffolding.

2.1.3 Number of storey

This is much related to the average storey height of a building, however its self refers to how high or the number of levels in terms of vertical construction. In reference to this constructing of tall buildings, it generally affects four major building elements significantly because of the number of storeys of high-rise building i.e. frame structure construction, external wall (curtain walling in most case for office buildings), lift installation and fire protection (as in services), all mainly vertical elements of the building.

Previous studies like [15] pointed out that construction cost generally falls as the number of storeys increases. In the United States, for [16], on the elements comprising of eight office buildings from 8 to 75 storeys on a hypothetical site. It was found out that generally unit building cost tended to rise moderately with building height. Meanwhile for [18], he found out that, minus the lower floors, the unit office building cost was almost constant building height was varied.

In addition to the ones stated above, other designs identified in the previous studies over the years included: Mechanical and Electrical Services; Circulation Space; Grouping of Buildings; Column Spacing; Building Size; Sharing walls; Floor spans and Constructability.

In summary, with this discovery of these design tools called design variables in use during design modelling and modification, this was handy in current survey study of Indonesian construction industry. This was put to a research questionnaire survey to establish among the available design variables which were in use in Indonesia. This was highly viewed as basis for further development to solve problem of inefficient and poor designs identified by department PU.

2.2 Theory of construction costs

The term cost in general terms can be defined as the value of any currency given in to obtain a product, or service, to expend labour and use equipment and tools or to operate a business. For the case of construction it was defined by [17] in their book titled, 'Cost Modelling', as; the cost of the contract incurred by the client. Meanwhile, the source for construction costs were generally grouped into two basic sources according to [18] and [19], i.e. the owner-designer costs i.e. ones that come up as a result of, owner's requirements and the design, and the contractors' and subcontractors' costs, through the competitive market and their own organizations.

3 Methodology

This research study took an approach of exploring the literature of previous studies; and then later surveyed of the practices by consultants and views of clients in the building industry. It should be noted that, this research was developed as an extension of the previous studies like [11, 20], hence, it was mainly focused on enhancing on these previous studies, basing on their discoveries about design variables, helping to establish if it was the same conditions that exist within the Indonesian industry. In line with that, sample study area was Surakarta and areas around it.

Thirty participants (30), consisting of 23 design consultants and 7 clients were sampled, basing on Roscoe in his book Research Methods for Business [21] about determination of sample respondents. The research data collection tool used i.e. questionnaire was designed basing on objectives of the research, hence, it was divided into two sections, firstly being about the practice of design variables and the second one about value engineering in the industry. Meanwhile, the sampling technique applied was stratified purposeful sampling, which enabled to use information rich cases for in depth study depending on their characteristics [22].

The data collected from the questionnaire for easy analysis, was done using Microsoft excel. This stated with data validity and reliability test using descriptive analysis technique i.e. formula of Product Moment rxy, which was then compared with 5 % error standard table. This descriptive analysis statistics method, analyses data by using descriptions or illustrations of the data collected from the questionnaire [23].

Furthermore, under analysis it included the responses that were received from the survey participants being tabulated and analyzed individually. This was done with usage of the following formula as suggested by [24]:

$$\text{Relative importance Index} = \left\{\frac{\sum_{i=1}^{5} a_i\, x_i}{5\sum_{i=1}^{5} x_i}\ x\ 100\%\right\} \qquad (1)$$

Where a_i = constant expressing the weight given to i;
x_i = variable expressing the frequency of each response for; i= 1, 2,3,4,5.

Finally, the findings were displayed by tabulation or using bar charts and pie charts, etc.

4 Results and discussion

The data reliability and validity test as seen in Table 1 obtained a product moment r_{xy} of 0.531 which lies in between 0.40 and 0.60 according to [21] guide, this meant that the connection between the responses given is enough.

Meanwhile, the test for correlation significance of product moment for N = 19 respondents, at a significance level of 5% shows that obtained r_{xy} 0.531 is greater than standard one from table i.e. r_t = 0.456, hence H_o is rejected and alternative H_1 is accepted.

Hence, the data collected was reliable as validity the test showed correlation between many answers obtained.

Table 1. Product moment r_{xy} computation.

N	r_{xy} Product Moment	Condition	Level of Connection	Product Moment r_t	Conclusion
19	0.531	0.40 -0.60	Connection is Enough	for 5% = 0.456	Ho rejected alternative H1 accepted

4.1 Building design variables used for modification in application of VE

On the building design variables as applied in decision making to modify designs so as to have value for money, the findings as shown in Fig. 1, were that in the Indonesian construction, building shape and its complexity is mostly as indicated by importance index of 83.2 and 82.1 respectively, this means design consultants in Indonesian like any other building industry ought to pay to much attention on the determinants of these parameters and their impact of any on costs. Therefore, this affirms findings in previous researches under review like [7, 25, 26] among others, who also said building shape variable with its complexity as the most influential design variables.

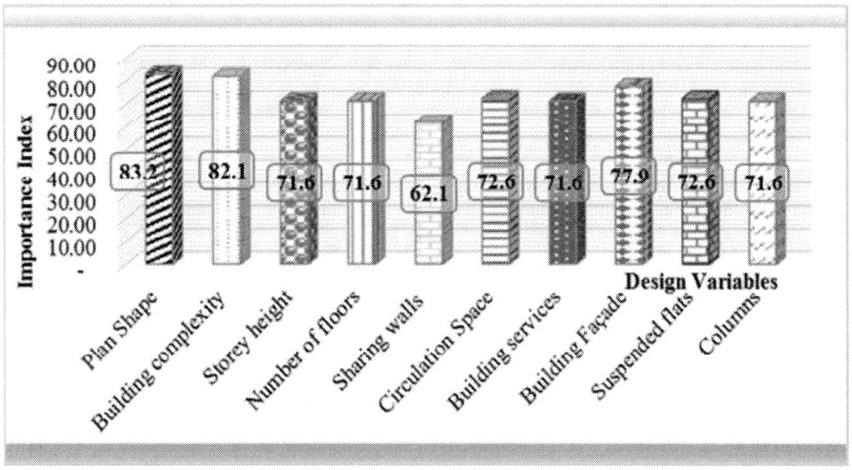

Fig. 1. Variables/parameters used for modifying designs.

This basically explains why these two parameters on construction costs i.e. plan shape layout and complexity are also becoming popular of recent, with the fact that many building projects, especially hotels and commercial buildings having a big portion of its their outer walls as glazed area or aluminium cladding. Hence, attention increasing, optimizing the shape. As it may go a long way in saving a great deal of these new expensive building materials.

4.1.1 Underling factors considered while adjusting designs

Thereafter, their factor considerations underlying application of some were also sought for. The findings obtained were as indicated in proceeding Tables 2 and 3 plus in Fig. 2.

For the building plan shape, however much as it may be and its complexity as seen in figure1, as the most common used design parameters during modelling and adjustments of designs to save costs, they did not work in isolation but had other factors influencing them i.e. which shape to use also depended on the following as seen in Table 2.

Table 2. Factors affecting choice of shape of a building.

	Factors	Importance Index	Ranking
A	Shape of building site	85,3	2
B	External characteristics of building membrane and building structure	86,3	1
C	Building Function	80,0	3
D	Symbolism	68,4	6
E	Structure in relation to light control	75,8	4
F	Cost saving of the building shape	72,9	5

As can be seen in Table 2, the building plan layout mainly depended on how external characteristics of a building membrane are and the shape of the available building site where it was to be constructed, as represented by importance index of 86.3 and 85.3 respectively. This means that the design architect should make through survey of the building site so as to make a design which is buildable on the available plan lay out and fits with in resultant shape membrane.

Meanwhile, with the underling factors for a particular building height, it's mainly ease of excess with in the building spaces and function use of a building by importance index of 82.1 and 80.1 respectively where most influential.

Table 3. Factors affecting choice of building height.

	Factor	Importance Index	Ranking
A	Natural wind circulation control	74,7	6
B	Easy of excess in the building	82,1	1
C	Considerations of standard height	75,8	5
D	Cost saving of height	76,8	4
E	Functional use of space	80,0	2
F	Cooling system to be used	77,9	3

As for the design variable of number of storeys, similar to building height it was established that also non-economic considerations were given more attention as indicated in Fig. 2, i.e. building rules and regulations followed by quality and type of soil at indices of 86.3 and 83.5 respectively were more determinant factors. Under this design variable, when it came aspect of economy i.e. to do with how the raise in height influences cost of construction, it was rated as number 3 with index of 61.1. In this case therefore, there is tendency of interplay of non-economic factors, such as regulations and tendency to follow standards, closely followed by economic factors.

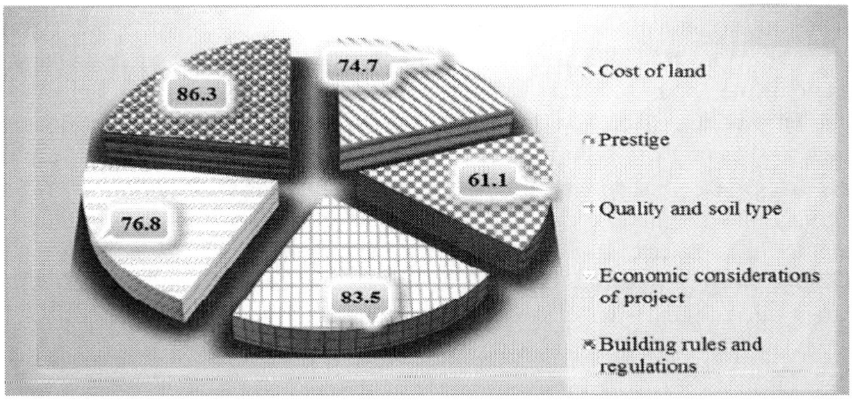

Fig. 2. Factors affecting choice for number of floors to use.

4.2 Consequences of applying design variables as part of VE practice

Basing on the study discoveries on consequences felt are as shown in figure 3 and 4, the concept of applying value engineering is the way to go, because also respondents surveyed agreed with previous researchers like [4] that this practice leads to outputting designs which are cost, and time effective as well as meeting quality specifications represented by 81.1 importance index. Secondly, design cost optimization can be attained with use of this approach during designing represented by index of 80.0.

Meanwhile, on the negative note once this practice is not applied with the growing complexities in designs, it could impact in a way that leads to material wastages and loss of confidence and consequently customers for the design consultant involved as shown by severity index of 80.0 each.

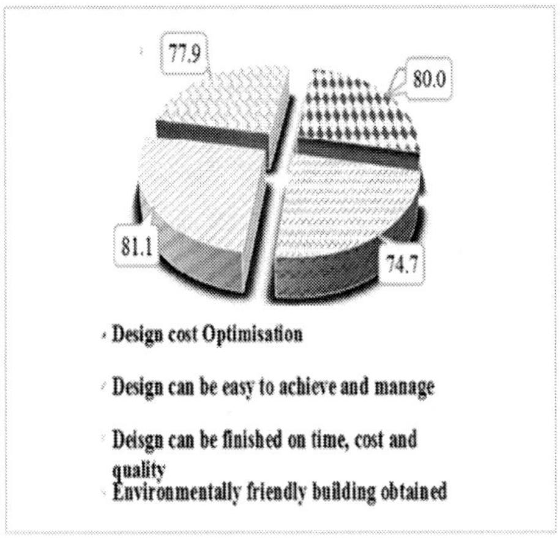

Fig. 3. Positive benefits due use of design variables.

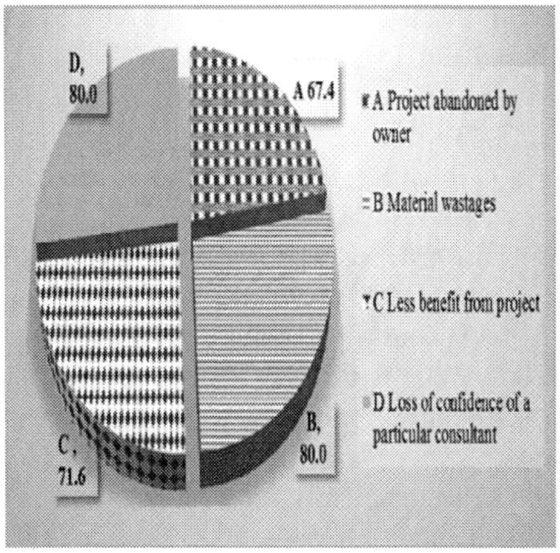

Fig. 4. Negative consequences due to over designing.

5 Conclusion

i) The top most five design variables often used by the architect consultants to modify building designs to meet client demands on costs identified in the survey were; building plane shape, building complexity, building façade, circulation space, and suspended flats.

ii) On the other hand the decision on which building shape plan to use, as the most common design variable, depended on a number of factors, like external characteristics of building membrane, building site shape.
iii) Value engineering as an approach which when used by building practitioners would increase efficiency and reduce unnecessary costs during designs, by delivering design plans that can be finished on time, cost and on quality, hence enabling smooth delivery of projects and profitability for all parties involved . This would solve challenge of PU which led to issuance of by law, *Nomor: 06/PRT/M/2008*

Lastly, the author recommends further research on the topic focusing on establishing to what extent each of the design variables influence the building costs of any project.

References

[1] D.T. Beeston, *A future for cost modelling. Building cost modelling and computers*, E FN Spon, 15–24, (1987)
[2] H. Zainuddin, CONPLAN. T.I.M.E Research Institution, Department of Survey, Univ. of Salford, UK, (1997)
[3] Rochmanhadi, *Pengantar dan dasar-dasar teknik penilaian (value engineering)*, Yayasan, Gema Aproteknika, Semarang, (1992)
[4] S. Palmer, A. Kelly, J. Male, Approximate of value engineering in Construction in U.S.A, J. Cons. Eng. Manag, **10**(1061), 324–328 (1996)
[5] G.P.A. Landow, *A Ruskin quotation?*, (2000). Available: http://www.victorianweb.org/ authors/ruskin/quotation.html (accessed 20 July 2015)
[6] S.A. Pena, W. Parshell, *Problem seeking: An architectural programming primer*, 4th ed. New York: Wiley, (2001)
[7] R.M. Ashworth, A. Skitmore, Accur, in Estim, Chart. Inst. Build., (1983)
[8] D. Morton, R. Jagger, *Design and the economics of building,* Spon Press, (1995)
[9] M. Fortune, C. Lees, The relat, perform, of new & trad, cost mod, in strat, advice and clients, R. Inst. Chart. Surv., (1994)
[10] A. H. Ashworth, *Willis' practice & procedure quantity surveyors*, Black, Scie, Ltd, (2004)
[11] A.D. Ibrahim, Effect of change in building, layout on the unit construction, cost of residual build, Quant. Surv., **7**(1), (2007)
[12] J. Ferry, D. Brandon, P. Ferry, *Cost planning of buildings,* 7th eds., Blackwell Science, (1999)
[13] E. Kouskoulas, V. Koehn, Predes, Cost-Estim, Func, for Build, J. Constr. Div. ASCE, **100**(12), 589–604 (1974)
[14] A. Ibrahim, applic, of reg, anal, for ass, eff, of var, in build, plan shape on unit cons, cost, Samaru J. Inf. Stud., **4**(2) 20 –26 (2004)
[15] N.G.Flanagan, The relation between construction, price & height, Chart. Surv. Build. Quant. Surv. Q., **5**(4), 68–71 (1978)
[16] J. Clark, W. Kingston, The skysc: A study in the economic height of modern office, Build, Am. Inst. Steel Constr., New York, (1930)
[17] V.K. Skitmore, R.M. Marston, *Cost modelling*, London: E & FN Spon, (1999)
[18] P.S. Ferry, D. Brandon, *Cost planning of buildings*, Oxford: Blackwell Scie. Public, (1991)
[19] D.J. Lowe, M.W. Emsley, Harding, relation between total construction cost & design related variable, J. Financ. Manag. Prop. Constr., **12**(1), 11–23 (2007)
[20] M.I. Sabri, Critical succes factor, Konsgres, Bang, Ged, Pada PT.X, Univ. Indo, (2011)

[21] Sugiyono, *Metode penelitian bisnis (pendekatan kuantitatif, kualitatif,* dan *R&D)*, Bandung Alf., (2009)
[22] B.C. Basheka, Public procurement skills requirement framework for local government systems in Uganda: Perceptions from professionals, (2010)
[23] Sugiyanto, *Hand-out meta-analysis,*Yogyakarta, UGM, (2004)
[24] A.A. Bubshait, A.A. Al-Musaid, Owner involvement in construction projects in Saudi Arabia, J. Manag. Eng., **8**(2), 176–185 (1992)
[25] K. Allsopp, An arch, view, Archit. J., **177**(4), 29 (1983)
[26] M. Coombs, Buildability: A delegate's view, Archit. J., **177**(5), 47 (1983)

Impact of Moistened Bio-insulation on Whole Building Energy Use

Eshrar Latif[1,*], *Devapriya* Chitral Wijeyesekera[2], and *Shaghayegh* Mohammad[3]

[1] Welsh School of Architecture, Cardiff University, Cardiff, United Kingdom
[2] Faculty of Civil and Environmental Engineering, Universiti Tun Hussein Onn Malaysia, Johor, Malaysia
[3] Department of Architecture and Civil Engineering University of Bath, Bath, United Kingdom

Abstract. One of the key properties of hemp insulation is its moisture adsorption capacity. Adsorption of moisture can increase both thermal conductivity and heat capacity of the insulation. The current study focuses on the effect of moisture induced thermal mass of installed hemp insulation on the whole building energy use. Hygrothermal and thermal simulations were performed using the CIBSE TRY weather data of Edinburgh and Birmingham with the aid of following simulation tools: WUFI and IES. Following simplified building types were considered: building-1 with dry hemp wall and loft insulations, building-2 with moistened hemp wall and loft insulation and building-3 with stone wool insulation. It was observed that the overall conditioning load of building-1 was 1.2 to 2.3% higher than building-2 and 3. However, during the summer season, the cooling load of building-2 was 3-7.5% lower than the other buildings. It implies that, moistened insulation can potentially mitigate the effect of increasing cooling degree days induced by global warming.

1 Introduction

Construction materials comprises 40-50% of worldwide annual material use [1] and the building industry accounts for the second highest consumption of raw materials , after food production.. Environmental impact of material consumption is manifold including resource depletion, waste production, disposal at the end of lifecycle, consumption of water and energy and emission of pollutants [2]. The seven core principles that defines the sustainability of construction are reducing resource consumption, reusing resources, using recyclable resources, protecting nature, eliminating toxics, applying life cycle costing, and focusing on quality [3]. One study shows that [4] a correct choice of building materials can reduce almost 30% CO2 emissions. Therefore, it is imperative that construction materials should be used responsibly for the greater good of the environment.

Natural materials from both renewable and non-renewable resources have generally lower embodied energy and toxicity than man-made materials as they require less processing [5]. Amongst these, the materials that are derived from renewable sources can

*Corresponding author: latife@cardiff.ac.uk

be considered as sustainable and green materials [6]. Plants can be a source of renewable and green materials if cultivated and harvested on a sustainable basis [7]. Furthermore, if produced locally, plant-based building materials will require less energy during transportation. Morel et al. [8]finds that it is possible to reduce the building energy by 215% through the use of local materials. Additionally, plant based materials work as a carbon sink during their growing period and they are biodegradable at the end of their life cycle. For all these attributes, plant based can be considered as material with low carbon [4] and low ecological footprint [7].

Plant fibres and shives can be used to manufacture thermal insulation taking advantage of their pore structure, elasticity and moisture and pollutant adsorption capacity[9, 7]. Wheat, rye, barley, oats, elephant grass, cotton, Hemp, maize, reed, flax and coconut are some of these plant-based materials used in building industry [10].

However, there is widespread concern that bio-insulations may perform poorly if they adsorb a lot of moisture. On the contrary, there is also belief that moistened bio-insulations might have increased thermal mass in highly insulated buildings which could be useful at certain boundary conditions. The Insulations used in the walls and loft of a "Passivhaus" are thicker than the insulation incorporated in a conventional house to meet the fabric heat loss requirements set for the Passivhaus. If hemp insulation is used in a vapour open construction with the thickness of insulation resembling that of a Passivhaus, it can be assumed that the thicker hemp insulation will work as a hygric and thermal mass and will influence the amount of heating and cooling load of the building. This assumption was tested for the weather conditions of Edinburgh and Birmingham using hygrothermal and thermal simulation tools.

2 Materials and methods

Hemp-2 and stone wool insulations were selected for the study. The summarised material properties are provided in Table 1 below. Notably, the mean density and declared Specific Heat Capacity of Hemp-2 are nearly double that of Stone wool while the measured mean thickness of Stone wool is double that of Hemp-2. However, the declared Dry Thermal Conductivities for both have very similar values.

Table 1. Summary Properties of the Insulations.

Materials	Measured Mean dry Density (kg/m^3)	Measured Mean Thickness (mm)	Constituents	Declared Specific Heat Capacity (J/Kg.K)	Declared Dry Thermal Conductivity (W/m.K)
Stone Wool	23	100-110	Amphibolite, about 6% lime stone, about 9% calcium oxide, resin	850	0.038
Hemp-2	50	55	85% hemp fibres, 10-12% bi-component fibres and 3-5% soda.	1600	0.038

Two different numerical tools were used in this study. These are: (i) The WUFI software, and (ii) The IES (Integrated Environmental Solution) Virtual Environment software. The WUFI software [11] has been used for hygrothermal simulation. The IES virtual environment is a software package consisting of a number of energy related simulation tools to address dynamic thermal properties, air flow, lighting, lifecycle analysis, computational fluid dynamics etc. in relation to buildings. The dynamic thermal simulation was performed by the ApacheSim software tool with solar shading analysis done by the SunCast and air flow by infiltration and ventilation calculated by the Macroflow tool.All of these are packages within the IES software. The IES uses the transient heat balance equations in its algorithm. The further details of the governing equation can be found in the website of IES [12].

To determine the effect of moisture adsorption by hemp-2 insulation on the whole house energy use in Edinburgh, the following method was used:

- Hygrothermal simulation was carried out in the WUFI software for the weather condition in Edinburgh. CIBSE TRY weather data were used in the simulation.
- From the analysed data of water content in the hemp-2 insulation, average water content in the insulation was determined. The water content was determined as 7.6 kg/m^3.
- From average water content of the insulation, the density and the heat capacity of the (hemp insulation and water) matrix was determined. The modified density of the moistened insulation was determined as 57.6 kg/m^3 and the modified heat capacity was calculated as 1943 J/kg.K based on the heat capacity of water as 4200J/kg ^{o}K.
- For the purpose of simplification, similar water content data was used for Birmingham during the input of hemp-2 material data in the IES software.

The instrumented wall section and the moisture content variations in the insulation are shown in Fig.1 and Fig. 2, respectively. The buildings incorporated dry hemp-2, moistened hemp-2 and stone wool insulations are henceforth called as building-1, building-2 and building-3, respectively.

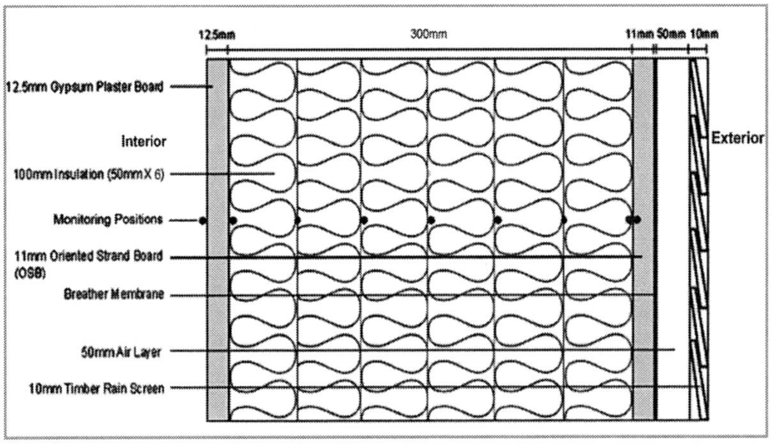

Fig. 1. The wall section.

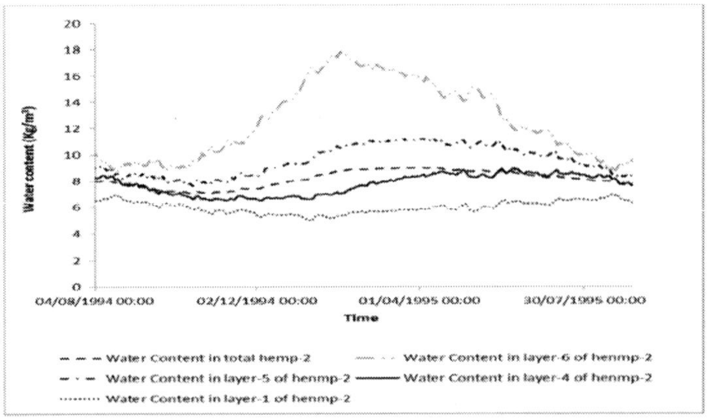

Fig. 2. Moisture content variations in hemp-2 insulation.

Dynamic thermal simulation was run in the IES Virtual Environment 2012 for three notional houses incorporating dry hemp, moistened hemp and stone wool insulation for the weather condition in Edinburgh and Birmingham. The simulation was run for one year with a preconditioning period of sixty days.

The plan and the IES software image of the buildings are shown in Fig. 3 and Fig. 4, respectively. As shown in Fig. 3, the notional building has two windows in the south wall with each window having the dimension of dimension of (2 X 2) m. There is a door in the north wall. The height of the building is 7 m, the ground floor is 4 m and the loft space is 3 m.

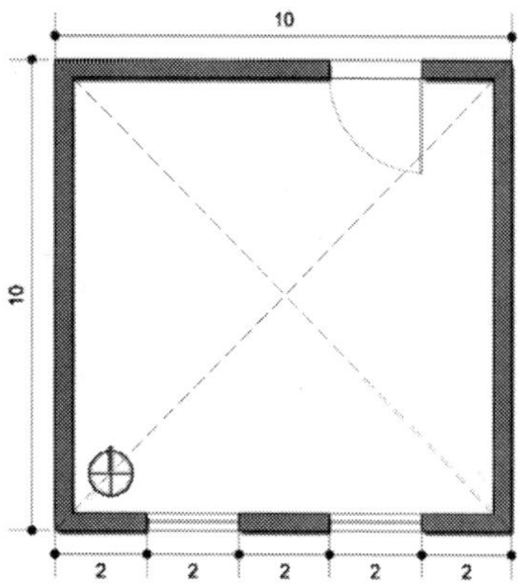

Fig. 3. The plan of the notional building for the IES software.

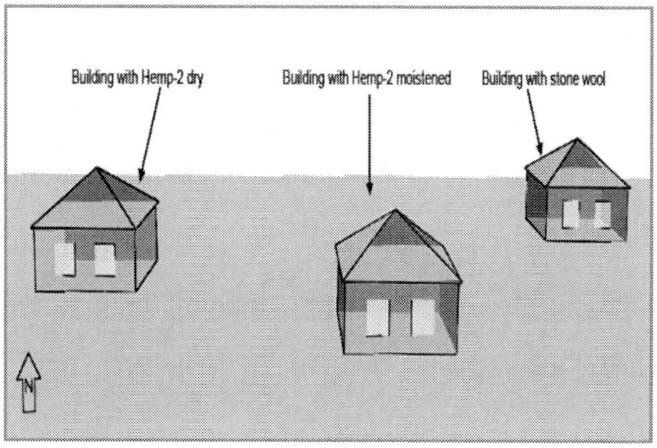

Fig. 4. The IES software image of the simulated buildings.

3 Results and discussion

Fig. 5 to Fig. 7 show the space conditioning sensible loads for the selected buildings for the weather conditions in Edinburgh and Birmingham. It can be observed that, in Edinburgh, the space conditioning sensible load of building-2 is 2.3% and 2.1% higher than that of building-1 and building-3, respectively. In Birmingham, the space conditioning sensible load of building-2 is 2.27% and 1.2% higher than that of building-1 and building-3, respectively. However, a closer observation of the monthly space conditioning loads revealed that, in Edinburgh, in summer time of June, July and August, the space conditioning sensible load of building-2 is lower than that of building-3 by 4% and higher than that of building-1 by 1.97%. In Birmingham, for the summer time of June, July and August, the space conditioning sensible load of building-2 is lower than that of building-3 by 7.5% and higher than that of building-1 by 1.75%. Thus, the moistened hemp-2 insulation is performing better than the stone wool insulation during the summer months. It can be assumed that, the heat capacity of the moistened hemp-2 insulation can be fully utilized when cooling degree days increase due to the global warming.

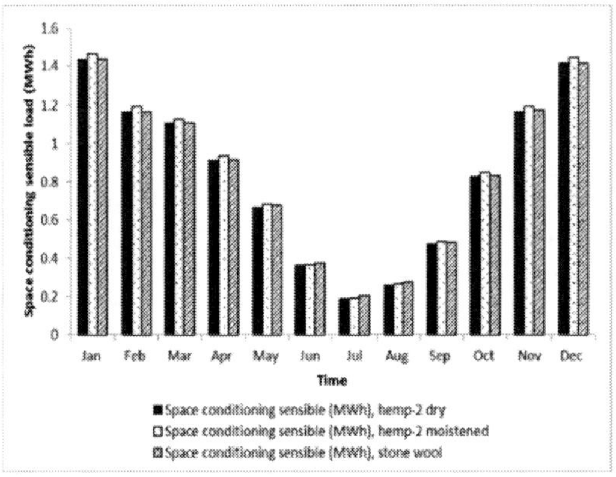

Fig. 5. The monthly space conditioning sensible loads of the selected buildings in Edinburgh.

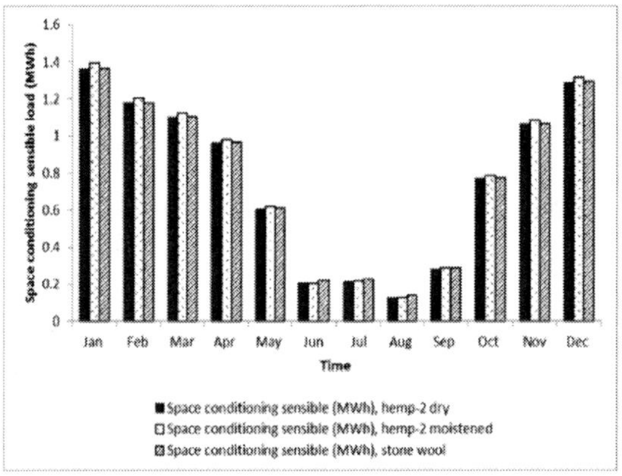

Fig. 6. The monthly space conditioning sensible loads of the selected buildings in Birmingham.

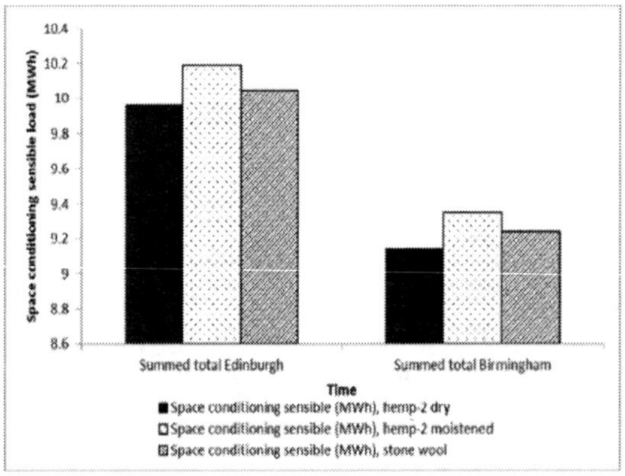

Fig. 7. The yearly space conditioning sensible loads of the selected buildings in Edinburg and Birmingham, respectively.

4 Conclusions

The objective of the study was to examine the effect of thermal mass of moistened hemp-2 insulation on whole building energy use. It was found that, the total space conditioning load of building-2 (with moistened hemp-2 insulation) was higher than that of building-1 (with dry hemp-2 insulation) and building-3 (with stone wool insulation) by 2.3% and 2.1%, respectively, in Edinburgh and 2.27% and 1.2%, respectively in Birmingham. However, it was also found that during the summer months of June, July and August the cumulative space conditioning load of building-2 was lower than that in building-3 by 3% and 7.5% for Edinburgh and Birmingham, respectively. Based on this result, it can be said that moistened hemp insulations will be more effective in saving energy when cooling degree-days increase as a result of the global warming.

References

[1] J.B. Storey, *Construction materials stewardship: The status quo in selected countries: CIB W115, Centre for Building Performance Research*, Victoria University of Wellington, (2008)

[2] R. Yao, *Design and management of sustainable built environments*, Springer, (2013)

[3] C.J. Kibert, and K. Grosskopf, Radical sustainable construction: envisioning next-generation green buildings, J. of Land Use & Environmental Law, **23**(1), 145, (2007)

[4] M.J. González, and J. García Navarro, Assessment of the decrease of CO2 emissions in the construction field through the selection of materials: Practical case study of three houses of low environmental impact, Building and Environment, **41**(7), 902-909, (2006)

[5] J.J. Kim, and B. Rigdon, *Sustainable architecture module: Qualities, use, and examples of sustainable building materials*, National Pollution Prevention Center for Higher Education, (1998)

[6] R. Spiegel, and D. Meadows, *Green building material: a guide to product selection and specification*, 3rd ed., John Wiley, Hoboken, N.J., (2012)

[7] B. Berge, *The ecology of building materials*, 2nd ed., Elsevier/Architectural Press, Amsterdam; Oxford, (2009)

[8] J.C. Morel, A. Mesbaha, M. Oggerob, and P. Walker, Building houses with local materials means to drastically reduce the environmental impact of construction, Building and Environment, **36**, 1119–1126, (2001)

[9] E. Latif, M. Lawrence, A. Shea, and P. Walker, Moisture buffer potential of experimental wall assemblies incorporating formulated hemp-lime, Building and Environment, **93** ,199-209. (2015)

[10] A. Cripps, *Crops in construction handbook*, CIRIA, London, (2004)

[11] H.M. Künzel, *Simultaneous heat and moisture transport in building components: One- and two-dimensional calculation using simple parameters,* Fraunhofer Institute of Building Physics, Fraunhofer Institute of Building Physics, Stuttgard, (1995)

[12] IES, *ApacheSim* Calculation M*ethods* (2016), Retrieved on October 1, 2016 from http://www.iesve.com/downloads/help/Thermal/Reference/ApacheSimCalculationMethods.pdf.

DOI: 10.1051/matecconf/201710303021

Acoustical Benefits of Plenum Window as Facade Device – A Parametric Study

Yean Ghing Tong[1,*], and *Shiu-keung* Tang[2]

[1]Jamilus Research Centre, Universiti Tun Hussein Onn Malaysia, 86400 Parit Raja, Johor, Malaysia
[2]Department of Building Services Engineering, Hong Kong Polytechnic University, Hong Kong, China

Abstract. Rapid population growth and economic development have led to serious noise pollution in densely populated cities. As the noise level keeps on increasing, opening windows for natural ventilation has become nearly not possible, especially in urbanized residential areas. Plenum window is the modification of a partially opened double glazing window system to the staggered inlet and outlet window openings which can provide better sound protection to the building and at the same time can allow certain degree of natural ventilation across it. A series of full-scale laboratory measurements were carried out in order find out the relationship between the important parameters and the acoustical performance of plenum window. From the measurement results, plenum window can attenuated at least 10 dBA traffic noise without treatment inside the window system. It is found that window opening sizes and overlapping width have significant effects on the acoustical protection of plenum window.

1 Introduction

Traffic noise is the major noise pollutant that has affected surrounding populations in densely populated cities. The situation becomes severe in the cities where large numbers of residential dwellings are required to build alongside the main traffic networks to satisfy the housing demands of the communities. When buildings are situated close to traffic roads, environment noise from transportation system becomes source of nuisance in the city. Many surveys were carried out over the century to evaluate the effects of traffic noise exposure to the nearby residents. This environment noise is found to be the most critical effects on human sleep disturbances in several studies [1-4] and could lead to health problems [5].

Roadside barriers are the common structures used in urban areas to reduce human noise exposure due to the noisy road networks in the residential areas [6]. Many studies of roadside barrier including experimental and theoretical have been carried out over the past few decades. Efforts have been made in changing geometries and materials of the barriers in order to improve the performance of barriers [7-9]. However, this measure is not cost-

* Corresponding author: ygtong@uthm.edu.my

effective and there are some cases where this massive structure cannot be built due to site constraints.

An alternative noise mitigation form is vital to deal with this serious problem. Treatment on the devices that attached on façade could be better solutions to deals with noise from transportation system because traffic noise mainly propagates from outdoor into the indoor environment through building facades. Besides the external walls, large portions of building façades are covered by windows which are designed to admit natural daylight and ventilation into the buildings. This makes window become a weak point of the facade because window constitutes the primary path of traffic noise enters into built environment. Natural ventilation and noise control are two conflicting issues. When natural ventilation is provided across the facade, windows will become the main noise transmission path for external noise ingress. As a results, residents living in the city with high traffic noise levels prefer to close all the windows and use air conditioning system. The use of mechanical ventilation increases electricity demand, energy consumption, environmental problems and against the principle of sustainability.

Recently, the concept of plenum chamber in duct silencing [10] has been proposed to be used in the window system [11]. The inlet and outlet openings of the proposed window system were designed in the zigzag configuration to block direct sound path from the outdoor to the indoor environment. This staggered inlet and outlet opening window system, referred as plenum window hereafter, has been attracted many researchers as it can provide high sound insulation while allowing natural ventilation across it. Plenum window proved to be effective on providing acoustical protection to the residential buildings compared to the conventional window system [12, 13]. In this paper, full scale laboratory measurements were carried out to investigate the effects of various plenum window configuration parameters to the acoustical performance of this facade device.

2 Measurement setup

2.1 Test chamber

All the measurements were carried out inside acoustic testing chamber of the Department of Building Services Engineering, The Hong Kong Polytechnic University. The test chambers were two coupled but isolated chambers originally used for the ISO 10140-2 tests [14] for sound transmission loss of building materials. The bigger chamber, usually used as source room, had a volume of about 240 m^3 and a height of 5 m, while the smaller one, usually used as the receiver room, had a volume of ~84 m^3. The source room was converted into a semi-echoic chamber by putting up 2-inch thick fibreglass curtains at about 1 m away from all walls and ceiling. This setup was made similar to the Kang's study [15].

2.2 Sound source

Traffic noise is regarded as a line source [16]. In this study, a 5m long line source consisted of 25 six-inch aperture loudspeakers was adopted to simulate road traffic noise. A constant magnitude of white noise signals were supplied by Brüel & Kjær 1405 Noise Generator, which was connected to a power amplifier to drive the loudspeakers. The loudspeaker array was placed on floor and located at 5 m horizontally away from the tested window.

2.3 Tested window configurations

Two different window systems, conventional casement window and plenum window were tested in the present study. The plenum window is composed of two layers of window systems where the outer layer is casement window and fixed glazed window while the inner layer is a movable sliding glass pane as shown in Fig. 1. This window design allows either direct opening or offset opening configurations by adjusting the position of the inner layer of the sliding glass pane. The window system was remained in the offset opening condition throughout the test. Two different configurations of outer side window, the single-leaf and the double-leaves casement windows were included in this study.

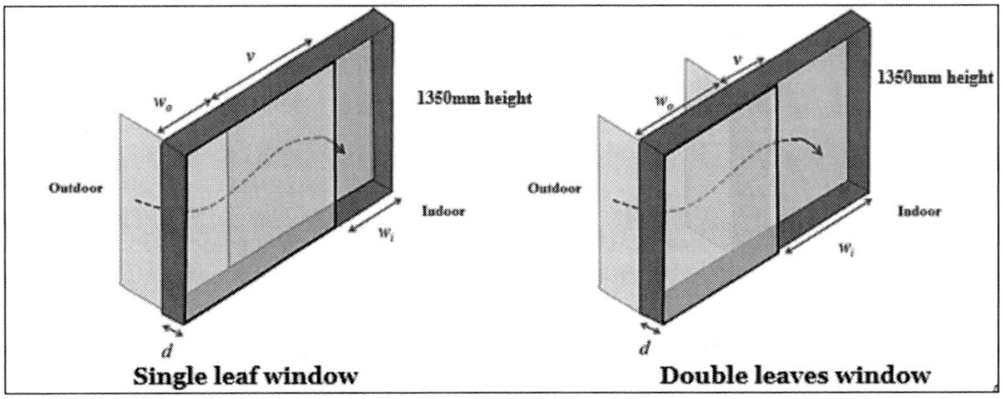

Fig. 1. Plenum window schematics. (left) Single leaf window; (right) Double leaves window

The performance of plenum window under different combinations of inner side window opening size (w_i), outer side window opening size (w_o), gap width (d) and overlapping length (v) with and without lining of sound absorption materials as shown in Table 1 were investigated. Fibreglass was lined inside the plenum cavity (top and both sides) which fitted the gap width (d) of the window system for the cases that plenum window were treated by sound absorption materials. From Table 1, eight different configurations of plenum windows were to compare with a single set of conventional casement window. The total length of the windows depended on the window opening sizes and overlapping length. The heights of all tested windows were 1350 mm. All windows glass panes were made of 6 mm thick single glazing with aluminium frames.

2.4 Measurement procedure

A total of nine Brüel & Kjær Type 4935 microphones were used in the receiver room used to capture the transmitted sound level. These microphones were spanned over the whole volume of the receiver room (at least 1 m away from walls) as shown in Fig. 2. On the other hand, the number of microphones used at the source room was depended on the size of the tested window start from three, nine and twelve microphones as shown in Fig. 3. The microphones at the source room side were placed at 1 m horizontally from the tested windows to estimate the average sound pressure level falling onto the tested specimens. All microphone signals were recorded simultaneously by Brüel & Kjær 3506D PULSE system. Three measurements were recorded for each setting which each measurement lasted for 30 seconds. The results presented later in this paper are the corresponding average values.

Since the objective of the present study is to estimate the acoustical protection of the plenum window in the presence of traffic noise, the normalized traffic noise spectrum depicted in the standard EN 1793-3 [17] was adopted. This normalized traffic noise spectrum was used in the estimation of the acoustical benefit of plenum window in term of a single rating.

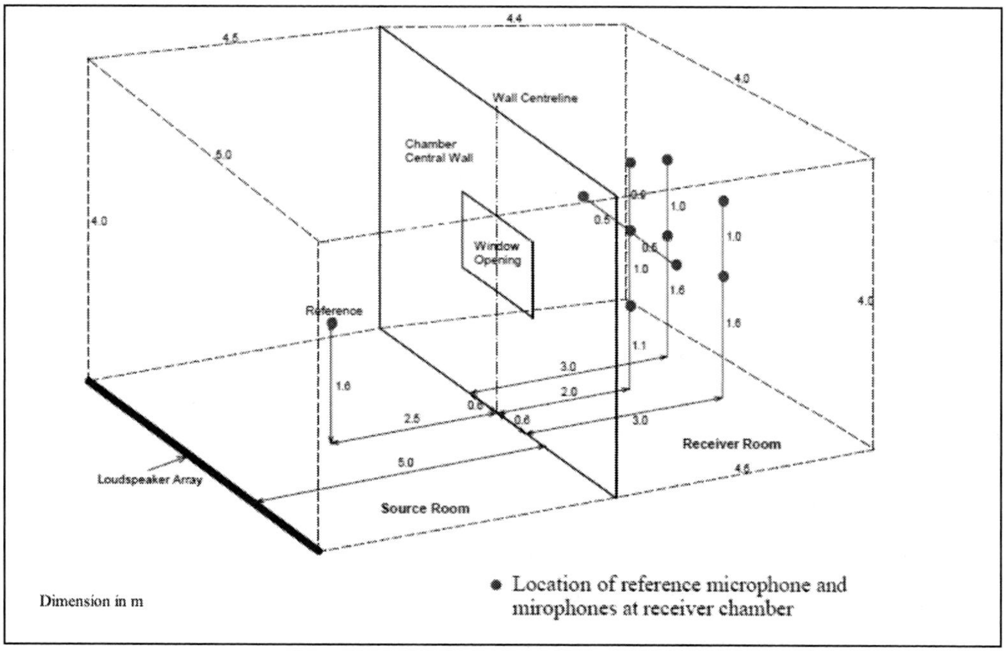

Fig. 2. Location of reference microphones and microphones at receiver room.

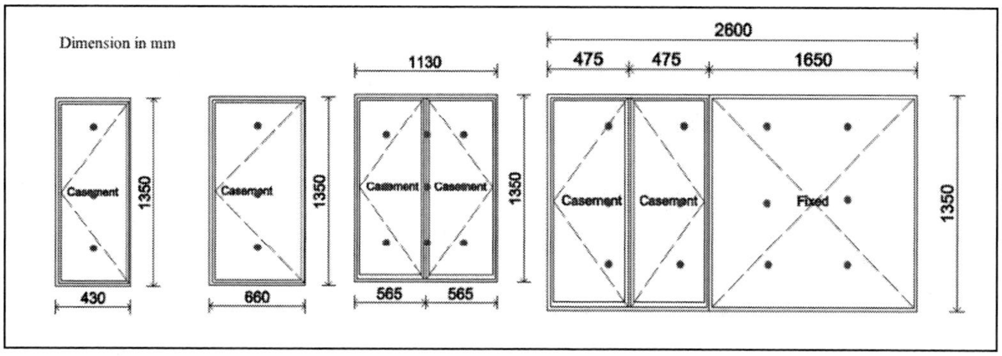

Fig. 3. Different location of microphone for different size of window at the source rooms.

Transmission loss (*TL*) was obtained from the differences of band noise levels between the source room and receiver room. This estimation evaluated the acoustical benefits offered by plenum window itself without make the comparison with the conventional casement window. The single rating TL can be expressed as follow.

$$\mathrm{T_i} = SPL_{i,source} - SPL_{\mathrm{i,\,receiver}} \tag{1}$$

$$TL = 10\log_{10}\left(\frac{\sum_{i=1}^{18} 10^{0.1(N_i - T_i)}}{\sum_{i=1}^{18} 10^{0.1T_i}}\right) \tag{2}$$

where i represents the ith one-third octave band data, from 100 Hz to 5 kHz, $SPL_{i,source}$ the average band level obtain in the source room, $SPL_{i,receiver}$ the average band level obtain in the receiver chamber, N_i the normalized noise band level.

Another acoustical benefits that estimated in the present study is insertion loss (IL_A) which is obtained from the differences of band noise levels inside the receiver room after replacing a conventional casement window by a plenum window. A single rating of insertion loss (IL_A) of plenum window when exposed to traffic noise can be defined as followed expression.

$$\mathrm{R_i} = SPL_{\mathrm{i,conventional}} - SPL_{\mathrm{i,Plenum}} \quad (3)$$

$$IL_A = 10\log_{10}\left(\frac{\sum_{i=1}^{18} 10^{0.1(N_i - R_i)}}{\sum_{i=1}^{18} 10^{0.1N_i}}\right) \quad (4)$$

3 Measurement results

Table 1. Plenum window tested scenarios and insertion losses

Scenario	Casement window width w	Plenum window w_o, w_i	v	d	Fibreglass	TL (dBA)	IL_A (dBA)
L1	430 (Single leaf)	320	100	145	Yes	17.2	9.8
L2					No	15.6	7.5
L3				205	Yes	18.2	13.0
L4					No	16.7	10.3
L5			700	145	Yes	19.0	14.2
L6					No	16.9	10.8
L7				205	Yes	18.9	13.2
L8					No	16.7	9.6
L9	660 (Single leaf)	600	100	145	Yes	17.4	13.0
L10					No	16.2	11.2
L11				205	Yes	17.0	11.2
L12					No	15.3	8.8
L13			700	145	Yes	10.3	14.2
L14					No	16.9	11.6
L15				205	Yes	18.8	13.5
L16					No	16.1	9.7
L17	1130 (Double leaves)	950	100	145	Yes	15.9	13.9
L18					No	14.6	11.6
L19				205	Yes	15.5	11.8
L20					No	13.9	9.4
L21			700	145	Yes	17.3	15.2
L22					No	15.8	12.7
L23				205	Yes	17.2	13.8
L24					No	15.1	11.2

3.1 Transmission loss (*TL*)

The results of laboratory measurements of 24 configurations plenum windows are depicted in Table 1. From the table, it show at least 10 dBA of traffic noise can be attenuated by using plenum window system, without any treatment on it. The maximum sound transmission loss obtained from the study was about 19 dBA. As expected, plenum window lined with sound absorption materials provided higher sound insulation to the indoor environment for all studied scenarios. For the same plenum window configuration, treatment on the plenum window system can increased about 2 dBA of *TL*.

In general, *TL* increased when the overlapping width of the plenum window increased especially for the larger opening sizes. This attenuation probably due to the multiple reflections of transmitted noise inside the plenum cavity before it ingress into indoor environment. For the smaller opening sizes (w_i or w_o= 320 mm), wider the overlapping width of the window only provide acoustic benefits to the plenum window that has smaller gap width (d=145mm).

When the opening sizes of the plenum window increased, it is noticed that the TL are slightly decreased. About 1.5 to 2 dBA of reduction of *TL* when the opening sizes of the plenum window expanded from 320mm to 950mm. The reduction of sound attenuation probably due to the facts that plenum window system received more sound waves propagated into the plenum cavity.

The results from the experiments suggested the acoustical transmission loss is not significantly affected by changing the gap width of the plenum window for all range of studied opening sizes and overlapping width.

3.2 Insertion losses (IL_A)

From the Table 1, one can observe that, for the same opening size, plenum window with larger overlapping width and smaller air gap width results in higher insertion loss (IL_A). The highest acoustic benefit is obtained when the opening and overlapping width of the window are larger (w_i/w_o = 950 mm; v = 700 mm), but smaller air gap width (145 mm). The maximum IL_A achieved is about 15 dBA. Similar to *TL*, plenum window with sound absorption material provides better sound insulation compared to the plenum window without any treatment. With fiberglass lined inside the cavity, an average of 2.7 dBA increase in term of IL_A.

It is noticed that increasing the air gap width of plenum window when other parameters are kept unchanged will decrease the IL_A except for the case of small opening (w_i or w_o) and small overlapping (v). The same trend of IL_A on the effect of air gap width (d) can be observed for the plenum windows lined with sound absorption material. The reduction of IL_A appears to decrease when the overlapping length (v) is increased. This is due to the fact that the mean propagation path for the outdoor noise to travel into the receiver room via plenum cavity become shorter when the overlapping length (v) of plenum window is reduced. For the case with small opening (w_i or w_o) and small overlapping (v), the window performance is sensitive to the change of air gap width (d) where about 3 dBA reduction can be noticed when d is reduced by 60 mm.

The total length (L) of the plenum window depends on the opening sizes (w) and overlapping length (v). Fig. 4 shows the effects of w/L ratio on the insertion loss IL_A. Smaller gap width (d) results in higher IL_A compared to the larger d cases except at the w/L=0.43. It is noticed that the IL_A are slightly increased when the w/L is increased from ~ 0.24 to ~ 0.32 and more rapid increment can be observed when the ratio is further increased to 0.37. However, the IL_A decreases quickly when the ratio is further increased to ~0.43. For the case of w/L= ~ 0.43, the window opening sizes and the overlapping length are small. Faster increase of insertion losses is observed for *w/L*> 0.43 for the smaller gap width (*d*) while the insertion losses keep droping for the cases of larger *d*. The insertion loss of the larger *d* then increase after *w/L*= ~ 0.46. An average of 2.5 dBA and ~ 3 dBA insertion loss can be obtained when sound absorption lined in the plenum window with *d* = 145 mm and 205 mm, respectively.

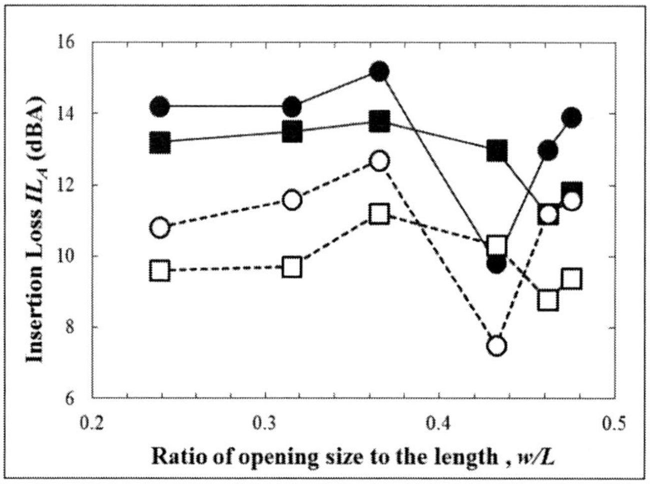

• : *d* = 145 mm; ▪ : *d* =205 mm; Closed symbols : with sound absorption; Opened symbols : without sound absorption.

Fig. 4. Effects of the w/L on insertion losses IL_A

4 Conclusions

The investigations of various combinations of full-scale plenum windows were carried out inside the two coupled but isolated chambers. A total of 24 different configurations of plenum windows with various opening sizes (*w*), gap widths (*d*) and overlapping lengths (*v*) with and without sound absorption materials lined inside the plenum cavity were tested. Two acoustical benefits are presented which are transmission loss *TL* and insertion losses of IL_A in the present study. IL_A is the difference of noise levels inside the receiver room when the conventional casement window is replaced to a plenum window, while *TL* is the difference sound levels before entering plenum window and after passing through plenum window.

From the results, it suggested that plenum window is an effective noise screening devices which can be used to attenuate traffic noise. The maximum of *TL* offers by plenum window is about 19 dBA. *TL* increased with the increasing of overlapping width (*v*) of plenum window or length of the window. However, increasing of opening sizes of the window system will resulting of reduction of *TL*. The changing of air gap width is found to be insignificantly effects the performance of plenum window in term of *TL*.

Higher insertion loss can be obtained for the window with larger overlapping (*v*) but smaller gap width (*d*) when the opening sizes of the plenum are fixed. For these configurations, increasing the opening sizes will enhance the acoustical benefit of the plenum window. The maximum IL_A offers by plenum window is about 15 dBA. The insertion loss tends to increase with increasing window span, decrease with increasing gap width and decreasing overlapping length in general.

References

[1] D.W. Robinson, *An outline guide to criteria noise for the limitation of urban, No. 1112*. Ministry of Technology, London, United Kingdom, (1970)
[2] A. Skånberg and E. Öhrström, Sleep disturbances from road traffic noise: A comparison between laboratory and field settings, J. Sound Vib., **290**(1-2), 3-16 (2006)
[3] E. Öhrström, A. Skånberg, H. Svensson and A. Gidlöf-Gunnarsson, Effects of road traffic noise and the benefit of access to quietness, J. Sound Vib., **295**(1-2), 40-59 (2006)
[4] E. Öhrström, E. Hadzibajramovic, M. Holmes and H. Svensson, Effects of road traffic noise on sleep: Studies on children and adults, J. Environ. Psychol., **26**(2), 116-126 (2006)
[5] W. Babisch, Road traffic noise and cardiovascular risk, Noise Health, **10**(38), 27-33 (2008)
[6] I. Ekici and H. Bougdah, A review of research on environmental noise barriers, Building Acoustics, **10**(4), 289-323 (2003)
[7] G.R. Watts, D. H. Crombie and D. C. Hothersall, Acoustic performance of new designs of traffic noise barriers: Full scale tests, J. Sound Vib., **177**(3), 289-305 (1994)
[8] M. Auerbach, A. Bockstedteb and O.V Estorff, Numerical and experimental investigations of noise barriers with helmholtz resonators, *Proc. of Noise-Con 2010*, Maryland, United States, (2010)
[9] F. Koussa, J. Defrance, P. Jean and P. Blanc-Benon, Acoustic performance of gabions noise barriers: Numerical and experimental approaches, Appl. Acoust., **74**(1), 189-197 (2013)
[10] R.J. Wells, Acoustical plenum chambers, Noise Control, **4**(4), 9-15 (1958)
[11] J. Kang and M.V. Brocklesby, Feasibility of applying micro-perforated absorbers in acoustic window systems, Appl. Acoust., **66**(6), 669-689 (2005)
[12] Y.G. Tong and S.K. Tang, Plenum window insertion loss in the presence of a line source -A scale model study, J. Acoust. Soc. Am., **133**(3), 1458-1467 (2013)
[13] Y.G. Tong, S.K. Tang, J. Kang, A. Fung and M.K.L. Yeung, Full scale field study of sound transmission across plenum windows, Appl. Acoust., **89**, 244-253 (2015)
[14] BS EN ISO 10140-2, *Acoustics - Laboratory measurement of sound insulation of building elements - Part 2: Measurement of airborne sound insulation*, British Standard Institution, London, (2009)
[15] J. Kang, An acoustic window system with optimum ventilation and daylighting performance, Noise Vib. Worldw., **37**(11), 9-17 (2006)
[16] Department of Transport Welsh Office, *Calculation of road traffic noise London* London, Her Majesty's Stationery Office, London, United Kingdom, (1988)
[17] BS EN 1793-3, *Road traffic noise reducing devices - Test method for determining - Part 3: Normalized traffic noise spectrum*, British Standard Institution, London, (1998)

Awareness of the Installation the Lightning Protection System (LPS) by Using Structural Bonding Method in Malaysia

Mustaqqim Abdul Rahim[1,3,*], *Shahiron* Shahidan[2], *Abdul Naser* Abdul Ghani[3], *Muhammad Arkam* Che Munaaim[1], *Zuhayr* Md. Ghazaly[1], *Syahirah* Ismail[1], and *Isham* Ismail[2]

[1] School of Environmental Engineering, Universiti Malaysia Perlis, Malaysia
[2] Jamilus Research Center, Faculty of Civil and Environmental Engieering, Universiti Tun Hussein Onn, Malaysia
[3] School of Housing Building and Planning, Universiti Sains Malaysia, Malaysia

Abstract. Structural Bonding Method (SBM) is one type of Lightning Protection System (LPS), design to protect human, structures, contents inside structures, electrical equipment, transmission lines and other from the lightning flash. Besides, SBM is a standard LPS that comply with technical standards or codes of practice or called as conventional Lightning Protection System. In order to know the level of the Awareness of the Installation LPS by using SBM in the building among Civil Engineering Consultants, conducting survey need to be done. This paper presents the Research Design and Research Strategy in conducting the survey. It is explaining about the way before conducting the survey which are determine the population of sample (Consultant Company at Northern Region Area), samples of respondents (Civil Engineer at Consultant Office with the number of sample is 40), data collecting process, structure of the questionnaire form and the way in analysis the data. After the analysis the data, the result of the level of awareness in the Installation of LPS by using SBM are consider as moderate level.

1 Introduction

Lightning is one of the natural phenomena, being a flow of electrical current between the earth and storm clouds, occurs as moving charges of positive and negative polarity build up in the atmosphere during a storm [1]. The protection against lightning had been increasingly used in the building to protect the building from the direct lightning impact. One of the methods is to embed the lightning protection cable in the concrete structure [2]. When Benjamin Franklin first experimented with electric charges in the 1700s using a kite, a key, and some string, he originally proposed that lightning rods could reduce or eliminate lightning by relieving the imbalance between clouds and the ground. However, he later realized that if the conductive metal rod was struck by lightning, then it worked to safely conduct lightning to the ground. In other words, the initial confusion was an issue of

* Corresponding author: mustaqqim@unimap.edu.my

prevention versus attraction. It turns out that Benjamin Franklin was correct, and prevention is indeed an option for lightning protection but the technology would not be available for another 200 years [3].

The lightning protection system (LPS) can be defined as a complete system used to reduce physical damage due to lightning flashes to the structure and lightning caused surges on surges and data lines. Before the installation of Lightning Protection System (LPS), the user or consumer must make sure the LPS is comply with the standard due to the legal and safe installation. The lightning protection system divided into two types which are conventional and unconventional LPS. Conventional LPS is a standard LPS that is comply with standard while unconventional LPS are a non-standard LPS which are not comply with standard. Example, the air terminals associated with the unconventional LPS is the early streamer emission (ESE) and the charge transfer system (CTS).

In 2005, the International Conference on Lightning Protection (ICLP) issued a warning that the use of the unconventional LPS presents a danger to the end users [4]. Structural Bonding Method (SBM) is a type of Lightning Protection System that complies with the standard. SBM consist of three components which are Air Termination System, Down-Conductor System and lastly is Earth Termination System [5]. For the structural bonding system is the method where the LPS cable concealed within the reinforced concrete structure. In this method, the air termination system will be connected with the lightning protection system embedded in the reinforcement steel bars in the concrete column, ground beam and pad footing before discharge to the earth [6]. The objective of this research is to determine the level of awareness of the installation of LPS by using Structural Bonding Method (SBM) among Civil Engineer Consultant in Northern Malaysia.

2 Materials and methods

2.1 Population of sample and respondents

This survey was conducted at the Consultant Company in Perlis, Kedah and Pulau Pinang. The Survey was focus on the Civil Engineer at Consultant Company around the Northern Region Area. The targeted respondents are the Civil Engineer from Consultant Company. There will be around Forty-five (40) Civil Engineer selected from Northern Region Area.

Table 1. Population sample

State	District	No. of Respondent
Perlis	Kangar	5
	Arau	5
Kedah	Alor Setar	15
	Sungai Petani	10
Pulau Pinang	Perai	5
Total		**40**

In deciding the number of sample size, there are two ways proposed which are by using the calculation (used formula) and using the table provided by Krejcie and Morgan. The formula demonstrates the ways in determining number of sample size.

$$S = \frac{P(1-P)}{\frac{A^2}{Z^2} + \frac{P(1-P)}{N}} \quad (1)$$

where S = Sample size that determine, N = Number of population, A = Accuracy that needed (sampling error), P = Percentage of population that interested. Commonly, expected is 50 %, Z = Exact in distribution of sampling. Value of exact commonly used, (1.96 = 95 %, 1.64 = 90 %)

2.2 Data collection and analysis

In this research, Questionnaires will be used as the method in collecting the Primary data. Then, for the Secondary data in the other hand is the data already have been collected by someone else and have been passed through the statistical process. The data were obtained from the journal, related articles, thesis, books, organizational records or statistics and information from the internet.

2.3 Structure of questionnaire

Questionnaires design can be categories into three components which are to decide the questions to be asked, select the type of each question and finally, design of the overall layout of the questionnaire. The questionnaire was printed documents containing summarization about the Structural Bonding Method (SBM), instructions answering the questions, and lastly is to get answers from the respondents. In this research, the questionnaire form consist of two (2) sections which are sections I (Demographic of Respondents) and II (Awareness).

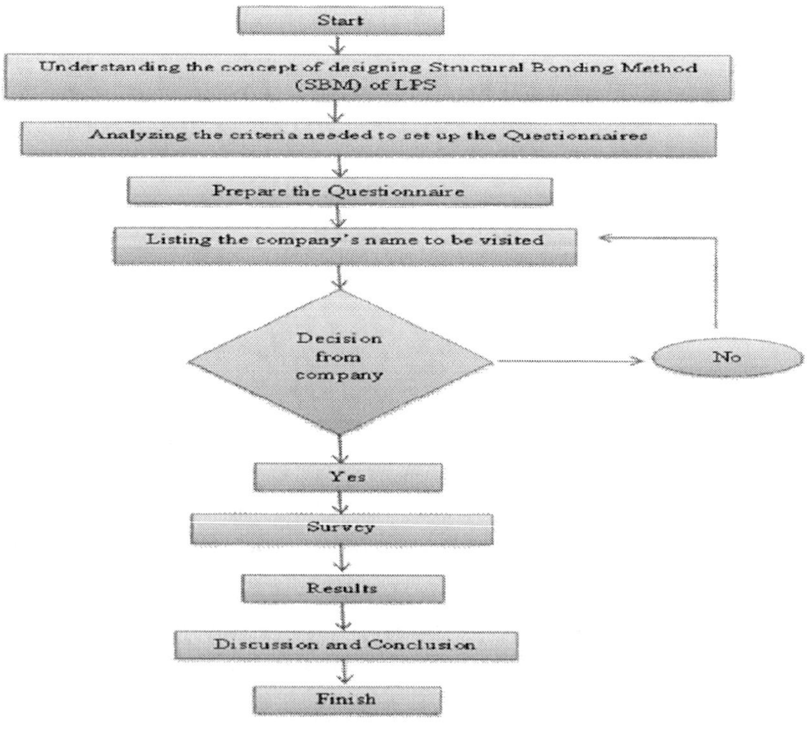

Fig. 1. Research framework

3 Results and discussions

3.1 Demographic of respondents

Table 2 shows the summary of the analysis of respondent's demographic information represents the major attributes of the selected Civil Engineering Consultants that used in this research.

Table 2. Demographic of respondents

Demographic of Respondents	Result
1. Gender of Respondents	Female (63.33 %)
2. Age of respondents	26 to 40 years old (50 %)
3. Respondents working in the Company	More than 5 years (43.3 %)
4. Respondents Experience	Less than 5 years (40 %)
5. Highest Education Level of Respondents	Degree Level (86.7 %)
6. Total Project Company involved	More than 40 Projects (53.3 %)
7. Values of Project	More than RM 10,000,000 (60 %)
8. Types of Project	Government building (33.3 %)
9. Sources of Information toward SBM	Internet and IEM (30 %)

3.2 Awareness of the installation of LPS by using SBM on company

Comprises of four (4) questions with the name of SWLS1 until SWLS4. Three (3) choices of answers are given which are Yes, No and Partially. Table 3 demonstrate the explanation of questions.

Table 3. Explanation of questions (part 1)

Number of Question	Question
SWLS1	Your company used Structural Bonding Method (SBM)?
SWLS2	Your company frequently uses Structural Bonding Method (SBM)?
SWLS3	You can differentiate between Structural Bonding Method (SBM) and other methods of Lightning Protection System (LPS)?
SWLS4	You participate during the designing of Structural Bonding Method (SBM) with Electrical Engineers?

Based on the Fig. 2, the percentage of the respondents are saying no with the question SWLS1 are 90 % and saying yes with the percentage of 10 %. In the question SWLS2, 83.3 % and 10 % of the respondent are saying no and partially. Only 6.7 % saying yes with the SWLS2 question. Besides that, for the question SWLS3, 46.7 % of the respondents are saying partially and 33.3 % saying no with the question. The respondents saying yes are 20 %. Lastly in the question SWLS4, the respondents saying no are 80 % and saying partially are 13.3 %. Then, the respondents saying yes in the question SWLS4 are 6.7 %.

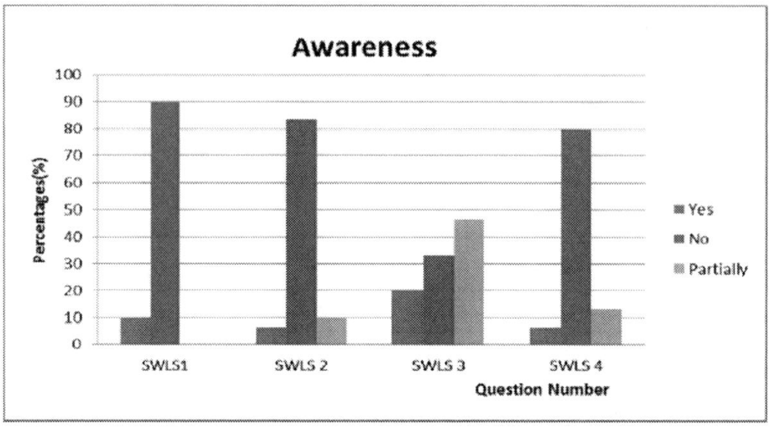

Fig. 2. Percentage of the awareness of installation of LPS by using SBM on company

3.3 Self-awareness of the installation of LPS by using SBM

This part consists of six (6) questions with the name of SWLS5 until SWLS10. Five (5) options of answers are given which is Absolutely Aware, Very Aware, Moderately aware, Slightly Aware and lastly is Not Aware. Table 4 shows the explanation of questions for Part 2.

Table 4. Explanation of questions (part 2)

Number of Question	Question
SWLS5	Installation of the Lightning Protection System (LPS)
SWLS6	Installation of the Structural Bonding Method (SBM)
SWLS7	Air Termination System of Structural Bonding Method (SBM)
SWLS8	Down Conductor System of Structural Bonding Method (SBM)
SWLS9	Earth Termination System of Structural Bonding Method (SBM)
SWLS10	The installation of down conductor embedded in the reinforcement

Based on the Fig. 3, the percentage of the respondents moderately aware with the question SWLS5 are 43.3 % and 26.7 % are slightly aware with that question. Besides that, 13.3 % of the respondents are absolutely aware and not aware with the question SWLS5. Then, 3.3 % of respondents are very aware of that question. For the question SWLS6, 53.3 % of the respondents are slightly aware and 23.3 % of the respondents are moderately aware with the question. Only 16.7 % of the respondents are not aware and 6.7 % are absolutely aware with the question SWLS6. In the question SWLS7, 40 % and 26.7 % of the respondents are moderately aware and slightly aware. The percentages of the respondents that are very aware with that question are 23.3 % and 10 % are not aware. In addition, for the question SWLS8, 43.3 % and 33.3 % of the respondents are slightly aware and not aware with the question. The 16.7 % of the respondents were moderately aware and 6.7 % were very aware with the question. For the question SWLS9, 60 % and 26.7 % of the respondents are slightly aware and not aware with the question. Then, 10 % of the respondents are moderately aware and 3.3 % of the respondents absolutely aware with the question SWLS9. Finally in the question SWLS10, 36.7 % and 33.3 % of the respondents are slightly aware and moderately aware. Besides that, 26.7 % and 3.3 % of the respondents are not aware and absolutely aware with the question SWLS10.

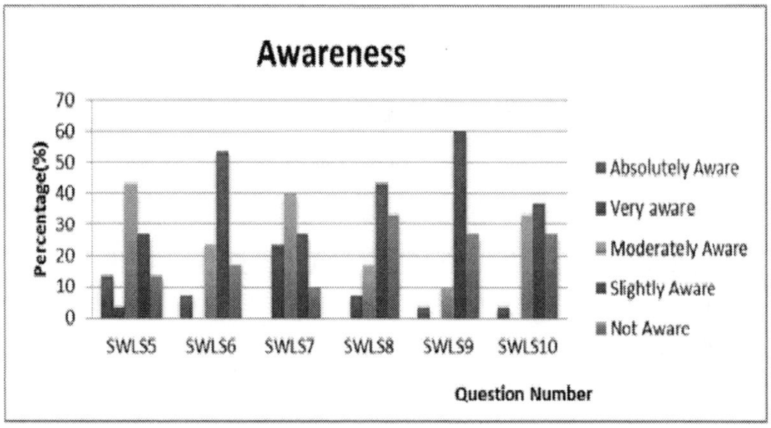

Fig. 3. Percentage of the self awareness of installation of LPS by using SBM

4 Conclusions

Based on the result achieved, the levels of Awareness of the Installation of Lightning Protection System (LPS) by using Structural Bonding Method (SBM) among Civil Engineers are moderate in the Northern Region of Malaysia. According to the feedback, the respondents are slightly aware with the Installation of Lightning Protection System (LPS) by using Structural Bonding Method (SBM) because of some factor such as company. Most of the respondents are agree that, the companies are not using Structural Bonding Method (SBM) in the Lightning Protection System (LPS). Thus it affects the level of awareness.

This research was funded by Fundamental Research Grants Scheme (9003–00484). The authors wish to extend appreciation to Ministry of High Education Malaysia, School of Environmental Engineering University Malaysia Perlis, School of Housing Building & Planning University Sains Malaysia and Department of Civil Engineering, University Tun Hussein Onn Malaysia.

References

[1] U.C. Foya, *Protection of structures against lightning*, MSc Thesis, Middle East Technical University, Ankara, Turkey, (2004)
[2] M. Abdul Rahim, A.N. Abdul Ghani and M.A. Che Munaaim, The temperature effect to the concrete due to the single strike of high electric current, Materials Science Forum, 857, 337-341 (2016)
[3] E.P. Krider, *Benjamin Franklin and lightning rods*, Physic Today, **59**(1), 42-48 (2006)
[4] Z.A. Hartono and I. Robiah, Conventional and unconventional lightning air terminal: An update, J. of Association of Consulting Engineers Malaysia, **First Quarter**, 1-5 (2007)
[5] BS EN 62305-3:2011, *Protection against lightning. Physical damage to structures and life hazard*, British Standards Institution, London, United Kingdom, (2011)
[6] M. Abdul Rahim, A.N. Abdul Ghani and M.A. Che Munaaim, Lightning protection system in Malaysia: Materials selection for down conductor, Jurnal Teknologi, 78, 5 (2016)

MATEC Web of Conferences **103**, 04001 (2017) DOI: 10.1051/matecconf/201710304001
ISCEE 2016

Environmental Degradation: A Review on the Potential Impact of River Morphology

Awang Nasrizal Awang Ali[1*], *Junaidah* Ariffin[2], *Mohd. Adib* Mohd. Razi[3] and *Jazuri A.*[2]

[1]Faculty of Civil Engineering, Universiti Teknologi MARA, 94300 Kota Samarahan, Sarawak, Malaysia
[2]Faculty of Civil Engineering, Universiti Teknologi MARA, 40450 Shah Alam, Selangor, Malaysia
[3]Faculty of Civil Eng. and Env., Universiti Tun Hussein Onn, 86400 Batu Pahat, Johor, Malaysia

Abstract. River morphology involves the lateral migration of matters deposited by flowing water in the river channel across its floodplain. This is driven by the erosion along the river banks and point bar deposition over time. This paper presents a review on river morphology studies and its potential impact to the society. The reviewed studies include mathematical models and computer simulation such as FLUVIAL-11 and RVR Meander Package that are significant to illustrate a continuous research development on channel adjustment. The findings also shows that a lot more area can still be explored to aid the fundamental of understanding river morphology and that East Malaysia will provide a good platform for the researchers to investigate the lateral migration of a river due to its diversity environment.

1 Introduction

As time passed, rivers may change its shape laterally or vertically. This phenomenon needs thorough investigation as any excessive development may also contribute to the changes [1]. Environmental degradation is a deterioration of the environment through the depletion of natural resources which leads to the destruction of ecosystems and wildlife extinction. The process might occur naturally but activities such as uncontrolled deforestation, improper agronomic practices and rapid urbanization will make the scenario worsen [2]. The interaction between river channel and the floodplain landscape is known as river morphology. Since the river is dynamic, they constantly changing in both space and time. River morphology can be categorized into degradation, widening and aggradation [3]. The morphology processes are as summarized in Table 1 and illustrated in Fig. 1.

Table 1. River morphology process.

Morphology Process	Caused By	Description
Degradation	Increased flow	Lowering bed elevation
Aggradation	Decreased flow	Increasing bed elevation
Widening	Increased flow	Increasing channel width

* Corresponding author: awang295@sarawak.uitm.edu.my

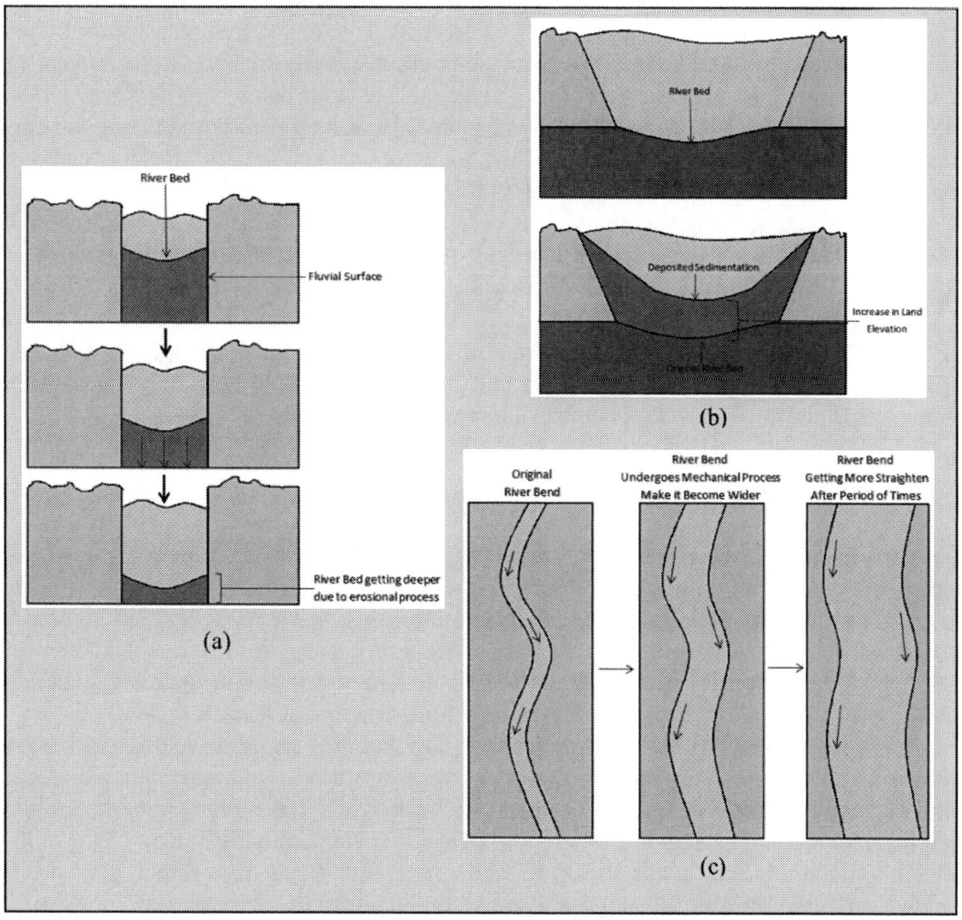

Fig. 1. River morphology process. (a) Degradation. (b) Aggradation. (c) Widening [3].

Degradation is the lowering of river bed elevation which can happen both at the downstream and also at the upstream. At the downstream, degradation can be related to the changes in the independent river channel variables such as increase in river flow, decrease in size of river bed material and decrease in river bed material discharge whereas for the upstream, it can be related to increase in the river slope which can occur as a result of natural river behaviour or human activities. Widening may occur by erosion of one of both river banks and leads to the increasing in the channel width. It is usually associated with flow acceleration due to a decreasing cross-sectional area coupled with deflection around growing bars. Aggradation is the accumulation of sediment in river channel, causing increase of the river bed elevation. It happens when the rate of sedimentation is greater than the rate of sediment transportation and when there is a decrease of bed slope [4].

2 Review on recent progress in established studies

Migration in a river is a process of river movement in which the water flow to erode the river outer bank and the sediments will be deposited on the opposite site of inner bank. Every year, there are a lot of cases being reported where properties or agriculture land which is located nearby to the river bank collapse. This tangible loss incurred to some extent is unavoidable which causing economic losses, damages and fatality.

The rate of lateral migration is unnoticeable through observations over a short period of time. In the past, time sequential of aerial photographs approximately 20 to 30 year apart have been used to measure the lateral migration rates of river bend [5]. The development of mathematical model and formula to measure the rate of lateral migration over the decade has contribute a lot to understand this lateral migration phenomenon. Studies had shown that there are a lot of factors in river migration which needs to be assessed and taken into consideration. Deep understanding on this occurrence is important as it can help researchers and local authorities to prepare and provide sufficient planning on the mitigation as a solution to the problems related to river migration.

Among the studies, there are significant numbers of findings related to river migration throughout the years. Most of the findings improve the way to assess and investigate the process and rate of river migration with some guidelines on action need to be taken. Table 2 shows the summary of studies from previous researchers since year 1984.

3 River dynamics: where the water goes

Nowadays, most of the engineers are concerned with the problem related to erosion and sediment in the river including slope stability, water quality, stream restoration and changes to the physicality of the river [14]. The purpose to understand river, both in their natural setting as well as how they respond to human induced changes in a watershed is to predict what changes will occur to the channel in response to alterations in watershed conditions and how these changes will impact the society, infrastructure and aquatic lives.

The process begins with the water coming down from the uppermost streams in the river network. It will flow down the concave-upward graded profile of water path and the high potential energy from the water flowing downward will have the ability to erode the bedrock and making a stream of water. The steeper the slope, the faster flow of the velocity will be and more energy of the stream to erode materials. When it is raining, the water will be drained from a higher elevation and cause drainage basin to occur with larger erosion problem. Once there is a drainage basin, there will be separation of the drainage basin from one another. Due to erosion, gully and rill will eventually formed up. When the water is flowing down, the potential energy in the water will be converted into kinetic energy. This will distinct the shape and the roughness of the channel. River morphology will occur when there is velocity in it. Velocity in the stream is controlled by the shape and the roughness of the channel. A smoother and narrower channel will produce larger velocity, whereas rougher and broader channel will produce smaller velocity. In a condition where there is a large velocity, erosion and deposition will occur more frequently. The larger the velocity, the more load a stream can carry. The result of these processes will leads to the formation of loads on the river bed. Sediments in rivers are transported either as bedload which is the coarser fragments or suspended load which is the finer fragments. There is also dissolved load, which are soluble products of chemical weathering process that dissolve in the stream. Generally, most material is transported as the bedload. It moves along the bed of stream by rolling, bouncing or sliding. These particles are too heavy to be carried by the flowing water. When sediment is light enough to become suspended in the water, it is carried by the stream as suspended load. Suspended load makes some streams appear to be cloudy [15].

The load is carried along by various processes. Traction is where the boulders or stones are rolled or slides along the river bed by the force of water. Suspension involves very small particles such as silt, clay and fine sands. As they float, these small particles are held up by the water and carried along the river. Then there is saltation, a combination of traction and suspension. This is where the velocity of the water lifted fragments like tiny pebbles off the river bed and they bounce along. Lastly, transportation in solution is when dissolved materials are carried in the water.

Table 2. Summary of studies from previous researchers outside Malaysia.

Year	Authors	Brief Related Research Work
1984	H. H. Chang	• Erosion on the river bank is caused by the removal of sediment on the bed of the river channel, causing the bank to collapse [6]. • Used FLUVIAL-11 to produce mathematical model of river channel. • Indicated the importance of lateral migration rate as one of the input data in modelling a river channel changes.
1984	Hickin and Nanson	• Conducted a series of investigation for 23 rivers in Canada and concentrated on channel-bend migration rates and control factors. • Found that maximum migration rates per unit width increase with increasing bend curvature. However the results may only be applied to the river with similarity in terms of hydrology and geomorphology.
1989	A. J. Odgaard	• Tested the analytical model to describe the platform development in terms of lateral and downstream migration rates. • The model is able to predict the rate, direction, magnitude and location of near-bank bed scour. However the model is non-applicable for channels with large curvatures [7].
1991	Ramasamy et al.	• Determined the river migration by observing the migratory signatures in Western India using Landsat photographs of 1:1 000 000 scale. • Explained that shifting of parent rivers greatly influenced and disorganized their tributary systems and that study on migration is important when selecting sites for dams, reservoirs and bridges [8].
1992	Alan D. H.	• Simulate the floodplain deposition model by assumes that the deposition rates decrease with distance from the closest channel and with increasing floodplain elevation. • Model combining flow and bed topography in meandering streams will provide realistic migration of simulated channels [9].
2001	Spitz et al.	• Emphasized on enhancing empirical databases to develop predictive model for bankline erodibility and floodplain characteristics. • A proper channel migration assessment are required as they may result in excess bridge pier and abutment scour, threats to bridge approaches and other highway infrastructure, worsened debris problems, and obstructed conveyance through bridge openings [10].
2006	T. J. Randle	• Simulates channel migration model by computing bank erosion as a function of sediment transport capacity, radius of channel curvature and the bank materials properties acting to resist the erosion including vegetation, large wood debris, cohesion and armouring. • Unlike previous one, this model is able to predict future channel alignments and make simulation for different hydrologic condition [11].
2008	Heo et al.	• Predicted the meandering channel migration by utilizing GIS. The result is satisfied. The longer the orthophotos-recorded, the better accuracy of the result is obtained. • The GIS approach offers simplicity and effectiveness. A specific tool such as satellite specifically used to monitor river migration may provide a better result [12].
2013	Ashraf and Liu	• Used RVR Meander package to relate migration rate to vertically averaged near-bank velocity through the use of coefficient of bank erosion (E). • The migration coefficient needs to be calibrated against historic channel centreline where the time period is chosen in such a way that it can capture large erosion events. Each section of the river bank have different coefficient value and the more coefficient used for different type of bank, the more accurate the result is [13].

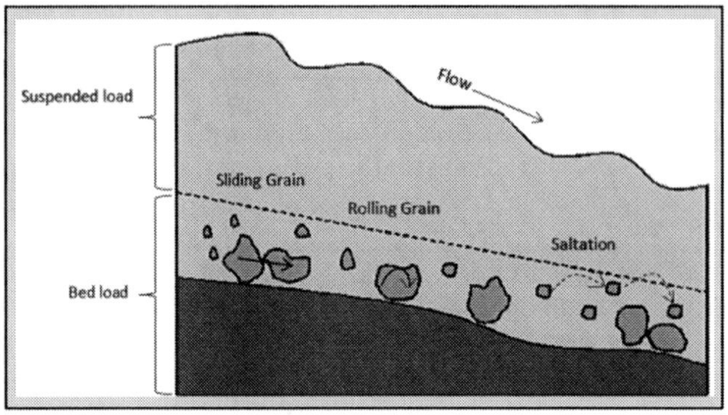

Fig. 2. Sediment motion in river morphology.

4 Potential impacts to the environment and society

The river morphology process that happens over time will have impact to the surrounding ecosystem and also the water quality. The suspended sediment increases with higher streamflow discharge. Development of landuse may accelerate surface soils erosion, gully erosion or soil mass movement along stream channels. When they are not controlled, the amount of suspended sediment might increase excessively and will affect other downstream uses of water such as hydroelectric pump, turbines and reservoir capacity. High levels of suspended sediment increase turbidity level which reduces light penetration in water and therefore can adversely affect the aquatic habitat.

As suburbanization progresses, the stream channels will have to handle risk of more frequent flood peaks. There will be changes in erosion and deposition of sediments and also to the structures in the area as it will be not very safe to live nearby the area. For example, Atrak River in North-East of Iran have shown that the rate of replacement of the river bank has reached 3.5m in year 2000 from only 0.36m in year 1956. The erosion processes that downcutting the surface or the sides of the stream will also slowly taking away the support of the trees and plants itself. In a certain period of time, landslide might occur and large woody debris will flow into the stream. Due to its large sizes, it will not be transported or affected by the speed of the water, thus making it sediments at certain point. From this, flood may occur as the discharge of the water at the downstream is lessened [16].

In rural areas, streams or rivers are their only source of daily water supply. The sediments and inputs of the erosion will have direct impact towards the physical properties of the water and the cleanliness of the water itself. This is hazardous to human health as the erosion of soil in river degrades the quality of the water. The changes of the river behaviour will also affect human daily activities such as farming due to high amount of water flow that rapidly change or even the severe erosion of riverbank. For example, 90.13% of respondents from Shantipur, India lost their land due to riverbank erosion over the past 40 years [17]. These show the significant of river morphology in engineering studies.

5 Study progress in Malaysia

Malaysia, a country which is located in Southeast Asia is rich with geographical value including rivers. The two distinct parts of this country are Peninsular Malaysia and East Malaysia. Malaysia climate is categorized as equatorial as it is located near the equator. The total land area of Malaysia is approximately 329,750 km^2. From the total land area, 1,200 km^2 is made up of water such as lakes and rivers. In total, there are 3,000 river basins with

189 of main river basins of more than 80 km², 74 from Peninsular Malaysia and 135 from East Malaysia (Sabah and Sarawak). The geographical location provides protection to the country from most of the major natural disasters such as earthquake, volcanoes eruption, tsunamis and typhoons. However, with the average rainfall of over 2,500-3,500mm per year, Malaysia tropical climate opens the country to the risk of floods, landslides and droughts problem [18].

Flooding issue has been a major concern due to uncontrolled rapid development which leads to the increase of impervious areas [19, 20]. With that, there will be more surface runoff and sediment buildup in the rivers system and overflowing water from the river banks. These will not only causing losses in terms of money, but also further damage to the economic structure and development in the area [21, 22].

As shown in Table 3, a few studies related to river morphology in Malaysia have been conducted, mostly in Peninsular Malaysia. However, there are still a lot to be explored especially the rivers in East Malaysia as both Sabah and Sarawak are known to be diverse in its surrounding environment.

Table 3. Summary of studies from previous researchers in Malaysia.

Year	Authors	Brief Related Research Work
2005	Kiat et al.	• The FLUVIAL-12 an erodible-boundary model which simulates inter-related changes in channel-bed profile, width variation and changes in bed topography was used. Good agreements were obtained for both water level and bed profiles between the measured data and predicted results as it shows the sediment size and channel geometry in Kulim River, Kedah changed significantly.
2013	Gasim et al.	• Floods occurred yearly at Pahang River especially during northeast monsoon and expected to be stimulated by the inconsistent condition of width and depth along the river which create sedimentation and meandering characteristic. The authors measured river drainage capacity, hydraulic parameters and estimation of flow discharge.
2015	Chang et al.	• The December 2014 flood in Pahang river basin affect not only to the economy but also the river morphology itself. Bank erosion, widened cross sections and sediment deposition can be easily identified after the flood recedes. Sediment size distributions were investigated.
2015	Khan et al.	• The most intense geomorphological process at the river bank is the process of erosion. Floods play a key role in landform evolution of an area Kelantan from Kemubu to Kuala Besar. The results show a significant change in the length of the cross section which suggest that the geomorphological processes play a key role in carving and shaping the river banks during the floods.

For example, Putatan River, a tributary which is located in Sabah was realigned in year 2004 to reduce flooding at the confluence with parent Moyog River at Penampang. However, detailed hydraulic investigation was not undertaken [23]. Now, the original Putatan River is no longer functioning as drainage path. It is ecologically degraded and there are rapid encroachment of housing and development into the river reserve. There is also an extensive sedimentation in the new channel, which now has limited its capacity.

Moyog River is geomorphologically unstable. Past attempts to train the river and protect infrastructure and property have been unsuccessful. One river in north Sabah, Benkoka River in Pitas experienced 15 m lateral movement per year due to large flood occurrence, with the whole villages requiring relocation. Such a large scale of relocation is unacceptable within the dense Penampang and Putatan area. Thus, this will make an interesting case study area for researchers to conduct their research in East Malaysia.

6 Conclusion

As a conclusion, having deeper understanding on river geomorphology is crucial especially on the lateral migration for the engineers. Both public and local authorities should also need to be aware of this phenomenon as many of the construction projects tend to be developed near to river banks. By assessing the rate of migration, prediction and planning for future development can be decided properly. Any structures that spanning across a river channel can be further enhance by providing a proper technique and detailed assessment of the migration rate of the river channel [24].

Usually, any river banks that have migrate and affect the structures or land used activity nearby need to be reviewed with proper river restoration methodologies [25]. Developers seldom conduct detailed hydraulic investigation as it requires time and cost. They will only investigate the water level and how often the flood happened in their construction site. It requires a lot of work and the construction proposal usually does not include detailed hydraulic investigation.

Over the years, all the breakthrough findings on lateral migration contribute a lot to the understanding of this phenomenon. The adaptation of modern technologies in recent centuries brought up the study to a whole new level [26]. The analytical model integrated with GIS that has been developed has a high potential for application and is more interactive where users are able to simulate and observe the potential impact in the affected areas. Knowledge on sediments transport is very important as it can be applied to determine the magnitude of erosion or deposition and the time and distance over which it will occur.

The authors wish to thanks the Ministry of Education, Malaysia for funding the research project through the Research Acculturation Grant Scheme [Ref. No. 600-RMI/RAGS 5/3 (180/2014)] and, the Research Management Unit (RMU), Universiti Teknologi MARA (UiTM) for the administrative support.

References

[1] G. Kondolf, Hungry Water: Effects of Dams and Gravel Mining On River Channels, Environmental Management. **21**(4), 533-551, (1997)

[2] M. Yamani, A. Goorabi, and J. Dowlati, , The Effect of Human Activities on River Bank Stability. American J. of Environmental Sciences, **7**, (3), 244-247. (2011)

[3] V. J. Galay, Causes of River Bed Degradation. Water Resources Research. **19**(5), 1057-1090, (2010)

[4] U. R. Mugade and J. B. Sapkale, Influence of Aggradation and Degradation on River Channels: A Review. Int. J. of Engineering and Technical Research, **3**(6), 209-212, (2015)

[5] E. J. Hickin, and G. C. Nanson, Lateral Migration Rates of River Bends. Hydraulic Engineering, 1557-1567, (1984)

[6] H. H. Chang,., Modeling Of River Channel Changes, Hydraulic Engineering, 157-172, (1984)

[7] A. J. Odgaard, River-Meander Model II: Application. Hydraulic Engineering, 1451-1464, (1989)

[8] S. M. Ramasamy, P. C. Bakliwali, and R. P. Verma, Remote Sensing and River Migration in Western India. International Journal of Remote Sensing, **12**(12), 2597-2609, (1991)

[9] D. H. Alan., Modeling Channel Migration and Floodplain Sedimentation in Meandering Streams. Lowland Floodplain Rivers: Geomorphological Perspectives, 1-41, (1992)

[10] W. Spitz, P. Lagasse, S. Schumm and L. Zevenbegen, A Methodology for Predicting Channel Migration. Wetlands Engineering & River Restoration, (2001)

[11] T. J. Randle, Channel Migration Model for Meandering Rivers. *Eight Federal Interagency Sedimentation Conf.*, (2006)

[12] J. Heo, T. A. Duc, , H. S., Cho and S.U. Choi, Characterization and Prediction of Meandering Channel Migration In The GIS Environment: A Case Study Of The Sabine River In The USA. Environment Monitoring and Assessment, 155-165, (2008)

[13] F. U. Ashraf, and X. Liu, River Meandering Prediction: Case Studies for Four Rivers in Texas. *World Environmental and Water Resources Congress 2013*, (2013)

[14] D. Rosgen, *Applied River Morphology*. 2nd Ed. Fort Collins, Wildland Hydrology, (1996)

[15] D. H. Richard, Dynamic Process-Response Model of River Channel Development. Earth Surface Processes 1979, **4**(1), 59-72. (1978)

[16] K. D. Tuhin, K. H. Sushil, D. G. Ivy and S. Sayanti, River Bank Erosion Induced Human Displacement And Its Consequences. Living Reviews in Landscape Research, **8**(3), 1-35. (2014)

[17] S. Chatterjee, and B. Mistri, Impact of River Bank Erosion on Human Life: A Case Study in Shantipur Block, Nadia District, West Bengal. Int. J. of Humanities and Social Science Invention, **2**(8), 2319-7722, (2013)

[18] G. D. Sani, B. G. Muhd, E. T. Mohd, and G. A. Musa, Floods in Malaysia: Historical Reviews, Causes, Effects And Mitigations Approach. Int. J. of Interdisciplinary Research and Innovations, **2**(4), 59-65. (2014)

[19] C. C. Kiat, A. A. Ghani, N. A. Zakaria, R. Abdullah, Sediment Transport In Kulim River, Malaysia. *XXXI IAHR Congress*, (2005)

[20] M. B. Gasim, M. E. Toriman, M. Idris, P. I. Lun, M. K. A. Kamarudin, A. A. N. Azlina, M. Mokhtar and S. A. Mastura, S. River Flow Conditions and Dynamic State Analysis of Pahang River. American J. of Applied Science. **10**(1), 42-57, (2013)

[21] C. K. Chang, A. A .Ghani and N. A. Zakaria, *Sediment Transport in River Due To a Major Flood Event: Case Study of Sungai Pahang* December 2014 Flood. Int. Conf. on Water Resources, (2015)

[22] M. M. A. Khan, N. A. Shaari, D. A. Nazaruddin, and H. E. Mansoor, Flood-Induced River Disruption: Geomorphic Imprints and Topographic Effects in Kelantan River Catchment from Kemubu To Kuala Besar, Kelantan, Malaysia. Int. J. of Environmental, Chemical Ecological, Geological and Geophysical Engineering. **9**(1), 10-14 (2015)

[23] B. Caddis, W. Hong, and C. Nielsen, The Challenges of Floodplain Management In Malaysian Borneo. *Floodplain Management Conf.*, (2011)

[24] W. Yord, J. R. Richardson, and V. J. Zink, Redesign and Rehabilitation of Railroad Bridges with Advanced Adverse Lateral Migration, *World Environmental and Water Resources Congress*, (2009)

[25] D. C. Baird, and C. C. Klumpp, Review of River Restoration Methodologies, *World Environmental and Water Resources Congress* (2012)

[26] J. Conaway, and T. Brabets, Copper River Channel Migration and Its Effects On The Copper River Highway, *Environmental and Water Resources Congress*, (2008).

Effective International Knowledge Exchange to Rehabilitate Rivers in Urban Delta's with RDM approach

Floris Boogaard[1,2*] , *Eric* de Boer[1], *Rick* Heikoop [3,4], and *Marius* Palsma[5]

[1] Hanze University of Applied Sciences Groningen, Zernikeplein 11, Groningen, The Netherlands
[2] TAUW BV, Zekeringstraat 43, PO 20748, Amsterdam, The Netherlands
[3] Rotterdam University of Applied Sciences Rotterdam, G.J. de Jonghweg 4-6, Rotterdam, The Netherlands
[4] RDM Centre of Expertise Rotterdam, Heijplaatstraat 23, Rotterdam, The Netherlands
[5] Waterwegen consultancy, Gasthuiskamp 23,7203 BH Zutphen, The Netherlands

Abstract. In coastal and low-lying vulnerable delta cities, such as the Rotterdam Metropolitan region and Metropolitan Manila, there has been an increase in vulnerability of urban flooding and water quality problems. New approaches address the current and future effects of climate change and increase urban quality, reduce vulnerability and increase water quality. New arrangements for international cooperation are set up in order to address these problems in the project 'IWASTO' ('to correct and to prevent') where several organisations from the Philippines and The Netherlands join forces to rehabilitate rivers and introduce new innovations for water quality monitoring and waste collection using the 'Research Design and Manufacaturing (RDM)' approach. One of the concrete results from this RDM approach in 2016 is the development and implementation of the 'Longganisa', a waste collection tool that was designed by a Dutch-Philippine team. In addition, the waterquality was measured and mapped with innovating methods using apps in order to locate and prioritize highly polluted areas. In the near future measures will be implemented to rehabilitate the rivers in Manila and a knowledge exchange forum will be set up to share the best management practices on international scale in countries as Indonesia, India and China.

1 Introduction

Urbanisation usually leads to degradation of the urban dense area by e.g. soil sealing, air, water and soil pollution with the result that humans are exposed to floods, contaminants and loss of biodiversity. Some of the biggest challenges are to bring down city temperatures, increase biodiversity, fight water shortages and protect properties from damage by flooding within small budgets and timeframes [1]. In 2016 the entire Cebu province was under a state of calamity due to drought [2] and Philippine temperatures reached dangerous levels [3].

* Corresponding author: floris@noorderruimte.nl

Fig. 1. Degradation of waterways in Manila

Given the worldwide increase in urbanization, and the impact of urban storm water and wastewater on both humans and aquatic ecosystems, the management of our water systems is a critically important challenge [4]. Mega cities, such as metropolitan Manila typically face rapid expansion and uncontrolled urban sprawl, resulting in the lack of infrastructure and wastewater treatment facilities. Urbanization also increases the variety and quantities of pollutants found in downstream receiving waters [5]. Metropolitan Manila is an example of a metropolis where rapid urban growth resulted in a river system that is biologically dead [6], as depicted in Fig. 1.

The quality of the river system in Metro Manila seems to be a status quo, which can not be changed. However, if we look at other similar cases in Europe such as the Amsterdam canals with UNESCO world heritage status [7], we see that these canals were not suitable for swimming decades ago [8,9], but in 2016 2966 swimmers participated in the Amsterdam City Swim in the Amsterdam canals.

The European experiences with achieving this improvement in water quality is applied to Manila, using the so called 'RDM-approach'. In this article the implementation of this international study is described and recommendations for the rehabilitaion process of the river system in Metro Manila are given as 'lessons learnt' for upscaling this method to countries as Indonesia, India and China.

2 Method

In the international knowledge exchange on rehabilitation of rivers, several research methods are applied such as: literature review, participatory waterquality monitoring, surveys and engagement of several stakeholders in interdiciplinair workshops. All information has been mapped on interactive international tools as www.climatescan.nl (Fig. 2) showing videos and photos of problems and solutions in Manila and other locations in the world. Visualisation of the fieldtrips, serious gaming, apps, articles and interactive (field)presentations in several workshops are also used as international knowledge exchange and engagement of stakeholders.

The main goal of the involved partners is to rehabilitate rivers and to become a showcase for river rehabilitation for other delta cities around the world. A network is being set up with other projects as the India Ganges riverfront project (Kanpur) and 'Transit oriented development & smart micro city' in China [10] and partners in Indonesia in order to exchange knowledge on solutions to improve the waterquality in a cost effective integrated way. The RDM approach is applied for establishing this international multidisciplinary group of experts and end users. The RDM method is based on the Community of Practice theory described by Wenger in 2001 [11]. Wenger identified three required components to organise interaction on a shared topic and to facilitate incremental change:

1. There needs to be a *domain*
2. There needs to be a *community*.
3. The members are *practitioners*.

In addition to these components triple helix networks elements are added to the model. The triple helix institutional actors are typically **U**niversity-**I**ndustry-**G**overnment (**U-I-G**), which can create synergy and enhance innovative capabilities of countries [12] and even more so if these networks could evolve into a quadruple helix which includes **C**ivil organisations (U-I-G-**C**) [13]. The Port of Rotterdam in which the RDM campus is located is a strong supporter of new innovate networks to enhance sustainability in Rotterdam's City Ports Area [14]. The research 'The impact of Triple Helix agents on entrepreneurial innovations'performance' [15] concludes that the positive effects of government subsidies is only evidenced when enterprises collaborate with universities.

The IWASTO project is set up to be able to facilitate change and to organise interaction between partners on the shared interest of river rehabilitation, the project started by the implementation of the RDM approach.

2.1 IWASTO

IWASTO, which means to correct and to prevent in Tagaloc, is a collaborative project of the Rotary Club of Makati Pasong Tamo, Department of Interior and Local Government, several Dutch (applied) universities and companies. The goal of IWASTO is to rehabilitate and sustain the integrity of the decaying and heavily polluted Pateros River for the sake of ecological balance and disease and flood prevention in the barangay areas that are most affected during natural and man-made calamities. The Pateros River traverses 11 barangays and affects the lives of around two million people: the local residents of the barangays of Pateros, Taguig, Pasig and Makati, and their respective business districts that line the banks of the Pateros River (Fig. 2).

Fig. 2. Pateros river case as shown on knowledge exchange platform www.climatescan.nl [16].

The first IWASTO actions that are planned [7]:

- Stakeholder analysis
- Assessment of the local situation
- Identifications of ambitions
- Cleaning up solid waste that floats in the river
- Initiate behavioural changes to prevent river pollution
- Improving the drainage system
- Education: public awareness (waste management, circular economy)
- Implementing innovating nature based solutions
- Monitoring of water quality by means of low-cost sensors in the river
- Evaluation & governance

The multiannual programme "IWASTO" aims to bring together Philippine and Dutch expertise in the fields of spatial planning, urbanization, mobility, spatial economics, civil engineering, water management, social science and other fields, in order to develop a sustainable vision for Manila and other cities. A 'meet in the middle strategy' is used which is a unique combination of a 'bottom up' and 'top down approach'. Following the mapping stage locations are prioritized and detailed solutions are designed to face the multiple challenges, such as innovating wastewater treatment and sustainable urban drainage systems on street level.

2.2 The RDM-approach

The RDM approach is widely accepted in The Netherlands and mostly located in the presence of (applied) universities as 'BuildingG' and 'EnTranCe' in Groningen, 'Watercampus' in Leeuwarden, 'YesDelft' in Delft and 'RDM campus' in Rotterdam. Within the Rotterdam RDM Centre of Expertise (RDM CoE) educational institutes, research centres and companies cooperate to improve technical education, create new knowledge and accelerate innovations that are necessary for the port and city of Rotterdam. RDM CoE is a Centre of Expertise of Rotterdam University of Applied Sciences recognized by the Ministry of Education, Culture and Science that has been set up on the campus of RDM Rotterdam, as part of a public-private partnership with the Port of Rotterdam Authority, the City of Rotterdam and other organizations. RDM CoE is the innovation engine for port and city. RDM Centre of Expertise focuses on wicked challenges that are difficult to address with multi stakeholder involvement. This cooperation takes place in innovative configurations such as Communities of Practice (CoP), where private firms and government agencies work together on applied research in collaboration with knowledge institutes [17]. Communities of Practise have been set-up in the fields of Maritime & Offshore, Logistics & Mobility, Energy & Process Technology, New Manufacturing Industry and Circular & Floating Construction.

According to Wenger [11], Communities of Practice can be defined, in part, as a process of social learning that occurs when people who have a common interest in a subject or area collaborate over an extended period, sharing ideas and strategies, determine solutions, and build innovations. Wenger gives a simple definition: "Communities of practice are groups of people who share a concern or a passion for something they do and learn how to do it better as they interact regularly".

A CoP is distinguished by three required components:
1. There needs to be a *domain*
2. There needs to be a *community.*
3. The members are *practitioners.*

1. There needs to be a *domain.* A CoP needs a shared domain of interest. Membership implied commitment to the domain and willingness to collaborate with the partners in this specific domain.

2. There needs to be a *community.* A necessary component is that members of a specific domain interact in shared activities, help each other, and share information with each other. They build relationships that enable them to learn from each other. There needs to be people who interact and learn together in order for a CoP to be formed.

The third requirement is that the members are *practitioners.* They develop a shared repertoire of resources which can include stories, helpful tools, experiences, stories, ways of handling typical problems, etc. This kind of interaction needs to be developed over time. Informal conversations held by people with the same passion and interests and a variety of professions and disciplines (including graduate students) help people share and develop a set of cases and stories that can become a shared repertoire for their practice, whether they realize it or not. Communities develop their practice through a variety of methods, including: problem solving, requests for information, seeking the experiences of others, reusing assets, coordination and synergy, discussing developments, visiting other members, mapping knowledge and identifying gaps [11].

The RDM-approach entails the acceleration of innovation in the Rotterdam metropolitan area which includes the largest port of Europe, by using a Communities of Practices method and in addition to that a structured approach towards problem solving within the triple helix collaboration in a living lab context. The approach focuses on applied research and applied solutions based on actual problems in the Rotterdam area. Knowledge is created by the involvement of both the private and public sector and knowledge institutes (U-I-G). The RDM campus and RDM Centre of expertise provide for the *tangible facilities* and the creation of *intangible favourable conditions* where partners can meet and come together, namely the Innovation Dock. The Innovation Dock is a 23.000 m^2 former shipyard which has been transformed into a state of the art testing facility (height 20 meters), see Fig 3.

Fig. 3. RDM campus, a former shipyard transformed into a knowledge centre where public, private and knowledge institutes work together on innovations [17].

The presence of high-tech prototypes and testing facilities and the collaboration of many innovative companies and start-ups at the Innovation dock make it a breeding ground for innovations. The continuous presence of students of different disciplines enhance the innovative atmosphere. Companies that wish to locate in Innovation Dock sign a contract which includes the condition that the companies involve students in their operations. In this way entrepreneurial thinking and real-life problem solving structurally slips into the head and mind of a new generation of students.

This triple helix approach (see Fig. 4) is materialised in the formulation of specific assignments from the private and public sector which are brought into the curricula and practice oriented research projects of the different courses (86 full time courses) at Rotterdam University. This approach helps to bring the 'real world' into the undergraduate education and reduces the distance to the labour market of young professionals.

The Living Lab Approach focuses on co-producing knowledge while recognising that the immense urban and societal challenges we are facing demand concerted collaborative efforts across sectors as well as between disciplines [18]. Co-production involves stakeholders understanding each other's contexts; in order to work together to frame research that delivers solutions that are more effective. While the range of initiatives that call themselves living labs is diverse, three core characteristics distinguish them from other approaches such as internships or practicums [19]: they comprise a geographically or institutionally bounded space, they conduct intentional experiments that make social and/or material alterations, and they incorporate an explicit element of iterative learning.

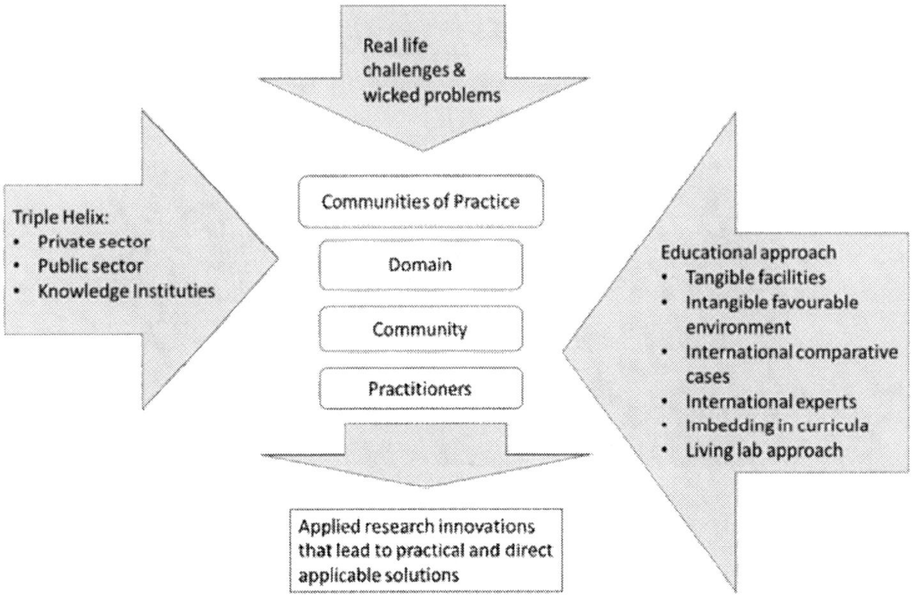

Fig. 4. RDM approach

3 Results

3.1 Case study Manila "IWASTO"

The 'IWASTO' initiative officially started with the signing of a Memorandum of Agreement (MOA) in November 2015 in the presense of: the mayors of Makati City, Pasig City, Taguig City and Pateros, the Rotary club of Makati Pasong Tamo, Department of Interior and Local government, The Netherlands Embassy, Rotterdam University of Applied Sciences, Groningen University of Applied Sciences, Tauw Group and small medium enterprises WaterWegen and INDYMO. The MOA embodies the commitment and support to rehabilitate and sustain the integrity of the decaying and heavily polluted rivers for the sake of ecological balance.

If we look closer at the IWASTO case, we see an immense problem, with multi stakeholders, which can be classified as a so called 'wicked problem,' the upper arrow in Fig. 4. The IWASTO case involves public stakeholders, namely the different mayors and barangay captains that signed the MOA. If we look at the left arrow in Fig. 4, we see that private partners are represented by the Dutch private partners Tauw Group., WaterWegen and Indymo. Looking at the knowledge institutes, we conclude that 2 international universities (Hanze University of Applied Sciences Groningen and Rotterdam University of Applied Sciences) are involved in the project. University of the Philippines participates in the project after establishing a Living Lab Water Philippines – Netherlands for which an MOU was signed during the second IWASTO conference in April 2016, and an MOU was signed with University of the Philippines to participate in the IWASTO Pasig river rehabilitation project.

At the RDM-campus a two-day event was organised by the Delta Talent Academy in the Netherlands, last September 2016 [20] The focus of the 2-day event was finding solutions for water quality improvement in Manila. 75 students participated representing 14 different universities in the Netherlands and 28 studies, worked together on the Manilla

DOI: 10.1051/matecconf/201710304002

case and pitched multiple innovative, creative, technical and social solutions for water quality problems. In the RDM-strategy the multidisciplinary approach is an important asset.

A clear Community of Practice (CoP) is established: looking at the three components of the CoP, in the centre of Fig. 4, the *domain* of the IWASTO case is the river rehabilitation challenge. River rehabilitation is a clear domain with many different aspects such as wastewater treatment, illegal discharge of water, solid waste pollution, illegal residents along the river, and lack of enforcement of rules and regulations.
The *community* is present. Partners are defined in the IWASTO MOA. Different mayors and barangay captains committed themselves to IWASTO. As the project is evolving, different partners join the community.

The number of *practitioners* is increasing. As more students are involved in this project, more studies and research results are generated. The presences of Dutch and Philippine students in the community along the Pateros river generates more attention in both the Philippines and the Netherlands. University of the Philippines (UP) was involved in the IWASTO challenge, which took place last April 2016. More UP students and professors are involved in IWASTO, and more will be involved as the project evolves and upscaled to other regions in the Philippines as Davao and Cebu in 2017.

The Living Lab characteristics are also clearly present, since the location of the Pasig river is geographically bounded. The ongoing activities that are conducted make social and material alterations, and there is an explicit element of iterative learning and concrete results.

3.2 Results IWASTO challenge 2016

April 2016, the IWASTO challenge took place in Manila with 15 Dutch and Philippine students, and 5 Dutch and Philippine experts. Together they worked on locating and mapping specific pollution sources (Fig. 5).

Fig. 5.a) Direct discharge of untreated wastewater in the Pateros river **5b)** the highly polluted rivers are being used as swimming water.

In order to locate the best locations for innovate solutions to address the water pollution new innovating tools were developed to monitor waterquality to analyse parameters as phosphate, nitrate, iron, electric conductivity, temperature and ph (Fig 6).

 DOI: 10.1051/matecconf/201710304002

Fig. 6. a) new innovating tools to map waterquality in Manila with an app. 6b) Knowledge exchange on innovating waterquality measurments with PRRC (Pasig River Rehabilitation Committee).

Researchers have taken samples of the waterquality at several places in Manila which have been analysed using apps on smartphones and being mapped on open source maps (Fig. 7). In the near future, more test results and parameters will be analysed and uploaded real time with the smartphone application. In a fieldworkshop with the Pasig River Rehabilitation Committee, measurements with apps and grap samples are simulatiniously taken. The results from the test strips and labresults will be compared to judge the cost effectiveness of this new method for several organisations. All the results are geolocated on a map and more results can be uploaded, accessible by anyone with the smartphone application. At places that are highly polluted underwaterdrones with sensors will be applied to make detailed 3D scans of the water near polutions sources as sewer outlets and industrial areas [21].

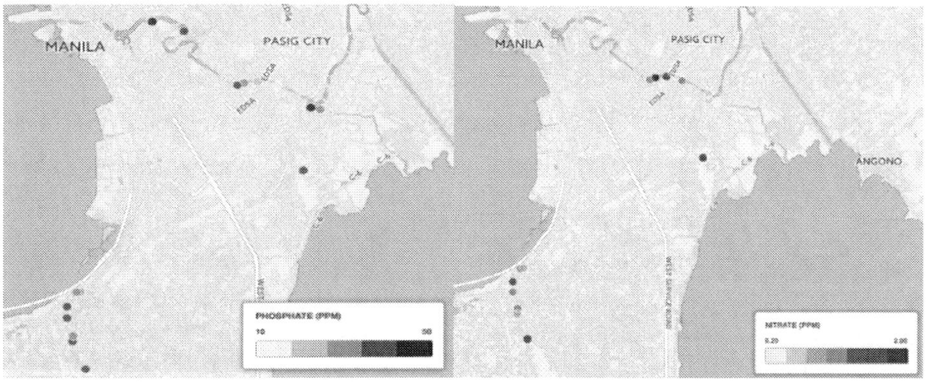

Fig. 7. Results of waterquality measurements in Manila (left: Phosphate, Right: Nitrate).

Waste management tool 'Longganisa'

An innovative design was developed by Dutch-Philippine students teams to catch solid waste from the rivers, the 'Longganisa' (sausage), shown in Fig 8. A floating 'sausage' accumulates the waste in the river by the flow of water. The waste is being gathered in the area of a segredation station where the floating waste (such as plastic) will be used to make products as chairs for schools and floating modulair platforms that can be used as bridges or boats. At highly polluted locations, the longganisa will also be constructed with filters to remove micropolutants from the water.

Fig. 8. Design of innovative 'Longganisa' to remove floating waste from rivers. 9b: the Longganisa constructed November 2016 (in the back is the old way of collecting trash from the river with a boat).

4 Discussion

The awareness increases with the number of *practitioners*. The IWASTO project has achieved the involvement of practitioners, but it is important to stay critical and continuously look for improvement and develop a joint vision. Looking at the right arrow in Fig. 4, we see that the criteria for a new educational approach are not yet fully met. The *Tangible facilities*, such as a campus or location were research can take place in an innovative environment, is not yet present in the area.

The intangible enabling environment, where practitioners and the community can meet and share ideas is also not present at the location. *International comparative cases* are present and Dutch practitioners and experts are present as well to share experiences from other cases. The case is also imbedded in the *curricula* of the Water Management course at Rotterdam and Groningen applied Universities, which enhances the number of practitioners. However, it is important that these kind of wicked problems are also embedded in the curricula of local universities. For the development of curricula that meet the current needs of the (global) society these kind of wicked problems should be included in the research agenda of universities.

The living lab is also present, although not yet completely. The Living Lab Approach at Rotterdam is characterised by mixed teams of both national students (from the country where the case is located) and international students. The Living Lab Approach needs more attention and will evolve in coming years so that the iterative learning aspect will be addressed.

5 Conclusion

The set-up of a Community of Practice (CoP) was effectively established and the 3 requirements were present:

1. The *domain* of the IWASTO case is the river rehabilitation challenge with its multidisciplinary challenges of wastewater treatment, discharge of water, solid waste pollution, illegal residents along the river, lack of enforcement of rules and regulations.
2. The *community* is present. Partners are defined in the IWASTO MOA. As the project is evolving, different partners join the community.
3. The awareness increased with the number of *practitioners* and actions.

Important actions in this project and their concrete results are:

- 2016 IWASTO project: identification of stakeholders, signing MOA between stakeholders in Manila, raising awareness in both the Netherlands and the Philippines, improve commitment. An innovative design was developed by Dutch-Philippine students teams to catch solid waste from the rivers, the 'Longganisa', shown in Fig. 8.
- Delta Talent academy organised at RDM campus Rotterdam: 75 students participated, raising awareness and getting involved
- RDM Manila pressure cooker November 2016 with 15 Dutch students and a number of Philippine students.
- A project proposal was written which will help the Community of Practice to successfully develop, and to monitor progress in the cleaning of Pateros river.
- The successful implementation of the CoP resulted in interest in many countries such as India, Indonesia and China, and the upscaling and implementation is expected in the upcoming years.
- New monitoringmethods are developed (Fig. 6)
- Waterquality is measured and mapped (Fig. 7)
- Tailor made solutions are designed and implemented at highly polluted areas for micropolutants and waste management (Fig. 8).

6 Recommendations

International knowledge exchange with the lessons learned from countries such as the Netherlands will help other countries as the Philippines to avoid costly investments and to implement their strategies cost effectively in the limited space of urban dense areas in a short timeframe. Lessons learned that will be addressed are:

- The 'Meet in the middle' strategy as a combination of 'top down' and 'bottom up' engagement of all stakeholders and show quick concrete results that are needed to keep people involved at different decision levels.
- When solutions are designed, all stakeholders should be involved and attention should be given to the maintenance efforts of solutions to guarantee their efficiency over time. The triple Helix ensures that education, research and entrepreneurship are involved for the design, implementation and evaluation of solutions and so enhance the sustainable and innovative approach.
- When practitioners from other parts of the world visit a certain location where a challenge or problem is observed, they often get very much motivated and inspired to continue further research and studies on this specific topic. The challenge that they face becomes a motivator for pursuing further studies and an international career.

This study would not have been possible without the funding and collaboration within the projects WaterCoG, INXCES and the the long-term support we have received from the several stakeholders from The Netherlands and the Philippines.

References

[1] F. Boogaard, Climate adaptation from Groningen to Mae Phaem. Groningen: Kenniscentrum NoorderRuimte Hanzehogeschool Groningen, The Netherlands, (2016)

[2] Dale Israel , *Entire Cebu province under state of calamity due to drought*, CNN, (April 13, 2016)

[3] Janvic Mateo, *Philippine temperature hits dangerous levels*, The Philippine Star, (April 14, 2016)

[4] T.D. Fletcher, H. Andrieu, and P. Hamel, Understanding, management and modelling of urban hydrology and its consequences for receiving waters; a state of the art. Advances in Water Resources., **51**, 261–279, (2013)

[5] B.E. Hatt, T.D. Fletcher, C.J. Walsh and S.L. Taylor, The influence of urban density and drainage infrastructure on the concentrations and loads of pollutants in small streams. Environ. Manag., **34**, 112–124, (2004)

[6] Pasig River Rehabilitation Commission. Annual Report 2014, Pasig river Rehabilitation commission (2014), Retrieved on September 25, 2016, from http://www.prrc.gov.ph/images/pdf/Transparent_Government/accomplishments/2014_ar.pdf (2016)

[7] F.C. Boogaard, Z. Vojinovic and R. Heikoop, Storm Water and Wastewater Management for Improving Water Quality, *Proc. - Global Water Safety Conf.*, Philippines, (April 25-29, 2016)

[8] Schets FM and v. W. The microbiological quality of the water in the Amsterdam canals. Bilhoven: RIVM Report (2007)

[9] Amsterdam City Swim. (2016), Retrieved on September 25, 2016 from Amsterdam city swim: https://www.amsterdamcityswim.nl/

[10] T. Venhoeven., Transit-Oriented Development & Smart Micro City 公交导向的城市在中国与荷兰的发展, Beijing, (November , 2016)

[11] E. Wenger, Communities of Practice. International Encyclopedia of the Social @ Behavioral Sciences, 2339-2342, (2001)

[12] Y. H. Lee, and Y. Kim, Analyzing interaction in R&D networks using the Triple Helix method: Evidence from Industrial R&D programs in the Korean government. Technological Forecasting and Social Change, 93-105 (2016)

[13] L. Leydesdorff, The Triple Helix, Quadruple Helix, and an N-tuple of Helices, (2010), Retrieved on November 14, from http://www.leydesdorff.net/ntuple/ntuple.pdf

[14] N. Frantzeskaki, Wittmayer, J. and Loorbach, D., The role of partnerships in 'realising' urban sustainability in Rotterdam's City Ports Area, the Netherlands. Journal of Cleaner Production, 406-417 (2014)

[15] M. Guerrero, and D. Urbano, The impact of Triple Helix agents on entrepreneurial innovations'performance: An inside look at entreprises located in an emerging economy. Technological Forecasting & Social Change ,(2016)

[16] J. Tipping, F. Boogaard, R. Jaeger, A. Duffy, T. Klomp and M. Manenschijn, Climatescan.nl: *The development of a web-based map application to encourage knowledge-sharing of climate-proofing and urban resilient projects*, International water week 2015, Amsterdam (2015)

[17] RDM Centre of Expertise. RDM Centre of Expertise, (2016), Retrieved on September 25,2016 from http://www.rdmcoe.nl/

[18] J. Evans, R. Jones, A. Karvonen, L. Millard and J. Wendler, Living labs and co-production: university campuses, Current Opinion in Environmental Sustainability, **16**, 1-6, (2015)

[19] J. Evans and A. Karvonen, Give Me a Laboratory and I Will Lower Your Carbon Footprint!' — Urban Laboratories and the Governance of Low-Carbon Futures. Int. J. of Urban and Regional Researc, 413–430 (2014)

[20] Delta Talent, Delta Talent Academy in volle gang, (2016), Retrieved on September 30, 2016, from Delta talent: http://www.deltatalent.nl/nieuws/delta-talent-academy-volle-gang

[21] W. Schampers, *3D scan gives new dynamic insights in waterquality* (in Dutch: 3Dscan geeft dynamisch inzicht in waterkwaliteit Groningse vijvers), vakblad riolering (November, 2016)

Frequency Analysis of Annual Maximum Flood for Segamat River

Noor Suraya Romali[1, 2, *] and *Zulkifli* Yusop[1]

[1]Faculty of Civil Engineering, Universiti Teknologi Malaysia, Skudai, Johor, Malaysia
[2]Faculty of Civil Engineering and Earth Resources, Universiti Malaysia Pahang, Gambang, Malaysia

Abstract. Several major floods had occurred in the last few decades in Segamat, causing extensive damage to properties and harm local community. For the purpose of flood risk management, this study estimated the average recurrence interval (ARI) and peak flows associated with the ARI based on the distributions of annual peak flow. The flood frequency analysis was performed for flood series data of Segamat River, at Sg. Segamat gauging station (Site 2528414) for the years 1960 – 2011. Five distribution models, namely Generalized Pareto, Generalized Extreme Value, Log-Pearson 3, Log-Normal (3P) and Weibull (3P) were tested for the 52 years flood series data. The goodness of fit test (GOF) of Kolmogorov-Smirnov (K-S) was used to evaluate and estimate the best-fitted distribution. The results obtained using Generalized Pareto distribution provided the best fit, followed by Generalized Extreme Value, Log-Pearson 3, Log-Normal (3P), and the least for Weibull (3P). The estimated peak flows for Segamat River for 50, 100 and 200 ARIs are 1362.2 m^3/s, 1914 m^3/s and 2642 m^3/s respectively. Results can be useful as a reference for further/future flood risk assessment works in the study area.

1 Introduction

Nowadays, a higher frequency of extreme rainfall is expected to occur more frequently due to the climate change phenomenon [1]. Flood had causes tremendous damages to properties [2] and may lead to the loss of human life [3]. In Malaysia, flood occur annually and affected an approximate area of 29,720 km^2, involving more than 4.915 million people and causing up to RM 915 million damage yearly [4]. Efforts have been made by researchers and local authorities to reduce the risk and mitigate the impact of flooding. Flood modelling has been used in flood mitigation to estimate floods associated with return periods of interest, which is called design flood. Design flood is essential in the flood plain management, development and planning controls, and in the design of hydraulic structures [5]. In Malaysia, a 100 year ARI has been used as a practice for designing hydraulic structures. However, recently this standard has been extended to 200 years return period [6].

* Corresponding author: suraya@ump.edu.my

Flood frequency analysis is the most direct method for determining design flood [5]. The purpose of flood frequency analysis is to estimate the return period associated with a given flood magnitude. It shows the relationship between the magnitude of an event and the frequency with which that event is exceeded [7]. Furthermore, the catchment characteristics, water availability and possible extreme hydrological conditions like floods and droughts at various locations of any river system may be illustrated through the flood frequency analysis [8]. The flood frequency analysis primarily uses observed annual maximum flood data at a gauging station to estimate flood magnitude [9]. A long period of recorded flood data is required for this purpose, and a statistical distribution method is needed [6].

Numerous probability distribution models have been used in flood frequency studies, such as log-Pearson 3 [3, 5], Generalized Extreme Value [2, 5], Generalized Pareto [8], log-normal (3P) [8] and Weibull [7]. The selection of appropriate probability distribution and associated parameter procedure is important in flood frequency analysis to avoid under- or over-estimation of design floods [5]. Hence, this paper is aimed at determining the most appropriate probability distribution model that could provide the hydrological frequency i.e. ARI and peak flows of the study area.

2 Model formulations

2.1 Study area and data

Segamat River is located in the southern part of Peninsular Malaysia at 102° 49" East and 2° 30.5' North, with a length of 23 km. The average width of Segamat River is 40 m and is 14 m above sea level. About 70% of the Segamat river watershed is classified as hilly with elevation up to 1000 meters above the mean sea level (msl), and the rest (30%) is undulating with little swamp. Segamat River is a tributary of Sungai Muar that flows through the Segamat town. The data used in the flood frequency analysis were 52 annual maximum flows of Sg. Segamat gauging station for the years 1960 until 2011. These data were provided by the Department of Irrigation and Drainage of Malaysia (DID). The location of Sg. Segamat gauging station (Site 2528414) is shown in Fig. 1.

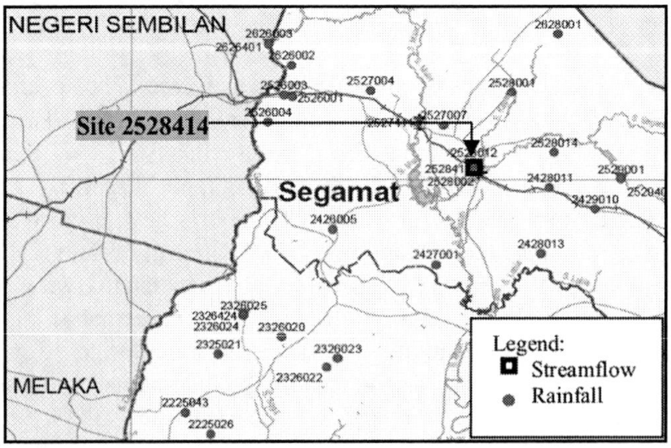

Fig. 1. The location of Sg. Segamat gauging station (Site 2528414)[10].

2.2 Flood Frequency Models

The purpose of flood frequency analysis is to extract information from a flow record to estimate the relationship between flows and return periods. Three different models i.e. annual maximum series (AM) model, partial duration series (PD) or peaks over the threshold (POT) model, and time series (TS) model could be considered for this purpose [11]. A numbers of probability distribution such as Generalized Extreme Value (GEV), Log-Pearson, Log-Normal, Gumbell, Weibull, and Generalized Pareto had been utilised in flood frequency studies worldwide. In order to determine whether the distribution model could fit the data properly, Goodness-of-Fit test such as Kolmogorov-Smirinov, Anderson-Darling, and Chi squared tests can be used [12].

Singo et al. [13] had adopted AM model in their study. 50 years annual maximum flow data from 8 stations were used to analyse flood frequencies in the Luvuvhu River Catchment in Limpopo province, South Africa. The result showed that Gumbel and Log Pearson type III distributions provided the best fit in the extreme value analysis. Rahman et al. [5] used a large annual maximum flood data set to select best probability distributions for at-site flood frequency analysis in Australia. They identified Log Pearson type III, GEV and Generalized Pareto as the top-three best fitted distributions. A frequency analysis study using POT flood data was conducted by Guru and Jha [8]. Comparison was made with another analysis using AM flood data where Generalized Pareto and Log Normal (3P) showed the best result for AM and PO flood data series respectively.

Study by Mohd Daud et al. [14] found that GEV was the most suitable distribution for annual maximum rainfall in Peninsular Malaysia. The analysis was done using annual maximum rainfall series for several time resolutions obtained from 17 recording rain gauges that are located all over the peninsular. Meanwhile, GEV and Generalized Logistic distribution are identified as the best fitted distribution for frequency analysis using annual flood data from more than 23 gauged river basin in Sarawak, Malaysia [15].

3 Methodology

Fig. 2 shows the general methodology adopted in this study. The first stage is the estimation of annual maximum stream flow that based from the flow historical data for certain years [6]. Then 52 selected flow data from the year 1960 until 2011 were analyzed using EasyFit Software to determine the distribution models that can best fit the data. EasyFit Software is a data analyzer and simulation software which is capable to fit and simulate statistical distributions with sample data, choose the best model, and then use the obtained result of analysis to provide better decisions [11].

3.1 Probability distributions, parameter estimation methods and Goodness of fit (GOF)

In this study, the annual maximum series (AM) model was adopted where only the peak flow in each water year is considered. Five different probability distributions i.e. Generalized Pareto, Generalized Extreme Value (GEV), Log-Pearson 3, Log-Normal (3P) and Weibull (3P) are considered for comparison. The selections of the distribution models are based on the previous studies where most of these have been used and recommended in various countries.

In EasyFit software, different parameter estimation methods are used for different probability distributions. Table 1 list the method uses for the five selected probability distributions. Method of L-moments is used for Generalized Pareto and GEV. Whereas, maximum likelihood method is used for Log-Normal (3P) and Weibull (3P), and method of moments is used for Log-Pearson 3. The most commonly adopted GOF tests are Kolmogorov-Smirinov (KS), Anderson-Darling, and Chi squared test. However, KS test is found to be the most used GOF test [12]. Hence KS test is applied in this study to determine whether the distribution is fitted to the data or not. K-S at 5% level of significant ($p<0.05$) was used to define the best fit ranking [6].

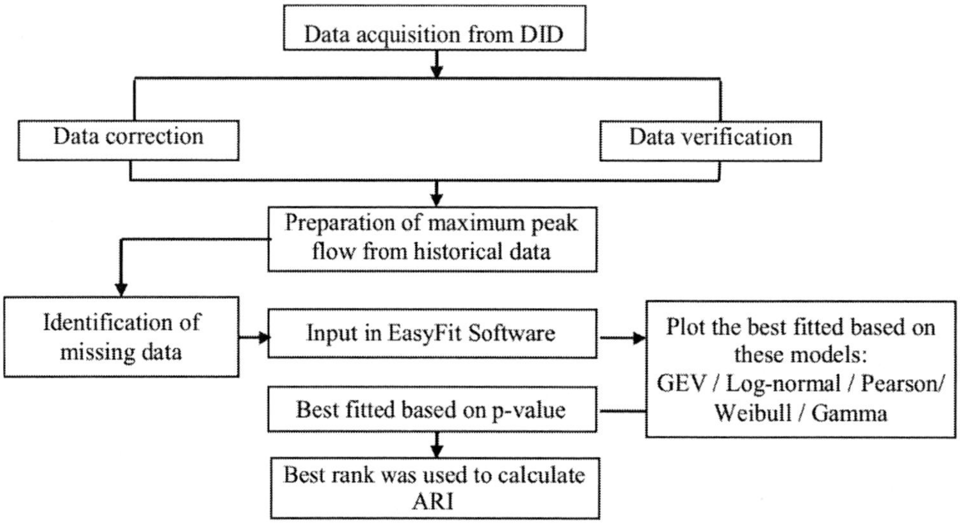

Fig. 2. The general methodology for flood frequency analysis and the determination of Average Recurrence Interval (ARI)

Table 1. Parameter estimation methods applied in this study.

Probability distribution	Parameter estimation method
Generalized Extreme Value, Generalized Pareto	Method of L-moments
Log-Pearson 3	Method of moments
Lognormal (3P), Weibull (3P)	Maximum likelihood method

3.2 Quantile estimation of Generalized Pareto

After the parameters of a distribution are estimated, quantile estimates (X_T) which correspond to different return periods may be computed [11]. For the case of Generalized Pareto distribution, the distribution function $F = F(x)$ is given by Equation (1) [11]:

$$F(x) = 1 - \left[1 - \frac{k}{\alpha}(x - \varepsilon)\right]^{1/k} \tag{1}$$

Using the inverse form of Equation (1), $x = x(F)$ and $F = 1 - (1/T)$, the T-year quantile (X_T) for Generalized Pareto distribution is given by Equation (2):

$$X_T = \varepsilon + \frac{\alpha}{k}\left[1 - T^{-k}\right] \tag{2}$$

4 Results and discussion

The annual flood variation for the respective years is shown in Fig. 3. The highest flow was recorded in 1983 which is 1615.5 m^3/s, while the minimum flow of 3.3 m^3/s was recorded in the year 1989. The average flow for the 52 years was 234.37 m^3/s. Four major flood events, labelled as 1, 2, 3 and 4 in Fig. 3 occurred in year 1969, 1979, 1983 and 2007 with peak flow more than 1000 m^3/s. A large flood had occurred in Segamat for years 2007 and 2011, which had caused tremendous damages and disruptions to local communities.

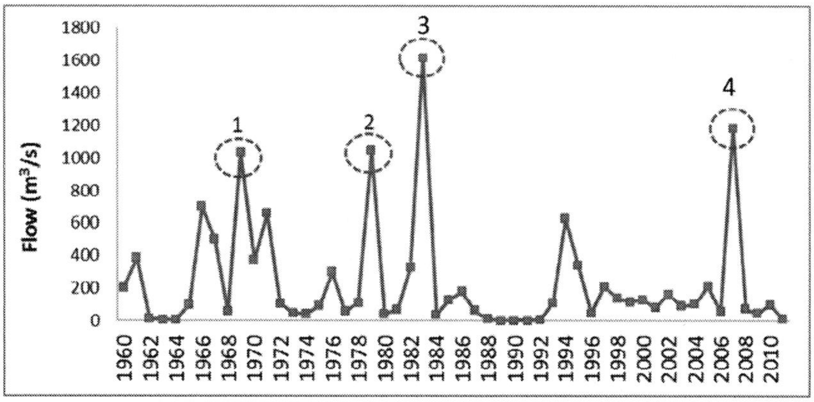

Fig. 3. The annual peak flow at Sg. Segamat gauging station from July 1960 to June 2011

4.1 Goodness of fit test result

The best parameter estimates generated from the EasyFit Software for the five distribution models are displayed in Table 2. Parameters (α, k) represent shape parameters, while (σ, β) and (μ, γ) representing the continuous scale parameters and continuous location parameters respectively. Table 3 ranks the performance of various cumulative density based on the K-S GOF tests. Generalized Pareto shows the best performance, followed by Generalized Extreme Value, Log-Pearson 3, Log-Normal (3P), and the least for Weibull (3P). The ranking is based on the p-value. A p-value closer to one indicates a better-fit distribution. The highest p-value is 0.83277 for the Generalized Pareto and the lowest is 0.29474 for Weibull (3P).

Table 2. Fitting results for probability distribution of annual flood

	Distribution	Parameters
1	Generalized Extreme Value	k=0.50376 σ=101.87 μ=75.491
2	Generalized Pareto	k=0.3988 σ=145.27 μ=-7.2636
3	Log-Pearson 3	α=23.428 β=-0.31913 γ=11.998
4	Lognormal (3P)	σ=1.5441 μ=4.512 γ=0.24503
5	Weibull (3P)	α=0.62873 β=172.5 γ=3.3

Table 3. Fitting results for probability distribution of annual flood

Distribution	Kolmogorov Smirnov	
	P	Rank
Generalized Pareto	0.83277	1
Generalized Extreme Value	0.67816	2
Log-Pearson 3	0.67300	3
Log-Normal (3P)	0.45227	4
Weibull (3P)	0.29474	5

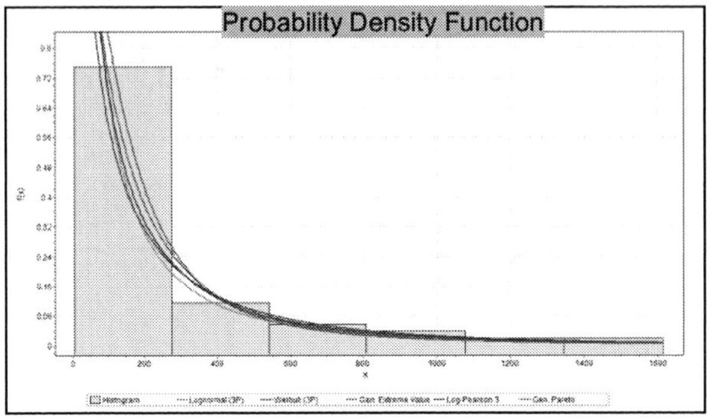

(a)

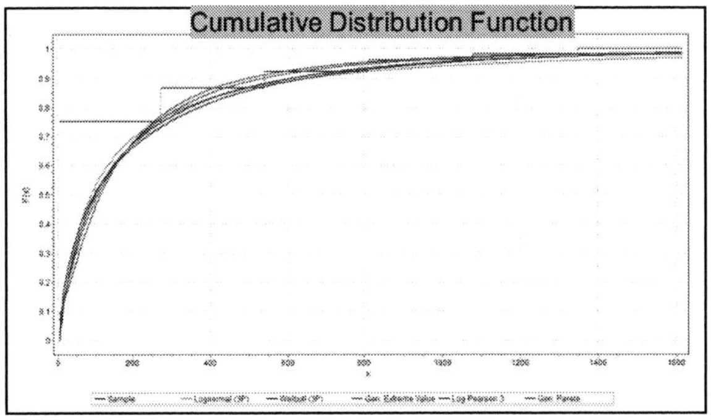

(b)

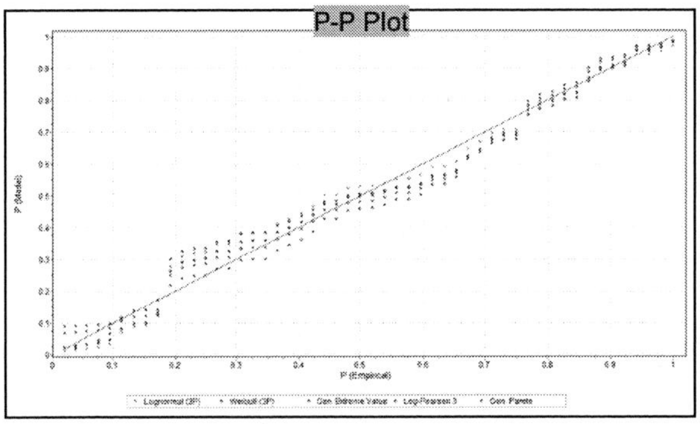

(c)

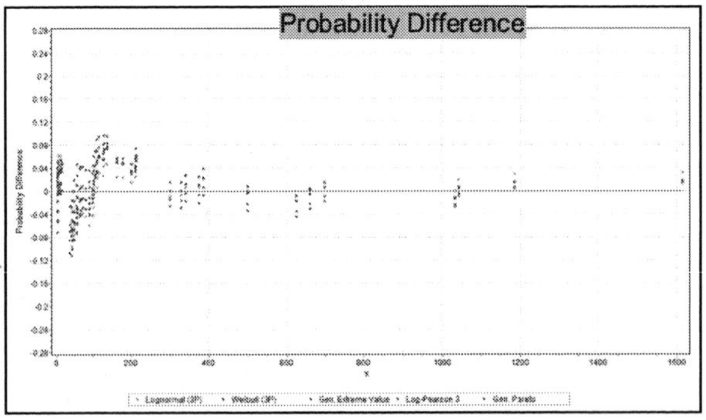

(d)

Fig. 4. a) Probability density function, b) Cumulative distribution function c) Probability-probability plot and d) Probability difference plot for the five distributions.

The Probability Density Functions (PDFs) for the five distribution models; Generalized Pareto (green), Generalized Extreme Value (blue), Log-Pearson 3 (purple), Log-Normal (3P) (orange) and Weibull (3P) (dark green) are shown in Fig. 4a. The Cumulative Distribution Function (CDF) in Fig. 4b shows the non-exceedance probability for a given magnitude. The P-P plot (Fig. 4c) is a graph of the empirical CDF values against the theoretical CDF values. The distribution that has the most number of points close to the line represents the best fitted distribution model. Through all the patterns shown in Fig. 4a to 4d, the best fitted is Generalized Pareto (GP) distribution. The second most chosen best fitting distribution is Generalized Extreme Value (GEV), followed by Log-Pearson 3, Log-Normal (3P), and the least for Weibull (3P). These result is differ from the finding by Mohd Daud et al. [14], which stated that GEV is the most suitable distribution for annual maximum rainfall in Peninsular Malaysia. However, it is in agreement with those obtained by Dan'azumi et al. [16] where in their study, GP is found to be the most suitable distribution for modelling the hourly rainfall intensity in Peninsular Malaysia. Furthermore, this findings are in accord with another study by Wan Zin et al. [17], indicating that GP is the most frequently selected fitting distribution of annual maximum rainfall in Peninsular Malaysia based on LQ-moment methods, together with Generalized Logistic distribution.

4.2 Peak Flow Estimation

Peak flows corresponding to return periods of 2, 5, 10, 50, 100 and 200 years were estimated using the best-fitted distribution models i.e. Generalized Pareto, as shown in Table 4. The estimated flows for 50, 100 and 200 years ARI are 1362.2 m^3/s, 1914 m^3/s and 2642.0 m^3/s respectively.

Table 4. Fitting results for probability distribution of annual flood

Return period, T (years)	Probability, P (%)	Flood discharge, Q (m^3/s)
2	50	108.7
5	20	320.6
10	10	540.9
25	4	943.5
50	2	1362.2
100	1	1914.2
200	0.5	2642.0

5 Conclusion

A total of five probability distributions namely Generalized Pareto, Generalized Extreme Value, Log-Pearson 3, Log-Normal (3P) and Weibull (3P) were tested using 52 annual flow data of Sg. Segamat gauging station to identify the best distribution model that fit the annual flood of Segamat River. This study found that Generalized Pareto distribution provided the best-fit, followed by Generalized Extreme Value, Log-Pearson 3, Log-Normal (3P), and the least for Weibull (3P). Peak flow of Segamat River for 50, 100 and 200 years ARI are estimated as 1362.2 m^3/s, 1914 m^3/s and 2642.0 m^3/s respectively. This information is useful for the flood risk management where the ARI and estimated flow values may be used to generate future flood risk mapping.

The authors would like to acknowledge Universiti Malaysia Pahang and Universiti Teknologi Malaysia for the financial support through research grant Institut Inovasi Strategik Johor (IISJ) vot

15H00 and Department of Irrigation and Drainage (DID) Malaysia for providing data and relevant information.

References

[1] M. Ren, B. Wang, Q. Liang and G. Fu, Classified real-time flood forecasting by coupling fuzzy clustering and neural network. Int. J. of Sediment Research, **25**(2), 134–148 (2010)

[2] L. Chang, C. Lin and M. Su, Application of geographic weighted regression to establish flood-damage functions reflecting spatial variation. Water SA, **34**(2)**,** 209–216 (2008)

[3] S. N. Jonkman, Global perspectives on loss of human life caused by floods. Natural Hazards, **34**(2), 151–175 (2005)

[4] DID Malaysia, *DID Manual: Flood management*, **1**(2009)

[5] A. S. Rahman, A. Rahman, M. A. Zaman, K. Haddad, A. Ahsan and M. Imteaz, A study on selection of probability distributions for at-site flood frequency analysis in Australia. Natural Hazards, **69**(3), 1803–1813(2013)

[6] A. Z. Ismail, Z. Yusop and Z. Yusof, Comparison of flood distribution models for Johor River basin. Jurnal Teknologi (Science and Engineering), **72**(1), 1–6 (2015)

[7] O.S. Selaman, S. Said and F. J. Putuhena, Flood frequency analysis for Sarawak using Weibull, Gringorten and L-Moments Formula. The Intitutions of Engineers, Malaysia, **68** (1), 43–52 (2007)

[8] N. Guru and R. Jha, Flood frequency analysis of Tel Basin of Mahanadi river system , India using annual maximum and POT flood data. Aquatic Procedia ,**4**, 427–434, (2015)

[9] B. P. Parida, R. K. Kachroo and D. B. Shrestha, Regional flood frequency analysis of Mahi-Sabarmati Basin (Subzone 3-a) using Index Flood Procedure with L-Moments. Water Resources Management, **12**(1), (1998)

[10] Department of Irrigation and Drainage of Malaysia (2016) ,Location of Sg. Segamat Gauging Station, Retrieve on September 26, 2016 from http://h2o.water.gov.my/

[11] A. Ramachandra Rao, K. H. Hamed. *Flood Frequency analysis*, CRC Press,(2000)

[12] H. Mehrannia and A. Pakgohar, Using Easy Fit Software For Goodness-Of-Fit Test and Data Generation, IJMA, **5**(1), 118–124, (2014)

[13] L. R. Singo, P. M. Kundu, J. O. Odiyo, F. I. Mathivha and T. R. Nkuna, Flood frequency analysis of annual maximum stream flows for Luvuvhu River Catchment , Limpopo Province , South Africa. In *16th SANCIAHS Hydrology Symposium*, 1-3 October 2012 at the University of Pretoria, South Africa, (2012)

[14] Z. Mohd Daud, A. H. Mohd Kassim, M. N. Mohd Desa and V. T. V. Nguyen, Statistical analysis of at-site extreme rainfall processes in Peninsular Malaysia. FRIEND 2002-Regional Hydrology: Bridging the Gap between Research and Practice, *(Proceedings of the Fourth International FRIEND Conference held at Cape Town, South Africa),* IAHS Publication, **No. 274**, 61–68, (2002)

[15] Y. H. Lim and L.M. Lye, Regional flood estimation for ungauged basins in Sarawak, Malaysia, Hydrological Sciences J., **48** (1), 79–94, (2003)

[16] S. Dan'azumi, S. Shamsudin, A. Azmi, Modeling the distribution of rainfall intensity using hourly data, American J. of Envi. Sciences, **6**(3), 238-243, (2010)

[17] W. Z. Wan Zin, A. A. Jemain and K. Ibrahim, The best fitting distribution of annual

maximum rainfall in Peninsular Malaysia based on methods of L-moment and LQ-moment, Theory Appl Climatol, **96**:. 337-334, (2009)

Flood Prediction at The Northern Region of UAE

Mohamed Elhakeem[1,*]

[1]Abu Dhabi University, Civil Engineering Department, P.O. Box 59911 Abu Dhabi, UAE

Abstract. Frequent flooding has been recently observed in the northern part of United Arab Emirates, particularly Ras Al Khaimah region (e.g., 2010, 2011 and 2016 floods). These frequent flood events raised the need for accurate estimates of surface runoff and possibly flooded areas in this region. Flooding is prevalent in this valley-coastal region due to the surrounding mountain stream network, which is characterized by flash floods from high precipitation amounts with high intensities between December and March. Most of the streams in this region have no hydraulic control structures at the outlets to regulate their flows. A hydrologic study was conducted at this urbanized valley-coastal area to identify the flood magnitudes and possible flooded areas using a number of geospatial, hydrologic and hydrodynamic models, namely GIS, HEC-SSP, Win-TR-20 and FESWMS. The study identified the flood magnitudes and possible flooded areas from large floods of return periods vary from 50 to 500-year. FESWMS simulations showed that the flooded area increases for the 500-yr return period compared to the lower ones. The water depth ranges on average from 0.5 to 8.0 m. Due to the natural slope of the simulated coastal area, higher depth was predicted, in general, close to the shoreline, while lower depth was predicted near the mountains.

1 Introduction

Frequent flooding has been recently observed in the northern part of United Arab Emirates (UAE), particularly Ras Al Khaimah region (e.g., 2010, 2011 and 2016 floods). These frequent flood events raised the need for accurate estimates of surface runoff and possibly flooded areas in this region. Flooding is prevalent in this valley-coastal region due to the surrounding mountain stream network, which is characterized by flash floods from high precipitation amounts with high intensities between December and March. Most of the streams in this region have no hydraulic control structures at the outlets to regulate their flows. A hydrologic study was conducted at this urbanized valley-coastal area to identify the flood magnitudes and possible flooded areas using a number of geospatial, hydrologic and hydrodynamic models, namely GIS, HEC-SSP, Win-TR-20 and FESWMS.

* Corresponding author: mohamed.elhakeem@adu.ac.ae

2 Study site

The hydrologic study was conducted in an urbanized valley-coastal area located in Ras Al Khaimah region, United Arab Emirates (Fig. 1). The study area is about 176 km^2 and is bounded with mountains from all sides except the north-west region, which is open to the Arabian Gulf Sea. A number of small towns and villages namely, Al Rams, Seih Al Birare, Qussaidat, Salhiyah, Fahlain, is located in the west side of the valley region. The boundary of the valley region is marked with red line in Fig. 1. On the eastern boundary of the valley, there are three main channels (Khors) namely, Wadi Biah, Wadi Al Beh, Wadi Najab, which contribute large amounts of surface runoff from their mountain watersheds to the valley-coastal region during severe storms. The three main channels are marked with orange lines in Fig. 1. Wadi Al Beh and Wadi Najab main channels end with large funnel-shape sediment deposits in the valley-coastal region. The only exit for the surface runoff water from valley-coastal area is the north-west coastal area, which delivers the water to the Arabian Gulf Sea, marked with the red line at location number 11.

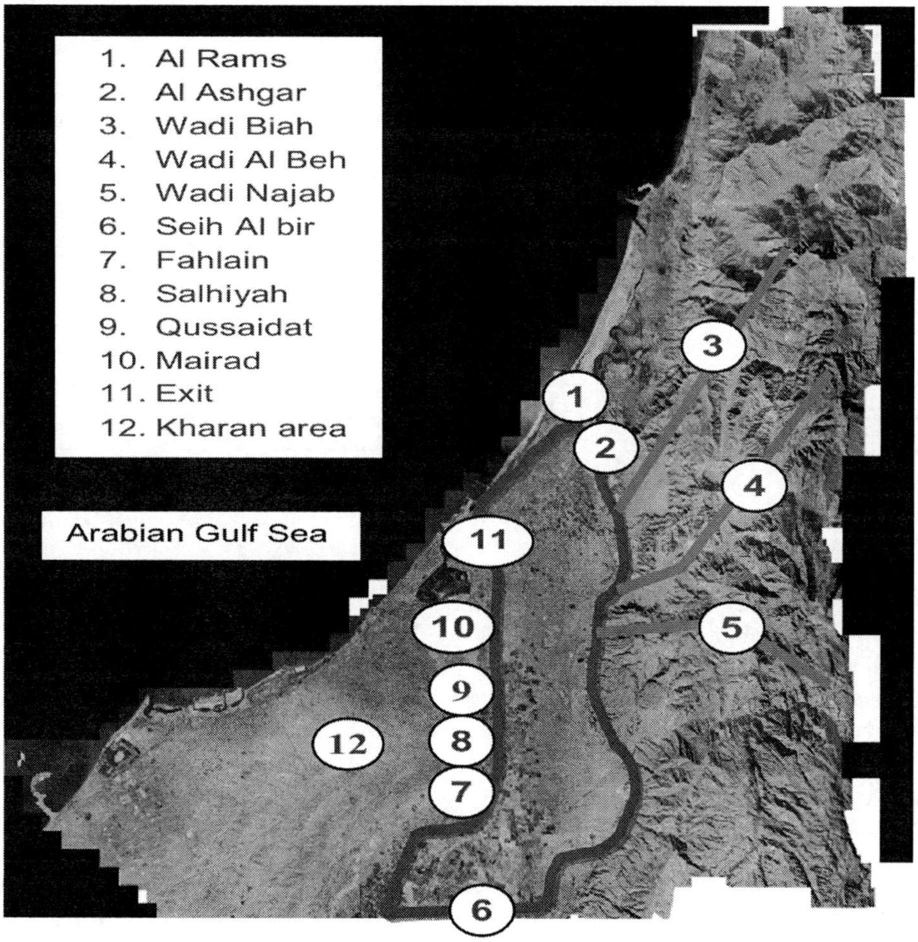

Fig 1. Aerial map of the study site location.

3 Results

3.1 HEC-SSP results

Prior studies have indicated that individual events are important because the effectiveness of flood mitigating practices is best examined on single storm events [1]. Therefore, the daily rainfall data of Ras Al Khaimah for the period from 1984 – 2014 (31 year) were used in the Statistical Software Package (HEC-SSP) analysis to estimate the extreme events and their expected return period [2]. Table 1 shows the return period of different daily peak rainfall events. The extreme events (> 50 yr return period) were used in delineating of the flood areas of the study site.

Table 1. Return period of daily peak rainfall events of Ras Al Khaimah (1984 – 2014).

Return period (yr.)	2	5	10	50	100	200	500
Rainfall (mm)	35	80	130	237	285	340	390

3.2 Win-TR-20 results

The effective rainfall model Win-TR-20 [3, 4] was used to predict the surface runoff volumes and peak flow rates from the rainfall data obtained from HEC-SSP. Table 2 provides the peak runoff flow rates obtained from Win-TR-20 for the 3 watersheds contributing to the study area, namely, Wadi Biah, Wadi Al Beh, and Wadi Najab.

Table 2. Peak runoff flow rate of the 3 watersheds contributing to the study site under different return period.

Return period (yr.)	2	5	10	50	100	200	500
Watershed	Peak runoff flow rate (m^3/s)						
Wadi Biah	0	1.4	17.7	107.6	162.4	232.2	301.8
Wadi Al Beh	0	14.3	143.2	989.3	1524.9	2186.4	2833.1
Wadi Najab	0	13.4	109.4	675.5	1035.4	1462.8	1869.5

3.3 FESWMS results

Due to the complex topography of the study area (Fig. 1), the 2D-hydrodynamic Finite Element Surface Water Modeling System (FESWMS) was used in the determining the flooded areas. FESWMS solves the differential forms of the continuity and the momentum equations in the stream wise and transverse directions using the Galerkin method of weighted residuals, providing water depth and depth-averaged velocity magnitude in x and y directions at each node in the grid [5]. The governing equations are written in the conservative form hence, the momentum is conserved along the streamline and the model is capable of capturing shock effects [6]. The conservative form was chosen because of its robustness in solving critical and transcritical flow fields under both low and high flow conditions by allowing dry-elements to exist within the computational mesh [6]. FESWMS solves the following equations simultaneously:

$$\frac{\partial z_w}{\partial t} + \frac{\partial q_1}{\partial x} + \frac{\partial q_2}{\partial y} - q_m = 0 \quad (1)$$

$$\frac{\partial q_1}{\partial t}+\frac{\partial}{\partial x}\left(\frac{q_1^2}{d}+\frac{1}{2}gd^2\right)+\frac{\partial}{\partial y}\left(\frac{q_1 q_2}{d}\right)+gd\frac{dz_b}{\partial x}+$$
$$gn^2\frac{q_1\sqrt{q_1^2+q_2^2}}{d^{7/3}}\sqrt{1+\left(\frac{\partial z_b}{\partial x}\right)^2+\left(\frac{\partial z_b}{\partial y}\right)^2}-2d\varepsilon_{xx}\frac{\partial^2\overline{u}}{\partial x^2}-\varepsilon_{xy}\frac{\partial}{\partial y}\left(\frac{\partial\overline{u}}{\partial x}+\frac{\partial\overline{v}}{\partial y}\right)=0 \quad (2)$$

$$\frac{\partial q_2}{\partial t}+\frac{\partial}{\partial x}\left(\frac{q_1 q_2}{d}\right)+\frac{\partial}{\partial y}\left(\frac{q_2^2}{d}+\frac{1}{2}gd^2\right)+gd\frac{dz_b}{\partial y}+$$
$$gn^2\frac{q_2\sqrt{q_1^2+q_2^2}}{d^{7/3}}\sqrt{1+\left(\frac{\partial z_b}{\partial x}\right)^2+\left(\frac{\partial z_b}{\partial y}\right)^2}-2d\varepsilon_{yy}\frac{\partial^2\overline{v}}{\partial y^2}-\varepsilon_{yx}\frac{\partial}{\partial x}\left(\frac{\partial\overline{u}}{\partial x}+\frac{\partial\overline{v}}{\partial y}\right)=0 \quad (3)$$

where, (1) is the continuity and (2) and (3) are the momentum equations in x and y directions, respectively. In the equations, t is time (s), d is water depth (m), ρ is water density (kg/m^3), g is the acceleration due to gravity (m/s^2), n is Manning's coefficient of roughness, z_w and z_b are water surface elevation and bed elevation above certain datum (m), q_1 and q_2 are the unit discharge fluxes (m^2/s) defined as $\overline{u}_d$ and $\overline{v}_d$, respectively, $\overline{u}$ and $\overline{v}$ (m/s) are the depth-averaged velocities of an element in the stream wise and transverse directions, respectively, q_m is the resultant inflow or outflow from that element (m/s), v_{xx} and v_{yy} are the normal components of the eddy viscosity (m^2/s) in the x and y directions, respectively, and v_{xy} and v_{yx} are the shear components of the eddy viscosity (m^2/s) applied to the $x-y$ plane.

FESWMS inputs are the Manning's coefficient of roughness n, and the eddy viscosity v. Model inputs must be assigned correctly to represent the physical processes occurring in the model reach and to produce accurate model predictions. Manning's n is an empirical coefficient that accounts for the total flow resistance caused by flow interaction with the boundary [7]. FESWMS utilizes Manning's n to account for momentum loss due to bed-shear, which may vary significantly in a stream reach in accordance to bed-bathymetry and roughness. The second input variable used by FESWMS is the eddy viscosity, v. Eddy viscosity accounts for flow resistance due to the internal shear stresses, or the Reynolds' stresses of the fluid incorporating the added energy dissipation due to turbulence in the flow [8]. Therefore, eddy viscosity is not a physical property of the fluid, but rather a turbulent characteristic of the flow. For isotropic flows in prismatic channels, a single eddy viscosity value is sufficient to describe the turbulence flow characteristics within the modelled reach.

The variation in the water depth of the coastal area was estimated from FESWMS with average Manning's value of 0.065 and eddy viscosity of 10 m^2/s for all the simulations [5]. Fig. 2 shows the predicted water depth (flooding) from FESWMS simulations for different return periods. As expected, it can be seen that the flooded area increases for the 500-yr return period compared to the lower ones. Fig. 2 shows that the water depth ranges on average from 0.5 to 8.0 m. Due to the natural slope of the simulated coastal area, higher depth was predicted, in general, close to the shoreline, while lower depth was predicted near the mountains.

DOI: 10.1051/matecconf/201710304004

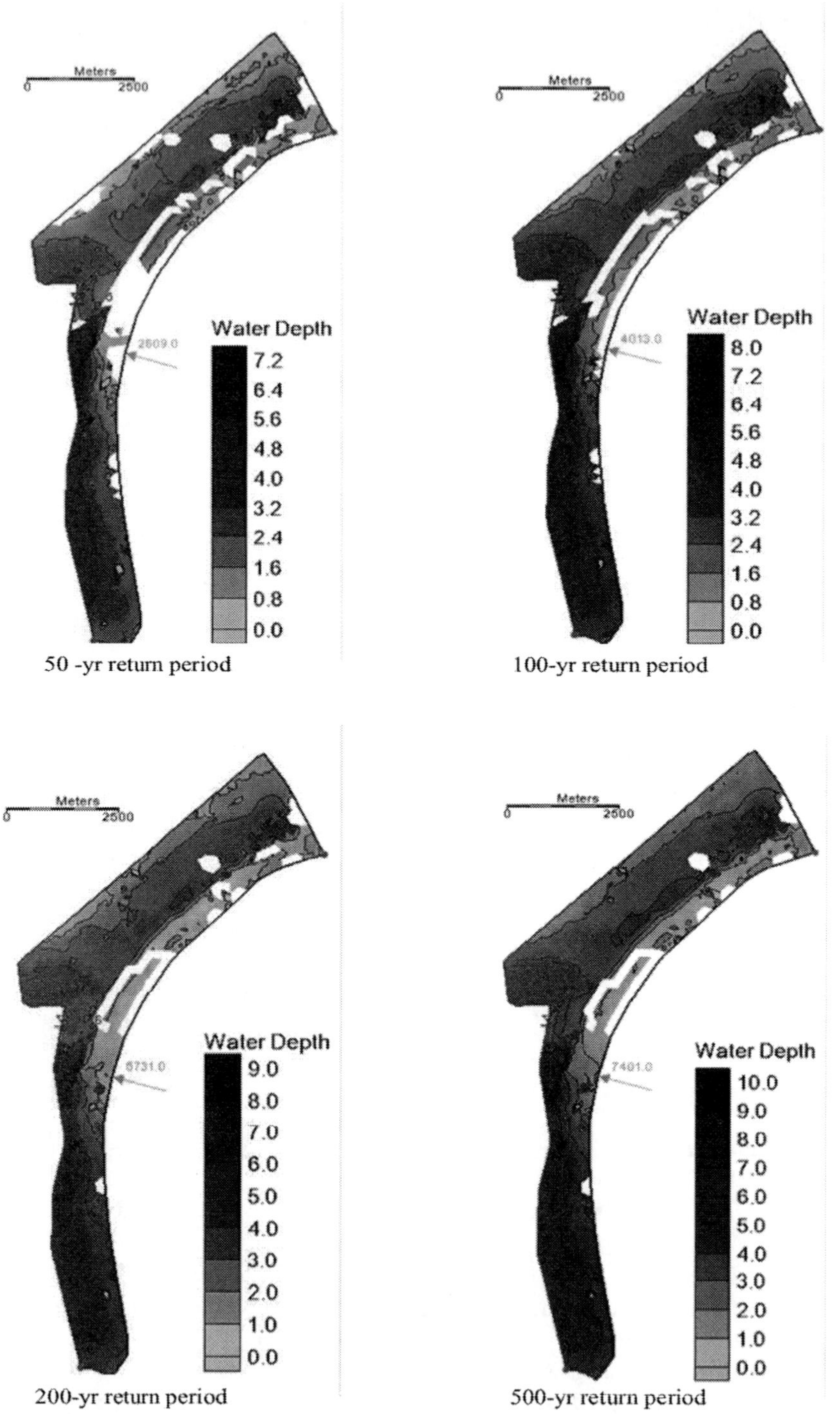

Fig 2. Predicted water depth at the coastal area under different return periods.

4 Conclusion

This study presents a hydrologic study performed in an urbanized valley-coastal area located in Ras Al Khaimah region, United Arab Emirates. Runoff volumes and peak flow rates were quantified at study site using established hydrologic modeling techniques and software, namely HEC-SSP, Win-TR-20 and FESWMS. The study aimed at identifying the flood magnitudes and possible flooded areas from large floods of return periods vary from 50 to 500-year. FESWMS simulations showed that the flooded area increases for the 500-yr return period compared to the lower ones. The water depth ranges on average from 0.5 to 8.0 m. Due to the natural slope of the simulated coastal area, higher depth was predicted, in general, close to the shoreline, while lower depth was predicted near the mountains.

References

[1] M. Elhakeem and A.N. Papanicolaou, Estimation of the runoff curve number via direct rainfall simulator measurements in the State of Iowa, USA, Water Resources Management, **23**(12), 2455-2473 (2009)

[2] U.S. Army Corps of Engineers, *HEC-SSP User's manual* (2007)

[3] Natural Resource Conservation Service (NRCS), *WinTR-20 User's Guide* (2004)

[4] R.H. McCuen, *Hydrologic analysis and design*, Prentice Hall, New York (2003)

[5] D. Froelich, *User's manual for FESWMS Flo2DH* (2002)

[6] M.H. Chaudhry, *Open channel flow*, Springer, New York (2008)

[7] D.M. Hicks and P.D. Mason, *Roughness characteristics of New Zealand Rivers*, New Zealand DSIR Marine and Freshwater Resources Survey, (1991)

[8] A.N. Papanicolaou and R. Hilldale, Turbulence characteristics in gradual channel transition, J. of Engineering Mechanics, **128**(9), 948-960 (2002)

Large Eddy Simulation of Flows Associated with Offshore Oil and Gas Pipeline

Z. Nizamani[1,*] and*A. Nakayama* [1]

[1]Department of Environmental Engineering, Universiti Tunku Abdul Rahman (UTAR), P.C. 31900, Kampar, Perak Darul Ridzuan, Malaysia

Abstract. Fluid structure interaction (FSI) applications are of wide range from offshore fixed and floating structures to offshore pipelines. Reynolds Averaged Navier Stoke (RANS) solution has limitation for unsteady and turbulent flow modelling. A possible approach is Large Eddy Simulation (LES) and it is applied to flows past a circular cylinder located far above, near and on a flat seabed. The Reynolds number considered is based on the real situation off Malaysia Coast and is sub-critical around 105. Hydrodynamic quantities in terms of mean pressure are predicted and vortex shedding mechanism is evaluated. The results are validated by comparing the simulation and experimental previous studies.

1 Introduction

One of the most significant problems in ocean engineering is flow around a circular cylinder. This flow represents flow around bluff body which carries significant interest in many engineering applications like hydrodynamic loading on marine risers, jacket legs, marine pipeline, FPSO and tension leg platforms moorings [1,2]. The design current velocity being used in offshore Malaysia varies for three different regions of Malaysia for operational and extreme conditions and it lies between 0.3 to 0.81 m/sec respectively.

The Reynolds number based on the pipe diameter and the mean current velocity varies from $8.1\text{x}10^4$ to $2.1\text{x}10^5$ in this region which indicates that the flow is in high subcritical range or near critical. The experimental work of obtaining detailed flow quantities at such high Reynolds number is very difficult and there are only few cases where it has been reported in literature [3]. This variation of Reynolds number makes it necessary to evaluate the flow around the pipe in the actual range. In this study the flow around and in the near wake of circular pipe is analysed with the expected range of the Reynolds number and other parameters such as the distance to the seabed. The grid system used in the numerical analysis is fixed rectangular mesh in Cartesian coordinates. The results could be enhanced if grid resolution is increased. Boundary geometries are of common shape. We assume that the bottom boundary is not consisted of loose materials like mud sediment. In the next section fundamental equations of motion and numerical models that are used in the analysis

*Corresponding author: zafarullah@utar.edu.my

are discussed. Then a LES method with a wall model is described. Its accuracy is verified by comparing with two cases of experimental data for the uniform flow. The results show that LES captures correctly the boundary layer separation and mean pressure distribution prediction at high Reynolds number of 10^5.

2 Previous work

Flow surrounding a circular cylinder has been studied by researchers for its flow phenomena and engineering applications in designing jacket legs, mooring lines of spar platforms, Floating production system and operations (FPSO), pipelines etc. Recent studies on flow around circular cylinder in uniform flow at high Reynolds number are reported by [4 - 12]. Due to the nature of many issues, there still remain many aspects of the flow that need further investigation [10]. LES is a best method available currently for determining turbulent flow characteristics. There are some advantages using fixed rectangular grids which present flow in natural and complex areas as compared to unstructured grid (boundary) following a mesh system or mesh free method. Conventional Computational Fluid Dynamics (CFD) method of RANS or advanced methods such as DNS or LES are based on basic governing equation of Fluid Dynamics. Incompressible viscous flow of Newtonian fluid is considered in this study.

2.1 Large Eddy Simulation (LES)

In this method large eddies are computed (resolved) and smallest sub-gird-scale eddies are model[11]. This is based on Kolmogorov's theory on turbulence which makes the assumption that large eddies of the flow are dependent on the flow geometry, whereas smaller eddies are self-similar and have a universal character. Large eddies are solved explicitly and effect of smaller eddies is modelled on large eddies. In this model large scale motions of the flow are evaluated similar to DNS method and effect of smaller universal scales on sub-grid scale model are modified. In this method it can be assumed that large scale is solved using DNS and small scale as RANS method.

Boundary conditions are very important in RANS and LES methods for all applications whereas initial conditions have secondary importance in these methods. This condition plays significant role in wall bounded flows of higher order quality but it comes with high computational costs. Wall boundary conditions are used in fluid flows of practical applications. The wall changes the fluid dynamics in an important manner this is due to the fact that in the near wall region, the sharp gradient for flow velocity will create large shear stresses as compared to far away regions from wall boundary. This is due to large viscosity stresses in the near wall region. Near wall flow under turbulence loading is still under study because study in this region involves a rapid change of turbulent time and length scales. CFD modelling requires wall boundary conditions to be properly defined for flow.

2.2 Computational Model

In this case an offshore oil and gas pipeline is selected for the analysis of LES model. LES model can be significantly The LES model for offshore pipeline has been discussed by some of the authors [1,12]. The LES model for circular cylinder general has been performed by [13,14]. Spatial and time wise scale of flow are very large and spacing of grid points and time step are made possible to resolve dissipation scales by dividing flow region into cubes. The order of cube is smaller in length for large scale flow and a fraction of turbulent fluctuation can be resolved to get mean flow and turbulent stresses. Filtered flow of this scale can represent overall characteristics of simulated flow.

2.2.1 Governing Equations

Filtered equations of motion and continuity for spatially filtered velocity components (u,v,w) and pressure p are shown in Equation 1,

$$\frac{\partial u}{\partial t}+\frac{\partial u^2}{\partial x}+\frac{\partial uv}{\partial y}+\frac{\partial uw}{\partial z}=-\frac{1}{\rho}\frac{\partial p}{\partial x}+\frac{\partial}{\partial x}\left(v\frac{\partial u}{\partial x}+\tau_{xx}\right)+\frac{\partial}{\partial y}\left(v\frac{\partial u}{\partial y}+\tau_{yx}\right)+\frac{\partial}{\partial z}\left(v\frac{\partial u}{\partial z}+\tau_{zx}\right)$$

$$\frac{\partial v}{\partial t}+\frac{\partial uv}{\partial x}+\frac{\partial v^2}{\partial y}+\frac{\partial vw}{\partial z}=-\frac{1}{\rho}\frac{\partial p}{\partial y}+\frac{\partial}{\partial x}\left(v\frac{\partial v}{\partial x}+\tau_{xy}\right)+\frac{\partial}{\partial y}\left(v\frac{\partial v}{\partial y}+\tau_{yy}\right)+\frac{\partial}{\partial z}\left(v\frac{\partial v}{\partial z}+\tau_{zy}\right)$$

$$\frac{\partial w}{\partial t}+\frac{\partial uw}{\partial x}+\frac{\partial vw}{\partial y}+\frac{\partial w^2}{\partial z}=-\frac{1}{\rho}\frac{\partial p}{\partial z}-g+\frac{\partial}{\partial x}\left(v\frac{\partial w}{\partial x}+\tau_{xz}\right)+\frac{\partial}{\partial y}\left(v\frac{\partial w}{\partial y}+\tau_{yz}\right)+\frac{\partial}{\partial z}\left(v\frac{\partial w}{\partial z}+\tau_{zz}\right) \quad (1)$$

Where,

x, y, z	=	Cartesian coordinates with *z* positive vertically upward
g	=	Gravitational acceleration
τ_{xy}, etc	=	Sub-grid scale stresses
ρ	=	Density of water
v	=	Kinematic viscosity of water

2.2.2 Subgrid and Wall model

The solution of equation of motion and sub-grid scale effects are very important specially flow near solid surfaces like wall region. Near wall region creates turbulence which gives resistance to flow, the bursting eddies in this region are largely responsible for this large resistance. The standard Smagorinsky eddy viscosity is used for the sub-grid stresses within the flow away from solid boundaries. Wall model is formulated such that it relates near wall flow velocity to wall shear stress on local and instantaneous conditions. The model which is used should reduce the sub-grid eddy viscosity or dynamically determined eddy viscosity coefficient should be used. The Cartesian shear stress components on the wall are shown in terms of velocity in Equation 2,

$$\tau_{xy}=C_d\rho V_1 v_1, \quad \tau_{xz}=C_d\rho V_1 w_1$$
$$\tau_{yx}=C_d\rho V_1 u_1, \quad \tau_{yz}=C_d\rho V_1 w_1$$
$$\tau_{zx}=C_d\rho V_1 u_1, \quad \tau_{yz}=C_d\rho V_1 v_1 \quad (2)$$

Where,

V_1		$=\sqrt{{u_1}^2+{v_1}^2+{w_1}^2}$
C_d	=	Model resistance coefficient
u_1, v_1, w_1	=	Velocity components at $z = z_1$

The symmetry conditions are required for stresses within flow which is specific for flow near solid wall and Equation 2 does not satisfy it. Thus here mean velocity and mean wall stress equations are used,

$$C_d = \left[\frac{A\ln(z_1 u_\tau)}{v} + B\right]^{-2}, \frac{k_s u_\tau}{v} \le 100\ and\ \frac{z_1 u_\tau}{v} > 10 \tag{3}$$

Where,

A, B, C	=	Model coefficient
z_1	=	Distance from wall
u_τ	=	Friction velocity
k_s	=	Roughness height

$$\left[\frac{z_1 u_\tau}{v}\right]^{-2}, \frac{k_s u_\tau}{v} \le 100, and\ \frac{z_1 u_\tau}{v} \le 10 \tag{4}$$

$$C_d = \left[A\ ln\frac{z_1}{k} + C\right]^{-2}, \quad \frac{k_s u_\tau}{v} > 100 \tag{5}$$

where A=2.5, B=5.2 ,C=8.5.

Equations (2-5) are solved iteratively as a "wall function method" used in two equation turbulence models.

3 Computation and results

As an example of application of the method described above, calculation of the flow past a circular cylinder is presented here. In addition to the basic flow past a cylinder in a uniform flow perpendicular to the cylinder axis of infinite extent shown in Fig.1 (a), the flow when the cylinder is placed near a solid surface as shown in Fig.1 (b) such as the case of a pipeline installed near the seabed, has been calculated. The latter is a subject of recent research in offshore engineering and calculations using RANS method has been conducted by [15] which can be compared to the present calculation method.

The region of calculation extends 8 cylinder diameters D upstream, 16D downstream, 8D above and below the cylinder normal to the flow direction and 2D in the direction of the cylinder axis for the case of uniform flow. This calculation region is resolved by the rectangular mesh of 127 x 95 x 16 grids with varying spacing made smallest near the cylinder. The minimum spacing is 0.625D so there are 16 grid points across the diameter of the cylinder. For the case of the cylinder near sea bed, the gap G between the cylinder and the seabed is 0.35D as [15] has showed that the effects of the wall become significant. In this case the boundary layer on the seabed develops from the beginning of the calculation region, since a uniform flow is assumed at the upstream end of the calculation region. This corresponds to a case where the pipeline is installed in an area where there is a mound of some height rather than on a flat seabed in open sea, in which case the bottom boundary layer would have developed to the full depth of the sea. The grid spacing in the vertical direction near the bed is taken small but the total number of the grid points is almost the same. The Reynolds number based on D for both calculations is 105 so that the flow is supercritical and the boundary layer on the cylinder becomes turbulent before separating. Fig. 2 shows the calculated flow fields at an instant t=60 $d/U0$ at which time the flow assumed uniform at initial t = 0 has fully developed. The figures show only part of the flow near the cylinder so the details can be seen. The symmetric vortex shedding pattern is seen in the basic uniform flow case without a wall. In the second case where a solid wall exists, the flow appears as a mixture of the wake behind a cylinder and the flow over an object on a surface. The third case deal with cylinder is resting on bed and flow is passing above the object. There are undulating wake in the area downstream of about 5D but in the region closer to the cylinder, the overall flow is retarded more and the stagnation point on the upstream surface of the cylinder is lower than the center of the cylinder. The flow in the gap is accelerated but for this size of the gap the decrease of the pressure is not significant and overall an upward lift is generated. This average flow effects and the reduced intensity of the vortex shedding appear the main influence of the bed.

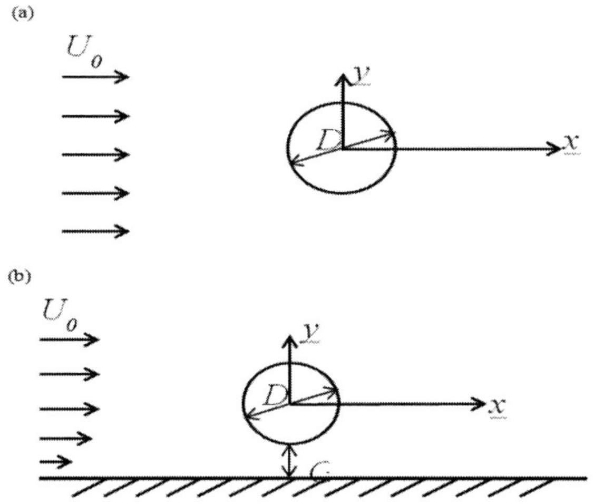

Fig.1. Sample calculation cases, flow past a circular cylinder in uniform flow and near a solid wall.

3.1 Hydrodynamic forces

There are a number of hydrodynamic effects that may become important in the design and installation of a pipeline. From the present numerical analysis, we have evaluated the time averaged and fluctuating pressure distributions around the circular cylinder for the three cases of the relative position from the seabed. Fig. 3 shows the distribution of the mean pressure coefficient $C_p=(P-P_0)/(\rho/2)U_0^2$as the function of the azimuthal angle θdefined by the inset in Fig. 3(b). Where G is the gap between the surface of circular cylinder and the plane boundary and D is the diameter of pipe P is the mean pressure and P0 is its value in the undisturbed flow away from the cylinder. The case in the uniform flow without a seabed is the well-studied case and the present results which is done for the Reynolds number typical of the real situation of a pipe line of $1.0x10^5$ are compared with the available experimental results at similar Reynolds number and that at lower Reynolds number of 3900.

The present results follow the experiments on the front surface up to about θ=60° and are between the two experiments. The point where the pressure becomes constant is repeated to the boundary-layer separation. The present results imply the separation a little upstream of the experimental point but the results are generally acceptable. Fig. 3(b) and Fig. 3(c) show the pressure distributions when the cylinder is placed closed to a seabed and right on it. It is seen that the pressure on the upper side of the cylinder decreases and the lower side increases as the pipe is placed near a seabed. On the other hand the pressure on the downstream half of the cylinder surface decreases due to the acceleration through the gap between the cylinder and the seabed. These wall proximity effects imply increased upward force and overturning clockwise moment. When the cylinder is placed on the seabed, the gap is completely closed and no acceleration and pressure decrease occur and the up-lifting force acts mostly on the upstream half of the pipe. These dramatic changes in the pressure distribution and the resulting force imply close examination of the flow effect should be made. The fluctuating pressure distributions are shown in Fig. 4 in terms of the root-mean-square C_{prms} of the fluctuating pressure coefficient. The fluctuation level is seen to increase as the cylinder is placed near the seabed. The present LES method can be used to examine these important hydrodynamic effects. The above example is only one application of the present simulation method and can be used in various other situations of the flow past objects.

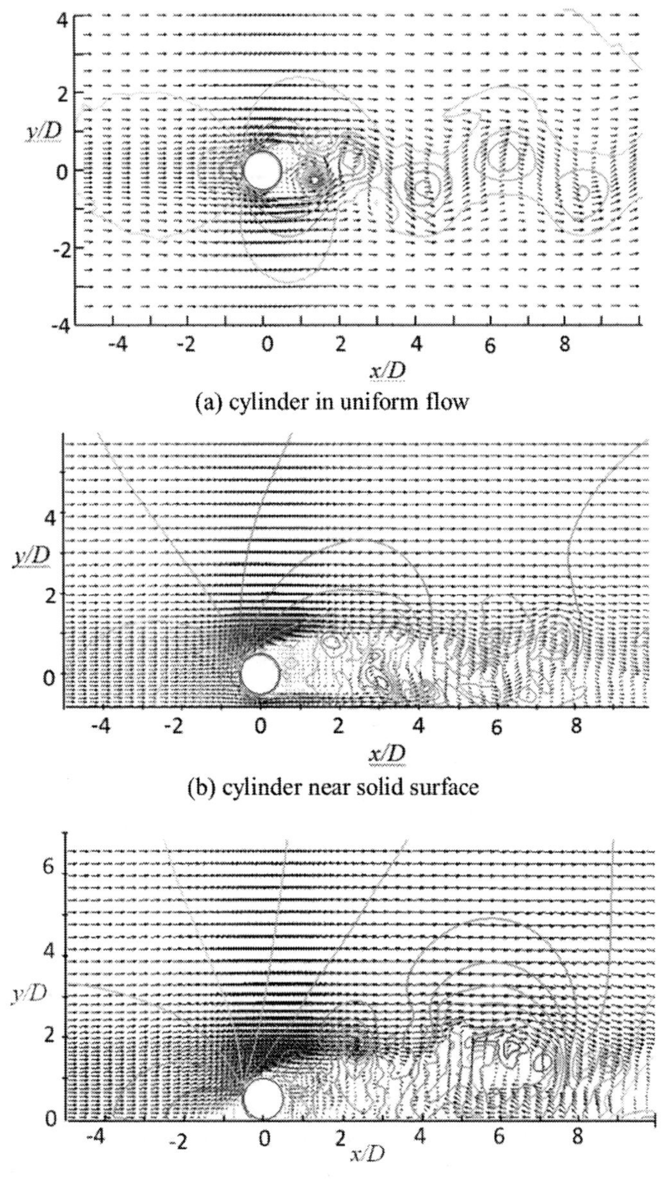

(a) cylinder in uniform flow

(b) cylinder near solid surface

(c) Cylinder at the mud level

Fig. 2. Simulation of offshore pipeline

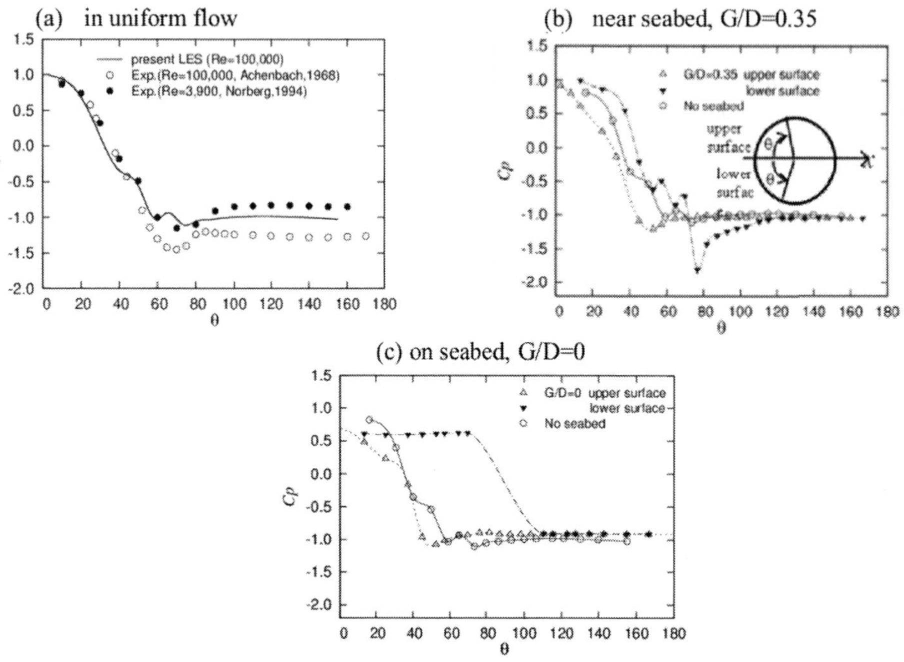

Fig.3. Mean pressure distributions on circular cylinderplaced away from, near and on the seabed

4 Conclusions

The results show that the effects of the seabed on the flow characteristics and the loads on the pipe are significant. In the case of without a wall symmetric vortex shedding pattern is seen in the basic uniform flow. The case where a solid wall exists, the flow appears as a mixture of the wake behind a cylinder and the flow over an object on a surface. There are undulating wake in the area downstream of about 5D but in the region closer to the cylinder, the overall flow is retarded more and the stagnation point on the upstream surface of the cylinder is lower than the center of the cylinder.

The flow in the gap is accelerated but for this size of the gap the decrease of the pressure is not significant and overall an upward lift is generated. This average flow effects and the reduced intensity of the vortex shedding appear the main influence of the bed.The results follow the experimental results up to θ= 60° and remain between two experimental results. The study shows separation a little upstream of the experimental point but the results are generally acceptable. When the pipe is placed near a seabed pressure on the upper side of the cylinder decreases and the lower side increases. Pressure decreases on the downstream half of the cylinder due to the acceleration through the gap. The present LES method can be used to examine these important hydrodynamic effects.It was expected that the vortex shedding is weakened since the wall or the seabed makes the wake flow non-symmetric and the generation of the vortices of alternating directions is naturally reduced. This is what is seen from the present results. More significant result which was not anticipated is that the pressure distribution on the cylinder changes significantly and suddenly as the distance to the wall is reduced. When the gap between the cylinder and the wall is larger than 3.5D, the pressure distribution is not much different from that in the uniform flow without walls, with slight decrease on the upper side and increase on the lower side causing an upward lift force. As the gap is narrowed from 3.5D to zero, the pressure on the front half on the lower side of the cylinder becomes very large, as large as stagnation pressure. It creates large lift and drag forces. This sudden change in the loads is very important to consider in the general design of pipelines. It is like a sudden large force an object experiences when it plugs a flow near intake or a drain, and can be very dangerous unless a precaution is taken to avoid a sudden approach to a wall.

References

[1] M. C. Ong, T. Utnes, L. E. Holmedal, D. Myrhaug, and B. Pettersen, Numerical simulation of flow around a smooth circular cylinder at very high Reynolds numbers. Marine Structures, **22**,142-153, (2009)

[2] M. Breuer, Large eddy simulation of the subcritical flow past a circular cylinder: numerical and modeling aspects. Int. J. for Numerical Methods in Fluids, **28**,1281-1302, (1998)

[3] C. Norberg, Experimental investigation of the flow around a circular cylinder: influence of aspect ratio. J. Fluid Mech , **258**, 287-316, (1994)

[4] P. Catalano, M.Wang, G. Iaccarino and P. Moin, Numerical simulation of the flow around a circular cylinder at high Reynolds numbers. Int. J. of Heat and Fluid Flow, **24**,463-469, (2003)

[5] M.C. Ong, T. Utnes, L.E. Holmedal, , D. Myrhaugand B. Pettersen, , Numerical simulation of flow around a circular cylinder close to a flat seabed at high Reynolds numbers using a k–ε model. Coastal Engineering, **57**, 931-947, (2010)

[6] J. Johansson, M.P. Nielsen andL.O. Nielsen, Uniform flow around a circular cylinder in the subcritical range - using the self-induced angular moment method turbulence model. *The 8th Asia-Pacific Conference on Wind Engineering*, Chennai, India, (2013)

[7] F.Tremblay, M. Manhart, and R. Friedrich, *LES of Flow around a Circular Cylinder at a Subcritical Reynolds Number with Cartesian Grids*. In: FRIEDRICH, R. & RODI, W. (eds.) Advances in LES of Complex Flows. Springer Netherlands, (2002)

[8] E. Achenbach, Distribution of local pressure and skin friction around a circular cylinder in cross-flow up to Re=5x106. J. Fluid Mech., **34**, 625-639, (1968)

[9] S. Li, B.W. Karney and G. Liu, FSI research in pipeline systems – A review of the literature. J. of Fluids and Structures, **57**,277-297, (2015)

[10] W. Lin, C. Lin, S. Hsieh and S. Dey, Flow Characteristics around a Circular Cylinder Placed Horizontally above a Plane Boundary. J.of Engineering. Mechanics, **135**, 697-716, (2009)

[11] A. Nakayama and S.N. Vengadesan, On the influence of numerical schemes and subgrid–stress models on large eddy simulation of turbulent flow past a square cylinder. Int. J. for Numeric. Methods in Fluids, **38**, 227–253, (2002)

[12] P. M. Abrahamsen, M. C. Ong, B. Pettersen and D. Myrhaug, Large Eddy Simulations of flow around a smooth circular cylinder in a uniform current in the subcritical flow regime. Ocean Engineering, **77**,61-73, (2014)

[13] A. Kulyakhtin, O. Shipilova and M. Muskulus, Numerical simulation of droplet impingement and flow around a cylinder using RANS and LES models. J. of Fluids and Structures, **48**, 280-294, (2014)

[14] I. Rodríguez , O. Lehmkuhl, J. Chiva, R. Borrell, and A. Oliva, On the flow past a circular cylinder from critical to super-critical Reynolds numbers: Wake topology and vortex shedding. Int. J. of Heat and Fluid Flow, (2015)

[15] D. Liang and L. Cheng, Numerical modeling of flow and scour below a pipeline in currents: Part I. Flow simulation. Coastal Engineering, **52**, 25-42, (2005)

Solar Trap for Banana Drying Method

S. Musa[1,*], *I.* Ibrahim[2] , *A. S.* Mohd[1], *M. M.* Shahabuddin[1] and *S.* Sulaiman[3]

[1]Faculty of Civil and Environmental Engineering, Universiti Tun Hussien Onn Malaysia,86400, Batu Pahat, Johor, Malaysia
[2]Faculty of Engineering Technology, Universiti Tun Hussien Onn Malaysia,86400, Batu Pahat, Johor, Malaysia
[3]Jabatan Kerja Raya Kluang, 86000 Kluang, Johor, Malaysia

Abstract. Food drying methods nowadays are mostly in high use of electricity and fuel which lead to high operational cost. This has resulted in a waste of energy and money due to the use of modern tools requires significant costs for implementation. Meanwhile, the traditional food drying process only uses sun rays in their process, where the process is far more efficient than the modern drying method. In this study, the test was conducted to determine the trapped solar heat energy requirements for the process of drying foods such as agricultural products, particularly bananas. The solar trap test by using solar trap container was carried out include determining the thermal energy requirement for drying, preparing equipment (solar trap container) to trap solar energy, handling and drying tests on samples of bananas. The percentage amount of water removal and energy required for the drying process was found to be 48% and 134 J. The results of this study can determine that solar trap drying method is easier, quicker and more effective than the usual method of drying because it use natural solar energy. Several proposals have been suggested for improvement for future study, such as controlling the solar trap air in the container, replacing the trap solar wall with a darker color, examining the floors slope so that more solar traps collected and installing a small hose on the base of the container so that the water evaporated in the solar trap may exit through the route.

1 Introduction

Solar energy is a technology to obtain useful energy from sunlight. Solar energy has been used in many traditional technologies for centuries and has been widely used in the absence of other energy supplies such as in remote areas and in space. Solar energy is now used in several applications such as hot water heating, heating of buildings and food. It is also used for electricity generation like fotovolik, engine heating and desalination of sea water. Its use becomes famous when awareness of the environmental costs and limited supplies of other energy sources such as fossil fuels become more pronounced [1].

* Corresponding author: sabariah@uthm.edu.my

Solar energy is very important to our daily lives. The use of solar energy can also save electricity consumption especially when other resources are lacking or absent. Therefore, various methods have been developed to expand the use of solar energy, especially in the method of drying ood and agricultural products. Among them is the decisive requirement of thermal energy for drying. Then, setting up equipment to trap solar energy, operating and carrying out tests and taking measurements of the drying removal of water in the food and amount of energy required [2].

Parit Raja, Batu Pahat, Johor is a very famous village as a result of making dried banana which only uses sunlight as a means of drying. This study is to identify the quantity of solar energy derived from the sun and collected in an air tight confine vessel that are used for drying of agricultural products. This study was conducted to test the effectiveness of using solar traps drying method than the usual drying methods without the use of solar trapping. The scope of this study is to determine the thermal energy needed for drying purposes, preparing for solar trapping equipment, conducting and testing solar traps and also measuring the removal of water from the food and energy needs.

2 Related Studies

Based on study [3], it found that evaporation is the conversion of water from liquid to gas. About 80% of all evaporation is from the ocean, with the remaining 20% coming from inland water and vegetation. Wind carries evaporated water over the world, affecting air humidity around the world.

However, heat (energy) is necessary for evaporation to occur. Energy is used to break the bonds that hold water molecules together, which is why water easily evaporates at the boiling point of (212°F, 100°C) but evaporates more slowly at the freezing point. Net evaporation occurs when the rate of evaporation exceeds the rate of condensation. Saturation exists when these two processes rates are equal, at which point the relative humidity is 100%. Condensation, the opposite of evaporation, occurs when saturated air is cooled below the dew point (the temperature at which air must be cooled at constant pressure to become fully saturated with water), such as the outside of a glass of ice water. In fact, the process of evaporation removes heat from the environment, which causes water to evaporate from your skin and becomes cool [4].

However, the drying process is divided into three categories: the imposition of air and air under atmospheric pressure, vacuum drying and freeze drying. In the air and the imposition of air under atmospheric pressure, the heat is transferred from the food through the hot air from hot or dry surfaces. Water vapour is removed through the air. In the vacuum drying, the evaporation of water occurs more easily at lower pressure compared when the pressure is higher. Heat transfer in vacuum drying generally occurs by conduction, sometimes through radiation. During freeze-drying, water vapour sublimed from the frozen food. Food structure is better maintained in this condition. Suitable temperature and pressure must be established in the dryer to ensure that the sublimation is accepted [5].

3 Materials and Methods

3.1 Determining thermal heat requirements for drying purpose

Determination of thermal energy is required for drying food stuffs and agricultural products to ensure that the energy require disappropriate and sufficient. The study of solar energy resources has been conducted and it was found that Malaysia is a country that receives

DOI: 10.1051/matecconf/201710304006

sunshine in a very long period of over 10 hours a day. In addition, the study of materials that was used to create a heat trap in the container.

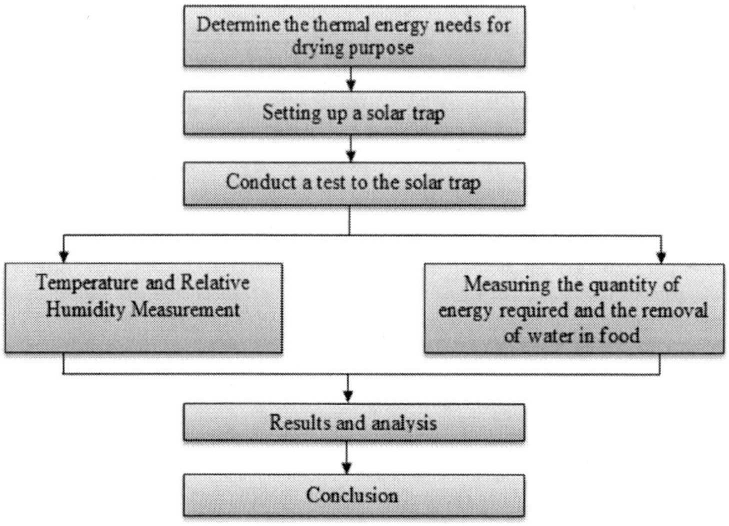

Fig. 1. Flow of banana drying process

3.2 Setting up a solar trap

A solar trap container as shown in Fig. 2 was prepared after a study was carried out to determine the thermal energy needs by using thermometer investigation. A container with the size of 2250 cm^3 heat trap was created for drying food stuffs and agricultural products. Among the features found on the solar trap are:

1. The surface of the transparent glass on the upper surface of the solar trap.
2. Insulation white polystyrene containers found in solar panels on each wall of the solar trap.
3. The surface of the solar trap coated with a layer of black water translucent paper.
4. There is a thermometer on the mirror surface on the inside of the solar trap.

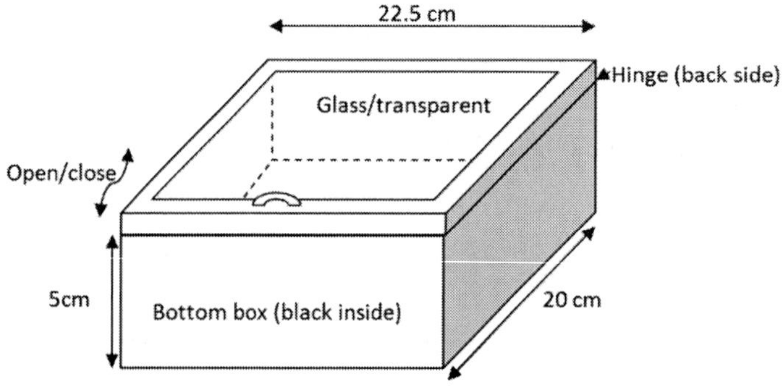

Fig. 2. Solar trap container for banana drying

3.3 Conducting solar trap testing

The operation and testing of solar trap was performed after the completion of the solar trap container. This study was conducted approximately in less than two months to measure temperature, relative humidity, the amount of energy required and also the removal of water in food depending on weather conditions in the surrounding area. This study should be done during the day as at night there is no sunlight.

3.4 Temperature and relative humidity measurement

Measurement of the level of relative humidity in solartrap container was done manually by taking the wet bulb and dry bulb temperature data every one hour using a thermometer. The study was conducted for a week depending on weather conditions. Meanwhile, the solartrap container is placed in open spaces necessarily far from the shadow of trees or buildings thatcan block sunlight from penetrating the solar trap container. This study was conducted to determine that the temperature and relative humidity in the solar trap is not influenced by ambient temperature. This comparison was done to prove that the temperature in the solar trap is higher than the ambient temperature as well as more effective for drying foods such as dried bananas.

3.5 Measurement quantity of energy required and water removal in food

Measurement of water removal in food and the energy needed was carried out during the testing of the solar trap. This measurement is done by collecting data in the solar trap to examine the quantity of solar energy from the sun that can be collected by a solar trap container for drying agricultural products purpose.

Energy measurement is also carried out by using ordinary measuring temperatures by taking into account the initial temperature and the temperature after drying for a day was applied to food stuffs. Meanwhile, water removal measurement was done with regard to the original quantity of water and quantity of water after the drying process on food stuffs using a syringe or weighing machine.

Water removal is done by collecting data of the specimen's dryness to test the dryness rate in solartrap. Then, the specimen was replaced with food such as bananas after the dryness test was successful. Next, the solar trap drying method is compared to the usual method without solar trap drying.

4 Results

4.1 Relative humidity and temperature in solar trap

Data of dry bulb temperature and relative humidity in the solar trap that has been recorded and analyzed based on the relationship for both parameters. The process has the potential to dry foods such as bananas as shown in Fig. 3. This method is intended to ensure that the relative humidity affect the temperature of the solar trap to prove the effectiveness of the drying process. If relative humidity is lower, the temperature is higher because of expansion process. The Equation 1 was generated as

$$RHs = 103.02 - 0.0017Ts^2 - 0.6808Ts \quad (1)$$

where;
RHs = Relative humidity of solar trap (%)
Ts = Temparature of solartrap (°C)

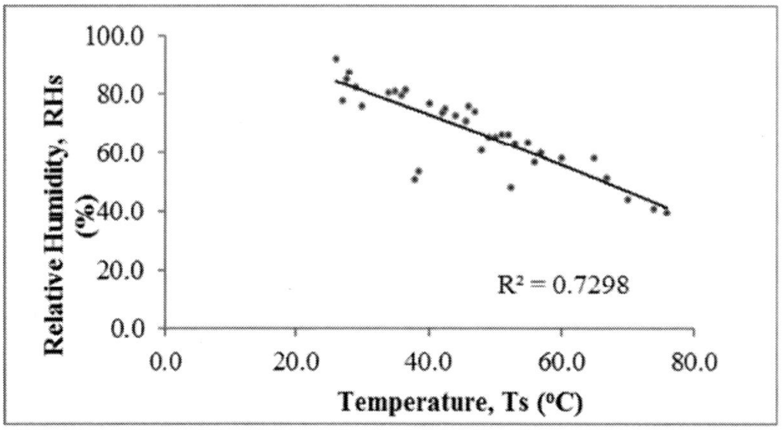

Fig. 3. Humidity and temperature in solartrap

4.2 Relative humidity solar trap and ambient

Fig. 4 shows the percentage of relative humidity of solar trap and ambient relative humidity more clearly. Ambient relative humidity percentages recorded the highest results compared to solartrap. This provesthat the former solar trap has a higherpotential and is very suitable for drying foods such as bananas than the usual drying methods due to being independent from the surrounding area.

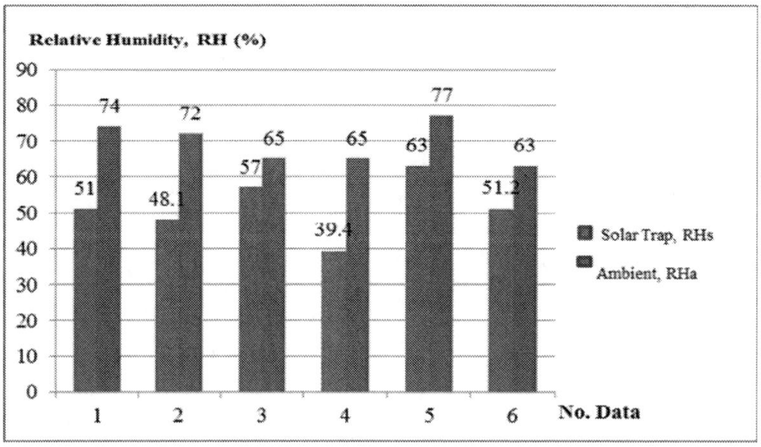

Fig. 4. Percentage reduction of relative humidity between solar trap and ambient

4.3 Temperature between solar trap and ambient

The variations in the temperature between the ambient and solar trap was shown in Fig. 5. This comparison was done to prove that the temperature in the solar trap is higher than the ambient temperature as well as more effective for drying foods such as dried bananas. This is due to several factor including ambient temperature of hot and cold air, the relative humidity was variable, wind speed and angle of sunlight either it is protected or unprotected. The relationship for each others was produced as Equation 2.

$$Ts = 3.4282Ta - 53.611 \qquad (2)$$

where;
Ts = Solar trap temperature (°C)
Ta = Ambient Temperature (°C)

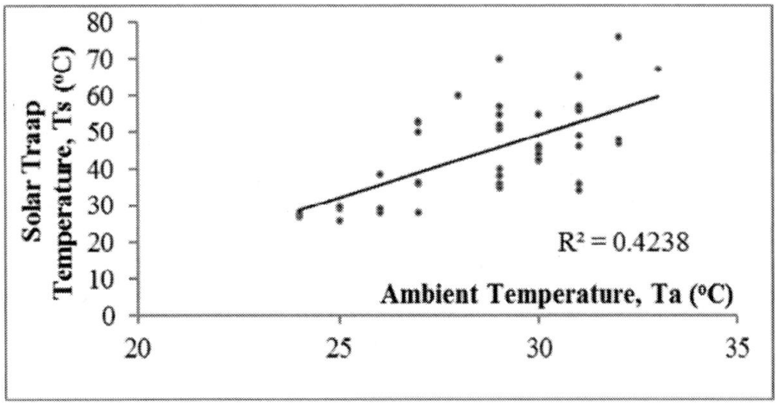

Fig. 5. Variation temperature in solar trap and ambient

4.4 Energy required in solar trap

The highest value for the average solar energy trapped in the solar trap is 31015.1 J/m^3, while the maximum energy that is trapped by solar traps was 49019.4 J/m^3. The average temperature change was considered was 16°C. This is due to weather conditions in Malaysia that are erratic and the temperature conditions are constantly changing according to the surrounding area. In conclusion, the average energy trapped in the trap solar is high and enough to dry foods such as dried bananas only one day. Table 1 shows the average energy trapped in the solar trap.

Table 1. Average energy trapped in solar trap

Date	Average solar trap temperature, Ts_{ave} (°C)	Average Ambient Temperature, Ta_{ave} (°C)	Average air density, p_{ave} (kg/m^3)	Average air mass, m (kg)	Change in temperature, ΔT (°C)	Energy trapped, E_{ave1} (J)	Energy trapped in air, E_{ave2} (J/m^3)
19/3/2015	36	30	1.106	0.0025	6	15.5	6907.3
1/4/2015	38	28	1.106	0.0025	11	26.9	11976.3
2/4/2015	45	28	1.106	0.0025	18	44.5	19774.9
3/4/2015	44	29	1.106	0.0025	15	37.1	16472.4
4/4/2015	56	28	1.106	0.0025	28	69.8	31015.1
5/4/2015	41	27	1.106	0.0025	14	35.1	15597.1
6/4/2015	48	30	1.106	0.0025	18	45.0	19983.8
				Average, ΔT_{ave}	16		

4.5 Dryness process

One dried banana only needs 134 J energy of drying process for 9 hours a day. While the average solar energy trapped in the trap is 31015.1 J/m^3. Approximately 200 to 300 bananas can be dried within one day compared to the normal drying, which takes three days at least. The data comparison between the sample and bananas are that bananas require lower energy and water removal rate for the drying process compared to the samples used within a day (Table 2). Therefore, the process of assessing the quantity of energy from the sun collected in a container for solar traps drying of agricultural products such as dried bananas had been achieved.

Table 2. Comparison between sample and banana

Comparison	Sample	Banana
Rate of water removal, MC (%)	60	48
Quantity of water removed (ml)	3.0	2.5
Mass of water removed, m (kg)	0.003	0.0035
Specific heat capacity, C_p (KJ/kg °C)	4.179	3.35
Energy required, E (J)	201	134

5 Conclusion

Based on the study that was conducted, dried banana needs only 134 J energy and drying process for 8 to 10 hours a day, while the average solar energy trapped in the solar trap is 31015.1 J/m^3. Therefore, the energy accumulated in the solar trap is sufficient to dry the banana. It can be estimated that about 200 to 300 bananas can be dried within one day compared to the usual drying methods which consume at least three days for the drying process.

This proves that the solar trap drying method is time, costs and energy efficient compared to drying methods without solar trap. It caused by the drying time and characteristics a drying place of foodstuffs . If bananas can be dried in large quantities in only one day, then this saves cost and time while using energy more efficiently and increasing banana production by 300%.

In conclusion, this study can provide convenience to the surrounding communities to perform an easier, faster and more effective method of drying food. In addition, drying using solar trap is able to maintain the quality of bananas from insect and animal intrusion because food are typically exposed to the environment during the drying process without the use of solar traps.

References

[1] S. Zekai, *Solar energy fundamentals and modeling techniques solar energy devices and collectors*, Springer, 239-252 (2008)

[2] P.K. Jauhri, *Sumber tenaga suria*, Encyclopaedia of resource geography, Anmol Publications PVT. LTD, New Delhi, (2005)

[3] Department of atmospheric sciences (DAS), *Urbana champaign,* University of Illinoi, (2010)

[4] H. Perlman, *The world's water*, (2014), Retrieved on Mar 6, 2014 from http://water.usgs.gov/edu/earthwherewater.html

DOI: 10.1051/matecconf/201710304006

[5] R. L. Earle and M. D. Earle, *Drying*, Unit operation in food processing, Chap. 7, The New Zealand institute of food science and technology Inc. (NZIFST), (1983)

Forecasting of Groundwater Level using Artificial Neural Network by incorporating river recharge and river bank infiltration

Mohd Khairul Nizar Shamsuddin [1,*], *Faradiella* Mohd Kusin[2], *Wan Nor Azmin* Sulaiman[2], *Mohammad Firuz* Ramli[2], *Mohamad Faizal* Tajul Baharuddin[3], *Mohd Shalahuddin* Adnan[3]

[1]Hydrogeology Research Centre, National Hydraulic Research Institute of Malaysia, Lot 5377, Jalan Putra Permai, 43300 Seri Kembangan, Selangor, Malaysia
[2]Faculty of Environmental Studies, Universiti Putra Malaysia, 43400 Serdang, Selangor, Malaysia
[3]Department of Environmental and Water Resources, Faculty of Civil and Environmental Engineering, Universiti Tun Hussein Onn Malaysia, Johor

Abstract. Groundwater tables forecasting during implemented river bank infiltration (RBI) method is important to identify adequate storage of groundwater aquifer for water supply purposes. This study illustrates the development and application of artificial neural networks (ANNs) to predict groundwater tables in two vertical wells located in confined aquifer adjacent to the Langat River. ANN model was used in this study is based on the long period forecasting of daily groundwater tables. ANN models were carried out to predict groundwater tables for 1 day ahead at two different geological materials. The input to the ANN models consider of daily rainfall, river stage, water level, stream flow rate, temperature and groundwater level. Two different type of ANNs structure were used to predict the fluctuation of groundwater tables and compared the best forecasting values. The performance of different models structure of the ANN is used to identify the fluctuation of the groundwater table and provide acceptable predictions. Dynamics prediction and time series of the system can be implemented in two possible ways of modelling. The coefficient correlation (R), Mean Square Error (MSE), Root Mean Square Error (RMSE) and coefficient determination (R^2) were chosen as the selection criteria of the best model. The statistical values for DW1 are 0.8649, 0.0356, 0.01, and 0.748 respectively. While for DW2 the statistical values are 0.7392, 0.0781, 0.0139, and 0.546 respectively. Based on these results, it clearly shows that accurate predictions can be achieved with time series 1-day ahead of forecasting groundwater table and the interaction between river and aquifer can be examine. The findings of the study can be used to assist policy marker to manage groundwater resources by using RBI method.

[*]Corresponding author: nizar@nahrim.gov.my

1 Introduction

River Bank Infiltration (RBI) is a natural filter process to improve the drinking water quality obtained from surface water flow through aquifer and mixture with groundwater. Surface waters are infiltrated through the aquifer media in the pumping wells during pumping activities and subsequently influence river and groundwater level [1-7] stated that river and aquifer interactions are exist when waters from vertical and horizontal wells that are located in alluvial adjacent to rivers/lakes are pumped. It must be managed in an integrated way to provide efficient water supplies and also concerns over the conservation of the natural environment. Changing of river water levels and groundwater level will effect flow and contaminant transport through infiltration in aquifer media [1]. The dynamic fluctuation of the river water levels during pumping activities does not only influenced by soil sediments clogging of the riverbed but also other factors such as climate, hydrogeological properties and pumping rates [5-6]. Therefore, groundwater tables monitoring is essential to be conducted for understanding river–aquifer interactions. Groundwater aquifer located closed to river have direct influence on river water level. At distances more than a few hundred meters from the river, the groundwater table depth is generally several meters higher than the river water level and subsequently does not effect to river water levels. In addition, groundwater tables are also not affected immediately by river water depth increase due to very low groundwater velocity. On the other hand, the other factors such as precipitation and stream flow can influence quickly the river water depth.

Network design and monitoring of groundwater tables depends on hydrogeological conditions and available logistical resources. Although mathematical and conceptual models are the main tools for representing hydrological variables and understanding the physical processes in a system but it has practical limitations [8]. When the data is not sufficient and accurate prediction is required, artificial neural network (ANN) models can be a good option [8]. The ability of ANNs to identify the relationship of hydrological variables patterns makes ANNs sufficiently to solve complex hydrologic problems [9]. Recently, ANNs have been successfully applied for identifying the temporal data to calculate groundwater level [10].[11-15] has developed an ANN model for hydrology and hydrogeology.

The aim of this study is to look of the applicability of ANNs in forecasting groundwater tables in a site at Jenderam Hilir, Selangor, Malaysia. The Jenderam Hilir is where a pilot project to develop a better understanding of sustainable water resources, and introduce RBI in Malaysia is located. This site was chosen due to the high water demand in the area and groundwater is seen as one of the source with very high potential to be developed as supplementary sources to meet the high public water supply demand. The applicability of the methodology is demonstrated when using ANN training algorithms namely Lavenberg-Marquardt (LM) algorithm for predicting groundwater tables in a Jenderam Hilir located in the tropical humid countries. The major focus of the current study is to investigate the potential of ANN approach in groundwater modeling focus on water level fluctuation and to predict the relationship between river and aquifer of the study area.

2 Materials and methods

2.1 Study area

The study area is located in a flat area of Langat River Basin at the confluence of Semenyih Rivers, Langat Rivers and a small Jenderam Hilir River flow to Langat River in the area.

These two rivers are the main source of raw water for Selangor state. This area is a former tin mining area in which there are three major ponds (ponds A, B and C) which are interconnected (Fig. 1). These three ponds function as storage to increase the capacity of raw water at the water intake. The study area is bounded by the hilly areas in the north and east, and by the sea in the southwest. The maximum elevation is 297 m and the minimum is 7.981 m, above mean sea level. The exact location is between latitude 2^0 53' 28.56" N and 2^0 53' 39.75" N, and longitude 101^0 42' 03.78" E and 101^0 44' 14.58"E, covering an area of 10km2. The daily average temperatures vary from 27-30^{0}C. This area is experiencing two monsoon periods; Northeast monsoon from October to January (wet season) and the Southwest monsoon from May to September (dry season). The annual precipitation in 2015 is 2,013 mm and 70-80% of annual precipitation is concentrated in November and December. The mean stream flow is 47.13 m^3/s and the highest river water level is about 5.64 m. The daily average of rainfall, stream flow, temperature, river water level were obtained from the Drainage Irrigation Department, Malaysia (DID) and groundwater levels were measured from the two pumping test wells, DW1 and DW2. The data (rainfall, stream flow, temperature, river water level, groundwater level) used for this study are from February 2015 to March 2016.

2.2 Design of ANN, Training Algorithms and Feed forward neural network (FNN)

A ANN model is characterised by its architecture that represents the pattern of connection between nodes, its method of determining the connection weights and the activation function [16] and will be used to reproduce groundwater fluctuations for long periods using the observed time series data within twelve months of monitoring period (February 2015 to March 2016). In order to perform the system identification, the neural model is first trained to perform 1-day a-head predictions of the groundwater table using previously observed groundwater level. Once this autoregressive model has been developed, simulations were carried out by feeding back its output as the simulation times increases. In this study feed forward neural network architecture has been used and coupled with ANN training algorithms and Levenberg–Marquardt (LM) algorithm. These algorithms have been used to identify a suitable procedure which performs to predict daily groundwater levels over the study area. Feed forward neural networks architecture and the corresponding learning algorithm can be viewed as a generalisation of the popular least-mean-square (LMS) algorithm [17]. A multilayer perceptron network consists of an input layer, one or more hidden layers of computation nodes, and an output layer. Fig. 2 shows a typical feed forward network with one hidden layer consisting of three nodes, four input neurons and one output. The input signal propagates through the network in a forward direction, layer by layer. The various parameters, which affect the performance of network, are number of hidden nodes, number of hidden layers, learning rate and momentum factor. The ANN architecture consist of an input layer with three nodes, a hidden layer with varying nodes and one output layer with single node, thus leading to a multi input single output network. In the present study, the 406 input nodes represent initial groundwater levels at the two sites, daily rainfall, average daily river stage, average daily stream flow and daily temperature (Fig. 3). The two output nodes represent groundwater levels at the two sites in the next time step (i.e., 1 day a head). The input data is the groundwater level during time 1 day ahead, 't-4', 't-3', 't-2', 't-1' and 't' and output data is groundwater level during time 't'. The structure of the neural network was determined by trial and error. The optimal number of nodes in the hidden layer and the stopping criteria were optimised by trial and error for obtaining accurate output. The activation function of the hidden layer and output layer was set as log-sigmoid transfer function as this proved by trial and error to be the best

DOI: 10.1051/matecconf/201710304007

among a set of other options. In this study, supervised type of learning with a batch mode of data feeding was used in ANN modelling by available data. The data length of 406 data sets has been dividing into 70% for training and 30% for testing. The entire ANN modelling was performed by using Neural Works Predict software and EXCEL.

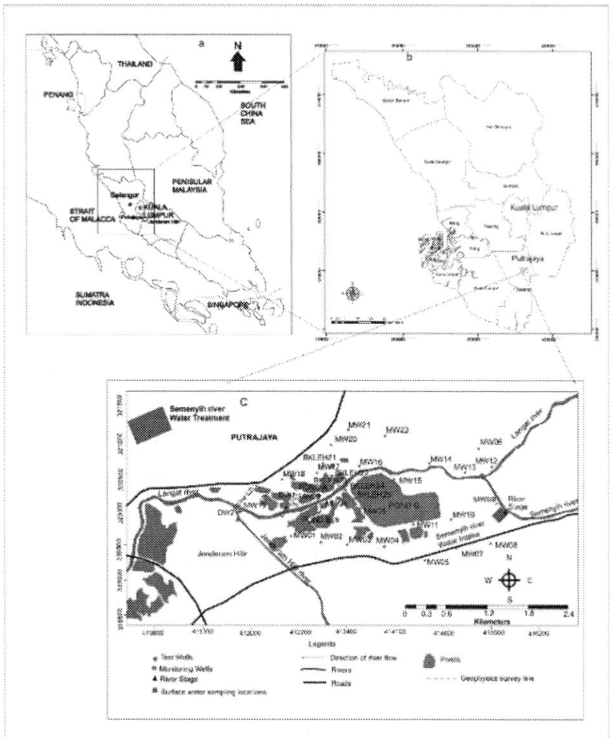

Fig. 1. Location of study area. a. Location of Selangor (box) within the Peninsular Malaysia. b. Location of Jenderam Hilir (shaded) in Selangor. c. Details of the study area including the locations of the Langat River, Semenyih River, monitoring wells and pumping wells.

2.2.1 Dynamic and Time series prediction

The temperature and groundwater level at time *t* at the DW1 and DW2, is a function of the past groundwater level at real time, the rainfall, stream flow and river water level from the Dengkil station. The lag is still at one day. In the time series prediction, time series are a sequence of number where in this case they are a sequence of groundwater level form the DW1 and DW2 in daily series or at time, (t) day). The groundwater level at time t, is a function of the past groundwater level at time t-1, t-2,t-3 and t-4 as shown below;

$$Y= f[\ y\ (t1), y(t-2), y(t-3), (y(t-4)] \quad (1)$$

Where y(t) is the models prediction of groundwater level form in the BI site at time t, y(t-1), y-2), y-3) and y(t-4).

2.2.2 *Evaluation of ANN efficiency*

The predictive performance of ANN are measured by four efficiency terms; the correlation coefficient (R); the mean error (ME), i.e the systematic difference between the predicted and measured values; the mean square error (MSE); and the root mean square error (RMSE). The ANN responses are more precise if R, MSE, RMSE, and ME are found to be close to 1, 0, 0 and 0, respectively.

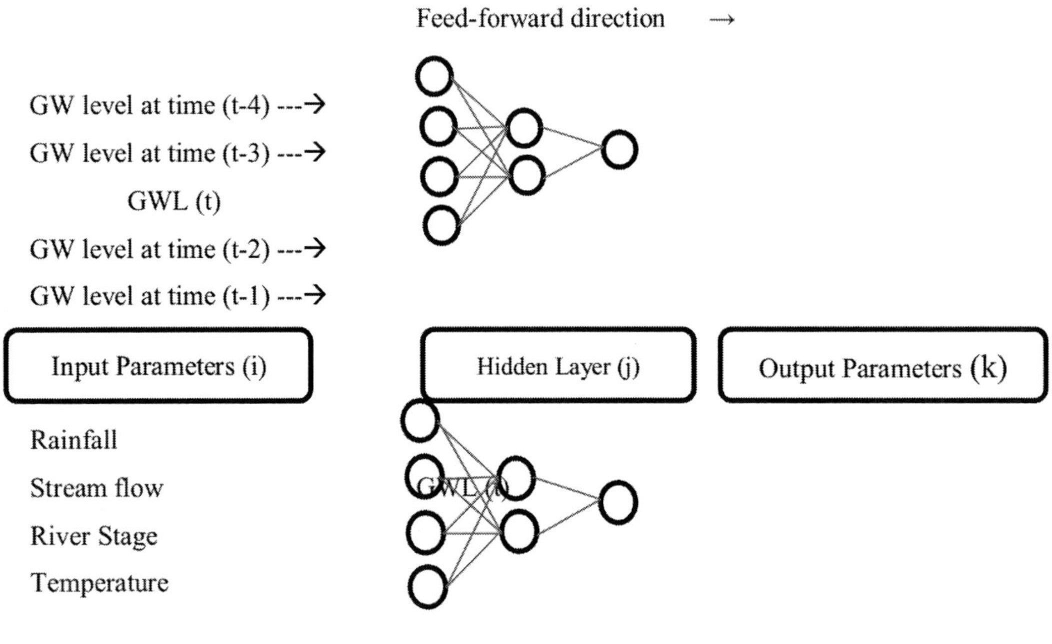

Fig. 2. Typical feed forward neural network of the study

2.2.3 Correlation Coefficient (R)

The correlation coefficient is a commonly used statistical parameter and provides information on the strength of linear relationship between the observed and predicted by the compute value. The value of R close to 1.0 indicates good model performance and can be calculated using equation below:

$$R = \frac{\sum_{i=1}^{n}(x_{obs} - \hat{x}_{obs})(x_{pre} - \hat{x}_{pre})}{\sum_{i=1}^{n}(x_{obs} - \hat{x}_{obs})^2 \sum_{i=1}^{n}(x_{pre} - \hat{x}_{pre})^2} \qquad (2)$$

Where, X_{obs}= observed groundwater $x\hat{}_{obs}$ observed levels=mean of X_{obs}, X_{pre}=predicted groundwater level, X_{pre}-= mean of X_{pre} and n =the number of data set used for evaluation.

2.2.4 Mean square Error (MSE) and Root Mean Square Error (RMSE)

For every data point, take the difference of the corresponding estimated values, and square the values. Then add up all those values for all data points, and divide by the number of points. The squaring is done so negative values do not cancel positive values. Smaller MSE indicates better prediction of the data. The MSE has the units squared of the parameter estimated.

$$MSE = \frac{1}{n}\sum_{i=1}^{n}(x_{obs} - x_{pre})\frac{1}{n}\sum_{i=1}^{n}(x_{abs} - x_{pre}) \quad (3)$$

It is probably the most easily interpreted statistic, since it has the same units as the parameter estimated. The RMSE is thus the difference, on average, of an observed data and the estimated data.

$$RMSE=\sqrt{\frac{\sum_{i=1}^{n}(x_{abs}-x_{pre})^2}{n}} \quad (4)$$

Different neural networks architectures were developed in order to establish a relationship between the input and output. All the networks were of the feed forward type. The network architectures were trained by varying the number of hidden layer and then by varying number of neurons in each hidden layer.

2.3 Coefficient Determination (R^2) and Residual error (RE)

The coefficient of determination R^2 (or sometimes r^2) is another measure of how well the least squares equation performs as a predictor of y.

$$\hat{y}= b_0 + b_1x \quad (5)$$

The R^2,is useful because it gives the proportion of the variance (fluctuation) of one variable that is predictable from the other variable. It is a measure that allows us to determine how certain one can be in making predictions from a graph. The higher the R^2, the more useful of the model, R^2 takes on values between 0 and 1. The residual error in the results is given by:

$$E_i = y_{obs} - \hat{y}_{pred} \quad (6)$$

Where y_{obs} is the observed and $\hat{y}_{pre}$ the predicted groundwater level. The percentage error of a variable is given by:

$$Ei = \frac{y_{obs} - \hat{y}_{pre}}{y_{abs}}\ x\ 100\% \quad (7)$$

3 Results and discussion

In the present study, the ANN model was designed to predict groundwater levels in two test wells with 1 day a-head time using a set of suitable input parameters. The input parameters for the ANN model were decided by considering the parameters potentially to affecting the groundwater level. A cross correlation analysis between the water levels in the test wells at various lags suggested that Lag 1 correlation is highly significant in the water level time series in all the two test wells. To examine the effect of rainfall, river stage, surface flow and temperature on groundwater, daily groundwater levels were plotted (Fig. 3). It shows that groundwater levels are generally higher on rainy days which indicate that rainfall is a paramount parameter that influences groundwater levels compared to river stage (Fig. 3) and other temperature parameters (Fig.4). In a semi-confined aquifer, apart from rainfall, surface water flow is another parameter that can influence the recharge to groundwater in BI sites (see Fig. 3). Therefore, daily surface water flow was also considered as one of the input parameters for the ANN model.The ANN models were trained with four input nodes for dynamic series model (Model 1) and five input nodes for time-series model (Model 2). The data were divided into two sets namely training and testing data. The calculated performance statistics for both models are shown in Table 1.Base on Table 1, the coefficient determination, R2 for the DW1 and DW2 for Model 1 is about 0.748 and 0.546 and for Model 2 is 0.640 and 0.602 respectively (Fig. 4 and 5). For a perfect predictor, the coefficient determination should be +1 or -1. In general the definition of R tells us that 100 R^2 is the percentage of the total variation of the predicted values which is explained by,or is due to their relationship with actual values. This is an important measure of the relationship between two variables; beyond this it permits valid comparisons of the strength of several relationships. Based on this value, indicating that the Model 1 and 2 is good even though the R^2 value is not really reaching 1.The correlation coefficient was also high on Model 1 for the DW1 and DW2 which is 0.8649 and 0.7392 and also model 2 which is 0.8006 and 0.7765 respectively. Fig.5a, positive (measure of direction) correlation or direct relationship indicates that a high score on the one variable is associated with a low score on the second variable. A negative correlation or inverse relationship indicates that a high score on one variable is associated with a low score on the second variable. The magnitude of the correlation coefficient indicates the strength of the relationship between the two variables. This magnitude can vary from 0.00 to 1.00. The closer the correlation coefficient is to either -1.00 or +1.00 the stronger the relationship. The stronger relationship between two variables is related produced the better prediction. This phenomenon can be seen in Fig. 4 and 5 for Model 1 and Model 2 respectively. Model 1 and 2 gives a clear view that the predicted and the actual process output are related which mean that the prediction output is quite near to the actual data which is the model 1 and 2 and it is a good correlation. As shown in Fig. 7 and 8, that the actual and predicted output for Model 1 and 2 is reasonably good and the groundwater level is always being recharged by river during the high flow period but during the low flow period the groundwater is recharging to the river. The prediction model is reasonably good in showing the relationship between river and groundwater. The residue analysis was also carried out in this study. The residual analysis is very useful in helping us to identify the performance of both models. The analysis for the residual is presented in Fig. 5. It can be seen that the residual for Model 1 and 2 is more consistent where output is near to zero even though in some samples the residual is high (Fig. 6 and 7). It shows that model output or prediction always near to the actual data. From the residual analysis, it determines that for the negative values, river gains from the aquifer and if river losses it shows the positive value. The statistical adequacies of the developed models for 1-day ahead forecasts for DW1 and DW2 test wells are summarized in Table 1. It is observed from Table 1, that the model performance is good and the models have forecasted the water levels with reasonable accuracy in terms of all statistical indices during calibration and validation. The correlation statistics that evaluates the linear correlation between the observed and the computed water table is consistent. The RMSE statistic, which is a measure of residual variance that shows the global goodness of fit between the computed and observed water levels, is very good as is evidenced by a low value (< 0.4 m) during both training and testing. While a 1-day ahead forecasts at higher lead time are required for efficient planning of RBI method or conjunctive use. A further analysis was performed by using the forecasted water levels at the DW1 and DW2 test wells as input to the models. An analysis that evaluates the input sensitivity to the model predictions was carried out by developing two models; one using rainfall, surface water level, temperature, stream flow and rainfall as input model. The results were found to be not very accurate and may be because the recharge time for water to reach the groundwater is quite high compared to the lag period considered in the model. The results presented are mainly in the form of percentages with 95% confidence intervals. The lower and upper limits of groundwater levels with 95% confidence interval are shown in the Fig.8. This Fig. 8 shows there is a decline and rise of groundwater table elevations for the entire period and the model compares well with observations with the highest groundwater level in the early February 2015 and 2016.

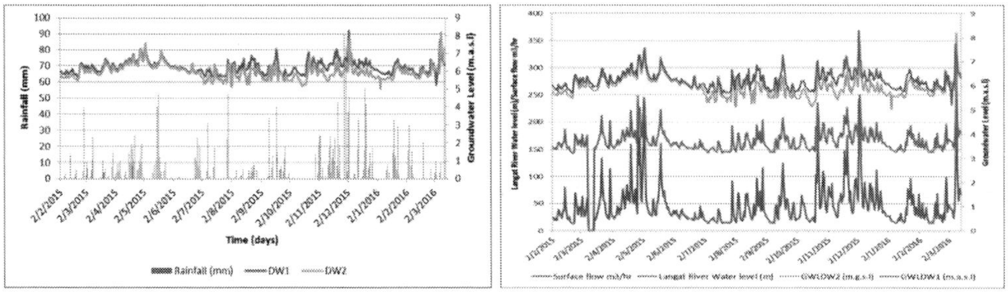

Fig. 3. (a) Daily groundwater level fluctuations and (b) Well hydrographs at sites DW1 and DW2 with river stage and surface water flow hydrograph at Jenderam Hilir

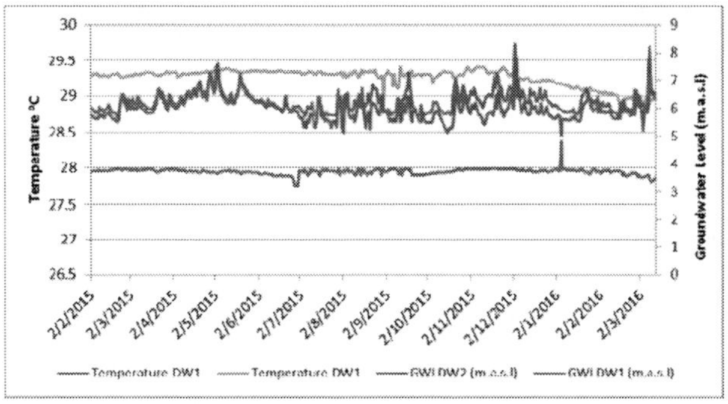

Fig. 4. Daily groundwater level fluctuations with groundwater temperature

Table 1. Performance statistic for model 1 and model 2.

	Model 1		**Model 2**	
	DW1	DW2	DW1	DW2
R	0.8649	0.7392	0.8006	0.7765
MSE	0.0356	0.0781	0.0513	0.0682
RMSE	0.01	0.0139	0.0113	0.013
R^2	0.748	0.546	0.64	0.602

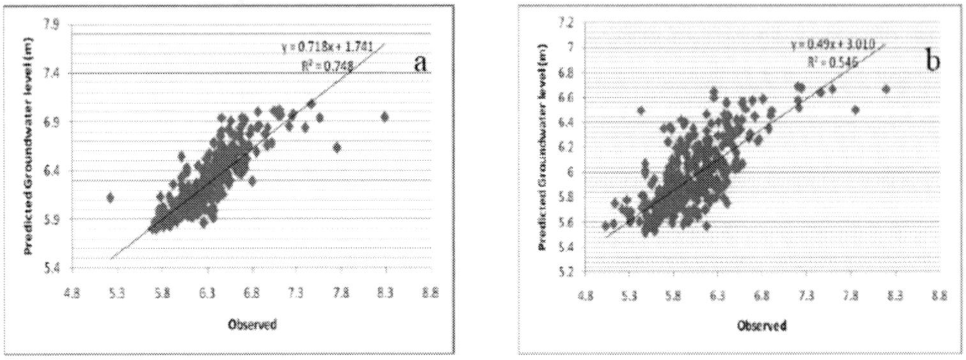

Fig. 5. Scatter plot for observation and predicted value for DW1 (a) and DW2 (b) in Model 1

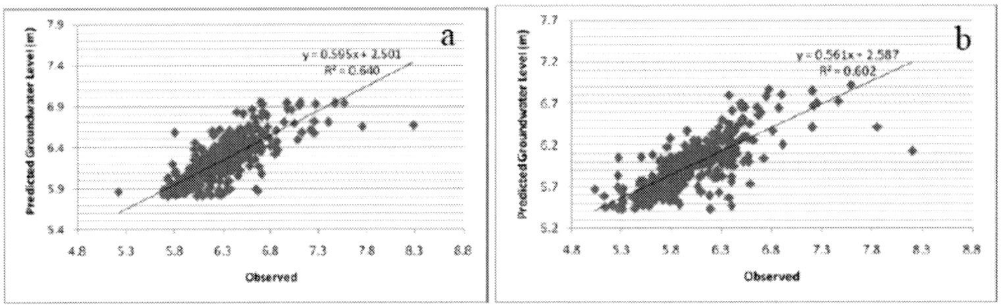

Fig. 6. Scatter plot for observation and predicted value for DW1 (a) and DW2 (b) in Model 2

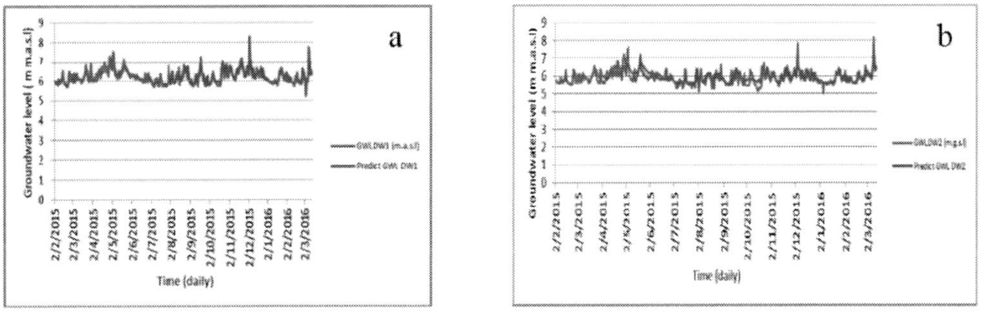

Fig. 7. Groundwater level prediction for DW1 (a) and DW2 (b) in a dynamics model

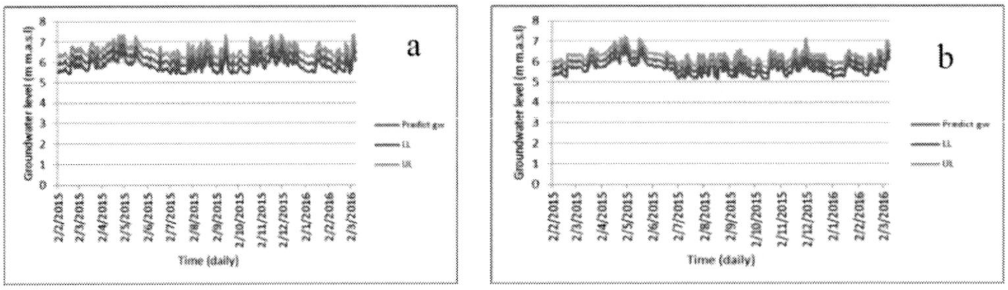

Fig. 8. The lower and upper limits are also shown in Model (DW1 (a) and DW2 (b)

4 Conclusion

In this study, the potential of neural network computing technique for forecasting groundwater level was investigated by developing ANN models for a shallow aquifer in Jenderam Hilir, Selangor in Malaysia. The result indicates that the capability of neural networks models in modelling of daily groundwater level using rainfall, temperature, stream flow and river water level data as inputs as well as the past value of the groundwater level. The two modelling techniques applied show that both models can perform well. This can be explained by referring to the correlation coefficient analysis, coefficient determination and RMSE. It was noted that the contribution of past values of groundwater level was very important in neural networks modelling as well as the value of the rainfall, temperature, stream flow and river water level. This was not observed in the time series modelling techniques where it only use past values of the groundwater level. This was the main reason, both the dynamic and time series modelling techniques was performed in this study. Therefore the appropriate techniques to model the groundwater level in study is by using the values of rainfall, temperature, stream flow and river water level as well as the values of groundwater level. A significant advantage of this model is that is can be provide good predictions with limitations of groundwater table record.

References

[1] J. Schubert, Hydraulic aspects of riverbank filtration-field studies. Journal of Hydrology, **266**, 145–161, (2000)

[2] R.A. Sheets, R.A. Darner and B.L. Whitteberry, Lag times of bank filtration at a well field, Cincinnati, OH, USA. J. of Hydrology, **266**, 162–174, (2002)

[3] P.J. Dillon, M. Miller, H. Fallowfield and J. Hutson, The potential of riverbank filtration for drinking water supplies in relation to microsystin removal in brackish aquifers. J. Hydrology, **266**, 209–221, (2002)

[4] C. Ray, G. Melin and R.B. Linsky, *Riverbank Filtration Improving Source-Water Quality*. Kluwer Academic Publishers, Dordrecht. (2002)

[5] J.Z. Wang, S.A. Hubbs and R. Song, *Evaluation of Riverbank Filtration as a Drinking Water Treatment Process*. AWWA Research Foundation and American Water Works Association, USA, (2002)

[6] B. Wett, H. Jarosch and K. Ingerle, Flood induced infiltration affecting a bank filtrate well at the River Enns, Austria. J. of Hydrology, **266**, 222–234, (2002)

[7] M.K.N. Shamsuddin, W.N.A. Sulaiman and S. Suratman, Hydrogeol J., **22**, 543. (2014)

[8] Nourani V and Mano A.. Semi-distributed flood runoff model at the sub continental scale for southwestern Iran. Hydrological Processes, **21**, 3173–3180, (2007)

[9] E.A. Coppola, C.F. McLane, M.M. Poulton, Szidarovszky F. and Magelky R.D.. Predicting conductance due to upconing using neural networks, Ground Water **43**: 827–836, (2005a)

[10] Lallahem S., Mania J., Hani A. and Najjar Y. On the use of neural networks to evaluate groundwater levels in fractured media, J. of Hydrology, **307**, 92–111, (2005)

[11] Coppola E.A., Rana A.J., Poulton M. M., Szidarovszky F. and Uhl V.W.. A neural network model for predicting aquifer water level elevations. Ground Water, **43**, 231–241, (2005b)

[12] Tayfur G., Swiatek D., Wita A. and Singh V.P.. Case study: finite element method and artificial neural network models for flow through Jeziorsko earth dam in Poland. J. of Hydraulic Engineering, **131**, 431–440, (2005)

[13] Gobindraju R.S. and Ramachandra Rao. A., *Artificial neural network in hydrology*. Kluwer, Dordrecht (2000)

[14] Coulibaly P, Anctil F., Aravena R., Bobee B., Artificial neural network modeling of water table depth fluctuations. Water Resources Research, **37**(4), 885–896, (2001)

[15] Coulibaly P., Anctil F. and Bobee B. Daily reservoir inflow forecasting using artificial neural networks with stopped training approach. J. of Hydrology, **230**, 244–257, (2000)

[16] Fausett L, *Fundamentals of neural networks*. Prentice Hall, Englewood Cliffs. . (1994)

[17] Haykin S. *Neural networks, a comprehensive foundation*, 2nd edn. Prentice Hall, Englewood Cliffs. (1999).

Modeling Flood Disasters: Issues Concerning Data for 2D Numerical Models

Ahmad Fikri Abdullah[3,*], *Zoran* Vojinovic [1,2], and *Vorawit* Meesuk[1]

[1]UNESCO-IHE, Westvest 7, 2611 AX Delft, The Netherlands
[2]University of Exeter, EX4 4QF, UK
[3]Department of Biological and Agricultural Engineering, Faculty of Engineering, Universiti Putra Malaysia, 43400 Serdang, Selangor, Malaysia

Abstract. Flood disasters are one of the worst natural disasters and they occur almost daily. Correspondingly, city managers are nowadays increasingly investing in data collection and modelling activities. Typically, flows in pipes and channels have been modelled using a one-dimensional (1D) approach. Also, if the flood flows on the ground are confined to streets and curbs, the use of 1D/1D modelling approach (a.k.a. as dual drainage concept) may be a feasible way forward. However, if the flows are two-dimensional with complex interactions taking place through surface/sub-surface links, then the required approach would be the so-called 1D/2D modelling approach. The ever-increasing power of computers is now allowing a 3D fluid dynamics analysis of overflow and manhole structures, and in time these may be coupled with 1D models that may be used in situations where flow characteristics require with higher dimensionality. With the use of 2D models, the importance of terrain data has become a growing issue. The current paper discusses some of the key modelling approaches and makes particular reference to the terrain data processing and collection which can either improve or deteriorate their predictive capability.

1 Introduction

The world has been overwhelmed with growing number of natural disasters where floods are most dominating the records. With the increase in flood disasters, the process of preparing for such disasters requires investments in data and models. In doing this, collection and processing of terrain data is almost equally important as instantiation of numerical models. By acquiring reliable data and setting up appropriate models, the city managers are able to study floods and to simulate various drivers (e.g., urbanization, population growth or climate change) and effects of rehabilitation measures that they may find acceptable to stakeholders [1-15].

As described in [16-19], the models used for modelling urban floods have a sound physical basis of shallow-water equations and their accuracy will depend on the quality of the data used for analysis, complexity of equations as well as the knowledge and experience

* Corresponding author: ahmadfikri@upm.edu.my

of the modeler. Traditionally, one-dimensional (1D) approach has been used widely as it has given a good support to hydraulic engineering in design and system rehabilitation projects. Correspondingly, even nowadays, many researchers and practitioners favor such modelling approach over other approaches. This is due to the fact that 1D models are easy to instantiate, operate and maintain. In case that 1D models are incapable of capturing the required physics some researchers have attempted to combine these models with data-driven models.

With the developments in computational power, 1D-2D approaches are increasingly gaining more attention [20-23]. A possibility of how to combine data-driven and physically based models for the purpose of modelling urban drainage flows can be found in [24] and [25].The accuracy of 2D models that can be reliably used in modelling flood disaster scenarios depends on the quality of terrain data, bathymetry and drainage network layouts [26-28]. In this paper, we discuss some of the key issues concerning data for 2D numerical models.

2 Method

2.1 Modelling Approaches

In urban flood modelling, we are typically concerned with the flood water interacting with urban environment in space and time. This objective leads us to consider different model complexities in space (one-dimensional, two-dimensional or three-dimensional) with a corresponding number of mathematical equations that describe the physical laws. Therefore, our focus is on how the geometry of physical environment affects flows at different points in the fluid, and how the changes in the geometry affect the propagation of flows. As mentioned earlier, if the flows are confined to pipes and channels a properly instantiated 1D model may be considered fit for purpose. As soon as the water spills from the drainage systems (or if it doesn't even enter the drainage system) and starts flowing on the ground our modeling efforts would require either the use of 1D/1D or 1D/2D modelling approaches.

The flows across urban floodplains are highly complex due to the presence of irregular features which may force flood flow paths to go in different directions. Developing models to reliably describe urban flood processes is a challenging task. The slopes of urban floodplains can often change which in turn can cause change in flow regimes. In such situations, the use of commercial modelling tools, which mainly use modified subcritical flow algorithms, may become an issue. It is often the case that head losses, which can form around irregular geometry of urban features, are particularly difficult to model and take into account [19]. Such issues drive the need to use 1D and 2D modelling systems which have a good predictive capability in modelling floods in urban areas; see, for example, [29-32].

Despite the advances in 1D/2D modelling algorithms, the predictive capability of 1D/2D models also depends on the quality of data and its associated processing techniques. In this paper, we investigate some of the key issues that relate to the terrain data for flood modeling applications. In our work, we apply different digital terrain models (DTMs) with different resolutions and different geometry of features as they were collected form LiDAR surveys. We applied these data sets for setting up 2D models. The resulting coupled 1D/2D model results are then mapped onto GIS maps to analyse the differences.

2.2 Terrain data collection and processing for flood modelling

A digital terrain model (DTM) term refers to a topographic map of an area, which contains terrain elevations and its characteristics. Unlike the DTM, the digital elevation model (DEM) term refers to the elevations of any surface, which can not necessarily be only the land (e.g., water surface, etc.). The DTM data can be acquired from different techniques. This is usually done through traditional ground surveys, aerial stereo photography and light detection and ranging (LiDAR) surveys.

Aerial Stereo Photography is one of the oldest remote sensing techniques. Satellite stereo imagery is yet another remote sensing technique, which combines images from two satellites, acquired from two different positions. Interferometric synthetic aperture radar (IfSAR) is a radar technique that can be applied in remote sensing. IfSAR mapping is a process of generating 3D maps from airborne IfSAR system, see for example [33]. The experience to date is that the coarser the IFSAR spatial resolution is, the more difficult it will be to produce the terrain elevation data [34].

Airborne Laser Scanning (ALS) or Light Detection and Ranging (LiDAR) is yet another remote sensing technique that is used to obtain elevation data from a particular area. LiDAR is based on an optical remote sensing technology to measure elevation points. LiDAR is a surveying method, which can give us accurate x and y coordinates, and z (height) positions. It can generate terrain data to a particular resolution. The LiDAR scanning provides a large number of elevation points which are spread across an area. The complexities that are inherent in processing such large amounts of data have called for better processing of raw data in order to deliver a more reliable set of points. Typically, operations that are referred to as thinning, filtering and interpolation are part of this process. Recently, a number of algorithms have been developed to enhance the processing of data that are acquired from LiDAR surveys (Fig. 1).

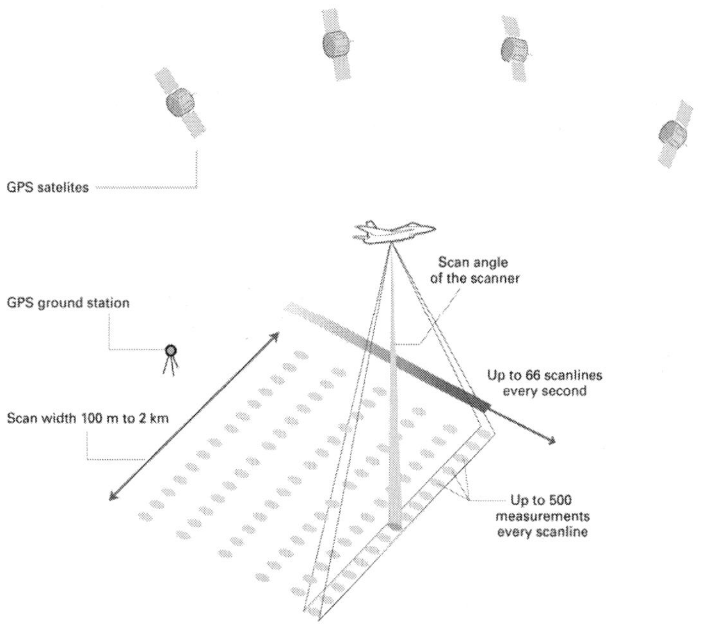

Fig. 1. LiDAR survey [19]. A large number of data points is collected.

The demand in LiDAR technology has increasing. Terrestrial LiDAR also known as terrestrial laser scanning (TLS) is establishing as an option to fulfill the demand. TLS is a ground based method to measure the position and create three dimensional information about the objects. It has the ability to collect very high accuracy point cloud from ground based platforms to model an object or scene. This TLS combines with a highly accurate differential GPS like RTK GPS, enables the user to register the georeference. TLS is a technique that provides detailed information for small-scale area [35]. TLS is often used for geomorphological analysis because it allows a fast and detailed 3D reconstruction of the sample object [36].

3 Results

3.1 Issues concerning DTM resolution

It is often assumed that a more detailed 2D model (or a 2D model with higher resolution) will be able to provide better distribution of flows over a particular urban floodplain. It is important to note that higher resolution of 2D data will inevitably increase computational time and mapping of such results will also take more considerable time. Correspondingly, coarsening the data will reduce model simulation time, but it will also reduce the level of details around urban features. Therefore, there is always a trade-off between the model simulation time and the level of details that can be captured within the domain of a 2D model.

In general terms, the work to date has shown that the results from 2D models with different DTM resolution can produce significant difference in flow depths [20] (Fig. 2).

Fig. 2. Impacts of different DTM resolutions on model results (5m and 10m DTM resolutions);

3.2 Issues concerning filtering algorithms

In addition to the issues related to DTM resolution, there are also issues associated with the efficacy of LiDAR filtering algorithms [31-32]. The work to date has shown that different filtering algorithms can produce different DTM data sets which in turn can cause different results from 2D models. Some of the most commonly used algorithms in the practice are considered in the present paper. These algorithms are: the morphological algorithm and a TIN based algorithm.

DOI: 10.1051/matecconf/201710304008

3.3 Flood maps with different filter algorithms

The morphological algorithm points over a particular height range and the underlying processing used in this algorithm are dilation and erosion.

For the TIN based algorithm, the nearby points are only used in the processing. This algorithm is capable of successfully removing small urban features but it does not work satisfactorily in removing some larger features.

Fig. 3 depicts the 1D/2D model results for which the data was processed with the morphological algorithm (left) and TIN based algorithm (right).

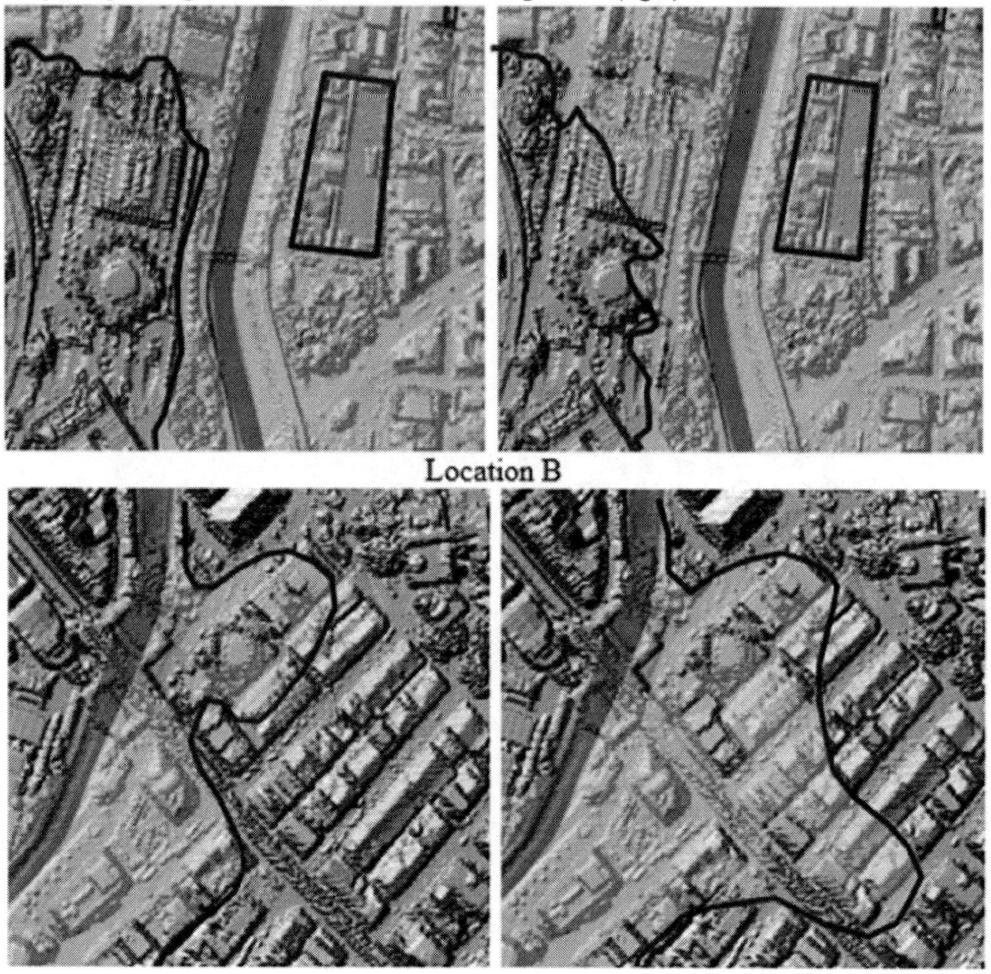

Fig. 3. Example of 1D/2D model results where different filtering algorithms are used in setting up 2D models (morphological algorithms – left; TIN based algorithms – right);

3.4 Fusion of top-view LiDAR and ground-view SfM data

Careful application of filtering algorithms can help in identification and representation of features that can be detected from top-view LiDAR data. However, if the features are hidden beneath vegetation and structures, the LiDAR filtering algorithms cannot detect such features. In this regard, the so-called Structure from Motion (SfM) technique can be used effectively to compensate for this shortcoming, [37]. This technique can be easily implemented by incorporating overlapping photos and video scenes.

With fusion of top-view LiDAR and ground-view SfM data it is possible to detect and take into account small urban features which may have substantial influence on propagation of flood flows. Fig. 4 depicts the results with SfM-LiDAR DTM (left) and conventional LiDAR DTM (right), see also [24].

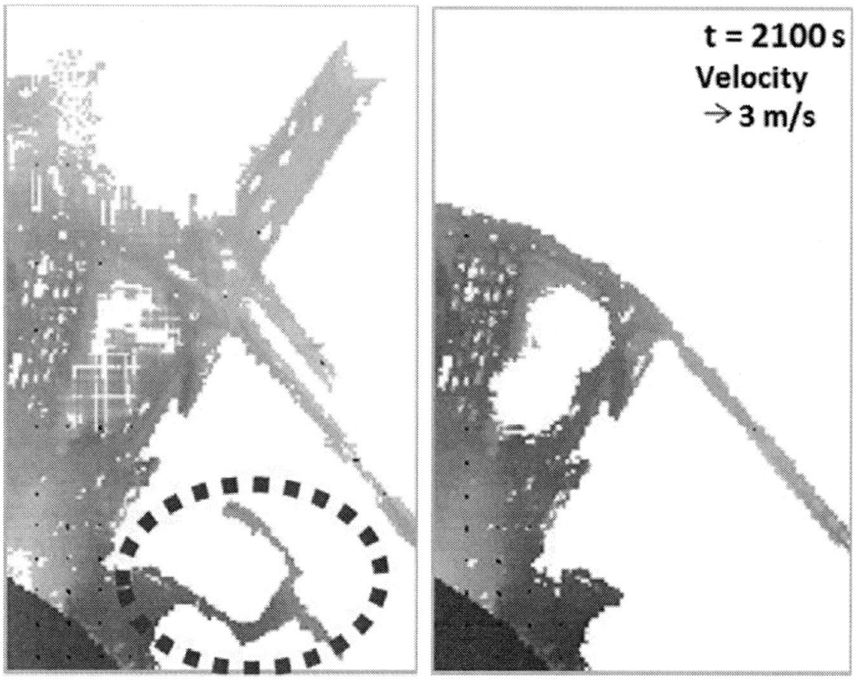

Fig. 4. Flood simulation results with SfM-LiDAR DTM (left) and conventional LiDAR DTM (right). The differences are found to be substantial and the 2D model based on SfM-LiDAR DTM gave more realistic results (the results were compared with the eye witness records from a flood event).

4 Conclusion

The present paper describes the differences that can be incurred due to different DTM data sets when processed at different grid resolutions and with different LiDAR filtering algorithms. In the present work, several LiDAR filtering algorithms were compared and it was found that none of them is fully reliable in capturing some important urban features.

Furthermore, the present work also considered the fusion of LiDAR data with Structure from Motion (SfM) data which proved to be capable of detecting urban features that are hidden beneath vegetation and other urban structures. When compared to the conventional DTM data, the results generated with SfM-LiDAR DTM were much more close to reality and on the ground measurements.

Although numerical models are invaluable for modelling floods and flood-related disasters, the overall analysis of results presented here suggests that straightforward application of LiDAR data may restrict the predictive capability of such models. Hence, incorporation of urban features within the DTM through appropriate filtering algorithms as well as with the fusion of ground-view SfM data can lead to a significant improvement of model results.

References

[1] Z. Vojinovic, D.P. Solomatine, R.K. Price, Dynamic Least-Cost Optimisation of Wastewater System Remedial Works Requirements, Water Sci Technol, **54**(6-7), 467-475 (2006b)

[2] W. J. Barreto, Z. Vojinovic, R. Price, D. Solomatine, Approches to multi-objective multi-tier optimisation in Urban Drainage Planning, *7th International Conference on Hydroinformatics,* Nice, France, (2006)

[3] Z. Vojinovic, J. van Teeffelen, An Integrated Stormwater Management Approach for Small Islands in Tropical Climates, Urban Water J., **4**(3), 211-231 (2007)

[4] R.K. Price, Z. Vojinovic, Urban Flood Disaster Management, Urban Water J., **5**(3), 259-276 (2008)

[5] Z. Vojinovic, A. Sanchez, W.J. Barreto, Optimising sewer system rehabilitation strategies between flooding, overflow emissions and investment costs, *11th International Conference on Urban Drainage*, Edinburg, Scotland, (2008)

[6] W. Barreto, Z. Vojinovic, R.K. Price, D.P. Solomatine, Multi-tier Modelling of Urban Drainage Systems: Muti-objective Optimization and Parallel Computing, *11th International Conference on Urban Drainage*, Edinburg, Scotland, (2008)

[7] A. Mynett, Z. and Vojinovic, Hydroinformatics in multi-colours – part red: Urban Flood and Disaster Management, J. Hydroinform., **11**(3-4) 166-180 (2009)

[8] W. Barreto, Z. Vojinovic, R. Price, D. Solomatine, Multiobjective evolutionary approach to rehabilitation of urban drainage systems. J. Water Resources Planning Management., **136**(5), 547-554 (2009)

[9] S. Seyoum, Z. Vojinovic, R.K Price, Urban pluvial flood modeling: development and application, *9th International Conference on Hydroinformatics*, Tianjin, China, (2010)

[10] A. Jonoski, L. Alfonso, A. Almoradie, I. Popescu, S.J. van Andel, Z. Vojinovic, Mobile phone applications in the water domain, Environ Eng Manag. J., **11**(5) (2012)

[11] D.S. Kumar, D.S Arya, Z. Vojinovic, Modeling of urban growth dynamics and its impact on surface runoff characteristics, Comput Environ Urban Syst., **41**, 124-135 (2013)

[12] Z. Vojinovic, S. Sahlu, S. Seyoum, A. Sanchez, H. Matungulu, Z. Kapelan, D. Savic, Multi-objective rehabilitation of urban drainage systems under uncertainties, J. Hydroinform, **16**(5) 1044-1061 (2014)

[13] A. Sanchez, N. Medina, Z. Vojinovic, R. Price, An integrated cellular automata evolutionary-based approach for evaluating future scenarios and the expansion of urban drainage networks, J. Hydroinform., **16**(2), 319-340 (2014)

[14] R. Singh, D.S. Arya, A. Taxak, Z. Vojinovic, Impact of climate change on Rainfall Intensity-Duration-Frequency (IDF) Curves in Roorkee, Water Resour Manag (2016)

[15] A. Alves, A. Sanchez, Z. Vojinovic, S. Seyoum, M. Babel, D. Brdjanovic, Evolutionary and holistic assessment of green-grey infrastructure for CSO reduction, Water **8**(9), 402 (2016)

[16] M.B. Abbott, B.M. Tumwesigye, Z. Vojinovic, The fifth generation of modelling in Hydroinformatics, *7th International Conference on Hydroinformatics,* Nice, France, (2006)

[17] M.B. Abbott, Z. Vojinovic, Applications of Numerical Modelling in Hydroinformatics, J Hydroinform, **11**(3-4), 308-319 (2009)

[18] R.K. Price, Z. Vojinovic, *Urban Hydroinformatics: Data, Models and Decision Support for Integrated Urban Water Management,* IWA Publishing, (2011)

[19] Z. Vojinovic, M.B. Abbott, *Flood Risk and Social Justice: From Quantitative to Qualitative Flood Risk Assessment and Mitigation,* IWA Publishing, (2012)

[20] Z. Vojinovic, S.D. Seyoum, J.M. Mwalwaka, R.K. Price, Effects of Model Schematization, Geometry and Parameter Values on Urban Flood Modelling, Water Sci Technol, **63**(3), 462-467 (2011)

[21] S.D. Seyoum, Z. Vojinovic, R.K Price, S. Weesakul, Coupled 1d and non-inertia 2d flood inundation model for simulation of urban pluvial flooding, J. Hydraul. Eng.-ASCE, **138**(1), 23-34 (2012)

[22] Z. Vojinovic, D. Tutulic, On the use of 1D and coupled 1D-2D approaches for assessment of flood damages in urban areas, Urban Water J., **6(3)**, 183–199 (2009)

[23] Z. Vojinovic, S. Seyoum, M.H. Salum, R.K Price, A.F. Fikri, Y. Abebe, Modelling floods in urban areas and representation of buildings with a method based on adjusted conveyance and storage characteristics, J. Hydroinform, **15**(4), 1150–1168 (2012)

[24] Z. Vojinovic, V. Kecman, V. Babovic, A Hybrid Approach for Modelling Wet Weather Response in Wastewater Systems, J. Hydraul. Eng.-ASCE, **129**(6), 511-521 (2003)

[25] Z. Vojinovic, A Complementary Modelling Approach to Manage Uncertainty of Computationally Expensive Models, Water Sci Technol., **56**(8), 1-9 (2007)

[26] Z. Vojinovic, Y. A. Abebe, R. Ranasinghe, A. Vacher, P. Martens, D. J. Mandl, S. W. Frye, E. van Ettinger, R. de Zeeuw, A machine learning approach for estimation of shallow water depths from optical satellite images and sonar measurements, .J Hydroinform, **15**(4), 1408-1424 (2013)

[27] Vojinovic, Z., ~~2015~~, Flood Risk: The Holistic Perspective, From Integrated to Interactive Planning for Flood Resilience, IWA Publishing, London, United Kingdom, (2015)

[28] Z. Vojinovic, M. Hammond, D. Golub, S. Hirunsalee, S. Weesakul, V. Meesuk, ... & M. Abbott, Holistic approach to flood risk assessment in areas with cultural heritage: a practical application in Ayutthaya, Thailand. Natural Hazards, **81**(1), 589-616. (2016)

[29] Z. Vojinovic, B. Bonillo, K. Chitranjan, R. Price, Modelling flow transitions at street junctions with 1D and 2D models, *7th International Conference on Hydroinformatics,* Nice, France (2006)

[30] A. Abdullah, A. Rahman, Z. Vojinovic, LiDAR filtering algorithms for urban flood application: Review on current algorithms and filters test, *Laserscanning09,* Paris, France, **38**, 30-36 (2009)

[31] A.F. Abdullah, Z. Vojinovic, R.K Price, A Methodology for Processing Raw LIDAR Data to Support 1D/2D Urban Flood Modelling Framework, J. Hydroinform., **4**(1), 75-92 (2011a)

[32] A.F. Abdullah, Z. Vojinovic, R.K and Price, Improved Methodology for Processing Raw LIDAR Data to Support Urban Flood Modelling - Accounting for Elevated Roads and Bridges, J. Hydroinform, **14**(2) 253-269 (2011)

[33] X. Li, A.B. Baker, T. Hutt, Accuracy of Airborne IfSAR mapping, *Proceedings of FIG XXII International Congress,* Washington, United State, 19-26 (2002)

[34] M. Holopainen, V. Kankare, M. Vastaranta, X. Liang, Y. Lin, M. Vaaja, ... P. Alho, Tree mapping using airborne, terrestrial and mobile laser scanning - A case study in a heterogeneous urban forest, Urban For Urban Gree, **12**(4), 546–553 (2013)

[35] N.A. Muhadi, A.F. Abdullah, M.S.M. Kassim, Quantification of terrestrial laser scanner (TLS) elevation accuracy in oil palm plantation for IFSAR improvement, *IOP Conference Series: Earth and Environmental Science,* **37** (1), 012042 (2016)

[36] P. Glira, C. Briese, N. Pfeifer, J. Dusik, L. Hilger, F. Neugirg, Accuracy analysis of height difference models derived from terrestrial laser scanning point clouds, *EGU General Assembly Conference Abstracts,* Vienna, Austria, **16**, 15987 (2014)

[37] V. Meesuk, Z. Vojinovic, A. Mynett, A.F. Abdullah, Urban flood modelling combining top-view LiDAR data with ground-view SfM observations, Adv Water Resour., **75**, 105–117 (2015)

Estimating Agricultural Losses using Flood Modeling for Rural Area

Nur Atirah Muhadi [3,*], *Ahmad Fikri* Abdullah [3], and *Zoran* Vojinovic [1,2]

[1] UNESCO-IHE, Westvest 7, 2611 AX Delft, The Netherlands
[2] University of Exeter, EX4 4QF, UK
[3] Department of Biological and Agricultural Engineering, Faculty of Engineering, Universiti Putra Malaysia, 43400 Serdang, Selangor, Malaysia

Abstract. Flooding is the most significant natural hazard in Malaysia in terms of population affected, frequency, flood extent, flood duration and social economic damage. Flooding causes loss of lives, injuries, property damage and leave some economic damage to the country especially when it occurs in a rural area where the main income is dependent on agricultural area. This study focused on flooding in oil palm plantations, rubber plantations and fruits and vegetables area. InfoWorks ICM was used to develop a flood model to study the impact of flooding and to mitigate the floods using a retention pond. Later, Geographical Information System (GIS) together with the flood model were used for the analysis on flood damage assessment and management of flood risk. The estimated total damage for three different flood event; 10 ARI, 50 ARI and 100 ARI involved millions of ringgits. In reducing the flood impact along the Selangor River, retention pond was suggested, modeled and tested. By constructing retention pond, flood extents in agricultural area were reduced significantly by 60.49% for 10 ARI, 45.39% for 50 ARI and 46.54% for 100 ARI.

1 Introduction

Flood is one of the major severe natural catastrophes experiencing in many countries around the world and the researchers and practitioners are challenged with urbanization, population growth and climate change and finding optimal flood protection and system rehabilitation strategies [1-8]. The process of preparing for such disasters requires investments in data and models. However the benefits of numerical models and data will depend on a number of issues such as the accuracy of the data and uncertainties [9-20]. The basic causes of flood in Malaysia are the incidence of heavy monsoon rainfall and the large concentration of runoff, which exceeds river capacity [21, 22]. Agricultural lands located in catchment areas have the highest probability to involve in flood events especially for those reside in low land area [23]. Flood damages can be categorized into tangible and intangible damages. Tangible damage can be calculated in monetory terms meanwhile intangible cannot be easily measured in monetory terms [24-26].

* Corresponding author: tiramuhadi@gmail.com

In disaster management strategies, several steps are needed in order to identify risks and making decision on the most appropriate measures with risk assessment is claimed to be the important part of the process [27]. Mitigation is another step required in disaster management strategies. The aim of flood mitigation is to reduce the long term impacts of disasters on a nation or community. Flood mitigation refers to any structural or non-structural measures taken to reduce the flooding impacts. Bustami and collegues [28] carried out a study for modeling of flood mitigation structures in Sarawak River sub-basin. It is reported that retention ponds provide the most efficient in mitigating the flood condition of the study area. In order to study the impact of flooding, a flood model is developed. Flood model will help in estimating the flood inundation for rural area. To meet the purpose, InfoWorks Integrated Catchment Modeling (ICM) was used to develop a comprehensive flood model that covers main river and floodplain. Results from the model was then used to carry out relevant analysis on flood damage assessment. In addition, InfoWorks ICM is also used to run simulation on certain scenarios on mitigation options. Besides, Geographical Information System (GIS) in combination with appropriate flood models provide ideal tools for the analysis and management of flood risk. GIS overlays of the land use and the flood extent are used to study the impact of flooding in the study area.

2 Method

2.1 Study area and data collection

Fig. 1. Location of the study area

This study has been carried out in a rural area located in Selangor river basin. However, only several parts of the Selangor River are involved which are Kuala Selangor, Pasangan and Bestari Jaya. Fig. 1 shows the location of the study area. There are two types of data required for this study, which are spatial data and non-spatial data. Spatial data include topographic data and land use data while non-spatial data includes rainfall intensity and river water level. In order to develop a good InfoWorks ICM model, good quality of GIS data of the catchment and good quality of digital elevation models (DEMs) data to represent natural depressions and embankments on the ground are required. Terrain is analyzed to avoid lands with steep slopes above permitted limits while regional watershed studies are performed to identify flood-prone areas. In this study, IFSAR data were used in developing the DEM for flood model. IFSAR data are best used in sparsely vegetated areas with relatively low slopes and away from urban areas. IFSAR data are appropriate for watershed-scale analysis. Because of its wider swath and greater flying speeds, IFSAR are more appropriate and more cost effective for large area applications. IFSAR is used in creating elevation models of the terrain from airborne platforms at superior levels of spatial detail typically 1-5 meters sample spacing and accuracy which is 15 cm – 3 meters RMSE vertical [29].

3 Results

In a report of Water Resources Planning Institute of Vietnam, it is suggested to use a 10 ARI as a flood standard for agricultural flood protection activities [30]. In addition, 100 ARI has been universally used to describe a reasonable flood protection level. Therefore, this study used three flood scenarios which are 10 ARI, 50 ARI and 100 ARI to assess the impacts of floods on agriculture in the Selangor river basin area. The model predicts the flood depth and flood extent for the 10 ARI, 50 ARI and 100 ARI scenarios and these are presented in Fig. 3 for 10 ARI flood depth. Table 1 represents the total inundated area during the flood events. The model predicts the areas of flood extents and flood depths for each of the flood scenarios. Hence, the proportions of the flood depth were calculated for each flood scenario. As a result of the flood model simulations, more than 40% of the total flood extent was covered by 1.2 m flood depth for each 10 ARI, 50 ARI and 100 ARI. Only area that will be flooded by 0.5 to 1.2 m of floods will be calculated during the analysis. The 0.001m depth is not a severe flood thus it is not accounted as flood for this study. Table 1 also shows the total inundated area based on flood depth for each flood event.

Table 1. Total inundated area based on flood depth

Flood Depth (m)	Total Inundated Area (ha)		
	10 ARI	50 ARI	100 ARI
0.001	1477.41	2204.07	2336.78
0.5	**1866.42**	**2792.51**	**3074.15**
1.2	**2502.52**	**4176.99**	**4767.46**
Total	5846.35	9173.57	10178.39

The flood risk maps resulted from the flood model was overlaid with the land use maps to assess the impact of 10 ARI, 50 ARI and 100 ARI in the study area. Fig. 2 shows the results for three various return periods which are overlaid with land use map. Different return periods have different flood extents. Among all types of land used, the model estimates that the greatest land use impacted by the floods is the agricultural land.

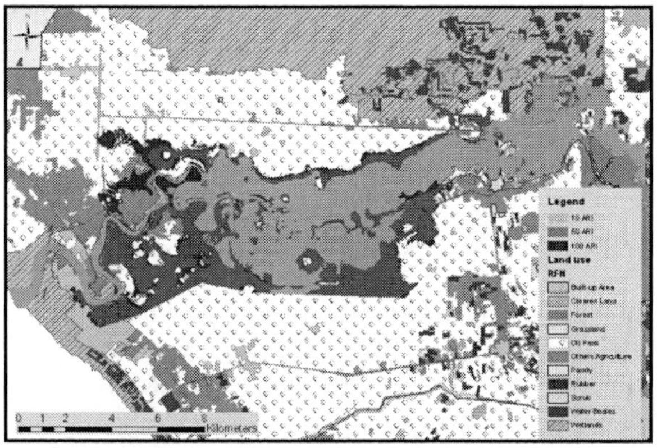

Fig. 2. The results for 10 ARI, 50 ARI and 100 ARI

3.1 Floods in Agricultural Sector

Based on agricultural cycle, excessive rainfall is detrimental to crops as yield is significantly affected, especially in oil palm and rubber plantations. Table 2 shows the estimated agricultural land that the model predicts to be affected.

Table 2. Estimated area of agricultural land affected by flood scenarios

Flood Scenario	Flooded area (ha)		
	10 ARI	**50 ARI**	**100 ARI**
Oil palm	1759.97	3642.32	4249.62
Rubber	39.68	70.41	79.39
Other agriculture	260.76	576.19	657.19
Total flooded in agricultural area	**2060.41**	**4288.92**	**4986.20**
Total flooded area	5846.35	9173.57	10178.40

The predicted flooded in agricultural areas for the 10 ARI, 50 ARI and 100 Ari scenarios are estimated to be 2060.41 ha, 4288.92 ha and 4986.2 ha respectively, which accounts for 47%, 62% and 64% respectively, of the total inundated area. The highest area affected by the floods is oil palm plantations, followed by the mixture of various types of agricultural crop and rubber plantations.

3.2 Estimation of Direct Flood Losses in Oil Palm Plantation

In order to simplify the calculation of losses in oil palm plantations, this study assumed that all oil palm trees were in the same maturity and they were in the early stage of growth. Besides, this study postulated that once the oil palm plantations are flooded, all the oil palm trees are damaged and cannot be harvested, so the worst case scenario is chosen. The average monthly Fresh Fruit Bunch (FFB) yields reported from Malaysian Palm Oil Board (MPOB) for March 2014 in Selangor state was 1.39 tonnes/hectare [31]. The reference price for FFB at the mill gate on March 18, 2014 is 634 RM/tonne [32]. This reference price was chosen because the simulation of flood model was carried out on March 18, 2014. The total loss on oil palm plantations for three flood events involved millions of ringgits. Table 3 shows the total area affected, total FFB and total price loss for each return period.

Table 3. The losses in oil palm plantations for each return period

Flood Event (years ARI)	Total Area (ha)	Total FFB (tonnes)	Total Price Loss (million RM)
10	1759.97	2446.36	1.55
50	3642.32	5062.82	3.21
100	4249.62	5906.97	3.75

3.3 Estimation of Direct Flood Losses in Rubber Plantation

Department of Statistics reported in the Monthly Rubber Statistics Malaysia [33] that the productivity of natural rubber in February 2014 was recorded at 129.1 kilogram per hectare. The prices of the rubber at the farm gate on March 18, 2014 were 510 cents/kg for low ammonia rubber and 605 cents/kg for rubber which contains high ammonia [33]. This study assumed that all the rubber trees in the study areas were in the same age and they will damage during floods. These assumptions are important in order to simplify the calculations of the rubber losses. Table 4 shows the total losses in rubber plantations during floods.

Table 4. The total losses in rubber plantations during floods

Flood Event (years ARI)	Total Area (ha)	Total NR (kg)	Total Price (thousands RM)	
			Low	High
10	39.68	5122.69	26.13	30.99
50	70.41	9089.93	46.36	54.99
100	79.39	10249.25	52.27	62.01

3.4 Flood Mitigation Structure in the Study Area

Mitigation is a strategy to reduce the long term impacts of disasters face by a nation or community. Flood mitigation refers to any structural or non-structural measures taken to reduce the flooding impacts. The flood mitigation structure proposed in this study was a retention pond. The retention pond was chosen because it is used to collect storm water runoff for the purpose of controlling of this runoff and it has been proved to be the effective mitigation option to reduce flood. The retention pond is located at an appropriate location along the Selangor River. It is located at 101°24'23.839"E and 3°25'57.396"N. The retention pond was built in a lower area so that water can easily flow from higher to lower land and go into the pond. This decision was made in order to reduce cost.

By building the retention pond in lowland, the cost of making sure the water flows go into the pond is reduced. Furthermore, the pond was built in an abandoned mine area which is located in Bestari Jaya. This location was chosen to ensure that the location of building the pond does not involve other residents. If the location of the retention pond has land owners, the cost of building the retention pond increases because the authorities have to pay for compensation for building on people's lands. The area of the retention pond is 341.5 ha. The depth of the retention pond is 8 meters depth. The size of the retention pond was determined based on the volume for 10 ARI, 50 ARI and 100 ARI. The retention pond can store only 2732 ha-m or 27,000,000 m^3 of water. Thus, this pond cannot store all the excessive water flows during the 10 ARI, 50 ARI and 100 ARI because the total volumes for these three various return periods are 3936.24 ha-m, 6408.65 ha-m and 7258.03 ha-m respectively. However, it can help in reducing the impacts of flood as it reduces the flood depths and flood extents. After the input of the retention pond was made, the simulations of the model were re-run to study the scenarios of the flood after being mitigated. Table 5 shows the area that will be flooded before and after flood mitigation and percentage reduced after flood mitigation is done.

Table 5. The percentages reduced in the study area after using flood mitigation structure

Flood Event (years ARI)	Area Before Mitigation (ha)	Area After Mitigation (ha)	Percentage Reduced (%)	Volume Before Mitigation (ha)	Volume After Mitigation (ha)	Percentage Reduced (%)
10	4368.94	2642.93	60.49	3936.24	2574.87	65.41
50	6969.5	3163.09	45.39	6408.65	3086	48.15
100	7841.61	3649.29	46.54	7258.03	3464.17	47.73

From the table, it is clearly shown that the retention pond helps in reducing the flood extent in the study area. Besides the reducing in flood extent, retention pond can also reduce the flood impact in term of flood depth. Table 6 also represents the decreasing in volume after applying flood mitigation condition in the study area. The decreasing in volume is due to the decreasing in flood depth and flood extent. The flood depth of the flooded area may not reduce in term of the flood height, but it reduces in term of flood coverage. To assess the flood damage more clearly, the flood maps were overlaid with the land use map. After the flood mitigation, the estimated flooded area in the agricultural area has been reduced. The floods cover oil palm area, rubber area, others agriculture area, built-up area, wetlands, water bodies and a few scrub, cleared land, and grassland.

3.5 Floods in Agricultural Sector after Mitigation

Even though the flooded area in agricultural land is reduced, the agricultural plantations are still affected by the floods. Table 6 shows the estimated agricultural land that the model predicts to be affected. The most affected area during flooding event is oil palm plantations, followed by the mixture of various types of agricultural crops and rubber plantation. These results are the same as the results before applying flood mitigation.

Table 6. Estimated area of agricultural land affected by flood scenarios after flood mitigation is done

Flood Scenario	Flooded area (ha)			Total Losses (million RM)		
	10 ARI	50 ARI	100 ARI	10 ARI	50 ARI	100 ARI
Oil palm	620.7	903.12	1193.58	0.55	0.80	1.05
Rubber	15.35	23.42	32.09	0.010	0.015	0.021
Other agricultural	73.45	95.23	145.16			
Total flooded in agricultural area	709.50	1021.77	1370.83			

4 Conclusion

This study presented an overview of flood mapping and GIS technology in order to determine the flood losses on agricultural land along the Selangor River. The impacts were determined in term of the price losses when the areas are flooded by three different flood event; 10 ARI, 50 ARI and 100 ARI. The flood damage during each flood events was calculated in the oil palm plantation area and rubber plantations. The total flooded area of agricultural land for 10 ARI was 2060.41 ha, 4288.92 ha for 50 ARI and 4986.20 ha for 100 ARI. The estimated total damage for these three return periods involved millions of ringgits. The calculated damage in agricultural area for 10 ARI was nearly RM2, 510,000 meanwhile for 50 ARI was RM5, 320,000. The highest damage was RM6, 150,000 during 100 ARI. An appropriate flood mitigation structure which was retention pond, in reducing the flood scenarios along the Selangor River was suggested, modeled and tested. The effects of the retention ponds were analyzed and evaluated for 10 ARI, 50 ARI and 100 ARI. By implementing retention pond, flood damage in agricultural area were reduced significantly by 67.33% for 10 ARI, 78.33% for 50 ARI and 74.14% for 100 ARI. It was found that the flood mitigation benefits in reducing the total flood damage in agricultural area for more than three million ringgits. This value is just an underestimation of the actual flood mitigation benefits as there are some assumptions made during the process of calculating the flood damage. Besides reducing the impacts of the floods in the study area, the flood mitigation can help in reducing the losses that country may face and the losses of the farmers' income. Other measures that can be used to overcome flood problem are by improving flood warning system so that people can have enough time to take action when flood occurs and by preserving wetlands and forestry which can act as sponges when flood overflows.

References

[1] W. Barreto, Z. Vojinovic, R. Price and D. Solomatine, Multi objective evolutionary approach to rehabilitation of urban drainage systems. J. water resour. plan. manage., **136**, 547-554, (2009)

[2] D.S. Kumar, D.S Arya and Z. Vojinovic, Modeling of urban growth dynamics and its impact on surface runoff characteristics, Comput Environ Urban Syst., **41**, 124-135, (2013)

[3] A. Mynett and Z. Vojinovic, Hydroinformatics in multi-colours – part red: Urban Flood and Disaster Management, J. Hydroinform, **11**, 166-180, (2009)

[4] Z. Vojinovic and J. van Teeffelen, An Integrated Stormwater Management Approach for Small Islands in Tropical Climates, Urban Water J., **4**, 211 – 231, (2007)

[5] Z. Vojinovic, D.P. Solomatine and R.K. Price, Dynamic Least-Cost Optimisation of Wastewater System Remedial Works Requirements, Water Sci Technol., **54**, 467-475, (2006)

[6] Z. Vojinovic, S. Sahlu, S. Seyoum, A. Sanchez, H. Matungulu, Z. Kapelan and D. Savic, Multi-objective rehabilitation of urban drainage systems under uncertainties, J. Hydroinform, **16**, 1044-1061, (2014)

[7] A. Sanchez, N. Medina, Z. Vojinovic and R. Price, An integrated cellular automata evolutionary-based approach for evaluating future scenarios and the expansion of urban drainage networks, J, Hydroinform, **16**, 319–340, (2014)

[8] Z. Vojinovic and A. Sanchez, Optimising Sewer System Rehabilitation Strategies between Flooding, Overflow Emissions and Investment Costs, *11th Int. Conf. on Urban Drainage, Edinburgh, Scotland*, (2008)

[9] A. Abdullah, A. Rahman and Z. Vojinovic, LiDAR filtering algorithms for urban flood application: Review on current algorithms and filters test. *Laserscanning* 09, **38**, 30-36, (2009)

[10] A.F. Abdullah, Z. Vojinovic and R.K Price, A Methodology for Processing Raw LIDAR Data to Support 1D/2D Urban Flood Modelling Framework, J. Hydroinform, **4**, 75-92,(2011)

[11] A.F. Abdullah, Z. Vojinovic, R.K and Price, Improved Methodology for Processing Raw LIDAR Data to Support Urban Flood Modelling - Accounting for Elevated Roads and Bridges, J. Hydroinform, **14**, 253-269, (2011)

[12] V. Meesuk, Z. Vojinovic, A. Mynett and A.F. Abdullah, Urban flood modelling combining top-view LiDAR data with ground-view SfM observations, Adv Water Resour **75**, 105–117, (2015)

[13] S.D. Seyoum, Z. Vojinovic, R.K Price and S. Weesakul, Coupled 1d and non-inertia 2d flood inundation model for simulation of urban pluvial flooding, J. Hydraul. Eng.-ASCE, **138**, 23-34, (2012)

[14] Z. Vojinovic, S. Seyoum, M.H. Salum, R.K Price, A.F. Fikri and Y. Abebe, Modelling floods in urban areas and representation of buildings with a method based on adjusted conveyance and storage characteristics, J. Hydroinform, **15**, 1150–1168, (2012)

[15] Z. Vojinovic and D. Tutulic, On the use of 1D and coupled 1D-2D approaches for assessment of flood damages in urban areas, Urban Water J., **6**, 183–199, (2009)

[16] Z. Vojinovic and M.B. Abbott, Flood Risk and Social Justice: From Quantitative to Qualitative, Flood Risk Assessment and Mitigation, **2**, (2012)

[17] Z. Vojinovic, M. Hammond, D. Golub, S. Hirunsalee, S. Weesakul, V. Meesuk and M. Abbott, Holistic approach to flood risk assessment in areas with cultural heritage: a practical application in Ayutthaya, Thailand. Natural Hazards, **81**, 589-616, (2016)

[18] Z. Vojinovic, S.D. Seyoum, J.M. Mwalwaka and R.K. Price, Effects of Model Schematization, Geometry and Parameter Values on Urban Flood Modelling, Water Sci Technol., **63**, 462-467, (2011)

[19] Z. Vojinovic, B. Bonillo, K. Chitranjan and R. Price, Modelling flow transitions at street junctions with 1D and 2D models, *7th Int. Conf. on Hydroinformatics,* (2006)

[20] M.B. Abbott, B.M. Tumwesigye and Z. Vojinovic, The fifth generation of modelling in Hydroinformatics, *7th Int. Conf. on Hydroinformatics,* (2006)

[21] K. H. Ho, A. H. Ghazali and S. F. Chong, Calibration and evaluation of modified tank model (flood forecasting model) for Kelantan river basin, Malaysia, *Proc. of the water Eng. Conf. Malaysia*, 23-24, (2002)

[22] M. E. Toriman and M. N. Shukor, An Analysis of Rainfall Interception on the selection Experimental Plot of Pangkor Hill Reserved Forest. J. of Wildlife and Parks, **22(2),** 169-178, (2007).

[23] A.H. Thieken, V. Ackermann, F. Elmer, H. Kreibich, B. Kuhlmann, U. Kunert and J. Schwarz, Methods for the evaluation of direct and indirect flood losses, *Proc. of the 4th Int. Sym. on Flood Defence: Managing Flood Risk, Reliability and Vulnerability, Toronto, ON, Canada,* **68**, 110, (2008)

[24] E. Genovese, A methodological approach to land use-based flood damage assessment in urban areas: Prague case study. *Technical EUR Reports, EUR, 22497*, (2006)

[25] L. C. P. M. Stuyt, J. E. A. Reinders, E. E. Van Der Hoek, E. G. M. Hermans, M. M. De Keizer and J. Icke, The Environmental Impact of Flooding of The Dutch 'Delta-Metropole', Flood Risk Management in Europe, 107-129, (2007)

[26] Z. Vojinovic, J. D. W. Ediriweera and A. Fikri, An approach to the model-based spatial assessment of damages caused by urban floods, *11th Int. Conf. on Urban Drainage,* **31** (2008).

[27] R. K. Price and Z. Vojinovic, Urban flood disaster management, Urban Water J., **5**, 259-276, (2008)

[28] R. Bustami, C. Bong, D. Mah, A. Hamzah and M. Patrick, Modeling of Flood Mitigation Structures for Sarawak River Sub-basin Using InfoWorks River Simulation (RS), World Acad Sci Eng Technol., 14-18, (2009)

[29] B. Mercer, Comparing LIDAR and IFSAR: What can you expect, *Proc. of Photogrammetric Week,* (2001)

[30] Institute of Water Resources Planning (IWRP), *Review and update the flood prevention plan for central provinces: Vu Gia-Thu Bon river*, Institute of Water Resources Planning, Hanoi, (2001)

[31] Malaysian Palm Oil Board (MPOB), MPOB Monthly Prices of Palm Oil Products Traded In March 2014 (RM/TONNE), Retrieved on April 7, 2014 from http://www.mpoc.org.my/

[32] Malaysian Palm Oil Board (MPOB), Monthly FFB Yield (Tonnes/Hectare), Retrieved on April 7, 2014 from http://www.mpoc.org.my

[33] Malaysian Rubber Board. Natural Rubber Statistics 2014, Retrieved on April 7, 2014 from http://www.lgm.gov.my

Modelling Technique for the Assessment of the Sub-Soil Drain for Groundwater Seepage Remediation

Mohamad Faizal Tajul Baharuddin[1,2,*], *Saiful Azhar* Ahmad Tajuddin[1], *Mohd Hazreek* Zainal Abidin[1], and *Aziman* Madun[1]

[1]Research Center for Soft Soils, Faculty of Civil and Environmental Engineering, Universiti Tun Hussein Onn Malaysia, 86400 Batu Pahat, Johor, Malaysia.
[2]Dept. of Water and Environmental Engineering, Faculty of Civil and Environmental Engineering, Universiti Tun Hussein Onn Malaysia, 86400 Batu Pahat, Johor, Malaysia

Abstract. Groundwater simulation technique was carried out for examining the performance of sub-soil drain at problematic site area. Sub-soil drain was proposed as one of solution for groundwater seepage occurred at the slope face by reducing groundwater table at Taman Botani Park Kuala Lumpur. The simulation technique used Modular Three-Dimensional Finite Difference Groundwater Flow (MODFLOW) software. In transient conditions, the results of simulation showed that heads increases surpass 1 to 2 m from the elevation level of the slope area that caused groundwater seepage on slope face. This study attempt to decrease the heads increase surpass by using different sub-soil drain size in simulation technique. The sub-soil drain capable to decline the heads ranges of 1 to 2 m.

1 Introduction

Groundwater seepage occurs on slope face may induced slope failures in the future if any relevant action is not taken. This seepage is triggered by existence of high pore water pressure in soil profiles. High pore water pressure measurements always related to groundwater tables observed at site and required a huge number of piezometers installation [1-2]. These works are involved intensive time and cost, and incomplete data coverage. The numerical groundwater flow model simulation technique will be able to overcome the limitation of the above method [3-5].This study used groundwater simulation technique for examining groundwater seepage appears at problematic slope face in Botani Park Kuala Lumpur. In addition, sub-soil drain approach was used in attempt to solve groundwater seepage by carried out groundwater modelling technique.

*Corresponding author: mdfaizal@uthm.edu.my

 DOI: 10.1051/matecconf/201710304010

2 Materials and Method

Panggung Anniversary of Botani Park in Kuala Lumpur was chosen as the study area to assess the ability of groundwater flow simulation technique in identifying the cause of seepage groundwater occurrence (Fig. 1 and 2). The establishment of conceptual model in groundwater simulation technique is important to set-up groundwater flow systems in study area. The model normally presented in the form of a block diagram or a cross section [6]. A few assumptions were made on conceptual model especially on the aquitard (sandy silt) layers and hydraulic properties. The model only considers the north and south of Panggung Anniversary as boundary condition with applied constant head boundary (Fig. 3). The observed water levels in the study were used to adjust the hydrogeology parameters applied in this model. Medium to coarse grained sand with dominant silt material formed the aquitard soil layer [7]. Aquitard thickness ranging from 13 to 25 m presumed as unconfined layer. The bedrock encountered at depths ranging from 13.0 to 25.0 m consists of inter-bedded shale and sandstone. Hydraulic conductivity ranging from 1.48 x 10^{-7} to 3.50 x10^{-8} cm/sec with an average of 9.25 x 10^{-8} cm/sec was used in the study for this area.

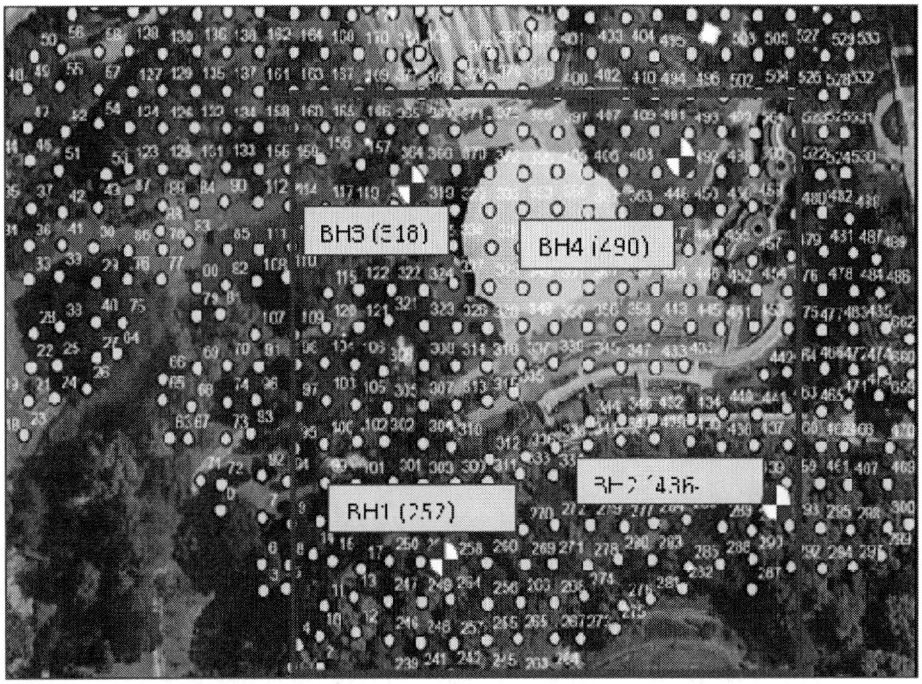

Fig. 1. The boundary locations of study area, boreholes location and surface elevation

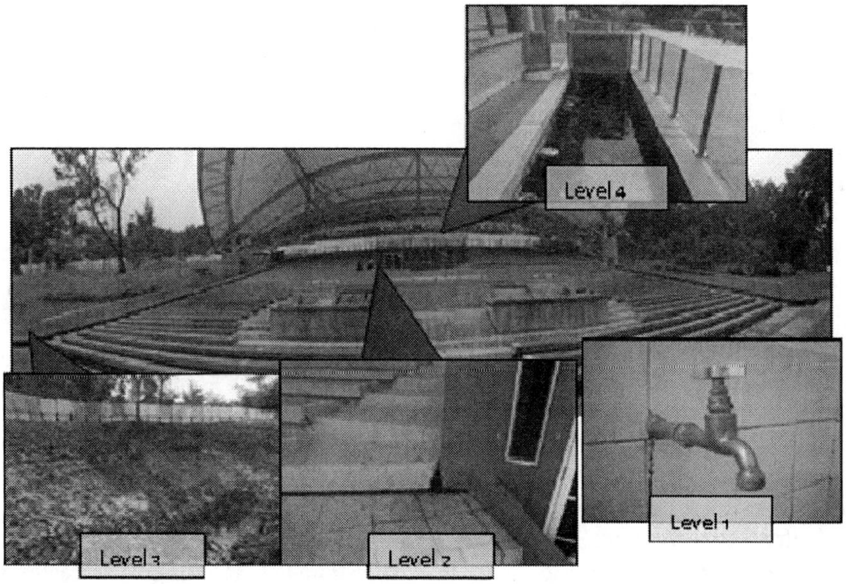

Fig. 2. Groundwater seepage locations at site

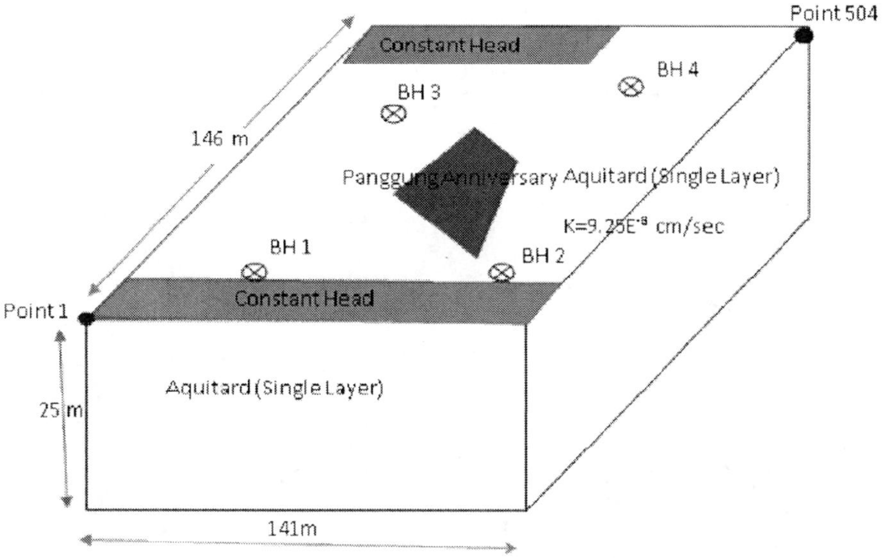

Fig. 3. Numerical and conceptual model at site.

2.1 Numerical model setup

Visual MODFLOW was used in creating three-dimensional representation of the site. In able to accelerate and facilitate the development of the MODFLOW model, Visual MODFLOW (Version 2003.3.1.0.85) was used to compile the input files. The three-dimensional model was constructed as a 141 m x 146 m area in the x and y directions (east–west and north–south directions, respectively) (Fig. 4). Total station theodolite was used to obtain the topography data from field survey (Fig. 1). The topography model (the first layer) was extracted from the elevation data for the wells (BH1-BH4). Surface elevations of the site was constructed by the digital elevation model. A total station theodolite was utilized to determine the relative elevations of water level monitoring points. The hydraulic heads in the model used mean sea level as the reference point (rectified skew orthomorphic). This permitted the assignment of the piezometer screens in the model at the terminated borehole depth.

2.2 Model parameters

The analysis data of permeability tests from BH2 and BH3 was used in estimating the hydraulic conductivities of the model layer. The dynamic viscosity value was used for estimating hydraulic condivities is about 8.01^{-4} Pa·s at 30°C. As the data were not enough to determine heterogeneity layer in this area, a single layer model was presumed to have uniform hydraulic properties. The data from previous studies [8-11] (Table 1) was utilized to estimate the model parameters data such as specific yields (Sy), specific storage (Ss), total porosity and effective porosity. Annual precipitation value of 5% and 20% of was used as an estimate of recharge for unavailable data [8]. The model calibration and validation processes has been carried out and recharge was found to be 10% of the average annual precipitation. 10% of total annual rainfall at open area was estimated as the recharge parameter of surface water into groundwater system [8]. Recharge parameter with zero value was used in closed area because there is no infiltration or recharge occur due to the concrete coverage of the Panggung Anniversary stage. Evapotranspiration values was estimated using 1.44 times by total annual transpiration at open area and zero value of evapotranspiration was imposed in closed area [8].

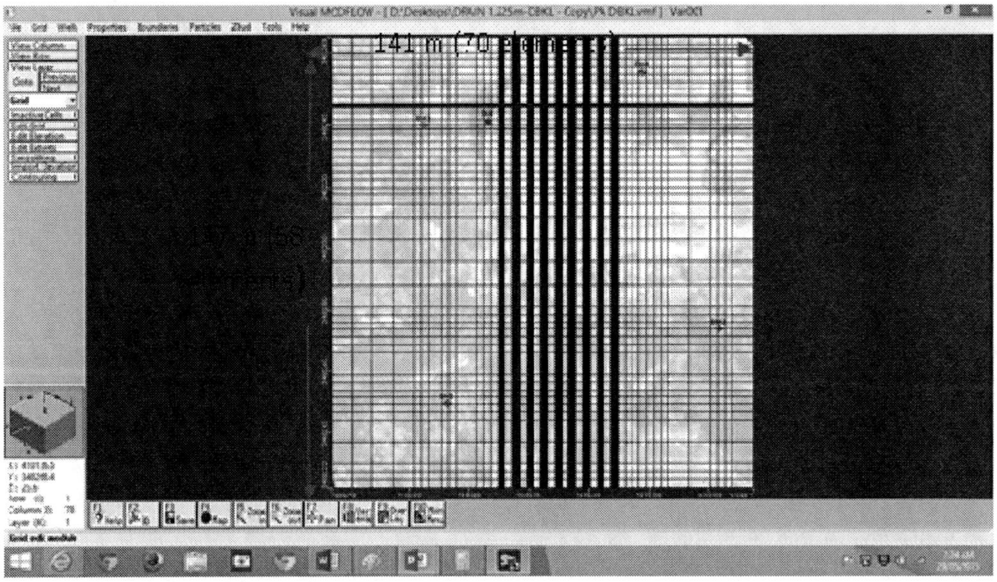

Fig. 4. Dimension of the grid mesh refine and area

2.3 Boundary Conditions

Northern and southern edges of the modelled area which are bounded by groundwater head of BH1-BH2 (south) and BH3-BH4 (north) were assigned with constant head boundaries. The heads in the model was referred at mean sea level datum. This depth was on average 5 m below the ground surface. The highest heads with an average head level of 48 m was located at the south area meanwhile the lowest heads with an average 38 m was located at the south area. DRAIN Package was applied as the model setup for the remediation technique in Visual MODFLOW software.

Table 1. Input for model parameters

Parameters	Value
K_x (cm/sec)	9.25e-8 (average value from hydraulic conductivity test at BH2 and BH3)
K_y (cm/sec)	9.25e-8 (average value from hydraulic conductivity test at BH2 and BH3)
K_z (cm/sec)	1e-9 [9]
S_s	1e-5 [10]
S_y	0.8 [11]
Total Porosity	0.3 [11]
Effective Porosity	0.15 [11]
Recharge (mm/year)	220 (10% of Annual Rainfall at barren area and 0 at Panggung Anniversary [8]
Evapotranspiration (mm/year)	500 and 0 at Panggung Anniversary [8]
Conductance (m^2/day)	0.00094 – 0.00562

2.4 Transient groundwater flow simulation and calibration

A partial differential equation was used as the equation for groundwater flow through the porous media:

$$\frac{\partial}{\partial x}\left(K_x\frac{\partial h}{\partial x}\right)+\frac{\partial}{\partial y}\left(K_y\frac{\partial h}{\partial y}\right)+\frac{\partial}{\partial z}\left(K_z\frac{\partial h}{\partial z}\right)=S_S\frac{\partial h}{\partial t}-W \qquad (1)$$

where K_x, K_y, and K_z are the hydraulic conductivities in the three orthogonal directions (m/s or ft/day), h is the head that drives the flow or the saturated thickness of the aquifer (m or ft),W is the volumetric flux per unit volume and represents the source/sink term for water or withdrawal (m^3/s or ft^3/day), Ss is the specific storage capacity of the porous medium (dimensionless), and t is the time (s or day). When the water withdraws, –W = R, where R is a general sink/source term, it is defined to be intrinsically positive to represent recharge (defines the volume of inflow to the system per unit volume of aquifer per unit of time (ft^3/ft^3/day or m^3/m^3/s)). The Visual MODFLOW model was run in transient state subsequently then calibrated to hydraulic heads recorded on a weekly basis from 16 April 2015 to 7 May 2015 at four standpipes (BH1-BH4) for calibration. An analysis of the observed versus computed water levels for each transient model run was conducted to determine the accuracy of the simulation.

3 Result and discussion

3.1 Calibration Groundwater Model

The computed difference between the simulated and the observed hydraulic heads was obtained from the assessment of the model fit. In this model the indication of a well calibrated model was shown by the estimated average of mean absolute error (MAE) of 0.5 m, the root mean square (RMS) of 1.2 m, and 12% of the normalized root mean square error (NRMS). As a result, the calibrated model was able to be use to simulate water levels and water level fluctuations satisfactorily to appraise the options of remediation methodology for groundwater seepage problem.

3.2 Sub-Soil Drain

In order to overcome a groundwater seepage problem at levels 2 and 4, remediation method using sub-soil drain was proposed. Two options on the sub-soil drain dimensional design were recommended taking into consideration of the availability of the materials in market.

Option 1:- 0.15 m (diameter), 40 m (length), 0.60 m (depth)
Option 2:- 0.10 m (diameter), 40 m (length), 0.60 m (depth)

Option 1 and 2 sub-soil drain showed heads result in ranges between 41 and 44 m at level 2 and 4 (Fig. 5 and 6), respectively. Drawdown is the difference in the head between the initial head (without sub-drain) and the head with sub-drain (option 1) showed difference of 1-2 m. It was suggested that option 1 and 2 to be use in the remediation of the groundwater seepage. This solution can reduced the heads to the optimal values to inhibit settlement of the stage structure.

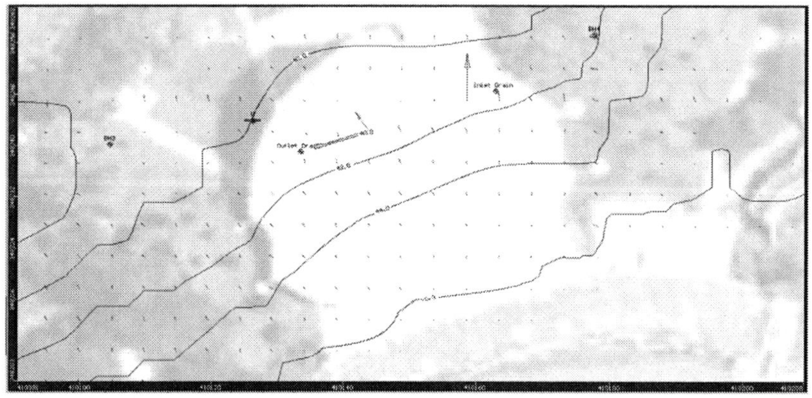

Fig. 5. Head results for Option 1

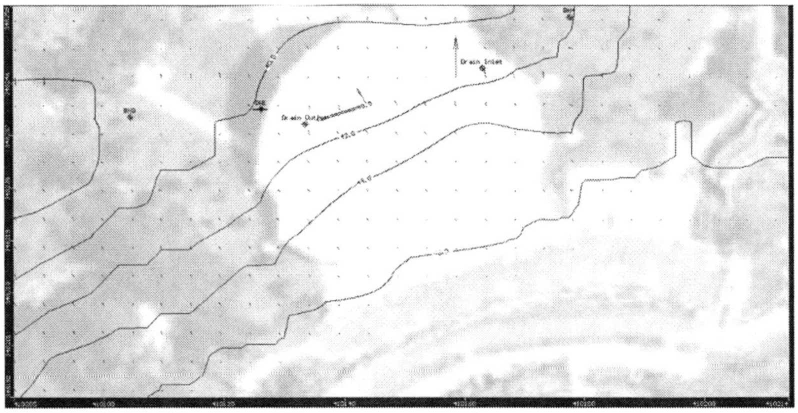

Fig. 6. Head results for Option 2

4 Conclusion

This study has showed that the numerical simulation groundwater flow (Visual MODFLOW), has succeeded in determining the cause of groundwater seepage problem in the study area. The elevated groundwater heads compared to the elevation levels of the area was identified as the cause for the groundwater seepage occurrence at Panggung Anniversary. The groundwater seepage caused by the increase of heads level can be overcome by reducing the head levels between 1 – 2 m. In order to reduce heads level, the installation of sub-drain with dimensional design of 0.10 to 0.15 m (diameter), 40 m (length) and using 0.60 m depth can be used. In order to prevent any structure settlement from occurring, the safest method to be use is the dimensional design remediation with sub-drain soil.

This project was supported by IKRAM Engineering Services Sdn Bhd. Many thanks also goes to all research members for their tremendous guidance and cooperation.

References

[1] M. F. Tajul Baharuddin, S. Taib, R. Hashim, M. H, Zainal Abidin, and M. F, Ishak, Time-lapse resistivity investigation of salinity changes at an ex-promontory land: a case study of Carey Island, Selangor, Malaysia, Environ. Monit. Assess., **180**, 345-369, (2011)

[2] M. H, Zainal Abidin., M. F. Tajul Baharuddin, M.H. Zawawi, N. Md. Ali, A. Madun, and S.A. Ahmad Tajudin. Groundwater Seepage Mapping using Electrical Resistivity Imaging, Applied Mechanics and Materials, 1524-1534, (2015)

[3] M. F. Tajul Baharuddin, S. Taib and R. Hashim. Electrical imaging resistivity study at the coastal area of Sungai Besar, J. of Applied Sciences, **9**(16), 2897-2906, (2009)

[4] M. F. Tajul Baharuddin, S. Taib, R. Hashim, M. H, Zainal Abidin, and M.A. Rashid, Evaluating freshwater lens morphology affected by seawater intrusion using chemistry-resistivity integrated technique: a case study of two different land covers in Carey Island, Malaysia, Environ Earth Sci., **69**, 2779–2797, (2013)

[5] M. F. T. Baharuddin,S. A. A. Tajudin, M. H. Z. Abidin and N. A, Yusoff, Simulation of Sub-Drains Performance Using Visual MODFLOW for Slope Water Seepage Problem *IOP Conf. Series: Materials Science and Engineering*, 136 (2016)

[6] J. Bear and A.H.-D. Cheng, *Seawater intrusion. In Modeling Groundwater Flow and Contaminant Transport*, Springer Science &Business Media, 593–636 (2010)

[7] IKRAM, Seepage Problem and Groundwater Studies at Panggung Anniversary, Taman Botani, Kuala Lumpur, IKRAM Preliminary Report, (2015)

[8] Waterloo, Visual MODFLOWv 4/1 User's Manual. Waterloo Hydrogeologic Inc., Waterloo, (2005)

[9] J. Bear, (Eds.). *Hydraulics of groundwater*, 384-435. New York, USA: McGraw-Hill (1979)

[10] C.W. Fetter, (Eds),*Applied Hydrogeology (4th Ed)*, New Jersey, USA: Prentice Hall Inc., 1-598,(2002)

[11] C.R. Fitts, (Eds.),*Ground Water Science*, London: Academic Press, (2003)

The Hydrological Performance of Lightweight Green Roofs Made From Recycled Waste Materials As the Drainage Layer

Nurul Shahadahtul Afizah Asman[1,*], *Salinah* Dullah[1], *Janice Lynn* Ayog[1], *Abdul Karim* Mirasa[1], and *Harimi* Djamila[1]

[1]Faculty of Engineering, Universiti Malaysia Sabah

Abstract. Green roofs can be used for promoting infiltration and provide temporary storage spaces. Hence, in urban stormwater structural design, the investigation of the hydrological performance investigation is often required. Thus, this paper presents the results of a hydrological investigation in term of peak flow reduction and green roof's weight using 0, 2, and 6% slope for three specimens drainage layer in green roofs. Three types of recycled waste are selected for each test bed which is rubber crumbs, palm oil shell, and polyfoam. Another test bed without a drainage layer as a control. The result indicates that rubber crumbs can be used as a stormwater control and runoff reduction while ensuring a good drainage and aeration of the substrate and roofs. From the results obtained shows that rubber crumbs are suitable as a drainage layer and a proposed slope of 6% are suitable for lightweight green roofs.

1 Introduction

In Malaysia, sustainable issues have become critical issues. Also, even though the average rainfall in Malaysia per year is 3500 mm, problems in the stormwater such as flash flood, water shortage, and water pollution still occurs. As stated by Ayub et al. [1], the conventional method in stormwater management such as using the concrete drain in a new development area cannot fulfill the expectation in controlling the volume and quality of runoff water. So, a green roof is a solution for sustainable problems. For other developed countries, green roof applications are widely used. However, in Malaysia green roof can be considered the relatively new concept and the less of application since only a few buildings in Malaysia practice the green roof construction on buildings.

Due to some sustainable practices, a recycled waste material for a green roof is introduced for the drainage layer in the green roof. This drainage layer is important since it must able to retain water during raining and ensuring a good drainage system and the aeration of the top soil and plant roots. A roof slope is believed can affect the water retention capacity. Thus the paper presenting a green roof system are tested under three different slope which is 0, 2 and 6% slope. The weight of designing a green roofs system is

[*] Corresponding author: nurulshahadahtul@gmail.com

necessary. Thus, the issues of the live load in wet and dry conditions are considered in determining the lightweight green roofs.

Objectives of this paper are to share the preliminary result found regarding the drainage layer used in green roofs using three different recycled waste materials for intensive green roofs in attenuation runoff and as lightweight green roofs in simulated rainfall.

2 Research background

As stated by Berndtsson [2], there are two types of green roofs which are intensive and extensive. The extensive green roof is the focus of this research due to the wide application and the weight which is lighter than the intensive green roof.

Green roof systems consist of consists of waterproofing layer, drainage layer (40 mm), a filter layer, substrate layer or a topsoil (50 mm) and growing medium [3]. Extensive green roofs are different from intensive because it is using smaller plants, less in maintenance and are established with thin substrate layers.

Chow, Bakar and Deck [4] in their research mention that it have been proven that it green roofs can produce many benefits such as stormwater mitigation, biodiversity, and habitat, save energy, air quality, aesthetics and quality of life, increased roof membrane durability, noise reduction and providing new amenity spaces.

According to the survey done by Ismail et al. [5] in Malaysia there are some obstacles of the application of green roofs, and it was proven that several factors were identified. The factors are past failure, challenging and high cost, complicated and hard to maintain, limited local expertise, lack of scientific research, fear of unknown risk, higher cost of materials supply, no design standards and guidelines, and believe that green roof is susceptible to fire.

Research on using recycled materials utilized for a drainage layer in a green roof are done by some researcher due to some sustainable practices to be applied in a green roof's design. Perez *et al.* and Rincon *et al.* [6-7] in their research are testing the use of rubber crumbs as a drainage layer by focusing on thermal reduction and hydrological performance. This recycled waste material can contribute to reducing waste materials production thus lessen the use of raw materials. Shahid *et al.* [8] also in their research mention that palm oil clinker can be used as the drainage layer because it has an excellent ability to drain the excess water and no effect on plant development. Roof slope hence is necessary for green roofs. The green roof slope is one of the factors that can affect green roof water retention capacity and the runoff [2].

As stated by Chow et al. [4], a good knowledge of engineering is essential in designing a green roof since all the critical aspects of design must be included such as the plant selection, environmental at the region and the weight of the system itself.

3 Study area and methodology

The study carries out in hydraulics laboratory of Faculty of Engineering, Universiti Malaysia Sabah. Test beds are placed under the rainfall simulator as shown in Fig. 1. Table 1 shows three test bed represent different recycled waste and one test bed as a control bed. The weight of the green roofs for dry conditions are measured using floor scale on 7-day antecedent dry conditions of the green roofs while the weights for wet conditions are collected after rainfall simulations have ended.

The inflow from rainfall simulator are for 15 minutes continuously. The value for outflow was taken for one hour continuously after the simulator started with five minutes interval. Three types recycled waste materials are used as the drainage layer as shown in Fig. 2.

Table 1. Materials used for control specimen and drainage layer.

Specimen				
Materials	Control	Test Bed 1	Test Bed 2	Test Bed 3
	Waterproofing layer	Waterproofing layer with Palm Oil Shell	Waterproofing layer with Polyfoam	Waterproofing layer with Rubber Crumps

Fig. 1. Testing under a rainfall simulator.

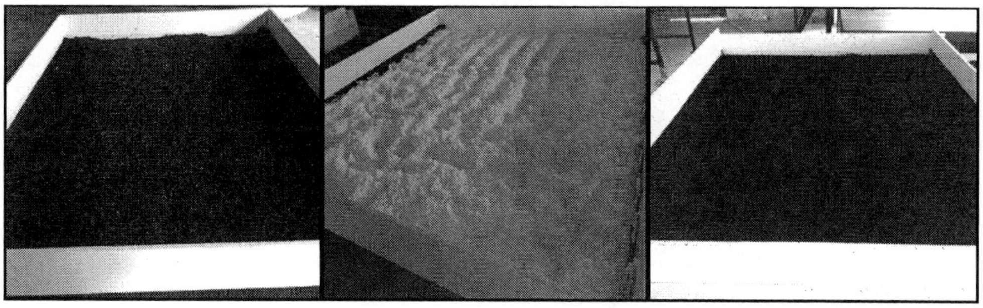

A B C

Fig.2. Drainage Layer used in the Green Roofs: A. Palm Oil Shell, B. Polyfoam, C. Rubber Crumbs.

4 Experimental results

4.1 Peak runoff

Three types of drainage layer are used in the testing, and four simulated rainfall has been done. Fig. 3 show the peak flow comparison among drainage layer using recycled waste materials on 0, 2, and 6% slope. Rubber crumbs have the lowest peak flow in all experiments data with the different slope in the range 17.67 to 26.00 cm^3/min. From the data recorded, the rubber crumbs have the highest flow reduction and able to attenuate peak flow. Second waste materials that have better hydrological performance in attenuation of peak flow is polyfoam. It gives the range of peak flow from 23.33 to 39.00 cm^3/min. Palm oil shells show the highest peak flow compare to other waste. The maximum peak flow for this waste is 43.33 cm^3/min whereas the lowest is 36.67 cm^3/min.

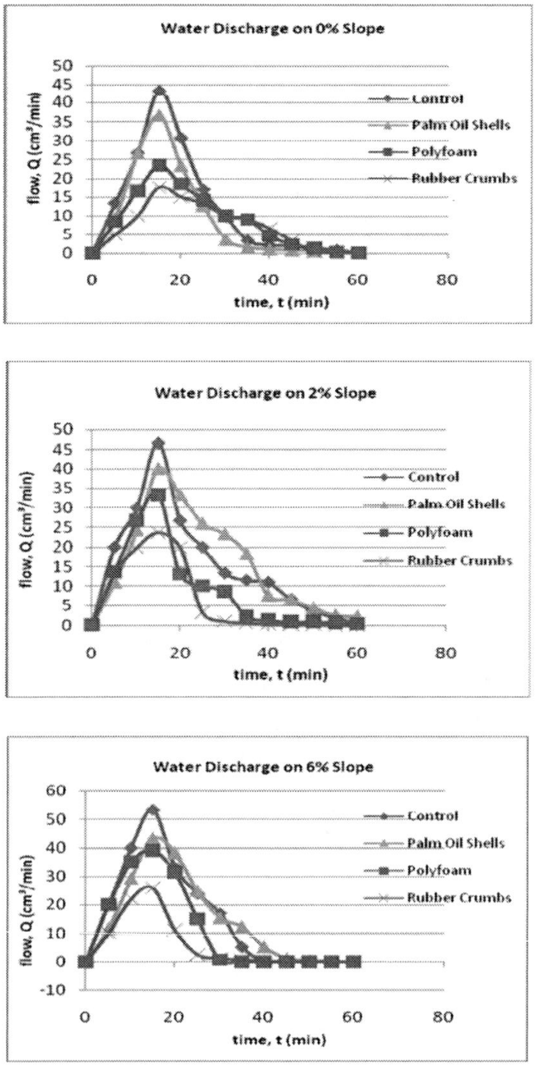

Fig.3. Peak flow comparison among drainage.

4.2 Weight of green roofs

In this study, all wasted materials studied have different weights. However, the most effective drainage layer is which gives the lowest weight. Fig. 4 and 5 show the details result from this study. Polyfoam has the lighter weight with 3.09 kg/m^2 increased weight on 0% slope. After the slope are increased, the weight of sample increased for the wet weight (0% slope: 3.51 kg/m^2, 2% slope: 1.53 kg/m^2, 6% slope: 1.28 kg/m^2). Followed by rubber crumbs (0% slope: 3.51 kg/m^2, 2% slope: 1.53 kg/m^2, 6% slope: 1.28 kg/m^2). The heaviest waste materials is palm oil shells (0% slope: 28.75 kg/m^2, 2% slope: 25.17 kg/m^2, 6% slope: 22.95 kg/m^2). From the figure also can be seen that the weight of the sample decreases as the slope are increased by 2 and 6% in slope. As shown in the results are believed due to the higher slope used to help water discharge and retention in the green roofs are reduced.

According to Table 2, for both 2% and 6% slope of green roofs, polyfoam have the highest reduction in weight where more than half percent of reduction which are 56.44% for 2% slope and 63.37 for 6%. Then, followed by rubber crumbs (2% slope:15.91%, 6% slope: 17.13%) and palm oil shells (2% slope:12.33%, 6% slope: 20.07%), respectively. For all the recycled waste materials, 6% of slope have the higher reduction in weight.

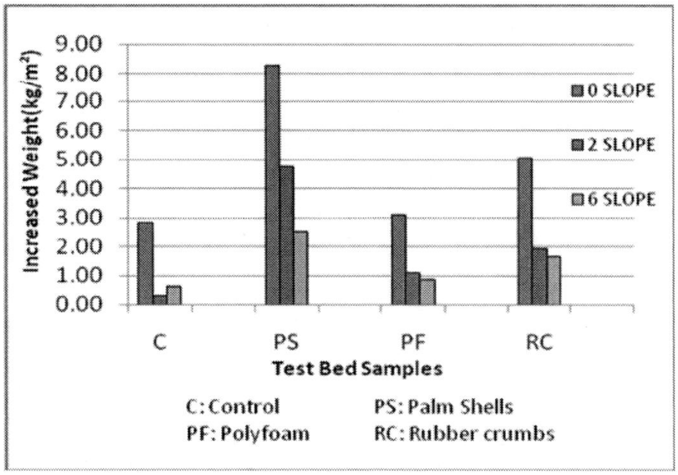

Fig.4. Increased Weight of Green Roofs for Drainage Layer with 0, 2 and 6 slope

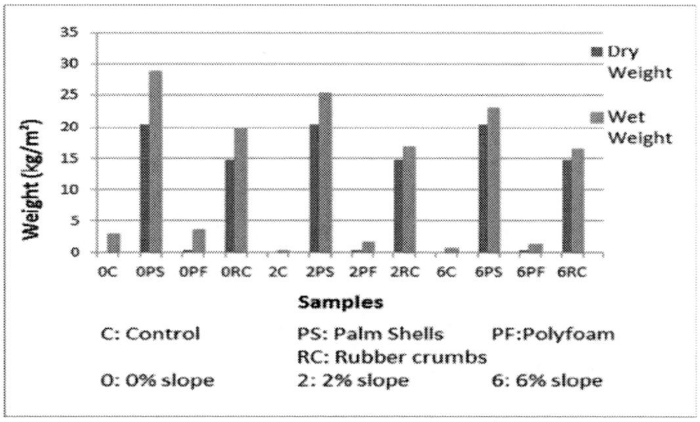

Fig.5. Weight of Green Roofs for dry and wet weight

Table 2. Reduction in weight of green roofs for drainage layer

Samples	Weight Reduction (%)	Weight Reduction (%)
	Slope : 2%	Slope : 6%
Control	89.02	78.05
Palm Oil Shells	12.33	20.07
Polyfoam	56.44	63.37
Rubber crumbs	15.91	17.13

5 Conclusions

Based on capabilities in peak flow attenuation and lighter green roofs, this study found the rubber crumbs suitable as green roof drainage layer beside this waste able to retain water when it rains, suitable for filtration while ensuring good drainage and aeration of the substrate and roots. The usage of rubber crumbs as waste materials in green roofs layer are suitable for stormwater management where it can reduce stormwater runoff, so there is less water directed into the storm drains. Second wastes material that is suggested for a green roof is polyfoam. The proposed slope for the green roofs is 6% since it ensures that large amounts of water are not retained on the roof, this implies a reduction in weight of the green roofs system. However, further details study for the green roofs with a combination of natural fibres with the same intensity of rainfall is needed.

This research was supported by University Malaysia Sabah under the Grant RAG0061-TK-2014. The authors of this paper also would like to thank the Faculty of Engineering, University Malaysia Sabah for providing the hydraulics lab for laboratory work.

References

[1] K.R. Ayub, A.A. Ghani, N. A, Zakaria and S. S. Efficiency of intensive green roof in high-intensity rainfall for stormwater treatment: selection of vegetations, *E-proceedings of the 36th IAHR World Congress*, **36**(1), 6730-6735, (2015)

[2] J. C. Berndtsson, Green roof performance towards management of runoff water quantity and quality: A review. Ecological Engineering, **36**(4), 351–360, (2010)

[3] N. Shahadahtul, S. Dullah and J. L. Ayog, The hydrological performance investigation of light weight green roofs made from natural fibres and recycle waste materials for stormwater runoff mitigation: A review, J. of Scientific Research and Development, **2**(12), 1-6, (2015)

[4] M. F. Chow, M. F. A. Bakar and R. Deck, A Review on the Development and Challenges of Green Roof Systems in Malaysia, **10** (1), 16–20, (2016)

[5] Z. Ismail, H. A. Aziz, N. M. Nasir, M. Zafrullah, M. Taib and S. Alam, Obstacles To Adopt Green Roof In Malaysia, (Chuser), 357–361, (2012)

[6] G. Pérez, A. Vila, L. Rincón, C. Solé and L. F. Cabeza, Use of rubber crumbs as drainage layer in green roofs as potential energy improvement material, Applied Energy, **97**, 347–354, (2012)

[7] L. Rincón, J. Coma, G. Pérez, A. Castell, D. Boer and L. F. Cabeza, Environmental performance of recycled rubber as drainage layer in extensive green roofs, A comparative Life Cycle Assessment. Building and Environment, **74**, 22–30, (2014)

[8] K. A. Shahid, M. A. Sulaiman, N. Ghazali, W. A. Rahman and A. Zukri, Palm Oil Clinker as Drainage Layer in Green Roof System under Malaysia Climatic Conditions, J. of Engineering and Technology, **5**(2), 27-37, (2014)

High Resolution Decision Maps for Urban Planning: A Combined Analysis of Urban Flooding and Thermal Stress Potential In Asia and Europe

Floris Boogaard [1,2*], *Zoran* Vojinovic[5], *Yu*-Cheng Chen[3], *Jeroen* Kluck [2,4] *Tzu*-Ping Lin[3]

[1]Hanze University of Applied Sciences Groningen, Zernikeplein 7, P.O. Box 30030, Groningen, The Netherlands
[2]Tauw bv, PoBox 133, 7400 AC Deventer, The Netherlands
[3]National Cheng Kung University, Daxue Rd, Tainan City, Taiwan 701, Taiwan
[4]Amsterdam University of Applied Sciences, Weesperzijde 190, 1097 DZ Amsterdam, The Netherlands
[5] UNESCO-IHE, Westvest 7, 2611 AX Delft, The Netherlands.

Abstract. Urban flooding and thermal stress have become key issues for many cities around the world. With the continuing effects of climate change, these two issues will become more acute and will add to the serious problems already experienced in dense urban areas. Therefore, the sectors of public health and disaster management are in the need of tools that can assess the vulnerability to floods and thermal stress. The present paper deals with the combination of innovative tools to address this challenge. Three cities in different climatic regions with various urban contexts have been selected as the pilot areas to demonstrate these tools. These cities are Tainan (Taiwan), Ayutthaya (Thailand) and Groningen (Netherlands). For these cities, flood maps and heat stress maps were developed and used for the comparison analysis. The flood maps produced indicate vulnerable low-lying areas, whereas thermal stress maps indicate open, unshaded areas where high Physiological Equivalent Temperature (PET) values (thermal comfort) can be expected. The work to date indicates the potential of combining two different kinds of maps to identify and analyse the problem areas. These maps could be further improved and used by urban planners and other stakeholders to assess the resilience and well-being of cities. The work presented shows that the combined analysis of such maps also has a strong potential to be used for the analysis of other challenges in urban dense areas such as air and water pollution, immobility and noise disturbance.

1 Introduction

Stormwater flooding and thermal stress have become two very important issues for most urbanised areas. Municipalities and authorities in charge of water management, urban

* Corresponding author: Floris@noorderruimte.nl

planning and building construction need insight into the risks of the storm water flooding and thermal stress. This is required in order to develop cost effective mitigating measures, preferably at a street or even house level.

Due to the climate change, the two issues of flooding and thermal stress are expected to increase in frequency and impact in the near future, especially for dense urban areas. For example, increasing temperatures are resulting in higher urban heat stress causing cities to be unpleasant and unhealthy places to live and work. Extreme heat waves ultimately result in a significant increase in the urban mortality rate due to heat stress [1].

It is therefore essential that public health and disaster management sectors adequately assess the vulnerability to stormwater flooding and thermal stress. Since both issues are expected to involve similar measures in the urban planning and design, it is logical to try to analyse and solve these issues simultancously. The interaction of the two issues therefore needs to be further investigated.

To achieve this goal, three cities in different climate regions with varying urban contexts were selected as pilot areas; Tainan in Taiwan, Ayutthaya in Thailand and Groningen in The Netherlands. Maps of stormwater flooding and thermal stress for those case studies have been compared in this study.

2 Method

2.1 Study areas

Tainan is a highly developed city (22°59'N, 120°11'E) in the tropical south of Taiwan. The total area of Tainan is 175.6 km^2, including 6 administrative districts and a population of almost 2 milion people. Groningen (53°13'N, 6°34'E) has a population just over of 200.000. It is the largest city in the north of the Netherlands with total area of 83.69 km^2. Ayutthaya city (14°21′ N, 100°35′E) covers an area of 14.84 km^2 with a population around 70.000.

Fig. 1 shows the average rainfall andtemperature of Tainan and Groningen. The comparisons show that the average temperature of each month varies between 18-30°C in Tainan, and between 2-16°C in Groningen. Concerning the precipitation, Tainan has the largest rainfall during June to August due to the typhoon season. The monthly rainfall in Groningen is lower than the peak in Tainan, and is more constant over the year. Ayutthaya has a tropical monsoon climate and the highest average temperature of any city in the world. Temperatures in Ayutthaya regularly remain well above 30°C throughout the year. The humidity levels are also high during this period and you can expect short spells of rainy weather, with frequent afternoon showers and occasional thunderstorms and monsoons [2].

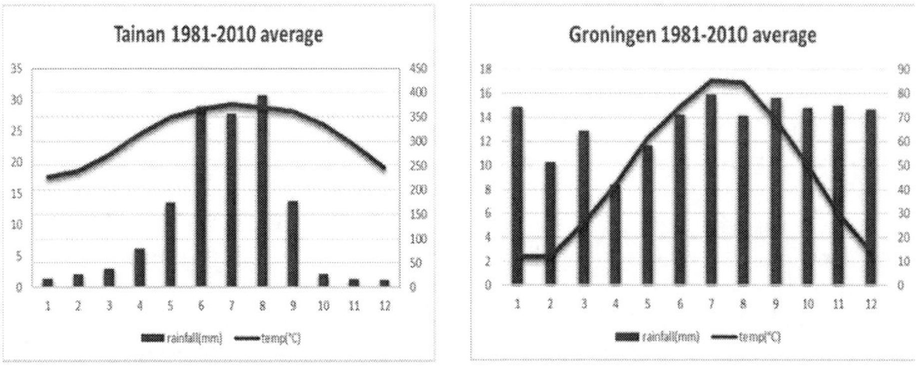

Fig. 1 Climate data in Tainan and Groningen

2.2 Basic information collection

The basis for the flood modelling in the Groningen case is an accurate DEM (Digital Elevation Model) of the urban area with a high resolution (4 points/m^2), maps of pervious and impervious surfaces and surface water. In addition, the heat stresss maps use information of land use including aerial photographs that allow green areas and trees to be mapped.

For Tainan an accurate DEM is unfortunately not available. Therefor a low resolution DEM (30 m resolution) was used in this study, validated by mobile field measured elevation data. The basic urban development information to estimate thermal stress potential was collected from the government and included the land use, population density, total floor area and height of building. The meteorological data in Tainan were derived from the central weather bureau. These data were used to set up the initial conditions for thermal stress calculation, and further make the verification of simulation results. In the case of Ayutthaya (Thailand), a DEM with 20 m resolution was applied. [3].

2.3 Spatial resolution of the mapping

The spatial resolution for the three cases differ greatly. This difference is due to the fact that a compromise had to be made between the size of the study area, the spatial resolution and the accuracy of modeling. In general, the larger the study area the simpler or less accurate the quick scan models need to be (physical simplifications and coarser resolutions) because of computational limitations. The continued improvement of the computational capabilities of computers and DEMs with high resolutions means it is now possible to make more accurate models with high resolutions. However to keep computational times limited, it is best to simplify the modeling of the physical processes. Comparing the results of the quick-scan assessments with results of more complete and more accurate models (for a limited area) shows that if the quick scans have sufficient detail they give valuable results.

2.4 Flood mapping

For the flood map analysis in Groningen, the Calamity Levels of Urban Drainage Systems (CLOUDS) was used. This is 'quick-scan' method to simulate storm water floodings [4]. CLOUDS is based on the assumption that for a cloud burst (> 60 mm/h) most of the stormwater will flow and stay above ground. CLOUDS visualizes the streamlines and the depth of stormwater in depressions where water will accumulate. The quick-scan is based on only readily available data. The most important is an accurate DEM, which is freely available for the whole of the Netherlands (AHN2). With 9 points per square meter and a vertical accuracy of several centimeters this provides an insight in the surface elevation. Other data used are: GIS-maps with infrastructure and houses, the TOP10 (open data topographic data for the roads and waterways) and the BAG (open dataset with all buildings (2d) of the Netherland). Combining the elevation model, the dataset with buildings and aerial photographs a 3d model of the city is constructed to get a better overview of the outcomes of the model.

The model was shown in the 3D virtual reality theater on a cylindrical screen using 6 HD projectors to project an image with a resolution of roughly 5000x1800. To display this model on a big screen a special 3D viewer was used, based on the open source OpenSceneGraph 3D toolkit. The software was running simultaneously on 7 PC's, one master PC for the control of the model and 6 slave PC's to drive the projectors.

The resulting maps show the expected water depths for cloudbursts expected once per 100 years (for The Netherlands 60 in one hour).

For Ayutthaya (Thailand), the flood modelling work involved the use of one-dimensional and two-dimensional (1D/2D) models. With such models, it was possible to explore the floodwater flows that are generated along roads, through fences and around buildings and to simulate and also optimize the consequent effects of any remedial works requirements [5-7]. The coupled 1D/2D models are invaluable for this purpose. However, there is a great body of literature which suggests that great care is needed in the data collection and processing as well as the need to use an appropriate numerical modelling technique [8-16]. The DEM resolution used in the Ayutthaya work was 20x20 m. The model and maps produced were used for communication with local stakeholders and selection of remedial and flood protection works [3].

In Taiwan, the flooding information is achieved by the flooding potential map offered by the government. The map is based on the precipitation depth in various return periods, highlighting the potential flooding area based on a certain precipitation scenarios.

2.5 Thermal stress maps

Two types of models have been used to assess the thermal stress. A quick-scan GIS-based thermal stress map for large scale (city scale), and a prognostic model EVNI-met for small scale (city quarter).

The quick-scan GIS-based thermal stress maps have been developed in The Netherlands in order to give a quick insight in possible thermal stress locations in a city. It is based on accurate DEM and the assumption that for a quick insight of thermal stress some rough simplifications of the actual physical processes can be made. The maps give an estimate of the maximum PET (physiological equivalent temperature) during a heat wave. Such maps also have been made for Groningen and Bangkok [17].

ENVI-met software is typically used to accurately simulate the climate conditions in urban areas. The model had been adopted to conduct research in many of countries. It is much more elaborate than the GIS method and attempts to take into account all important processes. In this study, ENVI-met is used to make a thermal stress map for a part of the city centre of Groningen and a part of Tainan (Fig. 5).

The thermal maps for the Dutch and Asian cases are used to compare the differences in simulation results between different climates zones [16]. It gives insight in the differences in hot spot distribution and thermal stress levels between Taiwan, Thailand and the Netherlands. Next to this the comparison of the GIS tool and the ENVI-met tool is used investigate the limitations and advantages of the different tools.

3 Results

3.1 Flood maps

The flood map of Tainan was made for the scenario of a three days heavy rain of over 1200 mm. The map shows that, the place with a lower elevation and a higher impervious pavement is vulnerable to flooding. However some places with low elevations but high pervious pavement ratios will be more safe when facing the flood related issues (e.g. parks and schools).

The flood map of Groningen shows inundated points in the center that are known to flood during heavy stormwater events with a maximum level about 50 cm. Social media posts, photos and videos from citizens were gathered by researchers to verify these points

(Fig. 2) and have been mapped on www.climatescan.nl [18]. Based on the insight in the vulnerable flooding spots (Fig. 3) the municipality has planned to implement some sustainable urban drainage systems (SUDS) to minimize these floods. The municipality is advised to reduce the vulnerability to flooding by reshaping some streets during reconstruction of the area. Those streets will be designed in such a way that they can convey the excess storm water without causing flood problems. The changes in Groningen and other cities as Amsterdam [4] will be carried out during a planned retrofitting of the urban spaces [16-17]. This way the roads themselves will act as a cost effective drainage system by discharging stormwater to the surface water bodies, preventing the flooding of houses. Similar plans have been developed using the results of the flood modelling in Ayutthaya (Fig. 4).

Fig. 2. Floodmap Groningen (The Netherlands)

Fig. 3. Combined flood and heatstress map Groningen

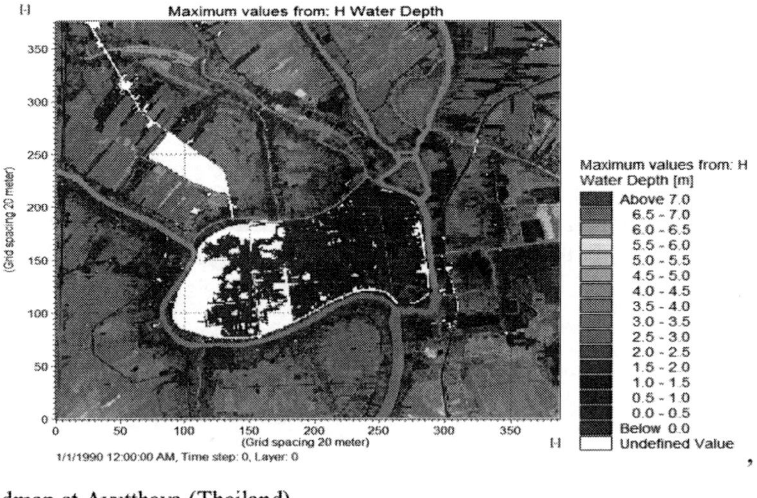

Fig. 4. Floodmap at Ayutthaya (Thailand)

3.2 Thermal stress map

Thermalhermal stress maps have been used to identify hot spots within the cities. PET is a indicator for thermal stress. PET-values in the hotter Asian cases are of course much higher than in the Dutch case due to higher air temperatures and radiation values [16]. The levels at which people perceive heat stress is also different. For The Netherlands there are no guidelines with maximum allowable PET-values. The maps are therefore mainly used to get attention for the issue of heat stress and to see the differences within the city.

The maps revealed that Tainan generally has high thermal stress (Fig. 5). Spaces with more greenery and pervious pavement like parks, have much lower thermal stress, than dense areas with high buildings and more impervious pavement, e.g. commercial districts show significantly more extreme thermal stress in summer time.

The thermal stress map of Groningen shows hotspots (Fig. 3 and 7) in the center where there is little green space or water bodies. The Noorderplantsoen, a park on the left top of figure 3, shows lower temperatures in the shades of the trees and near the water that flows through this green area. Climate measurements in these areas were taken at the same time as interviewing users of the public space. These results generally verified the quick-scan model results. An observation was that during the heat wave in the summer of 2015, inhabitants of the city avoided open paved shopping streets and marketplaces and instead retreated to the green park areas.

Although different models have been applied to predict the thermal stress in these three case studies, it is clear that the hot spots in most cases appear in the high density areas with impervious pavement, wheras the cool spots appear in the green areas and next to water bodies [16]. The comparative results demonstrate that the land use and land cover are highly related to thermal condition in the focus areas.

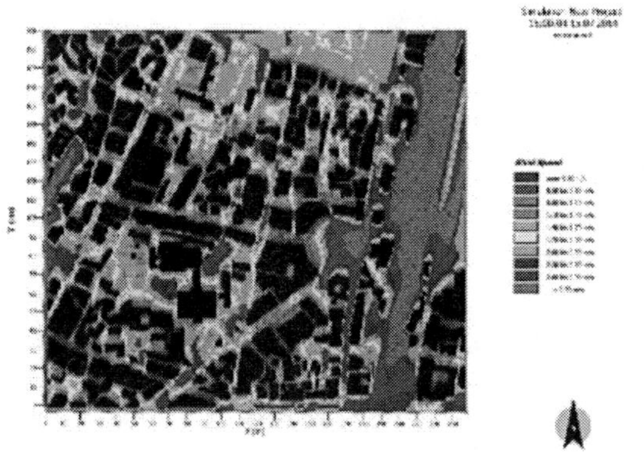

Fig. 5. Heatstress Tainan (Taiwan)

Fig. 6. Heat stress map in Ayutthaya (Thailand)

3.3 Combined approach for reduction of heat stress and flooding

GIS maps for heat stress and urban flooding are easily combined into one map showing which urban locations are vulnerable to the effect of climate change (Fig. 3). Measures to prevent or reduce flooding and to reduce urban heat can often be combined and carried out at the same time. This is especially true if above ground solutions are chosen to reduce flooding. Solutions like providing more green spaces in general help to abate both flooding and heat stress, since they increase evaporation and provide place to temporarily store excess storm water. The models and literature show that trees are the most effective in reducing heat stress, mainly because they provide shade.

3.4 3-D visualization

Combining the elevation model, the dataset with buildings and aerial photographs a 3D model of the city is created to get an improved overview of the outcomes of the model. The model was shown in the 3D virtual reality theater on a cylindrical screen using 6 HD projectors to project an image with a resolution of roughly 5000x1800. The software was running simultaneously on 7 PC's, one master PC for the control of the model and 6 slave PC's to drive the projectors [19]. After the flood modelling, the heat stress was visualized in 3D (Fig. 7). These visualisations and interactive communication tools can be used to encourage knowledge-sharing of climate-proofing and urban resilient projects [18].

Fig. 7. Heatstress 3D after 3D floodmodel, Groningen (The Netherlands).

4 Conclusions

GIS maps for heat stress and urban flooding are easily combined into one map showing which urban locations are vulnerable to the effect of climate change. Measures to prevent flooding and to reduce urban heat can often be combined and carried out at the same time. These combined maps are ideal quick-scan tools for urban planners who need to make early decisions quickly. In the three cities of this study, such combined mapping tools have proven invaluable in the decision-making process and it is envisaged that they will have similar successes in other cities the world over.

In Groningen these maps have been an important input for masterclasses on climateadaptation in Groningen and other cities in the Netherlands. It raised awareness on the need to implement measures to tackle floodings and heatstress and has led to the implementation of various sustainable urban drainage systems. This experience will be applied in the other regions of the world as Thailand and Taiwan.

Challenges and further developments associated with combined thermal stress and floods and other climate related issues in dense urban areas are continuously growing. The datasets are getting bigger and stakeholders are becoming more demanding and require faster andmore visual results. The technology using DEMs (digital elevation maps) is becoming more common and improved, both with a higher accuracy and a higher resolution. The accurate comparison of the results of the different models is still to be carried out. This will highlight to what level a quick-scan mapping is sufficient and when a more accurate and detailed modelling is needed. The choice between coarse and a fine scale modelling will vary within each situation, depending on the needs. The more accurate and complete models lack the ability to perform on a large scale. This can be tackled by the use of high performance computing facilities and/or by running the model on high performance clusters.

This study would not have been possible without the funding and collaboration within the projects climateproof cities (in Dutch: de klimaatbestendige stad), WaterCoG, INXCES, PEARL and the the long-term support we have received from the several stakeholders in the cities from The Netherlands, Thailand and Taiwan.

References

[1] J. Garssen, C. Harmsen and J. de Beer, The effect of the summer 2003 heat wave on mortality in the Netherlands. Eurosurveillance, **10**(7), 165-168, (2005)

[2] Worldweather online, (2016), Retrived on September 20, 2016 from http://www.worldweatheronline.com/bangkok-weather-averages/krung-thep/th.aspx

[3] Z. Vojinovic, D. Golub, M. Hammond, S. Hirunsalee, S. Weesakul, V. Meesuk, NP. Medina, M. Abbott, A. Sanchez, S. Kumara, 2015, Holistic approach to flood risk assessment in urban areas with cultural heritage: a practical application in Ayutthaya, Thailand, Natural Hazards, Springer, (2015)

[4] J. Kluck, F.C. Boogaard, D. Goedbloed, M. Claassen, Storm Water Flooding Amsterdam, from a quick Scan analyses to an action plan, International Waterweek 2015, Amsterdam, (2015)

[5] A. Mynett, and Z. Vojinovic, Hydroinformatics in multi-colours – part red: Urban Flood and Disaster Management, Journal of Hydroinformatics, **11**(3-4), 166-180, (2009)

[6] W. Barreto, Z.Vojinovic, R.K. Price and D.P. Solomatine, A Multi-objective Evolutionary Approach for Rehabilitation of Urban Drainage Systems, J. of Water Resources Planning and Management, **136**(5), 547-554, (2010)

[7] Z. Vojinovic, S. Sahlu, S. Seyoum, A. Sanchez, H. Matungulu, Z. Kapelan and D. Savic, Multi-objective rehabilitation of urban drainage systems under uncertainties, J. of Hydroinformatics, **16**(5), 1044-1061, (2014)

[8] A.F. Abdullah, Z. Vojinovic, and R.K. Price, Improved Methodology for Processing Raw LIDAR Data to Support Urban Flood Modelling - Accounting for Elevated Roads and Bridges, J. of Hydroinformatics, **14**(2), 253-269, (2011)

[9] A.F. Abdullah, Z. Vojinovic and R.K. Price, A Methodology for Processing Raw LIDAR Data to Support 1D/2D Urban Flood Modelling Framework, J. of Hydroinformatics, **14**(1), 75-92, (2011)

[10] V. Meesuk, Z. Vojinovic, A. Mynett and A.F., Abdullah, Urban flood modelling combining top-view LiDAR data with ground-view SfM observations, Advances in Water Resources, 75, 105-117, (2015)

[11] Seyoum, S.D., Vojinovic, Z., Price, R.K., and Weesakul, S., Coupled 1d and non-inertia 2d flood inundation model for simulation of urban pluvial flooding, ASCE Journal of Hydraulic Engineering, ASCE, 138(1), 23-34 (2012)

[12] Z. Vojinovic, S. Seyoum, M.H. Salum, R.K. Price, A.F. Fikri, and Y. Abebe, Modelling floods in urban areas and representation of buildings with a method based on adjusted conveyance and storage characteristics, Journal of Hydroinformatics, **15**(4), 1150-1168, (2012)

[13] Z. Vojinovic, and D. Tutulic, On the use of 1D and coupled 1D-2D approaches for assessment of flood damages in urban areas, Urban Water Journal, **6**(3), 183-199, (2009)

[14] Z. Vojinovic, S.D. Seyoum, J.M. Mwalwaka and R.K. Price, Effects of Model Schematization, Geometry and Parameter Values on Urban Flood Modelling, Water Science and Technology, **63**(3), 462-467, (2011)

[15] D. S. Kumar, D.S. Arya and Z. Vojinovic, Modelling of urban growth dynamics and its impact on surface runoff characteristics, Computers, Environment and Urban Systems, September 2013, 41, 124-135, (2014). Disaster Risk Management - urban flooding, heatstress & earthquakes [video]. Retrieved on September 20, 2016 from https://www.youtube.com/watch?v=mzFahS9RJPA

[16] F, Boogaard, Y. Chen, J. Kluck, S. Yang, L. van der Meulen, T. Lin, The integration of storm water flooding and thermal stress potential in Tainan (Taiwan) and Groningen (The Netherlands), Lyon NOVATECH, (2016)

[17] J. Kluck, P. Bosch, B. Heusinkveld, P. van der Wal, Quick scan tool for heat stress:How to help the urban planner, poster ECCA Copenhagen, (2015)

[18] J. Tipping, F. Boogaard, R. Jaeger, A. Duffy, T. Klomp, M. Manenschijn, Climatescan.nl: the development of a web-based map application to encourage knowledge-sharing of climate-proofing and urban resilient projects, International waterweek 2015, Amsterdam, (2015)

[19] M. Verlaat, L. van der Meulen, G. Schoof, F. Boogaard, J. Kluck, Disaster Risk Management: Urban Flooding and heatstress, Geomatics Workbooks n° 12 – FOSS4G Europe Como 2015, (2015)

Determination of Correlation for Extreme Metocean Variables

Zafarullah Nizamani [1,*], *Lim* Woan Yih [1] , *M.M.A* Wahab[2], and *Zahiraniza* Mustaffa[2]

[1]Department of Environmental Engineering, Universiti Tunku Abdul Rahman (UTAR), P.C. 31900, Kampar, Perak Darul Ridzuan, Malaysia
[2]Department of Civil Engineering, Universiti Teknologi PETRONAS, Malaysia, 32610 Bandar Seri Iskandar, Perak Darul Ridzuan, Malaysia

Abstract. Metocean environmental load includes wind, wave and currents. Offshore structures are designed for two environmental load design conditions i.e. extreme and operational load conditions of environmental loads are evaluated. The ccorrelation between load variables using Joint probability distribution, Pearson correlation coefficient and Spearman's rank correlation coefficients methods in Peninsular Malaysia (PM), Sabah and Sarawak are computed. Joint probability distribution method is considered as a reliable method among three different methods to determine the relationship between load variables. The PM has good correlation between the wind-wave and wave-current; Sabah has both strong relationships of wind-wave and wind-current with 50 year return period; Sarawak has good correlation between wind and current in both 50 years and 100 years return period. Since Sabah has good correlation between the associated load variables, no matter in 50 years or 100 years of return period of load combination. Thus, method 1 of ISO 19901-1, specimen provides guideline for metocean loading conditions, can be adopted for design for offshore structure in Sabah. However, due to weak correlations in PM and Sarawak, this method cannot be applied and method 2, which is current practice in offshore industry, should continueto be used.

1 Introduction

Offshore Malaysia is surrounded by various seas such as the South China Sea, Straits of Malacca and Sulu Sea. Offshore structures are designed using different load combinations based on operating and extreme conditions and whichever provides maximum stresses that load combination is considered to be governing the design. Extreme events like typhoon, storm and tsunami may occur once during the design life of offshore jacket platform. In South East Asia, typhoon is the extreme event, with a return period of 100 years and can damage the structures. Thus, to confirm the structural reliability of the jacket platformsduring its design life, operational 1-10 year and extreme loads 50-100 year data are used. Typhoons Vamei as well as tropical storm Greg is considered as the extreme

[*]Corresponding author: zafarullah@utar.edu.my

events which has occurred in past in Malaysia and they still may recur at any time once again. ISO 19901 recommends three different methods to analyse and determine the stresses in the jacket[1]. Correlation is evaluated between wind, wave and current using three methods of analysis, i.e. Joint probability distribution, Pearson correlation coefficient and Spearman's rank correlation coefficient methods.When all three variables occur concurrently, in the same direction and with same return period, they will cause maximum stresses. This will cause a huge impact on the jackets. On the contrary, if there is no correlation between variables, then we need to analyse the data to determine the relation between the variables with different return period.

2Literature Review

The wind, wave and current are considered to be random variables and may differ with time[1-9]. The jacket platforms are designed for extreme load conditions this is to make sure its structural reliability throughout its design life [10]. During the actual life of the jacket, extreme events may or may not occur during its operating conditions, but still there is low probability of its occurrence exists, with the impacts on the structure [11]. ISO 19901-1 suggests that if possible before actual design of offshore structure is undertaken, the joint probability of occurrence of uncertain metocean variables must be evaluated [1]. The design practice followed in Malaysia is based on return period of 100 year loads without giving any consideration for joint probability and the structure designed with this procedure will be safe but not economical. The joint probability analysis includes the determination of relationship between random variables which not only show their relationship, but also their distribution [12]. The knowledge about the joint probability distribution is an added benefit to the design engineer in considering the metocean loads on jacket structure. ISO 19901-1, recommends, that jackets to be analysed against long term return period metocean data typically 100 years return period metocean data should be considered. The ISO 19901-1 requires one of the three methods, should be adopted for the evaluation of environmental load parameters. 1) 100 year return period of wave height with associated period, wind or/and current speeds should be used. 2) The second method is being applied by design companies in Malaysia i.e. 100 year wave height, wave period, wind and current speed should be considered. 3) Any combination of wave height, period, wind and current speed that results in a) the global extreme environmental load acting on the offshore jacket, b) relevant global response of the jacket resulting from base shear or overturning moment.

For joint probability distribution method marginal and conditional distributions are evaluated. There are three techniques for the determination of the correlation. joint probability distribution method, Pearson correlation coefficient technique and Spearman's rank correlation coefficient. The Pearson correlation coefficient uses the given data for analysis. Spearman's rank correlation coefficient technique, before determining the correlation, ranks the given,data first.Correlation defines the relationship between the wind, wave and current, whereas the correlation coefficient shows the strength of a correlation. It needs to be evaluated here, whether the variables are related to each other or not, along with the strength of the relationship. The variables are considered to be related, if a variable (X) indicates the influence on the second variable (Y)[13].

Correlation coefficient range lies between -1.0 to 1.0 and is dimensionless [13-14]. Positive correlation indicates that the value of the variable is as high as 1.0. When one variable increases in value and another decreases it is considered to be negatively correlated variable. If the value of correlation coefficient is ±1.0 it is considered to be good correlation. A correlation coefficient of zero indicates that there is no linear relationship between the independent variables [15]. In case of difference among the values, it is recommended to take the lesser value.

The correlation tests are conducted to find the correlation between the metocean parameters for the dependent and independent variables. This is to describe the characteristics of the variables by classifying the exact association among variables [16]. The first stage is to determine the relationship between variables, whether they can be considered as dependent or independent. Subsequently, before the correlation test is carried out, the measurement scale for each variable needs to be determined as well as the number of variables involved. The final stage is to do the variable analysis using their relationship. Joint probability distribution is a study of considering the relationship between the variables together rather than taking their values separately. Joint distribution demonstrates that in what way two variables are associated and distributed [12]. Probability distributions are used for the determination of structural reliability. The probability distribution is determined for each variable, with information on mean, variance and standard deviation [12]. Joint probability distribution is advantageous for structural reliability analysis (SRA) because it helps to forecast of limit state function for structural damage [3]. The physics of offshore metocean parameters indicate that these variables should undergo multivariate analysis as compared with the values indicated by a single variable [10]. The advantage of using joint distribution is to determine the complex measured data [17]. The main difference between the Pearson and Spearman methods is Spearman uses ranked data instead of original data to conduct the research [14-15].

3 Methodology

In this study, metocean data, provided by government agency i.e. PETRONAS from three regions of Malaysia i.e. Peninsular Malaysia, Sabah and Sarawak is used to be analysed. The available data is a processed data and it is in shape of 1, 10, 50 and 100 years. A research was conducted where three statistical variables were considered independent; in this study a joint distribution is being considered [18]. In this study,50 and 100 year return perioddata is analysed to determine their co-relation. The available data belongs to 20 from PM, 8 from Sabah and 18 from Sarawak respectively. The hypothesis are built as shown in Figure 1 which shows that when the associated load variables have a positive relationship, all the variables will be acting in the same direction and are parallel. Thus, the loads produced, by use of this method, can provide severe case with regard to the forces on the structure. This will lead to the higher design loads being determined and thus uneconomical structure will result. This is because, to safeguard the safety of the structure, the resistance of the structure should be higher. This is achieved by using the highest grade of steel or enlarging the cross sectional area of the component. However, in case of associated variables have negative relationship, i.e. acting in the opposite direction, less force will be acting on the jacket and this can produce jacket which is not only safe but also economical.

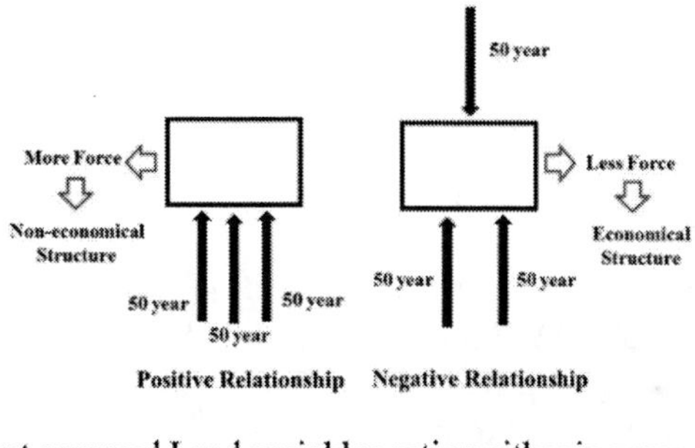

Fig. 1. Effect of different assumed Load variables acting either in same or opposite directions.

3.1 Joint Probability Distribution

Joint probability distribution is used to determine the probability distribution between two uncertain metocean random variables as shown in Equation (1) by X and Y. Marginal distribution of X given that Y is determined.

$$P(X = x \, and \, Y = y) = \frac{P(X = x \mid Y = y)}{P(Y = y)} \tag{1}$$

Here, X and Y are two different load variables. Equations (2-4) are used to determine mean (μ), variance (σ^2) and standard deviation (σ) respectively.

$$\mu = E(X) = \sum_{i=1}^{k} v_i \, p(v_i) \tag{2}$$

$$\sigma^2 = Var(X) = \sum_{i=1}^{k} (v_i - E(X))^2 \, p(v_i) \tag{3}$$

$$\sigma = Stdev(X) = \sqrt{Var(X)} \tag{4}$$

Here, x_i is random load variable and $p(x_i)$ is the random variable probability. The relationship between two random variables is determined by using the equations (5-6) by using the covariance and correlation.

$$Cov(X,Y) = \sum_{i=1}^{k} (x_i - E(X)) \, (y_i - E(Y)) \, p(x_i, y_i) \tag{5}$$

$$Corr(X,Y) = \frac{Cov(X,Y)}{Stdev(X) \times Stdev(Y)} \tag{6}$$

3.2 Pearson Correlation Coefficient

In the Pearson method the, xy, x^2, and y^2 are used to estimate the correlation coefficient. Here, x and y represent two different load variables. The correlation coefficient, r_p is determined by using Equation (7).

$$Correlation \, (r_p) = \frac{[(n \times \sum xy) - (\sum x \times \sum y)]}{\sqrt{[n((\sum x^2) - (\sum x)^2] \times [n(\sum y^2) - (\sum y)^2]}} \tag{7}$$

Here, *n*is a total set of data, *xy*has multiplied the*x* value with *y* value, x^2 is square of *x* value, y^2 is the square of *y* value, Σx is the sum of *x* values , Σy is the sum of *y* values, Σxy is the sum of the product of *xy*, Σx^2 is the sum of x^2 , Σy^2 is the sum of y^2.

3.3 Spearman's Rank Correlation Coefficient

For Spearman's method, the values of x and y in separate columns would be ranked by using Excel formula RANK.AVG(x) and RANK.AVG(y) respectively. The correlation coefficient (rs) is determined next and that is calculated by using Excel formula CORREL (Rank(x), Rank(y)). Here, x and y are two different load variables.

DOI: 10.1051/matecconf/201710304013

4 Result and Discussion

Correlation coefficient strength between the uncertain metocean variables in three regions of offshore Malaysia is determined using three different methods of correlation. Table 1 and Fig. 2 are based on the data of 50 year return period. It shows that a positive correlation exists between wind and wave using joint distribution, Pearson and Spearman's methods, i.e. the coefficients found are 0.820, 0.849 and 0.859 respectively. Table 1 and Fig. 3 show that the relationship between mean wind speed and current velocity is good in both joint distributions and Pearson's method, but the relationship in Spearman's method is found to be not strong. Table 1 shows that, in joint distribution, correlation coefficient between wave height and current velocity of surface is less than 0.5, while for Pearson and Spearman is around 0.2. Fig. 4 shows that the relationship for Joint distribution is average and positive between wave and current; but, for Pearson and Spearman is a weak and positive.

Table 1. Correlation Coefficients for Sabah Region.

Methods	Wind and Wave	Wind and Current	Wave and Current	Wind and Wave	Wind and Current	Wave and Current
	50 year	50 year	50 year	100 year	100 year	100 year
Joint Probability Distribution	0.820	0.649	0.406	0.483	0.617	0.520
Pearson Correlation	0.849	0.649	0.163	0.379	0.617	0.436
Spearman's Rank Correlation	0.859	0.270	0.163	0.220	0.015	0.325

Table 2 and Fig. 1 also shows the correlation coefficient for 100-year return period. It is found to be an average and positive between wind and wave for joint distribution.The

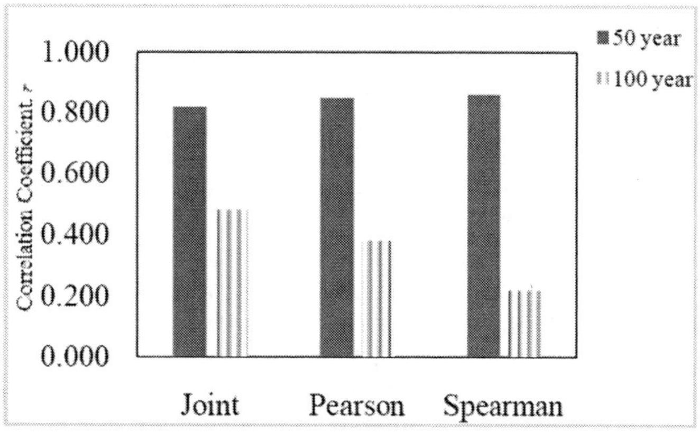

Fig. 2. Correlation Coefficient for Wind and Wave Sabah.

DOI: 10.1051/matecconf/201710304013

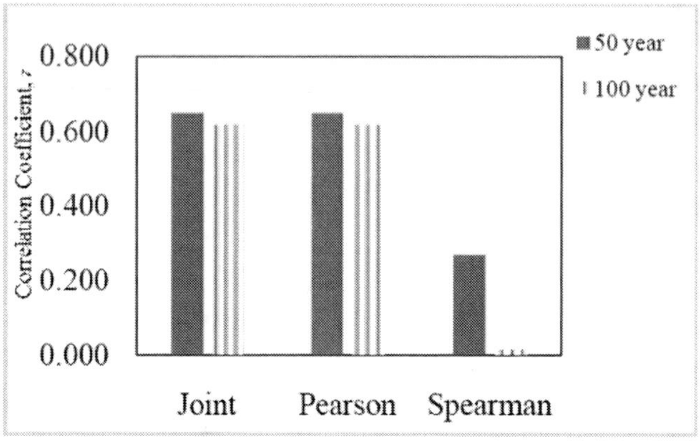

Fig. 3. Correlation Coefficient for Wind and Current _Sabah.

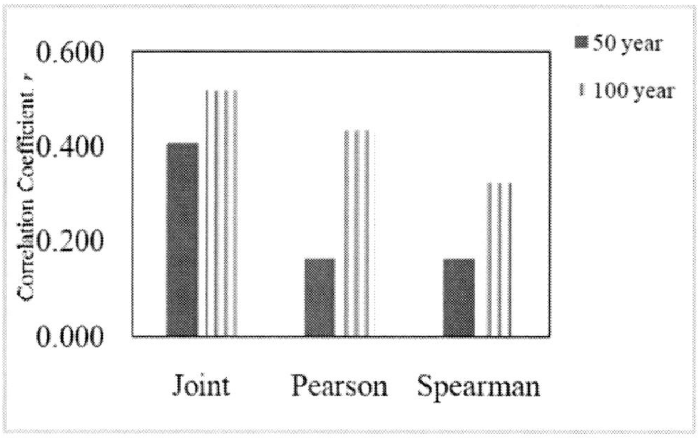

Fig. 4. Correlation Coefficient for Wave and Current _ Sabah.

Coefficient between the wind speed and wave height for Pearson and Spearman shows a weak relationship. Table 2 and Fig. 3 shows that the correlation coefficient is 0.617 between wind speed and current in Joint distribution. The Pearson and Spearman methods showed a correlation for wind and current as 0.617 and 0.015 respectively. Thus, it can be concluded that the relationship for wind and current using Joint distribution and Pearson is strong and positive. The same trend is found between wave and current as shown in Fig. 4. The Table 1 and Fig. 4 show that the correlation coefficient between wave and current. The relationship for both Joint distribution and Pearson is average and positively related and weak for Spearman between wave and current. In conclusion, for return periods of 50 years, we can recommend that, Sabah has strong and positive relationships between wind-wave and wind-current. For return periods of 100 years, it has a strong and positive relationship between wind and current. This could be due to the influence of the tropical Storm Greg and data collected was during that period. The maximum difference between Joint distribution and Pearson is in a range of 0.0-0.244 respectively, whereas for Joint and Spearman it is in range of 0.015-0.602 respectively. The results show that the difference between the joint distribution and the Pearson is smaller as compared tojoint distribution and Spearman. The reason for the difference between joint and Spearman is due to Spearman uses rank data [16], which is not the case when we use joint and Pearson method. However, as design of offshore structures follow[1]metocean specifications, which recommends joint probability distributions method to be used for metocean variables for the correlation. Thus, the joint probability distribution takes the precedence when there is an inconsistency among other methods.

Table 2 shows that Sabah has a strong correlation of all variables for return periods of 50 and 100 years. Thus, it can be concluded that jacket platforms for Sabah can be designed using the associated variables for return periods of 50 and 100. Furthermore, Method 1 of ISO 19901-1 can be applied to the design of jackets in Sabah. Whereas, it cannot be used in the PM and Sarawak in terms of associated variables for return periods of 50 and 100 years because of their weak correlations as shown in Table 2. Therefore, Method 2 of ISO 19901-1 which is the present practice in Malaysia should continue to be applied for PM and Sarawak until new data is made available. Table 2 shows negative co-relationship between two load variables, in some regions regardless of good or weak correlation, i.e. PM and Sabah obtained good, but the negative relationship correlation between wind and current for 50 year return period.

Table 2. Correlations for 50 and 100 Year Return Periods

Return Period	Correlation between variables	Good		Weak	
		Positive	Negative	Positive	Negative
50 year	Wind and Wave	PM, Sabah	-	Sarawak	-
	Wind and Current	Sabah, Sarawak	-	-	PM
	Wave and Current	-	PM, Sabah	Sarawak	-
100 year	Wind and Wave	-	Sabah	PM, Sarawak	-
	Wind and Current	Sabah, Sarawak	-	PM	-
	Wave and Current	Sabah	-	Sarawak	PM

5 Conclusion

The correlation between wind-wave and wave-current is strong in Peninsular Malaysia. Similarly correlation between wind-wave and wind-current is strong in Sabah for return periods of 50 years. Sabah has shown a strong relationship between wind and current for return periods 50 and 100 years. For return periods of 50 and 100 years, the metocean data for wind and current Sarawak provided good correlation. The highest difference in correlation coefficients in Joint probability distribution and Pearson is 0.244. The difference of minimum and maximum correlation coefficient for Joint probability distribution and Spearman are 0.015 and 0.602 respectively. It can be concluded that for associated variables Sabah has shown the best correlation for both return periods and ISO method 1 should be used for design of jacket platforms. Whereas PM and Sarawak regions are considered to have a weak overall correlation and therefore, ISO method 2, should to be continued for PM and Sarawak.

The first author would like to acknowledge the research grant IPSR/RMC/UTARRF/2014-C2/Z01 provided by the UniversitiTunku Abdul Rahman, Malaysia.

References

[1] ISO 19901-1 Petroleum and Natural Gas Industries – Specific Requirements for Offshore Structures – Part 1: Metocean Design and Operating Considerations, (2005)

[2] E.M.Bitner-Gregersen, K.C. Ewans and M.C. Johnson, Some Uncertainties Associated with Wind and Wave Description and Their Importance for Engineering Applications. Ocean Engineering, 86, 11-25, (2014)

[3] E. M.Bitner-Gregersen, Joint Met-Ocean Description for Design and Operations of Marine Structures. Applied Ocean Research, **51**, 279-292, (2015

[4] K. Burry,*Statistical Distribution in Engineering*. New York: The Press Syndicate of the University of Cambridge, (1999)
[5] S. K.Chakrabarti, *Hydrodynamics of offshore structures*. United Kingdom: WIT Press, (1986)
[6] C.P. Chang, C.H. Liu and C.H.Kuo, Typhoon Vamei: An Equatorial Tropical Cyclone Formation. Geophysical Research Letters, **30**(3), 50, (2003)
[7] K.Ewansand and P.Jonathan, Evaluating Environmental Joint Extremes for the Offshore Industry using the Conditional Extremes Model. Journal of Marine System, 130, 124-130, (2014)
[8] H.Mallahzadeh, G.Najafian, M.K.A Husain and N.I.M.Zaki, Derivation of Probability Distribution of Extreme Offshore Structural Response due to Combined Wave and Current Loading from the Probability Distribution of Extreme Surface Elevations. *7th International ASRANet Conference*, (2014)
[9] Q. Yiquan, Z. Zhixu and S.Ping, , Extreme Wind, Wave and Current in Deep Water of South China Sea. International Journal of Offshore and Polar Engineering, **20**(1), 18-23, (2010)
[10] P. Jonathan, K. Ewans and G. Forristall, Statistical Estimation of Extreme Ocean Environments: The Requirement for Modelling Directionality and Other Covariate Effects. Ocean Engineering, **35**(11-12), 1211-1225, (2008)
[11] P. Jonathanand K. Ewans, Statistical Modelling of Extreme Ocean Environments for Marine Design: A Review. Ocean Engineering, **62**, 91-109, (2013)
[12] S.C. Albright, W.L. Winston and C.J. Zappe, *Data Analysis & Decision Making with Microsoft Excel*, 3rd ed. Mason, USA: Thomson South-Western, (2012)
[13] N.S, Chok, *Pearson's versus Spearman's and Kendall's Correlation Coefficients for Continuous Data.* Master's Thesis, University of Pittsburgh, (2010)
[14] D.J. Booth,*A First Course in Statistics.* Second Edition. DP Publication Ltd.,London, UK, (1992)
[15] J. Haukeand T. Kossowski, Comparison of Values of Pearson's and Spearman's Correlation Coefficient on the Same Sets of Data. Questions Geographicae, **30**(2), 87-93, (2011)
[16] C.Y Piaw, *Mastering Research Statistics. Malaysia*: McGraw Hill Education, New York, United States,(2013)
[17] J.N. Sharma, S. Tryggested, & J. Bian, A Comprehensive Wind, Wave and Current Measurement Program in the South China Sea. Coastal Engineering 1996, 354-367, (1997)
[18] V.J. Kurian, M.S. Liew and Z. Nizamani, Statistical Modelling of Environmental Load Uncertainty for Jacket Platforms in Malaysia, *IEEE Colloquium on Humanities, Science and Engineering Research (CHUSER 2012)*, Malaysia, (2012)

Application of Linear Moments and Uncertainty Analysis to Extreme Rainfall Events in Sabah

Shu Lim Chuah[1,*], *Janice Lynn* Ayog[1], and *Nurmin* Bolong[1]

[1]Universiti Malaysia Sabah, Faculty of Engineering, Jalan UMS, 88400 Kota Kinabalu, Sabah, Malaysia

Abstract. Linear moments (LM) has been applied in extreme rainfall study for several countries, including China, United States of America, and Peninsular Malaysia. In this study, the LM procedures were applied to extreme rainfall data corresponding to locations provided in Malaysia Urban Stromwater Manual (MSMA) to derive new design rainfalls. Different record lengths were considered to assess the changes in design rainfall, and Monte Carlo simulations were carried out to compute confidence interval of the derived design rainfalls. Based on the Goodness-of-Fit (GoF) test results, the Generalized Extreme Value (GEV) probability distribution was chosen to derive the design rainfalls. The updated design rainfalls at all four locations showed significant reduction at design rainfalls of 50-year ARI and above. The difference of the design rainfalls from shorter record lengths with respect to the full record length and the confidence intervals do not necessarily reduce with a longer record. In hypothetical cases where 100-yr ARI rainfall was added, the increase in design rainfalls did not exceed the upper bound of the confidence intervals. The derived confidence intervals hence allow for better risk assessment, and should be considered in the design of critical structures, i.e. dams.

1 Introduction

Design rainfalls are often applied in the design of hydraulic structures, such as monsoon drains and detention ponds. The design rainfalls for Sabah in both MSMA [1] and HP No. 26 [2] are currently based on rainfall data prior to 1980. The main aim of this study is to update the design rainfalls in Sabah using LM procedures. Secondary aims in this study are assessing the differences in the design rainfall and confidence intervals derived from different record lengths, and whether the derived confidence interval was sufficient to accommodate for the presence of outliers.

2 Literature Review

The first stage in extreme rainfall studies is the compilation of extreme rainfall events, using either the Annual Maxima (AM) or Peak-Over-Threshold (POT) approach. The POT approach ensures that other extreme rainfall events apart from the annual maxima are

* Corresponding author: chuahshulim@gmail.com

included [3, 4], but suffers from the subjective selection of a suitable threshold [5, 6] and potential correlation between extreme events [7]. In contrast, the AM approach is simple, hence remaining popular in extreme rainfall studies [7].

Common probability distributions considered in extreme rainfall studies include Gumbel, Generalized Extreme Value (GEV), Generalized Pareto (GPA), and Pearson III (P3). For the AM series, the GEV is the limiting distribution [8], with Gumbel being a special case of GEV. Similarly, the PoT series converge to GPA [9]. P3 has been considered as the most suitable distribution in several extreme rainfall studies [10, 11]. The distributions are often fitted to the rainfall data by estimating the parameters via statistical procedures, such as the Maximum Likelihood Estimation and Linear Moments (LM).

The LM procedures were first published by Hosking [12], and are generally viewed as robust to outliers as there is no squaring and cubing of the variables involved [13, 14]. The LM procedures produce only small bias in small samples [15]. The LM procedures have since been applied in several countries, including USA, China, and Malaysia [3, 16, 17, 18, 19]. The design rainfalls in Peninsular Malaysia were updated using the LM procedures [20].

GoF tests are used to assess the validity of the fitted distributions in representing the extreme rainfall events. The k-s test and AD tests perform better than the Chi-Squared [21], with the AD deemed as more powerful as it places more emphasis on tail end of the distribution [22]. The critical values for the k-s and AD tests change based on the theoretical probability distribution considered and number of parameters being estimated, with various works [23-26] published on this matter. The GoF has been by used to determine the best-fitting probability distribution in studies where multiple distributions are considered [27, 28]. GoF test results should however be treated with caution, as the results depends not only on the PD choice, but also the fitting procedures [4].

In recent years, several extreme event deemed highly improbable based on standard design rainfall procedures had occurred, such as the 2003 storm at the Panama Canal [8], and the 1999 storm at Venezuela [29]. This indicates on the importance of allowing for uncertainty inherent in the derivation of design rainfalls to be incorporated via the construction of confidence intervals. The construction of confidence intervals would allow for the design risks to be properly assessed [12].

3 Method and Material

Rainfall data from four (4) stations highlighted in Fig. 1 were considered in this study due to close proximity to locations of design rainfall in MSMA [1], with record lengths listed in Table 1 considered. The 'Base' scenario represents the full data record available.

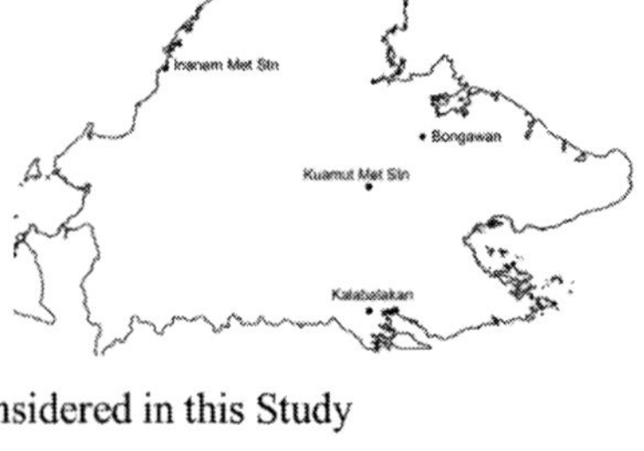

Fig. 1. Rainfall Stations Considered in this Study

Table 1. Rainfall Record and Sample Size Considered for each Scenario

Scenarios	Inanam, Kuamut, Bongawan		Kalabalakan	
	Record Period	**Sample Size**	**Record Period**	**Sample Size**
1	1985 – 1995	11	1987 – 1995	9
2	1985 – 2000	16	1987 – 2000	14
3	1985 – 2005	21	1987 – 2005	19
4	1985 – 2010	26	1987 – 2010	24
Base	1985 – 2015	31	1987 – 2015	29

The maximum annual 1-hour rainfall was extracted from the existing rainfall records, with L-moments and ratios computed using equations by Hosking [30]. L-moment estimators were applied for probability distribution fitting, considering the Gumbel, GEV, P3, GPA, Normal, and Exponential (Exp) probability distributions. The k-s and AD tests were applied to assess the GoF of the fitted probability distributions for each scenario. The k-stest statistic, D*, and AD test statistic, A*, were computed using equations published by Stephens and others [23-26], as listed below:

$$D^* = n^{1/2}\max[\{(i/n\text{-}\ F(x_i)\},\ \{F(x_i) - (i-1)/n\}],\ i = 1, 2, 3....n, \quad (1)$$

$$A^2 = \text{-}[\Sigma^n_{i\text{-}1}(2i - 1)\{\ln F(x_i) + \ln(1 - F(x_{n+1\text{-}i}))\}]/n - n,\ \text{i} = 1, 2, 3...n \quad (2)$$

Normal Distribution: $A^* = (1 + 4/n + 25/n^2)$ (3)

Exponential Distribution: $A^* = (1 + 0.6/n)$ (4)

Gumbel, GEV, P3 Distributions: $A^* = (1 + 0.2/n^{1/2})$ (5)

where n is the sample size, F(x) is the cumulative distribution function of the probability distribution, and x_i is the ordered data.

For the purposes of GoF, the P3 distribution was treated an extreme value distribution. Critical values published by Stephens and others [23-26] were referred to in this study, and listed in Tables 2 and 3. The distribution with the lowest test statistic value was deemed as the best fit. The design rainfall at 10, 50, and 100-years ARI were then extracted from the best-fit probability distribution, with the 95% confidence intervals constructed using Monte Carlo procedures outlined by Silva et al. [31]. The relative difference of the design rainfalls in various scenarios relative to the 'Base' scenario and the root mean square error (RMSE) of the Monte Carlo simulations were subsequently calculated and assessed.

Table 2. Critical Values for D* and A*

Probability Distribution	Critical Value for D*	Critical Value for A*	Critical Value for A*
Normal	0.895	0.787	0.787
Exp	1.094	1.341	1.341
Gumbel, GEV, P3, GPA (D* only)	0.819 @ n = 10; 0.843 @ n = 20; 0.860 @ n = 50	0.757	0.757

Note: n = sample size

Table 3. Critical Values for A^2 (GPA Only)

Shape Parameter, k	-0.9	-0.5	-0.2	-0.1	0	0.1	0.2	0.3	0.4	0.5
Critical Value	0.771	0.830	0.903	0.935	0.974	1.020	1.074	1.140	1.221	1.321

To further assess the confidence intervals, a 100-year ARI event derived from the fitted probability distribution was added to the original data series, and alternative 'outlier' design rainfalls derived at 10, 50, and 100-year ARI derived via re-fitting the same probability distribution based on the modified data series. The 'outlier' design rainfalls were then compared against the upper bound of the confidence intervals. All the procedures above were carried out using Microsoft Excel, a software capable of carrying out statistical analysis.

4 Results and Discussion

4.1 Goodness-of-Fit

A summary of the best fit for each scenario and station based on GoF tests is listed in Table 4, ranked in an ascending order. The GEV distribution provides the overall best performance in both k-s and AD tests, and was subsequently used in the derivation of the design rainfalls and confidence intervals. The Exp distribution was the worst performing, followed by the GPA distribution, with over half deemed as bad fits in the AD test. Although the Normal distribution shows good performance in the k-s test, its overall ranking falls behind both Gumbel and P3 in the AD test, indicating a lower fit for the Normal distribution at the tail ends. The P3 distribution slightly outperforms the Gumbel distribution, but has more bad fits in the AD test.

4.2 Design Rainfalls and Upper Bounds

The design rainfalls and the upper bounds based on the existing rainfall records are listed in columns 'Ori', and 'UB in Table 5. The 'outlier' design rainfalls based on the addition of the 100-yr ARI event into the data series is listed in column 'Out' in the same table. The derived design rainfalls are similar to the MSMA design rainfalls only at 10-yr ARI, becoming lower at higher ARI's of 50-year and above. The 'outlier' design rainfalls are higher due to the addition of the 100-year ARI event into the data series. The upper bounds were sufficient to cover for the 'outlier' design rainfalls at all rainfall stations.

4.3 Changes in Design Rainfall and Uncertainty

Changes in the design rainfall and the inherent uncertainty of the design rainfalls were quantified using the relative difference and RMSE, as listed in Table 6. The relative differences compared to the base scenario are generally not significant, with only several cases exceeding 10% at Inanam (scenario 3), and Kalabalakan (scenarios 1 & 2). There is no clear trend of reduction in both the relative difference and RMSE in relation to an increase in record length, with different trends occurring at each station. It may be surmised that longer record lengths do not necessarily lead to a reduction in design rainfall changes and its inherent uncertainty.

Table 4. Gof Ranking based on k-s and AD tests

Scenarios	Ranking based on Probability Distribution											
	k-s						AD					
	G	GEV	P3	GPA	Nor	Exp	G	GEV	P3	GPA	Nor	Exp
Inanam												
1	2	1	3	4	5	6	4	3	2	1	5	6
2	1	2	4	6	3	5	2	1	3	Fail	4	6
3	2	1	3	5	4	6	1	2	4	Fail	3	6
4	3	2	4	6	5	1	2	1	3	5	Fail	4
Base	2	1	3	5	6	4	2	1	3	4	Fail	6
Kuamut												
1	2	3	5	4	1	6	1	2	3	Fail	4	6
2	1	2	4	5	3	6	2	1	3	4	5	Fail
3	1	2	3	4	5	6	2	1	3	Fail	Fail	6
4	1	2	3	5	4	6	1	2	Fail	Fail	Fail	Fail
Base	2	3	4	5	1	Fail	1	2	3	Fail	Fail	Fail
Kalabalakan												
1	4	3	2	5	1	6	5	2	1	4	3	Fail
2	5	1	2	3	4	6	4	2	1	Fail	3	Fail
3	5	1	2	3	4	6	Fail	2	1	Fail	3	Fail
4	5	2	1	4	3	6	3	1	Fail	Fail	2	Fail
Base	Fail	2	1	4	3	6	3	1	Fail	Fail	2	Fail
Bongawan												
Base	5	3	1	4	2	6	Fail	2	Fail	Fail	1	Fail
2	5	3	4	1	2	6	5	2	3	1	4	6
3	5	4	3	2	1	6	5	2	3	1	4	Fail
4	5	4	3	1	2	6	4	3	2	1	5	6
Base	5	3	1	4	2	6	Fail	2	Fail	Fail	1	Fail

Table 5. Design Rainfalls

Scenarios	**Design Rainfall**								
	10-yr ARI			**50-yr ARI**			**100-yr ARI**		
	Ori	**Out**	**UB**	**Ori**	**Out**	**UB**	**Ori**	**Out**	**UB**
Inanam (Compared against Kota Kinabalu in MSMA)									
MSMA	91.9			124.1					
1	86.8	97.6	107.9	110.0	140.7	167.3	120.7	164.7	212.0
2	91.9	101.2	113.5	120.5	147.5	179.9	134.2	173.1	223.6
3	97.4	105.1	116.5	127.5	149.5	181.7	141.7	173.0	225.1
4	94.3	99.7	111.0	119.6	134.6	161.5	131.2	152.2	194.1
Base	93.5	97.7	104.7	115.5	126.2	147.0	124.8	139.4	172.1
Kuamut (Compared against Kuamut in MSMA)									
MSMA	88			106			116		
1	87.8	91.0	93.3	92.4	97.2	103.6	93.5	98.8	109.5
2	85.9	89.8	94.3	95.7	104.2	113.0	99.1	109.9	122.5
3	89.1	92.5	96.7	100.0	107.4	115.8	103.9	113.2	125.8
4	89.8	93.3	99.4	105.1	114.0	128.8	111.5	123.4	146.0
Base	89.4	91.6	95.3	98.6	102.5	109.2	101.3	105.9	115.0
Kalabalakan (Compared against Tawau in MSMA)									
MSMA	78								
1	78.9	89.6	95.8	97.2	126.0	146.8	105.0	145.0	181.9
2	73.9	80.4	88.0	89.3	105.1	122.7	95.4	116.5	142.6
3	81.3	87.9	98.1	103.7	121.3	149.0	113.5	138.0	178.0
4	85.1	91.2	101.8	111.3	128.0	153.1	123.2	146.6	187.4
Base	82.4	87.4	96.7	108.3	122.1	147.7	120.3	139.7	178.7
Bongawan (Compared against Sandakan in MSMA)									
MSMA	103			130					
1	94.7	102.9	110.1	108.6	125.4	138.4	113.3	134.1	158.5
2	93.5	96.7	100.1	100.3	105.3	112.3	102.0	107.6	118.2
3	91.5	94.0	97.2	98.3	102.3	108.5	100.1	104.7	114.0
4	94.9	97.4	101.2	103.9	108.4	115.3	106.5	111.8	121.9
Base	95.8	98.2	103.4	105.8	110.0	118.8	108.8	113.8	126.0

Table 6. Relative Difference and RMSE in Design Rainfalls

Scenarios	**Relative Difference (RD) and RMSE for Various ARI**					
	10-yr		**50-yr**		**100-yr**	
	RD	**RMSE**	**RD**	**RMSE**	**RD**	**RMSE**
Inanam						
1	7.1	9.72	4.7	23.94	3.3	35.75
2	1.7	9.73	4.4	24.78	7.5	36.81
3	4.2	8.84	11	23.03	14	34.47
4	0.9	7.23	3.6	17.85	5.1	25.51
Base		5.89		13.67		19.13
Kuamut						
1	1.8	3.39	6.3	5.27	7.7	6.64
2	4	4.42	3	8.05	2.2	10.41
3	0.3	4.11	1.5	7.65	2.6	10.00
4	0.4	4.71	6.7	10.21	10	14.04
Base		3.10		5.14		6.36
Kalabalakan						
1	4.3	8.89	10	20.63	13	30.06
2	10	6.56	18	14.30	21	19.90
3	1.3	7.58	4.3	17.80	5.6	25.35
4	3.2	7.28	2.8	11.46	2.4	13.27
Base		6.99		17.05		24.54
Bongawan						
1	1.2	7.68	2.7	13.89	4.1	18.63
2	2.5	3.65	5.2	5.76	6.3	7.14
3	4.5	3.10	7.1	4.86	8	5.99
4	1	3.38	1.8	5.51	2.1	6.87
Base		3.48		5.79		7.19

5 Conclusion

The updated design rainfalls allow for engineering structures to be designed accordingly with more recent data, extending the range to beyond 50-year ARI, which was previously not available in the existing design rainfalls. Derivation of design rainfalls using the L-moments procedure may be applied to even record lengths of 10 years, with generally no significant difference to the design rainfalls derived at longer record lengths. Although the inherent uncertainty in design rainfall does not necessarily reduce with an increasing record length, the full record length should still be used as it is the best representation of the rainfall characteristics.

The design rainfalls become much higher when outliers are added to the existing data series, highlighting the dangers of carrying out engineering design without considering the inherent uncertainty of the design rainfalls. It is recommended for confidence level bands to be constructed in the development of design rainfall curves. The upper bound of these design rainfalls may then be applied for critical engineering structures, such as dams.

Deep gratitude is given to the Malaysian Government, with this study being financially supported under the Malaysian Ministry of Education Research Acculturation Collaborative Effort (RACE) Grant No. RACE0020-TK-2015.

References

[1] Manual Saliran Mesra Alam, Chapter 13: Design Rainfall, 14, (2001)

[2] K.S. Low, M.F. Haji Mahmood and S. Salleh, *Hydrological Procedure No. 26*, Estimation of Design Rainstorm in Sabah and Sarawak, (1983)

[3] Z. Li, C. Li, Z. Xu and X. Zhou, Frequency analysis of precipitation extremes in Heihe River basin based on generalized Pareto distribution, Stoch. Environ. Res. Risk Assess, **28**, 1709-1721, (2014)

[4] D. Huard, A. Mailhot and S. Duchesne, Bayesian estimation of intensity-duration-frequency curves and of the return period associated to a given rainfall event, Stoch. Environ. Res. Risk Assess., **24**, 337, (2010)

[5] F.J. Acero, J.A. Garcia and M.C. Gallego, Peaks-over-threshold study of trends in extreme rainfall over the Iberian Peninsula, Journal of Climate, **24**, 1090, (2011)

[6] A. Mailhot, S. Lachance-Cloutier, G. Talbot and A-C. Favre, Regional estimates of intense rainfall based on the peak-over-threshold (POT) approach, Journal of Hydrology, **476**, 188, (2013)

[7] A. Ben-Zvi, Rainfall intensity-duration-frequency curves from large samples, *11th International Conference on Urban Drainage, Edinburgh, Scotland, UK*, (2008)

[8] E. Shamir, K.P. Georgakakos and M.J. Murphy Jr., Frequency analysis of the 7-8 December 2010 extreme precipitation in the Panama Canal watershed, Journal of Hydrology, **480**, 140, (2013)

[9] H. Van de Vyver and G.R. Demaree, Construction of intensity-duration-frequency (IDF) curves for precipitation at Lubumbashi, Congo, under the hypothesis of adequate data, Hydrological Sciences Journal, **55**(4), 556-557, (2010)

[10] S. Sarkar, N.K. Goel and B.S. Mathur, Development of isopluvial map using L-moment approach for Tehri-Gahrwal Himalaya, Stoch. Environ. Res. Risk Assess., **24**, 414, (2010)

[11] M.M. Rashid, S.B. Faruque and J.B. Alam, Modelling of short duration rainfall intensity duration frequency (SDR-IDF) equation for Sylhet City in Bangladesh, ARPN Journal Science of Technology, **2**(2), 92-95, (2012)

[12] J.R.M. Hosking, L-moments analysis and estimation of distributions using linear combination of order statistics, J. of the Royal Statistical Society Series B, **52**(1), (1990)

[13] D. Koutsoyiannis and G. Baloutos, Analysis of a long record of annual maximum rainfall in Athens, Greece, and design rainfall inferences, Natural Hazards, **29**, 37, (2000)

[14] S.S. Eslamian and H. Feizi, Maximum monthly rainfall analysis using L-moments for an arid region in Isfahan province, Iran, J. of Applied Meterology and Climatology, **46**, 495, (2007)

[15] R.V. Ramana, B. Chakaravorty, N.R. Samal, N.G. Pandey and P. Mani, Development of intensity duration frequency curves using L-moment and GIS technique, J. of Applied Hydrology, **XXI No.1 & 2**, 88-100, (2008)

[16] J.P. Raiford, N.M. Aziz, A.A. Khan and D.N. Powell, Rainfall depth-duration-frequency relationships for South Carolina, North Carolina, and Georgia, American Journal of Environmental Science, **3**(2), 78 – 84, (2007)

[17] P.S. Chu, X. Zhao, Y. Ruan and M. Grubbs, Extreme rainfall events in the Hawaiian Islands, J. of Applied Metereology and Climatology, **48**, 502-516 (2009)

[18] M.D. Norlida, I. Abutan, R. Abdullah, A.S. Yahaya, O. Sazali, M.D. Mohd Nor and M.S. Lariyah, Intensity-duration-frequency estimation using generalized Pareto distribution for urban area in a tropical region, *12nd International Conference on Urban Drainage, Porto Alegre/Brazil,* (2011)

[19] S.C. Liew, S.V. Raghavan and S-Y. Liong, Development of intensity-duration-frequency curves at ungauged sites: risk management under changing climate, Geosciences Letter, **1**(8), (2014)

[20] Reviewed and updated the Hydrological Procedure No.1 (Estimation of Design Rainstorm in Peninsular Malaysia), Volume I – Main Report, 32, (2010)

[21] N.Z. Abidin, M.B. Adam and H. Midi, The goodness-of-fit test for Gumbel distribution: a comparative study, Matematika, **28**(1), 35, (2012)

[22] A. Ben-Zvi and B. Azmon, Joint use of L-moment diagram and goodness-of-fit test: a case study of diverse series, Journal of Hydrology, **198**, 246, (1997)

[23] M.A. Stephens, EDF statistics for goodness of fit and some somparisons, J. of the American Statistical Association, **69**(347), 732, (1974)

[24] M.A. Stephens, Goodness of fit for the Extreme Value Distribution, Biometrika, **64**(3), 587, (1977)

[25] M. Chandra, N.D. Singpurwalla and M.A. Stephens, Kolmogorov Statistics for tests of fit for the Extreme Value and Weibull Distributions, J. of the American Statistical Association, **76**(375), 730, (1981)

[26] V. Choulakian and M.A. Stephens, Goodness-of-Fit tests for the Generalized Pareto Distribution, Technometrics, **43**(4), 480, (2001)

[27] B. Mohyment, G.R. Demaree and D.N. Faka, Establishment of IDF-curves for precipitation in the tropical area of Central Africa – comparison of techniques and results, Natural Hazards and Earth System Sciences, **4**, 375-387, (2004)

[28] O.O. Olofintoye, B.F. Sule and A.W. Salami, Best-fit probability distribution model for peak daily rainfall of selected cities in Nigeria, New York Science Journal, **2**(3), (2009)

[29] S. Coles, L.R. Pericchi and S. Sisson, A fully probabilistic approach to extreme rainfall modelling, J. of Hydrology, **273**, 35-50, (2003)

[30] J.R.M Hosking and J.R. Wallis, *Regional frequency analysis: an approach based on L-moments*, Cambridge University Press, (1997)

[31] A.T. Silva, M.M. Portela, J. Baez and M. Naghettini, Construction of confidence intervals for extreme rainfall quantiles, *Risk Analysis VIII*, 299, (2012)

Environmental Effect of Tidal Bore Propagation in Kampar River

Mubarak[1,*], *Albertus* Sulaiman[2] , and Efriyeldi[1]

[1]Department of Marine Sciences, Universitas Riau, Indonesia
[2]Geostech Laboratory BPPT, Kawasan Puspiptek Serpong, Indonesia.

Abstract. The environmental effect of the tidal bore propagation in kampar river is investigated. Propagation of the tidal bore satisfy Korteweg and de Vries (KdV) equation where the environmental effect will be given as a perturbation term of the KdV equation. The environmental efects taken into considérations are river discharge and bottom sediment. The result show that a river discharge that flow opposite to wave motion will give the damping effect that is decreasing of the wave velocity and the amplitude. The bottom sediments will give the damping effect so that a fine bottom sediment would cause the waves propagte slower than coarse sediment.

1 Introduction

Tidal bore is very unique phenomena. This is a huge wave with the amplitude up to 5m propagate from the mouth of estuary to the upstream with very fast velocity. The wave is called tsunami in the river. There only less than twenty of estuary have tidal bore phenomena in the world. One of them occure in Kampar River, Sumatra Indonesia. The waves ussually occure in November every years. This is generated by interaction of storm surge from the South china sea and river discharge at low tide condition. This wave have a local name the Bono and propagate with the velocity about more than 10m/s.

The Kampar river is a convergent alluvial estuaries. The most of studies tidal bore in the avlluvial estuaries have been focused on a small-scale bore processes i.e the propagation zone. In this scale, the environmental effect can not be ignored. The tidal bore will be influenced by the environmental condition such as sediment resuspension, bed erosion and score [1-3]. Another hand, the environment can influence the tidal bore propagation through the damping effect or breaking of the tidal bore. For example the river discharge damps the tidal bore propagation in alluvial estuary [4]. Another observation showed that the suspended sediment concentration (SSC) will increase when the tidal bore passage [5]. When the bore propagate into the upstream , the free surface increase and flow velocity decrease rapidly under the bore front to satisfy the conservation of mass and energy. The present of bottom sediment will break the conservation of energy then the damping effect may occur. In the paper, the environmental effect such as river discharge and bottom

* Corresponding author: mubarak@unri.ac.id

sediment on the tidal bore propagation are investigated. The study of environmental effect of this wave can help to identify dangerous region and can be used to river management plan. The paper is organized as follow, the physics of study of Kampar river and methodology are described in Sec-2. The tidal bore propagation and its environmental effect will be developed in Sec-3 and followed by the discussion of its simulation. The paper will be ended by a conclusion.

2 Methodology

Methodology used in the research is the simulations with analytical modeling supplied with the field observation.

2.1. Fields Observation

First of all, the morphology and a tidal bore observation of Kampar river will be highlighted. The Kampar river is a big river in Riau province Indonesia. It have 24.548 km^2 area with the depth is around 7.8m and the average depth is 413m. The river also used as main transportation with a groove of voyage crowded. From geomorphological point of view, the river is part of alluvial estuary that have trumpert form (converent) where this is a prerequest for the exixtence of a tidal bore. The tidal amplitude is around 2.1 to 4.5 m and propagate until 229km along. The study area is depicted in Fig.1.

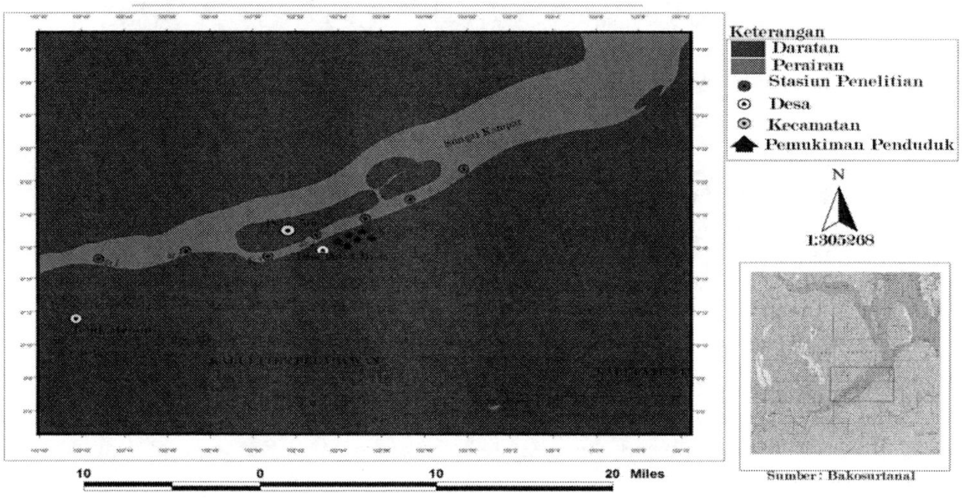

Fig.1. The study area. The red dot represent an observation point. The river water is represented by blue color.

The Kampar river have high river discharge and the interaction with the storm surge from the South china sea will generate a high amplitude of a tidal bore. The tidal bore in Kampar river ussualy occuring in November and December every years. The tidal bore wavefront can be seen in Fig.2. When the tidal bore passage, the local area of the river will be flooding due to run up of the waves. This is depicted in Fig.3. The picture describe a flooding during the tidal bore passage. The location is a small village called Teluk Meranti The run up can spread as far as 500m from the river. At the sometime, the water calor is brown and very turbid. The animal form the river such as fish, snake and the other can just took to the land.

Fig.2. Tidal bore of the Kampar river on November, 2014 (courtesy: Wisnu A. Martono).

Fig.3. The flooding in the side of river when the tidal bore passage.

The measurement of bottom sedimen is depicted in Fig.4. The dominant bottom sedimen is mud sediment. The previous studies showed that the suspended sediment will resuspend and increase the erosion when the tidal bore passage [4]. For example, the sediment transport due to the tidal wave propagation created some smal island in the river and bank erosion near Muda island and mouth of Sekap river [5]. The measurement suspended sediment concentration (SSC) transport induced by tidal bore in the See reiver showed that SSC increase greater than 30g/L during the tidal ore passage. After the tidal bore passage, then the SSC gradually increase over the whole water column with the amplitude is about 2m [5].

2.2. Analytical Modeling

Tidal bore in Kampar river appear in several form, sometime in the single pulse (undular bore) form and many pulse form. The first pulse have the amplitude higher than the succeses wave. The wave is not only propagate the energy but also the mass, so that this can transport the sediment. When the Bono propagate, there will be the silting-up of the bottom topography in many area of watershed especially in around Muda island and the estuary of Serkap river. Beside the sediment transport, the wave also cause the changing of the water colomn so that it can influence the river ecosystem. It is importance to study the effect of tidal bore propagation related to the environmental effect.

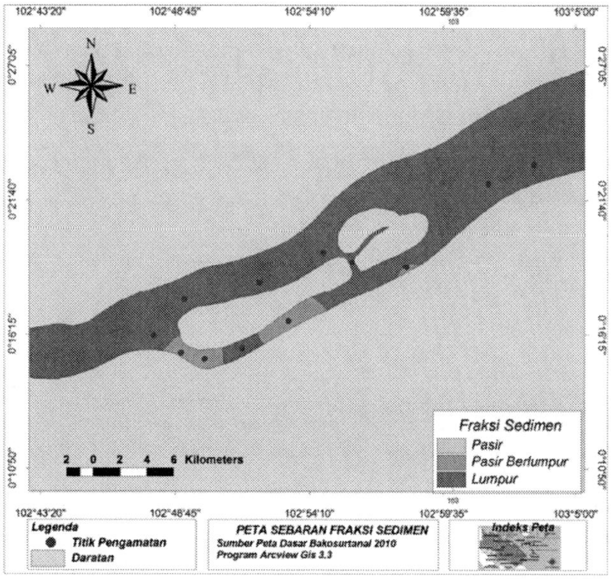

Fig.4. Botttom sediment observed June,16 to June, 19 2016. The bottom sediment type is represented by grey color. The light grey color is sand, the medium grey is muddy sand and the heavy grey is a mud sediment.

The tidal bore is a nonlinear phenomena. This wave is a balance between nonlinear and dispersion effect. The most appropriate model for description of the tidal bore propagation is the Korteweg and De Vries (KdV) equation. The derivation from this equation from shallow water model is very complicated. For simplicity we assume that the environment will give a perturbation for the tidal bore waves. This mean that the KdV equation should taking into account the environmental background. This phenomena can be described by the perturbed KdV equation as follow [8-10],

$$\frac{\partial \eta}{\partial t} + (gh)^{\frac{1}{2}}\left(1 + \tfrac{3}{2}\sqrt{gh}\,\frac{\eta}{h}\right)\frac{\partial \eta}{\partial x} + \frac{h^2}{6}(gh)^{\frac{1}{2}}\frac{\partial^3 \eta}{\partial x^3} + F(x)\eta = 0 \qquad (1)$$

where η tidal bore amplitude, g acceleration of gravitation, h is a water depth, x and t are spatial and temporal coordinate respectively and $F(x)$ is an environmental effect. The environmental effect of tidal bore propagation be investigated by giving the special form at $F(x)$ in the Equation (1). The single soliton solution will be used in this investigation. The first let we discuss the river discharge. The environmental effect such as river discharge also have significant contribution to damping effect of the tidal bore propagation [4]. The study showed that the fresh water discharge can no longer be neglected and this damp of tidal wave propagation significantly. The river discharge can seen as a constant perturbation so that in the Equation (1), we take $F(x)=\mu$, where μ is a constant parameter associated with river discharge. The single soliton solution of Equation (1) is given by [11],

$$\eta(x,t) = 8\eta_0^2 \exp\left[-\tfrac{8}{9}\mu\alpha\left(\frac{2\alpha}{3\beta}\right)^{\frac{1}{2}} t\right]\sec h^2(\varsigma) \qquad (2)$$

Where

$$\zeta = \eta_0 \exp\left[\frac{-4}{9}\mu\alpha\left(\frac{2\alpha}{3\beta}\right)^{1/2} t\right]\left[\left(\frac{2\alpha}{3\beta}\right)^{1/2}(x - Vt) - \xi\right] \tag{3}$$

$$\xi = 3\mu\eta_0^2\left(1 - \exp\left[\frac{-8}{9}\mu\alpha\left(\frac{2\alpha}{3\beta}\right)^{1/2} t\right]\right) . \tag{4}$$

By an assumption that the bottom sediment act as a constant force then the solution of Equation (1) is given by,

$$\eta(x,t) = 2\eta_0^2 e^{\left(-\frac{3}{4}F_0 t\right)} \sec h^2\left[\frac{(x + V_d t)}{\Delta}\right] \tag{5}$$

$$V_d = \left(1 - \frac{\eta_0}{2h}\right)\sqrt{gh}\left(1 - e^{\left(-\frac{3}{4}F_d t\right)}\right) \quad ; \quad \Delta = \sqrt{\frac{4h^3}{3\eta_0} e^{\frac{2}{3}F_0 t}} \tag{6}$$

3 Result and Discussion

We will discuss two kind of the environmental effect, the river discharge and the sediment resuspension. In this paper, the tidal bore will be modeled as a single soliton. If tidal bore propagate from right (mouth of estuary) to left (upstream) and the river discharge flow from the upstream to mouth of estuary then the traveling wave parameters shoulfd be in the form of $x+Vt$. The tidal bore propagation with the varying of river discharges are depicted in Fig.5.

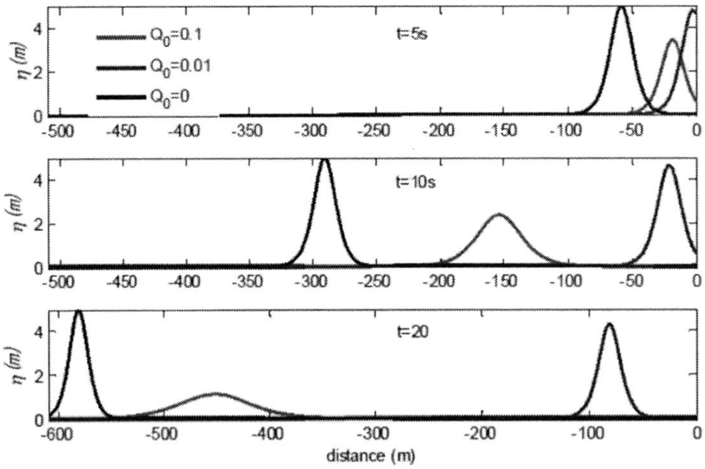

Fig.5. Simulation of tidal bore propagation with the varying river dischage.

This figure show various river effect on tidal bore propagation. This result show that the high discharge with propagate opposite with the propagation of tidal bore will damp the wave. The discharge will decrease the velocity and the amplitude of the tidal bore. The wave with high discharge will propagate faster but decrease the amplitude quickly compare with the lower discharge effect. The result also confirm by Bennenton [4].

The effect of bottom sediment respect to tidal bore propagtion can be studied through a definition of the bottom shear stress. The ottom stress ussualy define as $\tau = C_D\, \rho\, |V_b| V_b$, where C_D is a drag coefficient and V_b is bottom orbital velocity. Numerical study showed that the tidal bore will be damped due to the muddy bottom sediment and the wave dissipation causes significant downwave diffraction effects [12]. In this paper, we discuss the effect of bottom topography to the tidal bore propagation. Simulation of the solution by using various constant damping are depicted in Fig.5. The result show that the bottom sediment will give the damping effct of the tidal bore propagation. The tidal bore which propagate in the more dense bottom sediment will move faster than the thin sediment but the wave form more quickly decay. The bottom sediment in Kampar river compose of sand so that the damping effect is little paid influence on the wave propagation. This condition can explain that why the tidal bore on Kampar river can propagate over very long distance more than 200km.

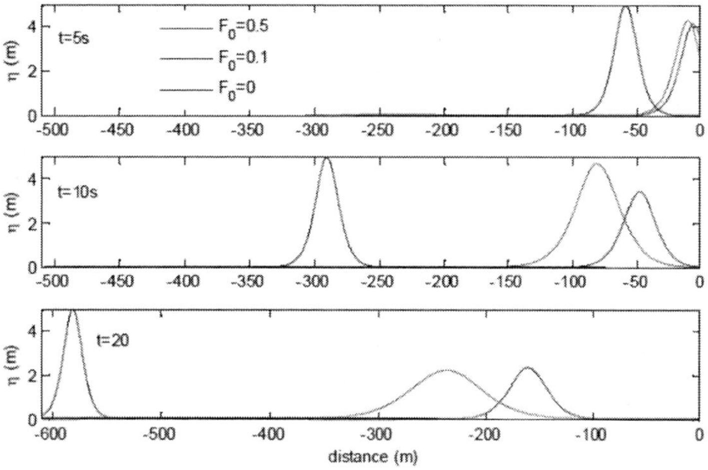

Fig.6. The tidal bore propagation with bottom sediment effect. The mud effect is modelled by assuming constant force. The 0.1 is represented by the blue color and 0.5 that is represented by red color.

4 Conclusion

Environmental effect of tidal bore propagation in the Kampar river is investigated. Two environmental effect such as river discharge and bottom sediment were studied. The result show that the high discharge with propagate opposite with the tidal bore propagation will damp the wave. The discharge will decrease the velocity and the amplitude of the tidal bore. The sediment effect show that the bottom sediment will give the damping effect of the tidal bore propagation. The simulation showed that the bore propagate in the dense bottom sediment will move faster than the thin sediment but the wave form more quickly decay.

This research founded by Universitas Riau in fiskal years 2016.

References

[1] Chanson, H., Environmental, Ecological and Cultural Impacts of Tidal Bores, Benaks, Bonos and Burros. *Proc. International Workshop on Environmental Hydraulics IWEH09, Theoretical, Experimental and Computational Solutions,* Valencia, Spain, 29–30 Oct., Editor P.A. Lopez-Jimenez et al., Invited keynote lecture, 20 pages, (2009)

[2] Chanson, H., Current knowledge in hydraulic jumps and related phenomena. A survey of experimental results, European Journal of Mechanics Fluids **28** (2): 191–210, (2009)

[3] Chanson, H., Momentum considerations in hydraulic jumps and bores, J. Irrigation and Drainage Engineering, ASCE,**138**(4), 382, (2012)

[4] Bonennton, P., Andrea G. Filippini., L. Arpaia., N. Bonneton, and M. Ricchiuto, On nonlinear shoaling properties of enhanced Boussinesq models, J. Estuarine, Coastal and Shelf Science, 172, 121-127, (2016)

[5] Furgerot, L., D. Mouaze., B. Tessier., L. Perez., S. Haquin., P. Weill and A. Crave, Sediment transport induced by tidal bores. An estimation from suspended matter measurements in the Sée River (Mont-Saint-Michel Bay, northwestern France), Comptes Rendus Geoscience, **348** (6), 432-441, (2016)

[6] Chanson, H., Tidal bore in the North Branch of the Changjiang Estuary, *Proc. International Conference on Estuaries and Coast*, Hanzhou, China, (2003)

[7] Yulistianto, B., Fenomena Gelombang Pasang Bono di Muara Sungai Kampar, Dinamika TEKNIK SIPIL, **9**(1), 19 – 26, (2009)

[8] Dingermans, M, *Water Waves Propagation over uneven Bottoms*, World Scientific, Singapore, (1997)

[9] Kundu, A., *Tsunami and Nonlinear Waves*, Springer Verlag, Berlin, (2007)

[10] Sulaiman, A., On the Internal soliton propagation over slope shelf topography , *Int. Theoretical Physics and Nonlienar Phenomena Conf,* Jakarta (2016)

[11] Sulaiman,A and M. Sadly, On Nonlinear Internal waves Propagation in the Narrow Strait of the Indonesian Seas, *The 2nd Indonesia Japan Joint Scientific Symposium*, Jakarta, (2006)

[12] Kaihatu,J.M., A. Sheremet, K.T. Holland., A model for the propagation of nonlinear surface waves over viscous muds, Coastal Engineering, 54, 752-764, (2007)

DOI: 10.1051/matecconf/201710304016

Application of Simple Dynamical System to Study the Hydrological Dynamics of a Mookervat Catchment Area

Agus Salim[1,*], *Thamzil* Las[1], *Nur* Aida[1], *Silvan* Erusani[1], *Albertus* Sulaiman[2]

[1]Faculty of Science and Technology, Universitas Islam Negeri Syarif Hidayatullah, Indonesia
[2]Geostech Laboratory BPPT,Kawasan Puspiptek Serpong, Indonesia.

Abstract. The paper study the characteristic of catchment behavior by using a simple first-order nonlinear dynamical system. The study area was located in a part of Jakarta Urban Lake was called Mookervart catchment area with some part of them are artifical river. The aims of the paper are to discover a new method to describe and to predict the dynamical of Mookervart river system by using a parameter simplify run off and rainfall. The average of rainfall was taken as long12 year showed that rainfall in Mookervart River under 500 mm/month which called as a "wet month". Mookervart river system was caused by 2 season: wet and dry month. The dry month starting from month of May up to October and wet month from November to April. In 2010 the wet month more longest than other year is caused by La Nina effect. By using the α=3 then the equation has two solution, this is mean that the solution is not unique. When the value of parameter increase then there are many kind of solution. This condition is called bifurcation. The rainfall on 8th days shown that increasing and decreasing from 15th day and then stabilize from 20th day to 100th day. Run off on the 18th days shown that increasing optimally and stabilize from 18th day to 100th day. The part or 50% of rainfall can be absorp to the land and others can be change to the runoff form. The application this model into the observation data of Mookervart catchment area is still developed. This is a restriction of a parameter value in our system that proofed that the behavior of the solution will be highlighted.

1 Introduction

Water fluxes in catchment area is very complicated phenomena which controlled by various physical processes and material properties. The recent advances technology such as satellite remote sensing, computer technology and geographic information system including digital terrain model an digital elevation model can be used to handle this complicated problem. The spatial problem canbe solved by these technology but the catchments dynamics is time depending and prediction the behavior in the future is still unresolved [1].

To describe time dependent of the complex phenomena, scientist use a physical parameter and investigate it by using a model. For example, the mayor theme in catchment

*Corresponding author: agus_salim@uinjkt.ac.id

hydrology has been the interpretation of streamflow variations interms of the drainage behavior of hillslope or channel storage elements [2]. In this paper we propose a model that the complex catchment area can be studied by a simple model that is a nonlinear differential equation. This idea is proposed by Kirchner [3]. He showed that one class of catchment can be characterized as simple first-order nonlinear differential dynamical systems. This approach can be used to reconstructing precipitation and evapotranspiration where direct measurements are unavailable.

2 Methodology

2.1 Field Site and Data

The location of research was located in Mookervart Catchment area.The Mookervart catchment area is the region surrounding the Mookervart river. The part of the Mookervart river i.e. close to the mouth of estuary is artificial river. This river have the average 4.5m depth and 32m and 14m width for the upper and the bottom respectively. The river can be storage of the discharge is about 125m^3/s. The area surrounding the river was used as a urban, industry and factory. The Mookervart cacthment area (67 km^2) is part of DKI Jaya province with the position is 106,38',15" to 106,44',50" E and 6,10',20" to 6,09',10"N. The catchment have water input from the Cisadane river and the output in West DKI beach. The tropical climate is dominant in this catchment area and have high rainfall. The topography of this catchment is relatively flat with the rank 0-25 m. The cathment area is depicted in Fig.1.

Mookevart river was initially used to be a river transportation and irrigation facility. In 1981 Mookervart river was designated to accommodate the discharge water capacity of 125m^3/s, with the following dimensions: wide cross-section over the river 32 m, a depth of 4.5 m and a base width of the river cross-section is 14 m. The current conditions of the Mookervart river right now has wide cross-section span over the river varies between 10 m - 33 m with a depth of cross-section between 1.5 m - 3 m and the width of the river varies between 9 m - 18 m. The banks of the Mookervart river left side directly bordered to the residential population, regional industry, and factories are less structured, while the right side of the Mookervart river is Daan Mogot Road.

Data and Materials Equipment such as: (i) Jakarta Map administration (Source: Coordinating Agency for Land Survey and Agroclimate (Bakosurtanal)), (ii) Data rainfall of Jakarta in 2002-2014 (Source: Meteorological and Geophysics Agency (BMG) Pondok Betung), (iii) Hydrology Map of Group Land and Land Cover (source: Map watersheds that flow into Jakarta, 2014) and (iv)Flow and water level data of the Mookervart River. (Source: Office of Public Works Jakarta)

The scope of the study includes : the forecast value of the catch water runoff generated around Mookervart River in Jakarta based method.The basic model of catchment-scale hydrology is the conservation of mass equation [1].

DOI: 10.1051/matecconf/201710304016

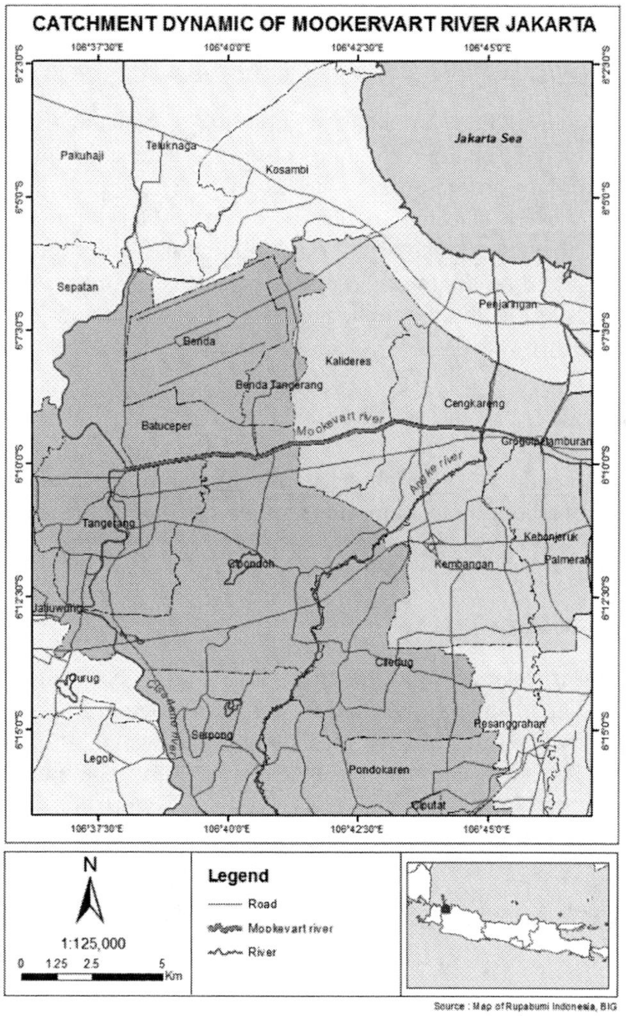

Fig. 1. The Mookervart catchment area.

2.2 Reaserch Methods

The study consisted of three phases. The first stage is a map overlay watershed delineation Mookervart with administrative boundaries Jakarta. The second stage is to make a map of the rainfall area of Jakarta and surrounding areas, the process of overlaying maps and rainfall, calculation of watershed surface (Q) resulting from each use of land in Jakarta and overlay maps with the maximum discharge of Jakarta and final stage using Cachment Model based on simple Nonlinier Dynamical System.

2.3 Methods of Analysis

2.3.1 Remote Sensing Technique

The Development and evolution of the lake will be studied by using satellite images such as Landsat TM. The remote sensing technique is used to acquire the lake area reduction and water quality deterioration information of urban lakes. The appropriate classification rule sand retrieval model were selected to obtain the important information regarding lake dynamics, such as the lake boundaries, land use classification, lake temperature and chlorophyll content based on the multi-source remote sensing images (MODIS, SPOT etc). The phenomena such as the abnormal change of the urban lake and track the development trend will be obtained.

2.3.2 Observation

In this research we conduct 1 year observation with sampling interval one weak for the parameters organic carbon, grab sampling, water quality, time series of temperature, salinity, pH etc.

2.3.3 Environmental Modeling

We use general mass balance equation to quantifying carbon budget in the lake and surrounding area. The model will be used to studies biological and/or chemical processes normally associated with water quality such as nutrient cycling studies; algal succession studies; food-web investigations; pathogen dynamics investigations; suspended solids studies; particulate carbon and toxic metal studies. Based on this simulation and comparison with observation data we can quantify carbon budget in Urban Lake.

2.4 Catchment Model based on Simple Nonlinear Dynamical System

The basic model of catchment-scale hydrology is the conservation of mass equation [1],

$$\frac{dS}{dt} = P - E - Q \qquad (1)$$

where S is the volume fo water storage in the catchment area (mm depth), P is the rate of praticipation, E is the evapotranspiration and Q is the discharge. All the function S,P,E and Q are function of times. In this paper we use a conceptual model. This model has two criterias: first, the structure of the model is specified prior to any modelling undertaken and the second not all parameters have a direct physical interpretation [4]. Further, we assume that the complex behaviour of a catchment area can be described by using a coupled nonlinear differential equation. This model has been used to predict the behavior of climate system such as El Nino and Sothern Oscillation. In the hydrological science, several scientist have discovered that a low dimensional model such as rainfall (R) and runoff (S) can be used to predict the behavior of hydrological processes. The method is called chaotic dynamics [5]. This system is sensitive to initial condition and there is no random effect involved [6].

In this paper we proposed dynamical model of run-off and rainfall as follow,

$$\frac{dR}{dt} = \mu R - vRS \tag{2}$$

$$\frac{dS}{dt} = \alpha S - \gamma S^3 + \lambda RS \tag{3}$$

The system in sensitive to initial condition where this mean that the long range prediction is imposible. There are three properties of the system: steady state or stable, limit cycle and chaotic behavior.

4 Result and Discussion

Three phase have been used for making GIS Map in this study. The first stage is a map overlay between watershed delineation *Mookervart River*map and administrative boundaries of Jakarta map. The second stage is mapping construction of the rainfall area of Jakarta and surrounding areas, and thencalculation of surface discharge (Q) which result from each land use in Jakarta. The final stage is overlaying map of Jakarta district with maximum discharge of Jakarta.

Data Processing using Mathlab (The MathWorks, Inc. 3 Apple Hill Drive Natick, MA 01760-2098). Such as: importing and exporting data, plotting data, missing data, inconsistent data, filter data, smooth data with convolution, detrending data and descriptif statistic.The rainfall data form 2002 to 2014 is depicted Fig.2.

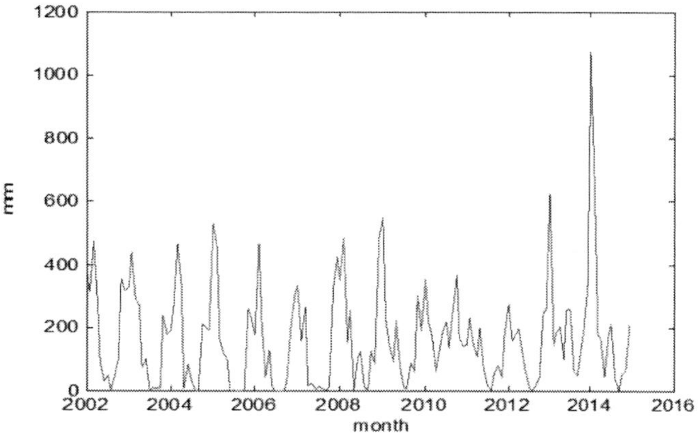

Fig. 2. The rainfall data from 2002 to 2014.

From Fig. 2 we can see that the average of rainfall was taken as long12 year showed that rainfall in Mookervart River under 500 mm/month which can we say "wet month". Mookervart river system was caused by 2 season: wet and dry month. The dry month starting from month of May up to October.From November up to April is wet month (Fig.2).From Fig. 2 we can see that in 2010 the wet month more longes than other year. In 2014 the wet season more than other year is caused by La Nina effect.

Now, we describe the nonlinear dynamics of Mookervart catchment area. The simplest model describe by only one variale. Let we describe the Equation (1) or example, the runoff equation with no rainfall can be modeled by the equation as follow,

$$\frac{dS}{dt} = \alpha S\left(1 - S^2\right) \tag{4}$$

This is called evolution equation. The solution of the equation depend on the parameter α is depicted in Fig.3.

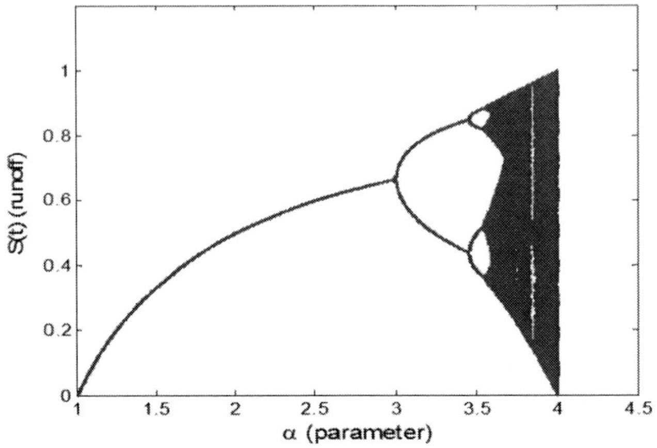

Fig.3. The bifurcation diagram of Equation (4).

The solution of a differential equation.

The solution of a differential equation has a unique solution if it well behavior (one-one mapping). The figure show that at the α=3 the equation has two solution, this is mean that the solution is not unique. When the value of parameter increase then there are many kind of solution. This condition is called bifurcation. This is a restriction of a parameter value in our system. The solution of Equation (4) is depicted in Fig. 4.

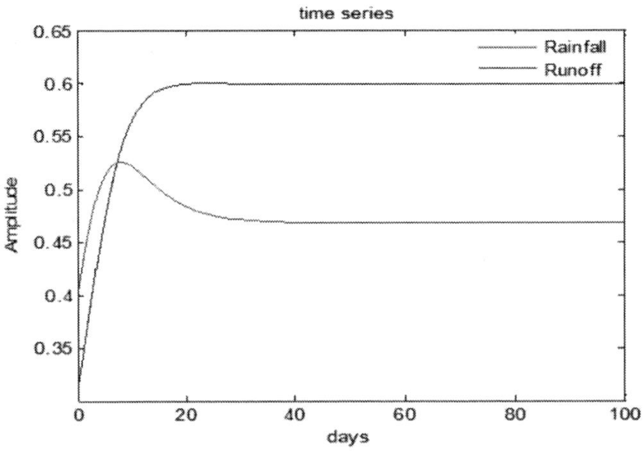

Fig.4. Solution of Equation 4 with parameter value are μ=0.14; ν =0.5; α=0.12; γ =0.4; λ=0.05;

From Fig. 4, we can see that rainfall on 8th days shown that incrising and decresing from 15th day and then stabilize from 20th day to 100th day. Run off on the 18th days shown that incrising optimally and stabilize from 18th day to 100th day. The part or 50% of rainfall can be absorp to the land and others can be change to the run off form. The application this model into the observation data of Mookervart catchment area is still ini progress.

5 Conclusions

The proposal a nonlinear dynamical system for Mookevart catchment area is breafly given. We proposed a nonlinear dynamical system describing Mookervart catchment area. The bifurcation analysis is given and numerical solution of equation is also described. The simulation show that the runoff profile is depend on the rainfall input.

This research founded by the grant of LPM of Syarif Hidayatullah State Islamic University 2016.

References

[1] W. Brutsaert, *Hydrology, An Introduction*, Cambridge University Press, (2005)

[2] A. Rinaldo, A. Marani and R. Rigon, Geomorphological Dispersion, Water. Resour, Res. **27**, 513-523, (1991)

[3] J.W. Kircher, Catchments as Simple Dynamical Systems: Catchment Characterization, Rainfall-Runoff Modelingand Doing Hydrology Backward, Water Resources research, **45**, (2009)

[4] I.G. Pechilivandis, B.M. Jackson, N.R. Mcintyre and H.S. Wheather, Catchment Scale Hydrological Modeling: AReview of Model Types, Calibration Approachs and Uncertainty Analysis Methods in the Context of Recent Development in Technology and Application, Global NEST J., **13**(3), 193-124, (2011)

[5] D. Koutsoyiannis, On the Quest for Chaotic Attractors in Hydrological Processes, Hydrological sciences, J. des Sceinces Hydrologiques, **51** (6), 1065-1092,(2006).

[6] S. Wiggin, *Introduction to Applied Nonlinear Dynamical System and Chaos*, Springer Verlag, Berlin, 820, (2003)

MATEC Web of Conferences **103**, 04017 (2017) DOI: 10.1051/matecconf/201710304017
ISCEE 2016

Evaluation on Flow Discharge of Grassed Swale in Lowland Area

Nurhazirah Mustaffa[1], *Noor Aliza* Ahmad[1,*], and *Mohd Adib* Mohammad Razi[1]

[1]Faculty of Civil and Environmental Engineering, Department of Water and Environmental Engineering, Universiti Tun Hussein Onn Malaysia, 86400 Parit Raja, Batu Pahat, Johor

Abstract. Grassed swale is an open vegetated channel designed specifically in attenuating stormwater runoff to decrease the velocity, to reduce the peak flows, and minimize the causes of flood. Therefore, the fundamental of this study is to evaluate the flow discharge of swale in Universiti Tun Hussein Onn Malaysia (UTHM), which has flat land surface area. There are two sites of study were involved to assess the performance of swale as stormwater quantity control, named as swale 1 and swale 2. Data collection was conducted on 100 meters of length for each swale. The velocity of swale was measured thrice by using a current meter according to the six-tenths depth method, after a rainfall event. The discharge of drainage area in UTHM was determined by the Rational Method (Q_{peak}), and the discharge of swales (Q_{swale}) was evaluated by the Mean-Section Method. Manning's roughness coefficient and the infiltration rate were also determined in order to describe the characteristics of swale, which contributing factors for the effectiveness of swale. The results shown that Q_{swale} is greater than Q_{peak} at swale 1 and swale 2, which according to the Second Edition of MSMA, the swales are efficient as stormwater quantity control in preventing flash flood at the campus area of UTHM.

1 Introduction

A rapid urbanization over the past few decades creates the need for construction of extensive storm drainage facilities [1]. Swales are one of the sustainable means to decrease the velocity, reduce the peak flows, and minimize the causes of flood, which applicable in many urban settings such as parking lots, roadways, commercial and light industrial facilities, and residential developments [2]. A grassed swale is a potential solution to transport stormwater from impervious surfaces by slowing down the flow and infiltrates into soils, and gives a significant improvement as a stormwater quantity and quality control over the traditional drainage ditch [3].

There are several factors that contributed to the ineffectiveness of urban drainage systems, such as topography, soil type, and substandard drainage design. In the case of Batu Pahat, Johor, some of the contributing factors include the lowland area where the land surface is flat and only 1 to 2 meters above sea level. Besides, the top soil is dominated by

[*]Corresponding author: aliza@uthm.edu.my

soft clay and peat soils, which caused the low infiltration rate [4]. In the end of 2006, Batu Pahat has been hit by severe flood, which was called as "banjir termenung". Universiti Tun Hussein Onn Malaysia (UTHM) was also affected by this flood where the old campus has been partially sunk in water. Even though UTHM has provided a detention pond, however, it is inefficiently functioned to reduce the peak flow rate as the elevation gabs is small between inflow and outflow, which caused back water from the main drain or river flows back into the drainage area [5].

Thus, the fundamental of this study is to find out whether grassed swale is one of the sustainable means to overcome flash flood. The performance of swale was assessed in terms of water quantity control by evaluating the flow discharge of swale. There are two sites of swales were involved in the campus area of UTHM, named as swale 1 and swale 2, which shown in Fig. 1. This study was carried out mainly referred to the Second Edition of Urban Stormwater Management Manual for Malaysia (MSMA).

Fig 1. Sites of swales in UTHM campus as swale 1 and swale 2.

2 Preliminary works

The sub-catchment or drainage areas of swales were measured via Google Earth application according to each site of swale, which then retrieved through Earth Point application to determine the drainage areas. The drainage area is based on the slope of overland flow, wherever a possibility of surface runoff would flow into the swale during a rainfall event. According to Fig. 1, the highlighted polygons are showing the drainage areas for swale 1 and swale 2. The area of swale 1 is 0.76 hectares, meanwhile the area of swale 2 is 0.62 hectares. The landuses were identified within a sub-catchment, where both sites have mixed landuses of grass cover and roadways. Therefore, Equation 1 is used to produce an average runoff coefficient (C_{avg}) for each site of swale, where C is the runoff coefficient of segment and A is the area of segment in hectares (ha).

$$C_{avg} = \frac{\sum CA}{\sum A} \quad (1)$$

The lengths of swale for both swale 1 and swale 2 are 100 meters, and divided into three sections (Section A, Section B, and Section C). The distance from Section A to Section B and Section B to Section C is 50 meters respectively. Four meters of a tagging line were stringed transversely across each section of the swale. Three points of verticals or subsections were established on the tagging line, labeled as left, center, and right, to define the location for wading rod during the current metering work. These points were divided at the center point of tagging line and distributed the left and right points corresponding to the bottom width of swale where the most shallow flow depth of swale would occur after a rainfall event. This tagging line was also facilitated for the levelling work to define the surface profile of swale.

3 Flow discharge measurement

Discharge or flow rate is expressed as Q. The flow discharge conveyed in the grassed swale is determined under fully developed conditions. Swales should be designed to convey the peak flow of the 10-year storm without overtopping. The swales were studied by determining the discharge of drainage areas (Q_{peak}) in UTHM evaluated by the Rational Method and the discharge of swales (Q_{swale}) evaluated by the Mean-Section Method. Q_{swale} must be greater than Q_{peak}, which means the swales in UTHM are able to accommodate the precipitation amount received by each swale.

3.1 Mean-Section Method

Mean-Section Method is used for calculating the discharge rate in the cross-sectional area of flow. To cope with the vertical distribution of velocity, measurements should be made at different depths in the water profile. This study has applied the depth of observation set at 0.6 of the total depth to measure the mean velocity, as the depth of water within the swale is less than 75 cm [6]. The equipment used to measure the velocity of swale in this study is a 1-D electromagnetic current meter (AEM1-D). The flow velocity of swale was measured three times at each vertical by waded the rod into the stream for several seconds to obtain the value as appeared on the display unit. During the measurement, the rod needs to be held in a vertical position, and the sensor of current meter must be opposition to the flow direction (facing the upper stream).

3.2 Rational Method

Malaysia and many other countries have frequently used the Rational Method as a technique for estimating the runoff peak [7]. This has agreed by Goyen et al. [8] to apply Rational Method for designing urban drainage networks in small catchments. The hydrology of grass swales can be characterized and compared using the Rational formula and the corresponding runoff coefficient [9]. The Rational Method is expressed as in Equation 2, where Q is the flow discharge, C is the runoff coefficient, i is the average rainfall intensity, and A is the drainage area.

$$Q = \frac{CiA}{360} \tag{2}$$

General procedures for estimating peak flow using the Rational Method, first must select design ARI [10]. Average Recurrence Interval (ARI) is the average length of time between rain events that exceeds the same magnitude, volume, or duration [11]. In this study, swales are categorized as a minor system which has 10-year ARI. The average rainfall intensity (i) for this study is determined by the empirical equation to minimize error in estimating the rainfall intensity values from the IDF curves [7], expressed in Equation 3,

$$i = \frac{\lambda T^{k}}{(d+\theta)^{\eta}} \tag{3}$$

where the fitting constants of λ, κ, θ, and η were adopted from the rainfall station of Sembrong Dam (Site 1931003) as according to the Second Edition of MSMA in Table 2.B1. Meanwhile, T is the ARI and d is the storm duration.

4 Contributing factors

Besides evaluating the flow discharge of each swale, the roughness coefficient and infiltration rate of swale were also determined in order to describe the characteristics of swale, which contributing factors in the effectiveness of swale. Roughness coefficients represent the resistance to flood flows in channels and flood plains [12]. Equation 4 is used to determine the Manning's roughness coefficient (n), where Q is the flow discharge, A is the cross-sectional area, R is the hydraulic radius, and S is the longitudinal slope. These hydraulic parameters can be obtained based on the plotted surface profile of each swale.

$$n = \frac{A}{Q} R^{2/3} S^{1/2} \tag{4}$$

The infiltration capacity of the soil indicates the maximum infiltration rate at a certain moment. The infiltration rate of each swale was determined as the campus of UTHM was claimed have high groundwater table and low infiltration rate. As noted by David [13], the swales are expected to have a decreasing capacity for infiltration storage, with the maximum infiltration rate decays asymptotically towards the saturated hydraulic conductivity of the soil. The occurrences of surface flow, storage, and eventual discharge within the swale were resulted once the rain exceeds the maximum infiltration rate. The infiltration test was carried out by using the double ring infiltrometer, which consists of two concentric stainless steel inner and outer rings. As according to ASTM D3385-03 standard test method, the rings were driven into the soil and partially filled with water. The drop of water level in the inner ring is measured over time to obtain the infiltration rate. This rate becomes constant when the saturated infiltration rate for the soil being measured has been reached. The infiltration rate (I) equation expressed as in Equation 5, where the sum of infiltration in millimeters divided by the accumulation of time in hours.

$$I = \frac{\sum Infiltration}{\sum time} \tag{5}$$

5 Results and discussion

Data collection for swale 1 and swale 2 were carried out on three days, 25 October 2015, 16 November 2015, and 25 November 2015. The bottom widths at each section of swale are constant values, meanwhile the top widths are varied in line with the flow depth. The measurement of flow depth and velocity has found that the average flow depths for swale 1 (0.195 m – 0.390 m) are higher than swale 2 (0.120 m – 0.236 m). As well as the mean velocities of swale, swale 1 (0.088 m/s – 0.094 m/s) has the higher ranges of velocities than swale 2 (0.065 m/s – 0.078 m/s). The measurements of flow discharge and roughness coefficient were evaluated for each section of swale on the certain days, where the results are shown in Table 1.

Table 1. The flow discharges and roughness coefficients of both swales on three days.

Date	Swale 1		Swale 2	
	Flow discharge, *Q* (m^3/s)	**Manning's, *n***	**Flow discharge, *Q* (m^3/s)**	**Manning's, *n***
25/10/2015	0.0207–0.0318	0.129–0.176	0.0066–0.0110	0.121–0.141
16/11/2015	0.0181–0.0351	0.120–0.181	0.0065–0.0101	0.121–0.144
25/112015	0.0167–0.0284	0.112–0.151	0.0050–0.0060	0.099–0.129

From previous study by Ali et al. [14] have compared the surface roughness of swale between a gravel bed and smooth bed, where the gravel bed has a slightly higher coefficient than the smooth bed. It shows that the rough the surface, the higher the coefficient will occur. However, the results on Table 1 have shown otherwise, where the values of Q were increased as the n values increased. In this case, previous study by Stagge [9] has claimed that at high-flow depths, water is not slowed by grass in the swale but allowing sedimentation and too high to undergo filtration. The results from Table 1 were also found similar to the previous study by Rhee et al. [15], where when the flow depth is greater than the grass height, the submergence will occur and the roughness started to increase as the velocity increased. These findings have explained why the flow discharges in this study did not decrease as the swale bed gets rougher.

As stated by Temple et al. [16], in case of large flows, thickness of the boundary zone by vegetative growth tends to be a minimum, thus the passing flow becomes negligible and the roughness coefficient tends to converge to a constant. This has seconded by Wu et al. [17], as the submergence starts to occur in the swale, the roughness coefficient tends to remain constant or rise. He also added that heights, types, and conditions of the vegetation are the factors that contribute to the changes in flow characteristics of the swale and its roughness coefficient. This can be concluded that the variations of roughness coefficients occurred at each section of swales as shown in Table 1 are due to the characteristics of grass growth. There is a possibility that there are differences in heights, types, and conditions of the grass in each section of swale even in a same swale. However, this study did not perform an evaluation on the characteristics of grass.

The calculation of flow discharge on swales (Q_{swale}) and flow discharge on drainage areas (Q_{peak}) are shown in Table 2 and Table 3 respectively. The channel areas in Table 2 were determined by applying the maximum flow depth of each swale. In evaluating the Q_{swale}, the n values were according to Table 2.3 as in the Second Edition of MSMA, so that the n values can be coincided with the estimation of time of concentration (t_c). In this case, the Q_{swale} was evaluated by using the Equation 4, in terms of Q.

Table 2. Calculation of Q_{swale}.

Site of study	Channel area, *A* (m^2)	Wetted perimeter, *P* (m)	Hydraulic radius, *R* (m)	Manning's, *n*	Q_{swale} (m^3/s)
Swale 1	1.415	3.632	0.390	0.050	0.477
Swale 2	1.099	3.581	0.307	0.050	0.316

Table 3. Calculation of Q_{peak}.

Site of study	Runoff coefficient, *C* Minor system (≤ 10 year ARI)	Average rainfall intensity, *i* (mm/hr)	Drainage area, *A* (ha)	Q_{peak} (m^3/s)
Swale 1	0.83	162.85	0.76	0.285
Swale 2	0.61	146.45	0.62	0.154

The Q_{peak} in Table 3 was evaluated by using the Equation 2. Based on Table 2 and Table 3, the comparisons between Q_{swale} and Q_{peak} can be made. The results shown that Q_{swale} is greater than Q_{peak} at swale 1 and swale 2, which according to the Second Edition of MSMA, the swales are efficient as stormwater quantity control in preventing flash flood. The capability of swales at the campus area of UTHM in collecting and conveying the stormwater runoff are in the range of 48.7% – 59.7%. In terms of reducing the flow discharge, previous studies have found significant results of the effectiveness of swale. Stagge [9] has compared to the direct runoff and found that the swales reduced the peak flow by an average 50% - 53%. Based on flow attenuation for swale surface, Ainan et al. [18] have found the percentage reduction of peak flow in the range of 28.9% - 55.9%. In field studies by Wu et al. [19] and supported by swale hydrologic models [20, 21] have noted a percentage reduction of peak flows in the range of 10% - 20%. This percentage reduction of peak flow has proven the swale is capable in controlling the stormwater quantity.

Table 4. Summary of swale characteristics.

Criteria	Site of study	
	Swale 1	Swale 2
Drainage area, A (ha)	0.76	0.62
Channel length, L (m)	100	100
Longitudinal slope, S_o (m/m)	0.001	0.001
Channel area, A (m^2)	1.415	1.099
Side slope, z	1.7	2.3
Maximum flow depth, y (m)	0.63	0.47
Infiltration rate, I (mm/hr)	21.14	4.67
Time of concentration, t_c (min)	18.9	22.8
Percentage of impervious area (%)	77.6	38.7
Land surface	Poorly grassed	Average grassed

Hydrologic soil group (HSG) is expressed in terms of four groups (A, B, C, and D) based on soil's infiltration rate defined by Soil Conservation Service [22]. According to Table 4, swale 1 has the infiltration rate of 21.14 mm/hr, which categorized the soil in Group A. Group A soils have low runoff potential and high infiltration rates even when thoroughly wetted and consist chiefly of deep, well to excessively drained sands or gravels. Meanwhile, swale 2 has the infiltration rates of 4.67 mm/hr, which categorized the soil in Group B. Group B soils have moderate infiltration rates when thoroughly wetted and consist chiefly of moderately deep to deep, moderately well to well drained soils with moderately fine to moderately coarse textures. From the results, it shows that swale 2 has low rate of water transmission compared to swale 1.

6 Conclusions

This study has carried out flow discharge measurements of grassed swales in UTHM campus, which took place at swale 1 within the area of 0.76 ha and swale 2 within the area of 0.62 ha. Based on the analyses, swale 1 and swale 2 are an effective grassed swale in conveying stormwater runoff ($Q_{swale} > Q_{peak}$). The swales are efficient in preventing flash flood with the average rainfall intensity (i) of 146.45 mm/hr – 162.85 mm/hr. The results showed the flow discharges (Q) are increased by the increase of roughness coefficients (n). This can be concluded that the n values started to be constant or rise once the submergence occurs in the swale, where the flow depth is greater than the grass height. The heights, types, and conditions of the grass covers are the factors that contribute to the changes in flow characteristics of the swale and its roughness coefficient. In this study, both swales have 0.001 of longitudinal slope (S_o), however, differ in side slope (z) where swale 1 is 1.7 and swale 2 is 2.3. The n values at swale 2 have resulted slightly less than the n values at swale 1. This has shown that the channel irregularity of swales also may have contributed to the variations of n. This study can be extended to determine other contributing factors to the variations of flow discharges such as the types of vegetation or the rainfall intensity.

The authors would like acknowledgement the Ministry of Higher Education Malaysia and Research, Innovation, Commercialization, and Consultancy Management Office (ORICC) of Universiti Tun Hussein Onn Malaysia (UTHM) for financially under FRGS Fund (Vot 1414) for supporting this study.

References

[1] T.A. Saybet, *Storm Water Management*, Wiley, John & Sons, 179-191, (2006)

[2] City of Indianapolis, *Stormwater Design and Specification Manual*, Chapter 4.7 Swales, 104-112, Retrieved on September 16, 2016 from http:www.cityoffortwayne.org/utilities/images/stories/designman/ 4.7%20swales.pdf (2009)

[3] D. Stevens,*For Three Weeks I Owned the University of Illinois*, Lulu.com, 135-136, (2011)

[4] L.A.Q. Tunji, A.A.A. Latiff, D. Tjanjanto and S. Akib, *The Effectiveness of Groundwater Recharges Well to Mitigate Flood*, Int. J. of the Physical Sciences, **6** (1), 8-14, (2011)

[5] D. Tjanjanto, S. Musa and M.B. Ridzuan, *A Study on Artificial Recharge Well as a Part of Drainage System and Water Supply in UTHM*, Nat. Seminar on Envir. Develop. & Sustainability: Ecological, Economical & Social Aspects, 106-111, (2008)

[6] M.A. Khan, K. Mahmood and G.V. Skogerboe, *Current Meter Discharge Measurements for Steady and Unsteady Flow Conditions in Irrigation Channels, Report No. T-7, Int. Irrigation Management Institute*, Pakistan, (1997)

[7] Dept. of Irrigation and Drainage Malaysia (DID), *Urban Stormwater Management Manual for Malaysia Second Edition*, Government of Malaysia, (2012)

[8] A. Goyen, B.C. Phillips and S. Pathirajas, *Urban Rational Method Review*, Australia Rainfall and Runoff Revision Project 13, Engineers Australia, Canberra, (2014)

[9] J.H. Stagge, *Field Evaluation of Hydrologic and Water Quality Benefits of Grass Swales for Managing Highway Runoff*, Thesis of Master of Science, Faculty of Graduate School, University of Maryland, College Park, (2006)

[10] Dept. of Irrigation and Drainage Malaysia (DID), *Urban Stormwater Management Manual for Malaysia*, Government of Malaysia, (2000)

[11] V.T. Chow, *Handbook of Applied Hydrology*, McGraw-Hill Book, New York, (1964)

[12] G.J. Jr. Arcement and V.R. Schneider, *Guide for Selecting Manning's Roughness Coefficients for Natural Channels and Flood Plains*, U.S. Geological Survey Water Supply Paper 2339, United States Government Printing Office, Washington, (1989)

[13] A.P. Davis, Field Performance Of Bioretention: Hydrology Impacts, J. Hydrologic Engineering, ASCE, **13** (2), 90-95, (2011)

[14] Z.M. Ali and N.A. Saib, Influence of Bed Roughness in Open Channel, *Proc. of Int. Seminar on Application of Science Mathematics*, (2011)

[15] D.S. Rhee, H. Woo, B.A. Kwon and H.K. Ahn, Hydraulic Resistance of Some Selected Vegetation in Open Channel Flows, River. Res. Applic. **24**, 673-687, (2008)

[16] D.M. Temple, K.M. Robinson, R.M. Ahring and A.G. Davis, *Stability Design of Grass-lined Open Channels*, Agricultural Handbook, 667, USDA: Washington, D.C., (1987)

[17] F. Wu, H.W. Shen and Y. Chou, Variations of Roughness Coefficients for Unsubmerged and Submerged Vegetation, J. of Hydra. Eng. ,**125** (9), 239-245, (1999)

[18] A. Ainan, N.A. Zakaria, A.A. Ghani, R. Abdullah, L.M. Sidek, M.F. Yusof and L.P. Wong, Peak Flow Attenuation Using Ecological Swale and Dry Pond, Advances in Hydro-Science and Engineering, **VI**, (2003)

[19] J.S. Wu, C.J. Allan, W.L. Saunders and J.B. Evett, Characterization and Pollutant Loading Estimation for Highway Runoff, J. Envir. Eng., ASCE, **124**(7), 584–592, (1998)

[20] A. Delectic, Performance of Grass Filters Used for Stormwater Treatment – A Field and Modeling Study, J. Hydrology, **317** (3-4), 168–182, (2006)

[21] D. Ackerman and E.D. Stein, Evaluating the Effectiveness of Best Management Practices Using Dynamic Modeling, J. Envir. Eng, ASCE **134** (8), 628–639, (2008)

[22] R. Cronshey, R. H. McCuen, N. Miller, W. Rawls, S. Robbins, D. Woodward, J. Chenoweth, S. Hamilton, W. Merkel, R. Rallison, H. Richardson, W. Styner and T. Seeman, *Urban Hydrology for Small Watersheds*, 210-VI-TR-55, Second Ed, (1986)

AUTHOR INDEX

AUTHOR INDEX

AUTHOR INDEX

AUTHOR INDEX

AUTHOR INDEX

AUTHOR INDEX

International Symposium on Civil and Environmental Engineering (ISCEE 2016)

MATEC Web of Conferences Volume 103 (2017)

Melaka, Malaysia
5 - 6 December 2016

Part 2 of 2

Editors:

Zainorizuan Mohd Jaini
Lee Yee Yong
Alvin John Lim Meng Siang
Mohamad Hanifi Othman
Siti Nazahiyah Rahmat
Mohd Shalahuddin Adnan

ISBN: 978-1-5108-3893-2

Printed from e-media with permission by:

Curran Associates, Inc.
57 Morehouse Lane
Red Hook, NY 12571

Some format issues inherent in the e-media version may also appear in this print version.

Printed by Curran Associates, Inc. (2017)

For additional information, please contact EDP Sciences – Web of Conferences at the address below.

EDP Sciences – Web of Conferences
17, Avenue du Hoggar
Parc d'Activité de Courtabœuf
BP 112
F-91944 Les Ulis Cedex A
France

Phone: +33 (0) 1 69 18 75 75
Fax: +33 (0) 1 69 28 84 91

contact-edps@webofconferences.org

Additional copies of this publication are available from:

Curran Associates, Inc.
57 Morehouse Lane
Red Hook, NY 12571 USA
Phone: 845-758-0400
Fax: 845-758-2633
Email: curran@proceedings.com
Web: www.proceedings.com

TABLE OF CONTENTS

PART 1

PART 2

The Soil-Root Strength Performance of *Alternanthera Ficoidea* and *Zoysia Japonica* as Green Roof Vegetation

Muhamad Firdaurs Abdullah[1], *Hartini* Kasmin[2,*], *Nor Azizi* Yusoff[1], *Safwan* Mustafar[2], *Shazren* Salimun[2], *Zaidi* Embong[1] and *Hanim* Ahmad[3]

[1] Research Centre for Soft Soil (RECESS), Universiti Tun Hussein Onn Malaysia, 86400 Parit Raja, Batu Pahat, Johor, Malaysia
[2] Micropollutant Research Centre (MPRC), Faculty of Civil and Environmental Engineering, Universiti Tun Hussein Onn Malaysia, 86400 Parit Raja, Batu Pahat, Johor, Malaysia
[3] Horticulture Research Centre, MARDI Headquarters, Serdang, P.O. Box 12301, 50774 Kuala Lumpur

Abstract. The rise of awareness on environmentalism has demanded that all parties involved in built environment to implement green technology in their construction projects. Great care must be taken when designing a green roof system including the selection of plants and appropriate substrates. This study was performed to investigate the soil-root composite strength of two types of green roof vegetation (*A. Ficoidea* and *Z. Japonica*) at different growth periods for up to 6 months. Both plants were planted in six plastic plots (45 cm × 29 cm × 13 cm) containing a mixture of perlite, vermiculite and organic soil. Every two months, a series of direct shear tests were conducted on a sample from each species to determine the root-soil shear strength. The tests continued until the 6th month. The average results showed that *Z. Japonica* had higher soil-root shear strength (49.1 kPa) compared to *A. Ficoidea* after two months of growth. In the 4th month however, *A. Ficoidea* managed to surpass *Z. Japonica* (28.7 kPa versus 18.5 kPa) in terms of shear strength. However, their average peak shear strength decreased sharply compared to the previous month. Lastly, in six months, *A. Ficoidea* sustained a higher average peak soil shear strength (56.5 kPa) compared to *Z. Japonica* (14.3 kPa). Therefore, it can be concluded that *A. Ficoidea* may offer a better soil reinforcement than Z. japonica and thus it could potentially be a good choice of green roof vegetation.

1 Introduction

Green roof is a green space which consists of layers of soil mediums, vegetation and drainage system. The concept of green roof nowadays has become increasingly popular as it brings many benefits towards the environment and helps promote a sustainable lifestyle

* Corresponding author: hartini@uthm.edu.my

[1]. The increase in the usage of green roof covers can potentially transform an urban environment into a cooler urban area.

Vegetation has been the basis of slope stabilization since ancient times. In more recent times, the roles of vegetation in some specific geotechnical processes have been recognized. Vegetation influences slope stability indirectly through its effect on the soil moisture regime. It will be able to intercept rainfall and draw water from the soil via evapotranspiration [2]. The depth of growing media is typically between 10 and 30 cm although some implementations (referred to as intensive green roofs) have deeper soils capable of sustaining large shrubs and even trees [3]. Fauzi et al, [4] reported the findings by Rowe [5] which stated that plant selection for different locations and climate requires a test on a myriad of plant species in order to evaluate their suitability for green roofs. A variety of plant species was used by Malaysian green roof researchers such as Johari [6] who has identified 3 family names of plants that can survive in Malaysian tropical climate which are *Crassulaceae*, *Portulaceae* and *Neoregelia*. On the other hand, Ismail [7] used *S.podophyllum, I. batatas, I. horsfalliae/J. Sambac* and *I. pes-caprae* for thermal studies. Furthermore, Yusoff [8] used *Axonopus Compressus* and evaluated its soil-root strength performance. Therefore, these studies suggested that more species can be tested for their suitability as green roof vegetation.

Storm water runoff is considered as the main concern of green roofs [4,9,10]. During heavy rainfalls in Malaysia, the direct impact of rain drops on green roofs can be regarded as the main force causing the separation of soil particles. This situation will happen when the bond between the particles of soil structure weakens and this could lead to soil erosion [11]. As a result, green roof strength may decrease. Hence, the presence of vegetation might reduce soil erosion through rainfall interception as well as enhance the infiltration and extraction of soil moisture via plant transpiration [12].

Therefore, this study aims to investigate dynamic soil shear strength development of two types of vegetation based on the green roof vegetation growth rate. It is important to identify which species of grass or herbaceous plants are able to perform better in terms of soil reinforcement (soil shear strength). An experiment was designed to assess soil-root shear strength of two plants within stipulated months so that a suitable candidate for green roof vegetation can be identified.

2 Materials and methods

2.1 Plot preparation and substrate materials

The experiment was conducted at the Research Centre for Soft Soil (RECESS), UTHM. The two plant species, *Alternanthera Ficoidea* (*A. Ficoidea*) and *Zoysia Japonica* (*Z. Japonica*), were used in the experiment and were allowed to grow up to 6 months. Both plant species were chosen for this experiment due to their popularity for landscape purposes. The plants were planted in March 2016 in six rectangular plastic plots (45 cm (W) × 29 cm (L) × 13 cm (H)); three plots for each plant contained a mixture of organic soil, perlite and vermiculite by the weight ratio of 2:3:5 respectively (Fig. 1(a)). The mixture has a maximum water capacity of 50% and 0.00433 cm/s of soil permeability [13]. The mixture may potentially be an ideal substrate as recommended by Beattie and Berghage [14] because it comprises of lightweight, well-drained material and adequate water and nutrient holding capacity. The substrate depth was 80 mm. All the plots were placed in the shade to prevent exposure to direct sunlight.

2.2 Plot maintenance

The plants were irrigated to excess once a day to ensure that they obtain enough water to survive. However, the frequency of watering the plants is usually influenced by several factors. Among the factors are weather, age of plants, species and soil types. Besides, plant supplements such as NPK fertilizer (with the ratio of 21% for each nutrient) was applied in the beginning of the treatment. This will ensure the successful growth of the plants.

Fig. 1. Plot preparation; (a) substrate materials, (b) A. *Ficoidea* and (c) *Z. Japonica*

2.3 Species sampling

2.3.1 Soil water content measurement

Prior to shear testing, to ensure that the soil water content of both plots (*A. Ficoidea* and *Z. Japonica*) is about the same, the plots were poured with excess water until the soil was fully saturated. Then, the plots were left in a dark room for 24 hours before the sampling process began. This was done to minimize the effect of soil suction on shear testing and to maintain soil moisture reading each time the test was performed.

2.3.2 Core sampling

Every two months, three individual *A. Ficoidea* plants were sampled from the plot by penetrating a square soil core (all sides equal to 8 cm as shown in Fig. 2(a)) diagonally across the plot [15]. The same sampling method was employed to retrieve three soil samples from the *Z. Japonica* plots (Fig. 2(b)). A portion of soil from those three samples per plot was taken and weighed before being placed in an oven at 105°C for 24 hours to obtain soil moisture content. Then the sample was transferred into a sample ring with a dimension of 6 cm × 6 cm × 2.5 cm. After skimming, the soil sample ready to be transferred to a direct shear apparatus.

2.4 Direct shear tests

Drained direct shear tests were carried out using Shear Trac II machine (Geocomp Inc, USA) available at RECESS, UTHM. It is a fully automatic direct shear testing machine that controls the whole process beginning from the consolidation phase until the end of the shear phase. An automatic data logging system of the machine will be able to download all

DOI: 10.1051/matecconf/201710304018

required data from the tensiometer, displacement transducer and load cell to a computer using software named 'SHEAR'. Soil samples from the ring with 2.5 cm thickness were inserted into the shear box carefully by using a special pressing tool (Fig. 2(c)) so that it would minimize the disturbance to the soil sample.

Prior to the direct shear test, the sample will go through a consolidation phase with a small vertical stress of 1 kPa applied on it. Such a little amount of stress was used after taking into consideration the shallow depth of the samples in the pots (less than 8 cm). Thus, it is important to adapt to the actual situation on the ground near to the surface. After the consolidation graph shows a constant result (Fig. 3), the direct shear test will commence. The test was carried out with a shearing rate of 1.0 mm/minute and a maximum displacement of 20 mm (33.3% strain).

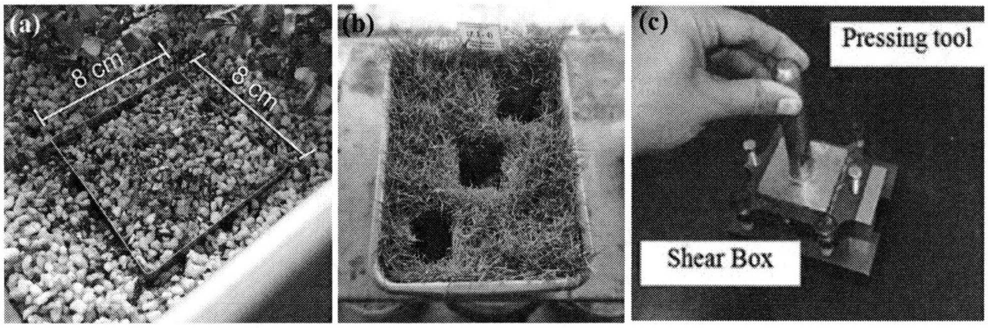

Fig. 2. Sampling process: (a) a square soil core, (b) three soil samples of *Z. Japonica* retrieved diagonally across the plot and (c) soil sample transferred into a shear box [16]

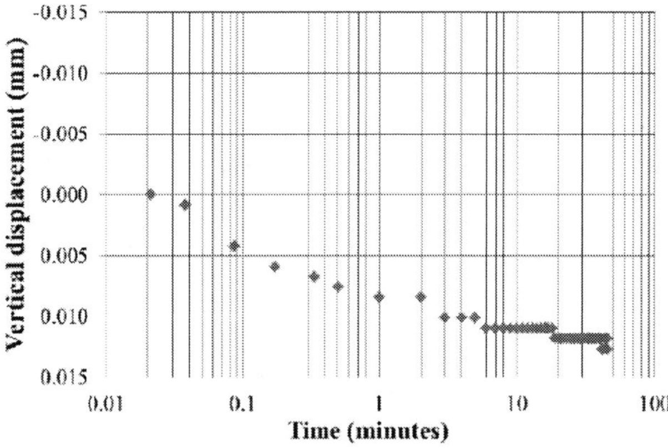

Fig. 3. Sample was consolidated with a vertical stress of 1 kPa until it reached a constant value

2.5 Above-ground biomass measurements

During the core sampling process, all the upper parts of plants including stems and leaves that were confined inside the square soil core (8 cm × 8 cm) were collected and kept in sealed plastic bags. The vegetation samples were later weighed and dried in an oven at 70°C for approximately two days until a constant sample weight indicated that all available water has been removed. The dried vegetation was weighed to a precision of 0.001g, yielding an estimate of above-ground biomass for every shear strength measurement. The biomass values were calculated by dividing the dry weight of vegetation by soil core area.

DOI: 10.1051/matecconf/201710304018

3 Results and discussion

3.1 Drained direct shear test

Fig. 4(a) shows the results of the direct shear tests of *A. Ficoidea* and *Z. Japonica* soil samples in the 2nd month which are 63, 33 and 13.5 kPa for *A. Ficoidea* samples 1, 2 and 3 respectively. The average peak shear strength for *A. Ficoidea* in the 2nd month was 37 kPa, denoted by a bold red dotted line in that figure. Peak shear strength in the 2nd for *Z. Japonica*'s soil samples 1, 2 and 3 were 62, 32 and 53 kPa respectively. Their average value is indicated by a bold green dotted line with a peak value equals 49 kPa, which is larger than *A. Ficoidea* of the same age.

After 4 months of growth, compared to the results for the past two months, the average soil shear strength had slightly decreased to 28.7 kPa and 18.5 kPa for *A. Ficoidea* and *Z. Japonica* respectively.

Fig. 4(b)). This time, *A. Ficoidea* has higher soil shear strength than *Z. Japonica*. Its shear stress values (sample 1 to 3) ranged from 24 to 33 kPa, which are higher than *Z. Japonica* (13 to 28 kPa). The decrease in shear strength for both species might be due to increasing soil moisture content and decreasing above ground biomass when compared to readings for the past two months.

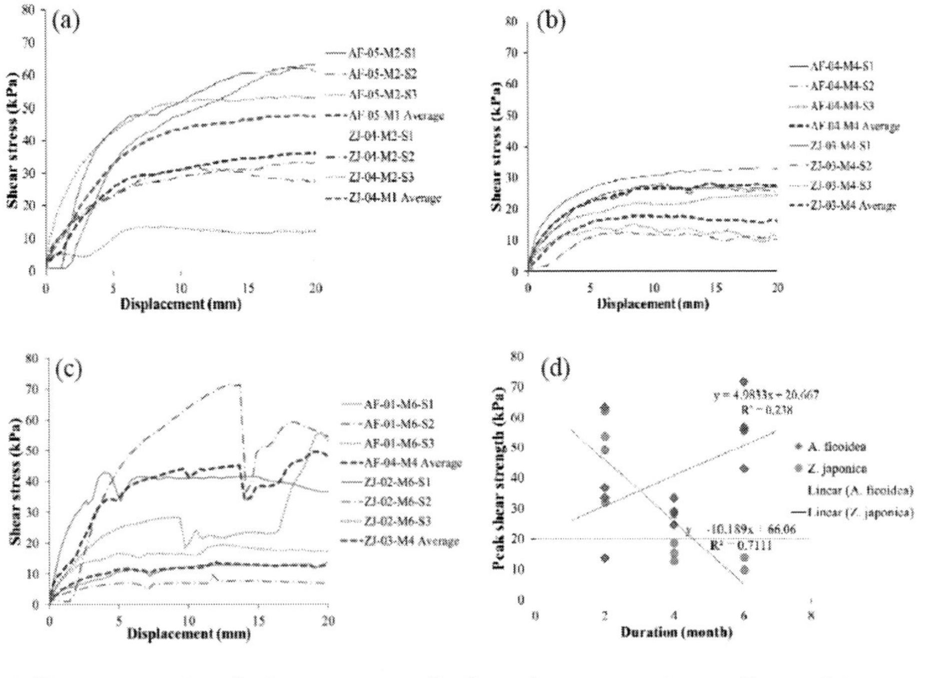

Fig. 4. Shear stress against displacement curves for direct shear tests on three replicates of *A. Ficoidea* and *Z. Japonica* of growth period; (a) month 2, (b) month 4 and (c) month 6. (d) Peak shear strength values of both plant species for the whole growth period (month 2, 4 and 6)

Fig. 4(c) shows the result of direct shear testing of *A. Ficoidea* and *Z. Japonica* after 6 months of growth. Surprisingly, the shear stress of *A. Ficoidea* is observed to be higher than *Z. Japonica* for all its samples although its readings were quite erratic. *Z. Japonica* seems to perform poorly as shown in the previous test (month 4), even its current average shear stress (14 kPa) was the lowest throughout the study. On the other hand, the average

 DOI: 10.1051/matecconf/201710304018

shear stress of *A. Ficoidea* was 57 kPa, which is the highest average shear strength recorded for the study. Based on

Fig. 4(c), the average of the shear stress curve of *A. Ficoidea* increased drastically at the early stage of the test (less than 3 mm displacement). It also has two distinct peak values at 14 and 19 mm displacement of its average shear curve. According to [17], the first peak value is due to maximum lateral root resistance whereby the second peak value is the maximum mobilized tap root resistance. This might be contributed by the resistance of both lateral and tap roots.

Overall, *A. Ficoidea* was able to improve soil shear strength better than *Z. Japonica* (Fig. 4(d)). This is proven by the increasing trendline with R^2 equals 0.238 of *A. Ficoidea* in the figure. Low R^2 value is due to a high variation in soil shear strength values recorded in the 2nd month and the 6th month. Meanwhile, *Z. Japonica* shows a decline in its shear strength throughout the study, indicated by a negative trendline with R^2 equals 0.7111. Despite its bad performance, it has a high R^2 value indicating a low variation in shear strength measured.

3.2 Soil moisture content

Fig.(a) shows the soil moisture content of *A. Ficoidea* and *Z. Japonica* for every soil sample tested every 2 months. There is no general trend observed for *A. Ficoidea* but *Z. Japonica*'s moisture content decreased with time (month). The readings for the moisture content in the 2nd month for *A. Ficoidea* were varied substantially (large standard error). However, the researcher managed to maintain soil moisture content in a range between 120 to 220 %.

3.3 Plant biomass

The above-ground biomass production of *A. Ficoidea* was observed to be higher than *Z. Japonica* for all months except the 2nd month (Fig. 5(b)). This is due to very little foliage of *A. Ficoidea* during its juvenile age. Meanwhile, *Z. Japonica* shows a slight decrease in biomass with an average value of around 400 g m^{-2}. There was no below-ground biomass as the mass of roots beneath the ground was too small and the root screening process was tedious.

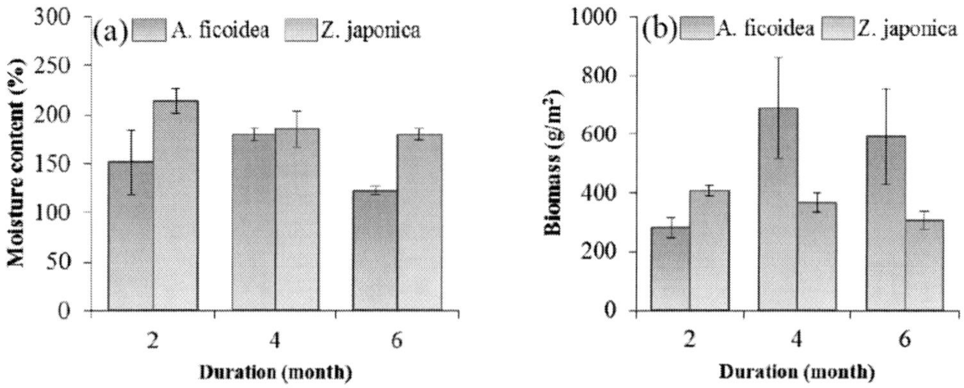

Fig. 5 (a) Soil moisture content against planting months and (b) Above-ground biomass of two plants after 2, 4 and 6 months. Vertical bars indicate standard error

3.4 Relationship between shear strength, moisture content and biomass

The results reveal correlations between shear strength and other parameters. It can be observed that the peak shear strength of *A. Ficoidea* correlates negatively ($R^2 = 0.667$) with moisture content (Fig. 6(a)). However, the relationship between peak shear strength of *Z. Japonica* is unusual. Normally, when soil moisture content decreases, the shear strength of soil will increase. As discussed by [18], saturation of soil will reduce the shear strength due to buoyancy reduction in normal force and it could also destroy the capillarity and apparent cohesion of soil.

Based on Fig. 6(b), there is no clear relationship between the peak shear strength of *A. Ficoidea* and above-ground biomass due to very low R^2 value (0.0176). In addition, the vertical trendline with high gradient ($R^2 = 0.548$) implies that above ground biomass of *Z. Japonica* does not have influence on its peak shear strength throughout its growth period. It should be noted that stable soil is primarily due to high biomass, which likely resulted in high soil water loss through plant transpiration [18]. However, it seems that the rule does not apply in this situation.

Fig.7 shows the relationship between soil moisture content and above-ground biomass for *A. Ficoidea* and *Z. Japonica.* Similarly, there is no clear relationship between moisture content of *A. Ficoidea* and its above-ground biomass due to very low R^2 value (0.0392). Besides, it is observed that the effect of above ground biomass towards moisture content is unusual, particularly on *Z. Japonica*. Generally, the increase in above-ground biomass will help to decrease soil moisture content. High biomass particularly above the ground would demand a large amount of water absorption by the root to produce soil-plant-atmosphere continuum (SPAC) [19]. Furthermore, 99% of water lost is due to transpiration and only 1% is due to evaporation [20].

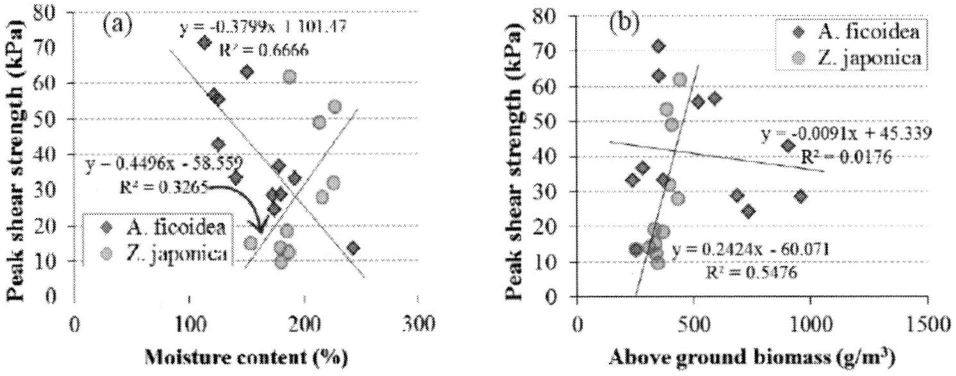

Fig. 6. Relationship between shear strength and other parameters; (a) moisture content and (b) above-ground biomass

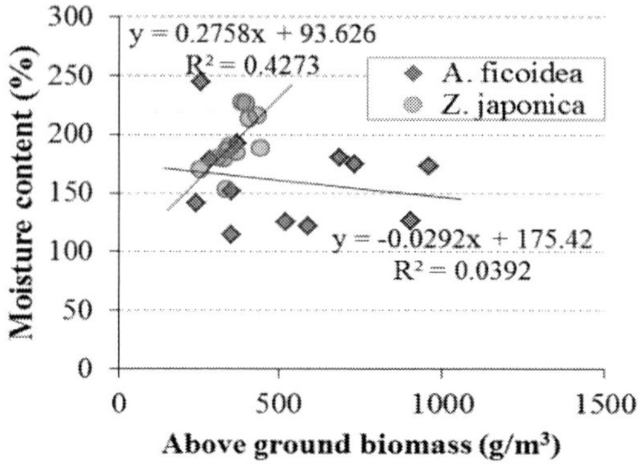

Fig. 7. Relationship between moisture content and above-ground biomass

4 Conclusion

Therefore, it can be concluded that the soil-root shear strength of *A. Ficoidea* has increased over the study period. It also performed better than *Z. Japonica* in terms of soil shear strength, removal of soil moisture and biomass production. Hence, *A. Ficoidea* is an ideal species for the purpose of green roof vegetation compared to *Z. Japonica*. However, further study is required to accurately investigate the correlation between shear strength, biomass and moisture content of both plants as the current results are not entirely convincing.

The materials used in this paper are based on work supported by Universiti Tun Hussein Onn Malaysia (UTHM) under the ERGS (E052) grant and advice from MARDI. Any findings, opinions, conclusions or recommendations expressed in this paper are those of the corresponding authors and do not necessarily reflect those of UTHM and MARDI.

References

[1] C.M. Fai, M. F. Abu Bakar, M.A.A. Roslan, F.A. Fadzailah, M.F.Z. Idrus, N.F. Ismail, L.M. Sidek and H. Basri, Hydrological performance of native plant species within extensive green roof system in Malaysia, ARPN J. of Eng. and Appl. Sci., **10**(15), 6419-6423 (2015)

[2] F.H. Ali and N. Osman, Shear strength of a soil containing vegetation roots, Soils and Foundation, **48**(4), 587-596 (2008)

[3] D.J. Sailor, A green roof model for building energy simulation programs, Energy and Buildings **40**(8), 1466-1478 (2008)

[4] M.A.Fauzi, N.A Malek and J. Othman, Evaluation of green roof system for green building projects in malaysia. World Acad. of Sci., Eng. and Tech., **74**(2), 124-130 (2013)

[5] D.B. Rowe, Green roofs as a means of pollution abatement, Environmental Pollution, **159**(8-9), 2100-2110 (2011)

[6] J. Johari, M.H. Rasidi and I. Said, Planting technology of green roof for building in tropical cities: a review on plant selection, 5th South East Asian Tech. Univ. Consort. Symp. Universiti Teknologi Malaysia (2011)

[7] A. Ismail, M.H.A. Samad and A.M.A. Rahman, The investigation of green roof and white roof cooling potential on single storey residential building in the malaysian climate, World Acad. of Sci., Eng. and Tech., **5**(4), 168-176 (2011)

[8] N.A. Yusoff, M.N. Ramli, T.N.T. Chik, H. Ahmad, M.F. Abdullah, H. Kasmin and Z. Embong, Soil-roots strength performance of extensive green roof by using Axonopus compressus, IOP Conf. Series: Mat. Sci. and Eng., **136**(1), 1-6 (2016)

[9] K. Getter, D.B. Rowe and J.A. Andresen. Quantifying the effect of slope on extensive green roof stormwater retention. J. of Eco. Eng., **31**(4), 225-231 (2007).

[10] N.D. VanWoert, D.B. Rowe, J.A. Andresen, C.L. Rugh, R.T. Fernandez and L. Xiao, Green roof stormwater retention, J. Env. Quality, **34**(3), 1036-1044 (2005)

[11] N.A. Yusoff, M.F.M. Fauzi, F. Abdullah, The study of soil-roots strength performance by using Pennisetum setaceum grass, *The 3rd Inter. Conf. on Civil and Env. Eng. for Sustain, IConCEES 2015*, EDP Sciences, **47,** 1-7 (2016)

[12] M. Saifuddin and N. Osman, Evaluation of hydro-mechanical properties and root architecture of plants for soil reinforcement, Current Science, **107**(5), 845-852 (2014)

[13] H. Kasmin, S.N.U. Munir, N.S. Razak, N.A. Yusoff, R. Kassim, H. Ahmad and Z. Sayuti, Exploration on the potential of locally waste-sourced green roof substrate mixture, *The Inter. Conf. on Mat. Eng. and Appl. (ICMEA)*, Hong Kong, China, (2016)

[14] D.J. Beattie and R. Berghage, Green roof media characteristics: the basic, *Proc. of 2nd North American Green Roof Conference: Greening Rooftops for Sustainable Communities,* Oregon, United States of America, (2004).

[15] N. Osman, S. Barakbah, Ecological Engineering, **37,** 139-147 (2011)

[16] M.S. Zainordin, N.A. Yusoff, T.N.T. Chik, M.A.H.A. Manap, Z. Sayuti, M.F. Abdullah, The study of soil-roots strength performance of soil slope by using Guinea grass, J. Appl. Mech. and Materials, **802,** 10-15 (2015)

[17] M.N. Abdullah, N. Osman and F.H. Ali, Soil-root shear strength properties of some slope plants, Sains Malaysiana, **40**(10), 1065-1073 (2011)

[18] N. Osman and S.S. Barakbah, Parameters to predict slope stability - Soil water and root profiles, Eco. Eng. **28,** 90-95 (2006)

[19] B. Huang and P.S. Nobel, Root hydraulic conductivity and its components, with emphasis on desert succulents. J. Agron, **86**(5), 767-774 (1994)

[20] A.H. Hazlifah, *Tanaman campuran - diversiti untuk kestabilan*, BSc Thesis, University of Malaya, Kuala Lumpur, Malaysia, (1995)

Flood Risk Assessment on Selected Critical Infrastructure in Kota Marudu Town, Sabah, Malaysia

Janice Lynn Ayog[1,2*], *Felix* Tongkul[2], *Abdul Karim* Mirasa[1,2], *Rodeano* Roslee[2], and *Salinah* Dullah[1]

[1]Faculty of Engineering, Universiti Malaysia Sabah, Jln UMS, 88400 Kota Kinabalu, Sabah, Malaysia
[2]Natural Disaster Research Centre, Universiti Malaysia Sabah, Jln UMS, 88400 Kota Kinabalu, Sabah, Malaysia

Abstract. This study investigates the risk of flood on selected critical infrastructure in a flood-prone catchment in Sabah, Malaysia. Kota Marudu, located in the Bandau floodplain, one of the Sabah's northern water catchments, was selected as the study site due to its frequent flood occurrence and large floodplain coverage. Two of its largest rivers, namely Sungai Bongon and Sungai Bandau, tends to flood during rainy season and cause temporary displacements of thousands of people living in the floodplain. A total of 362 respondents participated in the questionnaire survey in order to gather information on historical flood occurrence. Three flood depth groups were determined, which are 1) less than 0.3 meter, 2) 0.3 - 0.6 meter and 3) more than 0.6 meter, while three categories of critical infrastructure were defined, namely transportation system, communication system and buildings. It is found that the transportation system encounters the most severe impact as flood inundation increases, where 92% of the respondents believe that the transportation access should be abandoned when flood depth is more than 0.6m. The findings of this study will be used for detailed risk assessment, specifically on the vulnerability of the critical infrastructures to flood in this floodplain.

1 Introduction

The world community faces many risks from flooding as the pattern of floods across all continents have been changing, becoming more frequent, intense and unpredictable for local communities. Flooding is the most devastating type of natural disasters that hits human lives around the world. When flood occur in developing nations, they can effectively wipe out the entire investments in infrastructure and the personal wealth of many people, not to forget loss of lives, physical injuries, illness and psychological trauma that result from this disaster. Climate change may exacerbate the damage caused by extreme events, such as floods [1]. Critical infrastructure, which comprises of

* Corresponding author: jay@ums.edu.my

transportation and communication system, water and power lines and public institutions including schools, halls and health clinics, plays important roles as it includes the basic facilities, services and installations needed for the functioning of a community or society. Besides daily needs, it is also crucially important on the occasions of natural disasters, which refer to floods in this study, as the infrastructure will be used for safe road access for evacuations, to offer temporary refuge for the displaced victims, to provide medical treatments for injuries and many more. In fact, one of the essential parts in post-disaster reconstruction is rebuilding facilities and infrastructures for those affected, which includes critical infrastructures [2].

Flood events in the past revealed the need to safeguard infrastructure from destructive threats and prompted the need to work on preparedness, response and recovery plans associated to infrastructure safety. In order to gain the capacity to be prepared to avoid and manage hazardous flood situations, it is fundamental to understand and analyse the vulnerability of critical infrastructure and associated buildings. Through the vulnerability assessment, effective measurements for flooding can be improved and thus minimizes the damaging impacts done on critical infrastructure which in return minimizes the disruption done to the community as critical infrastructure is essential to public health and safety, nation's security as well as economic vitality.

1.1 Flood Risk, Hazard and Vulnerability

It is fundamentally important to understand the nature of flood risk in order to be able to future-proof human lives as well as economic damages and losses against the rapidly changing circumstances of the twenty-first century. Kates and Kasperson [3] considered that there are three distinct steps for risk assessment which include the identification of hazards, the estimation of risks and the evaluation of the associated consequences of the derived risk. However, Ologunorisa and Abuwua [4] believed that post-audits of all risk exercises should be considered as another additional step in order for sound risk management to occur.

Risk can be expressed mathematically as shown in Equation (1):

$$Risk = Hazard\ x\ Vulnerability\ x\ Exposure \quad (1)$$

Based on Equation (1), hazard is the potentially harmful event magnitude refers to both physical and statistical dimension of inundation such as the return period, flow velocity, extent and depth of a flood [5]. Past experience and records are taken into account to determine hazard. Exposure is the intensity, frequency and degree of human or assets exposed to an agent. The exploitations of flood prone areas due to the tremendously increased world's population have induced more exposure of people, property and infrastructure to the susceptibility of inundation. In the event of an inundation with no building contained in that flooded area, risk again does not exist owing to the fact that exposure is absent.

Vulnerability in the context of flood, is the extent to which a system is susceptible to floods due to exposure, a perturbation, in conjunction with its capacity or incapacity to be resilient, which is to cope, to recover or to adapt [6].

Linnekamp et al. [7] performed a research assessing household vulnerability to flooding as a result of climate change in two low elevation coastal zone cities in Caribbean. The authors intended to examine the perceptions of household of flood risks and considered vulnerability of households as the combined results of social-economic in entitlements and exposure to flooding and extreme rainfall. Both survey and interview were carried out during the research. The cities were compared in terms of their exposure, resistance and adaptive capacity to flooding. Balica et al. [8] has carried out a study in establishing a flood vulnerability index of 9 different coastal cities around the world, which took into account the hydro-geological, social, economic and politico-administrative components.

1.2 Background of Study Area & Research Problem

Malaysia, due to its geographical location with 189 water basins and an average rainfall over 2000mm per year, is prone to suffer from floods and flash floods especially during cyclical monsoon cold surge episodes which are characterized by extreme rainfall from roughly November to February [9]. Sabah is not excluded from the appalling effects of flooding. Nine districts in Sabah are identified as flood-prone regions according to the report made by the Department of Irrigation & Drainage Sabah in 1996. Kota Marudu is listed as one of them with approximately 8.0 square kilometers of flood-prone area.

Kota Marudu is unique as it is shaped like a flat-floored amphitheater, where most areas upstream are a part of the Kinabalu National Park (in which Mount Tambayukon, the third highest mountain in Malaysia is located). Due to this natural priviledge, vast areas of the upper catchment are intended to be gazetted as a water protection area. This should have resolved the flooding problems, however, illegal logging upstream and the issues of land acquisition by the natives (of which overlapped with the proposed water protection zone) have challenged the effort to effectively mitigate the flood woes. While these issues need to be tackled in one way or another, an alternative approach needed to be looked into to mitigate the annual flooding cases; hence, by developing flood vulnerability index, the critical infrastructures needed for the emergency and evacuation plans of the flood victims can be safeguarded.

Kota Marudu suffered from frequent occurrences of inundations which happen at least once a year. The flooding occurrences were contributed by the heavy flows of Sungai Bongon and Sungai Bandau, which are the main rivers in Kota Marudu. The critical infrastructures in Kota Marudu play very important roles in the flood events. During these seasonal, but extreme high flows, a huge number of villagers are displaced and evacuated to the nearest temporary relief centres and refuges, such as schools or halls. There were 2,027 people from 448 families in 16 villages were forced to abandon their homes and moved to nine temporary relief centres due to floods. The critical situations were further exacerbated by the inundation of road system as vehicles were no longer able to safely evacuate victims to a secure place. Emergency response teams swiftly took speed boats as alternatives to promptly send the victims to temporary shelters.

It is apparent that the safety of critical infrastructures is vital for flood emergency approaches in Kota Marudu. Owing to this, the vulnerability of critical infrastructures has to be analysed and assessed to ensure of its effectiveness for flood emergency plan in Kota Marudu. International concern about the disastrous flooding events has risen as the associated economic damage is significant. Hence, many countries have adopted the approaches of flood risk management as stimulated by severe catastrophic floods around the world [10]. The relentless upward global trend in vulnerability to flooding, resulted losses and threats from potential impacts of climate change on flood frequency represent the supporting factors that contribute to the emerging of flood risk management.

2 Research Methodology

This study intends to investigate the perception of the local residents pertaining to the risk of selected critical infrastructure to floods in Kota Marudu, Sabah, Malaysia. A questionnaire-based survey was done to gather the responses of the locals on the risk of flood inundation of these infrastructure based on a range of flood depths. The local community's perception on the flood damage, especially those who have encountered flood almost every year will shed unique insight of what the actual damage could have been, in order for a detailed vulnerability assessment of the affected critical infrastructure can be done accurately.

In the questionnaire survey, 362 respondents of the local residents were gathered through convenient sampling. Information regarding the vulnerability of existing critical infrastructures in Kota Marudu was acquired effectively through the knowledge and past experience of the local residents in the study area. The questionnaire was concentrated on the vulnerability of the three general critical infrastructures - transportation system, communication system and buildings. Three range of flood depths were determined, namely 1) less than 0.3 meter, 2) 0.3 – 0.6 meter and 3) more than 0.6 meter. A structured non-disguised questionnaire was prepared where questions were presented in pre-arranged order and the respondents were well informed of the purpose of the survey. The targeted survey respondents are those in the age of 15 and above.

The questionnaire consists of two parts. The first part consists of two demographic questions and four flood related questions. The second part required the respondents to assess the flood risk of the selected critical infrastructure based on three range of flood depths. Table 1 to Table 3 provide the descriptions of the flood risk based on the previously-mentioned three critical infrastructure categories.

Table 1. Flood risk and impact on the transportation system.

Risk	**Availability of Transportation**	**Description**
Severe	Cannot travel outdoor	Transportation access abandoned
Critical	Very hard to travel outdoor	Very limited transportation access
Moderate	Hard to travel outdoor	Limited transportation access
Light	Minor difficulties to travel outdoor	Slightly transportation access
None	Free to travel outdoor	Full transportation access

Table 2. Flood risk and impact on the communication system.

Risk	**Availability of Transportation**	**Description**
Severe	Cannot be contacted (both phone and face-to-face)	No communication access
Critical	Very limited contact (via phone only)	Very limited communication access
Moderate	Limited contact (face-to-face with major difficulties)	Limited communication access
Light	Good contact (face-to-face with minor difficulties)	Slightly limited communication access
None	Not affected	Full communication access

Table 3. Flood risk and impact on the buildings.

Risk	**Description**
Severe	More than 50% of the structure destroyed
Critical	30% to 50% of the structure destroyed
Moderate	5% to 30% of the structure destroyed
Light	Less than 5% of the structure destroyed
None	Structure not affected by flood

3 Results and Discussion

Fig. 1 shows the distribution of the respondents based on the age groups. It can be seen that the respondents participated in the questionnaire were distributed rather evenly – each age group consists of 20-30% of the total number of respondents. Majority of the respondents are aged between 45-65 years old (30%), while the least number of representative of age group is between 26-45 years old (21%). Fig. 2 shows the distribution of respondents based on occupation. It is found that 42%, which is the majority of the respondents, were farmers while the second largest group is the student group.

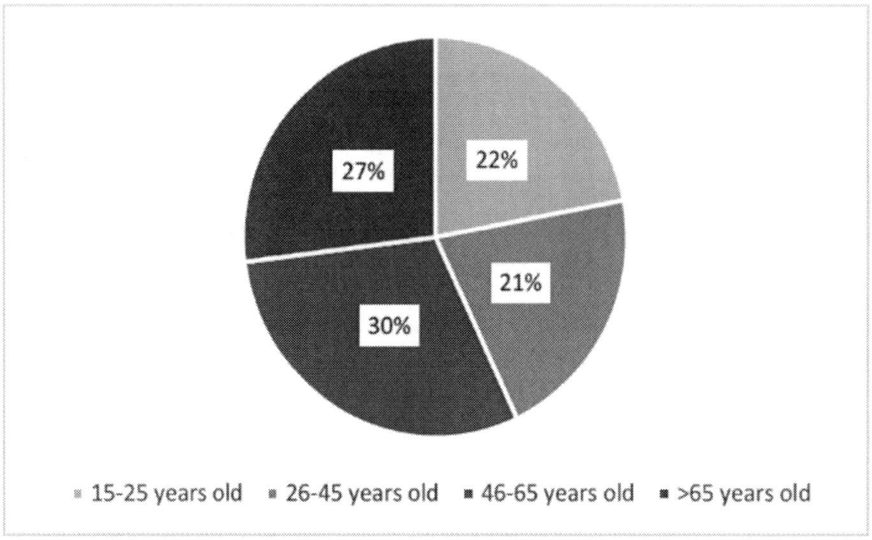

Fig. 1. The distribution of respondents based on the respondents' age group.

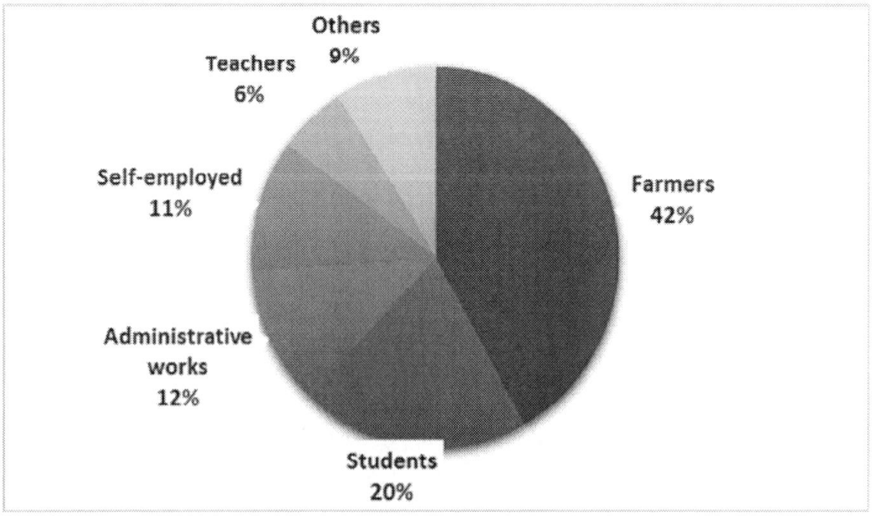

Fig. 2. The distribution of respondents based on the respondents' occupation.

Fig. 3 shows the responses on flood risk for transportation system. This study reveals that at a flood depth of 0.3 m, about 79.3% of the respondents has a perception that the

DOI: 10.1051/matecconf/201710304019

impact is light. At a flood depth between 0.3 m to 0.6 m, approximately 69.9% of the respondents perceived that the flood impact is at a moderate level, while a majority of the respondents (91.7%) think that flood could cause severe impact. This finding shows a trend where flood impact increases with flood depth.

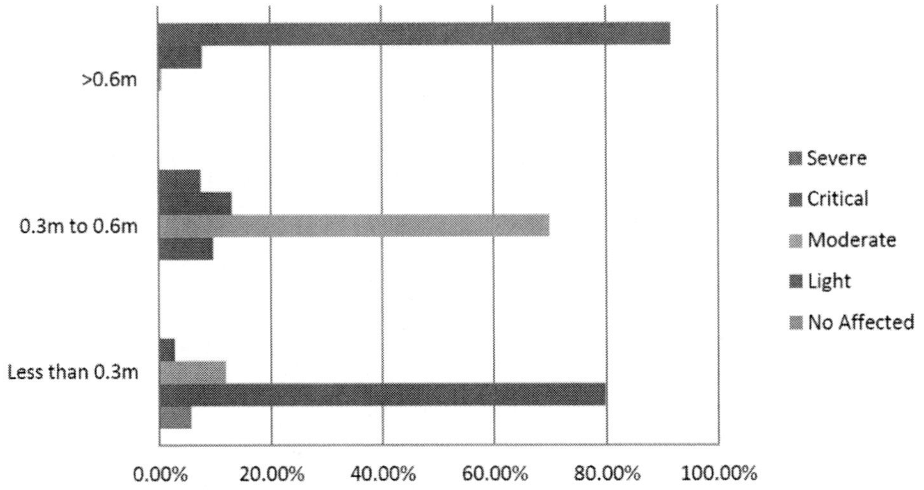

Fig. 3. Responses on flood risk for transportation system.

Different flood depth causes varying impact to the transportation system, which has direct exposure to the impact of flood. According to Heberger et al. [11], an increase of 1.4 m in sea level will cause double mileage under inundation, which indicates that transportation is affected due to flood. In Semarang, Indonesia, Marfai et al. [12] found that an increase of flood level about 0.5 m will cause disruption of accessibility and road.

Communication is another aspect which is frequently affected by flood. This aspect has been incorporated into this study and the result revealed that the effect of flood on the communication system in Kota Marudu increases with inundation. From Fig. 4, at flood depth less than 0.3 m, most of the respondent feels that the impact on communication is light, meaning that less flood impact is experienced. This situation changed as the flood depth increases. At flood depth between 0.3 m to 0.6 m, 52.3 % of total respondent perceived that flood impact is at moderate level. The impact raised to critical stage when the flood level is more than 0.6 m, as expressed by approximately 55.5 % of the total respondents. As with transportation, it can be observed that communication system is greatly affected by flood depth.

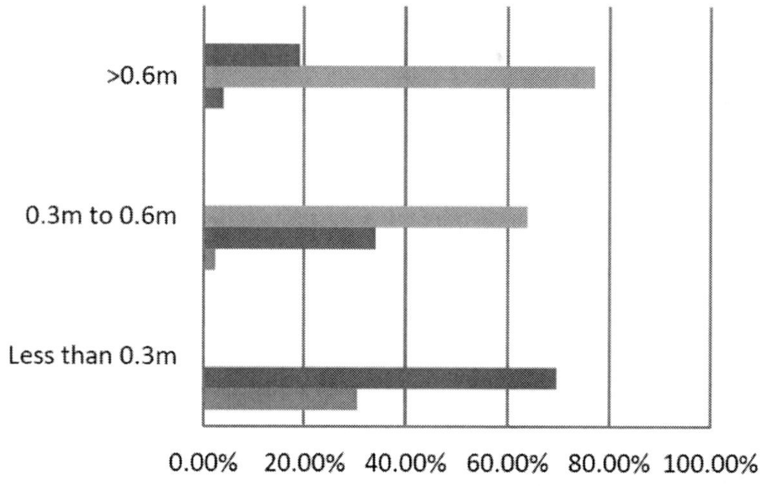

Fig. 4. Responses on flood risk for communication system.

Fig. 5 shows the response on flood risk on buildings. Building damages with its relation to flood is another aspect was investigated in this study. It is found that at a flood depth of 0.0 m to 0.6 m, building damage were perceived as light, while a flood depth beyond 0.6 m is thought to cause moderate impact to the building damage.

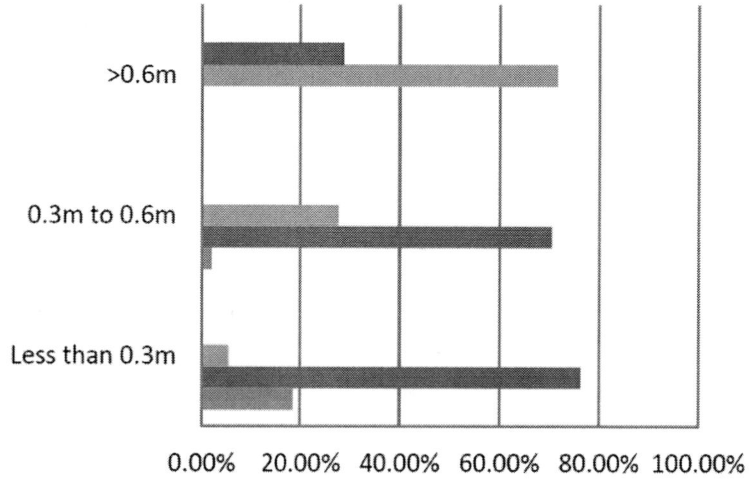

Fig. 5. Responses on flood risk for buildings.

According to Tu and Tingsanchali [13], most building is not much affected by flood depth less than 0.24 m and only slightly affected between 0.5 m to 1.5 m of depth. Higher impact is experienced for flood depth between 1.5 m to 3.0 m. Heberger et al. [11] states that the property in bay area is particularly vulnerable to flood impact due to sea level rise which could cause coastal floods. The location of Kota Marudu town, which is located at Marudu Bay may share the same flooding characteristics as reported by Heberger et al. [11].

4 Conclusion

In this study, the severity of flood impact on three social aspects namely transportation system, communication system and building damages were investigated. The responses gathered indicate that as the flood depths increases, the impact of flood gradually becomes critical. Transportation system encounters the most severe impact, where transportation access will be abandoned when the flood depth reaches above 0.6 m. This shows that transportation system is very sensitive to floods. Theoretically, flood issues such as these would require structurally and/or non-structurally reinforcing the critical infrastructures studied in this research, however, due to the complex nature of flood and as the results are solely based on the perception of the respondents alone, further analysis will be done through the development of flood models to corroborate the accuracy of the findings in this study and suggest appropriate solution to mitigate the impacts.

This study is financially supported by the Malaysian Ministry of Education Fundamental Research Grant Scheme (FRGS) Grant No. FRG0421-TK-1/2015. The authors would like to thank Mr. Vun Siew Fui, Ms. Chin Su Vun and Ms. Nicole Lee Siew Len for their significant contributions in this research.

References

[1] A. W. Salami, A. A. Mohammad and O. G. Okeola, Evaluation of climate change impact on runoff in the Kainji Lake Basin using Artificial Neural Network Model (ANN), Malaysian J. of Civil Eng, **26 (1)**, 35-50, (2014)

[2] C. E. Farahdilla, S. A. Mansur and A. Abdullah, Common issues in post-disaster reconstruction, Malaysian J. of Civil Eng, **25 (2)**, 267-276, (2013)

[3] R. W. Kates and J. X. Kasperson, Comparative risk analysis of technological hazards (a review), *Proceedings of National Academy of Science*, **80**, 7027-7038, (1983)

[4] T. E. Ologunorisa and M. J. Abuwua, Flood risk assessment: A review, J. of App. Sciences & Environmental Management, **9(1)**, 57-63, (2005)

[5] H. Apel, G. T. Aronica, H. Kreibich and A. H. Thieken, Flood risk analyses – How detailed do we need to be?, Natural Hazards, **49 (1)**, 79-98, (2009)

[6] S. F. Balica, N. Douben and N. G. Wright, Flood vulnerability indices at varying spatial scale, Water Sci. & Tech., **60 (10)**, 2571-2580, (2009)

[7] F. Linnekamp, A. Koedam and I.S.A. Baud, Household vulnerability to climate change: Examining perceptions of households of flood risks in Georgetown and Paramaribo, Habitat International, **35 (3)**, 447-456, (2011)

[8] S. F. Balica, N. G. Wright, F. van der Meulen, A flood vulnerability index for coastal cities and its use in assessing climate change impacts, Natural Hazards, **64 (1)**, 73-105, (2012)

[9] L. Juneng, F. T. Tangang, C. J. C. Reason, Numerical case study of an extreme event during 9-11 December 2004 over the east coast of Peninsular Malaysia, Met. and Atm. Physics, **98 (1-2)**, 81-98, (2007)

[10] J. W. Hall and E.C. Penning-Rowsell, Setting the Scene for Flood Risk Management, Flood Risk Sci. and Management, 1-16, (2011)

[11] M. Heberger, H. Cooley, P. Herrera, P. H. Gleick and E. Moore, Potential Impacts of Increased Coastal Flooding in California Due to Sea-level Rise, Climate Change, **109**, 229-249, (2011)

[12] M. A. Marfai, L. King, J. Sartohadi, S. Sudrajat, S. R. Budiani and F. Yulianto, The Impact of Tidal Flooding On A Coastal Community in Semarang, Indonesia, The Environmentalist, **28 (3)**, 237-248, (2008)

[13] V.U.T. Tu and T. Tingsanchali, Flood hazard and risk assessment of Hoang Long River basin, Vietnam, *Proceedings of the International MIKE by DHI Conference*, (2010)

DOI: 10.1051/matecconf/201710304020

Rainwater Harvesting (RWH) and Groundwater Potential as Alternatives Water Resources in Malaysia: A Review

Syafiqa Ayob[1] , and *Siti Nazahiyah* Rahmat [2*]

[1]Postgraduate Student, Faculty of Civil and Environment Engineering, Universiti Tun Hussein Onn Malaysia, 86400, Batu Pahat, Johor, Malaysia.
[2]Senior Lecturer, Faculty of Civil and Environment Engineering, Universiti Tun Hussein Onn Malaysia, 86400, Batu Pahat, Johor, Malaysia.

Abstract. Socio-economic growth at present exerting pressure on the capacity of water supply. Water has been harvested worldwide since antiquity, thus, it is essential to manage the potential available of natural water supplies in the catchment to save water for dry periods when the supply-demand gap is comparatively high. The overall aim of this paper is to present a general overview of rainwater harvesting system and groundwater, and their potential applications for daily use. Many countries have implement the rainwater harvesting system as a sustainable approach for supplementing their public water supply. In Malaysia, the guidelines for rainwater harvesting implementations was first created in 1999 and developed by time to time to attract users. By using the system, the main findings reveals that it has overcome several of water problem issues. Groundwater is another alternative water supply that can be counted. Research has found that the availability of groundwater in few states in Malaysiawas in freshwater category based on its hardness level results. Thus, conclusions and recommendations are drawn where the gaps exist and more research needs to be focussed.

1 Introduction

Demand on water resources has been increasing rapidly and the issue of supplying adequate water to meet social needs is one of the most significant challenges. Adopting the concept of conservation of water resources could help to cope with the global water shortage as well as mitigate flood[1]. Rainwater Harvesting (RWH) and groundwater are the sources that can be explored to meet the water shortage problem.

RWH has been the main source of water supplies since decades ago. It has been improved from time to time until it becomes more sophisticated to users. Previously, jar collection system was used and later converted to underground storage tank complete with a mesh filter for a better water quality results. There are several advantages of RWH system in which have been identified by a number of researchers[2-4]. To name a few, the system

*Corresponding author: nazahiya@uthm.edu.my

can be used in aquifer replenishment, to delay the surface water runoff which is very useful to prevent the occurrence of flash flood sand erosion control. Generally, RWH is a process where rainwater will be collected, stored and used, similar as treated water. Table 1 shows the guidelines and principle for rainwater harvesting implementation in Malaysia since 1999.

Table 1. Guidelines and principle for rainwater harvesting implementation[6].

No	Guidelines	Year
1	Guidelines for Installing a Rainwater Collection and Utilization System Ministry of Housing and Local Government	1999
2	Rainwater Harvesting: Guidebook on Planning and Design Department of Irrigation and Drainage Malaysia	2009
3	Guideline on Eco-Efficiency in Water Infrastructure for public Buildings in Malaysia	2011
4	Urban Storm water Management Manual for Malaysia, MSMA 2nd Edition	2012
5	Panduan Pelaksanaan Inisiatif Pembangunan Kejiranan Hijau e-Sistem Pengumpulan dan Penggunaan Semula Air Hujan	2012
6	Garis Panduan Perancangan Kejiranan Hijau	2012
7	Garis Panduan Sistem Pengumpulan dan Penggunaan Air Hujan	2013
8	Urban Storm water Management e Part 6: Rainwater Harvesting, MS2526-6:2014	2014

To meet the increasing water demand, the development of distant water resources is not always the best option. Groundwater can be an alternative water source to meet water demand especially in areas where the surface water resources have been developed to their limits. However, groundwater has not become a favorite because it is very costly and the quantification of water that can be extracted is different in every state depending on the geo-technological condition of the area [7].

2 Water Issues in Malaysia

In Malaysia, the source of water for daily use comes from treated water. Due to the growth of population and expansion of industrial and agricultural, the demand for water has increased manifold. The problems arise when the water is not sufficient to meet the demand or polluted [8]. Vast development in various sectors resulted in such impacts on high demand of clean water in some areas. In addition, climate change may be one of the contributors to the water supply problem[9-10]. It changes the availability, quantity and quality of the water resources, which will have impact on the whole cycle of water supply. Nowadays, Malaysia and other Southeast Asian countries receive very high intensities of rainfall during monsoon season[11]. In December 2014, the villagers in Kelantan and some parts of Terengganu faced water supply disruption after a water supply plant affected by the devastating floods. Tube wells were then built to supply clean water[12].

Apart from that, another event that is often associated with the shortage of water is El Nino. El Nino is a natural phenomenon occurs in Pacific Ocean when warm waters of the western coast of South America replace the colder nutrient rich waters and cause impacts on the weather patterns such increase in temperature between 0.5°C and 2°C and reduce the amount of rainfall [13]. Fig.1 shows the impact of El Nino in Malaysia (in circle) in 2015.The country experienced dry and warm conditions from June to August which had affected the water resources utilisation in some areas. This paper will shed light on how the RWH system and groundwater represent the best solutions among the numerous natural water supply sources available.

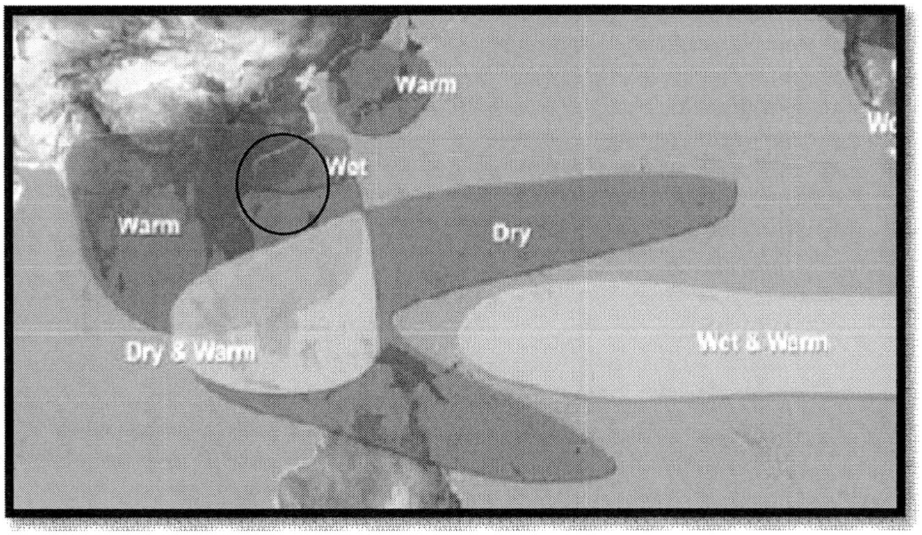

Fig.1. The impact of El Nino from June until August, 2015

3 Rainwater harvesting and groundwater utilisation around the world

3.1 Rainwater Harvesting

3.1.1 South Asia - India

Rainwater harvesting has become one of the recent trends as a sustainable approach for supplementing public water supply as well as for serving several other purposes. To control the wastage of water is one of the benefits of this system.RWH has been practiced as early as 4500 B.C. by the people of the United region. While the early water harvesting techniques used natural materials, 20thcenturytechnology has made it possible to use artificial means for increasing runoff from precipitation. Most of the cases in India, RWH systems being developed due to inadequate and unreliable water supply. People are resorting to rooftop water harvesting, which is the most convenient and economical water supply system. For instance, Aizawl city, located at the North East India has more than 10,000 rainwater harvesting tanks in individual houses [14]. In 2005, Vishwanath[15] reported the declination in ground water levels as well as the effects of pollution with nitrates which poses a threat to their colony. Two small tanks were built to collect rainwater and act as percolation tanks to recharge the aquifer. In addition, the second tank also accumulates the water recycled from sewage that has been treated physically and biologically. Both of these constructed tanks since then have supplied approximately about 2000 houses to their neighbourhood.

3.1.2 South East Asia- Japan, Thailand, Taiwan

In Japan, the RWH techniques have been used both in public and private levels. The main sumo wrestling stadium, Kokugikan, rainwater runoff from the roof of $8400m^2$redirectedinto underground tanks $1000m^3$ for non potable purposes[16]. Following the example of Kokugikan, new public facilities have begun to introduce a system of rain water usage in Tokyo. To meet the community demand, a simple and unique rainwater utilisation to utilise rainwater collected from the roof of private houses has been set up by local residents in the Mukojima district of Tokyo. To date, about 750 private and public buildings in Tokyo have introduce drain water collection and utilisation systems[14].

RWH techniques introduced by Thailand government called '*Thai Jar Programme*' is shown in Fig. 2. The Government of Thailand initiated this programme in the 1980s with the aim to provide an alternative and supplementary water supply in rural areas with emphasis on self sufficiency and conservation[17]. A study on rainwater quality was conducted in north-east Thailand and concluded that rainwater is the safest and most economical source of drinking water in the region with no major health implications to users, as compared to the highly contaminated conventional water sources[17].In Taiwan, under its Ministry of Economic Affairs, a registered agency named Water Resources Bureau (WRB) has been developed. WRB came out with an idea in giving priority to the agricultural sector and subsidies to install a RWH system for gardening. This project has benefited more than 3000 farmers[14].

Fig. 2. Construction of 2,000-litre cement rainwater jar

3.1.3 Malaysia

The water crisis that occurred around 1998 is a starting move in implementing the RWH system in Malaysia. The 'Guidelines for Installing Rainwater Collection and Utilization System' 1999 was a preliminary approach of rainwater harvesting policy issued by the government of Malaysia[18]. The main purpose of the guidelines is to reduce the dependence on treated water and provides convenient buffer in times of emergency or a shortfall in the water supply. Construction of mini dams or rainwater tanks in urban area has been proposed[19]. In 2004, the Ministry of Housing and Local Government prepared another paper to the National Water Resources Council to encourage government buildings to install a rainwater collection system[20]. In 2006, Ministry of Energy, Water and Telecommunication (KTAK) came out with two new water related laws; Water Services Industry Act 2006 and Water Services Commission Act2006. However, RWH system is still not well received by Malaysian, especially in the residential areas, due to lack of awareness and they are not exposed to the advantages of using this system.

3.2 Groundwater

Ground water is basically a renewable resource, however, the volume of water stored may vary greatly from place to place depending on physiography, climate, hydrogeology and rate of groundwater withdrawal used[21]. May and Mazlan [22]reported that approximately 18,184 m^3 of water per day is obtained from the aquifer to supply to the steel factory near Langat river in Malaysia. As surface water sources fail to meet the rising demands of water supply in urban areas, groundwater reserves are being tapped and over-exploited resulting into decline in ground water levels and deterioration of ground water quality. This precarious situation needs to be rectified by immediately recharging the depleted aquifers[23].Based on Segal[24], groundwater quantification can be determined by using these three (3) methods which are groundwater model, pumping test and flow model. Groundwater recharge is a category of planned potable reuse. The two main types of groundwater recharge are called surface spreading and direct injection. The difference between direct injection and surface spreading is the way of water being collected and treated. For spreading surface, the water will be stagnant in the reclamation plant and eventually seep through the soils layers such as rock, gravel and clay to the earthen basins. This kind of recharge allow the recycled water to undergo further physical, biological, and chemical purification as it filters through the soil and ultimately becomes part of the groundwater system. For direct injection method, the collected water goes through additional treatment process. For example, water that is near the sea will be treated to desalinate the salt in water. The treated water is then re-injected into the ground. In Nepal, Yi et al.[11] have found that their study area has shallow groundwater level, thus they have come up with an idea to recharge the groundwater incorporate with rainwater harvesting system. While in other cases, they have come out with the idea to overcome the flat area problems with a technique called artificial recharge. This technique is a simple concept in which water is stored in subsurface permeable aquifers when water is plentiful and being extracted during times of peak.

3.2.1 Potential Groundwater in Malaysia

Groundwater in Malaysia is an important source which can meet the increasing demand of fresh water for a variety of uses [25]. Though groundwater has been used for centuries, usage is limited to shallow unconfined aquifers using excavate dwells [26]. In Malaysia, deep tube wells in the rough sand aquifer was first developed in the past 30years for water supply of coastal cities such as Kota Bharu [27]. Along with the water shortage problems in Selangor that arise each year, government has taken the initiative to encourage many groundwater development projects. The categories of potential groundwater resources are listed in Table 2[28].

The pH value in the north side of Kelantan, Malaysia was slightly acidic ranged from 6.2 to 6.8 and also had low content of chloride [29]. Another successful study by Sapari et al. [30]conducted in Malacca and Selangor, where identification of quality and quantity in fractured met a sedimentary rocks found that the groundwater was in freshwater category based on its hardness level results. This study then concluded that the availability groundwater has benefited the development of industries in the west coast of Peninsular Malaysia. This resource has potential to be further exploited for water supply particularly to remote areas.

Table 2. Categories of potential groundwater resources[28].

Type of Soils	Features
• Alluvium (sand and gravel)	· Most productive
• Limestone	· Found in developed areas
• Fractured sandstone	· Has equivalent metamorphic
• Fractured igneous rock	· Least productive

3.2.2 Rainwater for recharging groundwater

Rainwater recharge methods widely used in country which receives tons of rainfall[16, 27–30]. However, the decision whether to store or recharge water depends also on the rainfall pattern of a particular region where in average total quantity of rain water that can be used from the rooftop collection in normal monsoon season is about 70% to 90%[23, 33]. India is the best example in implementing this technique. In Mumbai, at least three quarters of the city's surface covered in hard concrete by way of buildings and road[35]. In2011, Water Aid report discussed the importance of managing grain water in order to recharge groundwater sources in Nepal. The aim of this report was to investigate the potential recharge zone, volumetric agumentation and the water quality of groundwater after being recharged. It was found that some areas can work as a reservoir where rainwater can be stored in monsoon and extracted in dry season. The water quality on analysis of physical, chemical and biological parameters assured safe with the reference of National Drinking Water Quality Standards (NDWQS) except for nitrate and fecal which was observed more in monsoon after harvesting than in pre-monsoon[36]. Apart from that, Bhavsar and Bhalge [37] claimed that the recharge techniques can be the best solution in the areas that having saline groundwater or water containing fluorides. In addition, Abdul et al., [38]did indicated that the recharge well system will b successful when it is adopted to mitigate flood on site based on the physical model that has been developed.

4 Conclusions and Recommendations

Demand on water resources in Malaysia witnesses a substantial increase due to development, population increase, and global weather change. Adopting the concept of sustainability and conservation of water resources can help to cope with the global water shortage. Malaysia is blessed with an ample supply of water because of abundant rains. The only problem is lack of effective water resources management. Therefore, any attempt to manage water resources will have to take into account both the management of excess water (floods) and insufficient water (droughts). Based on this review the following conclusions can be drawn:

1. Water harvesting means making optimum use of rainwater to attain self-sufficiency in water supply, without being dependent on remote water sources.
2. In Malaysia, extended periods of droughts could cause water supply to fall short of water demands in states supporting large scale agriculture for rice production. This RWH system can be adopted to store the excessive water during wet season for later use (drought season).
3. Several water basins have shown high potential of groundwater. A comprehensive study needs to be carried out to explore potential area of groundwater in Malaysia for alternative to clean water resources.

4. Many researchers have adopted rainwater harvesting techniques in different parts of the country to improve the groundwater quality. The only thing is that it has not been done systematically in all places. In Malaysia, this integrated approach has not been comprehensively studied and it is worthwhile to further investigate.

This study was supported by Fundamental Research Grant Scheme (FRGS) (Vote 1523).

References

[1] M.S. Adnan, N.C. Ali, Y. Erfen, S.N. Rahmat, M.A.M. Razi and S. Musa, Analysis the Impact of Bridges Existance for the Segamat River Using Infowork RS, *IOP Conf. Ser. Mater. Sci. Eng.*, **136**(1), (2016)

[2] M. Nasir, N. Othman, K. Azlan, M.F. Omar and E.H. Omar, Effectiveness of Rainwater Harvesting System At Office, *Proc. 14th Int. Conf. Rain Water Catchment Syst. PWTC, KL*, (2009)

[3] K.E. Lee, M. Mokhtar, M. Mohd Hanafiah, A. Abdul-Halim and J. Badusah, Rainwater harvesting as an alternative water resource in Malaysia: Potential, policies and development, J. Clean. Prod, **126**, 218–222, (2015)

[4] S. Man, N. M. Hashim, A.H. Ahmad, M.T. Khin and N.S. Sidek, Kebolehupayaan sistem penuaian hujan sebagai bekalan air alternatif di Malaysia: Suatu penelitian awal, *Geogr. Malaysian J. Soc. Sp.*, **10**(6), 97–104, (2014)

[5] T.A. Mohammed, M.J. Megat Mohd Noorand A.H. Ghazali, Study on Potential Uses of Rainwater Harvesting in Urban Areas,*Proc. Colloq. Rainwater Util.*, (2007)

[6] Y.M.A. Sharifuddin, S.S. Mohamed, N.H. Adnan and K.A. Musa,Kajian Kuantiti Dan Kualiti Air Hujan Sistem Penuaian Air Hujan (SPAH) Di Rumah Panjang Bair Betong,*Inst. Penyelid. Hidraul. Kebangs. Malaysia*, **1**, 1–8, (2015)

[7] N. Weng, A Comparative Study Of Water Georgetown-Malaysia And Pattaya, Iranian J. Environ. Health Sci. Eng., **3**, (2006)

[8] S.N. Rahmat, A.Z. Mohd Ali, M.H. Wan Ibrahim, N.A. Alias, Oil and grease (O&G) removal from commercial kitchen waste water usingcarbonised grass as a key media, *MATEC Web of Conferences*, **87**(01010), (2017)

[9] S.N. Rahmat, N. Jayasuriya, M. Bhuiyan and M.S. Adnan, Temporal Trends of Discrete Extreme Events - A Case Study, *IOP Conf. Ser. Mater. Sci. Eng.*, **136**(1), (2016)

[10] S. N. Rahmat, N. Jayasuriya and M. A. Bhuiyan, Short-term droughts forecast using Markov chain model in Victoria, Australia,(2016)

[11] Y. Yi, L. Billa and A. Singh, Geoscience Frontiers Effect of climate change on seasonal monsoon in Asia and its impact on the variability of monsoon rainfall in Southeast Asia, *Geosci. Front.*, **6**(6), 817–823, (2015)

[12] A.F. Rahim, Rakyat Kelantan gembira selepas 25 tahun 'menderita' hadapi masalah air,*Utusan Online*, Kelantan, (2016)

[13] J. Saimon, El Nino: The Good, Bad and the Ugly of 2015, (2015)

[14] P.S Mathur and M. Prashad, *Rainwater Harvesting and Utilisation*. UN-Habitat, United Nations Centre for Human Settlements, Book 2: Beneficiries & Capacity Builders,Nairobi, Kenya, (2005)

[15] S. Vishwanath, Domestic Rainwater Harvesting Some applications in Bangalore, India, 1–5, (2005)

[16] K.F.A. Lo, Multi-Faceted Use of Rainwater Harvesting To Combat Water Problems,*XI IRCSA Conf.*, (1991)

[17] T.V. Luong, *A construction manual for cement rainwater Jars and Tanks*, Bangkok,Thailand, (2002)

[18] H.Y. Meng, F.T. Yong, N.M. Salleh, H. S. Seng, A.H.Md.Kassim, C. P. Siang, T.S. Tiat and K.Nantakumar, *Guidelines for Installing A Rainwater Collection and Utilization System*, Malaysia,(1999)

[19] H. Shawahid, A.R. Suhaimi, M.K. Rasyikah, S.A. Jamaluddin, Y.F. Huang and M.Farah, Policies and incentives for rainwater harvesting in Malaysia,*Rainwater Utilization Colloquium*, (2007)

[20] A.I.C. Ani, Rainwater Harvesting System Evaluation: A Resident Experience In Sabah,Malaysia, *Int. Symp. Dev. Econ. Commonalities Among Divers.*, 26–39, (2009).

[21] P. Madhnure and S.K. Jain,Additional Ground Water Storage Potential for Artificial Recharge in Phreatic Aquifers of Yavatmal District, Maharashtra, India,*Natl. Semin. Rainwater Harvest. Water Manag.*, 91–100, (2006)

[22] R.May and N.S.B. Mazlan, Numerical simulation of the effect of heavy groundwater abstraction on groundwater–surface water interaction in Langat Basin, Selangor, Malaysia, Enviromental Earth Sci., **71**(3), 1239–1248, (2014)

[23] W. Struckmeier, Y. Rubin and J.A.A. Jones, *Groundwater - reservoir for a thirsty planet?*, 2nd Edition, The Netherlands: International Year of Planet Earth, (2005)

[24] D. Segal, *Singapore's Water Trade with Malaysia and Alternatives*, John F.Kenedy School of Government, Harvard University,Singapore,(2004)

[25] A.A.F. Mohamed, W.Z. Wan Yaacob, M.R. Taha and Samsudin, Groundwater and Soil Vulnerability in the Langat Basin Malaysia, Eur. J. Sci. Res., **4**, 628–635, (2009)

[26] N.K. Ang, Hydrologic Framework and Groundwater Resources Utilisation in Peninsula Malaysia,*Proc. Groundwater Forum. No. 4, Geological Society of Malaysia*, 1.1–1.5 (1994)

[27] B. Sofner, *Groundwater Monitoring and Groundwater Protection in North Kelantan and Perlis, Malaysia.* Hannover : Fed. Inst. of Geosciences and Natural Resources, (1989)

[28] E. Fahnline, The production and remediation of malaysia's groundwater resources,*Ensearch Sustainability and Environmental Management Conf. and Exhibition*, October, (2013)

[29] A.R. Samsudin, A. Haryono, U. Hamzah and A.G. Rafek, Salinity mapping of coastal groundwater aquifers using hydrogeochemical and geophysical methods: a case study from north Kelantan, Malaysia,Environ. Geol., **55**(8), 1737–1743, (2008)

[30] N. Sapari, J. Raja Zainariah and H. Raja Azie, Quantity and Quality of Groundwater in Fractured Metasedimentary Rocks of the West Coast of Peninsular Malaysia,**40**(6), 537–542, (2011)

[31] A. Stringer and K. Nash, Design of Rainwater Harvesting Systems in Oklahoma,**5**, (2016)

[32] B. M. Mati, M. Malesu and A. Oduor, Promoting rainwater harvesting eastern and southern Africa, *World Agroforestry Centre, Nairobi, Kenya*, (2005)

[33] L.K. Bisoyi, Rain Water Harvesting – An ultimate need in 21st Century,*Natl. Semin. Rainwater Harvest. Water Manag.*, 105–110, (2006)

[34] Y.A. Reddy, Water Harvesting : Limitations in Implementation,*Natl. Semin. Rainwater Harvest. Water Manag.*, 70–77, (2006)

[35] S.H. Deshmukh and R.B. Magar, The Scope of Rainwater Harvesting in Urban Areas, *Natl. Semin. Rainwater Harvest. Water Manag.*, (2006)

[36] S.D. Shrestha, K. Rajbhandari, D.B. Singtan and S. Mishra, Rainwater harvesting for recharging shallow groundwater, *A WaterAid Nepal*, **41**, (2011)

[37] C. Bhavsar and P. Bhalge, Roof Top Rain Water Harvest- A Long Lasting Solution to Drive away the Need of Water Tankers, *Natl. Semin. Rainwater Harvest. Water Manag.*, (2006)

[38] L. Abdul, Q. Tunji, A.A.A.Latiff, D. Tjahjanto, and S. Akib, The effectiveness of groundwater recharges well to mitigate flood, **6**(1), 8–14, (2011)

Investigation of Infiltration Rate onSoft Claythat Affecting the Ponding Time

Mohd Shalahuddin Adnan [1,*], *Mohd Aliff* Mohd Anuar[2], *Muhammad* Nda[2] and *Siti Nazahiyah* Rahmat[1]

[1]Micropollutant Research Centre, Department of Water and Environment Engineering, Universiti Tun Hussein Onn Malaysia (UTHM), 86400 Parit Raja, Johor, Malaysia

[2]Department of Water and Environment Engineering, Universiti Tun Hussein Onn Malaysia (UTHM), 86400 Parit Raja, Johor, Malaysia

Abstract. Malaysia is a country that has abundant rainfall that depends on the seasonal such as monsoon, where,certain areas in peninsular area are expected to receive heavy rain during this monsoon season. This abundant of rain could lead to massive flood if no mitigation actions were taken. One of the factor that contribute to the flood occurrences is type of soils. Type of soils play an important role in determining the rate of infiltration. Thus, this study was conducted to investigate the infiltration rate that effecting the ponding time at the Universiti Tun Hussein Onn Malaysia (UTHM). This area is located at flat topography with thick layer of clay soil beneath the ground surface that indicated low infiltration capacity rate. This condition could causeponding of water that could lead to high surface runoff that could cause flood.This paper was aimed at understanding the effect of soil clay layer on infiltration rate and ponding time. Infiltration test was conducted at four (4) points to make a spatial analysis around the study area. The soil was classified based on soil classification system (USCS), while, ponding time was calculated based on Horton model. Based on the results, the average infiltration rate is ranging from 0.004 mm/s to 0.076 mm/s among the selected locations.Soil samples were sieve and thedominant soil comprises of minimal well graded silty clay and clayey sand which is more than 50% content sand and more than 12% content fines. Lastly, time of ponding was calculated by using Horton model and it shows that the ponding time started between 0.51h to 1.0h. This information could serve a vital information on determining the mitigation measures to tackle a flood problem in this area.

1 Introduction

Infiltration is one of the important processes in hydrologic cycle where this process can be translate into a process that water seeps into the ground through soil originate from water on the surface. The infiltration capacity primarily determine by the surface cover and properties of underlying soil. Infiltration capacity is much influenced by the underlying

[*]Corresponding author: shalahudin@uthm.edu.my

soil, where, a compacted soil like clay has a low infiltration rate compared to loose soil like sand. Normally clay soil has solid connection between the particles that reduce the pore size that contribute to the less space between the pore. This condition has led to decrease in infiltration rate. A study by Selim[1], discovered that the increasing of clay percentage has led to decreasing in soil infiltration. Infiltration is influenced by the hydraulic properties of soil, rainfall intensity and water content distribution with depth [2].

Undeveloped land has very little surface runoff, where, most of the rainfall soaks into the top soil and travel slowly through the soil mantle, as interflow to the stream, lakes or estuary.However, as the watershed developed and the land is covered over with an impervious surface (e.g. roads, parking lots, roofs, driveways and sidewalks) most of the rainfall is transformed into surface runoff. This will result in tremendous increasing in surface runoff and create the unbalance situation in hydrological cycle. Normally, it will resulted in the changes of proportion of evapotranspiration, runoff and infiltration [3].Accurate description of infiltration and post-ponding runoff remains a fundamental problem in hydrology. Water added to an unsaturated soil will be absorbed until it ponds on the surface.

There are several studies have been conducted related with infiltration and ponding time. Based on their study, there are several factors that affecting the infiltration rate such as soil texture and structure, conditions at soil surface, soil moisture content, type of vegetative covers, soil temperature, human activities on soil surface, soil density and biological crusts [4]. Owing to that, many researchers have contributed to the extensive study on the determining of infiltration rate as well as the ponding time in exploring the relationship among these factors. The prediction of ponding time is crucial as the time condensation approximation (TCA) remains a widespread practical tool. TCA basically assumes that at ponding there is a unique relationship between the cumulative infiltration and the flux, independent of previous history.In order to estimate the portion of precipitation that enters the soil and that which becomes overland flow, calculations require estimates of ponding time [5,2].

The understanding of soil properties as well the soil strata is very important before any development being implemented. In addition, this information will be used as a design criteria. Even though, many studies were conducted in determined the infiltration rate and capacity, but, low number of study was conducted to determine the ponding time. The ponding time could provide an information such as time taken for water to start Thus, In Malaysia, the conventional drainage systems that constructed using concrete arewidely used. This practice brings a significant impact on the environment as a whole where, concrete drains tend to cause flooding as runoff is being conveyed to watercourses directly in shorter duration and have less infiltration into the ground. Besides, the conventional drainage systems are highly exposed to clogging problems. As a result, water cannot flow properly to the catchment area. When heavy rain occurs, it can result in high water volume and spills out of the drains and then cause flooding[6].Thus, in this study the effect of soil types on infiltration rate as well as the determination of ponding time was conducted which could be used as guidance for the emergency response.

2 Methodologies

2.1 Study area

This study was conducted at Universiti Tun Onn Malaysia (UTHM) campus which is located in western part of Malaysia with coordinates 1.8529° N, 103.0866° E within Parit Raja, Batu Pahat, Johor (Fig. 1). The average rainfall distribution in this area is ranging between 2250 – 2600 mm/year [6].Since the expansion of this campus, the coverage of impermeable layer was increased compared to last 10 years. The development area consists of schools blocks, administration blocks, hostels and also infrastructures component such as roads, covered sidewalk, open sidewalk and many more.

Fig. 1. Location of study area located at Universiti Tun Hussein Onn Malaysia.

2.2 Sampling Method

Data collection is an essential part in completing this study. In this study, data collection was separated into two (2) which are primary and secondary data collection. For primary data collection, the infiltration rate, soil sampling, sieve analysis and rainfall were collected at the site by using appropriate apparatus and methods. On the other hand, for secondary data such as historical rainfall data, borelog report, infiltration test results, land use information, and topography were retrieved from respected local authority, library and journals [7].

Before the experimental was conducted, informationof the study areawas obtained to determine the suitable methods that will be applied. Based on preliminary study, sampling points were determined, while, the hydrological data such as precipitation data was derived from Department of Irrigation and Drainage (DID), topography dataand land use data of the study area were collected from University Facilities Centre.

2.2.1 Soil sampling and Infiltration test

Further exploration of the subsoil condition within the study area was made to provide additional data for further analysis. This includes soil sampling and infiltration tests where six locations were randomly selected. Moreover, previous borelog data was referred as reference to establish the soil profile by combing with the latest soil sampling.

For infiltration test, double ring infiltrometer instrument was used to determine the infiltration rate. The rate of infiltration is determined as the amount of water per surface area and time unit that penetrates the soil. This rate can be calculated on the basis of the measuring results and the Law of Darcy. The measurements exclusively take place in the inner ring through which the water runs virtually vertically.From the data collection, some early indication of soil type of the area can be determined.On the other hand, soil samples collected from borehole was brought back to laboratory to undergone sieve analysis and plasticity index tests.

2.2.2 Laboratory test

All of the soil samples taken from the field test were taken to the laboratory for further soil analysis. The Particle Size Distribution for fine and course-grained soil test was carried out in accordance with the procedures in BS 1377. The soil description is based on unified soil classification system (USCS).

2.2.3 Determination of ponding time

Infiltration, time to ponding, and runoff generation areconsidered as they manifest in three soils simulated to havebeen exposed to a natural rainfall event with highly variableintensity. The specific focus here is put on ponding time considering (1) the effect of the time interval for averagingrainfall intensity data and (2) the effect of the soil surfacesealing.The rainfall intensitywas used to calculate the ponding time using Horton Model Method. Equation 1 and2 were utilized for the purpose of ponding time calculation.

Based on Horton Model, the calculation of ponding time can be divided into two (2) conditions as follows;

Continuous water ponding

$$f_{(t)} = f_c t + \frac{f_o - f_c}{k}\left(1 - e^{-kt}\right) \tag{1}$$

Non continuous water ponding

$$f = \left[\frac{f_c}{k}\ln\left(f_o - f_c\right) + \frac{f_o}{k}\right] - \frac{f_c}{k} \tag{2}$$

Where;

f_o = initial infiltration rate,
f_c = asymtptotic infiltration rate
k = decay constant
f = potential infiltration rate
t = time

3 Results and Discussion

3.1 Soil type and Infiltration rate

In order to determine the soil strata underneath the land surface, soil sampling were conducted at the same site where infiltration rate was conducted. On the other hand, comparison with available borelog was made. Based on the sieve analyses, the dominant soil type which less than 10 meters depth consist of mainly well graded silty, clay, and clayey sand. This result showed a good agreement with the borelog data.

Based on the infiltration test, the infiltration rate and infiltration capacity were constructed as shown in Fig 2. The range of infiltration rate obtained for this study area ranges between 0.0044mm/s to 0.1167mm/s (Table 1). Based on this table, the value of infiltration for each station in varies due to different properties of each location. Station 1 shows the highest values of infiltration while the lowest value was recorded at Station 2. By referring to soil sampling and the available borelog data, the soil at station 2 and Station 6 consist of compacted soil which has less pore compared to non-compacted soil. A study byGregory et al.[8] has express that the value of runoff was increased not only due to the increase in impervious area associated with development but also due to compacted soil.

Table 1. Infiltration rate (mm/s) for each station

	Station 1	Station 2	Station 3	Station 4	Station 5	Station 6
Maximum	0.1167	0.0056	0.0167	0.0167	0.0167	0.0083
Minimum	0.0515	0.0044	0.0083	0.0093	0.0071	0.0051
Average	0.0795	0.0050	0.0098	0.0129	0.0085	0.0057

Fig.2(a) shows the graph of infiltration rate against time for station 1. Based on this graph, it clearly shows that infiltration rate was slowly decreased within time. The line stated that the water was slowly decreased infiltrated into the soil based on time. The range of infiltration rate at this station is from 0.0515 mm/s to 0.1167 mm/s.Moreover, the infiltration capacity was calculated as well to determine the maximum rate at which water can be absorbed by a soil. For station 1, high infiltration capacity was observed which show a sharp increment from the beginning of measurement until it become consistent and stagnant at value of 72 mm which take about 0.42 hours. Even though, the dominant soil in this area is clay, but, the land cover, soil protection and human activities contributes to the infiltration rate. High infiltrationrates were a consequence of improved soil properties, which can provide a high protectionfor soils against degradation or erosion. Therefore, the infiltration rate can reflect the level ofsoil protection[9].

Fig.2(b) shows the graph of infiltration rate against time for station 2. Based on this graph, it clearly shows that the infiltration rate is very low and took a longer time to infiltrate into the soil. This condition is due to the soil condition as this site where the top soil consist of compacted soil that lead to less void space between the soil particle. Even the experiment time was extended for a longer duration, the infiltration rate is still low. While, the infiltration capacity also revealed the same results where the values of infiltration capacity is low in the range of 15 – 18 mm at 0.694 hr.This shows that the infiltration capacity is significantly affected by the soil types as well as the behavior of the top soil. A study by Dejun et al. [10]has concluded that the sequence of soil infiltration capacity is coarse > fine >very fine > medium > medium fine, consistent to the sequence of soil saturated hydraulicconductivity Ks. The value of infiltration capacity is significantly affected by the pore size and void ratio of the soil particles.

On the other hand, the soil condition also plays an important role that affecting the value of infiltration rate and infiltration capacity. If the experimental was conducted after the rainfall event, the value of infiltration rate will be significantly low due to low infiltration capacity. During this time, the soil condition in saturated condition where the water still there and this greatly affected the infiltration rate. The infiltration rate is affected by the sediment (soil) texture and structure, the condition of the sediment surface, the distribution of soil moisture or soil moisture tension, the chemical and physical nature of the water, the head of the applied water, the depth to ground water , the length of time of application of water, biological activity, the temperature of the water and the type of equipment or method used [11].

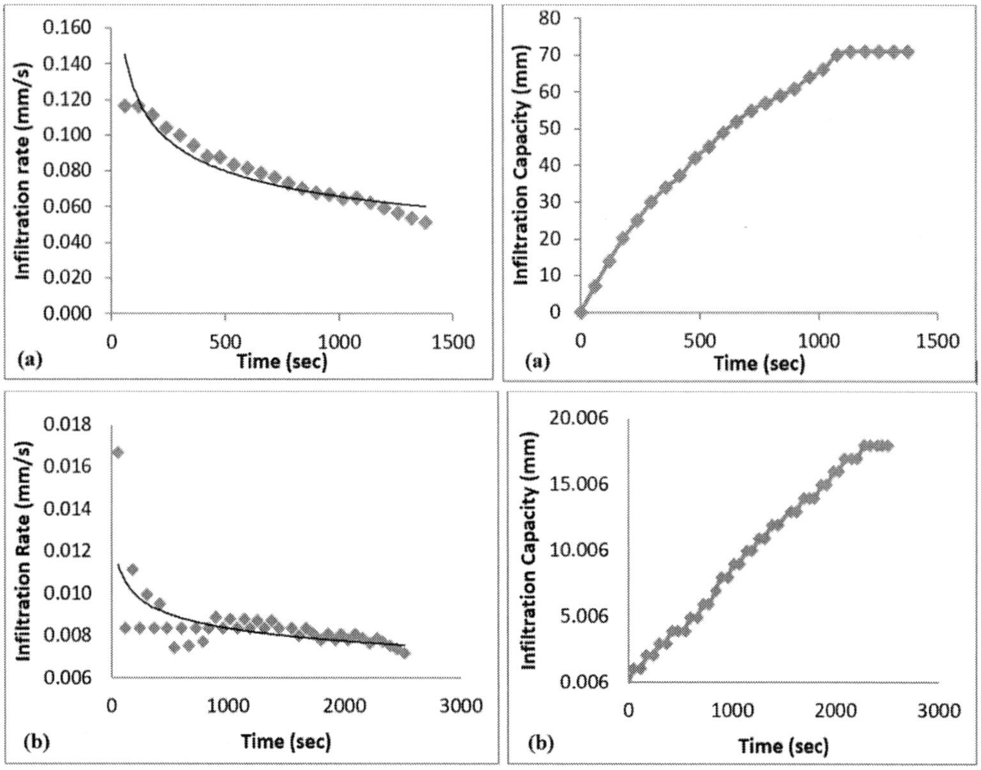

Fig. 2. Infiltration rate and capacity curves for **(a)** station 1 and **(b)**station 2

3.2 Ponding time

The ponding time corresponding with the infiltration rate, the increasing or decreasing of infiltration rate will significantly influenced the ponding time. Based on all six sampling stations, the ponding time is shows a good agreement with the infiltration rate results. Out of six stations, only Station 1 shows no ponding occurred with the same rainfall intensity used (Table 2). The ponding occurrences could be estimated when the value of runoff is increasing. The increasing of runoff values show that the soil is fully saturated and no infiltration occurred.

For Station 1, there is no ponding was observed since the value of runoff is zero. However, this result can be further improved by extending the time as well as the changes of rainfall intensity. The differences in rainfall intensity could produce different results since the pattern of runoff will be difference.The runoff value is ranging between 9.82 mm to 57.6mm.

Based on the calculated results, water ponding was observed to occur during precipitation and the beginning time recorded is 0.30hr. This value is corresponding with the soil type and rainfall intensity. The averaging time interval of rainfall relatively produced accurate estimation of time to peak that influenced the ponding time. The short duration of time to peak could tell us that value of runoff tend be increased [2]. In their study, they also mentioned about the soil type that play an important role in determining the ponding time.

Table 2. Summary of ponding occurrences

Time (min)	Station 1	Station 2	Station 3	Station 4	Station 5	Station 6	condition
0	-	-	-	-	-	-	-
10	Not	Not	Not	Not	Not	Not	i <fp
20	Not	Occur	Not	Not	Not	Not	i <fp
30	Not	Occur	Occur	Occur	Occur	Occur	i <fp
40	Not	Occur	Occur	Occur	Occur	Occur	i <fp
50	Not	Occur	Occur	Occur	Occur	Occur	i <fp
60	Not	Occur	Occur	Occur	Occur	Occur	i <fp

4 Conclusion

UTHM surrounded with pavement area and located at the low lying area with low infiltration capacity which makes this are prone to flood problem. In this study, the determination of soil is so important which could provide an important information on the soil types as well as the soil properties. Based on sieve analysis, it clearly shows that the dominant soil types in this study area comprises a minimal of well graded silty clay or clayey sand with the infiltration rate ranging in between 0.004 mm/s to 0.076 mm/s. Moreover, based on Horton model, the beginnings of ponding time start ranging from 0.30hr to 1.0hr while the total runoffs are ranging in from 9.82 mm to 57.6mm. On the other hand, suggestion to overcome the problem can be suggested to improve a drainage and retention basin to avoid critical water ponding during heavy rain.

This paper was funded by Universiti Tun Hussein Onn Malaysia (UTHM), Malaysia, The Ministry of Higher Education Malaysia.

References

[1] T. Selim, The effect of land use on soil infiltration rate in a heavy clay soil in Egypt, Vatten, **67**(1998), 161–166, (2011)

[2] S. Assouline, J. S. Selker and J. Y. Parlange, A simple accurate method to predict time of ponding under variable intensity rainfall,Water Resour. Res., **43**(3), 1–10, (2007)

[3] M. F. Yusof, N. A. Zakaria, A. A. Ghani, R. Abdullah and C. C. Kiat, Infiltration Study for Urban Soil: Case Studies – Butterworth and Engineering Campus, Universiti Sains Malaysia,*Proc. Congr. Assoc. Hydraul. Res.*, **1**, 1154–1162, (2005)

[4] O. S. Mangala, P. Toppo and S. Ghoshal, Study of Infiltration Capacity of Different Soils,Int. J. of Trend in Research and Development, **3(2)**,388–390, (2016)

[5] M.C. Donoso and N.M. VargasHim-Gonzalez, *Water Interactions with Energy, Environment, Food and Agriculture: Water Resources for Agriculture and Food Production*, **I**. (2009)

[6] M.S. Adnan, E. Yuliarahmadila, C.A. Norfathiah and H. Kasmin, Flood simulation using rainfall-runoff for Segamat River Basin,*Proc. of the Int. Conf. on Civil, Architectural, Structural and Constructional Eng.*, Dong-A University, Busan, South Korea, 369–373, (2016)

[7] S. Musa, N.A. Zakaria, D. Tjahjanto and L. Sai, The Potential of Recharge Well System in Flat Area with Low Infiltration Rate,*International Conference on Water*

Resources (ICWR 2009), 1-4, (2009)

[8] J. Gregory, M. Dukes, P. Jones and G. Miller, Effect of urban soil compaction on infiltration rate, J. Soil Water Conserv., **61**(3), 117–124, (2006)

[9] R. Alhassoun, *Studies on factors affecting the infiltration capacity of agricultural soils*, PhD thesis, Universität Carolo-Wilhelmina,(2009)

[10] Y. Dejun, B. Zhengfu, L. Shaogang, M. Shouguo and X. Jibing, Study on Modelling Infiltration Responses of Typical Top Soils Based on Infiltration Models,J. Residuals Sci. Technol., **12**(1), 123–131, (2015)

[11] A.I. Johnson, *A Field Method for Measurement of Infiltration*,Geol. Surv. Water-Supply Paper 1544–F, USGS, US, 27, (1991)

Sustainable Hydraulic Barrier Design Technologies for Effective Infrastructure Engineering

Devapriya Chitral Wijeyesekera[1,*], *Eric* Wooi Kee Loh[2] and *Alvin* John Lim Meng Siang[3]

[1,3]Faculty of Civil & Environmental Engineering, Universiti Tun Hussein Onn Malaysia, Johor, Malaysia
[2]Faculty of Built Environment, Linton University College, Negeri Sembilan, Malaysia

Abstract. Migration of liquids lead to embarrassing post construction scenarios such as that of leaks from roofs, potable water leaking from water tanks/ reservoirs, rising damp in walls with groundwater seeping into basement structures, leakage of water from ornamental lakes and ponds or leachate leakage into the environment from MSW landfill sites. Such failures demand immediate and expensive maintenance. A stringent control on structural and waterproof stability is deemed necessary for long term service life of structures and in particular underground and near surface structures. On a micro scale and over a longer time scale, the phenomenon of rising dampness occurs in older buildings with the groundwater rising up through walls, floors and masonry via capillary action. Even slower rates of contaminant fluid migration occur through landfill base liners. In this paper a variety of hydraulic barrier technologies is critically discussed against a backdrop of relevant case studies. The choice of an appropriate hydraulic barrier technology for a given scenario will depend also on the sustainability, financial affordability and subjective aesthetics.

1 Introduction

World Health Organisation, [1] defines moisture migration problem as "any visible, measurable or perceived outcome caused by excess moisture presence, indoor climate problems or problems of durability in building assemblies caused by various scenarios of water leaks". Impermeable barriers significantly impede such undesirable migration of fluids in a variety of sectors in infrastructure engineering. Water leaks through inherent weaknesses arising within the otherwise integral nature of pipes and roofs are embarrassingly visible effects causing constant concerns which are exasperated during heavy rainfall or excessive water pressure. On the other hand, a wet leaking basement or a poorly engineered landfill may with time manifest the undesirable consequences of mould growth / smell of dampness and pollution of the ground water regime respectively. Resource intensive and time consuming underground constructions demand a sustainable and long term service life. Furthermore, essential urban infrastructure such as the London

* Corresponding author: dcwijey@gmail.com

or Paris Metro is loathed to become out of service even temporarily be it even for routine maintenance and repair [2]. Increase in acidity of the environment in vicinity of tunnel (decreasing pH) increasingly corrode metal and concrete in underground structures, and also generate increasing amounts of soluble aluminium and iron conditions in the soil [3]. Adverse ground conditions such as old landfill sites require checks not only on water migration but also the ingress of noxious landfill gases [4]. Building and Construction Authority [5] reported that the weakest links in ensuring water tightness are in the precast joints; joints between precast elements themselves and also precast element joints with other fittings [6]. Table 1 illustrates the wide variety of documented standard hydraulic barrier technologies [7]. The sustainability and efficacy of the barrier technology/ product is dependent on the appropriateness for the onsite climate and the workmanship.

Table 1. Hydraulic Barrier Types (modified [7])

Type of barrier	Description	Application
Bonded sheet membranes	Bitumen-based, sheet membranes can be either: a) Cold-applied (self-adhesive); or b) Hot applied ("torch-on" or bonded using a hot melt bitumen adhesive). Composite sheet membranes BS 743: 1970, Class A; or BS 8747	Can be applied externally or sandwiched.
Liquid applied membranes	There are many types of liquid applied membranes, which include one or two part systems.	Can be applied externally or sandwiched.
Geosynthetic (bentonite) clay liners	These comprise of bentonite with a single or dual 'carrier' material, typically of geotextile or high-density polyethylene. There are two principal forms: dry granulated / powdered bentonite or pre-hydrated bentonite.	Can be applied externally or sandwiched.
Mastic asphalt membranes	These are applied in three coats as a hot liquid. BS EN12970	Can be applied externally or sandwiched.
Cementitious Crystallization slurries and powders	These are applied as coatings to surfaces of concrete walls and slabs or a solution or powder added to concrete.	Can be applied internally or externally.
Cementitious multi-coat renders, toppings and coatings	These are generally applied in multi-coats or slurries and are resistant to liquid water but allow some water vapour penetration.	Can be applied internally or externally.

2 Gravity Induced Water Migration

Thatched roofs (Fig. 1) are greatly admired architectural splendour and an outcome of past 'Biotechnology'. Water migrations in any leaking roof is gravity induced and result from; leaking capping / ridging (cracking of the cement mix; absence of expansion joints, punctured plastic sheeting), leaking valley sections (premature rotting from moisture concentrations and delayed drying), Wind damaged corners (thin weakly compacted thatching unable to restrain lifting by wind), Inadequately compacted thatch (require 35-50 kg//m^3 density to reduce fire risk, Animal or bird damage (deter the animals having access to the roof) and alike.

Fig. 1. Thatched Roofs – (modified [8])

Commercial and residential buildings are prone to face problems associated with water leakage; defective roofs, wet area water permeation (balconies, toilets, kitchens), internal / embedded pipe bursts or leaks from the pipe joints, and external wall infiltration by rainwater. Such undesirable leaks cannot be hidden for long as they manifest themselves; damaged ceilings /floors, dampened ceilings / walls and mould or fungi growth (Fig. 2).

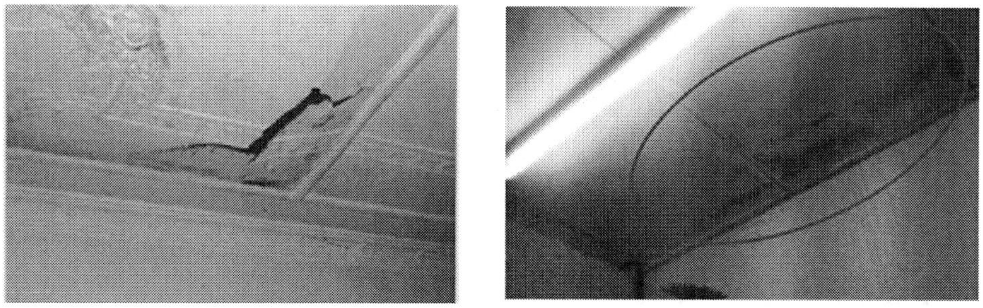

Fig. 2. Damaged Ceilings and Mould growth with Leaking Roofs – (modified [9])

Several generic types of waterproofing materials used for protecting concrete structures, such as cement-based, epoxy resin, polyurethane resin, acrylic resin, and silane /siloxane. These vary in terms of their price, durability performance, and method of application. The environment and the climatic conditions have a pronounced effect on the life span of such sealants. An initial comparative study of the efficiency of commercial waterproofing materials for roofs in Malaysia and UK was carried out by Roslan Talib et al. [10]. Summary of their research on waterproofing products are presented in Table 2.

3 Geosynthetic Clay Liners as Contaminant Barriers

Imposition of stringent environmental regulations of not contaminating previously untouched 'green-field' sites or removing the legacy of previous human occupations in the so called "brown-field" or derelict lands necessitate the adoption of sustainable contaminant barriers. A 1994 geo environmental case study investigation observed that a granulated bentonite Geosynthetic Clay Liner (granular/powder type GCL) "failed" in preventing the ingress of water into an old Victorian London vault. This construction was during the time of pioneering and emerging development of GCLs. An internal visual inspection of the vault roof showed that the GCL was not performing as a hydraulic barrier. Model tests were planned and carried out on the GCL in a tank apparatus developed to mimic the site conditions. Together with the support lip, sodium bentonite paste was extruded to give a receptive waterproof medium for the framed GCL. The GCL was attached to the frame using rivets to simulate a membrane of infinite area. A drainage layer of washed gravel was placed directly onto the membrane and the permeant was connected via pipes to a drainage trough. The base was given a 10-degree pitch to ensure positive drainage, with the gravel. thickness decreasing from 100mm down to 50mm to give a horizontal top surface.

Table 2. List of Waterproofing products [modified 9]

Company details	Year of Establishment & Logo	Product application	Climate
Bostik Paris, France Worldwide market	1889-	Sealants and adhesive	Temperate & Tropical all year
Soprema, Straspourg, France Worldwide market	1908 -	Liquid membrane	Temperate & Tropical all year
Sika AG Group Baar, Switzerland Worldwide market	1910 -	Epoxy coating (Outdoor); Cementitious (indoor wet areas / outdoor use to be rendered on top)	Temperate & Tropical all year
Axter Paris, France Worldwide market	1922-	Torch applied membrane: anti root with bituminous membrane	Temperate & Tropical all year
Fosroc Ltd. London, UK and Dubai UAE. Worldwide market	1940-	Construction sealants and, Wet areas cementitious	Temperate & Tropical all year
GE Sealants Hountsville, NC, USA Worldwide market	1940-	Construction sealants Silicone sealants	Temperate & Tropical all year
Shell Malaysia Kuala Lumpur, Malaysia Malaysia market only	1964-	Shell Flintkote for internal wet area waterproofing as well as for flat concrete roof and metal roofs	Tropical climate
Atlas Industry Sdn Bhd (Taiko Group) Selangor, Malaysia market only	1968-	Coating primer. Modified bituminous compound emulsions. Synthetic resin.	Tropical climate
Bitumat Dammam, KSA Mid-East, Africa, Asia, S America	1975-	Bitumen's torch membrane	Tropical climate
Polycell Sdn Bhd (Tyga Co) Selangor, Malaysia Malaysia market only	1980-	Foam grout	Tropical Climate
Axel Chemie Sdn Bhd Kuala Lumpur, Malaysia Malaysia market only	1986-	Roof coating for metal, clay and concrete flat roof	Tropical climate
Quicseal Pte Ltd Singapore Malaysia China , Phillipines, South Korea	1988-	Liquid based waterproofing system. Silicone sealant	Temperate & Tropical all year
ESTOP Sdn Bhd Shah Alam, Selangor Malaysia, Phillipines,S Korea Indonesia, Singapore, Hong Kong, China	1990-	Waterproofing system, coating, water stop, joint filler, sealants, concrete repair and grout	Temperate & Tropical all year
Bina Paint Marketing Sdn Bhd Selangor Malaysia Malaysia market only	1992-	Elastometric waterproofing coating	Tropical climate

Approximately 20 litres of water was showered onto the rye grass turf over a period of 5 days, in an attempt to simulate average rainfall. The water that penetrated the membrane upon initial wetting was 0.7 litres. After the test period of 8 months, the model was dismantled. The GCL's upper geotextile was then carefully cut away to expose the swollen bentonite (Fig. 3). X-ray diffraction analysis revealed that the bentonite was still a Na^+ rich montmorillonite with a moisture content of 27.7%. The hydraulic conductivity remained in the same order of magnitude of 1 x 10^{-11}m/s as was being achieved with standard hydraulic conductivity tests. There being no visual desiccation cracks or variation in X ray diffraction studies, assured that the GCL ought to have acted as an efficient barrier system.

However there was onsite evidence that the soil had been fertilised with an N-P-K (8-8-16) fertiliser, which consisted of calcium ions. Samples of the fertiliser were obtained and the GCL retested with a quantity of 500ml of N-P-K added to a 600 x 600 mm sample in the field simulation test apparatus. Water was again showered and the simulation repeated for the next 5 days to correspond to the rainfall of this time period. After five days the new GCL sample was exposed and is was clear from a visual inspection that the material had failed as a barrier system (Fig.4). The formation of a distinct crack pattern was clearly visible on the membrane. The material was then sampled for moisture content and x-ray diffraction. It was found that the moisture content had remained at 27.7% but the x-ray diffraction analysis revealed that the clay had undergone cation exchange resulting in a conversion to a Ca^{2+} (calcium) rich montmorillonite. Hence, X ray diffraction studies confirmed that the fertiliser with calcium ions caused the failure of the GCL in this case.

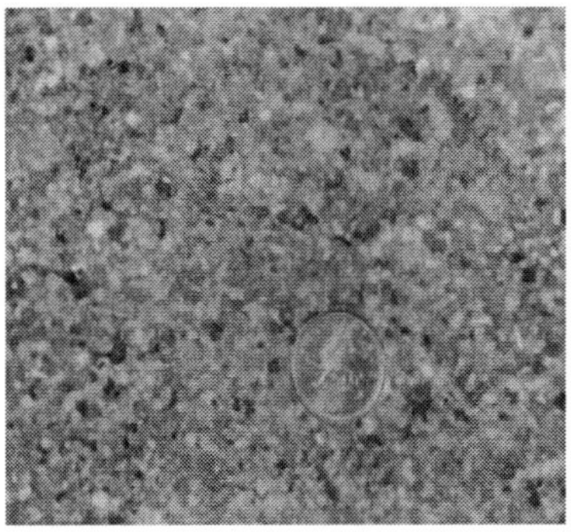

Fig. 3. Exposed bentonite showing swelling

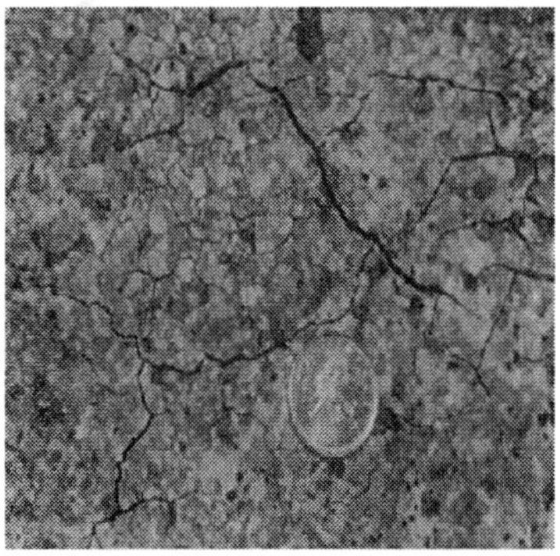

Fig. 4. Formation of cracks through cation exchange

3.1 Long Term Testing of Geosynthetic Clay Liners

Bentonite-based liners have gained acceptance for use as hydraulic barriers in industrial and municipal waste containment and sealing applications (k< 10^{-11} m/s). The current environmental monitoring practice adopted for measuring hydraulic conductivities are short-term laboratory tests with a modest testing routine. These fail to reflect the long-term dynamic changes in hydraulic conductivity performance of the bentonite in them. Interaction of bentonite with high electrolyte concentrations of high cation valence present in hostile geo-environment significantly increases the hydraulic conductivity of the bentonite through the reduction in the double layer thickness consequent to cation exchange. This defeats the prime objective of the field application of contaminant barrier as a sealing application intended to be sustainable perpetually or at least for 100 years.

A long term assessment was carried out on a polymer hydrated and vacuum extruded bentonite clay mat with non-standard liquids in order to examine the variation in hydraulic conductivity over time. This testing approach superseded the current short-term conventional laboratory testing and focused specifically on the long term hydraulic performance of GCLs when subjected to permeation that are representative of actual and commonly met hostile geo-environment conditions (Fig. 5).The overburden stress on the specimen was made up of a 300mm thick selected material appropriate to the respective field condition. Each specimen was placed between two porous plates and a free flow condition with a hydraulic head of 25 kPa maintained at the base and applied to the leachate at the testing times. Fig. 6 illustrates the variation in hydraulic conductivity (observed via a modified constant head test) during the reporting period of ten years. There is evidence of a marginal increase in hydraulic conductivity with time, but it is not detrimental to the performance criteria, even after 10 years' time.

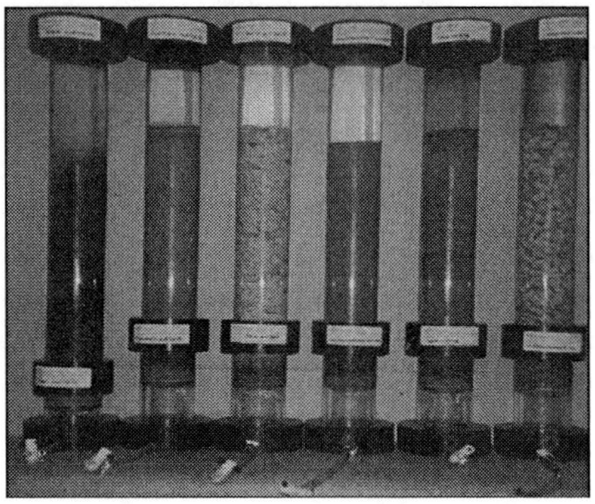

Fig. 5. Long-term Test Cells of Prehydrated GCL when permeated with non-standard liquids. (1) Mine Water, (2) Contaminated Water, (3) Rain water with Limestone Chip, (4) Sea Water, (5) Synthetic Leachate, (6) Transformer Oil

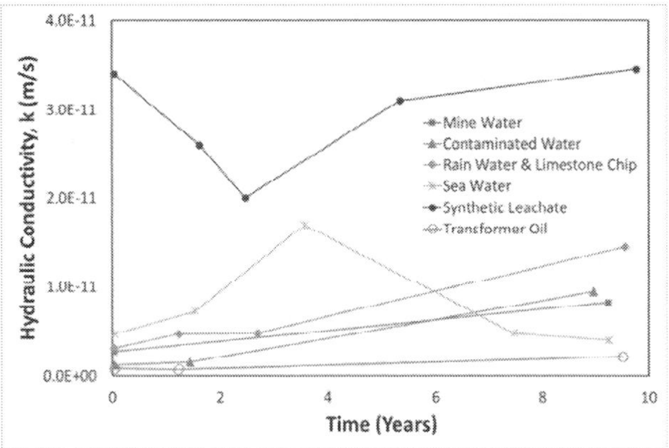

Fig. 6. Long term hydraulic conductivity results for Prehydrated GCL when permeated with non-standard liquids

4 Corrosion Control of Concrete Structures

Results of Al-Dulaijan et al. [11] showed that the adhesion of all the epoxy-resin- based coatings, to the concrete substrate, was better than that of the acrylic- resin-based surface coatings. The adhesion values of the resin-based surface coatings ranged from 1.25 to 2.03 MPa. Their extended investigations [12] on the performance of four cement-based surface coatings reported that the adhesion of the tested coatings ranged between 1.10 and 1.60 MPa. Effective waterproofing parameters include porosity, pore volume, permeability / hydraulic conductivity and more. The chloride permeability [13] was also evaluated and the values ranged between 552 and 1113 C. Pfeiffer and Scali [14] indicated that epoxies, methyl methacrylates, urethanes and silane provide an excellent barrier against the ingress of chloride ions into concrete cubes after 21 days of continuous immersion in 15% NaCl solution. They investigated the effect of concrete surface treatments on several properties of concrete including water absorption, initiation of reinforcement corrosion and cracking which resulted in the reduction in water absorption [15]. Precast Concrete (PC) framing systems fit snugly within the Industrialised Building System (IBS). Its full adoption is limited by water migration problems, where leakages occur through the joints of the precast elements. Hence, Lachimpandy et al. [16] proposed the alternative adoption of a mixed/ hybrid of In-situ Reinforced Concrete (IRC) and PC as an amalgamation of the conventional construction and IBS. In such instances, reliable solutions of sustainable moisture barrier technology in wet areas such as toilets will help reduce or eliminate use of IRC method of construction. PC joints are sealed to prevent ingress of water and air into building. PC joints can be sealed as open-drained, face-sealed and compression-sealed [17]. Of all the many types of sealant materials available, polyurethane (single and multi-component) and single component silicone are two (2) most commonly used sealant materials for sealing joints between PC components [18]. Sealant does not bond properly on wet and porous surfaces like concrete and usually needs base primers as adhesive to avoid premature failure [19, 20].

5 Rising Dampness in Buildings

Flooded basements and foundations can seriously weaken the structure and foster toxic mould growth--the newest threat to residential indoor air quality. Dynamics of water table movements are slow and seasonal. If basement leaks show up after a heavy rainfall or snowmelt, the cause is far easier to be noticed for repair. The exterior areas of basement walls must be coated with a high-grade waterproofing system since old fashioned paint on black mastic is not sustainable and fails too often.

Application of a variety of vertical membrane technologies prove promising to overcome difficult ground conditions to guarantee sustainable waterproofing of the structure for the building's lifetime. Water creates such high pressures that sealing cracks will not prevent water from leaking into the house if the soil around the house is saturated. It is important to seal concrete block walls on the outside and the inner surface of block walls need to be sealed with paint or waterproofing material to retard capillary action and help prevent radon entry. The common liquid injection is time consuming as each hole needed to be injected for several minutes in order for the correct amount of fluid to be injected.

6 Conclusion

This paper highlighted a variety of scenarios where migration of water (and fluids) needs to be appropriate remedied to rectify unsatisfactory hydraulic barrier problems. The sustainable effectiveness of the barrier system must be evaluated, (be it with appropriate and unconventional test methods), against its interaction with the onsite environment. Short term, conventional but unrealistic testing methods need to be carefully evaluated.

References

[1] World Health Organization (WHO), *Guidelines for Indoor air Quality: Dampness and Mould*, (2009)

[2] V. Novotny and P. Brown, *Cities of the future: towards integrated sustainable water and landscape management*, London, IWA Publishing, (2007)

[3] N. Ghafari, Corrosion control in underground concrete structures using a double waterproofing shield system (DWS), Int. J. of Mining Sci. and Tech., **23**, 603-611, (2013)

[4] D.C. Wijeyesekera, K. O'Connor and D.E. Salmon, Design & performance of a compacted clay barrier through a landfill, Engineering Geology, Special issue on Geoenvironmental Engineering, **60**(1-4), 295-306, (2001)

[5] Building Construction Authority, Singapore, Design of external walls, (2014), Retrieved on August 15, 2014 from http://www.bca.gov.sg,

[6] K.A.M. Kamar, Z.A. Hamid, M.K. Ghani and A.H. Rahim, *Industrial Building Systems: Current shortcomings and the vital role of R&D.* 2nd Quarter Master Building Publication, (2007)

[7] BS 8102: 2009, *Code of Preactice for protection of below ground structures against water from the ground*, British Standards Institution, (2009)

[8] Images for Leaking Thatched Roofs , 2016), Retrieved on October 1, 2016 from http:// www.morganasphalte.co.uk

[9] Quality Roof Leakage Repairs, (2016), Retrieved on October 1, 2016 from http://bina-waterproof.com/workingsecene.html

[10] R. Talib, D. Boyd, S. Hayhow, A.A. Ghafar and M. Suleiman, Investigating effective waterproofing materials in preventing roof leaking; initial comparative study: Malaysia, U.K. Science Direct Procedia Manufacturing, **2**,419-427, (2015)

[11] S.U. Al-Dulaijan, M. Maslehuddin, M.M. Al-Zahrani, A.M. Sharif, E.A. Al-Juraifani, and S.H. Al-Idi, Performance evaluation of resin based surface coatings. In: Deterioration and Repair of Reinforced Concrete in the Arabian Gulf, *Proc. of 6th ACI Int. Conf., Bahrain*, 345–62, (2000)

[12] S.U. Al-Dulaijan, M. Maslehuddin, M.M. Al-Zahrani, A.M. Sharif, E.A. Al-Juraifani, S.H. Al-Idi and M. Al-Mehthel, Performance evaluation of cement-based surface coatings. In: Repair, Rehabilitation, and Maintenance of Concrete Structures, and Innovations in Design and Construction, *Proc. of Fourth ACI Int. Conf.*, SP-193-20, Seoul, Korea, 321–35, (2000)

[13] American Society for Testing and Materials. ASTM C1202-91, *Standard test method for electrical indication of concrete ability to resist chloride ion penetration.* ASTM Standards, vol. 4.02, Philadelphia, (1991).

[14] D.W. Pfeiffer and M.J. Scali, *Concrete sealing for protection of bridge structures. National Cooperative Highway Research Program (NCHRP 244)*, Transportation Research board, Washington DC,(December , 1981)

[15] ASTM C642A. *American Society for Testing and Materials. ASTM C642-92, Standard test method for specific gravity, absorption, and voids in hardened concrete. vol. 4.08*, ASTM Standards, , Philadelphia, (1992).

[16] S.K. Lachimpandi, J.J. Pereira, M.R. Taha and M. Mokhtar, Construction Waste Minimisation comparing Conventional and Precast Construction (Mixed System and IBS) methods in High Rise Buildings: A Malaysian Case Study, Resources, Conservation and Recycling, **68,** 96-1-3, (2012)

[17] NPCCA, *National Precast Concrete Association Australia, Precast Concrete Hand Book*, (2012)

[18] SIKA Malaysia, *Technical Brochures*, (2010-2014)

[19] CPPCI, *Canadian Precast-Prestressed Concrete Institute, Architectural Precast Concrete Sealant Joint Guide*, (2010)

[20] N.A. Abd Rahman, Z.M. Zainordin, S. Ahmad and N. Mahat, Maximization of IBS Elements at Wet Areas in solving Leaking Problems and Promoting Better Quality Control, Procedia Social and Behavioral Sciences, **202,** 424-435, (2015)

Potential Study of Water Extraction from Selected Plants

S. Musa[1,*],*N.A. A.* Abu Mansor[1],*S. S.* Bujang[1], *N. A. F.* Mohd Kamil[1], *M. S.* Adnan[1]and *R.* Hamdan[2]

[1]Faculty of Civil and Environmental Engineering, Universiti Tun Hussein Onn Malaysia,86400, Batu Pahat, Johor, Malaysia
[2]Faculty of Engineering Technology, Universiti Tun Hussein Onn Malaysia,86400, Batu Pahat, Johor, Malaysia

Abstract. Water is absorbed by the roots of a plant and transported subsequently as a liquid to all parts of the plant before being released into the atmosphere as transpiration. In this study, seven(7)selected plant species collected from urban, rural and forested areas were studied and characterized. The water was collected using transparent plastic bag that being tied to the tree branches. Then, the vapouris water trapped inside the plastic bag and through the condensation process, it become water droplets. Water quality parameters such as temperature, pH value, DO, turbidity, colour, magnesium, calcium, nitrate and chloride were analyzed. The analysis was compared to drinking water quality standard set by the Ministry of Health Malaysia. Based on the results, it shows that banana leaf has a higher rate of water extraction compared to others. Thus, the plant can be categorised as a helpful guide for emergency use of water or as an alternative source to survival.

1 Introduction

Generally, the inclusion of water in the human body is determined by stimulation of thirst. When the volume of water lost from the body is higher than the volume of water consumed, the osmotic pressure of the extra cellular fluid (ECF) will increase. Anti diuretic hormone (ADH) is released from the hypothalamus gland osmoreceptors pituitary when activated[1]. The increase in both the osmotic pressure of the extracellular fluid and antidiuretic hormone generate the urge of thirst in humans [2]. Lack of water intake into the human body can cause diseases such as dehydration and hyperthermia. The disease can lead to death if not treated immediately [3].

There are many sources of water on earth that can be used as a source of clean drinking water such as rivers, lakes and rainwater [4]. However, these resources are not always available in all circumstances, for example, in the area of human settlements far from clean water sources or in situations where travellers in the woods are lack of clean water to drink[5]. In a desperate situation, alternative water source should be considered so that

*Corresponding author: sabariah@uthm.edu.my

people can still survive and to avoid health problems caused by lack of water in the body. One method that can be used is extracting water from plants.

Chand and Chang [6] stated that, the body of an adult human contains 60% of water, whilea new born baby is composed of more water (75%) than that of elderly person (50%). Also, the more muscular a body more water is contained. All our vital organs contain different amount of water, namely the brain, the lungs, the heart, the liver and the kidneys contain a large quantity of water between 65%-85% depending on the organ while bones contain less water (but still 31%) [3].

There are several areas in Malaysia that still suffer the deficiency of clean water supply such as Ranau, Sabah [4]. This problem has led to some of the residents in the area to use water of unknown level of cleanliness as drinking water [7]. The same problem exists in forests. Travelling over long distance of forest tracks may have difficulty to get clean water to survive during the journey.

One of the survival method is collecting water from plants by means of its transpiration process. However, there is no study to indicate whether such a water is drinkable[5, 9]. This study is meant to compare the characteristics of the water collected from the transpiration process of plants with drinking water quality standards set by the Ministry of Health Malaysia[8].

2 Extracting Transpiration of Plants

Transpiration bags are often used in the laboratory to coax plant transpiration. It can also be used for survival as emergency supplies to collect water through the transpiration process. This method starts with incorporating as many leaf of a plant into the bag and then ties it tightly to its branch tip. Water vapour is expected to evaporate through the plant leaf surface and condensewithin the boundary of the plastic bag before falling into the edges as drops of water. Bags should be transparent so that the light can penetrate into the bagtoincreasethe rate of plant transpiration [7].Some of the forest nomadic tribes are known to use the same method to collect water from plants. However, how safe the water for drinking by humans remains unknown.

2.1 Soil Inspection and Location

Soil inspection or geotechnical inspection is crucial in understanding the physical properties of soil and the rocks beneath the ground. Type of soils, soil moisture and groundwater content were analyzed to determine their properties as well as the characateristics. The water that derived from plant were compared with groundwater to determine the relationship between these two water to understand the effect of type soil and groundwater on the extraced water. Various tests are done to explore the sub-surface and surface characteristics of soil to estimate the plant-available concentrations of plant nutrients [3].

In this study, three (3) areas were chosen to identify the effects of water quality at difference locations and conditions, that covers urban area (Universiti Putra Malaysia, 43100 Serdanga, Selangor), rural area(Kg Serom 6, Sungai Mati, 84400 Muar, Johor) and forest area (Hutan Rekreasi Ayer Keroh Melaka, Lebuhraya Ayer Keroh, 75450 Melaka)(Fig. 1). These areas were selected because these area are known as a residential area that possible could apply this method to extract the water for their usage. The soil and water conditions are also analysed[10]. This is to identify any adverse effect in thewater extracted from the plants that may be due to the surrounding soils. Soil type and moisture content are recorded for each station [11].

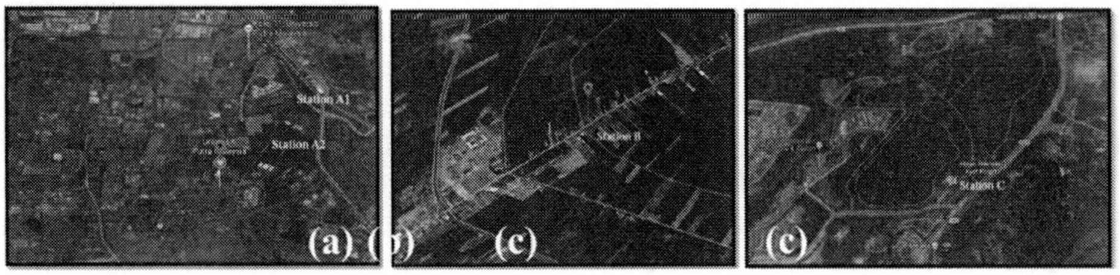

Fig. 1. (a) Urban area-UPM (b) rural area – Kg Serom 6 (c) forest area- Melaka

2.2 Type of Plants

The selection of plant at different locations is based on physical characteristics such as height and quantity of leaves. All selected plants should have thick bushes of leaves. This feature is emphasized in order to facilitate the work of water sampling to be carried out in which dense and shady leaves allow a plastic bag to be tied to a tree branch quickly and easily.

There are fruits plants, shrubs, flowers and poisonoustrees. These plants are selected because the plants grow well in the 3 areas [12]. These plants are also selected for its popularity and easily available in the vicinity of homes throughout Malaysia. This study will analyze the characteristics of the water parameters to be collected from the seven types. There are rambutan (*Nephelium lappaceum*), banana (*Musa Paradisiaca*), principal soursop (*Annona muricata L.*), flowers bells (*Allamanda cathartica*), shrimp clip (*Heliconias*), drasina (*Dracaena fragrans*), and flowerbeds setup (*Bauhinia Kockiana*).

2.3 Water Sampling

The use of transparent plastic bag for the purpose of water sampling has been justified at the beginning of Section 2.The size of the plastic bag is 25.4 cm x 40.64 cm for plants with small-size foliage such as rambutan, flowers bell, soursop, flowerbeds setup, drasina and heliconia. While plants with large-size foliage such as banana, the size of the plastic bag used is 50.8 cm x 91.44 cm (Fig.2).

Fig. 2. Water sampling method

 DOI: 10.1051/matecconf/201710304023

Due to the larger size surface is uneven, the methods used to measure the surface area of rambutan, flowers bell, soursoup and flowering plants are setup using the graphical method. Through this method, the shape of the leaf surface will be drawn on graph paper, then the total area will be determined by calculating the total area of each plot in the graph. Fig. 3 shows how to calculate the surface area of leaf using the graphical method.

However, this method cannot be applied to such a large leaf such as banana leaf, drasina and heliconia. The large leaf surface area will measure using a measuring tape. The length and width of the leaf was analysed by usingtrapizodial estimation. Fig. 3 shows the method of measuring the length and width of leaves using a measuring tape. Numerous methods have been employed from time to time for the determination of leaf areas, but it was found impracticable to use any of these methods without modification [13].

Moreover, in terms of number of leaf that are involved, it can be seen that the trees have small leaf like flowers bell requires more number of leaf that need to be loaded into the plastic in order to produce water transpiration than trees that have a size large leaf such as banana leaf that only requires a single leaf transpiration of water to produce a sufficient sample for the purposes of this study. The time required to collect water samples for this sampling isone-day collection.

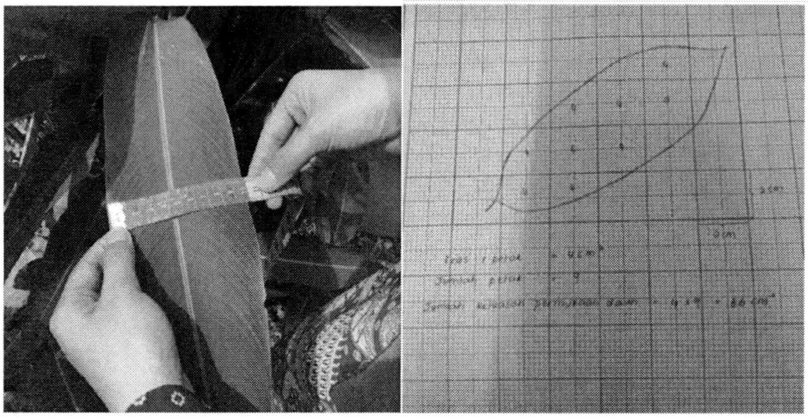

Fig. 3. Measurement of leaf surface area

With 60 minutes recording intervals, record the transpiration difference, average solar radiation and average saturated vapor pressure to analyze the model and formulate. The measuring intervals are the fitted measurement time for transpiration model of fitted plants.

The quantity of water should also be taken into account in terms of dependence on the light, the surface area of leaf and number of leaf [13]. Stomatal dimensions were determined with an ocular micrometer with the leaves attached to the plant [11, 14].High of temperature and the larger surface area of leaf or leaf number, will increase the total volume of water collected. Beside, if the weather is cloudy and less sunlight, volume of water also decrease in time. Transpiration process plant will be higher in the presence of abundant sunshine compared with less sunlight conditions [15].

2.4 Water Quality Assessment

Water quality is the whole process of evaluation of physical, chemical and biological water properties relating to the quality of natural, human impact and aims use [16], particularly use can affect human health. The guideline water quality standards and beverages is constantly monitored and updated so that is always in line with the current changes [5, 9].

Theoretically shows that the vapor through stomata process produces more cleanness and pure water extraction without any treatment. It also confidentially gives the good potential for water drinkable for tested plants. This investigation may identify that the effects of soil, water content, toxic rubber and safe plants where drinkable water without any filtration.

Collector water from leave was analyzed based on drinking water standard through selected parameters such as pH, DO, turbidity, chloride, fluoride, nitrate, sulphate and magnesium. All testing was handled at water quality laboratory by pH and DO meters, thermometer, turbidity meter, Ion Chromatography (IC) machine and Atomic absorption spectrometer (AAS) machine. Groundwater and water leaves samples were analyzed and compared respectively.

In this study, the chemicals contained in the samples will collect one-day data sample (from 9.30 am to 5.30 pm) [17] to identify the quality of water of transpiration of seven (7) types of plant chosen from three (3) different regions. Samples was measured by using ion chromatography,atomic absorption spectrometer instrument and others apparatus such as temperature, turbidity, DO, pH, colour, magnesium, calcium, nitrate and chloride [11,14,18]. The comparison to classify water sample for each plants will analyze and standardize as a guideline manual.

3 Results and Discussions

3.1 Water Quantity

Plant leaf surface area affects the total volume of water that can be collected. The leaf that have a larger surface area such as banana leaf can produce more water transpiration in a time compared to small leaf surface area.

From preliminary studies have showed that all the trees transpiration rates by 8 hours period. A banana tree was higher collector of water samples, followed by flowers bells and rambutan trees (Table 1). Shrimp clip tree is the lowest. It showed that the rate of transpiration for a species depends on the characteristics of the plants it self other than environmental factors.

Table 1. Average transpiration rate collection

Type of plant	**Average transpiration x (10^{-6}) (l/s)**
Banana	5.5
Principal soursop	3.4
Shrimp clip	1.8
Flowerbeds setup	3.3
Drasina	3.1
Flowers bells	4.1
Rambutan	3.6

3.2 Water Quality

According to preliminary study, the water extracted from all plants have good potential for drinking water and safe. It also showed that water quality was evaluated based on characteristics of plant for example; flower bells shows less secure features based on the content of the gummy that does not dissolve in the liquid. However, all the trees shows high turbidity values except drasina and soursop. This situation shows that the plant will produce a wide range of different water quality.

4 Conclusions

Through this transpiration method, maintain the number of flora that many on this earth for the well-being of all the object slighted up. In addition, the process of transpiration can continue to keep the air moist and helps atmosphere be able to reduce the temperature of the heat. In this study, the awareness study can be created about the importance of healthy drinking water quality and plant transpiration process itself. It is important not only to the water cycle on a rural and urban or global level it down acts as a solution to the survival in the forest looking for a source of water during emergencies. The existences of these plants have a very important role not only in the hydrological cycle but it can also serve as the urban landscape and give supply oxygen to peoples and living things. This feature is in keeping with the importance of this study in which the study are conduct as an alternative to the public drinking water if embattled clean water supply in the future.

Utmost appreciation and deepest gratitude to the UTHM hydro team of this study.

References

[1] O.V. De Sousa, G.D. Vechio Viera, J.D.J. De-Pinho, C.H. Yamamoto and M.S. Alves, Antinociceptive and anti-inflammatory activities of the ethanol extract of Annona muricata L. leaves in animal models. *Int. J. of Molecular Sciences*, **11(5)**, 2067-2078, (2010)

[2] N. W. Chan, Issues and challenges in water governance in Malaysia, *Journal of environmental health science and engineering*, **6**(3), 143-148, (2009)

[3] E. Jequier, and F. Constant, Water as an essential nutrient: the physiological basis of hydration, J. of clinical nutrition, **64**, 115-123, (2010)

[4] M. A. Fulazzaky, T. W. Seong, and M. I. Masirin, Assessment of Water Quality Status for the Selangor River in Malaysia. *Water Air Soil Pollution*, **205(1-4)**, 63-77, (2009)

[5] N. F. Gary, *Drinking water quality*. 2nd ed. United States of America: Cambridge University Press, (2008)

[6] C. Chang, and J. Chang, Study on Simple Transpiration Model and Fitted Measurement Time for Dominant Plant at Mudstone Area in Taiwan, *12th ISCO Conference*, Beijing, 244-250, (2002)

[7] H. Perlman, (2016) Retrieved on May 3, 2016 from http://water.usgs.gov/edu/watercycletranspiration.html

[8] T. Sterling, *Transpiration-water movement through plants*. Department of Entomology, Plant Pathology and weed science, (2004)

[9] F. R. Spellman, *The drinking water handbook*. $\mathbf{2^{nd}}$ed. London: CRC press, (2012)

[10] C. Maria, Singureanu., H. Denisa, and B. Erszebet, Heliconias- novelties and applicability in floral art, J. of horticulture, forestry and Biotechnology, **18**(1), 22-27, (2014)

[11] E. Steudle, How does water get through roots?, J. of Experimental Botany, **49**(322), 775-788, (1998)

[12] M. Rindam, T. N. Long Seman, R. Sulaiman, and R. Rashid, Proses sejatpeluhan tumbuhan terpilih sebagai kaedah mendapatkan dan menangani masalah sumber air, Malaysia, Malaysian J. of Society and Space, **6**(2), 10 - 19, (2010)

[13] W. L. C. Muenscher, A Study of the Relation of Transpiration to the Size and Number of Stomata, American J. of Botany, **2**(9), 487-504, (2016)

[14] W. M. Atwell, and N. B. Stout, Transpiration Rates under Controlled Environment : Species, Humidity, and Available Water as Variables, Ohio J. of Science. **62**(1), 18-26, (1962)

[15] T. E. Dawson, S. S. Burgess, K. P. Tu, R. S. Oliveira, L. S. Santiago, J. B. Fisher, A. R. Ambrose, Nighttime transpiration in woody plants from contrasting ecosystems. Tree Physiology, **27**(4), (2007)

[16] Food and Agriculture Organization of the United Nations (2014). Retrieved on June 20, 2016 from http://www.fao.org/world/regional/rap/index.asp

[17] A. E. Uduak, and S. U. Esther, Comparative phytochemical screening and nutritional potentials of the stems, leaves and flowers of allamanda cathartica (apocynaceae). Int. J. of science and technology, **4**(6), 248-253, (2015)

[18] M. T. Bashir, S. Ali, A., Bashir, Health effects from exposure to sulphate and chlorides in drinking water. Pakistan J. of medical and health sciences, **6**(3), 648 - 652, (2012).

Simulation of Saltwater Intrusion in Coastal Aquifer of Kg. Salang, Tioman Island, Pahang, Malaysia

Azrul Normi Idris[1,*], *Ahmad* Zaharin Aris[2], *Tahoora* Sheikhy Narany[2], *Sarva* Praveena Mangala[3], *Saim* Suratman[1], *Ismail* Tawnie[1], *Mohd* Khairul Nizar Shamsuddin[1] and *Anuar* Sefie[1]

[1]Hydrogeology Research Center, National Hydraulic Research Institute of Malaysia (NAHRIM), Lot 5377, Jalan Putra Permai, 43300 Seri Kembangan, Selangor, Malaysia
[2]Environmental Forensics Research Centre (ENFORCE), Faculty of Environmental Studies, 43400 UPM, Serdang, Selangor, Malaysia
[3]Department of Environmental and Occupational Health, Faculty of Medical and Health Science, 43400 UPM, Serdang, Selangor

Abstract. Numerical models are capable of simulating various groundwater scenarios and relate it towards groundwater management. A mesh based density dependent flow model, FEFLOW is used to simulate groundwater flow and transport for a coastal island aquifer in Kg. Salang, Tioman Island, Malaysia. FEFLOW is designed to simulate 2D and 3D, variable density groundwater flow and multi-species transport. The impacts of pumping and recharge rates represented by three different groundwater scenarios, which were investigated by means of hydraulic heads, TDS concentrations and water balance components. Scenario A showed the standard saturated groundwater flow and the steady state fluid flow. Over pumping and inconsistency in recharge rate are the stresses shown in Scenarios B and C. Scenario B involved in the maximum pumping rate of 96m^3 per day and recharge rate of 300mm per year has shown a drawdown of 1.5 m. Scenario C showed the extreme pumping rate of 1000m^3 per day and without recharge has shown a decrease in groundwater levels of 6.3 m.and groundwater storage (50%). Scenario B presented the most promising finding compared with Scenario C. Highest hydraulic heads, lowest mass concentration and positive groundwater storage (2578.6m^3/day) were obtained in Scenario B. Additionally, and further progress is needed in obtaining the water usage data from each part to determine the best pumping rate. A sustainable groundwater management plan is crucial to maintain the natural resources and social benefits as well as to protect the ecological balance.

* Corresponding author: azrul@nahrim.gov.my

1 Introduction

The fresh water of small tropical islands is often very limited and mostly depend on recharge quantity and storage capability of groundwater resources [1]. Under natural conditions, there is an equilibrium between saltwater and freshwater. Seaward freshwater gradient gets reduced and even sometimes may be reversed to landward direction due to the excessive pumping activities. This leads to aggressive saltwater intrusion from the sea contaminating inland fresh water aquifers to a large extent which may take several years to get remediated [2]. The impact of climate change such as recharge rate and human pressure pumping rate on freshwater/groundwater have been controlling the shape and thickness of the freshwater lens [3]. To address this issue effectively, density dependent groundwater model is required to track the movement of the solute in coastal aquifers [4].

Previously study in Kg. Salang using MODFLOW code for the groundwater storage study for alluvial aquifer in Tioman Island [5]. The study is to compare the pumping rate and the storage capacities in each of the village in Tioman Island. The advantage of FEFLOW software is the vast amounts of data generated from numerous sources in a variety of formats. Therefore, the objective of the present work is to use the FEFLOW code to develop a solute transport model for a coastal aquifer in Kg. Salang, Tioman Island, Malaysia and form the preliminary data for exploring the possibilities of further groundwater development for the region.

A 3D density-dependent transient flow and transport model has been implemented in order to evaluate the intrusion of seawater into the coastal freshwater aquifer [6]. In an unconfined aquifer that contacts the sea at the shoreline or seaward, the freshwater, which is less dense than seawater floats as a lens-shaped layer on top of seawater and the weight of the overlying freshwater depresses the seawater below sea level [7]. Generally, freshwater recharge in the aquifers moves down gradient and eventually discharges to low-lying coastal areas and into the sea which in turn can decrease or even reverse the seaward flow so that seawater moves landward into the freshwater aquifer known as seawater intrusion. The interface between the salty groundwater below and fresh groundwater above is a transition zone of gradually mixing fresh and salt waters. The location of the transition zone depends on several natural and human-made conditions: the relative densities of seawater and freshwater, the tides, the pumping from wells, the rate of groundwater recharge and the hydraulic features of the aquifer. According to the Ghyben-Herzberg Principle [8], salt water rises 40 meters for every 1 meter of freshwater depression and forms a cone of ascension.

An attempt has been made in the coastal island aquifers of Kg. Salang, Tioman Island to assess the ingress of saline water intrusion, where significant quantities of groundwater are drawn for various utilities like tourism industry and domestic utilities using numerical groundwater flow and solute transport model developed with FEFLOW code, using hydrogeological and hydrochemical data. The main objective of this study is the development of a saltwater intrusion prediction tool for the prevention of water pollution in tourism coastal regions.

2 Methodology

A FEFLOW model code was employed for the finite element subsurface flow and transport simulation system. For the purpose of modelling saltwater intrusion, three scenarios were used and compared for the mass concentration, groundwater flow, water table and groundwater storage capability. The model boundary was delineated by catchment area of Salang River.

2.1 Case Study

2.1.1 Site description

The study area is located in the North West of, Tioman Island, Malaysia (104°9'15"E – 104°9'38"E and 2°52'10"E – 2°52'40"E) covering an area of about 350,000 m^2 as shown Fig. 1. The subsoil of the coastal plain surrounding the site is characterized by alluvial sands, shells and corals. Geology of Kg. Salang is characterized by two different rock type which is granite in the south and pyroxene hornfel in the north. Alluvium is found overlying the granite in the southern part of Kg. Salang with about 4 to 30 m thick of medium to coarse sand with trace of corals and shells. The study area is underlain by unconfined aquifer comprising mainly of about 30.0 m thick medium to coarse sand confined to a narrow valley leading to the coast. These hydrogeological factors such as the permeability and porosity of the rocks and sediments have major influence on the distribution of groundwater on an island. Surface water resources prevail only on islands with relatively low permeability. Groundwater resources are most abundant on small islands with moderate to high permeability and porosity. Size, shape and topography of a small island influence the occurrence of both surface and groundwater resources. Since the area is located 50 metres from the coastline, groundwater flow and hydro-geochemistry is strongly influenced by seawater intrusion.

There are fifteen numbers of well scatter around the study areas. The wells were built in different method of drilling. The alluvial thickness is about 4 to 29 m thick from the bore log recorded. The groundwater flow from South east to the north west of the study area.

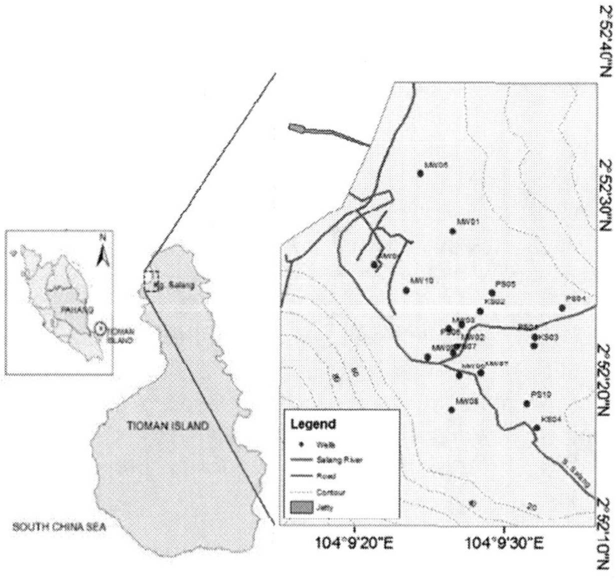

Fig. 1. Location map of wells in the study area

2.1.2 Density-dependent flow approach

In the low lying area of Kg Salang, Tioman Island from the hilly side to the shoreline was selected as study area (Fig. 1). Observation and abstraction wells are located on the low lying area (5 wells are currently operated and observed). There are different method of drilling were used to install the wells (washed boring, hand auger, jetting and hand dug caisson) and were distributed across the study area. Water level and mass concentration were observed and collected from April 2013 to February 2014 bi-monthly from the wells.

In the study, the hydraulic conductivity of horizontal and vertical direction was taken from the soil sample for the hydraulic conductivity and porosity analysis. The unconfined layer divided into 2 layer and 3 slices based on the hydrogeological information. Properties of Kg Salang, Tioman Island (Table 2) were assigned to the active model mesh. A TDS mass concentration of 35,000 mg/L was assigned for constant concentration and 0 mg/L for recharge concentration. In the study, the model calibration is achieved through trial and error approach by adjusting the value of conductivities and porosity until calculated values match the observed values. This method has been done in various groundwater model studies [9].

2.1.3 Model development

The groundwater flow has been simulated using the FEFLOW code [10], a modular three dimensional finite element groundwater flow model. The horizontal discretization of the unconfined aquifer was implemented using a triangular finite element mesh consisting of 111,378 mesh nodes and 148,234 mesh elements. The 3D numerical domain shows thickness of 38 m (from 0.5 to 35 m.s.l.) and the mesh boundary condition as in Fig. 2. It includes 2 layers (6 m of thickness) and 3 slices (or surfaces). The first and the last slices, represent the top and the bottom of the aquifer respectively, which have been produced using field data and kriging interpolation method. The model requires freshwater hydraulic heads as input, thus an equivalent hydraulic head needs to be calculated when applying boundary conditions and when comparing model results to field data [10]. A first type flow boundary condition was set along the coastline. For the first layer the hydraulic head at coastal area was set to zero and for the rest of the layers it was set as equivalent freshwater head, assuming a saltwater density of 1,025 kg/m^3 Pumping wells were defined as 4th kind flow boundary conditions and their screens were set at the second layer. An average monthly infiltration rate was also included as an inflow parameter on the top layer of the model. A first type mass boundary condition was imposed along the coastline with a constant value of 35 kg/L of TDS concentration. The porosity was estimated at 0.3 and the longitudinal and transverse dispersivities were 20 and 2 m, respectively, based on model calibration. The model was run in transient mode for a period of 365 days.

DOI: 10.1051/matecconf/201710304024

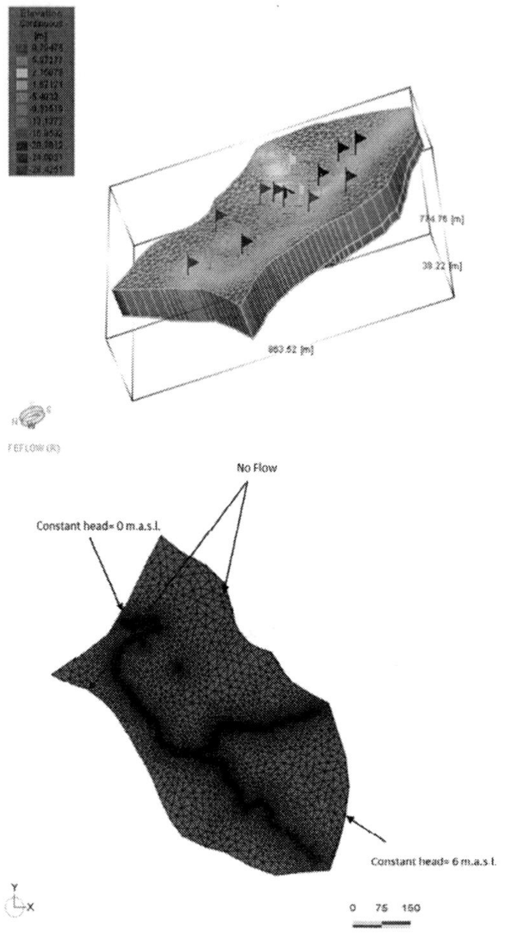

Fig. 2. Two layers and three slices of domain boundary and elevation of the study area (left), Mesh and boundary condition in the study area (right).

Constant values of hydraulic conductivity, derived from field data, characterize the whole domain: K_x=K_y=4.5x10-4 m/s, K_z=1.8x10-5 m/s. These values were evaluated in the sensitivity analysis. The system is influenced by the following physical boundaries: the sea (inflow-outflow), the up-stream hydraulic head (inflow), the pumping wells of the hydraulic barrier and the water supply wells (outflow). Recharged in the study area was calculated by a modified version of Chaturvedi [11] for tropical regions based on water level fluctuation and rainfall depths as given in Equation 1,

$$R=1.35[(P-14)]^{0.5} \qquad (1)$$

where, R is the net recharge and P is the annual precipitation in inches. The amount of recharge coefficient is defined as the ratio of recharge to effective rainfall [12]. The range of recharged varied from 215 to 308 mm/year. Fig. 3 show the flow chart of FEFLOW that including a conceptual model, finite element mesh (FEM), problem setting, model properties, simulation scenarios and output of model. The flow boundary conditions were assigned as specified in the previous paragraph; the hydraulic head was assigned to the sea considering the difference of density between sea-water and freshwater (density ratio). The TDS transport boundary conditions assumes the following constant concentration of TDS up gradient constant concentration of TDS is 0 mg/L; down gradient constant concentration

DOI: 10.1051/matecconf/201710304024

of TDS: is 35 kg/L. The static water table has been given as initial head. A 0 mg/L initial concentration of TDS has been assigned.

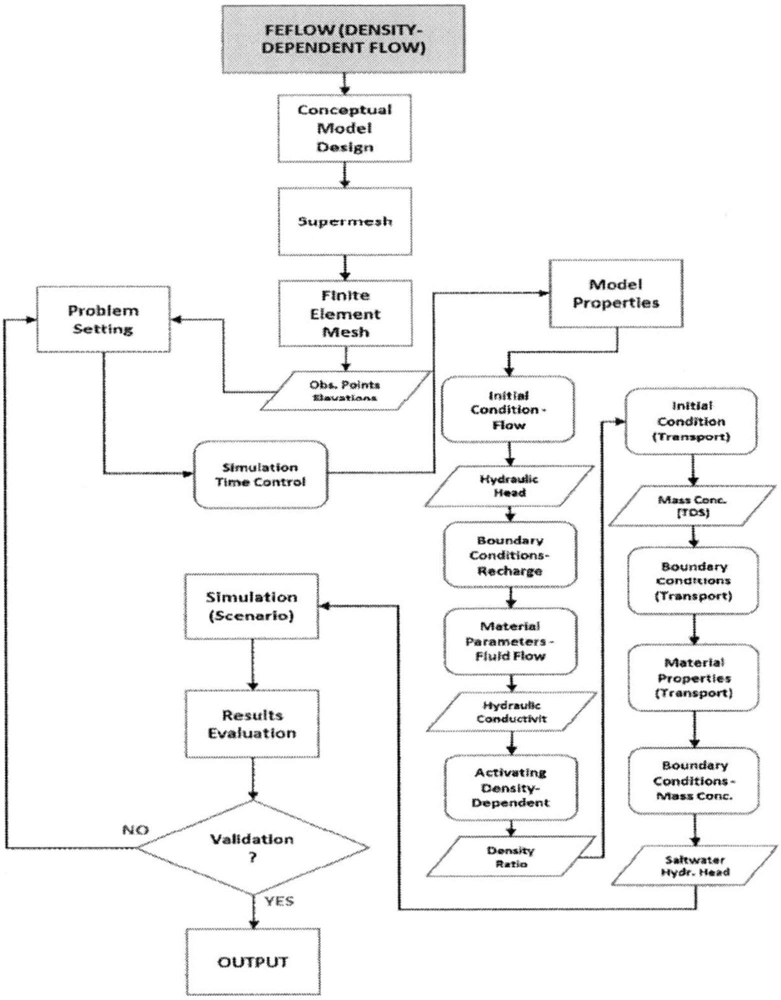

Fig. 3. FEFLOW Numerical modeling procedures flow chart

2.2 Boundary conditions and materials properties of flow and mass

The summary of the model physical parameters and data input for the boundary condition and properties were shown in Table 1. One the model was calibrated; it was run for verify the various scenarios where the predictive simulations were performed to minimize over-abstraction in the unconfined aquifer (Table 2). The simulations were done to predict the aquifer response toward the pumping activities (pumping rate) by human pressures on the freshwater resources. The scenarios will be focusing in the pumping rate to meet the needs of water supply for domestic and tourism industries in the study area.

Table 1. Parameters and model boundaries input data

Parameter	**Value**
Physical Parameters	
Grid/Mesh	
Element type	Triangle prism
Mesh element	148,234
Mesh Node	111,378
Number of layer	1
Number of slice	2
Total model boundary areas	343,298 m2
Materials Properties and Boundary Conditions (BC)	
Hydraulic Conductivity, Kx,Ky,Kz	4.5x10-4 m/s , 4.5x10-4, 1.8x10-5 m/s
Specific storage (Fluid flux)	0.002
Total Porosity	0.35
Longitude dispersivity	1 m
Tranverse dispersity	0.1 m
Density ratio	1,025 kg/m3
Constant head	0 m
Recharge concentration	0 mg/L
Constant head concentration	35,000 mg/L
Initial concentration	Based on April 2013
Initial head	Based on April 2013

Table 2. Three scenarios for the simulation of saltwater intrusion.

Simulation	**Description**
Scenario A	In the steady state of fluid and transport groundwater flow (without pumping)
Scenario B	The recharge rate of 300 mm per year and maximum pumping rate $96m^3$/day
Scenario C	The recharge rate of 0mm per year and maximum pumping rate of $1000m^3$/day.(transient of 365 days simulation time).

3 Results and Discussions

3.1 Calibration and Sensitivity Analysis

The sensitivity analysis was performed in order to understand the uncertainties caused by the estimates of the aquifer parameters, stresses and boundary conditions. Hydraulic head and dispersivity were varied during the sensitivity analysis for hydraulic conductivity as shown in Fig. 4. The calibration procedure consists in modifying the aquifer parameters and the boundary conditions in order to obtain a model suitable to simulate the groundwater flow and the mass concentration distribution, within the calibration target. In the calibration process, a trial and error procedure has been carried out in order to minimize the difference between the simulated and the measured levels/concentrations (residuals).

The transient model development has been based on the 3D steady-state flow model and takes into account the anthropogenic pressure due to pumping wells on the aquifer. Because of the complexity of the site, a 2D model has been implemented along a section perpendicular to the coast. This approach allowed estimating the longitudinal and

DOI: 10.1051/matecconf/201710304024

transverse dispersivity of the total dissolved salt (TDS) in the aquifer. These parameters are usually site-specific, obtained by tracer tests. Since tracer tests were not performed, this step is important to calibrate the aquifer dispersivity. Scenario A and scenario B show slightly drop of water table varies from 0.01 to 0.04m and scenario C drop from 0.05 to 0.22m (Fig. 5). This was due to the abstraction of 1,000m3/day from well, SW03. Table 3 shows the ability of storage after three different scenarios with pumping. Scenario B is the optimum pumping rate with small amount of groundwater abstraction with average of 1.5 m drawdown.

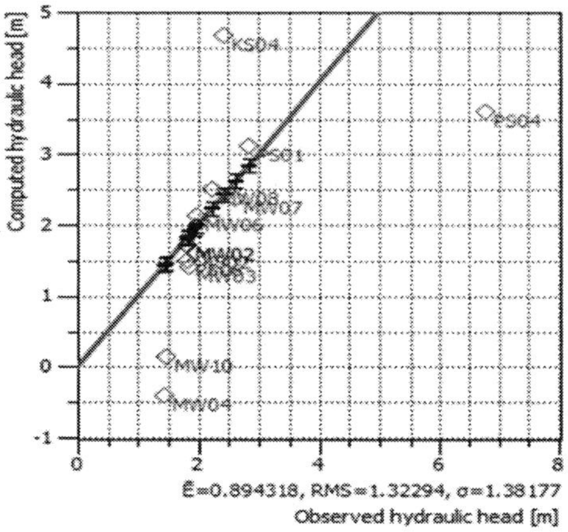

Fig. 4. Comparison of measured and computed hydraulic head values.

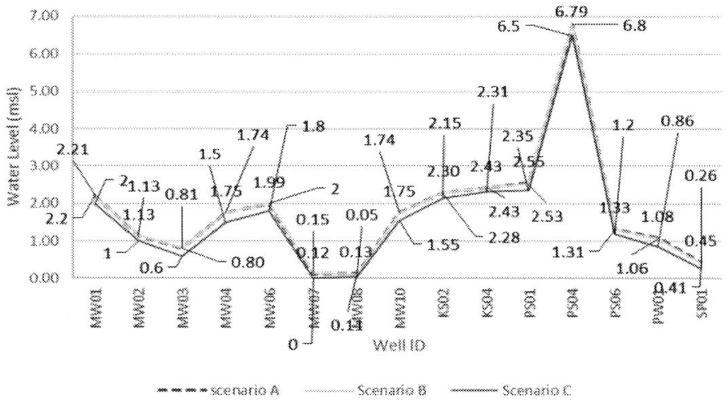

Fig. 5. Decline in ground water levels due to the different scenarios.

Table 3. Simulation after the scenarios of the capture storage.

Scenario	Storage Capture (m^3/d)	Groundwater Abstraction
Scenario A	2,983.5	0
Scenario B	2,578.6	96
Scenario C	1,788.3	1000

Maximum salinity was found 50 m from the coast line. Distance of the concentration were then measured are found that scenario C reach 1,000 mg/L concentration at 225 m from the coast.

Table 4. Movement of mass concentration into the aquifer based on the simulation.

Scenario	Maximum salinity (20-50)m from	Distance from coast (m) with
Scenario A	680	30
Scenario B	1,850	60
Scenario C	10,000	225

4 Conclusions

Seawater intrusion may further advance inland if withdrawals of groundwater by builders, resorts and others tourism activities continue to increase in the coming years. Groundwater quality need to be continuously be monitored near the coastal area, within 100m from the coast. The effect of seawater intrusion on fresh groundwater in coastal areas requires thorough examination, especially when the withdrawal for water supply in the coastal areas increases. An attempt has been made in the present study area to demarcate the saline influences by hydrochemistry and modeling approach.

References

[1] A.Z. Aris, M.H. Abdullah, and B. Musta. *Hydrochemical Analysis on Groundwater in Shallow Aquifer of Manukan and Mabul Island, Malaysia*. International Association of Hydrological Science Red Book (IAHS) Publication, **319**, 387-394, (2008)

[2] S.M. Praveena, M.H. Abdullah, A.Z. Aris, L.C. Yik and K. Bidin. Numerical Modelling of Seawater Intrusion in Manukan Island's Aquifer. World Applied Sciences J., **14**, 1-7, (2011)

[3] S.M. Praveena and A.Z. Aris. Groundwater Resources Assessment Using Numerical Model: A Case Study in Low-Lying Coastal Area. Int. J. Enviroment. Science. Technology, **7(1)**, 135-146, (2010)

[4] J. Lin, J.B Snodsmith, C. Zheng, and J. Wu. A Modeling Study of Seawater Intrusion in Alabama Gulf Coast, USA. Environmental Geology, **57**, 119–130, (2009)

[5] N.A. Ahmad Abdul Ghani, *Groundwater Study for Alluvial Aquifer in Tioman Island.* MSc. Thesis. Universiti Teknologi Malaysia, 187, (2004).

[6] P. Aitchison-Earl, M. Ettema, C. Hanson, S. Hayward, R. Larking, R. Sanders, and A. Veltman. *Coastal Aquifer Saltwater Intrusion Assessment Guidelines*. Report No. R04/18, (2003)

[7] E.D.P. Perera, K. Jinno, A. Tsutsumi, and Y. Hiroshiro. *Development and Verification of A Three Dimensional Density Dependent Solute Transport Model for Seawater Intrusion.* Memoirs of the Faculty of Engineering, Kyushu University. **68**, 93-106 (2008)

[8] J. Bear. *Seawater Intrusion in Coastal Aquifers: Concepts, Methods and Practices.* Boston Mass: Kluwer Academic, (1999)

[9] R. Rejani, M.K. Jha, S.N. Panda and R. Mull. Simulation Modelling for Efficient Groundwater Management in Balasore Coastal Basin, India. Water Resources. Management., **22(1)**, 23-50 (2008)

[10] H.J.G. Diersch. *FEFLOW: Finite Element Modelling of Flow, Mass and Heat Transport in Porous and Fractured Media.* Springer Heidelberg New York Dordrecht London (2014)

[11] R.S. Chaturvedi. *A Note on The Investigation of Groundwater Resources in Western Districts of Uttar Pradesh*. In Anuual Report. U.P. Irrigation Research Institute., 86-122, (1973)

[12] B.D.R. Misstear, I. Brown, and D. Daly. A Methodology for Making Initial Estimates of Groundwater Recharge From Groundwater Vulnerability Mapping. Hydrogeology J., **17(2)**, 275-285, (2009)

Characteristic and Prediction of Carbon Monoxide Concentration using Time Series Analysis in Selected Urban Area in Malaysia

Hazrul Abdul Hamid[1,*], *Ahmad Zia Ul-Saufie* Mohd Japeri[2] , and *Hasfazilah* Ahmat[2]

[1]Faculty of Civil and Enviromental Engineering, Universiti Tun Hussein Onn Malaysia
[2]Faculty of Computer Sciences and Mathematics, Universiti Teknologi MARA

Abstract. Carbon monoxide (CO) is a poisonous, colorless, odourless and tasteless gas. The main source of carbon monoxide is from motor vehicles and carbon monoxide levels in residential areas closely reflect the traffic density. Prediction of carbon monoxide is important to give an early warning to sufferer of respiratory problems and also can help the related authorities to be more prepared to prevent and take suitable action to overcome the problem. This research was carried out using secondary data from Department of Environment Malaysia from 2013 to 2014. The main objectives of this research is to understand the characteristic of CO concentration and also to find the most suitable time series model to predict the CO concentration in Bachang, Melaka and Kuala Terengganu. Based on the lowest AIC value and several error measure, the results show that ARMA (1,1) is the most appropriate model to predict CO concentration level in Bachang, Melaka while ARMA (1,2) is the most suitable model with smallest error to predict the CO concentration level for residential area in Kuala Terengganu.

1 Introduction

Air pollution is one of the important issues to deal with. Air pollution that happened today is from the result of a natural cycle phenomenon in which a very large impact on health in particular and economic development [1]. In Malaysia, Department of Environment (DoE) is responsible in monitoring the level of air pollutants and there are 52 monitoring locations belong to Department of Environment. There are 5 main pollutant that are significant and concerned by Department of Environmental Malaysia which is ground level ozone (O_3), carbon monoxide (CO), nitrogen dioxide (NO_2), sulfur dioxide (SO_2), and fine dust with aerodynamic diameter less than 10 micrometer (PM_{10}).

Carbon monoxide is one of the main pollutants in residential area since the main source of carbon monoxide is from motor vehicles [2]. Generally, Carbon Monoxide (CO) is a by-product of incomplete combustion of Hydrocarbon Fuels. Two kinds of health effects are attributed to CO, very long term and short term. Long periods of exposure to high

* Corresponding author: hazrul@uthm.edu.my

concentration of CO can raise the level of carboxyhemoglobin in the blood. Due to the high level of carboxyhemoglobin in blood, blood loses the capability to absorb oxygen. Due to this the amount of oxygen which is carried by blood to different parts of body is reduced [3]. Short periods of exposure to very high concentrations of CO can precipitate or hasten the death of individuals with severely impaired heart or respiratory functions. This hazardous particular substances can harm us by disturb our healthy body system. Every part of our body needs oxygen. Air is made up of approximately 20.9% oxygen and 79% nitrogen. The hemoglobin within our blood carries oxygen to every cell in our body. The hippocampus is the part of the brain that is involved in memory forming, organizing and storing. It is a limbic system structure that is particularly important in forming new memories and connecting emotions and senses, such as smell and sound, to memories [3].

Dangerous of carbon monoxide worsening because it can causes health effects to patients suffering respiratory problems. This is due it prevents the body from using oxygen as well as potential cancer patients [4]. Therefore prediction of carbon monoxide is important, especially as initial steps by give an early warning sufferers of respiratory problems such as asthma, wheezing as a precaution as well as early warning for us or the related authorities to be more prepared to prevent and take suitable action to overcome the problem. The main focus in this research is to understand the characteristic of CO concentration and also to find the most suitable time series model to predict the CO concentration in Bachang, Melaka and Kuala Terengganu.

2 Data and study area

Bachang and Kuala Terengganu are categorised by Department of Environment Malaysia as an urban area located in Melaka and Terengganu, respectively. Previously, this two monitoring station was classified as a residential area [5]. Geographically, Melaka state is situated on the western coast of peninsular Malaysia with the coordinates is latitude 02° 12.78'N of the equator and longitude 102° 14.05'E of the prime meridian. Terengganu state is situated on the northwest coast of peninsular Malaysia. The coordinates is 05° 20.23'N latitude and 103° 09.45'E longitude. Hourly average of CO concentration in this two location for 2013 and 2014 was obtained from the Department of Environment Malaysia and was measured in part per million (ppm) using carbon monoxide gas detector at the continuous air quality monitorind stations. Monitoring data was computed by averaging direct measurement from the monitoring sites on a yearly basis and cross-reference with Malaysia Ambient Air Quality Guidelines as shown in Table 1.

Table 1. Malaysia Ambient Air Quality Guideline for Carbon Monoxide

Averaging Time	ppm	mg/g^3
1 Hour	30.0	35.0
8 Hours	9.0	10.0

3 Procedure of time series analysis

3.1 Stationarity

Unit root test which is Augmented Dickey-Fuller (ADF) test was used to test the stationarity of data since this is the most commonly used test in time series analysis to determine the stationarity of any data set. The ADF test uses the following Equation (1):

$$ADF = \alpha_0 + p_1 y_{t-1} + \sum_{j=2}^{p-1} \beta_j \nabla y_{t-j} + e_t \quad (1)$$

where α_0 is drift components and e_t is error term [6]. The hypothesis to determine the stationary of the data is as follows:

H_o : the time series data is non-stationary
H_1 : the time series data is stationary

The null hypothesis will be rejected if the ADF value is greater than the critical value.

3.2 Model Identification and Estimation

Each time series model were identify part by part starting from Moving Average (MA), Autoregressive (AR), and Autoregressive Moving Average (ARMA). In this study Akaike Infomation Criterion was used as medium of lag length which is to identify the compatibility of error from those process come out either too small or large of error. In estimation period, each model is directly involved and process of its own equation. Their combination will create a new value that will be used in the next stage of analysis. Among the time series model used is :

i. Moving Average (MA) Process
ii. Autoregressive (AR) Process
iii. Autoregressive Moving Average (ARMA) or Autoregressive Integrated Moving Average (ARIMA) Process

The moving average process of order q is denoted MA(q) and defined by

$$X_t = \sum_{s=0}^{q} \theta s \in_{t-s} \quad (2)$$

where $\theta_1, \ldots, \theta_q$ are fixed constants, $\theta_0 = 1$, and $\{\in_t\}$ is a sequence of independent (or uncorrelated) random variables with mean 0 and variance σ^2 [7].

The autoregressive process of order p is denoted AR(p), and defined by the following Equation [7].

$$X_t = \sum_{r=1}^{p} \phi_r X_{t-r} + \in_t \quad (3)$$

where $\phi_1, \ldots, \phi_r$ are fixed constants and $\{\in_t\}$ is a sequence of independent (or uncorrelated) random variables with mean 0 and variance σ^2.

The autoregressive moving average process, ARMA(p, q), is combination of AR(p) model and MA(q) model which is can defined by

$$X_t = \sum_{r=1}^{p} \phi_r X_{t-r} = \sum_{s=0}^{q} \theta s \in_{t-s} \quad (4)$$

where again $\{\in t\}$ is white noise. This process is stationary for appropriate ϕ, θ.

3.3 Validation

To judge the developed model, three error measures were used. The root mean square error method is the most common indicator. For a good estimator, the RMSE value must approach zero. Therefore, a smaller RMSE value means that the model is more appropriate by [8]. It can be defined by :

$$RMSE = \sqrt{\frac{1}{N}\sum_{i=1}^{N}(P_i - O_i)^2} \tag{5}$$

The normalized absolute error is a more sensitive measure of residual error than RMSE [8]. It is defined as

$$NAE = \frac{\sum_{i=1}^{N}|P_i - O_i|}{\sum_{i=1}^{N}O_i} \tag{6}$$

Mean absolute percentage error is a relative error measure that uses absolute values to keep the positive and negative errors from canceling one another out and uses relative errors to enable you to compare forecast accuracy between time-series models. The formula for calculating the MAPE is as follows :

$$MAPE = \frac{\sum_{i=1}^{N}\left|\frac{(O_i - P_i)}{O_i}\right|}{N} \times 100 \tag{9}$$

where N is the number of monitoring records, O_i is the observed monitoring records and P_i is the predicted monitoring records [9].

4 Results and discussions

This section will discuss the results that have been obtained from the analysis of CO concentrations in Bachang, Melaka and Kuala Terengganu. Time series plot and descriptive statistics are used to get a general overview and compare the level of CO concentration at both monitoring stations before carried out the procedure for prediction.

4.1 Time Series Plot and Descriptive Statistics

Acquisition data for CO concentration level in Bachang, Melaka and Kuala Terengganu show respectively the time series plotting. Fig.1 and Fig.2 show the time series plot for Bachang and Kuala Terengganu respectively. Missing values were replaced using the mean top bottom method. Descriptive statistics for both monitoring site are shown in Table 2. Basically, the mean and mode of CO concentration at both monitoring sites are almost the same and both set of data are positively skewed, meaning that most of recorded data are lower than median.

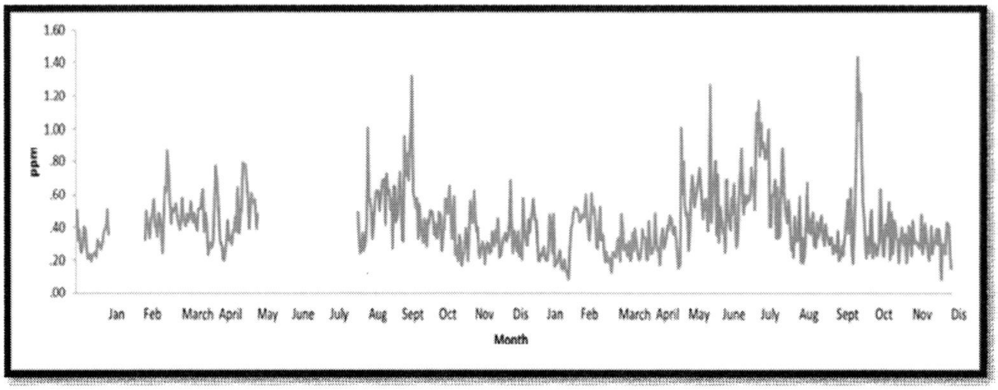

Fig. 1. Time Series plotting of CO concentration in Bachang, Melaka with missing value

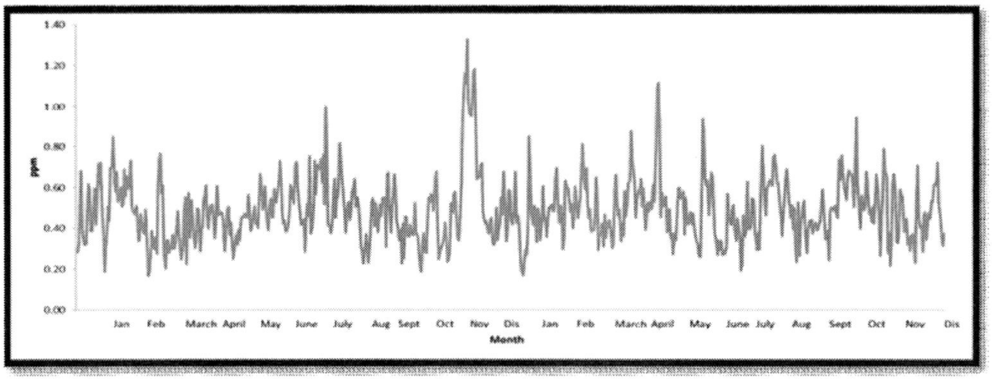

Fig. 2. Time Series plotting of CO concentration in Kuala Terengganu.

Table 2. Characteristics of CO concentration level in Bacang and Kuala Terengganu

	Bachang, Melaka	**Kuala Terengganu**
Mean	0.428	0.488
Median	0.410	0.470
Mode	1.430	1.330
Standard deviation	0.178	0.159
Skewness	1.463	1.202

4.2 Time Series Analysis

Results from Augmented Dickey Fuller test shows that the monitoring record of CO concentration from 2013 to 2014 are stationary for both monitoring stations with ADF statistic for Bachang, Melaka is -4.562 and ADF statistic for Kuala Terengganu is -4.106. Since both p-value are less than 0.05, the time series data can be concluded as stationary.

Autoregressive model (AR model), Moving Average model (MA model) and Autoregressive Moving Average model (ARMA model) were considered to find the most

DOI: 10.1051/matecconf/201710305001

appropriate model for these set of data by comparing the Akaike Info Criterion (AIC) statistic values. The small AIC statistic values indicate the most appropriate model with smallest error. For Bachang, Melaka, the most suitable model to make the prediction with the smallest AIC statistic value is ARMA (1, 1) where the AIC statistic value is -1.2462 while for Kuala Terengganu, ARMA (1,2) is the best choice based on the smallest AIC statistic value, -1.5151 compared to others.

Time series equation based on the best time series model for Bachang, Melaka is as follows:

$$X_t = 0.429 - 0.836X_{t-1} - 0.314e_{t-1} + e_t \tag{10}$$

For Kuala Terengganu, the time series equation based on the best time series model is

$$X_t = 0.489 + 0.747X_{t-1} - 0.023e_{t-1} - 0.077e_{t-2} + e_t \tag{11}$$

For the validation, 14 days ahead was predicted and the performance indicators for the time series model are summarized in Table 3. For the purpose of seeing more clearly, the 14 days observed and predicted of CO concentration using the most appropriate time series model were plotted in Fig.3 and Fig.4 for CO concentration in Bachang and Kuala Terengganu respectively.

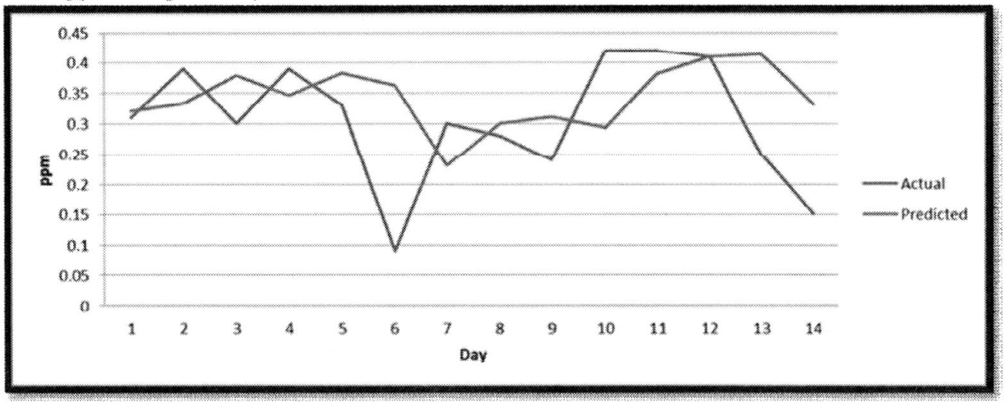

Fig. 3. Actual and predicted of CO concentration using ARMA (1, 1) for Bachang, Melaka.

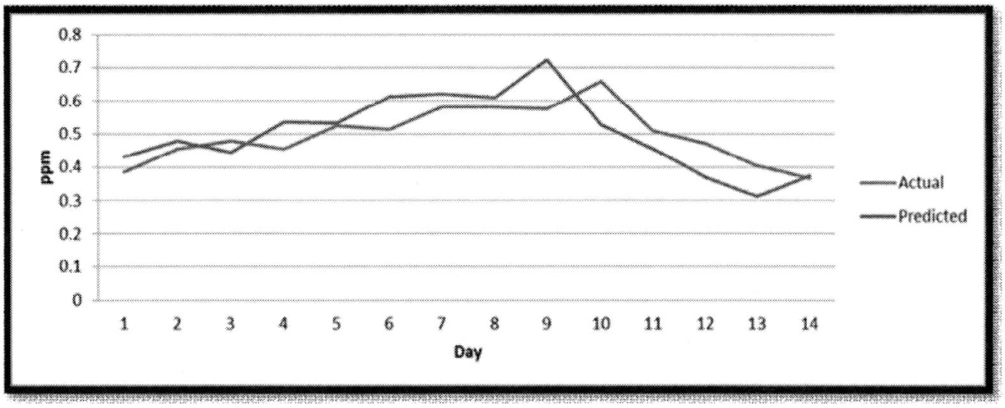

Fig. 4. Actual and predicted of CO concentration using ARMA (1, 2) for Kuala Terengganu

Table 3. Performance indicator

Performance Indicator (Error measure)			
	RMSE	**NAE**	**MAPE**
Bachang, Melaka	0.1119	0.0845	17.074
Kuala Terengganu	0.0773	0.0641	13.131

5 Conclusion

Set of secondary data from DOE were analyzed by varying the descriptive statistics and the result shows that mean value of CO concentration in Bachang, Melaka is slightly lower than Kuala Terengganu but for the skewness of CO concentration show, both urban areas have a positive values, indicated that data were skewed to the right which is means most of data are lower than median. To predict the CO concentration level using time series analysis by choosing the appropriate model for both monitoring station located in urban area, only autoregressive (AR), moving average (MA) and autoregressive moving average (ARMA) were considered since both sets of data are stationary. The quality and reliability of the developed models were validated using AIC and three performance indicators (NAE, RMSE, MAPE) and the result of this study show that ARMA (1, 1) is the most suitable model to predict the CO concentration for urban area in Bachang, Melaka. Similarly, another area which is for Kuala Terengganu, ARMA (1, 2) was best fitted to predict the CO concentration level.

References

[1] Janet C., Neidellf M. and Schmieder J.F, Air Pollution and Infant Health: Lesson from New Jersey, J. of Health Economic, **28**, 688-703, (2009)

[2] Kayode S.J. and Feyisayo K.,Air Pollution by carbon Monoxide (CO) poisonous gas in Lagos area Southwestern Nigeria, Atmospheric and Climatte Science, **3**, 510-514 (2013)

[3] Akinyemi M.L and Usikali M.R., Investigation of carbon monoxide concentration from anthropogenic sources in Lagos, Nigeria. Int. J. of Physical Sciences, **8**(21), 1128-1132. (2013)

[4] Peter A., Jack A. and Maureen A., *Healthy Home Issues : Carbon Monoxide. U.S Department of Housing and Urban Development*, Version 3, December 2005 (2015)

[5] Department of Environment Malaysia, *Malaysia Environmental Quality Report 2004. Kuala Lumpur : Ministry of Science, Technology and Environment*, Malaysia (2005)

[6] Abdel-Aziz A. and Frey, H.C., Development of hourly probabilistic utility NOx emission inventories using time series techniques : Part I – Univariate approach. Atmospheric Environment, **37**, 5379 – 5389A (2003)

[7] Madsen H., *Time Series Analysis*, Chapman & Hall/CRC texts in statistical science series, London (1955)

[8] Maxim V.S, Adriaan B., Nataliya L.S, Anton P.T., Timur A.J. and Valery A.K., A survey of forecast Error Measures, World Applied Sciences J., **244**, 171-176 (2013)

[9] Abdul Hamid H., Yahaya A.S., Ramli N.A., Ul-Saufie A.Z., Yasin M.N., Short-term prediction of PM_{10} concentrations using seasonal time-series analysis. *MATEC Web of Conference* , **47**, 05001, (2016)

Study on Composition and Generation of Food Waste in Makanan Ringan Mas Industry

Aeslina Abd Kadir [1,*], *Nur Wahidah* Azhari [1] *and Siti Noratifah* Jamaludin[1]

[1]Department of Environmental and Water Resources, Faculty of Civil and Environmental Engineering, Universiti Tun Hussein Onn Malaysia, Johor

Abstract. Food waste management is a major problem for most food premises in Malaysia. This study was conducted at one of the Small Medium Industries (SMIs) called Makanan Ringan Mas Industry that is located in Parit Kuari Darat, Parit Raja, Johor. This premise generates food waste almost every day including processed food waste (chips and coconut candy) and raw food waste (banana peels, tapioca peels, breadfruit peels and grated coconut). The objective of the study was to determine the waste generation and composition of food waste generated by the premise. Food waste collected from the premise once a week and tested for moisture content and density. The results demonstrated that Makanan Ringan Mas Industry generated more raw food waste compared to processed food waste. Banana peels recorded the highest amount at 27.15kg per month compared to other food waste. To conclude, banana peels were found to be the highest component in the food waste composition by Makanan Ringan Mas Industry whereas breadfruit peels were found to be the lowest.

1 Introduction

Waste generation and management have become an issue in most developing countries. Increasing solid waste generation has had a negative impact on human health and the natural environment. Furthermore, the increasing human population and economic growth in developing countries have also contributed to waste generation diversity [1]. In Malaysia, organic waste is the highest amount of waste generated, which is 50% by weight [2].

Organic waste includes kitchen waste, food leftovers, rotten fruit, vegetables and peels, straw and hay, leaves and garden trimmings, crop residues, rags, papers, animal excreta, bone and leather [3]. 60% of organic waste are usually generated by households [4] and this waste is disposed at landfills. The degradation of organic matter without oxygen in landfills will produce acids that affect other waste thus creating toxic pollutants known as leachate. Leachate will seep through ground water and contribute to ground water pollution [5].

Furthermore, the success in managing waste is a strong indicator of the ability of the authority within the city to work together to solve major urban environment problems. In

* Corresponding author: aeslina@uthm.edu.my

this research, Makanan Ringan Mas Industry is one of the industries that focuses on food production. From the observation, this industry mainly generates food waste. The problem facied by SMIs is improper waste management. Waste dumping into the ditch as well as open burning adds to atmospheric and hydrologic pollution load and gives a negative impact to the environment.

This paper highlights the amount of solid waste generated by Makanan Ringan Mas Industry at Parit Kuari Darat in eight months. Overall, food waste generation is something that needs to be monitored because improper disposal is the major source of surface and ground water contamination as well as air pollution.

2 Literature reviews

Solid waste can be defined as materials that are unwanted such as household garbage, food waste, yard waste and demolition or construction debris. In general, domestic waste generated are food waste, paper, cardboard and plastics. On the other hand, commercial waste mainly consists of wood, glass, hazardous waste, e-waste whereas agricultural waste usually consists of husks, cotton-stalks, coconut-shell waste from agricultural activities [6]. Generation of municipal solid waste in Malaysia is increasing due to human population growth, rapid urbanization, economic growth and its multicultural society [7]. Waste can be defined as any crap materials or other unwanted surplus or rejected products that arise from human activities [8].

In Malaysia, solid waste management is under the responsibility of the public sector [9]. It can be divided into three zones which are Southern Zone by SWM Environment Sdn Bhd. (Johor, Melaka and Negeri Sembilan), Northern Zone by Environment Idaman (Kedah and Perlis) and Central and Eastern zone by Alam Flora (Wilayah Persekutuan Kuala Lumpur/Putrajaya and Pahang).

Currently, human population contributes to the highest amount of municipal solid waste [10, 11]. It is reported that the composition of waste generation by Kuala Lumpur as the capital city of Malaysia with high population is food waste (57%), followed by mix paper (17%) mix plastic (15%) and other [12]. Most studies in the cities of developing countries showed that the solid waste composition is high in organic materials and contains less metal and plastic.

Other literature, by Mohan et al. [13], revealed that solid waste composition in Kathmandu, Nepal consists of about 71% organic materials, 12% plastic, 7.5% papers, 5% construction waste and 1% hazardous waste. Another study in Tulsipur, Nepal showed that household waste generation is made up of 46% organic wastes, 11% construction debris, 10% plastics, 7% glass, 6 paper, 5% metals and 5% rubber and leather [14]. On the other hand, a study in China showed that solid wastes consist of 57% organic waste [15]. Furthermore, Farouhar and Hristovski [16] who studied waste generation in Kabul, Afghanistan also found that about 70% of waste generated is organic materials. Moreover, data from previous research mostly also concluded that organic waste is the main waste generated compared to other types of waste. In addition, according to EWWR [17], food waste can be defined as any food; inedible parts of food; removed from the food supply chain to be recovered or disposed. It found that household waste had the highest production of food waste (42%), followed by manufacturing (39%), food services (14%) and wholesale and retail (5%).

3 Methodology

3.1 Sampling area

The study was carried out in one of the SMIs in Parit Raja which is Makanan Ringan Mas Industry in Parit Kuari Darat (Fig. 1). It is located about 13.5 km from Universiti Tun Hussein Onn Malaysia (UTHM). This industry operates 24 hours a day and the production depends on the customer demands. The industry is one of medium scale industries that focuses on food production such as variety of chips, *dodol*, coconut candy and others.

Fig. 1. Makanan Ringan Mas Industry location.

3.2 Sampling method

3.2.1 Collection, composition and classification

The waste collection was carried out once a week, for a period of 8 months. The waste was collected from the industry by car and unloaded at the Micropollutant Research Centre (MPRC) Laboratory. The waste generated is mainly organic material; food waste. The food waste was divided into processed food waste and raw food waste. Processed food waste are chips and coconut candy while raw food waste are tapioca peel, banana peel and grated coconut.

3.2.2 Physical Characteristics

Density, particle size and moisture content are some of the main physical characteristics of waste [18, 19]. Moisture content and density are two physical characteristics that were tested.

i. Moisture content:

The moisture sample for each waste was taken by dividing it into 1000g each and recorded as wet weight of the sample. Then, the wet sample was dried to a constant weight at a temperature at 105°C using an oven at the Geotechnic Laboratory, UTHM. The sample was allowed to be cooled and then the weight was measured and recorded as dry weight of the sample. The moisture content of each type of waste was calculated according to Equation (1) [20] and expressed in percentage (%).

$$\textit{Moisture Content (\%)} = \frac{\textit{wet sample (g)} - \textit{dry sample (g)}}{\textit{wet sample (g}} \quad (1)$$

ii. Density:

Food waste collected from the industry was filled in an a cylindrical container. The container was stamped 3 times in order to fill the spaces in the container with the food waste. The density was calculated according to Equation (2)[21].

$$\text{Density (kg/m}^3\text{)} = \frac{\text{weight (kg)}}{\text{volume (m}^3\text{)}} \quad (2)$$

4 Result And Discussion

4.1 Food Waste Generation in Small and Medium Industries (SMIs)

The quantity and rate of food waste generation in the industry were estimated. The mass-based compositions of the characterized food waste in the SMI studied, namely Makanan Ringan Mas Industry, are presented below.

4.1.1 Composition of Food waste from the SMIs

The results from the food waste collection for 8 months comprised the values of the mass-based composition of food waste generated by the industry. The total quantity of waste generated from the industry and the percentage composition of the wastes are presented in Table 1 and Fig. 2 respectively. Fig. 2 revealed that banana peel waste was the highest proportion of waste generated (27%) from the industry. Tapioca peels made up the second highest proportion (25%) of the wastes followed by grated coconut (16%). Banana and tapioca peels made up highest amount of overall waste as the consumption of bananas and tapioca in the industry is more than other types of food. The industry runs a chips business focusing mainly on banana and tapioca chips. Therefore, banana and tapioca peels are the major types of waste generated by this industry. Other food waste generated includes; chips 14%, coconut candy 10%, and breadfruit peels 8%. Although breadfruit is also used in the industry as an ingredient for making chips, it is not always available and in season. Thus, the amount of breadfruit peels generated is limited. The total quantity of waste generated from Makanan Ringan Mas Industry is estimated at 101.82 kg/month and the total density of the food waste is 32303.04 kg/m^3 per month.

Table 1. Composition of Food waste from the SMIs.

Food waste	Total Quantity of Food waste (kg/month)	Total Food Waste Density (kg/m^3)/month
Chips	14.02	4054.64
Coconut Candy	10.05	7384.63
Grated Coconut	16.86	6929.39
Banana Peels	27.15	6929.39
Tapioca Peels	25.46	6692.92
Breadfruit Peels	8.28	2791.07
Total	**101.82**	**32303.04**

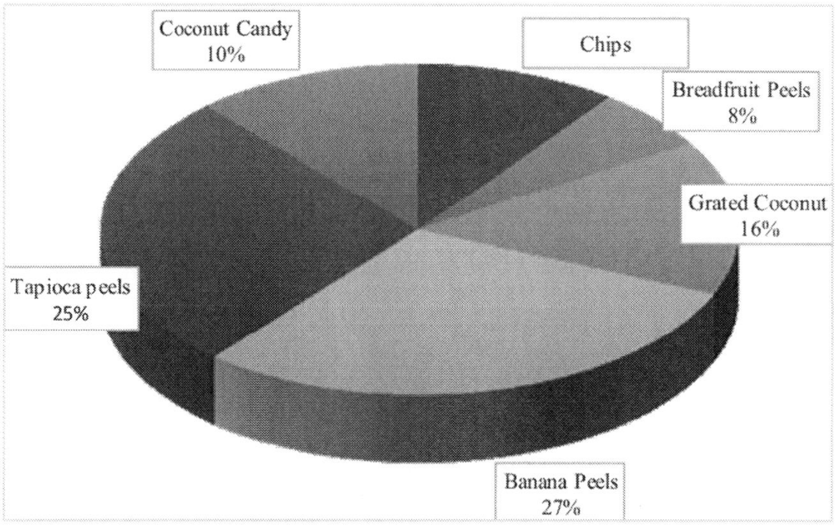

Fig. 2. Composition of Food waste from the SMIs.

4.1.2 Composition of Different Types of Food Waste

Table 2 shows the composition of the food waste between raw food waste (grated coconut, banana peels, tapioca peels and breadfruit peels) and processed food waste (chips and coconut candy). Raw food waste is produced from the fruit peeling process in the production of chips and is also known as unavoidable food waste. On the other hand, processed food waste is waste that results from the food packaging process where food with defects will be thrown away and become waste. This is also known as avoidable food waste. In this study, the total quantity of raw food waste is 77.75 kg/month with total a density of 20863.77 kg/m^3 per month, while total quantity of processed food waste is 24.07 kg/month with a total density of 11439.27 kg/m^3 per month. The results from Fig. 3 show that the total percentage of raw food waste is 74% which is higher than the processed food waste which is only 26%.

Table 2. Composition of Different Types of Food Waste by the SMIs.

Types of Food waste	Total Quantity of Waste (kg/month)	Total Waste Density (kg/m^3)/month
Raw Food waste	77.75	20863.77
Processed Food waste	24.07	11439.27
Total	**104.9**	**32303.04**

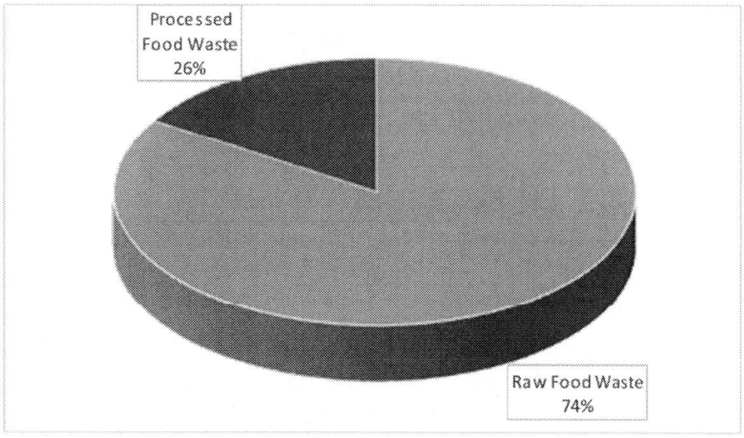

Fig. 3. Composition of Different Types of Food waste.

4.1.3 Moisture content of food waste from SMIs

Table 3 show the percentage of moisture content of food waste produced by the industries. Raw food waste such as grated coconut, banana peels, tapioca peels and breadfruit peels have a higher moisture content percentage compared to processed food waste such as chips and coconut candy. Banana peel has the highest moisture content at 86.04%, while chips have the lowest moisture content at 0.82%. From the results, we can see that the moisture content in the waste can also affect the total quantity of the food waste itself.

Table 3. Moisture content of food waste.

Food Waste	Moisture content (%)
Chips	0.82
Coconut Candy	5.90
Grated Coconut	53.42
Banana Peels	86.04
Tapioca Peels	83.00
Breadfruit Peels	83.50

5 Summary

Overall, the intention of this study was to estimate the food waste generated from Makanan Ringan Mas industry that is one of the small and medium industries (SMIs) in Parit Raja, Johor. The food waste produced by the industry are chips, coconut candy, grated coconut, banana peels, tapioca peels and breadfruit peels. From the results, the highest amount of food waste produced by Makanan Ringan Mas Industry consisted of banana peels and tapioca peels which are in the raw food waste category. The results also showed that the total quantity of food waste is 101.82 kg per month with a total density of 32303.04 kg/m^3 per month. Banana peels, breadfruit peels and tapioca peels have the highest moisture content compared to others which are also in the raw food waste category. The food waste generation from the industry should be managed to prevent improper disposal methods. Other than that, alternative low cost disposal methods should be encouraged as organic waste could be used again as fertilizers. This will help protect our environment and extend the lifespan of landfills.

References

[1] Shahabi. H., Khezri. S., Ahmad. B., and Zabihi. H, Application on Artificial Neutral Network in Prediction of Municipal Solid Waste Generation (Case Study: Saqqez City in Kurdistan Province), World Applied Sciences J., **20**(2), 336-343, (2012)

[2] Abd Jalil. Md, Sustainable Development in Malaysia: A Case Study on Household Waste Management, J. of Sustainable Development, **3**(3), 91-102, (2010)

[3] Inge Lardinois, Organic Waste. Technology Transfer for Development Waste Consultants. Series 1. (1993)

[4] Zero Waste. Fact Sheet. (2015), Retrieved on October 3, 2016 from http://www.hccrems.com.au/hccrems/media/RESOURCES/Education/OrganicsComposting.pdf

[5] Composting Fact Sheet (2009), Retrieved on October 3, 2016 from http://www.cleanup.org.au/PDF/au/ cleanupaustralia_factsheet_composting.pdf

[6] Ikemike Dolfina. O, Effective Solid Waste Management: A Panacea to Disease Prevention and Healthy Environment in Bayelsea State, Nigeria, International Journal Academic Research in Education and Review. **3**(3), 65-75 (2015)

[7] Chua, K.H., Mat Sahid, E.J., and Leong, Y.P, Sustainable Municipal Solid Waste Management and GHG Abatement in Malaysia, **4**(2), 1-8 (2011)

[8] Environmental Protection Agency (1996) (EPA), Retrieved on October 5, 2016, from https://www.epa.ie/pubs/reports/waste/wastecharacterisation/EPA_municipal_waste_characterisation.pdf

[9] Saat. S.A. and Ali. N, Analyzing the Sustainability of Solid Waste Policy in Malaysia using the Ecological Modernization Theory, Malaysian J. of Society and Space. **10**(6), 118-125. (2014)

[10] Johari. A., Alkali. H., Hashim. H., Ahmed. S.I., Mat. R. Municipal Solid Waste Management and Potential Revenue from Recycling in Malaysia. **8**(4), 37-49. (2014)

[11] PSN and IMechaE. (2013), Retrieved on September 18, 2016 from http://populationsustainability.org/global-food-waste-must-be-addressed-to-feed-growing-world-population/

[12] Saeed. M.O, Hassan. M.N., and Mujeebu. M.A. Development of Municipal Solid Waste Generation and Recyclable Components Rate of Kuala Lumpur: Perspective Study. Int. Conferences on Environment (2008)

[13] Mohan. B.D., Christopher. R. P., Micheal. A.U., Kenneth. G. G., J.M Reddy. Municipal Solid Waste Generation in Kathmandu, Nepal. J. of Environmental Management. **92**, 240-249. (2011)

[14] Mohan. B.D., Micheal. A. U., and Shashidhar. B. Characterization, Generation and Management of Household Solid Waste in Tulsipar, Nepal. Habitat International. **40**, 65-72, (2013)

[15] Chen. X., Geng. Y., and Fujita. T, An Overview of Municipal Solid Waste Management in China. Waste Management J., **30**, 716-724. (2010)

[16] Farouhar. A., Hristovski. K.D, Characterization of The Municipal Solid Waste Stream in Kabul, Afghanistan. Habitat Int., **36**, 406-413. (2012)

[17] EWWR (2014). Retrieved on October 6, 2016 from http://www.ewwr.eu/docs/PTD/PTDs_ 2014_stop_food_waste_info.pdf

[18] Tae- Kyung Park, Jae- Ram Park, Gook- Yong Jung, Sung- Jin Bae, Dong- Hoon Lee, Analysis on moisture content of mixed MSW considering distribution characteristics of physical composition, Int. Conference on Sustainable Energy and Environment. 508-511. (2014)

[19] Chandrappa and Das. *Book Chapter 2: Waste Quantities and Characteristics.* Solid Waste Management, Environmental Science and Engineering. (2012)

[20] Salami. L., Susu. A.A, Characterisation Study of Solid Wastes: A Case of Lagos State. Int. J. Applied Science and Technology, **1**(3), 47-52, (2011)

[21] European Standard. Solid materials – Determination of Bulk Density. (2005) Retrieved from https://www.ecn.nl/docs/society/horizontal/hor_desk_32_annex1.pdf

Reuse of Ablution Water to Improve Peat Soil Characteristics for Ornamental Landscape Plants Cultivation

Radin Maya Saphira Radin Mohamed[*,1], *Al-Gheethi* AA[1], *Junita Abdul Rahman*[1], *Amir Hashim Mohd Kassim*[1] *and Sabariah Musa*[1]

[1]Micro-pollution Research Centre (MPRC), Department of Water and Environmental Engineering, Faculty of Civil & Environmental Engineering, Universiti Tun Hussein Onn Malaysia, 86400 Parit Raja, Batu Pahat, Johor, Malaysia

Abstract: The present study aimed to reuse of ablution water for washing peat soil in order to reduce the concentrations of heavy metals in these soils which might effect negatively on the plant growth. The washing process design was similar to horizontal subsurface flow constructed wetlands (HSSFCW) consist of layers of peat and sand soil and surrounded by gravel on both sides. *Strelitzia* sp. was used to detect the presence negative effect of the washing process on the morphological characteristics of the plants. The physical and chemical characteristics of ablution water was examined before and after the washing process by using Inductively Couple Plasma- Mass Spectrometer (ICP-MS) Atomic Absorption Spectrometer (AAS). The characteristics of peat soil before and after the washing process were examined by using X-Ray Fluorescence (XRF). The results revealed that the percentage of FeO_3in peat soil reduced from 45.80 to 1.01%. The percentage of SiO_2 in sand soil dropped from 87.7 to 67.10%. Parameters of ablution water resulted from the washing process which including Biological Oxygen Demand (BOD_5) and heavy metals have increased but still within the standard limits for the disposal of ablution water into the environment. No atrophy was observed in *Strelitzia* sp. leaves, indicating the ability of plant to grow normally. It can be concluded that the utilization of ablution water in the washing of peat soil has improve the characteristics of the soil without effect on their organic constitutes.

1 Introduction

Peat soil is a class of soils which has between 50 and 75% of organic matter and formed in natural environments due to the slow decay of materials [1]. It represent 8% of the total area (about 3 million hectares) in Malaysia [2].Peat is considered as problematic soils in the view of design parameter by the geotechnical engineers because its engineering characteristic are inferior to those of the other soft soils which make it unsuitable for

* Corresponding author: maya@uthm.edu.my

construction in its natural stage. Peats are found to contain high organic matter and are generally associated with poor strength characteristics, large deformation, high compressibility, and high magnitude and rates of creep [3].

Constitutes of peat soil which including dissolved nutrients and organic matter as well as macro-elements are more benefit for the plant growth. However, the high concentrations of organic matter is associated with the lowering in pH values to less than pH 4 which might affect negatively on the plant growth and crop production. Moreover, the high concentration of inorganic matter such as presence of heavy metals might lead to be accumulated in the plant tissue and then transmitted into animal and human. Peat soil has a pH ranged from pH 3.5 to 5 [4], in which the toxicity of heavy metal become more effect on the plant growth. For instance, nickel ions with 0.5 to 1 mg L^{-1} of concentrations are toxic to a number of plants at. However, the toxicity it reduced at neutral or alkaline pH [5].The forest trees has a potential to grow at low pH values, but most of crop and ornamental plants grow best with optimum pH between pH 5.5 and 7.5. The adjustment of soil pH might be carried out by using lime which lead to increase the pH to alkaline values [6]. Nonetheless, the main chelating of peat soil lies in the presence of heavy metals which are not reduced by using the lime. In order to reduce the metal ions, some authors suggested to use phytoremediation process in which a heavy metal tolerant plant species are applied for cleaning polluted soils [7,8], but this technique take long time depend on the generation time of the plants. Other authors have investigated washing of metal contaminated garden soil using EDTA [9,10]. However, the utilization of chemical treatment of soil might associated with secondary productions with adverse effect on the soil characteristics. The safe method for washing of contaminated soil might be by using water. So far, the high expensive for water resource limit the utilization of this technique. Moreover, the ablution water might represent an alternative source of water used for the washing process. The water quality parameters of ablution water including Biological Oxygen Demand (BOD), Chemical Oxygen Demand (COD) and are within the standards limits of Environmental Quality Act 1974for the disposal, and thus it is usually flowed directly to the drain. Hence, these wastes might be used with an efficient for the peat soil washing and this is the focus of the current work which aimed to examine a design similar in the structure to the horizontal subsurface flow constructed wetlands (HSSFCW) and consisted of two layers including peat and sand soil and surrounded by gravel on both sides. This design will be used for cultivation of ornamental landscape plants which can be used in parks and public gardens as well as on the sides of public roads. The characteristics of peat soil which included heavy metals and acidity will be improved by the washing process using ablution water to be more applicable for the cultivation purpose and this emphasize the novelty of the work.

2 Materials and methods

2.1 Sampling

Peat soil were collected from Parit Raja, Johor Malaysia and then placed in a structure similar to HSSFCW design with 36×22×23 cm of size and mixed with the sand soil. The ablution water (10 L) were collected from UTHM Mosque and then transported into the laboratory. The physical and chemical characteristics of the ablution water were determined according to APHA [11] and then stored in a storage tank until the utilization for the washing process.

 DOI: 10.1051/matecconf/201710305003

2.2 Peat soil washing system

In order to wash peat soil, one layer of peat soil and two layers of sand soil were placed in horizontal form HSSFCW (Fig. 1). The volume of the container used for HSSFCW was 541.10 cm^3 fill-up with 8.125 kg of peat (2.708 kg) and sand soil (5.416 kg). Two HSSFCW container were used one without plant (Reactor A) (Fig. 1) and the second one with plant (Reactor B) (Fig. 2). Reactor B with *Strelitzia*sp. was used to detect the presence or absence negative effect of the washing process on the morphological characteristics of the plants. The ablution water was pumped from the storage tank through the layers with fixed flow rate (4.22 mL/ min) and the infiltration process conducted for 5 days. The flowrate and HRT used here was adjusted to be more appropriate to be used in the filed study. Reactor B has similar function to Reactor A, but as a result for using *Strelitzia*sp. the reactor B was supported on both side with gravel to be more appropriate to cultivate of *Strelitzia*sp.

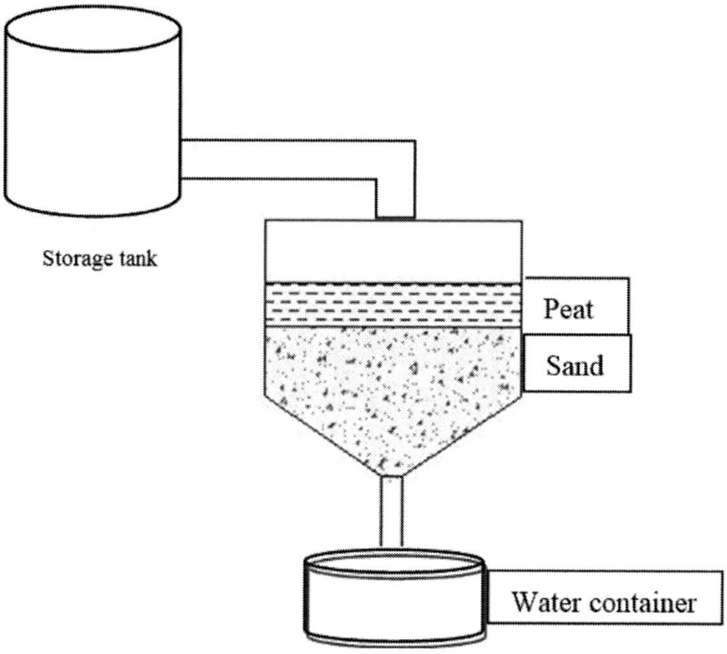

Fig. 1. Reactor A, washing system unit

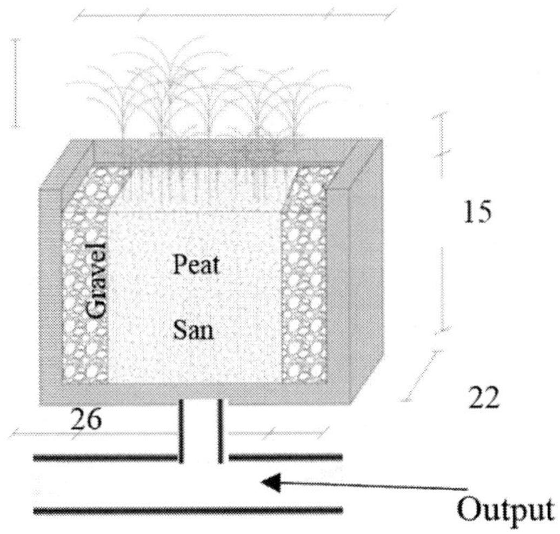

Fig. 2. Schematic diagram of Reactor B

2.3 Analytical methods

The characteristics of influent and effluent of HSSFCW including, BOD_5, turbidity, electrical conductivity (EC), pH, sodium adsorption ratio (SAR), sodium (Na^+), chloride (Cl^-), total coliform (TC) were conducted according to APHA (2005). The heavy metals and trace elements concentrations including cadmium (Cd), chromium (Cr), copper (Cu), manganese (Mn) and zinc (Zn) were carried out by using Inductively Couple Plasma- Mass Spectrometer (ICP-MS) while Iron (Fe) and magnesium (Mg) were determined by using Atomic Absorption Spectrometer (AAS) according to APHA [11]. These parameters were tested on 3,4 and 5th day of the washing process. The characteristics of peat soil before and after the washing process were examined by using X-Ray Fluorescence (XRF).

3 Results and discussions

The characteristics of peat soil before and after washing process conducted using X-Ray Fluorescence technique revealed that the peat soil has higher concentrations of Fe_2O_3 andK_2O than the sand soil, in contrast, sand soil has higher concentrations of SiO_2 than peat soil (Table 1). Other heavy metals found in peat soil included Al_2O_3 (7.88%), ZnO (6.03%), K_2O (5.15%), MgO (0.59%), and MnO (0.13%). The presence of SiO_2 (22.30%) in peat soil indicate to that the peat soil contains detectable amounts of sand soils.

Table 1. Chemical composition in peat and sand soil before and after washing process

Formula	Before washing		After washing	
	Peat (%)	Sand (%)	Reactor A (%)	Reactor B (%)
CO_2	0.10	0.10	0.10	0.10
Fe_2O_3	45.80	0.44	1.01	0.80
SiO_2	22.30	87.70	67.10	70.90
Al_2O_3	7.88	10.80	7.32	7.50
ZnO	6.03	< DL	< DL	< DL
K_2O	5.15	0.27	0.47	0.38
SO_3	5.12	< DL	0.69	0.68
CaO	4.87	< DL	1.21	0.56
Cl	0.85	< DL	0.14	0.16
P_2O_5	0.76	< DL	< DL	< DL
MgO	0.59	< DL	< DL	< DL
TiO_2	0.30	0.35	0.51	0.52
Br	0.17	< DL	< DL	< DL
MnO	0.13	< DL	< DL	< DL

*DL(Detection limits)

The washing system conducted in this study exhibited high efficiency in the reduction of the parameters of the peat soil.The composition of peat soils have changed significantly after the washing process. The percentage of FeO_3 in peat soil reduced from 45.80 to 1.01 and 0.8% in reactor A and B, respectively.The concentrations of K_2O, SO_3, CaO, Cl, decreased significantly, ZnO, P_2O_5, MgO, Br and MnO reduced to less than the detection limits. These findings confirm that the potential of ablution water to improve of peat soils. Some of heavy metals in the soil are necessary for plant growth because they acts as a trace elements and cofactors for several enzymes. For instance,Zn^{2+} is a cofactor in some essential enzymes in animals and plants, Cu^{2+} is an important metal for many biological functions, but at 0.1 to 1.0 mg L^{-1} is toxic to a number of plants [5]. Nonetheless, some of metals are toxics Pb^{2+} is a toxic element, and no useful biological function of lead has been reported [12, 13, 14,15]. Akrivos et al. [16] indicated that zinc (Zn^{2+}), copper (Cu^{2+}), nickel (Ni^{2+}) and chromium (Cr^{3+}) effects on plant growth and crop yields and are known to be cumulative toxins, while lead (Pb^{2+}), cadmium (Cd^{2+}) and mercury (Hg^{2+}) are toxic to the animal or human digest of the plants and might be mutagen causing cancer [17]. Hence, the reduction of the metal ions in the peat soil might limit the health risk. The health risk associated with the heavy metals lie in their accumulation in the human bodies and form strong toxic complexes which makes them too dangerous for any physiological function [12].

In order to confirm the reduction in parameters of peat soil, the characteristics of ablution water before and after the washing process were determined (Table 2). It can be noted that the parameters of these water have changed significantly. The concentrations of BOD_5 increased from 2 mg L^{-1} to 6.5 mg L^{-1} after 5 days of the washing process of peat soil. pH of the water has decreased from 8.65 to 5.4, indicating to migration proton ions from the peat soil to the ablution water. However, the concentrations of BOD_5 in the output water was not high and still within the standards limits of wastewater discharge.

Table 2. Chemical and physical parameters of ablution water before and after the utilization in the washing process of the peat soil

Parameter	Input ablution water	Output ablution water		
		3 days	4 days	5 days
BOD_5	2 mg/L	4.7 mg/L	6.1 mg/L	6.5 mg/L
Turbidity	7 NTU	13 NTU	14 NTU	17 NTU
pH	8.65	5.5	5.46	5.4
SAR	0.46meq/L	0.48meq/L	0.60meq/L	0.69meq/L
Cl	16.40mg/L	19.44 mg/L	19.85 mg/L	20.02 mg/L
EC	38.1µS/cm	42.4 µS/cm	48.5 µS/cm	49.2 µS/cm
E. coli	27 CFU/ 100mL	82 CFU/100mL	96 CFU/100mL	115 CFU/100mL

SAR (Sodium adsorption ratio), Biological Oxygen Demand (BOD), Chlorine (Cl), Electrical Conductivity (EC).

Sodium adsorption ratio (SAR) indicate to the concentrations of solids dissolved in the soil which including Na, Ca and Mg, presence of high concentrations for these metals decrease the ability of soil to form stable aggregates and loss of the soil structure. Besides, this situation lead to reduce the permeability and infiltration of the soil to infiltrate water for crop production. However, these metal play important role as micronutrient and macronutrient for the plant growth. Ca^{2+} ions are necessary to transport other nutrient in the plant and magnesium is needed for photosynthesis process [18]. Mg^{2+} is an essential element in biological systems because it is present in every cell type in every organism [19]. Moreover, Magnesium nutrient needed in plant with concentrations between 6 and 12% [20]. In washing process conducted in this study has reduced the SAR concentrations to be within the range required for plant growth. The high level of Cl in soil might deteriorate the soil properties and affect yields of different crops [21].

The results for concentrations of heavy metals in the ablution water revealed that some metal ion concentrations increased in the output ablution water at 3 Day HRT and then decreased at 5 Day HRT, while others were varied during the washing process period (Table 3). These variation might be related to the nature of heavy metals in the soil as free ions in the water contents or bounded to the organic matter. The concentrations of Cd^{2+} increased from 0.23 ppb in the ablution water used to 0.4 ppb at 3 Day HRT, and then reduced to 0.04 ppb at 5 Day HRT. Similar findings were recorded for Cr and Zn. These results revealed that the washing process of the peat soil with the ablution water was efficient in reduce of these metals to the minimum concentrations. Cd and Cr are a toxic elements, therefore, the reduction of these metal from the soil might support the plant growth. The concentrations of Cu and Mn increased at 3 Day HRT, and then decreased at 4 Day HRT, thereafter increased again at 5 Day HRT, these variations might be due to the degradation process which take place in the peat soil as a results of microbial activities. The degradation process lead to release some bounded metals on the organic compound into the surrounded medium.

Table 3. Heavy metals concentrations in the ablution water before and after utilization in the washing process

Metal ions	Metal ion concentration (ppb)					
	3 Day HRT		4 Day HRT		5 Day HRT	
	Input ablution water	**Output ablution water**	**Input ablution water**	**Output ablution water**	**Input ablution water**	**Output ablution water**
Cd	0.231	0.401	0.131	0.214	0.021	0.045
Cr	2.319	2.586	2.772	1.984	0.887	0.766
Cu	15.591	14.663	3.601	11.878	11.667	37.469
Mn	2.783	44.031	5.328	4.947	2.972	15.102
Zn	10.569	33.187	6.166	8.422	12.806	8.331
Fe	0.040	0.223	0.051	0.346	0.046	0.297
Mg	0.927	0.838	0.878	0.786	0.860	0.840

The morphological characteristics of *Strelitzia*sp. indicated that it's grow normally no more changes in their morphology. The leaves were between 30 and 200 cm long and between 10 and 80 cm broad. No atrophy was observed in the plant leaves, indicating the ability of plant to grow normally. Moreover, the *Strelitzia*sp. plant was suitable to be planted in the soil pH in range of 5 to 7. It has to mention that the current work might provide an alternative method to cultivation of ornamental landscape plant. By applying the peat soil and ablution water, no more fertilizers are required because the peat soil is very rich with the organic matters and elements necessary for plant growth. The ablution water would be used to irrigate the plants during the dry season, therefore, it will be an alternative for the storm water. The application of ablution water for the irrigation would lead to reduce the heavy metal load.

4 Conclusion

The findings in the present work revealed the efficiency of ablution water in the washing of peat soil and reduction the concentrations of heavy metals. The utilization of ablution water in the washing of peat soil might is an alternative methods for chemical and biological techniques. The ablution water has dual role for reduction of heavy metals contents and irrigation process. More critical analysis including the field scale experiments might provide more details on the efficiency this study.

The authors gratefully acknowledge Ministry of Higher Education of Malaysia for the research project financial support under fundamental research grant scheme (FRGS) vot No. 1574 and prototype research grant scheme (PRGS) vot G004.

References

[1] B. Kalantari, A. Prasad, A. Study of the effect of various curing techniques on the strength of stabilized peat. Transportation Geotechnics, **1** (3), 119-128, (2014)

[2] B. K. Huat, *Organic and Peat Soils Engineering*. Serdang, University Putra Malaysia, Malaysia, (2004)

[3] S. Kazemian, A. Prasad, B. B. Huat, M. Barghchi, A.State of art review of peat: Geotechnical engineering perspective. Int. J. Phys. Sci., **6**(8), 1974-1981, (2011)

 DOI: 10.1051/matecconf/201710305003

[4] J. Abd Rahman, C. M. Chan, A preliminary study of the threshold limit for cementation of peat at different decomposition levels. *Zaytoonah University Int. Engineering Conf. on Design and Innovation in Sustainability 2014 (ZEC Infrastructure 2014).* Amman, Jordan, (2014)

[5] FAO, *Irrigation in the Middle East region in figures. Aquastat Survey.* Edited by Frenken, K. Land and Water Division. Food and Agriculture Organization. FAO Water Reports **34,** (2008)

[6] N. P. Anderson, J. M. Hart, D. M. Sullivan, D. A Horneck, G. J. Pirelli, N. W. Christensen, *Applying lime to raise soil pH for crop production* (western Oregon). Corvallis, Or.: Extension Service, Oregon State University, (2013)

[7] A.L. Salido, K.L. Hastly, J.M. Lim, D.J Butcher, Phytoremediation of arsenic and lead in contaminated soils using Chinese brake ferns (*Pterisvittata*) and Indian mustard (*Brassica juncea*). Int. J. Phytoremed., **5**(2), 89–103, (2003)

[8] M. I. Lone, Z. L. He, P. J. Stoffella, X. E. Yang, Phytoremediation of heavy metal polluted soils and water: progresses and perspectives. J. Zhejiang Univ. Sci. B, **9**(3), 210-220, (2008)

[9] D.Voglar, D. Lestan, Pilot-scale washing of metal contaminated garden soil using EDTA. J. Hazard. Mat., **215**, 32-39, (2012)

[10] D.,Voglar, D. Lestan, Chelant soil-washing technology for metal-contaminated soil.Environ. Technol., **35**(11), 1389-1400, (2014)

[11] APHA, American Public Health Association, Inc. *Standard methods for the examination of water and wastewater* (21st Ed.). New York. USA, (2005)

[12] A.A. Al-Gheethi, I. Norli, J. Lalung, T. Azieda, M. O. Ab. Kadir, Reduction of faecal indicators and elimination of pathogens from sewage treated effluents by heat treatment. Casp. J. Appl. Sci. Res., **2** (2): 29-45, (2013)

[13] Al-Gheethi, A.A., Lalung, J., Efaq, A.N., Bala, J.D.,Norli, I. Removal of heavy metals and β-lactam antibiotics from sewage treated effluent by bacteria. Clean Technol. Environ Policy, **17** (8), 2101-2123, (2015)

[14] N.M. Jais, Mohamed R.M.S.R., A.A. Al-Gheethi, H.M. Amir, Dual role of phycoremediation of wet market wastewater for nutrients and heavy metals removal and microalgae biomass production. Clean Technol Environ Policy. (2016)

[15] Adel A.S. Banana, R.M.S Mohamed, A. A. S. Al-Gheethi, Mercury pollution for marine environment at Farwa Island, Libya. J. of Environ. Health Sci. Eng., **14**(5), (2016)

[16] J. Akrivos, D. Mamais, K. Katsara, A.D. Andreadakis, Agricultural utilization of lime treated sewage sludge. J. Water Sci. Technol., **42**,203-210 (2000)

[17] Z.L. He., Yang, X.E. Stoffella, P.J. Trace elements in agroecosystems and impacts on the environment. J. Trace Elements in Med. Biol., **19**, 125-140 (2005)

[18] Barker, Allen V., and David J. Pilbeam, eds.*Handbook of plant nutrition.* CRC press, (2015)

[19] J. E. Lusk, Williams, R. J. P., Kennedy, E. P. Magnesium and the growth of *Escherichia coli.* J. Biol. Chem., **243**(10), 2618-2624 (1968)

[20] Graham, Ellis R. Explanation of theory and methods of soil testing. (1959)

[21] B. B. Huat, Kazemian, S., Prasad, A.,Barghchi, M. State of an art review of peat: General perspective. Int. J.Phy. Sci., **6**(8), 1988-1996 (2011)

Effectiveness of Indoor Plant to Reduce CO_2 in Indoor Environment

Mohd Mahathir Suhaimi [1,*], *A.M* Leman[1], Azizi Afandi[2], *Azian Hariri*[2], *Ahmad Fu'ad* Idris[2], *S.N. Mohd* Dzulkifli [3], *and Paran* Gani[3]

[1] Faculty of Engineering Technology, Universiti Tun Hussein Onn Malaysia, 86400 Parit Raja, Batu Pahat, Johor, Malaysia
[2] Faculty of Mechanical and Manufacturing Engineering, UTHM, 86400, Parit Raja, Batu Pahat,Johor Malaysia
[3] Faculty of Civil and Environmental Engineering, Universiti Tun Hussein Onn Malaysia, 86400 Parit Raja, Batu Pahat, Johor, Malaysia

Abstract. Modern country strongly emphasizes on indoor air quality (IAQ) because it can effect on human health and productivity. Numerous efforts were performed to make sure that sustainability of IAQ is guaranteed. In the last 4th decade, researchers discover that indoor plants have abilities to reduce indoor air pollution. Generally, plants, carbon dioxide (CO_2), light, and temperature involve in the photosynthesis process. This paper intends to study the effectiveness of seven indoor plants (Anthurium, Dumb Cane, Golden Pothos, Kadaka Fern, Prayer Plant, Spider Plant, and Syngonium) to reduce CO_2 with different light level. This study was conducted in one cubic meter of chamber, and each plant was put into the chamber individually with CO_2 concentration in the chamber is set at 1000±50ppm, and light intensities is set at 300 and 700 lux, while temperature were fixed at 25 ± 1^0C. Based on the results, only the Spider Plant was not able to absorb CO_2 during the test at 300 lux of light intensity. Meanwhile, Prayer Plant performed well when tested at 300 or 700 lux of light intensity compare to other investigates plants. This study can conclude that light intensity play an important role for the plant to absorb CO_2 effectively. All the indoor plants absorbed more CO_2, when the light intensity is increased.

1 Introduction

IAQ issues are common in Malaysia. However there is lack of study, data and local regulation contributing toward health problem especially at Malaysian non-industrial sector [1]. IAQ must be preserved wisely because around 90% of the time people spent are at the indoor environment [2, 3]. Air quality significantly effect on human health and attributed

* Corresponding author: mohd_mahathir@rocketmail.com

the quality of life [4]. Thus, IAQ is essential to living environment [5].

There are many sources of pollutants like total volatile organic compounds (TVOC), even carcinogen (e.g. arsenic, benzene, formaldehyde, and vinyl chloride) that may have in the building which cause deterioration of IAQ [6]. In fact, CO_2 also contribute to indoor air pollution [7]. Generally, CO_2 come from human breathing during respiration process [8]. CO_2 is a IAQ problem when the concentration is increased above threshold limit [9]. At meantime, CO_2 limits that have been set in the building is below than 1000 ppm [10]. Effects of CO_2 on human are include respiratory and heart problems [7]

CO_2 gas is needed by plants to grow, survive, and propagate [11]. In other words, the more CO_2, the more the plant growth become effective [12]. Plants need CO_2 during photosynthesis process, where at the same time also react over the light and water, then produce the glucose (food for plants itself) and release the oxygen [13]. Without plants, humans simply could not exist, because sources of oxygen which is supplied by plants [14].

Human must take along plants to support their life in term of health especially at indoor spaces [15]. Extensive studies had been done on abilities of indoor plant to enhance IAQ for human benefit [16 – 21]. It is one of alternative method to sustain IAQ using non-mechanical or advance technology, and at the same time reduce cost.

Anyhow, plants originally come from outside the building, and of course the environment is different compare to the indoor in term of lighting and temperature. Light intensity and temperature is two factors that can effect plant growth [22 – 25]. Without the proper level of light and temperature that imposed on plant, growing proses will stunted [26, 16]. Nevertheless, certain plants are able to assimilated with different environment to keep on survive [27, 28]. Thus, there is need to conduct study to understand the effect of light intensity toward the effectiveness to reduce CO_2 by indoor plants.

2 Materials and Methods

Plants selected in this study is based on recommendation by former researchers [29, 30]. However, only seven plants were selected based on the availability of the plants at the local area of the study. The age for all indoor plants is one year, and has grown in pot of 17cm diameter and 25cm height with pot mix ration of 2:2:1 (garden soil, compost, and perlite). Indoor acclimated process has been carried out to all investigated plants by putting in indoor environment (25 ± 1^0C) for at least two weeks as suggested by Romanowska and Cavallero [31, 32]. Before the test in chamber is carried out, all indoor plants are watered every day and fertilized once a week.

The level of light intensity during acclimated process is 300 lux (minimum light intensity that specified by Industrial Code of Practice [10]. All plants were not be watered the day before the experimental conducted to avoid disruption value of relative humidity along experimental process [33]. Fig. 1 show all plants selected in this study. Main guideline methods for this study are based on report published by Horticulture Australia in 2011. Nevertheless, this study also refer others former researchers [34 – 38] as a guidance, to ensure accuracy results of the study.

Before the experiment is carried out, all plant selected were assimilated with indoor environment for two months. Assimilation process is important to familiarize the plant life in the building with different levels of temperature and light, as is done by other researchers [39]. During the experiment, chamber was seal with adhesive foam-plastic insulation tape to ensure no leaking. In this chamber consist of small fan (5V) for air circulation, and white fluorescent bulb for the plant live. The two white fluorescent bulbs placed inside the chamber, about 18 cm from the center of the plant. Meanwhile, value of light intensity is based on recommendation by Burchett [40]. Apart from that, a portable IRGA TSI IAQ meter also used to monitor CO_2 concentration in chamber, and was set to record CO_2 reading at 5 minute intervals.

Fig.1. Plants selected that involve in this study

All indoor plants will be tested individually in small scale of chamber (one cubic meter). The concentration of light was taken from minimum and maximum rate (300 & 700 lux) of Malaysian Standard [41]. The temperature during the study is follow to Industry Code of Practice and Prime Minister Office [10, 42], where the reading is 25±1^{0}C. Conversion process of inorganic carbon to organic compounds by plants was made under CO_2 concentration of 1000±10 ppm, because it is limit concentration permitted by ASHRAE [43]. Repeated experiment (three times at different plant) has been performed on every plants species, and time duration for every experiments take about eight hours.

3 Result and Discussion

Fig. 2 to Fig. 8 shows the results of CO_2 reduction by each of the investigated plants. Each of graph contain of two findings based on different light intensity (300 lux and 700 lux) that are imposed on every type of plant. CO_2 reading in the chamber is fluctuated, because apart of plants absorb CO_2 during photosynthesis, it also release CO_2 during respiration [26]. To facilitate the comparison between CO_2 reading by all plants at the chamber, correlation and fitting line by regression is used. The correlation analysis found that all data have strong relation to declare that CO_2 concentration is proportional with time, where the coefficient value is more than 0.5. Meanwhile, regression of fitting line analysis for all data have a statistically significant (all p-values is less than 0.05) [44].

Table 1 shows the total results of this study. According to the experiment that was conducted during 300 lux of light intensity, only Spider Plant is not capable to reduce CO_2 concentrations. While, other plants results that using the same light level (300 lux) shows that Prayer Plant is the most effective plant to reduce CO_2 followed by Symgonium , Kadaka Fern, Golden Pothos, Dumb Cane, and Anthurium with percent of CO_2 reductions is 7.0%, 6.72%, 6.5%, 6.1%, 5.5% and 2.2%. Respectively CO_2 that are reduced from 1000±10 ppm is 71.67 ppm (Prayer Plant), 66.67 ppm (Syngonium), 64.60 ppm (Kadaka Fern), 60.67 ppm (Golden Pothos), 55.4 ppm (Dumb Cane), and 23.67 ppm (Anthurium).

During all plants were tested with 700 lux light intensity shows that all plants are capable to reduce CO_2 concentration, however Spider Plant only reduce the CO_2 concentration with little amount of 0.1% (1.02 ppm) as shown in Fig. 7. At this light intensity, once again the Prayer Plant is the most effective plant to reduce CO_2 concentration followed by Kadaka Fern, Dumb Cane, Syngonium, Golden Pothos, Anthurium and Spider Plant with total reduction at 14.40% (154.63 ppm), 12.48% (123.33 ppm), 11.1% (111.33 ppm), 10.08% (104.00 ppm), 10.03% (101.33 ppm), 9.72% (100.67 ppm) and 0.1% (0.35 ppm) respectively.

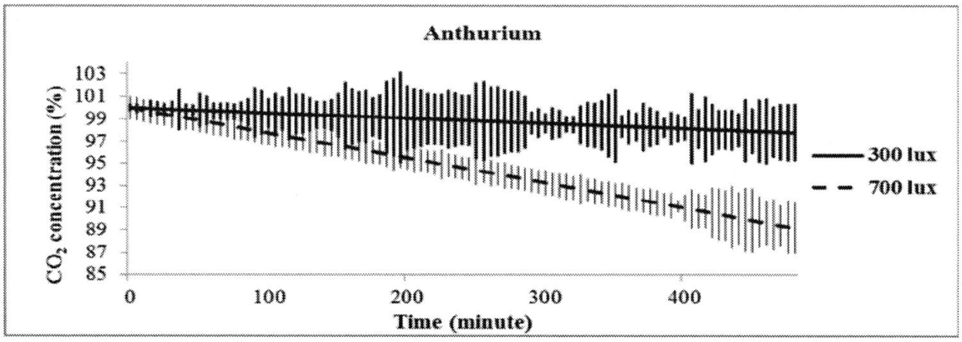

Fig. 2. Graph of CO_2 reduction by Anthurium using 300 lux and 700 lux light intensity

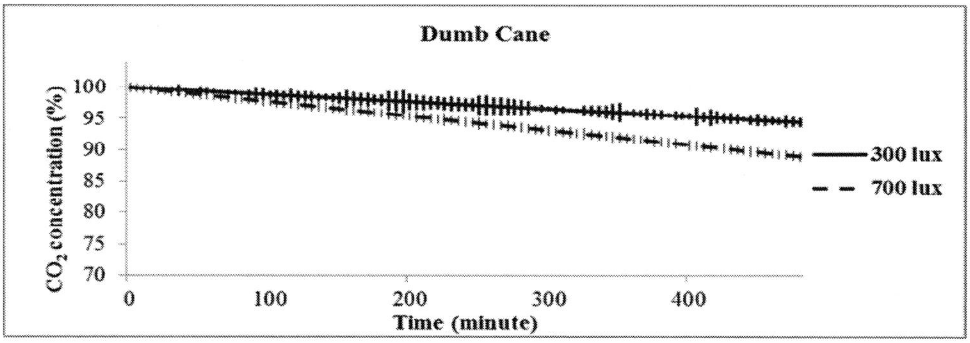

Fig. 3. Graph of CO_2 reduction by Dumb Cane using 300 lux and 700 lux light intensity

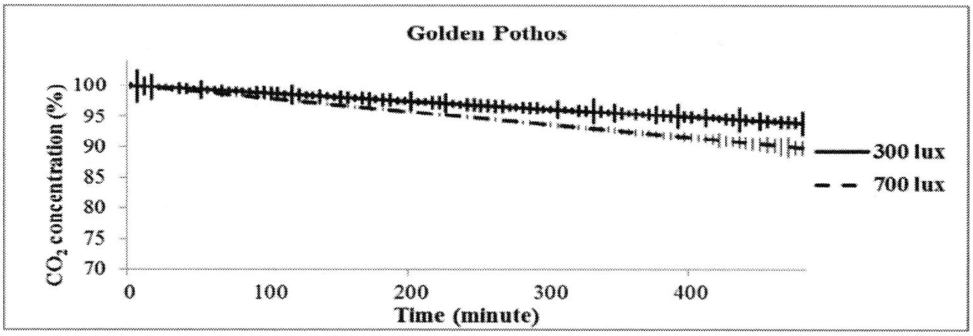

Fig. 4. Graph of CO_2 reduction by Golden Pothos using 300 lux and 700 lux light intensity

According to the all graphs (Fig. 2 to Fig. 8), CO_2 reduction is increased with the increased of the intensity of light. The investigated indoor plants absorb more CO_2 when the light intensity is increased. The results of this study is same agreement with others former researchers like David R. Holding, Torpy, and Wolverton [16, 26, 29]. Even Spider Plant did not show its capability to reduce CO_2 when exposed to 300 lux of light intensity, but CO_2 absorption still occurred when the light intensity increased to 700 lux; as shown in Table 1.

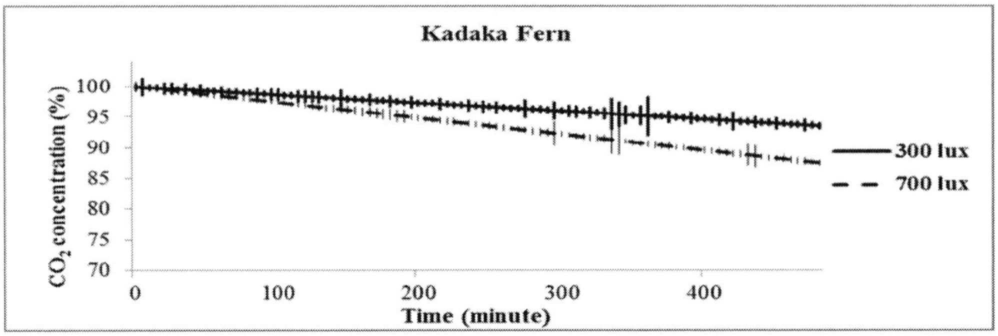

Fig. 5. Graph of CO_2 reduction by Kadaka Fern using 300 lux and 700 lux light intensity

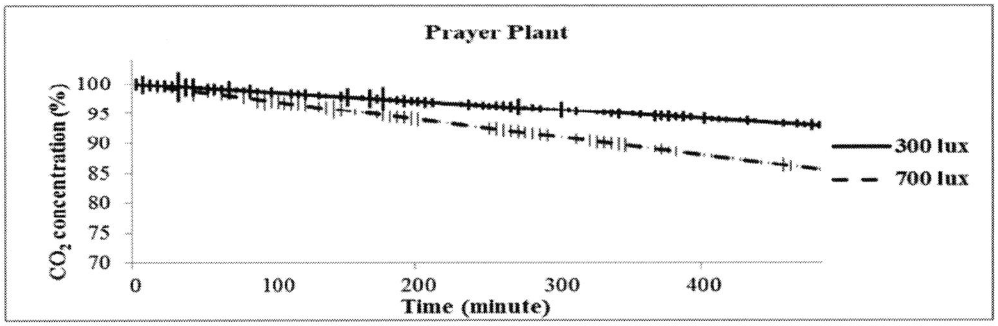

Fig. 6. Graph of CO_2 reduction by Prayer Plant using 300 lux and 700 lux light intensity

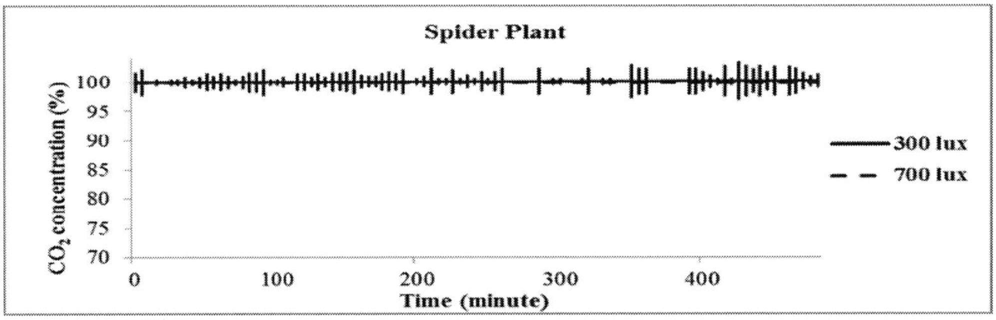

Fig. 7. Graph of CO_2 reduction by Spider Plant using 300 lux and 700 lux light intensity

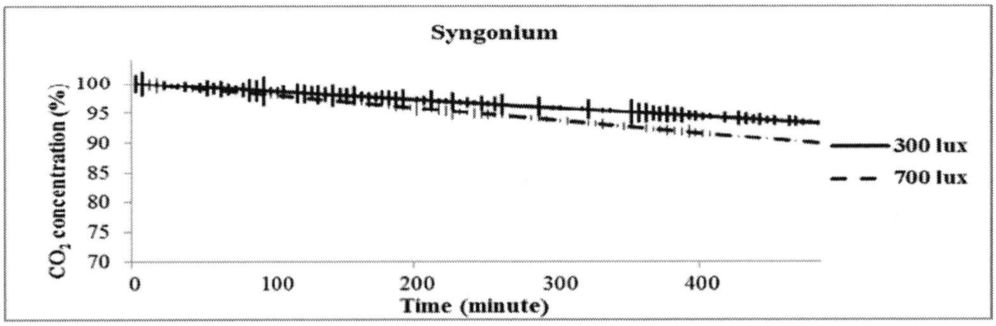

Fig. 8. Graph of CO_2 reduction by Syngonium using 300 lux and 700 lux light intensity

The most interesting in this study is the rate of CO_2 absorption by Anthurium, because the difference percentage of CO_2 absorption when the light level is changed from 300 lux to 700 lux is the highest value compare to others (7.52%). Even though, Anthurium just gave a moderate performance, but it may act as a great agent for CO_2 reduction when the right light imposed on this plant. Meanwhile, others plants like Dumb Cane, Golden Pothos, Kadaka Fern, Prayer Plant, and Syngonium also reduce more CO_2 when lighting level increased as show in Table 1.

Table 1. Rate of CO2 reduction by plants based on diffrent lighting

Plants	Light intensity	Percentage of CO_2 reduction (%)	Total CO_2 reduction (ppm)	Mean of standard deviation
Anthurium	300 lux	2.2	18.30	0.45
	700 lux	10.80	101.00	1.26
Dumb Cane	300 lux	5.50	55.4	0.45
	700 lux	11.10	111.33	0.18
Golden Pothos	300 lux	6.10	60.67	1.10
	700 lux	10.03	101.33	0.63
Kadaka Fern	300 lux	6.50	64.6	1.13
	700 lux	12.48	123.3	0.84
Prayer Plant	300 lux	7.00	71.67	0.64
	700 lux	14.40	154.63	0.62
Spider Plant	300 lux	*0.20	*0.67	1.32
	700 lux	0.10	1.02	1.17
Syngonium	300 lux	6.72	64.67	0.98
	700 lux	10.08	104.00	0.73

*Percentage of CO_2 increased

Table 1 also show the standard deviation (SD) and it is a measure to quantify the amount of variation or dispersion of a set of data values [45]. A low SD indicates that the data points tend to be close to the mean of the set, while a high standard deviation indicates that the data points are spread out over a wider range of values [46]. Based on results, when all plants were tested (three plants for every species) using 300 lux, SD for Anthurium is the highest (1.9), while Prayer Plant have a smallest value (0.64). Besides that, SD for all plants that are used 700 lux found that the Anthurium also get the highest value where the value is 1.26, but this value is small if compare with value SD for Anthurium that are tested with 300 lux. In the case of Anthurium, lighting level seems play an important role to make sure that all plants in the same species acting at the same rate to absorb CO_2. Meanwhile, Syngonium get a smallest value (0.45) when 700 lux was imposed on plant during the experiment.

4 Conclusions

The finding of this study indicated that level of light intensity greatly influenced indoor plants to absorb CO_2 during the experiment. This study reveals that only Spider Plant not able to absorb CO_2 during the test using 300 lux of light intensity. Even light intensity was increase to 700 lux, Spider Plant just reduce small amount of CO_2 (0.35 ppm). It can be concluded that Spider Plant not suitable to use at indoor to reduce CO_2. However, Prayer Plant has total reduction of CO_2 being the highest, whether using 300 lux or 700 lux in this study. Nevertheless, standard deviation for Prayer Plant both 300 lux and 700 lux is quite small, that's mean data points for Prayer Plant tend to be close to the average of the set. Prayer Plant is good for commercialized and real-siuation study, because all finding may easy to expect based on SD value of Prayer Plant.

References

[1] S. H. Ismail, B. Deros, and A. M. Leman, Indoor Air Quality Issues For Non-Industrial Work Place, *IJRRAS*, **5**, 235–244, (2010)

[2] R. Jiang, Effects of Indoor Particulate Matter Pollution from Biomass Fuels Burning : A Case Study in Six Shenyang Households , Northeastern China in Summer Season, Bull. Yale Top. Resour. Inst., **26**, 47–55, (2007)

[3] Mahathir, Leman, and H. Shafii, Monitoring of Indoor Plants Application for Indoor Air Quality Improvement : Proposed Study, *3rd Scientific Conference on Occupational Safety and Health-Sci-Cosh 2014*, Skudai, Malaysia, (2014)

[4] J. Namieśnik, T. Górecki, B. Kozdroń-Zabiega ła, and J. Łukasiak, Indoor air quality (IAQ), pollutants, their sources and concentration levels, Build. Environ., **27** (3), 339–356, (1992)

[5] Y. Su and C. Lin, CO_2 Purify Effect on Improvement of Indoor Air Quality (IAQ) through Indoor Vertical Greening, *Proceedings of the World Congress on Engineering*, London, UK, (2013)

[6] T. Birgitta, Bert, *Helmut, Effects of Indoor Air Pollution on Human Health*, European Concerted Action, Environment Institute, Luxembourg City, Luxembourg, (1991)

[7] S. A. Rice , Human health risk assessment of CO2: survivors of acute high-level exposure and population sensitive to prolonged low-level exposure, *Third Annu. Conf. Carbon Sequestration*, Grass Valley, USA (2004)

[8] P. N. Bierwirth, *Effects of rising carbon dioxide levels on human health via breathing toxicity - A critical issue that remains unapprehended*, (2015), Retrieved on May 21, 2015 from https://www.researchgate.net/directory/publications

[9] R. Prill, *Measuring Carbon Dioxide Inside Buildings – Why is it Important ?*, Energy Proram, WSU, Washington, USA, (2013)

[10] ICOP, *Industry Code Of Practice On Indoor Air Quality*, Kementerian Keselamatn Pekerjaan dan Kesihatan Sumber Manusia, Kuala Lumpur, Malaysia (2010)

[11] B. P. Driessen, *Carbon Dioxide : The Gas of Life*, Committee For A Constructive Tomorrow, Washington, USA, (2013)

[12] J. Robertson, *Carbon Dioxide Feeds the World*, Robertson Science, Queensland, Australia (2011)

[13] D. Whiting, *Plant Physiology : Photosynthesis , Respiration , and Transpiration*, Colorado Master Gardener, Colorado State University, Colorado State, USA (2014)

[14] William Hopkins, *Photosynthesis and respiration*, Chelsea House, New York, USA, (2006)

[15] B.C. Wolverton, *Interior Landscape Plants For Indoor Air Pollution Abatement*, Final Report September, NASA, Washington, USA, (1989)

[16] F. R. Torpy, P. J. Irga, and M. D. Burchett, Profiling indoor plants for the amelioration of high CO2 concentrations, Urban For. Urban Green, **13**(2), 227–233 (2014)

[17] Z. Wang, *Dynamic Botanical Filtration System For Indoor Air Purification*, PhD Thesis, Syracuse University, New York, USA, (2011)

[18] U. Shome, *Comparative efficiencies of indoor plant-based biofilters in removal of selected volatile organic compounds*, Master Thesis, The University of Guelph, Ontario, Canada, (2004).

[19] H. Kim, J. Yang, J. Lee, J. Park, K. Kim, B. Lim, G. Lee, S. Lee, D. Shin, and Y. Lim, House-Plant Placement For Indoor Air Purification And Health Benefits On Asthmatics, Environ. Health Toxicol., **29**, 1–8 (2014)

[20] R. a. Wood, M. D. Burchett, R. Alquezar, R. L. Orwell, J. Tarran, and F. Torpy, The Potted-Plant Microcosm Substantially Reduces Indoor Air VOC Pollution: I. Office Field-Study, Water. Air. Soil Pollut., **175**(1–4), 163–180 (2006)

[21] Z. Wang, J. Pei, and J. S. Zhang, Experimental Investigation of the Formaldehyde Removal Mechanisms in a Dynamic Botanical Filtration System for Indoor Air Purification, J. Hazard. Mater., **7**, 1–22 (2014)

[22] J. P. Allewalt, M. M. Bateson, N. P. Revsbech, K. Slack, and D. M. Ward, Effect of temperature and light on growth of and photosynthesis by Synechococcus isolates typical of those predominating in the octopus spring microbial mat community of Yellowstone National Park., Appl. Environ. Microbiol., **72**(1), 544–550 (2006)

[23] R. V. Ribeiro, E. C. Machado, and R. F. De Oliveira, Temperature response of photosynthesis and its interaction with light intensity in sweet orange leaf discs under non-photorespiratory condition, Ciência e Agrotecnologia, **30**(4), 670–678 (2006)

[24] C. Hew, G. Krotkov, and D. T. Canvin, Effects of Temperature on Photosynthesis and CO_2 Evolution in Light and Darkness by Green Leaves, Plant Physiol, **44**, 671–677 (1969)

[25] B. V Pennisi, *Growing Indoor Plants with Success Table of Contents*. Extension Floriculture Specialist, Georgia, (2012)

[26] A. M. S. David R. Holding, *Plant Growth Processes : Transpiration , Photosynthesis , and Respiration*, EC1268 ed, University of Nebraska–Lincoln Extension, United States, (2013)

[27] J. Sun, X. Sui, H. Huang, S. Wang, Y. Wei, and Z. Zhang, Low Light Stress Down-Regulated Rubisco Gene Expression and Photosynthetic Capacity During Cucumber (Cucumis sativus L.) Leaf Development, J. Integr. Agric., **13**(5), 997–1007 (2014)

[28] E. Weston, K. Thorogood, G. Vinti, and E. López-Juez, Light quantity controls leaf-cell and chloroplast development in Arabidopsis thaliana wild type and blue-light-perception mutants., Planta, **211**(6), 807–815 (2000)

[29] B. C. Wolverton, *How to grow fresh air*, Weidenfeld & Nicolson, London, UK, (1996)

[30] K. Omar-Hor, *1001 plants in singapore*, National Parks Board, Singapore, (2006)

[31] A. Drozak and E. Romanowska, Acclimation of mesophyll and bundle sheath chloroplasts of maize to different irradiances during growth, Biochim. Biophys. Acta - Bioenerg., **1757**(11), 1539–1546 (2006)

[32] L. Cavallero, L. Galetti, D. López, J. McCargo, and I. Barberis, Morphological variation of the leaves of Aechmea distichantha Lem. plants from contrasting habitats of a Chaco forest: a trade-off between leaf area and mechanical, Rev. Bras. Biociencias, **9**(8400), 455–464 (2011)

[33] K. J. Kim, M. J. Kil, J. S. Song, E. H. Yoo, K.-C. Son, and S. J. Kays, Efficiency of Volatile Formaldehyde Removal by Indoor Plants: Contribution of Aerial Plant Parts versus the Root Zone, J. Am. Soc. Hortic. Sci., **133**(4), 521–526 (2008)

[34] Y.-J. Liu, Y.-J. Mu, Y.-G. Zhu, H. Ding, and N. Crystal Arens, Which ornamental plant species effectively remove benzene from indoor air?, Atmos. Environ., **41**(3), 650–654 (2007)

[35] B.C. Wolverton, Plants And Soil Microorganisms Removal of Formaldehyde, Xylene, and Ammonia from the Indoor Environment, J. Mississippi Acad. Sci., **38**(2), 11–15 (1993)

[36] Z. Xu, N. Qin, J. Wang, and H. Tong, Formaldehyde biofiltration as affected by spider plant., Bioresour. Technol., **101**(18), 6930–6940 (2010)

[37] D. Llewellyn and M. Dixon, *Comprehensive Biotechnology*, Elsevier B.V., Guelph, Canada, (2011)

[38] A. Aydogan and L. D. Montoya, Formaldehyde removal by common indoor plant species and various growing media, Atmos. Environ., **45**(16), 2675–2682 (2011)

[39] G. Forti, Biochimica et Biophysica Acta The role of respiration in the activation of photosynthesis upon illumination of dark adapted Chlamydomonas reinhardtii, BBA - Bioenerg., **1777**(11), 1449–1454 (2008)

[40] M. Burchett, *Indoor-plant technology for health and environmental sustainability*, Horticulture Australia, Sydney, Australia (2011)

[41] Malaysian Standard 1525, *Code of Practice on Energy Efficiency and Use of Renewable Energy for Non-Residential Buildings*, Department Of Standard Malaysia, Kuala Lumpur, Malaysia (2007).

[42] Malaysian Prime Minister Department, *Surat Perkeliling Am Bilangan 2 Tahun 2014*. Kerajaan Malaysia, Putrajaya, Malaysia, (2014)

[43] ASHRAE, ASHRAE Standard, *Thermal Environmental Conditions for Human Occupancy*, ASHRAE, Atlanta, USA (2010)

[44] K. M. Hasan, Simple Linear Regression, Am. J. Neuroradiol., **31**(9), 21–45 (2010)

[45] G. Smith, *Standard deviations : flawed assumptions, tortured data, and other ways to lie with statistics*, The Overlook Press, New York, USA, (2014)

[46] M. Centre, *Variance and standard deviation*, (2003), Retrieved on June 28, 2015 from http://www.lboro.ac.uk/media/wwwlboroacuk/content/mlsc/downloads/var_stand_deviat_group.pdf.

The Effectiveness of Antifungal Controlling *Aspergillus Niger* Growth on Plasterboard

Umi Kalthsom Parjo[1] , *Norshuhaila* Mohamed Sunar[1,*], *Abdul Mutalib* Leman[1], *Er Chin* Ming[2], *Paran* Gani[2], and *Menega* Subramaniam[2]

[1]Faculty of Engineering Technology, Universiti Tun Hussein Onn Malaysia, 86400 Parit Raja, Batu Pahat, Johor, Malaysia
[2]Faculty of Civil and Environmental Engineering, Universiti Tun Hussein Onn Malaysia, 86400 Parit Raja, Batu Pahat, Johor, Malaysia

Abstract. Good indoor environmental quality is desired for a healthy indoor environment. The microbial growth under indoor environments contribute to the poor indoor environmental quality that can cause various of health problems. In this study, the applications of three types of antifungals to prevent microbial migration, subsequent growth and bio-deterioration of the substrates. The aim of this research was to evaluate the coating-bio resistance in remediation of indoor fungal using three types of antifungals with different types of wall finishing materials. The treatment was exposed to optimum temperature and relative humidity at 30°C and 90% respectively. The potassium sorbate, zinc salicylate and calcium benzoate are tested against *Aspergillus niger* which is collected from indoor rooms. This study has revealed the growth of *A. niger* are more affected by the potassium sorbate on thick wallpaper, which is the percentage growth are 47%.

1 Introduction

The indoor air quality (IAQ) in any building can be undermined by microbial contaminants (like fungal and bacteria), chemicals (such as carbon monoxide and formaldehyde), and allergens that can create health effects to the building's occupants [1]. These indoor microbial relates to high humidity has become common issues in tropical regions such as Singapore, Thailand, and Malaysia due to the construction building design, engineering and facility engineering. In Malaysia, the Department of Occupational Safety and Health (DOSH) under the Ministry of Human Resources launched the Industry Code of Practice on IAQ (ICOP-IAQ) [2], to identify the parameters distributed to the indoor air contamination. The good IAQ is needed for a healthy indoor work environment. Therefore, antifungal applied in this study was proposed to remediate the indoor fungal which is long lasting and safe method. The aim of this study is to evaluate the coating-bio

* Corresponding author: shuhaila@uthm.edu.my

resistance in remediation of indoor fungal using potassium sorbate (PS), calcium benzoate (CB) and zincsalicylate (ZS) on plasterboard covering with wall finishing which was exposed to optimum temperature and relative humidity. There were limited studies on IAQ have been carried out in a tropical climate [3–6]. In Malaysia, several studies were conducted based on the area of office, educational and institutional buildings such as in newly open hospital, building commissioning, non-industrial workplace and university's clinic [4,5][7,8]. In addition, the similar studies have also been conducted at secondary school which is cause from the fungal DNA. However, there is without proper treatment to remediate *A. niger* on new institutional building university in Malaysia. Thus, the research on different materials and wall finishing is very important as a fundamental element for characterization of indoor fungal growth problem especially at tropical climate and health effects in human.

2 Materials and Methods

2.1 Building materials

The interior finish plays a critical role in the risk management of fungal growth in the indoor environment. This works involved test methods for assessing the fungal resistance of finishing materials for the indoor application (including coatings, plasters, plastics such as silicon caulking and wood panel products). In this study, two types of paints and wallpaper were applied to plasterboard named as acrylic paint (AP), glycerol-based paint (GBP), thin wallpaper (THIN) and thick wallpaper (THICK).

2.2 Antimicrobial activity

The indoor fungus were isolated and identified based on the colony morphology. The samples were enumerated for viable cells in the microbiology laboratory. The incubation temperature was set at 37 ^{o}C for five days. The *A. niger* used were obtained from contaminated indoor coating rooms [9,10]. The 0.005% Tween 20 ((w/v), 0.85% sodium chloride (w/v) and 200 ml distilled water were used to prepare the spores suspension that have been transferred from old to new subculture [11]. Then, the concentration of spores was adjusted to 0.3 to 0.5 x 10^6 spores/ml concentration which were enumerated using the Neubauer chamber

2.3 Sample preparation- coating bio-resistance test

The plasterboards were cut in square size (5 cm X 5 cm) and cleaned with 70% ethanol and dried in laboratory atmosphere. Two types of paint (AP and GBP) were supplemented with 0.03% (w/v) of PS, CB and ZS were applied on plasterboard surfaces. On the wallpaper tested, two different types of wallpaper (THIN and THICK) were glued onto the supports. All the samples were left to dry about two days in the laboratory prior irradiated with UV-lamp for 40 minutes to prevent any contamination. The substrates were placed in a petri dish with medium extract agar (MEA). Each test was carried out in triplicate with the treatment samples and one sample as a control. Then, a 100µl of the spore suspension was placed on the each substrate by a homogenous surface, incubated at 30 ^{o}C ±1°C and a relative humidity (RH) of 90% ±2% in a plant growth chamber (Thermo Stable SWGC-450). The samples were observed at internal 3 days up to the day 27. A visual assessment using naked eye was made according to American Society for Testing and Materials

ASTMD 5590-00 standard method following scale: 0 (none), <10 % (1-trace growth), 10-30% (2-light growth), 30-60% (3-moderate growth) and 60-100% (4-heavy growth) [12].

3 Result and discussion

3.1 Coating bio-resistance test

Fig. 1 shows the percentage of *A. niger* growth on different wall finishing treatment by PS antifungals for 27days.The observation started on day 3 until day 27 of incubation period. The growth of *A. niger* on AP surface, increased from 9% to 99% at the end of incubation days. Meanwhile, *A. niger* extremely grown on the substrates surfaces of glycerol-based paint GBP which started from 73% and increased to 100% after the 12th days. The growth on THIN substrates surfaces was observed in day 3, by 11% and increase to 64%, while the *A. niger* growth on THICK started from 8% until 47%.

In addition, the *A. niger* growth on plasterboard treated by CB can be seen started after three days incubation time. The growth percentage for THIN and THICK was lower which were 10% and 9% respectively. Meanwhile, for AP and GBP the growth are 61% and 68% respectively after 9 days incubated.

The ZS was employed to reduce the *A. niger* growth on the THIN and THICK surfaces other than AP and GBP wall finishing. However, the percentage of growth was increased daily. The *A. niger* slower growth at the 3 days observation for THIN and THICK, which were 20% and 19% respectively. In contrast to AP and GBP were reached up to 64% and 81% respectively. All wall finishing are observed as a heavy growth of the *A. niger* on the plasterboards substrates. According to ASTM D5900-00 standard test method, the growths are 4 scales rating.

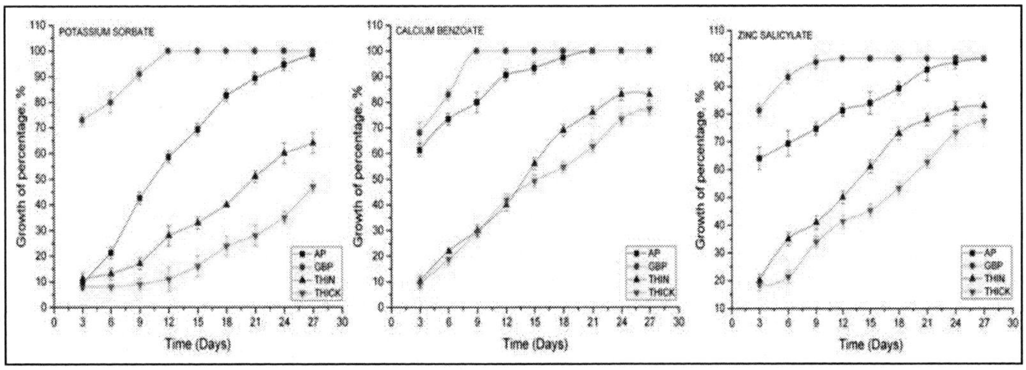

Fig 1. Observation result of fungal growth percentages within 27 days on plasterboards with different types of wall finishing and different antifungals application used

In summary, the plasterboards have shown their different characteristics to react with antifungals applied on different covering surfaces (Fig. 2 to 4). The antifungals to be more activate against *A.niger* on the THICK and THIN, but opposite occurred on AP and GBP. In addition, the potential of the antifungals to remediate indoor fungal are depending on wall finishing surfaces. The wallpaper surfaces had less mould growth compared to the painted surfaces. According to Singh & Chittenden [13], the paints contain organic components that serve as a nutrient source to fungal growth. On the other hand, behind the wallpaper layers may also encourage the fungal growth since they may provide an excellent nutrient source and water activity. Thus, the plasterboard has shown two different conditions; without treatment shows rapidly fungal infestation at 30 °C and 90% relative humidity (RH) demonstrating potentials use as a source of nutrients. At the same condition the plasterboard treated with antifungals revealed slower fungal growth on the treated wallpaper. In addition, in wet conditions, plasterboard gives nutritional conditions to provide the fungal growth quickly [14,15]. A study by Parjo et al [16]stated that celluse-based materials are more sensitive to contamination in comparison to inorganic materials specifically for gypsum, mortar, concrete and others. However, the fungal growth on the building materials occurs when the relative humidity seek a value 85% for plasterboard [17–19]. They also revealed that the influence of the temperature and incubation time also causes the fungal growth.

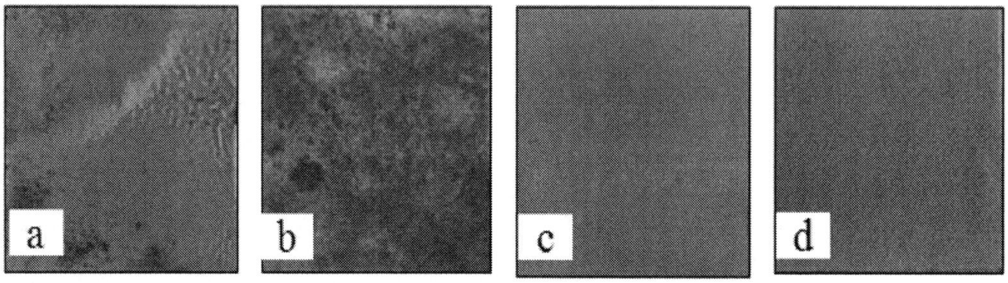

Fig 2. Treated plasterboard with PS on different on different covering after 27 days a)AP, b) GBP, c) THIN d) THICK

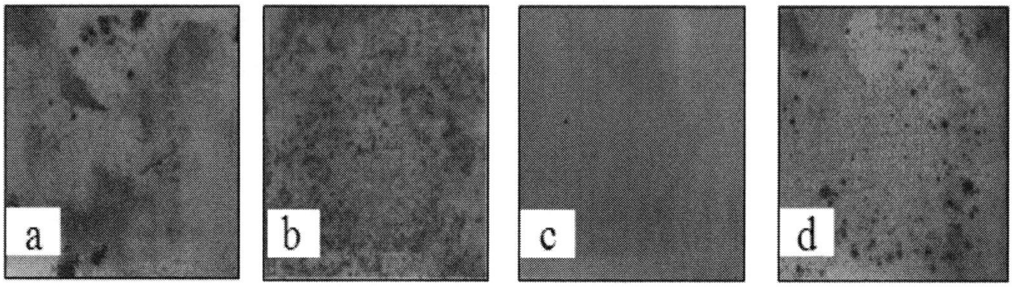

Fig 3. Plasterboard substrates treated treatment by CB with covering after 27 days incubation a) AP b) GBP, c) THIN d) THICK

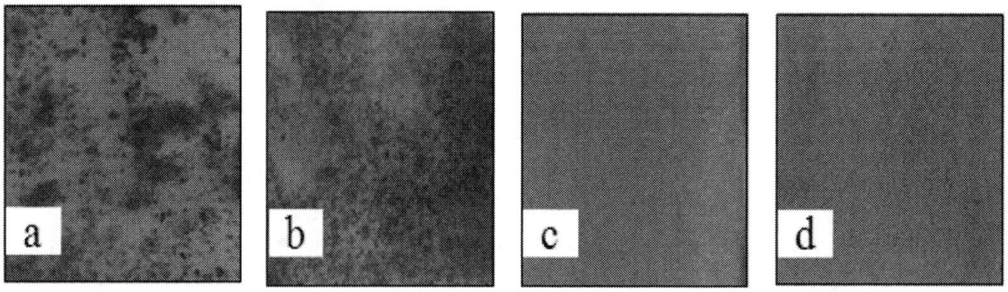

Fig 4. Treated plasterboard (ZS) with different coverings after 27 days a) AP b) GBP, c) THIN d) THICK

4 Conclusions

The composite building materials such as plasterboards are mostly used in building constructions. These materials exposed to fungal growth and may cause lung infection and illness to human. This study uses an environmental friendly antifungal in order to improve the durability of composite product. The indoor air quality must be enhanced to reduce the indoor quality problem in institution's building. The result presented in this study shows that the addition of PS, CB and ZS inhibits the growth of *A. niger* in certain wall finishing was succeed. The *A. niger* growth on THICK are slower until 27 days of observed days compare than AP and GBP wall finishing. Thus, for the conclusion, the use of comparisons was successful, as well as different wall finishing. The use of PS has shown the inhibition of *A. niger* on THICK. On the other hand, the ability of the antifungals capacity as an agent to reduce the fungus should be studied including others building materials.

The authors wish thanks to Universiti Tun Hussein Onn Malaysia, MyBrain15, Centre for Graduate Studies UTHM for sponsorship of this research. The authors would also like to acknowledge all participants involved in this research.

References

[1] G. Adam and H. Duncan, Development of a sensitive and rapid method for the measurement of total microbial activity using fluorescein diacetate (FDA) in a range of soils, Soil Biol. Biochem., **33**, 943–951, (2001)

[2] DOSH. (2010, August 30). Industry Code of Practice on Indoor air Quality, Mechanical Engineering Branch, Retrieved on May 5, 2013, from Mechanical Engineering Branch Official Website: http://www.fnreng.com/images/pdf/IEQGuidelineJKR.pdfl. Code, O. Practice, O.

[3] U.K. Parjo, N.M. Sunar, A.M. Leman, N.I. a. Ideris, P. Gani, Q. Emparan and C.M. Er, Indoor Fungal Treatment by Using Potassium Sorbate as Bio-Resistance Coating for Different Plasterboard Wall Finishings, Appl. Mech. Mater., **773-774**, 1116–1120, (2015)

[4] W.A. Rahman, H. Rosli, S.N. Baharuddin and B. Salleh, Incidence and remediation of fungi in a sick building in Malaysia: A case study, Aerobiologia (Bologna)., **28,** 275–283, (2012)

[5] I. Siti Hamimah, D. Baba and L. Abd Mutalib, Indoor Air Quality Issues for Non-Industrial Work Place, Int. J. Res. Rev. Appl. Sci., **5** ,235–244,(2010)

[6] K.S. Lian, P. Sri, N. Inangda, A. Ramly and P.S.N. Inangda, Sources of Indoor Air Quality Problem in a New Hospital in Malaysia, 6th Int. Conf. Indoor Air Qual. Vent. Energy Conserv. Build. Sendai, Japan. ,(2007)

[7] S. Kumar, S. Bakhda and W. Wc, Fungal remediation and protective antimicrobial treatment of a ten storey grossly contaminated hospital, **1**, 7, (2010)

[8] N. Mohd, R. Noraini, A.M. Leman, A. Sayuti and Z. Abidin, Building and Indoor Environment : A Study on Three Stages of A New Building Commisioning, **3,** 63–68, (2013)

[9] C.M. Er, N.M. Sunar, A.M. Leman, N. Othman, Q. Emparan, U.K. Parjo, P. Gani, N. a. Jamal and N.A. Ideris, The Evaluation of Indoor Microbial Air Quality in Two New Commissioning Higher Educational Buildings in Johor, Malaysia, Appl. Mech. Mater., **773-774,** 1068–1072, (2015)

[10] N. Bellotti, L. Salvatore, C. Deyá, M.T. Del Panno, B. del Amo and R. Romagnoli, The application of bioactive compounds from the food industry to control mold growth in indoor waterborne coatings, Colloids Surfaces B Biointerfaces., **104,** 140–144, (2013)

[11] U.K. Parjo, N.M. Sunar, A.M. Leman and C.M. Er, Effect of Fungal Growth on the Surface of Painted Plasterboards, Adv. Environ. Biol., **9,** 15–19, (2015)

[12] American Society for Testing and Materials, *Standard Test Method for Determining the Resistance of Paint films and Related Coatings to Fungal Defacement by Accelerated Four-Week Agar Plate Assay.* United States: ASTM D5590-00, (2010)

[13] T. Singh and C. Chittenden, Efficacy of essential oil extracts in inhibiting mould growth on panel products, Build. Environ., **45**, 2336–2342, (2010)

[14] T. Verdier, M. Coutand, A. Bertron and C. Roques, A review of indoor microbial growth across building materials and sampling and analysis methods, Build. Environ. **80**, 136–149, (2014)

[15] U.K. Parjo, N.M. Sunar, A.M. Leman, N.I. a. Ideris, P. Gani, Q. Emparan and C.M. Er, Coating Bio-Resistance Test of Different Wall Finishing for Isolated Indoor Fungal Treatment by Using Potassium Sorbate Biocide on Wood, Appl. Mech. Mater., **773-774**, 1384–1388, (2015)

[16] U.K. Parjo, N. M. Sunar, A.M. Leman, P.Gani and Q. Emparan, In-Vitro Indoor Fungal Treatment on Wood by using Potassium Sorbate As Biocide, In: *3rd Sci. Conf. Occup. Saf. Heal. Sci-Cosh 2014,* **12**(1), 151–154, (June 2015)

[17] P. Johansson, T. Svensson and A. Ekstrand-Tobin, Validation of critical moisture conditions for mould growth on building materials, Building and Environment, **62**, 201-209, (2013)

[18] P. Johansson, A. Ekstrand-Tobin, T. Svensson and G. Bok, Laboratory study to determine the critical moisture level for mould growth on building materials, Int. Biodeterior. Biodegrad., **7**,323–32, (2012)

[19] S. Bonetta, S. Mosso, S. Sampò and E. Carraro, Assessment of microbiological indoor air quality in an Italian office building equipped with an HVAC system, Environ. Monit. Assess., **161**, 473–483,(2010)

Greenhouse Gas Emission Estimation for Selected Crops and Rice Cultivation in Urban and Non-Urban Areas of Sarawak

N.Azimah. Bahrum[1,*], *M.* A. Malek [1,2], and *C.S.*Tan[3]

[1]Department of Civil Engineering, Universiti Tenaga Nasional, Jalan IKRAM-UNITEN, 43000 Kajang, Selangor, Malaysia
[2]Centre for Renewable Energy (CRE), Universiti Tenaga Nasional, Jalan IKRAM-UNITEN, 43000 Kajang, Selangor, Malaysia
[3]Institute of Energy Policy and Research (IEPRe), Universiti Tenaga Nasional, Jalan IKRAM UNITEN, 43000 Kajang, Selangor, Malaysia

Abstract. This study presents an updated estimation of annual Greenhouse Gas (GHG) emission from crop and rice cultivations from year 2008 to 2012 in Sarawak. This includes GHG estimation from rice cultivations, biomass and soils organic. The estimation of GHG emission in this study is conducted in accordance to 2006 IPCC guidelines for Agriculture, Forestry and Other Land Use (AFOLU) sector, where uncertainty analyses are incorporated. This study also presents findings at urban and non-urban areas of Sarawak and its comparisons. It is found that, the change in biomass for both urban and non-urban areas in Sarawak increases yearly with significant changes in biomass at urban areas at 160.29×10^3 tons C yr^{-1} as compared to non-urban areas at 45.90×10^3 tons C yr^{-1}. It is also found that, the annual carbon loss from cultivated organic soil at urban areas is higher at 6.17×10^6 tons C yr^{-1} in Miri. Total CH_4 emission from wet paddy in non-urban areas is found to be higher than urban areas at 19.01 Gg CH_4 yr^{-1}. This study would unable relevant authorities to use the values of GHG emission estimated in this study for the purpose of analysis on vulnerability, adaption and mitigation.

1 Introduction

Malaysia has signed the United Nations Framework Convention on Climate Change (UNFCCC) on 9 June 1993 and ratified it on 13 July 1994. Even though, Malaysia has no official obligations toward reducing emission of greenhouse gases (GHG) under the Kyoto Protocol as a signatory to the UNFCCC, Malaysia has an obligation to report its GHG inventory to UNFCCC [1]. Gases such as methane (CH_4) and nitrous oxide (N_2O) are the main two gases in agriculture sector besides carbon dioxide (CO_2). GHG is being estimated as part of the GHG inventory preparation and 2006 IPCC guidelines are used.

* Corresponding author: azimah.bahrum@gmail.com

It covers four (5) main sectors; energy, waste, industrial processes, land use, land use change and forestry (LULUCF) and Agriculture, Forestry and Other Land Use (AFOLU) sector [8]. The GHGs productions and global warming phenomena do not only impact the environment and economy, it also leaves conspicuous effects on human health. This study focuses on GHG emission emitted from cropland sector at urban (Kuching, Miri and Sibu) and non-urban (Limbang, Sri Aman and Betong) areas in Sarawak, Malaysia.

2 Backgrounds

Sarawak is one of the major agriculture producers in Malaysia. The total area classification for agriculture sector is 1,779,190 hectare. This study shows GHG emission from palm oil plantations at 68.2% and rice cultivations at 9.16% which are pertinent agricultural products in Sarawak. Based on review from Sarawak Gross Domestic Product (GDP) in 2012, agriculture sector is the second output of Sarawak by 15.1% [7]. Hence, it is a large potential source of GHG emission in Sarawak. GHG emissions may contribute to global warming phenomena due to its increment in atmospheric concentrations.

Greenhouse gas (GHG) is a gas in the atmosphere that absorbs and emits radiation within thermal infrared range. This process is fundamental because of the greenhouse effect. The greenhouse gases of concerned are carbon dioxide (CO_2), nitrous oxide (N_2O), and methane (CH_4). Formation of greenhouse gases from precursor gases is considered as an indirect emission. Activities in AFOLU such as management of croplands, forests, grasslands and wetlands contributed to anthropogenic GHG emission. In Malaysia, statistics showed that 3% of its GHG emission is from agriculture sector [2]. In order to observe this situation, monitoring efforts in this sector need to be established.

This study aims to estimate the amount and trending of GHG emissions from urban areas of Sarawak namely Kuching, Miri and Sibu as well as GHG emissions from non-urban areas namely Limbang, Sri Aman and Betong for cropland sector at the period of 2008 – 2012 years using 2006 IPCC guidelines. These areas are categorized based on several factors namely, gross domestic product, population size, infrastructural facilities, industrial and economy developments, standard of living and education level.

3 Methodology

Being a developing country, data collection for GHG emission on AFOLU is still at an infant stage in Malaysia. Through data exploration conducted, the state of Sarawak was found to have no regional specific data available, therefore Tier 1 methodology approach in 2006 IPCC guidelines was utilised.

This study uses the Tier 1 methodology of 2006 IPCC (Intergovernmental Panel on Climate Change) Guidelines in selected crops and rice cultivation has been applied to estimate (CO_2) and (CH_4) emissions in Sarawak, within the years from 2008 to 2012 and that gas emission inventory has been developed in this study, based on volume 4, 2006 IPCC Guidelines. The selected crops grown in Sarawak are paddy (wet cultivation), oil palm, rubber, coconut and pepper. Data collection is an important stage in any area of study. In this study emission of CO_2 are estimated from biomass cropland and CH_4 from rice cultivation.

4 Estimation of GHG Emission

Estimation of annual greenhouse gas emission from Cropland remaining Cropland includes biomass, soils organic matter and methane emissions from rice cultivation. In this study, the values GHG emissions are calculated using formula's provided in 2006 IPCC.

4.1 Annual Change in Carbon Stocks in Biomass, ΔC_B (tons C yr^{-1})

Changes in carbon for cropland biomass are estimated from annual rates of biomass gain and loss. The annual change of carbon stocks in biomass calculation for each sub-category cropland is shown in Equation 1, 2006 IPCC guidelines;

$$\Delta C_B = \Delta C_G - \Delta C_L \tag{1}$$

where;

ΔC_G = Annual increase in carbon stocks, tons C yr-[1] which annual area [3] of cropland multiply biomass accumulation rates, G [4].

ΔC_L = Annual decrease in carbon stocks due to biomass loss, tons C yr^{-1} which area of harvest is assumed to consist of 10% of total annual area [3] of cropland multiply biomass carbon loss, L under climate region category of tropical, moist [4].

4.2 Annual Change in Carbon Stocks in Mineral Soils, $\Delta C_{Mineral}$ (tons C yr^{-1})

Soils with high carbon content are called organic soils. Other soils with less carbon content are generally referred to as mineral soils [5]. Calculations on annual change in carbon stocks at each sub-category of cropland are conducted based on Equation 2, 2006 IPCC;

$$\Delta C_{mineral} = \frac{(SOC_o - SOC_{(0-T)}}{D} \tag{2}$$

where;

SOC_0 = Soil organic carbon stock in the last year of an inventory time period, tonnes C.

$SOC_{(0-T)}$ = Soil organic carbon stock at the beginning of the inventory time period, tonnes C.

D = Time dependence of stock change factors.

4.3 Annual Carbon Loss from Cultivated Organic Soils, $L_{Organic}$ (tons C yr^{-1})

Soils with high carbon content are called organic soils. Organic soils are common in wetland soils [5]. Calculations on annual carbon loss from cultivated organic soils at each sub-category of cropland are conducted based on Equation 3, 2006 IPCC;

$$L_{organic} = \sum (A \; x \; EF) \tag{3}$$

where;

A = Area of cropland, ha [1].

EF = Emission factor for climate, tons C yr^{-1} for cultivated organic soils under climatic temperature regime category of tropical/ sub-tropical used is 20.0 - Table 5.6 [4].

4.4 Methane emission from rice cultivation, $CH_{4\,Rice}$ (Gg CH_4 yr^{-1})

The annual amount of CH_4 emitted from a given area of rice and duration of crops grown, water regimes before and during cultivation period, soil type, temperature, and rice cultivar. Estimation on annual methane emissions from rice cultivation is based on Equation 4, 2006 IPCC guidelines;

$$CH_{4Rice} = \sum EF_i \;x\; t\; x\; A\; x 10^{-6} \tag{4}$$

where;

EF_i = A daily emission factor, kg CH_4 ha^{-1} day^{-1} [4].
t = Cultivation period of rice, day is assumed to eight (8) month or 240 days.
A = Annual harvested area of rice, ha yr^{-1} [3].

4.5 Uncertainty Analyses

In this study, the uncertainty analyses calculation is assumed to use a normal distribution, with a 95% confidence interval (CI) of 30% from Mean of Tier 1. The emission factors used in the uncertainty analyses were extracted from 2001 IPCC guidelines [6]. The uncertainty analyses conducted is based on Equation 5 and Equation 6.

$$U_{total} = \sqrt{U_1^2 + U_2^2 + \dots + U_n^2} \tag{5}$$

where;

U_{total} = The percentage uncertainty in the product of the quantities (half the 95 percent confidence interval divided by the total and expressed as a percentage).
U_1 = The percentage uncertainties associated with each of the quantities.

$$U_{total} = \sqrt{\frac{(U_1.x_1)^2 + (U_2.x_2)^2 + \dots\dots + (U_n.x_n)^2}{|x_1 + x_2 + \dots\dots + x_n|}} \tag{6}$$

where;
U_{total} = The percentage uncertainty in the sum of the quantities.
x_1 and U_1= The uncertain quantities and the percentage uncertainties associated with them, respectively.

5 Results

5.1 Carbon Dioxide (CO_2)

CO_2 are determined from estimating changes in carbon stocks for biomass cropland from year 2008 till 2012 as shown in Fig. 1. In this study, the change in biomass is calculated for perennial crops in urban (Kuching, Miri and Sibu) and non-urban (Limbang, Sri Aman and

Betong) areas of Sarawak. The selected crops in Sarawak are pepper, coconut, oil palm, paddy and rubber. It is found that, the change in biomass for both urban and non-urban areas increases yearly. The change in biomass at urban areas is more significant as compared to non-urban areas. Miri which is categorized as an urban area shows an amount of carbon stocks in biomass at 160.29×10^3 tons C yr^{-1} in 2012. Limbang which is categorized as non-urban area exhibited small changes for carbon stocks in biomass at 13.03×10^3 tons C yr^{-1} in year 2012.

In this study, the total uncertainties trend on CO_2 emission from biomass cropland is approximately 5.87% for Kuching, 2.20% for Miri, 16.73% for Sibu, 9.48% for Limbang, 21.13% for Sri Aman and 25.17% for Betong. The emission factors used in the uncertainty analyses were extracted from 2006 IPCC guidelines.

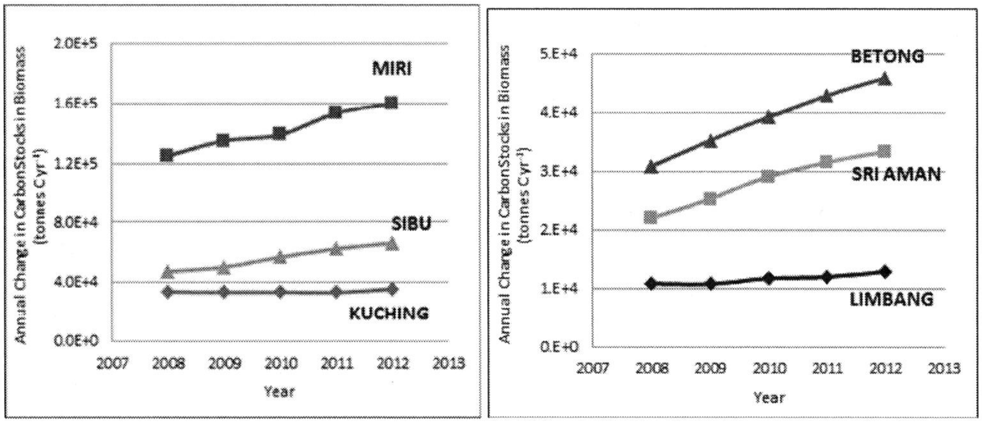

(a) (b)

Fig. 1. Total annual change for carbon stocks in biomass (tons C $yr-^1$) from perennial crops at (a) urban and (b) non-urban areas of Sarawak for year 2008 – 2012.

There have two types of soils organic, which is mineral soils and organic soils. For this study, types of soil organic for oil palm in Sarawak are under organic soils. The Fig. 2 showed the result of annual carbon loss of cultivated organic soil from year 2008 till 2012. In order to assign an annual emission factor to estimate the losses of C, in accordance to the area of drained and managed organic soils under each climate type, it is multiplied by the associated emission factor to derive an estimate of annual CO_2 emissions. In this study, it is found that, the annual carbon loss from cultivated organic soil for palm oil at urban areas is higher where the annual carbon loss in Miri is at 6.17×10^6 tons C yr^{-1} in year 2012. The annual carbon loss from cultivated organic soil at non-urban area is found to be lower, where carbon loss in Limbang at year 2012 is found to be at 343.01×10^3 tons C yr^{-1}.

Other soils with less carbon content are generally referred to as mineral soils. The selected crops in Sarawak are pepper, coconut, rubber and paddy. Fig. 3 shows the total value annual changes in carbon stocks for mineral soils of perennial crops in Sarawak. It is found that from year 2008 to 2012 the net change in mineral soils for perennial crops at Kuching, Miri and Sibu respectively is 14692.74 tonnes C yr^{-1}, 2770.39 tonnes C yr^{-1} and 6497.46 tonnes C $yr^{-}1$. In this study, it is found that, the net change in mineral soils for perennial crops at Kuching is higher than Miri and Sibu. While, it is found at Limbang, Sri Aman and Betong respectively is 1250.72 tonnes C yr^{-1}, 2953.72 tonnes C yr^{-1} and 10991.82 tonnes C yr^{-1} for non-urban areas. In this study, it is found that, the net change in mineral soils for perennial crops at Betong is higher than Limbang and Sri Aman.

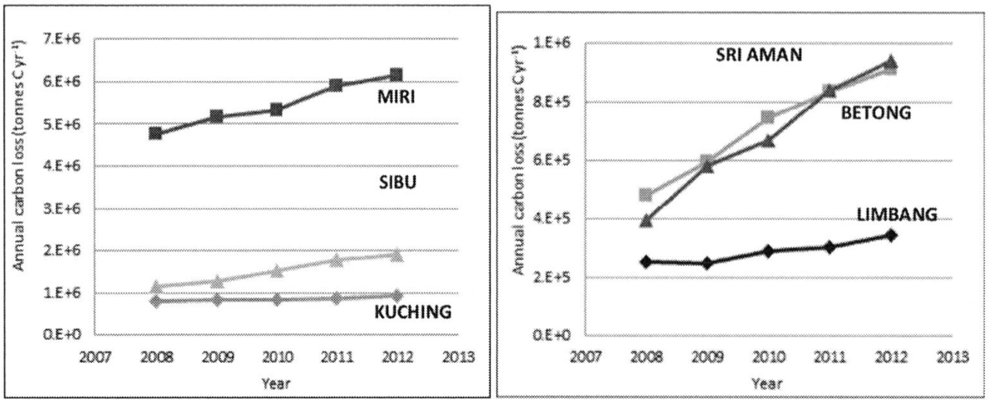

(a) (b)

Fig. 2. Total annual carbon loss from cultivated organic soils, $L_{Organic}$ (tonnes C yr^{-1}) for palm oil in (a) urban and (b) non-urban areas of Sarawak for year 2008 – 2012.

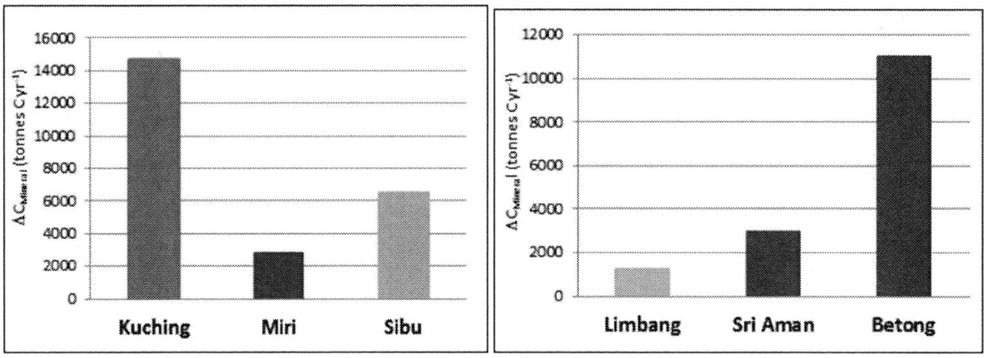

(a) (b)

Fig. 3. Total annual change in carbon stocks for mineral soils, $\Delta C_{Mineral}$ (tonnes C yr^{-1}) of perennial crop in (a) urban and (b) non-urban areas of Sarawak for year 2008 – 2012.

In this study, the total uncertainties trend on CO_2 emission from soils organic is approximately 1.70% for Kuching, 1.83% for Miri, 4.72% for Sibu, 3.88% for Limbang, 2.70% for Sri Aman and 9.92% for Betong. 2006 IPCC guidelines showed a low uncertainty factor for estimation of area under different cropping systems at less than 10% (<10%) [4].

5.2 Methane (CH_4)

Estimation of CH_4 emission is obtained from rice cultivation practiced in Sarawak for wet paddy. Wet paddy needs sufficient amount of water and requires stagnant water conditions. In Sarawak, the cultivation period for rice conditions is 240 days for wet paddy. Based on calculations conducted in this study, it is estimated that the production from wet paddy in urban area for 2012 at Kuching, Miri and Sibu respectively is 0.60 Gg CH_4 yr^{-1}, 1.59 Gg CH_4 yr^{-1} and 1.77 Gg CH_4 yr^{-1}. While, for non-urban area for 2012 at Limbang, Sri Aman and Betong respectively is 0.53 Gg CH_4 yr^{-1}, 1.91 Gg CH_4 yr^{-1} and 1.25 Gg CH_4 yr^{-1}.

Based on the results obtained in this study, it can be concluded the emission from 2008 to 2012 in urban areas has shown increasing trend but in non-urban areas has shown decreasing trend. Fig. 4 illustrates the CH_4 emission from wet paddy for both urban and non-urban areas in Sarawak for year 2008-2012.

In this study, the total uncertainties trend on CH_4 emission from rice is approximately 10.41% for Kuching, 2.28% for Miri, 7.77% for Sibu, 2.94% for Limbang, 3.22% for Sri Aman and 2.24% for Betong. It is found that the total uncertainties trend in GHG emissions from cropland for both urban and non-urban areas in Sarawak are less than 50%.

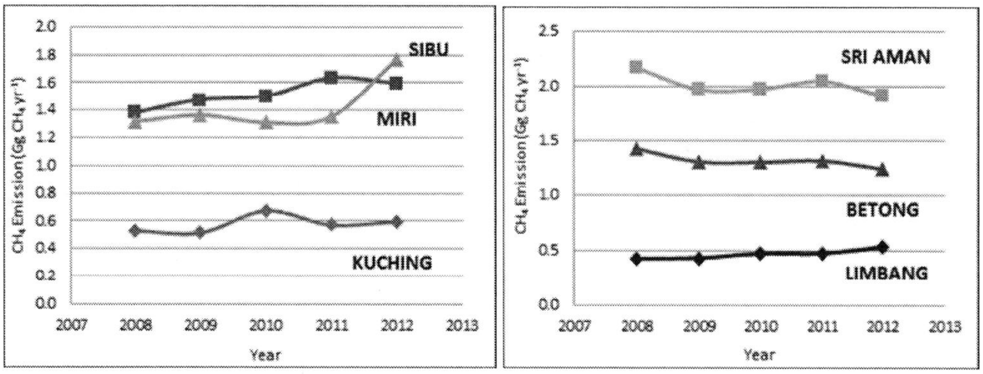

(a) (b)

Fig. 4. Total annual CH_4 emission from rice cultivation (Gg CH_4 yr^{-1}) for wet paddy in (a) urban and (b) non-urban areas at Sarawak for year 2008 – 2012.

6 Conclusions and Discussions

The study determines GHG emission from cropland at both urban and non-urban areas in Sarawak. The total carbon released from cropland heavily depended on the crop types. In order to determine the amount of emission from perennial crops, estimation on changes of carbon stock in biomass cropland is conducted. As illustrated in Fig. 1, the trend of changes in biomass carbon stock from year 2008 till 2012 for perennial crops for urban area (Miri) in Sarawak, shows total oil palm are the highest emission of CO_2 at 682.644×10^3 tons C yr^{-1}. It is reported that oil palm plantations are being greatly expanded, largely in Sarawak and Borneo Island [6].

The annual amount of CH_4 emitted from rice cultivation depends on the actual area of paddy being cultivated, number and duration of paddy grown, water regimes before and during cultivation period, as well as organic and inorganic soil amendments practiced [7]. For Sarawak, in the period of 2008 till 2012, it is found that total wet paddy contributed to the highest emission of CH_4 is 1.91 Gg CH_4 yr^{-1} at Sri Aman. While, hill paddy produces almost negligible amount of CH_4 emission. Hill paddy or upland rice fields, are not flooded, therefore this type of paddy do not produce significant quantities of CH_4.

Finally, the results presented in this study showed the emissions from CO_2 and CH_4. These gases trap heat energy from the sun. GHG that exist in the atmospheric help maintain and stabilize the earth temperature. However, if the temperature of these gases increases excessively, it will cause global warming. Reduction of emissions from these gases cannot be done dramatically. Methods in managing land and growing crops need to be harmonized, for example by reducing the amount of nitrogen in fertilizing crop. Increase usage of nitrogen can lead to higher N_2O emission. The increase of pasture quality to enhance animal productivity can reduce the amount of CH_4 emitted per unit of animal product. Practicing organic farming technique could result in less CH_4 being emitted to the atmosphere. In order to observe this situation, the GHG estimation established in this study can therefore be used for monitoring purposes.

The authors would like to thank UNITEN R&D Sdn. Bhd for Grant U-EN-CR-14-12 that enabled collaboration which leads to the development of this paper. Any of the opinions, findings and conclusions expressed in this paper are those of the authors and do not reflect the views of UNITEN R&D Sdn. Bhd.

References

[1] NC2, *Malaysia's Second National Communication (NC2) to the UNFCCC*, Ministry of Natural Resources and Environment, Malaysia, (2011)

[2] Sarawak Fact and Figures, State Planning Unit, Chief Minister's Department, *Sarawak Agriculture Statistics (2008 - 2012)*, from http://www.spu.sarawak.gov.my

[3] Department of Agriculture, Sarawak, *Sarawak Agriculture Statistics (2008-2012)*, Retrieved on June 25, 2016, from http://www.doa.sarawak.gov.my/modules/web/index.php

[4] IPCC, *2006 IPCC Guidelines for National Greenhouse Gas Inventories*, Prepared by the National Greenhouse Gas Inventories Programme, Eggleston H.S., Buendia L., Miwa K., Ngara T. and Tanabe K. (eds).Published: IGES, Japan, (2006)

[5] BUR, *Biennial Update Report to the UNFCCC*, Ministry of Natural Resources and Environment, Malaysia, (2015)

[6] IPCC, *Climate Change 2001*: The Scientific Basis - Contribution of Working Group I to the Third Assessment Report of the Intergovernmental Panel on Climate Change (IPCC). Cambridge, UK, Cambridge University Press, (2001)

[7] H. Neue, R. L. Sass, *Trace Gas Emissions from Rice Fields*, Int. J. of Environmental Science Research, **48**, 119-147, (1994)

[8] IPCC, *Revised 1996 IPCC Guidelines for National Greenhouse Gas Inventories*: **1-3**. Houghton, J.T., Meira Filho, L.G., Lim, B., Tréanton, K., Mamaty, I., Bonduki, Y., Griggs, D.J. and Callander, B.A. (Eds). Intergovernmental Panel on Climate Change (IPCC), IPCC/OECD/IEA, Paris, France, (1996)

Methane emissions from paddy cultivation and livestock farming in Sarawak, Malaysia

E.K. Peng[1,*], *M.*A. Malek[1,2] , and *C.S.*Tan[3]

[1]Department of Civil Engineering, Universiti Tenaga Nasional, Jalan IKRAM-UNITEN, 43000 Kajang, Selangor, Malaysia
[2]Centre for Renewable Energy (CRE), Universiti Tenaga Nasional, Jalan IKRAM-UNITEN, 43000 Kajang, Sekangor, Malaysia
[3]Institute of Energy Policy and Research (IEPRe), Universiti Tenaga Nasional, Jalan IKRAM-UNITEN, 43000 Kajang, Selangor, Malaysia

Abstract. In this study, implementation of Tier 1 methodology of 2006 IPCC (Intergovernmental Panel on Climate Change) Guidelines in paddy cultivation and livestock farming has been applied to estimate methane emissions in Sarawak, Malaysia within the years from 1998 to 2009. Methane emission inventory has been developed in this study, based on volume 4, 2006 IPCC Guidelines. Based on cultivation area and livestock population data as input to Tier 1 methodology, variations in paddy cultivation area and amount of livestock has been identified as the main contributor to emissions of methane. Methane emissions increased from 1.61 to 1.72 Gg CH_4/year during 1998 to 1999. Based on results obtained, the outcomes show that there would be a significant drop of methane emission from buffalo and sheep. Although there are gain and loss in emissions from enteric fermentation, drastic reduction is observed from 0.65 Gg CH_4/year in 1998 to 0.44 Gg CH_4/year in 2009 as well as 0.05 Gg CH_4/year to 0.02 Gg CH_4/year for buffalo and sheep respectively. Simultaneously, methane emissions from manure management of buffalo has decreased from 0.024 Gg CH_4/year in 1998 to 0.016 Gg CH_4/year in 2009 while for sheep, its emission from manure management dropped from 0.002 Gg CH_4/year in 1998 to 0.0007 Gg CH_4/year in 2009. Overall emission from paddy cultivation can be considered in upward trend due to gain from 1998 at 1.61 Gg CH_4/year to 1.67 Gg CH_4/year in 2009. As an addition, significant rise in methane emission by 0.24 Gg CH_4/year from 2000 to 2006 as well as 0.1 Gg CH_4/year from 2007 to 2009 show momentum gaining in enteric fermentation of cattle. It also indicates future increment in methane emission from cattle which coherently affects the state's emission level. As for emissions from manure management, emissions from cattle, goat and deer are gaining momentum in Sarawak.

[*] Corresponding author: pengek@gmail.com

1 Introduction

The agriculture sector consists of different variety of activities that may emit greenhouse gases (GHG) directly or indirectly. This includes rice-based production systems with manure management [1]. In Asia, areas of rice cultivation are gaining momentum to meet the food requirement of increasing population. Consequently, methane (CH_4) emission were higher than in other continents [2].

Both the domestic enteric fermentation and manure management are considered key categories in agriculture sector for both the initial and second national communication of Malaysia. Even though Malaysia was a net sink as reported in second national communication (NC2), increase in methane emission by 13% from initial NC to NC2 indicated emission has been increased throughout the years [3]. The highest source of methane emission in Malaysia is contributed by landfill which made up of 47% of the total emission. Meanwhile methane emission from rice production and enteric fermentation in domestic livestock contributed 4% and 2% of the total emission, respectively. Malaysia has ratified the Kyoto Protocol in September 2002, and as a developing country, it has been categorised into a non-Annex 1 members. This means, Malaysia is not obligated to reduce GHG emissions by any specific percentage. Nevertheless, during the Copenhagen Summit 2009, Malaysia Prime Minister has announced a conditional voluntarily target of 40% reduction in CO_2 intensity of Malaysian GDP by 2020 from a 2005 baseline, subjected to the availability of technology and finance from the Annex I countries.

The conditional voluntarily is applied for all sectors in the country namely energy, industrial processes, agriculture, land use, land use change and forestry (LULUCF) and waste. This paper aims to provide methane emissions from paddy cultivation and livestock farming in Sarawak. The reason methane is chosen is because the gas constituted 23.5% of the total GHG emissions in Malaysia, which is second largest contributor after carbon dioxide that contributed 75.1% of the total [3].

2 Overview of paddy cultivation and livestock farming in Sarawak

The ensuing section proves that rice cultivations and livestock farming are one of the main source of GHG emission in Malaysia, mainly in the form of methane (CH_4). As mentioned, emission from rice production and enteric fermentation in domestic livestock contributed 4% and 2% respectively in the total methane emission of Malaysia.

The percentage areas for each land use in Sarawak are shown in Fig. 1. The total land use area is recorded as 12 million hectares. It can be concluded that dry forest land contributed to the highest percentage area in Sarawak, followed by land use area of wet and hill paddy, other crops cultivations and swamp forests. Both wet and hill paddy cultivations occupied large portions of land use area and thus, contributed as major source of agricultural on GHG emission in Sarawak. Table 1 exhibits estimation of selected livestock population in Sarawak from 1998 to 2009.

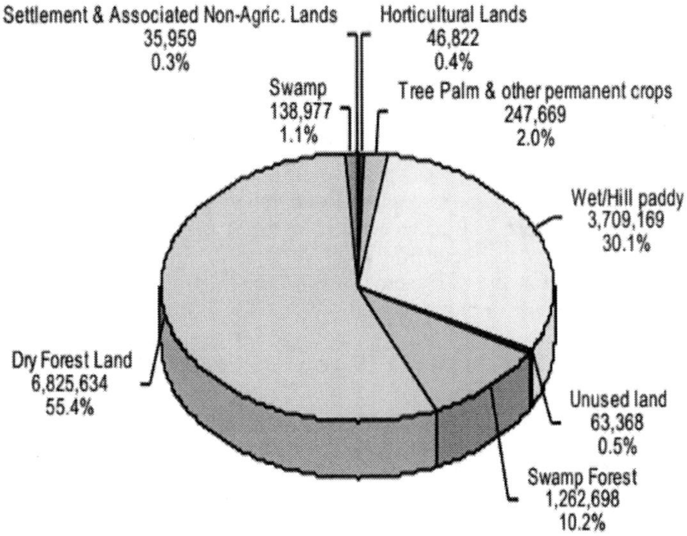

Fig. 1. Area and land use in Sarawak [4, 6].

Table 1. Number of livestock in Sarawak [5, 6].

Year	**1998**	**2002**	**2006**	**2009**
Cattle	8495	9532	12198	13983
Buffalo	11787	10924	8765	7973
Sheep	10189	5374	3408	3553
Goat	10214	9078	11146	14656
Deer	1384	3068	4434	1489

3 Material and methods

Tier 1 methodology of [7] is employed in this study due to difficulties in moving to higher tier methodology resulting from lack of detailed on local data and emission factors [8]. The origin of data used here are from Sarawak's Chief Minister's Department, State Planning Unit, and Publication of Sarawak Facts and Figures. Sarawak is one of the states in Malaysia, situated next to Sabah. It is located at the Eastern Region of Malaysia.

Rice fields in Sarawak are categorized as hill and wet paddy. However, hill paddy are not treated as source of methane emission due to the fact that they are not flooded for adequate time. Hence, only wet paddy is classified as one of the major sources of methane emission. As for livestock, by considering the changes in livestock population during year 1998-2009, a framework capable of estimating CH_4 emissions from enteric fermentation and manure management systems based on [7] was established in this study. This study aims to develop a GHG inventory which focuses on CH_4 emissions from livestock farming of cattle, buffalo, sheep, goat and deer plus wet paddy cultivation during the period of 1998-2009 in Sarawak.

Similar issues has been addressed and published, but not on GHG emissions specifically from rearing of cattle, buffalo, sheep, goat and deer as well as wet paddy cultivation in environment of Sarawak. Being a developing country, Sarawak is still struggling on Tier 1 methodology in estimating its GHG emissions.

3.1 Methodology of emission calculation from wet paddy cultivation

The 2006 IPCC Guidelines proposed a model to obtain yearly emissions based upon the baseline emission factor EF_c:

$$EF_i = EF_c \cdot SF_w \cdot SF_p \cdot SF_o \cdot SF_{s,r} \tag{1}$$

where: EFi: adjusted daily emission factor for a particular harvested area, EF_c: baseline emission factor for continuously flooded fields without organic amendments; SF_w: scaling factor to account for the differences in water regime during the cultivation period; SF_p: scaling factor to account for the differences in water regime in the pre-season before the cultivation period; SF_o: scaling factor should vary for both type and amount of organic amendment applied; SFs,r: scaling factor for soil type, rice cultivar, etc., if available.

A default average baseline emission (EF_c) of 1.30 kg-CH_4.ha^{-1}.d^{-1} is suggested in IPCC 2006 Guidelines with an error range of 0.80-2.20. All scaling factors affecting EF_c were extracted from 2006 IPCC Guidelines based on local crop and water management practices in Sarawak: rain-fed conditions (it is assumed that there is drought prone; non-flooded pre-season of more than 180 days; with farm yard manure as organic amendment). SF_o is the scaling factor reflecting both the type and amount of organic matter applied. Equation (2) determines SFo (IPCC, 2006):

$$SFo = (1 + ROA_i \cdot CFOA_i)^{0.59} \tag{2}$$

where: SFo: scaling factor for both type and amount of organic amendment applied; ROA_i :application rate of organic amendment i, in dry weight for straw and fresh weight for others (in tonne ha-1); $CFOA_i$:conversion factor for organic amendment i (in terms of its relative effect with respect to straw applied shortly before cultivation) [7]

Due to data constraint, in this study it is assumed that application rate of organic amendment is equal to one (1) as ROA_i. On the other hand, the scaling factors used for wet paddy cultivation are:

SFw : 0.25;SFp : 0.68; ROA_i : 1.0;$CFOA_i$: 0.14;ROA_i : 1; SFs,r : 1

Equation 3 is used to estimate CH_4 emission from rain-fed conditions:

$$CH_4\ Rice = \sum i\ [(EF_i \bullet t_i \bullet A_i \bullet 10^\wedge 6)] \tag{3}$$

In this study, it is assumed that the cultivation period of paddy in Sarawak is 4 months due to one (1) planting schemes per year practiced. Besides that, it is also assumed that only one (1), namely i condition of ecosystem, water regime, type of organic amendments are practiced for the whole year. These assumptions are made since Sarawak does not experience four (4) seasons throughout the year.

3.2 Methodology of emission calculation from enteric fermentation

Formula used for calculating CH_4 emissions from enteric fermentation is shown in Equation (4). In developing countries, the recommended value for enteric fermentation emission factor is presented in Table 2 and Table 3 for cattle and other livestock respectively.

$$Emissions = EF \cdot [(N/10^\wedge 6)] \tag{4}$$

where: Emissions = methane emissions from Enteric Fermentation, Gg CH_4 yr^{-1} ; EF = emission factor for the defined livestock population, kg CH_4 $head^{-1}$ yr^{-1}; N = number of head of livestock species in the country,

Table 2. Tier 1 enteric fermentation emission factor for cattle (kg CH_4 $head^{-1}$ yr^{-1}) [7].

Regional characteristics	Cattle category	Emission factor[1,2]	Comments
Asia: Small commercialised dairy sector. Most cattle are multi-purpose, providing draft power and some milk within farming regions. Small grazing population. Cattle of all types are smaller than those found in most other regions.	Dairy	68	Average milk production of 1,650 kg $head^{-1}$ yr^{-1}
	Other Cattle	47	Includes multi-purpose cows, bulls, and young
[1] IPCC Expert Group, values represent averages within region, where applicable the use of more specific regional milk production data is encouraged.			
[2] The following assumptions have been made in deriving these values: i) mature weights of animals have been used; ii) cows have been assumed to be non-lactating as lactation levels were low and, iii) the mix of bulls and castrates among "males" was undetermined as Cfi value for castrates was not specified.			

Table 3. Tier 1 enteric fermentation emission factors for buffalo, sheep, goat and deer (kg CH_4 $head^{-1}$ yr^{-1}) [7].

Livestock	*Developing countries
Buffalo	55
Sheep	5
Goats	5
Deer	20
*All estimates have an uncertainty of ±30-50% Source: Emission factor for buffalo from Gibbs and Johnson (1993); Deer from Clark et al., 2003.	

Note that the emission factors chosen are based on factors similar to condition in Sarawak. Also, emission factor of other cattle is chosen in this study rather than dairy cattle due to the fact that not all cattle in Sarawak are for the purpose of milk and beef yielding but the cattle are mainly consist of mixture of young, multi-purpose cows and bulls.

3.3 Methodology of emission calculation from manure management

Equation (5) shows calculation of CH_4 emission from manure management while Table 4 shows manure management methane emission factors by temperature for cattle and buffalo. Meanwhile, Table 5 shows manure management methane emission factors by temperature for sheep, goat and deer.

$$CH_{4Manure}=\sum[(EF \cdot N)/(10^{\wedge}6)] \quad (5)$$

where: $CH_{4Manure}$ = CH_4 emissions from manure management, for a defined population, Gg CH_4 yr^{-1}; EF = emission factor for the defined livestock population, kg CH_4 $head^{-1}$ yr^{-1}; N = the number of head of livestock species in the country

Table 4. Manure management methane emission factors by temperature for cattle and buffalo (kg CH_4 $head^{-1}$ yr^{-1}) [7].

Regional characteristics	Livestock species	CH_4 emission factors by average annual temperature (^0C)
		Warm
		$\geq$28
Asia: About half of cattle manure is used for fuel with the remainder managed in dry systems. Buffalo manure is managed in drylots and deposited in pastures and ranges.	Dairy Cows	31
	Other Cattle	1
	Buffalo	2

Table 5. Manure management methane emission factors by temperature for sheep, goat and deer (kg CH_4 $head^{-1}$ yr^{-1}) [4].

Livestock	CH_4 emission factor by average annual temperature (^0C)
	Warm (>25)
Sheep	0.20
Goats	0.22
Deer[a]	0.22
The uncertainty in these emission factors is ±30%. [a] Sneath *et al.* (1997)	

For cattle and buffalo, manure management methane emission factors are obtained from warm temperature, which is $\geq 28^0C$. This is justified by the average temperature in Sarawak which is 32-33^0C Emission factor of other cattle is chosen in this study.

4 Results and discussions

The annual emission of methane from paddy cultivation together with enteric fermentation and manure management in Sarawak is illustrated in Fig. 2 to 4 respectively. Unit for methane emission is Giga gram (Gg) CH_4/year.

DOI: 10.1051/matecconf/201710305007

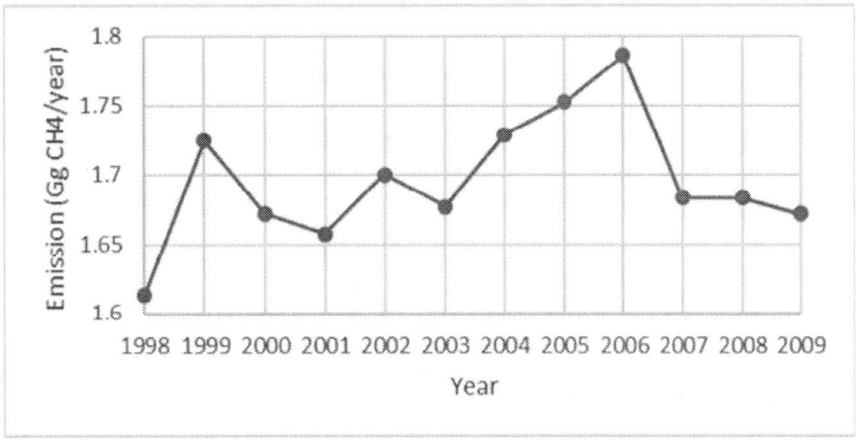

Fig. 2. Annual methane emission from paddy cultivation in Sarawak.

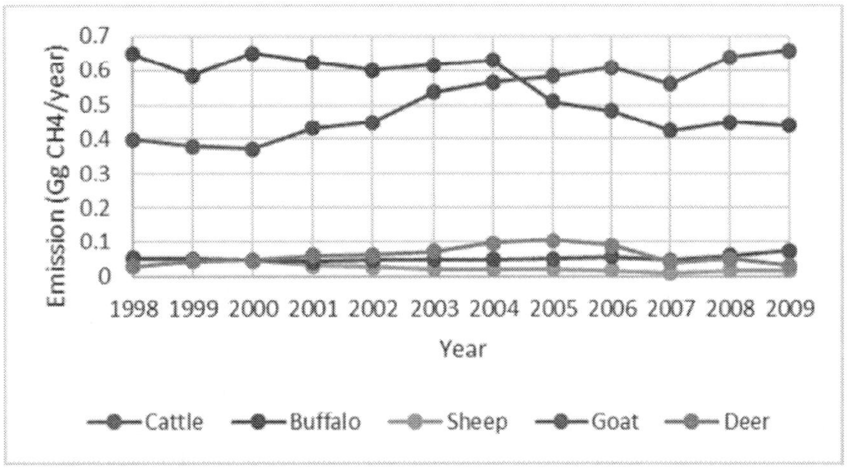

Fig. 3. Annual methane emission from enteric fermentation of livestock in Sarawak.

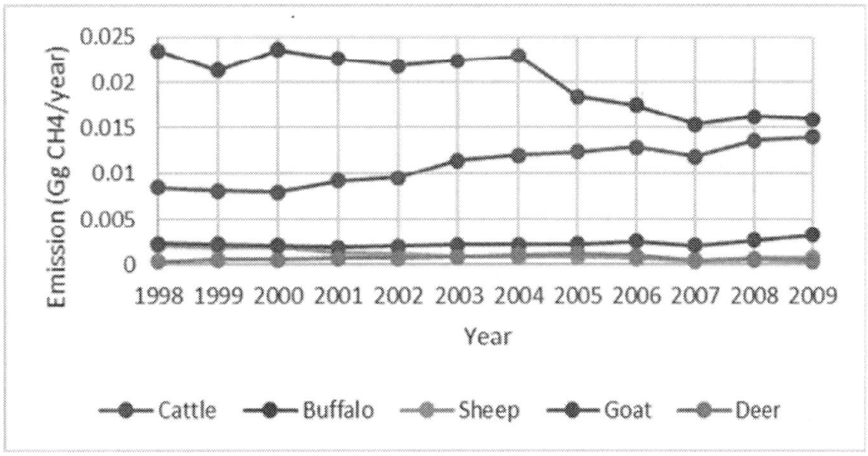

Fig. 4. Annual methane emission from manure management of livestock in Sarawak.

4.1 Cultivation of wet paddy

Fig. 2 shows the annual methane emission from paddy cultivation between 1998 and 2009. Practice in Sarawak is such that all animal waste is applied back to the field. The range of CH_4 emissions from wet paddy is 1.61 to 1.79 Gg CH_4/year as shown in this figure. The graph exhibits an incremental trend from 1998 at (1.61 Gg CH_4/year) till 1999 (1.72 Gg CH_4/year). The abrupt increasing trend from 2003 at 1.68 Gg CH_4/year to year 2006 at 1.79 Gg CH_4/year shows an alarming rate of CH_4 emissions. However, the emission from 2006 to 2009 has shown decreasing trend.

The upward emission trend is explained by changes in cultivation areas of paddy. Based on the results obtained in this study, it can be concluded that the gain in area will lead to a rise in CH_4 emissions. Simultaneously, [1] reported that a minor increase rate in the area for rice cultivation from 0.8% to 1% will emit an additional 0.10 tons/year of methane.

Mitigation options suggested in NC2 mentioned that water management infrastructures need to be tailored towards alternating irrigating and draining once or several times during the growing and fallow seasons. With the proper set up of water control infrastructures and good water management especially in the main granary areas, methane emissions from rice cultivation can be reduced [3].

4.2 Livestock - cattle

Fig. 3 presents annual methane emission from enteric fermentation of livestock between 1998 and 2009. Based on the figure, it is observed that methane emission from enteric fermentation of cattle increases in upward trend through the overall study duration. The trend increases steadily from 0.56 Gg CH_4/year in 2007 to 0.66 Gg CH_4/year in 2009. Chances are the number of population for cattle have increased via breeding due to huge demand. The trend is predicted to increase after year 2009 if and only if the demand continues to rise. As evidenced from Malaysia's NC2, GHG emissions will increase proportionally to the number of livestock kept. Technologies on reducing emissions from enteric fermentation are presently not economically feasible especially when Malaysia is importing most of its concentrate feed [3].

Fig. 4 demonstrates annual time series of methane emission from manure management of livestock between 1998 and 2009. From the figure, range of emission for cattle varies from 0.008 Gg CH_4/year to 0.014 Gg CH_4/year. It is shown that the overall emission increases with increase in population during 1998-2009. It is estimated that the emission after 2009 will continue to increase as the amount of cattle determines the amount of manure produced, thus affecting the emission. According to [10], the amount of methane released is related to the composition of manure. The more the amount of cattle, the higher the amount of methane generated due to increased manure produced.

4.3 Livestock - buffalo

From Fig. 3, methane emission from enteric fermentation of buffaloes in Sarawak showed overall decreasing trend. Abrupt decrease is from 0.63 Gg CH_4/year in 2004 to 0.42 Gg CH_4/year in 2007. The decreasing trend will continue to fall after year 2009 when the population of buffaloes are lesser. In this case, uncertainty may occur due to lack of detailed local data obtained. This is further proven by [11] who stated that emissions calculated using Tier 1 default IPCC emission factor were about 9.65 to 10.95% greater than emissions utilizing Tier 2 methodology of IPCC. Therefore collaboration and cooperation with data providers must be increased in order to improve data disaggregation.

From Fig. 4, methane emission from manure management for buffaloes varies from 0.015 to 0.024 Gg CH_4/year throughout year 1998-2009. This is due to the lessened population of buffaloes in Sarawak that brings to reduced amount of manure produced, thus lesser the methane released.

4.4 Livestock - sheep

Based on methane emission from enteric fermentation of sheep in Fig. 3, there is an abrupt decrease in methane emission starting from year 1998 at 0.051 Gg CH_4/year to year 2007 at 0.009 Gg CH_4/year. This is compatible with the average value obtained from Table 2 which demonstrated a decreasing trend throughout the years. Chances are the emission in the future after year 2009 will continue to decrease or maintain at the same level as in year 2009 due to low growth rate of sheep resulting from lower demand in the market in Sarawak. This is proven by reduction of sheep amount from year 1998 at 10189 to year 2007 at 1912.

For methane emission from manure management as shown in Fig. 4, the range of emission during 1998-2009 for sheep varies from 0.0006 to 0.002 Gg CH_4/year. The emission trend is probably reducing after year 2009. This is due to the way of feeding the sheep which will affect the portion of manure that decompose anaerobically. It is recommend that the manure of sheep shall be composted via aerobic composting in order to suppress emissions of methane.

4.5 Livestock - goat

Based on methane emission from enteric fermentation of goats in Sarawak, the trend of methane emission from enteric fermentation exhibited overall increasing trend from 0.05 to 0.06 Gg CH_4/year. The trend of methane emission will probably continue to increase after year 2009, depending on the goats' population. Similar to sheep, the factor that affect methane emission from enteric fermentation include population of goat kept. As shown in Table 1, quantity of goat has increased from year 2006 at 11,146 to year 2009 at 14,656, resulting in gain in emission. One also noticed that the default Tier 1 enteric fermentation emission factors provided by [4] is higher for developed nations compared to Malaysia. The variations in per head emissions originated from the amount and quality of feed intake in Sarawak.

For methane emission from manure management in Fig. 4, the range of emission for goat varies from 0.002 to 0.003 Gg CH_4/year. The use of composted manure in agricultural soils should be encouraged as it is a food source for soil microbes. Interactions between the manure and the microbes will increase the content of organic matter making the soil more fertile and sustainable for crop production while at the same time reducing GHG emissions both from manure decomposition and fertiliser usage [3].

4.6 Livestock - deer

Based on methane emission from enteric fermentation of farmed deer in Sarawak, overall increasing trend is spotted from 0.04 to 0.05 Gg CH_4/year during year 1998-2009. The methane emission from enteric fermentation after year 2009 will probably be reduced, as a result of decreasing population of deer. The slight increase in methane emission of enteric fermentation from 1998 till 2005 is due to ruminant livestock characteristics of the deer. Besides, default emission factors from 2006 IPCC Guidelines were used instead of detailed local emission factors. Therefore, it is suggested that research and development of local emission factors must be enhanced [8].

For methane emission from manure management, the range of emission for farmed deer varies from 0.0005 to 0.0006 Gg CH_4/year during year 1998-2009. As mentioned, composted manure brings various benefits. With the increase in prices of chemical fertilisers, the demand for composted manure is increasing, making it a good opportunity for farmers to earn some extra income. Adoption of this technology is easy and reduction of CH_4 emissions from deer manure can be achieved.

In summary, methane emission from wet paddy cultivation demonstrated upward trend from 1998 to 2009. Simultaneously, emission of methane from both enteric fermentation and manure management of cattle, goat and deer showed a raise in emission, while methane emission from enteric fermentation of buffalo and sheep experienced significant drop in emission. This is due to lower quality feed intake in Sarawak conditions. As for the methane emission from manure management, not much changes has been observed for buffalo except for that of emission from sheep which exhibited abrupt emission reduction. Last but not least, developing countries are expected to invest in rapid growth of animal production and consumption, which is much more different than developed countries. Hence emission from livestock shall be monitored [12].

5 Conclusion and recommendation

This study provides a tool that is capable of indicating the trend of methane emissions from paddy cultivation, enteric fermentation and manure management from livestock farming in Sarawak during the period of 1998-2009. Based on the emissions from paddy cultivation, it can be concluded that the emissions of methane from wet paddy varies depending upon the cultivation area. This is because of the varying area of cultivation and different cultivation practices. In addition, the geographic conditions of Sarawak allow for once per year of wet paddy cultivation. Hence, due to the importance of paddy cultivations for food security, it is recommended that the Malaysian Agricultural Research and Development Institute (MARDI) to install various apparatus on the paddy fields in Sarawak, in order to capture actual methane emissions. It is also recommended that MARDI to innovate paddy genetics that could emit less methane while providing more yield in the paddy fields.

In conclusion, even if we in this developing country faces many doubts on activity data collected, methods of data processing, methods of data and information presentation to the public, it is our responsibilities to pursue on these GHG estimation in order to determine the quantity of actual GHG being emitted by human activities such as crop cultivation, livestock farming and land use change so that Malaysia's voluntary indicator of up to a 40% reduction in the GHG emissions intensity of GDP compared to 2005 values by 2020 will become a reality as announced by the Malaysia Prime Minister at the UNFCCC COP 15 in Copenhagen. CH_4 which is produced in continuously flooded rice fields can be substantially reduced with more precise water management; and proper manure management can reduce CH_4 emissions, produce bio-fertilisers and facilitate the capture of biogas to serve as an energy source. It is hoped that the mitigation measures will reduce methane emissions thus reducing the total GHG footprint in Sarawak. Timely transfer of environmentally friendly and sound technologies and access to adequate financing under the UNFCCC (United Nations Framework Convention on Climate Change) will ensure Malaysia's achievement of the lower emissions per unit GDP scenario.

The authors would like to thank UNITEN R&D Sdn. Bhd for Grant U-EN-CR-14-12 that enables collaboration between them which leads to the development of this paper. Any of the opinions, findings and conclusions expressed in this paper are those of the authors and do not necessarily reflect the views of UNITEN R&D Sdn. Bhd.

References

[1] J. Maclean, B. Hardy, G. Hettel, *Rice Almanac, 4th edition: Source Book for One of the Most Important Economic Activities on Earth*, International Rice Research Institute, Laguna, Philippines, (2013)

[2] M. Cao, K. Gregson, S. Marshall, J.B. Dent, O.W. Heal, Global methane emissions from rice paddies, Chemosphere, **33**(5), 879-897, (1996)

[3] Ministry of Natural Resources and Environment, Malaysia, Malaysia's Second National Communication (NC2) to the UNFCCC, (2011), Retrieved on August 18, 2016 from http:// http://unfccc.int/resource/docs/natc/malnc2.pdf

[4] Land and Survey Department, Sarawak, Area and land use in Sarawak, (2013), Retrieved on May 15, 2016 from http://www.landsurvey.sarawak.gov.my/modules/web/index.php

[5] Department of Agriculture, Sarawak, Sarawak Agriculture Statistics, (2010), Retrieved on June 20, 2016 from http://www.doa.sarawak.gov.my/modules/web/index.php

[6] Sarawak Agriculture Statistics, *Sarawak Facts and Figures*, State Planning Unit, Chief Minister's Department, Sarawak, Malaysia, (2010)

[7] H.S. Eggleston, L. Buendia, K. Miwa, T. Ngara, K. Tanabe, The Intergovernmental Panel on Climate Change, *2006 IPCC Guidelines for National Greenhouse Gas Inventories, Volume 4: Agriculture, Forestry and Other Land Use*, Institute for Global Environmental Strategies, Kanagawa, Japan, (2006)

[8] Ministry of Natural Resources and Environment, Malaysia, Biennial Update Report (BUR) to the UNFCCC, (2015), Retrieved on July 5, 2016 from http://www.nre.gov.my/sites/TNC-BUR/Documents/BUR_Malaysia_Final.pdf/

[9] S. Anand, R.P. Dahiya, V. Talyan, P. Vrat, Investigations of methane emissions from rice cultivation in Indian context, Environment Int., **31**(4), 469-482, (2005)

[10] R.O. Yusuf, Z.Z. Noor, A.H. Abba, M.A. Hassan, M.F. Din, Methane emission by sectors: A comprehensive review of emission sources and mitigation methods, Renewable and Sustainable Energy Reviews, **16**(7), 5059-5070, (2012)

[11] A.K. Patra, Trends and projected estimates of GHG emissions from Indian livestock in comparisons with GHG emissions from world and developing countries, Asian-Australasian J. of Animal Sciences, **27**(4), 592-599, (2014)

[12] V. Sejian, R. Lal, J. Lakritz, Measurement and prediction of enteric methane emission, Int. J. Biometeorol, **55**, 1-16, (2011)

Potential Retrofit of Existing Sultan Ismail Parliament Building, Johor to Green Building

Mohd Norazam Yasin [1,*] *Abd Halid* Abdullah [1], *Hazrul* Abdul Hamid [1], and *Nur Shahirah Nadhilah* Wahab [2]

[1] Faculty of Civil and Environmental Engineering (FKAAS), Universiti Tun Hussein Onn Malaysia, Batu Pahat, 86400 Johor, Malaysia
[2] Faculty of Civil Engineering (FKA), Universiti Teknologi Malaysia, Johor Bahrut, 81310 Johor, Malaysia

Abstract. The concept of retrofitting an existing building towards green building was introduced to move towards greener development. This paper investigates the benefits of retrofitting an existing building to green building. The potential retrofit of the existing Sultan Ismail Parliament building was also determined and later the elements of the building to be retrofitted were proposed. The Sultan Ismail Parliament building is one of the newest development that serves as the Johor state administrative center under Iskandar Malaysia. Therefore, this paper focuses on whether the existing building has fulfilled all of the green building elements. A set of questionnaire was distributed to the targeted respondents who were the engineers, architects, technicians, staffs and those who were involved in the development of the Sultan Ismail Parliament building. A total of 60 respondents was involved. Based on the analyzed data, it shows that the existing building has fulfilled only one of the green building elements which is the Indoor Environmental Quality. The items that scored 3.5 and above mean index under this element include the daylight glare is not bothering, the ventilation in the building is adequate, inspections are done regularly, smoking regulation is implemented and the number of openings of windows and doors is sufficient. While it shows that other elements were only partially fulfilled. The criteria that achieved below 3.5 mean index values is considered to be the potential elements that could be used for future retrofitting initiative.

1 Introduction

The construction industry is one of the biggest industries in the world today. Global warming, climate change, soil erosion, air pollution and water pollution are some of the main problems arise due to the construction industry. Not only that it affects the environment, it also will affect the human health, comfort and productivity. Later, the concept of sustainable construction was introduced into the construction industry to promote the healthier construction environment to the people and to the nature itself.

[*] Corresponding author: norazam@uthm.edu.my

Malaysia is also one of the countries that are moving towards greener development. The Green Building concept was introduced. It serves as the solution which involves examining the interaction between the building elements in order to optimize the energy and the environmental performance (1). Hence, green building was introduced and established to promote the awareness of constructing sustainable to reduce the impact of construction. However, some of the building has not yet fulfilled the green building elements since some of them are still unaware of the importance of green building in a construction. The new Sultan Ismail Parliament building which is located at Kota Iskandar, Johor is taken into the main consideration in this study. The building claims to promote sustainability, reducing energy cost and increasing long term usage. Therefore, this paper is intended to study the benefit of retrofitting an existing building to green building, to determine whether the existing Sultan Ismail Parliament building has yet fulfilled all of the green building elements and to propose significant elements that could be retrofitted.

2 Green building and retrofitting

According to a report done by the Willmott Dixon Group, the construction industries uses 45% to 50% of global energy, nearly 50% of worldwide water usage and around 60% of total raw material use. It was also said that this industry contributes to 23% of air pollution, 50% of climate change gases, 40% of pollution of drinking water and 50% of landfill wastes. The development of new technologies means that building retrofits can allow for significant reductions in energy and water usage (2). In order to reduce the impact to the environment, the concept of sustainable construction was introduced and implemented since a few decades ago. Green Buildings should be designed and operated to reduce the overall environmental impact of the building and surrounding area (3). The concept of green development was introduced later and the terms such as the 'green building', 'green home' and 'green construction' are all connected to increase the building performance, to enhance the energy, water and resource efficiency, to protect the occupants health and comfort, to reduce waste and reduce negative impact to the environment. According to Green Building Index, a green building focusses to increase the efficiency of resource use, to reduce the building impact on human health and the environment during the building's lifecycle (4). There are six main green building elements which are being focused on including the energy efficiency, indoor environmental quality, sustainable site planning and management, materials and resources, water efficiency and innovation. All of these elements are important as they serve as the guideline to be used to achieve the green building strategy (5).

This green building concept, however, was only introduced a few years ago. Some of the buildings were built long before that such as the heritage buildings and there are also some of the new buildings have not followed the green building strategy and thus did not fulfilled the concept. Therefore, the retrofitting concept was later introduced. Retrofitting is one way to remodel or to modify an existing building towards green building. Retrofitting is not only improves the performance of the existing buildings, but it also reduces the carbon footprint to the environment. According to Garnaut (6), sustainable retrofit helps to lower the greenhouse emissions, lowers the energy usage and provides healthier workplace (7). Retrofitting can be done on any types of building and it is depending on the owner to determine what is the main objective and the scope of the retrofitting. To successfully achieve the retrofit, several strategies were introduced to ease the process. This includes the retrofit timing and duration, integrated planning, design and documentation, setting a retrofit budget, incorporating sustainability and replacing existing building system (8).

3 Sultan Ismail Parliament Building

The Sultan Ismail Parliament building is located at Kota Iskandar, Johor Bahru. The Iskandar Malaysia, Figure 1 was established under the Ninth Malaysian Plan as one of the regional development corridors. The Kota Iskandar was developed by the Cahaya Jauhar Sdn. Bhd and the key plan were to fulfill the sustainable concept. It was the first fully integrated solid waste collection and disposal system, the first non-intrusive security plan with direct link to the police and the first district cooling system outside of Kuala Lumpur (9). The building was constructed completely in the year 2008 and is now serving as the Johor state Assembly Hall. The land area is approximately 6.67 hectares and it was inspired by the Moorish-Andalusian and Johor-Malay design. The main building consists of five blocks and four stories each. There are also libraries, archives, operation rooms, banquet halls, and three themed courtyards.

Fig.1 Sultan Ismail Parliament Building.

4 Methodology

The methods used to achieve the objectives of the study consist of three phases. The first phase was identifying the research title, the background study, the problem statement, and the objectives of the study and the scope of study. The second phase was to study the benefits of retrofitting the existing building to green building by intensive research through past studies. The questionnaire survey was done to determine the potential of Sultan Ismail Parliament building. The questionnaire was designed into three parts, A, B and C. Section A was the demographic information of the respondents where the questions such as the gender, age, profession and work experience were asked. Respondents were required to answer by ticking the suitable answer provided. As for section B, the questions involves the respondent's opinion on the green building elements that have been implemented in the parliament building. Section C was asking about the respondent's agreement to suggest the best additional elements that can be implemented to support initiative in the future for the Sultan Ismail Parliament building. The respondents were required to answer by following the Likert Scale of agreement which have five main measures; strongly disagree, disagree, neither agree nor disagree, agree and strongly agree. A set of questionnaire was distributed to the developer of the Sultan Ismail Parliament building and to the staffs who are currently working at the building itself. A total of 60 respondents was involved in this study. The third phase was the analysis of data and the discussion were made. The data collected from the survey were analysed by using the mean index method. The data will also be analysed by using Spearman's correlation analysis to test the relationship of two variables.

5 Results and discussion

5.1 The potential of Sultan Ismail Parliament Building towards green building initiative

The Figure 2 shows the mean index for the Energy Efficiency element. Based on the figure, it shows that the most efficient sub-elements that has been implemented with a mean index of 4.02 is the number of lightbulbs installed is adequate and not excessive. This is because the lightbulbs were installed mostly in the working area where it needs sufficient lighting. Furthermore, this building enhances the natural lighting by designing skylights and openings that allows natural lighting into the building. Hence, this ensures that the artificial lightings were installed according to the requirement. While the lowest mean index is 1.82 which is the building is implementing a solar energy system. It shows that it is not agreeable that the building has yet implemented the solar energy system. Next, Figure 3 shows the Indoor Environmental Quality element. Based on the results, it shows that all of the elements have achieved more than 3.5 mean index, which could be concluded that the element of indoor environmental quality is fulfilled. The mean index for the Sustainable Site Planning and Management is shown in Figure 4 and it shows that most of the sub-elements scored higher than 3.5 which means it is acceptable with the highest score of 4.0 for the number of parking provided is sufficient. However, one of the sub-elements, the statement roof material used is suitable so the solar heat is not greatly transferred into the building scored 3.35 which means it is not yet fully fulfilled the element criteria since the building is currently using the conventional concrete roof tiles. Next, Figure 5 shows Materials and Resources element. The figure shows that all of the sub-elements have mean index lower than 3.5, which mean all of the sub-elements have not yet meet the requirements for this element. Figure 6 next shows the Water Efficiency element and it also shows that this element has not yet fulfilled since all of the sub-elements scored less than the 3.5 mean index. The last element which is the Innovation can be seen in the Figure 7 also shows that all the sub-elements score less than 3.5. Therefore, in this case the data could show that the only element that the Sultan Ismail Parliament building has fully achieved is the Indoor Environmental Quality. The element of Energy Efficiency and Sustainable Site Planning and Management are partially fulfilled. While the Materials and Resources, Water Efficiency and Innovation have not yet fulfill the green building strategy. Thus, all of the sub-elements which scored less than 3.5 mean index should be taken into consideration for the use of retrofitting in the future.

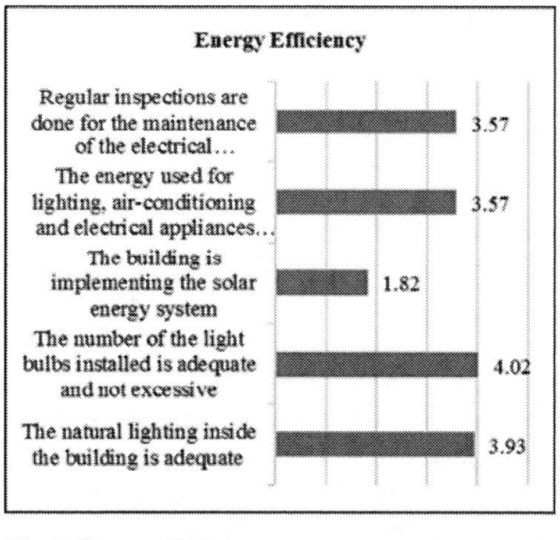

Fig.2. Energy Efficiency.

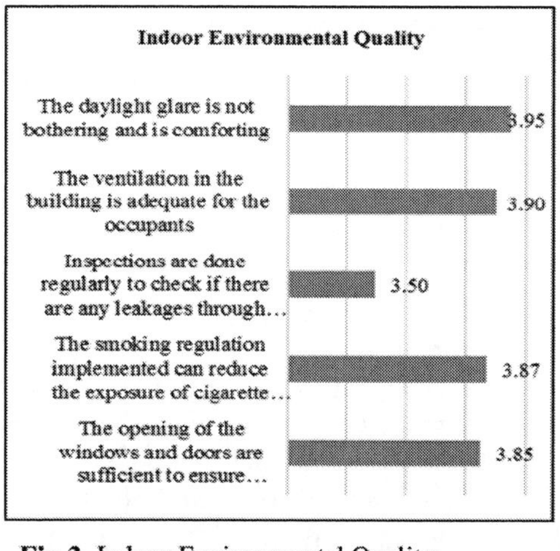

Fig.3. Indoor Environmental Quality.

DOI: 10.1051/matecconf/201710305008

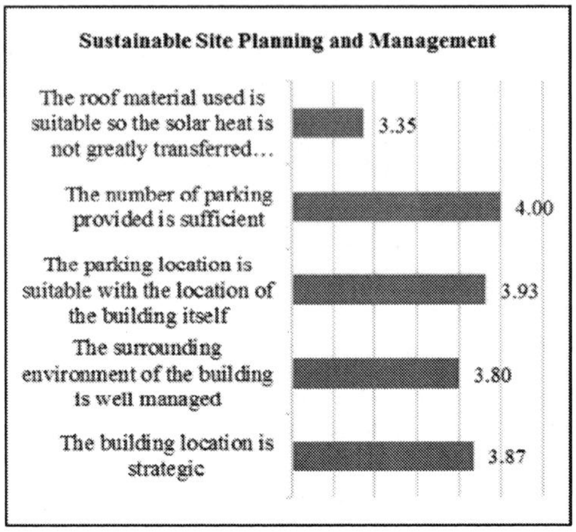

Fig.4. Sustainable Site Planning and Management.

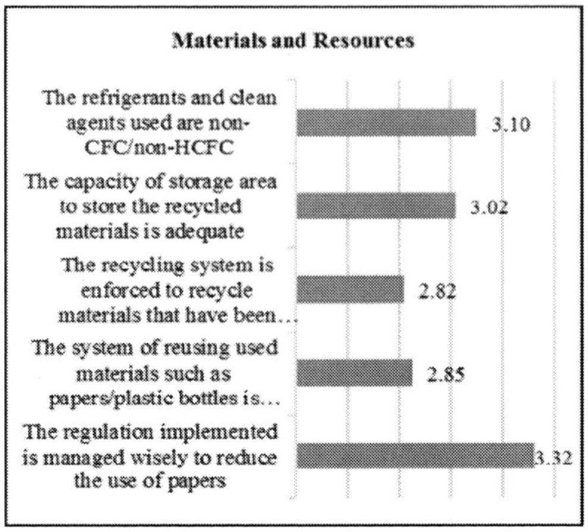

Fig.5. Materials and Resources.

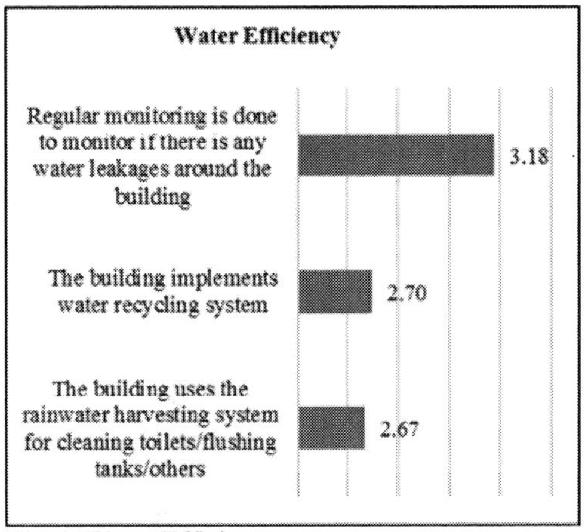

Fig.6. Water Efficiency.

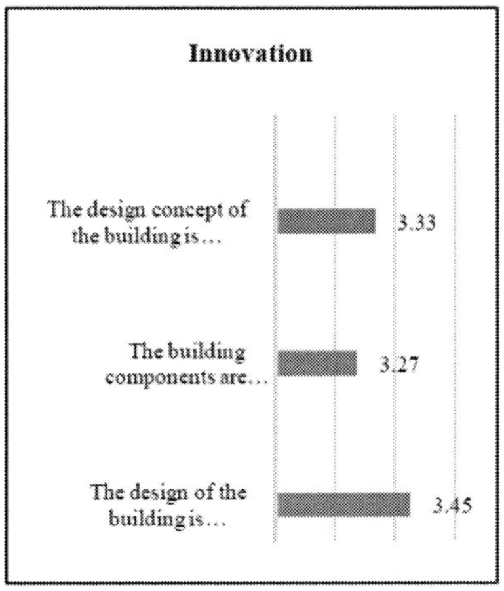

Fig.7. Innovation.

5.2 To propose retrofit for existing Sultan Ismail Parliament Building to green building

Based on Figure 8, all the items that have been listed to be proposed were scored above 3.5 mean index which shows that all the items are acceptable to be used later. The highest ranked item is to encourage the recycling and other waste management practices with mean 4.23. In workplaces such as the offices are encouraged to practice recycling and also the combination of recycling and waste prevention programs. It promotes not only recycling but also to reduce paper consumption and to reuse office supplies. The second item is to encourage the use of renewable energy technologies with mean index of 4.13. It shows that the respondents are agreeing that renewable energy could be implemented in the building. Renewable energy is essentially harnessing the power of the sun and wind. For example, the initiative that could be taken is by replacing a part of the conventional roof tiles to Photovoltaic (PV) or solar cells where the PV modules could convert sunlight into direct current. The third item is to install motion sensor for lighting system with a mean index of

4.12. A motion sensor can either replace conventional lighting switches or to be placed on an existing lighting fixture. The motion sensing lighting system is ideal for lighting areas that only being used briefly or as long as someone is there. The light will automatically switched off after a few minutes after there are no motion detected. Thus, electricity wastage can be prevented. While the lowest ranked is to install shading to reduce sunlight glare with a mean index of 3.53.

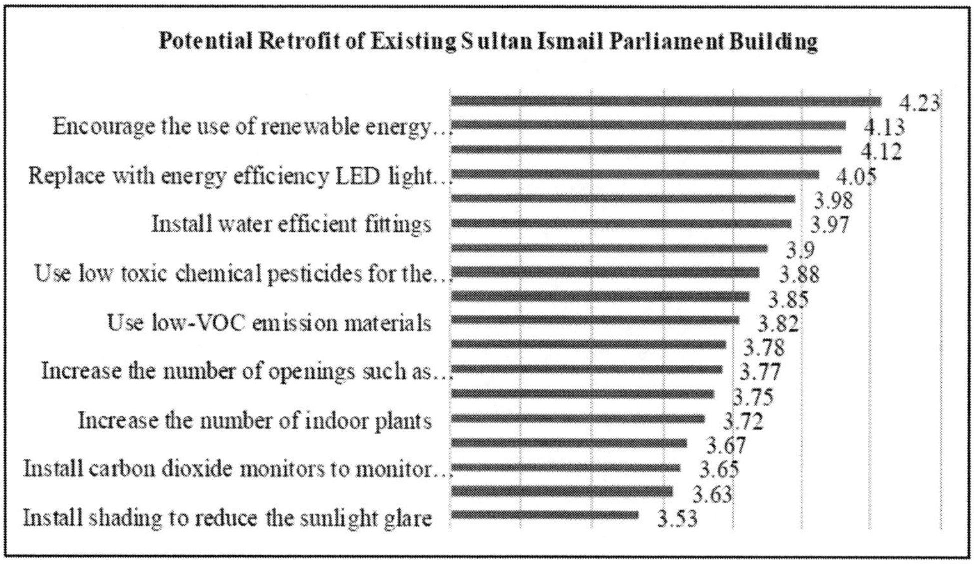

Fig.8. Potential Retrofit of Existing Building to Green Building.

5.3 Correlation Analysis

Spearman's Correlation Coefficient analysis was done to test the strength of the associations between two variables.

Table 1. Correlation of Two Variables (Dependent Variable: Energy Efficiency)

Independent Variable	(Spearman Correlation - r_s) Energy Efficiency	P (Significant Level)
Indoor environmental quality	0.572	0.0001
Sustainable site planning and management	0.406	0.0010
Materials and resources	0.301	0.0200
Water efficiency	0.461	0.0001
Innovation	0.493	0.0001

Table 1 shows the strength of the associations for Energy Efficiency with other elements could be considered as associated. Based on the table above, the strongest correlation is between Energy Efficiency and Indoor Environmental Quality with rs value of 0.572 which means that Energy Efficiency is moderately dependent to the Indoor Environmental Quality element. While the association between Energy Efficiency and other elements are weak and moderate. The significant level for all the elements is less than 0.05, which means the values are acceptable. Hence, it shows that there is a relevant correlation between Energy Efficiency with all other elements in the green building concept.

6 Conclusion

The study shows that the importance of retrofitting. This includes the reduction of energy consumption, reduction of greenhouse gas emission, improves working environment, increases the usage of renewable energy and reduce cost of water used. The study also shows the potential retrofit that could be applied to the Sultan Ismail Parliament building has been identified. The result shows that the building has only fully fulfilled one of the green building elements which is the 'Indoor Environmental Quality'. While, other elements should be considered for later use of retrofitting. Among the unfulfilled elements, it shows that waste management should be taken into account to improve the building materials and resources management. This study also identified that the most relevant steps that could be proposed for retrofit are to encourage the recycling and other waste management practices, to encourage the use of renewable energy technologies and to install motion sensor or auto sensor for lighting system.

References

[1] R. Zakaria, K. S. Foo, R. Mohammad Zin and S. Zolfagharian, Potential Retrofitting of Existing Campus Buildings to Green Buildings. J. of Applied Mechanics and Materials, **178-181,** 42- 45, (2012)

[2] M.C. Lee K. M, *Building and Environment*, 238- 244, (2012)

[3] M. N. Yasin, A. H. Abdullah, R. Yunus, H. Abdul Khalid, and N. S. N. Wahab, Potential Retrofit of Universiti Tun Hussein Onn Malaysia Existing Building towards Green Building, *ICTMBE Conference*, (2015)

[4] Malaysia Green Building (GBI) Index, (2015), Retrieved on June 3, 2016 from http//www.greenbuildingindex.org,

[5] Ministry of Energy, Water and Green Technology Malaysia, (2014), Retrieved on June 3, 2016 from http://www.kettha.gov.my

[6] Garnaut, R, *The Garnaut Climate Change Review.* Final Report. Cambridge University Press, Melbourne, Australia. ISBN: 9780521744447, (2007)

[7] L. Tobias and G. Vavaroustos, *Retrofitting office buildings to be green and energy-efficient*, Urban Land Institute, (2009)

[8] Kota Iskandar Website, (2016), Retrieved on September 3, 2016 from http//www.kotaiskandar.com,

[9] Cahaya Jauhar, (2015), Retrieved on September 3, 2016 from http://cahayajauhar.com,

Life Cycle Assessment of Producing Electricity in Thailand: A Case Study of Natural Gas Power Plant

Parnuwat Usapein[1,*] , and *Orathai* Chavalparit[2,3]

[1]Rattanakosin College for Sustainable Energy and Environment, Rajamangala University of Technology Rattanakosin, Nakorn Pathom 73170, Thailand
[2]Research Unit of Environmental Management and Sustainable Industry (EMSI), Faculty of Engineering, Chulalongkorn University, Phayathai Road, Patumwan, Bangkok 10330, Thailand
[3]Department of Environmental Engineering, Faculty of Engineering, Chulalongkorn University, Phayathai Road, Patumwan, Bangkok 10330, Thailand

Abstract. Environmental impacts from natural gas power plant in Thailand was investigated in this study. The objective was to identify the hotspot of environmental impact from electricity production and the allocation of emissions from power plant was studied. All stressors to environment were collected for annual natural gas power plant operation. The allocation of environmental load between electricity and steam was done by WRI/WBCSD method. Based on the annual power plant operation, the highest of environmental impact was fuel combustion, followed by natural gas extraction, and chemical reagent. After allocation, the result found that 1 kWh of electricity generated 0.425 $kgCO_2eq$ and 1 ton of steam generated 225 $kgCO_2eq$. When compared based on 1GJ of energy product, the result showed that the environmental impact of electricity is higher than steam product. To improve the environmental performance, it should be focused on the fuel combustion, for example, increasing the efficiency of gas turbine, and using low sulphur content of natural gas. This result can be used as guideline for stakeholder who engage with the environmental impact from power plant; furthermore, it can be useful for policy maker to understand the allocation method between electricity and steam products.

1 Introduction

Electricity is an important energy which help us to facilitate our daily life. To produce electricity, however, it consisted of various processes and required a fuel source to produce power. At present, the electric power industry can be mainly divided into two main types such as renewable energy power plant (solar, wind, biomass) and fossil fuel power plant

* Corresponding author: parnuwat.usa@rmutr.ac.th

(natural gas, coal, heavy oil). Main fuel for electricity production in Thailand was natural gas (69.22%), followed by coal (19.10%), hydro (8.65%), renewable energy (2.24%), and others (0.79%) [1]. Natural gas is fuel type that has high efficiency of combustion which can result in reducing environmental burden. Thailand, as one of the countries that has own natural gas resources, therefore selected natural gas as main fuel for country's development. In Thailand, Electricity Generating Authority of Thailand (EGAT) is a main operator in producing, supplying, and distributing electricity to every part of the country. EGAT is presently 43 power plants with capacity of 16,376 MW, which comprised of 3 thermal, 6 combined cycle, 22 hydro, 8 renewable, and 4 diesel power plants [2]. In addition to the electricity that produced by EGAT, EGAT has purchased the electricity from 11 of Independent Power Producers (IPP) with total capacity of 13,166 MW, and from Small Power Producers (SPP) with total capacity of 3,726.6 MW. Furthermore, Thailand also purchased electricity from neighbouring countries (Laos, and Malaysia) with total capacity of 2,404 MW. IPP and SPP are private companies and play an important role for electricity production in Thailand with 32% of total capacity production. Most of IPP and SPP used natural gas as main fuel for producing electricity with strict regulation on emissions to the environment. However, there are no evidence about life cycle inventory from IPP and SPP in Thailand to assess the environmental impact.

Life Cycle Assessment (LCA) can be used as a tool for identifying the environmental impact through cradle-to-grave (i.e. from raw material acquisition through material processing, distribution, use, and disposal). Energy policy makers can use this tool for supporting the decision making in the national level. There are some studies that apply LCA with power plant and compared the environmental impact of electricity production between combined-cycle and thermal technology in Thailand. They suggested that, in addition to natural gas feedstock, the other feedstocks and power production technologies should be further investigated for helping in decision making. Other studies also investigate the environmental impact from power plant but most of them concerns only carbon dioxide [3, 4].

Therefore, the objective of this study was to identify the hotspot of environmental impact from SPP power plants. In addition, this study evaluates the allocation of environmental burden between products from power plant (i.e. electricity and steam), not only carbon dioxide but also in every pollutant that emitted from power plant. The result of this study was expected to be beneficial for power plant in case of supporting data of environmental load in each product.

2 Materials and Methods

In this study, the framework of life cycle assessment was operated according to ISO 14040:2006 standard [5]. The step of normalization and weighting score were not considered in this study. Detail in each step can be described as follows.

2.1 Goal and scope of the study

The objective of this study was to evaluate the environmental impact of SPP power plant producing electricity and steam. The scope of this study was to provide the life cycle

DOI: 10.1051/matecconf/201710305009

inventory of natural gas power plant in both material balances and heat balance to allocate the environmental impact in each of the product (electricity and steam).

2.1.1 System boundaries

The scope of determining the environmental impact was started from the cradle-to-gate of electricity production. The stage of natural gas production, natural gas transmission, gas separation plant, and power generation at case-study site were included in this study, as shown in Fig. 1. A Combined Heat and Power (CHP) Plant was selected as a case study. In this process, the multiple forms of energy (i.e. electricity and steam) were generated at the same time from the same input energy supply. To produce the electricity, natural gas was combusted in the gas turbine-powered generator. An exhaust gas (as a result of the combustion of natural gas) was sent to Heat Recovery Steam Generator (HRSG) unit to make high-pressure steam and fed to steam turbine to produce electricity. From steam turbine, there are low-pressure steam left over which can feed to an industrial process.

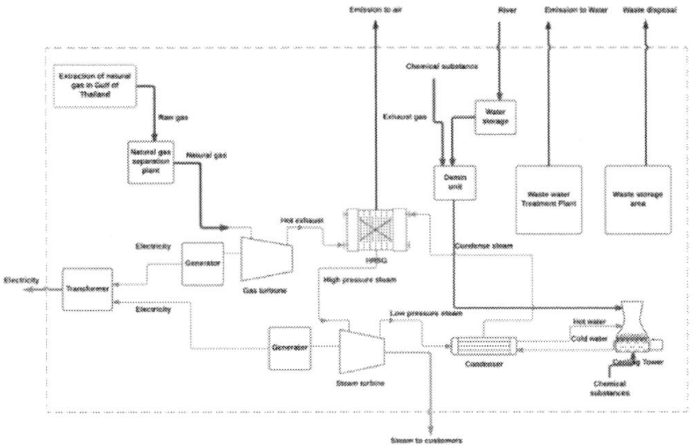

Fig. 1. The boundary system of life cycle assessment for producing electricity and steam production

2.1.2 Functional unit

This study, all energy and material used and environmental impacts were collected in the period of one year operation of natural gas power plant. The environmental impact will be allocated to1 kWh of electricity and 1 ton of steam production.

2.1.3 Allocation method

In this study, the multi-input of material and multi-output of product were determined which was required to allocation. The direct and indirect emissions those generated from raw material acquisition and at natural gas power plant were combined and allocated to the electricity and steam by using the Efficiency method as shown in Equation 1 and 2 [6].

$$E_H = \frac{\frac{H}{e_H}}{\frac{H}{e_H} + \frac{P}{e_P}} x E_T \tag{1}$$

$$E_P = E_T - E_H \tag{2}$$

where E_H is the emission allocated to steam production, H is the steam output (energy), e_H is the assumed efficiency of steam production, P is the delivered electricity generation (energy), e_P is the assumed efficiency of electricity generation, E_T is the total direct emissions of the CHP system, and E_P is the emission allocated to electricity production. With this method, the emissions generated from CHP plant were allocated based on the separate efficiencies of electricity and steam products.

2.2 Life cycle inventory

The data collected in each stage of the cradle-to-gate are as follows ;

(a) Raw material acquisition of natural gas: The data of natural gas extraction, separation, and transmission were based on the work [7].

(b) Direct emission at natural gas power plant : The data were collected from power plant case study. Emission data on SO_2, NO_x, and Particulate Matter (PM) were collected from the CEM database in the period of January 2014 to December 2016. The data from CEM were obtained as on hourly average records. CO_2 from natural gas combustion were calculated based on the composition according to the methodology that suggested by American Petroleum Institute 2009 [8].

Based on the composition of natural gas, CO_2 was estimated by using Equation 3,

$$E_{CO_2} = FC \ x \frac{1}{molar\ vol.conversion} x MW_{mixture} \ x \ Wt\%C_{mixture} \ x \frac{44}{12} \tag{3}$$

where E_{CO2} refers to mass emissions of CO_2, FC is the fuel consumed (scf), molar volume conversion is the conversion from molar volume to mass (379.3 scf/lbmole or 23.685 m^3/kgmole), $MW_{mixture}$ refers to molecular weight of mixture, and 44/12 is the stoichiometric conversion of C to CO_2. All data related with the quantity of raw material, energy, electricity and steam products, waste generation, were collected by using questionnaire form. The result of LCI was shown in Table 1.

Table 1. Overall LCI data collected from power plant case study

Mass/Energy Category	Total input	Unit	Reference
Natural gas	5,615,021	MMBTU	at site
Water	1,098,153	m^3	
Auxiliary process			
Diesel for system backup generator	1,200	liter	
Diesel for fire pump	400	liter	
Diesel for forklift	125	liter	
Chemicals input			
SF_6	1.00E-04	kg	at site
H_2SO_4	240	tons	EIA report (2555)
NaOH	144	tons	
NaOCl	35	tons	
Cooling water corrosion inhibitor	5	tons	
Oxygen scavenger	1.5	tons	
Na_3PO_4	5	tons	
Feed water corrosion inhibitor	1	tons	
Cooling water dispersent	6	tons	
Cooling water biocide	1.5	tons	
Scale inhibitor	1.5	tons	
Sodium bisulfate	6	tons	
Closed cooling water circuit corrosion inhibitor	1	tons	
Output			
Emission into air			
SO_2	21,030	$kgSO_2$	EIA report (2555)
NO_x	15,1361	$kgNO_x$	
CO_2	324,686,938	$kgCO_2$	
Particular Matter	40,460	kgPM	EIA report (2555)
Emission into water			
BOD	6.09	mg/l	at site
COD	52	mg/l	
Suspended solid	25	mg/l	
Dissolved solid	1,646	mg/l	
Waste			
Municipal waste	7,850	kg	at site
Non-hazardous waste	1,650	kg	
Hazardous waste	480	kg	

2.3 Life Cycle Impact Assessment (LCIA)

The environmental impact from life cycle inventory was calculated using The International Reference Life Cycle Data System (ILCD) method. ILCD method supports the analysis of emission into air, water, and soil; in addition, it also determines the impact of natural resources consumed in terms of their effect to different impacts [9]. This method determines sixteen impacts such as climate change, ozone depletion, human toxicity (non-cancer effects), human toxicity (cancer effects), particulate matter (PM), ionizing radiation (HH), ionizing radiation E (interim), photochemical ozone formation (POF), acidification, terrestrial eutrophication, freshwater ecotoxicity, land use, water resource depletion, and mineral, fossil & resource depletion. However, the impacts of land use and water resource depletion were not included in this study.

2.4 Interpretation

The results of impact assessment were interpreted to identify the hotspot of environmental impact from life cycle of electricity production. Measures to reduce environmental impact were suggested and also discussed on the allocation methodology.

3 Result and discussion

3.1 Characterization impacts

Fig. 2 shows the result of 14 impact characterizations for one year of operation of the natural gas power plant. Each impact category was affected by different stages of producing electricity. It can be seen that the emission of exhaust gas was a hotspot of climate change, particular matter, photochemical ozone formation, acidification, terrestrial and marine eutrophication.

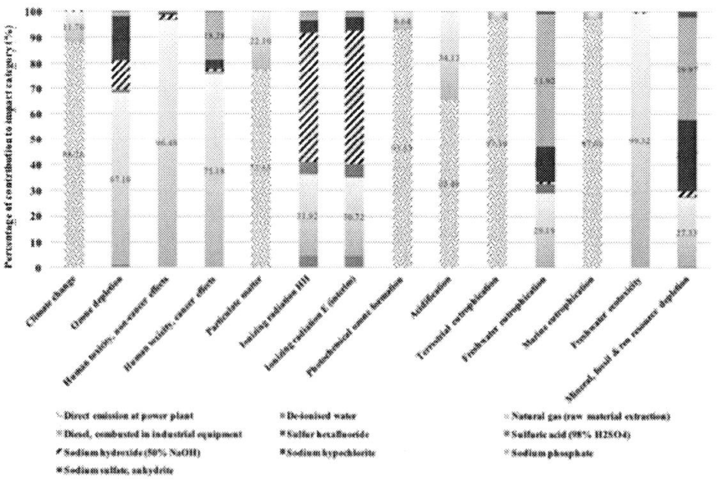

Fig. 2. The characterised result of 14 impacts for annual natural gas power plant

The main air pollutants was generated from natural gas combustion which depended on the composition of natural gas. CO_2 from fuel combustion was the major cause in climate change, while high sulfur content in natural gas will more affect to acidification impact. In case of POF, the major cause was generated from NO_x emission by fuel combustion. Although the power plant installed Dry Low NO_x Combustion Technology, NO_x is still dominant in this impact. The major of ozone depletion effect was generated from the refrigerant used in the natural gas extraction and separation. Also, the heavy metal was largely emitted in this section which was effect to the human toxicity category.

3.2 Allocation of Emissions

Due to the products from power plant consisted of more than one product (i.e. electricity and steam), it is therefore necessary allocating the environmental impact to reflect the environmental burden in each of product. The calculation of emission allocation in each category was done by using equation 1 and 2. The efficiency of electricity generation and steam production were 53% and 80%, respectively. With the total of 727,696 MWh and 259,200 tons-steam, the result of allocation was shown in Table 2. To determine the effect of allocation, this study compares the environmental impact in each product based on 1 GJ of energy product as shown in Fig. 3. After allocation, the environmental impact of electricity will be higher than steam product. This is because the efficiency of producing electricity has lower than steam production; it means that electricity used energy to produce more than steam. In this case study, the efficiency of producing electricity was about 50% which mainly based on the efficiency of turbine. Compared with steam production, the efficiency will be based on the HRSG performance which is normally about 80 - 85% of energy input [10].

Table 2. Allocation of environmental impact from electricity production

Impact category	Unit	Impact/kWh	Impact/ton steam
Climate change	kg CO_2 eq	4.25E-01	2.25E+02
Ozone depletion	kg CFC-11 eq	1.04E-10	5.50E-08
Human toxicity, non-cancer effects	CTUh	5.89E-10	3.11E-07
Human toxicity, cancer effects	CTUh	1.22E-11	6.43E-09
Particulate matter	kg PM2.5 eq	2.23E-05	1.18E-02
Ionizing radiation HH	kBq U235 eq	2.95E-05	1.56E-02
Ionizing radiation E (interim)	CTUe	2.82E-10	1.49E-07
Photochemical ozone formation	kg NMVOC eq	2.45E-04	1.29E-01
Acidification	molc H+ eq	3.17E-04	1.68E-01
Terrestrial eutrophication	molc N eq	9.88E-04	5.22E-01
Freshwater eutrophication	kg P eq	1.59E-08	8.39E-06
Marine eutrophication	kg N eq	9.03E-05	4.78E-02
Freshwater ecotoxicity	CTUe	1.28E-02	6.76E+00
Mineral, fossil & resource depletion	kg Sb eq	1.10E-08	5.83E-06

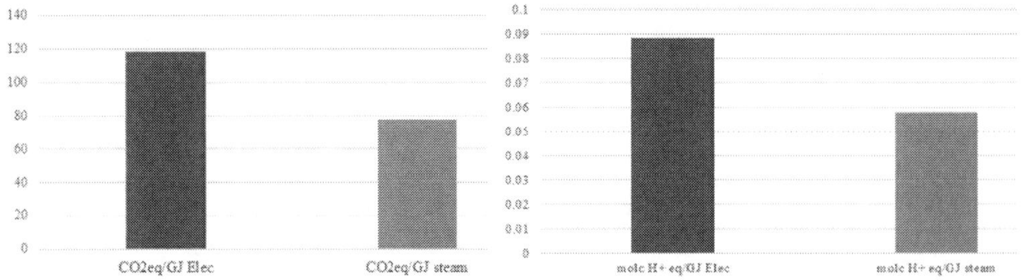

(a) Climate change (b) Acidification

Fig. 3. The comparison of allocating climate change (a) and acidification (b) impacts between electricity and steam product from power plant.

4 Conclusion

This study investigated the environmental impact of electricity production from natural gas power plant in Thailand. The boundary of assessment was cradle to gate. From the result, it can be concluded that the hotspot of environmental impact was fuel combustion. To improve the environmental performance, it should be suggested that low sulphur-content of natural gas could be playing an important role for the reduction of environmental impact. After allocated the emissions, the environmental impact of electricity is higher than steam, approximately 50% of every impact. For future investigation, it should be further study about the comparison of the environmental impact between different fuel types of power plant. The results presented in this study is expected to serve as a guideline for stakeholder who engage with the environmental impact from power plant, and for helping policy maker to understand the allocation method between electricity and steam products.

This research was funded by the Ratchadapisek Sompoch Endowment Fund (2016), Chulalongkorn University (CU-59-002-IC). The authors would like to acknowledge the factory staff for coordinating with the power plants.

References

[1] Electricity Generating Authority of Thailand (EGAT). The Proportion of Fuel Consumption for Electricity Production , (2015), Retrieved on August 15, 2016 from http://www.egat.co.th/

[2] Electricity Generating Authority of Thailand (EGAT). Electricity Capacity in Thailand, (2016), Retrieved on August 20, 2016 from http://www.egat.co.th/

[3] J. Kamsamrong, and C. Sorapipatana, An assessment of energy security in Thailand's power generation. Sustainable Energy Technologies and Assessments, **7**, 45-54, (2014)

[4] A. Pattanapongchai, and B. Limmeechokchai. *Energy Sources, Part B: Economics, Planning, and Policy,* **10**(3) (2014)

[5] ISO14040. *Environmental management—life cycle assessment—Principles and framework.*, 33-46, (2006)

[6] WRI/WBCSD. *Allocation of GHG Emissions from a Combined Heat and Power (CHP) Plant. Guide to calculation worksheets,* **1**, (2006)

[7] K. Phumpradab, S.H. Gheewala, and M. Sagisaka, Life cycle assessment of natural gas power plants in Thailand. Int. J. Life Cycle Ass **14**(4), 354 – 363, (2009)

[8] API, *Compendium of greenhouse gas emissions methodologies for the oil and natural gas industry*, (2009)

[9] JRC, *International Reference Life Cycle Data System in Recommendations for Life Cycle Impact Assessment in the European context-based on existing environmental impact assessment models and factors*, (2011)

[10] EPA, *Catalog of CHP Technologies, Section3. Technology Characterization – Combustion Turbines*, (2015), Retrieved on November 16, 2016 from https://www.epa.gov/

Assessment of Water Quality Index and Heavy Metal Contamination in Active and Abandoned Iron Ore Mining Sites in Pahang, Malaysia

Zafira Madzin[1,*], *Faradiella Mohd* Kusin[1,2], *Fasihah Mohd* Yusof [1], and *Siti Nurjaliah* Muhammad[1]

[1]Department of Environmental Sciences, Faculty of Environmental Studies, Universiti Putra Malaysia, 43400 UPM Serdang, Malaysia
[2]Environmental Forensics Research Centre (ENFORCE), Faculty of Environmental Studies, Universiti Putra Malaysia, 43400 UPM Serdang, Malaysia

Abstract. The composition of heavy metals in water and surface soils of iron ore mining sites were investigated to evaluate on the potential occurrence of heavy metal contamination. Physico-chemical characteristics of the waters were also investigated to determine the current status of water quality index (WQI) of the sites. Samples of water and surface soils of active mine (Kuala Lipis) and abandoned mine (Bukit Ibam) in Pahang were collected at four locations, respectively. The physico-chemical parameters measured for WQI were pH, dissolved oxygen, biological oxygen demand (BOD), chemical oxygen demand (COD), suspended solids (SS), and ammoniacal nitrogen (AN). The water quality parameters were classified according to the Department of Environment (DOE) water quality classification. The study revealed that most of the sites in Bukit Ibam and Kuala Lipis were categorized as clean to slightly polluted. On the other hand, heavy metal analysis in water showed that aluminium and manganese level in both sites have exceeded the allowable limits for raw and treated water standards by the Ministry of Health. For heavy metal compositions in soils showed most of the heavy metal concentrations were below the recommended guideline values except for lead, arsenic, zinc and copper.

1 Introduction

Pahang was reported as the top producer for most minerals in Malaysia [1]. However, there are several abandoned and active mining sites in Pahang that are not properly managed and could possibly become the source for water pollution. Mining industry has become a major economic contribution to the world. Several minerals such as iron ore, gold ore, sand ore and bauxite are known as the main source of energy and materials. From metal to non-metallurgic minerals, both are gradually extracted and exploited to meet the economic demands. Hence, this industry may also bring undesirable impacts to the world especially on the environmental aspects. Over exploitation of these minerals has cause

*Corresponding author: faradiella@upm.edu.my

excessive solid waste and wastewaters to remain untreated and discard hazardous substances to the biodiversity [2-6]. These toxic discharges which are the heavy metals and other trace elements can be uncontrollably percolated and penetrated to the soil and creating water and groundwater pollutions. A long-term exposure to heavy metals can be carcinogenic, affecting central and peripheral nervous system and circulatory effects [7]. High concentrations of heavy metals can cause potential risk to human health as heavy metal is easy to accumulate into the living tissues [8].

One of the serious problem that could arise from heavy metal contamination in mining industries is acid mine drainage (AMD). Acid mine drainage have resulted in serious pollution problem on surface or groundwater as well as in soil. AMD is acidic water that has high metal content formed by the reaction between oxygen and water in sulphide bearing minerals [9]. Hence, the objective of this research is to investigate the composition of heavy metals in water and surface soils of iron ore mining sites in Kuala Lipis and Bukit Ibam.

2 Materials and method

2.1 Sampling sites

The study was conducted at Kuala Lipis and Bukit Ibam where both are iron-ore mining sites. Four sampling locations were identified in Kuala Lipis while another four sampling sites were in Bukit Ibam. These iron mines are one of the top producers of iron in the state of Pahang as early as in the 1960s [10]. The Bukit Ibam abandoned mines were actively mined in 1960 and the iron mining activities have left hectares of abandoned ponds which have no buffer zone and the water can easily overflow to the river nearby during raining season. The iron ore mines in Kuala Lipis are still being mined actively in discrete areas but heavily connected to water stream (rivers). The sampling was carried out in June 2015 at location sites as shown in Fig. 1. Most of iron mining sites consist of Paleozoic and Mesozoic sedimentary rock sequence, which are dominated with limestone and shale then followed by volcanic and pyroclastic rocks [11].

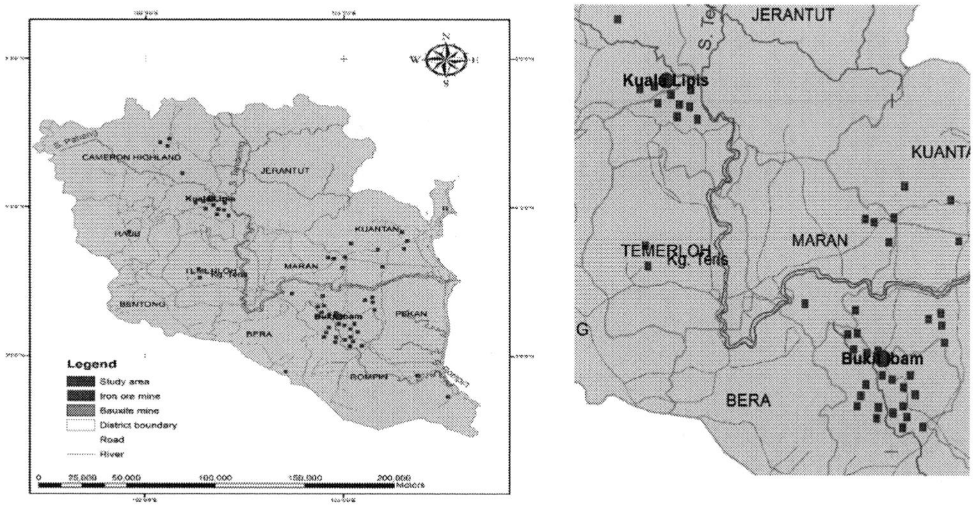

Fig. 1. Map of mining study sites of Bukit Ibam, Kuala Lipis (Mineral and Geoscience Department Malaysia, 2014)

2.2 Sampling of water and surface soil

2.2.1 Sampling of surface soil and water

Water samples were collected with great care to prevent contamination. Water samples for BOD, COD, TSS and AN were collected at site and brought back to the laboratory for analysis. Samples for COD and AN were acidified with H_2SO_4 until pH<2, whilst samples for BOD were analysed within 48 hours after sampling. Water pH and DO were measured on-site using Ultrameter 6P and a DO meter, respectively. Water samples for metal analysis were collected using 150ml clean polyethylene container and were preserved with 1% v/v HNO_3 until pH < 2. The water samples were filtered upon arrival to the laboratory and were analyzed using ICP-OES for metal elements (Cu, Fe, Mn, Pb, Zn, Cd, As, Cr, Ni and Al).

Samples of surface soils were collected at both active and abandoned iron mining sites. At each site, the soil were scooped into a clean polyethylene bags and then sealed and labelled. The sample of soils were kept in an icebox and brought back to the laboratory for further analysis. In the laboratory, all of the soil samples were oven dried for 3 hours at 105°c, pulverized into fine powder using mortar and pestle and then sieved using 2mm test sieve [12]. The preparation of samples was conducted accordingly to the USEPA (USEPA, 1996) method for heavy metal analysis. To extract heavy metal, acid digestion technique was used where 1.0g of dry soil samples were digested with aqua regia extraction method, using hydrochloric acid and nitric acid (ratio of 3:1) [10]. Then, the samples were analyzed for heavy metal (Fe, Pb, Cr, Cd, Zn, Cu, Mn, As, Cd, Cr and Ni) using inductively coupled plasma-optical emission spectrometer (ICP-OES).

2.2.2 Water quality parameters assessment

The Department of Environment (DOE) uses water quality index (WQI) to measure the water quality status in Malaysia. WQI serves as an indicator to discover the pollution levels and environmental basis in designation of classes for the beneficial use of National Water Quality Standards (NWQS) in Malaysia. It consists of series of sub-indices which are then compared to the given rating curve, combined and resulting in the final index [13]. The WQI index range from 0-100 where Class III, 0-59 (polluted), Class II, 60-80 (slighty polluted), and Class I, 81-100 (clean). The sub-indices that were specified to calculate WQI are the concentration of biological oxygen demand (BOD), chemical oxygen demand (COD), dissolved oxygen (DO), ammoniacal nitrogen (AN), total suspended solids (TSS) and pH [14,15]. Measurements of pH and DO were taken on-site whilst BOD, COD, TSS and AN were determined in the laboratory, and the WQI were calculated using equation 1[16];

$$WQI = (0.22 * SIDO) + (0.19 * SIBOD) + (0.16 * SICOD) + (0.15 * SIAN) + (0.16 * SISS) + 0.12 * SIPH) \quad (1)$$

3 Results and discussion

3.1 water quality assessment

For WQI assessment, the DOE-WQI was calculated for all sampling locations. The results of the WQI were shown in Table 1. From the observation, it is found that the lowest WQI was at S2 in Kuala Lipis (67). In Kuala Lipis, the WQI calculated was found in the order of S2<S4<S3<S1 (lowest to highest). S1 is a discharge just outside the mining area. S2 is a stream around the mining site. S3 is another discharge near S2 while S4 is a discharge that is close to resident area. S1 was classified as class I (clean) while S2, S3 and S4 were in class II which indicate that the water is clean but need further treatment if used as a raw water alternative (NWQS).

While in Bukit Ibam, the DOE-WQI resulted in clean status for all the four sites. In order of arrangement from lowest to highest, WQI was found in the order of S4<S1<S3<S2. The highest value is 90 which is at S2 (a pond near mining site) and then S3 which is a discharge around the mining area. S1 is the river point nearest the iron mine and S4 is another discharge at the mining area. The Bukit Ibam mine site was the biggest iron mining site in Malaya during 1960s and was one of the largest mining site in Southeast Asia [10]. A study on the water quality status in Bukit Ibam found that the pH was acidic [9] and there is a potential of AMD (acid mine drainage) problem. Heavy metal contamination arising from mining industries has given tremendous harmful risk to the environment [17]. The clean status of water quality may be due to the fact that the mine has been left for more than 40 years and hence the heavy metal concentration have been decreased and segregated away through the surface and groundwater streams but also could be settled down and penetrated into the sediments [18].

Additionally, further water quality evaluation is required as WQI does not include major and trace element geochemistry to understand the behaviour of heavy metals in water [13]. It was concluded that there was no apparent contamination in the water quality for both mining sites observed as all the readings are below the standards given by the MOH and NWQS. Table 2 shows the heavy metal concentrations in water of both mining sites. From the concentration values in Kuala Lipis, all of the parameters are under the standard guidelines except for the value of Mn in both S1 and S2. Mn is a common metalloid found in iron ore mines [19] and due to the alkaline condition of the water pH, the Mn may not dissolved easily and remain in the water [20]. Such water are in concern because the Mn tends to stay in solution, and can be carried to downstreams and eventually into domestic waters. Whilst in Bukit Ibam, all parameters are still under standard limit except for Al and Mn. Al is common in many ores including iron ore. Al is very hard to reduce and has been contributed to a lot of aquatic related pollutions [21].

3.2 Sediment analysis

Analyzing heavy metal concentrations in sediments/soils are an important reference indicator to water quality [22,23]. The values are compared with the UK Environment Agency standards (2008) and Canadian Counsils of Ministers of Environment (2011), due to the fact that Malaysia does not have its own standards for contaminated land. Table 2 shows the heavy metal compositions of the surface soils. Notwithstanding the WQI classification, it is interesting to note that in Kuala Lipis, all stations showed most of the heavy metal concentrations were below the recommended guideline values except for lead (Pb), arsenic (As) and copper (Cu). As illustrated in Table 3, S1 showed Pb, Cu and As which exceed the recommended values. S2 showed high content of Cu and As while S3 showed high concentrations of Pb and Cu. Whilst in BukitIbam abandoned mine, although the WQI status showed clean status, the heavy metal concentrations in the surface soils are high for all stations mainly Pb, Cu, Zn and As. As the mining sites are located at the oil palm plantation, apart from mining activities, the contribution of Zn in the agricultural soils might have contributed to the concentration of Cu in the soil. [21]. The concentrations or As has exceeded the allowable limit where the concentrations were the highest at S3 (632.5mg/kg), 268.5mg/kg (S1), 194.5mg/kg in S2 and 54mg/kg in S4, compared to 5.9

mg/kg of the allowable limit by the Canadian Councils of Ministers of Environment (2011). The arsenic-containing minerals from the iron mine possibly mobilized during the mining activities and deposited to the sediments or soils and are still there even after 40 years. Therefore, it was found that several of these metals were at high concentration level and hence careful consideration has to be given upon potential remediation strategies.

With regard the effect on human health, heavy metal such as Pb, in mines and smelters could lead to acute Pb poisoning where the symptoms are headache, abdominal pain and lead encephalopathy. Research found that a long term exposure of Pb could cause memory deterioration and diminished intellectual capacity [24], whereby too much Cu in liver cells is harmful and lead to liver damage and other symptoms that give serious on health. As is a well distributed metalloid occurring in metal rocks and also has been taken up in daily foods. Despite that, inorganic arsenic is acutely toxic and large exposure to arsenic could also give a harmful effect to the human bodies which gives severe disturbance to the nervous and gastrointestinal systems, and also gangrenous changes (black foot disease) as reported in Taiwan [25].

Table 1. Physico-chemical data of the waters

Location	pH	DO (mg/L)	TSS (mg/L)	BOD (mg/L)	COD (mg/L)	AN (mg/L)	WQI	Class	Status
Kuala Lipis (iron ore mining)									
S1-Kuala Lipis discharge	7.9	3.76	0.014	6.7	0.67	0.2	81	I	Clean
S2- Kuala LipisdischargeII	7.88	4.11	0.011	5.4	148.33	0.15	67	II	Slightly polluted
S3- Kuala LipisdischargeIII	7.86	2.75	0.022	3.6	2	0.2	78	II	Slightly polluted
S4- Kuala Lipis discharge	7.99	2.3	0.111	4.7	6.67	0.2	72	II	Slightly polluted
Bukit Ibam (abandoned iron ore mine)									
S1- Bukit Ibam (river)	6.99	5.31	0.035	5.4	2.33	0.25	85	I	Clean
S2- Bukit Ibam abandoned mine pond	6.45	4.94	0.028	3.5	1.33	0.11	90	I	Clean
S3- Bukit Ibam abandoned mine II discharge	7.96	5.25	0.004	4.6	3.33	0.42	86	I	Clean
S4- Bukit Ibam abandoned discharge	7.76	5.02	0.068	4.7	7	0.38	84	I	Clean
MOH (2009) (untreated raw water)	5.5-9.0	**	**	**	**	**			
MOH (2009) (treated water)	6.5-9.0	**	**	**	**	**			
WHO (2009) & USEPA (2009)	6.8-9.2	**	**	**	**	**			
EU drinking water standards (EU directive 98/83/EC	6.5-9.5	**	**	**	**	**			

#Data presented are mean data (*n*=) for all parameters
**not specified

 DOI: 10.1051/matecconf/201710305010

Table 2. Heavy metal concentrations in water

Location	Cu	Fe	Mn	Pb	Zn	Cd	Cr	Ni	Al
				mg/L					
Kuala Lipis (iron ore mining)									
S1-Kuala Lipis discharge	0.003	1.299	0.297	<LOD	0.008	<LOD	<LOD	<LOD	0.13
S2- Kuala LipisdischargeII	0.002	0.377	0.77	<LOD	0.005	<LOD	<LOD	<LOD	0.084
S3- Kuala LipisdischargeIII	0.005	0.286	0.106	<LOD	0.005	<LOD	<LOD	<LOD	0.079
S4- Kuala Lipis discharge	0.006	0.111	0.01	<LOD	0.003	<LOD	<LOD	<LOD	0.16
Bukit Ibam (iron ore mining)									
S1- Bukit Ibam (river)	0.025	0.104	1.027	<LOD	0.642	0.001	<LOD	<LOD	0.345
S2- Bukit Ibam abandoned mine	0.002	0.069	0.068	<LOD	0.044	0.001	<LOD	<LOD	0.19
S3- Bukit Ibam abandoned mine II	0.011	0.039	0.287	<LOD	0.017	0.002	<LOD	<LOD	0.089
S4- Bukit Ibam abandoned (sand mine)	0.006	0.039	0.104	0.005	0.008	<LOD	<LOD	<LOD	0.054
MOH (2009) (untreated raw water)	1.00	1.00	0.20	0.10	5.00	0.005	0.05	**	**
MOH (2009) (treated water)	1.00	0.30	0.10	0.05	5.00	0.005	0.05	**	0.2
WHO (2009) & USEPA (2009)	1.00	0.30	0.05	0.05	5.00	0.01	0.05	0.02	0.2
EU drinking water standards (EU directive 98/83/EC	2.00	0.25	0.05	0.01	**	0.005	0.05	0.02	

#Data presented are mean data (*n*=) for all parameters
**not specified

Table 3. Heavy metal composition of the surface soils

Station	Pb	Cu	Zn	Mn	Fe	As	Cd	Cr	Ni
					(mg/kg)				
Kuala Lipis (iron ore mining)									
S1-Kuala Lipis discharge	63.5	97	106	2127.7	128550	20.5	0.37	7.89	1.45
S2- Kuala LipisdischargeII	32	165.5	116	2213	127350	46.5	0.42	6.94	2.99
S3- Kuala LipisdischargeIII	72.5	67.5	93	2737	68200	5	0.063	9.34	4.36
Bukit Ibam (abandoned iron ore mine)									
S1- Bukit Ibam (river)	288.5	1798.5	1512.5	4478	134500	268.5	0.029	11.07	4.33
S2- Bukit Ibam abandoned mine	2195	1482.5	1542.5	4341	119600	194.5	0.06	0.24	7.96
S3- Bukit Ibam abandoned mine II	344.5	2845	1691	5285	211050	632.5	0.04	5.96	1.34
S4- Bukit Ibam abandoned (sand mine)	96.5	67.5	1682	920.5	75900	54	0.042	2.47	1.99
Canadian Council of Ministers of Environment (2001)	35	35.7	123	**	**	5.9	0.6	37.3	18
UK Environment Agency (2008)									
TEL – Threshold Effects Level	35	36.7	123	**	**	5.9	0.596	37.3	18
PEL – Predicted Effects Level	91.3	197	315	**	**	17	3.53	**	**
Background				400					

#Data presented are mean data (*n*=) for all parameters
**not specified

4 Conclusion

In this study, the water quality index have been calculated for both active and abandoned mining sites studied in Pahang. According to DOE-WQI classification, the WQI of the sites range from Class I (clean) to Class II (slightly polluted) for both Kuala Lipis (active) and Bukit Ibam (abandoned) mining sites. Although the water quality shows no apparent contamination, further water quality evaluation were conducted to measure the heavy metal concentration. From the heavy metal compositions in water, Mn and Al parameters have exceeded the allowable limits for raw water in both sites respectively. On the other hand, it was found that the heavy metal concentrations in surface soils are below the recommended guidelines values except for Cu, Pb, Zn and As in most of the sites. Despite the findings from this study, it is recommended that further studies and monitoring should be conducted to assess long term impact of these heavy metals on the quality of river ecosystem and on human health.

The authors would like to give a special acknowledgement to the State of Pahang Government for giving permission to enter the site. The preliminary results presented here are only for the sake of learning and the results here are not certified by any registered analytical chemists.

References

[1] A. Ab Majid, H. Mohamed Shaharudin, S. Alias, E. Adnan, A. I. Ali Hassan, M Z. Ali, Malaysia mining industry by Minerals and geoscience department Malaysia,**81**,(2013)
[2] M. Benvenuti, I. Mascaro, P. Lattanzi, P. Parrini and G. Tanelli, Minewaste dumps and heavy metal pollution in abandoned mining District of Boccheggiano Southern Tuscany, Italy. Environmental Geology,**30**, 238-243,(1997)
[3] R. Asklundand B. Eldvall, *Contamination of water resources in Tarkwa mining area of Ghana. Department of Engineering Geology*, Lund University, Sweden, (2005).
[4] S. Wang andC.N Mulligan, Occurrence of Arsenic contamination in Canada: Sources behavior and distribution. Science of the total Environment, 366, 701-721,(2005)
[5] M. C. Navorro, C. Perez-Sirvent, M.J. Martinez-Sainchez, J. Vidal, P.J. Tovar and J. Bech Abandoned mine sites as a source of contamination by heavy metals: A Case study in a Semi-Arid Zone. J. of Geochemical Exploration, **96**, 183-193,(2007)
[6] P., Schwab, D.Zhu, and M.K.Banks, Heavy metal leaching from mine tailings as affected by organic amendments. **98**, 2935-2941,(2007)
[7] Z. Madzin, M. F. Shai-in, F. M. Kusin, Comparing heavy metal mobility in active and abandoned mining sites at Bestari Jaya, Selangor, **30**, 232-237,(2015)
[8] W.S. Wan Ngah and M.A. Hanafiah,Removal of heavy metal ions from wastewater by chemically modified plant wastes as adsorbents: a review, Bioresource Technology, **99**(10), 3935-3948,(2007)
[9] Kumar, A., Mining contractors and engineers, **33**, (2013).
[10] S. M. Diami, F. M. Kusin, Z. Madzin, Potential ecological and human health risk of heavy metals in surface soils associated with iron ore mining in Pahang, Malaysia. Env. Sci.Pollut. Res.,(2016)
[11] W. Dzick and R.Carlson, Monument Mining Limited: Mengapur Project, Pahang State, Malaysia. Project No.V1165, Amended Technical Report, Vancouver Canada, 75,(2012)
[12] W. Z, Wan Yaacob and N.S. Mohd Pauzi, Acid mine drainage and heavy metals contamination at abandoned and active mine sites in Pahang, Bulletin of the Geological Society of Malaysia, **55**, 15 - 20, (2009)
[13] F. M. Kusin, M. S. M Zahar, S.N. Muhammad, N. D. Mohamad, Z.M. Zin, S.M. Sharif, Hybrid off-river augmentation system as an alternative raw water resource: the hydrogeochemistry of abandoned mining ponds, Environ. Earth Sci.,**75**(3), 1-15, (2016)
[14] N,Mustapha,*Indices for water quality in a river*, Master's thesis, Asian Institute of Technology, Bangkok, (1981)
[15] M.A. Haque, Y.F. Huang, T.S. Lee, Seberang Perai rice scheme irrigation water quality assessment. J. Institution of Engineers Malaysia,**71**, 42-49,(2010).
[16] Department of Environment Malaysia(DOE), Interim National Water Quality Standards for Malaysia. Kuala Lumpur, (2008)
[17] D.B. Johnson, K.B. Hallberg, Acid mine drainage remediation options: a review. Science of the total Environment, 338, 3-14,(2005)
[18] L,B.Clarke, Coal Mining and Water Quality. London7 IEA Coal Research, **99**, (1995)

[19] K.B Krauskopt, Separation of manganese from iron in sedimentary process,**12(**61), 84, (1957)
[20] B.G. Lottermoser, *Mine wastes, characterization, treatment and environmental impacts*, Springer, Berlin, Germany,108, (2007)
[21] B.O. Rosseland, T.D. Eldhuset, M. Staurnes Environmental effects of aluminium, Environ. Geochem. Health, **12**(1-2), 17-27, (1990)
[22] Y.Z. Chen, H. Yang, Z.K. Zhang, M.Z.Qin, F. Jin, J.J. Lu, Application of equilibrium partitioningapproach to the derivation of sediment quality guidelines for metals in Dianchi Lake. Pedosphere, **17**(3), 284–294,(2007)
[23] F.M. Kusin, A review of the importance of hydraulic residence time on improved design of mine water treatment systems, World Appl. Sci., **26**(10), 1316-1322, (2013)
[24] F.M.K usin, S.N. Muhammad, M.S.Md.Zahar, M. Zafira, Integrated river basin management: incorporating the use of abandoned mining pool and implication on water quality status. Desalin Water Treat. **57**(60), (2016)
[25] L. Jarup, Hazards of heavy metal contamination. Bri Med Bull, **68**, 167–182, (2003)

DOI: 10.1051/matecconf/201710305011

Groundwater Quality and Soil Characterization: A Case Study at Bukit Tembaga, Kuala Nerang, Kedah

Nor Izzah Abdul Aziz[1] and *Norzila* Othman[2,*]

[1]Department of Water and Environmental Engineering,Faculty of Civil and Environmental Engineering,UniversitiTun Hussein Onn Malaysia.
[2]Micropollutant Research Centre, Faculty of Civil and Environmental Engineering, UniversitiTun Hussein Onn Malaysia.

Abstract. Soil test can be used toanalyze soil characterization. The character of soils can reflect the quality of underground water. Nowadays groundwater can be used as an alternative to drinking water. The mineral contain in soil will give an added value to the quality of groundwater.This study extracted groundwater from lot 1981, Bukit Tembaga, Kuala Nerang Kedah which was in Padang Terap district. Generally this area is classified as hilly area and surrounded by a rubber plantation. The objective of this paper is to analyze characterization of the soil in term of oxide compound, groundwater quality and correlation between chemical concentration and groundwater quality. Chemical concentration in soil was analyzed using XRF. Groundwater quality was tested by using APHA standard 2012. The parameters for groundwater sourcewere compared with the Groundwater Standard.The correlation coefficientvalue is calculated using Pearson's method. The results indicated that certain metals present in the soil and groundwater samples. The negative value of correlation coefficientproved that high concentration of metals in soil resulted to low concentration of metals in groundwater.

1 Introduction

Soil comprises of all the materials in the surface layer of the earth's crust that are loose enough to be moved by spade or shovel [1]. Soil is substantially different from one place to another place in the world. The changes of the soil are caused by chemical, physical and biological processes [2]. These processes can be intensified by the action of the climate, relief, parent material, living organisms and time, which are the main factors of soil formation [3].

The transformation of rock into soil wasdesignated as soil formation. Soil forming factors are climates, organisms, parent material, topography and time [4]. Every soil originally formed from parent material, a deposit at the earth's surface. The material could have been bedrock that weathered in place or smaller materials carried by flooding rivers,

*Corresponding author: norzila@uthm.edu.my

moving glaciers, or blowing winds. Over time, sun, water, wind, ice, and living creatures help transform, or change, the parent material into soil [1].

In the soil formation, some major elements exist depend on nature of the process. These elements including silicon (Si), aluminium (Al), calcium (Ca) , Magnesium (Mg), Iron (Fe), Sodium (Na), Pottasium (K), Manganese (Mn) , Phosphorus (P) and titanium in soils which are mainly release from parent mostly overlain by igneous or sedimentary rocks [5].Therefore soil composition and formationwill give effect to land use and local culture indirectly.

Groundwater is the water found underground in the cracks and spaces in soil, sand and rock. It is stored in and moves slowly through geologic formations of soil, sand and rocks called aquifers [6].Groundwater and source of groundwater quality can be effected by type of soil, human activity, climate relief and time[7-10].Underground water sources are an alternative to the drinking water supply in Malaysia. Despite that, not all underground water can be used as sources of drinking water supplies. Groundwater quality problem are typically associated with high hardness, high salinity and elevated concentration of iron, manganese, ammonium, fluoride and occasionally nitrate and arsenic [11].The contamination of groundwater by heavy metals was triggered as threats to public health [12-13]. Groundwater and soil pollution in Malaysia for the pastfew yearshas not been identified as key environmental issue in Malaysia as compare to many water pollution problems [13-14]. This is true since not many cases of environmental and human health incidents have been reported. However with increasing demand for agricultural andalternative drinking water, groundwater and soil vulnerability has become an important environmental and human health issues [7].

The objective of this paper is to analyze characterization of the soil compound and groundwaterparameter.The correlation between oxide metal from soil and in the groundwaterwill be calculated.

2 Study area

The study area was located at lot 1981, Bukit Tembaga Utara, MukimTemak, Kuala Nerang Kedah which is in Padang Terap District as shown in Fig. 1. According to Mineral and Geoscience Department of Malaysia, this area is characterized by the presence of the following major soil types namely ereniterock and argillite. Geology of the area is underlain by rocks of Triassic age (230 – 250 million years ago) in Semanggol Formation. Generally this area is classified as hilly area. The western part of this area consists of hose layers of shale rock, mudstone, siltstone andalluvial sediments [8].This area is surrounding by rubber plantation. Although rubber can be planted on variety of soil types, but it grows best on deep (minimum 2m), well drained and well aerated, permeable soils of loamy or sandy clay texture (clay content >20%)[15].

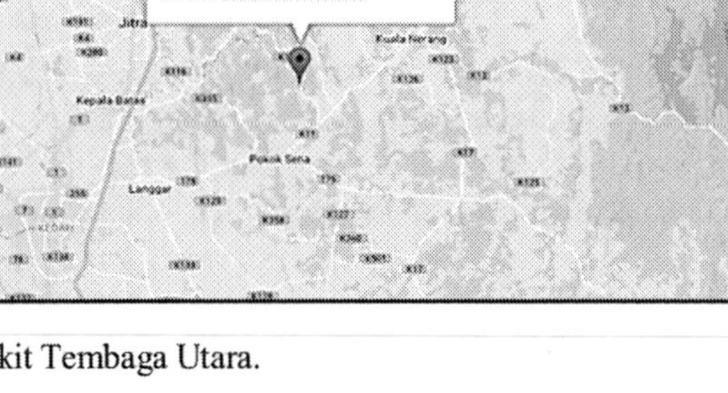

Fig.1. Map of lot 1981, Bukit Tembaga Utara.

3 Methodology

3.1 Soil sampling

Random sampling was used in soil sampling. Refer to EPA 1984 and EPA 1989 for a random number table and guidelines on selecting sample coordinates. The arbitrary selection of sample locations requires each sample location to be chosen independently so that results in all locations within the area of concern have an equal chance of being selected [15-16]as in Fig. 2.

This composited was comprised of 10cores of soil from sampling area 250ft^2. Every cores wastaken at the depth for each station of 1.2m and 2.4m, respectively. Sample A namely for the depth 1.2 m and sample B for the depth of 2.4m [16].

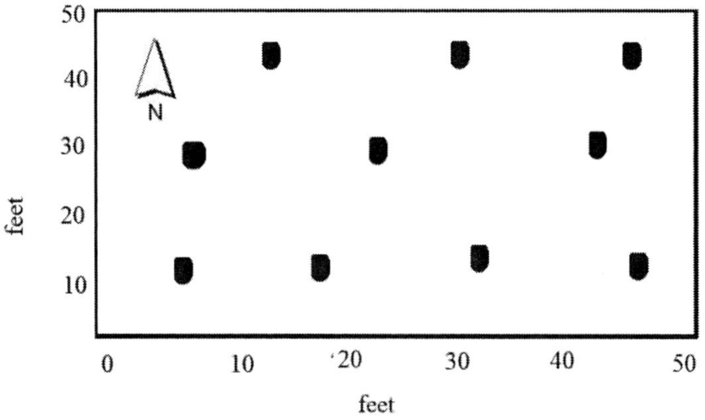

Fig.2. Random Soil sampling area and distribution [16].

Soil sample was taken from the surface to the depth of 15cm in cropping situation. The slices are placed in a clean plastic pail then mix thoroughly. The mixer was spread out on clean paper to dry in open air then crushed and ground to powder form before being analyzed by a X-Ray Fluorescent (XRF) machine [10].AnX- Ray Fluorescent Analysis is to determine the chemical composition of the soil. Prior to analysis 6.0g of soil were mixed homogeneously 4.0g binder wax and compressed into pallet using PE-Man Hydraulic press. Three samples were used for each depth per station.

3.2 Groundwater Sampling

Groundwater sampling have directed towards evaluating water quality of aquifers for water supply purpose[13].A total of 16groundwater samples were collected by grab sampling within 4 weeks from May to June 2016 [8]. Samples of groundwater were collected from groundwater tube well located atlot 1981, Bukit Tembaga Utara, MukimTemak, Kuala Nerang Kedah. The groundwater is located at foothill of Bukit Tembagaon the contour of 20m to 30m.The well is 112m depth and 200m radius andsurrounded by rubber tree plantation[8]. All sampling bottles were acid washed prior for sampling[13].

Groundwater was allowed to run for 30 minutes in order to remove stagnant water and allow representative groundwater to be sampled [9]. Sample bottle were fully filled with water samples to ensure the elimination of air bubbles and organic particulate matter entrapped in the bottles. Groundwater samples were filtered through 0.45μm membrane cellulose nitrate filter paper using glass filtration unit [14]. The filtrates were acidified with

HNO3 [15]. Then water samples were stored at 4°C before further analysis. Parameter analyzed as tabulated in Table 1.

Table 1. Method used for water parameter test [9].

Parameter	Unit	Method
Turbidity	NTU	APHA 2130B
pH	-	APHA 4500+H
Chloride	mg/L	In house method based on APHA 4110B
Magnesium (Mg)	mg/L	APHA 3120B
Sodium (Na)	mg/L	APHA 3120B
Total dissolved Solid	mg/L	APHA 2540C
Pottasium (K)	mg/L	APHA 3120B
Calcium (Ca)	mg/L	APHA 3120B
Silica	mg/L	APHA 4500 – SiO2 C

4 Results and Discussion

4.1 Soil Analysis

The soil sampleswere enriched with high concentration of Silicon dioxide(SiO_2) (43.90%) and Aluminium Oxide(Al_2O_3) (12.10%). High concentration of $SiO_2$and Al_2O_3were contributed by clay minerals such as vermivulate and illite [16]. Another factor influence the soil characteristic is due to the study areaunderlain by rocks of Triassic age (230 – 250 million years ago) in Semanggol formation, the chemical content in the soil is influenced by the type of rock especially parent material [1]. Other metals like iron, titanium, magnesium and zirconium present in low concentration. The chemical content in the soil at twodifferent depths were not significantly different.The result of all oxide metals as in Table 2.

Table 2. Chemical concentration in soil at lot 1981, Bukit Tembaga Utara.

Parameter	Sample A Concentration (%)	Sample B Concentration (%)
Silicon dioxide (SiO_2)	43.90	48.00
Aluminium Oxide (Al_2O_3)	12.10	13.00
Carbon (C)	10.00	10.00
Potassium oxide (K_2O)	2.35	3.15
Iron Oxide (Fe_2O_3)	2.33	1.79
Titanium Oxide (TiO_2)	0.70	0.70
Magnesium Oxide (MgO)	0.39	0.57
Zirconium dioxide (ZrO_2)	0.24	0.14
LOI (loss of ignition)	27.99	22.65

*Note:
1. Sample A is for 1.2m depth
2. Sample B is for 2.4m depth

4.2 Groundwater Analysis

Analysis of groundwater samples as tabulate in Table 3.The result of the analysis groundwater (raw water) is compared with Groundwater Standard and Ministry of Health Standard.According to groundwater standard the value of the groundwater in the range of the standard but certain essential minerals are present in low concentration.

The Ministry of Health Standard for raw water quality criteriawas adopted to ensure the quality of groundwater is applicable as source of drinking water after being treated [17].It is noted that the groundwater has high possibility to be used as drinking water due to the present of some mineral such as calcium, magnesium, zinc, iron, sodium, potassium, chloride, nitrates, sulphates and fluoride as essential mineral to human health [18-20]. Nevertheless, the values of minerals are still low compare to the values following MOH standard. Therefore, an extensive treatment is needed to further increase the minerals content in the groundwater. Adopted of sorption technology in removing and increasing essential elements is among choice of technology for groundwater treatment [21].

Table 3. Result for groundwater from Bukit Tembaga.

Parameters	Unit	Groundwater for study area	Ground water Quality Standard [raw water criteria]	MOH Drinking Water standard
Microbiological Escherichi Coli	MPN/ 100ml	absent	0/100mL	0/100mL
Physical				5
Turbidity	NTU	98.23	1000	15
Colour	TCU	10	300	1000
Total dissolved solid (TDS)	mg/L	208	1500	n.a
Dissolved Oxygen (DO)	mg/L	9.50	n.a	
Chemical				
pH	mg/L	6.23	5.5 – 9.0	6.5-9.0
Alkalinity		157	n.a	n.a
Chloride (Cl)		2.63	250	250
Sulphate (SO_4)		0.75	250	250
Nitrate(NO_3)		2.26	10	10
Fluoride(F)		0.10	1.5	0.6
Potassium(K)		4.12	200	200
Sodium (Na)		8.62	200	200
Bromide (Br)		0.41	n.a	n.a
Calcium (Ca)		31.84	n.a	n.a
Magnesium (Mg)		8.87	150	150
Zinc (Zn)		0.13	3	3
Silicon (Si)		8.11	n.a	n.a
Iron (Fe)		ND(<0.001)	1.0	0.3
Arsenic (As)		ND(<0.001)	0.01	0.01
Stronium (Sr)		0.06	n.a	n.a

*Note

1. ND – not detected by the instrument less than the minimum detection limit reported
2. n.a– Not available

Correlation is a statistical tool that helps to measure and analyze the degree of relationship between chemical concentration in soil and raw water parameter quantity.Pearson's method was used to find the correlationcoefficient (r) as in Table 4.

Table 4. The correlation between soil chemical and groundwater (raw water) content.

Formula	Soil (%)	Raw water (mg/L)	r
Silicon dioxide (SiO_2)	45.95	0.75	**-0.2752**
Potassium oxide (K_2O)	2.75	4.12	
Magnesium Oxide (MgO)	0.48	8.87	

The value of correlation coefficient is -0.2752, and it was consider weak correlation value. The present of high concentration of oxide metal in the soil resulted to low concentration the ions in the groundwater.

5 Conclusion

It could be concluded that the soil at the study area contain major and minor element of metals. The concentration of the major and minor element oxides has been notice with change in depth. Results of groundwater quality show that all the valuesare within the Ministry of Health Drinking for Groundwater Standard except for turbidity and Calcium (Ca). Significant metals that present in soil also available in groundwater analyzed.

This research was supported by Office for Research, Innovation, Commercialization, Consulting Management (ORICC), UniversitiTun Hussein Onn Malaysia (UTHM) under project vot; U569 and Post graduate Centre.

References

[1] S.K Sanjay,*Core principles of soil mechanis*. ICE publishing, London, (2014)

[2] J.Mark and J. Breitenstein, The relationship between soil and water: How soil amendments and compost can aid in salmon recovery, King Country department of Natural Resources. Washington. (1999)

[3] W. Chestworth. *Encylopedia of soil science*.Springer publishing (2008)

[4] A.S Vladychenskiy. Genesis of solis and factors of the soil formation, Environment Structure and Function: Earth System,**23–33**,(2004).

[5] H. Jenny and R. Amundson, *Factor of soil formation: A system of quantitative pedology*,McGrawHill, New York, (2005)

[6] S. N Zakir, L. Ali, S. A Khattak, Variation in major element oxide with time in the soils of Peshawar Basin: Their comparison with the normal agriculture soil,Journal of Himalayan Earth Sciences, **46**(2), 35–48, (2013)

[7] R. Frank and J.E.D Spellman,*The drinking water handbook, USA*: Taylor & Francisgroup, (2012)

[8] A.F Mohamed, W. Y. W. Yaacob, M.R Taha and A.R Samsudin,Groundwater and soil vulnerability in the Langat Basin Malaysia,European Journal of Scientific Research, **27**, 628-635, (2009)

[9] M. Mustaza and M. Muhammad.Perakuan jenis tanah dan punca air bagi Mukim Temak, Padang Terap ,Jabatan Mineral dan Geosains Malaysia, (2015)

[10] Water Environment& APHA, Standard Methods for the Examination of Water and Wastewater Part 1000, (2012).

[11] B. Musta, H. Fitria, W.Soehady,S. Tahir. Geochemical characterization of volcanic soils from Tawau, Sabah, Bulletin of the Geological Society of Malaysia, **54**, 33–36 (2008).
[12] N.Othman and S. Asharuddin.CucumisMelo Rind as Biosorbent to Remove Fe(II) and Mn(II) From Synthetic Groundwater Solution, Advanced Material Research, **795**,260-265, (2013)
[13] N. Othman, S. Mohd-Asharuddin, M-F-H Azizul Rahman. An overview of fruit waste as sustainable adsorbent for heavy metal removal,Applied Mechanics and Material, **389**,29-35, (2013)
[14] J.M.C Weaver, L. Cave, A.S Talma,Groundwater sampling,Water Research Commission, Republic of Soth Africa,(2007)
[15] H. Jenny,*Factors of soil formation*,Dovers publication New York, (2005)
[16] W, Verheye, Growth and production of rubber, Encylopedia of life support system (EOLSS), (2010)
[17] US, EPA. Soil Sampling Standard Operating Procedure: August 1997 Achive Document, (1997)
[18] Ministry Of Health, Malaysia: Kuala Lumpur. (2004)
[19] K.Kabilan, K.Mushilan and R.Sivaneswari,Study on the contents of bottled mineral water of various brands in Malaysia,J. of Natural Sciences Research, **2**(4) ,(2012)
[20] N. Othman, A. Che-Azhar and Suhaimi, A, Zinc removal using honewdew rind,Advanced Mechanics and Material, **680**, 150-153, (2014)
[21] N. Othman, Y.S. Kueh, M-F-H Aziziul-Rahman and R. Hamdan.Watermelon rind a potential adsorbent for zinc removal,Applied Mechanics and Material ,**680,** 146-149. (2014)

Acceptance of the Application of Sustainable Concept in Malaysia's Waste Management Infrastructure - Landfill

Tey Jia Sin[1,*], *Goh* Kai Chen[1], and *Peniel* Ang Soon Ern[2]

[1]Faculty of Technology Management and Business, Universiti Tun Hussein Onn Malaysia, Batu Pahat, Johor, Malaysia
[2]Faculty of Engineering Technology, Universiti Tun Hussein Onn Malaysia, Batu Pahat, Johor, Malaysia

Abstract. Sustainable concept has been a topic that is of broad and current interest in applying in many different areas of development. A sustainable concept encompasses three main criteria, namely environmental, social, and economic. This sustainable concept has usually been defined as the ability to develop something to meet the present needs yet it does not compromise the ability of future generation to meet their needs. Whereas waste management plays an important role in managing the municipal solid wastes produced. The landfill has been widely used in the disposal of solid wastes. However, the negative impacts caused by landfill have led to the need for application of sustainable concept. The acceptance of sustainable concept by stakeholders will determine the possibility of moving into sustainable waste management. This study aims at assessing the of acceptance by the stakeholders in the application of the sustainable concept into landfill. The research was conducted by reviewing of the past similar researches and through interview with ten waste management stakeholders. The results obtained show that there are some limitations in terms of understanding on the concept of sustainability and the willingness in accepting on this concept. Further action should taken in and moving towards sustainable waste management.

1 Introduction

The tremendous growth in Malaysian economy has led to the growth in the population along with the great influx of foreign workforce to cities. This phenomenon gives result an increase amount of wastes generated [1]. The increase of the production of municipal solid waste has brought to the need of having a proper waste management system. Besides, solid waste is already a significant problem for the Malaysian government [2].

Most of the municipal solid wastes produced in Malaysia are being disposed in landfill. This is due to the fact that some of the municipal solid wastes composition is not suitable to be burned and Malaysia is yet to have such technology to handle all different kinds of

* Corresponding author: jiasin.star@gmail.com

wastes. If the present rate of solid waste production goes on without effective supervision and disposal methods, there will be a substantial negative impact on the quality of our environment [2]. This implies that a waste management system that is sustainable is required. Through the application of sustainable concept into the waste management system, it could help to manage the solid wastes in a way that could minimise the impact to the environment, social, and economy.

2 Literature review

With all the negative impacts that landfill may cause, it is essential to tackle and provide a better solution for it through the application of sustainable concept. Besides, environmental problems caused by waste management have increased the pressure on waste managers, planners, and regulators. These push them to implement sustainable waste management in their projects [3] as well as selecting the sustainable option in managing waste. In this section, effort is made to search into the related literature in order to review the principle of sustainable development, acceptance as well as waste management infrastructure in Malaysia in relation to landfill

2.1 Principle of sustainable development

There are several ways to define the concept of sustainability. One of them is sustainability is the ability in sustaining the earth's resources while it is being exploited. The areas involved in sustainability are environmental responsibility, social awareness, and economic profitability [4]. In this context, sustainable concept is the idea of maintaining current resources for our future generation although the resources are being used for different purposes. This concept has slowly gained recognition by different agency and promotion in order to protect current resources from being exhaustively exploited. Thus, it is good to apply it into the management of municipal solid wastes.

Sustainable solid waste management aims to offer a chance to prevent waste through designs based on the full life cycle of the product, similar to natural cycles, which function without producing waste [5]. One of the researchers also supported that there is a need to work towards a sustainable waste management system, which requires environmental, institutional, financial, economics, and social sustainability [6].

There are many drivers affecting sustainability and the impact varies from one country to another, depending on political, socio-economic and environmental factors. The scenario is more complicated in Asia due to rapid industrialization and urbanization and changing waste composition and generation rates. Waste management technologies are generally conservative in Asia and the need for integrated waste management is imperative [7].

Sustainable landfill sites are becoming difficult to find as urban areas expand. Also, individuals are not willing to accept the implementation of a new landfill site near them because of concerns about smell, litter, pollution, pets, and the reduction in the value of their homes. There are large costs involved in providing conveniently located and environmentally responsible landfill facilities [5].

As a whole, the implementation of sustainable concept is about to bring more advantages to the waste management industry than the disadvantages. It should be accepted widely by the waste management stakeholders and applied in the daily decision of disposing waste through landfill.

2.2 Acceptance

Several researches have defined acceptance in different manners. Attitude, knowledge and support are the main factors to impinge on the acceptance of sustainable agriculture among farmers, but studies on the same scenario are lacking for Malaysia [8]. Furthermore, from a knowledge perspective, we can distinguish the transition in knowledge approaches for every type of agriculture, from conventional model to a sustainable one. Through this definition, though, Malaysia still lacking of the similar study yet the attitude, knowledge and support possess by the stakeholder can be a measure of the acceptance towards sustainable.

On the other hand, the acceptance is also being defined as the reflection of the behaviour that enables or promotes the use of a technology, rather than inhibits or demotes the use of it. Support can be expressed in proclaiming the technology (for example because of its environmental benefits, or purchasing and using the technology [9].

The lack of definition of acceptance as well as a model or framework in which to understand acceptance, and the drives behind it present a significant barrier both for understanding and working with acceptance. Since acceptance if developed by the individual it can only be based on the individual's perception of the system [10]. As a whole, the degree of acceptance towards sustainable concept can be identified through the individual attitude, knowledge possession and support that apply in the real practical work. An individual who is willing to accept will be shown through his action and attitude.

Hence, in this paper, the degree of acceptance towards the sustainable concept into waste management is based on the individual perception and the knowledge possess by the individual. It will be discussed in the following sections.

2.3 Waste management infrastructure in Malaysia – landfill

The major challenge in the urban areas, especially in the fast growing cities and towns is the management of solid waste. The reason behind it is the ever increasing population which lead to the increment of waste generation [11]. This is supported by Noor et al. [12] who also highlighted that the increasing population and rapid urbanization in Malaysia directly influence a municipal solid waste generation. This showed the needs to improve waste management system in order to contain the increasing amount of solid wastes.

Samsudin et al. [13] stated that there are about 230 municipal waste disposal sites in operation in Malaysia. However, most of the municipal waste disposal sites are open dumping sites, whereby the disposal of solid wastes are not operating according to the guideline and the sites are not well managed. Hence, the increasing quantities of wastes have made the efficient management of solid waste in Malaysia more important.

One of the main waste management technique adopted by Malaysia in handling the disposal of the solid waste is through landfill. However, although landfill appears to be capable of managing most of the solid wastes produced, but it undeniably has caused many negative impacts as well. Among the negative impacts are production of pollutants which include noise, odours, unsightliness, and air pollution through the heavy vehicle movement. It will contaminate the land and cause the land unsuitable for other usage purposes in the near future [14, 15]. In short, it brings negative social impacts towards the residents who stay near to the landfill sites. Both leachates and biogas that are the pollutant generated from the landfill. Biogas such as methane is generated from landfill and is released to the atmosphere is capable of causing global warming. The volume of biogas is estimated to be 23 times greater than that of the same volume of carbon dioxide gas [16]. All these negative impacts could be reduced by the implementation of sustainable concept into the construction of landfill, its operation and selecting the landfill sites.

DOI: 10.1051/matecconf/201710305012

Some authorities in Malaysia could not select a land for disposal site as the land is getting scarce and high cost is required in land acquisition [17]. The sites far from the source of waste generation increase transfer costs and require additional investments for infrastructure as well. Moreover, the solid waste is also often thrown directly into the waterways and rivers, or dumped on the roadside. Especially in the countryside, the amount of waste dumped openly is not known, without any environmental perspective plans. The reasons for these problems are lack of finance, land acquisition problems, and insufficient waste- collection and transfer system [5]. Through the stakeholders who possess the sustainable concept, a good landfill site that consists of sustainable criteria can be obtained.

3 Research methodology

This paper employed literature review and qualitative study, which is semi- structured interview to gather a more in- depth information from waste management industry. A total of ten waste management stakeholders who involved in the evaluation of a landfill site from the government sectors are interviewed.

The aim of this research is to identify the acceptance towards the implementation of the sustainable concept into waste management system. The concept of sustainability is desired to be implemented in the consideration of a suitable land to turn it into a landfill site. Thus, the perspective of each respondent which represents each government sector is to be determined in this research as well. The semi structured interview can provide the information for the development of a sustainable waste management.

3.1 Profile of respondents

The list of respondents is shown in Table 1. Among the departments are National Solid Waste Management Department, local authorities, State Economic Planning Unit, Public Work Department, Department of Environment, Town and Country Planning Department, Department of Irrigation and Drainage, Land Department, Solid Waste Management and Public Cleansing Corporation and a private company who works on managing solid waste.

Table 1. Respondents' Profile

No.	Position	Department
R1	Assistant Director	Department of National Solid Waste and Management
R2	Assistant Director	Solid Waste and Public Cleansing Management
R3	Assistant Director	Department of Environment
R4	Assistant Director	Town and Country Planning Department
R5	Assistant Director	State Economic Planning Unit
R6	Head of Assistant Director	Department of Irrigation and Drainage
R7	Assistant Engineer	Public Work Department
R8	Deputy	Local Authorities
R9	Assistant Administrative	Land and Survey Department
R10	Project Engineer	Private Company

Most of the respondents are professionals which are in high positions of making decision. Most of them are from different background which inferably contributed a vast variety of reliable information. The interview questions have been sent out to all the respondents prior to the interview session conducted.

4 Results and discussion

This research is aims in finding out the acceptance of sustainability concept in waste management industry by the waste management stakeholders. As stated in related literature as earlier reviewed, sustainability is the ability to sustain the earth's resources while it is being exploited. The areas involved in sustainability are environmental responsibility, social awareness, and economic profitability [4]. It can be clearly seen that there are two parties from each department that do not possess the whole understanding of sustainable concept.

It has been found out that all the respondents agreed to the importance of sustainability in the landfill site selection. It is also encouraging to obtain the information that all of the respondents seeing the importance to have the concept of sustainability. However, not all the government departments have applied it into the process of evaluation of a site to be used as landfill (see Fig. 1).

Based on Fig. 1, it can be observed that there are 90% of the respondents from each department that supported and implemented the concept of sustainability into the landfill site selection. However, through the data collected from the respondents, most of them are not applying the whole aspects of sustainability aspects into the landfill site selection. Some of the views of the respondents are presented as follows:

Yes, in this department, we did emphasize on sustainable concept. For example they set up buffer zones whereby they need to plan many trees. (R3)

Yes, it is very important. However, for Department of Environment, the environment and social aspects have always been emphasized in the evaluation of landfill site selection. (R5)

Fig. 1. The Application of Sustainable Concept into Landfill Site Selection.

It can be observed that most of the respondents claimed to apply the concept of sustainability in the process of site selection. Among the respondents are R1, R2, R3, R4, R5, and R6, with each representing different departments. Conversely, most of them are only able to take into consideration on environmental aspects and some social aspects. This also implies that there is a strong urge to further emphasize the sustainable concept and to help all the related parties to get the full understanding of sustainability aspects.

From Fig. 1, that other 10% is on the Public Work Department. As the respondents stated:

I do agree that the concept of sustainability is important, however, it is not implemented in this department because the Public Work Department is focusing on the road and access only. (R7)

The Public Work Department's focus on the road and access has pushes them found that it is not necessary to integrate sustainable concepts into landfill site selection. It is important to help them to be aware of working as a whole unit in the landfill site selection and the importance of how each department chooses to contribute the best idea in the landfill site selection.

On the other hand, R1 and R9 expressed that they do not understand the sustainable concept before this interview session, while R4 mistakenly thought that the sustainable concept is focused only on the environmental criteria and R2 regarded that sustainability is all about the separation of the solid waste at its source. However, all these respondents were willing to accept and consider applying this concept into the management of solid wastes after listening to what sustainability concept is all about.

However, Respondents 6, 8 and 10 showed the understanding and willingness to accept the application of the sustainable concept into landfill. These small part of respondents showed that there are still some waste management stakeholders whom possessed the understanding and good response to the sustainability concept. Respondent 8 expressed the sustainable concept as:

Sustainable concept is a way to plan for future usage. Also, it is environmental friendly (R8)

From the data obtained, it can be concluded that the concept of sustainability is yet to be widely comprehended by all the waste management stakeholders. Mohamed [18] also supported that there is a lack of knowledge in the sustainable concept in the local governments and that it would in turns cause several issues such as improper public sanitation among others. Although the importance of sustainability has been agreed and acknowledged by the respondents. Most of the respondents still need to be educated in order for them to realise the benefits of adopting sustainable concept into their real life project. The conclusion can be drawn that the basic knowledge of sustainability has not been fully captured and applied into all the relevant parties and their job scope. Hence, it strengthens the needs to promote sustainable concept into waste management system, landfill.

5 Conclusion

Sustainable concept is getting more attention and being used in different industries. Due to the increasing number of municipal solid wastes in a country, it led to the need in improving the waste management industry. This can be done by implementing the concept of sustainability in the operation of waste management system. Therefore, it requires the acceptance by the waste management stakeholders in order to be implemented in Malaysia waste management industry. However, through the research, it is found out that the acceptance of the sustainable concept in Malaysia at this stage is still considerably low. Moreover, the results have reflected that their understanding towards the concept of sustainability is also low. From the study, it reinforces the need of imparting the concept of sustainability to the waste management stakeholders as to improve the whole waste management industry.

The authors express grateful acknowledgement to the industry stakeholders for their valuable contributions towards the success of this research. This research was partly sponsored by the Centre for Graduates Studies UTHM.

References

[1] S. Kathirvale, M. N. Muhd Yunus, K. Sopian, and A. H. Samsuddin, Energy potential from municipal solid waste in Malaysia, *Renewable energy,* **29**, 559-567, (2004)

[2] S. A. Haron, L. Paim, and N. Yahaya, Towards sustainable consumption: an examination of environmental knowledge among Malaysians, *International Journal of Consumer Studies,* **29**, 426-436, (2005)

[3] R. Minciardi, M. Paolucci, M. Robba, and R. Sacile, A Multi-objective Approach for solid waste management, in *Proc. iEMSs*, 205-210, (2002)

[4] M. Pitt, M. Tucker, M. Riley, and J. Longden, Towards sustainable construction: promotion and best practices, Construction Innovation: Information, Process, Management, **9**, 201-224, (2009)

[5] U. N. Ngoc and H. Schnitzer, Sustainable solutions for solid waste management in Southeast Asian countries, Waste Management, **29**, 1982-1995, (2009)

[6] S. Rathi, Alternative approaches for better municipal solid waste management in Mumbai, India, Waste Management, **26**, 1192-1200, (2006)

[7] P. Agamuthu, S. Fauziah, K. Khidzir, and A. N. Aiza, Sustainable waste management-Asian perspectives, in *Proceedings of the international conference on sustainable solid waste management*, **15**, (2007)

[8] A. Azman, J. L. D'Silva, B. A. Samah, N. Man, and H. A. M. Shaffril, Relationship between attitude, knowledge, and support towards the acceptance of sustainable agriculture among contract farmers in Malaysia, Asian Social Science, **9**, 99, (2013)

[9] N. M. Huijts, E. Molin, and L. Steg, Psychological factors influencing sustainable energy technology acceptance: A review-based comprehensive framework, Renewable and Sustainable Energy Reviews, **16**, 525-531, (2012)

[10] E. Adell, The concept of acceptance, in *ICTCT-workshop,* (2008).

[11] A. Johari, S. I. Ahmed, H. Hashim, H. Alkali, and M. Ramli, Economic and environmental benefits of landfill gas from municipal solid waste in Malaysia, Renewable and Sustainable Energy Reviews, **16**, 2907-2912, (2012)

[12] Z. Z. Noor, R. O. Yusuf, A. H. Abba, M. A. Abu Hassan, and M. F. Mohd Din, An overview for energy recovery from municipal solid wastes in Malaysia scenario, Renewable and Sustainable Energy Reviews, **20**, 378-384, (2013)

[13] A. R. Samsudin, B.-e. E. A. Rahim, W. Z. W. Yaacob, and U. Hamzah, Mapping of contamination plumes at municipal solid waste disposal sites using geoelectric imaging technique: Case studies in Malaysia, J. of Spatial Hydrology, **6**, (2006)

[14] N. P. Cheremisinoff, *Handbook of Solid Waste Management and Waste Minimization Technologies,* First ed. United States of America: Butterworth-Heinemann, Elsevier Science, (2003)

[15] S. El Haggar, *Sustainable Industrial Design and Waste Management: Cradle-To-Cradle for Sustainable Development*: Academic Press, (2010)

[16] S. Renou, J. G. Givaudan, S. Poulain, F. Dirassouyan, and P. Moulin, Landfill leachate treatment: Review and opportunity, J. of hazardous materials, **150**, 468-493, (2008)

[17] Z. Sakawi, Municipal solid waste management in Malaysia: Solution for sustainable waste management, J. of Applied Sciences in Environmental Sanitation, **6**, 29-38, (2011)

[18] S. Mohamed, Preliminary concept of knowledge transfer practices towards sustainable development for local governments, *In: MUCET 2015*, 11-13 October 2015, Johor (2015)

MATEC Web of Conferences **103**, 05013 (2017) DOI: 10.1051/matecconf/201710305013
ISCEE 2016

Noroviruses Surrogate Detection Using Loop-Mediated Isothermal Amplification (LAMP), Conventional RT-PCR and Quantitative RT-PCR in Sg. Tekala and Sg. Gabai Stream Water

Dzulaikha Khairuddin[1,3,*], *Nurul Yuziana* Mohd Yusof[2], *Jan Maizatulriah* Jani[3], and *Marfiah* Ab. Wahid[3*]

[1]Malaysia Genome Institute, Jalan Bangi, 43000 Kajang Malaysia
[2]School of Biosciences, Faculty of Science and Technology, Universiti Kebangsaan Malaysia, 43600 UKM Bangi, Malaysia
[3]Department of Water Resources and Environmental Sustainability, Faculty of Civil Engineering, Universiti Teknologi MARA, 40450 Shah Alam, Malaysia

Abstract. A preliminary study was carried out in order to evaluate a reverse transcriptase loop-mediated isothermal amplification (RT-LAMP) assay for detecting Noroviruses (NoV) RNA surrogates as external standard for NoV.The detection limit of the NoV RT-LAMP assay was observed to be 22 copies/μL. This RT-LAMP assay sensitivity was comparablewith the quantitative reverse transcriptase-Polymerase Chain Reaction (PCR) and shown to be 10-fold more sensitive than end-point conventional reverse transcriptase-PCR (RT-PCR). The NoV RT-LAMP assay showed high specificity to NoV targeted gene when specificity test was completed with no cross-reactivity with other 17 environmental strains. The assay also was performed with 11 spiked recreational stream water randomly picked from two recreational areas in Hulu Langat Malaysia, Sg. Tekala and Sg. Gabai. The RT-LAMP assay is simpler compared to the conventional PCR and real-time PCR, which in optimum isothermal temperature of 63°C, the amplification can be completed in 40 minutes. Results of spiked recreational stream water samples suggested that the NoV RT-LAMP assay can be used as monitoring tool for NoV surveillance in recreational stream water.

1 Introduction

There are not many surveillance of NoV in Malaysia have been carried out in water and environment area except very few studies had been performed in food application[1,2]. The lack of information could be contributed to misdiagnose of severe diarrhea symptoms of NoV to other gastroenteritis bacterial such as *Shigella*sp. and *Escherichia coli* and other enteric viruses like Rotavirus and Adenovirus [3].It is suggested that the infection of NoV occurred from the activities of swimming, canoeing or other recreational activities relating

* Corresponding author: adawiyahdanial@yahoo.com

to the fecal contaminated water [4–6]. For decades, fecal indicator bacteria (FIB) such as *E.coli* and enterococci has been widely used as standard procedure to evaluate fecal borne pathogen contamination in water. Although this technique is being used for water quality criteria [7], it has limitation in protecting the public from recreational waterborne disease especially viruses infection. The requirement of 18 to 24 hours of virus incubation post-exposure prior to issue advisory on infectivity may increased the greater risk of infection[8,9]. Another drawback of using FIB as indicator is that these bacteria has been reported to continuously grow in the environment, leads to a bias in assessment especially in tropical weather like Malaysia[10–12]. Human enteric viruses such as enteroviruses, NoV and adenoviruses which directly associated with human diseases found to be more reliable indicators for fecal contaminated water [13,14].These enteric viruses have the ability to survive for longer period of time and reflect their existence in real number since they cannot replicates without permissive host organism in water [13,15].

Although, the public health risks assessment of NoV in recreational water is very important, the need on the advanced diagnostic tool and high analytical skill in determining the availability of NoV in water samples has concealed the perceptive on the real existence, concentration and behaviours of NoV in aqueous environment[16]. PCR and RT-PCR method is time saving and able to detect non cultivable viruses or slow-growing viruses. However, the downside of these methods is the need of bulky, expensive thermo cycler which usually inaccessible in a remote area associated with waterfalls and streams.These lacks of accessible leads the development of loop-mediated isothermal amplification (LAMP) which employs simpler method for DNA amplification; isothermal condition around 63 to 65°C and directly visualized after 30to 60 minutes incubation.Since no cultivation activity involved like FIB enumeration method [7], this molecular methodhas the advantage to serve as a faster and safer platform for monitoring NoV contamination in recreational water with simplified risk consideration for recreational water contamination management.In this study, recreational stream waters were spiked with NoV to mimic the real detection of NoVusing PCR, real-time RT-PCR and RT-LAMP, and the performances of these methods for NoV detection in environmental sample were compared.

2 Materials and methods

2.1 Laboratory strains and specificity test

Specificity test for the primers used in this study were validated using 15environmental DNA strains as described in our previous study [17]: *Burkholderia cepacia, B. thailandensis, B. pseudomallei, Bacillus subtilis, B. macerans, B. circulans, B. megaterium, Staphylococcus* sp., *S. epiderminis, S. heamolyticus, E. coli* BL21, *E. coli* Nova Blue, *Shigella* sp., *Salmonella* sp. and *Shinella granuli.* All strains, positive control (DNA plasmid of NoV) and negative control (steril dH_2O) were assessed simultaneously in optimized LAMP condition.

2.2 NoV surrogates construction and transcription *in-vitro*

The external standard for NoV was constructed synthetically in plasmid (Integrated DNA Technology, USA) by inserting specific target genes sequences for NoVd etection.The molecular weights of nucleotide base pairs of the products recombinant plasmid are 207bp and the sequences are:

5′GTGGTATGGATTTTTACGTGCCCAGGCAAGAGCCAATGTTCAGATGGATGAGATTC
TCAGATCTGAGCACGTGGGAGGGCGATCGCAATCTGGCTCCCAGTTTTGTGAATGAAGATG

GCGTCGAATGACGCCAACCCATCTGATGGGTCCGCAGCCAACCTCGTCCCAGAGGTCAACA ATGAGGTTATGGCTTTGGAGCCCGTTGTC-3'. The target gene was previously being used as NoV specific marker [18]. Briefly, a total of 400 complete and partial RNA-dependent RNA polymerase and capsid protein gene sequences of NoV from Genbank database were aligned using Vector NTI Advance 11 (Invitrogen, Auckland, New Zealand). The most conserved region among these genes was inserted into pIDT-SMART-NoV vector and sequencing-verified (Integrated DNA technology, USA). This template was further used as template for *in-vitro* transcription using the Ribo-MAX™ Large Scale RNA production System-T7 (Promega, Madison, USA)as described by the manufacturer's protocol. The transcribed RNA was then purified with QIAamp Viral RNA Mini Kit (Qiagen, Hilden, Germany) and measured using Nanodrop 1000 (Thermo Scientific, USA). The reading of RNA absorbencies were then converted to molecular copies according to Krieg formula [19] and serially 10-fold diluted from 10^{-1} to 10^{-8} for further use.

2.3 Primers design for RT-LAMP of NoV

The oligonucleotides were designed using LAMP primer designing software Primer Explorer 4.0 (http://primerexplorer.jp/e/) according to the conserved region of target gene inserted to the pIDT-Smart-NoV vector. List of primers suggested by the software were evaluated as described by Notomi *et al.*[20]. The primers set consist of six primers; two outer primers (B3 and F3), two inner primers (BIP and FIP) and two loop primers (LF and LB) that recognizing eight target sequence. The B3 and F3 oligonucleotide primers used for the RT-LAMP of NoV were also used for quantitative RT-PCR and conventional RT-PCR. The details of the selected primers sequences are shown in Table 1.

Table 1 Sequences of primer set used for Noroviruses surrogate RT-LAMP, RT-PCR and qRT-PCR

Assays	Primer	Sequence (5'-3')	Reference
RT-LAMP/ RT-PCR/ qRT-PCR	F3	GTGGTATGGATTTTTACGTGCC	[18]
RT-LAMP/ RT-PCR/ qRT-PCR	B3	GACAACGGGCTCCAAAGC	
RT-LAMP	FIP	CAGATTGCGATCGCCCTCCCGAG CCAATGTTCAGATGGATGA	
RT-LAMP	BIP	TGAAGATGGCGTCGAATGACGCA ACCTCATTGTTGACCTCTGG	
RT-LAMP	LF	CGTGCTCAGATCTGAGAATC	
RT-LAMP	LB	TCCGCAGCCAACCTCGT	

2.4 RT-LAMP reaction and detection

The RT-LAMP reaction mixture consist of 1.0 µL of 10x ThermoPol Reaction Buffer, 0.4 µL of 8U/µL*Bst* DNA polymerase (New England Biolabs), 0.4 µL of 200U/µL Superscript III Reverse Transcriptase (Invitrogen, USA), 1.0 µL of 10mM dNTP mix (Invitrogen, USA), 1.6 µL of 5M Betaine (Sigma, USA), 1.2 µL of 50mM $MgCl_2$ (Invitrogen, USA), 0.4 µLof 5 µM each primers F3 and B3 (Integrated DNA Technology, USA), 0.4 µL of 40 µM each primers FIP and BIP, 0.4µL of 20µM each LF and LB primers and 0.8 µL of target RNA/DNA. The reaction mixture was incubated at 63°C for 45 min and heated at 80°C for 5 min to terminate the reaction.

The RT-LAMP reaction were observed visually for colour changing from soft orange to fluorescent green in the presence of LAMP amplified products. Alternatively, the RT-LAMP also been observed by electrophoresis in 2.5% agarose gel (Fermentas , USA) added with 1:10000 SYBR® safe (Bio Rad, California, USA) staining dye and viewed in UV-transilluminator (Gelcompany, San Francisco, USA) for the present of ladder-like pattern.

2.5 Quantitative RT-PCR and RT-PCR

The cDNA of RNA transcript was produced using Superscripts III first-strand synthesis system (Invitrogen, USA). The reaction was done in one reaction tube with two steps. The first step was done with mixture of 1.0μL of 50ng/μL random hexamers, 1.0 μL of 10mM dNTP mix and 8.0 μL of total RNA and incubated at 65°C for 5 min, and cold down in ice for 1 min. Next, 2.0 μL of 10x RT buffer, 4.0μL of 25 mM $MgCl_2$, 2.0 μL of 0.1 MDTT, 1 μL of 40 U/μL RNAse OUT and 1 μL of 200U/ μL Superscript III reverse transcriptase were added. The cDNA were synthesized at 25°C for 10 min, 50°C for 50 min and terminated at 85°C for 5 min. The product was stored in -20°C for further usage.

RT-PCR reaction mixture was performed with the NoV specific primers targeting the same region as RT-LAMP using the B3 and F3 primers (Table 1). The reaction mixture was carried out in a total volume 10 μL containing 5 μL of 2X Gotaq® Green Mastermix (Promega, USA), 1 μL of 10 μM F3 and B3 primers, 1 μL of cDNA template and 2 μL of dH_2O. The tube were incubated in thermal cycler (MJ Research,MA, USA) using programme; 94°C for 5 min, 30 cycles of 94°C for 45 s, 55°C for 30 s and 72 °C for 30s and a final extension cycles at 72°C for 5 min. The expected PCR product was 207bp.

The quantitative RT-PCR was performed in Mastercycler®quantitative PCR system (Eppendorf,USA) using SYBR Green I based real-time PCR amplification system. The amplification was carried out in a 25 μL reaction using SsoFast™EvaGreen® Supermix (BioRad, California, USA). The reaction mixture consist of 12.5 μL of SsoFast™ EvaGreen® Supermix, 1.0 μL of each F3 and B3 primers (10μM), 2.0 μL of cDNA template and 8.5 μL of dH_2O. The cycler was programmed to95°C for 30 s, followed by 40 cycles of 95°C for 5 s, 55°C for 10 s and 72°C for 10s. The melting curve analysis was performed after the amplification cycles finished verifying the targeted products.

2.6 Application of the RT-LAMP and RT-PCR in spiked water samples

Aliquots from 11water samplesrandomly collected from two recreational streams located in the Hulu Langat province (Sg. Tekala and Sg. Gabai) were used as spiked water samples.NoV surrogates (RNAs)from the stock solution were measured based on theabsorbance at 260 nm and 10 μL of 55 ng/μL of NoV surrogates were added in 10% aliquots of the pelleted water samples continue with the RNA extraction. All extracted RNA from 11 spiked water samples were used for RT-LAMP and RT-PCR assays.

3 Results

3.1 RT-LAMP amplification and specificity test

The reaction mixture of NoV RNA surrogates RT-LAMP was incubated at 63°C for 45 min. Although a positive reaction could be detected after 20 to 25 min incubation, the incubation time was prolonged to ensure the optimum reaction occurs during incubation.Positive RT-LAMP amplification can be detected due to the presence of fluorescent detection reagent. The observation made by colour changes from light orange to fluorescent green and presence of a ladder-like pattern on 2.5% agarose gel electrophoresis under UV-transilluminator. No positive reactions were found in all other environmental species *Burkholderia cepacia*, *B. thailandensis*, *B. pseudomallei*, *Bacillus subtilis*, *B.*

macerans, *B. circulans*, *B. megaterium*, *Staphylococcus* sp., *S .epiderminis*, *S. heamolyticus*, *E. coli* BL21, *E. coli* Nova Blue, *Salmonella* sp., *Shigella* sp. and *Shinella granuli*(data not shown).

3.2 Detection Limit of the RT-LAMP assay

The detection limit of the assay was determined by copy number of the 10-fold dilution of RNA transcripts. RNA concentration measured based on the absorbance at 260 nm was 55ng/μL and the transcribed length of RNA is 448 bp. The NoV RNA initial concentration was calculated and converted to copy numbers to get 2.17x 10^8copies μL^{-1} based on the formula by Krieg [25]. The lowest of detection (LOD) of the RT-LAMP assay was estimated to be ~22 copies μL^{-1} (Fig. 1B). The last positive results from the copy number dilution was defined as the detection limit (only samples with three replicates tested positives was considered positive (3/3))

3.3 Evaluation of the RT-LAMP, RT-PCR and quantitative RT-PCR

Standard curve of the 10-fold serially diluted RNA transcripts of NoV were tested in parallel with RT-PCR and quantitative RT-PCR. The detection limits of both assays were 220 copies and 22 copies, respectively (Fig. 1). The amplification of quantitative RT-PCR, conventional RT-PCR and RT-LAMP assays showed a linear plots against range of input RNA transcripts concentration generating coefficient of correlation for NoV surrogates detection R^2= 0.995 (Fig. 2A). Amplification and dissociation plots showed in melting curve analysis proved the primers were specific to the target gene sequence where all the amplified products have same melting temperature at 85.9°C (Fig. 2B).

The sensitivity results of NoV RT-LAMP and quantitative RT-PCR were both equal and 10-fold more sensitive than conventional RT-PCR on agarose gel electrophoresis.However, the RT-LAMP has the advantage over quantitative RT-PCR for its visibility by naked eye. The total end result of the RT-LAMP amplification was completed in 45 min faster than that quantitative RT-PCR (2.0 h) and conventional RT-PCR (2.5h).

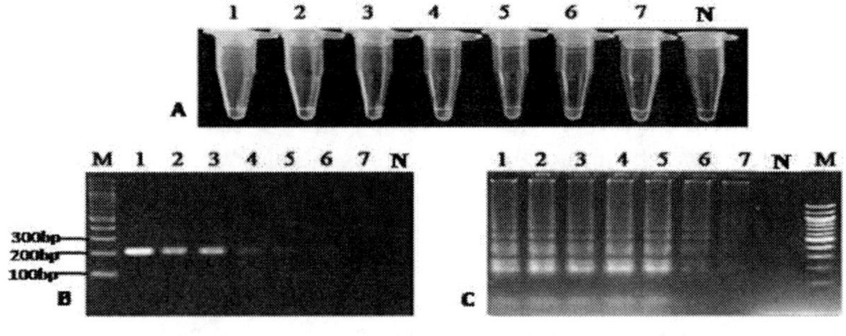

Fig. 1 Limit of detection of NoV RT-LAMP assay (positive results will show ladder-like amplicons, while negative results shows no amplification) was examined against the concentration of NoV surrogates ranging from 2.17 x 10^7 to ~22 copies/μL in 10-fold serial dilutions (lane M; 100bp marker, lane 1; 2.17 x 10^7, lane 2; 2.17x 10^6, lane 3; 2.17x 10^5, lane 4; 2.17x 10^4, lane 5; 2.17x 10^3, lane 6; 2.17x 10^2, lane 7; 2.17x 10^1, lane N; non-template control).(**A**) Color changes of the RT-LAMP reaction stained with fluorescent detection reagent showed the detection limit is 22 copies/μL.(**B**) Gel electrophoresis of PCR product (207bp) from conventional RT-PCR showed a 10-fold less sensitive then (**C**) the RT-LAMP method electrophoresis results in 2.5% agarose gels.

DOI: 10.1051/matecconf/201710305013

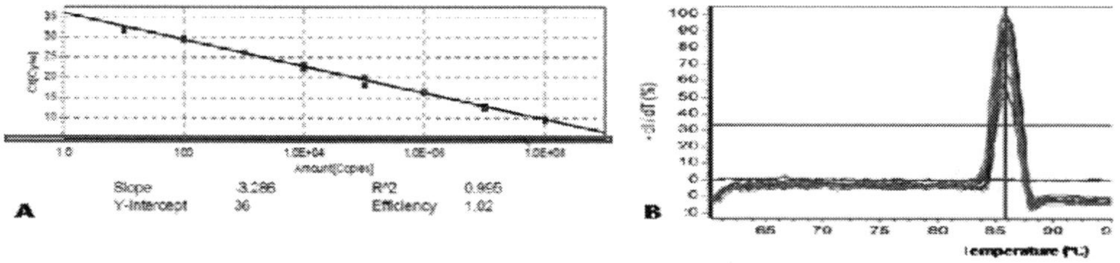

Fig.2 The amplification by quantitative RT-PCR (**A**) A graph plotted by real-time PCR system showed the coefficient of correlation for range of cDNA concentration input was R^2= 0.995. (**B**) Melting curves analysis showed amplification and dissociation of all PCR products happened at the same temperature of 85.9°C. A single target gene was amplified from the NoV surrogates.

3.4 Application of NoV RT-LAMP in spiked sample

The RT-LAMP assay was further evaluated with spiked water samples randomly picked from Hulu Langat province. Using the RT-LAMP and conventional RT-PCR the samples were examined for the presence of Noroviruses surrogates spiked in it. The results found that the RT-LAMP assays showed all positive (100%) for the presence of Noroviruses surrogates while the conventional RT-PCR showed only 8 (72.3%) out of 11 spiked of positive samples (Table 3).The gel electrophoresis analysis for RT-PCR assays and eyes observation of the RT-LAMP assays are shown in Fig. 3.

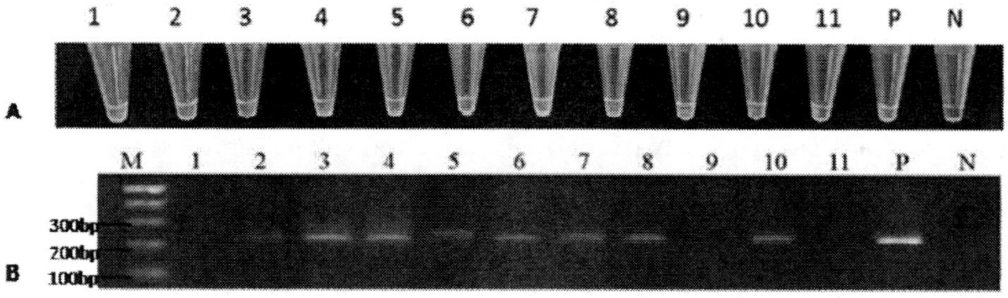

Fig. 3 RT-PCR and RT-LAMP assaysfor NoV surrogates detection in spiked recreational stream water samples. Results in lane 1 to 11 (stream water samples from lane M; 100bp marker, lane 1;Tekala 1, lane 2;Tekala 2, lane 3; Tekala 3, lane 4; Tekala 4, lane 5; Tekala 5, lane 6; Gabai 1, lane 7; Gabai 2, lane 8; Gabai 3, lane 9; Gabai 4, lane 10; Gabai 5, lane 11; Gabai 6, lane P; positive control, lane N; non-template control) showed that (**A**) RT-LAMP assay was able to detect NoV*in-vitro* transcribed surrogates spiked in all recreational stream water samples. (**B**) Conventional RT-PCR assay was unable to detect the NoV surrogates from sample 1, sample 9 and sample 11 and showed amplified products (207 bp) only in other 8 samples.

Table 2 Results of spiked water samples tested with RT-LAMP and RT-PCR

Water type	RT-LAMP assay	RT-PCR assay
Stream water tested	11	11
Positive/tested	11/11	8/11
(%)	(100%)	(72.3%)

4 Discussion

Noroviruses are shed in the faces of infected individuals and have high possibility to exist in very low concentration in water environment [21]. In recreational water, the risk for contamination may associate with overt fecal accidents or direct contact of contaminated bodies in the water [5].As a major causative agents of nonbacterial gastroenteritis in many countries, a rapid detection on the occurrence of NoV in recreational water is very much needed[1,3,5]. A study based on human volunteer showed the NoV infectious dose may be as low as 10to 100 viruses particles [6]emphasizes the need on highly sensitive method to detect their prevalence in water. NoV is usually detected in 3to 4h by RT-PCR method[22]. In latest innovation in DNA detection by amplification techniques, loop-mediated isothermal amplification (LAMP) has gain attention for it versatility in shorten the DNA amplification time to around one hour with results visible by naked eye using fluorescent dye. In this study, the RT-LAMP assay was demonstrated for the detection of NoV in comparison with the quantitative RT-PCR and conventional RT-PCR to evaluate the performance of each assay. The RT-LAMP has shown higher sensitivity with detection limit as low as 22 copies of NoV surrogates in 10μL reaction mixture with fluorescent dye. Detection of target NoV surrogates by RT-LAMP when compared to detection by quantitative RT-PCR was equivalent or more sensitive. The specificity test of the RT-LAMP, quantitative RT-PCR and conventional RT-PCR assays showed no-cross reactivity towards other 15 water and environmental bacteria strains available in our laboratories. Previously, the primers have been tested to be specific to NoV GII amongst other viruses including rotavirus, astrovirus, adenovirus, and other calicivirus and sapovirus[18].

Although DNA plasmid has been used to replace the RNA templates for assay optimization purpose in some study [23] (for the reason of RNA molecules are instable *in-vitro*) however, the DNA plasmid is not fully representative the RNA viruses such as NoV.In this study, the DNA plasmid was transcribed *in-vitro* which would served as better NoV surrogates. The sensitivity test and spiked recreational stream water sample was done using the *in-vitro* transcribed RNA. This form of surrogates would provide comparative reference to other laboratories doing the similar research.The use of fluorescent detection reagent for visualizing f RT-LAMP results simplifies the need of agarose gel electrophoresis. The step for opening the reaction tube also would introduce risk of contamination to the handler. Despite higher sensitivity thequantitative RT-PCR assay demands higher cost commitment which off benefit to the unequipped laboratoriesand therefore provide advantage toRT-LAMP assay for routine detection of NoV isolated from recreational stream water even on-site application.

5 Conclusion

The feasibility of RT-LAMP assay as method for detection of NoV surrogates in stream water should have a great potential for surveillance of NoV in Malaysia to improve the capacity for early prevention management.

References

[1] C. Tuan Zainazor, M.S. Noor Hidayah, L.C. Chai, Tunung, F. Mohamad Ghazali, R. Son, The scenario of norovirus contamination in food and food handlers, J. Microbiol. Biotechnol. , **20**, 229–237 (2010)

[2] T.C. Tuan Zainazor, L. Afsah-Hejri, M.S. Noor Hidayah, M.R. Noor Eliza, M.D. Naziehah, J.Y.H. Tang, Noorlis, N.A. Elexson, L.C. Chai, F.M. Ghazali, Y.K. Cheah, R. Son, Assessment of noroviruses in selected ulam from local market, Int. Food Res. J. , **19**, 877–

882 (2012).

[3] M.F.C. Beersma, F.H.A. Sukhrie, J. Bogerman, L. Verhoef, M.M. Melo, A.G. Vonk and M. Koopmans, Unrecognized Norovirus Infections in Health Care Institutions and Their Clinical Impact, J. Clin. Microbiol., **50**, 3040–3045 (2012)

[4] T.J. Wade, R.L. Calderon, E. Sams, M. Beach, K.P. Brenner, A.H. Williams and A.P. Dufour, Rapidly measured indicators of recreational water quality are predictive of swimming-associated gastrointestinal illness, Environ. Health Perspect., **114**, 24–28 (2006)

[5] C.J.P.A. Hoebe, H. Vennema, A. Maria De, R. Husman, Y.T.H.P. Van Duynhoven and C. Hoebe, Norovirus Outbreak among Primary Schoolchildren Who Had Played in a Recreational Water Fountain, J. Infect. Dis., **189**, 699–705 (2004)

[6] S. a Schaub and R.K. Oshiro, Public health concerns about caliciviruses as waterborne contaminants., J. Infect. Dis. , **181 Suppl**, S374–S380 (2000)

[7] U. Epa, Criteria for Bacteria -1986 Bacteriological Ambient Water Quality Criteria for Marine and Fresh Recreational Waters, (1986)

[8] S. Dorevitch, N.J. Ashbolt, C.M. Ferguson, R. Fujioka, C.D. McGee, J.A. Soller and R.L. Whitman, Meeting report: Knowledge and gaps in developing microbial criteria for inland recreational waters, Environ. Health Perspect., **118**, 871–876 (2010)

[9] J.H. Kim and S.B. Grant, Public mis-notification of coastal water quality: A probabilistic evaluation of posting errors at Huntington Beach, California, Environ. Sci. Technol. , **38**, 2497–2504 (2004)

[10] Z. Liang, Z. He, C.A. Powell and P.J. Stoffella, Survival of Escherichia coli in soil with modified microbial community composition, Soil Biol. Biochem., **43**, 1591–1599 (2011)

[11] T.R. Desmarais, H.M. Solo-Gabriele, C.J. Palmer, Influence of soil on fecal indicator organisms in a tidally influenced subtropical environment, Appl. Environ. Microbiol., **68**, 1165–1172 (2002)

[12] E.J. Viau, K.D. Goodwin, K.M. Yamahara, B.A. Layton, L.M. Sassoubre, S.L. Burns, H.I. Tong, S.H.C. Wong, Y. Lu and A.B. Boehm, Bacterial pathogens in Hawaiian coastal streams-Associations with fecal indicators, land cover, and water quality, Water Res., **45**, 3279–3290 (2011)

[13] S. Jiang, R. Noble and W. Chu, Human Adenoviruses and Coliphages in Urban Runoff-Impacted Coastal Waters of Southern California Human Adenoviruses and Coliphages in Urban Runoff- Impacted Coastal Waters of Southern California, Appl. Environ. Microbiol., **67**, 179–184 (2001)

[14] T.-T. Fong and E.K. Lipp, Enteric viruses of humans and animals in aquatic environments: Health risks, detection, and potential water quality assessment tools, Microbiol. Mol. Biol. Rev., **69**, 357–371 (2005)

[15] J.L. Melnick, C.P. Gerba and G. Berg, The ecology of enteroviruses in natural waters, C R C Crit. Rev. Environ. Control., **10**, 65–93 (1980)

[16] Y. Suzuki, S. Narimatsu, T. Furukawa, A. Iwakiri, M. Miura, S. Yamamoto and H. Katayama, Comparison of real-time reverse-transcription loop-mediated isothermal amplification and real-time reverse-transcription polymerase chain reaction for detection of noroviruses in municipal wastewater, J. Biosci. Bioeng., **112**, 369–372 (2011)

[17] D. Khairuddin, M.A. Wahid, N.Y.M. Yusof and J.M. Jani, Adaptation of NoV LAMP Primers by PCR for Highly Sensitive Detection of Noroviruses in Water, *Proc. of the Int. Symp. on Flood Research and Management,* Springer, 287–294 (2016).

[18] J. Luo, Z. Xu, K. Nie, X. Ding, L. Guan, J. Wang, Y. Xian, X. Wu and X. Ma, Visual Detection of Norovirus Genogroup II by Reverse Transcription Loop-Mediated Isothermal Amplification with Hydroxynaphthol Blue Dye, Food Environ. Virol., **6**, 196–201 (2014)

[19] P.A. Krieg, A laboratory guide to RNA. New York: Wiley-Liss, Willy-Liss, New York,

1996.

[20] T. Notomi, H. Okayama, H. Masubuchi, T. Yonekawa, K. Watanabe, N. Amino and T. Hase, Loop-mediated isothermal amplification of DNA, Nucleic Acids Res., **28**, (2000)

[21] M.R. Karim, F.W. Pontius and M.W. LeChevallier, Detection of noroviruses in water -- summary of an international workshop, J. Infect. Dis., **189**, 21–28 (2004)

[22] S. Fukuda, S. Takao, M. Kuwayama, Y. Shimazu and K. Miyazaki, Rapid Detection of Norovirus from Fecal Specimens by Real-Time Reverse Transcription – Loop-Mediated Isothermal Amplification Assay, J. Clin. Microbiol., **44**, 1376–81 (2006)

[23] H. Wei, J. Zeng, C. Deng, C. Zheng, X. Zhang, D. Ma and Y. Yi, A novel method of real-time reverse-transcription loop-mediated isothermal amplification developed for rapid and quantitative detection of human astrovirus, J. Virol. Methods., **188**, 126–131 (2013)

DOI: 10.1051/matecconf/201710305014

Evaluation of physical, chemical and heavy metal concentration of food waste composting.

Aeslina Abdul Kadir[1,*], *Nur Wahidah* Azhari[2] , and *Siti Noratifah* Jamaludin[3]

[1] Senior Lecturer, Department of Environmental Engineering and Water Resources, Universiti Tun Hussein Onn Malaysia (UTHM) 86400 Parit Raja, Batu Pahat, Johor, Malaysia

[2,3] Department of Environmental Engineering and Water Resources, Universiti Tun Hussein Onn Malaysia (UTHM) 86400 Parit Raja, Batu Pahat, Johor, Malaysia

Abstract: In this study, food waste composting with rice husk and coconut fibre as compost medium were carried out. Two types of different fermentation liquids were prepared which were fermented liquid (banana peel) and fermented liquid from fermented soybeans. During the composting process, a compost samples for a twenty week duration at an interval time of two weeks. Among the physico-chemical parameters that were tested were temperature, moisture content, pH value, Total Nitrogen, Total Phosphorous, Potassium and Total Organic Carbon and Carbon Nitrogen ratio. Heavy metals such as copper, cadmium, lead, nickel and arsenic were observed and analysed. From this study, it was found that, the temperature increased during the thermophilic phase while there was gradually increase of Total Nitrogen, Total Phosphorous and Potassium from the beginning till the end of the composting process. It was also found that the total organic carbon (TOC) and the carbon nitrogen ratio decreased significantly during the decomposition process. Traces amounts of heavy metals were also detected and remains below the standard Malaysian Environmental regulations. It was concluded that, the composting process was faster with processed food waste followed by combination of processed food waste and raw. Raw food waste were demonstrated the lowest degradation rate.

1 Introduction

Environmental Quality Act 1974 describe , waste included any matter prescribed to be scheduled waste or any matter, whether in a solid, semi-solid or liquid form or in the form of gas or vapour which is emitted, discharged or deposited in the environment in such volume, composition or manner as to cause pollution. This poor solid waste management lead to the environmental pollution such as leachate [1,2]. The challengers to overcome the municipal solid waste (MSW) generation are location and treatment of the waste. Landfill and open dumping is main disposal method practice in Malaysia. Landfilling of mixed waste is the worst solution of all as biological, chemical and physical degradation processes. Furthermore, the high amount of food waste in landfill contributes the foul odor,

* Corresponding author: aeslinaabdulkadir@gmail.com

toxic leachate, emission of greenhouse gasses and vermin infestation [3]. In addition, according to Hesnawi and Mohamed [24] when food waste decomposes under the low oxygen conditions that are typically encountered in landfills, it produces methane, a harmful greenhouse gas that is twenty one times more contaminated than carbon dioxide, and it might contaminate groundwater

Therefore, composting is one of the potential waste management elements to divert organic waste generated to landfill [4]. The presence of a high percentage of biodegradable organic matter, acceptable moisture content and C:N ratio in the waste and factors that influence the composting process [5]. Production of organic fertilizer from food wastes is an effective environmental sanitation and appropriate resource recovery strategy. Nutrient value of compost varies widely, depending upon the nature of feedstock composted [6].

2 Materials and Method

2.1 Study area

In this study, the waste was collected from small medium scale industry, Makanan Ringan Mas (MASi) at Parit Kuari Darat, Parit Raja, Johor. This industry waste mainly. mainly dominated by food waste. The food waste can be divided into two categories which are processed food waste and raw food waste. The processed food wastes consist of chips and candy, while the main raw food wastes that is generated are fruit peeling which are tapioca peel, banana peel and coconut husk. The average quantity of organic waste generated by this industry is in between 101.82 kg/month. The improper management of waste generated has been practiced and contributed to unhygienic conditions. Therefore, in this study, backyard composting food waste was carried out as alternatives to solid waste management due to its low cost and easy to operate

2.1.1Compost preparation

In this study, two composting medium namely, rice husk and coconut fibre were mixed and prepared as the composting medium to compost of food waste (processed of food waste, raw of food waste and combination processed and raw of food waste). Composting was carried out at Micropollutant Research Centre (MPRC), UTHM. Two type of fermentation liquid was prepared which were fermented banana peel (S*) and fermented soybeans (C*) and eight reactors are prepared as in Table 1. Aeration of each reactors was accomplished through periodic turning at every 7-day. During composting process, the samples was collected through the degradation stages and placed into a close container. The samples were brought from week 1 to week 20 to the laboratory for physical-chemical characteristics and heavy metal concentrations.

2.1.2 Physico-Chemical Analysis

Temperature and moisture content value were monitored throughout the composting process. Temperature was measured by using thermometer whilst moisture content was conducted by using gravimetric method. Compost samples (10g) was dried in an oven at 105 °C for 24 h. Changes in pH was determined immediately after transportation into the laboratory and measured by using a ph meter using a method described by Dashan et al. [7] and was measured by pH meter. On the other hand, the total carbon was determined by Shimadzu SSM-5000 with TOC-Vcph analyzer by Method 5310 B.As for nutrient content, total nitrogen content was measured by Micro- Kjeldahl method, where available of total Phosphorous was extracted by digestion with nitric acid according to the method that was carried out by Webb and Adeloju [8]. Samples were analysed by Westco SmartChem 140 automated wet chemistry discrete analyser. As for potassium and heavy metal, the samples were extracted by nitric acid digestion according to Fanta & Ade [9] and analysed by Atomic Absorption Spectrometer (AAS) and Inductively Coupled Plasma Mass Spectrometry (ICP-MS).

Table 1. Reactors prepared in the experiment work

	C*+ *Rice Husk*	
C1	: Control	: -
C2	: Processed food waste (Coconut Candy & chips)	: 250 g
C3	: Raw food waste (Banana peel, tapioca peel & grated coconut)	: 250 g
C4	: Processed and Raw food waste (Coconut Candy, chips, Banana peel, tapioca peel &grated coconut)	: 250 g
	S*+ Coconut Fibre	
S1	: Control	: -
S2	: Processed food waste (Coconut Candy & chips)	: 250 g
S3	: Raw Food waste (Banana peel, tapioca peel & grated coconut)	: 250 g
S4	: Processed and Raw food waste (Coconut Candy, chips, Banana peel, tapioca peel &grated coconut)	: 250 g

3 Result

3.1 Temperature

In composting, heating is important. It is indicates that, microbial activity take place to degrade the organic matter [10]. According to Venglovsky et al. [11] it was suggested, that the microbial evaluation actually steadily and maximize around 35 and 40 °C during the composting process while according to Boulter et al. [12], when it reached 40 to 45 °C, it represent thermophilic phase. During thermophilic phase, the presence of weed seeds and pathogens in the compost are killed, due to hemicellulose and lignins was degraded.

In this study, these fluctuation of temperature reading, Fig. 1 may be explained by the differences of waste was placed into the reactor supported by Pagans et al. [13] where the combination of organic compounds. Which organic matter is mainly composed easily degradable and energetic compounds such as protein and fats and presents of large amount of organic nitrogen. At early early stage of composting the temperature is low due to less water content. It was slight increases in temperature after keep turning for aeration and after adding water. At the end of composting process, both treatments shows the temperatures tend to decrease where the compost was considered nearing mature phase. However, for reactor C3 and S3, both treatments had a quite low because the excess of moisture, cause slow the activity of microbial activity. The fluctuation of this temperature was same as previous studies by Abdul Kadir et al. [14], where it rise from the initial mesophilic phase to the thermophilic phase.

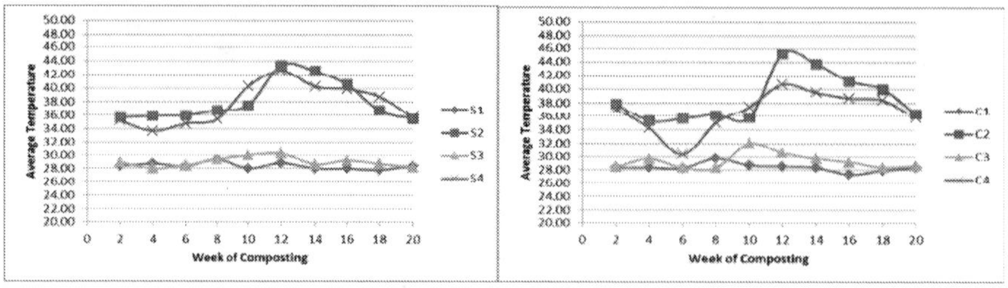

Fig. 1. Temperature by week in the reactor

3.2 Moisture Content

At the beginning, of the composting the moisture content for all reactors was low due to addition of food wastes, as shown in Fig. 2. In order to control the moisture content, water addiction was mad due to the moisture content of the food waste was low compare to the raw of food waste, before it was placed into each reactor. The waste was watered to ensure that the waste optimum moisture content was maintained in each reactor between 45-65 % moisture content were recorded. Middle of composting process, both composting medium, the moisture content was declined during week 5 and 6 for composting material reactor C2,C3,S2 and S3.It may due to heat produces at thermophilic stage by actives of the microbial. It was supported by Shymala & Belagali [15], that found the moisture content value were decreased during the composting period due to degradation of waste and heat generated by microbial activity.

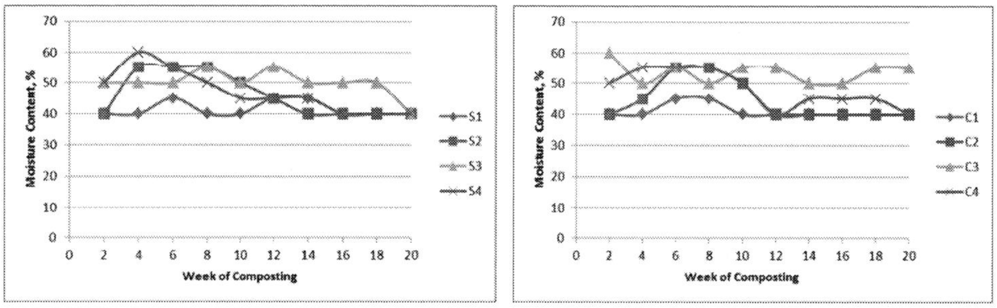

Fig. 2. Moisture content by week in the reactor

3.3 pH value

According to Arslan et al.[16], pH value was important in evaluating the microbial environment. This is also supported by Hubbe et al. [17] that, action of microbes can be found in the variation of pH during composting and the best for competent biological decomposition of organic waste also depend on pH [18]. The optimum pH for the degradation of organic waste in between of 6 to 7.5 [12,19]. Fig. 3 shows that, both treatments have similarity results which, were it acidic in early of composting. Then the pH, increased to between 7 and 8 in the Week 3. The results indicates that the thermophilic phase has occurred and at the end of the composting the pH have been neutral.

DOI: 10.1051/matecconf/201710305014

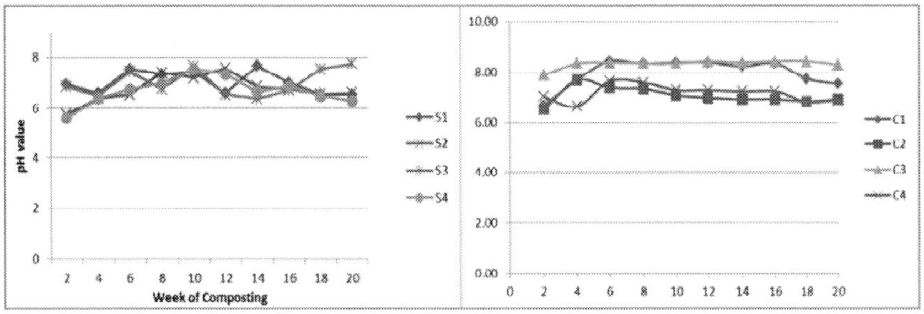

Fig. 3. pH value by week in the reactor

3.4 Total Organic Carbon

The amount of carbon found in an organic caompound is called total organic carbon. In this study, the graph shows that, decomposition of food waste by microorganisms during the composting period was decrease at the end of composting. The initial compost for C4 are 25.19% and 5.49% at end of composting. While for S4, the intial reading are 30.31% and decreased at the end of composting with 8.07%. Total carbon concentration declined slightly for both compost as shown in Fig. 4. According to Jusoh et al. [20] carbon is sources of energy for microorganisms and transformed to carbon dioxide during metabolism process of the cells, while according to Ch'ing et al. [21], carbon was reduce at the end of composting due to active microbial cellulytic degradation and microbial proliferation which immobilize Nitrogen.

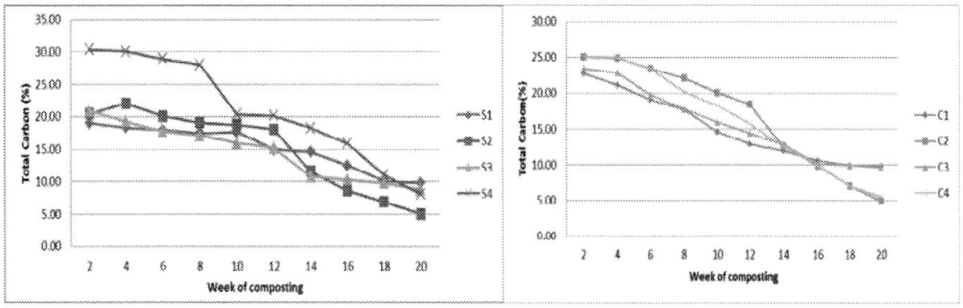

Fig. 4. Total Carbon by week in the reactor

3.5 Total Nitrogen

In this study, the concentration of total nitrogen all the reactors demonstrate increase value with composting time. As shown in Fig. 5, it was recorded that, C2 and S2 contain high N with 0.22% and 0.21% respectively. Compare with other previous researcher, concentration of total nitrogen, obtained is a quite low for this type of compost. This might due to ammonia volatilization and elevated temperature during composting process [22]. The concentration of total nitrogen, that retrieved from previous researchers varied from 1.00 to 2.45%. In this study, lower total nitrogen was recorded, however the results are similar recorded by Gautam et al. [5] which is between 0.03 to 0.07%.

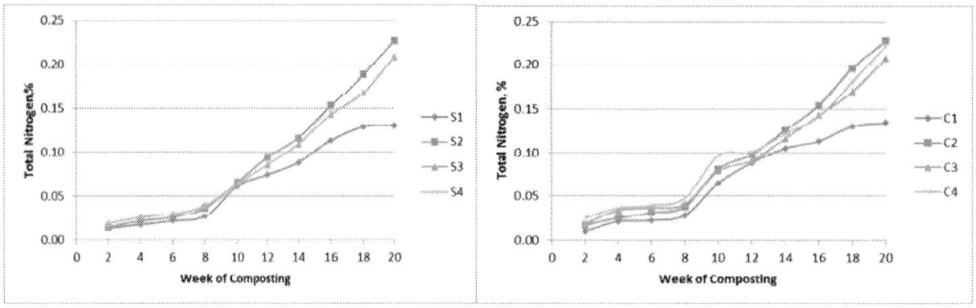

Fig. 5. Total Nitrogen by week in the reactor

3.6 Total Phosphorous

Fig. 6 shows values of phosphorous along 20 week of composting process. C4 (2.047 ppm to 11.615 ppm) and S4 (3.308 ppm to 11.418 ppm) reactors demonstrated increment every week compare to reactor C2,C3,S2 and S3. This is also discussed, the increase of phosphorous may be due to decreases of water solubility with humification so that phosphorous solubility during the decomposition was subjected to further immobilization factor [23,15]

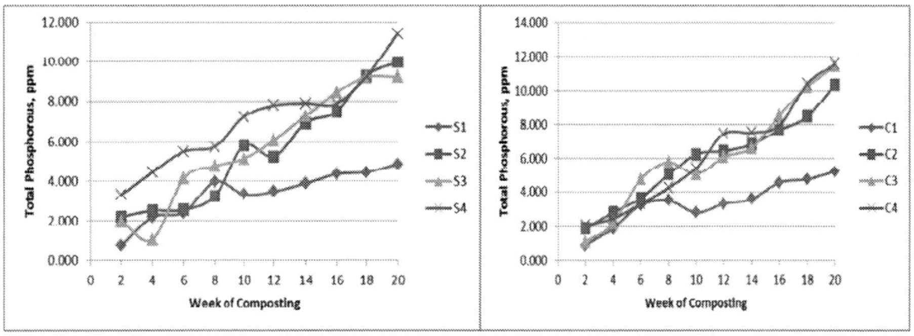

Fig. 6. Phosphorous versus week

3.7 Potassium

From Fig. 7, the values of potassium were observed increase at the end of composting process. At the end of the decomposition process, the value measured was obtained for reactors C2 633.40 ppm, S2 666.95 ppm, C3 516.00 ppm, S3 582.21 ppm, C4 645.55 ppm and S4 606.65 ppm. Factors that contribute to increase the potassium value, are the decomposting medium used such as, coconut fibre and rice husk. Where this compost medium can absorb and maintain the moisture content. It supported by Syamala and Belagali [15] reported that, the raw material use is factor that influences of potassium increased, due to characteristics where allowed the absorption of moisture content and maintain structural integrity and porosity.

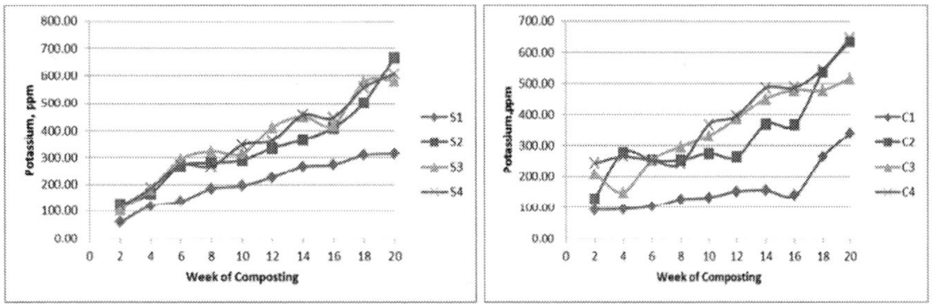

Fig. 7. Potassium versus week

3.8 Heavy Metal

Compost usually contains heavy metal based on their initial raw materials. In this study, heavy metals such as copper, cadmium, arsenic, nickel and lead concentration were measured in all reactor and compared to standard value retrieved from 9,26. Trace elements of heavy metal were found in all of the reactors as shown in Table 2. The value is in accordance with the usage of a small amount of is essential for plant growth and vice versa.

Table 2. Average heavy metal concentration along composting process

	S1	*C1*	*S2*	*C2*	*S3*	*C3*	*S4*	*C4*	*Standard Value (mg/kg)*
	(ppb)								
Cu	167.86	133.48	172.57	187.75	190.59	206.2	182.87	193.08	1500[a,b]
Cd	3.58	3.26	3.73	2.39	2.59	3.32	3.96	3.25	35[a]-39[b]
As	144.31	138.96	123.35	122.83	154.24	139.2	141.1	141.83	41[a,b]
Ni	91.3	140.11	117.4	152.54	106.51	117.79	113.65	140.08	420[b]
Pb	157.8	167.11	170.12	172.47	199.76	116.48	211.6	145.81	300[a,b]

a=Retrieved from [15] ; b=Retrieved from [24]

3.9 C/N ratio

Fig. 8 shows the decrease in C/N value in both treatments, it may be due to the mineralization of organic matter [20], on the other hand, according to Norbu [25] reduction in the C/N ratio was due to the degradation of organic matter. In this study, it was slow of decrease in C/N ratio. At week 10, both composting reactor C2 and S2 was 21.49 and 22.07 while the other reactor was between 25 to 45. It can be concluded that, when food waste is fed into the reactor the C/N slowly decrease. The degradation of waste is quite low. It can be compared with the control reactor. The C/N ratio value can be faster decrease due to addition of sludge [25], otherwise many researchers reported that, by adding waste materials as animal waste, bagasse or garden waste can improved the C/N ratio and enhance the decompose process.

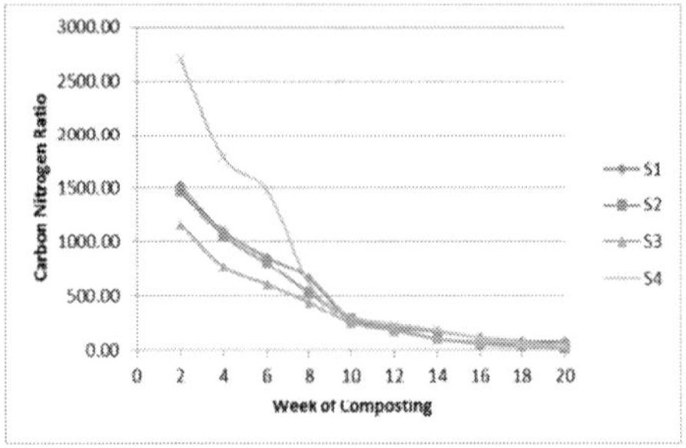

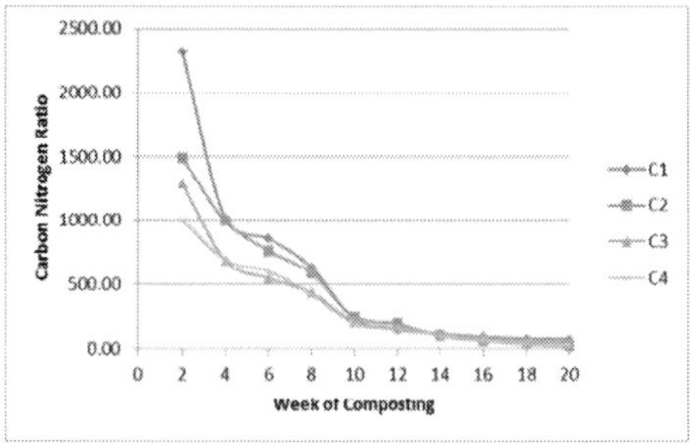

Fig. 8. Carbon Nitrogen ratio (C/N) value versus weeks

4 Conclusion

In this study, composting method was conducted as one of the feasible alternative method for reducing food waste generated at Makanan Ringan MAS. The results obtained proved that the composting method could produce acceptable value of nitrogen, phosphorous and potassium (NPK) and the heavy metal were also found to be in desirable limits. Therefore the organic compost produce could be used as fertilizers.

References

[1] S.N.S. Ismail and L.A. Manaf, The challenge of future landfill: A case study of Malaysia, J. of Toxicology and Environmental Health Sci., **5(6)**, 86-96, (2013)

[2] M.D.M. Samsudin, and M.M. Don, Municipal Solid Waste Management in Malaysia: Current Practices, Challenges and Prospect. J. Technology, **62(1)**, 95-101, (2013)

[3] A.A. Hamid, A. Ahmad, M.H. Ibrahim and N.N.N.A. Rahman, Food Waste Management in Malaysia- Current situation and future management options. J. of Industrial Research & Tech., **2(1)**, 36-39, (2012)

[4] R. Hassan, F.A.A. Majid and A.A. Rahman, The Potential Of Implementing Food Waste Composting At Source Using Biosense Scheme. J. of Research in Applied Natural and Social Sciences, **3**(4), 55-66, (2015)

[5] S. Gautam, P. Bundela, A. Pandey, M. Awasthi and S. Sarsaiya, Composting of municipal solid waste of Jabalpur city, Global J. of Env. Research, **4**(1), 43-46, (2010)

[6] R.A.Z. Taleb, A.T. Rakad and R.A. Abdullah, Changes in Compost Physical and Chemical Properties during Aerobic Decomposition. Int. J. of Current Microbiology and Applied Sciences, **3**(10), 479-486, (2014)

[7] H. Dahshan, A.M.M. Abd-Elall, M.A. Abd-El-Kader and A.M. Megahed, Field-scale management and evaluation of recycled cattle manure solids from livestock in Nile Delta ecosystem. African J. of Agricultural Research, **8**(23), 2829-2836, (2013)

[8] B. Webb and S.B. Adeloju, Evaluation of some wet digestions methods for reliable determination of total phosphorus in Australian soils. Microchemical J. **111**, 47-52 (2013)

[9] G.M. Fanta and A.G. Ede, Analyzing Selected Heavy Metals Content of Compost used in Ethiopia, Int. J. of Innovative and Applied Research, **2**(9), 5-11, (2014)

[10] A. Hassen, K. Belguith, N. Jedidi, A. Cherif, M. Cherif and A. Boudabous, Microbial characterization during composting of municipal solid waste. Bioresource Technology, **80**(3), 217-225, (2001)

[11] J. Venglovsky, N. Sasakova, M. Vargova, Z. Pacajova, I. Placha, M. Petrovsky and D. Harichova, Evolution of temperature and chemical parameters during composting of the pig slurry solid fraction amended with natural zeolite. Bioresource technology, **96**(2), 181-189, (2005)

[12] J. Boulter, G. Boland and J. Trevors, Compost: a study of the development process and end-product potential for suppression of turfgrass disease. World J. of Microbiology and Biotechnology, **16**(2), 115-134, (2001)

[13] E. Pagans, R, Barrena, X. Font and A. Sánchez, Ammonia emissions from the composting of different organic wastes. Dependency on process temperature. Journal Chemosphere, **62**(9), 1534-1542, (2006)

[14] A. Abdul Kadir, N. Rahman and N. Azhari, The Utilization of Banana Peel in the Fermentation Liquid in Food Waste Composting. *IOP Conference Series: Materials Science and Engineering*, **136**(1), (2016)

[15] D.C. Shymala and S.L. Belagali, Studies on Variations in Physico-Chemical and Biological Characteristics at Different Maturity Stages of Municipal Solid Waste Compost. Int. J. of Environment Sciences, **2**(4), 1984-1997, (2012)

[16] E.I. Arslan, A. Ünlü and M. Topal, Determination of the effect of aeration rate on composting of vegetable–fruit wastes. CLEAN–Soil, Air, Water, **39**(11), 1014-1021, (2001)

[15] M.A. Hubbe, M. Nazhad and C. Sánchez, Composting as a way to convert cellulosic biomass and organic waste into high-value soil amendments: A review. BioResources, **5**(4), 2808-2854, (2010)

[16] K. Atalia, D. Buha, K. Bhavsar and N. Shah, A Review on Composting of Municipal Solid Waste. IOSR J. of Environmental Science, Toxicology and Food Technology (IOSR-JESTFT), 2319-2402, (2015)

[17] M. Gómez-Brandón, C. Lazcano and J. Domínguez, The evaluation of stability and maturity during the composting of cattle manure. Chemosphere, **70**(3), 436-444, (2008)

[18] M.L.C. Jusoh, L.A. Manaf, and P.A. Latiff, Composting of rice straw with effective microorganisms (EM) and its influence on compost quality. Iranian J. of Env. Health Science & Engineering, **10**(1), (2013)

[19] H.Y. Ch'ng, O.H. Ahmed, S. Kassim and N.M.A. Majid, Co-composting of pineapple leaves and chicken manure slurry. Int. J. of Recycling of Organic Waste in Agriculture, **2**(1), 1-8, (2013)

[20] G. Civeira, Influence of municipal solid waste compost on soil properties and plant reestablishment in peri-urban environments, (2011)

[21] T. Janakiram and K. Sridevi, Conversion of waste into wealth: A study in solid waste management. J. of Chemistry, **7**(4), 1340-1345, (2010)

[22] J.E. Montoya, T.M. Waliczek and M.L. Abbott, Large scale composting as a means of managing water hyacinth (Eichhornia crassipes). Invasive Plant Science and Management, **6**(2), 243-249, (2013)

[23] T. Norbu, *Pretreatment of municipal solid waste by windrow composting and vermicomposting,* A thesis for the degree of Master of Science, 1-103, (2002)

[24] Hesnawi, R. M., and Mohamed, R. A, Effect of organic waste source on methane production during thermophilic digestion process. Int. J. of Environmental Science and Development, **4**(4), 435. (2013)

Study on NPK Performance in Food Waste Composting by Using Agricultural Fermentation

Siti Noratifah Jamaludin[1], *Aeslina* Abdul Kadir[2,*], *and Nur Wahidah* Azhari[1]

[1]Department of Environmental Engineering and Water Resources, Universiti Tun Hussein Onn Malaysia (UTHM) 86400 Parit Raja, Batu Pahat, Johor, Malaysia.
[2]Associate Professor, Department of Environmental Engineering and Water Resources, Universiti Tun Hussein Onn Malaysia (UTHM) 86400 Parit Raja, Batu Pahat, Johor, Malaysia.

Abstract: Food waste represents almost 60% from the total municipal solid waste disposed in the landfill. Alternative disposal method for food waste could be conducted by using composting method. In this study, investigationon food waste composting by using agricultural fermentation was carried out to find out the performance of the compost. Two types of compost were produced which were commercial compost and research compost and total of 8 reactors were tested during this study. Research compost consist of coconut fiber (decomposing medium) and the combination of salt and breadfruit peels as the fermentation liquid, while rice husk was used as decomposing medium for commercial compost along with fermented soybeanand brown sugar as fermentation liquid. Physical and chemical parameters which are temperature, pH value, moisture content, Total Nitrogen (N), Total Phosphorus (P) and Potassium (K) concentration were determined. Based on the results of 20 weeks composting, the overall temperature range from 27 °C to 45 °C which shown the active phase for composting occurred. On the other hand, during the period of composting, most of the pH value in each reactor is above 5 and approximately at neutral. This shown that the microbial respiration in the composting reactor was inhibited and had approached the mature phase. As for NPK content, Total Nitrogen value range from 98 ppm to 2268 ppm for commercial compost, while 84 ppm to 2240 ppm for research compost. Total Phosphorus has the values of0.871 ppm to 11.615 ppm for commercial compost and 1.785 ppm to 14.143 ppm for research compost. On the other hand, result for potassium is from 91.85 ppm to 645.55 ppm for commercial compost and from 133.95 ppm to 686.2 ppm for research compost. As a conclusion from the results obtained, the compost in this study is sufficient to be use for agricultural purposes and the best performance of NPK value was demonstrated by Reactor C2 from research compost.

[*]Correspondingauthor: aeslinaabdulkadir@gmail.com

1 Introduction

The issue of food waste are now starting to become a regional concern even for developed countries that are more likely to find the best management methods along with other solid waste. Countries such as Korea and Japan has been practicing the separation of food waste from other solid waste due to the breakdown of food waste, lack of landfill space, transport issue and disposal of food waste problem that also occurs in many combustion processes [1].

In Malaysia, the solution is sought to solve the problem of solid waste management based on the concept of environmentally acceptable to the public. There isstill a non-existentof food waste separation in Malaysia but the solutions for this matter need to be resolved immediately because food waste accounts for 65% of municipal solid waste [2]. The total generation of food waste is a major cause of many problems associated with the landfill, such as odor, leachability of toxic, and greenhouse gas emissions [3].

Food waste recycling in Malaysia is rarely implemented. There are several options for implementing the method of recycling food waste. One of the option were implemented by few countries including Germany and Britain [4] that perform the anaerobic decomposition by nurture methane gas to be used as fuel. Another option is the composting method.

Composting are the process of decomposition and stabilization of organic materials in biology under controlled conditions to produce a final product that can be applied to land without adverse effect to the environment. Microorganisms play an important role in this process. It decompose organic materials to make it a more stable material called compost.

The purpose of this research is to determine the performance of nutrient which are Total Nitrogen (N), Total Phosphorus (P) and Potassium (K) by utilizing agricultural fermentation liquid (research compost) and commercial fermentation liquid (commercial compost)also determine the temperature and the pH of the compost.

2 Materials and Method

2.1 Preparation of fermentation liquid

There are two types of fermentation liquid prepared in this study which are using bread fruit peels and salt for research fermentation liquid and using fermented soybean and brown sugar for commercial fermentation liquid. By using Takakura Composting Method three liters of water was mixed with 250 gram salt and bread fruit peel or 250 gram brown sugar and fermented soybean. Fermentation should take no longer than 7 days. It should be done around 3-5 days. During the fermentation process, the lid of the bottle should be loosened once a day to remove the gas that trapped in the bottle. The sweet and sour smell produced indicates that the fermentation liquid has been successfully conducted.

2.2 Preparation of decomposing medium

There are also two types of decomposing medium prepared in this study which are using rice husk with soil (commercial decomposing medium) and coconut fiber with soil (research decomposing medium). The ratio for the decomposing medium should be 2 to 1 with soil and rice husk or soil and coconut fiber also using Takakura Composting Method. After mixing the medium with the right ratio, the prepared fermentation liquid was added to the medium with the right amount and then homogenously mixed. The mixture should able to form without liquid dripping.

2.3 Preparation of the reactors

Eight reactors were set up, with 4 reactors each for commercial compost and research compost. For each set, one reactor acted as control while the rest were fed with 500 grams of different types of food waste which are classified as processed food waste, raw food waste and combination of processed and raw food waste. Table 1 shows the feeding of each reactor use in this study.

Table 1. Reactors with different types of food waste.

Compost Type	Reactor	Classification food waste	Types of food waste
Commercial Compost	X1	Control	-
	A1	Processed food waste	Candy + chips
	B1	Raw food waste	Banana peel + tapioca peel + bread fruit peel + coconut husk
	C1	Processed food waste + Raw food waste	Candy + chips + tapioca peel + banana peel + bread fruit peel + coconut husk
Study Compost	X2	Control	-
	A2	Processed food waste	Candy + chips
	B2	Raw food waste	Banana peel + tapioca peel + bread fruit peel + coconut husk
	C2	Processed food waste + Raw food waste	Candy + chips + tapioca peel + banana peel + bread fruit peel + coconut husk

2.4 Parameters Measurement

Total Kjeldahl Nitrogen (TKN) was used to analyze the concentration value of total nitrogen (N) and Discrete Analyzer (DA) was used to analyze total phosphorus (P) concentration. On the other hand, Atomic Absorption Spectrophotometer (AAS) was used to get the concentration of potassium (K). According to method in APHA Water and Wastewater, pH meter was used to measure the pH value of the reactor every week whether it is acid or alkali and thermometer was used to measure temperature everyday.

3 Result and Discussions

3.1 Temperature

Temperature is the main parameter that controls microbial activity during the composting process. The temperature of each reactor were measured before disposing any food waste into the reactor. The highest temperature determine in this study average between 42˚C to 45˚C. The increase of temperature during the composting procedure is due to the high temperature generated by the microorganisms through respiration activities and decomposition of organic materials. The temperature of composted matter determines the rate of many biological processes and plays a basic role in microorganism succession understood as a change in the quantitative and qualitative composition of microorganism population [5].

Temperature changes during composting process occurs in three stages: the mesophilic, termophilic and curing stages (cooling stage) [6]. In the termophilic stage, microbes are active until the compost reaches a certain stage of decomposition. Then, the temperature will drop to the initial temperature during the cooling stage. The temperature decreased during the composting process for each reactor in commercial and research compost. The beginning of the composting process can be observed in the mesophilic phase, and in a few weeks, the temperature increased until it almost reached to the termophilic phase. When the temperature reaches 40 °C to 50 °C, both phase microorganism takes place [7]. It is the active phase of composting. Based on the temperature the most active reactor was reactor A1 and A2 that contain process of food waste (chips and coconut candy) for both commercial and research compost. This explains the findings that the microorganisms can withstand one extreme environmental factor, or which is high temperature[8].

Based on the temperature at week 20, every reactor had almost achieved the curing stage. The compost maturation phase begins when the compost temperature falls to the ambient air temperature. Fig. 1 shows the temperature value for both commercial and research compost.

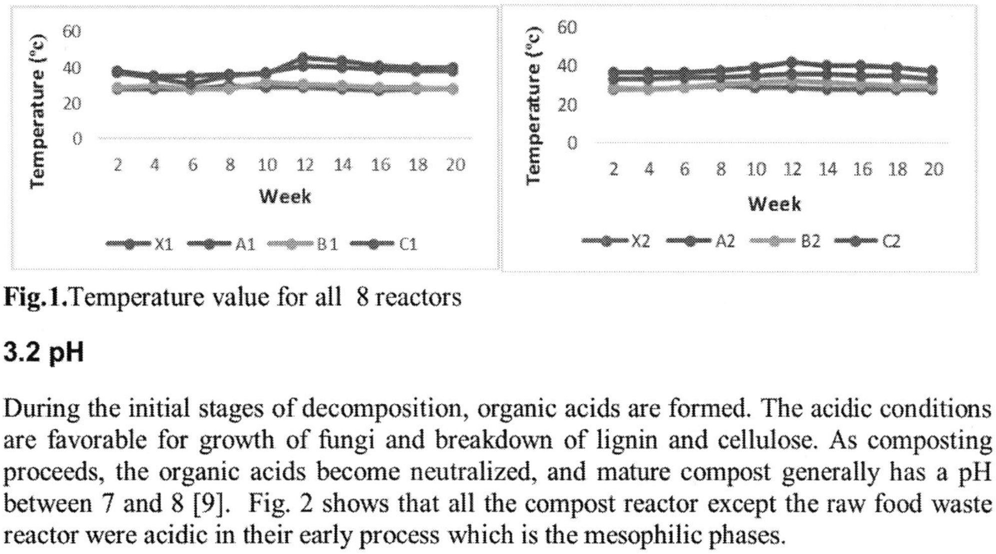

Fig.1. Temperature value for all 8 reactors

3.2 pH

During the initial stages of decomposition, organic acids are formed. The acidic conditions are favorable for growth of fungi and breakdown of lignin and cellulose. As composting proceeds, the organic acids become neutralized, and mature compost generally has a pH between 7 and 8 [9]. Fig. 2 shows that all the compost reactor except the raw food waste reactor were acidic in their early process which is the mesophilic phases.

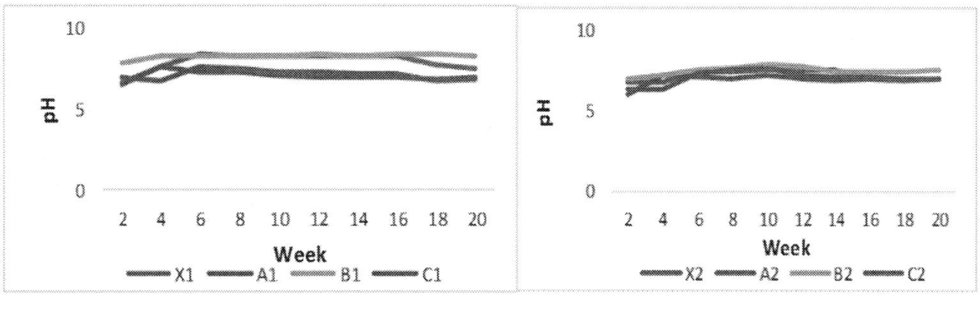

Fig.2. pH value for all 8 reactors.

In general, in the composting process, the pH falls below neutral in the beginning due to the formation of organic acid and later rise above neutral because the acids are consumed[10]. The pH increased around in the 6th week, when the composting process entered its thermophilic phase and slightly decreased in some reactor and become almost neutral (7 to 8) during the 8th week, when the compost reached its mature state. Moreover, Cooperband [11] stated that optimum pH value for composting (matured) is between 7 and 8. In this study, food waste that used in each reactors does not effect much in term of pH because the control reactor has almost the same trend line with other sample reactors.

3.3 Total Nitrogen Concentration

In all the reactors, total nitrogen increased slightly during the first four weeks and it increased gradually until week 20 in all compost reactors except for control reactor in both type of compost. Total nitrogen began to rise due to loss of dry mass in terms of carbon dioxide emissions, water loss by evaporation and activities of nitrogen fixing bacteria. The increased in total nitrogen content after 20 days of composting which indicated it might have been due to net loss of dry mass in terms of carbon dioxide and water loss by evaporation caused by heat evolved during oxidation of organic carbon. The same condition was observed by Ajay and Azmi [12]. From Fig. 3, the graph show sthat the concentration of the total nitrogen is increased gradually week by week especially at week 8 for all reactors. Reactor X1 and X2 which the control compost for both commercial and research compost do not increased as much as other reactors due to the lack of nutrient in the reactor itself.The highest concentration demonstrated is at the last week for all reactors because the total nitrogen fixing bacteria might also have contributed to the increase in the total nitrogen in the later stage of composting [13].

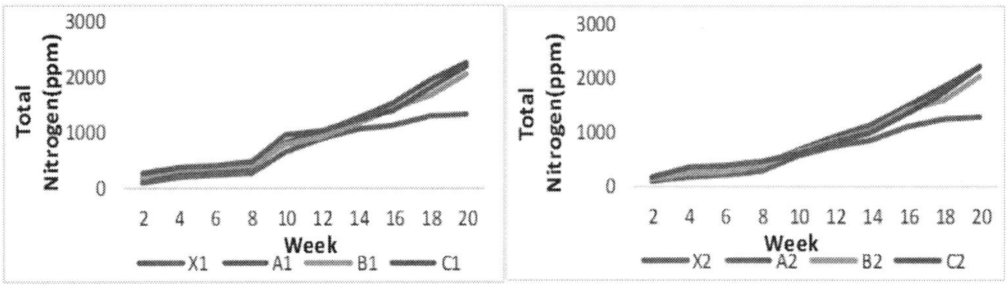

Fig.3. Total Nitrogen for all 8 reactors.

3.4 Total Phosphorus Concentration

In general, the trends of total phosphorus concentration in the reactors are slightly increase week by week. Total phosphorus content in Reactor B gradually increased during the composting process as shown in Fig. 4. The highest concentration was obtained in Reactor B2 and the lowest concentration was demonstrated in Reactor X1. This is because the increase in total phosphorus during composting was possibly caused by concentration effect arising from the higher rate of carbon loss that occurs when organic matter is decomposed[14].

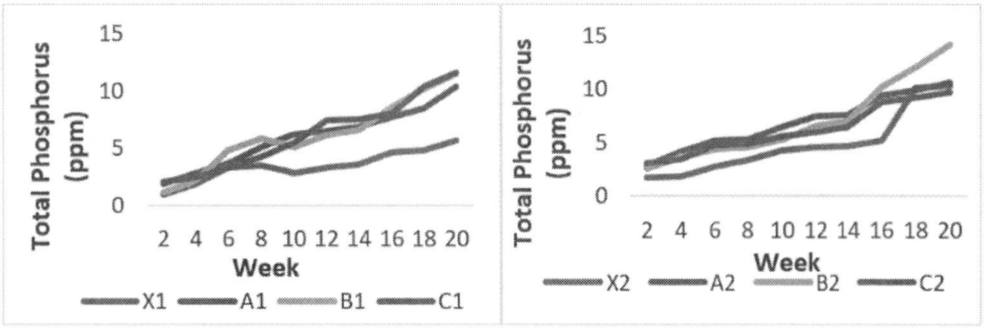

Fig.4. Total Phosphorus value for all 8 reactors.

3.5 Potassium Concentration

In this study, the potassium content is increased week by week but not constant. The highest concentration is in C2 which contain all the food waste in thi sstudy. From Fig. 5, the graph line show that the potassium value is not constantly increasing due to the interruption of microorganisms in the composta. Razzak [15] also claim that the values decreased and increased on a weekly basis and may be unstable due to the activity of microorganisms present in compost that also need nutrients. Therefore, the activity of microorganisms, which also need nutrients at certain times may cause the concentration of compost before maturity is not stable and declined. Most of the concentrations of all compost is growing very week.

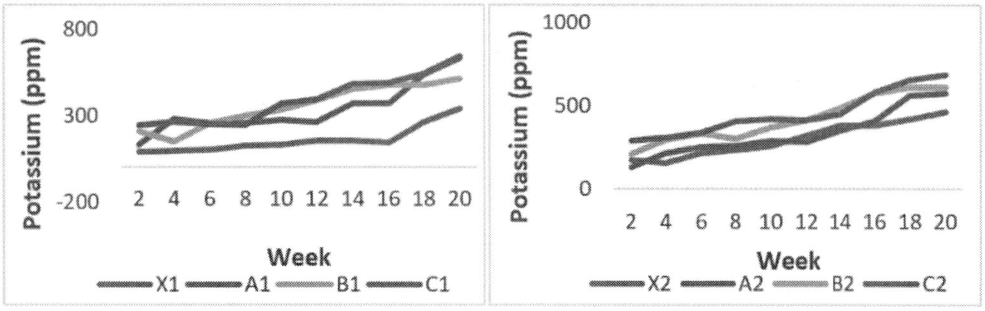

Fig.5. Potassium value for all 8 reactors.

4 Conclusion

In this composting study, temperature and pH values during the composting process were found to reach the level of the proposed range with 42°C to 45°C and 6.0 to 8.5respectively.Total nitrogen value were 98 ppm to 2268 ppm for commercial compost and 84 ppm to 2240 ppm for research compost. As for total phosphorus the result obtained is 0.871 ppm to 11.615 ppm for commercial compost and 1.785 ppm to 14.143 ppm for research compost. On the other hand, result for potassium is 91.85 ppm to 645.55 ppm for commercial compost and 133.95 ppm to 686.2 ppm for research compost. Reactor C2showed the most effective composting process in this composting study. The concentration of total nitrogen increased gradually between each reactor by week. On the other hand, the concentration of total phosphorus and potassium increased weekly but fluctuated and research compost show higher concentration compare to commercial compost reactors. The different concentration indicated that the presence of different types of food waste added will differ the amount of NPK due to microbial activity. There is a potential for the compost in this study to be used for agricultural purpose since the nutrient performance which is the NPK is increasing week by week.

This paper was partly sponsored by the Centre for Graduate Studies UTHM.

References

[1] M.H. Kim and J.W. Kim, Comparison through a LCA evaluation analysis of food waste disposal options from the perspective of global warming and resource recovery. Science of the Total Environment, **408** (19), 3998-4006(2010)

[2] M.O. Saeed, M.N. Hassan and M.A. Mujeebu,Assessment of municipal solid waste generation and recycle materials potential in Kuala Lumpur, Malaysia. Waste Management, **29**, 2209-2213,(2009)

[3] S.H. Lee, K.I. Choi, M. Osakoand D.J.I. Jong-In,Evaluation of environmental burdens caused by changes of food waste management systems in Seoul, Korea. Science of The Total Environment, **387** (1-3) 42-53,(2007)

[4] A.D. Knipe, *The management of household food waste.* Environmental Research and Consultancy. West Morden, United Kingdom, (2005)

[5] J.D. Wolna-Maruwka and A. Sawicka,Effect of temperature on the number of selected microorganism groups and enzymatic activityof sewage sludge composted with different additions in cybernetic.Agron. Res.,**7**(2), 875–890,(2009)

[6] E. Yenie,*Kelembapan bahan dan suhu kompos sebagai parameter yang mempengaruhi proses pengomposan pada unit pengomposan rumbai.* Jurusan Teknik Kimia Fakultas Teknik Universitas Riau, Pekanbaru, 28293, (2008)

[7] J. Peigné and P. Girardin,Environmental Impacts of Farm-Scale Composting Practices.Water, Air, Soil Pollut.,**153**(1–4), 45–68, (2004)

[8] C. Sundberg, D. Yu, I. Franke-Whittle, S. Kauppi, S. Smårs, H. Insam, M. Romantschukand H. Jönsson,Effects of pH and microbial composition on odour in food waste composting.Waste Manag., **33**(1), 204–211, (2013)

[9] K. Nakasaki, H. Yaguchi, Y. Sasaki and H. Kubota, Effects of pH Control on Composting of Garbage.Waste Management Resource. **11**(2), 117–125, (1993)

[10] C. Sundberg, S. Smårs and H. Jönsson,Low pH as an inhibiting factor in the transition from mesophilic to thermophilic phase in composting.Bioresour. Technol., **95**(2), 145–150,(2004)

[11] L. Cooperband, *Sustainable use of by products in land management.* In Bartels, J.M., and W.A. Dick (eds.) Land application of agricultural, industrial and municipal by products. SSSA Book Series 6. p. 215- 235. Soil Science Society of America SSSA, Madison, Wisconsin, USA, (2000)

[12] S.K. Ajay and A.A. Kazmi, Rotary Drum Composting of Mixed Organic Waste based on Different C/N Ratios. Proc. of the Int. Conf. on Sustainable Solid Waste Management, Chennai, India. 258-265,(2007)

[13] A. Abdul Kadir, S.N.M. Ismail and S.N. Jamaludin, Food Waste Composting Study from Makanan Ringan Mas. *IOP Conf. Series: Materials Science and Engineering,* **136,** (2016)

[14]C. Lin,A negative-pressure aeration system for composting food wastes.Bioresour. Technol., **99**(16), 7651–7656, (2008)

[15]L.A. Razzak, Nutrition And Growth Of Bacteria, (2010) Retrieved on August 10, 2016 from http://www.uobabylon.edu.iq/eprints/publication_10_1062_180.pdf,

Carbon Dioxide (CO_2) Sequestration In Bio-Concrete, An Overview

A. Faisal Alshalif[1], *J.M.Irwan*[1*], *N. Othman*[2], *M.M.Zamer*[1] *and L.H. Anneza*[1]

[1]Jamilus Research Centre for Sustainable Construction (JRC), Faculty of Civil and Environmental Engineering, UniversitiTun Hussein Onn Malaysia, 86400, Parit Raja, Johor, Malaysia
[2]Micro-Pollutant Research Centre (MPRC), Faculty of Civil and Environmental Engineering, University Tun Hussein Onn Malaysia, 86400, Parit Raja, Johor, Malaysia

Abstract. The emission of CO_2 into atmosphere which has increased rapidly in the last years has led to global warming. Therefore, in order to overcome the negative impacts on human and environment, the researchers focused mainly on the reduction and stabilization of CO_2which represent the main contributor in the increasing global warming. The natural capturing and conversion of CO_2 from atmosphere is taken place by biological, chemical and physical processes. However, these processes need long time to cause a significant reduction in CO_2. Recently, scientists shifted to use green technologies that aimed to produce concrete with high potential to adsorb CO_2 in order to accelerate the reduction of CO_2. In the present review the potential of bio-concrete to sequestrate CO_2 based on carbonation process and as a function of carbonic anhydrase (CA) is highlighted. The factors affecting CO_2 sequestration in concrete and bacterial species are discussed. It is evident from the literatures, that the new trends to use bio-concrete might contribute in the reduction of CO_2 and enhance the strength of non-reinforced concrete.

1 Introduction

Concentration of Carbon dioxide and other gasses in the atmosphereare increasing due to anthropogenic activities. The increase of CO_2 had caused intensive phenomena such as; climatic changes, global warming and change in rainfall and rise of sea level [1]. CO_2 emissions in atmosphere come from different sources, mainly from the combustion of fossil fuels used in power generation, transportation, industrial processes and residential and commercial buildings. In precisely, industrial process such as cement manufacturing is considered as the major source of CO_2 in atmosphere. Whereas, cement manufacturing is responsible for 88% of the anthropogenic CO_2 emissions while land-use change, primarily deforestation, is accountable for the remaining 12% [2]. The above discussion pointed out that, there is a need to reduce the atmospheric concentration of CO_2 to be below of the current levelin future [3].

[1]Corresponding author: irwan@uthm.edu.my

In recent years, there are many researches [1-4], with different technologies interest on reducing of atmospheric CO_2 concentration and limiting the emission of CO_2 from differencegenerated sources. based on previous studies, sequestration of CO_2 can be done using different technologies such as; chemical absorption, physical separation, membrane separation and biological fixation [5-7]. Biological sequestration of CO_2 can naturally convert CO_2 to biominerals including carbonate minerals such as calcium carbonate ($CaCO_3$) and magnesium carbonate through bio-mineralization process [8]. In this case, bacterial carbon anhydrate (CA) used as bio-mimetically for CO_2 capture and conversion. CA has the ability to accelerate hydration of CO_2 to HCO_3^-, which can be precipitate in the form of carbonate as $CaCO_3$ [4,9,10]. On the other hand, CO_2 penetrates and sequestrates into concrete by reacting with Portland cement in the presence of moisture to form $CaCO_3$. The rate of carbonation process in concrete depend on several factors whereas, concentration of CO_2 considered as the most factor affect carbonation process as illustrated in section 3 in this manuscript [11-12].

Nowadays, concrete technology used different types of bacteria such as *Ureolytic, Bacillus sphaericus, Shewanella and Bacillus pasteurii, pseudomonas aeruginosa* as natural additive material to concrete in producing new type of concrete namely bio-concrete. This type of concrete has the ability to accelerate the precipitation process of $CaCO_3$ in concrete pores [13-16]. Bio-concrete incorporating ureolytic bacteria such as *Bacillus sphaericus*proved that, theused of urease helps in mineralization of $CaCO_3$and hydrolyzing urea that present in the environment. In line with that, $CO_2$is released from urea and combines with calcium ions resulting in deposition of $CaCO_3$ in the form of calcite[17]. The objective of this paper is to highlight the potential of bio-concrete to sequestration CO_2. This paper will provide understanding on the use of bacteria as CO2 sequestration.

2 Carbon dioxide sequestration in concrete

Sequestration and capture of $CO_2$are considered as a new technology that can reduce and eliminate of CO_2 emissions. Most of previous studies focused on sequestration of CO_2 using geological sequestration conducted by injecting CO_2 into deep underground rock formation, or capture CO_2 by using power plants and biological sequestration [18-19].
Nowadays, many studies in concrete technology were conducted to reduce CO_2 emission by reducing the amount of cement used in concrete mix and replaced it by other materials such as fly ash, silica fume and waste ash of palm oil, which produced different types of green concrete can be used in construction rather than conventional concrete [20-23].
On the other hand, sequestration of CO_2 from atmosphere in concrete and lime can be achieved via carbonation process. Carbonation occurred on the surface of concrete though CO_2 penetration into the pores inside concrete. CO_2 penetration into concrete pores depend on different factors such as; CO_2 concentration and relative humidity, the availability of $Ca(OH)_2$ and water, the additional materials that replace cement [24]. Carbonation is famed as a neutralizing process, which occurred as a results of the chemical reaction of $Ca(OH)_2$ and calcium-silicate-hydrate (C-S-H) with CO_2 to form calcium carbonate$CaCO_3$ and H_2O [25].

$$Ca(OH)_2 + CO_2 \rightarrow CaCO_3 + H_2O \qquad (1)$$

Pade & Guimaraes[26] mentioned that, carbon dioxide production can be re-capture in concrete by chemical reaction to calcination in the form of carbonation. Furthermore, carbonation can improve the surface hardness strength and durability of concrete by pore refinement of the cement paste matrix. However, carbonation is useful in non-reinforced concrete. It can lead to steel corrosion in reinforced concrete by reduced pH of carbonated cement paste due to carbonation [26].

Previous study by Pade & Guimaraes [26] approved that, carbonation process in concrete is generally very slow especially in normal concrete. Therefore, there is an interest in accelerating carbonation processespecially in non-reinforcement concretealong primary life applications and long term life of concrete in order to make the life cycle of concrete more carbon neutral with respect to the calcination reaction. In order to achieve the purpose, factors affecting carbonation process in concrete must be studied, if there is an interesting in accelerating carbonation in the concrete [24].

3 Factors affecting carbonation process in concrete

There are many factors that affect the acceleration process of carbonation into concrete such as; concentration of CO_2, concrete porosity, types of material, water cement ratio, relative humidity and degree of hydration of concrete materials.

Yoon et al [27] approve that, CO_2 concentration plays an important role on the carbonation depth of concrete. At the same time, the different of water cement ratio(W/C) and microclimatic conditioncontributed affectively of carbonation of concrete structures. The results show that, the carbonation in concrete increased with increased of CO_2 concentration and with high W/Ccompared to low W/C.

According to the study by Chang & Chan [28], three distinct regions of carbonated zone in concrete were identified; fully carbonated, partly carbonated and noncarbonated zones. Concrete with the fully carbonated zone is identified with the degree of carbonation greater than 50% and pH< 9.0. Furthermore, the degree of carbonation in the partly carbonated zone lies between 0% and 50% when pH is between 9.0 and 11.5. The non-carbonated zone is where the test specimen shows no sign of carbonation. The change of pH and concentration of CO_2 play an important role on penetration of carbonation into concrete. Therefore, carbonation of the concrete is considered as deterioration mechanism, because it help to decrease the pH of the concrete. Concrete with the lower pH causes the reinforcement steel to lose its electrochemical protection, so that corrosion may initiate and progress. For this reason, carbonation is not good for reinforced concrete. However, carbonation can be helpful for non-reinforced concrete [29]. Many of researches mentioned that, the relative humidity with surrounding environment around 50% to 70% can help to increase carbonation of concrete [11,12,29]. Furthermore, concrete with high internal moisture content shows a low rate of carbonation because the diffusion of CO_2 becomes difficult when the pores in concrete is saturated by water. Carbonation rate also decreases at lower internal moisture level due to insufficient water in pores [4].

4 Bacteria in concrete

The used of bacteria as additional material in concrete produce new type of concrete namely bio-concrete. The most popularbacteria used in concrete are presented in Table 1. Bacteria can play an important role in concrete by microbial mineral precipitation due to metabolic activities of some specific microorganisms in concrete. Whereas, metabolic activities of some microorganisms improve the overall behavior of concrete specially, compressive strength and concrete penetration which in role reduce the environmental factors that affect concrete which may lead to concrete deterioration. These improvements were due to the precipitation of calcium carbonate due to presence of bacteria in bio-concrete. In addition, bacteria plays an important role in bio-concrete to improve the deterioration of pores materials and enhance the durability of building materials especially sand, also healing of porosity and cracks in the concrete [15,30].

Table 1. Types of bacteria used in bio-concrete for different applications.

Researcher	Application	Types of bacteria
Bang et al. [13]	Microbial concrete as crack healer	*Bacillus Pasteurii* *Pseudomonas Aeruginosa*
Ghosh et al. [14]	To increase strength in concrete mixture	*Shewanella*
Achal et al. [37]	Enhanced urease and calcite production	*S. pasteurii*
Al-Thawadi. [16]	Strength enhancement of sand	*Ureolytic, Bacillus*
Wiktor & Jonkers. [15]	Self-healing concrete	*Bacillus alkalinitrilicus*
Pacheco-Torgal & Labrincha.[38]	Microbial concrete as surface treatment	*Bacillus Sphaericus*
Nosouhian et al. [39]	Enhancement of concrete properties	*B. subtilis*

There are some factors affecting the growth of bacteria such as growth medium, temperature, high pH value and anaerobic conditions [31]. The factors restricting the growth of bacteria in bio-concrete were extreme pH and anaerobic conditions.

According toTittelboom et al [17],when bacteria are used to heal pores in concrete, the major hindering factor is the highly alkaline pH of the concrete which restricting the growth of bacteria. Therefore, bacteria cells immobilized in silica gel to protect them from the high pH in concrete. Other authors had already suggested the use of polymer encapsulation [13].

The study by Wiktor & Jonkers [15] found that anaerobic condition restricting the growth of bacteria. However, the denitrification is the biological reduction of nitrogenous oxides to gaseous products during anaerobic (no oxygen) bacterial growth.

Al-Thawadi[16] approved that, ureolytic bacteria (*Bacillus)* have the ability to precipitate calcium carbonation which used in many application such as removing contaminates and surface coating of monumental estuaries to improve the mechanical properties of the concrete especially sandy material.Furthermore, *Bacillus* bacteria has a great innovation to hydrolyzed urea to generate adenosine triphosphate (ATP) using the efflux of ammonium ions through ATP-synthase which in role produce carbonate. Whereas, the presence of calcium ions with excess amount, conformed that calcium carbonate will precipitate. Availability of calcium carbonate is known as bio-grout or bio-cementation used as superior to the chemical cemented sandy material in terms of reduced penetration and insignificant decrease in porosity, also used for the purpose of strength formation and resistance of weathering factors.

Tittelboom et al [17] mentioned that, urease can be used to catalyzes the hydrolysis of urea ($CO(NH_2)_2$) into ammonium (NH_4^+) and carbonate (CO_3^{2-}). Whereas, urea is hydrolyzed intracellular in carbonate and ammonia (Eq.2). Carbonate spontaneously hydrolyses to form ammonia and carbonic acid (Eq.3).These products from bicarbonate and ammonium and hydroxide ions (Eqs.4-5). Which in role caused increase in pH, which can shifts the bicarbonate equilibrium, resulting in the formation of carbonate ions (Eq. 6).

$$CO(NH_2)_2 + H_2O \rightarrow NH_2COOH + NH_3 \quad (2)$$
$$NH_2COOH + H_2O \rightarrow NH_3 + H_2CO_3 \quad (3)$$
$$H_2CO_3 \leftrightarrow HCO_3^- + H^+ \quad (4)$$
$$2NH_3 + 2H_2O \leftrightarrow 2NH_4^+ + 2OH^- \quad (5)$$
$$HCO_3^- + H^+ + 2NH_4^+ + 2OH^- \leftrightarrow CO_3^{2-} + 2NH_4^+ + 2H_2O \quad (6)$$

When bacteria cell wall charged negatively, the bacteria draw cations from the environment, including Ca^{2+}, to deposit on their cell surface. The Ca^{2+} -ions react with CO_3^{2-} -ions, which in role lead to precipitation process of $CaCO_3$ at the cell surface (Eqs.7-8).

$$Ca^{2+} + Cell \rightarrow Cell\text{-}Ca^{2+} \quad (7)$$
$$Cell\text{-}Ca^{2+} + CO_3^{2-} \rightarrow Cell\text{-}CaCO_3\downarrow \quad (8)$$

5 Influence of bacterial carbonic anhydrase on accelerate bio-sequestration of CO_2

In the global effort to eliminate CO_2 emission, several technologies are being deliberated. Bio-sequestration of CO_2 is one of the most important technology applied to accelerate sequestration process.

In order to accelerate the process of biological sequestration of CO_2 under a conducive environment, which progressesthrough the hydration of CO_2, an enzyme known namely carbonic anhydrase (CA) is employed [32,33]. CA is a metallo-enzyme containing zinc (Zn^{+2}) metal ion in its active site, encoded by almost all organisms including eukaryotes and prokaryotes, and catalyzes the reversible hydration of carbon dioxide (CO_2) into bicarbonate ions (HCO_3^-) which in role has the ability to promote $CaCO_3$ [32]. CA can be extracted from different bacteria such as; *Ureolytic bacteria, Bovine CA, Bacillus cereus and Citrobacter freundii* for sequestration purpose of CO_2 [4,10,34-35].

Chen et al [34] & Ramanan et al [35] mentioned that, the precipitation process of carbonate minerals is naturally very slow. Therefore, the used of CA can play an important role to accelerate the reaction and the formation of carbonate under a conducive environment as presented in Eqs 9-13. $CO_{2\,(g)}$ diffuse through the solution, producing $CO_{2(aq)}$, $HCO_3^-{}_{(aq)}$and $CO_3^{2-}{}_{(aq)}$, which could be used for precipitation reaction. While, the reaction started by releasing of bicarbonate ion into the reaction medium by hydration of CO_2[13,35]. At the end of this reactions solid calcium carbonate ($CaCO_3$) is produced in the presence of calcium ion.

$$CO_{2\,(g)} \Leftrightarrow CO_{2\,(aq)} \quad (9)$$
$$CO_{2\,(aq)} + H_2O \Leftrightarrow HCO_3^- + H^+ \quad (10)$$
$$HCO_3^- \Leftrightarrow H^+ + CO_3^{2-} \quad (11)$$
$$Ca^{2+} + CO_3^{2-} \rightarrow CaCO_3\downarrow \quad (12)$$
$$Cell\text{-}Ca^{2+} + CO_3^{2-} \rightarrow Cell\text{-}\ CaCO_3\downarrow \quad (13)$$

According to reaction series for the precipitation process of $CaCO_3$ (Reaction 10-12), the hydration of CO_2 to form carbonate ion and proton (Reaction 10) forwards with an extremely slow pace and pace is the rate limiting step. When the biocatalyst CA is employed to catalyze the hydration of CO_2 (Reaction 10), resulting dramatically increases in the rate of hydration which in role make the reaction much faster. It can be seen clearly that, the CA enzyme play vital role in making the reaction much faster, which can help in accelerate the process of CO_2 sequestration [36].

6 Concluding remarks

$CO_2$is considered as one of the main factors affecting environmental and global warming challenges. Many researches were conducted to minimize emissions of CO_2 and to capture CO_2 in the atmosphere. Carbonation of concrete can be considered as one of CO_2 sequestration process but it is very slow. Furthermore, concrete carbonation caused steel corrosion in structural reinforced concrete. Therefore, carbonation can be conducted in non-reinforced concrete to avoid steel corrosion. Accelerating of carbonation process in non-reinforced concrete tobe one of the vital attribution in CO_2sequestration. the used of bacteria in concrete can help in precipitation of $CaCO_3$ process which used bacteria enzyme and atoms of CO_2 during the chemical reaction. Previous studies had pointed out that many bacteria had showed possibility in improving concrete properties and self-healing. Therefore, it is very important to study the role of bacteria enzyme in carbonation process and the ability to accelerate the sequestration process of CO_2 into bio-concrete.

This study has been granted by the Ministry of Education (MOE) under Fundamental Research Grant Scheme (FRGS Vot. 1577) and supported by UniversitiTunHussien Onn Malaysia (UTHM).

References

[1] C. Karthikeyan, S. Rajeswari, S. Maruthamuthu, K. Subramanian, and G. Rajagopal, Biogenic ammonia for CO_2apturing and electrochemical conversion into bicarbonate and formate, J. CO_2 Util., **6**, 53–61, (2014)

[2] C. Le Quere, M. R. Raupach, J. G. Canadell and G. Marl, Trends in the sources and sinks of carbon dioxide. Nature Geosciences 2-12, 831–836, (2009)

[3] H. Siikavirta, R. Zevenhoven and J. Kohlmann, CO_2 capture, storage and utilisation in Finland, TechnolClim.,(2002)

[4] I.G. Kim, B.H. Jo, D.G. Kang, C.S. Kim, Y.S. Choi and H.J. Cha, Biomineralization-based conversion of carbon dioxide to calcium carbonate using recombinant carbonic anhydrase, Chemosphere., **87**(10), 1091–1096, (2012)

[5] J.D. Figueroa, T. Fout, S. Plasynski, H. McIlvried and R.D. Srivastava, J. Int, Greenhouse Gas Control 2, 9–20 , (2008)

[6] K.M.K.Yu, I. Curcic, J. Gabriel and S.C.E.Tsang, Recent advances in CO_2 capture and utilization, ChemSus Chem., **1**(11), 893-899 (2008)

[7] C.K. Ahn, H.W. Lee, Y.S Chang, K. Han, J.Y. Kim, C.H. Rhee, H.D. Chun, M.W. Lee and J.M. Park, Int. J. Greenhouse Gas Control, **5**, 1606–1613 (2011)

[8] S. Mann, Biomineralization: *Principles and Concepts in Bioinorganic Materials Chemistry*, Oxford University Press, New York, (2001)

[9] M.J. Aguilar, Urine as a CO_2 absorbent, J. Hazard Mater, **213–214**, 502–4, (2012)

[10] C. Karthikeyan, S. Rajeswari, S. Maruthamuthu, K. Subramanian and G. Rajagopal, Biogenic ammonia for CO_2capturing and electrochemical conversion into bicarbonate and formate, J. CO_2Util., **6**, 53–61, (2014)

[11] AM. Neville, JJ. Brooks, *Concrete technology*, Longman Group UK Ltd, (1993)

[12] AM. Neville, Properties of concrete, Fourth Edition, Longman Group UKLtd, (1995)

[13] S. Bang, J.K. Galinat and V. Ramakrishnan, Calcite precipitation induced by polyurethane-immobilized Bacillus pasteurii, Enzyme and microbial technology, **28**(4), 404-409, (2001)

[14] A. F. Alshalif, J.M. Irwan, N. Othman and L.H. Anneza, Isolation of Sulphate Reduction Bacteria (SRB) to Improve Compress Strength and Water Penetration of Bio-Concrete, *In MATEC Web of Conferences*, **47**, EDP Sciences, (2016)

[15] J.M. Irwan, L.H. Anneza, N. Othman and A. Faisal. Alshalif, Compressive Strength and Water Penetration of Concrete with Enterococcus faecalis and Calcium Lactate, In Key Engineering Materials, **705**, 345-349, Trans Tech Publications, (2016)

[16] J.M. Irwan, L.H. Anneza, N. Othman, A.F. Alshalif, M.M. Zamer and T. Teddy, Calcium Lactate addition in Bioconcrete: Effect on Compressive strength and Water penetration, *In MATEC Web of Conferences*, **78**, 01027, EDP Sciences, (2016)

[17] K. Van Tittelboom, N. De Belie, W. De Muynck and W. Verstraete, Use of bacteria to repair cracks in concrete, Cement and Concrete Research., **40**(1), 157-166, (2010)

[18] G.M. Bond, N. Liu, A. Abel, B.J. McPherson and J. Stringer, Biomimetic Sequestration of CO_2 in Carbonate Form: Role of Produced Waters and Other Brines, Prepr, Pap-Am. Chem. Soc, Div. Fuel Chem, **49**(1), 420, (2004)

[19] W.D. Gunter, E.H. Perkins and T.J. McCann, Aquifer disposal of CO_2 rich gases: reaction design for added capacity, Energy Conversion and Management, **34**, 941–948, (1993)

[20] E.E. Berry, R.T. Hemmings, M.H. Zhang, B.J. Cornelius and D.M. Golden, Hydration in high-volume fly ash concrete binders. Materials J., **91**(4), 382-389, (1994)

[21] R.D. Hooton, Influence of silica fume replacement of cement on physical properties and resistance to sulfate attack, freezing and thawing, and alkali-silica reactivity, materials J., **90**(2), 143-151, (1993)

[22] W. Tangchirapat, T. Saeting, C. Jaturapitakkul, K. Kiattikomol and A. Siripanichgorn, Use of waste ash from palm oil industry in concrete, Waste Management, **27**(1), 81-88, (2007)

[23] C.J. Kibert, *Sustainable construction: green building design and deli*very, John Wiley & Sons, (2016)

[24] F.M. Lea, Some Special Cements and Cement Properties, The Chemistry of Cement and Concrete, Chemical Publishing Company, NY, USA, 544-547, (1971)

[25] B. Johannesson and P. Utgenannt, Microstructural changes caused by carbonation of cement mortar, Cem, Concr, Res **31**, 925–93, (2001)

[26] C. Pade, M. Guimaraes, The CO_2 uptake of concrete in a 100 year perspective. Cement and Concrete Research, **37**, 1348–1356, (2007)
[27] I.S. Yoon, O. Çopuroğlu and K.B. Park, Effect of global climatic change on carbonation progress of concrete, Atmos, Environ, **41**(34), 7274–7285, (2007)
[28] C. Chang and J. Chen, The experimental investigation of concrete carbonation depth, **36**, 1760–1767, (2004)
[29] C.D. Atis, Carbonation-porosity-strength model for fly ash concrete, Journal of Materials in Civil Engineering, **16**(1), 91-94, (2004)
[30] N.K. Dhami, A. Mukherjee, S.M. Reddy and J. Seto, Biofilm and microbial applications in biomineralized concrete, Intech open access publisher, (2012)
[31] J.M. Irwan, A. F. Alshalif, N. Othman and A.L. Hll, Effect of ureolytic bacteria on compressive strength and water permeability on bio-concrete. *International Conference on Civil, Architectural, Structural and Constructional Engineering*, August 21-23, (2015)
[32] G.M. Bond, N. Liu, B.J. McPherson, A. Abel and J. Stringer, Laboratory scale pilot of enzyme catalyzed CO_2 sequestration with produced waters as cation source, netl proceedings paper (2004), Retrieved on September 2, 2016 from www.netl.doe.gov/publications/proceedings/04/carbon seq/175.pdf,
[33] N. Favre, M.L. Christ, A.C. Pierre, Biocatalytic capture of CO_2 with carbonic anhydrase and its transformation to solid carbonate, J. of Molecular Catalysis B: Enzymatic, **60**, 163–170, (2009)
[34] W. Li, W.S. Chen, P.P. Zhou and L.J. Yu, Influence of enzyme concentration on bio-sequestration of CO_2 in carbonate form using bacterial carbonic anhydrase, Chem. Eng. J, **232**, 149–156, (Oct, 2013)
[35] R. Ramanan, K. Kannan, S.D. Sivanesan, S. Mudliar, S. Kaur, A.K. Tripathi and T. Chakrabarti, Bio-sequestration of carbon dioxide using carbonic anhydrase enzyme purified from Citrobacterfreundii, World J. Microbiol, Biotechnol., **25**(6), 981–987, (2009)
[36] A.Y. Shekh, K. Krishnamurthi, S.N. Mudliar, R.R. Yadav, A.B. Fulke, S.S. Devi and T. Chakrabarti, Recent Advancements in Carbonic Anhydrase–Driven Processes for CO_2 Sequestration, Minireview,Crit. Rev, Environ, Sci, Technol., 42(14), 1419–1440, (2012)
[37] V. Achal, A. Mukherjee, P.C. Basu and M.S. Reddy, Lactose mother liquor as an alternative nutrient source for microbial concrete production by Sporosarcina pasteurii, J. of industrial microbiology & biotechnology, **36**(3), 433-438, (2009)
[38] F. Pacheco-Torgal and J.A. Labrincha, Biotech cementitious materials: Some aspects of an innovative approach for concrete with enhanced durability, Construction and Building Materials, **40**, 1136-114, (2013)
[39] F. Nosouhian, D. Mostofinejad and H. Hasheminejad, Influence of bio-deposition treatment on concrete durability in a sulphate environment, Biosystems Engineering, **133**, 141-152, (2015)

Efficiency of *Moringa oleifera* Seeds for Treatment of Laundry Wastewater

Al-Gheethi AA[1], *Mohamed* RMSR[1*], *Wurochekke* AA[1], *Nurulainee* NR[1], *Mas Rahayu* J[1], *Amir Hashim* MK[1]

[1]Micro-pollution Research Centre (MPRC), Department of Water and Environmental Engineering, Faculty of Civil & Environmental Engineering, Universiti Tun Hussein Onn Malaysia,86400 Parit Raja, Batu Pahat, Johor, Malaysia

Abstract: Laundry wastewater has simple characteristics in which the detergents compounds are the main constitutes. But these compounds have adverse effects on the aquatic organisms in the natural water bodies which received these wastes without treatment. Few studies were conducted on these wastes because it represent a small part of the total wastewater generated from different human activities. Moreover, the coagulation process for laundry wastewater might be effective to remove of detergents compounds. Therefore, in the present study, the efficiency of coagulation process by using chemical (ferrous sulphate) and natural coagulants (*Moringa oleifera* seeds) were investigated. The raw laundry wastewater samples were collected from laundromat located at Taman Universiti, Parit Raja. The characteristics of these wastes were determined and then the wastewater was subjected for the treatment process consisted of three units including aeration, coagulation and sedimentation process. The chemical and natural coagulants were used with four dosage (30, 60, 90 and 120 mg L^{-1}) and the coagulation process was carried out at room temperature (25±2°C) for one hour. The results revealed that the laundry wastewater have high concentrations of turbidity (57.8-68.1 NTU) and Chemical Oxygen Demand (COD) (423-450 mg L^{-1}) with pH value between 7.96 and 8.37. *M. oleifera* seeds exhibited high efficiency for removal of turbidity (83.63%) with 120 mg L^{-1} of dosage, while 30 mg L^{-1} of $FeSO_4$ was the best for removal of COD (54.18%). However, both parameters still more than Standard B for wastewater disposal suggesting the need to increase the period of coagulation process with *M. oleifera* seeds or to subject of the treated effluents for a secondary coagulation process with natural coagulant products to improve the characteristics of laundry wastewater without a secondary products as that generated with the chemical coagulants.

1 Introduction

Detergent compounds are classes of the organic micro-pollutants (OMPs) in the laundry wastewater. The disposal of these compounds into the natural water bodies lead to

[1] Corresponding author: maya@uthm.edu.my

aesthetic losses due to the foam which can also cause adverse effects on ecosystems and biodiversity [1,2]. Detergents compounds in the laundry wastewater are biodegradable compounds and they have a short persistence time in the environment. Nevertheless, the main hazard for these compounds are associated with their concentrations which might be very active and poisonous directly for the organisms within a short exposure time. The indirect effect of the detergent for aquatic life lies by lowering the surface tension of the water. Besides, the chemical composition of the laundry wastewater such as high chemical oxygen demand (COD) and total suspension solids (TSS) as well as pH might change the natural water characteristics and induce the occurrence eutrophication phenomenon. Therefore, these parameters have to be reduced in the laundry wastewater to the minimum concentrations. Among several technologies used for the treatment and reduction of organic compounds in the wastewater, the coagulation process represent the most potent technique for improving the characteristics of wastewater before the disposal into the environment. The coagulants substances used might be chemicals such as ferrous sulphate or natural such as *Moringa oleifera*. Nonetheless, natural coagulant have more advantages than chemical substances in term of toxicity and efficiency as well as the cost. The toxicity of chemical coagulants have reported in the literature, the correlation between the availability of ferrous sulphate (above 200 mg L^{-1}) in the treated water and Alzheimer's and neuro-degenerative diseases has revealed by Bhatia et al. [3].

Moringa oleifera is known as a tropical plant with high contents of oil and water soluble molecules in their seeds [4]. *M. oleifera* seeds are used in the coagulation process due to the biodegradable characterize of their compounds. One advantages more of *M. oleifera* is the biological activity against bacteria which has demonstrated by the authors. *M. oleifera* has the potential to reduce cloudiness, clay and bacterial content of water in comparison to aluminium salts. Therefore, it represent an alternative and future technology for the wastewater treatment. Two process included absorption and neutralization of the colloidal charges are explained the mechanism of coagulation with *M. oleifera*. *M. oleifera* is non-toxic and biodegradable substrates, eco-friendly and unlike alum, has no effect on the pH and conductivity of the treated wastewater. Moreover, less amounts of sludge are generated when *M. oleifera* are used in the coagulation of wastewater in comparison to alum [5].

M. oleifera seeds extracts has exhibited an efficiency in the reduction of surfactant such as sodium lauryl sulphate from aqueous solutions, where 80% of the reduction has recorded [1]. It has been reduced TSS by 92%, total coliform by 89.6% and faecal coliform by 98.1% with 10 mL L^{-1} (v/v) of seeds concentrations [6]. However, the maximum reduction recorded for BOD and COD were 32 and 48% respectively. *M. oleifera* seeds have used for treatment different types of wastewater such as tapioca starch wastewater. Suhartini et al. [7] used a system consisted of two-stage clarifier tanks filled with sand or coconut fiber and *M. oleifera* seeds. The system occurred high reduction of BOD, COD and TSS to meet the Indonesian standards limits of tapioca starch wastewater discharge. Laundry wastewater has a different composition to the domestic and agriculture wastewater. These differences lies the quality of the chemical characteristics rather than their quantity. Therefore, the potential of natural coagulant for the treatment of these wastes need to be investigated in order to give an evaluation for their role in the reduction of physical and chemical parameters in these wastes before the final disposal into the environment and this is the focus of the current work.

This study aimed to examine the efficiency of *M. oleifera* seeds as coagulant of laundry wastewater in comparison to the commercial coagulant such as ferrous sulphate. The potential of *M. oleifera* seeds in improving the quality of laundry wastewater was evaluated based on the reduction of COD, TSS, turbidity and pH which represent the main parameters of the laundry wastewater.

2 Materials and methods

2.1 Study area and sampling

The laundry wastewater samples (5 L) were collected from laundromat located at Taman Universiti, Parit Raja (Latitude 1°51'3.12"N and longitude 103° 4'29.63"E). The laundromat was selected because it's a self-service laundromat where different types of detergents with different dosages are used and thus represent a good model to test the efficiency of natural coagulants. The samples were collected from discharge point (Fig. 1) in a plastic bottle (5 L) and then transported into the laboratory in ice box and subjected for the chemical and physical tests within two hours to prevent the degradation of detergents compounds. The characteristics of laundry wastewater including pH, turbidity, COD and TSS were determined according to APHA [8].

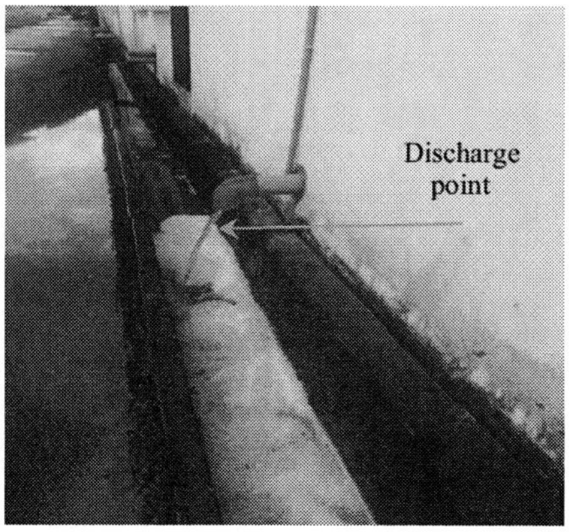

Fig. 1. Laundry wastewater discharge point

2.1.1 Preparation of M. oleifera seeds and ferrous sulphate

M. oleifera seeds were obtained from a local market at Parit Raja. The seeds were dried in order to improve level of polyelectrolyte presented in the kernels [9]. Thereafter, they were grounded into fine particles less than 0.20 mm of size and dried again at 100°C for 24 hrs [10]. The dried seeds were milled into a fine powder by using domestic blender and then passed through 50-mesh sieve according to coagulant-method [3]. Ferrous sulphate ($FeSO_4.7H_2O$) salts (CAS: 7782-63-0, Fisher Scientific) was supplied from Tai`an Ming Chen Chemical Import and Export Co. Ltd, China.

2.2 Coagulation studies

The raw laundry wastewater were treated by using a treatment system consisted of three units included aeration, coagulation and sedimentation tank. The aeration process was carried out in order to remove carbon dioxide and oxidize soluble iron and manganese as well as to reduce hydrogen sulphide and ammonia. The coagulation process were performed with four concentrations of *M. oleifera* seeds and ferrous sulphate in order to determine the optimal concentrations required to achieve high removal percentage of the main parameters of laundry wastewater. The sedimentation tank was used to separation solid materials resulted from the coagulation process and clarified wastewater.

A fixed volume of (1 L) of laundry wastewater sample was stored in the aeration tank (25×17×18 cm) at room temperature (25±2 °C) for 30 min. The wastewater was pumped into the coagulation tank using a water pump (Shanda), 120V. The coagulation tank made up of High-density polyethylene (HDPE) transparent bottle (14 cm×20 cm×14 cm) and Direct Current motor (DC Geared Motor, DC12V) with 1.1 watt, 103 rpm, 410 mA, 127.4 mN and 8 mm of shaft diameter [10]. Four concentrations [30, 60, 90 and 120 mg L^{-1} (w/v)] of *M. oleifera* seeds and ferrous sulphate were investigated separately. The wastewater sample was mixed with each concentration for one hour. At the end of the coagulation process, the wastewater was pumped slowly into sedimentation tank to settle for one hour. The supernatant generated from the sedimentation tank was used to determine pH, turbidity, COD and TSS according to APHA [8]. The control samples were conducted without chemical or natural coagulant. The removal efficiency was calculated according to the following equation;

$$Removal\ (\%) - \frac{c_i - c_f}{c_i} \times 100 Removal\ (\%) = \frac{c_i - c_f}{c_i} \times 100 \quad (1)$$

where, C_i is the initial parameter concentration (mg L^{-1}), C_f is the remaining parameter concentration (mg L^{-1})

3 Results and discussions

The characteristics of laundry wastewater demonstrated that these wastes have high concentrations of COD (423-450 mg L^{-1}) and turbidity (57.8-68.1 NTU) more than the standards limits recommended by Environmental Quality Act, 1974 (sewage and Industrial effluents, Regulation 2009). These standards are used for sewage and industrial wastewater, but it can be used for discharging laundry wastewater due to absence a specific regulation of these wastes. The high concentrations of COD might be due to the presence of detergents compounds, which lead to increase of chemical reaction level between organic substances. Laundry wastewater has high impurities due to the presence of sand and particles, oil and grease, surfactants, detergent and phosphates [11]. Despite, TSS and pH values were within the standards range, pH of these waste was more than pH 7.5, which might be related to the presence detergents compounds that include wide range of amphoteric, anionic and cationic compounds. Laundry water contains detectable concentrations of TSS, salts, nutrients, organic matter and pathogens raised from the clothes, laundry detergents and fabric softener residues [12]. Moreover, the detergents are the most abundant organic chemicals in these wastes [13]. Therefore, removal of these compounds is required before the discharge into the environment.

The efficiency of chemical and natural coagulant in the removal of TSS is presented in Fig. 2 a. TSS was within the standard limits of wastewater disposal regulated by Environmental Quality Act, 1974 (sewage and Industrial effluents, Regulation 2009). Moreover, *M. oleifera* seeds and ferrous sulphate exhibited an efficiency to reduce it by 73 and 78% at 90 and 120 mg L^{-1} of both coagulants respectively. It has revealed that *M. oleifera* seeds contain significant quantities of proteins with high molecular weight and water soluble protein, which carries a positive charge to the solution. These flocculating protein works as a clarifying agent of turbid water [4].

M. oleifera seeds and ferrous sulphate have recorded a significant reduction in the turbidity in comparison to the control (without coagulant) (Fig. 2 b). The maximum removal was noted with the high concentrations (120 mg L^{-1}) of both coagulant.

However, *M. oleifera* seeds exhibited high efficiency in the turbidity removal than ferrous sulphate (83.63 *vs.* 59.02%). The efficiency of *M. oleifera* seeds in removal of turbidity might be related to the presence of water soluble cationic proteins which indicate for electronic charge and enhance coagulation and flocculation [14]. In contrast, $FeSO_4$ exhibited more efficiency in the removal of COD, the reduction was 54.18% with 30 mg L^{-1} for $FeSO_4$, and 42.72% at 90 mg L^{-1} of *M. oleifera* (Fig. 2 c). These findings might be due to the nature of chemical reaction between $FeSO_4$ and organic constitutes in the wastewater which might occur more than that between *M. oleifera* components and the organic compounds. However, it has to mention that the reduction of COD with the addition of $FeSO_4$ take place due to the chelating process that bind to the organic matter with $FeSO_4$. This process relied on the equilibrium between the substrates during the coagulation process. The high concentrations of $FeSO_4$ has decreased the removal; percentage of COD, due to the decreases in the affinity of organic compounds to adsorb of Fe ions. Besides, coagulation process with chemical chelating agents is accomplished with the generation of sludge. On the other hand, the efficiency *M. oleifera* seeds in the reduction of COD take place by the degradation of organic matter by using their compounds which have the degradation potential. The high ability of *M. oleifera* to degrade the organic compounds relate to the presence of protein in the *M. oleifera.* The binding of proteins to the organic compounds facility their consumption by the microorganism as carbon sources by using hydrolyses enzymes. But the degradation process need more time to cause a significant reduction in COD. In the present study the coagulation process was conducted for one hour which might be sufficient for the chelating process but not for the degradation process. It has demonstrated that even with high concentrations of *M. oleifera* seeds the reduction in COD might still limited [15]. Nonetheless, it can be indicated that the *M. oleifera* still better than $FeSO_4$ for reducing COD, because less sludge is produced with the natural coagulants as revealed by previous studies in literature [16] and the produced sludge is a degradable substances and has no toxicity characteristics. The absence of toxicity of produced sludge related to the nature of *M. oleifera* seeds which have no toxicity and they are used for the medical applications.

In all case, the results recorded here indicated that the final concentrations of COD and turbidity still more than the standard A and B required for the disposal of wastewater (Table 1). These findings indicate that the treatment system has to be conducted for more than one hour rather than to increase the concentrations of coagulants. This is because the increasing in these concentrations correlated with the reduction in pH value the minimum pH value recorded was pH 5.77 with 120 mg L^{-1} of both *M. oleifera* and $FeSO_4$, this value is most close to the minimum pH required by Standard B (pH 5.5), therefore, the addition more coagulant dose might lead to decrease pH value to less than the standards. Hence, in order to achieve more reduction of COD and turbidity the coagulation period might be extended to more than one hour. However, the results for the coagulation with $FeSO_4$ noted that the maximum reduction of COD was observed with 30 mg L^{-1} while the increasing of concentrations to more than 30 mg L^{-1} associated with decreases in the removal percentage which indicating that the increasing of coagulation period might effect negatively on the chemical coagulation. In contrast the increasing of removal percentage with increasing of *M. oleifera* dose indicating that the increasing of time might also achieve high efficiency based on the concept of equilibrium between substrates and produces. The alternative way to achieve more reduction in both COD and turbidity is by perform a secondary coagulation process of supernatant results from the first coagulation process. The increasing in the coagulant dose or settling time might achieve the reduction levels required for the disposal process while, the adjusting pH of laundry wastewater might be unacceptable in terms of reduction of chemical additives for the treatment of wastes.

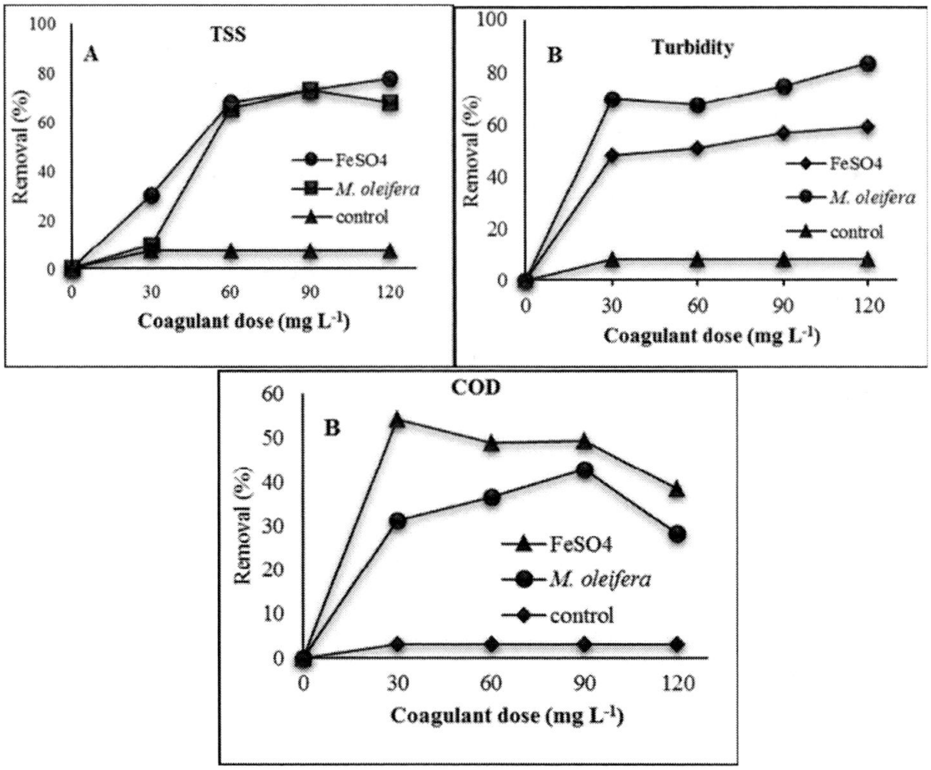

Fig. 2 Removal parentage of the main laundry wastewater during the coagulation process using chemical and natural coagulant; A) TSS; B) Turbidity; C) COD

Table 1. Characteristics of laundry wastewater coagulated with different doses of *M. oleifera* and $FeSO_4$

Dose (mg L^{-1})	pH		Turbidity (NTU)		COD (mg L^{-1})		TSS (mg L^{-1})	
	M. oleifera	$FeSO_4$	*M. oleifera*	$FeSO_4$	*M. oleifera*	$FeSO_4$	*M. oleifera*	$FeSO_4$
0	7.96-8.37		57.8-68.1		423-450		0.38-0.44	
30	6.14	6.42	19	32.5	300	200	0.37	0.25
60	7.01	6.02	20.5	30.9	277	223	0.14	0.13
90	6.52	6.09	15.8	27.3	250	222	0.11	0.11
120	5.77	5.7	10.3	25.8	313	269	0.13	0.09
Standards (A)	6-9		<5		50		50	
Standards (B)	5.5-9		50		100		100	

COD (Chemical Oxygen Demand); TSS (Total suspension solids)

4 Conclusions

The results of COD, turbidity, pH and TSS indicate that the laundry wastewater have a different characteristics due to the presence of detergent compounds which lead to increase COD and pH values. *M. oleifera* seeds exhibited high efficiency for the removal of turbidity but not for the COD in comparison to $FeSO_4$. However, *M. oleifera* seeds is eco-friendly since the sludge generated is consist of degradable substances. Both COD and turbidity still more than the standards A and B for the disposal of wastewater, but the increase the period of coagulation with *M. oleifera* seeds or the subjection for a secondary coagulation process might contribute in their reduction of these parameters to be within the standards limits. Nevertheless, these suggestion need to be investigated to confirm their effectiveness.

The authors gratefully acknowledge Research, Innovation, Commercialization, Consultancy Management (ORICC), UTHM for Financial support of this work

References

[1] J. Beltrán-Heredia and J. Sánchez-Martín, Removal of sodium lauryl sulphate by coagulation/flocculation with *Moringa oleifera* seed extract. J. Hazard Materials, **164**(2), 713-719, (2009)

[2] J. K. Braga and M. B. Varesche, Commercial laundry water characterisation. Am J Anal Chem., **5**(1), 8, (2014)

[3] S. Bhatia, Z. Othman and A. L. Ahmad, Coagulation–flocculation process for POME treatment using *Moringa oleifera* seeds extract: optimization studies. Chem. Eng. J., **133**(1), 205-212, (2007)

[4] K. A. Ghebremichael, K. R. Gunaratna, H. Henriksson, H. Brumer and A. Dalhammar, simple purification and activity assay of the coagulant protein from *Moringa oleifera* seed. Water Res., **39**(11), 2338-2344, (2005)

[5] A. Ndabigengesere, K.S. Narasiah and B. G. Talbot, Active agents and mechanism of coagulation of turbid waters using *Moringa oleifera.* Water Res., **29**(2), 703-710, (1995)

[6] M. A. Ashmawy, M. S. Moussa, A. K. Ghoneim and A. Tammam, Enhancing the efficiency of primary sedimentation in wastewater treatment plants with the application of *Moringa oliefera* Seeds and Quicklime. J Am. Sci., **8** (2), (2012)

[8] S. Suhartini, N. Hidayat and E. Rosaliana, Influence of powdered Moringa oleifera seeds and natural filter media on the characteristics of tapioca starch wastewater. Int. J. Rec. Org. Waste Agr. **2**(1), 1-11 (2013)

[9] APHA, American Public Health Association, Inc. Standard methods for the examination of water and wastewater (21st Ed.). New York. USA, (2005)

[10] R. Nirmala and M. V. Jadhav, Enhancing filtrate quality of turbid water incorporating seeds of *Strychnos potatorum*, pads of *Cactus opuntia* and mucilage extracted from the fruits of *cocciniaindica* as coagulants. J. Environ. Res. Dev. **7** (2), 668-674 (2012)

[11] A. A. Al-Gheethi, R. M. Mohamed, M. Afaiz Ab. Rahman, J. Mas Rahayu and H.K. Amir, Treatment of wastewater from car washes using natural coagulation and filtration system, *Int. Conf. On Sustainable Environment & Water Research (ICSEWR 2015),* 25-26 Oct. 2015, Johor Baru, Malaysia, (2015)

[12] R.S. Mandac, B. Bogunović, D. Žagar and J. Faganeli, Riverine impact on the thermohaline properties, turbidity and suspended solids in a shallow bay (Bay of Koper, northern Adriatic Sea). Acta Adriatica, **55** (2), 195 – 212, (2014)

[13] E. R. Howard, R. Misra, Loch, and N. Le-Minh. *Laundry grey water potential impact on Toowoomba soils*-final report. National Centre for Engineering in Agriculture Publication 1001420/2, USQ, Toowoomba, Australia. (2005)

[14] M. Abu-Zreig, R.P. Rudra and W. T. Dickinson, Effect of application of surfactants on hydraulic properties of soils. Biosys. Eng., **84**(3), 363-372 (2003)

[15] K. H. Lee, K. Y. Park, S. K. Khanal and J. W. Lee, Effects of household detergent on anaerobic fermentation of kitchen wastewater from food waste disposer. J. Hazard. Mat., **244**, 39-45 (2013)

[16] N. Muralimohan, T. Palanisamy and M.N. Vimaladevi, Experimental Study on removal efficiency of blended coagulants in textile wastewater treatment. IMPACT: Int. J. of Research in Engineering & Technology., **2**, 2347-4599, (2014)

[17] R.M. Mohamed, A.I. Kutty, N. Mariam, M. Kassim, A. Hashim, Efficiency of using commercial and natural coagulants in treating car wash wastewater treatment. Aus. J. Basic Appl. Sci., **8**(16), 227-234, (2014)

Application of Dual Coagulant (Alum + Barley) in Removing Colour from Leachate

Nur Shaylinda Mohd Zin[1,*], *Zulazimie* Azraff Zulkapli[1]

[1] Faculty of Civil and Environmental Engineering, Universiti Tun Hussein Onn Malaysia, 86400 Parit Raja, Batu Pahat, Johor, Malaysia

Abstract. Coagulation/ flocculation is one of the treatment method for highly polluted leachate. One of the main affecting factor for this process is the coagulant used. Coagulant is divided into natural and chemical coagulant. In the current study, Alum (chemical coagulant) and barley (natural coagulant) were used as dual coagulant. The aim of this study is to examine the effectiveness of dual coagulant made from alum and barley in removing colour from the effluent of Simpang Renggam landfill leachate aeration lagoon through coagulation/flocculation method. Coagulation/flocculation process with single alum coagulant, single barley coagulant and dual coagulant (alum+barley) were examined by evaluating the optimum values of pH and dose. Optimum dose and pH for alum and barley as single coagulant were; 3 g/L & pH 5; 0.8 g/L & pH 6. Higher removal of colour was recorded for alum compared to barley. Application of alum and barley as dual coagulant had higher colour removal than alum and barley as single coagulant. The optimum pH and dose for dual coagulant were at pH 6, 3.0 g/L of alum and 0.8 g/L of barley respectively. However, at pH 6, 2 g/L alum and 1.6 g/L barley, the removal of colour was similar to alum at 3 g/L. It can be concluded that barley as coagulant aid able to reduce 33 % usage of alum at par removals of colour. Thus, the dual coagulant consist of alum and barley has the potential to be applied as a coagulant for leachate treatment.

1 Introduction

Increased number of population had causing enormous pressure on the local environment especially in solid waste management. The fast development in industrial sector and improved standard of living produces huge volume of solid waste. In Malaysia, it is a common practice that most of solid waste are disposed by using landfilling method. However the negative outcomes of landfilling is the production of leachate. Leachate is characterised as a highly polluted liquid waste that contains high value of polluted matters such as ammonia, chemical oxygen demand, suspended solid, colour and etc. [1]. Thus, proper treatment of leachate is crucial to maintain a healthy surrounding for living organism.

* Corresponding author: nursha@uthm.edu.my

There are many methods that can be used to treat this highly contaminated liquid. Basically the treatment of leachate consists of physical, chemical or biological methods. The selection of this treatment method is depending on the maturity of the leachate. One of the treatment methods used is coagulation, which is suitable for intermediate and old leachate [2]. Operation of coagulation is simple yet effective and required small area [3]. One of the main factors that affect coagulation effectiveness is coagulant type. Coagulant used consists of chemical (alum, ferric chloride, poly aluminium chloride) and natural (moringa oleifera, starch, cactus, guar gam) coagulant [4]. It is well known that chemical coagulant is better than natural coagulant. However application of chemical coagulant is exposed to after used effects such as Alzheimer and neuro-degenerative diseases [5]. Furthermore, the cost involved in coagulation is quite expensive for chemical coagulant. Thus, there's a need in finding less expensive and less health effect coagulant. Application of dual coagulant made from natural and chemical coagulant is able to reduce the dose of chemical and reduce the toxicological effect [4]. Successful application of combination coagulant made from natural and chemical coagulant for leachate treatment were proven by Ong, Ramli, Kamaluddin, and Zin et al, [6-9].

Barley is recognised as one of the major crops in daily diets of people [10]. Barley commonly used as animal feed, production of alcoholic beverages and food for human. Barley grain contains starch, fat and fibre components [11]. The starch in barley indicates good coagulant properties and known as natural polymer coagulant [4]. This natural polymer coagulant contains mixture of amylose and amylopectin [12]. The polymer chain structure of the starch, especially the amylopectin improved the coagulation/flocculation ability through bridging mechanism between the branched of the polymer chain [13-14]. However, application of natural polymer as main coagulant is limited by its efficiency in treating high strength wastewater. Thus, the addition of barley as a coagulant aid is seen able to increase the coagulation of the dual coagulant in this study.

Current treatment of Simpang Renggam landfill (SRL) involved aerated lagoon. However, the current treatment is not sufficient to treat the leachate. In order to improve the effluent quality of SRL, this study aimed to examine the effectiveness of dual coagulant made from alum and barley as a coagulant for Simpang Renggam landfill aeration lagoon effluent. The effectiveness of this dual coagulant was measured by determining the percentage removal of colour from SRL aeration lagoon effluent. The optimization of pH and coagulant dose were tested in this study.

2 Methodology

2.1 Study area and sampling

Samples of leachate were collected from Simpang Renggam landfill site (SRL) located in Simpang Renggam, Johor, Malaysia. SRL is located at latitude 10 53'41.64" North and longitude 1030 22'34.68" East in Kluang district. The treatment process of leachate in SRL is an aeration lagoon. The leachate samples were taken at the last outlet after leachate was treated by aerated lagoon. Sampling and analytical procedures were conducted in accordance with the American Public Health Association (APHA) Standard Method [15]. Age of SRL is 10 years old and is categorised as an old leachate.

2.2 Preparation of coagulant

The alum ($Al_2\ (S0_4)^3$) (3 % concentration) was used as the primary coagulant, while barley (1 % concentration) as coagulant aid. Preparation of barley as coagulant was based on modification of Okuda et.al study [16]. Dry barley seeds were blended using dry blender. Then, the barley powder was sieved and weighed according to concentration needed (1% concentration). The barley powder were then added in the beaker containing distilled water. The beaker were stirred using magnetic stirrer and heated up to 90°C for 30 minutes. After that, the solution was used as the coagulant aid.

2.3 Jar test

Coagulation tests were carried out using a jar test (Flocculator JTL5) apparatus. For this study, only pH and dose were considered in jar test process. While the factors listed in Table 1 were kept constant. The dose of alum and barley was calculated in g/L. NaOH and HCL were used to adjust the pH of leachate. A 1 L beaker was filled with 500 mL of effluent sample of SRL aeration lagoon, followed by the addition of alum under rapid stirring and addition of barley at the start of slow mixing. Then, the leachate was allowed to settle. Finally, 50 mL supernatant was collected 3 cm from the surface of the leachate for analysis of the color removal. The control sample was tested in each optimization test. The removal percentage was calculated based on the following formula:

$$\text{Percentage removal} = \frac{(C_i - C_f) \times 100}{C_i} \qquad (1)$$

where C_i (mg/L) and C_f (mg/L) are the initial and final parameter concentration respective.

Table 1. Constant jar test factors.

Factors	Value
Rapid mixing (rpm)	200
Rapid mixing duration (minutes)	4
Slow mixing (rpm)	30
Slow mixing duration (minutes)	15
Settling time (minutes)	30

Efficiency of dual coagulant (alum+barley) was analysed based on percentage removal of colour from SRL leachate. At the same time, comparison of removal percentage of colour between single alum coagulant, single barley coagulant and dual coagulant was also made.

3 Results and discussions

3.1 Effect of dose

In this study, colour efficiency of leachate effluent treated by single alum coagulant, single barley coagulant and dual coagulant were compared. As shown in Fig. 1, the colour removal ability of barley is the lowest indicates barley as single coagulant is not effective in removing colour. While alum, as single coagulant records 88% colour removal at 3.5 g/L. Application of dual coagulant increased the removal ability. Alum 3 g/L+barley and barley 1.6 g/L+alum, show higher trends compared to single barley and alum. Thus, addition of barley improved the coagulation process. Barley as coagulant aid, produce better strength of floc and increased the weight of the floc and finally increased the colour removal ability [4]. In comparison with Alum 3 g/L+barley and barley 1.6 g/L+alum, alum 3 g/L+barley is better as higher colour removals are recorded. At 3 g/L alum and 0.8 g/L barley, 98 % colour removal was achieved. At 3 g/L alum as single coagulant, just 85% colour removal was obtained. Therefore, addition of barley increased the colour removal ability. Reduction of alum dose at 1.6 g/L barley, achieved slightly low colour removals. However, based on Fig. 1 lower alum dose from 2-2.5 g/L with 1.6 g/L barley, has better colour removal compared to alum as single coagulant. Dual coagulant consist of alum 2 g/l and barley 1.6 g/L, achieved 85% colour removal with alum dose less than 33% to alum (3 g/L) as single coagulant. Based on the results mentioned above, the addition of barley is able to reduce the dose of alum used and improve the coagulation ability. Besides, reduction dose of alum used indicate possible reduction cost of treatment and toxicity effect of the chemical coagulant.

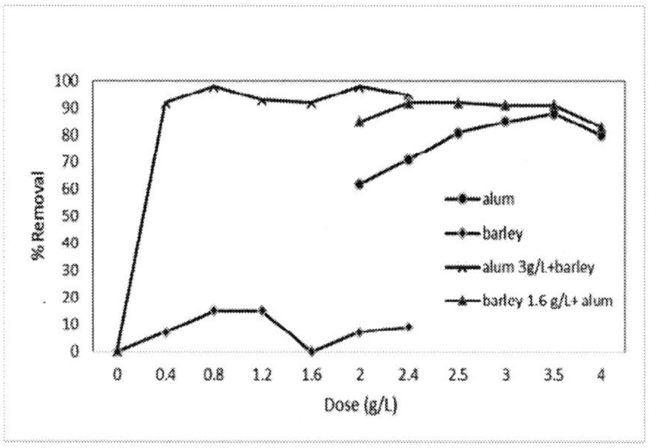

Fig. 1. Effect of dose on color removal.

3.2 Effect of pH

The subsequence experiments were designed to find the optimal pH of dual coagulant (alum 2.5 g/L+barley 1.6 g/L) and compare with single alum (3g/L) and single barley (0.8 g/L). Ability of barley as single coagulant is still the lowest compared to alum and alum+barley. Decolourization of leachate under the influence of pH at acidic condition, resulted on similar trend for alum, barley and alum+barley. However alum, showed higher removal compared to barley and alum+barley at pH range 3-5. The amount of alum added was more than amount of barley, so most of the coagulation optimum condition will favoured the optimum condition of alum as a single coagulant. Thus, the difference of percentage removal ability of alum and alum+barley were not too far and this is in line with Chen et. al (2015) study [17]. However, at pH 6 alum+barley, recorded slightly higher removal compared to alum and this indicate a point where the pH condition is suitable for alum and barley as dual coagulant. Optimum pH for maximum colour removal efficiencies of alum+barley, alum and barley were at 6, 5 and 5, respectively. The colour removal recorded at this optimum pH for alum+barley, alum and barley were 92%, 93% and 22%, respectively. Lower removals are recorded at alkaline condition for all the coagulant tested as shown in Fig. 2.This instances was in good agreement with the previous finding [9]. The results above indicate that dual coagulant and single alum performed better in acidic condition.

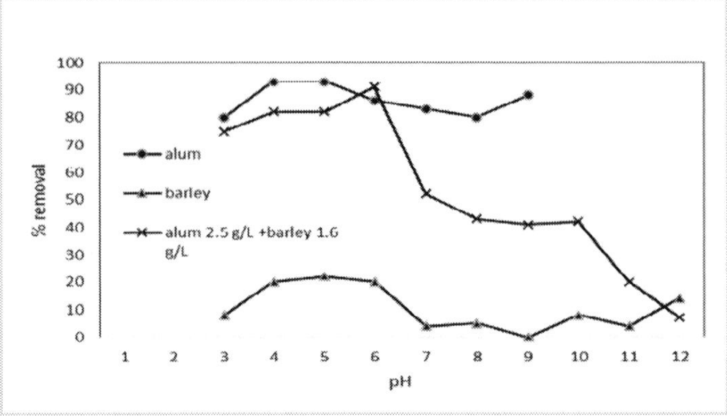

Fig. 2. Effect of pH on color removal.

4 Conclusions

Single alum coagulant, single barley coagulant and dual coagulant (alum+barley) were comparatively investigated in terms of colour removals under the influence of dose and pH. Higher removal of colour was achieved by dual coagulant compared to single alum and barley under the influence of dose. All three coagulants, performed better at acidic condition. However, the best performance was recorded by dual coagulant (alum+barley) at pH 6, alum 3 g/L and barley 0.8 g/L with 98% colour removal. The best color removal of single alum at pH 4, dose 3 g/L was 93 %. Thus, barley as coagulant aid is able to improve coagulating ability of alum at better pH and produced higher removal of colour.

The authors would like to gratefully acknowledge Ministry of Higher Education of Malaysia for the research financial support under Fundamental Research Grant Scheme (FRGS) vot NO.1570

References

[1] N. S. M. Zin, H. Abdul Aziz, M. N. Adlan, A. Ariffin, M. S. Yusoff, I. Dahlan, A comparative study of Matang and Kuala Sembeling landfills leachate characteristics. Applied Mechanics and Materials, **361**, 776-781, (2013)

[2] S. Ghafari, H.A. Aziz, M.J. Bashir, The use of poly-aluminum chloride and alum for the treatment of partially stabilized leachate: A comparative study. Desalination, **257**(1), 110-116, (2010)

[3] L.M. Rui, Z. Daud, A.A.A. Latif. Treatment of Leachate by Coagulation-Flocculation using different Coagulants and Polymer: A Review. International Journal on Advanced Science, Engineering and Information Technology, **2.2,** 114-117, (2012)

[4] N.S.M. Zin, *Treatment of Partially Stabilized landfill Leachate using composite coagulant derived from Prehydrolized Iron and Tapioca Starch*, PhD. Thesis, Universiti Sains Malaysia, (2015)

[5] S. Bhatia, Z. Othman, A.L. Ahmad, Coagulation–flocculation process for POME treatment using Moringa oleifera seeds extract: optimization studies. Chemical Engineering J., **133** (1), 205-212, (2007)

[6] N. Kamaludin, *Kajian penyingkiran kekeruhan, sulfat, nitrat, klorida, cadmium, kuprum dan zink dari air dan air sisa mengunakan kanji semulajadi.* Master thesis. University Sains Malaysia, (2005)

[7] R.Y. Ramli, *The use of tapioca starch as coagulant and coagulant aid in leachate treatment*, Degree Thesis, Universiti Sains Malaysia, (2011)

[8] S.L. Ong, *A study of landfill leachate treatment by PACL, Sago starch and Tapioca starch, Hymenocallis Liriosme and Aloe Vera as coagulant*, Master Thesis, Universiti Sains Malaysia, (2010)

[9] N.S.M. Zin, H.A. Aziz, M.N. Adlan, A. Ariffin, M.S. Yusoff, I. Dahlan, Treatability Study of Partially Stabilized Leachate by Composite Coagulant (Prehydrolyzed Iron and Tapioca Flour), Int. J. of Scientific Research in Knowledge **2** (7), 313, (2014)

[10] Tsuji, Hideaki, Masumi Kimoto, and Yasuo Natori. Allergens in major crops. Nutrition Research, **21** (6), 925-934, (2001)

[11] L.D. Campbell, R. J. Boila, S. C. Stothers, Variation in the chemical composition and test weight of barley and wheat grain grown at selected locations throughout Manitoba, Canadian journal of animal science, **75** (2), 239-246, (1995)

[12] Jankowski, A. Jeffrey, T. Calvin, T. Tobison, Starch/cationic polymer combinations as coagulants for the mining industry, U.S. Patent, **No. 6**,042,732, (2000)

[13] Pérez, Serge, E. Bertoft, The molecular structures of starch components and their contribution to the architecture of starch granules: A comprehensive review, Starch-Stärke, **62** (8), 389-420, (2010)

[14] J. Bratby, *Coagulation and flocculation in water and wastewater treatment* (IWA publishing, 2006)

[15] APHA, AWWA. WEF *Standard methods for the examination of water and wastewater* (American Public Health Association, American Water Works Association, and Water Environment Federation, 2005)

[16] Okuda, Tetsuji, Aloysius U. Baes, Wataru Nishijima, and Mitsumasa Okada. Improvement of extraction method of coagulation active components from Moringa oleifera seed, Water research, **33** (15), 3373-3378, (1999)

[17] Chen, Ting, B. Gao, and Q. Yue. Effect of dosing method and pH on color removal performance and floc aggregation of polyferric chloride–polyamine dual-coagulant in synthetic dyeing wastewater treatment, Colloids and Surfaces A: Physicochemical and Engineering Aspects, **355** (1), 121-129, (2010)

Performance of Vetiver Grass (*Vetiveria zizanioides*) for Phytoremediation of Contaminated Water

Sharifah Nur Munirah Syed Hasan[1], *Faradiella* Mohd Kusin[1*], *Ashton Lim* Sue Lee[1], *Tony Austin* Ukang[1], *Ferdaus* Mohamat Yusuff [1], and *Zelina* Zaiton Ibrahim[1]

[1]Department of Environmental Sciences, Faculty of Environmental Studies, Universiti Putra Malaysia, 43400 UPM Serdang, Selangor, Malaysia

Abstract. In tolerance towards metal uptake, there is a need to evaluate the performance of vetiver grass for metal removal to reduce water impurity. This study was aimed to evaluate contaminant removal by vetiver grass at varying root length and plant density and determine the metal uptake in vetiver plant biomass. Pollutant uptake of vetiver grass was conducted in laboratory experiment and heavy metal analysis was done using acid digestion and Atomic Absorption Spectrometry. Findings indicated that the removal of heavy metal was decreased in seven days of the experiment where iron shows the highest percentage (96%; 0.42 ppm) of removal due to iron is highly required for growth of vetiver grass. Removal rate of heavy metals in water by vetiver grass is ranked in the order of Fe>Zn>Pb>Mn>Cu. Results also demonstrated greater removal of heavy metals (Cu, Fe, Mn, Pb, Zn) at greater root length and higher density of vetiver grass because it increased the surface area for metal absorption by plant root into vetiver plant from contaminated water. However, findings indicated that accumulation of heavy metals in plant biomass was higher in vetiver shoot than in root due to metal translocation from root to the shoot. Therefore, the findings have shown effective performance of vetiver grass for metal removal in the phytoremediation of contaminated water.

1 Introduction

Vetiver grass (*Vetiveria zizanioides*) has been identified as the potential plant to be used in the phytoremediation treatment. Vetiver grass (VG) is a tall (1–2 m), fast-growing, perennial tussock grass. It has a long (3–4 m), massive and complex root system, which can penetrate to the deeper layers of the soil [1-4]. The dense high fiber, fine root system that forms an enormous volume accelerates habitat for many symbiotic bacteria, and fungi made the ecology system naturally contributes to the pollutant uptake. Furthermore, VG has ability to absorb large amounts of environmental pollutants such as heavy metals,

* Corresponding author: faradiella@upm.edu.my

pesticides/herbicides and petroleum hydrocarbon without effect its growth [5]. Through these characteristics, VG can be used for purification of polluted sites and normalised the eutrophic areas.

In addition, VG also involve in translocation process to uptake the contaminant by using plants roots which act as phytoextraction [4, 7]. The rate of phytoextraction depends on the root depth in soil or medium because root acts as the metal accumulator [4, 8]. This application is almost similar to rhizofiltration but it is normally used on groundwater, wastewater and surface water as hydroponic treatment; because the root can come in contact with water [9, 10]. The process occurs during the adsorption of contaminants onto root surface or plant root absorption. Thus, the plant root is responsible to accumulate the heavy metal with the application of plant translocation.

Another unique characteristic of VG is that it has high tolerance to elevated concentrations of heavy metals such as As, Cd, Cu, Cr, Pb, Hg, Ni, Se and Zn in the soil [1, 11, 12]. Besides, VG has high tolerance and adaptability to extreme edaphic conditions such as soil with high acidity and alkalinity. It can thrive under a wide pH range and adapt to saline, sodic and magnesic conditions, and aluminium and manganese toxicities [7, 11, 13]. VG also can survive between pH 3.3 to 9.5 at nutrient-adequate condition and saline soil with high electrical conductivity up to 4.0 [1, 10, 13, 14]. Therefore, the outstanding morphological and physiological features of VG make it suitable choice for water treatment purposes.

Lack of previous research discover the potential of VG in removal of heavy metal for phytoremediation especially in polluted running water. In this study, treatment using phytoremediation of VG was applied in removing or controlling various types of pollutants from contaminated water. The objectives of this study are to evaluate contaminant removal by VG at different root length and plant density and also to determine the metal uptake in vetiver plant biomass.

2 Materials and methods

2.1 Testing of vetiver grass performance

2.1.1 Experimental works

The experimental works were carried out as a platform for more complex analytical conditions on vegetative, physical and physico-chemical parameter measurements. The conditions at actual site was simulated in the laboratory and therefore experiment was undertaken in seven days of period under controlled environment through continuous flow experiment (Fig. 1). The laboratory experiment of VG performance on pollutant uptake were carried out by planting VG at different root length (10 cm, 20 cm, 30 cm) and plant density (4 tillers, 10 tillers) that has been trimmed accordingly to given root length and density in hydroponic system. However, the tillers that were too young was ignored and not counted as a tiller.

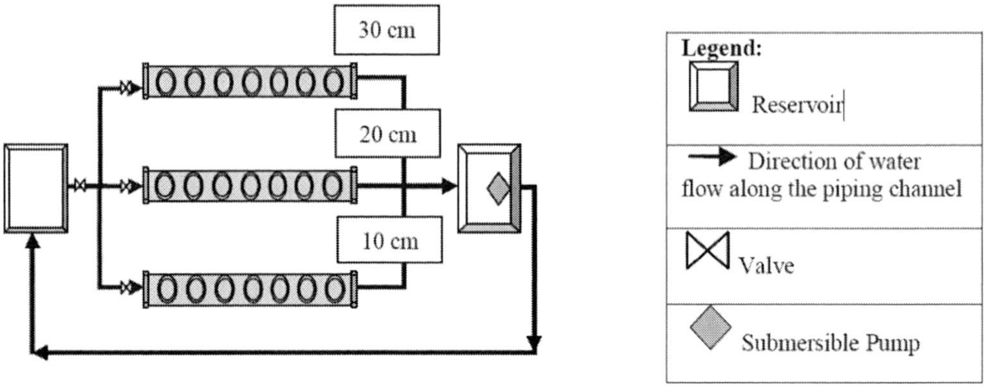

Fig. 1. Example of a set of hydroponic system (plan view).

2.2 Performance of vetiver grass for phytoremediation technique

2.2.1 Vetiver grass planting

The plants (vetiver grass) were obtained from local suppliers and were typically delivered as young vetiver in boxes (200 tillers of VG in each box). For the purpose of young vetiver planting in water medium, the soil is washed off in order for the plant to adapt to the environment prior to the planting. The plants were kept afloat in the container with the crowns above the water surface (Fig. 2).

Fig. 2. (a) Young plants in the box **(b)** Plants being soaked in a pail of water to loosen up the soil before washing **(c)** Make sure to keep the crown above water surface.

Note: Make sure that the crown is not soaked in the water as it would damage the crown, which in turn, killing the plants.

Then, the VGs were transferred to the containers for plant acclimatization and growth of the root length. Hydroponic solution of (N6: P3: K10 + Ca + Mg) was added one week after the plant acclimatization in tap water to assist in the growing process.

2.3 Heavy metal uptake by vetiver grass in static condition

2.3.1 Water sampling

The experiment was conducted for 10 days to observe the effect of root length and plant density on contaminant removal. The initial concentration was set to be 10 ppm for all five heavy metals which are copper (Cu), iron (Fe), manganese (Mn), lead (Pb) and zinc (Zn) based on the worst case scenario of heavy metal occurrence found in Malaysia [15, 16, 17].

The water sampling was conducted every day which involved in-situ and ex-situ. The in-situ parameters such as pH, temperature, electrical conductivity (EC), and salinity were recorded, while the ex-situ parameter was done through the analysis of heavy metals uptake by VG in water and plant biomass such as root and shoot. 50-mL of water sample was taken, placed in a centrifuge tube and were filtered using 0.45μm cellulose acetate filter paper to remove large biological materials. Then, the water samples were acidified to pH < 2 by addition of 0.7-ml of nitric acid (HNO_3) Suprapur 65%.

2.3.2 Plant harvesting

Plants biomass was separated in two samples which contain of shoots and roots. The vetiver shoots and roots were dried at 70 – 85°C in the oven to a constant weight and the total harvestable dry biomass production was then determined via dry weight of each sample. Then, 0.1 g of shoots and roots are grinded using mortar and pestle and was digested with 10 mL of 65% HNO_3 (AR grade; BDH). The acid digestion was prepared using a digestion block at 90°C and it was then cooled at room temperature. The digestate sample was filtered using 0.45μm cellulose acetate filter paper by syringe filter. Digested vetiver samples were analyzed using Atomic Absorption Spectrometry (AAS model AA-6800 Shimadzu) to determine concentration of heavy metal in shoots and roots.

2.4 Data analysis

The removal efficiency of each heavy metal was calculated to determine the potential of heavy metal uptake of VG based on following equation [19]:

$$\%\ \text{Removal efficiency} = \frac{(C\,ini - C\,fin)}{C\,ini} \times 100$$

where C_{ini} represent the initial concentration of metal content, while C_{fin} signify the final concentration of metal content.

3 Results and discussion

3.1 Heavy metal removal by vetiver grass

General trend shows that the concentration of all pollutant (Cu, Fe, Mn, Pb, Zn) were decreased from day one up to day seven of the experiment. Fig. 3 shows that the highest percent of metal removal was iron (96%; 0.42 ppm), followed by zinc (75%; 2.55 ppm), lead (50%; 4.97 ppm), manganese (33%; 6.71 ppm) and the least removal was copper (25%; 7.47 ppm) in seven days of period. The findings were in line with previous studies that iron (Fe) had high removal compared to other heavy metals [7, 18]. This is because the growth of VG requires high uptake of Fe and other macronutrient such as nitrogen (N), phosphorus (P) and potassium (K) by the root and also for photosynthesis reactions [5, 19]. Within seven days, uptake of heavy metal are differ due to its high tolerance towards elevated concentrations of heavy metal. It is suggested that VG required a longer time (more than a week) to absorb other heavy metals such as Pb, Mn and Cu that are not essential for vetiver growth. This indicated that VG have great ability in removing heavy metal within a short days of period. Thus, it can be seen that the percentage removal of heavy metal is ranked in the order: Fe>Zn>Pb> Mn>Cu.

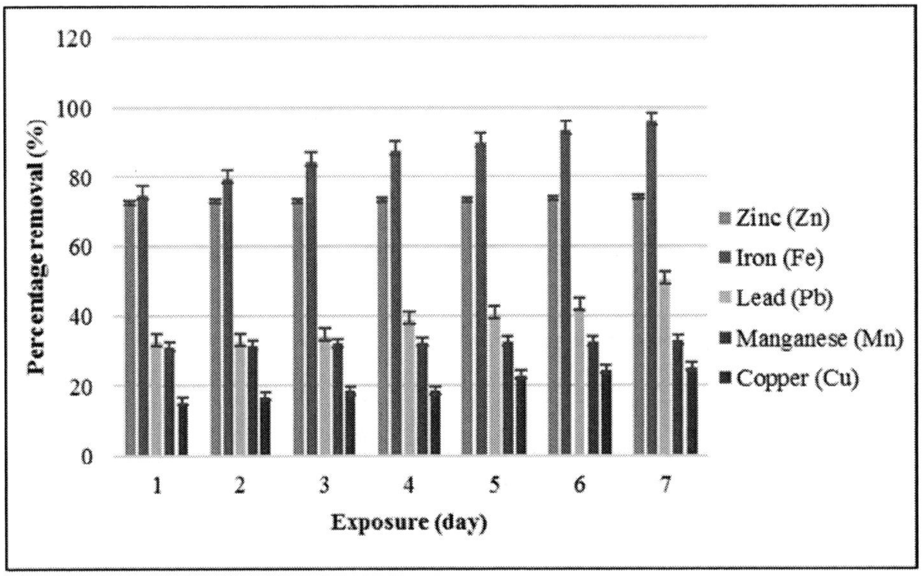

Fig. 3. Percentage of heavy metals removal (%) by vetiver grass within 7 days of experiment.

In general, greater removal of all heavy metal was found at greater root length and higher density with different metal concentration. Results show that most of heavy metals were being removed at a very fast rate for concentration less than 2 ppm and Fe has the highest removal at both four vetiver tillers (Fig. 4a) and 10 vetiver tillers (Fig. 4b). Findings indicated that large number of vetiver tillers used can increase the removal of heavy metals as the concentration rises. When density (number of vetiver tillers) increases, removal of heavy metal also increases. By the end of four days, most of the heavy metals were already taken up by the plants. This might due to higher surface area from many plants that can absorb large amount of heavy metals from metal concentration [19, 20]. The removal among heavy metals in water regardless of their density can be ranked in the order of Fe>Zn>Pb>Cu>Mn. Therefore, number of vetiver tillers used is one of the important factors that has significant role in reducing heavy metal contents especially in metal-contaminated water.

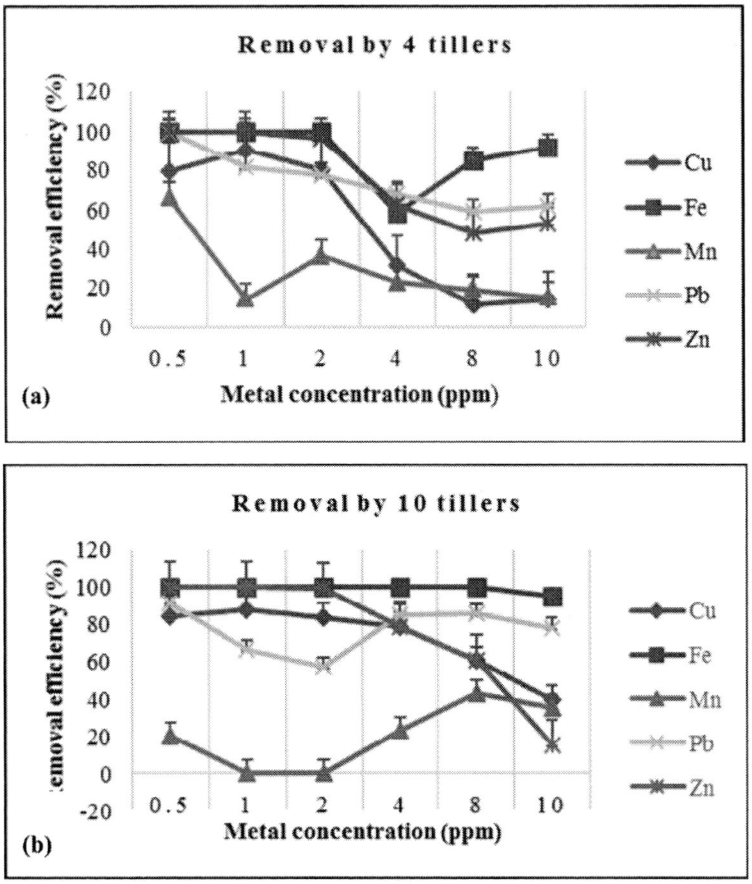

Fig. 4. Heavy metals (Cu, Fe, Mn, Pb, Zn) removal after four days by 4 vetiver tillers **(a)** and 10 vetiver tillers **(b)**

3.2 Metal accumulation in plant biomass

Overall, the plant biomass accumulation is notably higher at greater root length, while higher metal accumulation was found in shoot compared to the root (Fig. 5). The metals uptake by plant biomass in root was found higher at 30 cm of root length (Fig.5a). Findings indicated that the longer the root length, the higher accumulation of metals. This is because greater root length has higher surface area for metal absorption by the root into the plants from polluted water [19, 20]. Besides, its complex and massive vetiver root system encouraged the potential of high amount of metals uptake at different root length [1-4]. Thus, the vetiver root length is a good metal-accumulator due to its ability in taking up excessive amount of pollutant contents in polluted water.

However, result shows that metal accumulation was higher in shoot than in root (Fig. 5b). Current findings contradict the previous research where plant root uptake high amount of metal contents compared to the shoot [18, 21]. This is because of metals translocation from root to shoot which results in high accumulation of heavy metals in the shoot [4, 5, 8]. This suggests that plant root can uptake high metal contents during the early stage of the experiment and resulting in high accumulation of metals occurred at plant shoot. Therefore, current findings have shown the unique physiological features of VG in tolerance towards metal-accumulation and translocation of heavy metals from water into plant biomass at different root lengths.

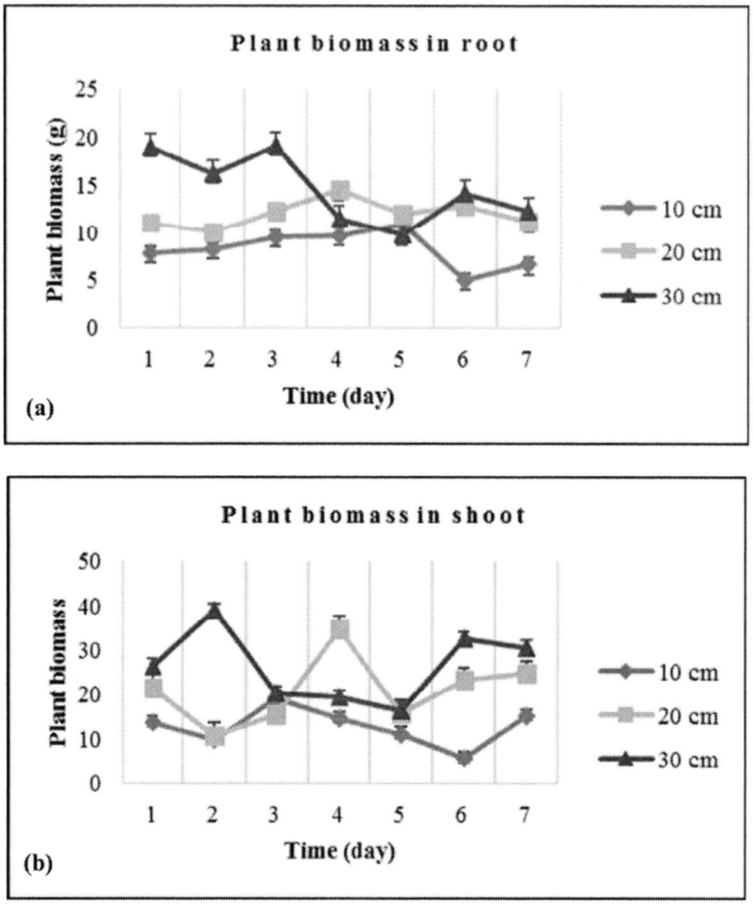

Fig. 5. Plant biomass in root **(a)** and shoot **(b)** at different root lengths.

4 Conclusion

Vetiver grass is a good accumulator of heavy metals in phytoremediation due to great ability in absorption of metals in water except for Mn and Zn. All the heavy metal elements have shown decreasing trends of metal concentration at different root length and density. Longer plant root length and higher vetiver density increased the uptake of heavy metals in polluted water. Metal accumulation occurred at both vetiver root and shoot but greater in shoot due to translocation of heavy metals from root to shoot.

The research was supported by Humid Tropic Centre (HTC), Department of Irrigation and Drainage (DID) Malaysia, Humibox (M) Sdn. Bhd. as a supplier for the vetiver grass and Faculty of Environmental Studies, Universiti Putra Malaysia for providing facilities.

References

[1] L.T. Danh, P. Truong, R. Mammucari, T. Tran and N. Foster, Vetiver grass, *Vetiveria zizanioides*: a choice plant for phytoremediation of heavy metals and organic wastes, Int. J. of Phytoremediation, **11**, 664–691 (2009)

[2] N.M.R. Pichai, R. Samjiamjiaras and H. Thammanoon, Vetiver and its multifold applications, Asian Infrastruct. Res. Rev., **3**, 1–4 (2001)

[3] P.A. Dalton, R.J. Smith and P.N.V. Truong, Vetiver grass hedges for erosion control on a cropped flood plain: hedge hydraulics, Agr. Water Manage., **31**, 91–104 (1996)

[4] P.N. Truong, Vetiver grass technology for environmental protection, *Proc. of the Second Int. Vetiver Conf.,* Bangkok, Thailand, (2000)

[5] N. Roongtanakiat and P. Chairoj, Uptake potential of some heavy metals by vetiver grass, Kasersart J. (Nat. Sci.), **35**, 46–50 (2001)

[6] D.E. Salt, M. Blaylock, N.P.B.A. Kumar, V. Dushenkov, B.D. Ensley, I. Chet and I. Raskin, Phytoremediation: a novel strategy for the removal of toxic metals from the environment using plants, Biotechnology, **13**(5), 174–468 (1995)

[7] P.N. Truong, Y.K. Foong, M. Guthrie and Y.T. Hung, Phytoremediation of heavy metal contaminated soils and water using vetiver grass, Environ. Bioengineering, **11,** 233–275 (2010)

[8] S. Kidney, Phytoremediation may take root in Brownfields, The Brownfields Report, **2**(14), 1–6 (1997)

[9] National Risk Management Research Laboratory (NRMRL), Environment Protection Agency (2000)

[10] F.M. Kusin, A review of the importance of hydraulic residence time on improved design of mine water treatment systems, World Appl. Sci., **26**(10), 1316–1322 (2013)

[11] N. Chomchalow, The role of vetiver in controlling water quantity and treating water quality: an overview with special reference to Thailand, *The Fourth Thai Nat. Conf. on Vetiver,* Bangkok, Thailand, AU J.T., **6**(3), 145–161 (2003)

[12] W.S. Shu, Exploring the potential utilization of vetiver in treating acid mine drainage (AMD), *Proc. of the Third Int. Vetiver Conf.,* Guangzhou, China, (2003)

[13] F.M. Kusin, A.P. Jarvis and C.J. Gandy, Hydraulic performance and iron removal in wetlands and lagoons treating ferruginous coal mine waters, Wetlands, **34**(3), 555–564 (2014)

[14] R. Webb, The Permaculture Research Institute, Vetiver grass – a hedge against erosion, (2009), Retrieved on June 15, 2015 from http://permaculturenews.org

[15] M.V. Prasanna, S.M. Praveena, S. Chidambaram, R. Nagarajan and A. Elayaraja, Evaluation of water quality pollution indices for heavy metal contamination monitoring: a case study from Curtin Lake, Miri City, East Malaysia, Environ. Earth Sci., **67**, 1987–2001 (2012)

[16] H. Hatar, S.A. Rahim, W.M. Razi and F.K. Sahrani, Heavy metals content in acid mine drainage at abandoned and active mining area, *AIP Conf. Proc.,* **1571**, 641–646 (2013)

[17] T. Hadibarata, F. Abdullah, A.R.M. Yusoff, R. Ismail, S. Azman and N. Adnan, Correlation study between land use, water quality, and heavy metals (Cd, Pb, and Zn) content in water and green lipped mussels *Perna viridis* (Linnaeus.) at the Johor Strait, Water Air Soil Pollut., **223**, 3125–3136 (2012)

[18] N. Roongtanakiat, S. Tangruangkiat and R. Meesat, Utilization of vetiver grass (*Vetiveria zizanioides*) for removal of heavy metals from industrial wastewaters, Science Asia, **33**, 397–403 (2007)

[19] N. Darajeh, A. Idris, P. Truong, A.A. Aziz, R.A Bakar and H.C. Man, Phytoremediation potential of vetiver system technology for improving the quality of palm oil mill effluent, Adv. Mater. Sc. Eng., **4** (2014)

[20] I. Raskin, R.D. Smith and D.E. Salt, Phytoremediation of metals: using plants to remove pollutants from the environment, Curr. Opin. Biotechnol., **8**, 221–226 (1997)

[21] S.S. Andra, R. Datta, D. Sarkar, S.K.M. Saminathan, C.P. Mullens and S.B.H. Bach, Analysis of phytochelatin complexes in the lead tolerant vetiver grass [*Vetiveria zizanioides* (L.)] using liquid chromatography and mass spectrometry, Environ. Pollut., **157**, 2173–2183 (2009)

[22] S. Ladislas, C. Gérente, F. Chazarenc, J. Brisson and Y. Andrès, Performances of two macrophytes species in floating treatment wetlands for cadmium, nickel, and zinc removal from urban stormwater runoff, Water, Air, Soil Pollut., **224**, 1408 (2013)

Analysis of Mass Transfer Resistance for Adsorption of Phosphate onto Industrial Waste Materials in Plug-flow column

Mohd Hairul Khamidun[1,*], *Mohammad Ashraf* Abdul Rahman[2]

[1]Faculty of Civil and Environmental Engineering Universiti Tun Hussein Onn Malaysia, Batu Pahat, Johor, Malaysia
[2]Faculty of Engineering Technology, Universiti Tun Hussein Onn Malaysia, Batu Pahat, Johor, Malaysia

Abstract. The understanding resistance of mass transfer for adsorption of phosphate (PO_4^{3-}) onto industrial waste such as clay brick and ceramic as an adsorbent is important to verify. This study presented the removal of PO_4^{3-} from a synthetic solution by adsorbing onto the clay brick and ceramic wastes in a plug flow column (PFC). The experimental results showed that increasing the flow rate was decreases the breakthrough time. The mass transfer factor (MTF) models were used to study the behaviours of breakthrough curve and to determine the resistance of mass transfer. The MTF models verified that the resistance of mass transfer could be dependent on porous diffusion until the percentage of outflow reaches 72% for clay brick waste and 86% for ceramic waste, even though film mass transfer can play a minor role in controlling the movement of PO_4^{3-} from the bulk water to film zone. The results of MTF models coefficient could be indicated to increase the capacity of clay brick and ceramic wastes to adsorb PO_4^{3-} from synthetic solution, it needs to develop the porosity of these adsorbents by either physical or chemical modification.

1 Introduction

Adsorption is the accumulation of a substance in solution with a suitable interface. The solute that accumulates or adsorbs at the interface is called an adsorbate; the solid that is adsorbedly the solute is called the adsorbent. The adsorbate is said to be adsorbed onto the adsorbent when it is attached at the active sites of the adsorbent surface, and it is said to be desorbed when it is released from the adsorbent into solution. The mechanisms of adsorption consist of three successive steps in the transfer of solutes onto adsorbents in liquid, such that: (1) external mass transfer is the mechanism to transport an adsorbate from the bulk liquid to film zone exterior the grain immediately connecting the pores, (2) internal mass transfer is mechanism to diffuse an adsorbate from the film zone towards acceptor sites of the adsorbent, and (3) fixation is mechanism to attach the adsorbate onto acceptor

* Corresponding author: mhairulk@gmail.com

sites of the adsorbent interior the grain [1]. The adsorption process is one of the purification technologies and can be utilized to remove pollutants from the water. This method requires simple operation and design, and draws up a small amount of sludge [2].

Nowadays, many researchers have been studied the use of waste materials as a low cost adsorbent such cockle shell, fish scale, waste coal, solid biomass waste, food waste, newspaper waste and coffee industrial waste to remove pollutants from waters [3-7]. In the previous studies, the adsorption of PO_4^{3-} from synthetic solution onto used brick as reported by Jia *et al.*, [8], can achieve 0.90, 0.97, 1.09 and 1.29 mg P g^{-1} of adsorption capacity for concentration of 5, 10, 25 and 50 mg L^{-1}, respectively. The use of ceramic waste for the removal of PO_4^{3-} from phosphorus solution to having approximately 250 μg g^{-1} when feeding with an initial concentration of 2000 μg L^{-1} [9]. Even if many adsorption models have been used to investigate the ability of using certain adsorbents to improve the adsorption performance, the use of the empirical models for scrutinizing the mass transfer factors of adsorption of PO_4^{3-} onto clay brick and ceramic wastes from synthetic solution is still not fully understood. The objective of this study is to verify application of the modified mass transfer factor models and determine the variations of global mass transfer, external mass transfer and internal mass transfer coefficients for the adsorption of PO_4^{3-} onto clay brick and ceramic wastes from synthetic solution in PFC.

2 Materials and methods

2.1 Preparation of adsorbents

Two types of industrial waste were used in this study as adsorbent i.e. clay brick and ceramic wastes. The clay brick waste was collected from a Materials Engineering Research Laboratory, Universiti Teknologi Malaysia. The ceramic waste was obtained from ceramic factory in Ayer Hitam, Johor state, Malaysia. All these adsorbents were crushed and passed through a sieve fraction of 1.18 mm and retained 0.85 mm mesh; thus the mean size of adsorbents was around 1.02 mm. Approximately 2 kg of these adsorbents were washed with distilled water and dried at 103°C for 24 h in an oven. The dried adsorbents were used as adsorbent for the adsorption of PO_4^{3-} from synthetic solution in PFC. Surface morphology of the raw adsorbents was then characterised using a scanning electron microscope (SEM, Zeiss Supra 35 V). The Fourier Transform Infrared (FTIR) Spectroscopy (Perkin-Elmer Spectrum One FTIR Spectrometer) analyses was performed in the range 600-4000 cm^{-1} to investigate the surface functional groups at the surface of the adsorbents.

2.2 Preparation of synthetic solution

Synthetic solutions with concentration of 100 mg L^{-1} of PO_4^{3-} were prepared by dissolving a 0.4394 g of KH_2PO_4 (analytic grade), respectively, into 1 L of deionised water in a volumetric flask. The synthetic solution was successively diluted with deionised water to obtain the predetermined concentration of PO_4^{3-}. The pH of the synthetic solutions was set up at 6.5 for all the experiments.

2.3 Plug-flow column reactor

The mass transfer resistance mechanism and hydrodynamics breakthrough curve for PO_4^{3-} were determined through continuous feeding of synthetic solution into PFC. A lab-scale PFC was set up, which consisted of a peristaltic pump to control the flow rate and columns

DOI: 10.1051/matecconf/201710306004

filled with adsorbents (see Fig. 1). Samples were collected at the influent tank and at the outlet of the columns before analyzing the concentration of PO_4^{3-}. The bed height of adsorbent filled in the column is 65 cm and the flow rate are 1.6, 2.8 and 3.9 mL min^{-1}.

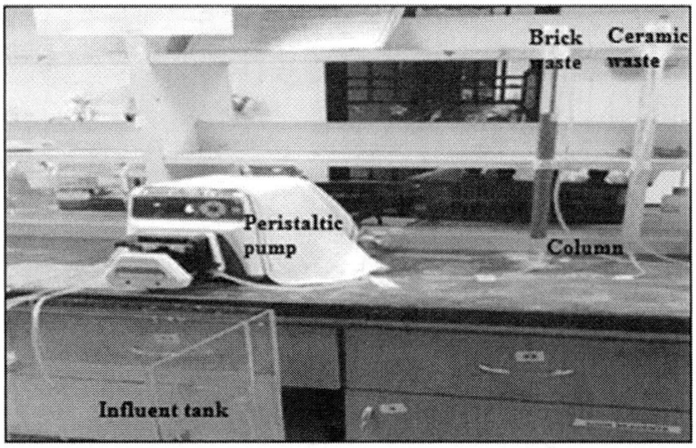

Fig. 1. Laboratory scale plug-flow column system

2.4 Mass transfer factor models

The mass transfer factors such as global, external and internal mass transfer can be determined by using empirical models developed by Fulazzaky [1, 10]. The modified mass transfer factor models were developed to allow the integration of knowledge concerning the considered phenomena of global, external and internal mass transfer. The use of the modified mass transfer models is able to determine the resistance of mass transfer for the adsorption of PO_4^{3-} present in water onto adsorbents in PFC. The models are expressed in Equation (1) as follows

$$ln\left(\frac{C_i}{C_e}\right) = [k_L a]_g \times e^{-\beta \times ln(q)} \times t \tag{1}$$

where C_i is the PO_4^{3-} concentration at influent the column (mg L^{-1}), C_e is the concentration of PO_4^{3-} effluent at the column (mg L^{-1}), $[k_L a]_g$ is the global mass transfer factor (h^{-1}), β is the absorbent – adsorbent affinity parameter (g h mg^{-1}); q is the cumulative quantity of the solute to be adsorbed onto adsorbents (mg g^{-1}); and t is the accumulated time (h)
The mathematical deduction of equation (1) can be simplified as:

$$\ln(q) = \alpha \ln(t) + \mathrm{B}\ln(q) = m\ln(t) + B \tag{2}$$

$$\text{where,} \quad \alpha = 1/\beta \tag{3}$$

and B is the intercept of the plot ln(q) versus ln(t):

$$B = \frac{\ln([k_L a]_g) - \ln\left\{ln\left(\frac{C_i}{C_e}\right)\right\}}{\beta} \tag{4}$$

DOI: 10.1051/matecconf/201710306004

The value of $[k_La]_g$ can be determined using Equation (1) since the index B and the parameter β has been verified on a straight line of a plot of $\ln(q)$ versus $\ln(t)$. Equation (5) permits the computation of the film mass transfer factor (h^{-1}) $[k_La]_f$, and Equation (6) can be used to determine the porous diffusion factor (h^{-1}) $[k_La]_d$.

$$[k_La]_f = [k_La]_g \times e^{-\beta \times ln(q)} \quad (5)$$

$$[k_La]_d = [k_La]_g - [k_La]_f \quad (6)$$

3 Results and discussions

3.1 Characteristics of adsorbents

Fig. 2 shows the FTIR spectra of unmodified clay brick (Fig.2A) and ceramic (Fig.2B) wastes. FTIR spectra of the clay brick waste exhibited bands at 1750, 1620 1042, 747 and 632 cm^{-1} (see Fig. 2A). The bands in range 1620 – 1750 cm^{-1} are attributed to H-O-H bending [11]. The asymmetric stretch of Al-O are observed at bands 1042 cm^{-1} [12]. The bands at 747 and 632 cm^{-1} indicated the presence of mineral quartz and Si–O–Al bond, respectively [13]. Fig. 2(B) is FTIR spectrums of the ceramic waste. The analysis of spectrums indicated sharp peak at 1580 cm^{-1} is due to scattering from atmospheric oxygen [14]. The peaks in the region of 1600 – 1800 cm^{-1} are associated with the acetate groups [15]. The broad peak around 1100 cm^{-1} is characteristics of asymmetrical Si–O–Si stretching [16].

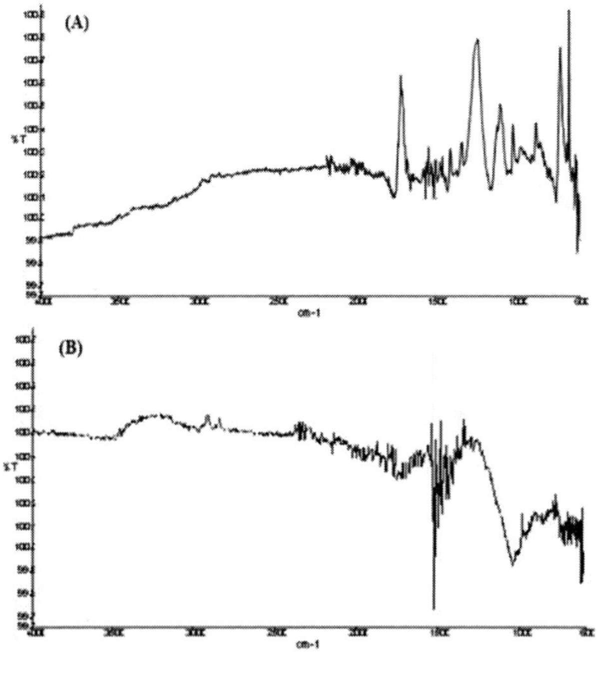

Fig. 2. FTIR spectra of **(A)** brick waste and **(B)** ceramic waste

The surface morphology of brick and ceramic wastes was investigated by SEM analysis and the SEM images are shown in Fig. 3. Fig. 3(A) shows that the SEM image of 1000 times magnification are drusy texture with micro porosity of the brick waste surface.

 DOI: 10.1051/matecconf/201710306004

Fig. 3(B), the microscopic image of ceramic waste, shows quite smooth and wavy ripple morphology on the surface.

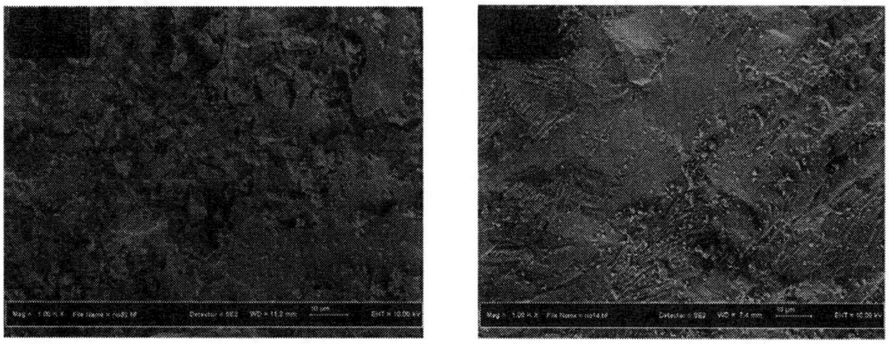

Fig. 3. SEM photomicrograph of (**A**) clay brick waste and (**B**) ceramic waste at overview 1000 times magnification

3.2 Plug-flow column study

Fig. 4 shows the variations of C_e/C_i versus t for the adsorption of PO_4^{3-} onto clay brick waste (Fig. 4(A)) and ceramic waste (Fig.4(B)) from synthetic solution in continuous PFC. The figure shows that the use of brick waste to adsorb PO_4^{3-} could be more efficient than that the use of ceramic waste. As shown in Fig. 4, the adsorption efficiency decreases with increasing the flow rate. The decrease in adsorption efficiency due to fact that the increased flow rate may decrease the residence time of PO_4^{3-} to attach onto the surface of adsorbents. The higher adsorption of PO_4^{3-} onto clay brick waste due to H-O-H, Al-O and Si–O–Al functional groups on the surface could ion exchange and precipitate with PO_4^{3-} [17], [18]. The analyses of mass transfer factor models were performed for the adsorption of PO_4^{3-} in PFC at flow rate of 1.6 mL min^{-1}.

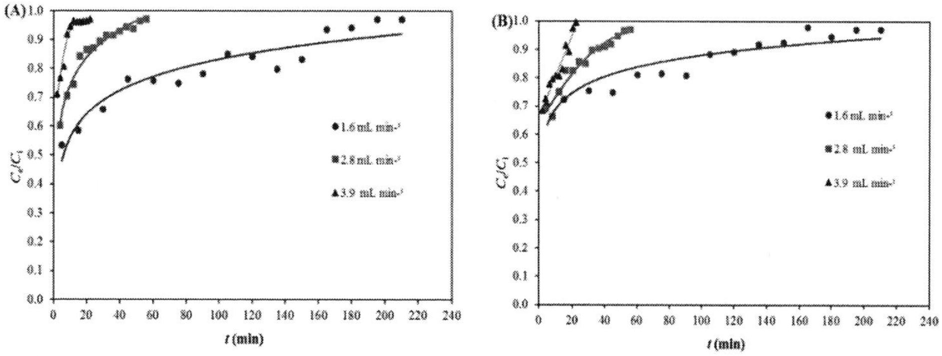

Fig. 4. Breakthrough curves for the adsorption of PO_4^{3-} from synthetic solution onto (**A**) brick and (**B**) ceramic wastes.

3.3 Application of Mass Transfer Factor Models

A linear regression analysis of using the experimental data shows a good correlation for all the curves of plotting ln(q) versus ln(t) in Equation (2) for flow rate of 1.6 mL min^{-1} ($R^2 > 0.976$; see Table 1), meaning that the use of the parameters B and β can be used to determine the mass transfer potential and adsorbate–adsorbent affinity for the adsorption of PO_4^{3-} from synthetic solution onto clay brick and ceramic wastes. The experimental data validation (Table 1) shows that the value of *B* and *β* for clay brick waste is higher than ceramic waste could be due to a high PO_4^{3-} concentration in the stagnant film on the water side of the interface, affecting the value of driving force reduces because of the difference in PO_4^{3-} concentration between the bulk water and film zone lowers [19].

Table 1. Results of determining the values of β and *B* for clay brick and ceramic wastes in plug-flow column

Adsorbent	*β* (g h/mg)	*B* (mg/g)	R^2
Clay brick waste	1.3358	-9.913	0.988
Ceramic waste	1.2758	-10.38	0.978

Using Equation (4) permits us to determine the variations of [*k*L*a*]g pursuant to the percentage of outflow if the parameters *B* and β were verified. Fig. 5 shows the variations of [*k*L*a*]g decreased progressively from a high to low mass transfer potential with increasing of the percentage of outflow. The variation of [*k*L*a*]g value for clay brick waste packed-bed column is quite high when comparing to ceramic waste packed-bed column as shown in Fig. 5. This could be due to high adsorbate–adsorbent affinity of clay brick waste to attract PO_4^{3-} from bulk water when passing through packed-bed column [20].

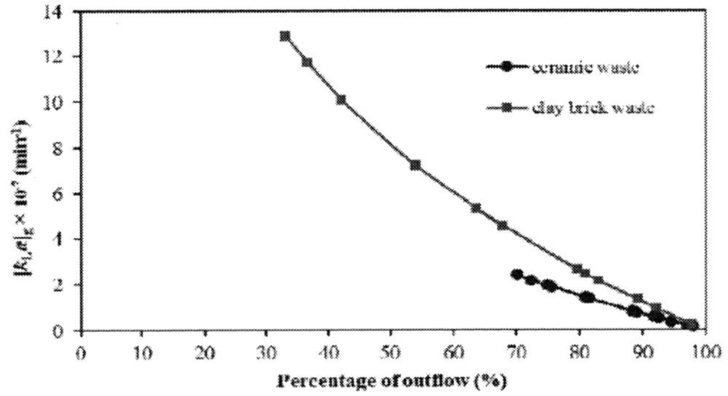

Fig. 5. Variations of $[k_La]_g$ pursuant to the percentage of outflow for clay brick waste (**red line**) and ceramic waste (**black line**) in plug-flow column

As can be seen in Fig. 6 the variations of $[k_La]_f$ rapidly decrease, counterbalanced by the variations of $[k_La]_d$ rapidly increase when the percentages of outflow are still less than 63% for clay brick waste and 76% for ceramic waste. After passing a 63% and 76% of outflow for clay brick waste and ceramic waste, respectively; the variations of $[k_La]_f$ slowly decrease but the variations of $[k_La]_d$ slowly increase until percentage of outflow reach at approximately 72% and 86% for clay brick waste and ceramic waste, respectively. Based on the variations of $[k_La]_f$ and $[k_La]_d$, it would be concluded that the resistance of mass transfer for the adsorption of PO_4^{3-} onto clay brick and ceramics wastes is dependent on porous diffusion before a breakthrough occurs and then to continue until it reaches at approximately 63% and 76% of the outflow for clay brick waste and ceramic waste, respectively. In spite of the transport of PO_4^{3-} from the bulk water to film zone and then to acceptor sites within the pores is very slow, the resistance of mass transfer could be dependent on both film mass transfer and porous diffusion when the percentages of outflow are greater than 72% and 86% for clay brick waste and ceramic waste, respectively.

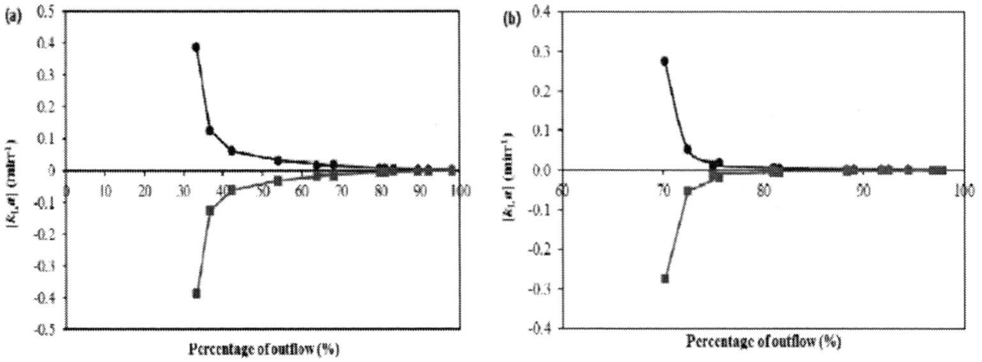

Fig. 6. Variations of $[k_L a]_f$ (black line) and $[kLa]d$ (red line) pursuant to the percentage of outflow for (a) clay brick waste and (b) ceramic waste in plug-flow column.

4 Conclusion

As a conclusion, the adsorption of PO_4^{3-} onto clay brick and ceramic wastes might allow precipitation occurred by a chemisorption process while the van der Waals forces would be quite difficult due to the porosity of the adsorbents very small (confirmed by SEM image). Therefore, in order to increase the capacity of these adsorbents to adsorb PO_4^{3-} from waters, it needs to develop the porosity of clay brick and ceramic wastes by either physical or chemical modification. In addition, the resistance of mass transfer for the adsorptions of PO_4^{3-} onto clay brick and ceramic wastes from synthetic solution processing in a PFC was verified to contribute to the development of such porous materials for improving separation performance.

The author thank the Universiti Tun Hussien Onn Malaysia, for short term grant under grant no Vot U635.

References

[1] M. A. Fulazzaky, Determining the resistance of mass transfer for adsorption of the surfactants onto granular activated carbons from hydrodynamic column, Chem. Eng. J., **166**(3), 832–840, (Feb.,2011)

[2] R. Boopathy, S. Karthikeyan, A. B. Mandal, and G. Sekaran, Adsorption of ammonium ion by coconut shell-activated carbon from aqueous solution: kinetic, isotherm, and thermodynamic studies., Environ. Sci. Pollut. Res. Int., **20**(1), 533–542, (Jan, 2013)

[3] B. G. Tsyntsarski, B. N. Petrova, T. K. Budinova, N. V Petrov, and D. K. Teodosiev, Removal of phenol from contaminated water by activated carbon produced from waste coal material, Bulg. Chem. Commun., **46**(2), 353–361, (2014)

[4] M. Shanmugaprakash, V. Sivakumar, and M. Manimaran, Batch and Dynamics Modeling of the Biosorption of Cr (VI) from Aqueous Solutions by Solid Biomass Waste from the Biodiesel Production, Environ. Prog. Sustain. Energy, **33**(2), 342–352, (2014)

[5] A. Shehzad, M. J. K. Bashir, S. Sethupathi, and J.-W. Lim, An overview of heavily polluted landfill leachate treatment using food waste as an alternative and renewable source of activated carbon, Process Saf. Environ. Prot., **98**, 309–318, (2015)

[6] M. H. Dehghani, D. Sanaei, I. Ali, and A. Bhatnagar, Removal of chromium(VI) from aqueous solution using treated waste newspaper as a low-cost adsorbent: Kinetic modeling and isotherm studies, J. Mol. Liq., **215**, 671–679, (2016)

[7] N. Zayadi and N. Othman, Characterization and Optimization of Heavy Metals Biosorption by Fish Scales, Adv. Mater. Res., **795**, 260–265, (2013)

[8] C. Jia, Y. Dai, J. Chang, C. Wu, Z. Wu, and W. Liang, Adsorption characteristics of used brick for phosphorus removal from phosphate solution, Desalin. Water Treat., **51**, 5886–5891, (2013)

[9] J. Wang, Y. Zhang, C. Feng, H. Wang, and L. Wang, Adsorption Capacity Comparison among Three Filter Media for Phosphorus, *2010 4th Int. Conf. Bioinforma. Biomed. Eng.*, 1–4, Jun. (2010)

[10] M. A. Fulazzaky, M. H. Khamidun, M. F. M. Din, and A. R. M. Yusoff, Adsorption of phosphate from domestic wastewater treatment plant effluent onto the laterites in a hydrodynamic column, Chem. Eng. J., **258**, 10–17, (2014)

[11] M. F. Zawrah, R. A. Gado, N. Feltin, S. Ducourtieux, and L. Devoille, Recycling and utilization assessment of waste fired clay bricks (Grog) with granulated blast-furnace slag for geopolymer production, Process Saf. Environ. Prot., **103**, 237–251, (2016)

[12] F. Kooli, L. Yan, R. Al-Faze, and A. Al-Sehimi, Removal enhancement of basic blue 41 by brick waste from an aqueous solution, Arab. J. Chem., **8**(3), 333–342, (2015)

[13] J. A. Calabria, W. L. Vasconcelos, and A. R. Boccaccini, Microstructure and chemical degradation of adobe and clay bricks, Ceram. Int., **35**, 665–671, (2009)

[14] T. Hirose, T. Fujino, T. Fan, and H. Endo, Effect of carbonization temperature on the structural changes of woodceramics impregnated with liquefied wood, Carbon N. Y., **40**, 761–765, (2002)

[15] M. A. Gulgun, M. H. Nguyen, and W. M. Kriven, Polymerized Organic – Inorganic Synthesis of Mixed Oxides, J. Am. Ceram. Soc., **82**(3), 556–560, (1999)

[16] K. Lin, W. Chang, T. Chang, C. Lee, and C. Lin, Recycling thin film transistor liquid crystal display (TFT-LCD) waste glass produced as glass – ceramics, J. Clean. Prod., **17**(16), 1499–1503, (2009)

[17] M. A. Fulazzaky, N. A. A. Salim, N. H. Abdullah, A. R. Mohd Yusoff, and E. Paul, Precipitation of iron-hydroxy-phosphate of added ferric iron from domestic wastewater by an alternating aerobic-anoxic process, Chem. Eng. J., **253**, 291–297, (2014)

[18] S. Sengupta and A. Pandit, Selective removal of phosphorus from wastewater combined with its recovery as a solid-phase fertilizer., Water Res., **45**(11), 3318–3330, (2011)

[19] M. H. Khamidun and M. A. Fulazzaky, Adsorption of phosphate from synthetic solution onto the limestone in a plug - flow column, Int. J. Adv. Appl. Sci., **2**(12), 7–13, (2015)

[20] M. H. Khamidun, M. A. Fulazzaky, M. F. Md Din, and A. R. Mohd Yusoff, Resistance of mass transfer , kinetic and isotherm study of ammonium removal by using a Hybrid Plug-Flow Column Reactor (HPFCR), *Proceedings of the 2013 International Conference on Frontier of Energy and Environment Engineering, ICFEEE 2013*, 555–559 (2014)

Thin Film Composite Membranes: Mechanical and Antifouling Properties

Norin Zamiah Kassim Shaari[1,*] , *Norazah* Abd Rahman[1] , *Nurul Aida* Sulaiman[1], *Ramlah* Mohd Tajuddin[2]

[1] Faculty of Chemical Engineering, Universiti Teknologi MARA (UITM), 40450 Shah Alam, Selangor Darul Ehsan, Malaysia
[2] Faculty of Civil Engineering, Universiti Teknologi MARA (UITM), 40450 Shah Alam, Selangor Darul Ehsan, Malaysia

Abstract. As compared to membranes produced from pure polymer or pure inorganic materials, a hybrid membrane possesses better mechanical and thermal properties. This paper reported on the effect of incorporating silica nano-precursor (tetraethylorthosilicate) as well as glycerol in the formulation of hybrid membrane on the mechanical properties and antifouling properties of the resultant thin film composite membranes. The mechanical properties were measured in terms of tensile strength, tensile strain and elastic modulus. Whereas for antifouling properties, it was evaluated through the measurements of relative flux decay (RFD) and relative flux recovery (RFR), along with the permeate flux rate, percentage glycerol permeated and NaCl rejection. Results showed that the presence of silica and glycerol in hybrid membrane's formulation had increased the tensile strength and elongation of the resultant membranes. In addition to that, the incorporation of glycerol has resulted in thin film composite with better antifouling properties as compared to the thin film composite with barrier layer from the pure polymer blend. Based on its performance, the fabricated thin film composite has a great potential to be used as a pathway for crude glycerol purification due to some advantages over the existing process that employ membrane.

1 Introduction

In the previous paper [1], it was reported that the incorporation of glycerol in the hybrid membrane formulation had increased the thermal stability and porosity of the resultant membrane with better performance in crude glycerol purification process as compared to that without it. In the preparation of hybrid membrane, sol-gel method was employed. The sol gel reaction involves the hydrolysis of nano-precursor such as silica and titania and condensation of the resulting hydroxyl groups to form a nanostructure. The nanometer-ordered inorganic minerals precipitate in situ regularly in the organic polymer matrix forming hydrogen bond between organic phase and inorganic phase, where this strong interaction between the organic polymer and inorganic mineral may result in a hybrid of

* Corresponding author: norinzamiah@salam.uitm.edu.my

markedly improved mechanical properties [2]. It is therefore, the purpose of this paper is to observe the degree of improvement on the mechanical properties as well as the effect of glycerol incorporation.

Mechanical property of the pressure-driven membranes is very important characteristic to evaluate the strength of the membranes in term of withstanding the trans-membrane pressure during filtration and cleaning such as back flushing, where the membranes have enough strength and not easily broken. The mechanical properties of a membrane usually estimated in terms of tensile strength, tensile strain and elastic modulus [3]. As for the hybrid membranes, the analysis was performed to evaluate the effect of cross-linking by silica nano precursor as well as the incorporation of glycerol on the mechanical strength. Besides the mechanical properties, this paper will report on the evaluation on antifouling properties of the resultant thin film composite membranes. Membrane separation process always suffers from fouling phenomenon which normally causes declining in flux as well as the rejection. In water treatment process, in order to maintain the flux, the operating pressure is increased that will lead to higher energy demand and product water cost [4].

There are three main types of foulants in the feed solution during membrane separation process such as inorganic (clays and mineral particles), biological (bacteria, fungi) and organic (oils and humics) [3]. Since the application of the thin film composite is for crude glycerol purification, this paper only focused on the organic foulant by using the actual crude glycerol solution with 75% additional water. The main mechanisms of fouling are adsorption of partially rejected matter within the membrane pores (pore constriction), plugging of the pores by particle which has similar size as the pores (pore blocking) and cake formation where the top of membrane surface is accumulated with the rejected particles [5]. Fouled membranes are traditionally cleaned using a combination of chemical and physical methods. Pore blocking and cake formation are reversible fouling and could be cleaned by physical hydraulic back flushing unlike pore constriction which is irreversible but could be cleaned by aggressive chemical cleaning process [4]. Besides that, the reversible fouling could be reduced by introducing and increased stirring near the membrane surface. Realizing the above negative impacts caused by fouling phenomenon, fouling resistant membranes are highly desirable for membrane separation process. In this study, the dead-end mode filtration was employed that would impose on severe fouling scenario as compared to cross-flow mode where the shearing force could reduce the fouling effect.

2 Materials and Methods

2.1 Materials

For the preparation of thin film composite membrane, the material were polyvinyl alcohol with degree of hydrolysis of 87-89% (molecular weight: 85000-124000 g/mol), polysulfone resin pellet (molecular weight: 44,000-53,000 g/mol), polyethylene glycol (molecular weight: 400 g/mol), pure glycerol, tetraethylorthosilicate with 99% purity as silica nano-precursor, and hydrochloric acid with 37% purity as the catalyst. All the chemicals were obtained from Sigma Aldrich, Malaysia. 1-methyl-2-pyrrolidone with purity of 99% was obtained from Merck, Malaysia and deionized water as solvent. Crude glycerol was obtained from one of the biodiesel manufacturers in Malaysia.

2.2 Methods

2.2.1 Preparation of micro porous support membrane

In the preparation of support membrane, phase inversion technique was employed. Polysulfone polymer solution was prepared by dissolving specific amount of polysulfone pellet into 1-methyl-2-pyrrolidone as the solvent to yield 11% wt/wt concentration of polymer solution. The solution was continuously stirred at 60°C until homogeneous solution was achieved. Then, the solution was left for several hours at room temperature for the removal of air bubbles. By using the casting machine and by adjusting the thickness to 45-50 µm, the polymer solution was cast onto the glass plate. The film was left in the ambient temperature for 30 seconds before it was immersed in the water as the coagulation medium for 1 hour. Then, the film was kept in the large amount of water for 24 hours. The membrane was subsequently cured at 45°C in the oven for 1 hour [6] before preparation of the thin film composite. Fig. 1 shows schematic diagram of the preparation of micro porous support membrane.

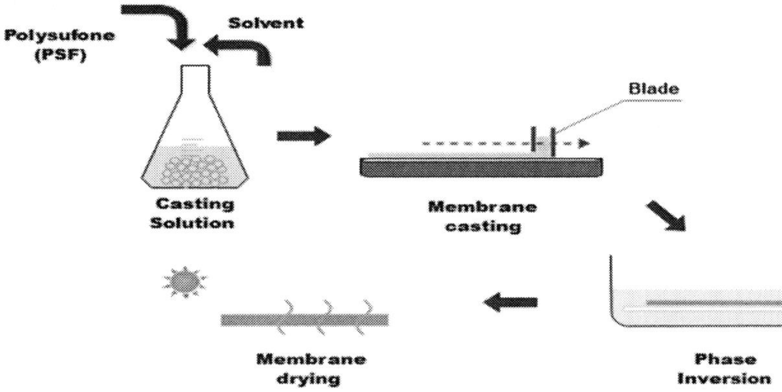

Fig. 1. Formulation of PSF membrane as base membrane (using phase inversion method)

2.2.2 Preparation of a hybrid membrane

In the preparation of a hybrid membrane, a sol-gel method was employed. The measured amount of polyvinyl alcohol (PVA) was blended with polyethylene glycol (PEG) with a weight ratio of 9:1 [7]. A measured amount of deionized water as solvent was added to the mixture and the solution were heated at 90°C until homogeneous solution was obtained. Glycerol acted as organic additive was added to the solution, which was then stirred at 90°C for 15 minutes. The mixture was left to cool down at room temperature. By using sol-gel method, tetraethylorthosilicate (TEOS) as nano precursor at fixed concentration was added at to the solution. Then, 1 ml of hydrochloric acid (37% purity) was added as the catalyst. The solution was continuously stirred for 10 hours at 30°C. The composition of PVA, PEG, glycerol and TEOS used in the preparation of the hybrid membrane for this study was shown in Table 1. Four formulations were prepared, which comprise of membrane from pure polymer blend of PVA/PEG, hybrid membranes incorporated with glycerol and TEOS 2.5% and 4% respectively and hybrid membrane with TEOS 4% but without the incorporation of glycerol.

Table 1. Formulation of membranes

Materials	Formulation			
	F1	F2	F3	F4
Blended polymer, PVA:PEG (weight ratio)	9:1	9:1	9:1	9:1
Glycerol (wt.% blended polymer)	0	10	10	0
TEOS (wt. % total polymer)	0	2.5	4	4
Total polymer + TEOS (wt.%)	5	5	5	5
Water (wt.%)	95	95	95	95

2.2.3 Preparation of thin film composite (TFC)

The polysulfone membrane was attached to a glass plate. A specific concentration of a hybrid membrane solution was poured onto the membrane and by using the glass rod, a thin layer of the hybrid membrane solution was coated onto the membrane. The thin film composite membrane was left for 24 hours at room temperature and subsequently cured in the oven at 45°C for one hour [6] before characterization and performance evaluation of the membrane. Fig. 2 below show the schematic diagrams on the preparation of hybrid membrane and thin film composite.

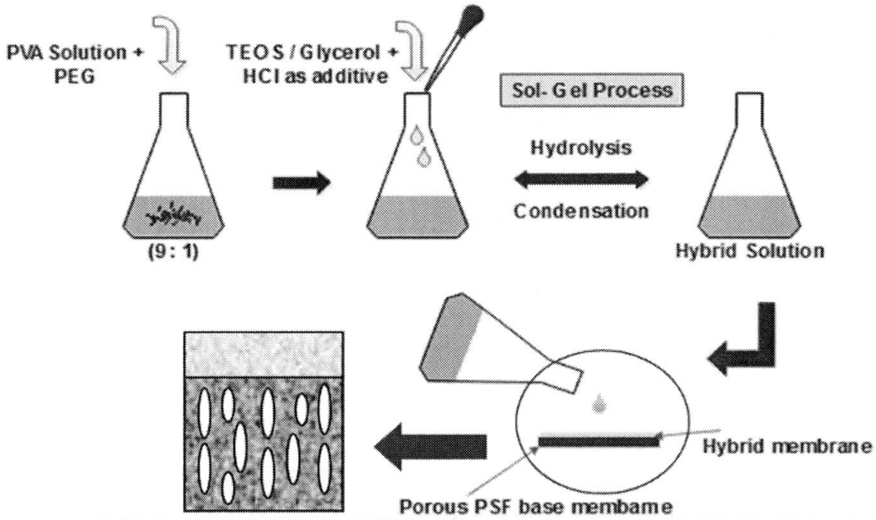

Fig. 2. Preparation of hybrid membrane (using sol-gel method) and thin film composite

2.3 Characterization

2.3.1 Scanning electron microscopy (SEM)

Cross-sectional of the thin film composite with hybrid membrane (formulation F3) coated on polysulfone membrane from 11 wt.% polymer was observed by using FESEM Supra 40 VP from Carl Zeiss SMT-Nano Technology Systems Division at 5 kV accelerating voltage. Prior to the analysis, the membrane was coated for 50 s with a 10-nm-thick layer of gold using a sputter coater (Bal-tec, Model: SCD 005). The test was conducted at the instrumentation laboratory, Faculty of Applied Science, UiTM Shah Alam.

2.3.2 Tensile strength measurement

Mechanical properties of the membranes were measured in the dry state at room temperature using a tensile strength instrument (Model: Instron) with 10 N load cell and it was conducted in the Instrumentation Laboratory, Faculty of Applied Science, UiTM Shah Alam. Hybrid membranes with a thickness < 100 μm were cut into rectangular shape (4 mm width and 100 mm length) for the measurements of tensile strength, strain and elastic modulus. The test method is according to ASTM D882, which is used for the determination of tensile properties of plastics in the form of thin sheeting with less than 1 mm thickness. This includes film which has been arbitrarily defined as sheeting having nominal thickness < 0.25 mm. From the stress-strain graph generated from the analysis, the tensile strengths, elongation and elastic modulus were determined. Tensile strength is the maximum stress the film can sustain before it actually fractures or amount of force necessary to pull a material apart. Strain refers to the amount the material will stretch before breaking. Elastic modulus is a measure of the force required to deform the film. It is a measure of the film stiffness. The cross head speed was fixed at 50 mm/min and the gauge length was set at 25 mm [8]. Five samples were tested and the average value was recorded. This test was conducted at instrumentation laboratory, faculty of applied science, UiTM Shah Alam.

2.3.3 Zeta potential

The particle size of the synthetic crude glycerol mixtures was measured by using a Zeta Potential Meter with Particle Sizer. Prior to the analysis, the parameters for the analysis and the characteristics of the sample were recorded in the Standard Operating Procedure (SOP) of the software by clicking on 'Start SOP' button. The record was saved. Before conducting the analysis, 'Measure' button followed by 'Start SOP from file' were pressed to retrieve the recorded SOP. Then the sample name was written. Before starting, the crude glycerol solution was prepared according to formulation as shown in Table 2. Then, a test cell was filled with the sample until certain level (the sample height must be below the maximum level). Then the test cell was placed into the closed container. The 'Start' button was pressed and the analysis was executed. The analysis was performed at three cycles. A result was reported in term of size distribution by intensity and an average size of the particles was recorded as Z-average size in nm. The result from the analysis was also shown in Table 2.

Table 2. Formulation of crude glycerol solution and the particle size

Compositions	Crude glycerol solutions
Additional water (wt.%)	75%
Crude glycerol (g)	200
Water (g)	150
Total solution (g)	350
Particle size (μm)	5.8

2.4 Antifouling performance

The antifouling experiment was performed by using a membrane test machine, which consists of a dead end stirred cell resting on a magnetic stir plate. Crude glycerol solution with 75 wt.% of additional water was employed as the organic foulant. For each thin film composite membrane, there are three stages involved in the evaluation of anti fouling performance. The first stage involved the permeation of distilled water through the thin film composites and the stabilized flux was recorded for 30 minutes (J_0) [3]. The second stage involved filtration of the crude glycerol solution through the TFC for 2 h and the flux was recorded as J_P. Then the TFC was cleaned by immersing it into 50 mL distilled water and being stirred at 200 rpm for 30 min. In the final stage, the pure water flux was recorded again for 30 min (J_1). For each experiment, the feed solution was placed in the feed chamber and was pressurized by using nitrogen gas to permeate through the thin film composite. The flux rate was calculated by dividing the flow rate of the permeate stream with the membrane area. For antifouling performance, two parameters were evaluated such as RFD (relative flux decay) and RFR (relative flux recovery) by the following equations:

$$RFD = [(J_0 - J_P)/ J_0] \text{ x } 100 \quad (1)$$

$$RFR = (J_1/ J_0) \text{ x } 100 \quad (2)$$

The compositions of the permeate stream in terms of glycerol content and NaCl concentration was determined by glycerol content determination method and by using ion meter respectively. Concentration of NaCl in the feed and permeate was determined by ion meter (Eutech PC 2700 Meter). The salt rejection was calculated by using Equation (3):

$$R(\%) = \left[1 - \frac{C_p}{C_f}\right] \times 100 \quad (3)$$

where R is the rejection, Cf and Cp are the concentrations of NaCl in the feed and permeate respectively.

Glycerol content determination was conducted according to standard method, ISO 2879-1975. The percentage permeation of glycerol (P) was calculated by using equation (4) where C_f and C_p are the concentrations of glycerol in the feed and permeate respectively. The process was repeated for three times by using another piece of the TFC and the average values were calculated.

$$P = \left[\frac{C_p}{C_f}\right] \times 100 \quad (4)$$

3 Results and discussion

Fig. 3 shows a cross sectional view of the TFC, where the presence of the hybrid membrane's layer on the base polysulfone membrane could be observed clearly. Based on the figure, no distinct pores could be observed from the hybrid membrane's layer. Obviously, the thin layer acts as the barrier layer, which dominates the transport resistance of the thin film composite [9] in which the rejection of NaCl and impurities will occur during the permeation of crude glycerol.

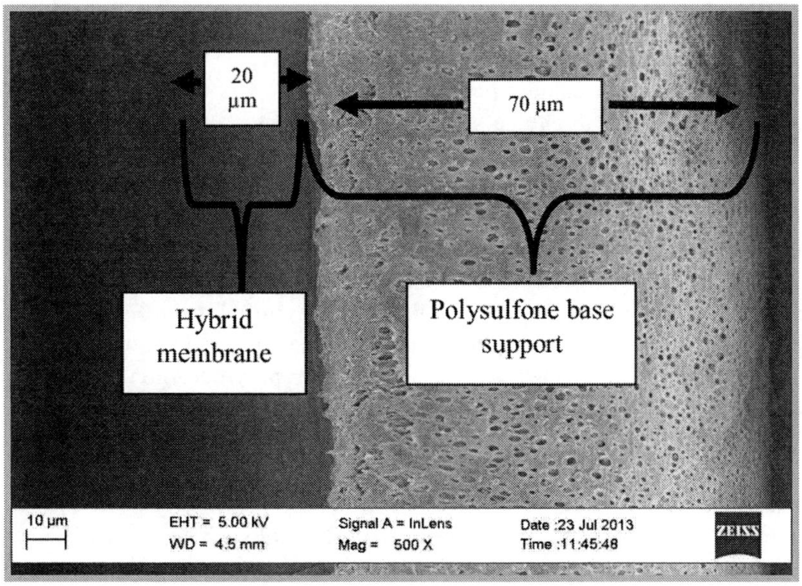

Fig. 3. Cross sectional view of TFC with F3 formulation

Based on Fig. 4, membranes from hybrid formulations, F3 (with glycerol) and F4 (without glycerol) exhibited higher tensile strength and strain as compared to membrane from pure polymer blend (F1). This result had ascertained the claim that cross-linking reaction between organic and inorganic materials resulted in better mechanical properties of the resultant hybrid membrane. From Table 3, it is clearly shown that the incorporation of glycerol in the pure polymer blend (2) has improved the tensile strength and strain of the membrane. As for the hybrid membranes, the mechanical properties of membrane F3, which consists of 10 wt.% glycerol improved significantly as compared to that without glycerol (F5). Based on the figure, it was obvious that a hybrid membrane incorporated with glycerol (F3) has higher tensile strength (30.22 kPa) and strain (6.20) as compared to that without it (F4) where it has tensile strength 15.98 kPa and strain 2.8. However, based on the elastic modulus values as shown in Table 3 which is determined from the slope of the initial linear portion of the stress vs strain curve, F4 has higher elastic modulus (113.78 MPa) as compared to F3 (48.82 MPa). Elastic modulus represents the stiffness of hybrid membrane, which is resistance to elastic strain [10] and further confirmed through lower elongation that is attributed to the reduction in the ductility of the membrane as stated by Dahe et al. [11]. The needs for good ductility of membrane is important to ensure easy manufacturing of membrane where it will not easily crack and can be easily molded into the required shape. As described by Zubir and Ismail [10], in the production of PTFE (Polytetrafluoroethylene) membrane, the targeted outcomes were membranes with high tensile strength and strain but with low elastic modulus for easy shaping of the membrane into a thin membrane with a uniform thickness.

Generally, the presence of silica in the hybrid membrane could increase the mean distance between polymer chains which causes the increase in stiffness of the polymer structure due to restricted segmental motion [12] that leads to increase in the elastic modulus. The presence of glycerol as the plasticizer has reduced the rigidity structure of the hybrid membrane that is observed through lowering the elastic modulus [13]. Besides that, synergistic effect of glycerol with the cross-linking network portrayed by organic-inorganic materials would substantiate the enhancement in the mechanical properties of the hybrid membrane, which was claimed by Liang et al. [14] in their research on the production of chitosan/polyvinyl alcohol gel films by using cross-linking method. Although the incorporation of glycerol has increased the elastic modulus of F3 but the value is marginal and it has been compensated with higher elongation.

Fig. 5 displays the flux change percentage over time from antifouling experiment. As shown in Table 4, all membranes portray high relative flux decay (RFD) due to dealing with high viscosity of feed solution as compared to pure water. However, the flux declining rate for F1 as displayed in Table 4 is higher as compared to other membranes, where membranes from F3 and F4 displaying the lowest RFD. However, all membranes portrayed comparable flux declining rate for filtration on crude glycerol solution. Based on Table 4, after undergoing cleaning process for 30 min, the pure water flux for TFC with barrier layer from F2 and F3 were almost recovered back to the initial values to give relative flux recovery (RFR) 95.51% and 82.41% respectively. Whereas for TFC with F1 and F4, lower RFR were measured that shows barrier layer from polymer blend as well as hybrid membrane with the absence of glycerol were easily fouled and it is not a reversible fouling. Higher RFR exhibited by F2 and F3 would justify the enhancement in hydrophilicity of the membrane's surface as a result of glycerol incorporation during the formulation

In term of the performance of all membranes on crude glycerol purification, TFC with hybrid membranes exhibited better permeation of glycerol with sufficient rejection of NaCl as displayed in Table 4. This situation is caused by the formation of hydrogen bonding between organic phase and inorganic phase during the sol-gel reaction [15], where this property would benefit the crude glycerol purification process as water soluble material could permeate through the membrane. TFC with F3 as the barrier layer shows the best performance in term of rejecting higher percentage of NaCl by allowing higher percentage of glycerol passing through the TFC.

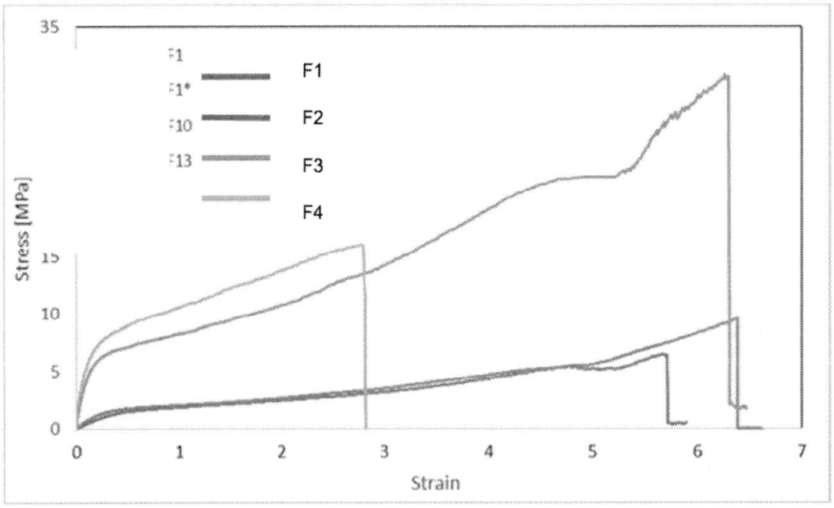

Fig. 4. Stress vs elongation for membranes from various formulations

Table 3. Mechanical properties of various membranes

Membrane	Tensile strength (MPa)	Break Strain	Elastic modulus (MPa)
F1	6.456	5.67	3.925
F2	13.36	6.39	21.28
F3	30.22	6.20	48.82
F4	15.98	2.80	113.78

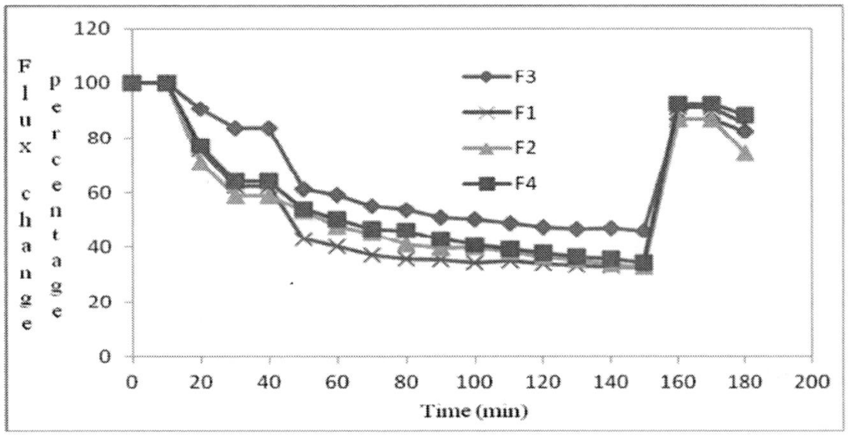

Fig. 5. Variation in the permeate flux of various TFC membranes during filtration of crude glycerol solution

Table 4. Performance of various membranes in term of antifouling behaviour and filtration

Formulation	J_0 (L/m^2.day)	J_P (L/m^2.day)	J_1 (L/m^2.day)	RFD (%)	RFR (%)	NaCl rejection (%)	Glycerol permeated (%)
F1	2484.38	52.56	1397.12	97.88	56.24	29.8	80.02
F2	2033.43	55.19	1942.13	97.28	95.51	23.01	90.74
F3	630.78	39.77	462.02	93.34	82.41	32.62	86.60
F4	1610.14	74.56	879.77	95.37	54.64	23	93.96

4 Conclusion

The presence of glycerol in the hybrid membrane formulation had reduced the rigidity structure of the resultant membrane that was observed through the improvement on the tensile strength and elongation with the reduction in the tensile modulus. Furthermore, the enhanced hydrophilicity of the membrane's surface had increased its antifouling behaviour which was displayed through higher relative flux recovery values. Based on its performance, the fabricated thin film composite has a great potential to be used as a pathway for crude glycerol purification due to some advantages over the existing process that employ membrane.

The author would like to thank Ministry of Higher Education for funding the research through Fundamental Research Grant Scheme (FRGS) No. 600-RMI/ST/FRGS 5/3/Fst (100/2010).

References

[1] K.S. Norin Zamiah and A.R. Norazah, Performance of thin film composite with hybrid membrane as the barrier layer for the separation process of synthesized crude glycerol, *Proc. of the 2013 IEEE Business Engineering and Industrial Applications Colloquium*, Bayview Hotel, Langkawi, (2013)

[2] S. Yano, K. Iwata, and K. Kurita, Physical properties and structure of organic-inorganic hybrid materials produced by sol-gel process, Material Science and Engineering., **6**(2-3) , 75-90, (1998)

[3] X.Zhu, H.E. Loo and R. Bai, A novel membrane showing both hydrophilic and oleophobic surface properties and its non-fouling performances for potential water treatment applications, J. of Membrane Science, **436**, 47-56, (2013)

[4] E.M.V. Hoek, A.K. Ghosh, X. Huang, M. Liong and J.I. Zink, Physical-chemical properties, separation performance, and fouling resistance of mixed-matrix ultrafiltration membranes, Desalination, **283**, 89-99, (2011)

[5] S. Kim, M. Marion, B.H. Jeong and E.M.V. Hoek, Crossflow membrane filtration of interacting nanoparticles suspensions, J. of Membrane Science, **283**, 89-99, (2011)

[6] G.J. Francisco, A. Chakma and X. Feng, Membranes comprising of alkanolamines incorporated into poly(vinyl alcohol) matrix for CO2/N2 separation, J. of Membrane Science, **303**, 54-63, (2007)

[7] L.Y.Ye, Q.L.Liu, Q.G. Zhang, A.M. Zhu, and G.B. Zhou, Pervaporation characteristics and structure of Poly(vinyl alcohol) / Poly(ethylene glycol) / Tetraethylorthosilicate hybrid membranes, J. of Applied Polymer Science, **105**, 3640-3648, (2007)

[8] N.A. Zubir and A.F. Ismail, Effect of sintering temperature on the morphology and mechanical properties of PTFE membranes as a base substrate for proton exchange membrane, Songklanakarin J. Sci. Technology, **24**, 823-831, (2002)

[9] H.J. Li, Y.M. Cao, J.J. Qin, X.M. Jie, T.H. Wang, J.H. Liu and Q.Yuan, Development and characterization of anti-fouling cellulose hollow fiber UF membranes for oil-water separation, J. of Membrane Science, **279**, 328-335, (2006)

[10] G.J. Dahe, R.S. Teotia and J. R. Bellare, The role of zeolite nanoparticles additive on morphology, mechanical properties and performance of polysulfone hollow fiber membranes," Chemical Engineering J., **197**, 398-406, (2012)

[11] G.J. Dahe, R.S. Teotia and J. R. Bellare, The role of zeolite nanoparticles additive on morphology, mechanical properties and performance of polysulfone hollow fiber membranes, Chemical Engineering J., **197**, 398-406,(2012)

[12] M. Sadeghi, M.A. Semsarzadeh and H. Moadel, Enhancement of the gas separation properties of polybenzimidazole (PBI) membrane by incorporation of silica nano particles, J. of Membrane Science, **331**, 21-30, (2009)

[13] J.C. Wang, P. Chen, L. Chen, K. Wang, H. Deng, F.Chen, Q. Zhang and Q. Fu, Preparation and properties of poly (vinylidene fluoride) nanocomposites blended with grapheme oxide coated silica hybrids, eXPRESS Polymer Letters, **6**(4), 299-307, (2012)

[14] S. Liang, L. Liu, Q. Huang and K.L. Yam, Preparation of single or double-network chitosan/poly(vinyl alcohol) gel films through selectively cross-linking method, Carbohydrate Polymers, **77**, 718-724, (2009).

[15] K. Xie, Y. Yu and Y. Shi, Synthesis and characterization of cellulose/silica hybrid materials with chemical crosslinking, Carbohydrate Polymers, **78**, 799-805, (2009)

Measurement of Water Quality Parameters for Before and After Maintenance Service in Water Filter System

Nuraida Shaharudin [1,*], *Nurfarhana* Suradi[1], *Nor Amani Filzah* Mohd Kamil[1]

[1]Water and Environmental Engineering Department, Faculty of Civil & Environmental Engineering UniversitiTun Hussein Onn Malaysia, BatuPahat, Johor, Malaysia.

Abstract. An adequate supply of safe drinking water is one of major ways to obtain healthy life. Water filter system is one way to improve the water quality. However, to maintain the performance of the system, it need to undergo the maintenance service. This study evaluate the requirement of maintenance service in water filter system. Water quality was measured before and after maintenance service. Parameters measured were pH, turbidity, residual chlorine, nitrate and heavy metals and these parameters were compared with National Drinking Water Quality Standards. Collection of data were involved three housing areas in Johor. The quality of drinking water from water filter system were analysed using pH meter, turbidity meter, DR6000 and Inductively Coupled Plasma-Mass Spectrometer. pH value was increased from 16.4% for before maintenance services to 30.7% for after maintenance service. Increment of removal percentage for turbidity, residual chlorine and nitrate after maintenance were 21.5, 13.6 and 26.7, respectively. This result shows that maintenance service enhance the performance of the system. However, less significant of maintenance service for enhance the removal of heavy metals which the increment of removal percentage in range 0.3 to 9.8. Only aluminium shows percentage removal for after maintenance with 92.8% lower compared to before maintenance service with 95.5%.

1 Introduction

Malaysians spend billions of ringgits each year on water filter system units that installed in their house. According to the survey National Consumer Water Quality January 1999, 38% of resident use a water filter system at home which is increase by 10% since 1995 [1]. By comparing to nowadays, the number of user may be increase double or triple. It is shown that most Malaysian used water filter system to improve on their tap water in whole Malaysia. [2] was conducted a study to compare water quality from water filter system and user perception. Mostly respondents were not satisfied with the water supply to their homes. The respondents also think that the water could bring adverse impact to their health if no water filter system installed in their homes. Finding from the study was similar with

* Corresponding author: noramani@uthm.edu.my

[3] which main problems with their tap water was colour, odour and taste. The study also reported that 85% of respondents used water filter system in their homes due to convenience and comfortable compare to boiling the water.

Some of the users rarely or do not conduct any maintenance service for their water filter system [2]. This is may be due to manufacturer only supply the water filter system to user without explaining the maintenance service that need to be conducted by user. User that had used the water filter system must carry out proper maintenance periodically as recommended by the supplier to ensure the quality and safety of the tap water. If no maintenance service was conducted, maybe other contaminants that flow into the system may increase the contaminant's concentration and effect the effluent water. [4] found that concentration of bacteria was 10,000 times higher in filtered water compare to ordinary tap water which may be due to growth or biofilm formation in the filter material. The filter material contains a mixture of ion exchange resins and silver impregnated activated carbon. From time to time, the filter maybe covered with contaminants and debris and affected the flow of water through the filter. When the water flow drastically reduced, the filter system may produce very high concentrations of contaminants.

Many studies were conducted to evaluate the performance of water filter system and these studies have shown high removal percentage [5-9]. [5] showed that reverse osmosis in water filter system removed total suspended solids and turbidity more than 99% and 98%, respectively, by using electrocoagulation as a media filter. Study conducted by [6], showed that water filter system based on ultrafiltration removed of 97%, 95% and 99% of arsenic, organic matter and turbidity, respectively, by using trickling filter with biosand filter. Moreover, [7] shown that the percentage removal for lead, barium, nickel and copper were more than 80%.The study use bench-top water filter and the media wasion exchange resin and Silverized granular activated charcoal. [8] shown the percentage removal for arsenic is 95% by using a ceramic filter as water filter system and the media is clay soil and rice bran. According to [9], the percentage removal for arsenic were 7.7– 50% by using modified homemade filter and the media was crushed brick. Even though the percentage removals were high and shows the good performance of water filter system, the performance after long duration is not emphasize. In addition, the quality of water after maintenance service was never measure.

In this study, the effectiveness of maintenance service on water filter system was analyzed. Result from the analysis was compared with National Drinking Water Quality Standards to ascertain the effluent water safe for drinking. Parameters selected for this study were pH, turbidity, chlorine, nitrate and heavy metals (arsenic, barium, cadmium, copper, chromium, ferum, lead, nickel, aluminium and selenium).

2 Material and method

This section consists of sampling areas and sampling handling and analysis of water quality parameters.

2.1 Sampling areas and sampling handling

Three housing areas were selected as site locations which are Taman Pura Kencana in Seri Gading, Kg. Melayu Majidee in Johor Bahru and Taman Perling in Johor Bahru. These selected housing areas can be categorized as terrace housing. These sampling areas were chosen due to the usage of similar water filter systems

Water sampling was carried out twice every two months of maintenance service in each site. Polyethylene bottles were used for collection of water samples which had been thoroughly washed and rinsed using distilled water to flush out any concealed pollutants.

For each sampling area, three samples from outlet before maintenance service and three samples from outlet after maintenance service were collected. Unfiltered water samples were collected from a water tap on site. Water samples were kept in low temperature at 4°C in the dark before testing. The samples were analysed within 48 hours for pH, turbidity, residual chlorine and nitrate. Meanwhile, for heavy metals the holding time for samples can be up to 6 months if properly preserved.

2.2 Analysis of Water Quality Parameter

Water samples collected from water filter system was tested at the Environmental Laboratory in Universiti Tun Hussein Onn Malaysia. pH was analysed by using pH meter onsite. Turbidity was tested by using turbidity meter. Residual chlorine and nitrate were analysed using UV-VIS spectrophotometer (DR6000) by using Method 8021 and Method 10206, respectively. Heavy metals, i.e., arsenic, barium, cadmium, copper, chromium, ferum, lead, nickel, aluminium and selenium were analysed by using Inductively Coupled Plasma-Mass Spectrometer (ICP-MS), Agilent 7900, United States.

3 Results and discussion

This section discussed concentrations and percentage removal of selected parameters at influent, effluent before and after maintenance service.

3.1 Concentrations of selected parameter at influent, effluent before and after maintenance service

Fig. 1 shows the pH value and concentrations of turbidity, chlorine, nitrate, arsenic, barium, cadmium, copper, chromium, ferum, lead, nickel, aluminium and selenium at influent, effluent before maintenance service and after maintenance service. Whereas, Table 1 shows the allowable limit in National Water Quality Standards [12]. pH value and concentration of all parameters obtained from the three housing areas were below limit stated in the National Drinking Water Quality Standards except for residual chlorine. pH value was increased from 5.93 to 6.9 after through the water filter before maintenance service. After maintenance service, the pH was increased higher, from 5.93 to 7.75 which makes water more alkaline.

The turbidity at the influent was much lower compare to the standard, i.e., 0.792 NTU which almost similar to [10] that obtain average turbidity for tap water was 0.86 NTU. This concentration complies with the recommended value of 1 NTU for safe drinking water using chlorine as disinfectants [10]. After passing through the water filter, concentration at the effluent was reduced to 0.433 NTU for before maintenance service. After maintenance service, the turbidity was reduced to 0.263 NTU, which clearer water was obtained due to neo sense filter services that help removes the fine particles or dust and other impurities in the water [11]. In addition, the presence of reverse osmosis (RO) membrane with 0.001 micron further reduced the contaminant.

Concentrations of residual chlorine at the effluent of water filter were much lower than the standard. However, these low concentrations may not bring adverse impact to user of the water filter system. The residual chlorine acts as a protection against additional microbial contamination [13]. The water filter system used in this study was designed for user drinking directly at the effluent outlet, which can be classified as indoor water filter system. Thus, additional microbial contamination may not occur. If the water filter system is used outdoor, then the system will not be suitable for contamination can easily occur. For any house in Malaysia, water from the mains is channelled into two parts, first directly to the kitchen sink and second is to an elevated water storage tank [14]. If the effluent water is directed to the elevated water storage tank and stored for a long time, the effluent may not safe for drinking due to probability of contamination. In addition, pH measured after maintenance with value 7.75 which higher than requirement for effective disinfection with chlorine [10]. Thus, residual chlorine was less significant in the water filter system in terms against protection of additional microbial contamination.

Concentration of nitrate was much lower at influent, 0.03 mg/L, compared to the standard. Concentration was reduced at the effluent before and after maintenance from 0.013 mg/L and 0.005 mg/Lrespectively due to reverse osmosis membrane that had physically separate impurities from the water, including nitrates [11]. For heavy metals, concentrations at influent were much lower compared to the standard, which in range 0.407 ppb to 75.82 ppb. Concentrations at the effluent were reduced in range 0.0121 ppb to 3.397 ppb for before maintenance service. After maintenance service, concentrations of heavy metals were much reduced in range 0.009 ppb to 0.598 ppb except for aluminium. Concentration of aluminium was increased from 3.397 ppb for before maintenance service to 5.464 ppb for after maintenance service.

pH value and concentration of all parameters for unfiltered water samples were below the limit stated in National Drinking Water Quality Standards. Thus, this finding shows that without water filter system, the water is safe for drinking. The water filter system was more reduced the contaminants concentrations and increased the quality of water.

Table 1. Allowable limit in National Drinking Water Quality Standards [12]

Parameters	pH	Turbidity	Residual chlorine	Nitrate	As	Ba	Cd	Cu	Cr	Fe	Pb	Ni	Al	Se
Unit		NTU	mg/L		ppb									
Standards [12]	6.5-9.0	5	0.2-0.5	10	10	700	3	1000	50	300	10	20	200	10

As: Arsenic; Ba: Barium; Cd: Cadmium; Cu: Copper; Cr: Chromium; Fe: Iron: Pb: Lead: Ni: Nickel; Al: Aluminium; Se: Selenium

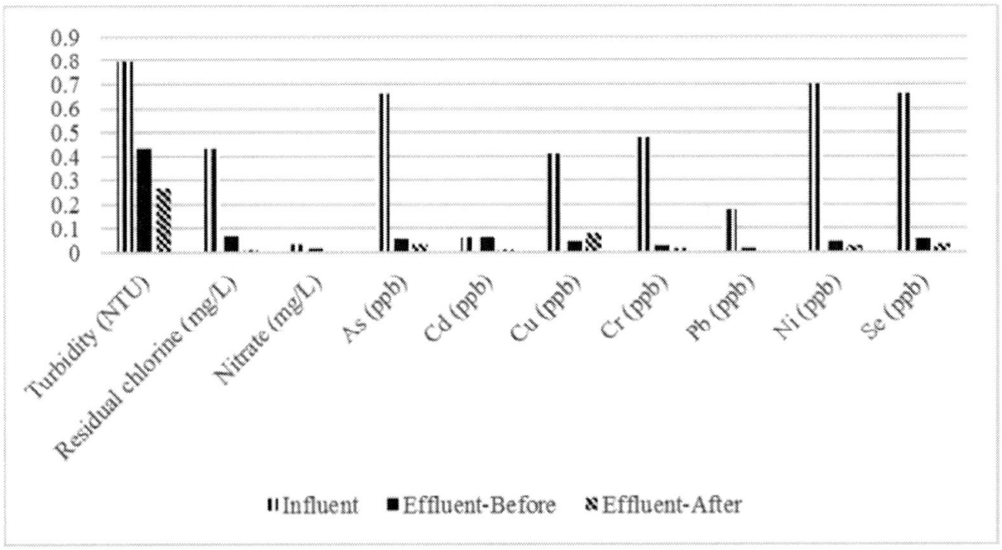

(a)

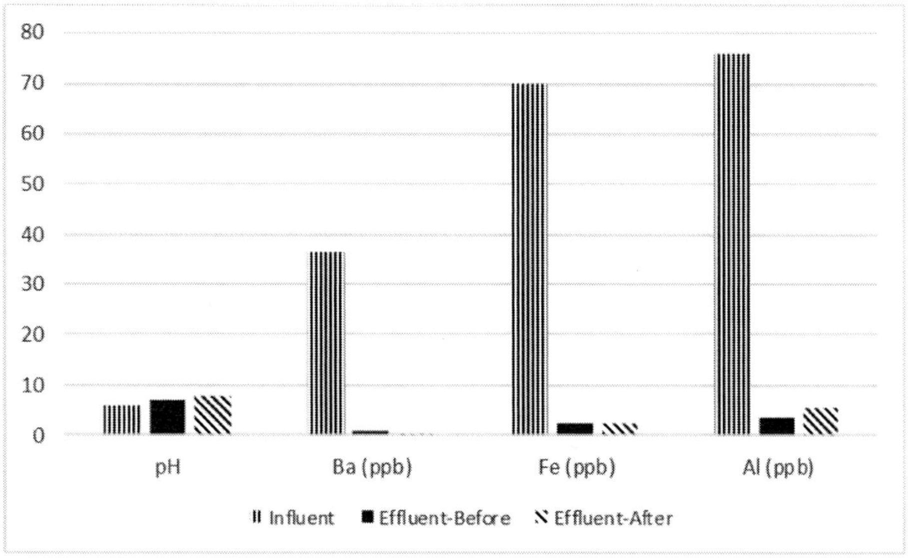

(b)

Fig. 1. Concentrations of pH, turbidity, chlorine, nitrate and heavy metals at influent, effluent before maintenance service and effluent after maintenance service (a & b).

3.2 Percentage of Removal of selected parameter before and after maintenance service

Fig. 2 shows the percentage of removal for before and after maintenance service of pH, turbidity, chlorine, nitrate, arsenic, barium, cadmium, copper, chromium, ferum, lead, nickel, aluminium and selenium before and after maintenance. The increase in pollutant removal shows that sanitation during maintenance service has enhanced the performance of water filter system except for aluminium. Nitrate shows the highest increment, 26.7 and Ferum shows the lowest increment, 0.4.

The sanitation work significantly to enhance the performance of water filter for pH, turbidity, residual chlorine and nitrate. There are three maintenance service provided by manufacturer, namely, sanitation, check-up and filter change [15]. In this study, sanitation was selected due to highest frequency of this service. There are six filters in the water filter system, namely, plus sediment filter, pre-carbon filter, RO membrane, post-carbon filter, fine filter and ceramic filter [11]. During sanitation, only plus sediment filter was backwashed with tap water. Other filter are changed within six months or a year after use following the schedule by manufacturer [15]. The plus sediment filter was use to remove sand, mud and rust [11]. Thus, this removal enhanced the performance of the water filter system to remove turbidity, residual chlorine and nitrate. In addition, pH also increase due to reduction of impurities during sanitation.

However, the sanitation has reduced the potential of aluminium removal. This may be due to tap water contains high concentrations of aluminium, 75.82 ppb as shown in Table 1. Thus, aluminium from tap water may flow into the water filter system and reduce the performance after maintenance service. In addition, the sanitation has less significant for removal of heavy metals. For heavy metals, changing of RO membrane may be more significant to the heavy metals removal. RO membrane was used to reduce heavy metals such as arsenic, barium, cadmium, selenium, radium, chromium and lead [11]

A study conducted in 2007 shows that better water quality was obtained for unfiltered water compared to filtered water [2] which contrast to this study. This is may be due to the filter not being serviced at all for that year. In 2007, clients are expected to conduct the maintenance service by themselves. By comparison nowadays, many suppliers of water filter system are providing the maintenance service in their package.

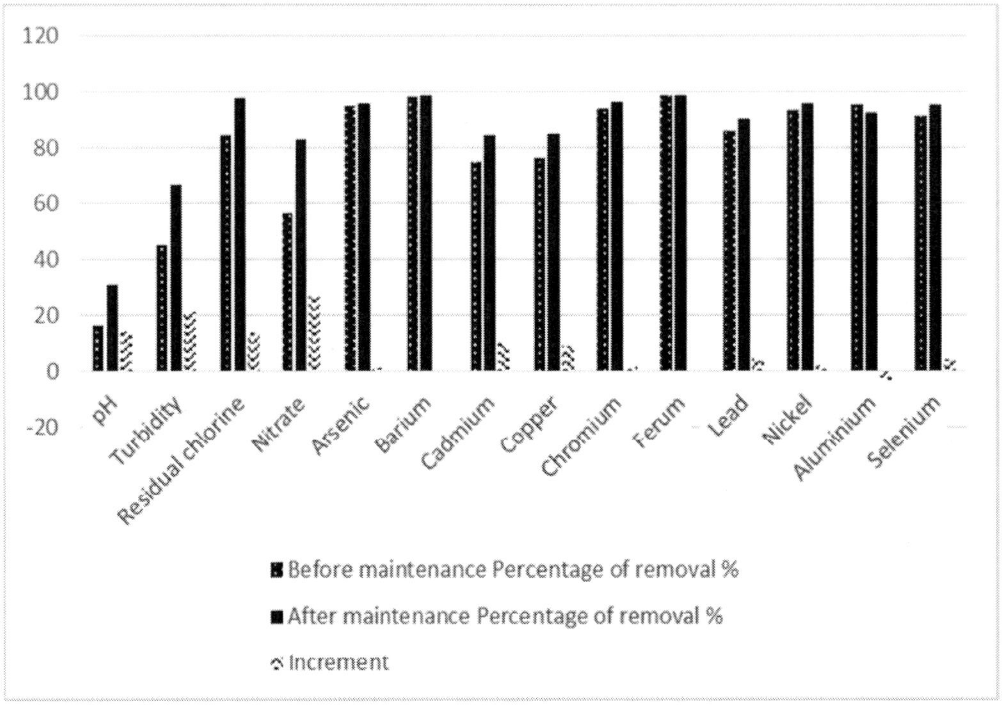

Fig. 2. Percentage of removal for before and after maintenance of pH, turbidity, chlorine, nitrate and heavy metals

4 Conclusion

As conclusion, concentrations for all parameters were below the standard which render the water is safe for drinking except for residual chlorine. However, the concentration of residual chlorine may bring adverse impact to user if directly use without storing the water first. All parameters show better removal after maintenance service except for aluminium. The maintenance service enhanced the performance of water filter system for pH, turbidity, residual chlorine and nitrate, but less significance for heavy metals. Thus, the maintenance service, namely, sanitation is required to maintain and enhance the performance of water filter system. It is recommended to measure the effectiveness for emerging pollutants found in pharmaceutical, pesticides, surfactants, industrial additives and solvents. In addition, it was suggested to measure the requirement of maintenance service for outdoor water filter system.

The author would like to thank the supplier of water filter system that had agree to conduct this study. Author also would like to thank for staff of maintenance service for maintenance information.

References

[1] K. H. Loh, H. Yaacob, H. Andan, S. Omar and M. Jamaludin, A Study of The Effect of Home Water Filtration Systems On Fluoride Content of Drinking Water in Johor, Malaysian Dental J., **33**(2), 8-13, (2011)

[2] R. Mat-Salleh, Water Quality, *Perception And Consumer's Satisfaction Towards Domestic Water Filters*, Doctoral dissertation, Universiti Putra Malaysia, (2007)

[3] Aini, M. S., Fakhrul-Razi, A., Mumtazah, O., and Chen, J. M. (2007). Malaysian households' drinking water practices: A case study. The Int.l J. of Sustainable Development & World Ecology, **14**(5), 503-510.

[4] F. D. Daschner, H. Rüden, R. Simon, and J. Clotten. Microbiological contamination of drinking water in a commercial household water filter system. European J. of Clinical Microbiology and Infectious Diseases, **15**(3), 233-237. (1996)

[5] K. Sadeddin, A. Naser, and A. Firas, Removal of turbidity and suspended solids by electro-coagulation to improve feed water quality of reverse osmosis plant. Desalination, **268**(1), 204-207. (2011).

[6] M. Maeng, J. Park, H. Lee, J. E. Tobiason, and S. Dockko. Sustainable technology of trickling biosand filter (TBSF) combined with rock media to reduce organic matters for drinking water. Desalination and Water Treatment, **57**(17), 7733-7741. (2016)

[7] B. L. Gulson, A. Sheehan, A. M. Giblin, M. Chiaradia, and B. Conradt. The efficiency of removal of lead and other elements from domestic drinking waters using a bench-top water filter system. Science of the total environment, **196**(3), 205-216. (1997)

[8] M. Shafiquzzaman, M. S. Azam, J. Nakajima, and Q. H. Bari. Investigation of arsenic removal performance by a simple iron removal ceramic filter in rural households of Bangladesh. Desalination, **265**(1), 60-66. (2011)

[9] R. Devi, E. Alemayehu, V. Singh, A. Kumar, and E. Mengistie. Removal of fluoride, arsenic and coliform bacteria by modified homemade filter media from drinking water. Bioresource technology, **99**(7), 2269-2274. (2008)

[10] Ong, C., Ibrahim, S., and Sen Gupta, B. (2007). A survey of tap water quality in Kuala Lumpur. Urban Water J, **4**(1), 29-41.

[11] C. S. Yen. U.S. *Patent No. 5,914,037.* Washington, DC: U.S. Patent and Trademark Office. (1999).

[12] Engineered Services Division Ministry of Health Malaysia. *National Drinking Water Quality Standard.* (2004)

[13] P. Payment. Poor efficacy of residual chlorine disinfectant in drinking water to inactivate waterborne pathogens in distribution systems. Canadian J. of Microbiology, **45**(8), 709-715. (1999)

[14] M. Shuhada. *Elak Guna Penapis Air*, Harian Metro. 6 Mei 2011.

[15] Interview with staff of maintenance service from water filter system company. (2015)

Green Approach in the Bio-removal of Heavy Metals from wastewaters

Paran Gani[1,*], *Norshuhaila* Mohamed Sunar[2], *Hazel* Matias-Peralta[3], *Ab Aziz* Abdul Latiff[1], *Umi Kalthsom* Parjo[2] and *Adeleke AbdulRahman* Oyekanmi[1]

[1]Faculty of Civil and Environmental Engineering, Universiti Tun Hussein Onn Malaysia, 86400 Parit Raja, Batu Pahat, Johor, Malaysia
[2]Faculty of Engineering Technology, Universiti Tun Hussein Onn Malaysia, 86400 Parit Raja, Batu Pahat, Johor, Malaysia
[3]Faculty of Science, Technology and Human Development, Universiti Tun Hussein Onn Malaysia, 86400 Parit Raja, Batu Pahat, Johor, Malaysia

Abstract. Cultivation of microalgae has been suggested as a green approach for a sustainable wastewater treatment especially heavy metal bioremediation. This study investigated the bio-removal of zinc (Zn), iron (Fe), cadmium (Cd) and manganese (Mn) from domestic wastewater (DW) and food processing wastewater (FW) using green microalgae, *Botryococcus* sp.. The total of five treatments represented by five different cell concentrations (1×10^3, 1×10^4, 1×10^5, 1×10^6 and 1×10^7 cells/mL) of *Botryococcus* sp. in the wastewaters medium. The results revealed high removal efficiency of Zn, Fe, Cd and Mn after 18 days of the culture compared to control (wastewaters without algae). In DW , Zn, Fe, Cd and Mn were successfully removed at the highest efficiencies up to 71.5%, 51.2%, 83.5% and 97.2%, respectively while in FW, the same metal concentrations were reduced by up to 64.4%, 53.3%, 52.9% and 26.7%, respectively. Overall, most of the algae cell concentrations tested were successfully reducing the metals contaminant presence in both wastewaters and provides a baseline for further phycoremediation coupled with biomass production.

1 Introduction

Rapid population growth and industrial development are expected to contribute extremely to the generation of waste, especially wastewater. Increasing in population normally will be accounted to the production of domestic wastewater which is highly contaminated and harmful to the environment. Similarly, the booming food industry in Malaysia is becoming one of the leading contributors of wastewater that pollutes the receiving water bodies [1]. Both domestic and food industry wastewaters contained a high amount of organic and inorganic nutrients which are suitable for growth of microalgae [2]. The growing microalgae can then be used to carry out the phycoremediation process [1, 2]. However,

* Corresponding author: parancgat@yahoo.com

these wastewaters contained not only nutrients but also harmful heavy metals that threaten the aquatic life and public health.

The heavy metals present in domestic wastewater (DW) is mainly contributed by pipe sedimentation, corrosion and domestic product usages such as medicine, caulking products, cleansing product cosmetics and others [4]. Meanwhile, metals concentration in the food wastewaters (FW) were mainly contributed by fertilisers and pesticides used during the production and growing stage of the fruit plants. Conventional techniques to remove heavy metal are often ineffective and very expensive. In addition, these methods are used to reduces heavy metal contaminations at low concentrations [5]. In fact, most of the conventional techniques provided incomplete metals removal, require high energy and used a significant amount of reagent, and had limited tolerance to pH changes [5]. Hence, the green approach has been introduced to address these problems. What is meant by green approach is to apply the cultivation technology of microalgae using wastewater and simultaneously reduce the concentration of hazardous metals indirectly [6]. This approach has a promising potential to contribute to achieving this goal in an environmentally friendly manner. Moreover, green technologies comprise low-cost, high-efficiency technique for heavy metals bio-removal from wastewaters [5–9]. However, the effectiveness of metal removal using this approach depends on microalgal species, environmental factors and microalgae concentrations [2, 10, 11]. As highlighted earlier, there has been few information exist on the potential of microalgae, *Botryococcus* sp. in the removal of heavy metals from wastewater. Previously, Onalo et al. [12] grew microalgae, *Botryococcus* sp. in heavy metal contaminated textile industry wastewater for the potential Cr, Cu, As and Cd bio-removal. They found that *Botryococcus* sp. effectively reduce Cr, Cu, As and Cd up to 94%, 45%, 9% and 2%, respectively. Soeprobowati and Hariyati [14] reported using *Chaetoceros calcitrans* to remove Pb, Cd, Cu and Cr from synthetic wastewater while Worku and Sahu [15] also used synthetic wastewater to phytoremediate the Cr, Ni, Fe and Hg using microalgae, *Synechocystis salina*. In 2012, Krustok and Nehrenheim [16] had demonstrated the cultivation of microalgae (species not specify) for the reduction of heavy metals present in domestic wastewater. Accordingly, the objective of the present work is to investigate the capability of *Botryococcus* sp. in bio-removal of selected heavy metal present in domestic wastewater and food processing wastewater with different cell concentrations under indoor culture condition.

2 Materials and Methods

2.1 Wastewaters sampling and analytical analysis

The domestic wastewater (DW) was obtained from wastewater treatment plant located in the main campus of Universiti Tun Hussein Onn Malaysia, Johor (N 01° 51' 55.224" E 103° 5' 21.183"). This plant receives daily wastewater generated from different activities in the campus area of which student's residential college and campus cafeterias contributed the most. Meanwhile, the food industry wastewater (FW) was collected from Azhar Food Manufacturing Industry located in Rengit, Johor (N 01° 41' 13.2" E 103° 07' 43.6"). The industries carry out processing activities and the production of snack foods such as potato chips, banana chips and cookies. Both wastewater samples were collected using acid washed sample bottles at the site and immediately transferred to the laboratory and preserved at temperatures below 4°C in a refrigerator [17]. The chemical oxygen demand (COD) and total phosphorus (TP) were measured using DR 6000 Spectrophotometer (Hach, USA) according to Method 8000 and Method 10127, respectively. Meanwhile, the biochemical oxygen demand (BOD) and total suspended solid (TSS) were analysed according to the standard method [17]. For the heavy metal (Zn, Fe, Cd and Mn) analysis, the inductively coupled plasma spectrometry (ICP-MS) was used.

2.2 Experimental design

The microalgae namely *Botryococcus* sp. used in this study was obtained from tropical rainforest located in the southern region of Peninsular Malaysia. The preparation of *Botryococcus* sp. inoculum follows Gani et al. [3]. This was followed by cell observation and cell concentration count using Neubauer haemocytometer chamber.

Indoor culture condition was employed in the current study. The experiments were conducted in an environmental chamber (Wisd – ThermoStable SWGC) with control temperature (33°C) light intensity (243 µmol $m^{-2}s^{-1}$) and 24:0 hours of photoperiod [6]. These factors were considered to be optimal for this microalgae species according to our preliminary biomass productivity examination (data not shown). There are five initial microalgae cell concentration were tested (1×10^3, 1×10^4, 1×10^5, 1×10^6, and 1×10^7cells/mL) excluding control sample [3].

The wastewater samples were filtered using a membrane filter (Whatman) with a 0.45µm pore size to remove other microorganisms and suspended solids before commencing the experiments. A total of 36 Erlenmeyer flasks (500 mL) were filled with 350 mL of prepared DW, and FW were used in this study. These 36 flasks correspond to 12 different treatments including control in triplicates. Each treatment represents initial concentrations of microalgae at the beginning of the experiment. The culture flasks were exposed to controlled indoor condition. The flasks were covered with sterile cotton plugs and repeatedly shaken four times daily to ensure homogenised cells in the wastewaters [18]. The experiment was carried out for 18 days in total. Total removal efficiencies calculated following to Equation 1.

$$\mathrm{Re}\,moval\,,\% = \frac{Initial\ \text{conc.}\ (ppb)\ \text{- Final conc.}\ (ppb)}{Initial\ \text{conc.}\ (ppb)} \times 100 \qquad (1)$$

2.3 Statistical analysis

Data analyses of average, mean differences, standard deviation, and significant different level (95%) for heavy metals removal were performed accordingly using analysis of variance (One-way ANOVA) SPSS 20. Basically, this statistical analysis served to test the mean difference of the dependent variable for three or more microalgae concentrations simultaneously.

3 Result and discussion

3.1 Characteristic of wastewaters

Wastewater characterization is important for determining the organic and inorganic nutrient supplements required for microalgae growth during bioremediation process. The FW showed higher COD, BOD and TSS compared to DW (Table 1). Similarly, heavy metals, zinc, cadmium and manganese were found higher in FW than DW (Table 1). The physicochemical parameters of both wastewaters were found to be above the limit set by the Environmental Quality Act of Malaysia 1974 [19] (Table 1). On the other hand, the heavy metal (Zn, Fe, Cd and Mn) contents of both DW and FW were still within the allowable limit (Table 1).

DOI: 10.1051/matecconf/201710306007

Table 1. Raw characteristics of domestic wastewater (DW) and food processing wastewater (FW) compared to effluent standard

Parameter	[a] Concentration		Effluent standard (Environmental Quality Act, 1974)	
	Domestic wastewater (DW)	Food processing wastewater (FW)	Standard A	Standard B
Physiochemical, mg/L				
Chemical oxygen demand (COD)	129.00±2.35	2838.62±793.46	50	100
Biochemical oxygen demand (BOD)	71.08±2.80	793.89±110.20	20	50
Total phosphorus (TP)	8.99±0.53	6.81±1.93	5	10
Total suspended solid (TSS)	2158.00±10.32	12418.67±8989.20	50	100
pH	6.96±0.31	5.86±0.75	6.0 – 9.0	5.5 – 9.0
Heavy metal, ppb				
Zinc, Zn	334.6±30.0	670.9±290.0	1000	1000
Ferum, Fe	891.5±30.0	685.2±120.0	1000	5000
Cadmium, Cd	0.7±0.00	1.3±0.00	10	20
Manganese, Mn	159.3±20.0	219.3±170.0	200	1000

* All unit in mg/L except for pH and heavy metal (ppb)
[a] All experiments conducted in replicates (n=3)

In general, the characteristics of the DW (Table 1) were highly variable but comparable to the range reported in previous studies. The concentration of COD and BOD were 129 mg/L and 71.08 mg/L, respectively; this concentration was different from that used in other research paper. For instance, Gani et al. [3] used untreated DW containing 76.1 mg/L of COD and 44 mg/L of BOD to cultivate green microalgae *Botryococcus* sp. Zhang et. al. [20] cultivated mixotrophic microalgae strain in DW containing 142 mg/L of COD while Mostafa et. al. [21] used DW containing 50 mg/L of COD and 15 mg/L of BOD to grow cyanobacteria and *Chlorella vulgaris*. Nevertheless, the wastewaters used in this study contained 8.99 mg/L of TP 6.96 of pH value, indicating an acceptable concentration compared to the effluent standard and suitable enough for microalgae cultivation [22].

In terms of metal elements, manganese shows higher concentration compared to effluent Standard A (Table 1), while the rest of the elements studied were within allowable limit as stated in the standard (Table 1). Although the selected metal elements studied were still within the range of standard effluent limit, however, they still need to be reduced considerably if not completely removed before discharged to the environment. The need for this action is because heavy metal have the capacity to bioaccumulate in living freshwater organisms habitating the receiving water bodies, hence a threat to public health [12]. Thereby, this study may allow the use of DW and FW for the development of *Botryococcus* sp. culture and simultaneously to perform the bio-removal process of metals.

3.2 Heavy metal bio-removal efficiency

The removal of zinc (Zn), ferum (Fe), cadmium (Cd) and manganese (Mn) using cultivation of microalgae, *Botryococcus* sp. from domestic wastewater (DW) and food processing wastewater (FW) with different cells concentration are successfully performed (Table 2 and Fig. 1). Statistically, there is a significant effect ($P<0.05$) of microalgae concentrations on the bio-removal of Zn, Fe, Cd and Mn from both DW and FW under indoor culture condition. The highest removal of Zn from DW and FW was 71.5% and 64.4%, respectively. Accordingly, Zn had been reduced from 156.67 ppb to 44.7 ppb for DW while 141 ppb to 50.23 ppb for FW at the concentration of 1×10^4 cell/mL and 1×10^6 cell/mL, respectively (Table 2). However, the reduction of Fe in DW was up to 51.2% (from 302.19 ppb to 147.59 ppb) at microalgal cell concentration of 1×10^5 cell/mL; while the presence of Fe in FW was able to be taken up about 53.4% at the concentration of 1×10^6 cell/mL (reduced from 455.33 ppb to 212.67 ppb). The Cd ion in DW was reduced from 0.34 ppb to 0.056 ppb which is equivalent to 83.5%. Meanwhile, Cd in FW reduced from 0.18 ppb to 0.085 ppb (52.9%). Other toxic such as Mn in DW was able to be removed optimally about 97.2% at the lowest microalgae cell concentration applied in which reduced from 112.33 ppb to 3.15 ppb. Whereas, in FW, Mn could be eliminated up to 26.7% only under indoor culture condition.

Table 2. The bio-removal efficiency of heavy metal from domestic wastewater (DW) and food processing wastewater (FW) using *Botryococcus* sp.

Microalgae cells concentration, cell/mL	Initial concentration, ppb		Final concentration, ppb		Removal efficiency, %	
	DW	FW	DW	FW	DW	FW
Zinc, Zn						
1×10^3	156.67± 3.06	141± 3.00	49.3±3.4	61.20±3.02	68.53±1.97	56.60±3.06
1×10^4			44.7±4.49	57.77±2.54	71.47±2.83	59.03±2.66
1×10^5	116.67± 3.51		33.83±3.82	58.37±2.10	71.00±2.42	58.61±2.33
1×10^6			48.7±3.04	50.23±3.02	58.26±3.13	64.37±2.89
1×10^7			63.93±2.61	60.70±2.52	45.20±0.59	56.95±2.70
Control			80.51±2.27	94.79±2.86	30.99±4.05	32.78±0.81
Ferum, Fe						
1×10^3	333.67± 3.51	455.33± 2.52	174.45±3.09	215±5.00	47.72±1.24	52.78±1.26
1×10^4			172.95±2.62	218.67±12	48.17±0.49	51.98±2.44
1×10^5	302.19± 3.02		147.59±2.51	218±13.89	51.16±0.34	52.13±2.84
1×10^6			192.45±3.09	212.7±11.0	36.31±0.70	53.29±2.43
1×10^7			166.19±3.38	300±6.00	45.00±0.86	34.11±1.56
Control			175.45±1.50	443.67±5.5	41.94±0.12	2.56±1.42
Cadmium, Cd						
1×10^3	0.48± 0.03	0.18± 0.002	0.2±0.0095	0.115±0.01	57.87±1.90	36.11±1.45
1×10^4			0.179±0.005	0.085±0.01	62.36±1.78	52.89±2.04
1×10^5	0.34± 0.04		0.056±0.004	0.10±0.004	83.52±2.69	43.89±1.63
1×10^6			0.093±0.004	0.103±0.00	72.48±3.18	42.78±2.30
1×10^7			0.1463±0.01	0.096±0.01	56.97±3.54	46.41±3.61
Control			0.206±0.004	0.152±0.01	38.71±7.78	15.37±4.48
Manganese, Mn						
1×10^3	112.33± 5.51	68.3± 1.75	3.15±1.23	50.13±2.01	97.20±1.14	26.60±1.08
1×10^4			3.66±1.52	50.33±2.52	96.74±1.22	26.31±4.63
1×10^5	55.17± 0.76		4.55±0.51	50.73±2.61	91.75±1.00	25.72±3.06
1×10^6			7.4±0.53	50.07±1.10	86.58±0.97	26.70±1.74
1×10^7			7.62±0.54	60.13±2.01	86.18±1.15	11.96±0.72
Control			18.65±0.57	66.13±2.73	66.20±1.48	3.17±1.92

*All experiments conducted in replicates (n=3)

The present study, however, makes several noteworthy contributions to the application of green approach in the removal of metals in wastewater using microalgae, *Botryococcus* sp. is highly influenced by the different microalgae concentration employed. Obviously, each metal element study required a different amount of microalgae cell to perform the bio-removal process effectively. These phenomena may be due to the difference species of microalgae used and also affected by the metal concentration itself in wastewater [23].

Overall, cultivation of *Botryococcus* sp. in contaminated DW for biomass production and simultaneously remove some metal elements (Zn, Fe, Cd, and Mn) was successfully performed. Previously, El-Sheekh et al. [9] found that freshwater microalgae *Chlorella vulgaris* was able to remove Zn, Mn, and Fe concentration in sewage water up to 64.96%, 100%, and 100%, respectively while *Chlorella salina* successfully reduced about 15.6-28.5%, 89.94-93.71%, and 97.24%, respectively after 10 days of treatment. In another study, *Chlorella vulgaris* also successfully reduced the concentration of Zn, Fe, and Mn up to 80.1%, 100%, and 100%, respectively from domestic secondary effluent [24].

In 2014, Onalo et al. [12] investigated the removal of Cd in textile wastewater using *Botryococcus* sp. only up to 2%. Meanwhile, Chan et al. [13] reported that *Chlorella vulgaris* and *Spirulina maxima* capable of reducing Zn concentration in DW up to 96.3% and 94.9%, respectively. Other than that, Hamouda et al. [25] studied removal of heavy metals in industrial wastewater using green microalgae and found that *Scenedesmus obliquus* able to reduce Cd about 70% in the light and dark culture condition. Therefore, heavy metals removal in wastewater mostly depending on the species of microalgae used in which has different accumulation affinities towards the tested elements. Moreover, it is also relying on the nature and charge of the cell wall polysaccharides of microalgae [9]. In fact, green microalgae cells cultivated in wastewater with high metal contents also reduced higher metal concentration [26]. In other words, the metal accumulate was independent on the external metal concentration [9].

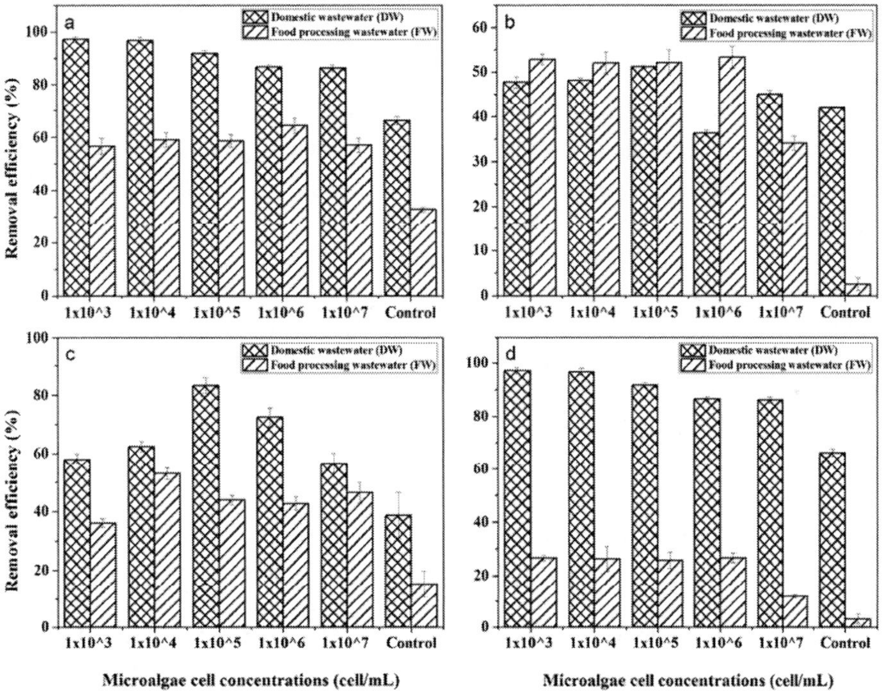

Fig. 1. The comparison of heavy metal removal efficiency between domestic wastewater and food processing wastewater; (a) Zinc, (b) Ferum, (c) Cadmium and (d) Manganese

4 Conclusions and recommendations

The purpose of the present study was to investigate the effect of microalgae; *Botryococcus* sp. cell concentration on the bio-removal of heavy metal from domestic wastewater and food processing wastewater under indoor culture condition. The capability of *Botryococcus* sp. in metals removal is clearly supported by the current findings. However, removal efficiencies are highly depending on the microalgae concentration applied. Most of all concentration employed were successfully removed Zn, Fe, Cd and Mn between 11.96% and 97.2%. More broadly, the investigation is also needed to determine the influence of the application of outdoor culture condition on the bio-removal of metal ions coupled with potential precious biomass production.

The financial contributions provided by Office for Research, Innovation, Commercialization and Consultancy Office (ORICC) of Universiti Tun Hussein Onn Malaysia under grant GPP (Vot B060), STG (1008) and grant FRGS (Vot 1476 and Vot 1479) is gratefully acknowledged. This paper was partly sponsored by the Centre for Graduate Studies UTHM.

References

[1] P. Gani, N. Mohamed, H. Matias-peralta and A.A.A. Latiff, Application of Phycoremediation Technology in the Treatment of Food Processing Wastewater by Freshwater Microalgae *Botryococcus* sp., J. Eng. Appl. Sci., **11**, 7288–7292 (2016)

[2] P. Gani, N.M. Sunar, H. Matias-Peralta, A.A.A. Latiff and S.F.Z.M. Fuzi, Growth of microalgae *Botryococcus* sp. in domestic wastewater and application of statistical analysis for the optimisation of flocculation using alum and chitosan, Prep. Biochem. Biotechnol., **In Press** (2016)

[3] P. Gani, N.M. Sunar, H. Matias-Peralta, A.A.A. Latiff and A.R.A. Razak, Influence of Initial Cell Concentrations on the Growth Rate and Biomass Productivity of Microalgae in Domestic Wastewater, Appl. Ecol. Environ. Res., **14,** 399–409 (2016)

[4] H. Ali, E. Khan and M.A. Sajad, Phytoremediation of heavy metals-Concepts and applications, Chemosphere, **91**, 869–881 (2013)

[5] K. Suresh Kumar, H.-U. Dahms, E.-J. Won, J.-S. Lee and K.-H. Shin, Microalgae - A promising tool for heavy metal remediation., Ecotoxicol. Environ. Saf., **113,** 329–352 (2015)

[6] P. Gani, N.M. Sunar, H. Matias-Peralta, S.S. Jamaian and A.A.A. Latiff, Effects of different culture conditions on the phycoremediation efficiency of domestic wastewater, J. Environ. Chem. Eng., **4,** 4744–4753 (2016)

[7] P. Gani, N.M. Sunar, H. Matias-peralta, A.A.A. Latiff, U. Kalthsom and A.R.A. Razak, Phycoremediation of Wastewaters and Potential Hydrocarbon from Microalgae: A Review, Adv. Environ. Biol., **9,** 1–8 (2015)

[8] I. Rawat, S.K. Gupta, A. Shriwastav, P. Singh, S. Kumari and F. Bux, Microalgae Applications in Wastewater Treatment, in: Y. Chisti (Ed.), Algae Biotechnol. Green Energy Technol., Springer International Publishing Switzerland, 249–268 (2016)

[9] M.M. El-Sheekh, A.A. Farghl, H.R. Galal and H.S. Bayoumi, Bioremediation of different types of polluted water using microalgae, Rend. Lincei., **22**(2), 401 - 410 (2015)

[10] S. Dwivedi, Bioremediation of Heavy Metal by Algae: Current and Future Perspective, J. Adv. Lab. Res. Biol. **III,** 195–199 (2012)

[11] K. Ben Chekroun and M. Baghour, The role of algae in phytoremediation of heavy

metals : A review, J. Mater. Environ. Sci., **4,** 873–880 (2013)

[12] J.I. Onalo, H.M.M. Peralta and N. Mohamed Sunar, Growth of Freshwater Microalga, *Botryococcus* sp. in Heavy Metal Contaminated Industrial Wastewater, J. Sci. Technol., **6**, 29–40 (2014)

[13] A. Chan, H. Salsali and E. McBean, Heavy Metal Removal (Copper and Zinc) in Secondary Effluent from Wastewater Treatment Plants by Microalgae, ACS Sustain. Chem. Eng., **2,** 130–137 (2014)

[14] T. Soeprobowati and R. Hariyati, Phycoremediation of Pb, Cd, Cu, and Cr by Chaetoceros calcitrans (Paulsen) Takano, Int. J. Adv. Chem. Eng. Biol. Sci., **1,** 37–40 (2014)

[15] A. Worku and O. Sahu, Reduction of Heavy Metal and Hardness from Ground Water by Algae, J. Appl. Environ. Microbiol., **2,** 86–89 (2014)

[16] O.M. Krustok I and Nehrenheim E, Cultivation of Microalgae for Potential Heavy Metal Reduction in a Wastewater Treatment Plant, in: Int. Conf. Appl. Energy, Suzhou, China (2012)

[17] APHA, *Standard Methods for the Examination of Water and Wastewater: American Public Health Association*, AWWA (American Water Works Association), Wasshington DC, USA, (2012)

[18] P. Gani, N.M. Sunar, H. Matias-Peralta, A.A.A. Latiff, I.T.K. Joo, U.K. Parjo, Q. Emparan and E.C. Ming, Phycoremediation of Dairy Wastewater by Using Green Microalgae : Botryococcus sp ., Appl. Mech. Mater., **773–774,** 1318–1323 (2015)

[19] Environmental Act Quality, Effluent Standards, Department of Environment, Ministry of Natural Resources & Environment, Malaysia (1974).

[20] T.-Y. Zhang, Y.-H. Wu, S. Zhu, F. Li and H.-Y. Hu, Isolation and heterotrophic cultivation of mixotrophic microalgae strains for domestic wastewater treatment and lipid production under dark condition., Bioresour. Technol., **149,** 586–595 (2013)

[21] S.S.M. Mostafa, E.A. Shalaby and G.I. Mahmoud, Cultivating Microalgae in Domestic Wastewater for Biodiesel Production, Not. Sci. Biol., **4,** 56–65 (2012)

[22] L. Creswell, Phytoplankton Culture for Aquaculture Feed, Florida, US (2010)

[23] K.V. Ajayan, M. Selvaraju, Heavy metal induced antioxidant defense system of greem microalgae and its effective role in phytoremeduation of tannery effluent, Pakistan J. Biol. Sci., **15**, 1056–1062 (2012)

[24] F. Gao, C. Li, Z. Yang, G. Zeng, J. Mu, M. Liu and W. Cui, Removal of nutrients, organic matter, and metal from domestic secondary effluent through microalgae cultivation in a membrane photobioreactor, J. Chem. Technol. Biotechnol., **In Press,** (2016)

[25] R.A. Hamouda, D.S. Yeheia, M.H. Hussein and H.A. Hamzah, Removal of Heavy Metals and Production of Bioethanol by Green Alga *Scenedesmus obliquus* Grown in Different Concentrations of Wastewater, Sains Malaysiana, **45,** 467–476 (2016)

[26] I. Priyadarshani, Sahu, D. and R. Biswajit, Microalgal bioremediation: Current practices and perspectives, J. Biochem. Technol., **3,** 299–304 (2011)

Preparation of Palm Oil Mill Effluent Sludge Biochar for the Treatment of Landfill Leachate

Mohammed J.K. Bashir[1]*, *Jing Wei Wong*[1], *Sumathi Sethupathi*[1], *Ng Choon Aun*[1], *Lim Jun Wei*[2]

[1]Department of Environmental Engineering, Faculty of Engineering and Green Technology (FEGT), Universiti Tunku Abdul Rahman, 31900 Kampar, Perak, Malaysia.
[2]Department of Fundamental and Applied Sciences, Universiti Teknologi PETRONAS, 32610 Seri Iskandar, Perak Darul Ridzuan, Malaysia

Abstract. In recent years, landfilling has emerged as the most widely employed method for municipal solid waste disposal especially in ASEAN countries. However, due to the presence of refractory compounds in stabilized landfill leachates which results in the inefficiency of biological treatment methods, other alternatives such as physico-chemical treatment methods are being look into and adsorption method seems promising. Nevertheless, the raw materials used for the preparation of adsorbents in the industry are usually non-renewable materials thus resulting in high operational costs. Therefore, present study evaluates the ability of palm oil mill effluent sludge (POMES) derived biochar in the treatment of chemical oxygen demand (COD), color and ammoniacal nitrogen (NH_3-N) from stabilized landfill leachate obtained from Sahom landfill located in Kampar, Perak, Malaysia using response surface methodology (RSM) technique. It was found that the removal efficiency of the leachate pollutants by POMES biochar is highly dependent on the production process such as temperature and holding time. In this study, 44.26% of COD, 70.20% of color and 30.56% of NH_3-N were removed from stabilized landfill leachate via optimized biocharat452°C temperature and 66 minutes of holding time.

1 Introduction

From time to time, Malaysia faces serious complications with its municipal solid waste (MSW) originated from various sources such as commercial, residential, agricultural and industrial. The trend of the waste produced annually is increasing at an alarming rate which signifies the need for MSW management and a compelling strategy to overcome the current situation. The scenario worsens with the continual increase in both economy and population growth which means more waste to be produced. The waste generation in Malaysia is expected to rocket up from 292 kg/capita in 2000 to 511 kg/capita in 2025 and at the time being, landfilling of waste is still the main waste disposal solution for Malaysia which accounts for up to 80% of total waste disposal methods [1]. In the year 2020, landfilling is

*Corresponding author: jkbashir@utar.edu.my

still expected to uphold about 65% of total waste disposal methods in the country according to the 10th Waste Management Policy of Malaysia Plan, 2010 to 2020. The popularity of using this method for waste disposal is due to certain advantages such as low cost and easy disposal, however the long-standing drawback associated with municipal landfill is the self-generation of highly contaminated leachate which is formed by the mixing of water sources with the wastes in the landfill either through an open landfill or the top cap of a complete landfill. This highly contaminated water may contain large quantities of ammonia, heavy metals, suspended solids, dissolved organic matter, xenobiotic organic substances, phenols and many more [2]. The high strength of COD and NH_3-N resulted from the decomposition of organic matters are considered the best way to characterize and analyse a leachate sample due to the fact that these are the pollutants that play the role for main chemical features of an environmental threat. As a result, stringent discharge standards have been imposed by the Malaysian government which leaves conventional treatments insufficient to fulfil the stipulated requirement. Therefore this leads to various leachate treatment techniques being employed worldwide.

In spite knowing the fact that the whole treatment process for landfill leachate is an extremely sophisticated procedure and requires high operational costs, leachate treatment is being done and widely used to minimize its environmental impacts.Different processes have been employed for example chemical oxidation, coagulation-flocculation, ion exchange, adsorption and electrochemical treatment. Of all, adsorption process is the most extensively used technique worldwide when it comes to the removal of recalcitrant organic compounds from landfill leachate [2]. Having certain significant properties such as highly porous structure, low acid/base reactivity, thermo-stability, high surface reactivity, relatively low cost involvement, adsorption process has emerged as the most promising technique in removing landfill leachate pollutants having abilities to attract pollutants onto and into its surface using chemical and physical interactions. Adsorbents that are commercially available in the market are expensive and mostly derive from non-renewable sources such as coal, so the use of conventional materials such as agricultural waste or biomasses that are readily available locally in large quantity such as bamboo, pecan shell, oil palm fibre and sugarcane bagasse could be the solution to this matter. These materials can be chemically and physically modified to become a new adsorbent with minimum expenditure.

Up to date, no study has been made on utilizing POMES-derived biochar for the treatment of highly contaminated landfill leachate. Therefore, due to the promising adsorption properties of biochar produced from biomass for example POMES, this study focuses on the preparation and effectiveness of using POMES biochar to remove the complex and highly contaminated pollutants from landfill leachate.

2 Materials and methods

2.1 Site characteristic and leachate sampling

The leachate samples were collected from Sahom landfill site, located in Kuala Dipang in Perak, Malaysia. The operation of Sahom landfill has begun ever since 1992 with a total landfilling area of approximately 37 acres. The anaerobic landfill mainly serves the waste collected within Kampar district and has recorded a population service to a total of 101,183 people with an average of 100 tonnes of MSW received daily. The collected leachate samples were instantaneously transported to the Environmental Laboratory and stored at 4°C in minimal light and air exposure condition in order to reduce the chemical and biological changes. The leachate was characterized according to the Standard Method for the Examination of Water and Wastewater [3]. In total, three samples were collected and characterised from October 2015-March 2017.For chemical oxygen demand (COD),ammoniacal-nitrogen (NH_3-N), color, Biochemical Oxygen Demand (BOD_5), turbidity, total suspended solids (SS) and pH as presented in Table 1. Based on leachate characteristics indicted in Table 1, Sahom landfill leachate is classified as stabilised leachate.

Table 1. Sahom landfill leachate characteristics.

Parameters	Average Value	Malaysian Discharge Standards for Leachate
Temperature	24.8°C	40°C
pH	8.06	6.0-9.0
Turbidity	207 NTU	-
BOD_5	37 mg/L	20 mg/L
COD	549 mg/L	400 mg/L
BOD_5/COD	0.067	0.05
Color	1020 Pt/Co	100 Pt/Co
Total Suspended Solids	95 mg/L	50 mg/L
NH_3-N	144 mg/L	5 mg/L

2.2 Char preparation

The palm oil mill effluent sludge (POMES) was collected from Tian Siang Oil Mill Sdn. Bhd. Located in Air Kuning, Perak, Malaysia. The POMES was obtained from the anaerobic digestion system and was evenly spread on 3 metal trays and air-dried for 3 days to remove bad odor and moisture. The POMES was later screened to the desired mesh size of 1-2mm and further dried in an oven at 105°C for 24 hours to remove the water content because presence of high moisture content in the sample will result in production of fuel gases and affect the production of biochar [4]. Then, the dried POMES were allowed to cool down before being stored in a dry bucket and sealed up to minimize moisture contamination and fungal infection.

The prepared precursor was later measured for weight and loaded in a stainless steel vertical tubular and placed in the tube furnace in the carbonization unit. The carbonization process was conducted under purified nitrogen (N_2) supply with the flow rate at 300ml/min using control valves. The range of pyrolysis temperature (300-600°C) and holding time (60-180 min) were taken on the basis of literature [5]. Carbonization process was carried out based on the experimental design by a standard RSM method.

2.3 Design of experiment (DOE)

Design-Expert ® Software Version 10.0 by Stat-Ease, Inc., Minneapolis, USA was used to generate a series of experiments to be conducted by the different combinations of experimental variables using a standard response surface methodology (RSM) technique known as central composite design (CCD). RSM utilizes the combination of statistical and mathematical workings to optimize and design an experiment based on numerous experimental variables with minimum amount of experimental runs and analyse the relationship between the dependent and independent variables [6].CCD was used to verify the effective parameters with definite number of experiments and analysed the interface between the parameters. By using CCD the experiment could be designed based on 2^n factorial runs with 2naxial runs and $n_c$center runs (replicate experimental runs), whereby n indicates the number of factors involved. In this study, POMES biochar preparation conditions were found to be important criteria affecting the characteristics of biochar produced. The two parameters under analysis include heating temperature and holding time. Therefore the full experimental matrix consists of 4 factorial points, 4 axial points with 5 replicates at the center points summing up to 13 experimental runs as denoted by using Equation 1:

$$N = 2^n + 2n + n_c = 2^2 + 2(2) + 5 = 13 \text{ runs} \quad (1)$$

In Equation 1, N is the total number of experimental runs, and n is the number of affecting factors.Numerous variables may affect the response of the system studied and it is practically impossible to identify and control the small contributions from each one. Therefore, the independent variables are coded as $-\alpha$, -1, 0, +1,$+\alpha$ interval which indicates the low, center and high level of each variableswhere α is the axial points from the center which makes the design rotatable.

2.4 Model fitting and statistical analysis

The analysis of variance (ANOVA) is a reliable way to evaluate the quality of a fitted model by examining the statistical significance of the equations as well as the degree of fitness of the equations. It depicts the co-relation between the actual results obtained with the equation which encapsulates all the responses and variable denoted in the equation. As replicates of the central point are made, it is possible to calculate the pure error associated with the repetitions since variation occurs due to pure error which could be defined as unpredictable changes and variables that occur during the experiment. In general, a model will be well fitted to the experimental data if it presents a significant regression and a non-significant lack of fit.

2.5 Batch equilibrium studies

The prepared biochar under respective variables was added into the leachate samples to identify the efficiency of the biochar. 0.2 g of biochar was added into a 250mL Erlenmeyer Flask with 50mL of leachate sample and was shaken for 3 hours with rotating speed of 200rpm using the orbital shaker (IKA KS 260 B S2, Germany). The mixture was then filtered using filter paper with pore size of 90μm. The filtrate was then tested for the concentration of COD, color, and NH_3-N to determine the removal efficiency of biochar. The removal efficiency of the color, COD, ammoniacal-nitrogen were identified using Equation 2. Each experiment was repeated for three times to obtain an average reading.

$$R(\%) = \frac{C_i - C_f}{C_i} \times 100 \quad (2)$$

In Equation 2, *R* is the removal efficiency in percentages and C_i and $C_f$are the initial and final concentrations of COD (mg/L), NH_3-N (mg/L) and color (PtCo).

3 Result and discussion

3.1 Leachate characteristics

The initial measurements recorded high values of COD, color and ammoniacal-nitrogen values. The BOD_5 value is relatively low and thus resulting in low BOD_5/COD ratio. Due to these characteristics, Sahom landfill leachate was thus classified as matured landfill leachate with pH value larger than 7.5, COD reading lower than 4000mg/L, BOD_5/COD ratio below 0.1 and heavy metal concentration lower than 2ppm. The organic components present in the stabilized landfill leachate were mostly non-biodegradable as they contain refractory compounds. As such, conventional biological treatment methods are no longer adequate and efficient to be utilized for Sahom landfill leachate treatment, therefore physicochemical process in particular carbon adsorption becomes one of the compatible alternative for stabilized landfill leachate treatment.

3.2 Preparation of POMES derive biochar using (RSM)

The responses (pollutant percentage removal) for respective experimental runs were examined and tabulated in Table 2. The equations with coded factors used by the software to estimate and maximize the removal efficiency of the dependent responses COD removal (Y_1), color removal (Y_2) and NH_3-N removal (Y_3) with pyrolysis temperature (X_1) and holding time (X_2) are given in Equations 3, 4 and 5:

$$Y_1 = 41.76 + 6.77*(X_1) - 0.90*(X_2) + 0.080*(X_1)*(X_2) - 17.20*(X_1^2) + 3.94*(X_2^2) \quad (3)$$

$$Y_2 = 69.29 + 1.88*(X_1) - 0.99*(X_2) + 0.075*(X_1)*(X_2) - 8.81*(X_1^2) + 2.17*(X_2^2) \quad (4)$$

$$Y_3 = 26.53 - 0.19*(X_1) - 1.41*(X_2) - 2.50*(X_1)*(X_2) - 4.62*(X_1^2) + 1.72(X_2^2) \quad (5)$$

Positive terms represent synergistic effect, whereas negative terms represent antagonistic effect [6]. The significance of quadratic models for all responses was examined using ANOVA and the results are presented in Table 3.

Table 2. Experimental Design Matrix coupled with respective parameter readings.

	Factors		Response (Removal Efficiency, %)		
Experimental Run	**Temperature (°C)**	**Holding Time (min)**	**COD, (mg/L)**	**Color, (PtCo)**	**NH3-N, (mg/L)**
1	300	60	22.94	61.83	25.56
2	300	180	21.65	59.94	22.56
3	375	120	30.93	65.32	24.32
4	450	90	42.78	70.14	26.91
5	450	120	43.40	69.90	26.32
6	450	120	42.53	69.57	27.82
7	450	120	43.40	69.90	26.05
8	450	120	42.53	69.57	26.37
9	450	120	43.30	69.90	27.46
10	450	150	39.18	68.16	26.25
11	525	120	40.46	67.49	25.67
12	600	60	35.64	65.37	24.81
13	600	180	34.67	63.78	21.80

Table 3. ANOVA analysis for response surface quadratic model ANOVA

	Source	Sum of Square	df *	Mean Square	F value	p-value	Remarks
COD	Model	678.09	5	135.62	32.37	0.0001	Significant
	Lack of Fit	28.62	3	9.54	53.63	0.0011	Significant
Color	Model	138.55	5	27.71	51.64	<0.0001	Significant
	Lack of Fit	3.63	3	1.21	36.99	0.0022	Significant
NH_3-N	Model	32.67	5	6.53	9.04	0.0058	Significant
	Lack of Fit	2.61	3	0.87	1.42	0.3615	Not Significant

Note: df*: Degree of freedom

3.3 Removal of pollutants by POMES derived biochar

The removal efficiencies of COD, color and NH_3-N in relation with pyrolysis temperature and holding time are as shown in Fig. 1 (a-c). The higher the temperature applied, the more complete is the carbonization process and therefore the surface area is increased as well as the development of pores within the biochar. When the temperature is high, the destruction of aliphatic alkyl as well as the ester groups takes place thus exposing the aromatic lignin core. This might be due to the escape of volatile substances inclusive of hemicellulose, cellulose and lignin from the biomass and forming channel structures during pyrolysis which in turn facilitate the increase of surface area and porous structure of biochar[7-8]. The pore-filling mechanism for organic pollutants is in relation to its adsorption capability which in turn is directly proportional to its microporous surface area. The greater the pore volumes and surface area promoted, the higher the adsorption for organic matters.

However, when the temperature is increased after the peak at 450^{0}C, the removal efficiency starts to drop. This is because the surface of the biochar has ruptured or sealed off due to the high temperature resulting in blockage of pores for adsorption [7]. Holding time does not play significant effect on the removal of organic pollutants but the removal efficiency is observed to decrease with increase of holding time. More time is allowed for the heat energy to break the surface bonds on the biochar thus reducing the functional groups available for interactions. However, noticing that the removal efficiency slightly increases when the holding time is approaching 180 minutes. At high temperatures, volatile organic matters will be removed from the biochar thus results in unblocking of the pores. Therefore, the slight increase in removal efficiency may be due to the fact that volatile matters are being removed from the pores when sufficient holding time is applied. Based on Fig. 1 (a-c), the constructed quadratic models readily fit with the experimental results.

The model fit summary of COD, color and NH_3-N is to determine the degree of fitness of respective responses. The goodness of fit was proven by the determination of coefficient R^2 of 0.9585, 0.9736 and 0.8659 for the COD, color and NH_3-N respectively. A good fit model should have R^2>0.8, thus this shows that the proposed models are fit [7]. R^2values closer to 1 will result in individual quadratic models to be statistically desirable, better fit and reasonable agreement to experimental data.The experimental conditions with the highest desirability were selected to be verified.

DOI: 10.1051/matecconf/201710306008

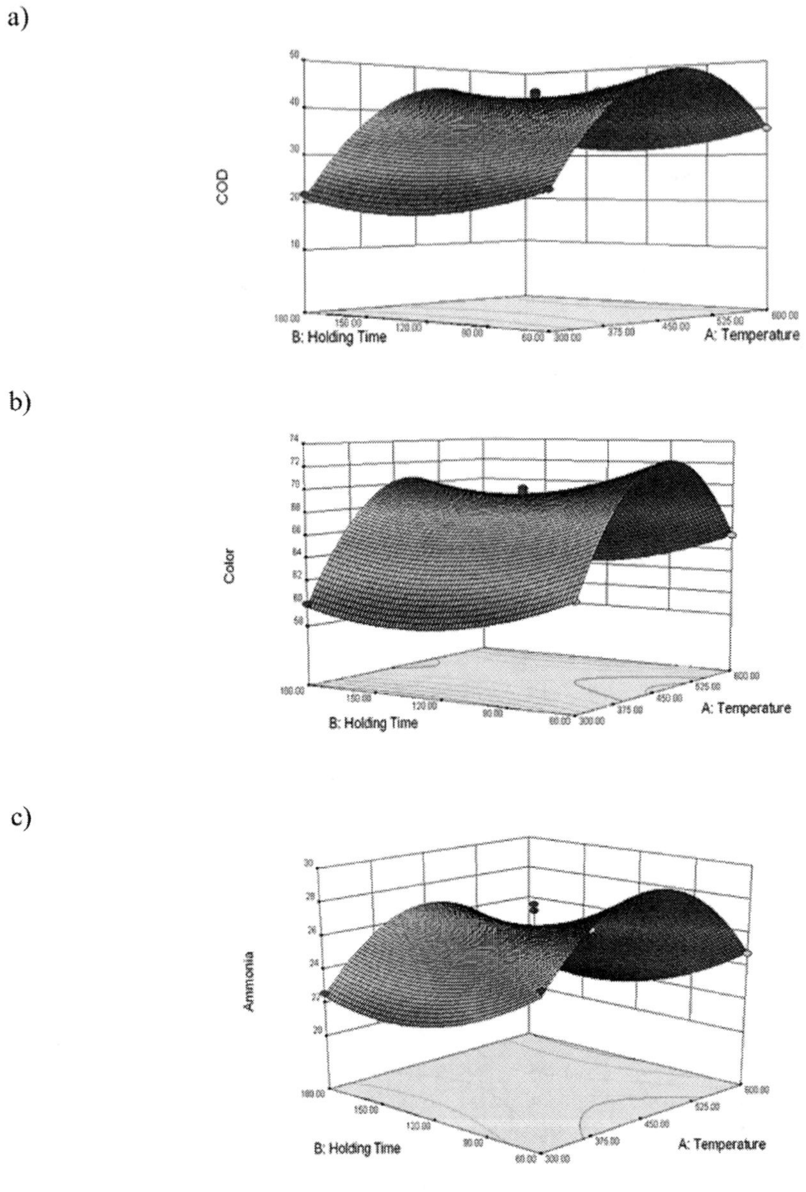

Fig.1. Three-dimensional response surface plot for removal efficiency of (a) COD, (b) color and (c) NH_3-N

3.4 Process optimization

Optimization was done simultaneously in this study by using Design-Expert ® Software 10.0. One solution was chosen out of 28 solutions proposed by the software by taking into consideration of the solution desirability and also taking into account the removal efficiency as well. The closer the solution desirability to 1.0 indicates the closer the predicted responses to the target requirements. The independent variables proposed were taken into consideration as well to help minimize operational cost but at the same time with high removal efficiency of the leachate pollutants. The experiment was repeated by using the optimized values for the independent variables to check and compare between the actual

 DOI: 10.1051/matecconf/201710306008

and predicted removal efficiency. Based on Table 4, only a small percentage error was observed between the actual and predicted results under optimum conditions. The optimum operational conditions of POMES biochar for leachate treatment were found to be at 452°C of pyrolysis temperature and 66 minutes of holding time with pollutant removal percentages for COD, color and NH_3-N of 44.26%, 70.20% and 30.56%.

Table 4. Predicted and experimental result of POMES biochar at optimum conditions

Response Parameter	Actual Removal Efficiency, %	Predicted Removal Efficiency, %	Percentage error,%
COD	44.26	45.89	3.55
Color	70.20	71.93	2.41
NH_3-N	30.56	29.22	4.38

4 Conclusions

In this study, complicated landfill leachate from Sahom landfill was treated by biochar produced from POMES from Tian Siang Oil Mill Sdn. Bhd. with the removal percentage of 44.26% of COD, 70.20% of color and 30.56% of NH_3-N under optimal conditions. The optimization process of biochar preparation was done by using RSM and the optimum conditions obtain were 452°C pyrolysis temperature and 66 minutes of holding time. It can be observed from the series of experiments done that the removal capability of biochar is greatly dependent on the method and parameters used to produce the biochar which directly and indirectly affects its porous structure. It can be concluded that the main factor affecting the adsorption efficiency of POMS biochar is the pyrolysis temperature as compared to holding time with only minor effects on its adsorption capability.

References

[1] S.N.S. Ismail andL.A. Manaf.The challenge of future landfill: A case study of Malaysia, J. Tox. Environ. Health Sci., **5**(6), 86-96,(2013)

[2] K.Y. Foo., B. H.Hameed,An overview of landfill leachate treatment via activated carbon adsorption process, J. Hazard. Mater,**171**, 54-60, (2009)

[3] American Public Health Association (APHA), *AWWA &WEF*, 21stEdition,(2005)

[4] D. Beneroso., J.M. Bermudez., A. Arenillas, J.A. Menendez ,Integrated microwave drying, pyrolysis and gasification for valorisation of organic wastes to syngas, Fuel, **132,** 20-26, (2014)

[5] S. L. Low, *Preparation, characterization and utilization of palm oil mill sludge biochar for copper removal.* Undergraduate Thesis,Universiti Tunku Abdul Rahman, (2015)

[6] N. B. Azmi, M.J.K. Bashir., S. Sethupathi and C. A. Ng., Anaerobic stabilized landfill leachate treatment using chemically activated sugarcane bagasse activated carbon: kinetic and equilibrium study, Desalination and WaterTreatment,**57**, 3916-3927, (2016)

[7] A. Shehzad.,M.J.K. Bashir., S. Sethupathi and J.W. Lim,An insight into the remediation of highly contaminated landfill leachate using sea mango based activated bio-char: optimization, isothermal and kinetic studies. Desalination and Water Treatment, **57**(47),22244-22257, (2016)

[8] M.B Ahmed Zhou, H.H. Ngo.and W. Guo, Insight into biochar properties and its cost analysis, Biomass and Bioenergy, **84**,76-86, (2016).

Optimization of pH in Ammonia, Heavy Metals, and Chlorophyll-a Removal by using Electric Arc Furnace Slag and Activated Carbon Cassava Peel

Nabilah Zayadi[1], *Shareen* Hwong Shiaw Ling[1], *Norzila* Othman[1,*] and *Rafidah* Hamdan[1]

[1]Department of Water and Environmental Engineering, Faculty of Civil and Environmental Engineering, University Tun Hussien Onn Malaysia

Abstract. The wastewater contain high concentrations of inorganic and organic pollutant and that could be toxic to aquatic life and may lead to toxic environment of the water bodies. The poolutants able to accumulate in living tissues, causing various illnesses. In this study, parameters that affected the removal of heavy metals, ammonia (NH_4-N) and chlorophyll-a were investigated in a synthetic water of mixed metals, NH_4-N, and alga synthetic solution. The removal was strongly affected by pH by using activated carbon cassava peel (CPAC) and Electric Arc Furnace Slag (EAFS) as media For CPAC, optimum condition for removing heavy metals, NH_4-N and chlorophyll-a can be achieved at pH 7.5 where removal up to 95% for heavy metals, 49.2% for NH_4-N and 56.8% for chlorophyll-a. While for EAFS, the optimum condition for removing heavy metals, NH_4-N and chlorophyll-a can be achieved at pH 7.5 where removal up to 93% for heavy metals, 77.7% for NH_4-N and 80.7% for chlorophyll-a. In order to obtain the maximum removal of heavy metals, NH_4-N and chlorophyll-a, two conditions obtaining CPAC and EAFS were considered. The optimum condition for removing heavy metals, NH_4-N and chlorophyll-a by 1/3 CPAC + 2/3 EAFS was pH 7.5 as well where removal up to 95% for heavy metals, 86.2% for NH_4-N and 90.9% for chlorophyll-a. By using 1/3 EAFS + 2/3 CPAC, the optimum removal of heavy metals, NH_4-N and chlorophyll-a was at pH 7.5 where removal up to 96% for heavy metals, 80.7% for NH_4-N and 86.4% for chlorophyll-a. This study highlighted that EAFS and CPAC peel are promising method in removing Zn, Al, Fe ions, NH_4-N and chlorophyll-a from aqueous solution.

1 Introduction

One of the major concerns of this century is the preservation of environmental quality. Water contamination and shortage of clean water is one of the most urgent problems for human activities. The water quality degenerated caused by enrichment of nutrients and heavy metals which help the growth of algae species and lastly cause eutrophication which

* Corresponding author: norzila@uthm.edu.my

defined as the excessive enrichment of waters by nutrients are widespread in lakes and reservoirs throughout the world [1]. In order to eliminate targeted heavy metals, ammonia and chlorophyll-a, biosorbent with most favorable and hold high composition of elements and functional groups were selected [2-4]. Maximum biosorption particularly occurs in weak acidic pH because of the involvement of carboxyl group and other acidic functional groups. These groups were responsible for binding metal cations through various mechanisms [5]. Moreover, the presence of Si, Ca, Mg, Fe, rich in functional groups and surface porosity of media were significant factors which attribute to facilitate heavy metals and nutrient removal [6]. Study by Nabilah *et al.,* [6] supports the potential use of cassava peel and steel slag through characterization of medias by XRF, FTIR, FESEM-EDX, XRD and TG analysis.

EAFS was used to remove phosphate and ammonia nitrogen from wastewater. EAFS has benefits such as loose structure, large pores, stable chemical composition, and it contains the free CaO, MgO and Fe and therefore effect filtration, adsorption, and precipitation on the pollutants (ammonia, chlorophyll-a and heavy metals) in the water [7, 8]. CPAC is an eventual cheap biosorbent for metal ions sequestration. Therefore, this paper will evaluate the potential of EAFS and CPAC in removing ammonia, heavy metal and chlorophyll-a.

2 Materials and Methods

2.1 Preparation of CPAC

Cassava peel was gathered at Kilang Kerepek Ahmad Shah, Parit Raja Darat, Johor. Raw fruit peels were soaked together with distilled water for extra clean up for a minimum of 24 hours and dried in an oven overnight. Cassava peel was pre-treated with 5% of nitric acid (HNO_3) for 6 hours to get rid of attached dust and soluble impurities from the surface. Next, it was oven-dried for 24 hours and continues to be for impregnation process. Prior to carbonization, the dried peel was impregnated using phosphoric acid 30% v/v for 6 hours. The impregnated peel then oven-dried at 110°C and kept for carbonization process [9]. Impregnated peel using phosphoric acid 30% (v/v) was used in the carbonization. This process was done in muffle furnace at temperature of 380°C with a heating rate of 5°C/min at 5 min holding time. The soaking duration at final temperature was 30 minutes. As the starting material is non-woody originates, the temperature-ramp was conducted at the rate of 5°C /min at 5 minutes holding time.

Finally, the activated samples were cooled to room temperature before it was taken out from the furnace [9]. The samples required to wash to reduce its pH and to eliminate ash from the surface of AC. The samples were washed batch wise at room temperature with hot water 70°C several times followed by cool distilled water until the filtrate reached approximately pH 6-7 [10]. After that, the washed samples were oven-dried at 85°C until constant weight [9]. CP-AC were then sieved to get particle size between 0.833 mm – 2.36 mm (8/20 mesh size). Finally, the samples were kept in tight bottle for analysis.

2.2 Preparation of EAFS

Electric Arc Furnace Slag (EAFS) were collected at Antara Steel Mills, Pasir Gudang, Johor. EAFS were used to remove total nutrient in this study. XRF and SEM-EDX analysis were performed to analyze chemical compositions and elemental distributions, respectively. EAFS were washed with tap water two times and then followed by distilled water once. This process was done in order to remove any dirt or particulate that adheres to the media before being used for batch study. The washed media were oven dried at 105°C for 24 hours and sieved to obtain 10 – 20 mm of particle size [11].

2.3 Experimental design in batch study for optimization of pH

All experiments were conducted using the batch technique by differing pH of mixed solution. Synthetic solution were prepared in mixed of Fe, Zn, Al ions, NH_4-N, and alga synthetic solution. The working condition of concentrated NH_4-N, heavy metals, and chlorophyll-a were selected based on results of Sembrong's water characterization. Each experiment was conducted in triplicate with three blank samples.

In this study, 100 ml of initial concentration (NH_4-N, heavy metals and alga) solution were prepared for each experiment. Biosorbent with aqueous solution was agitated at 125 rpm [12]. The samples were then centrifuged at 1400 rpm within five minutes. At the end of each experiment, the biosorbent that had been filtered through 0.45 µm filter paper will be kept in an airtight container. Finally, the supernatants were kept in airtight plastic bottles and analyzed to perform NH_4-N and heavy metals while chlorophyll-a analysis were run immediately or stored in amber glass covered with aluminium foil. Sampling, sample preservation and laboratory analysis were based on Standard Methods for the Examination of Water and Wastewater [13]. Heavy metals were analyzed by Inductively Coupled Plasma-Mass Spectrometer (ICP-MS), NH_4-N by distillation unit, and alga were measured as chlorophyll-a content .

2.4 Effect of pH on heavy metal ions, ammonia, and chlorophyll-a removal

In this study, pH (NH_4-N, Fe, Zn, Al, and alga) solution were optimized by using batch study. Table 1 shows that working range of pH in synthetic solution.

Table 1. Working range of pH in synthetic solution.

Biosorbent	Parameter Test	Constant Parameter						
	pH	Time (hr)	Dosage (g)	Initial Concentration (mg/L)				
				NH_4	Fe	Zn	Al	Algae (cell/ml)
CPAC	3.0, 3.5, 4.0, 4.5, 5.0, 5.5, 6.0, 6.5, 7.0, 7.5, 8.0, 8.5, 9.0	2	5	1.5	0.115	0.010	0.025	1000
EAFS	3.0, 3.5, 4.0, 4.5, 5.0, 5.5, 6.0, 6.5, 7.0, 7.5, 8.0, 8.5, 9.0	2	20	1.5	0.115	0.010	0.025	1000
1/3 CPAC + 2/3 EAFS	3.0, 3.5, 4.0, 4.5, 5.0, 5.5, 6.0, 6.5, 7.0, 7.5, 8.0, 8.5, 9.0	2	1.7 CPAC 13.3 EAFS	1.5	0.115	0.010	0.025	1000
1/3 EAFS+ 2/3 CPAC	3.0, 3.5, 4.0, 4.5, 5.0, 5.5, 6.0, 6.5, 7.0, 7.5, 8.0, 8.5, 9.0	2	6.7 EAFS 3.3CPAC	1.5	0.115	0.010	0.025	1000

2.5 Data analysis for biosorption efficiency

The removal efficiency for NH_4-N, heavy metals, and chlorophyll-a is determined using Equation (1). Equation 1 has been widely used and reported in previous studies on biosorption field [12, 14, 15].

$$\text{Percent removal} = \frac{C_0 - C_f}{C_0} \times 100\% \qquad (1)$$

Where:

C_0 = Initial concentration (μg/ L @ mg/L)
C_f = Final concentrations (μg/ L @ mg/L)

3 Result and Discussion

3.1 Effect of pH

The pH is a key factor that should be considered in the removal in heavy metals, NH_4-N and chlorophyll-a. The extent of removal of ions is strongly influenced by the pH of the solution. Figures below showing the relationship graph between pH and percentage removal of heavy metals, NH_4-N and chlorophyll-a by CPAC, EAFS and combination of CPAC and EAFS.

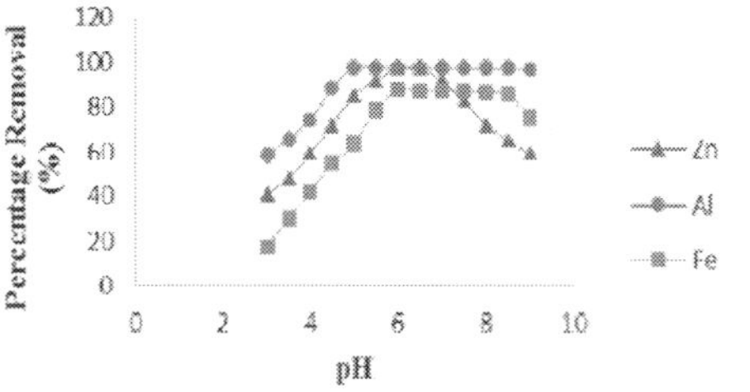

Fig. 1. Effect of pH on Zn, Al, and Fe ions removal by using CPAC

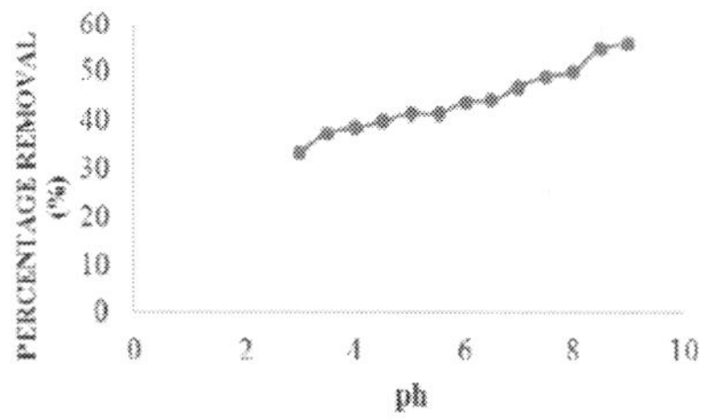

Fig. 2. Effect of pH on NH_4-N removal by using CPAC

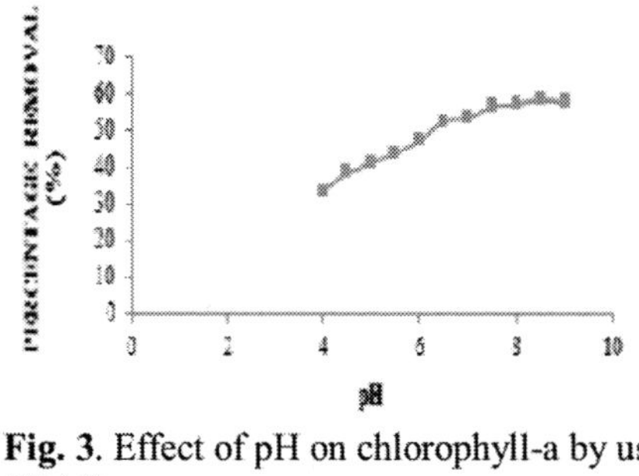

Fig. 3. Effect of pH on chlorophyll-a by using CPAC

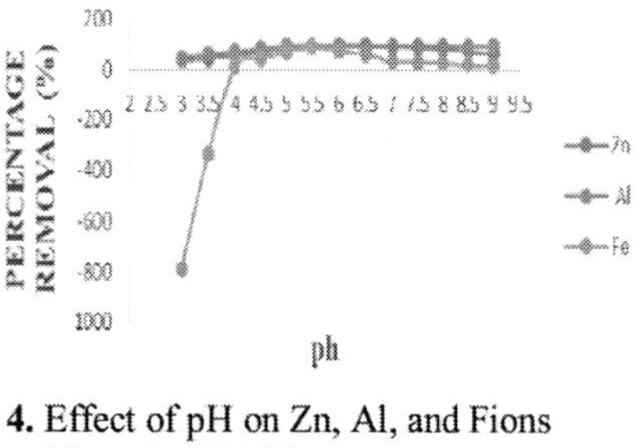

Fig. 4. Effect of pH on Zn, Al, and Fions removal by using EAFS

The effect of pH on removal of heavy metals, NH_4-N and chlorophyll-a by using CPAC were shown in Fig. 1, 2, and 3. The optimum pH was 7.5 that resulted in removal up to 95% for heavy metals, 49.2% for NH_4-N and 56.8% for chlorophyll-a. An increase of percentage removal above pH 7 could be due to the electrostatic repulsion between the negatively charged species and the biosorbent surface. Hence, it can be concluded that an effective removal of heavy metals from aqueous solution can be accompanied by using CPAC and were supported through characterization of cassava peel by XRF, FTIR, FESEM-EDX and TG analysis [6].

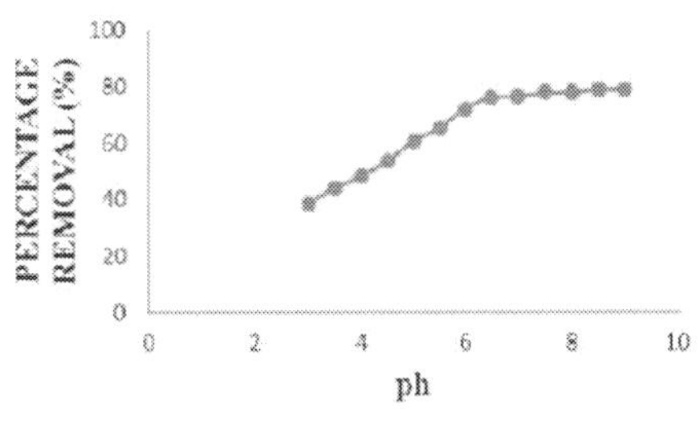

Fig. 5. Effect of pH on NH_4-N by using EAFS

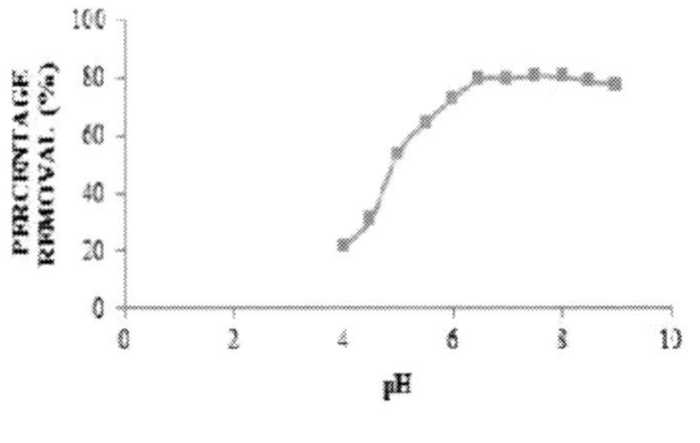

Fig. 6. Effect of pH on chlorophyll-a by using EAFS

Fig. 4, 5 and 6 show removal of metals, NH_4-N and chlorophyll-a by using EAFS. The optimum pH was also 7.5 which resulted to removal up to 93% for heavy metals, 77.7% for NH_4-N and 80.7% for chlorophyll-a. The results of the effect of pH on removal that the percentage removal by NH_4-N is strongly dependent on pH. The effect of pH on removal of heavy metals, NH_4-N and chlorophyll-a by using 1/3 CPAC + 2/3 EAFS and 1/3 EAFS + 2/3 CPAC were shown in Fig. 7, 8, 9 and Fig. 10, 11, 12, respectively.

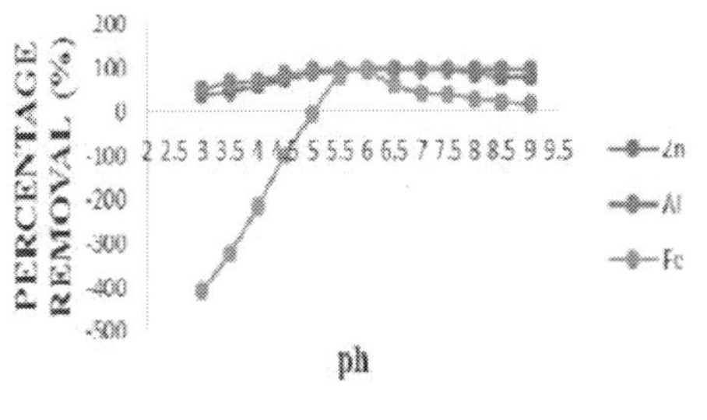

Fig. 7. Effect of pH on Zn, Al, and Fe ions by using 1/3 CPAC + 2/3 EAFS

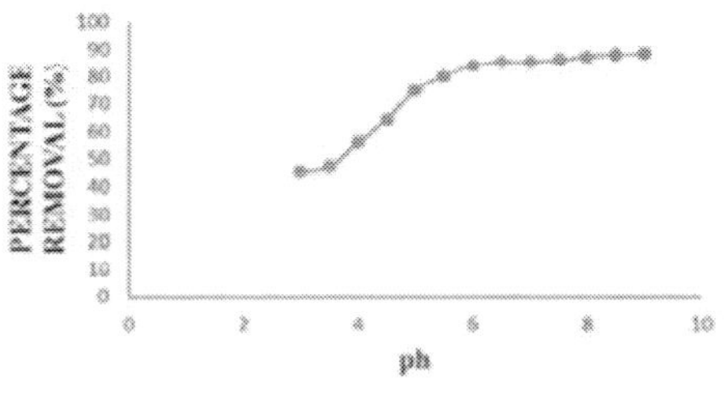

Fig. 8. Effect of pH on NH_4-N by using 1/3 CPAC + 2/3 EAFS

DOI: 10.1051/matecconf/201710306009

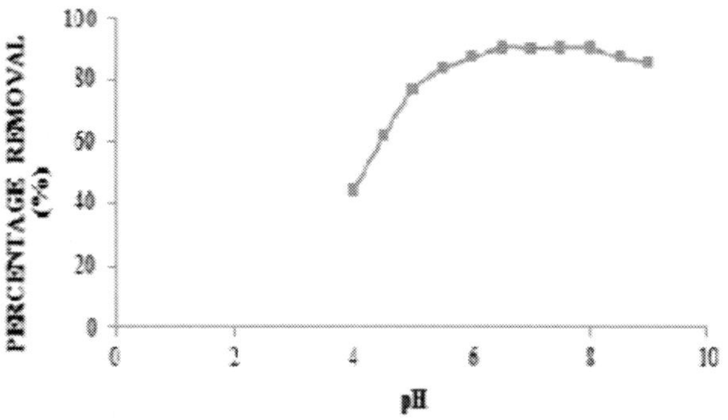

Fig. 9. Effect of pH on chlorophyll-a by using 1/3 CPAC + 2/3 EAFS

The change of surface characteristics can affect removal efficiency to some extent. The surface characteristics of EAFS, such as the surface charge resulted from proton transfers at the surface. The increase of OH− in solution could affect the electrostatic properties of the slag. The presence of –OH on the surface of slag leads to the formation of a new charged layer, making the slag surface being a comparable lower affinity to ammonia.

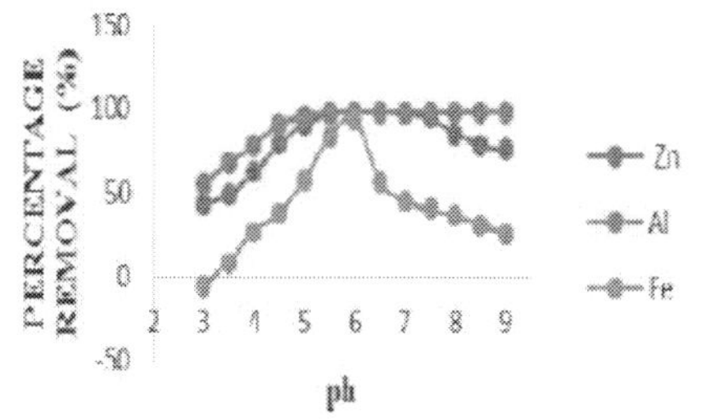

Fig. 10. Effect of ph on Zn, Al, and Fe ions by using 1/3 EAFS + 2/3 CPAC

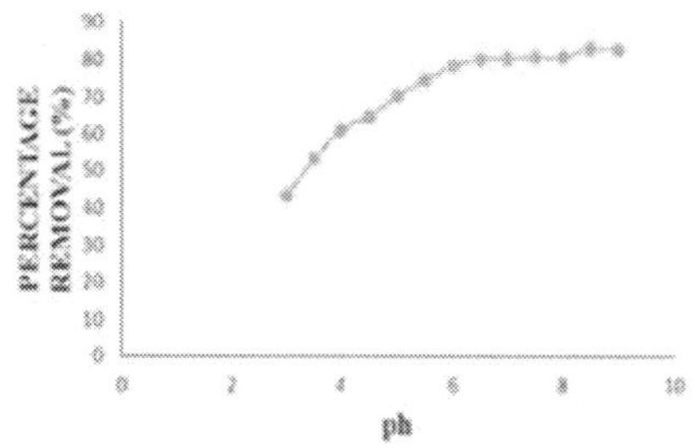

Fig. 11. Effect of pH on NH_4-N by using 1/3 EAFS + 2/3 CPAC

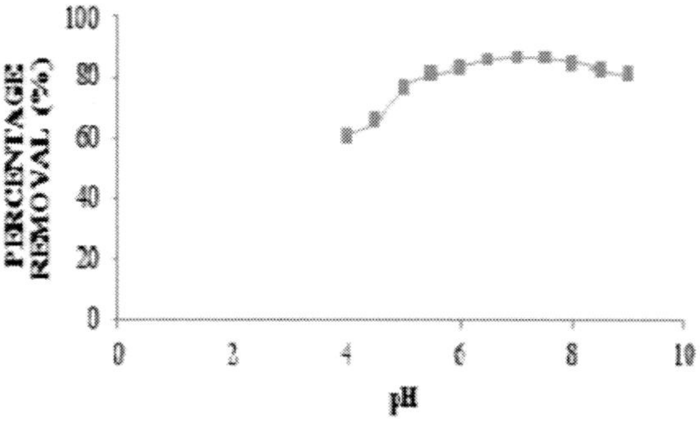

Fig. 12. Effect of pH on chlorophyll-a by using 1/3 EAFS + 2/3 CPAC

With an increase in pH, OH− competes strongly with ammonia for active sites, which affects the removal efficiencies. This therefore implies that removal is favored by high pH values which was pH 7.5.

4 Conclusions

Based on the result obtained, it is proven that the NH_4-N, heavy metals and chlorophyll-a in the water could be removed by using CPAC and EAFS. Overall, all of the optimum pH are in pH 7.5. Condition using 1/3 CPAC + 2/3 EAFS were slightly higher compared to other where removal up to 86.2% for NH_4-N and 90.9% for chlorophyll-a. Meanwhile, condition using 1/3 EAFS + 2/3 CPAC achieved slightly higher in heavy metals removal which is 96%. This study highlighted that EAFS and CPAC are promising method in removing Zn, Al, Fe ions, NH_4-N and chlorophyll-a from aqueous solution. The application of CPAC and EAFS were under range and they are not harmful to aquatic life.

This study was supported by the Fundamental Research Grant Scheme (FRGS Vot 1575) sponsored by Ministry of Higher Education (MOHE).

References

[1] V. H. Smith, Eutrophication of freshwater and coastal marine ecosystems a global problem, Environ Sci Pollut Res Int., **10**(2), 126-39, (2003)

[2] Othman, N., Kueh. S.Y., Azizul Rahman, M-F-H and Hamdan, R, Watermelon rind: a potential adsorbent for zinc removal, Applied mechanics and material, 680, 146-149, (2014)

[3] N. Zayadi and N.Othman, Characterization and optimization of heavy metals biosorption by fish scale, Advanced Material Research, 795, 360-365, (2013)

[4] C. G. Rocha, D. A. M. Zaia, R. V. D. S. Alfaya and A. A. D. S. Alfaya, Use of rice straw as biosorbent for removal of Cu(II), Zn(II), Cd(II) and Hg(II) ions in industrial effluents, J. Hazard. Mater, **166**(1), 383–388, (2009)

[5] N.Othman, A.Che-Azhar, and A. Suhaimi, Zinc removal using honeydew rind, Applied Mechanics and Material, 680, 150-153, (2014)

[6] N. Zayadi, N. Othman, R. Hamdan, A potential waste to be selected as media for metal and nutrient removal. *In IOP Conference Series: Materials Science and Engineering* **136**(1), 012058, (2015)

[7] L.Huang, Vertical undercurrent steel slag applied to artificial wetlands to deal with high ammonia nitrogen sewage test research, China University and Mining Technology, (2009)

[8] H. Motz, J. Geiseler, Products of steel slags : an oppurtunity to save natural resources, Waste Manage, **21**(3), 285-293, (2001)

[9] Z. M. Yunus, N. Othman, R. Hamdan, N. N. Ruslan, Characterization of phosphoric acid impregnated activated carbon produced from honey dew peel, Jurnal Teknologi. **76** (5), 15-19, (2015)

[10] R. G. D. Alvaro, R. Z. Elessandra, H. Elizabete, M. Fernande Aline, F. C. Cesar, Oxidation of fermented cassava starch using hydrogen peroxide, Carbohydr. Polym. **86**(1), 185-191, (2011)

[11] W-A. Wan-Mohamed, R. Hamdan, N. Othman, Study of the pH effects on the phosphorus removal mechanism in lab-scale electric arc furnace slag and limestone filters in synthetic wastewater. Applied Mechanics and Mater., 752-753, 277-282, (2015)

[12] K. Srividya and K. Mohanty, Biosorption of hexavalent chromium from aqueous solution by catla catla scales: Equilibrium and kinetics studies, Chem. Eng. J., **155**, 666-673, (2009)

[13] APHA., AWWA., WEF, *Standard Methods for the Examination of Water and Wastewater*. 22nd ed. Washington, DC: American Public Health Association, (2012)

[14] V. K. Jha, Y. Kameshima, A. Nakajima, K. Okada, Utilization of steel-making slag for the uptake of ammonium and phosphate ions from aqueous solution, J. Hazard. Mater. **156**(1-3), 156-162, (2008)

[15] M. H. Khamidun, M.A. Fullazaky, M.F.M. Din and A.R.M. Yusoff. Resistance of mass transfer, kinetic and isotherm study of ammonium removal by using a hybrid plug flow column reactor. Environment Energy and Sustainable Development-Sung, Koa & Chen, 555-559, (2014).

Preliminary Assessment of Growth Rates on Different Concentration of Microalgae Scenedesmus sp. in Industrial Meat Food Processing Wastewater

Nur Atikah Ahmad Latiffi [1], *Radin Maya Saphira Radin Mohamed* [1,*], *Najeeha Mohd Apandi* [1], *and Ramlah Mohd Tajuddin* [2]

[1]Faculty of Civil and Environmental Engineering, Universiti Tun Hussein Onn Malaysia, Johor, Malaysia
[2]Faculty of Civil Engineering, University Technology MARA, UiTM, Selangor, Malaysia

Abstract. This study is aimed to evaluate and access the growth rates and biomass productivity in different concentrations of microalgae *Scenedesmus sp.* using Industrial Meat Food Processing Wastewater as a media. The focus of this study is to determine the best concentrations of microalgae *Scenedesmus sp.* in raw wastewater in terms of kinetics of cells growth rates. The study verified that concentration of $1x10^6$ cells/ml of microalgae gives the highest specific growth rates of biomass at 0.4488 day-1 and 1720 cells/ml/day compare to the other concentrations, while the lowest occurred at concentration of $1x10^3$ cells/ml at 0.4108 day^{-1} and 14.9 cells/ml/day. The result shows the different concentration of microalgae *Scenedesmus sp.* culturing in Industrial Food Processing Wastewater influence the cells growth of biomass and the optimum were obtained at concentration of $1x10^6$ cells/ml which suggested use for Industrial Meat Food Processing Wastewater Treatment purposed. With this finding, it should be seemly to adopt and applied efficiently in treating the wastewater especially for *Scenedesmus sp.* type of microalgae.

1 Introduction

Microalgae are known as a survival microorganism which can grow and adapt well in an extreme environmental conditions such as in various type of wastewater. However, there are some species who can only survive in certain type of conditions [1]. There are more 50,000 species of microalgae has been found which varies from types, species, size and organism groups [2-4]. The structure of algae typically comprises of plastids chlorophyll which ensuing photosynthesis process that devours light as energy and converts inorganic substance into sugar [5-6]. This expressions indicates that microalgae has absorption ability indirectly can restore and reduce substances from wastewater [7].

[*] Corresponding author: maya@uthm.edu.my

Wastewater normally contains high amount of organic matter as well as inorganic nutrients and according to [8] cultivation of microalgae for biomass productions by using wastewater can significantly reduce the water usage up to 90%. This can be achieved by assimilating all the nutrients present in wastewater. Industrial Meat Food Processing wastewater is known to be highly biodegradable, non-toxic and rich in nitrogen and phosphorus elements. Its characteristics is differ comparing with typical municipal wastewater that managed by public treatment plants as it is difficult to predict due to the seasonal and demand nature of business [9]. Activities involving in the meat processing industries known to be high in water usage up to 62 Mm^3/year, as only a small amount from this total are fully utilized into the final products, subsequently the rest of wastewater been discharge into the water bodies [10] and according to [11], meat food processing facility itself are using clean water up to 40m^3/metric tons of meat products which is the highest consumers compare with other food and beverage industries by 24% [11-12]. All the steps involved in the process of production until packaging indirectly gives an impact to the environmental as the waste generates from all of these process will eventually initiate several environmental problems to human beings and aquatic organism as they both rely on water to sustained [9, 12]. Furthermore, nitrogen and phosphorus are an ideal and essentials nutrients that can stimulate a fast growth of algae in freshwater, wastewater streams from varies of sources such as from agricultural run-off, industrial, municipal wastewater and marine ecosystem [13-14]. Even though wastewater could possibly offer some essentials goods for microalgae cultivation activity, however, the raw wastewater compositions are differ compare to the culture medium prepared in the laboratory. Concentration of nitrogen and phosphorus as well as the existent wide range of other microorganism, presence of heavy metals and other toxics comes from wastewater may interfere the growth and uptake of nutrients by microalgae [15]. On the other hand, nitrogen presents in wastewater normally in ammonia forms which can restrain microalgae growth if high in concentration [1, 16, 17]. Therefore, the most significant key in selecting the best concentration of microalgae is it can strongly adapt and have high specific growth rates interpretation in selected wastewater.

2 Materials and methods

2.1 Microalgae *Scenedesmus sp.*

Three (3) steps of procedure on preparations of microalgae i.e. microalgae sampling, isolation and identification need to be done prior culturing. Microalgae of *Scenedesmus sp.* was obtained from plankton located in Endau Rompin National Park, Johor. This location is selected due to well preserved national park that has less environmental pollution which contributes to a better colonies and growth. Many previous studies by [18-19] have been used and applied microalgae from this location for treatments and growth studies that portrayed effective results and outcome. A sterilize glass bottle was used for collection and seal with cotton wool for preservations to allowed air flow and photosynthesis. A small amount of algae were isolated and isolation process was done under microscope by using platinum needles. The isolated microalgae then will be examined for initial identification and confirmation using NIKON Eclipse E600 microscope. For culturing, the isolated microalgae is cultivate in Basal Bold Medium, BBM [20] and later were place under natural condition of 12h: 12h intensity lights with temperature ±35°C for growth observations.

2.2 Wastewater Sampling Preparations

Sample was collected at one of small and medium food industries located in Johor. Sampling was done once a week for 8 weeks duration to get the average values and range. The collected wastewater was preserve in plastic bottle of Polyethylene Terephthalate (PET) and was stored at 4^0C without any acid preservation. Wastewater that was collected from the study site were filtered before inoculation process takes place to minimized all unwanted microorganism or bacteria that may exists in the wastewater.

2.3 Inoculations of different microalgae *Scenedesmus sp.* concentration for laboratory batch optimizations

Six (6) different concentration of microalgae *Scenedesmus sp.* was prepared and use in this batch study i.e. concentration $1x10^3$ cells/ml, $1x10^4$ cells/ml, $1x10^5$ cells/ml, $1x10^6$ cells/ml, and $1x10^7$ cells/ml respectively. The concentration was calculated using haemocytometer together with equations by Stephenson [21] i.e. $C_1V_1 = C_2V_2$ where by C_1 is the initial concentration of the stocks solutions, V_1 is the amount of stock solution required, C_2 is the concentration of the wanted stock concentration and V_2 is the volume of the stock solutions. This equation is suitable for solution, mixture and media as stated by Stephenson [21]. The growth rates data were collected and counted for every 24hours until the growth enter the death or decline stage [22].

3 Cells Concentration Counting

Microalgae cell concentration was determined through cell counting using haemocytometer [23-24]. The haemocytometer is a slide glass size 30 mm x 70 mm x 4 mm thick. It has 3 chambers and cells counts are performed at the center part. The specific growth rates of microalgae were calculated using Eq.1 derives formula from Andersen R.A. [24] i.e. μ / day $= (\text{Ln}(N_t/N_o)/(t_f - t_i))$. Where N_t is Final algae concentrations, N_0 is Initial algae concentrations and Δ_t are subtraction of Final time with the Initial time (day).

4 Data Analysis

4.1 Microalgae *Scenedesmus sp.* characteristics

Characteristics of microalgae *Scenedesmus sp.* is shown in Table 1 and its physical characteristics is shown in Fig. 1 by using SEM micrograph.

Table 1. Characteristics of microalgae *Scenedesmus sp.*

Parameters	Values
pH	6.38
Total Suspended Solid, TSS (mg/L)	375
Turbidity (NTU)	77.8
Sulfate, SO_4^{2-} (mg/L)	35
Dissolved Oxygen, DO (mg/L)	16.38
Total Nitrogen, TN (mg/L)	308.9
Total Phosphorus, TP (mg/L)	0.75
Total Organic Carbon, TOC (mg/L)	38.01
Ammonia Nitrogen, NH_3-N (mg/L)	1.95
Zinc, Zn (mg/L)	0.47
Copper, Cu (mg/L)	1.24

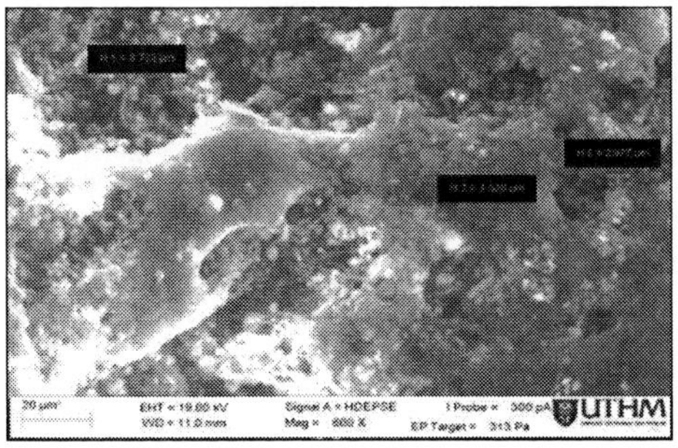

Fig. 1. SEM micrgraph (600X) of *Scenedesmus sp.*

4.2 Industrial Meat Food Processing Wastewater Characteristics

Wastewater characterization is significantly important as it is to determine the availability of nutrients in the wastewater for microalgae growth. Table 2 shows the physical and chemical parameter of meat processing wastewater in comparison with standard discharge effluent allowable by local authorities. Chemical Oxygen Demand (COD) generally defined as the total quantity of oxygen required for complete oxidation of organic compound to carbon dioxide and water and it is significantly to continually measure to determine the degree of pollution in an effluents [25]. Concentration of COD is 1643 mg/L, which is over the allowable limit permit by the local authorities. BOD_5 is known to be the basic of parameter in finding the degree of pollutions in water. As for Biological oxygen demand, (BOD), the concentration of the wastewater is 895 mg/L which also above the limits of standard A and Standard B. This wastewater also contains other parameter such as Total Nitrogen, Ammonia Nitrogen, Total Phosphorus, Orthophosphate, Total Organic Carbon, Total Suspended Solids as well as Total Dissolved Solids with concentration of 151 mg/L, 10 mg/L, 45 mg/L, 35 mg/L, 333 mg/L, 333 mg/L and 1120 mg/L respectively. All of the mentions parameters are way above limits line as indicate in the [26]. In terms of pH, the wastewater pH value is acceptable and fit for microalgae cultivation as refer to [27]. There have been few researchers who have done and shown some potentials in cultivating microalgae in other different types of raw wastewater [28-31], or phycoremediation and bio transform those pollutants to essentials biomass as well as cleaned the wastewater at the same time. Moreover, growth efficiency of the cells is subjected to several of environmental factors and nutrients availability. Therefore, this study is to utilized raw industrial meat food processing wastewater into cultivating the *scenedesmus sp.* for the growth cells efficiency.

4.3 Cell Growth of *Scenedesmus sp. in* Industrial Meat Food Processing Wastewater

The growth of *Scenedesmus sp.* cells curve in the industrial meat food processing wastewater are showing positive growing over time for all concentration except for $1x10^7$ cells/ml. A typical growth of curve for microorganism as stated by Shuler [22] consist of five (5) phase; the lag, exponential, deceleration, stationary and death or decline phase. As for these experimental growth (Fig. 2), cells of $1x10^3$ cells/ml, $1x10^4$ cells/ml, $1x10^5$ cells/ml and $1x10^6$ cells/ml shows an increasing of biomass concentration over time except for $1x10^7$ cells/ml. The results of growth are similar to the growth of *Botryococcus sp.*in domestic wastewater that has been reported by Gani [28], whereby, too much of populations of microalgae will cause suffocation of nutrients available in the wastewater to sustained the microalgae growth, resulting no increasing of biomass concentrations over time and based [22], at high concentration of microalgae in the medium, the growth rates will inhibited.

Table 2. Characteristics of the Industrial Meat Food Processing Wastewater used as growth media for *Scenedesmus sp.* Sampling date: January 2015 ~ February 2015.

Physiochemical parameters	Average Concentrations Mg/L	*Standard A	*Standard B
Biological Oxygen Demand, BOD	1070±350.913	20	50
Chemical Oxygen Demand, COD	2350±835.427	50	100
Total Nitrogen	317.22±171.147	-	-
Total Phosphorus	62.86±36.525	5	10
Orthophosphate	47.37±27.530	-	-
Total Organic Carbon	493.82±267.881	-	-
Total Suspended Solids, TSS	1400±299.100	50	100
pH	6.5-8.0	6.0 – 9.0	5.5 – 9.0

*EQA, 1974

Fig. 2 shows the comparison between the collected experimental data in comparing with mathematical modeling [32]. Each of specific growth rates of the concentrations can be calculated and obtained during the exponential phase of the microalgae growth by drawing a straight line that touched at least three or more points in Fig. 3 as accordingly to Andersen [24] methods. The growing curve between experimental data and modeling data is considerably different as during experiments works, lots of other factors such as environmental factors, equipment's error and human errors contribute during the analysis.

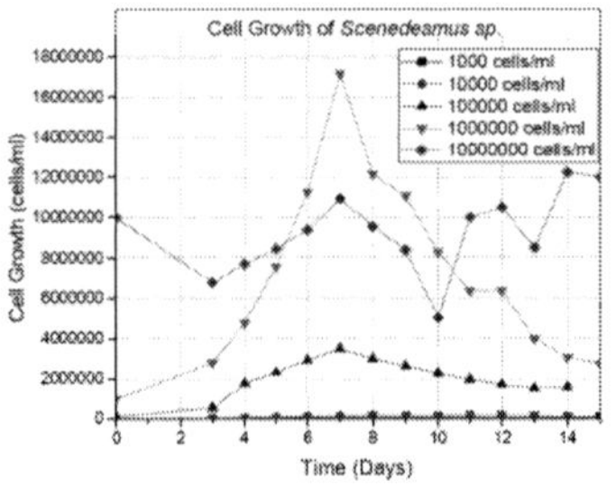

Fig. 2. Scenedesmus sp. growth curve in different concentrations.

 DOI: 10.1051/matecconf/201710306010

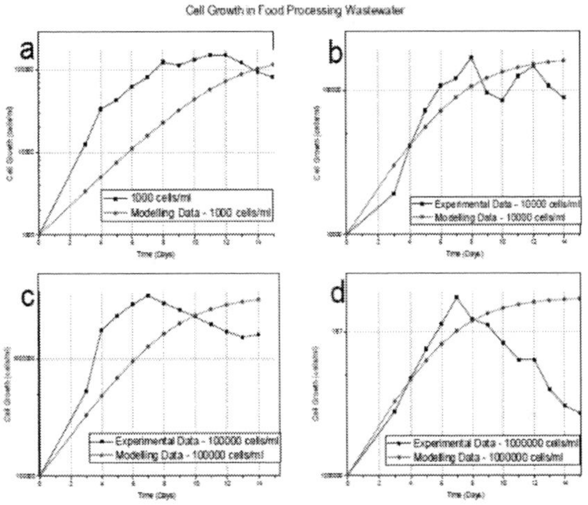

Fig. 3. a) Cell growth of Scenedesmus sp. for $1x10^3$ cells/ml concentration; b) Cell growth of Scenedesmus sp. for $1x10^4$ cells/ml concentrations; c) Cell growth of Scenedesmus sp. for $1x10^5$ cells/ml concentration; d) Cell growth of Scenedesmus sp. for $1x10^6$ cells/ml

Table 3. Comparison of growth parameters between experimental data and mathematical modelling for *Scenedesmus sp.* in different initial of concentrations.

Concentration (cells/ml)	Experimental Data				
	Max Specific Growth Rates μmax/day	**Division per day (Dd)**	**Doubling Time, td (day)**	**Biomass productivity (cells/ml/day) 10^4**	**R^2**
$1x10^3$	0.4108±0.0425	0.5927	1.687	14.9	0.9393
$1x10^4$	0.4125±0.0583	0.5951	1.680	16.8	0.9321
$1x10^5$	0.4250±0.0308	0.6131	1.631	348.0	0.8187
$1x10^6$	0.4483±0.0244	0.6468	1.546	1720.0	0.9967
$1x10^7$	0	0	0	0	0

* Experiments conducted in triplicate manner *(n =3)*

The specific growth rates of biomass increases with microalgae concentration until they reached a maximum value that associate with optimal concentrations, concentration $1x10^6$ cells/ml shows the highest specific growth of biomass increment followed by $1x10^5$ cells/ml, $1x10^4$ cells/ml and $1x10^3$ cells/ml with specific growth rates of 0.4483 day^{-1}, 0.4250 day^{-1}, 0.4125 day^{-1} and 0.4108 day^{-1} respectively. Beyond these concentrations, it can be considered as saturated phase where the growth rapidly decreased or no growth occurred [33]. As per $1x10^6$ cells/ml, the exponential curve starts from day 3rd until day 7th, similarly to $1x10^5$ cells/ml while $1x10^4$ cells/ml and $1x10^3$ cells/ml starts their exponential curve at day 3rd until day 8th correspondingly. In identifying specific growth rates, can be directly obtained by applying Equation (1). Biomass productivity on the other hand shows some significant increments for all concentrations, particularly for $1x10^6$ cells/ml with maximum biomass productivity and specific growth rates of 1720 cells/ml/day and 0.4483 day^{-1}. This results are slightly higher than the results obtained by [34] studies with specific growth rates of 0.103 day^{-1} when they cultivate *Scenedesmus sp.* in municipal wastewater and biomass productivity up to 50.75 mg/L/day (dried weight) [35]. The specific growth rate obtained in this study is slightly higher than study done by [36] with maximum specific growth rate of 0.17 day^{-1}, and biomass production of 253.45

mg/l in terms of dry weight by using BG-11 medium media during cultivation of *Scenedesmus SDEC-8.* Moreover, *chrollella SDEC 4* on the other hand obtained 0.18 day^{-1} of maximum specific growth rate with 157.30 mg/l in terms dry weight of biomass production. Different species will give different growth rates due to its algometric relationship between growth and cell size as well as metabolic process, resulting smaller size of species tends to grew faster than the bigger size [33], same goes with different media or medium used.

5 Conclusion

The series of experiments conducted here are aimed to access and verified the best concentration of microalgae *scenedesmus sp.* cultivating in Industrial Meat Food Processing Wastewater. The results shows concentration of $1x10^6$ cells/ml of microalgae *scenedesmus sp.* gives the highest specific growth rates biomass of 0.4488 day-1. Thus, from these findings, the most adequate population that suitable for sustaining the microalgae *scenedesmus sp.* growth in Industrial Meat Food Processing Wastewater is verified. This study is another small chapter on accessing the effectiveness of cultivating microalgae *Scenedesmus sp.* in industrial meat food processing wastewater for wastewater purification and biomass productivity, as the culturing microalgae in industrial meat food processing wastewater are highly dependable on wastewater nutrients availability as well as initial cell concentrations. Therefore, more research on microalgae *Scenedesmus sp.* would be essentials for phycoremediation and potentially biomass.

This research was funded under the FRGS (Vote 1435) and eScience Fund project no: 02-01-13-SF0135. The authors thank to any parties that involved in this project especially Universiti Tun Hussein Onn Malaysia for providing the equipment and research facilities to carry out this project.

References

[1] Y. Wu, H. Hu, Y. Yu, T. Zhang and S. Zhu, Microalgal species for sustainable biomass lipid production using wastewater as resource: A review, Renewable and Sustainable Energy, **336**, 75-688, (2014)

[2] A. Richmond, *Handbook of microalgal culture: biotechnology and applied phycology.* Blackwell Science Ltd; (2004)

[3] J. P. Maity, J. Bundschuh, C.-Y. Chen, and P. Bhattacharya., Microalgae for third generation biofuel production, mitigation of greenhouse gas emissions and wastewater treatment: Present and future perspectives - A mini review, Energy, **78**, 104–113, (2014)

[4] D. Hanelt, K. Bischof and K. Dunton, Life strategy, eco-physiology and ecology of seaweeds in polar waters, Reviews in Environmental Science and Biotechnology 2007, **6**(1), 95–126, (2007)

[5] A. Al-Darmaki, L. Govindrajan, S. Talebi, S. Al-Rajhi, T. Al-Barwani, Z. Al-Bulashi, Cultivation and characterization of microalgae for wastewater treatment. *Proceedings of the world congress on Eng.* **1**,(2012)

[6] D. S. Shekhawat, A. Bhatnagar, M. Bhatnagar, and J. Panwar, Potential of Treated Dairy Waste Water for the Cultivation of Algae and Waste Water Treatment by Algae, J. Environ. Res. Technol., **2**, 101–104, (2012)

[7] A. Worku and O. Sahu, Reduction of Heavy Metal and Hardness from Ground Water by Algae, J. Appl. Environ. Microbiol., **2**(3), 86–89 (2014)

[8] X.E. Yang, X. Wu, H.L. Hao, and Z.L. He, Mechanisms and assessment of water eutrophication. J. Zhejiang Univ. Sci. B. **9**(3), 197-209, (2009)

[9] Onet C., Characteristics of the Untreated Wastewater Produced By Food Industry Vol Xv, 709 – 714, University of Oradea- Faculty of Environmental Protection, (2010)

[10] E. Sroka, K. Wladysław, and B. Jolanta, Biological Treatment of Meat Industry Wastewater, Desalination **162**, 85-91, (2004)

[11] C. F. Bustillo-Lecompte and M. Mehrvar, Slaughterhouse wastewater characteristics, treatment, and management in the meat processing industry: A review on trends and advances, J. of Environmental Management, **161**, 287-302, (2015)

[12] W. Qasim and A. V. Mane, Characterization and treatment of selected food industrial effluents by coagulation and adsorption techniques, Water Resources and Industry, **4**, 1–12, (2013)

[13] J. Liu, Z. Sun, and H. Gerken, *Recent Advances in Microalgal Biotechnology. Potential Application of Microalgae Wastewater Treatment.* OMICS Group eBooks, (2014)

[14] V. H. Smith, G. D. Tilman, and J. C. Nekola,, Eutrophication : impacts of excess nutrient inputs on freshwater, marine, and terrestrial ecosystems, Env. Pollution, **100**, 179-196, (1999)

[15] Z-F, Su, X. Li, H-Y, Hu, Y-H, Wu, T, Noguchi, Culture of Scenedesmus sp LX1 in the modified effluent of a wastewater treatment plant of an electric factory by photo-membrane bioreactor. Bioresource Technology 1; **102**, 7627-32. (2011)

[16] T. Kallqvist, A. Svenson, Assessment of ammonia toxicity in tests with the microalga, Nephroselmispyriformis, Chlorophyta. Water Res., **37**, 477–84, (2003)

[17] X. Yuan, A. Kumar, AK. Sahu, and Sj. Ergas. Impact of ammonia concentration on Spirulina platensis growth in an airlift photobioreactor, Bioresour. Technol., **102**(3), 234-9, (2011)

[18] Gani, P. S. Sunar, N. M. M. A. Peralta, H. A. B. L. Latiff, and A. A. A. B. R. Razak, Influence of Initial Cell Concentrations on the Growth Rate And Biomass Productivity of Microalgae In Domestic Wastewater, **14**(2), 399-409, (2016)

[19] N.A.A. Latiffi, S. Radin, N.M. Apandi and H. Mohd, Application of Phycoremediation using Microalgae Scenedesmus sp. as Wastewater Treatment in Removal of Heavy Metals from Food Stall Wastewater, J. of Applied Mechanics and Materials, (773 – 774), 1168 -1172, (2014)

[20] Nichols Hw and Bold Hc., Hand Book of Physiological Methods, Growth media-Fresh Water, Cambridge University, London, U.K., 7-24, (1965)

[21] F. H. Stephenson, Calculations for Molecular Biology and Biotechnology, Elsevier's Science and Technology, Oxford, U.K., (2010)

[22] Mi. L. Shuler, Bioprocess Engineering-Basic Concepts (2nd Edition), Prentice Hall PTR, New Jersey, U.S.A., (2002)

[23] APHA (AMERICAN PUBLIC HEALTH ASSOCIATION), *Standard Methods for Examination of Water and Wastewater (21stedn.)*, American Public Health Association, Washington DC, U.S.A., (2012)

[24] R. A. Andersen, *Algae culturing technique,* Elsevier's Science and Technology, Oxford, U.K., (2005).

[25] E. A. Echiegu and J. T. Liberty, Effluents Characteristics of Some Selected Food Processing Industries in Enugu and Anambra States of Nigeria, J. of Environment and Eart Science, **3**(9), 46–54, (2013)

[26] Environmental Quality Act 1974, *Department of Environmental Malaysia: A Guide For Investor*, Putrajaya, Malaysia, (2010)

[27] L. Creswell, Phytoplankton Culture for Aquaculture Feed, Florida, US, (2010), Retrieved on August 23. 2016 from http://www2.ca.uky.edu/wkrec/PhytoplanktonAlgaeCulture.pdf

[28] P. Gani, N. M. Sunar, H. Matias-peralta, A. Aziz and A. Latiff, Application Of Phycoremediation Technology In Thetreatment Of Food Processing Wastewater By Freshwater Microalgae *Botryococcus Sp.,* Applied Ecology and Environmental Research, **11**(11), 7288–7292 (2016)

[29] T.-Y. Zhang, Y-H. Wu, S-F. Zhu, F-M. Li, H-Y. Ho, Isolation and heterotrophic cultivation of mixotrophic microalgae strains for domestic wastewater treatment and lipid production under dark condition, Bioresource Technology **149**, 586–9 (2013)

[30] Ş.S. Can, V. Demir, S.A. Korkmaz, E. Can, Treatment of domestic wastewater with Botryococcus braunii (Cholorophyceae), J. of Food, Agriculture and Environment, **11**(4&3) 3–5 (2013)

[31] M.-K. Ji, R-A.I. Abou-Shanab, S-H. Kim, E-S. Salama, S-H. Lee, A.N. Kabra, Y-S. Lee, S. Hong, B-H. Jeon, Cultivation of microalgae species in tertiary municipal wastewater supplemented with CO2 for nutrient removal and biomass production, Ecological Engineering **58**, 142–148, (2013)

[32] Z. Arbib, J. Ruiz, P. Álvarez-Díaz, C. Garrido-Pérez, and J. a. Perales, Capability of different microalgae species for phytoremediation processes: Wastewater tertiary treatment, CO_2 bio-fixation and low cost biofuels production, Water Research, **49**, 465– 474 (2014)

[33] R. Bouterfas, M. Belkoura, P. M. Abdellah, S, B. P., D. Narbonne, and T. Cedex, The effects of irradiance and photoperiod on the growth rate of three freshwater green algae isolated from a eutrophic lake Source of the organisms isolated from the eutrophic Takerkoust barrage ' s, **25**(3), 647–656, (2006)

[34] M. S. De Alva, V. M. Luna-pabello, E. Cadena, and E. Ortíz, Green microalga Scenedesmus acutus grown on municipal wastewater to couple nutrient removal with lipid accumulation for biodiesel production, Bio resource Technology**146**, 744–748, (2006)

[35] D.Y. Shin, H.U. Cho, J.C. Utomo, Y.-N. Choi, X. Xu and J.M. Park, Biodiesel production from Scenedesmus bijuga grown in anaerobically digested food wastewater effluent, Bio resource technology **184**, 215–21, (2015)

[36] M. Song, H. Pei, W. Hu, S. Zhang, G. Ma, L. Han, and Y. Ji, Identification and characterization of a freshwater microalga Scenedesmus SDEC-8 for nutrient removal and biodiesel production, Bio resource Technology, **162**, 129–135, (2014)

Protein and Lipid Content of Microalgae *Scenedesmus* sp. Biomass Grown in Wet Market Wastewater

Najeeha Mohd Apandi[1], *Radin Maya Saphira Radin Mohamed*[1,*], *Nur Atikah Ahmad Latiffi*[1], *Nur Farahanim Mohd Rozlan*[1], *and A.A.S. Al-Gheethi*[1]

[1]Department of Water and Environmental Engineering, Faculty of Civil and Environmental Engineering, Universiti Tun Hussein Onn Malaysia, 86400 Parit Raja, Batu Pahat, Johor, Malaysia

Abstract. Microalgae biomass is a type of organism that has various advantages and often used for variety of uses not only for wastewater treatment purposes but also serve as bio-based products development. Because of high in protein and lipid accumulation, this freshwater microalgae *Scenedesmus* sp. was selected for the present study. In this study, microalgae *Scenedesmus* sp. was grown in four concentrations of wet market wastewater (10%WM, 15%WM, 20%WM and 25%WM) in comparison with the Bold Basal Medium (BBM) as a control. The result shows that the highest protein content (50.72%) was in 10%WM followed by 15%WM, 20%WM, 25%WM and BBM. As for lipid content, the best concentration was obtained in the 25%WM with the value of 26.7% compared to other concentrations and BBM. Therefore, the result indicates that microalgae *Scenedesmus* sp. biomass cultured using wet market wastewater has a great potential in producing protein-rich and lipid content which significantly could be used as a fish feed product.

1 Introduction

An assortment of microalgae, including *Scenedesmus* sp. have been concentrated on for their conceivable viability as bioresources for application as fish feed, supplement, human consumption due to their nutritious compounds eg; protein, lipids, carbohydrates and other valuable trace element and additionally for the bioremediation of contaminated water [1-3]. In recent years, many researchers have studied the value potential of dual application of microalgae for wastewater treatment and biomass production [4, 5]. Unfortunately, wastewater from wet market contain nutrient and elements derived from the fresh foodstuff, waste scraps of poultry and fish entrails which consists of variety components such as nitrogen and phosphorus commonly discharge directly to the drainage without any treatment [6]. In Malaysia, the disposal of these wastewaters is commonly direct into the drainage and end up flowing into the river. This phenomenon escalates the pollution to obnoxious smell and filthy surrounding environment due to the wastewater discharge

* Corresponding author: maya@uthm.edu.my

freely. This issue may prompt an uncomfortable condition directly to the people and occupants at the wet market business area [7].

Basically, conventional wastewater treatment is highly expensive due to chemicals needed to treat the wastewater. However, biological treatment using microalgae as an alternative to reduce the pollutant with less expensive and at the same time valuable biomass microalgae can be produced. Microalgae based wastewater treatment is eco-friendly and offers the advantage of a cost effective way of nutrient removal and biomass production [8]. Since microalgae have the ability to reduce nitrogen and phosphorus levels in wastewater [9], instead of the nutrients being wasted; this will provide nutrients for the microalgae to grow which in turn become valuable biomass products. Since microalgae has been recognized as a very favourable source of biomass, this can be applied in order to treat wastewater efficiently without involve high production cost.

The idea of using microalgae as a source of finding an alternative fish feed replacement is not something new and it is being considered important as unconventional feed ingredients in replacement of high cost feed stuffs that has been increasing [10,11]. Microalgae has the promising substitution protein-rich and lipid sources for fish, where several of studies had investigate the effect of partial replacement of fish feed with microalgae *Scenedesmus* sp. [12, 13]. The objective of this study was to determine the protein and lipid content of *Scenedesmus* sp. cultivated in different concentration of wet market wastewater under outdoor culture condition. In addition, the morphological of *Scenedesmus* sp. after growing in wet market wastewater also characterised using Scanning Electron micrograph (SEM) accordingly.

2 Materials and methods

2.1 Cultivations of *Scenedesmus* sp. using wet market wastewater

Microalgae *Scenedesmus* sp. was obtained from the culture collection of Faculty of Science, Technology and Human Development, Tun Hussein Onn University of Malaysia. The inoculum of *Scenedesmus* sp. was maintained in Bold basal medium (BBM) prior to experiment. The experiment was set up using 2L of Erlenmeyer flask containing 1.5 L of wastewater and Bold Basal Media (BBM). BBM medium was used as control media. Wastewater was transported to the laboratory and stirred for about 30 minutes and allowed to settle for 1 hour. The supernatant was separated from the remaining suspended solids and diluted with distilled water into four concentrations (10%, 15%, 20% and 25%). Microalgae was cultivated under outdoor sunlight with continuous aeration at 3.2 L/min to ensure the distribution of nutrients using atmospheric air from an aquarium pump through air sparger. The experiment flask was inoculated with the same initial microalgae cells of (10^6 cells/ml) at the initial day of the experiment. The biomass was harvested immediately after 12th days by centrifugation at 4000 rpm for 5 min. The cultivation batches were carried out in duplicate.

2.2 Analysis of proximate composition

The measurement of protein, ash, fiber and moisture content was performed based on AOAC (2002) [14] methods. Lipid content was determined according to Bligh and Dyer [15] and Folch et al. [16]. All biomass samples were kept in an airtight container and were sent for further analysis of Scanning Electron Microscopy (SEM) for cell morphology characterization after cultivation.

3 Results and discussion

3.1 Protein and Lipid Content of *Scenedesmus* sp.

Table 1 shows the chemical composition of *Scenedesmus* sp. biomass which were cultivated in different types of media. The highest protein content was recorded by 10%WM followed by 15%WM, 20%WM, 25%WM and BBM respectively. Low protein content recorded in 25%WM with 37.3% while the highest protein content was recorded at 10%WM with 50.72%. It is to be found that the protein content increased along with the decreasing wet market wastewater ratio added to the cultures. On the other hand, the highest lipid content was found to be in 25%WM with a total value of 26.7% were significantly higher than those obtained from other concentration and BBM. As the concentration goes lower, the lipid content were decreased (Fig. 1). These results show that it may possibly due to micronutrient deficiency or some other stress factor was involved; higher nitrogen involvement achieved higher lipid productivity [17].

Table 1. Proximate composition (% of dry matter)

Composition	10%WM	15%WM	20%WM	25%WM	BBM
Crude potein (%)	50.72 ± 0.64	41.2 ± 0.23	38.6 ± 1.27	37.3 ± 0.10	35.0 ± 1.52
Crude lipid	8.0 ± 0.10	14.7 ± 0.17	19.9 ± 0.18	26.7 ± 0.56	15.1 ± 0.19
Ash (%)	3.1 ± 0.25	7.1 ± 0.04	9.2 ± 0.67	12.0 ± 0.48	10.2 ± 0.29
Moisture (%)	6.4 ± 0.26	8.5 ± 0.25	8.7 ± 0.13	9.0 ± 0.10	7.7 ± 0.29
Crude fiber (%)	6.91 ± 0.21	6.39 ± 0.19	6.72± 0.36	6.24± 0.34	6.50 ± 0.07

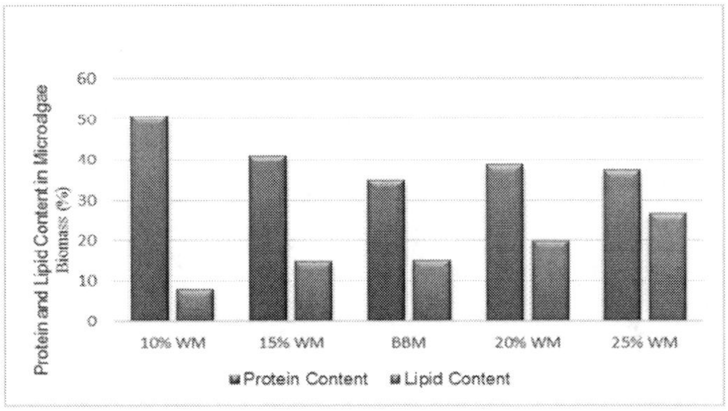

Fig. 1. Protein and lipid content

Protein content was found higher and the lipid content was lower in the 10%WM than in the 25%WM. This measurement indicates that stress conditions in 10%WM were higher than in the 25%WM in the cultures. The stress condition triggered due to the nutrient content of nitrogen and phosphorus as both a particular key for protein and lipid accumulation in microalgae cells. It is likely known that the aspects of this stress conditions are tightly linked to the available amount of nutrients; limitation of nitrogen, phosphorus, and the presence of other microorganisms [18]. Moreover, the trend value of protein and lipid were similar to the values reported by Markou *et al.* [19] who cultivated microalgae using slaughterhouse wastewater and revealed that this phenomenon could be explained to

DOI: 10.1051/matecconf/201710306011

the fact that the presence of higher nitrate and phosphorus concentration related to the changes in the protein and lipid content. Assuming further used for this microalgae *Scenedesmus* sp, as fish feed, this study is already fulfill the requirement of protein (40%-50%) and lipid (6%-13%) percentage needed for fish in the standard from Food and Agriculture Organization of The United Nations (FAO) [20]. Thus, this proximate quality of microalgae *Scenedesmus* sp. biomass cultivated from wet market wastewater in this study can be the ideal amount to supply in terms of protein and lipid based on the percentage amount to be consumed by fish. However, in order to simplify whether this microalgae *Scenedesmus* sp. biomass grown on wet market wastewater is harmless to be used as fish feed, thoroughly follow-up research is needed.

3.2 Microalgae cell characterization via SEM

The cell morphology of *Scenesdesmus* sp. that has been grown on five different of wet market wastewater media was obeserved under Scanning electron micrograph (SEM) asdepicted in Fig. 2. The SEM at the magnification of 1200× was used to characterize the biomass for both culture media (WM and BBM). It shows that wet microalgae *Scenedesmus* sp. biomass from each media was characterized by having different length and longer axis for all sample which ranged of 6mμ - 8μm to 2μm - 3μm. In fact, the individual microalgae cells were found to be in spherical and unicellular as typical colonies consisting of numerous cells.

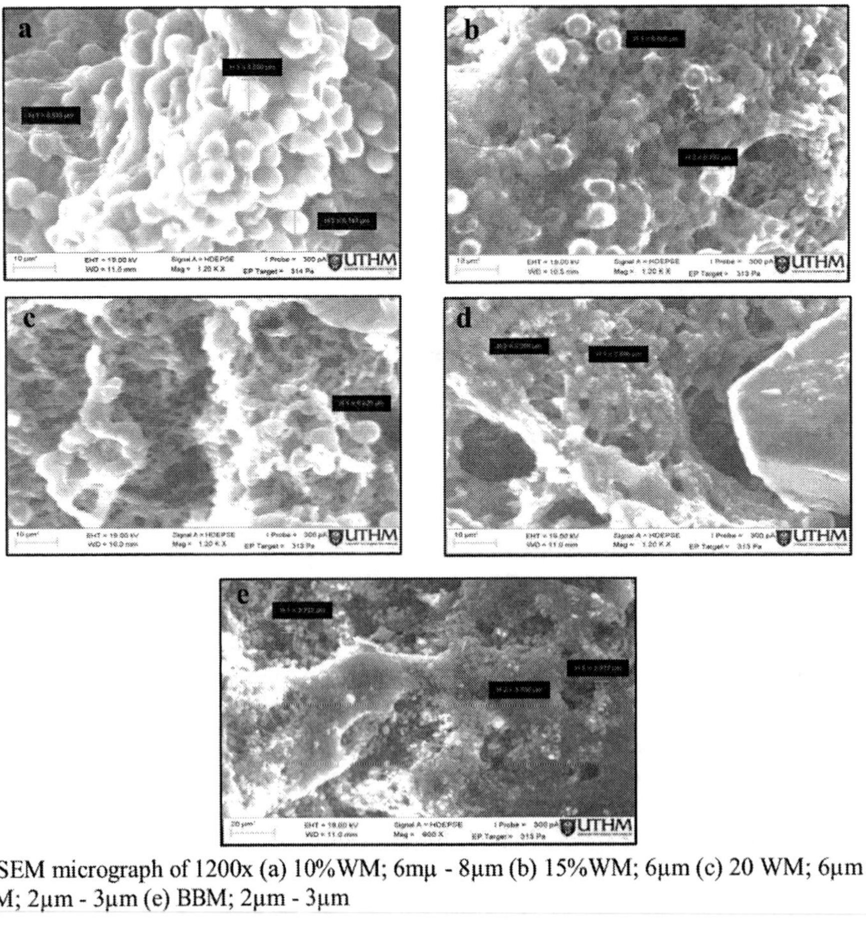

Fig. 2. SEM micrograph of 1200x (a) 10%WM; 6mμ - 8μm (b) 15%WM; 6μm (c) 20 WM; 6μm (d) 25%WM; 2μm - 3μm (e) BBM; 2μm - 3μm

4 Conclusion

This study shows that wet market wastewater can be used as an alternative media for the cultivation of microalgae *Scenedesmus* sp. protein and lipid content production. Among the four different concentrations of WM, 20% WM showed the best performance in respect of protein and lipid content. Besides, a tremendously high of protein content was found in 10%WM with 50.72% while high lipid content was found in 25%WM with 26.7%. This indicates that microalgae *Scenedesmus sp.* biomass was rich in protein but lower in lipid straightly can be a potential application for bio-based product such as fish feedstock. This research is also motivated by the increasing interest in application of microalgae biomass and the implementation of environmentally friendly tools in treating wastewater.

This research was funded under the PRGS (G004) and eScience Fund project no: 02-01-13-SF0135. Special thanks to Dr. Hazel Monica Matias-Peralta for providing the microalgae seed and any parties that involved in this project specifically University Tun Hussein Onn Malaysia for providing the equipment and research facilities to carry out this project. This paper was partly sponsored by the Centre for Graduate Studies UTHM.

References

[1] E.W. Becker, Micro-Algae as a Source of Protein, Biotechnology Advances, **25**(2), 207–10, (2007)

[2] V. Makareviciene, A. Vaida, S. Virginija, and K. Jurate, Cultivation of Microalgae Chlorella Sp. and Scenedesmus Sp. as a Potentional, Biofuel Feedstock **57**(3), 21-27, (2011)

[3] Spolaore, Pauline, Claire Joannis-Cassan, Elie Duran, and Arsene Isambert, Commercial Applications of Microalgae. Journal of bioscience and bioengineering **101**(2), 87–96, (2006)

[4] I. Rawat, K. Ranjith, R. T. Mutanda, and F. Bux, Dual role of microalgae: Phycoremediation of domestic wastewater and biomass production for sustainable biofuels production, Applied Energy, **88**(10), 3411-3424, (2011)

[5] A.C. Wilkie and W.W. Mulbry, Recovery of dairy nutrients by benthic freshwater algae, Bioresource Technology, **84**, 81-91, (2002)

[6] H. S. Plumber and B. H. Kiepper, Impact of Poultry Processing By-Products on Wastewater Generation, Treatment and Discharges, *Proc. of the 2011 Georgia Water Resources Conf.,* Univ. of Georgia , Athens, (2011)

[7] A. Izzuddin and Ghani, *Application of bioparticle and constructed wetlands in treating wet market wastewater*, Universiti Teknologi Malaysia, Undergraduate's Thesis, (2011)

[8] S. Sriram, and R. Seenivasan, Microalgae Cultivation in Wastewater for Nutrient Removal, J. of Algal Biomass Utilization **3**(2), 9-13, (2012)

[9] M. Adamsson, Potential Use of Human Urine by Greenhouse Culturing of Microalgae (Scenedesmus Acuminatus), Zooplankton (Daphnia Magna) and Tomatoes (Lycopersicon). **16**, 243-54, (2000)

[10] D. Van Thang, A. Faruq, S. R. Thomass-Hall, Q. Simon, N. Ekaterina and P. M. Schenk, High Protein- and High Lipid-Producing Microalgae from Northern Australia as Potential Feedstock for Animal Feed and Biodiesel, Frontiers in Bioengineering and Biotechnology, **3**, (2015)

[11] Tongsiri, Sudaporn, Kringsak Mang-amphan and Yuwadee Peerapornpisal, Effect of Replacing Fishmeal with Spirulina on Growth, Carcass Composition and Pigment of

the Mekong Giant Catfish, Asian Journal of Agricultural Sciences, **2**(3), 106-110, (2010)

[12] M. B. Tartiel, E. M. Ibrahim, and M.M. Zeinhom, Partial Replacement of Fish Meal With Dried Microalga (*Chlorella* Spp and *Scenedesmus* Spp) in Nile Tilapia (*Oreochromis Niloticus*) Diets, *Int. Symposium on Tilapia in Aquaculture*, (2008)

[13] Dawah, Aida M, *Efficency of Inoculating the Green Algae*, **11**(3), 115-25, (2007)

[14] A.O.A.C. Association of Official Analytical Chemists: *Official Methods of Analysis*, 18th Edition, Maryland, USA., (2005)

[15] E.G. Bligh, W.J. Dyer, *A* rapid method of lipid extraction and purification. Canadian J. of Biochemistry and Physiology, **37**, 911-917, (1959)

[16] A. L.-S, Folch, A simple method for isolation and purification of total lipids from animal tissues, J. Biol. Chem., **226**, 497-509, (1957)

[17] K. Chiu-Mei, C. Tsai-Yu, L. Tsung-Hsien, K. Chien-Ya, L. Jinn-Tsyy, C. Jo-Shu and L. Chih-Sheng, Cultivation of Chlorella sp. GD using piggery wastewater for biomass and lipid production, Bioresour. Techol., **194**, 326-33, (2015)

[18] Juneja, Ankita, Ruben Michael Ceballos, and Ganti S. Murthy, Effects of Environmental Factors and Nutrient Availability on the Biochemical Composition of Algae for Biofuels Production: A Review, Energies, **6**(9), 4607-38, (2013)

[19] Markou, Giorgos, Dimitris Iconomou, and Koenraad Muylaert, Applying Raw Poultry Litter Leachate for the Cultivation of Arthrospira Platensis and Chlorella Vulgaris. Algal Research, **13**, 79-84, (2016)

[20] Food and Agriculture Organization of the United Nations (FAO), (2013), *Aquaculture Feed and Fertilizer Resources Information System*, Retrieved on September 9, 2016 from http://www.fao.org/fishery/affris/species-profiles/nile-tilapia/faqs/en

A Chemical and Morphological Study of Cassava Peel: A Potential Waste as Coagulant Aid

Syazwani Mohd-Asharuddin[1], *Norzila* Othman[1,*], *Nur Shaylinda* Mohd Zin[1] and *Husnul Azan* Tajarudin[2]

[1]Faculty of Civil and Environmental Engineering, Universiti Tun Hussein Onn Malaysia, 64200 Batu Pahat, Johor, Malaysia
[2] School of Industrial Technology, Universiti Sains Malaysia, 11800 Penang, Malaysia

Abstract. This study investigates the chemical and morphological characteristics of cassava peel (CP) biomass as a potential coagulant aid for turbidity, heavy metals and microbial removal. FE-SEM micrograph shown the surface of the CP samples was covered with smooth and globular in shaped of bound starch granules. FTIR spectra demonstrated that carboxyl and hydroxyl groups were present in abundance. Whereas analysis by XRF spectrometry indicated the CP samples contain Fe_2O_3 and Al_2O_3 which might contribute to its coagulation ability. The features of CP obtained from this study promotes the feasibility of CP to be further developed and studied to produce effective coagulant aid as sustainable alternative to reduce the usage of chemical coagulants.

1 Introduction

The production of potable water from most raw water sources usually involves the process called coagulation and flocculation. This process plays a major role in surface water treatment by reducing turbidity, algae, colour, organic compound, clay particles as well as removing microbial content in the water which results in minimizing the risk of waterborne diseases [1].

Aluminium salts, ferric salts and synthetic polymers are the most common coagulant used in water treatment worldwide. These coagulants are often expensive and many developing countries can hardly afford the costs of imported chemicals for water and waste water treatment. Besides, many researches have raised concern regarding the health effect related to Alzheimer disease caused by residual aluminium ions in the treated waters [2]. Meanwhile, some synthetic organic polymers such as acrylamide are regarded to neurotoxicity and strong carcinogenic effect [1].

For these reasons, alternative approach has been taken to produce novel coagulants with comparable coagulation capability from natural materials that are renewable, abundantly

* Corresponding author: norzila@uthm.edu.my

available and relatively cost effective. Natural polymer coagulants such as chitosan, *Moringa oleifera* seeds, cactus mucilage and plant tannins are mainly composed of polysaccharides and proteins. These biopolymers shows promising potential to be a good choice of water coagulant not only due to their chemical composition but also they are usually available locally at low price and biodegradable [1, 3].

In view of the difference in coagulation capacity of various coagulants as well as considering techno-economic constrain, cassava peel (CP) that is regionally available industrial waste is studied to evaluate it suitability to be selected as coagulant aid for water treatment system. A large amount of CP waste is generated annually by small and medium scale industries. This has led to a new strategy of complete utilization of raw materials so that there will be little or no residue left that could pose pollution problems. Conversion of this industrial by-product into coagulant and adsorption material for water treatment will not only enhance their market value but also reducing waste generated by the producer.

Cassava peels typically have higher concentration of cyanogenic glucosides than the pulp which this characteristic makes the peel unsuitable as animal feed [4]. However it is considered safe to use this agricultural waste as an alternative adsorbent since CP also contains a cyanide detoxification enzyme (cyanoalanine synthase) which sufficiently fast to maintain cyanide at safe concentration [5]. A study also reported that the CP contains sugars in the form of polysaccharides such as starch and holocellulose. Pectin, cellulose and amino acids containing abundant of carboxyl, hydroxyl and amino groups which has significant capabilities for metals binding from aqueous solution as well as flocculating agent [6, 7].

Hence, this paper aims to investigate the characteristics of CP waste as an early study in developing an alternative coagulant aid for turbidity, heavy metals and microbial removal. The surface morphological study and chemical composition analysis were performed on three different sample of CP namely CP1 (whole peel waste consist of periderm, cortex and flesh), CP2 (cortex only) and CP3 (flesh only). The cross sectional of cassava tuber is shown in Fig. 1.

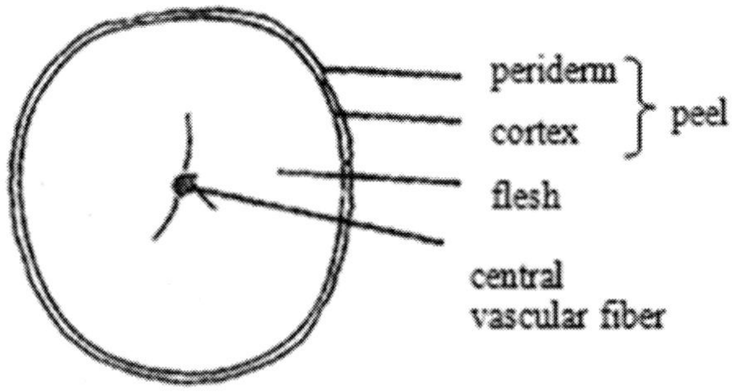

Fig. 1. Cross sectional of cassava tuber [8]

2 Materials and methods

2.1 Preparation of biomaterial

CP was collected as waste from Kilang Kerepek Ahmad Shah, Parit Raja Darat in Johor state, Malaysia. The raw CP was further processed to separate the sample into three type of sample consist of CP1 (periderm, cortex and flesh), CP2 (cortex only) and CP3 (flesh only). Then the CP samples were washed with tap water repeatedly to get rid of surface impurities. Then the peels were soaked in distilled water for further clean up at least for 24 hours followed by oven drying at 60°C for 24 hours to substantially remove its moisture content. The dried peels were then ground to fine powder using domestic blender and sieved to obtain desirable particle size of < 150μm [9, 10]. Final products were kept in air tight plastic container for further analysis.

2.2 Field Emission Scanning Electron Microscopy (FE-SEM)

This analysis was conducted to determine the surface morphology of the CP samples. Analysis was conducted at Microelectronics and Nanotechnology – Shamsuddin Research Centre (MINT-SRC), UTHM. The samples were mounted on the copper stub using an adhesive carbon tape. Then the samples were sputtered with platinum for 120 seconds at 30 mA under high vacuum until they were completely covered and ready to be used for the microscopic analysis [9]. The FE-SEM micrograph were observed using JEOL JSM-7600F Field Emission Scanning Electron Microscope (USA).

2.3 Fourier Transform Infrared (FTIR) spectroscopy

FTIR spectroscopy was done to identify surface functional group of native CP samples and performed using Perkin Elmer Spectrum 100 [9]. The adsorption bands were recorded at characteristics wave numbers between 600cm^{-1} and 4,000 cm^{-1}.

2.4 X-Ray Fluoresence (XRF) spectrometry

All CP samples were also subjected to XRF analysis to determine the elemental composition of the samples using Bruker AXS S4 Pioneer XRF spectrometer [9]. Prior to analysis, 8.0 g of pulverized samples were mixed homogeneously with 2.0 g of binder wax and compressed into pellet using PE-MAN Manual Hydraulic Press.

3 Results and discussions

3.1 Field Emission Scanning Electron Microscopy (FE-SEM)

The surface morphologies of the CP samples are depicted in Fig. 2-4. Based on the SEM micrographs at 500x magnification, it was observed that the CPs surface feature non-porous and heterogeneous characteristics. Similar observations were reported by other studies on SEM analysis on Cassava peel [9-11]. Smooth and globular in shape of bound starch granules were observed to mostly cover overall surface of CP1 and appeared to be in cluster on CP3 sample. While free unbound starch granules were observed in CP2. CP1 sample consist of the whole CP waste composed of the periderm, cortex and flesh layer which might be the reason of numerous appearance of starch granules on its surface.

Cassava tuberous roots are composed of a peel which contain about 10 – 20% of the tuberous root. The cork layer (periderm) covers 0.5 – 2.0% of the total tuberous root weight while the fleshy edible portion make up about 80 – 90% of the tuberous root. The roots are mainly consist of digestible carbohydrates and stored in starch granules. Cassava starch granules are composed of amylose (20%) and amylopectin (80%) [12]. A study on nutritional status of Cassava peel by Aderemi and Nworgu [13] showed that the protein content of the peels was 5.35%, cellulose and hemicellulose contents were 5.40% and 21.65% respectively.

Abdul-Aziz et. al [7] studied the sago starch as coagulant aid for heavy metals and turbidity removal. The research suggests that sago starch enhances the removal of positively charged heavy metals through charge neutralization mechanism. The starch attracts positively charged metals; destabilize the metals which cause them to agglomerate than precipitate.

Fig. 2. FE-SEM micrograph of CP1

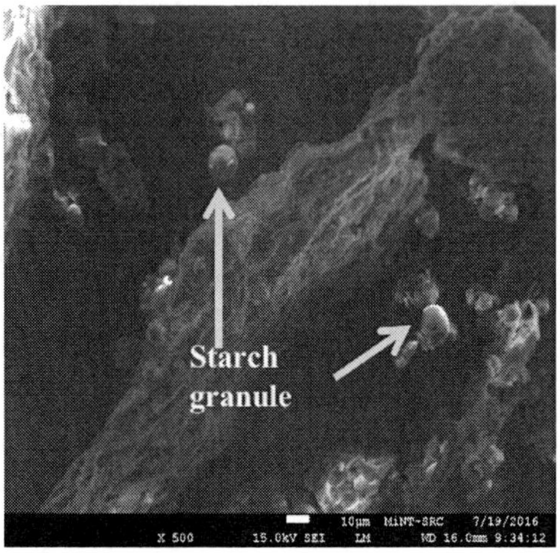

Fig. 3. FE-SEM micrograph of CP2

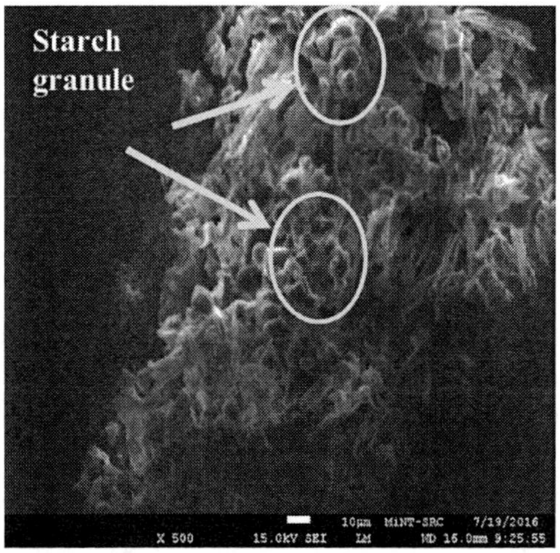

Fig. 4. FE-SEM micrograph of CP3

3.2 Fourier Transform Infrared (FTIR) spectroscopy

A comparison between the FTIR spectra of sample CP1, CP2 and CP3 is shown in Fig. 5. Peaks appearing in the FTIR spectra were assigned to various chemical groups and bonds in accordance with their respective wavenumbers (cm^{-1}).

The broad band observed between 3500 – 3200 cm^{-1} may be assigned to O-H groups of free hydroxyl groups and bonded O-H group in polymeric compounds such as alcohols, phenols and carboxylic acids presented in pectin, cellulose and lignin on the CPs. These bioactive components have been found to have antimicrobial properties [14]. The peak at wavenumber between 3000 – 2850 cm^{-1} may indicate the stretch of symmetric or asymmetric CH, CH_2, of aliphatic acids. Furthermore, the characteristic peak appeared between 1750 – 1680 cm^{-1} are demonstrating the stretching vibration of C=O bond of carboxyl groups. Whereas the peaks observed between `1375 – 1300 cm^{-1} may reflect the stretching vibration of ionic carboxylic groups of pectin. While the deep peaks between 1300 – 1000 cm^{-1} indicate the C-O stretching of –COOH. The results were in agreement with FTIR analysis done onto *Cucumis melo* peels [15], mango peels [16] and citrus peels [17] as shown in Table 1.

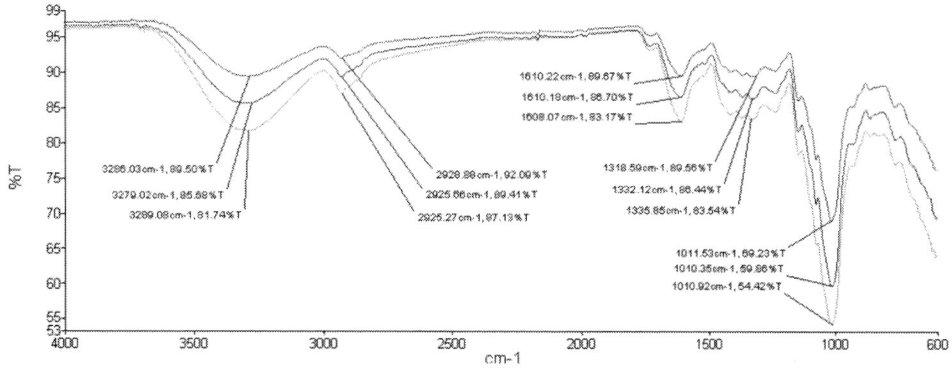

Fig. 5. FTIR spectra of CP1, CP2 and CP3

Table. 1. Functional groups identified by FTIR analysis

Agricultural Waste material	Wavelength Range (cm^{-1})	Identified groups
Cassava peel [this study]	3500 – 3200	O-H groups of free hydroxyl groups and bonded O-H in polymeric compounds such as alcohol, phenols and carboxylic acids
	3000 - 2850	Symmetric or assymetric stretching of CH, CH2 of aliphatic acids
	1750 - 1680	Stretching vibration of C=O bond of carboxyl groups
	1375 - 1300	Stretching vibration of ionic carboxylic groups
	1300 - 1000	C-O stretching of COOH
Cucumis melo rind [15]	3500 – 3200	Bonded hydroxyl groups (-OH) and (-NH) groups
	3000 – 2850	C-H stretching, -CH, $-CH_2$, $-CH_3$
	1750-1680	C=O carbonyl groups
	1640-1500	Carboxylate and carboxyl groups
	1300-1000	-C-O or –C-N bonds
Mango peel [16]	3500 – 3200	Bonded hydroxyl groups (-OH) and (-NH) groups
	3000 – 2850	C-H stretching, -CH, $-CH_2$, $-CH_3$
	1750-1680	C=O carbonyl groups
	1300-1000	-C-O or –C-N bonds
Citrus peel [17]	3000 – 2850	C-H stretching, -CH, $-CH_2$, $-CH_3$
	1750-1680	C=O carbonyl groups
	1640-1500	Carboxylate and carboxyl groups

Since cassava peel is principally made up of lignin and cellulose, it is well demonstrated from the FTIR spectra of each samples that carboxyl and hydroxyl groups were present in abundance especially in CP1 sample. These groups that constituted in biopolymers may function as proton donors; hence deprononated hydroxyl and carboxyl groups may be involved in the formation of alum-starch complex. The insoluble alum-starch salt may initially forms micro flocs (coagulation) which gradually grow to become macro flocs (flocculation). Over time, the macro flocs get settled and improve the water clarity [18].

Awang and Abdul-Aziz [19] suggest that carboxyl (C=O), hydroxyl (O–H) and amino or amide (–NH_2) groups as well as hydrogen bonding, which were the preferred groups for the flocculation process. The deprotonated carboxyl (–C=O-) groups also might serves as ion bridge or binding sites for divalent metal cations (Ca^{2+}, Mg^{2+}, etc.) at the particle surface of wastewater to induce the coagulation activity.

Another research on the feasibility of CP to remove Cu(II) and Ni(II) indicates that the presence of of hydroxyl, carboxyl and carboxylate group contribute to the removal of the metals up to 90% [20]. The inclusion of hydroxyl group in CP on metal binding also reported by Kokasih et. al [5]. Jorgetto et. al [21] suggested that the oxygen on each hydroxyl group acts as a strong Lewis base because of the presence of its vacant double electrons, and this hydroxyl group undergoes a complex coordination with metal Cu(II) which is electron deficient.

3.3 X-Ray Fluorescent (XRF) Spectrometry

XRF analysis results are presented in Table 2. The elements present in the CP that gave it the coagulant properties are Al_2O_3 and Fe_2O_3. Both elements are expected to give the CP the ability to aid the coagulation process via precipitation of the particles in the raw water.

Standard alum consisted of 17-18 wt% Al_2O_3 and 0.4-0.7 wt% Fe_2O_3. Furthermore, the presence of SiO_2 and CaO in CP might help in the addition of weight and size to particles in the raw water as in the case of activated silicate as a flocculation aid [22].

Moreover, SiO_2 was also found in the CP. Same observation was reported in a study of rice husk characterization [23] and watermelon rind [24]. The study proposed that element that may be responsible for sorption are carbon and silica. Rao et al. [25] suggested that when the presence of Si-OH bonds would also act as cation exchange centers. This could be due to the substitution of metal ions at the site produces net negative charge which enables adsorption of cations.

Table 2. Elemental composition of cassava peel

Formula	Weight (%)		
	CP1	CP2	CP3
C	0.10	0.10	0.10
K_2O	5.48	6.18	5.83
CaO	4.20	3.90	3.48
Fe_2O_3	1.54	0.33	0.41
SiO_2	0.87	0.23	0.40
SO_3	0.87	0.94	0.94
Al_2O_3	0.74	-	0.26
P_2O_5	-	-	0.42

4 Conclusions

The characterization of CP samples which were accomplished through the analysis of FE-SEM, FTIR and XRF show the potential of CP1 sample which composed of the whole cassava peel waste comprises of the periderm, cortex and flesh to be developed as coagulant aid. The presence of starch granules, functional groups such as hydroxyl, carboxyl and amines as well as Al_2O_3 and Fe_2O_3 are important properties which accredited to the coagulating agents. Modification and optimization studies will be conducted in the future to further evaluate the performance of CP as coagulant aid for turbidity, heavy metals and microbial removal.

This paper was partly sponsored by the Centre for Graduate Studies UTHM and Fundamental Research Grant Scheme (FRGS Vot. 1575) supported by Ministry of Higher Education (MOHE).

References

[1] S. Vara, Screening and evaluation of innate coagulants for water treatment: A sustainable approach, Int. J. of Energy and Environmental Engineering, **3**, 1-11, (2012)
[2] S.Y. Choy, K.M. Nagendra-Prasad, T.Y. Wu, M.E. Raghunandan, R.N. Ramanan, Utilization of plant-based natural coagulants as future alternatives towards sustainable water clarification, J. of Environmental Science, **26**, 2178-2189, (2014)
[3] B. Ramavandi, Treatment of water turbidity and bacteria by using a coagulant extracted from *Plantago ovate*, Water Resource Industry, **6**, 36-50, (2014)
[4] J. Kongkiattikajorn, B. Sornvoraweat, Comparative Study of Bioethanol Production from Cassava Peels by Monoculture and Co-Culture of Yeast, Kasetsart Journal, **45**, 268-274, (2011)
[5] A.N. Kosasih, J. Febrianto, J. Sunarso, Y.H. Ju, N. Indraswati, S. Ismadji, Sequestering of Cu(II) from aqueous solution using cassava peel (Manihot esculenta), Journal of Hazardous Materials, **180**, 366-374, (2010)
[6] G. Crini, Non-conventional low-cost adsorbents for dye removal: A review, Bioresource Technology, **97**, 1061-1085. (2006)
[7] A.A. Hamidi, A. Mohd-Nordin, M.D.M. Abdul, R. Sujatha, M.I. Mohd-Koflly, A. Mohd-Harmizi, Study on the anionic natural coagulant for heavy metals and turbidity removal in water at ph 7.5 and alum concentration 25mg/l – laboratory scale, Indian Journal of Engineering and Material Sciences, **7**, 195-199, (2000)
[8] Food and Agriculture Organization of the United Nations (FAO). Rome, Italy. (1998)
[9] A. Kurniawan, A.N. Kosasih, J.Febrianto, Y.H. Ju, J. Sunarso, N. Indraswati, S. Ismadji, Evaluation of cassava peel waste as lowcost biosorbent for Ni-sorption: Equilibrium, kinetics,thermodynamics and mechanism, Chemical Engineering Journal, **172**, 158-166, (2011)
[10] N. Zayadi, N. Othman., R. Hamdan, A potential waste to be selected as media for metal and nutrient removal, IOP Conference Series : Material Science and Engineering, **136**(1), 012051, (2016)
[11] I.R.M. Benesi, *Characterization of Malawian cassava germplasm for diversity, starch extraction and its native and modified properties*. University of The Free State, Bloemfontein, South Africa. PhD. Thesis. (2005)
[12] I.R.M., Benesi, Genotype x Environment interaction effects on native cassava starch quality and potential for starch use in the commercial sector, African Crop Science J., **12**, 205-216, (2004)

[13] F.A. Aderemi, F.C. Nworgu, Nutritional status of cassava peels and root sieviate biodegraded with *Aspergillus niger,* American-Eurasian J. of Agriculture and Environmental Science, **2**, 308-311, (2007)
[14] R.U. Maheswari, V. EuginAmala, Analyzing and Determining the Activity of Antimicrobial, Functional Group and Phytochemicals of *Cymbopogon citratus* using Well Diffusion, FT-IR and HPLC, Int. J. of Science and Research, **4**, 138-141, (2015)
[15] N. Othman, S. Mohd-Asharuddin, *Cucumis melo* rind as biosorbent to remove Fe(II) and Mn(II) from synthetic groundwater solution, Advanced Materials Research, **795**, 266-271, (2013)
[16] M. Iqbal, A. Saeed, S.I. Zafar, FTIR spectrophotometry, kinetics and adsorption isotherms modeling, ion exchange, and EDX analysis for understanding the mechanism of Cd^{2+} and Pb^{2+} removal by mango peel waste, Journal of Hazardous Materials, **164**, 161-171, (2009)
[17] S. Schiewer, S. Patil, Modeling the effect of pH on biosorption of heavy metals by citrus peels, J. of Hazardous Materials, **157**, 8-17, (2008)
[18] O.N. Abiola, Appraisal of cassava starch as coagulant aid in the alum coagulation of congo red from aqua system, Int. J. of Environmental Pollution and Solutions, **2**, 47-58, (2014)
[19] N.A. Awang, H. Abdul-Aziz, Hibiscus *rosa-sinensis* leaf extract as coagulant aid in leachate treatment, Applied Water Science, **2**, 293-298, (2012)
[20] K.N. Awokoya, O.J. Owoade, B.A. Moronkola, B.K. Oguntade, A.A. Ibikunle, O.C. Atewolara-Odule, S.A. Ogundare, Morphological characteristics of cassava peel and its effect on the adsorption of heavy metal ions from aqueous media, J. of Multidisciplinary Engineering Science and Technology, **3**, 5342-5348, (2016)
[21] A.D.O. Jorgetto, A.C.P, da-Silva, B.C.R.C. Barbosa, M.A. Utrera, and G.R. de-Castro, Cassava root husks as a sorbent material for the uptake and pre-concentration of cadmium(II) from aqueous media, Orbital: The Electronic J. of Chemistry, **5**, 206-212, (2013)
[22] I. Muhammad-Misau, A.W. Abdulkarim-Yusuf, Characterization of water melon seed used as water treatment coagulant, Journal of Advanced Studies in Agricultural, Biological and Environmental Sciences, **3**, 22-29, (2016)
[23] S. Noor-Syuhadah, H. Rohasliney, Rice husk as biosorbent: A Review, Health and the Environmental J., **3**, 1-7, (2012)
[24] N.Othman, S.Y Kueh, M-F-H Azizul Rahman, R, Hamdan,R. Watermelon rind: a potential adsorbent for zinc removal, Applied mechanics and material, **680**, 146-149, (2014)
[25] K.S. Rao, M. Mohapatra, S. Anand, P. Venkateswarlu, Review on cadmium removal from aqueous solutions, Int. J. of Engineering, Science and Technology, **2**, 81–103, (2010)

Application of Agricultural Wastes Activated Carbon for Dye Removal – An Overview

Amirza, M. A. R.[1], *Adib*, M. M. R.[1,*], and *Hamdan*, R.[2]

[1]Department of Water and Environmental Engineering, Faculty of Civil and Environmental Engineering, Universiti Tun Hussein Onn Malaysia, 86400 Batu Pahat, Johor, Malaysia.
[2]Department of Chemistry Engineering Technology, Faculty of Chemistry Engineering Technology, Universiti Tun Hussein Onn Malaysia, 86400 Batu Pahat, Johor, Malaysia.

Abstract. Dyes are an important class of pollutants and can even be identified by the human eyes. Disposal of dyes in precious water resources must be avoided especially those that are not easily biodegradable, however, and for that various treatment technologies are in use. Among various methods adsorption occupies a prominent place in dye removal. The growing demand for efficient and low-cost treatment methods and the importance of adsorption has given rise to agricultural waste. This review highlights and provides an overview of these activated carbons prepared by non-woody and woody materials and their application for dyes removal. From a comprehensive literature review, it was found that many researchers used non-woody material as activated carbons to removal dye contaminants.

1 Introduction

Saving water to rescue the planet from water polluted situation and to make the future of mankind being safe is what we need now. With the growth of mankind, society, science and technology our world is reaching to new high horizons but the cost which we are paying or will pay in near future is surely going to be too high. Among the consequences of this rapid growth is environmental disorder with a big pollution problem. Besides other needs the demand for water consumption ("Water for People Water for Life" United Nations World Water Development Report UNESCO) has increased tremendously with agricultural, industrial and domestic sectors consuming 70, 22 and 8% of the available fresh water, respectively and this has resulted in the generation of large amounts of wastewater [1] containing a number of 'pollutants'.

In recent years, properties of dyes solution as their visibility and recalcitrance generate problem for various industries such as manufacturing of fabrics, paper, silk, cotton, cosmetics and textile [2-16] in order to color their products [4, 6, 10, 17-28]. Wastewater contaminated by synthetic azo-dyes have complex aromatic molecular structure which make them inert and biodegradable [12, 14-15, 29-31] difficult when discharged into the environment [3, 5, 8, 10-11, 16, 26] if not being removed properly [13, 27, 32-34]. The

* Corresponding author: adib@uthm.edu.my

worldwide production of synthetic dyes is estimated at 7×10^5 to 1×10^6 tons dyestuff [28] produced annually with 10%-60% of dye-containing effluents are emitted into the environment without proper treatment [5, 9, 12, 22, 28, 30, 35]. Dyes contaminants from industrial disposal cause severe problems associated, as dyes given toxicity [12, 14-15, 27, 30-31] to the aquatic flora and fauna life that are present in the ecosystem [2-4, 8-11, 20-23, 25, 36]. The effluent of dyes may damaging the aesthetic nature of the environment [2, 4, 18, 23, 25, 27, 32, 35, 37-38] includes visible pollution, reduce photosynthetic activities and not readily biodegradable by the indigenous [2, 4-6, 8, 22, 25, 28, 31, 36, 38], even at low concentrations [15, 17, 19, 35]. As well as the potential mutagenic and carcinogenic effects of cancer [2, 3, 5, 9, 11, 15, 19-23, 28, 31, 36] after transformations or degradation yield compounds such as benzidine, toluidine and other aromatic amines [8, 25, 32, 35, 39-40] are indiscriminately. Where certain dyes were determined to have a direct effect on the newly formed micronuclei, chromosomal breakage and aneuploidy in human cells [19]. The dyes also can cause severe damage to human beings such as allergy dermatitis [3, 9, 36], skin irritation [3, 9, 26, 36], dysfunction of kidney [25], reproductive system, liver, brain and central nervous system [2, 20, 25].

Conventional techniques available for dye removal such as photo-catalytic degradation [3, 7-8, 15, 21, 25, 36], Fenton degradation, electrochemical degradation [3-4, 8, 14, 21-22, 26, 33, 36, 41], electrocoagulation [8, 11, 15, 26], liquid-liquid extraction [21], irradiation, biodegradation [6], photochemical [11], coagulation-flocculation [12, 16, 20, 27, 37], membrane filtration [6, 16, 21, 27, 29], ultrafiltration membrane [3, 15], ion exchange [29, 32] and persulfate oxidation [2, 16-18, 27, 40] is efficient and economically feasible but it also create secondary pollution notably spent catalyst and fouled membrane that requires further disposal [19]. However, these conventional methods have proven to display low effectiveness of dye removal inherent constraints due to their high cost [13, 29], high energy requirements and generation of hazardous by-products [2, 6, 32, 33, 37, 38] caused by their complex aromatic compounds [25].

In additional to already mentioned methods, the adsorption process has been widely being used for dyes removal. Adsorption techniques particularly by carbon adsorbents has emerged to be the best one of the most promising alternatives [3, 5-6, 9-12, 22-23, 27, 30] efficient technology, which finds diverse applications in several disciplines of purification and decontamination processes [34] for the removal of organic pollutants from wastewaters. Factors that affect the adsorption technique in terms of initial cost, simplicity of design, ease of operation, high removal efficiently, regeneration capacity and insensitivity to toxic substances [5, 18-19, 21-22, 25-26, 29-30, 40] have been found to be among the most effective and proven treatments of dye-loaded wastewater. The adsorptive properties of activated carbon depends on the surface area, good internal porous structure, high thermal stability, low acid/base reactivity and chemical structure of the surface [35, 38, 43].

The most important characteristics of an adsorbent are the quantity of adsorbate it can accumulate which is usually calculated from the adsorption isotherms. Dyes that are difficult to biological breakdown can often be removed by using the adsorbents. A good adsorbent should generally possess a porous structure (resulting in high surface area) and the time taken for adsorption equilibrium to be established should be as small as possible so that it can be used to remove dye wastes in lesser time.

2 Production of Activated Carbon from Agricultural Waste

Generally, activated carbons (ACs) is considered as the most commonly used adsorbent in water and wastewater treatment worldwide [16, 42]. The properties of ACs are a tasteless, solid, microcrystalline, non-graphitic form of black carbonaceous material with a porous structure [16]. ACs has been regarded as a unique and versatile adsorbent due to its extended surface area [26, 28-29, 31], micro porous structure [26], good chemical and mechanical stability, high adsorption capacity [26, 29]. The high degree of activated carbon surface reactivity are caused by heterogeneous porous structures [28-29] which make them perfect adsorbents [11, 15, 17-19, 21, 27, 32-33, 36, 43-46]. The ACs can be produced from a variety of natural and synthetic substances, where lignocelluloses materials being one of the most used precursors [42]. The composition of the lignocelluloses material regarding the cellulose, hemicellulose and lignin content determines, to some extent, the porosity development of the produced ACs [24, 28]. However, a major challenge of commercial available activated carbon (CAC) are still considered expensive materials for many countries due to the use of non-renewable and relatively expensive starting materials such as coal [2-7, 9, 12-13, 19, 22, 26-28, 30, 35-36, 38, 46] but poor regeneration limit its potential usage in the treatment system [2, 7, 11, 13, 27-28, 32, 37]. Commonly in market, commercial activated carbon (CAC) was sales in range of RM90.00/Kg, however for production of sugarcane bagasse activated carbon (SBAC) was in range of RM12.00/Kg. For that reason, production of SBAC using lower heating temperature (500°C) and using raw material from agricultural waste compared with CAC being used higher heating temperature (≥ 1000°C).

Therefore, ACs obtained from agricultural byproducts has the advantage of offering an effective [4, 27, 47,], low cost replacement for non-renewable coal-based granular activated carbon (GAC) [3, 8-11, 24, 27-29, 36] may provide that they have similar of better adsorption efficiency [25,43] such as solid pineapple waste [17], oil palm empty fruit bunches [44], *Thevetia peruviana* [18], cattail [46], peanut shell [40], rambutan (*Nephelium lappaceum*) [35], banana empty fruit bunch (BEFB) and *delonix regia* fruit pod (DRFP) [43], palm shell waste-based [37], coir pith [32], orange peel [38], sawdust and rice-husk [19]. The production of activated carbon prepared from agricultural waste can be revised based on non-woody materials (Table 1) and woody materials (Table 2).

Apparently, activated carbons are prepared by physical and chemical activation processes [15]. In physical activation also known as pyrolysis process, the precursor is being carbonized at high temperature and activated by passing CO_2 of steam under pressure to increase the porosity and surface area [8, 28]. Meanwhile, in chemical activation, both activation and carbonization process take place simultaneously in which the raw precursor is impregnated with activating agents and heated at desired temperature [5, 8, 25, 28, 48]. Chemical activation leads to both physical and chemical modifications on produce activated carbon. The agricultural wastes provide both woody and non-woody materials for adsorption methods replaced with commercial activated carbon. For woody materials composed lignocellulose only compared with non-woody materials composed sugar, starch, lignocellulose and oils [49]. Many authors used non-woody materials due to the composition of materials such as starch and lignocellulose which have high pore structure compared with woody materials.

Based on the review of Table 1 and Table 2 for materials woody and non-woody, it can be concluded that specific surface area of activated carbon produced depends significantly on the type of materials and activating agents used in the adsorption processes. Specific surface area was important parameters used in evaluating many capabilities of powders and porous materials such as activity, adsorptive performance and catalytic performance [50]. The factors affecting on the measured value of specific surface area of activated carbon such as degassing temperature, degassing time and adsorption point number.

Certain agricultural waste products have been tested and proposed for dye removal. Which agricultural waste adsorbent is better to being used? There is no direct answer to this question due to their each agricultural waste has its specific physical and chemical characteristics such as porosity, surface area and physical strength, as well as inherent advantages and disadvantages in wastewater treatment. However, it is clear from the present literature survey that agricultural waste adsorbents may have potential as readily available, inexpensive and effective sorbents. They also possess several other advantages that make them excellent materials for environmental purposes such as high surface area (Table 1 and 2).

Table 1. Non-woody materials of agricultural waste activated carbon.

Raw precursor	**Chemical used in impregnation method**	**Pollutants in dye removal**	**Specific surface area analysis (BET) (m^2/g)**	**Ref.**
Pomegranate peel	$ZnCl_2$, nitric acid solution	Dye direct blue 106	-	[2]
Orange peel	H_2SO_4	Dye direct navy blue 106	-	[3]
Coir pith	H_2SO_4	Dyes reactive orange 12, reactive red 2 and reactive blue 4	Coir pith carbon (598) and Granular coir pith carbon (557)	[12]
Rice husk	H_2SO_4	Dyes crystal violet, direct orange and magenta	98.27	[13]
Rice husk	-	Dyes congo red and magenta	-	[14]
Sugarcane bagasse	$ZnCl_2$	Basic dye	-	[15]
Hazelnut bagasse	$ZnCl_2$	Dye acid blue 350	1489	[16]
Solid pineapple waste biomass, SPWB (crown, leaves, stem)	$ZnCl_2$	Dye methylene blue (MB)	1002 (leaf), 955 (stem), 794 (crown)	[17]
Thevetia peruviana	Na_2SO_4, H_3PO_4, $ZnCl_2$, KOH, HCl, H_2SO_4, direct pyrolysis, dolomite, H_2SO_4 + H_2O_2	Dyes methylene blue (MB), Basic green 4, Acid violet 49, Reactive orange 4 and Direct blue 71	329.27 (pyrolysis), 205.63 (Na_2SO_4), 862.39 (H_3PO_4), 503.77 ($ZnCl_2$), 561.1 (KOH), 156.24 (HCl), 390.69 (H_2SO_4), 85.93 (Dolomite), 299.85 (H_2SO_4 + H_2O_2)	[18]
Sawdust and rice-husk	-	Dye acid yellow 36	Sawdust carbon was 516.3 while rice-husk was 272.5	[19]
Silk cotton hull, coconut tree sawdust, sago waste, maize cob, banana pith	H_2SO_4	Dyes were rhodamine-B, congo red, methylene blue, methyl violet, malachite green	-	[20]
Flamboyant pods	NaOH	Dyes acid yellow 6, acid yellow 23 and acid red 18	2854	[23]
Esparto grass	-	-	1122	[24]
Avocado peel	-	Dyes naphthol blue black and reactive black 5 and basic blue 41	53 (raw material) and 452 (after carbonization)	[25]

Table 1. Non-woody materials of agricultural waste activated carbon (*Cont...)*

Raw precursor	Chemical used in impregnation method	Pollutants in dye removal	Specific surface area analysis (BET) (m^2/g)	Ref.
Sugarcane bagasse pith	H_3PO_4 and $ZnCl_2$	Dye reactive orange	-	[27]
Grape processing waste	$ZnCl_2$	Dyes methylene blue and metanil yellow	1455	[28]
Apple pulp and peel	H_3PO_4	Dye methylene blue	1552 (apple peel) and 1103 (apple pulp)	[29]
Orange peel	H_3PO_4	Dyes methylene blue and rhodamine B	1090	[30]
Prickly pear peels, broccoli stems and white sapote seeds	H_3PO_4	Dyes acid blue 74, direct blue 80, basic blue 9 and basic violet 3	1025 (prickly pear peels), 1177 (broccoli stems) and 1043 (white sapote seeds)	[31]
Coir pith	$ZnCl_2$	Dyes acid brilliant blue, acid violet, methylene blue, rhodamine B, direct red 12B, congo red, procion red, procion orange	Precursor presence of $ZnCl_2$ was 910 and 167 for absence of $ZnCl_2$	[32]
Jerusalem artichoke stalk-based	$ZnCl_2$	Dyes methy orange and methylene blue	Jerusalem artichoke AC was 1632 compared to PAC was 1410 and 582 for GAC	[33]
Pomela skin	NaOH	Dyes methylene blue and acid blue 15	52 (Char) and 1335 (PSAC)	[34]
Rambutan (*Nephelium lappaceum*)	KOH	Dye acid yellow 17	971.54	[35]
Orange peel	H_2SO_4 + $NaHCO_3$	Dye methylene blue	-	[38]
Tomato processing solid waste	$ZnCl_2$	Dyes methylene blue and metanil yellow	1093	[41]
Date palm leaflets	KOH	Dye methylene blue	823	[42]
Banana empty fruit bunch (BEFB) and *Deloxia regia* fruit pod (DRFP)	H_3PO_4, KOH	Dye methylene blue (MB)	BEFB for 0.32 (untreated), 15.38 (H_3PO_4) and 1.04 (KOH) DRFP for 17.55 (untreated), 22.29 (H_3PO_4) and 0.01 (KOH)	[43]
Cattail	H_3PO_4	Dyes Neutral red and Malachite green	1279	[46]
Corncob	H_3PO_4	Dye methylene blue	1809	[47]

Table 1 represents some of the non-woody materials of agricultural waste of activated carbon reported. From the recent literature reviewed, adsorbents that stand out for high surface area are flamboyant pods (2854 m^2/g), corncob (1809 m^2/g), Jerusalem artichoke

(1632 m^2/g), apple peel (1552 m^2/g), hazelnut bagasse and grape waste with the surface area of (1489 m^2/g) and (1455 m^2/g). As usual, the solution pH is an important parameter in the adsorption process due to the interaction effect of the surface functional groups of the adsorbate and the adsorbent. Besides that, the properties of the activated carbon and of the dyes may indicate whether adsorption process is favorable in acid or base solutions.

[23] stated that at initial concentration of 500 mg/L, the maximum removal percentages of activated carbon from flamboyant pods were 99.88%, 99.87% and 98.97% for acid yellow 6, acid yellow 13 and acid red 18 dyes, respectively. Activated carbons prepared by Jerusalem artichoke stalk based mesoporous (MAC) were studied as adsorbents [33] for adsorption removal of anionic (methyl orange, MO) and cationic dyes (methylene blue, MB). It was observed that the amount of dyes adsorbed, q_t (mg/g) values increased from 170.2 mg/g to 223.7 mg/g with the pH value increased from 3.44 to 10.6 at initial 60 min for MO dye, but different condition for anionic dye (MO) stated that the q_t value decreased from 238.1 mg/g to 179.3 mg/g when pH was increased from 3.15 to 10.7 at initial 60 min. For instance, the MAC becomes negatively charged, which may produce a considerable high electrostatic attraction to simulate the MB moving toward the MAC surface, but when pH value was lower than pH_{pzc} (3.70) condition, the surface of MAC is positively charged by adsorbing H^+ ions which is attractive to the anionic dye due to electrostatic repulsion between MO and negative charges on the MAC surface. Therefore, the authors suggested that the electrostatic interaction attraction may enhance the adsorption rate but is not the primary adsorption mechanism between the cationic dye and the partially negative charge MAC surface.

Grape processing waste (GW) was studied by Saygili *et al.*, (2015) for the adsorption of methylene blue (MB) and metanil yellow (MY) dyes. The authors investigated the impregnation ratio of $ZnCl_2$ increased from 1:1 to 1:6 (GW/$ZnCl_2$), the surface area values also increased from 911 m^2/g to 1361 m^2/g, respectively. It is obvious that impregnated with chemical as an activating agent is very efficient in order to produced ACs with high surface area and porosity. Due to the chemical activation, the equilibrium data for both dye adsorptions onto grape processing waste activated carbon (GWAC), showing the maximum monolayer adsorption capacity of 417 mg/g for MB and 386 mg/g for MY. This may attributed to the fact that $ZnCl_2$ selectively extracted H and O away from the GW rather than ACs. Hence, this causes eventually to an increase in the surface area and porosity [51].

Due to their low cost and local availability, agricultural wastes such as Fox nutshell and Holm oak acorn are classified as woody materials (Table 2) and can produced activated carbon used as adsorbents for dye removal [10, 21]. Recently, [21] studied the removal of methylene blue and phenol onto activated carbon from Fox nutshell (FNAC) by chemical activation ($ZnCl_2$). The maximum % removal of MB was observed at pH 11. Similar results are reported for the adsorption of methylene blue on Jute fiber and wheat shells [52-53]. When the pH value decreased at pH 3, the authors found that the low adsorption rate of MB occurred on FNAC due to the positive charged on the surface, causing H^+ ions to compete effectively with MB cations. The equilibrium adsorption of MB was increased from 249.88 tom968.74 mg/g which indicates that the MB adsorption process onto FNAC was exothermic in nature.

[10] studied that the highest micropore percentages was obtained using KOH as an activating agent for activated carbon from Holk oak acorn. Further, in order to know the effect of chemical treatment and to improve its efficiency the authors also tested the potential of the adsorbent by treating it with H_3PO_4 and using $ZnCl_2$, found that activated carbons obtained by H_3PO_4 had a mesopore structure while those obtained from $ZnCl_2$ had a heterogeneous pore distribution consisting of micropores and mesopores. Similar to the findings of [54] preparation of activated carbons from sewage sludge via chemical activation using H_3PO_4 may contribute large mesoporous and some microporous being

observed for all activated carbons. The increase in surface area of ACs due to a phosphoric acid reagent has been explained in the reviews as H_3PO_4 aiding the thermal decomposition of the lignocelluloses material.

Table 2. Woody materials of agricultural waste activated carbon.

Raw precursor	Chemical used in impregnation method	Pollutants in dye removal	Specific surface area analysis (BET) (m^2/g)	Ref.
Coconut shell	H_3PO_4	Dye yellow (Tartrazine E102)	-	[4]
Cocoa shell	Red mud + lime + KOH + $Al(NO_3)_3$ + Na_2SO_4	Dye reactive violet 5	-	[5]
Macore fruit	NaOH	Dyes methylene blue and methyl orange	229.51	[6]
Cashew nut shell	KOH + TiO_2, KOH, TiO_2	Dyes brilliant green and methylene blue	-	[7]
Walnut and poplar woods	H_3PO_4	Dye acid red 18	-	[8]
Coconut shell	H_3PO_4	Dye reactive blue 19	-	[9]
Holm oak acorn	KOH, $ZnCl_2$, H_3PO_4, H_2SO_4	Dye disperse orange 30	968 (H_3PO_4, 550°C), 1305 ($ZnCl_2$, 750°C), 582 (KOH, 750°C)	[10]
Apricot stones	H_3PO_4 + HNO_3	Dyes methyl orange and methylene blue	359.40 (H_3PO_4 + HNO_3) and commercial activated carbon (510.57)	[11]
Fox nutshell	$ZnCl_2$	Dye methylene blue (MB)	2869	[21]
Spent coffee grounds	KOH	Acid orange 7 and methylene blue	704.23	[22]
Peanut sticks wood	HCl and HNO_3	Dye methylene blue	218.89	[26]
Olive stone	-	Dye methylene blue	-	[36]
Palm shell-waste based	NaOH	Dye methylene blue (MB)	731.50	[37]
Peanut shell	Conventional pyrolysis (P sample) and microwave irradiation followed by pyrolysis (MW-P sample)	Dyes Direct black 38 and reactive red 141	370.10 (P sample) and 395.80 (MW-P sample)	[40]

3 CONCLUSIONS

The majority of the reviewed studies concluded that a wide range of activated carbons prepared from agricultural waste as adsorbents of dye removal. It is worthwhile noting that the removal of dyes can be done by various materials such as non-woody and woody waste. However, there exists no such methodology of which activated carbons can successfully remove all types of dyes at low cost adsorbents. The agricultural waste is converted into value-added materials as activated carbons, which are generated in huge amounts annually. The modification of the surface chemistry of carbons is also an important influenced, although depending on the raw materials of activated carbon, both increases and decreases in surface area have been reported. It has been generally shown that many researchers using agricultural waste from non-woody materials compared to woody materials in term of removal dyes in water polluted. As a reason, non-woody materials is ease to be conducted and chemical composition of these agricultural waste have high ash content and fiber such as hemicellulose, cellulose and lignin which may reduce the mechanical strength and adsorption capacity of activated carbon.

As environmental regulations become stricter, the effectiveness and cost of treatment processes for dye becomes more significant. Instead, the dye removal data available in literature suggests that removal of dyes is possible by agricultural waste to a certain extent since some promising results are obtained in some of the studies. Therefore, it is understood that impregnation ratio of chemical activating agents significantly affect to porosity. With low impregnation ratio, the formation of tar is inhibited and the release of volatiles is promoted, producing more mesopores. Rather at higher impregnation ratio, the more swelling impregnated precursor materials and stronger release of volatiles in the activation process will be lead to the widening of pores: micropores formed are subsequently converted to mesopores [55]. The application of activated carbon in adsorption process was mainly depends on the surface chemistry and pore structure or porous carbons. If possible, develop modification of surface chemical activation method on activated carbon prepared by non-woody material should be provided. Various methods should be produced such as acidic treatment and alkaline treatment.

This paper was partly sponsored by the Centre for Graduate Studies UTHM. The author also would like to express gratitude to the Ministry of Education Malaysia under Postgraduate Research Grant Scheme (FRGS: Vot 1456), Department of Environment Malaysia (DOE) and Universiti Tun Hussein Onn Malaysia for the support in preparing this paper. Grateful acknowledgement also goes to all that are involved directly or indirectly to completing this paper.

References

[1] V.K. Gupta and Suhas, Application of low-cost adsorbents for dye removal: a review, J. of Environmental Management, **90,** 2313-2342, (2009)

[2] N.K. Amin, Removal of direct blue 106 dye from aqueous solution using new activated carbons developed from pomegranate peel: adsorption equilibrium and kinetics, J. of Hazardous Materials, **165,** 52-62, (2009)

[3] A. Khaled, A.E. Nemr, A. El-Sikaily and O. Abdelwahab, Removal of direct N blue 106 from artificial textile dye effluent using activated carbon from orange peel: adsorption isotherm and kinetic studies, J. of Hazardous Materials, **165,** 100-110, (2009)

[4] M. Jibril, J. Noraini, L.S. Poh and A.M. Evuti, Removal of color from waste water using Coconut Shell Activated Carbon (CSAC) and commercial Activated Carbon (CAC), J. teknologi (Sciences & Engineering), **60**, 15-19, (2013)

[5] M.C. Ribas, M.A. Adebayo, L.D.T. Prola, E.C. Lima, R. Cataluna, L.A. Feris, M.J. Puchana-Rosero, F.M. Machado, F.A. Pavan and T. Calvete, Comparison of a homemade cocoa shell activated carbon with commercial activated carbon for the removal of reactive violet 5 dye from aqueous solutions, Chemical Engineering J., **248,** 315-326, (2014)

[6] K.N. Aboua, Y.A. Yobouet, K.B. Yao, D.L. Gone and A. Trokourey, Investigation of dye adsorption onto activated carbon from the shells of Macore fruit, J. of Environmental Management, **156**, 10-14, (2015)

[7] S. Ragupathy, K. Raghu and P. Prabu, Synthesis and characterization of TiO_2 loaded cashew nut shell activated carbon and photo catalytic activity on BG and MB dyes under sunlight radiation, Spectrochimica Acta Part A: Molecular and Biomolecular Spectroscopy, **138**, 314-320, (2015)

[8] B. Heibati, S.R. Couto, M.A. Al-Ghouti, M. Asif, I. Tyagi, S. Agarwal and V.K. Gupta, Kinetics and thermodynamics of enhanced adsorption of the dye AR 18 using activated carbons prepared from walnut and poplar woods, J. of Molecular Liquids, **208**, 99-105, (2015)

[9] I.A. Umar, G. Abdulraheem, S. Bala, S. Muhammad and M. Abdullahi, Kinetics, equilibrium and thermodynamics studies of C. I. reactive blue 19 dye adsorption on coconut shell based activated carbon, International Biodeterioration and Biodegradation, **102**, 265-273, (2015)

[10] U.T. Un, F. Ates, N. Erginel, O. Ozcan and E. Oduncu, Adsorption of disperse orange 30 dye onto activated carbon derived from holm oak (*Quercus Ilex*) acorns: A 3^k factorial design and analysis, Journal of Environmental Management, **155**, 89-96 (2015)

[11] C. Djilani, R. Zaghdoudi, F. Djazi, B. Bouchekima, A. Lallam, A. Modarressi and M. Rogalski, Adsorption of dyes on activated carbon prepared from apricot stones and commercial activated carbon, J. of the Taiwan Institute of Chemical Engineers, **53**, 112-121, (2015)

[12] K. Santhy and P. Selvapathy, Removal of reactive dyes from wastewater by adsorption on coir pith activated carbon, Bioresource Technology, **97**, 1329-1336, (2006)

[13] V.K. Verma and A.K. Mishra, Kinetic and isotherm modeling of adsorption of dyes onto rice husk carbon, Global NEST J., **12**(2), 190-196, (2010)

[14] S. Saeed, S. Khan, S. Saeed and R. Khan, Removal of dyes from textile waste water using adsorption by activated carbon of rice husk, Int. J. of Innovation and Scientific Research, ISSN 2351-8014 **17**, 191-196, (2015)

[15] M.B. Fabon, G.J. Legaspi, K. Leyesa and M.C. Macawile, Removal of basic dye in water matrix using activated carbon from sugarcane bagasse, *Int. Conf. on Innovations in Engineering and Technology*, 198-201, (2013)

[16] H. Demiral, I. Demiral, B. Karabacakoglu and F. Tumsek, Adsorption of textile dye onto activated carbon prepared from industrial waste by $Zncl_2$ activation, J. Int. Environmental Application & Science, **3**(5), 381-389, (2008)

[17] M.N. Mahamad, M.A.A. Zaini and Z.A. Zakaria, Preparation and characterization of activated carbon from pineapple waste biomass for dye removal, Int. Biodeterioration and Biodegradation, **102**, 274-280, (2015)

[18] J.R. Baseri, P.N. Palanisamy and P. Sivakumar, Preparation and characterization of activated carbon from *Thevetia peruviana* for the removal of dyes from textile wastewater, Advances in Applied Science Research, **3**(1), 377-383, (2012)

[19] P.K. Malik, Use of activated carbons prepared from sawdust and rice-husk for adsorption of acid dyes: a case study of acid yellow 36, Dyes and Pigments, **56**, 239-249, (2003)

[20] K. Kadirvelu, M. Kavipriya, C. Karthika, M. Radhika, N. Vennilamani and S. Pattabhi, Utilization of various agricultural wastes for activated carbon preparation and application for the removal of dyes and metal ions from aqueous solutions, Bioresource Technology, **87**, 129-132, (2003)

[21] A. Kumar and H.M. Jena, Removal of methylene blue and phenol onto prepared activated carbon from Fox nutshell by chemical activation in batch and fixed-bed column, J. of Cleaner Production, **137**, 1246-1259, (2016)

[22] K.W. Jung, B.H. Choi, M.J. Hwang, T.U. Jeong and K.H. Ahn, Fabrication of granular activated carbons derived from spent coffee grounds by entrapment in calcium alginate beads for adsorption of acid orange 7 and methylene blue, Bioresource Technology, **219**, 185-195, (2016)

[23] A.M.M. Vargas, A.L. Cazetta, A.C. Martins, J.C.G. Moraes, E.E. Garcia, G.F. Gauze, W.F. Costa and V.C. Almeida, Kinetic and equilibrium studies: adsorption of food dyes acid yellow 6, acid yellow 23 and acid red 18 on activated carbon from flamboyant pods, Chemical Engineering J., **181-182**, 243-250, (2012)

[24] J.V. Nabais, C. Laginhas, M.M.L.R. Carrott, P.J.M. Carrott, J.E.C. Amoros and A.V.N. Gisbert, Surface and porous characterization of activated carbons made from a novel biomass precursor, the esparto grass, Appl. Surf. Sci, **265**, 919-924, (2013)

[25] C. Palma, L. Lioret, A. Puen, M. Tobar and E. Contreras, Production of carbonaceous material from avocado peel for its application as alternative adsorbent for dyes removal, Chinese J. of Chemical Engineering, **24**, 521-528, (2016)

[26] M. Ghaedi, A.G. Nasab, S. Khodadoust, M. Rajabi and S. Azizian, Application of activated carbon as adsorbents for efficient removal of methylene blue: kinetics and equilibrium study, J. of Industrial and Engineering Chemistry, **20**, 2317-2324, (2014)

[27] N.K. Amin, Removal of reactive dye from aqueous solutions by adsorption onto activated carbons prepared from sugarcane bagasse pith, Desalination, **223**, 152-161, (2008)

[28] H. Saygili, F. Guzel and Y. Onal, Conversion of grape industrial processing waste to activated carbon sorbent and its performance in cationic and anionic dyes adsorption, J. of Cleaner Production, **93**, 84-93, (2015)

[29] R.H. Hesas, A. Arami-niya, W.M.A. Wan Daud and J.N. Sahu, Preparation and characterization of activated carbon from apple waste by microwave-assisted phosphoric acid activation: application in methylene blue adsorption, Bioresources, **8**(2), 2950-2966, (2013)

[30] M.E. Fernandez, G.V. Nunell, P.R. Bonelli and A.L. Cukierman, Activated carbon developed from orange peels: batch and dynamic competitive adsorption of basic dyes, Industrial Crops and Products, **62**, 437-445, (2014)

[31] A.A.P. Cid, A.M.H. Gonzalez, M.S. Villanueva and A.B. Hernandez, Elimination of textile dyes using activated carbons prepared from vegetable residues and their characterization, J. of Environmental Management, **181**, 269-278, (2016)

[32] C. Namasivayam and D. Sangeetha, Recycling of agricultural solid waste, coir pith: removal of anions, heavy metals, organics and dyes from water by adsorption onto $ZnCl_2$ activated coir pith carbon, J. of Hazardous Materials, **B135**, 449-452, (2006)

[33] L. Yu and Y.M. Luo, The adsorption mechanism of anionic and cationic dyes by Jerusalem artichoke stalk-based mesoporous activated carbon, J. of Environmental Chemical Engineering, **2**, 220-229, (2014)

[34] K.Y. Foo and B.H. Hameed, Microwave assisted preparation of activated carbon from pomelo skin for the removal of anionic and cationic dyes, Chemical Engineering J., **173**, 385-390, (2011)

[35] V.O. Njoku, K.Y. Foo, M. Asif and B.H. Hameed, Preparation of activated carbons from rambutan (*Nephelium lappaceum*) peel by microwave-induced KOH activation for acid yellow 17 dye adsorption, Chemical Engineering J., **250**, 198-204, 2014)

[36] R. Hazzaa and M. Hussein, Adsorption of cationic dye from aqueous solution onto activated carbon prepared from olive stones, Environment Technology & Innovation, **4**, 36-51, (2015)

[37] K.T. Wong, N.C. Eu, S. Ibrahim, H. Kim, Y. Yoon and M. Jang, Recyclable magnetic-loaded palm shell-waste based activated carbon for the effective removal of methylene blue from aqueous solution, J. of Cleaner Production, **115**, 337-342, (2016)

[38] T. Teka and S. Enyew, Study on effect of different parameters on adsorption efficiency of low cost activated orange peels for the removal of methylene blue dye, International J. of Innovation and Scientific Research, ISSN 2351-8014, **8**, 106-111, (2014)

[39] B. Ratna and B.S. Padhi, Pollution due to synthetic dyes toxicity and carcinogenicity studies and remediation, Int. J. Environ. Sci, **3**, (2012)

[40] J. Georgin, G.L. Dotto, M.A. Mazutti and E.L. Foletto, Preparation of activated carbon from peanut shell by conventional pyrolysis and microwave irradiation-pyrolysis to remove organic dyes from aqueous solutions, J. of Environmental Chemical Engineering, **4**, 266-275, (2016)

[41] H. Saygili and F. Guzel, High surface area mesoporous activated carbon from tomato processing solid waste by zinc chloride activation: process optimization, characterization and dyes adsorption, J. of Cleaner Production, **113**, 995-1004, (2016)

[42] E.I. El-Shafey, S.N.F. Ali, S. Al-Busafi and H.A.J. Al-Lawati, Preparation and characterization of surface functionalized activated carbons from date palm leaflets and application for methylene blue removal, J. of Environmental Chemical Engineering, **4**, 2713-2724, (2016)

[43] P. Sugumaran, V.P. Susan, P. Ravichandran and S. Seshadri, Production and characterization of activated carbon from banana empty fruit bunch and *deloxic regia* fruit pod, J. of Sustainable Energy and Environment, **3**, 125-132, (2012)

[44] M.Z. Alam, E.S. Ameem, S.A. Muyibi and N.A. Kabbashi, The factors affecting the performance of activated carbon prepared from oil palm empty fruit bunches for adsorption of phenol, Chem. Eng. J, **155**, 191-198, (2009)

[45] M.J. Ahmed and S.K. Dhedan, Equilibrium isotherms and kinetics modeling of methylene blue adsorption on agricultural wastes-based activated carbons, Fluid Phase Equilib, **317**, 9-14, (2012)

[46] Q. Shi, J. Zhang, C. Zhang, C. Li, B. Zhang, W. Hu, J. Xu and R. Zhao, Preparation of activated carbon from cattail and its application for dyes removal, J. of Environmental Sciences, **22**(1), 91-97, (2010)

[47] G.Z. Zhu, X.L. Deng, M. Hou, K. Sun, Y.P. Zhang, P. Li and F.M. Liang, Comparative study on characterization and adsorption properties of activated carbons by phosphoric acid activation from corncob and its acid and alkaline hydrolysis residues, Fuel Processing Technology, **144**, 255-261, (2016)

[48] J. Hayashi, A. Kazehaya, K. Muroyama and A.P. Watkinson, Preparation of activated carbon from lignin by chemical activation, Carbon, **38**, 1873-1878, (2008)

[49] C.L. Williams, T.L. Westover, R.M. Emerson, J.S. Tumuluru and C. Li, Sources of biomass feedstock variability and the potential impact on biofuels production, Bioenergy Research, **9**(1), 1-14, (2016)

[50] J. Song, L. Wang and G. Song, Research on influence factors on determination of specific surface area of carbon material by N_2 adsorption method, J. of Applied Science and Engineering Innovation, **1**(1), 77-82, (2014)

[51] I. Ozdemir, M. Sahin, R. Orhan and M. Erdem, Preparation and characterization of activated carbon from grape stalk by zinc chloride activation, Fuel Process. Technol,**125**, 200-206, (2014)

[52] Y. Bulut and H. Aydin, A kinetics and thermodynamics study of methylene blue adsorption on wheat shells, Desalination, **194**, 259-267, (2006)

[53] S. Senthilkumaar, P. Varadarajan, K. Porkodi and C. Subbhuraam, Adsorption of methylene blue onto jute fiber carbon: kinetics and equilibrium studies, J. Colloid Interface Sci, **284**, 78-82, (2005)

[54] T. Boualem, A. Debab, A. Martinez de Yuso and M.T. Izquierdo, Activated carbons obtained from sewage sludge by chemical activation: gas-phase environmental applications, J. Environ. Manage, **140**, 145-215, (2014)

[55] J. Yang and K. Qiu, Development of high surface area mesoporous activated carbons from herb residues, Chem. Eng. J, **167**(1), 148-154, (2011).

Enhanced Freshwater Production Using Finned-Plate Air Gap Membrane Distillation (AGMD)

Mohammad Jabed Perves Bappy[1], *Rubina* Bahar[1,*], *Shaliza* Ibrahim[2] and *Tasnim Firdaus* Ariff[1]

[1]Department of Manufacturing and Materials Engineering, International Islamic University Malaysia, 50728 Kuala Lumpur, Malaysia

[2]Department of Civil Engineering, University of Malaya, 50603 Kuala Lumpur, Malaysia

Abstract. Air Gap membrane distillation (AGMD), a special type of energy efficient membrane distillation process, is a technology for producing freshwater from waste water. Having some benefits over other traditional processes, this method has been able to draw attention of researchers working in the field of freshwater production technologies. In this study, a basic AGMD system with flat coolant plate has been modified using a specially designed channelled coolant plate of portable size to observe its effect over the production rate and performance of the system. Attempt has been made to increase the amount of distillate flux by using the "fin effect" of the channelled coolant plate. A finned plate have been used instead of a flat coolant plate and experiments were conducted to compare the effect. Coolant temperature and feed temperature of the system have been varied from 10°C to 25°C and 40°C to 70°C respectively. Comparing the data, around 50% to 58% distillate enhancement has been observed for channelled coolant plate. Also, it was seen that the enhancement was higher for higher feed temperatures and coolant temperatures. With these findings, a better performing AGMD module has been introduced to mitigate the scarcity of freshwater.

1 Introduction

Membrane distillation (MD) is becoming one of the most popular technique of wastewater treatment in recent years which combines the benefits of thermal and membrane separation processes [1]. The process contains a hot aqueous solution or wastewater side which purifies itself by being separated through a very thin microporous hydrophobic membrane toward a cooler side with lower temperature. The hydrophobic nature of the membrane prevents penetration of the aqueous solution into the pores, resulting in a vapour-liquid interface at each pore entrance. The driving force in the MD process is the vapour pressure difference induced by the temperature difference across the hydrophobic membrane [2] . MD is somewhat a recent technique with its energy efficient nature and cost effectiveness

* Corresponding author: rbahar@iium.edu.my

which allow it to be installed anywhere in the world with minimum capital compared to other methods available in the market [3].

The common available methods of treating wastewater includes thermal distillation and reverse osmosis (RO). Thermal distillation requires special infrastructure to install and RO needs very high pressure to overcome the osmotic pressure. Also the feasibility of thermal distillation is decreasing for its high energy consumption as energy saving is becoming a big issue around the world. Reverse Osmosis in still a popular one but have limitations above 50°C where the membrane quality deteriorates significantly which degrades the performance of the system [4]. Membrane distillation fairly overcomes the limitation of thermal distillation and reverse osmosis. The requirement of very low grade energy and high customizability give it the added advantages over those processes [5].

The work aims to improve the overall production from a small scale wastewater treatment system by implementing channelled coolant plates instead of the traditionally used flat plates. The coolant plate is used in a special type of water purification system which is called Air Gap Membrane Distillation (AGMD) [5]. It is a type of MD process where the vapour from the waste/saline water travels through the hydrophobic membrane pores and gets condensed on the cooler side on a condensing (coolant) plate [6]. If the coolant plate can be designed and manufactured with channels, it would impose a "fin"-effect on the coolant side which subsequently would accelerate condensation and thus make more room for more vapour to come in. As AGMD units are compact and portable, and does not need high pressure/temperature [6-9], it seems to be a promising technology for water supply in the arid and remote areas with water scarcity. Improvement of this system by implementing channelled coolant plates will aid in the productivity and help to maintain future demand.

2 Experimental procedure and setup

2.1 Design

The 3-D design of the AGMD module has been done using Solidworks 2013 CAD software package. Dimensions have been chosen to maintain the portability of the system. The complete module have five basic parts; 1) Hot solution Chamber, 2) Membrane ramp, 3) Air gap frame, 4) Fined coolant plate and 5) Cold solution chamber. The dimension of the two chambers, hot and cold, was selected as 162.40 x 106.60 mm. A detailed drawing of the two symmetric chambers is shown in Fig. 1. The inlet 15 mm diameter pipe of the feed chamber was inclined at 60° angle with the base of the chambers. The angle was given so that the stream of circulating hot water can hit the membrane directly thus reducing polarization of temperature and concentration on the evaporating surface. Exit pipe diameter was 15 mm too with straight 90° angle with the chamber.

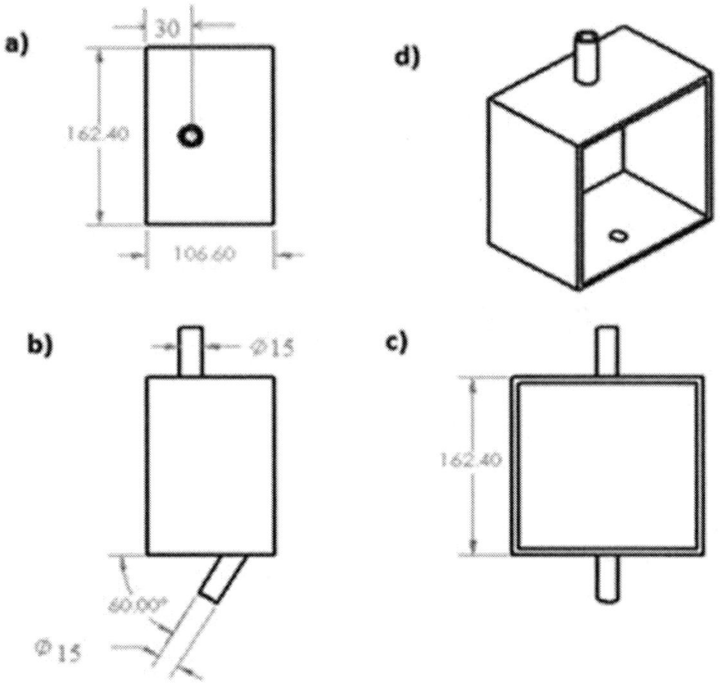

Fig. 1. Design of the hot and cold solution chamber a) top view, b) front view, c) side view and d) isometric view.

The study mainly focussed on implementing the finned coolant plate which was designed very carefully. 60 fins were put on the 165.33 mm*165.33 mm sized plate. Each fin's dimension is 30mm*5mm. The detail spacing of the fins are shown in Fig. 2 where front, top, side & isometric view of the finned plate is mentioned.

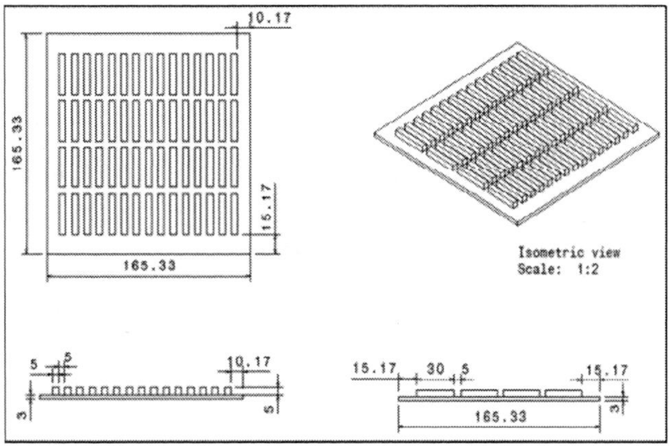

Fig .2. Design of the channeled coolant plate (front, top, side and isometric view).

The membrane ramp dimension have been shown in Fig. 3 where it can be seen that the rectangular plates dimension is 202*202 mm. Membrane will sit on the ramp in the middle. The inside hollow circle's diameter is 134 mm. The thickness of the ramp is 5mm. Four holes on the four corners of the ramp have been made to screw it up with the module with a diameter of 7 mm. Air gap frame is a rectangular frame which will provide the air gap for AGMD system.

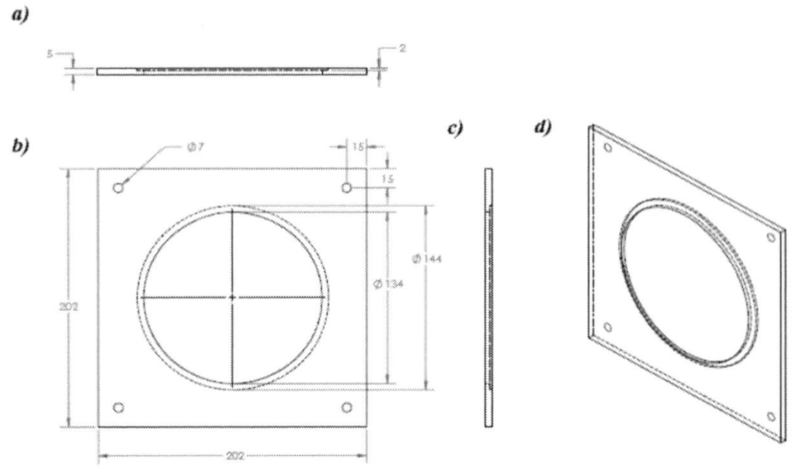

Fig. 3. Design of the membrane ramp of AGMD module a) top view, b) front view, c) right side view and d) isometric view.

The air gap was considered 5mm. And the overall design parameters are summarized in Table 1.

Table 1. Design parameters of the AGMD module (Rect.=Rectangular).

Hot & cold Chambers		**Membrane Ramp**		**Air gap frame**		**Finned Coolant plate**	
Shape	Rect. box	Shape	Rect. plate (mm)	Shape	Rect. hollow frame (mm)	Shape	Rect. finned plate
Height	162.40 mm	Height	202	Height	165.33	Height	165.33 mm
Width	162.60 mm	Width	202	Width	165.33	Width	165.33 mm
Length	106.60 mm	Hollow circle dia	134	Frame thickness	5	Fin height	30 mm
Inlet Dia	15 mm	Ramp area thickness	5	Material thickness	5	Fin width	5 mm
Outlet Dia	15 mm	Screw hole Dia	7	--	--	Number of fins	60
Inlet angle	60°	Material Thickness	5	--	--	Material thickness	3 mm

2.2 Materials

The material selection for fin included an intensive literature study to understand the suitability. A material with cost effectiveness, easy machinability and thermal effiectiveness was to be selected. Considering all the factors, aluminium was chosen to be used with significantly high thermal conductivity compared to other available materials [10]. PTFE hydrophobic membrane of 0.50 μm pore size and 150 μm thickness with 85% porosity was used from Fluoropore membrane series supplied by Merck Millipore.

Material for the other parts of the module was made mainly by grade 'A' polycarbonate sheets. Thickness of the sheet was 5mm. Dichloromethane was used to join the polycarbonate sheet to make the box sized chamber as it provide a weld like joint making a chemical bonding with the polycarbonate sheets which is very effective for high temperature operation. High quality silicon glue was used to make the chambers airtight. The overall 3-D design of the module can be seen in Fig. 4. The circulation pipes were made from reinforced PVC. Steel rings and water regulators were used to control the bending of the PVC pipe and water flow inside the water circulating pipe.

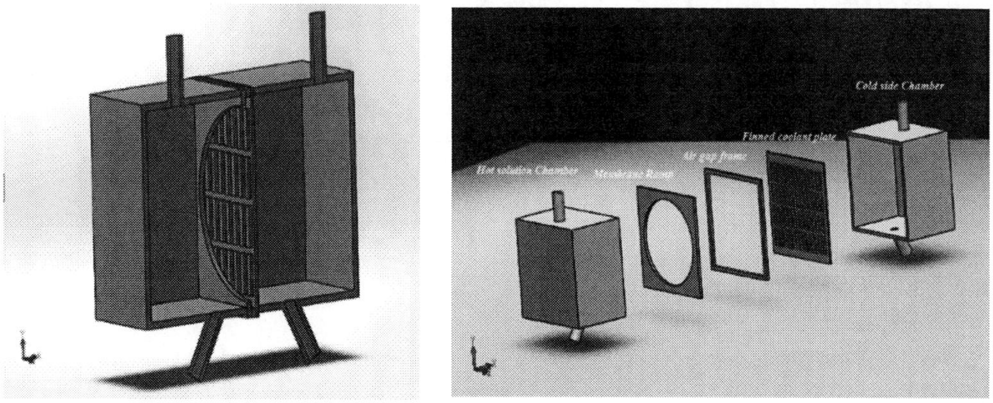

Fig. 4. The 3D design of the AGMD module using Solidworks 2013 software package.

2.3 Construction and experimental setup

For the construction of the whole AGMD setup, Polycarbonate sheets were cut very precisely with CNC milling machine according to the dimension from the design using G-codes. The cut pieces were joined to build the box using high quality epoxy glue and dichloromethane chemical additive. 2.5 mm screws have been used to combine all five parts into a compact chamber. The whole box was put on a steel frame base and all the PVC pipelines was joined later. The edges of the pipelines and chambers have been sealed to make it air and water tight using silicon glue.

The experimental setup is shown in Fig. 5 indicating the major parts including the schematic, overall experimental setup and the membrane holding module. Two sets of LAUDA Alpha RA8 Heating and cooling thermostats recirculated the flow of hot feed and coolant. The feed used was tap water mixed with food colour. The food colour was mixed to observe any immediate leakage in membrane. The distillate colour should be totally transparent and this was the major criteria during experiments. The temperature difference between hot and cold side drives the vapour to pass the membrane and travel through the air gap to condense on the coolant plate. As the PTFE hydrophobic membrane only allows the vapour to pass through the membrane, it completely separates any impurities on that water. The vapour get condensed on the plate and can be collected as distillate from the below of the chamber inside a graduated distillate collection beaker.

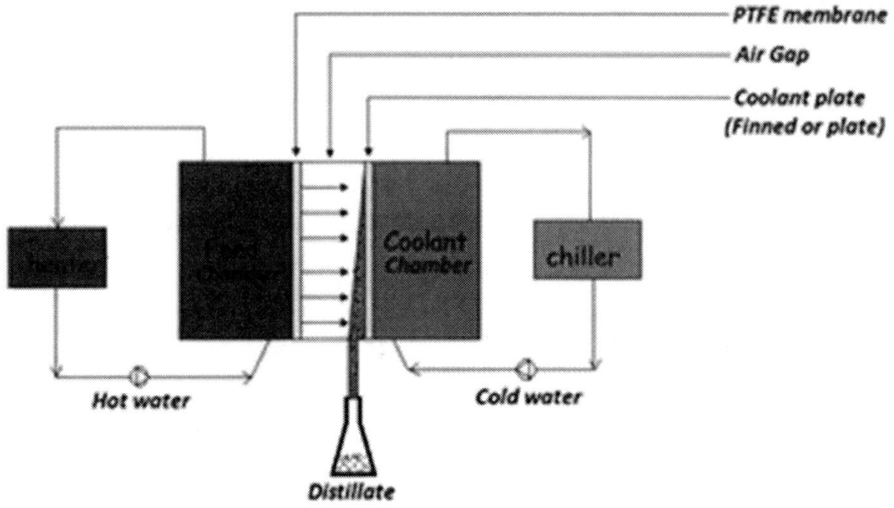

Fig. 5. Experimental setup of the AGMD module (a) Schematic diagram (b) Overall setup (c) Membrane Module

Data were taken varying the feed and coolant temperatures. Coolant temperature was varied from 10°C to 25°C with 5°C minor. On the other hand feed temperatures was varied from 40°C to 70°C with the same minor. Distillate was collected in every 10 minutes for every single data using a scaled measuring beaker. The flowrate was further lowered from the supply source(LAUDA recirculator) using a flowrate valve and was kept constant throughout the experiments. Temperatures have been measured with 0.1 mm thermocouple temperature probes with an accuracy of 0.5 °C.

3 Results and discussion

In the study, the amount of freshwater production was observed after using channelled coolant plate instead of a flat plate. The evaporation process on the surface of the membrane got tweaked and increased largely because of the increased condensation heat transfer surface of the fin.

The vapour coming out of the membrane pore found a bigger area to condense, thus in a way, the partial pressure difference was manipulated. The feed temperature is varied from 40°C to 70°C on this experiment where cold side temperature was varied from 10°C to 25°C. From Fig. 6 it can be observed that a significant enhancement of the distillate flux was found in channelled plat compared to flat plate for different coolant temperatures. It was also observed that, with higher feed temperature, the enhancement in production with finned plate was also increasing. Fig. 7 shows the percentile increase in freshwater production using the channelled coolant plate for various coolant temperatures. The effect of fins in producing freshwater was more dominant for increased coolant temperature as well.

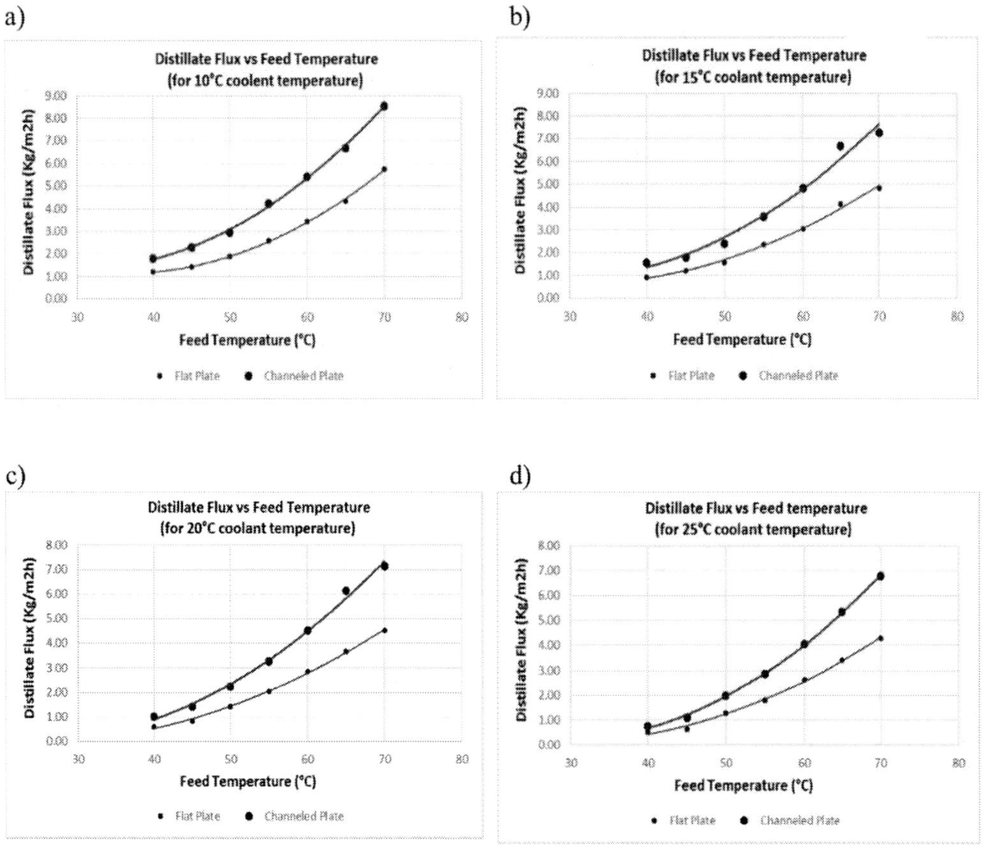

Fig. 6. Distillate Flux vs Feed temperature graph at a)10°C coolant side temperature, b) 15°C coolant side temperature, c) 20°C coolant side temperature and d) 25°C coolant side temperature.

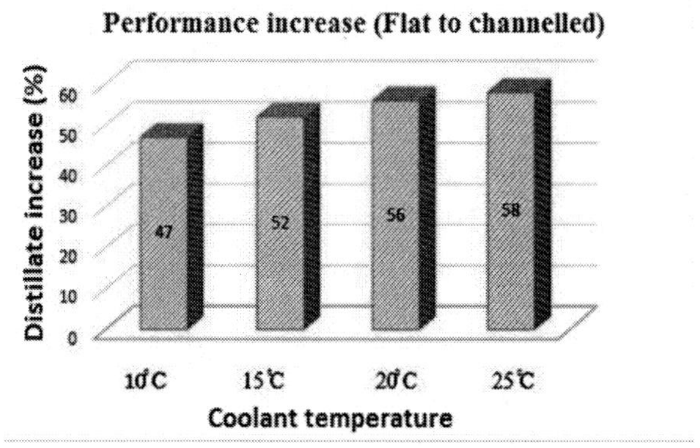

Fig. 7. Performance increase of the system from flat to channelled collant plate at various coolant temperatures.

4 Conclusion

Finned/channelled coolant plate has been able to increase the freshwater production rate significantly for a lab scale AGMD unit. After imposing 60 fins, the increased distillate production of 47%, 52%, 56% and 58% have been achieved respectively at 10°C, 15°C, 20°C and 25°C coolant side temperatures. This increment has been influenced by the trend in increasing feed and coolant temperatures. The distillate was collected through a sealed path inside a sealed container. However, some entrapment of the distillate droplets has been observed on the top surface of each fin; this condition can be eliminated with a circular fin. The increased performance of the module can supply more amount of fresh water per unit time which will make the system more cost effective and can be made usable in daily basis even in home environment with a compact design. The initial feed water used for this study was coloured tap water. In future, household waste water will be used as the feed after necessary pre-treatment.

The work was jointly supported by Fundamental Research Grant Scheme (FRGS-14-118-0359), IIUM Endowment fund (EDW B14-134-1019) and IIUM-UM joint grant (CG015-2014) by providing necessary financial and technical assistance for this project.

References

[1] A. Alklaibi and N. Lior, Membrane-distillation desalination: status and potential. Desalination, **171**,(2), (2005)

[2] S. Bouguecha, R. Chouikh and M. Dhahbi, Numerical study of the coupled heat and mass transfer in membrane distillation, Desalination, **152**(1), (2003).

[3] A. Alkhudhiri, N. Darwish and N. Hilal, Membrane distillation: a comprehensive review, Desalination, **287,** (2012).

[4] K.P. Lee, T.C. Arnot and D. Mattia, A review of reverse osmosis membrane materials for desalination—development to date and future potential, J. Membr. Sci., **370**(1), (2011).

[5] R. Tian, H. Gao, X.H. Yang, S.Y. Yan and S. Li, A new enhancement technique on air gap membrane distillation, Desalination, **332**(1), (2014).

[6] U. Dehesa-Carrasco, C. Pérez-Rábago and C. Arancibia-Bulnes, Experimental evaluation and modeling of internal temperatures in an air gap membrane distillation unit, Desalination, **326**,(2013).

[7] M.I. Ali, E.K. Summers and H.A. Arafat, Effects of membrane properties on water production cost in small scale membrane distillation systems, Desalination, **306,** (2012).

[8] A. Alkhudhiri, N. Darwish and N. Hilal, Treatment of saline solutions using air gap membrane distillation: experimental study, Desalination, **323,** (2013).

[9] Geng, H., et al., Experimental study of hollow fiber AGMD modules with energy recovery for high saline water desalination. Desalination, **344,** (2014).

[10] J.R. Davis and J.R. Davis, *Aluminum and aluminum alloys,* ASM international, (1993).

Factor Affecting Textile Dye Removal Using Adsorbent From Activated Carbon: A Review

Mohd Adib Mohammad Razi[1,*], *Mimi Nur Attahirah* Mohd Hishammudin[1] and *Rafidah* Hamdan[2]

[1]Faculty of Civil Engineering and Environmental, Universiti Tun Hussein Onn Malaysia, 86400 Batu Pahat, Johor
[2]Faculty of Engineering Technology, Uniersiti Tun Hussein Onn Malaysia, 86400 Batu Pahat, Johor

Abstract. Industrial company such as textile, leather, cosmetics, paper and plastic generated wastewater containing large amount of dye colour. The removal of dye materials are importance as the presence of this kind of pollutant influence the quality of water and makes it aesthetically unpleasant. As their chemical structures are complicated, it is difficult to treat dyes with municipal waste treatment operations. Even a small quantity of dye does cause high visibility and undesirability. There have been various treatment technique reviewed for the removal of dye in wastewater. However, these treatment process has made it to another expensive treatment method. This review focus on the application of adsorbent in dye removal from textile wastewater as the most economical and effective method, adsorption has become the most preferred method to remove dye. The review provides literature information about different basis materials used to produce activated carbon like agricultural waste and industrial waste as well as the operational parameters factors in term of contact time, adsorbent dosage, pH solution and initial dye concentration that will affect the process in removing textile dye. This review approach the low cost and environmental friendly adsorbent for replacing conventional activated carbon.

1 Introduction

Textile industries processes are among the most industrial that release coloured wastewater containing dye that become major environmental concern. Without a proper treatment, the disposal of dye wastewater can cause harm for the aquatic species and environment. [1]. Therefore, wastewater from textile industry has to be treated before being discharged into the environment. The textile industry is classified into three main categories as in Table 1. Dye is a coloured substance that been used widely in textile industries which provide bright and lasting colour when binding with fabric or surface. Dye was estimated to be used for thousands year back and about 180,000 year ago, colorant is believed to be use earliest by Neanderthal [2]. Synthetic dyes gained huge popularity in textile and dyeing process because of their stability to light, temperature, detergent and variety in colour [3]. In fact, it

[1]Corresponding author: adib@uthm.edu.my

has reached nearly 100,000 different types of dyes and annually produce over $7.0x10^5$ tonnes per year worldwide [3-8].

Table 1. Fibre and fabric categories for textile industries.

Fibres	Fabric	Type of dye and chemical	Ref.
Cellulose	Cotton, rayon, linen, ramie, hemp and lyocell	Reactive dyes (remazol, procion MX and cibaxron F). direct dyes (congo red, direct yellow 50 and direct brown 116), naphthol dyes (fast yellow GC, fast scarlet R and fast blue B) and indigo dyes (indigo white, tyrisn purple and indigo carmine)	[9]
Protein	Wool, angora, mohair, cashmere and silk	Acid dyes (azo dyes, triarylmethane dyes and anthraquinone dyes) and lanaset dyes (Blue 5G and Bordeaux B)	
Synthetic	Polyester, nylon, spandex, acetate, acrylic, ingeo and polyproplylene	Dispersed dyes (disperse yellow 218 and disperse navy 35), basic dyes (basic orange 37 and basic red 1) and direct dyes	

Loss of up to 40% dye during the dyeing process are the result of coloured wastewater that passed into the environment or treatment plants [10]. The coloured wastewater are unpleasant and heavily pollute the environment. Dyes present in industrial wastewater are carcinogenic, mutagenic, teratogenic and toxic [11]. In addition, most of dye leads to hazardous affects not only to animal and human health but also microorganisms [12-14]. Dye have complex aromatic molecular structures and stable towards heat and oxidizing agent [14,15]. Therefore, dyes removal is prime important issue for researchers. Recently, many different techniques such as biological treatment [16], nanofiltration, oxidation process, ion exchanges, ozonations, ultrafiltration, coagulation, etc. have been studied for the treatment of dyes from wastewater [12,17-19]. However, these processes are expensive and ineffective to treat huge range of wastewater unlike adsorption using activated carbon which is proven to be one of the most effective technique [20].

Adsorption process is among the famous process that been studied by researchers. Adsorption is effective in removing colour and soluble organic pollutants include organic dye, toxic chemical, phenol, pesticides and cyanides [18,21]. The most used adsorbent in removing colours and other pollutants include heavy metal from effluents is activated carbon. However, conventional activated carbon that been used today are expensive for a large scale of treatment [21-23]. Recently, the researchers has been focus on the development of low cost adsorbent for a full scale treatment [5,24,25]. Therefore, this review provide comprehensive information on the application of activated carbon for the treatment of dye and study the effect of adsorption factors towards dye removal.

2 Textile dye

Commercially, more than 100,000 dye are used and available in colour index today [19]. Dye are stable to light [26] and can be classified in several ways based on their structures, application method used or even by colour [27]. The common dyes applied in textile industries are classed as direct dye, acid dye, basic dye, mordant dye, azo dye, reactive dye, disperse dye, sulphur dye, vat dye [28]. Nowadays, the classification of dye in term of application used in dyeing process are often. In addition, dye can also be classified by their solubility in water [29] or else by their particle charge during dissolution such as anionic dyes (reactive, direct and acid dye), cationic dye (basic dye) and also non-ionic dye (disperse dye) [30]. In addition, anionic and cationic dye including in water soluble dye while non-ionic dye is in the categories of water insoluble dye.

Reactive dye been design for cellulose fiber and contain high colour of organic substance [31]. This reactive dye have high solubility in water and the most permanent dye among others dye [32]. As for direct dye from azo group, these dye provide dull colour and is lightfast as compared to reactive dye. Some azo direct dye may contain benzidine-based dye that could result an exposure to allergies, irritation, cancer risk and treating aquatic life if they are directly release without any treatment. Back then, benzidine-based dye are used to colour textile, leather, wood, paper, plastic, fur and cosmetic industries [33]. As for worldwide textile industries, dye are discharged more than 1000 tones/year into waste stream [5,34]. Consequently, these textile wastewater containing benzidine-based dye may cause major damage to ecosystem including loss of soil productivity, disturbing the oxygen transfers in water interface and also interrupt photosynthesis activity.

3 Classification of adsorbent

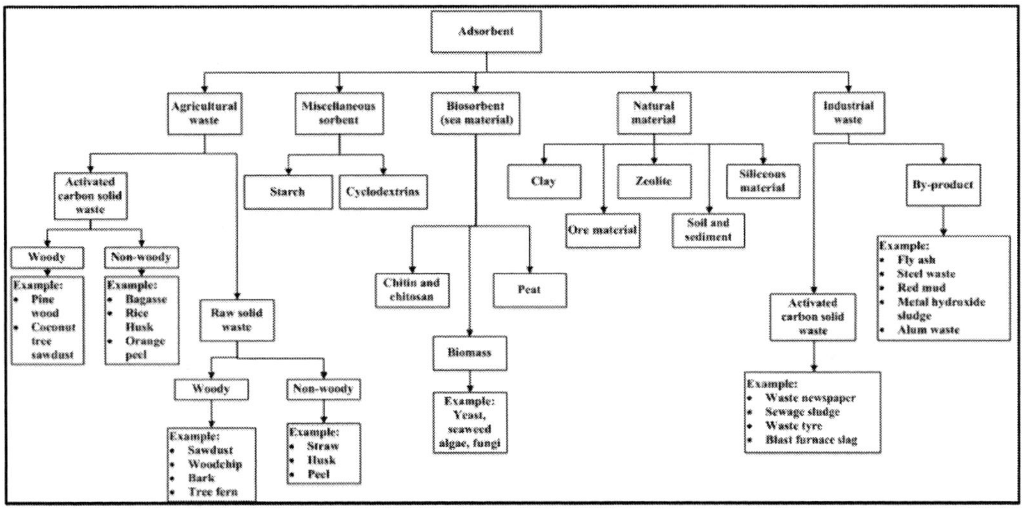

Fig. 1. Type of adsorbent materials used for wastewater treatment. [24,25]

Among several wastewater treatment process, adsorption have the potential in reducing water pollutants and dyes from textile industries. Adsorption is effective to be used in lowering dye concentration in the effluents [35-36]. Fig. 1 shown various type of adsorbent being studied in order to treat wastewater containing dye, pigments and other pollutants. Today, the most commonly adsorbent used for treatment is activated carbon [37] and applied for various water pollutant removal such as dye and heavy metal [35,38]. Activated carbon also known as solid sponge [39] is a carbon form by using either physical or chemical treatment [40]. Activated carbon have extended surface area, high capacity of adsorption, higher surface reactivity degree and also micro-pore structure which is suitable in eliminating dye from wastewater [41-43]. However, treating wastewater by using conventional activated carbon that is available in the market are expensive [1, 5, 31, 43-44] and its regeneration are even costly [26]. Therefore, various raw material have been examined to produce activated carbon and each presents different properties as some previous research of adsorbent shown in Table 2.

Table 2. Previous research on adsorbent in removing dye from wastewater.

Adsorbent	Type of dye	Max Adsorption Capacity	Ref.
Date pits activated carbon	Methylene blue	244 mg/g	[20]
Date pits activated carbon	Remazol yellow	173 mg/g	[20]
Nickel doped zinc sulphide activated carbon	Congo red	285.714 mg/g	[45]
Palladium nanoparticles activated carbon	Congo red	126.58 mg/g	[45]
Myrtus communis activated carbon	Congo red	19.231 mg/g	[1]
Pomegranate activated carbon	Congo red	10 mg/g	[1]
Argemone Mexicana	Direct red 81	6.9 x 10-5 mol/g	[46]
Grape marc-based activated carbon	Reactive Black 5	333 mg/g	[47]
TiO2/GMAC	Reactive Black 5	56 mg/g	[47]
Pomegranate peel activated carbon	Remazol Brilliant Blue R	370.86 mg/g	[32]
Bagasse fly ash	Malachite green	170.33 mg/g	[26]
Activated carbon commercial grade	Malachite green	8.27 mg/g	[26]
Activated carbon laboratory grade	Malachite green	42.18 mg/g	[26]
Phoenix dactylifera seeds	Congo red	61.72 mg/g	[14]
Sugarcane bagasse activated carbon	Methylene blue	198 mg/g	[48]
Spent tea leaves	Brilliant green, Porcion red	9.57 mg/g 4.8 mg/g	[49]
Jackfruit peels	Brilliant green, Porcion red	9.47 mg/g 6.12 mg/g	[49]
Rambutan peels	Brilliant green, Porcion red	9.64 mg/g 4.12 mg/g	[49]
Mangosteen peels	Brilliant green, Porcion red	9.27 mg/g 4.51 mg/g	[49]
Sugarcane bagasse activated carbon	Acid Blue 80	384.6 mg/g	[38]

3.1 Agricultural waste

Adsorbent produce from agricultural waste is cheap and can be divided onto two groups: i) activated carbon ii) solid waste which from raw agricultural waste and material waste from forest industries. Various studied by researchers has been conduct on raw and activated carbon agricultural waste [50] such as sugarcane bagasse [8,18,50-55], olive stone [56], date pits [20],langsat peel [34], apple waste [42], macadamia shell [57], rice husk [52], orange peel [58], prosopis juliflora plant [59], bamboo [60], corncob [41], coffee residue [61], pomegranate peel [1,32], coconut shell [11,62] etc. These agricultural waste have been investigated for various dye removal from aqueous solution such as basic dye [15,63-64], direct dye [11], benzidine-based dye [6], reactive dye [47,65] and others. Wan suraya et al. [44] have been studied the potential of sugarcane bagasse to be an alternative for conventional activated carbon. Sugarcane bagasse which is non-woody activated carbon are proven to be environmental friendly and low cost activated carbon and also efficient adsorption agent in removing dye and other pollutants from textile wastewater as compared to woody activated carbon [35,38,44,64]. Fig. 2 shown an example of structure surface for agricultural waste adsorbent. The use of agricultural waste in order to treat wastewater not only can solve environmental issue with inexpensive cost but also can reduce the agricultural waste in landfill as the waste can be reused.Agricultural waste is one of famous material to be used as activated carbon, however only few research are mention in this review as this review focusing on factors affect adsorption process.

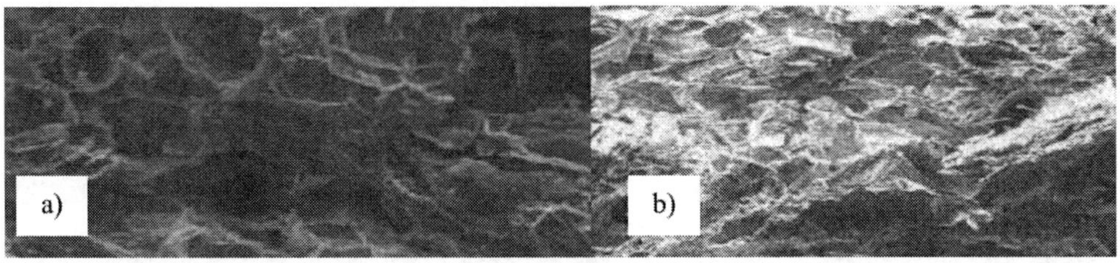

Fig. 2. Agricultural waste adsorbent SEM picture a) sugarcane bagasse activated carbon [38] b) raw sugarcane bagasse [38]

3.2 Industrial waste

Industrial waste also produce adsorbent that can be divided into two groups which are activated carbon solid waste and by-product. Sewage sludge is one of the adsorbent materials included in activated carbon solid waste categories. In addition, several sewage sludge based adsorbent has been used to remove pollutants [22,66-68]. These adsorbent can be produced using different activation process to adsorb metals and dye from wastewater [22]. Example of structure surface for industrial waste adsorbent can be shown as in Fig. 3.Research by Kacan and Kutahyali [67] analysed that the structure surface of sewage sludge prepared from chemical activation shortened the time of adsorption. As for industrial by product categories, fly ash and red mud is popular to be adsorbents materials [24,25]. Moreover, fly ash used in many countries even though it may contain some hazardous substance [25] because it is abundance and easily available material [36].There are lot of researcher also interest to studies on industrial waste but not all are listed in this review.

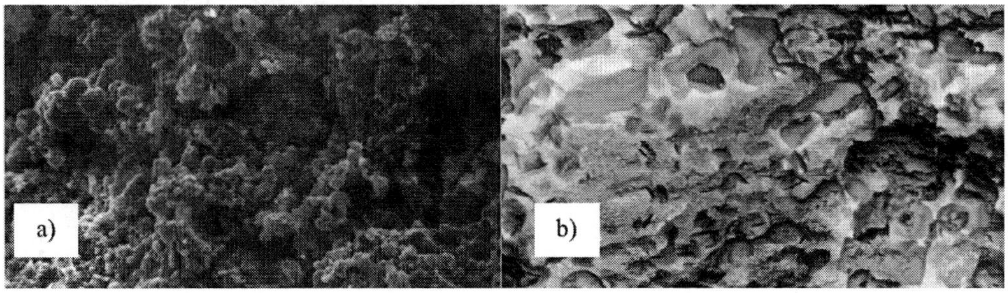

Fig. 3. Industrial waste adsorbent SEM picture a) sewage sludge activated carbon [67] b) usedtyre activated carbon [68]

3.3 Biosorbent

Biosorbent or sea material based adsorbent also has been studied by several researcher as an alternative material to treat wastewater [69-72]. Chitosan, algea and seafood waste are include in biosorbent materials [24] and example of surface structure for biosorbent is shown in Fig. 4. In Copello et al. [23] research, chitosan and chitin was used to remove three type of dye which is reactive dye (Reactive Black 5), basic dye (Basic Red 5 and Basic Violet 3) and acid dye (Acid Red 51). Other research explored the used of biomass of

Trametes Virsicolor in eliminating direct dye which is benzidine based dye [70]. The result shown that heat-treat biomass adsorb more dye effectively as compared to native biomass for both direct blue 1 and direct red 128 dye. The maximum biosorption capacities of heat-treat biomass recorded is 152.3 mg/g direct blue 1 and 225.4mg/g for direct red 128. Das and Charumathi [71] mention that yeast have attractive futures as compared to fungi and bacteria with the ability to resist unfavourable environment and can rapidly growth. Miao et al. [72] research using alligator weed as activated carbon which produced a rough and high surface area, resulting high adsorption capacity.As same as agricultural waste and industrial waste, biosorbent also listed in researcher favourite subject but only few are mention in this review.

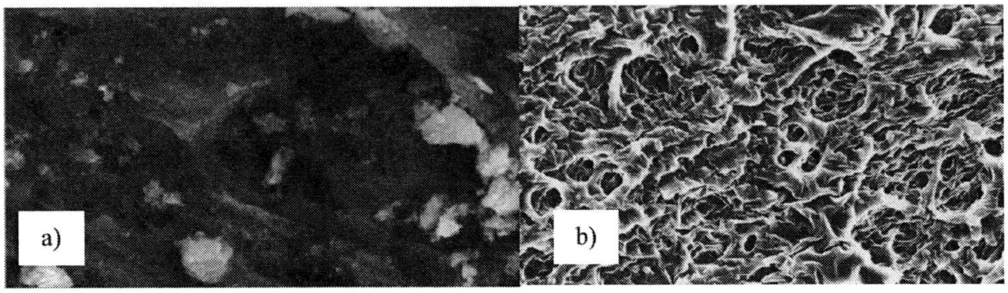

Fig. 4. Biosorbent SEM picture a) alligator weed activated carbon [71] b) fungus [70]

4 Factor affecting adsorption.

Table 3 shows the operational parameter factors that affecting adsorption process conducted by other researchers. Dye adsorption can be affected by various factor like contact time, dosage of adsorbent, pH and initial dye concentration. These factors optimisation can be refer to be applied on full scale treatment process for eliminating dye. Thus, these parameters will be discussed in this review.

Table 3. Review on factors affecting adsorption studied by previous researchers.

Type of adsorbent	**Type of removal**	**Operational parameter factors**						**Ref.**
		pH	**Contact time**	**Adsorbent concentration**	**Dosage**	**Initial dye concentration**	**Temperature**	
Raw bagasse and modified bagasse	Basic dye (malachite green)			✓	✓			[73]
Treated sugarcane bagasse and powdered activated carbon	Acid dye (acid red 2)	✓			✓	✓		[8]
Modified immobilized activated alumina	Reactive dye (reactive yellow)		✓	✓	✓			[74]
Fly ash bagasse and commercial activated carbon	Basic dye (malachite green)	✓	✓		✓	✓		[26]

Table 3. Review on factors affecting adsorption studied by previous researchers. (*Cont...*)

Type of adsorbent	Type of removal	Operational parameter factors						Ref.
		pH	Contact time	Adsorbent concentration	Dosage	Initial dye concentration	Temperature	
Cloth activated carbon	Arsenic	✓	✓	✓		✓		[75]
Green vegetable waste activated carbon	Copper II	✓	✓		✓			[76]
Orange peel activated carbon	Direct dye (direct navy blue 106)	✓	✓		✓	✓		[58]
Alligator weed activated carbon	Cephalexin	✓	✓		✓	✓	✓	[72]
Coconut shell activated carbon	Reactive dye (reactive blue 19)	✓	✓			✓		[31]
Walnut activated carbon and poplar activated carbon	Acid dye (acid red 18)	✓	✓		✓			[77]
Cryogenic grinding used tyre activated carbon	Cationic dye (methylene blue)	✓	✓		✓	✓		[68]
Commercial activated carbon	Reactive dye (reactive black 5)	✓	✓			✓	✓	[78]
Loofa sponge activated carbon	Cationic dye (methylene blue)	✓			✓	✓		[79]
Olive stone activated carbon	Cationic dye (methylene blue)	✓	✓			✓		[80]
Ficus carica bast activated carbon	Cationic dye (methylene blue)	✓	✓		✓	✓		[14]
Lansium domesticum peel	Cationic dye (methylene blue)	✓	✓		✓	✓		[34]
Raw wheat straw, raw sawdust and sugarcane activated carbon	Fluoride	✓	✓		✓	✓		[55]

 DOI: 10.1051/matecconf/201710306015

4.1 Effect of contact time

Generally, dye removal rate increase with an increase in contact time to a certain extent. Due to deposition of dyes on the available adsorption site on adsorbent material, any further increase in contact time will not increase the uptake [5]. At this point, the amount of dye desorbing form the adsorbent is in a state of dynamic equilibrium with the amount of dye being adsorbed onto the adsorbent. The time required to attain this state of equilibrium is termed the equilibrium time and the amount of dye adsorbed at the equilibrium time reflects the maximum adsorption capacity of the adsorbent under those operating conditions [34]. El-Sayed et al. [41] mention that contact time between adsorbent and adsorbate significantly affecting the performance of dye removal. As result in Fig. 5a, the removal was rapidly increase at the first 10 min but then slowly to obtain the equilibrium because of the strong attraction force between methylene blue dye and activated carbon. The time taken for corncob activated carbon with 400°C temperature of activation is 45 min, while for 500°C and 600°C longer time needed which is 120 min. The relation between direct blue 6 with contact time was studied in Khaled et al. [58] research. The time to reach equilibrium for orange peel activated carbon claim to be nearly 3h for maximum adsorption to obtained as in Fig. 5b, with pH at 2 and dye concentration at range 50-150 mg/L. The slow process was mention probably due to the slow of pore diffusion of solution ion into adsorbent bulk [58]. Based on theliterature information, the pattern of contact time effect for agricultural waste activated carbon are basically shown as in Fig. 5.

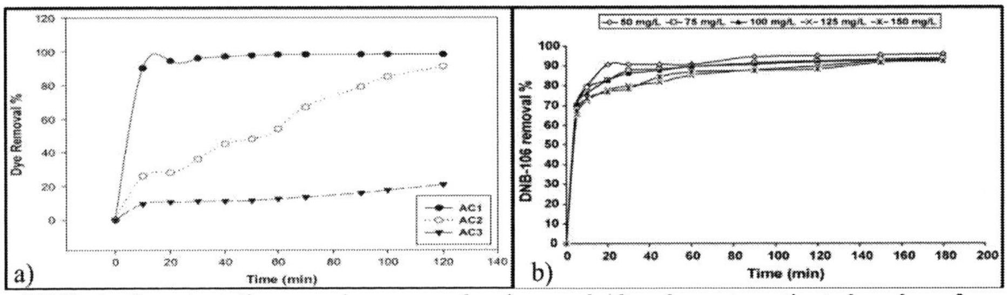

Fig. 5. Effect of contact time on dye removal using agricultural waste activated carbon from a) corncob [41] b) orange peel [58].

4.2 Effect of adsorbent dose

Dosage of adsorbent is one of important parameter in order to determine the adsorbent's capacity for a given amount of adsorbate at the operating conditions. In general, the increase in dye removal along with the increasing of adsorbent dose, where the amount of sorption site at the adsorbent surface will increase by increase of adsorbent dosage that will result on the increase of dye removal percentage [81]. In order to study the effect of adsorbent dose on the adsorption process, it can be carried out by preparing adsorbent-adsorbate solution with different amount of adsorbents added to fixed initial dye concentration then shaken together until reach the equilibrium time [82]. Sharma and Uma [83] carried out an experiment using different dose of adsorbent (0.40–0.60 g) in 50 ml of dye solution in order to find the effect on different rice husk activated carbon dosage to remove methylene blue. They reported that by increasing adsorbent dose from 0.40 to 0.60 g, the percentage of removal increased from 86.75 to 99.83%. Another researcher Heibati et al. [77] analysed the effect of adsorbent dose on the removal of acid red 18 (acid dye) by activated carbon from walnut (ACW) and poplar wood (ACP). The result shown as in Fig. 6a that the efficiency of dye removal increased as the adsorbent dosage increased up to 10 mg/g and after that remain almost constant for both activated carbon. In Amin N.K [18] research on bagasse pith activated carbon to remove reactive dye, the same condition of result being detected. As in Fig. 6b, the percentage in removing reactive orange dye increased with the increase of adsorbent dose whereby the result may due to increment of dosage provides more surface area which lead to more biding site for adsorbent. Based on the literature review,analysing the effect of adsorbent dose showed the adsorption ability toward dye using minimum dosage of adsorbent. The rise in dosage provide more site for dye saturation during adsorption. Referring Fig. 6a and 6b, woody activated carbon optimum dose obtain at 1g of adsorbent while non-woody activated carbon claimed the optimum dose to be at 15g of adsorbent. The activated carbon from agricultural waste reviewedusing minimum dose of adsorbent and have a great potential in remove dye.

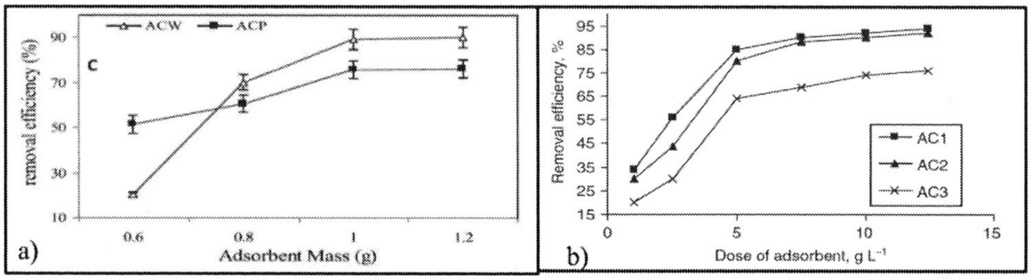

Fig. 6. Effect of adsorbent dose on dye removal using agricultural waste a) woody activated carbon [76] b) non-woody activated carbon [18]

4.3 Effect of pH

The pH factor is very important for dye adsorption in the adsorption process. A medium pH will control the electrostatic charges magnitude which are imparted by ionized dye molecules and resulting the adsorption rate vary with the pH of the medium used [68,84]. Generally, dye removal percentage for cationic dye will decrease at a low pH, while increment in dye removal percentage for ionic dyes adsorption. In contrast, cationic dye adsorption preferred for high solution of pH but as for anionic dye adsorption the efficiency will be lower [5]. As surface charge density decreases with an rising of pH, the electrostatic repulsion between the positively charged dye and adsorbent surface is lowered, which may result in an increment in the adsorption extent [4,85-86]. Yasin et al. [85] investigated that in condition of 28°C of temperature with 180 min contact time and 0.25 g / 25 ml dosage of adsorbent, the pattern as in Fig. 7a obtained for effect of pH on adsorption. The trends shown increment in methylene blue removal with the rising in pH, while they predict that at the low pH, the presence of H^+ ions competing with cations in dye for adsorption sites resulting lower adsorption [85]. Ahmad et al [32] reported that the effect of pH on the adsorption of remazol de by pomegranate activated carbon with the adsorption for the azo dye was initial at pH 3.6 and the maximum dye removal was at 94.36%. In other research, N.K. Amin [18] studied the adsorption of reactive dyes on activated carbon from sugarcane bagasse which result obtained that the dye adsorption was maximum at intial pH of 1 and increment in pH decrease the percentage of dye removal as shown in Fig. 7b.Based on literature review, the increment of removal for cationic dye at high pH due to the decrement of positive charge at the solution interface which resulting the adsorbent surface to be negatively charged. As the increment of anionic dye adsorption at lower pH as result of adsorbent surface appear to be positively charged due to positive charge increament in solution interface.Fig. 7 shows the evidence where the percentage of dye removal for activated carbon was affected by pH variation. The strong force of interaction either H^+ or OH^- ions between the activated carbon and the dye could influence the adsorption capacity [28].

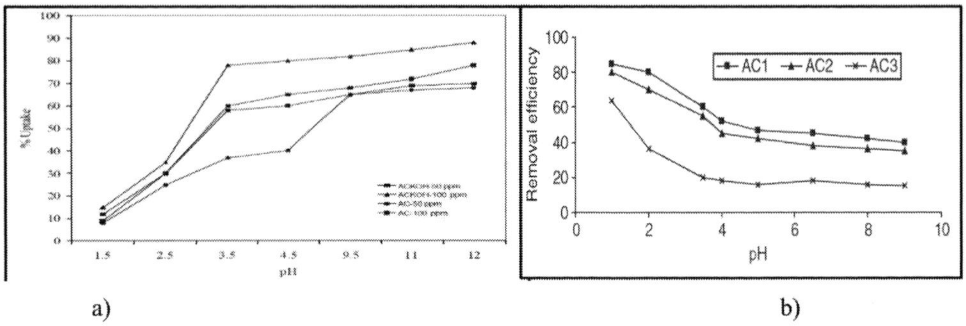

a) b)

Fig. 7. Effect of solution pH on removal of a) cationic dye [85] b) anionic dye [18] using agricultural waste activated carbon

4.4 Effect of initial concentration

A given mass of sorbent material can only adsorb a fixed amount of dye, hence the initial dye concentration of an effluent is one of important factor to studied [5]. The effect of dye initial concentration can be conducted by shaken adsorbent-adsorbate solution until equilibrium using fixed the adsorbent dosage with different initial dye concentration for different time intervals [82]. The effect of increment in dye initial concentration will increase the adsorbent loading capacity [81]. Amin N.K. [18] analysed the effect of initial dye concentration on removing reactive orange dye where the percentages of removal was noticed to be decrease. In research on Myrtus communis and pomegranate activated carbon [1], the different pattern was obtained. At concentration of 30 mg/L congo red, the percentage of removal was significantly increase with increment of congo red initial concentration. They concluded that by lowering the initial dye molecules, the adsorbent surface area will be low as well, hence the adsorption is depends on the initial concentration of dye [1]. Hazzaa and Hussein [80] reported that the increment of initial concentration enhance higher adsorption of dye as the driving force of mass transfer become large. Fig. 8b shown the effect of initial concentration of dye in Hazzaa and Hussein [80] research where the risingof 50 to 150 mg/L initial concentration, the methylene blue adsorbed by olive stone activated carbon (OSAC) increased from 4.8 to 12.4 mg/g at pH of 5, 25°C of temperature with 30 min contact time and 10 g/L dosage of adsorbent.From the literature review, it is mention that the immediate relation between the available surfaces site on adsorbents and the concentration of dye affect the initial concentration factors [68]. As the initial concentration increase, the available site on adsorbent surface area become fewer which causingincrement in dye being adsorbed. However, the removal dye percentage decreased[8] which been proved in Fig. 8.

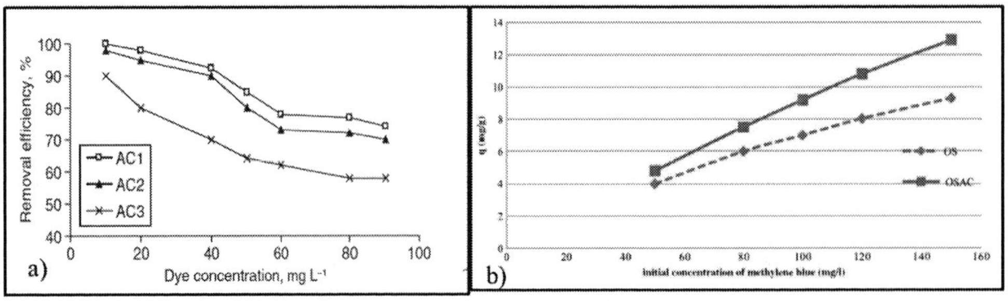

Fig. 8. Effect of initial concentration on removal of a) reactive dye [18] b) direct dye [80] by using agricultural activated carbon

5 Discussion

Literature information give evident that activated carbon is the most preferred adsorbent material to treat wastewater that containing dyes and pigments. In addition, activated carbon prepared from agricultural waste came to be promising in the case of sugarcane bagasse, pomegranate peel, coconut shell [31-32,44,87]. Several activation process also been studied to improve the efficiency in removing dyes and pollutants in wastewater [38,52,46,87-88]. However, most of the study referred in this review referring to lab-scale experiment.Based on literature information, only few study focus on comparing the adsorbent in removal of dye and pollutants [25]. These problems may depends on the performance of material used. These easily available agricultural waste may have a variable characteristic of each materials and each of the materials or adsorbent has its own feature in the treatment for wastewater that will influence the adsorbent properties parameter. As example sugarcane bagasse effective in removing acid dye [35], grape marc-based in removing reactive dye [47] and coconut shell offer effective removal of basic dye but a bit lower for direct dye [11].Yet, the variables and condition in adsorption process also influence the effectiveness of treatment including dose of adsorbent, contact time, adsorbent concentration, pH value and temperature. To ensure the adsorbent effectively can be used for industrial scale, further investigation on development of adsorption process for real condition of wastewater need to be conducted as the industrial effluents containing several pollutants simultaneously. Additional information on kinetic study and modelling is needed on this topic.Based on literature review, manyresearchers had beenstudy on adsorbent especially activated carbon. However they were focusing on the production of the activated carbon and those researches were not mentioned in this review. As for adsorption process study, environment condition is important as the effectiveness in removing dye depends to it. There a lot more factor that can affect the dye removal, however solution pH, contact time, initial concentration and adsorbent dose gave the major affect in percentage of dye remove from wastewater.Further research is necessary,especially concentrate on the characteristics of adsorbents including the average size of particles and activation time to standardize the activated carbon. Besides that, more investigation should be conducted on benzidine-based dye (direct black 38, direct brown 95 and direct blue 6) as these dye is harmful to human, animals and environment. In addition, interest studies should be focus in real wastewater as majority of previous studies used synthetic solution to predict the performance of adsorption process towards removal of dye.

6 Conclusion

A review of various type of activated carbon as adsorbent has been presented. The use of these activated carbon as adsorbent is recommended since they show a good potential in eliminating dyes from industrial wastewater, easily available, low cost and renewable. This paper presented the potential of adsorption process using activated carbon from agricultural waste to remove dye from textile wastewater. The factors effecting dye adsorption such as contact time, adsorbent dosage, pH and initial dye concentration has been discussed. Solution of pH turn to be the most important condition in adsorption process as for anionic dye, a low pH value are preferable in contrast for cationic dye where the suitable pH value is high. For the adsorbent dose, that the adsorption capacity increase along with the increment of adsorbent dosage due to the increase of available amount of sorption site. Is was also highlighted that the contact time between adsorbent and dye affecting the efficiency of dye removal where strong attraction force will shortened the time.As for the effect of dye initial concentration, increasing the initial concentration, enhance the increment of adsorbent surface area to adsorb dyes.Extensive studies in literature show that industrial waste and biosorbent activated carbon are among the less selective compound in removing dye. Furthermore, agricultural waste activated carbon showed effectiveness in removing dye such as basic dye and direct dye yet often being test for other pollutant such as copper II, fluoride and phenol. However, the raw material of activated carbon to be used are depends on the local sources available at low cost. According to the literature reviewed, these adsorption method using agricultural waste activated carbon have potential to be applied at full-scale wastewater treatment. These low cost adsorbent can used to replace the expensive adsorbents in the market nowadays.

References

[1] M. Ghaedi, H. Tavallali, M. Sharifi, S.N. Kokhdan, and A. Asghari, Preparation of low cost activated carbon from Myrtus Communis and pomegranate and their efficient application for removal of congo red from aqueous solution, Spectrochimica Acta Part A,**86**, 107-114 (2012)
[2] V.K. Gupta and Suhas, Application of low-cost adsorbents for dye removal – A review, J. Environ. Manag.,**90**, 2313-2342 (2009)
[3] S.R. Couto, Dye removal by immobilised fungi, Biotech. Adv.,**27**, 227-235 (2009)
[4] S.J. Allen, and B. Koumanova, Decolourisation of water/wastewater using adsorption (review), J. University Chem. Technol. Metallurgy,**40**(3), 175-192 (2005)
[5] K.S. Bharathi and S.T. Ramesh, Removal of dyes using agricultural waste as low-cost adsorbents: a review, Appl. Water Sci.,**3**, 773-790 (2013)
[6] M.A. Kamboh, A.A. Bhatti, I.B. Solangi, S.T.H. Sherazi and S. Memon, Adsorption of direct black-38 azo dye on *p-tert*-butylcalix[6]areneimmobilized material, Arabian J. Chem.,7, 125-131 (2014)
[7] S. Lalnunhlimi and V. Krishnaswamy, Decolorization of azo dyes (Direct Blue 151 and Direct Red 31) by moderately alkaliphilic bacterial consortium, Brazilian J. Microbiology,**47**, 39-46 (2016)
[8] S.Saiful Azhar, A. Ghaniey Liew, D. Suhardy, K. Farizul Hafiz and MD. Irfan Hatim, Dye removal form aqueous solution by using adsorption on treated sugarcane bagasse, American J. Appl. Sci.,**2**(11), 1499-1503 (2005)
[9] A.E. Ghaly, R. Ananthashankar, M. Alhattab and V.V. Ramakrishnan, Production, characterization and treatment of textile effluents: a critical review, J. Chem. Eng. Process. Technol.,**5**, 1-18 (2014)
[10] A. Rehorek and A. Plum, Characterization of sulfonated azo dyes and aromatic amines by pyrolysis gas chromatography/mass spectrometry, Anal. Bioanal. Chem.,**388**, 1653-1662 (2007)
[11] A.M. Aljeboree, A.N. Alshirifi and A.F. Alkaim, Kinetics and equilibrium study for the adsorption of textile dyes on coconut shell activated carbon, Arabian J. Chem. In Press, (2014)
[12] B. Noroozi, G.A. Sorial, H. Bahrami and M. Arami, Adsorption of binary mixtures of cationic dyes, Dyes Pigm.,**76**, 784-791 (2008)
[13] W. Lemlikchi, N. Drouiche, N. Belaicha, N. Oubagha, B. Baaziz and M.O. Mecherri, Kinetic study of the adsorption of textile dyes on synthetic hydroxyapatite in aqueous solution, J. Ind. Eng. Chem., **32**, 233-237 (2015)
[14] D. Pathania, A. Sharma and Z. Siddiqi, Removal of congo red dye from aqueous system using Pheonix dactylifera seeds, J. Molecular Liquids **219**, 359-367 (2016)
[15] A.-A. Pelaez-cid, A.-M. Herrera-Gonzalez, M. Salazar-Villanueva and A. Boutista-Hernandez, Elimination of textile dyes using activated carbons prepared from vegetable residues and their characterization, J. Environ. Manag., **181**, 269-278 (2016)

[16] B. Naresh, J. Jaydip, B. Prabhat and P. Rajkumar, Recent biological technologies for textile effluent treatment, Int. Res. J. Biol. Sci.,**2**(6), 77-82 (2013)
[17] M.S.I. Mozumder and M.A. Islam, Development of treatment technology for dye containing industrial wastewater, J. Sci. Rec.,**2**(3), 567-576 (2010)
[18] N.K. Amin, Removal of reactive dye from aqueous solutions by adsorption onto activated carbons prepared from sugarcane bagasse pith, Desalination,**223**, 152-161 (2008)
[19] P. Nigam, G. Armour, I.M. Banat, D. Singh and R. Marchant, Physical removal of textile dyes from effluents and solid-state fermentation of dye-adsorbed agricultural residues, Bioresour. Technol.,**72**, 219-226 (2000)
[20] S.S. Ashour, Kinetic and equilibrium adsorption of methylene blue and remazol dyes onto steam-activated carbons developed from date pits, J. Saudi Chem. Society,**14**, pp. 47-53 (2010)
[21] V.B. Upadhye and S.S. Joshi, Advances in wastewater treatment – A review, Int. J. Chemical Sci. Applications,**3**, 264-268 (2012)
[22] T.L. Silva, A. Ronix, O. Pezoti, L.S. Souza, P.K.T. Leandro, K.C. Bedin, K.K. Beltrame, A.L. Cazetta and V.C. Almeida, Mesoporous activated carbon from industrial laundry sewage sludge: adsorption studies of reactive dye remazol brilliant blue R, Chem. Eng. J.,**303**, 467-476 (2016)
[23] G.J. Copello, A.M. Mebert, M. Raineri, M.P. Pesenti and L.E. Diaz, Removal of dyes from water usning chitosan hydrogel/SiO2 and chitin hydrogel/SiO2 hybrid materials obtained by the sol-gel method, J. Hazard. Mater.,**186**, 932-939 (2011)
[24] S.D. Gisi, G. Lorano, M. Grassi and M. Notarnicola, Characteristics and adsorption capacities of low-cost sorbents for wastewater treatment: a review, Sustainable Mater. Technol.,**9**, 10-40 (2016)
[25] G. Crini, Non-conventional low-cost adsorbents for dye removl: A review, Bioresour. Technol.,**97**, 1061-1085 (2006)
[26] I.D. Mall, V.C. Srivastava, N.K. Agarwal and I.M. Mishra, Adsorptive removal of malachite green dye from aueous solution by bagasse fly ash and activated carbon-kinetic study and equilibrium isotherm analyses, Colloids Surf. A: Physicochem. Eng. Aspects,**264**, 17-28 (2005)
[27] E. Clarke and R. Anliker, Organic dyes and pigments, Handb. Environ. Chem.,**3**, 181-215 (1980)
[28] A. Demirbas, Agricultural based activated carbons for the removal of dyes from aqueous solutions: a review, J. Hazard. Mater.,**167**, 1-9 (2009)
[29] K.D. Belaid, S. Kacha, M. Kameche and Z. Derriche, Adsorption kinetics of some textile dyes onto granular activated carbon, J. Environ. Chemical Eng.,**1**, 496-503 (2013)
[30] M. Joshi, R. Bansal and R. Purwar, Colour removal from textile effluents, Indian J. Fibre Text. Res.,**29**, 239-259 (2004)
[31] U. Isah A., G. Abdulraheem, S. Bala, S. Muhammad and M. Abdullahi, Kinetics, equilibrium and thermodynamics studies of C.I. reactive blue 19 dye adsorption on coconut shell based activated carbon, Int. Biodeterioration Biodegradation,**102**, 265-273 (2015)
[32] M.A. Ahmad, N.A. Ahmad Puad and O.S. Bello, Kinetics, equilibrium and thermodynamics studies ofsynthetic dye removal using pomegranate peel activated carbon prepared by microwave-induced KOH activation, Water Resc. Ind.,**6**, 18-35 (2014)
[33] E.R. Bandala, M.A. Pelaez, A.J. Gracia-Lopez, M.J. Salgado and G. Moeller, Photocatalytic decolourisation of synthetic and real textile wastewater containing benzidine-based azo dyes, Chem. Eng. Process.,**47**, 169-176 (2008)

[34] M.A. Mohd Salleh, D. K. Mahmoud, N.A. Awang Abu, W.A. Wan Abdul Karim and A. Idris, Methylene blue adsorption from aqueous solution by langsat (lansium domesticum) peel, J. Purity, Utility Reaction Environ.,**1**, 472-495 (2012)
[35] W.T. Tsai, C.Y. Chang, M.C. Lin, S.F. Chien, H.F, Sun and M.F. Hsieh, Adsorption of acid dye onto activated carbons prepared from agricultural waste bagasse by ZnCl2 activation, Chemosphere,**45**, 51-58 (2001)
[36] S. Sivamani and B. Leena Grace, Removal of dyes from wastewater using adsorption – A review, Int. J. BioSci. Technol.,**2**, pp. 47-51 (2009)
[37] S. Babel and T.A. Kurniawan, Low-cost adsorbents for heavy metals uptake from contaminated water: A review, J. Hazard. Mater.,**97**, 219-243 (2003)
[38] M. Valix, W.H. Cheung and G. McKay, Preparation of activated carbon using low temperature carbonisation and physical activation of high ash raw bagasse for acid dye adsorption, Chemosphere,**56**, 493-501 (2004)
[39] R. Zanzi, X. Bai, P. Capdevila and E. Bjornbom, Pyrolysis of biomass in presence of steam for production of activated carbon, liquid and gaseous fuels, *Proc. of 6th world congress on chemical engineering, Melbourne, Australia*, 1-8 (2001)
[40] A.A. Mamun, Y.M. Ahmed, S.A. Muyibi, M.F.R. Al-Khatib, A.T. Jameel and M.A. Al-Saadi, Synthesis of carbon nanofibers on impregnated powdered activated carbon as cheap substrate, Arabian J. Chem.,**9**, 532-536 (2016)
[41] G.O. El-Sayed, M.M. Yehia and A.A. Asaad, Assessment of activated carbon prepared from corncob by chemical activation with phosphoric acid, Water Resc. Ind.,**7-8**, 66-75 (2014)
[42] R.H. Hesas, A. Arami-Niya, W.M.A. Wan Daud and J.N. Sahu, Preparation and characterization of activated carbon from apple waste by microwave-assisted phosphoric acid activation: Application in methylene blue adsorption, BioResour.,**8**(2), 2950-2966 (2013)
[43] H. Saygili, F. Guzel and Y. Onal, Conversion of grape industrial processing waste to activated carbon sorbent and its performance in cationic and anionic dyes adsorption, J. Cleaner Production,**93**, 84-93 (2015)
[44] R.W.S. Wan Suraya, M.R. Mohd Adib and H. Rafidah, Overview of acid optimization in impregnation method for sugarcane bagasse activated carbon production, Adv. Environ. Bio.,**9**(12) 1-5 (2015)
[45] K. Ahmadi, M. Ghaedi and A. Ansari, Comparison of nickel doped zinc sulphide and/or palladium nanoparticle loaded on activated carbon as efficient adsorbents for kinetic and equilibrium study of removal of congo red dye, Spectrochimica Acta Part A: Moecular and Biomolecular Spectroscopy,**136**, 1441-1449 (2015)
[46] S. Khamparia and D. Jaspal, Adsorptive removal of direct red 81 dye from aqueous solution onto Argemone Mexicana, Sustainable Environment Research (2016)
[47] H. Belayachi, B. Bestani, N. Benderdouche and M. Belhakem, The use of TiO2 immobilized into grape marc-based activated carbon for RB-5 azo dye photocatalytic degradation,Arabian J. Chem. In Press, (2015)
[48] C.X. Chen, B. Huang, T. Li and G.F. Wu,Preparation of phosphoric acid activated carbon from sugarcane bagasse by mechanochemical processing, Bioresour.,**7**(4), 5109-5116 (2012)
[49] N. Mohd Nor, T. Hadibarata, Z. Yusop and Z. Mat Lazim, Removal of brilliant green and procionred dyes from aqueous solution by adsorption using selected agricultural wastes, J. Teknologi (Sciences & Engineering),**74**, 117-122 (2015)
[50] M.A. Yahya, Z. Al-Qodah and C.W. Zanariah Ngah, Agricultural bio-waste materials as potential sustainable precursors used for activated carbon production: A review, Renewable Sustainable Energy Reviews,**46**, 218-235 (2015)

[51] Z. Zhang, L. Moghaddam, I.M. O'Hara and W.O.S. Doherty, Congo red adsorption by ball-milled sugarcanr bagasse, Chem. Eng. J.,**178**, 122-128 (2011)
[52] J.X. Yu, R.A. Chi, J. Guo, Y.F. Zhang, Z.G. Xu and C.Q. Xiao, Desorption and photodegradation of methylene blue from modified sugarcane bagasse surface by acid TiO2 hydrosol, Appl. Surf. Sci.,**258**, 4085-4090 (2012)
[53] S.L. Pandharipande, Y.D. Urunkar and A. Singh, Characterization and adsorption studies of activated carbon prepared from rice husk, sugarcane bagasse and saw dust, Int. J. Adv. Eng. Technol.,**3**(3), 60-62 (2012)
[54] H.D.S.S. Karunarathneand B.M.W.P.K. Amarasinghe, Fixed bed adsorption column studies for the removal of aqueous phenol from activated carbon prepared from sugarcane bagasse, Energy Procedia,**34**, 83-90 (2013)
[55] A.K. Yadav, R. Abbassi, A. Gupta and M. Dadashzadeh, Removal of floride from aqueous solution and groundwater by wheat straw, sawdust and activated bagasse carbon of sugarcane, Ecological Eng.,**52**, 211-218 (2013)
[56] S.M. Yakout and G. Sharaf El-Deen, Characterization of activated carbon prepared by phosphoric acid activation of olive stones, Arabian J. Chem, **9**, S1155-S1162, (2016)
[57] N. Dejang, O. Somprasit and S. Chindaruksa, A preparation of atctivated carbon from macadamia shell by microwave irradiation activation, Energy Procedia,**79**, 727-732 (2015)
[58] A. Khaled, A. El-Nemr, A. El-Sikaily and O. Abdelwahab, Removal of direct N blue 106 from artificial textile dye effluent using activated carbon from orange peel: adsorption isotherm and kinetic studies, J. Hazard. Mater.,**165**, 100-110 (2009)
[59] A.J. Ahamed and K.R. Ahamed, Preparation and characterization of activated carbon from the Prosopis juliflora plant, Asian J. Chem.,**20**(3), 1702-1706 (2008)
[60] A.A. Ahmad and B.H. Hameed, Effect of preparation conditions of activated carbon from bamboo waste for real textile wastewater, J. Hazard. Mater. **173**, 487-493 (2010)
[61] S.M. Lamine, C. Ridha, H.M. Mahfoud, C. Mouad, B. Lotfi and A.H. Al-Dujaili, Chemical activation of an activated carbon prepared from coffee residue, Energy Procedia **50**, 393-400 (2014)
[62] N. Arena, J. Lee and R. Clift, Life cycle assessment of activated carbon production from coconut shells, J. Cleaner Production **125**, 68-77 (2016)
[63] K.A.G. Gusmao, L.V.A. Gurgel, T.M.S. Melo and L.F. Gil, Application of succinylated sugarcane bagasse as adsorbents to remove methylene blue and gentian violet from aqueous solutions – kinetic and equilibrium studies, Dyes Pigm.,**92**, 967-974 (2012)
[64] Y.S. Ho, W.T. Chiu and C.C. Wang, Regression analysis for the sorption isotherms of basic dyes on sugarcane dust, Bioresour. Technol.,**96**, 1285-1291 (2005)
[65] Z. Mat Lazim, N.S. Zulkifli, T. Hadibarata and Z. Yusop, Removal of cresol red and reactive black 5 dyes by using spent tea leaves and sugarcane bagasse powder, J. Teknologi (Sciences & Engineering),**74**, 147-151 (2015)
[66] E. Kacan, Optimum BET surface ares for activated carbon produced from textile sewage sludges and its application as dye removal, J. Environ. Manag.,**166**, 116-123 (2016)
[67] E. Kacan and C. Kutahyali, Adsroption of strontium from aqueous solution using activated carbon produced from textile sewage sludges, J. Anal. Appl. Pyrolysis,**97**, 149-157 (2012)
[68] I. Ould Brahim, M. Belmedani, A. Belgacem, H. Hadoun and Z. Sadaoui, Discoloration of azo dye solutions by adsorption on activated carbon prepared from the cryogenic grinding of used tires, Chem. Eng. Transactions, **38**, 121-126 (2014)

[69] Y. Gao, S. Xu, Q. Y. Y. Wu and B. Gao, Chemical preparation of crab shell-based activated carbon with superior adsorption performance for dye removal from wastewater, J. Taiwan Institute Chem. Engineers,**61**, 327-335 (2016)
[70] G. Bayramoglu and M.Y. Arica, Biosorption of benzidine based textile dyes 'Direct Blue 1 and Direct Red 128" using native and heat-treated biomass of Trametes versicolor, J. Hazard. Mater.,**143**, 135-143 (2007)
[71] N. Das and D. Charumathi, Remediation of synthetic dyes from wastewater using yeast – An overview, Indian J. Biotech.,**11**, 369-380 (2012)
[72] M.S. Miao, Q. Liu, L. Shu, Z. Wang, Y.Z. Liu and Q. Kong, Removal of cephalexin from effluent by activated carbon prepared from alligator weed: kinetics, isotherms and thermodynamic analyses,Process Safety and Environmental Protection (2016)
[73] H. Tahir, M. Sultan, N. Akhtar, U. Hameed and T. Abid, Application of natural and modified sugarcane bagasse for the removal of dye from aqueous solution, J. Saudi Chem Society (2012)
[74] A. Wasti and M.A. Awan, J. Assoc. Arab Uni. Basic Appl. Sci.,**20**, 26-31 (2016)
[75] L. L. Matovic, N.S. Vukelic, U.D. Jovanovic, K.R. Kumric, J.B. Krstic, B.M. Babic and A.B. Dukic, Mechnochemically improved surface properties of activated carbon cloth for the removal of As(V) from aqueous solutions, Arabian J. Chem (2016)
[76] M.I. Sabela, K. Kunene, S. Kanchi, N.M. Xhakaza, A. Bathinapatla, P. Mdluli, D. Sharma and K. Bisetty, Removal of copper (II) from wastewater using green vegetable waste derived activated carbon: An approach to equilibrium and kinetic study, Arabian J. Chem. (2016)
[77] B. Heibati, S. Rodriguez-Couto, M.A. Al-Ghouti, M. Asif, I. Tyagi, S. Agarwal and V.K. Gupta, Kinetics and thermodynamics of enhanced adsorption of the dye AR 18 using activated carbons prepared from walnut and poplar woods, J. Molecular Liquid,**208**, 99-105 (2015)
[78] D.A. Giannakoudakis, G.Z. Kyzas, A. Avranas and N.K. Lazaridis, Multi-parametric adsorption effects of the reactive dye removal with commercial activated carbons, J. Molecular Liquids,**213**, 381-389 (2016)
[79] H. Cherifi, B. Fatiha and H. Salah, Kinetic studies on the adsorption of the methylene blue onto vegetal fiber activated carbons, Appl. Surf. Sci.,**282**, 52-59 (2013)
[80] R. Hazzaa and M. Hussein, Adsorption of cationic dye from aqueous solution onto activated carbon prepared from olive stones, Environ. Technol. Innovation,**4**, 36-51 (2015)
[81] A.A. Adeyemo, I.O. Adeoye and O.S. Bello, Adsorption of dyes using different types of clay: a review, Appl. Water Sci. (2015)
[82] M.A. Mohd Salleh, D.K. Mahmoud, W.A. Wan Abdul Karim and A. Idris, Cationic and anionic dye adsorption by agricultural solid wastes: A comprehensive review, Desalination,**280**, 1-13 (2011)
[83] Y.C. Sharma and S. Uma, Optimization of parameters for adsorption of methylene blue on a low-cost activated carbon, J. Chem. Eng. Data,**55**, 435-439 (2010)
[84] U.J. Etim, S.A. Umoren and U.M. Eduok, Coconut coir dust as a low cost adsorbent for the removal of cationic dye from aqueous solution, J. Saudi Chem. Society (2012)
[85] E. Yilmaz, S. Memon and M. Yilmaz, Removal of direct azo dyes and aromatic amines from aqueous solution using two β-cyclodecxtrin-based polymers, J. Hazard. Mater.,**174**, 592-597 (2010)
[86] Y. Yasin, M.Z. Hussein and F. Ahmad, Adsorption of methylene blue onto treated activated carbon, Malaysian J. Anal. Sci. **11**(11), 400-406 (2007)
[87] Z. Nadzirah, H. Nor Haslina and M.R. Mohd Adib, Studies on the preparation of activated carbon sugarcane bagasse on removal of chemical oxygen demand, alkalinity and oil and grease of car wash wastewater, Adv. Environ. Biology, **9**(12), 15-21 (2015)

[88] K.A. Adegoke and O.S. Bello, Dye sequestration using agricultural wastes as adsorbents, Water Resour. Ind.,**12**, 8-24 (2015)

Applications of Natural Coagulants to Treat Wastewater – A Review

Vicky Kumar[1], *Norzila* Othman[2,*], and *Syazwani* Asharuddin[1]

[1]Department of Water and Environment, Faculty of Civil and Environmental Engineering, UniversitiTun Hussein Onn Malaysia, 86400 Parit Raja, BatuPahat, Johor, Malaysia
[2]Micro-Pollutant Research Centre (MPRC), Faculty of Civil and Environmental Engineering, University Tun Hussein Onn Malaysia, 86400, Parit Raja, Johor, Malaysia

Abstract. The natural water falls from the mountain is merging into the oceans. This water is preserved by humans that are consumed for agriculture, industrial, and municipal use. This water become wastewater after different usage, and finally, completes the hydrological cycle. The water becomes wastewater due to population growth, urbanization, industrialization, sewage from household, institutions, hospitals, industries and etc. Wastewater can be destructive for the public because it contains a variety of organic and inorganic substances, biological substances, toxic inorganic compounds and the presence of toxic materials. The coagulant chemicals and its associated products are resourceful but these may change the characteristics of water in terms of physical and chemical characteristics, this make matters worse in the disposal of sludge. An option of natural polymer can be used in water and wastewater in this review. The natural polymers are most efficient that provide several benefits such as; prolific, exempt from physical and chemical changes from the treated water.

1 Introduction

Water is a key substance in all natural and human activities. It regenerates the shape of oceans, seas, rivers, lakes and forests, becoming part of hydrological cycle that is important for the development of ecosystems and human life [1]. However since water have been considered as basic need of life, current misuse of water coupled with growing population size, industrialization, change in climate, and urbanization has increased the shrink in cleaning the water reserves [2]. The need for clean water had trigged more comprehensive research in water and wastewater. Heavy metals released into environment that is active in development and rapid growth of industrialization that affect bionetwork. Industries such as; metal coating, excavation, and battery production, consequence of profound metals to ecosystem [3]. It has been discussed that in the developing countries more than 1.6 million people are using the unhygienic water & among them most of the people suffers from diarrhoea and other water related diseases [4].

*Corresponding author: norzila@uthm.edu.my

It is described that developing countries are paying a high cost for water treatment by importing the chemicals [5]. There are several methods are utilized to treat the water, depends on the quality of the wastewater. Wastewater is the large seasonal variation that is associated with problems like turbidity, due to solubilize matter and results to more compactable treatment [6]. The physical, chemical, and biological characteristics can be altered by dissolving impurities, which is comprised of organic compounds, gases, minerals and E-coli these effect can be controlled by the concentration and composition of chemical reactions among pollutants [7].

The proper treatment of wastewater resulted to accept that of effluents discharged. Once the effluents enters in environment such river it may health effect such as eye infections, skin diseases, as well as worm infections by hygienic practice that can result the major improvements in health conditions [8]. These effect may be due to pollutant namely suspended solids & turbidity of the wastewater by using coagulation-flocculation.

Wastewater treatment techniques that are widely used are chemical precipitation, lime coagulation, ion exchange, reverse osmosis and solvent extraction [9].Coagulants are used that added to the water to withdraw the forces that stabilizes the colloidal particles and causing the particles to suspend in the water [10]. Once the coagulant is introduced in the water, the individual colloids must aggregate and grow bigger so that the impurities can be settled down at the bottom of the beaker and separated from the water suspension (as shown in table 1). Various types of coagulants show potential application in treating water and wastewater. It ranges from chemical to non-chemical coagulant. The coagulant also could be synthetic material or natural coagulant with the properties of coagulant having +ve charge, these positive charge proteins would bind to the -ve charged particles in the solution that cause turbidity. water and wastewater treatment is the removal of suspended and colloidal particles, untreated matter, microbes and other substances that are deleterious to life, in search of lowest cost deployment, operation, maintenance, and reduced environmental impacts to the contiguous [11]. Coagulants are characterised according to the wastewater properties that relates to enhance the efficiency of treatment process to achieve required quality of water on standards (as shown in Table 1).Coagulants normally in form of natural (as shown is Table 2) & synthetic (as in Table 3).Both coagulants aim to remove pollutant in form of physical (solids & turbidity) or chemical (BOD & COD). The coagulants also present with advantages various among them.

Table 1. Factors Affecting Coagulation

Coagulant Applications	Effective Characteristics	Natural Water Properties
Coagulants Extraction Coagulants Solubility Coagulants Dosage Charge on Particles Basicity of a coagulant	Settling time Turbulence 1. Rapid Mixing 2. Slow Mixing Coagulant adds quantity Particles type	pH Alkalinity Availability of Bacteria's Presence of Elements (Cl, Na, Mn, Si, Fl, NH3, Fe) Total dissolved solids Suspended Solids Temperature Turbidity Dissolved Oxygen

The impact of pH and coagulant doses on the coagulation procedure was contemplated in order to streamline comparing to the best evacuation of turbidity. The optimum dosage of pH will lead to the optimum conditions of JAR test.Coagulation affects the performance of the other stages of the treatment, favouring microbiological quality of the final product increasing the lifetime of filters and reducing the final cost of the treated water [11]. Coupled coagulation, flocculation step constitutes conditions of time and agitation where the particles from colliding and coagulating destabilized by forming flakes could be eliminated by sedimentation.

Table 2. Natural coagulants Efficiency

Natural Coagulants	Turbidity	COD
CicerAretinum	81.20%	90.00%
Moringa Oleifera	82.02%	83.33%
Cactus	78.54%	80.65%

Table 3. Synthetic/Inorganic Coagulants Advantages and Disadvantages

Name	Advantages	Disadvantages
Aluminiumsulphate (Alum) $Al_2\ (SO_4)_3\cdot\ _{18}H_2O$	Easy to handle and apply; most commonly used; produces less sludge than lime; most effective between pH 6.5 and 7.5	Adds dissolved solids (salts) to waster; effective over a limited pH range.
Sodium Aluminate $Na_2Al_2O_4$	Effective in hard waters; small dosage usually needed	Often used with alum; high cost; ineffective in soft waters
Polyaluminium Chloride (PAC) $Al_{13}(OH)_{20}(SO)_4.Cl_{15}$	In some applications, Floc, formed is more dense and faster settling than alum	Not commonly used; little full scale data compared to other aluminium derivatives
Ferric Sulfate $Fe_2(SO_4)_3$	Effective between pH 4-6 and 8.8-9.2	Ads dissolved solids(salts) to water; usually need to add alkalinity
Ferric Chloride $FeCl_3.6H_2O$	Effective between pH 4 and 11	Adds dissolved solids (salts) to water; consumes twice as much alkalinity as alum
Ferrous Sufate (Copperas) $FeSO_4\ .7H_2O$	Not as pH sensitive as lime	Ads dissolved solids(salts) to water; usually need to add alkalinity
Lime $Ca(OH)_2$	Commonly used; very effective; may not add salts to effluent	pH dependent; produces large quantities of sludge; overdose can result in poor effluent quality

2 Applications of natural coagulants to treat wastewater

The natural coagulants are used in wastewater treatments include microbial polysaccharides, starches, gelatin galactomannans, cellulose derivatives, chitosan, glues, and alginate. Coagulants which carry natural characteristics supposed to be harmless for human health, whereas existence of aluminium zest may provoke neurology & pathology diseases [11]. Natural coagulants are mixed with some artificial coagulants that are consumed as coagulant aid, their effectiveness as the key coagulant remains stays at early stages. The process of treatment in these coagulants composed of molecules bridging, adsorption, and charge balancing. Natural coagulants are capable for wastewater treatment following effluents discharge standard.

3 Non-plant-based coagulants

Polymeric coagulants preserve cationic, anionic, or non-ionic, in which sometimes two are communally described as polyelectrolytes. Generally, the polymers used in wastewater treatment are artificial, the natural polyelectrolyte is persistent. Previous studies regarding natural coagulants is taking place among those are referred as 'polyelectrolytes' although many of these studies did not essentially carry out in-depth to determine chemical characterization and their ionic activity. Naturally, coagulants are polysaccharides or proteins. Polymers categorized as non-ionic due to lack of charged interactions, there may be interactions among partially charged groups of the polymer and a solvent within a solution may contain including –OH along its chain [12]. Chitosan is a natural polysaccharide with many useful lineaments such as; hydrophilicity, biocompatibility, and the capability of adsorbing number of metal ions because of its amino groups. Chitosan is an example cellulose-like biopolymer generally distributed in nature, mainly in marine invertebrates, insects, yeasts, and fungi.

Chitosan is the acetyl group and derivative of chitin, a natural polysaccharide found primarily in the arthropods and some fungi. Chitosan contains behaviour like non-toxic, biodegradable, linear cationic polymer of high molecular weight. Moreover, chitin extraction does not cause any flap to the ecosystem, it gains all advantages provided by polysaccharides, in vision of it as the source of chitosan [13].

Alginates are plentiful in nature, having both capsular polysaccharides in soil bacteria and structural component in aquatic brown algae (phaeophyceae) that encompass up to 40% of dry matter. Alginates are linked to their ability to keep water viscosifying, their gelling, and stabilizing properties. Alginates is the simply polysaccharide, it naturally contains carboxyl groups and various abilities for functional materials.

Components of actinobacteria contain infinite sources of new compound and hold an essential arrangement due to their variety and proved ability to create new bioactive compound [14]. Oladoja in [15] gives comprehensive review of research summary involving actinobacteria coagulants. The research of biocoagulating activity using Cellulomonas and Streptomyces species were evaluated individually in synthetic turbid water of kaolin suspension showed significant flocculating activity [16]. The chemical analyses revealed that the presence of polysaccharides and protein with neutral sugar, amino sugar and uranic acids which might contributing to its high flocculating activity.

4 Plant-based coagulants (PBC)

The PBC are broadly utilized for the purification of contaminated water that is in less urbanized, because they seems to be less carrying cost treated coagulates as compared to artificial. PBC coagulants are assumed to treat water showing low-to-medium turbidity range (50–500) NTU. It is unforeseen that a complete decisive analysis of existing PBC is still imaginary in this 21st century. The significance of PBC to ecosystem in the end yield to an instant area for researcher in strengthening the investigation to find inherent resources [17]. Among PCB described namely, Nirmali seeds, Moringa oleifera, Tannin and Cactus.

Anionic polyelectrolytes are polymers of nirmalii seed whose particles get faded in water by through inter particle bridging. The seed extracts contains –COOH and –OH groups that are able increase the coagulation competency because lipids, carbohydrates and alkaloids. Galactomannani and galactanii are mixture of polysaccharide division that are extracted from Strychnospotatorum seeds which is capable in reducing turbidity up to 80%. In all aspects, the galactomannans are prepared up 1,4-linked β-d-mannopyranosyli residuum mien end α-d-galactopyranosyl units associated with 0–6 point of some manner residue. Nirmali seeds and its associated coagulation process yet not been studied,[18] the presence of abundant amount of –OH groups along chains of galactomannan and galactan provides weakly but profuse adsorption that eventually lead to the preceding coagulant interparticle bridging effect more research can be carried out in this aspects [18].

Moringa oleifera, is a non-hazardous (low concentrated) stifling plant whose seeds contain an edible oil and water soluble substance. It is widely known that this plant can be utilized for beneficial purposes. Moringa is most often consuming as provisions and remedial resource limited to urban areas [19]. The most recommended coagulating agents with 12–14 kDa mass and about 10-11 isoelectric point (pI) are dimeric cationic proteins. Adsorption and charge neutralization its main coagulation method. Cationic protein having pI greater than 9.6 and mass, no less than 6.5 kDa. Conversely, Okuda argue that active elements from an aqueous extraction of salt that is not a polysaccharide or protein, actually they are organic polyelectrolyte having molecular mass of 3.0kDA [20]. It is also discussed that bivalent cations are highly regarded (e.g. Ca2+ and Mg2+) that raise the effect of moringa oleifera.

It is known that interface provides strong adsorption in system, it also neutralized particles that may occur charge problem [21]. Accumulation of cations can enhance the coagulant competency. Natural materials that are achieve from polyphenolic substructure are commonly known as tannin, such as Acacia, Castanea, or Schinopsis extracted from bark and wood. Applications of tannin are widely studies to treat water containing nitrogen atom with a lone pair and carbolic acid groups. Tannin gives 3meq/g charge density, since cationic in environment there is a single tertiary amine group per monomer [22]. Tannis molecular weight ranges from 100 to 10,000 that is generally tanning agent in lather industry. However, such studies still can be explored in future [23]. Furthermore, it also defines that tannin is impartial coagulant as aid that can be used to treat the water and wastewater treatment. It is general comprehension phenolic groups stabilized via resonance. This gives attention that tannin structure is capable for coagulation.

Most known cactus genus treating water ‘nopal’ in Mexico and ‘prickly pear’ in North America. This type of cactus is related with nutritional food sources & remedial characteristics. Beside opuntia cactus latifaria is widely used natural coagulant. The opuntia high coagulation ability recognized to the presence of mucilage assumes as complex carbohydrate and sticky, storage of these compounds in internal cactus and surface pads has high water retention capability. Mucilage in cactus opuntia has fiber such as l-rhamanosei, l-arabinose, d-galactose, d-xyloose, and galacturonic acid, predominantly during a bridging coagulation method solution particulates carry out indirectly contact from the cactus genus [24]. The most capable coagulation and active element is galacturonic acid, it is well-known 50% of turbidity removal [25].

Another option of plant based coagulant using unexploded waste such as cassava peel. Fresh cassava peels have three main efficiencies: spread very rapidly, contain phytates, and huge amounts of cyanogenic glycosides. To shrink cyanogenic & phytate substance to maintain their natural value, altered methods are adopted in sinking cyanogenic glycoside like sun-drying, ensiling, and soaking + sun-drying. Peels are chopped to obtain good quality 2 cm for easy compression, and sagging for two days to decrease moisture substance from 70-75% to about 40% [25]. After 21 days cassava peel silage was observe as light brown in appearance, with satisfying odour. The pH recorded as 4.4, and fungal growth was negligible.

5 Mechanism of coagulation by natural coagulants

The mechanism of coagulation established an idea that there may be interaction between the polymer and the dissolved particles in a solution due to the fact that natural polymer contain numerous charged functional groups residing in their polysaccharide chain such as

–OH, –COOH, and –NH. In general processing coagulant involves four steps, such as bridging mechanisms, charge neutralization, double layer compression, and sweep-floc mechanism [26]. Natural coagulants are commonly composed of the combination of several macromolecules such as carbohydrates, protein and lipids. In many cases, the major building blocks are the polymer of polysaccharides and amino acids [27].

In many practical cases, cationic polyelectrolytes are the most promising flocculants to undermine contaminated particles that are -ve charged. Electrostatic interaction provides strong adsorptions that neutralize the particles surface and potentially cause charge reversal. Moreover, flocculation may take place due to the reduction of particles surface charge and decreased electrical repulsion between them [21]. Many studies carried out on polyelectrolyte of maximum charge density is efficient simply as, at a given dosage, they have good ability to deliver further charge to the particle shell. Moreover, when maximum charge density polyelectrolyte absorbed on -ve surfaces with a fairly less density of potential sites, another prospect arises called 'electrostatic patch mechanism' [21]. The adsorbed polyelectrolyte chains thus type of charge alternating positive and negative charges which give particle attraction and hence flocculation occurred. Coagulating agents in moringa oleifera are cationic proteins with molecular mass of 12-14kDa and iso-electric point (Pi) lies under 10 - 11. The researchers also conclude that the main mechanisms that govern the coagulation activity are adsorption and charge neutralization.

The bridging mechanism has been recognized by many studies using various coagulants like moringa oleifera remove, sago starch and cassia seed gum. Long chain polymers absorbed on particles can have head and tails entering several techniques into the solution. For a bridging mechanism, it is necessary to have much space that can attach divisions of polymer chains to engross other particles. Therefore, importance of most favourable amount for bridging mechanism is suitable [21]. Previous study has follow that the microbe extract of Nirmali contain carbohydrate, lipids, and alkaloids having –COOH and –OH groups which enhance its coagulating potential. Polysaccharide particles extracts from Nirmali seed contained galactomannan and galactan which is competent in turbidity removal up to 80% of kaolin solution. Abundant adsorption sites of –OH beside the chains of galactomannan and galactan that constituted in nirmali seed extracts lead to inter particle bridging effect.

Generally, sweep flocculation gives significantly particle elimination than just particle destabilization like in charge neutralization mechanism due to superior rate of aggregation as the result of increased solids concentration. Sweep flocculation is used to trap suspended solids that are harmful for the mesh composition [28]. Same mechanism was proposed, on the treatment of turbid water by chitosan. The study showed that sweeping mechanism dominates the kaolinites removal when the pH in the range of 7.0 to 8.5 [29].

6 Processing Steps of Natural Coagulants

The first processing phase is widely and essentials for conjugal application for utilization. In this step plant is being cut, sliced, chop and peel for drying purpose by manual or mechanical pulverization to avoid undesired parts. Conventional method involves aeration and following pulp of plant parts into particles are then used by local area due to lack of such equipment. In the second processing phase, organic and/or alcoholic solvents extraction is used to remove the active agents. However, water&salt solution (NaCl) extraction [30] also used. The previous research pointed out that by using NaCl solution extraction with lower dosages as 7.4 times higher to eliminate kaolinite turbidity. The author also endorsed improved the substance in proteins whereas substance increases protein anion and ability of solution. Third process phase involves dialysis, lyophilization, ion-exchange and precipitation to treat water as viable refining methods for*Moringa* *oleifera*stepped for treatment of higher volume of turbid water [31]. Such methods are not much actively used over other PBCs; therefore it shows opportunity for researcher to explore this area.

7 Conclusion

Natural coagulants are obtained from several natural sources, when functional as coagulants primary or auxiliary coagulation/flocculation present as feasible and economical alternatives. There are two types of coagulants are found natural and synthetic. Function of coagulants in wastewater treatment is to remove various parameters. Coagulants are present in twoform, first is plant based coagulants (PBC) and second is non-plant based coagulants. The preparation of natural coagulants is consisting over three steps for the substitution of the dosage of the coagulant in treating wastewater.The treatment through bio coagulants signifies to a vital development in viable environment for better worth of eco-system particularly for less urbanised area. An attempt is using eco-friendly coagulant as a natural coagulant for the process of coagulation to treat wastewater. New coagulant processing technique such as composite polymerization and impregnation method can be incorporated in producing coagulants with enhanced capability. This review highlighted that many potential advantages in using natural coagulants from various sources of plants, animal or biomass.

This research was supported by UniversitiTun Hussein Onn Malaysia (UTHM), Malaysia, The Ministry of Higher Education Malaysia through Fundamental Research Grant Scheme (FRGS), Vot. 1575.

References

[1] J.D.P Theodoro, G.F Lenz, R.F Zara, R. Bergamasco, Coagulants and Natural Polymers: Perspectives for the Treatment of Water, Plastic and PolymerTechnology (PAPT), **2**, 55-62 (2013)

[2] A.D, BENETTI, "Water reuse: issues, technologies, and applications" Sanitary and Environmental Engineering", **13**(3), 247-248, (2008)

[3] N.Othman, S.M.Asharuddin, "Cucumismelorind as biosorbent to remove Fe (II) and Mn (II) fromsyntheticgroundwater solution, " In Adv. MaterialsResearch", **795**, 266-271 (2013)

[4] R. Sowmeyan, J. Santhosh, and R. Latha, Effectiveness of herbs in community water treatment, Int: Research J. of Biochemistry and Bioinformatics, **1**, 297-303, (2011)

[5] K.A.Ghebremichael, K. R. Gunaratna, H.Henriksson, H.Brumer, G.Dalhammar,A simple purification and activity assay of the coagulant protein from Moringa oleiferaseed, Water research, **39**, 2338-2344 (2005)

[6] G. L.McConnachie, G. K. Folkard, M. A. Mtawali, J. P. Sutherland, Field trials of appropriate hydraulic flocculation processes, Water research, **33**,1425-1434 (1999)

[7] A .Sethupathy, An Experimental Investigation Of Alum And Moringa Oleifera Seed In Water Treatment, Int. J. of Adv. Res.,, **3**, 515-518 (2015)

[8] K.A Yongabi, "Biocoagulants for water and waste water purification: a review,Int: review of chemical engineering, **2**, 444-458 (2010)

[9] R.Azizul, M.F, Mohd-Suhaimi, N.Othman, Biosorption of Pb(II) and Zn(II) in Synthetic Waste Water by Watermelon Rind (Citrulluslanatus), AppiedMechanics and Materials, **465**, 906-910 (2014)

[10] A.I. Cissouma, F. Tounkara, M. Nikoo, N. Yang, and X. Xu. Physico-Chemical Properties and Antioxidant Activity of Roselle Seeds Extracts.,Advance J. of Food Science and Technology, **5**(11), 1483-1489 (2013)

[11] R.Rajendran, M. Abirami, P. Prabhavathi, P. Premasudha, B. Kanimozhi, A. Manikandan, Biological treatment of drinking water by chitosan based nanocomposites. African J. of Biotechnology,**14**(11), 930-936 (2015)

[12] M A AHusan, L S Hui, Z.N Zainura, Removal of Boron from Industrial wastewater by chitosan via chemical precipitation, J. Chemistry Nat Res Engineering, **4**, 1-11(2009)

[13] A. Montembaulti, V.Christophe, and D.Alain., Biomaterials, Physico-chemical studies of the gelation of chitosan in a hydro alcoholic medium, Biomaterials, **26**, 933–943 (2005)

[14] M.E Mabrouk, Production of bioflocculant by the marine actinomycete Nocardiopsisaegyptia sp. Nov, life sciences J. **11**, 27-35 (2014)

[15] N.A. Oladoja, Headway on natural polymeric coagulants in water and wastewater treatment operations, J. of Water Process Engineering, **6**, 174–192 (2015)

[16] U.U Nwodo, G. Ezekiel, V. Leonard, K.O, Mabinya, K.Rumbold, C.O.Lawrence, I.Anthony, Bioflocculant production by a consortium of Streptomyces and Cellulomonas species and media optimization via surface response model.Colloids and Surfaces B: Biointerfaces, **116**, 257-264 (2014)

[17] M.J. Pearse, Historical use and future development of chemicals for solid–liquid separation in the mineral processing industry., Mineral Engineering, **16**(2), 103-108 (2003)

[18] C.Y.Yin, Emerging usage of plant-based coagulants for water and wastewater treatment.Process Biochemistry, **45**(9), 1437-1444 (2009)

[19] F.Anwar, S Nahid Zafar, and Umer Rashid, Characterization of Moringa oleifera seed oil from drought and irrigated regions of Punjab, Pakistan. Grasas Y Aceites, **57**(2), 160-168 (2006)

[20] Okuda, Tetsuji, U.B. Aloysius, N. Wataru, O.Mitsumasa, Isolation and characterization of coagulant extracted from Moringa oleifera seed by salt solution. Water research, **35**(2), 405-410 (2001)

[21] B.Bolto, J.Gregory, Organic polyelectrolytes in water treatment." Water research, **41**(11), 2301-2324 (2007)

[22] ÖZACAR, MAHMUT, I.A.ŞENGİL, The use of tannins from turkish acorns (valonia) in water treatment as a coagulant and coagulant aid,Turkish J. of Engineering and Environmental Sciences, **26**(3), 255-264 (2002)

[23] Graham, Nigel, F.Gang, G.Fowler, M.Watts, Characterisation and coagulation performance of a tannin-based cationic polymer: A preliminary assessment. Colloids and surfaces A: Physicochemical and engineering aspects, **327**(1), 9-16 (2008)

[24] C.Saenz, E. Sepúlveda, B.Matsuhiro, Opuntia spp mucilage's: a functional component with industrial perspectives. J. of Arid Environments, **57**(3), 275-290 (2004)

[25] S.M.Miller, E.J.Fugate, V.O.Craver, J.A.Smith, J.B.Zimmerman, Toward understanding the efficacy and mechanism of Opuntia spp. as a natural coagulant for potential application in water treatment. Environmental science & technology, **42**(12), 4274-4279 (2008)

[26] Renault, Francois, B.Sancey, P-M.Badot, G.Crini, Chitosan for coagulation /flocculation processes–an eco-friendly approach. European Polymer J., **45**(5), 1337-1348. (2009)

[27] M.Adinolfi. M.M.Corsaro, R.Lanzetta, M.Parrilli, G.Folkard,W.Grant, J.Sutherland, Composition of the coagulant polysaccharide fraction from Strychnospotatorum seeds. Carbohydrate research, **263**(1) 103-110 (1994)

[28] M.B. Sciban, M.T.Klasnja, J.Lj.Stojimirovic. Investigation of coagulation activity of natural coagulants from seeds of different leguminose species. Acta PeriodicaTechnologica (Serbia and Montenegro) APTEFF, **36**, 1-266 (2005)

[29] B. Bina, M. Mehdinejad, M. Nikaeen, H.M. Attar, Effectiveness of chitosan as natural coagulant aid in treating turbid waters. J. of Environmental Health Science & Engineering, **6**(4), 247-252 (2009)

[30] Adamafio, N. Ayikailey, M. Sakyiamah, J. Tettey, Fermentation in cassava (ManihotesculentaCrantz) pulp juice improves nutritive value of cassava peel. African J. of Biochemistry Research, **4** (3), 51-56 (2010)

[31] B. Lukuyu, I.Okikie, A.J Duncan, M. Beveridge, M.Blummel, Use of cassava in livestock and aquaculture feeding programs. Int. Laboratory for Research on Animal Di, **25**, 1-83 (2014)

Organics Characteristics of Sludge from a Full-Scale Anaerobic Digester Treating Domestic Mixed Sewage Sludge

Roslinda Seswoya[1,*], and *Ahmad Tarmizi* Abdul Karim[1]

[1]Micropollutant Research Centre, Faculty of Civil and Environmental Engineering, Universiti Tun Hussein Onn Malaysia, Parit Raja, Johor.

Abstract. Sewage sludge, normally in form of mixed sewage sludge is treated using anaerobic digester worldwide. In Malaysia, sewage sludge was categorized as domestic sewage sludge since sewage treatment plant treats only domestic sewage. The complex organic compounds in form of carbohydrates and proteins are transformed to methane during anaerobic digestion. The characteristics of complex organic compounds in domestic mixed sewage sludge are needed to assess the energy recovery form digesting domestic mixed sewage sludge. Besides that, it is common to use anaerobic biomass from existing anaerobic digester for the new setup of the anaerobic reactor. Therefore, this study was outlined to study the characteristics of domestic mixed sewage sludge and anaerobic biomass, particularly on the complex organic compounds. The complex organic compounds measured were carbohydrates and proteins. The higher complex organic solubilisation as a result of thermal pre-treatment was proven to improve the methane production. Therefore, in this study, the impact of low thermal pre-treatment in improving the organics solubilisation was assessed too. Low thermal pre-treatment at 70°C and 90°C at various treatment time were applied to the domestic mixed sewage sludge. The results indicated that the domestic sewage sludge and anaerobic biomass from a full-scale anaerobic digester contained complex organic compounds; existed mostly in form of particulate as shown by the low value of soluble to total ratio. Besides that, the low thermal treatment at 70°C and 90°C increased the organics solubilisation. Protein solubilisation was observed exceeded 8% after being treated for 20 min at both thermal treatments. However, the impact of low thermal treatment was better at 90°C, in which higher solubilisation was observed at longer treatment time.

1 Introduction

Anaerobic digestion is a biological process to degrade the organic pollution with the existence of several types of anaerobic microorganisms, in which the biogas is produced. The biogas has a high calorific value and is considered as a renewable energy source [1].

* Corresponding author: roslinda@uthm.edu,my

The airtight tanks are used for anaerobic digestion. Many types of the organic material can be digested, except for stable woody materials. The lignin in the stable wood is hard to be degraded by the anaerobic micro-organisms lignin [1]. Substrate to anaerobic digester varies including landfill leachate, and sewage sludge [2], [3].

Asia Pacific Economic Cooperation (APEC) economies (including Malaysia) have made commitments to promote renewable power generation and have implemented a variety of policies to support wider adoption of renewable technologies [4]. This energy policy will shape the region's' energy future. According to Kumaran [5], if entire sewerage system in Malaysia adopts anaerobic digestion, the methane (renewable energy) from sewage sludge is estimated as 1,132,512.50 m^3 CH_4 /day and the energy production is 4.75 GW.h/day. Unfortunately, only few sewage treatment in Malaysia adopting the anaerobic digestion. Anaerobic digester in Jelutong Sewage Treatment Plant (STP) was producing an average of 800 m^3/day biogas, used for lighting within the plant compounds, odor control system and lift for the digester equipment building, consequently complying with the sustainability principle [6]. It is a huge economical loss to neglect application of anaerobic digester for sewage treatment because this technology able to reduce sludge volume; by reducing of about 30% of the organic load (part of the solid organic matter is converted into gas) besides producing renewable energy (biogas) [7].

Despite the advantages of anaerobic digestion as mentioned above, the long retention time (typical 20–30 days) and less organic reduction were the major drawbacks of this anaerobic digestion. Appels [1] reported that organics reduction from sewage sludge through anaerobic digestion is limited to about 50% even after 20 days residence time. During the anaerobic digestion of complex organic matter, the hydrolysis is always the rate-limiting state. Therefore the optimization of the anaerobic digestion process strongly depends on the increase in hydrolysis efficiency [1, 8]. Moreover, the good contact between substrate and biomass is a crucial key in hydrolysis [9].

Pre-treatment is a method to improve hydrolysis, in which sewage sludge disintegration is achieved before entering the digester. In comparison with other methods of pre-treatment, thermal pre-treatment has been considered to be a more promising technique [10, 11]. Most of the pre-treatments have only been applied to sewage secondary sludge (or waste activated sludge, WAS), due to the difficulty of sewage secondary sludge digestibility compared to sewage primary sludge. According to Jones [12], the gas yield (volume gas / volume sludge) from the digestion of WAS at 10 days solids retention time (SRT) is approximately four-fold lesser than what was observed from the digestion of primary sludge. Low thermal pre-treatment ($< 100^oC$) was able to increase the solids and organic solubilisation of sewage sludge [13-16]. The higher organic soluble enhance the hydrolysis step consequently increased methane production [13, 17].

There are nine (9) options for energy recovery from sewage sludge (from the organic compounds in sewage sludge) including anaerobic digestion of sewage sludge, production of biofuels from sewage sludge, and direct production of electricity from sewage sludge in microbial fuel cells [18]. Biogas production is directly influenced by the type of sludge and the operating conditions of the digester. For a mixture of primary and secondary sludge the gas production amounts to approximately 1 m^3 of biogas/kg of organic solids biodegraded [4]. It is necessary to measure the composition of the sewage sludge to assess the options for energy recovery from sewage sludge as a sole substrate, particularly on the complex organic concentration. The pathway of the transformation of complex organic to biogas is established [19]. Therefore, the aim of this study was to characterize the domestic mixed sewage sludge and anaerobic biomass from the full-scale anaerobic digester particularly on the organic compounds, in which very few studies on this subject were reported. In addition, this study also investigated the impact of low thermal pre-treatment on the domestic mixed sewage sludge characteristics.

2 Material and Methods

2.1 Sample Collection

Two types of sample were taken from the full-scale local anaerobic digester treating the domestic mixed sewage sludge located in Malaysia. The domestic mixed sewage sludge (DMS), which goes directly into the anaerobic digester and anaerobic biomass or anaerobically digested sludge (AnDS) were collected at each sampling day. AnDS was taken from the anaerobic digester itself. The sludge was stored at 4°C in the laboratory before use.

2.2 Preparation for pre-thermally treated domestic mixed sewage sludge (DMS)

After being stored at 4°C for a short time (< 24 hours), the low thermal pre-treatment was applied to the DMS to investigate the effects of low thermal pre-treatment, particularly on organics solubilisation. Low thermal pre-treatment is expected to increase the organic solubilisation of domestic mixed sewage sludge.

The thermal pre-treatments were carried out using a thermostatic water bath operated at a temperature of 70°C and 90°C respectively. The procedures for preparing the pre-thermally treated DMS were described by Roslinda [20]. According to Salsabil, [21], the organic solubilisation (S_X in %), can be calculated as below:

$$S_x = \left(\frac{x_s - x_{so}}{x_{PO}} \right) x100\% \quad (1)$$

where,

X = represents either COD, proteins or carbohydrates concentration.
X_S = soluble concentration
X_{SO} = initial soluble concentration
X_{PO} = initial particulate concentration

2.3 Analytical method

Untreated DMS, pre-thermally treated DMS, and AnDS were analyzed for chemical oxygen demand, carbohydrate, and proteins. The analytical method for that parameter was described previously by Seswoya [22]. The sample for determination of the soluble concentration of chemical oxygen demand (COD), carbohydrates, and proteins were firstly centrifuged at 10000 rpm for 10 min, then filtered through a cellulose acetate filter of 0.45μm pore size prior to analysis [23, 24].

DR 6000 UV-Vis spectrophotometer, DRB200 Reactor (digester) and COD vial (TNT plus 822) were used for total and soluble COD determination, based on Reactor Digestion Method which was approved by USEPA.

3 Results and Discussion

3.1 AnDS Characteristics

Table 1 showed the ratios of soluble to particulate of each measured parameters. The ratios of soluble COD to total COD (SCOD/TCOD) were low. This was also observed for the

AnDS originated from lab-scale continuous stirred tank reactor operated at mesophilic temperature (37^{0}C and 35^{0}C) treating sewage sludge (mixed and waste activated sludge) [25,26,15,16]. AnDS also contained carbohydrates and proteins. AnDS from this study having higher protein than carbohydrate, which is opposite to what was observed by Lu [26] Astals [25] in which AnDS the in their study originated from lab-scale continuous stirred tank reactor. The lower concentration of soluble form of COD, carbohydrates, and proteins showed that the organics mostly existed in the form of a particulate. This was also observed by Astals [25] and Lu [26] previously.

Table 1. Soluble to total ratio of measured organic compounds in untreated AnDS

Parameters	Sampling date	
	27/7/15	1/9/15
SCOD / TCOD	0.14	0.09
SProteins / TProtein	0.05	0.04
SCarbohydrate / TCarbohydrate	0.03	0.05

The data presented in the table were calculated from the average value from three sample

3.2 DMS Characteristics

The ratio of SCOD/TCOD ranged from 0.07 to 0.11. In previous work of Astals [25], three different municipal mixed sewage sludge were having SCOD/TCOD equals to 0.07. SCOD/TCOD value which is low as 0.04 was also observed from municipal mixed sewage sludge [27]. Proteins are higher in DMS than carbohydrates. The similar observation was observed when the municipal mixed sewage sludge and municipal secondary sewage sludge were studied [10, 21]. However, this organic compounds existed in the form of particulate as shown by the low value of soluble to particulate ratios.

Table 2. Soluble to total ratio of measured organic compounds in untreated DMS

Parameters	Sampling date	
	27/7/15	1/9/15
SCOD / TCOD	0.11	0.07
SProteins / TProteins	0.03	0.02
SCarbohydrates / TCarbohydrates	0.05	0.01

The data presented in the table were calculated from the average value from three samples

3.3 Pre-thermally treated DMS Characteristics

Thermal treatment on DMS was performed for two different temperatures and several treatments period. An overview of the results is shown in Table 3. The thermal treatment released the monitored component to the water phase as indicated by the improved soluble to the total ratio as compared to the ratio from raw DMS at each sampling. The soluble COD was higher at 90°C compared to 70°C, which is almost twofold, although the treatment period was 20 min. Generally, the soluble COD and soluble carbohydrate were greater at the longer treatment time regardless the temperature. However, the highest values were observed at pre-treatment of 90°C.

Applying Equation (1) to appropriate data (data not presented), the COD, proteins, and carbohydrates solubilisation were calculated as shown in Table 4. Fig. 1 showed the graphical illustration of organics solubilisation affected by the low thermal pre-treatment. At both temperatures (70°C and 90°C), the solubilisation of COD and complex organics compounds were observed at each temperature. The gradually increased solubilisation was consistently observed for proteins at both pre-treatment conditions. Meanwhile, the carbohydrate solubilisation fluctuated at pre-treatment of 70°C. The remarkable changes of carbohydrate solubilisation were observed at pre-treatment at 90°C compared to thermal treatment at 70°C. The Amadori compounds or melanoidin as a result of the reactions of soluble carbohydrate with themselves or soluble protein was not observed in this study due to the pre-thermally DMS supernatant remained transparent and did not turn brown after treatment to form the Amadori compounds or melanoidins is not observed.

Table 3. Organic characteristics of pre-thermally treated DMS

Parameters	Sampling date: 27/7/15			Sampling date: 1/9/15		
	DMS 7020	DMS 7040	DMS 7060	DMS 9020	DMS 9040	DMS 9060
Total COD, TCOD (g/L)	9.61 ± 0.05	8.83 ± 0.11	7.35 ± 0.00	16.30 ± 0.07	20.60 ± 0.00	17.78 ± 0.10
Soluble COD, SCOD (g/L)	1.95 ± 0.07	1.84 ± 0.00	1.57 ± 0.00	3.32 ± 0.01	3.48 ± 0.01	3.69 ± 0.00
SCOD / TCOD	0.20	0.21	0.21	0.20	0.17	0.21
Total Proteins, TProteins (mg /L)	3850.33 ± 124.03	3138.33 ± 45.79	2625.00 ± 91.02	2532.50 ± 1.73	3007. 00 ± 38.16	2257.33 ± 33.39
Soluble Proteins SProtein (mg /L)	467.80 ± 2.13	548.97 ± 6.32	545.65 ± 6.03	457. 63 ± 2.91	500.52 ± 0.06	556.48 ± 2.95
SProteins / TProteins	0.12	0.17	0.21	0.18	0.17	0.25
Total Carbohydrates, TCarbohydrates (mg /L)	281.83 ± 6.53	271.17 ± 6.81	48.67 ± 0.29	780.50 ± 7.81	1246.00 ± 23.26	917.00 ± 3.28
Soluble Carbohydrates, SCarbohydrates, (mg /L)	8.62 ± 0.41	38.75 ± 0.33	23.00 ± 0.33	125.65 ± 3.96	128.02 ± 0.56	136.57 ± 2.24
SCarbohydrates / TCarbohydrates	0.03	0.14	0.47	0.16	0.10	0.15

1) DMS 7020 is referring to the domestic mixed sewage sludge treated at 70°C for 20 minutes. Using this description all the sample were labeled accordingly.
2) DMS 9020 is referring to the domestic mixed sewage sludge treated at 90°C for 20 minutes. Using this description all the sample were labeled accordingly.
3) 9.61 ± 0.05 = average ± stdev.
4) The coefficient of variation (CV) for each data is less than 5%. (CV < 5%)

At treatment time of 60 min at 90°C, protein solubilisation has exceeded 14%. Meanwhile at similar treatment time at 70°C, protein solubilisation reduced almost 5%. The COD solubilisation seem to be low at the temperature of 70°C even after the treatment time increased to 60 min. This was not observed for COD solubilisation of thermal treatment at 90°C. On contrary, the COD solubilisation increased gradually over the treatment time.

Table 4. Organics solubilisation

Sampling Date		27/7/15			1/9/15		
Sample ID		**DMS 7020**	**DMS 7040**	**DMS 7060**	**DMS 9020**	**DMS 9040**	**DSS 9060**
Organic Solubilisation (%)	COD	8.90	7.77	5.01	7.55	8.26	9.18
	Carbohydrates	-2.82	4.37	0.61	8.75	8.94	9.62
	Proteins	8.40	10.34	10.26	12.49	13.80	15.52

1) DMS 7020 is referring to the domestic mixed sewage sludge treated at 70°C for 20 minutes. Using this description all the sample were labeled accordingly.
2) DMS 9020 is referring to the domestic mixed sewage sludge treated at 90°C for 20 minutes. Using this description all the sample were labeled accordingly.
3) All date presented in the table are calculated using average data.
4) The solubilisation in negative is showing that the soluble concentration of monitored parameter is less than what was observed from the non-treated sample.

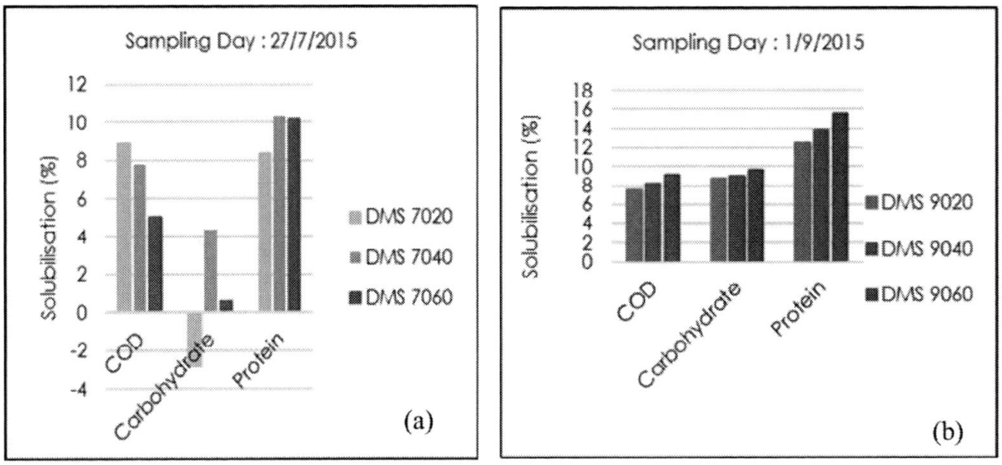

Fig. 1. Solubilisation Changes due to low thermal pre-treatment a) 70°C and b) 90°C

4 Conclusions

The present paper studied the characteristics of AnDS and DMS originated from full-scale anaerobic digester treating domestic mixed sewage sludge. It was seen that the AnDS also contained carbohydrates and protein, less than what was observed in DMS. Generally, the treatment operations and conditions would influence the characteristics of DMS and AnDS. The paper also presented the studies of the application of low temperatures (70°C and 90°C) as a pre-treatment method, to improve the hydrolysis prior to anaerobic digestion of domestic mixed sewage sludge. Generally, the organic solubilisation was observed after 20 min treatment time at each temperature except for carbohydrate. However, the organic compounds are solubilised efficiently during thermal pre-treatment at 90°C. As for thermal treatment at 90°C, more soluble organic compounds released at the higher treatment time. This finding suggested that the thermal pre-treatment at low temperature able to improve the organic solubilisation. However, in future work, pre-thermally treated DMS (with improved organic solubilisation) will acted as a substrate for anaerobic digestion, in which the energy recovery will be estimated.

The authors wish to express their gratitude to Universiti Tun Hussein Onn Malaysia (UTHM) for providing financial support and to R&D and Innovation Unit, Indah Water Konsortium (IWK) for their technical support.

References

[1] L. Appels, J. Baeyens, J. Degrève, and R. Dewil, Principles and Potential of The Anaerobic Digestion of Waste-Activated Sludge, *Prog. Energy Combust. Sci.*, **34**(6), 755–781, (2008)

[2] M. B. Ridzuan, Z. Daud, A. A. Abdul Latiff, Z. Ahmad and Z. Zakariah, Landfill Leachate Treatment Using Hybrid Up-Flow Anaerobic Sludge Blanket (HUASB) Reactor, Int. J. Integr. Eng., **5**(1), pp. 8–13, (2013)

[3] C. Arnaiz, J. C. Gutierrez and J. Lebrato, Biomass stabilization in the anaerobic digestion of wastewater sludges, Bioresour. Technol., **97**(10), 1179–1184, (2006)

[4] A.S. Muhamad Izham, A. Chrisnawan, M. Brown-Santirso, E.T. Gelindon, A. Kabalinskiy, L.M. Juan Roberto, T. Otsuki and M.O. Sinocruz, *APEC Energy Demand And Supply Outlook 6th Edition*, Vol. 1, Asia Pacific Energy Research Centre, Tokyo, (2016)

[5] P. Kumaran, D. Hephzibah, R. Sivasankari, N. Saifuddin and A.H. Shamsuddin, A review on industrial scale anaerobic digestion systems deployment in Malaysia: Opportunities and challenges, Renew. Sustain. Energy Rev., **56**, 929–940, (2016)

[6] Indah Water Konsortium Sdn Bhd, "*Cleaning the Unseen For a Sustainable Future*, Sustainable Report 2012-2013, (2013)

[7] E. Farno, J.C. Baudez, R. Parthasarathy and N. Eshtiaghi, Rheological characterisation of thermally-treated anaerobic digested sludge: impact of temperature and thermal history, Water Res., **56**, 156–61, (2014)

[8] W. Lv, F.L. Schanbacher and Z. Yu, Putting Microbes to Work in Sequence: Recent Advances in Temperature-Phased Anaerobic Digestion Processes, Bioresour. Technol., **101**(24), 9409–14, (2010)

[9] I. Angelidaki and W. Sanders, Assessment of the anaerobic biodegradability of macropollutants, Rev. Environ. Sci. Bio/Technology, **3**(2), 117–129, (2004)

[10] S. Ponsá, I. Ferrer, F. Vázquez and X. Font, Optimization of the hydrolytic-acidogenic anaerobic digestion stage (55 degrees C) of sewage sludge: influence of pH and solid content., Water Res., **42**(14), 3972–80, (2008)

[11] W.J. Park and J.H. Ahn, Effects of Microwave Pretreatment on Mesophilic Anaerobic Digestion for Mixture of Primary and Secondary Sludges Compared with Thermal Pretreatment, Environ. Eng. Res., **16**(2), 103–109,(2011)

[12] R. Jones, W. Parker, Z. Khan, S. Murthy and M. Rupke, Characterization of sludges for predicting anaerobic digester performance., Wat. Sci.Tech, **57**(5), 721–726, (2008)

[13] L. Appels, J. Degrève, B. Van der Bruggen, J. Van Impe and R. Dewil, Influence of low temperature thermal pre-treatment on sludge solubilisation, heavy metal release and anaerobic digestion, Bioresour. Technol., **101**(15), 5743–8, (2010)

[14] I. A. Nges and J. Liu, Effects of anaerobic pre-treatment on the degradation of dewatered-sewage sludge, Renew. Energy, **34**(7), 1795–1800, (2009)

[15] I. Ferrer, S. Ponsá, F. Vázquez and X. Font, Increasing biogas production by thermal (70°C) sludge pre-treatment prior to thermophilic anaerobic digestion, Biochem. Eng. J., **42**(2), 186–192, (2008)

[16] I. V Skiadas, H. N. Gavala, J. Lu and B. K. Ahring, Thermal pre-treatment of primary and secondary sludge at 70 degrees C prior to anaerobic digestion., Water Sci. Technol., **52**(1-2), 161–166, (2005)

[17] Y. Yan, H. Chen, W. Xu, Q. He and Q. Zhou, Enhancement of Biochemical Methane Potential from Excess Sludge with Low Organic Content by Mild Thermal Pretreatment, Biochem. Eng. J., **70**, 127–134, (2013)

[18] W. Rulkens, Sewage Sludge as a Biomass Resource for the Production of Energy : Overview and Assessment of the Various Options, Energy and Fuels, **22**, 9–15, (2008)

[19] W. Gujer and A. J. B. Zehnder, Conversion Process of Anaerobic Digestion, Wat.Sci.Tech, **15**, 127–167, (1983)

[20] R. Seswoya and A. T. Abdul Karim, Low Thermal Pretreatment as Method for Increasing the Bioavailability of Organic Matters in Domestic Mixed Sludge, *Int. Conf. on Civil and Env. Eng. for Sustanability (IConCESS 2015)*, (2016)

[21] M. R. Salsabil, J. Laurent, M. Casellas and C. Dagot, Techno-economic evaluation of thermal treatment, ozonation and sonication for the reduction of wastewater biomass volume before aerobic or anaerobic digestion, J. Hazard. Mater., **174**, 323–33, (2010)

[22] R. Seswoya and A.T. Abdul Karim, Characterization Of Organic Matter Solubilisation For Domestic Sewage Sludge Before And After Thermal Pretreatment At 70 0 C, Int. J. Appl. Eng. Res., **10**(95), 21–25, (2015)

[23] C. A. Wilson and J. T. Novak, Hydrolysis of Macromolecular Components of Primary and Secondary Wastewater Sludge by Thermal Hydrolytic Pretreatment, Water Res., **43**(18), 4489–98, (2009)

[24] H. Yi, Y. Han and Y. Zhuo, Effect of Combined Pretreatment of Waste Activated Sludge for Anaerobic Digestion Process, Procedia Environ. Sci., **18**, 716–721, (2013)

[25] S. Astals, M. Esteban-Gutiérrez, T. Fernández-Arévalo, E. Aymerich, J. L. García-Heras and J. Mata-Alvarez, Anaerobic digestion of seven different sewage sludges: A biodegradability and modelling study, Water Res., **47**(16), 1–11, (2013)

[26] X. Lu, G. Zhen, Y. Liu, T. Hojo, A. L. Estrada and Y.-Y. Li, Long-term effect of the antibiotic cefalexin on methane production during waste activated sludge anaerobic digestion, Bioresour. Technol., **169**, 644–51, (2014)

[27] A. S. Ucisik and M. Henze, Biological hydrolysis and acidification of sludge under anaerobic conditions: the effect of sludge type and origin on the production and composition of volatile fatty acids, Water Res., **42**(14), 3729–38, (2008)

Comparisons Study of Phosphate Removal in Unaerated and Aerated High Calcium Steel Slag Filter System of Different pH Feed

Siti Zu Nurain Ahmad[1], Hamdan R.[1,*], *Wan Afnizan* Wan Mohamed[2], N. Othman[2], *Nur Shaylinda* Mohd Zin[2], S. Musa[2]

[1]Department of Civil Engineering Technology, Faculty of Engineering Technology, UniversitiTun Hussein Onn Malaysia, 86400 Parit Raja, BatuPahat, Johor, Malaysia
[2]Department of Water and Environmental Engineering, Faculty of Civil and Environmental Engineering, UniversitiTun Hussein Onn Malaysia, 86400 Parit Raja, BatuPahat, Johor, Malaysia

Abstract. Excess phosphorus in water body will lead to eutrophication. This study investigated the phosphate removal efficiencies of unaerated and aerated filter systems using high composition of Calcium (Ca) steel slag as the filter media at different pH values of the wastewater influents. Lab-scale filters were developed using 25 mg/L synthetic wastewater and weekly sampling was done to monitor the phosphate removal efficiencies together with the concentration of metals (Calcium (Ca) and Magnesium (Mg). The results show that both unaerated and aerated systems have excellent phosphate removal efficiency at all acidic, neutral and alkaline pH feed, though unaerated systems removed slightly better compared to aerated systems; 76-98% and 69-97% respectively. The dominant phosphate removal mechanism for aerated systems was adsorption, meanwhilefor unaerated systems; both adsorption and precipitation for acidic and neutral pH, whileprecipitation was more dominant at basic pH. The performance of unaerated systems are slightly better compared to aerated systems, however, aerated systems are recommended to be applied when simultaneous removal of nutrients (phosphorus and nitrogen) are concerned.

1 Introduction

There is a dual concern regarding nutrients, since they contribute to unfavorable health and environmental effects, thus their control is imperative, but the same nutrients are essential for the performance of the natural biological treatment systems[1].The most significant effect of phosphorus in the environment is eutrophication or algal bloom. This process will promote a great deal of plant and animal biomass production, which eventually leads to succession in aquatic ecosystems [2]. Since 1950s, development of technologies of phosphorus removal has been studied in response to the issue of eutrophication. Since then,

[*]Correspondingauthor: rafidahh@uthm.edu.my

scientists had realized the need to reduce the amount of phosphorus allowed to enter the water body [3].

Nowadays, natural wastewater treatment process such as waste stabilization ponds (WSP), constructed wetlands (CW) and rock filters (RF) are very outstanding alternatives for phosphorus removal.. The utilization of low-cost and easily available materials as the filter media has been widely demonstrated by previous studies for their reliability and contribution in the removal of phosphate including fly ash [4], limestone[5], iron oxide tailings [6], blast furnace slag [7], basic oxygen furnace slag and electric arc furnace slag [8].

Steel slag has shown great phosphate removal efficiency as filter media for RF. One of the types of steel slag is an electric arc furnace (EAF) slag which is a by-product of melting recycled scrap in an electric arc furnace during the steel-making [9]. The significant contribution of steel slag in phosphate removal is generally due to high concentration of metal (Ca, Fe, Al, Mg) oxides [10]. However, the chemical compositions of all steel slags are not the same due to the variations in the grade of steel products and type of furnaces used during the process of steel-making [10]. Nevertheless, adsorption and precipitation with metal oxides are the main phosphate removal mechanisms in rock filters [11].

Performance of phosphorus removal was affectedby parameters include pH, temperature, concentration of metallic salts, HRTv and many more [12]. Of all these, pH values of the influents are known to contribute significant effects on phosphorus removal. The mechanism of phosphate removal can be differentdue to the different chemical composition of the steel slag. Very high percentage of CaO of the steel slag providesa significant rate of dissolution of Ca, causing formation of Ca phosphate, which were formed at high pH value and the precipitation dominates the removal mechanism. On the other hand, a high percentage ofFe_2O_3 of the steel slag prefers adsorption onto Fe oxides forthe phosphate removal mechanism at acidic and neutral pH ranges [13]. Besides that, precipitation might occur resulting from the dissolution of Fe from Fe oxides [14]. This is due to thefact that Fe ions tend to dissolve at acidic condition, which can be related to the precipitation of Fe phosphates meanwhile Ca ions tend to dissolve at basic condition which was related to Ca phosphate formation.

Unaerated system and aerated system differ in the introduction of oxygen into the systems during the aeration process. Aerated system provides aerobic condition to the system during the aeration process. Under aerobic conditions, Ca and Al adsorb onto the surface of adsorbent which can be the adsorption site for phosphate removal whereas phosphate are precipitated out with Fe ions [15]. Besides that, high dissolved oxygen in aerated conditions causes much more carbon dioxide to be stripped to the atmosphere thus produce little carbonic acid. This results in the increase in pH levels in the system [12]. Therefore, this study was carried out to compare the difference performance of aerated and unaerated high Ca steel slag systems at different pH values for its phosphate removal efficiency.

2 Experimental

2.1 Materials

The high Ca electric arc furnace (EAF) steel slag was obtained from Pasir Gudang, Johor, Malaysia. The steel slag was characterizedusing X-ray Fluorescence Spectroscopy (XRF) to obtain its chemical composition. The XRF results obtained were shown in Table 1. For preparation of XRF samples, the solid steel slag was crushed manually using a hammer and the crushed samples were collected and sieved through a 63 micrometer test sieve (British

Standard sieve BS410/1986) using a shaker, Endecotts Lombard Rd. London, model Swl93BR, England. Meanwhile,the steel slag samples in the range of sizes of 9.5 mm to 20 mm were accepted for use as filter mediain the column study. The samplespre-treatmentincludewashing up twice the steel slag with tap water followed by distilled water before they were dried up in the oven for 24 hours at 105°C.

Table 1. The XRF Results for Chemical Composition of Steel Slag

Compounds	**Percentage (%)**
CaO	49.5
SiO_2	18
MgO	4.32
SO_3	2.58
Al_2O_3	2.23
Fe_2O_3	1.64
MnO	0.38
TiO_2	0.22

2.2 Filter Column Study

To investigate the differences between unaerated and aerated systems on the phosphate removal efficiency using high Ca steel slag as the adsorbents, lab-scale vertical column filters study was developed. Twelve column filters with different pH values of the influents (pH 3, 5, 7, 9, 11 and Control) for unaerated and aerated systems with the inner diameter of 150 mm, 6 mm thickness and total height of 400 mm made from Perspex materials were set up as shown in Fig. 1. The column filters for Control systems used distilled water as the feed. The pH values of all different pH systems were adjusted using 2N NaOH and 2N H_2SO_4.

The pH values were monitored and adjusted frequently in order to minimize its fluctuation, especially for very low and very high pH values. HANNA HI-99310 portable pH meter was used to measure the pH and temperature. The influent was set to 25 mg/L synthetic wastewater (typical concentration range of phosphorus in wastewater in Malaysia) which was prepared by dissolving KH_2PO_4 salts (as the phosphate source) into distilled water. Synthetic wastewater feed was flowed via gravity from the influent tank located at the higher level and connected to the influent point which was located at the bottom of each column filters. The effluents of the samples were collected at the upper layer valves of each column filter. The hydraulic loading rate (HLR) used for the systems was set at 0.60 m^3/m^3.d following Johnson [15].

Meanwhile, aerated system was developed by connecting the tube from the air pump to the base of the column filters. The column filters were continuously fed with synthetic wastewater andweekly sampling was done for the influents and the effluents of each filter to test for pH, orthophosphate concentration and metals (Ca and Mg) concentration.

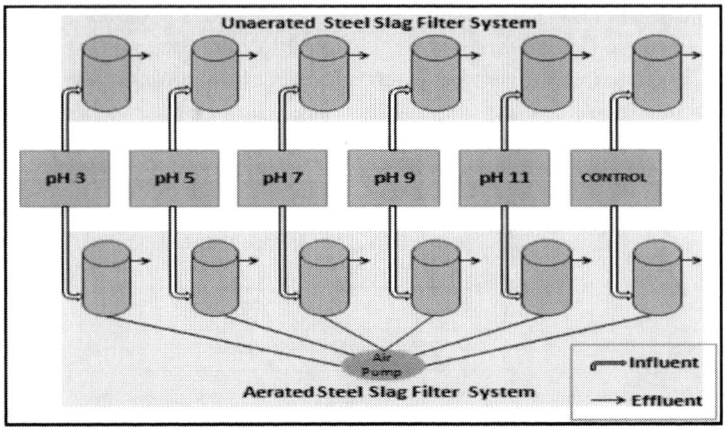

Fig. 1. Schematic diagram of lab-scale vertical steel slag column filters system

2.3 Analysis Methods

The samples were analyzed for the concentration of orthophosphate following the ammonium molybdatespectrometric method using WESTCO Discrete Analyzer, model Smartchem 200, France. The phosphate removal efficiency was obtained by determining the removal percentage of the phosphate. Meanwhile, the concentration of Ca ions and Mg ions were analyzedfollowing3111B Standard Methods for Examination of Water and Wastewater; atomic absorption spectrometric (AAS) method; [10] using model Perkin Elmer AAnalyst 100.

3 Results and Discussion

3.1 Phosphate removal efficiencies of unaerated and aerated high Calcium steel slag filter systems

Fig. 2 (a-e) illustrates the phosphate removal efficiencies of high Ca steel slag for unaerated and aerated filters at pH values of 3, 5, 7, 9 and 11 respectively. Generally, similar trends can be detected for all of the different pH systems; unaerated systems performances were slightly better compared to aerated systems. Nevertheless, aerated systems' performance can still be considered as efficient in removing phosphate since their percentage removal is around 69-97% while unaerated systems removed 76-98%.

Fig. 2(a) shows the removal efficiency of phosphate at pH 3where very similar trends of phosphate removal efficiencies were shown with slightly better performance of unaerated pH3 system compared to aerated pH 3 systems. Both systems showed excellent removal of phosphate between 77-97%. The highest percentage removal was shown during day 21 sampling at 97.9%. Meanwhile, systems at pH 5 also showed similar behaviour which can be seen in Fig. 2(b); highest removal for the first four weeks then started to decline; 77-97% and 74-97% for unaerated and aerated system respectively.

In Fig. 2(c), it can be seen that the removal efficiency of phosphate for unaereated pH 7 system was 76-98% whereas aerated pH 7 system was 71-95%. Unaerated system performed slightly better in removing phosphate compared to aerated system. Same goes to pH 9 and pH 11 systems, unaerated systems perform better; 77-97% for pH 9 and 76-97% for pH 11. The aerated systems followed closely behind; 69-96% for pH 9 and 67-91% for pH 11.

DOI: 10.1051/matecconf/201710306018

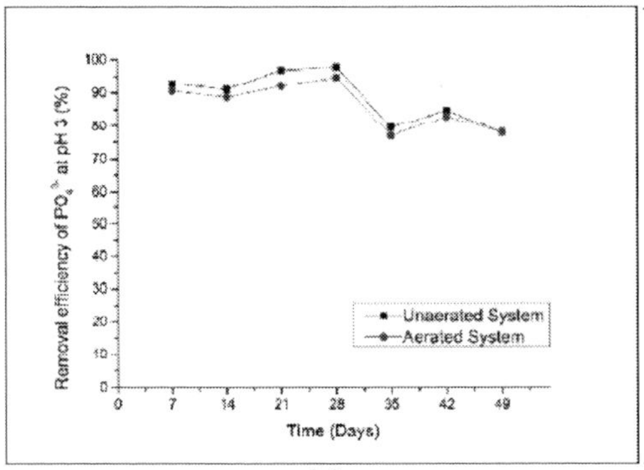

(a)

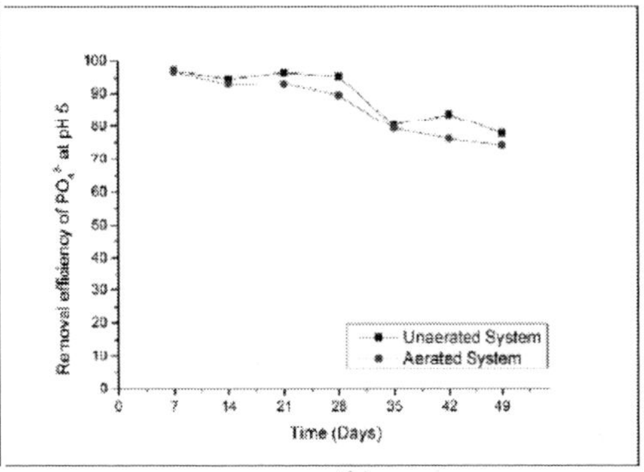

(b)

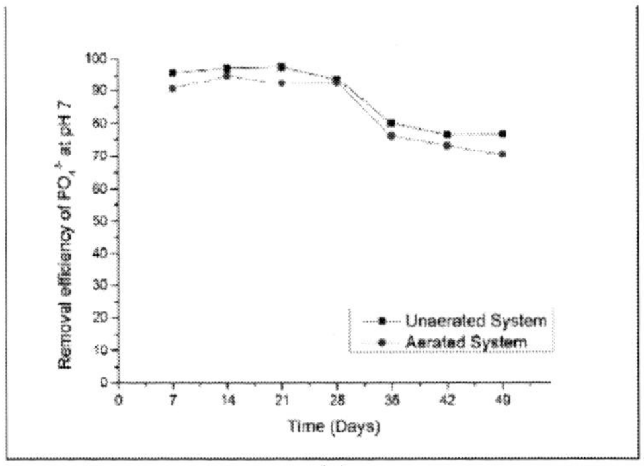

(c)

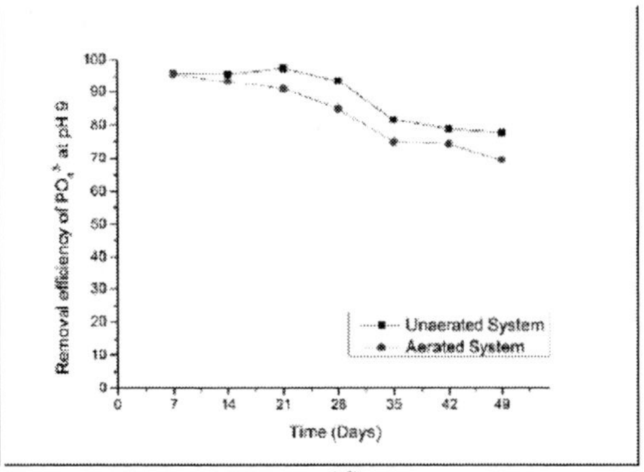

(d)

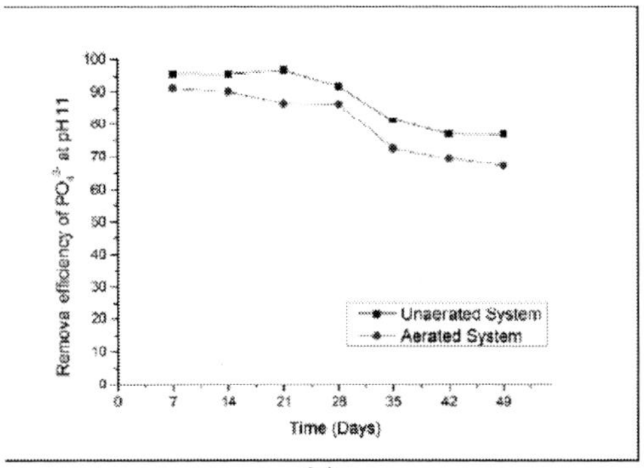

(e)

Fig. 2. (a-e) represents removal efficiency of phosphate from the filter systems of: (a) pH 3 systems; (b) pH 5 systems; (c) pH 7 systems; (d) pH 9 systems; (e) pH 11 systems.

All of the systems illustrated that the phosphate removal efficiencies are the best during the first four weeks before they started to decrease slightly and then consistent. During the first four weeks, many surface sites for removal mechanism were available. Therefore, they performed the best. However, after that, the surface sites for reaction had decreased a little due to the precipitates formed as the results of the removal causing certain surface sites to be unavailable for further reaction. This is true for unaerated systems because the systems were not disturbed; no aeration which allowed precipitates to form. On the other hand, aerated systems did not show a significant amount of precipitates inside the filters though the removal of phosphate still happened. This indicated that the removal mechanism followed adsorption mechanism rather than precipitation. However, the sampling duration should be extended some more to see the exact trends following after.

Overall, unaerated systems performed slightly better in phosphate removal compared to aerated systems. All of the different pH systems, either acidic, neutral or alkaline, has high phosphate removal efficiencyand this could be very advantageous in the removal of phosphate at different pH feed for wastewater treatment technology.

3.2 The effects of pH related to the concentration of Ca, and Mg ions in effluents and the phosphate removal mechanism

Fig. 3(a-f) illustrated the concentration of Ca and Mg in effluents at pH values of 3, 5, 7, 9, 11 and Control for both unaerated and aerated systems. At pH 3 as shown in Fig. 3(a), concentrations of Ca in effluents are significantly greater in unaerated systems compared to aerated system. This showed that the significant role of Ca ions in the phosphate removal mechanism via precipitation for unaerated system.

Similar trends can be seen in pH 5, 7, 9 and Control systems as seen in Fig.3 (b), (c), (d) and (f) respectively. The concentration of Ca in unaerated systems was obviously greater compared to aerated systems. In unaerated systems, visible white precipitates formed could be seen especially at the inlet of the filters, which were thought to be Calcium-phosphate precipitates either hydroxyapatite (HAP) or its precursors. The amount of white precipitates formed was greater for basic pH systems (pH 9 and pH 11 unaerated systems). This showed that the phosphate removal mechanism for unaerated systems followed both precipitation and adsorption mechanisms, only the significance of precipitation were greater and ruled the pathway at basic pH compared to acidic pH[14].

Meanwhile, aerated systems caused less concentration of Ca in effluents. This was actually contradicted to the assumptions made that aeration causes greater rate of dissolution of the metal ions inside the steel slag due to vibration. Physically, only a small amount of cloudy sludge can be seen in the filters and no visible white precipitates were formed. Nevertheless, their performance in removing phosphate was considered as excellent, comparable to unaerated systems. Therefore, adsorption might be more significant phosphate removal mechanism in aerated systems since no white precipitates were visible in the filters. The adsorption follows inner-sphere ligand exchange mechanism [15] where they reacted by forming monodentate and bidentate complexes which were determined by how they were coordinated [16].

Meanwhile, the concentrations of Mg are very small for all of the systems, less than 1 mg/L. Little can be obtained from here since the trends are very similar for both unaerated and aerated systems, with slightly higher concentration of Mg for unaerated systems. However, this shows that Mg ions have a less significant contribution compared to Ca ions in the removal of phosphate.

The performance of unaerated systems are slightly better compared to aerated systems. Nevertheless, since the importance of aeration is very crucial when removing nitrogen from wastewater; therefore when the desired results are to remove both nutrients simultaneously, therefore aerated systems' performance can be considered as an excellent choice.

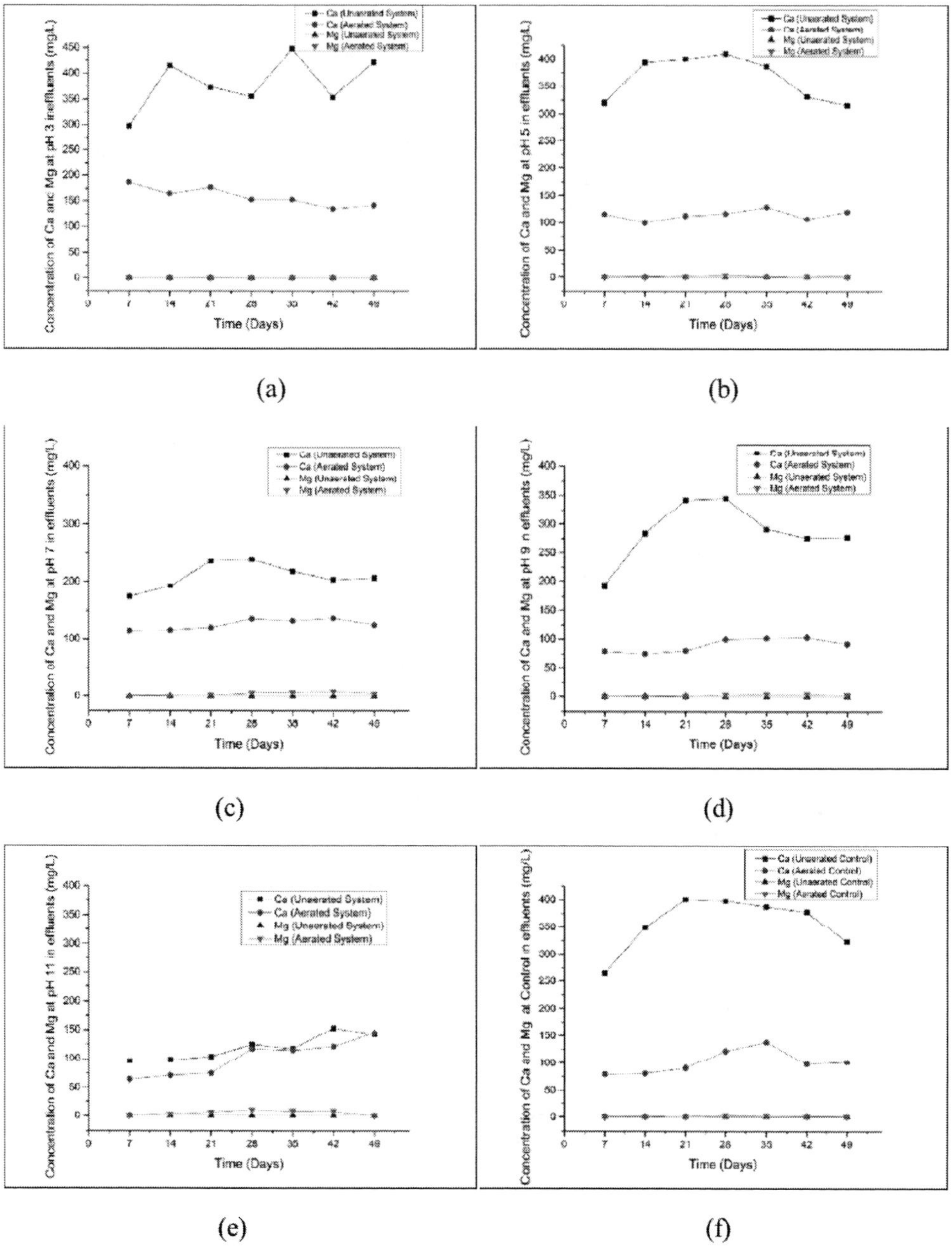

Fig. 3. (a-f) represents the concentration of Ca and Mg in effluents at systems of: (a) pH 3 systems; (b) pH 5 systems; (c) pH 7 systems; (d) pH 9 systems; (e) pH 11 systems; (f) Control systems

4 Conclusions

This study showed that both unaerated and aerated, high Ca steel slag filter systems were very efficient in removing phosphate at all acidic, neutral and basic pH influents. The percentages of phosphate removal efficiency for unaerated systems were found to be around 76-98%, while aerated systems removed 69-97% of phosphate. Ca concentrations in effluents are higher for unaerated systems compared to aerated systems, meanwhile Mg concentrations are similar for all the systems. The dominant phosphate removal mechanism followed adsorption for aerated systems and both adsorption and precipitation for unaerated systems; where at acidic and neutral pH, both adsorption and precipitation ruled the mechanism, meanwhile at basic pH, precipitation was more dominant.

We are grateful for the fundings provided by Ministry of Higher Education and University Tun Hussein Onn Malaysia; ERGS VOT.E031 and GPPS VOT U513.

References

[1] S.C. Reed, *Nat.systems for waste management and treatment*, 2nd ed. New York:McGraw-Hill,(1995)

[2] S. Manahan, *Fundamentals of Enviro. Chem.*, 3rd ed. Boca Raton, Florida:Taylor&Francis/CRC Press,(2009)

[3] G.K. Morse, S.W. Brett, J.A. Guy andJ.N.U. Lester, Review : Phosphorusremoval and recovery technologies. The Sci. of the Tot. Env., **212**, 69–81,(1998)

[4] Y.Z. Li, C.J. Liu, Z.K.Luan, X.J. Peng, C.L. Zhu, Z.Y. Chen, Z.G. Zhang, J.H. Fan and Z.P. Jia, Phosphate RemovalFromAqueous Solution UsingRaw and ActivatedRedMud and Fly Ash. J. Hazard. Mater. **137,** 374-383, (2006)

[5] L. Johansson-Westholm, Substrates for PhosphorusRemoval–potentialBenefits for On-site WastewaterTreatement, Water Resources. **40**(1), 23-36. (2006)

[6] I. Blanco, P. Molle, L. Miera, G. Ansola, Basic OxygenFurnaceSteelSlagAggregates for PhosphorusTreatment : Evaluation of ItsPotential Use as a Substrate in ConstructedWetlands, Water Research, **89**, 355-365 (2016)

[7] C. Barca, C. Gerente, D. Meyer, F. Chazarenc, Y. Andres,Phosphate RemovalfromSynthetic and Real WastewaterUsingSteelSlagsProduced in Europe, Water Resources. **89**, 1-6 (2012)

[8] C. Barca, S. Troesch, D. Meyer, P. Drissen, Y. Andres, F. Chazarenc, SteelSlagFilters to Upgrade PhosphorusRemoval in ConstructedWetlands: TwoYears of Field Experiments, Env. Sci. Tech.,**47**, 549-556 (2013)

[9] C. Barca, D. Meyer, M. Liira, P.Drissen, Y. Coemau, Y. Andres, F. Chazarenc, SteelSlagFilters to Upgrade PhosphorusRemoval in WastewaterTreatment Plants: RemovalMechanisms and Performance, Eco. Eng., **68**, 214-222 (2014)

[10] R. Hamdan, D. Mara, Study of In-filterPhosphorusRemovalMechanisms in an Aerated Blast FurnaceSlag, Int. J. of Research in Eng. and Tech., **2**(8), 130-136 (2013)

[11] C. Han, Z. Wang, W. Yang, Q. Wu, H. Yang, X. Xue, Effects of pH on PhosphorusRemovalCapacities of Basic OxygenFurnaceSlag, Eco. Eng.,**89**,1-6 (2016)

[12] P. Wifert, P.S. Kumar, L. Korving, G. Witkamp, M.C.M. Loodsrecht, The Relevance of Phosphorus and IronChemistry to the Recovery of PhosphorusfromWastewater: A Review, Env. Sci. and Tech., **49**, 99400-9414 (2015)

[13] M.Scholz, *WetlandsSystems to Control UrbanRunoff*, Elsevier, Oxford, United Kingdom, (2006)

[14] S.E. Johnson, R.H. Loeppert, Role of OrganicAcids I Phosphate MobilizationfromIronOxides, SoilSci. Soc. Am. J., **70**, 222-234 (2006)

[15] B. Zhong, R. Stanforth, S. Wu, J.P. Chen, Proton Interaction in Phosphate Adsorption onto Goethite, J. of ColloidalInterfacialSci., **308**, 40-48 (2007)

DOI: 10.1051/matecconf/201710306018

[16] A. Klimeski, W.J. Chardon, E. Turtola, R. Uusitola, Potential and Limitations of Phosphate Retention Media in Water Protection: A Process-basedReview of Laboratory and Field-scale Tests, Agr. and Food Sci., **21**, 206-223 (2012)

Potential of Copper Electrodes in Electrocoagulation Process for Glyphosate Herbicide Removal

Rabiatuladawiyah Danial[1,*], *Luqman Chuah* Abdullah[2], and *Shafreeza* Sobri[1]

[1]Department of Chemical and Environmental Engineering, Faculty of Engineering, Universiti Putra Malaysia, 43400 Serdang, Selangor, Malaysia
[2]Institute of Tropical Forestry and Forest Product, Universiti Putra Malaysia, 43400 Serdang, Malaysia

Abstract. Herbicide has been widely used since it was first marketed. Consumption of water contains glyphosate herbicide can bring adverse effects to human health. This study was intended to evaluate the performance of electrocoagulation process using copper electrodes for glyphosate removal in aqueous solution. Synthetic glyphosate powder has been diluted in deionized water. An electrocoagulation tank of 500 mL with two copper plates electrodes, same in dimensions were setup. The effects of continuous variables of initial glyphosate concentration, electrocoagulation time and current density, were discussed in detail. Production of copper cations showed an ability to neutralize negatively charged particles, then, encouraged to bind together to form aggregates of flocs composed of a combination of glyphosate and copper hydroxide. This study revealed that electrocoagulation process using copper electrodes is reliable by indicating 46.69% glyphosate removal. Finally, it can be concluded that electrocoagulation process using copper electrodes is efficient for glyphosate removal from aqueous environments.

1 Introduction

Herbicide has been widely used and application of herbicide in agriculture has further extended with increased usage for crops. In Malaysia, 172 herbicide products containing glyphosate with concentration range from 70% to 95%. Glyphosate, N-(phosphonomethyl)-glycine ($C_3O_5H_8PN$), which is a non-selective herbicide, has a function to control plants, including grasses, broad-leaves weeds, sedges, and woody plants [1,2]. Department of Environment Malaysia reported that many types of herbicides were sold in order to improve the quantity and quality of agricultural productions. This scenario indeed caused a very large amount of herbicide used. The widespread and enormous usage of herbicide can create a potential source of pollution and poses a significant threat to contamination of both surface water and underground water. It is due to discharge of herbicide wastes from the point or diffused sources, such as agricultural runoff. Additionally, consumption of water

* Corresponding author: adawiyahdanial@yahoo.com

contains glyphosate can bring adverse effects to human health in short-term and long-term [3]. Therefore, a treatment needs to be introduced to remove glyphosate from water.

Electrocoagulation process is susceptible to produce flocs of higher size and density, which facilitates pollutant removal by sedimentation in an aqueous solution. It was reported that electrocoagulation process has been widely applied to treat wastewater such as cutting oil emulsion wastewater [4] and textile wastewater [5]. In addition, researchers claimed that electrocoagulation process also successfully removed heavy metals from wastewater such as mercury (III) [6], phosphate [7,8], arsenic [9], fluoride [10] and antimony [11]. Electrocoagulation has advantages of minimum chemical addition and low maintenance. Electrocoagulation was selected because it has low operating cost compared to other treatments such as bioremediation, ozonation, advanced oxidation and nanofiltration. Copper was chosen as electrode plates in this study because it easy to be ionized in electrolysis system and its sludge would have high potential to be recycled.

This study handles the electrocoagulation process of a glyphosate aqueous solution in batch mode on a laboratory scale. It is mainly focused on the potential of using copper electrodes. The objective of this study is to investigate the impact of operating parameters which are initial glyphosate concentration, electrocoagulation time and current density; on the percentage removal of glyphosate. This study also comprehensively discusses the mechanism of electrocoagulation in removing glyphosate.

2 Methodology

Methods of this study include experimental setup and determination of glyphosate percentage removal.

2.1 Experimental setup

The basic setup of the electrocoagulation experiment comprised of a glass cylinder as electrocoagulation cell using a 500 mL beaker (Pyrex, England) with two copper electrode plates (Dingzhou Jinlong Metal Production Co. Ltd., China) of the same dimensions of (50 mm x 50 mm x 3 mm) with a total effective surface area of 50 cm^2, as anode and cathode placed in parallel. Fig. 1 shows a schematic diagram of experimental setup of this study. This laboratory scale experiment was set up in batch mode. Each batch contained 400 mL of feed. Glyphosate aqueous solution was prepared by dissolving white fine glyphosate powder (95% grade, Hebei Eminent Chemical Technology Co., Ltd.) within an electrolyte. Different amounts of glyphosate powder were added to deionized water (pH 6.7 ± 0.2, conductivity 1.0 μS/cm), depending on the required initial glyphosate concentration of 25 mg/L, 50 mg/L, 75 mg/L, 100 mg/L and 125 mg/L.

The conductivity of the solution was fixed by adding 400 mg of sodium chloride (NaCl). One plate was connected to the positive and the other plate connected to the negative outlet of a direct current (DC) power supply (HY 3003-2, Digimess, China). The DC power supply was then switched on to start the electrocoagulation process and a constant direct current was applied. The DC was adjusted to give the desired current density of 2 mA/cm^2, 4 mA/cm^2, 6 mA/cm^2, 8 mA/cm^2, 10 mA/cm^2 and 12 mA/cm^2, respectively, which is defined as the ratio of the applied direct current to the total effective surface area of the electrodes.

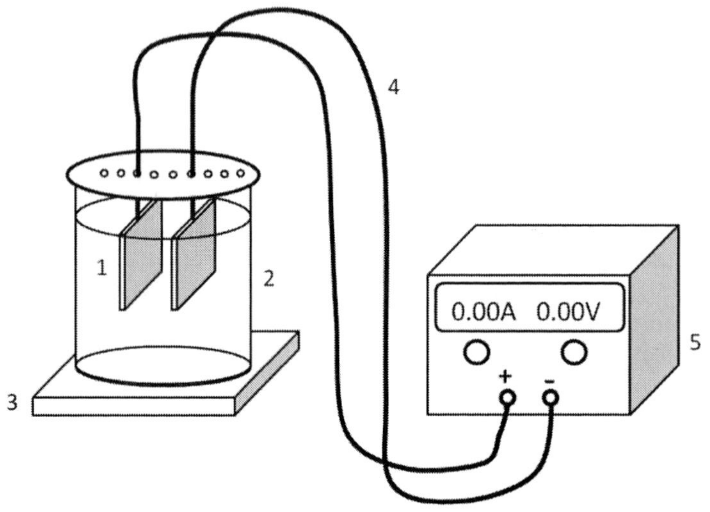

Indicators:
1. Electrodes plates (both copper)
2. Plexiglass beaker reactor (electrochemical cell)
3. Platform board
4. Wire
5. Digital DC power supply

Fig. 1. Schematic diagram of experimental setup for electrocoagulation process.

The time of electrocoagulation was varied over 10 min, 20 min, 30 min, 40 min, 50 min, 60 min and 70 min, respectively. When the DC power supply was switched off, the solution was left for 4 hours in order to allow all flocs to settle down at the bottom of the electrocoagulation cell until the treated aqueous solution exhibited as clear solution. The clear treated solution was then pumped into a filtration chamber in order to filtrate by filter paper with 0.45 μm pore (Whatman, UK) and the filtrate concentration was then measured. The settled sludge was dried at 100 °C using oven (UF55, Memmert, Germany) for two hours and then was allowed to cool in a desiccator (VT144, Joan Lab, China) for one hour.

2.2 Determination of glyphosate percentage removal

According to the glyphosate molecular structure, glyphosate has a phosphate functional group. A chemical reaction procedure to detect phosphate in aqueous solution was adopted to determine the existence of glyphosate. A basic mathematical equation used on a regular basis in the wastewater industry is percentage of removal [12]. The glyphosate removal was measured in terms of percentage removal which is defined as;

$$Glyphosate\ removal\ (\%) = \frac{C_o - C}{C_o} x\,100 \tag{1}$$

where C_0 is the initial glyphosate concentration before electrocoagulation (mg/L) and C is the residual glyphosate concentration after sedimentation (mg/L) at a given electrocoagulation time (t), respectively.

3 Mechanism of electrocoagulation

This section discussed about mechanism of electrocoagulation process in removing glyphosate and its proposed concept. Normally, four main processes occur serially during electrocoagulation processes which are electrolytic reactions at the electrode surfaces, the formation of coagulants in the aqueous phase, adsorption of soluble or colloidal pollutants on coagulants, and removal by sedimentation or floatation [13]. Thus, glyphosate removal from this study also underwent the same processes. Electrolytic reactions at electrode surfaces are the most important process in order to form copper hydroxide as a coagulant in aqueous medium. Coagulation of glyphosate and copper hydroxide were then formed flocs and precipitated as sludge.

3.1 Electrochemical reactions

According to Equations (2) – (8), the oxidation and reduction process mechanisms occurred simultaneously during electrocoagulation. The electrocoagulation process using copper electrodes produced copper cations and copper hydroxide in situ as a sorbent or a coagulant to remove the glyphosate. The electrochemical reaction with copper as the electrodes was summarized as follows:

Anodic electrochemical dissolution:

$$Cu(s) \rightarrow Cu^{2+}(aq) + 2e^{-} \qquad (2)$$

Oxidation of copper ions:

$$Cu^{2+}(aq) + H^{+}(aq) + \frac{1}{4}O_2(g) \rightarrow Cu^{2+}(aq) + \frac{1}{2}H_2O(l) \qquad (3)$$

Hydrolysis reaction:

$$Cu^{2+}(aq) + 2H_2O(l) \rightarrow Cu(OH)_2(s) + 2H^{+}(aq) \qquad (4)$$

Cathodic electrochemical reaction:

$$2H^{+}(aq) + 2e^{-} \rightarrow H_2(g) \qquad (5)$$

Overall reaction:

$$2Cu(s) + 3H_2O(l) + \frac{1}{2}O_2(g) \rightarrow 2Cu(OH)_2(s) + H_2(g) \qquad (6)$$

3.2 Glyphosate bonds break out

Glyphosate ($C_3O_5H_8PN$) bonds break out by the surrounding heat that exists from electrodes in an electrocoagulation cell during electrocoagulation time was as follows.

$$C_3H_8NO_5P(s) \rightarrow COH_5PNO^{-}(aq) + C_2H_2O_3(l) + H^{+}(aq) \qquad (7)$$

$$C_3H_8NO_5P(s) + H_2O(l) + H_2(g) \rightarrow CO_2(g) + CH_5N(l) + CH_2PO_4(l) + \frac{5}{2}H_2(g) \qquad (8)$$

4 Results and discussion

The analysis was evaluated based on the main aim which is the percentage removal of glyphosate.

4.1 Effect of initial glyphosate concentration

The initial glyphosate concentration was one of the important parameters that have a high impact on the percentage removal of glyphosate [14]. Fig. 2 shows that when the initial glyphosate concentration was increased, the percentage removal of glyphosate decreased. The effect of glyphosate concentration from 0 to 125 mg/L on the removal efficiency has been investigated. It was observed that the increase in concentration for the same electrocoagulation time of 60 min and current density of 6 mA/cm^2 resulted in decreasing removal efficiency. The performance of the electrocoagulation process was slowed down when the initial glyphosate concentration approached 100 mg/L. The highest percentage removal was at an initial glyphosate of 25 mg/L. However, at 100 mg/L the percentage removal reached equilibrium. At 100 mg/L initial glyphosate concentration, 46.68% removal was indicated. The removal efficiency was limited by the amount of copper hydroxide produced.

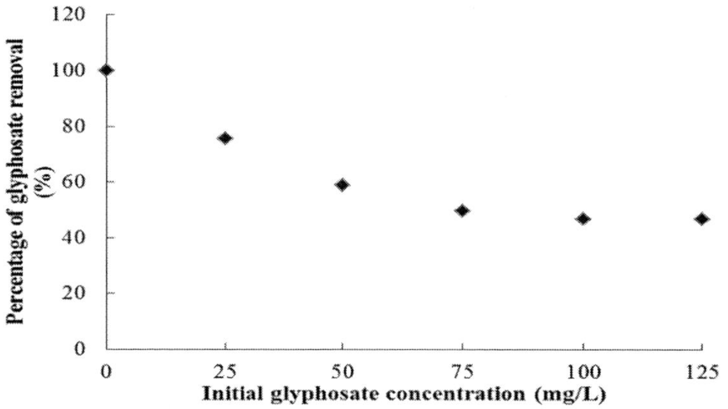

Fig. 2. Effect of initial glyphosate concentration on percentage removal of glyphosate (set up electrocoagulation condition: 60 min electrocoagulation time, 0.4 g NaCl added, 6 mA/cm2 current density, 6 cm distance between electrodes).

4.2 Effect of electrocoagulation time

The duration is the one of the most important parameters in the process of electrocoagulation. It requires an adequate period of time for the process to generate copper hydroxide, hence, to complete the coagulation of the impurities. Two reactions occur simultaneously during the electrocoagulation time. These reactions are the electrode deposition reaction and the flocs formation reaction. Both reactions occur when the current supply is turned on and end when the current is turned off. Normal process time ranges from 15 min to 75 min are for the highest possible removal of different copper ions [2]. In the present work, the electrocoagulation time used to remove glyphosate was from 10 min to 60 min. Fig. 3 shows that the highest removal percentage was 46.69%, with a duration of 60 min. The percentage removals indicated in these results are in very good agreement with the results reported in the study of [14], in which removal percentage increases with an increase in electrocoagulation time. No further increase or decrease in the removal percentage was observed when the electrocoagulation time was increased beyond 60 min.

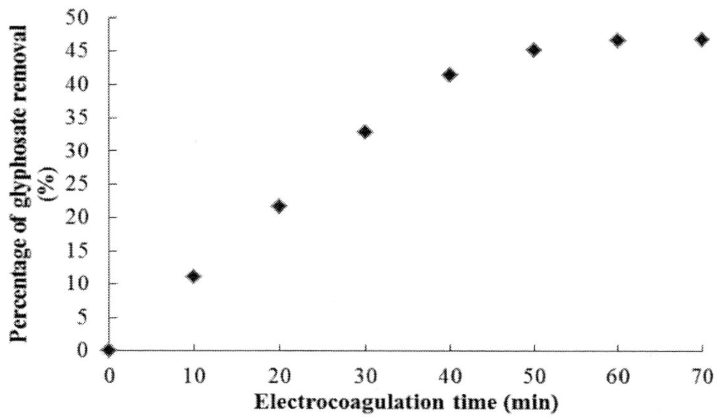

Fig. 3. Effect of electrocoagulation time on percentage removal of glyphosate (set up electrocoagulation condition: 100 mg/L initial glyphosate concentration, 0.4 g NaCl added, 6 mA/cm2 current density, 6 cm distance between electrodes).

4.3 Effect of current density

Current density is defined as the current applied per unit surface area of the electrode. In all electrochemical processes as well as the electrocoagulation process, the current density is an important parameter to control the reaction rate within the cell. In this study, the selected range of applied current density was from 0 to 12 mA/cm^2. Generally, a greater current density results in greater removal of solutes, but this may not be true beyond a certain value of current density [15-18]. However, when too large a current is used, there is a greater chance of electrical energy wasted and it is noticed when the solution became warm. Fig. 4 shows that the highest percentage removal was at 6 mA/cm^2 in which 46.69% removal. Direct current (DC) induces an impermeable oxide layer on the cathode due to oxidation which prevents effective current transfer between the anode and cathode. Therefore, the efficiency of the electrocoagulation process might decline when it reaches the optimum current density.

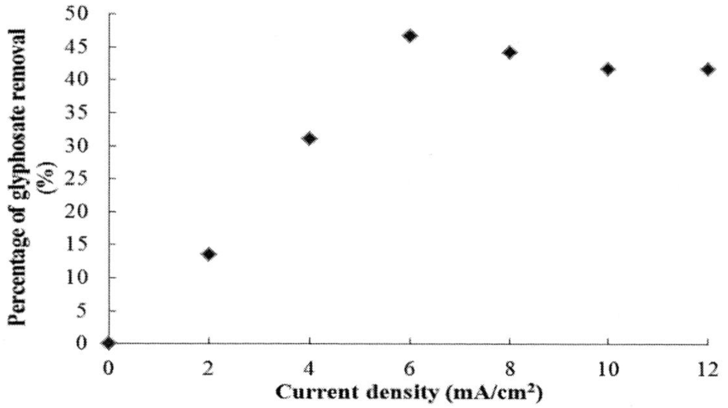

Fig. 4. Effect of current density on percentage removal of glyphosate (set up electrocoagulation condition: 60 min electrocoagulation time, 100 mg/L initial glyphosate concentration, 0.4 g NaCl added, 6 cm distance between electrodes).

4.4 Sludge recovery

The purpose of drying the sludge is to make it easy to store or to decompose. Sludge can be a big issue for the health of the environment if it is produced in large amounts and it is not able to be recycled. Several compounds detected might have a potential to be recycled. Copper sludge can be sent to a copper factory for isolation and further recovery [19]. Copper sludge can also be reused and recycled since the cost of copper mining is very high nowadays. The copper sludge can be useful since it is proposed to mix with cement to harden a concrete of building walls.

5 Conclusion

The electrocoagulation process using copper has been proven to be effective in removing glyphosate. Hence, by removing glyphosate, it reduces the amount of glyphosate in the water before the water is safe to be consumed. The impacts of the operating parameters on glyphosate removal, such as electrocoagulation time, initial concentration, current density, have been discussed in detail. Electrocoagulation using copper electrodes indicated a high percentage removal of glyphosate (46.69%). Electrocoagulation is suitable for application only for a medium or small sized community compared to a huge community. Application to a large wastewater treatment plant would not be economical or cost-effective.

The authors thank Faculty of Engineering, Universiti Putra Malaysia and Ministry of Science and Technology Malaysia (MOSTI) for the grant and financial aids for this project achievement.

References

[1] S.O. Duke, S.B. Powles, Glyphosate: A once-in-a-century herbicide: Mini-review. Pest Manag. Sci., **64**, 319–325 (2008)

[2] G. Chen, Electrochemical technologies in wastewater treatment. Sep. Purif. Technol., **38**, 11–41 (2004)

[3] P.K. Holt, G.W. Barton and C.A. Mitchell, The future for electrocoagulation as localised water treatment technology, Chemosphere, **59**, 355–367 (2005)

[4] K. Bensadok, S. Benammar, F. Lapicque and G. Nezzal, Electrocoagulation of cutting oil emulsions using aluminium plate electrodes. J. Hazard Mater., **152**, 423–430 (2008)

[5] I. Zongo, A. Hama, J. Wéthé, G. Valentin, J. Leclerc, G. Paternotte and F. Lapicque, Electrocoagulation for the treatment of textile wastewaters with Al or Fe electrodes: Compared variations of COD levels, turbidity and absorbance. J. Hazard. Mater., **169**, 70–76 (2009)

[6] C.P. Nanseu-njiki, S. Raoul, P. Claude, A. Darchen and E. Ngameni, Mercury (II) removal from water by electrocoagulation using aluminium and iron electrodes. J. Hazard. Mater., **168**, 1430–1436 (2009)

[7] A. Attour, M. Touati, M. Tlili, M.B. Amor, F. Lapicque, and J. Leclerc, Influence of operating parameters on phosphate removal from water by electrocoagulation using aluminum electrodes. Sep. Purif. Technol., **123**, 124–129 (2014)

[8] S. Vasudevan, J. Lakshmi, J. Jayaraj and G. Sozhan, Remediation of phosphate-contaminated water by electrocoagulation with aluminium, aluminium alloy and mild steel anodes. J. Hazard. Mater., **164**, 1480–1486 (2009)

[9] E. Lacasa, P. Ca, C. Sáez, F.J. Fernández and M.A. Rodrigo, Removal of arsenic by iron and aluminium electrochemically assisted coagulation. Sep. Purif. Technol., **79**, 15–19 (2011)

[10] N. Drouiche, S. Aoudj, M. Hecini, N. Ghaffour, H. Lounici and N. Mameri, Study on the treatment of photovoltaic wastewater using electrocoagulation: Fluoride removal with aluminium electrodes-Characteristics of products. J. Hazard. Mater., **169**(1-3), 65–9 (2009)

[11] J. Zhu, F. Wu, X. Pan, J. Guo and D. Wen, Removal of antimony from antimony mine flotation wastewater by electrocoagulation with aluminum electrodes. J. Envi. Sci, **23**(7), 1066–1071 (2011)

[12] N. Daneshvar, A. Oladegaragoze and N. Djafarzadeh, Decolorization of basic dye solutions by electrocoagulation: an investigation of the effect of operational parameters. J. Hazard. Mater., **B129**(1-3), 116–22 (2006)

[13] M.F. Ni'am, F. Othman, J. Sohaili and Z. Fauzia, Electrocoagulation technique in enhancing COD and suspended solids removal to improve wastewater quality. Water Sci. Technol., **56**(7), 47–53 (2007)

[14] S.A. Abdel-Gawad, A.M. Baraka, K.A. Omran and M.M. Mokhtar, Removal of some pesticides from the simulated waste water by electrocoagulation method using iron electrodes. Int. J. Electrochem. Sci., **7**, 6654–6665 (2012)

[15] A. Aleboyeh, N. Daneshvar and M.B. Kasiri, Optimization of C.I. Acid Red 14 azo dye removal by electrocoagulation batch process with response surface methodology. Chem. Eng. Process.: Process Intensification, **47**(5), 827–832 (2008)

[16] M. Kobya, E. Demirbas, A. Dedeli and M.T. Sensoy, Treatment of rinse water from zinc phosphate coating by batch and continuous electrocoagulation processes. J. Hazard. Mater., **173**(1-3), 326–334 (2010)

[17] M. Saleem, A.A. Bukhari, M.N. Akram, Electrocoagulation for the treatment of wastewater for reuse in irrigation and plantation. J. Basic Appl. Sci., **7**(1), 11–20 (2011)

[18] M. Tir and N. Moulai-Mostefa, Optimization of oil removal from oily wastewater by electrocoagulation using response surface method. J. Hazard. Mater., **158**(1), 107–115 (2008)

[19] J. Rodriguez, S. Stopić and B. Friedrich, Continuous electrocoagulation treatment of wastewater from copper production. World Metall., **60**(2), 81–87 (2007)

Adsorption of Heavy Metal from Palm Oil Mill Effluent on the Mixed Media Used For the Preparation of Composite Adsorbent

Adeleke Abdulrahman Oyekanmi [1,2,*], *Ab Aziz* Abdul Latiff[1,2], *Zawawi* Daud[1,2], *NurFalilah* Mat Daud[1], and *Paran* Gani[1]

[1]Department of Water and Environmental Engineering, Faculty of Civil and Environmental Engineering, Universiti Tun Hussein Onn Malaysia, 86400, Parit Raja Batu Pahat, Johor, Malaysia
[2]Team of Researchers in Integrated Solid Waste Management (TRISM). Faculty of Civil and Environmental Engineering, Universiti Tun Hussein Onn Malaysia, 86400, Parit Raja Batu Pahat, Johor, Malaysia

Abstract. The adsorption of lead ion from palm oil mill effluent produced directly from the mill was investigated using mixed adsorbents that were used to prepare the composite. Experiments were carried out under predetermined conditions of pH, shaking speed, contact time and particle size. Equilibrium study was carried out to determine the adsorption capacity of the mixed media. The reduction of the lead ion was effective on the adsorbent materials. The Temkin and Freundlich models were applied to describe the adsorption pattern on the mixed media. The experimental data fitted well to the Temkin isotherm with a correlation coefficient of 1, this was used to indicate the level of interaction of the adsorbent-adsorbate and also the energy utilized in the adsorption process. The hydrophobic behaviour of the activated coconutshell and cow bone carbons was observed for average contact angle of 105° and 95° for the coconut shell and cow bone respectively. The influence of the hydrophobic materials in the mixed media and the zeolite assisted in the ion exchange and in the adsorption of the heavy metal.

1 Introduction

Palm oil mill effluent (POME) is well known as a very high strengthwastewater. The discharge of POME from the processing mill without effective treatment may result to the deterioration of water bodies, other sources of waste containing pollutants are from domestic, urban and other industrial sectors [1]. Some of the pollutants include chemicals, sewage, fertilizer, pesticide and heavy metals. The presence of these pollutants in the aquatic ecosystem has severe damaging effect on the health of the environment, human and the aquatic habitat [2].

Heavy metal in water source isa big challenge threatening the ecosystem for researchers. The presence of heavy metals causes health disorder [3]. Several methods have

*Correspondingauthor: abdulkan2000@yahoo.com

been reported for the removal of heavy metals of POME from water bodies. POME is the major wastewater in many Asian countries such as Malaysia and Indonesia but the treatment of the effluent mostly has been reported using conventional means.Some of these methods are ion exchange, chemical precipitation, membrane filtration and photo-catalysis.Adsorption offers economic benefits interms of cost, efficiency interms of treatment, and produces high quality effluents better than other conventional methods. Also, adsorption materials have great potentials of regeneration by desorption process. The process of regeneration requires low maintenance cost, high treatment efficiency and ease of operation [4]. Adsorption study is widely conducted using single material in the past, such as the use of cow bones [5], banana peel [6], maize cob [7], peat [8], kaoline [9] etc.Due to the high strength nature of POME, previous researches on adsorption have been conducted as the secondary treatment after majortreatment from the aerobic pond or before discharge. [10-12].

In recent study of adsorption, adsorbent materials are identified on the basis of their efficiency; usually two or more materials are combined with suitable binder to develop a new composite adsorbentto enhance adsorption. In the present work, batch adsorption of lead was investigated as primary treatment of POME using the mixed media that was used in the preparation of the composite. The materials were activated cowbone powder, activated coconut shell carbon and zeolite. The ordinary Portland cement (OPC) was used as binder because of its good pozolonic properties. The objective was to investigate the effectiveness of the mixed media for the reduction of lead ion in raw palm oil mill effluent (POME).

2 Materials and Method

2.1 Preparation of the Adsorbent Materials

The locally sourced materials for the preparation of composite were selected on the basis of their hydrophobic and hydrophilic behaviour in fluid. This is to classify materials on the basis of the effectiveness to adsorb polar and non-polar pollutants in the adsorbate. The process is usuallydetermined by the contact angle measurement. The materials chosen for this work were coconut shell carbon, cow bone carbon and zeolite. The coconut shells were obtained from the factory of Kian Hoe Johor, Malaysia.

The materials were washed thoroughly, dried and crushed. A predetermined weight of 500g of the dried sample of coconut shellwas obtained at 110°C for 2h. The carbonization of the crushed sample was obtained at 250°C to form char, this was followed by activation process at 850°C for 2h under the flow of CO_2, and this was achieved at heating rate of 5°C. The bulk density of 0.57g/ml was obtained and used for the composite. Fresh cow bones were obtained in the local market in Parit Raja Johor. The method of preparation and characterization of the activated cow bone having bulk density measured at0.63g/ml was in our previous work [13]. Also, the natural zeolite used for this work was from Clinoptiloliteorigin has the bulk density of 0.97g/ml, the materialwas purchased at 3.7RM/Kg from local suppliers. The bulk densities of the starting materials were obtained using the method in[14]. Also, the OPC provided the binding effect at different adsorbent:OPC binding ratios. The materials were thoroughly mixed together using a mechanical mixer and the particle size of 150μm was used for the bulk densities used present study, this was obtained using the ceramic ball grinder. The adsorbent: OPC ratio used is presented in Table 1.

The surface morphology of the mixed materials (adsorbent:OPC) was achieved using the scanning electron microscopy at 500x magnification. Prior to the investigation, adsorbent material was plated with gold. The bulk density of the adsorbent: OPC used was 0.768g/ml. The elemental composition of the mixed media was obtained using the energy dispersive x-ray spectroscopy (EDX INCA-Oxford High Wycombe, UK). The result is presented in Table 2.

2.2 Adsorption Experiment

The sorption capacity of the composite was determined by the surface contact of the different mixed ratios of the contributing materialsin Table 1, the materials combined were selected from the contact angle measurement in Fig.1.The combined materials were obtained from the bulk densities of the prepared adsorbents. The surface characterization of mixed media is illustrated in Fig.2.

Table 1. Adsorbent: OPC ratio

% Adsorbent	90	80	70	60	50	40	30	20	10
Adsorbent(g)	21.79	19.37	16.944	14.524	12.103	9.682	7.262	4.841	2.4206
CBP	5.106	4.54	3.97	3.404	2.8365	2.2692	1.7021	1.135	0.5673
CAC	8.511	7.57	6.62	5.674	4.7275	3.7820	2.8369	1.189	0.9456
ZEO	8.171	7.27	6.354	5.447	4.5386	3.6308	2.723	1.8154	1.5129
% Binder	10	20	30	40	50	60	70	80	90
Binder (g)	0.5	1.0	1.5	2.0	2.5	3.0	3.5	4.0	4.5

The batch study was conducted using 250ml Erlenmeyer flasks at fixed condition of 150rpm shaking speed, 105 minutes contact time and at neutral pH condition. The adjustmentof the pH was done using 0.1M HCl and 0.1M NaOH. After the completion of each contact time, filtration using 0.45 µm membrane was conducted to separate the adsorbate from the solutions. The filtered samples were analysed using the inductively coupled plasma mass spectrometer ICP-MS ELAN 9000. All experiments were conducted in triplicate. The equilibrium liquid phase concentration was calculated as:

$$Q_e = \frac{C_o - C_e}{W} v \tag{1}$$

where$Q_e$is the equilibriumuptake (mg/g), V is the volume of solution (L), $C_o$is the initial concentration of lead(mg/L), $C_e$is the final concentration of lead(mg/L).

2.3 Adsorbent Experiment

In this study, two adsorption isotherms were adopted, the Freundlich isotherm to investigate the heterogeneity of the surface of the adsorbent on the adsorbate and the Temkin isotherm, to express the adsorbent-adsorbate relationship and the heat energy used in the adsorption process.

3 Results and Discussion

The adsorption of lead was conducted on mixed materials used to prepare the composite adsorbent. The materials selected were of hydrophobic and hydrophilic characteristics in the adsorbate. The measurement of the contact angles of these materials using water drop can be shown in Fig. 1. An adsorbent material is hydrophobic if the contact angle is greater than 90° [15]. It is highly hydrophobic if is greater than 120° [16]. Also when a material has contact angle less than 90°, it is considered as hydrophilic [17].

Fig.1. Contact angle 105°, 95°, 21° for coconut shell, cowbone and zeolite respectively.

The surface morphology was obtained at 20.0kv using a magnification image of 500x, the image analysis was done simultaneously with EDX analysis in Table 2. The surface characterization is shown in Fig. 2. The microstructure of the prepared media appears smooth except in cases of cracks and holes experienced on some surfaces. The percentage of carbon in the mixed material was reasonably high, this may be as a result of the presence of the hydrophobic coconut shell and cow bones. The presence of carbon on the available sites for sorption provided the attachment of the lead metal on the mixed media.

The adsorption of lead on the mixed media was conducted on fresh samples of POME for primary treatment. The adsorptionstudywas conducted at different adsorbent: binder ratio in Table 1.

Fig.2. Surface Characterization of adsorbent

The concentration of the metal uptake was obtained after the completion of each shaking time and filtration process of the treated samples using 0.45 micro filter membranes. The batch isotherm study on the experimental data showed that the model fitted better to the Temkin isotherm with a correlation coefficient (R^2) of 1.00, the Temkin constants of BT served as the expression of slope and KT (L/g) as the expression of the intercept. The value of BT was 2570 and KT was 1.4527L/g. The Temkin isotherm showed that the adsorption of lead ion decreases linearly on the adsorbate due to the adsorbent-adsorbate interactions.

DOI: 10.1051/matecconf/201710306020

Table 2. Elemental composition of the mixed Adsorbent.

Element	Mass%	At%
C_k	28.62	39.52
O_k	48.39	50.15
Na_k	1.56	1.12
Mg_k	1.07	0.73
P_k	2.14	1.15
Ca_k	16.85	6.97
Cu_k	1.38	0.37
Total		100.00

The Temkin constant (BT) describes the variation of the adsorption energy, the model is based on the principle that the decrease of the heat energy of adsorption tends to be linear than logarithmic [18]. If adsorption of BT >1, the adsorption process is exothermic, also the adsorption process is endothermic if BT <1 In this study, the value of BT exhibited an exothermic behaviour of the process of adsorption[19]. This suggests that there is an electrostatic interaction and the heterogenous pore spaces of the coconut shell activated carbon and the cow bone activated carbon played a very significant role. The Freundlich isotherm is widely used to describe the adsorption process on heterogeneous surface. The Freundlich model showed R^2 of 0.9828, the slope 1/n was 0.71598 and intercept K_F of 0.4514 was acheived. The slope 1/n describes the surface heterogeneity or the adsorption intensity and K_F shows the adsorption capacity. The value of 1/n is known to be in the range of 0 to 1. A value closer to zero indicates favourable adsorption and a heterogeneous surface. From the slope of the Freundlich isotherm, it can be said that the mixed media have high heterogeneous surface in terms of the slope of the isotherm. The adsorption process was noticeably high at a higher ratio of adsorbent to minimal OPC application, this was due to the difference in the concentration between the bulk solution and the solid-liquid interface. The zeolite in the composite provided the adsorption sites and enhanced the interaction of the lead ion with the sites[20].The result of the adsorption isotherm study is presented in Fig. 3 and 4:

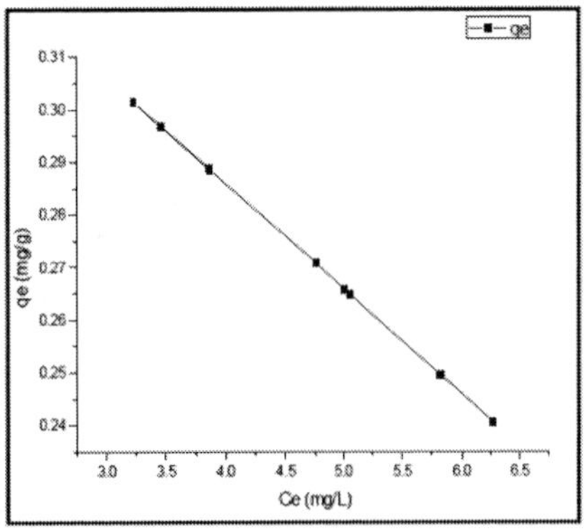

Fig.3. the Temkin Isotherm

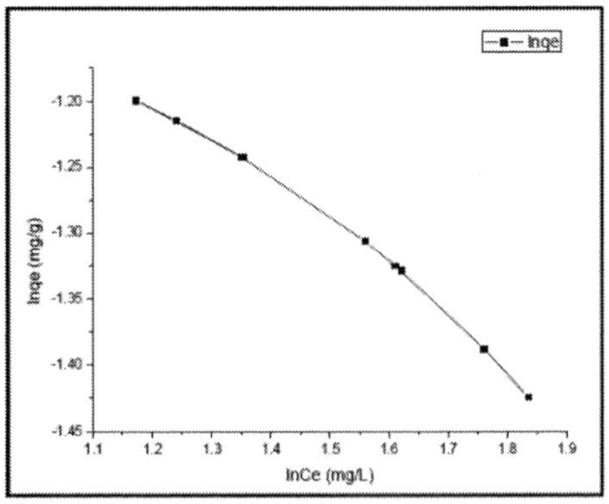

Fig.4. Freundlich Isotherm

4 Conclusions

The materials used for the preparation of the composite adsorbent in this work showed effective reduction of lead ion from POME. The result indicated that high carbon content of the adsorbent, the hydrophobic behaviour of the coconut shell activated carbon and cow bone carbon favoured the attachment of the metal ion on the surface of the sorbent. The isotherm model fitted better to the Temkin model. This was used to determine the extent of the adsorbent-adsorbate interaction and the energy involved in the adsorption process. The adsorbent materials showed great potential for a new composite adsorbent which can be well suited for the effective reduction of heavy metals in other high strength wastewater.

This research was supported by Universiti Tun Hussein Onn Malaysia (UTHM), Malaysia, The Ministry of Higher Education Malaysia through Fundamental Research Grant Scheme (FRGS), Vot. 1571.

References

[1] N. N. N. Daud and A. A. Syazwani. Preliminary assessment of lakes water quality status at campus area in selangor, malaysia,**131**, 279-292,(2016)

[2] K. Iwahori, J. I.Watanabe, Y.Tani, H, Seyama and N.Miyata. Removal of heavy metal cations by biogenic magnetite nanoparticles produced in Fe (III)-reducing microbial enrichment cultures. J. of bioscience and bioengineering, **117**(3), 333-335, (2014).

[3] M. Matouq, N. Jildeh, M. Qtaishat, M. Hindiyeh and M. Q. Al Syouf. The adsorption kinetics and modeling for heavy metals removal from wastewater by Moringa pods. J. of Environmental Chemical Engineering, **3**(2), 775-784. (2015).

[4] M. Kobya, E. Demirbas, and F. Ulu. Evaluation of operating parameters with respect to charge loading on the removal efficiency of arsenic from potable water by electrocoagulation. J. of Environmental Chemical Engineering, **4**(2), 1484-1494, (2016).

[5] A. R. O Adeleke, A. A.L. Aziz, Z. Daud, B. Ridzuan, M. Daud and N. Falilah, Remediation of raw wastewater of palm oil mill using activated cow bone powder through batch adsorption. In Key Engineering Materials, **705**, 380-384, (2016).

[6] J. Anwar, U. Sfafique, M. Salman, A. Dar and S. Anwar. Removal of Pb (II) and Cd (II) from water by absorption on peels of banana, Bioresource Technology, **10**(6), 1752-1755, (2010).

[7] K. Y. Foo,Value-added utilization of maize cobs waste as an environmental friendly solution for the innovative treatment of carbofuran, Process Safety and Environmental Protection, **100**, 295-304, (2016).

[8] H. I. Chieng, L. B.Lim, and N. Priyantha, Sorption characteristics of peat from Brunei Darussalam for the removal of rhodamine B dye from aqueous solution: adsorption isotherms, thermodynamics, kinetics and regeneration studies, Desalination and Water Treatment, **55**(3), 664-677, (2015).

[9] E. I. Unuabonah, B. I. Olu-Owolabi and K. O. Adebowale. Competitive adsorption of metal ions onto goethite–humic acid-modified kaolinite clay. Int. J. of Environmental Science and Technology, **13**(4), 1043-1054, (2016).

[10] R.R. Mohammed and M.F. Chong. Treatment and decolorization of biologically treated palm oil mill effluent (POME) using banana peel as novel biosorbent. J. of Env.Management.**132**, 237-249, (2014).

[11] A.L., Ahmad, S. Sumathi and B.H. Hameed, Adsorption of residual oil from palm oil mill effluent using powder and flake chitosan: Equilibrium and kinetic studies. Water Research, **39**(12), 2483-2494, (2005).

[12] M.A. Shavendi, Z, Haddadian, M.H.S Ismail, N., Abdullahi and Z.Z. Abidin, Removal of Fe(III), Mn (III) and Zn (II) from Palm Oil Mill Effluent (POME) by natural zeolite. J. of the Taiwan Institute of Chemical Engineers, **43**(5), 750-759, (2012).

[13] A. Latiff, A.Aziz, O. Adeleke AbdulRahman, Z. Daud, M. B. Ridzuan, M. Daud, and N. Falilah, Batch adsorption of manganese from palm oil mill effluent onto activated cow bone powder, ARPN J. of Engineering and Applied Science, **11**(4), 2627-2631, (2015).

[14] C. H. Chia, A. Downie, and P.Munroe, *Characteristics of biochar: physical and structural properties, Biochar for environmental management: science and technology*, Earthscan Books Ltd, London, 89-109, (2015).

[15] T. Monroe, S. Bhaduri, A.Koch and K. Lant, Making Clays Water Resistant: New Technology That Minimizes Formation Damage. In *SPE European Formation Damage Conference and Exhibition*, Society of Petroleum Engineers, (2015).

[16] M. R, Hibbs, L. K. McGrath, S. Kang, A. Adout, S. J. Altman, M. Elimelech and C. J. Cornelius, Designing a biocidal reverse osmosis membrane coating: Synthesis and biofouling properties, Desalination, **380,** 52-59, (2016).

[17] J. Zhao, H. Lai, Z. Lyu, Y. Jiang, K. Xie, X. Wang and J. Liu. Hydrophilic Hierarchical Nitrogen-Doped Carbon Nanocages for Ultrahigh Supercapacitive Performance, Advanced materials, **27**(23), 3541-3545, (2015).

[18] X.j., Hu, J.S., Wang., Y.G., Liu, G.M., Zeng, Z.I., Bao. A. W. X-Zeng and F. L. Chen. Adsorption of chromium (VI) by ethylene diamine modified cross linked magnetic chitosan resin, Isotherm, kinetics and thermodynamics, J. of Harzardous material, **185**, 306-314, (2011).

[19] A. C. Martins, O. Pezoti, A. L. Cazetta, K. C. Bedin, D. A.Yamazaki, G. F. Bandoch and V. C.Almeida. Removal of tetracycline by NaOH-activated carbon produced from macadamia nutshells: Kinetic and equilibrium studies, Chemical Engineering J., **260**, 291-299, (2015).

[20] T. Motsi, N.A., Rowson and M.J.H. Simmons, Adsorption of heavy metals from acid mine drainage by naturalzeolite, Int.J. of Mineral Processing, **92**, 42-48, (2009).

Preliminary Study on Remediation of Contaminated Clay Soil Using Cement and Sugarcane Bagasse

Mohamad Azim Mohammad Azmi [1], *Saiful Azhar* Ahmad Tajudin[2,*], *Shahiron* Shahidan[3], *Md Asrul* Nasid Masrom[4], and *Ain Nabila* Abdul Talib[5]

[1,2,5]Research Centre for Soft Soil (RECESS),University Tun Hussein Onn Malaysia, 86400 Johor, Malaysia
[1,2,3]Faculty of Civil and Environmental Engineering,University Tun Hussein Onn Malaysia, 86400 Johor, Malaysia
[4]Faculty of Technology, Management and Bussiness, University Tun Hussein Onn Malaysia, 86400 Johor, Malaysia

Abstract. Disposals of agricultural waste in a large volume have an extremely harmful effect on the environment if they are ineffectively treated. To solve this, several appropriate methods has been identified to produce new recycling technique of utilizing the agricultural waste. In this study, the feasibility of using sugarcane bagasse (SCB) as partial replacement binder with cement to stabilized and solidified (S/S) the contaminated soil are investigated. This paper focused on the strength and the leaching characteristic of lead (Pb) contaminated soil treated with SCB and cement. Two tests, namely the Unconfined Compressive Strength (UCS) test and Toxicity Characteristic Leaching Procedure (TCLP) were employed to measure the strength and leaching performance of S/S samples. The experimental results demonstrated that the UCS at 28 days was in a range of 0.82 to 5.66 MPa for sample containing cement and SCB compare to 0.29 MPa of control mix at the same age. The concentration of Pb in the leachates was within the limits specified by US EPA as low as 2.11 mg/L in 28 days. This shows that, the quality of S/S sample containing cement and SCB significantly improve the strength development as well as effectively in reducing the Pb leachability. Based on the finding, SCB could be useful as cheaper, easy available alternatives binder for the treatment of contaminated soil.

1 Introduction

Soil that is contaminated by heavy metals is a worldwide problem that requires urgent attention. Heavy metals in soil may cause threat to human health either through accidental soil ingestion, breathing in contaminated soil dust particles or by ingesting polluted drinking water or farm products tainted by contaminated soil [1]. On the other hand, heavy metal contaminated soil is mainly due to the subsequent migration of leachate form and

* Corresponding author: saifulaz@uthm.edu.my

within the landfill waste cell. According to fauziah et al. [2], the natural process taking place within the boundaries of the waste cells accelerate the process of heavy metal leaching from the waste components sources of heavy metal within a landfill system. The released of heavy metal into the adjacent environment is a serious environmental concern and a threat to public health and safety [3].

Kamari et al., [4] claimed that heavy metal behavior in soil and biological effects are caused by their presence in increased concentrations. Heavy metal behavior is in fact largely determined by the processes of metal released from the solid phase into the soil solution and factors influencing the chemical forms of metal in soil. The contamination of soil by heavy metal, particularly lead (Pb), is a common problem throughout the world [5]. Lead has been reported as a metal that affect the human nervous system, blood vessels and kidneys. Currently, the governments are trying to minimize the lead impact that produce adverse effects on the mental and physical development of humans, even at the lower exposure level [6]. Therefore, remediation of contaminated soil is of great importance for both engineers and researchers. The stabilization/Solidification (S/S) remediation method has been employed as a potential technology with the addition of a binding agent to reduce the movement of hazardous waste elements due to its low cost, availability and versatility [7].

2 Materials and methods

2.1 Binder system and soil contaminated process

Clay used for this study was collected at a location near the Research Center for Soft Clay (RECESS), University Tun Hussein Onn Malaysia (UTHM). The soil sample was dried in an oven at 105°C for 24 hours at the laboratory. After undergoing a drying process for 24 hours, the soil was crushed using a rubber hammer before it was reduced to 2 mm in size by a grinder machine. The soil particles which pass the 2 mm sieve were stored in polyethylene plastic. The initial concentration of Pb would be 500 ppm. The selection of this percentage would be to test the worst case of the contaminated soil. Type II ordinary Portland cement (OPC) was used as the main binder together with treated sugarcane bagasse (SCB) in order to treat the lead contaminated soil. The SCB was collected from sugarcane juice hawker located around Parit Raja, Johor with the SCB underwent a treatment process under acidic conditions to remove cellulose fibers and to liberate the lignin. The SCB was boiled with 0.1 M HCl for 45 minutes and the residue was washed to rid it of sugars and hydrolysis products. This procedure was repeated 3 to 4 times until the filtrate became colorless. Finally, the SCB was dried in the oven at a temperature of 105 °C for 24 hours to remove all moisture content before being grinded into particles measuring 90μm in size.

The raw materials were mixed by using a small mixer to ensure the homogenous of the S/S sample. Then, S/S sample then compacted in a split mould to form a sample of 38 mm in diameter and 76 mm in height. A specially designed miniature hand compacting tool was used to compact the mixture in 4 layers, 50 blows each. The extruded specimens were then wrapped and stored for 7, 14 and 28 days. Table 1 shows the mix design of S/S sample used in this study.

Table 1. Sample mix proportions (by weight)

Binder			Total weight of the sample (g)	w/b	Symbol
Clay soil (%)	OPC (%)	SCB (%)			
95	5	-	160	0.4	C95O5
95	2.5	2.5	160	0.4	C95O2.5S2.5
90	10	-	160	0.4	C90O10
90	5	5	160	0.4	C90O5S5
85	15	-	160	0.4	C85O15
85	7.5	7.5	160	0.4	C85O7.5S7.5
100	-	-	160	0.4	control

2.2 Unconfined compressive strength (UCS)

The UCS test has been carried out by using the Geocomp LoadTrac II system available in RECESS UTHM. Testing was conducted according to BS1377-7:1990. The samples were extruded from the moulds, leveled, measured for length and diameter, weighed and subjected to uniaxial compression test (UCT) at a constant rate of strain of 1% per minute. Before testing, the loading device was adjusted carefully to ensure the upper platen just makes contact with the sample. Care was taken to ensure that both ends of the sample were as flat as possible to minimize bedding error during the test, especially the stiffer samples. The tested samples than collected and dried in the oven at 105oC for 24 hours before being crushed to pass through a 1 mm sieve prior to TCLP testing.

2.3 Toxicity characteristic leaching procedure (TCLP)

A TCLP test was conducted by adding a fixed volume of TCLP leachant No. 2 (5.7 ml/l glacial acetic acid aqueous solution at pH 2.88) to 20 times the weight of the sample (20:1). The extraction period for the sample lasted 18 hours under rotary agitation at 30 rpm. The sample was then filtered using Grade GF/F 0.7-μm glass fiber filter paper and the pH was measured at the end of the process. The sample was filtered within 2 h of the 18-h extraction period to ensure method reproducibility [8]. This procedure differs from the EPA TCLP [9] method in that one-tenth (1:10) of the amount of the sample was used. Apart from the modified step, all another aspect of the method remained the after filtration, metal concentrations in the leachate was determined via Perkin-Elmer AA800 Atomic Absorption Spectrometer (AAS).

3 Result and discussion

3.1 Strength development of cement-SCB

Compressive strength is an important factor as it determines the load that can be carried by a material. The strength of the mixture depends on the additive added. The results of the strength of S/S are presented in Fig. 1. It could be seen that sample with 100% OPC (sample C95O5, C90O10 and C85O15) has a highest compressive strength that is 0.73 Mpa, 2.79 Mpa and 5.66 Mpa respectively at 28 curing days. While, control sample shows the lowers strength with only 0.29 Mpa at 28 days. Sample with treated SCB performs an intermediate compressive strength ranging from 0.35 Mpa to 1.39 Mpa. When SCB added as a partial replacement in OPC, the strength becomes weaker compared to the sample without SCB. This is contributed by cement percentage ranging 5% to 20% are routinely used to stabilized and solidified contaminated soil. Low cement content causes the individual soil particles to be inadequately coated. However, it has a tendency to form a rigid form of waste [10].

Additionally, the figure also presents the UCS of S/S lead spike soil sample at 7, 14 and 28 days. It is obviously shown that a sample containing 100% OPC has the highest strength than another sample. This evidently shows the effect of using OPC as additive in the contaminated sample. It can be concluded that hydration time also effect the strength development. All samples become more harden for 28 days [11-14]. This is caused by the slow formation of calcium hydrated silicates (C-S-H) which resulted from the reaction of anhydrous-silicate constituents of cement with water [15][16]. However, even the UCS of sample with SCB are not high compared to sample without SCB, the result is still exceeded the minimum landfill disposal limit stated by US EPA of 0.35 Mpa (350kPa) at 28 curing days.

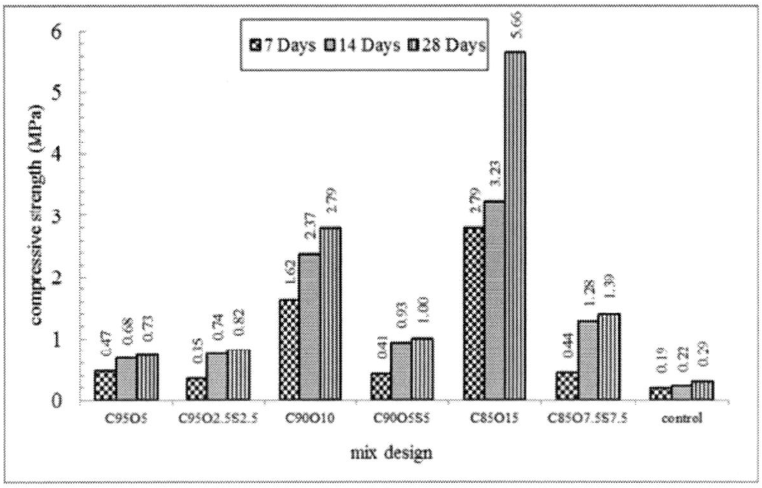

Fig 1. Strength development of S/S sample

3.2 Leachability of the Pb

In this study, Toxicity Characteristic Leaching Procedure TCLP test has been conducted to investigate the leaching characteristic that is suitable for organic waste or waste containing high sulfate compounds. This testing produced leachate that is similar to the leachate obtained from a sanitary landfill site and found to deplete various types of metal [17]. The Fig. 2 shows the Pb leachability at 7, 14 and 28 days. The figures demonstrate that the concentration of Pb in S/S sample depends on two main factors which are mix ratio and curing days. The higher reduction is showed by sample C85O7.7S7.5 with Pb concentration 6.23 mg/L, 4.26 mg/L and 2.11 mg/L at 7, 14 and 28 days respectively. The small reduction are showed by control sample with 236.11 mg/L, 183.19 mg/L and 93 mg/L at 7, 14 and 28 days respectively. This scenario is expected because the leachability of heavy metals in S/S method depends on the hydration days. It is proved by Antemir et al. [18] which stated that the long-time hydration of S/S sample has been decreased the leachability in groundwater. This is also supported by Wang et al. [19] which study the leachability of heavy metals at 17 years old S/S site and found that, the slow bonding reaction happen between metals and binder.

Apart from that, the figure also indicates that sample containing SCB has leached out least amount of Pb in final leachate. In a nutshell, the research has found that SCB as a fibre material, when incorporated with OPC has been successful adsorbed the Pb in S/S method.

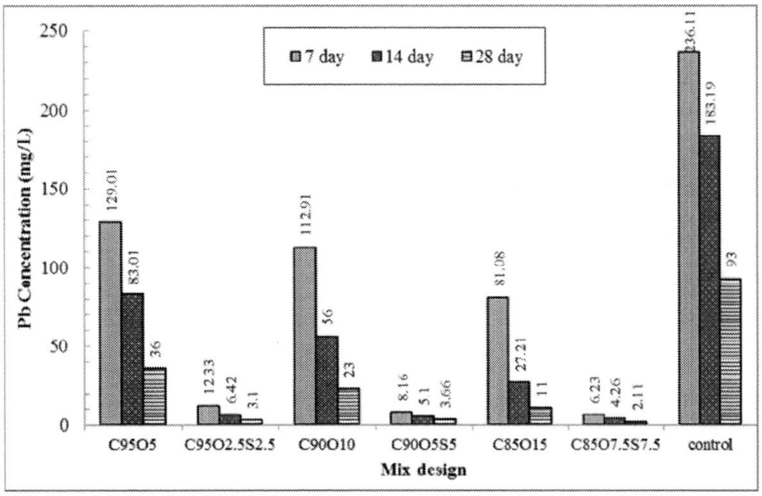

Fig 2. Leachability of S/S sample

4 Conclusion

From the study, it was found that the use of SCB is not substantial enough for obtaining the highest strength compared to a sample containing only OPC. However, all samples exceeded the landfill disposal limit of 0.34 Nmm^2 or 340 kPa. The UCS at 28 days was in a range of 0.82 to 5.66 MPa for sample containing cement and SCB compare to 0.29 MPa of control mix at the same age. Despite that, the TCLP extraction has been successful in reducing the concentrate of Pb below the standard limit set by US EPA. The concentration of Pb in soil samples containing OPC and SCB remain low. For example, a sample containing 15% OPC leached out the Pb at 11 mg/L at 28 days of curing. The better result showed by a sample containing 7.5% OPC+ 7.5% SCB that is 2.11 mg/L. This significant reduction indicates that the use of SCB as a partial replacement is possible to reduce the concentration of Pb in the final leachate. In the same way, all sample containing SCB as additive materials to OPC obtained the US EPA limit (< 5mg/L).

The authors would like to thank the Ministry of Higher Education and Universiti Tun Hussein Onn, Malaysia for supporting this research under the Exploratory Research Grant Scheme (ERGS) Vot. 40.

References

[1] P. Yin and L. Shi, Remediation of Cd, Pb, and Cu-Contaminated Agricultural Soil Using Three Modified Industrial By-products, Water, Air, Soil Pollu, An Inter.J. Enviton.Poll., **225** 11,2194 (2014)

[2] S. H. Fauziah, M. N. Izzati, and P. Agamuthu, Toxicity on Anabas Testudineus: A Case Study of Sanitary Landfill Leachate, Procedia Environ. Sci., **18**, 14–19 (2013)

[3] A. T. S. Azhar, M. A. M. Azmi & A N A Talib, Stabilization/Solidification Remediation Method for Contaminated Soil: A Review, IOP Conference Series: Materials Science and Engineering, **136** 1-6 (2016)

[4] A. Kamari, I. D. Pulford, and J. S. J. Hargreaves, Binding of heavy metal contaminants onto chitosans--an evaluation for remediation of metal contaminated soil and water, J. Environ. Manage., 92 **10**, 2675–82, (2011)

[5] C. E. Halim, J. a Scott, R. Amal, S. a Short, D. Beydoun, G. Low, and J. Cattle, Evaluating the applicability of regulatory leaching tests for assessing the hazards of Pb-contaminated soils, J. Hazard Mater., **120** 1–3, 101–11 (2005)

[6] J. Aslam, S. A. Khan, and S. H. Khan, Heavy metals contamination in roadside soil near different traffic signals in Dubai, United Arab Emirates, J. Saudi Chem. Soc., **17** 3, 315–319 (2013)

[7] A. T. S. Azhar, N. S. Nordin, A. Marto, A. Madun, M.H.Z. Abidin, I. Jefferson, M. A. M. Azmi, "The monitoring and cementation behavior of electrokinetic stabilisation technique on Batu Pahat marine clay, Inter. J. GEOMATE **11** (4), 2581-2588 (2016)

[8] M. W. A. Raouf and H. G. Nowier, Assessment of Fossil Fuel Fly Ash Formulations in the Immobilization of Hazardous Wastes, J. Environ. Eng., 499–507, (2004).

[9] M. T. Montanes, R. Sanchez-Tovar, and M. S. Roux, The effectiveness of the stabilization/solidification process on the leachability and toxicity of the tannery sludge chromium, J. Environ. Manage., **143**, 71–9, (2014)

[10]R. Malviya and R. Chaudhary, Factors affecting hazardous waste solidification/stabilization: a review, J. Hazard. Mater.J. Hazard. Mater., **137** 1, 267–76 (2006)

[11] A T S Azhar, A. Marto, M. A. M. Azmi, A. Madun & M. H. Z. Abidin, Utilization of sugarcane Bagasse ash for stabilization/solidification of lead-contaminated soils, Jurnal Teknologi **77**:11, 119–125 (2015)

[12] A. T. S. Azhar, M. A. M Azmi , S. Shahidan, M.H.Z. Abidin, A. Madun, Relationship of physical parameters in Pb-contaminated by stabilization/solidification method, MATEC Web of Conferences **47**, 1-6 (2016)

[13] A .T. S. Azhar, A T A Nabila, M S Nurshuhaila, M Z N Shaylinda and M A M Azim, Electromigration of Contaminated Soil by Electro- Bioremediation Technique, IOP Conference Series: Materials Science and Engineering, **136** 1-6 (2016)

[14] A. T. S. Azhar, A T A Nabila, M S Nurshuhaila, E Zaidi, M. A. M Azmi and S M S Farhana, Assessment and Comparison of Electrokinetic and Electrokinetic-bioremediation Techniques for Mercury Contaminated Soil, IOP Conference Series: Materials Science and Engineering, **160** 1-9 (2016)

[15] A T S Azhar , A T A Nabila, M S Nurshuhaila , E Zaidi , M A M Azim and A M F Zahin, Novel Technique to improve the pH of Acidic Barren Soil using Electrokinetic-bioremediation with the application of Vetiver Grass, IOP Conference Series: Materials Science and Engineering, **160** 1-9 (2016)

[16] A T S Azhar, M A M Azim, M Aziman and A T A Nabila Leachability of Arsenic (As) Contaminated Landfill Soil Stabilised by Cement and Bagasse Ash. IOP Conference Series: Materials Science and Engineering, **160** 1-9 (2016)

[17] K. Nemati, N. K. Abu Bakar, M. R. Abas, and E. Sobhanzadeh, Speciation of heavy metals by modified BCR sequential extraction procedure in different depths of sediments from Sungai Buloh, Selangor, Malaysia, J. Hazard. Mater., **192** 1, 402–10 (2011)

[18] A. Antemir, C. D. Hills, P. J. Carey, K. H. Gardner, E. R. Bates, and A. K. Crumbie, Long-term performance of aged waste forms treated by stabilization/solidification, J. Hazard. Mater., **181**, 1–3, 65–73, (2010)

[19] F. Wang, H. Wang, F. Jin, and A. Al-Tabbaa, The performance of blended conventional and novel binders in the in-situ stabilisation/solidification of a contaminated site soil, J. Hazard. Mater., **285C**, 46–52 (2014)

Evaluation of Unknown Tube Well Depth Using Electrical Resistivity Method

Mohd Hazreek Zainal Abidin[1,2,*], *Aziman* Madun[1,2], *Saiful Azhar* Ahmad Tajudin[1,2], *Mohamad Faizal* Tajul Baharuddin[1,2], *Mohd Fairus* Yusof[1,2], *Muhammad Nizam* Zakaria[1,2], *Siti Nazahiyah* Rahmat[1]

[1]Faculty of Civil and Environmental Engineering, Universiti Tun Hussein Onn Malaysia, 86400 Batu Pahat Johor, Malaysia
[2]Research Center for Soft Soil, Universiti Tun Hussein Onn Malaysia, 86400Batu Pahat Johor, Malaysia

Abstract. Electrical resistivity method has increasingly adopted in engineering, environmental, mining and archaeological studies. Systematic and proper studies of unknown civil engineering structure evaluation particularly on tube well depth was rarely being established. Conventionally, camera test or string with weight approach has been used to evaluate unknown tube well depth thus exposed to several restriction due to its expensive and time consuming. Hence, this study focused on evaluation of unknown tube well depth using indirect test with particular reference to electrical resistivity method (ERM).A single spread line of electrical resistivity survey was performed using ABEM SAS 4000 equipment set based on Wenner and Pole-dipole array in line with the tube well position. Electrical resistivity raw data was processed using RES2DINV software producing electrical resistivity tomography (ERT) of the subsurface profile studied. Then, electrical resistivity value (ERV) obtained from RES2DINV analyses (ERT) was extracted and analysed using plotted graph (depth versus ERV) specifically at tube well position based on electrical resistivity spread line performed. It was found that both array have shown some good similarity results in term of tube well depth (20 m) thus able to verify the result interpreted. Both array have shown some good similarity of ERV representing groundwater (ERV = 10 – 100 Ωm) and soil with water (ERV > 100 Ωm) at depth of 0 – 20 m and >20 m respectively. All those interpretation have shown good agreement based on verification thru established ERV of earth materials references, geological map and nearest available boreholes data. Hence, this study has shown that the application of ERM was applicable in evaluation of unknown tube well depth which efficient in term of cost, time and environmental sustainable.

[*] Corresponding author: hazreek@uthm.edu.my

1 Introduction

Civil engineering subsurface mapping involve several sub-disciplines including environmental engineering, geotechnical engineering, earthquake engineering, water resources engineering, wastewater engineering and construction engineering. Several civil engineering underground structure (e.g. tube well, foundation, subsoil drain, etc.) was commonly being investigated due to forensic assessment. Common forensic assessment involve the investigation related to unknown or failure structure. In other words, forensic assessment is the investigation of materials, products, structures or components that lost, fail or do not operate or function as intended, which may cause some negative impact afterward. Forensic engineering normally investigate and collect data related to the materials, products, structures or components that lost or failed such as inspections, collecting evidence, measurements, developing models, obtaining prototype products, and performing experiments.

A tube well is a type of water well with specific design which being bored into an underground aquifer. The main purpose of tube well is to extract groundwater for several application in domestic and industrial sector such as agriculture, aquaculture, livestock, factory, mining, etc. From past experienced, proper and systematic forensic studies related to unknown tube well depth identification was rarely established. Unknown tube well depth was commonly occurred due to the long abandon project or loss of record. Conventional approach to measure unknown tube well depth was commonly based on informal method particularly by string and weight to measure the well depth. his approach may expose for some argument due to its efficiency of result accuracy and reliability.

Nowadays, geophysical method has increasingly adopted in most of the subsurface profile mapping due to its efficiency in term of cost, time and data coverage. Moreover, the nature of indirect or surface technique (non-destructive method) used in geophysical data acquisition has contribute benefit with our environment particularly in reducing site damageability thus sustainable to our environment. However, the successful performance of geophysical method may relative to fundamental physical constraint (e.g. penetration, resolution, and signal to-noise ratio) [1-2].Geophysical method used the physics concept to characterize the subsurface profile. Common, physics properties used including resistivity, wave, density, magnetic susceptibility, etc. Then, specific geophysical equipment has been invented based on those physics properties (e.g. electrical resistivity, seismic, ground penetrating radar, gravity, magnetic, etc.).Geophysical method has widely being applied in several areas related to forensic assessment [3-6], environmental [7-11], engineering [12-14], mining [15-17] and archaeological [18-20] studies .Consequently, this study was performed to establish the application of electrical resistivity method for unknown tube well depth evaluation which being verified using established properties of earth materials resistivity, geological map and borehole data.

2 Methodology

2.1 Study area and geological setting

This study was performed at hydro-meteorology station in UTHM, Johor as shown in Fig.1. Generally, site topography was flat which surrounded by private factory, university buildings (laboratory, class and offices) and golf course. General geology of Peninsular Malaysia area has been well documented [21]. Based on geological map shown in Fig.2, this study area consists of unconsolidated deposits from clay and silt (marine). In general, the present of this type of materials will exhibits soft soil phenomenon due to its high water content derived from high water table of lowland areas. Moreover, high rainfall intensity which was commonly occurred in tropical climate country as Malaysia also has contribute to the increment of water table thus promoting to the soft soil phenomenon in study areas. Previous borehole results obtained from PPH, UTHM also found that this area consist of thick clay and silt [22] thus suitable for groundwater carriage and storage. However, fine particle of soils such as clay and silt may associate with low hydraulic conductivity which may influence the effectiveness of groundwater recharge and quality extracted from the existing tube well. Hence, several testing such as pumping and water quality test need to be performed thoroughly in order to obtain comprehensive results and conclusion. Based on site observations, soft soil condition related to wet clay and silt geomaterials can easily found in study area. These evident can be used as an indicator regarding the composition of subsurface profile that possibly consist of homogeneous soft soil geomaterials.

Fig. 1. Study area

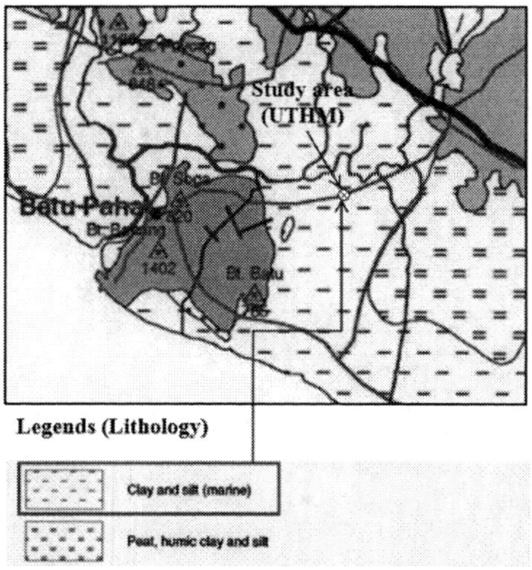

Fig. 2. Geology of study area [21]

2.2 Data acquisition and processing

Electrical resistivity imaging (ERI) was performed using the ABEM Terrameter SAS 4000, combined with ES 10-64 electrode selector. A fixed single line of electrical resistivity survey was performed across the target of interest (tube well unknown depth) zone. Testing configuration was based on Wenner and Pole-dipole array using four (4) resistivity land cables, sixty one (61) numbers of inline electrode plus one remote electrode and sixty five (65) numbers of jumper cable. Equal electrode spacing of 3.0 m was used for all 61 electrodes producing total electrical resistivity spread line length of 240 m. Field arrangement and alignment of spread line of the electrical resistivity imaging was given in Fig. 3 and 4 respectively. Wenner and Pole-dipole array was performed in this study due to verification purposes representing low and high intensity of electrical resistivity data. Furthermore, Pole-dipole array was able to provide greater depth of subsurface profiles within limited spaced area during the resistivity data acquisition. Raw data obtained from data acquisition was processed using commercialize RES2DINV software of [23] to provide an inverse model that approximates the actual subsurface structure. The inversion algorithm of RES2DINV was used to process the data, as proposed by [24] in order to obtain the 2-D resistivity section (Electrical resistivity tomography, ERT). Finally, electrical resistivity value (ERV) obtained from RES2DINV analysis was extracted and plotted with depth specifically at horizontal distance (x) of 121.5 m (based on tube well position from the first electrode within the spread line performed in Northsouth direction). Schematic diagram explaining the extraction of ERV from ERT was given in Fig. 5.

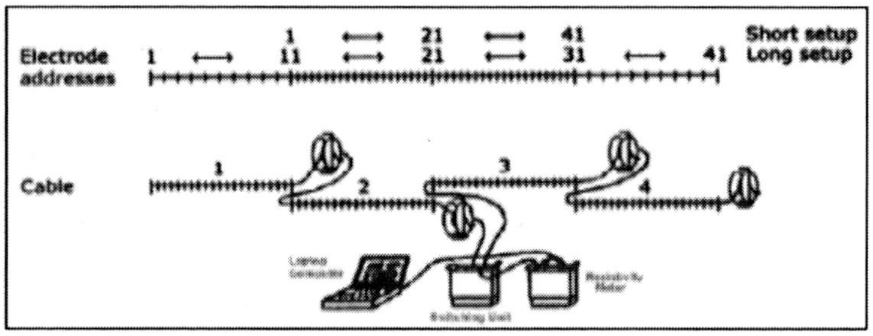

Fig. 3. Electrical resistivity imaging (ERI) field arrangement performed at hydro-meteorology station, UTHM Johor [25]

Fig. 4. Alignment (Northsouth) of electrical resistivity (spread lines) performed at hydro-meteorology station, UTHM Johor

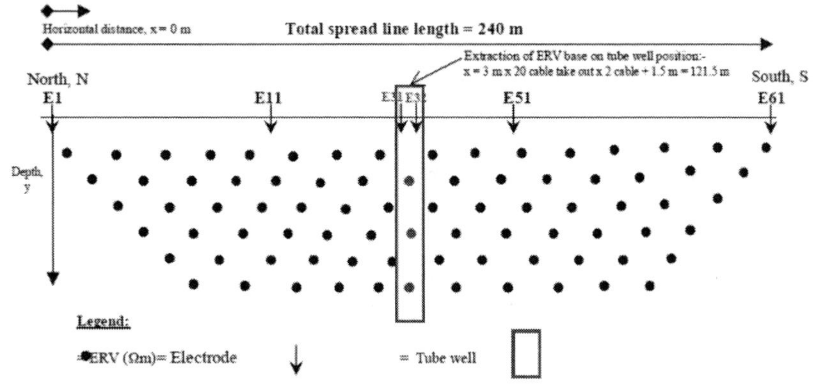

Fig. 5. Extraction of ERV from ERT pseudosection

3 Results and discussions

Basically, electrical resistivity value (ERV) was determined by measuring the potential difference at points on the ground surface which caused the propagation of direct current through the subsurface [26]. Traditionally, four (4) electrodes related to two (2) outer current electrodes and two (2) inner electrodes representing current and potential electrode was used to measure the resistivity of earth materials as shown in Fig. 6. Resistivity of earth material will be plotted at the middle of both potential electrode (between P1 and P2) after current being injected thru the subsurface profile using current electrode. This traditional four electrode system of resistivity technique was popularly known as vertical electrical sounding (VES) and constant separation traversing (CST) subjected to depth and lateral resistivity measurement. The limitation of traditional VES and CST technique was due to time consuming and limited data coverage (1-D) especially when working in a large and difficult study area. Nowadays, electrical resistivity technique has evolved due to the rapid development of high technology and sophisticated innovative equipment. The development of high technology and sophisticated electronics devices has invent the electrical resistivity method from VES and CST to imaging technique. Electrical resistivity imaging (ERI) technique permit to performed the data acquisition (field measurement) of resistivity method in two dimensional (2-D) and three dimensional (3-D) perspectives resulting an electrical resistivity tomography (ERT). Basic configuration of electrical resistivity imaging (ERI) was given in Fig. 7. Hence, field measurement was able to be performed quickly, low cost, large data coverage and sustainable to study area.

All results of resistivity method was presented based on Wenner array (Table 1, Fig. 8 and 10) and Pole-dipole array (Table 2, Fig. 9 and 11). ERV obtained in Table 1 and 2 was originally extracted from ERT based on Fig. 10 and 11 as explained in section 2.2. Variation of ERV with depth obtained at this particular point (x = 121.5 m) will be an indicator of the tube well depth identification. Then, ERV data series from Table 1 and 2 was plotted with depth in order to analysed and conclude depth of tube well studied (Fig. 10 and 11). According to Fig. 10 and 11, it was found that length of tube well was approximately interpreted as 20 m depth from the ground level. Fig. 8 and 10 has demonstrate clearly that the ERV from depth of 0 to 20 m was varied around 1 – 100 Ωm representing groundwater. Then, ERV starts to increase at depth of 20 m and over by > 100 Ωm representing soil with water. In addition, ERT given in Figure 10 and 11 has shown that the subsurface profile compose of thick and homogeneous of soft layer representing marine clay and silt. Stiffness of soil profile has been dominated by weak soil which gradually increased of its stiffness with depth. From soil mineral composition perspectives, fine particles such as clayey and silty soil experience higher minerals with particular reference to kaolinite, illite, montmorillonite and vermiculite which allows the ease of the current propagation in the soil tested thus produces low resistivity value in contrast with coarse soil (sand and gravel). Furthermore as reported by Abidin et al. [5], resistivity data exhibits a low value for a fine soil such as clayey and silty while the coarser soil such as sand and gravel will produce a higher resistivity value. This ERT interpretation was verified thru UTHM's borehole report [22] which stated that soil profile in UTHM consist of thick (> 30 m depth) and homogeneous fine soil (clay and silt).

Geophysical methods are incapable to stand alone in order to provide solutions to any particular problems [27-28]. Hence, strong verification from established data was important to support the electrical resistivity result interpreted. Previous original document of tube well studied (detail drawing) was missing and still in searching progress since this project has been abandon several years ago. Moreover, the original researcher leader of this tube well project was already resign and move to another agency sector. Since there are no other supportive information or data to verify above finding, those tube well depth interpretation was verified thru established ERV references based on ERV of groundwater and soil. Moreover, two (2) types of array (Wenner and Schlumberger) performed in this study was found to be in good agreement (consistency of ERV vs Depth given in Fig. 12 and 14) thus able to verify the result reliability interpreted. Based on Table 1 and 2, dissimilarity of ERV was obtained due to the different array used. Dissimilarity of ERV relative to different array performed was commonly occurred even it has been performed exactly at same particular spread line position/alignment/location together with similar environment and ground condition. One of the main factors contributed to this phenomenon was due to the geometry factor, K derived from different types of array established [29-30]. Hence, the value of ERV calculated was different based on K factor applied in every measurement due to its dissimilarity of current propagation for ERV determination. Geometry factor, K describes the geometry of the electrode configuration used in data acquisition. Apparent resistivity (ρa) is ERV estimated based on half-pace geometry assumption which refers to the field ERV. According to [31], apparent resistivity will be equal to the true resistivity provided the current and configuration was applied over the homogeneous isotropic ground. Field ERV was determined using Wenner and Pole-dipole array with a geometry factor as given in final Equation 1 and 2 which is derived from basic Equation 3 and 4. Wenner and Pole-dipole geometry factor, K used in this study was derived from Equation 4 based on basic four electrode system of measurement. The schematic diagram of field resistivity configuration used was given in Fig.12 and Fig. 13 while the schematic diagram for the four electrode system is given in Fig. 14.

Electrical resistivity for groundwater and clays was varied from 10 – 100 Ωm [31]. Meanwhile, ERV for clayey and silty soil was varied from 1 – 100 Ωm [31] and 200 – 500 Ωm [32] respectively subjected to several factors such as degree of saturation, basic physical properties of soil, salt concentration, etc. The fluctuation of ERV at tube well and soil layer may influenced by variation of physical and chemical content present within the subsurface profile studied. As reported by [21], study area composed of clay and silt from marine environment thus may influenced the consistency of ERV due to the possibility of ancient seawater trap. Samsudin et al. [33], have reported that ancient seawater that was trap within the sediment long time ago may influence the groundwater characteristic. Resistivity value was highly influenced by pore fluid and grain matrix of geomaterials [34]. According to Abidin et al [35-38], ERV of geomaterials was largely influenced by the variations of basic physical properties of soil such as grain size fraction, moisture content and density. As reported by [39], a soil's ERV generally varies inversely proportional to the water content and dissolved ion concentration as clayey soil exhibit high dissolved ion concentration, wet clayey soils have lowest resistivity of all soil materials while coarse, dry sand and gravel deposits and massive bedded and hard bedrocks have the highest ERV. Furthermore a decrease of ERV was results from an increase of metal ions or inorganic elements in geomaterials [40].

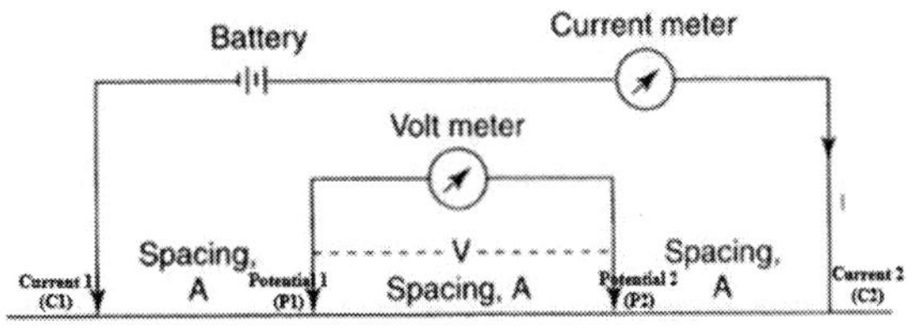

Fig. 6. Basic configuration of VES and CST in electrical resistivity method [26]

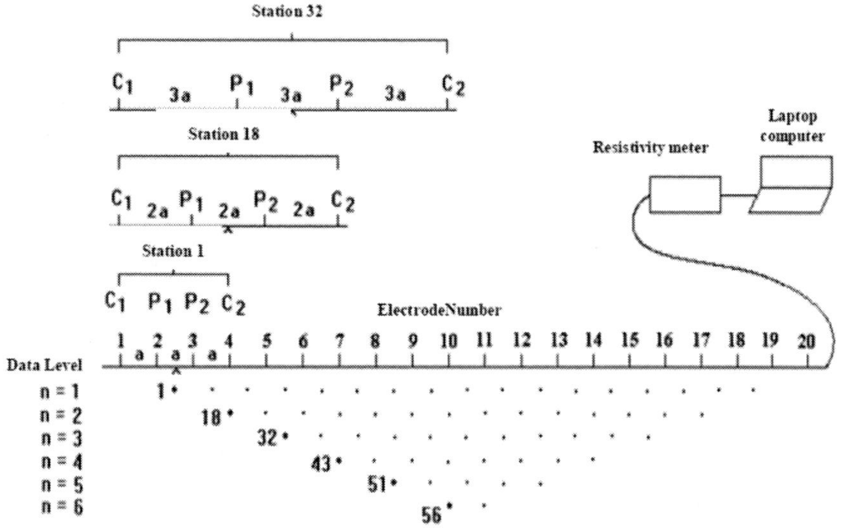

Fig. 7. Basic configuration of 2-D ERI in electrical resistivity method and the sequence of measurements used to build up a pseudosection [41]

Table 1. ERV data series at tube well position (Wenner array).

Wenner array (x = 121.5 m)	
Depth (m)	ERV (Ohm.m)
0.75	6.285
2.25	6.285
3.825	6.466
5.557	8.309
7.463	9.103
9.56	10.077
11.866	14.115
14.402	28.958
17.192	56.416
20.261	87.305
23.638	108.04
27.351	117.92
31.437	121.72
40.873	124.58
46.311	309.84

DOI: 10.1051/matecconf/201710307002

Table 2. ERV data series at tube well position (Pole-dipole array).

Pole-dipole array (x = 121.5 m)	
Depth (m)	ERV (Ohm.m)
0.75	63.354
2.25	38.039
3.825	32.758
5.557	31.728
7.463	6.683
9.56	1.838
11.866	1.838
14.402	19.362
17.192	108.73
20.261	187.8
23.638	204.11
27.351	205.47
31.437	206.34
35.93	207.03
40.873	207.64
46.311	208.16
58.871	209.03
66.108	209.37

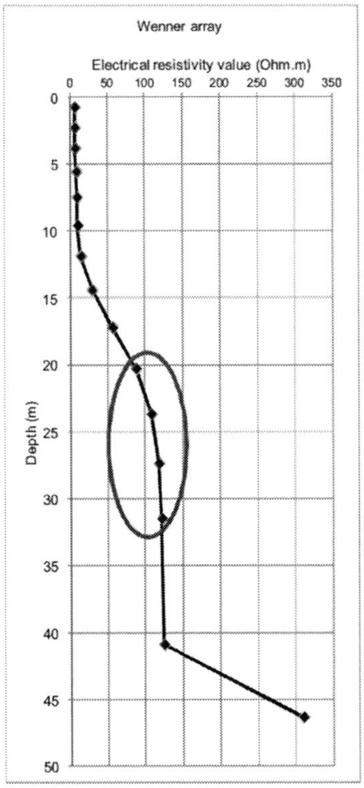

Fig. 8. ERV vs depth (Wenner array)

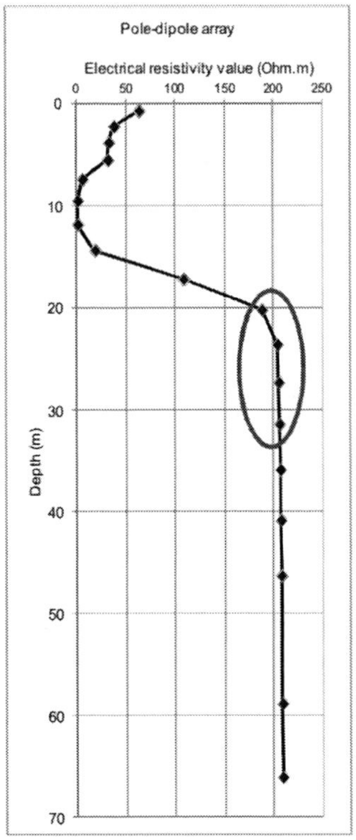

Fig. 9. ERV vs depth (Pole-dipole array)

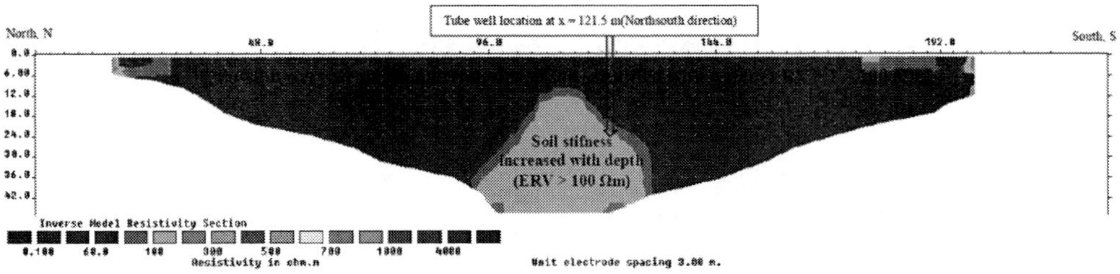

Fig.10. Electrical resistivity tomography (ERT) based on Wenner array (Northsouth direction)

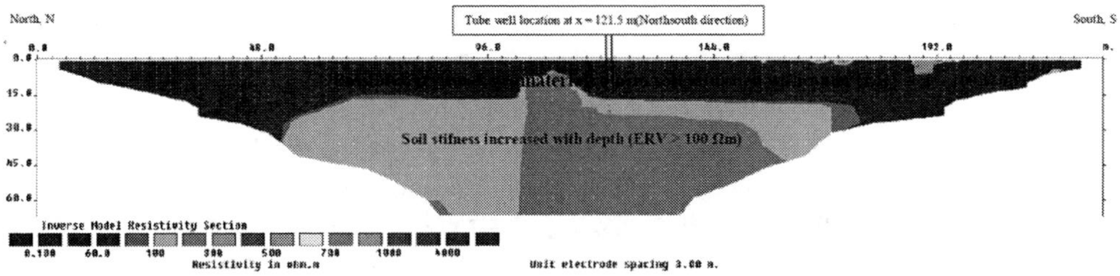

Fig. 11. Electrical resistivity tomography (ERT) based on Pole-dipole array (Northsouth direction)

$$\rho_a = 2\pi a * R \quad (1)$$

where R is a resistance term given by R=ΔV/I

$$\rho_a = ((2\pi ab)/(b-a)) * R \quad (2)$$

where R is a resistance term given by R=ΔV/I

$$\rho_a = K * R \quad (3)$$

where R is a resistance term given by R=ΔV/I, K is geometry factor based on pole-dipole electrode configuration

$$\rho_a = ((2\pi\Delta V)/I)) * ((1/(1/r1 - 1/r2) - (1/r3 - 1/r4)) \quad (4)$$

where $K = ((1/(1/r1 - 1/r2) - (1/r3 - 1/r4))$

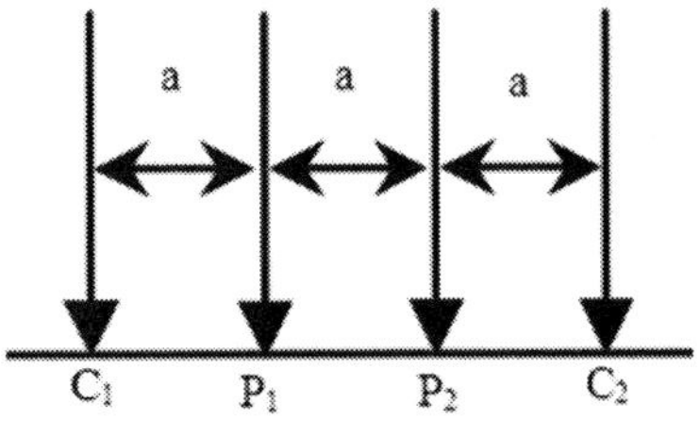

Fig. 12. Wenner electrode array arrangement

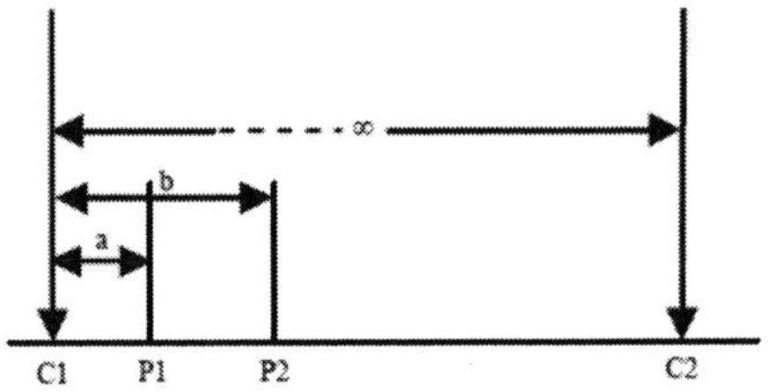

Fig. 13. Pole-dipole electrode array arrangement

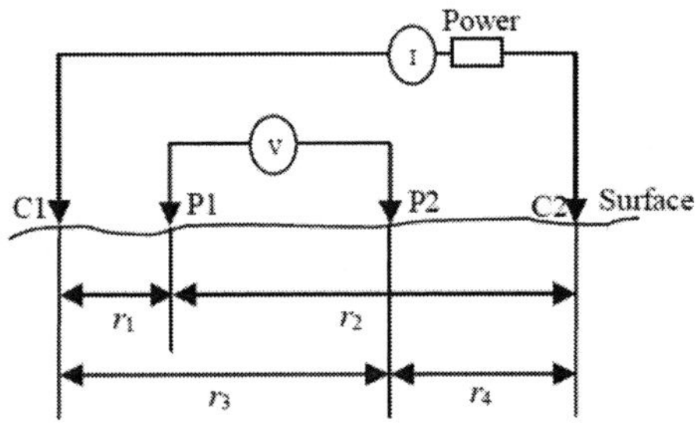

Fig. 14. Four electrodes arrangement on the surface of homogeneous isotropic ground of resistivity

4 Conclusion

The problematic of unknown tube well depth (20 m) was successfully being investigated using electrical resistivity method. The geometry and electrical resistivity anomaly distribution has been determined by analyzing electrical resistivity data obtained along the tube well location and the result has shown a good correlation based on previous references. This finding has proved that this approach was applicable to investigate the tube well depth particularly involving unknown tube well depth. Electrical resistivity method was successfully mapped the ground profile which able to extend the surface information observed during the physical mapping. The geometry of tube well can be easily recognized based on variations of ERV obtained. This geophysical method is suitable for our sustainable ground investigation since it can reduce time, money and compliment others conventional method. The application of ERI in conjunction with other records (borehole data, geological information, established resistivity values of geomaterials, etc) was applicable being applied to the evaluation of the unknown tube well depth due to its ability to differentiate properties of earth materials.

This work was funded by Universiti Tun Hussein Onn Malaysia (IGSP of Vote U258 and FRGS of Vote 1455). Many thank are due to all research members for their tremendous work and cooperation.

References

[1] H.J. Mauritsch, W. Seiberl, R. Arndt, A. Römer, K. Schneiderbauer, G. P. Sendlhofer, Geophysical investigations of large landslides in the Carnic Region of Southern Austria, Engineering Geology, **56**, 373-388 (2000)
[2] D.M. McCann, A. Forster, Reconnaissance geophysical methods in landslide investigations, Engineering Geology, **29**, 59-78 (1990)
[3] Z.A.M. Hazreek, A.T.S. Azhar, M. Aziman, S.M.S.A. Fauzan, J.M. Ikhwan, M.A.N. Aishah, Forensic assessment on ground instability using electrical resistivity imaging (ERI), *The 2nd Int. Conf. on Rheology and Modelling of Materials*, Miskolc-Lillafüred, Hungary, (2015)
[4] M.H. Z. Abidin, R. Saad, F. Ahmad, D.C. Wijeyesekera, M.F.T. Baharuddin, Seismic refraction investigation on near surface landslides at the Kundasang area in Sabah, Malaysia, Procedia Engineering, **50**, 516-531 (2012)
[5] M.H.Z. Abidin, R. Saad, F. Ahmad, D.C. Wijeyesekera, M.F.T. Baharuddin, Integral analysis of geoelectrical (resistivity) and geotechnical (SPT) data in slope stability assessment, Academic J. of Science, **1**(2), 305-316 (2012)
[6] M.H.Z. Abidin, M.F.. Baharuddin, A.F. Kamarudin, Application of geoelectrical method in subsurface profile forensic study, *Proc. of Malaysian Technical Universities Conf. on Engineering and Technology*, Melaka, Malaysia, (2010)
[7] A. Abdulrahman, M. Nawawi, R. Saad, A.S. Abu-Rizaiza, M.S. Yusoff, A.E. Khalil, K.S. Ishola, Characterization of active and closed landfill sites using 2D resistivity/IP imaging: Case studies in Penang, Malaysia, Environmental Earth Sciences,**75**, 347, (2016)
[8] K. Parial, A. Biswas, S. Agrahari, S.P. Sharma, D. Sengupta, Identification of contaminated zones using direct current resistivity surveys in and around ash ponds near Kolaghat Thermal Power Plant, West Bengal, India, Int. J. of Geology & Earth Sciences, **1**(2), 55-64 (2015)
[9] M.F.T. Baharuddin, A.R. Othman, S. Taib, R. Hashim, M.H. Zainal Abidin, M.A. Radzuan, Evaluating freshwater lens morphology affected by seawater intrusion using chemistry-resistivity integrated technique: A case study of two different land covers in Carey Island, Malaysia, Environmental Earth Sciences, **69**, (2013)
[10] M.F.T. Baharuddin, S. Taib, R. Hashim, M.H. Z. Abidin, Assessment of seawater intrusion to the agricultural sustainability at the coastal area of Carey Island, Selangor, Malaysia, Arabian J. of Geosciences, **6**, 3909-3928 (2013)
[11] M.H. Loke, Electrical imaging surveys for environmental and engineering Studies, A Practical Guide to 2-D and 3-D Surveys, 1-67 (1999)
[12] R. Arjwech, M. E. Everett, Application of 2D Electrical Resistivity Tomography to Engineering Projects: Three Case Studies, Songklanakarin J. Science and Technology, **37**(6), 675-681, (2015)

[13] A. Madun, I. Jefferson, K.Y. Foo, D.N. Chapman, M.G. Culshaw, P.R. Atkins, Characterization and quality control of stone columns using surface wave testing, Canadian Geotechnical J., **49**, 1357-1368 (2012)
[14] R. Saad, N.M. Muztaza, E.T. Mohamad, The 2D electrical resistivity tomography (ERT) study for civil and geotechnical engineering purposes, Electronic J. of Geotechechnical Engineering, **16**, 1537-1545 (2011)
[15] M. Zarrocaa, R. Linaresa, P.C. Velásquez-Lópezb, C. Roquéc, R. Rodríguezd, Application of electrical resistivity imaging (ERI) to a tailings dam project for artisanal and small-scale gold mining in Zaruma-Portovelo, Ecuador, J. of Applied Geophysics, **113**, 103–113 (2015)
[16] S.K. Nasab , A. Hojat, A. Kamkar-Rouhani, H.A. Javar, S. Maknooni, Successful use of geoelectrical surveys in area 3 of the Gol-e-Gohar Iron Ore Mine, Iran, Mine Water and the Environment, **30**(3), 208-215 (2011)
[17] M.N. Tijani, O.O. Osinowo, O. Ogedengbe, Mapping of sub-surface fracture systems using integrated electrical resistivity profiling and VLF-EM methods: A case study of suspected gold mineralization, Materials and Geoenvironment, **56**(4), 415–436 (2009)
[18] J.M. Terrón, V. Mayoral, J.A Salgado, F.A. Galea, V.H.C. Odriozola, P. Mateos, A. Pizzo, Use of soil apparent electrical resistivity contact sensors for the extensive study of archaeological sites, Archaeological Prospection, **22**(4), 269–281 (2015)
[19] A. Hawamdeh, R. Jaradat, Z. Alsaad, Integrated application of geophysical methods for investigation of the Al-Berktain archaeological site in the City of Jerash, Jordan, Environmental Earth Sciences,**73**(7), 3665–3674 (2015)
[20] M. Marchetti, V. Sapia, A. Garello, D.D. Rita, A. Venuti, Geology and geophysics at the Archeological Park of Vulci (Central Italy), Annals of Geophysics, **57**, (2014)
[21] M.G.D. Malaysia, *Geological Map of Peninsular Malaysia*, 8th Ed.,(1985)
[22] T. Gerudi, *Cadangan Pembinaan Kolej Universiti Tun Hussein Onn*, Laporan Penyiasatan Tapak, (2002)
[23] M.H. Loke, I. Acworth, T Dahlin, A comparison of smooth and blocky inversion methods 2-D electrical imaging surveys, Exploration Geophysics, **34**(3), 182–187 (2013)
[24] M.H. Loke, R.D. Barker, Rapid least squares inversion of apparent resistivity pseudosection using a Quasi-Newton Method, Geophysics Prospective,**44**(1), 131- 152 (1996)
[25] ABEM, *Terrameter SAS 4000/SAS 1000 Instruction Manual*, ABEM Instrument AB, Sweden, (2010)
[26] H.R. Burger, A.F. Sheehan, C.H. Jones, *Introduction to Applied Geophysics* , W.W. Norton & Company, (2006)
[27] S. G. C. Fraiha, J. B. C. Silva, Factor analysis of ambiguity in geophysics, Geophysics., **59**(7), 1083-1091(1994)
[28] R.C.Benson, L. Yuhr, R.D. Kaufmann, Some considerations for selection and successful application of surface geophysical methods, *Proc. the 3rd Int. Conf. on Applied Geophysics*, Orlando, Florida (2003)
[29] M.H.Z. Abidin, R. Saad, D.C. Wijeyesekera, F. Ahmad, N.A. Ismail, the influence of electrical resistivity array on its soil electrical resistivity value, Applied Mechanics and Materials, **510**, 185-192 (2014)
[30] M.H.Z. Abidin, R. Saad, D.C. Wijeyesekera, F. Ahmad, The influence of electrical resistivity array on its soil electrical resistivity value, Electronic J. of Geotechnical Engineering, **18**, 5643-5653 (2014)
[31] W.M. Telford, L.P. Geldart, R.E. Sheriff, *Applied Geophysics,* Cambridge University Press, Cambridge, (1990)

[32] T.S. Lee, *Slope Stability and Stabilization Methods: Geologic Site Investigation in chapter 4*, John Wiley & Sons. Inc., (2002)
[33] A.R. Samsudin, A. Haryono, U. Hamzah, A.G. Rafek, Salinity mapping of coastal groundwater aquifers using hydrogeochemical and geophysical methods: A case study from North Kelantan, Malaysia, Environmental Geology, **55**(8), 1737-1743 (2008)
[34] D.H. Griffiths, R.F. King, *Applied Geophysics for Geologist and Engineers: The Element of Geophysical Prospecting*, Pergamon Press, (1981)
[35] Z.A.M. Hazreek, S. Rosli, W.D. Chitral, F. Ahmad, A.T.S. Azhar, M. Aziman, B. Ismail, Soil identification using field electrical resistivity method, *J. of Physics: Conf. Series*, **622**, 1, (2015)
[36] Z.A.M. Hazreek, M. Aziman, A.T.S. Azhar, F. Fauziah, S. Rosli, The behaviour of laboratory soil electrical resistivity value under basic soil properties influences, *IOP Conf. Series: Earth and Environmental Sciences*, **23**, 1, (2015)
[37] Z.A.M. Hazreek, F. Fauziah, D.C. Wijeyesekera, R. Saad, The influence of basic physical properties of soil on its electrical resistivity value under loose and dense condition, *J. of Physics: Conf.* Series, **495**, 1, (2014)
[38] M.H.Z. Abidin, R. Saad, F. Ahmad, D.C. Wijeyesekera, M.F.T. Baharuddin, Correlation analysis between field electrical resistivity value (ERV) and basic geotechnical properties (BGP), Soil Mechanics and Foundation Engineering, **51**(3), 117-125 (2014)
[39] C. Liu, J.B. Evett, *Soils and Foundation*, Pearson Prentice Hall, (2008)
[40] Y. Jung, Y. Lee, H. Ha, application of electrical resistivity imaging techniques to civil and environmental problems, Use of Geophysical Methods in Construction, (2000)
[41] R.D. Barker, Electrical imaging and its application in engineering investigations, Geological Society, London, Engineering Geology Special Publications, **12**, 37-43(1997)

Visual Inspection of Historical Buildings Using Micro UAV

Masiri Kaamin[1,*], *Noor Azyyati* Idris[2], *Saifullizan* Mohd Bukari[2], *Zaurin* Ali[2], *Norhafiza* Samion[1], and *Mustaffa* Anjang Ahmad[2]

[1]Pusat Pengajian Diploma, Universiti Tun Hussien Onn Malaysia, 86400 Batu Pahat, Johor
[2]Smart Driving Research Center (SDRC), Fakulti Kejuruteraan Awam dan Alam Sekitar, Universiti Tun Hussien Onn Malaysia,86400 Batu Pahat, Johor

Abstract. The historical building is one of the country's heritage that needs to be taken care of failures in design and structure. Damage or dilapidation study that conducted by Conservation of Historical Building Guide Lines that practices for documentation any problems and defects in buildings. The stages to be followed to determine the damage was shooting photos for every element studied. Photo has been taken directly or by using stair for higher parts of structures. Inspection of the historical building should be emphasized to ensure that the building is in a safe condition for long periods. This research aims is to investigate the historical building defects by using Unmanned Aerial Vehicle (UAV) to identify the type of defects occur at the exterior structure of museum. The case study is building of Perak Museum that located at Taiping, Perak. The process involved in this study are planning and preparation before the flight, data acquisition using UAV, image processing and analyzing data from video to image. The result of analysis, a total numbers of 91 locations have been identified and documented suffer from defects or faults such as frame decayed, vegetation, perforated pipe and crack. All defects information obtained suing UAV visual inspection can help the museum maintainance team for conservation and maintainace work.

1 Introduction

The historical building is an old structure and has its own unique aesthetics value in addition to the specific functions which caused the building ware gazette as historical buildings. Important historical research carried out for the scan and get a true picture of the historical appearance of the building and the environmental conditions during the era of the building was built [1]. In Malaysia, historical buildings are different form one another in terms of architectural style, function, structure and building defects which require a comprehensive approach to building conservation and systematic [2]. The inspection conducted at Perak Museum as case study for this research. There was several type of defect to the structure of the museum such as frame decayed, vegetation, perforated pipe

* Corresponding author: masiri@uthm.edu.my

and crack. Previously, the inspection is usually done from very limited naked eye observation. UAV has being used as alternative way for collecting data visually for the exterior structure.

2 Research background

Survey the state of the building and damage known as dilapidation study. It is a process to identify and record the condition or degree of disability of the building. This study is a practice of documenting the problems and defects of the building is run [3]. The study is used to record and document, a thorough investigation is necessary to determine the condition of the building, the defect and the cause. Documentation in the form of digital and photography need to detect the circumstances and the nature of the materials of the existing building. Perak Museum was selected as study area because the history value and the first building undergoing conservation work.

According to J. Unger et.al , the high resolution camera attached to the UAV showed good results even under non-optimal flight conditions and confirmed that already visual recording methods provide valuable information for infrastructural inspection purposes [4]. Also the image post-processing has to be improved by reducing manual workflow by appropriate image stitching and mosaicking software, ideally with an integrated crack detection feature based on the results shown which still requires to be developed.

Due to management requirements in buildings, the risk of personal injury and damage to the infrastructure needs to be carried out as low as that may be. Redundancy factor therefore to position structures and parts of the complete system, then the reliability of flight important function, is a matter of paramount importance. Other major aspect was upon request to the characteristics of stable hovering to ensure planned and detailed examination of the damage

2.1 Unmanned aerial vehicle

Micro Unmanned Aerial Vehicle (UAV) equipped with a camera that offers the possibility to map the different areas quickly and with high flexibility compares to classic aerial photography. UAV also known as the Autonomous Navigation system because it is fly by using Global Positioning System (GPS) and support by the telemetry systems at the Ground Control Station (GCS) to monitor and navigate the UAV when capturing the data [5]. UAVs are becoming popular for use low budget and use in large scale mapping [6]. DJI Phantom FC40 used for this research as alternative ways for collecting data visually. Micro UAV able to fly less than 1 hour, limited load factor and optimum level of image quality that is useful between 800 feet to 1200 feet and an image resolution of 6cm to 15cm per pixel [7].

3 Building inspection

For building inspection of Perak Museum, UAV flight has been used for collecting data. This study followed several steps as the methodology. There are planning and preparation before fly, data collection using UAV, image processing from video to image and analysis of study.

Fig. 1 shows the UAV that has been used in this research. DJI Phantom FC40 is equipped with camera for recording image or videos. The recording will display through the smart phone that have install FC40 software. It comes with its own built in DJI designed smart camera with 720p/30fps HD video support, and iOS or Android app control. Using a 2.4G Wi-Fi connection, it provides a full FPV experience at up to 100m.[8] It's the ideal choice for those getting started with aerial cinematography. The software will be downloaded at Google Play to connect with the camera. The software will function when the smart phone and the camera were connected through the Wi-Fi application.

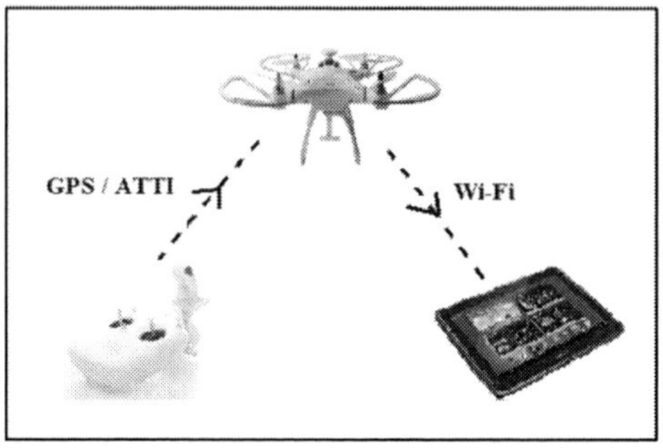

Fig. 1. Connection between remote control, DJI Phantom FC40 & smartphone.

3.1 Planning and preparation

Rigorous planning prior flight made to avoid things that are not expected to determine the distance between UAV and building must be less or equal to 2m [9] (Fig. 2). The direction of UAV line of flight in vertically [10] as shows in Fig. 3. The direction of flight starting from the bottom moves upward until the top of the building or at the roof and moves to the right side and slowly downward until reaches the ground. This steps repeat until the end of the building.

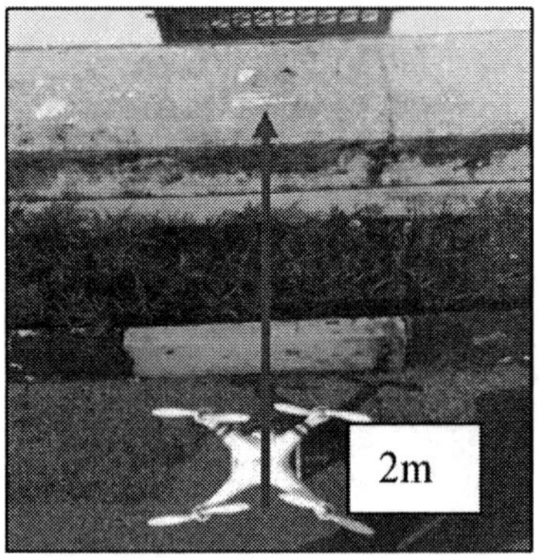

Fig. 2. Distance between UAV and building

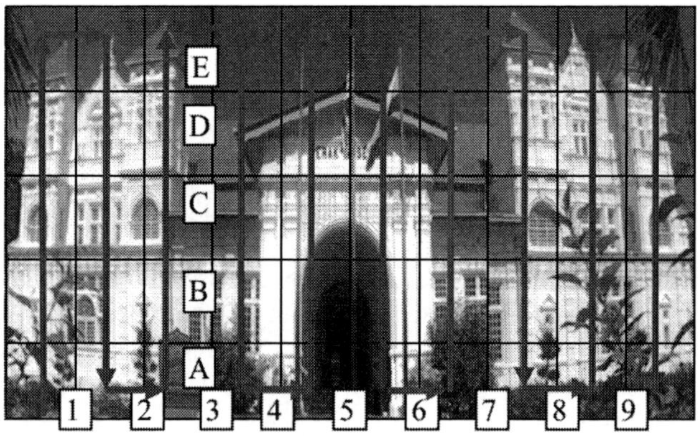

Fig.3. Direction of flight

3.2 Data collection

Recording image and video from UAV camera is the method of collecting data. Because of the museum is bigger, the study divided the data collection into 3 phase. There are right side, front and left side. To conduct this research, at least need 2 persons involve while flying the UAV. One of them will control the remote of UAV and another one person need to monitor the smartphone and give the direction to the controller. There are 26 videos taken as the data for this research to show the whole building and to analysis the video.

DOI: 10.1051/matecconf/201710307003

Table 1 shows total number of data collecting on site and Fig. 4 shows the system were setup and data collection conducted.

Table 1. Data collection

Parts of Building	Video	Image	Editing Image
Right	3	38	26
Front	9	105	28
Left	10	101	42
Behind	4	16	8
Total	26	260	104

Fig. 4. System setup and Data Collection

3.3 Image processing

From the video that have been taken from the site, extra process needs to be done into the image version as the effective way to shows the defect for the building. The step that involves in this process is from the video then play in the VLC Media Player and "Take Snapshot" for defect detects and open the image in Adobe Photoshop. From Adobe Photoshop, the image will be zooming in the display clearly the defect and the brightness and contrast will be adjusted to enhance the image result. The total of image that has been edited by using Adobe Photoshop is 104 pictures. From the edited image, there are 91 defect has been detected for the whole building and Table 2 shows, 3 image has identify the type of defect and location of defect.

DOI: 10.1051/matecconf/201710307003

Table 2. Image processing

No.	Location & Type of Defect	Image Editing Method Used to Identify Defect	Image Defect	Zoom Image
1.	Right Side Frame decayed	Ink Outline Method		
2.	Front Side Fungus	Trace Contour Method		
3.	Left Side Gutter rusted and rotted	Trace Contour Method		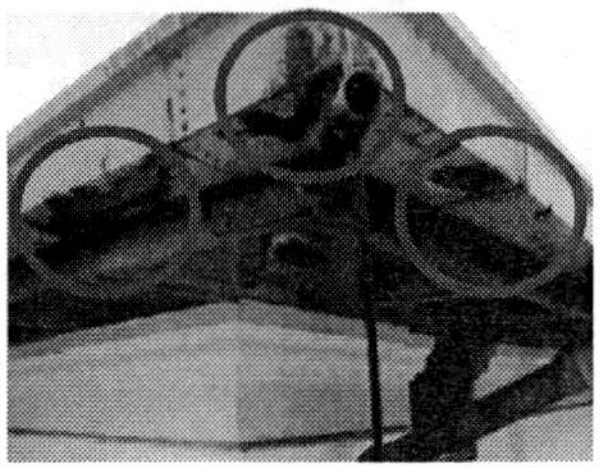

3.4 Analysis of study

Through the image processing technique using Adobe Photoshop the total number of defects that occur can be identify. The defect has been choose for analysis is based on the repetition occurring. This stage will help to analysis the defect in different way from image processing that just zoom the image. Adobe Photoshop has been use to shows clearly the defect occur. Fig. 5 shows the image editing technique using 'Ink Outlines' method for frame decayed. The black color at the image after editing provide an indicator of broken or missing frame.

Fig. 6 show defect of fungus that occur at the front side of the building. For this defect, the research use Gradient Map and Trace Contour to make the defect shown clearly. It need couple of editing because of the first editing do not get excellent result and the image with colored dots representing fungus on the surface of the building. Fig. 7 show defect of gutters rusted and rotted. By using Brush Strokes Accented Edges, the rusted and rotted part visible clearly and the circle represent the defect.

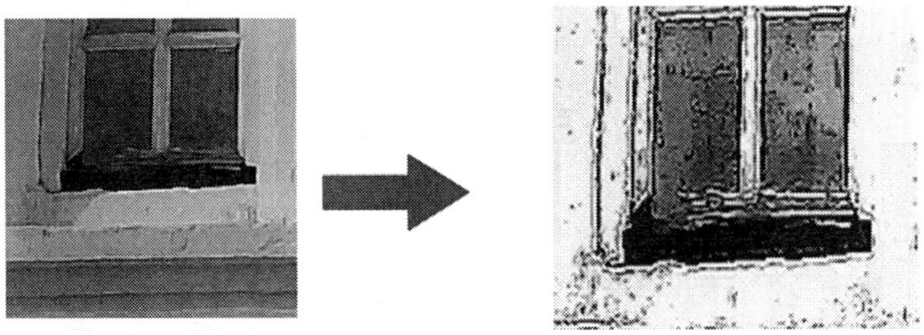

Origional shape After using 'Ink Outlines' method

Fig. 5. Analysis for frame decayed

Original image Edit using Gradient Map Edit using Trace Contour

Fig. 6. Analysis for fungus.

Before edit using Adobe Photoshop After using 'Brush Strokes Accented Edges'

Fig. 7. Analysis for gutter rusted and rotted.

During the process of improvement of images, more than one an editing process to ensure that all the effects of the defect becomes clearer. For example, the process of brush strokes is more suitable to improve the image of the line as the image of the gutter.

4 Results and discussions

From image processing, the total number of defect has been can be identified. The Table 3 shows the data of defect occur at Perak Museum by the section of the building. From the table, the type of defect and the total defect can be recognized and categorized according to the section of the building.

Table 3. Type and total defect by section.

Section	Type of defect	Total defect	Total
Right	Frame decayed	2	5
	Vegetation	1	
	Perforated pipe	1	
	Discoloration	1	
Front	Frame decayed	4	32
	Vegetation	7	
	Discoloration	1	
	Peeling paint	3	
	Clogged gutters	1	
	Fungus	5	
	Dirt	3	
	Broken mirror	2	
	Gutters rusted and rotted	2	
	Cleft, Crack	2	
	Eroded surface	1	
	Faded paint	1	
Left	Frame decayed	3	44
	Vegetation	2	
	Peeling paint	6	
	Fungus	9	
	Dirt	1	
	Gutters rusted and rotted	4	
	Cleft, Crack	5	
	Eroded surface	5	
	Faded paint	3	
	Moldy pipeline	4	
	Roof broken/dirty	2	
Back	Frame decayed	1	10
	Discoloration	1	
	Peeling paint	3	
	Fungus	4	
	Wall rupture	1	

From table 2, the top 7 defect occur at museum building is fungus, peeling paint, frame decayed, vegetation, cleft and crack, gutters rusted and rotted and eroded surface.

5 Conclusions

By using unmanned aerial vehicle (UAV) to visual inspection of the building is the alternative technique to know the defect occur at exterior of the building. Research has been prove by using the UAVs is visual recording provide information for inspection of the building. Besides that, image processing and analysis will shows in detail the defect occur and the severity of the defect for the building. Also, inspection of building using UAV will help to minimize the cost of operation, reduce time and friendly user.

DOI: 10.1051/matecconf/201710307003

This research was supported supported by Short Term Grant, U385 UTHM. We thank our colleagues from PPD and FKAAS who provided insight and expertise that greatly assisted the research, although they may not agree with all of the interpretations/conclusions of this paper.

References

[1] Jabatan Warisan Negara, *Garis Panduan Pemuliharaan Bangunan Warisan*, Jabatan Warisan Negara, Kementerian Pelancongan Dan Kebudayaan Malaysia, Kuala Lumpur, Malaysia (2016)

[2] A Ghafar Ahmad, Understanding common building defects: the dilapidation survey report. Majalah Arkitek, **16**(1), 19-21 (2004).

[3] A. Ghafar Ahmad, *Rangka Kerja Pemuliharaan Bangunan Warisan di Malaysia,* (2006), Retrieved on October 15, 2016 from http://www.hbp.usm.my/ conservation/ Sejarah@Msia/Rangka%20Kerja.htm

[4] J. Unger, M. Reich & C. Heipke, UAV-based photogrammetry: monitoring of a building zone. The International Archives of Photogrammetry, Remote Sensing and Spatial Information Sciences, 40(5), 601-606 (2014).

[5] A.M. Samad, N. Kamarulzaman, M.A. Hamdani, T.A. Mastor, The potential of Unmanned Aerial Vehicle (UAV) for civilian and maping application. *Proc. Of 3rd International Conference on System Engineering and Technology*, Shah Alam, Malaysia, (2013)

[6] N. Darwin, A. Ahmad, and O. Zainon, The potential of unmanned aerial vehicle for large scale mapping of coastal area. IOP Conference Series: Earth and Environmental Science 18(1), 1-6. (2014)

[7] I. Ibrahim, Penggunaan UAV bagi applikasi awam., (2009) Retrieved on October 15, 2016 from http://gisondemand.blogspot.my/2009/08/penggunan-uav-bagi-applikasi-awam.html

[8] Masiri Kaamin, Misbahul Muneer Abd Rahman, Norhayati Ngadiman, Aslila Abd Kadir, Ahmad Hakimi Mat Nor, Nor Baizura Hamid, and Mardiha Mokhtar, Visual inspection of historical buildings using Unmanned Aerial Vehicles (UAV): A case study of Sultan Abu Bakar Mosque, Johor, Advanced Science Letters, 22(9), 2160-2163 (2016)

[9] Masiri Kaamin, Nurunnazifah Abd Aziz, Saifullizan Mohd Bukari, Zaurin Ali, Norhafiza Samion, Aslila Abd Kadir, and Norhayati Ngadiman. Kaedah pemeriksaan bangunan tinggi menggunakan pesawat tanpa pemandu (UAV) Jurnal Teknologi **78**(5-10), 83-89 (2016)

[10] C. Eschmann, C.M. Kuo, C.H. Kuo, and C. Boller, Unmanned aircraft systems for remote building inspection and monitoring. *Proc. of the 6th European workshop on structural health monitoring*, DGZIP, Germany (2012)

The Behaviour of Remolded Batu Pahat Soft Clay with Different OCR Values under Cyclic Loading

Alvin John Lim Meng Siang [1,*], and *Muhammad Arif Azraei* Izzudini[1]

[1]Faculty of Civil and Environmental Engineering, Universiti Tun Hussein Onn Malaysia, 86400 Parit Raja, Johor, Malaysia

Abstract: The Batu Pahat Soft Clay, (BPSC) of Universiti Tun Hussein Onn Malaysia, (UTHM) were low in shear strength, bearing capacity, and suffer large settlements when subjected to loading. They undergo varies of dynamic cyclic loadings during their design lifetime and the response was typically more complex, requiring engineers to investigate the dynamic behaviour of soils thoroughly in the laboratory. The objectives of this research were to simulate remolded Batu Pahat Soft Clay (RBPSC) samples with different $\sigma_{v'}$, to study the dynamic shear modulus, G and damping ratio, D of (RBPSC) under cyclic loading with different f, OCRs, $\sigma_{v'}$ and $\sigma_{c'}$ and to analyses the correlation between G and D of the (RBPSC) with the OCR values under cyclic loading. As result, the 100 x 50 mm of remoulded samples are succeed simulated by using large strain consolidation apparatus of 50, 80 and 100 kPa of pre consolidation stress, contain moisture ranging within 42 to 55%. The series of remolded consolidated undrained dynamic cyclic triaxial test were ran under OCR of 1, 2, 3.85 and 4. Hence, it could be concluded that the G decreased when increasing of f and OCRs subjected to increasing axial strain while the G corresponding to each $\sigma_{v'}$ increases slightly as the $\sigma_{v'}$ becomes higher. The D shows minor increased when increasing of the f, OCRs and $\sigma_{v'}$ when subjected to increasing axial strain. Thus, input of parameter G and D can be review as technical values to key design structure on top of soil layer.

1 Introduction

The soils are often vulnerable to undrained cyclic loadings such as earthquakes, wind, wave loading and many more. These cyclic loadings have a wide range of amplitude, frequency, and duration. Similarly, response of soils to cyclic loading also varies widely and depends on various factors. Therefore, cyclic behaviour of clay is a matter of interest in geotechnical engineering practice and has been studied extensively, [15]. Present, development of city has become a major critical impact in Batu Pahat because of the development and modernisation. The problem occurred during post construction is the high settlement rate,

[*] Corresponding author: alvin@uthm.edu.my

which reduce the strength of workability and shortens the design life of structure. In addition, Soft Clay are well known for their low strength and high compressibility. Usually, due to sedimentary process on different environment, both physical and engineering properties of the clays (namely void ratio, water content, grain size distribution, compressibility, permeability and strength) show a significant variation. Furthermore, they exhibit high compressibility (including an important secondary consolidation), reduced strength, low permeability and compactness, and consequently low quality for construction, [4].

2 Problem on soil due to cyclic loading

The soil deposits in many geotechnical engineering projects undergo dynamic cyclic loadings during their design lifetime. These repeated loadings may be due to environmental factors, such as seismic activity and ocean storms, or human activities, such as passing traffic and vibrating machinery installed on a structure or site. Importantly, the soil response generated by these dynamic cyclic loadings is typically more complex than that considered when conducting static analyses, requiring engineers to investigate the dynamic behaviour of soils in the laboratory, as well as in the field. The dynamic loading which are shear modulus and damping ratio of soil are expressed as a function of cyclic stress - strain amplitude and depend on various parameters such as effective vertical consolidation stress, OCR/Stress history, loading frequency and etc. They was important parameter to analyses the soil failure behaviour under specific cyclic loading circumstances. Various researchers investigated dynamic loading in various soils and the influence of the above mentioned factors. The hysteresis loops can be considered in order to calculate the secant shear modulus, G and damping ratio, D from the slope and the total area of the loop respectively. [11], observed an increase in shear modulus values with loading frequency from tests on Leda clay but drop at 1.0 Hz, so they are not significant. The shear modulus corresponding to each effective confining stress increases slightly as the confining stress becomes higher. The damping ratio decreases as confining stress increases. The shear modulus decreases with increasing loading frequency. Moreover, when the loading frequency increases the damping ratio increases, [15]. However, for greater strain levels, the dependency of D and G on the loading frequency becomes less clear; moreover, both the shape of cycle and the average shear stiffness evolve progressively with the number of cycles imposed, [19]. The site-specific ground response analysis was performed using the modulus reduction and damping curves established for the site and the results are compared to those obtained using conventional practice.

The 'SHAKE' type site response analysis is currently the most common means of estimating site specific response during an earthquake. The modulus reduction and damping curves are key input to such an analysis, in addition to the soil profile and ground motion stress time history. According to related studies, the over consolidation state is an important effect for soil liquefaction potential. If a soil mass has experienced stresses higher than its current state, it is an over consolidated soil (OCR > 1). Over consolidated soils, OC have fewer settlements due to external loadings as compared with normally consolidated soils, NC, [15]. Moreover, normally consolidated soil specimens yield highest peak strength. The peak strength of specimens decreases with an increase in OCRs. As OCR increases, the initial tangent modulus of clean sand decreases due to the lower current effective consolidation stress and consequently, the soil specimens become weaker. However, the OCR seems to have negligible effect on the initial tangent modulus of soil specimens with clay fines content, [16].This paper aims to explain in creating remolded (BPSC) samples with different pre consolidation pressures, to study the dynamic shear modulus, G and damping ratio, D of (RBPSC) under cyclic loading with different f, OCRs, σ_v' and σ_c' and to analyses the relationship between dynamic shear modulus and damping ratio of the remolded (BPSC) with different OCR values under cyclic loading.

3 Materials for testing

The sampling of Batu Pahat Soft Clay, (BPSC) was taken at Research Centre of Soft Soil Universiti Tun Hussein Onn Malaysia, (RECESS UTHM) at depth of 1.5 meter using backhoe. Table 1 below shows the geotechnical properties of BPSC. The sample conducting the geotechnical properties was disturbed sample.

Table 1. Batu Pahat Soft clay soil characteristics

Soil Characteristic – Batu Pahat Soft Clay (RECESS)	
Liquid Limit, (%)	69.00
Plastic limit, (%)	34.46
Plastic Index, (%)	34.54
Natural moisture content, (%)	87
Dry unit weight, γd, (kN/m^3)	10.28
Specific Gravity	2.700
Particles (%)	10.2 clay, 80.1 silt, 9 sand and 0.7 gravel

4 Test equipment and procedures

Method use for pre consolidation preparing remolded sample is using large strain consolidation apparatus. First the sample is dry in the oven over 1 day period and grinder into smaller size particles using crusher machine. Then, the sample of Batu Pahat Soft Clay, BPSC is mix with certain amount of water until it's become slurry using mixer machine and pour into cylinder mould diameter of 12.2 cm and being applied load of 50, 80 and 100 kpa on top. After 2 to 3 weeks consolidated, the samples are extrude out from the cylinder mould using pvc pipe size of 100 x 50 mm using machinery extruder and wrap with plastic wrapper to avoid loss of moisture content and stored into plastic container under room temperature shows in Fig. 1.

Fig. 1. Process of preparing sample

The pre - consolidation stress of 50, 80 and 100 kpa are chosen which are low, medium and high value because to represent different depth of sampling at the field but the samples conducting in laboratory method. The maximum load can be applied using large strain consolidation is 100 kpa. On top of that, the standard for dynamic cyclic triaxial test will follow ASTM D3999-11 for load control method and BS 1377 – 8: 1990 for Saturation, B –

check and Consolidation Stages. In this research, the frequency selected are 0.1 Hz (wave action), 0.5 Hz (wind action) and 1 Hz (earthquake). The total number of samples are 27 will be tested using GDS Enterprise Level Dynamic Triaxial Testing System (ELDYN) show in Fig. 2 and Table 2.

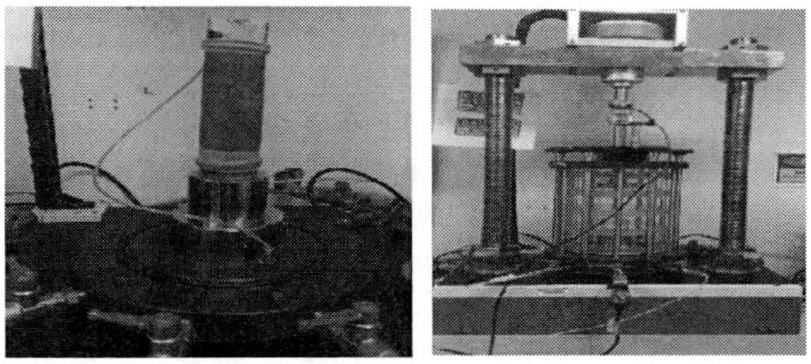

Fig. 2. GDS Enterprise Level Dynamic Triaxial Testing System (ELDYN)

Table 2. Total number of samples

σc' (kpa)	Load (kg)	σv' (kpa)	Depth (m)	OCR value	OCR status	f (Hz)	No. of cycles	No. of samples
50	60	13	1.187	3.85	OC	0.1, 0.5, 1	100	3
		25	1.811	2	OC	0.1, 0.5, 1	100	3
		50	3.112	1	NC	0.1, 0.5, 1	100	3
80	95	20	1.551	4	OC	0.1, 0.5, 1	100	3
		40	2.592	2	OC	0.1, 0.5, 1	100	3
		80	4.673	1	NC	0.1, 0.5, 1	100	3
100	120	25	1.811	4	OC	0.1, 0.5, 1	100	3
		50	3.112	2	OC	0.1, 0.5, 1	100	3
		100	5.713	1	NC	0.1, 0.5, 1	100	3
Total numbers of sample								27

5 Results

5.1 Effect of loading frequency on shear modulus and damping ratio

The Fig. 3 and 4 show the relationships between the shear modulus and the damping ratio with axial strain due to subjected loading frequencies of 0.1 Hz, 0.5 Hz and 1.0 Hz. It can also be seen in Fig. 1 that at given loading frequency, shear modulus decreases with increasing axial strain and loading frequency. The shear modulus of 0.1 Hz start from 79.141 kpa down to 10.011 kpa whereas 0.5 Hz start from 96.541 kpa down to 9.011 kpa and 1.0 Hz start from 97.777 kpa down to 8.012 kpa which is the lowest value recorded in situation one. Meanwhile Fig. 2 shows that the damping ratio increases with increasing axial strain and loading frequency. The peak damping ratio of 0.1 Hz is 0.116 (lowest), 0.5 Hz is 0.121 (second highest) and 1.0 Hz is 0.131 (highest) in situation two. This trend was similar as Thammathiwat. A, and Chim – oye. W, 2006, whom found that the shear modulus decreased when loading frequencies increased while the damping ratio increased when loading frequencies increased undergone increasing in axial strain, [14]. This happen because frequency is the number of occurrences of a repeating event per unit time or cycle per time which is, in this research all the samples undergo 100 cycle of dynamic cyclic loading. The frequency loading of 0.1 Hz is noted as wave action which run 1000 seconds per cycle, 0.5 Hz is noted as wind action which run 200 seconds per cycle and 1.0 Hz is noted as earthquake action which run 100 seconds per cycle, hereafter simulated in the laboratory work using set of dynamic cyclic triaxial apparatus. This can be said that, the value of shear modulus and damping ratio are related on the loading frequency within period of time under specific number of cycle and vice versa or also can be said increase in loading frequency with time, more lower or higher value recorded in both situation. In real work at site, the value of shear modulus (stiffness) and damping ratio (decrease in the amplitude of an oscillation as a result of energy being drained from the system to overcome frictional or other resistive forces) can be used as technical value to construct structure above soil to withstand these prediction loading frequencies action.

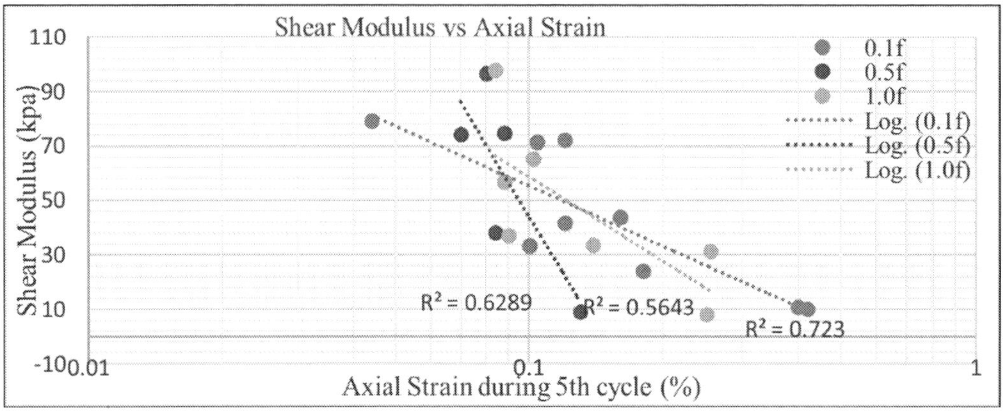

Fig. 3. Relationship between shear modulus and axial strain under different loading frequencies

5.2 Effect of over – consolidation ratio, OCR on shear modulus and damping ratio

The Fig. 5 and 6 show the relationships between the shear modulus and the damping ratio with axial strain due to subjected over - consolidation ratio, OCR of 1, 2, 3.85 and 4. It can also be seen in Fig. 3 that at given OCR, shear modulus decreases with increasing axial strain and Fig. 4 that at given OCR, damping ratio increases with increasing axial strain. Moreover, normally consolidated, NC = 1 shows the highest shear modulus value started from 97.777 kpa down to 41.667 kpa, follow by over – consolidated, OC = 2 which started from 74.747 kpa and down to 24.073 kpa, OC = 4 which started from 69.767 kpa down to 9.011 kpa and OC = 3.85 shows the lowest value which started from 38. 097 kpa down to 8.012 kpa. This happen because in this research, NC = 1 has effective vertical stress of (σ_v' = 50, 80 and 100kpa) which 100kpa is the highest, while OC = 2 has effective vertical stress of (σ_v' = 25, 40 and 50kpa) which 50kpa is the second highest, OC = 4 has effective vertical stress of (σ_v' = 20 and 25kpa) which 25kpa is the third highest and lastly OC = 3.85 has effective vertical stress of (σ_v' = 13kpa) which 13kpa is the lowest effective vertical stress.

This trend was similar with Thian [16], whom found that the stiffness of soil decreased when OCRs of soil was increased. The pre - consolidation stress, σ'_c used in this research were 50 kpa, 80 kpa and 100 kpa. This result can be said that the NC is stiffer than OC but must referring to the effective vertical stress value. The consolidation is any process which involves a decrease in water content of saturated soil without replacement of water by air.

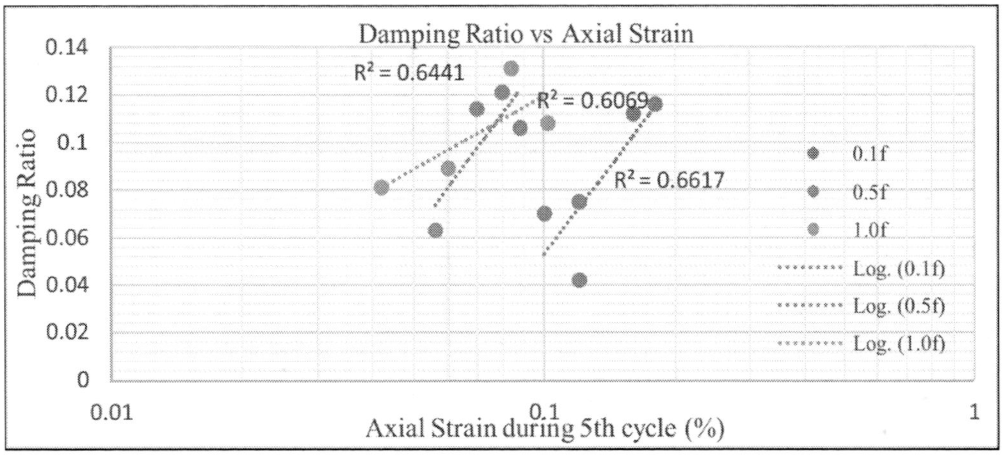

Fig. 4. Relationship between damping ratio and axial strain under different loading frequencies

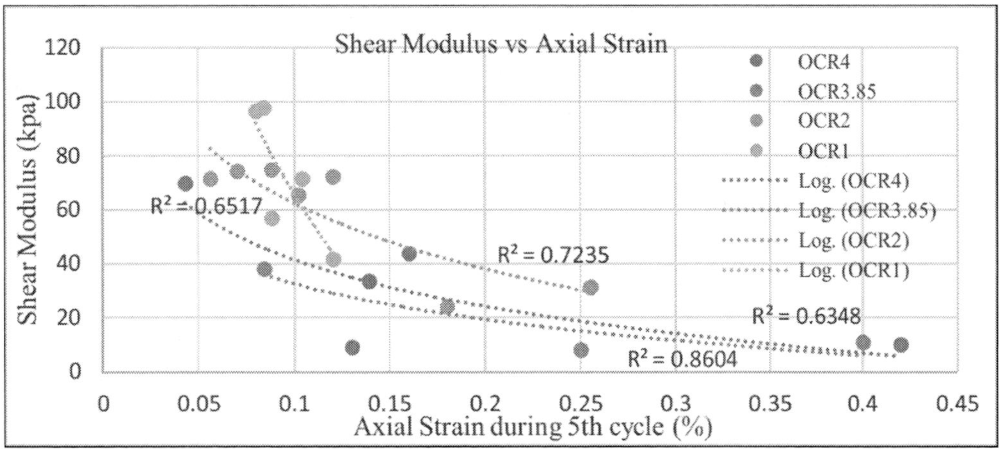

Fig. 5. Relationship between shear modulus and axial strain under different OCR

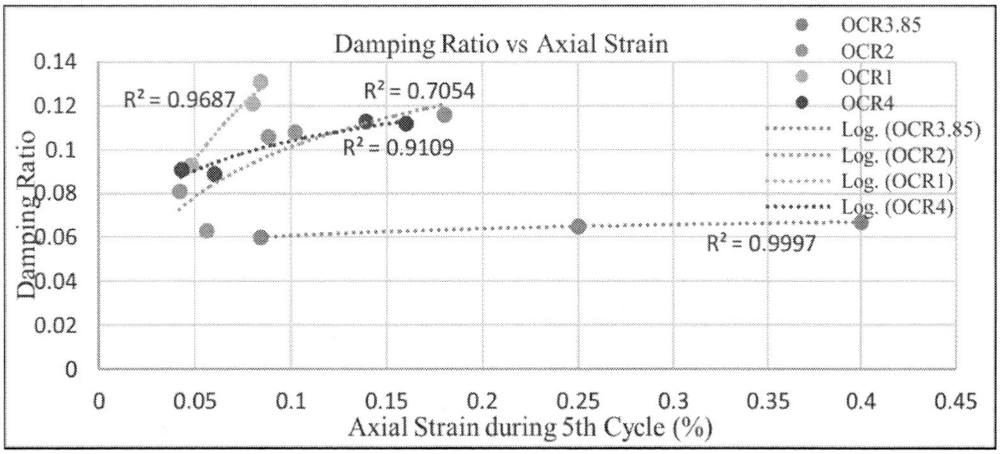

Fig. 6. Relationship between damping ratio and axial strain under different OCR

When stress is removed from a consolidated soil, the soil will rebound, regaining some of the volume it had lost in the consolidation process whereas when stress reapplied, the soil will consolidate again. This happen may be due to geologic activity or human activity. The soil which had its load removed is considered to be over consolidated. The over - consolidation ratio, OCR is defined as the highest stress experienced divided by the current stress. On the other hand, if the cohesive soil is over consolidated, there would be no significant settlement when a load is applied to the soil, [17]. Therefore, the significance and importance of OCR value can be used as technical parameter measurement indicator on the stress history and current stress applied on top of soil when construction take placed at real site condition.

5.3 Effect of effective vertical consolidation stress/confining stress on shear modulus and damping ratio

The Fig. 7 and 8 show the relationships between the shear modulus and the damping ratio with axial strain due to subjected effective vertical consolidation stress, (σ_v' = 13, 20, 25, 40, 50, 80 and 100kpa). It can also be seen in the Fig. 5 that at given σ_v', shear modulus decreases with increasing axial strain and decreased in σ_v' and the Figure 6 shows that at given σ_v', damping ratio increases with increasing axial strain and σ_v'. Furthermore, at (σ_v' = 13 kpa, the peak shear modulus, G recorded was 38.097 kpa which was lowest), at (σ_v' = 20 kpa, G = 40.139 kpa), at (σ_v' = 25 kpa, G = 71.43 kpa), at (σ_v' = 40 kpa, G = 74.231 kpa), at (σ_v' = 50 kpa, G = 74.747 kpa), at (σ_v' = 80 kpa, G = 79.141 kpa) and (σ_v' = 100 kpa, G = 97.777 kpa which was highest).

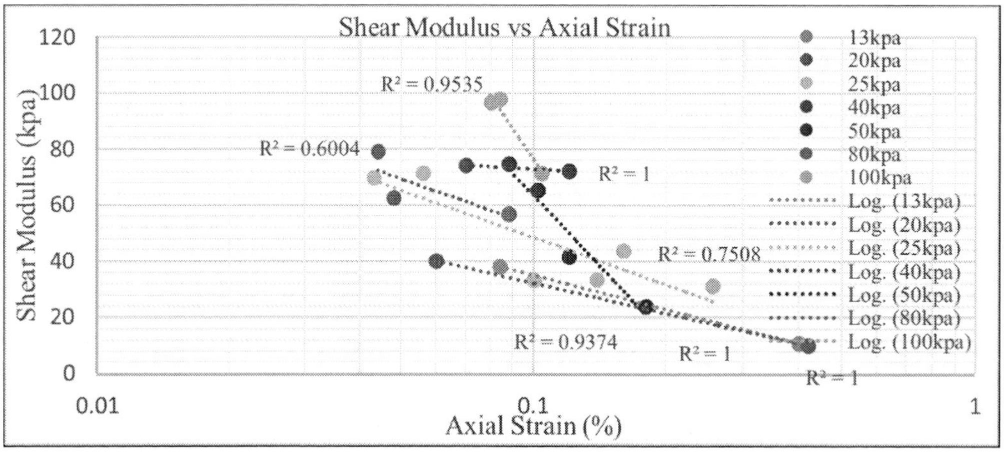

Fig. 7. Relationship between shear modulus and axial strain under different σ_v'

However, the differences between peak shear modulus and smallest value of each σ_v' can't be taken as significant argument because each σ_v' had slightly differ reduction. In contra, the damping ratio, D at (σ_v' = 13 kpa, the peak D recorded was 0.067 which was lowest), at (σ_v' = 20 kpa, D = 0.09), at (σ_v' = 25 kpa, D = 0.113), at (σ_v' = 40 kpa, D = 0.114), at (σ_v' = 50 kpa, D = 0.116), at (σ_v' = 80 kpa, D = 0.121) and (σ_v' = 100 kpa, D = 0.131 which was highest). This trend result was similar as previous journal, Thammathiwat. A, and Chim – oye. W, 2006, whom found that the shear modulus decreased when effective vertical consolidation stress increased while the damping ratio increased when effective vertical consolidation stress increased subjected to increasing in axial strain, [14]. Hence, it

 DOI: 10.1051/matecconf/201710307004

can conclude that the effective vertical stress, σ_v' value play high role in term of factors affecting the cyclic behaviour of soft clay under specific circumstances of cyclic loading. The shear modulus, G and damping ratio, D are not only needed for earthquake analysis but also for the design of machine foundations, wind power plant foundations, fast transportation systems, impacts and blast resistant structures and etc. [2].

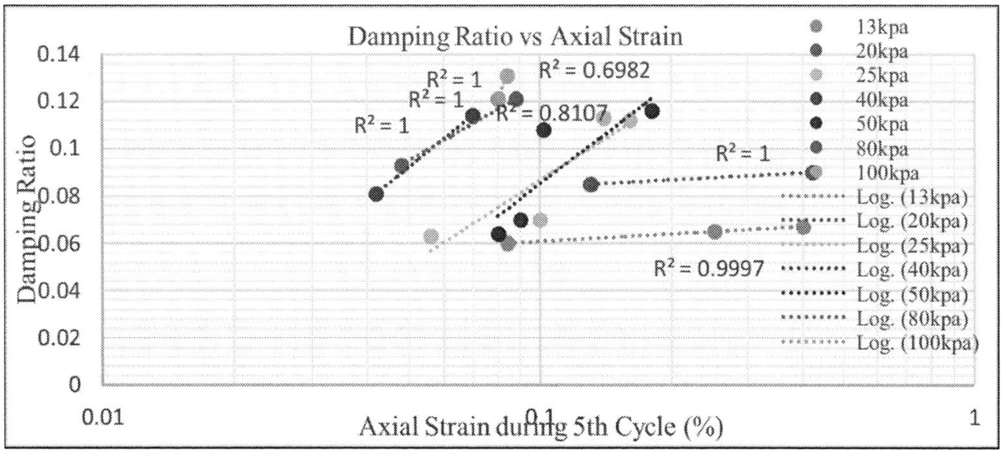

Fig. 8. Relationship between damping ratio and axial strain under different σ_v'

6 Conclusion

From the results of dynamic cyclic triaxial tests to investigate the cyclic strength which were shear modulus and damping ratio of remolded Batu Pahat Soft Clay, BPSC, the following conclusions may be drawn:

i) The 100 x 50 mm of remoulded samples are succeed simulated by using large strain consolidation apparatus of 50, 80 and 100 kPa of pre consolidation stress, contain moisture ranging within 42 to 55%.
ii) The shear modulus decreased with increasing of axial strain for a given loading frequencies but damping ratio increased with increasing of loading frequencies and axial strain.
iii) The shear modulus corresponding to each effective vertical consolidation stress, $\sigma_{v'}$ increases slightly as the $\sigma_{v'}$ becomes higher. The damping ratio increases as $\sigma_{v'}$ increases.
iv) The peak strength of specimens decreases with an increase in OCRs but referring to its effective vertical consolidation stress respectively. Normally consolidated, NC of soil specimens yield highest peak strength compare to OC.
v) At given OCR, shear modulus decreases with increasing axial strain and at given OCR, damping ratio increases with increasing axial strain.
vi) The damping ratio increasing by slightly margin on increasing axial strain under specific loading frequencies, effective vertical consolidation stresses and OCRs.

References

[1] K.R. Devi, Cyclic triaxial tests on kolkata clay, Int. J. of Engineering Research and Technology, **2**(7), 439 – 443, (2013)

[2] N. Dibaji, W. Arghavan, Damping ratio of mixed gravel and clay, Int. of J. Science and Technology, **32**(B5), 501–518, (2008)
[3] A. Erken, M.B.C. Ülker, The post cyclic shear strength of fine grained soils, *14th World Conference on Earthquake Engineering*, Beijing, (2008)
[4] M.H. Ho, C.M. Chan, Some mechanical properties of cement Stabilized Malaysian Soft Clay, Int. J. of Civil, Environmental, Structural, Construction and Architectural Engineering, **5**(2), 76 – 83, (2011)
[5] L. Hu, J. Ding, Mechanical behavior of marine clay under wave loading, International J. of Offshore and Polar Engineering, **20**(1), 5381, (2010)
[6] Y. Hu, H. Kempfert, Some important aspects in evaluating cyclic triaxial test on clayey soil, *14th Southeast Asian Geotechnical Conference,* Hong Kong, (2001)
[7] M. Jiang, Stiffness degradation of soft marine clay under uniaxial cyclic loading, European J. Geotechnical Engineering,**17**, 887 – 3895, (2012)
[8] H.M. Mohamad, Post-cyclic behaviour of soil – A critical review, *8th Malaysian Technical Universities Conf. on Engineering and Technology*, Melaka, (2014)
[9] J. Ni, B. Indraratna, X.Y. Geng, C. Rujikiatkamjorn, The effect of the strain rate on soft soil behaviour under cyclic loading, *11th Australia - New Zealand Conf. on Geomechanics*: *Ground Engineering in a Changing World,* Melbourne, (2012)
[10]C. Pasten, H. Shin, J.C. Santamarina, Long-term foundation response to repetitive loading, J. of Geotechnical and Geoenvironmental Engineering, **04013036**, 1-11, (2013)
[11]K.K. Rasmussen, *An Investigation of Monotonic and Cyclic Behaviour of Leda Clay*, Master Thesis, University of Western Ontario, Canada, (2012)
[12]M. Shahin, R. Loh, H. Nikraz, Some observations on the behavior of soft clay under undrained cyclic loading, J. of Geo Engineering, **6**(2), 109 – 112, (2011)
[13]H. Soltani-Jigheh, A. Soroush, Cyclic behavior of mixed clayey soils, Int. J. of Civil Engineering, **8**(2), 99-106, (2010)
[14]A. Thammathiwat, W. Chim-Oye, Behavior of strength and pore pressure of soft bangkok clay under cyclic loading stress, Int. J. on the Biology of Stress, **9**(4), 21- 28, (2004)
[15]T. Theenathayarl, *Behaviour of sensitive leda clay under simple shear loading*. Master Thesis, University of Peradeniya, Peradeniya, Sri Lanka, (2015)
[16]S.Y. Thian, Stress history effect on mining sand with fines contents, Int. Journal of Geomatics and Geoscience, **2**(1), 1- 10, (2011)
[17]S.Y. Thian, C.Y. Lee, Effect of OCR on cyclic shear strength degradation of marine clay, Int. J. of Civil and Structural Engineering Research, **4**(1), 280- 285, (2016)
[18]P. Tsai, N. Ni, Effects of types of additives on dynamic properties of cement stabilized soils, Int. J. of Applied Science and Engineering, **10**(2), 131-144, (2012)
[19]C. Zambelli, A. D'onofrio, F.S.D. Magistris, Dependency of the mechanical behaviour of granular soils on loading frequency: Experimental results and constitutive modelling, *Soil Stress Strain Behaviour Measurement Modeling and Analysis Geotechnical Symposium Proc.,* Rome, (2006)

DOI: 10.1051/matecconf/201710307005

Numerical Evaluation on the Different Shapes of Gravelly Sand Columns to Increase the Loading Capacity of Soft Clay

Sif Allah Meghzili[1,2], *Aziman* Madun[1,2,*], *Saiful Azhar Ahmad* Tajudin[1,2], *Mohd Fairus* Yusof[1,2], and *Mohd Ashraf Mohamad* Ismail[3]

[1] Universiti Tun Hussein Onn Malaysia, Batu Pahat, Johor, Malaysia
[2] Research Centre for Soft Soil, University of Tun Hussein Onn, Malaysia
[3]University Sains Malaysia

Abstract. Improvement on soft clay by the installation of stone column is one of the most popular methods followed worldwide. Different analytical and numerical solutions have already been developed for understanding the load transfer mechanism of soft soil reinforced with stone column. This study investigated a bearing capacity of the gravelly sand column, installed in soft clay bed at 15kpa of undrained shear strength. The column variable of length and diameter ratio at 7, 8 and 9 were evaluated. On top of that, the combination of two diameters in single column was tested and the uniform diameter was used as a control. In the numerical analysis, Mohr-coulomb model was adopted in the idealization of the behaviour of the gravelly sand column and soft clay materials. The results revealed that the optimum design that gave the highest loading capacity of the combination l1=l2 of column diameter was the length and diameter ratio of 8.

1 Introduction

There are a numerous amount of methods for ground improvement were developed throughout the world. Amongst the wide range of ground improvement methods is stone column, which is often considered as cost-effective and environmentally friendly comparing to other types of stiffening column techniques [1]. The stone column technique of ground improvement has been proven to be successful in improving slope stability of both embankments and natural slopes, via increasing ground bearing capacity and accelerating the period of settlement. The stone column length has been always designed to be reached at stiff layer [2]. In cases of very deep weak soil, which has shear strength below 30 kPa, the application of stone column is not economics [3]. In recent years, the stiffening column technique has been adopted the floating stone column, when the soft ground is too deep. The bearing capacity was remained constant for the floating stone column at length to diameter ratio (L/D) of 10 [4-6]. According to Sivakumar et al [7] and Tekin and Ergun [8], the effective length to diameter ratio of the column was between 7 and 9. However, longer

*Corresponding author: aziman@uthm.edu.my

column may be needed to reduce the settlement. The ultimate bearing capacity, settlement, and general stability are the parameters to identify the stone column performance.

Numerical analysis enables the resolution of complex problems where the combination of the analytical solutions and experimental measurements can be established. The numerical models were developed using the finite element analysis software such as plaxis. Regardless of the development of more selected constitutive models for soil, the Mohr–Coulomb yield criterion remains a popular choice for geotechnical analysis due to its ease and simplicity to be used by practising engineers [9]. However, Mohr-Coulomb model is the most common constitutive laws used in finite element methods [10]. Al-Saidi [11] studied the behaviour of stone column using the axisymmetric single stone column using (unit cell) concept in the finite element programs. The result showed the failure depth, where the bulging zone occurred between 2 to 3 times the diameter. Gunduz [12] used the plaxis software and adopted a hardening soil and Mohr-coulomb models to analyse the deformation for 50 years period. Both model analyses showed similar results. Shien [13] used Mohr-coulomb model which indicated that the plastic zone was at 0.6 times of the length. Mats [14] indicated that the Mohr-Coulomb material model is inadequate to predict the primary consolidation of soft clay, and thus needs to refer to actual deformation measurements. Meanwhile, Ruben [15] investigated the behaviour of single stone columns via adopting Mohr-Coulomb and modifying Drucker-Prager models which resulted the column diameter influenced the loading capacity and the optimum length to diameter ratio was between 7 and 9. Therefore, this study has considered the design of the floating stone column at length to diameter ratio between 7 and 9. On top of that, the combination of two diameters in single column was tested and the uniform diameter was used as a control. In the numerical analysis, Mohr-coulomb drained and undrained model materials were adopted in the idealization of the behaviour of the gravelly sand column and soft clay materials.

2 Methodology

The compression modelling was performed using the plaxis 2-dimensional program. The model was based on the scaled-down at 25 times similar to Kelly [5]. The model of 84 mm wide and 450 mm high represents the unit cell (De) containing a soft clay as shown in Fig 1(a). It was deemed to represent a field scenario where gravelly sand column was installed in a triangular pattern (De=1.05S) where the spacing (S) at 80 mm. The uniform diameter of the column as a control shape was installed at the centre of the soft clay with a diameter (D) at 44 mm to achieve a replacement ratio in the average (as) between 0.25 and 0.50.

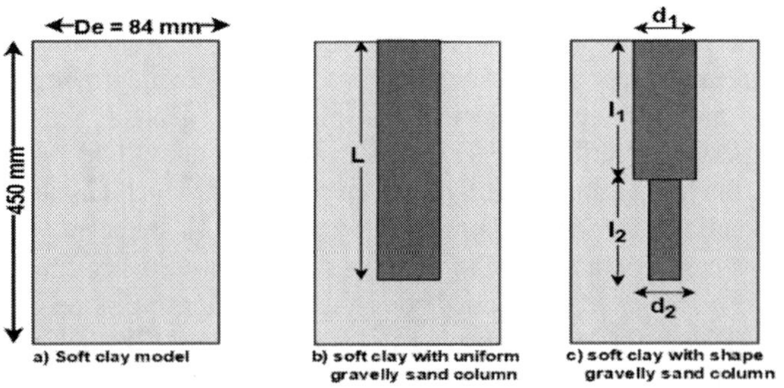

Fig. 1. Geometry of models

The various lengths (L) of the column were adopted at 308, 352 and 396 mm according to the ratio of length and diameter (L/D) between 7, 8 and 9 respectively as shown in Fig. 1(b). The combination of two column diameters (d1 and d2) and lengths (l1 and l2) was installed at the centre of the soft clay as shown in Fig. 1 (c). The first diameter (d1) is similar with control shape and for the second column diameter (d2) is smaller at 44 mm and 20mm respectively. There were 9 variables adopted at first diameter column lengths (l1) of 102.6, 154, 205.4, 117.4, 176, 234.6, 132, 198, 264 mm and second diameter column lengths (l2) at 205.4, 154, 102.6, 234.6, 176, 117.4, 264, 198, 132 mm. Table 1 and 2 summarize the parameters adopted in the numerical analysis.

Table 1. Uniform column diameter

No	Length (L), mm	Diameter (D), mm	Ratio (L:D)
1	308	44	7:1
2	352		8:1
3	396		9:1

Table 2. Combination of 2 columns diameter

No	Length (l1), mm	Length (l2), mm	Ratio (l1:l2)	Total length, (L) mm	Diameter (d1), mm	Diameter (d2), mm	Ratio (L:d1)
1	102.6	205.4	1:2	308	44	20	7:1
2	154	154	1:1				
3	205.4	102.6	2:1				
4	117.4	234.6	1:2	352			8:1
5	176	176	1:1				
6	234.6	117.4	2:1				
7	132	264	1:2	396			9:1
8	198	198	1:1				
9	264	132	2:1				

3Material properties

Kaolin from Associated Kaolin Industries was the main material used to form a soft clay bed. This material was used to ensure a high degree of sample control and reduce variability throughout the test program. Its properties were determined during the laboratory tests as summarized in Table 3. The material for constructing the column was obtained by sieving a fine quarry aggregate in the laboratory. The suitable grain sizes were based on the scaled-down prototype grain diameter at 25. For this study, a prototype of the bottom feed technique of constructing sand columns was used with a typical grain size diameter of 10 to 40 mm and thus, the scaled-down grain size diameter was between 0.4 mm and 1.6 mm. Properties of the gravelly sand were determined during the laboratory tests and summarized in Table 3.

4 Constitutive model

The Mohr - coulomb model is widely used in finite element analysis of geotechnical applications due to its simplicity and sufficient accuracy. The Mohr-coulomb model for drained and undrained condition was adopted to model the behaviour of the gravelly sand and soft clay materials respectively. Meanwhile, an Elasto-perfectly plastic behaviour with the Mohr-coulomb failure criterion was adopted for the material response under loading conditions. An elasto-perfectly plastic behaviour with the Mohr-Coulomb failure criterion was adopted for the material response under loading conditions. In terms of the principal stresses (σ1 and σ3), the failure envelope of the Mohr-Coulomb model is defined using (1).

$$(\sigma 1 - \sigma 3) + (\sigma 1 + \sigma 3) \sin\varphi - 2c \cos \varphi = 0 \quad (1)$$

where c and φ are the cohesion and the angle of internal friction of the material respectively.

Table 3. Properties of soft clay and gravelly sand materials

Properties	Unit	Soft clay	Gravelly sand
Specific gravity	---	2.65	2.70
Moisture content	%	42	9
Undrained shear strength	kPa	15	N.G
Cohesion	N/mm2	0.00887	0.000376
Angle of internal friction	o	31.2	47.9
Dilatancy angle	o	10.4	15.97
Permeability	mm/s	$7e^{-6}$	0.012
Young modulus	N/mm2	1.8	40
Poisson ratio	---	0.35	0.30
Unit weight saturation	N/mm3	$1.738\ e^{-5}$	$2.138\ e^{-5}$
Unit weight unsaturation	N/mm3	$1.224e^{-5}$	$2.04\ e^{-5}$

5 Finite element analysis

5.1 Program used

Plaxis as the most used software packages in geotechnical engineering was employed for finite element analysis. Plaxis 2D is a two dimensional finite element program interface but currently restricted to 2D plane strain and axisymmetric problems. The program can deal with undrained and drained consolidation analysis of two dimensions (with axisymmetric loading).

5.2 Finite element geometry, boundary conditions

The problem of axially loaded sand column surrounded with soft clay is an axisymmetric one, because the materials are homogenous. Axisymmetric finite element analysis was carried out for the assessment of the load settlement of the soft clay and gravelly sand column. Due to symmetry, only half of the axisymmetric problem is considered. The boundary conditions of the axisymmetric problem domain are shear free with no radial movement at the lateral sides and prevent the bottom boundary from both radial and vertical movements as shown in Fig. 2. Fifteen nodes elements are used to model the soft clay and gravelly sand column. An example of the domain meshes generated from the analyses is shown in Fig. 3.

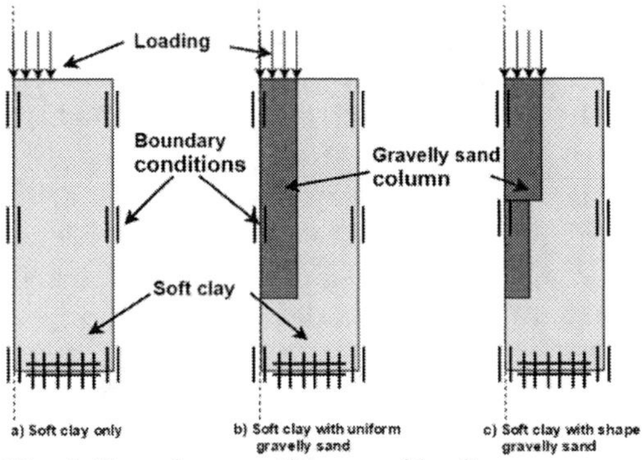

Fig. 2. Boundary conditions and loading

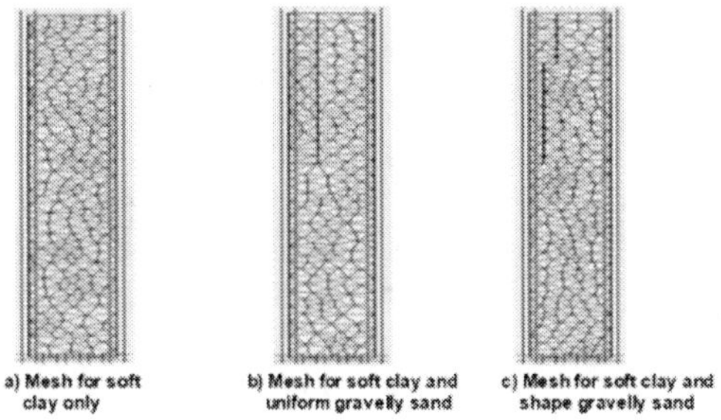

Fig. 3. Mesh of the models

6 Results and discussion

Fig. 4 shows the relation between the load and the displacement for three cases of soft clay only, uniform column diameter, and the combination of two column diameters (l1 and l2) when the ratio of length and diameter (L/D) is 7.

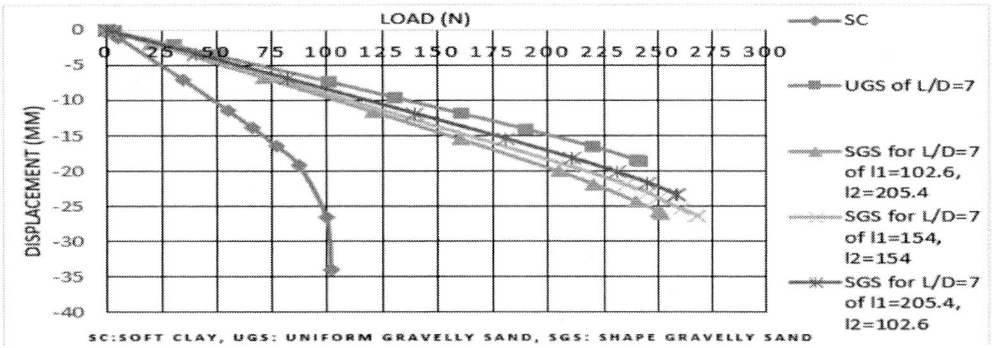

Fig. 4. Load vs. displacement for column at ratio (L/D=7) 7:1

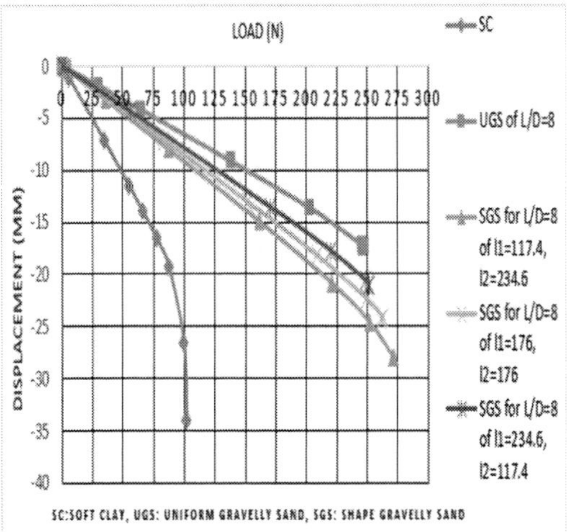

Fig. 5. Load vs. displacement for the column at ratio (L/D) 8:1

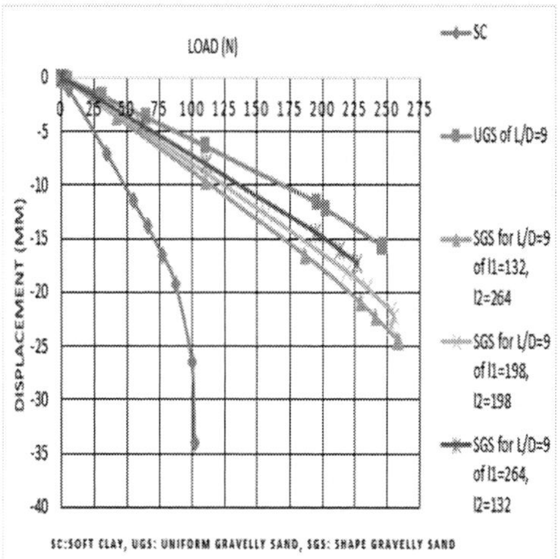

Fig. 6. Load vs. displacement for the column at ratio (L/D) 9:1

The loading capacity increased more than twice with the column installation at uniform and a combination of two diameters. The highest loading capacity for column length at 308 mm was the uniform column diameter and a combination of two column diameters (l1 and l2) at ratio 1:1 of 268.506 N.

In the case of the column length at 352 mm and diameter 44 mm at a ratio (L/D) 8:1 as shown in Fig. 5, the loading capacity increased more than twice with the column installation at uniform and a combination of two diameters. The highest loading capacity for column length of 352 mm was the combination of two column diameters (l1 and l2) at ratio 1:2 of 271.543 N.

In the case of the column length at 396 mm and diameter 44 mm at a ratio (L/D) 9:1 as shown in Fig. 6, the loading capacity increased more than twice with the column installation at uniform and a combination of two diameters. The highest loading capacity for column length of 396 mm was the combination of two column diameters (l1 and l2) at ratio 1:2 of 258.865N.

7 Conclusion

From the finite element analysis, the following conclusions can be drawn:

- The loading capacity without and with the column is more than twice, indicating that column is able to improve the bearing capacity of soft clay.
- The loading capacity of the uniform column and the combination of two column diameters are comparatively equal. However, in many cases, the highest loading capacity was a combination of two column diameters.
- The column ratio length and diameter (L/D= 8) and the combination of two column diameters (l1 and l2) at ratio 1:2 (L/D=8 of l1=176, l2=176) have the highest loading capacity of 271.543N.
- There is unbeneficial to increase the length to diameter ratio more than 8.

The combination of two column diameters, where the larger column diameter at the top is practical to be implemented and able to reduce the volume of aggregate to build the column. It is recommended to explore the laboratory and numerical modelling of bulging and punching for the different column diameters.

The authors would like to thank the University Tun Hussein Onn Malaysia (UTHM), the Office for Research, Innovation, Commercialization and Consultancy Management (ORICC) for their financial support on GIPS phase 1/2015 vot. U270.

References

[1] E. Venere, *Ground-improvement methods might protect against earthquakes*, Purdue University News, (March 19, 2014)

[2] H.J. Priebe, *The design of vibro replacement*, Proceedings of the Institution of Civil Engineers, Ground engineering, Technical paper GT 037-13 E 31–37, (1995)

[3] A. Salahi, H. Niroumand, K.A. Kassim, *Evaluation of stone columns versus liquefaction phenomenon*, Department of Civil Engineering, Faculty of Engineering, Islamic A. University, Tehran, Iran, (2015)

[4] J.A Black, V. Sivakumar, M.R Madhav, G.A Hamill, *Reinforced stone columns in weak deposits laboratory model study*, Journal of Geotechnical and Geoenvironmental Engineering, **133**(9), 1154-1161, (2007)

[5] K. Micheál, *Numerical modelling of small groups of stone columns*, College of Engineering and Informatics, National University of Ireland, Galway, (2012)

[6] P. Kelly, *Soil structure interaction and group mechanics of vibrated stone column foundation*, Department of civil and structural engineering, University of Sheffield, (2014)
[7] V. Sivakumar, D. Glynn, J.A Black, M. Neill, A laboratory model study of the performance of vibrated stone columns in soft clay, *Proceedings 14th European Conference on Soil Mechanics and Geotechnical Engineering*, Madrid, (2007)
[8] M. Tekin, M.U. Ergun, A model study of strains under footings supported by floating and end-bearing granular columns, *18th ICSMGE*, Paris, (2013)
[9] A.J. Abbo , A.V. Lyamin, S.W. Sloan, J.P. Hambleton, *A C2 continuous approximation to the Mohr–Coulomb yield surface centre for geotechnical and materials modelling*, University of Newcastle, Callaghan, (2011)
[10] H.S. Yu, *A closed-form solution of stiffness matrix for Tresca and Mohr-Coulomb plasticity models*, The University of Newcastle, Australia, (2003)
[11] A.A. Al-Saidi, *The use of stone columns to improve soft soils*, MSc Thesis, University of Baghdad, Iraq, (2000)
[12] B. Gunduz, *Analysis of settlements of test embankments during 50 years - A Comparison between field measurements and numerical analysis*, Lund University, Sweden, (2010)
[13] N.K. Shien, *Numerical study of floating stone columns*, National University of Singapore, Singapore, (2013)
[14] M. Kahlström, *Plaxis 2D comparison of Mohr-Coulomb and soft soil material models*, Luleå University of Technology, Sweden, (2013)
[15] R.A. Gnandji, *Experimental and numerical analyses of the behavior of rammed stone columns installed in a South African Soft Soil*, University of Cape Town, South Africa, (2014).

DOI: 10.1051/matecconf/201710307006

Microzonation Analysis of Cohesionless and Cohesive Soil

Choy Soon Tan[1,*], *Yong Chang* Koh[1] , and *Aminaton* Marto[2]

[1]Faculty of Engineering, UCSI University, 56000 Cheras, Kuala Lumpur, Malaysia
[2]Faculty of Civil Engineering, Universiti Teknologi Malaysia, 81310 UTM Skudai, Malaysia

Abstract. Urban seismic risk is a continuous worldwide issue, numerous researchers are putting great effort in dealing with how to minimise the level of the threat. The only way to minimise the social and economic consequences caused but the seismic risk is through comprehensive earthquake scenario analysis such as ground response analysis. This paper intends to examine the characteristic of shear wave velocity and peak ground acceleration on cohesionless and cohesiveness soil. In order to examine the characteristic of shear wave velocity and peak ground acceleration on cohesionless and cohesiveness soil, ground response analysis was performed using Nonlinear Earthquake Site Response Analysis (NERA) and Equivalent-linear Earthquake Site Response Analysis (EERA). The value of ground acceleration was initially high at bedrock and vanishes during the propagation process. It is thus, the measured acceleration at surface is therefore much lower as compare to at bedrock. Result shows that seismic waves can travel faster in harder soil as compared to softer soil. Cohesive soil contributes more to the shaking amplification than cohesionless soil such as sand and harder soil. This is known as local site effect. The typical example is the Mexico Earthquake that happened in 1985. As conclusion, peak ground acceleration for cohesive soil is higher than in cohesionless soil.

1 Introduction

Urban seismic risk is a continuous worldwide issue, numerous researchers are putting great effort in dealing with how to minimise the level of the threat. Earthquake as a natural hazard is inevitable, more than billions of people throughout the world are subjected to this significant hazard. The only way to minimise the social and economic consequences caused but the seismic risk is through comprehensive earthquake scenario analysis such as ground response analysis. The occurrence of the earthquake is due to the release of seismic energy in the form of waveform from earth crust. The transmission of the wave form is largely depending on the density and elasticity. Figure 1 illustrates the seismic wave propagation from the epicentre to the local site. The seismic wave could be either amplified or de-amplified depending on the local site effects which include soil type, soil density, thickness and etc [1]. Most researchers [2-4] found that the shear wave velocity (Vs) is higher when

* Corresponding author: cstan@ucsiuniversity.edu.my

the density is higher. Ansal et al. [5] and Cox et al. [6] stated that the shear wave velocity in clay soil is lower as compared to as in sand. For the thickness of soil, researchers [7-10] stated that the shear wave velocity is increasing when the soil thickness is increased.

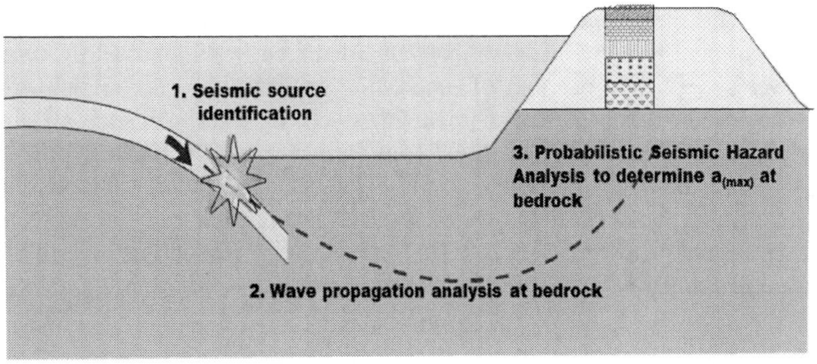

Fig. 1. Wave propagation from bedrock to surface. [11]

In general, studies have shown that there are two important geologic characteristics which affect the levels of ground shaking during an earthquake which are the softness of the ground, and the thickness of sediments above hard bedrock [12]. For the soil which is soft and thick, the shaking or amplification of the waves produced by an earthquake will be greater. This is because seismic waves can travel faster through hard rock as compared to softer soil. In soft soil, the waves slow down and their amplitude increases which then causes the shaking to be stronger. Thus, this paper intends to examine the characteristic of shear wave velocity and peak ground acceleration on cohesionless and cohesiveness soil.

2 Methodology

In order to examine the characteristic of shear wave velocity and peak ground acceleration on cohesionless and cohesiveness soil, ground response analysis was performed using Nonlinear Earthquake Site Response Analysis (NERA) and Equivalent-linear Earthquake Site Response Analysis (EERA).

There are two main input parameters was needed for the analysis which are the geotechnical data and earthquake data. The geotechnical data was extracted from the site investigation report. A total number of four site investigation reports to represent the soil condition of Johor Bahru were obtained. Altogether 25 borehole logs were used to represent the soil profile that consist mainly cohesionless soil (sand) and cohesive soil (clay). There are a few important information shall be extracted from the site investigation report and to be input to the NERA and EERA software. These inputs are includes Standard Penetration Blow Count (SPT-N), unit weight, soil thickness, soil type. Using correlation proposed by Bowles [5], the unit weigh for each soil layer was estimated. It is important to note that the software required the input of shear wave velocity, which was not included in the site investigation report. Thus, empirical formula that established by Marto et al [13, 14] that represent both type of cohesionless and cohesive soil was used, as follows:

$$V_s = 77.13\ \mathrm{N}^{0.377} \tag{1}$$

In which Vs = shear wave velocity
N = standard penetration count

As the main interest of this work is on the characteristics of ground response parameters on different soil type, the default earthquake data was use in the analysis procedure. There are a few ground motion parameters could be obtained as the output of the software such as acceleration, amplification factor, Fourier spectrum, response spectrum and so on. However, the main output parameter that will be extracted for the comparison purposes in this study is the peak ground acceleration that calculated by the software. Acceleration is the rate of velocity change per unit of time while peak ground acceleration is the largest value of acceleration that will be recorded for a seismic event. In addition, the value of amplification was also been used for the comparison purpose. According to Kokusho [15], the amplification behaviour of the seismic wave that occurs between the ground surface and bedrock is depending on several factors such as soil layer, shear wave velocity, soil density and the damping characteristics of layer. In general, soft soil such as cohesive soil would amplify more as compare to cohesionless soil during the wave propagation process.

3 Results and discussion

This particular section summarised the analysis results. Table 1 and 2 show the example of the geotechnical data of two different site investigation report, mainly cohesive soil and cohesionless soil. It can be seen that the borehole in Table 1 consist mainly clay, which is cohesive soil while Table 2 consist mainly sand, which is cohesionless soil. Based on all these geotechnical data, the ground response analysis was carried out using NERA and EERA. The following subsection will summarise and discuss the ground response parameters such as shear wave velocity, peak ground acceleration and amplification obtained from the analysis. In addition, the comparison between the data obtained from NERA and EERA will be done.

Table 1. Typical geotechnical data of borehole at Taman Century.

Layer Number	Soil Type	Thickness (m)	Unit weight (kN/m^3)	Vs (m/sec)	SPT-N Value
1	Sand	1.5	14.14	77.13	1
2	Sand	1.5	14.14	130.08	4
3	Clay	1.5	17.28	116.71	3
4	Clay	1.5	20.42	183.75	10
5	Clay	1.5	20.42	202.85	13
6	Clay	1.5	17.28	130.08	4
7	Clay	1.5	17.28	116.71	3
8	Clay	1.5	18.85	151.56	6
9	Clay	1.5	18.85	141.49	5
10	Clay	1.5	18.85	160.63	7
11	Clay	1.5	20.42	176.59	9
12	Clay	1.5	20.42	229.33	18
13	Clay	1.5	20.42	238.62	20
14	Sand	1.5	20.42	259.57	25
15	Sand	1.5	20.42	303.95	38
16	Clay	1.5	20.42	337.08	50
17	Clay	1.5	20.42	337.08	50
18	Clay	1.5	20.42	337.08	50
19	Clay	1.5	20.42	337.08	50

Table 2. Typical geotechnical data of borehole at Taman Bayu Puteri.

Layer Number	Soil Type	Thickness (m)	Unit weight (kN/m^3)	Vs (m/sec)	SPT-N Value
1	Sand	1.5	18.85	160.63	7
2	Sand	1.5	17.28	100.16	2
3	Clay	1.5	17.28	130.08	4
4	Clay	1.5	18.85	77.13	1
5	Clay	1.5	20.42	238.62	20
6	Clay	1.5	20.42	183.75	10
7	Clay	1.5	18.85	337.08	50
8	Sand	1.5	18.85	337.08	50
9	Sand	1.5	18.85	337.08	50
10	Sand	1.5	18.85	337.08	50
11	Sand	1.5	18.85	337.08	50

3.1 Shear wave velocity

From Table 1 and 2, it can be seen that the increase of unit weight will increase the value of shear wave velocity. Fig. 2 shows an example of the unit weight and shear wave velocity for the same borehole. In general, the shear wave velocity is increased as the unit weight is increased. This is because as the unit weight is high, meaning the soil is in much compacted state, thus waves can travel faster within the particle. At the same time, the increase of SPT-N value will also cause the increment of the shear wave velocity. This result was also been observed by Maheswari et al. [16]. By way of illustration from Table 1, layer 15 which has a blow counts of 38 achieved a lower shear wave velocity at 303.95 m/s as to be compared to layer 16 which has a blow counts of 50 at 337.08 m/s despite of both having the same unit weight of 20.42 kN/m^3.

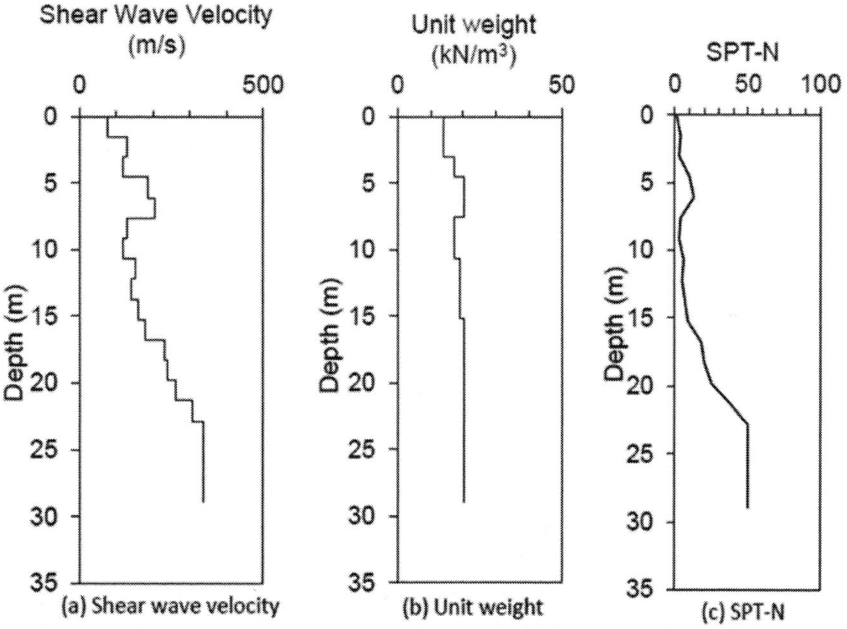

Fig. 2. Value of unit weight and shear wave velocity of borehole at Taman Century.

3.2 Peak ground acceleration

The bedrock acceleration and surface acceleration for all boreholes for all four projects' site investigation in Johor was examined through NERA. Table 3 summarised the range of ground and surface acceleration computed for each of the site investigation report obtained from NERA. From the table, it can be seen that the acceleration was reduced for all of the four sites. The value of ground acceleration was initially high at bedrock and vanishes during the propagation process. It is thus, the measured acceleration at surface is therefore much lower as compare to at bedrock.

Table 3. Results of peak ground acceleration.

Project Site	Range of bedrock acceleration (g)	Range of surface acceleration (g)
Taman Century	0.119-0.183	0.053-0.061
Tanjung Piai	0.016-0.078	0.013-0.056
Indaputra	0.0.09-0.123	0.052-0.071
Taman Bayu Puteri	0.119-0.183	0.053-0.061

To look into more specifically, Table 4 summarised the value of acceleration for the site investigation report of Taman Century. From the table, it shows that the bedrock acceleration for the borehole 3 is dropped from 0.183 g to 0.056 g at surface. This is the highest drop among the five boreholes. On the other hand, borehole 5 has the lowest difference in acceleration which has the value of 0.061 g. This is because seismic waves can travel faster in harder soil as compared to softer soil which is stated from Kumar [12]. In soft soil, the waves slow down and their amplitude increases which causes the shaking to be stronger. The borehole 3 consist mainly the cohesive soil (clay) while borehole 5 consist mainly sand which is cohesionless. In general, cohesionless soil is harder as compare to cohesive soil.

Table 4. Results of peak ground acceleration for Taman Century.

Borehole No.	Bedrock Acceleration (g)	Surface Acceleration (g)	Difference, Δa (g)
1	0.158	0.057	0.101
2	0.158	0.053	0.105
3	0.183	0.056	0.127
4	0.163	0.061	0.102
5	0.119	0.058	0.061

3.3 Amplification factor

The factor of amplification can be defined as the ratio between the peak horizontal acceleration at ground surface and the peak horizontal acceleration at bedrock. Table 5 summarised the range of amplification factor computed for each of the site investigation report obtained from NERA. In general, the higher the surface acceleration, higher the amplification factors the soil will be. The highest amplification factor can be observed for the site of Indaputra while the lowest is at Taman Century. For the range, Taman century have bigger range of the value for the amplification factor as compare to Indaputra. This is because the soil profile of Taman Century consist both cohesionless and cohesive soil. Indaputra site consist only the clay. Thus the range of amplification factor for Indaputra is relatively small as compared to Taman Century. However, it is at the highest value side of it. To explain this observation, Kamaruddin [17] stated that cohesive soil contributes more to the shaking amplification than cohesionless soil such as sand and harder soil. This is known as local site effect. The typical example is the Mexico Earthquake that happened in 1985 [1].

Table 5. Results of amplification factor.

Project Site	Range of Amplification Factor, A_F
Taman Century	0.134-0.691
Tanjung Piai	0.546-0.629
Indaputra	0.536-0.732
Taman Bayu Puteri	0.146-0.556

From Table 5, it can be seen that for this project, the range of the amplification factor is from 0.546g to 0.629g. Also, from the table, it can be seen that borehole number 4 produces the highest amplification factor as the area while borehole number 2 produces the least amplification factor. Being said, a higher acceleration will cause the amplification factor to increase.

3.4 Comparison between NERA and EERA

In order to compare the difference between NERA and EERA, the site investigation report of Taman Century was selected for the analysis. Despite EERA and NERA are having similar system of the software under Microsoft Excel interface, the both software delivers a very similar but not entirely same results. It is necessary to get certain ground response parameters from the software in order to compare the frequency content on the investigated area and also to have a better understanding on the behaviour of the soil on the investigated area. Therefore, amplitude Fourier spectra and response spectra were plotted to observe the difference from both of the software.

3.4.1 Amplitude fourier spectra

Fig. 3 shows the Fourier spectra graph obtained through both NERA and EERA software. In the graph, it can be seen that the Fourier amplitude obtained through EERA is slightly higher as compared to the spectra obtained through NERA. Through EERA, it produced a value of Fourier amplitude of 0.0607 gs while NERA produced a value of Fourier amplitude of 0.0479 gs where both achieved at the frequency of 1.46 Hz.

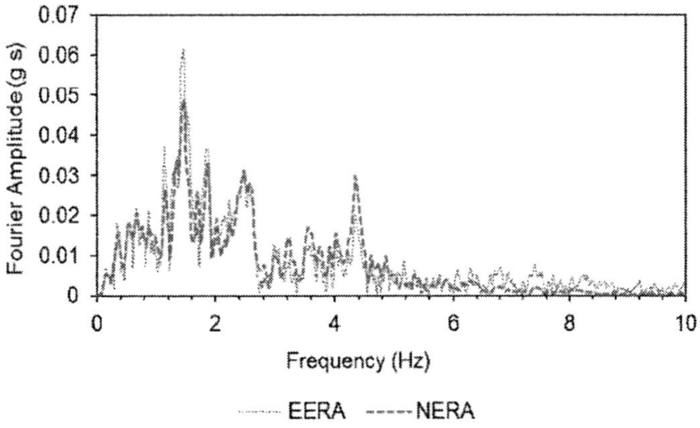

Fig. 3. Amplitude Fourier Spectra comparison from EERA and NERA software.

3.4.2 Response spectra

By plotting a graph on the response spectra from NERA and EERA software, response spectra is measureable. Just like amplitude Fourier spectra, EERA software, too, achieved a generally higher value as compared to NERA software. From Fig. 4, it can be observed that EERA software obtained 0.48 g while NERA software obtained 0.36 g at the same period of time of 0.67 s.

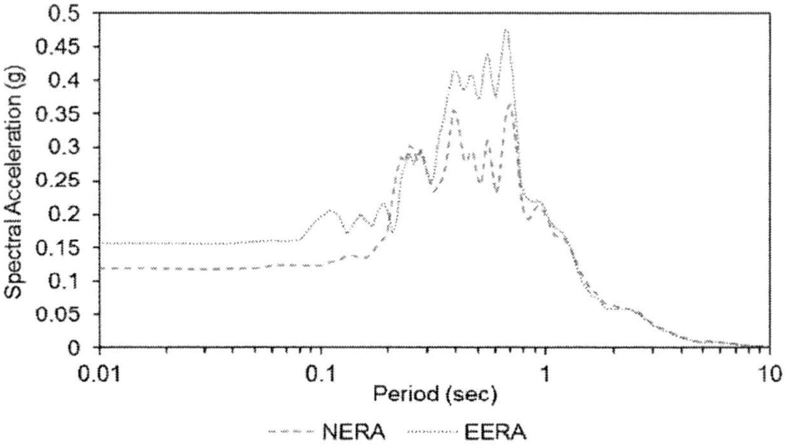

Fig. 4. Response spectra comparison from EERA and NERA software.

4 Conclusion

From the study, the following conclusions could be drawn which includes:

i) Shear wave velocity in cohesive soil is higher than cohesionless soil
ii) Seismic wave will be more amplified compared to cohesionless soil
iii) Peak ground acceleration for cohesive soil is higher than in cohesionless soil

The authors gratefully acknowledged the financial supports by the Ministry of Higher Education (MOHE) Malaysia, under the Fundamental Research Grant Scheme (UTM Vote No.: 4F849).

References

[1] S.L. Kramer, *Geotechnical Earthquake Engineering*, Prentice-Hall, New Jersey, USA (1996)
[2] A. Aksinenko, T. Hofstetter, V. Giller, A. Shvartsburg, D. Giller, I. Dan, M. Kalmanovich, Site effect and microzonation of the coastal plain from Northern Akko to Nahariyya for the Assessment of Earthquake Hazard, **No. 533/545**, 10 (2010)
[3] H.K.E.D. Moustafa, A.E.K.A. El-Aal, S. Salem, Microzonation of El-Fayoum, New City, Egypt, 64-83 (2013)
[4] P. Teves-Costa, I.M. Almeida, P.L. Silva, Microzonation of Lisbon: 1-D theoretical approach, 2579-2596 (2002)
[5] A. Ansal, G. Tönük, A. Kurtulucs, Microzonation for urban planning, 133-152 (2009)
[6] B.R. Cox, J. Bachhuber, E. Rathke, C.M. Wood, R. Dulberg, A. Kottke, S. M. Olson, Shear wave velocity and geology-based seismic microzonation of Port-au-Prince, Haiti, **27**(S1), 67-92 (2011)

[7] M.L. Sharma, J.P. Narayan, K.S. Rao, Seismic microzonation of Delhi region in India, **2043** (2004)
[8] M.S. Fnais, A.M. Al-Amri, K. Abdelrahman, Seismic microzonation and site effect response of Damman and Al Khobar cities in Eastern Saudi Arabia, 1-191 (2012)
[9] R. Motamed, A. Ghalandarzadeh, I. Tawhata, S.H. Tabatabaei, Seismic microzonation and damage assessment of Bam City, Southeastern Iran, **11**(1), 110-132 (2007)
[10] Z. Wang, A technical note on seismic microzonation in the Central United States, **117**(2), 749-756 (2008)
[11] M.A.Adnan, A. Hendriyawan, M. Irsyam, Microzonation maps for Kuala Lumpur and Putrajaya, Malaysian Journal of Civil Engineering, **23**(1), 63-85 (2011)
[12] K. Kumar, *Basic Geotechnical Earthquake Engineering*, 49-61 (2008)
[13] A. Marto, C.S. Tan, F. Kasim, M. Suhatril, A correlation of shear wave velocity and standard penetration resistance, Electronic Journal of Geotechnical Engineering, **18**, 463-471 (2013)
[14] A. Marto, C.S. Tan, T.K. Leong, Universal correlation of shear wave velocity and standard penetration resistance, Electronic Journal of Geotechnical Engineering, **18**, 2727-2738 (2013)
[15] T. Kokusho, K. Sato, Site amplification formula for seismic zonation based on downhole array records during strong earthquakes, 12-17 (2008)
[16] R. Uma Maheswari, A. Boominathan, G.R. Dodagoudar, Use of surface waves in statistical correlations of shear wave velocity and penetration resistance of Chennai soils, Geotechnical and Geology Engineering, **28**(2) ,119-137 (2010)
[17] A.F. Kamarudin, Development of microzonation map of southern industrial and logistic clusters, **80** (2008)

Complex Permittivity to Detect Lead in Soil with Various Salt Forms and Combinations

Samer Al Martini[1,*], and *Julie Q.* Shang[2]

[1] Department of Civil Engineering, College of Engineering, Abu Dhabi University, Abu Dhabi, UAE
[2] Department of Civil Engineering, Western University, London, Ontario, Canada

Abstract. The response of soil-water systems to an external electric field can be described as polarization and conduction. Polarization represents the electric charges stored in a material, whereas conduction is the ability of a material to conduct free charges. The combined effects of polarization and conduction can be expressed in terms of complex permittivity. Since the complex permittivity of a soil-water system is a function of soil properties, this research was undertaken to detect soil contamination using the complex permittivity measurements. The system used in this investigation to measure the complex permittivity of a lead contaminated soil consisted of an Automatic Network Analyzer (ANA), a set of coaxial cables, a personal computer (PC) for data processing and a coaxial sample holder. Two sample holders were first used: long holder and short holder. It was found that the short holder can give reliable and close measurement to the long holder. The short holder was later adopted. Soil samples were prepared in the lab and mixed with lead nitrate solutions and with hybrid blend of lead nitrate and lead chloride solutions. The relationship between the complex permittivity and lead concentration was investigated. A chemical analysis of the pore fluids for all lead contaminated soil samples was conducted to trace the presence of lead in soil. It was found that the complex permittivity is affected by lead concentration in soil, as the real part decreased and the imaginary part increased with an increase in lead concentration. Lead was not detected in the soil pore water for the soil tested, which suggested that lead was immobilized by precipitation and the adsorption mechanism.

1 Introduction

The last century witnessed vast industrial revolution that resulted in global environmental contamination problems. Many industries all over the world, and especially in developed countries, generate wastewater daily, which contains high levels of metals. In the past, at the start of the industrial revolution, any damage that heavy metals could have caused to the environment was not known. Many industries discharged their wastewaters directly to rivers and lakes causing pollution of water and soil. Eventually, the contamination is transferred to humans, wildlife and vegetation. The environmental problem has become an

* Corresponding author: samer.almartini@adu.ac.ae

important issue in the world. Strict environmental regulations have been established in order to protect the environment from pollution. Researchers recently focused on research contamination issues and monitoring systems [1- 4].

Heavy metals in general and lead in particular are the most dangerous contaminants in soil. Heavy metals usually have very low mobility due to their strong affinity for the solid phase [5]. Due to its low mobility, when introduced to soil, lead usually accumulates in the upper layer of the ground surface where vegetables and grass grow. This highlights the danger of lead contamination in several ways including the pathway from vegetation to humans, animals and direct contact [6].

It is essential that lead contamination be monitored. Monitoring includes gathering information about the extent of pollution in the area, so that the environment can be protected controlled, and lead levels maintained within the allowed limits using government regulations and standards. Many countries have put standards and regulations of heavy metals and lead in place. What remains to be seen is how these regulations can be followed and controlled.

The first step is to provide a comprehensive monitoring system which most countries do not have. This is due to the lack of a means of lead monitoring and because this issue has not been given high priority. Consequently, the level of heavy metals contamination, and lead specifically, is not known precisely [7]. It has now become urgent that monitoring systems be developed and installed beginning with regulations and ending with the development of a means to measure lead levels in the ambient environment. The monitoring system includes off situ and in situ systems. Many companies have taken important steps to develop lead monitoring systems. Extensive research work has been done and reported in the literature to achieve this objective.

A soil is a three-phase system (i.e. solids, liquid and gas). Each component of soil has a different dielectric behaviour and thus different dielectric constant values. As a result, the complex permittivity of the soil water mixture is a function of the dielectric properties of all its constituents. Water is present in the soil as free water and adsorbed water. The microwave aquametry studies the interaction between the soil-water system and the incident electromagnetic waves [8]. The complex permittivity of soil was investigated by several researchers to monitor soil contamination [1, 2, 9-11]. The complex permittivity (ε^*) consists of two terms real part and imaginary part.

In the course of the current research, the soil complex permittivity method is investigated and evaluated for detecting lead contamination in soil. The system used to measure the soil complex permittivity consists of coaxial cable, sample holder, analyzer, and computer. The soil is placed in a coaxial holder to measure its complex permittivity. Because several researchers have had success, when using the coaxial holder to characterize soil properties [12, 13], this holder was used in this investigation. Two lead salts (PbCl2, Pb(NO3)2) at different concentrations were mixed with soil. The lead concentrations were selected based on Ontario Ministry of Environment and Energy [11] for soil (200 ppm – 1000ppm). The lowest concentration was chosen according to the MOEE Guideline [14] for residential areas, while the highest concentration was based on the allowable lead concentration for industrial areas. The objective of this paper was to investigate the relationship between soil complex permittivity and lead concentrations in various forms, which could be used as an indicator to detect soil contamination with lead.

2 Experimental program

The Halton Till soil was used in this investigation. This soil is a brown to gray clayey silt with illite, chlorite, and smectite being the major clay minerals. The non-clay constituents include quartz, carbonates, and feldspar [8]. Lead nitrate salt (Pb (NO3)2) was dissolved in water at five different concentrations. In addition, multiple salt solutions were prepared using two lead salts (($Pb(NO_3)_2$) and ($PbCl_2$)) to study the possibility of tracing lead contamination in various forms. Lead solutions were mixed with soil sample.

The soil was placed in the oven at a temperature of 105 oC for 24 hours. The soil was then pulverized and sieved through sieve #4. The pulverized soil was mixed with distilled water or lead solutions with a moisture content of 19±1% (the mass ratio of water to dry soil). The wet soil was placed in a plastic bag and sealed to prevent evaporation, and it was left overnight. A small specimen was taken from the bag for the moisture content measurement. The empty sample cell was weighed and its weight was recorded. The soil sample was compacted in the cell in three layers with three blows for each layer using a special apparatus designed by Scholte [13]. A Proctor standard hammer was employed to compact the tested soil with a weight of 24.5N and at a fall height of 0.305 providing 76.4 KJ/m3 compacting energy per fall [13]. The compacted sample cell was weighed and the mass was recorded. The bulk density of the sample was calculated.

The system used to measure the complex permittivity of a lead contaminated soil consists of an Automatic Network Analyzer (ANA), a set of coaxial cables, a personal computer (PC) for data processing and a coaxial sample holder (see Fig. 1). The soil sample was installed in the holder for complex permittivity measurements. The measurements were performed quickly to minimize any drying effect so that the measured complex permittivity reflects the real soil conditions. It took experienced personnel less than one minute to perform one measurement.

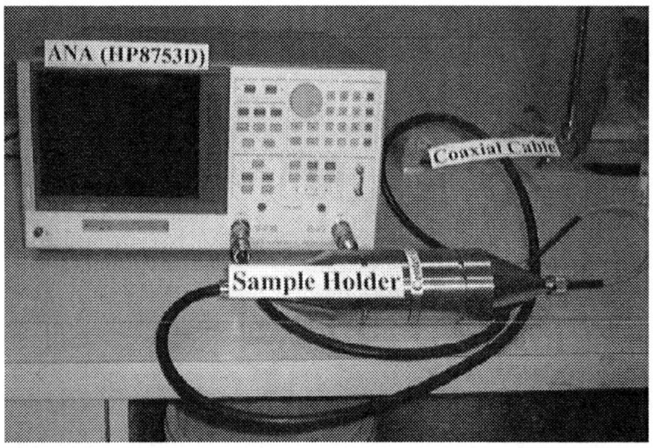

Fig. 1. Complex permittivity measurement system

3 Results and discussion

3.1 Complex permittivity of soil sample mixed with distilled water

Fig. 2 shows six repetitive measurements of the complex permittivity for a soil mixed with distilled water. Fig. 2(a) shows results for real part and Fig. 2 (b) shows results for the imaginary part. These Fig.s show that the six results are highly consistent, which indicates that the system can produce reliable repetitive measurements. The longitudinal resonance is observed at the frequency where a discontinuity in the real and imaginary parts of the complex permittivity occurs. The resonance frequency occurs at a range between 910 and 930 MHz.

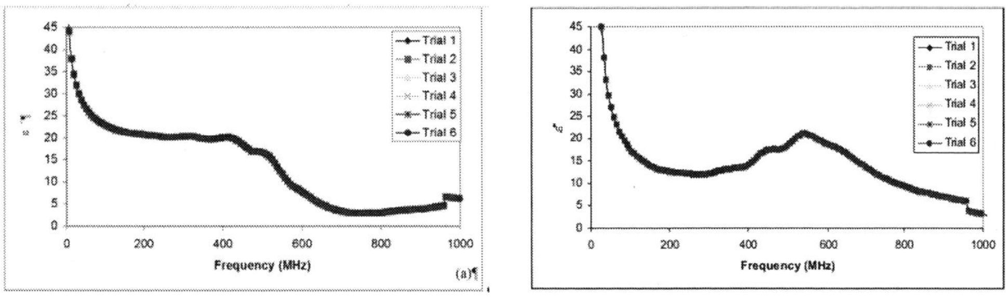

(a) Real part   (b) Imaginary part

Fig. 2. Repeatability of complex permittivity measurements for soil mixed with distilled water using short holder

3.2 Complex permittivity of soil samples mixed with lead nitrate solutions

Five soil samples of Halton Till were prepared with five different concentrations of lead nitrate solution (c=0.8, 1.6, 4.5, 8.1, 14.3 g/l). The complex permittivity of soil samples were measured in both short holder (l = 261 mm) and long holder (l = 321mm). The results for both real and imaginary parts of the complex permittivity were plotted over frequencies 0.3 to 1000 MHz for each compacted soil sample. Fig. 3 (a) shows the real part (long-short holders) of complex permittivity for soil samples mixed with lead nitrate solutions (c=8.1, g/l), and Fig. 3(b) shows the imaginary part for the same soil sample. As shown in Fig. 3(a), the relative permittivity in low frequency was high, which is attributed to the polarization of the electrical double layer [15]. When the polarization of the adsorbed electrical double layer can no longer change quickly enough to reach equilibrium with an applied external alternating electric field, dielectric dispersion occurs [13]. The dielectric dispersion begins where the permittivity starts to drop. As indicated by Scholte [13], the dielectric dispersion usually occurs in frequencies of 200 to 500 MHz for Halton Till. The dielectric dispersion was noticed at frequencies of 395 and 418 MHz for long and short holders, respectively. The relative permittivity was approximately constant over the frequency range 200 to 400 MHz for both long and short holders. This permittivity is known as the static dielectric constant (ε_s). After these frequencies (the dielectric dispersion) the relative permittivity decreased until it reached the lowest values (ε'_{∞}).These Fig.s show that the dielectric dispersion for the short holder occurred at higher frequencies than for the long holder. The loss factor was lower in the short holder than in the long holder (Fig. 3(b)).

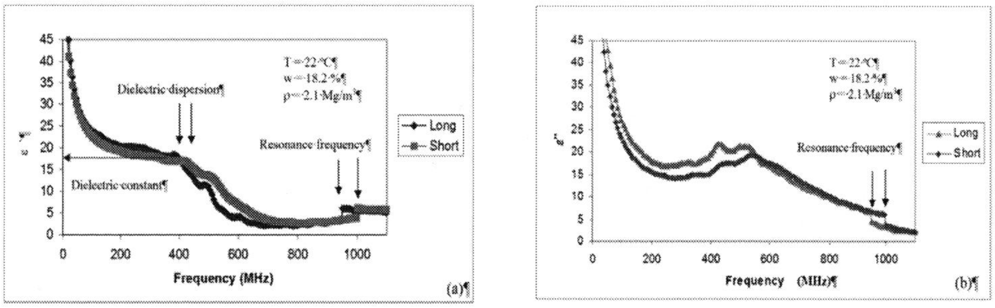

(a) Real part   (b) Imaginary part

Fig. 3. Permittivity measurements of soil sample mixed with lead nitrate solution (c=8.1 g/l) for long and short holders

Fig. 4(a) shows the relationship between the dielectric constant and lead nitrate concentrations for the short holder. This Fig. illustrates that the dielectric constant decreased linearly with an increase of lead nitrate concentration. Fig. 4(b) shows the relationship between the loss factor and the lead nitrate concentrations when the short holder was used. This Fig. shows that the loss factor increased with an increase of lead nitrate concentration. This indicates that using the short holder for measuring the complex permittivity can distinguish the change in lead nitrate concentrations in soil.

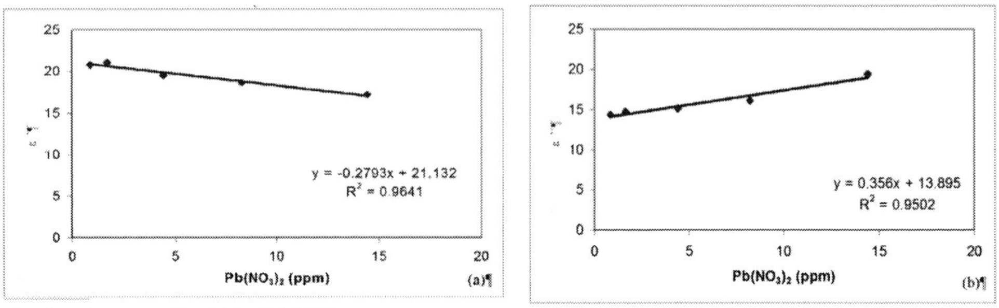

(a) Real part at f = 200-400 MHz (b) Imaginary part at f = 150-250 MHz

Fig. 4. The relationship between the complex permittivity and lead nitrate concentrations (short holder)

3.3 Complex permittivity of soil mixed with PbCl2+Pb(NO3)2 solutions

Fig. 5 shows results for soil sample mixed with $PbCl_2$+$Pb(NO_3)_2$ solutions. Fig. 5a presents the relationship between the lead concentrations and the real relative permittivity at the dispersion frequency (320MHz). As shown in this Fig. 5, the real relative permittivities decreased when salt concentrations increased. The relationship was linear with $R^2 = 0.729$. Fig. 5b presents the relationship between the imaginary part and lead concentrations at a frequency of 200MHz. They are also linearly related. The imaginary part increased with an increase in lead salt with $R^2 = 0.82$. Fig. 5c shows the relationship between the imaginary and relative real permittivity. They are linearly related with $R^2 = 0.82$.

4 Summary and conclusions

Lead contaminated soil samples were prepared using aqueous solutions for two salts (lead nitrate and lead chloride). These two salts were chosen based on their solubility in water. Lead nitrate has very high water solubility and lead chloride has moderate water solubility. In order to detect lead contamination in soil in various forms, aqueous solution with blends of lead nitrate and lead chloride were prepared in five different concentration and mixed with soil. The complex permittivity of the lead contaminated soil samples was measured in order to study the capability of the complex permittivity method for detecting lead in soil water systems. The analysis of the pore fluid chemistry for the lead contaminated soil samples tested herein showed the absence of lead in the squeezed pore water. This suggests that lead is immobilized and retained on the soil particles by the mechanisms of adsorption, precipitation and exchange.

It was observed that complex permittivities are sensitive to changes in lead salt concentrations in soil. In general, the relative permittivity at dielectric dispersion frequency decreased, whereas, the loss factor at 200 MHz increased with an increase in lead concentration. This indicates that complex permittivity can be used to monitor lead concentration in soil. The results also indicate that this method can be used as a fast and nondestructive method to detect lead in contaminated sites.

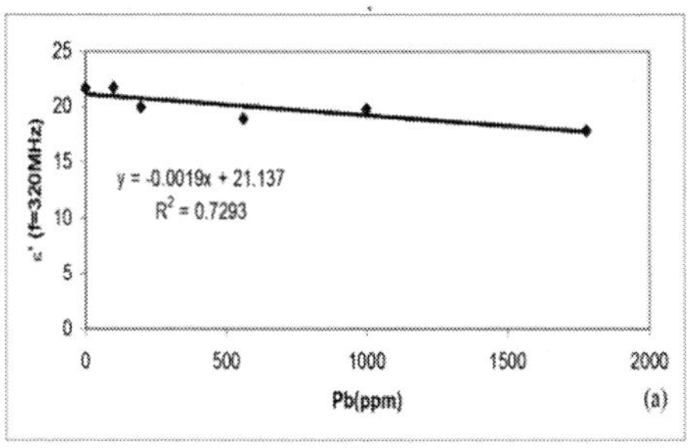

a) ε' at dielectric dispersion (f=320 MHz) versus concentration

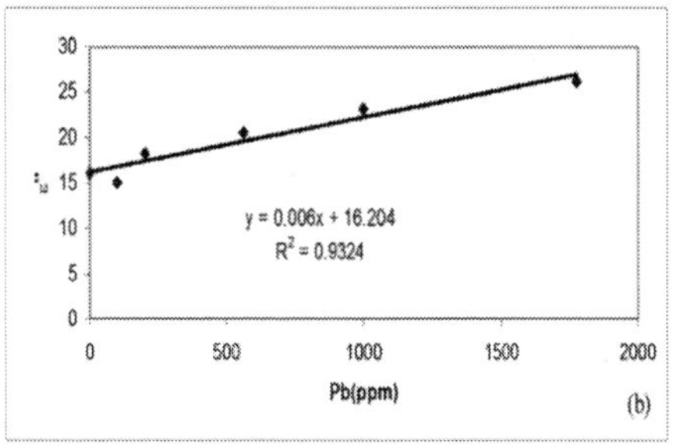

b) ε'' at (f=200 MHz) versus concentration

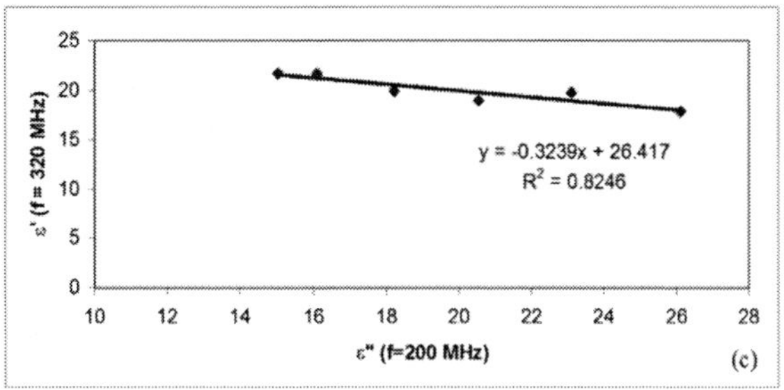

c) ε' at dielectric dispersion (f=320 MHz) versus ε'' at (f=200 MHz)

Fig. 5. Halton Till mixed with lead chloride and lead nitrate ($PbCl_2$+ $Pb(NO_3)_2$) complex permittivity relationships

References

[1] Q.J. Shang, K.R. Rowe, W. J. Scholte, Soil characterization using static conductivity and permittivity, *Canadian Geotechnical Conference*, Calgary, Alberta, 968-975 (2001)

[2] Q.J. Shang, W.J. Scholte, K.R. Rowe , Multiple linear regression of complex permittivity of halton till at frequency range from 200 MHz to 400 MHz, Subsurface Sensing Technologies and Applications, **1**(3), 337-356 (2000)

[3] S. Al-Martini, Q.J. Shang, Complex permittivity measurement system for detection of soil contamination, Alhosn University Journal of Engineering and Applied Science **4**(1), 73-88 (2011)

[4] S. Al-Martini, Q.J. Shang, Complex permittivity of lead in contaminated soils, *Proc. of GeoMontreal 2013*, Montreal, Canada, (2013)

[5] A. Kabata-Pendias, *Trace Elements in Soils and Plants*, 3rd ed., CRC Press, (2001)

[6] R. Wainner, T. Harmon, R. Miziolek, L. Mc Nesby, D. French, Analysis of environmental lead contamination, Spectrochimica Acta Part, B56, 777-793 (2001)

[7] A. Karivelil, *Kinetics of Lead(II) Biosorption by Calcium Alginate Biopolymers*, M.E.Sc. Thesis, The University of Western Ontario, (1995)

[8] A. Kraszewski, *Microwave Aquametry*, IEEE Press, New York, (1997)

[9] Q.J. Shang, K.Y. Lo, I. I. Inculet, Polarization and conduction of clay- water electrical systems, Journal of Geotechnical Engineering, **121**(3), 243-248 (1995)

[10] N.R. Peplinski, F.T. Ulaby, M.C. Dobson, Dielectric properties of soils in the 0.3-1.3 GHz. range, IEEE Transactions on Geoscience and Remote Science, **33**(3), 803-807 (1995)
[11] R.K. Rowe, J. Q. Shang, Y. Xie, Complex permittivity measurement system for detecting soil contamination, Canadian Geotechnical Journal, **38**, 498-506 (2001)
[12] J.A. Umana, *Measurement of dielectric permittivity of pavement materials*, M.E.Sc Thesis, University of Western Ontario, (1998)
[13] J.W. Scholte, *The complex permittivity of compacted halton till*, M.E.Sc Thesis, University of Western Ontario, (1999)
[14] MOE, *Guideline for use at contaminated sites in Ontario*, Queen's Printer for Ontario, PIBS 3161E01, (1996)
[15] J.Q. Shang, J.A. Umana, F.M. Bartlett, J.R. Rossiter, Measurement of complex permittivity of asphalt pavement materials, Journal of Transportation Engineering, **125**(4), 347-356 (1999)

Scale Effect on Mode of Failure and Strength of Offset Rock Joints

Omer Mughieda[1,*], *Kenan* Hazirbaba[2], and *Osama* Mohamed[1]

[1]Department of Civil Engineering, Abu Dhabi University, P.O. Box 59911, Abu Dhabi, UAE
[2]Department of Civil Infrastructure and Environmental Engineering, Khalifa University Abu Dhabi, UAE

Abstract. As a sustainable construction material, the use of rock has increased significantly. In this experimental study, the scale effect on failure mechanisms and compressive strength of rock blocks was investigated. Samples of rock with non-persistent offset joints were subjected to uniaxial loading. The angle of orientation of the rock bridge with respect to the applied axial load and the size of the block were studied. Two different block sizes, having dimensions of (63.5 x 28 x 20.3) cm and (30.5 x 15.24 x 10) cm, were tested. The joint inclination angle was maintained at 22.5° in both cases. Also, degree of persistence was kept constant at 0.3 for all tested blocks. However, the offset angle which connects the inner tips of the joints was changed from 30°-90° with an increment of 15°. The results showed a reduction in strength with increasing the size of the sample. This reduction is becoming more significant as the bridge inclination angle increases. This behavior is due to the fact that as the bridge inclination angle increases the mode of failure shifted from shear to tension mode which is more dependent on the size of sample due to the presence of more micro flaws. No effect of block size was noticed on mode of failure for the tested blocks.

1 Introduction

The use of rock in the built-environment has increased significantly as rocks emerge as sustainable construction material. This trend requires engineers to carefully examine the behaviour of rock under loading. Specifically, factors affecting the strength of rock and contributing to failure need to be evaluated and quantified for design/application purposes. The scale effect on strength of rock mass with persistent rock joints has been studied previously by many researchers [1, 2] and they indicated the existent of size effect on the strength of rock mass

The exact behavior of non-persistent rock joints at failure under uniaxial or biaxial loading condition is not well understood due to the interaction between the intact rock segments and the joint segments. The interaction between the intact rock and the joints depends on many parameters such as the geometry of the joint and bridge segments, the slope of the joint segment, the stiffness of the rock among others [3-10].

* Corresponding author: omer.mughieda@adu.ac.ae

Generally, the process of failure begins with the initiation of cracks perpendicular to the joint segment which then becomes parallel to the principal loading axis. These cracks are called wing cracks. Wing cracks are followed by another type of cracks called secondary cracks that initiate from the inner joint segment tips and propagate toward each other and meet across the intact rock segment causing the coalescence of the joint segments leading eventually to failure of the rock. Propagation and coalescence of the secondary cracks are very quick compared to that of the wing cracks.

This current work is a continuation to the previous work that has been done by the first author and his co-workers. Mughieda and Alzoubi [11] have studied the mode of failure of offset non-persistent rock joints while the joint inclination angle was kept at 45^0 in all blocks tested and Mughieda and Khawaldeh [12, 13] have studied the scale effect on fracture mechanisms and strength of blocks with non offset (planar) rock joints with bridge inclination angles of 67.5^o and 45. It is the first time to investigate the behaviour of offset non-persistent rock joints with joint inclination angle of 22.5^o.

2 Specimen preparation and testing

2.1 Specimen preparation

In this study the model material used was made of mixing silica sand, ordinary Portland cement type I and water in 72%, 16%, and 12% proportion by weight respectively. In order to obtain grain to grain contact that cause low tensile strength and high friction angle, high sand content and low cement content were used.

The friction angle of the model material determined from direct shear tests was 33°. The ratio of unconfined compressive strength to the tensile strength for material was 12 and the ratio of secant modulus to the unconfined compressive strength was 1100.

The procedure developed by Jamil [4] for preparing open non-persistent joints was used in this research with some modification. The sample preparation mold and set-up is shown in Fig. 1. Each sample was cured for 28 days prior to testing. It is important to note that wooden boards were placed on all sides of the sample and secured during handling of the samples to avoid splitting of the jointed sample. A typical sample dimensions are shown in Fig. 2.

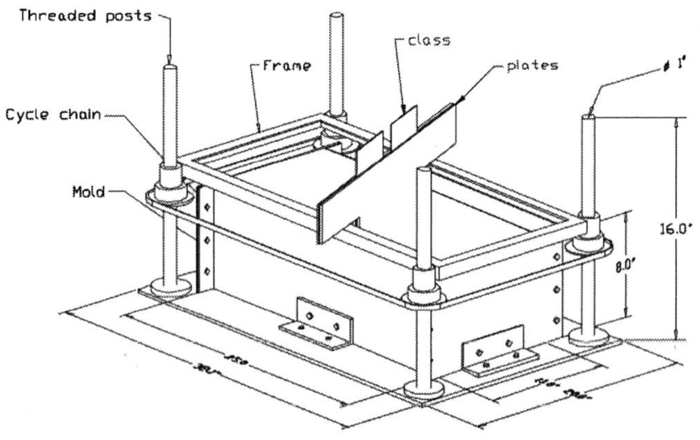

Fig. 1. Sample preparation mold and set-up for making open joints

DOI: 10.1051/matecconf/201710307008

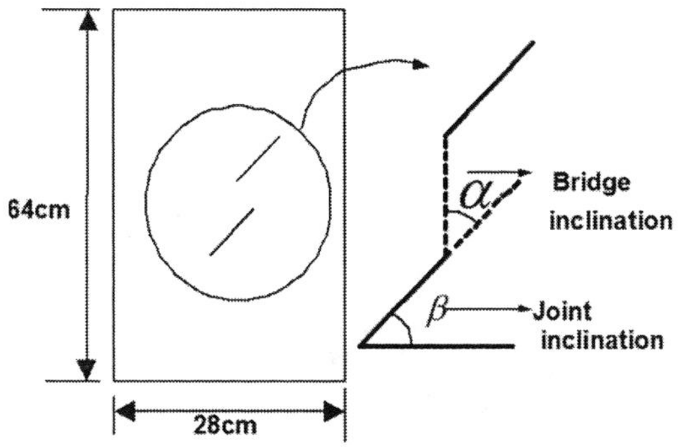

Fig. 2. Geometry of the specimens and pre-existing cracks

2.2 Test machine, instrumentation and recording

The tests were performed using a 2000kN universal compression machine. The load applied on the sample was measured by load cells installed on the testing machine. The total deformation and the displacements in the vicinity of the joint, were measured by Linear Variable Differential Transducers (LVDT) with a maximum range ±5mm (sensitivity ±0.001mm). Readings of the Load cells and LVDTs were recorded and collected by using a Hewlett Packard data acquisition system.

3 Results and discussion

3.1 Mode of failure

3.1.1 Slightly offset joints (β=22.5°, α=30°)

The failure started with the initiation and propagation of the wing crack the wing first and then the secondary cracks initiated at the tips of the joints and propagated to coalesce at a point in the bridge causing sample failure. Crushed and pulverized materials have been noticed on the failure surface. Shear displacement was noticed at the tips of the ligament between the joints. No scale effect was noticed on mode failure. The failed specimen and sketch of the wing and secondary cracks are shown in Fig. 3.

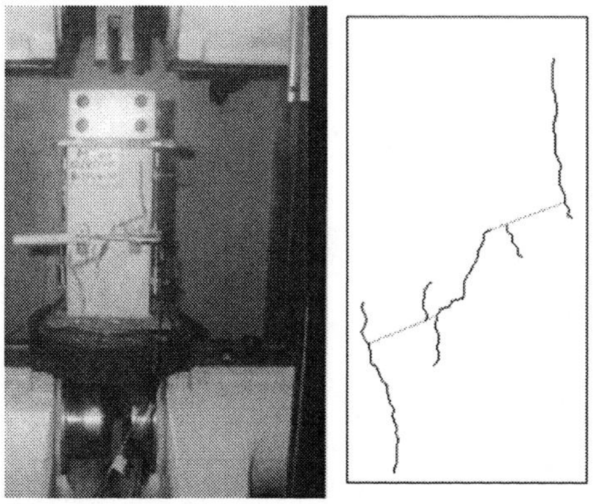

Fig. 3. Crack growth in 22.5°-30° flaws

 DOI: 10.1051/matecconf/201710307008

3.1.2 Offset joints (β= 22.5°, α= 45°)

In these blocks wing cracks propagated from the external flaw tips only and kept growing till the boundary of the block. Coalescence occurred through the growth of secondary crack matching through the flaws. The mode of failure in both sizes was identical. However size shows slight effect on the strength of the joint. The failed specimen and sketch of the wing and secondary cracks are shown in Fig. 4.

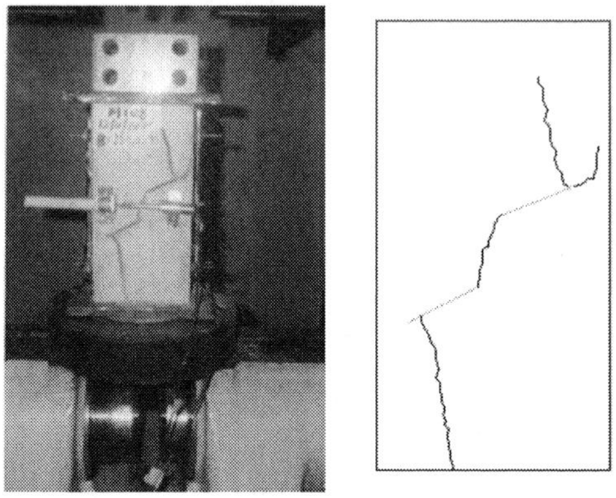

Fig. 4. Crack growth in 22.5°-45° flaws

3.1.3 Offset joints (β= 22.5°, α= 60°)

In both block sizes the failure initiated by wing cracks propagated from the external flaw tips only, extended to a certain distance from the tips of the joints and opened. Coalescence occurred through the growth of secondary crack matching through the flaws. No scale effect was noticed on mode of failure while strength has affected by the block size. The failed specimen and sketch of the wing and secondary cracks are shown in Fig. 5.

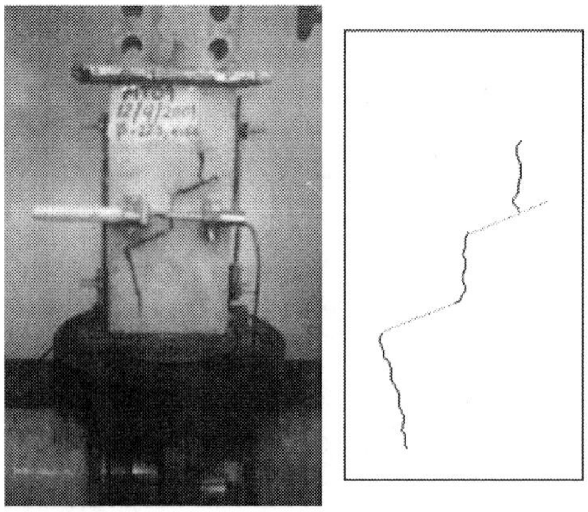

Fig. 5. Crack growth in 22.5°-60° flaws

3.1.4 Overlapped joints (β= 22.5°, α= 90°)

In these blocks failure started as wing cracks propagated from the external and internal flaw tips, they extended and opened. Coalescence occurred through two wing cracks matching through the flaws. The failed specimen and sketch of the wing and secondary cracks are

DOI: 10.1051/matecconf/201710307008

shown in Fig. 6. The results of mode of failure of all tested samples are summarized in Table 1.

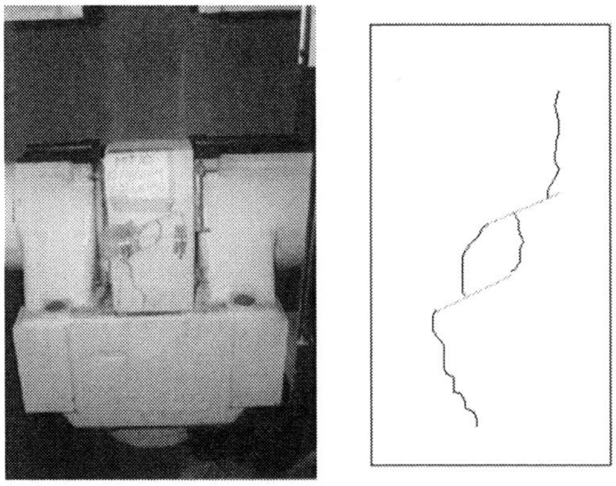

Fig. 6. Crack growth in 22.5°-90° flaws

Table 1. Mode of block failure with different bridge inclinations

	30°	45°	60°	90°
Large sample	Tensile + Shear failure	Tensile + Shear failure	Tensile failure	Tensile failure
Small sample	Tensile + Shear failure	Tensile + Shear failure	Tensile failure	Tensile failure

3.2 Effect of scale on strength

The current research showed that the scale has slight effect on the strength on the offset non persistent rock joints. Also, the results showed that this effect increased as the bridge inclination angle increases. This behaviour is understood as the mode of failure changes from shear to tension failure as bridge inclination angle increases. As the size of the block increases the intensity of micro flaws also increases which resulted in decreasing the tensile strength of the rock. To eliminate the effect of variation in curing time for the samples, variation in mixes for the samples, and material sources, strength of each sample was normalized by dividing it by the average unconfined compressive strength (f'_c) of the cylinders for each sample. The results of normalized uniaxial compressive strength of the large samples and small samples tested are shown in Table 2 and Table 3 respectively.

Table 2. Uniaxial compressive strength (MPa) of large samples

Large sample	**30°**	**45°**	**60°**	**90°**
σ_{1f} **(MPa)**	11	5	6	7
σ_{1f} / f_c'	0.92	0.68	0.66	0.58

Table 3. Uniaxial compressive strength (MPa) of small samples

Small Sample	**30°**	**45°**	**60°**	**90°**
σ_{1f} **(MPa)**	13.76	7.46	12.52	12.4
σ_{1f}/f_c'	0.98	0.918	0.84	0.81

4 Conclusions

The following conclusions may be drawn from the presented research:

i) Mode of failure of Rock Bridge varies from shear to tensile failure as the bridge inclination increases. For slightly offset joint ($\alpha = 30^o$ or 45°) the failure was due to combined shear and tension failure while it was purely tension for overlapped joints ($\alpha = 90^o$).

ii) Wing cracks (Tensile cracks) grew first, and were followed by secondary cracks (Shear cracks) for all the geometries tested, and for both sizes.

iii) Size has small or no effect on the mode of bridge failure.

iv) Strength is a scale dependent property. As the size of sample increases the strength decreases. The influence of scale becomes more significant as the bridge inclination angle increases.

References

[1] S.C. Bandis, A. Lumsden and N. Barton, Experimental studies of scale effects on the shear behavior of rock joints, Int. J. of Rock Mechanics and Mining Sciences, **18**, 1-21, (1981)

[2] T. Blejwas and F. Hansen, Scale in the shear behavior of joints in welded tuff, *Proc. Loen Conf. Rock Joints*, Norway, (1990)

[3] O. Reyes and H.H. Einstein, *Stochastic and centrifuge modeling of jointed rock, Part I-fracturing of jointed rock*, Final report submitted to the Air Force Office of Scientific Research and Air Force Engineering Services Center, (1991)

[4] S.M. Jamil, *Strength of non-persistent rock joints*, PhD Thesis, University of Illinois at Urbana-Champaign, IL, United States, (1992)

[5] S. Boatang, *Mechanics of fractures and inverting bridges in hard rocks*, Doctoral Thesis, Royal Institute of Technology, (1993)

[6] O. Mughieda, *Failure mechanisms and strength of non-persistent rock joints*, PhD Thesis, University of Illinois at Urbana-Champaign, IL, United States, (1997)

[7] H.C. Wong Robina and K.T. Chau, Crack coalescence in a rock-like material containing two cracks, Int. J. of Rock Mechanics and Mining Sciences, **35**(2),147-164, (1998)

[8] H. Park and A. Bobet, Crack coalescence in specimens with open and closed flaws: A comparison, Int. J. of Rock Mechanics and Mining Sciences, **46**(2), 819-829, (2009)

[9] Z. Xiao-Ping, L.N.Y. Wong, Cracking processes in rock-like material containing a single flaw under uniaxial compression: a numerical study based on parallel bonded-particle model approach, Rock Mechanics and Rock Engineering, **45**(5), 711-737, (2012)

[10] J. Jin, P. Cao and J. Liu, The influence of bridge length on failure patterns of specimens containing two coplanar cracks under uniaxial compression load, Applied Mechanics and Materials,**711**, 410-413, (2015)

[11]O. Mughieda and A. Alzo'ubi, Fracture mechanisms of offset rock joints-A laboratory investigations, Geotechnical and Geological Engineering J., **22**, 545-562, (2004)
[12]O. Mughieda and I. Khawaldeh, Scale effect on engineering properties of open non-persistent rock joints under uniaxial loading, *RockMEC'2004-VIIth Regional Rock Mechanics Symposium*, Sivas, Turkey, (2004)
[13]O. Mughieda and I. khawaledh, Scale effect on fracture mechanisms of non-persistent rock joint under uniaxial loading, Geotechnical Engineering J. of Southeast Geotechnical Society, **37**(2), (2006)

Correlation of JKR Probe with Undrained Shear Strength

Mohd Fairus Yusof[1,2,*], *Md Nor Faliq* Abd Khalid[2], *Saiful Azhar* Ahmad Tajudin[1,2], *Aziman* Madun[1,2], and *Mohd Hazreek* Zainal Abidin[1,2]

[1]Research Center for Soft Soil (RECESS), UTHM, 86400 Parit Raja, Batu Pahat, Johor, Malaysia
[2]Faculty of Civil and Environmental Engineering, UTHM, 86400 Parit Raja, Batu Pahat, Johor, Malaysia

Abstract. The JKR probe test is a popular field test in Malaysia, and is especially used in soft soil areas to determine bearing capacity. However the correlation of undrained shear strength (S_u) and the JKR probe test is not available at this moment. This study was performed to develop an empirical correlation of JKR probe and undrained shear strength of soft to stiff soil. In order to produce the correlation, the JKR probe test and the vane shear test have been performed at Universiti Tun Hussein Onn Malaysia (UTHM). Six JKR probe tests and 12 field vane shear tests were performed in close proximity to obtain good correlation. The results show that undrained shear strength and JKR probe can be correlated as S_u=18$J^{0.3}$. The proposed correlations will serve as a reference for engineers in estimating undrained shear strength, especially in low-risk construction design.

1 Introduction

The JKR probe is a simple site investigation procedure that is relatively inexpensive. The probe is often used in Malaysia, especially for preliminary site investigation, to assess the subsoil layer and the bearing capacity of soils. It can also be used effectively to identify soft or weak layer materials or slip plane failure [1]. The use of the probe test may reduce the number of bore holes required and reduce the overall cost of the project.

JKR probe is slightly different with the Mackintosh probe in terms of cone angle, diameter of rods and couplings, hammer weight and height of hammer free fall, as shown in Table 1. However, for practical application, the results of JKR Probe and Mackintosh Probe may be considered equivalent [1]. The bearing capacity of the soil can be simply obtained by using a correlation graph, as shown in Fig. 1.

2 Correlations of Mackintosh and JKR Probe

The correlation of JKR probe test with undrained shear strength has not been addressed in detail in the literature. However, there have been several correlations proposed from

* Corresponding author: fairus@uthm.edu.my

Mackintosh Probe test [2,3]. The detail of correlations of the undrained shear strength, S_u (kN/m^2) and Mackintosh probe (M) is shown in Table 2.

Table 1. Difference between JKR probe and Mackintosh probe

Equipment	Cone angle (°)	Diameter of rods (mm)	Diameter of coupling (mm)	Weight of hammer (kg)	Height of hammer fall (mm)
JKR Probe	60	12	22	5.0	280
Mackintosh Probe	30	13	24	4.5	300

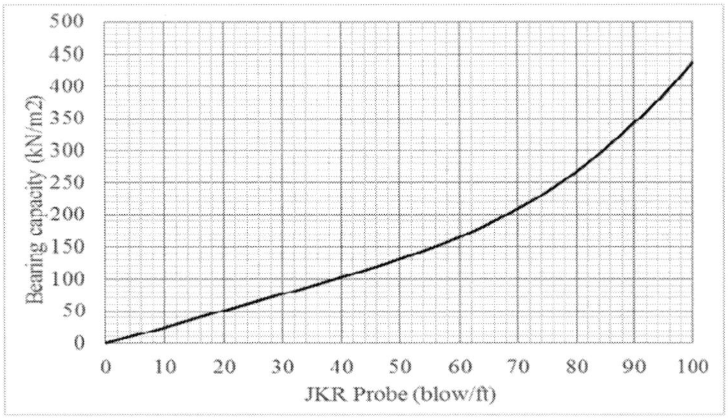

Fig. 1. Estimation chart of bearing capacity

Table 2. Previous correlations between S_u – M (mackintosh probe)

Researcher	Soil types	Location	Equations	Remarks
Hossain & Ali [2]	Soft Clay	Saudi Arabia	S_u (kPa)= K.M	K = 1.59-2.04
	Alluvium	Khozestan province, South of Iran	$S_u = 2.5M$	from vane shear, UU and UCT
Fakher et al. [3]	Alluvium and soft clay	Saudi Arabia and Iran	$S_u = 0.90\ M^{0.96}$	from N and M correlation and Equation 1
	Very soft soil		$S_u = 0.64M + 15.35$	
Fatnanta et al. [6]	Soft soil	Indonesia	$S_u = 0.94M+10.88$	vane shear & Mackintosh probe
	Medium Stiff soil		$S_u = 0.63M+19.75$	

M represents the number of blows for 100 mm penetration of Mackintosh Probe. The first correlation was proposed by Hossain & Ali [2] in 1990 for soft clay in Sabkha, Saudi Arabia, using field vane shear and Mackintosh probe. Later, Fakher et al. [3] proposed two equations. The first equation was derived from the results of a vane shear test, unconsolidated undrained test (UU) and unconfined compression test (UCT) with a Mackintosh probe. The second equation is based on the correlation of standard penetration test (N) and Mackintosh probe (M) by combining data from Sabtan & Shehata [4]. They proposed a new relationship as follows:

DOI: 10.1051/matecconf/201710307009

$$N = 0.15M^{0.96} \tag{1}$$

By adopting Terzaghi et al. [5] equation ($S_u = 6N$), they proposed a correlation of S_u and (M) as:

$$S_u = 0.90\ M^{0.96} \tag{2}$$

In 2013, Fatnanta et al. [6] proposed a correlation based on field vane shear test and Mackintosh probe test performed at Pekan Baru, Indonesia. They categorized correlations for very soft soil, soft soil, and medium stiff soil. The correlation of very soft soil is questionable because the undrained shear strength of very soft soil is higher than that of soft soil. Furthermore, the proposed equations are not valid if M is equal to zero.

3 Data collection

To develop the correlation of Su and M, the JKR probe and the vane shear tests were performed at the Research Centre for Soft Soil (RECESS) in Universiti Tun Hussein Onn Malaysia (UTHM). There are six points for the JKR probe test with maximum depth of 6.9 m, while 12 points of vane shear test have been performed in close proximity (less than 0.3 m) to ensure a good correlation obtained as shown in Fig. 2. The number of blows and undrained shear strength for each 300 mm penetration were recorded for JKR probe test. Additionally, a vane shear test was performed at every 0.6 m depth. The vane shear test was pushed statically to the desired depth and torque was applied at a rate of 6° per minute.

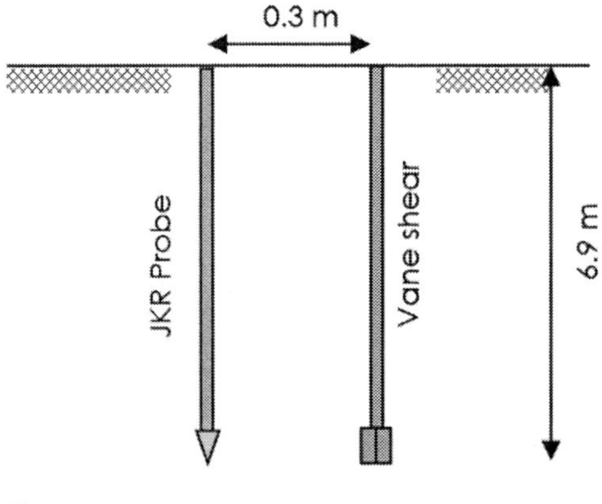

Fig. 2. Details of field testing

4 Results and discussions

4.1 JKR probe test and vane shear test results

The example results of the JKR probe test and vane shear test for point 1 and 2 are shown in Fig. 3 and 4. The results show a good pattern of number of blows and undrained shear strength. This increases with increasing soil depth due to increasing of stress in soil. From Fig. 3 and 4, it is clear that the hard layer exists from the ground surface to a depth of 1 m. This result is due to present of fill material above the existing soft soil layer. These unsaturated fill soils are located above the ground water table, so their behaviors are

 DOI: 10.1051/matecconf/201710307009

influenced by the soil water characteristic curve (SWCC). The SWCC of residual soil or fill material in Malaysia has been discussed in detail by Mohd Fairus *et al.* [7].

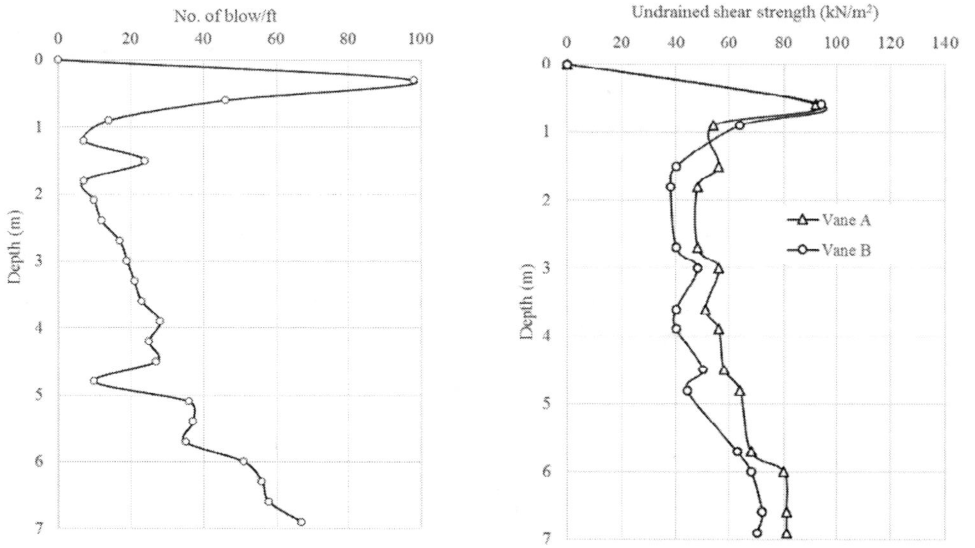

Fig. 3. Number of blow and undrained shear strength versus depth for point 1

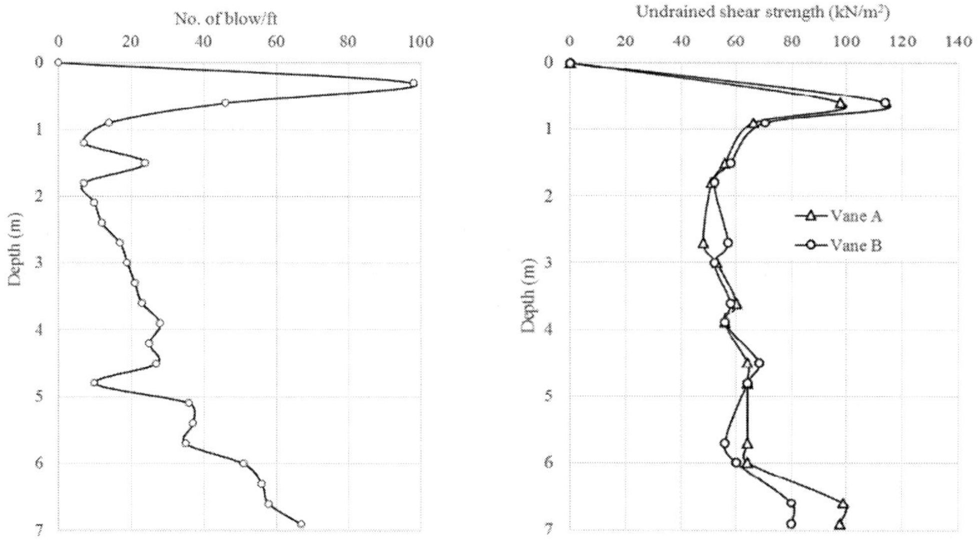

Fig. 4. Number of blow and undrained shear strength versus depth for point 2

4.2 Proposed correlation of JKR probe (J) and undrained shear strength (S_u)

The number of data points obtained from the JKR probe test and the vane shear test were 138 and 168 respectively. The graph of undrained shear strength, S_u versus the number of blows (J-Value) was constructed to develop the correlation.

A graph of undrained shear strength versus the number of blows is shown in Fig. 5. The power correlation was developed and the undrained shear strength could correlate with number of blows using the simplified equation as follows:

 DOI: 10.1051/matecconf/201710307009

$$S_u = 18J^{0.3} \tag{3}$$

where S_u is the undrained shear strength (kN/m^2) and J is the number of JKR probe blow. Referring to the coefficient of determination, $R^2 = 0.5027$ for the proposed equation, it can be categorized as a moderately good correlation [8].

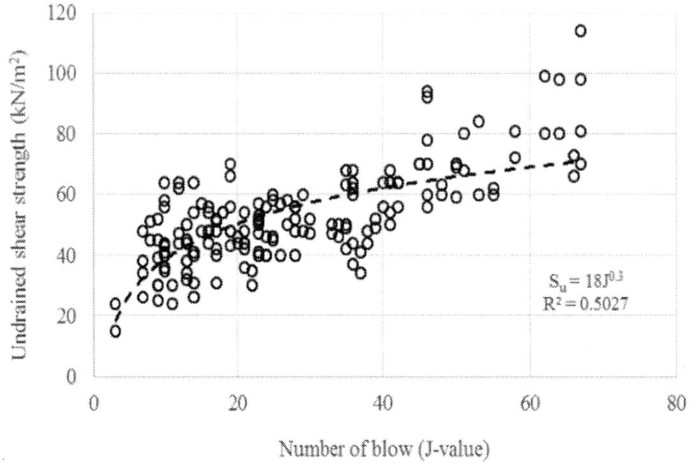

Fig. 5. Correlation of undrained shear strength and J-value

4.3 Parametric studies of proposed correlation

A parametric study was conducted to verify the proposed equation and make comparisons with past researchers. Fig. 6 contains a graph of undrained shear strength versus the number of blows from Mackintosh and JKR probe based on proposed equation. For practical applications, the results from Mackintosh and JKR probe are assumed to be identical.

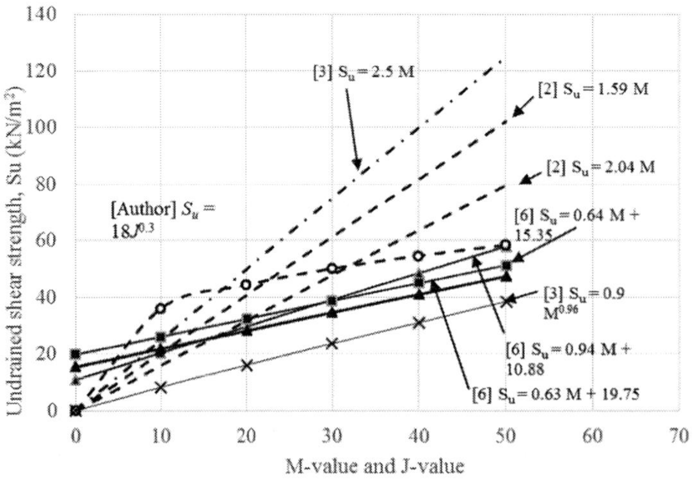

Fig. 6. Comparison of S_u and M or J relationship from proposed equation

5 Conclusion

This study provides a correlation of JKR probe test and undrained shear strength of soft soil. From the study, the undrained shear strength can be predicted as $Su = 18J^{0.3}$.

References

[1] S.S. Gue, Y.C. Tan, Subsurface investigation and interpretation of test results for foundation design in soft clay, *Seminar on Ground Improvement-Soft Clay,* UTM, Kuala Lumpur, 1-23 (2000)
[2] D. Hossain, K.M. Ai, Mackintosh-vs-vane estimation of undrained shear strength correlation for a Sabkha Clay of Saudi Arabia, Quarterly J. of Eng. Geology and Hydrogeology, **24**, 269-272 (1990)
[3] A. Fakher, M. Khodaparast, C.J.F.P. Jones, The use of the mackintosh probe for site investigation in soft soils, Quarterly J. of Eng. Geology and Hydrogeology, **39**, 189-196 (2006)
[4] A.A. Sabtan, W.M. Shehata, Mackintosh probe as an exploration tool, Bulletin of the International Association of Engineering Geology, Paris, **50**, 89-94 (1994)
[5] K. Terzaghi, R.B. Peck, G. Desri, *Soil Mechanics in Engineering Practice,* 2nd ed., John Wiley and Sons, Chichester, (1967)
[6] F. Fatnanta, S.A. Nugroho and H. Rosyida, Determination of shear strength based on mackintosh probe test (in Indonesia language), *Konferensi Nasional Teknik Sipil 7*, Universitas Sebelas Maret, Surakarta, 175-182, (2013)
[7] M.F. Yusof, A.S. Setapa, S.A.A. Tajudin, A. Madun, M.H.Z. Abidin, A. Marto, The soil water characteristic curve of unsaturated tropical residual soil, IOP Conferences Series: Material Science and Engineering, **136**, (2016)
[8] A. Marto, *Volumetric compression of a silt under periodic loading*, PhD Thesis, University of Bradford, England (1996)

Multiple Regression Analysis of Unconfined Compression Strength of Mine Tailings Matrices

Ali A. Mahmood[1,*], and *Maria* Elektorowicz[2]

[1] Department of Civil Engineering, University College of Technology Sarawak, 868 Persiaran Brooke, 96000 Sibu, Sarawak, Malaysia
[2] Department of Building Civil and Environmental Engineering, Concordia University, Montreal, Quebec H3G 1M8, Canada

Abstract. As part of a novel approach of sustainable development of mine tailings, experimental and numerical analysis is carried out on newly formulated tailings matrices. Several physical characteristic tests are carried out including the unconfined compression strength test to ascertain the integrity of these matrices when subjected to loading. The current paper attempts a multiple regression analysis of the unconfined compressive strength test results of these matrices to investigate the most pertinent factors affecting their strength. Results of this analysis showed that the suggested equation is reasonably applicable to the range of binder combinations used.

1 Introduction

Mine tailings are the waste materials resulting from the mining industry; they typically contain all other hazardous materials except the extracted material, left out from the processing of the ore. Mine tailings are usually disposed of in large impoundments surrounded by dams constructed not far away from the tailings mine. However, this classical method of disposal was the reason behind several severe tailings dam failures [1, 2, 3]. Hence, to lessen the harmful effects from the storage and failure of tailings dams and to attempt to economically benefit from these tailings in a sustainable manner, a new approach of managing these tailings was deployed and tested by several researchers. Researchers such as Demers and Haile [4], Zou and Sahito [5], Celik et al. [6] Roy at al. [7], Swami et al. [8], Onuaguluchi and Eren [9], Das et al. [10], Ugama et al. [11] and Sabat et al. [12] attempted to investigate the applicability of using tailings as substitutes to traditional building materials. While others like Belem and Benzaazoua [13], Fall et al. [14], Fall et al. [15], Benzaazoua et al. [16], Ercikdi et al. [17], Fall and Pokharel [18], Ercikdi et al. [19], Helinski et al. [20] and Helinski et al. [21] investigated the use of different mine tailings in the most traditional manner by using them as cemented paste backfill. The tailings, in many of these studies, have been mixed with cement to solidify and increase their effectiveness. However, cement is relatively expensive and hence the

[*] Corresponding author: ali.mahmood@ucts.edu.my

need to add pozzolanic materials such as fly ash and slag to the tailings product to reduce cost and utilize the final product in a sustainable manner.

Recently there has been a new development in the additives sector with the advent of Calsifrit™. Calsifrit™ is a special cement that was developed by a Quebec Based company by the name of NovaFrit International [22]. It is a totally amorphous material, which is a matrix of calcium and sodium fluoro-aluminosilicate. It is a homogenous solid substance possessing high reactivity potential that shows cementitious properties when finely ground. No previous research study had been involved with testing the compression strength of mine tailings mixed with Calsifrit. Therefore, a comprehensive experimental and numerical research program was devised to formulate new tailings matrices and assess their suitability as road base materials in cold regions [23]. As part of this research program, this is an attempt to investigate computationally, using multiple regression analysis, the effect Portland cement, slag, fly ash and Calsifrit have on the unconfined compressive strength of these mine tailings matrices.

2 Preparation of cylindrical specimens

Cylindrical specimens of tailings matrices measuring (44 diameter x 74 height) mm were molded for the purpose of testing in accordance with ASTM D 4842 [24]. The binders used were the following: cement, Calsifrit, fly ash and slag. These cylindrical specimens were, subsequently, cured in the moisture chamber for 28 days in the case of wetting and drying, and for 43 days for the freezing and thawing testing. Unconfined compression tests were conducted on these mine tailings matrices after subjecting them to 12 cycles of drying at 60 °C and wetting by being submerged in room temperature distilled water. The freezing/thawing samples were compressed after 12 cycles of freezing at -20 °C in a freezing cabinet and thawing in distilled water at room temperature (22 °C). When performing the unconfined compression tests, speed of descent of the platen was kept constant at 70 mm/min. Cylindrical specimens of the tailings matrices were subjected to wetting/drying and freezing/thawing weathering resistance tests as further explained in Mahmood [23].

3 The predictive equation

In the modeling of engineered systems a recurrent problem may arise that where the dependent variable or variables (Y) are found to depend on a combination of independent variables (x_1, x_2,….., x_i) without the availability of a known physical mechanism describing this dependence. According to conventional statistical analysis such dependence can be best described using the method of multiple regression analysis. Such analysis can be based on fitting linear or polynomial equations, as described by Rumsey [25].

Multiple regression models can be used, both, to generate predictions based on previously unseen data and to explain observational relationships. In multiple regression the general form of the model is the following:

$$Y = \beta_o + \beta_1 X_1 + \beta_2 X_2 + \ldots + \beta_k X_k + \varepsilon \tag{1}$$

where β_o is the Y- intercept, β_i (where i=1 to k) represents the true slope of the regression surface in the X_i direction, and ε is the error term. The regression equation can be solved using the method of least squares, which generates the estimates for each of the beta (β) coefficients. These estimates are represented by b_i.

The UCS data for the wetting/drying and freezing/thawing experiments were fitted using design expert®, which is a commercial software program used for the design of these experiments. Multiple regression analysis was used to fit the data together with the objective of finding a relevant predictive equation that fits the binders used with the UCS results.

Table 1 shows the data being fitted into this proposed model. Binder combinations were used to represent the factors of regression and the UCS was used as the response. Percentages of fly ash/binder, slag/binder, Calsifrit/binder, cement/binder, binder/tailings and type of weathering were used as the factors in developing this model. However, only the factors indicating the percentages of Calsifrit/binder, slag/binder and binder/tailings were found to be statistically relevant. As for the other 3 independent variables, since their effect was found to be statistically weak on the response variable (UCS), it was thought better not to include them in this model to avoid overfitting and to get a general view of the response. In the future, more models should be attempted using the same independent variables, to further ascertain this.

Table 1. The data used in the regression model

Run No.	Calsifrit/ binder ratio (%)	Fly ash/binder ratio (%)	Slag/ Binder ratio (%)	Cement/ Binder ratio (%)	Binder/ Tailings ratio (%)	Weathering(1: wetting/ drying, 2: freezing/ thawing)	UCS (N/ mm^2)
1	0.00	0.00	0.00	100.00	50.00	1	24.43
2	10.00	0.00	0.00	90.00	50.00	1	25.65
3	25.00	0.00	0.00	75.00	50.00	1	25.24
4	0.00	25.00	0.00	75.00	50.00	1	22.97
5	10.00	15.00	0.00	75.00	50.00	1	26.22
6	20.00	5.00	0.00	75.00	50.00	1	25.83
7	0.00	0.00	25.00	75.00	50.00	1	28.62
8	10.00	0.00	15.00	75.00	50.00	1	27.9
9	20.00	0.00	5.00	75.00	50.00	1	29.47
10	0.00	0.00	0.00	100.00	37.50	1	13.5
11	10.00	0.00	0.00	90.00	37.50	1	16.41
12	25.00	0.00	0.00	75.00	37.50	1	19.42
13	0.00	25.00	0.00	75.00	37.50	1	16.91
14	10.00	15.00	0.00	75.00	37.50	1	19.3
15	20.00	5.00	0.00	75.00	37.50	1	21.79
16	0.00	0.00	25.00	75.00	37.50	1	19.79
17	10.00	0.00	15.00	75.00	37.50	1	20.09
18	20.00	0.00	5.00	75.00	37.50	1	17.5
19	0.00	0.00	0.00	100.00	50.00	2	23.51
20	10.00	0.00	0.00	90.00	50.00	2	26.31
21	25.00	0.00	0.00	75.00	50.00	2	27.1
22	20.00	5.00	0.00	75.00	50.00	2	28.00

23	0.00	0.00	25.00	75.00	50.00	2	25.7
24	10.00	0.00	15.00	75.00	50.00	2	27.83
25	20.00	0.00	5.00	75.00	50.00	2	24.88
26	0.00	0.00	0.00	100.00	37.50	2	16.57
27	10.00	0.00	0.00	90.00	37.50	2	20.94
28	25.00	0.00	0.00	75.00	37.50	2	23.33
29	0.00	25.00	0.00	75.00	37.50	2	17.78
30	10.00	15.00	0.00	75.00	37.50	2	16.71
31	20.00	5.00	0.00	75.00	37.50	2	16.65
32	0.00	0.00	25.00	75.00	37.50	2	16.35
33	10.00	0.00	15.00	75.00	37.50	2	16.74
34	20.00	0.00	5.00	75.00	37.50	2	20.11

The final equation, based on this analysis, describing the dependency of UCS on binder combinations was found to be the following:

$$UCS = -6.856 + 0.14A + 0.092C + 0.6198E \qquad (2)$$

where A: Calsifrit/binder ratio (%),C: slag/binder ratio (%), E: binder/tailings ratio (%).

It was found during the analysis that the effect of cement could not be associated directly with UCS. Its effect was evident in biner/tailings ratio (E), as this ratio includes cement as the main binder constituent. Table 2 shows the Analysis of Variance (ANOVA) calculated for this model. The P values were found to be much less than the required 0.05 indicating that significance level has been reached.

Table 2. The Analysis of Variance of the regression model

The Source	Sum of Squares (SS)	df (degrees of freedom)	Mean Square	F - value	P - value
Model	580.11	3	193.37	58.93	< 0.0001
A-A	47.27	1	47.27	14.41	0.0007
C-C	18.76	1	18.76	5.72	0.0233
E-E	506.2	1	506.2	154.25	< 0.0001
Residual	98.45	30	3.28		

Equation (2) is specifically applicable to the range of binder combinations and the two types of weathering used. If different binder combinations and ratios are required, beyond the range used here, it is best to base a new equation on the new binder combinations. This is advised to retain a high numerical accuracy and closer proximity to experimental results.

The mathematical difference between the observed (y_i) and model predicted (y'_i) values of the dependent variable, for a certain given set of independent variables, specifically y_i-y'_i, is known as the prediction error or residual. In order for any mathematical model to be valid, four assumptions regarding the model residuals must be met [26]:

- Residuals are normally distributed
- Residuals have a mean of zero
- Residuals have a constant variance
- Residuals are independent

One key assumption that must be fulfilled in order for any model to be valid is that the residuals are normally distributed. In order for this assumption to hold true a plot of the residuals must look like a random sample from a normal bell-shaped distribution. Histograms and normal scores plots are considered useful tools for detecting the normalcy of residuals. In the case of the latter, a straight line plot should be expected. Another assumption made in the development of the statistical model is that all data points are collected with equal precision. In order for this assumption to hold true, a plot of the model residuals against model predictions should be free of obvious trends. The third assumption concerning these residuals relates to their independence with respect to time and the values of the model variables. For this assumption to be true, plots of residuals in time order, as well as plots of residuals against the values of the dependent and independent variables, must be free of any obvious trend. The fourth and final assumption suggests that a good model fit will have residuals that are scattered around a mean of zero [26].

The first condition to meet is that the residuals must have a normal distribution with a mean zero. The software Design-Expert© [27] is used to draw the statistical figures shown below. Figure 1 below shows how well the residuals match a normal distribution. The residuals fall on a straight line, which means that the normality condition was met.

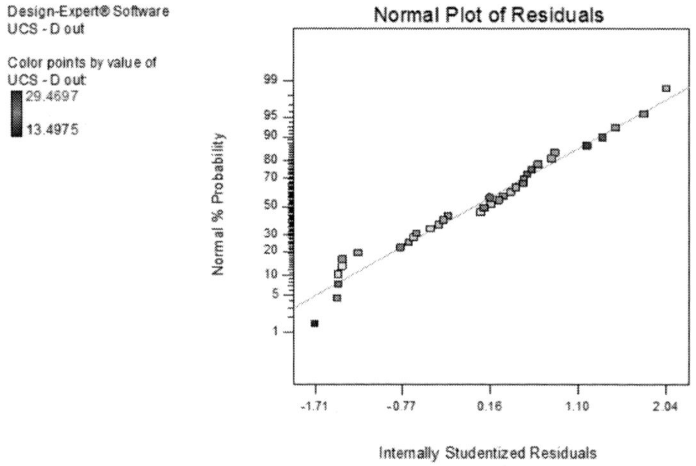

Fig. 1. Normal probability plot of model residuals

Fig. 2 shows a plot of the model residuals versus predicted UCS values. The horizontal line across that plot is at zero as a marker. Residuals were centered around zero in a manner that has no predictable pattern, with the same amount of variability around the horizontal line that crosses at zero from left to right. This plot shows more residuals scattered around zero, where the middle lump would be on a standard normal distribution plot, and fewer and fewer residuals as we move away from zero. Due to all of this, this plot confirms the normal distribution criteria. The residuals average out at the zero line, which indicates that the mean of zero condition holds for these residuals.

Fig. 2 shows no change in the amount of spread in the residuals around the horizontal zero line as we move from left to right or vice versa. This clearly indicates that condition three; the residuals have constant variance, has been met. Fig. 3 that shows the residuals plotted by observation order indicates no discernible pattern. This shows that condition four; the residuals are independent, has been met.

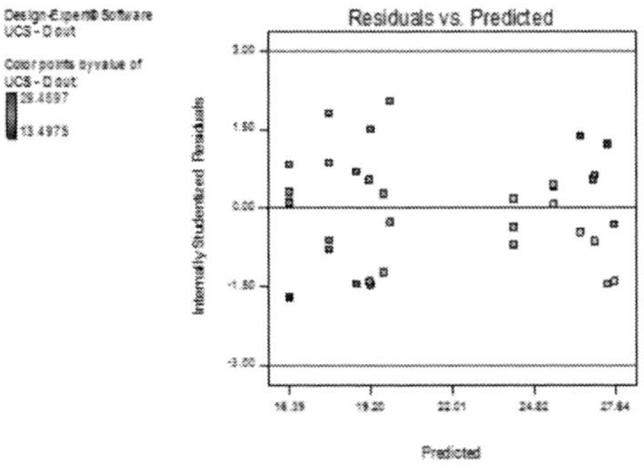

Fig. 2. The residual vs. predicted UCS values

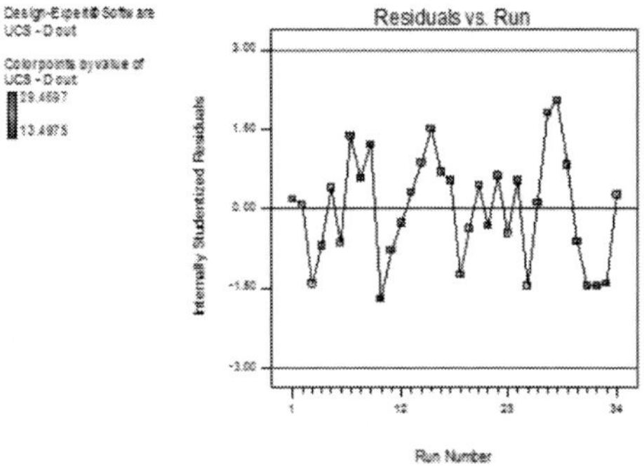

Fig. 3. The residuals plotted by observation order

Plots of the model residuals against each of the independent variables were found to be free of trends as can be seen in Fig. 4 to 9. They also seem to cluster symmetrically around the middle line. This shows that the suggested model is correct and the assumption of linearity is accurate.

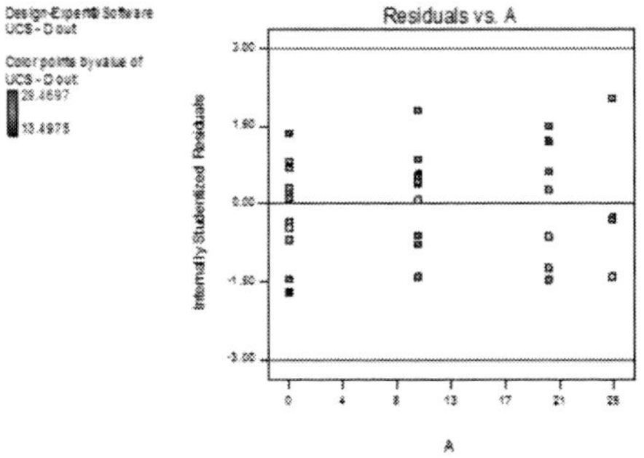

Fig. 4. The residuals vs. Calsifrit/binder ratio

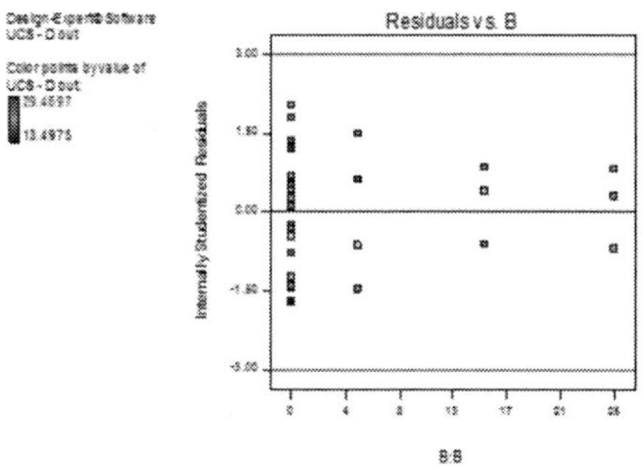

Fig. 5. The residuals vs. fly ash/binder ratio

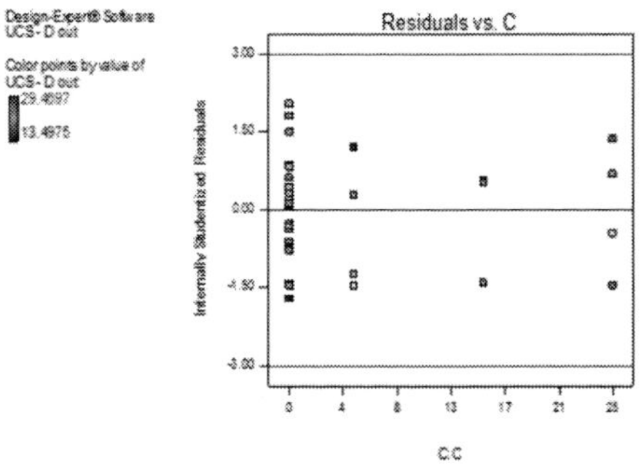

Fig. 6. The residuals vs. slag/binder ratio

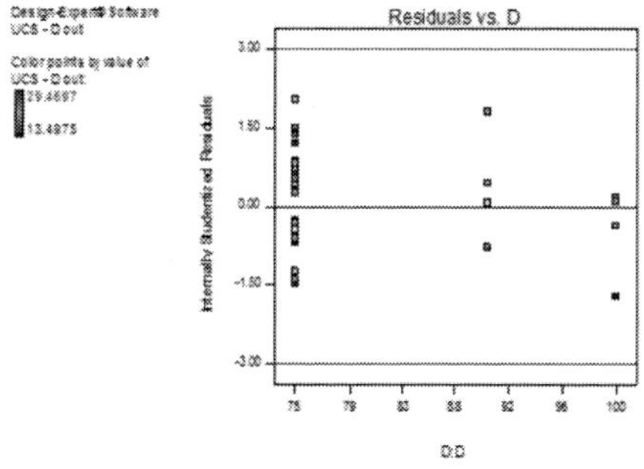

Fig. 7. The residuals vs. cement/binder ratio

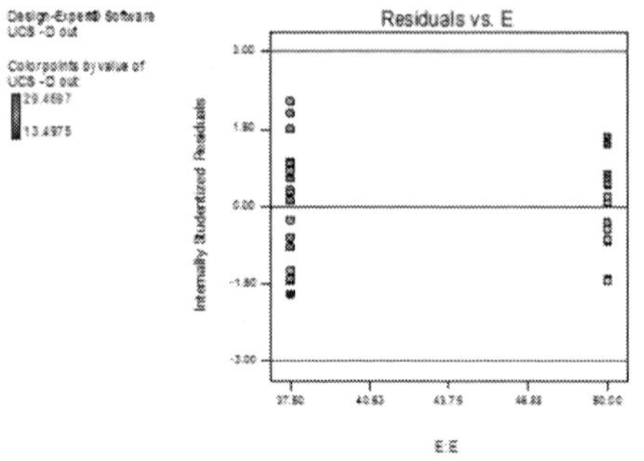

Fig. 8. The residuals vs. binder/tailings ratio

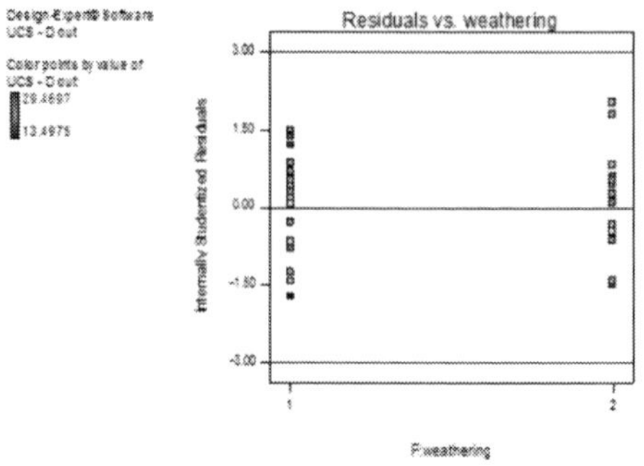

Fig. 9. The residuals vs. weathering type; 1 for wetting/drying and 2 for freezing/thawing

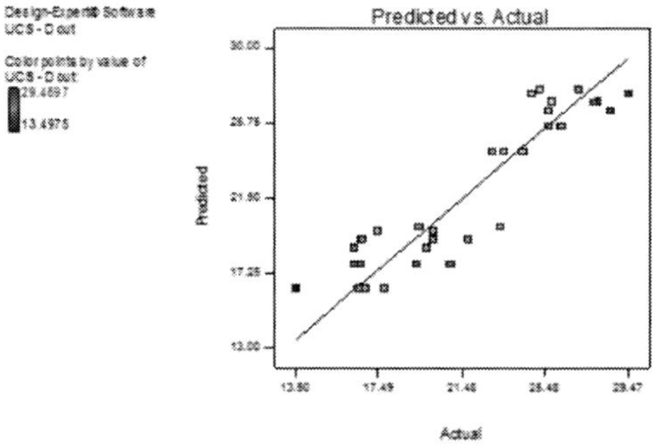

Fig. 10. The predicted vs. actual UCS values

Fig. 10 shows a good correlation between predicted versus actual values. However, more tests need to be conducted using the same independent variables suggested here to fine tune the model and get a more accurate representation of these independent variables. Calsifrit was shown to a relevant factor in the unconfined compression equation. The latter finding was further confirmed by experimental results [23].

4 Conclusion

The above indicates that the multiple regression model suggested here is reasonable and can be used for the prediction of the unconfined compressive strength of mine tailings matrices, based on the variables used. Calsifrit, was shown to be a relevant factor in the unconfined compression equation. The proposed Equation applicable to the range of binder combinations and the 2 types of weathering used. If different binder combinations are required, beyond the range used above, it is best to base a new equation on the new binder combinations. This is done in order to retain a high numerical accuracy and closer proximity to experimental results. The proposed Equation needs to be validated with more experimental data using the same independent variables.

References

[1] A. B. Fourie and G. Papageorgiou. Defining an appropriate steady state line for Merriespruit gold tailings, Canadian Geotechnical J., **38**(4), 695-706 (2001)

[2] A. B. Fourie, G.E. Blight and G. Papageorgiou. Static liquefaction as a possible explanation for the Merriespruit tailings dam failure, Canadian Geotechnical J., **38**(4), 707-719 (2001)
[3] *World Information Service on Energy-Uranium Project*, website information: http://www.wise-uranium.org/index.html, retrieved in (2012)
[4] B. Demers and G. Haile, Management of tailings stabilized by lime and cement at Canadian electrolytic zinc, Valleyfield, Quebec, *Proceedings of Sudbury 2003 Mining and the Environment*, Sudbury, Ontario (2003)
[5] D.H. Zou and W. Sahito, Suitability of mine tailings for shotcrete as a ground support, Canadian J. of Civil Engineering, **31**(4), 632–636 (2004)
[6] O. Celik, I.Y. Elbeyli and S. Piskin, Utilization of gold tailings as an additive in Portland cement, Waste Management and Research, **24**(3), 215-224 (2006)
[7] S. Roy, G.R. Adhikari and R.N. Gupta, Use of gold mill tailings in making bricks: A feasibility study, Waste Management and Research, **25**(5), 475–482 (2007)
[8] R.K. Swami, N.K.S. Pundhir and S. Mathur. Kimberlite tailings a road construction material, J. of the Transportation Research Board, **2**(1989), 131–134 (2007)
[9] O. Onuaguluchiand and Ö.Eren. Cement mixtures containing copper tailings as an additive: durability properties. Materials Research, **15**(6), 1029-1036 (2012)
[10] S.K. Das, J. Ghosh, A.K. Mandal, N. Singh and S. Gupta, Iron ore tailing: A waste material used in ceramic tile compositions as alternative source of raw materials, Transactions of the Indian Ceramic Society. **71**(1), 21-24 (2012)
[11] T.I. Ugama, S.P. Ejeh and D.Y. Amartey, Effect of tailings on the properties of concrete, IISTE, **6** (10), (2014)
[12] V. Sabat, M. Shaikh, M. Kanap, M. Chaudhari, S. Suryawanshi and K. Knadgouda, Use of iron ore tailings as a construction material, Int. J. of Conceptions on Mechanical and Civil Engineering, **3**(2), (2015)
[13] T. Belem and M. Benzaazoua, An overview on the use of paste backfill technology as a ground support method in cut-and-fill mines, *Proc. of the 5th Int. Symp. on Ground support in Mining and Underground Construction*,Tayler and Francis, London, 637-650 (2004)
[14] M. Fall, M. Benzaazoua and S. Ouellet. Experimental characterization of the influence of tailings fineness and density on the quality of cemented paste backfill. Minerals Engineering, **18**(1), 41–44 (2005)
[15] M. Fall, M. Benzaazoua and E.G. Saa, Mix proportioning of underground cemented tailings backfill, Tunnelling and Underground Space Technology, **23**(1), 80–90 (2008)
[16] M. Benzaazoua, B. Bussiere, I. Demers, M. Aubertin, E. Fried and A. Blier, Integrated mine tailings management by combining environmental desulphurization and cemented paste backfill: Application to mine Doyon, Quebec, Canada, Minerals Engineering, **21**(4), 330–340 (2008)
[17] B. Ercikdi, A. Kesimal, F. Cihangir, H. Deveci and I. Alp, Cemented paste backfill of sulphide-rich tailings: importance of binder type and dosage, Cement and Concrete Composites, **31**(4), 268–274 (2009)
[18] M. Fall and M. Pokharel, Coupled effects of sulphate and temperature on the strength development of cemented tailings backfills: Portland cement-paste backfill, Cement and Concrete Composites, **32**(10), 819–828 (2010)
[19] B. Ercikdi, F. Cihangir, A. Kesimal, H. Deveci and I. Alp. Effect of natural pozzolans as mineral admixture on the performance of cemented-paste backfill of sulphide-rich tailings. Waste Management and Research, **28**(5), 430–435 (2010)

[20] M. Helinski, M. Fahey and A. Fourie. Coupled two-dimensional finite element modeling of mine backfilling with cemented tailings, Canadian Geotechnical J., **47**(11), 1187-1200 (2010)
[21] M. Helinski, M. Fahey and A. Fourie, Behavior of cemented paste backfill in two mine stopes: measurements and modelling, J. of Geotechnical and Geo-environmental Eng., **137**(2), 171-182 (2011)
[22] NovaFrit International, 1200 Garnier Street, Ville Ste-Catherine, Quebec, J0L 1E0 (2006)
[23] A.A. Mahmood, *Experimental and computational assessment of tailings binder matrices for construction purposes in cold regions*, PhD Thesis, Concordia University, Canada (2012)
[24] ASTM D 4842-90, *Standard test method for determining the resistance of solid wastes to freezing and thawing*, Annual Book of ASTM Standards, American Society for Testing and Materials, Philadelphia, (1996)
[25] D. Rumsey, *Intermediate statistics for dummies*, John Wiley and Sons, New Jersey, (2007)
[26] C.W. Baxter, D.W. Smith and S.J. Stanley, A comparison of artificial neural networks and multiple regression methods for the analysis of pilot scale data, J. of Environmental Engineering and Science, **3**(S1), S45-S58 (2004)
[27] Stat-Ease. Stat-Ease, Inc. 2021 E. Hennepin Avenue, Suite 480 Minneapolis, MN 55413-2726, USA (2008).

Understanding and Identifying Natural Hazard for Bandung City Preparedness and Mitigation against Natural Disaster

Paulus P. Raharjo[1,*]

[1]Geotechnical Engineering, Universitas Katolik Parahyangan, Bandung, Indonesia

Abstract. Understanding Natural Hazard is an important aspect for the City Planning and Implementation of Policy and Development. The Local Government of Bandung City is developing a map of Hazard and Disasters based on the Natural Condition, Geography and Existing Condition. This paper discuss the scenario of the natural and main emphasis on geological hazard in the City of Bandung related to the Mitigation and Disaster Management as an input that should be adopted by the Local Government. The City is surrounded by active volcanoes and a number of faults that might cause natural disasters including earthquakes, volcano eruptions, flood and landslides. on the other hand the city developement can be directed towards the mitigation and risk reduction against these hazards.

1 Introduction

In most city development, hazard and disaster have become important aspects to be considered. Much of the data has been collected throgh research and investigations, and many has been due occurance in the past. One of the challenge of the city is due to uncontrolable development which were not well planned and many of them have caused disasters, which could not be avoided. However understanding and awareness of the hazard will help a better plan for future development of the city.

Bandung is the capital of the West Java province with population of about 2.5 millions. Due to the fact of the surrounding mountain in boundary of the city, future developement will face challenges for instance to build houses and infrastructures on difficult soil condition such as deep soft deposit at the lowland of Bandung area or on the slope which might not be stable. In fact the city is recently advised by a team consisting of experts from unversities, government research centers and practical engineers for buildings and infrastructures permit before implemented. However the regulations are still not well understood.

This paper discuss some evidence and data which shows aspects of the geological hazard which may affect the develoment and to be considered as strategy in the management of the disaster as outlined by the Indonesian Law no 24/2007. According to

* Corresponding author: rahardjo.paulus@gmail.com

the law, the disaster management should be focussed on the preparedness to reduce the risk of disasters rather than the management during or after disasters.

The understanding of common people on the disaster management so far has been focussed on action during and after disaster which is very weak for the mitigation purposes. Hence the mitigation plan should be socialized among planners, educators, developers, stake holders and designers and education on natural disaster should be part of the instruction for students of all level since elementary school to universities.

1.1 Development and master plan of Bandung City

Bandung City is located at high elevation of 675-1050 m above sea level with areas of 16.729,650 Ha consisting of 30 Kecamatan and 151 Kelurahan. With 2.5 million of population, Bandung City is the third largest city in Indonesia, after Jakarta and Surabaya. Fig. 1, shows the Bandung City Area with boundaries of Kabupaten Bandung Barat and Kabupaten Bandung at the north, the City of Cimahi at the west boundary, and Kabupaten Bandung at the east and south boundaries.

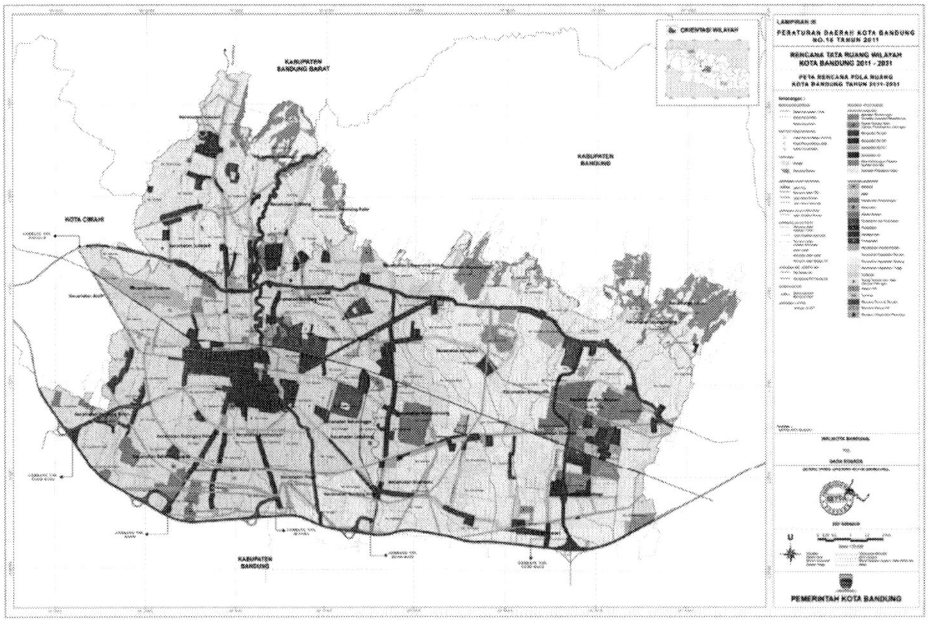

Fig. 1. The city of Bandung map (source : Government of Bandung city [5])

1.2 The underlying law for preparedness and mitigation of Bandung City

The impact of natural disasters are fatalities, the loss of people's belonging including houses, the loss of jobs, the damages of infrastrutures, difficulies in the supply of food, electricity, communication and transportation. All of those will cause economic loss, chaos, trauma, backward of the city development etc.

The term "natural disaster' refers to those disasters that are triggered by natural phenomena, however the term can be misleading because it implies that the disasters are solely a result of natural hazards – when in fact, human endevour are a major contributing factor in creating a disaster. The complete realm of disaster related activities is "disaster management'. Generally people tend to think of disaster management only in terms of post disaster actions taken by relief and reconstruction officials; yet disaster management covers a much more broader scope, and many modern disaster managers may find themselves far more involved in pre-disaster activities than in post-disasters response. Disaster management can be defined as the range of activities designed to maintain control over disaster and emergency situations and to provide a framework for helping at risk persons to avoid or recover from the impact of the disaster. Disaster management deals with situations that occur prior to, during and after the disasters to reduce or avoid the human, physical and economic lossses suffered by individuals, by the society and by the country at large, to reduce personal suffering and to speed the disaster recovery.

The underlying Indonesian law covers the preparedness and the management of disaster as shown at UU No. 24/2007 about the response to disaster and reduce risk due to natural disaster. The Law No. 26/ 2007 on the Land Use (Penataan Ruang) and the City Regulation Perda Kota Bandung No. 18/2011 on the Development of the City of Bandung 2011-2031.

Indonesian law UU RI no. 24 and 26 year 2007 changed paradigm of disaster management into management of disaster focusing on 'preparedness" for mitigation before they occur. Land use planning that consider hazard and disaster aspects are best to be implemented to reduce the negative effect of disaster.

2 The geology of Bandung City

Million years ago, Bandung was land with are influenced by tectonic activities and notrh of Bandung are mainly lahar formations. At north of Bandung City, the Tangkuban Perahu volcano is an active volcano in the close proximity to Bandung City and combined with the existence of Lembang Fault which is still active and many people live near Lembang Fault. Fig. 3 shows the location of Lembang Fault. At the south of Bandung City there are Patuha Volcano and Malabar Volcano, even though not reported as active volcanoes however they are registerd as being active.

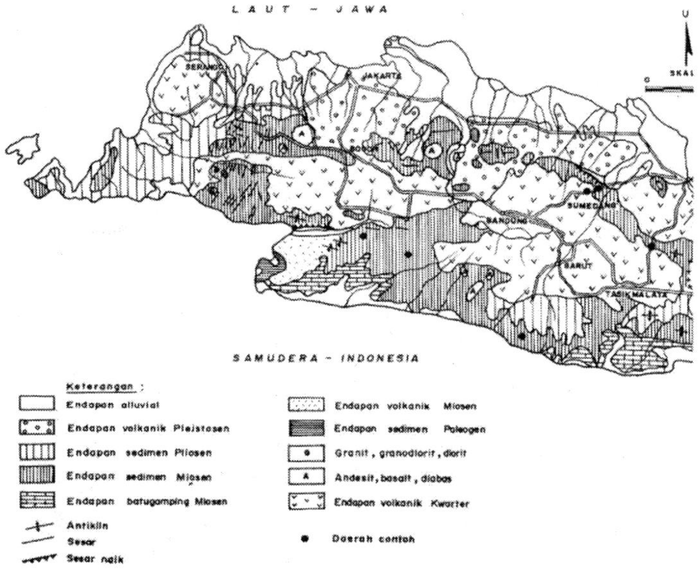

Fig. 2. The geology map of West Java (source : Direktorat Geologi Bandung [5])

.

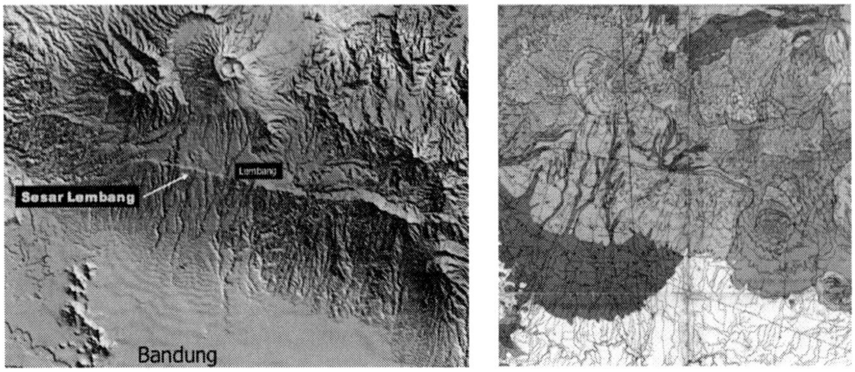

Fig. 3. Lembang fault and northern Bandung geology (source : Direktorat Geologi Bandung [5])

However south of Bandung is a big difference, this area is mainly lowland that was originated from two ancient big lakes. Fig. 1 shows the schenario of the lakes. They were very deep (more than 30m) and hence all sediments into the lakes create soft layers. The geotechnical hazard is any buildings in these areas will be prone to settlement or land subsidence which will damage the houses and infrastructures. Many of the people in Bandung not realizing this fact and many of the development failed. Moreover Bandung City is planning to develop new territory south of Bandung which is known as Bandung Technopolis. This development will be affected by the soil condition.

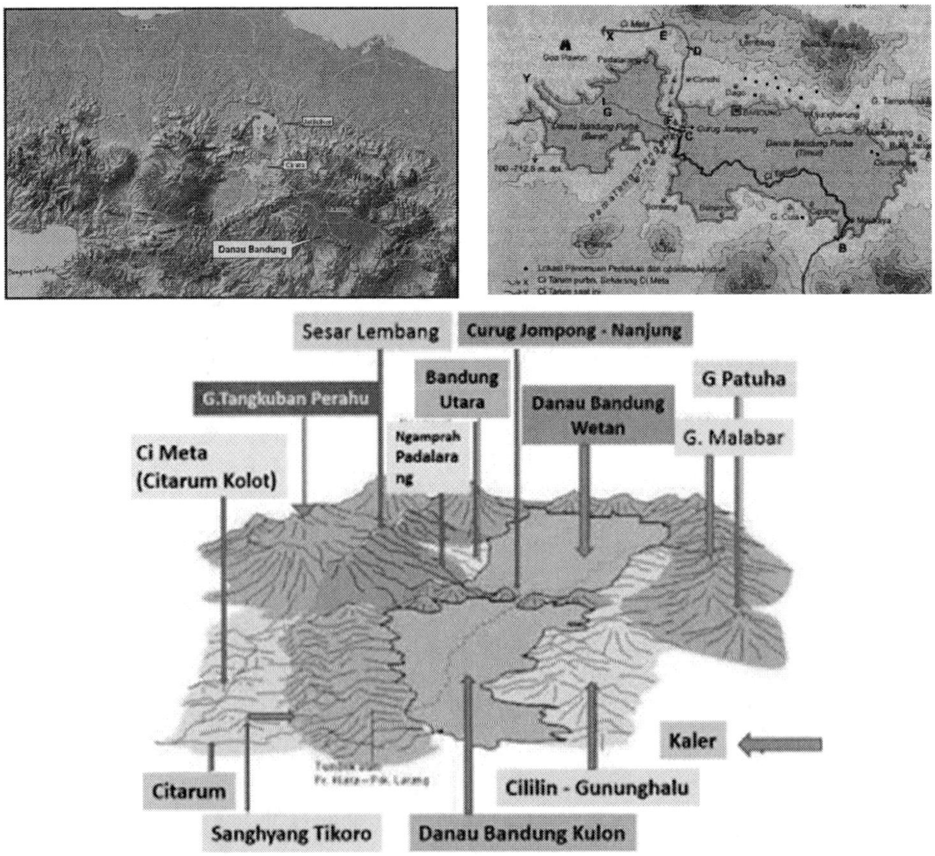

Fig. 4. Bandung soft soils originated from two large ancient lakes

3 The natural hazard of Bandung City and mitigation plan

Geohazard in Bandung City includes Earthquakes where Cimandiri Fault and Lembang Fault are very near to the city, Landslides specially north of Bandung, however landslides could also be due to man made slopes such as north of Bandung City. The main reason would be the existence of colluvial which people might not aware of or non engineered fill material, Volcano Eruption (there are several active volcanoes surrounding Bandung City. One of the most active volcanos are Galunggung which is located near Tasikmalaya and Tangkuban Perahu which is located north of Bandung.

The soft soils condition south of Bandung will be worst during earthquakes. Hence building regulation at this area must be very stricht. Moreover, this area is law land with the main river Citarum flowing the water from east to west. The low elevation and the fact that Bandung is like a cup cause south Bandung to be frequently flooded.

Flood frequently occured in combination with landslides, specially because of the high rain intensity, then land subsidence commonly occured at Bandung south due to very specific deep soft soils originated by sedimentation through the ancient Bandung Lake as described in the previous section.

There is already plan for evacuation at Bandung City based on Local government regulation (Peraturan Daerah Kota Bandung) No. 18 year 2011 on the Land Use at Bandung City (Peraturan Daerah tentang Rencana Tata Ruang Wilayah Kota Bandung Tahun 2011 – 2031, pasal 58). Development Plan for Evacuation Area of Bandung City against Disaster:

i) Development of flood evacuation plan is directed at Taman Tegallega at Kecamatan Regol and the Main Stadium Gelora Bandung Lautan Api at Kecamatan Gedebage
ii) Development of Landslide evacuation plan is directed at Taman Gasibu and Sasana Budaya Ganesha located at Kecamatan Bandung Wetan and Taman Pacuan Kuda (Kecamatan Arcamanik)
iii) Development of parks in the scale of community called Taman Rukun Tetangga (RT), on the scale of Rukun Warga (RW), sports area, or open public space for evacuation points in the residential area
iv) Development for evacuation during earthquakes by taking advantage of open public space like alun alun or soccer field, open area at public buildings like schools and other area at the city scale
v) Development for evacuation due to fires at the parks at Rukun Tetangga and Rukun Warga scales, sport halls and public open space

4 Conclusions summary

i) There is urgent need for socialisation of natural hazard to the Bandung City Community and how to live in harmony with the nature.
ii) There is need to develop hazard map and the scenario of disaster and prepare evacuation plan, muster points the location for help for every region.
iii) There is need for coordination plan both the public sector and the role of the private party for Bandung City Mitigation Plan.
iv) It is necessary to enforce law specially for infrastructure and buildings, they need to follow the hazard map and the regulation for save structures.
v) Master Plan and Development Plan of Bandung City must follow the hazard map.
vi) Education of Hazard and How to live with nature is a necessity for curriculum since elementary school to university students for better understanding of the hazard and mitigation.

References

[1] A. Tohari, *Geologi Cekungan Bandung, FDG ilmu kebumian untuk mitigasi bencana di Wilayah Cekungan Bandung*, Forum Grup Diskusi (FGD) Cekungan Bandung, Lipi Geoteknologi, (2015)

[2] A. Sunarwibowo, *Tata ruang Kota Bandung*, Badan Perencanaan Pembangunan Daerah Kota Bandung, (2015)

[3] A. Taruna, *Mitigasi bencana gempa bumi di sekitar Sesar Lembang*, Laporan Departemen Teknik Tambang, Institut Teknologi Bandung, (2012)

[4] B. Brahmantyo, *Review geologi Cekungan Bandung*, Forum Grup Diskusi (FGD) Cekungan Bandung, Lipi Geoteknologi, (2015)

[5] Direktorat Geologi Bandung, Peta Geologi Kota Bandung

[6] P.P. Rahardjo, Mempersiapkan Kota Bandung menghadapi bencana alam, *Seminar Bandung Kota Teknopolis*, Universitas Katolik Parahyangan, (2016)

Physico-Chemical Properties of Kaolin-Organic Acid

S.W. Yeo[1], *Felix N. L.* Ling[1,*], *V. S.* Low[2], and *K. L.* Toh[2]

[1]Research Center for Soft Soil (RECESS), Universiti Tun Hussein Onn Malaysia, 86400 Parit Raja, Batu Pahat, Johor, Malaysia
[2]Faculty of Civil & Environmental Engineering., Universiti Tun Hussein Onn Malaysia, 86400 Parit Raja, Batu Pahat, Johor, Malaysia

Abstract. Soil with more than 20% of organic content is classified as organic soil in Malaysia. Contents of organic soil consist of different types of organic and inorganic matter. Each type of organic matter has its own characteristic and its effect on the properties of the soil is different. Hence, a good understanding on the effect of specific organic and inorganic matter on the physico-chemical characteristic of organic soils can serve as a guide for predicting the properties of organic soils. The main objective is to unveil the effect of organic acid on the physico-chemical properties of soil. Artificial organic soil (kaolin mixed with organic acid) was utilized in order to minimize the geochemical variability of studied soil. The organic acid which consists of humic acid and fulvic acid was extracted from highly humificated plant–based compost. The effect of organic acid on the physico-chemical properties of soil was determined by varying the concentration of organic acid. The specific gravity, Atterberg limits, pH, bulk chemical composition and the functional group of kaolin-organic acid were determined. It was found that the plasticity index, specific gravity and pH value were decreased with lowered concentration of organic acid. However, the liquid limits and plastic limits were found to be increased with the concentration decrement of organic acid. The analysis of XRF on the bulk chemical composition and analysis of FTIR spectra on the functional group of artificial organic soils with different concentration have confirmed little geochemical variability between samples.

1 Introduction

Malaysia, which situated at the tropical belt is rich in peat soil. Maintenance of road construction over peat soil deposit faced great challenges and cost millions of funding on it. Besides that, the most commonly used soil replacement method and resting structures on piles over peat deposits were also reported to be not social-economic sustainable [1].

Peat is generally consists of fragmented organic materials that formed in wetlands which is termed as soil organic matter. The contents of peat varies with location due to the factors such as degree of humification, temperature and origin of fiber [2]. However, it was

*Corresponding author: felix@uthm.edu.my

reported that the peat properties of Hokkaido peat that had been studied has many similarities when compared with Malaysia peat particularly with Johor peat. Both peat soils were reported to be hemic peat with high ash contents and lightly acidic [3]. Hence, it motivated more studies to generalize the behavior of peat and organic soil through artificial soil [4, 5, 6].

Basically, soil organic matter is derived from the soil biomass, which consists of both living and dead organic matter [7]. Thus, soil organic matter can be classified into two major categories which are living organic matter and non-living organic matter. In general, soil biota such as bacteria, fungi and algae, and others fresh undecomposed animal or plant debris were categorized as living organic matter (a minor fraction of soil organic matter). The major faction of soil organic matter, which is non-living organic matter, can be defined as the plant or animals at different stages of decomposition and transformation.

Non-living organic matter can be further divided into humified and non-humified organic matter depending on its degree of humification. Humic substances which consist of fulvic acid, humic acid and humin are categorized as humified organic matter [4]. Studies found that highly humified matter can coagulate the soil particles resulting in bigger particles which may reduce the compatibility of soil [8] increases the elasticity and deformation resistance of the aggregates and thus increases the stability and strength of aggregates [9]. However, it was reported that the settlement of embankments over peat deposit with high organic content remains excessively large and continued for many years [10,11]. The findings of the previous researcher [8, 9, 10] showed that the effect of organic matter was different with the contents of organic matter. Hence, it emerged a need to enhance the understanding of effect of various organic matters on the physico-chemical properties of soil.

2 Materials and methods

Artificial organic soils were used in this study for the purpose to minimize the geochemical variability that normally encountered in natural organic soils. Commercial processed kaolin grade S300 was used as base soil. Kaolin grade S300 which marketed by Kaolin (M) Sdn. Bhd. was reported to be dominant silt sized with small contents of sand and clay sized particles [12]. Organic acid utilized in this study was extracted from highly humificated plant–based compost. The organic acid was determined to consist of 0.37% humic acid and 46.60% fulvic acid [13]. A total 15% of organic acid based on dry weight of kaolin were mixed with kaolin to form artificial organic soils. Organic acid was prepared in four different dilution factors with distilled water in order to study the effect of organic acid on the properties of soil. The dilution factor of organic acid was named based on its ratio to the distilled water which was 1:0, 1:1, 1:2 and 1:3 respectively. Table 1 summarizes the percentage of each composition of artificial organic soil samples that utilized in this study.

Table 1. Composition of artificial organic soils.

Material	**Percentage for each component (%)**			
	Soil 1	**Soil 2**	**Soil 3**	**Soil 4**
Kaolin S300	85	85	85	85
Organicacid	15	7.5	5	3.7
Distilled water	0	7.5	10	11.3

Physical properties of the artificial organic soils namely Atterberg limits (liquid limit, and plastic limit), and specific gravity were determined. The liquid limit of the soil samples were determined by using cone penetrometer method based on standard specification [14]. Whereas, the plastic limit of soil samples were determined by using modified cone penetrometer method [15]. The procedure is same as the liquid limit and the only differences was the mass of the cone. Modified cone penetromenter method utilizing a cone with 240g weight, 3 times heavier than standard liquid limit cone with same geometry and penetration depth for plastic limit test. Small pycnometer method was used to determine the specific gravity of soils based on standard specification [14].

The pH of the artificial organic soils was determined by using electrometric method with reference to standard specification [16]. A total amount of 30g dry soil was mixed with 75mL of distilled water in a 100mL beaker for pH determination. The X-ray fluorescence (XRF) and Fourier Transform Infrared Spectroscopy (FTIR) were utilized to determine the geochemical composition of artificial organic soil samples with various dilution factors by determining its bulk chemical composition and functional group.

3 Results and discussion

Table 2 shows the physico-chemical properties test that had been carried out for kaolin-organic acid in different concentration. Based on the data collected, the differences between use of pure organic acid and after diluted can be seen. Soil samples with less concentration organic acid shows higher value in liquid limit and plastic limit compared to the soils with more concentrated organic acid. Fig. 1 shows the plasticity range were determined for the kaolin-organic acid with different concentration. The differences between the two boundaries were defined as plasticity index where the upper and lower boundary is element for liquid limit and plastic limit. Hence, with the plasticity index value, soils samples are classified as intermediate plasticity silt based on plasticity chart. The specific gravity of soil samples decreasing when the concentration of organic acid used are less concentrated. The average specific gravity of kaolin S300 was determined as 2.6 [17]. As for this study, the used of kaolin as the clay material to mixed with organic acid have produced the specific gravity between the ranges of 2.6 to 2.7 that can be accepted as well. From previous researcher [18], the plastic limit and liquid limit for kaolin-humic acid both shows decreasing when humic acid content increases. In contrast, the Atterberg limit for kaolin-organic acid were shown increasing when the dilution factor increasing. In the same paper reported that the specific gravity for kaolin-humic acid decreased when the humic acid content increasing. Simultaneously, when the used of organic acid is less concentrated, the specific gravity of soil samples were slightly decreasing.

Table 2. Physico-chemical properties for kaolin – organic acid in different concentration.

Physico-chemicalPropertiesTesting	Percentage for each component (%)			
	Soil 1	Soil 2	Soil 3	Soil 4
Liquidlimit (%)	36.8	37.2	37.5	38.0
Plastic limit (%)	33.1	33.6	35.0	35.7
Plasticity index (%)	3.7	3.6	2.5	2.3
Specificgravity	2.80	2.66	2.66	2.64
pH	6.68	6.66	6.35	6.18

For chemical properties test, through the pH testing can know that, the pH value for kaolin was 4.74 and the organic acid had pH value of 7.74. Overall, all the soil samples had the pH value in the range of 6.18 to 6.68 which is in between the pure kaolin and organic acid. The pH value of soil samples shows more acidic when distilled water is added for dilution purpose. XRF test are carried out and the results are summarized as below Table 3. Theoretically, magnesium oxide (MgO) and potassium oxide (K2O) content will increase the soil pH. However, the pH result in this study did not comply with this condition. This may due to the characteristic of organic acid that are mostly soluble either in acid and base environment. Fig. 2 shows the FTIR spectra for kaolin-organic acid in different concentration. All the samples exhibits sharp bands in the region of wavelength 1100-1000 and 990-945, which can attributed to Si-O-C groups.

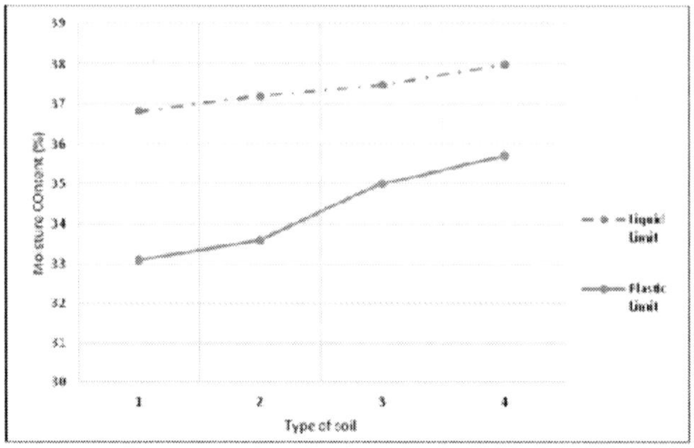

Fig. 1. Effect of concentration on plasticity of kaolin-organic acid.

Table 3. XRF result for kaolin – organic acid in different dilution factor

Formula	**Sample**			
	Soil 1	**Soil 2**	**Soil 3**	**Soil 4**
SiO_2	48.30%	49.80%	50.80%	50.40%
Al_2O_3	11.70%	17.40%	17.80%	16.30%
K_2O *	3.10%	3.62%	3.57%	3.56%
C	1.00%	1.06%	1.11%	1.00%
Fe_2O_3	0.76%	1.00%	1.00%	0.97%
MgO *	0.64%	0.82%	0.83%	0.87%
TiO_2	0.60%	0.74%	0.77%	0.66%
ZrO_2	0.35%	0.56%	0.59%	0.46%
Percentage**	0.056	0.0592	0.058	0.060

*Alkaline and alkali earthelements
**Percentage of alkaline & alkali earth elements over total inorganic element

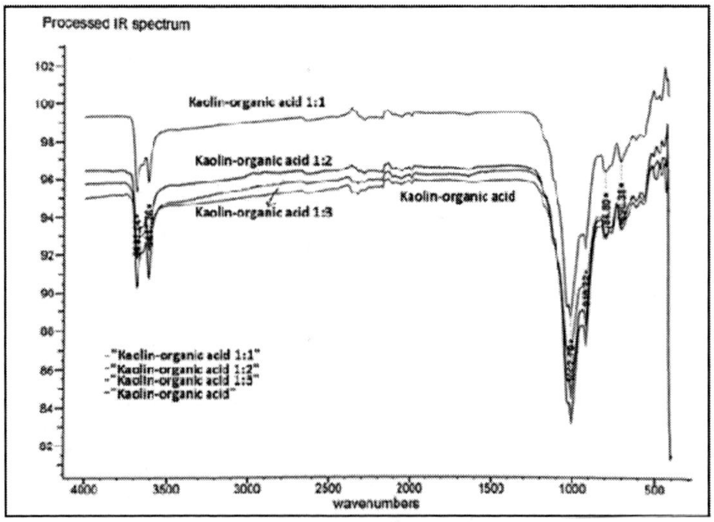

Fig. 2. FTIR Spectra of kaolin-organic acid in different concentration.

4 Conclusion

Throughout this study, the effect of concentration on the physico-chemical characteristic had been identified. When the concentration of organic acid become less concentrated, the physical and chemical characteristic such as liquid limit and plastic limit shows increasing but the plasticity index decreased when dilution factor of organic acid increased. However, the pH value shows more acidic when the concentration for organic acid used are decreased. Furthermore, all the soil samples are mainly composed by silicon oxide based on the XRF and FTIR result. Overall, the results obtained throughout this study will be benefit for the understanding on basic characteristic of soil samples as a reference for further experiment to determine the strength and improvement in real construction environment.

The authors wish to express grateful and thankful appreciate to the financial support from University Tun Hussein Onn Malaysia through the Centre for Graduate Studies and Office for Research, Innovation, Commercialization and Consultancy Management (ORICC). The authors also thanks all laboratory technicians in Research Centre for Soft Soil (RECESS), Department of Infrastructure and Geomatic Engineering, Department of Water and Environmental Engineering of Faculty of Civil and Environmental Engineering, Universiti Tun Hussein Onn Malaysia. Their advice, knowledge and technical support had made this research possible.

References

[1] M.K. Abu Talib, *Effectiveness of sugarcane bagasse ash (SCBA) utilization in peat stabilization,* PhD Thesis, Kyushu University, Fukuoka, (2016)

[2] S. Kazemian, B.B.K. Huat, A. Prasad, M. Barghchi, A state of art review of peat: Geotechnical engineering perspective, Int. J. of the Phys. Sciences, **6**(8), 1974-1981 (2011)

[3] M.K. Abu Taib, Y. Noriyuki, An overview on Japan and Malaysia peat relating to geotechnical characteristic, Int. J. of Integrated Eng., **6**(1), 1-7 (2014)

[4] K. Kujala, M. Makikyro, O. Lehto, Effect of humus on the binding reaction in stabilized soils, *Proc of the Int. Conf on Ground Improvement and Geosystems, Tokyo,* 415-420 (1996)

[5] H. Tremblay, J. Duchesne, J. Locat, S. Leroueil, Influence of the nature of organic compounds on fine soil stabilization with cement, Can. Geotech. J., **39**, 535-546 (2002)

[6] C. Modmoltin, J. Lu, K. Onitsuka, Influence of humic acid and salt concentration on lime-stabilized ariake clays and microstructure research, Chinese J. of Geotech. Eng. **26**(2), 281-286 (2004)

[7] K.H. Tan, *Humic Matter in Soil and the Environment. Principles and Controversies,* Marcel Dekker Inc., New York, (2003)

[8] R. Sugita, Y. Marumo, Screening of soil evidence by a combination of simple techniques: validity of particle size distribution, Forensic Sci. Int., **122**, 155-158 (2001)

[9] B.D. Soane, The role of organic matter in soil compactibility: A review of some practical aspects, Soil and Tillage Res., **16** (1-2), 179-201 (1990)

[10] Y. Duraisamy, B.K. Huat, A.A. Aziz, Compressibility behavior of tropical peat reinforced with cement columns, American J. of App. Sci., **4** (10), 768-773 (2007).

[11] A. Zainorabidin, S.H. Mansor, Investigation on the shear strength characteristic at Malaysian peat, ARPN J. of Eng. and App. Sci., **11**(3), 1600-1606 (2016)

[12] F.N.L. Ling, K.A. Kassim, A.T. Abdul Karim, Size distribution analysis of kaolin using laser diffraction technique, Adv. Materials Research, **341-342**, 108-112 (2012)

[13] M.I. Jaimi, *Test Report of Sampel LMB03*, Report No. BK 35/14, Forest Research Institute Malaysia (FRIM), Kepong, Selangor, (2014)
[14] BS1377-2, *Methods of test for soils for civil engineering purposes. Classification tests*, British Standards, London, (1990)
[15] A.S.A Rashid, K.A. Kassim, A. Katimo, M.N. Norhazilan, Determination of plastic limit of soil using Modified Methods, Malaysian J. of Civil Eng. **20**(2), 295-305 (2008)
[16] BS1377-3, *Methods of test for soils for civil engineering purposes. Classification tests*, British Standards, London, (1990)
[17] F.N.L. Ling, *Physico-chemical and microstructure of artificial soils stabilised with lime-zeolite*, PhD Thesis, Universiti Teknologi Malaysia, Johore, (2016)
[18] F. Pakir, A.T. Abdul Karim, F.N.L. Ling, K.A. Kassim, Effect of humic acid on geochemistry properties of kaolin, Adv. Materials Research, **701**, 310-313 (2013)

Highly Organic Soil Stabilization by Using Sugarcane Bagasse Ash (SCBA)

Mohd Khaidir Abu Talib [1,*], and *Yasufuku* Noriyuki[2]

[1] Department of Infrastructure and Geomatic Engineering, Faculty of Civil and Environmental Engineering, Universiti Tun Hussein Onn Malaysia, 86400 Batu Pahat, Johor, Malaysia
[2] Geotechnical Engineering Laboratory, Department of Civil Engineering, Faculty of Engineering, Kyushu University, 744 Motooka, Nishu-ku 819-0395 Fukuoka, Japan

Abstract. The study objective is to develop alternative binders that are environment friendly by utilizing sugarcane bagasse ash (SCBA) in the organic soil stabilization. Together with SCBA, Ordinary Portland Cement (OPC), calcium chloride ($CaCl_2$) and silica sand (K7) were used as additives to stabilize the peat. In obtaining the optimal mix design, specimens of stabilized peat were tested in unconfined compression. It was found that stabilized peat comprising 20% and 5% (PCB1-20 and PCB2-5) partial replacement of OPC with SCBA 1 and SCBA 2 attain the maximum unconfined compressive strength (UCS) and discovered greater than UCS of peat-cement (PC) specimen. At the optimal mix design, the UCS of the stabilized peat specimens increased with increasing of curing time, preloading rate, OPC and K7 dosage. For PCB1-20 mixture, inclusion of a minimum OPC of 300kg/m^3 and K7 of 500kg/m^3 along with curing under 20kPa pressure is recommendable for the peat stabilization to be effective. However for PCB2-5, it suggested to use more OPC and K7 dosage or alternatively increase the preloading during curing to 40kPa in order to achieve target UCS. It can be concluded that SCBA 1 has better quality than SCBA 2 in peat stabilization especially the contribution made by its fine particle size.

1 Introduction

As a result of high organic content, presence of humic acid and less solid particles in peat soil, the proper hydration of stabilized peat could be hindered [1-5]. That means the large amount of cement is needed in order to neutralize the acids or otherwise the process of the soil stabilization remains retarded. However, adding a large quantity of cement into the peat is absolutely an unfriendly and uneconomical solution to peat ground improvement considering the fact that the peat ground is covers a wide area, the rising cost of cement and its transportation to the site [6]. Cement is responsible for about 5% – 8% of global carbon oxide, CO_2 emissions and expected to grow 0.8 to 1.2% per year until may reach 4.4 billion tonnes of productions in 2050 [7].

* Corresponding author: khaidir@uthm.edu.my

DOI: 10.1051/matecconf/201710307013

The productions of sugarcane is world number one commodities with amount around 1.9 billion tonnes in 2013[8] and produced a lot of bagasse. Bagasse is the residue left after the crushing of sugar cane for juice extraction and on average about 32% of bagasse is produced from every tonne of sugar cane. Bagasse is burnt to generate power required for diverse activities in the factory and finally leave the bagasse ash as a waste. Increasing concern of disposal of bagasse residual creates interest to explore the potential application of this material. Increasing concern of disposal of bagasse residual creates interest to explore the potential application of this material [9]. The sugarcane industry is still seeking solutions to dispose of the wastes generated by the sugar and alcohol production processes. This ash is used as fertilizer in the plantations, but it does not have adequate mineral nutrients for this purpose [10]. Currently, there is no evidence in the current literature of its use in soil stabilization especially for organic soil.

Consequently, it could be something beneficial to develop alternative binders that are environment friendly and contribute towards sustainable management. The utilization of SCBA in the stabilization of peat soil can be an interesting idea and appears to be promising alternative when considering issues of energy consumption and pollution. The objective of this research works is to evaluate effectiveness factors of two type of SCBA inclusion in peat stabilization on its strength.

2 Experimental studies

The materials that had been used in this study are peat soil, Ordinary Portland Cement (OPC), calcium chloride ($CaCl_2$), silica sand and Sugarcane Bagasse Ash (SCBA). Peat samples were obtained from Sapporo in the Hokkaido region, Japan. Two types of SCBA samples namely SCBA 1 and SCBA 2 were directly brought from Shinko Sugar Industry Co., Ltd, Kagoshima prefecture in Kyushu, Japan. The well graded silica sand so called K7 type was prepared as a filler to increase the solid particles and enhance the filling effect of the stabilized peat. The entire laboratory test regulation and standards that had been implemented in this study was adopted from American Society for Testing and Materials (ASTM) [11] standards. The UCS tests were conducted at all samples with the aim to clarify the stabilized peat strength improvement. The whole mix designs of stabilized peat for laboratory testing are shown in Table 1. The mix designs of stabilized peat were formulated in term of binder composition and dosage. Each binder dosage was determined based on the bulk density of peat at its average natural water content. After all materials were mixed with untreated peat by using a laboratory mixer at approximately 10 minutes, the stabilized specimens were place into the mould by filling and tamping in about five equal layers. Each cylinder mould has a size of 60 mm internal diameter and 300 mm height. The samples then immersed in water for curing at identified duration under initial pressure by using air pressure to simulate the surcharge pressure on the stabilized soil at site. After curing durations complete, the cylinder tube is removed and the test specimen is cut to the required size for testing. Towards the degree of improvement evaluation, the established parameters of the stabilized soil should be compared to those of untreated peat.

3 Results and discussions

3.1 Materials properties

Basic properties of the untreated peat and the percentages of chemical compounds of the materials were exposes in Table 2. With the intention of ensure the reliability of results in Table 2, the basic properties ranges for typical Hokkaido peat were collected from published data [12-14] and had been used for guidance purpose. In this study, it can be understood that the peat has fiber content between 33% and 67%, ash content exceed 15% and pH value at 5.46. Therefore, this peat can be classified as hemic peat with high ash content and moderate acidic [11]. With the high water content, liquid limit, void ratio, compression index accompanied by low shear strength, the studied peat demonstrated the high compressibility and instability characteristics.

Table 1. Laboratory mix designs

No. test	Type of test	Curing duration (days)	Initial pressure (kPa)	Cement dosage (kg/m^3)	Silica sand dosage (kg/m^3)	Binder compositions @ acronym
1	Unconfined Compressive Strength, UCS	7	20	300	500	100% OPC @PC 95%OPC:5%SCBA @ PCB5 90%OPC:10%SCBA @ PCB10 85%OPC:15%SCBA @ PCB15 80%OPC:20%SCBA @ PCB20 75%OPC:25%SCBA @ PCB25
2		7, 14, 21, 28, 60	20	300	500	The optimum binder composition from test (1)
3		7	0, 20, 40, 60, 80, 100	300	500	
4		7	20	100, 150, 200, 250, 300	500	
5		7	20	300	0, 100, 200, 300, 400, 500	

Table 2. Peat soil basic properties

Untreated Hokkaido peat soil properties	Results	Published data
Natural water content, %	580	115 - 1570
Ash content, %	16.79	2 - 80
Organic content , %	83.21	20 - 98
Bulk unit weight (kN/m^3)	10.57	7.1 - 19.7
Specific gravity, G_s	1.67	1.04 - 2.63
Fiber content, %	43	42 - 86.9
Acidity ,pH	5.46	-
Liquid Limit, %	375	-
Void ratio, e_o	9.65	2 - 21
Compression Index, C_c	4.89	0.3 - 14
Preconsoilidation pressure, σ'_c (kPa)	25	-
Unconfined compressive strength, q_u (kPa)	13.6	5- 40

The particle size distribution of the peat and other materials are tabulated in Fig. 1. The particle size distribution curves for the soils were obtained by dry sieve analysis. In addition to the normal sieve test method, the finer fraction was analyzed using a laser diffraction method (SALD). As can be seen, Hokkaido peat consists of 90% of the soil that finer than 4.75mm, and about 2% is finer than 75μm. Most of K7 particles sizes are 0.05 to 0.3mm in range. The SCBA 1 and SCBA 2 particles consist about 80% and 15% finer than 0.045mm (passing No. 325 sieve) respectively. This result also indicates that SCBA 1 is finer than SCBA 2 with average particle size are (D_{50}) around 18 and 260 μm individually.

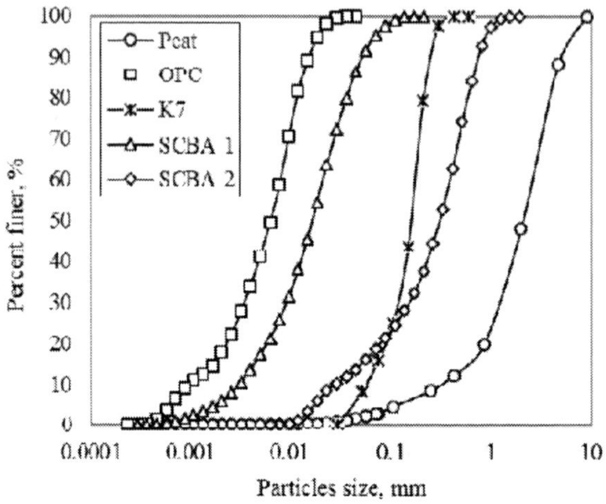

Fig. 1. Particle size distribution of peat and materials

3.2 Effect of partial replacement of OPC with SCBA on the strength characteristic

Fig. 2 shows the experimental results of the effect of both SCBA percentage replacements on the UCS of the stabilized peat. An optimal UCS of the stabilized soil was evaluated based on the results of UCS on the specimens of stabilized peat with partial replacement of the cement with SCBA that varies from 5% to 25% as shown in Table 1. The UCS ratio, q_u/q_o means the ratio between unstabilized and stabilized peat compressive strength. It can be detected that the test specimen with 20% partial replacement of OPC with SCBA 1 (PCB1-20) has the highest UCS of 387kPa and was discovered to be about 30 times better than unstabilized peat (P) and approximately 1.2 times greater than UCS of cement stabilized (PC) specimen. On the other hand, SCBA 2 mixture achieves the optimum at only 5% of cement replacement (PCB2-5) and recorded slight increment compared to PC specimen. As stated by ASTM-D4609 [15], an increase in UCS of 345kPa (adopted as minimum UCS target in this study) or more must be reached for a treatment to be considered effective.

The main reasons why PC mixture gave lower strength than optimum mixture of SCBA 1 and SCBA 2 are because peat soil consists of high organic content and less solid particles. Organic soils can retard or prevent the proper hydration of cement in soil mixtures and become insufficient to provide the required function for peat stabilization. Matched to clay and silt, peat has a considerably lower content of clay particles that can enter into secondary pozzolanic reactions [1, 2]. The combination of humic acid with calcium ions produced in cement hydration makes it difficult for the calcium crystallization, which is responsible for the increase of peat soil-cement mixture strength to take place [5]. The fact that the test specimen of PCB1-20 and PCB2-5 has the UCS may be explained by a condition whereby it has achieved an optimal effect of hydration reaction. By means of the inclusion of pozzolan such as SCBA in the soil-cement mixture, hydration of cement is accelerated when the pozzolan reacts with calcium hydroxide and water to form more secondary calcium silicate hydrate (CSH) along with calcium alumina silicate hydrates (CASH). This is probable because the pozzolan which contains extra silica and alumina that activated by cement is able to counterbalance the acid and create an alkaline atmosphere that boosts the secondary pozzolanic reaction within the cemented soil. Additional CSH and CASH densify the stabilized peat, thereby further enhancing its strength [6]. Frequently, the pozzolanic effect depends not only on the pozzolanic reaction but also on the physical or filler effect of the smaller particles in the mixture [16]. The positive result indicates that the optimal mix design can be effectively applied to stabilize the peat in such a way that the fine particles of pozzolan fill up the pore spaces of the cemented soil, thus closely packing, reinforcing and strengthening its matrix as the hydration and pozzolanic products are formed during cement hydrolysis [17]. In this assessment, it can be understood that SCBA 1 has better quality than SCBA 2 in peat stabilization especially the contribution made by its fine particle size.

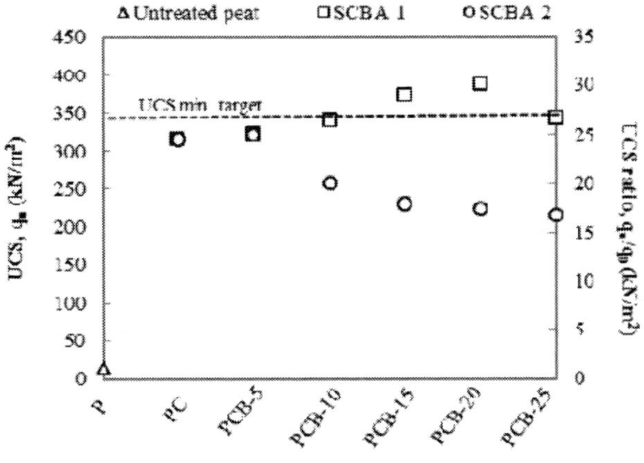

Fig. 2. Effect of SCBA percentage replacement on the UCS of the stabilized peat

3.3 Effect of curing duration and preloading on the optimum PCB mixtures

It can be observed from Fig. 3(a) that the UCS of both optimum PCB mixture specimens increased while increasing the duration of curing. The UCS of test specimens increased progressively at the curing time in water of 7, 14, 21, 28 and 60 days. It is evident from the findings that the duration of curing influenced the UCS of test specimens. The minimum UCS achieved at 7 and 21 days at the respective PCB1-20 and PCB2-5. Both optimum PCB mixtures show the slow increment compare to PC mixtures and overall PCB 1 mixtures demonstrates the higher positive results. This results shows the good agreement with the fact that cement hydration (PC mixtures) is normally rapid and effective at first month but almost stop or complete after that duration while pozzolan reaction can be occurred continuously until several month or even years [6, 18, 19]. Therefore it is plainly indicate that the quality of selected SCBA is very important and should be prudently examined in order to get long term strength gain.

The results in Fig. 3(b) shows the positive and linear improvement when the initial loading during curing rise. It observed that preloading gave the significant increment even only subjected the pressure as small as 20kpa. For PCB2-5, it is suggested to use at least 40kPa preloading to the aim of minimum strength. The strength of stabilized peat is pronouncedly affected by the application of initial load shortly after mixing with binder. Particularly in stabilizing peat, preloading in the field has been regarded necessary in order to create a more homogeneous stabilized mass of peat. Besides that, preloading provides a trafficable bed for the continuous stabilization of adjacent areas, thereby considerably improve the strength of stabilized peat. The void spaces between the binder grains and the solid soil particles in peat would be reduced by the compression that occurs under preloading. [3].

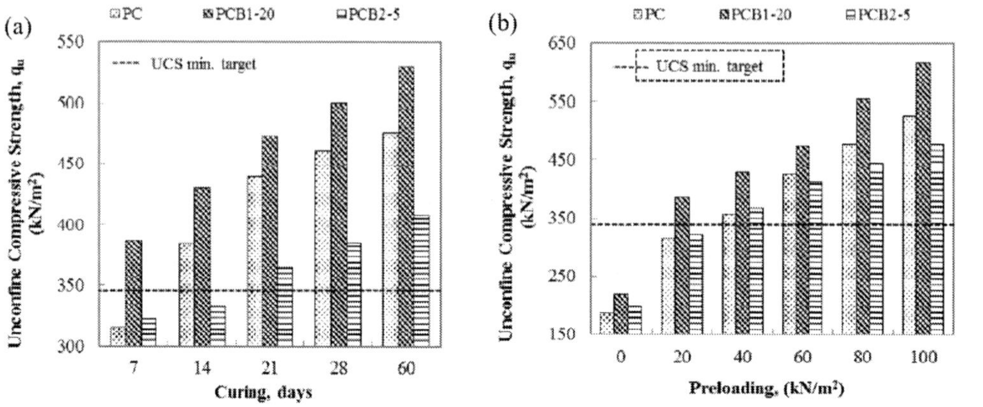

Fig. 3. Effect of optimum PCB mixture on the UCS at various (a) curing durations, and (b) preloading

3.4 Effect of OPC and K7 dosage on the optimum PCB mixtures

Fig. 4(a) and Fig. 4(b) show the results of the effect of the OPC and K7 dosage to the UCS of optimum PCB mixtures. It can be seen that strength gain gradually increases with OPC dosage and it's more obvious after the amount exceed 250kg/m^3. It is revealed that at satisfactory OPC quantity, the acid in the peat could be neutralized. This also imply that below the threshold binder dosage, stabilization of the peat remains hindered due to insufficient binder to induce cement hydration and pozzolanic reactions in the test specimens. There was a progressive increment of the UCS when the dosage of K7 of the test specimens increases. This suggests that well graded K7 had a positive effect on the development of the UCS in such a way that it increased the density and reduced the void of the test specimens. Because K7 is chemically inert, it did not take part in the cement hydration process. However, it provided solid particles for the binder to bind and form a load bearing stabilized soil. For PCB1-20 mixture, inclusion of a minimum OPC dosage of 300 kg/m^3 and K7 dosage of 500 kg/m^3 along with curing under 20kPa pressure is recommendable for the peat stabilization to be effective. However for PCB2-5 mixture, it suggested to use more OPC and K7 dosage to achieve minimum strength target.

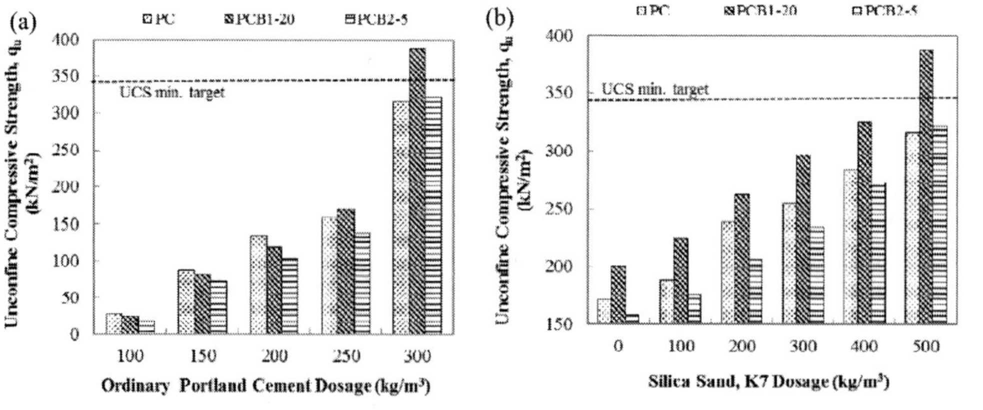

Fig. 4. Effect of optimum PCB mixture on the UCS at various (a) OPC dosage, and (b) K7 dosage

4 Conclusions

Highly organic or peat soil stabilization by utilizing the SCBA waste as a partial cement replacement has been studied. It can be summarized from the experimental outcomes that SCBA has made a substantial effect on the strength properties of the stabilized peat. Based on the results of the laboratory investigation, the following concluding comments are made:

i) Untreated peat in this study can be classified as hemic peat with high ash content and moderate acidic. With the high water content, high liquid limit and low shear strength, this soil demonstrated the high compressibility and instability characteristics.
ii) It was observed that PCB1-20 has the highest UCS of 387kPa and was discovered to be about 30 times better than unstabilized peat (P) and approximately 1.2 times greater than UCS of PC mixture. On the other hand, SCBA 2 mixture achieves the optimum at only 5% of cement replacement (PCB2-5) and recorded slight increment compared to PC specimen.
iii) The UCS of the stabilized peat specimens at the optimal mix design increased significantly with increasing of curing time, preloading during curing, OPC dosage and K7 dosage.
iv) For PCB1-20 mixture, inclusion of a minimum OPC dosage of 300 kg/m^3 and K7 dosage of 500 kg/m^3 along with curing under 20kPa pressure is recommendable for the peat stabilization to be effective. However for PCB2-5 mixture, it suggested to use more OPC and K7 dosage or alternatively increase the preloading during curing to 40kPa in order to achieve minimum strength target.
v) It can be concluded that SCBA 1 has better quality than SCBA 2 in peat stabilization especially the contribution made by its fine particle size. The quality of selected SCBA is very important and should be prudently examined in order to get long term strength gain.

Overall, it is suggested that the obtained optimum mix design can be applied to stabilize the peat layer in order to support road embankment for rural or agricultural road network.

References

[1] S. Hebib, E.R. Farrells, Some experiences on the stabilization of Irish peats of paper, Canadian Geotechnical Journal, **40**(1), 107-120 (2003)
[2] M. Janz and S.E. Johansson, *The function of different binding agents in deep stabilization of report*, Swedish Deep Stabilization Research Center, (2002)
[3] H. Ahnberg, *Strength of stabilised soils - A laboratory study on clays and organic soils stabilised with different types of binder of report*, Swedish Deep Stabilization Research Centre, (2006)
[4] K. Axelsson, S.E. Johansson, R. Anderson, *Stabilization of organic soils by cement and puzzolanic reactions- feasibility study of report*, Swedish Deep Stabilization Research Centre, (2002)
[5] H. Chen, Q. Wangs, The behaviour of organic matter in the process of soft soil stabilization using cement of paper, Bulletin of Engineering Geology and the Environment, **65**(4), 445-448 (2006)
[6] L.S. Wong, R. Hashim, F.H. Alis, A review on experimental investigations of peat stabilization of paper, Australian Journal of Basic and Applied Sciences, 3(4), 3537-3552 (2009)
[7] World Business Council for Sustainable Development, *International Energy Agency, Cement Technology Roadmap 2009*, (2009)
[8] *Food and agricultural commodities production of paper*, (2012)

[9] S.C. Lee, M. Mariattis, The effect of bagasse fibers obtained (from rind and pith component) on the properties of unsaturated polyester composites of paper, Materials Letters, **62**(16), 2253-2256 (2008)
[10] A. Sales, S.A. Limas, Use of Brazilian sugarcane bagasse ash in concrete as sand replacement of paper, Waste Management, **30**(6), 1114-1122 (2010)
[11] American Society of Testing Materials, *Annual Book of ASTM Standards, Section Four: Construction*, 04.08, (2000)
[12] S. Noto, *Peat Engineering Handbook, Civil Engineering Research Institute*, Hokkaido Development Agency, Japan. 1-35 (1991)
[13] S. Hamamoto, Effects of moisture content and shrinkage on soil-thermal properties for peat soils in Japan, *International Conference on Sustainable Built Environment*, Kandy, (2010)
[14] H. Hayashi, S. Nishimoto, M. Takahashis, Field performance of PVD combined with reinforced embankment on peaty ground of paper, Soils and Foundation, Japanese Geotechnical Society, **51**(1), 191-201, (2011)
[15] ASTM-D4609s, *Standard Guide for Evaluating Effectiveness of Admixtures for SoilStabilization of paper*, American Society for Testing and Materials Standards, Pennsylvania, (2005)
[16] G.C. Isaia, A.L.G. Gastaldini, R. Moraess, Physical and pozzolanic action of mineral additions on the mechanical strength of high-performance concrete of paper, Cement and Concrete Composites, **25**(1), 69-76 (2003)
[17] L.S. Wong, R. Hashim, F. Alis, Utilization of sodium bentonite to maximize the filler and pozzolanic effects of stabilized peat of paper, Engineering Geology, **152**(1), 56-66 (2013)
[18] EuroSoilStab, *Development of Design and Construction Methods to Stabilize Soft Organic Soils, Design Guide Soft Soil Stabilization*, Project No. BE 96-3177, Industrial and Materials Technologies Programme, European Commission, (2002)
[19] F.O. Allu, Ramboll, *Mass Stabilization Manual*, Orimattila, Findand, (2005)

Stabilization of Peat Soil by Using Envirotac

Putri Nur Farah Izzati Norazam[1], *Ismail* Bakar [1,*], *Alvin John* Lim Meng Siang[1], and *Herman* Shah Herman[1]

[1] Faculty of Civil and Environment Engineering, University Tun Hussein Onn Malaysia, 86400 Parit Raja, Johor, Malaysia

Abstract. Peat soil is wet and soft. It is also known as a very soft and difficult soil with high water and organic content, low shear strength, low bearing capacity and their mechanical chemical and biological properties degenerate with time. The characteristics and geotechnical properties of peat soil are mainly related to its high moisture content and high organic content respectively. This is very challenging to geotechnical engineers and construction industry as peat soil characteristic can cause excessive settlement.Soil stabilisation is the process of improving the engineering properties of the soil which will then in return make it more stable. As a result of soil stabilization, the bearing capacity and the strength of the soil will increase with an added bonus of water tightness, resistance to wash out and other properties which will also be improved. In this research, Envirotac polymer is used as the soil stabilizer which is in a form of liquid. This study is done in order to investigate the optimum mix proportion of Envirotac as a liquid stabilizer on Peat soil. The objective of this study was to determine the strength effects of 15%, 30% and 45% of Envirotec on peat sample and to investigate the shrinkage effect of stabilizing peat soil by using Envirotac. A mould sample with a diameter of 50mm and a height of 100mm was prepared. Samples were then tested after 7, 14 and 21 days. The physical properties tests conducted for this research are the Atteberg limits and moisture content test, while the mechanical test was Unconfined Compressive strength (UCS) test. The tests were carried out according to BS 1377: 1990. From the result, 30% Envirotac on peat sample has higher strength from control sample and more economical compare to use 45% Envirotac which had greater shrinkage effect

1 Introduction

Soil stabilization is the recent improvement to a certain desired properties in soil to make it stable and useful for a specific purpose especially construction aspects. Dating back from previous stabilization techniques cannot provide an economical and environmental friendly solutions. In this stabilization method, soil are bonded together using liquid stabilizer. This research will focus on establishing peat profile at Gamuda Development Land Jalan Sri Cheeding, Jenjarom Dengkil,Selangor. This is an industrial based laboratory research

[*] Corresponding author: bismail@uthm.edu.my

DOI: 10.1051/matecconf/201710307014

project which is mainly related to experimental works in the usage of Envirotac soil stabilizer to increase the strength of peat soil on site. It is because Envirotac can increase the economical functions and sustain the environment aspects in the investigations. However, when volume of water is reduce to a shrinkage limit, the linear shrinkage of Envirotac peat soil may decrease in one dimension on a soil mass, commonly expressed as a percentage of the original dimension. It was clearly determined according to BS 1377: Part 2: 1990 [1].

2 Methodology

The determination of the material and testing of the peat soil sample with suitable are important to obtained data for analysis purposes. Undisturbed soil is used as control sample of the research. This research has conducted physical test and mechanical test. For physical test is atterberg limit, meanwhile mechanical test is unconfined compressive strength (ucs) and linear shrinkage test.

2.1 Envirotac

Envirotac is a unique soil stabilizer, dust and erosion control product. Envirotac commonly in liquid form and it is white colour. During drying process, it binds the soil particles together by forming a clear plastic and resin bond. Besides, envirotac also can be use effectively on in-situ material such as sand soil,laterite soil, sily and. Envirotac also confirmed as non-toxic, non-hazardous, environmentally safe dust control which is concentrated liquid format and finally cured in solid state. However, it is still considered as new material in this field, research on this soil stabilizers is yet to be discovered. Fig. 1 shows Envirotac, soil stabilizer in liquid form.

Fig. 1. Envirotac soil stabilizer in liquid form

2.2 Soil Sample preparation

Preparation of soil sample for this study were strictly complied to [1] and ASTM D1997 Standards. This study are focus on several testing to identify the physical and chemical characteristics of the sample. Intially, the peat soil sample will be mixed with different percentages of Envirotac and disttiled water. This sample undergo several test to determine its effectiveness and performance of stabilised peat soil. Table 1 shows sample informations.

Table 1. Samples of information

Type of Soil	Additives	Number of samples		
		7 days	14 days	21 days
Peat Soil	Blank	3	3	3
	15% Envirotac	3	3	3
	30% Envirotac	3	3	3
	45% Envirotac	3	3	3

2.3 Atterberg limit

Atterberg limit is used to determine the water content corresponding to different state of soil. The testing that conducted for this study is moisture content (MC), liquid limit (LL) and the plastic limit (PL). It has physical meanings for remoulded fine grained mineral soils and they correlate with many fundamental soil parameters used in design and construction practice [4]. LL test is repeated with same soil which is peat soil with addition of distilled water, meanwhile PL test is cannot be determine if peat soil is cohesion less [5].

2.4 Unconfined compressive strength (UCS) test

The purpose of this test is to determine the soil strength for civil engineering purposes according to Clause BS 1377: 1990 Part 7. Prepare a cylindrical sample which is then subject to increasing axial compression until failure occurs. UCS is the highest load that could be attain per unit area on a cylindrical sample tested in compression. Fig. 2 shows UCS Test equipment.

Fig.2 UCS test equipment

2.5 Linear shrinkage test

Linear shrinkage test is conducted using standard semi-circular linear shrinkage moulds of 25 mm diameter x 140 mm length in accordance to BS 1377:1990 (Test 5). The test aimed to examine the reduction in linear shrinkage of stabilized peat in comparison to that untreated peat. Fig. 3 shows linear shrinkage test of the sample.

Fig. 3. Linear shrinkage test of the sample

3 Result

The experiment was conducted in Research Center Soft Soil (RECESS), Universiti Tun Hussein Onn Malaysia. The result are shown clearly from the analysis of graphs. Unicofined compressive strength between the control specimens and specimens containing substituents, and the shrinkage effect of the envirotac into the peat soil was analyzed.

3.1 Alterbert limit test

For this test, the moisture content of the soil is 237 %. From the data, it can be classified as peat soil due to it is range 200% - 700% [3]. Table 2 shows moisture content of the soil sample. Next, the liquid limit test was conducted based on BS1377: Part 2: 1990: 4.3 The liquid limit percentages of the peat soil sample is 264% while the plastic limit percentages is 139%. The data obtained shows that liquid limit and plastic limit is high and can be categorized as peat [2],[6]. Table 3(a) shows plastic limit of the soil sample, Table 3(b) shows liquid limit of the soil sample and Figure 4 shows the liquid limit of the soil sample.

Table 2. Moisture content of the soil sample

Container No	1	2	3
Wet soil + container (g), m2	55.54	55.31	55.18
Dry soil + container (g), m3	20.73	20.63	20.91
Container (g), m1	6.12	6.27	6.32
Moisture Loss (g), m2-m3	34.81	34.68	34.27
Dry soil (g), m3-m1	14.61	14.36	14.68
Moisture Content(%)	238.3	241.5	233.4
Average Moisture Content (%)	237		

Table 3(a). Plastic limit of the soil sample

Test Number	1	2
Mass of Can + Moist Soil, M_{cws} (g)	20.76	20.63
Mass of Can + Dry Soil, M_{cs}, (g)	15.28	12.05
Mass of Can M_c (g)	5.35	5.86
Mass of Dry Soil, M_s (g)	6.46	6.19
Mass of water, M_w (g)	8.95	8.58
Water content,w (%)	138.5	138.6

Table 3(b). Liquid limit of the soil sample

Test Number	1		2		3		4	
Dial Gauge Reading (Start) (mm)	0	0	0	0	0	0	0	0
Dial Gauge Reading (End) (mm)	15.1	15.2	18.0	18.1	21.1	21.2	24.3	24.2
Cone Penetration (mm)	15.1	15.2	18.0	18.1	21.1	21.2	24.3	24.2
Average penetration (mm)	15.15		18.05		21.15		24.25	
Can Number	A		B		C		D	
Mass of Can + Moist Soil, M_{cws} (g)	41.03		41.04		40.24		40.52	
Mass of Can + Dry Soil, M_{cs}, (g)	15.28		15.23		14.82		14.99	
Mass of Can M_c (g)	5.42		5.42		5.23		5.43	
Mass of Dry Soil, M_s (g)	9.86		9.81		9.59		9.56	
Mass of water, M_w (g)	25.75		25.81		25.42		25.53	
Water content,w (%)	261.2		263.1		265.1		267.1	

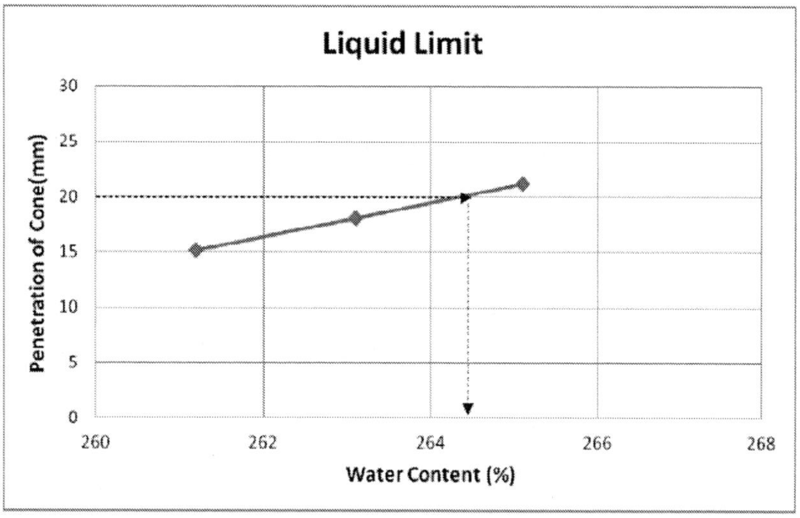

Fig. 4. Liquid limit of the soil sample

3.2 Analysis of unconfined compressive strength test

UCS test was performed on the cured peat soil sample with different percentages of stabilizer. In this research, the standard curing (air curing technique) were used during sample preparation. From the data, the strength of the peat soil with Envirotac was increases as the percentage Envirotac in the peat soil sample increases. It is because the addictive was homogenously bond together with the peat soil sample. Table 4 shows result of the peat soil sample with the Envirotac meanwhile Fig. 5 shows unconfined compressive strength of peat soil with Envirotac curing time.

Table 4. Result of the peat soil sample with Envirotac.

Type of Soil	Additives	Unconfined Compressive Strength (kPa)		
		7 days	14 days	21 days
Peat Soil	Blank	42.46	44.84	46.31
	15% Envirotac	175.81	405.53	501.27
	30% Envirotac	241.87	537.34	561.72
	45% Envirotac	422.95	560.55	573.89

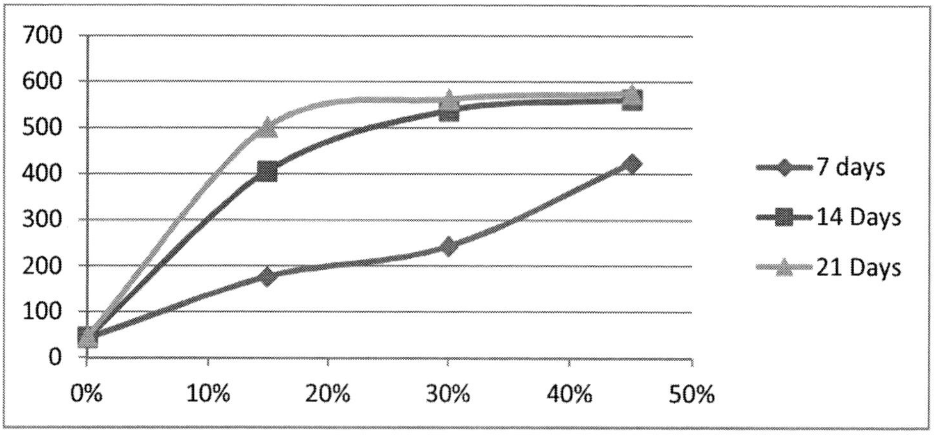

Fig. 5. Unconfined compressive strength of peat soil sample with Envirotac curing time.

3.3 Analysis of linear shrinkage test

This test aimed to determine the reduction in linear shrinkage of stabilized peat in comparison to that untreated sample, treated with polymer which is Envirotac. From the result, the addition of Envirotac into soil sample make it more compact. Besides, it is also homogeneous.

Table 5. Linear Shrinkage of the Sample

Sample	Length of mould, L (mm)	Longitudinal shrinkage of the specimen, Ls (mm)	Linear Shrinkage, LS (%)
Peat	140.22	118.80	15.28
Peat + 15% Envirotac	140.51	116.34	17.20
Peat + 30% Envirotac	140.43	115.10	18.04
Peat + 45% Envirotac	140.33	110.64	21.16

4 Conclusion

As conclusion, the addition of the Envirotac into peat soil can increase the strength of peat soil sample. However, the shrinkage of the peat soil will increase as the addictive react with it. The best percentage of the Envirotac in the peat soil is 30%. The strength of 45% of the Envirotac in the peat soil is the highest. However, it is suffered a lot of shrinkage rather than 30% of the Envirotac in the peat soil. Besides, the strength of 30% and 45% Envirotac after 21 days curing time is almost similar. From here, it can be more costly for treat the peat soil.

Next, this research still new in this field and a lot of improvement must be carry out near future. As the recommendation, Envirotac must be mix with other material to control the shrinkage of the peat soil. High Density Polyethylene (HDPE) is can be use one of the solution to control the shrinkage of the sample with Envirotac.

References

[1] BS 1377, *Methods of test for soils for civil engineering purposes*, British Standard Institution, London, (1990)
[2] B.B.K. Huat, S. Kazemian, A. Asadi, A. Prasad, Geotecnics of organic soils, CRC Press/Balkema, London, (2014)
[3] B. Huat, *Organic and peat soils engineering*, International Standard Book Number: 983-2871-08-5, (2004)
[4] P. Kolay, Peat stabilization using gypsum and fly ash, UNIMAS E-Journal of Civil Engineering, 1(2), (2010)
[5] P. Kolay, H. Sii, S.Taib, Tropical peat soil stabilization using class F pond ash from coal fired power plant, International Journal of Civil and Environmental Engineering **3**(2), (2011)
[6] S.N.M. Razali, I. Bakar, A. Zainorabidin, Behaviour of peat soil in instrumented physical model studies, Procedia Eng., **53**, 145–155 (2013)

Critique on the Conceptual Developments of Innovative Practices with Lightweight Fills

Devapriya Chitral Wijeyesekera[1,*], *Nurvin* Ratha[1], and *Alvin* John Lim Meng Siang[1]

[1] Faculty of Civil and Environmental Engineering, Universiti Tun Hussein Onn Malaysia, Johor, Malaysia

Abstract. Conceptual developments are underpinned by the intellectual process of constant reviewing and developing innovative research ideas into realistic design. There exists no one quick fix, fit for purpose solution to ensure stability of structures being constructed on a wide range of demanding ground conditions and variety of project constraints. Routine highway design and construction technology conformingly adopt only conventional fill methods and thus fail even to address adequately the less favoured conditions in challenging situations: a high initial construction cost arising from a full replacement of existing unfavourable soil conditions or an unsustainable construction design leading to a condition of rapid disintegration requiring regular expensive and disruptive highway maintenance activity. Undulating road embankment surfaces result from the transference of heavy self-weight of the embankment fill on yielding and non-uniform subgrade such as in peat soils. The Manitoba Water Services Board [1] specifies four classes of fills (viz; common, compacted common, compacted select granular backfill and unshrinkable backfill) for backfilling of pipeline trenches. Both conventional and alternatively sustainable backfills are discussed in this paper with new and appropriate technology opening doors with other tangible benefits for innovative and cost saving outcomes for differing construction scenarios.

1 Introduction

Sustainable infrastructure developments are vital socio-economic pathways that foster a better quality of life to all and in particular, for the less fortunate of a developing country's population. Competitiveness and the sustainability of the developed products and the technologies are additional parameters of information for the critique of conceptual design. Methods include the performance specifications method, quality function deployment (QFD), and morphological charts. Means include brainstorming, synectics, analogies, benchmarking, and reverse engineering.

Cost effective infrastructure construction is necessarily vital for the economic health of any country. This must be deemed a priority particularly in developing countries. Nearly linear infrastructures such as highways, rural roads, utility pipelines (and others such as power lines) desire to be post construction maintenance free. Daud et al [2] emphasised

* Corresponding author: dcwijey@gmail.com

that the 'green road' approach to be the most appropriate for rural road construction in developing countries and be set in place through environmental friendly construction practice with optimum utilisation of local resources together with simple (self-help) technologies. Historically, the first documented road construction dates to 4000BC (stone paved roads in Ur, Middle East and Roman Corduroy Roads in Glastonbury, UK) and 3000 BC (brick paving in India) [3]. Emerging infrastructure projects such as routing of new highways demand construction to inevitably take place on less favourable soils, such as on soft yielding soils or on even cavernous rocks that require embankment fills.

Cracking of road pavements and sidewalks (Fig. 1(a) and 1(b)) are clear indicators of failure initiations and the presence of inherently imminent problems. Learning must extend beyond recognising a problem, to seek a better understanding of the cause of failure and recommend pragmatic, sustainable and lasting solutions. The conceptual development phase for any lasting solutions can be reverberantly time-consuming, depending on the level of expertise demanded for the investigation. Lightweight fills being the researched area, current engineering problems faced in transportation infrastructures and the potential methods to address those challenges is highlighted in this paper.

2 Road pavement cracking as early failure indicators

Roads with heavier loads need to be designed differently than the lower volume roads. Furthermore, paved roads flex excessively through fatigue in the spring (monsoonal weather, under surface seepage or flood situations) when it is softened through saturation. Some common types of pavement cracking are fatigue/ alligator cracking, longitudinal cracking, transverse cracking, block cracking, slippage cracking, reflective cracking and edge cracking [4] (Fig. 1). Sub grade softening occur through ingress of water into existing cracks, standing water on a road without a crown, piping of ground water or from leaking buried pipes. Longitudinal cracking parallel to the laydown direction are a type of fatigue cracking, but can also result from the lateral squeezing of the softened particulate subgrade through and into the unsupported side drains. Alligator cracking is characterised by interconnected / interlaced cracking in the asphalt layer (similar to that on an alligator hide). Crack sizes range between 150 mm and 300 mm in size and numerous factors such as sub-base failure, poor drainage or repeated over loading can be the cause.

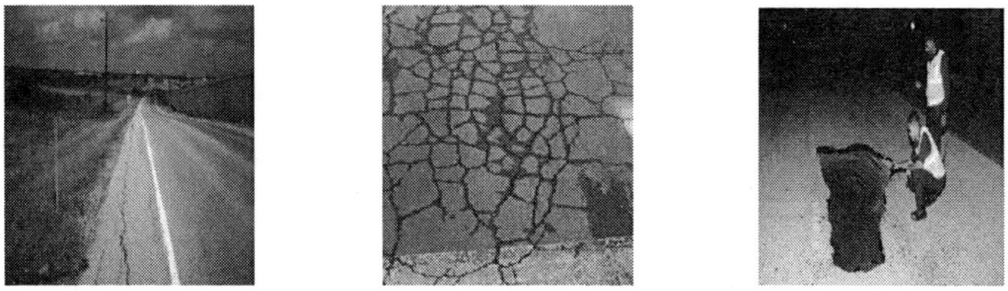

(a) (b) (c)

Fig. 1. Few examples of Road Pavement failures (a) Longitudinal Cracking (b) Alligator Cracking (c) Road collapse - Taman Kinara [5]

Fig. 1(c) echoes a report in New Strait Times, 2011 [5] of a not so uncommon road collapse that occurred catastrophically in Section 5, Taman Kinara, Batu 9, Jalan Puchong leaving a gaping sinkhole (6m deep and 4m wide). This is a characteristic sinkhole failure in the upper Silurian marble rock of the calcareous rock formation in Kuala Lumpur. Such failures emphasise the need for careful ground investigations even in Highway design and construction over presumed sound and "solid" rock geology.

2.1 Maintenance on the approach roads to Parit Yaani roadbridge

On a hotly pursued research note, Fig. 2 gives location (coordinates 1.876205, 103.016594; on J125 - Jalan Parit Yaani/Seri Bengkal Kampung Sengkuang) where recurrent and costly maintenance of road pavement distress is being done. During 2010/ 2013, the road bridge over a minor tributary of Sungai Simpang Kanan was constructed on piled foundations driven into soft marine clay. The approach road to the reinforced concrete bridge has since been repaired six times (Nov 2013, Apr 2014, Sept 2014, Feb 2015, Aug 2015 and Feb 2016). Each time fresh conventional particulate fill (sometimes with reinforcing geomembranes) were placed. Fig. 2(c) and 2(d) taken now in October 2016 show the development of new transverse linear cracking at the crest of the approach road fills, resulting from continued self-weight settlement of the underlying soft marine clay.

(a) Google View of the Bridge site (c) and (d) Current views the Bridge approach

(b) View of the Bridge approach after 2013 repair (d)

Fig. 2. Views of J125 - Jalan Parit Yaani/ Seri Bengkal Kampung Sengkuang

2.2 Gravel roads – Extremes of problems

On the other hand, unpaved rural roads pose their own suite of problems. As examples, poor drainage and construction can result in wash boarding effects (Fig. 3(a)) [6]. In dry weather conditions, unsealed gravel roads cause dust (Fig. 3(b)) which is an aggravating problem to the people who live nearby and to the wider environment. Chlorides of Calcium, Magnesium and even Sodium are being used as dust suppressant stabilisers.

3 Conventional and natural fill material

An embankment fill is a volume of earthen material that is placed and compacted for the purpose of raising the grade of a roadway (or railway) above the level of the existing ground surface, filling in a hole or depression. They are constructed of materials that usually consist of soil, including aggregate, rock or crushed paving material [8] and are

DOI: 10.1051/matecconf/201710307015

used as bases in rural road construction or as a sub-base for flexible or rigid pavements, and are also used (bound or unbound) in the construction industry as trench backfills. Department of Transport (UK) specifies 2 types of granular materials (Type 1 and Type 2) for Road and Bridge Works. Type 1 is a well graded aggregate that is restricted to crushed rock, crushed concrete or well burnt non-plastic colliery shale (Fig. 4). Material that is finer than 425μm should necessarily be non-plastic. Type 2 additionally includes natural sand and gravel and the plasticity index of the material finer than 425μm is limited to 6%. For use as a subbase it must have a minimum California Bearing Ratio (CBR) of 20%, and needs to be protected from wet conditions.

(a) (b)

Fig. 3. Corrugation (wash boarding) in gravel roads having inadequate crown [6], and dust hazards from gravel roads during dry weather [7]

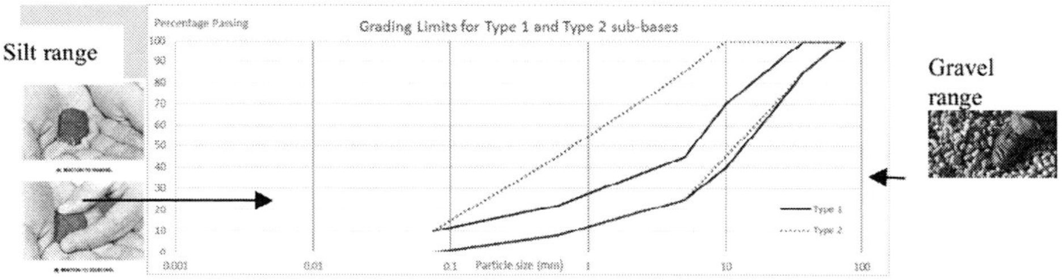

Fig. 4. Grading Curve Limits for Type 1 and Type 2 fill lying between silt and gravel

The durability of a pavement depends on the underlying subgrade soil (Fig. 5(a)). It shows that anything finer than sand gives a poor support with a clear decrease in CBR values. Fig. 5(b) [9] indicates the JKR guidelines for the selection of alternative pavement structures for low volume (< 1 m ESAL) roads.

4 Alternative lightweight fill materials

Wijeyesekera et. al. [11] describes the use of artificial aggregates (manufactured from Municipal Solid Waste) as an innovative, improved and sustainable trench backfill and bedding material for buried plastic pipes. Large quantities of postconsumer glass can be recycled and are being reused as crushed aggregate filler in mortars [12]. Table 1 is an overview of different alternative fills for road subgrades that are being pursued in contemporary research.

Subgrade Soils for Design	Unified Soil Classifications	Load Support and Drainage Characteristics	Modulus of Subgrade Reaction (k), psi/inch	Resilient Modulus (M_R), psi	CBR Range
Crushed Stone	GW, GP, and GU	Excellent support and drainage characteristics with no frost potential	220 to 250	Greater than 5700	30 to 80
Gravel	GW, GP, and GU	Excellent support and drainage characteristics with very slight frost potential	200 to 220	4500 to 5700	30 to 80
Silty gravel	GW-GM, GP-GM, and GM	Good support and fair drainage, characteristics with moderate frost potential	150 to 200	4000 to 5700	20 to 60
Sand	SW, SP, GP-GM, and GM	Good support and excellent drainage characteristics with very slight frost potential	150 to 200	4000 to 5700	10 to 40
Silty sand	SM, non-plastic (NP), and >35% silt (minus #200)	Poor support and poor drainage with very high frost potential	100 to 150	2700 to 4000	5 to 30
Silty sand	SM, Plasticity Index (PI) <10, and <35 % silt	Poor support and fair to poor drainage with moderate to high frost potential	100 to 150	2700 to 4000	5 to 20
Silt	ML, >50% silt, liquid limit <40, and PI <10	Poor support and impervious drainage with very high frost value	50 to 100	1000 to 2700	1 to 15
Clay	CL, liquid limit >40 and PI >10	Very poor support and impervious drainage with high frost potential	50 to 100	1000 to 2700	1 to 15

(a)

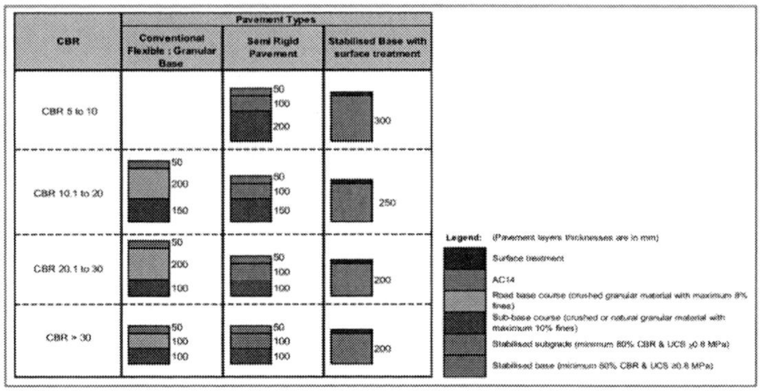

(b)

Fig. 5. (a) Suitability as subgrade [9] (b) JKR guide for selection of alternative pavement structure [10]

Table 1. Overview of some emerging alternative sustainable fills.

Alternative Composite Material & Production Method	**Functional use, Technical Properties & Approximate Costs**
Foamed recycled glass [13]; Low weight foamed recycled glass aggregates of up to 40 mm in size	***Non-structural fills in embankments, Retaining wall backfill and pipe bedding;*** Gap graded materia ; Max. Dry density = 2.84 kN/ m^3 ; LA Abrasion = 94%
Recycled Asphalt Pavement(RAP) – Fly ash(FA) Geopolymer [14] A 30 % FA mixture and a 50:50 ratio of NaOH and Na2SiO3 is used at40 C curing temperature	***Stabilized pavement base course material;*** Compressive strength= 5.3 MPa Fly Ash 1 tonne – Price insignificant. Transporation cost and safety certificate required due to hazardous nature.
Cement-treated sand and expanded polystyrene (EPS) beads [15] EPS beads are heated with steam to expand 50 times its original size. Cement binds sand and EPS.	***Backfill material for use in highway embankments in soft soil areas*** Compression Strength = 800 KPa Unit Weight = 10 kN/m^3
Clays Modified with Recycled Crumb Rubber [16] Kaolin clay and bentonite were used as soils for modification. 2-4 % of rubber crumbs give maximum strength.	***Soil modification*** Max dry density = 1296 kg/m^3 Max. Strength = 482 kPa
Blast furnace slag [17] The cooled slag is transported to a crushing and screening plant for further processing into various products including aggregates.	***As fill material for subgrade, sub-base or base.*** Particle Density= 2,550 — 2,650 kg/m^3 LA abrasion= 37 — 43% OMC= 8 — 12 % ; RM 50.10 /metric tonne
Geofoam [18] Polymerization of styrene monomer is followed by impregnation of the polymerized polystyrene beads with a blowing agent. Styrene monomer and water is charged to the reaction kettle equipped with an agitator.	**As fill material for highways and embankments on peat or soft soil.** Min. density= 15 kg m^3 ; Compressive stress=60 – 110 kPa Flexural strength= 60 – 300 kPa Shear strength= 80 – 130 kPa ; Tensile strength= 110 – 290kPa 1.2m x 1.8m x 0.6m ~RM 300.0

5 Alternative and lightweight fill technology

Wijeyesekera et al [19] designed the compacted clay surround barrier for the Heathrow Express Tunnel (portion passing through an old landfill site) to prevent methane and other environmentally hazardous material from the landfill penetrating the box tunnel in the form of gas, solid or as them dissolved in the landfill leachate. They adopted a sustainable approach of reusing (with strict moisture conditioning) the London clay being excavated from a nearby shaft for Terminal 4 construction. The three main acceptance criteria for the compacted clay surround (Fig. 6) were; to achieve a low permeability in the clay fill of less than 10^{-9} m/s, swelling pressures to be less than the horizontal stress caused by the overburden, and the fill to be flexible in accepting a degree of differential movement without fissuring or cracking. As shown in Fig. 6, this was achieved with strict compaction and moisture control of three specific layers of fill at plastic limit (PL) of 1.1PL, 1.2PL and 1.3PL. Additional environmental gas venting and post construction deformation monitoring provisions are also shown in Fig. 6.

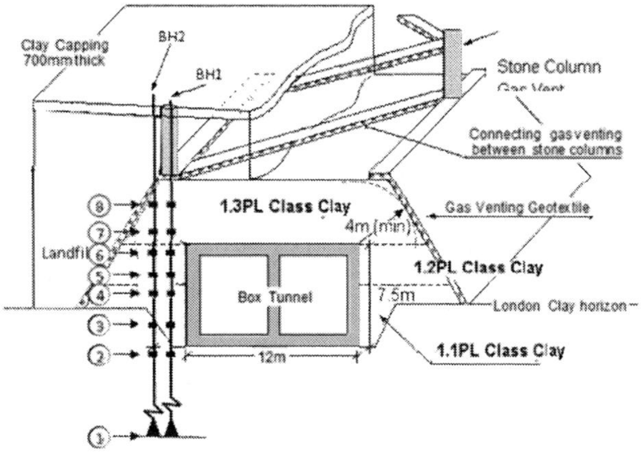

Fig. 6. Compacted Clay Liner (Fill) surround for Heathrow Express Tunnel [19]

Further technology to mitigate settlement has been the intuitive adoption of bio engineering concepts in using natural material (tree trunks- logs or bamboo) to reinforce the embankments. Fig. 7(a) illustrates a 2500 BC Babylonian technology[20], where hewn tree trunks were laid on yielding subgrade to form a permeable and stiffened / reinforced subbase (but lacking monolithic rigidity found in reinforced concrete).

Somewhat similar bio engineering concept is further seen in a case study reported by Priyadarshee et al [21] who reviewed construction of a low-volume semi rigid pavement with the provision of a grid of bamboo as soil reinforcement(A bamboo mesh 250x250mm placed on 100 mm wooden spacers). Such practice is being popularly used in current construction with the additional use of geofabrics separating the reinforced bamboo cages and interspersed with sand layers (Fig. 7(b)).

Environmentally friendly adoption of used tyres as tyre bales or mats (Fig. 8(a) and 8(b)) over soft ground demonstrates the need for a stiffer but lightweight infill to the subgrade or embankment. Tyre bales measure approximately 1.3x1.55x0.8 m. Tyre mat manufacture is more labour intensive and its length can be custom made while the width (tyre circumference) and the thickness (tyre width) are fixed with the split tyres held together by steel rods. Fig. 8(c) [22] illustrates the some of the potential applications in highway structures. As seen in Fig. 8(d), geotextiles are used as appropriate surrounds to restrict the movements of fines between the regulating sand layers and the ground.

DOI: 10.1051/matecconf/201710307015

Fig. 7. Technology with a bio engineering concept; (a) Early road construction in Ur – adopted from [20]; (b) Bamboo grid and cell reinforcement – construction site in Selangor, photographed by author.

Blocks of expanded Polystyrene (EPS) have been used extensively in Malaysia to remedy the settlement in approach road embankments (Bridge abutments). EPS is popularly used in the form of a lightweight (< 50 kg/m^3, dry density) block material, to mitigate settlement by reducing gravitational self-loading.

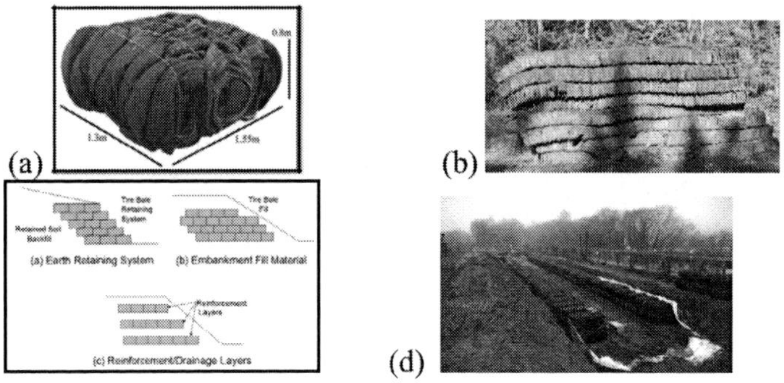

Fig. 8. Sustainable, environmentally appropriate fill technology; (a) Tyre bales [23] (b) Tyre mat – photographed by author (c) Proposed uses of tyre bales [22] (d) Construction site [22]

The very low density of the EPS makes it vulnerable to high buoyancy forces [24]. EPS is a combustible material and there are instances of failures caused by fire.

6 Conclusions

A variety of lightweight fill materials and emerging technologies have been studied and presented in this paper circumspectly. The importance of the fill being lightweight was highlighted through examples of various alternative fill technologies in practise. Furthermore, by using composite and existing lightweight materials as alternative fill material, it contributes in achieving the necessary balance between economics and sustainability, thus an effective replacement for the conventional fill.

References

[1] The Manitoba Water Service Board, *Standard Construction Specifications Section 022180*, Pipe Excavation, Bedding and Backfill, (2013)
[2] D. Daud, P.C. Tan, M. Lee, D.Q.Wu, Green and quality rural roads in Malaysia, *Proc. 7th Int. Conf. on Road and Airfield Pavement Technology*, 1-8 (2011)
[3] R.S. Kirby, *Engineering in History*, Pubs. Courier Corporation, (1990)

[4] New Strait Times, 27th September (2011)
[5] U.S. Department of Transportation FDHA, *Gravel Roads – Maintenance and Design Manual,* South Dakota Local Transport Assistance Program, (2000)
[6] New Zealand Research Council, *Dust – Nuisance Issues*, Retrieved on October 2016 www.nrc.govt.nz/environment/air/issues/dust_nuisance
[7] Federal Highways Authority, *Gravel Roads*, FHWA-RD-97-148, US Department of Transportation, (1997)
[8] American Pavement Concrete Association, *Asphalt Roads*, Asphalt Paving Association, Ohio, (1996)
[9] Jabatan Kerja Raya, *Pavement Design Guide*, Kuala Lumpur, (2012)
[10] D.C. Wijeyesekera, D. Newport, K. Edwards, Manufactured Aggregate as improved backfill material for plastic pipes, *Plastic Pipes XII*, Italy, (2004)
[11] V. Corinaldesi, A. Nardinocchi, J. Donnini, Reuse of recycled glass in mortar manufacturing, European Journal of Environmental and Civil Engineering, (2016)
[12] A. Arulrajah, M.M. Disfani, F. Maghoolpilehrood, S.Horpibulsuk, A.M. Udonchal, Imteaz, Y.J. Du, Engineering and environmental properties of foamed recycled glass as a lightweight engineering material, Journal of Cleaner Production, **94**, 369-375 (2015)
[13] M. Menglim Hoy, S. Horpibulsuk, A. Arulrajah, Strength development of recycled asphalt pavement – Fly ash geopoly mer as a road construction material, Construction and Building Materials, **117**, 209–219 (2016)
[14] F. Wang, L. Miao, A proposed lightweight fill for embankments using cement-treated Yangzi River sand and expanded polystyrene (EPS) beads, Bull Eng Geol Environ, **68**, 517–5(2009)
[15] B. Beena Ajmera, B. Tiwari, J. Koirala, Geotechnical properties of clays modified with recycled crumb rubber, Geotechnical and Structural Engineering Congress, ASCE, (2016)
[16] Australasian Slag Association, *A Guide to the Use of Slag in Roads*, (2002)
[17] A.F. Elragi, *Selected engineering properties and applications of EPS geofoam*, ProQuest Disssertations and Theses, (2000)
[18] D.C. Wijeyesekera, K. O'Connor, D.E. Salmon, Design and performance of a compacted clay barrier through a landfill, Engineering Geology, **60-4**, 295-306 (2001)
[19] Information retrieved from http://sunerianshakespeare.com/mediac/450-0/.jpg (2011)
[20] A. Priyadarshee, A.K. Chhotu, V. Kumar, Use of bamboo in low volume rigid pavement as reinforced material: A review. Journal of Civil Engineering and Environmental Technology, 1(3), 14-16 (2014)
[21] B. Freilich, J.G Zornberg, *Mechanical Properties of Tire Bales for Highway Applications*, Technical Report for Texas Department of Transportation and the Federal Highway Administration, (2009).
[22] C. Mackenzie, T. Saareketo, *Use of recycled tyre bales in a lightweight road embankment over peat*, Research Report THE B 871, Road Scanners, (2003)
[23] J.S. Horvath, *Geofoam fills and the non-issue of buoyancy. Lessons learned from failure involving geofoam in roads and embankments*, Report No. CE/GE -99-1 Civil Engineering Manhattan College, (1999)

Exterior Noise Due to Interaction of Tire-Thermoplastic Transverse Rumble Strips

Zaiton Haron[1], *Mohd Hanifi* Othman[2,*], *Khairulzan* Yahya[1], *Mohd Rosli* Hainin[3], *Kamarudin* Ambak[2], and *Nadirah* Darus[1]

[1]Department of Structure and Materials, Faculty of Civil Engineering, Universiti Teknologi Malaysia, 81310 Skudai, Johor, Malaysia
[2]Smart Driving Research Centre (SDRC), Faculty of Civil and Environmental Engineering, Universiti Tun Hussein Onn Malaysia, 86400, Parit Raja, Johor, Malaysia
[3]Department of Geotechnics and Transportation, Faculty of Civil Engineering, Universiti Teknologi Malaysia, 81310 Skudai, Johor, Malaysia

Abstract. Transverse rumble strips (TRS) are a common choice to reduce vehicle speed and increase driver alertness on roadways. However, there is a potential trade-off using them on rural roadway due to the noise problem created when vehicles go over the strips. The present study investigated the noise level, spectral analysis and the possible noise generation mechanism when the TRS hit by a vehicle. Ten-raised-rumbler (RR) and three-layer-overlapped (TLO) were selected in this study as they have received complaints from the public. Results showed that RR generated relatively higher noise and impulse at low speed, and increased sound level in each octave band. Based on these results, RR may irritate human ears even when the vehicle travels at low speed. It was found that RR increased all noise generation mechanism of tire-pavement interaction whilst for the TLO increased structural resonance, sidewall and surface texture vibration.

1 Introduction

Motor vehicle accidents are the second most frequent cause of death in the entire world. Around 1.2 million people are killed each year on the road and 50 million more are injured [1]. Speeding and careless driving are the two main causes of accidents, contributing 32.8 and 28.2 percent respectively to the total number of accidents in Malaysia [2]. Speeding also contributed to about 30 percent of fatal crashes in the United States [3]. Alternative measures consisting of a road layout and its associated features including transverse rumble strips which are able to subconsciously inform drivers of upcoming road conditions were introduced. Transverse rumble strips (TRS) made of a groove or raised lateral pattern are introduced to reduce vehicle speed and increase driver alertness on the roadway. Generally, TRS around the world are diverse in terms of configuration, dimensions, colour, and profile. TRS is intended to give an audible, visual and tactile cue of when an operational decision point is approaching [4]. TRS vertically deflects the wheels of a vehicle driving over them and produces both noise and vibration [5].

* Corresponding author: hanifi@uthm.edu.my

As in Malaysia, thermoplastic TRS is one of the most frequently used forms. It is classified as a raised rumble strip and it has a thickness in the range 3–7mm [6]. Thermoplastic TRS are commonly applied as they contribute several factors including a readiness for immediate use, high durability, good retro reflectivity and relatively low cost. In order to increase the visibility through retro reflectivity elements, glass beads are also intermixed into the thermoplastic and partially embedded onto the surface of the marking binder material [7]. Studies of TRS made of thermoplastic have shown that it is able to reduce vehicle speed and road accidents [8]. According to [9], thermoplastic TRS with narrow strips are able to increase the road noise level, LAeq by around 2–4 dBA. In addition, TRS noise is claimed to be in the pattern of impulse noise which is the most annoying to people [9, 10]. Malaysia is a developing country, and is presently suffering from an increase in noise pollution arising from traffic noise. The problem is compounded with TRS which are widely used for reasons of safety, despite the fact many premises are built alongside the roads. Malaysia has also not yet developed a practical policy to cope this situation.

The primary purpose of the present study, therefore, was to assess the characteristics of noise generated by roads with the installation of thermoplastic transverse rumble strips. Two typical TRS profiles widely used in Malaysia were selected and the anticipated noise mechanism generation for the typical TRS were obtained to facilitate the local authorities in the selection of the reliable installation of TRS. This specific characteristic should be given close attention since many premises are located close to the road.

1.1 Tire-pavement noise generation

Sound generated by interaction between tire and pavement play a significant role when the vehicle speed exceeding 30km/h. according to previous research sound could be generated by the followings mechanisms; tire tread block impact occurred at 600Hz–900Hz [11], surface texture impact was dominant at 700Hz–1300Hz [12], tire belt/carcass vibration is the dominant mechanism for the frequency around 500 Hz [13], side wall vibration occurred at a frequency <1000 Hz [11, 14], and the air pumping mechanism was at a frequency range between 500Hz–3kHz [15] or above 1 KHz [14].

An adhesion mechanism is produced as the tyre rolls on the surface of the pavement. The release of the block produces tangential or radial vibration and yields a carcass response and side wall vibration. If the tyres hit the contact surface with a high angle of attack compared with the road surface, the tangential and radial vibration become higher. Air pumping occurs when the tread blocks enter the contact patch and trap air between the pavement and tyre which is then compressed and pumped out of the tread block. According to [14], the macrotexture and megatexture of the road surface play significant roles in noise generation. Increases in megatexture to 80 mm will increase the tyre noise in the low frequency of less than 1 KHz and when the macrotexture increases to 3 mm the tyre noise in frequency greater than 1 KHz decreases due to air pumping. The megatexture is responsible for radial and tangential vibration of the tyre, which causes the mechanism in low and high frequency ranges.

The authors consider that the interaction of the tire and TRS have similarities with the interaction of the tire and road. Thus, an experimental method was conducted based on the controlled pass by method test procedure, derived from the Coast-By Method [12, 16] for the selected sites.

2 Materials and methods

Two sites installed with typical thermoplastic rumble strips were selected. Site 1 was installed with the saw blade pattern or ten-raised-rumbler (RR) profile, and site 2 with the three-layer-overlapped profile (TLO) (Fig. 1).

The Coast-By Method was used as it can directly measure tire-pavement noise from individual vehicles in free-flowing and essentially constant-speed traffic. To avoid the effect of the test vehicle type and the conditions of the test site [12] the same test vehicle was used throughout the experiments at site 1 and site 2. For each site, the test vehicle was ran on the roadway using four levels of speed of 30 km/h, 50 km/h, 70 km/h and 90 km/h. The specification of each profile can be seen in Fig. 1, and both roads are considered straight in order to meet the requirements of ISO 13325 [16], with the condition of the pavement and TRS surfaces being good and dry.

The test vehicle was a typical passenger car used on the road in Malaysia, which is a 2005 model Perodua Myvi with a weight capacity of 950 kg. This vehicle is classified as a light vehicle since the gross vehicle weight (GVW) is less than 4500 kg. The tire size is 175/65 R14. The tire pressure was fixed at 250 kPa for all measurements. A type-1 integrating-averaging Pulsar Model 33 sound level meter was used for the measurement. The sound level meter was calibrated using a Model 105 acoustic calibrator before and after the measurements. The microphone was equipped with a foam windscreen to reduce the effect of wind noise on the data. A multifunctional anemometer was used for measuring the wind speed and air temperature, and an IR thermometer was used to measure TRS and pavement temperature.

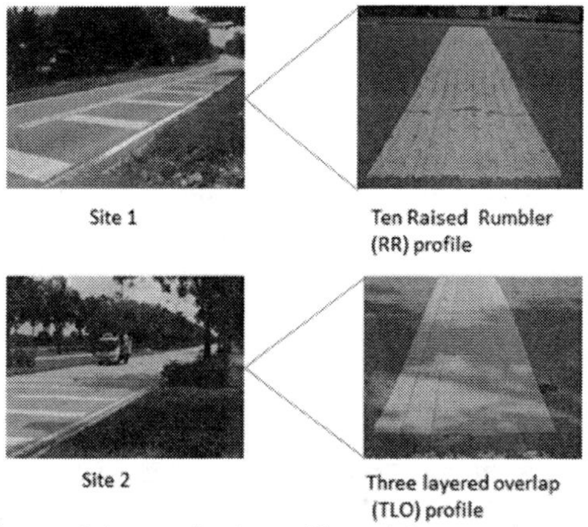

(a) Types of thermoplastic profile and the experimental sites

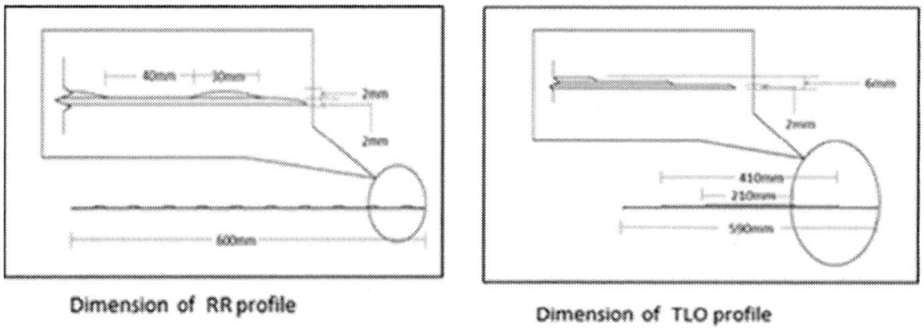

(b) Dimensions of profiles

Fig. 1. Profile of thermoplastic TRS and their dimensions

Each site has two points of measurement (Fig. 2). Point 1 is located in the middle of a set of TRS. At this location the readings were taken 4 times, i.e. once for each speed. At Point 2 the measurements were carried out at 300 m from Point 1 to avoid the effect of TRS. This position is sufficient because it is larger than the value proposed by the Transportation Association of Canada which states that the noise due to the grooved surface of the TRS will be dissipated at 100 m [5]. At Point 2 the same experiments were repeated without TRS. The sound level meter was installed at 7.5m from the middle of the test lane and the microphone was at a height of 1.5 m. The sound measurements were carried out using fast time weighting (fast) and A-frequency weighting, impulse time weighting (impulse) and A-frequency weighting, and spectral mode analysis of the one-third octave band were measured.

The parameters considered in the comparison between the sound level for the roadway with and without TRS were maximum measured LAFmax, and maximum measured LAImax. LAFmax, and LAImax were the maximum reading when the test car was 7.5 m from the sound level meter. Readings for wind speed and temperature of the air, TRS and pavement surface were taken before and after the measurements. Surface temperature correction was applied for every reading, as required by ISO 13325 [16]. LAFmax and LAImax were then corrected using Equation 1 to obtain LAF and LAI, and instead of pavement, in this study TRS temperature was taken as the surface temperature. To avoid external noise from other vehicles that can lead to data bias, the measurement process was carried out between 12:00am and 3:00 am.

$$LAF = LAFmax + K(20 - t) \quad (1)$$

where: LAF = is the corrected measured LAFmax, K = is the coefficient equal to −0.03dB (A-weighted)/°C when the measured test surface temperature is >20°C and −0.06dB (A-weighted)/°C when the measured test surface temperature is less than 20°C, t = surface temperature

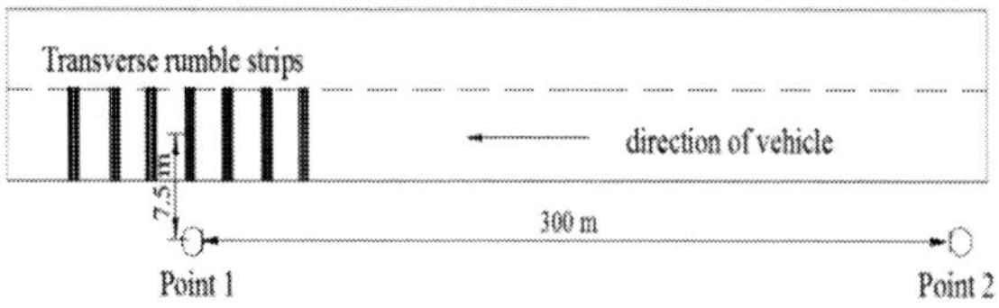

Fig. 2. Location of the sound level meter

The spectrum analysis for when the TRS was hit by vehicles were analysed and compared with previous findings from tire/road interaction frequency spectra analysis conducted by Kim et al. [11], Sandberg and Ejsmont [12], Keulen and Duškov [15] and Wayson [14].

3 Results and analysis

3.1 Relationship between sound level and vehicle speed

Fig. 3 presents the variation in maximum LAF generated when a vehicle hit RR and TLO, in comparison with and without TRS. Generally, for both RR and TLO profiles, it was clearly observed that faster vehicle speeds generate a higher noise. The sound pressure levels have been corrected based on the TRS surface temperature using Equation 1 as the pavement, and the TRS temperatures were 29.2°C and 29.6 °C, respectively, at site 1 and 30.3°C and 31.3°C, respectively, at site 2. Both locations had the same background noise level (45 dBA). The readings were valid as air temperatures were within 5°C to 40°C and the measured wind speed at the microphone height was less than 5 m/s, as recommended by ISO 13325 [16] to avoid sound refraction that will affect the readings [17].

Fig. 4 shows in detail the increment in noise level with and without both of the TRS profiles and a comparison between these profiles. It is seen that RR increased in noise level by 2.4 dB(A) (3.8%), while TLO only increased by 0.9 dB(A) (3.2%) for the single vehicle speed of 30 km/h. At a vehicle speed of 50 km/h the RR profiles exhibited a similar trend with a higher increase in noise level. However at 70 km/h the TLO exhibits a higher increase than RR of 4.5 dB(A) (5.7%), and at a vehicle speed of 90km/h, both profiles generated a similar increase. This finding shows that the RR is able to produce a just noticeable difference at 70 km/h and 90 km/h. Thus the installation of RR or TLO is able to exhibit noise changes that can be detected by the local residents near the roadway.

Fig. 5 presents a comparison between the LAF and LAI generated from the road installed with either profile. It can be clearly seen that LAI for each speed and each profile exhibit a higher sound pressure level. The increase in Fig. 6 shows the (LAI-LAF) values for RR of 3 dBA for a speed 70 km/h and below and reduce afterwards. However the trend for the TLO fluctuates. This result indicates that the noise has an impulsive character. In comparison with the road without RR and TLO, the (LAI-LAF) has differences of up to 1 dBA. This finding shows that the thermoplastic RR and TLO are similar with to pavement groove types of TRS and roads with metal cleats [18,19].

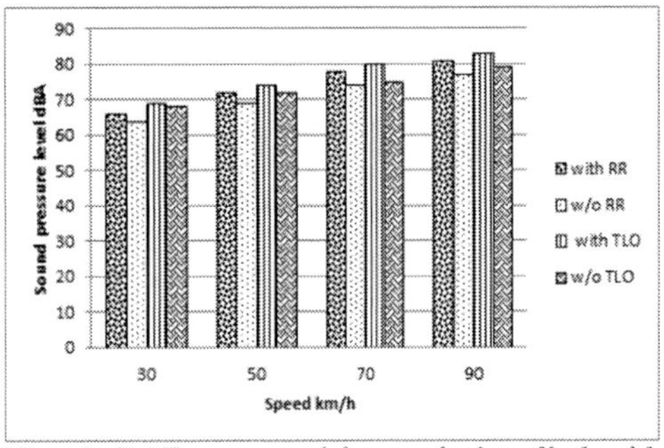

Fig. 3. LAF generated by roads installed with and without RR and a road installed with and without TLO

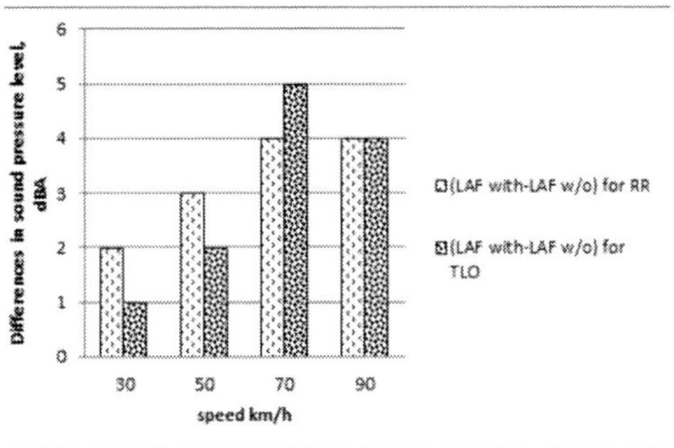

Fig. 4. Differences in LAF between roads installed with and without RR and road installed with and without TLO

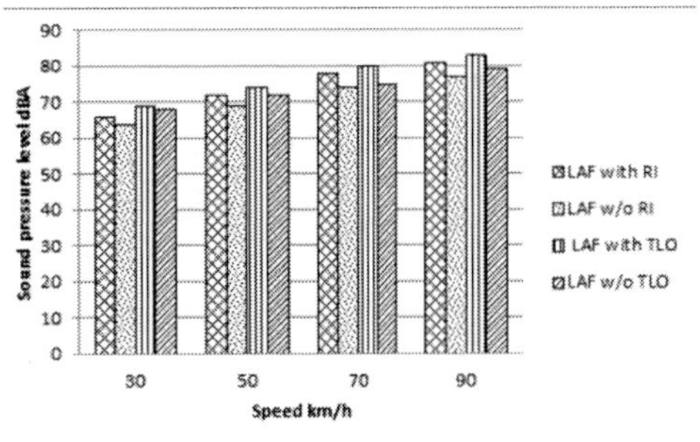

Fig.5. LAF and LAI generated by roads installed with and without RR and road installed with and without TLO

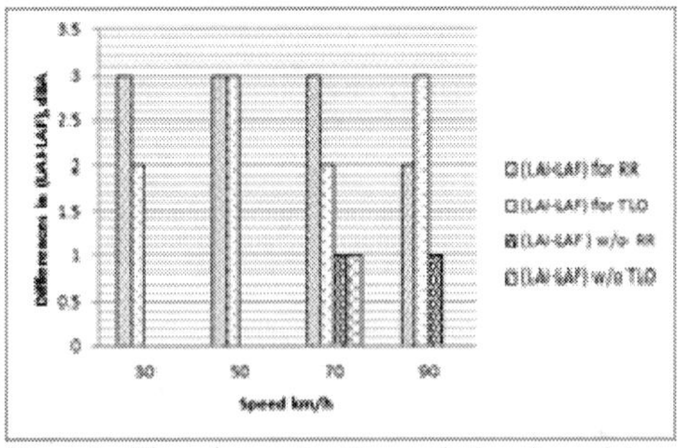

Fig. 6. The determination of the impulsive character

3.2 Frequency spectrum analysis and anticipated tire-TRS mechanism of noise generation

The frequency spectrum of noise at the moment the tire hits the RR and TLO can be seen in Fig. 7(a-d) and Fig. 8(a-d), respectively. This reading was the maximum reading when the test car was 7.5 m from the sound level meter. The RR profile tends to have a more uniform increase in all of the frequencies. The anticipated mechanism was air pumping, side wall vibration, and surface and tire impact. Air pumping as a result of the findings of Keulen and Duškov [15] was dominant for the frequency range between 1–3kHz. Air pumping occurs when the tire is rolling and a volume of air is enclosed in the contact patch within the grooved line between the raised lines and pores constituted by the tread pattern grooves and the surface texture. Air is compressed and pressed away at the front of the contact patch and expanded and sucked into the cavities at the rear. The variations in surface texture and tread pattern grooves, the latter of which will be rapidly squeezed, produce variations in air flow over time. This generates vibrations in the surrounding air and therefore constitutes a source of sound.

Alternatively, sidewall vibration, surface and texture tread block, according to Kim et al. [11] and Sandberg and Ejsmont [12] were found for frequencies less than 1000Hz, 700–1300 Hz and 600–900 Hz, respectively. Side wall vibration is caused by a collision between the tread blocks and TRS. As the tire interacts with the macro-texture of the RR (2 mm), the tread vibrations are transported to the side wall which acts as a 'sounding board' and radiates sound. Surface texture impact is generated when the tire tread enters the contact zone at certain angles relative to the road, the tire tread displacement itself contains a radial and tangential component. TRS with a macro-texture of 2mm gives a higher attack angle and generates more gradual displacement. Tire tread impact is represented by the radial vibration which is the sudden displacement of tread element, in relation to its original position in the rotating tire, when it impacts on the TRS surface. This may explain why the ten-raised-rumbler profile produced a more uniform increment in sound level for each frequency since it was dominated by these three mechanisms. In addition, the RR profile has a relatively high height of unevenness (2 mm), width (30 mm) and spacing (40 mm). The asperity height unevenness and the asperity spacing are defined as important parameters that govern tire vibration noise.

On the other hand, the generation of noise from TLO is governed by structural resonance, surface texture vibration, and side wall vibration. In Fig. 8, TLO tends to have a significant increase in sound level at low frequencies below 315 Hz. Based on the finding by Bolton and Kwon [13], structural resonance (Fig. 9) is the dominant mechanism at low frequencies. Structural resonance is represented by tire belt/carcass vibration [12]. The overall thickness of the TLO is relatively thick at 6 mm which may be a key factor to this type of mechanism. However, according to Perisse [19], the increment in sound level occurs at a higher frequency as the speed is increased. This can be seen in Fig. 8(c) and (d) for a vehicle speed of 70 km/h and 90 km/h. TLO also induced more dominant surface texture vibrations, which is combination of radial (Fig. 10) and tangential vibration (Fig. 11) [12]. Frequency is quite dominant in the 250 Hz to 2 kHz range at speeds of 70km/h and 90km/h, which may be due to this mechanism since Sandberg and Ejsmont [12] state that this mechanism is dominant at 700–1300Hz. Sidewall vibration involve dominant frequency <1000kHz [11]. Sidewall vibration (Fig. 12) induces collision between the tread blocks and the road [15], and in this case the tread block and TRS. As the TLO is relatively thick the collision becomes stronger and results in higher sound pressure levels for frequencies of less than 1000 Hz.

DOI: 10.1051/matecconf/201710307016

3.3 Effect of TRS on noise exposure

From the above findings, it is notable that TRS increase the noise level by up to 5 dBA, and both profiles have impulsive characteristics. Furthermore, although TRS increases the noise level by a relatively small amount, their psychological impact may be greater because of the impulsive impact. In the work of Bendtsen et al. [9], TRS generated impulsive and suggested that a 5 dB penalty for impulsive noise in order to predict the community response. The saw blade pattern or RR increases the sound pressure level at high frequency (1 to 4 KHz) which enable the human ear most sensitive with while the TLO mainly increases the sound pressure level in the low frequency which cannot be heard by the human ear. Thus, based on the impulsive nature and frequency spectral analysis, it is recommended that the saw blade pattern or RR can only be installed on roads that are located away from sensitive premises.

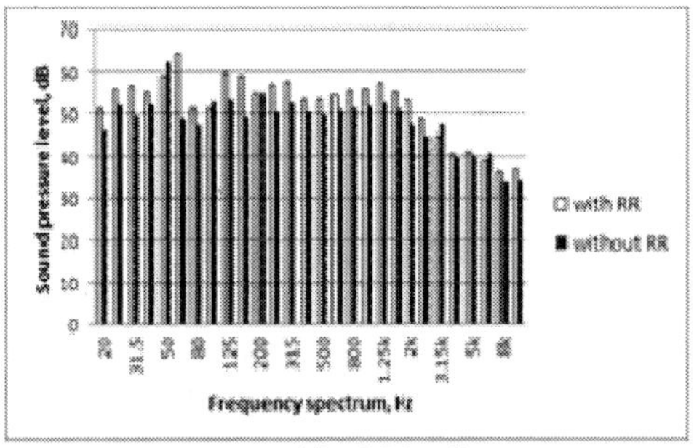

(a) 30 km/h

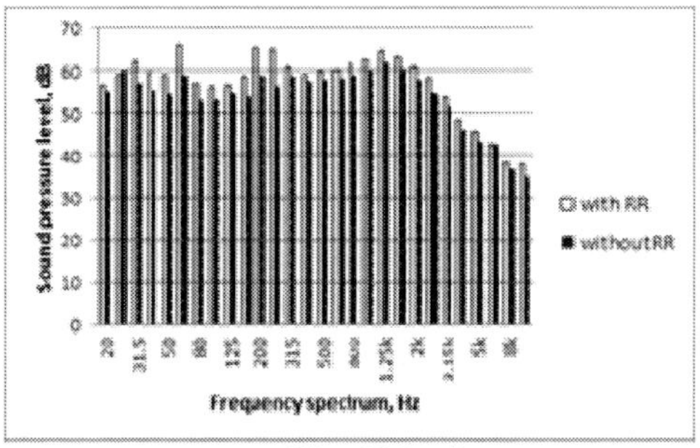

(b) 50 km/h

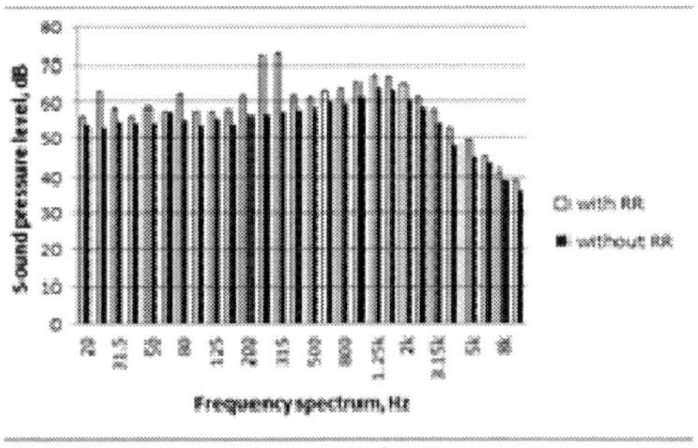

(c) 70 km/h

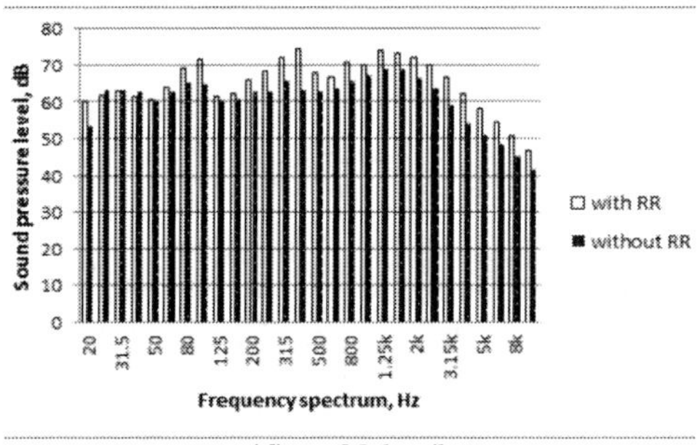

(d) 90 km/h

Fig. 7. Spectral frequency analysis of each vehicle's speed for roads with and without RR

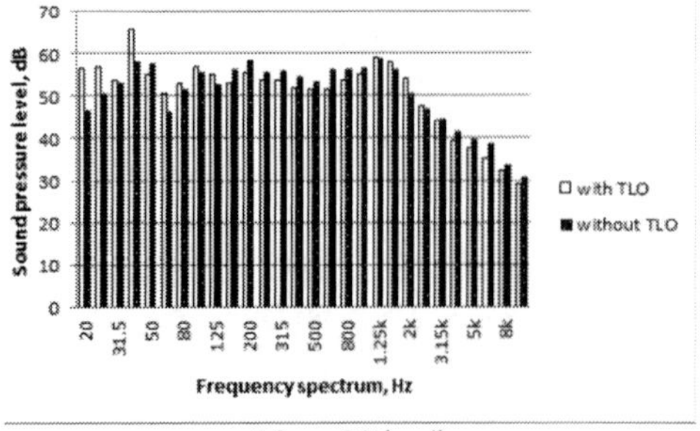

(a) 30 km/h

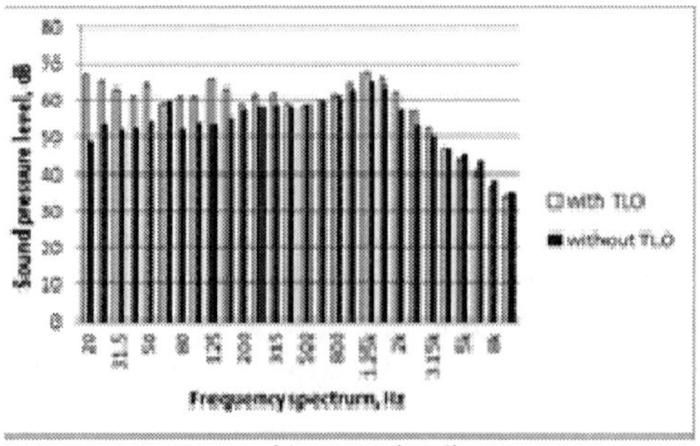

(b) 50 km/h

DOI: 10.1051/matecconf/201710307016

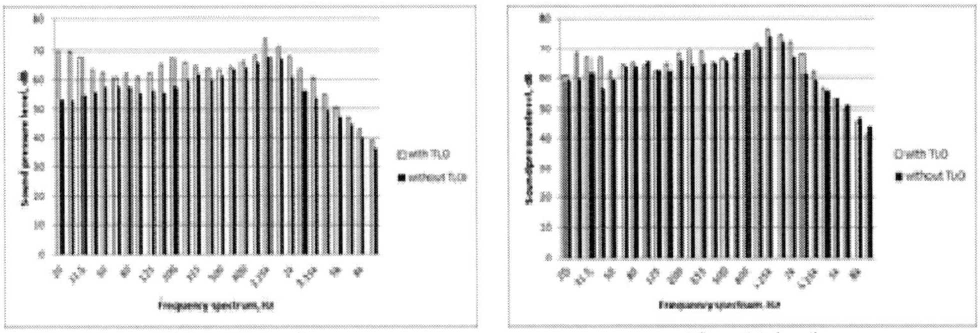

(c) 70 km/h (d) 90 km/h

Fig. 8. Spectral frequency analysis of each vehicle's speed for roads with and without TLO

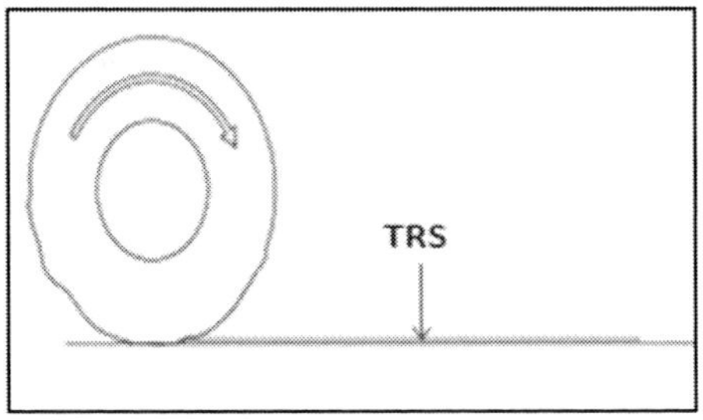

Fig. 9. Structural resonance - tire belt/carcass vibration

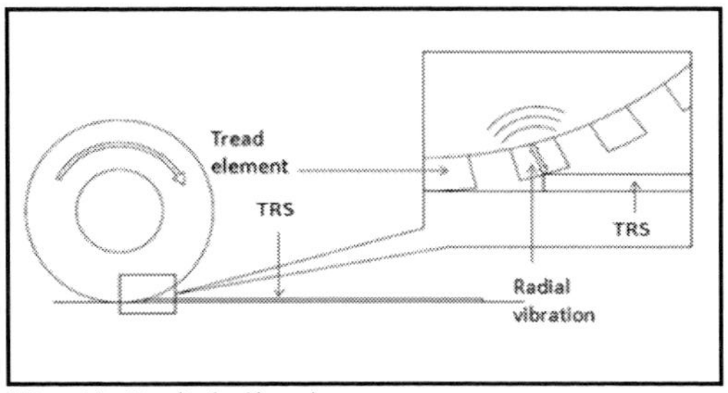

Fig. 10. Radial vibration

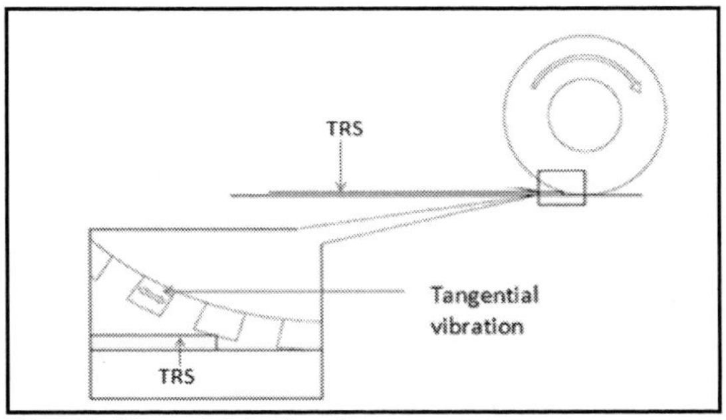

Fig. 11. Tangential vibration

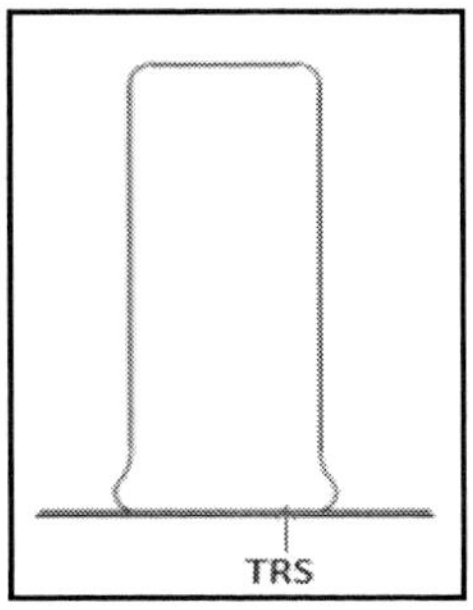

Fig. 12. Sidewall vibration

4 Conclusion

This study revealed the characteristics of the noise of tire-TRS interaction for two types of typical profile widely used in Malaysia. The results showed that the ten-raised-rumbler profile and three-layer-overlapped profile increased the noise level and have impulsive characteristics. The ten-raised-rumbler is capable of generating a relatively high noise at low speed, compared to the three-layer-overlapped system. Frequency spectral analysis showed that the ten-raised-rumbler profile produces a uniform increase in sound level for each octave band, and that the three-layer-overlapped profile has dominant noise at low level frequency, which is at about <315 Hz. Thus, the ten-raised-rumbler tends to irritate human ears. Based on this result and on the previous findings in the literature, the dominant generating mechanisms for the ten-raised-rumbler profile noise are tire tread vibration, surface texture vibration, 'air pumping', and slight structural resonance and sidewall vibration. Meanwhile for the three-layer-overlapped profile it is structural resonance, sidewall vibration and surface texture vibration. However, the mechanism is just anticipated speculation based on a comparison of spectral frequency with the previous literature. It should be noted that the frequency spectra for each mechanism is also influenced by tire design, tire pressure, and road design. These findings hopefully will provide a contribution to the development of a new approach in the manipulation of TRS profiles that can provide optimal auditory impact to road users without compromising the comfort of road users and generating excessive noise that may disturb surrounding localities.

This work is supported by the Fundamental Research Grant Scheme (FRGS-R.K130000.7840.4F653) financed by the Ministry of Education, Malaysia.

References

[1] D. Shinar, *Traffic Safety and Human Behaviour*, Elsevier, Amsterdam, (2007)
[2] W.K. Ng, P. Selva, *OSH profile in the transport sector in particular commuting hazard*, (2003), Retrieved 5 February 2011. from Malaysia Trade Union Congress: http://www.mtuc.org.my/osh_profile_transport.htm
[3] B.J. Katz, *Peripheral Transverse Pavement Markings for Speed Control*, PhD Thesis, Virginia Polytechnic Institute and State University, (2007)
[4] T.D. Thompson, M.W. Burris, P.J. Carlson, Speed changes due to transverse rumble strips on approaches to high-speed stop-controlled intersections, Transportation Research Record: Journal of the Transportation Research Board, **1973**, 1–9, (2006)
[5] G. Bahar, T. Erwin, M. MacKay, A. Smiley, S. Tighe, *Best Practice Guidelines for the Design and Application of Transverse Rumble Strips*, Transportation Association of Canada, Ottawa, (2005)
[6] Ministry of Work, *Traffic Calming Guidelines*, Highway Planning Unit, Kuala Lumpur, (2002)
[7] C.A. Lopez, *Pavement Marking Handbook*, Texas Department of Transportation, Austin, (2004)
[8] P. Liu, J. Huang, W. Wang, C. Xu, Effects of transverse rumble strips on safety of pedestrian crosswalks on rural roads in China, Accident Analysis and Prevention, **43**(6), 1947– 1954 (2011)
[9] H. Bendtsen, J. Haberl, U. Sandberg, G. Watts, *SILVIA Deliverable 12: Traffic management and noise reducing pavements Recommendations on additional noise reducing measures*, European Commission, Denmark, (2004)
[10] T. Yano, A. Kobayashi, Effects of duration of repeated impulsive sounds on noisiness, Environment International, **16**(4-6), 547-554 (1990)
[11] B.S. Kim, G.J. Kim, L.T. Lee, The identification of sound generating mechanisms of tyres, Applied Acoustics, **68**(1), 114-133 (2007)
[12] U. Sandberg, J.A. Ejsmont, *Tyre/Road Noise Reference Book*, INFORMEX Ejsmont and Sandberg Handelbolag, Kisa, (2002)
[13] J.S. Bolton, H.S. Kwon, *Nearfield acoustical holography applied to sound radiation from tires*, Purdue University, Indiana, (1998)
[14] R. Wayson, *Relationship Between Pavement Surface Texture and Highway Traffic Noise- A Synthesis of Highway Practice*, Transportation Research Board, Washington, (1998)
[15] W.V. Keulen, M. Duškov, *Inventory study of basic knowledge on tyre/road noise*, Road and Hydraulic Engineering Division of Rijkswaterstaat, Delft, (2005)
[16] ISO 13325, *Tyres - Coast-by methods for measurement of tyre-to-road sound emission*, International Organization for Standardization, Geneva, (2003)

[17] M.J. Crocker, *Theory of Sound - Predictions and Measurement*, In M.J. Crocker (ed.) Handbook of Noise and Vibration Control, John Wiley, New Jersey, (2007)
[18] D.E. Karkle, M.J. Rys, E.R. Russell, Centerline rumble strips: Study of external noise, Journal of Transportation Engineering, **137**(5), 311-318 (2011)
[19] J. Perisse, A study of radial vibrations of rolling tyre for the road noise characterisation, Mechanical Systems and Signal Processing, **16**(6), 1043–1058 (2002)

Correlation Between P-wave Velocity and Strength Index for Shale to Predict Uniaxial Compressive Strength Value

H. Awang[1], *N. R.* Ahmad Rashidi[1], *M.* Yusof[1,*], and *K.* Mohammad[2]

[1]Institute for Infrastructure Engineering & Sustainable Management, Universiti Teknologi MARA, 40450 Shah Alam, Selangor, Malaysia
[2]Faculty of Civil Engineering, Universiti Teknologi MARA, 40450 Shah Alam, Selangor, Malaysia

Abstract. Seismic refraction survey is a non destructive method used in site investigation to identify the seismic velocity subsurface strata. Although it is widely known, the reliability of the result is still doubtable for some reason as well as due to an engineer's ignorant, which insist on using conventional method rather than new advanced method causing the lack of usage in geophysical method for testing. This study aims to produce a correlation between P-wave velocity value and point load strength index value for shale. Both field and laboratory tests were carried out. In order to obtain the P-wave value, seismic refraction method was conducted as a field test at Precint 4, Putrajaya, Malaysia to achieve the P-wave velocity value of the shale bed. Ten samples of shale were collected from the field and laboratory tests were conducted. The tests are divided into three sections, namely non-destructive laboratory test, physical properties test and mechanical properties test. Ultrasonic Velocity Test via PUNDIT test was conducted as non-destructive laboratory test to acknowledge the P-wave velocity value in laboratory. Both field and laboratory P-wave velocity value were then compared and the result delivers are reliable due to it is within the range. For physical properties test, the rock density and porosity were acknowledged. Meanwhile, Point Load Test was conducted as mechanical properties. Correlation for both P-wave velocity value and point load strength value were achieved via producing an empirical relationship as the end result. Prediction of uniaxial compressive strength (UCS) value was made via converting the point load strength value to UCS value using a correlation. By acknowledging this empirical relationship, it shows that geophysical methods are able to produce a reliable result. Hence more and widely used of geophysical method will be profound in the future.

1 Introduction

Prior to site investigation purposed, most of the methods used are destructive, consumed more time and high cost. Besides that, the results are discrete which is only single point of

* Corresponding author: haryatiawang@yahoo.co.uk

DOI: 10.1051/matecconf/201710307017

data could be obtained. Underground profile produced from this method is from combining a few numbers of results of the holes points. A continuous profile shows the soil and rock layer by assuming the layer is similar from one point to another. However, the reliability of the result is still doubtable due to misinterpret of the profile between the points. Due to this, geophysical methods are now becoming popular in producing the subsurface profile especially in differentiating the soil and rock layers. Many studies have been conducted to prove the reliability of the geophysical test results as well as correlations between the geophysics and strength parameters to promote this advanced non-destructive method without the needs to use the conventional methods.

This study is about conducting a geophysical method called as seismic refraction as a non-destructive test method to obtain the subsurface profile of a site. Besides that, a laboratory test using pulse ultrasonic no-destructives test (PUNDIT) and other strength tests on samples collected from site were also conducted. The objectives of this study is to compare between the velocity of wave called as primary wave velocity (P-wave velocity, v_p) measured from field and laboratory. The other objective is to correlate between the P-wave velocity and point load strength for the purpose of estimating the uniaxial compressive strength (UCS) of the rock samples. The study was conducted at Precint 4, Putrajaya, Malaysia Fig. 1. This site lies on Kenny Hill's Formations that contain of shale, phylite, slate and sandstones.

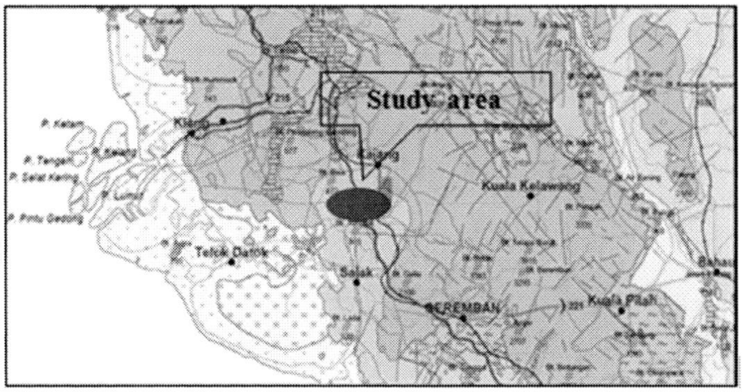

Fig. 1. Location of study area

2 Theories and previous study

This study applied seismic refraction method test that measures travel times of seismic waves throught ground layers via receivers at known-points along the ground surface. Basically seismic refraction technique is a measurement of time taken for the wave to travel from one location to another. As end result it produces P-wave velocity, and this value is then recorded for analysis. In recent years, the use of coconut fibre as reinforcement on concrete has received high attention because of renewability, low density, high specific properties and low cost.

In engineering applications, the used of P-wave velocity value is proven to be important due to the parameters of Poisson's ratio and Young's modulus. The relationship between velocity (v_p) and Young's modulus (E) is address as,

$$v_p = \sqrt{\frac{E}{\rho}} \tag{1}$$

where ρ is density of material; *E* and v_p have been define as above. The use of these parameters could be used to calculate foundation settlement on rock according to [1] as in Equation (2).

$$S = \frac{\pi}{2}\left[\frac{q\left(1-v^2\right)R}{E}\right](Ig) \qquad (2)$$

where *S,q,E, Ig, R* and *v* is for settlement, uniform load per area, Young's modulus, depth reduction factor, radius of the pile and Poisson's ratio accordingly. This equation shows the relation between Poisson's ratio, Young's modulus and P-wave velocity. By substituting Equation (1) into (2) the settlement of foundation is calculated as:

$$S = \frac{\pi}{2}\left[\frac{q\left(1-v^2\right)R}{\rho v_p^{\,2}}\right](Ig) \qquad (3)$$

The physical and mechanical properties of rock are essential to predict the behavior of the rock mass. It can be achieved via conducting laboratory test on specimens collected from the field. [2] found that among the rock mass properties involved are the rock type, strength, and degree of alteration, structure, fabric abrasiveness, moisture content and the seismic velocity.

Uniaxial compressive strength (UCS) value is an important engineering parameter needed in construction design due to the rock strength. The prediction model is developed by converting the value from the result achieved by PLT to UCS value using a correlation factor. [3] stated that the correlation factor to convert the result achieve from PLT to UCS as in Equation (4).

$$UCS = 14.10 Is_{50} + 4.50 \qquad (4)$$

[4] produced a prediction model between uniaxial compressive strength (UCS) and Young's modulus (*E*) acknowledged as :

$$E = 0.3752UCS + 4.428 \qquad (5)$$

3 Experimental studies

3.1 Field test

Test begins with conducting a seismic field test at Precint 4, Putrajaya. A total of 24 geophones were pinned into the ground along the site with a total length of 115m with interval 5 m between the geophones. The seismic energy source was triggered by 8 kg hammer shot on steel plate at the ground surface. As the wave travel in refracted path the travel times is recorded by the geophones as receivers. The velocity of the wave is determined by calculating the time taken for each distance of the receiver points. Offset were set up at 0.5 m for both end sides. Seismometer was placed in the middle of the layout between geophones 12 and 13. The line was named as SL1. Fig. 2 shows the field setup for the seismic refraction method.

DOI: 10.1051/matecconf/201710307017

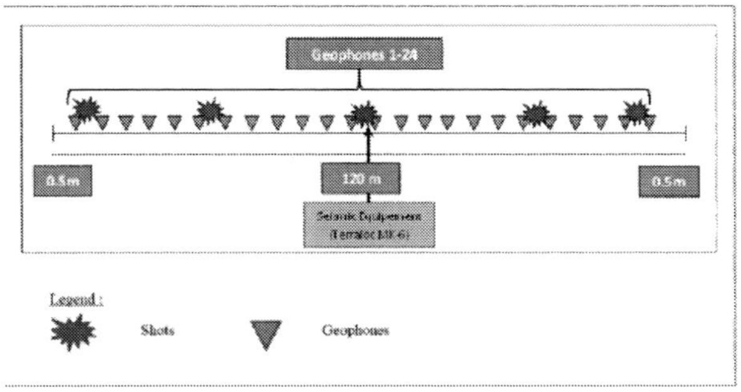

Fig. 2. Field setup for the seismic refraction method (SL1)

3.2 Laboratory tests

For laboratory experimental study, ten irregular specimens were collected for PUNDIT and point load test (PLT). Density and porosity tests on the other hand were conducted to acknowledge the physical properties of the rock. Samples were collected from the location of field testing as the used to conduct laboratory testing in broken cored form. It represent characteristics and properties of rock material of the field. Samples received were in irregular condition due to weak condition of the rock which is soft and easily to break into parts. Therefore, to conduct the test, the best sample need to be chosen. The sample was chosen based on their conditions which is no fractured occurring at the specimens. Apart from that the sample dimension also were considered to ensure that the specimen size are eligible for laboratory tests.

4 Results and discussion

The results of field test is presented as a seismic tomogram profile as in Fig. 3. The profile was interpreted based on seismic velocity index by [5] demonstrates geological sections along the seismic refraction survey line consist of varieties of velocity. The upper layer (L1) consists of overburden soil with the P-wave velocity value ranging between 100ms^{-1} and 1000 ms^{-1} which the thickness is approximately 9m. Layer 2 (L2) produced seismic velocity value of 2000ms^{-1} to 3800ms^{-1} with 21m in thickness. The bottom layer (L3) has velocity more than 4000 ms^{-1} with the thickness value is considered to be approximately 2m. The seismic velocity of each layer is determined in range as summaried in Table 1.

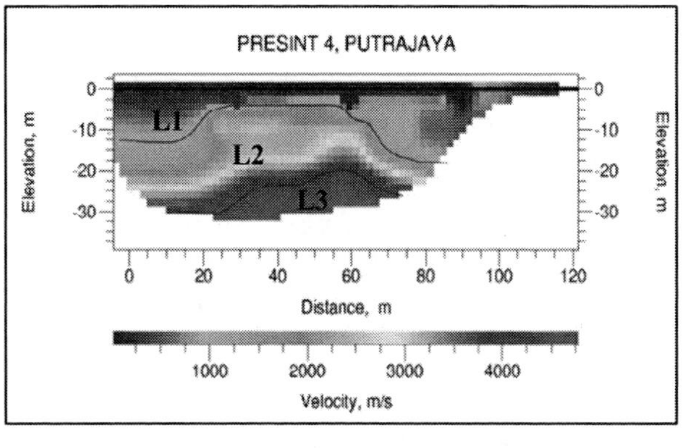

Fig. 3. P-wave velocity model produced

Table 1. Typical primary (vp) velocities for different material

Layer	v_p (ms^{-1})	Description
1	100 - 1000	Overburden soil
2	2000 - 3800	Shale
3	>4000	Sandstone

Result achieved via conducting the laboratory tests for porosity, density and Pulse wave value is tabulated in Table 2.

Table 2. Results of density, porosity , P-velocity value , PLT and predicted UCS & Young's modulus

Sample No	Density (kg/m^3)	Porosity (%)	Distance (mm)	Time (ηs)	Velocity (ms^{-1})	Load (P) (kN)	Is(50)	Predicted UCS (MPa)	E (GPa)
1	2620	0.47	70	27.5	2545	4.70	1.87	30.87	16.01
2	2341	0.52	69	22.7	3040	6.10	2.26	36.37	18.07
3	2523	0.54	73	27.0	2704	5.50	2.17	35.10	17.60
4	2418	0.48	63	21.5	2930	5.80	2.31	37.07	18.34
5	2489	0.59	63	21.0	3000	5.70	2.25	36.23	18.02
6	2335	0.51	71	24.5	2898	5.20	2.03	33.12	16.85
7	2532	0.56	70	26.0	2692	5.00	1.99	32.56	16.64
8	2612	0.49	69	22.5	3067	5.50	2.19	35.38	17.70
9	2308	0.52	67	23.5	2851	5.90	2.28	36.65	18.18
10	2471	0.51	68	25.0	2720	4.90	1.93	30.87	16.01

For physical properties, as shown in Table 2 the density value ranging from 2308 kg/m^3 to 2620 kg/m^3. When compared to [6] the density value of shale ranging from 2300kg/m^3 to 2700kg/m^3. This showed that the result of this study is reliable. For porosity value results from this study is ranging from 0.47% - 0.56%. This finding was compared with a study conducted by [7], where the porosity value of shale ranging from 0.50% to0.90%. Hence it showed that the result achieved by this study is reliable.

Based on the Table 4 the point load streng index, $Is_{(50)}$ value for all samples are from 1.87 MPa ~ 2.28 MPa in average. According to [3], the $Is_{(50)}$ value of shale ranging 0.06 MPa~ 2.56 MPa. The $Is_{(50)}$ obtained in this study is reliable that showed a good agreement with previous study.

P-wave velocity value has been recorded for both field and laboratory test. Field test consists of two layers which the results ranging between 2000 - 4000 ms^{-1}. Meanwhile, for laboratory test consist of ten samples and the results achieve ranging between 2545 ms^{-1} – 3067 ms^{-1}. The average P-wave velocity value for field test is 2900 ms^{-1} meanwhile for laboratory test were 2845 ms^{-1}. The differences between two values were 55.3 ms-1. Nevertheless, the value is still reliable to be interpreted as shale. According to [5] the P-wave velocity value for shale ranging from 2000 ms^{-1} to 4100 ms^{-1}. The P-wave velocity value from both field and laboratory test were compared (Fig. 4). This comparision was made to identify that the P-wave velocity values obtainded in the laboratory represented the rock mass. The result shows that the range of P-wave measured in the labortaory accomodated to the field value.

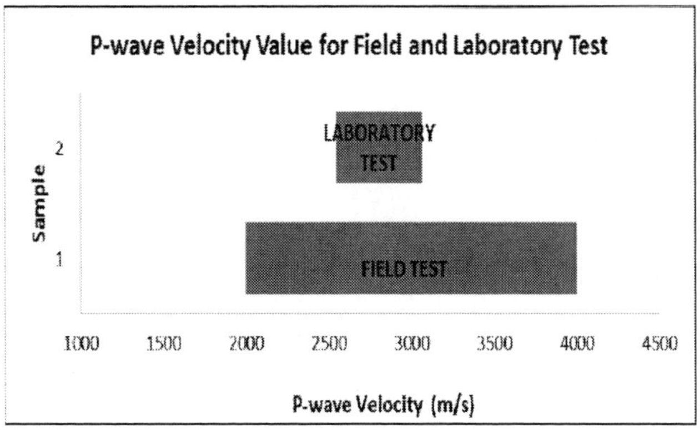

Fig. 4. P-wave velocity value for field and laboratory test

The relationship between P- wave velocity value and point load strength index value Is $_{(50)}$ shows a good correlation on both parameters (Fig. 5). This can be acknowledged that the point load strength index value can be predicted by knowing the P-wave valueof the rock either from seismic field test or by PUNDIT test.

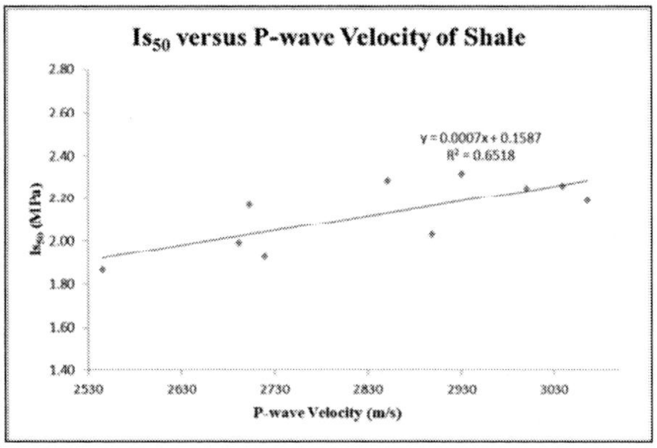

Fig. 5. Correlation between P-wave velocity value and point load strength index value

Referring to the Table 2, it shows that after estimating UCS values using a correlation as per Equation (2) , the UCS value obtained ranging between 30.87 MPa ~ 36.65 MPa. According to [3] the value for UCS for shale in Kenny Hill Formation is ranging between 5.33 MPa ~ 35.23 MPa. Hence the value obtained in this study is considered reliable as compared to previous study. As the UCS value is an important parameter in engineering to measures the rock strength, therefore, the correlation between P-wave velocity and UCS value can be acknowledged without the need of conducting UCS test. This means that by knowing the P-wave velocity value based on seismic refraction test, we can estimate the UCS value . Resulting from the correlation as in Fig. 6 an empirical relationship between UCS and P-wave velocity value is developed. The empirical relationship equation produces is stated in Equation (6).

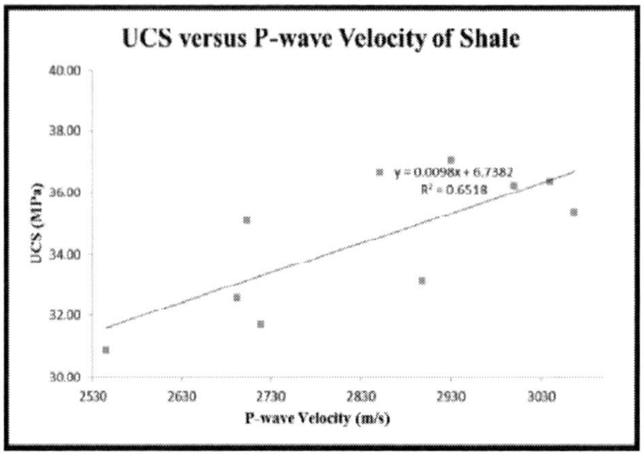

Fig. 6. Correlation between P-wave velocity value and UCS

$$UCS = 0.0098\ Vp + 6.7382 \quad (6)$$

In general from this study that was guided by [8], P-wave velocity value was achieved via seismic refraction method as a field test, then the data were analyzed using OPTIM software. Ultrasonic pulse velocity (PUNDIT) test on the other hand also produce P-wave velocity value, but it was conducted in the laboratory. Meanwhile Point Load Test (PLT) was conducted to identify the rock's mechanical properties. Both PUNDIT and point load strength index value was achieved using samples collected at the same location where the seismic field test conducted. A correlation has been developed between P-wave velocity test and point load strength index value and furthermore predictions of uniaxial compressive strength test was produced.

Based on the empirical relationship, it shows that to find the UCS value of shale this empirical equation can be used. This equation is achieved based on the regression value produced from the graph. By acknowledging this empirical relationship, it is now easier to identify UCS value in without the needs of physical testing on samples as well as borehole and sample collecting procedures. Due to seismic test that produces P-wave velocity value which performed on the ground surface, we can acknowledge the compressive strength index value that is important for design factor. Both UCS and E value are one of the most important parameters in rock engineering properties. Many equations can be produced via these parameters for example settlement, (S) equation as per stated in Equation (4) by [1]. By knowing E value, settlement (S) of foundation on rock value can be calculated. Refering to the Table 2 it shows that the Young's Modulus (E) value ranging between 16.01 GPa ~ 18.34 GPa matched to [4].

5 Conclusions

This study is pointing out the reliability of using geophysical method prior to site investigation. The other aim is to predict uniaxial compressive strength by using P-wave velocity and point load strength index. The results revealed that a correlation between P-wave velocity value and point load test strength index value can be acknowledged via an empirical relationship equation. Settlement equation could be determined from P-wave velocity as well as Young's modulus value. In a nutshell, this study is beneficial to engineering development and more of such studies are needed in future due to the contribution in sustainable type of living,

References

[1] T.Y. Irfan, G.E., Powell, *Foundation design of Caissons on granitic and volcanic rocks*, Geo Report No.8., Geotechnical Engineering Office, Civil Engineering Department, Hong Kong. (1995)
[2] T. Bozdag, Indirect rippability assessment of coal measure rocks, Doctoral Dissertation, METU, Ankara, (1988)
[3] Z. Mohamed, K. Mohamed, H. Awang, Empirical strength models, elastic modulus and stiffness of weathered sandstone and shale as a composite rock, Electronic Journal of Geotechnical Engineering, 0730, (2007)
[4] C. Sachpazis, Correlating Schmidt hardness with compressive strength and Young's Modulus of carbonated rocks, Bull, Eng. Geol. Environ., **42**,75-83 (1990)
[5] H. Kohen, The temperature dependence of seismic waves in ice, Journal of Glaciology, **13**(67), 144-147 (1974)
[6] J.H. Schön, *Physical properties of rock*, Handbook of Petroleum Exploration and Production, Elsevier, USA, (2015)
[7] H.S. Poelchau, D.R. Baker, T. Hantschel, B. Horsfield, B.,Wygrala, Basin simulation and design of the conceptual basin model, Petroleum and Basin Evaluation, Springer-Verlag, New York, 3-70 (1997)
[8] International Society of Rock Mechanics Rock Characterization, *Testing and Monitoring: Suggested Methods For Determining the Point Load Strength*, Pergamon Press, Oxford (2007)

 DOI: 10.1051/matecconf/201710308001

Driver Anger Scale (DAS) Among Car Drivers: How Serious Are They?

Kamarudin Ambak[1,*], *Shuhada* Shamsudini[1], *Basil* David Daniel[1], and *Ahmad Raqib* Abd Ghani[1]

[1]Smart Driving Research Center, Faculty of Civil and Environmental Engineering, Universiti Tun Hussein Onn Malaysia, 86400 Batu Pahat, Johor, Malaysia

Abstract. Nowadays, anger while driving on the road is a crucial issue in Malaysia. The anger of driver may lead to traffic crashes. Road accident issue is an important agenda in every country. Driver behaviour and attitude such as impatience, hasty and hot-tempered on the road has become one of the causes in road accidents. Uncontrolled anger will affect the behavior of oneself during driving and could cause an individual to display aggressive attitude. Therefore, this study is aimed to identify contributing factor in anger of drivers and evaluate the Driver Anger Scale (DAS) in Batu Pahat, Johor. In addition, this study also analyzes the relationship between driver anger and the factor of DAS. Cross sectional study method was conducted in this study by distributing 250 questionnaires to car drivers. The data collectied was analyzed by descriptive, chi-square test, correlation and regression analysis using *Statistical Package for Social Sciences* (SPSS) version 20.0. The findings show that the discourtesy was reported as the most dominant factor contributing to driver anger. As a recommendation, to overcome driver anger issue, the authorities should prepare an action plan, promote and nurture public to practice good driving behaviour. Moreover, curriculum and driver training lesson should be improved to create better attitude among drivers. Drivers also should change their driving attitude by becoming safe drivers and exhibiting defensive driving skills on the road.

1 Introduction

Anger is an emotion that is commonly experienced while driving but angry drivers present a problem on the roads. Driver anger was conceptualised as a personality trait related to an individual's underlying predisposition for anger, but specific to road situations [1]. This is because driver anger has been identified as a predictor of traffic crashes and precursor to behaviors that increase the likely hood of traffic crashed [2]. Furthermore, research found that those individuals who reported high levels of anger are more likely to be outwardly aggressive [3]. Therefore driver anger also leads to aggressive driving behaviours [4]. Angry drivers, however, can become aggressive and tend to incorporate a higher level of risk-taking into their driving style [5]. Studies have shown that significant relationship exist

---

* Corresponding author: kamardin@uthm.edu.my

between driving anger and crash related conditions such as losing control of their vehicle, losing concentration, speeding, tailgating, near misses and moving violations (tickets)[4, 6]. While driving, high anger drivers experience more anger triggers, frequent and intense anger, hostile thinking, aggression, risky behavior, and some crash related conditions than low anger drivers [7].

For these reasons, it is important to understand what makes drivers become angry and how serious their anger is in that situation. Many researchers often use the Driving Anger Scale (DAS) was developed by Deffenbacher [1] to measure individual's "trait driving anger" which is their propensity to become angry across a number of different driving situations. DAS produced six factors relating to the types of situations likely to provoke anger which consists: discourtesy, hostile gestures, traffic obstruction, slow driving, police presence and illegal driving. These six factors have been reported to show generally positive correlations with each other, which indicate a general trend for driving anger in addition to independent reactions to different types of situation [1,8]. Hence, the objectives of this study were to identify which situations are major contributing factors that cause the anger of drivers and evaluate the Driver Anger Scale (DAS). Another objective for this study was to analyze the relationship between driver anger with the factors of DAS.

2 Material and methods

This survey was carried out in district of Batu Pahat, Johor. Batu Pahat was chosen as the location of study because it represents the region where the highest number of road accidents were recorded in Johor [9].

2.1 Instrument and measurement

This study was conducted using the cross sectional study method. Questionnaires were used distribute to car drivers in Batu Pahat and interviews were conducted with respondents. The questionnaire was developed based on previous studies as in [1,10] and divided into three parts. In the first part, background information of respondents was collected. This included gender, age, race, status of marriage, and occupation of respondent. The second part of the questionnaire involves the experience of driving among respondents, which included type of driving license, duration of driving experience, frequency of driving, distance of driving, traffic accident history, traffic ticket history and frequency of drivers feeling angry while driving. The last part of the questionnaire is Driving Anger Scale (DAS) section, which describes situations that provoke driver anger. Participants were asked to rate how angry they would become if they came across each situation. Ratings were given based on a five point Likert scale (1=not angry, 2= slightly angry, 3 = angry, 4=very angry 5=extremely angry). A pilot study was conducted to determine the reliability of the questionnaire. 25 sets of questionnaire were distributed to respondents and the study yieled an Alpha Cronbach test value of 0.907, which showed that questionnaire was highly reliable.

2.2 Sample size and location of study

The sample size of this study included 250 respondents randomly selected among car drivers in Batu Pahat, Johor and these involved drivers who were willing to complete the survey and be interviewed. The duration for data collection was two months, beginning from March 2015 and ending in May 2015.

2.3 Statistical analysis tool

The Statistical Package for the Social Sciences (SPSS) version 20.0 was used to analyse the dataas analyze through the application of statistical methods such as descriptive analysis, chi-square test, correlation test and multiple regression analysis.

3 Results and discussions

A total of 250 respondents completed the questionnaire. The following subsections detail the statistical analyses conducted on the data gathered.

3.1 Descriptive analysis

From this study, there were 250 respondents who drove cars in Batu Pahat, Johor. Majority of them were male (52.4%). Most of them were aged between 26 to 35 years old (28.8%). Besides that, more than half of the respondent were married (58.8%) and over a quarter of the work in the government sector (26.4%). Most drivers have full car driving license (91.2%) and 28.8% of the respondent have driving experience 15 years and more. Furthermore, large number of repondent (69.2%) drove their cars everyday, with 38% of them driving for a distance of between 11km and 20 km. 42.8% of them admitted to receiving traffic ticket between one and five times and while 60% of them were involved in road accident (Table 1).

Table 1. Respondent Demographic

Respondents	Frequency	Percentage (%)
1. Gender		
Male	131	52
Female	119	48
2. Age (years)		
Below 25	72	29
26-35	51	20
36-45	55	22
46 and above	72	29
3. Profesion		
Self-employed	42	19
Government Sector	66	26
Private Sector	47	19
Student	60	24
Others (unemployed)	35	14
4. Licenses		
Competent (Class D)	228	91
Probationary (Class P)	19	19
Learner (Class)	3	1
5. Experience (years)		
Less than 5	50	20
5-10	62	25
11-15	66	26
15 above	72	29

6. Driving distance (km)		
Less than 2	24	10
2-10	84	34
11-20	95	38
20 above	47	19
7. Traffic Summons		
1 time	107	43
2-5 times	89	36
6-10 times	48	19
Over 10 times	6	2
8. Accident Involved		
Yes	151	60
No	99	40

3.2 Driver anger scale (DAS) analysis

For the purpose of identifying the major factors that contribute to driver anger while driving, the average mean score and standard deviation of frequently angry driver and factors of Driving Anger Scale (DAS) showed in Table 2. The average score mean for frequently angry driver was rated 4.25 (SD = 1.15). It can be said that the most frequently angry drivers' score were more than 4 out of 7 to that of quite often angry drievr, seeing that the mean score were closer to the score of always angry driver. From the six factors causing driving anger, discourtesy was rated as the most anger provoking with an average score mean of 3.76, (SD =.64). The situations that were found to intensely provoke driver anger weres "Someone did not give signal to change lane" and "Someone takes the car parking spot you were waiting for". This results were statistically similar to the subscale mean obtained for Malaysia sample [9]. Furthermore, the factor that was found to be least provoking was police presence, which had the lowest average anger rating (M=2.08, SD=0.78) of the six factors. This finding is inline with a previous study that reported, the presence of police did not evoke any appreciable amounts of anger amongst their sample of Turkish and Malaysian drivers [11,12].

Table 2. Average means and standard deviation of driver anger and DAS factors

Sub scale factor	No. item	Average mean score	Standard deviation (SD)
Frequently angry driver	1	4.25	1.15
Discourtesy	4	3.76	0.64
Hostile gesture	4	3.44	0.97
Slow driving	4	3.30	0.69
Police presence	4	2.08	0.78
Traffic obstruction	4	3.17	0.66
Illegal driving	4	3.30	0.73

Table 3 shows that summary of driving anger by age, gender, driving experience, driving distance, traffic crash involvement and receiving traffic ticket. By using chi- square test, it was found that age, driving experience, driving distance, traffic crash involvement and receiving traffic ticket were significant ($p<0.05$) influenced driving anger. It can be concluded that age, driving experience, driving distance, traffic crash invovlement and receiving traffic ticket have relative order with DAS.

The results of chi-square test aalso found that young drivers (57.7%) were more likely to be angry than older drivers. This findings is comparable to a previous study that showed young drivers are more likely to display anger and aggression [13,14]. Moreover, less experienced drivers (47.7%) were more likely to be angry than more experienced drivers. Drivers driving for more than 10 km (54.2%) was also found to cause a driver to be more likely angry while driving. These result can linked to a previous study that found some relationship between annual mileage and anger [13]. Besides that, this study also found that drivers who had prior experience of being involved in traffic crashes (60.9%) and those who had received traffic tickets (53.4%) were more likely to be angry. These results are in line with a previous research [8], which stated that drivers who reported minor losses of control and those who reported receiving a recent traffic ticket had reliably higher tendencies to become angry.

Table 3. Summary chi-square test result for descriptive variable with DAS

Descriptive variable	Chi squre (χ^2)	df	Significance ($p<0.05$)	Result
Age	9.172	1	0.002	Yes
Gender	0.227	1	0.634	No
Driving experience	5.058	1	0.025	Yes
Driving distance	5.104	1	0.024	Yes
Involved traffic crashed	25.532	1	0.000	Yes
Traffic ticket	5.315	1	0.021	Yes

3.3 Correlation analysis

The correlation coefficient is a statistical value that measures the strength of relationship between one or more variables [13]. Table 4 shows the results of the for correlation between driver anger and the DAS factors.

Table 4. Correlation coefficient for DAS factors

	Driver anger	Discourtesy	Hostile gesture	Slow driving	Police presence	Traffic obstruction	Illegal driving
Driver anger	1						
Discourtesy	.470**	1					
Hostile gesture	.253**	.478**	1				
Slow driving	.227**	.391**	.449**	1			
Police presence	.016	.097	.220**	.317**	1		
Traffic obstruction	.249**	.276**	.215**	.483**	.440**	1	
Illegal driving	-.003	.048	.046	.096	.184**	.306**	1

**. Correlation is significant at the 0.01 level (2-tailed).

It was found that discourtesy wase highly correlated with driver anger, showing a positive linear relationship and significant ($p<0.01$). Meanwhile, hostile gesture, slow driving and traffic obstruction were found to be positively correlated and significant ($p<0.01$) with driver anger. It is can be concluded that four out of six factors of DAS had strong and significant relationship with driver anger. Based on a previous study the DAS was found to be related to trait anger scale with a strong correlation coefficient [10,13]. Another research, Yasak and Esivol [11] stated that the correlations between anger symptoms and the six factors of DAS have positive correlation and were significant among the factor DAS, except for illegal driving.

3.4 Multiple Regression Analysis

Table 5 shows the model summary of multiple regression analysis conducted for driver anger. Model 1, yeilded a R^2 of .250 which means that a 25% change in the independent variable (factor of DAS) due to the dependent variable (driver anger).

Table 5. Summary of multiple regression toward driver anger and factor of DAS

Model	R	R^2	Adjusted R square	Standard Error of the Estimate
1	.500[a]	.250	.232	1.005

Table 6 shows that the summary of ANOVA analysis for driver anger. It was found that a significant relationship ($p<0.05$) exists between the independent variable (factor of DAS) with the dependent variable (driver anger).

Table 6. Summary of ANOVA analysis for driver anger with factor of DAS

Model		Sum of square	df	Mean	F	Sig.
	Regression	81.832	6	13.639	13.511	.000[b]
1	Residual	245.292	243	1.009		
	Total	327.124	249			

Table 7 show the results of a regression analysis conducted on the factor of DAS as the independent variable and driver anger as the dependent variable. It was determined that the factor of discourtesy and traffic obstruction have a significant relationships ($p<0.05$) with driver anger. For this model, the discourtesy factor showed the highest beta value ($\beta= .412$, $t= 6.257$, $p< .05$) which concludes that the discourtesy factor exerts the greatest influence on a driver become angry while driving.

4 Conclusion

In conclusion, this study has found that the major contributing factors to driver anger were discourtesy, hostile gesture, slow driving, illegal driving and traffic obstruction.The, the relevant authorities should be give more attention to these driver misbehaviours in order to curtail driver anger or road rage. Apart from that, this study also found that young drivers, less experienced drivers, long distance driving, drivers with prior involve in road accidents and drivers who have received traffic ticket were more likely feel angry while driving. In addition, discourtesy was found to have positive linear relationship and was significantly influential towards driver anger. Discourtesy was also found to be the greatest contributing factor in causing drivers to become angry while driving. In order to overcome this issue, the authorities and enforcement teams should be more focused on minimising discourtesy behaviour among drivers. Campaign via mass media can be instrumental in highlighting the types of good driving behaviour, advocating tolerance in driving style and showing the negative impacts of engaging in discourteous driving behaviour.

Table 7. Regression coefficient

Model		Unstandardized coefficient		Standardized coefficient	t	Sig.
		B	Std. Error	Beta		
	(Constant)	.918	.495		1.853	.065
1	Discourtesy	.738	.118	.412	6.257	.000
	Hostile gesture	.053	.080	.045	0.660	.510
	Slow driving	-.016	.117	-.010	-0.141	.888
	Traffic obstruction	.344	.124	.197	2.774	.006
	Illegal driving	-.101	.092	-.065	-1.103	.271
	Police presence	-.155	.093	-.105	-1.663	.098

The authors would like to thank Universiti Tun Hussein Onn Malaysia (UTHM) for funding this study under the Multidisciplinary Research Grant (MDR) Vot U090 and Smart Driving Research Center (SDRC), Faculty of Civil and Environmental Engineering for support and providing facilities to accomplish the study.

References

[1] J.L. Deffenbacher, E.R. Oetting, R.S. Lynch, Development of a driver anger scale, Psychological Reports, **74**, 83–91 (1994)

[2] A.N. Stephens, J.A. Groeger, Anger-congruent behaviour transfers across driving situations, Cognition and Emotion, **25**, 1423–1438 (2011)

[3] R. Lawton, A. Nutter, A comparison of reported levels and expression of anger in everyday and driving situations, British Journal of Psychology, **93**, 407–423 (2002)

[4] J.L. Deffenbacher, R.S. Lynch, L.B. Filetti, E.R. Dahlen, E.R. Oetting, Anger, aggression, risky behaviour, and crash-related outcomes in three groups of drivers Behav. Res. Ther, **41**, 333–349 (2003)

[5] J.L. Deffenbacher, R.S. Lynch, E.R. Oetting, D.A. Yingling, Driving anger: Correlates and a test of state-trait theory, Personality and Individual Differences, **31**, 1321–31 (2001)

[6] T. Lajunen, D. Parker, Are aggressive people aggressive drivers? A study of the relationship between self-reported general aggressiveness, driver anger and aggressive driving, Accident Anal. Prev., **33**, 243–255 (2001)

[7] M.J.M. Sullman, A.N. Stephens, D. Kuzu, The expression of driving anger amongst Turkish taxi drivers, Accident Analysis Prevention, **56**, 42–50 (2013)

[8] J.L. Deffenbacher, Angry drivers: Characteristics and clinical interventions, Revista Mexicana de Psicologia, **26** (1), 5-16, (2009)

[9] Polis Diraja Malaysia, *Statistik kemalangan jalan raya 2011 di daerah Batu Pahat*, Batu Pahat, Balai Polis Trafik Batu Pahat, (2011)

[10] M.J.M. Sullman, A.N. Stephens, M. Yong, Driving anger in Malaysia, Accident Analysis and Prevention, **71**, 1-9 (2014)

[11] Y. Yasak, B. Esiyol, Anger amongst Turkish drivers: Driving anger scale and its adapted long and short version, Safety Science, **47**, 138–144 (2009)

[12] K. Ambak, R. Ismail, R.A. Abdullah, A.A. Latiff, M.E. Sanik, Application of technology acceptance model in predicting behavioral intention to use safety helmet reminder system, Research Journal of Applied Sciences, Engineering and Technology, **5** (3), 881-888 (2013)

[13] M.J.M. Sullman, A.N. Stephens, A comparison of the driving anger scale and the propensity for angry driving scale, Accident Analysis Prevention, **58**, 88–96 (2013)

[14] M.B. Ahmed, K. Ambak, A. Raqib, N.S. Sukor, Helmet usage among adolescent in rural road from the extended theory of planned behavior, Journal of Applied Sciences, **13**, 161-166 (2013)

Accident Analysis and Highway Safety

Noorliyana Omar[1,*], *Joewono* Prasetijo[1], *Basil* David Daniel[1], and *Mohd Asrul Effendi* Abdullah[2]

[1]Department of Infrastructure and Geomatic Engineering, Universiti Tun Hussein Onn Malaysia, Johor, Malaysia
[2]Department of Statistics and Mathematics, Universiti Tun Hussein Onn Malaysia, Johor, Malaysia

Abstract. Since 2010, Federal Route FT050 (Jalan Batu Pahat-Kluang) has undergone many changes, including the improvement of geometric features (i.e., construction of median, dedicated U-turns and additional lanes) and upgrading the quality of the road surface. Unfortunately, even with these enhancements, accidents continue to occur along this route. This study covered both accident analysis and blackspot study. Accident point weightage was used to identify blackspot locations. The results reveal hazardous road locations and blackspot ranking along the route.

1 Introduction

Road accidents have been recognised as one of the major problems in the world. In addition, it has major impacts on the society, economy and progress of a country. Road accidents occur when traffic conflicts between vehicular movements which can cause delay and traffic congestion [1]. Rohani [2] mentioned that human error, traffic, vehicle and road conditions, and the surrounding environment have been determined as key factors that influence driving behaviour, which can in turn contribute to road accidents.

Road accidents increase relatively from year to year even though the Malaysian government has carried out many programs. The increment is proportionate to the growth in population, economic development, industrialisation and motorisation in the country [3]. Furthermore, road accidents are the leading cause of death in Malaysia [4].

There are many road accidents reported every day. In 2013, Malaysia recorded 477,204 accidents, which include 6,915 deaths. This resulted in an average of 19 deaths from road accidents every day. Deaths per 10,000 registered vehicles was 2.9, which is high in comparison to developed countries that register 2.0 deaths or less per 10,000 registered vehicles [5].

Fig. 1 shows that in 2013, federal roads in Malaysia had the highest number of road accidents compared to other road categories. Federal Route 50 (FT050) is a four-lane road that runs from Batu Pahat to Kluang (Fig. 2). The road has many access points, almost one per kilometre. The road also has a high density of private driveways and access to properties. 1420 accidents were recorded along this stretch in 2013 alone [6]. So far, at least

*Corresponding author: liyana@uthm.edu.my

five studies have been conducted along this road [6][7][8][9][10]. Seven locations at FT050 were identified as accident blackspots, which are KM posts 2, 4, 7, 23, 48, 57 and 58 [5]. The accident rate at this road is among the highest compared to other federal roads in the area.

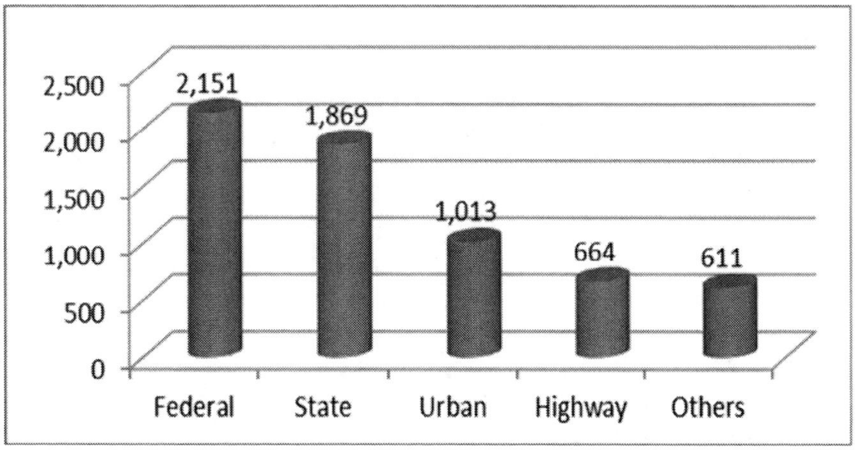

Fig. 1. Accidents based on road types in 2013 [5].

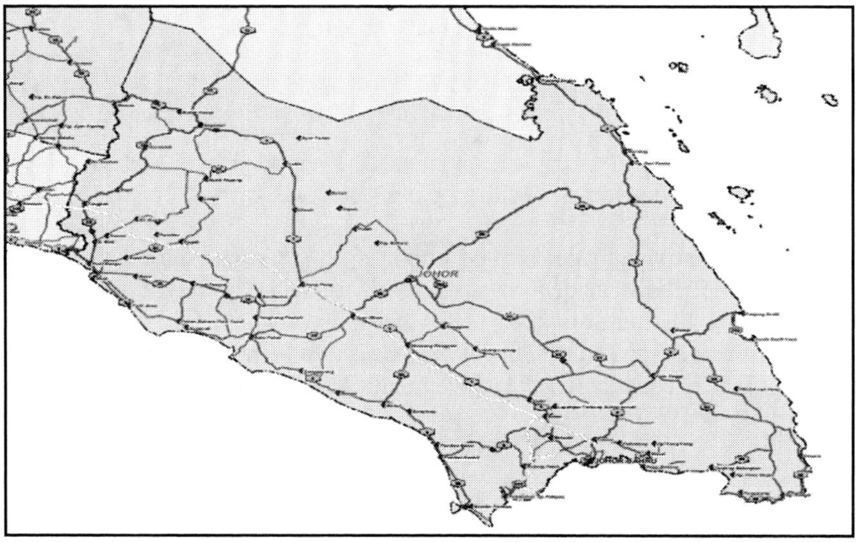

Fig. 2. Study location at Federal Route FT05.

In order to achieve a "developed country" status, Malaysia needs to minimise road accidents and increase the safety level for road users. The Malaysian Institute of Road Safety (MIROS) was established in order to monitor and enhance road safety in Malaysia. It acts as the main organiser in developing young minds and educating people about road safety. In addition, the institute, through their advanced Research and Development unit, focuses on how to minimise road accidents through road standards and design.

The location of this study is along Federal Route 50 (FT050), one of the black spot areas in Johor [11]. Based on the road accident statistics, the accidents involved cars, lorries, bicycles, pedestrians and mostly motorcyclists [8].

2 Aims

The aim of this study is to analyse the accident data in order to identify blackspots along Federal Route 50.

3 Method

The methodology adopted for this study involves identification of the study area, accident data collection from the police station and blackspot identification.

3.1 Data collection

Accident data was collected from PDRM covering all police recorded accidents. The data included; number of fatalities, number of seriousness of the injuries (i.e., serious and slight injury) and number of accidents that did not involve any injury. A five-year accident period was used, starting from 2010 until 2014.

3.2 Analysis of accident data

The raw data were collected from PDRM as mentioned in section 3.1 and analyzed to determine the accident pattern at the studied area. The analysis consist of accidents severities by year, accidents by section and black spot location.

3.3 Blackspot identification

An accident blackspot refers to a section of a road regarded as a high-risk location for vehicle crashes. Identification of accident blackspots was carried out using the accident point weightage method.

The accident point weightage formula was used to determine the ranking of accidents for every kilometre. Each accident severity level has its own weightage. For accidents involving fatalities, the numbers need to be multiplied by 6.0. For serious injury, slight injury and not involving any injury, the numbers need to be multiplied by 3.0, 0.8 and 0.2, respectively [12]. The equation for the above mentioned method is shown in Equation (1).

$$APW = X_1(6.0) + X_2(3.0) + X_3(0.8) + X_4(0.2) \quad (1)$$

where, X_1 = Number of fatalities, X_2 = Number of serious injuries, X_3 = Number of slight injuries, and X_4 = Number of damages only

4 Results and analysis

This study was designed to analyse accident data and identify accident blackspot. The following sections report the results of this study.

4.1 Accident severity by year

Data on accident severity was analysed from year 2010-2014. The results can be seen in Fig. 3.

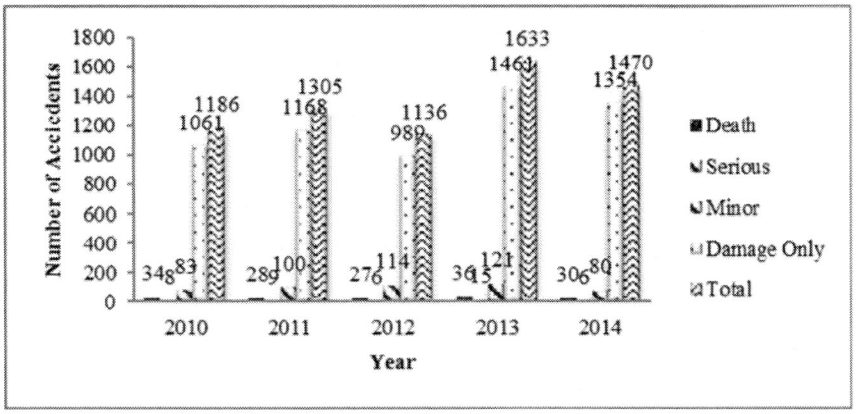

Fig. 3. Accident severity from 2010 – 2014.

As observed from Figure 3, the accidents increased from 2010 to 2011. From only 1186 accidents in 2010, the number increased to 1305 in 2011. In 2012, there was a slight decrease in the number of total accidents. This is because of the upgrading process of this road which causes drivers to decrease their speed along the work zone. However, the total number of accidents increased from 1136 to 1633 from 2012 to 2013. This may be due to a sharp increase in total vehicle registration as reported by the Road Transportation Department (2015) [13]. In 2014, the number of accidents decreased to 1470. The highest number of recorded accidents between 2010 and 2014 occurred in the year 2013, where a total of 1633 accidents were recorded.

4.2 Accidents by section

From the accident severity data by year, this study focused on the highest number of accidents based on section location. The results can be seen in Fig. 4.

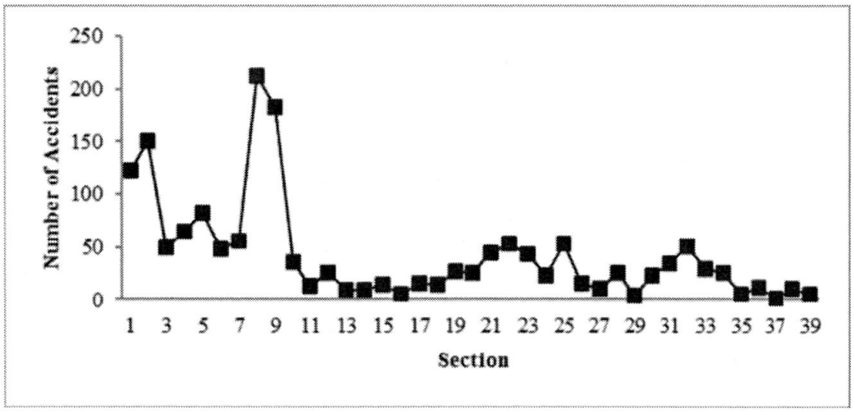

Fig. 4. Total Number of Accidents by Section in 2013.

The highest numbers of accidents was recorded in 2013 at KM 8 (Fig. 4). Land use and vertical alignment are amongst the factors that contribute to this figure. Along this section, there are several buildings such as Pejabat Kaji Cuaca, Maktab Perguruan Tun Hussein Onn and Kolej Vokasional Batu Pahat. The type and functions of buildings along the road have a considerable influence on accident risk [14].

The highest number of fatalities occurred in 2010 at KM 10. The high number of fatalities was contributed by two factors, which are land use [14] and road alignment. At this location, there are a number of factories. Furthermore, traffic conflict occurred due to the presence of many access points and road alignment [15].

The highest number of serious accidents occurred in 2013 at KM 9. Since this section is near Pejabat Kaji Cuaca, Maktab Perguruan Tun Hussein Onn and Kolej Vokasional Batu Pahat, the contributing factors to the high accident rate are land use and type of terrain. As mentioned before, type and functions of building along the road has a considerable influence on the accident risk [14].

The highest number of minor accidents was recorded in 2012 at KM 32. The main factor that caused minor accidents is land use. At this location, there are shop lots along the road. More traffic conflict occurred due to the presence of many access points [15].

The highest "damage only" accidents were recorded in 2014 at KM 1. The main factors that caused minor accidents are land use and road alignment (roundabout). At this location, there are roundabouts and shop lots along the road. More traffic conflict occurred due to the presence of many access points and roundabouts [15]. Traffic conflict is "an observable situation in which two or more road users approach each other in space and time for such an extent that there is a risk of collision if their movement remain unchanged" [16].

4.3 Black spot

Black Spot is the simplest way of ranking sites and it is recommended to be used in Malaysia [7]. All that is needed to be done is to shortlist the ranking of the accident blackspot area. From the analysis, the ranking shows that the highest Accident Point Weightage value from each section analysis of the traffic accident record by PDRM.

Fig. 5 shows the top ten accident ranking from 2010 to 2014. In 2010 and 2011, Sharp/ROXY factory was placed in the first ranking. As mentioned previously, any accident which occurred along the road was due to land use [14]. Besides that, many pedestrians use this road as reported by [14, 17, 18].

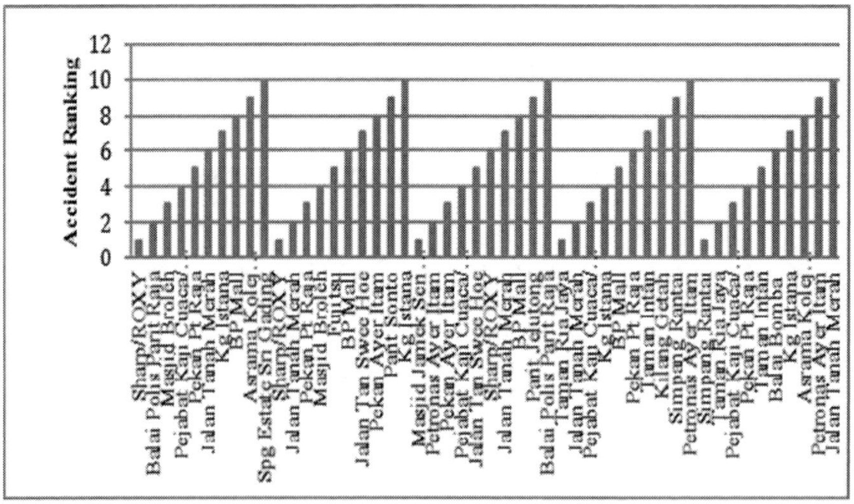

Fig. 5. Accident Ranking by Location from year 2010-2014.

Masjid Jamek Seri Beroleh has the top ranking in 2012. A petrol station is located beside this place. These locations are parallel to this route. There is driveway from the main road to this place. Therefore, this situation can create conflict at this access point and affect the number of accidents [17].

In 2013, Taman Ria Jaya was ranked 1st. Since this section is near Pejabat Kaji Cuaca, Maktab Perguruan Tun Hussein Onn and Kolej Vokasional Batu Pahat, the contributing factors that make it at the top of the list are land use and type of terrain. Type and functions of building along the road have a considerable influence on the accident risk [14].

Ranked 1st was Simpang Rantai in 2014. This location is an unsignalized intersection with a roundabout thus this section had many traffic conflicts. Traffic control will affect the number of accidents [17, 19, 20].

5 Conclusions

From this accident analysis, the following conclusions can be drawn:

i) Since the upgrade of FT050 in 2010, the number of accidents has increased instead of producing a safer route for the community.
ii) The authorities and government should take this issue seriously and introduce precautionary steps to prevent more accidents from happening along this route.
iii) This study found that accident data fluctuates from year to year.
iv) This study had also established the accident point weightage to rank the accident blackspots along the FT050 stretch (KM1.0 – KM38.0) to determine hazardous road sections.
v) The result shows that KM 1 – KM 10 and KM 21 – KM 23 are the most dangerous stretches along FT050. This result was supported by data obtained by the Royal Malaysian Police [5].
vi) Further research should be conducted in order to improve information, safety awareness and propose actions for the public or the authorities.

The authors gratefully acknowledge the Ministry of Higher Education Malaysia (MOHE) for providing financial support for the authors and to Universiti Tun Hussein Onn Malaysia (UTHM) for providing the resources needed to complete this paper.

References

[1] J. Prasetijo, W. Razzaly, N. Wu, K. Ambak, M.E. Sanik, M.M. Rohani, and H. Ahmad, Capacity Analysis of priority intersections with flare under mixed traffic conditions, Procedia - Soc. Behav. Sci., **138**, 660–670 (2014)
[2] M. Rohani, R. Buhari, B.D. Daniel, K. Ambak, N.A. Sukor, and F.S. Asmah, Car driving behaviour on road curves: A study case in Universiti Tun Hussein Onn Malaysia, Appl. Mech. Mater., **10**, 990–995, (2015)
[3] W.N.W. Manan, *Accident Prediction Model at Un-Signalized Intersections Using Multiple Regression Method*, Master Thesis, Universiti Tun Hussein Onn Malaysia, (2011)
[4] J. Prasetijo, S. Nur, F. Mohd, K. Ambak, and B.D. Daniel, Investigation of the impact on different type of toll plaza crash barriers by numerical approach, **11**, 3690–3696, (2016)
[5] RMP, Royal Malaysia Police.pdf, (2015)
[6] F. Mustakim, and B. D. Daniel, Accident investigation, blackspot treatment and accident prediction model at federal route FT50 Batu Pahat-Air Hitam, *The 4th National Technical Post Graduate Symposium*, (2006)

[7] F. Mustakim, A. Aziz, and A. Samad, Blackspot study and accident prediction model using multiple liner regression, *Advancing and Integrating Construction Education, Research and Practice*, 121–136 (2008)
[8] A. Raqib, A. Ghani, M.E. Sanik, R. Aida, and M. Mokhtar, Comparison of accident prediction model, *Int. Semin. Appl. Sci. Math.*, (2011)
[9] F. Mustakim, and M. Fujita, Development of accident predictive model for rural roadway, Int. J. of Civil, Archit. Struct. Constr. Eng., **5**, 46–51 (2011)
[10] J. Prasetijo, and Z.F. Zainal, Development of continuous speed profile using GPS at Johor Federal Roads F0050, *MATEC Web Conf.*, **24**, 0–5 (2016)
[11] B.Y.C. Tan, *Batu Pahat district records the highest increase in road accident fatalities,* The Star, 23–25, (2012)
[12] B.N.P. Alireza Hadayeghi, and Amer S. Shalaby, TRB 2003 Annual Meeting CD-ROM Paper revised from original submittal, (2003)
[13] JPJ, Jumlah Pendaftaran Lain-lain Kenderaan Mengikut Tahun.pdf, (2015)
[14] P. Greibe, Accident prediction models for urban roads, Accident Analysis and Prevention, **35**, 273–285 (2003)
[15] R. Fu, Y. Guo, W. Yuan, H. Feng, and Y. Ma, The correlation between gradients of descending roads and accident rates, Saf. Sci., **49**, 416–423 (2011)
[16] K. El-Basyouny, and T. Sayed, Safety performance functions using traffic conflicts, Saf. Sci., **51**, 160–164 (2013)
[17] A. Ahmed, A. F. M. Sadullah, and A. S. Yahya, Accident analysis using count data for unsignalized intersections in Malaysia, Procedia Eng., 77, 45–52 (2014)
[18] M. Simonc, Road accidents in Slovenia involving a pedestrian , cyclist or motorcyclist and a car, *Accid. Anal. Prev.*, **33**, 147–156 (2001)
[19] S. Daniels, T. Brijs, E. Nuyts, and G. Wets, Extended prediction models for crashes at roundabouts, Saf. Sci., **49**, 198–207 (2011)
[20] B. De Brabander, and L. Vereeck, Safety effects of roundabouts in Flanders: Signal type, speed limits and vulnerable road users., Accid. Anal. Prev., **39**, 591–599 (2007)

Crossing Behaviour of Pedestrians Along Urban Streets in Malaysia

Walid A. Al Bargi[1], *Basil* David Daniel[1,*], *Joewono* Prasetijo[1], *Munzilah* Md Rohani[1], and *Siti Naquiyah* Mohamad Nor[1]

[1]Department of Infrastructure And Geomatic Engineering, Faculty of Civil and Environmental Engineering, Universiti Tun Hussein Onn Malaysia, 86400 Batu Pahat, Johor, Malaysia

Abstract. Road crossings are considered as an unavoidable part of walking in which the desirable route of pedestrians interacts with vehicles. These interactions may expose the pedestrians to risks or delays. In Malaysia, road accident statistics show that pedestrian casualties are fairly high. Inappropriate gap acceptance when pedestrians cross roads is a main contributing element to this situation. In this context, the purpose of this study was to develop realistic models for pedestrian road crossing behaviour using the regression technique for mid-block street crossing. A choice model was produced to capture the decision making process of pedestrians whereas rejected or accepted vehicular gaps was based on the discrete choice theory. Gap acceptance data under real mix traffic conditions was collected using video camera on a typical unsignalised two lane one way urban street section in the city center of Kuala Lumpur, Malaysia. The lognormal regression model developed for the crossing behaviour model shows that traffic speed, pedestrian waiting time, gender, crossing distance, age group, frequency of attempts and pedestrian number are the significant factors which are able to predict 77.0% of variance or changes in accepted gap size at 0.05 significance level. Higher traffic speed, lower waiting time, being a male, wider crossing distance, older age group, lower frequency of attempts and higher number of pedestrian were found to influence pedestrians to accept a bigger gap size. The binary logistic regression developed for the crossing choice model was found to be influenced by traffic speed, driver yield, pedestrian number and age group. Furthermore, lower traffic speed, willingness of drivers to slow down, more pedestrian crossings at the same time and a younger age group lead to a higher chance or probability of crossing roads. The model was validated again using 100 isolated samples and an accuracy of 98% was obtained compared to the calibrated model which yielded an accuracy of 98.9%.

1 Introduction

The concept of "pedestrian" can have many different meanings. The Concise Oxford dictionary defines it as "a person walking in the street and not travelling in a vehicle".

[*]Corresponding author: basil@uthm.edu.my

Crossing facility users are always at risk when they make their daily trips. Therefore, crossing facilities, roadway designs and traffic control devices must consider factors of safety to protect pedestrians [1]. In developing countries, pedestrians face difficulty when they are attempting to cross the street due to the drivers' behaviour towards them. Pedestrians usually have to wait for a long time before finding a volunteer who stops for them to allow them to cross safely otherwise they are forced to wait until there are no more vehicles. This phenomenon may be caused by a lack of awareness on the rule of the right of way in such situations [2]. Pedestrians' choice of suitable gaps in vehicular stream flow depends on their demographics, pedestrian and vehicle behavioural characteristics and street geometry [3]. The identification of safe gaps between passing cars when crossing a road is a task most pedestrians accomplish successfully on a daily basis [4]. In the previous decades, pedestrian behaviour has been placed in the spotlight. Many research studies incorporating different aspects of the minimum gap from vehicles that is accepted by pedestrians who intend to cross streets at mid-blocks have been performed by various researchers at different durations across the world [5]–[8]. The space between the pedestrians and motor vehicles appear to affect the most minimum gap accepted by pedestrians. Moreover, an increase in traffic volume leads to smaller accepted gaps. These gaps are often described by estimating the means of liner regression modeling or by probability distributions. Indicatively, it can be declared that the minimum accepted gap has been estimated to be 2 seconds while the mean accepted gap has been estimated to be 8 seconds [17].

2 Literature overview

In the past, many studies investigating the behaviour of pedestrians, particularly those crossing the street, have been done. Several factors which influence pedestrian behaviour have been identified. These factors can be categorised as pedestrians, road environment and traffic factors. Delays due to long waiting time significantly affect pedestrian behaviour. Pedestrians have a higher tendency to jaywalk after a long waiting time. They lose their patience while waiting to accept safe gaps. Instead of waiting for safer gaps (greater loss of time), a pedestrian may opt to use rolling gaps across several lanes [10], [11]. The time spent waiting for safer gaps depends on whether the pedestrian is alone or in groups. Pedestrians tend to cross illegally (cross on red) if a member of the group initiates the illegal cross. From observations made, male pedestrians were more likely to cross on red compared to females [12]. In addition to this, the number of pedestrians waiting in a group also affects the behaviour of pedestrians in that group. Smaller groups tend to make more illegal crosses compared to larger ones [13]. In accepting gaps to cross, each pedestrian has a different perception on choosing the safest gap. Their physical characteristics affect their movement, i.e. walking speed. Taller pedestrians accept smaller gaps compared to shorter pedestrians due to them generally being able to walk faster [1].

The distance between vehicles and pedestrians has a significant effect on the acceptance of safe gaps to cross [7], Interestingly, male pedestrians made the most inaccurate choices where they accepted smaller gaps to cross compared to their female counterparts [14]. Pedestrians' decisions on crossing the road, when to stop or wait, and where to cross the road are very complex. The decision is influenced by many factors such as comfort level, the convenience to cross and safety. Pedestrian flow, traffic volume, footpath width, road width, surface condition and walkway obstructions were found to significantly affect pedestrian comfort and safety [15].

Pedestrian behaviour is not constant. It often changes according to the surroundings. Nevertheless, pedestrians naturally adapt to their surroundings [16]. The mean accepted gap for pedestrian to make a cross is estimated to be 8 seconds [17].

3 Methodology

Jalan Sultan Ismail, located in the center of Kuala Lumpur, Malaysia, has been chosen due to the considerable volume of vehicles and pedestrians. It was also observed that the selected location shows great interaction between traffic and pedestrians. The data was collected using a high resolution video camera. Data analysis was conducted to compute the descriptive statistics including standard error, t-value, and p-value, under Multiple Linear Regression and Binary Logic. Basic steps and function under Multiple Linear Regression and Binary Logic were used for the data analysis. However, pedestrian crossing behaviour was recorded in real traffic situations. Factors considered include pedestrian waiting time, traffic waiting time, gap sizes rejected or accepted by pedestrians, the related number of crossing attempts and each vehicle's speed. In addition to pedestrian individual characteristics such as gender, it is significant to indicate that illegal parking on Jalan Sultan Ismail occurred frequently and the existence of illegally parked vehicles was videotaped during the data collection. Vehicle types in the traffic stream are classified as Malaysian standard cars, motorcycles, vans/medium vehicles and heavy vehicles.

4 Data analysis

This section explains the structure of analysis as well as the variables involved in the study. The data of study was obtained through the recording of road crossings and the descriptive analysis of the observed variables. Observed variables such as traffic speed, waiting time, frequency of attempt, driver yield behaviour, crossing distance of the road, size of vehicle, occurrence of illegal parking, pedestrian gender, age group, etc. will be discussed. Next, the road crossing behaviour of pedestrians was analysed and modeled using lognormal regression model. The regression model was able to explain the accepted gap size for road crossing based on the various predictors as mentioned above. Subsequently, the pedestrian crossing behaviour was analysed in order to develop a crossing choice model using binary logistic regression with 78% of the sample for model calibration and 22% of the sample for model validation. In addition, the relationship between predictors and dependent variables (log gap size and gap acceptance) was presented visually to ease understanding.

Basically, there are 13 variables recorded as presented in Table 1,in which 2 variables are dependent variables and 11 variables are independent variables or predictors. Based on the table, gap size is a continuous dependent variable that measures the gap between two incoming cars at the crossing spot in seconds while gap acceptance is a discrete dependent variable that records whether a pedestrian chooses to cross the road or not (0: Reject; 1: Accept). Traffic speed is a continuous independent variable that measures the speed of a vehicle at the crosswalk area in km/h. The time pedestrians spend waiting at the crossing area until they cross the road is measured in seconds. Frequency of attempt is a continuous independent variable that records the number of attempts made by pedestrians until they cross the road. Next, driver yield is a discrete independent variable that shows whether the driver of an approaching car slows down (assign a value of 1) or does not slow down (assign a value of 0). The crossing distance of a road is a continuous independent variable measured in meter unit. The accepted gap size of respondents may be influenced by the crossing distance. In terms of vehicle size, which is an independent variable with a discrete scale, a value of 0 is assigned for small vehicles and a value of 1 is assigned for big vehicles. Illegal parking is another discrete independent variable that records the presence of illegally parked cars along the road side with a value of 0 when there is no illegally parked car and a value of 1 if there is an illegally parked car. Other independent discrete variables include pedestrian gender (0: Male, 1: Female), gap type which measures the distance between pedestrians and the gap (0: Near, 1: Far) and pedestrian age (1: Young, 2:

Middle, 3: Elder). Lastly, pedestrian number basically refers to the number of pedestrians waiting in a group to cross. It was noticed that these factors are considered to be the most significant ones influencing pedestrian crossing behaviour at mid-block, based on to the literature (9).

Table 1. List of variables

Variable	Type	Unit	Description
Gap Size	Continuous	sec	Gap between two incoming cars at the crossing spot
Gap Acceptance	Discrete	0: Reject 1: Accept	Gap acceptance refers to whether a pedestrian chooses to cross the road
Traffic Speed	Continuous	km/h	Speed of vehicle at the crosswalk area
Pedestrian waiting time	Continuous	sec	Time spent waiting at the crossing area for a suitable gap
Frequency of attempt	Continuous	Count	Number of attempts made by a pedestrian before accepting the vehicular gap
Driver Yield	Discrete	0: No 1: Yes	Driver reduces speed when pedestrian is trying to cross
Crossing Distance	Continuous	m	The crosswalk width that pedestrians need to cross
Vehicle size	Discrete	0: Small 1: Big	Size of vehicle
Illegal Parking	Discrete	0: No 1: Yes	Presence of illegally parked cars
Pedestrian Gender	Discrete	0: Male 1: Female	Gender of pedestrian
Pedestrian Number	Continuous	Count	Number of pedestrians in a group
Gap Type	Discrete	0: Near 1: Far	Distance between pedestrians and the gap
Age	Discrete	1: Young 2: Middle 3: Elders	Visual appearance of pedestrians

4.1 Descriptive analysis of variables

In this study, 448 samples were collected at Jalan Sultan Ismail, Kuala Lumpur and the descriptive analysis for both discrete and continuous variables are presented in Table 2 and Table 3. Based on Table 2, 77.2% of the sample crossed the road by accepting the gap, and about 62.1% of drivers slowed down their vehicle speed. In addition, about 31% of the pedestrian road crossings had illegally parked cars. In terms of pedestrian gender, the majority of them are males, accounting for 61% of the total sample. Results showed that 41% of the sample consists of middle-aged people, 34% are young pedestrians while 25% consists of the elderly. Results also showed that 55.4% of pedestrians cross the road when the gap is near while 44.6% of them choose to cross the road while the gap is still far from them. Lastly, 91.5% of the vehicles observed were small vehicles and only 8.5% were big vehicles. As far as continuous variables are concerned, results illustrated their mean value, standard deviation, minimum and maximum values. For gap size, the mean time is 8.91 seconds with a range of 1 to 30 seconds while traffic speed had a mean of 36.5 km/h with a range of 23 to 55 km/h. Furthermore, pedestrian waiting time was estimated to be 8.87 secs with a range between 1 to 60 seconds. Frequency of attempt was 0.22 with a range between

0 to 2 times while the pedestrian number showed a mean value of 1.36 with a range between 1 to 8 pedestrians.

Table 2. Descriptive statistic for discrete variables

Variables	Frequency	Percent (%)
Gap Acceptance		
Reject	102	22.8
Accept	346	77.2
Driver Yield		
No	170	37.9
Yes	278	62.1
Illegal Parking		
No	308	68.8
Yes	140	31.3
Pedestrian Gender		
Male	273	60.9
Female	175	39.1
Gap Type		
Near	248	55.4
Far	200	44.6
Age Group		
Young	151	33.7
Middle	183	40.8
Elder	114	25.4
Vehicle size		
Small vehicle	410	91.5
Large Vehicle	38	8.5

Table 3. Descriptive statistic for continuous variables

Variable	Mean	S.D.	Minimum	Maximum
log_gap	0.90	0.204	0.0	1.5
Gap Size	8.91	4.344	1.0	30.0
Traffic Speed	36.50	8.765	23.0	55.0
Pedestrian waiting time	8.87	5.022	1.0	60.0
Frequency of attempt	0.22	0.458	0.0	2.0
Crossing Distance	6.77	0.878	6.3	8.3
Pedestrian Number	1.36	0.782	1.0	8.0

4.2 Pedestrian behavioural model

The multiple linear regression (MLR) model was used to evaluate the accepted gap size for pedestrians. A lognormal regression model was chosen due to the normal distribution assumption for the logarithm of the dependent variable which will likely be able to fit the logarithm of the gaps. Prior to running the test, some samples were filtered out (those who do not cross) since the model was meant to evaluate the accepted gap size. In other words, only samples that accepted the gap (crossed the road) were used, in line with the purpose to examine the factors that affect the accepted gap size of pedestrians.

The lognormal regression model was conducted using the Stepwise method, leaving only truly significant variables and the results are summarised in Table 4. Based on the table, F value of 161.352 with a p value close to 0 show that the model is significant. Moreover, an R square value of 0.770 indicates that 7 predictors were able to explain 77.0% of variance in log gap size. First, the variance inflation factor (VIF) value was lower than 10, suggesting that there are no multicollinearity issues for the model. Based on the table, pedestrian waiting time and frequency of attempt showed negative regression weight, indicating that both variables showed a reverse relationship with accepted gap size. Such findings suggest that the longer the pedestrian waiting time and the more attempts taken, the shorter the gap size that pedestrians willing to take. Referring to the B value of -0.010 and -0.028, a unit increase in waiting time and pedestrian attempt will result in the log accepted gap size decreasing by 0.01 second and 0.028 seconds respectively. In short, the longer the pedestrian waiting time or the more the pedestrian attempts, the more likely they will take the risk by accepting a shorter gap size.

Apart from that, this study discovered that traffic speed, pedestrian gender, crossing distance, age group and pedestrian number showed a positive impact on the log accepted gap size with a positive regression weight, B. Such positive coefficients indicate that an increase in the predictor will increase the log accepted gap size as well. For continuous variables, each unit increase in the continuous predictor will increase the pedestrian accepted log gap size by 0.013 seconds (traffic speed), 0.036 seconds (crossing distance), and 0.016 seconds (pedestrian number) respectively. As for discrete or categorical variables (refer back to the coding), a positive coefficient in pedestrian gender suggests that female pedestrians are more prone to accept longer gap size compared to male pedestrians. Besides, a positive coefficient of age groups indicates that older pedestrians tend to accept a longer gap size.

In regression modelling, it is important to assess the relationship strength of predictors to understand their impact on the dependent variable (log gap size). The absolute value of standardised regression weight (BETA) is examined and it was found that traffic speed and pedestrian waiting time showed a relatively strong relationship with log gap size with absolute BETA values of 0.373 and 0.247 respectively. The strength of relationship is then followed by pedestrian gender and age groups which have relatively normal strength or impact towards log gap size. As for frequency of attempt and pedestrian number, both are relatively weak in terms of relationship or impact towards the log gap size. However, it was observed that lognormal regression assumes a normal distribution for the logarithm of the dependent variables. The general model framework is given below:

$$\text{Log} - \text{Gap} = \beta 0 + \beta 1\, \text{X1} + \beta 2\, \text{X2} + \beta 3\, \text{X3} + \ldots\ldots\ldots\ldots\ldots + \beta n\, \text{Xn} \quad (1)$$

where Log-Gap= logarithm of accepted gaps; Xi-n= explanatory variables; β1-n= are estimated parameters from the model; β0= constant

Table 4. Regression weight statistics

Variable	**B**	**Beta**	**t**	**p-value**	**Tolerance**	**VIF**
(Constant)	0.261		4.83	0.000		
Traffic Speed*	0.013	0.374	10.33	0.000	0.519	1.928
Pedestrian waiting time*	-0.01	-0.248	-7.013	0.000	0.547	1.828
Pedestrian Gender	0.074	0.189	5.663	0.000	0.61	1.64
Crossing Distance*	0.036	0.172	5.056	0.000	0.587	1.702
Age Group	0.041	0.12	3.633	0.000	0.626	1.596
Frequency of attempt*	-0.027	-0.07	-2.003	0.046	0.552	1.811
Pedestrian Number*	0.016	0.067	2.025	0.044	0.628	1.592

DV: Log gap, $R^2 = 0.770$, $F(7,338) = 161.352$, p value ≈ 0.000, *continuous variable

The final model is shown below:

$$\text{Log gap size} = 0.261 + 0.013\,TS - 0.01\,PWT + 0.074\,PG + 0.036\,CD + 0.041\,AG - 0.027\,FA + 0.016\,PN \quad (2)$$

To better illustrate the relationship, the relationship of particular predictors was modelled using the regression model obtained earlier while other variables were held constant with their corresponding mean values (Table 5). For example, if the study illustrates the relationship of traffic speed and gap size, the mean values of other variables in Table 5 were substituted into model Equation 1. The regression equations for each variable is presented in Table 6.

Table 5. Mean statistic for variables

Variables	Mean
Traffic Speed	32.77
Pedestrian waiting time	10.37
Pedestrian Gender	0.45
Crossing Distance	6.89
Age Group	1.61
Frequency of attempt	0.29
Pedestrian Number	1.36

Table 6. Relationship between independent variables and the dependent variable (gap size)

Independent variables	Relationship
Traffic speed (km/h)	y = 0.013x + 0.5415
Waiting Time (s)	y = -0.01x + 1.0506
Gender (0: Male, 1: Female)	y = 0.074x + 0.9138
Crossing Distance (m)	y = 0.036x + 0.6988
Age (1: Young, 2: Middle, 3: Elder)	y = 0.042x + 0.8793
Frequency of attempt (count)	y = -0.028x + 0.955
Pedestrian number (count)	y = 0.016x + 0.9251

4.3 Choice behaviour model

A Binary Logit Model was employed to analyse the choice opportunities for pedestrian road crossing behaviour at an uncontrolled mid-block location. It is important to state that the data was divided into approximately 78% (n=348) and 22% (n=100) randomly, where 78% of data was used to analyse and calibrate the Binary Logit model while22% of the data was utilised to validate the prediction model.

The probability of selecting an alternative (accept/reject) is based on a linear combination function (utility function) expressed as:

$$Ui = C + \beta i1X1 + \beta i2X2 + \beta i3X3 + \beta i4X4 + \beta i5X5 \ldots\ldots + \beta inXn \quad (3)$$

where Ui = the utility of choosing alternative i, i = the alternative (accept/reject), n = number of independent variables, C = constant, and ß = coefficients.

The utility of alternative 'i' has to be transformed into a probability in order to predict whether a particular alternative will be chosen or not. The probability of choosing alternative 'i' is, in this case, crossing the road which is then calculated using the following function:

$$P(i) = \exp(Ui) \;/\; [1 + \exp(Ui)] \tag{4}$$

Based on the result, the final Binary Logit Model can be written as:

$$Ui = 48.265 - 0.853TS + 0.7442\,DY + 2.829\,PN - 7.153\,AG \tag{5}$$

From the regression weight results in Table 6, it seems that traffic speed and age group produced a negative coefficient in terms of the decision of road crossing or gap acceptance (0: reject, 1: accept). The negative B value shows that if vehicles approach with higher speed, pedestrians are more likely to reject the gap or chose not to cross compared to vehicles with lower traffic speed. For age groups (1: young, 2: middle, 3: elder), a negative B value means that pedestrians who looked older are more likely to reject a gap compared to pedestrians who look younger or more energetic. It was found that driver yield and pedestrian number (in the group) have a significant positive effect on pedestrian crossing choice. Besides that, driver yield (0: no, 1: yes) occurs when drivers slow down their vehicle and a positive B value indicates that if a driver slows down, pedestrians are more likely to cross the street.

Similarly, a positive B value in terms of pedestrian number shows that if pedestrians move in a larger group, they are more likely to cross the street compared to those crossing alone. This could be due to the fear factor when doing it alone or the fact that drivers tend to slow down for a larger group of pedestrians, which then promote higher gap acceptance. Looking at the exponential B value, it seems that if drivers were to decrease their speed by 1 unit (1km/h), the probability of pedestrians crossing the road increases by 2.3 times. Also, if the drivers were to slow down (driver yield), the probability of pedestrians crossing the road increases by about 1700 times, which is a very significant predictor in the Crossing Choice Model. On the other hand, an increase of each person in the pedestrian group will increase the gap acceptance probability by about 17 times. Lastly, if the pedestrian age were to decrease, for instance, from the elderly group to the middle-aged group, the chance that a pedestrian will cross or accept the gap increases by about 1000 times.

With the calibrated model, this study validates the model by calculating the probability of crossing as follows:

$$p(i) = \frac{e^{48.265 - 0.853TS + 0.7442\,DY + 2.829\,PN - 7.153\,AG}}{1 + e^{8.265 - 0.853\,TS + 0.7442\,DY + 2.829\,PN - 7.153\,AG}} \tag{6}$$

If the calculated probability of crossing is greater than 0.5, the study will then classify or predict that the sample chooses to cross the road. As such, the predictive accuracy of the model is summarised in Table 8. Based on the table, the calibrated model showed an overall accuracy of 98.9% while the validation model showed an accuracy of 98% which show that the model can be used to predict the crossing choice of pedestrians based on 4 significant predictors namely traffic speed (of vehicle), driver yield (slow down), pedestrian number (number in group crossing), and physical age group (by observation).

While altering the predictors' one at a time, the remaining predictors were held constant at a value of 37 km/h for traffic speed, 0 for driver yield, 1 for pedestrian number and 3 for age groups. Based on Table 7, it is not hard to see that traffic speed and age group

DOI: 10.1051/matecconf/201710308003

have a reverse impact on the crossing probability while driver yield and pedestrian number have a direct impact towards the pedestrian crossing probability.

Table 7. Regression weight for binary logistic regression

Variable	B	S.E.	Wald	df	P value	Exp(B)
Traffic speed	-0.853	0.319	7.138	1	0.008	0.426
Driver yield	7.442	3.102	5.758	1	0.016	1706.601
Pedestrian number	2.829	1.267	4.987	1	0.026	16.936
Age group	-7.153	3.093	5.349	1	0.021	0.001
Constant	48.265	19.056	6.415	1	0.011	9.14625E+20

Cox & Snell R Square = 0.632, Nagelkerke R Square = 0.977

Table 8. Model accuracy statistic

			Gap Acceptance (p)		Correct %	
			Reject	Accept	Partial	Total
Gap Acceptance (o)	Calibrated	Reject	73	2	97.3	98.9
		Accept	2	271	99.3	
	Validated	Reject	25	2	92.6	98.0
		Accept	0	73	100	

5 Summary

This study found that there were basically seven factors that influence the pedestrian behaviour at Jalan Sultan Ismail, Kuala Lumpur, in terms of accepted gap size in seconds. In fact, traffic speed, pedestrian waiting time, pedestrian gender, crossing distance, age group, frequency of attempts and pedestrian number were found to have significant impact on accepted gap size among pedestrians. Pedestrian waiting time and frequency of attempt are the two variables that showed a negative relationship with log gap size while the rest showed a positive relationship. In short, a pedestrian who is female and experiences higher traffic speed, lesser waiting time, longer crossing distance, older age, lesser attempts and crosses in a larger group will be more likely to accept longer gap size.

In terms of crossing choice, traffic speed, driver yield, pedestrian number and age were found to influence pedestrian choice. Results also showed that lower traffic speed, driver yield (slow down), crossing in larger groups and younger pedestrians will promote a higher probability of crossing the road.

The authors would like to acknowledge the Ministry of Higher Education, Malaysia for providing financial support to this research under the Fundamental Research Grant Scheme (Project Reference No. FRGS/2/2014/TK07/UTHM/02/1; Vot 1485).

References

[1] B.H. Goh, K. Subramaniam, Y.T.Wai, A.A.Mohamed, A. Ali, Pedestrian crossing speed: Tthe case of Malaysia, International Journal for Traffic and Transport Engineering 2, **34**, 323-332 (2012)

[2] N.I. Ibrahim, F.A. Kidwai, M.R. Karim, Motorists and pedestrian interaction at unsignalised pedestrian crossing, *Proceedings of the Eastern Asia society for transportation studies*, **5**, 120-125 (2005)

[3] B.R. Kadali, P. Vedagiri, Modelling pedestrian road crossing behaviour under mixed traffic condition, Eur. Transp. - Trasp. Eur., **55**, 1-17 (2013)
[4] T. Petzoldt, On the relationship between pedestrian gap acceptance and time to arrival estimates, Accid. Anal. Prev., **2**, 127-133 (2014)
[5] D. Sun, S.V.S.K. Ukkusuri, R.F. Benekohal, S.T. Waller, Modeling of motorist-pedestrian interaction at uncontrolled mid-block crosswalks, **51**, 61801 (2002)
[6] T. Wang, J. Wu, P. Zheng, M. McDonald, Study of pedestrians' gap acceptance behavior when they jaywalk outside crossing facilities, *Intelligent Transportation Systems (ITSC), 2010 13th International IEEE Conference*, (2010)
[7] J.A. Oxley, E. Ihsen, B.N. Fildes, J.L. Charlton, R.H. Day, Crossing roads safely: An experimental study of age differences in gap selection by pedestrians, Accident Analysis & Prevention, **37**, 962-971 (2005)
[8] R. Rastogi, S. Chandra, J. Vamsheedhar, V.R. Das, Parametric study of pedestrian speeds at midblock crossings, Journal of Urban Planning and Development, **137**, 381-389 (2011)
[9] G. Yannis, E. Papadimitriou, A. Theofilatos, Pedestrian gap acceptance for mid-block street crossing, Transportation Planning and Technology, **36**, 450-462 (2013)
[10] M. Brewer, K. Fitzpatrick, J. Whitacre, D. Lord, Exploration of pedestrian gap-acceptance behavior at selected locations, Transportation Research Record: Journal of the Transportation Research Board, **1982**, 132–140 (2006)
[11] B.R. Kadali, P. Vedagiri, Effect of vehicular lanes on pedestrian gap acceptance behaviour, Procedia-Social and Behavioral Sciences, **104**, 678-687 (2013)
[12] R. Lobjois, V. Cavallo, Age-related differences in street-crossing decisions: The effects of vehicle speed and time constraints on gap selection in an estimation task, Accident Analysis & Prevention, **39**, 934-943 (2007)
[13] C. Holland, R. Hill, Gender differences in factors predicting unsafe crossing decisions in adult pedestrians across the lifespan: A simulation study, Accident Analysis & Prevention, **42**, 1097-1106 (2010)
[14] M.M. Ishaque, R.B. Noland, Behavioural issues in pedestrian speed choice and street crossing behaviour: A review, Transport Reviews, **28**, 61-85 (2008)
[15] B.D. Daniel, S.N.M. Nor, M.M. Rohani, J. Prasetijo, M.Y. Aman, K. Ambak, Pedestrian footpath level of service (FOOT-LOS) model for Johor Bahru, MATEC Web of Conferences, **47**, EDP Sciences, (2016)
[16] T. Rosenbloom, Crossing at a red light: Behaviour of individuals and groups, Transportation Research Part F: Traffic Psychology and Behaviour, **12**, 389-394 (2009)
[17] E. Papadimitriou, G. Yannis, J. Golias, A critical assessment of pedestrian behaviour models, Transportation Research Part F: Traffic Psychology and Behaviour, **12**, 242-255 (2009)

Comparison of Traffic Speed Before, During and After "Banci Lalu Lintas" at Federal Road ft005

R. Rahman[1,*], *M. S.* Nemmang[1], *N. A.* Nazimuddin[1], and *Hazrul* Abdul Hamid[1]

[1]Department of Infrastructure and Geomatic Engineering, Universiti Tun Hussein Onn Malaysia, Johor, Malaysia

Abstract. Traffic engineering uses engineering methods and techniques to achieve the safe and time efficient movement of people and goods on roadways and it depends on traffic flow. The three main parameters of a traffic flow are volume, speed and density. Speed is an important transportation consideration because it relates to safety, time, comfort, convenience, and economics. This study is to show the difference of traffic speed for before, during and after a primary traffic survey called "Banci Lalu Lintas". This study also is conducted at Federal Road FT005 with collaboration of Jabatan Kerja Raya (JKR). In achieving the goal for this study, traffic speed is recorded by using two methods which are manual method and Automatic Traffic Count (ATC). For before and after the survey, manual method is used and the data is collected for 15 minutes, while during the survey, ATC is used in collecting data for 24 hours per day in a week. The data obtained where the mean speed is recorded and is compared as well as analyzed between three categories which are before, during and after "Banci Lalu Lintas" and using statistical analysis. In result, the speed of vehicles for during the survey is the lowest compared to before and after survey where the differential percentage are 6.68% and 23.64% for before – during and during – after "Banci Lalu Lintas". The study concluded that drivers tend to decrease their vehicles speed when there is an event or unexpected conditions on the road. The result is important for future development and safety of road in Malaysia.

1 Introduction

Seeing people rushing to run errands, and going to work during the day are normal sight in this modern era. To cope with time, majority of people chose to drive their own vehicle rather than using public transportation or walking. This is because, most people are not very satisfied with the services of public transportation as mentioned in [1]. Basically, speed is an important transportation consideration because it relates to safety, time, comfort, convenience and economics. This is because, most of the drivers have different perceptions of appropriate operating speed on the roads as suggested in [2]. Other than that, in [3] that studies the accidents behaviour stated that the increase of speed and access point would trigger an increase in the frequency of road accidents. Therefore, the transportation system

* Corresponding author: raha@uthm.edu.my

offers a good speed, and majority of them using it as much as possible, then the best enemy of speed which is congestion will occur. A road system should be able to occupy the number of vehicles with the road condition so that drivers will maintain their speed and congestions can be reduced.

In Traffic Engineering, traffic flow is the study of the movement of vehicles at a given point in a given time. To better represent traffic flow, there are three vital elements that should be considered in traffic flow which are speed, density and volume. In this study, it is concentrate more on speed. In [4] defined as a rate of motion expressed as distance per unit of time, generally as Kilometres per hour (km/hr). At the point where the measurement of speed is taken, various vehicles that passes the point will be travelling at their own different speeds. Therefore, this study used 2 types of method in collecting data which are manual traffic count and automatic traffic count (ATC). ATC data is used for a variety of purposes to support transport planning and design considerations. It is a highly cost effective method of collecting large volumes of a week or more of classified and/ or speed data. ATC are normally identified as two rubber tubes laid across the carriageway linked to a road side recorder box. These tubes can measure both speed and number of vehicle movements and in some cases the weight of vehicles passing over. Besides that, manual traffic count needed a human effort in collecting data. In the end of this studies, the data collected have been compared which are before, during and after survey to get the result which are the speed differences on that three (3) times. Thus, the average speed of vehicles is recorded because it is almost impossible to track every vehicle speed on roadway.

2 Literature review

Speed is a primary factor in all modes of transportation and it is very important factor in the geometric design of roads. According to [5], the speed of vehicles on a road depends on the capabilities of the drivers and their vehicles, upon general conditions such as the physical characteristics of the road, the weather, the presence of other vehicles and the legal speed limitations.

The available sight distance on a roadway should be sufficiently long to enable a vehicle traveling at or a near the design speed to stop before reaching a stationary object in its path. To allow drivers to perceive, react, and safely stop, a minimum stopping sight distance must be available. In [6] stated that stopping sight distance can be defined as the following sum of two distances, reaction distance, where the distance traversed by the vehicle from the instant the driver sees an object necessitating a stop to the instant the brakes are applied. Braking distance, where the distance needed to stop the vehicle from the instant brake application begins.

Speed has been identified as a key risk factor in road traffic injuries, influencing both the risk of a road crash as well as the severity of the injuries that result from crashes [7]. Excess and inappropriate speed are responsible for a high proportion of the mortality and morbidity that result from road crashes. According to [8], speed reduction will be occurring due to the changing of road segment or have some event on the road segment and it will lead to the traffic shock wave.

3 Method and data collection

This study was conducted at KM24 Jalan Batu Pahat-Pontian FT005 as shown in Fig.1. Data is collected for before, during and after “Banci Lalu Lintas”. For before and after, manual method is used while for during, Automatic Traffic Count is used as shown in Fig. 2. For manual method which are manual observation and using manual counter, the data is collected for 15 minutes in the evening based on previous year of "Banci Lalu Lintas" traffic volume data as the peak hour determination. For the other method which is by using ATC, the data is collected within 24 hours for one-week period and the determination of specific time for peak hour can be obtained by taking the highest traffic volume from the survey data for during "Banci Lalu Lintas". Fig 3(a) shows the visualization of equipment setup and Fig. 3(b) show the length between port A and port B for the installation of ATC. Both of this method have collected traffic speed that passed through the road but the best. Later the data collected were analyzed to compare by using statistically analysis.

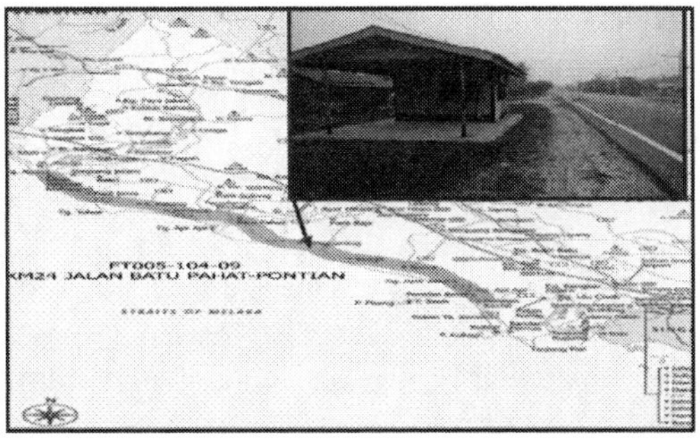

Fig. 1. Site Location at KM24 Jalan Batu Pahat – Pontian and Census Station JR112.

Fig. 2. Pneumatic Road Tube Installation.

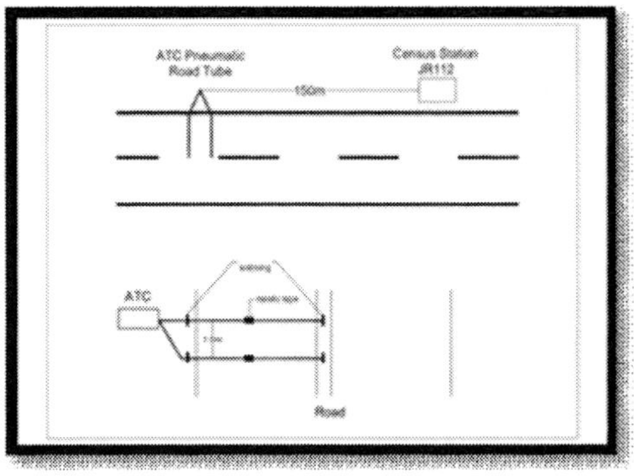

Fig. 3(a). Layout Plan for ATC and Manual Method Speed Data Collection

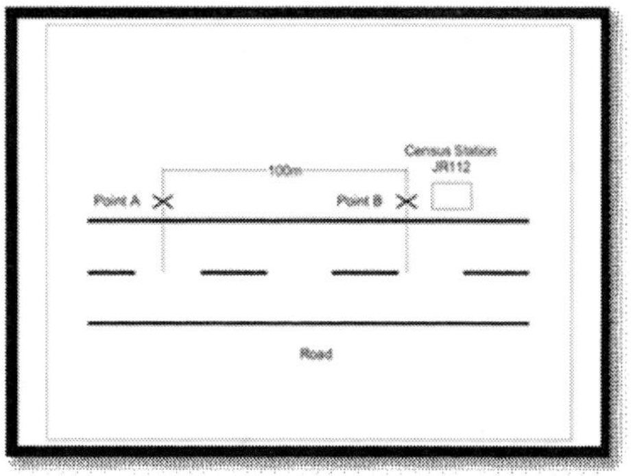

Fig. 3(b). Length between port A and port B for the installation of ATC

4 Result and analysis

In result, this study will indicate the difference of traffic speed for before, during and after "Banci Lalu Lintas" as in Table 1.

Table 1: Traffic volume and speed for before, during and after survey

Categories	Before Survey	During Survey	After Survey
Average Speed (km/hr)	71	66	84
Traffic Volume	94	131	86

While in Fig. 4 below illustrated that traffic speed during "Banci Lalu Lintas" is the lowest than the other two categories where the percentage are 6.68% and 23.64% for before – during and during – after "Banci Lalu Lintas". This has proved that drivers will reduce their speed when they sees any event beside the roads which are people wearing the safety jacket and focusing on the vehicle movement on the road.

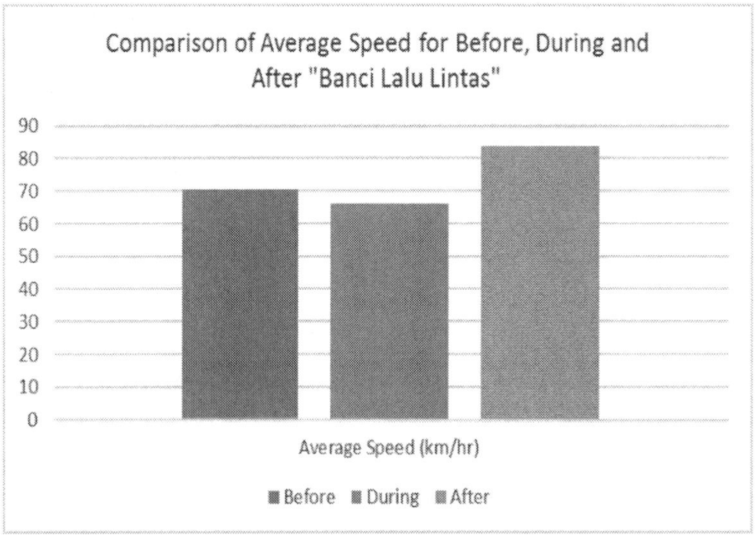

Fig. 4. Comparison of average speed for before, during and after survey

As in Table 2, for the During-After and Before-After "Banci Lalu Lintas", P-value calculated is 0.0002 and 0.0004 which both of these value are < 0.05, thus the null hypothesis is rejected which "there are significant differences". While for the Before-During "Banci Lalu Lintas", P-value calculated is > 0.05, thus the null hypothesis is accepted. From this result, there are differences in the traffic average speed data collection between the During-After and Before-After "Banci Lalu Lintas", unlike Before-During "Banci Lalu Lintas".

However, "Banci Lalu Lintas" only conducted for 16 hours per day in a week, starting at 6am until 10pm, after 16 hours, drivers no longer aware about their speed which leads them to increase the speed since there is no one in the census station or any other traffic speed traps. It has been shown in Table 3 and Fig. 5 that the vehicles speed for after 16 hours, Wednesday 6th April 2016.

Table 2. P-value for two-sample t-test in respect to categories of "Banci Lalu Lintas"

Categories of "Banci Lalu Lintas"	**P – value**
Before – During	0.91
During – After	0.0002
Before – After	0.0004

Table 3. Traffic mean speed at 11pm to 5am for Tuesday – Wednesday.

Tuesday, April 5, 2016						
Time	**Traffic volume**	**Mean speed**	**85th percentile speed**	**Minimum speed**	**Maximum speed**	**Mean speed exceeding 90**
2300	139	74	94	28.4	133	105
Wednesday, April 6, 2016						
0000	81	78.9	98.3	47.2	128.4	103
0100	42	75.1	89.6	30.5	130.4	109.4
0200	44	75.2	89.3	24.3	129.3	110
0300	23	72.8	95.4	26.8	125.6	106.8
0400	17	75.2	96.1	38.3	110.4	100.6
0500	52	79.7	99.7	43.1	132.1	105.6

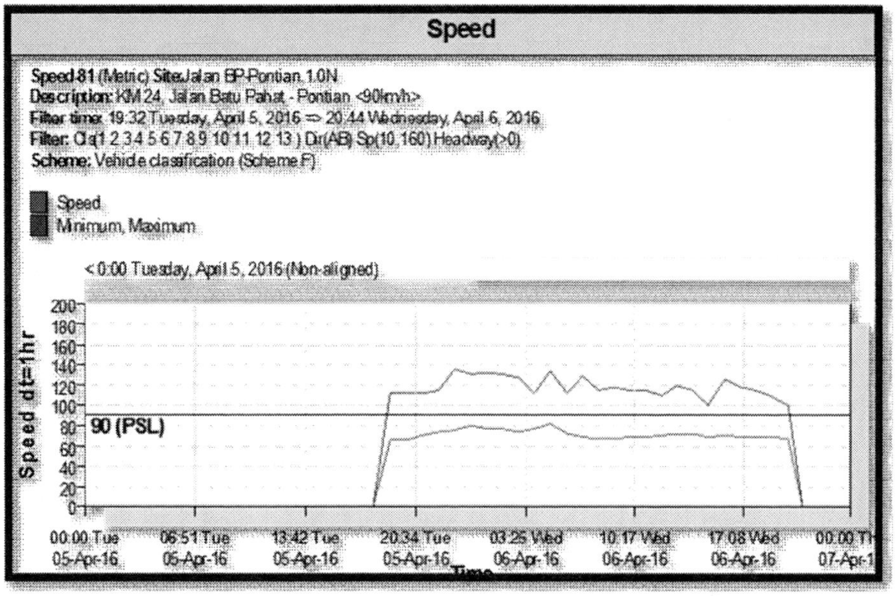

Fig. 5. Traffic speed after "Banci Lalu Lintas" conducted for Tuesday – Wednesday

The result can be concluded that the study has achieved the two objectives. The traffic speed will change either increase or decrease due to the event/activity at the location. From the data obtained, the speed is decreased during the survey while before and after the survey, it increased. Traffic speeding will leads to accident which can be proved by the past years data accident. Fig. 6 below is the graph of accidents and casualty at FT005 from 2007 to May, 2010 as stated in [9]. It indicated that the total number of accidents recorded at this route is 2774 cases as it increased every year. Thus, it is shown that the number of cases is high for within 3 years. Therefore, it shows that the driver were tend to speeding during driving in this road.

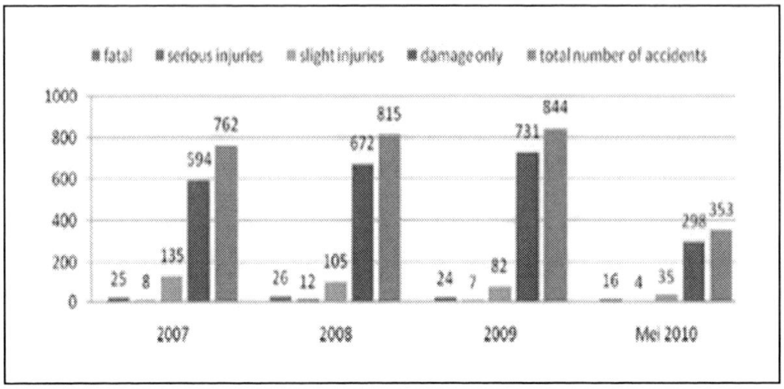

Fig. 6. Accidents and Casualty at FT005 for 2007 until May, 2010 (Source: [10])

5 Conclusion

This study is considered successful since the two objectives has been achieved as there are differences of traffic speed for before, during and after "Banci Lalu Lintas" which is conducted at KM24 Jalan Batu Pahat – Pontian within a week observation by using ATC and 15 minutes by using manual method. The results have been proven that during "Banci Lalu Lintas", the speed of vehicles decreased compared to before and after "Banci Lalu Lintas" where the percentage are 6.68% and 23.64% for before – during and during – after "Banci Lalu Lintas". In conclusion, research obtained that drivers were aware of their speed when there is an event or unexpected conditions for instance "Banci Lalu Lintas" that make them following the rules.

This research is supported by the Short Term Grant (STG) of Universiti Tun Hussein Onn Malaysia (UTHM) in Vot U341.

References

[1] R. Ismail, M.H. Hafezi, R.M. Nor, and K. Ambak, Passengers preference and satisfaction of public transport in Malaysia Sustainable Urban Transport Research Centre (SUTRA), Department of Civil and Structural, Aust. J. Basic Appl. Sci., **6**, 410–416 (2012)

[2] B.D. Daniel, N. Alan, and G. Koorey, Investigating speed patterns and estimating speed on traffic-calmed streets, *IPENZ Transp. Gr. Conf. Auckl. March*, (2011).

[3] M.I. Masirin, W.A. Al-bargi, J. Prasetijo, and B.D. Daniel, Road accident analysis: A case study of federal route, MATEC Web Conf., **4**, 1–7 (2016)

[4] Transportation Research Board, National Research Council, *Highway Capacity Manual*, Washington, (2000)

[5] R. Branch, *Arahan Teknik Jalan 8/86 (Pindaan 2015), A Guide on Geometric Design of Roads*, Public Works Department Malaysia, Kuala Lumpur, (2015)

[6] American Association of State Highway and Transportation Officials, *A Policy on Geometric Design of Highways and Streets*, Washington, (1994)

[7] M. Idrus, M. Masirin, and N.A. Mohamad, Analysis of road infrastructural audit along jalan Batu Pahat-Kluang Malaysia : A case study, *International Integrated Engineering*

Summit, 1–7 (2014)
[8] R. Rahman, Extent of traffic kinematic waves and queuing caused by midblock u-turn facilities, *Proceedings of the ITRN2012*, (2012)
[9] C. Norbalkish, I. Yusof, M. Erwan, *Report of Prediction of Accident Trend at Two-Lane Federal, 178-190,*Universiti Tun Hussein Onn Malaysia, (2011)
[10] Royal Malaysian Police, *Accidents and Casualty at FT005 for 2007 until May*, Kuala Lumpur, (2010)

Safer Roads: Comparisons Between Road Assessment Program and Composite Road Safety Index Method

Intan Suhana Mohd Razelan[1,*]

[1]Universiti Malaysia Pahang, 26600, Pekan, Pahang

Abstract. In most countries, crash statistics have becoming very crucial in evaluating road's safety level. In Malaysia, these data are very important in deciding crash-prone areas known as black spot where specific road improvements plan will be proposed. However due to the unavailability of reliable crash data in many developing countries, appropriate road maintenance measures are facing great troubles. In light of that, several proactive methods in defining road's safety level such as Road Assessment Program (RAP) have emerged. This research aim to compare two proactive methods that have been tested in Malaysian roads ; road assessment program and road environment risk index which was developed based on composite index theory in defining road's safety level. Composite road environment risk index was combining several crucial environment indicators, assigning weight and aggregating the individual index together to form a single value representing the road's safety level. Based on the results, it can be concluded that both road assessment program and composite road environment risk index are contradicted in six different ways such as type of speed used, type of analysis used and their final outcomes. However, with an aim to promote safer roads, these two methods can be used concurrently as the outcomes in both methods seems to fulfil each other's gap very well.

1 Introduction

In country like Malaysia who is still at its early phase of instilling road safety educations, having new system that is capable to elaborate the most crucial aspect of the road can be seen as a start for a better road traffic system. Apart from that, development of road safety index in Malaysia was also seen as a proactive measure in identifications of problematic road sections that require urgent treatment [1]. Malaysia has experience very high number of crash where most of the road safety evaluation is based on the crash record (reactive measure). Based on this data, lots of programmes have been initiated to help educating Malaysian road users. However those programmes seem to be insufficient in reducing the number of crashes in Malaysian roads and transforming the people's behaviour and perceptions towards road safety [2]. In an attempt to overcome this issue, a proactive

* Corresponding author: intan@ump.edu.my

method of composite road safety index targeting at the environment factor of the road was developed to proactively determine risky road sections before crashes was initiated.

2 Concept of composite road safety index for road safety evaluation

According to [3], development of road safety index may not possibly show the whole road safety picture in a country but at least this index should be able to show most of the picture. Crash were suspected to be contributed by various factors so examining a single factor at one time is seems to be insufficient. A single system that can capture all those factors in-whole and can be further used to elaborate the process leading to crash as well as suggesting a countermeasure is believed to be the best solutions. In compliance to that, European Transport Safety Council (ETSC) has come out with an idea of road safety performance indicator where the outcome is a single value. According to [4], safety performance index (SPI) was defined as a measure to reflect the operational conditions of the road traffic system, which influence the system's safety performance. The main characteristic of the safety performance index is its ability in giving out a more complete picture of the road safety status and can also point the emergence of developing problems at an early stage, before these problems show up in the form of accidents [5]. In short, safety performance index has been recognized to have below credibility:

- To reflect the current safety conditions of a road traffic system (i.e. they are considered not necessarily in the context of a specific safety measure, but in the context of specific safety problems or safety gaps)
- To measure the influence of various safety interventions
- To compare different road traffic systems (e.g. countries, regions, etc).

Realizing the needs to combine as much information as possible from current road conditions in mitigating correct crash factors, a concept of composite index was introduced. Here, each indicator developed from the targeted road aspect was given specific weight and were later combined to produce single value which represented the road safety status for the selected area. As indicated by [3], the philosophy of combining performance outcomes into one single quantitative value, so called "composite index", is not a new one. Although the development of composite index is seen as a challenging matter to road safety researcher [6], this index is desperately needed as an easy and understandable tools for policy makers as well as the public to measure the road safety status [3]. In line with that, many composite indices have been developed internationally and used across fields to indicate a progress or achievement between comparable entities such as countries.

2.1 Road assessment program (RAP) method

International road assessment program or better known as iRAP is a continuation from similar successful road assessment programmed carried out in United States of America (usRAP), European Countries (EuroRAP) and Australia (AusRAP) to name a few. Main outcomes generated from this road assessment programmed were the risk mapping and the performance tracking protocols where both of the protocols were aims at producing safer road networks focusing at developing countries. According to iRAP report produced in year 2011, risk mapping is a map produced from sets of crash data used to show the sections of road where the risk level was high. Performance tracking on the other hand is a protocol used to identify consistently high risk roads and the most improved roads so that benchmarking between the high risk road and the improved roads can be done.

Within the iRAP procedures, detailed visual inspections of road's infrastructure elements that were known to have impact on the likelihood of a crash were carried out using drive-through and video based inspection. The results from the inspections and video recordings were used to calculate star rating for every 100 m road sections using Road Protection Score (RPS) method. RPS is a unit-less measurement represented in either 5-star for safest infrastructure conditions and 1-star to the riskiest infrastructure conditions.

RPS are calculated by focusing on three different road infrastructure factors chosen as the most critical road infrastructure factors; the median, the side area and the junction areas. It is important to note that the Star Rating represents the safety risk faced by an individual road user if they (or the traffic around them) are travelling within the speed limit. Traffic flow and estimation of actual speeds are not included in the calculation.

Fig. 1 shows a car occupant star rating along most Malaysian Expressway and Federal roads and specifically along Federal Road 3 start from Kuantan to Kuala Rompin, Pahang. Calculation of star rating for this stretch of road shows that more than 50% of roads are rated as having a 2 star (red colour) infrastructures conditions.

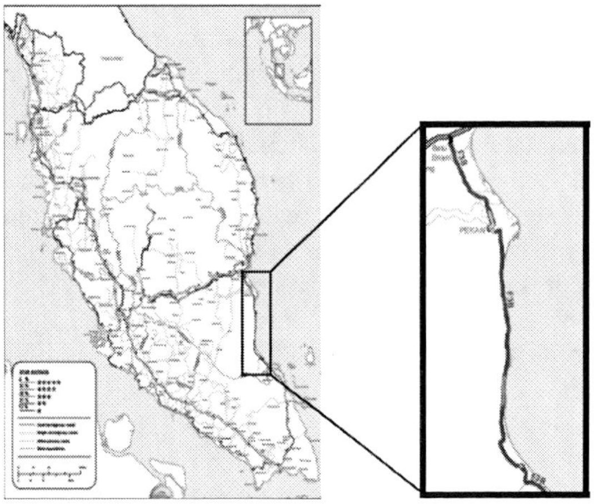

Fig. 1. Car Occupant's Star Rating for Malaysia Roads (iRAP Malaysia Results, 2009)

Major upgrading works have been done in year 2011 to this particular road section where part of road from Batu Enam to Pekan for approximately about 45km length has been upgraded to dual carriageway road with concrete median to separate opposing traffic flow as shown in Fig. 2. Nevertheless, based on crash data Royal Malaysia Police Force in year 2012, numbers of crashes in this road are still soaring up to 18 crashes/500 m road length.

Fig. 2. Current Condition of Road between Batu Enam and Pekan

3 Road environment safety evaluation based on composite road safety index

A stretch of road within Federal Road 3 has been chosen as the research area where the principle of composite road safety index was tested. This road starts at Skudai Town and ends at Rantau Panjang, Kelantan with 739 km in total length. The stretch of road adopted as study area is between KM 183 and KM 318 making a total of 135 km length road. Road environment factor was chosen as the main research theme since the complexity of road environment has been proven to be highly related in producing high risk to drivers [7] especially in developing countries like Malaysia who has a very unique road environment factors that need to be further explored [8]. Fourteen indicators comprised of 12 numerical indicators and 2 categorical indicators as shown in Table 1 below were used to represent the local road environment conditions of this area.

Table 1. Selected road environment indicators

Road Environment Indicators
Number of Motorcycles/section
Number of Pedestrian /section
Percentage of Poor Roadway Condition
Number of Lanes/section
Percentage of Unpaved Shoulder/section
Percentage of Poor Road Delineation/section
Percentage of Roadside Development/section
Number of non-Signalized Access/section
Number of Signalized Intersection/section
Percentage of Road Without Median
Lane Changing Opportunity
Non-Traversable Object at Roadside
Number of Heavy Vehicles/section
Actual Driving Speed/section

The criteria in developing the road environment risk indicators were done by adapting the criteria proposed for the development of road safety performance indicators prepared by SafetyNet through its report entitled 'State of the Art Report on Road Safety Performance Indicators, 2005 and thesis written by [3]. The criteria set – up for constructing the road environment risk indicators are as listed below.

- Indicators must be designed to be highly accurate in explaining the road environment dimension.
- Indicators must have clear objectives, clear measurement methods and expectations.
- Each indicator is possible to thoroughly explain the outcomes expected to be derived from the indicators.
- Each indicator must be quantifiable in a sense that possible and reasonable score could be obtained.
- The indicators should express an ordinal relation with road risk (slight changes in score will gives very clear changes on road risk level).

Indicators can be measured in terms of a percentage, a rate, an exact values or even a level of qualitative statements such as very good, good and poor [3]. In comparions with the development of RAP, the selection of each indicator/attribute is somewhat different. While RAP's attributes are focusing on developing a scale showing how well road protects their user from death or serious injuries once the accident occurs, indicators in composite road safety index is measuring the risk generated from the selected indicators where the indicators must reflect their level of risk towards drivers

After conducting several preliminary analyses to test the suitability of each environment indicator to be included in the development of the composite road environment safety index, four indicators have been excluded from the dataset. The balance of 10 indicators have been combined together and produced a Kaiser-Meyer-Olkin (KMO) value of 0.760 and Bartlett's significance value of 0.0001 indicating that this combination has an ability to produce good road environment risk index. Principal component analysis (PCA) has been chosen as the extraction method in the development of the composite index. Results from several iterations using PCA and Varimax with Kaiser normalization as extraction and rotation method respectively, shows that those indicators could be further classified into three different groups representing three different environment characteristics. Fig. 3, 4 and 5 show a composite index value of each section within the study area according to its characteristic.

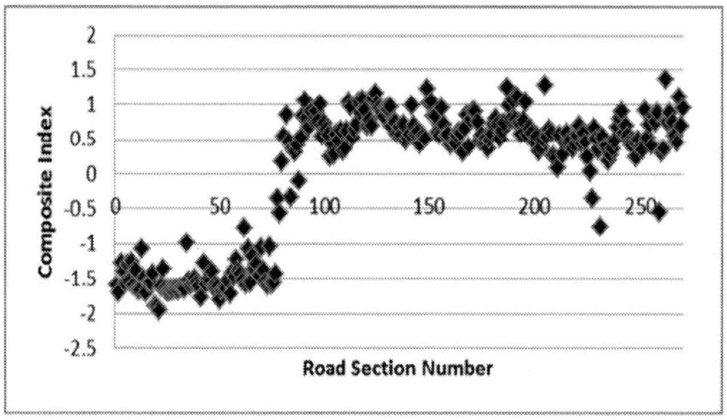

Fig. 3. Road Section's Composite Index based on Infrastructure Environment Risk Factor: Federal Road 3

Fig. 3 clearly shows the trend of risk generated from road infrastructure environment of this study area. This study area is a double carriageway lane with concrete median for the first 45 km of the road and the rest of the road is a multilane road. The difference in its infrastructure environment was clearly illustrated in this figure where areas equipped with raised median to separate the opposing traffic have significantly lower risk compared to areas where no raised median is built.

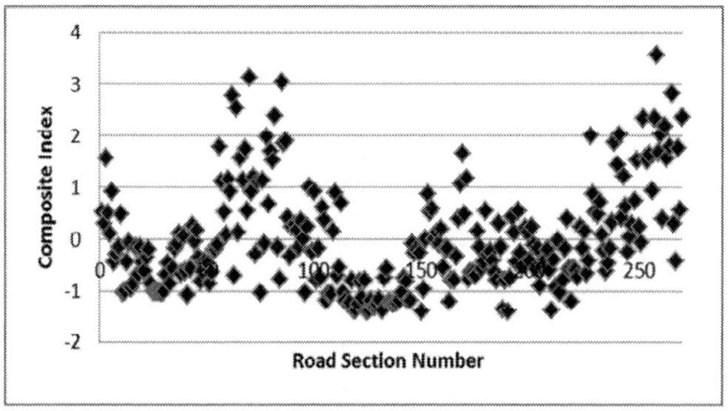

Fig. 4. Road Section's Composite Index based on Roadside Environmnet Risk Factor: Federal Road 3

Fig. 4 shows the composite index value of the road sections based on roadside environment risk factor. A clear fluctuating trend can be observed where the risks generated from roadside environment are quite high in certain locations and low in certain locations. This study area is encompassing two major towns i.e. Pekan and Kuala Rompin where the roadside activities in these two towns are very high. So it could be early concluded that the two major peaks (highest risk) as depicted in above figure are actually representing these two towns.

Fig. 5 shows the composite index values calculated from operational environment risk factor of this area. The composite index trend shows that the composite values are extremely high in several sections indicating that the risks of operational environment in those particular sections are also very high. It is also observed that numbers of sections with low composites values are higher compared to sections with high composite index values.

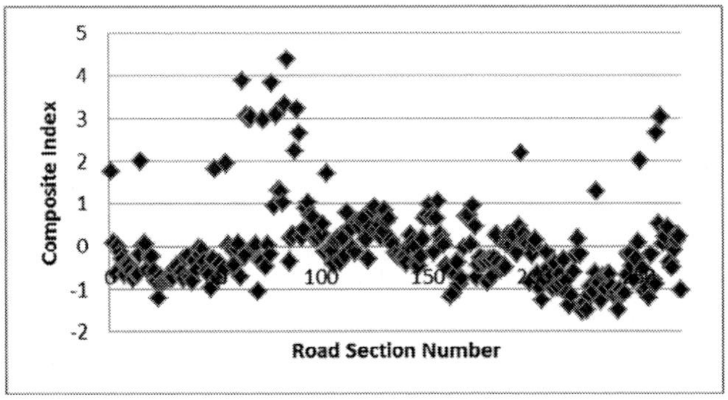

Fig. 5. Road Section's Composite Index based on Operational Environmnet Risk Factor: Federal Road 3

Based on these figures, the composite index values of the continuous road sections would generally indicates the trend of road environment risk along the study area where in general, comparison between high risk sections with low risk sections can be directly detected.

4 Composite road safety index vs. road assessment program

Based on the brief introductions of both methods above, it can be clearly observed that both composite road environment risk index produced from this research and the concept of star rating produced by iRAP team shared the same aim; providing safer road to the road users. Table 2 shows the comparisons between composite road safety index and iRAP's procedures in producing safer roads.

Final outcomes of road environment risk index is a single value called composite index for each road sections which were used to rank each section risk level so that the riskiest sections could be correctly identified. iRAP's final outcomes are values calculated from 3 main road infrastructure aspects translated into 1-5 star rating relates to the risk faced by individual member of each road user group (motorcyclist, car occupant etc).

Based on the Table 2, it was concluded that there were six substantial differences in the procedures of safety status evaluations between the composite road environment risk index and iRAP's procedure. The differences were mainly on the main frameworks which involved the main parts of the procedures. However, one of the important advantages that iRAP procedures had was the ability of the star rating in fulfilling the basic needs of a road rating theory; the calculation of risk based on different types of road users i.e. the car occupant, motorcyclist, bicyclist and pedestrians.

It is generally acknowledged that different type of road users experienced different source of risk while travelling on the road. Risks faced by car drivers and risk faced by motorcyclist for example is totally different. While car drivers is safely prohibited in an hard shell of steel, motorcyclist on the other hand have to face risk coming from all angle thus have a very limited opportunity to protect themselves from being injured or fatal [9-12].

Table 2. Comparisons of Final Outcomes from Composite Road Safety Index and iRAP's Star Rating

Composite Road Safety Index	Star Rating : International Road Assessment Programme
Measuring the risk level of the road sections.	Measuring the safety performance of a road network
Targeting on specific road environment attributes that known to instantly generate risks towards car drivers and triggering crash.	Covering the aspects of road infrastructures that known to influence likelihood of crash and severity
The road environment indicators were weighted and individual index of each indicator was calculated based on statistical method.	The road attributes were scored based on judgments of the selected road inspectors from video recordings of the road areas.
The weight of each indicator was calculated based on the exclusive behavior / characteristics of the indicators itself on each road network.	The weight of each road infrastructure aspects was pre-determined based on the level of risk produced by each road infrastructure aspects
Actual speed was used throughout the composite index calculation so that the actual risk experienced by the road users was presented.	Posted speed limit was used in determination of the risk resulting to final outcomes was only valid if only the road users is travelling within the speed limit.
Final outcomes were single values called composite index for each road sections which were used to rank each section risk level so that the riskiest sections could be correctly identified.	Final outcomes are values of calculated from 3 main road infrastructure aspects translated into 1-5 star rating relates to the risk faced by individual member of each road user group (motorcyclist, car occupant etc).

Since this is the first research to test the applicability of composite index in evaluating the road section's risks based on prevailing road environments condition in developing countries, only one type of road users were taken into considerations. Considering huge numbers of cars used by most road users, car drivers was taken as the main issues despite motorcycles who is also one of the most commonly used vehicles in developing countries especially in South East Asia countries. But, since motorcycles are quite uncommon in other developing countries like Chile, Costa Rica and most of South African countries [13], car drivers was mostly suitable to use as preliminary subject of this research. This would allow for wider usage of the research outcomes.

Other than that, final outcomes derived from both methods were also different. While composite road environment risk index presented the risk level of each road sections in terms of the composite index value, iRAP procedures was presenting the safest road infrastructure design through star rating received by each road networks with 5-star represents the safest road infrastructure design for the prevailing speed limits. This has come to a conclusions that 5-star given to the road infrastructure built in federal road with a speed limit of 90 km/h is incomparable with 5-star given to the road infrastructures designed for expressway having a speed limit of 110 km/h. Taking road barrier for example, the safest design of road barriers for motorcyclist traveling on federal road with speed limit of 90km/h would not necessarily be given 5-star if the same design was applied within an expressway with speed limit of 110 km/h. Better barrier's design must be introduced to accommodate the needs to have the safest barriers for motorcycles at prevailing speed limit of 110 km/h. This is the challenges that must be faced by the road improvements departments since they have to design range of road infrastructures aspects for every type of road having different speed limits.

Crashes always occurs in an event of speeding [14], so by only studying the safest road infrastructure for road users who are travelling within the speed limit is insufficient to represent the real road situations especially if the main aim of the assessment program is to reduce the number of crashes and providing safer roads in general. Moreover, as indicated by [15], most road users doesn't always aware of the posted speed limits during their driving period whereas the environment of the roads is always being used as indicators in defining their prevailing speeds. So, it was clear that designing a system that highly depended on the speed limits was insufficient in evaluating the exact situations of the roads.

5 Conclusion

In overcoming this issue and in an attempt to ensure that the developed composite index is presenting actual conditions of the roads, the theory of naturalistic driving method was applied in this study. Naturalistic driving method was defined as data collection method where the research subject (car drivers in this case) would drive its own car, under its normal conditions, without any descriptions or interventions from the data collection procedures and the drivers was allows to drives just like they normally do [16]. The outcomes generated using this method was very helpful especially in defining the relationship between road users, road, traffic and the environment under driver's normal conditions. Hence, this efforts were seems to be appropriate in filling in the missing part in iRAP method.

References

[1] N. Daud, K. Ibrahim, K. Sopian, Hierarchical bayesian approach for ranking of accident blackspots with reference to cost of accidents, *2nd WSEAS Int. Conf. on Computer Engineering and Applications (CEA'08) Acapulco*, Mexico, (2008)

[2] S.H. Musthar, M. Muda, B.M. Adji, M.R. Karim, Audience involvement in Malaysian road safety campaigns among young adults, *Proceeding of the Eastern Asia Society for Transportation Studies*, **9**, (2013)

[3] A.H. Ghazwan, *Road Safety Development Index*, Dissertation Number: 1100, ISBN: 978-91-85715-04-6, Linkoping University, (2007)

[4] European Transport Research Council, *Transport Safety Performance Indicators*, ISBN: 90-76024-11-1 2, (2001)

[5] A.S. Hakkert, V. Gitelman, *Road Safety Performance Indicators: Manual*, Deliverable D3.8 of the EU FP6 Project SafetyNet, (2007)

[6] E. Hermans, F. Van den Bossche, G. Wets, Combining road safety information in a performance index, Accident Analysis and Prevention, **40**, 1337–13 (2008)

[7] C.M. Rudin-brown, J. Edquist, M.G. Lenné, Effects of driving experience and sensation-seeking on drivers' adaptation to road environment complexity. Safety Science, **62**,121–129 (2014)

[8] L. Rogers, H. Hashim, *Star Rating Road Improvements: 2009 Malaysia Black Spot Programme*, International Road Assessment Programme (iRAP), Report Number 504.10, (2011)

[9] H. Hussain, *Development of Capacity and Level of Service for Uninterrupted Exclusive Motorcycles Lanes in Malaysia*, PhD Thesis, Universiti Putra Malaysia, (2006)

[10] L.B. Meuleners, A.H. Lee, C. Haworth, Road environment, crash type and hospitalisation of bicyclists and motorcyclists presented to Emergency Departments in Western Australia, Accident Analysis and Prevention, **39**, 1222–1235 (2007)

[11] S.M. Rifaat, R. Tay, A. De Barros, Severity of motorcycle crashes in Calgary, Accident Analysis and Prevention, **49**, 44–49 (2012)

[12] M.G. Masuri, K.A.M. Isa, M.P.M. Tahir, Children, youth and road environment: Road traffic accident, Procedia - Social and Behavioral Sciences, **38**, 213–218 (2012)

[13] iRAP and EuroRAP, *Crash Rate-Star Rating Comparisons, Review of Available Evidence*, Working Paper Number 504.2, (2011)

[14] B. Fildes, S. Lee, *The Speed Review: Road Environment, Behaviors, Speed Limit, Enforcements and Crashes*, Monash University Accident Research Center, A report prepared for Road Safety Bureau, Roads and Traffic Authority of New South Wales and Federal Office of Road Safety Department of Transport and Communications, (1993)

[15] S. Daniels, J. Vanrie, A. Dreesen, T. Brijs, Additional road markings as an indication of speed limits: results of a field experiment and a driving simulator study, Accident Analysis and Prevention, **42**, 953–960 (2010)

[16] SWOV, *Fact Sheet on Naturalistic Driving: Observing Everyday Driving Behaviour*, Leidschendam,1–53 (2010)

Comparison Between the Effect of Motorcycle Safety Campaign Using Animated Video and Live Action Video-A Preparation Stage

Leksmono Suryo Putranto[1,*], and Rostiana[2]

[1]University of Tarumanagara, Department of Civil Engineering, Jalan Letjen S. Parman 1, Jakarta 11440, Indonesia
[2]Univiversity of Tarumanagara, Faculty of Psychology, Jalan Letjen S. Parman 1, Jakarta 11440, Indonesia

Abstract. Hypothetically, traffic safety campaign using animated video and live action video will get different result. This paper discuss the preparation stage of a study to compare between the effect of motoryccle safety campaign using animated video and live action video.There were some activities on this stage. Firstly was the selection of three live action videos from nine live action videos available from previous study. This was followed by the development of thre animated videos based on these three live action videos. There will be also development of completely new three animated videos. These three brand new videos were developed in animated version because they will be too dangerous to create live action videos of this version, e.g. to describe motorcyclist risky behaviour such as motorcycle travelling in opposite direction. A hundred students from four universities in Indonesia will be asked to observed different type of videos and fill some questionnaires to compare between the effect of motoryccle safety campaign using animated video and live action video.

1 Introduction

As a result of uncontrolled development of Indonesian cities land use and lack of seamless public transport system, severe congestions were happened daily. People should keep conducting their duties and therefore they need to travel every day. Driving a car will trap people in long traffic queues. Therefore they purchase motorcycle which perceived to travel them faster to the diestination with the ability to use small gap to avoid congestion [1].

In authors previous research the effect of various motorcyclist safety campaign media (audio, still picture and video) to the attitude change was evaluated. Although all type of campaign media have similar short term positive effects on attitudes change, the effect of video was slightly higher [1]. Persumably, video can be effectively change motorcycle rider behaviours into better attitudes because it can deliver stronger messages as a combined effect of moving pictures, audio messages, text messages, animations, sound effects and musical illustration supporting each other.

* Corresponding author: lexy_putranto@yahoo.co.id

As some campaign messages (such as motorycle traveling in opposite direction) are to dangerous to be developed in live action videos, the authors decided to develop such campaign messages in animated videos. This present research will compare the effect of watching animated and live action motorcycle safety videos to motorcyclists attitudes.

2 Related works

Putranto and Rostiana [1] conducted a research on 107 student respondents from University of Udayana (Denpasar), University of Tanjungpura (Pontianak), Institut Teknologi Sepuluh Nopember (Surabaya) and University of Andalas (Padang). All 22 motorcycling related behaviours daily conducted by the respondents can be changed into better attitudes after observing pictures and/ or hearing audios and/ or watching videos and/ or watching testimonial videos. All changes were positive changes and statistically significant at 0.05 significant level

Furthermore Putranto and Rostiana [1] stated that V9, a three minutes long video containing various messages can change 11 motorcycling related behaviours into better motorcycling related attitudes. Meanwhile P2, a picture about children helmet can not change any motorcycling related behaviour into better attitude. It can also be seen that the power of pictures to change motorcycling related behaviours into better motorcycling related attitudes were limited, whilst the power of audios, videos and testimonial videos to change motorcycling related behaviours into better motorcycling related attitudes were quite many. As stated before, video can be effectively change motorcycle rider behaviours into better attitudes because of its combined effect of different elements supporting each other.

Colclough [2] defined animation as the art of drawing sketches and objects and then showcasing them in series of frames, making it look like the images are moving. Animated videos tend to used in any from of explainer videos used to educate and inform the viewers. On the contrary, live action video is a recording of either still or moving objects. Live action videos can refer to both cinematography that does not make use of animations. Animated videos tent to be less expensive. However costs can still amount due to all scripting, illustration, animatoins, editing and sound effects. Animation however might take longer time to develop if it is developed from scratch. Therefore the use of library of pre made one can speed up the process. Creating human emotion using animation is not easy, but one advantage of using animation is it can explain tricky idea in faster manner than live action video.

3 Material and method

This paper is prepared based on the third year of a three years research on Indonesian motorcyclists risky behaviour. In the first year research, an Indonesian Motorcycle Rider Behaviour Questionnaire (MRBQ) was developed. The Indonesian MRBQ was consists of 38 question items categorized into 6 subscales, i.e. traffic errors, control errors, speed violations, traffic violations, safety violations and stunts [3].

In the second year research, effect of different type of traffic safety campaign media to motorcyclists attitude change was observed and reported in [4] and [1]. There were several different campaign media observes, i.e. still pictures in the form of poster, audios in the form of radio advertisement, short live action videos in the form of television/ movie adversitsement and testimonial videos in the form of television/ movie advertisements.

The third year is still about the effect of different type of traffic safety campaign media to motorcyclists atitude change, but specifically on the difference of live action video and animated video efffects. The third year research methodology is presented in Fig 1.

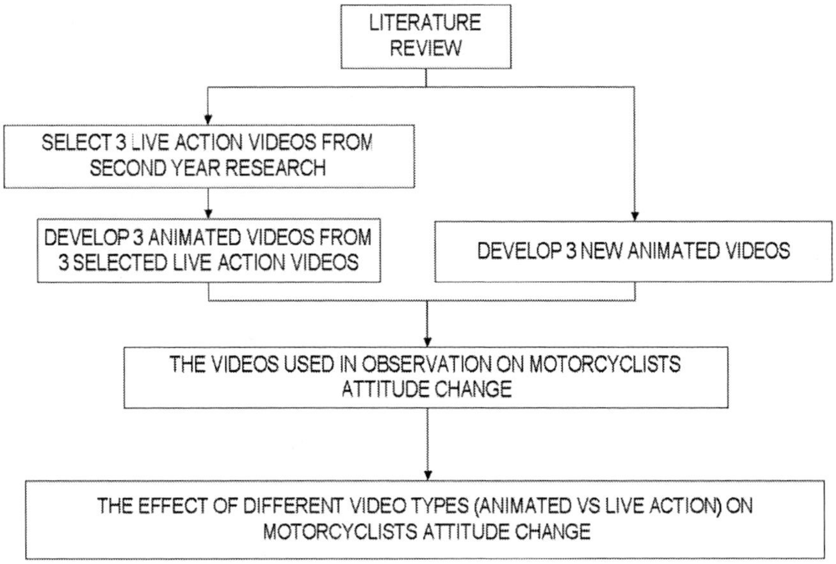

Fig. 1. Third year research methodology

In the second year research, nine live action videos were observed by the respondents. The theme of those videos were:

- Helmet for Children
- Overloaded Motorcycle
- Riding Safety Equipments
- My First Helmet
- Standard Helmet
- Risky Rides
- Never Come Back (Underage Rider)
- For Love (Riding Safety Euipments)
- Practical Motorcycling Guidance

In this third year research, the author select three of them based on two considerations, i.e.:

- Variation on campaign message delivered.
- Complication of actualizing the scene in live action video, e.g. risky actions

After considering the above criteria, the three chosen videos were:

- Helmet for Children
- Overloaded Motorcycle
- Never Come Back (Underage Rider)

The following were the synopsis of the selected videos:

- Helmet for Children (30 seconds):
  A father carried his little daughter with his motorcycle. The mother has remined her husband to equipped the child with helmet and jacket but he neglected her and argued that it will be a short trip only. Few seconds later there was a motorcycle

accident suffered by the father and daughter. The kid was unconscious after her had hit the road pavement, as she did not wear a helmet.

- Overloaded Motorcycle (30 seconds):
  A fruit trader has just received an order to deliver some fruits. He thought that if he brought as many fruits as he can at once, he will get substantial benefit. Her wife disagree when she saw her husband trying to brought every single ordered fruits at once. The fruit trader shouted to other road users to move aside and give him a way. As a result the he lost his balance and the fruits fell apart.
- Never Come Back (Underage Rider) (60 seconds):
  A father was so proud of his son academic achievement in a junior high school. So he gave his son a motorcycle as a present. The young boy rode the motorcycle and almost hit other road users until finally he hit a car. The video ended with a scene showing that his son who badly injured should use a wheelchair.

When transfering these live action videos into animated videos certain aspects were considered, among others:

- Although it is imposible to recreate every single details in live action videos into animated scenes, but all efforts should be done to recreate the scenes as complete as possible.
- To develop smooth movement of each characters in the animated videos are too costly and time consuming, but any simplification of movement should guarantee that no awkward movement of any main characters in the videos are allowed
- Dialects of the main characters should be as similar as possible with the live action videos to avoid differences in cultural senses.

Therefore animated videos were produced by professionals in animation. Their services include the followings:

- The production of story board. This is more comprehensive compare to a story line. Story board consists of serial of cartoon pictures that can be used to describe the story in multimedia softwares [5].
- The choice of 2D and 3D assests. It is time consuming for animator to create 2D and 3D assests from scratch. Usually they use either available free 2D and 3D libraries or the paid ones.
- Animating.
- Motion Graphics, i.e. pieces of digital footage or animation which create the illusion of motion or rotation, and are usually combined with audio for use in multimedia projects.
- Composing.
- Voice Over, i.e. a production technique where a voice—that is not part of the narrative)—is used in a radio,television production, filmmaking, theatre, or other presentations. The voice-over is read from a script and may be spoken by someone who appears elsewhere in the production or by a specialist voice talent.
- Editing and Scoring as a part of post production, to finalize the animated video.

Besides recreating the three live action videos from the second year research, a fully new animated videos will also be produced. Again, these videos would be too dangerous if actualized in live action videos. The followings are the synopsis of the videos:

- Motorcyle Flowing in the Opposite Directions (60 seconds). A woman was riding a motorcycle in the opposite of the determined direction in a four lanes-two ways devided road. She did that because she live faraway (more than 2 km) from the nearest u-turn. Moreover after that, she should make another u-turn located 3.5 km from the first u-turn. Her daily destination was ony 1 km from her residence. There was no police officer on duty nearby the site of this traffic law violation. At that time (8 a.m.) the weather was fine and the traffic flow was rather high but still flowing. Suddenly a car was emerge from a small side road perpendicular to the main road. The visibility of this access was insufficient due to the existence of 2 m high solid wall. As a result the woman was hit and squashed by the car.

- Motorcycle with Overloaded Passengers and Luggages (30 seconds). A man carried his wife and their five years old son on a trip to their hometown to celebrate a yearly religious event. They brought many luggages by installing a wooden board at the back of the morcycle. When they almost reach their destination, the man felt exhausted, so he accelarated the motorcycle and reaching very hgh speed. It was around noon. As he became more exhausted, he lost his balance and as a result they fell into paddy field in the roadside.
- Motorcycle flowing on Sidewalk and Pedestrian Bridge (30 detik). A young man rode a motorcyle in a heavy traffic of an afaternoon peak hour. To avoid congestion he rode the mororcycle on a sidewalk and when he intended to make a u-turn he utilized a pedestrian bridge. Shortly after leaving the pedestrian bridge, he rode the motorcycle in the sidewalk in a quite high speed. As his vision was obstructed by the pedestrian bridege piers, it was to late for him to realize that there were groups of a highschool students walking on the opposite direction of the sidewalk. As a result he hit two of the students until they are heavily injured.

Based on the previous experience in [1] sample size of about a hundred students were sufficient for the purpose of the study. As the respondents were all undergradute students with similar age and similar cognitive level, increase the number of respondents might be meaningless. Therefore, in this study about a hundred students from 4 universities in Indonesia will be asked to:

i) Fill selected items in Indonesian MRBQ.
ii) Watch 3 live action videos and then fill another qustionnare to identify whether the live action videos are able to generate attitude change.
iii) Watch 3 animated videos (that recreate scenes in live action videos) and then fill another qustionnare to identify whether the animated videos are able to generate attitude change.
iv) Watch 3 brand new animated videos and then fill another qustionnare to identify whether the brand new animated videos are able to generate attitude change.
v) Provide evaluation on the difference of the 3 live action videos and 3 animated videos that recreate scenes in live action videos in terms of (a) the power to deliver overall message of the traffic safety campaign theme The power to deliver overall message of the traffic safety campaign theme (b) the missing scenes (c) The missing main characters (d) The inappropriate animated scenes to recreate live action scenes (e) Inappropriateness of main characters dialects (f) Inappropriateness of musical illustration.
vi) Provide evaluation on 3 brand new animated videos when compared to their synopsis in terms of (a) the power to deliver overall message of the traffic safety campaign theme (b) the missing scenes (c) the missing main characters (d) inappropriateness of main characters dialects (e) Inappropriateness of musical illustration.

A Likert Scale of 1 to 5 is to be used to both the selected itmes of Indonesian MRBQ and other questionnaires to identify any possible attitude change after watching different type of videos. T-test of mean difference between responses from baseline survey (using the selected items of Indonesian MRBQ) and after intervention survey (after watching different type of videos) will be conducted with 0.05 significant level.

4 Future works

The animated videos production was still on progress, so the main survey has not been conducted yet. However in our hypothesis the general expected result will be (i) As found in the second year research, a short term attitude change will be shown by most of respondents after watching the 3 live action videos. (ii) Similar results might happen after watching the 3 animated videos that recreate 3 live action videos. However compare to point number 1 the attitude change might be marginally lower in terms of mean diffierence of responses. (iii) Similar results might also happen after watching 3 brand new animated videos. However compare to point number 2 the attitude change might be marginally lower in terms of mean diffierence of responses. (iv) The respondents might satisfy with the power of animated videos to deliver overall message of the traffic safety campaign. However they might critically pointed out some inappropriateness. The defficiencies might more apparent in 3 brand new animated videos as the scenes and the characters should be created almost from scratch.

5 Concluding remarks

This is stil an ongoing research. The author hope to finish the research by the end of 2016. Final findings will be reported in further publication. Preliminary analysis shows that short term attitude change were shown by most respondents after watching live action videos. Similar effect was found after watching animated videos. I acknowledge the Ministry of Research, Technology and Higher Education who provide the fund for this three years research on risky behaviour of Indonesian motorcyclist

References

[1] L.S. Putranto, Rostiana. *Proceeding of 10th Asia Pacific Conference on Transportation and Environment,* (2016) (to be published soon)
[2] L. Colclough, *Animated video vs live action video: Which one shoud you choose?* (2016), retrieve on 20 September 2016 from https://www.moovly.com/blog/animated-video-vs-live-action-video-which-one-should-you-choose
[3] L.S. Putranto, N.L.P.E. Setyarini, Rostiana, R. Bunawan, *Proceeding of 17th International Symposium of Indonesian Inter-University Transportation Studies Forum*, **2**(1), 1038-1044 (2014)
[4] L.S. Putranto, Rostiana, *Proceeding of Conference of ASEAN Road Safety*, 369-373 (2015)
[5] D. Prasetyo, *Pengertian Storyboard*, (2016), retrieve on 20 September 2016 from http://www.davidprasetyo.com/2015/11/ pengertian-storyboard.html (2015)

Evaluation of Exclusive Stopping Space for Motorcycle at Signalized Intersections in Large Cities in Indonesia

A. Caroline Sutandi[1,*], and *Dennis Dwika* Siregar[1]

[1]Parahyangan Catholic University, Jalan Ciumbuleuit No.94, Bandung 40141, West Java, Indonesia

Abstract. Proportion of motorcycles in large cities in Indonesia as a developing country is around 60%-71% of all vehicles and increases 19%-37%every year. Data from Indonesia Statistics Centre Bureau indicated that number of motorcycle increase from 13,563,017 in 2000 to 92,976,240 in 2014. This condition leads worse traffic congestion in large cities and more specific at signalized intersections. Motorcycles usually stop everywhere including among cars and in front of stop-line at signalized intersection during red time period. One of solutions to reduce the poor traffic congestion is implementation of exclusive stopping space (ESS) for motorcycle. Manual of ESS for motorcycle in Indonesia is used as a guidance of implementation. The aim of this study is to determine signalized intersections that need ESS for motorcycle, determine the type, and then evaluate the implementation. This study is specific, important, and crucial to do because usually exclusive stopping space in developed countries is for bicycle. Case study is carried out at all signalized intersections in large city Bandung in Indonesia. Results of study indicated that ESS for motorcycle is needed with 52.2%-100%effectiveness. Furthermore, recommendation that could be provided is more specific ESS guidance for existing variation of lane widths.

1 Introduction

Traffic congestion usually occurs in large cities in developing countries including in Indonesia. This condition is worse because number of motorcycle in the large cities in Indonesia is around 60% up to 71% of number of all vehicles and increase 19% up to 37% per year [1]. They stop everywhere among cars and even in front of stop-line at signalized intersections during red time periods. This condition blocks traffic flow from other leg intersections during next green time period. One of solutions to reduce the poor traffic congestion is implementation of exclusive stopping space (ESS) for motorcycle.

The aim of this study is to determine the signalized intersections that need ESS for motorcycle, determine the type of ESS for motorcycle, and then evaluate the implementation. This study is very specific, important, and crucial to do because usually exclusive stopping space in developed countries is for bicycles. Manual of ESS for

* Corresponding author: caroline@unpar.ac.id

motorcycle in Indonesia is used as a guidance of implementation. Case study is carried out at all signalized intersections in large city Bandung in Indonesia. Results of this study are not only beneficial for large city Bandung and other large cities in Indonesia, but also beneficial for other large cities in other developing countries that have similar traffic and geometric conditions.

2 Exclusive stopping space (ESS) for motorcycle

Exclusive stopping space (ESS) for motorcycle is a space facility for motorcycle while stopping at signalized intersection during red time in large cities. The exclusive area is located between stop line at signalized intersection and stop line for vehicle with 4 wheels or more.

2.1 Requirements for ESS for motorcycle implementation

There are some requirements that have to be fulfilled by signalized intersections in order to implement ESS for motorcycle. The requirements are as follow [1]:

- General requirement i.e. size of signalized intersection is medium (10m-14m) and large (> 15m) [2].
- Geometric requirements. Each leg intersection has at least 2 lanes straight through only. Left lane for left turn on red (LTOR) in not included. A number of intersection types that fulfil the geometric requirement are presented in Fig. 1, Fig. 2, and Fig. 3. Another geometric requirement is that lane width is 3.5m. With this lane width, motorcycle can move among vehicles to the ESS for motorcycle. The position of motorcycles among cars is presented in Fig. 4.
- Traffic Requirements. If number of motorcycle stop at leg intersection with 2 lanes at least 30 motorcycles or at leg intersection with 3 lanes at least 45 motorcycles, during red time period, therefore ESS for motorcycle is needed. If leg intersection has more than 3 lanes, it needs ESS for motorcycle if there are at least 15 motorcycles stop during red time period. The example is presented in Table 5.

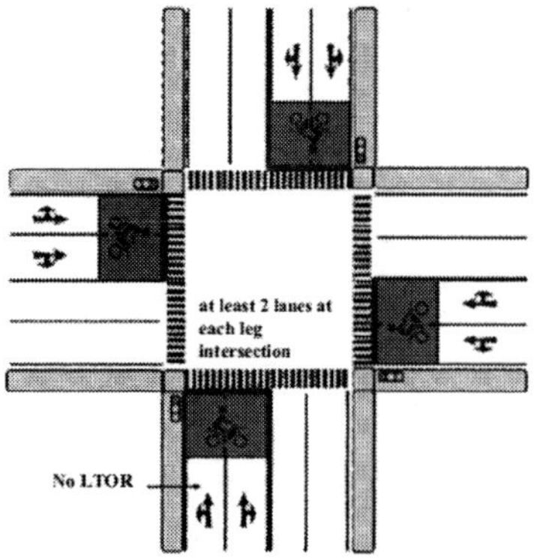

Fig. 1. Location of ESS for motorcycle at signalized intersection without LTOR and without traffic island [1]

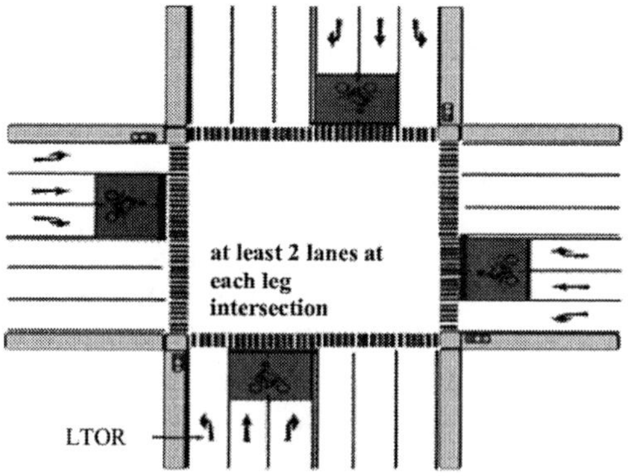

Fig. 2. Location of ESS for motorcycle at signalized intersection with LTOR and without traffic island [1]

DOI: 10.1051/matecconf/201710308007

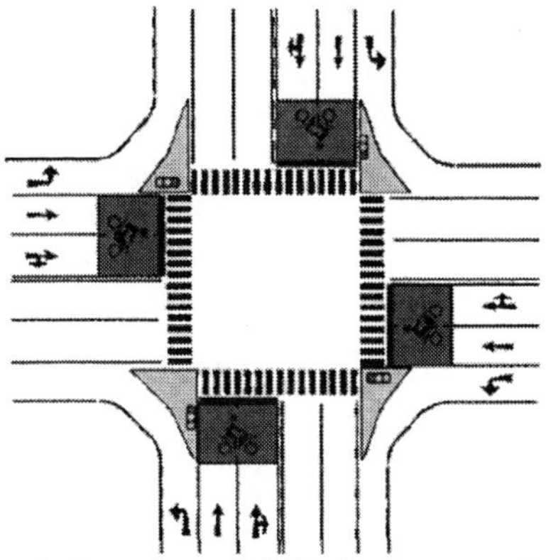

Fig. 3. Location of ESS for motorcycle at signalized intersection with LTOR and with traffic island [1]

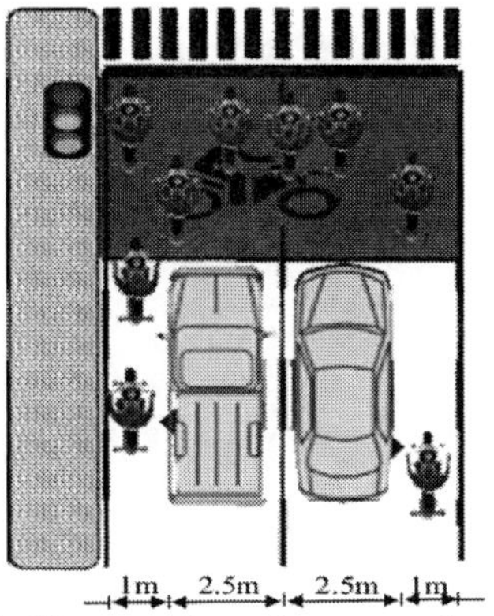

Fig. 4. The position of motorcycle among cars [1]

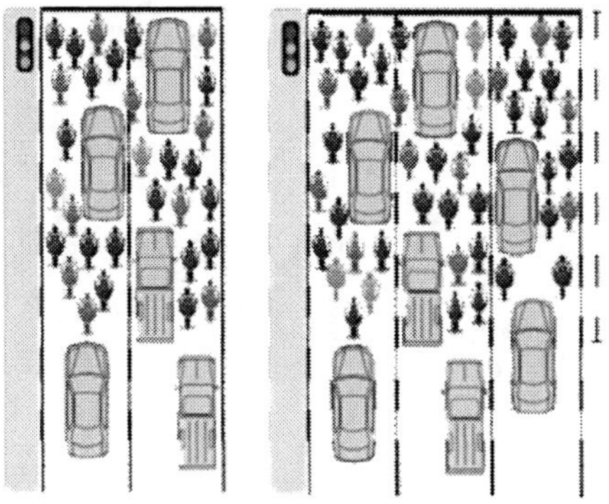

Fig. 5. Examples of leg intersections with 2 and 3 lanes that need ESS for motorcycle [1]

2.2 Survey before implementation of ESS for motorcycle

Before implementation of ESS for motorcycle, survey for counting the number of motorcycle at each leg intersection at each signalized intersection during red time period is needed. Steps of survey are as follow [1]:

- Survey is conducted 3 times, in the morning, at noon, and in the afternoon.
- Survey duration minimum is during 10 phases per time, so that total minimum survey duration is during 30 phases per day.
- Survey technique can be manual i.e. using traffic counter or survey form or semiautomatic using video camera.
- Summary of number of motorcycle stop at each leg intersection at signalized intersection will determine whether the leg intersection needs ESS for motorcycle or not. Furthermore, is to determine the type of ESS for motorcycle.

2.3 Types of ESS for motorcycle

Type of ESS for motorcycle depends on number and size of motorcycle at each leg intersection. General size of motorcycle used in Indonesia is 0.8m x 2.0m (1.6m^2), the largest kind with cylinder of 110-125cc. Usually there is 2 types of ESS for motorcycle i.e. Square type and P type. More detail explanation is as follow [1].

2.3.1 Square type of ESS for motorcycle

Square type of ESS for motorcycle is used if number of motorcycle on each lane at leg intersection is almost the same. Length of Square type of ESS for motorcycle is between 8m-12m depend on number of motorcycle stops during red time period. While width of Square type of ESS for motorcycle is depend on the lane width. Square type of ESS for motorcycle is presented in Fig. 6.

2.3.2 P type of ESS for motorcycle

P type of ESS for motorcycle is used if proportion of motorcycle on left lane at leg intersection is 60%higher for leg intersection with 2 lanes and 70%higher for leg intersection with 3 lanes, than other lanes. Length of P type of ESS for motorcycle is between 12m-16m depend on number of motorcycle stops during red time period. While width of Square type of ESS for motorcycle is depend on the lane width. P type of ESS for motorcycle is also presented in Fig. 6.

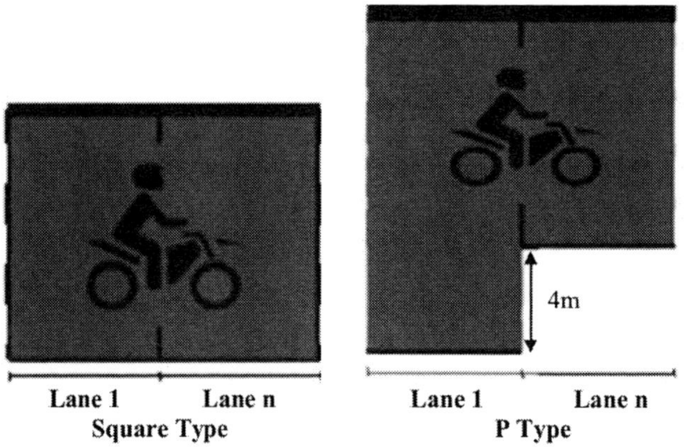

Fig. 6. Square type and P type of ESS for motorcycle [1]

3 Methodology

Methodology of determination, implementation, and evaluation of ESS for motorcycle in large cities in Indonesia is presented in Fig. 7. Besides manual that are used, traffic regulations also used as legal formal foundation for implementation i.e. law of Republic of Indonesia No.38 year 2004 about Road and law of Republic of Indonesia No.22 year 2009 about traffic and road transport.

4 Data of SSE for motorcycle

Field data is collected from all signalized intersection in large city Bandung in Indonesia. There are 87 working signalized intersections of 148 signalized intersections in Bandung. Location of the signalised intersections is presented in Fig. 8 [3, 4, 5]. 39 of 87 signalized intersections have already implemented ESS for motorcycle and will be evaluated. 48 of 87 signalized intersections have not already implemented ESS for motorcycle and will be determine whether they need implementation of ESS for motorcycle or not.

DOI: 10.1051/matecconf/201710308007

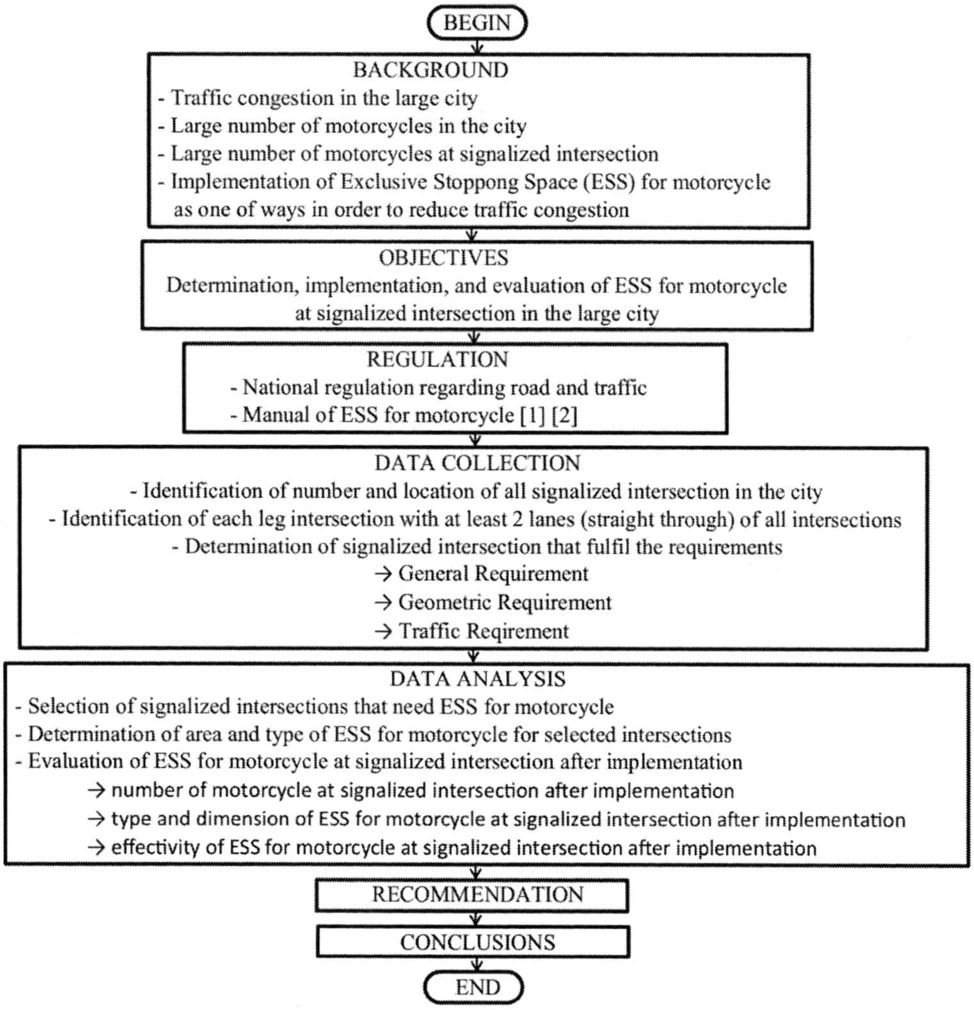

Fig. 7. Methodology of determination, implementation, and evaluation of ESS for motorcycle in large cities in Indonesia

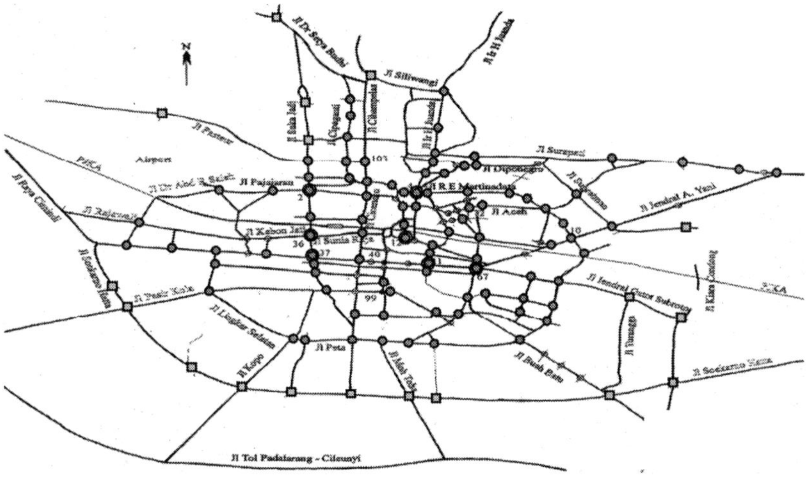

Fig. 8. Location of signalised intersections in Bandung, Indonesia [3, 4, 5, 6]

4.1 Determination and implementation of ESS for motorcycle

Based on size of intersections, the 48 signalized intersections that have not already implemented ESS for motorcycle consist of 7 large intersections (> 15m), 30 medium intersections (10m-14m), and 11 small intersections (6m-9m).

Furthermore, signalized intersections that fulfil general requirement (large and medium intersection), geometric requirements (leg intersection with 2 straight through lanes without LTOR, and lane width is 3.5m) and traffic requirement (30 motorcycles stop at leg intersection with 2 lanes and 45 motorcycles stop at leg intersection with 3 lanes, during red time period) is only 4 intersections. Characteristics and type of ESS for motorcycle of the 4 intersections are presented in Table 1 and Fig. 9 up to Fig. 12.

Table 1. Characteristicsand type of ESS for motorcycle of the 4 intersections.

Signalized Intersection			Leg intersection that fulfil all requirements		Average number of motorcycles during red time period	Proportion of motorcycle on left lane (%)	Type of ESS for motor cycle
No	Name	Size	Approach	Number and width of lane			
1	Lingkar Selatan - Otto Iskandar Dinata	Large	North	3: 3.6m, 3.5m, 3.5m	60	80.12	**P**
2	Dipatiukur - Siliwangi	Large	North	3: 3.5m, 3.5m, 3.5m	42	54.18	**Square**
3	RE.Martadinata - Ir.H.Juanda	Medium	North	2: 3.5m, 3.5m	36	65.90	**P**
4	Lingkar Selatan - Sriwijaya	Medium	West	3: 3.6m, 3.5m, 3.5m	52	54.57	**Square**

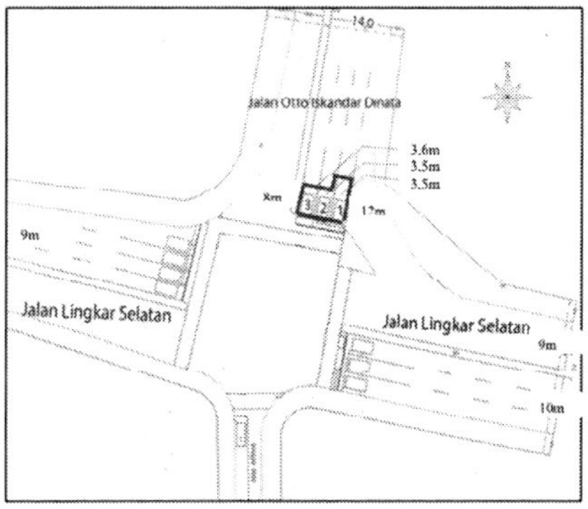

Fig. 9. Lingkar Selatan-Otto Iskandar Dinata large intersection in Bandung [2, 6]

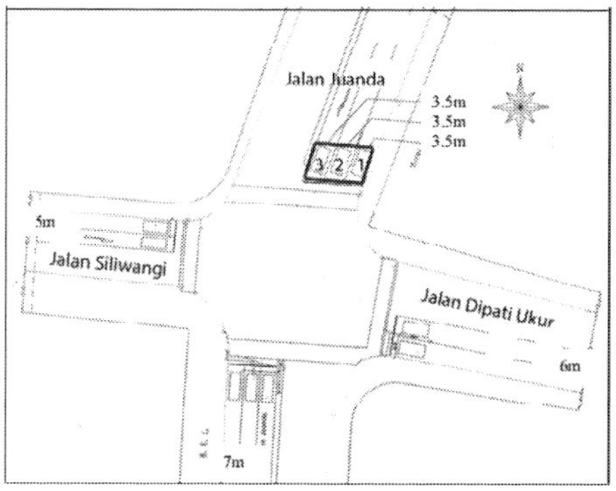

Fig. 10. Dipatiukur-Siliwangi large intersection in Bandung [2, 6]

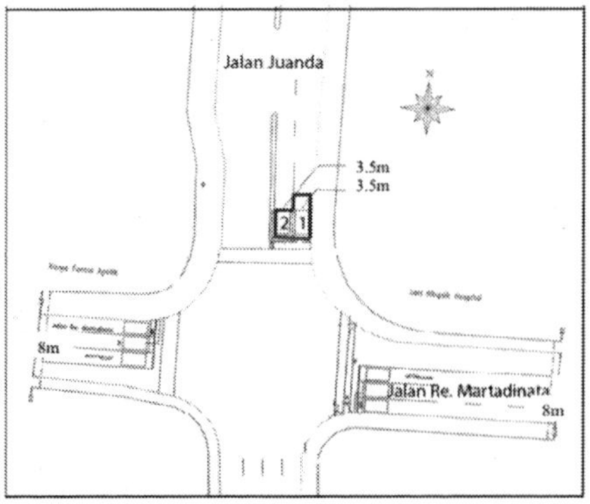

Fig. 11. RE.Martadinata-Ir.H.Juanda medium intersection in Bandung [2, 6]

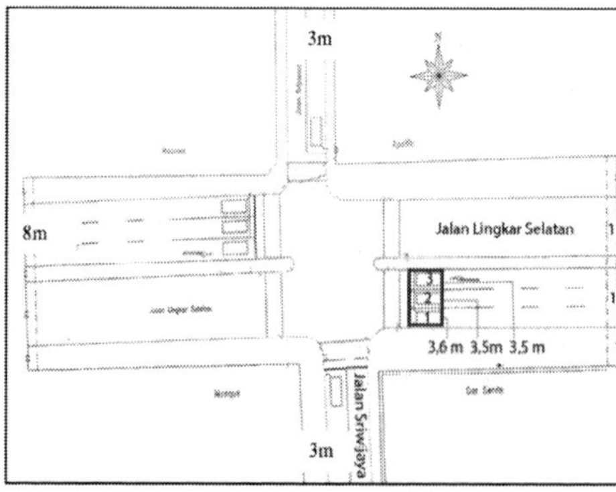

Fig. 12. Lingkar Selatan-Sriwijaya medium intersection in Bandung [2, 6]

4.2 Evaluation of implemented ESS for motorcycle

Evaluation is going to conduct on implemented ESS for motorcycle in Bandung, Indonesia. 9 most congested signalized intersections of 39 signalized intersections that have already implemented ESS for motorcycle are evaluated. Criteria of evaluation are proportion of motorcycle on left lane (%), capacity of ESS for motorcycle of all lanes (N1), and number of motorcycle stop during red time period (N2).

Furthermore, based on proportion of motorcycle on left lane (%), existing type of ESS is evaluated, whether still square type or need to change to P type. Moreover, based on capacity of ESS for motorcycle of all lanes (N1) and number of motorcycle stop during red time period (N2) then effectiveness is determined by divide N2 with N1 in percentage. Fields data and evaluation of the 9 signalized intersections are presented in Table 2 [7].

Table 2. Evaluation of implemented ESS for motorcycle.

Signalized Intersection			Existing condition of ESS for motorcycle [7]				ESS type evaluation *)	Effectiveness (N2/N1)x100% ≤50% → Low 51%-79% → Moderate ≥80% → High
No	Name	Number of leg intersection with ESS	Type	Proportion of motorcycle on left lane (%)	Capacity of all lanes –N1 (motorcycle)	Number of motorcycle stop during red time period N2		
1	Jendral Sudirman – Otto Iskandar Dinata	1 with 2 lanes	Square	52.30	92	80	Square	89
2	Jendral Sudirman – Jamika	2 with 2 lanes	Square	47.10; 54.70	46; 37	24; 22	Square	52.2; 59.5
3	Jamika – Peta	1 with 3 lanes	Square	54.70	70	47	Square	67.10
4	Soekarno Hatta – Kiara Condong	3 with 3 lanes	Square	48.20; 67.90; 57.40	116; 70; 46	95; 69; 36	Square	81.90; 98.60; 78.30
5	Soekarno Hatta – Buah Batu	2 with 2 lanes	Square	52.60; 59.60	46; 56	34; 52	P	73.90; 92.9
6	Soekarno Hatta – Kopo	1 with 2 lanes	Square	43.30	46	45	Square	97.80
7	Lembong – Tamblong	1 with 2 lanes	Square	49.80	46	46	Square	100
8	Tamblong – Naripan	1 with 3 lanes	Square	66.50	70	62	Square	88.60
9	Veteran – Sunda	2 with 3 lanes	Square	65.90; 50.80	56; 37	54; 33	Square	96.40; 89.20

***) = P type if 60%proportion (2 lanes) 70% proportion (3 lanes)**

4.3 Analysis of implementation of ESS for motorcycle

Table 1 shows that 2 of 4 signalized intersections need Square type of motorcycle and 2 other need P type. Furthermore, although the 2 intersection only need Square type, but proportion of motorcycle on left lane (%) is high i.e. 54.18% and 54.57%. regular monitoring is needed to collect newest field data of increasing number of motorcycle during red time period. So that P type of ESS for motorcycle can be implemented soon. This is important because not changing Square type to P type on time as needed, can cause many motorcycle stop among cars that can block traffic flow and increase traffic congestion.

Table 2 shows that based on existing condition, effectiveness of using ESS for motorcycle is between moderate and high (52.2%-100%). Although ESS for motorcycle occupy space of other vehicles (disadvantage), but motorcycles located on exclusive area that will not block traffic flow and can reduce traffic congestion (advantages). This condition indicated that ESS for motorcycle in the city is needed and also need to be monitor regularly and consistently

5 Recommendation

Based on existing conditions, recommendations that can be provided are as follow:

- Revised manual of ESS for motorcycle in Indonesia so that calculation and implementation also involve lanes with lane width less than 3.5m. This is important because number of motorcycle stop on lanes with lane width less than 3.5m during red time period, in the city are, is high.
- Further studies are evaluation whether traffic parameter at intersection for example queue length, number of queue, and number of stops are better or not after implementation of ESS for motorcycle and also relationship between signalized intersection capacity and traffic congestion before and after implementation of ESS for motorcycle.
- Socialization to the society regarding benefit of using ESS for motorcycle in reducing traffic congestion at signalized intersections and also in the city.

6 Conclusion

Innovation of ESS for motorcycle is specific for cities and urban areas with high proportion of motorcycle like Indonesia. Moderate to high effectiveness value i.e. 52.20%-100% indicated that ESS for motorcycle is needed. The most important thing is that benefit of the result study for Bandung, other large cities in Indonesia, and can be implemented to other cities in other developing countries that have similar traffic and geometric conditions.

References

[1] Road and Bridge Research and Development Centre, *Guidance of Exclusive Stopping Space for Motorcycle*, 4th Designed, Ministry of Public Works, Republic of Indonesia (2012)

[2] Directorate General of Bina Jalan Kota, *Indonesian Highway Capacity Manual 1997*, Directorate General of Bina Marga, Ministry of Public Work, Republic of Indonesia, (1997)

[3] AWA Plessey, *Bandung Area Traffic Control, Final System Design*, Directorate General of Land Transport, Ministry of Communications, Republic of Indonesia, (1996)

[4] A.C. Sutandi, *Advance Traffic Control Systems, Performance Evaluation in Developing Country*, Lambert Academic Publishing, (2010)

[5] A.C. Sutandi, Evaluation of the impacts of VMS on traffic performance measure in an urban area in Indonesia, Civil Engineering Dimension, Journal of civil Engineering Science and Application, **10**(1), 28 – 34 (2008)

[6] D.D. Siregar, *Choosing Signalized Intersections in Bandung City that Need ESS*, Thesis, Parahyangan Catholic University, Bandung, (2016)

[7] M. Cahyaningtyas, *Evaluation of ESS Implementation at Intersections in Bandung*, Thesis, Parahyangan Catholic University, Bandung, (2014)

Analysis of Speeding Behaviour During Approaching the U-Turn Facility Road Segment Based On Driving Simulation Test

M. S. Nemmang[1], *R.* Rahman[1,*], *M. M.* Rohani[1], *N.* Mashros[2], and *J.* Md Diah[3]

[1]Department of Infrastructure and Geomatic, University Tun Hussein Onn Malaysia, Batu Pahat
[2]Faculty of Civil Engineering, Universiti Teknologi Malaysia (UTM), Johor Bahru, Johor, Malaysia
[3]Faculty of Civil Engineering, Universiti Teknologi Mara, Shah Alam, Selangor 40450, Malaysia

Abstract. The main purose of this study is to analysis the speeding behavior during approaching the U-turn facility road segment based on driving simulation test. In previous studies, it has been shown that speeding behavior is a complex problem that requires a full understanding of drivers' attitudes and beliefs. Such understanding is needed to improve the speeding behavior of drivers that comes from the effective design and interventions. Therefore, this study will analyse the speeding behavior in approaching the U-turn facility as a key variable in this study which may affect the collision result. Totally 50 participants were recruited in this experiments based on driving simulator. The scenarios and road environment in the driving simulator were based on site location at FT050 Jalan Batu Pahat – Kluang, Johor. Results shows significant increase in speed of up to 40.01% more than the speed limit at the approaching the U-turn facility road segment. The paper concluded that speeding behavior during approaching the U-turn facility road segment based on driving simulation test will trigger significant increasing of speed.

1 Introduction

This study is conducted to find out the driver speeding behavior in the area of U-turn at FT050 form Batu Pahat to Kluang. FT050 has six (6) midblock U-turn facility in total from Batu Pahat to Parit Raja with only two (2) U-turn facility available for heavy vehicles. The speed profile is an important tool for road safety especially in the evaluations and in design process of existing toad because it can be used for safety considerations. The data recorded by Royal Malaysia Police in 2012 [1] showed that accident rate were increasing every year where in year 2012 the number of accident was 62,316 with 1,073 fatal and location for this study which is in Batu Pahat, where it is the second highest in Johor the number of accident recorded was 7,439 with 176 fatal recorded which is the highest in Johor during that year. Sadly, in 2015 it keeps increasing with the 197 fatal. Abdul Manan [2] and Abdul Manan & Várhelyi [3] also mentioned that Batu Pahat – Kluang FT050 road recorded 1000 accidents and 36 deaths since 2006. This study was using the driving simulator to collect the data.

* Corresponding author: raha@uthm.edu.my

However, when drivers driving in actual driving situations rear-end risk still exists. The development of driving simulator with character safer driving environment, easier to use, and lower cost makes the experiment for data collection in a more convenient way. Studies using driving simulator have been conducted in many fields such as using driving simulator in model development and to examine the effect of intersection and as stated by Md Diah et al. [4] and Nemmang and Rahman [5].

Speed choice is one of the most studied driver behaviors. One reason may be because speed is strongly related to accident risk. Therefore, the studies on speeding behavior are important in this road especially in approaching the U-turn facility because the speed limit in the U-turn area are 60km/h where is at the roadway before and approaching U-turn segment was 90km/hr the reduction and increasing speed can lead the accident as studied by Rahman and Ben-Edigbe [6]. Studies of speeding behavior have been conducted in many aspects. Mehmood, [7] studies the speeding behavior used radar gun to measure the spot speed of randomly selected vehicles from three major arterial roads in order to assess the current extent of the speeding problem in Al Ain. The results indicated that speeding behavior contributing to 37% of severe road crashes. While Montella et al. [8] studies speeding behavior in rural intersections using driving simulator and the results showed that that intersections significantly affect drivers' speed behavior.

Some researchers also study about the travel speed at the multilane. Rahman and Ben-Edigbe [6] trigger significant travel speed reduction in median openings zone facilities irrespective of their traffic conflicts minimization merits. While Daniel et al. [9] studies about the speed patterns on traffic calmed. They provided that driver choice of speed not only when traversing the devices but also as they move towards and away from the devices that affected into the effects of traffic calming devices on driver behavior. Mashros et al. [10] explore about the mean speed at the dry and rainfall condition. The research successfully finds out the speed reduction in dry and rainfall condition. Moreover, Rohani et al. [11] assess the impact of business and non-business hours on speed and acceleration of car drivers. The researcher divided the road into four (4) segment and find out the speed in every segment during the business hour and non-business hour. Speed study have been familiar for the researchers in order to reduce the number of accident and find out the driver behavior in choosing the speed. Usually, mean speed will be used to find out the average speed of driver tend to drive in certain area such as in U-turn area. Mean speed is calculated as the sum of all speeds divided by the number of speed observations and a measure of the central tendency of the data. This speed is believed as an indicator of average travel speed. Therefore, this studies find out the speed behavior of drivers in order to minimize the probability of accident especially in U-turn area.

2 Methodology

In this paper, the methodology in collecting the data is divided into 3 steps.

Step 1: Participant

The process begins with a recruited a total of 20 participants to use the driving simulator as a sample. During the selection of the participants, gender (10 males vs 10 females). The participants were given time to practice the driving simulator to make sure they are familiar with the road and condition. The most important thing this group of people should have a valid Malaysian driving license and must in 19 years old and above. In order to ensure the availability of the data, participants were excluded for sickness and for the sleep debt which may influence a normal driving.

Step 2: Apparatus

The apparatus in this experiment was conducted by a fixed- basis driving simulator in University Teknologi Mara (UITM) by using Automotive Simulator for Driver's Behavior and Competency Evaluation (ASIS). The driving simulator (Fig. 1) give the participant drive like driving a real car on a real road because it can simulate driving scenarios and various kinds of road traffic environments. The vehicle cockpit is based on the true steering simulation system, accelerated and brake pedal system, real instrument control interface, vibration and noise simulation system and the vehicle is located in a 180 degree rearview that can display virtual road environment.

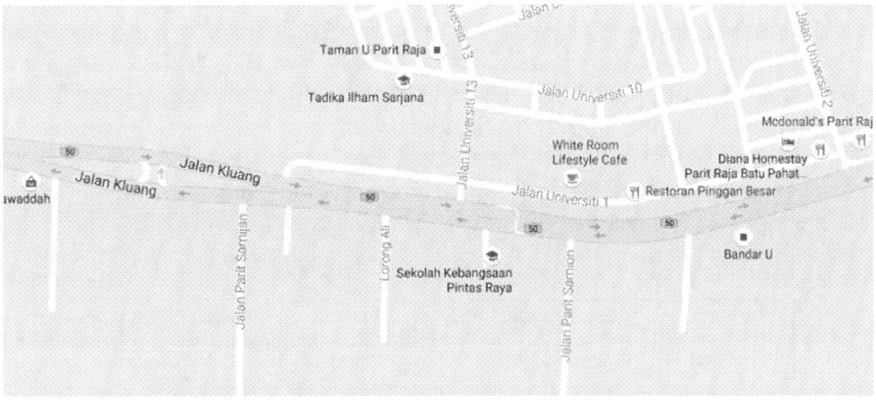

Fig. 1. Area of research study at FT050

Fig. 2. Driving simulator screen simulation

Step 3: Scenarios

The purpose of this study is to study the drivers speeding behavior in approaching the U-turn facility. Thus the scenarios or road environment are based on the road environment of FT050 as shown in Fig. 2. To meet the requirements of designing scenarios and simulating virtual traffic environment, a set of software as shown in Fig. 3 are adopted. Other than that, the roads design was dual carriageway highway with a median that give participants choose the best lane and speed for their driving. The road and the U-turn area development has a midblock in the middle of the road. The road providing the U-turn signboard which is 150m before the U-turn area and at the U-turn itself that given the driver warning and guide that U-turn area at the front. The length of the road was 700m which is 500m before the U-turn area and 200m after the U-turn. It gives the driver a chance to get momentum on the speed of the vehicle in approaching the U-turn area. Therefore, driver will be driving with their own intentions without any disruptions from the road design.

3 Results and findings

3.1 Analysis of drivers speeding behavior

In this paper, the analysis of driver behavior focuses on the speeding behavior of the driver in the U-turn area. Table 1 presents the mean speed of driver driving from the driving simulation based on the road environment of FT050 from Batu Pahat to Kluang at the U-turn area of KM15. The speed limit in the area of U-turn area which is 150m from the U-turn is 60km/h. From the Fig. 3, the histogram shows that the driver behavior tends to speeding in the U-turn area where the graph nearly uniforms from all the drivers. The highest average speed is 91.95km/h which is 31.95km/h more than the speed limit. The lowest average speed comes from the driver 19 with speed 62.87km/h.

Table 1. Mean speed of driver

Drivers	Mean Speed (km/h)	Drivers	Mean Speed (km/h)
Driver 1	86.88	Driver 11	78.15
Driver 2	88.17	Driver 12	85.93
Driver 3	85.29	Driver 13	90.41
Driver 4	91.95	Driver 14	87.22
Driver 5	90.60	Driver 15	87.26
Driver 6	83.16	Driver 16	91.52
Driver 7	86.55	Driver 17	62.91
Driver 8	88.06	Driver 18	83.99
Driver 9	87.72	Driver 19	62.87
Driver 10	87.72	Driver 20	74.83

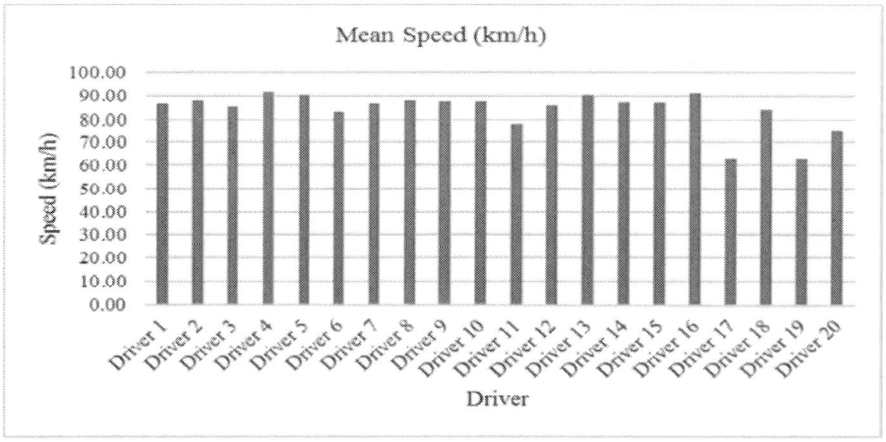

Fig. 3. Mean speed histogram

Graphical view of the speeding behavior of the driver shown in Fig. 4. It shows that the speeding behavior in the area of U-turn nearly closed between all the drivers as shown in the graph line for speed differences that formed nearly flat line from driver 1 to driver 16. Only driver 16 and 19 have speed nearly 60km/h that follows the speed limit of the area. It shows that drivers tend to speeding in this area where it can bring into an accident.

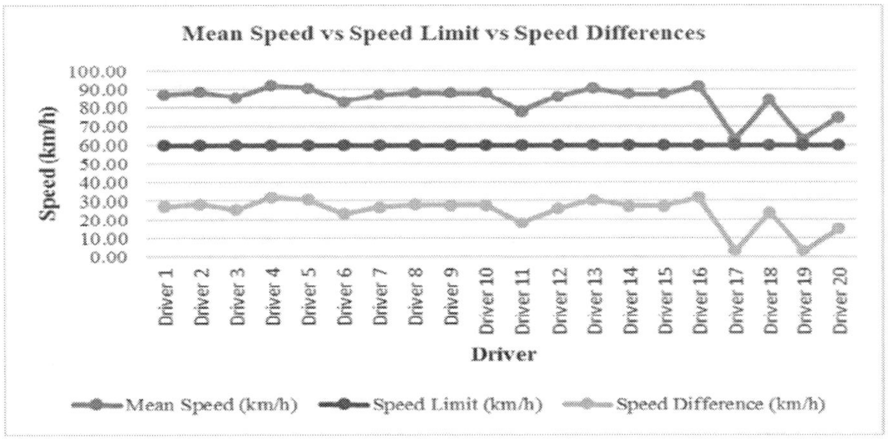

Fig. 4. Graphical view of mean speed vs speed limit vs speed differences

The summary of average speeds is presented in Table 2. Result shown that average speeds in the U-turn area increasing in 40.01% which is the average speed is 84.06km/h. T-test was performed on these speed data and the result is given in Table 3. Result indicated that the speed increasing in average speed was found to be statistically significant.

Table 2. Descriptive analysis of speeding

	N	Mean	Std. Deviation	Std. Error Mean
Mean_Speed	20	84.0595	8.32670	1.86191
Speed_Differences	20	24.0595	8.32670	1.86191

Table 3. T-test for speeding behavior

	Test Value = 0					
					95% Confidence Interval of the Difference	
	t	df	Sig. (2-tailed)	Mean Difference	Lower	Upper
Mean_Speed	45.147	19	.000	84.05950	80.1625	87.9565
Speed_Differences	12.922	19	.000	24.05950	20.1625	27.9565

3.2 Speeding behavior based on gender

Descriptive statistics on gender was shown in Table 4. It was found that male drivers seemed to have a higher speed than the female drivers. It shows that mean speed for male drivers is 89 km/h while female drivers has a mean speed 79 km/h which is about 10 km/h lower than male drivers. The significance T-test as shown in Table 5 shows that speeding behavior between male and female drivers has a significant differences where the P value is 0.005 which is lower than 0.05.

Table 4. The descriptive statistics of speeding behavior by gender

Gender	N	Mean	Std. Deviation
Male	10	88.9340	1.95528
Female	10	79.1850	9.47347
Total	20	84.0595	8.32670

Table 5. The significance T-test of speeding behavior by gender

			Sum of Squares	df	Mean Square	F	Sig.
Mean_Speed * Gender	Between Groups	(Combined)	475.215	1	475.215	10.157	.005
	Within Groups		842.128	18	46.785		
	Total		1317.343	19			

4 Conclusion

Research findings suggested that in general, speeding behavior of the driver in the area of the U-turn not affected by the signboard and the speed limit. Even though all the drivers have been given warning by all the signboard, drivers were ignored all the signboard where the research found that drivers were drive up to 40.01% significantly faster than the speed limit. It shows that, drivers tend to speeding without notices the signboard and speed limit that may cause to an accident with the U-turn vehicles. Other than that, speeding behavior based on gender shows that male drivers tend to speeding where it is higher than a female driver when approaching the U-turn facility road segment.

This research is supported by the GPPS of Universiti Tun Hussein Onn Malaysia (UTHM) in Vot U439.

References

[1] Polis Diraja Malaysia, *Laporan Perangkaan Kemalangan Jalan Raya Malaysia,* Malaysia, (2012)
[2] M.N. Abdul Manan, *Motorcycle Crashes in Malaysia - Status, Risk Factors and Interventions*, MIROS, (2015b)
[3] M.N. Abdul Manan, and A. Várhelyi, Motorcycle fatalities in Malaysia. *IATSS Research, **36***(1), 30–39, (2012)
[4] J. Md Diah, M.Y. Abdul Rahman, M.A. Adnan, I. Atan, Weaving section flow model at the weaving area of Malaysian conventional roundabout, Journal of Transportation Engineering, **136**(8), 782–792 (2010)
[5] M.S. Nemmang, R. Rahman, An Overview of vehicles lane changing model development in approaching at u-turn facility road segment, Jurnal Teknologi, **7**, 59–66 (2016)
[6] R. Rahman, J. Ben-edigbe, Impact of multilane median openings zone on travel speed,Technology Journal, *4*, 15–20 (2015)
[7] A. Mehmood, Determinants of speeding behavior of drivers in Al Ain (United Arab Emirates), Journal of Transportation Engineering, **135**, 721–729 (2009)

[8] A. Montella, V. Claudio, F. Mauriello, drivers' speed behaviour at rural intersections: Simulator experiment and real world monitoring, Road Safety and Simulation, 1–14 (2011)
[9] B.D. Daniel, N. Alan, G. Koorey, Investigating speed patterns and estimating speed on traffic-calmed streets, *Proc. of the IPENZ Transportation Group Conference*, Auckland, (2011)
[10] N. Mashros, R. Rahman, J. Ben-Edigbe, Exploration of sample quantiles and mean speed distributions, *International Conference on Statistics in Science, Business and Engineering: Empowering Decision Making with Statistical Sciences*, ICSSBE 2012 – Proceedings, 158–161, (2012)
[11] M. Rohani, B.D. Daniel, M.Y. Aman, J. Prasetijo, A.A. Mustafa, R. Rahman, Business hours vs non - business hours in business zone: Impact on speed and acceleration, Jurnal Teknologi, **4**, 71–77 (2016)

Speed-Flow Relationship at Basic Segment of Federal Highway

Mohd Erwan Sanik[1,2,*], *Joewono* Prasetijo[2,3], and *Ahmad Sofian* Azmi[3]

[1]Department of Civil Engineering, Centre for Diploma Studies, Universiti Tun Hussein Onn Malaysia, 86400 Parit Raja, Johor, Malaysia
[2]Smart Driving Research Centre, Universiti Tun Hussein Onn Malaysia, 86400 Parit Raja, Johor, Malaysia
[3]Department of Infrastructure and Geomatic Engineering, Faculty of Civil and Environmental Engineering, Universiti Tun Hussein Onn Malaysia, 86400 Parit Raja, Johor, Malaysia

Abstract. Theoretically, at many traffic facility during peak hour, speed-flow relationship is only appeared as a normal flow, represented by half of speed-flow curve. Forced flow will only occur during severe overcrowding resulted in drivers trying to use all the space in a way so that the amount of overcapacity. The purpose of this study is to analyse the relationship between speed-flow traffic flows along a basic segment. The objectives of this study is to analyse the time series of speed and flow at basic segment of A550 and A551 study location for 24 hours in a month as well as to determine the maximum flow and critical speed through the speed-flow models. Data obtained from this study were analysed using video recording and image processing techniques. In order to achieve these objectives, the Federal Highway 2 was selected as location for the study. Results show that the speed – flow diagram in one month on both A550 and A551 obtain the maximum flows and critical speeds of 4200 and 5320 veh/hr/ln and 55 and 60 km/hr, respectively.

1 Introduction

Congestion usually relates to an excess of vehicles on a portion of roadway at a particular time resulting in speeds that are slower sometimes or much slower than normal or free flow speeds [1]. When the capacity of a highway section is exceeded, breakdown occurs where speeds dropped and vehicles crowded together.

Massive daily traffic congestion along the Federal Highway 2 is common especially early in the morning and late afternoon. This congestion create a condition of forced flow which can be interpreted using the speed-flow relationship [2]. The objectives of this study is to analyse the time series of speed and flow at basic segment of A550 and A551 study location for 24 hours in a month as well as to determine the maximum flow and critical speed through the speed-flow models.

* Corresponding author: erwans@uthm.edu.my

 DOI: 10.1051/matecconf/201710308009

2 Methodology

2.1 Study framework

Fig. 1 shows the flowchart of study as framework of this study.

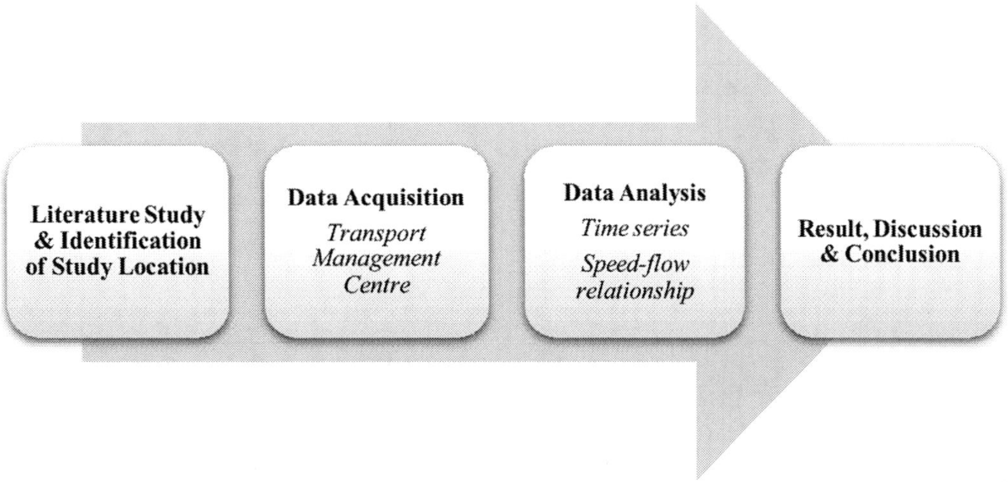

Fig. 1. Framework of study

2.1.1 Study location and data acquisition

This study was carried out at basic segment of Federal Highway 2 near Kompleks Astana. Secondary data were used in this study. Data were obtained from the Transport Management Centre (TMC). Two automatic incident devices were used to record and extract data at study locations, namely A550 and A551.

2.1.2 Data analysis

Data were plotted into time-series graph for both location. The speed and flow data were analysed to develop speed-flow relationship of both locations. The maximum flow and critical speed can be determined from the graph.

3 Result and discussion

All data for speed and flow rate in month of January 2012 was plotted as graph in time-series over the period of time for each location. The primary and secondary x-axis on the left was represented by speed while on the right was flow rate value, respectively.

3.1 Time series analysis

Fig. 2 and Fig. 3 show examples of time series of speed and flow on January 1st at A550 and A551.

DOI: 10.1051/matecconf/201710308009

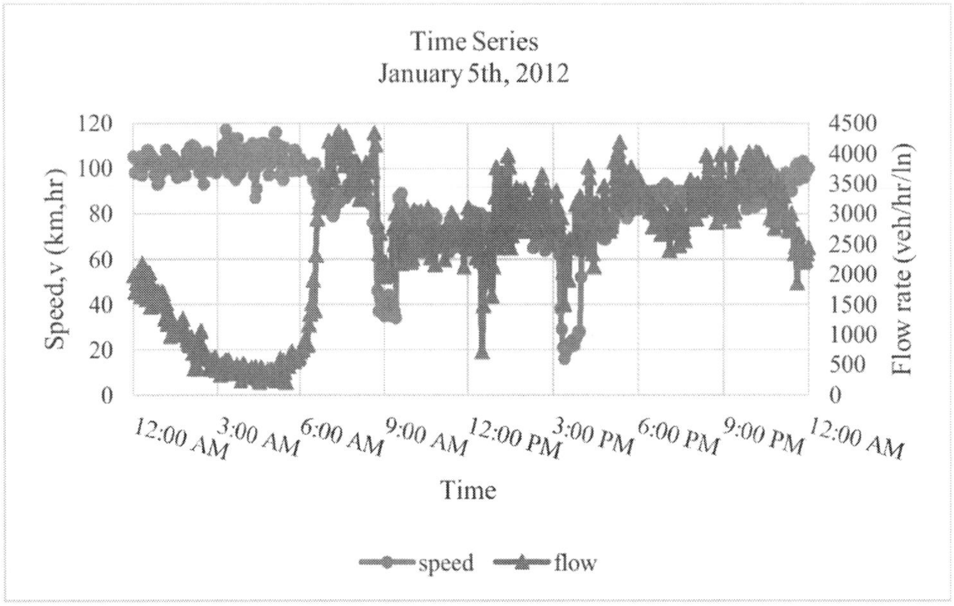

Fig. 2. Time series of speed and flow on January 1st 2012 at A550

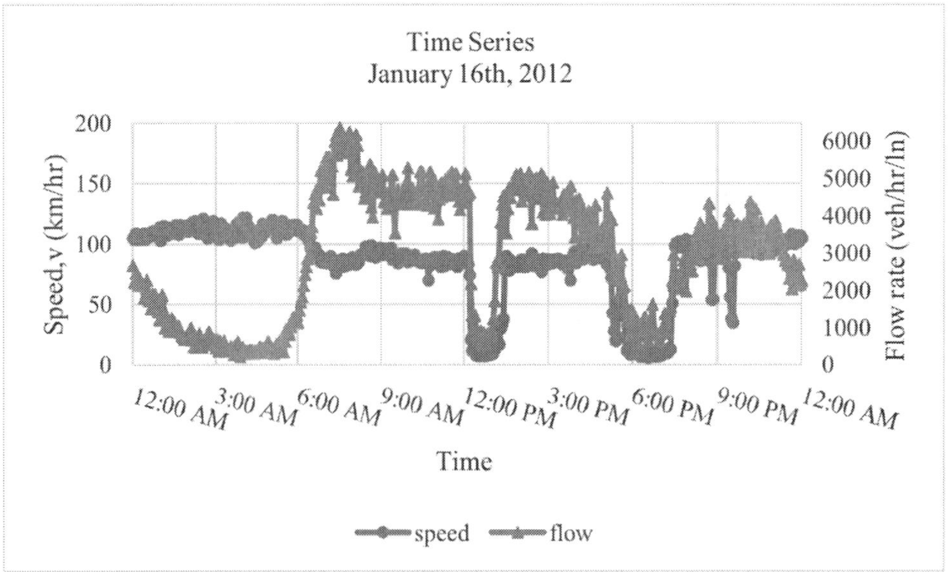

Fig. 3. Time series of speed and flow on January 1st 2012 at A551

Referring to Fig. 2, flow rate of A550 was started low at 40 to 60 veh/hr at 12.00 AM and kept on decreasing before suddenly increasing at 5.00 AM up to 4500 veh/hr. The flow rate was then kept on highly fluctuating until 9.00 AM. However, at this morning rush hour, the speed of vehicles was start to reduce suddenly at 9.00 AM. This may due to long stretch of basic segment. The speed of vehicles was again reduce tremendously after 3.00 PM. This may due to industrial employees finish and change shift around this time and start to travel home.

Referring to Fig. 3, several rush hours can be noticed however speeds were inaccurately reflected to the flow rate pattern. This condition may due to unexpected event on study locations that was not recorded by TMC.

3.2 Speed-flow model

Fig. 4 and Fig. 5 show the speed-flow relationship of AID 550 and AID 551 throughout the entire month of January 2012, respectively.

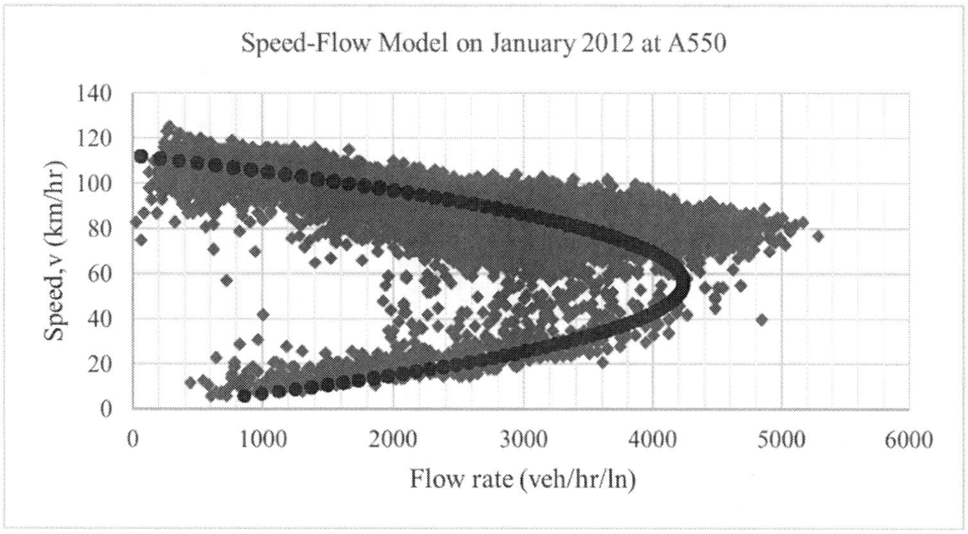

Fig. 4. Speed-flow relationship of AID 550 throughout the entire month of January 2012

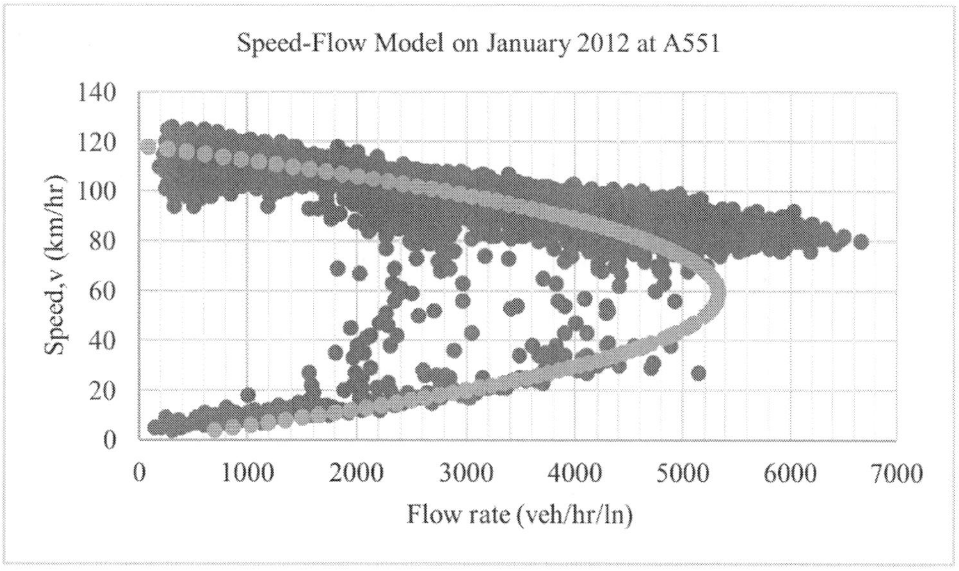

Fig. 5. Speed-flow relationship of AID 551 throughout the entire month of January 2012

Referring to Fig. 4 and Fig. 5, the estimated maximum flow for AID 550 location is 4200 veh/km/ln and the critical speed is 55 km/hr while for AID 551 location, the maximum flow is 5320 veh/hr/ln and the critical speed is 60 km/hr. The average curves to estimate maximum flow were developed using Greenshields approach [3].

4 Conclusion

The objectives of the study were successfully achieved and can be concluded that theoretically, during off-peak hour, the speed-flow relationship will only appear as a normal flow, represented by half of speed-flow curve. The forced flow will only happen during heavy congested while driver try to utilize all space on road until the volume exceeded capacity. At basic segment, normal flow may likely occur at all time unless in the presence of traffic accident occurrence or maintenance works at one of the lanes. By looking at the speed – flow diagram in one month on both AID 550 and AID 551 the maximum flows obtained were 4200 and 5320 veh/hr/ln and 55 and 60 km/hr, respectively. The estimated maximum flow values are considerably high if compare to standard capacity of highway which is 2400 veh/hr/ln.

References

[1] T.V. Mathew, *Freeway Operations, Transportation Systems Engineering*, IIT Bombay, (2014)
[2] M. Lorenz, L. Elefteriadou, *A Probabilistic Approach to Defining Freeway Capacity and Breakdown*, The Pennsylvania Transportation Institute, (2000)
[3] B.D. Greenshields, *A Study of Traffic Capacity*, Highway Research Board, (1935)

The Study of Relationship Between Pedestrian and Safety based on the Theory of Planned Behaviour at Batu Pahat, Johor

M. A. M. Bilema[1], *M. M.* Haurula[1], and *R.* Rahman[1,*]

[1]Department of Highway and Traffic Engineering, Faculty of Civil and Environmental Engineering, Universiti Tun Hussein Onn Malaysia, 86400 Parit Raja, Johor, Malaysia

Abstract. Pedestrian are among those who were contributed to the increase of road accidents and identified as vulnerable rod user in the country. Various aspects and factors that lead to the accident among the pedestrian and other road users are needed to be studied, so that the safety all road user should be improved. This study was conducted to identify the characteristics of user demographic in terms of age, gender, employment and others. In addition, the characteristics and the perception of pedestrians were also evaluated in order to achieve the objectives of this study. The observational study was carried out to the users on the usage of pedestrian facilities. The study was focused in the city area of Batu Pahat. The type of pedestrian facilities included pedestrian walkaways, pedestrian bridges and zebra crossings. A total of 200 sets of survey questions were distributed among respondents and analyzed using Statistical Package for Social Science (SPSS) based on the behavioral model called the multiple regression analysis. The correlation analysis shows that attitudes have a significant effect on the safety of pedestrians and other factors do not show significant results. However, the two-step regression analysis demonstrates that the intention has a strong relationship with behavior on pedestrian safety. In order to enhance the government efforts to mitigate road accidents, it is efficient to use an education as a medium which helps to change the attitude of road users in the country.

1 Introduction

Pedestrian is one of the road users that contributes about second highest in the case of road accidents in the world. In 1999, 119,000 of pedestrians were involved in road accidents while 6552 from them was died. In Asian country included Hong Kong and Korea, it is about 70% from road accidents involved the pedestrians and 50% was recorded died. The

* Corresponding author: raha@uthm.edu.my

total of road accidents involved pedestrian recorded in Middle-East country show increasing 30% every year [1].

Most of the pedestrians that involved in the road accidents include children, elderly and disabled peoples. Based on the report of World Healthy Organization (WHO), the groups of elderly citizens are more risks to danger compared with the younger citizens as the road users. The probabilities of pedestrians at the age of 65 years old and above will be killed is 71% compared with the pedestrians age in the range of 15 to 35 years old which is only 20% [2].

In Malaysia, the case of road accidents involved pedestrians is always increasing at every year as much as the increasing the number of vehicles. Besides that, the case of road accidents involved pedestrians is recorded higher at the city area compared with rural area where the death rate was recorded about 562 cases per year [3]. One of the reasons is the facilities provided for the pedestrians in Malaysia still not fulfill the safety requirement besides the behavior of road users itself [4]. Thus, good condition of pedestrian facilities and good behavior of road users may help to increase the safety for both.

2 Literature review

Discuss the previous study that related to the relationship between the pedestrians behavior towards safety and other factors influenced. The main topic that will be discussed is about the psychological and behavior, definition of pedestrian, safety of pedestrian, facilities of pedestrian and Theory of Planned Behavior.

2.1 Psychological and behavior

Psychology known as the set of academic, clinical and industrial that related with the explanation and prediction of behavior, thinking, emotion, motivation, personal relation, ability and pathology [5]. The behavior and changes behavior of pedestrian are one of the main aspects in the safety of pedestrian itself. The concept of human behavior can be stated as the one of the idea that related to the emotion with influencing by one action class to another certain situation class [6]. It is the reaction depends to the situation. Moreover, behavior also known as the endurance that will be changed and learned without existed.

2.2 Pedestrian

A pedestrian is a person traveling on foot, whether walking or running. In some communities, those traveling using tiny wheels such as roller skates, skateboards, and scooters, as well as wheelchair users [7]. Pedestrian also known as the most high risk users compared with the other class of road users. This is because, pedestrian is more exposed to danger when using the road.

2.3 Safety of pedestrian

In Malaysia, there are some organization involved to improve the safety of pedestrian which is IKRAM, JKJR and MIROS. This is because, the amount of road accident including the pedestrian is recorded higher in the urban area compared with the sub-urban area even though the rate of safety is already upgraded. According to the previous study, there are some important points to suggest the facilities for safety of pedestrians included flow of vehicles and pedestrians, average of crossing time (gap and acceptance of pedestrian) and previous accident report at the suggestion crossing area [8].

2.4 Facilities of pedestrian

Pedestrian crossing system is the most important facilities needed in the area such as school, shopping complex, offices, residential and industrial area. This system needed is to reduce the traffic conflict between vehicles and safety for pedestrian itself. There are some system for pedestrian crossing practiced in Malaysia that included Pelican Crossing, Overpass Crossing and Zebra Crossing [9].

2.5 Theory of Planned Behavior

It is one of the theories about relationship between the believing and behavior. The behavior is the most important factors to influence the person to take the action either positive action or negative action [10]. Individual evaluation towards the action taken will be produced two types of believing which is favorable and un-favorable. If a person found out that the action taken give the good feedback, he/she will going through for it but if it is not, he/she will not further the action should be taken. Regarding to the previous study, it was stated that the human behavior give affects the person's intention in establishing a pedestrian wants to cross the road safely or harmful according to the particular circumstances even though the good facility for pedestrians is provided [11].

3 Methods

This study is using the Theory of Planned Behavior as the model to determine the behavior of pedestrian towards to the road safety. The components involved as follows:

3.1 Types of study

There are two types of study involved which is field work and library research. Field work included to find the information, overview and data surveyed by using survey form provided from the theory of planned behavior (TPB) as sown in Fig. 1. This form will be distributed to the respondent to collect the data according to the pedestrian behavior. Meanwhile, library research involved the activities to find the information from the journal, previous research, thesis, references books, newspaper and internet. All the sources are to perform the objectives of this study.

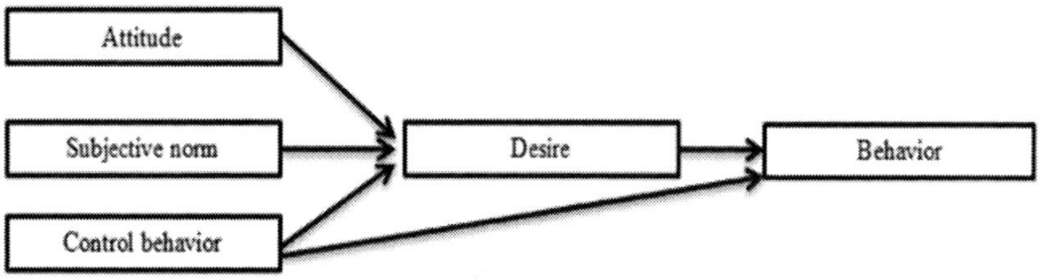

Fig. 1. Model of planned behavior towards behavior with pedestrian safety [10]

3.2 Design of study

To achieve the objectives of this study, there are some steps should be followed in the design of study which is included identify objectives and research problems, data collection, data analysis, final results, conclusion and recommendation of research.

3.3 Instrument of study

The instrument of study used is questionnaire. A set of survey questions are provided to achieve the objectives of study. Questionnaires will be distributed to the respondents who are pedestrians in Batu Pahat where a concentration involves a lot of walking activity among road users. The questionnaire consists of three parts, namely Part A, Part B and Part C.

Part A related to respondents' personal information and demographic data such as your gender, race, age, state of life, occupation, level of education and many others. Part B is involves a pedestrian destination, distance, time, walking, etc. part C is related to the perception of pedestrians where the questions are based on the variables and factors to be taken into account in the theory of planned behavior.

3.4 Scale of measurement

The scale of measurement in this study will show the level of agreement from the pedestrian based on the questions provided by survey form. The scale started with 1 to 5 which is totally disagreeing to totally agree. Besides that, the size of samples is collected in randomly. In this study, the population of drivers in Batu Pahat is estimated about 10,000 peoples. However, there is no accurate information to state the total number of drivers in Batu Pahat. Thus, this study used a 200 of sample size to be distributed to the respondents in the study area by refering from previous studies.

3.5 Area of study

The site selection factor is very important to collect the data. The area chosen for this study are in Batu Pahat where the focus area in the shopping center, school or residential area. This is because the area was involved a lot of pedestrian activities and movements besides facilities provided such as zebra crossing.

3.6 Data analysis

Method of data analysis is the most important to interpret the data collection to become reliable and used conclusion. Thus, statistic method is used by the researcher to understand and describe the relationships between variables in research. There are some analysis involved in this activity which is descriptive analysis, minimum score, trustworthiness analysis and correlation analysis that shown in Table 1 until Table 4 while regression analysis follows the Equation (1). Those analysis works should be analyzed with properly and used the right method to overcome the strong results from this study.

Equation (1) shows the basic formula for the regression:

$$\mathrm{Y} = a + \sum bi\ Xi \tag{1}$$

where Y is score predictor on the dependent variable, a is refer to Y when $X = 0$, bi is the regression coefficient for the predictor I and Xi is scores on predictor i (independent variables).

Table 1. Difference between descriptive statistics and inferences [13]

Statistic	Purpose	Statistic
Descriptive	- Describes the characteristics of variables - Used for drawing the conclusions about the numerical data / numbered - Not too common from the sample to the population	Frequency, mean, mode, median, range, standard deviation, variance, proportion, percentage, percentage change, rates, ratios, normal distribution etc.
Inferences	- Describe the relationship between variables - Describe the characteristics of a selected sample of the population - The characteristics of the sample population in general	t-test, ANOVA, Khi Square etc.

Table 2. Interpretation and measurement of tendency level based on mean scores [14]

Mean score	Interpretation	Tendency level
1.00 – 2.33	Totally disagree	Low
2.34 – 3.66	Agree	Medium
3.67 – 5.00	Totally agree	High

Table 3. Level measurement of alpha Cronbach (α) [13]

Value of Alpha Cronbach (α)	Level measurement
>0.9	Excellent
>0.8	Good
>0.7	Accepted
>0.6	Questioned
>0.5	Incomplete
>0.4	Rejected

Table 4. Differences in the interpretation of the coefficients and the strength of the relationship Davis and Guilford [15]

Davis		Guilford	
Coefficient	**The strength of the relationship**	**Coefficient**	**The strength of the relationship**
Above 0.70	Very strong	0.90 – 1.00	Very strong
0.50 – 0.69	Strong	0.70 – 0.90	Strong
0.30 – 0.49	Moderate strong	0.40 – 0.70	Moderate
0.10 – 0.29	Weak/not strong	0.20 – 0.40	Weak
0.01 – 0.09	Can be ignored	Below 0.20	Very weak

4 Results and discussions

This part will describe the results of data collection through questionnaires that were distributed to 200 respondents in the area of Batu Pahat, Johor. The analysis involved as follows:

4.1 Demographic of respondents

Table 5 shows the information of respondent demographic regarding to the questionnaire made in this study.

Table 5. Information of respondent demographic

Item	Frequency	Percentage (%)
Gender		
- Male	120	60
- Female	80	40
Age		
- Below 20	60	30
- 20 – 25	100	50
- 26 – 30	20	10
- Above 30	20	10
Type of occupation		
- Government	20	10
- Private	60	30
- Self-employed	40	20
- Student	80	40
Level of education		
- Primary school	3	1.5
- Secondary school	59	29.5
- Training institution	72	36
- IPT	66	33
Experience in using facility?		
- None	20	10
- Sometimes	60	30
- Mostly used	40	20
- Always	40	20
Experience involving in road accident?		
- Yes	40	20
- No	160	80

4.2 Characteristic of pedestrians

Table 6 shows the information of pedestrian characteristic based on their purpose and need regarding to this study. Respondents who intended to work, social activities and personal matter shows the percentage of 10%.

Table 6. Information of pedestrian characteristic

Item	Frequency	Percentage (%)
Purposes of pedestrian		
- Work	20	10
- Social activity	20	10
- Personal matter	20	10
- Shopping	80	40
- School/education	60	30
Distance (meter)		
- Below 100	40	20
- 100 – 200	80	40
- 200 – 300	20	10
- 300 – 350	20	10
- Above 350	40	20
Time (minute)		
- Below 10	80	40
- 10 – 20	80	40
- 21 – 30	20	10
- 31- 45	10	5
- Above 45	10	5
Alternative transportation		
- Public	40	20
- Private car	70	35
- Motorcycle	80	40
- Bicycle	10	5

While who would like to go to store or shopping center shows 40% and 30% to the school or educational institution. According to the distance taken by pedestrians, it is shows that 20% respondents are walking less than 100m and more than 350m. Meanwhile, who is walking about 100m to 200m shows 40%. It is only 10% respondents that walking about 200m until 300m and 300m until 350m.

For the time taken by pedestrians to go to the destination, it is about 40% from the respondents achieved less than 10 minutes and in the range 10 minutes to 20 minutes. Then, respondents that achieved the time taken about 21 minutes to 30 minutes is about 10% while it is only 5% respondents that achieved time taken in the range 31 minutes to 45 minutes and more than 45 minutes in this survey. Regarding to the percentage of alternative transportation, it is shows that 20% from the respondents use the public transport while 30% uses private car. The highest percentages can be seen that the most respondents use the motorcycle which is about 40% while it is only 5% that use a bicycle.

4.3 Minimum score

The value of minimum score in this study was summarized in Table 7 where it is based on the theory of planned behavior regarding to the feedback from the pedestrian.

Table 7. Summarization of minimum score

Variables	No. of items	Average of min score	Results
Attitude	4	4.26	Agree
Subjective norms	5	3.17	Agree
Control behavior	4	3.94	Agree
Desire	3	3.90	Agree
Behavior	2	4.75	Agree

4.4 Analysis of trustworthiness

The level of trustworthiness is depend on the theory of alpha Cronbach where the value of alpha will produces the good results and accepted as shown in Table 8.

Table 8. The value of alpha Cronbach

Variables	No. of items	Value of alpha Cronbach	Results
Attitude	4	0.7	Accepted
Subjective norms	5	0.7	Accepted
Control behavior	4	0.6	Accepted
Desire	3	0.6	Accepted
Behavior	2	0.8	Good

4.5 Analysis of correlation

From the analysis, the value of correlation between subjective norms and attitude of pedestrian shows vice versa. It is means the subjective norm is not influencing the attitude of pedestrian according to the safety. Besides that, the control behavior and desire towards behavior shows the same result as mention before. It is can be explained in the situation when a pedestrian show a good attitude and polite when crossing the street, pedestrian will be in safe.

5 Conclusions

Based on the main objective of this study is to identify the behavior of pedestrian safety is seen to have achieved the objectives. This is because the results from the analysis show the significance of the factors involved. Attitudes, subjective norms, behavioral control and the desire to influence the behavior of pedestrians on the road such as attitude in a hurry to pass, the influence of family and close friends, the sensitivity of the surrounding state and others. Meanwhile, the objective of the study to apply the theory of planned behavior is achieved when the analysis was a relationship between the variables involved on the behavior of pedestrian safety in urban area such as Batu Pahat, Johor.

References

[1] C.A. O'Flaherty, M.H. Parkinson, Movement on city centre footway, Traffic Engineering and Control, **13**, 434 – 438 (2006)
[2] World Health Organization, *World Report on Traffic and Injury Prevention*, Geneva, (2011)
[3] Jabatan Keselamatan Jalan Raya, *Laporan Kemalangan Jalan Raya,* Kuala Lumpur, (2007)
[4] Malaysian Institute of Road Safety Research, *Laporan Kemalangan Jalan Raya 2011*, Kuala Lumpur (2011)
[5] R. Lancaster, R. Ward, Management of work and age related road safety, Entec UK Limited for The Health and Safety Executive and Scottish Executive, (2002)
[6] H.C. Triandis, E.R. Hall, R.B. Ewen, Member heterogeneity and dyadic creativity, Human Relations, **18**(1),33 – 55 (1965)
[7] Austroads, *Road Safety Audit*, Austroads Publication, Sydney, (2009)
[8] H. Davies, M.A. Winett, Why do Pedestrian Accidents Happen?, *PTRC Proceedings*, **P365**, 315 – 324 (2012)
[9] M.R. Ibrahim, M.R. Karim, F.A. Kidwai, The Effect of digital count-down display on signalized junction performance, American Journal of Applied Sciences, **5**(5), 497-482 (2008)
[10] I. Ajzen, The theory of planned behaviour, Organizational Behavior and Human Decision Processes, **50**, 179 – 211 (1991)
[11] D. Evans, P. Norman, Understanding pedestrians road crossing decisions: An application of the theory of planned behaviour, Health Education Research, **13**(4), 418 – 489 (2003)
[12] M.O. Azuan, *Analisis Tingkah Laku dan Sikap Pemandu Kenderaan Kecil Terhadap Kelajuan*, Thesis, Universiti Tun Hussein Onn Malaysia, (2012)
[13] Y.P. Chua, *Kaedah Penyelidikan – Kaedah dan Statistik Penyelidikan – Buku 1*, McGraw Hill Education, Kuala Lumpur, (2006)
[14] Norhayati, *Persepsi Pelajar Terhadap Sikap, Pengetahuan dan Kemahiran dalam Mata Pelajaran Vokasional (MPV) di Sekolah Menengah Akademik di Batu Pahat, Johor*, Master Thesis, Kolej Universiti Teknologi Tun Hussein Onn, (2005)
[15] R.D. *Yockey, SPSS Demystified. A Step-By-Step Guide to Successful Data Analysis*, Pearson Education, (2008)

Enhancing Road Safety Behaviour Using a Psychological and Spiritual Approaches

Mohd Tarmizi Ghous[1], *Kamarudin* Ambak[1,*], *Ryan White* Hamsaari[1], and *Ahmad Sharifuddin* Mustapha[2]

[1]Smart Driving Research Center, Faculty of Civil and Environmental Engineering, Universiti Tun Hussein Onn Malaysia, 86400 Batu Pahat, Johor, Malaysia
[2]Department of Islamic and Social Science Studies, Faculty of Science, Technology and Human Development, Universiti Tun Hussein Onn Malaysia, 86400 Batu Pahat, Johor, Malaysia

Abstract. Main causes of accident is due to driver itself that is influenced by their bad attitude while driving. Human attitude is closely related to the human psychology. Apart from that, spiritual aspect also influence human attitude. Hence, this study carried out to improve driver safety using a new approach through psychology and spiritual factors. Objectives of this study are to identify then analyze factors of psychological and spiritual that contribute towards safety driving. A self-administered questionnaire were distributed among 256 respondents from various type of background. An analysis descriptive statistics show demographic and experience of respondents. Chi-square analysis showed only education level and traffic summon are significant to safety driving. Furthermore, correlation analysis shows psychological factors has strong linear relationship on attitude of drivers towards safety driving while spiritual factor, the perception of the spiritual and practices, both have a strong relationship to safety driving. Regression analysis demonstrates boths psychological and spiritual factors have strong evidence and significant relationship with safety driving. Thus, it can be identified that spiritual psychological factors encourage drivers to drive more safely and reduce road accidents. Therefore, this study propose useful guidelines to related agencies in order to enhance safety among drivers to be able drive safely on the road.

1 Introduction

Accidents problem occurred in Malaysia is one of the most important agenda seriously discussed and various methods were carried out to overcome this problem. Accidents that occurred resulted loss of valuable lives, injury and property [1]. Transport system is divided into three main components which are physical environment, vehicles and road users. There have been many improvements that have been carried out to reduce accidents, especially on the physical environment and the vehicle but less improvement in road users itself. In accordance with population growth, the demand for vehicle use and transportation are also

* Corresponding author: kanardin@uthm.edu.my

increased and it was undoubtedly resulting more accidents. Behavior of road users is one of the factors contributing majorly in road accidents involving death or injury.

In a nutshell, users fail to implement some important aspects before driving such as not wearing a safety belt, breaking the speed limit, taking drugs or under the influence of alcohol and not wearing a helmet correctly [2, 3, 4]. Human behavior relates to human internal characteristic such as psychological factors. Psychology is the study of human behavior that was made systematically and scientifically. It is a discipline that studies scientifically human or animal behavior, especially thoughts, attitudes, feelings, instincts, potential, intelligence, needs and human responses to stimuli and social environment [5].

In other perspective, spiritual factors also influence human attitude and behavior. In expressing the concept of spiritual or religious from previous study, mention the impact of religious practice would indicate a person can control or push yourself , relieve stress and shows concern about life and death [6]. Moreover, driver bringing spiritual element in the road such as displaying good behavior, patient, do not rush, humble as recommended in the religion can also reduce accidents [7]. In State of Johor was recorded the highest mortality rate compared to other states for 1128 deaths. This state also contributed most accidents case and increased from the previous year's total of 1073 deaths [8].

Therefore, the need for in-depth study to improve road safety through the study of the psychological and spiritual factors approach and recommend appropriate action. This study was conducting surveys for quantitative data regarding the psychological and spiritual to guide people from different backgrounds, age also in employment among car drivers. This study focusing in determine the factors from psychological and spiritual that contributing to safety driving. Hence, data of this study was collected from driver around Batu Pahat, Johor.

2 Material and methods

2.1 Instrument and measurement

Instrument and measurement of this study used quantitative data by collected data through questionnaire from respondent which is the designation of questionnaire referred to the Driver's Behavior Questionnaire [9, 10] for psychological elements and for spiritual questionnaire design from *Religious Orientation Scale* (ROS) [11]. Questionnaire consist of demography question relating to the personal notice of respondents such as age , gender , marital status and others.This questionnaire also contents question for individual perception of psychology and spiritual using likert scale 1 to 5. Other part from questionnaire is open ended asked about respondent opinion in enhancing driving safety. Data was analyzed by using Service Package for Science Sosial (SPSS).

2.2 Sample size and location of study

Sample size was determined by referring to the number of population in Batu Pahat. Sample size was determined from [12] which is based on the statistics of population in Batu Pahat 2010 of 3,230,44 then obtained the highest number was 384. This study archive 256 of respondent around Batu Pahat which is 67% from 384 number of respondent. This study was conducted around city of Batu Pahat, Johor.

2.3 Analysis tool

For analysis method, this study uses a static descriptive analysis method for demography to determine the mean score of respondent. To determine the significant between demography to the factor of psychology and spiritual, Chi-Square analysis was used. In order to determine which factor in psychological and spiritual that contributed to the safety driving, Correlation and Multiple Regression were used.

3 Results and discussions

3.1 Descriptive analysis

Based on finding from 256 respondents, (79%) 203 are males and the rest 53 (21%) are females. Most of the respondents are by the age of 20 – 25 years old with a total of 118 (46%) respondent and followed by group age of 30 years old and above with a total of 67(26%) respondents. From 256 respondent, 71 respondents works in private sector. This finding also shows the highest number of respondents have SPM/STPM with (38%) 96 respondent and second was have Bachelor, Master or PHD with (35%) 91. Most of the respondents are Muslims and Malay race. As for the driving licences, most of them posseses competent (Class D) licenses. The demographic part also asked respondent about experiences in driving, involved in accident and traffic summons. Result shows respondent in 2- 5 years' experience is the highest with 84 (33%) and 146 (57%) respondent have no experience involved in accident while (53%) 135 were involved in traffic summons.

Table 1. Respondent Demographic

Respondents	Frequency	Percentage (%)
1. Gender		
Male	203	79
Female	53	21
2. Age		
Below 20 year	12	5
20-25 year	118	46
26-30 year	59	23
30 year above	67	26
3. Profesion		
Self-employed	38	15
Government Sector	33	13
Private Sector	71	28
Student	88	34
Others	26	10
4. Education		
Diploma	65	25
Bachelor/Master/PHD	91	35
SPM/STPM	96	38
5. Religion		
Muslim	243	95
Non – Muslim	13	5

7. Licenses		
Competent (Class D)	207	81
Probationary (Class P)	27	11
Learner (Class L)	5	2
No Licenses	17	6
8. Experience		
Less than 2 years	38	14
2-5 years	84	33
6-9 years	73	29
10 years above	61	24
9. Accident Involved		
Yes	110	43
No	146	57
10. Traffic Summon		
Yes	135	53
No	121	47

3.2 Chi-square analysis

Table 2 shows the result of chi – square analysis between factor of psychology and spiritual with background of respondent demography towards safety driving. Based on the result, significant *p* value less than 0.05 indicated to gender and subjected to the traffic summon which is this two variable for psychological was associated to safety driving. However, all variable for spiritual factors had *p* more than 0.05 shows that there had no associated to safety driving.

Table 2. Chi - Square analysis

	Psychology			Spiritual		
	X^2	P	Result	X^2	P	Result
Age and safety driving	0.023	0.880	Not Significant	2.000	0.157	Not Significant
Gender and safety driving	0.104	0.747	Significant	0.537	0.464	Not Significant
Education and safety driving	5.356	0.021	Significant	1.128	0.288	Not Significant
Experience and safety Driving	1.037	0.309	Not Significant	2.095	0.148	Not Significant
Summon and safety driving	2.758	0.097	Significant	1.127	0.288	Not Significant

3.3 Correlation analysis

Table 3 shows the result of correlation analysis between psychological factors as variable and safety driving as constant variable. Based on result, psychology factor have a weak positive linear relationship with value r = 0.24 while variable for driver's behavior have a moderate negative linear relationship with r = - 0.274 meaning that driver's behavior have significant relationship with safety driving. For other factors variable obtained strong positive linear relationship and significant *p* value to safety driving. While in spiritual factors, all of independent variable have strong positive linear relationship to dependent variable safety driving as shown in Table 4.

Table 3. Correlation for Psychological factors towards safety driving

	Safety driving	**Psychology factors**	**Drivers behavior**	**Other factors**
Safety driving	1			
Psychology factors	0.243*	1		
Drivers attitude	-0.274**	-0.612 **	1	
Other factors	0.479**	-0.002	-0.549**	1

** Correlation is significant at the 0.01level (2-tailed)*
*** Correlation is significant at the 0.05level (2-tailed)*

Table 4. Correlation for Spiritual factors towards safety driving

	Safety Driving	**Driver's Perception**	**Driver's Practices**
Safety Driving	1		
Driver's Perception	0.542**	1	
Driver's Practices	0.537**	0.345**	1

** Correlation is significant at the 0.01level (2-tailed)*
*** Correlation is significant at the 0.05level (2-tailed)*

3.4 Multiple regression analysis

Table 5 shows the summarize results of multiple regression psychological factors were performed using SPSS which is R squared of 32 % of variables contribute towards driving safety. While Table 6 shows the results of multiple regression spiritual factors were R squared of 43 % of variables contribute towards driving safety.

Table 5. Summarize R square for psychological factors

Model	R	R Square	Adjusted R Square	Std error of the estimate
1	0.563[a]	0.317[a]	0.307	0.5676286

Table 6. Summarize R square for spiritual factors

Model	R	R Square	Adjusted R Square	Std error of the estimate
1	0.658[a]	0.433	0.428	0.5231901

Table 7 shows that the summary result from variance analysis (ANNOVA) for safety driving as dependent variable to independent variable in psychology and spiritual factors. Based on the result, value of $p = 0.000$ obtained shows that there is a significant relationship between safety driving and factor from psychological and spiritual. From this result shows that the psychological and spiritual aspect have a strong impact to drivers towards safety driving.

Table 7. Summary of ANNOVA analysis

	Model	Sum of Squares	Df	Mean Squares	f	Sig
Psychology	Regression	31.005	3	10.352	32.128	0.000
	Residual	67.018	252	0.322		
	Total	98.073	255			
Spiritual	Regression	43.068	2	21.534	78.670	0.000
	Residual	56.388	253	0.274		
	Total	99.456	255			

From Table 8 shows the summary result of coefficient multiple regression also where all of factors obtained $p < 0.05$ value which is all of factor from psychology and spiritual have strong relationship and affect drivers towards safety driving as dependent variable. Value of ß obtained from table also shows the factors was affected to another factor and support objectives of this study to determine the factors from psychology and spiritual that contribute to safety driving.

Table 8. Summary of Coefficient Multiple Regression for Psychological and Spiritual factors

	Model	Unstandard Coefficient		Standard Coefficient	t	sig
		B	Std. Error	Beta		
Psychology	Safety Driving	3.962	0.398	-	9.946	0.000
	Psychology factors	-0.291	0.057	-0.373	-5.142	0.000
	Drivers behavior	-0.186	0.067	-0.203	-2.799	0.006
	Other factors	0.342	0.043	0.458	7.969	0.000
Spiritual	Safety Driving	1.535	0.227	-	6.758	0.000
	Driver's Perception	0.446	0.062	0.405	7.247	0.000
	Driver's Practices	0.298	0.042	0.397	7.112	0.000

4 Conclusion

In conclusion, the driving safety is closely associated with the driver itself and then it is the same to previous studies which indicates that the driver itself is the main cause in contributing to accident occurred and it corresponds with the study carried out by [13], which stated that human behavior itself nowadays contributes to accident with a high percentage of almost 80%. Based on the findings, it can be concluded that the needs of education and learning on the psychological and spiritual and same as [14], human behavior is not something that is available since birth but must be learned, honed and developed through experience. Since these factors have strong evidence which is indicating a strong relationship as a factor to drive safely thereby avoiding accidents where it would lead to so much loss to many parties, especially to oneself if involving injury or death.

The authors would like to thanks Universiti Tun Hussein Onn Malaysia (UTHM) for funding this study under the Multidisciplinary Research Grant (MDR) Vot U090 and Smart Driving Research Center (SDRC), Faculty of Civil and Environmental Engineering for support and providing facilities to accomplish the study.

References

[1] T. Rothengatter, T. Elander, Psychological aspects of road user behaviour, Applied Psychology, **2**, 8-9 (1997)
[2] International Federation of Red Cross and Red Crescent, *Practical Guide on Road Safety*, Geneva, (2007)
[3] K. Ambak, I. Rozmi, R. Atiq, S. Foad, Do Malaysian motorcyclists concern to safety helmet usage? A cross-sectional survey, Journal of Applied Sciences, **11**(3), 555-560 (2011)
[4] M.B. Ahmed, K. Ambak, A.R.A. Ghani, N.S. Sukor, Helmet usage among adolescents in rural road from the extended theory of planned behaviour, Journal of Applied Sciences **13**(1),161-166 (2013)
[5] A. Yahaya, *Psikologi Sosial Alam Remaja*, PTS Profesional Publishing, Selangor, (2005)
[6] M. Argyle, B.Beit-Hallahmi, *The Psychology of Religious Behavior, Belief and Experience*, Routledge, London, (1997)
[7] A. Abdullah, *Tangani Kemalangan Jalan Raya Melalui Pendekatan Syariat Islam*, Pemikir No. 44, Utusan Karya, 125 139 (April-June 2006)
[8] Polis Diraja Malaysia, *Road Accidents Statistical Reports Malaysia 2012*, Cawangan Bukit Aman, Kuala Lumpur, (2012)
[9] J. Reason, A. Manstead, S. Stradling, J. Baxter, K. Campbell, Errors and violations on the roads: A real distinction?. Ergonomics, **33**(10-11), 1315-1332 (1990)
[10] N. Rosli, K. Ambak, B.D. Danial, J. Prasetijo, Penerimaan sistem penguatkuasaan automatik (AES) Dalam kalangan pemandu menggunakan model penerimaan teknologi, Jurnal Teknologi, **75**(1),1-6 (2015)
[11] G.W. Allport, J.M. Ross, Personal religious orientation and prejudice, Journal of Personality and Social Psychology, **5**, 432-443 (1967)
[12] R.V. Krejcie, D.W. Morgan, Determining sample size for research activities, Educational and Psychological Measurement, **30**, 607-610 (1970)
[13] R. Lancaster, R. Ward, The *Contribution Of Individual Factors To Driving Behavior: Implications For Managing Work-Related Road Safety* ,Entec UK Limited, (2002)
[14] A.K. Lim, *Hubungan Sikap Pengguna dengan Keterlibatan dalam Kemalangan Jalan Raya. Kajian Kes di Negeri Melaka*, Master Thesis, Universiti Sains Malaysia, Penang, (2007).

Road Fatality Model Based on Over-Dispersion Data Along Federal Route F0050

Wan Zahidah Musa[1,*], *Joewono* Prasetijo[1], and *Zaffan Farhana* Zainal[1]

[1] Smart Driving Research Center (SDRC), Faculty of Civil and Environmental Engineering, Universiti Tun Hussein Onn Malaysia, 86400 Batu Pahat, Johor, Malaysia

Abstract. According to The World Health Ranking 2011 has ranked Malaysia as 20th in its list of countries with the most deaths caused by road accidents. Road accidents also have been identified as the prime cause of death in Malaysia after the heart disease, stroke, influenza and pneumonia. To date, previous researches from Malaysian Institute of Road Safety (MIROS) have reported that averages of 18 people were killed on Malaysian road daily. There are many kinds of models that have been developed in modelling the circumstance of accidents. The most widely applied was Poisson and Negative Binomial regression models while Zero-inflated Poisson and Zero-inflated Negative Binomial are the modification of Poisson and Negative Binomials regression models. This study interested to focus on road F0050 as statistic data from Royal Malaysian Police 2014 list F0050 as one of the high accident road in Malaysia from kilometre 0 until kilometre 58. R programming was chose to analyse the relationship between road fatality and its factor (annual average daily traffic (AADT), speed, shoulder width, lane width). Negative binomial and Zero-inflated negative binomial (ZINB) were shown to be preferred modelling methods for this study. Significant positive relationships were also identified between road fatality and annual average daily traffic (AADT) and lane width. This relationship can be a helpful support to the decision making of accident management for road F0050.

1 Introduction

Traffic accidents have been recognized as one of the major causes for human and economic losses both in developed and developing countries. In 2014, Malaysia launched a new road safety plan, aligned with the Global Decade of Action for Road Safety which aims at having the predicted number of fatalities by 2020 [1]. Most recent safety data from Royal Malaysian Police found that there were 6674 road deaths in 2014 which were 3.5% less compared to 2013. It has been conclusively shown that this is the largest reduction in the number of road fatalities in the past 10 years [2]. Traffic accidents represent a significant cost for the Malaysian society which based on a willingness-to-pay estimation, traffic crashes cost each year around 1.6% of the national gross domestic product. In a following-

* Corresponding author: wanzahidahmusa@yahoo.com

 DOI: 10.1051/matecconf/201710308012

up study, [3, 4] found that the statistical value of life in Malaysia is RM 1.3 million, around EUR 330 000.

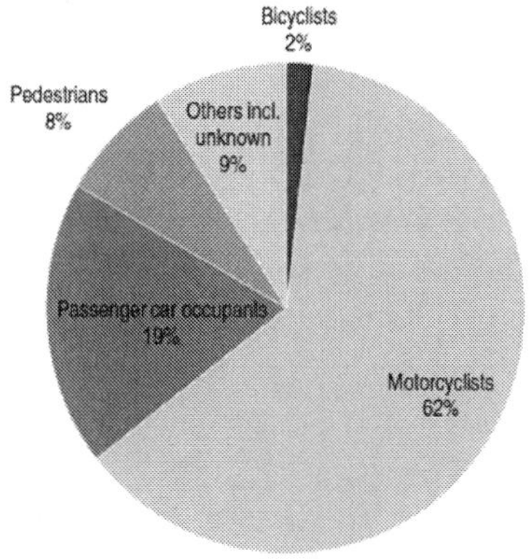

Fig. 1. Road fatalities figures by road user group (total fatalities) 2014 [ITF, 2015].

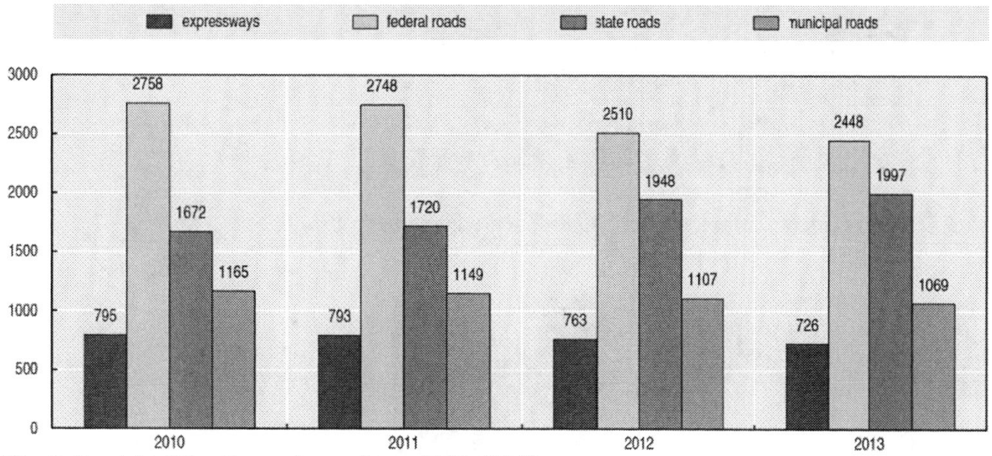

Fig. 2. Road fatalities figures by road type [ITF, 2015]

Zero-Inflated models has a pivotal role in predicting probability of crashes and identify the most relevant contributory factors at the hazardous sections as this models can handle an over-dispersion data and high excess zero properly [5-15]. Malaysia has it owns plan and target that set by the Global Decade of Action for Road Safety 2011-2020 to reduce the predicted number of road deaths in 2020 by 50%. MIROS present a comprehensive review on the absence of a comprehensive road safety programme, there would be 8760 road fatalities in 2015 and 10716 in the year 2020 based on their 2015 annual report. This shows an effective reduction of 22% in the number of road fatalities when compared to 2010 level.

Table 1. Fatalities target [ITF.2015]

	2015	**2020**
Predicted number of fatalities (business as usual)	8760	10716
Fatalities targets with the implementation of the plan	6570	5358

Over dispersion in Poisson models occurs when the response variance is greater than the mean [16]. The main issue in need of future effort is the reliability test of some statistical models (Poisson, Negative Binomial and Zero Inflated) in yielding better results in traffic safety. Although these types of model have been well coordinated in many other fields such as economics and sociology, it is only less than 10 years when traffic safety analysts first reported relevant results in well-known publications. Especially, while most applications are limited to types of vehicles, types of crashes, varying-intercept model, and varying-slope model deserves more effort because of the complexity in model estimation and result interpretation. Therefore, this study is the new approach of quantifying road safety performance at the heterogeneous traffic conditions at route F0050.

2 Method

2.1 Accidents

Fig. 3. Road accidents in front of traffic light UTHM at kilometre 19

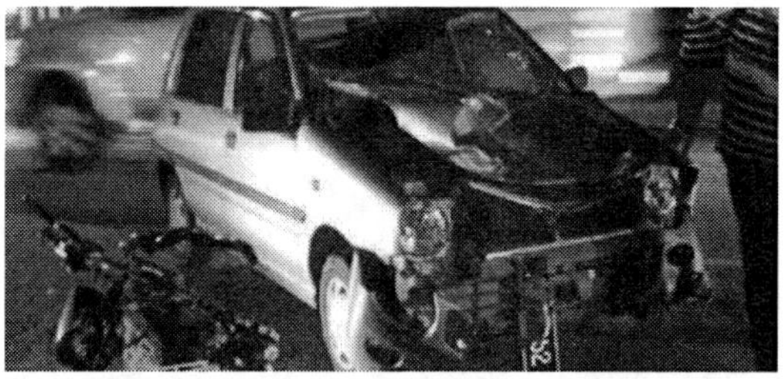

Fig. 4. Road accidents in front of shell gas station: Parit Raja at kilometre 12

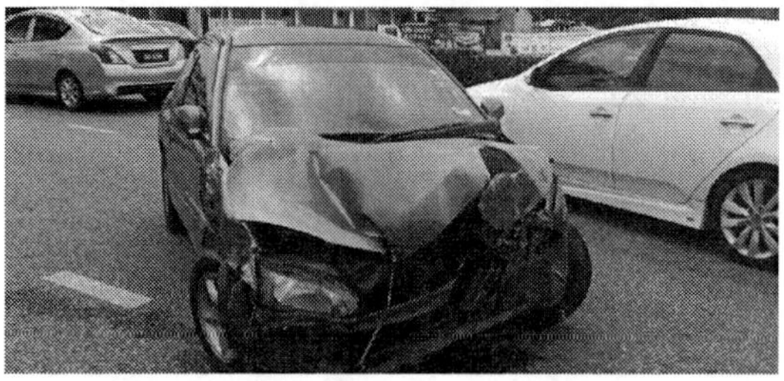

Fig. 5. Road accidents in front of Batu Pahat Mall at kilometre 4

Up to now, there are three types of road accident that applied in Malaysia which was fatal, serious injured and slight/minor injuries. Road fatality was defined as death resulting from a road crash within 30 days after the crash. While serious injured is when a person injured as a result of a road crash. Lastly, a slight injury is any injury that does not fit under death or serious injury. This paper attempts to show an interest only to road fatality and investigates the relationship with the factors that determine the fatality.

2.2 Study Location: Federal Route F0050

2.2.1 Route F0050: Batu Pahat-Ayer Hitam-Kluang

Fig. 6. Hotspot accident locations at kilometre 23

According to [2], F0050 was listed among the highest accident road in Malaysia. As shown in Figure 6 and Figure 7, there are the hotspots locations of accident happen. F0050 is a four lane two ways divided road that runs from Batu Pahat through Ayer Hitam to Kluang. Most of the hazardous locations have high access points and annual average daily traffic(AADT).

Fig. 7. Hotspots location of accident at kilometre 22

2.3 Stochastic Regression Models

2.3.1Poisson, Negative Binomial, Zero-inflated Poisson, and Zero-inflated Negative Binomial

The data of crashes was under characteristic of discrete, sporadic and random. Stochastic regression models were applied to describe the occurrence of accidents instead of deterministic models [11]. The most widely applied was Poisson and Negative Binomial regression models while Zero-inflated Poisson and Zero-inflated Negative Binomial are the modification of Poisson and Negative Binomials regression models. They were proposed to handle over dispersion and excess zeros in the data set.

DOI: 10.1051/matecconf/201710308012

3 Results and discussion

Table 2. Variables in the final model

Independent Variable	Location	Dependent Variable
LN AADT, SP, SW, LW	F0050	Fatal/Yr/Km

Table 3. Results of Pr(>|z|) modeling accidents for model

	P2	NB2	ZIP2	ZINB2
LN.AADT	0.842	0.963	0.686	0.961
SP	0.116	0.240	0.097	0.223
SW	0.244	0.319	0.257	0.320
LW	0.704	0.765	0.461	0.768

Signif. codes: 0 '***' 0.001 '**' 0.01 '*' 0.05 '.' 0.1 ' ' 1

Table 4. Summary statistics of model

Variable	Variance	Mean	S.D	Min	Max
Fatal /Yr/Km	5.8135	3.254	2.4111	0	10
LN AADT	-	9.469	0.4611	7.91	9.81
SP	-	80.15	24.5998	36.12	140.87
SW	-	1.475	0.5409	0.50	2.80
LW	-	3.521	0.4211	2.50	3.75

Table 5. Correlation summary F0050 for model

Variable	F	LN.AADT	SP	SW	LW
F	1				
LN.AADT	-0.003	1			
SP	-0.148	0.176	1		
SW	-0.113	-0.085	-0.041	1	
LW	-0.022	-0.088	0.429	-0.103	1

Table 6. Over-dispersion test for model

Location	Dispersion Test (Poisson)
F0050	Over dispersion test data: poisson z = 2.5283, p-value = 0.005731 alternative hypothesis: true dispersion is greater than 1 sample estimates: dispersion 1.72433

Table 7. Estimation results for model

	P2	NB2	ZIP2	ZINB2
LN.AADT	0.032	0.010	-0.073	0.010
SP	-0.005	-0.005	-0.005	-0.005
SW	-0.167	-0.185	-0.166	-0.185
LW	0.074	0.077	0.147	0.078

Table 8. Results of model evaluation and comparison model

	P	NB	ZIP	ZINB
LL	-133.57	-127.69	-131	-127.7
K	4	4	4	4
N	58	58	58	58
AIC	275.14	267.39	270	263.4
BIC	274.19	262.43	269.05	262.45
Theta	-	4.46	-	4.464

Table 9. Comparison for best performance for model

	P	NB	ZIP	ZINB
AIC	-	**	*	***
BIC	-	***	*	**
Vuong	-	***	*	**
LL	-	***	*	**

Criteria: '***' (Best & Fit), '**' (Fit), '*'(Average), '-'(Low)

Stochastic regression model was used to analysis all of the collected data which was Poisson model, Negative Binomial model, Zero-inflated Poisson and Zero-inflated Negative Binomial. Table 7 above illustrates the estimation results of each model that was analyzed using R programming. In order to see the performance of the model the estimation value of variables has been examined. Models with logical algebraic signs of the variables were selected. As shown in Table 7 provides the summary of how the dependent variable associates with independent variables. Interestingly as can be seen, mostly from all model that the variables speed (SP) and shoulder width (SW) have negative association with fatality crashes. While for the variables AADT (LN.AADT) and lane width (LW) have majorly positive association with the accident. Taken together, these results suggest that there is preliminary evidence that in route F0050 with too many access points with increase of average annual daily traffic (AADT) have more contribution to fatality crashes. This proved by the most hazardous sections along this route was at main point by local people mostly around business and residential area such at kilometre 2,4,7,19,22,23,48,57 and 58. High AADT leads to lower speeds and trigger driver to take over the front car and this increase the accidents in federal route F0050. It is interesting to note that in most study found that lane width (LW) is negative association with fatality but incompatible with result found in Table 7. Lane width (LW) increases the likelihood of fatality in route F0050, although when lane width (LW) of lanes increases the traffic flows gets more convenient but the traffic conflicts and driving maneuvers such as lane changing increase and hence increases the probability of accident happened. As comparison of the shoulder width, a negative sign of shoulder width suggests that more is the shoulder width less will be the accident and it is logically acceptable.

Table 8 presents the results of model evaluation and comparison between the models. The best log-likehood has the maximum value. In Table 8, negative binomial (NB) has the higher value which was -127.69 followed by zero-inflated negative binomial (ZINB) (-127.7), zero-inflated Poisson (ZIP) (-131) and Poisson (P) (-133.57). While for the akaike information criteria (AIC), the lesser the fittest. Zero-inflated negative binomial (ZINB) has the least akaike information criteria (AIC) value (263.4); followed by negative binomial (NB) (267.39), zero-inflated Poisson (ZIP) (270) and Poisson (P) (275.14). Comparisons of best performance of model are set out in Table 9. Data from this table can be compared with the results from Table 8 which shows the comparison to choose the best and fittest performance of model. What is interesting in this data that negative binomial (NB) has the most points of criteria for the best and fittest model followed by zero-inflated negative binomial (ZINB), zero-inflated Poisson (ZIP) and lastly Poisson (P). Although negative binomial (NB), has the most points, the different value of NB performance and ZINB performance in this model just 0.01. Interestingly, these summarised that for this data was being under over-dispersion but not have high excess zero. In the final part for these studies both NB and ZINB model was choose as the best model.

4 Conclusions

This study set out to determine a model that can properly handle the highly stochastic nature of crash event and finding the most relevant contributory factors at the hazardous road sections at Johor federal route. In general, Zero-inflated regression and negative binomial was found to determine an empirical association between fatality road accidents, route geometric and traffic parameters. The results suggest that valuable insight into the underlying association between probability factors and vehicle crash.

The author would like to thanks the Ministry of Education Malaysia in funding this study under the Fundamental Research Grant Scheme (FRGS) Vot. 1452. Further thanks are also for Smart Driving Research Center (SDRC), Faculty of Civil and Environmental Engineering, Universiti Tun Hussein Onn Malaysia (UTHM) for support and providing facilities to accomplish the study.

References

[1] OECD/ITF, *Road Safety Annual Report 2015*, OECD Publishing, Paris, (2015)
[2] RMP, *Statistical Report on Road Accidents in Malaysia*, (2014)
[3] G. Nor, M. Fauzi, Value of life of Malaysian motorists : Estimates from a nationwide survey, Journal of the Eastern Asia Society For Transportation Studies, **5**, (2003)
[4] M. Yusoff, Malaysian value of fatal and non-fatal injury due to road accident : The willingness to pay using conjoint analysis study, *Proceedings of the Eastern Asia Society for Transportation Studies*, **8**, (2011)
[5] C. Pudprommarat, M. Khamkong, P. Bookkamana, Zero-inflated poisson regression in road accidents on major road in the North of Thailand, *IRCNSA Proceedings*, 323-330 (2005)
[6] A.H. Lee, M.R. Stevenson, K. Wang, K.W. Yau, Modeling young driver motor vehicles crashes: Data with extra zeros, Accident Analysis and Prevention, **34**, 515-521 (2002)
[7] S.P. Miaou, The relationship between truck accidents and geometric design of road sections: Poisson versus negative binomial regressions, Accident Analysis and Prevention **26** , 471-482 (1994)
[8] E. Ayati, E. Abbasi, Modelling accidents on mashad urban highways, Open Journal of Safety Science and Technology, **4**, 22-35 (2014)
[9] M. Asrul, N.N. Naing, Mortality in hospitalized human immunodefiency virus (HIV) and tuberculosis (TB) infection patients: Zero-inflated negative binomial death rate (ZINBDR) models, Applied Mathematical Sciences, **7**, 37-47 (2013)
[10] A.K. Sharma, V.S. Landge, Zero inflated negative binomial for modelling heavy vehicle crash rate on Indian rural highway, International Journal of Advance in Engineering & Technology, **5**(2), 292-301 (2013)
[11] N.J. Garber, L. Wu, *Stochastic Models Relating Crash Probabilities with Geometric and Corresponding Traffic Characteristics Data*, Research Report No. UVACTS-5-15-74, (2001)

[12] M.H. Pour, J. Prasetijo, A.S.Yahaya, S.M.R. Ghadiri, Modeling vehicle-pedetrian crashes with excess zero along Malaysia Federal Roads, *Procedia-Social and Behavioral Sciences 53* , 1218-1227 (2012)
[13] D. Lord, S.P. Washington, J.N. Ivan, Further notes on the application of zero-inflated models in highway safety, Accident Analysis and Prevention **39**, 53-57 (2007)
[14] J. Prasetijo, W.Z.Musa, Modeling zero-inflated regression of road accidents at Johor Federal Road F001, MATEC Web of Conferences, **47**, 03001, (2016)
[15] J. Prasetijo, W.Z.Musa, Z.F. Zainal, Identification of hazardous road sections using over dispersion-excess zero data of vehicle accident at Johor Federal Roads, *Proceedings International Conference on Engineering of Traumagara*, **1**(1), (2015)
[16] J.M. Hilbe, *Negative Binomial Regresion*, 2nd ed., *California Institute of Technology and Arizona State University*, (2011)

Road Service Performance Based On Integrated Road Design Consistency (IC) Along Federal Road F0023

Zaffan Farhana Zainal[1] , *Joewono* Prasetijo[1,*], and *Wan Zahidah* Musa[1]

[1] Smart Driving Research Center (SDRC), Faculty Civil and Environmental Engineering, Universiti Tun Hussein Onn Malaysia 86400 Parit Raja, Johor

Abstract. Road accidents are one of the world's largest public health and injury prevention problems. In Malaysia, the west coast area of Malaysia been stated as the highest motorcycle fatalities and road accidents are one of the factors that cause of death and injuries in this country. The most common fatal accident is between a motorcycle and passenger car.The most of the fatal accidents happened on Federal roads with 44 fatal accidents reported, which is equal to 29%. Lacks of road geometric designs consistency where the drivers make mistakes errors due to the road geometric features causes the accident kept rising in Malaysia. Hence, models are based on operating speed to calculate design consistency of road. The profiles were obtained by continuous speed profile using GPS data. The continuous operating speed profile models were plotted based on operating speed model (85th percentile). The study was conduct at F0023 from km 16 until km 20. The purpose of design consistency is to know the relationship between the operating speed and elements of geometric design on the road. As a result, the integrated design consistency motorcycle and cars along a segment at F0023, the threshold shows poor design quality for motorcycles and cars.

1 Introduction

Road accidents are one of the world's largest public health and injury prevention problems. About one million people are killed, three millions are severely disabled for life and thirty millions are injured in road accident according to WHO. In Asian, about 400 00 people are killed on the roads annually. Malaysia is one of the developing countries that facing road accidents in Asian and road accidents are one of the factors that cause of death and injuries in this country. In Malaysia, the west coast area of Malaysia been stated as the highest motorcycle fatalities due to the highest number of registered motorcycles and population. Johor was ranked the highest with 27 deaths recorded followed by Selangor (23) and Perak (22). In 2010, the federal government estimated that the number of deaths on motorcycles was 30 times the number of death in cars, per miles travels [1]. The most common fatal accident is between a motorcycle and passenger car. Generally fatal accidents happened on

* Corresponding author: joewono@uthm.edu.my

Federal roads with 44 fatal accidents reported, which is equal to 29%. Most of the road accidents were attributed to motorcyclists with 92 cases reported, which is equivalent to 61% of the 150 accidents recorded. According to [2], in early investigation shows that motorcycle accidents are 17 times more dangerous than driving a passenger car.

Accidents usually result from a combination between 4 elements; the driver, the road, the vehicle and the environment [3]. Previous study stated the causes of road accident are faulty vehicles, uneven roads, careless/reckless driving, speeding, drunk driving, inadequate sleep, alcohol and many more. [4] Accidents keep rising in Malaysia, because lacks of road geometric designs consistency where the drivers make mistakes errors due to the road geometric features. If the consistency are poor, the potential for accident happen are highest [5, 6].

2 Design consistency estimation

Most previous study have developed model of consistency but are less known. One of pervious study have put forth a new model of consistency based on data through continuous speed profile and relate the speed profile with the geometric variables of the road segment and with accident statistics for the segment [7]. Design consistency are defines as the relationship between the geometric characteristics of a highway with the conditions the driver expects when drive. Models are based on operating speed to calculate design consistency of road. Operating speed are defined as the speed selected by the drivers when no restricted by other users or speed under free flow conditions and normally represented by the 85th percentile speed, V_{85}. The operating speed profile were developed by of two types of operating speed models; one for horizontal curves and other for tangents. The profiles were obtained by considering continuous speed profile using GPS devices and continuous speed profiles are used to determine the global speed variation along a road segment and determining a single consistency value for the whole road segment.

According [8] purposed new method to check design consistency. Parameters for this model are the bounded area between the profile and the average speed along a two-lane highway segment. As a design consistency increased, crash rates will shown to decrease significantly. The corresponding equations are

$$R_a = \frac{(\Sigma\, a_i)}{L} \tag{1}$$

Where;$R_a$is relative area (m/sec) measure consistency, $\sum a_i$ is sum of i areas bounded between the speed profile and the average operating speed (m^2/sec), L is entire segment length (m).

$$\sigma = \left\{\frac{(V_j - V_{avg})}{n}\right\}^{0.5} \tag{2}$$

Where; Σ is standard deviation of operating speeds (km/h), V_j is operating speed along the jth geometric element (tangent/curve) (km/h), V_{avg} is average weighted (by length) operating speed along a segment (km/h), n is number of geometric elements along a segment (km/h).

$$V_{avg} = \frac{\Sigma V_i L_i}{\Sigma L_i} \tag{3}$$

Where; V_{avg} is average weighted (by length) operating speed along a highway segment (km/h), V_i is operating speed (km/h),L_i is length of segment (km). The model of consistency (C) and the model integrated design consistency is [8]:

 DOI: 10.1051/matecconf/201710308013

$$C = 2.808.exp^{(-0.278.R_a.\sigma)} \quad (4)$$

Where; C is basic consistency of a highway segment, R_a is normalized area bounded by the average speed profile of cars/ motorcycle and the average operating speed (m/sec), σ is standard deviation of cars speeds (m/sec).

$$C = 2.808.exp^{(-0.278.R_a.\sigma)} \quad (5)$$

Where; IC is integrated consistency of road sections, R_a is normalized bounded area by speed profile of cars /motorcycle (m/sec), σ is standard deviation of car speeds/ motorcycle, ACT is normalized bounded area between the speed profile of cars and motorcycle (m/sec). To determination threshold for design consistency proposed by [8] is:

Table 1. Design-consistency quality

Design-consistency quality $C = 2.808.exp^{(-0.278.Ra.\sigma)}$		
Good	Acceptable	Poor
C > 2 (m/s)	1< C ≤ 2 (m/s)	C ≤ 1 (m/s)

3 Research methodology

In this study, continuous speed profile was used rather than collecting using spot speeds. GPS DG-200 was attached with the vehicle to record data along a segment. Continuous speed data was collected at the study site in direction of traffic flow. The purpose of the study is to estimate value of index the integrated road design consistency (IC) of motorcycles and cars at F0023 from km16 to km20.

3.1 Site collection

The data was collected through the test driver with the same vehicle for motorcycle and car. The study site was selected along km16 until km20 at F0023, Jalan Muar – Segamat, Johor. Fig. 1 shows location km16 – km20 (f0023).

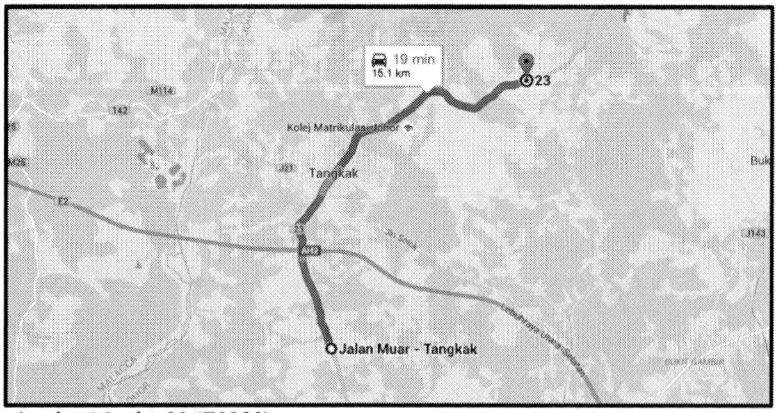

Fig. 1. Location km16 – km20 (F0023)

3.2 Primary data collection

Speed data in this study known as a primary data collection. Many years, a GPS technology has been used to collect traffic data for transportation studies. By previous studies, most of them using automated traffic counters or radar speed gun measurement to measure speed along the segment [9].Therefore, the continuous speed profile data was using new method analysis to develop road design consistency profile of motorcycle and cars by using global positioning system (GPS). By using GPS, the continuous speed profile data was using new method analysis to develop road design consistency profile of motorcycle and cars and determined value of index the integrated road design consistency (IC). 100 test drivers and riders were instructed to drivers and riders along the selected segment by using the same vehicle and GPS (DG- 200) was placed into moving vehicles; cars and motorcycles. To perform data collection (speed data), only two persons, a driver and a GPS recorder were needed. Speed data from signalized intersection been remove about 300 m to avoid the effect of traffic control devices on vehicle speeds [10]. Fig. 2 shows GPS DG-200 that be used in this study.

Fig. 2. GPS DG-200

3.3 Secondary data collection

Accident data, traffic data and road features are been included as secondary data collection. Accident data collection was collected from data police that includes accident type, location and number of vehicles involving. Traffic data collection includes average annual daily traffic (AADT). Road Features include roads and road characteristics such as length of tangent, length of curve, shoulder width, lane width and access point. The data was collected from authorities or site visit

4 Results and discussion

By using statistical software (Minitab), analysis was carried out to develop operating speed model. ACAD Software analysis was using to developed design consistency model.

4.1 Development operating speed model tangent and curve (85th percentile)

Profile was plotted based using continuous speed data obtained through GPS DG-200. The models are developed based on the new method of speed profiles analysis. Model prediction of 85th percentile speed for tangent and curve are using linear regression method to developed prediction operating speed model for motorcycles and cars.

- Motorcycles

$$V_{85t} = 61.2 + 0.41\, L_t + 11.5\, SW + 0.55\, AP \qquad R^2 = 0.43 \qquad (6)$$

$$V_{85c} = 66.0 + 5.7\, L_c + 6.8\, SW + 0.43\, AP \qquad R^2 = 0.34 \qquad (7)$$

- Cars

$$V_{85t} = 40.5 + 6.46\, L_t + 28.9SW - 1.41\, AP \qquad R^2 = 0.94 \qquad (8)$$

$$V_{85c} = 74.6 - 34.8\, L_c - 9.8\, SW + 6.48\, AP \qquad R^2 = 0.88 \qquad (9)$$

Where; V_{85t} = Operating speed of tangent, V_{85c} = operating speed of curve, L_t= length of tangent, L_c = length of curve, SW = shoulder width, AP = access point

4.2 Development design consistency

The continuous operating speed profile models were plotted based on operating speed model (85th percentile). To developed road design consistency model, ACAD software was used to developed road design consistency model. Plots of operating speed model profile (85th percentile) along the length are called as speed profile for this study. The consistency was calculated for each segment from the speed profile. Fig. 3 shows speed profile F0023 of motorcycles at km21 – km25. Fig. 4 shows speed profile F0023 of cars at km21 – km 25.

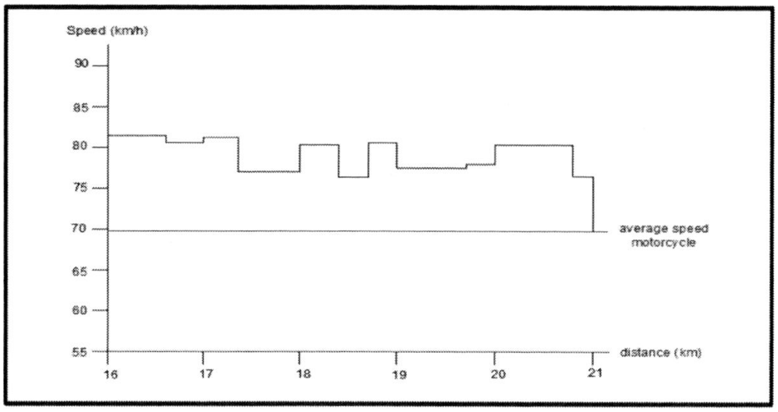

Fig 3. Speed profile F0023 of motorcycles (KM16 – KM20)

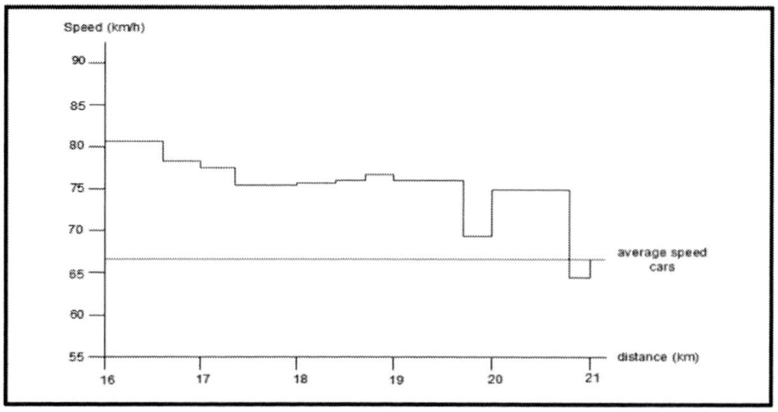

Fig 4. Speed profile F0023 of cars (KM16 – KM20)

The consistency index was estimated based on speed profiles. To estimate consistency index of eq. (4) two parameters were considered. First the value of standard deviation, σ was obtained using the operating speed. Second the bounded area between the profile and the average speed. Table 2 shows design consistency motorcycles at km16 – km20 per km. Table 3 shows design consistency cars at km16 – km20 per km. The design consistency model for motorcycle and cars shows the threshold of design-consistency quality is poor design. This is based on road features conditions in this area for motorcycles and cars.

Table 2. Design consistency motorcycle (KM16 –KM20) per km

KM	Design Consistency	Threshold
16	0.16	poor
17	0.30	poor
18	0.25	poor
19	0.38	poor
20	0.23	poor

Table 3. Design consistency cars (KM16 –KM20) per km

KM	Design Consistency	Threshold
16	0.11	poor
17	0.27	poor
18	0.28	poor
19	0.46	poor
20	0.51	poor

4.3 Integrated design consistency model

The model shows the integrated design consistency model between speed profile of cars and motorcycle. Fig. 5 shows speed profile F0023 between motorcycles and cars at km16 – km 20. The integrated design consistency between the speed profiles motorcycles and cars can be determined through normalized bounded area between the speed profile of cars and motorcycle. Table 4 shows integrated design consistency between motorcycles and cars at km 16 – km 20 per km.

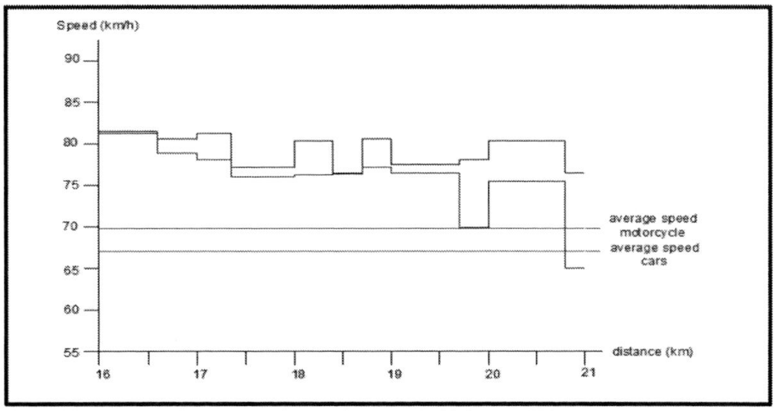

Fig 5. Speed profile F0023 of motorcycles and cars (KM16 – KM20)

The integrated design consistency motorcycle and cars along a segment, the threshold shows poor design for both type vehicles. According to the data police statistics for 5 years (2010 – 2014) and based on APW shows km16 is a high accident area. The road design at km 16 gave influence of the road driving in this area based on threshold design consistency. Other than, for whole road km 16 until km 20 at F0023 has poor design. There may be several factors that affect the design such as access points that located in this area. F0023 (km16 – km20) the area has no median, the main road of the residents in this area, industrial area and estates area along the segment. It shows, accident occur are not based on speed fault but the factor of surrounding area and driver behavior.

Table 4. Integrated Design consistency motorcycles and cars (KM16 – KM20) per km

KM	Integrated Design Consistency	Threshold	APW
16	0.11	poor	35
17	0.27	poor	15.4
18	0.27	poor	14.2
19	0.45	poor	16.4
20	0.48	poor	6.8

5 Conclusions

Most of the model developed using the spot speed data from equipment such as radar gun. GPS data loggers were used to develop the continuous speed profiles. Furthermore, design consistency profile of motorcycles - cars were developed by 85th percentile operating speed model of motorcycle and cars. In Malaysia, the design consistency models have not been used to evaluate geometric design at rural road, previously. The purpose of design consistency is to know the relationship between the operating speed and elements of geometric design on the road.

Design consistency was influenced the driver to drive in a safe condition. The design consistency model was developed based on two independent; the bounded area between the profile and the average speed and the standard deviation of speeds with results in Integrated Design Consistency (IC). The integrated design consistency models were based on normalized bounded area between the speed profile of cars and motorcycle and design consistency.

It can be concluded from the study that integrated design consistency (IC) motorcycles - cars shows that the poor design consistency with the higher potential accident occurs. Even though, the accident was not high according to APW, but the road design shows poor design consistency along a segment.

The authors would like to thanks to the Universiti Tun Hussein Onn Malaysia in funding this study under the Geran Kontrak UTHM Vot. U395. Further thanks are also for Smart Driving Research Center (SDRC), Faculty of Civil and Environmental Engineering, Universiti Tun Hussein Onn Malaysia (UTHM) for support and providing facilities to accomplish the study.

References

[1] D.G. Lee, *Motorcycle vs Car Accident Statistics*, (2012). Reached at September 5, 2016, from http://motorcycleaccidentlawyerpa.com/motorcycle-vs-car-accident-statistics/
[2] U. Radin, Updates of road safety status in Malaysia, IATSS Research, **29**, 78-80 (2005)
[3] H. Farah, A. Polus, M. Cohen, Multivariate analyses for infrastructure-based crash-prediction models for rural highways, Road and Transport Research: A J. of Australian and New Zealand Research and Practice, **16**(4), 26 (2007)

[4] A. Kareem, Review of global menace of road accidents with special reference to Malaysia-a social perspective, Malaysian Journal of Medical Sciences **10**(2), (2003)
[5] R. Lamm, B. Psarianos, T. Mailaender, *Highway design and traffic safety engineering handbook*, New York, (1999)
[6] J. De Oña, L. Garach, Accidents prediction model based on speed reduction on Spanish two-lane rural highways, Procedia-Social and Behavioral Sciences, **53**, 1010-1018 (2012)
[7] L.G. Morcillo, F.J.C. Poyo, M.P. Fernández, J. de Oña, Measurement of road consistency on two-lane rural highways in Granada (Spain), Procedia-Social and Behavioral Sciences, **162**, 237-242 (2014)
[8] C. Mattar-Habib, A. Polus, H. Farah, Further evaluation of the relationship between enhanced consistency model and safety of two-lane rural roads in Israel and Germany, EJTIR, **4**(8), 320-332 (2008)
[9] J. Prasetijo, Z.F. Zainal, W.Z. Musa, Speed profile based on design consistency, *Proc. Int. Conf. on Engineering of Tarumanagara*, Jakarta, Indonesia, (2015)
[10]J. Prasetijo, Z.F. Zainal, Development of continuous speed profile using GPS at Johor Federal Roads F0050, MATEC Web of Conferences, **47**, (2016)

Analysis of Pedestrian Gap Acceptance and Crossing Decision in Kuala Lumpur

Siti Naquiyah Mohamad Nor[1], *Basil* David Daniel[1,*], *Rizati* Hamidun[2], *Walid* A Al Bargi[1], *Munzilah* Md Rohani[1], *Joewono* Prasetijo[3], *Mohamad Yusri* Aman[1], and *Kamarudin* Ambak[3]

[1]Faculty of Civil and Environmental Engineering, Universiti Tun Hussein Onn Malaysia (UTHM), Johor, Malaysia
[2]Malaysian Institute of Road Safety Research, (MIROS), Selangor, Malaysia
[3]Smart Driving Research Centre (SDRC), UTHM, Johor, Malaysia

Abstract: Pedestrians are most vulnerable of all road users. This research aims to investigate and model pedestrian road crossing behaviour at crossing facilities. In particular, they have two aspects of pedestrians crossing behaviour are examined, namely the size of traffic gaps acceptance by pedestrians and the decision of pedestrians either to cross the road or not. A fields survey was carried out at six crossing facilities which from a zebra crossing at midblock. In this survey, the data were recorded in real traffic condition using video recorder. Determine the associations between characteristics of pedestrians, crossing facilities and vehicular traffic through on-site observations of pedestrian behaviour. This data will analysis using statistical analysis which is multiple regression and binary logit regression method. It is hope that through this research, the model of pedestrian gap acceptance and pedestrian crossing decision can be reached and what are the indicators that pedestrians look for when accepting gaps to cross the road.

1 Introduction

A pedestrian is defined as people who go on foot or who utilize assistive devices to facilitate them to walk. Walking is one of the most usual ways of moving for each person. Each type of mode transportation used also involves some by movement on foot. Walking can be categorized as one of the main modes of transport and it is healthier for human society [1]. Consequently, the pedestrian is one of the key elements in the system traffic, transportation arrangement, specially in urban traffic. The growth in pedestrian in Malaysia is parallel with the increment in population rapidly in Malaysia. In relation to this, the increase in the number of the pedestrian is also the reason of the increasing number of a traffic accident. A pedestrian was the most vulnerable when on the road. There has high traffic congestion in Malaysia, particularly in Kuala Lumpur. This situation can lead to conflicts between the movement of vehicles and pedestrians. This conflict sometimes can make accidents occurred on the pedestrian.

[*] Corresponding author: basil@uthm.edu.my

2 Literature review

There are many different factors that affect the gap acceptance behaviour such as those relating to the site or location where the maneuvers take place, the conditions at the time of action and driver or vehicle involved. A gap in traffic is the space and time between vehicles and pedestrian. Some people might be can acceptable a gap in traffic, but some people might not be. Whether or not a gap is acceptable depends on the person's level of risk acceptance, how much the person trusts that the drivers will stop, and the person's perception of how long the gap is that perception may not be correct (the gap might be much longer or shorter than the person thinks it is).

The time spent waiting for safer gaps depend on whether the pedestrian is alone or in a group. Pedestrians tend to cross illegally (cross on red) if a member of the group initiates the illegal cross. In accepting gaps to cross, each pedestrian has a different perception on choosing the safest gap. Their physical characteristics affect their movement, i.e. walking speed. Taller pedestrians accept smaller gaps compared to shorter pedestrians due to them generally being able to walk faster [2]. The distance between the vehicles and the pedestrian have a significant effect on accepting safe gaps to cross [3]. Male pedestrians made the most inappropriate choices where they accepted smaller gaps to cross compared to their female counterparts discovered by [1]. Younger-old and older-old pedestrian group was also found to cross more slowly and accepts larger gaps than young pedestrian [4].

Previously study was found that the gap acceptance can find using some parameter such as with traffic condition and with the vehicular and pedestrian characteristic. Based on the gap acceptance theory researchers have developed mathematical models to represent the gap acceptance behavior of pedestrian with the help of pedestrian demographic characteristics such as gender and age [5], types of vehicle and waiting time [6], effect of rolling gap [7], effect of parked vehicles [8], non-compliant road crossing behavior at midblock locations [9], [10] and non-compliant road crossing behavior at signalized intersection [11].

3 Research methods

A field survey obtained the information about pedestrian crossing behavior. The selection of sites for pedestrian gap acceptance based on the area selected has the highest pedestrian count of 100 pedestrian/hour or more. The choices of sites also consist of the shopping complex, job center, public transport or education center. Six sites were chosen to observe pedestrian behavior in the area of Kuala Lumpur which is Lorong Gombak, Jalan Esfahan, Jalan Tun Perak, Jalan Pudu, Jalan Melaka and Jalan Merah Cagar. Manual counts were carried out to determine the volume of both pedestrians and vehicles, while a speed meter was used to measure vehicular speed. Video recording was recorded pedestrian who actually cross the road either using the crosswalk or not within 50 meters to know pedestrian flows and pedestrian behavior. The pedestrian that crosses a road at a distance more than 50 meters were not counted as samples for this study. Figure 1 shows an example of crossing facilities in Kuala Lumpur.

The process of street crossing can be excused by the utility maximization theory that pedestrians want to select the most satisfactory facilities and locations to cross the street. As a result, pedestrians make the utmost utility. It would be fair to assume that pedestrians' most satisfactory decisions are dependent on the location and the type of crossing facility. For instance, as shown in Figure 2, there are crossing facilities in the country. If a pedestrian's origin is in zone 4 and the destination is in zone 2, there are two potential routes to cross the street. The first path is to use the crosswalk and the second route is to use the overpass. Therefore, the existence of crossing facility may change pedestrians' crossing behavior and also induce a traffic violation because pedestrians cannot cross the street at the desired location.

Fig. 1. Example crossing facilities in Kuala Lumpur

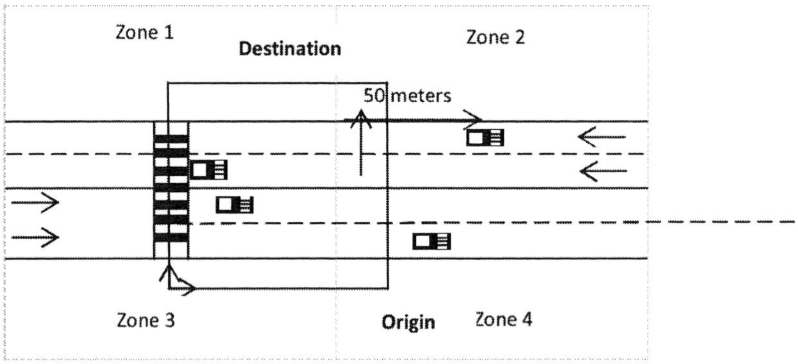

Fig. 2. Illustration of Crossing Route

4 Model framework

Analysis data is one of the processes to develop a model. All the data that have been collected has been analyzing for accuracy and suitability. Before beginning the analysis, there are several elements that should be recognized. This is very important to make sure the process will not affect for developing the model. A factor that must have to the identification of variables that to be included in the analysis are continuous variables and discrete variable. Data from video recording focused on the pedestrian crossing on the road either use crossing facilities or not use. The collected variables are shown in Table 1.

Table 1. The Collected Variables

Variable	Type of Variable	Unit or Code	Description
Gap size	Continuous	Time in sec	Time gap between two vehicles with reference to crosswalk point
Waiting time	Continuous	Time in sec	Time spent at the curb or median for suitable gap
Vehicle speed	Continuous	m/s	Speed of the vehicle at crosswalk area
Traffic volume	Continuous	vehicle/hour	Volume of vehicle in crosswalk
Pedestrian speed	Continuous	m/sec	The speed of the pedestrian while crossing the road
Pedestrian Volume	Continuous	pedestrian/hour	Volume of pedestrian in crosswalk
Gender	Discrete	0: Women 1: Man 2: Both	Male or female
Type of vehicle	Discrete	0: Motorcycle 1: Car 2: MPV 3: Lorry and bus	Type of vehicle
Pedestrian platoon	Discrete	0: Single 1: Two 2: More than two	Number of pedestrian in group
Gap acceptance	Discrete	0: Rejected 1: Accepted	Whether pedestrian accepting gap or rejecting

4.1 Model pedestrian gap acceptance

After several of the trials using SPSS software on different independent variables, the final model which has the best statistical results is developed. The gap acceptance was considered as the dependent variable and the remaining variables are considered as independent variables.

$$\begin{aligned} \textit{Log-Gap} = 34.663 + 0.529(WS) - 0.271(WT) - 1.477(G) - 3.012\,(J) - 0.252(VS) \\ + 0.263(TV) - 0.627(RW) \end{aligned} \quad (1)$$

Where; Log-Gap = Logarithm of accepted gap, WS = Walking Speed, WT = Waiting Time, G = Gender, J = Jaywalk, VS = Vehicle Speed, TV = Type of Vehicle, RW = Road Width. Based on the analysis has shown the p-value is less than α which is 0.05. Therefore, the comparison shows that the null hypothesis (H_0) rejected. The value of H_0 shows that there are an influence between gap acceptance and the factors. The descriptive statistics of multiple linear regression analysis test, t-value and p-value are summarized in Table 2. Reported t-value and p-value are statistical test values of each independent variable.

4.2 Model pedestrian crossing decision

The purpose of modeling pedestrian's crossing decision is to develop a linear function of the selected independent variable. In this study, there are 8 possible factors has been determined to influence pedestrian's crossing decisions which are gender, walking speed, waiting time, gap size, conformity psychology, jaywalking, vehicle speed, type of vehicle and road width.

The model will provide the probability of a pedestrian either to accept or reject the street. Table 4.5 summarizes the descriptive statistic binary logit model.

Table 2. Summaries the descriptive

Model		Coefficients	*t*-value	*p*-value
1	(Constant)	34.663	22.141	0.000
	WS	.529	2.305	0.021
	WT	-.271	-5.383	0.000
	G	-1.477	-2.127	0.034
	J	-3.012	-4.542	0.000
	VS	-.252	-16.574	0.000
	TV	.263	3.785	0.000
	RW	-.627	-3.391	0.001

Table 3. Summaries the descriptive statistic Binary Logit Model

		B	S.E.	Sig.	Exp(B)
Step 1[a]	Walking Speed	-.059	.051	.245	.943
	Waiting Time	.010	.011	.363	1.010
	Gap Size	-.012	.006	.028	.988
	Jaywalk	-.084	.145	.561	.919
	Vehicle Speed	-.033	.040	.001	.876
	Type of vehicle	.020	.019	.286	1.021
	Road Width	.003	.004	.436	1.003
	Constant	2.749	.416	.000	15.620

Based on table 4.5, the *p*-value is used to determine the variables that statistically fit the model. A significant value for each other factor that suitable for decision model must be less than 0.1 (p-value<0.1).

After the analysis, it is found out that gap size and vehicle speed are significant variables for the model whereby the significance value is less than 0.1. The other factors are considered insignificant due to significant value more than 0.1. Thus, the model is fitted by the method of maximum likelihood to the data as shown in Equation (2).

$$z_i = 2.749 - 0.012(GS)_i - 0.033(VS)_i \tag{2}$$

The probability for a pedestrian to cross the street based on gap size and vehicle speed may be explained by a binary logit regression model in the form as shown in Equation (3).

$$P(U) = \frac{e^{(2.749-0.012(GS)-0.033(VS))}}{1+e^{(2.749-0.012(GS)-0.033(VS))}} \tag{3}$$

5 Analysis and discussion

Most of the previous study gets a different gap size. The mean Malaysia pedestrian gap size at the crosswalk in second for man, women and both are 4.1, 5.4 and 4.9 respectively (Table 4). This value is higher than a study from [1] which for women is 4.876 second and man 3.422 second. It shows that women select larger gap than men. Difference for both categories. It can explain when a particular pedestrian is surrounded by several other pedestrians, he/she may feel protected by others and then may act more aggressively [10].

Table 4. The minimum accepted gap size for gender group

Gender	Gap Size (sec)		
	Minimum	Maximum	Mean
Man	3.5	5.5	4.1
Women	4.6	8.0	5.4
Both	4.1	7.4	4.9

Fig. 3 shows the possibilities of a pedestrian's crossing decision according to the increasing speed of the vehicle and mean of gap size. According to figure 3, as speed increases, the possibility for a pedestrian to cross is decreased. When the vehicle speed is more 90 km/h with the mean of traffic gap 3 seconds, pedestrian chooses not to cross.

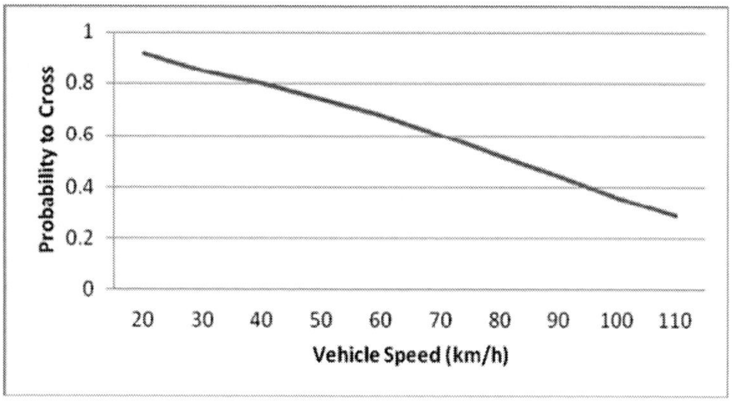

Fig. 3. Probability for a pedestrian crossing decision against vehicle speed

Fig. 4 shows the possibilities of pedestrian's crossing decision based on increasing gap size with the mean of vehicle speed. Based on the graph, it can be concluded that as gap size increases, the possibility of a pedestrian to cross is higher. When smaller gap size reduces, the possibility of a pedestrian to cross the road is lower.

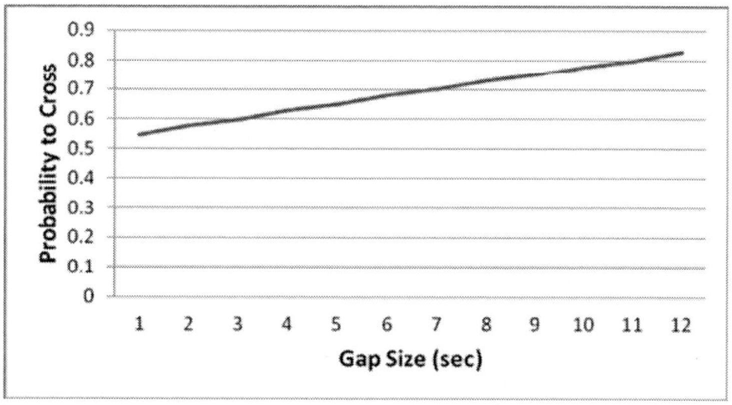

Fig. 4. Probability of a pedestrian's crossing decision against gap size

6 Conclusion

This study analyzed the gap acceptance and crossing decision in Kuala Lumpur. Information was obtained through surveying and observation pedestrian at crossing facilities at 6 sites. The following conclusions are drawn based on the result:

i) A multiple regression model was developed in order to examine the variable that affected by the size of traffic gaps accepted by pedestrians. It was found that the gap size depends on walking speed, waiting time, gender, jaywalk, vehicle speed, type of vehicle and road width.

ii) A binary logit regression model was also developed in order to examine the variable that will affect to pedestrian when making the decision either to cross the street or not. The result showed that the decision depends on gap size and vehicle speed.

iii) The minimum gap size for man, women and both are 3.5, 4.6 and 4.1. It showed that man has a lower gap size because in the human nature there is an obvious difference between men and women at term of how aggressive they are.

iv) Probability to cross when vehicle speed was lower is higher, while the probability to cross the road when the gap size fewer is lower.

The findings of this survey show that pedestrian`s gap accepted is significantly influenced not just by the vehicular traffic condition but also the road environment affect. Moreover, it was found that pedestrian crossing decisions are strongly connected with vehicle speed from the incoming vehicle and gap size because pedestrian`s choose not to cross if the vehicle speed is higher and the gap size were smaller.

The authors would like to acknowledge the financial support of the Ministry of Higher Education Malaysia under the Fundamental Research Grant Scheme (Project Reference No. FRGS/2/2014/TK07/UTHM/02/1/1485).

References

[1] B.R. Kadali and V. Perumal, Pedestrians' gap acceptance behavior at mid-block location, Int. J. of Engineering and Technology, **4**, 158–161 (2012)

[2] B.H. Goh, K. Subramaniam, Y.T. Wai and A.A. Mohamed, Pedestrian crossing speed : the case of malaysia, Int. J. of Traffic and Transport Engineering, **2**, 323–332 (2012)

[3] J.A. Oxley, E. Ihsen, B.N. Fildes, J.L. Charlton and R.H. Day, Crossing roads safely: An experimental study of age differences in gap selection by pedestrians, Accident Analysis and Prevention, **37**, 962–971(2005)

[4] A. Dommes, V. Cavallo, J.B. Dubuisson, I. Tournier and F. Vienne, Crossing a two-way street: Comparison of young and old pedestrians, J. of Safety Research, **50**, 27–34 (2014)

[5] A. Jain, A. Gupta and R. Rastogi, Pedestrian crossing behaviour analysis at intersections, Int. J. of Traffic and Transportation Engineering, **4**, 103–116 (2014)

[6] D. Sun, S.V. Ukkusuri and R.F. Benekohal, Modeling of motorist-pedestrian interaction at uncontrolled mid-block crosswalks, Transportation Research Record, Urbana, 51, 61801(2002)

[7] M. Brewer, K. Fitzpatrick, J. Whitacre and D. Lord, Exploration of pedestrian gap-acceptance behavior at selected locations. J. of the Transportation Research Board, **1982**, 132–140 (2006)

[8] G. Yannis, E. Papadimitriou and A. Theofilatos, Pedestrian gap acceptance for mid-block street crossing. Transportation Planning and Technology, **36,** 450–462 (2013)

[9] C. Cherry, B. Donlon, X. Yan, S.E. Moore and J. Xiong, Illegal mid-block pedestrian crossings in China: gap acceptance, conflict and crossing path analysis, Int. J. of Injury Control and Safety Promotion, 1–11 (2011)
[10] T. Wang, J. Wu, P. Zheng and M. McDonald, Study of pedestrians' gap acceptance behavior when they jaywalk outside crossing facilities, *13th Int. IEEE Conference*, 1295–1300 (2010).
[11] J.M. King, D.W. Soole and A. Ghafourian, Illegal pedestrian crossing at signalized intersections: Incidence and relative risk, Accident Analysis and Prevention, **41**, 485–490 (2009)

Comparison of Bus Driving Control Based on Working Shift Hour and Impact on Bus Fuel Consumption

M. M. Rohani [1,*], *B. D.* Daniel [1], *M.Y.* Aman[1], *J.* Prasetijo[1], *R.* Buhari[1], *N. N.* Mohd Nawi[1], *N. S. A.* Sukor[2], and *S. A.* Hasan[3]

[1]Faculty of Civil and Environmental Engineering, Universiti Tun Hussein Onn Malaysia, 86400 Batu Pahat, Johor, Malaysia
[2]School of Civil Engineering, Universiti Sains Malaysia
[3]Faculty of Civil Engineering, Universiti Teknologi Malaysia

Abstract. This study is aimed to investigate and compare how the work schedule of bus drivers in urban areas can have impact on bus drivers' control on speed and acceleration. This study also assesses the influence of bus driving behavior on fuel consumption. The work schedule assessed in this research is morning shift (6.00 am to 3.00 pm), afternoon shift (3.00 pm to midnight) and direct shift (6.00 am to midnight). The result from this study indicated that for all driving shifts, drivers spent more than 70% of their driving time for accelerating and braking. Drivers were also significantly driving faster and applying sharper acceleration when accelerating and braking during morning shift. The variation of bus driving behavior between different working shift hours were found to influence bus fuel consumption significantly.

1 Introduction

There are some evidence that working shift hour can directly influence driver behavior. This can be seen in Rospa [1] study on driver fatigue and road accidents. Researcher suggested that drivers who are working on early morning (12 am) and in the afternoon have higher accidents risks as drivers are prone to fell asleep. Horne and Reyney [2] supported the statement in their research about sleepy driver. They suggested that driving for a long period is boring and will cause sleepiness. Besides that, the researchers suggested that the driver will be 50 times sleepier and have 3 times more risk to accident at 2 am compared to 10 am. Results from 102 operators who were involved as a sample in Howarth [3] study demonstrated that there are different sleep length between drivers who work split shift (two periods of work within one day) and straight shift (non-split). The study about investigation of sleep length and fatigue of transit bus operators while working with different work schedules found split shift drivers sleep less on working days compared to straight shift drivers. From that, Howart suggested drivers who work on split shift are more exposed to mental fatigue compared to drivers who work on straight shift.

* Corresponding author: munzilah@uthm.edu.my

The study carried out towards transit bus driver in Florida also found that drivers who work in split shift tend to get fatigue compared to straight shift due to inadequate rest [4]. Researchers performed the analysis using Fatigue Audit Interdynamics (FAID) to relate between fatigue and accident (divided into red, yellow, and green) found the result is 56.48% (in red) which shows fatigue is related to accidents. Apart from that, the increasing higher risk of accidents at peak hours (at night) multiplied 10 times from daytime. Therefore, working shift influences the driver's performance and safety. A previous study has shown that fatigue, increases as working time increases, with associated increase in accidents. Past research about the factors of fatigue and bus accident suggested that working schedule is a cause of accidents as irregular working hours led to fatigue and causing accidents [5]. Besides that, unsuitable working schedule also caused lack of rest or sleep to the bus drivers and will affect their driving performance.

Since it is known that working shift hour can have impact on driver behavior, it is no doubt that driving behavior is the main factor impacting bus fuel consumption. For example, study carried out in Southampton by Rohani [6] on bus driver issues and fuel consumption suggested that fuel consumption is sensitive towards acceleration change. The change of driving behavior from aggressive to normal and from normal to economic provides the fuel saving of 20% to 40%. However, the fuel consumption increases to 67% when the driver accelerates from rest. According to Mierlo e al. [7], the changing of driving behavior from aggressive to pro-aggressive can reduce the fuel consumption of 5% to 25% while Vlieger et al. [8] suggested that aggressive driving behavior increases the fuel consumption until 40% but it also depends on road conditions and vehicle technology.

Another study that showed the impact of driving behavior on fuel consumption was carried out in Belgium [9]. Researchers reported that the average of fuel consumption before and after attending driving course have shown a reduction at about 5.8%. The difference in fuel usage was also determined by the change of driving behavior based on economic driving tips such as gear changing after approaching 2000 rpm,using steady speed and reducing gear change especially during long journeys.

2 Research approach

The four research buses used in this study were lower floor single decker Scania. It has a Euro 3 engine with six automatic gear transmissions. The bus weight with and without load are 11000kg and 18000kg respectively. Global Positioning System (GPS) (GLOBALSAT DG-200 model) device was used to collect real speed profile, bus position, date and time from research buses (Fig. 1). The devices were mobile based device and powered by an internal battery. Data were stored in its internal data logger at 1 Hz, allowing up to 24 hours of continuous data collection. The GPS was placed on board on each research bus without the knowledge of the drivers so that the drivers will drive the bus based on their actual behaviour.

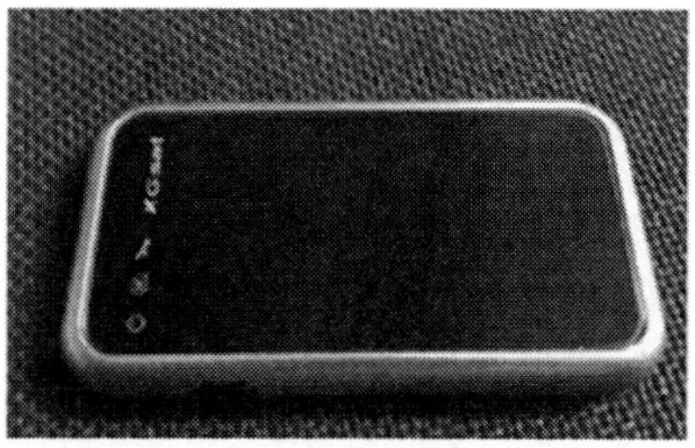

Fig. 1. Global Positioning System (GPS) device used in this study

Eleven bus drivers from private bus company participated in this study and were asked to drive the bus based on their working shifts hours. Three different working shifts implemented by the bus company are morning shift, afternoon shift and full day shift. The duration hours for these working shifts are between 6.00 am to 3.00 pm (9 hours), 3.00 pm to 12.00 am (9 hours) and 6.00 am to 12.00 am (18 hours) respectively. In the middle of the drivers' working period, they were given break time to ensure that they have sufficient rest so that they are to be able to drive the bus safely.

Route selected for this research connecting center of Putrajaya and Cyberjaya. During the period of data collection, the buses were driven continuously based on bus trip schedule for both directions from Putrajaya to Cyberjaya and from Cyberjaya to Putrajaya by drivers. Each bus driving trip (for both directions) took between 30 to 45 minutes. As presented in Table 1 total of 549 driving profile were observed within 2 weeks of data collection.

Table 1. Number of driving profile observed based on working shift period

Bus Route	Working Shift period	Number of driving profile observed
Putrajaya to Cyberjaya	Morning	113
	Afternoon	81
	Full day	71
Cyberjaya to Putrajaya	Morning	120
	Afternoon	74
	Full day	90
Total		549

3 Research data

Data were downloaded from GPS datalogger during the end of the bus daily schedule. Since the GPS recorded the data continuously and unable to split the data according to research needs, this process was conducted manually using Microsoft Excel software and Google Maps. In addition, the GPS recorded the time based on Coordinated Universal Time (UTC). Therefore, the research data need to be adjusted based on Malaysia time. It should be noted that, the hour differences between UTC time and Malaysia time is 8 hours. Further to the data processing, the acceleration data used in the analysis were calculated from the speed change recorded by GPS devices.

An additional software: SIDRA TRIP was used in this study to estimate bus fuel consumption based on driving profile and the bus' specific parameters such as engine size, bus load, etc. To ensure the fuel estimation is correct for further analysis, the estimated fuel data were compared with daily fuel filled recorded in log book. As the result from comparison test have shown that SIDRA TRIP fuel estimation was not significantly different from fuel used recorded in the log book, the estimated fuel consumption were used for further analysis.

4 Analysis and results

4.1 Research data

In present study, the driving styles of the bus driver were experimentally monitored to investigate the variation of the speed and acceleration control. Further to the investigation, fuel consumption impact from driving styles has been determined. It is based on real time measurement from four research buses driven by eleven bus drivers. Dynamic variables of speed, acceleration and estimated fuel consumption have been analysed. Each data was categorized based on driving phases of idling, cruising, accelerating and braking. It should be noted that the term 'accelerating' and 'braking' used in this study refers to any data with positive and negative acceleration respectively. The overall proportion of driving phases derived from the entire data are as presented in Fig. 2. It shows that more than 70% of bus driving activity in city environment was for accelerating and braking. These driving phases are the major cause of higher fuel consumption (Fig. 3).

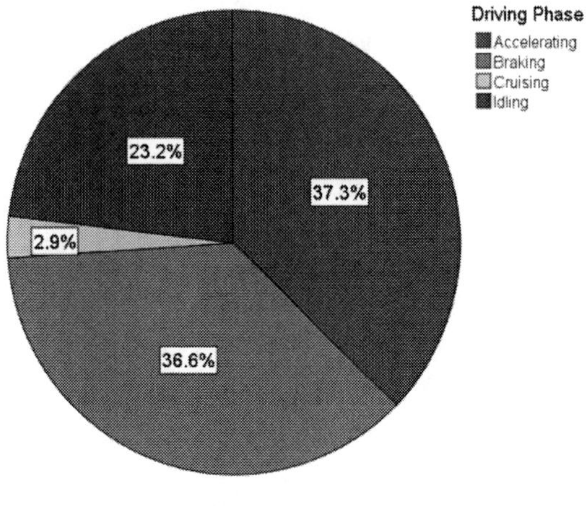

Fig. 2. Proportion of driving phase in the research data

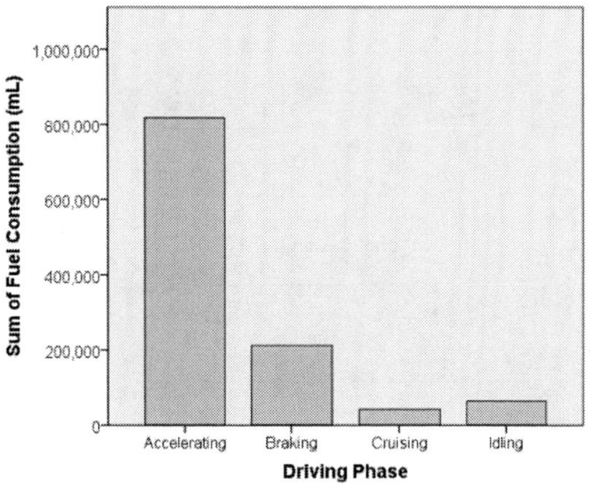

Fig. 3. Sum of fuel consumption according to different driving phase calculated from research data

In Fig. 4 to Fig. 6, the distribution of speed, acceleration and fuel consumption measures are shown. The observed speed peaked at lower speed range of 0 – 2 km/h is normal for bus driving in city environment due to stop and go activities; as the drivers are require to stop at signalized intersection for red light, or due to traffic interruption within the road segment. Furthermore, since the bus serves passengers in town areas, passenger boarding and alighting activities at dedicated bus stop not only influnce the speed but also the acceleration data's distribution. As presented in the graph, the acceleration calculated from the speed change, range between -3.94 to 2.69 m/s^2. The peak acceleration was observed at lower acceleration ranged at 0 – 0.2 m/s^2. The distribution of speed and

DOI: 10.1051/matecconf/201710308015

acceleration captured from the data are also found to influence 30% and 63% of bus fuel consumption respectively.

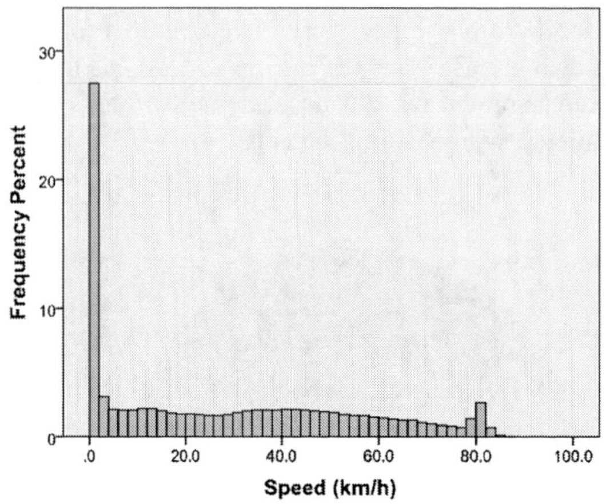

Fig. 4. Distribution of bus speed

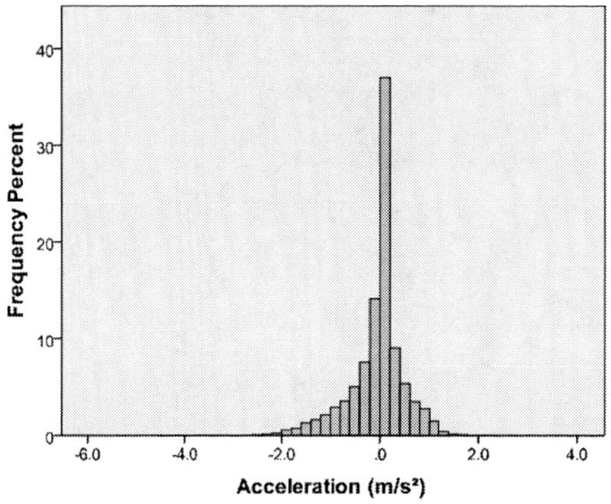

Fig. 5. Distribution of bus acceleration

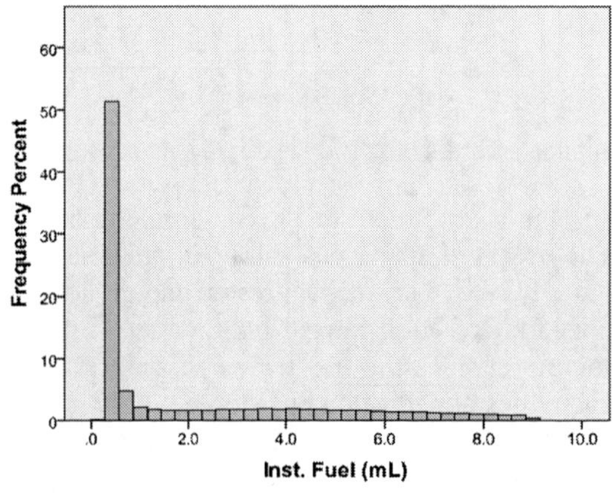

Fig. 6. Distribution of bus fuel consumption

4.2 Comparison of acceleration and deceleration control of bus drivers based on working hour shift

One of the aims of this study is to compare the variation of bus speed and acceleration control during the three different working hour shifts as mentioned in the research approach. Fig. 7 presents the proportion of driving phases spent by drivers who work at different working shift hour. The figure exhibits that the driving phase proportion between the drivers who work at different working hour shift are not significantly different. It was shown that, in the city environment, bus drivers were dealing with majority of accelerating and braking (more than 70%) control. Beside unable to maintain the constant speed, the bus drivers were also found to spend more than 20% of their journey time for idling. Traffic and road geometric factors are generally seen as strong factors that influence how the drivers control their driving.

To further investigate, the analysis was conducted to examine whether the drivers control the speed and acceleration differently at different working shift hour. The overall result from correlation test indicated that, speed, and acceleration control and driving hour shift have low correlation to each other (R2 < 0.1). However, the mean and maximum speed and acceleration reached by the drivers as presented in Table 3 suggested that mean acceleration rate when accelerating is higher among drivers who drove in direct shift (0.339 ms^{-2}). As for braking behavior, the overall data showed that, mean of deceleration rate when drivers braking was higher in morning shift (-0.550 ms^{-2}).

The significant difference of bus driving control is not only shown when drivers were accelerating and braking, but was also captured in bus speed control (as shown in Table 2). As presented in Table 3, mean speed is higher during morning shift driving (29.48 km/h) followed with direct shift (27.969 km/h). Specifically, drivers who work for direct shift were also found to apply 4% (accelerating) and 8% (braking) higher acceleration rate during morning shift.

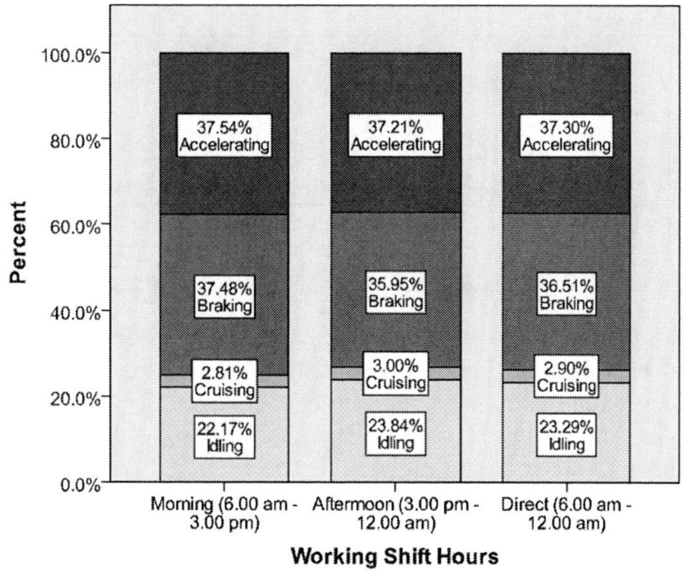

Fig. 7. Proportion of driving phases spent indifferent working shift period

Table 2. Comparison test for acceleration and speed control between different working shift hour

		Sum of Squares	df	Mean Square	F	Sig.
Acceleration for Accelerating	Between Groups	1.144	2	.572	4.680	.009
	Within Groups	22696.969	185680	.122		
	Total	22698.114	185682			
Acceleration for Braking	Between Groups	33.316	2	16.658	56.833	.000
	Within Groups	53347.408	182009	.293		
	Total	53380.724	182011			
Speed	Between Groups	467137.180	2	233568.590	343.122	.000
	Within Groups	339293850.589	498437	680.716		
	Total	339760987.769	498439			

Table 3. Mean and maximum speed and acceleration control at different working shift hour

	Accelerating		Braking		Speed (km/h)		
	Mean	Max	Mean	Max	Mean	Max	Min
Morning (6.00 am - 3.00 pm)	.337	2.167	-.550	-4.403	29.481	86.500	.000
Afterrnoon (3.00 pm - 12.00 am)	.333	2.694	-.516	-5.375	26.934	86.700	.000
Direct (6.00 am - 12.00 am)	.339	2.389	-.541	-3.875	27.969	87.900	.000

Table 4. Mean of acceleration and speed of the driver who work for direct shift

	Driving control		
	Accelerating	Braking	Speed (km/h)
Morning	.3453	-.5586	28.1690
Afternoon	.3310	-.5162	27.6995

Fig. 8 shows how drivers dealtwith acceleration at specific speed range while accelerating and braking. The figure shows that at speed lower than 30km/h, the 85th percentile acceleration of drivers who work in the morning shift were found to be sharper than the afternoon shift. At higher speed, drivers accelerating and braking behavior were considerably similar for all working shift hour.

4.3 Impact of bus driving variation on fuel consumption

The variations of speed and acceleration control of bus driving in the city areas contributed to disparity of bus fuel usage. As presented in Fig. 2 and Fig. 3, majority of bus fuel were used for accelerating. The accelerating behavior consumed 72% of the total bus fuel usage for all working shift hour. Besides, although drivers spent almost equal duration for braking and accelerating, the analysis conducted suggested that only 19% of fuel usage is used for braking (Table 5). Overall result also shows that, lesser fuel is consumed for cruising and idling.

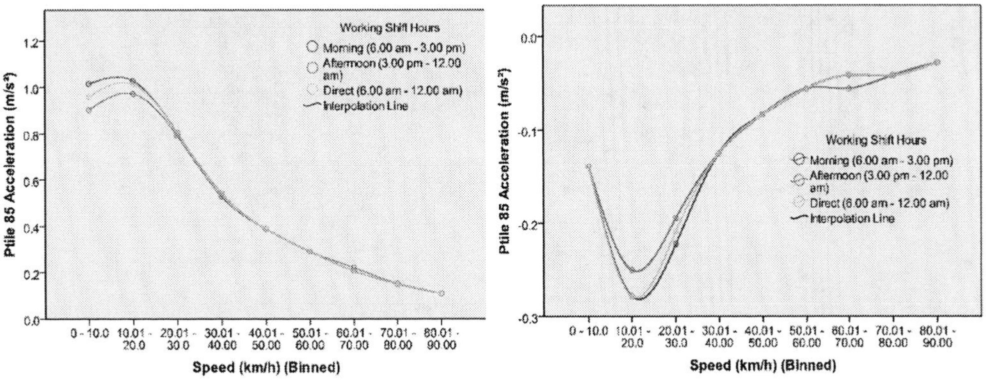

(a) Accelerating (b) Braking

Fig. 8. Comparison of acceleration applied by drivers at specific speed range during accelerating and braking

Table 5. Proportion of bus fuel consumption based on different driving phase and Working shift hour

	Accelerating	Braking	Cruising	Idling
Morning (6.00 am - 3.00 pm)	72%	19%	4%	5%
Afternoon (3.00 pm - 12.00 am)	72%	18%	4%	6%
Direct (6.00 am - 12.00 am)	72%	19%	4%	6%

This study evaluates the correlation between speed, acceleration and fuel consumption. Non-linear relationship was investigated. Table 6 shows the correlation coefficients. As presented, positive and good correlation can be seen between speed, acceleration and fuel consumption. A scatterplot demonstrated in Fig. 9 summarises the results.

Table 6. Correlation test between speed, acceleration and fuel consumption

		Acceleration (m/s^2)	Inst. Fuel (mL)	Speed (km/h)
Acceleration (m/s^2)	Correlation	1	.539	-.072
	Sig. (2-tailed)		.000	.000
	N	498441	498441	498441
Inst. Fuel (mL)	Correlation	.539	1	.477
	Sig. (2-tailed)	.000		.000
	N	498441	498441	498441
Speed (km/h)	Correlation	-.072	.477	1
	Sig. (2-tailed)	.000	.000	
	N	498441	498441	498441

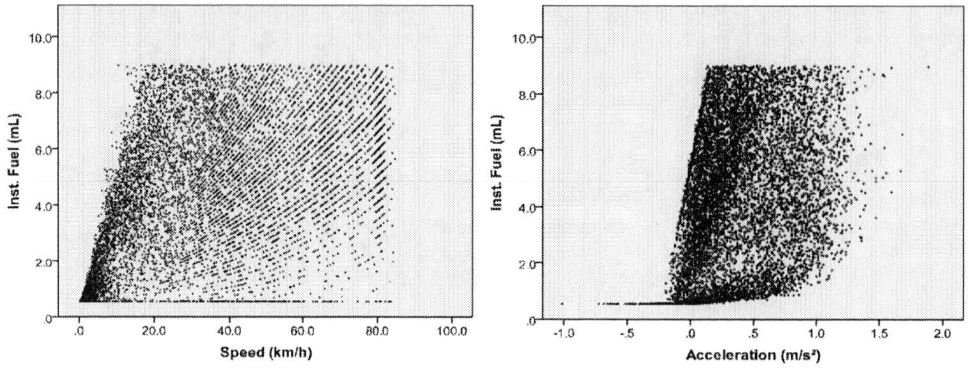

(a) (b)

Fig. 9. Relationship between (a) speed and Instantaneous fuel consumption and (b) acceleration and instantaneous fuel consumption

Further investigation on bus fuel consumption has been conducted to compare the differences of fuel used between morning and afternoon shifts. The result presented in Table 7 and Table 8 shows that, the drivers who drove during morning shift consumed 4% significantly higher fuel than drivers who drove during afternoon shift. This expected result, is consistent with drivers' negotiation on speed and acceleration as presented previously in this paper which showed, morning shift driversapplied higher speed and sharper acceleration rate.

Specifically, Fig. 10 presents the mean of instantaneous fuel consumption at every 10km/h speed bin. Overall data exhibit that fuel consumption increases as speed increases. In addition, it can be highlighted that, sharp increase in fuel consumption occurred at speed range 0 – 20km/h. However, within the 20.01 km/h to 50 km/h speed range, the fuel consumption is likely to remain steady. The figure also shows that, within the speed range of 50.01 km/h to 80 km/h, the fuel increased approximately linearly with the increase in speed.

Table 7. Comparison of bus fuel consumption between morning and afternoon shift hour

t		14.232
df		498390.384
Sig. (2-tailed)		.000
Mean Difference		.0981
Std. Error Difference		.0069
95% Confidence Interval of the Difference	Lower	.0846
	Upper	.1117

Table 8. Descriptive analysis of bus fuel consumption for morning and afternoon shift hour

Working shift hour	N	Mean of fuel consumption (ml/s)	Std. Deviation	Std. Error Mean
Morning (6.00 am - 3.00 pm)	254034	2.3279	2.4700	.00
Afternoon (3.00 pm - 12.00 am)	244406	2.2298	2.3997	.00

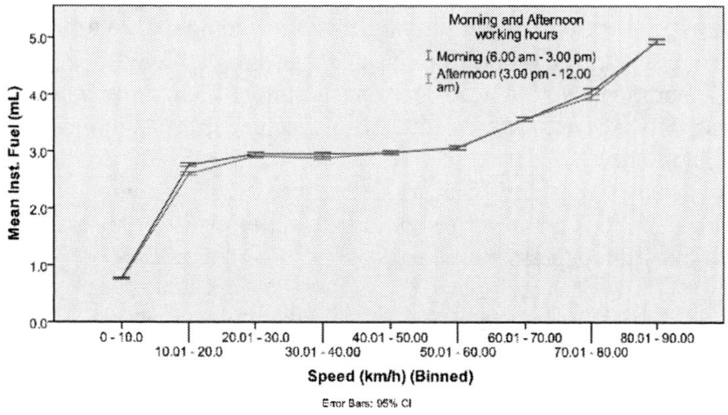

Fig. 10. Mean of instantaneous fuel consumption vs. Speed

Table 9. Mean of acceleration at different speed range

Speed range	Accelerating		Braking	
	Morning shift	Afternoon shift	Morning shift	Afternoon shift
0 - 20km/h	.5673	.5161	-.6944	-.7487
20.01 – 50 km/h	.3410	.3323	-.5163	-.5537
50.01 - 90 km/h	.0889	.0828	-.2471	-.2635

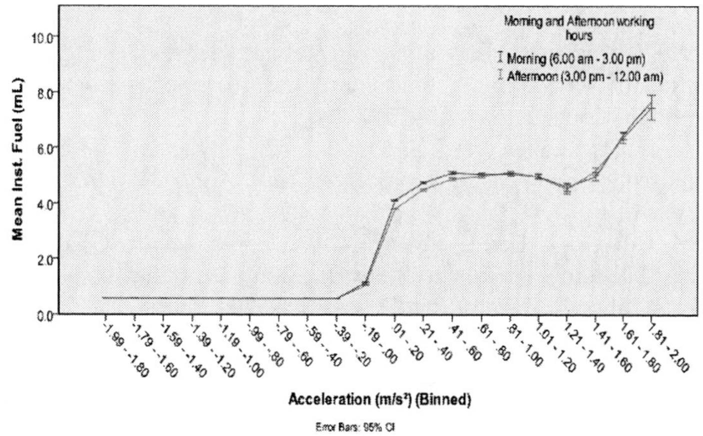

Fig. 11. Mean of instantaneous fuel consumption vs. acceleration

The analysis on acceleration impact on fuel consumption is also assessed in this study. Fig. 11 shows that the bus fuel consumption rate increased as the acceleration increases when drivers increased the acceleration rate (>0 ms^{-2}). While as drivers reduce the speed (decelerating), minimum fuel is consumed.

5 Conclusion

The results from this study found that there are differences in bus driving behavior and working shift. The bus drivers used high speed and high acceleration in the morning shift. The drivers also used sharp braking in the morning shift compared to afternoon shift. This is related to the busy traffic condition in the morning as everyone isrushing to the workplace. This condition also impacts fuel consumption. This study indicates that for all driving shifts, the bus drivers were found to consume more than 70% fuel while accelerating and braking.

As a conclusion, working shift has an influence on bus driving behavior and fuel consumption. Therefore, bus companies should take action with these issues. Further study should be carried out with express buses.

We are grateful to Siti Maisarah Che Hasan, Universiti Tun Hussein Onn Malaysia for her contribution to this study and the Ministry of Higher Education, Malaysia for funding and supporting this research project under grant code FRGS 1451

References

[1] Rospa, *Driver Fatigue and Road Accidents*, The Royal Society for the Prevention of Accidents, (2001)

[2] J. Horne, L. Reyner, Vehicle accidents related to sleep: A review, Occupational and Environmental Medicine, **56**(5), 289-94 (1999)

[3] H.D. Howarth, *An Investigation of Sleep And Fatigue In Transit Bus Operators on Different Work Schedules*, (2002)

[4] T. Sando, E. Mtoi, R. Moses, *Potential Causes of Driver Fatigue : A Study on Transit Bus Operators In Florida*, (2010)

[5] A.D.N.M. Abdullah, H. L. Von, Factors of fatigue and bus accident, *International Conference on Innovativation, Management and Service*, (2011)

[6] M.M. Rohani, *Bus Driving Behaviour and Fuel Consumption*, PhD Thesis, University of Southampton, (2012)

[7] J. Van Mierlo, G. Maggetto, E. Van De Burgwal, R. Gense, Driving style and traffic measures-influence on vehicle emissions and fuel consumption, Journal of Automobile Engineering, **218**(1), 43–50 (2004)

[8] I. De Vlieger, D. De Keukeleere, J.G. Kretzschmar, Environmental effects of driving behaviour and congestion related to passenger cars, Atmos. Environ., **34**(27), 4649–4655, (2000)

[9] B. Beusen, S. Broekx, T. Denys, C. Beckx, B. Degraeuwe, M. Gijsbers, L.L. Panis, Using on-board logging devices to study the long-term impact of an eco-driving course, Transportation Research - Part D, **14**, 514–520 (2009)

Differences of Driving Experience and Gender on Traffic Offences Among Malaysian Motorists

Sharifah Liew[1,*], *Rizati* Hamidun[1], and *Nor Fadilah* Mohd Soid[1]

[1]Malaysian Institute of Road Safety Research, Kajang, Malaysia

Abstract. This study attempt to examine the influence of driving experience and gender on different traffic offences. Questionnaire survey was conducted to measure the frequency of committing eleven types of traffic. Data collected were analysed using t-test and Chi-square. Results indicate that male drivers have high score in speeding, red light running, tailgating and using emergency lane. Significant differences emerged between driving experiences in speeding, red light running, dangerously cutting into traffic, using hand phone while driving/riding and driving/riding with noncompliance specification vehicle.

1 Introduction

Road accidents is known as a concern to public health, resulted about 1.2 million deaths worldwide [1]. Proposing traffic regulation is one of the countermeasure taken aimed to reduce the frequency and severity of road accident. Ignore to the traffic regulation is among the human behavioural factor that may lead to the road accident [2].

Traffic offences such as speeding, red light running, tailgating and using emergency would trouble other road users and might end up with an accident. Speeding would contribute to more severe accident due to high impact [3], yet difficult to handle since it is a widespread phenomenon [4]. The offence of speeding have found to be associated with driver impulsivity [5]. Traffic offences somehow related to the expression of driving anger [6].

Relationship of traffic offences and driving experience had been studied by previous researcher. For example, red light running violation through simulation experiment examined by Wu et al. [7] for taxi drivers and non-professional drivers. Results indicated that taxi drivers prone to violate traffic light as compared to the non-professional drivers. In other studies, effect of fog to driving performance for professional and non-professional drivers had been conducted by [8] using simulation experiment.

Although several studies have focus on the driving experience, study on the differences of driving experience on traffic offences is still limited. Thus, this study explores the differences of driving experience toward accident involvement, summon received and 10 types of traffic offences and gender.

* Corresponding author: sharifahliew@miros.gov.my

2 Methodology

2.1 Data collection

This study was conducted using semi-quantitative approach and was carried out by using self-administered questionnaire to collect data on the road users' aberrant behaviour. The survey was conducted in north and south of PLUS expressway and selected area in Klang Valley. The target population for this study comprised of road users who own a valid driving license. Total of 300 respondents were involved in this study, 177 (59.8%) respondents were motorist and 119 (40.2%) respondents were motorcyclists. The respondents were randomly selected from two zone Klang Valley and expressway heading to north and south as shown in Table 1 below.

Table 1. Survey locations

Zone	Location
Klang Valley	Tesco Klang
	Tesco Ampang
Expressway heading to north and south	Genting Sempah
	RnR Rawang
	RnR Seremban

2.1.1 Respondents

A total of 300 respondents participated in the study, 56% of them are male and 44% of them females. The majority of the respondents came from the age group of 26 to 35 years old (33.3%), 36 to 45 years old (20.7%) & more than 46 years old (25.7%); there were only 20.3% of respondents who were less than 25 years old. All respondents have driving licences with varies period in driving experiences. The majority of respondents has 3-10 years of driving experience. About 73.1% of the respondents were married. In terms of race, 71.6% were Malays, 13.9% Chinese and 14.5% Indian. More than half of the respondents had higher than tertiary education level. Among the respondents, 59.8 % were car drivers while the rest were motorcyclists.

2.1.2 Questionnaires

The questionnaires for this study were designed to gather information related to the respondents' demographic characteristics, accident and summon experience and their involvement in the traffic offences. The questionnaire consisted of three parts. Part A contained the respondents' demographic information such as age, driving experience, education, race, religion; Part B asking about an involvement of respondent in an accident and summon received. Part C is about the frequency of committing different types offences by respondents. Offences measured are: (1) speeding; (2) red light running; (3) overtaking at double lines; (4) queue jumping; (5) illegal use of emergency lane using mobile phone while riding/driving; (6) dangerously cutting into traffic; (7) using hand phone while driving / riding; (8) not wearing seatbelt / helmet; (9) driving/riding with expired road tax; (10) driving/riding with noncompliance specification vehicle and (11) dangerously cutting into traffic. The frequency of committing the 11 traffic offences was measured using score from 0 (never) to 10 (most of the time).

2.2 Data analysis

Data collected has been coded and analysed using SPSS. Descriptive statistics such as mean and percentage was performed to describe the overall of aberrant behaviour and the perception towards traffic enforcement visibility. While inferential statistics such as the Chi-square, T-test and ANOVA was performed to achieve the objective of the research which were to examine the differences between aberrant behaviour and selected demographic factors.

3 Results

Table 2 shows the non-parametric correlation between the demographic variables and the accident involvement of the respondents. It shows that there is a significant relation between the gender and their accidents involvement.

Table 2. Non-parametric test for accident involvement and demographic details

Demographic Details	Accident Involvement		Chi-Square	P-value
	Yes (n)	No (n)		
Gender			12.262	0.000
Male	69	99		
Female	29	103		
Driving Experience			4.667	0.198
Less than 2	5	23		
3-10	35	75		
11-20	35	54		
More than 20	23	50		

Table 3 presents the non-parametric test between the demographic variables and the summon received by the respondents. It is found that there is a significant relation between gender and the summons received by the individuals.

Table 3. Non-parametric test for summon received and demographic details

Demographic Details	Summon Received		Chi-Square	P-value
	Yes (n)	No (n)		
Gender			14.110	0.000
Male	78	87		
Female	34	97		
Driving Experience			7.374	0.061
Less than 2	6	22		
3-10	41	69		
11-20	32	57		
More than 20	36	37		

Table 4 shows the means for both sex for the different measures of traffic offences studied. The t-values found are presented along with their statistical meaning. By comparison, the men showed higher score in speeding, red light running, illegal use of emergency lane and tailgating significantly.

Table 4. Mean and sex comparisons of the measures of traffic offences

Traffic Offense	Male (Mean)	Female (Mean)	t	p-value
Speeding	2.70	1.82	2.878	0.004*
Red light running	1.67	1.06	2.447	0.015*
Overtaking at double line	1.30	0.89	1.792	0.074
Queue jumping	1.41	1.34	0.267	0.790
Illegal use of emergency lane	1.33	0.86	1.965	0.050*
Dangerously cutting into traffic	1.88	1.47	1.403	0.162
Not wearing seatbelt/helmet	1.82	1.47	1.216	0.225
Using hand phone while driving/riding	1.43	1.25	0.765	0.228
Tailgating	1.39	0.77	2.930	0.004*
Driving/riding with expired road tax	0.67	0.54	0.698	0.486
Driving/riding with noncompliance specification vehicle	0.59	0.39	1.125	0.262

Result on the comparison between driving experience and the traffic offences is indicated in Table 5. It is observed that experience has significant differences in the area of speeding, red light running, dangerously cutting into traffic, using hand phone while driving/riding and driving/riding with noncompliance specification vehicle.

Table 5. Comparisons between driving experience and the traffic offences

Traffic Offense	F	p-value
Speeding	2.952	0.033*
Red light running	3.471	0.017*
Overtaking at double line	0.945	0.419
Queue jumping	0.193	0.901
Illegal use of emergency lane	0.746	0.525
Dangerously cutting into traffic	4.874	0.003*
Not wearing seatbelt / helmet	2.118	0.098
Using hand phone while driving / riding	3.299	0.021*
Tailgating	0.549	0.649
Driving/riding with expired road tax	2.065	0.105
Driving/riding with noncompliance specification vehicle	2.803	0.040*

4 Discussions and conclusion

The purpose of the study was to explore the differences of driving experience and gender in accident involvement, summon received and involvement in traffic offences. Analysis showed gender differences in accident involvement, experience of being summonsed and several traffic offences including speeding, red light running, illegal use of emergency lane and tailgating.

Comparing gender, male is more frequent to be involved in accident. These result is supported by [9-11]. A possible explanation for this might be due to the fact that male drivers has more number of kilometres driven as compared to female drivers and number of female drivers is significantly lower than the number of male drivers. However, it is an undisputable fact that male and female have different behaviour. Sudden reactions of males to events, their nervousness as well as their desire to "prove" themselves via their driving skills create a basis for accidents [12].

In regard to the receiving summon by gender, male drivers have received more summons as compared to female drivers. This finding were supported by [6, 13, 14] whereby number of fines and accident were greater among male drivers and higher tendency to traffic violations. It seems possible that this result is due to dangerous driving behaviour are highest in males [15, 16].

Furthermore, male drivers have significantly higher score in speeding, red light running, illegal use of emergency lane and tailgating among eleven traffic measures. This finding indicates that male drivers are more prone to risky driving behaviour as compared to female drivers, which in line with findings by [17]. Several studies regarding gender difference in traffic offences have shown that peer influence leads to increased speeding behaviour among male drivers [18]. One interesting finding by [19] where males and females have difference perception and preference of speeding.

Besides that, men tailgated more than women and tailgating occurred more during rush-hour and when there was a traffic officer present [20]. The scenario is a norm in Malaysia as public had to travelled long distance to their workplace and the addition of registered vehicle does not seem to decreasing. Unless the service of public transportation could be improved, the possibility of publics shifting their mode of transport will increase.

With respect to the findings related to the effect of driving experience on traffic offences, high experienced driver was more prone to commit traffic offences such as speeding, red light running, dangerously cutting into traffic, using hand phone while driving/riding and driving/riding with noncompliance specification vehicle. These findings are in agreement with those obtained by [14, 21]. It seems possible that these results are due to drivers became more sensitive to situational cues that they were likely to behave in a manner that was congruent with the surrounding situation and individuals [22]. For example, experienced drivers drove faster in clear visibility than novice drivers, yet they reduced their speed more in reduced visibility so that both groups drove at the same speed in simulated fog [8]. Even in some extent, these drivers can predict just by observing the enforcement officer whether or not they will be stopped when you violating traffic rules. This is where the role of perception of being caught (POBC) is important in enhancing traffic obligation.

Based on the findings, a special need to plan proper intervention to specific target groups and traffic violations are important in behaviour modifications and towards the reduction of national accident rate. Nevertheless, the authorities and stakeholder must have the same vision and committed in order to solved our nation's concern which in this context of road safety issues.

References

[1] World Health Organization, *Global status on road report 2015*, Geneva, (2015)

[2] E. Petridou, M. Moustaki, Human factors in the causation of road traffic crashes, European J. of Epidemiology, **16**(9), 819–26 (2000)

[3] N.N. Sze, S.C. Wong, Diagnostic analysis of the logistic model for pedestrian injury severity in traffic crashes, Accident Analysis and Prevention, **39**(6), 1267–78 (2007)

[4] R. Elvik, Why some road safety problems are more difficult to solve than others,

Accident Analysis and Prevention, **42**(4), 1089–1096 (2010)
[5] P. Bıçaksız, T. Özkan, Impulsivity and driver behaviors, offences and accident involvement: A systematic review, Transportation Research Part F: Traffic Psychology and Behaviour, **38**, 194–223 (2016)
[6] B. González-iglesias, J.A. Gómez-fraguela, M.Á. Luengo-martín, Driving anger and traffic violations: Gender differences, Transportation Research Part F: Traffic Psychology and Behaviour, **15**, 404–412 (2012)
[7] J. Wu, X. Yan, E. Radwan, Discrepancy analysis of driving performance of taxi drivers and non-professional drivers for red-light running violation and crash avoidance at intersections, Accident Analysis and Prevention, **91**, 1–9 (2016)
[8] X. Li, X. Yan, S.C. Wong, Effects of fog, driver experience and gender on driving behavior on S-curved road segments, Accident Analysis and Prevention, **77**, 91–104 (2015)
[9] E. Jiménez-Mejías, C.A. Prieto, V. Martínez-Ruiz, J.D.D.L. Del Castillo, P. Lardelli-Claret, J.J. Jiménez-Moleón, Gender-related differences in distances travelled, driving behaviour and traffic accidents among university students, Transportation Research Part F: Traffic Psychology and Behaviour, **27**, 81–89 (2014)
[10] A.H. Al-balbissi, A.H. Al-balbissi, Role of gender in road accidents, Traffic Injury Prevention, **4**(1), 64-73 (2003)
[11] M. Karacasu, A. Er, An analysis on distribution of traffic faults in accidents, based on driver's age and gender: Eskisehir case, Procedia-Social and Behavioral Sciences, **20**, 776–785 (2011)
[12] Social Issues Research Center, *Sex differences in driving and insurance risk*, Oxford, United Kingdom, (2004)
[13] J. Rudjanakanoknad, P. Prarom, S. Panwai, Attitudes of drivers towards speed enforcement measures on Bangkok expressways, Procedia-Social and Behavioral Sciences, **48**, 222–233 (2012)
[14] H.S. Lonczak, C. Neighbors, D.M. Donovan, Predicting risky and angry driving as a function of gender, Accident Analysis and Prevention, **39**(3), 536–545 (2007)
[15] D. Iliescu, P. Sârbescu, The relationship of dangerous driving with traffic offenses: A study on an adapted measure of dangerous driving, Accident Analysis and Prevention, **51**, 33–41 (2013)
[16] D. Yagil, Gender and age-related differences in attitudes toward traffic laws and traffic violations, Transportation Research Part F: Traffic Psychology and Behaviour, **1**(2), 123–135 (1998)
[17] S. Oltedal, T. Rundmo, The effects of personality and gender on risky driving behaviour and accident involvement, Safety Science, **44**(7), 621–628 (2006)
[18] M. Moller, S. Haustein, Peer influence on speeding behaviour among male drivers aged 18 and 28, Accident Analysis and Prevention, **64**, 92–99 (2014)
[19] C. Horvath, I. Lewis, B. Watson, The beliefs which motivate young male and female drivers to speed: A comparison of low and high intenders, Accident Analysis and Prevention, **45**, 334–341, (2012)
[20] N.M. Monteiro, S.K. Balogun, M. Kote, K. Tlhabano, Stationary tailgating in Gaborone, Botswana: The influence of gender, time of day, type of vehicle and presence of traffic officer, IATSS Research, **38** (2), 157–163 (2015)
[21] Y. Hoffman, T. Rosenbloom, Driving experience moderates the effect of implicit versus explicit threat priming on hazard perception test, Accident Analysis and Prevention, **92**, 82–88 (2016)
[22] Y. Xu, Y. Li, L. Jiang, The effects of situational factors and impulsiveness on drivers' intentions to violate traffic rules: Difference of driving experience, Accident Analysis and Prevention, **62**, 54–62 (2014)

Energy Consumption of Insulated Material Using Thermal Effect Analysis

M. A. Fadzil[1,*], *M. A.* Norliyati [1], *M. A.* Hilmi[1], *A. R.* Ridzuan[1], *M. H.* Wan Ibrahim[2], and *R. Z.* Assrul [1]

[1]Institut of Infrastructure Engineering and Sustainable Management (IIESM), Faculty of Civil Engineering, Universiti Teknologi MARA, 40450 Shah Alam, Selangor, Malaysia
[2]Faculty of Civil and Environmental Engineering, Universiti Tun Hussein Onn Malaysia, 86400 Parit Raja, Johor, Malaysia

Abstract. Wall is one of the structures elements that resist direct heat from the atmosphere. Modification on several structures is relevance to reduce filtrate thermal movement on wall. Insulation material seems to be suitable to be implemented since its purpose meets the heat resistance requirement. Insulation material applied as to generate positive impact in energy saving through reduction in total building energy consumption. Fiberglass is one of the insulation materials that can be used to insulate a space from heat and sound. Fiberglass is flammable insulation material with R Value rated of R-2.9 to R-3.8 which meets the requirement in minimizing heat transfer. Finite element software, ABAQUS v6.13 employed for analyze non insulated wall and other insulated wall with different wall thicknesses. The several calculations related to overall heat movement, total energy consumption per unit area of wall, life cycle cost analysis and determination of optimal insulation thickness is calculated due to show the potential of the implementation in minimize heat transfer and generate potential energy saving in building operation. It is hoped that the study can contribute to better understanding on the potential building wall retrofitting works in increasing building serviceability and creating potential benefits for building owner.

1 Introduction

A constructed building relatively contribute to the effect of global warming, as a result, there are an issues in building energy consumption and carbon emission [1]. Based on the geological location and condition, Malaysia is experiencing warm and humid climate condition throughout a year. Due to that condition, Malaysia building having an impact which is high intensity of solar radiation and high daily air temperature [2] world climate change is considered to have strong effects on building needs and requirement as the cooling needs are related to temperature conditions and weather variations. [3].
Refer to the previous study, most of the buildings prefer to assign maximum cooling load to accommodate the comfort level [4]. Therefore, the energy demand especially in cooling

*Corresponding author: fadiil2013@yahoo.com

DOI: 10.1051/matecconf/201710308017

load has increase and currently burdens the building owner. Furthermore, these directly will give great impact towards air pollutant and carbon emission in reflected heat and heat generated from the cooling devices.

It is estimated that buildings consume at least 40% of energy and 16% of water used annually [5]. Energy growth in most develop countries has accounted that about 13.6% of total energy consumption and 48% of electricity consumption consumed in commercial and residential buildings [6]. More than 40% of the energy consumed by Malaysian buildings can be reduced by proper energy efficiency practice and application of suitable sustainable technology to all building in Malaysia [7]. This is next lead to the existence of MS 1525 2007 which is the Energy Efficiency code of practice as a guideline for energy efficient measure in Malaysia buildings.

Modification and renovation can be made by installation of thermal insulation material. Thermal insulation is applicable in facing the climatic problem and to reduce the air pollution forms from heat reflect and air conditioning operation. In reflected to the R-Value of the importance of thermal insulation material where an insulated material consists of high resistance value. Majority of the material R-Value of the wall is attributable to the insulation. R-value is influenced by the surface material that installed as wall surface including wall installed with insulation surfaces [8]. The main purpose of insulation material is to retards the heat movement on two different surfaces with different temperature, less electricity or energy is required the desired inside temperature [9].

An analysis tool such as ABAQUS v6.13 was used in this study. Finite element method is a numerical method to solve an integral or differential equation. How it works is by predicting the piecewise continuous function for the solution and obtaining the parameters of the function to minimize the error in the solution. The intention of this study is to determine heat movement inside the specimen to generate the different between two specimens with and without insulation material.

2 Analytical studies

A complete analysis using ABAQUS v6.13 requires a description of the material, the model configuration, temperature on both outside and inside, duration of surface to be thermally tested, and boundary condition. The modeling analyzed based on 3m x 6m wall panel as shown in Fig. 1.

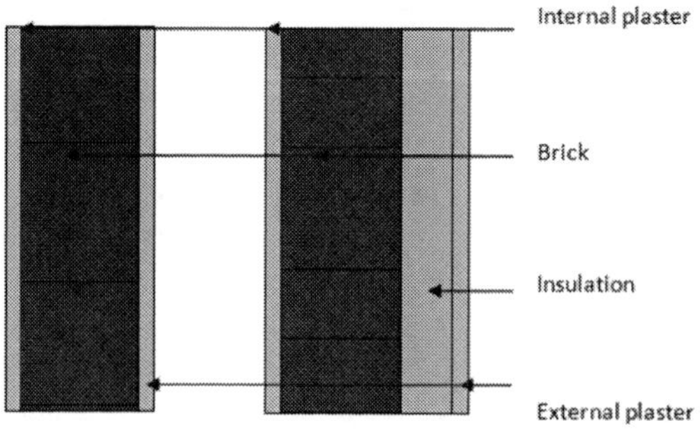

Fig. 1. Test Set up

Data such as absolute zero, Stefan Boltzmann constant, emissivity, material thermal conductivity, and specific heat of all related material will influence the result accuracy. Each data is important in thermal analysis work. Therefore, data such as surface contact between two materials is important in analysis since it will influence result of the analysis.

In finite element analysis the structural component is divided into detail shape of meshing forms. The smaller the meshing is made, the more accurate data possessed. Meshing work is done individually on every material. However, there were limitation numbers in meshing element where it will affect the analysis process.

For this study, analysis is conducted on four different samples which are non insulated wall and another 3 insulated wall with 25mm, 50mm and 75mm thicknesses. Detail of material properties used in analysis are shown in Table 1.

Table 1. Material properties

Properties	Cement Plaster	Clay Brick	Fiberglass (Insulation Material)
Density (ρ), (kg/m^3)	1860	2243	16
Thermal Conductivity (k), (W/mK)	0.72	0.60	0.04
Specific Heat (c), (J/kgK)	840	900	1000
Emissivity (ε)	0.54	0.45	0.75

Heat is applied from both external and internal wall surface due to replace heat from outside heat radiation and room heat temperature. The indoor air kept at a fixed design temperature T_i.

While on the external wall surface is subjected to an outdoor air temperature T_o. During the analysis, wall is exposed to heat in 25200 sec since normally wall is exposed to extreme heat radiation from 11 am to 6 pm. The emissivity, ε of the material is important due to response towards heat. Heat applied on external wall surface is 37°C while 21°C applied on internal wall surface to response as room temperature. Heat transfer coefficient which combine both convection and radiation effects are considered for both inside and outside wall surfaces with assumption of the surrounding surfaces to be at the same temperature as the ambient air.

First analysis is conducted between non insulated wall and insulated wall due for the purpose to define differential heat transmission characteristic and to determine the response of insulation material towards thermal activity. Then, analysis is continued with different insulation thickness. The purpose of the analysis is to determine the heat transfer and heat release differential influenced from the thickness adjustment.

Result of heat transfer is analyzed and illustrated in heat contour and temperature on each heat transfer phase which segregated with different colour from maximum temperature, red to the minimum temperature represent by blue colour contour as shown in Fig. 2 and Fig. 3.

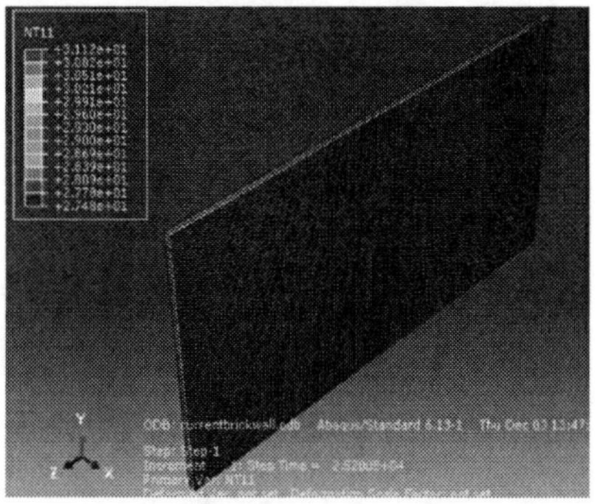

Fig. 2. Non isolated material

 DOI: 10.1051/matecconf/201710308017

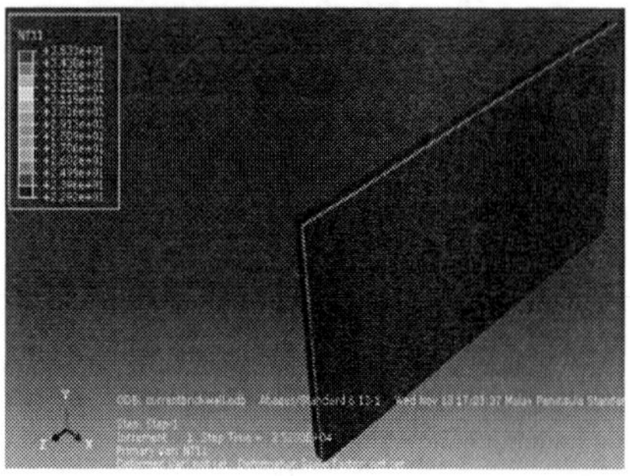

Fig. 3. Isolated material

From the results obtained through the analysis, data will be examined mathematically towards its impact on energy consumption, and it reflection on energy savings. Lastly, the determination of Profit Worth Factor, PWF is measured to identify the effectiveness of the implementation investment on N period of time. From the PWF data measured, the optimal thickness which relevance to be implemented is obtained due to identify the relevance of implementation towards the investment made.

3 Results and discussions

3.1 Full finite element analysis result

From the analysis result, it is shown that heat flow is reducing by installation of insulation material. Refer to Table 2, the temperature start to reduce at the fiberglass portion.

Table 2. Thermal reaction on wall

Insulation Thickness	Thermal Contour Activity
Non Insulated wall	
25mm Insulation	
50mm insulation	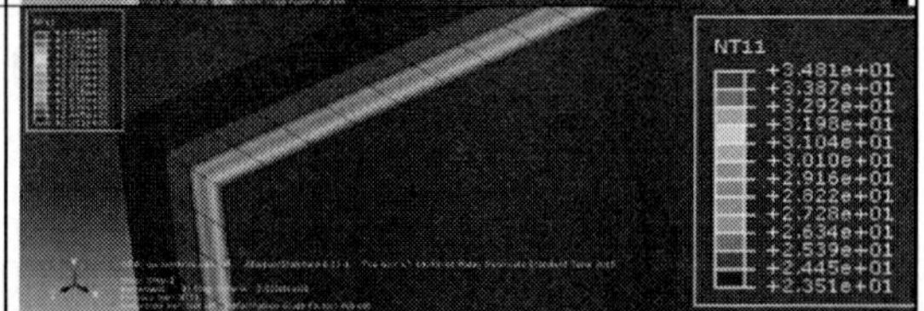
75mm insulation	

DOI: 10.1051/matecconf/201710308017

Other than that, it is shown that there were different heats flow contour characteristic throughout the analysis result for all 4 type of walls.

3.2 Thermal effect between non insulated and insulated wall

Resulting from the analysis as shown in Fig. 2 and Fig. 3, there were different thermal contour shapes illustrated from the simulation. Thermal activity that react on non insulated wall shown that the thermal movement is equal from external plaster layer to internal plaster layer while the insulated walls contour shown that heat transfer activity is slow and constant starting form insulation layer and continued to external layer. The heat stored on the last layer of the insulated wall is lesser than non insulated wall. During nighttimes, less energy is required for cooling purposes where the heat released on the internal area is lower than temperature release in common non insulated wall. Fig. 4 and Fig. 5 show contour illustration for non isolated wall and isolated wall respectively.

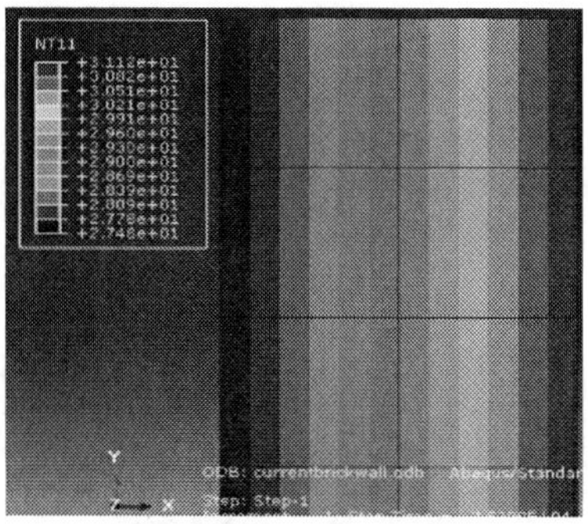

Fig. 4. Non insulated wall

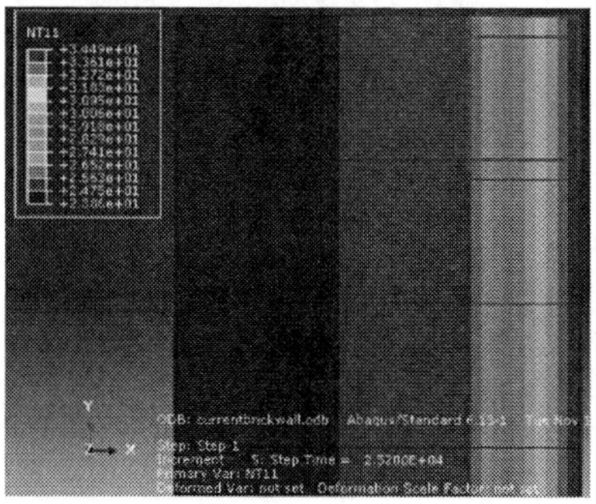

Fig. 5 Insulated wall

Fig. 6 shows that the heat flow of insulated wall start to reduce at insulation layer and the heat flow released to the brick and internal layer are lower than temperature of non insulated wall. Non insulated wall is having a constant heat reduction and the percentage of temperature reduce is smaller than insulated wall. The introduction of insulation can effectively reduce energy cooling demand and peak load of the cooling system. By adding insulation on wall structure energy cooling load demand and peak load will decrease by 12%.

DOI: 10.1051/matecconf/201710308017

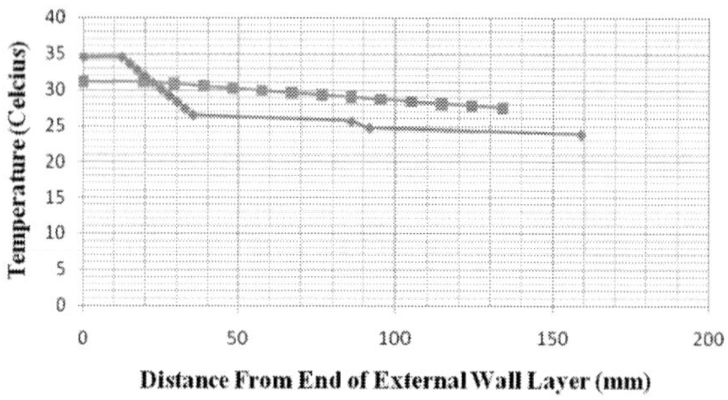

Fig. 6 Graph of temperature versus wall thickness for non insulated and insulated wall

3.3 Thermal radiation impact on walls

Fig. 7 shows the percentage temperature versus insulation thickness graph of the walls reaction after having a natural outside temperature of 37°C on the external surface. From the graph, it shown that by introduction of insulation layer, the heat stored on the external plaster layer is higher when the thickness of insulation increase. In these cases, result shows that heat radiated or reflected to the environment is decreases parallel to the increment of fibreglass insulation layer thickness.

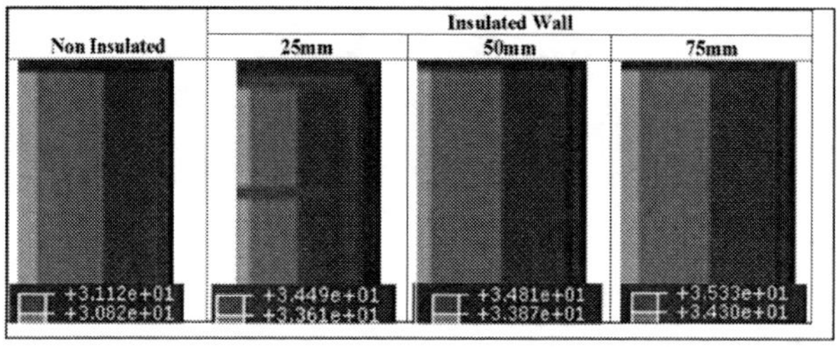

Fig. 7 Heat Radiation on external wall

Based on Fig. 7 and Fig. 8, it is shown that without installation of insulation material, a wall surface can contribute 16% of heat that being reflected back to the atmosphere. While by the adding thermal insulation layer, it is shown that the thermal radiation rates slowly decrease by the increment of insulation thicknesses. By implementing 25mm insulation layer, the heat radiation percentage decrease 56% from non insulated wall in total heat radiation percentage of 7%. The percentage is continuously decreased to 6% on 50mm insulation thickness and 5% on 75mm insulation thickness installation. This show the effectiveness of insulation walls which will react on slowing the thermal movement through the wall.

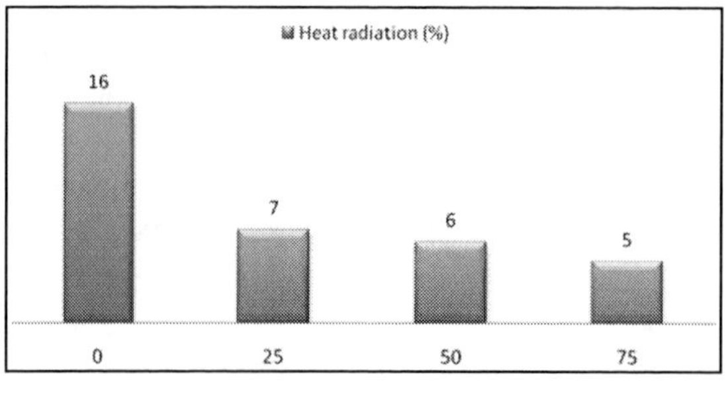

Fig. 8. Heat radiation percentages

3.4 Thermal Conduction impact on internal area

From the simulation, result of the heat stored at two last layer shown that the temperature reading four all four material is different as shown in Table 3. Temperature recorded in the last layer is 27.48°C. Meanwhile the temperature recorded at all three others insulated walls shown the temperature start decreasing by the increase of insulation thickness. On 25 mm insulation thicknesses, results shown that the temperature is 23.86°C, on 50mm thickness, temperature recorded from the simulation is 23.51°C and lastly, at the thickness of 75mm, temperature stored at the both brick and internal layer is 22.92°C. From these results, it is shown that insulation will give benefit in term of reducing the heat release on the internal area which needs a comfort environment and at these cases, to human thermal comfort level is at 24°C.

Table 3. Thermal conduction temperature

Type of wall	Temperature °C
Non Insulated	27.48
25mm Insulated Wall	23.86
50mm Insulated Wall	23.51
75mm Insulated Wall	22.92

Table 4 Thermal Reaction Characteristic

Type of Wall	Simulation Visualization	Result description
Non insulated	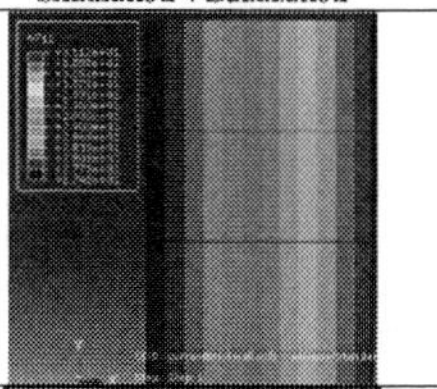	• thickness of dark blue region equal to others temperature region
25mm insulated wall	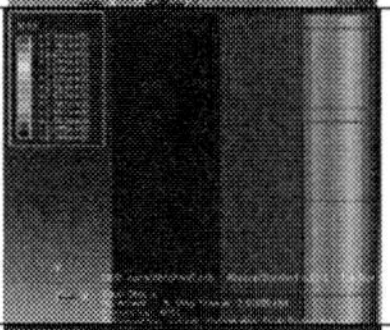	• Blue region region wider up to 55% of the clay brick thickness
50mm insulated wall		• Dark blue region covering up to 65% of clay brick thickness
75mm insulated wall		• The temperature starts to be consistent at 22.92°C at 80% of clay brick layer. •

Based on the results illustrated generated from the simulation, it review the different thermal reaction on all four examined walls. From Table 4, it can be described that fibreglass insulation material reacts on reducing the heat movement from outside to inside area. In non insulated walls, the dark blue region thickness is almost equal to others region. By adding fibreglass as thermal insulation material the trend of heat movement shows a different contour where the blue or cool temperature region is thicker than other region. The blue cool region starts to cover the clay brick layer. This lastly support the fact on chapter two which described insulation materials is installed as a median to reduce the heat flow and reducing the heat released to the internal area [10] that resulting a major impact on cooling load demand reduction.

Figure 9 below shows the graph of temperature reaction on all walls analyzed during the study. It show different heat flow recorded from both non insulated and insulated wall. Changing the fibreglass thermal insulation of the external wall from 25mm to other 50mm and 75mm reduces the cooling energy demand and peak load by 14% and 16% respectively. It can be seen that the ticker the insulation the greater reduction in both cooling energy demand and cooling load demand.

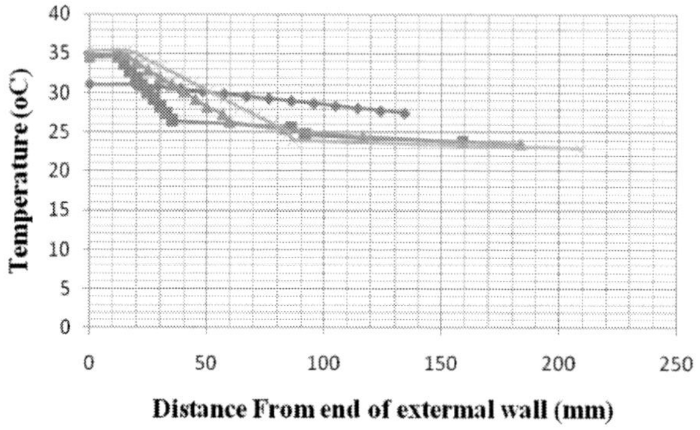

Fig. 9. Graph Thermal temperature reactions on walls

3.5 Annual energy cost for unit wall area

In this study, annual energy cost calculated by considering data such as fuel cost, Cf in RM/kg and its heating value *Hu* in J/kg. From the calculation, annual energy cost per unit area for all four walls is tabulated in Figure 10 below. From previous data description, the annual energy cost is influenced by the thickness of insulation. The thicker it goes, the lower energy cost resulted in the calculation. This calculation leads to the effectiveness of insulation layer installation in promoting energy saving opportunities in buildings wall. Annual energy cost recorded about 50% cost reduction in increment of insulation thickness layer. It is shown in 25mm thickness, the rate of annual energy cost per unit area are RM12.49/m^2 and the cost drop to RM 7.17/unit wall area after addition of 50% insulation thickness to 50mm and lastly at 75mm insulation thickness, the annual energy cost is drop to RM 5.03 per unit wall area. From the simulation, result of the heat stored at two last layer shown that the temperature reading four all four material is different as shown in Table 3.

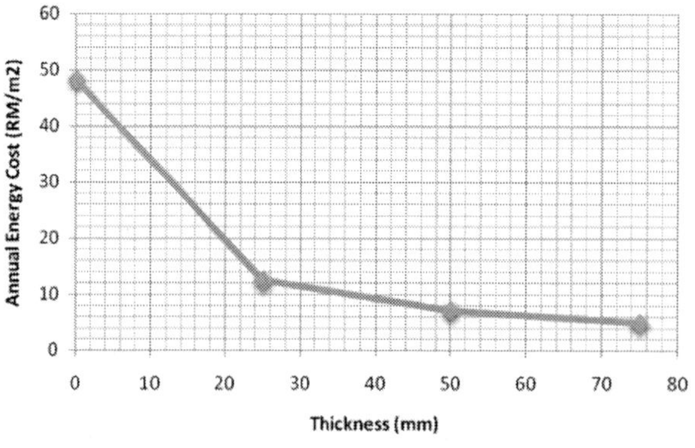

Fig. 10. Graph of Annual energy cost versus insulation thickness

3.6 Energy saving

Refer to Table 5 and Fig. 10, total energy consumption (EA) shows that the total energy consumption has been reduced starting from 25mm Insulation wall at 40.74kWh energy consumption, and continuously reduce to 40.53kWh and 40.33kWh at both 50mm insulation wall and 75mm insulation wall. By reduction of energy consumption, less energy demand is required to accommodate cooling load for the internal area.

Table 5. Total energy consumption

Thickness (mm)	0	25	50	75
ADH	4.5	4.5	4.5	4.5
q	26.67	26.53	26.39	26.26
COP	2.93	2.93	2.93	2.93
EA (kWh)	40.96	40.74	40.53	40.33

Result is be fortified as shown in Fig. 11and Fig. 12 where the energy consumed per air conditioning system and annual energy consumption per air-conditioning system reduce at every walls with insulation materials. Even though the cost involve in implementing insulation wall is high, both graph shown that the guaranteed returns of the investment will give the benefits to the house owner.

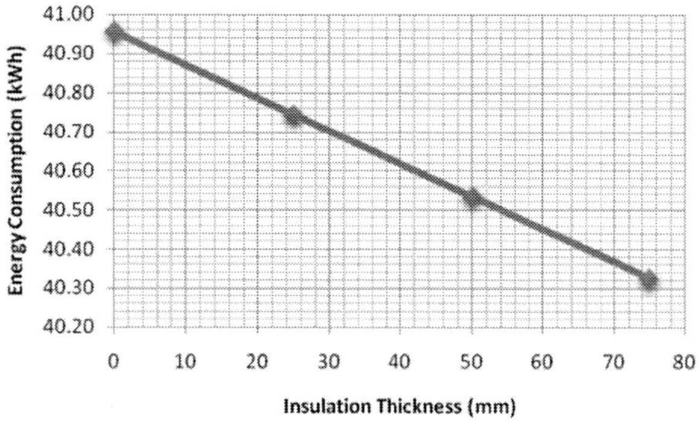

Fig. 11. Graph total energy consumption versus insulation thickness

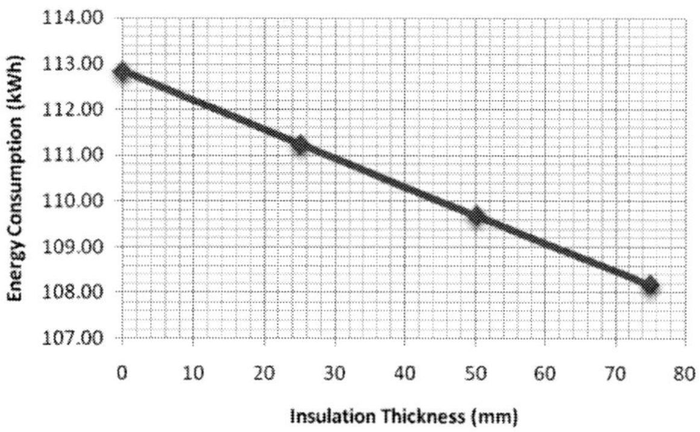

Fig. 12. Graph annual energy consumed per air conditioning unit versus insulation thickness

3.7 Life cycle analysis of implementation and optimum thickness layer

In this study, life cycle cost employed as to determine the relevance of the investment for long term period. Life cycle cost study is computes the cost that reflected by parameters such as heating cost over the lifetime of buildings, the total cost of heating throughout N years where 10 years is estimated and evaluated by using present worth factor (PWF) in calculation of PWF, current interest rate and inflation rate data is important to monitoring the relevance of this investment. In this study, PWF value calculated is 5.04. This PWF data will next influence the determination of insulation cost and yet the optimum thickness that found relevance to applied during N years' time.

The optimal thickness represent the optimal investment that worth to invest and can generate both saving and provide thermal heat reduction in order to reduce cooling demand on internal spaces. In this study, the optimal thickness measured that suits PWF at 10 years duration is 110 mm.

4 Conclusions

From this analytical analysis, the following conclusion can be drawn:

i) Temperature reduces in implementation of thermal insulation material. Heat release to the internal area will be lower than non-insulated wall. Insulation given an impact 13.17% heat reduction.
ii) Heat movement inside structure body is different between non insulated and insulated wall.
iii) Addition of insulation thickness will give continuous positive increasing heat reduction that directly will reduce more cooling demand and cooling load demand.
iv) The annual energy cost per unit area of the wall having 50% cut off after insulate the wall with fibreglass material.
v) The energy consumption used in cooling purposes is reduce on thermal insulation material.
vi) Optimum insulation thickness best to be installed is 110mm.

We would like to thanks to the Universiti Teknologi MARA, Malaysia for computer laboratory fascilities. The first and second author are supported by Ministry of Higher Education, Malaysia for supporting the research. (References 600-RMI/FRGS 5/3 (0120/2016)

References

[1] A. Ismail, M.H. Abdul Samad, A.M.A. Rahman, F.S. Yeok, Cooling potentials and CO2 uptake of Ipomoea Pescaprae installed on the flat roof of a single storey residential building in Malaysia, Procedia - Social and Behavioural Sciences, **35**, 361-368 (2012)

[2] N.A.M. Al-Tamimi, S.F. Syed Fadzil, Thermal performance analysis for ventilated and unventilated glazed rooms in Malaysia (comparing simulated and field data), **20**, 222-229 (2011)

[3] K. Papakostas, T. Mavromatis, N. Kyriakis, Impact of the ambient temperature rise on the energy consumption for heating and cooling in residential buildings of Greece, **35**, 1376-1379 (2010)

[4] T. Kubota, S. Jeong, D.H.C. Toe, Energy consumption and air conditioning usage in residential buildings of Malaysia, Journal of international Development and Cooperation, **17**, 61-69 (2011)

[5] D.M. Roodman, N. Lenssen, A building revolution: How ecology and health concerns are tranforming construction, Worldwatch Paper 124 (2005)

[6] A. Al - Mofleh, S. Taib, Analysis of sectoral energy conservation In Malaysia, Energy, **34**,733-739 (2009)
[7] A. Azni Zain, Integrating sustainable energy in buildings: A case study in Malaysia, *FAU Conference*, Copenhagen, 78-91 (2008)
[8] A. Sami, Al-Sanea, M.F. Zedan, Improving thermal performance of building walls by optimizing insulation layer distribution and thickness for same thermal mass, Applied Energy, **88**, 3113-3124 (2011)
[9] Paul Dunham Close, Thermal insulation of buildings: A Treatise on the principles of thermal building insulation for dwellings and other structures, **01**, 1-3 (2007)
[10] E.F. Strother, W.C. Turner, Thermal Insulation Building Guide, 2-322(1990)

Future Parking Demand at Rail Stations in Klang Valley

Phooi Wai Ho[1,*], *Seyed Mohammadreza* Ghadiri[1], and *Premkumar* Rajagopal[1]

[1]School of Science and Engineering, Transportation and Logistics Department, Malaysia University of Science and Technology, Petaling Jaya, Malaysia

Abstract. Klang Valley, Malaysia is currently undergoing a massive development of rail transportation system expansion where the current integrated rail transit system will see new extensions of two MRT lines and an LRT line by year 2020. By year 2017, the first MRT line will be ready to run with 31 new rail stations connected to the current passenger rail network. The existing Park and Ride facilities in Klang Valley are commonly known as being unable to sufficiently cater for the current parking space demand. Therefore, with the expansion of many additional rail stations which are rapidly under construction, there are doubts that the future parking space at rail stations will be able to accommodate the sudden rise of rail passengers. Although the authorities are increasing parking bays at various locations, will the future parking demand at rail stations be sufficient? This paper studies the factors influencing parking demand in terms of population, car ownership, new car registrations and passenger rail ridership and estimating the future parking demand using Linear Regression method. Result shows that the forecasted parking demand at rail stations after the implementation of the first new MRT system in 2017 is 2.7 times more than in 2014.

1 Introduction

Klang Valley, consists of cities and towns in Kuala Lumpur, the Selangor state and Putrajaya and is home to 7.8 million population and growing [1]. With the addition of around 500,000 new vehicle registrations every year, this contributes to the enormous traffic congestion in Klang Valley [2].

Public transportation such as rail, buses and taxis play an important role to reduce traffic congestion in Klang Valley. The Malaysia Land Public Transport Commission (SPAD) has set out an integrated 20 year plan to transform land public transportation in the region to respond to local needs and aspirations [3]. This includes the massive development of rail transportation system expansion where the current integrated rail transit system will see new extensions of two Mass Rapid Transit (MRT) lines and an Light Rapid Transit (LRT) line, with a total of 91 new rail stations by year 2020 [4], [5]. This is indeed in line with Malaysia's aim to increase public transportation modal share from 18.1% in 2014 to 40%

* Corresponding author: phooiwai@gmail.com

by year 2030 [6] and to achieve transportation sustainability and a high income and developed nation by year 2020, with a world class rail transit system [2].

2 Park and ride facilities in Klang Valley

Park and Ride (P&R) facilities in Klang Valley allow rail passengers to drive and park their cars to get on the passenger trains such as the LRT and MRT. It is crucial for P&R facilities to facilitate the first mile travel to enable the transfer of rail passengers to the trains along the rail network in order to increase rail ridership, thus the public transportation modal share. The existing P&R facilities in Klang Valley are commonly known for having parking space constraint problem and is unable to sufficiently cater for the current parking space demand.

2.1 Parking space constraint at park and ride facilities

Illegal and indiscriminate parking at P&R facilities indicates the lack of parking space at rail stations. These scenarios of parking space constraint can be further analyzed and confirmed with a parking utilization survey, where in such surveys, the parking occupancy rate represents the parking demand of the facility [7].

Parking space constraint at P&R facilities in Klang Valley can be traced from as far back as year 2008 where many vehicle double parking cases were seen at the Rawang P&R facility, in which 94% of the facility's regular users suggested that more parking spaces should be provided [8].

In 2009, cars were noticed parking illegally by the road side in the residential area causing traffic congestion and on plots of land where high tension pylons were located near the Kelana Jaya LRT station. Similar cases were also noticed at the Taman Bahagia and Taman Paramount LRT stations [9]. A survey conducted in 2014 to examine the parking usage at the Kelana Jaya LRT station showed a high 92% parking occupancy rate [10].

The concern of parking space constraint at LRT stations was highlighted again in 2013 where hundreds of rail passengers daily parked at undesignated areas near LRT stations, causing inconvenience and risking damage to their own vehicles. Stretches of cars were found parked at every nook and corner that fits and on grassy roads between houses in residential areas. At the Sentul LRT station, vehicles were parked at the parking area at almost double its capacity. Few parking bays at the Bangsar LRT station, which are on a first-come, first-served basis, were often taken up by those working nearby [11].

At the Putrajaya P&R facility, vehicles were noticed to have parked illegally at the surface parking with 95% parking occupancy rate [12]. Similarly, the multi-storey parking recorded an average 90% occupancy and can go up to 100% during weekends [13]. The Gombak P&R facility had reached 100% occupancy rate in 2013, a month after its official opening [14].

Note that some of the above data were taken in 2008, 2009 and 2013 which means that the most recent parking occupancy rates may be more than what were founded before. In actual fact, there could be many more rail stations facing the same problem. This problem has been a deterrent factor for many urbanites in the Klang Valley who cite it as a reason they are reluctant to travel by train to their destinations [6], [11]. Closer to the completion of the new MRT, there were parties who raised concerns about the possible illegal parking after the opening of the MRT station, due to the lack of parking space [15].

2.2 Parking concern after the implementation of new MRT line

In the near future, by year 2017, the first MRT line will be ready to run with 31 brand new rail stations connected to the current passenger rail network. Considering that the current parking space capacity is already an existing problem, with the expansion of the new MRT which are rapidly under construction, there are doubts whether the future parking space at P&R facilities will be able to accommodate the sudden rise of rail passengers. The lack of parking space at P&R facilities after the implementation of the new rail lines, may cause massive traffic congestion [16], [17], thus air pollution [18], [19] and the limitation of rail transit system ridership [20]–[23] for the Klang Valley integrated rail transit system.

The success of P&R facilities are confirmed in attracting rail passengers with users reported high levels of satisfaction [24]. Therefore, it is necessary to provide more parking space to encourage the usage of rail transportation. However, although the authorities are looking at finding a solution, they are challenged with the lack of land to provide more parking bays [11]. How many parking bays at P&R facilities are enough to cater for the future in 2017, after the implementation of the first new MRT line in Klang Valley? This depends on the future parking demand.

2.3 Factors influencing parking demand at rail stations in Klang Valley

Parking demand refers to the amount of parking that would be used at a particular time, place and price. It is a critical factor in evaluating parking problems and solutions. Parking demand is affected by vehicle ownership, among all other behavioral factors such as trip rates, mode split, parking duration and type of trip, and factors of other travel modes such as the quality of alternative travel, fuel and road pricing [25].

In Klang Valley, vehicle ownership is an exceptional factor and plays a significant role in influencing parking demand as there are around 500,000 new vehicle registrations added to the region every year [2]. Malaysia also tops the list as the country with third highest car ownership in 2014. 88% of Malaysian car owners also intend to upgrade their vehicle when they are financially able [26]. With car ownership in the growing trend, the parking demand in Klang Valley, especially at rail stations, is expected to increase.

Besides car ownership, the growth of rail passenger ridership due to the massive rail development in Klang Valley is expected to increase parking demand at rail stations. At the same time, population growth in Klang Valley is also expected to increase rail passenger ridership, thus parking demand at rail stations [23], [27]–[29]. It is projected that Malaysia's population will grow from around 30.6 million to 38.6 million in 2040 [1], [30].

2.4 Estimating parking demand at rail stations in Klang Valley

It is important to estimate the parking demand in 2017 to gauge the parking bays needed as there will be a sudden rise of rail passengers after the 31 new rail stations are in operations. This is to prepare for the future parking space needs and to ensure the success of the integrated rail transit system in Klang Valley.

3 Methodology

This study estimates parking demand in 2017 at rail stations in Klang Valley by forecasting using the Least Squares Regression method or popularly known as Linear Regression method. Factors influencing parking demand specifically population, car ownership and rail passenger ridership are forecasted in 2017 to estimate the parking demand in 2017.

4 Results and discussion

4.1 Population in Malaysia and Klang Valley

Historical data shows that the population in Malaysia and Klang Valley have been increasing with a similar upward trend since 2011 until 2014 as shown in Table 1. The total population increases with an average growth rate of 1.7% and 2.0% annually in Malaysia and Klang Valley respectively during these period. The average increase in Klang Valley is higher Malaysia which shows that Klang Valley is one of the preferred location to live in Malaysia.

Table 1. Population in Malaysia and Klang Valley [31-34].

Region/Year	2011	2012	2013	2014
Malaysia	29,062,000	29,510,000	30,213,700	30,597,900
Annual Growth Rate		*1.5%*	*2.4%*	*1.3%*
Klang Valley	7,370,800	7,482,800	7,707,800	7,829,100
Annual Growth Rate		*1.5%*	*3.0%*	*1.6%*

By 2017, the forecast of the total population in Klang Valley is 8.3 million people as shown in Fig. 1. This is approximately 6.2% increase from the population in Klang Valley from 2014. With the future population increase, parking demand at rail stations is also expected to increase.

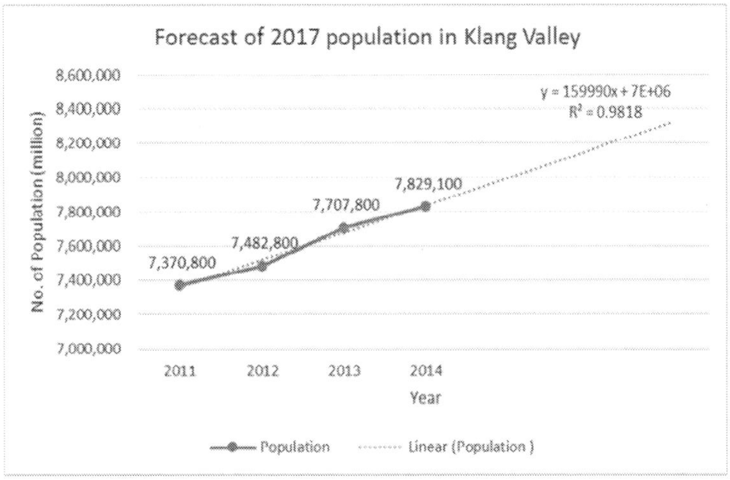

Fig. 1. Forecast of 2017 population in Klang Valley [31-34]

4.2 Car ownership in Malaysia and Klang Valley

Registered cars in Table 2 shows the number of new and old cars registered in Malaysia and Klang Valley, which also represents car ownership. 43% of the total registered cars in Malaysia, almost half is located in Klang Valley. Therefore, parking demand will be exceptionally high in this region, especially after the implementation of the new MRT line.

Historical data shows an upward trend for car ownership in Malaysia and Klang Valley from 2011 until 2014. The gradual increase of the number of cars is at an average annual rate of 5.8% and 4.6% respectively in Malaysia and Klang Valley. For the same period, the car registrations growth rate is higher than the population growth rate in Malaysia and Klang Valley. This shows that the transportation in Malaysia is currently unsustainable.

Table 2. Registered cars in Malaysia and Klang Valley [31]–[34].

Region/Year	2011	2012	2013	2014
Malaysia	9,324,276	9,952,515	10,535,575	11,028,296
Annual Growth Rate		*6.7%*	*5.9%*	*4.7%*
Klang Valley	4,114,759	4,385,120	4,479,562	4,701,751
Annual Growth Rate		*6.6%*	*2.2%*	*5.0%*

By 2017, the forecast of the car ownership in Klang Valley is 5,255 thousand of cars as shown in Fig. 2. This is approximately an 11.8% increase from 2014. With the increase of car ownership in Klang Valley, parking demand at rail stations is also expected to increase.

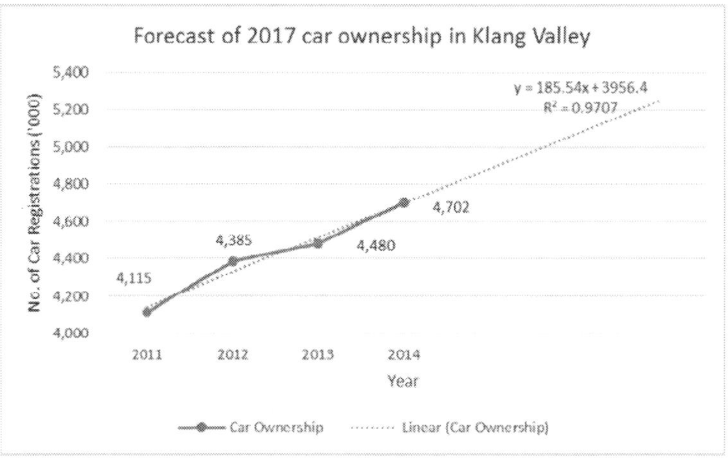

Fig. 2. Forecast of 2017 car ownership in Klang Valley [31-34]

4.3 New car registrations in Malaysia and Klang Valley

The new car registrations represents only the new cars registered in Malaysia and Klang Valley, with a similar trend as shown in Table 3. The numbers picked up from 2011 until 2012, thereafter have dropped in 2013 and reduced further in 2014.

Table 3. New car registrations in Malaysia and Klang Valley [31-34]

Region/Year	2011	2012	2013	2014
Malaysia	594,610	628,239	583,060	492,721
Klang Valley	255,460	267,689	262,818	222,189

Although new car registrations in Malaysia and Klang Valley have dropped from 2012 until 2014, the total number of car ownership were gradually increasing. This only shows that Malaysians were not purchasing as many new cars as before, but they still rely on their cars as a mode of transport by purchasing second hand cars or retaining their existing cars.

4.4 Ridership for integrated rail transit system in Klang Valley

Table 4 shows the annual ridership for all passenger rail services in Klang Valley where the new MRT will be built and integrated into. The total ridership for these rail services shows a strong upward trend from 2011 until 2014 with an average growth rate of 6.2% annually. By 2017, the forecast of annual passenger rail ridership is 257 million based on the current existing number of rail stations. However, the number of rail passengers will definitely surpass this figure when the new MRT line commences in 2017.

Table 4. Annual ridership for passenger rail services [1, 35]

Passenger Rail Services/Year	2011	2012	2013	2014
Putra LRT	68,398,561	71,574,675	78,702,931	81,971,322
Star LRT	53,568,672	56,809,978	60,207,397	63,270,431
KL Monorail	24,200,299	24,113,242	25,437,623	24,303,466
KLIA Express	1,581,476	1,649,410	2,062,223	2,928,302
KTM Commuter	35,598,901	34,847,254	43,941,777	46,956,723
Total Passengers	**183,347,909**	**188,994,559**	**210,351,951**	**219,430,244**
Annual Growth Rate		*3.1%*	*11.3%*	*4.3%*

According to SPAD, the expected ridership of the new MRT line is 445,000 passengers per day [3]. Table 5 shows the projected rail ridership in 2017 including the new MRT line. There will be a total of 555 million of annual passenger rail ridership with an approximate 153% increase from 2014. With the tremendous increase of rail passengers in 2017, parking demand will certainly shift up to a new level.

Table 5. Forecast of 2017 annual passenger rail ridership [3]

Passenger Rail Services	Daily Ridership	Annual Ridership
LRT Kelana Jaya	496,000	181,040,000
LRT Ampang	352,000	128,480,000
KL Monorail	115,000	41,975,000
KLIA Express	16,500	6,022,500
KTM Commuter	95,000	34,675,000
MRT Line 1	445,000	162,425,000
Total Passengers	**1,519,500**	**554,617,500**

4.5 Car parking bays demand in Klang Valley for year 2017

To project the car parking demand at rail stations in Klang Valley for 2017, the car ownership for rail passengers, which directly influences the car parking demand is estimated for 2017 and 2014. The rail passengers' car ownership per total rail passengers (cars per person) is equal to the total car ownership per total population (cars per person) in Klang Valley, with the estimation that the average car ownership in Klang Valley is the same for all population as per Equation (1).

$$\frac{Q}{R} = \frac{O}{P} \tag{1}$$

where O = Car ownership in Klang Valley, P = Population in Klang Valley, Q = Rail passengers' car ownership, R = Daily rail ridership .

The Rail Passengers' Car Ownership (RPCO) can be derived as per Equation (2).

$$Q = \frac{O}{P} \times R \tag{2}$$

Based on Equation (2), the forecasted RPCO for Klang Valley is 960,055 and 361,037 in total for 2017 and 2014 respectively. With this, the ratio of RPCO in Klang Valley for 2017 to 2014 is as per Equation (3).

$$\frac{Q_{KV-2017}}{Q_{KV-2014}} = 2.7 \tag{3}$$

Referring to the result of Equation (3), the RPCO for 2017 has increased tremendously by 2.7 times compared to 2014. Therefore, based on all other factors affecting the use of cars for rail passengers remain unchanged, the car parking bays demand in Klang Valley for 2017 is 2.7 times more than the number of car parking bays provided in Klang Valley for 2014.

5 Conclusion

Based on the above forecast, the future parking demand at rail stations after the implementation of the first new MRT line is 2.7 times more than in 2014. The fact that Klang Valley is currently facing parking space constraint at rail stations should be seriously looked into. Measures should be engaged to ensure that there will be sufficient parking bays available to those planning to commute the train in the near future. It is crucial especially for passengers who will be commuting the train for the first time and are planning to change their travel mode to train for long term. If the authorities are unable to react to the situation promptly, the unpleasant experiences such as time and resources wastage and traffic congestion, will be part of normalization, which will further result to underutilize of rail transit system. Whereas, sufficient parking bays will encourage rail ridership, increase public transportation modal share thus achieving transportation sustainability.

This study was funded by the Malaysian Ministry of Higher Education through the MyBrain15 MyPhD scholarship award. Authors are indebted to the anonymous reviewers for their valuable comments and suggestions.

References

[1] Department of Statistics, *Malaysia Statistical Handbook 2015*, Malaysia, (2015)
[2] Government Transformation Programme, *Improving Urban Public Transport, National Key Results Areas*, GTP Annual Report 2014, Performance Management and Delivery Unit (PEMANDU), (2014)
[3] Land Public Transport Commission, *Greater KL/Klang Valley Land Public Transport Master Plan*, Malaysia, (2013)
[4] Prasarana Malaysia Berhad, LRT 3, (2016), Retrieved on July 23, 2016 from http://www.lrt3.com.my
[5] Mass Rapid Transit Corporation, The Klang Valley MRT, (2016), Retrieved on July 23, 2016 from http://www.mymrt.com.my/
[6] Land Public Transport Commission, *Land Public Transport Commission (SPAD) Annual Report 2013*, Malaysia, (2013)
[7] U.A. Khalid, A.A. Kadar Hamsa, Parking utilization pattern of park and ride facility at public transportation terminals, *Proceedings of the Eastern Asia Society for Transportation Studies*, **9**, 16 (2013)

[8] N. Abdul Hamid, J. Mohamad, M.R. Karim, Travel behaviour of the park and ride users and the factors influencing the demand for the use of the park and ride facility, *EASTS International Symposium on Sustainable Transportation incorporating Malaysian Universities Transport Research Forum Conference 2008*, 1–13 (2008)
[9] F. Aziz, *Parking Issue at Station*, The Star (October 16, 2009)
[10] A.A.K. Hamsa, S.A.A.S Adnan, U.A. Khalid, Analysis of parking usage at the park and ride facility in Klang Valley, Malaysia, WIT Transactions on the Built Environment, **138**, 179–193 (2014)
[11] V. Nair, *Commuters Face Daily Struggle to Find Space at LRT Stations' Parking Lot*, The Star (July 29, 2013)
[12] S.A.A. Syed Adnan, A.A. Kadar Hamsa, Evaluating the parking demand at park and ride facility at Putrajaya public transportation terminal, *Proceedings of the Eastern Asia Society for Transportation Studies*, **9**, 14 (2013)
[13] N. Azhar Zainal, *Putrajaya Bus Service Has Far To Go*, The Star, (March 11, 2015)
[14] S. Tamrin, *LRT and Monorail Ridership to Reach the Two Billion Mark in May*, The Sun Daily, (April 30, 2015)
[15] The Star, *MRT Station Welcomed with Some Concerns*, (November 17, 2015)
[16] H. Shang, W. Lin, H. Huang, Empirical study of parking problem on university campus, J. Transp. Syst. Eng. Inf. Technol., **7**, 135–140 (2007)
[17] W. Yan-Ling, W. Xin, Z. Ming-chun, Current situation and analysis of parking problem in Beijing, Procedia Eng., **137,** 777–785 (2016)
[18] A.M. Abdullah, M.A.A. Samah, Y.J. Tham, An overview of the air pollution trend in Klang Valley, Malaysia, Open Environ. Sci., **6**, 13–19 (2012)
[19] S. Masahina, R. Afroz, J. Duasa, N.A. Mohamed, Framework to estimate the willingness to pay of household for airquality improvement: A case study in Klang Valley, Malaysia, Int. J. Sustain. Dev., **4**, 11–16 (2012)
[20] M. Brons, M. Givoni, P. Rietveld, Access to railway stations and its potential in increasing rail use, Transp. Res. Part A Policy Pract., **43**, 136–149 (2009)
[21] J. Zhao, W. Deng, Y. Song, Y. Zhu, What influences Metro station ridership in China? Insights from Nanjing, Cities, **35**, 114–124 (2013)
[22] D. Merriman, How many parking spaces does it take to create one additional transit passenger?, Reg. Sci. Urban Econ., **28**, 565–584 (1998)
[23] B.D. Taylor, C.N.Y. Fink, Explaining transit ridership: What has the evidence shown?, Transp. Lett., **5**, 15–26 (2013)
[24] G. Parkhurst, Park and ride: Could it lead to an increase in car traffic?, Transp. Policy, **2**, 15–23 (1995)
[25] Victoria Transport Policy Institute, *Parking Evaluation*, (2012), Retrieved on February 28, 2012 from http://www.vtpi.org/tdm/tdm73.htm
[26] Nielsen, *Rising Middle Class Will Drive Global Automotive Demand in the Coming Two Years*, (2014), Retrieved on January 30, 2016 from http://www.nielsen.com/my/en/press-room/2014/rising-middle-class-will-drive-global-automotive-demand.html
[27] A. Yoh, P. Haas, B. Taylor, Understanding transit ridership growth: Case studies of successful transit systems in the 1990s, Transp. Res. Rec., **1835**, 111–120 (2003)
[28] B.D. Taylor, D. Miller, H. Iseki, C. Fink, Nature and/or nurture? Analyzing the determinants of transit ridership across US urbanized areas, Transp. Res. Part A Policy Pract., **43**, 60–77 (2009)
[29] H. Sung, K. Choi, S. Lee, S. Cheon, Exploring the impacts of land use by service coverage and station-level accessibility on rail transit ridership, J. Transp. Geogr., **36**, 134–140 (2014)
[30] M. Chi, *Population to Hit 30 Million This Week, Statistics Department Says*, Malay

Mail Online, (February 26, 2014)
[31] Department of Statistics, *Social Statistical Bulletin Malaysia 2015*, Malaysia, (2015)
[32] Department of Statistics, *Social Statistical Bulletin Malaysia 2014*, Malaysia, (2014)
[33] Department of Statistics, *Social Statistical Bulletin Malaysia 2013*, Malaysia, (2013)
[34] Department of Statistics, *Social Statistical Bulletin Malaysia 2012*, Malaysia, (2012)
[35] Department of Statistics, *Malaysia Statistical Handbook 2014*, Malaysia, (2014)

Stability and Volumetric Properties of Asphalt Mixture Containing Waste Plastic

Siti Aminah Abd Kader[1], *Ramadhansyah* Putra Jaya[2,3,*], *Haryati* Yaacob[4], *Mohd Rosli* Hainin[5], *Norhidayah* Abdul Hassan[6], *Mohd Haziman* Wan Ibrahim[7], *Abdullahi* Ali Mohamed[8], and *Ichwana*[9]

[1,2,4,5,6]Faculty of Civil Engineering, Department of Geotechnics and Transportation, Universiti Teknologi Malaysia, 81310 UTM Skudai, Johor Bahru, Malaysia
[3]Faculty of Engineering, Department of Civil and Environmental Engineering, Universiti Teknologi PETRONAS, 32610 Seri Iskandar, Perak Darul Ridzuan, Malaysia
[7]Faculty of Civil and Environmental Engineering, Universiti Tun Hussein Onn Malaysia, 86400 Batu Pahat, Johor, Malaysia
[8]Faculty of Engineering, The University of Nottingham Malaysia Campus, Jalan Broga, 43500 Semenyih, Selangor Darul Ehsan, Malaysia
[9]Faculty of Agriculture, Department of Agricultural Engineering, Universitas Syiah Kuala, 23111 Darussalam, Banda Aceh, Indonesia

Abstract. The objectives of this study are to determine the optimum bitumen content (OBC) for every percentage added of waste plastics in asphalt mixtures and to investigate the stability properties of the asphalt mixtures containing waste plastic. Marshall stability and flow values along with density, air voids in total mix, voids in mineral aggregate, and voids filled with bitumen were determined to obtain OBC at different percentages of waste plastic, i.e., 4%, 6%, 8%, and 10% by weight of bitumen as additive. Results showed that the OBC for the plastic-modified asphalt mixtures at 4%, 6%, 8%, and 10% are 4.98, 5.44, 5.48, and 5.14, respectively. On the other hand, the controlled specimen's shows better volumetric properties compared to plastic mixes. However, 4% additional of waste plastic indicated better stability than controlled specimen.

1 Introduction

Plastic is used regularly in today's lifestyle, and its disposal is a great problem [1]. Plastic is a non-biodegradable product; thus, it causes environmental pollution and problems [2]. The issue now is to reduce plastic waste by transforming it into something beneficial. Given that the plastics used in PET bottles, disposal glasses, handbags, and covers of various appliances soften up at 160°C [3], a solution is to use waste plastic in such a manner that it gets coated over by an aggregate via heating (140–160°C) [4]. Asphalt mixtures, commonly called asphalt concrete in the United States, are composite materials commonly used in surface roads, parking lots, and airports. Asphalt concrete consists of aggregate and asphalt/bitumen laid in layers and then compacted. Bitumen acts as binder to bind the

* Corresponding author: ramadhansyah@utm.my

aggregates together. One of the main problems of polyethylene is that without special treatment it is not readily biodegradable and thus accumulates. Over 500 billion to 1 trillion plastic bags are consumed worldwide. Plastic waste is a serious problem that can lead to many other problems [5]. A recent study had found one more way to reduce the amount of plastic waste by using it in asphalt mixtures [6]. Waste plastic bags can be used as a modifier of hot-mix asphalt [7]. Costa *et al.* [8] reported that utilization of waste plastic in flexible pavement has been established since 2012. A study was carried out by using plastic waste of 0%–12% by weight of bitumen. As a result, the stability, strength, fatigue life, and other characteristics of the asphalt mixture were improved even under adverse water-logging conditions. Kumar and Garg [9] recycled waste plastics bags in bituminous mix by using 1.2–3 mm particle size. They found that the stability and flow were improved. It can be concluded that the added waste plastic can strongly improve the binder properties, provide safer road, and reduce the maintenance cost by increasing the stiffness of the bitumen and by improving its temperature susceptibility. The process is environment friendly. Thus, these processes are socially highly relevant and give better infrastructure.

2 Materials and experimental procedures

2.1 Raw materials preparation

The binder used in this study is bitumen 60/70. Bitumen were heated in the oven till fluid conditions, then plastic were prepared to be added gradually to the bitumen and blended for three to five minutes. Mixing was continued for a certain time to produce homogenous mixtures. Three specimens were prepared for each percentage of plastic. After Specimens have been compacted using Marshall Hammer, then were tested for stability, flow, VTM, VMA, and VFB. Furthermore, granite aggregates used in this investigation is nearly always massive, hard and. The gradation test (Fig. 1) was conducted following the procedure outlined in BS EN [10] and JKR [11].

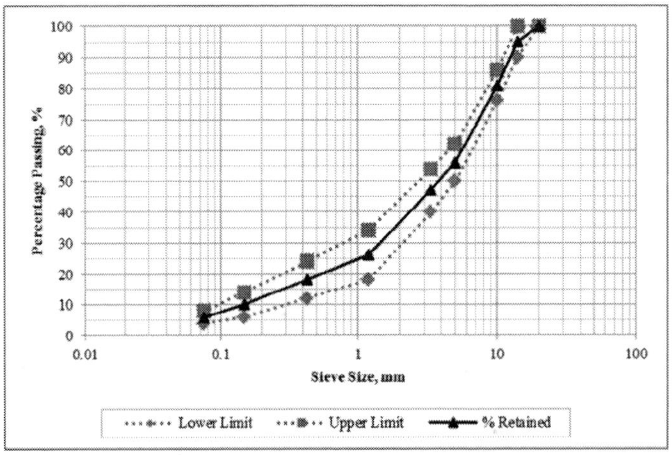

Fig. 1. Aggregate gradation of asphalt mix [11]

2.2 Marshall mix design

In the laboratory, the aggregate, binder and plastic were respectively mixed and compacted at 180±0.5°C. The mixes were compacted with 75 blows on each side with the standard

Marshall hammer to avoid disintegration of materials. After compaction, the specimens were removed from the molds and allowed to cool down.

2.3 Stability test

The stability value of each test specimen shall then be determined in accordance with BS EN [12]. In the stability test, the specimens were prepared with the specified temperature by immersing in a water bath at a temperature of 60°C±1°C for a period of 45 minutes. It was then placed in the Marshall Stability testing machine and loaded at a constant rate of deformation of 50.8 mm/minute until the maximum load was reached.

3 Result and discussion

3.1 OBC determination

According to Ramadhansyah *et al.* [13], the OBC of the controlled sample (AC14) is 5.1%. However, the OBC values for 4%, 6%, 8%, and 10% plastic added are shown in Table 1. The OBC increases for 4%, 6%, and 8% and then decreases for 10%. OBC has a relationship with the stability value. Stability value also increases up to a certain amount of plastic added, and further addition of plastic beyond this limit decreases the stability.

Table 1. OBC at different percentage plastic

% Plastic	OBC
4.0	4.98
6.0	5.44
8.0	5.48
10.0	5.14

3.1.1 Stability

Fig. 2 shows the relationship between stability and percentage of plastic. The stability of 4% plastic is the highest (i.e., 184 kN), and it slightly decreases with the increase of plastic addition. A large gap between the stability of the conventional asphalt mixture and the plastic modified asphaltic concrete is observed (i.e., more than 100%). Stability verifies the performance of the asphalt mixture under loading. The results showed that plastic can increase the stability of the asphalt mixture up to certain percentages and after that it will decrease.

3.1.2 Flow

Fig. 3 shows the effects of flow with different percentage of plastic in asphalt mixture. Flow is related to stability and shows the flexibility of the mix. Flow for the conventional asphalt mixture is 2.85 mm [14], which is the lowest. Flow of the modified asphalt mixture is higher than the flow of the conventional asphalt mixture. The difference between the modified and conventional asphalt mixture varies from 19% to 37%. The highest flow is 3.9 mm, which is when 10% plastic was added.

DOI: 10.1051/matecconf/201710309002

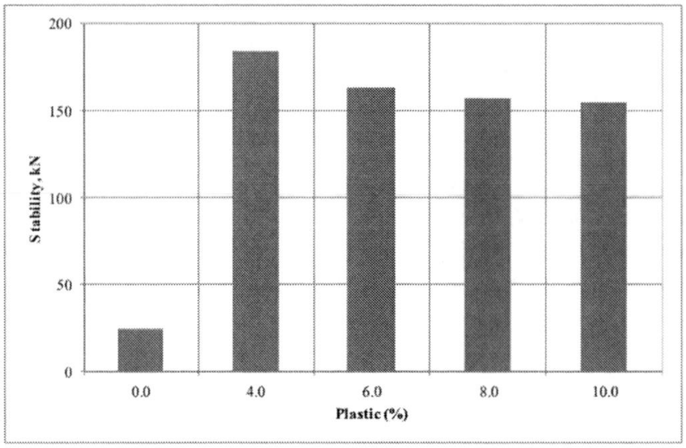

Fig. 2. Stability of asphalt mixture at different percentage plastic

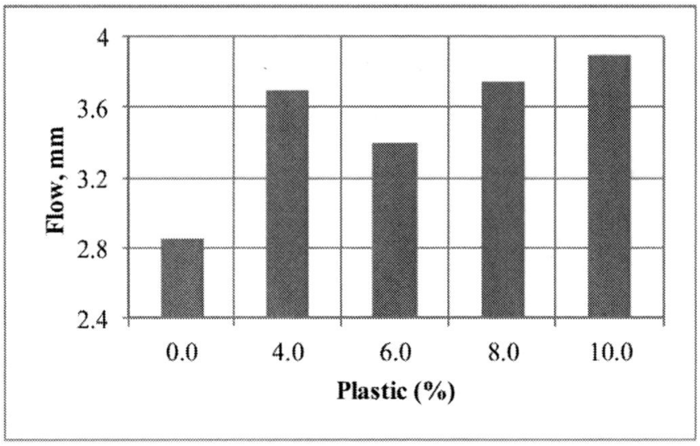

Fig. 3. Flow of asphalt mixture at different percentage plastic

3.1.3 Stiffness

Fig. 4 shows the effect of different percentage of plastic to the stiffness. The stiffness of the modified asphalt mixture increases (4% and 6%) but decreases for 8% and 10%. Therefore, plastic can increase the stiffness of the asphalt mixture up to certain percentages and then decrease after it reaches the optimum value. The conventional asphalt mixture has the lowest stiffness, that is, 8.53kN/mm. Stiffness is an important parameter to measure the rutting. A higher stiffness pavement is good. However, a too high stiffness can lead to fatigue cracking [15].

3.2 Volumetric properties

The volumetric properties are reflected on the volume of bitumen and aggregates required to produce a mixture of desired properties. Mixture volumetric properties are important for long-term performance and durability of pavements.

3.2.1 Density

Density is the mass per unit volume. The densities of asphalt mixture at different percentage of plastic are shown in Fig. 5. The density of the asphalt mixture decreases with the increase of percentage plastic. Conventional asphalt mixture has density of 2.349 g/cm^3. Plastic-modified asphalt mixture with 4% plastic has the highest density, that is, 2.500 g/cm^3, which is 6% higher than the conventional asphalt. As the percentages of plastic increase, the volume of the asphalt mixture increased. Therefore, the density decreased.

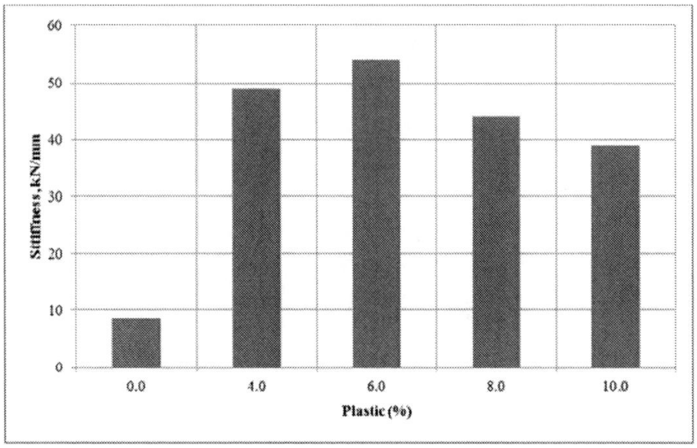

Fig. 4. Stiffness of asphalt mixture at different percentage plastic

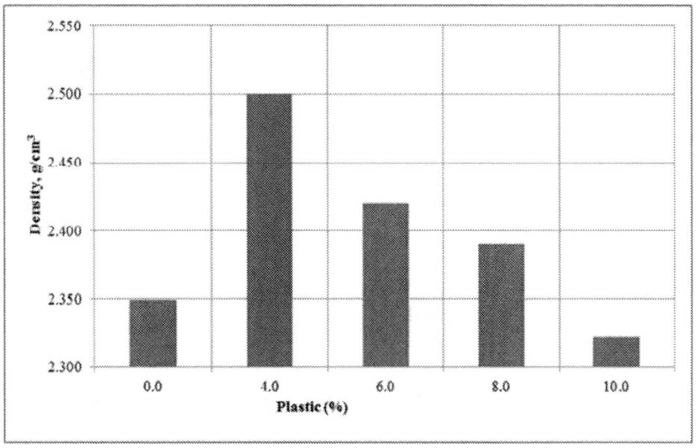

Fig. 5. Density of asphalt mixture at different percentage plastic

3.2.2 Voids in Total Mix

Fig. 6 shows the effect of plastic added on voids in total mix. The VTM of the conventional asphalt mixture is 3.6%, whereas the VTM of the plastic-modified asphalt mixture varies from 4.0% to 7.0%. According to JKR [11], the range of VTM is from 3% to 5%. Based on the result presented in the figure, the VTM values for 4% and 8% plastic are out of range, whereas the VTM for 6% and 10% plastic are in the range. VTM is also known as air voids and is the ratio of the volume of voids in a compacted mixture to the total compacted mixture volume. Voids can be reduced by conducting compaction properly. The compactor machine needs to be maintained regularly to make sure that it works well during the compaction process.

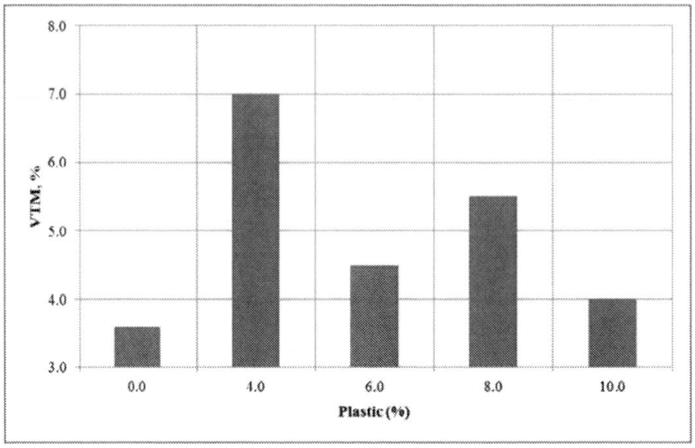

Fig. 6. VTM of asphalt mixture at different percentage plastic

3.2.3 Voids Filled with Bitumen

The voids filled with bitumen (VFA) are plotted against the percentage of plastic addition in Fig. 7. Voids filled with asphalt binder is the percent of voids in mineral aggregate (VMA) that is filled with asphalt cement. VFA for the conventional asphalt mixture is 76%, and it is quite similar to VFA for 10% plastic, that is, 73%. VFA for 4% plastic is the highest (i.e., 140%). The VFA for 6% and 8% are constant (i.e., 97% and 95%, respectively), but still out of range of the JKR [11] specification that is in between 70% to 80%.

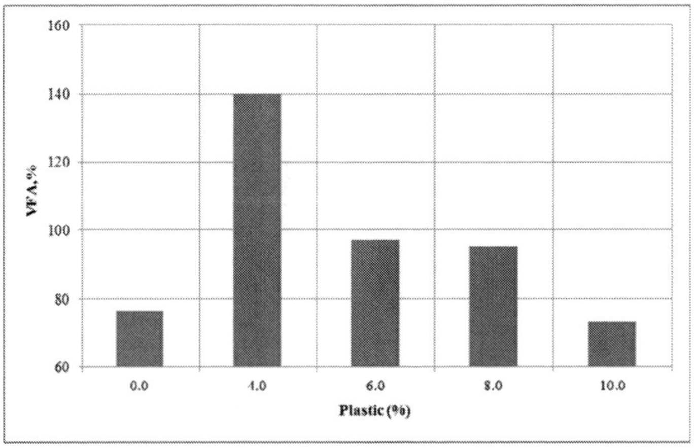

Fig. 7. VTM of asphalt mixture at different percentage plastic

4 Conclusion

Based on the result and analysis, it can be seen that the OBC of asphalt mixture containing waste plastic increases (4% - 8%) and then decreases at 10% added. In comparison between plastic percentage levels, 4% waste plastic showed excellent stability. Furthermore, the results revealed that controlled specimen's shows better volumetric properties compared to waste plastic mixes. However, 4% added of waste plastic indicated better performance than controlled specimen.

The support provided by Malaysian Ministry of Higher Education (MOHE) and Universiti Teknologi Malaysia (UTM) in the form of a research grant number Q.J130000.2522.11H58 for this study is very much appreciated.

References

[1] E.A. Asamany, M.D. Gibson, M.J. Pegg, Evaluating the potential of waste plastics as fuel in cement kilns using bench-scale emissions analysis, Fuel, **193**(1), 178-186 (2017)
[2] T. Sri Devi Kumari, A.J.J. Jebaraj, T.A. Raj, D. Jeyakumar, T.P. Kumar, A kish graphitic lithium-insertion anode material obtained from non-biodegradable plastic waste, Energy, **95**, 483-493 (2016)
[3] B.T. Ashwini Manjunath, Partial replacement of e-plastic waste as coarse-aggregate in concrete, Procedia Environmental Sciences, **35**, 731-739 (2016)
[4] M. Naskar, T.K. Chaki, K.S. Reddy, Effect of waste plastic as modifier on thermal stability and degradation kinetics of bitumen/waste plastics blend, Thermochimica Acta, **509**(1–2), 128–134 (2010)
[5] S. Köfteci, P. Ahmedzade, B. Kultayev, Performance evaluation of bitumen modified by various types of waste plastics, Construction and Building Materials, **73**, 592–602 (2014)
[6] Johnson Kwabena Appiah, Victor Nana Berko-Boateng, Trinity Ama Tagbor, Use of waste plastic materials for road construction in Ghana, Case Studies in Construction Materials, **6**, 1–7 (2017)
[7] M. Garcia-Morales, P. Partal, F.J. Navarro, C. Gallegos, Effect of waste polymer addition on the rheology of modified bitumen, Fuel, **85**, 936–943 (2006)
[8] L.M.B. Costa, H.M.R.D. Silva, J.R.M. Oliveiran, S.R.M. Fernandes, Incorporation of waste plastic in asphalt binders to improve their performance in the pavement, Int. J. Pavement Research and Technology, **6**(4), 457–464 (2013)
[9] P. Kumar, R. Garg, Rheology of waste plastic fibre-modified bitumen, Int. J. Pavement Engineering, **12**(5), 449–459 (2011)
[10] BS EN 933-1:2012, *Tests for geometrical properties of aggregates. Determination of particle size distribution, sieving method*, BSI, London, (2012)
[11] JKR/SPJ/2008-S4, *Standard specification for road works*, Malaysian Public Works Department, Kuala Lumpur, (2008)
[12] BS EN 12697-34:2012, *Bituminous mixtures. Test methods for hot mix asphalt. Marshall test*, BSI, London, (2012)
[13] P.J. Ramadhansyah, M. Nurfatin Aqeela, J. Siti Nur Amiera, M. Norhafizah, A.H. Norhidayah, S.J. Dewi, Use of coconut shell from agriculture waste as fine aggregate in asphaltic concrete, ARPN J. Engineering and Applied Sciences, **11**(12), 7457-7462 (2016)
[14] M. Norhafizah, P.J. Ramadhansyah, J. Siti Nur Amiera, M. Nurfatin Aqeela, A.H. Norhidayah, M.R. Hainin, C.W. Che Norazman, The effect of coconut shell on engineering properties of porous asphalt mixture, J. Teknologi, **78**(7-2), 127–132 (2016)
[15] S.N.A. Jeffry, R.P. Jaya, N. Manap, N.A. Miron, N.A. Hassan, The influence of coconut shell as coarse aggregates in asphalt mixture, Key Engineering Materials, **700**, 227-237 (2016)

Performance Evaluation of Conventional Hot Mix and Porous Asphalt Under Accelerated Loading

Ludfi Djakfar[1,*], Harimurti[1], *Yulvi* Zaika[1], and *Ketut* Sugiarto[1]

[1] Department of Civil Engineering, Universitas Brawijaya, Jl. M.T. Haryono 167 Malang, Indonesia

Abstract. The purpose of the study is to evaluate and compare the performance of conventional hot mix asphalt and porous hot mix asphalt under accelerated loading. To achieve the objective, a test section was constructed and tested with a loading device previously developed. The pavement material conforms to the Indonesian Directorate General of Highway. While the porous surface conforms to the Australian standard for asphalt porous. The thickness of the test section was prepared to simulate the real condition of the pavement structure. The load, equivalent to the standard 8,160 kg axle load was applied for 1,000 passes. For every 100 passes, stresses and deformation were recorded. The result shows that there is some consistency between theory and field results in terms of stresses and deformation. Conventional surface course seems to perform better in deformation/rutting. Future research should be directed to improve the performance of the device and also the performance of the porous asphalt.

1 Introduction

As a country positioned in the equatorial lines, Indonesia experiences heavy rain almost 6 months in a year. This condition has caused some water to present in the pavement surface, which has some detrimental effect either to the pavement itself or to the road users since it may cause accidents due to hydroplaning. The majority of the road in Indonesia was constructed using conventional hot mix asphalt (HMA). While many studies and experience showed that porous HMA has satisfactory performance in dealing with splash water and hydroplaning, the Highway Agency seems to be unconvinced with these facts. Very few porous HMA have ever been constructed.

Many researches have been conducted to evaluate the performance of the porous HMA. Most of the efforts, however, are still in lab-based evaluation. Therefore, research should be directed to evaluate the performance of pavement using field data. Previous efforts have been conducted. Louisiana Transportation Research Center has constructed an Accelerate Loading Facilities (ALF) in order to evaluate the performance of several types of pavement materials [1]. Bonaquist et al. [2] used the ALF to evaluate the effects of tire pressure on flexible pavement response and performance. Sebaaly et al. [3] used data from previous ALF research to evaluate relationships between surface cracking and the structural capacity

* Corresponding author: Ldjakfar@ub.ac.id

of both thin and thick pavements. Kadar [4] used the ALF test results to assess the relative performance of a variety of asphalt surface types for pavement rehabilitation. Johnson-Clark et al. [5] employed the ALF to investigate the effectiveness of a geotextile reinforced seal between the subgrade and gravel layers to rehabilitate low volume roads.

Several researchers have also developed and utilized accelerated pavement testing to investigate the performance of some pavement materials. Tang et al [6] used Model Mobile Load Simulator (MMLS3) to investigate the use of geogrids for subgrade stabilization in flexible pavements. Al Qadi et al. [7] investigated the effect of various tire configurations on Geogrid-reinforced low-volume flexible pavement using the mobile Accelerated Testing

Loading Assembly (ATLAS) housed at University of Illinois at Urbana-Champaign.
One hindrance of using existing equipment is its cost and its availability, which may become the main reason for the highway department particularly in developing countries to hesitate to use it. Therefore, simpler and modest equipment need to be developed for the purpose of mobile accelerated testing of pavement.

2 Objective of study

The objective of the study is to evaluate the performance of conventional and porous asphalt when subjected to accelerated loading.

3 Methods

To achieve the objective, two test sections: (1) conventional and (2) porous sections were constructed and tested. Fig. 1 shows the scheme and dimension of the test section. The thickness of base and subbase section was prepared simulating the common dimension encountered in the field, while the thickness of the surface was 3 cm. The reason of using 3 cm thick of surface material in the test section is to ensure that the strain gauge installed beneath each layer can respond and read the signal caused of the applied load. Since this is the preliminary testing, the researchers would like to make sure that the strain gauges could function as intended, by using a thinner surface course.

3.1 Materials

Materials used to construct the test section consist of Class A for base and Class B for subbase and conform to the Indonesian Directorate General of Highway (IDGH) Specifications. Tables 1 and 2 present the material characteristics for base and surface course. The conventional HMA surface course was designed to conform to the Specification requirement of the IDGH Specs, while porous HMA surface course was designed based on porous asphalt specification with 8% Gilsonite additive, since previous research [8] showed that adding the Gilsonite additive increases the stability up to 800 kg without significantly reducing its permeability capability, as shown in Table 2. The pavement was constructed on the sandy soil subgrade with 10% CBR. Fig.2 presents the test section construction.

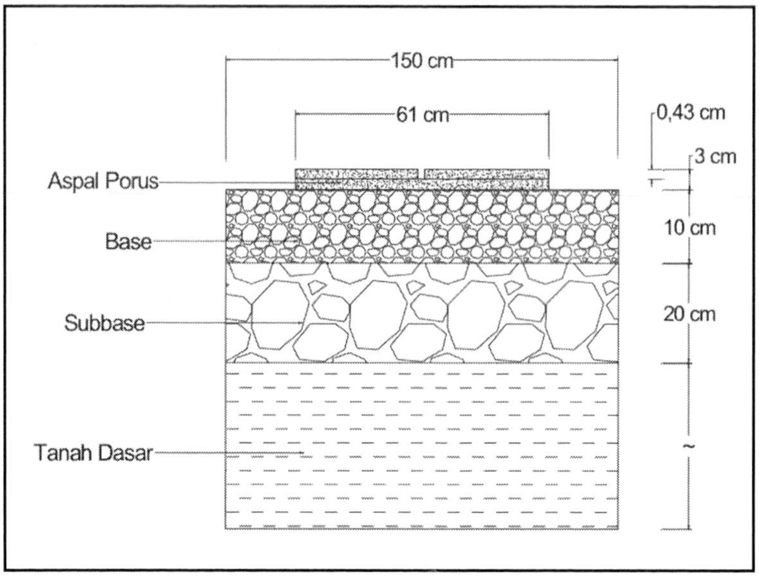

Fig. 1. Dimension of pavement structure constructed at the test section

Table 1. Unit weight and CBR of the subgrade, base and subbase materials

Layer	Unit Weight (gr/cm^3)		CBR (%)
	Laboratory	Field	
Subbase	1.80	1.83	75
Base	1.79	1.69	82
Subgrade	1.57		15

Table 2. Marshall characteristics of the surface course

Marshall Characteristics	**Conventional HMA**	**Porous HMA**
VIM	3%	8,1 %
Stability	1500 kg	805 kg
Flow	2,5 mm	3,8 mm
MQ	550 kg/mm	219 kg/mm

Fig. 2. Construction of test section

3.2 Installation of the strain gauge

To record the strain and stress occurring in the pavement due to load, a set of strain gauge was installed between the surface and base course as shown in Fig. 3, and spaced at 40 cm as shown in Fig. 3b.

3.3 Loading device

To simulate the traffic load, a loading device was developed as shown in Fig. 4. The device can run at 2.7 km/h. The model test tire footprint is 1 cm x 1 cm while the applied load is 17 kg. It is based on the assumption that the model should produce stress on the pavement similar with that produced by standard axle load (8160 kg) and the truck tire footprint (11 x 11 cm). The first set of the research using this device has been published elsewhere [9].

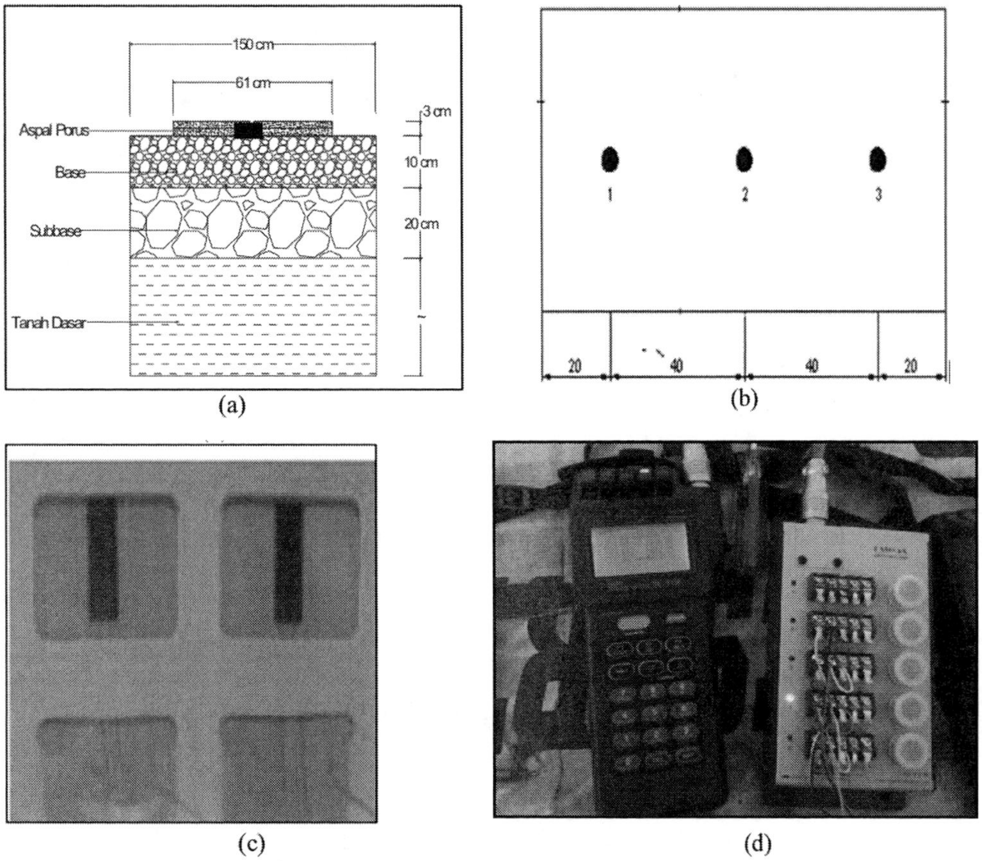

Fig. 3. Installation of strain gauge. (a) cross section, (b) location of strain gauge, (c) type of strain gauge, (d) strain meter

Fig. 4. Loading device

4 Results

Fig. 5 and 6 present the stresses recorded below the surface course for conventional and porous HMA, respectively. The measurement was made when the moving load was on top of the strain gauge. Fig. 7 present the deformation recorded at the surface course.
Fig. 5 shows that the stress recorded at the 3 cm, 5 cm, and 20 cm from the surface generally shows constant over time, although at some it tends to be lower. It can therefore be interpreted that with constant loading, the stress induced by the load remains constant in the pavement structure. It also shows the relative consistency of the loading and recording device.

The test result also shows some peak at stress recorded at every 50 passes. At every 50 passes, the loading was stopped temporarily to allow the deformation readings. The researchers also suspected that this peak was due to the nature of the pavement material elasticity, where after the load, it needs some time to get back to its original position.

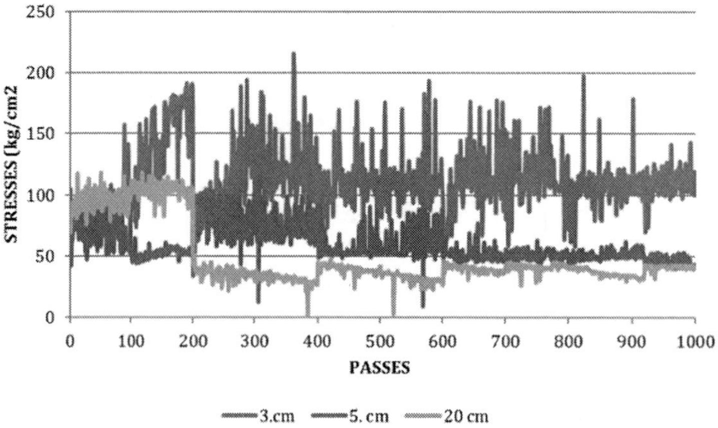

Fig. 5. Stress vs passes for conventional HMA course

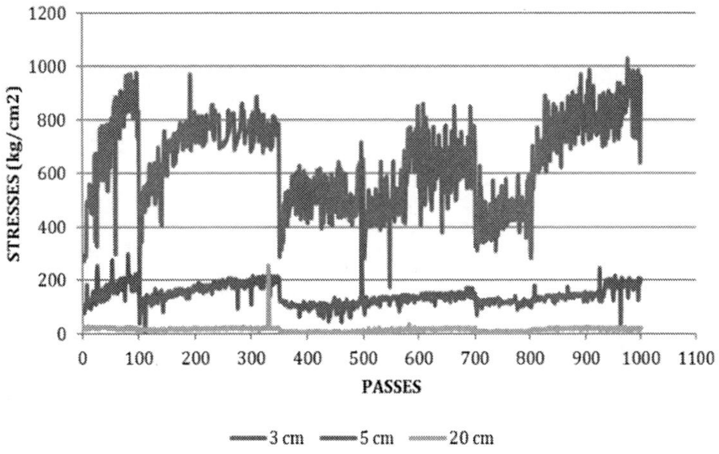

Fig. 6. Stress vs passes for porous HMA course

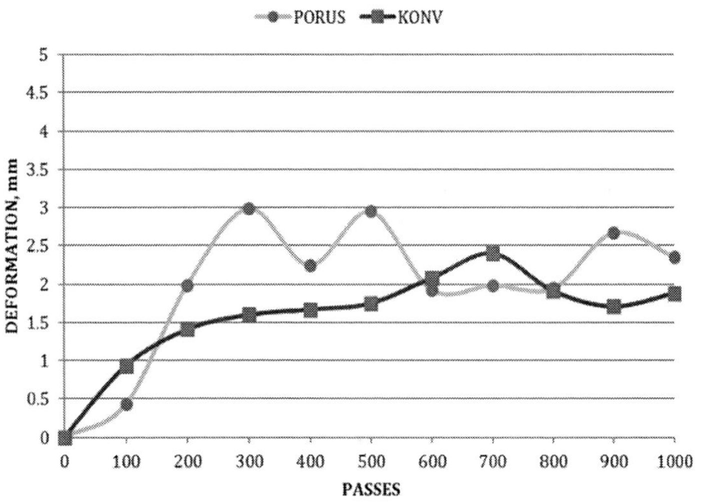

Fig. 7. Deformation comparison between conventional and porous HMA

Fig. 7 shows the deformation recorded from the test section. It is recorded every 100 passes. The figure shows that the conventional HMA structure tends to rut lesser than that of porous structure. It is due to the structure of the conventional HMA has less void compared to porous HMA, which may cause the structure has better rutting performance.

5. Conclusion and recommendation

Based on the results from the test section, the following conclusion can be drawn:

i) At low passes (up to 1000 passes) the equipment developed in this research has shown a good result in evaluating the stress and deformation occurred in the pavement.
ii) The stress occurring in both conventional and porous pavement tends to constant at the constant loading.

iii) The porous structure has slightly lower performance compared to conventional in terms of rutting.
iv) The result presented in this paper is a preliminary one, further enhancement of the equipment and longer testing period is needed in order to better simulate the field behavior of pavement.

References

[1] L. Djakfar, F.L. Roberts, *Performance Prediction of Louisiana ALF Test Sections, Transportation Research Record 1597*, Transportation Research Board,Washington, (2000)
[2] R. Bonaquist, R. Surdahl and W. Mogawer, *Effect of Tire Pressure on Flexible Pavement Response and Performance*, Transportation Research Record 1227, Transportation Research Board, Washington, 97-106 (1989)
[3] P. Sebaaly, N. Tabatabaee, R. Bonaquist and D. Anderson, *Evaluating Structural Damage of Flexible Pavements Using Cracking and Falling Weight Deflectometer Data*, Transportation Research Record 1227, Transportation Research Board, Washington, 115-127 (1989)
[4] P. Kadar, *The Performance of Overlay Treatment and Modified Binders Under Accelerated Full Scale Loading- The Callington ALF Trial,* Australian Road Research Board, Victoria, (1990)
[5] J.R. Johnson-Clark, K.G. Sharp and P.D. Walter, *The Performance of Pavements with Geotextile Reinforced Seals: The Brewarrina, N.S.W. ALF Trial*, Research Report 241. Australian Road Research Board, Victoria, (1993)
[6] M.S. Tang, Medeiros and G.R. Chehab, Numerical modeling and laboratory testing of geogrid-reinforced flexible pavements, efficient transportation and pavement systems, *Proceedings of the 4th International Gulf Conference on Roads*, Doha, 10–13, (2008)
[7] I.L. Al Qadi, S.H. Dessouky and E. Tutumluer, Geogrid-reinforced low-volume flexible pavement response to loadings by various tire configurations. Efficient Transportation and Pavement Systems, *Proceedings of the 4th International Gulf Conference on Roads*, Doha, (2008)
[8] N. Ramadhan, and R.R. Burhanuddin, *Effect of Gilsonite Additive on the Marshal Characteristics of Porous Pavement Mix*, Thesis, Universitas Brawijaya, (2014)
[9] L. Djakfar, Harimurti and Y. Zaika, Preliminary assessment of the newly developed mobile accelerated pavement testing, *Proceeding of the 9th International Conference on Road and Airfield Pavement Technology*, Dalian, China, (10 – 12 August 2015)

Volumetric Analysis and Performance of Hot Mix Asphalt with Variable Rap Content

Ahmad Kamil Arshad[1,2,*], *Masyita* Mohammad[2] , *Ekarizan* Shaffie[2] , *Wardati* Hashim[2] , and *A. G.* Abdul Halim[3]

[1]Institute for Infrastructure Engineering and Sustainability Management (IIESM), Universiti Teknologi MARA, Shah Alam, Selangor, Malaysia
[2]Faculty of Civil Engineering, Universiti Teknologi MARA, Shah Alam, Selangor, Malaysia
[3]Faculty of Civil and Environmental Engineering, Universiti Tun Hussein Onn, Malaysia

Abstract. Incorporating Reclaimed Asphalt Pavement (RAP) to the asphalt concrete mixture for highway construction offer many benefits including energy consumption, conservation of natural resources and preservation of the environment to associated emissions. This paper presents a study on performance of Hot Mix Asphalt with variable RAP content. The study is carried out to evaluate the Marshall Properties and Performance of RAP-Asphalt mixes using conventional asphaltic concrete mix AC14. Marshall Mix Design Method was used to produce control mix (0% RAP) and RAP-Asphalt mixes samples which consist of 15% RAP, 25% RAP and 35% RAP in accordance with Specifications for Road Works of Public Works Department, Malaysia. The Marshall Properties analysis was performed to ensure compliance with Marshall Requirements, The resilient modulus test was performed to measure the stiffness of the mixes while Modified Lottman test was conducted to evaluate the moisture susceptibility of these mixes. The results obtained showed that there were no substantial difference in Marshall Properties, moisture susceptibility and indirect tensile strength between RAP-Asphalt mixes with the control mix. The test results indicated that recycled mixes performed as good as the performance of conventional HMA in terms of moisture susceptibility and resilient modulus. It is recommended that further research be carried out for asphalt mixes containing more than 35% of RAP material.

1 Introduction

Hot Mix Asphalt (HMA) is a dense graded mix that is commonly used as road surfacing intended for general use. If HMA is properly designed, it will provide impermeable characteristic which allow the water to run away from the pavement's surface. However, over time, pavement distresses inhibit the optimal function of roads and contribute to traffic accident. Damaged pavements are considered one of the largest contributing factors to fatal motor vehicle crashes [1]. Therefore, there is a need to rehabilitate the pavement and in most cases, remove the asphalt surfacing material to be replaced by a new asphalt overlay.

* Corresponding author: drahmadkamil@salam.uitm.edu.my

Approximately about 80% of asphalt pavement removed during highway rehabilitation or reconstruction is recycled back into new hot mix asphalt (HMA), disposing of only 20% through normal waste stream channels [2]. Recycling of asphalt pavement is an alternative technology developed to rehabilitate and or replace pavement structure suffering from pavement deformation and structural damage. However, although recycling is beneficial in most cases by reducing the consumption of virgin materials, the performance of the highway pavement should not be compromised for cost reduction [3]. It has been accepted that RAP can be a feasible constituent in HMA pavements and if properly designed and constructed, HMA mixtures with RAP can perform as well as conventional mixtures [4]. Some studies indicates that incorporating low percentages of RAP (up to 15% RAP) have no significant changes in the performances of mixes [5].

Studies in Europe and the United States have concluded that over 80% of the recycled material is reused in the construction of roads, but regulations are still strict in allowing the inclusion of RAP in proportions ranging between 5 and 50% for production of new HMA mixtures [6]. Previous researchers stated that RAP replacement at proportions below 50% is feasible to produce new HMA mixtures, obtaining satisfactory results in the mechanical properties. Likewise, the susceptibility to moisture damage was low with the Tensile Strength Ratio (TSR) values close to 95%[7][8]. In addition, the HMA mixtures with RAP replacement increased in 50% the Indirect Tensile Strength (ITS) compared to the HMA mixtures with only virgin materials [9].

In addition, RAP replacement at proportions above 50% is feasible to produce new HMA mixtures, obtaining satisfactory results in the mechanical properties [10]. Partial blending of RAP binder occurs in the hot mix asphalt when RAP is mixed with virgin mineral materials and virgin binder. The amount of RAP has been limited to certain portion because the degree of blending between the RAP and the virgin materials is not known. If the weight by percent of all materials and their recycling technology are properly selected, the composition and properties of RAP HMA are improved.

Research on pavement recycling and RAP asphalt mixtures in Malaysia is very limited. Few research have been carried out on mixes containing RAP exceeding 25% as higher amount of RAP produced variability in Marshall Properties and performance of HMA. The objective of this study is to evaluate the performance of HMA mixtures with variable RAP content (15%, 25% and 35% RAP), in particular the Marshall properties and performance in terms of resilient modulus and moisture susceptibility.

2 Methodology

This study is based on laboratory experimental work. The samples were prepared using the Marshall Test method in accordance with ASTM D1559. The mix design used AC14 aggregate gradation in accordance with PWD Malaysia's Specification for Road Works [11], with the range of binder content from 4.0% to 6.0%. A total of 15 samples were prepared for the control mix (100% virgin aggregates) and the asphaltic concrete mixes with RAP contents of 15%, 25% and 35%. All the samples were compacted with 75 blows on both sides of the samples using Marshall Compactor. Volumetric properties and analysis were carried out to determine the optimum binder content (OBC) for control mix and recycled mix. The optimum bitumen content value was used for the performance tests including resilient modulus and moisture susceptibility. Moisture susceptibility test was performed in accordance to the procedure described in AASHTO T 283. Samples for the test were prepared at 7 + 0.5 percent air voids. Resilient modulus test was performed in accordance to ASTM D 4132. Testing temperatures of 25 and 35°C were used and loading of 1300N with load pulse of 0.1 seconds and rest period of 0.9 seconds were set.

3 Result and discussion

The initial laboratory tests have been conducted to determine physical properties of the materials used. marshall mix design was conducted to determine the marshall properties and the optimum binder content (obc) for control mix and rap-asphalt mixes. the results of moisture susceptibility test and resilient modulus test as performance tests were then compared between the control mix and rap mixes.

3.1 Comparison of RAP aggregate and virgin aggregate properties

RAP aggregate properties assist in evaluating the suitability of the aggregates to be used in road construction. The basic properties tests that were conducted for both virgin and RAP aggregates consist of Los Angeles abrasion test, aggregate impact test and specific gravity. The detailed test results for both types of aggregate are shown in Table 1.

For this study, based on the results as shown in Table 1, the LAAV, AIV, Specific Gravity for RAP aggregate is higher than virgin aggregate. This indicates that the strength of the recycled aggregate is lower than virgin aggregate. However, the value complied with PWD Malaysia's specification requirements which must not exceed 25% for LAAV and AIV. This is due to the fact that the recycled aggregate was exposed to traffic loading for a long period that reduced the strength of aggregate, unlike the virgin aggregate which has not been exposed to traffic. Since both types of aggregate complied with the specification requirements, the aggregate is suitable to be used for the asphalt mixtures preparation.

Table 1. Basic properties of virgin aggregate and RAP aggregate

Parameter	Virgin Aggregate	Reclaimed Asphalt Aggregate (RAP)	PWD Malaysia's Specification requirement
Los Angeles Abrasion Value (%)	24.00	24.91	< 25
Aggregate Impact Value (%)	14.99	19.98	< 25
Bulk Specific Gravity, Gsb (g/cm^3)	2.648	2.625	-
Apparent Specific Gravity (g/cm^3)	2.705	2.711	-
Water Absorption (%)	0.626	1.063	< 2.0

3.2 Marshall properties

Marshall Test was conducted to determine the volumetric properties of mixture. The volumetric properties determined included bulk specific gravity, air void, voids in mix (VMA), stability and flow. The Optimum Binder Content (OBC) for the control and RAP-Asphalt mixes were determined from the individual plots of bulk density, voids in total mix, voids in mix, flow and stability versus percent asphalt content.

Mix properties for virgin materials and varying different percentage of RAP showed that there are significant effect on the Marshall Properties with the inclusion of RAP in the asphalt mixes. Marshall Test result showed that the Marshall flow, VFB and Bulk SG increased as the RAP content increased. However, Marshall Stability value and voids in mix showed a decreasing trend with increasing RAP content. The trends are shown in Fig. 1 to Fig. 5 below. This result could be attributed to the fact that the virgin bitumen failed to rejuvenate the RAP binder as its percentage increases. However, the results and analysis showed that there is not much variation between conventional mixes with RAP-Asphalt mixes and all the mixes satisfy the requirements of the Specifications for Marshall Test.

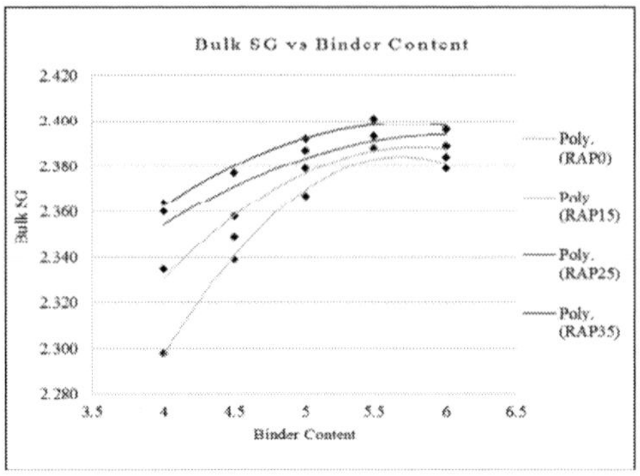

Fig. 1. Bulk specific gravity versus binder content for different RAP contents

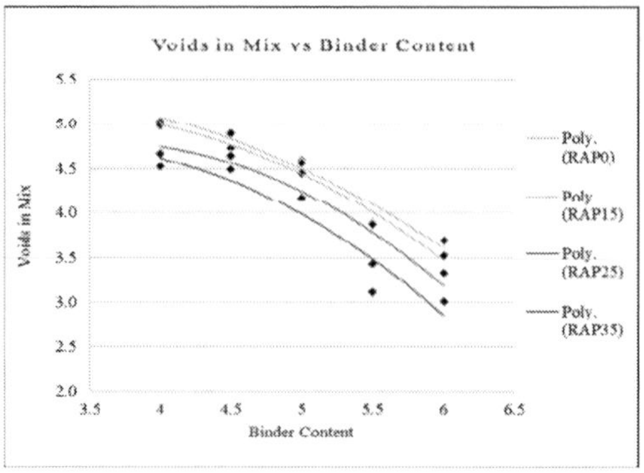

Fig. 2. Voids in mix versus binder content for different RAP contents

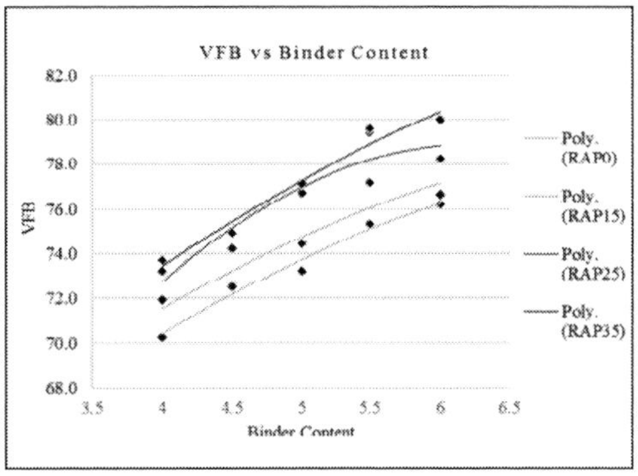

Fig. 3. VFB versus binder content for different RAP contents

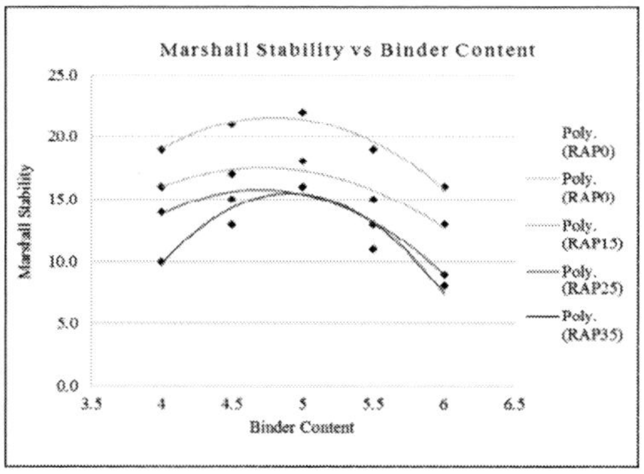

Fig. 4. Marshall stability versus binder content for different RAP contents

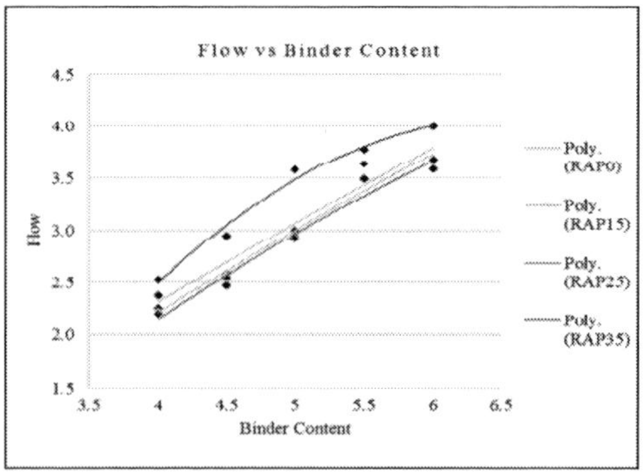

Fig. 5. Flow versus binder content for different RAP contents

3.3 Marshall properties and determination of optimum binder content

The OBC for control mix and RAP-Asphalt mixes were taken as the arithmetic mean of binder from the five smooth curves plotted and the value was checked with the limits set in Public Works Department specification. Table 2 showed the Optimum Binder Content (OBC) for the control and RAP-Asphalt mixes. The OBC decreased as the RAP content increased. The OBC for control mix, RAP15 and RAP25 and RAP35 are 5.34 %, 5.22 %,

5.10 % and 4.88 % respectively. The result obtained showed that there were no substantial differences in Marshall Properties.

Table 2. Mix Design selection criteria

Properties	Control Mix	RAP15	RAP25	RAP35	PWD Specification
OBC (%)	5.34	5.22	5.10	4.88	-
Bulk S.G	2.388	2.393	2.399	2.401	-
VIM (%)	4	4	4	4	3-5
VFB (%)	75.33	75.20	75.20	75.10	70-80
Stability (kN)	22	18	16	18	>8
Flow (mm)	3	2.95	2.92	2.95	2-4

3.4 Indirect tensile strength

The results of indirect tensile strength (ITS) for control mix and RAP-Asphalt mixes are shown in Fig. 7. It is noticed that the increase of RAP in the asphalt mix reduces the tensile strength values compared with the control mix. The results shows that the ITS values of wet condition and dry condition decreased with increasing RAP content. The graph's trend showed that the use of relatively higher RAP content reduces the moisture resistance. Since RAP aggregate is weaker compared to virgin aggregate, some particles could have been crushed during the compaction and fractured surface is exposed to water. Good quality aggregate is important to reduce the moisture damage in mix. The values of Control Sample, RAP15, RAP25 and RAP35 reduced by 4.29 %, 12.16 %, 16.13 %, and 20.0 % respectively, after the wet conditioning process. The highest value of ITS in dry condition is 0.74 MPa for RAP15 and 0.67 MPa in wet condition for Control sample.

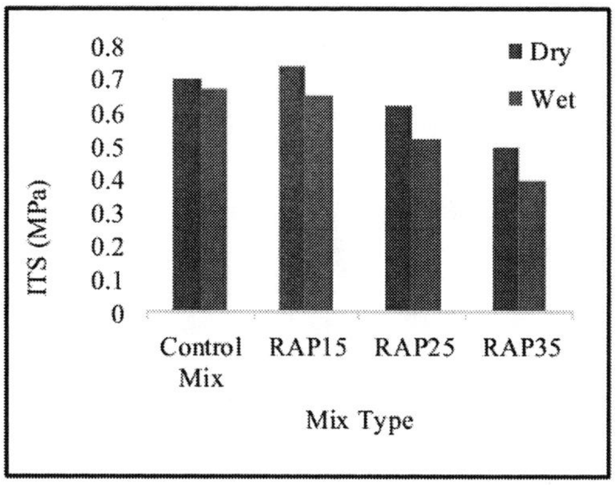

Fig. 6. Indirect Tensile Strength (ITS) for variable RAP content

The Tensile Strength Ratio (TSR) values for control mix and RAP-Asphalt mixes are shown in Fig. 7. From the results, it shows that the TSR values decreased with increasing of RAP content. The highest TSR value is 95.96 % for Control mix while the lowest TSR value is 80.14 % for RAP35. The results indicates that Control Mix to be less susceptible to moisture damage compared to RAP-Asphalt mixes. However, all the mixes have the potential to resist moisture damage since the TSR values obtained are more than 80%.

DOI: 10.1051/matecconf/201710309004

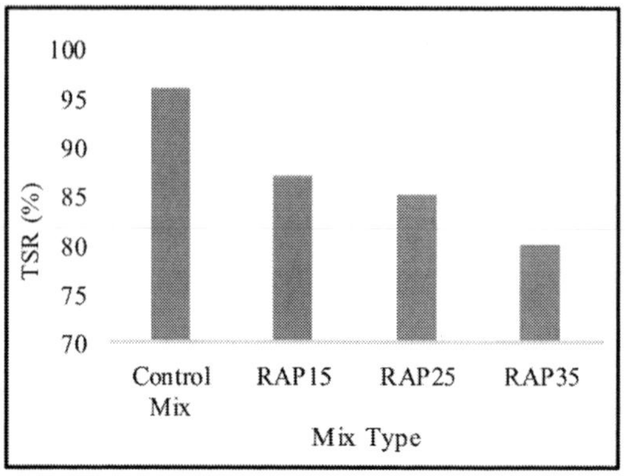

Fig. 7. Tensile Strength Ratio (TSR) for variable RAP Content

3.5 Resilient modulus of RAP mixtures

The results of resilient modulus test for different RAP contents at temperatures of 25 °C and 35°C are shown in Fig. 8. The resilient modulus for all mixes at temperature 25°C is higher for all mixes than at temperature of 35°C. The results also showed that resilient modulus of the mixes decreases with increasing RAP content. The control sample has highest resilient modulus under both testing temperature compared to samples with RAP content. The resilient modulus of control sample at temperatures of 25 °C and 35 °C are 3953 MPa and 2842 MPa respectively. This is followed by RAP15 with values of 3784 MPa and 1991 MPa at temperatures of 25°C and 35°C respectively. The results showed that RAP15 has the highest resilient modulus compared to other samples with RAP content.

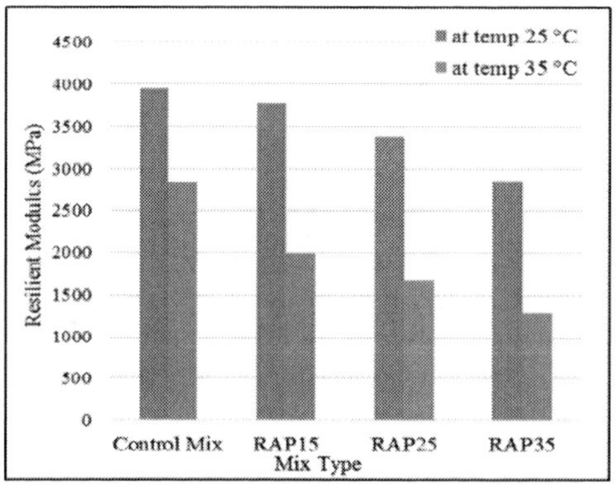

Fig. 8. Resilient modulus for variable RAP content at temperature of 25°C and 35°C

Table 3 shows the percentage of resilient modulus reduction for different RAP contents at temperature 0f 25°C to 35°C. The results in Table 3 show that the resilient modulus decreased significantly when the temperature increased from 25°C to 35°C. The resilient modulus decreased approximately 28.12 % for control mix while the resilient modulus for RAP15, RAP25 and RAP35 decreased 47.39 %, 50.75 % and 54.91 % respectively. RAP35 having highest reduction of resilient modulus from temperature 25°C to 35°C. This finding suggested that the percentage reduction of resilient modulus increases with increasing RAP in the mixture.

Table 3. Percentage of reduction in resilient modulus for variable RAP content from temperature of 25°C to 35°

Mixture Type	Resilient Modulus at 25°C (MPa)	Resilient Modulus at 35°C (MPa)	Percentage Reduction of Resilient Modulus (%)
Control Mix	3953.01	2841.59	28.12
RAP15	3784.09	1990.92	47.39
RAP25	3386.34	1667.75	50.75
RAP35	2858.48	1288.84	54.91

4 Conclusion

From this study, the following can be concluded:

i. The laboratory results on the physical properties of virgin aggregate and RAP shows that the aggregates used in this study complied with the requirements in the Public Works Department specifications. The recovered aggregate from RAP has lower strength compared to virgin aggregate. This is due to the RAP material which is exposed to the environment and traffic for a certain period of time.

ii. The OBC of control mix, RAP15, RAP25 and RAP35 showed a decreasing trend with increasing RAP content. The OBC of control mix, RAP15, RAP25 and RAP35 are 5.34 %, 5.22 %, 5.10 % and 4.8 % respectively. This shows that partial blending of aged binder occurs but the effect is not as significant because of the low percentage inclusion of RAP proportion in mixes. The result indicated that use of lower percentage of RAP in mix does not significantly affect the properties of the mix.

iii. Moisture susceptibility test and resilient test were performed to evaluate the performance of recycled mixes. The ITS and TSR value of the mixture decreases as the RAP content increases under both conditions. RAP-Asphalt mixtures satisfied the minimum requirement value for the moisture susceptibility test.The resilient modulus of the RAP-Asphalt mixtures decreased with increasing RAP content and temperature.

Based on the findings of the study, it can be concluded that it is possible to design acceptable quality asphaltic concrete mixes with RAP that meets the required volumetric and desired performance criteria. As the performance of RAP mixes satisfied the minimum requirements of PWD Malaysia, the use of recycled mixes should be encouraged as the use of RAP likely to be cheaper than normal mix. The results of this study indicate that the addition of RAP to HMA is feasible as the performance is comparable to conventional asphaltic concrete mixes. The use of higher percentages of RAP content requires a detailed study and investigation to be carried out to ensure the RAP suitability and compliance with the specification requirements.

The authors would like to express gratitude to the Institute for Engineering and Sustainability Management (IIESM) and the Faculty of Civil Engineering, Universiti Teknologi MARA (UiTM) for the support given to make this study possible.

References

[1] H. Fares, K. Shahata, E. Elwakil, Modeling the performance of pavement in cold weather conditions, Structure and Infrastructure Engineering, **8**(11), 1067-1079 (2010)

[2] A. Carter, *Development of A Non-Solvent Based Test Method for Evaluating Reclaimed Asphalt Pavement Mixes*, PhD Thesis, Auburn University, Alabama, (2004)

[3] L. Mohammad, I. Negulescu, Z. Wu, C. Daranga, W. Daly, investigation of the use of recycled polymer modified asphalt binder in asphalt concrete pavements, Journal Associated of Asphalt Pavement Technology, **72,** 551-594 (2003)
[4] B. Huang, X. Shu, D. Vukosavljevic, Laboratory investigation of cracking resistance of hot-mix asphalt field mixtures containing screened reclaimed asphalt pavement, ASCE Journal of Materials in Civil Engineering, **23**(11), 1535-1543 (2011)
[5] H.R. Paul, Evaluation of recycled projects for performance, Proceedings of The Association of Asphalt Paving Technologists, **65**, 231-254 (1996)
[6] W. Mengqi, W. Haifang, W. Muhunthan, N. Kalehiwot, Influence of RAP content on the air void distribution, permeability and moduli of the base layer in recycled asphalt pavements, *Proceedings of the 91st Transportation Research Board Meeting*, TRB (2012)
[7] C. Celauro, C. Bernardo, B. Gabriele, Production of innovative, recycled and high performance asphalt for road pavements, Journal of Resource, Conserve and Recycle, **54**, 337-347 (2010)
[8] P. Shirodkar, Y. Mehta, A. Nolan, K. Sonpal, A. Norton, A study to determine the degree of partial blending of reclaimed asphalt pavement (RAP) binder for high RAP hot mix asphalt, Journal of Construction Building and Material, **25**, 150-155 (2011)
[9] A.E.A. El-Maaty, A.I. Elmohr, Characterization of recycled asphalt pavement (RAP) for use in flexible pavement, American Journal of Engineering and Applied Sciences, **8**(2), 233–248 (2015)
[10] K. Vislavicius, H. Sivilevicius, Effect of reclaimed asphalt pavement gradation variation on the homogeneity of recycled hot-mix asphalt, Journal of Civil and Mechanical Engineering, (2013)
[11] Public Works Department of Malaysia, *Standard Specification for Road Works - Section 4: Flexible Pavement (JKR/SPJ/2008-S4)*, Malaysia, (2008)

Multiple Stress Creep and Recovery of Nanosilica Modified Asphalt Binder

Ahmad Kamil Arshad[1,2,*], *Mohamad Saifullah* Samsudin[2], *Khairil Azman* Masri[2], *Mohamed Rehan* Karim[3], and *A. G.* Abdul Halim[4]

[1]Institute for Infrastructure Engineering and Sustainability Management (IIESM) Universiti Teknologi Mara, Shah Alam, Selangor, Malaysia
[2]Faculty of Civil Engineering, Universiti Teknologi MARA, Shah Alam, Selangor, Malaysia
[3]Faculty of Engineering, Universiti of Malaya, 50603 Kuala Lumpur, Malaysia
[4]Faculty of Civil and Environmental Engineering, Universiti Tun Hussein Onn, Malaysia

Abstract. The main objective of this study is to evaluate the effect of nanosilica (NS) concentration on the rutting parameter of based asphalt binder using the multiple stress creep and recovery (MSCR) test. NS with concentrations between 1% to 5% (1% increment) was used to modify asphalt binder. The nanosilica modified asphalt binder (NSMB) were aged using rolling thin film oven (RTFO) before tested. MSCR test was conducted at 64^{0}C with two stress level (100 Pa and 3200 Pa). The results found that the addition of NS reduced the accumulative strain of NSMB. Besides that, non-recoverable creep compliance decreased and recovery strain increased, which indicates an improvement of rutting resistance and elasticity of the binder, respectively. The grade of asphalt binder also improved from heavy (H) grade to extreme (E) grade. Overall, NSMB 2% show the best improvement in terms of Jnr and %R, thus NSMB 2% considered as a optimum percentage.

1 Introduction

Permanent deformation or rutting is one of the main problems in asphalt pavement; it is caused by repeated loading from heavy traffic loading which results in progressive accumulation of permanent deformation under repetitive tire pressures [1]. Rutting significantly affect the pavement performance and reduce its service life. The resistance of asphalt pavement to rutting depends on many factors. One of the main importance factor is the stiffness of the asphalt binder. Therefore, it is necessary to modify asphalt binder in order to improve it stiffness. The use of nanotechnology in asphalt binder modification promises better result in improving asphalt binder stiffness. Various nanomaterials have been used to modify asphalt binder such as nanosilica (NS).

NS has been widely used in polymer, concrete and asphalt binder as inorganic filler to improve the properties of polymeric, mechanical and bituminous materials. NS is an inorganic material which has a maximum dimension about 30nm. NS possesses good properties such as large surface area, good dispersal ability, strong adsorption, high

* Corresponding author: drahmadkamil@salam.uitm.edu.my

DOI: 10.1051/matecconf/201710309005

chemical clarity and excellent stability [2]. Yao et al (2013) have used NS to modify asphalt binder. The addition of NS (4% and 6%) in the base asphalt binder decrease the viscosity value of base asphalt binder as well as anti-aging property. Besides that, the dynamic modulus of NS modified asphalt mixture is higher than the control asphalt mixture by an average 30%. NS also gives high contribution in increasing the value of flow number where it increases the flow number by an average of 300%. This value indicates that it has more possibilities to resist the permanent deformation at high temperature compared to control binder [3,4]

Previously, the rutting parameter (G*/sinδ) from dynamic shear rheometer (DSR) test was used to evaluate the rutting performance of asphalt binder. As the DSR test used low stress level (about 10%), thus it may underestimate the performance of modified binder. This is because at low strain level, the modifier may not be fully activated. Recently, the performance grade (PG) asphalt binder specification introduced the multiple stress creep and recovery test (MSCR) to indicate the rutting performance of the asphalt binder. The MSCR test was developed by D'Angelo et al (2007) to replace the dynamic oscillatory shear test at high temperature [5]. The non-recoverable creep compliance (J_{nr}) in MSCR test has higher correlation with rutting performance of asphalt binder compared to the rutting parameter (G*/sinδ) as reported by Federal Highway Administration (2011) and Asphalt Institute (2012)[6]. According to the AASHTO TP70 standard, the test consists of 10 cycles of creep and recovery. Each cycles consist of 1s loading and 9s recovery upon two different stress level started with low stress level (100 Pa) and followed by high stress level (3200 Pa)[7]. The use of two different stress levels in the MSCR test correlated with varrying traffic loading levels on the pavement surface. Fig. 1 show the typical creep and recovery in MSCR test.

The main objective of this study was to evaluate the high temperature rutting performance of NSMB using MSCR. The effect of NS on J_{nr} and %R of base asphalt binder will be explored as an indicator of rutting performance. Maximum NS percentage for asphalt binder also will also be determined.

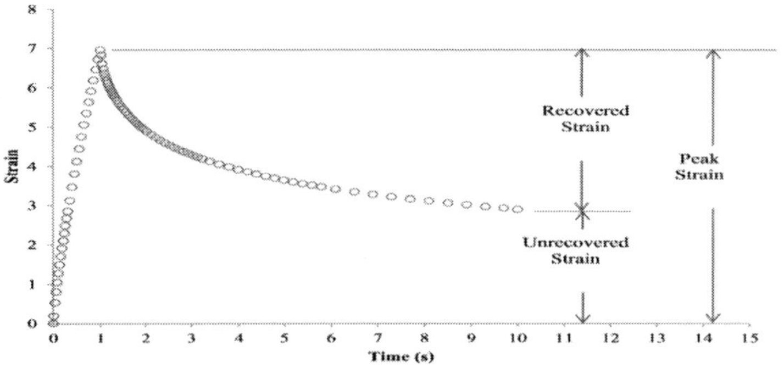

Fig. 1. Typical creep and recovery in MSCR test

2 Environmental program

2.1 Materials and sample preparation

The study used 60/70 Penetration Grade (60/70 PEN) of asphalt binder. Then, the binder was modified with NS with average size of 10-15 nm. Table 1 shows the properties of asphalt binder and NS. Asphalt binder was blended with 1% to 5% NS (1% increment) by weight of the virgin asphalt binder. The modification of asphalt binder was conducted by mechanical mixer. Asphalt binder was heated up to 160°C until it achieved the processing viscosity. Cylindrical container was filled with about 400g of hot asphalt and placed on a hot plate. The temperature of hot plate was set to 160°C to maintain the viscosity of asphalt binder during mixing. NS was gradually added into the hot asphalt binder while stirring with the mechanical steel stirrer. The speed of stirrer was set to 2000 rpm. The mixing process continued for one hour in order to achieve uniform dispersion of NS. The modified binder was placed in the oven to remove the bubbles, before preparing the test sample. To facilitate in referring each sample, nanosilica modified asphalt binder (NSMB) were named using following abbreviations: NSMB 0%, NSMB 1%, NSMB 2%, NSMB 3%, NSMB 4% and NSMB 5%.

Table 1. Properties of asphalt binder and NS

Materials	Properties	value
Asphalt binder	Softening point (°C)	48-56
	Penetration (0.1 mm)	60-70
	Ductility (cm)	>100
	Flash point (°C)	331
Nanosilica	Appearance	Slight milky transparent to translucent liquid
	SiO_2 (wt%)	30±1%
	Na_2O (wt%)	0.5%
	pH (20°C)	8.5-10.5
	Density (20°C, g/cm3)	1.19-1.22
	Particle size (nm)	10-15

2.2 Short term aging

The RTFO equipment was used to simulate short term aging according to ASTM D2872. In this test, the hot binder was poured into a cylindrical glass bottle with a 35 g each. The glass bottles which contained binder were placed horizontally in a rotating carriage slots in the oven. Then, the carriage was rotated at a rate of 15 rpm for 85 minutes in the oven at temperature of 163°C. Throughout the test, the opening glass bottles containing the binders were exposed to the air pressure at 4000 ml/min.

2.3 Multiple stress creep and recovery

Multiple stress creep and recovery (MSCR) test was carried out on RTFO aged sample at temperature 64°C using DSR equipment in accordance with AASHTO TP70. Two different stress levels were chosen, 100 Pa and 3200 Pa. Two continuous sets of ten loading cycle of repeated creep and recovery testing was applied. The test used 1s creep load followed by 9s recovery for each cycle. The main parameters achieved from MSCR are recoverable strain (R) and non-recoverable creep compliance (J_{nr}). R value provides an indication of the presence of elastic response and stress dependency of the binder, while J_{nr} was defined as the ratio of the non-recovered strain to the stress applied during MSCR testing. R and J_{nr} were calculated using the following equations:

$$R = \frac{\varepsilon_p - \varepsilon_u}{\varepsilon_p} \times 100\% \qquad (1)$$

$$J_{nr} = \frac{\varepsilon u}{\sigma} \tag{2}$$

where, Ɛp represents the peak strain, Ɛu represents the unrecovered strain and σ represents the stress level. The average of R of 10 cycles at 100 Pa and 3200 Pa are calculated and expressed as R100 and R3200 respectively, while the average of J_{nr} at 100 Pa and 3200 Pa are expressed as $J_{nr}100$ and $J_{nr}3200$. The stress sensitivity parameter, Rdiff and J_{nr}-diff are calculated using the following equation;

$$Rdiff = \frac{R100 - R3200}{R100} \times 100 \tag{3}$$

$$J_{nr}\text{-}diff = \frac{jnr3200 - jnr100}{jnr100} \times 100 \tag{4}$$

3 Result and discussion

3.1 Strain

Fig. 2 shows the typical strain outputs from MSCR testing of NSMB at 64^{0}C with two stress levels. The first 100 second represents low stress level (100 Pa) and next 100 second represents a high stress level (3200 Pa). Fig. 3 represents one cycle of MSCR of NSMB at 100 Pa. The MSCR test result included two phase which is creep phase and recovery phase to complete one cycle. As can be seen from the Fig. 3, in the one-second creep phase, the strain is going up under loading. In the nine-second recovery phase, the strain recovered when the loading was removed. In the recovery phase, the strain recovered immediately at the beginning but the recovery rates decreased with time. Also, the strain and recovery rate was very high at the beginning of the cycle, but the rate decreased with time as shown in Fig. 2. This MSCR test result is a reflection of the visco-elastic-plastic property of asphalt binder; when the load is removed, the elastic strain recovered quickly, while the viscous strain recovered slowly [8]. Comparing the accumulated strain by the stress level, it can be observed that the accumulated strain at low stress level is lower than at high stress level. This indicates that, with the increasing stress level, the accumulated strain also increased. In terms of the effect of NS on the MSCR results, it seems that the accumulated strain is significantly reduced compared to base asphalt binder. This indicates that NS improve the stiffness of the asphalt binder at high service temperature. It can be concluded that NSMB2% has lower accumulated strain compared to other percentages of NSMB.

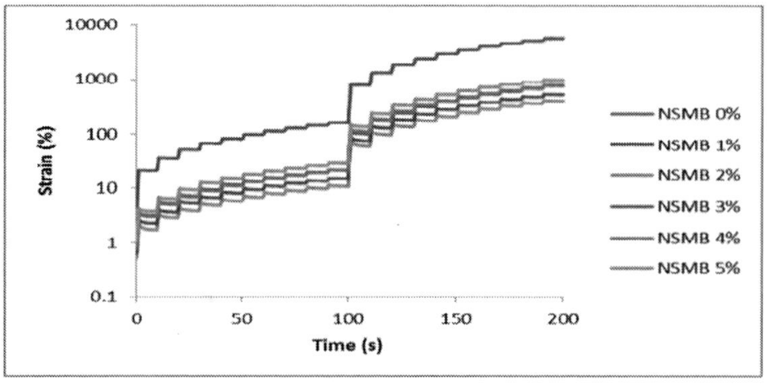

Fig. 2. Strain versus time of MSCR testing for ten cycle

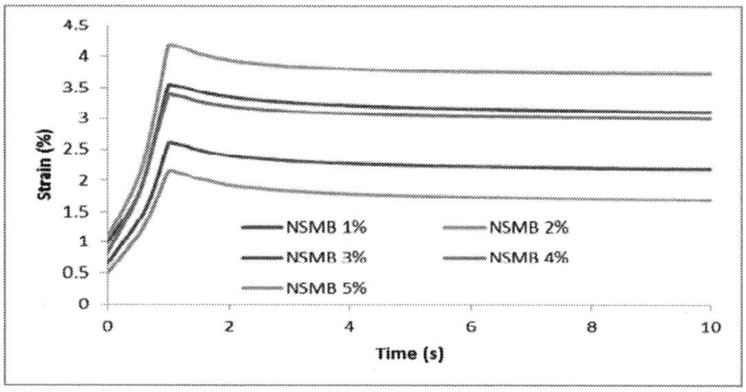

Fig. 3. Strain versus time of MSCR testing for one cycle at 100 Pa

3.2 Nonrecoverable creep compliance (J_{nr})

The J_{nr} parameter was proposed to measure the binder contribution to mixture rutting performance, where lower value of J_{nr} indicates better resistance to rutting performance and higher value of J_{nr} indicates poor resistance to rutting performance. Fig. 4 shows the J_{nr} results for NSMB at stress levels of 100 Pa and 3200 Pa. It was observed that the trends of the bar chart are similar for both stress levels. Also, J_{nr} increased slightly as the level of stress changed from 100 Pa to 3200 Pa. The changes of J_{nr} were more pronounced for NSMB0%. The addition of nanosilica into asphalt binder also reduced the value of J_{nr} compared with the base asphalt binder. For stress level of 100 Pa, the J_{nr} of NSMB 0%, NSMB 1%, NSMB 2%, NSMB 3%, NSMB 4% and NSMB 5% were 87.0%, 93.4%, 91.0%, 87.0%, and 82.5% lower than that of the base asphalt binder, respectively. For stress level of 3200 Pa, the J_{nr} of NSMB 0%, NSMB 1%, NSMB 2%, NSMB 3%, NSMB 4% and NSMB 5% were 86.8%, 93.1%, 90.9%, 86.6% and 83.6% lower than that of the base asphalt binder, respectively. The value of percentage difference between NSMB and base asphalt binder proved that addition of nanosilica reduced the value of J_{nr} and indicate that rutting resistance of asphalt binder improved.

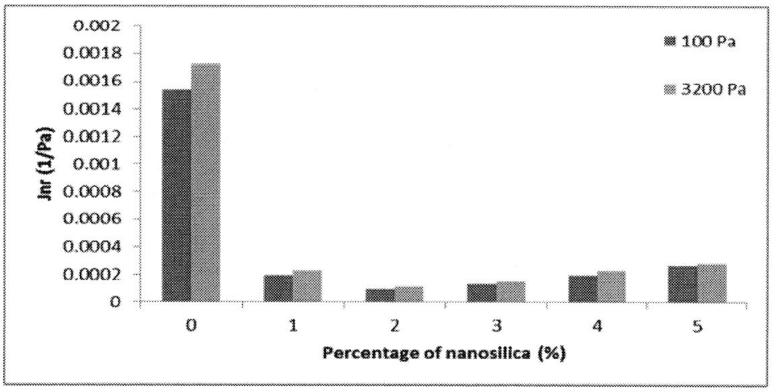

Fig. 4. Nonrecoverable creep compliance (J_{nr}) at difference stress level of NSMB

3.3 Percentage recoverable strain (%R)

The percentage recoverable strain is used to determine the elastic behavior of asphalt binder under loading. It is an important parameter to evaluate the ability of asphalt binder to recover after deformation. The higher percentage of recoverable strain of asphalt binder indicates the lower susceptibility to permanent deformation. Percentage recoverable strain showed an inverse relationship with nonrecoverable creep compliance. Fig. 5 shows the percent recoverable strain at stress levels of 100 Pa and 3200 Pa. As it can be seen that the NSMB have a higher value of percent recoverable strain compared to the control binder. This indicates that the addition of small percentages of nanosilica improves elastic recovery of asphalt binder and improves its rutting resistance. By comparing the results at two stress levels, the value slightly decreased at 3200 Pa compared to the value at 100 Pa. For stress level of 100 Pa, the %R of NSMB 0%, NSMB 1%, NSMB 2%, NSMB 3%, NSMB 4% and NSMB 5% were 80.6%, 89.0%, 85.0%, 78.8%, and 77.9% higher than that of the base asphalt binder, respectively. For stress level of 3200 Pa, the Jnr of NSMB 0%, NSMB 1%, NSMB 2%, NSMB 3%, NSMB 4% and NSMB 5% were 94.1%, 96.8%, 95.6%, 92.6% and 92.2% higher than that of the base asphalt binder, respectively. Saboo and Kumar (2015) reported that if binder has at least 20% recovery, irrespective of any stress level or temperature, it will indicate the presence of delayed elastic behavior. By comparing with this study, the addition of 2% of NS passed the minimum strain recovery [9].

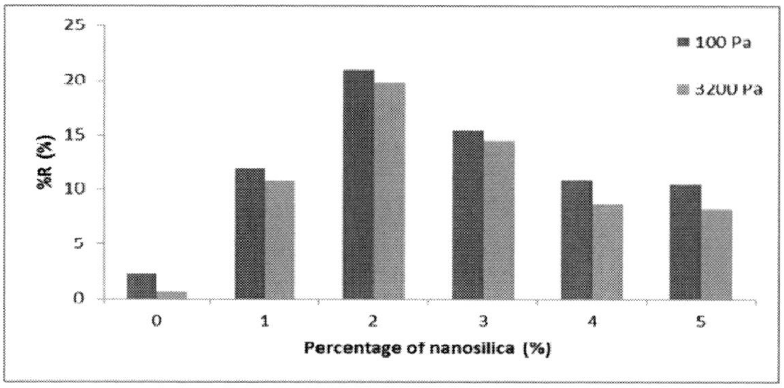

Fig. 5. Percent recoverable strain (%R) at difference stress level of NSMB

3.4 Stress sensitivity

Percent difference of non-recoverable creep compliance (J_{nr}-diff) and creep recovery (Rdiff) is a measure of the sensitivity of the asphalt binder to the increasing stress level. Lower values of J_{nr}diff and Rdiff indicate less stress sensitivity of the asphalt binder. Fig. 6 shows the stress sensitivity of J_{nr}diff and Rdiff for NSMB. According to Superpave MSCR procedure [2], the maximum value for J_{nr}diff are 75% at the high PG grade and the loading-unloading times of 1-9s. From Fig. 6, it can be seen that, at high percentages of nanosilica shows high sensitivity compared to low percentages of NS. All percentages of NSMB complied with the specification. J_{nr} of NSMB2% had the highest stress sensitivity at high service temperature. In contrast, %R exhibit less sensitivity at high service temperature against rutting.

3.5 Relationship between Jnr and %R

Fig. 7 shows the relationship between %R and J_{nr} at different stress levels of NSMB. This graph was divided into 4 grades in accordance to AASHTO TP70 which are standard (S) J_{nr} < 4.5 1/kPa, heavy (H) J_{nr} < 2.0 1/kPa, very heavy (V) J_{nr} < 1.0 1/kPa, and extreme (E) J_{nr} < 0.5 1/kPa based on the value of J_{nr}. As can be seen from the graph, the standard (S) grade was neglected because there was no value in that grade. From Fig. 7 and Table 2, it was observed that the base asphalt binder can be classified as standard (S) and heavy (H) grade while NSMB can be classified for all grades. This proved that NSMB improved the rutting performance and improved the grade of the asphalt pavement.

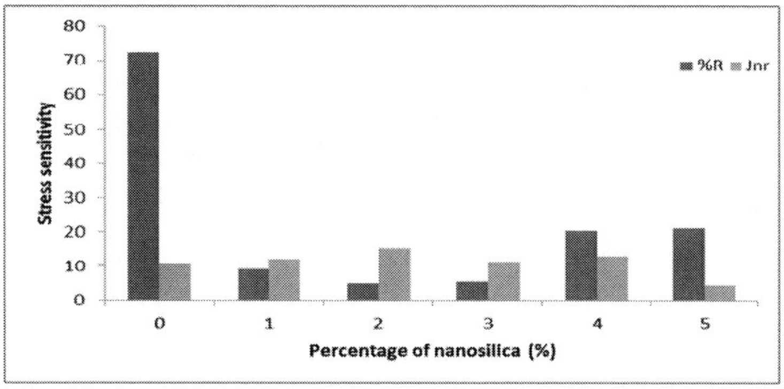

Fig. 6. Stress sensitivity of %R and J_{nr} of NSMB

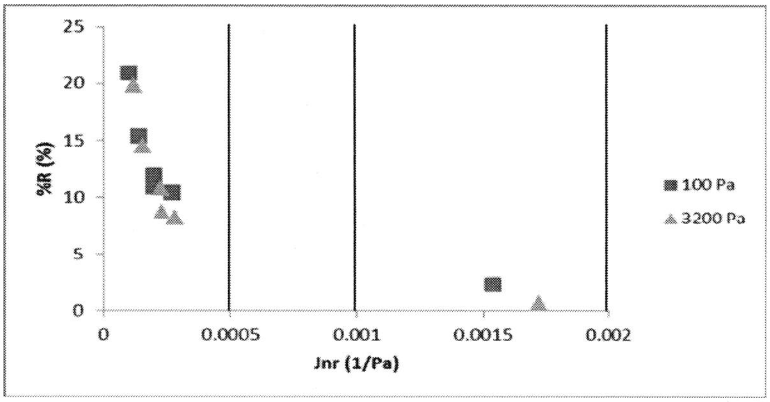

Fig. 7. Relationship between J_{nr} and %R of NSMB

Table 2. Suitability of NSMB for pavement grade at 64^{0}C

Type of grade	J_{nr} maximum (kPa)	100 Pa						3200 Pa					
		0%	1%	2%	3%	4%	5%	0%	1%	2%	3%	4%	5%
S	4	Y	Y	Y	Y	Y	Y	Y	Y	Y	Y	Y	Y
H	2	Y	Y	Y	Y	Y	Y	Y	Y	Y	Y	Y	Y
V	1	N	Y	Y	Y	Y	Y	N	Y	Y	Y	Y	Y
E	0.5	N	Y	Y	Y	Y	Y	N	Y	Y	Y	Y	Y

(Y=Suitable, N=Not suitable)

4 Conclusion

This study investigated the effect of the NS on the rutting performance of asphalt binder using MSCR. By adding small percentages of NS in the base asphalt binder, the strain values decreased significantly. Besides that, the non-recoverable creep compliance (J_{nr}) also decreased, while percentage recovery increased. Thus, this indicates that high temperature stability and elasticity of NSMB improved. The grade of the asphalt binder also improved from Heavy (H) to Extreme (E). It was found that NSMB 2% has the greatest performance against rutting, compared to other NSMB.

The authors expressed utmost gratitude to FRGS Research Grant: FRGS/1/2015/TK08/UIYM/02/3 from Ministry of Higher Education, Malaysia for financial support which enables this paper to be written. The authors also would like to thank to Department of Civil Engineering, Faculty of Engineering, University of Malaya, 50603 Kuala Lumpur, MALAYSIA for permitting the use of their equipment.

References

[1] S. Tayfur, H. Ozen, and A. Aksoy, Investigation of rutting performance of asphalt mixtures containing polymer modifiers, J. Constr. Build. Mater., **21**, 328-337 (2007)

[2] T.H. Hsieh, A.J. Kinloch, A.C. Taylor, and S. Sprenger, The effect of silica nanoparticles and carbon nanotubes on the toughness of a thermosetting epoxy polymer, Journal of Applied Polymer Science, **119**, 2135-2142 (2011)

[3] H. Yao, Z. You, L. Li, Lee, and S.W. Goh, Evaluation of the master curves for complex shear modulus for nano-modified asphalt binder, CICTP, 3399-3414 (2012)

[4] H. Yao, Z. You, L. Li, C.H. Lee, D. Wingard, Y.K. Yap, X. Shi, and S.W. Goh, Rheological properties and chemical bonding of asphalt modified with nanosilica, Journal of Materials in Civil Engineering **25**, 1619-1630 (2013)

[5] J. D'Angelo, R. Kluttz, R. Dongre, K. Stephens, and L. Zanzotto, Revision of the Superpave high temperature binder specification: The multiple stress creep recovery test, J. Assoc. Asphalt Paving Technol., **76**, 123-162 (2007)

[6] U.S Department of Transportation Federal Highway Administration, *The multiple stress creep and recovery (MSCR) procedure*, FHWA-HIF-11-038, (2011)

[7] AASHTO TP70-10 Multiple Stress Creep Recovery (MSCR), *Test of Asphalt Binder Using a Dynamic Shear Rheometer (DSR),* American Association of State and Highway Transportation Officials, (2013)

[8] X. Yang, and Z. You, High temperature evaluation of bio-oil modified asphalt binder using the DSR and MSCR tests, Construction and Building Materials, **76**, 380-387 (2015)

[9] N. Saboo, and P. Kumar, A study on creep and recovery behavior of asphalt binder, Construction and Building Material, **96**, 632-640 (2015)

Performance of Hot Asphalt Mixtures Containing Plastic Bottles as Additive

Hakeem Jan[1,*], *Mohamad Yusri* Aman[2], *Sheraz* Khan[3], and *Fazal* Karim[4]

[1,2] Faculty of Civil and Environmental Engineering, Universiti Tun Hussein Onn Malaysia, 86400 Parit Raja, Johor, Malaysia
[3,4]Faculty of Civil Engineering, SUIT, 25000 Peshawar Pakistan

Abstract: This study focuses on evaluating the resistance of polymer modified asphalt mixes and the role played by asphalt in the realm of construction is undeniably important. Addition of polymers(PB) as additives to asphalt helps to improve the strength and water repellent property of the mix and as well as helps environment in various ways and at the same time, analyzing its lower maintenance activities and service life is most important. The use of inexpensive polymers, in this case, waste polymers has without any doubt proven to be the most convenient way of reducing the cost of construction and at the same time maintaining quality. The main resolve for this research was to establish the effects of the use of plastic bottles on hot asphalt and its mixtures. In order to put this into perspective, varying percentages of asphalt mixtures were calculated and subjected to laboratory tests. The two-factor variance analysis (ANOVA) was conducted to determine the significance at various confidence limits. The results indicate that the inclusion of Polyethylene Terephthalate (PET) had a particularly substantial effect on the properties of asphalt. Consequently, it can encourage the re-utilization of waste in the manufacturing industry in an ecologically friendly and cost-effective way.

1 Introduction

Asphalt is the product of a mixture of binders, aggregates and filler. It is used both for construction and maintenance of the road, pavements and sometimes even parking areas. Polyethylene Terephthalate is a non-biodegradable, polymer. It is half-crystalline thermoplastic in nature. For this reason, it is considered to be a polyester material. It forms the greater percentage of most plastic wastes in any modern municipality. Plastics are famous the world over. Large volumes of waste polyethylene terephthalate are manufactured mostly by European nations. They come in various shapes and forms including, molding, bottles, fibers as well as sheets. Polyethylene Terephthalate comes in handy in the food industry thanks to its superlative nature. Bottles and cans are used as packaging containers. However, Extensive generation of wastes would bring disastrous effects to the ecosystem especially when disposed of in landfills.

* Corresponding author: khyberpass08@gmail.org

According to [1], asphalt mixtures' service life has been on the decrease over the last decade. They attribute this decline in quality and longevity to the considerable increase in the number of heavy vehicle on roads. The most affected types of vehicles are vans and trucks. In addition to that, the performance of asphalt is influenced by the atmospheric temperature. The reason for this behavior is because asphalt is visco-elastic in nature [1]. For this is a reason, it is greatly affected by fluctuations in temperature. Naturally, high temperatures induce softer or fewer viscose binders which in turn affect the asphalt's performance in a negative way.

In order to avoid these complications and come up with a better mixture that has a prolonged service life, most road engineers have resorted to modifying asphalt mixtures. They achieve this feat by using additives namely; polymers and fibers.In their investigation, [2] also confirms that the modification of asphalt using polymers provides a vital stance upon which the designing of a flexible pavement is inclined. Majorly, this contributes towards increasing the pavements service length or reducing its layer or the thickness of its base.

What is more, it has been found that polymers affect bitumen. Not only does it increase the stiffness of asphalt, but it also ensures enhanced temperature vulnerability. [3] posits that increased stiffness guarantees better altered rutting resistance in hot weather. Additionally, they suggest that bitumen can be relatively used with a softer base with a view to achieving better results during cool temperatures. It is also revealed that polymers possess better bonding agents along with interconnected properties. Polymers also act as an additive for asphalt concrete mixtures [4]. Consequently, it serves as an aggregate material majorly used for coating. These kinds of coatings enhance the toughness of an aggregate surface. In this way, superior engineering properties in the production of the mixtures of asphalt are guaranteed.

There exist a number of polymers that are often used in the modification of bitumen. They are subdivided into a couple of groups following their behaviors when subjected to heat [4]. These groups include; thermosetting and thermoplastic polymers. On one hand, thermosetting polymers, that is elastomers and resins are cured by the use of heat in order to produce a three-dimensional structure bearing crosslinking bonds which offer due resistance to polymers matrix tempering. On the other hand, thermoplastic polymers are susceptible to heat; hence their shapes can be easily altered and redesigned into other forms. [4], confirms that the significant difference between the two is that the latter is reversible while the former is irreversible. As it seems, pavement defects such as rutting in high temperatures, cracking in areas experiencing low temperatures and so on, are as a result of vehicle loading and perhaps most importantly asphalt concrete's inability to resist changes in temperature.

Scientists and researchers the world over have put themselves to task to explore the recycling of waste material to the advantage of the environment as well as the economy. Most importantly they attempt to look into the possibility of the re-use of solid waste material in the construction of roads. According to [2] the best way to get rid of the excessive solid waste accumulating urban and industrial areas is to implement the re-use of waste material. This initiative would have a significant impact on both the environment and the economy in some ways.

To begin with, it would help in reducing the over-reliance on natural resources saving them from exhaustion in the process. Secondly, it would keep environmental pollution by wastes generated from urban and industrial areas in check. Thirdly, it helps in saving money as well as energy [2]. Generally, in this plastic study bottle which is in the actual sense a form of polyethylene terephthalate as an asphalt mixture additive was investigated. The study evaluates the performance of hot asphalt by using plastic bottles as additives [3].

2 Materials and methodology

In this experiment, polyethylene was used in the modification of asphalt mixtures. Black gravels were collected and pulverized into rough aggregate for performance evaluation of polyethylene on hot asphalt mixtures. The resultant fine dust was used as a filler material. It was acquired from the aggregate mixture of Domar. As the stone was being crushed, a large amount of dust was produced. Some of it was trapped in the pulverized rock. The dust was eventually mixed with the sand in a bid to make certain the maximum utilization of stone dust.

This experiment incorporated the use of domestic waste as well as low-density bags polyethylene bags. After cleaning the waste polyethylene, they were eventually shredded into tiny particles of 2-3 mm for the preparation of recycled polyethylene. [5] recommended a gravity of 0.94 and a melting temperature of 115°C. These conditions were imperative for the preparation of the polyethylene. It would enhance the acquisition of better results.

As dictated by AASHTO, samples of bitumen were tested to assess different properties of asphalt, commonly referred to as fire and flash points, softening points as well as penetration value [6]. Mainly, the properties of bitumen binder were maintained within the penetrating specification of asphalt grade 80/100. Specimens of bitumen were then prepared for evaluating the Marshall stability which is concomitant with the mixtures of asphalt and waste polyethylene. These preparations were guided by test formula by Bruce Marshall that was improved by the Engineer Corps of the United States [4].

Aggregates of approximately 1200 g were for that reason used in the preparation of this specimen which was 101.6 mm in diameter with the breadth of around 63.5 mm of unadulterated bitumen. Moreover, the average amount of bitumen relevant in the preparation of the mixture was ascertained by this method [7]. It is equally imperative to note that the Marshall Stability was utilized in the prediction of the performance degree of stability of the mixture, the capacity of weight carrying as well as the permanent deformation resistance. The same goes for the flow test. A total of nine samples were used in the process of determining the content of optimum asphalt binder with different proportions of road [7].

The already crushed black stone was further, manually broken down into 25.00 mm downgraded pieces. As dictated by [5], the rough aggregate needed to be sieved of using a 2.36 mm sieve. As a result, a superlative aggregate that would pass through 2.36 mm but is retained by a sieve of 0.075 mm was obtained from the coarse aggregate sand. It is recommended that a non-plastic sieve of not more than 0.075 mm should be used to meet the mineral filler's requirements. According to the authors, a constantly graded bituminous aggregate is essential in the evaluation of the asphalt concrete mix that is improved using additives on behavior. On the other hand, the latter needs even aggregate assortment gradation for it to attain a dense mix with a free-air controlled ideal content for the construction of steady as well as long-lasting supple pavement.

3 Results

This investigation evaluated the performance of hot asphalt with regards to flash, fire, bitumen penetration as well as softening points. Samples of different ground polymers (plastics) were mixed with asphalt. The results that were obtained were presented in details in the subsequent section. The results revealed an inverse variation in the comparison of polymer added in bitumen and the value of penetration. The highest possible value of penetration of 85 was achieved with the addition of a polymer of 0% whereas; the lowest penetration value of zero was obtained with the addition of a polymer of 16%. Other results received included; penetration value of 70 with an addition of a polymer of 2%, penetration value of 64 with an addition of a polymer of 4%, penetration value of 57 with an addition of a polymer of 6%, , penetration value of 46 with an addition of a polymer of 8%, penetration value of 38 with an addition of a polymer of 10%,, penetration value of 24 with an addition of a polymer of 12%, penetration value of 13 with an addition of a polymer of 14%.

Table 1. Comparison between polymer added (%) and penetration value.

Polymer added in Bitumen (%)	Penetration Value
0	85
2	70
4	64
6	57
8	46
10	38
12	24
14	13
16	0

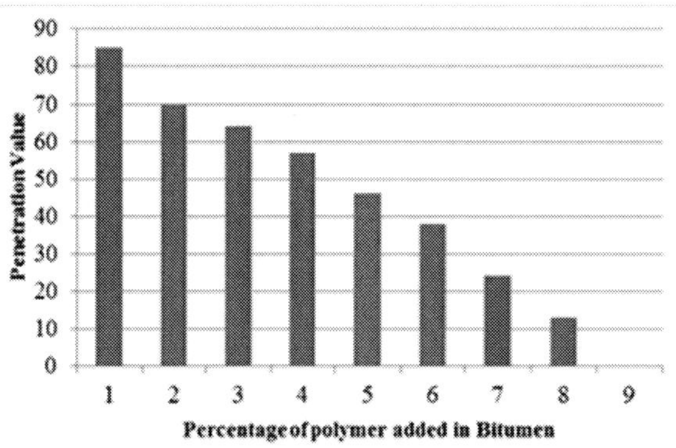

Fig. 1. Comparison between polymers added (%) and penetration value.

When comparing the additional polymer and their resultant points of softening, the research revealed that these factors were directly interrelated with one other. For instance, zero percent unadulterated Bitumen, had a corresponding softening point of 55 C°, Bitumen 57 C° with polymer of 1%, Bitumen at 61C° with polymer of 2%, Bitumen at 65 C° with polymer of 3% and finally bitumen at 68 C° with polymer of 4% (Table and Figure 2). The mean softening point was thus found to be ± SD = 27.980 and Std. Error = 9.327.

Table 2. Comparison between Polymers added (%) and Softening Point.

Polymer added in Bitumen (%)	Softening Point (C°)
0	55
1	57
2	61
3	65
4	68

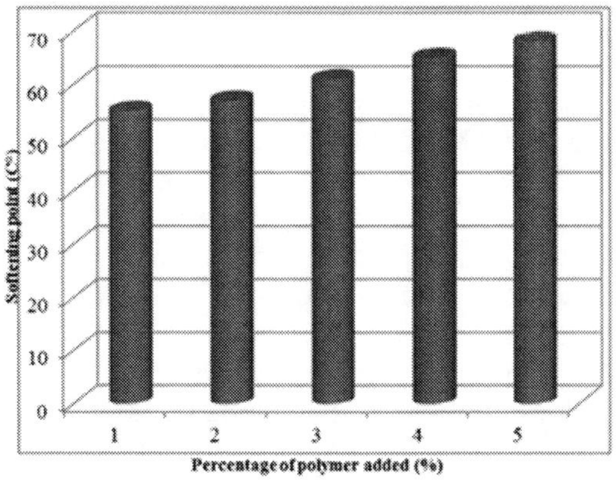

Fig. 2. Comparison between Polymers added (%) and Softening Point.

When comparing the various percentages of fire point, flash point, and plastics added, the outcomes revealed that unadulterated bitumen possessed flash and fire points of 246 C° and 238 C° respectively. The flash and fire points were 252 C° and 241 C° for an additional polymer of 1%, while the flash and fire points were 278 C° and 268 C° for an additional polymer of 1% respectively. The mean flash point was 260.8 ± SD = 11.649 and Std. Error = 5.210. Additionally, mean fire point was 250.8 ± SD = 12.194 and Std. Error = 5.453. (Table and Fig. 3).

Table 3. Percentage of polymer added, fire point and flash point

Polymer added in Bitumen (%)	Fire point (C°)	Flash Point(C°)
0	238	246
1	241	252
2	250	263
3	257	269
4	268	274

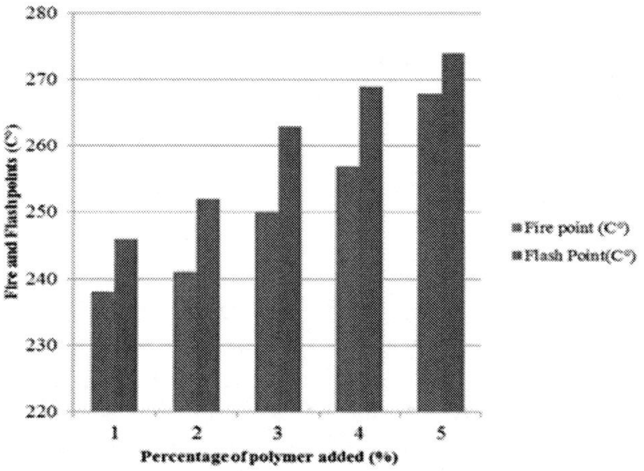

Fig. 3. Percentage of polymer added, fire point and flash point

4 Discussions

Bitumen is often used in constructing supple pavements by tying the aggregates together by way of layering the aggregates. Moreover, the material endorses the high point as well as the life service of the pavement. On the other hand, following its poor characteristics, for instance, its resistance to water, bitumen is universally bettered through the modification of its properties of rheology by mixing it with artificial polymers, for example, rubber and plastics [7]. Since the use of waste plastic with bitumen is similar to using polymer modified bitumen, this test used discarded plastic to assess the performance of hot asphalt with regards to flash, fire, bitumen penetration as well as softening points.

This research established a marked decline in the penetration value with the subsequent increase in the content of polymer Bitumen. This without any doubt suggests the necessity for a tougher modifying binder. These results are in favor of what was reported until this time by [8] which posited that the standards of penetration flexibility and enhanced bitumen lessened with the increase in the content of plastic up to 12% in weight. In different study carried out by [9], performance appraisal of bitumen mixtures, (15%) bitumen and (8%) plastic were likened to a blend of typical bituminous concrete. The outcomes showed enhanced properties, for example, retained immovability to incidental ductile strength and rutting with polymer modified bitumen concrete. The authors maintained that the modified mixtures are impervious to lasting deformation in raised temperatures.

Moreover, the plastics that are used in the process of bitumen modification can as well be classified into the above groups. The most commonly used thermoplastic polyester is referred to as polyethylene terephthalate (PET). As stated earlier in the paper it is also known as 'polyester' [1]. It hardly goes without noticing that the name – polyester – causes mix-up due to the similarities it shares with polyester resins which belong to the group of thermosetting polymers. By and large, polyethylene terephthalate could be labeled as a crystal clear polymer with a decent perfunctory property as well as dimensional constancy in dissimilar load. According to [2], this polymer boasts of good barrier properties on gas as well as excellent resistance to chemicals. This characteristic offers an explanation for its prevalent use in the making of bottles, electrical components, and thermally stabilized films and in the manufacturing of textile.

According to [5], there has been a general increase in the use of asphalt in the construction of road pavements, especially in binding aggregates. Even so, the main disadvantage is that the asphalt coating stratum is always vulnerable to adverse temperatures [6]. It experiences rutting during high temperatures, fatigue during medium temperatures and cracking during low temperatures. This validates the act of modifying asphalt mixtures with a view to ensuring enhanced application. Improved road properties may be realized through the use of apposite starting rudimentary oil while making individual meticulous refinery process [1]. However, in both cases, it is quite challenging to realize.

5 Conclusion

This experiment revealed that discarded plastic materials could be used effectively in the modification bitumen for the construction of ductile pavements. It is easily achieved through the process of mixing treated waste plastic of different quantities by weight of asphalt to in a bid to guarantee considerable enhancement in the Marshall stability, strength, as well as other associated characteristics of the improved bituminous concrete with a view to attain durability and improved performance of roads and pavements with negligible bitumen utility. Furthermore, the procedure is kind to the surroundings. Nonetheless, this investigation vouches for additional studies in order to establish the precise quantities of the mixtures that provide the best results.

References

[1] T.B. Moghaddam, M. Soltani and M.R. Karim, Evaluation of permanent deformation characteristics of unmodified and Polyethylene Terephthalate modified asphalt mixtures using dynamic creep test, Materials & Design, **53**, 317-324 (2014)
[2] E. Ahmadinia, M. Zargar, M.R. Karim, M. Abdelaziz and P. Shafigh, Using waste plastic bottles as additive for stone mastic asphalt, Materials & Design, **32**(10), 4844-4849 (2011)
[3] M.T. Awwad and L. Shbeeb, The use of polyethylene in hot asphalt mixtures, American Journal of Applied Sciences, **4**(6), 390-396 (2007)
[4] A. Gawande, G.S. Zamre, V.C. Renge, G.R. Harsakale and S. Tayde, Utilization of waste plastic in asphalting of roads a review, Sci. Revs. Chem. Commun., **2**(2), 147-157 (2012)
[5] M.N. Rahman, M. Ahmeduzzaman, M.A. Sobhan and T.U. Ahmed. Performance evaluation of waste polyethylene and pvc on hot asphalt mixtures, American J. of Civil Engineering and Architecture, **1**(5), 97-102 (2013)
[6] A. Jamshid, M.O. Hamzah and Z. You, Performance of warm mix asphalt containing Sasobit®: State-of-the-art, Construction and Building Materials, **38**, 530-553 (2013)
[7] G. Sarang, B.M. Lekha, G. Krishna and A.U.R. Shankar, comparison of stone matrix asphalt mixtures with polymer-modified bitumen and shredded waste plastics, Road Materials and Pavement Design, 1-13, (2015)
[8] C.E.G. Justo and A. Veeraragavan, *Utilization of waste plastic bags in bituminous mix for improved performance of roads*, Centre for Transportation Engineering, Bangalore University, Bangalore, India, (2002)
[9] S. Rashid, T.A. Khan, Sangita, D.K. Sharma, Performance evaluation of waste plastic/polymer modified bituminous concrete mixes, Journal of Scientific and Industrial Research, **68**(11), 975-979 (2009)

Plastic Bottles Waste Utilization as Modifier for Asphalt Mixture Production

Hakeem Jan[1,*], *Mohamad* Yusri Aman[1], and *Fazal* Karim[2]

[1]Faculty of Civil and Environmental Engineering, Universiti Tun Hussein Onn Malaysia, 86400 Parit Raja, Johor, Malaysia
[2] Faculty of Civil Engineering, SUIT, 25000 Peshawar, Pakistan

Abstract. Plastic Bottles was used as the polymeric waste to investigate performance of asphalt mixture Aggregates obtained from Margalla, Burhan and Karak quarries. 12 samples were prepared for conventional asphalt mixtures and 48 samples were prepared for PB modified asphalt mixture of each quarries at various proportions of PB waste. The PB used for modification according to wet process are 15%, 20%, 25% and 30% by weight of Optimum Bitumen Content (OBC). OBC of 4.2 % was concluded for conventional asphalt mixtures. The stability and flow values of the conventional and modified Asphalt Mixture were compared. The average Stability of the modified Margalla asphalt mixtures when 15% PB was used was much higher as compared to conventional asphalt mixtures. But when PB was used beyond 15%, the Marshall stability showed a decreasing trend for Margalla aggregates, increasing trend for Karak aggregates and decreasing trend for Burhan aggregates. This decline in stability is attributed to a decline in interlocking of aggregates due to lubricating effect. The corresponding flow for the Modified asphalt mixtures first showed a decreasing trend for Margalla aggregates at 15% PB modification but beyond 15%, an increasing trend in flow as compared to conventional asphalt mixtures The decrease in flow or increase in Marshall Stability is attributed to improvement in interlocking and decline in flow or stability is attributed to a decline in interlocking offered by binder and PB coated aggregate particles in modified asphalt.

1 Introduction

Bitumen when used in conjunction with polymers would lead to many improvements in the asphalt mixture properties such as improvement in rut resistance, fatigue resistance, resistance to thermal cracking, resistance to stripping and decline in temperature susceptibility. The above significant features of polymer modified bitumen (PMB) make it possible to replace pure bitumen by PMB in many paving and maintenance applications. Performances of PMB by itself depend on various parameters such as polymer-bitumen characteristics, compatibility and mixing conditions [1]. The disposing of Plastic waste originated from the plastic bottles has become a major part of the

* Corresponding author: khyberpass08@gmail.org

world's solid waste management problems. Several attempts were made to recycle these polymeric wastes for useful engineering purposes in asphalt pavement industry in order to get rid of their disposing problems. The polymeric wastes can be used as additive as well as partial replacement of bitumen in asphalt mixtures to improve performance of asphalt pavements [2]. Certain polymers are significantly used as a binder modifier in asphalt mixtures as a performance accelerator particularly for rut resistance and resistance to thermal cracking in asphalt pavements [3] Polypropylene was also subjected to thermal degradation during pyrolysis in wet basis where polypropylene was mixed with the asphalt in various proportions ranging from 1, 3, 5 and 7% by weight of asphalt binder at temperature 160 ± 5 °C for 5 min. It was concluded that the tensile strength ratio for different mixtures containing polypropylene was greater by 85% than the conventional mixtures. The penetration index values recorded in the test indicate that polypropylene reduced the temperature susceptibility of the asphalt and therefore, there would not be any problem for mixture when exposed to moisture. The increase in Marshall Stability and decrease in flow value was also recorded [4].An improvement in Marshall Stability with specified flow was found for the pavement base and surface mix design when modified with plastic wastes bags as a particle having size of 1.2mm-3mm up to 2.5% by weight[5]. It has already been proved that the permanent deformation of asphalt mixture can be reduced by modifying asphalt with recycled Polyethylene Terephthalate (PET). PET is widely used in water and soft drink bottles and it's commonly recycled[6] The performance of conventional and plastic polymer (PP) modified asphalt mixture was compared and concluded that waste PP modified bituminous mixes are expected to be more durable and have an improved performance in field[7]. It has already been reported that plastic modified asphalt concrete has improved rutting resistance and resistance against low-temperature cracking as compared to the conventional asphalt mixtures. The field tests proved that plastic wastes used as an additive not only enhance the serviceable life of the roads but also solve the disposing of plastic bottles waste [8] It has also been concluded that the service life of pavement decreases with the increase in number and frequency of traffic but particularly heavy traffic such as trucks when having low number of axles greatly affect the design life. So, there are a number of options available to improve asphalt mixture performance such as to build thicker sections or to use additives as modifier in asphalt mixture [9]. Plastic is available in various forms such as packaging, protecting, agriculture, toys, domestic utensils which constitute waste when their useful life is over [10].Different forms of wastes like plastic bottles, Nylon, polythene bags, polyvinylchloride which need to be recycled for engineering purpose not only to get rid of these growing wastes but to recycle them for useful purpose in pavement industry. Asphalt pavement performance is affected by several factors such as the type and amount of binder, aggregate and additive in the asphalt mix. The field performance of asphalt mixtures can be enhanced due to utilization of various types of additives such as Polymers, latex, fibers and many chemical additives [11].The use of additives in asphalt mixture would be very economical and time saving as compared to having thicker sections in order to improve pavement performance and control its deterioration. Unfortunately, plastic is non-biodegradable material which would remain in the environment even when buried in the earth crust for a long period of time thereby causing environmental pollution. So, therefore, some innovative and sustainable approaches are required to be applied to reuse these growing quantities of wastes. The best solution for this problem is to recycle these wastes for adequate engineering products [12]. The performance of plastic bottles modified asphalt mixture was studied in terms of Marshall Stability and flow for both the soaked and unsoaked conditions.

The two grades of bitumen 60/70 and 80/100 modified with six different percentages of plastic bottles ranging from 2% to12% with an increment of 2% were used to prepare asphalt mixtures. The unsoaked samples gradually showed 15% increase in stability for 60/70 grade bitumen and 10% increase in stability for 80/100 grade bitumen respectively but, for the soaked conditions, the samples showed a gradual decrease in stability for both the 60/70 and 80/100 grades bitumen[13]Due to plenty use of plastic in our lives in various forms such as packaging bottles , Cups , tyres, plastic furniture and toys , medical devices which lead to a wide spread generation of plastic wastes which require a proper end-of-life management. These are non-biodegradable materials and cost much to be disposed. So, it is a good option to re-use it for useful engineering products [14]. PVC and waste polyethylene were used in conjunction with asphalt mixture as modifier in order to check the performance as well as the design criteria of modified asphalt mixtures. A remarkable improvement in performance in terms of Marshall Stability, Stiffness and voids characteristics was found for the modified asphalt mixture when modified with polyethylene up to 10% and up to 7.5% PVC in asphalt mixture particularly in warmer region [15].

2 Methodology

The test procedure and preparation of samples were conducted as per the procedure described in ASTM D1559. The test procedure and preparation of samples were conducted as per the procedure described in ASTM D1559.Aggregates used in this study were obtained from three different locations (Margalla, Karak and Burhan) hills. The aggregates were provided as a mixture of different sizes of particles which were then sieved in the suit laboratory to the required sizes and then mixed according to the trial gradation. Bitumen was collected from local Attock Refinery of grade 60/70 is used as a binder to prepare asphalt mix then prepared modified Asphalt Mixtures using waste plastic bottles then testing the samples. The following flow chart shows the detailed approach for the study program.

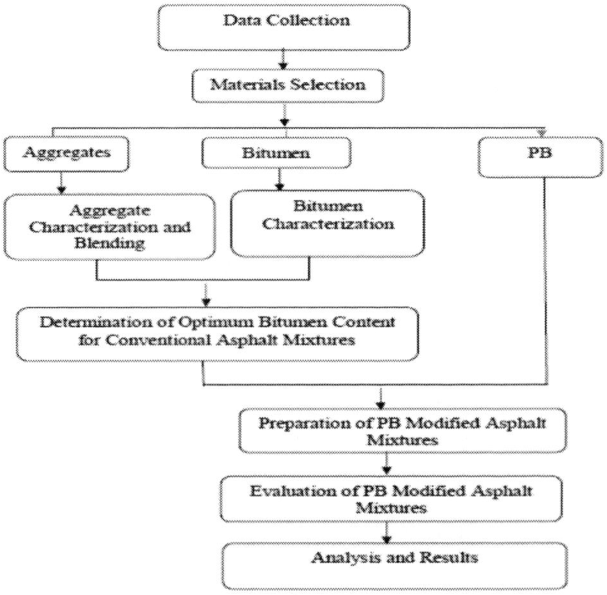

Fig. 1. Flow chart showing laboratory testing procedure

2.1 Material properties

2.1.1 Bitumen properties

60/70 penetration grade bitumen was collected from local Attock Refinery and is used as a binder to prepare asphalt mix. The important properties of bitumen used are given below.

2.1.2 Physical properties of bitumen

Table 1. Properties of bitumen

Property	Value
Penetration at 250C (mm)	66.8
Softening Point (0C)	48.62
Specific Gravity	1.01

2.1.3 Physical properties of aggregate

The aggregate samples were collected from three local quarries named Margalla, Karak and Burhan. The important properties of the aggregates used are given below

Table 2. Properties of aggregates

Property	Margalla	Karak	Burhan
Abrasion (%)	19.46	21.8	21
Flakiness Index (%)	2	4.5	3.7
Elongation Index (%)	11	12.7	14
Bulk Specific Gravity	2.6	2.53	2.49

2.2. Preparation and testing of asphalt mixtures

This study was based on the laboratory work conducted in the Highway and Transportation Engineering laboratory of SUIT Peshawar. The conventional and modified asphalt mixtures were prepared according to the Marshall procedure as per ASTM D 1559-89. The coarse and fine aggregate were mixed according to the adopted gradation described as per the procedure of ASTM. In order to prepare Conventional asphalt mixtures, the Marshall mould having a dia of 4 inch, thickness of 2.5 inch and volume of 1200 gram was filled with conventional and modified asphalt mixture and compacted with a compactive effort of 75 blows by the standard Marshall hammer on both sides of the Marshall briquettes. The hardened samples were extracted from the mould by sample extruder after twenty four hours the samples were kept in water at 600C for one hour before testing. Marshall Tester was then used for finding Stability and Flow of the samples. 4.2 % Optimum Binder Content (OBC) was concluded for the conventional unmodified asphalt mixtures by using Marshall Test. Different percentages of the PBW ranging from 0, 15, 20,25 and 30% of the weight of OBC was heated at 1600C along with bitumen before adding aggregate to get a modified binder. The modified binder was added to the pre-heated asphalt mixture and thoroughly mixing was done mechanically till the color and consistency of the mixture appeared to be uniform. The mixture was then poured in to a pre-heated Marshall mold and the samples were compacted by Marshall Compactor for heavy traffic and were kept over night for cooling to room temperature.

3 Results and data analysis

A comparison among asphalt mixture performance when plastic bottles waste was used as additive in asphalt mixture prepared in aggregates obtained from Margalla, Karak and Burhan quarries is described below.

The result presented in table 3 reflects the stability and flow values for the conventional asphalt mixture prepared in specified three types of aggregates at an Optimum Binder Content (OBC) of 4.2 %. The average stability of the unmodified asphalt mixtures prepared in Margalla aggregates is maximum, the Burhan is medium and the Karak is minimum. Where as the corresponding flow values has an increasing trend for the specified three types of asphalt mixtures. This decreasing trend in stability or increasing trend in flow for Margalla, Burhan and Karak samples is attributed to a decreasing trend in interlocking or cohesion offered by binder among the grains of the samples. The physical properties indicate that the flakiness index , elongation index, abrasion values are low for Margalla aggregates, medium for Karak aggregates and maximum for Burhan aggregates which is the main reason for the decreasing trend in stability and increasing trend in flow for Margalla , Burhan and Karak aggregates. It also shows that the Margalla aggregate are Superior to be used in asphalt mixture production.

Table 4 and Fig 3 give the Stability and Flow values for the 15 % PB waste modified asphalt mixture for the three types of aggregates. It was concluded that 15% modification with PB is Optimum in terms of Marshall Stability and flow. The average Stability of the modified Margalla asphalt mixture when 15% PB were used is 13.8 % higher as compared to conventional asphalt mixtures. Whereas the improvement in Karak sample at 15% PB modification is just 1 % but there is fall in the stability values by 3.5 % in Burhan sample as compared to conventional unmodified asphalt mixtures. The corresponding flow values at 15 % PB modification shows a decreasing trend for Margalla and Karak but increasing trend for Burhan.

Table 5, 6, 7 and Fig 4, 5,6 shows that when PB were used beyond 15%, the Marshall stability showed a decreasing trend. This decline in stability is attributed to a decline in interlocking of aggregates. The corresponding flow for the Modified asphalt mixtures first showed a decreasing trend for Margalla aggregates when 15% PB was used but beyond 15%, an increasing trend in flow as compared to conventional asphalt mixtures has been shown. The decrease in flow or increase in Marshall Stability is attributed to improvement in interlocking and decline in flow or stability is attributed to a decline in interlocking offered by binder and 'PB'coated aggregate particles.

Table 8 shows that Margalla samples give 13.8 % higher strength at 15 % PB modification as compared to conventional mix but the stability then declines at the remaining percentages of 20, 25 and 30. It also shows that there is little increase in stability of the Karak at 20 and 25 % modification but it declines at 15 and 30 % modification. The Burhan samples gradually show a decreasing trend at all percentages of modification. So, the effect of 15 % PB modification of asphalt mix is very significant and recommended for construction.

Table 9 shows that Margallah samples give decrease in the average flow at 15 % PB modification as compared to conventional mix but the flow then increase at the remaining percentages of 20, 25 and 30.

The he karak samples there was decrease at 20 % as compared to conventional mix but the flow then increase at the remaining percentages 15, 25 and 30.The Burhan samples give decrease the average flow at 15 % PB modification as compared to conventional mix but the flow then increase at the remaining percentages 20, 25 and 30.

Table 3. Stability and Flow values for conventional unmodified asphalt mixtures at optimum bitumen content of 4.2 %, Stability (Kg), Flow (mm)

Sample	Margalla	Karak	Burhan
01	1651	1639	1672
02	1583	1473	1562
03	1693	1540	1628
04	1678	1488	1599
Avg.	1651.25	1535	1615.25

Sample	Margalla	Karak	Burhan
01	2.5	2.8	2.9
02	2.9	3.1	3.1
03	2.4	2.7	2.7
04	2.6	2.9	2.6
Avg.	2.6	2.87	2.82

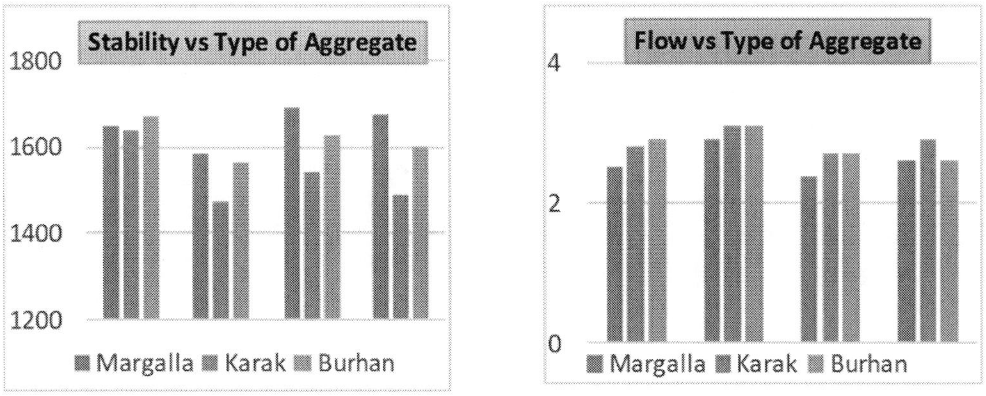

Fig. 2. Stability and Flow values for conventional unmodified asphalt mixtures, Stability (Kg), Flow (mm)

Table 4. Stability and Flow values for the 15 % PB waste modified asphalt mixtures, Stability (Kg), Flow (mm)

Sample	Margalla	Karak	Burhan
01	1892	1510	1540
02	1808	1553	1518
03	1913	1562	1577
04	1903	1571	1594
Avg.	1879	1549	1557.25

Sample	Margalla	Karak	Burhan
01	2.2	2.7	2.6
02	2.4	2.1	3.1
03	2.0	2.5	2.5
04	2.1	2.2	2.0
Avg.	2.17	2.37	2.55

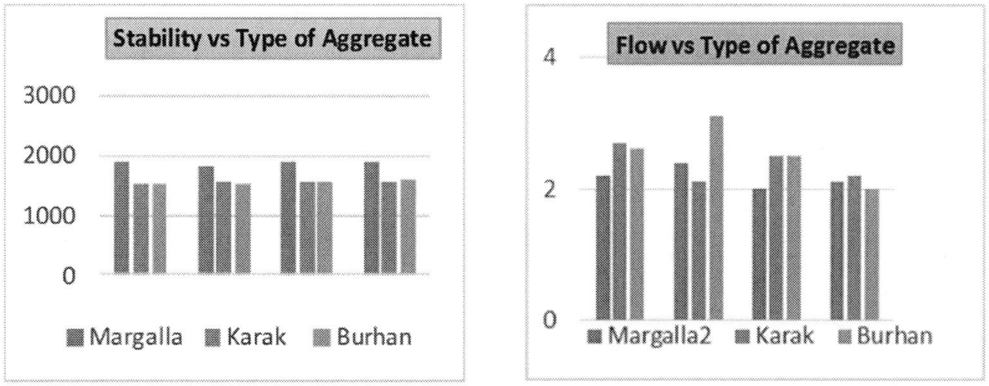

Fig. 3. Stability and Flow values for the 15 % PB waste modified asphalt mixtures, Stability (Kg), Flow (mm)

Table 5. Stability and Flow values for the 20 % PB waste modified asphalt mixtures, Stability (Kg), Flow (mm)

Sample	Margalla	Karak	Burhan
01	1660	1503	1342
02	1650	1546	1298
03	1694	1523	1386
04	1688	1582	1378
Avg.	1673	1538.5	1351

Sample	Margalla	Karak	Burhan
01	2.4	2.4	3.1
02	2.5	2.2	3.3
03	2.7	2.1	3.0
04	2.6	2.0	2.9
Avg.	2.55	2.17	3.07

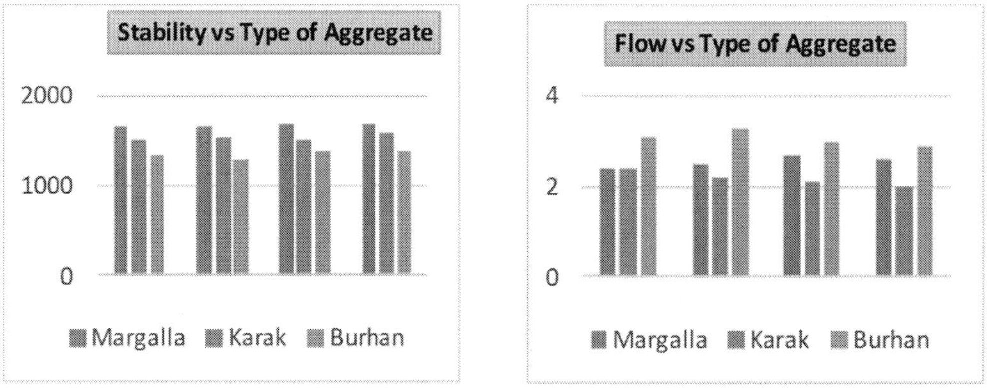

Fig. 4. Stability and Flow values for the 20 % PB waste modified asphalt mixtures, Stability (Kg), Flow (mm)

Table 6. Stability and Flow values for the 25 % PB waste modified asphalt mixtures, Stability (Kg), Flow (mm)

Sample	Margalla	Karak	Burhan
01	1452	1607	1225
02	1518	1510	1254
03	1540	1556	1183
04	1531	1613	1144
Avg.	1510.25	1571.5	1201.5

Sample	Margalla	Karak	Burhan
01	2.9	2.3	3.4
02	2.6	2.7	3.2
03	2.5	2.4	3.7
04	2.6	2.3	3.5
Avg.	2.65	2.42	3.45

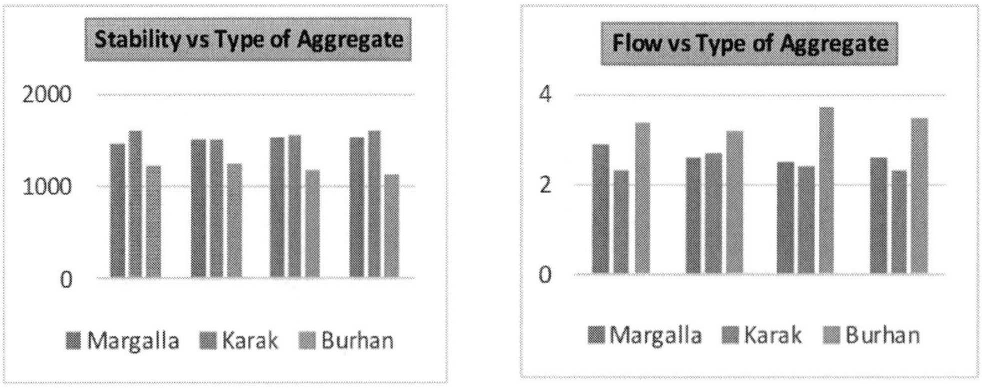

Fig. 5. Stability and Flow values for the 25 % PB waste modified asphalt mixtures, Stability (Kg), Flow (mm)

Table 7. Stability and Flow values for the 30 % PB waste modified asphalt mixtures, Stability (Kg), Flow (mm)

Sample	Margalla	Karak	Burhan
01	1386	1496	1035
02	1364	1373	1077
03	1378	1387	1122
04	1293	1403	1143
Avg.	1355.25	1414.75	1094.25

Sample	Margalla	Karak	Burhan
01	3.0	3.1	3.4
02	3.1	3.3	3.2
03	2.9	3.0	3.0
04	3.3	2.8	2.9
Avg.	3.0	3.05	3.12

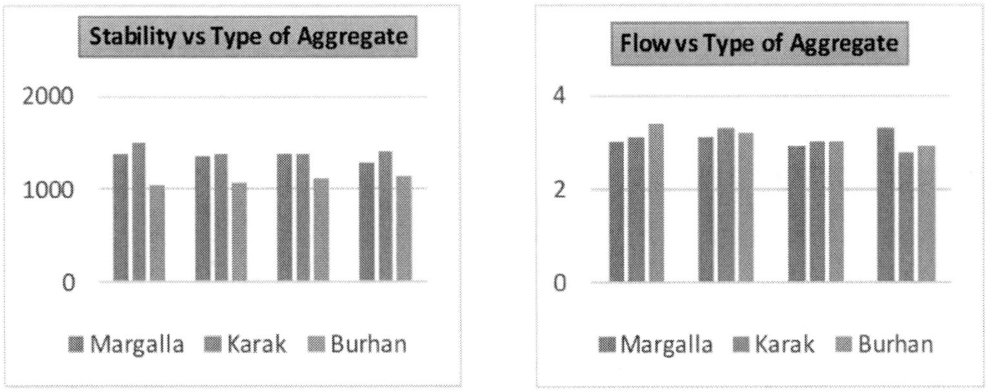

Fig. 6. Stability and Flow values for the 30 % PB waste modified asphalt mixtures, Stability (Kg), Flow (mm)

Table 8. Comparison of the Average Stability values for various percentages of PB waste modified asphalt mixtures, Stability (Kg), Flow (mm)

Sample	Conventional	15 %PB	20 % PB	25 % PB	30 % PB
Margalla	1651.25	1879	1673	1510	1355.25
Karak	1535	1549	1538.5	1576.25	1414.75
Burhan	1615.25	1557.2	1351	1201.5	1094.25

Table 9. Comparison of the Average Flow values for various percentages of PB waste modified asphalt mixtures, Stability (Kg), Flow (mm)

Sample	Conventional	15 % PB	20 % PB	25 % PB	30 %PB
Margalla	2.6	2.17	2.5	2.65	3
Karak	2.87	2.37	2.17	2.42	3.1
Burhan	2.82	2.55	3.0	3.45	3.12

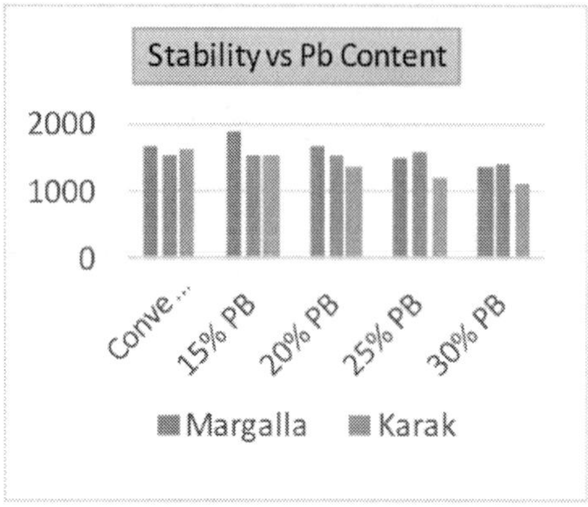

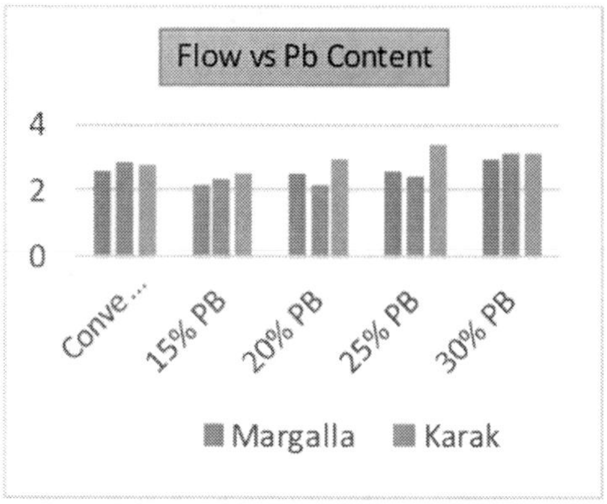

Fig. 7. Comparison of the Average Stability and Flow values for various percentages of PB waste modified asphalt mixtures, Stability (Kg), Flow (mm)

4 Conclusions

The experimental work while comparing the performance of Conventional Asphalt Mixtures and PB Modified asphalt mixtures prepared in aggregates obtained from Margalla, Karak and Burhan quarries give the following outcomes:

- Plastic Bottles can be conveniently used as modifier for asphalt mixture production as a sustainable management of plastic wastes to avoid disposing problems.
- The Optimum Bitumen Content was found to be 4.2% and the Optimum Plastic Bottles waste was found to be 15 % by weight of bitumen for the better
- performance in terms of Marshall stability with specified flow of asphalt mixturesThe average Stability of the modified Margalla Asphalt Mixtures when 15% PB waste was used is 13.8 % higher as compared to conventional asphalt mixtures. But when PB waste was used beyond 15%, the Marshall stability showed a continuous decreasing trend.
- The Margalla and Karak aggregates give higher stability at 15 % PB waste modification as compared to the asphalt mixture prepared in Burhan aggregates.
- There is a gradual increase in stability of asphalt mixture prepared in Karak aggregates at 15%, 20% and 25% PB modification but, a sudden fall has been observed in stability of asphalt mixture at 30 % PB modification.
- A continuous fall in stability of asphalt mixture prepared in Burhan aggregates at 15%, 20%, 25% and 30% PB modification has been observed.
- Margalla aggregate are recommended to be used at 15% PB modification for good performance of asphalt mixtures.

The authors of this study acknowledge the center of graduate studies(CGS) UTHM Malaysia that contributed to the success of this research work to support the technical and financial.

References

[1] Y. Becker, M. Mendez and Rodriguez, Polymer modified asphalt, Vision Tecnologica, **9**(1), (2001)
[2] S. Bose and S. Raju, Utilization of plastic waste in bituminous concrete mixes, Roads and Pavements, **3**, (2004)

[3] M. Awwad and Shabeeb, The use of polyethylene in hot asphalt mixtures, American Journal of Applied Sciences, **4**(6), 390-396 (2007)

[4] A. Al-Hadidy and Tan, The effect of plastomers polymer type and concentration on asphalt and moisture damage of sma mixtures, Al-Rafidain Engineering Journal, **19**(5), (2011)

[5] Q. Adnan and I. Mansoor, utilization of plastic waste bags in itummous mix for improved performance of roads, Better Road, (2008)

[6] T.F. Chen, *Evaluation of rutting performance on hot mix asphalt modified with plastic bottles*, B.Sc Thesis, University of Technology of Malaysia, Johor Bahru, (2009).

[7] S. Rashid, T.A. Khan, Sangita, D.K. Sharma, Performance evaluation of waste plastic/polymer modified bituminous concrete mixes, Journal of Scientific and Industrial Research, **68**(11), 975-979 (2009)

[8] V.S. Punith and A. Veeraraghavan, Laboratory fatigue studies on bituminous concrete mixed utilizing waster shredded plastic modifier, *Proc. of 21st ARRB Transport Research (ARRB)*, **38**(5), (2010)

[9] T.B. Moghaddam, M.R. Karim and Mahrez, A review on fatigue and rutting performance of asphalt mixes, Scientific Research and Essays, **6**(4), 670-682 (2011)

[10] A.F. Abdalqader, *Landfills Needs Assessment In Gaza Strip And Sites Selection Using Gis*, M.Sc Thesis, Islamic University of Gaza, Palestine (2011)

[11] Taih, The effect of additives in hot asphalt mixtures, Journal of Engineering and Development, **15**(3), (2011)

[12] V. Swami, A. Jirge, K. Patil, S. Patil and K. Salokhe, Use of waste plastic in construction of bituminous road, International Journal of Engineering Science and Technology, **4**(5), 2352-2355, (2012)

[13] K.S. Prasad, K.R. Manjunath and K.V.R Prasad, Study on marshall stability properties of BC mix used in road construction by adding plastic waste bottles, IOSR Journal of Mechanical and Civil Engineering, **2**(2), 12-23 (2012)

[14] Plastic Common Wastes & Materials US EPA [Online]. Available (2013), www.epa.gov/osw/conserve/material/plastic.htm

[15] R. Md. Nobinur, Performance evaluation of waste polyethylene and PVC on hot asphalt mixtures, American Journal of Civil Engineering and Architecture, 97-102, (2013)

Parking Space Occupancy at Rail Stations in Klang Valley

Phooi Wai Ho[1,*] *S. M. Sabri* Ismail[2] , and *Premkumar* Rajagopal[3]

[1]School of Science and Engineering, Transportation and Logistics Department, Malaysia University of Science and Technology, Petaling Jaya, Malaysia
[2]Research and Development, PRAISE, Prasarana Malaysia Berhad, Kuala Lumpur, Malaysia
[3]School of Science and Engineering, Transportation and Logistics Department, Malaysia University of Science and Technology, Petaling Jaya, Malaysia

Abstract. The development of Klang Valley Integrated Rapid Transit system in Klang Valley, Malaysia has been quickly gaining momentum during the recent years. There will be two new MRT lines (MRT Line 1 and MRT Line 2) and one new LRT line (LRT Line 3) extended from the current integrated rail transit system by year 2020 with more than 90 new rail stations. With the substantial addition of potential rail passengers, there are doubts whether the existing Park and Ride facilities in Klang Valley are able to accommodate the future parking space demand at rail stations. This research studies the parking occupancy at various Park and Ride facilities in Klang Valley namely Taman Jaya, Asia Jaya, Taman Paramount, Taman Bahagia and Kelana Jaya by applying the non-conventional method utilizing Google Earth imageries. Results showed that the parking occupancy rate at these LRT stations were 100% or more before the commencement of LRT extension (Kelana Jaya and Ampang Lines) in 2016 and in the range of 36% to 100% after the commencement of LRT extension due to the additionally built car parks and changes in parking pattern with dispersed passenger traffic.

1 Introduction

The Klang Valley Integrated Rail Transit system in Klang Valley, Malaysia comprises of four existing rail transit line namely, KTM Commuter, Kuala Lumpur (KL) Monorail, KLIA Express Line and Light Rail Transit (LRT) [1]. During the initial stage of this research, the construction of the LRT Extension was underway with the extension of Kelana Jaya and Ampang lines, with additional 13 stations each [1]. At the same time, the construction of the new MRT Line 1 is also underway with 31 rail stations. Thereafter, the construction of MRT Line 2 and LRT Line 3 will commence with 35 and 25 rail stations respectively [2], [3]. The MRT Line 1 is scheduled for completion in July 2017 while both the MRT Line 2 and LRT Line 3 are scheduled for completion in year 2020 [2], [3].

These huge expansion of MRT and LRT projects will be extended into outlying suburbs and upgraded to a full-fledged integrated rail transit system. This is part of the blueprint of

* Corresponding author: phooiwai@gmail.com

the Malaysia Land Public Transport Commission (SPAD), which has set out an integrated 20 year plan to transform land public transportation in the region to respond to local needs and aspirations [4], which will see a total of 91 new rail stations by year 2020 [2], [3]. This is indeed in line with Malaysia's aim to increase public transportation modal share from 18.1% in 2014 to 40% by year 2030 [5] and to achieve transportation sustainability and a high income and developed nation by year 2020, with a world class rail transit system [6].

2 Historical data of parking occupancy at Park and Ride facilities

Researches have been conducted to study the parking occupancy at Park and Ride (P&R) facilities namely at Putrajaya Sentral, Shah Alam, Seremban, Kelana Jaya, Gombak and Rawang rail stations. The parking occupancy rates range from 90% to 100% as shown in Table 1 [7]–[12]. Parking occupancy rates above 85% showed parking problems at car parks where the efficiency flow of cars coming and going will be affected [13]. The high parking occupancy rates also showed that there were parking space constraint at these rail stations. The most recent parking occupancy rates are probably more than before due to the increase of population, car registrations and rail ridership throughout the years [14-17].

Table 1. Parking occupancy at Park and Ride facilities [7-12].

Park and Ride Facility	Parking Occupancy Rate
ERL Putrajaya (Surface parking)	95%
ERL Putrajaya (Multi-storey parking)	90%
KTM Shah Alam	95%
KTM Seremban	95%
LRT2 Kelana Jaya	92%
LRT2 Gombak	100%
KTM Rawang	100%

With the rapid expansion of the new rail transit projects, there are doubts whether the future parking space at P&R facilities will be able to accommodate the sudden rise of rail passengers. In the case if there is lack of parking space at P&R facilities after the implementation of the new rail lines, this may cause massive traffic congestion [18], [19], thus air pollution [20], [21] and the limitation of rail transit system ridership [22]–[25] for the Klang Valley integrated rail transit system. Therefore, it is crucial to examine the parking occupancy at P&R facilities to be able to plan for the future parking demand.

3 Parking space occupancy at Park and Ride facilities

This research studied the parking occupancy at various P&R facilities in Klang Valley along the Kelana Jaya LRT Line namely Taman Jaya, Asia Jaya, Taman Paramount, Taman Bahagia and Kelana Jaya by applying the non-conventional method utilizing Google Earth imageries. The parking data for these facilities were taken from Google Earth images on Monday, 2nd November 2015 between 8.35am to 5.47pm on a normal working weekday. For images that were blocked by trees, buildings or other structures, an on-site survey and observation were conducted on Tuesday, 2nd February 2016 between 9.00am to 11.00am.

3.1 Taman Jaya Park and Ride facilities

The Taman Jaya LRT station, KJ20, is situated along Persiaran Barat, Section 52, Petaling Jaya and is set around the neighbourhood of Section 9, 10 and 11, Petaling Jaya. The P&R facilities are located next to the LRT station (Reserved Car Park) and along Lorong Sultan not far away from the LRT station (Car Park 1).

The Taman Jaya Reserved Car Park is dedicated for the special needs. From Google Earth image, the car park was partially blocked by trees. Based on a site survey, there were 4 car parking bays and all of them were occupied. The parking occupancy rate was 100% with no parking charges. Other observed parking areas at the car park were for taxis, buses and cars pick-up and drop-off. There were 11 car parking bays for cars pick-up and drop-off, 3 of which were occupied. However, these car parking bays were not for P&R purpose.

The Taman Jaya Car Park 1, is a common car park managed by Majlis Bandaraya Petaling Jaya (MBPJ), the local municipal council. From Google Earth image, there were 59 parking bays and all of them were occupied. The parking occupancy rate was 100%. The parking fee for Taman Jaya Car Park 1 was RM0.60 per hour, a standard for all MBPJ bays in town. These car parking bays do not only cater for LRT passengers. In fact, these parking bays were shared for all premises nearby the LRT station such as A&W restaurant, Kelab PJ Sports and Recreation Club, Shah's Village Hotel, PJ Palms Sports Centre and a local food court.

Besides the MBPJ managed car park, there were also other private car parking operators operating along the road with their own private parking space. For example, the valet parking operated by PJ Palms Sports Centre charges RM3.00 per hour, RM5.00 per 3 hours or RM15.00 per 24 hours. There were also other commercial car parks opposite the LRT station such as Yayasan Selangor (Open Site 1 & 2), Yayasan Selangor (Petaling Jaya), Amcorp Trade Centre and PJX PJ Exchange. The Yayasan Selangor (Open Site 1 & 2) charges RM5.00 for 2 hours and RM4.00 for subsequent hours. Yayasan Selangor (Petaling Jaya) charges RM4.00 for the 1st hour and RM3.00 for subsequent hours. Amcorp Trade Centre charges RM3.00 for the first 3 hours and RM2.00 for subsequent hours. Lastly, PJX PJ Exchange charges RM3.00 for the 1st hour and RM2.00 for subsequent hours.

The parking occupancy rate of 100% for the Taman Jaya Reserved Car Park is expected because there were only 4 parking bays. Moreover, there were no parking charges. Likewise, the parking occupancy rate of 100% for the Taman Jaya Car Park 1 is also expected because not only does this car park caters for LRT passengers but also for visitors at other nearby premises. Furthermore, the other nearby commercial car parks were charging much more than the MBPJ bays. The cheapest commercial car park is Amcorp Trade Centre with parking fee RM3.00 for the first 3 hours and RM2.00 for subsequent hours, while other commercial car parks charges up to RM4.00 for the first hour and RM3.00 for subsequent hours, which makes it reluctant for LRT passengers to park at these commercial car parks. The parking situation at the Taman Jaya LRT station is similar to the scenario at the Gombak LRT station in 2015 where the parking occupancy rate was 100% [10].

Table 2. Parking occupancy at Taman Jaya Park and Ride facilities.

Taman Jaya Park and Ride	No. of Car Parking Bays	No. of Occupied Car Parking Bays	Parking Occupancy Rate
Reserved Car Park	4	4	100%
Car Park 1	59	59	100%
Total Average	*63*	*63*	*100%*

3.2 Asia Jaya Park and Ride facilities

The Asia Jaya LRT station, KJ21, is situated Off Jalan Utara and Jalan 51A/223. It is set around the industrial and commercial districts of Section 51A & Section 52, and is near the residential neighbourhood of Section 8 and Section 14, Petaling Jaya. The P&R facilities are located next (Reserved Car Park and Car Park 1) and behind (Car Park 2) the LRT station.

The Asia Jaya Reserved Car Park is dedicated for staff of Rapid KL, subsidiary of Prasarana. From Google Earth image, the car park was mostly blocked by the LRT station. Based on a site survey, there were 49 cars at the car park squeezed into 44 of the car parking bays. Cars were seen double parking, parked next to and in front of one another. The parking occupancy rate was 111% with no parking charges.

From Google Earth image, the Asia Jaya Car Park 1 has 9 car parking bays and all of them were occupied with 1 car was seen parking illegally at the end of the parking bays. The parking occupancy rate was 111% with no parking charges.

The Asia Jaya Car Park 2 was built in 2014 to curb the parking constraint problem at the LRT station. The parking fee is RM3.00 per entry. Officially, there was a total of 607 parking bays catered for the needs. However, Google Earth image indicated that the total cars parked was 632 numbers, 25 more cars than the allocated bays. Cars were also seen to be parking at every nook and corners possible. However, 3 empty parking bays were found. Since 4 cars were seen driving at the car park, the 3 empty parking bays will most probably also be parked. Due to the aggressive parking pattern at the car park and the number of cars parked were more than the actual allocated parking bays, the parking occupancy rate was considered 104% even though there was 3 parking bays spotted empty. The parking occupancy rate was 104%.

Besides the car park managed by Prasarana and MBPJ, there were also other private car parks around the vicinity of the Asia Jaya LRT station with the nearest private car park located just 100 meters away from the entrance of the LRT station. The car park was always full during normal working days with cars starting to park from as early as 6.00am until 9.00am to reach its full parking occupancy of about 250 car parking bays. This car park also catered for office workers nearby and it charges RM6.00 per entry. Other commercial car parks within 300 meters from the LRT station, such as Menara Axis, Crystal Plaza and Menara LYL charges RM3.00 for the 1st hour and RM2.00 for subsequent hours on Mondays to Fridays during office hours.

The parking occupancy rate of 111% at the Asia Jaya Reserved Car Park is expected because not only does this car park caters for the personnel permanently based at this LRT station. It also caters for other visiting Prasarana's personnel from other locations. Likewise, the parking occupancy rate of 111% for the Asia Jaya Car Park 1 is also expected because there were only 9 bays. Most of all, there were no parking charges for these parking bays and they were also the nearest to the LRT station. The same goes to the Asia Jaya Car Park 2 where the parking occupancy rate of 104% is expected because based on the Google Earth image in December 2014, it has already been long shown to have full parking occupancy. The high parking fee charges at the nearby commercial car parks may drive office workers to park at the Asia Jaya Car Park 2 instead. The cheapest commercial car park charges RM6.00 per entry but it was always full. Therefore, the next alternative is RM3.00 for the 1st hour and RM2.00 for the subsequent hours. This makes the parking fee at the Car Park 2 looks attractive with only RM3.00 per entry.

The situation at the Reserved Car Park and Car Park 1 were similar to the scenario at the Putrajaya P&R facility where double parking and illegal parking were observed [26]. Also, the situation at the Car Park 2 were similar to the scenario at the Rawang P&R facility with double parking and 100% parking occupancy rate [12].

Table 3. Parking occupancy at Asia Jaya Park and Ride facilities.

Asia Jaya Park and Ride	No. of Car Parking Bays	No. of Occupied Car Parking Bays	Parking Occupancy Rate
Reserved Car Park	44	49	111%
Car Park 1	9	10	111%
Car Park 2	607	632	104%
Total Average	*660*	*691*	*105%*

3.3 Taman Paramount Park and Ride facilities

The Taman Paramount LRT station, KJ22, is situated at the edge of Jalan 21/1. It is set near the neighbourhood of Section 14, 20, 21 and 22 as well as Kampung Tunku in Section SS1, Petaling Jaya. The P&R facility is located behind the LRT station (Car Park 1). The Taman Paramount Car Park 1 was able to accommodate up to 87 cars and all of them were fully occupied from as early as 7.00am. There were also 3 cars double parking in the parking area. The parking occupancy rate was 100% with parking fee of RM3.00 per entry.

The parking occupancy rate of 100% at the Taman Paramount Car Park 1 is expected because this scenario can clearly be seen on Google Earth image since 2013. The situation of parking space constraint at the Taman Paramount LRT station is similar to the scenario at the same station back in 2013, where double parking was observed [27].

Table 4. Parking occupancy at Taman Paramount Park and Ride facilities.

Taman Paramount Park and Ride	No. of Car Parking Bays	No. of Occupied Car Parking Bays	Parking Occupancy Rate
Car Park 1	87	87	100%

3.4 Taman Bahagia Park and Ride facilities

The Taman Bahagia LRT station, KJ23, is situated along Jalan SS2/3 near the LDP interchange at Taman Megah and is set near the neighbourhood of SS2, SS3, SS4 and SS23. The P&R facilities are located directly next to the LRT station (Reserved Car Park) and at the back of the LRT station (Car Park 1).

The Taman Bahagia Reserved Car Park is dedicated for the staff of Prasarana. From Google Earth image, the car park was partially blocked by the LRT track. Based on a site survey, there were 6 cars squeezed into a small area catered for 4 cars. Cars were seen double parking in front of the row of car parks. The parking occupancy rate was 150% with no parking charges. Meanwhile, the road just outside of the LRT station was also parked with cars. Furthermore, the narrow lane near the residential area behind the LRT station was parked with cars on both sides of the lane. Unfortunately, these were not proper car parking bays. However, due to the lack of parking space, these cars just parked there anyway.

The Taman Bahagia Car Park 1 is an empty space car park managed by MBPJ with no parking fee imposed yet. From Google Earth image, there were a total of 81 cars parked at the parking bays and 79 of the parking bays were occupied. Some cars were seen parking right up to the entrance of the car park. The parking occupancy rate was 98%.

The parking occupancy rate of 150% at the Taman Bahagia Reserved Car Park is expected because not only does this car park caters for the personnel permanently based at this LRT station. It also caters for other visiting Prasarana's personnel from other locations. Moreover, the double parking, illegal and indiscriminate parking around the LRT station up to the nearby neighbourhood showed that there was need for more parking. Likewise, the parking occupancy rate of 98% for Car Park 1 is also expected because that was the only empty space near the LRT station. Most of all, there were no parking charges at this car park. The situation at the Taman Bahagia Reserved Car Park and Car Park 1 were similar to the scenario at the same LRT station back in 2009 where illegal parking were seen around the LRT station [28].

Table 5. Parking occupancy at Taman Bahagia Park and Ride facilities.

Taman Bahagia Park and Ride	No. of Car Parking Bays	No. of Occupied Car Parking Bays	Parking Occupancy Rate
Reserved Car Park	4	6	150%
Car Park 1	81	79	98%
Total Average	*85*	*85*	*100%*

3.5 Kelana Jaya Park and Ride facilities

The Kelana Jaya LRT station, KJ24, is situated along the busy Damansara Puchong Expressway. It is set near the neighbourhood of SS2, Kelana Jaya (SS4 and SS5) and Taman Mayang (SS25). The P&R facilities are located beside (Reserved Car Park and Car Park 1) and behind (Car Park 2) the LRT station.

The Kelana Jaya Reserved Car Park is dedicated for staff of Prasarana. From Google Earth image, the car park was partially blocked by LRT tracks. Based on a site survey, there were 6 cars at the reserved car park where all of the parking bays were occupied. The parking occupancy rate was 100% with no parking charges.

From Google Earth image, the Kelana Jaya Car Park 1 was partially blocked by LRT tracks and trees. However, it can be seen that the car park was fully occupied with double parking. Based on a site survey, there were 322 cars squeezed into all 282 parking bays. The parking occupancy rate was 114% with parking fee of RM4.00 per entry.

The Kelana Jaya Car Park 2 was partially blocked by LRT tracks and a high tension pylon on Google Earth image. However, it can be seen that the car park was fully occupied. Based on a site survey, all 241 parking bays were occupied with 4 cars parked outside of the entrance of the car park. The parking occupancy rate was 102% with parking fee of RM4.00 per entry.

Besides the car parks managed by Prasarana, there were 2 private car parks operating near the LRT station which charges RM4.00 per entry. The entrance to these 2 car parks was from Jalan SS4C/1. From Google Earth image, Private Car Park 1 was almost fully occupied with little space at the far end of the car park, while Private Car Park 2 was fully occupied.

Table 6. Parking occupancy at Kelana Jaya Park and Ride facilities.

Kelana Jaya Park and Ride	No. of Car Parking Bays	No. of Occupied Car Parking Bays	Parking Occupancy Rate
Reserved Car Park	6	6	100%
Car Park 1	282	322	114%
Car Park 2	241	245	102%
Total Average	*529*	*573*	*108%*

The parking occupancy rate of 100% at the Kelana Jaya Reserved Car Park is expected because this car park caters for Prasarana's personnel permanently based at this LRT station and from other locations. Likewise, the parking occupancy rate of 114% and 102% respectively for the Kelana Jaya Car Park 1 and Kelana Jaya Car Park 2 were also expected because Kelana Jaya LRT station is the terminus station along the Kelana Jaya LRT Line. Based on Google Earth image on 10th May 2013, it has already been long shown to have full parking occupancy. The situation at the Reserved Car Park, Car Park 1 and Car Park 2 were similar to the scenario at the same LRT station back in 2009 where illegal parking were seen around the LRT station [28].

4 Parking space occupancy after the implementation of LRT extension

By 30th June 2016, all of the new rails stations on Ampang and Kelana Jaya LRT extension have commenced operations [29]. There were also additional car parks built at some of the LRT stations under the scope of works of this research, namely Taman Paramount, Taman Bahagia and Kelana Jaya LRT stations. Therefore, after the commencement of the LRT Extension, there were changes in parking patterns at the P&R facilities. Thus, additional research was conducted at P&R facilities at the 5 LRT stations from 29th August 2016 till 8th September 2016 on a normal working weekday between 8.30am to 11.00am.

The average parking occupancy rate at Taman Jaya, Asia Jaya, Taman Paramount, Taman Bahagia and Kelana Jaya Park and Ride facilities after the commencement of the LRT extension were 100%, 83%, 49%, 36% and 79% respectively as shown in Table 7.

Table 7. Parking occupancy at Park and Ride facilities after the commencement of LRT extension.

Park and Ride Facility†	Reserved Car Park	Car Park 1	Car Park 2	Car Park 3	Total Average
Taman Jaya	100% (4)	100% (59)	-	-	100%
Asia Jaya	100% (44)	100% (9)	81% (492)	-	83%
Taman Paramount	-	100% (87)	36% (123)	-	49%
Taman Bahagia	100% (6)	100% (104)	0% (0)	Pending	36%
Kelana Jaya	100% (6)	114% (321)	38% (91)	Pending	79%

5 Conclusion

The average parking occupancy rate at Taman Jaya, Asia Jaya, Taman Paramount, Taman Bahagia and Kelana Jaya P&R facilities were 100%, 105%, 100%, 100% and 108% respectively, which showed that there were parking space constraint at the LRT stations, where parking occupancy rates above 85% showed parking problems at car parks where the efficiency flow of cars coming and going will be affected [13].

After the commencement of LRT extension, the parking space constraint at the Taman Jaya LRT station remains, while Asia Jaya, Taman Paramount, Taman Bahagia and Kelana Jaya LRT stations do not show any parking constraint anymore. For Taman Bahagia and Taman Paramount, the occupancy drop were due to new car park built at the stations. However, the increase of the total number of occupied parking bays were obvious, from 85 to 110 bays at Taman Bahagia and from 87 to 210 bays at Taman Paramount. For Asia Jaya and Kelana Jaya, the occupancy drop were possibly due to more car parks built at the new LRT extension rail stations such as Ara Damansara, Glenmarie, USJ21, Alam Megah, Subang Alam and Putra Heights. Therefore, parking pattern has changed with dispersed passenger traffic. Even though the occupancy rates were 36% and 79% respectively at Taman Bahagia and Kelana Jaya LRT station, the authorities are still building car parks to cater for the future parking demand after the implementation of the upcoming new rail transit lines.

It is expected that the remainder of the parking space and the new car parks would be easily filled up after the commencement of the additional 91 rail stations. The success of P&R facilities shown by the increase of car parking bay users are confirmed in attracting rail passengers with users reported high levels of satisfaction [30]. Therefore, it is necessary to provide more parking space to encourage the usage of rail transportation.

References

[1] Prasarana Malaysia Berhad, *Integrated Rail Transit Map*, (2016), Retrieved on July 23, 2016 from http://www.myrapid.com.my/rail/klang-valley-integrated-rail-transit-map
[2] Mass Rapid Transit Corporation, *The Klang Valley MRT*, (2016), Retrieved on July 23, 2016 from http://www.mymrt.com.my/
[3] Prasarana Malaysia Berhad, *LRT 3*, (2016), Retrieved on July 23, 2016 from http://www.lrt3.com.my
[4] Land Public Transport Commission, *Greater KL/Klang Valley Land Public Transport Master Plan*, Malaysia, (2013)
[5] Land Public Transport Commission, *Land Public Transport Commission (SPAD) Annual Report 2013*, Malaysia, (2013)
[6] Government Transformation Programme, *Improving Urban Public Transport, National Key Results Areas*, GTP Annual Report 2014, Performance Management and Delivery Unit, (2014)
[7] S.A.A.S Adnan and A.A.K.H Hamsa, Evaluating the parking demand at Park and Ride facility at Putrajaya public transportation terminal, *Proceedings of the Eastern Asia Society for Transportation Studies*, Taipei, (2013)
[8] N. Abdul Hamid, Utilization patterns of park and ride facilities among Kuala Lumpur commuters, Transportation (Amst)., **36**, 295–307 (2009)
[9] U.A. Khalid and A.A.K. Hamsa, Parking utilization pattern of Park and Ride facility at public transportation terminals, *Proceedings of the Eastern Asia Society for Transportation Studies*, Taipei, (2013)
[10] S. Tamrin, *LRT and monorail ridership to reach the two billion mark in May*, The Sun Daily, (April 30, 2015)
[11] N.A. Zainal, *Putrajaya bus service has far to go*, The Star, (March 11, 2015)
[12] N.A. Hamid, J. Mohamad and M.R. Karim, Travel behaviour of the Park and Ride users and the factors influencing the demand for the use of the Park and Ride facility, *EASTS International Symposium on Sustainable Transportation incorporating Malaysian Universities Transport Research Forum Conference 2008 (MUTRFC08)*, Johor Bahru, (2008)
[13] J. Feit, *Extended hours and low rates: Parking data in context*, Seattle Met, (February 16, 2015)
[14] Department of Statistics, *Social Statistical Bulletin Malaysia 2014*, Malaysia, (2014)
[15] Department of Statistics, *Social Statistical Bulletin Malaysia 2013*, Malaysia, (2013)
[16] Department of Statistics, *Social Statistical Bulletin Malaysia 2012*, Malaysia, (2012)
[17] Department of Statistics, *Social Statistical Bulletin Malaysia 2015*, Malaysia, (2015)
[18] H. Shang, W. Lin, and H. Huang, Empirical study of parking problem on university campus, J. Transp. Syst. Eng. Inf. Technol., **7**, 135–140 (2007)
[19] W. Yan-ling, W. Xin, and Z. Ming-chun, Current situation and analysis of parking problem in Beijing. Procedia Eng., **137**, 777–785 (2016)
[20] A.M. Abdullah, M.A.A. Samah and Y.J. Tham, An overview of the air pollution trend in Klang Valley, Malaysia. Open Environ. Sci., **6**, 13–19 (2012)
[21] S. Masahina, R. Afroz, J. Duasa and N.A. Mohamed, Framework to estimate the willingness to pay of household for airquality improvement: A case study in Klang

Valley, Malaysia. Int. J. Sustain. Dev., **4**, 11–16 (2012)
[22] M. Brons, M. Givoni and P. Rietveld, Access to railway stations and its potential in increasing rail use, Transp. Res. Part A Policy Pract., **43**, 136–149 (2009)
[23] J. Zhao, W. Deng, Y. Song and Y. Zhu. What influences Metro station ridership in China? Insights from Nanjing, Cities, **35**, 114–124 (2013)
[24] D. Merriman, How many parking spaces does it take to create one additional transit passenger?, Reg. Sci. Urban Econ., **28**, 565–584 (1998)
[25] B.D. Taylor and C.N.Y. Fink, Explaining transit ridership: What has the evidence shown?,Transp. Lett., **5**, 15–26 (2013)
[26] A.A.K Hamsa, S.A.A.S Adnan and U.A. Khalid, Analysis of parking usage at the park and ride facility in Klang Valley, Malaysia, WIT Transactions on The Built Environment, **138**, 179–193 (2014)
[27] V. Nair, *Commuters face daily struggle to find space at LRT stations' parking lot*, The Star, (July 29, 2013)
[28] F. Aziz, *Parking issue at station*, The Star, (October 16, 2009)
[29] N. Zulkifli, *Fifteen new LRT stations in Klang Valley begin operations today*, New Straits Times Online, (June 30, 2016)
[30] G. Parkhurst, Park and ride: Could it lead to an increase in car traffic?, Transp. Policy, **2**, 15–23 (1995)

Review on Operational Performance of Employee Shuttle Service in Industrial Areas of Jababeka-1 Cikarang, West Java

Achmad Wicaksono [1,*], *Panji* Pasa Pratama[1], *Harnen* Sulistio[1] , and *Rahayu* Kusumaningrum[1]

[1] Department of Civil Engineering, Faculty of Engineering, Universitas Brawijaya, Indonesia

Abstract. Bekasi has one of the biggest industrial areas in Indonesia called Industrial Area of Jababeka 1 Cikarang, which has provided facility of employee "shuttle" transport for the employees. However, the employee shuttle service has not been able to accommodate all employees in the industrial areas especially in Industrial Area of Jababeka 1 Cikarang. This study aims to investigate the existing condition of operational performance of employee shuttle service, knowing the satisfaction and the necessity level of the service based on users' perception, and provides recommendation for the improvement of employee shuttle service in the Industrial Area of Jababeka 1 Cikarang. This study uses Importance Performance Analysis (IPA) method and multiple linear regression models. The final regression model is $Y = 2.034 + 1.621X_2 + 1.039X_4 + 1.311X_5 + 0.829\ X_6 + 1.046X_7 + 1.877X_8 + 1.619X_{10}$. This results shows that there are 3 variables which influence the performance of employee shuttle service (R^2= 0. 934) including X_8 variable (Sanitary Facility/Dust Bin) and X_2 variable (Vehicle Driving Skills), and X_{10} variable (Age of the Vehicle).

1 Introduction

Industrial areas in Indonesia are spread out almost in all regions and islands in this country. Those industrial areas generally have a large land, many of company and employees, production, and flow of goods (either export or domestic). Bekasi is one of regencies in West Java Province where the capital city is in Cikarang. Economy in Bekasi is sustained by agriculture, trade, and industry. Nowadays, there are a lot of industries in Bekasi which include seven biggest industrial areas located in Cikarang. Those industrial areas are Jababeka, MM 2100, Delta Mas, Lippo Cikarang, Hyundai, EJIP, and Bekasi Fajar.

Industrial area of Jababeka is the biggest industrial area in Bekasi especially in Cikarang with an area of approximately 1,570 hectares in which there are more than 1,500 local and multinational companies. The growth of industrial area in Bekasi has made on the increase of the employment and the number of employees, either originally from Bekasi or come from outside of Bekasi. The existing industrial area of Jababeka 1 Cikarang, which is the

* Corresponding author: wicaksono1968@yahoo.com

biggest industrial area, also shows that there is an increased number of employees. This will result in the increase of trips in Bekasi which directly impacts on the transportation activity in this area. Although there is a public transport terminal as the main entrance to the industrial areas of Jababeka 1 Cikarang, but the public transport in this terminal does not giving a service to all regions in industrial area of Jababeka 1 Cikarang, because it passes only on one road. Therefore, the existing public transport does not serve the employees who need transportation to reach their destination.

In reality, the employee shuttle service has not been able to accommodate all employees in the industrial area of Jababeka 1 Cikarang. There are still a lot of employees who use private vehicles such as car or motorbike. Based on that problem, a further study is needed to investigate the operational performance of the existing employee shuttle service in order to reach all the industrial area of Jababeka 1 Cikarang.

This study has several ojectives as: (i) To evaluate the existing operational performance of employee shuttle service in the industrial area of Jababeka 1 Cikarang, (ii) To determine the satisfaction and importance level of some employee shuttle service characteristics according to user's perception using Important Performance Analysis (IPA) method, and (iii) To give some recommendations in order to improve the employee shuttle service in the industrial area of Jababeka 1 Cikarang based on the result of multiple linear regression analysis.

2 Literature review

Employee shuttle service is a transport service for people whose an employee and utilizing car for public passengers or bus [1]. This special mode of transport is a transport mode which has its origin and/or destination to serve public passengers commuter, especially employees commuter. Employee shuttle service is implemented in such a way which serves from and to workplace with several passenger pick-up points, that scheduled, should not be stopped at the terminal, using yellow license plate car, with direct or indirect payments. There are several previous studies on this employee shuttle service, these are:

i) Satisfaction Level of Public Transport Users in DIY [2]. The paper shows that the public transport service in Yogyakarta is generally bad as it is indicated by low level of the compliance. Factors that influence the satisfaction level are safety, security, and comfort.
ii) Prospect of Employee Shuttle Service - A case study in Government Office of Central Java Province [3]. The result shows that the use and development of employee shuttle service for government employee of Central Java Province are influenced by the level of income, education class, family size, vehicle ownership, and travel time.
iii) Modeling on Guided Busway Needs Using Structural Equation Modeling [4]. The journal shows that travel origin and destination, travel patterns, willingness to pay and ability to pay have positive and significant impact on the Guided Busway Needs.
iv) Influencing Factors of School Bus Service Quality [5]. The analysis shows that there are 4 variables that has to be improved, these are: facilities, infrastructures, teachers, and employees.
v) Preliminary Study of School Bus Operation (Case Study: Bus Halokes Malang) [6]. The paper shows that the operational performance of the school bus is good enough. Based on the user's perception there are three important variables that should be improved i.e: comfort, safety and reliability.

3 Methods

3.1 Data collection

There are several methods of data collection that used in this study:

i) Inventory Survey on Transport Infrastructures
To determine the transport infrastructures in the industrial area of Jababeka 1 Cikarang

ii) Static Survey of Employee shuttle service
To determine the number of employee shuttle service operated in the industrial area of Jababeka 1 Cikarang

iii) Important Performance Analysis (IPA) Method Questionnaire
To determine the users' perception toward the performance of the employee shuttle service in the industrial area of Jababeka 1 Cikarang

3.2 Sampling

Based on the calculation result by using Slovin formula for determining the number of samples, there is a need of 400 respondents for IPA method questionnaire survey.

3.3 Analysis method

Analysis method used in this study was:

i) Descriptive Analysis on Operational Performance of Employee shuttle service
To determine the operational performance of employee shuttle service in the industrial area of Jababeka 1 Cikarang

ii) Important Performance Analysis (IPA) Method
To determine the users' perception toward the employee shuttle service in industrial area of Jababeka 1 Cikarang in which the result was plotted in a Cartesian diagram.

iii) Multiple Linear Regression Analysis
To obtain a multiple linear regression equation to understand the relation of each variables in affecting the employee shuttle service performance.

4 Finding and discussion

4.1 Existing condition on operational performance of employee shuttle service in industrial area of Jababeka 1 Cikarang

Based on the data obtained from Customer Service Division of PT. Jababeka Tbk, there were 844 companies and 249,925 employees in the industrial area of Jababeka 1 Cikarang. There are 88 employee shuttle services which passing through Gate 1, 19 employee shuttle services passing through Gate 7, and 13 employees shuttle services passing through Gate 9. Therefore, the total number of employee shuttle services operated in industrial area of Jababeka 1 Cikarang is 120 vehicles. There are nine routes of employee shuttle service servicing in the industrial area of Jababeka 1 Cikarang. Three routes through arterial streets including Route D (Cibarusah Destination), Route E (Cibitung, Tambun, Bekasi Timur destination), and Route H (Kerawang Destination-By Pass). Besides the three routes, some employee shuttle services are operated through tollroads.

Table 1. Data of Bus company operated in Cikarang-1 Industrial Area

No	Bus Company	Fleets			Total vehicles each company
		Gate 9	Gate 1	Gate 7	
1	Aneka Jaya	0	0	1	**1**
2	Bina Transport	1	2	0	**3**
3	Hiba Utama	1	24	1	**26**
4	GMS	1	2	0	**3**
5	JM	0	0	1	**1**
6	Parahyangan	4	20	7	**31**
7	Putra Regu	0	2	0	**2**
8	Restu	0	4	0	**4**
9	SLJU	6	34	9	**49**
Total Each Door		**13**	**88**	**19**	
Total		**120**			**120**

4.2 The result of importance performance analysis (IPA) methods

Based on the questionnaires which have been analyzed, it is obtained that the score of the average rate of importance is 3.44 and performance level is 3.02. The Cartesian diagram is obtained from the scores and the data from questionnaire using IPA method with the number of respondents of around 400. This diagram consists of four quadrants including quadrant I (main priority), quadrant II (keep achievement), quadrant III (redundant), and quadrant IV (low priority) which abscissa and ordinate lines that form the quadrant derived from the average value of the item aspect and level of performance (X) and the level of interest (Y). Cartesian quadrant position indicator is shown on Fig. 1. The indicators of each quadrant are described in Table 2.

Table 2. The indicator of quadrant

1	**Quadrant I main priority (high importances but low performance)** This indicator consists of 10 items: (1) *item 4* - use of a window film glass of vehicle, (2) *item 7* - driver's driving skills, (3) *item 11* - safety equipment such as fire extinguisher, glass breaker, and a flashlight, (4) *item 12* - aspect of emergency health facility such as first aid kit, (5) *item 13* - information emergency response with the availability of sticker shows contact to complaints and emergency situation, (6) *item 19* - passenger's seat belt, (7) *item 24* - facility to regulate air room temperature/AC, (8) *item 25* - hygiene facility in the form of dust-bin, (9) *item 26* - smoking ban stickers, and the last (10) *item 31* - age of vehicle.
2	**Quadrant II, maintain achievement (high importance and high performance)** This indicator consists of 3 items: (1) *item 2* - the identity of the vehicle users, (2) *item 29* - time between vehicles (headway) and (3) *item 30* – total number of fleet in operation.
3	**Quadrant III, excessive (high performance but low importance)** This indicator consists of 9 items: (1) *item 5* - emergency/dangerous lights signs and the position of that light shall be on top of vehicle, (2) *item 8* - attitude and behavior of the driver, who must obey the rule and friendly, (3) *item 9* - physical condition of the driver, (4) *item 10* - working time of the driver, (5) *item 14* -

	facility of handrail for standing passengers, (6) *item 15* - embarking and disembarking facility for passenger, (7) *item 16* - the vehicle's tire, (8) *item 20* - disembarking passenger's signs, and (9) *item 21* - electricity for audio visual facility.
4	**Quadrant IV low priority (low importance and low performance)** This indicator consists of 9 items: (1) *item 1* - identity of the vehicle; (2) *item 3* - vehicle interior lighting; (3) *item 6* - knowledge of the driver; (4) *item 17* – availability of rail curtain on window; (5) *item 18* - speed limiter instrument; (6) *item 22* - Depo and maintenance of the vehicle (pool); (7) *item 23* - vehicle's horsepower; (8) *item 27* - information transport services and (9) *item 28* - bus stop sign at the bus stop.

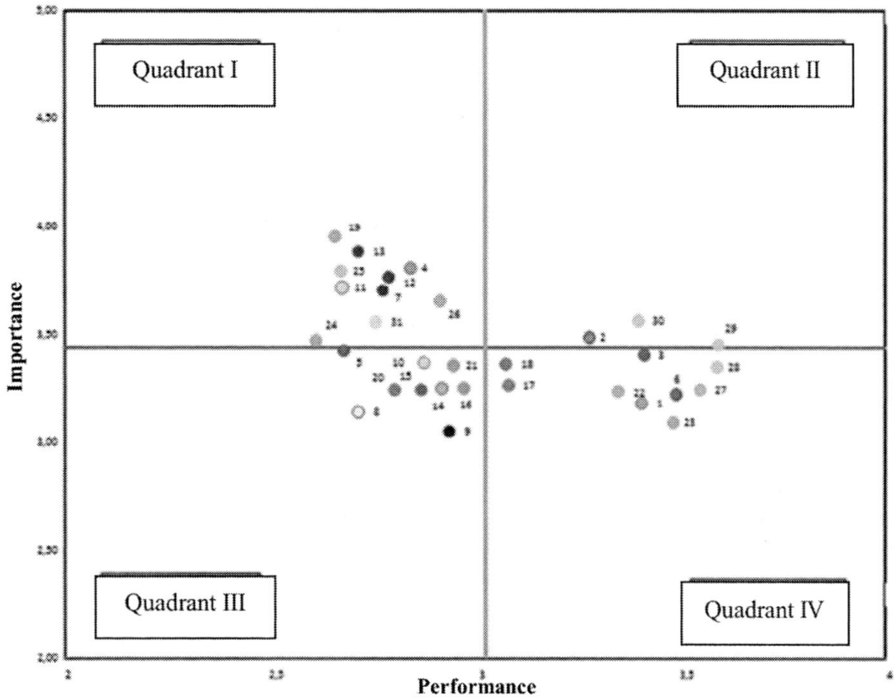

Fig. 1. Cartesian diagram of IPA.

4.3 Regression Analysis

According to the analysis of Importance Performance Analysis (IPA) method there are 10 variables included to the quadrant I but once the researcher obtain the data from the analysis of correlation between variables there are only seven variables that will be used in multiple linear regression analysis, these are: (i) Driver's Driving Skills (X_2), (ii) First Aid Kit (X_4), (iii) Emergency Response Information (Sticker Complaints Emergency) (X_5), (iv) Seat Belt (X_6), (v) Air Room Temperature Facility (AC) (X_7), (vi) Hygene facility (Dust-Bin), and (Vii) Age of Vehicles (X_{10})

Based on the results of multiple linear regression analysis which had been done with the help of statistical program, it then obtained multiple linear regression equation with R^2 value is 0.934. So it can be expressed mathematically in the following equation:

$$Y = 2.034 + 1{,}621X_2 + 1{,}039X_4 + 1{,}311X_5 + 0{,}829X_6 + 1{,}046X_7 + 1{,}877X_8 + 1{,}619X_{10} \quad (1)$$

where:
Y = Performance of Employee shuttle service
X_2 = Driver's Driving Skill
X_4 = First Aid Kit
X_5 = Emergency Response Information (Sticker Complaints Emergency)
X_6 = Seat Belt
X_7 = Air Room Temperature Facility (AC)
X_8 = Hygiene facility (Dust-Bin)
X_{10} = Age of Vehicle

From the equation above it can be concluded that there are three things that mostly influence the employee shuttle service based on the value of the coefficient, these are variable X_8 Hygiene Facility (Dust-Bin), variable X_2 (Driver's Driving Skill) , and variable X_3 (Vehicles' Age).

5 Conclusions and suggestion

5.1 Conclusion

Based on the result of analysis and discussion, it can be concluded as follows:

i) Existing Condition on operational performance of employee shuttle services are as follows:
   a. The employee shuttle service operate in the industrial area of Jababeka 1 Cikarang were giving a service to some routine commuters. The total number of employee shuttle service which operated in the industrial area of Jababeka 1 Cikarang was 120 fleets which only serve 14 industrial companies.
   b. There are nine routes of employee shuttle service which operated in the industrial area of Jababeka 1 Cikarang. There were three routes through arterial streets including Route D (Cibarusah Destination) , Route E (Cibitung, Tambun, Bekasi Timur destination), and Route H (Kerawang Destination-By Pass). Besides the three routes, some employee shuttle services are operated through toll roads.

ii) Based on IPA Analysis, there are 10 variables in Quadrant I as the main priority (high importance but low performance). It consist of 10 items including (1) item 4-sunglass windows film, (2) item 7 – Driver's driving skills, (3) item 11- safety equipment such as fire extinguisher, glass breaker, and a flashlight, (4) item 12 - health facility such as first aid kit, (5) item 13 - emergency response information in the form of complaints emergency sticker, (6) item 19 - seat belt, (7) item 24 – air room temperature facility -AC, (8) item 25 - hygiene facility (dust-bin), (9) item 26 - smoking ban stickers and the last (10) item 31 - age of vehicle.

iii) Based on the regression analysis of 10 variables handling priority. There are only seven variables that influence the employee shuttle service including Driver's Driving Skill (X_2), Health Facilities/First Aid Kit (X_4), Emergency Response Information /Sticker Complaints Emergency (X_5), Seat Belt (X_6), Air Room Temperature Facilities (AC) (X_7), Hygiene Facility/Dust-Bin (X_8) , and Age of Vehicle (X_{10}) with a multiple linear regression equation :
   $Y = 2.034 + 1{,}621X_2 + 1{,}039X_4 + 1{,}311X_5 + 0{,}829X_6 + 1{,}046X_7 + 1{,}877X_8 + 1{,}619X_{10}$
   From the equation, it can be concluded that there are three variables that mostly influence the employee shuttle service based on the value of the coefficient, these are the variable X_8 (Hygiene Facility/Dust Bin), variable X_2 (Driver's Driving Skill), and variable X_{10} (Age of Vehicle).

5.2 Suggestion

Related to the topic of this study, there are several suggestions for the government of Bekasi regency and the future researchers.

i) For the Bekasi' government and the district manager of industrial area of Jababeka 1 Cikarang, it is needed to carry out cooperation to organize the employee shuttle service and to see feedback from the service users.
ii) For the Ministry of Transport, it is needed to develop the minimum service standards for employees transport.
iii) For further researcher, it can be used as guidance to study in other industrial area as the further research area.

References

[1] Ministry of Transport, *Traffic and road-based public transport act No 22 Year 2009*, Indonesia, (2009)
[2] Z. Idris, Study on public transport passenger satisfaction at Yogjakarta Province, J. Dinamika Teknik Sipil, **9**(2), 189-196 (2009) (in Bahasa Indonesia)
[3] Supoyo, *Prospect of empoyees shuttle service a case study at central office of Center Java Province*, Master Thesis, Universitas Diponegoro, Semarang, Indonesia, (2004) (in Bahasa Indonesia)
[4] D. Supriyatno, H. Sulistio, L. Djakfar, and A. Wicaksono, Guided-busway needs modeling using structural equation model, J. Transportasi FSTPT, **12**(2), 143-152 (2012) (in Bahasa Indonesia)
[5] Nurtania, and A. Saddad, Factors affecting quality of service, J. Administrasi Pembangunan, **2**(1), 93-98 (2013) (in Bahasa Indonesia)
[6] T. Prasetyo, L. Djakfar and S. Abusini, Preliminary study of school bus operation (Case study: Bus Halokes Malang), Proc. *of The 18th FSTPT International Symposium,* Lampung, Indonesia, (2015)

The Effect of Deck Width Addition Toward Stability of Cable Stayed Bridge: Case Study of Siak Sri Indrapura Bridge, Riau

Bambang Supriyadi[1,*], *Suprapto* Siswosukarto[1], *Algazt A*. Masagala[2], *Imanuel Elvis S.* Hadjoh[3]

[1]Civil Engineering and Environment Department, Universitas Gadjah Mada, Yogyakarta
[2]Technology University of Yogyakarta
[3]Department of Civil and Environmental Engineering, Universitas Gadjah Mada, Yogyakarta

Abstract. An aeroelastic interaction always take places in long-span bridge, i.e. the dynamic relationship between wind forces (aerodynamic) and bridge structural motion (dynamics). Wind forces may lead to serviceability and safety problems, and may even cause instability of the whole bridge structure due to the flexibility nature of long span bridge structures.The slimer girder plate will make the structure more unstable as can be seen from the occurrence of lateral deformation. This paper presents the results analysis of cable stayed bridge that has total span of 640 m, two planes configuration of harp-typed hanger cable, the A shape pylons span along 320 m (main span),and side span of 160m.Structural analysis was conducted to determine the optimum ratio between width and spans of cable stayed bridge. The bridge was modeled with various width of 7 to 22 m. Structural modelling was conducted using SAP2000 software to analyze the structural stability of cable stayed bridge under wind load at speed of 35 m/s.The influence of wind loads to the cable stayed bridge stability can be seen based on the bridge deck deformation at the Y-axis (U_2), in which for the width of the bridge deck of 7 m, 8 m, 9 m and 10 m, the deformation of U2 are 0.26 m; 0.17 m; 0.12 m and 0.10 m, respectively. Meanwhile, for bridge's width of 11-22 m, the value of U2 axis deformation is relatively constant between 0.08 m to 0.07 m. This finding suggests that the ratio between width and length of bridge greatly affect the stability of the cable stayed bridge. Cable-stayed bridge, especially for concrete bridge, with two plain system having optimum ratio of width and length show no sign of an aerodynamic instability when fulfills the requirements of $B \geq L / 3$.

1 Introduction

Cable stayed bridge is one type of long span bridge whose main structure is girder and is held by one or more cable that transfer the load from the girder to the tower or pylon. Cable-stayed bridge is effectiveto be used in medium and long-span (300 m - 1600 m). This

*Corresponding author: bbsupri@tsipil.ugm.ac.id

type of bridge has an advantage in neglecting the use of many pillars which can interfere the mobility activities of the underneath bridge.

However, there is a problem in long-span bridge structure associated with wind load that could result in a significant influence on the structural stability of bridge or even catastrophic failure of the whole bridge. Research studies have shown that long span bridge should be designed as such so that it can withstand static as well as dynamic loads resulted from an interaction between the action of wind load and dynamic behavior of structural motion of the bridge. The effect of wind load can lead to serviceability and safety problems due to the nature of structural flexibility of long span bridge.

Nowdays, design and construction of long span bridge are very much influenced by the collapse of the old Taccoma Narrow bridge on 7 November 1940. The bridge failed and collapsed due to wind at the speed of 42 mph speed (18.8 m/s). The blowing wind on to the bridge caused the bridge deck to wobble violently side to side, twisted and then fell down. Further investigation revealed that the main principal to cause the bridge to collapse is the extent of an excessive flexibility of the bridge deck that initiate drag and lift, or flutter. Therefore, the modern design of long-span bridge requires the fulfillment of stability requirements against flutter as one important aspect in design.

Many researches study have devoted their work to investigate the influence of the deck characteristic, such as shape and model, but less attention was given on the effect of deck width to the dynamic behavior of long span bridge. The purpose of this research is to studythe influence of deck width variation of cable stayed bridge on the bridge structural stability. The research covers the study on the bridge's structure behavior resulted from live load for both static and dynamic due to wind, the investigation of natural frequency of structures and resonancepattern variation, the evaluation of bridge deck deformation of cable stayed bridge and the study to determine the optimum span to width ratio of bridge that still meet the aerodynamics stability requirements of the long bridge under wind load.

2 Literatures review

Suangga and Wiryana [1] conducted a research on Flutter Analysis of Old Tacoma Narrows Bridge. The researcher analyzed bridge model that was similar to the original design of the old bridge that made it the third longest bridge in the United States. The distance between pylons was 2,800 ft (853.44 m) with the height of 425 ft (129.54 m) and had steel girder plate as main structure. The stabilityequation for 3 dimension model of bridge considering the actions of aerodynamic forces is as follows:

$$[M]\ddot{u} + [K]u = F_{ae} \tag{1}$$

The result finding revealed that the principal cause of the bridge's instability is the extent of excessive flexibility of the bridge girder. The girder flexibility was very high either in vertical or tortional directions. It is because the thickness of the girder is so thin and the width of the bridge deck is so small compared to the span of the bridge.

Fig. 1 shows that at the wind speed about 1.4 ft/s (0.43 m/s), at the testing scale, the first torsion asymmetrical resonance pattern arises. But, as the wind speed increases, the vibration resonance muffled. At the wind speed of about 2.75 ft/s (0.83 m/s), the same resonance pattern arises again, but at this time, the resonance was getting stronger that finally made the bridge collapsed. This was also the case for vertical resonance pattern but the effect was not serious to cause the bridge to fail [1-3]. From *flutter* analysis in 3 dimensions of the old Taccoma Narrow bridge, shown in Fig. 2, it was found that flutter failure phenomenon occurred at the wind speed of 14 m/s. Therefore based on this analysis, it can be concluded that the cause of the bridge to collapse was due to the dynamic action of

 DOI: 10.1051/matecconf/201710309010

wind forces. Meanwhile, based on information in the field, it was known that the wind speed at the time (7 November 1940) when the bridge collapse wasat 42 mph (18.8 m/s).

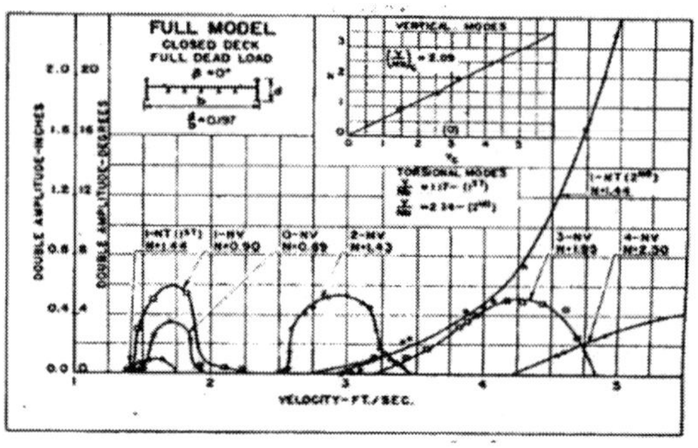

Fig. 1. Result of testing of Tacoma Bridge model.

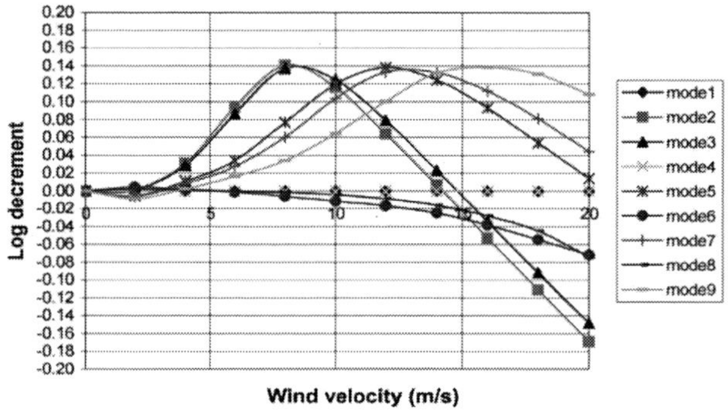

Fig. 2. Result of *flutter* analysis of the old Tacoma Narrow bridge [1]

Cable stayed bridge's span is generally quite large so that the wind's effect needs special attention in stability analysis of the bridge. The wind tends to influence the tortional ossilation as well as the flexural movement of stucture, and the change of angle of incidence to the girder will change the magnitude of lift. It is well known that girder flexibility greatly influences the bridge stability that make many researchers devoted their work to account for this aspect. One of aspect affecting the girder flexibility is bridge deck. Abdel-Aziz. A. et all [4] carried out study on the influence of bridge deck section shape of suspension bridge by Numerical Analysis. Various shape of bridge decks were investigated and the results show that the shape of the bridge deck contribute significantly to the potential cause of failure of long span bridge. Bai et al. [5] studied the aerodynamic measures impact on flutter stability of steel truss suspension bridge. Many arrangements of stabilizers on the steel truss deck bridge were introduced and was found to greatly influence the flutter characteristic of the bridge. Result of experiment conducted by Kirrenko, 1967 [in 6] shows that change of moment inertia of a girder has insignificant effects on the stress in the cable, and bending stress in the girder and *pylon*. Although the effect of deck bridge shape on the dynamic behavior of long span bridge received so much attention, but the influence of the bridge deck width variation is less studied.

For concrete *cable stayed* bridge with cable system and planes at the two end corners, it will not show an aerodynamic instability when it full fills the following requirement:

$$B \geq 10H \text{ or } B \geq L/30 \quad (2)$$

where:-$B = bridge's\ width, H = bridge's\ height, and\ L = main\ span$

According to Podolny and Scalm [7] the height of pylon is determined by several factors, such as: cable system type and aesthetic comparison of the height of pylon and span length. Therefore, it is recommended to have proportion value of the longest span and the highest pylon between 0.19 – 0.25. The height of pylon is the function of panel's length [6] that can be written as follows.

$$\frac{H}{L} = \frac{1{,}4\,a}{n + 1{,}3a} \tag{3}$$

According to Nobuaki [8] the available span (to determine the length of every span) has close correlation with configuration and anchorage method of cable. Some actual cases of three symmetrical spans bridges has ratio of side span/middle span (l/L) about 0.4.

The design code for loading used in this research refers to Indonesian standard code of loading for bridges [9]. In general, the design load is should be determined to account for several type of load as follows:

Dead Load from self-weight and and Super Imposed Dead Load.

a) Live load plan in the form of "T" Load, "D" Load, as well as live load from pedestrian.
b) Environmental Loads, namely the wind load, earthquake load, loads due to temperature effect, load due to soil pressure, and water pressure effects.
c) Other loads, i.e. loads due to collission, execution load etc.

Table 1. Plan Wind velocity Value (Vw)

Limit condition	Location	
	Until 5 km from beach	>5km from beach
Service Performance	30 m/s	25 m/s
Ultimit	35 m/s	30 m/s

According to design load of Indonesian code for Earthquake Load for Bridges [10], there are many ways to analyses earthquake load. *Time history* linear dynamic analysisis very appropriate to be used for structural analysis on irregular shape of structures under earthquake. Considering difficulties in estimating earth movement due to earthquake, the earthquake input is determined based on simulation. In this analysis, the record of land movement due to earthquake is taken from the accelerogram of *El-Centro N-S*earthquake recorded on 15 May 1940.

3 Research method

In this research, stability analysis was carried out on the *cable stayed* bridge with total bridge span of 640 m. The bridge model has sharp type suspension cable with two planes configuration,an A-shaped *pylon* with span between *pylons* (*main spane*) of 320 m, *side span* 160 m, and variation of bridge's width from total deck width of 7 to 22 m were introduced. The deck shape of the bridge is concrete box girder. The bridge pylon have A shape and is made of concrete. The analysis of modells were carried out using *SAP2000* software to analysis the stability of bridgeunderwind at speed of 35 m/s. Based on the result analysis, it is expected to be able to determine the optimum span to width ratioof the long span bridge. The illustration of technical designof the bridge model is presented in Fig. 3, Fig. 4 and Fig. 5.

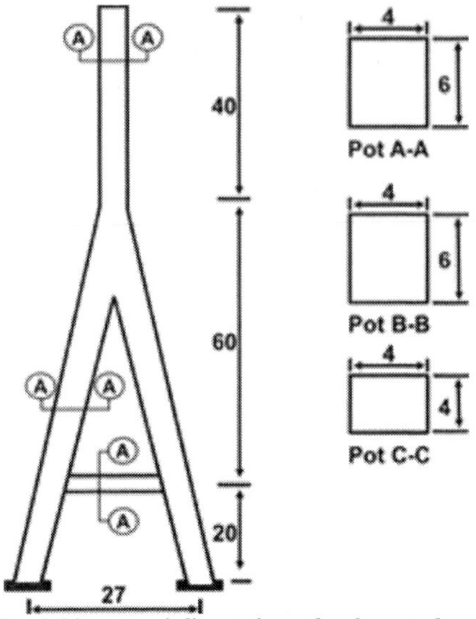

Fig. 3. Shape and dimension of *pylon girder*

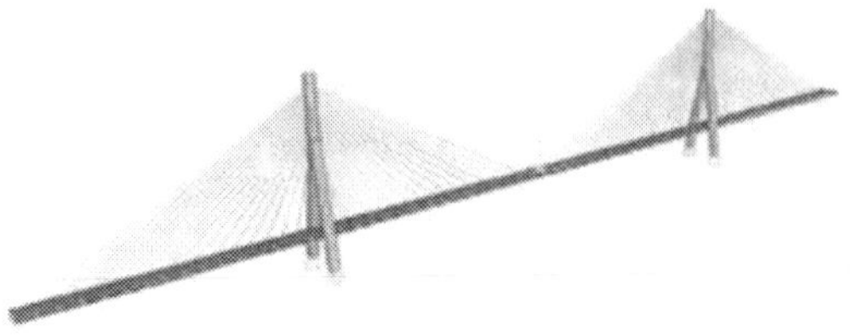

Fig. 4. Modelling of *cable stayed* bridge

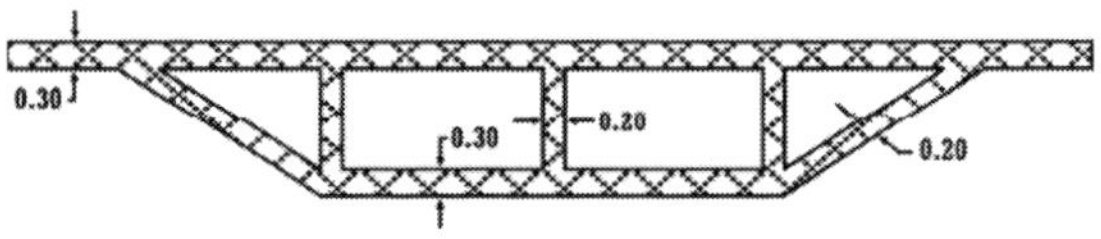

Fig. 5. Cross-section of b*ox girder*

3.1 Longitudinal girder consisted of high strength concrete box girder

The suspension cablebridge was modelled on SAP2000 program as *cable*element that capable to sustain only axial forces in tension. Therefore, the *stiffness properties* for moment, shearing forces and torsion for cable were denoted to be zero. Meanwhile, for *pylon* and box girder were modelled in rigid 3D frame. The materials specifications used for structural modelling in SAP 2000 are as follows:

a) Concrete for the main structure (girder and *pylon*) using high quality concrete of K-500that is equal to f'c = 41,5 MPa.
b) Steel for frame has yield strength (fy) of 400 MPa.
c) The suspension cable type is *Low Relaxation Strand* fabricated by VSL with specification according ASTM A 416-74 has fy = 1670 MPa and Fu = 1860 MPa.
d) Pretention cable for concrete is *Low Relaxation Strand* type from VSL with specification of Type ASTM A 416-74 by fy = 1.670 MPa and Fu = 1.860 MPa.

4 Basic design

4.1 Initial dimension

According to Walther [11], the length of panel for *cable-stayed* bridge with concrete deck is about 5 m – 10 m, so the length of each bridge panel being analyzed was taken as 10 m. Podolny [6] recommends that the height of pylon is determined based on the proportion between the longest span and the pylon's height of about 0.19 – 0.25. In this case, the bridge's model height of pylon is 60 m and the length of *main span* is 320 m, therefore, the ratio of pylon height to bridge span (H/L) is 60/320 = 0.19 (this value is in accordance with the recommendation of 0.19 – 0.25).

4.2 Length and width of the bridge

Based on equation proposed by Leonhardt[12], ratio between the bridge's width (B) and the girder's height (H) isB $\geq$ 10 H, while ratio between the bridge's width (B) and the main

 DOI: 10.1051/matecconf/201710309010

span length of the bridge (L) is B ≥ L/30. Therefore, based on the *Leonhardt's* equation, all the pre-assumptions for the adjustment of the bridge's width (B) toward the girder's height (H) = 1.5 m and the main span length (L) = 320 m can be seen in Fig. 6.

4.3 Design guidelines

The guidelines used in analysis of *cable-stayed* bridge in this study refer to:

a) Indonesian Concrete Structure Design for Bridges (RSNI T-12-2004) [13]
b) Indonesian Standards of Earthquake Resistance Design for Bridges (RSNI T-12-2004)[10]
c) Indonesia Standards of Loading for Bridges (SNI T-02-2005)[9]
d) For cable selection, size and type of cable refer to VSL specification (VSL-1984)[14]

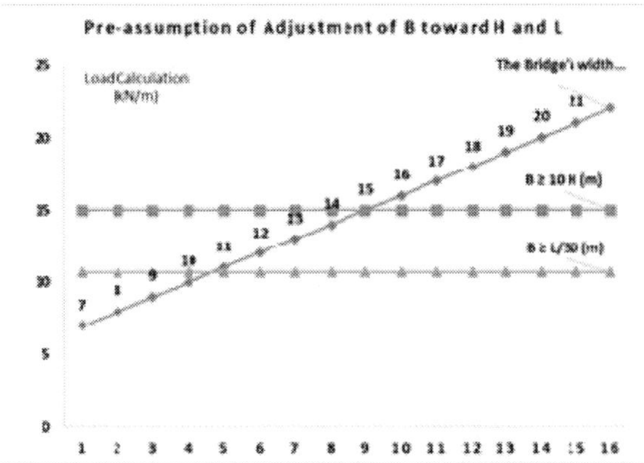

Fig. 6. Pre-assumption of adjustment of B toward H and L

4.4 Structural modelling

The structural modelling of *cable stayed* bridge using SAP2000program was carried out in three dimensional analysis. The modelling of each element is different depending on the type of element.The following Fig. 4 shows the *cable stayed* bridge modelling with width of 17 m using SAP2000 program. The bridge modelling and analysis was carried out for each width variations of bridge width, from bridge width of 7 m to 22m. Results of SAP2000 yields some mechanical characteristic of bridge, such as natural frequencys, internal forces, as well as structural deformation of the *cable stayed* bridge models.

5 Results and discussions

5.1 Load combination

Loading combination was determined according to Indonesian standards load for bridges taking into account the limit condition of service performance of bridge. The load was determined considering each variations of bridge's width under investigated. The result of loading calculation is presented in the following Fig.7, 8 and 9, for dead load (DL), live load (LL) and wind load (EQ), respectively.

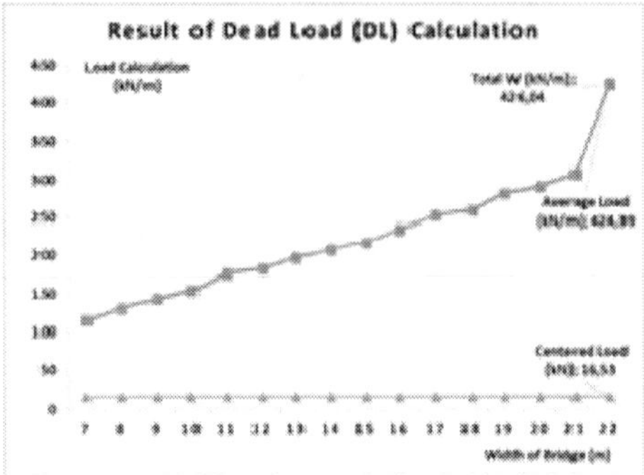

Fig. 7. Result of Dead Load (DL)

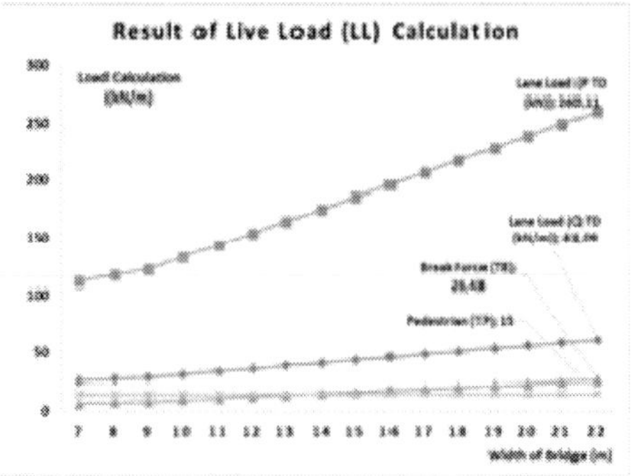

Fig. 8. Result of Live Load (LL)

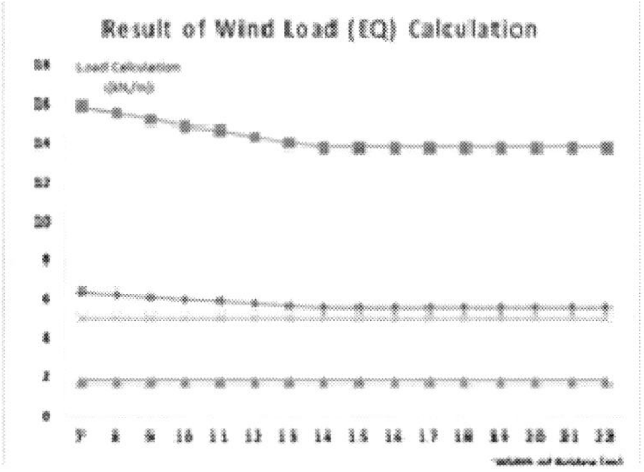

Fig. 9. Result of Wind Load (EQ)

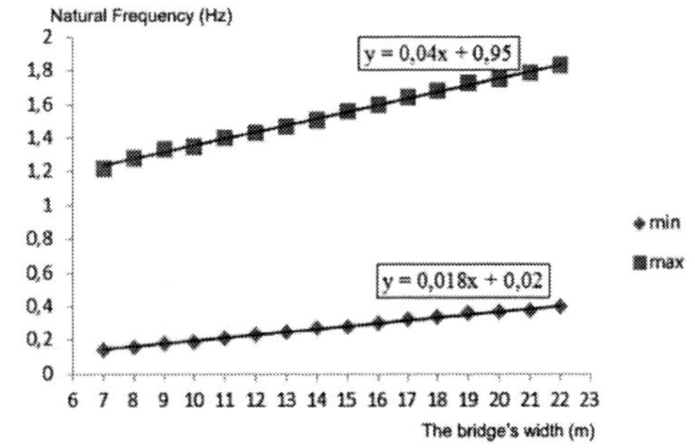

Fig. 10. Bridge's Natural Frequency

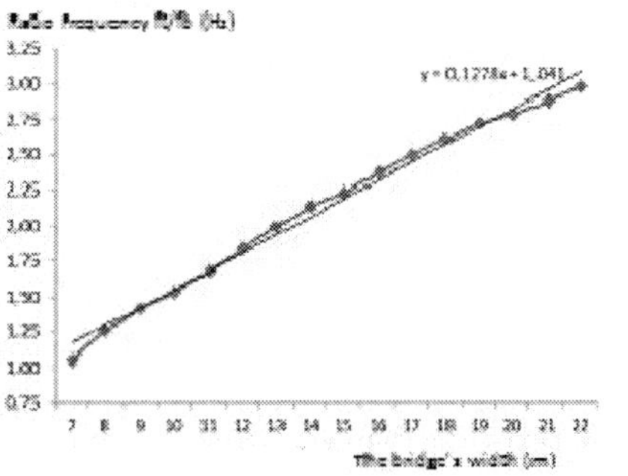

Fig. 11. Ratio of f_t/f_b

From Fig. 10, it can be seen that for the same span of bridge, the wider deck of *cable stayed* bridge will result in larger flexural frequency (f_b). The rate of increase in natural frequency of the bridge is somewhat larger than the rate of increase in bridge width as shown by the magnitude of the line gradient (0.04 for frequency and 0.018 for bridge width). Similar phenomenon was also observed for flexural (f_b) and torsional (f_t) frequencies where both frequencies increase with the increase in bridge width. For example, for the bridge width of 15 m, the f_b value is bigger, more than 1.00 Hz. This is also the case for torsional frequency (f_t), the wider bridge will give larger f_t. For the bridge width of more than 14 m, the magnitude of f_t value is also larger, more than 2.00 Hz (Fig. 11). The rate of increase in both flexural and torsional frequencies is nearly linear. This finding shows that the increase in bridge width will increase the incrtia of the bridge thereby increasing the stiffness of the bridge structures.

5.2 Suspension cable stress

From the result analysis of suspension cable, it is shown that for wider bridge width, the axial forces developed within cable will also be bigger. This is due to the fact that the wider bridge width will result in heavier self-weight and service load of the bridge. For example, for bridge width of 7 m, the calculation show that the minimum cable force (P) is 7,378.91 kN, while for bridge width of 22 mhave the minimum value (P) of 18,014.88 kN (Fig. 12). The increase in magnitude of forces in cables seem to be linear with the increase in bridge width.

5.3 Girder

Girder was modelled as a 3D *frame* element. The girder was designed as a *box girder*which has panel's length of 10 m and height of 1.5 m for all variations of the bridge's width. Results of analysis show that the wider of the bridge's width will result in greater internal forces within the girder. This is easy to understand since as the width increase so as the load in the structure due to the larger number of road lane (Fig. 13). The rate increase of forces is somewhat larger than the increase in bridge width for normal, shear, flexural and torsional forces as shown by the magnitude of lines gradient for each forces. For torsional forces, the rate increase is even greater (exponential) for wider width bridge than the other forces. This is a phenomenon that need to be put into special consideration in designing of long bridge.

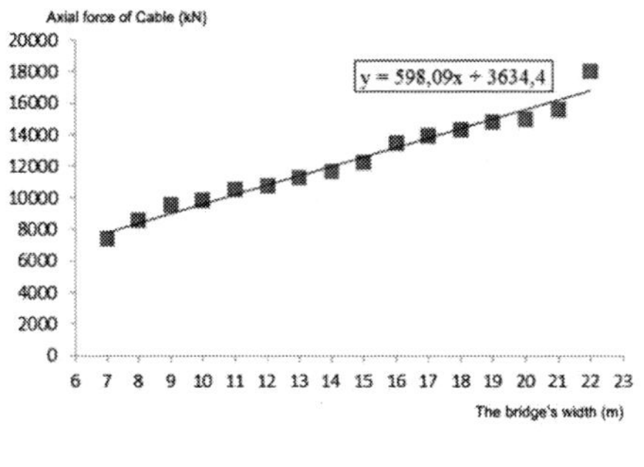

Fig.12. Axial force of cable

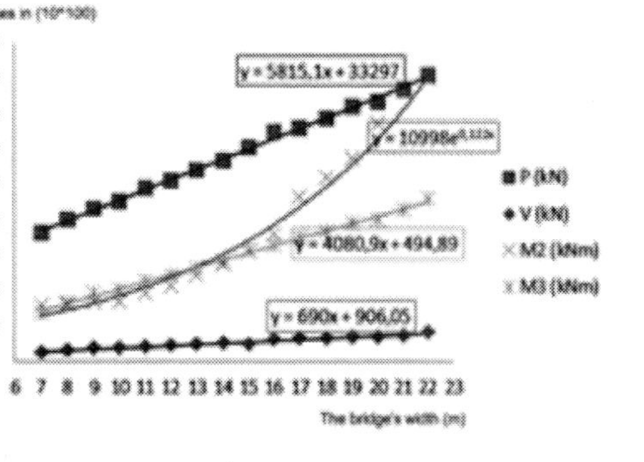

Fig.13. Forces in girder

5.4 Pylon

The A shape p*ylon* was modelled as 3D frame having dimension of 400/600 cmand connection beam size of 400x400 cm for all variation of the bridge's deck width. The pylon acts like portaland is capable to sustain axial forces (P), shear forces (V), torsion (T) and moment (M).The total height of pylon is 120 m measured from pedestal until the end of pylon. Based on the results of analysis it is shown that the bigger the dimension the bridge's width, the forces developed within the pylon will also be larger. The reason for this is due to heavier load acting on the bridge resulted from the heavier self-weight (Fig. 14). The rate increase of shear, flexural and torsional forces is nearly proportional. But, the rate increase in axial forces is somewhat greater than the other forces.

5.5 Pylon deformation

Pylon maydeform in three directions, i.e. X-axis (U_1), Y-axis (U_2) and Z-axis (U_3). The biggest deformation occurs at the top end of pylon.The pylon deformation at X-axis (U_1) shows that the wider of the bridge's width yields larger bridge deformation at X direction; for bridge's width of 7 m, U_1 = 0.19 m, while for the bridge's width of 22 m, the value of U_1 = 0.33 m. This is due to the fact that the position of suspension cable that sustains the girder's load at the X axis. For pylon deformation at Y axis (U_2), the magnitude is not really different for the variation of the bridge's width. For bridge's width of 7 m, U_2 = 0.15 m, for bridge's width of 22 m, U_2 = 0.10 m. This is due to the less flexibility of the bridge system as the bridge width as a result of significant increase in its inertia. The increase in stiffness thereby reduce deformation at Y-direction. For deformation at Z (U_3), no significant magnitude of deformation was observed. For bridge's width of 7 m, U_3 = 0.03 m, and for bridge's width of 22 m, U_3 = 0.05 m. The reason for this is due to rigidity nature of concrete pylon (Fig. 15).

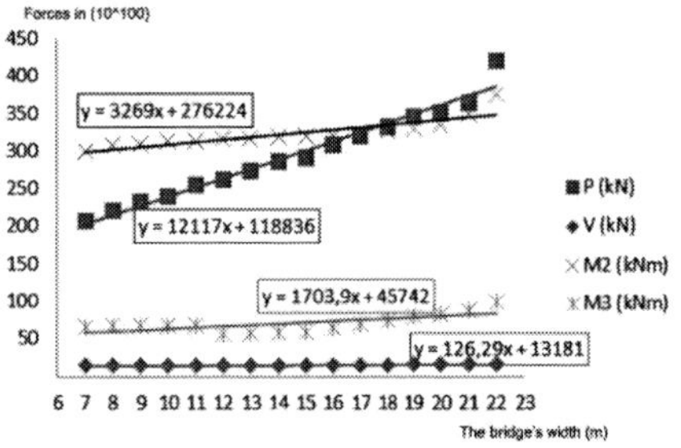

Fig. 14. Forces in pylon

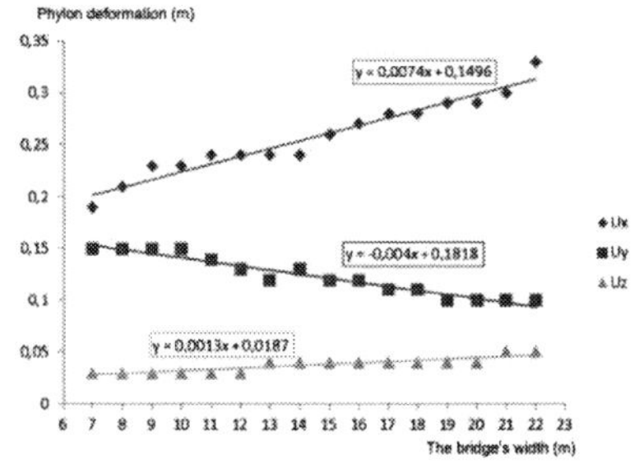

Fig. 15. Deformation for pylon

5.6 Bridge deck deformation

The Bridge deck is part of the first structure that supports loads. So, the stability of the Bridge deck structure greatly affect the stability of the overall bridgestructure as shown by magnitude of deformation toward either X axis (U_1), Y axis (U_2), or Z axis (U_3), as presented in Fig. 16.

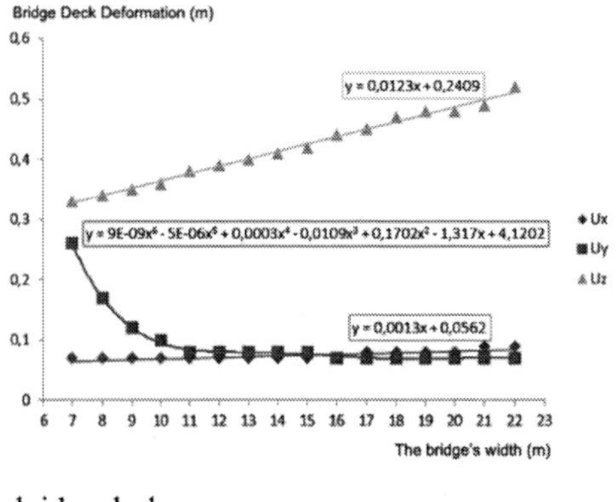

Fig. 16. Deformation of the bridge deck

At X-axis, the *deck* deformation is nearly similar for all variation of bridge's width. For example, for the bridge width of 7 m and 22m, the magnitude of $U_1$and U2 are both 0.09 m. This is due to the similar magnitude of stiffness and hence strength of the *box girder*.Meanwhile, at Y-axis (U_2), the value depends on the predominant load acting Ydirection, namely the wind load acting on side plane of the bridge deck. Thus, the wider the bridge deck, the structure of the bridge will be more stable under aerodynamic affect due to the wind. The improved of bridge stability at Y-direction is resulted from the significant increase in girder inertia. Based on *Leonhardt's* requirement, the bridge's width of 7 m to 10 m is considered as unsafe under aerodynamic effect. This is in accordance with the results of SAP2000 analysis, i.e. U_2 value in the bridge's width of 7 m, 8 m, 9 m, and 10 m are 0.26 m; 0.17 m; 0.12 m; 0.10 m, respectively. Meanwhile, U_2 value for the bridge's width of 11 to 22 m, the magnitude relatively stable that is 0.08 m to0.07 m, respectively.

For pylon deformation at Z-axis (U_3), the value depends on load combination acting atZ axis, namely it self-weight of the deck as well as service load. For wider bridge deck, the U3will be greater. For example, for bridge's deck width of 7 m, U3= 0.33 m, while for bridge's deck of 22 m, U3 = 0.52 m. The rate increase in U3 is somewhat greater than the rate increase in deck width as shown by the magnitude of the line gradient in Fig. 16. Based on deflection requirements of AISC (δ max = 0.89 m) and AASHTO (δ maks = 0,80 m), all variations of the bridge's deck width still meet the requirements, when viewed from Z deformation.

6 Conclusions

Based on the results of analysis of *cable stayed* bridge considering bridge's deck widths variation of 7 m to 22 m, it can be concluded as follows.

a. The wider of bridge deck width will increase the inertia and so as the stiffness of the bridge girder that result in a greater flexural (fb) and torsional (ft) frequencies. The rate of increase in natural frequency of the bridge is somewhat larger than the rate of increase in bridge width.
b. The wider bridge deck width result in larger forces developed within cables and girders. For the case of girder, the rate increase of forces is greater than the increase in bridge's width for normal, shear, flexural and torsional forces. An exception for torsional forces where it increases exponentially.
c. The bigger the bridge's width, the forces as well as deformation developed within the pylon will also be larger.
d. The bridge deck deformation at X-axis is nearly similiar for all variation of bridge width. At Y-axis, the structure if more stable due to the significant increase in girder stiffness int this direction.

7 Further research

Further research will be devoted to find the optimum value of angle considering the aerodynamic effects necessary to improve the servicability. Further research will also employ a more special software specially design for long bridge.

References

[1] M. Suangga and A. Wiryana, Flutter analysis of old tacoma narrow bridge, Jurnal Dinamika Teknik Sipil (available in Indonesian), **11**, (2008)
[2] M. Suangga and S. Mahatmaputra, Flutter analysis software prototype for long span-bridge, Jurnal Dinamika Teknik Sipil (available in Indonesian), **12**, 40 – 49 (2009)
[3] B. Supriyadi and A.S. Muntohar, *Jembatan*, (available in Indonesian), Beta Offset, Yogyakarta, (2007)

[4] A. Abdel-Aziz and W.A. Attia, Aeroelastic investigation of different deck sections for suspension bridges by numerical analysis, International Journal of Engineering and Innovative Technology (IJEIT), **4**(12), (2015)
[5] H. Bai, S. Huang and J. Li, Study of the aerodynamic measures impact on flutter stability of steel truss suspension bridge, *The Eighth Asia-Pacific Conference on Wind Engineering*, Chennai, India, (2013)
[6] M.S. Troitsky, *Cable Stayed Bridges Theory and Design*, BSP Professional Book, London, (1997)
[7] Podolny and J.B. Scalm, *Construction and Design of Cable Stayed Bridges*, John Wiley & Sons Inc., New York, (1976)
[8] F. Nobuaki, *Long Span Bridges Engineering*, BIPRAN Bina Marga, Jakarta, (1990)
[9] National Standard Board, *SNI T – 02 – 2005, Standard Code of Loading for Bridg*es, Jakarta,(2005)
[10] National Standard Board, *RSNI T – 12 – 2004, Seismic Resistance Design for Bri*dge, Jakarta, (2004)
[11] R. Walther, *Cable Stayed Bridges*, Thomas Telford, London, (1988)
[12] I. Zarkasi and S. Rosliansjah, *Perkembangan Akhir Jembatan Cable Stayed,* Makalah Pada Konferensi Reginal Teknik Jalan (KRTJ), (available in Indonesian), Padang (1995)
[13] National Standard Board, , *RSNI T – 12 – 2004, Reinforced Concrete Design for Bridge*, Jakarta, (2004)
[14] Anonim, *VSL Stay Cables for Cable-Stayed Bridges*, VLS International LTD, Berne, Switzeland, (1984)

Alighting and Boarding Time Model of Passengers at a LRT Station in Kuala Lumpur

Peay San Hor[1], and *Mohd Masirin* Mohd Idrus[1,*]

[1]Faculty of Civil and Environmental Engineering, Universiti Tun Hussein Onn Malaysia, 86400 Batu Pahat, Malaysia

Abstract. A research was conducted to study the factors affecting the alighting and boarding rate of passengers and establish a prediction model for alighting and boarding time of passengers for a passenger rail service in Malaysia. Data was collected at the KL Sentral LRT station during the morning and evening peak hours for a period of 5 working days. Results show that passenger behaviour, passenger volume, crowdedness in train and mixture of flow has significant effects on the alighting and boarding time though mixture of flow is not significant in the prediction model produced due to the passenger behaviour at the platform.

1 Introduction

As rail transportation is emerging to be the preferred choice for public transport in Kuala Lumpur, the quality of service provided by the rail operators is most important for the public. Apparently, dwell time is a key parameter of system performance, service reliability and quality in any mode of public transportation [1, 2]. Over the years, more researches had been conducted about train dwell time. A review by Hor & Mohd Masirin on past dwell time models revealed that the volume of alighting, boarding and standing through-passengers had the most influence on dwell time models [3].

One of the earliest known train dwell time models was developed by Wirasinghe & Szplett for Calgary LRT [4]. They produced three separate linear equations based on the percentage of boarders for a train. The authors found out that not only the volume of passengers boarding and alighting affected the dwell time, but the percentage of boarding and alighting passengers in the flow also had significant effect on dwell time, thus revealing the effect of a mixed flow.

This was soon followed by a non-linear model developed by Weston, which revealed that the dwell time increased with the volume of alighters and boarders at a declining rate, indicating a crowding effect [5]. In 2007, Harris & Anderson applied the same model to estimate dwell times of metros and suburban railways around the world. They concluded that Weston's formula seemed to be valid around the world, and the overall structure of the approach appeared sound [6]. A non-linear dwell time model was also developed by Puong for the Massachusetts Bay Transport Authority (MBTA) in Boston revealed that the presence of standing through-passengers had an effect on the dwell time, specifically the

[*] Corresponding author: masiri@uthm.edu.my

boarding time if there were more than 3 of standing through-passengers. Yet, Puong also found out that this had no or only a minor effect on the alighting process [1].

Lam, et al. developed a generalised linear model that included a fixed time for doors opening and closing, and the number of boarding and alighting passengers per train for the Hong Kong MTR system. They concluded that the dwell time did not increase infinitely with the increase of passengers due to the maximum allowable dwell time of trains being governed by the train headway [7].

Douglas established a model to predict train dwell time for RailCorp in Sydney [8]. Similar to Weston's model [5], he used a power function of 0.7 for both alighting and boarding because it gave a better fit than a linear function. Mixed flow of passengers and crowding effects caused by standing through-passengers in the model were included in the model by multiplying the number of standing passengers with the combined total of boarding and alighting passengers.

The studies mentioned above are related to dwell time, which include the time passengers alight and board the train, the time before the train doors are closed, and the time the train waits to depart with the doors closed [8]. The most critical part of the dwell time that relates to passenger movement is the alighting and boarding time but this varies from door to door due to the difference of volume of passengers alighting and boarding at each door and the crowdedness in the vehicle [2]. In Malaysia, there is little or even no published work about the alighting and boarding time for a door for an urban rail transit system. Therefore, this research aims to study the factors affecting the alighting and boarding time of passengers and establish a prediction model for alighting and boarding time of passengers for an urban rail transit system in Malaysia.

2 Data collection

The KL Sentral station, which is located on the Kelana Jaya LRT Line is a major interchange station that connects with the various rail routes in Kuala Lumpur. It has one of the highest passenger flows during the morning and evening peak hours. According to the statistics obtained from Public Land Transport Commission (SPAD) in May 2015, the arrival and departure ridership at KL Sentral exceeded 10,000 and 5,000 respectively in the morning peak hours from 07:00 to 09:00. In the evening peak hours (17:00 to 19:00), the arrival and departure ridership at KL Sentral were around 7,000 and 10,000 respectively. Therefore, it was chosen as the site for this study.

The movement of passengers alighting and boarding at one of the doors of each train was observed throughout the observation period. A total of 115 and 99 observations were recorded in the morning and evening peak respectively. The data collection was conducted from 07:30 to 08:30 during the morning peak and from 17:30 to 18:30 during the evening peak hours for a period of 5 working days at the KL Sentral LRT station.

Observation was conducted using a camera at the platform. The field data obtained from this observation were: a) Alighting passengers, b) Boarding passengers, c) Alighting and boarding times, and d) Crowdedness in the vehicle.The volume of alighting and boarding passengers and the time required for the whole process were obtained directly from the recordings whilst the crowdedness in the vehicle was estimated by observing the standing through-passengers in it.

On average, the dwell time for all trains in the morning and evening peak hours at KL Sentral was 50 s if there was no external interruption. The dwell times of trains at a station is fixed as the Kelana Jaya LRT Line is a fully automated train service. The dwell time of a train is taken to be from the moment it stopped at the platform to the moment it left the platform. This included the time required for passengers to alight and board the train, the time before the train doors were closed, and the time the train waits to depart with the doors closed [8].

For the purpose of this study, the time recorded during the observation is the alighting and boarding time for a selected door of each train, limited to clustered alighting and boarding only. In this study, a passenger is assumed to be part of a cluster if the time interval in the alighting and boarding process between his predecessor and himself is 3 seconds at the maximum. The time was measured from the moment the first person passed through the door (usually an alighter) till the moment the last person passed through it (usually a boarder).

To estimate the crowdedness of the train at each door, the fixed-route passenger load LOS [9] was used. Table 1 describes the standing passenger area for each LOS category, used mainly for vehicles designed to have most passengers standing. With the total standing area of the train at each door calculated using the measurements provided by the operator, the passenger load LOS at each door was estimated.

Table 1. Fixed-route passenger load LOS [9].

LOS	Standing passenger area (m^2/pass)	Standing passenger(s) per each m^2 (pass/m^2)
A	>1.00	<1.00
B	0.76-1.00	1.32-1.00
C	0.51-0.75	1.33-1.99
D	0.36-0.50	2.00-2.85
E	0.20-0.35	2.86-5.00
F	<0.20	>5.00

3 Results and discussion

3.1 Passenger behaviour

From the observation conducted at the platform, it was noticed that there were more boarding passengers in the morning peak whilst more alighting passengers were sighted in the evening peak. This had an effect on the movement of passengers at the platform.

In the morning peak, where there were more boarding passengers, the passengers waiting for the train's arrival queued up in two separate lines at each door. This created a narrow channel in between the two lines, allowing the alighting passengers to enter the platform without being obstructed when they alighted, as can be seen in Fig. 1. Most of the time, the boarding passengers observed would allow all alighting passengers to alight first before they started boarding, except for a few instances where passengers began to board when they noticed that there was enough space in the train for them to move in.

Fig.1. Passengers queuing prior to boarding

In the evening peak, where there were more alighting passengers than boarding passengers, most of the boarding passengers queued in two separate lines at each door, as those observed in the morning, albeit shorter ones due to the lesser number of passengers. There were some instances where the boarding passengers did not form a queue and were standing in the keep-out zone, as shown in Fig. 2. This interrupted the flow of the alighting passengers, thus increasing the time required for alighting and boarding.

Fig. 2. Passengers waiting at keep-out zone

3.2 Passenger volume

The time used for alighting and boarding was measured for a selected door of each train. Fig. 3 shows that the increase in volume of passengers had increased the time required for alighting and boarding. Yet, it was observed that the volume of passengers and time required for boarding and alighting did not increase infinitely, but was bound by the capacity of the train and the dwell time of the train at the platform. Most of the observations show that the time required for alighting and boarding was below 30 s, except for two observations which had an alighting and boarding time of 31 s and 37 s respectively. In terms of the volume of alighting and boarding passengers, a maximum of 51 passengers was recorded during the observation period.

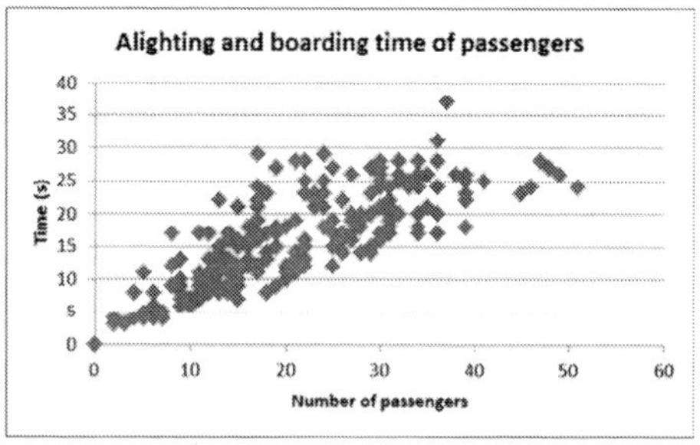

Fig. 3. Relationship between alighting and boarding time of passengers with the passenger volume

3.3 Train crowdedness

Crowdedness in the train was observed and analysed to study its effect upon passenger flow rate. The train crowdedness in this study is defined by the level of service (LOS) in the vehicle for a selected door observed.

As can be seen from Fig.4, the crowdedness of a train increased from LOS "C" to LOS "F", it was found that the passenger flow rate decreased significantly from a mean of 1.37 to 0.52 pass/s. This is a reduction of more than 60% as compared to LOS "A", "B" and "C". Similarly, the range of the passenger flow rate also became smaller as the train became more crowded. When the number of standing through-passengers increases in the train, the

DOI: 10.1051/matecconf/201710309011

amount of space available in the train reduces (refer to Table 1). This also means that the space available for the passengers to move when alighting and boarding the train is limited and causes the passenger flow rate to decrease significantly.

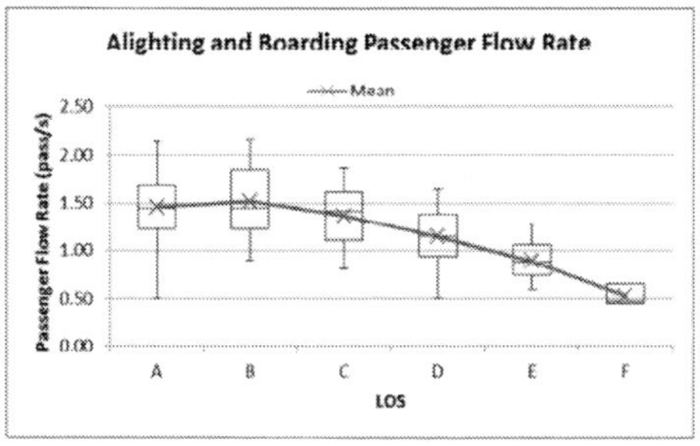

Fig. 4. Relationship between alighting and boarding passenger flow rate with LOS

3.4 Mixture of flow

Mixture of flow in the passenger flow is described as the difference between the number of alighters and boarders at a door of a train. Referring to Fig. 5, the overall passenger flow rate is lowest when the number of alighters and boarders are similar (high mixture of alighters and boarders). In contrast, the overall passenger flow rate is highest when there is a low mixture of alighters and boarders i.e. consisting mostly alighters or boarders only.

When there is a high mixture of alighting and boarding passengers, passengers are bound to "collide" with each other more, thus reducing the rate of alighting and boarding. Besides, passengers usually move in groups with a specific motivation: alighting or boarding. Thus, passengers with different motivations have an effect on the alighting and boarding rate [10].

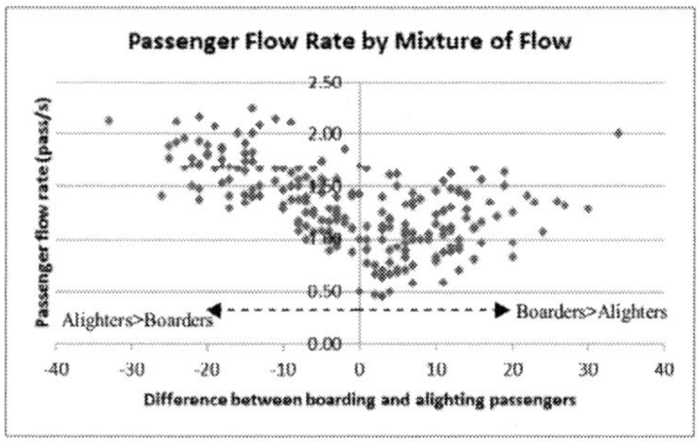

Fig. 5. Passenger flow rate by mixture of flow

3.5 Model

Using the regression analysis, a non-linear model to predict alighting and boarding time was produced. Initial variables included in the analysis were number of alighters, number of boarders, LOS in the train and percentage of alighters and boarders. The regression analysis removed variables that were not significant and as a result, Equation (1) was produced.

$$\text{Time} = 1.127\,A^{0.7} + 1.937\,B^{0.7} + B \times Cr \quad (1)$$

with A = number of alighters, B = number of boarders, Cr = crowdedness index based on the LOS in the train (refer to Table 2).

Table 2. Crowdedness index

LOS	Cr
A, B, C	0
D	0.122
E	0.372
F	0.790

This model does not include a constant like many other dwell time models [1, 4, 5, 7, 8] as this is an alighting and boarding time model. It does not take into consideration the dispatching time and unused time because they are independent of the passenger movement. Thus, this model considers only the time affected by the movement of passengers, which is the alighting and boarding time.

Based on the model produced, it is estimated that the time a passenger requires for boarding is more than that of alighting. Though this is in contrast to the models by Lam, et al. [7] and Douglas [8], this finding is similar to that of Puong's [1], who also found that boarding requires a longer time than alighting. This difference is acceptable because the behaviour of passengers at these stations studied were different.

If passengers at the platform stand in the keep-out zone, they would become a resistance to the alighting passengers, thus significantly increasing the time required to alight the train. In this study, it was observed that passengers at the platform had queued up and did not stand in the keep-out zone at most times. This allowed passengers to alight the train without obstruction and resistance. A power function of 0.7 was adapted from the model by Douglas [8] and this produced a slight improvement as compared to a linear function for both alighting and boarding times.

This model also reveals that when the LOS in the train is in the categories of A, B or C, it does not have any effect on the boarding time. But when the crowdedness in the train increases beyond LOS C, this would result in the increase of the boarding time, thus increasing the overall alighting and boarding time. It is also noted that the crowdedness in the train has no or very minimal effect on the alighting time, similar to the finding by Puong [1].

In the analysis, the variable for mixed flow effect is found to be insignificant and thus removed from the model. This could be due to the reason that the observation was conducted during the peak hours, where most boarding passengers had queued up and did not obstruct the passengers who alighted from the train. Most of the boarding passengers were also found to practice the *alight first board later* system. This eliminates the effect of mixed flow and thus found to be not significant in the model produced.

4 Conclusion

Dwell time at station is a key indicator of system performance, service reliability and quality for passenger rail services [1, 2]. Among the components of the dwell time, the alighting and boarding time is most related to the passenger movement at a platform. From this study at the KL Sentral LRT station, passenger behaviour has an effect on the passenger flow rate, with non-queuing passengers at the platform and those standing in the keep-out zone reducing the passenger flow rate. The increase in the passenger volume also increases the alighting and boarding time, though not infinitely as it is bound by the train capacity and the dwell time of the train at the platform. Crowdedness in the train reduces the passenger flow rate significantly as it increased from LOS "C" to LOS "F". Yet it has no effect when the LOS is LOS "A" to LOS "C". In the event of a mixed flow, the alighting and boarding rates of passengers are lowest when there is a high mixture of alighters and boarders. The prediction model to estimate the alighting and boarding time of passengers produced using the regression analysis is significant with the inclusion of volume passengers and the LOS as variables. The effect of crowdedness is accounted for in this model but is not so for the effect of mixed flow. This is due to the passenger behaviour at the platform. Further study with different conditions i.e. during non-peak hours and other stations is recommended to further investigate the effects of the factors mentioned in this paper.

This paper was sponsored by the Office for Research, Innovation, Commercialization and Consultancy Management (ORICC), UTHM.

References

[1] A. Puong, Dwell time model and analysis for the MBTA red line, Massachusetts Inst. Technol. Res. Memo, 1–10 (2000)
[2] W. Chu, X. Zhang, J. Chen, and B. Xu, An ELM-Based approach for estimating train dwell time in urban rail traffic, Math. Probl. Eng., **2015**, 1–9 (2015)
[3] P.S. Hor, and M.I. Mohd Masirin, Train dwell time models for rail passenger service, MATEC Web Conf., **47**, 3005, (2016)
[4] D. Szplett, and S.C. Wirasinghe, An investigation of passenger interchange and train standing time at LRT stations: (i) Alighting, boarding and platform distribution of passengers, J. Adv. Transp., **18**, 1–12 (1984)
[5] J.G. Weston, *Train service model-technical guide*, London Underground Operational Research Note 89/18, (1989)
[6] N.G. Harris, and R.J. Anderson, An international comparison of urban rail boarding and alighting rates, Proc. Inst. Mech. Eng. Part F J. Rail Rapid Transit, **221**, 521–526 (2007)
[7] W.H.K. Lam, C.Y. Cheung, and Y.F. Poon, A study of train dwelling time at the hong kong mass transit railway system, J. Adv. Transp., **32**, 285–295 (1998)
[8] N. Douglas, *Modelling Train and Passenger Capacity*, Sydney, (2012)
[9] TCRP, *Report 100: Transit Capacity and Quality of Service Manual*, Washington, (2003)
[10] Q. Zhang, B. Han, and D. Li, Modeling and simulation of passenger alighting and boarding movement in Beijing metro stations, Transp. Res. Part C Emerg. Technol., **16**, 635–649 (2008)

Existing Noise Level at Railway Stations in Malaysia

Shahiron Shahidan[1,*], *Mohamad Zulkhairi* Md Maarof[2], *Nurul Izzati* Raihan Ramzi Hannan[3], and *Noorwirdawati* Ali[4]

[1, 3, 4]Jamilus Research Centre, Faculty of Civil and Environmental Engineering, Universiti Tun Hussein Onn Malaysia.
[2]Faculty of Engineering Technology, Universiti Tun Hussein Onn Malaysia, 86400 Batu Pahat, Johor, Malaysia.

Abstract. Railway transportation known as one of the most environmental friendly transportation mode. However, the significance problems of railway transportation are noise pollution and negatively impact the wellbeing of the whole community. Unfortunately, there has been lack of public awareness about the noise level produce by the railway transportation in Malaysia. This study investigates the noise level produced by railway transportation in Malaysia specifically by Keretapi Tanah Melayu Berhad (KTMB). Methods of collecting existing noise level at railway stations in Malaysia are briefly discussed in this study. The finding indicates that the noise level produced by the railway transportation in Malaysia which is by KTMB is considered as dangerous to human being and also exceed the noise limit that has been assigned by Department of Environment Ministry of Natural Resources and Environment of Malaysia. A better noise barrier and improved material should be developed to mitigate the existing noise level produced by railway transportations in Malaysia.

1 Introduction

Unnecessary noise is the most cruel abuse of care which can be inflicted on either the sick or the well [1]. Plus, the noise pollution also reported as a primary cause of sleep deprivation [1]. Recent investigations indicate that railway noise leads to significant sleep fragmentation and cardiovascular activations during sleep and to subjective distress as well as long-term effects of prolonged exposure to noise [2,3][24]. The World Health Organization (WHO) estimates that more than 1 million healthy life years are lost every year in western Europe due to noise exposure [4]. Railway noise causes much nuisance to the residents near the railway tracks [5][27]. With the growth of cities, societies is increasingly questioning these problems and people living in the proximities of railway tracks consider noise the most serious environmental problem [6]. Research on cardiovascular disease has specifically explored the hypothesis that exposure to transportation noise increases the risk for ischemic heart disease (IHD) [7]. Night-time

[*] Corresponding author: shahiron@uthm.edu.my

freight train noise and vibration also can accelerate heart rate during sleep which may in turn be linked to cardiovascular disease (CVD) []. The difference in dose-response relationships for railway and road traffic noises between western and eastern countries was ascribed to the distance between railways and houses, the position of the balcony and the attitudes of the residents toward the source of the noise [9-11].

The events corresponding to noise sources, different from the transit noise but associated anyway with the rail such as manoeuvring, loading and unloading, truck movements, braking, squeals, whistles, arrivals and departures of trains, speakers, passengers, internal works, generators, bells and crossings were labelled as "unconventional noise sources" whilst the rail transit and the passage of freight trains were separately identified [11]. To reduce this noise, different measures could be implemented in different places such as at the source, between the source and the receiver and at the receiver [12][26]. Rail transport which is ecologically and economically and significantly accepted, faces the demand to guarantee on such values of noise emission that do not exceed the permitted noise limits monitored in tracks for different speed zones which standard, speed and high speed track [13]. Noise load of the rail transport can be reduced by different building and technical noise arrangements which can generally can be divided into active and passive [13]. Active noise arrangements are such modifications of noise sources which involve certain modifications on the railway vehicles and railway tracks while the passive noise arrangements are such measures that absorbing the noise already emitted to the surrounding such as noise barrier, noise tunnels and noise windows [13].

However, in order to mitigate the effect of noise pollution cause by railway transportation, the existing noise level produced by KTMB train services should be measured. A good knowledge of the nature and relative strengths of the various sources of noise is a fundamental requirement if railway noise is to be understood and moreover to be reduced [14]. The basic reason for investigating source mechanism is that, in many cases, due to system considerations, simple shielding of the sources is not possible [14][25].

The Malaysian Department of Environment had published a guidance on acceptable noise limits for various types of land use and human actitivities . This is to provide a guideline for the acceptance criteria for quantitative assessment of noise to define disturbance or otherwise for any new developments or projects. Table 1 represents the limiting sound level guideline for railway transportations.

Table 1. Limiting sound level for railway including transits for new development and re-alignments

Receiving Land Use Category	Day time 7.00am-10pm	Night time 10.00pm-7.00am	Lmax (Day & Night)
Noise Sensitive Areas Low Density Residential Areas	60 dBA	50 dBA	75 dBA
Suburban And Urban Residential Areas	65 dBA	60 dBA	80 dBA
Commercial, Business	70 dBA	65 dBA	80 dBA
Industrial	75 dBA	65 dBA	NA

Furthermore, for reasons of efficiency in terms of reduction, the depth of knowledge necessary for each source has to be considered, with respect to the state of the art in the subject [14][15]. One of the solutions to reduce noise is by using noise barrier. However, the function of the noise barrier is only as noise reducer but not completely block the sound annoyance [15][16]. The noise level produced by the KTMB train services must be measured to be characterized and analysed either it is in acceptance level or considered dangerous to living things.

The aim of this paper is to investigate the existing noise level produced by all KTMB railway services such as intercity train services, komuter services, Electric Train Services (ETS) and freight train.

2 Data collections method

2.1 Sound Level Meter (SLM)

The data collections being conducted using Sound Level Meter (SLM) type of ISO-TECH SLM-1352N. The SLM was located at two different locations which are as near as possible to the operating trains within permissible area along the train platform to measure the maximum noise level produced by the operating trains and also located away from the operating train where the possible barriers are exist such as trees, building and wall. Fig. 1 displays the closest distance of SLM located to the noise source of ETS at Gemas Railway Station while Fig. 2 shows the location of SLM to the closest of diesel engine noise source available on locomotive diesel train and Figure 3 represent the SLM being located as close as possible to the pantograph noise available on ETS.

Fig.1 : Location of SLM near Electric Train Service (ETS)

Fig.2 : Location of SLM near engine of KTM Intercity Diesel Train

Fig.3 : Location of SLM near pantograph of the ETS

The measurement process was conducted during the day and night to be compared with the noise regulation that had been published by the Department of Environment Ministry of Natural Resources and Environment Malaysia [17]. This is to determine either the existing noise level produce by KTMB train services is considered safe for human or not.

2.2 Selection of study location

There are four preferred study locations which are Kluang Railway Station, Segamat Railway Station, Gemas Railway Station and Kuala Lumpur Railway station. The selected study locations were based on the different criteria which involves new train and old train line. The old system single track still being used from Gemas to the south of west Malaysia which end at Johor Bahru Sentral before entering Singapore. The construction of a new double track system from Gemas to Johor Bahru Sentral is still in plan and does not started yet. It involves only old train systems which are locomotive intercity diesel train and freight train. However, a new and modern Electric Train Services (ETS) available from Gemas towards the north of west Malaysia. Thus, this noise level measurement involves both old and new train system provided by KTMB train services. Figure 4 show KTMB intercity train services with yellow crossed mark shows the locations of noise level measurements being conducted which are Kuala Lumpur Railway Station, Gemas Railway Station, Segamat Railway Station and Kluang Railway Station.

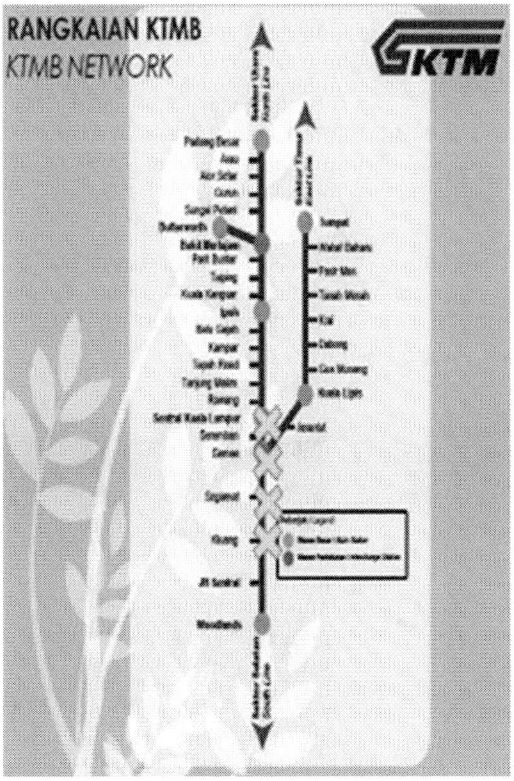

Fig.4 : Roadmap of KTMB intercity train services

3 Result and discussion

3.1 Different locations

The noise levels recorded are varies from each stations. There are several factors influence the noise level produced at the railway stations such as the design of the station and surrounding area of each particular stations. Table 1 reveals the maximum and minimum noise level recorded at each railway stations involve in this study.

Table 2. Noise level recorded at different railway stations

Location / Noise level	Kluang (dB)	Segamat (dB)	Gemas (dB)	Kuala Lumpur (dB)
Minimum	53.1	46.4	43.2	57.9
Maximum	89.6	85.4	93.9	90.4

All these stations are open structure without enclosure except for the Kuala Lumpur Railway station which has fully enclosure along the stations and platform. Table 1 shows that the highest noise level recorded is at Gemas Railway Station. This is because the

Sound Level Meter (SLM) is able to be located as close as possible to the train noise sources such as the diesel engine or the rail track.

Kuala Lumpur Railway station recorded the highest ambient noise without the presence of the trains. This is possibly due to reverberation phenomenon where it is a result of multiple reflection of sound wave. A sound wave in an enclosed or semi-enclosed environment will be broken up as it is bounced back and forth among the reflecting surfaces as in this case the noise produced by the human activities, station announcement through the loudspeaker upon train arrival and departure and also noise from building air conditioning system available at the station. As a result, it contributed to the significant increase of the ambient noise compared to the other ambient noise measured at other railway stations which typically range between 45 to 55 dBA without the presence of the train. Table 1 reveals that Kluang Railway Station recorded a high minimum noise levels compared to other railway station despite of having an open sructure of railway station. This is because the present of local food restaurant at the railway station produce a higher noise level from human activities. Fig. 6 indicates the Kuala Lumpur Railway Station with enclosed roof top structure.

Fig. 5. Kuala Lumpur Railway Station with enclosed roof top structure

3.2 Different types of trains

Different types of trains produce different levels of noises. The highest level of noise for KTM train services is recorded from the intercity diesel train and freight train. The other two types of KTM train services which are KTM Komuter and ETS produces a lower noise level compared to the old diesel train. Table 2 represents the comparison of noise level and percentage difference compared to permissible noise level according to the guideline of Department of Environment provided by Ministry of Natural Resources and Environment Malaysia . The oldest train used by the KTMB is the diesel locomotives operate for KTM intercity passenger services and KTM Kargo services. Currently, the diesel locomotives used by KTM are 19 class, 23 class, 24 class, 25 class, 26 class, 29 class and YDM4 which Y stood for metre gauge, D for diesel and M for mixed loads.

All KTM locomotive types of train use diesel and produce a high noise level during operation either during carrying passengers or loads for freight train. KTM Komuter is first introduced in 1995 as Malaysia's first electric train services. It provides all-stop local commuter rail services across KTM's electrified double track sectors primarily linking the suburbs to and from the city centre. The electric powered KTM Komuter produce low noise level compared to the old one KTM diesel locomotives. It divided into two operating sector which are northern sector and Klang Valley sector.

Table 3. Percentage difference of different types of trains

Type of trains	Freight train	Locomotive Diesel Train	KTM Komuter	Electric Train Services (ETS)
Noise levels measured	93.9	93.9	79.9	72.2
Permissible noise level	65	65	70	70
Percentage of difference %	44.46%	44.46%	14.14%	3.14%

Table 2 revealed that the intercity diesel locomotive for passenger and freight train are the only KTM train services which recorded the highest and maximum noise level that SLM able to record which is 93.9 dBA. The latest KTM train services is Electric Train Services (ETS) which produced the lowest noise level measured. It run at a maximum speed of 160km/h and the operational speed is up to 140km/h on electrified double track sectors between Gemas and Padang Besar. Table 2 also indicated ETS recorded the lowest maximum noise level compared to other trains services provided by the KTM.

3.3 Different distances of noise receiver

The sound level meter (SLM) recorded a different values of noise levels at different distances. Table 3 display the comparison of noise levels produced by the KTM train services at different distances of noise receiver. It shows that the highest noise level recorded is the closest to the noise source which is within 1m. The second highest noise level recorded is 5m from the noise level and the third highest and the lowest noise level recorded is within 10m and 15m from the noise source respectively.

Table 4. Noise levels produced at various distances

Distance	**1 m**	**5 m**	**10 m**	**15 m**
Noise level (dBA)	93.9	93.9	77.8	57.9
Permissible noise level (dBA)	60	60	60	60
Percentage of difference %	56.5	56.5	29.6	-3.626

Fig. 6. SLM located at different platform 15m away from train passing-by

Fig. 7. SLM located away from train noise with the presence of physical barriers

It shows that the higher the distances between the noise source and the receiver, the lower the noise level received by the receiver. The noise level measured decrease when the SLM device is located far away from the railway track or the noise source. Fig. 6 shows the location of the SLM which is located within 15 m away from the noise source while Figure 9 represents the SLM located within 10 m away from the train noise with the presence of physical barriers such as trees and bushes. This shows that increase in distance between noise source and the receiver or if there is any presence of barrier between noise source and the receiver will significantly reduce the noise level received by the receiver [18,19].

3.4 Day vs night

The Department of Environment Ministry of Natural Resources and Environment Malaysia had provided an Environmental Noise Limits and Controls as guidance for various types of land use and human activities. The guideline as shown in Table 5 reveals the noise acceptance criteria for quantitative assessment of noise to define disturbance or otherwise. Table 5 reveals that the Department of Environment had provided a strict noise level of acceptance criteria during the night since the nocturnal exposure to a traffic or transportation noise constitute a real probe for sleep and quality of life [20][23]. Thus, the percentage difference of exceeding noise level produced by KTM train services is higher during night time. The selection of receiving land use category is based on where the data collection had been conducted during day and night which are at Segamat and Kuala Lumpur. The reason behind the selected study locations is to compare the noise level produced at the low residential area which is at Segamat with the noise levels produced at the commercial or business area which is in Kuala Lumpur. Nocturnal noise disturbance mainly concern detrimental effects on cardiovascular reactivity and sleep architecture with the loss of the most critical sleep stages involved in physical and metal recovery [21][22].

Table 5. Limiting sound level for railway including transits for new development and re-alignments

Receiving Land Use Category	Day time 7.00am-10pm	Night time 10.00pm-7.00am
Noise Sensitive Areas Low Density Residential Areas	60 dBA	50 dBA
Commercial, Business	70 dBA	65 dBA

Table 6. Noise level recorded during day and night

Location \ Condition	Recorded Noise Level During Day Time (dBA)	Percentage Difference During Day Time (%)	Recorded Noise Level During Night Time (dBA)	Percentage Difference During Night Time (%)
Segamat	93.9	56.5	93.9	87.8
Kuala Lumpur	83.8	19.71	89.6	37.85

4 Conclusions

This study presents the result of existing noise level available at the different locations of KTMB railway stations which are Kuala Lumpur Railway Station, Gemas Railway Station, Segamat Railway Station and Kluang Railway Station. Based on the result of the study the following conclusion may be presented:

i) The old conventional diesel locomotive train produces higher noise level.
ii) Increase in distance between the noise source and the receiver will significantly reduce the noise level receive by the receiver.

iii) The presence of the barrier between the noise source and the receiver will reduce the noise level received by the receiver.
iv) The ambient noise in enclosure structure is higher compared to the open air structure.
v) The noise produced by the new and fully electrified train which is ETS produce lower noise compared to other type of KTM train Services such as diesel powered locomotive train.

The authors would like to express their deepest appreciation to the Ministry of Higher Education Malaysia, Universiti Tun Hussein Onn Malaysia for funding the research under FRGS grant vot 1576.

References

[1] R. Khaiwal, T. Singh, J.P. Tripathy, S. Mor, S. Munjal, B. Patro, and N. Panda, Assessment of noise pollution in and around a sensitive zone in North India and its non-auditory impacts, Sci. Total Environ., **566–567**, 981–987 (2016)
[2] S. Shahidan, N.M. Bunnori, N. Md Nor, and S.R. Basri, Damage severity evaluation on reinforced concrete beam by means of acoustic emission signal and intensity analysis, *2011 IEEE Symposium on Industrial Electronics and Applications*, 337–341, (2011)
[3] G.J.L. Micheli, and S. Farné, Urban railway traffic noise: Looking for the minimum cost for the whole community, Appl. Acoust., **113**, 121–131 (2016)
[4] M. Gouveia, M. Borges, J. Costa, and A.V. Carneiro, Burden of disease from hypercholesterolemia in Portugal, Rev. Port. Cardiol., **23**, 255–270 (2004)
[5] N. Ozkurt, D. Sari, A. Akdag, M. Kutukoglu, and A. Gurarslan, Modeling of noise pollution and estimated human exposure around Istanbul Atatürk Airport in Turkey, Sci. Total Environ., **482–483**, 486–492 (2014)
[6] F. Bunn, and P.H.T. Zannin, Assessment of railway noise in an urban setting, Appl. Acoust., **104**, 16–23 (2016)
[7] D. Vienneau, C. Schindler, L. Perez, N. Probst-Hensch, and M. Röösli, The relationship between transportation noise exposure and ischemic heart disease: A meta-analysis, Environ. Res., **138**, 372–380 (2015)
[8] M.G. Smith, I. Croy, M. Ögren, O. Hammar, and K.P. Waye, Freight trains, nocturnal vibration and noise, and their physiological effects during sleep, *11th Int. Congr. Noise as a Public Heal. Probl.*, 1–8 (2014)
[9] P. Lercher, B. de Greve, D. Botteldooren, and J. Rüdisser, A comparison of regional noise-annoyance-curves in alpine areas with the European standard curves, *Proc. 9th Congr. Int. Comm. Biol. Eff. noise (ICBEN 2008)*, 562–570 (2008)
[10] T. Morihara, T. Sato, and T. Yano, Comparison of dose-response relationships between railway and road traffic noises: The moderating effect of distance, J. Sound Vib., **277**, 559–565 (2004)
[11] G. Licitra, L. Fredianelli, D. Petri, and M. Angela, Science of the total environment annoyance evaluation due to overall railway noise and vibration in Pisa urban areas, Sci. Total Environ., (2015)
[12] A. Tumavičė, A. Laurinavičius, A. Jagniatinskis, and A. Vaitkus, Environmental noise mitigation measures for Lithuanian Railway Network, Transp. Res. Procedia, **14**, 2704–2713 (2016)
[13] A. Pultznerov, and L. Izvolt, Structural modifications, elements and equipment for railway noise reduction, Procedia Eng., **91**, 274–279 (2014)
[14] C. Talotte, P.E. Gautier, D.J. Thompson, and C. Hanson, Identification, modelling and reduction potential of railway noise sources: A critical survey, J. Sound Vib., **267**, 447–468 (2003)

[15] H. Halim, R. Abdullah, A. Abdullah, A. Ali, and M. J. Mohd, Effectiveness of existing noise barriers : Comparison between vegetation, concrete hollow block, and panel concrete, Procedia Environ. Sci., **30**, 217–221 (2015)
[16] S. Shahidan, H.B. Koh, A.M.S. Alansi, and L.Y. Loon, Strength development and water permeability of engineered biomass aggregate pervious concrete, MATEC Web Conf., **47**, 2–7 (2016)
[17] The Planning Guidelines for Environmental Noise Limits and Control, Vol. 1, Department of Environment Ministry of Natural and Environment, Malaysia, (2016)
[18] S. Shahidan, R. Pullin, K.M. Holford, M.B.N, and N. Nor, Quantitative evaluation of the relationship between tensile crack and shear movement in concrete beams, Adv. Mater. Res., **626**, 355–359 (2013)
[19] M. Abdul Rahim, N. M. Ibrahim, Z. Idris, Z. M. Ghazaly, S. Shahidan, N. L. Rahim, L. A. Sofri, and N. F. Isa, "Properties of Concrete with Different Percentange of the Rice Husk Ash (RHA) as Partial Cement Replacement," Mater. Sci. Forum, vol. **803**, pp. 288–293, (2014).
[20] P. Tassi, O. Rohmer, A. Bonnefond, F. Margiocchi, F. Poisson, and S. Schimchowitsch, Long term exposure to nocturnal railway noise produces chronic signs of cognitive deficits and diurnal sleepiness, J. Environ. Psychol., **33**, 45–52 (2013)
[21] D. Halperin, Environmental noise and sleep disturbances: A threat to health?, Sleep Sci., **7**, 209–212 (2014)
[22] J.M. Irwan, M.M.K. Annas, A.K. Aeslina, R.M. Asyraf, and S.K. Faisal, Cracking propagation of reinforced concrete using polyethylene terephtalate (PET) bottles as fine aggregate, Advanced Materials Research, **911**, 474-478 (2014)
[23] M.H. Wan Ibrahim, N. Jamaludin, J.M. Irwan, P.J. Ramadhansyah, and A. Suraya Hani, Compressive and flexural strength of foamed concrete containing polyolefin fibers, Adv. Mater. Res., **911**, 489–493 (2014)
[24] N.I.R. Ramzi hannan, S. Shahidan, M.Z. Maarof, and N. Ali, Physical and chemical properties of coal bottom ash (CBA) from Tanjung Bin Power Plant, IOP Conf. Ser. Mater. Sci. Eng., **160**, 12056 (2016)
[25] S. Shahidan, S.S.M. Zuki, and N. Jamaluddin, Damage grading system for severity assessment on concrete structure, Case Stud. Constr. Mater., **5**, 79–86 (2016)
[26] S. Shahidan, S.S. Mohd Zuki, M.A. Rahim, and N. Ali, Amplitude distribution of emission wave for cracking process, MATEC Web Conf., **47**, 2–8 (2016)
[27] S. Shahidan, N.I.R.R. Hannan, M.Z. Maarof, and A.S. Leman, A comprehensive review on the effectiveness of existing noise barriers commonly used in the railway industry, MATEC Web Conf., **87**, 1–8 (2017)

Investigating the Feasibility of Using Jatropha Curcas Oil (JCO) as Bio Based Rejuvenator in Reclaimed Asphalt Pavement (RAP)

Kabiru Abdullahi Ahmad[1,2,*], *Mohd* Ezree Abdullah[1], *Norhidayah* Abdul Hassan[3], *Nura* Usman[1], and *Kamarudin* Ambak[1]

[1] Faculty of Civil and Environmental Engineering, Universiti Tun Hussein Onn Malaysia, 86400 Parit Raja, Batu Pahat, Johor, Malaysia
[2] Faculty of Engineering, Department of civil engineering, Bayero University Kano, 3011 Kano State Nigeria
[3] Faculty of Civil Engineering, Universiti Teknologi Malaysia, 81310 UTM Skudai, Johor, Malaysia

Abstract. Reclaimed asphalt pavement (RAP) usage has increased recently due to the decreasing supply of liquid asphalt and concerns over adverse health effects when making use of petroleum-based/chemical recycling agents which increased the need for bio-based recycling agents designed to return the RAP binder to its original state. The objective of this study is to investigate the possibility of using jatropha curcas oil (JCO), which is a non-edible oil (do not compete with the food chain) that cannot be used for nutritional purposes due to the presence of anti-nutritional factors, such as phorbol esters, as a bio-base rejuvenating agent for aged bitumen. The physical properties and storage stability of the bio-oil, original bitumen, aged bitumen and rejuvenated bitumen were measured. The results of the investigation indicated that the bio-oil have potential to rejuvenate aged bitumen to condition that resembled the original bitumen, the use of the bio-oil show benefits from both health and environmental perspectives. Also the rejuvenated bitumen was found to be very stable in term of storage.

1 Introduction

Bitumen is the glue that binds aggregates together due to its complex, viscoelastic, rheological and non-crystalline nature (black or dark brown in colour) composed of heavy hydrocarbons with low dielectric constant (ε'), which is substantially soluble in carbon disulphide (CS2). Bitumen is a low less material with loss tangent, tan δ ($\varepsilon''/\varepsilon'$) < 0.5, exhibits both adhesive and water-proofing characteristics, and microwave permittivity (dielectric constant, ε') value ranging from 2–7 depending on the grade of the bitumen and asphaltenes content of distilled products obtained from the oil refining process. Bitumen been a product obtained from the oil refining process. Though it is used as a

* Corresponding author: kabiruaahmad@gmail.com

binder for flexible pavement all over the world, it has severe environmental impacts. Bitumen is non-hazardous at room temperature but when heated at 165–200 C, it becomes toxic. Consequences, such as environmental degradation, depleting petroleum reserves and price-spiking, led researchers to explore alternative sources of obtaining binder for pavement. In the world market the price of petroleum is constantly increasing as the consumption level spikes. The price of petroleum crude oil was $61.95 per barrel in 2008 but in 2013 it had gone up to $97.98 per barrel [1, 2]. This in turn has led to an increase in the price of asphalt. Because of the rapid increasing demand for natural petroleum products such as bitumen and its unstable price and also it's severe effect on our environment as well as the health of pavement workers. Researchers, Engineers and pavement industries need to find new products and materials[3]. They must not just rely on traditional routes, but need ingenuity to develop new ideas to use and reuse resources in different ways, and to look for new solutions to old problems of high demand for liquid asphalt is increasing rapidly and petroleum oil reserves becoming depleted. [4].Researchers and pavement industries have been seeking of new technology and approaches to reduce the use of petroleum asphalt. Using alternative materials is one of the most effective and environmentally friendly ways to solve this problem. Some positive approaches were applied, such as the recycling of asphalt pavement materials. Over the years, considerable experience has been gained; today recycling methods are also considered for heavy trafficked roads, and the recycling ratio may be close to 100%. The use of RAP have increase remarkably, this in turn has led to an increase in demand of rejuvenators. [5, 6].Bio resources researchers and pavement engineers have investigated a range of different bio oils such as waste cooking oil, waste engine oil as rejuvenating agents designed to return the RAP binder to its original state[7]. The result of the investigation indicate that by utilizing waste engine oil (WEO) and waste cooking oil (WCO) as rejuvenating agents have revive the aged bitumen to a condition that resembles 60/70 and 80/100 penetration grade bitumen respectively[8, 9]. As a result of increasing emphasis on pavement sustainability and promoting the use of alternative resources as binder and also the success achieved in using WEO and WCO powered the idea of using jatropha curcas oil (JCO) which is a non-edible oil (do not compete with the food chain) that cannot be used for nutritional purposes due to the presence of anti-nutritional factors, such as phorbol esters as a bio-base rejuvenating agent for aged bitumen. The novelty of this paper is using jatropha curcas oil (JCO) as a bio-based recycling agent in aged bitumen. The main objective of this study was to investigate the possibility of utilizing JCO as a bio-based recycling agent through physical properties test and storage stability properties of asphalt binder. The potential use of non-edible bio-oil as a bio-base RAP rejuvenator is an attractive way to increase the longevity and enhance the performance of asphalt pavements.

2 Material and experimental procedure

2.1 Asphalt binder

In this study, artificial reclaimed bitumen was obtained by subjecting bitumen grade 60/70 to long-term aging by means of the Pressure Aging Vessel equipment. After the aging process was completed, the artificial reclaimed bitumen was tested to using penetration test to determine the group of the aged bitumen. The aged bitumen based on the penetration test was found to 40/50 penetration grade. The choice to use a laboratory aged bitumen rather than a bitumen directly extracted from milled material does not limit the general validity of this study as each extracted bitumen would have its own characteristics and could be considered in any case not exhaustive of the all possible conditions, analogously to the laboratory aged one. Moreover, the laboratory aged bitumen is always reproducible for further investigation or comparative purpose.

2.2 Bio-oil

The bio-oil used in this study is hot water extracted jatropha curcas oil (JCO) which is a non-edible oil that cannot be used for nutritional purposes due to the presence of anti-nutritional factors, such as phorbol esters. Before adding bio-oil into the control asphalt, the physiochemical properties of bio-oil were investigated.

2.3 Materials preparation

Aged bitumen 40/50 penetration grade asphalt was uniformly heated first in a temperature-regulated heating mantle and continuously stirred using a high shear mixer. Then, 0%, 1%, 3%, 5% bio-oil were added into aged bitumen 40/50 penetration grade asphalt. When test temperature reached 155 °C, and then blended for 25 min by high shear mixer with a speed of 600rpm to achieve a homogeneous mixing state [4, 8, 10, 11].

2.4 Experimental methods

Firstly, physiochemical properties investigation of the bio-oil (JCO) was carried out. Then, the physical properties of both the control and the rejuvenated aged bitumen were studied. Viscosity test, ductility, storage stability and temperature susceptibility in accordance with ASTM D36, D5, and D 4402, respectively.

3. Results and discussion

3.1 Physical properties of jatropha curcas oil

In this case Table 1 shows that jatropha curcas have low iodine below 100, pour point of 4°C and cloud point of 14°C, indicating that the oil is non-dry oil can perform satisfactorily even in cold climatic conditions. A low peroxide value, as seen increases the suitability of the oil. For a long-time storage due to a low level of oxidative and lipolytic activities. Also the lower the oil viscosity, the easier it is to pump and atomize, and achieve finer results. In addition, Oils with flash point above 66°C are considered as safe oils. With a flash point of 150°C, Jatropha curcas oil can prevent auto ignition and fire hazard at high temperatures during transportation and storage.

3.2 Chemical composition of jatropha curcas oil

Table 2 shows the presence of high amount of polyunsaturated fatty acids (linoleic acid) in JCO makes it a great potential for oleo chemical application Such as surface coating industries and bio lubricant base oil applications whereas the high amount of monounsaturated fatty acid can find an application as a biodiesel feed [12].

Table 1. Physical properties of Jatropha curcas seed oil.

Property	Value
Boiling point	124°C
pH	5.2
Free fatty acids	0.0718 mg KOH g–1 oil
Specific gravity	0.9186
Flash point	230°C
Cloud point	14°C
Saponification value	155 mg KOH g–1 oil
Peroxide value	7.20 meq g–1 oil
Iodine value	51.27 g 100 g–1 oil
Pour point	4°C
Density at 27°C	0.725 g cm–
Acid value	0.1428 mg KOH g–1 oil
Viscosity	8.2 cst

Table 2. Shows the chemical compositions of jatropha curcas oil

Fatty acid composition	(%)
Palmitic (C16:0)	11.3
Linoleic (C18:2)	47.3
Oleic (C18:1)	12.8
Stearic (C18:0)	17
Palmitoliec (C16:1)	0.7
Linolenic (C18:3)	0.2
Arachidic (C20:0)	0.2
Margaric (C17:0)	0.1
Myristic (C14:0)	0.1
Saturated	21.6
Monounsaturated	45.4
Palmitic (C16:0)	11.3

3.3 Physical properties of bitumen

Penetration and softening tests are an empirical tests used to evaluate the consistency of the binder i.e. to measure of the resistance offered by a fluid to continuous deformation when it is subjected to shearing stress in accordance to with ASTM D[13-15].

The physical properties of the rejuvenated binder are illustrated in Fig. 1 and 2 with a decrease in softening points and an increase in penetration values resulted from adding different percentages (0, 1%, 3%, and 5%) of JCO by weight of the aged bitumen will soften the bitumen.

According to the results which have been presented in Fig. 1, the addition of 3% JCO by weight of bitumen into the aged bitumen penetration grade 40/50 will increase the penetration value that resembles the original 80/100 bitumen with the penetration value of 85.5. While in Fig. 2 the softening point of the aged bitumen was decrease to a value that is equivalent to original 80/100 bitumen.

DOI: 10.1051/matecconf/201710309013

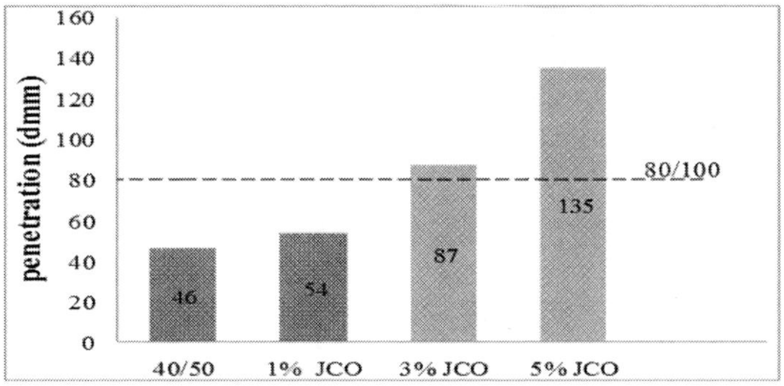

Fig. 1. Penetration results of 40/50 binder blended with JCO.

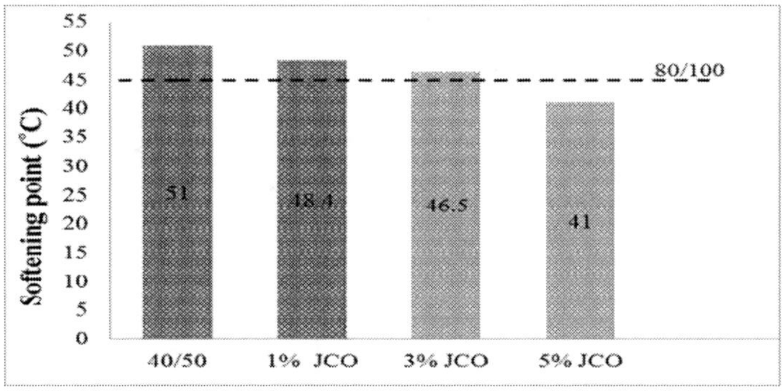

Fig. 2. Softening results of 40/50 binder blended with JCO.

3.3.1 Ductility test

Ductility test was used to evaluate the asphalt binder ductility by stretching a standard-sized briquette of asphalt binder to its breaking point in accordance to ASTM D113[16]. According to the results which have been presented in Fig. 3, it can be concluded that the ductility of the aged bitumen was observed to decrease with increase in the percentage of JCO.

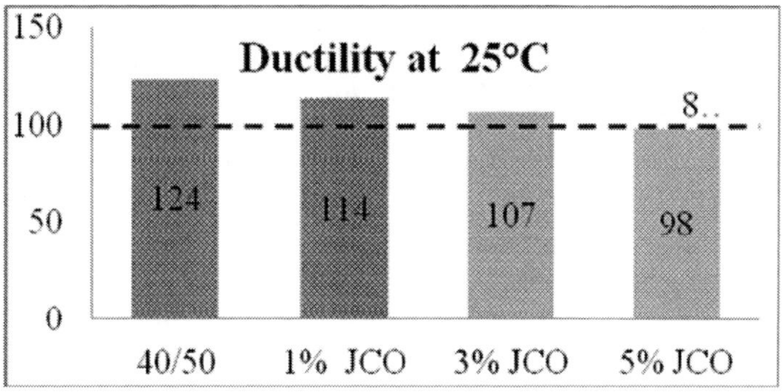

Fig. 3. Ductility results of 40/50 binder blended with JCO.

3.3.2 Storage stability

According to the results which have been presented in Fig. 4, it can be seen that the difference between softening point of the top and the bottom sections of aged bitumen

DOI: 10.1051/matecconf/201710309013

(40/50) was 0.2°C. Meanwhile, the result of aged bitumen rejuvenated with different percentages of JCO are 0.55°C (1%), 0.5°C (3%) and 0.45°C (5%). thus, It can be seen that all rejuvenated binder samples are less than 2.2°C and these could be regarded as storage stable blend.

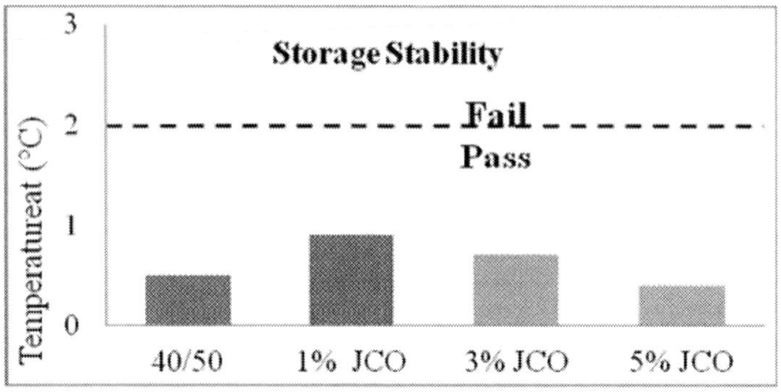

Fig. 4. Storage stability results of 40/50 binder blended with JCO.

3.3.3 Viscosity test

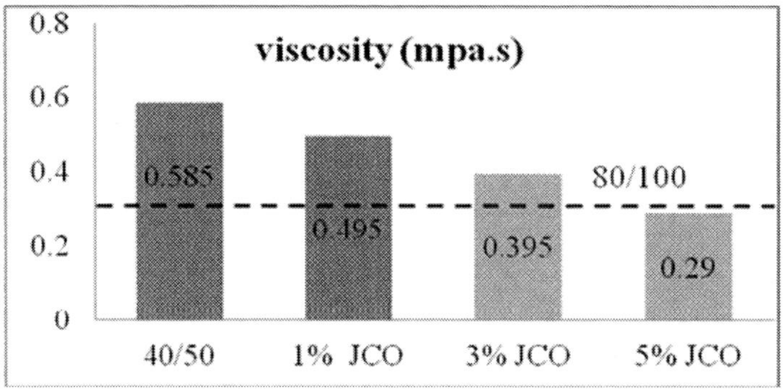

Fig. 5. Viscosity results of 40/50 binder blended with JCO.

According to (ASTM D4402) viscosity (degree of fluidity) grading, higher the grade, stiffer the bitumen[17]. Viscometer test was used to evaluate the difference in viscous behaviour's aged bitumen rejuvenated with different percentages of JCO. The temperatures selected for RV test was 135 °C which is usually applied to measure the workability of the bitumen, according to Superpave specifications. From the rotational viscosity plots shown in Fig. 3, it can be seen that the aged bitumen has the highest value. However, the viscosity value obtain on the addition of 5% JCO into the aged bitumen resembles the viscosity of original penetration 80/100 bitumen but 3% JCO shows a higher viscosity than original penetration 80/100 even though its lower than that of the aged bitumen (40/50).

3.3.4 Temperature susceptibility

Temperature susceptibility is defined as the change in the consistency parameter as a function of temperature. The temperature susceptibility of the modified asphalt binder samples was calculated in terms of penetration index (PI), using the results obtained from penetration and softening point tests according to proposed equation of shell bitumen handbook[18].

$$PI = \frac{1952 - 500\log(\text{pen at } 25°\text{C}) - 20(\text{sp})}{50\log(\text{pen at } 25°\text{C}) - \text{sp} - 120} \qquad (1)$$

The PI value of virgin bitumen reduces after aging. As illustrated in Fig. 5, PI value of aged bitumen increases with the addition of JCO. The higher the percentage added, the higher the PI value increases which indicate lower thermal susceptibility. Asphalt mixtures containing bitumen with higher PI are more resistant to low temperature cracking as well as permanent deformation (rutting)[13].

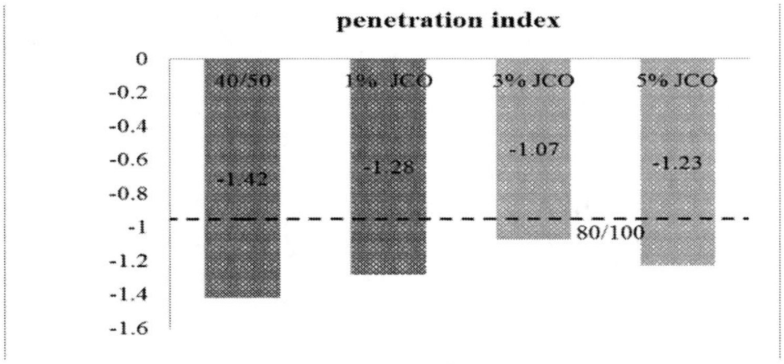

Fig. 6. Temperature susceptibility results of 40/50 binder blended with JCO

3.3.5 Dynamic shear rheometer

Fig. 7 is a plot of the G*/Sin (δ) parameters obtained from the DSR test results. These results are for the 40/50 binder that was then blended with 1%, 3% and 5% and JCO [16, 19]. The control binder, shown in blue (40/50) have an upward shift for all data points, and had an overall of PG 76, indicating the binder stiffness. The binder blend with 3% JCO performed almost identically to the original binder by reducing the PG to 64, as indicated by the G*/Sin (δ) less than the specification of 1000 Pa at 64°C. Statistically, it can be seen that the coefficient of determination (R^2 = 0.98) shows that the rejuvenated binder is 98% similar to original 80/100 bitumen. The addition of JCO into the aged bitumen led to a drop in high temperature penetration grade, which means the rutting resistance is moving toward the rejuvenated bitumen.

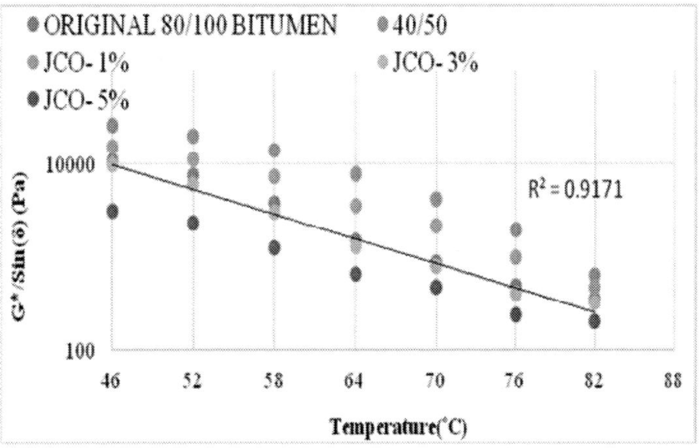

Fig. 7. DSR results of 40/50 binder blended with JCO

4 Conclusion

As mention earlier, the main purpose of this study is to investigate the feasibility of JCO as a bio-based recycling agent through physical tests and storage stability properties of bitumen. From these test results, the following can be concluded. (1) The increase in penetration grade due to aging corresponds to an increase in stiffness and rutting resistance. With subsequent percentages of oil added, that stiffening was reduced and ultimately the bitumen was softened.(2) Results showed that 3 - 4% JCO has the ability to counteract the stiffening of the 40/50 aged bitumen to a value that is equivalent to original 80/100 bitumen. (2) The increase in softening point and decrease in penetration of the aged bitumen shows the effectiveness of JCO in reducing age-hardening. (4) PI value increases which indicate lower thermal susceptibility. Asphalt mixtures containing asphalt binder with higher PI are more resistant to low temperature cracking as well as permanent deformation (rutting) (5) Storage stability tests indicated that the rejuvenated bitumen is very stable.(6) Statistically, it can be seen that the coefficient of determination ($R^2 = 0.91$) shows that the rejuvenated binder is similar to original 80/100 bitumen. Above all the result shows the potential of JCO as a bio-base rejuvenator for recycling bituminous mixtures in order to reduce the construction and maintenance cost of flexible pavement.

Special thanks to Office for Research, Innovation, Commercialization and Consultancy Management (ORICC) and centre for graduate studies Universiti Tun Hussein Onn Malaysia (UTHM) for funding this project.

References

[1] Bioenergy, Retrieved 2015. http://www.avellobioenergy.com /en/products/ bioasphalt_binder/

[2] M.M.A. Aziz, M.T. Rahman, M.R. Hainin, W.A.W.A. Bakar, An overview on alternative binders for flexible pavement, Construction and Building Materials, **84**, 315-319 (2015)

[3] N. Usman, M.I. Mohd Masirin, A.K. Abdullahi, W. Anwaruddin Ahmed, Reinforcement of asphalt concrete mixture using recycle polyethylene terephthalate fibre, Indian Journal of Science and Technology, **9**(46), (2016)

[4] J. Peralta, M.A. Raouf, S. Tang, R.C. Williams, Bio-renewable asphalt modifiers and asphalt substitutes, Sustainable Bioenergy and Bioproducts, Springer, 89-115 (2012)

[5] Z. You, J. Mills-Beale, E. Fini, S.W. Goh, B. Colbert, Evaluation of low-temperature binder properties of warm-mix asphalt, extracted and recovered RAP and RAS, and Bioasphalt,. Journal of Materials in Civil Engineering, (2011)

[6] M. Zaumanis, R.B. Mallick, R. Frank, 100% recycled hot mix asphalt: A review and analysis. Resources, Conservation and Recycling, **92**, 230-245 (2014)

[7] M. Zaumanis, R.B. Mallick, R. Frank, Use of rejuvenators for production of sustainable high content RAP hot mix asphalt, *XVIII International Baltic Road Conference*, Vilnius, (2013)

[8] M. Zargar, E. Ahmadinia, H. Asli, M. R. Karim, Investigation of the possibility of using waste cooking oil as a rejuvenating agent for aged bitumen, Journal of Hazardous Materials, **233**, 254-258 (2012)

[9] H. Asli, E. Ahmadinia, M. Zargar, M.R. Karim, Investigation on physical properties of waste cooking oil–Rejuvenated bitumen binder, Construction and Building Materials, **37**, 398-405 (2012)

[10] E.H. Fini, E.W. Kalberer, A. Shahbazi, M. Basti, Z. You, H. Ozer, Q. Aurangzeb, Chemical characterization of biobinder from swine manure: Sustainable modifier for asphalt binder, Journal of Materials in Civil Engineering, **23**(11), 1506-1513 (2011)
[11] H. Wen, S. Bhusal, B. Wen, Laboratory evaluation of waste cooking oil-based bioasphalt as an alternative binder for hot mix asphalt, Journal of Materials in Civil Engineering, **5**(10), 1432-1437 (2012)
[12] K.A. Ahmad, N. Abdul Hassan, K. Ambak, A. Musbah, N. Usman, S.K. Abu Bakar, Extraction techniques and industrial applications of jatropha curcas, Jurnal Teknologi, **78**, 53-69 (2016)
[13] M.E. Abdullah, K. Ahmad Zamhari, N. Nayan, M. Hermadi, M.R. Hainin, Storage stability and physical properties of asphalt modified with nanoclay and warm asphalt additives, (2011)
[14] ASTM-36, *Standard Test Method for Softening Point of Bitumen (Ring-and-Ball Apparatus)*, West Conshohocken, (2009)
[15] ASTM-D36, *Standard Test Method for Softening Point of Bitumen (Ring-and-Ball Apparatus)*, American Society for Testing and Materials, (2009)
[16] M.E. Abdullah, K.A. Zamhari, M.R. Hainin, E.A. Oluwasola, N.A. Hassan, N.I.M. Yusoff, Engineering properties of asphalt binders containing nanoclay and chemical warm-mix asphalt additives, Construction and Building Materials, **112**, 232-240 (2016)
[17] ASTM-D4402, *Standard Test Method for Viscosity Determinations of Asphalts at Elevated Temperatures Using a Rotational Viscometer*, American Society for Testing and Materials, (2002)
[18] J. Read, D. Whiteoak, *The shell bitumen Handbook*, Thomas Telford, (2003)
[19] ASTM-D7175, *Standard Test Method for Determining the Rheological Properties of Asphalt Binder Using a Dynamic Shear Rheometer (DSR)*, American Society for Testing and Materials, (2005)

An Exploration on the Developing of Expert System in Transport Engineering

Shuhairy Norhisham[1,*], *Amiruddin* Ismail[2], *Muhamad Nazri* Borhan[2], and *Herda Yati* Katman[1]

[1]Department of Civil Engineering, College of Engineering, Universiti Tenaga Nasional
[2]Department of Civil and Structural Engineering, Faculty Engineering and Built Environment, Universiti Kebangsaan Malaysia

Abstract. This paper aims to determine the expert system developed from past research observations. This overview of expert system indicates the use of system which have slight affects on the results. Planning of research design enables a researcher to decide on what system it should be conducted. Two main components categorized for this paper are knowledge base obtained from previous study and a comparative study between the expert system to get a summary of the best inference engine, knowledge base, benefit and understanding.

1 Introduction

The expert system is subjected to smart computers that basically solve any intriguing difficulty and the transport engineering area is matured for development. Generally, an expert system is about incorporating human expertise into computer program to solve any problems requiring a human expert [14]. In order to use the expert system, planning and operational should be firstly identified. The purpose for planning is to determine the margins of share for specific transport modes while the operational purpose is to provide the survey with parameter values from the existing state or for making a short-term forecast for these parameters [12]

To identify the characteristics for the expert system, [15] conducted a survey on the classification by function in four very broad groups. They were traffic management control; traffic impact and safety; highway design and planning; and highway management. All these boards were surveyed according to their functions.

Consequently, the ability of expert system to access regular updated regulations from commercial programs was one requirement for the expert system [8].

In other hand, developing expert system can assist and advice engineers in dealing with traffic safety problem. Hence, the gap between traffic safety and people (decision makers, engineers and students) would be filled [4].

* Corresponding author: shuhairy@uniten.edu.my

2 Expert system evaluation proses

Developing an expert system should combine with the evaluation process. The evaluation process employed for expert systems differs substantially from that used for traditional software projects and is much more problematic, mainly because of the nature of problems being solved by the expert system [14].

Verification, validation and evaluation are three processes that should be measured. Verification is the consistency, completeness and correctness of the software. Hence for software verification it totally determines if the system was built according to specifications. While validation is concerned with the quality of the conclusions. Thus it focuses on the effectiveness of system. A commonly borrowed phraseology to describe the difference between verification and validation is that verification is concerned with building the system right, whereas validation is concerned with building the right system.

Meanwhile evaluation process is concerned with user issues such as user acceptance and system usefulness. Therefore, with respect to all of above, validation is the cornerstone of system evaluation process and is the most difficult component to accomplish.

3 Expert system definition

Expert system is a computer system that emulates or acts in all respects, with a human expert decision making capabilities. Generally, the system contains knowledge base of accumulated experience and a set of rules for applying the knowledge base to each particular situation that is described to the program.

[12] defined the expert system as the creation of basic idea on the modal spilt against basis parameter for transport system demand and supply in the city. Meanwhile [14] described the expert system as a computer program that react to solve problems and require a human expert. Besides that, an expert system can solve problem requiring judgement and experience if the problem is complex or have inadequate information input.

On the other hand, [10] named the expert system as an advisory system which contains advice for user and assist problem solving that are normally solved by human expert. According to [4], expert system is also known as computer based programs that are developed to mimic and imitate problem solving processes along with the reasoning of human expert in different knowledge fields. Thus, this expert system slightly assist human in solving problem which require extensive knowledge or a long time.

3.1 Processes of building expert system

Expert system is currently used as support methods that are developed for best result. This paper is currently searching for process of building expert system knowledge, they include:

i) The knowledge engineer establishes a dialog with the human expert to elicit knowledge.
ii) The knowledge engineer codes the knowledge explicitly in the knowledge base.
iii) The expert evaluates the expert system and gives a critique to the knowledge engineer.

4 Application of expert system in transport engineering

4.1 Automated fare

The automated fare collection is introduced by using smart card technology and electronic payment system. [11] explored the essential to integrate effective information technology management in public transport strategic information technology planning. It involved how essential it is to integrate effective information technology management with the public transport strategic information technology. The reason why strategic information technology planning is needed is because the system is a core instrument for communication within organizations and to external stakeholders and customers. Basically, the function of automated fares is to ease the seamlessness of multi-modal passengers to travel. The automated fare collection could address many key issues especially with regards to fare collection faced by the public transport agency [11].

4.1.1 Advantage

With all the motivation and opportunities that drove the multi-modal automated fare collection solution, several rationales benefit from the system including [11]:

1. Reduce maintenance and operational costs
2. Increase transaction throughput for travelers
3. Reduce fraudulence activities. Innovations for ticketing and fare options
4. Flexible fare pricing

4.2 Geographic information system for transport (GIS-T)

[16] developed GIS as a computer-based system which could manage objects and their information with their spatial characteristics. Basically, GSI-T is a geographical information system for transport, created by combining the advantages of both GISs and TIS to store geographic information, visualize, plan, and analyze transport information. Then, [5] introduced GIS-T software as interconnected systems that consist of hardware, software, data, people organization and institutional arrangements for collecting, storing, analyzing and disseminating information about areas on earth that are used, influenced or affected by transport activity

The study aimed to evaluate bus delays based on some factors such as traffic congestion, passenger's waiting time and the bus line interaction [16].

In this study, the GIS-T provided a database of detailed information to support simulations. A digital map was coordinated to the system as a background map. The background map contained information on block, road, railway, water area and green area. Then, all the bus lines and bus stations were located on the digital map with the coordinates.

4.2.1 Advantage

The software is simple and has a convenient interface. It is also a direct representation of the network that is constructed on supported background pictures. Then, by providing the advanced signal control mechanism it makes simulations of crossroads intersection closer to reality. Other than that, the software usually customized results so that the user can control data collection speed by adjusting the sampling time. Besides that the software can seamless interface with the network constructed in VISUM would be imported into VISSIM to simulate. Meanwhile the simulated VISSIM network must be transferred into VISUM to get a detailed graphical representation like cobweb graphs. In addition, the urban traffic simulation accuracy is well improved because of the Innovative pedestrian simulation.

4.3 System Dynamics (SD)

System Dynamics (SD) is a methodology of wide applicability and pioneered the use of system concepts and computer simulation for complex problems analysis. This system has become a modelling style used by various disciplines.

This system provides a framework structure through a large easily accomodated scale system, Other than that, the socio-economic and demographic forecasts are obtained by using separate modelling techniques. Then, SD models also utilize easily available data which is mainly required to initialize the model for simulation run. Besides that, using SD for transport modelling, the way in which traditional submodel are linked to each other and to other subsystems, occur simultaneously rather than step by step. In addition, results of SD transport model are achieved through the dynamic, casual, feedback interactions of the model structural components.

4.4 Case-Based Reasoning/ Rule Based Reasoning

Case-Based Reasoning can be defined as the program that acts as solution to unsolved problems based on pre-existing solutions of a similar nature.

On the other hand, CBR is also known as brilliant techniques to solve human problem and new problems will be solved by recalling and adapting the solutions of similar past problems. CBR is also one of the famous and active research area since many companies have adopted the program.

Rule based reasoning is used to store and manipulate knowledge to interpret useful information. RBC is also known as simple patterns and an inference engine that searches for patterns in the rules that match patterns in the data.

- Function: CBR functions as Crew Scheduling Planning According, Crew Scheduling, Crew Attendance Management and Weekends Shift [9].
- Advantage: CBR and RBR give some benefits that include representing the rules neutrality that represent method with high comprehensibility level. Then, modularity reflects a rule with discrete knowledge unit that can be inserted into or removed from the knowledge base. Other than that, it provides compact representation of general knowledge [13].

4.5 Radioactive Materials Prototype

The Radioactive Materials Prototype is initially built to demonstrate the feasibility of building an expert system. The system is built according to module for transporting and packaging radioactive materials. Then the module is developed with respect to hazardous chemicals.

In order to build the system, the feasibility stage included in few steps. Firstly, the commercial software that was related to regulation access was analyzed. Secondly, the knowledge acquisition was obtained. Finally, the expert system prototype was developed [8].

The main function for this system is to ensure straight forward, consistent and error-free application towards regulations for hazardous materials transport [8].

4.6 Integrated Transportation Systems

Multilayer and Multilevel Space of Integration is a natural approach for transport system. ITS is an integration of a number of technologies consisting of information processing, communication, control and an electronic intelligent link between travelers, vehicles and infrastructure. This clearly shows that ITS offers problem solving for transport problems [1].

From the knowledge base (KB) aspect, the system application includes traffic management, supervision and control system that were presented as motivated by several important reasons. The traffic control problems were normally well-structured and formulated to be solved by purely algorithmic techniques [1].

ITS provide KB system with problem solving methods, mechanisms working with some degree of multiple tasking and problem domain tools for automatic generation and customization of dedicated knowledge [1].

4.6.1 Advantage

The system provides many benefits through the traditional system that result in provision of area-wide transport demand management. Other than that, it also gives area wide synchronized response to the full range of mobility needs. Development on the efficient multi-layered technical provide opportunity to multi modal planning [1].

4.7 E-SUTRA

E-SUTRA is an engineering software that is developed as a problem solving expert system. The expert system shell was developed for young and inexpert transport engineers and planners who are currently involved in transport demand management (TDM) strategy planning and implementation. As for the knowledge base model, the nine class module is made up of each deal with specific strategy [10].

This system functions as a decision support system that represent a teaching tool for transport engineers, planners, private developers and also government officials. The system has reliable advice that can efficiently decide on the Transport Demand Management strategic planning and implementation process. In other hand, this system provided knowledge, skills and experiences; because they can undertake a complex scenario analysis [10].

The expert system is user friendly and it does not require the user to have experience or knowledge on programming. This is because the user can smoothly navigate through nine different input screens before the system outputs its summary. On the other hand, the system help young and fresh engineers and planner in their decision making. Moreover, the system is very costly and time saving [10].

4.8 Intelligent Electronic Devices (IEDs)

[2] proposed IEC 61850-9-2 as a developer system that operates protection IEDs performance in terms of dependability, security, and operating speed. The system is an internationally recognized non-proprietary standard for power system substation communication networks and automation. The system currently detects the process bus with a reliability of power system protection. IEDs real protection was used in hardware in-the-loop set-up with the RTDS. It is important to use the actual process bus equipment when carrying out performance analysis to investigate and record the equipment performance [2].

The advantage of implementing this system is a reduction in the overall cost of substation protection, automation and control system because the cost of numerous copper conductor is reduced. Moreover, the system is safer to use and the IEC 61850-9-2 permits wide availability of measurements to individual protection IEDs on the digital communication bus and system reconfiguration can be done in less time with a minimum cost as compared to the conventional hardwired method. To summarize, the system benefits substations safety, utilize less copper, facilitate maintenance and easy reconfiguration, etc. while giving the same performance in terms of operational speed, security, and dependability available from conventional analogue inputs IEDs [2].

4.9 CALMSYS

CALMSYS is a web-based traffic calming expert system developed to assist engineers improve safety all road-users and residents in residential streets safety [4].

The system is divided into three sections namely Knowledge Acquisition, Selection of Building Tool and Knowledge Based. For Knowledge Acquisition, the expert system is obtained from human expertise in traffic calming and written sources of knowledge. Meanwhile for the Selection of Building tool, Microsoft Visual Basic.NET is widely used for developing expert system and the software is easy to learn with flexible language that enables coding and creates GUIs (graphical user inter- faces). For Knowledge Based, the system works independently but the collaboration between traffic calming strategies modules existed [4].

4.9.1 Advantage

The system is important because it classifies the strategies and solutions for traffic calming subject. The strategies are categorized in physical, nonphysical and combined categories. In addition, the system can assist engineers in finding proper traffic calming strategies for traffic safety problems. Moreover, by this system can help in better and more succeful handling with a computerized system rather than with human experts due to the presence of many data received [4].

5 Discussion and conclusion

In conclusion, the expert system has its own benefits: increased availability, reduced cost, reduced danger, performance, multiple expertise, increased reliability, explanation, fast response, steady, unemotional, and complete responses at all times, intelligent tutor and intelligent database.

The reviewed expert system in this study express the importance of applying modern technology in research design. In analyzing transport engineering, the most appropriate system must be introduced during the research study. The system can affect the achieved results. Several systems may be used but not all the system are appropriate for the research study. Therefore, in conclusion computer based simulation can provide decision-makers with strategies for intermodal terminals development.

References

[1] A. Adamski, Intelligent integrated transportation systems, *The 9th Meeting of the EURO WG Intermodality, Sustainability and ITS*, Bari, 565–570 (2002)

[2] A.C. Adewole, R. Tzoneva, Impact of IEC 61850-9-2 standard-based process bus on the operating performance of protection IEDS : Comparative study, *The 19th World Congress The International Federation of Automatic Control*, Cape Town, South Africa, 2245–2252 (2014)

[3] P. Chandra, P. Soni, R.K. Keshari, RFID-based ticketing for public transport system : Perspective megacity, Int. J. of Advance Research in Computer Science and Management Studies, **7782**, 187–191 (2014)

[4] A. Falamarzi, M.N. Borhan, R.A.O.K. Rahmat, Developing a web-based advisory expert system for implementing traffic calming strategies,The Scientific World J., (2014)

[5] D. R. Fletcher, *Geographic Information Systems for Transportation: A Look Forward*, Transportation in the New Millenium: State of the Art and Future Directions, (2000)

[6] V. Graudina, J. Grundspenkis, Technologies and multi-agent system architectures for transportation and logistics support: An overview, In *International Conference on Computer Systems and Technologies*, 1–6 (2005)

[7] E.J. Heikkila, T. Kim, J.M. II, *Incorporating Expert Geographic Information Systems Into Urban Land Use and Transportation Planning Models*, In Future Directions For EGIS: Application to Land Use and Transportation Planning, (1989)

[8] J.J. Ferrada, D. Micbelhaugh, R. Rawl, development of an expert system for transportation of hazardous and radioactive materials, *International Nuclear and Hazardous Waste Management*, (1994)

[9] T. Liu, J. Ma, W. Guan, Y. Song, P. Fu, Design and implementation of bus crew scheduling system using integrated case-based and rule-based reasoning, *2012 Fifth International Joint Conference on Computational Sciences and Optimization*, 475–479 (2012)

[10] R. Mansyur, R.A.O.K. Rahmat, A. Ismail, M.R. Kabit, Decision suport system for transport demand mangement : Object oriented approach using Kappa Pc 2 . 4 expert system shell, ARPN J. of Engineering and Applied Sciences, **6**(2), 73–81 (2012)

[11] N.M. Noor, A multimodal automated fare collection solution for facilitating strategic information technology planning of public transportation in Malaysia. *International Symposium on Information Technology*, 1–8, (2008)

[12] M.M. Popović, J.J. Jović, Concept of expert system for modal split in transportation planning, Yugoslav J. of Operations Research, **16**(1), 107–124 (2006)

[13] J. Prentzas, I. Hatzilygeroudis, Categorizing approaches combining rule-based and case-based reasoning, Expert Systems **24.2**, 1–34 (2007)
[14] G.S. Spring, Critical review of expert system validation in transportation, J. of the Transportation Research Board, (970106), 8–12 (1997)
[15] J.A. Wentworth, Expert systems in transportation, 24–29 (1993)
[16] L. Zhang, X. Ren, A two-factor evaluation of bus delays based on GIS-T database and simulation, (2010)
[17] M. Zouari, C. Diop, E. Exposito, Multilevel and coordinated self-management in autonomic systems based on service bus, J. of Universal Computer Science, **20**(3), 431–460 (2014)

Regional Financial CGE Model for Infrastructure Investment Policy

Atsushi Koike[1,*], and *Naoki* Segawa[1]

[1]Department of Civil Engineering, Kobe University, 1-1 Rokkodai Nadaku Kobe, Japan

Abstract. In this paper, regional financial computable general equilibrium (RFCGE) model was developed. The RFCGE model is a useful tool in the evaluation of regional fiscal policies. So far, computable general equilibrium (CGE) models have often been used to evaluate regional infrastructure investment policies. However, conventional CGE models do not consider financial assets, such as deposits, equities, and government bonds and decision-making mechanism of government. Therefore, applying CGE models might overestimate the benefit incidence of policies on the rural society. To confirm this theory, the RFCGE model was applied to regional infrastructure investment policy.

1 Introduction

Over the past decade, various tools have been used to evaluate local fiscal policies, one of the most powerful tools being the computable general equilibrium (CGE) model[1)]. CGE models make it possible to explicitly estimate the effects of policies. However, CGE models have a theoretical constraint in that they do not consider the financial economy. This means that CGE models do not consider the effects of financial assets, such as deposits, equities, and government bonds. For example, households use their assets not only for buying goods but also for buying equities, government bonds, and making deposits. In the real world, these financial assets may spill over from a local region to other regions through the financial market. Buying government bonds leads to spill-over effects. Further, if local firms borrow loans from banks in other regions, the payment of interest by the former leads to an outflow of assets. This means that the net wealth (assets of households) may spill over to other regions. Therefore, the benefit incidence on a local region owing to local fiscal policies might be overestimated by CGE models. Among policies, an infrastructure investment policy may have a major influence on the benefit incidence, because it involves a large amount of money.

In this paper, a regional financial computable general equilibrium (RFCGE) model was developed to include the financial economy in the CGE model. The RFCGE model is based on a financial computable general equilibrium (FCGE) model, which was developed by Lance Taylor (1990)[2)] and Mark Thissen (2000)[3)]. The FCGE model not only considers the real economy but also the financial economy. So far, the FCGE model has been used to

[*] Corresponding author: koike@lion.kobe-u.ac.jp

evaluate national policy. In this paper, the FCGE model has been applied to evaluate the effect of policy on a local region. Therefore, some endogenous variables in the FCGE model have been changed to exogenous variables. Further, the RFCGE model considers inter-regional trade and asset movement between regions. In this paper, infrastructure investment policy was analyzed using the RFCGE model.

2 Model

The RFCGE model developed in this study has been presented in this section. The model consists of four economic agents—firms, households, local banks, and other regions, and six markets—the goods market, labor market, loan market, stock market, government bond market, and deposit market. The goods and labor markets are real markets. The stock market, government bond market, and deposit market are financial markets.

2.1 Schematic

2.1.1 Real side of the model

Fig. 1 shows a schematic of the real side of the model.

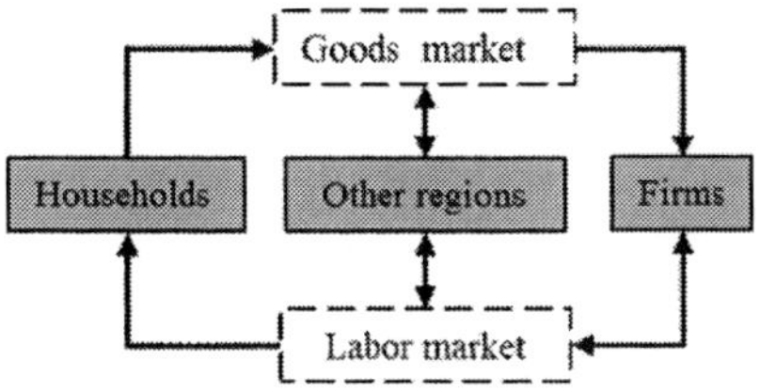

Fig. 1. Real-side schematic

2.1.2 Financial side of the model

Fig. 1 shows a schematic of the financial side of the model.

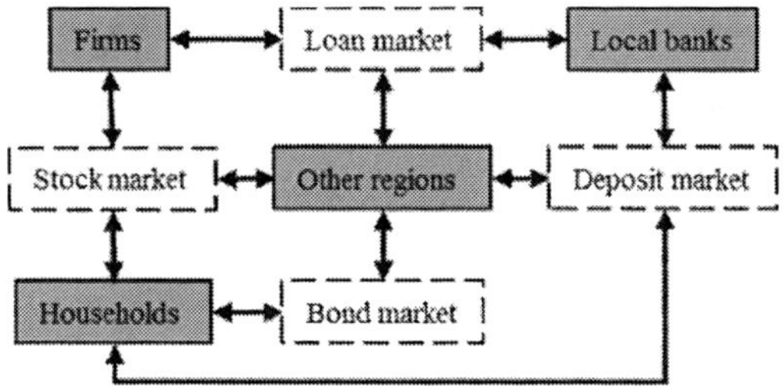

Fig. 2. Financial-side schematic

2.2 Economic agents

2.2.1 Firms

Three firms have been considered in the model: a service-sector firm, industry-sector firm, and an agricultural-sector firm. Firms receive income from the sale of goods, purchase intermediate inputs, pay direct taxes, make wage payments and investments, distribute dividends to households, and demand bank loans. The income that remains with firms after meeting their expenses is saved. Firms finance their expenditures through retained equities and borrowings from local banks as well as banks from other regions. Production is determined by the Cobb-Douglas production function with two inputs—labor and capital. In addition, production is assumed to be carried out with constant technological growth.

2.2.3 Households

The demand for goods by households is a result of the maximization of utility, subject to a budget constraint. The utility function of households is represented by equation (1). The budget constraint is determined by activity of households in the real economy. Households receive labor income from firms and firms in other regions and profit income (dividends) from firms. The remaining income is saved post the deduction of direct taxes.

Net wealth is held in the form of currencies, deposits, government bonds, equities, and savings (for the current period). Households divide their net wealth again to currencies, deposits, government bonds, and equities in the current period. Moreover, deposits and equities are classified on the basis of region as local and other regions. Equations (2)–(15) represent the abovementioned flow of assets.

$$U = C_1^{\alpha_1^{Dem}} C_2^{\alpha_2^{Dem}} C_3^{\alpha_3^{Dem}} \tag{1}$$

$$NW_h(t) = CURR_h(t-1) + DEP_h(t-1) + P_z(t)Z_h(t-1) + GB_h(t-1) + S_h(t) \tag{2}$$

U : Utility, C_i (i = 1 - 3): Consumption, α_i^{Dem} (i = 1 - 3): Share parameter, NW_h: Net wealth of households, $CURR_h$: Currencies owed by households, DEP_h : Household deposits, P_z : Equity price, Z_h : Household equity demand (Unit : Quantity), GB_h : Government bonds owned by households, S_h : Household savings, t : Current period, $t-1$: Previous period

$$q_h = A_1^h\left(i_d / \bar{i}_{dh}\right)^{\rho_h - 1} + A_2^h\left(r / \bar{r}_h\right)^{\rho_h - 1} + A_3^h\left(i_g / \bar{i}_{gh}\right)^{\rho_h - 1} + A_4^h \tag{3}$$

$$\phi_1^h = A_1^h \frac{\left(i_d / \bar{i}_{dh}\right)^{\rho_h - 1}}{q_h} \tag{4}$$

$$\phi_2^h = A_2^h \frac{\left(r / \bar{r}_h\right)^{\rho_h - 1}}{q_h} \tag{5}$$

$$\phi_3^h = A_3^h \frac{\left(i_g / \bar{i}_{gh}\right)^{\rho_h - 1}}{q_h} \tag{6}$$

$$\phi_4^h = A_4^h \frac{1}{q_h} \tag{7}$$

$$DEP_h = \phi_1^h NW_h(t) \tag{8}$$

$$ZZ_h = \phi_2^h NW_h(t) \tag{9}$$

$$GB_h = \phi_3^h NW_h(t), \tag{10}$$

$$CURR_h = \phi_4^h NW_h(t) \tag{11}$$

q_h : Mean return of portfolio, A_i^h : Distribution parameter in asset demand functions (I = 1–-4), i_d: Deposit interest rate, $\bar{i}_{dh}$: Deposit interest rate (standard), r_h: Equity interest rate, $\bar{r}_h$: Equity interest rate (standard), i_g: Government interest rate, $\bar{i}_{gh}$: Government interest rate (standard), σ_h: Elasticity of substitution, ϕ_i^h : Distribution ratio (i = 1–- 4).

$$DEP_h = DEP_h^L + DEP_h^O \tag{12}$$

$$DEP_h^L = \Delta^{DEP} DEP_h \tag{13}$$

$$ZZ_h = ZZ_h^L + ZZ_h^O \tag{14}$$

$$ZZ_h^L = \Delta^{ZZ} ZZ_h \tag{15}$$

DEP_h^L : Local region deposits, DEP_h^O : Other region deposits, Δ^{DEP} : Ratio of deposit, ZZ_h^L : Equity from local firms, ZZ_h^O : Equity from firms in other regions, Δ^{ZZ} : Equity ratio.

2.2.4 Local banks

Operating expenses such as labor costs are not taken into account. Moreover, local banks do not buy or sell goods. Funds of local banks come from deposits from households and remittances from other regions. Local banks provide loans to local firms. Because other regions require banks to hold funds in the form of reserves, some funds are held for that purpose.

2.2.5 Other regions

Other regions include consumers, producers, governments, and central banks. They are involved in the exchange of money with households and local firms through various markets.

2.3 Price and interest rate

The producer price, consumer price, and intermediate price are different for each of the three firms, while the equity price and wage rate are the same for all three firms. Thus, there are eleven different prices, all of them being endogenous variables.

There are four interest rates—deposit interest rate, equity interest rate, government bond interest rate, and bank loan interest rate. The bank loan interest rate is an endogenous variable.

2.4 Markets

Endogenous variables are determined by market equilibria. Each market equilibrium has been explained in this section.

2.4.1 Goods market

The goods market equilibrium is expressed by equation (16). Producer prices are determined from the goods market equilibrium equation. Consumer prices and intermediate prices are determined from producer prices. Consumer prices take into account direct taxes.

$$X_i = \sum_{j=1}^{3} X_{ji}^L + C_i + IN_i^L + IM_i, (i = 1,2,3) \tag{16}$$

X_i : Gross production, X_{ji}^L : Intermediate demand, IN_i^L : Infrastructure investment in a local region, IM_i : Net exports.

2.4.2 Labor market

The labor market equilibrium is expressed by equation (17). The wage rate is determined from the labor market equilibrium equation.

$$\sum_{i=1}^{3} L_i = L \tag{17}$$

L_i : Labor demand by firm i (i = 1 - 3), L: Labor supply.

2.4.3 Equity market

Equity price is determined using equation (18).

$$P_z = {ZZ_h}/{(Z_1 + Z_2 + Z_3)} \tag{18}$$

P_z : Equity price, ZZ_h : Household equity demand (Unit: Monetary), Z_i : Firm i's equity supply (Unit: Quantity).

2.4.4 Loan market

Bank loan interest is determined using equation (19). This equation indicates that the assets under management of local banks are equal to the loans demanded by firms.

$$QL_B = \sum_{i=1}^{3} BLN_i^L \tag{19}$$

QL_B : Funds of local banks BLN_i^L : Bank loan demand of firm i in a local region.

3 Social accounting matrix

The numbers in a CGE model must be consistent with the national income and the input-output accounting equations the model contains. A social accounting matrix (SAM) represents the flow of all economic transactions that take place within an economy. An SAM for a local region has been developed in this paper. The SAM is based on the input-output table for Nara prefecture[4)]. The exogenous variables in the model are determined by the numbers in the matrix, which have been provided in the appendix on the last page.

4 Comparative statics

This section comprises an analysis of the effect of infrastructure investment policy on a local region using comparative statics. Fig. 3 shows the result of comparative statics, which indicates that infrastructure investment policy increases the net wealth of households.

Moreover, the productivity of firms in the subsequent period improves, because the amount of funds held by the firms depends on the deposits and equity held by households. However, these comparative statics do not take into account the financial effect. The negative effect on the local region, taking the financial market into consideration, has been shown in the next subsection.

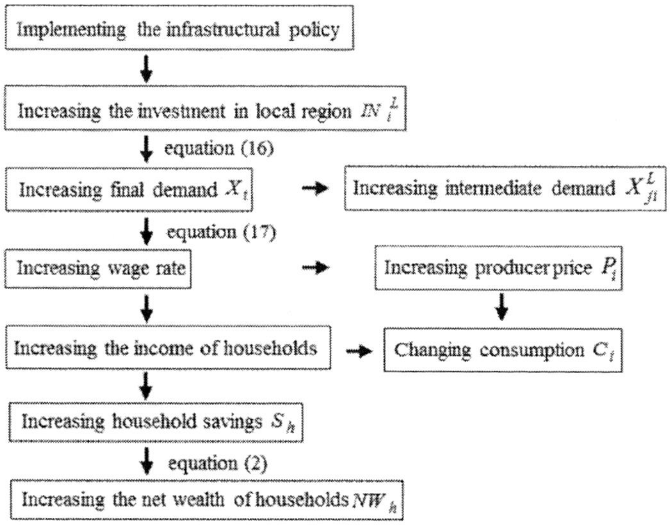

Fig. 3. Flowchart of comparative statics

4.1 Spill-over effects

In this subsection, the effects of the investment policy have been analyzed from the viewpoint of the financial economy. Fig. 3 shows that the investment policy results in increasing net wealth for households. However, there are some negative effects of the policy on the local region. These effects have been analyzed from two different viewpoints as follows:

The first viewpoint involves the outflow of households' assets. Households categorize their net wealth as deposits, government bonds, equities, and currencies. Further, deposits and equities are categorized into those from the local region and other regions. Further, funds raised by sale of government bonds to government bonds spills over to the government. The government redistributes these funds to other regions, which means that some of the increased net wealth spills over to other regions.

The second viewpoint involves the outflow of firm assets. Local firms borrow funds from local banks or banks from other regions, for which they pay interest to the banks. If the firms choose to borrow from banks from other regions, assets from the local region will spill over to other regions in the form of interest payments.

As discussed above, asset outflow can occur in two different ways. Thus, it can be observed that benefit incidence on the local region may be overestimated by CGE models owing to spill-over effects. High-cost policies, such as an infrastructure investment policy, in particular, result in such negative effects.

5 Calibration

In this section, the effect of an infrastructure investment policy on a local region has been calibrated using the RFCGE model. In this case study, we assumed a 20% increase in infrastructure investment in the local region. Because of space limitations, only a few results have been shown. Table 1 shows that net wealth increased by 100 billion yen after the policy was implemented. However, the total amount of financial assets in the other region ($DEP_h^O + ZZ_h^O + GB_h$) increased by 14 billion yen. This implies that an amount of 14 billion yen spilled over to the other region owing to the policy.

Table 2 shows that there was an increase in producer prices. This increase was because producer prices depend on the wage rate, which rose because of the policy. Further, the increase in producer prices resulted in a decreased demand for goods.

In table 3, EV refers to equivalent variation, which is a measure of economic welfare changes associated with changes in prices. In this paper, EV represents the effect of the policy on the real economy in the local region. In addition, ΔNW represents the effect of the policy on the financial economy in the local region. Thus, EV + ΔNW represents the overall effect of the policy on the local region. EV + ΔNW is -93. This is because the demand for goods decreased owing to an increase in the prices.

Table 1. Financial assets of households

	Before	After	After/Before
NW	8710	8810	1.012
DEP_h	6200	6099	0.984
DEP_h^L	4660	4584	0.984
DEP_h^O	1540	1515	0.984
ZZ_h	2046	2255	1.102
ZZ_h^L	1632	1799	1.102
ZZ_h^O	414	456	1.102
GB_h	206	203	0.984
$CURR_h$	258	254	0.984
(Unit: Billion yen)			

Table 2. Prices of and demand for goods

	Before	After	After/Before
P1	1.00	1.39	1.39
P2	1.00	1.37	1.37
P3	1.00	1.35	1.35
C1	27	24	0.90
C2	430	391	0.91
C3	2100	1948	0.93
(Unit: Billion yen)			

Table 3. Effect of policy

	Effect of policy
EV	-193.5
ΔNW	100.5
EV + ΔNW	-93.0
(Unit: Billion yen)	

6 Conclusion

In this paper, an RFCGE model was developed to solve the theoretical issue of CGE models. The effect of an infrastructure investment policy on a local region was assessed using the RFCGE model. As a result of calibration, the net wealth of households increased. However, a portion of the increased household assets spilled over to other regions. Moreover, producer prices increased owing to an increased wage rate. Further, the demand for goods decreased due to increased prices. Thus, EV+Δ NW had a negative value. This implies that infrastructure investment policy may, sometimes, result in negative effects on the local region. In conclusion, it is difficult to explicitly evaluate the effect of policy on the local region by merely using the CGE model. Therefore, it is important to evaluate infrastructure investment policy using the RFCGE model.

References

[1] Euijiune Kim et al., *Project Evaluation of Transportation Projects: An Application of Financial CGE Model*, http://ideas.repec.org/p/wiw/wiwrsa/ersa15, (2015)
[2] Lance Taylor, *Socially Relevant Policy Analysis: Structuralist Computable General Equilibrium Models for the Developing World*, MIT Press, 1-69, (1990)
[3] Mark Thissen, *Building Financial CGE Models: Data, Parameters, and the Role of Expectations. A Financial CGE Model for Egypt*, University of Groningen, (2000)
[4] *Input-Output Table of Nara Prefecture*, http://www.pref.nara.jp/16380.htm, (2011)

Appendix: Social Accounting Matrix

									Portfolio																												
Unit: Billion yen	Use of Incomes								Firm 1				Firm 2				Firm 3				Households				Local banks				Other regions								
	Production(1-3)			Capitalist(1-3)			House holds	Invest ment	Physical capital	Bank borrow ing	Other region borrow ing	Equity	Physical capital	Bank borrow ing	Other region borrow ing	Equity	Physic al capital	Bank borrow ing	Other region borrow ing	Equity	Curre ncy	Deposit	Equity	Govern ment bond	Loan	Reser ve	Deposit	Remit tance	Invest ment	Export /Impor t	Remit tance	Deposit	Govern ment bond	Equity	Loan	Re serve	Total
Initial Portfolio																																					
Capital									-23				-80				-300																				-5000
Loan										100				1000				2000							-3100												0
Reserve																										-100										100	0
Currency													0	0							-250																-250
Deposit													0									-4500					4500										0
Equity												100	0			500				1000			-1600														0
Other region																						-1500										1500					0
Other region																							-400											400			0
Other region											50				500				1000																-1550		0
Remittance																												200			-200						0
Government bond																								-200									200				0
Sales																																					
Production1	9	40	11				27	1																						-26							62
Production1	11	970	320				430	540																						-11							2,260
Production1	10	520	820				2100	86																						-756							2,780
Incomes																																					
Capitalist1	23																													15							38
Capitalist2		80																												230							310
Capitalist3			300																											250							550
Households	9	850	1329	12	120	250																								606							2,976
Other regions				4	40	125	159																														328
Flow of funds																																					
Firm1				22					-62	30	10	4																	-4								0
Firm2					150								-230	60	20	12													-12								0
Firm3						175											-335	90	30	16									24								0
Households							260														-8	-200	-46	-6													0
Local banks																									-180	-20	160	40									0
Other regions																														20	-40	40	6	14	-60	20	0
Asset balance																																					0
Loan										-130				-1060				-2090							3280												0
Equity												-104				-512				-1016			1632														0
Currency																					258																258
Reserve																										120										-120	0
Deposit																						4660					-4660										0
Other region deposit																						1540										-1540					0
Remittance																												-240			240						0
Government bond																								206									206				0
Other region equity																							414											-414			0
Other region loan											-60				-520				-1030																1610		0
Total	62	2,260	2,780	38	310	550	2976	627	-85	0	0	0	-310	0	0	0	-635	0	0	0	0	0	0	0	0	0	0	0	8	328	0	0	0	0	0		

Transportation Investments and Productivity Analysis of a Japanese Case Study

Atsushi Koike[1,*], and *Ryota* Okumura[1]

[1] Department of Civil Engineering, Kobe University, 1-1 Rokkodai Nadaku Kobe, Japan

Abstract. Over the last two decades, there have been several studies on productivity related to transportation investments in not only Japan but also several foreign countries. However, these prior studies did not achieve common results as they did not use a unified format for measuring the service level of the transportation network and using the dataset. Therefore, this study classifies previous studies from six points of view. These are analytical method, functional type, explained variable, service level of the transportation network, aggregation level, and estimation method. This study examines the differences in results caused by the diversity of analytical methods. In addition, this study estimates the production function using two analytical methods with Japanese data. Then, the relationship between the productivity and transport investments is revealed. From this result, several issues are confirmed.

1 Introduction

In the last two decades, there have been several studies about productivity related to transportation investments in not only Japan but also several foreign countries, because it is assumed that transportation investments have a significant effect on economic growth. Aschauer [1] is the first study that considered the relationship between aggregate productivity and government expenditures in public-sector capital stock including transportation investments. The study revealed that the public capital stock had most explanatory power for productivity. The method of statistically estimating the production function including the level of service was established in the United States. Later, the relationship between the level of service and productivity was studied using a similar method in Japan. However, most prior studies have not used a unified format for this issue because of the difference in the estimation process and dataset. For example, Kanemoto and Okawara [2] revealed that public investments had a negative effect on production in 1985. Yoshino et al. [3] revealed that the marginal productivity of public investments decreased from 1970 to 1993. (See Ejiri et al. [4], Tsukai et al. [5], and Hayashi [6] for more information.) Thus, it is necessary to establish a method of estimation of the production function considering the service of the transportation investments; it is also necessary to lead to a robust opinion.

[*] Email address: koike@lion.kobe-u.ac.jp

In contrast, OECD published a manual [7] on the method of measuring productivity. The purpose of this manual was to solve the problem of the difference in the estimation process. Thus, the change in productivity can be compared on an international level using this manual. However, this manual does not consider the formulation of the transportation service. In response, the purpose of this study is to establish the normal analytical method of the production function including the formulation of the transportation service. Therefore, this study indicates problems in each analytical method and analytical results using Japanese data.

2 Classification of prior studies

This chapter classifies the prior studies using six points (see Fig. 1.).

Analytical method	Functional type	Explained variable
• Time-series analysis • Cross-section analysis • Panel analysis	• Cobb-Douglas type • CES type • Translog type	• Gross output • Value-added
Service level of the transportation network	**Aggregation level**	**Estimation method**
• Amount of Investments • Accessibility	• Industry level • Region level	• One-step estimation • Two-step estimation

Fig. 1. Classification points

2.1 Analytical method

There are three types of analytical methods. The first method is time-series analysis, which involves regression analysis using the data with time change. For this, it is assumed that parameters, such as the social or industrial structure, are constant for the time change. It is not useful to use the time-series analysis for a long period.

The second analytical method is cross-section analysis, which involves regression analysis using the data with region change. For this, it is assumed that the parameters are constant for the region change. This analysis is used by remaining constant with the time-series analysis, and it needs data by region.

The third analytical method is panel analysis, which involves regression analysis using the data with both time and region change. Thus, panel analysis is a combination of time-series and cross-section analyses. For this analysis, it is assumed that the parameters are constant for both time and region change. Therefore, the assumption of panel analysis is stronger than that of both time-series and cross-section analyses.

2.2 Functional type

Three main functional types are used for estimating the production function. The first is the Cobb-Douglas type, which involves the assumption that the elasticity of substitution is 1. Therefore, this type is stronger than all other types and is mainly used for estimating the macro production function.

The second is the constant elasticity of substitution (CES) type, which involves the assumption that the elasticity of substitution is constant. Thus, this type is generalized from the Cobb-Douglas type.

The third is the translog type, which involves the assumption that the elasticity of substitution is flexible. Thus, this type is the most general production function and is often estimated simultaneously with the cost-share function because of too many parameters.

2.3 Explained variable

There are two types of explained variables. The first is gross output, which includes not only the final demand but also the intermediate input. Thus, it is necessary to acquire the transaction data of the intermediate products, but it is difficult to acquire this data by industry. This is why gross output is not used often for estimating the production function by industry.

The second explained variable is the value-added. This does not include the intermediate input, so it can be explained by only labor and capital input. The value-added is used more often than gross output.

2.4 Service level of the transportation network

There are two types of service level of the transportation network. The first is the amount of investments, which is the government expenditure for public investment. The amount of investments includes flow and stock, but the stock is used as an explanatory variable for estimating the production function. The amount of investments is calculated by the price base, so that this is easily treated together with other explanatory variables.

The second is accessibility, which is derived by the required time and cost. Thus, this index can reveal the effect of the transportation service. However, the accessibility is derived in many different forms, variables, and units. Therefore, this index itself indicates variety.

2.5 Aggregation level

The aggregation level includes the industry and region levels. First, this section explains the industrial aggregation. When industries are aggregated, the industrial movement cannot be observed. Thus, the change of productivity of dataset aggregated industries is lower than that with dataset by industry. In addition, it is difficult to acquire the necessary dataset by industry.

Second, this section explains the regional aggregation. This is an important problem to settle at the region level, because the region level relates to the transportation service level. Therefore, the region level decides the range of a transportation investment's effect, which could be at the macro, prefecture, municipality, or firm level. Moreover, it is difficult to acquire the necessary data at the low region level.

2.6 Estimation method

There are two types of estimation methods. The first is one-step estimation, which estimates the production function including the explanatory variable of the level of service. Thus, this estimation step is used more often than two-step estimation because of the simple calculation.

The second estimation method is two-step estimation. This method is used in Holl [8], and this section explains this method by using Holl's (2016) model. First, the production function not including the level of service is estimated as the following equation.

$$\ln Y_{it} = a_{it} + \beta_1 \ln K_{it} + \beta_2 \ln L_{it} + u_{it} \quad (1)$$

Here, Y_{it}: value-added of firm i at time t, L_{it}: labor input of firm i at time t, K_{it}: capital stock of firm i at time t, and u_{it}: error of firm i at time t.

Second, the following equation including the constant term and error of the estimated production function is estimated as follows.

$$a_{it} = \delta \ln HWDIS_{it} + u_{it} \quad (2)$$

Here, $HWDIS_{it}$: highway access distance of firm i at time t.

The effects of the transportation service are estimated from equation (2). Thus, this estimation method is more complex than one-step estimation because two regressions are necessary.

2.7 Classification of prior studies

This section classifies prior studies that estimated the production function including the level of service on the above-discussed six points. Table 1 shows the classification of the prior studies' results.

Table 1. Classification of prior studies*

	Analytical method	Functional type	Explained variable	Service level	Aggregation	Estimation method
Aschauer (1989)	T	C-D	G	I	1 industry 1 region	One
Takenaka and Ishikawa [9]	T	C-D	V	I	1 industry 1 region	Two
Asako and Sakamoto [10]	P	C-D	V	I	1 industry 42 regions	One
Mitsui and Inoue [11]	T	C-D	V	I	1 industry 1 region	One
Iwamoto et al. [12]	P	C-D	V	I	3 industries 42 regions	One
Yoshino and Nakano [13]	P	TL	G	I	1 industry 9 regions	One
Economic Planning Agency [14]	T	C-D	V	I	1 industry 2 regions	One
Kanemoto and Okawara (1996)	C	C-D	V	I	1 industry 456 regions	One
Yoshino et al. (1999)	T	C-D TL	V	I	1 industry 1 region	One
	P	TL	V	I	3 industries 11 regions	One
Kim and Hewings [15]	C	TL	V	Acc	16 industries 237 regions	One
Holl (2016)	C	C-D	V	Acc	Firm-level	Two

*Service level: Service level of the transportation network, T: Time-series analysis, C: Cross-section analysis, P: Panel analysis, C-D: Cobb-Douglas type, TL: Translog type, G: Gross output, V: Value-added, I: Amount of investments, Acc: Accessibility, One: One-step estimation, and Two: Two-step estimation

3 Analysis

This chapter explains the analytical method, dataset, and results. This study is composed of two analyses. The first is time-series analysis not including the level of service, and the second is panel analysis including the level of service.

3.1 First analysis

The first analysis is time-series analysis for 1970–2009 not including the level of service. Therefore, the estimated production function in this section is composed of labor input and private capital stock, which is the general production function. The production function is a derived Cobb-Douglas type and presented as follows:

$$\ln Y_t - \ln L_t = \alpha + \beta(\ln K_t - \ln L_t) + u_t \tag{3}$$

Here, Y: real value-added (million yen, price in 2000), L: the annual total actual hours worked (/1000), and K: real private capital stock (million yen, price in 2000). These variables are data estimated by Tokui et al. (2013) [16].

The productivity is derived by the estimated constant term and error. Thus, this index indicates the production efficiency not explained by labor and capital input.

$$P = \alpha + u_t \tag{4}$$

Here, P: productivity.

The productivity includes the effects of not only the transportation service but also the total level of service.

3.2 Second analysis

The second analysis is panel analysis for 1970–2009 of 47 regions including the level of service. This study uses the amount of investments as the level of service. Parameters are constant for time and region change in the panel analysis. However, it is not natural for parameters to remain constant for a long period, such as from 1970 to 2009. Therefore, this study divides the dataset every five years and analyzes the divided panel dataset. The estimated production function is a derived Cobb-Douglas type and presented as follows:

$$\ln Y_{tr} - \ln L_{tr} = \alpha + \beta(\ln K_{tr} - \ln L_{tr}) + \gamma \ln G_{tr} + u_{tr} \tag{5}$$

Here, G: net level of service stock (million yen, price in 2005). This variable is data estimated by Cabinet Office, Government of Japan (estimation of the level of service stock).

The marginal productivity of public investments by time and region is calculated from the estimated parameter as follows:

$$F_{tr} = \gamma \frac{Y_{tr}}{G_{tr}} \tag{6}$$

Here, F_{tr}: marginal productivity of the public investments of region r at time t.

The marginal productivity of public investments in all of Japan is derived as its weighted average by region.

$$F_t = \theta \cdot F_{tr} \tag{7}$$

$$\theta = \frac{Y_{tr}}{\sum_r Y_{tr}} \tag{8}$$

Here, F_t: marginal productivity of public investments in all of Japan at time t and θ: share of the real value-added.

The marginal productivity of public investments is the change in the value-added output resulting from increasing one more unit in the level of service input.

3.3 Results

This section indicates results from the first and second analyses as well as implications of the analytical results. Fig. 2 shows the productivity from the first analysis, marginal productivity of public investments in all of Japan from the second analysis, and real public investments (trillion yen, price in 2005) from the Cabinet Office, Government Of Japan (estimation of the level of service stock).

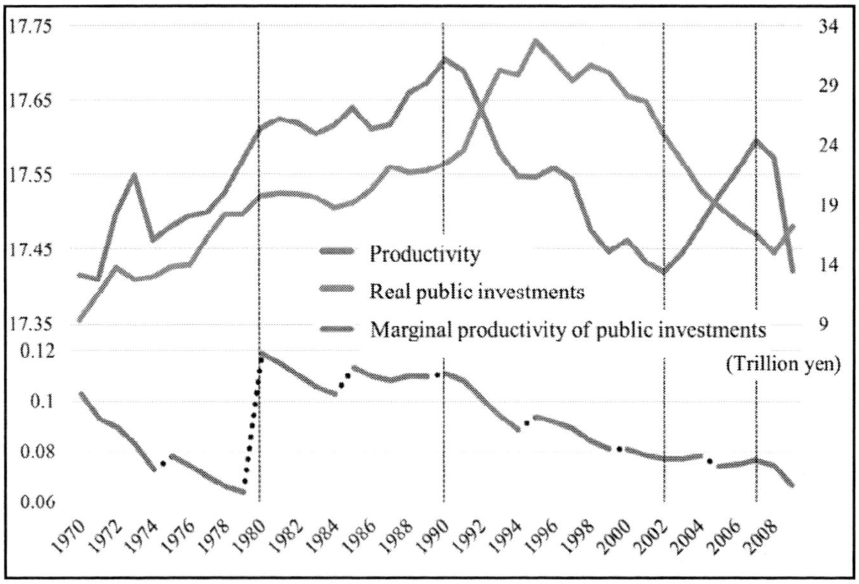

Fig. 2. Productivity, real public investments, and marginal productivity of public investments

In the 1970s, the marginal productivity of public investments decreased as productivity and real public investments increased. In the 1980s, the decline in the marginal productivity of public investments slowed down slightly. In the 1990s, namely after the burst of the economic bubble, the productivity declined dramatically. From 2002 to 2007, productivity increased as real public investments decreased. From 2007 to 2009, productivity declined because of the decline in the marginal productivity of public investments.

4 Conclusion

There are several studies about productivity related to the service level of the transportation network. However, a unified view on this issue is not revealed because of the diversity of analytical methods. This study classifies prior studies using six points of view to reveal this diversity. Each analytical method has individual assumptions and problems. Thus, it is necessary to develop a normal manual on productivity related to the service level of the transportation network. Therefore, this study uses Japanese dataset to conduct analyses using two methods and presents the analytical results. The results reveal that the relationship between productivity and public investments, which is used as the level of service in this study, is not constant for the time change.

Next, it is necessary to formulate the transportation investments, because this study does not use transportation investments but total public investments. That means that the results of this study include effects of not only the transportation service but also the level of service other than transportation. Thus, it is necessary to extract the transportation service's effects from the total level of service to reveal the relationship between productivity and transportation investments. Use of the accessibility derived by the required time and cost is considered the solution to this problem.

In addition, it is necessary to continue developing the analytical manual in order to indicate the robust truth about the relationship between productivity and transportation investments.

References

[1] D. A. Aschauer, Is public expenditure productive?, JME, **23**, 177-200 (1989)
[2] Y. Kanemoto, T. Okawara, Is Tokyo too large? An empirical analysis of city sizes in Japan, Central Research Institute of Electric Power Industry, **37**, 29-42 (1996)
[3] N. Yoshino, T. Nakajima, M. Nakahigashi, Shakaishihonnoseisanryokukoka, Nippon Hyoron Sha Co., Ltd., 11-88 (1999)
[4] R. Ejiri, M. Okumura, K. Kobayashi, The productivity of infrastructure and economic growth: A state-of-the-art review, JSCE, **688/IV-53**, 75-87 (2001)
[5] M. Tsukai, R. Ejiri, M. Okumura, K. Kobayashi, Productivity of infrastructure and spill over effects, JSCE, **716/IV-57**, 53-67 (2002)
[6] M. Hayashi, Shakaishihonnoseisanseitodojisei, ESRI, **21** (2002)
[7] P. Schreyer, Measuring Productivity - OECD Manual, Keio University Press (2009)
[8] A. Holl, Highways and productivity in manufacturing firms, JUE, **93**, 131-151 (2016)
[9] H. Takenaka, T. Ishikawa, Nihonnoshakaishihonsutokkutokyokyusaido, NLI Research Institute, **6**, 19-35 (1991)
[10] K. Asako, K. Sakamoto, Seifushihonnoseisanryokukoka, Japanese Ministry of Finance, IFMP, **26**, 97-102 (1993)
[11] K. Mitsui, J. Inoue, Shakaishihonnoseisanryokukoka, Nippon Hyoron Sha Co., Ltd., 43-65 (1995)
[12] Y. Iwamoto, S. Ouchi, T. Takeshita, T. Bessho, Shakaishihonnoseisanseitokokyotoshinochiikikambunseki, Japanese Ministry of Finance, IFMP, **41**, 27-52 (1996)
[13] N. Yoshino, H. Nakano, Kokyotoshinochiikikanhaibuntoseisanryokukoka, Japanese Ministry of Finance, IFMP, **41**, 41-51 (1996)
[14] Economic Planning Agency, Kaikakuhehonkakukidousurunihonkeizai, EWP, **1.5.**, 1-5 (1997)
[15] E. Kim, G. J. D. Hewings, An application of an integrated transport network - Multiregional CGE model to the calibration of synergy effects of highway investments ESR, **21**, 377-397 (2009)
[16] J. Tokui, K. Fukao, T. Makino, T. Miyagawa, N. Arai, S. Arai, T. Inui, K. Kawasaki, N. Kodama, N. Noguchi, R-JIPdetabesunokochikutochiikikanseisanseikakusano bunseki, RIETI, **13-J-037** (2013)

Short-run Economic Assessment of the Transportation Recovery Policy After an Earthquake

Atsushi Koike[1,*], and *Yoshinao* Miyamoto[1]

[1]Kobe University, Department of Civil Engineering, 1-1 Rokkodai-cho, Nada-ku, Kobe-shi, Hyogo 657-8501, Japan

Abstract. This paper develops a short-run economic damage assessment model. The model contains two sub models. One is the conventional traffic assignment model, which calculates traveling time under the damaged transportation infrastructure. The other is the economic damage assessment model, which determines the decrease in production level in the short run. This method facilitates the identification of critical infrastructure that could reduce the economic damage when the disaster occurs. As a case study, we applied this model to several recovery plans for transport facilities. The results suggest that the proposed comprehensive model should be considered as a prevention plan.

1 Introduction

There is a 70% possibility that a huge earthquake will hit the Nankai trough in Japan in the next 30 years. Thus, the economic damage assessment of earthquakes is required for effective recovery.

When the Great East Japan Earthquake occurred, there were several types of economic damage: first, "the middle/long-run damage," which is the damage to production facilities or customers; second, "the short-run damage," which is the shortage of labor or intermediate supply by the damage of transportation infrastructure. [1] Since the response just after the disaster influences the economic recovery both temporally and spatially, it might be important to be focused on the short-run damage. In addition, there is a possibility that the spatially uniform policy does not work well in some regions. To sum up, it is assumed that the comprehensive policy, considering the damage factor in each region, should be planned for the effective recovery.

There are many previous studies on the economic damage assessment (e.g., Koike et al. (2014), Tsuchiya et al. (2006) [2, 3]). However, they are focused on the middle/long-run effects, while there no studies on the analysis of the short-run effect.

This paper proposes an economic damage assessment model in the short-run. Additionally, the model is applied to the Chukyo metropolitan area, which is predicted to be the most seriously damaged by the Nankai Trough Earthquake. We analyze the

* E-mail address: koike@lion.kobe-u.ac.jp

economic damage and specify the main factors of damage in each industry of every region. In addition, we simulate two recovery policies on transportation facilities to develop the effective recovery policy.

2 Structure of the economic damage assessment model in the short-run

2.1 Outline of the model

Fig. 1 shows the whole structure of the economic damage assessment model in the short - run. This model contains two sub models, namely the transportation model to calculate traveling time and the economic damage assessment model to determine the decrease in products in the short-run. This model can analyze the factors that affect economic damage the most in each industry in every zone. If the earthquake damages the road or railway network, traveling time increases and it causes the limitation of labor supply, intermediate supply, or production supply. Finally, the decrease in production is determined by the production function with a Leontief technology.

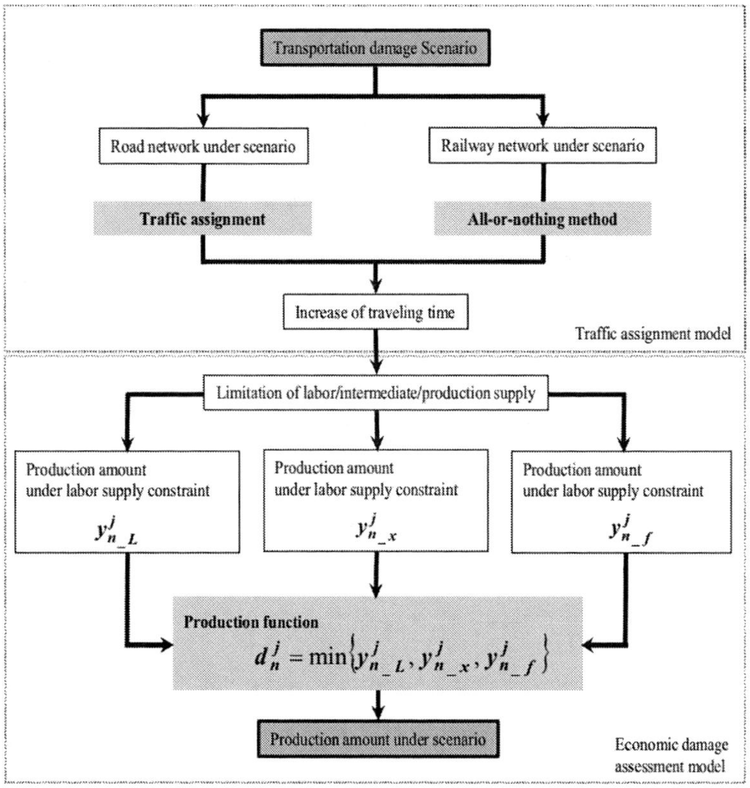

Fig. 1. Structure of the economic damage assessment model in the short-run

2.2 Transportation model

In the transportation model, traveling time is calculated by analyzing the traffic assignment using the origin-destination (OD) table and network data. We applied the incremental assignment method to the road network. Whereas the all-or-nothing method is applied to the railway network on the ground to ensure there is no congestion.

Generally, the traffic assignment analysis applies the four-step method. However, in this paper, the traffic assignment model is used on the assumption that the trip generation, destination, and modal split are not changed in the short-run. In addition, we apply the incremental assignment to the road network to reduce the calculation load.

2.3 Economic damage assessment model

2.3.1 Production function

The production amount at zone j of industry n is determined by the following: $y_{n\_L}^j$ means the production amount under constraint wherein only the labor cannot commute and the intermediate or production goods are supplied as usual. This equation shows that the production amount is determined as that under the most seriously damaged factor.

$$d_n^j = \min\{y_{n\_L}^j, y_{n\_x}^j, y_{n\_f}^j\} \tag{1}$$

where d_n^j: production amount; $y_{n\_L}^j$, $y_{n\_x}^j$, $y_{n\_f}^j$: production amount under damage factor constraint

The production amounts under the damage factor constraint are described as follows. In this paper, the production functions (Equation (1), (2), and (3)) are all specified with the Leontief technology since it is assumed that production factors in the short-run are not substituted.

$$y_{n\_L}^{j\,'} = \frac{L_j^{n\,'}}{a_{Ln}^j} \tag{2}$$

$$y_{n\_x}^{j\,'} = \min\left\{\frac{x_{1n}^{j\,'}}{a_{1n}^j}, \frac{x_{2n}^{j\,'}}{a_{2n}^j}, \ldots, \frac{x_{mn}^{j\,'}}{a_{mn}^j}\right\} \tag{3}$$

$$y_{n\_f}^{j\,'} = u_j^m \cdot y_n^j \tag{4}$$

where L_j^n: labor supply; a_{Ln}^j, a_{mn}^j: input coefficient; u_j^m: reduction rate of production amount; (A' means A under the scenario).

2.3.2 Limitation of labor/intermediate/production supply

In this paper, we assume that the increase of traveling time influences the working time or decision making of freight transportation. The reduction rate of labor supply s_j^L is defined as Equation (5) and the amount of labor supply is calculated by Equation (6). Next, we set the working time as 8 hours per day per worker and it is reduced by the increase in traveling time.

$$s_j^L = \frac{\sum_i (8 - 2E_{ij}^{car})N_{ij}^{car} + 8N_{ij}^{pub\,'}}{\sum_i 8N_{ij}^{car} + 8N_{ij}^{pub}} \tag{5}$$

$$L_j^{n}{}' = s_j^L \cdot L_j^n \quad (6)$$

where N_{ij}^{car},N_{ij}^{pub}: a number of commuting trips; E_{ij}^{car}: increase in traveling time

Moreover, the limitations of freight transportation are determined depending on the increase in the rate of traveling time. It is assumed that the firms decide to stop transport if it takes more than 1.5 times the usual traveling time. The reduction of transported intermediate goods is $I_{ij}^{m}{}'$ and the reduction rate of intermediate supply is calculated by Equation (7). Finally, the amount of intermediate supply under damaged by transportation is determined by Equation (9). Additionally, it is the same for production supply.

$$t_j^m = \frac{\sum_i I_{ij}^{m}{}'}{\sum_i I_{ij}^m} \quad (7)$$

$$u_j^m = \frac{\sum_k F_{jk}^{m}{}'}{\sum_k F_{jk}^m} \quad (8)$$

$$x_{mn}^{j}{}' = t_j^m \cdot x_{mn}^j \quad (9)$$

where I_{ij}^{m}: intermediate supply; F_{jk}^{m}: production supply.

2.4 Data and data sources

The model requires the following dataset shown in Table 1. In the transportation model, the required network data and velocity data is obtained from the Road Traffic Census. In addition, the OD matrix of commuting and freight transportation is developed by using the Chukyo PT Survey and Chukyo Logistics Survey data. In the economic damage assessment model, a dataset is required to develop the GDP data in each industry in every region. If there are some regions where GDP data is not available, it could be estimated based on the number of workers in that region. The input coefficients of labor and intermediate goods are derived from the Input-Output table.

Table 1. Data sources

	Data sources
Road network	Digital Road Map Road Traffic Census
Rail network	National Land Numerical Information
Commute OD matrix	Chukyo Urban Area Person Trip Survey
Freight OD matrix	Chukyo Urban Area Logistics Survey
GDP table	GDP Classified by Economic Activities Economic Census for Business Frame Economic Census for Business Activity Census of Manufactures Input-Output Table
Input coefficient	Input-Output Table

 DOI: 10.1051/matecconf/201710309017

3 Case study

3.1 Details on assumptions and scenario setting

3.1.1 Coverage area and transport infrastructure

The coverage area is Chukyo metropolitan area (Aichi Prefecture, the northern part of Mie prefecture and the southern part of Gifu prefecture), which is divided into 111 zones. Fig. 3 and Fig. 4 show the transport infrastructure in the coverage area. We have excluded the local streets and the Shinkansen line.

Fig. 2. Coverage area (111 ZONE)

Fig. 3. Road network

Fig. 4. Railway network

3.1.2 Setting the scenario

We applied the model to evaluate the damage of transport infrastructure in the Chukyo metropolitan area. We set the scenario wherein the road network and railway network cannot be used in a flood-assumed area. Fig. 5 shows that the traveling time increases considerably between the regions across the Kiso, Nagara and Ibo rivers.

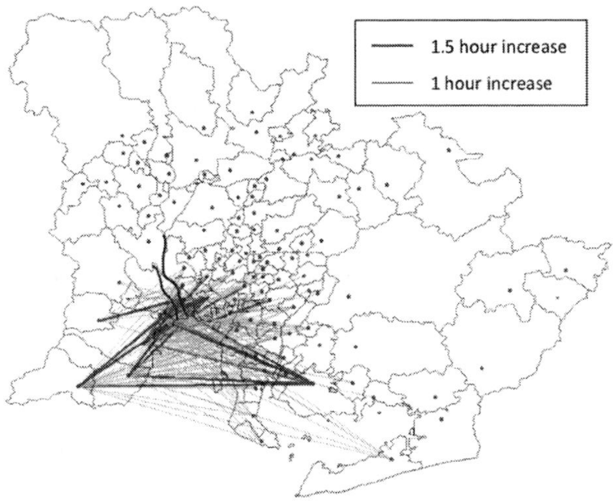

Fig. 5. Increase in traveling time

3.2 Outcomes of scenario

3.2.1 Economic damage of earthquake by region and industry

The total financial damage is calculated to be approximately 10 trillion yen/year, which constitutes 25% of GDP in the Chukyo metropolitan area.

From an industry perspective, the manufacturing, wholesale/retail and service industries are seriously damaged. The study focuses on the manufacturing industry because the purpose of this study is to analyze which factor, caused by preventing commuting or physical distribution, has the most significant effect on economic damage. Fig. 7 shows that the transport equipment industry is most seriously damaged from the manufacturing industry as there are several major car related companies (e.g., TOYOTA).

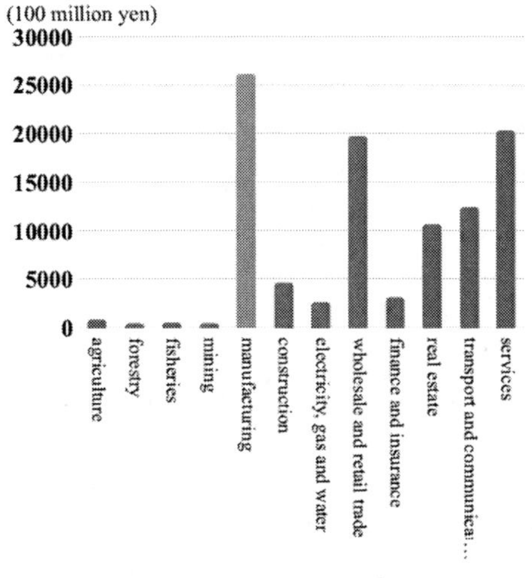

Fig. 6. Economic damage by industry

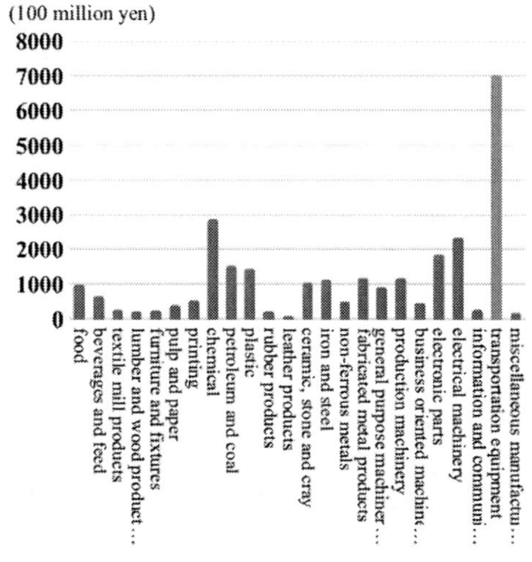

Fig. 7. Economic damage of manufacturing

3.2.2 Main damage factor by region (the transportation equipment industry)

Fig. 8 shows which factor has the most significant effect on economic damage by zone. Red signifies the limited supply of labor, blue signifies the limited intermediate supply, and green signifies the limited production supply.

In the coastal areas, the limited intermediate or production supply is the main damage factor. However, in the interior area, the limited labor supply is the main damage factor. Fig. 7 shows that the traveling time increases greatly in the coastal area and an unfavorable route around the rivers needs to be taken. Thus, it is assumed that intermediate or production goods are limited.

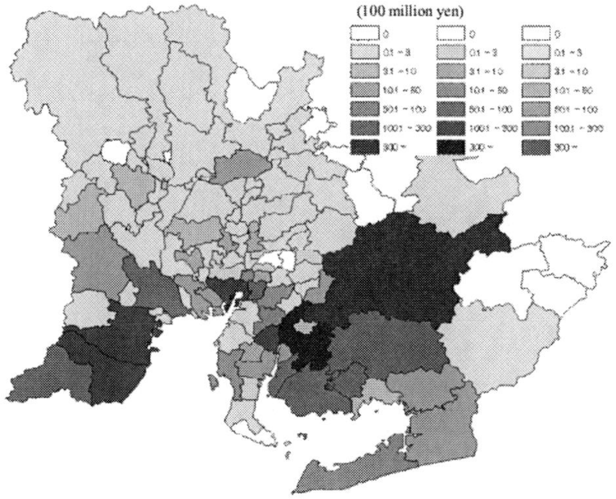

Fig. 8. Economic damage by region and the damage factor (transportation equipment industry)

3.2.3 Effects of recovery plan for transport facilities

We simulate two cases of transportation recovery policies in order to evaluate the effect of the economic damage reduction by the policy and propose an efficient policy.

Table 2 shows the economic damage after implementing the two transportation recovery policies. Across the whole industry, the effect of the economic damage reduction by policy 1 is about 1.1 trillion yen/year; in contrast, that of policy 2 is about 6.6 trillion yen/year. In the transport equipment industry, the effect on the recovery of the road network is much more effective than that of the railway network. This could be because railway network is only used for commuting, but road network has a multi-use (for commuting and freight transportation).

Table 2. Economic damage after each policy (100 million yen)

	Whole industry	**Transport equipment**
No recovery policies (Fig. 8)	96,948	6,948
Recovery of railway network (Fig. 9)	85,710	5,973
Recovery of road network (Fig. 10)	30,665	1,782

Fig. 9 shows the distribution of the damage factors after the recovery of the railway network. On comparing Fig. 8 and Fig. 9, we can see that the effect of the policy is not seen in regions where the main damage factor is the limitation of intermediate or production supply; although there is some effect in regions where the main damage factor is the limitation of labor supply. However, we also found that there are some regions, such as

Inabe-shi or Tahara-shi, where the shortage of labor supply is not largely eliminated even though the railway network has recovered. A similar result is observed with policy 2. In the coastal area, where the main damage factor is the limitation of intermediate or production supply, the economic damage due to the freight factor is largely reduced; however, there are regions such as Suzuka-shi, Nishio-shi, or Minato-ku where most of the economic damage still remains.

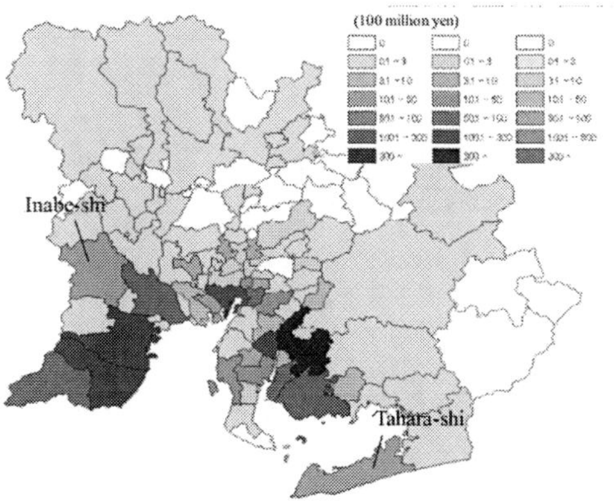

Fig. 9. Recovery of railway network

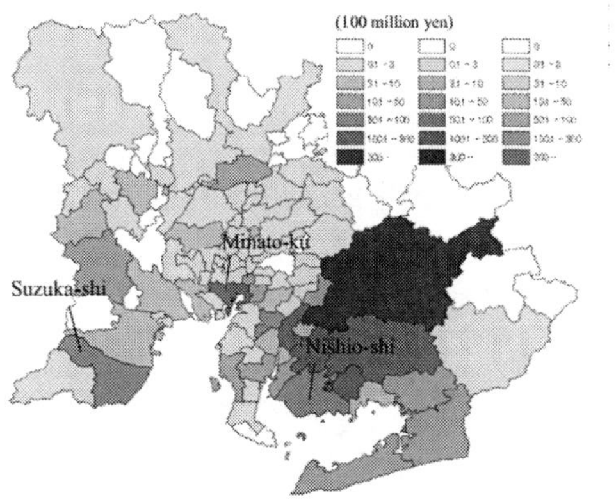

Fig. 10. Recovery of road network

4 Conclusion

We developed the economic damage assessment model in the short -run and applied the model to the Chukyo metropolitan area. We found the following results:

- The economic damage is mainly seen in the coastal areas where the traveling time increases largely.
- From an industry viewpoint, the transport equipment industry is most seriously damaged from the manufacturing industry.
- In the coastal areas, the limited intermediate or production supply is the main damaging factor. On the other hand, in the interior area, the limited labor supply is the main damage factor.
- The effect of the recovery of the road network is much more significant than that of the railway network.
- However, there are some regions where the effects of recovery policies are negligible even though only either one of the road network or railway network has recovered.

Thus, it is assumed that the comprehensive policy, considering the damage factor and the regional characteristics (including the development status of the transportation facilities or the industrial structure), should be planned.

References

[1] N. Hamaguchi, *Result and Discussion of Survey of Disaster-hit Firms from the East Japan Quake*, RIETI Policy Discussion Paper Series (2012) (in Japanese)

[2] A. Koike, T. Sasaki, Y. Sasaki, and K. Yamasaki, Evaluation of economic damage of Great East Japan Earthquake by SCGE model of municipality level, Journal of Japan Society of Civil Engineers, Ser. D3, **70**(5), 151-159 (2014) (in Japanese)

[3] S. Tsuchiya, Y. Tatano, and N. Okada, A framework of measuring economic losses from the Niigata-Chuetsu earthquake, Journal of Japan Society of Civil Engineers, Ser. D3, No. 23, No. 2 (2006) (in Japanese)

Effect of Waste Plastic as Bitumen Modified in Asphalt Mixture

Mohd Ezree Abdullah[1], *Siti Aminah* Abd Kader[2], *Ramadhansyah* Putra Jaya[3,4,*], *Haryati* Yaacob[5], *Norhidayah* Abdul Hassan[6], and *Che Norazman* Che Wan[7]

[1]Faculty of Civil and Environmental Engineering, Universiti Tun Hussein Onn Malaysia, 86400 Batu Pahat, Johor Bahru, Malaysia
[2,3,5,6]Faculty of Civil Engineering, Department of Geotechnics and Transportation, Universiti Teknologi Malaysia, 81310 UTM Skudai, Johor Bahru, Malaysia
[4]Faculty of Engineering, Department of Civil and Environmental Engineering, Universiti Teknologi PETRONAS, 32610 Seri Iskandar, Perak Darul Ridzuan, Malaysia
[7]Department of Civil Engineering, Politeknik Ungku Omar, 31400 Ipoh, Malaysia

Abstract. The objectives of this study are to investigate the engineering properties of the asphalt mixtures containing waste plastic at different percentages *i.e.* 4%, 6%, 8%, and 10% by weight of bitumen. The experimental tests performed in the study were stability, tensile strength, resilient modulus and dynamic creep test. Results showed that the mixture with 4% plastic has the highest stability (184kN). However, the stability slightly decreases with the increase of plastic additive. On the other hand, the highest tensile strength among the modified asphaltic concrete is 1049kPa (8% plastic added). The modified asphalt mixture with 8% plastic has the highest resilient modulus, which is 3422 MPa (25°C) and 494Mpa (40°C). Where the highest creep modulus recorded is 73.30Mpa at 8% plastic added. It can be concluded that the addition of 8% plastic gave the highest value properties of asphalt mixture. Finally, it can be said that 8% plastic is the optimum value adding.

1 Introduction

Plastic is used regularly in today's lifestyle, and its disposal is a great problem. Plastic is a non-biodegradable product; thus, it causes environmental pollution and problems. The issue now is to reduce plastic waste by transforming it into something beneficial. Given that the plastics used in PET bottles, disposal glasses, handbags, and covers of various appliances soften up at 160°C, a solution is to use waste plastic in such a manner that it gets coated over by an aggregate via heating (140–160°C). Plastic bags are made from one of three basic types of polyethylene (PE), that is, high-density polyethylene (HDPE), low density polyethylene (LDPE), and linear low-density polyethylene (LLDPE). The thick, silky shopping bags from malls are LLDPE; grocery bags are HDPE; and clothing bags from the dry cleaners are LDPE. Many types of PE are known, with most having the chemical

* Corresponding author: ramadhansyah@utm.my

formula $(C_2H_4)n$. Thus, PE is usually a mixture of similar organic compounds that differ in terms of the value of n. One of the main problems of polyethylene is that without special treatment it is not readily biodegradable and thus accumulates. Over 500 billion to 1 trillion plastic bags are consumed worldwide. Plastic waste is a serious problem that can lead to many other problems. It may be hard to totally remove plastic bag usage from our life. Nevertheless, we can manage it by following the common simple rule of 3R: reduce, reuse, and recycle. A recent study had found one more way to reduce the amount of plastic waste by using it in asphalt mixtures [1]. Waste plastic bags can be used as a modifier of hot-mix asphalt [2]. We review the possibility of incorporating waste plastic to improve the performance of asphalt mix. Köfteci [3] reported that utilization of waste plastic in flexible pavement has been established since 2012. Arabani and Pedram [4] reported that modifying the rheological properties of bitumen by mixing with synthetic polymers, such as rubber and plastics, is a common method to improve bitumen quality. Polymer-modified bitumen has better resistance to temperature and water. According to Naskara et. al [5], recycled waste plastics bags in bituminous mix by using 1.2–3 mm particle size in the surface and base mix design of up to 2.5% for improved performance of roads to improve the desired mechanical characteristics of a particular road mix. As the result, the stability and flow were improved. Using waste plastics (polyethylene terephalate) in asphalt concrete mixture can replace the mineral coarse aggregate of an equal size (2.36–4.75 mm) and resulted in an optimum value of 6.6% vol. This method led also to a reduction of bulk compacted mix density. Modification of asphalt with polymers is one of the best options to improve asphalt properties [6]. The added polymer can strongly improve the binder properties, provide safer road, and reduce the maintenance cost by increasing the stiffness of the bitumen and by improving its temperature susceptibility [7]. Modarres and Hamedi [8] reported that the use of recycled waste plastic in pavement asphalt represents a valuable outlet for such materials. The use of modified bitumen with the addition of processed waste plastic of about 5%–10% by weight of bitumen helps in substantially improving the Marshall stability, fatigue, life strength, and other required properties of bituminous concrete mix, which improves the durability and pavement performance with marginal saving in bitumen usage. The process is environment friendly. The use of waste plastics in the manufacture of roads and laminated roofing also helps to consume large quantity of waste plastics. Thus, these processes are socially highly relevant and give better infrastructure.

2 Materials and methods

2.1 Binder, aggregate and gradation

The binder used in this study was 60/70 PEN bitumen supplied by Chevron Malaysia. On the other hand, granite aggregates was used throughout this investigation. Prior to aggregate batching, the aggregates were washed, dried and sieved into their respective size ranges. The aggregate gradation conformed to JKR/SPJ/2008-S4 [9] specifications for asphaltic concrete as shown in Fig. 1.

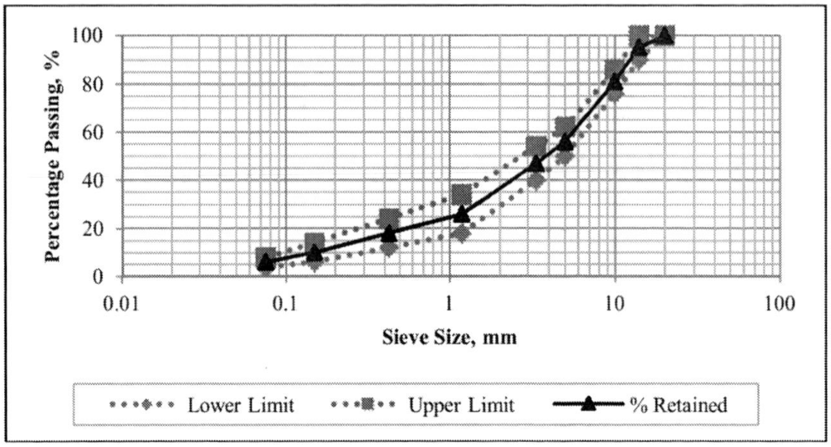

Fig. 1. Aggregate gradation of asphalt mix [9]

2.2 Mix preparation

Ordinary Portland cement and hydrated lime were used as filler in this investigation. The 60/70 bitumen was used as binder for mix preparation. The aggregates were first mixed into batches according to the designated gradations and weight. These batches were then heated in an oven at the designated mixing temperature for at least 4 hours before the mixing process. The mixing and compaction temperatures adopted were 180±0.5°C. The heated aggregate batches were then mixed with a specified amount of bitumen. The mixes were compacted with 75 blows on each side with the standard Marshall hammer. After compaction, the specimens were removed from the molds and allowed to cool overnight.

2.3 Marshall stability test

In the stability test, the specimens were prepared with the specified temperature by immersing in a water bath at a temperature of 60°C±1°C for a period of 45 minutes. It was then placed in the Marshall Stability testing machine and loaded at a constant rate of deformation of 50.8 mm/minute until the maximum load was reached. Test was conducted in accordance with ASTM D6927-06 [10]. The stability result was recorded on the Marshall testing machine in kN.

2.4 Indirect tensile strength test

The indirect tensile strength of bituminous mixtures is conducted by loading a cylindrical specimen across its vertical diametral plane at a specified rate of deformation and test temperature. The samples were conditioned at 25°C for 2 h. Afterward, diametral compressive loading was applied to the conditioned samples until failure occurred. The peak load at failure is recorded and used to calculate the IDT strength of the specimen. This test followed ASTM D6931-12 [11].

2.5 Resilient modulus test

A five pulse indirect tensile modulus test conforming to ASTM D4123 [12] test method was conducted using the Universal Testing Machine, MATTA. The specimen was tested at

25°C and 40°C after a 4-hour-conditioning period. In the test, a pulsed diametral loading force was applied to a specimen and the resulting total recoverable diametral strain was then measured. The test sequence consisted of the application of 150 conditioning pulses followed by 5 pulses where data acquisition took place. The test procedure was repeated by orientating the specimen at 90 degree.

2.6 Dynamic creep test

Dynamic creep test is the simplest method of assessing the resistance to permanent deformation or rutting. The specimens that were used in indirect dynamic creep test were the Marshall samples, which dimensions were similar to that of the specimen for indirect tensile resilient modulus test. The prepared samples were conditioned at 40°C for about 2 h to 3 h. Repeated load was applied for 3600 cycles for 40 min to 45 min. Test was conducted in accordance with BS EN 12697-25 [13].

3 Results and discussion

3.1 Stability

Fig. 2 shows the relationship between stability and percentage of plastic. The stability of 4% plastic is the highest (i.e., 184kN), and it slightly decreases with the increase of plastic addition. A large gap between the stability of the conventional asphalt mixture and the plastic modified asphaltic concrete is observed (i.e., more than 100%). Stability verifies the performance of the asphalt mixture under loading. The results showed that plastic can increase the stability of the asphalt mixture up to certain percentages and after that it will decrease.

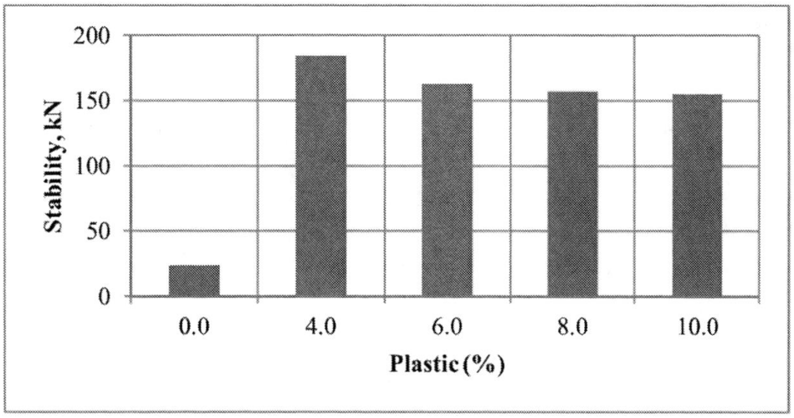

Fig. 2. Effect different percentage plastic to stability

3.2 Indirect tensile strength

Indirect tensile strength (ITS) test is used to measure the indirect tensile strength of the mix, which helps to find out the resistance of thermal cracking of the mixture. The ITS results of conventional and modified asphalt mixture are graphically presented in Fig. 3. From the bar chart, the tensile strength of modified asphalt mixture is lower than conventional asphalt mixture. The tensile strength for conventional asphalt mixture is 1095kPa. The highest tensile strength among the modified asphaltic concrete is 1049kPa with 8% plastic, which is 4% lower than the conventional asphalt mixture tensile strength. The lowest tensile strength is 844kPa with 4% plastic, which is 23% lower than the conventional asphalt mixture.

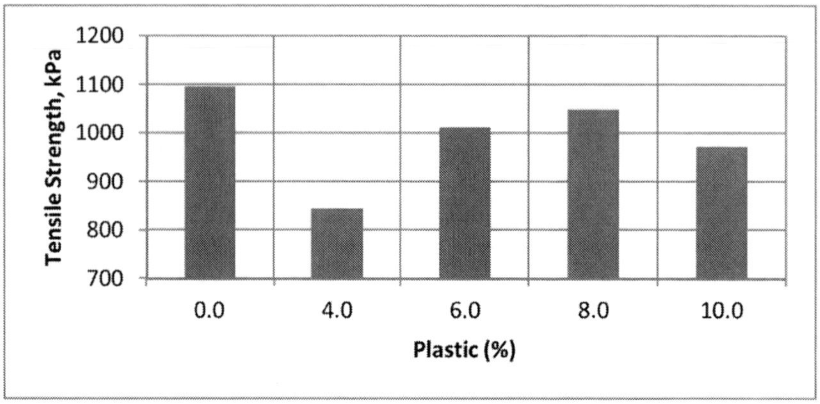

Fig. 3. Effect of different percentage plastic to tensile strength

3.3 Resilient modulus

Fig. 4 shows the results of different percentage plastic to resilient modulus at different temperatures. A similar trend is observed for both temperatures. The resilient modulus increases with the increase of the plastic until it reaches the optimum plastic content. The resilient modulus then decreases with further increase of plastic. Modified asphalt mixture with 8% plastic has the highest resilient modulus for both 25 and 40°C temperatures, which is almost double from the conventional asphalt mixture, that is, 3422MPa for 25°C and 494Mpa for 40°C. In addition, 4% plastic has the lowest resilient modulus, that is, 69% and 51% lower than the conventional asphalt mixture for 25 and 40°C temperatures.

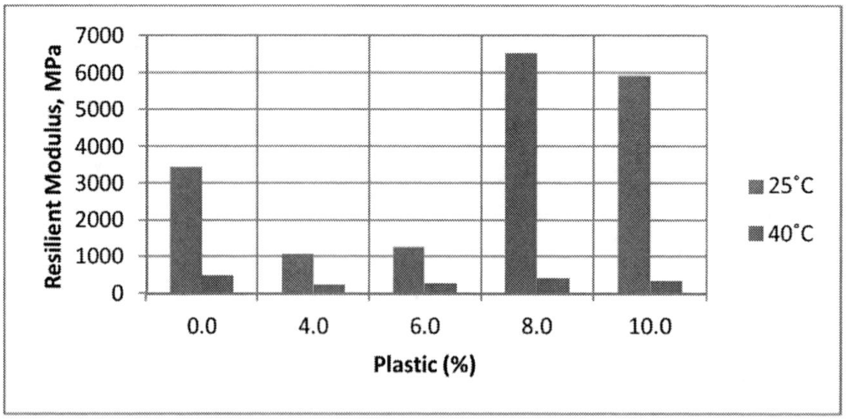

Fig. 4. Effect of different percentage plastic to resilient modulus at 25˚C and 40˚C

3.4 Dynamic creep

The dynamic creep test result is presented in Fig. 5. Dynamic creep test is repeated on the asphaltic concrete and causes the development of rutting. The bar chart in Figure 5 shows that the creep modulus increases with the increased percentage of plastic until 8% of plastic and then decreases. The highest creep modulus is 73.30 Mpa with 8% of plastic. The lowest creep modulus is 24.92 Mpa with 4% plastic, which is 5% lower than the conventional asphalt mixture. Recycling of plastic bags is useful to reduce permanent deformation in the form of rutting and reduce low temperature cracking of pavement surfacing.

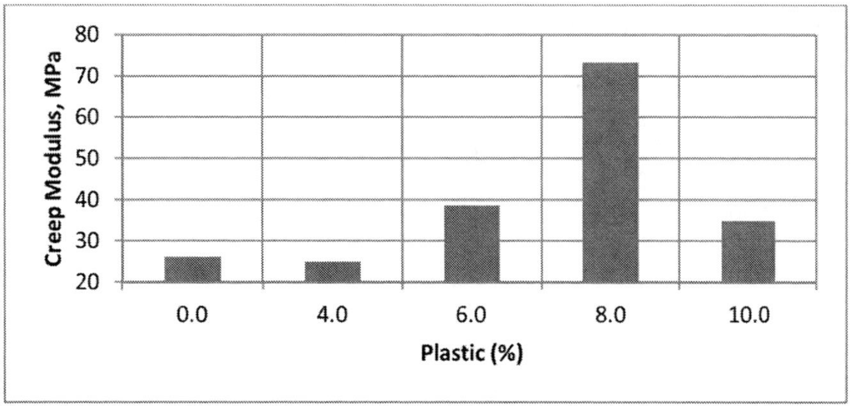

Fig. 5. Effect of different percentage of plastic to creep modulus

4 Conclusions

Based on the result and analysis:

(a) The highest tensile strength among the modified asphaltic concrete is 1049kPa with 8% plastic added.

(b) Modified asphalt mixture with 8% plastic added is the highest resilient modulus for both 25°C and 40°C temperatures which is 3422 MPa for 25°C and 494 Mpa for 40°C. Where the highest creep modulus is 73.301Mpa with 8% of plastic added.

(c) Since 8% plastics added give the highest value for tests, it can be conclude that 8% plastic added is the optimum value.

Research work presented in this paper was supported by Ministry of Higher Education Malaysia (MOHE), Fundamental Research Grant Scheme (research grant no. 1458), Universiti Tun Hussein Onn Malaysia and Universiti Teknologi Malaysia (research grant no. Q.J130000.2522.11H58), the financial assistance for this study is highly appreciated.

References

[1] M.R. Mohd Hasan, C. Baron, Y. Zhanping, J. Ali, A.H. Patricia and O.H. Meor, A simple treatment of electronic-waste plastics to produce asphalt binder additives with improved properties, Constr. Build. Mater., **110**, 79–88 (2016)

[2] I.M. Khan, K. Shahid, A.A. Majed and F.A. Feras, Asphalt design using recycled plastic and crumb-rubber waste for sustainable pavement construction, Procedia Eng., **145**, 1557–1564 (2016)

[3] S.Köfteci, A. Perviz and K. Baurzhan, Performance evaluation of bitumen modified by various types of waste plastics, Constr. Build. Mater., **73**, 592–602 (2014)

[4] M. Arabania and M. Pedram, Laboratory investigation of rutting and fatigue in glassphalt containing waste plastic bottles, Constr. Build. Mater., **116**, 378–383 (2016)

[5] M. Naskara, T.K. Chakia and K.S. Reddy, Effect of waste plastic as modifier on thermal stability and degradation kinetics of bitumen/waste plastics blend, Thermochimica Acta, **509**(1–2), 128–134 (2010)
[6] N. Abdul Hassan, G.D. Airey, N.I. Md. Yusoff, M.R. Hainin, R. Putrajaya, M.E. Abdullah and M.M.A. Aziz, Microstructural characterisation of dry mixed rubberised asphalt mixtures, Constr. Build. Mater., **82**, 173-183 (2015)
[7] M.E. Abdullah, K.A. Zamhari, R. Buhari, M.N. Nayan and M.R. Hainin, Short term and long term aging effects of asphalt binder modified with montmorillonite, Key Eng. Mater., **594-595**, 996-1002 (2014)
[8] A. Modarres and H. Hamedi, Effect of waste plastic bottles on the stiffness and fatigue properties of modified asphalt mixes, Mater. Design, **61**, 8-15 (2014)
[9] JKR/SPJ/2008-S4, *Standard Specification for Road Works Section 4: Flexible Pavement.*
[10] ASTM D6927, *Standard Test Method for Marshall Stability and Flow of Bituminous Mixtures*, Annual Book of ASTM Standards, West Conshohocken, (2006)
[11] ASTM D6931, *Standard Test Method for Indirect Tensile (IDT) Strength of Bituminous Mixtures*, Annual Book of ASTM Standards, West Conshohocken, (2012)
[12] ASTM D4123, *Standard Test Method for Indirect Tension Test for Resilient Modulus of Bituminous Mixtures*, Annual Book of ASTM Standards, West Conshohocken, (1995)
[13] BS EN 12697-25, *Bituminous Mixtures, Test Methods, Cyclic Compression Test*, British Standards Institution, London, (2016)

Characterization of Kaolin as Nano Material for High Quality Construction

M. A. Fadzil[1,*], *M. S.* Muhd Nurhasri[2], *M. A.* Norliyati [1], *M. S.* Hamidah[1], *M. H.* Wan Ibrahim[3], and R. Z. Assrul [1]

[1]Institut of Infrastructure Engineering and Sustainable Management(IIESM), Faculty of Civil Engineering, Universiti Teknologi MARA, 40450 Shah Alam, Selangor, Malaysia
[2]Faculty of Civil Engineering Universiti Teknologi MARA (UiTM), Pulau Pinang, Malaysia
[3]Faculty of Civil and Environmental Engineering, Universiti Tun Hussein Onn Malaysia, 86400 Parit Raja, Johor, Malaysia

Abstract. At the moment utilisation of nano technology in every aspect in human life were growing rapidly. In this research, a new nano material was produce from kaolin clay and compare to OPC in terms of surface analysis, particle sizing and micrograph image on new modification of kaolin clay particles. Kaolin clay was established in two processes which are before and after heat treatment. Apart from that, transformation of kaolin clay to nano material was monitor by using Field Emission Scanning Electron Microscope (FESEM) and new nano materials were formed. Those images were supported by X Ray Diffraction analysis (XRD), X Ray Fluorescence (XRF) and laser particle analyser to see the chemical composition and particle size for all specimens. A combination of rough, smooth and long section can be analysed. From this analysis a new develops nano materials can be achieved and can be utilised especially for construction purposes.

1 Introduction

Since late 50's, human trying to develop a new material than has a super-size reduction from millimetres to macron based material [1-3]. The idea came to develop from macron to micron started at the early 60's and micron based technology had improved a lot to manufactured small scale gadget or instrument and benefit to human and improved quality of life. Since then, human are trying to improve and enhance technology by inventing less than micro size materials. The idea came from Sobolev to reduce and creating micro to smaller size. From that nano size where introduced to world [3-5].

Kaolin clay or white clay, rich in kaolinite mineral and also classified as layered silicate mineral. Due to endothermic process or dehydroxylation process which involves heat, kaolin particles can easily transform into other shape or particles [6-8]. The reaction of endothermic process will creates metakaolin. In metakaolin formation, the new formed kaolin based product was very reactive. Metakaolin were popular in the ceramic industries and also as construction materials [9, 10]. Most of the researcher concludes that, metakaolin performs better resistance to water and enhance mechanical properties of concrete [11-13].

[*] Corresponding author: fadiil2013@yahoo.com

Metakaolin in concrete reduces water penetration and alkali silica reaction by acting as filler and increase hydration process. Apart of refining of voids, concrete structures strength of concrete can be increase due to densifying of voids and less voids created [14, 15].

Process of nano forming includes of bottom to up and top to down approach. Selection of that nano process depends on the availability, cost and expertise [16]. Bottom to up approach is consider as expensive but forming a uniform and smaller particles. On contrary, top to down approach provides less cost and easy process of nano forming particles but the size in not uniform [3-5].

In this research, forming of kaolin product as nano materials is not been done intensively. Raw kaolin was produced into metakaolin, nano kaolin and nano clay and technique applied was top to down for nano kaolin and bottom to up for nano clay. There is a quite number of research on nano kaolin but nano kaolin in this research uniquely produced by milling technique. Calcination also involves in transforming the not reactive kaolin based product into reactive kaolin such as metakaolin, nano metakaolin and nano metaclay. From this research a new process is produced from kaolin and the mesopores or microstructures of new nano materials can be discussed.

2 Methodology

2.1 Sample preparation

In this research all kaolin clay based will be compared to Ordinary Portland Cement (OPC). OPC for this research was supplied by Tasek Cement from Perak, Malaysia. Two types of kaolin clay kaolin clay will be analyse which is nano kaolin and nano clay. Although those kaolin based were the same, but the manufacturing process was different. Before that, kaolin clay was procured by AKI Kaolin (M) Sdn Bhd. in Selangor, Malaysia. Average size given by the manufactured was 10 to 5 micron. To transform kaolin to nano kaolin, high energy milling was the technique applied. Fritsch milling analyser, was the instrument used for transforming kaolin to nano kaolin. To prevent heat generate during milling, ball and jar for was from Zirconia. 15 zirconia balls with 2mm diameter were used. Speed of milling was fixed to 400 rpm and duration of milling also fixed to 24 hours. For nano clay, material was procured by Sigma Aldrich (M) Sdn Bhd.

All specimens will undergo secondary treatment. The process of secondary treatment was needed for all kaolin based sample to transform from a single phase structure into amorphous stage. The process of secondary treatment involves heat and also can be called as calcination. Calcination process will be done by using electric furnace. All kaolin sample will underwent calcination process which the temperature and duration was fixed to 700°C within 3 hours heated.

To performed analysis of micrograph image of all sample powders, Field Emission Scanning Electron Microscope (FESEM) was implemented. From FESEM analysis, shape and surface texture of all samples can be evaluated. As and addition, chemical composition of all sample powders evaluated by using Horiba Mesa 50 X Ray Fluorescene Analyzer (XRF) and Rigaku X Ray analytical instrument (XRD). Finally Fritsch laser particle analyser was used to analyse particle size of all powder sample. All testing regarding chemical analysis were done at Institute of Science (IOS), UiTM Malaysia.

3 Result and discussion

3.1 Surface and shape analysis

From Fig. 1 to 7 shows specimen in powder form for kaolin based material. Kaolin samples will be disgust on the formation before and after milling. Discussion also will include on the behaviour of kaolin samples before and after treatment. All kaolin samples will be compared to OPC in terms of shape and surface texture.

Image of OPC specimen can be seen in Fig. 1. At a glance, rough, irregular, angular shape and texture were the major formation in the molecule of OPC. Most of the rough surface in OPC structure was in amorphous state due to the process of calcination in the cement plant. During calcination silica and alumina combines and creating Tricalcium Silicate (C_3S) which contributes strength, Dicalcium Aluminate (C_2A) which controlling the setting phases during hydration and other minor compound. Angular and irregular texture that can be seen in OPC was the formation C_3S and C_2A. Most of the angularity formation in OPC structure was created by C_3S formation. For C_2A formation, represent by light whitish line along the outer ring of OPC particles. Rough surface in the OPC structures were created due to incomplete and non-uniform calcination process in cement plant. We can see the formation of tiny layering or mini island texture on the surface of bigger component. Those tiny layering or mini island forms will slowly been deceased during the hydration process which includes water. During hydration tiny spots contain of C_3S will react and dilute alumina in the outer ring and creates bonding between other OPC molecules. Although bonding between pastes will enhance strength, due to non-uniformity of OPC particles will creates minor voids between pastes. As a result, surface of newly hydrated product called Calcium Silica Hydrate (CSH) gel were not uniform.

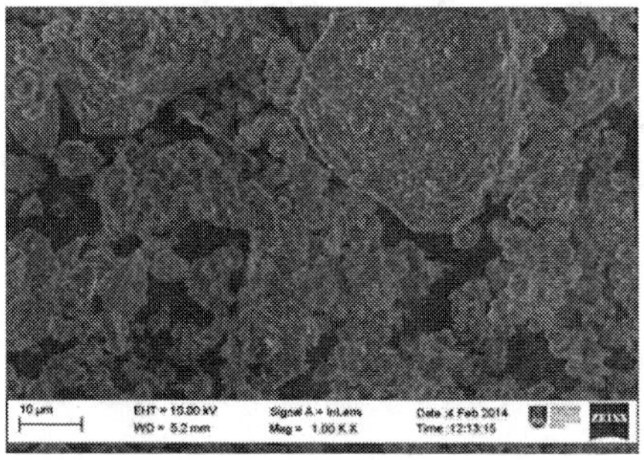

Fig. 1. OPC

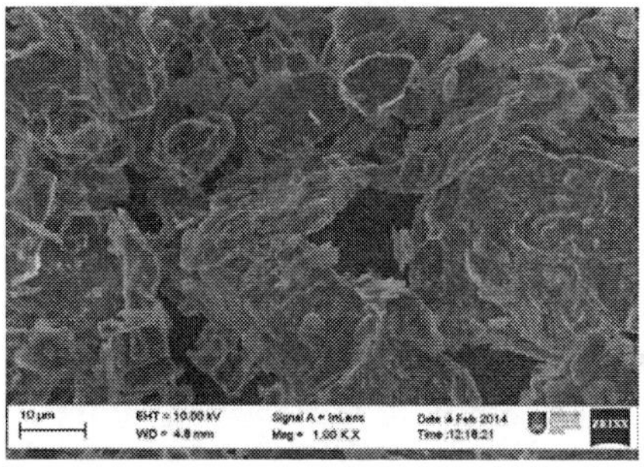

Fig. 2. Kaolin

A kaolin particle was portrayed by Fig. 2. Kaolin also known as white clay regards as high pozzolanic materials due to high content of silica and alumina compare to OPC. However the formation of kaolin particles were a lot different and unique compare to OPC. Shape of kaolin can be described as flaky and platy formation. A smooth surface of kaolin texture can be seen and shows a lot of different compare to OPC texture. The formation of kaolin particles were in layering or flakes system formation and close from one flake to another. As been mentioned earlier, silica and alumina was the major chemical occurs in kaolin. From that we can see that the different between angularities of kaolin flakes are more uniform and compare to sharp edges of OPC. More whitish layer forming the outer ring of kaolin particles represent by alumina shield system. Alumina line can be seen clearly in the flakes system of kaolin particles compare to OPC. During single crystal formation which can be seen in kaolin, alumina line performs a bonding effect from one to other flakes. In addition, alumina shield system creating and forming more surface area between kaolin particles and larger voids can be filled up by other kaolin flake. It was clear thus effect of micron based kaolin can be applied as filler in any solid matters such as concrete, soil treatment and water filtering system. In other words, kaolin can also perform as micro filler agent. To support this statement, in terms of packing formation in kaolin less empty voids can be compared to OPC. Due to micro size effect kaolin structure were in tidy and uniform layered component compare to OPC.

Metakaolin micrograph image can be seen from Fig. 3. The formation of metakaolin was from kaolin after calcination. Flaky and platy structure was the main shape texture for metakaolin. Obviously calcination process transforms kaolin into metakaolin that's involving oxidation in the kaolinite structure. Metakaolin surface were smoother compare to kaolin. Formation of metakaolin particles can be seen by bigger flakes were formed and separated from one another. Smooth surface of metakaolin were cause by dilution effect of alumina in metakaolin. Before calcined, alumina bond metakaolin flakes and provides layering system. Tiny spots or mini island of flakes were move apart due to calcination and from that surface of metakaolin were refined. Shape of metakaolin flakes reformed a neat angularity and sharp edges were reduced. Size of metakaolin flakes also are bigger compare to kaolin resulting more space and larger surface area were created. Increase in flakes size due to dilution effect of alumina component during calcination. As been mentioned earlier, alumina can be seen by whitish layer outside the metakaolin flakes. After calcined, whitish layer which represent alumina was obviously seen compare to kaolin.

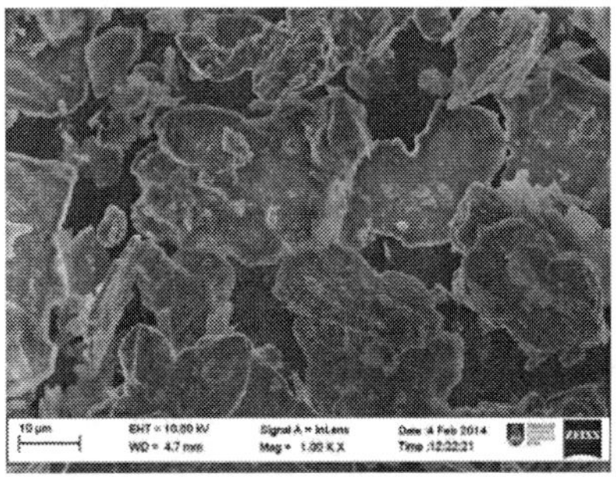

Fig. 3. Metakaolin

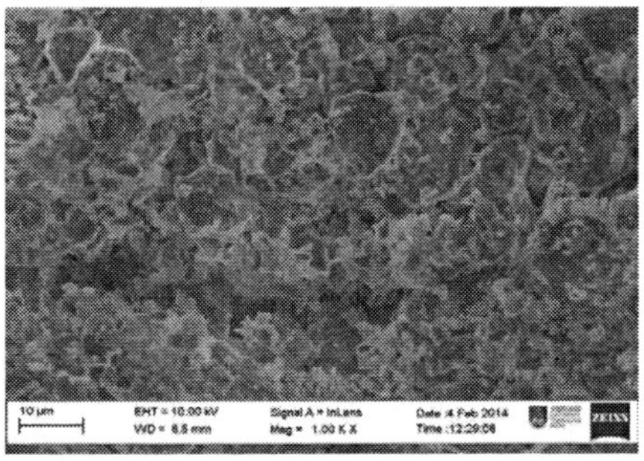

Fig. 4. Nano Kaolin

Nano kaolin was created from kaolin after undergo milling process. Fig. 4 shows the micrograph image of nano kaolin. Milling process proves to reduce the size of kaolin particles. In terms of shape and surface texture, nano kaolin shows a similar texture to kaolin. However due to milling the kaolin flakes has been reduced to nano particles and forming nano kaolin. Average size of nano kaolin recorded was less than 200 nm. As been stated earlier, shape of nano kaolin still maintaining the flaky, platey and tetrahedron formation also known as layering structure. Surface of nano kaolin also in smooth similar to kaolin but increase in the formation of tiny spots or mini island overlay from one another. Smaller particles in the nano kaolin structure resulting in increase in surface area. Since nano kaolin not undergo secondary treatment, alumina layer is not dilute and nano kaolin flakes maintaining the neat tetrahedron or layering flaky formation.

Nano metakaolin were formed from nano kaolin after undergo secondary treatment which is calcination. Fig. 5 portrait image of nano metakaolin. The analysis of shape and surface texture of nano metakaolin is similar to metakaolin only the different was size of particles was smaller. Due to calcination, tetrahedron or layering of flakes were dilute and increase surface area. Again increase in surface area due to the dilution of alumina resulting in fluidising the flakes and unlock the bonding of nano kaolin particles. Calcination process refines and densifies angularity and sharp edges of nano metakaolin flakes and form a uniform nano metakaolin structure. Whitish layer represent alumina, can be seen forming at the outer ring of nano metakaolin flakes. Surface of nano metakaolin were smoother compare to metakaolin due to uniform and incomplete calcination process due to smaller particles compare to metakaolin.

Fig. 6 shows micrograph image of nano clay. A different image can be seen through nano clay compare to OPC and kaolin. At a glance, nano clay shape consists of rounded and angular edges. Most of the particles were rounded or ball shape, only size of balls is not equal or uniform. Based on the rounded shape, nano clay can perform a ball bearing effect such as fly ash and silica fume. In addition, nano clay particles are freely to move or mobility to move during mixing with other elements and workable. Although in terms of mobility, movement or flow ability of nano clay is not freely like fly ash or silica fume. This occurs due to clay properties in kaolin performs absorption effect. Surface of nano clay can be considered smoother than OPC but not to kaolin and other samples. It seems like ball formation of nano clay particles in rounded shape overlay from one particle to another. In complete fluidity of nano clay due to no secondary treatment causing to the less smooth surface of nano clay. Uniformity of nano clay particles compare to other samples was due to processing stages by using bottom to up approach. Combination of chemical synthesis method during the production of nano clay help to modify the originality of kaolin particles from flakes particles into rounded or ball flocking formation structure.

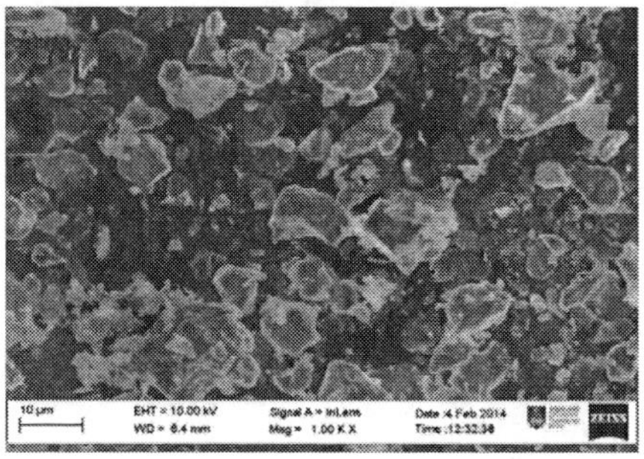

Fig. 5. Nano Metakaolin

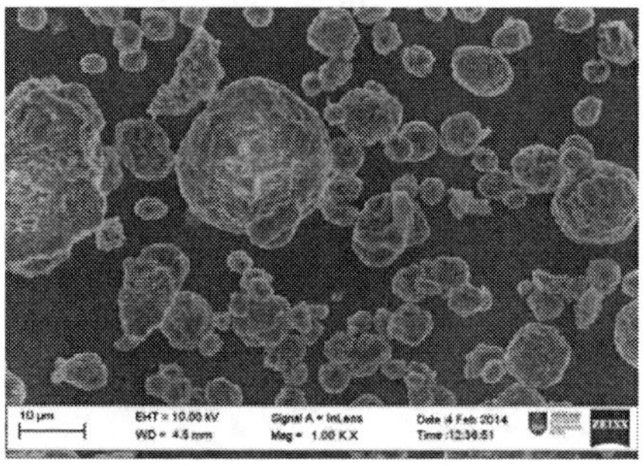

Fig. 6. Nano Clay

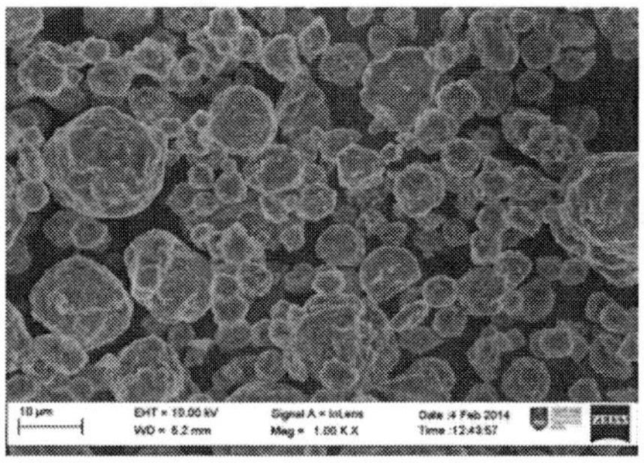

Fig.7. Nano Metaclay

Nano metaclay was created by secondary treatment which involves calcination of nano clay. Fig. 7 shows micrograph image of nano metaclay. Due to calcination, alumina content of nano clay fluidise thus creating dehydrolixation process and causing ball flocking formation. Apart from that, structure of nano clay tends to separate from one to another and smaller ball formation was created. From this situation, more surface area was created and can perform a good filler agent. Fluidise effect of alumina dilution can be seen clearly by whitish line colour outside the outer ring of nano metaclay. Those alumina layer were not seen in nano clay image and from that an obvious different can be seen before and after calcine. Smoother surface of nano metaclay also can be compare to nano clay due to fluidise effect. Smaller particles of nano metaclay were created after calcination and resulting in increase in surface area. In conclusion, calcination proves to change the ball flocking formation of nano metaclay into ultra size and smoother surface compare to nano clay.

3.2 Chemical composition

Table 1 shows the chemical composition of OPC, kaolin, metakolin, nano kaolin, nano metakaolin, nano clay and nano metaclay. The chemical composition were detected by using X-Raf Fluorenscence (XRF) technique. It can be seen that the content of SiO_2 for nano metakaolin was the highest as compared to that of OPC and other specimen. Almost 90% of silica is recorded in nano metakaolin thus confirmed the potential of strength increment when blending with OPC. The second goes to metakaolin around 75%, nano metaclay less than 70% and OPC records the lowest content of SiO_2 which less than 15%. It can be seen that after treatment from kaolin into metakaolin, nano metakaolin and nano metaclay, a slight increament of silica content were produced. It is believe due to liquidation of tetahedron structure from heat were the factor contributed to this phenomena. For Al_2O_3, nano metakaolin also record was the highest as compared to the other which is about 19%. Most of kaolin specimen records higher alumina with respect to OPC. Although transformation of kaolin into other specimen which includes secondary heat treatment, only a slight increament of alumina can be seen for all specimen. The alumina oxide will accelerate the hydration process and setting time. The content of CaO for nano metakaolin is very minimum which less than 1% which will reduce the formation of calcium hydroxide in cement paste during hydration. As a conclusion, nano metakaolin can be classified as a very reactive pozzolannic materials because the high content of SiO_2 and Al_2O_3.

Table 1. Chemical composition of kaolin clay specimen

Chemical	Mass Percentage (%)						
	OPC	Kaolin	Metakaolin	Nano Kaolin	Nano Metakaolin	Nano Clay	Nanao Metaclay
SiO_2	11.6	73.3	74.3	87.7	89.6	63.9	65.9
Al_2O_3	2.2	14.5	17.2	17.9	18.9	14.0	15.1
CaO	75.17	0.058	2.4	3.33	0.43	4.8	4.3
TiO_2	0.4	1.38	1.45	1.65	1.82	1.01	0.9
Fe_2O_3	5.38	1.99	1.81	2.2	2.0	13.41	11.4
K_2O	0.43	4.86	4.64	3.41	4.55	0.27	0.24

3.3 Particle size analysis

Table 2 and Fig. 8 shows the analysis for particle size for OPC, kaolin, metakaolin, nano kaolin, nano metakaolin, nano clay and nano metaclay respectively. From Table 1.2 we can see huge transformation from micron based into nano particles can be produced. Kaolin and OPC considered as micron based materials have a specific surface area less than 20000 m^2/kg. After calcination of kaolin, kaolin changed to metakaolin and the surface area has increased almost 30000 m^2/kg. the same pattern is seen for calcination sample which involves nano metakaolin and nano clay. Increase in surface area was recorded and the highest was nano metaclay which records around 170000 m^2/kg. Apart from that, nano process which involves milling and chemical also helps to reduce the surface area of kaolin and transform into nano kaolin. Effective size for kaolin also be influence by nano transforming process. From Table 2, size of kaolin can be reduced into smaller particles depends on the transforming process. Huge reduction of nano metakaolin particles was recorded which around 176 nm as compared to OPC and kaolin. Eventually for nano metaclay, record the lowest particle size as compared to other sample and size recorded is 146 nm.

Table 2. Particle size of kaolin clay specimen

Specimen	D10 (effective size) (nm)	D30 (nm)	D60 (nm)	Specific Surface Area (m^2/kg)
OPC	2433	6954	10972	12748
Kaolin	1612	5585	13325	18231
Metakaolin	925	2106	7282	27847
Nano kaolin	419	773	1704	63612
Nano Metakaolin	176	395	767	140792
Nano clay	269	487	556	98456
Nano metaclay	146	228	315	168987

Fig. 8 shows particle size distribution of OPC, kaolin, metakaolin, nano kaolin, nano metakaolin, nano clay and nano metaclay. Nano metakaolin record the highest particles less than 500 nm. For nano metakaolin, almost 85% records the size less than 500 nm as compared to other sample. Next highest goes to nano kaolin which the product of milling before secondary treatment. For nano metaclay almost 70% of particles are less than 500 nm.

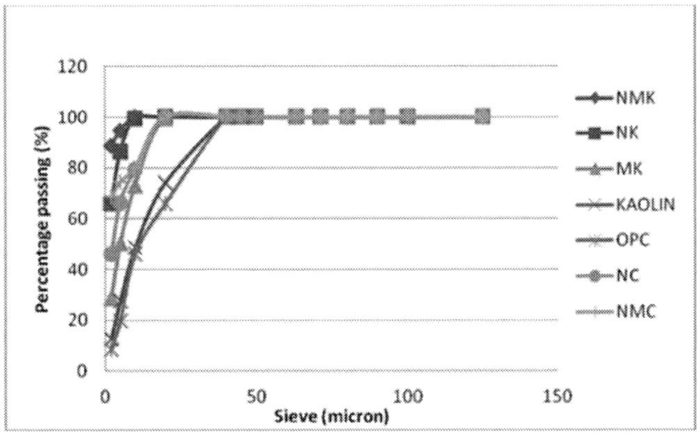

Fig. 8. Particle size distribution of seven (7) samples Of OPC, kaolin, metakaolin, nano kaolin, nano metakaolin, nano clay and nano metaclay

3.4 XRD analysis

Fig. 9 and 10 shows the result of XRD analysis of kaolin, metakaolin, nano kaolin, nano clay and nano metaclay. From Fig. 9 a different peak is seen regarding the process of calcination of kaolin specimen. Kaolinite formation of kaolin was reduced as been compared to metakaolin. This happen due to endothermic process during calcination creates a n amorphous structure of metakaolin. Major content of kaolinite are alumina and alumina and oxidation creates a better formation of alumina and silica. It can be seen by metakaolin where by a lower peak or counts as compared to kaolin. A different peak can be seen by nano kaolin and nano metakaolin in Fig. 9. Counts and peak of nano kaolin and nano metakaolin is higher as compared to OPC. Due to nano particles created, causing the peak detecting smaller particles as compared to kaolin. Another different between those specimens, we can see a better or refines peak and noise as compared of kaolin sample and after calcination. Due to oxidation, refining the crystal structures can be seen by a clear noise and more sharp peak by metakaolin and nano metakaolin.

Fig. 10 shows the XRD pattern for nano clay and nano metaclay. Montmorilite is the major compound in nano clay and in montmorilite contains of kaolinite, mica and silica. Due to oxidation, montmorilite peak will be reduced and amorphous structure created. Another examples can be seen on the formation of amorphous structure is the peak of nano metaclay is smoother as compared to nano clay.

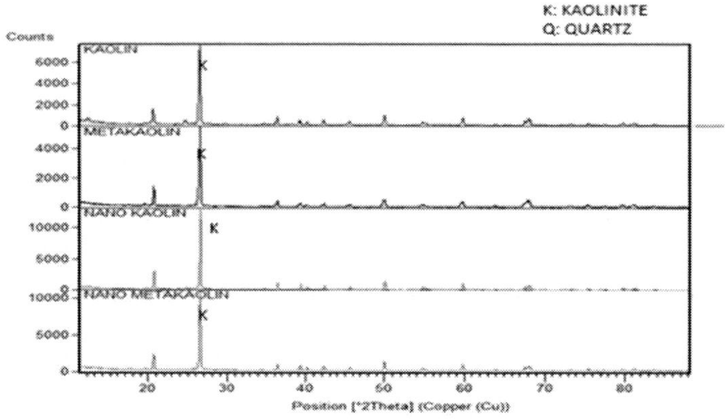

Fig. 9. XRD for kaolin,metakaolin, nano kaolin and nano metakaolin

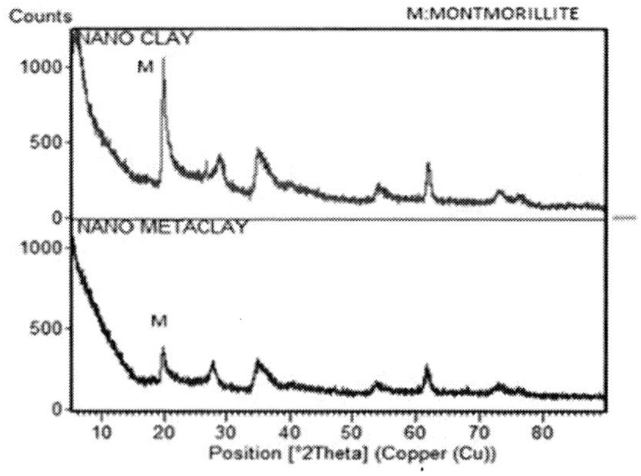

Fig. 10. XRD for nano clay and nano metaclay

4 Conclusion

From this research conclusions are drawn as follows:

i) High speed milling can be utilised for processing nano materials. Parameters namely types of milling, types of ball, numbers of ball, speed of rotation and duration of milling are the factors considered.
ii) There are differences in morphology between OPC, kaolin, metakaolin, nano kaolin and nano metakaolin. Calcination process has altered shape and surface texture of kaolin.

References

[1] N. Crainic, Nano-composites: A state of the art review, Key Engineering Materials, **230-232,** 656-660 (2002)
[2] A.H. Shekaria, Influence of nano particles on durability and mechanical properties of high performance concrete, Procedia Engineering, **14,** 3036–3041 (2011)
[3] K. Sobolev, I.F. Roman Hermosillo and L.M. Torres-Martínez, Nanomaterials and nanotechnology for high-performance cement composites, *Proceedings of ACI Session on Nanotechnology of Concrete: Recent Developments and Future Perspectives,* Denver, (2006)
[4] K. Sobolev, The development of a new method for the proportioning of high-performance concrete mixtures, Cement and Concrete Composites, **26**(7), 901-907 (2004)
[5] K. Sobolev and M. Ferrada-Guterrez, How nanotechnology can change the concrete world: Part 1, American Seramic Society Bulletin, **10**, 14-17 (2005)
[6] M.S. Morsy, H. Abbas and S.H. Alsayed, Behavior of blended cement mortars containing nano-metakaolin at elevated temperatures, Construction and Building Materials, **35**, 900–905 (2012)
[7] L. Dvorkin, A. Bezusyak, N. Lushinikova and Y. Ribakov, Using mathematical modeling for design of self compacting high strength concrete with metakaolin admixture, Construction and Building Materials, **37**, 851-864, (2012)
[8] S.M. Mansour, K. Bekkour and I. Messaoudene, Improvement of rheological behaviour of cement pastes by incorporating metakaolin, European J. of Scientific Research, **42**(3), 442-452 (2010)
[9] M.A. Megat Johari, S. Khabir and P. Rivard, Influence of supplementary cementitious materials on engineering properties of high strength concrete, Construction and Buildings Materials, **25**, 2639 – 2648 (2011)
[10] J. Bai, S. Wild, J.A. Ware and B.B. Sabir, Using neural networks to predict workability of concrete incorporating metakaolin and fly ash, Advances in Engineering Software, **34**(11), 663-669 (2003)
[11] B.B. Sabir, S. Wild and J. Bai, Metakaolin and calcined clays as pozzolans for concrete: A review,Cement and Concrete Composites, **23**(6), 441-454 (2001)
[12] A.M. Rashad, Metakaolin as cementitious material: History, scours, production and composition – A comprehensive overview, Construction and Building Materials, **41**, 303-318 (2013)
[13] A.M. Rashad, Alkali-activated metakaolin: A short guide for civil Engineer – An overview, Construction and Building Materials, **41**, 751-765 (2013)
[14] A. Nadeem, S.A. Memon and T.Y. Lo, Mechanical performance, durability, qualitative and quantitative analysis of microstructure of fly ash and metakaolin mortar at elevated temperatures, Construction and Building Materials, **38**, 338-347 (2013)
[15] R. Madandoust and S.Y. Mousavi, Fresh and hardened properties of self-compacting concrete containing metakaolin, Construction and Building Materials, **35**, 752-760, (2012)
[16] F. Sanchez and K. Sobolev, Nanotechnology in concrete – A review, Construction and Building Materials, **24**(11), 2060-2071 (2010)

AUTHOR INDEX

AUTHOR INDEX

AUTHOR INDEX

AUTHOR INDEX

AUTHOR INDEX

AUTHOR INDEX